2014 IEEE International Electron Devices Meeting

(IEDM 2014)

San Francisco, California, USA
15-17 December 2014

IEEE Catalog Number: CFP14IED-POD
ISBN: 978-1-4799-8002-4

Copyright © 2014 by the Institute of Electrical and Electronic Engineers, Inc
All Rights Reserved

Copyright and Reprint Permissions: Abstracting is permitted with credit to the source. Libraries are permitted to photocopy beyond the limit of U.S. copyright law for private use of patrons those articles in this volume that carry a code at the bottom of the first page, provided the per-copy fee indicated in the code is paid through Copyright Clearance Center, 222 Rosewood Drive, Danvers, MA 01923.

For other copying, reprint or republication permission, write to IEEE Copyrights Manager, IEEE Service Center, 445 Hoes Lane, Piscataway, NJ 08854. All rights reserved.

***This publication is a representation of what appears in the IEEE Digital Libraries. Some format issues inherent in the e-media version may also appear in this print version.**

IEEE Catalog Number: CFP14IED-POD
ISBN 13:   978-1-4799-8002-4

Additional Copies of This Publication Are Available From:

Curran Associates, Inc
57 Morehouse Lane
Red Hook, NY 12571 USA
Phone: (845) 758-0400
Fax: (845) 758-2633
E-mail: curran@proceedings.com
Web: www.proceedings.com

TABLE OF CONTENTS

1.1 SILICON CARBIDE POWER DEVICE DEVELOPMENT FOR INDUSTRIAL MARKETS................1
John W. Palmour

1.2 ATOMIC SCALE DEVICES: ADVANCEMENTS AND DIRECTIONS................9
Enrico Prati,Takahiro Shinada

1.3 RESEARCH INTO ADAS WITH DRIVING INTELLIGENCE FOR FUTURE INNOVATION................13
Hideo Inoue

2.1 SOCIAL IMPACT OF POWER SEMICONDUCTOR DEVICES (INVITED)................20
B. J. Baliga

2.2 SIC POWER DEVICES FOR HEV/EV AND A NOVEL SIC VERTICAL JFET................24
Tsuyoshi Ishikawa, Yasunori Tanaka, Tsutomu Yatsuo,Koji Yano

2.3 APPLICATION SPECIFIC TRADE-OFFS FOR WBG SIC, GAN AND HIGH END SI POWER SWITCH TECHNOLOGIES................28
R. Rupp, T. Laska, O. Häberlen, M. Treu

2.4 HIGH VOLTAGE SILICON BASED DEVICES FOR ENERGY EFFICIENT POWER DISTRIBUTION AND CONSUMPTION................32
A. Kopta

2.5 PROGRESS IN ULTRAHIGH-VOLTAGE SIC DEVICES FOR FUTURE POWER INFRASTRUCTURE................36
T. Kimoto, J. Suda, Y. Yonezawa, K. Asano, K. Fukuda, H. Okumura

2.6 600 V JEDEC-QUALIFIED HIGHLY RELIABLE GAN HEMTS ON SI SUBSTRATES................40
T. Kikkawa, T. Hosoda, K. Imanishi, K. Shono, K. Itabashi, T. Ogino, Y. Miyazaki, A. Mochizuki, K. Kiuchi, M. Kanamura, M. Kamiyama, S. Akiyama, S. Kawasaki, T. Maeda, Y. Asai, Y. Wu, K. Smith, J. Gritters, P. Smith, S. Chowdhury, D. Dunn, M. Aguilera, B. Swenson, R., L. McCarthy, L. Shen, J. McKay, H. Clement, J. Honea, S. Yea, D. Thor, R. Lal, U. Mishra, P. Parikh

2.7 POWER DEVICES ON BULK GALLIUM NITRIDE SUBSTRATES: AN OVERVIEW OF ARPA-E'S SWITCHES PROGRAM (INVITED)................44
Timothy D. Heidel, Pawel Gradzki

3.1 AN ENHANCED 16NM CMOS TECHNOLOGY FEATURING 2ND GENERATION FINFET TRANSISTORS AND ADVANCED CU/LOW-K INTERCONNECT FOR LOW POWER AND HIGH PERFORMANCE APPLICATIONS................48
S.-Y. Wu, C. Y. Lin, M. C. Chiang, J. J. Liaw, J. Y. Cheng, S. H. Yang, S. Z. Chang, M. Liang, T. Miyashita, C. H. Tsai, C. H. Chang, V. S. Chang, Y. K. Wu, J. H. Chen, H. F. Chen, S. Y. Chang, K. H. Pan, R. F. Tsui, C. H. Yao, K. C. Ting, T. Yamamoto, H. T. Huang, T. L. Lee, C. H. Lee, W. Chang, H. M. Lee, C. C. Chen, T. Chang, R. Chen, Y. H. Chiu, M. H. Tsai, S. M. Jang, K. S. Chen, Y. Ku

3.2 ANALOG CIRCUIT AND DEVICE INTERACTION IN HIGH-SPEED SERDES DESIGN IN 16NM FINFET CMOS TECHNOLOGY................52
Freeman Zhong, Ashutosh Sinha

3.3 16 NM FINFET HIGH-K/METAL-GATE 256-KBIT 6T SRAM MACROS WITH WORDLINE OVERDRIVEN ASSIST................56
Makoto Yabuuchi, Masao Morimoto, Yasumasa Tsukamoto, Shinji Tanaka, Koji Tanaka, Miki Tanaka, Koji Nii

3.4 DYNAMIC SINGLE-P-WELL SRAM BITCELL CHARACTERIZATION WITH BACK-BIAS ADJUSTMENT FOR OPTIMIZED WIDE-VOLTAGE-RANGE SRAM OPERATION IN 28NM UTBB FD-SOI................59
O. Thomas, B. Zimmer, S. O. Toh, L. Campolini, N. Planes, R. Ranica, P. Flatresse, B. Nikolic

3.5 DRAIN EXTENDED MOS DEVICE DESIGN FOR INTEGRATED RF PA IN 28NM CMOS WITH OPTIMIZED FOM AND ESD ROBUSTNESS................63
Ankur Gupta, Mayank Shrivastava, Maryam Shojaei Baghini, Dinesh Kumar Sharma, A. N. Chandorkar, Harald Gossner, V. Ramgopal Rao

3.6 HETEROGENEOUSLY INTEGRATED SUB-40NM LOW-POWER EPI-LIKE GE/SI MONOLITHIC 3D-IC WITH STACKED SIGEC AMBIENT LIGHT HARVESTER................67
Chang-Hong Shen, Jia-Min Shieh, Wen-Hsien Huang, Tsung-Ta Wu, Chien-Fu Chen, Ming-Hsuan Kao, Chih-Chao Yang, Chein-Din Lin, Hsing-Hsiang Wang, Tung-Ying Hsieh, Bo-Yuan Chen, Guo-Wei Huang, Meng-Fan Chang,Fu-Liang Yang

3.7 A 14NM LOGIC TECHNOLOGY FEATURING 2ND-GENERATION FINFET TRANSISTORS, AIR-GAPPED INTERCONNECTS, SELF-ALIGNED DOUBLE PATTERNING AND A 0.0588µM^2 SRAM CELL SIZE................71
S. Natarajan, M. Agostinelli, S. Akbar, M. Bost, A. Bowonder, V. Chikarmane, S. Chouksey, A. Dasgupta, K. Fischer, Q. Fu, T. Ghani, M. Giles, S. Govindaraju, R. Grover, W. Han, D. Hanken, E. Haralson, M. Haran, M. Heckscher, R. Heussner, P. Jain, R. James, R. Jhaveri, I. Jin, H. Kam, E. Karl, C. Kenyon, M. Liu, Y. Luo, R. Mehandru, S. Morarka, L. Neiberg, P. Packan, A. Paliwal, C. Parker, P. Patel, R. Patel, C. Pelto, L. Pipes, P. Plekhanov, M. Prince, S. Rajamani, J. Sandford, B. Sell, S. Sivakumar, P. Smith, B. Song, K. Tone, T. Troeger, J. Wiedemer, M. Yang, K. Zhang

3.8 HIGH PERFORMANCE 14NM SOI FINFET CMOS TECHNOLOGY WITH 0.0174µM² EMBEDDED DRAM AND 15 LEVELS OF CU METALLIZATION..74

C.-H. Lin, B. Greene, S. Narasimha, J. Cai, A. Bryant, C. Radens, V. Narayanan, B. Linder, H. Ho, A. Aiyar, E. Alpetkin, J.-J. An, M. Aquilino, R. Bao, V. Basker, N. Breil, M. Brodsky, W. Chang, L. Clevenger, D. Chidambarrao, C. Christiansen, D. Conklin, C. DeWan, H. Dong, L. Economikos, B. Engel, S. Fang, D. Ferrer, A. Friedman, A. Gabor, F. Guarin, X. Guan, M. Hasanuzzaman, J. Hong, D. Hoyos, B. Jagannathan, S. Jain, S.-J. Jeng, J. Johnson, B. Kannan, Y. Ke, B. Khan, B. Kim, S. Koswatta, A. Kumar, T. Kwon, U. Kwon, L. Lanzerotti, H.-K. Lee, W.-H. Lee, A. Levesque, W. Li, Z. Li, W. Liu, S. Mahajan, K. McStay, H. Nayfeh, W. Nicoll, G. Northrop, A. Ogino, C. Pei, S. Polvino, R. Ramachandran, Z. Ren, R. Robison, I. Saraf, V. Sardesai, S. Saudari, D. Schepis, C. Sheraw, S. Siddiqui, L. Song, K. Stein, C. Tran, H. Utomo, R. Vega, G. Wang, H. Wang, W. Wang, X. Wang, D. Wehelle-Gamage, E. Woodard, Y. Xu, Y. Yang, N. Zhan, K. Zhao, C. Zhu, K. Boyd, E. Engbrecht, K. Henson, E. Kaste, S. Krishnan, E. Maciejewski, H. Shang, N. Zamdmer, R. Divakaruni, J. Rice, S. Stiffler, P. Agnello

3.9 A 55 NM TRIPLE GATE OXIDE 9 METAL LAYERS SIGE BICMOS TECHNOLOGY FEATURING 320 GHZ F_T / 370 GHZ F_{MAX} HBT AND HIGH-Q MILLIMETER-WAVE PASSIVES..77

P. Chevalier, G. Avenier, G. Ribes, A. Montagné, E. Canderle, D. Céli, N. Derrier, C. Deglise, C. Durand, T. Quémerais, M. Buczko, D. Gloria, O. Robin, S. Petitdidier, Y. Campidelli, F. Abbate, M. Gros-Jean, L. Berthier, J. D. Chapon, F. Leverd, C. Jenny, C. Richard, O. Gourhant, C. De-Buttet, R. Beneyton, P. Maury, S. Joblot, L. Favennec, M. Guillermet, P. Brun, K. Courouble, K. Haxaire, G. Imbert, E. Gourvest, J. Cossalter, O. Saxod, C. Tavernier, F. Foussadier, B. Ramadout, R. Bianchini, C. Julien, D. Ney, J. Rosa, S. Haendler, Y. Carminati, B. Borot

4.1 MOS CAPACITOR DEEP TRENCH ISOLATION FOR CMOS IMAGE SENSORS..80

N. Ahmed, F. Roy, G.-N. Lu, B. Mamdy, J.-P. Carrere, A. Tournier, N. Virollet, C. Perrot, M. Rivoire, A. Seignard, D. Pellissier-Tanon, F. Leverd, B. Orlando

4.2 THREE-DIMENSIONAL INTEGRATED CMOS IMAGE SENSORS WITH PIXEL-PARALLEL A/D CONVERTERS FABRICATED BY DIRECT BONDING OF SOI LAYERS..84

Masahide Goto, Kei Hagiwara, Yoshinori Iguchi, Hiroshi Ohtake, Takuya Saraya, Masaharu Kobayashi, Eiji Higurashi, Hiroshi Toshiyoshi, Toshiro Hiramoto

4.3 HIGH SENSITIVITY IMAGE SENSOR OVERLAID WITH THIN-FILM CRYSTALLINE-SELENIUM-BASED HETEROJUNCTION PHOTODIODE..88

S. Imura, K. Kikuchi, K. Miyakawa, H. Ohtake, M. Kubota, T. Nakada, T. Okino, Y. Hirose, Y. Kato, N. Teranishi

4.4 9.74-THZ ELECTRONIC FAR-INFRARED DETECTION USING SCHOTTKY BARRIER DIODES IN CMOS..92

Zeshan Ahmad, Alvydas Lisauskas, Hartmut G. Roskos, Kenneth K. O

4.5 EXPERIMENTAL DEMONSTRATION OF A STACKED SOI MULTIBAND CHARGED-COUPLED DEVICE..96

Chu-En Chang, Julie D. Segal, Aaron J. Roodman, Christopher J. Kenney, Roger T. Howe

4.6 ENHANCED TIME DELAY INTEGRATION IMAGING USING EMBEDDED CCD IN CMOS TECHNOLOGY..100

Piet De Moor, Jo Robbelein, Luc Haspeslagh, Pierre Boulenc, Alper Ercan, Kyriaki Minoglou, Anne Lauwers, Koen De Munck, Maarten Rosmeulen

4.7 A SOLID STATE THIN FILM INCANDESCENT LIGHT EMITTING DEVICE..104

Yue Kuo

5.1 CONTACT RESISTANCE REDUCTION USING FERMI LEVEL DE-PINNING LAYER FOR MOS₂ FETS..108

Woojin Park, Yonghun Kim, Sang Kyung Lee, Ukjin Jung, Jin Ho Yang, Chunhum Cho, Yun Ji Kim, Sung Kwan Lim, In Seol Hwang, Han-Bo-Ram Lee, Byoung Hun Lee

5.2 TOWARDS HIGH-PERFORMANCE TWO-DIMENSIONAL BLACK PHOSPHORUS OPTOELECTRONIC DEVICES: THE ROLE OF METAL CONTACTS..112

Yexin Deng, Nathan J. Conrad, Zhe Luo, Han Liu, Xianfan Xu, Peide D. Ye

5.3 ATOMICALLY THIN GRAPHENE PLANE ELECTRODE FOR 3D RRAM..116

Joon Sohn, Seunghyun Lee, Zizhen Jiang, Hong-Yu Chen, H.-S. Philip Wong

5.4 GRAPHENE INDUCTORS FOR HIGH-FREQUENCY APPLICATIONS - DESIGN, FABRICATION, CHARACTERIZATION, AND STUDY OF SKIN EFFECT..120

Xiang Li, Jiahao Kang, Xuejun Xie, Wei Liu, Deblina Sarkar, Junfa Mao, Kaustav Banerjee

5.5 NANOPHOTONICS WITH TWO-DIMENSIONAL ATOMIC CRYSTALS..124

Thomas Mueller, Marco M. Furchi, Andreas Pospischil, Dmitry K. Polyushkin

5.6 BROADBAND 10GB/S GRAPHENE ELECTRO-ABSORPTION MODULATOR ON SILICON FOR CHIP-LEVEL OPTICAL INTERCONNECTS..128

Y. T. Hu, M. Pantouvaki, S. Brems, I. Asselberghs, C. Huyghebaert, M. Geisler, C. Alessandri, R. Baets, P. Absil, D. Van Thourhout, J. Van Campenhout

5.7 FAST VISIBLE-LIGHT PHOTOTRANSISTOR USING CVD-SYNTHESIZED LARGE-AREA BILAYER WSE₂..132

Pang-Shiuan Liu, Chang-Hsiao Chen, Wei-Ting Hsu, Chih-Pin Lin, Tzu-Ping Lin, Li-Jen Chi, Chao-Yuan Chang, Shih-Chieh Wu, Wen-Hao Chang, Lain-Jong Li, Tuo-Hung Hou

6.1 CONTROLLING OXYGEN VACANCIES IN DOPED OXIDE BASED CBRAM FOR IMPROVED MEMORY PERFORMANCES..136

G. Molas, E. Vianello, F. Dahmani, M. Barci, P. Blaise, J. Guy, A. Toffoli, M. Bernard, A. Roule, F. Pierre, C. Licitra, B. De Salvo, L. Perniola

6.2 PROCESS INTEGRATION OF A 27NM, 16GB CU RERAM..140

J. Zahurak, K. Miyata, M. Fischer, M. Balakrishnan, S. Chhajed, D. Wells, H. Li, A. Torsi, J. Lim, M. Korber, K. Nakazawa, S. Mayuzumi, M. Honda, S. Sills, S. Yasuda, A. Calderoni, B. Cook, G. Damarla, H.Tran, B. Wang, C. Cardon, K. Karda, J. Okuno, A. Johnson, T. Kunihiro, J. Sumino, M. Tsukamoto, K. Aratani, N. Ramaswamy, W. Otsuka, K. Prall

6.3 RESISTIVE MEMORIES FOR ULTRA-LOW-POWER EMBEDDED COMPUTING DESIGN..144

E. Vianello, O. Thomas, G. Molas, O. Turkyilmaz, N. Jovanovic, D. Garbin, G. Palma, M. Alayan, C. Nguyen, J. Coignus, B. Giraud, T. Benoist, M. Reyboz, A. Toffoli, C. Charpin, F. Clermidy, L. Perniola

6.4 POINT TWIN-BIT RRAM IN 3D INTERWEAVED CROSS-POINT ARRAY BY CU BEOL PROCESS 148
Yung-Wen Chin, Shu-En Chen, Min-Che Hsieh, Tzong-Sheng Chang, Chrong Jung Lin, Ya-Chin King

6.5 EXPERIMENTAL AND THEORETICAL UNDERSTANDING OF FORMING, SET AND RESET OPERATIONS IN CONDUCTIVE BRIDGE RAM (CBRAM) FOR MEMORY STACK OPTIMIZATION 152
J. Guy, G. Molas, P. Blaise, C. Carabasse, M. Bernard, A. Roule, G. Le Carval, V. Sousa, H. Grampeix, V. Delaye, A. Toffoli, J. Cluzel, P. Brianceau, O. Pollet, V. Balan, S. Barraud, O. Cueto, G. Ghibaudo, F. Clermidy, B. De Salvo, L. Perniola

6.6 ULTRATHIN (~2NM) HFO$_X$ AS THE FUNDAMENTAL RESISTIVE SWITCHING ELEMENT: THICKNESS SCALING LIMIT, STACK ENGINEERING AND 3D INTEGRATION 156
Liang Zhao, Zizhen Jiang, Hong-Yu Chen, Joon Sohn, Kye Okabe, Blanka Magyari-Köpe, H.-S. Philip Wong, Yoshio Nishi

6.7 3D-STACKABLE CROSSBAR RESISTIVE MEMORY BASED ON FIELD ASSISTED SUPERLINEAR THRESHOLD (FAST) SELECTOR 160
Sung Hyun Jo, Tanmay Kumar, Sundar Narayanan, Wei D. Lu, Hagop Nazarian

6.8 HIGH-DRIVE CURRENT (>1MA/CM²) AND HIGHLY NONLINEAR (>10³) TIN/AMORPHOUS-SILICON/TIN SCALABLE BIDIRECTIONAL SELECTOR WITH EXCELLENT RELIABILITY AND ITS VARIABILITY IMPACT ON THE 1S1R ARRAY PERFORMANCE 164
Leqi Zhang, Bogdan Govoreanu, Augusto Redolfi, Davide Crotti, Hubert Hody, Vasile Paraschiv, Stefan Cosemans, Christoph Adelmann, Thomas Witters, Sergiu Clima, Yang-Yin Chen, Paul Hendrickx, Dirk J. Wouters, Guido Groeseneken, Malgorzata Jurczak

7.1 IMPACT OF 3D INTEGRATION ON 7NM HIGH MOBILITY CHANNEL DEVICES OPERATING IN THE BALLISTIC REGIME 168
W. Guo, M. Choi, A. Rouhi, V. Moroz, G. Eneman, J. Mitard, L. Witters, G. Van der Plas, N. Collaert, G. Beyer, P. Absil, A. Thean, E. Beyne

7.2 A MOBILITY ENHANCEMENT STRATEGY FOR SUB-14NM POWER-EFFICIENT FDSOI TECHNOLOGIES 172
B. De Salvo, P. Morin, M. Pala, G. Ghibaudo, O. Rozeau, Q. Liu, A. Pofelski, S. Martini, M. Cassé, S. Pilorget, F. Allibert, F. Chafik, T. Poiroux, P. Scheer, R. G. Southwick, D. Chanemougame, L. Grenouillet, K. Cheng, F. Andrieu, S. Barraud, S. Maitrejean, E. Augendre, H. Kothari, N. Loubet, W. Kleemeier, M. Celik, O. Faynot, M. Vinet, R. Sampson, B. Doris

7.3 COUPLED MONTE CARLO SIMULATION OF TRANSIENT ELECTRON-PHONON TRANSPORT IN SMALL FETS 176
Y. Kamakura, I. N. Adisusilo, K. Kukita, G. Wakimura, S. Koba, H. Tsuchiya, N. Mori

7.4 MODELING AND OPTIMIZATION OF GROUP IV AND III-V FINFETS AND NANO-WIRES 180
Victor Moroz, Lee Smith, Joanne Huang, Munkang Choi, Terry Ma, Jie Liu, Yunqiang Zhang, Xi-Wei Lin, Jamil Kawa, Yves Saad

7.5 PERFORMANCE EVALUATION OF INGAAS, SI, AND GE NFINFETS BASED ON COUPLED 3D DRIFT-DIFFUSION/MULTISUBBAND BOLTZMANN TRANSPORT EQUATIONS SOLVER 184
Seonghoon Jin, Anh-Tuan Pham, Woosung Choi, Yutaka Nishizawa, Young-Tae Kim, Keun-Ho Lee, Youngkwan Park, Eun-Seung Jung

7.6 SIMULATION ANALYSIS OF III-V n-MOSFETS: CHANNEL MATERIALS, FERMI LEVEL PINNING AND BIAXIAL STRAIN 188
Enrico Caruso, Daniel Lizzit, Patrik Osgnach, David Esseni, Pierpaolo Palestri, Luca Selmi

7.7 ASSESSMENT OF HOLE MOBILITY IN STRAINED INSB, GASB AND INGASB BASED ULTRA-THIN BODY PMOSFETS WITH DIFFERENT SURFACE ORIENTATIONS 192
Pengying Chang, Xiaoyan Liu, Gang Du, Xing Zhang

8.1 TWO-DIMENSIONAL NANOELECTROMECHANICAL SYSTEMS (2D NEMS) VIA ATOMICALLY-THIN SEMICONDUCTING CRYSTALS VIBRATING AT RADIO FREQUENCIES 196
Philip X.-L. Feng, Zenghui Wang, Jaesung Lee, Rui Yang, Xuqian Zheng, Keliang He, Jie Shan

8.2 INTEGRATED ON-CHIP ENERGY STORAGE USING POROUS-SILICON ELECTROCHEMICAL CAPACITORS 200
D. S. Gardner, C. W. Holzwarth III, Y. Liu, S. B. Clendenning, W. Jin, B. K. Moon, C. Pint, Z. Chen, E. Hannah, R. Chen, C. P. Wang, C. Chen, E. Mäkilä, J. L. Gustafson

8.3 COMPREHENSIVE ANALYSIS OF DEFORMATION OF INTERFACIAL MICRO-NANO STRUCTURE BY APPLIED FORCE IN TRIBOELECTRIC ENERGY HARVESTER 204
Myeong-Lok Seol, Jin-Woo Han, Jong-Ho Woo, Dong-Il Moon, Jee-Yeon Kim, Yang-Kyu Choi

8.4 A HIGH EFFICIENCY FREQUENCY PRE-DEFINED FLOW-DRIVEN ENERGY HARVESTER DOMINATED BY ON-CHIP MODIFIED HELMHOLTZ RESONATING CAVITY 208
X. J. Mu, C. L. Sun, H. M. Ji, L. Y Siow, Q. X. Zhang, Y. Zhu, H. B. Yu, J. F. Tao, Y. D. Gu, D. L. Kwong

8.5 FABRICATION OF INTEGRATED MICROMETER PLATFORM FOR THERMOELECTRIC MEASUREMENTS 212
Maciej Haras, V. Lacatena, F. Morini, J.-F. Robillard, S. Monfray, T. Skotnicki, E. Dubois

8.6 ATOMIC SCALE ENGINEERING OF METAL-OXIDE-SEMICONDUCTOR PHOTOELECTRODES FOR ENERGY HARVESTING APPLICATION INTEGRATED WITH GRAPHENE AND EPITAXY SRTIO₃ 216
Li Ji, Martin D. McDaniel, Li Tao, Xiaohan Li, Agham B. Posadas, Yao-Feng Chang, Alexander A. Demkov, John G. Ekerdt, Deji Akinwande, Rodney S. Ruoff, Jack C. Lee, Edward T. Yu

9.1 FDSOI CMOS DEVICES FEATURING DUAL STRAINED CHANNEL AND THIN BOX EXTENDABLE TO THE 10NM NODE 219
Q. Liu, B. De Salvo, P. Morin, N. Loubet, S. Pilorget, F. Chafik, S. Maitrejean, E. Augendre, D. Chanemougame, S. Guillaumet, H. Kothari, F. Allibert, B. Lherron, B. Liu, Y. Escarabajal, K. Cheng, J. Kuss, M. Wang, R. Jung, S. Teehan, T. Levin, M. Sankarapandian, R. Johnson, J. Kanyandekwe, H. He, R. Venigalla, T. Yamashita, B. Haran, L. Grenouillet, M. Vinet, O. Weber, E. Josse, F. Boeuf, M. Haond, J.-L. Bataillon, W. Kleemeier, T. Skotnicki, M. Khare, O. Faynot, B. Doris, M. Celik, R. Sampson

9.2 FUTURE CHALLENGES AND OPPORTUNITIES FOR HETEROGENEOUS PROCESS TECHNOLOGY: TOWARDS THE THIN FILMS, ZERO INTRINSIC VARIABILITY DEVICES, ZERO POWER ERA 223

S. Deleonibus, O. Faynot, T. Ernst, M. Vinet, P. Batude, F. Andrieu, O. Weber, D. Cooper, F. Bertin, H. Moriceau, L. DiCioccio, T. Signamarcheix, M. Sanquer, X. Jehl, O. Cueto, H. Fanet, F. Martin, H. Okuno, F. Nemouchi, G. Poupon, Y. Lamy, D. Gasparutto, X. Baillin, L. Duraffourg, J. Arcamone, L. Perniola, B. De Salvo, E. Vianello, L. Hutin, C. Poulain, E. Beigne, R. Tiron, L. Pain, S. Tedesco, S. Barnola, N. Posseme, C. Le Royer, A. Villalon, R. Salot

9.3 FIRST EXPERIMENTAL DEMONSTRATION OF GE CMOS CIRCUITS 227

Heng Wu, Nathan Conrad, Wei Luo, Peide D. Ye

9.4 INALP-CAPPED (100) GE NFETS WITH 1.06 NM EOT: ACHIEVING RECORD HIGH PEAK MOBILITY AND FIRST INTEGRATION ON 300 MM SI SUBSTRATE 231

Xiao Gong, Qian Zhou, Man Hon Samuel Owen, Xin Xu, Dian Lei, Shu-Han Chen, Gene Tsai, Chao-Ching Cheng, You-Ru Lin, Cheng-Hsien Wu, Chih-Hsin Ko, Yee-Chia Yeo

9.5 GE N-CHANNEL FINFET WITH OPTIMIZED GATE STACK AND CONTACTS 235

M. J. H. van Dal, B. Duriez, G. Vellianitis, G. Doornbos, R. Oxland, M. Holland, A. Afzalian, Y. C. See, M. Passlack, C. H. Diaz

9.6 IN-SITU DOPED AND TENSILY STAINED GE JUNCTIONLESS GATE-ALL-AROUND NFETS ON SOI FEATURING I_{ON} = 828 $\mu A/\mu M$, I_{ON}/I_{OFF} ~ 1×10^5, DIBL= 16-54 MV/V, AND 1.4X EXTERNAL STRAIN ENHANCEMENT 239

I-Hsieh Wong, Yen-Ting Chen, Shih-Hsien Huang, Wen-Hsien Tu, Yu-Sheng Chen, Tai-Cheng Shieh, Tzu-Yao Lin, Huang-Siang Lan, C. W. Liu

9.7 EXPERIMENTAL REALIZATION OF COMPLEMENTARY P- AND N- TUNNEL FINFETS WITH SUBTHRESHOLD SLOPES OF LESS THAN 60 MV/DECADE AND VERY LOW (PA/μM) OFF-CURRENT ON A SI CMOS PLATFORM 243

Y. Morita, T. Mori, K. Fukuda, W. Mizubayashi, S. Migita, T. Matsukawa, K. Endo, S. O'uchi, Y. Liu, M. Masahara, H. Ota

10.1 JOT DEVICES AND THE QUANTA IMAGE SENSOR 247

Jiaju Ma, Donald Hondongwa, Eric R. Fossum

10.2 SPAD BASED IMAGE SENSORS 251

Edoardo Charbon

10.3 TOWARD 1GFPS: EVOLUTION OF ULTRA-HIGH-SPEED IMAGE SENSORS: ISIS, BSI, MULTI-COLLECTION GATES, AND 3D-STACKING 255

T. G. Etoh, V. T. S. Dao, K. Shimonomura, E. Charbon, C. Zhang, Y. Kamakura, T. Matsuoka

10.4 IMAGING WITH ORGANIC AND HYBRID PHOTODETECTORS (INVITED) 259

Sandro F. Tedde, Patric Büchele, Rene Fischer, Frank Steinbacher, Oliver Schmidt

10.5 A CMOS-COMPATIBLE, INTEGRATED APPROACH TO HYPER- AND MULTISPECTRAL IMAGING 261

Andy Lambrechts, Pilar Gonzalez, Bert Geelen, Philippe Soussan, Klaas Tack, Murali Jayapala

10.6 IMAGE SENSORS FOR HIGH-THROUGHPUT, MASSIVELY-PARALLEL DNA SEQUENCING: REQUIREMENTS AND ROADMAP 265

Annette Grot

10.7 HIGH PERFORMANCE SILICON IMAGING ARRAYS FOR COSMOLOGY, PLANETARY SCIENCES, & OTHER APPLICATIONS (INVITED) 267

Shouleh Nikzad, Michael E. Hoenk, John Hennessy, April D. Jewell, Alexander G. Carver, Todd J. Jones, Samuel L. Cheng, Timothy Goodsall, Charles Shapiro

10.8 DETECTING ELEMENTARY PARTICLES USING HYBRID PIXEL DETECTORS AT THE LHC AND BEYOND 271

Michael Campbell

11.1 EXTREMELY LOW ON-RESITANCE ENHANCEMENT-MODE GAN-BASED HFET USING GE-DOPED REGROWTH TECHNIQUE 275

Asamira Suzuki, Sonbeak Choe, Yasuhiro Yamada, Shuichi Nagai, Miori Hiraiwa, Nobuyuki Otsuka, Daisuke Ueda

11.2 WIDE TEMPERATURE (10K-700K) AND HIGH VOLTAGE (~1000V) OPERATION OF C-H DIAMOND MOSFETS FOR POWER ELECTRONICS APPLICATION 279

H. Kawarada, T. Yamada, D. Xu, H. Tsuboi, T. Saito, A. Hiraiwa

11.3 GAN-BASED GATE INJECTION TRANSISTORS FOR POWER SWITCHING APPLICATIONS 283

Tetsuzo Ueda, Hiroyuki Handa, Yusuke Kinoshita, Hidekazu Umeda, Shinji Ujita, Ryo Kajitani, Masahiro Ogawa, Kenichiro Tanaka, Tatsuo Morita, Satoshi Tamura, Hidetoshi Ishida, Masahiro Ishida

11.4 SCHOTTKY-ON-HETEROJUNCTION OPTOELECTRONIC FUNCTIONAL DEVICES REALIZED ON ALGAN/GAN-ON-SI PLATFORM 287

Baikui Li, Xi Tang, Qimeng Jiang, Yunyou Lu, Hanxing Wang, Jiannong Wang, Kevin J. Chen

11.5 THE SUPER-LATTICE CASTELLATED FIELD EFFECT TRANSISTOR (SLCFET): A NOVEL HIGH PERFORMANCE TRANSISTOR TOPOLOGY IDEAL FOR RF SWITCHING 291

Robert S. Howell, Eric J. Stewart, Ron Freitag, Justin Parke, Bettina Nechay, Harlan Cramer, Matthew King, Shalini Gupta, Jeffrey Hartman, Megan Snook, Ishan Wathuthanthri, Parrish Ralston, Karen Renaldo, H. George Henry, R. Chris Clarke

11.6 MIT VIRTUAL SOURCE GANFET-RF COMPACT MODEL FOR GAN HEMTS: FROM DEVICE PHYSICS TO RF FRONTEND CIRCUIT DESIGN AND VALIDATION 295

Ujwal Radhakrishna, Pilsoon Choi, Sushmit Goswami, Li-Shiuan Peh, Tomás Palacios, Dimitri Antoniadis

12.1 SCALING BREAKTHROUGH FOR ANALOG/DIGITAL CIRCUITS BY SUPPRESSING VARIABILITY AND LOW-FREQUENCY NOISE FOR FINFETS BY AMORPHOUS METAL GATE TECHNOLOGY 299

Takashi Matsukawa, Koichi Fukuda, Yongxun Liu, Junichi Tsukada, Hiromi Yamauchi, Yuki Ishikawa, Kazuhiko Endo, Shin-ichi O'uchi, Shinji Migita, Wataru Mizubayashi, Yukinori Morita, Hiroyuki Ota, Meishoku Masahara

12.2 A CIRCUIT LEVEL VARIABILITY PREDICTION OF BASIC LOGIC GATES IN ADVANCED TRIGATE CMOS TECHNOLOGY 303

E. R. Hsieh, C. M. Hung, T. Y. Wang, Steve S. Chung, R. M. Huang, C. T. Tsai, T. R. Yew

12.3 ULTRA-HIGH-Q AIR-CORE SLAB INDUCTORS FOR ON-CHIP POWER CONVERSION 307
Naigang Wang, David Goren, Eugene O'Sullivan, Xin Zhang, William J. Gallagher, Philipp Herget, Leland Chang

12.4 EFFICIENT WIRELESS POWER TRANSMISSION TECHNOLOGY BASED ON ABOVE-CMOS INTEGRATED (ACI) HIGH QUALITY INDUCTORS 311
Salahuddin Raju, Xing Li, Yan Lu, Chi-Ying Tsui, Wing-Hung Ki, Mansun Chan, C. Patrick Yue

12.5 A MAGNETIC TUNNEL JUNCTION BASED TRUE RANDOM NUMBER GENERATOR WITH CONDITIONAL PERTURB AND REAL-TIME OUTPUT PROBABILITY TRACKING 315
Won Ho Choi, Yang Lv, Jongyeon Kim, Abhishek Deshpande, Gyuseong Kang, Jian-Ping Wang, Chris H. Kim

12.6 DTMOS MODE AS AN EFFECTIVE SOLUTION OF RTN SUPPRESSION FOR ROBUST DEVICE/CIRCUIT CO-DESIGN 319
Shaofeng Guo, Ru Huang, Peng Hao, Mulong Luo, Pengpeng Ren, Jianping Wang, Weihai Bu, Jingang Wu, Waisum Wong, Scott Yu, Hanming Wu, Shiuh-Wuu Lee, Runsheng Wang, Yangyuan Wang

12.7 POLY PITCH AND STANDARD CELL CO-OPTIMIZATION BELOW 28NM 323
Marlin Frederick Jr.

13.1 NEM RELAY DESIGN FOR COMPACT, ULTRA-LOW-POWER DIGITAL LOGIC CIRCUITS 327
Tsu-Jae King Liu, Nuo Xu, I-Ru Chen, Chuang Qian, Jun Fujiki

13.2 HIGH I_{ON}/I_{OFF} GE-SOURCE ULTRATHIN BODY STRAINED-SOI TUNNEL FETS: IMPACT OF CHANNEL STRAIN, MOS INTERFACES AND BACK GATE ON THE ELECTRICAL PROPERTIES 331
Minsoo Kim, Yuki Wakabayashi, Ryosho Nakane, Masafumi Yokoyama, Mitsuru Takenaka, Shinichi Takagi

13.3 COMPREHENSIVE PERFORMANCE RE-ASSESSMENT OF TFETS WITH A NOVEL DESIGN BY GATE AND SOURCE ENGINEERING FROM DEVICE/CIRCUIT PERSPECTIVE 335
Qianqian Huang, Ru Huang, Chunlei Wu, Hao Zhu, Cheng Chen, Jiaxin Wang, Lingyi Guo, Runsheng Wang, Le Ye, Yangyuan Wang

13.4 A SCHOTTKY-BARRIER SILICON FINFET WITH 6.0 MV/DEC SUBTHRESHOLD SLOPE OVER 5 DECADES OF CURRENT 339
Jian Zhang, Michele De Marchi, Pierre-Emmanuel Gaillardon, Giovanni De Micheli

13.5 CAN PIEZOELECTRICITY LEAD TO NEGATIVE CAPACITANCE? 343
Justin C. Wong, Sayeef Salahuddin

13.6 SUB-60 MV/DECADE STEEP TRANSISTORS WITH COMPLIANT PIEZOELECTRIC GATE BARRIERS 347
Raj K. Jana, Arvind Ajoy, Gregory Snider, Debdeep Jena

14.1 PROGRESSIVE VS. ABRUPT RESET BEHAVIOR IN CONDUCTIVE BRIDGING DEVICES: A C-AFM TOMOGRAPHY STUDY 351
U. Celano, L. Goux, A. Belmonte, G. Giammaria, K. Opsomer, C. Detavernier, O. Richard, H. Bender, F. Irrera, M. Jurczak, W. Vandervorst

14.2 UNDERSTANDING THE IMPACT OF PROGRAMMING PULSES AND ELECTRODE MATERIALS ON THE ENDURANCE PROPERTIES OF SCALED TA_2O_5 RRAM CELLS 355
C. Y. Chen, L. Goux, A. Fantini, A. Redolfi, S. Clima, R. Degraeve, Y. Y. Chen, G. Groeseneken, M. Jurczak

14.3 PULSED CYCLING OPERATION AND ENDURANCE FAILURE OF METAL-OXIDE RESISTIVE (RRAM) 359
S. Balatti, S. Ambrogio, Z.-Q. Wang, S. Sills, A. Calderoni, N. Ramaswamy, D. Ielmini

14.4 IMPACT OF LOW-FREQUENCY NOISE ON READ DISTRIBUTIONS OF RESISTIVE SWITCHING MEMORY (RRAM) 363
S. Ambrogio, S. Balatti, V. McCaffrey, D. Wang, D. Ielmini

14.5 A NEW APPROACH FOR TRAP ANALYSIS OF VERTICAL NAND FLASH CELL USING RTN CHARACTERISTICS 367
Daewoong Kang, Changsub Lee, Sunghoi Hur, Duheon Song, Jeong-Hyuk Choi

14.6 THROUGH SILICON VIA (TSV) EFFECTS ON DEVICES IN CLOSE PROXIMITY- THE ROLE OF MOBILE ION PENETRATION -CHARACTERIZATION AND MITIGATION 371
C. Kothandaraman, S. Cohen, C. Parks, J. Golz, K. Tunga, S. Rosenblatt, J. Safran, C. Collins, W. Landers, J. Oakley, J. Liu, A. J. Martin, K. Petrarca, M. Farooq, T. L. Graves-Abe, N. Robson, S. S. Iyer

14.7 HIGHLY BENEFICIAL ORGANIC LINER WITH EXTREMELY LOW THERMAL STRESS FOR FINE CU-TSV IN 3D-INTEGRATION 374
M. Murugesan, T. Fukushima, J. C. Bea, Y. Sato, H. Hashimoto, K. W. Lee, M. Koyanagi

15.1 AN ULTRA-SENSITIVE RESISTIVE PRESSURE SENSOR BASED ON THE V-SHAPED FOAM-LIKE STRUCTURE OF LASER-SCRIBED GRAPHENE 378
He Tian, Yi Shu, Xue-Feng Wang, Mohammad Ali Mohammad, Cheng Li, Yi Yang, Tian-Ling Ren

15.2 LARGE-SCALE FABRICATION OF GRAPHENE-BASED ELECTRONIC AND MEMS DEVICES 382
Debin Wang, He Tian, Iñigo Martin-Fernandez, Yi Yang, Tian-Ling Ren, Yuegang Zhang

15.3 FLEXIBLE, TRANSPARENT SINGLE-LAYER GRAPHENE EARPHONE 386
He Tian, Yi Yang, Cheng Li, Mohammad Ali Mohammad, Tian-Ling Ren

15.4 A SEMICONDUCTOR BIO-ELECTRICAL PLATFORM WITH ADDRESSABLE THERMAL CONTROL CIRCUITS FOR ACCELERATED BIOASSAY DEVELOPMENT 390
T.-T. Chen, C.-H. Wen, J.-C. Huang, Y.-C. Peng, S. Liu, S.-H. Su, L.-H. Cheng, H.-C. Lai, T.-C. Liao, F.-L. Lai, C.-W. Cheng, C.-K. Yang, J.-H. Yang, Y.-J. Hsieh, E. Salm, B. Reddy, F. Tsui, Y.-S. Liu, R. Bashir, M. Chen

15.5 LABEL-FREE OPTICAL BIOCHEMICAL SENSOR REALIZED BY A NOVEL LOW-COST BULK-SILICON BASED CMOS-COMPATIBLE 3-DIMENSIONAL OPTOELECTRONIC IC (OEIC) PLATFORM 394
Junfeng Song, Xianshu Luo, Jack Sheng Kee, Chao Li, Guo-Qiang Lo

15.6 AN INTEGRATED TUNABLE LASER USING NANO-SILICON-PHOTONIC CIRCUITS 398
M. Ren, H. Cai, J. F. Tao, Y. D. Gu, K. Radhakrishnan, Z. C. Yang, D. L. Kwong, A. Q. Liu

16.1 FIRST DEMONSTRATION OF HIGH-GE-CONTENT STRAINED-SI$_{1-x}$GE$_x$ (X=0.5) ON INSULATOR PMOS FINFETS WITH HIGH HOLE MOBILITY AND AGGRESSIVELY SCALED FIN DIMENSIONS AND GATE LENGTHS FOR HIGH-PERFORMANCE APPLICATIONS 402

Pouya Hashemi, Karthik Balakrishnan, Sebastian U. Engelmann, John A. Ott, Ali Khakifirooz, Ashish Baraskar, Marinus Hopstaken, Joseph S. Newbury, Kevin K. Chan, Effendi Leobandung, Renee T. Mo, Dae-Gyu Park

16.2 DUAL-CHANNEL CMOS CO-INTEGRATION WITH SI NFET AND STRAINED-SIGE PFET IN NANOWIRE DEVICE ARCHITECTURE FEATURING SUB-15NM GATE LENGTH 406

P. Nguyen, S. Barraud, C. Tabone, L. Gaben, M. Cassé, F. Glowacki, J.-M. Hartmann, M.-P. Samson, V. Maffini-Alvaro, C. Vizioz, N. Bernier, C. Guedj, C. Mounet, O. Rozeau, A. Toffoli, F. Alain, D. Delprat, B.-Y. Nguyen, C. Mazuré, O. Faynot, M. Vinet

16.3 V$_{TH}$ ADJUSTABLE SELF-ALIGNED EMBEDDED SOURCE/DRAIN SI/GE NANOWIRE FETS AND DOPANT-FREE NVMS FOR 3D SEQUENTIALLY INTEGRATED CIRCUIT 410

Chih-Chao Yang, Jia-Min Shieh, Tung-Ying Hsieh, Wen-Hsien Huang, Hsing-Hsiang Wang, Chang-Hong Shen, Tsung-Ta Wu, Chun-Yuan Chen, Kuei-Shu Chang-Liao, Jung-Hau Shiu, Meng-Chyi Wu, Fu-Liang Yang

16.4 ENHANCEMENT MODE STRAINED (1.3%) GERMANIUM QUANTUM WELL FINFET (W$_{FIN}$=20NM) WITH HIGH MOBILITY (μ_{HOLE}=700 CM²/VS), LOW EOT (~0.7NM) ON BULK SILICON SUBSTRATE 414

A. Agrawal, M. Barth, G. B. Rayner Jr., Arun V. T., C. Eichfeld, G. Lavallee, S.-Y. Yu, X. Sang, S. Brookes, Y. Zheng, Y.-J. Lee, Y.-R. Lin, C.-H. Wu, C.-H. Ko, J. LeBeau, R. Engel-Herbert, S. E. Mohney, Y.-C. Yeo, S. Datta

16.5 FIRST DEMONSTRATION OF 15NM-WFIN INVERSION-MODE RELAXED-GERMANIUM N-FINFETS WITH SI-CAP FREE RMG AND NISIGE SOURCE/DRAIN 418

J. Mitard, L. Witters, H. Arimura, Y. Sasaki, A. P. Milenin, R. Loo, A. Hikavyy, G. Eneman, P. Lagrain, H. Mertens, S. Sioncke, C. Vrancken, H. Bender, K. Barla, N. Horiguchi, A. Mocuta, N. Collaert, A. V.-Y. Thean

16.6 HIGH-PERFORMANCE TRI-GATE POLY-GE JUNCTION-LESS P- AND N-MOSFETS FABRICATED BY FLASH LAMP ANNEALING PROCESS 422

K. Usuda, Y. Kamata, Y. Kamimuta, T. Mori, M. Koike, T. Tezuka

16.7 DEEP SUB-100 NM GE CMOS DEVICES ON SI WITH THE RECESSED S/D AND CHANNEL 426

Heng Wu, Wei Luo, Mengwei Si, Jingyun Zhang, Hong Zhou, Peide D. Ye

17.1 THE DYNAMICS OF SURFACE DONOR TRAPS IN ALGAN/GAN MISFETS USING TRANSIENT MEASUREMENTS AND TCAD MODELLING 430

Giorgia Longobardi, Florin Udrea, Stephen Sque, Jeroen Croon, Fred Hurkx, Jan Sonský

17.2 THERMALLY INDUCED THRESHOLD VOLTAGE INSTABILITY OF III-NITRIDE MIS-HEMTS AND MOSC-HEMTS: UNDERLYING MECHANISMS AND OPTIMIZATION SCHEMES 434

Shu Yang, Shenghou Liu, Cheng Liu, Zhikai Tang, Yunyou Lu, Kevin J. Chen

17.3 IMPACTS OF FLUORINE-TREATMENT ON E-MODE ALGAN/GAN MOS-HEMTS 438

X. Sun, Y. Zhang, K. S. Chang-Liao, T. Palacios, T. P. Ma

17.4 HIGH-TEMPERATURE LOW-DAMAGE GATE RECESS TECHNIQUE AND OZONE-ASSISTED ALD-GROWN AL$_2$O$_3$ GATE DIELECTRIC FOR HIGH-PERFORMANCE NORMALLY-OFF GAN MIS-HEMTS ...442

S. Huang, Q. Jiang, K. Wei, G. Liu, J. Zhang, X. Wang, Y. Zheng, B. Sun, C. Zhao, H. Liu, Z. Jin, X. Liu, H. Wang, S. Liu, Y. Lu, C. Liu, S. Yang, Z. Tang, J. Zhang, Y. Hao, K. J. Chen

17.5 TRAPPING AND HIGH FIELD RELATED ISSUES IN GAN POWER HEMTS 446

Gaudenzio Meneghesso, Matteo Meneghini, Alessandro Chini, Giovanni Verzellesi, Enrico Zanoni

17.6 CMOS-COMPATIBLE GAN-ON-SI FIELD-EFFECT TRANSISTORS FOR HIGH VOLTAGE POWER APPLICATIONS 450

Man Ho Kwan, K.-Y. Wong, Y. S. Lin, F. W. Yao, M. W. Tsai, Y.-C. Chang, P. C. Chen, R. Y. Su, C.-H. Wu, J. L. Yu, F. J. Yang, G. P. Lansbergen, H.-Y. Wu, M.-C. Lin, C. B. Wu, Y.-A. Lai, C.-W. Hsiung, P.-C. Liu, H.-C. Chiu, C.-M. Chen, C. Y. Yu, H. S. Lin, M.-H. Chang, S.-P. Wang, L. C. Chen, J. L. Tsai, H. C. Tuan, A. Kalnitsky

18.1 DEVICE AWARE HIGH-SPEED TRANSCEIVER DESIGN IN PLANAR AND FINFET TECHNOLOGIES 454

Ken Chang, Jafar Savoj, Parag Upadhyaya, Yohan Frans

18.2 MISMATCH IN HIGH-K METAL GATE PROCESS ANALOG DESIGN 458

A. Woo, H. Eberhart, Y. Li, A. Ito

18.3 CHALLENGES OF ANALOG AND I/O SCALING IN 10NM SOC TECHNOLOGY AND BEYOND 462

A. Wei, J. Singh, G. Bouche, M. Zaleski, R. Augur, B. Senapati, J. Stephens, I. Lin, M. Rashed, L. Yuan, J. Kye, Y. Woo, J. Zeng, H. Levinson, A. Wehbi, P. Hang, V. Ton-That, V. Kanagala, D. Yu, D. Blackwell, A. Beece, S. Gao, S. Thangaraju, R. Alapati, S. Samavedam

18.4 TECHNOLOGY PATHFINDERS FOR LOW COST AND HIGHLY INTEGRATED RF FRONT END MODULES 466

C. Raynaud

18.5 DIGITALLY-INTENSIVE RF TRANSCEIVERS IN HIGHLY SCALED CMOS 470

Chih-Ming Hung

18.6 CIRCUIT AND DEVICE INTERACTIONS FOR 3D INTEGRATION USING INDUCTIVE COUPLING 474

Tadahiro Kuroda

19.1 CO/NI BASED P-MTJ STACK FOR SUB-20NM HIGH DENSITY STAND ALONE AND HIGH PERFORMANCE EMBEDDED MEMORY APPLICATION 478

G. S. Kar, W. Kim, T. Tahmasebi, J. Swerts, S. Mertens, N. Heylen, T. Min

19.2 CHALLENGING ISSUES FOR TERRA-BIT-LEVEL PERPENDICULAR STT-MRAM 482

J. G. Park, T. H. Shim, K. S. Chae, D. Y. Lee, Y. Takemura, S. E. Lee, M. S. Jeon, J. U. Baek, S. O. Park, J. P. Hong

19.3 AREA DEPENDENCE OF THERMAL STABILITY FACTOR IN PERPENDICULAR STT-MRAM ANALYZED BY BI-DIRECTIONAL DATA FLIPPING MODEL 486

K. Tsunoda, M. Aoki, H. Noshiro, Y. Iba, S. Fukuda, C. Yoshida, Y. Yamazaki, A. Takahashi, A. Hatada, M. Nakabayashi, Y. Tsuzaki, T. Sugii

19.4 0.026μM² HIGH PERFORMANCE EMBEDDED DRAM IN 22NM TECHNOLOGY FOR SERVER AND SOC APPLICATIONS 490

C. Pei, G. Wang, M. Aquilino, N. Arnold, B. Chandra, W. Chang, X. Chen, W. Davies, K. Hawkins, D. Jaeger, J. B. Johnson, O.-J. Kwon, R. Krishnasamy, W. Kong, J. Liu, X. Li, B. Messenger, E. Nelson, K. Nummy, K. Onishi, D. Poindexter, S. Rombawa, C. Sheraw, T. Tzou, X. Wang, M. Yin, G. Freeman, T. Kirahata, E. Maciejewski, J. Norum, N. Robson, S. Narasimha, P. Parries, P. Agnello, R. Malik, S. S. Iyer

19.5 A NEW SAW-LIKE SELF-RECOVERY OF INTERFACE STATES IN NITRIDE-BASED MEMORY CELL 494

Yuh-Te Sung, Po-Yen Lin, Jim Chen, Tzong-Sheng Chang, Ya-Chin King, Chrong Jung Lin

19.6 A NOVEL DOUBLE-TRAPPING BE-SONOS CHARGE-TRAPPING NAND FLASH DEVICE TO OVERCOME THE ERASE SATURATION WITHOUT USING CURVATURE-INDUCED FIELD ENHANCEMENT EFFECT OR HIGH-K (HK)/METAL GATE (MG) MATERIALS 498

Hang-Ting Lue, Roger Lo, Chih-Chang Hsieh, Pei-Ying Du, Chih-Ping Chen, Tzu-Hsuan Hsu, Kuo-Ping Chang, Yen-Hao Shih, Chih-Yuan Lu

20.1 LOW-FREQUENCY NOISE AND RTN ON NEAR-BALLISTIC III-V GAA NANOWIRE MOSFETS 502

N. Conrad, M. Si, S. H. Shin, J. J. Gu, J. Zhang, M. A. Alam, P. D. Ye

20.2 RTN AND PBTI-INDUCED TIME-DEPENDENT VARIABILITY OF REPLACEMENT METAL-GATE HIGH-K INGAAS FINFETS 506

J. Franco, B. Kaczer, N. Waldron, Ph. J. Roussel, A. Alian, M. A. Pourghaderi, Z. Ji, T. Grasser, T. Kauerauf, S. Sioncke, N. Collaert, A. Thean, G. Groeseneken

20.3 DIRECT OBSERVATION OF SELF-HEATING IN III-V GATE-ALL-AROUND NANOWIRE MOSFETS 510

S. H. Shin, M. Masuduzzaman, M. A. Wahab, K. Maize, J. J. Gu, M. Si, A. Shakouri, P. D. Ye, M. A. Alam

20.4 GATED AND STI DEFINED ESD DIODES IN ADVANCED BULK FINFET TECHNOLOGIES 514

S.-H. Chen, D. Linten, J.-W. Lee, M. Scholz, G. Hellings, A. Sibaja-Hernandez, R. Boschke, M.-H. Song, Y. See, Guido Groeseneken, A. Thean

20.5 STUDY OF THE PIEZORESISTIVE PROPERTIES OF NMOS AND PMOS Ω-GATE SOI NANOWIRE TRANSISTORS: SCALABILITY EFFECTS AND HIGH STRESS LEVEL 518

J. Pelloux-Prayer, M. Cassé, S. Barraud, P. Nguyen, M. Koyama, Y.-M. Niquet, F. Triozon, I. Duchemin, A. Abisset, A. Idrissi-Eloudrhiri, S. Martinie, J.-L. Rouvière, H. Iwai, G. Reimbold

20.6 RELIABILITY CHALLENGES FOR THE 10NM NODE AND BEYOND 522

James H. Stathis, M. Wang, R. G. Southwick, E. Y. Wu, B. P. Linder, E. G. Liniger, G. Bonilla, H. Kothari

20.7 WILL RELIABILITY LIMIT MOORE'S LAW? (INVITED) 526

Anthony S. Oates

21.1 ON THE MICROSCOPIC STRUCTURE OF HOLE TRAPS IN PMOSFETS 530

T. Grasser, W. Goes, Y. Wimmer, F. Schanovsky, G. Rzepa, M. Waltl, K. Rott, H. Reisinger, V. V. Afanas'ev, A. Stesmans, A.-M. El-Sayed, A. L. Shluger

21.2 ANALYTICAL FORMULATION OF SIO₂-IL SCAVENGING IN HFO₂/SIO₂/SI GATE STACKS: A KEY IS THE SIO₂/SI INTERFACE REACTION 534

Xiuyan Li, Takeaki Yajima, Tomonori Nishimura, Kosuke Nagashio, Akira Toriumi

21.3 FIRST PRINCIPLES STUDY OF SIC/SIO₂ INTERFACES TOWARDS FUTURE POWER DEVICES 538

K. Shiraishi, K. Chokawa, H. Shirakawa, K. Endo, M. Araidai, K. Kamiya, H. Watanabe

21.4 NEW FRAMEWORK FOR THE RANDOM CHARGING/DISCHARGING OF OXIDE TRAPS IN HFO₂ GATE DIELECTRIC: AB-INITIO SIMULATION AND EXPERIMENTAL EVIDENCE 542

Jingwei Ji, Yingxin Qiu, Shaofeng Guo, Runsheng Wang, Pengpeng Ren, Peng Hao, Ru Huang

21.5 MICROSCOPIC UNDERSTANDING OF THE LOW RESISTANCE STATE RETENTION IN HFO₂ AND HFALO BASED RRAM 546

B. Traoré, P. Blaise, E. Vianello, H. Grampeix, A. Bonnevialle, E. Jalaguier, G. Molas, S. Jeannot, L. Perniola, B. De Salvo, Y. Nishi

22.1 NANOSYSTEMS MONOLITHICALLY INTEGRATED WITH CMOS: EMERGING APPLICATIONS AND TECHNOLOGIES 550

J. Arcamone, J. Philippe, G. Arndt, C. Dupré, M. Savoye, S. Hentz, T. Ernst, E. Colinet, L. Duraffourg, E. Ollier

22.2 A SELF-SUSTAINED NANOMECHANICAL THERMAL-PIEZORESISTIVE OSCILLATOR WITH ULTRA-LOW POWER CONSUMPTION 554

Kuan-Hsien Li, Cheng-Chi Chen, Ming-Huang Li, Sheng-Shian Li

22.3 OPTIMIZING THE CLOSE-TO-CARRIER PHASE NOISE OF MONOLITHIC CMOS-MEMS OSCILLATORS USING BIAS-DEPENDENT NONLINEARITY 558

Ming-Huang Li, Chao-Yu Chen, Chi-Hang Chin, Cheng-Syun Li, Sheng-Shian Li

22.4 HIGH PERFORMANCE POLYSILICON NANOWIRE NEMS FOR CMOS EMBEDDED NANOSENSORS 562

I. Ouerghi, J. Philippe, L. Duraffourg, L. Laurent, A. Testini, K. Benedetto, A. M. Charvet, V. Delaye, L. Masarotto, P. Scheiblin, C. Reita, K. Yckache, C. Ladner, W. Ludurczak, T. Ernst

22.5 INTEGRATION OF RF MEMS RESONATORS AND PHONONIC CRYSTALS FOR HIGH FREQUENCY APPLICATIONS WITH FREQUENCY-SELECTIVE HEAT MANAGEMENT AND EFFICIENT POWER HANDLING 566

Humberto Campanella, Nan Wang, Margarita Narducci, Jeffrey Bo Woon Soon, Chong Pei Ho, Chengkuo Lee, Alex Gu

22.6 A MONOLITHIC 9 DEGREE OF FREEDOM (DOF) CAPACITIVE INERTIAL MEMS PLATFORM 570

Ilker E. Ocak, Daw D. Cheam, Sanchitha N. Fernando, Angel T .H. Lin, Pushpapraj Singh, Jaibir Sharma, Geng L. Chua, Bangtao Chen, Alex Y. D. Gu, Navab Singh, Dim-Lee Kwong

25.1 NOVEL INTRINSIC AND EXTRINSIC ENGINEERING FOR HIGH-PERFORMANCE HIGH-DENSITY SELF-ALIGNED INGAAS MOSFETS: PRECISE CHANNEL THICKNESS CONTROL AND SUB-40-NM METAL CONTACTS 574
Jianqiang Lin, Dimitri A. Antoniadis, Jesús A. del Alamo

25.2 HIGH-PERFORMANCE III-V DEVICES FOR FUTURE LOGIC APPLICATIONS 578
D.-H. Kim, T.-W. Kim, R. H. Baek, P. D. Kirsch, W. Maszara, J. A. del Alamo, D. A. Antoniadis, M. Urteaga, B. Brar, H. M. Kwon, C.-S. Shin, W.-K. Park, Y.-D. Cho, S. H. Shin, D. H. Ko, K.-S. Seo

25.3 HIGH-PERFORMANCE CMOS-COMPATIBLE SELF-ALIGNED IN$_{0.53}$GA$_{0.47}$AS MOSFETS WITH GMSAT OVER 2200 µS/µM AT V$_{DD}$ = 0.5 V 582
Y. Sun, A. Majumdar, C.-W. Cheng, R. M. Martin, R. L. Bruce, J.-B. Yau, D. B. Farmer, Y. Zhu, M. Hopstaken, M. M. Frank, T. Ando, K.-T. Lee, J. Rozen, A. Basu, K.-T. Shiu, P. Kerber, D.-G. Park, V. Narayanan, R. T. Mo, D. K. Sadana, E. Leobandung

25.4 LOW POWER III-V INGAAS MOSFETS FEATURING INP RECESSED SOURCE/DRAIN SPACERS WITH I$_{ON}$=120 µA/µM AT I$_{OFF}$=1 NA/µM AND V$_{DS}$=0.5 V 586
C. Y. Huang, S. Lee, V. Chobpattana, S. Stemmer, A. C. Gossard, B. Thibeault, W. Mitchell, M. Rodwell

25.5 INGAAS/INAS HETEROJUNCTION VERTICAL NANOWIRE TUNNEL FETS FABRICATED BY A TOP-DOWN APPROACH 590
Xin Zhao, Alon Vardi, Jesús A. del Alamo

25.6 IN$_{0.17}$AL$_{0.83}$N/ALN/GAN TRIPLE T-SHAPE FIN-HEMTS WITH G$_m$=646 MS/MM, I$_{ON}$=1030 MA/MM, I$_{OFF}$=1.13 µA/MM, SS=82 MV/DEC AND DIBL=28 MV/V AT V$_D$=0.5 V 594
S. Arulkumaran, G. I. Ng, C. M. Manojkumar, K. Ranjan, K. L. Teo, O. F. Shoron, S. Rajan, S. B. Dolmanan, S. Tripathy

26.1 ULTRALOW POWER TRANSPONDER IN THIN FILM CIRCUIT TECHNOLOGY ON FOIL WITH SUB - 1V OPERATION VOLTAGE 598
Tung-Huei Ke, Kris Myny, Adrian Chasin, Robert Müller, Paul Heremans, Soeren Steudel

26.2 THIN-FILM HETEROJUNCTION FIELD-EFFECT TRANSISTORS FOR ULTIMATE VOLTAGE SCALING AND LOW-TEMPERATURE LARGE-AREA FABRICATION OF ACTIVE-MATRIX BACKPLANES 602
Bahman Hekmatshoar, Ali Afzali-Ardakani

26.3 HIGH PERFORMANCE METAL OXIDE TFT AND ITS APPLICATIONS FOR THIN FILM ELECTRONICS (INVITED) 606
Gang Yu, Chan-Long Shieh, Juergen Musolf, Fatt Foong, Tian Xiao, Guangming Wang, Kristoffer Ottosson

26.4 INTEGRATION OF SOLUTION-PROCESSED (7,5) SWCNTS WITH SPUTTERED AND SPRAY-COATED METAL OXIDES FOR FLEXIBLE COMPLEMENTARY INVERTERS 610
L. Petti, F. Bottacchi, N. Münzenrieder, H. Faber, G. Cantarella, C. Vogt, L. Büthe, I. Namal, F. Späth, T. Hertel, T. D. Anthopoulos, G. Tröster

26.5 SOLUTION-PROCESSED POLY-SI TFTS FABRICATED AT A MAXIMUM TEMPERATURE OF 150 C 614
M. Trifunovic, J. Zhang, M. van der Zwan, R. Ishihara

26.6 HIGH PERFORMANCE ULTRA-THIN BODY (2.4NM) POLY-SI JUNCTIONLESS THIN FILM TRANSISTORS WITH A TRENCH STRUCTURE 618
Mu-Shih Yeh, Yung-Chun Wu, Min-Hsin Wu, Yi-Ruei Jhan, Ming-Hsien Chung, Min-Feng Hung

26.7 PERFORMANCE ENHANCEMENT OF A NOVEL P-TYPE JUNCTIONLESS TRANSISTOR USING A HYBRID POLY-SI FIN CHANNEL 622
Ya-Chi Cheng, Hung-Bin Chen, Chi-Shen Shao, Jun-Ji Su, Yung-Chun Wu, Chun-Yen Chang, Ting-Chang Chang

27.1 WAFER LEVEL SYSTEM INTEGRATION FOR SIP 626
Douglas C. H. Yu

27.2 HIGH-PRECISION WAFER-LEVEL CU-CU BONDING FOR 3DICS 630
Masashi Okada, Isao Sugaya, Hajime Mitsuishi, Hidehiro Maeda, Toshimasa Shimoda, Shigeto Izumi, Hosei Nakahira, Kazuya Okamoto

27.3 A MANUFACTURABLE INTERPOSER MIM DECOUPLING CAPACITOR WITH ROBUST THIN HIGH-K DIELECTRIC FOR HETEROGENEOUS 3D IC COWOS WAFER LEVEL SYSTEM INTEGRATION 634
W. S. Liao, C. H. Chang, S. W. Huang, T. H. Liu, H. P. Hu, H. L. Lin, C. Y. Tsai, C. S. Tsai, H. C. Chu, C. Y. Pai, W. C. Chiang, S. Y. Hou, S. P. Jeng, Doug Yu

27.4 MONOLITHIC 3D INTEGRATION OF LOGIC AND MEMORY: CARBON NANOTUBE FETS, RESISTIVE RAM, AND SILICON FETS 638
Max M. Shulaker, Tony F. Wu, Asish Pal, Liang Zhao, Yoshio Nishi, Krishna Saraswat, H.-S. Philip Wong, Subhasish Mitra

27.5 NEW INSIGHTS ON BOTTOM LAYER THERMAL STABILITY AND LASER ANNEALING PROMISES FOR HIGH PERFORMANCE 3D VLSI 642
C. Fenouillet-Beranger, B. Mathieu, B. Previtali, M.-P. Samson, N. Rambal, V. Benevent, S. Kerdiles, J.-P. Barnes, D. Barge, P. Besson, R. Kachtouli, M. Cassé, X. Garros, A. Laurent, F. Nemouchi, K. Huet, I. Toqué-Trésonne, D. Lafond, H. Dansas, F. Aussenac, G. Druais, P. Perreau, E. Richard, S. Chhun, E. Petitprez, N. Guillot, F. Deprat, L. Pasini, L. Brunet, V. Lu, C. Reita, P. Batude, M. Vinet

27.6 FLEXIBLE HIGH-PERFORMANCE NONVOLATILE MEMORY BY TRANSFERRING GAA SILICON NANOWIRE SONOS ONTO A PLASTIC SUBSTRATE 646
Ji-Min Choi, Jin-Woo Han, Yang-Kyu Choi

28.1 LOW POWER AND HIGH DENSITY STT-MRAM FOR EMBEDDED CACHE MEMORY USING ADVANCED PERPENDICULAR MTJ INTEGRATIONS AND ASYMMETRIC COMPENSATION TECHNIQUES 650
K. Ikegami, H. Noguchi, C. Kamata, M. Amano, K. Abe, K. Kushida, E. Kitagawa, T. Ochiai, N. Shimomura, S. Itai, D. Saida, C. Tanaka, A. Kawasumi, H. Hara, J. Ito, S. Fujita

28.2 CHALLENGE OF MOS/MTJ-HYBRID NONVOLATILE LOGIC-IN-MEMORY ARCHITECTURE IN DARK-SILICON ERA 654
Takahiro Hanyu, Daisuke Suzuki, Akira Mochizuki, Masanori Natsui, Naoya Onizawa, Tadahiko Sugibayashi, Shoji Ikeda, Tetsuo Endoh, Hideo Ohno

28.3 TECHNOLOGY AND CIRCUIT OPTIMIZATION OF RESISTIVE RAM FOR LOW-POWER, REPRODUCIBLE OPERATION 657
D. C. Sekar, B. Bateman, U. Raghuram, S. Bowyer, Y. Bai, M. Calarrudo, P. Swab, J. Wu, S. Nguyen, N. Mishra, R. Meyer, M. Kellam, B. Haukness, C. Chevallier, H. Wu, H. Qian, F. Kreupl, G. Bronner

28.4 VARIABILITY-TOLERANT CONVOLUTIONAL NEURAL NETWORK FOR PATTERN RECOGNITION APPLICATIONS BASED ON OXRAM SYNAPSES 661
D. Garbin, O. Bichler, E. Vianello, Q. Rafhay, C. Gamrat, L. Perniola, G. Ghibaudo, B. De Salvo

28.5 3D SYNAPTIC ARCHITECTURE WITH ULTRALOW SUB-10 FJ ENERGY PER SPIKE FOR NEUROMORPHIC COMPUTATION 665
I-Ting Wang, Yen-Chuan Lin, Yu-Fen Wang, Chung-Wei Hsu, Tuo-Hung Hou

28.6 HIGHLY DEPENDABLE 3-D STACKED MULTICORE PROCESSOR SYSTEM MODULE FABRICATED USING RECONFIGURED MULTICHIP-ON-WAFER 3-D INTEGRATION TECHNOLOGY 669
K.-W. Lee, H. Hashimoto, M. Onishi, S. Konno, Y. Sato, C. Nagai, J.-C. Bea, M. Murugesan, T. Fukushima, T. Tanaka, M. Koyanagi

28.7 PAIRWISE COUPLED HYBRID VANADIUM DIOXIDE-MOSFET (HVFET) OSCILLATORS FOR NON-BOOLEAN ASSOCIATIVE COMPUTING 673
N. Shukla, A. Parihar, M. Cotter, M. Barth, X. Li, N. Chandramoorthy, H. Paik, D. G. Schlom, V. Narayanan, A. Raychowdhury, S. Datta

28.8 HYBRID CMOS/BEOL-NEMS TECHNOLOGY FOR ULTRA-LOW-POWER IC APPLICATIONS 677
Nuo Xu, Jeff Sun, I-Ru Chen, Louis Hutin, Yenhao Chen, Jun Fujiki, Chuang Qian, Tsu-Jae King Liu

29.1 OPTIMIZATION METRICS FOR PHASE CHANGE MEMORY (PCM) CELL ARCHITECTURES 681
M. Boniardi, A. Redaelli, C. Cupeta, F. Pellizzer, L. Crespi, G. D. D'Arrigo, A. L. Lacaita, G. Servalli

29.2 55-μA GE$_X$TE$_{1-X}$/SB$_2$TE$_3$ SUPERLATTICE TOPOLOGICAL-SWITCHING RANDOM ACCESS MEMORY (TRAM) AND STUDY OF ATOMIC ARRANGEMENT IN GE-TE AND SB-TE STRUCTURES 685
N. Takaura, T. Ohyanagi, M. Tai, M. Kinoshita, K. Akita, T. Morikawa, H. Shirakawa, M. Araidai, K. Shiraishi, Y. Saito, J. Tominaga

29.3 PHASE CHANGE MEMORY AND ITS INTENDED APPLICATIONS 689
Chung H. Lam

29.4 CAPACITY OPTIMIZATION OF EMERGING MEMORY SYSTEMS: A SHANNON-INSPIRED APPROACH TO DEVICE CHARACTERIZATION 693
Jesse H. Engel, S. Burc Eryilmaz, SangBum Kim, Matthew BrightSky, Chung Lam, Hsiang-Lan Lung, Bruno A. Olshausen, H.-S. Philip Wong

29.5 EXPERIMENTAL DEMONSTRATION AND TOLERANCING OF A LARGE-SCALE NEURAL NETWORK (165,000 SYNAPSES), USING PHASE-CHANGE MEMORY AS THE SYNAPTIC WEIGHT ELEMENT 697
G. W. Burr, R. M. Shelby, C. di Nolfo, J. W. Jang, R. S. Shenoy, P. Narayanan, K. Virwani, E.U. Giacometti, B. Kurdi, H. Hwang

29.6 STATISTICS OF SET TRANSITION IN PHASE CHANGE MEMORY (PCM) ARRAYS 701
M. Rizzi, N. Ciocchini, S. Caravati, M. Bernasconi, P. Fantini, D. Ielmini

29.7 CIRCUIT-LEVEL BENCHMARKING OF ACCESS DEVICES FOR RESISTIVE NONVOLATILE MEMORY ARRAYS 705
P. Narayanan, G. W. Burr, R. S. Shenoy, K. Virwani, B. Kurdi

29.8 A NOVEL INSPECTION AND ANNEALING PROCEDURE TO REJUVENATE PHASE CHANGE MEMORY FROM CYCLING-INDUCED DEGRADATIONS FOR STORAGE CLASS MEMORY APPLICATIONS 709
W. S. Khwa, J. Y. Wu, T. H. Su, H. P. Li, M. BrightSky, T. Y. Wang, T. H. Hsu, P. Y. Du, S. Kim, W. C. Chien, H. Y. Cheng, R. Cheek, E. K. Lai, Y. Zhu, M. H. Lee, M. F. Chang, H. L. Lung, C. Lam

30.1 SOURCE-INDUCED RDF OVERWHELMS RTN IN NANOWIRE TRANSISTOR: STATISTICAL ANALYSIS WITH FULL DEVICE EMC/MD SIMULATION ACCELERATED BY GPU COMPUTING 713
Akito Suzuki, Takefumi Kamioka, Yoshinari Kamakura, Kenji Ohmori, Keisaku Yamada, Takanobu Watanabe

30.2 PERSPECTIVE OF TUNNEL-FET FOR FUTURE LOW-POWER TECHNOLOGY NODES 717
A. S. Verhulst, D. Verreck, Q. Smets, K.-H. Kao, M. Van de Put, R. Rooyackers, B. Sorée, A. Vandooren, K. De Meyer, G. Groeseneken, M. M. Heyns, A. Mocuta, N. Collaert, A. V.-Y. Thean

30.3 PERFORMANCE EVALUATION OF MOS$_2$-WTE$_2$ VERTICAL TUNNELING TRANSISTOR USING REAL-SPACE QUANTUM SIMULATOR 721
Kai-Tak Lam, Gyungseon Seol, Jing Guo

30.4 AB-INITIO SIMULATIONS OF MOS$_2$ TRANSISTORS: FROM MOBILITY CALCULATION TO DEVICE PERFORMANCE EVALUATION 725
Aron Szabo, Reto Rhyner, Mathieu Luisier

30.5 PERFORMANCE EVALUATION AND DESIGN CONSIDERATIONS OF 2D SEMICONDUCTOR BASED FETS FOR SUB-10 NM VLSI 729
Wei Cao, Jiahao Kang, Deblina Sarkar, Wei Liu, Kaustav Banerjee

30.6 ATOMIC DISORDER SCATTERING IN EMERGING TRANSISTORS BY PARAMETER-FREE FIRST PRINCIPLE MODELING 733
Qing Shi, Lining Zhang, Yu Zhu, Lei Liu, Mansun Chan, Hong Guo

31.1 BIO-MEMS TOWARDS SINGLE-MOLECULAR CHARACTERIZATION 737
Hiroyuki Fujita

31.2 AN AC AND PHASE NANOWIRE SENSING FOR SITE-BINDING DETECTION 741
Marco Tartagni, Marco Crescentini, Michele Rossi, Hywel Morgan, Enrico Sangiorgi

31.3 MEMS FOR CELL MECHANOBIOLOGY 745
Beth L. Pruitt

31.4 ORGANIC ELECTROCHEMICAL TRANSISTORS FOR BIOMEMS APPLICATIONS 749
Dimitrios A. Koutsouras, Pierre Leleux, Marc Ramuz, Jonathan Rivnay, George G. Malliaras

**31.5 MULTIFUNCTIONAL SMART LAB-ON-A-TUBE (LOT) PROBE FOR MONITORING TRAUMATIC
BRAIN INJURY (TBI)** 753
Chunyan Li, Pei-Ming Wu, Zhizhen Wu, Nirjhar Bhattacharjee, Jed A. Hartings, Raj K. Narayan, Chong H. Ahn

31.6 SMALL SOFT SAFE MICROMACHINES FOR BIOMEDICAL APPLICATIONS 757
Satoshi Konishi

**31.7 BIO-INTEGRATED SYSTEMS WITH STRETCHABLE DESIGNS FOR SKIN-MOUNTED WEARABLE
HEALTH MONITORING** 761
Milan Raj, Pinghung Wei, Shyamal Patel, Xianyan Wang, Bryan McGrane, Lauren Klinker, Paolo DePetrillo, Roozbeh Ghaffari

**32.1 ELECTRICAL CHARACTERIZATION OF FINFETS WITH FINS FORMED BY DIRECTED SELF
ASSEMBLY AT 29 NM FIN PITCH USING A SELF-ALIGNED FIN CUSTOMIZATION SCHEME** 764
H. Tsai, H. Miyazoe, J. Chang, J. Pitera, C.-C. Liu, M. Brink, I. Lauer, J. Y. Cheng, S. Engelmann, J. Rozen, J. J. Bucchignano, D. P. Klaus, S. Dawes, L. Gignac, C. Breslin, E. A. Joseph, D. P. Sanders, M. E. Colburn, M. Guillorn

**32.2 HIGHLY RELIABLE CU INTERCONNECT STRATEGY FOR 10NM NODE LOGIC TECHNOLOGY
AND BEYOND** 768
R.-H. Kim, B. H. Kim, T. Matsuda, J. N. Kim, J. M. Baek, J. J. Lee, J. O. Cha, J. H. Hwang, S. Y. Yoo, K.-M. Chung, K. H. Park, J. K. Choi, E. B. Lee, S. D. Nam, Y. W. Cho, H. J. Choi, J. S. Kim, S. Y. Jung, D. H. Lee, I. S. Kim, D. W. Park, H. B. Lee, S. H. Ahn, S. H. Park, M.-C. Kim, B. U. Yoon, S. S. Paak, N.-I. Lee, J.-H. Ku, J. S. Yoon, H.-K. Kang, E. S. Jung

**32.3 A NEW HIGH-K/METAL GATE CMOS INTEGRATION SCHEME (DIFFUSION AND GATE
REPLACEMENT) SUPPRESSING GATE HEIGHT ASYMMETRY AND COMPATIBLE WITH HIGH-
THERMAL BUDGET MEMORY TECHNOLOGIES** 772
R. Ritzenthaler, T. Schram, A. Spessot, C. Caillat, M. Cho, E. Simoen, M. Aoulaiche, J. Albert, S. A. Chew, K. B. Noh, Y. Son, P. Fazan, N. Horiguchi, A. Thean

**32.4 ULTRA LOW CONTACT RESISTIVITY ($< 1 \times 10^{-8}$ Ω-CM^2) TO IN$_{0.53}$GA$_{0.47}$AS FIN SIDEWALL
(110)/(100) SURFACES: REALIZED WITH A VLSI PROCESSED III-V FIN TLM STRUCTURE
FABRICATED WITH III-V ON SI SUBSTRATES** 776
Rinus T. P. Lee, Y. Ohsawa, C. Huffman, Y. Trickett, G. Nakamura, C. Hatem, K. V. Rao, F. Khaja, R. Lin, K. Matthews, K. Dunn, A. Jensen, T. Karpowicz, Peter F. Nielsen, E. Stinzianni, A. Cordes, P. Y. Hung, D.-H. Kim, R. J. W. Hill, W.-Y. Loh, C. Hobbs

**32.5 DRAMATIC EFFECTS OF HYDROGEN-INDUCED OUT-DIFFUSION OF OXYGEN FROM GE
SURFACE ON JUNCTION LEAKAGE AS WELL AS ELECTRON MOBILITY IN N-CHANNEL GE
MOSFETS** 780
Choong Hyun Lee, Tomonori Nishimura, Cimang Lu, Shoichi Kabuyanagi, Akira Toriumi

32.6 EVOLUTION OF DIRECTED ION BEAMS FROM DOPING TO MATERIALS ENGINEERING 784
Anthony Renau

**32.7 A NOVEL JUNCTIONLESS FINFET STRUCTURE WITH SUB-5NM SHELL DOPING PROFILE BY
MOLECULAR MONOLAYER DOPING AND MICROWAVE ANNEALING** 788
Y.-J. Lee, T.-C. Cho, K.-H. Kao, P.-J. Sung, F.-K. Hsueh, P.-C. Huang, C.-T. Wu, S.-H. Hsu, W.-H. Huang, H.-C. Chen, Y. Li, M. I. Current, B. Hengstebeck, J. Marino, T. Büyüklimanli, J.-M. Shieh, T.-S. Chao, W.-F. Wu, W.-K. Yeh

**33.1 EXPERIMENTAL DEMONSTRATION OF FOUR-TERMINAL MAGNETIC LOGIC DEVICE WITH
SEPARATE READ- AND WRITE-PATHS** 792
D. M. Bromberg, M. T. Moneck, V. M. Sokalski, J. Zhu, L. Pileggi, J.-G. Zhu

**33.2 PERPENDICULAR-ANISOTROPY COFEB-MGO BASED MAGNETIC TUNNEL JUNCTIONS
SCALING DOWN TO 1X NM** 796
S. Ikeda, H. Sato, H. Honjo, E. C. I. Enobio, S. Ishikawa, M. Yamanouchi, S. Fukami, S. Kanai, F. Matsukura, T. Endoh, H. Ohno

**33.3 HIGH PERFORMANCE, EXCELLENT RELIABILITY MULTIFUNCTIONAL GRAPHENE OXIDE
DOPED MEMRISTOR ACHIEVED BY SELF-PROTECTIVE COMPLIANCE CURRENT STRUCTURE** 800
Kuan-Chang Chang, Rui Zhang, Ting-Chang Chang, Tsung-Ming Tsai, Tian-Jian Chu, Hsin-Lu Chen, Chih-Cheng Shih, Chih-Hung Pan, Yu-Ting Su, Pei-Jung Wu, Simon M. Sze

**33.4 REALIZING A TOPOLOGICAL-INSULATOR FIELD-EFFECT TRANSISTOR USING
IODOSTANNANANE** 804
William G. Vandenberghe, Massimo V. Fischetti

**33.5 HYBRID SI/TMD 2D ELECTRONIC DOUBLE CHANNELS FABRICATED USING SOLID CVD FEW-
LAYER-MOS$_2$ STACKING FOR V$_{TH}$ MATCHING AND CMOS-COMPATIBLE 3DFETS** 808
M.-C. Chen, C.-Y. Lin, K.-H. Li, L.-J. Li, C.-H. Chen, C.-H. Chung, M-D. Lee, Y.-J. Chen, Y.-F. Hou, C.-H. Lin, C.-C. Chen, B.-W. Wu, C.-S. Wu, I. Yang, Y.-J. Lee, W.-K. Yeh, T. Wang, F.-L. Yang, C. Hu

33.6 HIGH-PERFORMANCE CARBON NANOTUBE FIELD-EFFECT TRANSISTORS 812
Max M. Shulaker, Gregory Pitner, Gage Hills, Marta Giachino, H.-S. Philip Wong, Subhasish Mitra

**34.1 NEW INSIGHTS INTO THE DESIGN FOR END-OF-LIFE VARIABILITY OF NBTI IN SCALED HIGH-
K/METAL-GATE TECHNOLOGY FOR THE NANO-RELIABILITY ERA** 816
Pengpeng Ren, Runsheng Wang, Zhigang Ji, Peng Hao, Xiaobo Jiang, Shaofeng Guo, Mulong Luo, Meng Duan, Jian F. Zhang, Jianping Wang, Jinhua Liu, Weihai Bu, Jingang Wu, Waisum Wong, Shaofeng Yu, Hanming Wu, Shiuh-Wuu Lee, Nuo Xu, Ru Huang

34.2 NBTI OF GE PMOSFETS: UNDERSTANDING DEFECTS AND ENABLING LIFETIME PREDICTION 820
J. Ma, W. Zhang, J. F. Zhang, B. Benbakhti, Z. Ji, J. Mitard, J. Franco, B. Kaczer, G. Groeseneken

34.3 ACCURATE PREDICTION OF PBTI LIFETIME FOR N-TYPE FIN-CHANNEL TUNNEL FETS 824

W. Mizubayashi, T. Mori, K. Fukuda, Y. X. Liu, T. Matsukawa, Y. Ishikawa, K. Endo, S. O'uchi, J. Tsukada, H. Yamauchi, Y. Morita, S. Migita, H. Ota, M. Masahara

34.4 BTI RELIABILITY OF ADVANCED GATE STACKS FOR BEYOND-SILICON DEVICES: CHALLENGES AND OPPORTUNITIES 828

G. Groeseneken, J. Franco, M. Cho, B. Kaczer, M. Toledano-Luque, Ph. Roussel, T. Kauerauf, A. Alian, J. Mitard, H. Arimura, D. Lin, N. Waldron, S. Sioncke, L. Witters, H. Mertens, L.-Å. Ragnarsson, M. Heyns, N. Collaert, A. Thean, A. Steegen

34.5 NEW UNDERSTANDING OF STATE-LOSS IN COMPLEX RTN: STATISTICAL EXPERIMENTAL STUDY, TRAP INTERACTION MODELS, AND IMPACT ON CIRCUITS 832

Jibin Zou, Runsheng Wang, Shaofeng Guo, Mulong Luo, Zhuoqing Yu, Xiaobo Jiang, Pengpeng Ren, Jianping Wang, Jinhua Liu, Jingang Wu, Waisum Wong, Shaofeng Yu, Hanming Wu, Shiuh-Wuu Lee, Yangyuan Wang, Ru Huang

34.6 NEW OBSERVATIONS ON HOT CARRIER INDUCED DYNAMIC VARIATION IN NANO-SCALED SION/POLY, HK/MG AND FINFET DEVICES BASED ON ON-THE-FLY HCI TECHNIQUE: THE ROLE OF SINGLE TRAP INDUCED DEGRADATION 836

Changze Liu, Kyong Taek Lee, Sangwoo Pae, Jongwoo Park

34.7 MULTIPLE BREAKDOWN PHENOMENA AND MODELING FOR NON-UNIFORM DIELECTRIC SYSTEMS 840

Ernest Wu, Baozhen Li, James H. Stathis, Ravi Achanta

35.1 A PHYSICS-BASED COMPACT MODEL FOR FETS FROM DIFFUSIVE TO BALLISTIC CARRIER TRANSPORT REGIMES 844

Shaloo Rakheja, Mark Lundsrom, Dimitri Antoniadis

35.2 A PHYSICS-BASED RTN VARIABILITY MODEL FOR MOSFETS 848

Maurício Banaszeski da Silva, Hans Tuinhout, Adrie Zegers-van Duijnhoven, Gilson I. Wirth, Andries Scholten

35.3 EXPERIMENT AND MODEL FOR DEVIATION FROM PELGROM SCALING RELATION IN DEVICE WIDTH 852

Ning Lu, Jeffrey S. Brown, Rainer Thoma, Pooja M. Kotecha, Richard A. Wachnik

35.4 A COMPREHENSIVE PLATFORM FOR THERMAL STUDIES IN TSV-BASED 3D INTEGRATED CIRCUITS 856

P. M. Souaré, P. Coudrain, J. P. Colonna, V. Fiori, A. Farcy, F. de Crécy, A. Borbely, H. Ben-Jamaa, C. Laviron, S. Gallois-Garreignot, B. Giraud, N. Hotellier, R. Franiatte, S. Dumas, C. Chancel, J.-M. Rivière, J. Pruvost, S. Chéramy, C. Tavernier, J. Michailos, L. Le-Pailleur

35.5 A NEW SURFACE POTENTIAL-BASED COMPACT MODEL FOR A-IGZO TFTS IN RFID APPLICATIONS 860

Zhiwei Zong, Ling Li, Jin Jang, Zhigang Li, Nianduan Lu, Liwei Shang, Zhuoyu Ji, Ming Liu

35.6 PHYSICS-BASED FACTORIZATION OF MAGNETIC TUNNEL JUNCTIONS FOR MODELING AND CIRCUIT SIMULATION 864

Kerem Yunus Camsari, Samiran Ganguly, Deepanjan Datta, Supriyo Datta

Author Index

2014 International Electron Devices Meeting
TECHNICAL DIGEST

Papers have been printed without editing as received from the authors.

All opinions expressed in the digest are those of the authors and are not binding on The Institute of Electrical and Electronics Engineers, Inc.

Publication of a paper in this Digest is in no way intended to preclude publication of a fuller account of the paper elsewhere.

WELCOME FROM THE GENERAL CHAIR

On behalf of the entire IEDM committee, I would like to welcome you to the 2014 IEEE International Electron Devices Meeting to be held December 15-17, 2014 in San Francisco. The 60th annual IEDM continues to be the world's premier venue for presenting the latest breakthroughs in electronic device technologies. This year we have a strong collection of both contributed and invited papers that will be presented by industrial, academic leaders and students from around the world. Short summaries of all the papers are available on the IEDM web site, which we encourage everyone to visit – http://www.ieee-iedm.org/. We will continue to distribute an abbreviated digest at the meeting, along with electronic versions of the complete abstracts. This is the 3rd year that we are providing an official IEDM smartphone and tablet application that supports iPhone, iPad and Android platforms to help our attendees navigate the conference. The full digest will be available on the IEEE Xplore website and the DVD package offered by the IEEE Electron Devices Society after the conference.

Howard C.-H. Wang
General Chair

In addition to the many regular paper sessions - and the always informative and entertaining IEDM Luncheon on Tuesday –we are continuing the Entrepreneurs Lunch on Wednesday for a 3rd year.

For the Tuesday luncheon, we are fortunate to have Dr. T. J. Rodgers, founder and CEO of Cypress Semiconductor, as our speaker. For our Wednesday Entrepreneurs Lunch Series at IEDM, we are very excited to have Kathryn Kranen, who was the president and CEO of Jasper Design Automation, a profitable and fast-growing private company delivering formal verification solutions to semiconductor and systems companies. She will share her thoughts and experience in developing Jasper into a growing and profitable company, and how to build and lead a global team with diverse talents.

On Saturday afternoon, December 13, 2014, we will continue to offer our highly successful short tutorials that were introduced in 2011. These tutorials are directed to students and new hires, or anyone who wants to learn the basics of key electron device technologies. Three tracks run in parallel, for a total of six tutorial topics.

On Sunday December 14, 2014, two short courses will be offered: "Challenges of 7nm CMOS Technology", and "3D System Integration Technology". These courses have been organized and will be presented by internationally recognized researchers active in their respective areas of technology. The topics and instructors have been carefully chosen to have broad appeal to IEDM participants, and will include material suitable for both newcomers as well as experts in the field.

John Suehle
Technical
Program Chair

The Plenary Session on Monday morning will feature three invited talks: John Palmour from Cree Inc. will give a talk titled "SiC MOSFET Development for Industrial Markets," followed by Enrico Prati of Consiglio Nazionale delle Ricerche who will present a talk titled "Are 3D Atomic Printers Around the Corner?" The final talk titled "Research into ADAS with Driving Intelligence for Future Innovation" will be provided by Hideo Inoue from Toyota Motor Corporation, and will focus on the role electronics are poised to play in advanced driver assistance.

Given the rapid evolution of electron device technologies, IEDM has increased its focus on emerging areas by introducing multiple special "Focus Sessions" since 2013. This year IEDM will feature four Focus Sessions, with talks from leading experts in exciting new areas. The first focus session is hosted by the PC subcommittee on Monday afternoon and is titled: "Power Devices". The 2nd focus session is hosted by the DIS subcommittee on Tuesday morning titled: "Novel Imagers and Specialty Imaging Applications". The 3rd focus session is hosted by the CDI subcommittee on Tuesday afternoon and is titled: " Analog and Mixed Signal Circuit Device Interaction". The 4th focus session is hosted by the SMB subcommittee on Wednesday morning and is titled: "Devices for Biological Science, Medical Applications and Health Monitoring".

On Tuesday night, we will feature an interactive Panel session that promises to be engaging and popular. This year, the panel is titled: "60 Years of IEDM and Counting" and has been organized by Prof. Krishna Saraswat from Stanford University. The panelists will discuss the major technological innovations that are well documented in the past 60 years of IEDM, and will attempt to look into their crystal ball for predictions of what the future holds for electronics devices in the next 60 years.

On behalf of the IEDM, John Suehle, the Technical Program Chair, and Patrick Fay, the Technical Program Vice-Chair, I want to express my sincere appreciation to all of the IEDM authors and to each of the members of the IEDM committee. The IEDM committee members did an outstanding job in planning and organizing the 2014 conference. The authors really make the IEDM what it is; a forum for the presentation of the leading work in our field.

Patrick Fay
Technical Program
Vice Chair

The IEDM is sponsored by the IEEE Electron Devices Society. If you are not already an IEEE member, please consider joining this great institution which has played such an important role globally for over 120 years. More detailed information regarding the IEEE is available at the conference and on their website – http://www.ieee.org

It is again my great honor and pleasure to extend a warm welcome to everyone attending the 2014 IEEE International Electron Devices Meeting, and helping to celebrate our 60th year.

Howard C.-H. Wang
General Chair

AWARD PRESENTATIONS

PLENARY SESSION AWARD

Monday, December 15

2013 Roger A. Haken Best Student Paper Award

To: Umberto Celano, imec

For the paper entitled: "Conductive-AFM Tomography for 3D Filament Observation in Resistive Switching Devices"

EDS 2013 Paul Rappaport Award

To: Rui Zhang, Po-Chin Huang, Ju-Chin Lin, Noriyuki Taoka, Mitsuru Takenaka, and Shinichi Takagias

For the paper entitled: "High-Mobility Ge p- and n- MOSFETs with 0.7-nm EOT Using HfO2/Al2O3/GeOx/Ge Gate Stacks Fabricated by Plasma Postoxidation"

EDS 2013 George E. Smith Award

To: Takashi Ando, Eduard A. Cartier, John Bruley, Kisik Choi, Vijay Narayanan

For the paper entitled: "Origins of Effective Work Function Roll-Off Behavior for High-k Last Replacement Metal Gate Stacks"

and

Yong-Young Noh, Kang-Jun Baeg, Dongyoon Khim, Juhwan Kim, Dong-Yu Kim, Si-Woo Sung, Byung-Do Yang

For the paper entitled: "Flexible Complementary Logic Gates Using Inkjet-Printed Polymer Field-Effect Transistors"

2014 EDS Distinguished Service Award

To: Yuan Taur

"To recognize and honor outstanding service to the Electron Devices Society"

2014 EDS Education Award

To: Juin J. Liou

"For promoting and inspiring global education and learning in the field of electron devices"

2014 EDS J.J. Ebers Award

To: Joachim N. Burghartz, Institute for Microelectronics (IMS CHIPS)

"For contributions to integrated spiral inductors for wireless communication ICs and ultra-thin silicon devices for emerging flexible electronics."

2014 IEEE/EDS FELLOWS

This is a complete listing of the 2014 IEEE/EDS Fellows. Not all Fellows will be recognized at the 2014 IEDM.

Seiichi Aritome, SK Hynix Inc., Icheon-si, South Korea

Phaedon Avouris, IBM Research Corporation, Yorktown Heights, NY, USA

Richard Brown, University of Utah, Salt Lake City, UT, USA

Babu Chalamala, MEMC Electronic Materials, Inc., St. Peters, MO, USA

Edward Chang, National Chiao Tung University, Hsinchu, Taiwan

Jing Kevin Chen, Hong Kong University of Science and Technology, Kowloon, Hong Kong

Donald Disney, Avogy, Inc., Cupertino, CA, USA

Ichiro Fujimori, Broadcom Corporation, Irvine, CA, USA

Bruce Gurney, HGST, a subsidiary of Western Digital, San Jose, CA, USA

Kazunari Ishimaru, Memory Division, Toshiba Corporation, Yokohama, Japan

Byoungho Lee, Seoul National University, Seoul, South Korea

Kwyro Lee, Korea Advanced Institute of Science and Tech., Daejeon, South Korea

Zachary Lemnios, IBM Corporation, Yorktown Heights, NY, USA

Philip Mok, Hong Kong Univ. of Science & Technology, Hong Kong, China

Taiichi Otsuji, Tohoku University, Sendai, Japan

Daniel Radack, Institute of Defense Analyses, Kensington, MD, USA

Jean-Pierre Raskin, Universit Catholique de Louvain (UCL), Louvain-la-Neuve, Belgium

William Redman-White, University of Southampton, Hampshire, UK

Robert Reed, Vanderbilt University, Nashville, TN, USA

Mircea Stan, University of Virginia, Charlottesville, VA,USA

Jacobus Swart, State University of Campinas – UNICAMP, Campinas, Brazil

Jan Van Houdt, IMEC, Leuven, Belgium

Ya-Hong Xie, University of California, Los Angeles, CA, USA

Rui Yang, University of Oklahoma, Norman, OK, USA

IEDM LUNCHEON

Tuesday, December 16

2014 IEEE Andrew S. Grove Award

To: Sanjay Banerjee, University of Texas at Austin

"For contributions to column-IV MOSFETs and related materials processing."

LUNCHEON PRESENTATION

Dr. T.J. Rodgers, Founder, President and CEO, Cypress Semiconductor

Dr. T.J. Rodgers (TJR) is founder, president, and CEO of Cypress Semiconductor Corporation since 1982. TJR is a former chairman of the Semiconductor Industry Association (SIA) and SunPower Corporation (solar energy systems) and currently sits on the board of directors of high-technology companies, including, Bloom Energy (fuel cells), Deca Technologies (advanced packaging), AgigA Tech (nvRAMs) and Cypress Envirosystems (energy-saving systems). TJR holds 14 patents and has received numerous awards and recognitions. TJR has been inducted to the Silicon Valley Engineering Council Hall of Fame joining Silicon Valley icons such as Robert Noyce, Gordon Moore and Steve Wozniak. He was a member of the board of trustees of Dartmouth College, his alma mater. TJR has testified before Congress five times.

TJR was a Sloan scholar at Dartmouth, where he graduated in 1970 as Salutatorian with a double major in physics and chemistry and won the Townsend Prize as the top physics and chemistry student in his class. He attended Stanford University on a Hertz fellowship, earning a master's degree (1973) and a Ph.D. (1975) in electrical engineering. At Stanford, TJR invented, developed, and patented VMOS technology, which he sold for cash and royalties to American Microsystems Inc. (AMI). He managed the MOS memory design group at AMI from 1975 to 1980 before moving to Advanced Micro Devices (AMD), where he ran AMD's static RAM product group until 1982.

TJR founded Cypress back in 1982 focusing on making SRAM memory on CMOS technology that its scaling is governed by the Moore's Law based on Dennard's scaling rules. At that time computers were made with chips based on bipolar transistors and were water cooled to manage the thermals. Cypress attacked the bipolar products and made chips just as fast with five times less power in CMOS. As the semiconductor industry progressed based on Moore's Law, transistors got smaller, cheaper, and faster while they consumed less power. Consequently more transistors could be integrated on a chip. Today's chips can be viewed as systems. A key part of adding value to the semiconductor industry is to architect these systems to solve valuable problems. As an example, Cypress' programmable system on chip (PSoC) integrates processing cores with various kinds of memories, programmable logic, programmable analog and programmable bus such that it can turn into any system one wants. These programmable solutions then can interact with the analog real world around us while being linked with the internet. Two decades ago, this could only be achieved with 40 or so pieces of silicon on a printed circuit board.

These achievements are the consequence of the progress we have made in the semiconductor industry. Moving forward, understanding of systems and efficiently architecting these systems are critical. Increasing cost of process technology and challenges with Moore's Law are driving new integration schemes to create differentiated value for the systems required to efficiently move data on the internet and secure, store and retrieve data.

TJR has shared some of his views in various industry forums. Earlier in 2014, TJR discussed several key topics regarding the semiconductor industry and the future of disruptive innovation at a panel of industry CEOs and VCs. In fact, TJR has perpetuated a spirit of entrepreneurship by launching and incubating a series of autonomous businesses that have relied on the parent company for funding in much the same way as startup companies rely on venture capital. This "federation of autonomous subsidiaries" has delivered multiple successes including Cypress Microsystems, which lead to developing Cypress's PSoC and SunPower which was spun out of Cypress in 2008 for $2.6 billion.

Dr. Rodgers' rich illustrious career in our industry and his entrepreneurship spirit in the Silicon Valley provide an opportunity for the IEDM attendees to hear his perspectives at the 60th IEDM anniversary luncheon.

CONFERENCE HIGHLIGHTS

Date	Time	Room	Event
12/13	2:45 p.m. – 6:30 p.m.	Imperial Ballroom A/B/Plaza A	Tutorials
12/14	9:00 a.m. – 5:30 p.m.	Continental Ballroom 1-4/6-9	Short Courses
12/15	9:00 a.m. – 12:00 p.m.	Grand Ballroom B	Plenary Session
12/15	6:30 p.m. – 8:00 p.m.	Grand Ballroom B	Reception
12/16	12:20 p.m. – 2:00 p.m.	Grand Ballroom B	Tuesday Luncheon
12/16	8:00 p.m. – 10:00 p.m.	Continental Ballroom 4-6	Panel Session
12/17	12:30 p.m. – 1:30 p.m.	Plaza A/B	Wednesday Luncheon

Entrepreneurs Lunch at IEDM 2014!

Sponsored by IEDM and EDS Women in Engineering

Speaker: Kathryn Kranen, Former President and CEO, Jasper Design Automation

Wednesday, December 17
12:30 pm - 1:30 pm.

Kathryn Kranen is an experienced technology CEO, based in Silicon Valley, and is one of the most influential persons in the Electronic Design Automation industry's functional verification domain. Her specialties are pioneering new markets, monetizing software value, and building and leading excellent management teams leveraging diverse global talent.

Kathryn was president and CEO of Jasper Design Automation, a profitable and fast-growing private company delivering formal verification solutions to semiconductor and systems companies. Jasper won the Red Herring Top 100 award for private companies in 2013. Kathryn sold Jasper to Cadence Design Systems for $170MM in 2014.

Kathryn will share her thoughts and experience in turning around Jasper to a growing and profitable company and how to build and lead a global team with diverse talents.

Thuy Dao, of Freescale Semiconductor and the EDS Women in Engineering Committee Chair, will be the event moderator.

COMMITTEES

IEDM EXECUTIVE COMMITTEE

Howard C.-H. Wang
General Chair
TSMC
Hsinchu, Taiwan

John Suehle
Technical Program Chair
NIST
Gaithersburg, MD

Patrick Fay
Technical Program Vice Chair
University of Notre Dame
Notre Dame, IN

Ken Rim
Publications Chair
Qualcomm
San Diego, CA

Suman Datta
Publicity Chair
Penn State University
University Park, PA

Mariko Takayanagi
Publicity Vice Chair
Toshiba Corporation
Tokyo, Japan

Martin Giles
Short Course Chair
Intel Corporation
Hillsboto, OR

Satoru Yamada
Short Course Vice Chair
Samsung Electronics Co., Ltd.
Gyeonggi-Do, Korea

Stefan De Gendt
Tutorial Chair
IMEC
Leuven, Belgium

Mina Rais-Zadeh
Focus Session Chair
University of Michigan
Ann Harbor, MI

Michael Wu
Asian Arrangements Chair
TSMC
Hsinchu, Taiwan

Masakazu Kanechika
Asian Arrangements Co-Chair
Toyota
Aichi, Japan

Thomas Ernst
European Arrangements Chair
CEA-LETI
Grenoble, France

Tibor Grasser
European Arrangements Co-Chair
Vienna University of Technology
Vienna, Austria

Phyllis Mahoney
Conference Manager
Widerkehr and Associates
Gaithersburg, MD

Polly Mahoney
Conference Planner
Widerkehr and Associates
Gaithersburg, MD

Seated from left to right: Ken Rim, Publications Chair; Patrick Fay, Technical Program Vice Chair; Howard C.-H. Wang, General Chair; John Suehle, Technical Program Chair; Stefan De Gendt, Tutorial Chair

Standing from left to right (second row): Polly Mahoney, Conference Planner; Mina Rais-Zadeh, Focus Session Chair; Merlyne De Souza, Modeling and Simulation Chair; Daping Chu, Display and Imaging Systems Chair; Satoru Yamada, Short Course Vice Chair; Peter Moens, Power and Compound Semiconductor Devices Chair; Masakazu Kanechika, Asian Arrangements Co-Chair; Martin Giles, Short Course Chair; Tibor Grasser, European Arrangements Co-Chair; Thomas Ernst, European Arrangements Chair; Gianluca Boselli, Characterization, Reliability, and Yield Chair; Barbara De Salvo, Memory Technology Chair; Yee-Chia Yeo, Process and Manufacturing Technology Chair; Tian-Ling Ren, Sensors, MEMS, and BioMEMS; Phyllis Mahoney, Conference Manager

Standing from left to right (third row): Suman Datta, Publicity Chair; Mariko Takayanagi, Publicity Vice Chair; Taiichi Otsuji, Nano Technology Devices Chair; Chia-Hong Jan, Circuit and Device Interaction Chair

SUBCOMMITTEE ON CIRCUIT AND DEVICE INTERACTION

Chia-Hong Jan
Circuit and Device Interaction Chair
Intel
Portland, OR

François Andrieu
CEA-LETI
Grenoble, France

Shashank Ekbote
Qualcomm
San Diego, CA

Jan Hoentschel
GLOBALFOUNDRIES
Dresden, Germany

Ru Huang
Peking University
Beijing, China

Ali Keshavari
Cypress
San Jose, CA

Shouhei Kousai
Toshiba
Kawasaki, Japan

Yves Laplanche
ARM
Grenoble, France

Shignobu Maeda
Samsung
Gyeonggi-Do, Korea

Myunghee Na
IBM
Lagrangeville, NY

Peter Rickert
Texas Instruments
Richardson, TX

Robert Wu
Broadcom
USA

Shyh-Horng Yang
TSMC
Hsinchu, Taiwan

Scott Yu
SMIC
Shanghai, China

SUBCOMMITTEE ON CHARACTERIZATION, RELIABILITY, AND YIELD

Gianluca Boselli
Characterization, Reliability, and Yield Chair
Texas Instruments
Dallas, TX

Su Jin Ahn
Samsung
Gyeonggi-Do, Korea

Alain Bravaix
ISEN Toulon
Toulon, France

Mikael Casse
CEA/Leti
Grenoble, France

Francesco Driussi
University of Udine
Udine, Italy

Ahmad Ehteshamul Islam
Air Force Research Laboratory
Beavercreek, OH

Ben Kaczer
imec
Leuven, Belgium

Ziyuan Liu
Renesas
Kanagawa, Japan

Souvik Mahapatra
IIT Bombay
Mumbai, India

Yuichiro Mitani
Toshiba
Kanagawa, Japan

Rosana Rodriguez
University Autonoma Barcelona
Bellaterra, Spain

Wen-Jer Tsai
Macronix
Hsinchu, Taiwan

Ernest Wu
IBM
Essex Junction, VT

Jian Zhang
Liverpoll John Moores University
Liverpool, UK

SUBCOMMITTEE ON DISPLAY AND IMAGING SYSTEMS

Daping Chu
*Display and Imaging Systems
Chair*
University of Cambridge
Cambridge, UK

Piet De Moor
IMEC
Leuven, Belgium

David James Gundlach
NIST
Gaithersburg, MD

Ed Hudson
Jasper Displays
Santa Clara, CA

Ryoichi Ishihara
TU Delft
Delft, The Netherlands

Jack Luo
Zhejiang University
Hangzhou, China

Hisayo Momose
Toshiba
Kawasaki, Japan

Francois Roy
STMicroelectronics
Crolles, France

Daniel Smalley
Brigham Young University
Provo, UT

Taku Umebayashi
Sony
Kanagawa, Japan

Ching-Chun Wang
TSMC
Tainan City, Taiwan

SUBCOMMITTEE ON MODELING AND SIMULATION

Merlyne De Souza
Modeling and Simulation Chair
University of Sheffield
Sheffield, UK

Suman Banerjee
Texas Instruments
Dallas, TX

Katsumi Eikyu
Renesas
Hyogo, Japan

Hervé Jaouen
STMicroelectronics
Crolles, France

Jeffrey Johnson
IBM
Essex Junction, VT

Christoph Jungemann
RWTH Aacehn
Aachen, Germany

Roza Kotlyar
Intel

Luca Larcher
University of Modena
Reggio Emilia, Italy

Gengchiau Liang
National University of Singapore
Singapore

Blanka Magyari-Kope
Stanford
Stanford, CA

Andires Scholten
NXP
Eindhoven, Netherlands

Shigeyasu Uno
Ritsumeikan University
Shiga, Japan

Dmitri Veksler
Sematech
Albany, NY

Jeff Wu
TSMC
Hsinchu, Taiwan

SUBCOMMITTEE ON MEMORY TECHNOLOGY

Barbara De Salvo
Memory Technology Chair
CEA/LETI
Grenoble, France

Tetsuo Endoh
Tohoku University
Miyagi, Japan

Yoosang Hwang
Samsung
Gyeonggi-do, Korea

Fernanda Irrera
University of Rome
Rome, Italy

Tao-Cheng Lu
Macronix
Science Park, Taiwan

Giuseppina Puzzilli
Micron
Boise, ID

Pawan Singh
Spansion
Sunnyvale, CA

Sabina Spiga
CNR-IMM, Laboratorio MDM
Agrate Brianza, Italy

Joseph (Zhongze) Wang
Qualcomm
San Diego, CA

Toshitake Yaegashi
Toshiba
Yokkaichi, Japan

Joshua Yang
Hewlett-Packard Labs
Palo Alto, CA

Paola Zuliani
STMicroelectronics
Agrate Brianza, Italy

SUBCOMMITTEE ON NANO DEVICE TECHNOLOGY

Taiichi Otsuji
Nano Technology Devices Chair
Tohoku University
Aoba-ku, Japan

Joerg Appenzeller
Purdue University
West Lafayette, IN

Bernard Dieny
CEA Saclay
Grenoble, France

Elena Gnani
University of Bologna
Bologna, Italy

Max Lemme
Siegen University
Siegen, Germany

Chen-Hsin Lien
National Tsing Hua University
Hsinchu, Taiwan

Witek Maszara
GLOBALFOUNDRIES
Sunnyvale, CA

Kirsten Moselund
IBM Zurich
Rüschlikon, Switzerland

Matthias Passlack
TSMC
Leuven, Belgium

Ian Post
Intel Corp.
Hillsboro, OR

Alan Seabaugh
University of Notre Dame
Notre Dame, IN

Shinichi Takagi
University of Tokyo
Tokyo, Japan

Katsuhiro Tomioka
Hokkaido University
Hokkaido, Japan

SUBCOMMITTEE ON POWER AND COMPOUND SEMICONDUCTOR DEVICES

Peter Moens
*Power and Compound
Semiconductor Devices Chair*
ON Semiconductor
Oudenaarde, Belgium

Subramaniam Arulkumaran
Nanyang Technological University
Sinagpore

Kevin Chen
Hong Kong University of Science
and Technology
Hong Kong

Gilbert Dewey
Intel Corporation
Hillsboro, OR

Minghwei Hong
National Taiwan Univerity
Taipei, Taiwan

Matteo Meneghini
University of Padova
Padova, Italy

Miura Naruhisa
Mitsubishi Electric
Hyogo, Japan

Serge Oktyabrsky
SUNY CNSE
Albany, NY

Mikael Östling
KTH, Royal Institute of Technology
Kista, Sweden

Sameer Pendharkar
Texas Instruments
Allen, TX

David Sheridan
RFMD
Greensboro, NC

Keisuke Shinohara
HRL
Malibua, CA

Matthias Stecher
Infineon Technologies AG
Neubiberg, Germany

Michael Uren
University of Bristol
Bristol, UK

Niamh Waldron
IMEC
Leuven, Belgium

SUBCOMMITTEE ON PROCESS AND MANUFACTURING TECHNOLOGY

Yee-Chia Yeo
*Process and Manufacturing
Technology Chair*
TSMC
Hsinchu, Taiwan

Salih Muhsin Celik
STMicroelectronics
Albany, NY

Chorng-Ping Chang
Applied Materials
Santa Clara, CA

Jeff Hull
Micron
Boise, ID

Sangjin Hyun
Samsung Electronics
Gyeonggi-Do, Korea

Subhash Joshi
Intel Corp.
Hillsboro, OR

Jingfeng Kang
Peking University
Beijing, China

Carlos Mazure
Soitec
Bernin, France

Hiroshi Morioka
Fujitsu - Japan
Tokyo, Japan

Tsutomu Tezuka
Toshiba
Kawasaki, Japan

Dina Triyoso
GLOBALFOUNDRIES
Dresden, Germany

Anabela Veloso
IMEC
Leuven, Belgium

Maud Vinet
CEA/LETI
Grenoble, France

Jeff Xu
Qualcomm
San Diego, CA

SUBCOMMITTEE ON SENSORS, MEMS, AND BIOMEMS

Tian-Ling Ren
Sensors, MEMS, and BioMEMS Chair
Tsinghua University
Beijing, China

Nuria Barniol
University Autonoma de Barcelona
Bellaterra, Spain

Wolfgang Benecke
Fraunhofer ISIT
Itzehoe, Germany

Gary Fedder
Carnegie Mellon University
Pittsburgh, PA

Carlotta Guiducci
EPFL
Lausanne, Switzerland

Rainer Minixhofer
AMS
Unterpremstaetten, Austria

Yuji Miyahara
Tokyo Medical and Dental University
Tokyo, Japan

Siavash Pourkamali
University of Denver
Richardson, TX

Debbie Senesky
Stanford University
Stanford, CA

Kea Tiong Tang
National Tsing Hua University
Hsinchu, Taiwan

Dana Weinstein
Massachusetts Institute of Technology
Cambridge, MA

Fengnian Xia
Yale University
New Haven, CT

Jianbin Xu
The Chinese University of Hong Kong
Hong Kong

Gang Zhang
A*STAR
Singapore

Silicon Carbide Power Device Development for Industrial Markets

John W. Palmour
Cree, Inc.
4600 Silicon Drive, Durham, NC 27703, USA
John_Palmour@cree.com

Abstract— SiC power devices have the ability to greatly outperform their Silicon counterparts. SiC material quality and cost issues have largely been overcome, allowing SiC to start competing directly with more traditional Si devices. 150 mm substrates and epitaxy are now commercially available. Commercially released 4H-SiC MOSFETs with a specific on-resistance ($R_{ON,SP}$) of 5 mΩ·cm^2 for a 1200 V rating are now available, and research has further optimized the device design and fabrication processes to greatly expand the voltage ratings from 900 V up to 15 kV for a much wider range of high-power, high-frequency energy-conversion applications. Performance for voltage ratings from 900 V up to 15 kV have been achieved with a $R_{ON,SP}$ as low as 2.3 mΩ·cm^2 for a breakdown voltage (BV) of 1230 V and 900 V-rating, 2.7 mΩ·cm^2 for a BV of 1620 V and 1200 V-rating, 10.6 mΩ·cm^2 for a BV of 4160 V and 3300 V-rating, 123 mΩ·cm^2 for a BV of 12 kV and 10 kV-rating, and 208 mΩ·cm^2 for a BV of 15.5 kV and 15 kV-rating. All of these devices exhibit very high frequency switching performance over silicon IGBTs. For even higher voltages, bipolar devices in SiC have been demonstrated from 15 kV up to 27 kV. SiC GTOs have been shown up to 22 kV with 200 A capability. SiC n-IGBTs are reported up to 27 kV, with 20 A capability. This is the highest voltage semiconductor device reported to date.

I. INTRODUCTION

Compared to silicon, 4H-silicon carbide (4H-SiC) is a wide bandgap semiconductor that offers a factor of > 8 times higher electric breakdown field, which allows SiC power devices to have a much thinner drift region and higher doping for a given voltage rating than their silicon counterparts. The much thinner and higher doped drift region in SiC greatly reduces the on-resistance of the devices, which enables the use of a simple unipolar SiC power MOSFET device structure. Without the introduction of any complicated Super-Junction or bipolar device structures, SiC MOSFETs reported to date already significantly outperform any available Si-based power switches at the same voltage rating in terms of reduced switching losses and at least 5-10 times higher switching frequencies. As a result, one can fabricate unipolar power switches in SiC with voltage ratings more than 10 times higher than is feasible for Si unipolar devices.

Among the most commonly pursued switches, the SiC MOSFET offers the most desirable features in terms of device performance from a user's perspective, such as normally-off operation, low turn-off losses due to the lack of bipolar current tail, low conduction losses, and low gate charge. In early 2013, Cree commercially released its 2nd generation (C2M series), 25 mΩ and 80 mΩ, 1200 V, SiC MOSFETs, which have been used in inverters to provide an efficiency gain of > 2.5% over the all-Si solutions. In 2011, we also demonstrated 10 kV/10 A SiC power MOSFETs that can be switched efficiently at 20 kHz and above [1]. With further optimized device design and rapid advancement in SiC substrate and epitaxial material quality in recent years, the next-generation SiC power MOSFETs being reported here show even greater capability with a further reduction in on-state resistance and improved blocking performance, resulting in a further reduction in switching losses and fabrication cost at voltage ratings from 900 V up to 15 kV compared to our commercially released SiC MOSFETs. When compared to state-of-the-art silicon high-voltage devices, like IGBTs, Thyristors, and PiN diodes, with a blocking voltage close to 8 kV, SiC MOSFETs with voltage ratings at and above 10 kV can offer a significant improvement in system efficiency, switching speed, and power density due to much lower switching losses, a reduced number of components required in series as well as the number of levels utilized to achieve desired blocking voltage, resulting in a much simplified system design with improved overall reliability at the system level. The next-generation SiC power MOSFETs reported here will allow even further penetration of SiC MOSFETs into energy conversion systems not only at lower voltages, but will also enable entirely new topologies to be achieved at very high voltages. The performance improvements by utilizing the SiC MOSFETs over commercial Si bipolar switches in high-voltage switching converters will be also discussed.

II. STATUS OF SiC MATERIAL

A. Bulk Growth

Silicon Carbide bulk crystals are typically grown using a seeded sublimation process [2]. A polycrystalline SiC material source is heated to the range of 2200-2400°C, at which it sublimes into SiC vapor that then migrates to a single crystal SiC seed due to an induced temperature gradient. The crystal then grows out as more SiC vapor reaches the growth

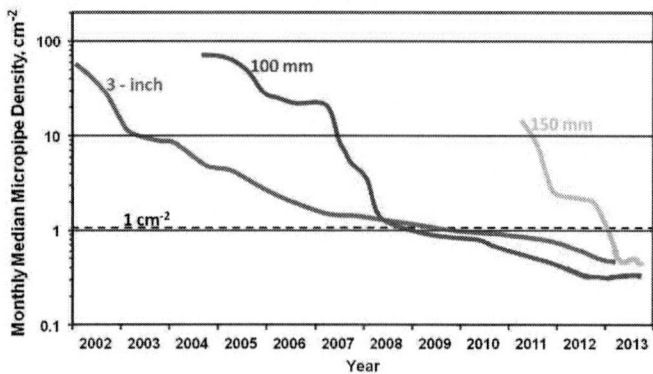

Fig. 1. Progress in SiC micropipe defect reduction over the last 14 years.

front on top of the seed. After the crystal growth process, the resulting boules are then typically sliced at 4 degrees off-axis in order to provide three-dimensional crystallographic topology for subsequent epitaxial growth.

SiC bulk crystals had long been plagued by a high density of defects termed "micropipes", which were the open core of a super screw dislocation that resulted in micron-sized holes that ran through the crystals. However, this defect has by and large been reduced to the point where they are no longer a major yield limiter for devices. The trend in micropipe reduction over the last 14 years is shown in Fig. 1. Micropipe densities are generally less than 0.5 cm^{-2} for all wafer sizes.

While 100 mm diameter SiC wafers are still the norm for device production today, 150 mm diameter wafers have been commercially available for the last 2 years from multiple sources [3], and there is a lot of activity by those with already existing 150 mm wafer fabs to migrate production to this larger size. A photograph of a 150 mm diameter n$^+$ SiC substrate is shown in Fig. 2, where the micropipe defects have been etched and highlighted, showing only two micropipes on the wafer, resulting in a density of less than 0.05 cm^{-2}.

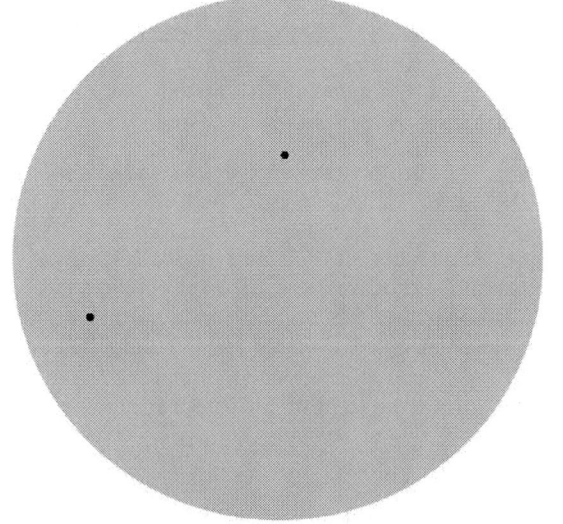

Fig. 2. A 150 mm diameter n$^+$ SiC substrate where the only two micropipe defects present have been highlighted, resulting in a density < 0.05 cm^{-2}.

B. Epitaxial Growth

SiC epitaxial growth can typically be achieved in two ways. The first and most common method is "warm wall epitaxy". However, for the growth of thicker epitaxial layers, "hot wall" epitaxy is often preferred. Again, most epitaxy for SiC device production is currently focused on 100 mm diameter wafers, but epitaxy is now well established on 150 mm diameter wafers as well [4].

The results achieved at Cree in warm wall epitaxy in a 6x150 mm reactor are shown in Fig. 3, which is a doping map for an R&D-best 150 mm-diameter epitaxial layer, intentionally n-type doped at ~5E15 cm^{-3}, with ~3.2% σ/mean doping uniformity. The doping was measured along a radius, perpendicular to the primary flat, at 7 evenly spaced points from the center, with a 9.3 mm edge exclusion. The doping uniformity is essentially unchanged at ~3.1% when including a point located only 5 mm from the wafer's edge. Due to the wafer rotating during growth, the doping around the wafer is also very symmetrical, resulting in a 1.9% σ/mean doping asymmetry. The average doping and thickness uniformity for this 6 wafer run was 3.9 and 1.6% σ/mean respectively.

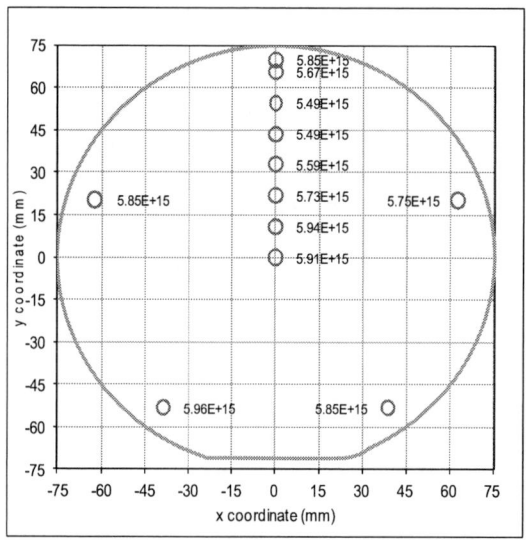

Fig. 3. CV doping map for an R&D best 150 mm-diameter epitaxial layer with ~3.1% σ/mean doping uniformity (5 mm edge exclusion). The average doping uniformity for this run was 3.9% σ/mean [4].

III. SiC MOSFETs

A. Design and Performance from 900V through 15 kV

Fig. 4 shows a schematic cross-section of a unit cell structure with the major resistive components of the 900 V to 15 kV SiC power MOSFETs. Details of the device fabrication can be found elsewhere [5]. The SiC MOSFETs reported in this work have a chip size and an active conducting area of 6.79 mm^2 and 3.41 mm^2 for the 900 V, 1200 V, and 1700 V-ratings, 46.72 mm^2 and 28.11 mm^2 for the 3300 V-rating, 65.61 mm^2 and 32 mm^2 for the 10 kV-rating, and 63 mm^2 and 32 mm^2 for the 15 kV-rating. The total on-resistance (R_{ON}) of the SiC MOSFETs consists of channel resistance (R_{ch}), JFET resistance (R_J), spreading resistance (R_{spd}), drift-layer resistance (R_d), and substrate resistance (R_{sub}), assuming that

978-1-4799-8002-4/14 $31.00 © 2014 IEEE

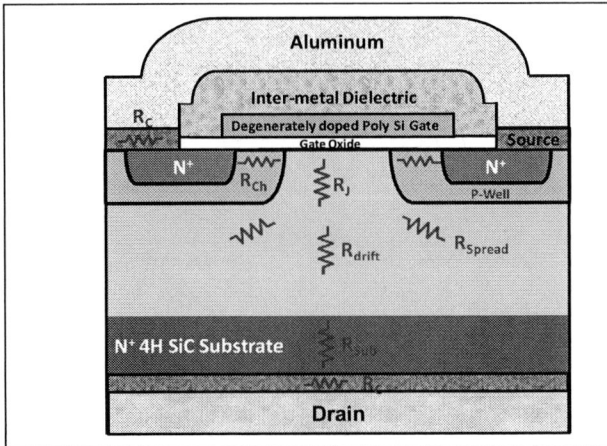

Fig. 4. Schematic cross-section of a unit cell showing resistive components of the 900 V to 15 kV SiC DMOSFETs.

contact resistance (R_C) to the source and backside drain regions are negligible. The limitation of the SiC MOSFET structure for the 900 V ~ 1700 V-ratings is in the MOS channel, which currently suffers from relatively low effective channel mobility (μ_{eff}). Because of the low μ_{eff}, a moderately high gate bias is needed to fully turn-on the device. At higher voltages, the SiC bulk resistance becomes dominant over the SiC MOS-channel resistance.

As shown in Fig. 5, despite the much lower inversion channel mobility than in Si MOS-based power switches, extremely low specific on-state resistance ($R_{ON,SP}$) has been achieved by further optimization of the device design in the recently developed 900 V, 1200 V, 1700 V SiC power MOSFETs. As the voltage rating increases to 3300 V and higher, the SiC MOSFET channel resistance becomes much less significant as compared to the SiC bulk resistance, therefore the total on-resistance is closer to its theoretical value for the MOSFETs with higher breakdown voltage. By using the planar SiC MOS channel structure, the electric fields within the active region can be designed to ensure maximum reliability during high-voltage operation as opposed to the use of trench structures, but still outperform most published $R_{ON,SP}$ values for SiC trench MOSFETs. We have now achieved DMOS performance in SiC power MOSFETs with a $R_{ON,SP}$ as low as 2.3 mΩ·cm^2 for a BV of 1230 V and 900 V-rating, 2.7 mΩ·cm^2 for a breakdown voltage of 1620 V and 1200 V-rating, 3.38 mΩ·cm^2 for a breakdown voltage of 1830 V and 1700 V-rating, 10.6 mΩ·cm^2 for a breakdown voltage of 4106 V and 3300 V-rating, 123 mΩ·cm^2 for a breakdown voltage of 12 kV and 10 kV-rating, and 208 mΩ·cm^2 for a breakdown voltage of 15.5 kV and 15 kV-rating.

Comparing to competing technologies for 12-13 A / 600 V-ratings described in Table 1, the $R_{DS,ON}$ of a 15 A / 900 V SiC MOSFET is almost one half of the published 600 V Si super-junction (SJ) MOSFETs and GaN cascode HEMTs at 25°C and less than one third at 150°C and 175°C [6], while the leakage current of the 600 V GaN HEMT and 600 V Si SJ-MOSFET are significantly higher than those of the 900 V SiC MOSFET at both 25°C and 150°C. Fig. 6 further illustrates that the $R_{DS,ON}$ of the 900 V SiC MOSFET has much less

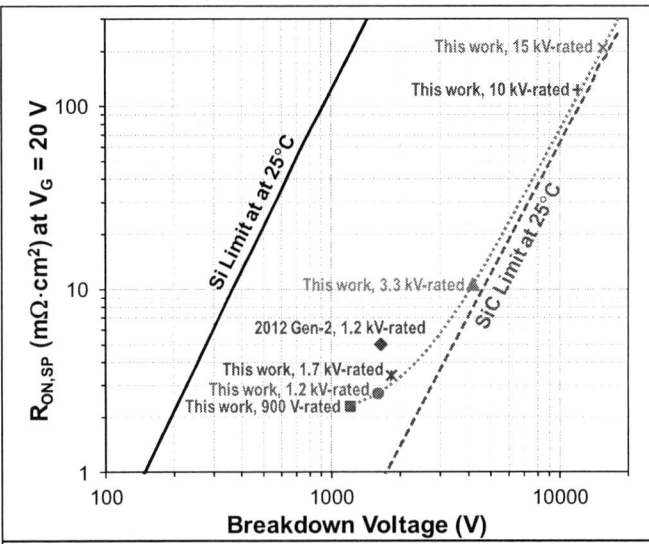

Fig. 5. Specific on-resistance, $R_{ON,SP}$ in mΩ·cm^2, of the next generation SiC DMOSFETs measured at gate bias of 20 V as a function of breakdown voltage at 25°C.

Table 1. Comparison of the 900 V/15 A SiC MOSFETs in this work to the 600 V/12 A GaN Cascode HEMT [6] and the 600 V/13 A Si SJ-MOSFET

Parameter	This Work 900 V / 15 A SiC MOSFET	600 V / 12 A GaN Cascode HEMT	600 V / 13 A Silicon SJ MOSFET
ON Resistance, $R_{DS,ON}$ (25°C)	78 mΩ	150 mΩ	150 mΩ
ON Resistance, $R_{DS,ON}$ (150°C)	100 mΩ		400 mΩ
ON Resistance, $R_{DS,ON}$ (175°C)	108 mΩ	330 mΩ	
I_{DSS} at V_{DS}=600V, (25°C)	50 nA	2.5 µA	1 µA
I_{DSS} at V_{DS}=600V, (150°C)	50 nA	20 µA	10 µA

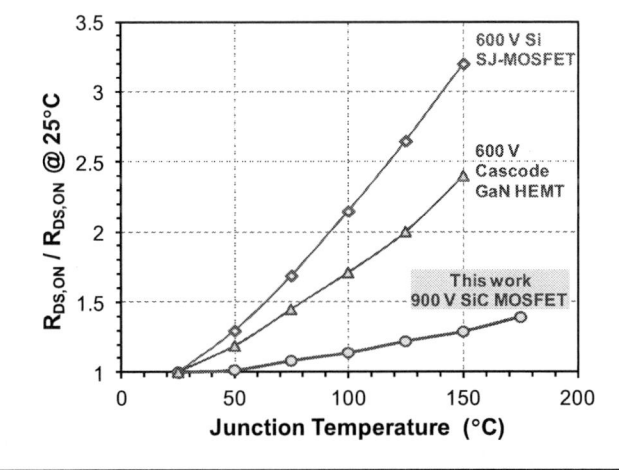

Fig. 6. Increase in $R_{DS,ON}$ as a function of temperature showing SiC advantage at high temperature over GaN HEMT [6] and Si Super-junction MOSFET data.

temperature dependence than those of the 600 V GaN cascode HEMT and the Si SJ-MOSFET up to 175°C.

Figs. 7&8 show the switching losses of the 1700 and 3300 V SiC MOSFETs at 25°C in an inductive load double-pulse switching setup. A rectifier built with C3D10170H 1700V

978-1-4799-8002-4/14 $31.00 © 2014 IEEE

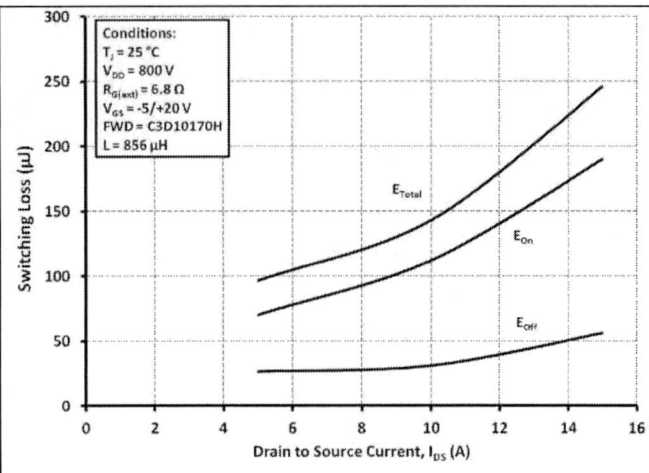

Fig. 7. Switching losses as a function of drain current of the 1700 V, 15 A SiC MOSFET at 800 V and 25°C.

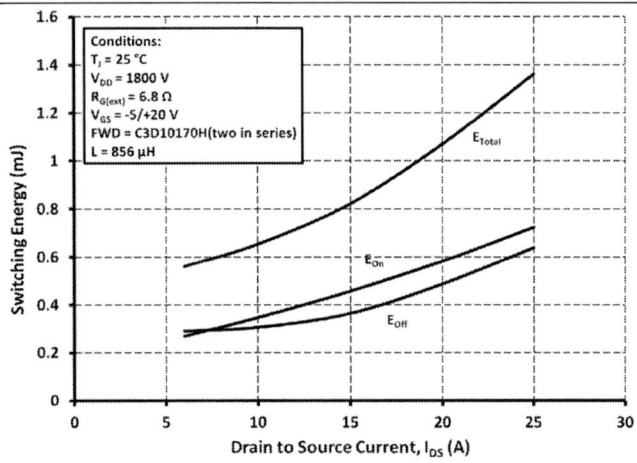

Fig. 8. Switching losses as a function of drain current of the 3300 V, 25 A SiC MOSFET at 1800 V and 25°C.

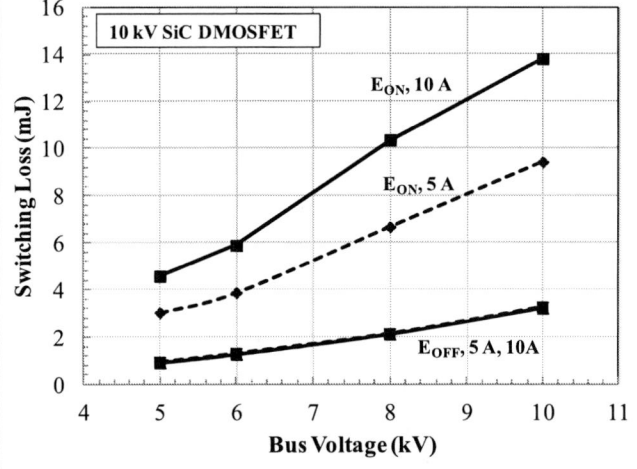

Fig. 9. Switching losses as a function of bus voltage for the 10 kV SiC MOSFET at 5 A and 10A at 25°C.

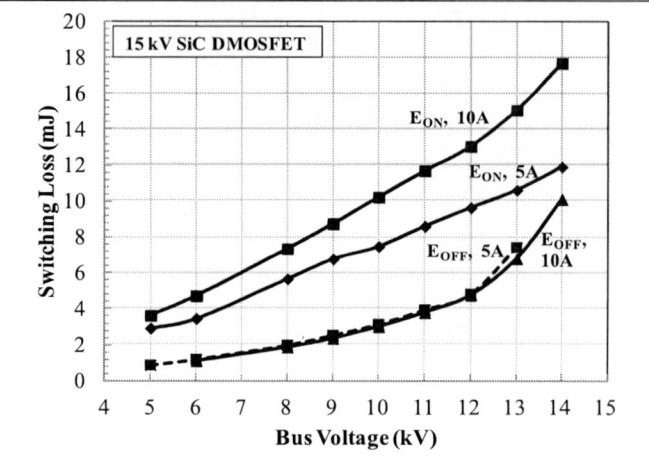

Fig. 10. Switching losses as a function of bus voltage for the 15 kV SiC MOSFET at 5 A and 10A at 25°C.

SiC JBS diodes connected in a parallel configuration was used as the freewheeling diode for the test. A supply voltage of 800 V to the 1700 V MOSFET and 1800 V to the 3300 V MOSFET and an inductor of 856 µH were used. A V_G of +20 V was used to turn on the SiC MOSFET and a V_G of -5 V was used to turn-off the MOSFET. Due to ~ 2x higher blocking voltage and about 7x larger active area, the total switching losses of the 3300 V MOSFET (1.35 mJ) is > 6x higher than that of the 1700 V MOSFET (245 µJ) when switched at about 50% of the rated voltage and 100% rated current.

The device performance of the 10 kV and 15 kV SiC MOSFETs were evaluated using a different high-voltage double-pulse switching set-up. A low capacitance air-core 14 mH inductor was used for the measurement. The inductor current is commutated by two 10 kV SiC JBS diodes connected in series on the high side of the double pulse circuit. The external gate resistance used in the set up was 6.7 Ω. Both the 10 kV and 15 kV SiC MOSFETs were measured in the same set-up. Unlike bipolar devices, SiC MOSFETs have no current tail, and hence have small switching losses in both hard switched and soft switched topologies. The turn-ON and turn-OFF energy losses of the 10 kV and 15 kV MOSFETs at 10A is shown in Figs. 9 & 10. The 10 kV SiC MOSFET was switched at 10 A and bus voltages from 5 kV to 10 kV, whereas the 15 kV MOSFET developed in this work was switched at 10 A and bus voltages from 5 kV to 14 kV. The total energy losses at room temperature is about 17 mJ for the 10 kV MOSFET switched at 10 kV, 10 A and about 27.5 mJ for the 15 kV MOSFET switched at 14 kV, 10 A.

Due to the superior SiC material properties and its unipolar characteristics, the SiC MOSFET at 10 kV and above exhibits incredibly high frequency switching performance over high-voltage Si power devices. Table 2 compares the device characteristics of high-voltage commercial 6.5 kV Silicon IGBTs [7], which have been widely used in various power converter applications, to the 10 kV and 15 kV SiC MOSFETs. The unipolar SiC MOSFETs show a nearly 30x reduction in switching losses over the 6.5 kV Silicon IGBTs. Fig. 11 illustrates that the high switching losses limit the operation of the silicon IGBT to the realm of a few kHz in a

978-1-4799-8002-4/14 $31.00 © 2014 IEEE

Table 2: Comparison of 10 kV, 15 kV SiC MOSFETs in this work to a commercial 6.5 kV Si IGBTs

Device	Max Tj (°C)	Rated Current	Die Size (mm²)	$V_{DS\,ON}$ @ Max T_J & Rated Current	E_{ON}	E_{OFF}
Si IGBT [7]	125	25 A	13.6 x 13.6	5.4 V	200 mJ @ 3.6 kV and 25 A	130 mJ @ 3.6 kV and 25A
10 kV/10 A SiC MOSFET	150	10 A	8.1 x 8.1	10.2 V	5.9 mJ @ 6 kV and 10A	1.3 mJ @ 6 kV and 10A
15 kV/10 A SiC MOSFET	150	10 A	8 x 8	16.3 V	8.9 mJ @ 8 kV and 10A	1.9 mJ @ 8 kV and 10A

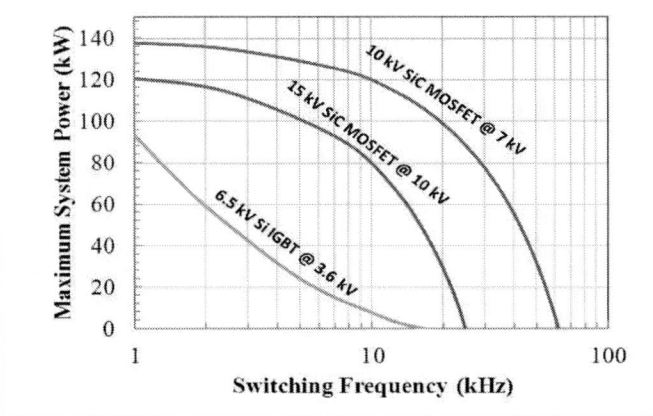

Fig. 11. Comparison of a 6.5 kV Si IGBT, 10 kV SiC MOSFET chip and 15 kV SiC MOSFET chip for a similar thermal environment.

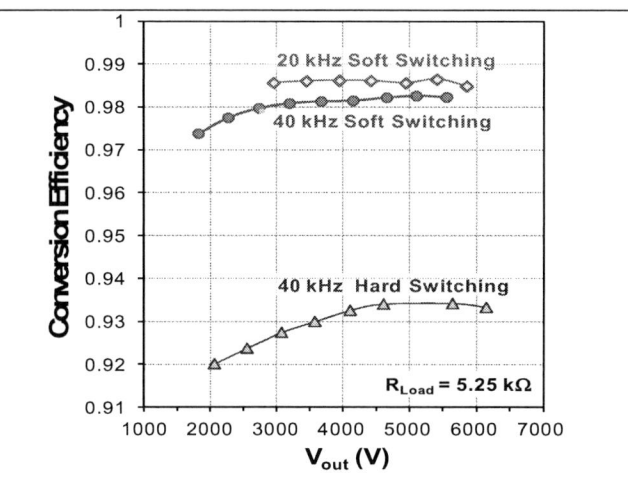

Fig. 12. Efficiency vs. output voltage of a boost converter using 15 kV SiC DMOSFET under 20 kHz and 40 kHz for both soft- and hard-switching conditions.

similar thermal environment, whereas the SiC MOSFETs can operate at tens of kHz with much lower losses and higher voltage blocking capability. As a result, these 10 kV and 15 kV SiC MOSFETs can deliver higher system power at frequencies in the kHz range.

Furthermore, a boost converter efficiency based on the 15 kV SiC MOSFET was evaluated as a function of the output voltage under 20 kHz and 40 kHz for both soft-switching (ZVS, zero-voltage-switch) and hard-switching (No ZVS) conditions, separately [8]. As shown in Fig. 12, when switched at ~ 6 kV and ~ 5 A, the 15 kV SiC MOSFET exhibits the capability of a very high conversion efficiency of 98.5% during the 20 kHz soft-switching, 98.2% during the 40 kHz soft-switching, and 93.2% during the 40 kHz hard-switching. These results are extremely promising for high power and high frequency applications that can significantly impact the system size, weight, and cost of the future advanced power conversion and transmission systems.

B. SiC MOSFET Reliability

One of the largest challenges in SiC MOSFET device reliability has been the threshold voltage (V_T) stability, which has been examined in numerous reports [9, 10]. As part of the development of the next generation SiC MOSFET, we have placed a great deal of emphasis on refining the device design and processing steps to achieve V_T stability. Figure 13 shows

the threshold voltage of 1200V rated MOSFETs during the course of a 1000 hour high temperature gate bias (HTGB) stress test at V_G = -15 V and T = 150°C. Threshold voltage was measured periodically *in situ* at 150°C. At 1000 hours, the average V_T shift for the 15 devices under stress was -50 mV, with a maximum shift of -90 mV. The maximum negative V_T shift occurs at about 4 hours into the stress (with a maximum shift of any device of -140 mV), and after that time, V_T begins shifting back in the positive direction. Figure 4 shows V_T shift under a +20V gate bias at 175°C, with a maximum V_T shift of 280 mV at 1000 hours of stress. The majority of the shift occurs within the first 5 to 10 hours, and the ΔV_T reverses rapidly upon removal of the positive bias. In a power switching application, since the MOSFETs are constantly switching from positive gate bias to zero or negative gate bias, V_T shift in actual usage is minimal.

Accelerated life testing is best used to project the long term reliability of devices. Fig. 15 shows the results of accelerated high temperature reverse blocking (HTRB) testing performed very close to avalanche in Cree's C2M 80 mΩ part at 150°C. Extrapolating back to a line voltage of 800 V yields a mean time to failure of 30 million hours at 150°C.

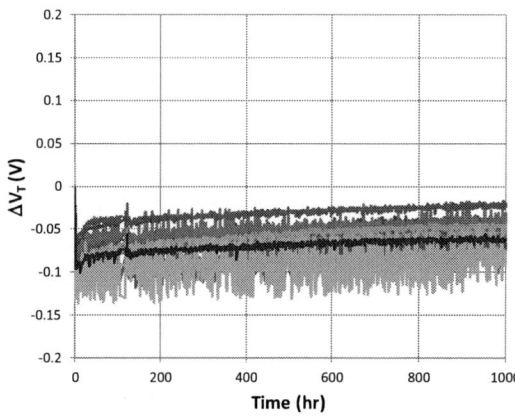

Fig. 13. Threshold voltage shift as a function of time for a 1000 hour HTGB stress at -15 V and 150°C.

978-1-4799-8002-4/14 $31.00 © 2014 IEEE

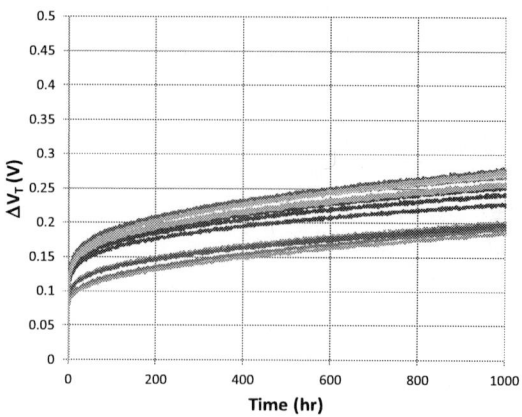

Fig. 14. Threshold voltage shift as a function of time for a 1000 hour HTGB stress at +20 V and 175°C.

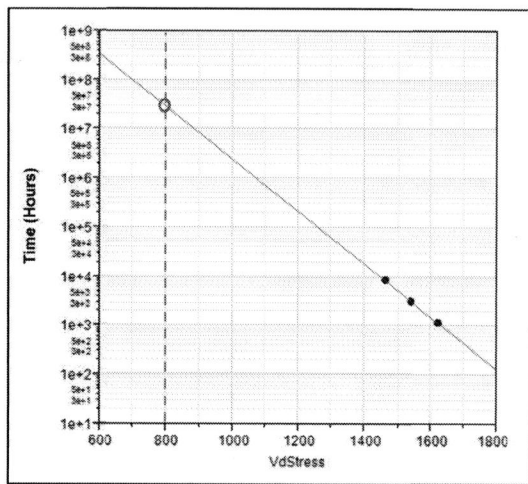

Fig. 15. Accelerated HTRB testing results at 150°C for Cree's 1200V 2nd generation (C2M) MOSFETs. Points show the MTTF as a function of very high bias, which can then be extrapolated back to lower bias. Data projects to a MTTF of 30 million hours using a line voltage of 800V at 150°C.

IV. ULTRA HIGH VOLTAGE SiC BIPOLAR DEVICES

Above 15 kV, the on-resistance of unipolar devices increases to the point where it is impractical from a cost and yield standpoint. This is similar to the case of Silicon between 600 V and 1200V, but at a much higher voltage due to the much higher breakdown electric field of SiC. Therefore, for devices above 15 kV it is best to focus on bipolar designs, so that one can utilize conductivity modulation in order to obtain a much higher current density than would be possible with a MOSFET. We have focused on two main types of devices for this very high voltage range. The first is SiC GTOs, which offer the highest current density due to current injection from both sides, and the SiC IGBT, which has less current injection, but has the advantage of a voltage controlled MOS gate, and is capable of higher frequency operation.

A. SiC Gate Turn-off Thyristors

Figure 16 shows a schematic cross-section view of a 22 kV SiC Gate Turn-off Thyristor (GTO) cell. A 2.5 μm thick p-type field-stop buffer layer was grown on an n^+ SiC substrate to achieve an asymmetrical structure. A 160 μm,

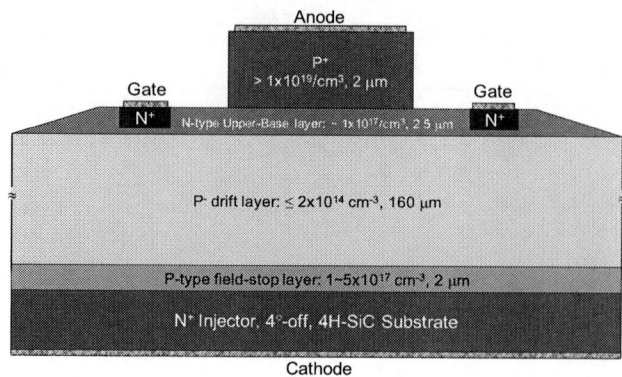

Fig. 16. Schematic cross-section of a 22 kV 4H-SiC GTO thyristor.

$2x10^{14}$/cm³ lightly doped, p-type lower-base region was then grown, followed by the growth of a 2.5 μm n-type upper-base region. The heavily doped, p-type anode layer of 2 μm thickness was grown as the final top layer. Details of the device fabrication can be found elsewhere [11, 12]. This 22 kV SiC GTO thyristor has a chip area of 2 cm² and an active conducting area of 0.53 cm².

Due to the very high-voltage ratings (> 20 kV) of the 4H-SiC GTO thyristors, the device was placed in a customized high-voltage metal-can package. The package was filled with Silicone Gel to prevent the device from arcing during the very high voltage testing. As shown in Fig. 17, the SiC GTO thyristor was able to block 22.1 kV at a leakage current of 1.76 μA from gate to cathode, which corresponds to an one-dimensional (1D) parallel-plate maximum electrical field of ~ 1.58 MV/cm at room-temperature (RT).

To measure the SiC GTO thyristor at high current injection levels (≥ 100 A/cm²), the forward characteristics of the device were evaluated using a Tektronix 371 curve tracer in pulse mode. As shown in Fig. 18, a very low $R_{ON,Diff}$ of 7.7 mΩ·cm² at 20°C and 7.6 mΩ·cm² at 150°C were measured at high current injection levels ≥ 100 A/cm², respectively. The slightly reduced $R_{ON,Diff}$ is likely a result of the slightly improved ambipolar carrier lifetime of this 22 kV, SiC GTO thyristor at 150°C. These results indicate that the SiC GTOs could be paralleled to achieve much greater current handling capability at elevated temperatures.

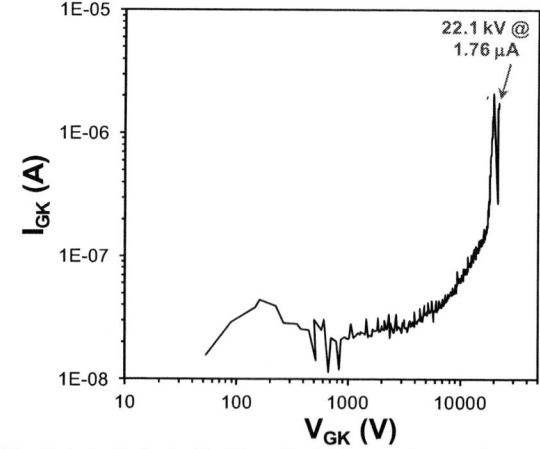

Fig. 17. Gate-to-Cathode blocking (V_{GK}) characteristics of a 2 cm², 22 kV 4H-SiC GTO thyristor at room temperature.

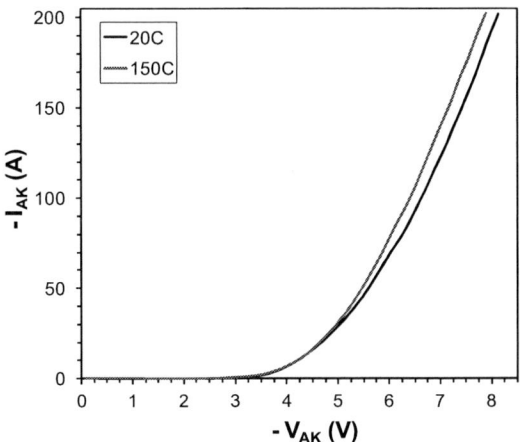

Fig. 18. Forward conduction characteristics of a 2 cm², 22 kV 4H-SiC GTO thyristor at temperatures at 20°C and 150°C, showing a differential $R_{ON,SP}$ of 7.7 mΩ·cm².at 20°C and 7.6 mΩ·cm² at 150°C at high current injection level ≥ 100 A/cm².

B. SiC Insulated Gate Bipolar Transistors

The high voltage device design that is easiest to control and has faster switching speed is the Insulated Gate Bipolar transistor (IGBT), due to its voltage controlled MOS gate. Just as is the case with silicon, we can take the same basic MOSFET process developed, and apply it to an IGBT by replacing the n-substrate with a p-type injection layer [13]. A simplified cross-section of the fabricated SiC n-IGBT is shown in Fig. 19. A deep punch-through design was selected for evaluation: (210 μm thick, 1×10^{14} cm⁻³ doped, $V_{PT} \approx 4$ kV). The drift region design results in a calculated parallel plane breakdown voltage greater than 24 kV [14]. The n-type field-stop buffer was approximately 1-2 μm thick, and doped sufficiently to avoid reach-through breakdown at the avalanche voltage. A 1.5 mm wide edge termination consisting of floating guard rings was used to alleviate high electric fields at the active region edge. Prior to device fabrication, a lifetime enhancement procedure consisting of 15 hours of thermal oxidation at 1300 C was performed to increase the as-grown drift ambipolar lifetime from less than 2 μs to more than 10 μs. Details of the device fabrication can be found in reference [15].

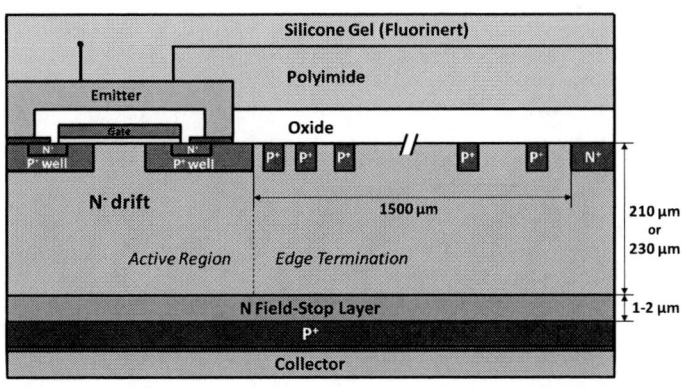

Fig. 19. Cross-section of fabricated 4H-SiC n-IGBTs including passivation scheme to achieve high blocking voltage.

The static forward blocking current-voltage characteristics for the best measured device is shown in Fig. 20, showing under 10 μA of leakage at 27.5 kV [16]. Device threshold voltages at 1 μA of current and V_C=10 V were greater than 2 V, guaranteeing normally-off operation. The static IV characteristics measured in pulse mode for the IGBT are shown in Fig. 21. A positive temperature coefficient is observed for these devices, indicating that they could be easily paralleled. At room temperature, the on-state forward voltage drop at 20 A for the n-IGBT with a 210 μm drift region was 11.7 V. The device has approximately the same on-state forward voltage drop with 20 V gate bias across the range of operating currents, indicating that the minority carrier lifetime is sufficient to guarantee conductivity modulation of the drift region.

The inductive load double-pulse turn-off switching waveforms at a bus voltage of 14 kV and load current of 20 A at 25 °C, 75 °C, and 125 °C are shown in Fig. 22. Two series 10 kV SiC JBS Schottky diodes with balance resistors were used as a freewheeling device; and the air-core load inductor used had a value of 13.8 mH. The current "notch" observed immediately after the initial voltage rise is a result of the diode capacitance; computations of current flow through the diode capacitance exactly match the observed current drop.

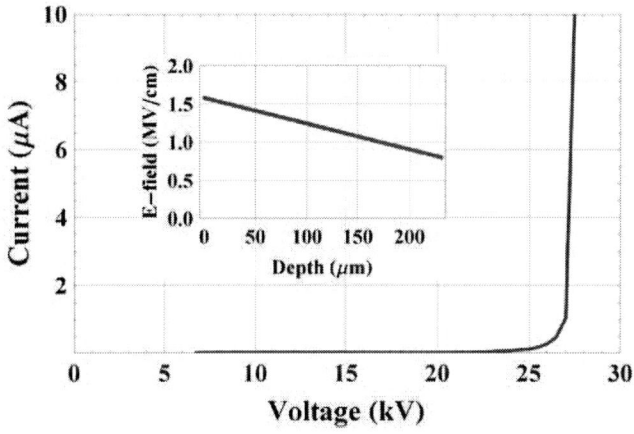

Fig. 20. Static blocking characteristics for n-IGBT with V_{GS}=0 V. Inset: Electric field with an applied voltage of 27.5 kV.

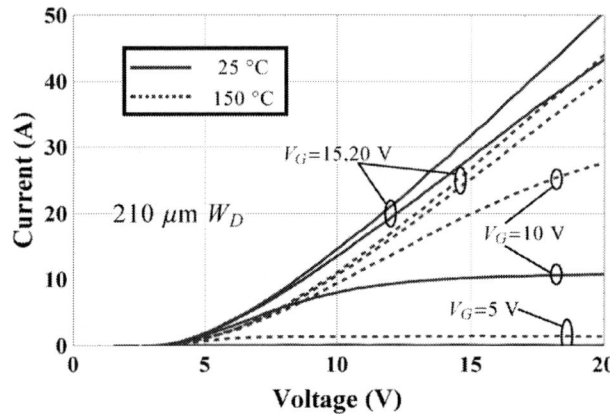

Fig. 21. Forward IV characteristics of SiC n-IGBT with deep punch – through structure at 25°C and at 150°C.

978-1-4799-8002-4/14 $31.00 © 2014 IEEE

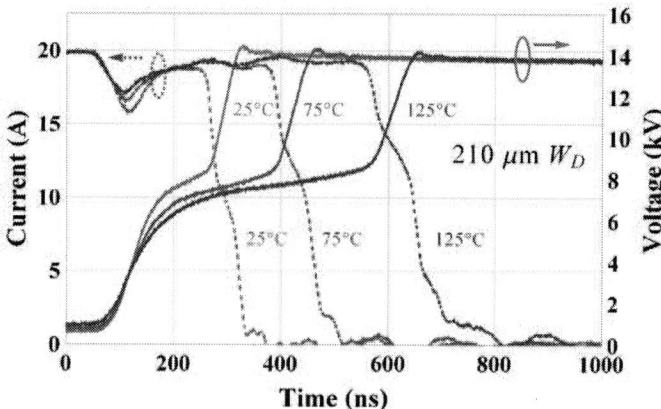

Fig. 22. 14 kV, 20 A double pulse switching waveforms for SiC n-IGBT with deep punch –through structure from 25°C up to 150°C.

A low impedance gate drive was used to switch the device as quickly as possible. As can be seen, increasing temperature leads to increased stored charge, and thus increased switching times and larger turn-off losses. The deep punch-through device has a fast transition to the bus voltage upon punch-through, which is desirable from a turn-off energy standpoint. The peak value of dv/dt for this deep-punch through device is 120 kV/µs, and occurs during the fast transition to the bus voltage after punch-through; this peak is comparable to the dv/dt of 100 kV/us that occurs immediately after switching.

In order to compare how these bipolar devices compare with the on-resistance of unipolar devices in SiC, we can compare the forward voltage drop for the rated current of these devices, including the built-in potential of the wide bandgap junction, to calculate an "effective" specific on-resistance for these high voltage devices. Figure 23 shows that the higher voltage SiC IGBTs fall roughly a factor of three below the theoretical line for a unipolar device. Due to the more efficient conductivity modulation allowed by the dual injection of the GTO structure, these devices fall more than an order of magnitude below the unipolar line, but with the requirement of more complex control circuitry.

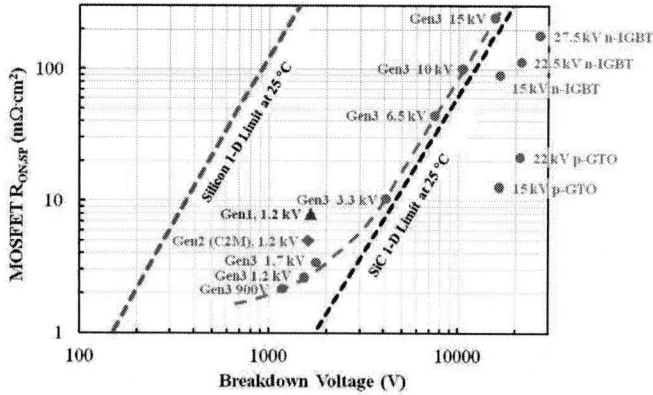

Fig. 23. Specific on-resistance, $R_{ON,SP}$ in mΩ·cm^2, of the next generation SiC DMOSFETs as was shown in Fig. 5 as a function of breakdown voltage at 25°C, along with the "effective" $R_{ON,SP}$ of high voltage bipolar SiC n-IGBTs and p-GTOs.

V. CONCLUSION

In summary, SiC technology has developed to the point where it can have a large impact on industrial markets. SiC substrates and epitaxy are now commercially available in 150 mm diameters with excellent crystal quality and uniformity. We have developed and demonstrated next generation SiC power MOSFETs with excellent performance over a wide range of voltage-ratings from 900 V up to 15 kV. By further optimizing device design and fabrication processes, these SiC MOSFETs show not only record low specific on-resistance but also exhibit very high switching frequency performance with extremely low switching losses over conventional Si power devices at the similar voltage ratings. The simple planar DMOS structure allows for very high reliability as demonstrated by Accelerated HTRB, and the threshold voltages are stable. At 10 kV and above, entirely new applications can be explored with 10-15 kV MOSFETs with extremely fast switching speeds. For even higher voltages, bipolar devices such as GTOs and IGBTs may be utilized. We have demonstrated 22 kV GTOs with 200 A capability. The highest voltage switching device demonstrated to date is a 27 kV SiC IGBT.

REFERENCES

[1] D. Grider, M. Das, A. Agarwal and J. Palmour, "10 kV/120 A SiC DMOSFET half H-bridge power modules for 1 MVA solid state power substation", IEEE Electric Ship Technologies Symp, 2010, pp 131-134.

[2] V. Tsvetkov, R. Glass, D. Henshall, D. Asbury and C.H. Carter, Jr., "SiC Seeded Boule Growth", Mater. Sci. Forum, vol. 264-268 (1998), pp. 3-8.

[3] http://www.cree.com/News-and-Events/Cree-News/Press Releases/2012/August/150mm-wafers, Cree Introduces 150-mm 4HN Silicon Carbide Epitaxial Wafers (2012)

[4] A.A. Burk, et al., "Latest SiC Epitaxial Layer Growth Results in a High Through-put 6 x 150 mm Warm-Wall Planetary Reactor" Mater. Sci. Forum, vol. 778-780 (2014), pp. 113-116.

[5] L. Cheng, et al, "High-Temperature Performance of 1200 V, 200 A 4H-SiC Power DMOSFETs", Mater. Sci. Forum, vol. 717-720 (2012), pp. 1065-1068, 2012.

[6] J.M. Briere, "GaN based power devices," RPI CFES Conf., (2013).

[7] 5SMX 12M6501 IGBT datasheet, www05.abb.com

[8] Gangyao Wang, "Design, Development and Control of 13kV Silicon-Carbide MOSFET based Solid State Transformer (SST)., Doctor of Philosophy Thesis, Dept. Elect. Engr, NC State University (2013).

[9] A. J. Lelis, D. Habersat, G. Lopez, J. M. McGarrity, F. B. McLean, and N. Goldsman, Mater. Sci. Forum, vol. 527–529 (2006), pp. 1317–1320.

[10] M. J. Tadjer, K. D. Hobart, E. A. Imhoff, and F. J. Kub, Mater. Sci. Forum, vol. 600-603 (2009), pp. 1147-1150.

[11] L. Cheng, et al., Proc. of IMAPS High Temperature Electronics Conference (HiTEC 2012), 149 – 153, (2012).

[12] L. Cheng, et al., Mat. Sci. Forum, vol. 740-742 (2013), 978-981, (2012).

[13] S-H. Ryu, et al., "Development of 15 kV 4H-SiC IGBTs", Mat. Sci. Forum, vol. 717-720, (2012) pp. 1135-1138.

[14] A. O. Konstantinov et al, App. Phys. Lett., vol. 71, pp 90-92, 1997

[15] E. Van Brunt, et al., "22 kV, 1 cm^2, 4H-SiC n-IGBTs with Improved Conductivity Modulation," Proc. 2014 26th Intnl. Symp. on Power Semi. Devices, June 15-19, 2014 Waikoloa, Hawaii. pp. 358-361.

[16] E. Van Brunt, L. Cheng, M.J. O'Loughlin, J. Richmond, V. Pala, J. Palmour, C.W. Tipton, and C. Scozzie, "27 kV, 20 A 4H-SiC n-IGBTs" to be published in Mat. Sci. Forum (2014).

978-1-4799-8002-4/14 $31.00 © 2014 IEEE

Atomic scale devices: advancements and directions

Enrico Prati[1] and Takahiro Shinada[2]

1. Istituto di Fotonica e Nanotecnologie, Consiglio Nazionale delle Ricerche, Piazza Leonardo da Vinci 32, I-20133 Milano, Italy

2. Center for Innovative Integrated Electronics System, Tohoku University 468-1 Aramaki Aza Aoba, Sendai, Miyagi, 980-0845, Japan

Abstract:

We review the theoretical and experimental advances in nanometric-scale devices and single atom systems. Few electron devices are currently obtained either by fabricating nanometric-scale semiconductor FinFETs and quantum dots, or by doping them with few impurity atoms. Devices of such size, originally realized by employing either pre-industrial or laboratory processes, are now being fabricated in commercial 14 nm node architecture. They have lead, starting from the 90's, to the observation of classical non-linear effects, to spin- and orbital-related quantum effects, manipulation of few qubits and to many-body quantum effects. As scaling of devices continues, the natural question is whether single atom and few electron devices will represent the ultimate scaled technology. We highlight high points and major constraints and limitations to state-of-the-art fabrication based on lithography and doping, and their possible integration with different methods such as self-assembly, inspired by biology and natural systems. *"At the atomic level, we have new kinds of forces and new kinds of possibilities, new kinds of effects. The problems of manufacture and reproduction of materials will be quite different. I am, as I said, inspired by the biological phenomena in which chemical forces are used in repetitious fashion to produce all kinds of weird effects (one of which is the author). R. Feynman, 1957"*

Introduction:

A variety of research experiments have highlighted the extended capabilities of nanoelectronic devices when approaching the limit of nanometric scale operations. Highly non-linear and purely quantum effects originally observed in III-V semiconductors (1), have been later reproduced in group IV heterostructures and FinFETs (2,3), carbon/graphene based devices (4), and one/few impurity atom systems (5,6). The driving forces towards scalable atomic scale devices and circuits are provided by both the need of controlling the geometric properties of devices at single lattice spacing for highly reproducible performances, and the employment of quantum mechanical properties associated to the wave behavior of single/few excess carriers, leading for instance to solid state quantum information control (7,8). The manipulation of number, spin and orbital state of excess electrons confined in nanoelectronic devices paved the way to a new generation of one-off devices with dramatically new properties. Even if reproducibility needed for tunnel devices has been achieved (9), according to Kelly himself (10), there is a limit around 3 nm due to the sample-to-sample variability which is intrinsic to modern forms of deposition including epitaxy, both e-beam and ultra-deep UV lithography, and precision etching, so one-off fabrication is possible, but manufacture is not. Unfortunately, variability of just 1 nm dramatically alters quantum transport (11). In the following we review the theoretical and experimental advances in nanometric scale devices and single atom systems, with special attention to issues connected to further miniaturization of devices. Next, a radically different approach is reviewed, consisting of bottom-up self-assembly and self-organization of nanostructures, which provide complementary methods towards further scaling beyond traditional large scale fabrication methods.

Nanometric scale devices: from FinFETs to quantum dots

The release of 14 nm FinFET 3D commercial technology at Q4 2014 represents the achievement of another milestone along the road of scaling of devices. In Broadwell architecture, Intel has integrated 14 nm channel transistor scale with 14 nm node interconnects. In parallel, 10 nm and 7 nm nodes are under development. According to IBM, the major challenges to develop the 7 nm node are density scaling (compared with 10 nm), advanced patterning (Extreme Ultraviolet Lithography - EUL) to enable aggressive Fin pitch scaling, channel material innovation for boosting performance and finally Middle-Of-Line (MOL) and Back-End-Of-Line (BEOL) via resistances. If we compare with existing state-of-the-art silicon quantum dots, such miniaturized size is sufficient for observing energy quantization effects in the density of electron states in the channel at either cryogenic (14 nm) or nearly room temperature (7 nm). Under such conditions, the system behaves more similarly to a quantum dot where Coulomb blockade takes place than standard transistors. When energy separation of the

978-1-4799-8002-4/14 $31.00 © 2014 IEEE

electron states, enhanced by the small size, overcomes thermal broadenindg kT, the device can be treated as an artificial atom, (1) with Hund-like orbital shells in sufficiently low disordered conditions, including valley sub-shells in silicon (12). The one-by-one filling of excess electrons is tuned by the control gate voltage which shifts its ground state to arbitrary value below the Fermi energy of the drain and source contacts. Therefore, a miniaturizaed system at the scale of tens of nm potentially behaves as an artificial atom and it is able to manifest highly non linear behaviour due to Coulomb blockade regime. Exclusion principles-related selection rules (13,14,15), room temperature Coulomb blockade (16,17), control of a single spin in silicon (8) as well as in graphene (18) and qubit operations based on multiple spin states (19), quantum turnstiles (20), valley-related transport in silicon (21) are among recent advacements in one-off devices. Tuneable strong coupling of excess electrons in a quantum dot with electrons in the contacts have been experimentally demonstrated in the past (22).

The capability of achieving control on individual excess electrons and on their quantum degrees of freedom should not be confused with the ability of controlling electrons at atomic level. Indeed, the size of the device is in the range 5-100 nm and it allows to create atomic like energy spectra with an enhanced Bohr radius. Such effects remain dominant at smaller scale, but top-down approaches are currently limited to ~3 nanometers scale for large scale fabrication.

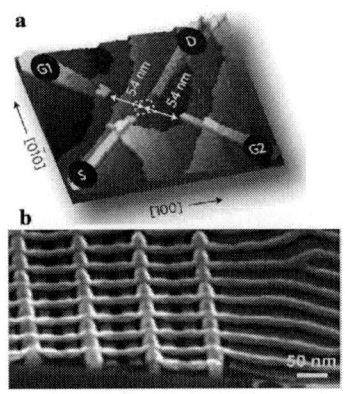

Figure 1 **a**. Single atom transistor fabricated with STM from (5). **b.** self assembled block-copolymer system from (45).

Atom based devices:

Recent studies have reported on quantum transport through single donor/acceptor systems, memory effects in coupled-donor systems, donor-band formation, metal-insulator transition from delocalized to localized states in the donor arrays, dopant-based applications such as turnstiles, memories,

photonic devices, or nanoscale pn junctions (6,23,24,25). Today, it is possible to adjust electrical properties by controlling individual dopants, which is referred to deterministic doping (26). Deterministic doping methods accelerate the understanding of single-atom-based phenomena and the device development. Such progresses are creating a more quantitative knowledgebase that will enable the design of future CMOS devices toward the single atom transistor limit.

The feasibility of single ion implantation (SII) has been verified (27), and SII is becoming a more and more important tool for enabling systematic studies of atom based devices (28,29,30,31,32,33). Such technology pursuits to place a specified number of desired dopants at designed locations precisely. Key challenges are to achieve single ion implantations with high spatial resolution and flexibility in dopant species, as well as 100% single dopant detection. One-by-one implantation of ions can be achieved by the detection of secondary electrons, photons, electron-hole pairs, changes in transistor channel currents, or direct imaging changes in surface topography. Dopant positioning errors, such as implantation spot size, straggling range, and diffusion and segregation during annealing, must be addressed for SII to achieve accuracy higher than 10 nm.

Atomic-scale fabrication platform has recently been established by dopant positioning with scanning tunneling microscope (STM) lithography, subsequent low-temperature activation and molecular beam epitaxy (5). First single-donor transistors with narrowest, lowest resistance conducting Si wires and, atomically abrupt delta-doping for shallow S/D contacts have been fabricated in both Si and Ge substrates. The method provides high stability and full activation in a high density n-type systems and low-temperature process. Potential benefits of the STM approach include: the ability to pattern with atomic precision in three dimensions (Fig. 1a); extremely high density, atomically planar and abrupt doping profiles; the investigation of novel device architectures; and applicability to other dopant sources/metals/organics. Atomic-scale control of dopant position have been demonstrated by STM and deterministic ion implantation for atom based devices beneficial to exploration of fundamental device limits and new functionalities, including quantum computing and 3D device architectures. Practical atomic-level control and cost-effective techniques are needed to move towards high-volume manufacturing.

Integration with self assembly:

In 2003, G. M. Whitesides and coworkers reviewed the potential use of self-assembly for the fabrication of nano-scale electronic and photonic devices (34). The four advantages offered by self-assembly consist of intrinsic parallelism, subnanometer precision, 3D architectures, and control by external forces and geometry. The methods were classified among self-assembled molecular monolayers, self-assembly in supramolecular chemistry, of nanocrystals and nanowires, of phase-separated block copolymers (BC), and colloidal. Molecular monolayers have been recently proposed for uniform doping by spike annealing (35). The method proved promising for doping of nanostructures, even if not for atomic scale control of dopants as they diffuse with part of the precursor used for self-assembly (36). The two-dimensional supramolecular chemistry provides interesting solutions to surface-based processes at sub-nanometer scale (37). A variety of self-assembled structures are possible, and host–guest chemistry is viable in the surface environment potentially facilitating the preparation of nanoscale devices. A simple approach to create a surface-based self assembled structure starts by a single molecule which creates an extended supramolecular network like 1,3,5- benzenetricarboxylic acid (38). Satisfying agreement has been demonstrated between STM images and Montecarlo simulations at sub-nanometer scale of self-assembly behavior of molecular building blocks into porous networks. Interplay of solvent, concentration, and temperature has been quantitatively clarified (39). For what concern nanostructures, strain- and defect-free faceted crystals forming space-filling arrays up to tens of micrometers in height have been grown by a mechanism of self-limited lateral growth by fast, low-temperature epitaxial growth of Ge and SiGe crystals onto micrometer-scale tall pillars etched into Si(001) substrates (40). Ordered SiGe islands grown on pit-patterned Si (001) substrates, compared to islands grown on flat substrates, show by atomic-force-microscopy-based nanotomography improved size, shape, and compositional homogeneity. The pyramids, domes, and barns have a base of 200 nm (41). Large scale self-assembly of long horizontal nanowires into orthogonally oriented bundles have been obtained during in situ annealing of a few monolayers of Ge on Si(001) (42). Coulomb blockade of holes has been observed by quantum transport characterization in controlled epitaxially grown of self-assembled SiGe nanocrystals, by a process that controls the size, composition and position of the nanocrystals. The quantum dots were operated in weak coupling regime with aluminum electrodes (43). Block copolymer self-assembly have been developed for fabricating regular arrays of silicon nanocrystals. A polystyrene-b-poly(methyl-methacrylate) (PS-b-PMMA) polymeric mixture has been employed to transfer the BC cylindrical organization to the material inclusions into the matrix (44). There, Si nanocrystals have been used as the active material to process, in the form of well-defined arrays over a macroscopic surface. The Si nanoparticle arrays onto sub-micrometric trenches are defined by graphoepitaxy, whose periodicity and topography affect the possibility of the conformal deposition of the ordered block polymeric mask. Three of the four advantages of Ref. (34) (with exception of the sub-nm precision) are granted by the registration and alignment of a monolayer of microdomains in self-assembled BC three-dimensional multilevel structures. A bilayer film of a cylindrical morphology BC, templated by an array of posts suitable functionalized, form a rich variety of three-dimensional structures consisting of cylinder arrays (Figure 1b) with controllable angles, bends, and junctions whose geometry is controlled by the template periodicity and arrangement (45), with feature size of ~18 nm. Minimum feature size reported is currently around 10 nm for PS-b-PMMA (46), and 8 nm for PS-b-PDMS (47).

Conclusions

Unless some revolutionary fabrication method will overcome the 3 nm limit of top-down methods, the integration of fabrication of solid state devices with bottom-up self-assembly methods seems to be the only viable way to achieve large scale fabrication with molecular scale control. Co-integration of different platforms may grant the best compromise between large scale fabrication and molecular scale features.

We gratefully acknowledge David N. Jamieson (University of Melbourne) for useful discussions and suggestions.

References:

.(1) R. Hanson et al., "Spins in few electron quantum dots", *Rev. Mod. Phys.*, vol. 79, pp. 1217-1265 (2007)

(2) H. W. Liu et al., "Pauli-spin-blockade transport through a silicon double quantum dot", *Phys. Rev. B*, vol. 77, pp. 073310 (2008)

(3) E. Prati et al., "Few Electron Limit of n-type Metal Oxide Semiconductor Single Electron Transistors", *Nanotechnology*, vol. 23, pp. 215204 (2012)

(4) P. Jarillo-Herrero, et al., "Electron-hole symmetry in a semiconducting carbon nanotube quantum dot", Nature, vol. 429, pp. 389-392 (2004)

(5) M. Fuechsle, et al., "A single-atom transistor", *Nature Nanotechnology*, vol. 19, pp. 242 (2012)

(6) E. Prati. M. Hori, F. Guagliardo, G. Ferrari, T. Shinada, "Anderson–Mott transition in arrays of a few dopant atoms in a silicon transistor", *Nature Nanotechnology*, vol. 7, pp. 443 (2012)

(7) B. M. Maune, et al. "Coherent singlet-triplet oscillations in a silicon-based double quantum dot", *Nature*, vol. 481, pp. 344 (2012)

(8) J. J. Pla, et al., "A single-atom electron spin qubit in silicon", *Nature*, vol. 489, pp. 541 (2012)

(9) C. Shao, J. Sexton, M. Missous, and M. J. Kelly, "Achieving reproducibility needed for manufacturing semiconductor tunnel devices," *Electronics Letters*, vol. 49, pp. 674-675 (2013)

(10) M. J. Kelly, "Intrinsic top-down unmanufacturability", *Nanotechnology*, vol. 22, pp. 245303 (2011)

(11) T. F. Watson, et al., "Transport in Asymmetrically Coupled Donor-Based Silicon Triple Quantum Dots", *Nano Letters*, vol. 14, pp. 1830-1835 (2014)

(12) M. De Michielis, E. Prati, M. Fanciulli, G. Fiori, G. Iannaccone, "Geometrical Effects on Valley-Orbital Filling Patterns in Silicon Quantum Dots for Robust Qubit Implementation", *Appl. Phys. Expr.*, vol. 5, pp. 1 (2012)

(13) C. W. J. Beenakker, "Theory of Coulomb-blockade oscillations in the conductance of a quantum dot", *Phys. Rev. B*, vol. 44, pp. 1646-1656 (1991)

(14) G. Yamahata, et al., "Magnetic field dependence of Pauli spin blockade: A window into the sources of spin relaxation in silicon quantum dots." *Physical Review B*, vol. 86, pp. 115322 (2012)

(15) E. Prati, "Valley blockade quantum switching in silicon nanostructures", *J. Nanosc. and Nanotech,*, vol. 11, pp. 8522-8526 (2011)

(16) S. J. Shin, et al., "Si-based ultrasmall multiswitching single-electron transistor operating at room-temperature", *Appl. Phys. Lett.*, vol. 97, pp. 103101 (2010)

(17) S. J. Shin et al., "Room-Temperature Charge Stability Modulated by Quantum Effects in a Nanoscale Silicon Island", *Nano Lett.*, vol. 11, pp. 1591-1597 (2011)

(18) B. Trauzettel, et al., "Spin qubits in graphene quantum dots", Nature Physics vol. 3, pp. 192 - 196 (2007)

(19) Z. Shi et al. "Fast coherent manipulation of three-electron states in a double quantum dot", *Nature Communications,* vol. 5 (2014)

(20) A. Rossi, et al. "An accurate single-electron pump based on a highly tunable silicon quantum dot,", *Nano Letters* (2014)

(21) G. P. Lansbergen, et al. "Lifetime-enhanced transport in silicon due to spin and valley blockade." Phys. Rev. Lett., vol. 107, pp. 136602 (2011)

(22) N. J. Craig, et al. "Tunable nonlocal spin control in a coupled-quantum dot system,", Science, vol. 304, pp. 565-567 (2004)

(23) M. Fuechsle, et al.,, "Spectroscopy of few-electron single-crystal silicon quantum dots", *Nature Nanotechnology*, vol. 5, pp. 502 (2010)

(24) T. Shinada, et al., "Quantum transport in deterministically implanted single-donors in Si FETs", *IEDM*, pp. 697 (2011)

(25) E. Hamid, et al., "Electron-tunneling operation of single-donor-atom transistors at elevated temperatures", *Phys. Rev. B*, vol. 87, pp. 085420 (2013)

(26) T. Shinada, et al.. "Enhancing semiconductor device performance using ordered dopant arrays." *Nature*, vol. 437, pp. 1128 (2005)

(27) International Technology Roadmap for Semiconductors, 2013 Edition, Emerging Research Chapter, 24.

(28) T. Matsukawa, T. Fukai, S. Suzuki, K. Hara, M. Koh, I. Ohdomari. "Development of single-ion implantation - Controllability of implanted ion number." *Appl. Surf. Sci.*, vol. 117, pp. 677 (1997)

(29) J. Meijer, et al., "Towards the implanting of ions and positioning of nanoparticles with nm spatial resolution", *Appl. Phys. A*, vol. 91, pp. 567 (2008)

(30) E. Bielejec, J. A. Seamons and M. S. Carroll. "Single ion implantation for single donor devices using Geiger mode detectors", *Nanotechnology*, vol. 21, pp. 085201 (2010)

(31) T. Shinada, et al., "Performance evaluation of MOSFETs with discrete dopant distribution by one-by-one doping method", *IEDM*, pp. 592 (2010)

(32) M. Ilg, et al., "Improved single ion implantation with scanning probe alignment", *J. Vac. Sci. & Technol. B*, vol. 30, pp. 06FD04 (2012)

(33) A. D. Alves, et al., "Controlled deterministic implantation by nanostencil lithography at the limit of ion-aperture straggling", *Nanotechnology*, vol. 24, pp. 145304, (2013)

(34) B. A. Parviz, D. Ryan, and G. M. Whitesides, "Using Self-Assembly for the Fabrication of Nano-Scale Electronic and Photonic Devices", *IEEE Trans. on Adv. Pack.*, vol. 26, pp. 233 (2003)

(35) J. C. Ho, et al., "Controlled nanoscale doping of semiconductors via molecular monolayers", *Nature Materials*, vol. 7, pp. 62-67 (2008)

(36) Y. Shimizu, et al., "Behavior of phosphorous and contaminants from molecular doping combined with a conventional spike annealing method", *Nanoscale*, vol. 6, pp. 706-710 (2014)

(37) A. G. Slater, et al. "Two-dimensional supramolecular chemistry on surfaces", *Chem. Sci.*, vol. 2, pp. 1440–1448 (2011)

(38) S. Griessl, M. Lackinger, M. Edelwirth, M. Hietschold and W. M. Heckl, *Single Mol.*, vol. 3, pp. 25 (2002)

(39) K. J. Adisoejoso, et al., "One building block, two different nanoporous self-assembled monolayers: A combined STM and Monte Carlo study", *ACS Nano*, vol. 6, pp. 897-903 (2012)

(40) C. V. Falub, et al., "Scaling hetero-epitaxy from layers to three-dimensional crystals", *Science*, vol. 335, pp. 1330 (2012)

(41) J. Zhang, A. Rastelli, O. Schmidt, G. Bauer, "Compositional evolution of SiGe islands on patterned Si (001) substrates", *Appl. Phys. Lett.*, vol. 97, pp. 203103/1-3 (2010)

(42) J. Zhang et al., "Self-organized evolution of Ge/Si(001) into intersecting bundles of horizontal nanowires during annealing", *Appl. Phys. Lett.*, vol. 103, pp. 083109 (2013)

(43) G. Katsaros, et al., "Hybrid superconductor-semiconductor devices made from self-assembled SiGe nanocrystals on silicon", *Nature Nanotechnology*, vol. 5, pp. 458 (2010)

(44) P. Pellegrino, et al., "Fabrication of well-ordered arrays of silicon nanocrystals using a block copolymer mask", *Physica Status Solidi (A)*, vol. 210, pp. 1477-1484 (2013)

(45) A. Tavakkoli et al., "Templating Three-Dimensional Self-Assembled Structures in Bilayer Block Copolymer Films", *Science*, vol. 336, pp. 1294 (2012)

(46) G. Seguini, et al., "Thermally induced self-assembly of cylindrical nanodomains in low molecular weight PS-b-PMMA thin films", *Nanotechnology*, vol. 25, pp. 045301 (2014)

(47) S.-M. Park, et al., "Sub-10 nm Nanofabrication via Nanoimprint Directed Self-Assembly of Block Copolymers", *ACS Nano* vol. 5, pp. 8523-8531 (2011)

Research into ADAS with Driving Intelligence for Future Innovation

Hideo Inoue
TOYOTA MOTOR CORPORATION, 1200 Mishuku, Susono, Shizuoka, 410-1193, JAPAN
Phone: (+81) 55-997-9094, Fax: (+81) 55-997-7884, E-mail: hideo_inoue@mail.toyota.co.jp

1. Abstract

Japan is facing up to the challenges of a rapidly aging society. In addition to measures to help vitalize this situation, the automotive sector is also working to resolve issues such as congestion and traffic accidents. With this background, this paper introduces two collaborative projects about intelligent vehicle research and development in Japan. The first project aims to develop an intelligent driving system to achieve a safe and secure traffic society for elderly drivers. The main purpose of this project is to realize an intelligent driving system incorporating an experienced driver model to help recover the deterioration in the recognition, decision-making, and operation capabilities of elderly drivers, and to achieve a significant improvement in road safety. The second is the Smart Traffic Flow Control Project. This project focuses on the fact that an advanced driver assistance system (ADAS) with driving intelligence has the potential not only to enhance safe and secure driving but also to reduce congestion. This paper uses these topics to describe perspectives about intelligent vehicle technologies.

2. Intelligent Driving System to Achieve Safe and Secure Traffic Society for Elderly Drivers Project

2-1. Summary

This project was adopted and started in 2010 as one of the technology innovation projects to vitalize Japan's aging society. The project is sponsored by the Japan Science and Technology Agency (JST). This project is being carried out based on a collaborative framework between universities and industry, including the Tokyo University of Agriculture and Technology (TUAT), the University of Tokyo, Toyota Central R&D Labs., Inc., and Toyota Motor Corporation. The main purpose of this project is to realize an intelligent driving system incorporating an experienced driver model to help recover the deterioration in the recognition, decision-making, and operation capabilities of elderly drivers. The vehicle control concept proposed in this project will be disseminated throughout the market as a part of ADAS, and the concept is applicable to intelligent driving systems designed for certain traffic circumstances in the near future. This paper describes the current activities and research results gained from the project studies.

2-2. Motivation and Objectives

Recent advances in passive and active safety technologies have greatly contributed to a remarkable decrease in traffic fatalities. However, the number of road accidents remains high according to national statistics [1]. By 2030, it is estimated that the ratio of people aged 65 years or older will reach one-third of the Japanese population, and that the ratio of people aged 75 years or older will reach one-fifth [2]. In Europe, the share of people aged 65 years or older in the total population is projected to increase from 17.1% to 30.0% between 2008 and 2060, an increase from 84.6 to 151.5 million [3]. Focusing on elderly drivers, it is also estimated that half of people aged 60 years will remain driving license holders. Based on these facts, accidents involving elderly drivers have been increasing every year owing to the declining physical capabilities of the elderly in terms of recognition, decision-making, and vehicle operation. For instance, the effective visibility range for elderly drivers is narrower than that of younger drivers. A driving test of a forward collision warning system using a large driving simulator found that many elderly drivers could not recognize the collision warning sound produced by a driver assistance system [4]. Moreover, it was also found that some drivers that did recognize the collision warning system sound could not brake in time to react to a critical situation, as shown in Fig. 1 [4]. These facts indicate that in critical situations, an advanced driver assistance system (ADAS) with driving intelligence may be a promising solution to prevent collisions involving elderly drivers. Moreover, this deterioration in driving capabilities reduces the self-confidence of elderly drivers. Some elderly people still need a vehicle as a means of mobility for active participation in society as well as to improve the quality of life [5]. To recover this deterioration in driving capabilities, this research project aims to considerably improve road safety by allowing autonomous driving control intervention in the last few seconds before an accident involving elderly drivers. Other requirements for this system include thorough acceptance by drivers and society for introducing this type of assistance technology. Some of the questions and research issues regarding the interaction of autonomous technologies with an older driver population are stated in the literature by Yang and Coughlin [6].

Intervention by active controls such as emergency braking and steering assist is regarded as a promising technical solution to facilitate safe driving by elderly drivers as well as to prevent accidents by compensating for human error with intelligent vehicle and robot technologies. Research has been conducted into intelligent driving system design, such as automatic lane keeping functions, automated headway distance control, and a combined steering and braking control system for forward object collision avoidance [7][8][9]. The target of this project is not to achieve driverless vehicle technology. With this intelligent driving technology, the driver is still in the control loop and has responsibility for driving. Moreover, when the assistance provided by intelligent driving technologies is able to take over major aspects of driving tasks, the dynamic balance of capabilities, authority, control, and responsibility between the driver and the driver assistance systems must be investigated and optimized to achieve good safety performance and acceptance [10].

This project aims to improve road safety by the intervention of intelligent driving controls, thereby preventing accidents at an early stage before the risk becomes imminent. The intelligent driving system is designed based on the key concept of experienced driver behavior modeling with potential hazard anticipation. In this project, thorough driver and social acceptance are also key system requirements. Anticipated project outcomes and social impacts are the vitalization of society in rural areas, more lively and active lifestyles for elderly people, the prevention of accidents caused by deterioration in driving capabilities, and the global marketing of new types of innovative active safety technology. Technology resulting from the project may also be able to improve the potential of innovation in Japan toward becoming a leading country for business solution providers.

2-3. Project Outline and Key Concept

The following key technologies are important to construct the above-mentioned intelligent driving system.
(1) High-precision road environment recognition technology (using cameras, LIDAR, radar, GPS/IMU, and the like)
(2) Digital maps and electronic horizon technology (topological maps and surroundings data)
(3) Intelligent driving technology (recognition, knowledge-based driver model, risk potential prediction)
(4) Driver diagnostics technology (driver model, driver acceptance)
(5) Optimized human-machine interface (HMI) technology.
Moreover, the extension of intelligent driving utilizing big data as well as connected vehicle technology utilizing vehicle-to-X communication (V2X) are also included in the development platform. In such intelligent driving applications, when the driver acts as a main controller and the system acts as a subordinate controller, it is important to investigate driver acceptance in terms of shared authority concerns as well as social acceptance when introducing the intelligent driving system in an actual vehicle. The effectiveness in real-world driving will also be verified by conducting field operational tests (FOT).
The research plan contains the following three development process stages.
Stage 1: Development of sensing/control devices and prototype vehicle completion
Stage 2: System improvement based on FOT and preparation for commercialization
Stage 3: Commercialization, system deployment, and further development toward global standardization

2-4. Conceptual Design of Intelligent Driving for the Elderly

A governors' association for the promotion and development of senior citizen-friendly vehicles was established in 2009 to study the requirements of new-type vehicles for senior citizens. The association gathered senior drivers' opinions relevant to daily driving, as well as opinions related to a concept hypothesis and the advanced vehicle functionalities. According to the questionnaire survey, it was found that most elderly drivers have a high motivation to maintain the habit of driving and that many elderly people also need individual mobility to improve the quality of life (Fig. 1).

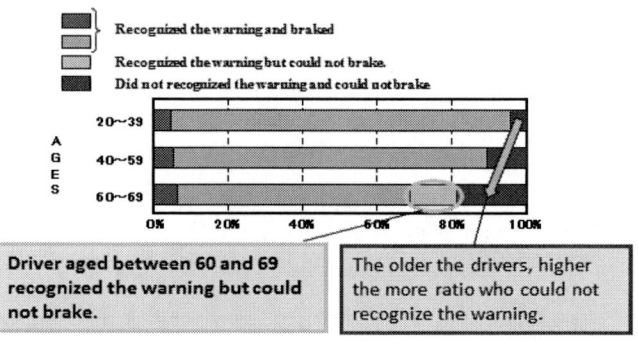

Fig.1 Elderly drivers' characteristics in driving simulator study

However, deterioration in driving capabilities reduces self-confidence and causes a fear of driving. Therefore, the developed intelligent driving system must be able to help recover the deterioration in driving capabilities and overcome the fear of driving. Table 1 summarizes the requirements of vehicle technologies to recover the deteriorated driving characteristics of elderly drivers.

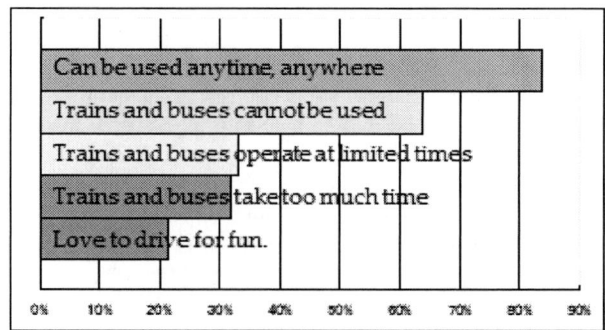

Fig.2 Questionnaire results collected from 10,000 elderly drivers in rural areas

Table 1 System requirements based on elderly drivers' characteristics

Elderly drivers' characteristics	System functionality requirements
When only a warning system is used, the ratio of elderly drivers cannot completely avoid a collision increases.	The system needs to provide assistance by vehicle control intervention such as autonomous braking.
The recognition ability survey found that people aged 65 years or older have a narrower effective field of view.	An environment perception and recognition function with a wide range and field of view is required.
Driving scenarios that are difficult for elderly drivers, such as reversing and parking, increase.	An emergency assistance function for pedal misapplication with object detection is necessary.
Elderly drivers still have high motivation to drive. The driving ability of elderly drivers is high due to their experience.	Although driverless vehicle technology is not needed, a shared driving configuration between the driver and the system must be developed. It is important to provide assistance adapted to the driver state.

2-5. Conceptual Design of Experienced Driver Model

This study analyzed the near-miss incident database constructed by the Smart Mobility Research Center of TUAT to obtain a list of critical driving scenarios containing the key parameter conditions of the surrounding environment and driver collision behavior characteristics. Currently, the database contains 89,000 items of crash-relevant-event data registered as of March 2014. Each data contains 15-second snapshot front-view video images, as well as data related to vehicle dynamics, vehicle location, and driver operation recorded by drive-recorders mounted on 200 taxis in Japan. Some drive-recorders also include the driver's face for detailed analysis of driving behavior. By using drive recorders, information such as vehicle velocity, as well as the acceleration and braking operations of drivers can be acquired. This project focused on the following main driving scenarios and 22 detailed scenarios were defined to develop the driver models and control functionalities.

(1) Forward vehicle collision avoidance
(2) Pedestrian collision avoidance
(3) Lane departure prevention
(4) Intersection head-on collision avoidance

As an example of the crash-relevant data analysis from the near-miss incident database, it was found that the current active safety systems cannot completely avoid a collision in certain circumstances due to the physical limitations of vehicle dynamics. For example, in a scenario in which a pedestrian appears from behind an object to cross a road, such as from behind a parked car on an urban road, current autonomous braking systems cannot stop the vehicle in time if the pedestrian suddenly appears at a very close distance to the driver's vehicle. In the real-world, drivers should decelerate when approaching an area with poor visibility and also prepare to brake to avoid potential collisions. This is the so-called hazard-anticipatory mechanism of expert drivers. Therefore, incorporating an expert driver model into a driver assistance system to reduce driving risk is a promising means to further reduce the number of accidents.

(a) Avoidance by braking

(b) Avoidance by braking and steering

Fig.3 Pedestrian collision avoidance test with combined braking and steering control system

Figure 3 shows the effectiveness of a collision avoidance system when a vehicle is approaching and passing a parked vehicle, and a pedestrian dash out from behind the vehicle. A combined braking and steering avoidance system based on a theoretical path planning algorithm was designed. To respond to crossing pedestrian scenarios including those where objects are blocking the view of the driver, a novel collision avoidance system based on potential field theory was designed by treating the object as a potentially hazardous area, and its effectiveness was verified using a test vehicle. Figure 4 shows an example of the path planning results after the application of potential field theory.

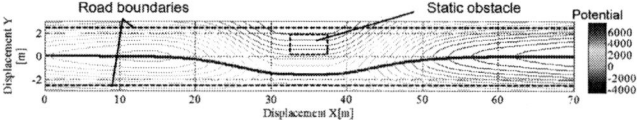

Fig.4 Potential field computational result for determining path trajectory and speed

2-6. Fundamental Design of Intelligent Driving System

The intelligent driving technology designed in this research project aims to enhance the performance of an ADAS and to fulfil the requirements obtained from the survey. The system must be able to avoid potential collisions caused by human error by direct vehicle control intervention when the risk becomes imminent. The intelligent driving system is composed of three main components: sensing, intelligence, and vehicle dynamics control. The control level can be adjusted with respect to the operational behavior of the human driver since the system is designed based on the driver-in-the-loop principle. Figure 5 shows a schematic diagram of the intelligent driving system, and Fig. 6 shows a block diagram of the main control system. The features of the intelligent driving system are as follows.

(1) The main part of the system is configured as a control-oriented driver model consisting of three-level modes: normal driving control, risk-potential based control, and emergency avoidance control.

(2) The system determines the appropriate assistance level by utilizing the risk potential prediction of experienced drivers.

(3) The system realizes the control performance taking into account the vehicle dynamic characteristics.

(4) The system is configured to be applicable to all sensors and hardware systems.

(5) The system can be applied to existing on-board chassis control systems.

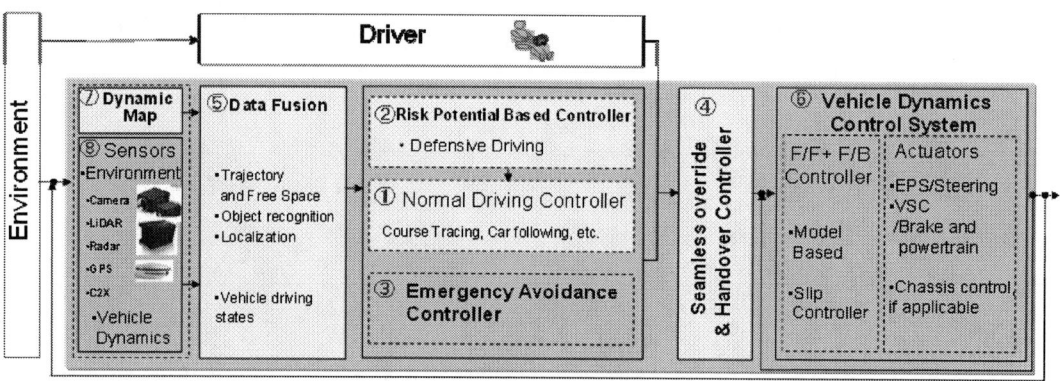

Fig.5 Structure of proposed intelligent driving system

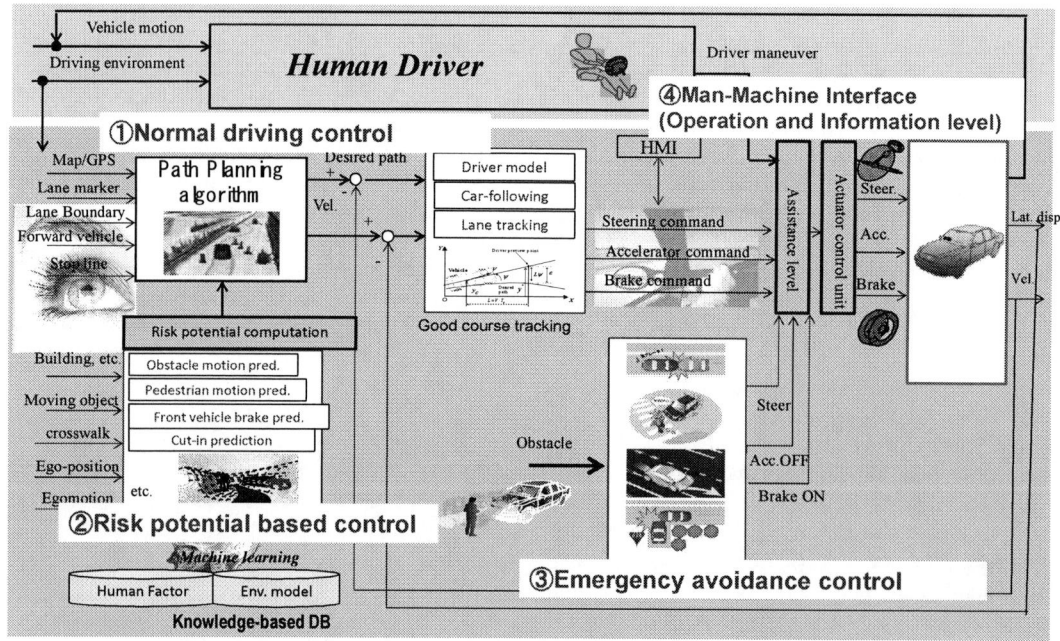

Fig.6 Main structure of an intelligent driving system composed of three-level controllers and HMI for human driver

2-7. Verification of Intelligent Driving System

A three-level driver model was implemented on the test vehicle as a controller to avoid collisions, and the effectiveness of the intelligent driving system was verified by test drives. The following scenario was used: the driver's vehicle needs to overtake a parked vehicle in the same lane and a pedestrian may dart out from behind the parked vehicle (Fig. 7). Experienced drivers predict the potential appearance of a pedestrian in advance and reduce speed to prepare for a worst-case scenario in which the pedestrian darts out. This experienced driver behavior is referred as defensive driving. Defensive driving behavior was modeled using potential field theory and the model was incorporated into the test vehicle. As a result, a level of autonomous driving equivalent to that of an experienced driver can be achieved. Figure 8 compares the collision avoidance performance of a conventional autonomous emergency braking (AEB) system and the proposed braking system. By incorporating the risk-potential based control in the driver model, the advanced braking is activated to reduce the vehicle speed when passing through an area containing objects that block the driver's view. Even though a pedestrian darts out from behind an object, a collision can be avoided effectively without high-G braking.

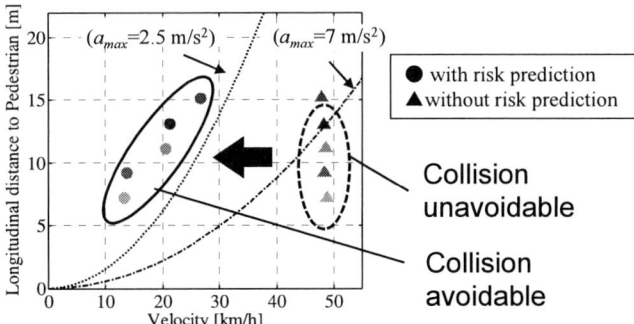

Fig. 7 Typical driving scene containing object that requires risk prediction

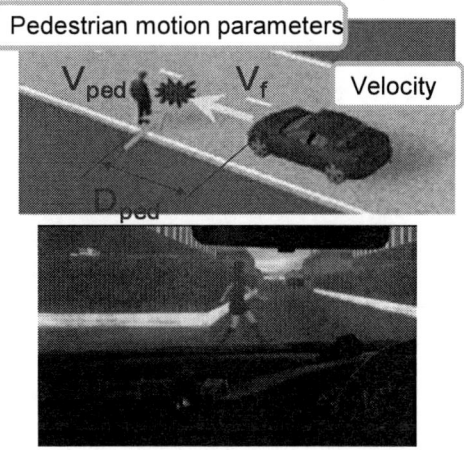

This is a typical case that requires a risk potential based control. Scenario modeling and collision avoidance control system design for other typical driving scenarios are being studied to extend the functionalities of the intelligent driving system.

2-8. Driver Acceptance Issues

Driving simulator studies and near-miss incident database analysis were conducted to investigate driver acceptance when the intelligent driving system is activated, and to identify the driving characteristics of drivers, especially elderly drivers, by focusing on the perception, recognition, decision-making (prediction), and operation capabilities. An anticipatory driving model with knowledge-based decision-making and a driving intention detection method were studied based on analysis of driving data collected from experienced drivers in the real-world. Over-trust and over-dependence in a system are key issues for introducing intelligent driving systems into a vehicle. Figure 9 shows the accident scenario used by the driving simulator developed by TUAT for investigating driver acceptance and HMI issues.

Fig.9 Accident scenario on TUAT driving simulator

2-9. Driving Environment Model and Database Construction by FOT

The framework of the FOT and the construction of driving intelligence utilizing big data are very important to enhance the performance of ASAS for accident prevention. It is necessary to improve the system based on real-world driving data using machine learning techniques, and to investigate human driver behavior adaptation using test vehicles equipped with a driver assistance or intelligent vehicle system to reveal changes in driver characteristics and the total driving performance in terms of safety. Figure 10 shows the scheme of the FOT conducted in this project, including the concept of global and local intelligent driving for enhancing the performance of the active safety vehicle.

As shown in Fig. 10, to build up the next-generation of driver assistance systems, it is necessary to integrate human factors with traffic environment data collected by

Fig. 8 Effectiveness of risk-potential based control algorithm on collision avoidance performance

978-1-4799-8002-4/14 $31.00 © 2014 IEEE   17

on-board, V2X, and vehicle-to-vehicle (V2V) sensors using data-fusion technology, and to evolve the intelligent driving model using machine learning techniques. The data fusion module functions in conjunction with environment perception, and recognition layer algorithms will be studied and synthesized. In the final stage, this project aims to create an innovative system architecture that takes into account the environment model of the finite preceding horizon with respect to the precise driver's vehicle location utilizing the horizon as well as dynamic maps. The constructed system must also be able to utilize knowledge-based information extracted from the driving database in real-time. Based on the research achievements

of this project, a test-bed-oriented framework will be created, aiming at the construction of an innovative platform for ADAS development and standardization.

3. Smart Traffic Flow Control Project

This section describes a second GIA collaborative project related to intelligent vehicle technologies. With the aim of developing a traffic smoothing system for expressways using adaptive cruise control (ACC), automakers formed a research consortium and started the Smart Traffic Flow Control Project in 2009. This project is targeting ways of reducing congestion at sag sections and tunnels, which currently accounts for the majority of congestion on expressways in Japan.

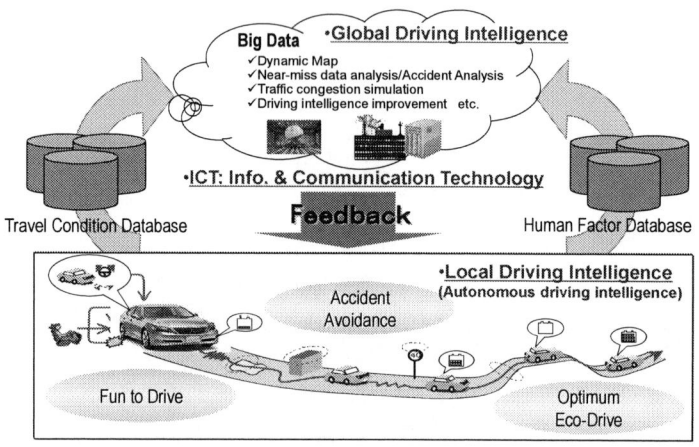

Fig.10 Vision of global and local intelligent driving system structure

Fig. 11 Scene from vehicle platoon tests in joint Automaker project

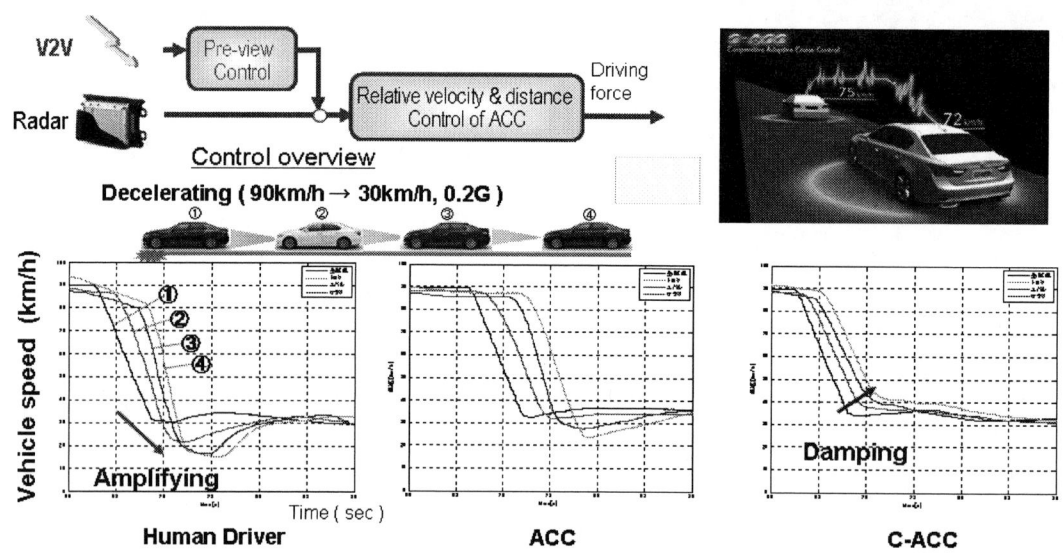

Fig.12 Study of following stability of ACC and C-ACC by research consortium

Congestion at sag sections is caused by drivers not noticing that the road is sloping upward, resulting in a gradual reduction in speed. At tunnels and similar locations, drivers tend to slow down, which causes congestion when these reductions in speed propagate to traffic approaching from behind. ACC has the potential to interrupt this propagation since it is capable of controlling both speed and vehicle-to-vehicle distance. This research consortium has identified the characteristics of ACC from the results of platoon driving tests (Fig. 11) and its effects on traffic flows using vehicle following simulations.

It has also predicted the congestion-reduction effect of ACC using data measured at sag sections in the Yamato district of the Tomei Expressway. This research consortium has also verified the effect of cooperative-ACC (C-ACC) systems, an advanced version of ACC that incorporates vehicle-to-vehicle (V2V) communication (Fig. 12).

In collaboration with the Ministry of Land, Infrastructure, Transport and Tourism (MLIT), the National Institute for Land and Infrastructure Management (NILIM), the University of Tokyo, and other institutions, demonstration runs using actual vehicles were carried out at the 2013 ITS World Congress in Tokyo and the effectiveness of C-ACC was disseminated widely. Based on these results, each automaker intends to refine its ACC systems, launch vehicles equipped with C-ACC, and further research and development into even more intelligent systems.

In addition, since these systems can be configured with the same architecture as the intelligent driving system shown in Figs. 5, 6, and 10, it should be possible to expand the applications of next-generation ADAS with driving intelligence.

4. Conclusions

This paper described the framework of an innovative driver assistance system utilizing the concept of intelligent driving and an experienced driver model that can potentially reduce accidents by recovering the deteriorated driving capabilities of elderly drivers. Current driver assistance systems can be categorized into three main types of development process.

(1) Enhancing safe driving and accident prevention
(2) Enhancing convenient and comfortable driving
(3) Fully-automated driving systems for driverless vehicle or center-based automatic transport systems

The final goal of this project is to utilize driving to enhance the quality of life and encourage active social behavior by elderly drivers through the various aspects of the intelligent driving technologies described in this paper. The development of intelligent driving system as the main focus of this project is an important innovative technology, and continuous efforts must be made to advance and deploy this technology in the near future.

References

[1] Traffic Bureau, National Police Agency, Statistics 2007 Road Accidents in Japan, International Association of Traffic and Safety Sciences (IATSS), (2008).

[2] Cabinet Office, Government of Japan, White Paper of the Aging Society in 2013.

[3] GOAL project website: http://www.goal-project.eu/

[4] Satomi, Y., Murano, T., Aga, M. and Yonekawa, T., A characteristic analysis of driving behavior to rear-end collision warning using a driving simulator, Proceedings of 18th JSME Conference on Transportation and Logistics (TRANSLOG), pp. 283-286 (2009).

[5] Raksincharoensak, P., Autonomous Driving System to Enhance Safe and Secured Traffic Society for Elderly Drivers, 18th World Congress on ITS (Invited talk), Special Interest Session No.30, Orlando, USA, (2011).

[6] Yang, J. and Coughlin, J.F., In-Vehicle Technology for Self-Driving Cars: Advantages and Challenges for Aging Drivers, International Journal of Automotive Technology, Vol.15. No.2, pp.333-340 (2014).

[7] Pongsathorn Raksincharoensak, Masao Nagai, Motoki Shino, Lane Keeping Control Strategy with Direct Yaw Moment Control Input by Considering Dynamics of Electric Vehicle, Vehicle System Dynamics Supplement Vol. 43, pp.192-201, 2006.

[8] Pongsathorn Raksincharoensak, Yuta Takimoto and Masao Nagai, Radar-Based Vehicle Following Control Algorithm of Micro-Scale Electric Vehicle, Proceedings of APAC07 (2007-01-3590), California, USA, (2007).

[9] Ryuzo Hayashi, Juzo Isogai, Pongsathorn Raksincharoensak and Masao Nagai, Autonomous Collision Avoidance System by Combined Control of Steering and Braking using Geometrically-Optimized Vehicular Trajectory, Vehicle System Dynamics, Vol. 50, Supplement, 151-168, (2012).

[10] Flemisch, F., Heesen, M., Hesse, T., Kelsch, J., Schieben, A. and Beller, J., Towards a dynamic balance between humans and automation, Cognition, Technology and Work - Special Issue on Human-automation Coagency, Volume 14 Issue 1, March 2012, pp. 3-18.

Social Impact of Power Semiconductor Devices
(Invited Paper)

B. J. Baliga, *IEEE Life Fellow*

North Carolina State University, Raleigh, NC 27695

Abstract: **Silicon IGBTs are now used in all the major sectors of our economy including transportation, consumer, lighting, industrial, medical, and renewable energy generation. The improved efficiency derived from IGBT-based automotive electronic ignition systems, adjustable speed motor drives, and compact fluorescent lamps has reduced gasoline consumption by over 1 Trillion gallons and electricity consumption by over 50,000 Terra-Watt-Hours. The social impact includes consumer cost savings of more than $ 15 Trillion and carbon dioxide emission reduction by over 75 Trillion pounds.**

Introduction

The social impact of power semiconductor devices is not fully appreciated because they are an embedded technology hidden from the eyes of society. These devices are used in all the major sectors of our economy improving the comfort and convenience of billions of people in the world. It is remarkable that at the same time this technology provides a huge cost savings for consumers because of the improved efficiency for the delivery and management of energy. The reduced energy consumption also provides an environment benefit.

The Insulated Gate Bipolar Transistor (IGBT)

The IGBT was first developed and commercialized at the General Electric Company in the early 1980s. This device combined the physics of operation of the existing bipolar transistor and the power MOSFET to generate much superior characteristics for power electronic applications. The high power handling capability of the IGBT device and the simplicity of its gate control circuit enabled substantial improvements in the size, weight, and cost of power electronic circuits for a wide range of applications. Within GE, it was immediately recognized that the IGBT would impact the small appliance (steam irons, toasters, mixers, food processors, etc.), large appliance (refrigerators, microwave ovens, washing machines, etc.), transportation (locomotive drives), factory automation (robotics), lighting (Triad-PAR, CFL), and medical (CT, MRI scanners) business sectors. Its applications have spread over the last 30 years to almost all sectors of the economy. This paper provides the analysis of the energy savings derived by using the IGBT in major sectors of our economy and the resulting social benefits.

Industrial Sector

Two-thirds of the electricity in the world is used for running

motors. Before the availability of IGBTs, pumps and fans were operated using induction motors with dampers to regulate the output. Although efficient at full output conditions, the efficiency of this approach drops to below 50% at light loads. The development of adjustable speed drives (ASDs) using IGBTs as switches allowed delivery of power with a variable frequency power source to the motor. The motor speed could be controlled without the use of dampers improving the efficiency by over 40%.

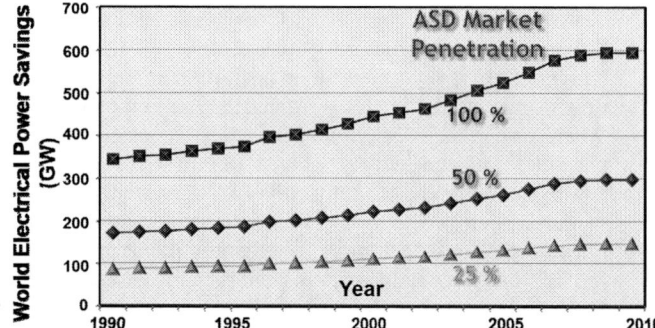

Fig. 1. Power Savings derived using IGBT-based Adjustable Speed Motor Drives.

The electricity consumption in the world has grown from 10,000 Terra Watt Hours in 1990 to 20,000 Terra Watt Hours in 2010. The electrical energy savings derived by using IGBT-based ASDs can be calculated based upon the 40% improvement in efficiency. The data is plotted in Fig. 1 for the time span from 1990 to 2010. The adoption of ASDs has grown rapidly in developed nations but more slowly in the underdeveloped countries because of its larger installation cost. The figure provides the energy savings for three penetration rates. It is estimated that the average world-wide penetration rate for ASDs is 50% during the last 20 years. Using this penetration rate and the 40% enhancement in efficiency derived using IGBT-based ASDs, a cumulative reduction in electricity consumption by 41,870 Terra Watt Hours is obtained during this period.

The electrical energy savings derived from ASDs provides consumers the benefits of cost savings. The price for electricity varies around the world and within each country as well. In the United States, the price for electricity is about $ 0.10 per kW-hr. However, it is much greater in Western Europe ranging from $ 0.31 kW-hr in Germany to $ 0.43 kW-hr in Demark. The average global cost for electricity is about $ 0.20 per kW-hr. Using this value, the cost savings derived by consumers is charted in Fig. 2 for the three ASD penetration rates. Assuming an average penetration rate of 50% for ASDs around the world, the cumulative cost savings during the period from 1990 to 2010 adds up to more than $ 8 Trillion.

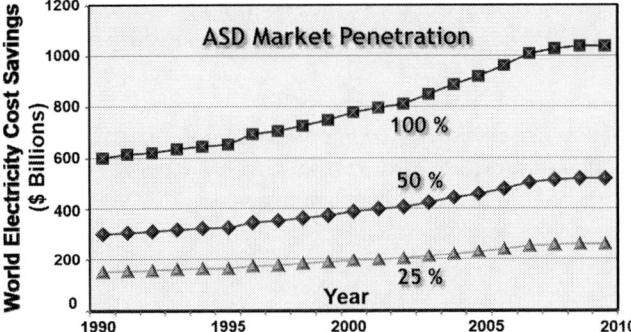

Fig. 2. Annual cost Savings derived using IGBT-based Adjustable Speed Motor Drives.

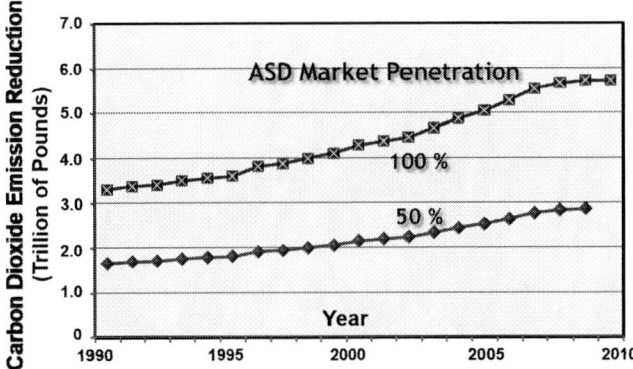

Fig. 3. Annual reduction in carbon dioxide emissions derived using IGBT-based Adjustable Speed Motor Drives.

Unfortunately, most (68%) of the electricity in the world is produced by burning fossil fuels such as coal and natural gas. The rest is produced using hydroelectric, nuclear, and more recently renewable energy sources such as solar and wind power. Coal fired power plants produce carbon dioxide emissions at the rate of 1.1 pounds per kW-hr of generated electricity. The carbon dioxide released from human activity, has been correlated to a rise in global temperature due to the green-house effect. In addition, the burning of fossil fuels introduces carcinogens into the atmosphere. Based up on the electricity savings derived by using IGBT-based ASDs, the reduction in carbon dioxide emission can be calculated and is shown in Fig. 3. Based up on a 50% global penetration rate for ASDs, the cumulative reduction in carbon dioxide emission during the period from 1990 to 2010 adds up to 46 Trillion pounds. As the penetration rate for ASDs grows to 100% around the world, twice this amount of carbon dioxide emission reduction can be anticipated.

Transportation Sector
Over the past century, society has globally become dependent on using cars and trucks operated using the internal combustion engine (ICE) with gasoline as the fuel for the transportation of people and merchandise. The ICE operates by using a spark plug to ignite a mixture of gasoline and air in the cylinders. Until the 1990s, the Kettering mechanical ignition system was used with the timing of the spark plug controlled using a rotating distributor. It was essential to periodically tune the distributor to maintain good timing for the spark plug to achieve good efficiency and avoid engine knocking problems. The Kettering system had poor accuracy for the timing and its drift due to mechanical wear degraded fuel efficiency.

The availability of rugged IGBTs that could tolerate the adverse conditions under the hood of a car enabled introduction of the electronic ignition system (EIS) in the late 1980s. In the electronic ignition system, the IGBT is used on the primary side of the ignition coil (actually a transformer) to regulate the current flow. A very high voltage is generated on the secondary side of the ignition coil to ignite the spark plug under the control of the IGBT. Billions of IGBTs have been sold by multiple manufacturers for this application since 1990.

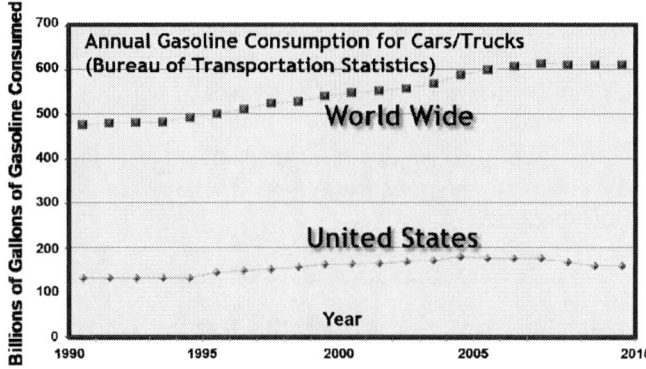

Fig. 4. Annual gasoline consumption in the United States and World Wide.

The precise electronic control of the spark with the EIS allows using a leaner fuel mixture which has improved the fuel efficiency by at least 10%. The fuel savings derived from this IGBT-based innovation can be determined by using the available data on gasoline consumption in the United States and around the world as shown in Fig. 4. Between 1990 and 2010, the gasoline utilization for automobile transportation ranges from 100 to 200 Billion gallons per year in the United States and from 500 to 600 Billion gallons per year World-Wide. Using a conservative 10% savings in fuel derived by IGBT-based EIS, a cumulative gasoline savings of 1.1 Trillion gallons has been achieved between 1990 and 2010.

The cost benefit to consumers from having to purchase less gasoline due to the improved efficiency derived from IGBT-based EIS can be quantified using data on the price of gasoline. The average price of unleaded regular gasoline in the United States between 1990 and 2010 is shown in Fig. 5. It

978-1-4799-8002-4/14 $31.00 © 2014 IEEE

remained fairly low around $ 1.30 until 2002 when it began increasing steadily (with a spike in 2008). The price for gasoline in Europe and Asia has been maintained at about 3.5 times greater than in the United States. Combining this price for gasoline with the gasoline consumption shown in Fig. 4 and utilizing a 10% reduction in fuel consumption due to the IGBT-based EIS yields a cumulative cost savings for World-Wide consumers of $ 5.65 Trillion from 1990 to 2010.

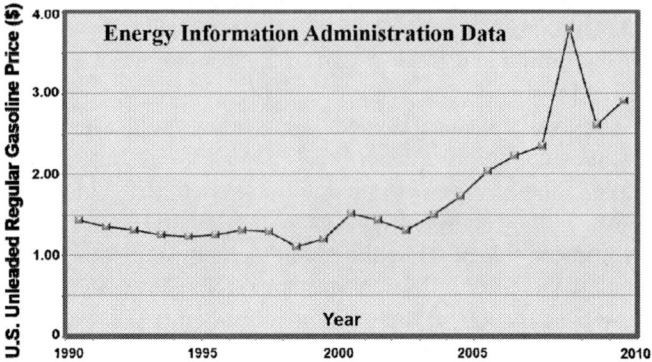

Fig. 5. Price for unleaded regular gasoline in the United States.

Burning one gallon of gasoline produces 19.4 pounds of carbon dioxide as well as emission of carcinogenic compounds. Consequently, the reduction of gasoline consumption derived by the development of the IGBT-based EIS has produced a huge annual reduction in carbon dioxide emission as shown in Fig. 6. The cumulative reduction in carbon dioxide emissions from this technology adds up to 22 Trillion pounds.

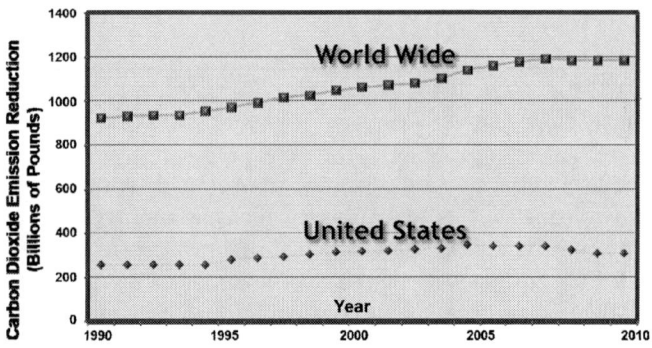

Fig. 6. Annual reduction in carbon dioxide emissions derived using IGBT-based Electronic Ignition System.

In addition, the introduction of electric and hybrid electric vehicles has reduced gasoline consumption especially in urban areas. Every EV and HEV utilizes IGBTs for control of the power delivered from the battery to the electric motors. In addition, IGBTs are employed in the regenerative braking systems to improve fuel efficiency even further. As the number of EVs and HEVs increases, the impact of this technology will become comparable if not greater than that discussed above.

Lighting Sector

The availability of lighting in homes and offices for 24 hours per day using electricity created a major enhancement in the quality of life and productivity for people around the globe. The development of the incandescent lamp, attributed to the work by Edison in the 19th century, has served mankind for over 100 years. Unfortunately, the incandescent lamp operates by heating a tungsten filament to 3000 °C with less than 5% of the energy being converted into visible light. In addition, it has a relatively short lifespan of 1000 hours.

A major enhancement in efficiency for lighting was achieved by the introduction of compact fluorescent lamps (CFLs) in 1990. Today, more than 14 Billion CFLs have been sold and are utilized around the world. A typical 60 W incandescent lamp can be replaced with a CFL that consumes only 15 W of power to produce the same amount of light. The challenges of miniaturization of the electronic ballast used to control the gas discharge in the CFL were overcome due to the ability of the IGBT to control large amounts of power with a small footprint, the ability to integrate its low power gate control circuit, and its high temperature operating capability.

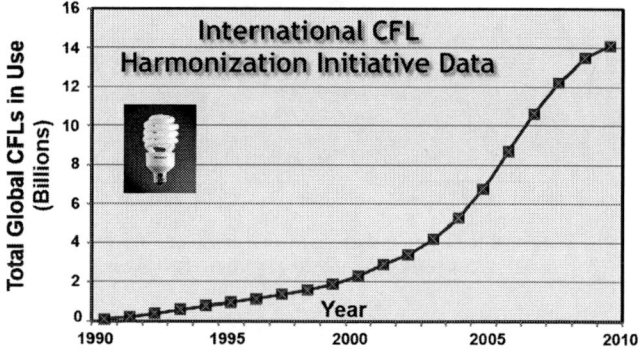

Fig. 7. Number of CFLs in use globally.

The analysis of the power savings derived from using IGBT-based CFLs requires determination of the number of lamps in use by consumers. This is not equal to the number of CFLs sold because of their finite lifespan. According to manufacturers, the lifespan for a CFL is typically 10,000 hours. The CFL must be replaced after 5 years of use if turned-on for 6 hours per day. It must therefore be assumed that all CFLs purchased by consumers after 1995 are first used to replace purchases made 5-years ago and the rest for replacing incandescent lamps. The number of CFLs in use globally is shown in Fig. 7 after taking this into account. This growth has been spurred by European and Asian initiatives to encourage the use of CFLs in order to reduce energy demands.

The power savings derived by the widespread global utilization of CFLs can be computed assuming that each 60-W CFL is used for 6 hours per day with 45 watts of reduced electricity consumption. The global power savings derived from IGBT-based CFLs has grown from 4.5 GW in 1990 to over 600 GW in 2010 as shown in Fig. 8.

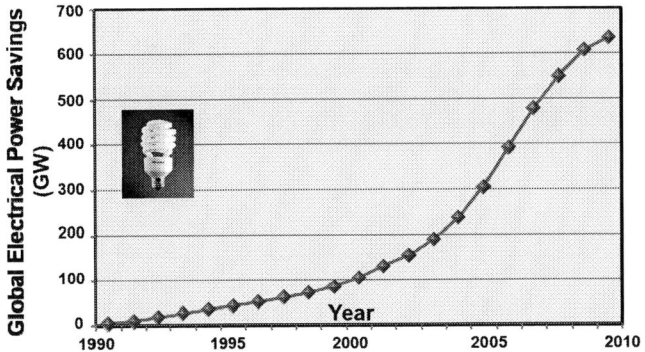

Fig. 8. Global power savings derived from IGBT-based CFLs.

The cost savings for consumers from the reduced electricity consumption using CFLs can be computed using the above power savings and an average world-wide cost of electricity of $ 0.20 per kW-hr. The annual cost savings have grown from $ 2 Billion in 1990 to $ 278 Billion in 2010. The cumulative cost savings derived by consumers from this IGBT-enabled technology adds up to $ 1.8 Trillion over this time span.

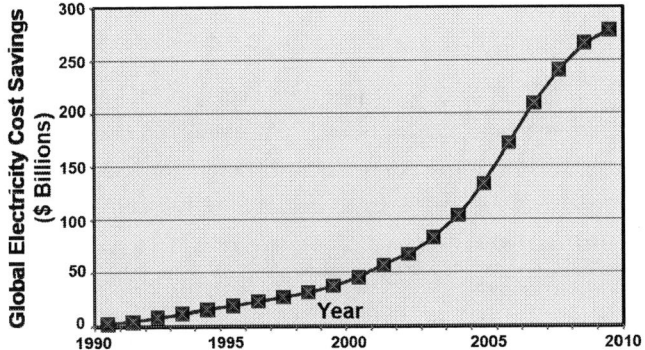

Fig. 9. Global annual consumer cost savings derived from IGBT-based CFLs.

The annual reduction in carbon dioxide emission derived from the electricity power savings can be computed using the information in Fig. 8 and an emission rate of 1.1 pounds per kW-hr from fossil fuels. The results are provided in Fig. 10. The cumulative carbon dioxide emission reduction derived from CFLs adds up to 10 Trillion pounds during this time span from 1990 to 2010.

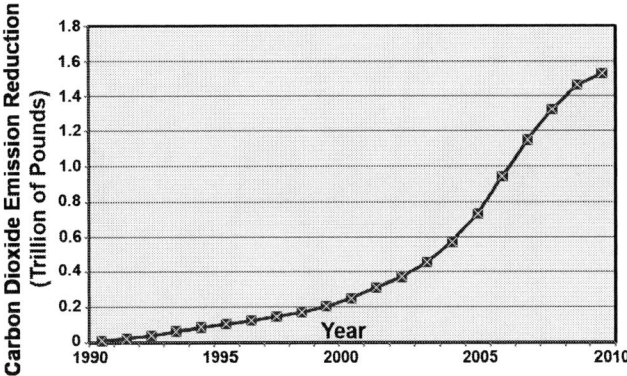

Fig. 10. Global annual carbon dioxide emission reduction derived from IGBT-based CFLs.

Utilities

Each one GigaWatt coal fired power plant for electricity generation requires an investment of $ 2 to 3 Billion by the utilities. The combined largest annual power savings (in 2010) derived from IGBT-enabled ASDs and CFLs is 1,200 GW. This reduction in power generation derived from IGBT-enabled ASDs and CFLs has been beneficial to the utility industry by eliminating the need for $ 2.8 Trillion in capital investments.

Conclusions

Power semiconductor devices, with the IGBT as a specific example, have enabled innovations in the industrial, transportation, and lighting sectors with a total electrical energy savings of over 50,000 Terra Watt Hours and gasoline savings of over 1 Trillion gallons during the last 20 years. This has saved consumers over $ 15 Trillion and reduced global carbon dioxide emissions by over 75 Trillion pounds. In addition, utilities have saved over $ 2.8 Trillion in coal-fired power plant investments because of the reduced demand for electricity.

The information used for the social and economic analysis provided in this paper is discussed in greater detail in a new book: "The IGBT: Device Physics, Design, and Applications" written by the author that will be published by Elsevier Press in 2015. This book will also contain an extensive list of references for the data used in this paper.

In the future, power devices made using wide band gap semiconductors (WBGS), such as Silicon Carbide and Gallium Arsenide, promise reduction of on-state and switching losses in the rectifiers and transistors used in applications. The availability of the silicon IGBT has already pushed the efficiency for most applications into the 90% range. The greatest impact of WBGS will be in raising the operating frequency of the power circuits to reduce the size, weight, and losses in the passive components.

978-1-4799-8002-4/14 $31.00 © 2014 IEEE

SiC Power Devices for HEV/EV and a Novel SiC Vertical JFET

Tsuyoshi Ishikawa, Yasunori Tanaka*, Tsutomu Yatsuo*, and Koji Yano**

Toyota Central R&D Labs., Inc., Nagakute, Aich, JAPAN
*National Institute of Advanced Industrial Science and Technology (AIST), Tsukuba, Ibaraki, JAPAN
**University of Yamanashi, Kofu, Yamanashi, JAPAN
Tel: +81-561-71-8104, Fax: +81-561-63-6042, Email: t-ishi@mosk.tytlabs.co.jp

Abstract

We propose a novel SiC VJFET with low feedback capacitance C_{rss}. A key feature of the proposed VJFET is the p^+ screen grid inserted between gate and drain electrode. The screen grid effectively reduces the C_{rss} by about 80% compared to a conventional VJFET. The lowest total power dissipation among existing SiC power devices can be achieved by the low C_{rss}. This new VJFET can be a promising candidate for a high-speed and low-loss SiC power device.

Introduction

Silicon carbide (SiC) is a promising material for power electronics because of its high breakdown voltage, high operating temperature and superior thermal conductivity. SiC high voltage (1000~2000 V) and low on-resistance unipolar switching devices (including Schottky barrier diodes, JFETs and MOSFETs) have become commercially available due to the recent improvement of crystal growth

and device processing technologies. Switching loss of these devices is considerably low compared with Si PN diodes and Si IGBTs because there is no reverse recovery and tail current in the case of SiC unipolar devices. Decrease in switching loss achieves high-frequency operation of power conversion system and extending the driving range of hybrid electric vehicles (HEVs) and electric vehicles (EVs). High frequency operation also leads to the miniaturization of the DC/DC converter components, especially passive components including inductor and capacitor (Fig. 1). SiC power devices will contribute to boost the performance of HEVs and EVs.

Issue of high speed operation for SiC devices

Reliability and long term stability issues in SiC MOSFETs (1, 2) give advantage to SiC VJFETs which do not require a gate insulator. However, it is difficult for conventional normally-off SiC VJFETs to achieve high-speed switching without gate-drive assist (3, 4) due to the large feedback capacitance C_{rss} and limited low gate-overdrive voltage V_G-V_{th} (< 2.5 V). SiC MOSFETs also suffer from large C_{rss}

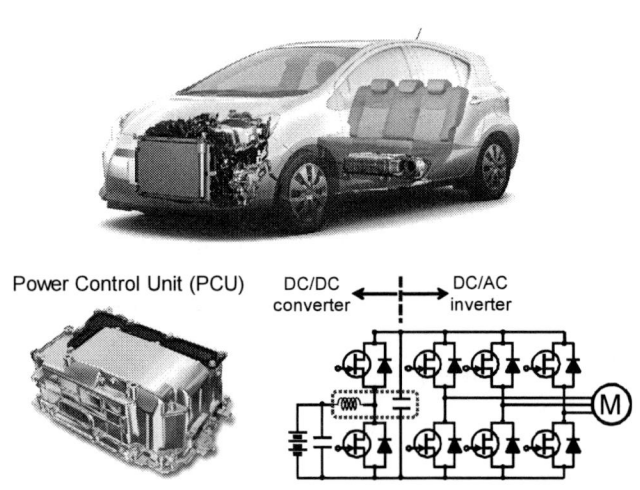

Figure 1 Power conversion system for hybrid electric vehicles (HEVs). Power control unit consists of DC/DC converters, DC/AC inverters, passive components and water cooler.

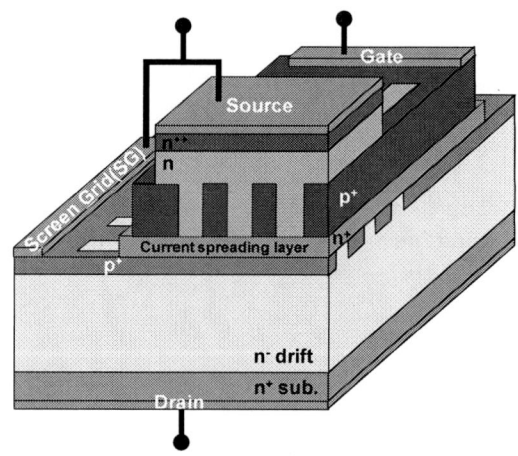

Figure 2 A schematic diagram of the proposed SiC VJFET structure (Screen Grid VJFET: SG-VJFET).

978-1-4799-8002-4/14 $31.00 © 2014 IEEE

because of the thin depletion layer width due to the high impurity concentration of the drift layer compared to Si-based devices with the same blocking voltage (5). This is a common issue for SiC power devices. Therefore, it is important for SiC power devices to develop a new device structure reducing C_{rss}. In this paper, we propose a novel normally-off SiC VJFET with low C_{rss} based on the SiC Buried Gate Static Induction Transistor (BGSIT) (6, 7).

Novel SiC VJFET with low-feedback capacitance

A. Device structure and simulation setup

A major feature of the new VJFET is the p+ screen grid, the fourth electrode, located between gate and drain (Fig. 2). This electrode is usually connected with the source electrode. It reduces the feedback capacitance C_{rss} due to the decrease in capacitive coupling between gate and drain. On the other hand, the screen grid restricts the current flow to narrow regions, so it leads to increase the on-resistance. In order to suppress this effect, current spreading layer, which is the n-type high impurity concentration layer, is introduced between p+ gate and screen grid. We call this new device Screen Grid VJFET (SG-VJFET).

In order to compare the performance of SG-VJFET with that of other SiC FET devices (Table 1) by using device simulator, the thickness and doping concentration of n⁻ drift layer and n+ substrate are common among three devices (n⁻ drift layer: 10 µm, 8×10^{15} cm⁻³. n+ substrate: 350 µm, 6×10^{18} cm⁻³). Device modeling of trench-type VJFET and trench-type MOSFET, was carried out by using device simulator based on published data (8, 9).

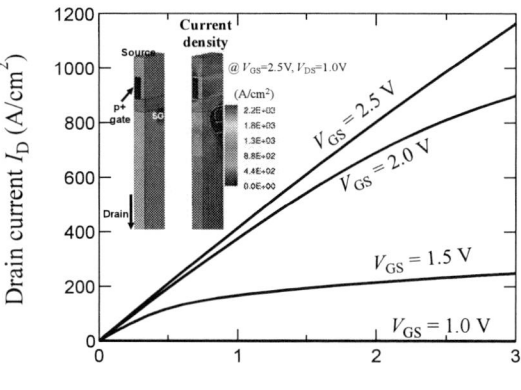

Figure 4 Simulated output characteristic (I_D vs. V_{DS}) of SG-VJFET at V_{GS} = 1.0~2.5 V.

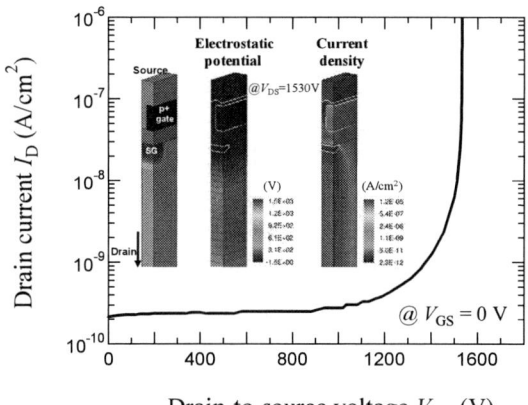

Figure 5 Simulated forward blocking characteristic of SG-VJFET at V_{GS} = 0 V.

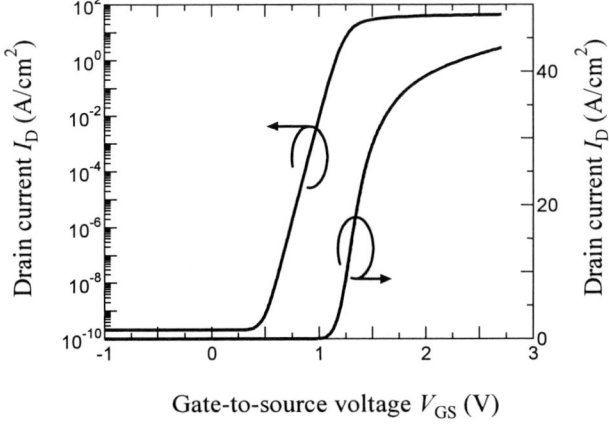

Figure 3 Simulated transfer characteristic (I_D vs. V_{GS}) of SG-VJFET.

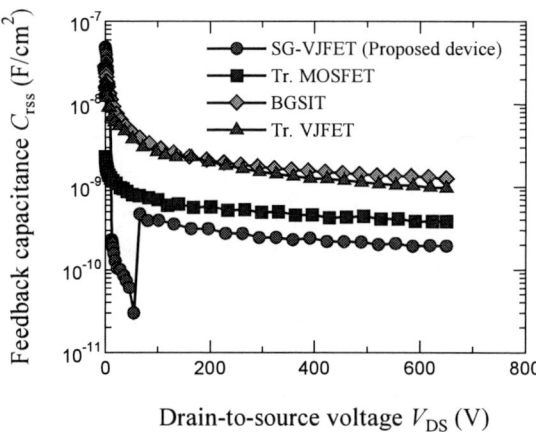

Figure 6 Comparison of simulated feedback capacitance C_{rss} among three different types of SiC switching devices.

Table 1 Summary of static characteristics for different types of SiC devices.

	(Proposed device) SG-VJFET	Tr. VJFET	Tr. MOSFET
Device structure			
On resistance R_{on} @ V_{DS}=1.0V	2.4 mΩ•cm² [*3]	3.3 mΩ•cm² [*3]	2.7 mΩ•cm² [*4]
Threshold volatage V_{th} @V_{DS}=0.1V, I_D=0.1A/cm²	+1.0 V	+1.0 V	+6.1 V
Feedback capacitance C_{rss} @V_{DS}=650V	0.19 nF/cm²	0.97 nF/cm²	0.38 nF/cm²

[*1] n⁻ drift layer: 8×10^{15} cm⁻³, 10 μm   [*3] @ V_G = 2.5 V
[*2] n⁺ substrate: 6×10^{18} cm⁻³, 350 μm   [*4] @ V_G = 15 V

B. Static characteristics

Static characteristics of SG-VJFET are calculated by using device simulator. Both normally-off (V_{th} = +1.0 V) and low on-resistance (R_{on} = 2.4 mΩ•cm²) are achieved due to the fine line and space p⁺ gate pattern (Fig. 3 and Fig. 4). As shown in Figure 5, blocking voltage > 1400 V has been realized even at zero gate bias V_{GS} = 0 V. Figure 6 shows the comparison of C_{rss} among three different SiC switching devices. The C_{rss} of SG-VJFET is the smallest and reduced by 80% compared to the conventional VJFETs.

C. Dynamic characteristics

To investigate the improvement in switching performance of SG-VJFET, an inductive load double pulse tests were performed on three devices (Active area: 0.25 cm²) which have the static characteristics shown in Table 1. A 650V DC voltage is supplied to an inductor of 100μH and a freewheeling SiC Schottky Barrier diode (Active area: 0.25 cm²). Due to the low C_{rss}, SG-VJFET is the highest switching speed and the lowest switching loss among three devices under the same di/dt condition (see Fig. 7 and Fig. 8). Supposing that the converter carrier frequency and duty ratio are 10 kHz and 0.5 respectively, the total power loss of the FETs is estimated from the results of static and dynamic characteristics (Fig. 9). The total loss of SG-VJFET is the lowest among existing SiC power devices. Based on this estimation, carrier frequency f_c of the DC/DC converter with the SG-VJFET can be increased by more than two times

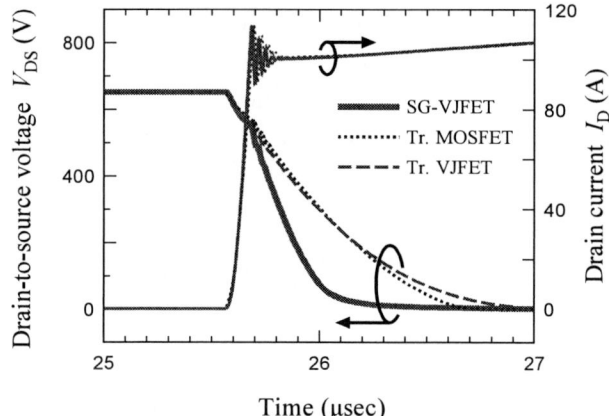

Figure 7 Simulated turn-on switching waveforms of three different devices. (V_{DD}=650V, I_D=100A)

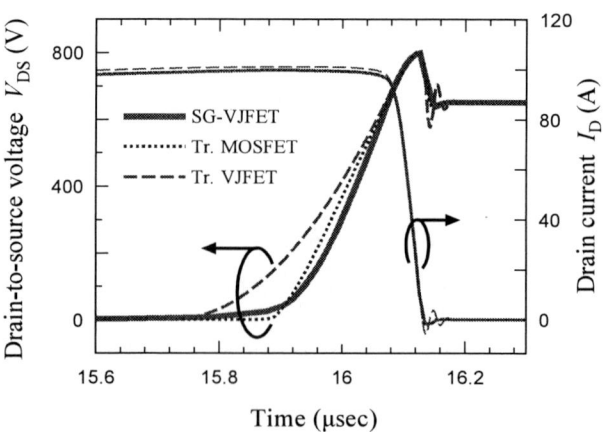

Figure 8 Simulated turn-off switching waveforms of three different devices. (V_{DD}=650V, I_D=100A)

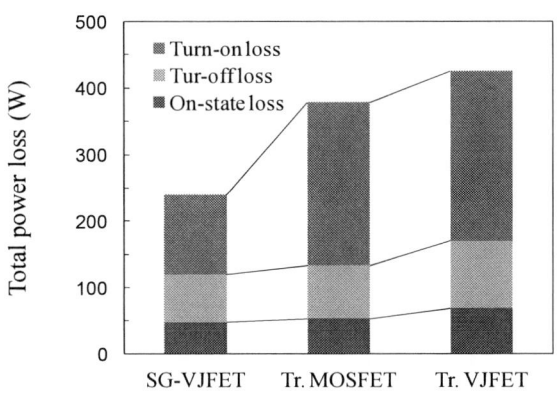

Figure 9 Comparison of FET total power loss among three different devices. (Switching current: 400A/cm²)

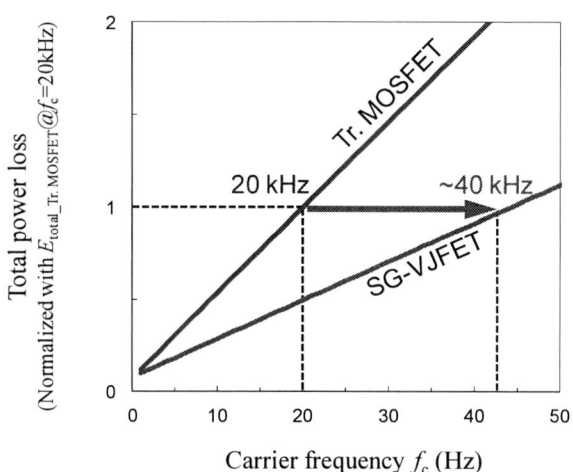

Figure 10 Dependence of total power loss on carrier frequency f_c. Each data is normalized with total loss of trench-gate MOSFET at $f_c = 20$ kHz.

compared to that using the trench-gate MOSFET to maintain the same dissipation (Fig. 10).

Conclusion

We propose a novel SiC vertical JFET (VJFET) with low feedback capacitance C_{rss}. A key feature of the proposed device is the p+ screen grid inserted between gate and drain electrode. The screen grid is effective to reduce the C_{rss} by 80% compared to conventional VJFETs. Due to the low C_{rss}, the total power loss of the SG-VJFET is the lowest among existing SiC power devices. From these results, SG-VJFET could be a promising candidate to improve the performance of power conversion systems for HEVs and EVs.

References

(1) R. Singh, "Reliability and performance limitations in SiC power devices", *Microelectronics Reliability*, vol. 46, pp. 713-730, 2006.

(2) Mrinal K. Das, Sarah K. Haney, Jim Richmond, Anthony Olmedo, Q. Jon Zhang, and Zoltan Ring, "SiC MOSFET Reliability Update", *Mat. Sci. Forum*, vols. 717-720, pp. 1073-1076, 2012.

(3) Katusmi Ishikawa, Hidekatsu Onose, Yasuo Onose, Takasumi Ooyanagi, Tomoyuki Someya, Natsuki Yokoyama, and Hiroshi Hozouji, "Normally-off SiC-JFET inverter with low-voltage control and a high-speed drive circuit", *Proceedings of 19th International Symposium on Power Semiconductor Devices & ICs* (ISPSD), pp. 217-220, 2007.

(4) Robin Kelley, Andrew Ritenour, David Sheridan, and Jeff Casady, "Improved two-stage DC-coupled gate driver for enhancement-mode SiC JFET", *Proceedings of Applied Power Electronics Conference and Exposition Annual IEEE Conference* (APEC), pp. 1838-1841, 2010.

(5) Tsuyoshi Funaki, Yuki Nakano, and Takashi Nakamura, "Comparative Study of SiC MOSFETs in High Voltage Switching Operation", *Mat. Sci. Forum*, vols. 717-720, pp. 1081-1084, 2012.

(6) Yasunori Tanaka, Mitsuo Okamoto, Akio Takatsuka, Kazuo Arai, Tsutomu Yatsuo, Koji Yano, and Masanobu Kasuga, "700-V 1.0-mΩ•cm2 Buried Gate SiC-SIT (SiC-BGSIT)", *IEEE Electron Device Lett.*, vol. 27, pp. 908-910, 2006.

(7) Akio Takatsuka, Yasunori Tanaka, Koji Yano, Tsutomu Yatsuo, and Kazuo Arai, "980 V, 33A Normally-Off 4H-SiC Buried Gate Static Induction Transistors", *Mat. Sci. Forum*, vols. 679-680, pp. 662-665, 2011.

(8) D.C. Sheridan, A. Ritenour, V. Bondarenko, P. Burks, J.B. Casady, "Record 2.8mΩ-cm2 1.9kV Enhancement-Mode SiC VJFETs", *Proceedings of 21st International Symposium on Power Semiconductor Devices & ICs* (ISPSD), pp. 335-338, 2009.

(9) A. Ritenour, D.C. Sheridan, V. Bondarenko, and J.B. Casady, "Performance of 15 mm2 1200 V Normally-Off SiC VJFETs with 120 A Saturation Current", *Mat. Sci. Forum*, vols. 645-648, pp. 937-940, 2010.

Application specific trade-offs for WBG SiC, GaN and high end Si power switch technologies

R. Rupp[1], T. Laska[1], O. Häberlen[2] & M. Treu[2]

[1] Infineon Technologies AG, Am Campeon 1-12, 85579 Neubiberg, Germany

[2] Infineon Technologies Austria AG, Siemensstrasse 2, 9500 Villach, Austria

Abstract

There is an increasing choice of power switches in the 600V to 1700V range for the application engineers. Besides the well-established Si SJ (Super Junction) MOSFETs and IGBTs now also silicon carbide (SiC) and latest gallium nitride (GaN) power switches are available for new designs. Complete new system optimizations are possible driven by totally different trade off options e.g. between static and dynamic losses and their temperature dependencies. In this paper we explain these trade-offs for the different device types and show the consequences based on some prominent sample applications.

Introduction

In the last decade the competition of power semiconductor switch technologies was enriched by two new members: SiC-FETs (MOSFETs and JFETs) and lateral GaN-HEMTs. Whereas the SiC devices convince with great performance but traditionally suffer from high wafer costs and wafer diameter limitations, GaN-HEMTs can be manufactured on large and cheap Si-wafers – but still have deficiencies with respect to ruggedness and require significant nominal voltage derating. In parallel to this new semiconductor switch solutions, the traditional Si-based technologies of IGBT (like Trench Fieldstop) and Super Junction MOSFETs (like CoolMOS) have improved continuously. Especially in the 600V to 1200V

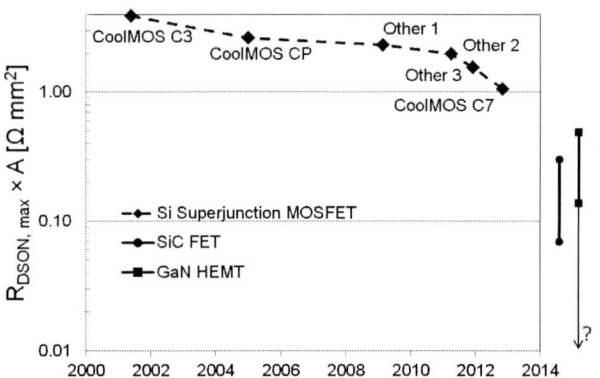

Fig. 1:
Trend of the chip resistivity $R_{DSON,max} \times A$ over time for best available 600V or 650V part and an estimated range for the first 2 to 3 generations of potential commercial SiC and GaN switches (see also [1]).

blocking range this results in a very competitive situation with no clear long term winner identified today. The race will be decided individually depending on the target application. This paper will deal – therefore – with some of these application-specific trade-offs and development trends related to the mentioned power switch technologies.

Competitive situation of power switch technologies

Super junction MOSFETs:

Fig. 1 shows the trend in $R_{DS,on} \times A$ development exemplary for the Infineon CoolMOS technology family. The value of $1\Omega mm^2$ is already achieved by reducing cell pitch and compensation accuracy. Modern SiC switch concepts promise a factor 5 lower on-resistance, but this is easily eaten up by the higher SiC area costs for 650V devices. In parallel also the E_{oss} values (energy stored in the output capacitance, Fig. 2a) of the most recent SJ Si devices were reduced by more than a factor of two now being pretty close to the SiC and GaN values [1]. Therefore, for hard switching applications like classical power factor correction there is very little benefit left for WBG switches. However for the same SJ Si devices there is stagnation with respect to Q_{oss} (charge stored in the output capacitance, Fig. 2b), the key parameter for so called resonant applications or hard switching bridges. Here both SiC and GaN will clearly outperform the Si MOS solutions [1] and get justification for a price premium. This is very important, because SJ devices and IGBTs are starting to be manufactured on a 300 mm wafer platform, paving the way for even lower manufacturing costs.

IGBTs:

By tailoring the local plasma density in sophisticated trench and tricky vertical structures the power density of 650V - 1700V IGBTs was also significantly improved in the last years, coming to new and improved $V_{CE,sat}$ vs. switching loss trade-offs. In many cases this approach is only hampered by the high resulting saturation current and the related low short circuit robustness. This is a restriction in various applications, however it can be improved by better power dissipation allowed by improved die attach (diffusion soldering, sintering) and thicker front side power metal serving as short-term heat sink. Also another drawback of the IGBT – the missing integrated free-wheeling

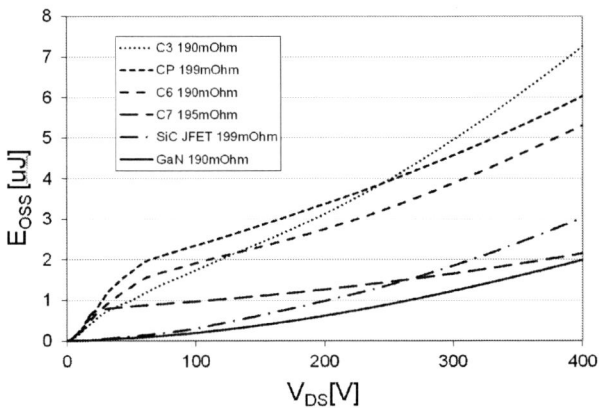

Fig. 2a:

Comparison of the energy stored in the output capacitance E_{OSS} for different CoolMOS devices and potential SiC and GaN devices.

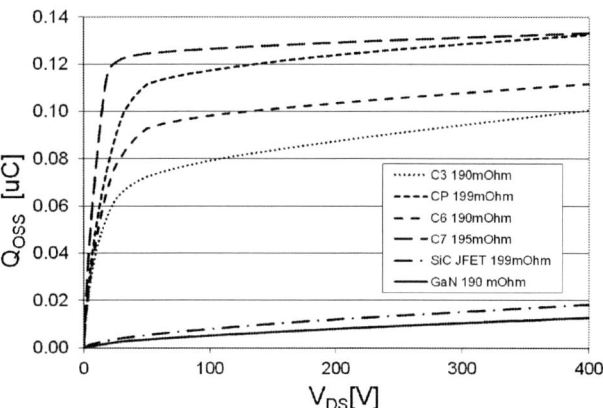

Fig. 2b:

Comparison of the charge stored in the output capacitance Q_{OSS} for different CoolMOS devices and potential SiC and GaN devices.

diode – is meanwhile addressed and quite encouraging results are available for the so called reverse-conducting IGBT (Fig. 3, [2]). With respect to manufacturing cost, the IGBT is clearly superior to all other power switch technologies and has the lowest T-dependence of conduction losses (Fig. 4). But even with all the recent performance improvements, the WBG switches are providing only ~1/10th of the switching losses, if the application allows high dV/dt and the assembly setup is low inductive.

Furthermore the switching losses of IGBTs as bipolar devices are T-dependent (Fig. 5) due to the continuously increasing minority carrier lifetime. They are rising by 40-50% between room temperature and 150°C for the classical field stop IGBTs and increase even more for devices with lifetime killing. However, when using a Reverse Conducting IGBT as shown in Fig. 3 and applying an active gate control during diode operation to reduce the minority carrier density (desaturation) before turning off the diode, a >50% reduction of the rectifying losses are achievable [3]. But this requires a more sophisticated, nonstandard gate control algorithm, which is a certain barrier for the market introduction of this device concept.

Therefore, whether an IGBT or one of the other devices has benefits in certain applications depends mainly on the ratio between static and dynamic losses and whether the circuit itself allows steep dI/dt and dV/dt gradients (e.g. has low parasitic loop inductance).

SiC FETs:

SiC diodes meanwhile have a commercial history of more than a decade and are a fixed part of high-end power supplies in server and telecom applications, whereas SiC switches are still in an early stage of market penetration. Reasons are challenges

concerning gate oxide quality and channel mobility which do not allow the full exploitation of the material's potential yet.

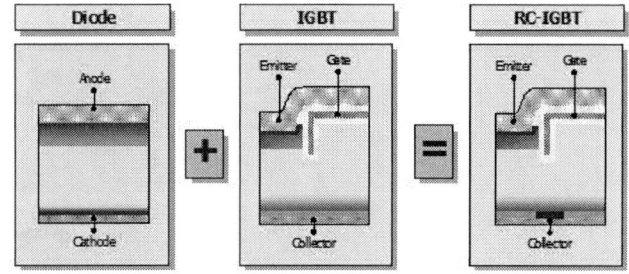

Fig. 3:

Reverse Conducting IGBT with integrated Free Wheeling Diode function. In case of a 1200V IGBT this diode functionality increases the chip area only by about 20%.

On the other hand, the average annual cost down of SiC base material was >12% over the last decade, allowing the SiC switches becoming more and more competitive. Especially in 1200V applications, where no super junction MOSFETs are available as competition, higher frequencies provide system advantages (higher power densities), efficiency is critical and a low loss body diode is required (Solar converters, UPS), SiC power switches are gaining a lot of interest. Fig. 6 shows an efficiency comparison of a solar converter using most recent SJ devices, GaN or SiC power switches (see also [4]): The SiC JFET [5] used here clearly outperforms the CoolMOS C7 over the whole output power range, and is in the same efficiency range as the GaN HEMT devices – but with a much better ruggedness and voltage safety margin.

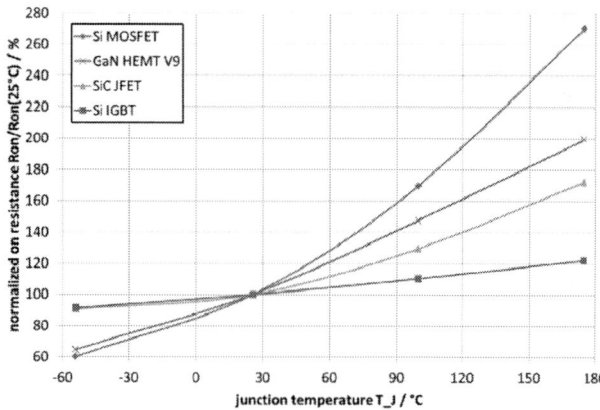

Fig. 4:

R_{on} over junction temperature (normalized to 25°C) for 650V power switches: GaN-HEMT (normally on), SiC-JFET, newest IGBT and CFD CoolMOS™

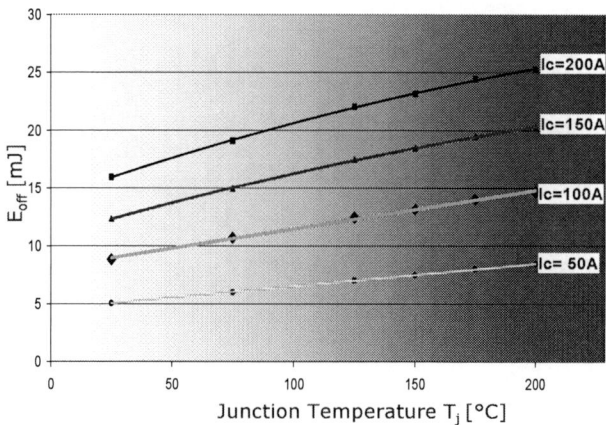

Fig. 5:

100A/1200V Trench Field Stop IGBT (IGBT 3) turn off losses up to 200°C

GaN HEMTs:

These devices are completely different from the traditional FETs as they are lateral, have very low gate and output charge, virtually zero reverse recovery losses, but show destructive breakdown witch current filamentation instead of avalanche. And they are based on hetero-epitaxial material growth, as only GaN-on-Si is cost competitive to the other technologies yet. The consequence of this approach is a very high defect density of the material requiring a significant voltage de-rating (use-voltage vs. breakdown voltage). The very low $R_{DS,on}$ of these devices allows high current densities, but on the other hand this is hampered by the low thermal capacitance and thermal conductivity of the GaN-on-Si structure. However extremely low switching losses

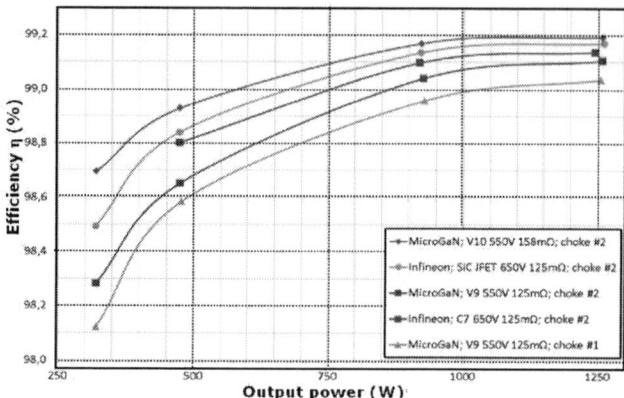

Fig. 6:

Efficiency results of a compact solar boost converter prototype as function of output power (U_{in}=300V, U_{out}= 350V, f_{sw}=200kHz) [4]

due to the high carrier mobility motivate even the development of new circuit designs for e.g. power supplies constructed around the specific set of properties of the GaN switches – especially as there is hope, that in the future such GaN switches will be even cheaper than CoolMOS in the 650V range. Furthermore these switching losses are independent from temperature, as they are pure capacitive losses – similar to Si SJ MOSFETs and SiC FETs. In contrast to this – unfortunately – the conduction losses show quite significant temperature dependence as depicted in Fig. 4 (80% increase between room temperature and 150°C). In combination with the limited thermal capacity and conductivity of the GaN-on-Si HEMTs this leads to a high risk of thermal runaway and to quite limited surge current and short circuit capability. In applications where this is important (e.g. drives or PFC after AC drop out) this may require a significant over-dimensioning of the HEMTs in comparison to other devices or the installation of protection circuits.

Outlook & Conclusion

Table 1 shows a simplified comparison of all the power switches covered in this article. The most interesting voltage class is around 600V – this is a market segment with very high volume, where all technologies compete. Neglecting the cost differences between the various technologies would lead to a clear champion: the SiC FET. This device offers the highest flexibility with respect to the various applications as good Q_{oss} and E_{oss} is combined with a very fast integrated body diode (low Q_{rr}), a high ruggedness concerning surge events and excellent thermal properties. But the high SiC material cost dims this picture and limits the market acceptance.

978-1-4799-8002-4/14 $31.00 © 2014 IEEE

On the other hand, the cost down of SiC base material of >12%/a is encouraging and very likely we will also see 200mm SiC base material at the end of this decade. Under this boundary conditions SiC power switches will continue to increase their market share in the next few years.

However it is still questionable, whether this will be accompanied by an increase of the maximum junction temperature to values of 200°C and above. Of course SiC allows such high T_j without losing blocking capability, but the related increase of static losses and the challenge of a significantly increased temperature cycling capability are serious obstacles [6].

In case of GaN the market penetration will depend strongly on application relevant demonstration of long term reliability, minimized de-rating of use-voltage vs. breakdown voltage and cost position. The last point is mainly related to the still quite high GaN epitaxial costs in comparison to the proposed potentially achievable values and the yield limiting epitaxial defect density. Finally, a real breakthrough for GaN based devices would be enabled by the availability of cheap GaN substrates in larger wafer diameters (150mm and 200mm) avoiding all the problems related to the hetero-structural approach. This would also allow GaN to stretch out to higher voltage classes beyond 650V, as limitations due to thermal expansion mismatch on hetero-substrates will vanish.

On the other hand both SJ and IGBT technologies are getting closer to their technological limits. There are still new ideas to further improve the trade-off between static and switching losses and keeping short circuit ruggedness, but they are fighting with the fact, that performance improvements are counterbalanced by increased processing costs coming with the more sophisticated processing and reduced process tolerances. The newly available 300 mm wafer process environment for such Si-based power switches helps with respect to this cost aspect – but will be probably the last significant productivity gain for Si based power electronics for the next decade.

In sum the race is getting tighter with respect to device cost and application specific differentiation is more and more important. In the opinion of the authors this will lead to a co-existence of all the mentioned power switch technologies in the coming years with possibly even fluctuating market shares depending on application trends and "carbon footprint reduction" driven legal regulations.

References

[1] M. Treu, E. Vecino, M. Pippan, O. Häberlen, G. Curatola, G. Deboy, M. Kutschak, U. Kirchner; Proc. IEDM 2012

[2] T. Laska, M. Münzer, R. Rupp, H. Rüthing; 1st Conference on Automotive Power Electronics, APE '06, Paris 2006

[3] D. Werber, F .Pfirsch, T. Gutt, V. Komarnitskyy, C. Schaeffer, T. Hunger, D. Domes; Proc. ISPSD 2014 pp. 35

[4] T. Stubbe, R. Mallwitz, W. Bergner, O. Häberlen, G. Pozzovivo, R. Rupp; Proc. CIPS 2014 pp. 139

[5] W. Bergner, R. Rupp, U. Kirchner, D. Kück; Proc. ICSCRM 2013 Mat. Sci. Forum Vols. 778-780 (2014) pp 871

[6] P. Friedrichs; Proc. ICSCRM 2013 Mat. Sci. Forum Vols. 778-780 (2014) pp 1104

Table 1: Summary of technology and performance parameters related to the power switch technologies covered in this paper.

Device	Typical blocking voltage range	Cost position	Maximum wafer diameter today	Main performance weakness today	Main advantage
Si-IGBT	400V - 6.5 kV	Very low cost	200mm-300mm (dep. on voltage class)	High switching losses, increasing with T	Excellent controllability of switching speed, nearly no increase of static losses with T
Si SJ	500V - 900V	Low cost	300mm	High Q_{oss} and high Q_{rr} of body diode	Very low E_{oss}
SiC FETs	600V – 10kV	High Cost	150mm	Increase of static losses with T	Low Q_{oss} and E_{oss} integrated body diode with low Q_{rr} and high ruggedness
GaN-on-Si HEMTs	30V – 600V	Moderate cost	150mm (200mm)	Missing ruggedness concerning overvoltage and overcurrent	Very low Q_{oss}, near zero Q_{rr}, integration capability

High voltage silicon based devices for energy efficient power distribution and consumption

A. Kopta

ABB Switzerland Ltd, Semiconductors, Fabrikstrasse 3, CH-5600 Lenzburg, Switzerland
Phone: +41 58 586 12 54, e-mail: arnost.kopta@ch.abb.com

Abstract

This paper gives an overview of future requirements and recent progress of silicon based semiconductor devices for very high power applications. The first part provides an outline of future trends in the areas of power transmission and power consumption and the resulting requirements on device design and performance. The second part elaborates on the recent advances of bipolar power devices and the corresponding packaging technologies used in these high power applications.

Introduction

In the future, ecological and political concerns will increase the efforts to both grow the level of regenerative energy sources, as well as the efficiency with which the energy is utilized. As a part of this change, efficient power semiconductor devices will play an increasingly important role to ensure low losses, high reliability and affordable energy prices. In this paper, the highest part of the power spectrum, where silicon based bipolar devices are utilized, will be discussed. In this power range, devices with blocking voltages of up to 10kV and current handling capabilities of several kA are used. The devices are typically employed in hard-switched topologies with switching frequencies of up to a few kHz, reaching turn-off power densities of more than $1MW/cm^2$. The general development efforts in this field are focused on increasing the power densities in order to lower the total system cost. On the semiconductor side, this transforms into reduced losses, higher operational temperatures, as well as higher device reliability. Due to the high currents, voltages and stray inductances, the devices have to have high safe operating areas and show controllable and soft switching behavior to prevent EMI problems.

Power distribution and grid stability

The transition to distributed regenerative power sources will completely change the demand on the power distribution grid. With increasing integration of wind and solar power, the energy production will take place in areas far away from the centers where it is consumed. Fig. 1 shows such a future scenario for Europe, where power needs to be transported with low losses over large distances. On other continents, the situation is similar and already today there is a big demand for transmission of bulk power generated in large hydro-power plants in remote areas to the densely populated areas where it is consumed. Due to the inherent intermittency of the most regenerative power sources, alongside with long distance transmission, additional measures to stabilize the grid and to store energy in times of overproduction will also become necessary in the future.

Today, most of the generated power is transported using 400kV High-Voltage Alternating Current (HVAC) lines. Over distances greater than about 300km, transmission with HVAC becomes economically unattractive due to the high reactive power consumption, the associated losses caused by

the reactive currents and ultimately by the stability of the transmission line. This problem can be overcome by changing to High Voltage Direct Current (HVDC) transmissions (Fig. 2). An HVDC system consist of two converter stations at each end of the transmission line. Depending on the current flow, one station always acts as an AC to DC rectifier and the second one as an inverter to transform the DC-current back to AC. The possible length of the transmission line is not limited by any fundamental stability criteria as in HVAC. HVDC systems with transmission lengths in excess of 2000km are in operation today. Depending on the converter technology, HVDC systems can be divided into two main categories. Line Commutated Converters (LCC) and self-commutated Voltage Source Converters (VSC).

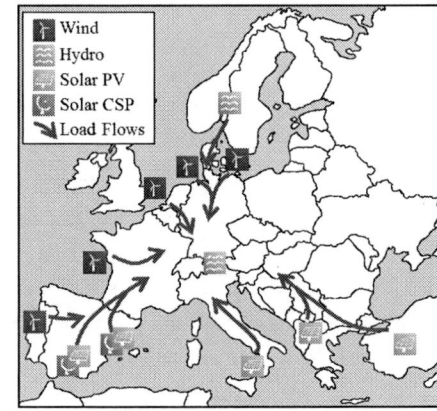

Fig 1: Future power flow in Europe. Power from renewable energy sources (solar and wind) will be transported over large distances.

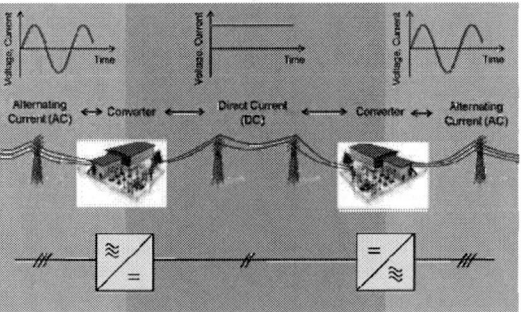

Fig. 2: HVDC transmission system. Depending on the system type, converters employing either line-commutated thyristors or self turn-off capable devices like the IGBTs as semiconductor switches are in use.

Line-commutated systems use Phase Control Thyristors (PCTs) as the main semiconductor switches. These systems are used when very high power transmissions are necessary, which is enabled by the inherently low losses of the converter topology and the utilized thyristor valves. The most recent systems are reaching transmitted powers of up to 8GW using DC-voltages of +/-800kV (1). HVDC systems operated at +/-1100kV with a transmitted power of more than 10GW are currently planned in China (2).

978-1-4799-8002-4/14 $31.00 © 2014 IEEE   32

One of the biggest drawbacks of line-commutated HVDC systems is that they require good quality AC-networks on both sides of the DC-line to commutate (turn-off) the thyristor valves. This limitation can be overcome by using voltage source converter systems with turn-off devices like IGBTs or integrated gate-commutated thyristors (IGCTs) as the main power switches. Such systems have additional features that are important to ensure the stability of grids with a high level of regenerative power sources. As there is no need for a strong AC-net, such systems can be used to connect large offshore wind-parks to land (3). A VSC-based system can fully control the reactive power and therefore be used to stabilize the AC networks it connects to. Today's VSC-based HVDC systems are reaching transmission levels of 1GW utilizing voltages of +/-320kV.

The first generation voltage source converters for HVDC where made using a two level topology, where individual IGBT modules were connected in series to reach the desired voltage level of the system. The introduction of the modular multilevel converter topology (4) has significantly affected the requirements on the power semiconductor devices in this field. Thanks to the many levels, the individual semiconductors only need to be switched at frequencies close to the fundamental output frequency of 50Hz (Fig. 3). As a consequence, the devices have to be optimized to have low conduction losses, whereas the switching losses are less important than in corresponding two level systems. The blocking voltage class of the utilized devices is given by a trade-off between losses and system cost (number of levels) and is typically in the 4.5kV range.

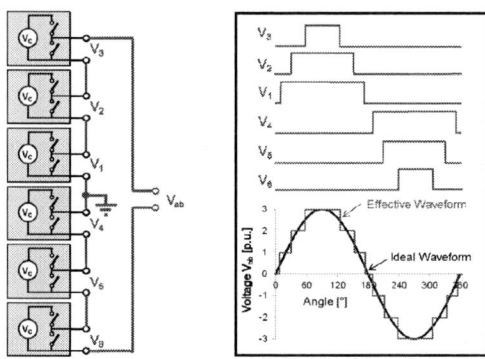

Fig. 3: Modular multi-level converter topology. This topology is used in HVDC systems and large motor drives. One advantage is the low switching frequency.

One of the main problems of grids with high levels of wind and solar power is how to deal with the high intermittency these energy sources have by nature. Energy needs to be stored in times of overproduction and released when production drops below demand. The storage systems needs to be powerful, efficient and have a fast response time. Currently, pumped hydro power is the only mature technology, which reasonably fulfills all these demands at an affordable price. Pumped hydro power using synchronous generators directly coupled to the grid is a well established technology. In order to increase the efficiency and response time, future pumped hydro power plants will however be realized using converter fed synchronous motors, enabling variable speeds and thus efficiently adjustable power levels. Such converters have output voltages of 10 to 20kV and output powers of up to 300MVA (6). Due to the high output voltages, the modular multilevel topology has also in this case shown to be the most efficient converter topology putting similar requirements on the employed power devices as VSC-based HVDC converters.

Efficient power consumption

Power semiconductor devices are today present in a large variety of applications where electrical energy is converted and consumed. The different applications range from switched mode power supplies for computers and charging devices to multi-megawatt motor drives. An example of a high power application is the variable speed drive used in traction vehicles. A typical high-end locomotive has a rated power of 6MW and contains up to 100 IGBT modules. This sums up to a total silicon area of about $0.7m^2$. The semiconductors are exposed to harsh environmental conditions with large temperature differences and high humidity. The chips as well as the package have to be designed for high reliability to meet the low failure rates and long lifetimes expected in these applications (up to 30 years).

Silicon based devices for high power applications

Modern silicon based devices for high power applications can be divided into two main categories. Line commutated turn-on devices like the thyristor and turn-off capable devices like IGBTs or IGCTs. In this section, the latest progress in these both areas will be outlined.

In order to support the development of new generations HVDC systems with voltage levels above 1000kV and transmission capabilities beyond 10GW, ABB Semiconductors has developed a new range of low loss phase controlled thyristors (PCTs) (5). In the HVDC converter, the individual devices are connected in series in several stacks to reach the desired voltage level of the system. Typically, around 200 thyristors need to be seriesed depending on the transmitted power and the trade-off between converter cost and semiconductor losses.

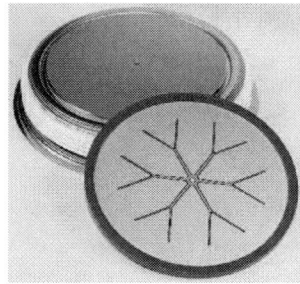

Fig. 4: Newest generation 8.5kV/5kA thyristor (PCT). The thyristor is made out of a single 150mm wafer.

Fig. 4 shows the newly developed PCT based on a 150mm wafer with a nominal blocking voltage of 8.5kV and a rated current of 5kA. The on-state voltage (V_T) is 1.75V at 5kA and a junction temperature of 90 °C. The power rating of the new device was increased by 20% compared to the previous generation by improving the diffusion profiles, the silicon resistivity and the lifetime profiles together with an improved silicon area utilization achieved by an optimized cathode design.

Ever since the invention of the IGBT, significant development efforts have been made to reduce the overall losses by enhancing the properties of the IGBT cell. The newest generation of IGBTs employs an enhanced trench structure (Fig. 5), which combines low losses with a controllable switching behavior (7), (8). This new IGBT design utilizes a combination of an N-enhancement layer underneath the P-well with an optimized trench cell layout. This further increases the electron-hole plasma concentration on the cathode side of the IGBT, reducing the conduction losses compared to current devices which, use only the trench or only the enhanced planar structure (9).

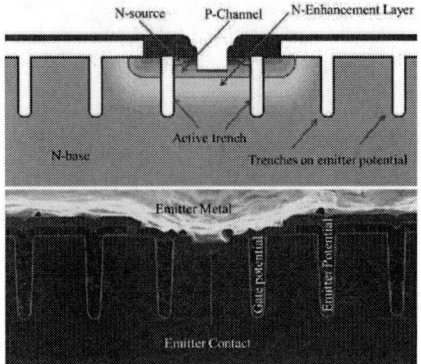

Fig. 5: Cross-section of the enhanced trench IGBT cell showing the N-enhancement layer (for low on-state losses) and the dummy trenches connected to emitter-potential (for good controllability).

Fig. 6 shows the trade-off curve comparison between the 3.3kV enhanced trench IGBT and the older device generation using an enhanced planar structure. Thanks to the significant reduction of the conduction losses, it was possible to increase the rating of the new IGBT by 20% to 75A. This correspond to a current density of 66A/cm² of active area.

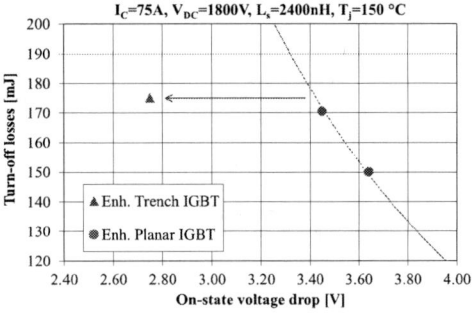

Fig 6: 3.3kV IGBT trade-off curve between on-state voltage drop and turn-off losses 3.3kV IGBT. Comparison between the enhanced trench and enhanced planar structures.

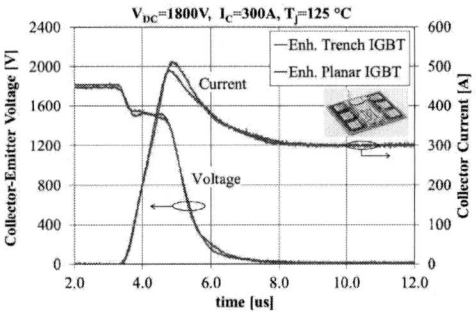

Fig 7: 3.3kV IGBT turn-on at nominal conditions (substrate level). Comparison between the enhanced trench and enhanced planar IGBTs. Thanks to the optimized emitter design, the enhanced trench IGBT achieves a good turn-on controllability comparable to the planar IGBT.

Until recently, one of the biggest drawbacks of trench IGBTs was the unsufficient turn-on controllability and increased requirements on the gate driving power caused by the inherently higher gate-collector capacitance compared to planar device concepts. A good IGBT turn-on controllability is essential in order to reach a good trade-off between diode reverse recovery performance (low losses without disturbing

oscillations) and IGBT turn-on losses. Through optimization of the dummy trench structure and the gate-collector capacitance, an improved turn-on performance very similar to the planar IGBT was achieved. Fig.7 shows the turn-on waveforms at nominal conditions comparing the enhanced trench IGBT with the enhanced planar structure. The measurements were done on substrate level with four IGBTs and two diodes in parallel.

A further performance step, which goes beyond the optimization of the IGBT cell, can be achieved by integrating the free-wheeling diode into the IGBT structure thus creating a reverse conducting device. In this way, the overall silicon utilization can be improved, increasing the total power density as explained in Fig. 8. The figure shows one phase-leg of a three-phase voltage source inverter as used in many standard drive applications. The basic switch consists of an IGBT and an anti-parallel free-wheeling current path, which in the normal case is a fast recovery diode chip. In operation, the current flow intermits between the IGBT and the diode. Using a standard two-chip approach, this means that the silicon area is poorly utilized as the current flows either in the IGBT or in the diode. Integrating the diode into the IGBT structure therefore allows a better utilization of the available footprint. The figure shows a standard high power module, which in the normal case contains 24 IGBTs and 12 diodes. Equipped with reverse conducting IGBTs, the same module footprint provides 36 chips for both IGBT and diode mode operation. The power density of such a module can in this way be increased by 10 to 40% depending on the application and the relative loading in IGBT and diode mode.

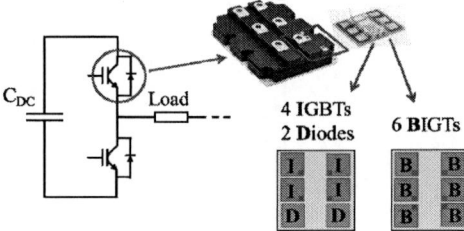

Fig. 8: One phase-leg of a three-phase inverter with IGBT modules as switches. Internally the module consists of 6 substrates connected in patrallel (see also Fig. 10). Conventional substrate are equiped with 4 IGBT and 2 diode chips, whereas the use of BIGTs allows 6 chips to be used for both function → better usage of the substrate and module footprints.

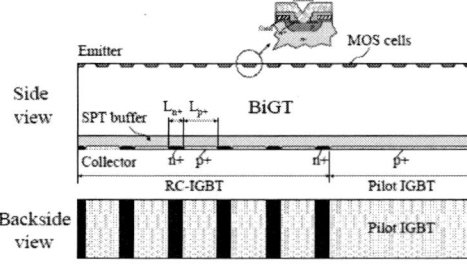

Fig. 9: Cross-section of an addvanced reverse conducting IGBT structure (BIGT). The device offers an increased power density through the integration of the IGBT and diode into a single device.

Fig. 9 shows a cross-section of such an integrated structure, referred to as the Bi-mode Insulated Gate Transistor (BIGT) (10). On the emitter side, the BIGT uses the same MOS-structure as standard IGBTs. On the collector side, the BIGT consists of a hybrid structure with a continuous central P+ area acting as a pilot IGBT and an outer region with intermitting P+ and N+ areas enabling the reverse conducting path. In IGBT conduction mode, the pilot IGBT part enables a good spreading of the electron-hole plasma into the outer reverse conducting regions. When current flow is

978-1-4799-8002-4/14 $31.00 © 2014 IEEE   34

reversed, the N+ regions take the role of the diode cathode, whereas the IGBT P+ channel acts as the anode. The main difficulty is to optimize the reverse recovery characteristics of the integrated diode without affecting the IGBT-mode performance. A good compromise was achieved by using a two step local lifetime-control of the P+ channel. The BIGT concept offers particularly large benefits in applications with low switching frequencies and in cases where the diode and the IGBT parts are not loaded equally. This especially holds for multilevel converter topologies, which makes the BIGT ideal for such applications.

Packaging

The packaging concept has a large influence on the device performance and reliability. For converter topologies where no series connection is necessary, the chips are packaged in insulated modules like the one shown in Fig. 10. The main challenge for the module design are the thermal expansion cycles during operation and the consequent wear out of bond wires and solder joints. Different methods to improve the thermal cycling capability like improved bond wire patterns and the usage of advanced joining techniques like low temperature bonding have been adopted to improve this capability (11), (12).

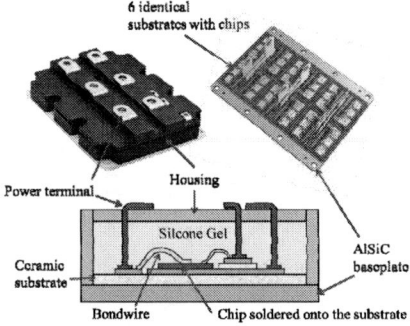

Fig. 10: IGBT Module with 24 IGBT and 12 antiparallel diode chips (140 x 190mm) in a single switch configuration. The current rating using the newest 3.3kV enhanced trench IGBTs (Fig. 5) is 1800A.

The modules are not hermetically sealed and in recent time considerable efforts have been made to increase the reliability when used under humid conditions. The weak spot is thereby the junction termination and passivation of the encapsulated chips. Due to the large widths of the high voltage terminations, there are large areas with high electric fields reaching outside of the semiconductor, requiring rugged encapsulation materials and termination designs to prevent premature breakdown failures. In order to assess the capabilities, new humidity tests with high voltages have been proposed and are currently under evaluation (13). In many cases the failure mechanisms are however not yet fully understood and intensive research efforts to improve the termination structures and passivation stacks are currently made.

IGCT

The IGCT is as of today the most powerful turn-off device, combining the low conduction losses of the thyristor structure with a highly rugged transistor-like turn-off. Fig. 11 shows the new reverse conducting IGCT, which was designed for very high power applications (14). The switch and the free-wheeling diode are accommodated on a single 150mm wafer. The device has a rated blocking voltage of 4.5kV and is able to turn off currents of up to 9.5kA as shown in Fig 12. Thanks to it's low losses and high current handling capabilities it can be used to design converters without the need of device paralleling, which makes it poss-

ible to design compact and cost efficient high voltage, high power converters using modular multilevel topologies like suggested for pumped hydro power storage.

Fig. 11: 4.5kV high-power reverse conducting IGCT. One IGCT and one diode integrated on a 150mm wafer shown together with the gate unit.

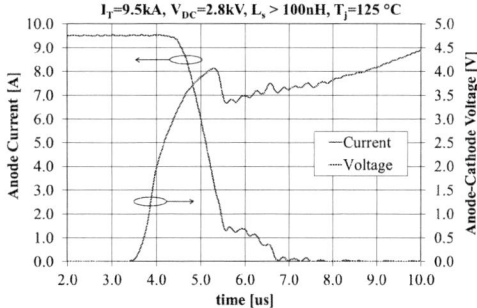

Fig. 12: 4.5kV IGCT turn-off waveforms

Conclusion

The future integration of renewal power sources will require the energy to be transported over large distances. HVDC transmission system are ideally suited to take over this task. In addition, energy storage systems with large capacities and fast response times will become necessary to smooth out the intermittency of the these energy sources. Pumped hydropower using variable speed drives is currently the most promising technology to take over this role. The introduction of the modular multilevel converter topology has changed the requirements on the semiconductors used in the highest power range, requiring devices optimized for low conduction losses.

References

(1) Thomas Freyhult, Mats Berglund, Åke Carlsson,"UHVDC Meeting the needs of the most demanding power transmission applications", www.abb.com/hvdc

(2) R. Montano, B. Jacobson, D. Wu, L. Arevalo, "Corridors of power", ABB Review, Special Report: 60 years of HVDC, 2014.

(3) U. Wijk, et al, "Dolwin1 – Further Achievements in HVDC Offshore connections", EWEA Offshore 2013.

(4) R. Marquardt, et al, "New Concept for High Voltage – Modular Multilevel Converter", PESC 2004

(5) J. Vobecký, T. Stiasny, V. Botan, K. Stiegler, U. Meier, "New Thyristor Platform for UHVDC (>1 MV) Transmission", PCIM 2014

(6) S. Linder, "Power Electronics: The Key Enabler of a Future with more than 20% Wind and Solar", ISPSD 2013

(7) M. Andenna, et al, "The Next Generation High Voltage IGBT Modules utilizing Enhanced-Trench ET-IGBTs and Field Charge Extraction FCE-Diodes", EPE 2014

(8) Y. Toyota, et al, "Novel 3.3-kV Advanced Trench HiGT with Low Loss and Low dv/dt Noise", ISPSD 2013

(9) M. Rahimo, et al, "Novel Enhanced-Planar IGBT Technology Rated up to 6.5kV for Lower Losses and Higher SOA Capability" ISPSD 2006.

(10) M. Rahimo, et al, "The Bi-mode Insulated Gate Transistor (BIGT) A Potential Technology for Higher Power Applications", ISPSD 2009

(11) J. Rudzki, et al, "Power Modules with Increased Power Density and Reliability Using Cu Wire Bonds on Sintered Metal Buffer Layers", CIPS 2014

(12) S. Hartmann, et al, "Packaging Technology Platform for Next Generation High Power IGBT Modules", PCIM 2014

(13) C. Zorn, N. Kaminski, "Temperature Humidity Bias (THB) Testing on IGBT Modules at High Bias Levels", CIPS 2014

(14) T. Wikström, et al, "The 150 mm RC-IGCT: a Device for the Highest Power Requirements", ISPSD 2014

Progress in Ultrahigh-Voltage SiC Devices for Future Power Infrastructure

T. Kimoto[1], J. Suda[1], Y. Yonezawa[2], K. Asano[3], K. Fukuda[2], and H. Okumura[2]

1) Department of Electronic Science and Eng., Kyoto University, A1-301 Katsura, Nishikyo, Kyoto 615-8510, Japan
2) Advanced Power Electronics Research Center, AIST, 1-1-1 Umezono, Tsukuba, Ibaraki 305-8568, Japan
3) Kansai Electric Power Co., Inc., 3-11-20 Nakoji, Amagasaki, Hyogo 661-0974, Japan

Abstract

UHV (> 15 kV) SiC PiN diodes and IGBTs with improved on-state performance are presented. Through enhancement of carrier lifetime and optimization of junction termination, a breakdown voltage over 26.9 kV and a differential on-resistance of 9.7 mΩcm^2 were achieved for PiN diodes. Flip-type n-channel IGBTs with a chip size of 8 mm × 8 mm exhibited a breakdown voltage over 16 kV, and 6.5 kV – 60 A switching at 250°C was demonstrated.

Introduction

Silicon carbide (SiC) has received increasing attention as an innovative semiconductor for high-power devices (1)-(3). Ultrahigh-voltage (UHV) and low-loss power devices are key components for future smart grids and high-voltage power supplies. A typical voltage of power distribution is 6.6-7.2 kV, where 13-15 kV power devices are required for constructing single-phase converters. Solid-state transformers are one of attractive applications (4). Since the specific on-resistance of UHV unipolar devices becomes very high (> 100 mΩcm^2) even with SiC, bipolar devices will be attractive, owing to the conductivity modulation effect (5)-(10). In this paper, recent progress in UHV (> 15 kV) SiC PiN diodes and insulated gate bipolar transistors (IGBTs) with improved on-state performance is presented.

Ultrahigh-voltage PiN diodes

SiC PiN diodes having five different thicknesses of n$^-$-layer (i-layer or voltage-blocking layer) from 48 to 268 ∞m were fabricated. The doping concentration of i-layer was almost fixed to (1-2)×10^{14} cm^{-3} (nitrogen-doped) for all the diodes. Fig. 1 shows a schematic structure of a fabricated SiC PiN diode. After growing a very thick, high-purity n-type epilayer, the carrier lifetime was enhanced by utilizing an original elimination process of carbon vacancy (thermal oxidation at high temperature) (11), which has been identified as the lifetime killer. Fig. 2 depicts the photoconductance decay curves measured for as-grown and thermally oxidized epilayers (thickness: 198 ∞m), indicating successful enhancement of the carrier lifetime from 1.8 ∞s to 21.6 ∞s. The p^+-anode was epitaxially grown after this process. To alleviate electric field crowding, combination of mesa and Al-implanted junction termination extension (JTE) structures was employed (5)-(7), detail of which is

described later. Near the mesa edge, 0.8 ∞m-deep JTE structures were formed by multistep Al$^+$ implantation at room temperature followed by annealing in Ar at 1650 °C for 20 min. The surface was passivated with a nitrided oxide and polyimide.

By the lifetime-enhancement process, the forward characteristics of PiN diodes were remarkably improved, as shown in Fig. 3, in agreement with a previous report (12). Fig. 4 shows the differential on-resistance versus the thickness of i-layer of SiC PiN diodes. In this figure, the experimental data are plotted with circles, and the dashed lines are the calculated on-resistances, taking account of carrier injection and recombination in the end regions (p$^+$ and n$^+$ regions) (13). By eliminating the carbon-vacancy defects throughout very thick (> 200 ∞m) layers, low on-resistances of 1.3-9.7 mΩcm^2 were attained for 7-35 kV-class diodes.

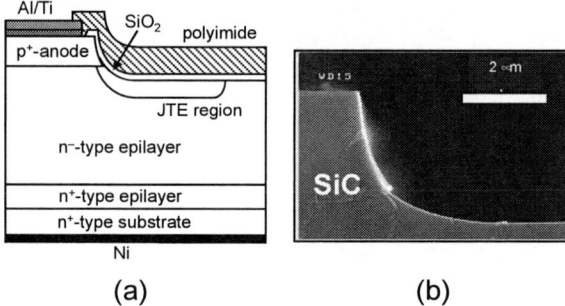

(a)   (b)

Fig. 1. (a) Schematic structure of a fabricated SiC PiN diode. The pn junction was epitaxially grown. The edge termination was achieved by combination of mesa and Al-implanted JTE. (b) Cross-sectional SEM image of a mesa region.

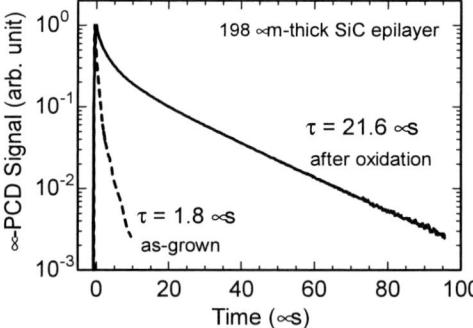

Fig. 2. Photoconductance decay curves measured for as-grown and thermally oxidized n-type SiC epilayers (thickness: 198 ∞m). By high-temperature oxidation, excess carbon atoms diffuse and fill carbon vacancy defects present in SiC, leading to remarkable enhancement of carrier lifetime.

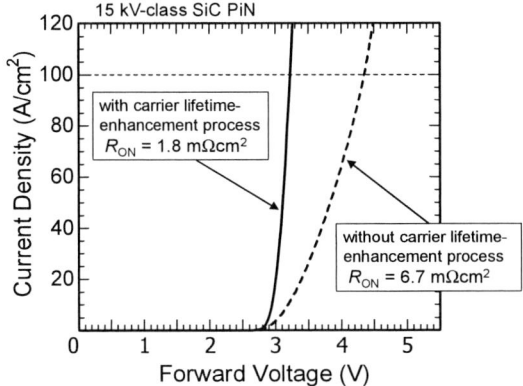

Fig. 3. Forward characteristics of fabricated 15 kV-class SiC PiN diodes with and without the lifetime-enhancement process shown in Fig. 2. The thickness and donor concentration of the n-type voltage blocking layer are 98 ∞m and 2×10^{14} cm^{-3}, respectively. The specific on-resistance was significantly improved from 6.7 mΩcm^2 to 1.8 mΩcm^2.

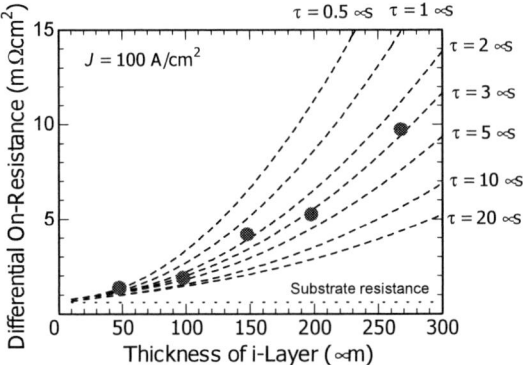

Fig. 4. Differential on-resistance versus the thickness of n^--type voltage-blocking layer of fabricated SiC PiN diodes. By eliminating the carbon-vacancy defects throughout very thick (> 200 ∞m) layers, low on-resistances of 1.3-9.7 mΩcm^2 (at 100 A/cm^2) were attained.

In this study, an improved space-modulated junction termination extension (SM-JTE) (14), where multiple rings with modulated width and spacing are embedded inside a RESURF-type JTE, was employed. The proposed JTE structure offers a much wider optimum window of JTE dose to obtain nearly ideal blocking voltage, compared with two-zone JTE, as shown in Fig. 5. Fig. 6 demonstrates the current density–voltage characteristics of a fabricated SiC PiN diode with a space-modulated JTE structure (total JTE length: 1050 ∞m). The thickness and donor concentration of i-layer were 268 ∞m and $(1-2)\times10^{14}$ cm^{-3}, respectively. The JTE dose was 1.8×10^{13} cm^{-2} for Dose1 and 4.5×10^{12} cm^{-2} for Dose2. The PiN diode did not exhibit breakdown up to 26.9 kV (a limit of the measurement setup). In such a UHV diode, a low differential on-resistance of 9.72 mΩcm^2 was achieved. Since the on-resistance of a Schottky barrier diode fabricated on the same wafer was as high as 460 mΩcm^2, the low on-resistance of this PiN diode indicates a remarkable effect of conductivity modulation owing to the enhanced carrier lifetime.

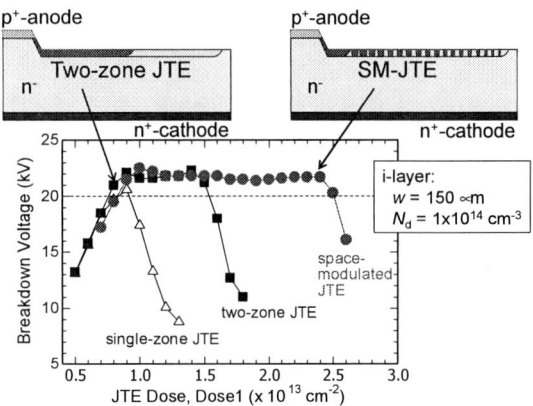

Fig. 5. JTE dose dependence of breakdown voltage for SiC PiN diodes with various 600 ∞m-long JTE structures (simulation). An optimum dose window for obtaining nearly ideal breakdown voltage can be remarkably enlarged for the improved space-modulated JTE shown in the figure. The thickness and donor concentration of the n-type voltage- blocking layer are 150 ∞m and 1×10^{14} cm^{-3}, respectively.

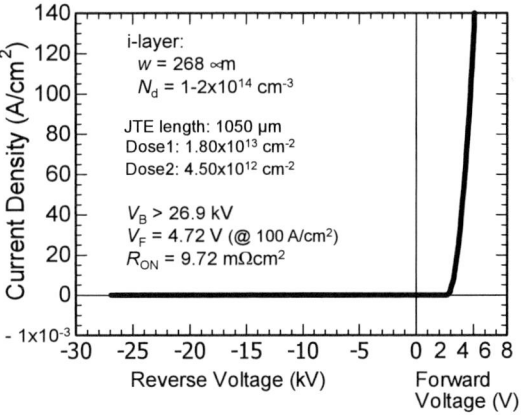

Fig. 6. Current density–voltage characteristics of a SiC PiN diode with a space-modulated JTE structure (total JTE length: 1050 ∞m). The thickness and donor concentration of the n^--type epilayer were 268 ∞m and $1-2\times10^{14}$ cm^{-3}, respectively. The diode did not exhibit breakdown up to 26.9 kV and showed a low differential on-resistance of 9.7 mΩcm^2.

Ultrahigh-voltage n-channel IGBTs

Flip-type n-channel implantation and epitaxial (IE)-IGBT on the 4H-SiC (000$\bar{1}$) C face was developed, the schematic structure of which is shown in Fig. 7 (15),(16). Taking account of the doping control capability in epitaxial growth, the n-type voltage-blocking layer (150 ∞m, 4×10^{14} cm^{-3}) and p^+-type collector (200 ∞m, 2×10^{14} cm^{-3}) were subsequently grown on the (0001) Si face. Then, the wafer was flipped and substrate was mechanically removed. After this process, the MOS structure was formed on the (000$\bar{1}$) C face. Note that high n-channel mobility can be obtained on the C face, compared with the Si face (17). Furthermore, by combination of implantation and epitaxial (IE) processes, the surface roughness caused by implantation and subsequent high-temperature annealing can be greatly reduced, leading to a high mobility in real devices (18). The gate oxide was formed dry oxidation followed by annealing in a hydrogen-rich ambient. The channel mobility of MOSFETs processed on the same wafer was 60-75 cm²/Vs.

A 2 ∝m-thick *n*-type charge-storage layer was introduced beneath the *p*-wells to improve the injected carrier profile and thereby to reduce the on-state resistance. The JFET width is another important parameter, which determines both on-state and off-state characteristics. After optimization by device simulation and experiments, the authors selected a rather narrow JFET width of 1.6 ∝m for obtaining high breakdown voltage while keeping low on-resistance. The cell pitch of fabricated IE-IGBTs was 14.8 ∝m. The chip size and active area were 8 mm × 8 mm and 0. 37 cm², respectively.

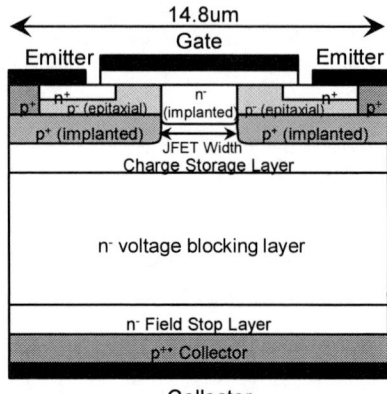

Fig. 7. Schematic structure of a fabricated flip-type *n*-channel SiC IE-IGBT cell (*n*-type voltage-blocking layer: 150 ∝m, 4×10¹⁴ cm⁻³). The MOS channel was formed on the C face by using an original implantation and epitaxial (IE) process.

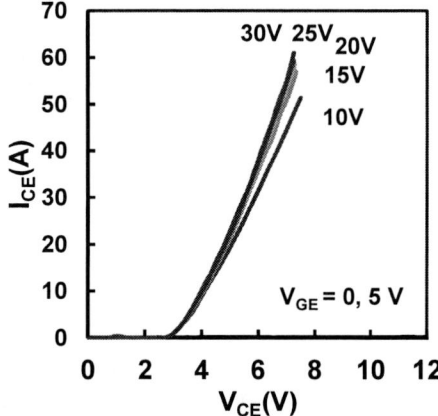

Fig. 8. On-state collector characteristics of a fabricated SiC IE-IGBT with an 8 mm × 8 mm chip size. The IGBT exhibited a low forward voltage drop of 4.8 V and differential on-resistance of 23 mΩcm² at 20 A.

The pulsed on-state characteristics of a fabricated IE-IGBT are shown in Fig. 8. At a gate voltage of 30 V (oxide field: 3 MV/cm), the forward voltage drop

was 4.8 V at 20 A (54 A/cm²) and 7.2 V at 60 A (162 A/cm²), respectively, with a differential on-resistance of 23 mΩcm². The temperature dependence of on-state characteristics is very small from room temperature (RT) to 250°C, as shown in Fig. 9. The decrease of bulk and channel mobilities at elevated temperature may be compensated by an enhanced carrier lifetime.

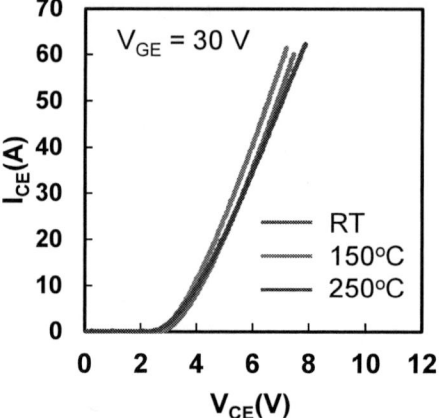

Fig. 9. Temperature dependence of on-state characteristics of a fabricated SiC IE-IGBT with an 8 mm × 8 mm chip size. The gate voltage was 30 V. The on-state performance did not degrade very much at elevated temperature up to 250 °C.

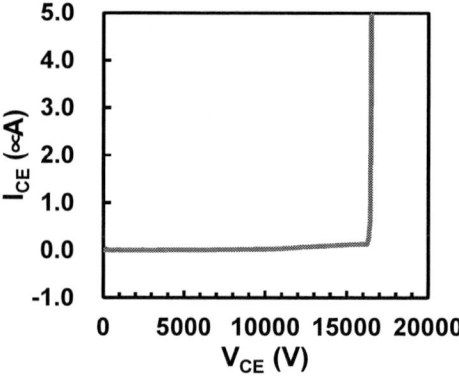

Fig. 10. Blocking characteristics of a fabricated IE-IGBT (8 mm × 8 mm) with zero gate bias, demonstrating a high blocking voltage over 16 kV.

The blocking voltage with zero gate bias exceeded 16 kV, as depicted in Fig. 10. The leakage current was as low as 10⁻⁸ A at 10 kV. The authors confirmed that epitaxially-induced extended defects such as a triangular defect and in-grown stacking faults severely affect the blocking performance.

In order to evaluate the switching characteristics of the IE-IGBT, ultrahigh-voltage power modules were assembled (19). The power module consisted of a tungsten base plate, a direct bonded copper (DBC) base with Si₃N₄, and a copper electrode. The DBC base and power devices were molded using a resin and packed in a module case.

978-1-4799-8002-4/14 $31.00 © 2014 IEEE

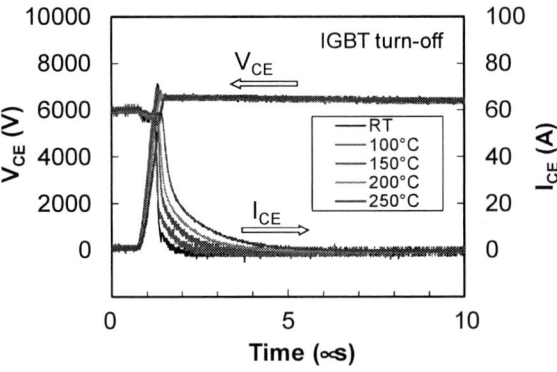

Fig. 11. Turn-off switching waveforms of the SiC IE-IGBT (8 mm × 8 mm) in the temperature range from RT to 250°C. In spite of UHV bipolar devices, fast turn-off (1-4 μs) could be obtained under the condition with V_{CE} = 6.5 kV and I_{CE} = 60 A.

The switching characteristics were characterized with the chopper circuit configuration. SiC PiN diode (5.3 mm × 5.3 mm) modules were separately used as freewheeling diodes in the circuit. A 300-Ω gate resistance was employed for both turn-on and turn-off switching. The maximum collector voltage (V_{CE}) was 6.5 kV, and V_{GEoff} = -10 V and V_{GEon} = +25 V were applied as the gate voltages, respectively. The turn-off switching waveforms of the fabricated IE-IGBT are shown in Fig. 11 in the temperature range from RT to 250°C. Smooth and rather fast turn-off waveforms were successfully obtained at V_{CE} = 6.5 kV and I_{CE} = 60 A. The turn-off energy was 36 mJ/pulse at RT and increased to 199 mJ/pulse at 250°C. The increase in the tail current may originate mainly from the prolonged carrier lifetime at elevated temperature.

Conclusions

Progress in UHV (> 15 kV) SiC PiN diodes and IGBTs for future power infrastructures was presented. Through enhancement of carrier lifetime and optimization of junction termination, a breakdown voltage over 26.9 kV and a differential on-resistance of 9.7 mΩcm^2 were achieved for a PiN diode having a 268 μm-thick i-layer. Flip-type n-channel IGBTs on (000$\bar{1}$) C face with a chip size of 8 mm × 8 mm were developed. The IGBTs having 150 μm-thick voltage-blocking layer exhibited a breakdown voltage over 16 kV and voltage drop of 4.8 V at 20 A. By using UHV IGBTs and PiN diodes, 6.5 kV – 60 A switching at 250°C (single device) was demonstrated.

Acknowledgments

The authors acknowledge Mr. H. Niwa, N. Kaji with Kyoto University and the project members in AIST for their contribution. This work was supported by the Funding Program for World-Leading Innovative R&D on Science and Technology (FIRST Program) from the Japan Society for the Promotion of Science. Part of this work has been implemented under a joint research project of the Tsukuba Power Electronics Constellations (TPEC). They also acknowledge Tokyo Electron Limited for supplying thick epitaxial wafers.

References

(1) M. Bhatnager and B.J. Baliga, "Comparison of 6H-SiC, 3C-SiC, and Si for power devices", *IEEE Trans. Electron Devices*, vol. 40, pp. 645-655, 1993.

(2) J.A. Cooper, Jr., M.R. Melloch, R. Singh, A. Agarwal, and J.W. Palmour, "Status and prospects for SiC power MOSFETs", *IEEE Trans. Electron Devices*, vol. 49, pp. 658-664, 2002.

(3) T. Kimoto and J.A. Cooper, *Fundamentals of Silicon Carbide Technology* (Wiley, 2014).

(4) J. Wang, A.Q. Huang, W. Sung, Y. Liu, and B.J. Baliga, "Development of 15-kV SiC IGBTs and their impact on utility applications", *IEEE Industrial Electronics Magazine*, June, 2009, pp.16-23.

(5) Y. Sugawara, D. Takayama, K. Asano, R. Singh, J. Palmour, and T. Hayashi, "12-19 kV 4H-SiC PiN diodes with low power loss", *Proc. Int. Symp. Power Semiconductor Devices and ICs*, 2001, pp. 27-30.

(6) R. Singh, J.A. Cooper, M.R. Melloch, T.P. Chow, and J.W. Palmour, "Large area, ultra-high voltage 4H-SiC p-i-n rectifiers", *IEEE Trans. Electron Devices*, vol. 49, pp.2308-2316, 2002.

(7) H. Niwa, J. Suda, and T. Kimoto, "21.7 kV 4H-SiC PiN diode with a space-modulated junction termination extension", *Appl. Phys. Express*, vol. 5, pp. 064001/1-3, 2012.

(8) S.H. Ryu et al., "Development of 15 kV 4H-SiC IGBTs", *Mater. Sci. Forum*, vol. 717-720, pp. 1135-1138, 2012.

(9) H. Miyake, T. Okuda, H. Niwa, T. Kimoto, and J. Suda, "21-kV SiC BJTs with space-modulated junction termination extension", *IEEE Electron Device Lett.*, vol. 33, pp. 1598-1600, 2012.

(10) L. Cheng et al., "Strategic overview of high-voltage SiC power device development aiming at global energy savings", *Mater. Sci. Forum*, vol. 778-780, pp. 1089-1095, 2014.

(11) T. Hiyoshi and T. Kimoto, "Reduction of deep levels and improvement of carrier lifetime in n-type 4H-SiC by thermal oxidation", *Appl. Phys. Exp.* vol. 2, pp. 041101/1-3, 2009.

(12) K. Nakayama, A. Tanaka, M. Nishimura, K. Asano, T. Miyazawa, M. Ito, and H. Tsuchida, "Characteristics of a 4H-SiC pin diode with carbon implantation/thermal oxidation," *IEEE Trans. Electron Devices*, vol. 59, pp. 895 - 901, 2012.

(13) J. Luts, H. Schlangenotto, U. Scheuermann, and R. De Doncker, *Semiconductor Power Devices* (Springer-Verlag, Berlin Heidelberg, 2011), Chapter 5.

(14) G. Feng, J. Suda, and T. Kimoto, "Space-modulated junction termination extension for ultrahigh-voltage p-i-n diodes in 4H-SiC", *IEEE Trans. Electron Devices*, vol. 59, pp. 414-418, 2012.

(15) X. Wang and J.A. Cooper, "High-voltage n-channel IGBTs on free-standing 4H-SiC epilayers", *IEEE Trans. Electron Devices*, vol. 57, pp. 511-515, 2010.

(16) Y. Yonezawa et al., "Low Vf and highly reliable 16 kV ultrahigh voltage SiC flip-type n-channel implantation and epitaxial IGBT", *Tech. Digest of Int. Electron Devices Meeting 2013*, 6.6.1-6.6.4, 2013.

(17) K. Fukuda, M. Kato, K. Kojima, and J. Senzaki, "Effect of gate oxidation method on electrical properties of metal-oxide-semiconductor field-effect transistors fabricated on 4H-SiC (000-1) face", *Appl. Phys. Lett.*, vol. 84, pp. 2088-2090, 2004.

(18) S. Harada, M. Kato, K. Suzuki, M. Okamoto, T. Yatsuo, K. Fukuda and K. Arai, "1.8 mΩcm^2, 10 A Power MOSFET in 4H-SiC," *Tech. Digest of Int. Electron Devices Meeting 2006*, pp. 903-906, 2006.

(19) T. Hayashi, T. Izumi, T. Hemmi, and K. Asano, "Insulating properties of package for ultrahigh-voltage, high-temperature devices", *Mater. Sci. Forum*, vols.740-742, pp.1036-1039, 2013.

600 V JEDEC-Qualified Highly Reliable GaN HEMTs on Si Substrates

Toshihide Kikkawa[1], Tsutomu Hosoda[1], Kenji Imanishi[1], Ken Shono[1], Kazuo Itabashi[1], Tsutomu Ogino[1], Yasumori Miyazaki[1], Akitoshi Mochizuki[1], Kenji Kiuchi[1], Masahito Kanamura[1], Masamichi Kamiyama[1], Shiniichi Akiyama[1], Susumu Kawasaki[1], Takeshi Maeda[1], Yoshimori Asai[1], YiFeng Wu[2], Kurt Smith[2], John Gritters[2], Peter Smith[2], Saurabh Chowdhury[2], Dixie Dunn[2], Martin Aguilera[2], Brian Swenson[2], Ron Birkhahn[2], Lee McCarthy[2], Likun Shen[2], Jim McKay[2], Heber Clement[2], Jim Honea[2], Sung Yea[2], Douglas Thor[2], Rakesh Lal[2], Umesh Mishra[2], and Primit Parikh[2]

[1]Transphorm Japan Inc., 3 Kogyodanchi, Monden-machi, Aizu-Wakamatsu, 965-8502, Japan
[2]Transphorm Inc., 75 Castilian Drive, Goleta, CA, 93117, U.S.A

Abstract

In this paper, we demonstrate 600 V highly reliable GaN high electron mobility transistors (HEMTs) on Si substrates. GaN on Si technologies are most important for the mass-production at the Si-LSI manufacturing facility. High breakdown voltage over 1500 V was confirmed with stable dynamic on-resistance (R_{ON}) using cascode configuration package. These GaN HEMT on Si based cascode packages have passed the qualification based on the standards of the Joint Electron Devices Engineering Council (JEDEC) (1-5) for the first time. High voltage acceleration test was performed up to 1150 V. Even considering most conservative failure mechanism, mean time to failure (MTTF) of over 1×10^7 hours at 600 V was predicted at 80°C. Additional conclusion is that conventional packages such as TO-220 are still suitable for high speed circuit application without using a specific gate driver. Ultimately GaN will significantly reduce conversion losses endemic in all areas of electricity conversion, ranging from power supplies to PV inverters to motion control to electric vehicles, enabling consumers, utilities and governments to contribute towards a more energy efficient world.

Introduction

In the past few years, GaN HEMTs have attracted much attention for power conversion. As shown in Fig. 1, there will be expanding markets for GaN HEMTs, especially for high frequency application with small size and high efficiency.

In general, GaN device researchers have encountered technological difficulty for normally-off operation which is required from the application side. In addition, large gate voltage swing capability compatible to Si-MOSFET is also essential to use a non-special gate driver in the circuit. In this paper, cascode-type normally-off devices are described with robust performance (2).

Experimental Data and Discussions

A. Device configuration

Fig. 2(a) shows the cascode configuration using a 600 V normally-on GaN HEMT and a low voltage Si-MOSFET. Using this configuration, normally-off operation was established and the same gate swing range as the Si-MOSFET is allowed with keeping GaN advanced performance. Fig. 2(b) shows the TO-220 package consisting of cascode configuration inside. Fig. 2 (c) shows the 600 V normally-on GaN HEMT cross sectional view. AlGaN/GaN epitaxial layers were grown on Si substrates by metal organic vapor phase epitaxy (MOVPE).

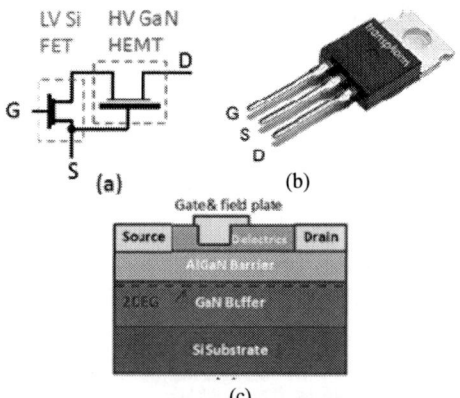

Fig. 2 (a) Cascode configuration using a 600V normally-on GaN HEMT and a low voltage Si-MOSFET. (b) TO-220 package consisting of cascode configuration inside. (c) 600 V normally-on GaN HEMT cross sectional view.

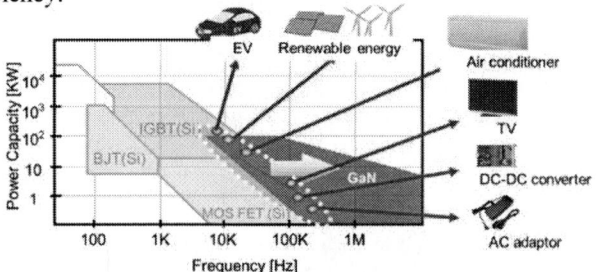

Fig. 1 Market for GaN HEMTs

978-1-4799-8002-4/14 $31.00 © 2014 IEEE

Fig. 3 (a) shows a 6-inch GaN on Si epitaxial wafer. No crack can be seen. Fig. 3 (b) shows a HEMT wafer which was processed at the Au-free 6 inch Si-CMOS compatible process line. In this line, no III-V Evaporator or liftoff is used. Sputter/etch process using standard CMOS Fab is applied.

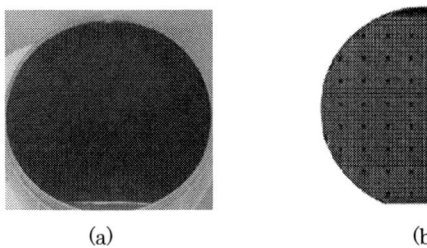

(a)   (b)

Fig. 3(a) 6-inch GaN on Si epitaxial wafer. (b) HEMT wafer which was processed at the Au-free 6 inch Si-CMOS compatible process line.

B. Device performance

Fig. 4 (a) shows the typical I_D-V_D characteristics of the GaN HEMT-based cascode package, TO-220 by using Tektronix 370-A curve tracer in pulsed mode (3). It is seen that, at V_{GS} = 2.5 V, the drain current is nearly zero while at V_{GS} = 8 V the device is fully on and carries 20 A of drain current with a voltage drop of 3 V, corresponding to 0.15 Ω static on-resistance. The Si FET and bond wires account for 10%, or 15 mΩ, of this total resistance.

Fig. 4 (b) shows the typical I_D-V_G transfer characteristics. Normally-off operation is available with no gate leakage. Additional measurements performed on the same devices revealed the gate threshold voltage at 1 mA drain current to be 2.3 V and the maximum pulsed drain current at V_{GS} = 8 V and V_{DS} = 10 V to be >60 A. The gate charge is 10 nC at V_{GS} = 8 V and I_{DS} = 8 A, which is practically independent of drain voltage from 100 V and above due to the very low gate drain capacitance (C_{GD}) in the cascode design.

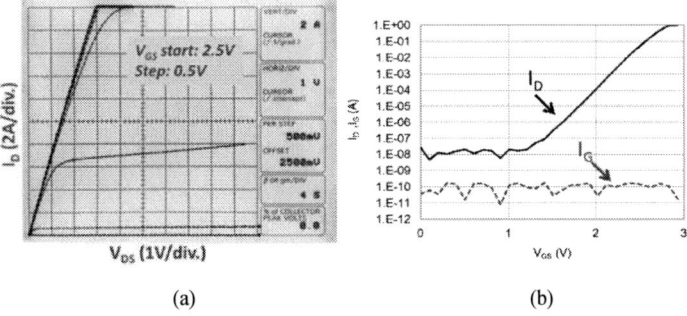

(a)   (b)

Fig. 4 (a) I_D-V_D characteristics and (b) I_D-V_G transfer characteristics at V_D of 1 V of the GaN HEMT-based cascode package device

Fig. 5 shows the cumulative probability plot of contact resistance (R_C) across an entire wafer using a process monitor dies at Au-free 6-inch Si-CMOS compatible line. Transmission line monitor (TLM) was used to estimate contact resistance. 133 wafers were characterized without any sorting. Uniform and extremely low R_C around 0.3 Ωmm was obtained reproducibly.

Fig. 6 shows the cumulative probability plot of R_{ON} across an entire wafer using a process monitor dies at Au-free 6-inch Si-CMOS compatible line. Liner behavior in the probability plot verifies that only one mechanism can be implied for determining R_{ON} with good controllability.

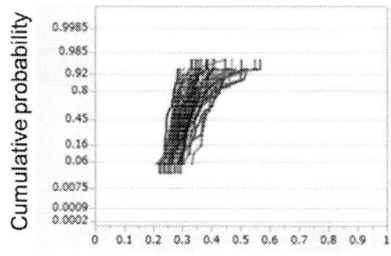

Fig. 5 Cumulative probability of ohmic contact resistance across an entire 6-inch wafer. Au-free 6-inch Si-CMOS compatible line was used. Transmission length monitor (TLM) was used to estimate contact resistance.

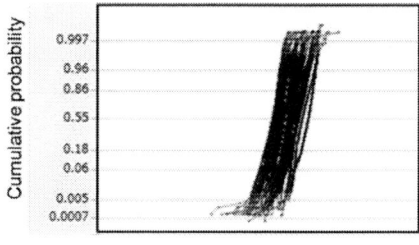

Fig. 6 Cumulative probability of R_{ON} across an entire 6-inch wafer. Au-free 6-inch Si-CMOS compatible line was used. 40 wafers data were shown.

Fig. 7 shows the typical off-state leakage measurements of GaN cascode package up to 1800 V. Over 1500 V breakdown voltage was confirmed. High temperature leakage is actually lower than Si Cool MOS and all our devices are rated for a spike voltage tolerance of 750 V. Contrary to the popular myth, dislocations do not impact the leakage current levels or reliability in otherwise well designed and optimally manufactured lateral GaN devices.

Dynamic R_{ON} characteristics or current collapse is one of the most important parameters to enter the power conversion device market. Dynamic R_{ON} of us-level was measured up to 1000 V as shown in Fig. 8. When tested in a resistive switching circuit from a blocking (off) state to full conduction at 5 A current, a voltage drop was detected through an oscilloscope at 0.5 µs after turning-ON. No degradation of R_{ON} was observed in Fig. 8. Temperature characteristics were also evaluated by the on-wafer measurement using 6-inch GaN HEMT on Si as shown in Fig. 9. No change was observed from room temperature (RT) to 150°C .

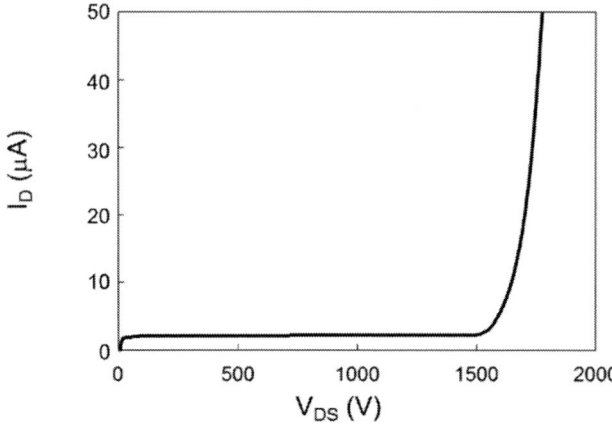

Fig. 7 Typical off-leakage characteristics of GaN cascode package.

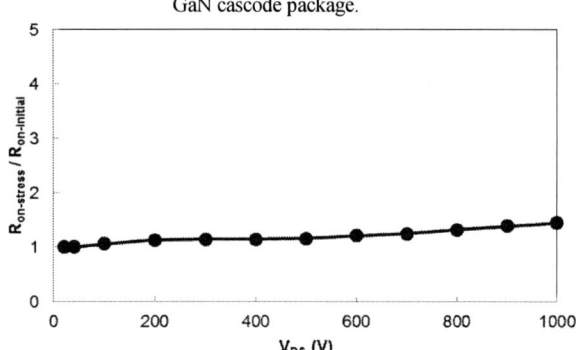

Fig. 8 Dynamic R_{ON} measurements of GaN cascode package up to 1000 V.

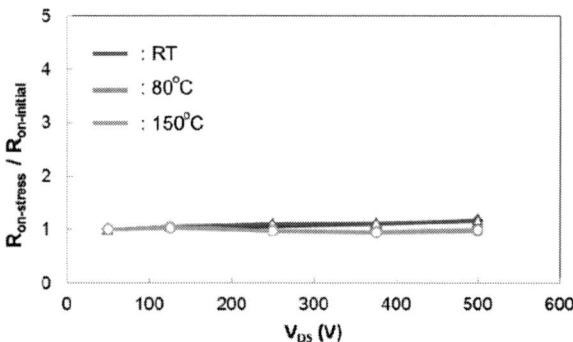

Fig. 9 Temperature characteristics of on-wafer dynamic R_{ON} of GaN-HEMT up to 500 V.

Table I is the summary of GaN cascode package compared with Si-MOSFET. The finished device in TO-220/TO247 and PQFN packages also features the Quiet-TabTM

package scheme, with the package base as a low-dV/dt source/drain terminal. Such a configuration allows 200% increase in switching speed with little compromise in induced EMI noises. The FOM of Ron*Qg ~1 nVs and Ron*Qrr ~8.5 nVs, is a significant enhancement already over the mature Si SJ MOSFETs (2).

TABLE I Summary of GaN cascode package compared with Si-MOSFET.

Symbol	Unit	Cool MOS IPP60R160P6	GaN-HEMT TPH3006PS	Test condition
V_{DSS}	V	600	600	
$R_{DS(ON)}$	ohm	0.16	0.15	T=25°C
Qg	nC	44	6.2	VDS =100 V VGS= 0-4.5 V, ID = 11 A
Qgd	nC	15	2.2	
Co(er)	pF	72	**56**	VGS=0 V, VDS=0 V to 480 V
Co(tr)	pF	313	**110**	
Qrr	nC	5300	54	IDS=9 A, VDD=480 V, di/dt =450 A/µs
trr	ns	350	30	

C. Reliability

Table II shows the JEDEC qualification measurement results (2). High temperature reverse bias (HTRB), Highly Accelerated temperature and humidity Stress Test (HAST), Temperature cycling test (T/C), Power cycling test (P/C), and High temperature storage test (HTSL) were shown in the Table. All JEDEC needs were satisfied with GaN HEMT on Si based cascode package for the first time.

Fig. 10 shows the box plot of HTRB operated at 150°C and 480 V when using GaN on 6-inch Si. No change of R_{ON} and leakage were observed after 1000 hours. Voltage acceleration test was also investigated to predict long-term reliability.

TABLE II Summary of GaN cascode package compared with Si-MOSFET.

item	conditions	duration	# of sample	results	judge
HTRB	480 V, 150°C	1000 hrs	77pcs*3Lot	0/231	Pass
HAST	100 V, 130°C,85%	96 hrs	77pcs*3Lot	0/231	Pass
T/C	-55°C to 150°C	1000 c	77pcs*3Lot	0/231	Pass
P/C	25°C to 150°C	5000 c	77pcs*3Lot	0/231	Pass
HTSL	150°C	1000 hrs	77pcs*3Lot	0/231	Pass

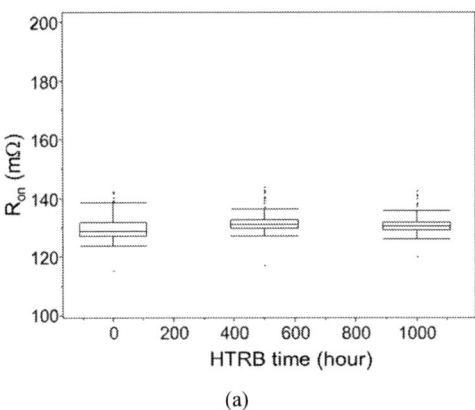

(a)

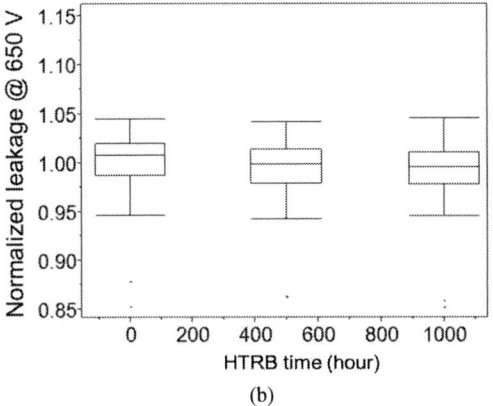

(b)

Fig. 10 Box plot of high temperature reverse bias (HTRB) test at 150°C under 480 V when GaN on 6-inch Si was used. (a) R_{ON}, (b) drain leakage at 650 V measurements of cascode package.

Fig. 11 shows high voltage off state (HVOS) testing. Over 1×10^7 hour lifetime was confirmed. To our knowledge, this is the first report of HVOS for any GaN power device. Further, through 1:2 400 Watt boost converter testing, we have also completed 3000 hour high temperature operation life (HTOL) testing with no degradation in any parameter (4,5).

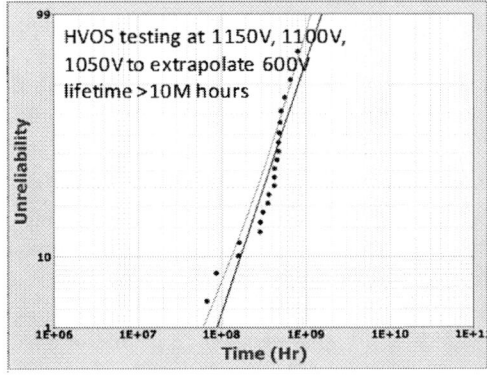

Fig. 11 HVOS testing demonstrating >1×10^7 hours lifetime at 80°C.

D. Circuit Efficiency

Fig. 12 (a) shows the all-in-one power supply, i.e., power factor control (PFC) with LLC converter, using GaN based cascode package device (3). Total size was decreased by 45% compared with Si-MOSFET based power supply. Maximum efficiency of over 95% was obtained at 200 kHz using GaN, as shown in Fig. 12 (b). Si-MOSFET based circuit showed the efficiency lower than 94% at 50-80 kHz.

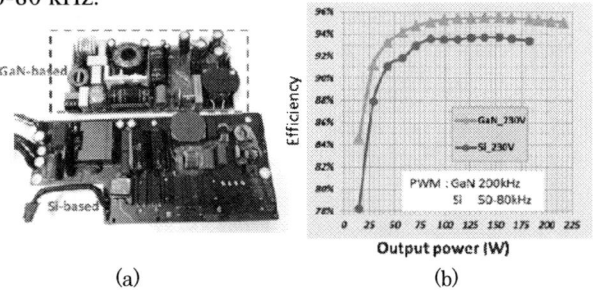

(a)   (b)

Fig. 12 (a) All-in-one power supply consisting of power factor control (PFC) and LLC converter, using GaN based cascode package device. (b) Efficiency as a function of output power. Efficiency of Si-MOSFET circuit was also shown.

Conclusion

Highly reliable cascode-package normally-off device consisting of GaN HEMT on Si was demonstrated. World's first JEDEC qualified data were shown with stable dynamic R_{ON}. 6-inch Au-free CMOS compatible process for GaN on Si fabrication was well demonstrated. Voltage accelerated test showed MTTF of over 1×10^7 hours at 600 V at 80°C. These results verify that 600 V GaN HEMT on Si reached the commercially ready status for new energy-saving society. In addition, conventional packages such as TO-220 are still suitable for high speed circuit application without using a specific gate driver.

References

(1) JEDEC JESD47I "Stress-Test-Driven Qualification of Integrated Circuit."

(2) Y.-F. Wu, J. Gritters, L. Shen, R.P. Smith, J. McKay, R. Barr and R. Birkhahn, "Performance and Robustness of First Generation 600-V GaN-on-Si Power Transistors," *The 1st IEEE Workshop on Wide Bandgap Power Devices and Applications (WiPDA)*, Oct. 27-29, Columbus, 2013, S1-002.

(3) Y.-F. Wu, J. Gritters, L. Shen, R. P. Smith, and B. Swenson, "kV-Class GaN-on-Si HEMTs Enabling 99% Efficiency Converter at 800 V and 100 kHz," IEEE Transactions on Power Electronics, Vol. 29, No. 6, June 2014, pp. 2634-2637.

(4) K. V. Smith and J. Haller, "High Field Reliability of GaN Power Devices," Reliability of Compound Semiconductors Workshop, Denver, May, 2014, S3.

(5) P. Parikh, Y. Wu, U. Mishra, L. Shen, R. Birkhahn, B. Swenson, J. Gritters, R. Barr, L. McCarthy, J. Honea, J. Yea, K. Smith, P. Smith, D. Dunn, J. McKay, H. Clement, T. Kikkawa, T. Hosoda, Y. Asai, K. Imanishi, and K. Shono, "Commercialization of 600V GaN HEMTs," 2014 SSDM, Tsukuba, Sep., 2014.

Power Devices on Bulk Gallium Nitride Substrates:
An Overview of ARPA-E's SWITCHES Program

Timothy D. Heidel

Advanced Research Projects Agency – Energy (ARPA-E)

U.S. Department of Energy

Washington, DC, USA

Pawel Gradzki

Booz Allen Hamilton

Washington, DC, USA

Abstract

Wide bandgap power semiconductor devices offer substantial energy efficiency opportunities in a wide range of applications. However, to date, relatively high cost has impeded the widespread adoption of these devices in many high volume applications. Recent progress in high quality bulk GaN substrates offers a new potential pathway to the development of novel vertical power semiconductor devices in gallium nitride. If successfully developed, these devices could offer a pathway to functional cost parity with silicon-based power devices at higher power levels. The Advanced Research Projects Agency-Energy (ARPA-E)'s recently launched SWITCHES program is targeting the development of bulk GaN, 1200 V, 100 A transistors and diodes. In this paper, we give an overview of the technical approaches within the program and discuss some of the major anticipated challenges.

Introduction

Power electronics are projected to play a significant and growing role in the delivery of electricity. It has been estimated that as much as 80% of electricity could pass through power electronics between generation and consumption by 2030 (1). (30% of electrical energy passes through power electronics converters today.) Technical advances in power electronics promise enormous energy efficiency gains throughout the United States economy (2).

Achieving high power conversion efficiency requires low-loss power semiconductor switches. Today's incumbent power semiconductor switch technology is silicon (Si) based MOSFETs, IGBTs and thyristors. However, silicon power semiconductor devices have several important limitations:

1. High Losses: The relatively low silicon bandgap (1.1 eV) and low critical electric field (30 V/µm) require high voltage devices to have substantial thickness. The large thickness translates to devices with high resistance and associated conduction losses.
2. Low Switching Frequency: Silicon high voltage power MOSFETs require large die areas to keep conduction losses low. Resulting high gate capacitance and gate charge produce large peak currents and losses at high switching frequencies. Silicon IGBTs have smaller die than MOSFETs due to utilization of minority carriers

and conductivity modulation, but the relatively long lifetime of minority carriers reduces the useful switching frequency range of IGBTs.
3. Poor High-Temperature Performance: The relatively low silicon bandgap also contributes to high intrinsic carrier concentrations in silicon-based devices, resulting in high leakage current at elevated temperatures. Temperature variation of the bipolar gain in IGBTs amplifies the leakage and limits the maximum junction temperature of many IGBTs to 125 °C.

New opportunities for higher efficiency have emerged with the development of wide bandgap (WBG) power semiconductor devices. WBG semiconductor-based devices are capable of low-loss operation at high voltages (> 1 kV to tens of kV), high frequencies (tens of kHz to tens of GHz), and high temperatures (>150°C). Power converters based on WBG devices can achieve both higher efficiency and higher gravimetric and volumetric power conversion densities. For example, in a recent demonstration, a 2kW motor driven by high frequency GaN devices resulted in an increase in efficiency of over 2% at full load and 8% at low load relative to the same motor being driven by Si IGBTs (3).

The Advanced Research Projects Agency-Energy (ARPA-E) has invested in WBG power semiconductor devices since 2010. ARPA-E's ADEPT (Agile Delivery of Electrical Power Technologies) program, funded several teams to develop a range of SiC and GaN devices and demonstrate their efficacy in power converters (4).

SiC and GaN have also made important commercial progress over the past decade with 1200 V SiC devices (5) and 600 V GaN (6) devices now qualified and commercially available. However, SiC and GaN device technology remains relatively immature relative to Si and currently carry a substantial cost premium, limiting their widespread adoption. The state of technology is such that many of the largest opportunities for increased energy efficiency and reduced energy-related emissions exist in extremely cost conscious industries, including markets for railway traction drives, automotive applications, and industrial motors. Cost for an equivalent functional performance at the device level remains a major barrier to the widespread adoption of WBG devices, despite opportunities for superior performance (including reductions in system costs). WBG devices will have to approach functional cost parity with Si power devices to gain widespread adoption.

In addition to high cost, most WBG discrete devices demonstrated to date have had relatively low current ratings. Traction, industrial, and grid applications advantage devices with high current ratings (>100 A) because implementation of low amperage devices in parallel increases power system complexity and cost.

In 2014 ARPA-E launched a new program entitled SWITCHES (Strategies for Wide Bandgap, Inexpensive Transistors for Controlling High-Efficiency Systems). The SWITCHES program is aiming to catalyze the development of high voltage (1200V+), high current (100A) single die power semiconductor devices that, upon ultimately reaching scale, have the potential to reach functional cost parity with silicon power transistors on a $/A basis, while also offering breakthrough relative performance (low losses, high switching frequencies, and high temperature operation). An overview of the project teams funded under the program can be found in Fig. 1.

Opportunities

Recent research results indicate that new materials advances, device architectures, and device fabrication processes could substantially accelerate progress towards WBG devices that achieve both higher current ratings and functional cost parity with silicon-based devices (7)(8)(9). These approaches have, as of yet, received relatively little attention from industry and the research community since they are perceived to be technically unproven and high risk.

Vertical device architectures for GaN power semiconductor transistors could substantially reduce cost and increase current densities (relative to die size). The dominant GaN device architecture today is the High Electron Mobility Transistor (HEMT) heterostructure depicted in Figure 2(a). The high electron mobility that is achieved in a 2D Electron Gas (2DEG) makes this device architecture appealing. However, substantial gate/drain lateral spacing must be maintained to allow for high breakdown voltages, reducing the effective current density (relative to die size) that can be achieved in these devices, increasing device cost per Ampere substantially for high voltage devices.

Recent progress in high quality bulk GaN substrates offers a new potential pathway to the development of vertical power semiconductor devices in gallium nitride (10)(11), as depicted in Figure 2(b). These devices could overcome the inherent limitations of lateral device structures and offer a pathway to functional cost parity with silicon-based power devices. As with vertical FET and IGBT technologies in silicon, it is expected that vertical devices will be able to achieve higher effective current densities and will enable improved thermal management.

There are several classes of unmet technological and scientific challenges that need to be solved in order to achieve high voltage, high current vertical GaN devices:

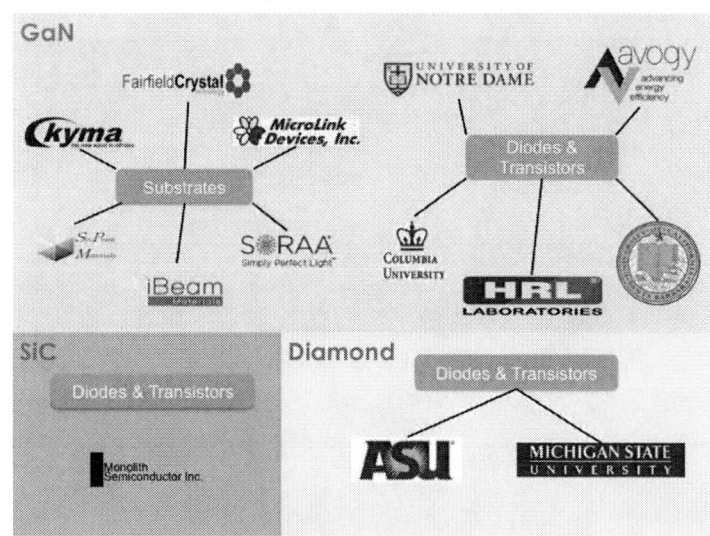

Figure 1: The ARPA-E SWITCHES program is funding 14 separate research teams in the United States. Only project lead organizations are shown here.

1. Substrates: Low defect density, large area bulk GaN substrates are required for many vertical device concepts. Today, GaN substrates are limited to small sizes and are very costly to produce. Many existing GaN substrate products have small domain structure unsuitable for high current power devices. High quality bulk GaN substrates with low defect densities are critical, as it has been shown that thermal conductivity of GaN is strongly dependent on material quality (12). New GaN substrate fabrication processes are needed, including new crystal growth and wafering approaches.

2. Device Structures and Fabrication Processes: New GaN device structures that are compatible with higher current ratings and low cost are needed. Improved approaches to fabricating buried p-type layers or improvements in metal-semiconductor contact formation on N-polar or semi-polar GaN crystal orientations are also needed. Device structures minimizing mask count could also reduce costs. Device structures that reduce thermal management challenges could also be important.

Technical Targets

Aggressive targets for performance and cost are set by ARPA-E so that technologies can rapidly surpass the current state-of-the-art and have disruptive impact. The superior performance of WBG devices, relative to the silicon switches they are replacing, will allow them to gain adoption in many applications before reaching functional cost parity. However, truly widespread adoption in high power, conservative, cost conscious applications, such as industrial motors and automotive, may require functional cost parity for the devices themselves.

The SWITCHES program is focused on the technical targets in Table 1. These targets are aimed at enabling high performance and widespread market adoption. The cost target

978-1-4799-8002-4/14 $31.00 © 2014 IEEE

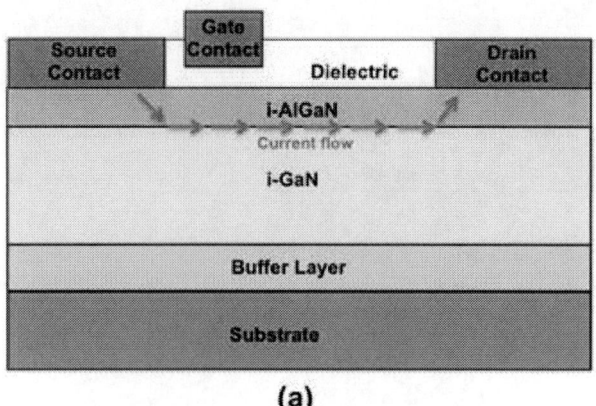

(a)

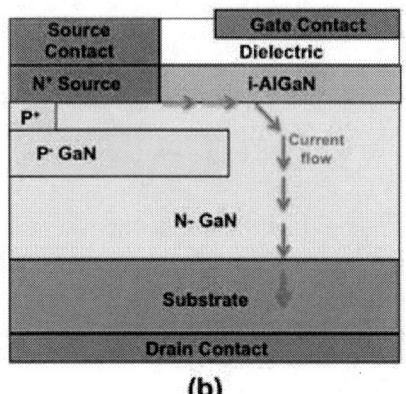

(b)

Figure 2: (a) Generalized device structure for AlGaN/GaN High Electron Mobility Transistor (HEMT) (b) Generalized device structure for vertical GaN transistor.

in particular was set based on an analysis of current 1200V silicon power device prices, see Fig 3. The cost target is intended to be a forward looking consideration of device costs, including packaging, assuming successful technology development and subsequent scaling to manufacturing.

The remaining technical targets describe device characteristics to be demonstrated experimentally by all research teams in the program by the end of the project period. These targets are consistent with the characteristics of devices that are required in high power applications, including many motor drive systems and electric vehicles. The dynamic performance target requires that all teams demonstrate that devices designed and fabricated under this program are suitable for high frequency switching applications.

SWITCHES GaN Projects

SWITCHES program GaN projects [1] are focused, generally, on two categories of innovations: (a) demonstration of low cost, high quality, freestanding GaN substrate growth techniques and (b) demonstration of vertical GaN diodes and transistors. As the SWITCHES program kicked-off earlier this year, there are few specific results available thus far. Our

---

[1] While the SWITCHES program is funding efforts in SiC, GaN, and diamond, we focus on GaN in this paper. The SWITCHES program features 14 separate R&D projects, 11 are focused on GaN devices.

TABLE 1
SWITCHES PROGRAM TECHNICAL TARGETS

Category	Value
Discrete Device Cost (Packaged)	<= $0.10 /A
Drain-Source Breakdown Voltage	>= 1200 V (V_{DSS} @ T_C = 25°C and V_{GS} = 0)
Continuous Drain Current Rating (Single Die)	>= 100 A (I_D @ T_C = 25°C and V_{GS} <= 20 V)
Operating Junction Temperature	-55 °C to 150 °C·
I_{off}/ I_{on} Ratio	> 10^6
V_{th} (not applicable to diodes)	> 2 V @ I_D = 5 mA
Dynamic Performance	Project must demonstrate device driving a hard switched boost (PFC) converter at f >= 40 kHz, Vout = 800 V, Imax = 50 A.
Specific R_{DSON}	< 3 mΩ·cm^2 @ V_{GS} = 15 V
Switching Loss $E_{ON}+E_{OFF}$	< 0.5 mJ @ 800 V and 50 A

intent here is to only introduce the objectives of each project at a relatively high level.

In the substrates area, several program teams are focusing on using either ammonothermal growth and/or hydride vapor phase epitaxy (HVPE) to demonstrate larger substrate diameters with lower defect densities. Soraa (Goleta, CA), for example, is working on developing scalable methods for the synthesis of large area, high quality seed crystals that would be capable of enabling bulk GaN boule growths up to 150mm in diameter via ammonothermal growth. Fairfield Crystal Technology (New Milford, CT) is exploring unique crystal growth processing conditions and reactor designs that may allow them to demonstrate 2" and 3" diameter, high quality single crystal GaN boules up to 1" long, grown at growth rates exceeding 200 µm/h. SixPoint Materials (Buellton, CA) and Kyma Technologies (Raleigh, NC) are attempting to combine HVPE growth with ammonothermal bulk seeds to enable high quality, low cost GaN boule production. Finally, Columbia University (New York, NY) and Microlink Devices (Niles, IL) are leading projects seeking to demonstrate scalable methods for epitaxial lift-off and substrate reuse. The Columbia University project is targeting substrate re-use up to 50 times using controlled spalling. The Microlink Devices project is attempting to develop 6" bulk GaN "superstrates" via tiling of high quality seeds. These superstrates are envisioned to provide a large-area, low dislocation density template for subsequent epitaxial growth. The team will employ epitaxial liftoff to enable superstrate reuse. Finally, iBeam Materials

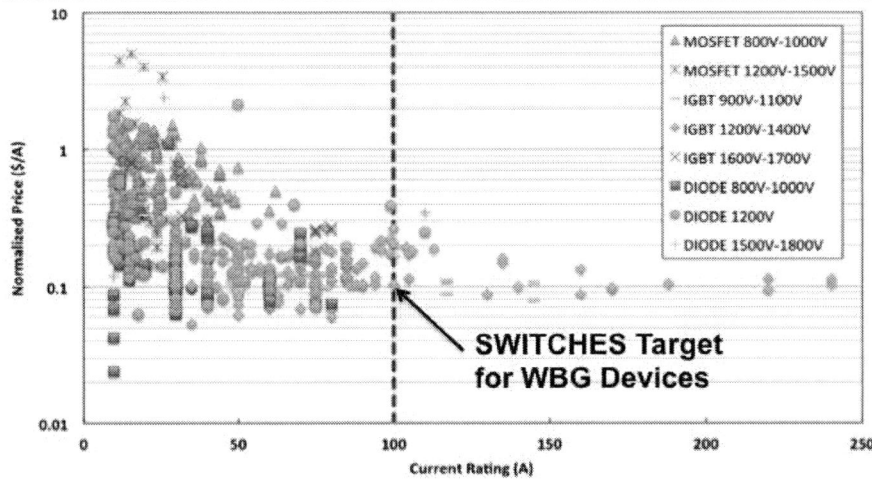

Figure 3: The price ($/A) of existing high-voltage silicon power semiconductor devices using data from the electronic components distributor Digi-Key. All in stock devices with relevant ratings and pricing information readily available on the Digi-Key website as of late May 2013 are included in the figure. Variability is associated primarily with different package technologies. This plot indicates that 1200V, 100A devices are typically priced between $0.10/A and $0.20/A today.

(Santa Fe, NM) is attempting to demonstrate the epitaxial growth of GaN on ion-beam assisted deposition (IBAD)-textured films. This approach, if successful, could enable continuous, roll-to-roll manufacturing of GaN transistors.

Teams in the program are also exploring a variety of novel vertical GaN device structures. These projects, led by Avogy (San Jose, CA), the University of California Santa Barbara (Santa Barbara, CA), the University of Notre Dame (Notre Dame, IN), HRL Laboratories (Malibu, CA), and Columbia University (New York, NY) are all seeking to advance upon devices previously demonstrated in the literature and meet the program targets. All of these teams are working to first demonstrate new device structures in simulation and then to actually fabricate working devices. We expect to see fundamental new contributions in many areas of GaN device fabrication including the development of reliable, efficient contact structures, strategies to demonstrate reliable p-doped GaN layers, and the development of high quality dielectric growth on GaN for MOS controlled, normally-off devices. All of the device-focused projects are expected to culminate in circuit demonstrations, in which vertical GaN transistors will switch at 800 V and 50 A in a boost converter at frequencies of at least 40 kHz to prove their suitability for use in high power converters.

Conclusion

The ARPA-E SWITCHES program, launched in 2014, is funding a portfolio of potentially transformational and disruptive GaN boule growth technologies, substrate fabrication methods, epitaxial growth techniques, and devices for a wide range of power electronics applications. Project teams are all focused on enabling the development of high voltage (1200V+), high current (100A) single die power semiconductor devices that have the potential to reach functional cost parity with silicon power transistors at scale. Successful projects will demonstrate devices operating in high power converters. We believe these research projects have promise to reduce the barriers to ubiquitous deployment of low-loss WBG power semiconductor devices.

References

(1) L.M. Tolbert, T. J. King, B. Ozpineci, J. B. Campbell, G. Muralidharan et al., "Power Electronics for Distributed Energy Systems and Transmission and Distribution Applications: Assessing the Technical Needs for Utility Applications." (Oak Ridge, TN: Oak Ridge National Laboratory, 2005)

(2) "SiC and GaN electronics: Where, when and how big?" *Compound Semiconductor*, July 27, 2012, http://www.compoundsemiconductor.net/csc/features-details.php?cat=feat ures&id=19735293

(3) Y-F. Wu D. Kebort, J. Guerrerol, S. Yea, J. Honea et al., "High-Frequency, GaN Diode-Free Motor Drive Inverter with Pure Sine Wave Output," *Power Transmission Engineering*, October 2012, 40-43, http://www.powertransmission.com/issues/1012/motor_drive_inverter.pdf

(4) "Agile Delivery of Electrical Power Technologies," ARPA-E, U.S. Department of Energy, accessed June 2, 2013, http://arpa-e.energy.gov/?q=arpa-e-programs/adept

(5) Cree, Inc., "Cree Announces Volume Production of Second-Generation SiC MOSFET, Bringing Significant Cost Savings to Power-Conversion Systems," news release, March 13, 2013, http://www.cree.com/news-and-events/cree-news/press-releases/2013/mar ch/2nd-gen-mosfet

(6) Transphorm, Inc., "Transphorm Releases First JEDEC-Qualified 600 Volt GaN on Silicon Power Devices," news release, March 14, 2013, http://www.transphormusa.com/news/transphorm-releases-first-jedec-qual ified-600-volt-gan-silicon-power-devices

(7) S. Chowdhury, M. H. Wong, B. L. Swenson, and U. K. Mishra, "CAVET on Bulk GaN Substrates Achieved With MBE-Regrown AlGaN/GaN Layers to Suppress Dispersion," *IEEE Electron Device Letters* 33, no. 1 (2012): 41-43, doi: 10.1109/LED.2011.2173456

(8) R. Yeluri, C. A. Hurni, S. Chowdhury, J. S. Speck, and U. K. Mishra., "Demonstration of Low ON-Resistance CAVETS with Ammonia MBE Grown Active p-GaN Layer as the Current Blocking Layer for High Power Applications." Paper presented at The Pacific Rim Meeting on Electrochemical and Solid-State Science, Honolulu, HI, October 2012.

(9) I.C. Kizilyalli, A. Edwards, D. Bour, H. Shah, H. Nie et al., "Vertical Devices In Bulk GaN Drive Diode Performance To Near-Theoretical Limits," *How2Power Today*, March 2013, http://www.how2power.com/newsletters/1303/articles/H2PToday1303_de sign_Avogy.pdf

(10) T. Uesugi and T. Kachi, "Which are the Future GaN Power Devices for Automotive Applications, Lateral Structures or Vertical Structures?," Paper presented at CS MANTECH Conference, Palm Springs, CA, May 2011.

(11) D. Disney, H. Nie, A. Edwards, D. Bour, H. Shah at al. "Vertical power diodes in bulk GaN," Paper presented at 25th International Symposium on Power Semiconductor Devices, Kanazawa, Japan, May 2013

(12) C. Mion, J. F. Muth, E. A. Preble, and D. Hanser, "Accurate dependence of gallium nitride thermal conductivity on dislocation density," *Applied Physics Letters* 89, 092123 (2006), doi: 10.1063/1.233597

An Enhanced 16nm CMOS Technology Featuring 2nd Generation FinFET Transistors and Advanced Cu/low-k Interconnect for Low Power and High Performance Applications

Shien-Yang Wu, C.Y. Lin, M.C. Chiang, J.J. Liaw, J.Y. Cheng, S.H. Yang, S.Z. Chang, M. Liang, T. Miyashita, C.H. Tsai, C.H. Chang, V.S. Chang, Y.K. Wu, J.H. Chen, H.F. Chen, S.Y. Chang, K.H. Pan, R.F. Tsui, C.H. Yao, K.C. Ting, T. Yamamoto, H.T. Huang T.L. Lee, C.H. Lee, W. Chang, H.M. Lee, C.C. Chen, T. Chang, R. Chen, Y.H. Chiu, M.H. Tsai, S. M. Jang, K.S. Chen, Y. Ku

168, Park Ave. 2, Hsinchu Science Park, Hsinchu, Taiwan, R.O.C., Email: shien-yang_wu@tsmc.com
Taiwan Semiconductor Manufacturing Company

Abstract

Advancing the state-of-the-art 16nm technology reported last year, an enhanced 16nm CMOS technology featuring the second generation FinFET transistors and advanced Cu/low-k interconnect is presented. Core devices are re-optimized to provide additional 15% speed boost or 30% power reduction. Device overdrive capability is also extended by 70mV through reliability enhancement. Superior 128Mb High Density (HD) SRAM Vccmin capability of 450mV is achieved with variability reduction for the first time. Metal capacitance reduction by ~9% is realized with advanced interconnect scheme to enable dynamic power saving.

Introduction

FinFETs with excellent electrostatic and short channel control enable low voltage operation, critical for next generation's low power and high performance applications [1-3]. However, the external parasitic capacitances distributed between fins, in addition to intrinsic device capacitance, increase the dynamic power of circuit operation. This paper presents an enhanced 16nm CMOS technology with effective capacitance reduction in both FinFET transistors and Cu/low-k interconnects for low power and high speed applications. In addition, device reliability is enhanced in order to support maximum operation voltage (Vmax) for additional speed boosts by overdriving the devices. DVFS (Dynamic Voltage Frequency Scaling) range up to 300mV for a product use condition at 85C for 10 year life-time can be realized with technology enhancements.

Process Architecture

Fin patterning and formation on bulk silicon with a 48nm fin pitch is realized using pitch-splitting technique where the fin width is determined by the sidewall thickness of a mandrel. Fin profile and gate profile are carefully co-optimized to balance among the needs to maintain excellent short channel control, to enhance drive current and to reduce parasitic capacitance of the devices. Poly-silicon deposition and gate patterning with a gate pitch of 90nm on the 3-dimensional fin structure is followed by high-K metal gate (HK/MG) RPG process. A dual-gate oxide flow is employed to support core and I/O devices. Gate height is carefully managed in order to provide an optimized combination of gate resistance and gate-to-source/drain (gate-to-S/D) capacitance. Raised source/drain with dual epitaxy process is used and optimized in order to mitigate source/drain (S/D) parasitic resistance. MEOL with tungsten (W) plug provides local routing connected to gate and source/drain. A M1 / Mx metal pitch of 64nm is enabled using advanced patterning scheme, whereas single patterning is adopted for metal pitches of 80nm/90nm and above. Advanced Cu/low-k interconnect process scheme is optimized to provide lower metal capacitance. A planar MiM with high-k dielectrics is also integrated to provide on-chip capacitance >15fF/um^2 for noise reduction.

Transistor Performance

The second generation FinFET transistors are introduced in the enhanced version of our 16nm CMOS technology. Figure of Merits (FOM) based on Inverter, NAND, and NOR circuitry with a fan-out of 3 (F.O.=3) illustrate a 15% speed gain or a 30% total power reduction over our previous reported work [1] as shown in Figure 1. Gate delay (CV/I) of an inverter Ring Oscillator (R.O.) with F.O.=3 measured at different voltages is shown in Figure 2. At 0.8V Vdd, the gate delay is reduced by ~20%. Reduction of gate delays becomes further enhanced as the voltage is further reduced. Superior electrostatic and short channel effect (SCE) of FinFETs are illustrated with competitive DIBL <40mV/V & sub-threshold swing <70mV/dec. [1-4] for core & I/O devices shown in Figure 3.

978-1-4799-8002-4/14 $31.00 © 2014 IEEE   48

These values are substantially better than our previously reported 28nm HKMG technology based on planar transistors [5]. Analog characteristics are further enhanced for both core and I/O devices. Figure 4 shows core device intrinsic gain (gm/gds) v.s. analog current (defined as the drain current measured at Vgs=Vt+200mV and Vds=0.5Vdd), whereas I/O device intrinsic gain as a function of gate length achieves significant improvement as shown in Figure 5. Furthermore, the RF cut-off frequency (f_T) and thermal noise have also been improved. Figure 6 shows the improvements in RF cut-off frequency (f_T) of the 2nd Gen. FinFET devices in this work over our previously reported works [1] and [5] measured at a bias of Vgs=0.5V and Vds=0.8V. Figure 7 compares thermal noise of the FinFET devices in this work against those in previous reported works, showing substantial improvement over 28HK/MG devices [5].

SRAM and Interconnect

High Density (HD) and High Current (HC) SRAM cells are further improved to provide speed gain with less standby-leakage current (Isb) over the previous work [1]. Figure 8 compares Isb and SRAM speed of this work with those reported in [1]. As can be seen, a greater than 25% speed gain is realized with lower Isb. This improvement allows the use of a 512 bits per bit-line scheme instead of a 256bits per bit-line scheme to reduce the periphery circuit size. As a result, the macro size for a Gb SRAM macro is reduced by 7~10%. The butterfly curves of the 0.07um^2 HD SRAM cell measured at different voltages are shown in Figure 9, where the excellent cell stability down to 0.4V is clearly demonstrated. The Vccmin of 128Mb SRAM is demonstrated in Figure 10. Vccmin down to 450mV with tight distribution can be clearly observed. In addition, Figure 11 shows the interconnect sheet resistance (Rs) and capacitance (C) of Mx metal with a pitch of 64nm. As can be seen, ~9% metal capacitance reduction has been achieved at the same metal sheet resistance. Via profile is re-optimized in order to maintain the similar Via resistance (Rc) values. Back-end-of-Line (BEOL) defectivity is monitored with a yield tile consists of long metal lines (290m) and 380 million of vias. Stable yields for BEOL as screened by metal line sheet resistance (Mx Rs), metal line leakage current (Mx LK) and Via resistance (Via Rc) are demonstrated in Figure 12.

Device Reliability

Device reliability of the second generation FinFET like time-dependent-dielectric-breakdown (TDDB), bias-temperature-instabilities (BTI) and hot carrier injection (HCI) are further improved and characterized to maximize the performance gain of the technology. With careful post high-k/MG thermal optimization, TDDB and BTI for both NMOS and PMOS are improved. Figure 13 shows the improved the Mean-time-to-failure (MTTF) of TDDB for NMOS and PMOS in this work versus those in the previous work [1], Figure 14 shows improved BTI with reduced threshold voltage (Vt) shifts under various stress voltages for core NMOS and PMOS. High quality I/O gate oxide/high-k dielectric enables good TDDB and BTI for I/O devices. In addition, hot carrier immunity is also addressed with junction doping profile optimization for both core and I/O devices. Figure 15 shows improved Vt shift from HCI as a function of different voltages (Vdd). Overall the maximum operation voltage (Vmax) of core devices is raised by 70mV through reliability enhancement. Figure 16 compares speed and leakage power of a representative ring oscillator (R.O.) at the nominal bias voltage, at the Vmax of the previous work [1] and at the Vmax of this work. The increase in Vmax of this technology is found to translate into a 10% speed increase over the speed at the maximum operation voltage of work [1] as illustrated in Figure 16.

Conclusion

An enhanced second generation 16nm FinFET CMOS foundry technology with lower power, higher performance and smaller SRAM macro size and lower SRAM Vccmin is developed based on improved capacitance, better SRAM speed, lower Isb and enhanced SRAM cell stability. Reliability robustness and variability reduction provide an additional 10% performance gain at the Vmax and wider manufacturing margins for volume production.

References

[1] S.-Y. Wu et. al., IEDM Tech. Dig., pp. 224-227, 2013
[2] K. Seo et al., VLSI Tech. Symp. pp.14-15, 2014
[3] C.-H. Jan et. al., IEDM Tech. Dig., pp. 44-47, 2012
[4] Q. Liu et al., IEDM Tech. Dig., pp. 228-231, 2013
[5] S. Yang et al., CICC, pp. 1-5, 2011

978-1-4799-8002-4/14 $31.00 © 2014 IEEE

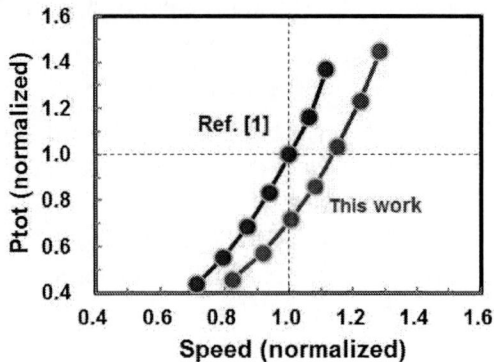

Fig. 1 FOM shows 15% speed gain or 30% power reduction.

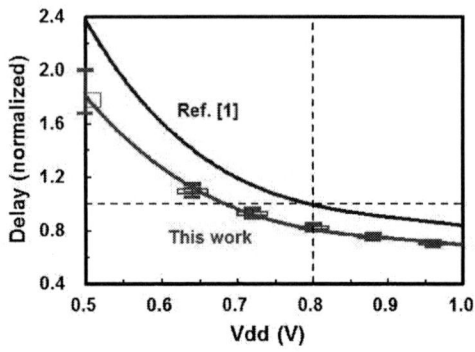

Fig. 2 CV/I of an inverter R.O. (F.O.=3) at different Vdd.

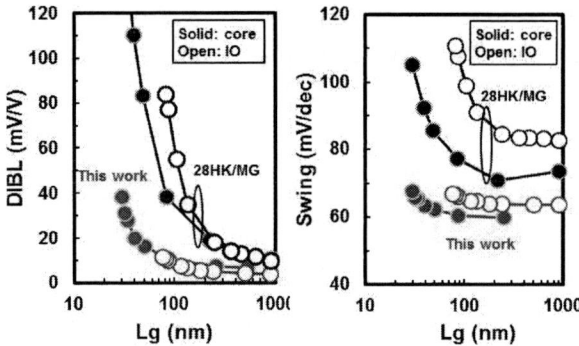

Fig. 3 Superior FinFET device DIBL and sub-threshold swing.

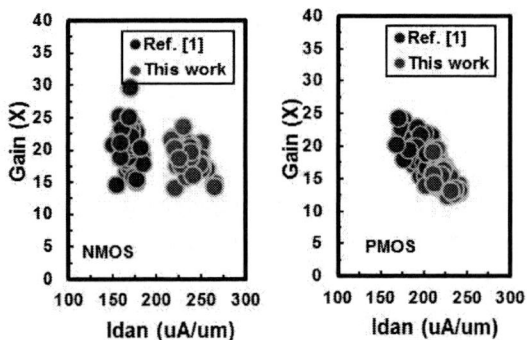

Fig. 4 Core device intrinsic gain (gm/gds) vs.drain current at Vgs=Vt+200mV and Vds=0.5*Vdd.

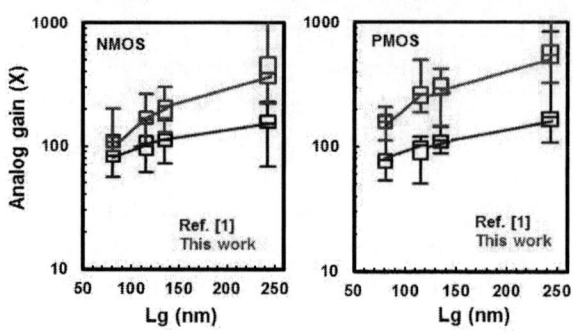

Fig. 5 Improved I/O device intrinsic gain (gm/gds) vs. Lg.

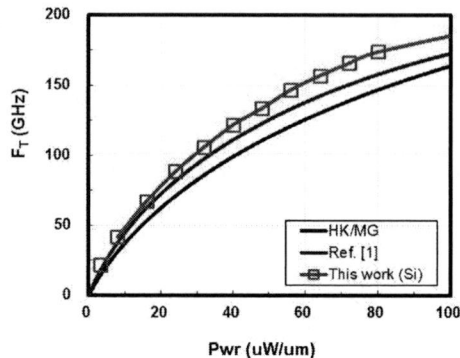

Fig. 6 NMOS RF cut-off frequency (f_T) vs. power at Vgs=0.5V and Vds=0.8V.

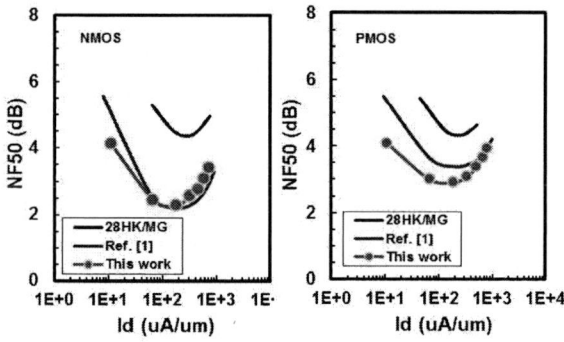

Fig. 7 FinFET devices achieve lower thermal noise than 28HK/MG.

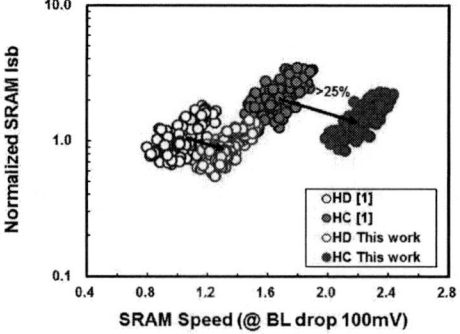

Fig. 8 HC/HD SRAM cell speed is boosted by >25% with less Isb.

978-1-4799-8002-4/14 $31.00 © 2014 IEEE

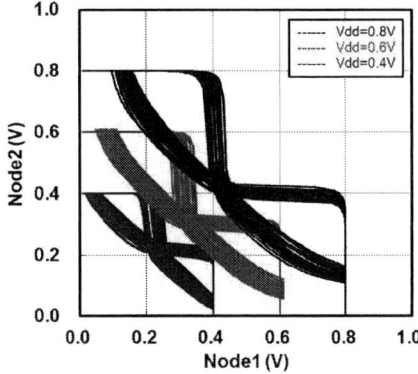

Fig.9 SNM of 0.07um² HD SRAM cell is illustrated down to 0.4V.

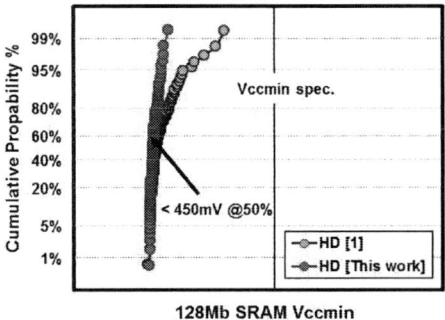

Fig. 10 128Mb HD SRAM Vccmin capability of 450mV is achieved.

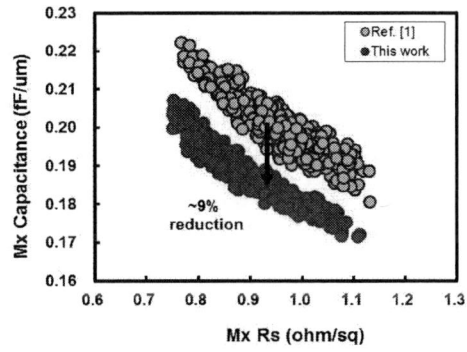

Fig.11 Metal capacitance is reduced by ~9% for Mx pitch of 64nm.

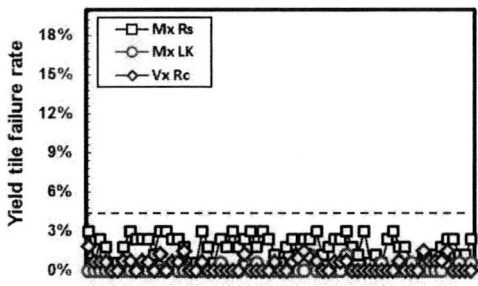

Fig. 12 Stable BEOL yield tile consists of metal Rs and via Rc.

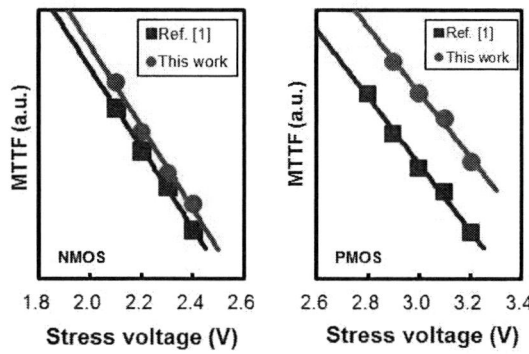

Fig. 13 Improved core NMOS and PMOS TDDB in this work.

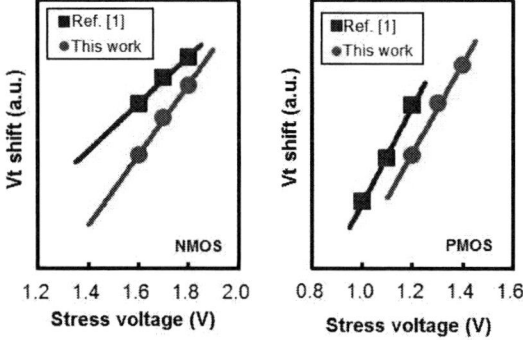

Fig. 14 Improved core device NBTI and PBTI in this work.

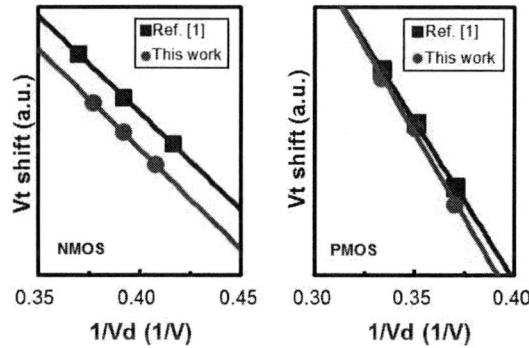

Fig. 15 I/O NMOS and PMOS HCI are improved by S/D junction opt.

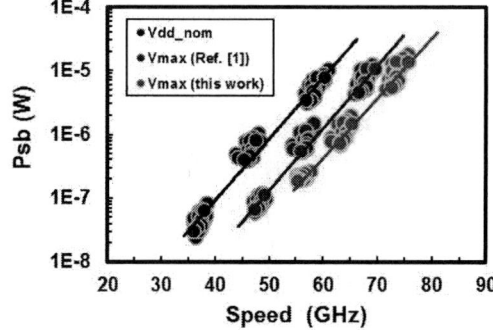

Fig. 16 The improved Vmax provides additional 10% R.O. speed gain.

978-1-4799-8002-4/14 $31.00 © 2014 IEEE   51

Analog Circuit and Device Interaction in High-Speed SerDes Design in 16nm FinFET CMOS Technology

Freeman Zhong, Ashutosh Sinha

Avago Technologies Inc, San Jose, CA 95131

Abstract

SerDes deals with data serialization, deserialization and channel equalization up to data rate of 28+Gb/s. Process technology and device characteristic greatly impacts architecture, circuit topology, and design merit of a SerDes. Several architecture choices, analog circuits, and techniques to mitigate undesired device characteristic in 16nm FinFET are discussed in this paper. With advanced CMOS technology and mitigation techniques, a prototype 28Gb/s SerDes was developed and demonstrated desired performance, power and die area.

Introduction

SerDes is one of the most critical components in communication systems. As data traffic increases exponentially and CMOS technology advances to 16nm in past 15 years, data rate of SerDes has increased from 1.25Gb/s to 28+Gb/s. As shown in Fig. 1, 2, 3, a serial link consists of a transmitter, channel, and receiver. It not only functions as data serialization from N to 1 bit, and deserialization from 1 to N bit, but also deals with signal integrity challenges, such as inter-symbol-interference, cross-talk, reflection and system jitter as data rate increases. Due to low-pass nature of a channel, the received eye is totally closed at a receiver. With channel equalization from both transmitter and receiver, the equalized eye is open enough for clock and data recovery as shown in Fig. 4. Many SerDes were designed and fabricated in 250nm to 28nm CMOS technology (1), (2), (3).

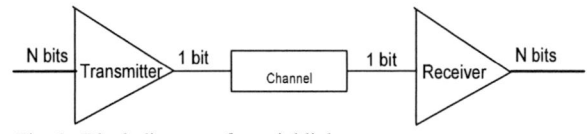

Fig. 1– Block diagram of a serial link

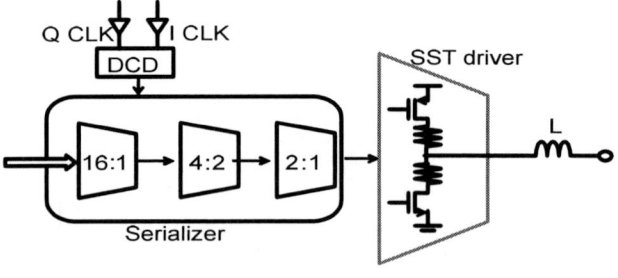

Fig. 2 – Block diagram of a transmitter

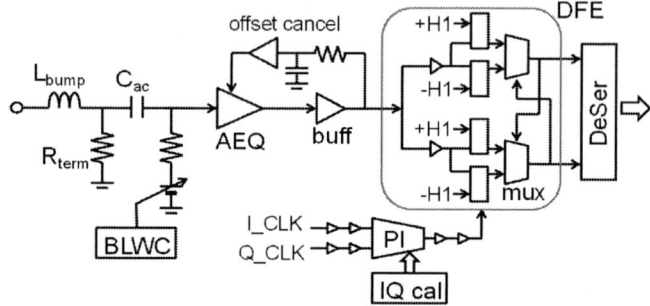

Figure 3 – Block diagram of a receiver

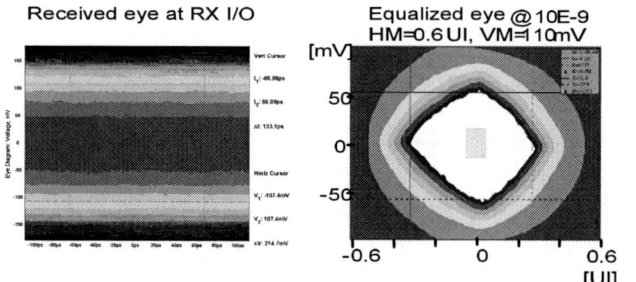

Fig. 4 – Received and equalized eyes at a receiver with 28Gb/s PRBS31 data traffic over a channel of 34dB loss

SerDes architecture and process technology interaction

Process technology greatly impacts the architecture, implementation, performance and power consumption of a SerDes. The critical process parameters, such as unity gain frequency, threshold voltage, leakage, and maximum supply voltage of devices, are dominant considerations for SerDes architecture choice. For example, a decision feedback equalizer (DFE) is a timing critical circuit that generates a feedback signal and adds it to the received signal within one unit interval (UI). For data rate below 16Gb/s, direct feedback DFE, as shown in Fig. 5, is an optimal implementation where the timing constraint of Tck-q + Tdac + Tsum + Tsetup < 1UI can be met in in 28nm/16nm process. However, when data rate increases to 28Gb/s where 1UI is only 35.7ps, the timing constraint of direct DFE is no longer met. A h1 loop unrolled DFE, shown in Fig. 6, was developed where the +/-h1 feedback signals are pre-calculated and added to received signal, and 2 speculative outputs are selected based on previous data decision, thus, new timing constraint of Tck-q + Tmux + Tsetup < 1UI can be met in 28nm/16nm CMOS.

As CMOS technology scales down to16nm, thin-oxide device operates at lower supply voltage that presents challenges of headroom and linearity to analog circuits. To tackle these

978-1-4799-8002-4/14 $31.00 © 2014 IEEE

challenges, simple circuit topologies without device stacking are used, and inverter-based analog circuits are one of these topologies. As shown in Fig 2, source serial terminated driver (SST), consisting of a CMOS inverter and 50 ohms termination resistor, delivers output differential swing the same as supply voltage to the driver with ¼ power consumption of its current mode logic (CML) counterpart. Another example is TAS-TIA-based analog front-end (AFE), shown in Fig 7, where an inverter is configured in close-loop with a compact inductor and feedback resistor, and forms a trans-impedance amplifier (TIA) with ultra-high bandwidth and dynamic range. Fig. 8 shows the measured AC transfer function of this AFE.

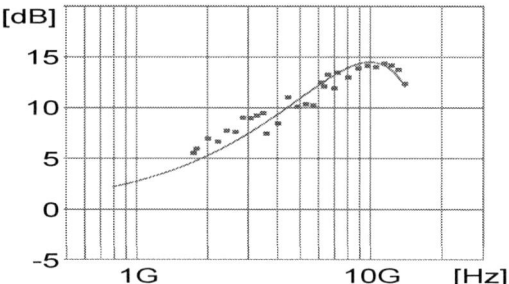

Fig. 8 – Measured transfer function of TAS-TIA AFE with 14dB boost at 12GHz

Analog circuit and device interaction

Device characteristics are fundamental considerations in analog circuit design. Compared with planar CMOS, 16nm FinFet devices are of low sub-threshold swing ratio, low threshold voltage and its variation, small drain induced barrier lowering effect, and body effect, low device noise, and similar Idsat of both PMOS and NMOS devices that are greatly beneficial to high performance analog designs (4). However, higher gate capacitance and limitation imposed by quantized channel width and length needs to be mitigated to gain full benefits of the process. Fig. 9, 10 show comparisons of device output impedance and gate capacitance between 28nm planar and 16nm FinFET devices. Taking the advantage of high output impedance, the intrinsic device gain Gm*Ro is 2-3X times compared to 28nm process, and device size can be reduced to achieve similar electrical and matching performance as 28nm's. However, the equivalent gate capacitance of a 16nm FinFET device is higher by ~20%, especially in Vgs around 0.3V, compared with a device of similar gm in 28nm that makes its unity gain frequency equal or lower than 28nm's.

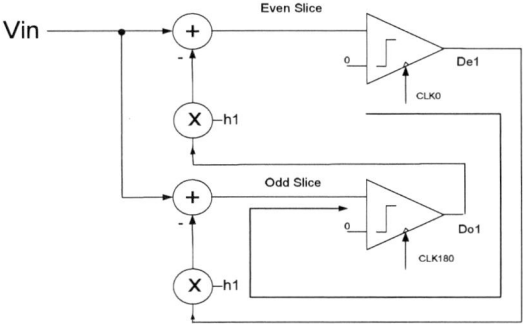

Fig. 5 – Block diagram of direct feedback h1 DFE implementation

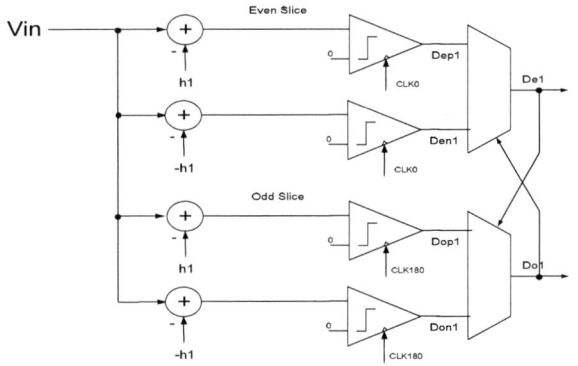

Fig. 6 – Block diagram of loop unrolled h1 DFE implementation

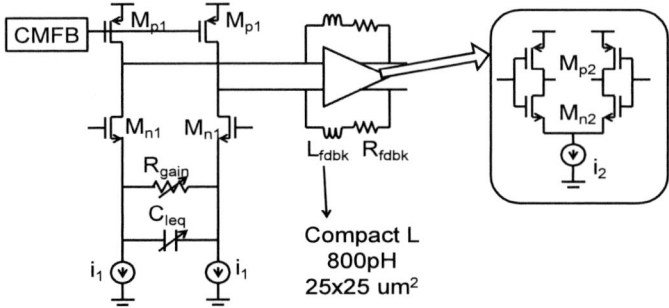

Fig. 7 – Block diagram of TAS-TIA based analog front-end

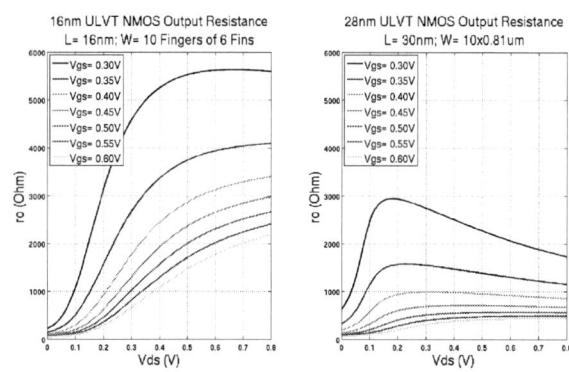

Fig. 9 – Comparison of output impedance of 16nm FinFET and 28nm NMOS

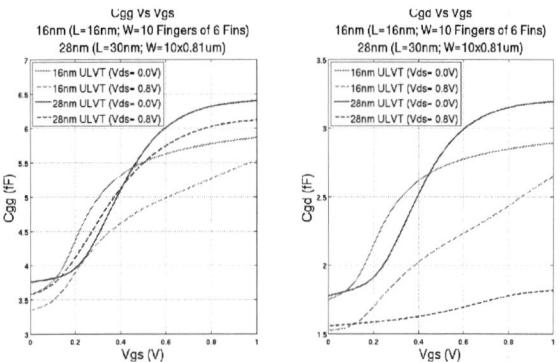

Fig. 10 – Comparison of gate capacitance of 16nm FinFET and 28nm NMOS

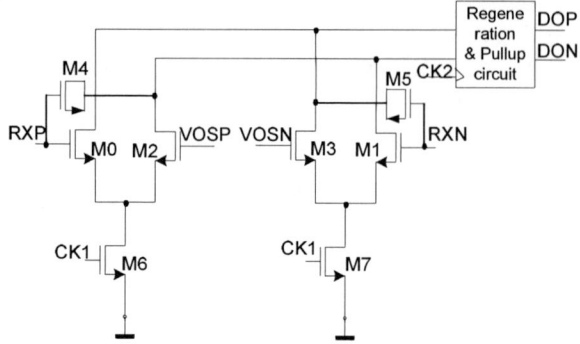

Fig. 12 – Block diagram of sense-amplifier based capture latch with low input gate capacitance and noise cancellation

To tackle high gate capacitance in analog circuit design, many circuit design techniques are used. First, device parasitic depends strongly on the layout, like fin number of each finger. A ring-based oscillator with different fin number was studied to identify relationship of oscillation frequency vs. fin number. As shown in Fig. 11, the oscillator reaches maximum frequency with fin number of 4. Hence, device of 2-5 fins are used for all high-speed signal paths balancing speed and layout area. Other circuit techniques are elaborated in design of a sense-amplifier based capture latch, as shown in Fig. 12. Equipped with wide range of offset cancellation circuit up to 120mV, a capture latch can use very small size for 2 differential pairs of M0&M2, and M1&M3 since feedback signals (VOSP, VOSN) is applied to M2&M3 to cancel out DC offset of 2 differential pairs. The input differential pair is configured as a frequency doubler where input gate capacitance of M0 and M1 is equal to $Cgs/2 + Cgd$ since the gates of M2 and M3 are AC ground. Due to the high Cgd in 16nm and its Miller effect, the effective Cgd increases significantly and puts heavy load on driving circuit. High Cgd also induces kick-back noise on its driver. A pair of feed-forward capacitors, M4 and M5, is used to reduce effective load and kick-back noise on the driver. With all these circuit techniques, the capture latch is capable to operate at 28Gb/s and has sensitivity better than 15mV over PVT.

Varactor is another critical device in analog circuit design, but exhibits asymmetrical C-V characteristic as shown in Fig. 13, thus, asymmetrical voltage-control oscillator (VCO) gain KVCO vs. control voltage in a LC oscillator, as shown in Fig. 14. A linearization circuit, as shown in Fig. 15, is used to shift the center point of C-V curve to zero voltage and symmetrical KVCO over control voltage is achieved, as shown in Fig. 16

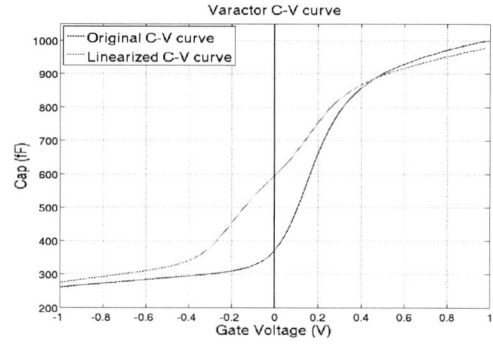

Fig. 13 – Intrinsic and linearized C-V curve of a varactor in 16nm FinFET

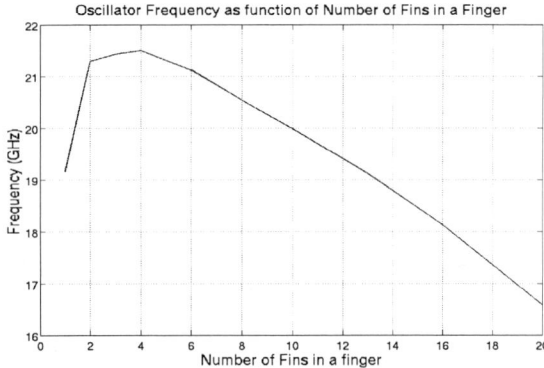

Fig. 11 – Oscillation frequency vs. number of fins

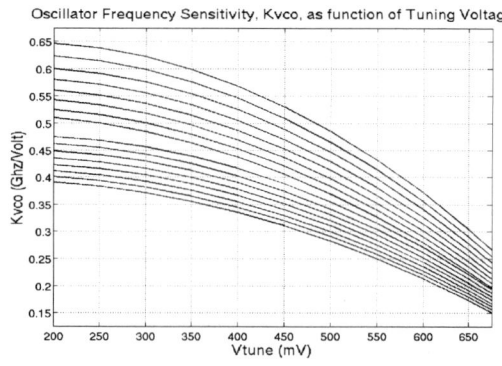

Fig. 14 – KVCO vs control voltage of LC-VCO with intrinsic varactors

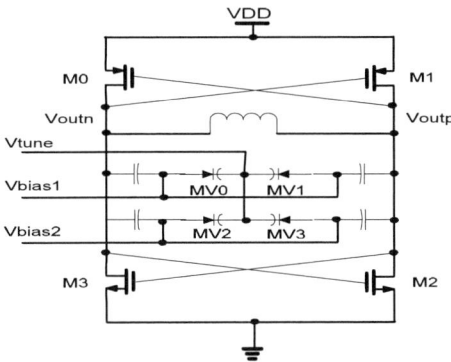

Fig. 15 – Block diagram of LC-VCO with linearized varactors

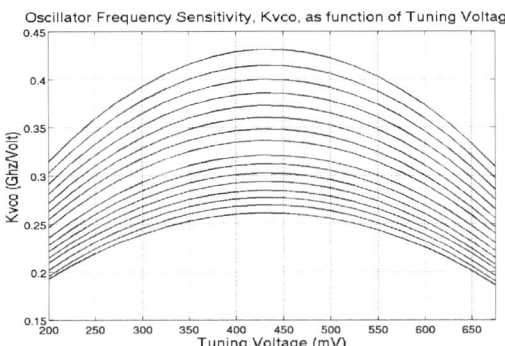

Fig. 16 – KVCO vs control voltage of LC-VCO with linearized varactor

Physical implementation is extremely important for analog designs, especially for high-speed SerDes since highly compact layout is desired to minimize routing parasitic resistance and capacitance. On the other hand, a device with dummy pattern and guard-ring surrounding it, as shown in Fig. 17, is recommended to achieve good device model to silicon correlation (minimize lithography and layout effects by containing the device surroundings), local & global device matching and noise isolation but makes layout less compact. To address this layout challenge, a layout methodology with shared dummy pattern and guard-ring over a group of devices can be used. With this methodology, most analog circuits of a SerDes achieve compact and small area, desired electrical performance, and good simulation-to-silicon correlation. Fig. 18 is the example of a CML frequency divider layout.

Fig. 17a, NMOS   Fig. 17b, NMOS w dummy   Fig. 17c, NMOS w dummy & guard-ring
Fig. 17, 16nm FinFET device layout patterns

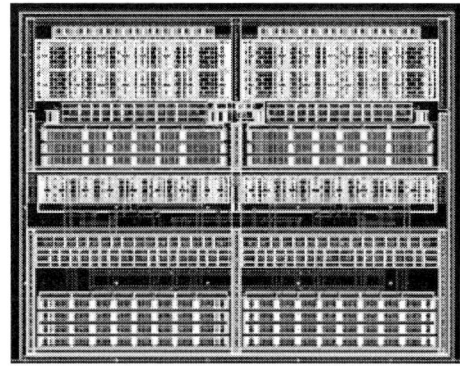

Fig. 18, Layout of CML frequency divider in 16nm FinFET CMOS

Conclusion

The architecture, circuit topologies, physical implementation, and design merit of a SerDes are substantially impacted by process technology and device characteristics. The 16nm FinFET CMOS technology offers many good features and device characteristics for high-speed SerDes designs compared with 28nm planar technology. Taking advantage of 16nm FinFET CMOS technology, a prototype 28Gb/s SerDes was developed and demonstrated desired performance, power and die area, even though many circuit techniques were required to mitigate undesired device characteristics.

References

(1) H. Kimura, P. Aziz, T. Jing, A. Sinha, R. Narayan, H. Gao, et al., "28Gb/s 560mW Multi-Standard SerDes with Single-Stage Analog Front-end and 14-Tap Decision-Feedback Equalizer in 28nm CMOS," in IEEE ISSCC Dig. Tech. Papers, 2014, pp. 38-39.
(2) J.F. Bulzacchelli, C. Menolfi, T.J. Beukema, D.W. Storaska, J. Hertle, D.R. Hanson, et al., "A 28-Gb/s 4-Tap FFE/15-Tap DFE Serial Link Transceiver in 32-nm SOI CMOS Technology," in IEEE ISSCC Dig. Tech. Papers, 2012, pp. 324-325.
(3) F. Zhong, S. Quan, W. Liu, P. Aziz, T. Jing, J. Dong, et at., "A 1.0625~14.025Gb/s Multi-Media Transceiver With Full-Rate Source-Series-Terminated Transmit Driver and Floating-Tap Decision-Feedback Equalizer in 40nm CMOS," IEEE J. Solid-State Circuits, vol. 46, no. 12, pp. 3126-3139, Dec., 2011.
(4) Shien-Yang Wu, C.Y. Lin, M.C. Chiang, J.J. Liaw, J.Y. Cheng, et al., "A 16nm FinFET CMOS Technology for Mobile SoC and Computing Applications" IEEE International Electron Devices Meeting (IEDM), 2013, pp. 9.1.1 – 9.1.4.

978-1-4799-8002-4/14 $31.00 © 2014 IEEE

16 nm FinFET High-k/Metal-gate 256-kbit 6T SRAM macros with Wordline Overdriven Assist

Makoto Yabuuchi, Masao Morimoto, Yasumasa Tsukamoto, Shinji Tanaka, Koji Tanaka, Miki Tanaka and Koji Nii

Renesas Electronics Corporation, Tokyo, 187-8588, Japan

Tel.: +81-42-328-5900, Fax: +81-42-327-8195, Email: {makoto.yabuuchi.ub, koji.nii.uj}@renesas.com

Abstract: We demonstrate 16 nm FinFET High-k/Metal-gate SRAM macros with a wordline (WL) overdriven read/write-assist circuit. Test-chip measurements confirm improved minimum operating voltage (V_{min}), standby leakage current, and access time compared to planar bulk CMOS. The proposed assist circuit improves V_{min} by 50 mV and improves read-access-time by more than 1.5 times in 256-kbit SRAM macros. Read current (I_{read}) dependence against the fin diffusion length was observed. An extra design guard-band is needed to provide a reliable operation margin.

16 nm FinFET High-k/Metal-gate 6T SRAM bitcells

6T SRAM bitcells are crucial for products in bulk FinFET CMOS technology as well as in planar bulk CMOS technology [1, 2]. **Fig. 1(a)** shows a schematic. Several layouts specified by the number of fins are presented in **Fig. 1(b)**. For example, 1-2-2 bitcell type stands for the combination of *one* pull-up (PU) PMOS fin, *two* pull-down (PD) NMOS fins and *two* pass-gate (PG) NMOS fins. **Fig. 2** presents a comparison of local variations of threshold voltage, sigma-Vt, for PU/PD/PG at 28 nm, 20 nm planar bulk, and 16 nm FinFET. FinFET achieves lower sigma-Vt attributable to the dopant-less channel and higher gate controllability, which contributes to enhancement of the read/write margin at lower supply voltage operation. In advanced technology nodes, SiGe enhancer technologies for PMOS are introduced [1, 2]. **Fig. 3** shows the alpha ratio (defined as I_{PU}/I_{PG} with on-current of PU (I_{PU}) and that of PG (I_{PG})). Applying SiGe increases the alpha ratio, with an accompanying decrease of the write-margin because of strong PU. **Fig. 4** presents a comparison of static noise margin (SNM) and write margin (WM) of 6T SRAM, confirming the degradation on WM in 16 nm FinFET with 1-1-1 bitcells. SNM degradation results from stronger PG and PD with fixed PU fin.

Wordline overdriven read/write-assist circuit

Fig. 5 presents various assist circuit techniques for enhancing WM [3–10]. The negative bitline technique [5] improves WM better than the VDD lowering technique does [4]. However, the power overhead increases, as pointed out in an earlier report [8]. Therefore, we propose the WL overdriven assist circuit to enhance not only WM but also read-access-time with minimum power and area overhead. The WL overdrive induces SNM degradation in read-operation or that of half-selected cells in write-operation. However, dynamic stability [11] alleviates SNM in AC operation. It can therefore compensate for this degradation such that the WL overdrive does not harm the SNM. **Fig. 6** presents a schematic of the proposed WL overdriven assist circuits. The supply voltage of VDDW (>VDD), which is provided by an external source or on-die DC–DC converter, is connected to the WL driver. **Fig. 7** shows timing waveforms for read/write-operation with and without assist. In high-speed mode, the WL is overdriven for both read and write cycles. Monte Carlo simulations of SNM and WM are carried out for 28 nm planar bulk CMOS bitcell and 16 nm FinFET (**Fig. 8**). The WL overdriven assist circuit decreased the SNM by 17%, but

improved the WM considerably by 99% compared to 28 nm without assist. I_{read} was also increased by 45%.

Test chip implementation and silicon measurement results

We designed and fabricated test chips using TSMC 16 nm FinFET bulk CMOS technology [2]. **Fig. 9** shows a test chip photograph and layout plots of 256-kbit SRAM macros with 1-2-2 and 1-3-3 bitcells. Measured V_{min} distributions of 256-kbit SRAM macros at -40°C and 125°C are presented in **Fig. 10**, achieving operation below 0.7 V with good distributions. The distributions of standby leakage current are depicted in **Fig. 11**. No dependence on stored data patterns was observed: all-0, all-1, and stripes. Therefore, no issues of variation arise from any double-patterning problems or mask misalignment. **Fig. 12** presents the relation between read access times and core MOS (standard Vt, SVT) on-current measured by the other DC test circuit, showing good correlation. Mean values are 1.03 ns and 1.04 ns, respectively, for 1-2-2 and 1-3-3 macros. **Figs. 13(a) and 13(b)** show plots of read-access-time *vs.* supply voltage for 256-kbit SRAM macros in high-speed mode (with assist) at -40°C. The mean value at 0.7 V is 648 ps (641 ps) for 1-2-2 (1-3-3) SRAM macro, which is over 1.5× faster than that of w/o assist. **Fig. 14** shows measured V_{min} distributions with/without WL overdriven assist at -40°C, showing that the proposed read/write-assist improves 50 mv (70 mV) of the mean value for 1-2-2 (1-3-3) macro, and no V_{min} degradation is observed. In the FinFET technology, the layout dependency effect (LDE) becomes significant compared to that of planar bulk CMOS. For example, the dependence of the fin length affects the on-current characteristics of FinFET [12]. To confirm the LDE in the SRAM bitcell array, we also implemented test structures to evaluate dependencies on I_{read} *vs.* the fin-diffusion length of PD/PG NMOSs in a column (**Fig. 15**). The measured data in the 128-row test structure show that I_{read} of the bitcells near the array edge is decreased (ca. 5%) compared to that at the center. No dependence was found in the short eight-row test structure. The measured data showed that the design guard-band of large SRAM array with many rows should be considered appropriately.

Conclusion

16 nm FinFET High-k/Metal-gate SRAM macros were designed and fabricated successfully. The proposed WL overdriven read/write-assist circuit improved V_{min} by 50 mV and improved over 1.5× read access time of 256-kbit SRAM macros.

Acknowledgment

We thank all TSMC staff for supporting test chip fabrication.

References

[1] C. Auth et al., VLSI Tech., pp. 131-132, 2012. [2] S.-Y. Wu et al., IEDM pp. 224-227, 2014. [3] M. Yamaoka et al., JSSC, Vol. 41, No. 3, pp. 705-711, 2006. [4] K. Zhang et al., JSSC, Vol. 41, No. 1, pp. 146-151, 2006. [5] K. Nii et al., VLSI Cir., pp. 212-213, 2008. [6] E. Karl et al., ISSCC, pp. 230-231, 2012. [7] J. Chang et al., ISSCC, pp. 316-317, 2012. [8] M. Yabuuchi et al., ISSCC, pp. 234-235, 2014. [9] T. Song et al., ISSCC, pp. 232-234, 2014. [10] Y.-H. Chen et al., ISSCC, pp. 238-239, 2014. [11] Y. Tsukamoto et al., CICC, pp. 1-4, 2011. [12] M. Garcia Bardon et al., VLSI Tech., 2013.

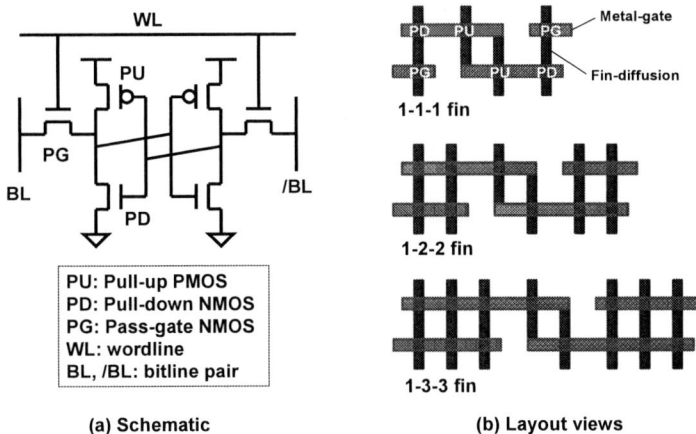

(a) Schematic **(b) Layout views**

Fig. 1: (a) Schematic of 6T SRAM bitcell. (b) Layout views of 6T SRAM bitcells with 1-1-1, 1-2-2, and 1-3-3 fin numbers.

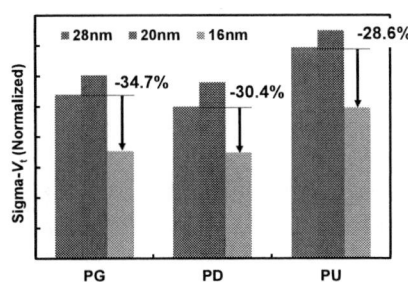

Fig. 2: Local V_t variations of pull-up (PU) PMOS, pull-down (PD) NMOS and pass-gate (PG) NMOS for 28 nm, 20 nm (planar bulk), and 16 nm (FinFET).

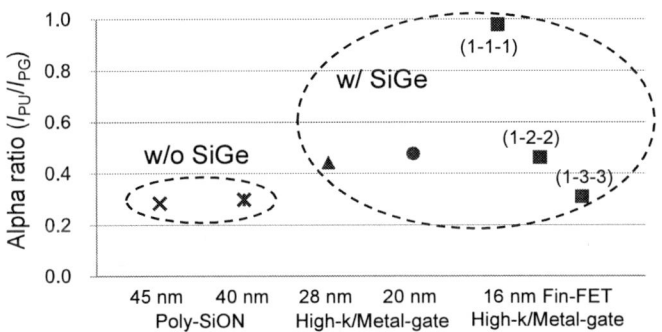

Fig. 3: Technology trends of alpha ratio (I_{PU}/I_{PG}) in 6T bitcells.

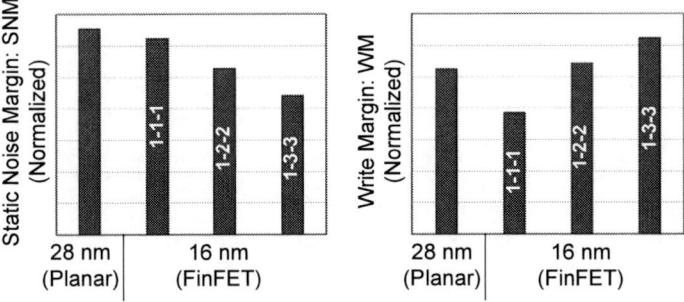

Fig. 4: Simulated SNM (left) and WM (right) of 6T SRAM bitcells for 28 nm planar bulk CMOS and 16 nm FinFET CMOS at the worst temperature/process-corner, and nominal VDD.

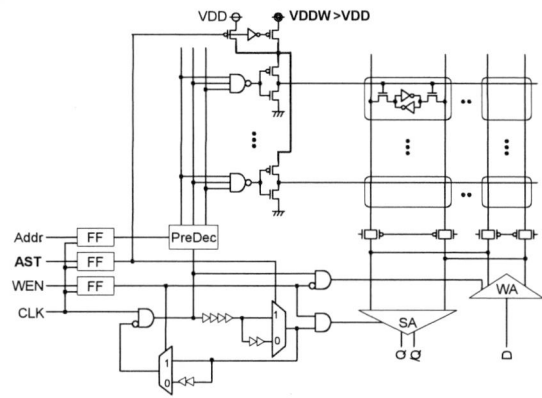

Scheme	VDD floating VDD lowering	Negative BL bias	WL lowering WL boost, overdriven		
			Lowering [5, 6, 8, 9]	Lowering & Boosted [7, 9]	Overdriven [8] / Overdriven (Proposed)
SNM	→ (no change)	→	+	+	→
WM	+ (improved)	++	–	+	++
Ta	→ (no change)	→	–	-	++

Fig. 5: Comparison of published and proposed assist circuit techniques for enhancing SNM, WM, and read-access time (*T*a).

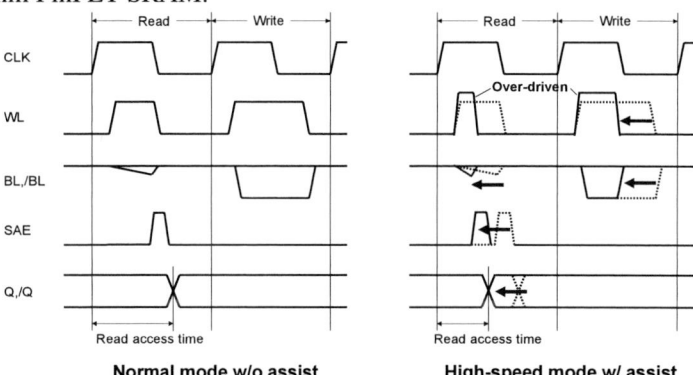

Fig. 6: Proposed WL overdriven read/write assist circuit for 16 nm FinFET SRAM.

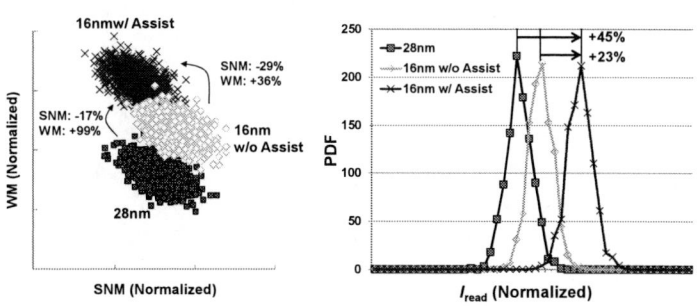

Normal mode w/o assist **High-speed mode w/ assist**

Fig. 7: Waveforms of read and write cycles for normal mode (w/o assist) and high-speed mode (with WL overdriven assist).

Fig. 8: Monte Carlo simulation results of SNM, WM, and I_{read} for 28 nm planar bulk CMOS and 16 nm FinFET with 1-2-2 bitcell.

978-1-4799-8002-4/14 $31.00 © 2014 IEEE

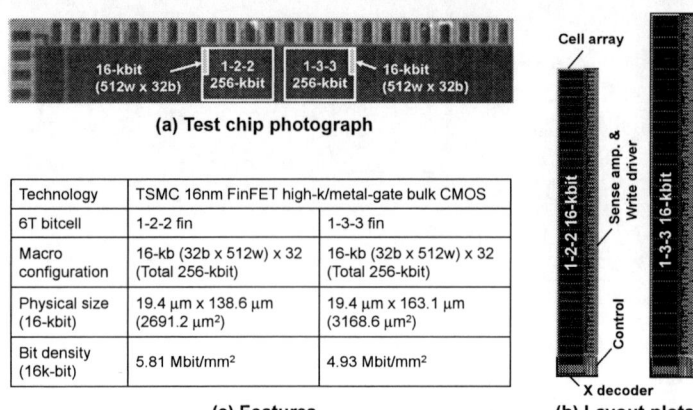

(a) Test chip photograph

Technology	TSMC 16nm FinFET high-k/metal-gate bulk CMOS	
6T bitcell	1-2-2 fin	1-3-3 fin
Macro configuration	16-kb (32b x 512w) x 32 (Total 256-kbit)	16-kb (32b x 512w) x 32 (Total 256-kbit)
Physical size (16-kbit)	19.4 μm x 138.6 μm (2691.2 μm²)	19.4 μm x 163.1 μm (3168.6 μm²)
Bit density (16k-bit)	5.81 Mbit/mm²	4.93 Mbit/mm²

(c) Features

(b) Layout plots

Fig. 9: (a) Test chip photograph. (b) Layout plots of 256-kbit SRAM macros. (c) Summary of features.

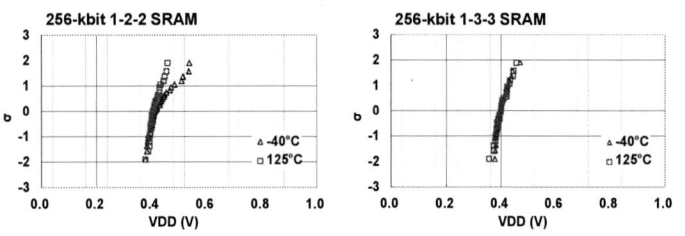

Fig. 10: Measured V_{min} distributions for 256-kbit SRAM macros with 1-2-2 and 1-3-3 bitcells at -40°C and 125°C, and 2-MHz clock cycle (test mode w/o WL-overdriven assist).

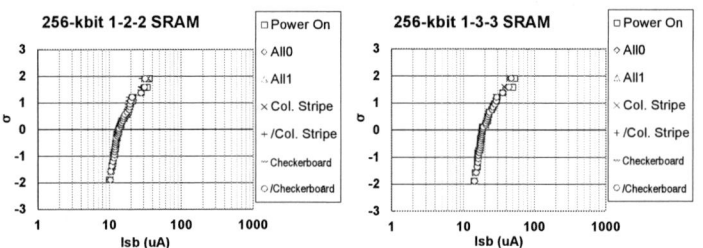

Fig. 11: Measured standby current (I_{sb}) distributions for 256-kbit SRAM macros with 1-2-2 and 1-3-3 bitcells at 25°C.

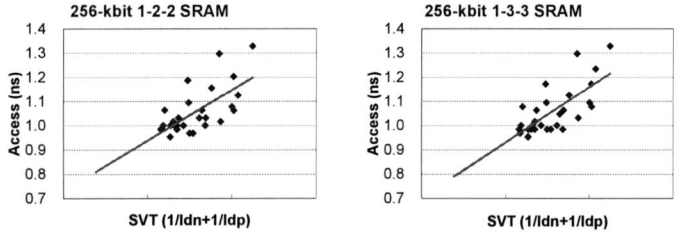

Fig. 12: Measured distributions of read access time for 256-kbit SRAM macros with 1-2-2 and 1-3-3 bitcells at 25°C and 2-MHz clock cycle (test mode w/o WL-overdriven assist). A good correlation exists with (1/Idn + 1/Idp) SVT core MOS distribution.

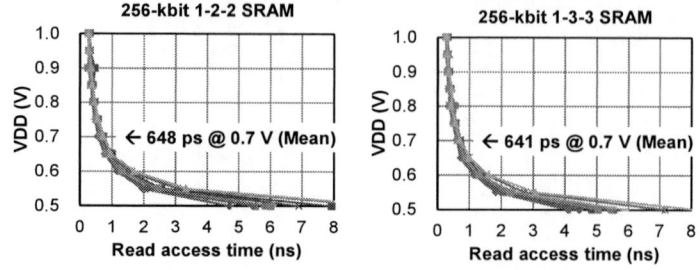

Fig. 13: Plots of read-access-time *vs.* supply voltage for 256-kbit SRAM macros with 1-2-2 and 1-3-3 bitcells under the high-speed mode (w/ WL overdriven assist) at -40°C.

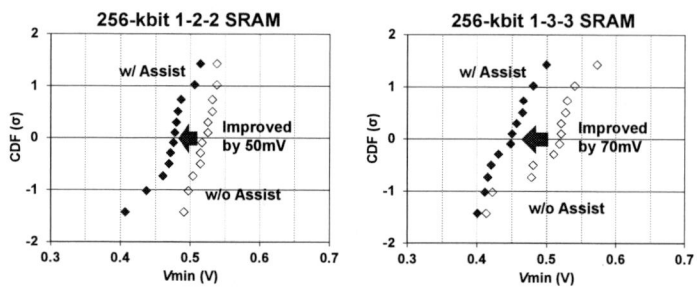

Fig. 14: Measured V_{min} distributions of 256-kbit SRAM macro with 1-2-2 bitcell with/without WL overdriven assist at -40°C.

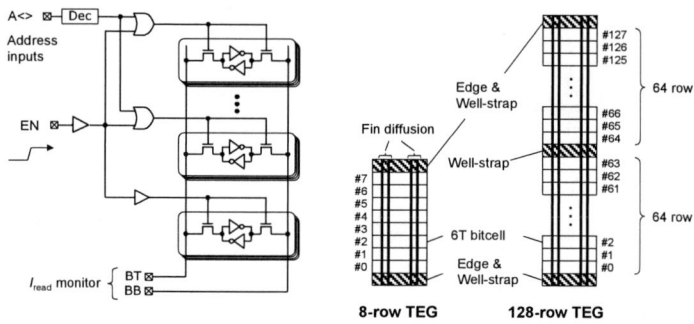

Fig. 15: Test structures for evaluating layout effect dependency (LDE) on I_{read} *vs.* length of fin-diffusion of PD/PG NMOSs in a column with short (8-row) and long (128-row) arrangements.

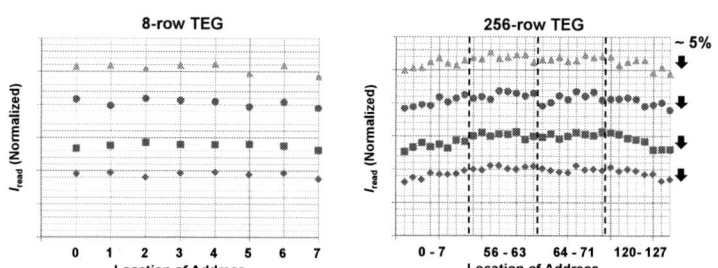

Fig. 16: Measurement results of LDE on I_{read} *vs.* length of fin-diffusion of PD/PG NMOSs in a column. Around 5% degradations of I_{read} are observed near the edge.

Dynamic Single-P-Well SRAM bitcell characterization with Back-Bias Adjustment for Optimized Wide-Voltage-Range SRAM Operation in 28nm UTBB FD-SOI

O. Thomas[1,2], B. Zimmer[1], S. O. Toh[1], L. Ciampolini[3], N. Planes[3], R. Ranica[3] P. Flatresse[3] and B. Nikolić[1]

[1] Berkeley Wireless Research Center, Berkeley, CA, United States, email : olivier.thomas@cea.fr – [2] CEA-LETI Minatec Campus, 38054 Grenoble Cedex 9, France, Crolles – [3] ST Microelectronics, 38926 Crolles, France

Abstract

This paper demonstrates the 28nm ultra-thin body and buried oxide (UTBB) FD-SOI high-density (0.120µm²) single p-well (SPW) bitcell architecture for the design of low-power wide voltage range systems enabled by back-bias adjustment. The results from a 140kb programmable dynamic SRAM characterization test module provide both information about location and cause of failures as well as power and performance by mimicking system operating conditions over a wide supply voltage range. A 410mV minimum operating voltage and less than 310mV data retention voltage with a leakage current close to 100fA/bitcell are measured. Improved bitcell read access time and write-ability through back-bias are demonstrated with less than 5% of stand-by power overhead.

Introduction

Voltage reduction and increased variability associated with technology scaling compromise margins necessary for robust SRAM operation. Ultra-thin body and buried oxide (UTBB) FD-SOI eliminates channel doping to lower intrinsic transistor variability [1], and offers multiple threshold voltages through selection of the well under the buried oxide [2-3] with an ability to lower SRAM operating voltage. Traditionally, SRAM functionality is evaluated through static voltage margins during the development phase and is monitored with built-in self-test (BIST) in production [4]. Dynamic metrics [5] more accurately characterize SRAM read and write margins than static ones. In scaled technologies dense SRAM may only be functional with dynamic stimuli [6]. This work demonstrates the dynamic performance of the high-density (0.120µm²) single-P-well (SPW) 6T bitcell architecture [7-8] for wide voltage range modern

wireless systems, using a complete SRAM dynamic defect and performance characterization module designed in an early production 28nm UTBB FD-SOI dual V_T CMOS process.

UTBB devices and Single-P-Well SRAM features

UTBB-FDSOI devices (Fig.1) are fabricated with a gate-first high-k metal-gate 28nm technology implementing an ultra-thin silicon layer (7nm) on top of a 25nm buried-oxide (BOX) [9]. The thin BOX isolation enables to apply efficient extended body biasing (V_B) without source-drain junction leakage and to adjust transistor V_T with well doping types providing low-V_T (LVT) and regular-V_T (RVT) devices (Fig.2) [2-3]. In the SPW design (Fig.3) [7], both PMOS and NMOS devices are placed over a common p-well, leading to LVT PMOS pull-up (PU) devices and RVT pull-down (PD) and pass-gate (PG) NMOS devices (Fig.4). The common PW is isolated from the grounded p-substrate by using a deep n-well (DNW) biased at the SRAM array supply voltage or higher. By changing V_{PW} (Fig.5), back-bias is applied to all SRAM transistors trading-off access time, power consumption and stability. Derived from high-density (0.120µm²) regular SRAM bitcell architecture, SPW design does not require process and footprint modifications (Fig.6) [8] and reduces scaling limitations caused by well-proximity and diffusion issues.

Test chip features

The test chip (Fig.7) provides a dynamic characterization module designed to evaluate SRAM bitcell retention-stability (RET), read-stability (RS), write-ability (WA), and read-access timing faults (RA) (Fig.8). The module (Fig.9) consists of a 140kb SPW SRAM macro clocked by an on-chip pulse generator (PG) and controlled by a programmable BIST. The SPW SRAM macro includes 4 arrays of 280 columns by 128 word-lines (WL). The BIST is a finite state machine with 3 paths (Fig.10). BIST setup data are loaded through a scan chain and can be scanned out for verification through the first path. The memory initialization is checked through the second path, while the third path is dedicated for the memory tests. Input and output data are scanned at low frequency (CLK_L) and the tests can be performed at up to 800MHz using an external clock reference (CLK_H). A synchronizer circuit avoids hazardous glitches for signal traversing from low (high) to high (low) clock frequency. The BIST supply voltage (V_{DDB}) is separated from the SRAM. The SRAM output data (Do<69:0>) are level-shifted (LV) to the BIST to extract errors (E<69:0>). While the BIST runs at a nominal voltage, the SRAM macro can be run at a lower voltage. The SRAM periphery (V_{DDP}) and array (V_{DDA}) supply voltages are also electrically independent for bitcell power consumption measurement. SRAM bit error rate (BER) evaluation consists of 3 main steps (Fig.11). It begins with a safe initialization (INIT) of the SRAM, continues with the selected test (RD, WR…) and ends with a safe check (CHECK). Before each step, the SRAM supply voltage and the clock frequency are set to fulfill the initialization, test and check conditions. The initialization, test and check are performed column-by-column to avoid half-selected bitcell failures in BER evaluation.

Experimental results and discussion

The characterization test module determines both the location and cause of failures for every bitcell in the array (Fig.12) by measuring independently the BER of each mode (RET, RA, RS, and WA) (Fig.13). By mimicking the SRAM operating conditions, it enables optimization of conflicting design requirements (i.e. WA vs. RS) and performance trade-offs (i.e. column height vs. RA) (Fig.14). The minimum SPW bitcell operating voltage (V_{MIN}) is RA limited for WL pulse widths lower than 10ns and WA limited for longer pulse widths, reaching static test results (Fig.15). To demonstrate the use of p-well back-bias voltage in a wide tuning range to improve SRAM performances and reduce standby-power, Fig.16-20 show how the RA and WA BER as well as the minimum data retention voltage (DRV) and the bitcell static power

are affected. Increasing V_{PW} reinforces NMOS and weakens PMOS (cf. Fig.5), and in turn, improves RA but harms DRV and WA. Thus for fast access time a positive back-biasing ($V_{PW}>0V$) can be used to improve RA (Fig.16), whereas for ultra-low operating voltage a V_{PW} close to 0 enables a V_{MIN} as low as 410mV. In sleep mode, V_{PW} is adjusted to minimize DRV and stand-by power. In negative back-bias ($V_{PW}<0$) the strength of the bitcell inverter devices is more balanced. A DRV lower than 310mV is achieved (Fig.17) allowing an ultra-low leakage current in average close to 100fA/bitcell (Fig.18). The bitcell static power depends on both PMOS and NMOS leakage currents. While positive (negative) back-biasing increases (reduces) the NMOS leakages it reduces (increases) the PMOS leakages. The minimum static current is obtained for V_{PW} in between -0.5V and 0V (Fig.19). However, it is worth to note that back-biasing adjustment has a low impact on the SPW bitcell static power due to the NMOS PMOS leakage compensation. Thus, it enables wide back-bias range adjustment to improve access time and bitcell robustness with a very low stand-by power overhead. The test module also helps to track bitcell mechanisms of failures. WA V_{MIN} is limited by the completion of the transition of the high logic level node driven by PMOS PU transistors and the discharge of the high logic level node driven by PG/PU current ratio (Fig.20) [7]. Completion and discharge failures are traded-off versus the PU strength. In negative back-biasing range WA failures are caused by the weak PG/PU ratio, while in positive back-bias range it originates from a weak PU completion explaining the optimum WA achieved at V_{PW} 0V.

Conclusion

The SPW bitcell architecture combined with a wide back bias range in UTBB FDSOI enables both low operating voltage and fast access time. The low sensitivity of the bitcell I_{OFF} current to changes in body bias enables access time improvement through V_B adjustment. In sleep mode, V_B is set to minimize the bitcell standby leakage current and the data retention voltage. A V_{MIN} of 410mV and a DRV lower than 310mV with a leakage current close to 100fA/bitcell are demonstrated. Key causes of dynamic bitcell failures are analyzed, providing early feedback for both design decisions and process improvement, scalable to finer process geometries.

978-1-4799-8002-4/14 $31.00 © 2014 IEEE

References:

[1] K. Cheng et al., p. 49, IEDM'09.
[2] O. Weber, et al., p. 58, IEDM'10.
[3] J-P. Noel, et al., Vol. 58, No. 8 p. 2473, TED'11.
[4] Y.-H. Chen, et al., Vol. 47, p. 969, JSSCC'12.
[5] S.O. Toh, et al., p. 35, VLSI'10.
[6] Y. Morita, et al., p. 37, VLSI'10.
[7] O. Thomas, et al., p. 1, SOI-Conference'12.
[8] R. Ranica, et al., p. 210, VLSI'13.
[9] N. Planes, et al., p. 133, VLSI'12.

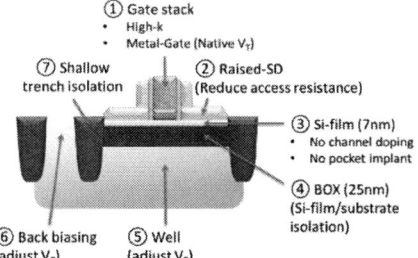

Fig.1: UTBB FD-SOI cross section view. Device V_T is set by the gate stack and adjusted by well doping type [2-3]. Wide V_T tuning is feasible by back-biasing changing the well bias.

	NMOS		PMOS	
V_T	Well type	V_{BS} (V)	Well type	$|V_{BS}|$ (V)
RVT	P	0	N	0
LVT	N	0	P	V_{DD}

RVT: Regular-V_T
LVT: Low-V_T

Fig.2: UTBB FD-SOI V_T definition.

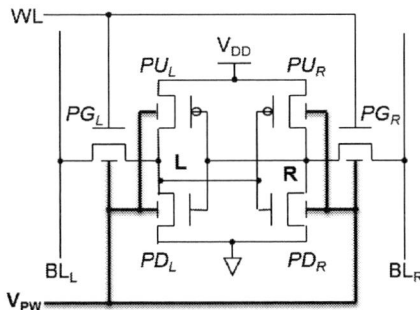

Fig.3: SPW bitcell schematic. Back gate of NMOS (PG, PD) and PMOS (PU) devices are electrically connected by the common P-well.

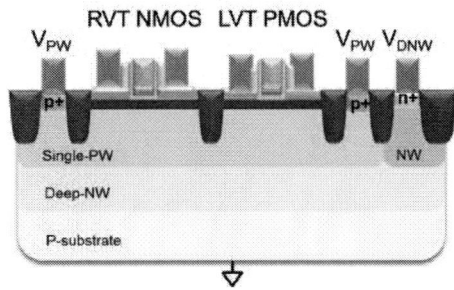

Fig.4: SPW bitcell device cross section view. DNW isolates the PW from the p-substrate enabling wide voltage range PW back biasing.

Back bias (BB)	PD, PG	PU
$V_{PW} = 0V$ (default)	No BB	No BB
$V_{PW} > 0V$	FBB	RBB
$V_{PW} < 0V$	RBB	FBB

FBB: Forward-BB → $|V_T|$ reduces
RBB: Reverse-BB → $|V_T|$ increases

Fig.5: SPW bitcell device V_T shift and back bias mode versus V_{PW} considering V_T definition in Fig.2.

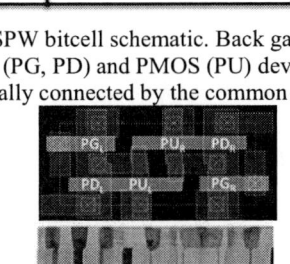
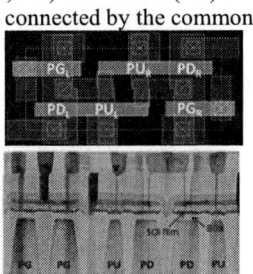

Fig.6: High density (0.120μm²) SPW bitcell layout and TEM cross section [8]. 28nm High-k metal-gate technology implementing 7nm Si-film relying on 25nm BOX thickness [9].

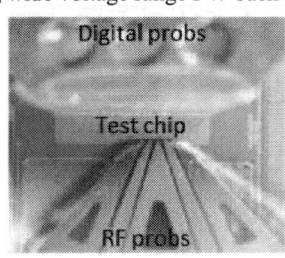

Fig.7: Test chip photograph. Implemented to be testable on wafer during the process development. 25 digital and 2 RF probes are used to bring in and output low and fast signals.

Metric	Definition
RET	Bitcell ability to retain data in retention
RS	Bitcell ability to retain data versus read access time
RA	Bitcell ability to discharge bitline versus read access time
WA	Bitcell ability to flip data versus write access time

Fig.8: SRAM bitcell metric definitions testable with the test chip.

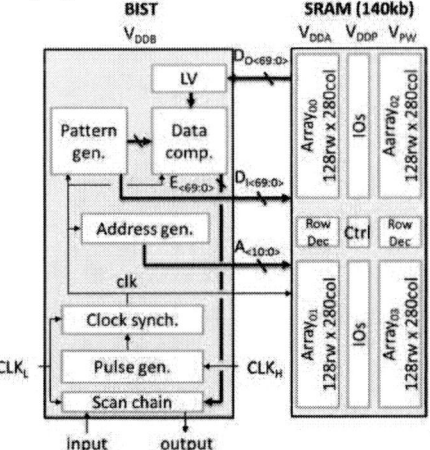

Fig.9: Dynamic characterization module architecture, including a 140kb SPW SRAM macro clocked by an on-chip pulse generator and controlled by a programmable BIST. Input and output are scanned in and out at low clock frequency (CLK_L) while tests can be performed at higher frequency (CLK_H).

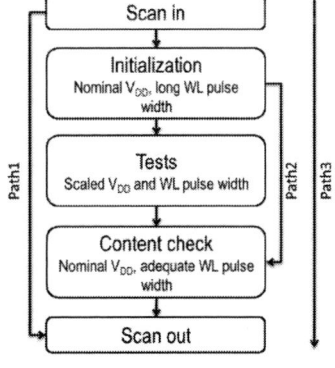

Fig.10: BIST finite state machine:
Path1 – Scan chain test
Path2 – SRAM initialization test
Path3 – SRAM tests

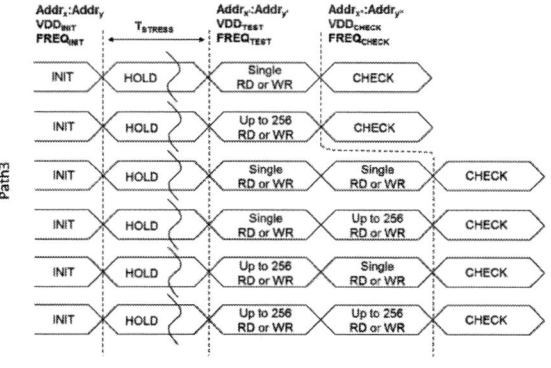

Fig.11: SRAM access patterns for evaluating retention, read, write and timing SRAM failure rate, while considering multiple cycle and electrical stress.

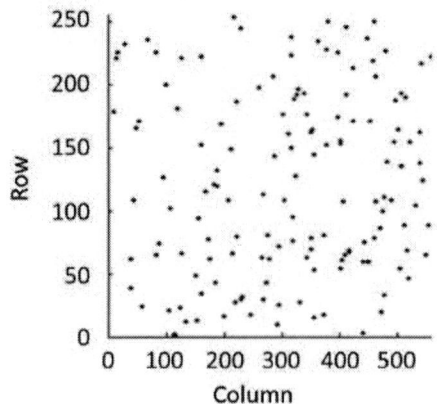

Fig.12: Memory map of RS errors measured at V_{DD}=300mV and 5000ns word-line pulse width. The random distribution of errors confirms that errors measured are caused by the bitcells and not the periphery.

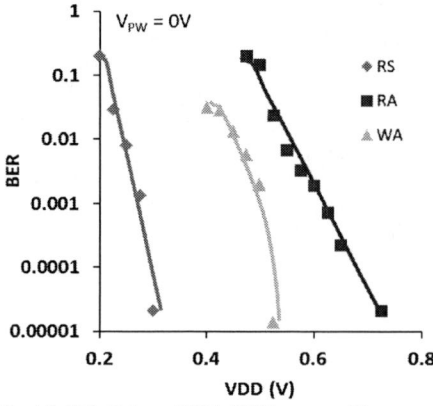

Fig.13: RS, RA and WA BER versus V_{DD} considering a word-line pulse width of 2.5ns. Each metric is measured independently column by column to avoid false error measurements caused by column interleaving.

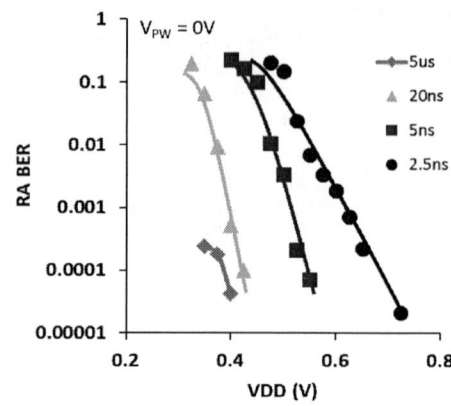

Fig.14: RA BER versus V_{DD} for various word-line pulse widths.

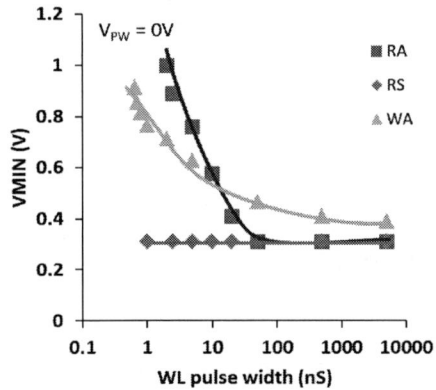

Fig.15: RS, WA, RA V_{MIN} versus WL pulse width. For WL pulse width longer than 10ns V_{MIN} is limited by WA, while for shorter pulse width V_{MIN} becomes limited by RA. RS is not the primary V_{MIN} limiting condition.

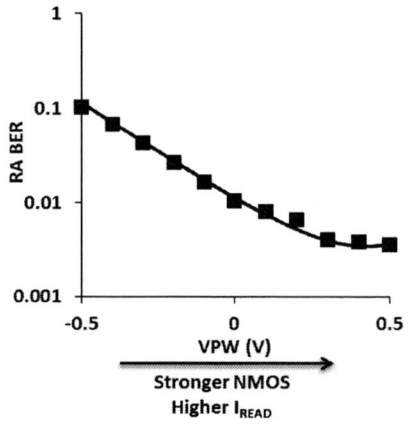

Fig.16: RA BER versus V_{PW} (V_{DD}=500mV, WL pulse width=5ns). V_{PW} increase forward biases the PD PG NMOS devices leading to an over one decade decrease of RA BER due to the increase of bitcell read current enabling faster access time.

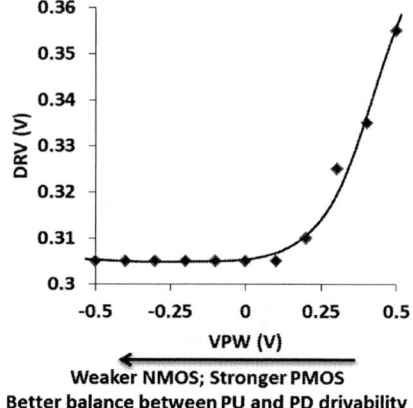

Fig.17: DRV versus V_{PW}. Decreasing V_{PW} balances NMOS and PMOS device strength improving bitcell retention voltage margin. Positive back-bias weakens PU degrading the retention of high logic state, increasing DRV.

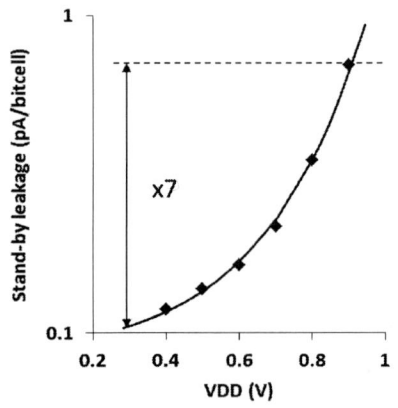

Fig.18: Bitcell stand-by current versus V_{DD} measured at ambient temperature.

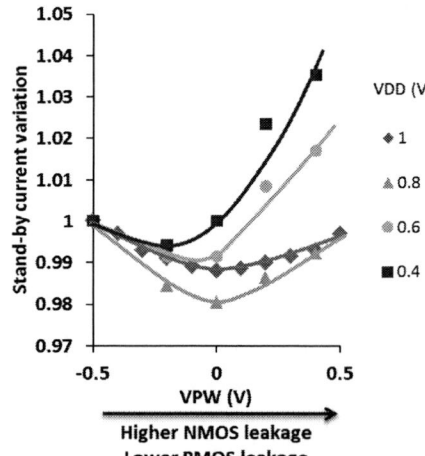

Fig.19: Bitcell stand-by current variation versus V_{PW}. Low impact of V_{PW} on stand-by current due to the leakage current compensation between PMOS and NMOS. Lowest stand-by current is reached when -250mV<V_{PW}<0V.

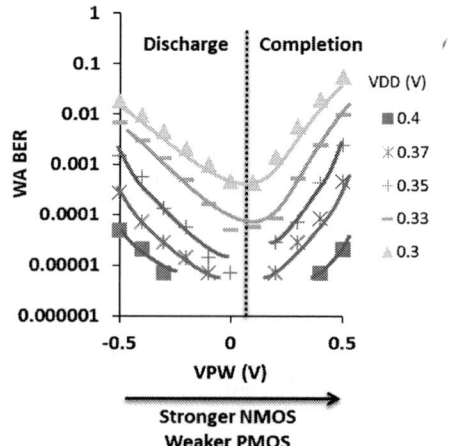

Fig.20: WA BER versus V_{PW} (WL pulse width =5000ns). When V_{PW} < 0, the weak PG/PU ratio causes discharge failures, while for V_{PW} > 0 PU completion fails. Measurements agree with simulation results in [7].

978-1-4799-8002-4/14 $31.00 © 2014 IEEE

Drain Extended MOS Device Design for Integrated RF PA in 28nm CMOS with Optimized FoM and ESD Robustness

Ankur Gupta[1], Mayank Shrivastava[2], Maryam Shojaei Baghini[1], Dinesh Kumar Sharma[1], A. N. Chandorkar[1], Harald Gossner[3] and V. Ramgopal Rao[1]

[1]CEN, EE Department, Indian Institute of Technology Bombay, India (email: agupta@ee.iitb.ac.in), [2]Department of ESE, Indian Institute of Science Bangalore, India (email: mayank@dese.iisc.ernet.in), [3]Intel Corp., Platform Engineering Group, Neubiberg, Munich, Germany (email: harald.gossner@intel.com)

Abstract

This paper explores drain extended MOS (DeMOS) device design guidelines for an area scaled, ESD robust integrated radio frequency power amplifier (RF PA) for advanced system-on-chip applications in 28nm node CMOS. Simultaneous improvement of device-circuit performance and ESD robustness is discussed for the first time. By device design optimization a 45% increase in gain and 25% in power-added efficiency of RF PA at 1GHz, and 5× improvements in ESD robustness are reported experimentally.

Introduction

Advanced system-on-chip (SoC) concepts push to integrate all the functionalities on the same Si Die including high power radio frequency power amplifiers (RF PA) and power management modules [1]-[2]. State-of-the-art high power RF PA is realized by discrete III-V devices [3]. However, integration of these device types on Si substrate is not trivial. Therefore, for CMOS based advanced SoC, a Si based solution has to be explored. Drain extended NMOS (DeNMOS) device is a potential candidate for such applications in advance CMOS nodes [4]-[5]. The challenge for implementation of an integrated RF PA into a sub-65nm node CMOS is low breakdown voltage, sensitivity towards ESD stress and nonlinear device characteristics. This paper discloses DeMOS device design guidelines for designing an integrated high performance RF PA with an excellent ESD robustness.

Experiments and Results

DeNMOS device (Fig. 1) was fabricated in 28nm CMOS technology node, using process and design rules developed for low power devices. Total fabricated electrical width of the device was 1.6mm. In order to study device's suitability for RF PA application, the RF circuit was mounted on a test board. Schematic of the PA circuit is shown in Fig. 2a along with the biasing and matching networks. A capacitor ladder is used at the DC supplies in order to de-couple the RF noise generated by the circuit from DC. Fig. 2b shows the photograph of the fabricated PA circuit prepared on a low loss laminate. Small signal s-parameter measurements were conducted on die, to design the matching and biasing network for the RF PA (Fig. 3). The DeNMOS device's gate was biased slightly above the threshold voltage in order to operate the PA in class-AB mode. Output power of the PA was

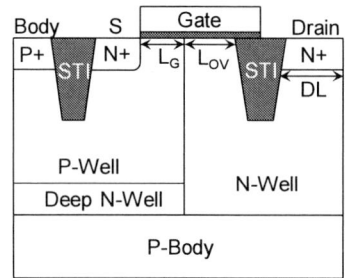

Fig. 1: Cross-sectional view of the DeNMOS device fabricated in 28nm CMOS node. Figure depicts a single finger structure; however, it is realized in a two finger configuration on silicon with sharing of N-well between two fingers.

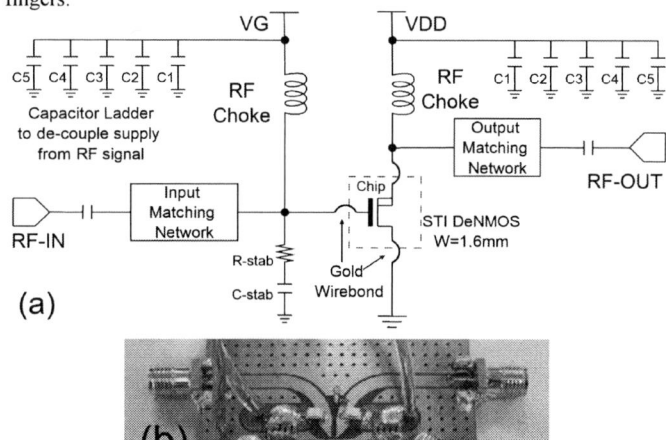

Fig. 2: (a) RF PA circuit, which is used as a vehicle to analyze RF PA performance of the advance CMOS node DeNMOS devices manufactured in this work. (b) Board level implementation of RF PA for device-circuit co-design and RF PA performance analysis. A low loss laminate of thickness = 0.8mm (TanD = 0.0004), dielectric constant = 2.2 and copper cladding of 70μm was used for board level implementation. Electrical connections from die to PCB were made using 25μm gold wire. Die is pasted using thermal conductive epoxy for proper thermal transport as junction temperature rise can be a serious issue to the structure [6].

measured for a 50Ω load, as shown in Fig. 4. RF PA using the standard DeNMOS device offers an output power of 23.9dBm at 1-dB compression point (P_{1-dB}), which corresponds to a power density of 0.16 W/mm, at 1 GHz. Measured RF gain of the PA circuit is 10.8dB, peak drain-efficiency is 46% and power added-efficiency (PAE) is 40.2% at 1-dB compression point. RF distortion produced by

978-1-4799-8002-4/14 $31.00 © 2014 IEEE

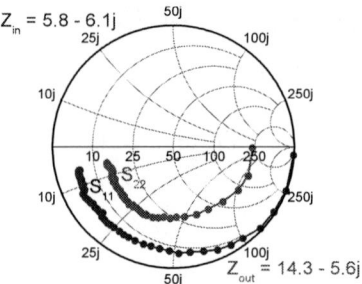

Fig. 3: S-parameter measurement results (S_{11} & S_{22}) for the 1.6mm (electrical width) standard DeNMOS device, which are used for designing matching network for RF PA (Z_{in} and Z_{out} are shown for 1GHz).

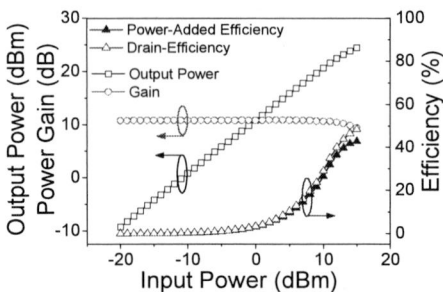

Fig. 4: Performance of the implemented DeNMOS based RF PA, extracted experimentally using class-AB mode of PA operation and tested at 1GHz.

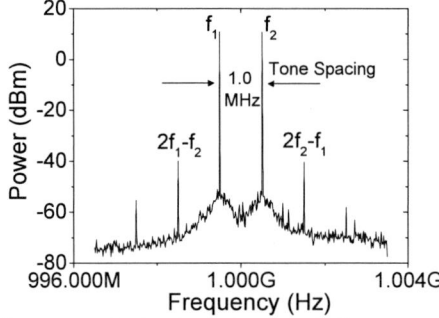

Fig. 5: Circuit response to two-tone signal excitation measured at 0dBm input power. The spectrum shows two fundamental tones (f_1 and f_2) along with two 3rd order harmonics ($2f_1$-f_2 and $2f_2$-f_1) generated due to the non-linearity in the device.

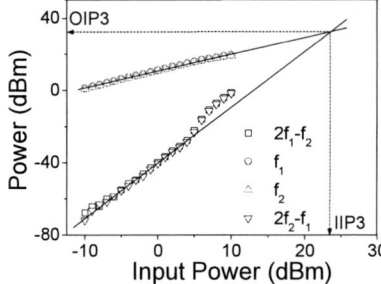

Fig. 6: Two-tone output power measurements of the standard DeNMOS based RF PA design as a function of input power. Both the plots were extrapolated to calculate IIP3 and OIP3.

the device is very critical for deeply scaled technologies [7]. In order to explore linearity of the device for RF applications two-tone measurements were performed. Two sinusoids (f_1 and f_2) centered at 1GHz with a tone spacing of 1MHz were applied at the circuit input port. Fig. 5 shows the output power spectrum and the third harmonic frequency ($2f_1$-f_2 and

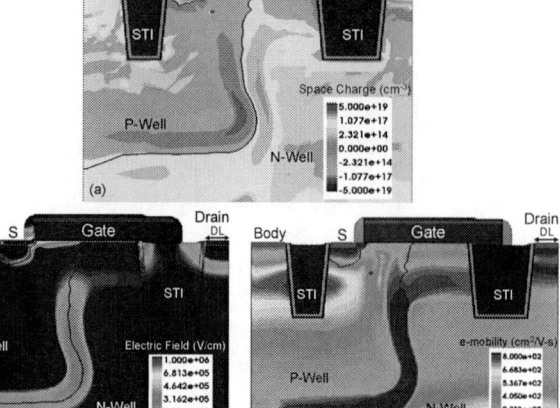

Fig. 7: (a) 2D Space charge contour across the standard DeNMOS device at high drain current (V_G=2V, V_D=5V). Figure depicts significant amount of space charge modulation at high drain current, which may be a possible cause of mitigated device performance and early quasi saturation. (b) 2D electric field contour across the standard DeNMOS device at high drain currents. Figure depicts very high electric field underneath the drain diffusion and in the gate-to-N-well overlap region. High electric field in the drift region is attributed to an early space charge modulation leading to localized field distribution. (c) 2D electron mobility contour across the standard DeNMOS device at high drain current. Figure depicts significant electron (majority carrier) mobility reduction in the drift region, especially under the drain diffusion (or drain N+) and Gate-to-Nwell overlap region. Mobility reduction is attributed to very high electric fields in these regions due to an early space charge modulation.

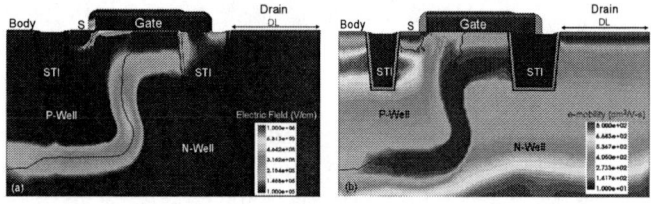

Fig. 8: 2D (a) electric field and (b) electron mobility contour across the modified DeNMOS device (DL= 5× of min. allowed) at high drain currents. Figure depicts lowering of electric field under the drain diffusion region, however increased electric field in the Gate-to-Nwell overlap region. Lowering of electric field is attributed to absence of space charge modulation underneath the drain diffusion, which results in an improved electron mobility in this region.

$2f_2$-f_1) power spectrum generated due to the device's non-linearity. Two-tone measurements (Fig. 5 & 6) show an excellent inter-modulation distortion (IMD3) level of -72dBm at -10dBm of input power. Fundamental and harmonic powers were extrapolated to obtain the input and output third order intercept points (IIP3 and OIP3 respectively). RF PA offers 37dBm of output 3rd order intercept point (OIP3), which is >10dB higher than P_{1-dB}.

Quasi-Saturation, Device Design & RF PA Performance

To optimize DeMOS devices, quasi saturation leading to g_m drop needs to be reduced. Device TCAD simulations were performed under high current condition. Fig.7a shows an early space charge modulation, which leads to localized electric field peaking underneath the drain diffusion (i.e.

978-1-4799-8002-4/14 $31.00 © 2014 IEEE

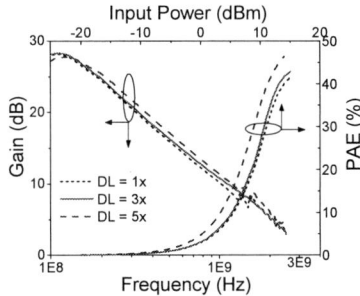

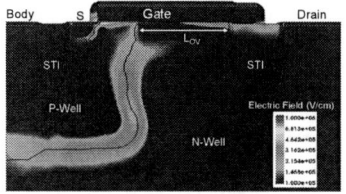

Fig. 12: Lowering of the electric field in the gate-to-Nwell overlap region by engineering the overlap length and thereby doping profile.

Fig. 9: Measured small signal gain and power added efficiency of the RF PA circuits realized using DeNMOS devices of different drain diffusion lengths (DL). A clear increment in both Gain and PAE can be seen after drain engineering.

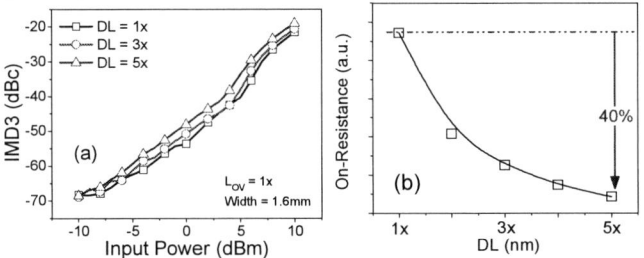

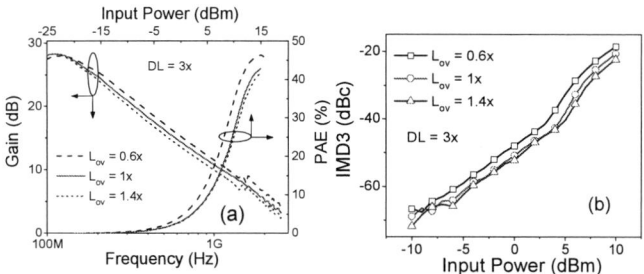

Fig. 13: (a) Measured small signal gain and power added efficiency of the RF PA circuits realized using DeNMOS devices of different gate-to-Nwell overlap length (L_{OV}) and thereby doping profile. (b) IMD3 level of the gate-to-Nwell overlap engineered device based RF PA.

Fig. 10: (a) IMD3 level of the drain engineered devices with respect to two tone input power. IMD3 levels are almost unchanged. (b) On-resistance of drain engineered devices. Figure depicts 40% reduction in the on-resistance of the device, which is a significant factor for power management (or switching) applications.

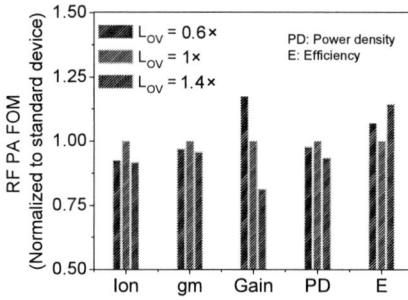

Fig. 14: Gate-to-Nwell overlap engineered device and implemented RF PA FoM.

In order to avoid electric field peaking in the gate-to-Nwell overlap region, the overlap length (L_{OV}) was increased. Although the electric field peaking was mitigated by an increased L_{OV}, (Fig. 12), the RF PA characteristics and FoM show a non-linear trend (Fig. 13 & 14). This can be attributed to the trade-off between Miller/overlap capacitance, non-linearity in the trans-conductance, drain parasitic capacitance and drift region resistance.

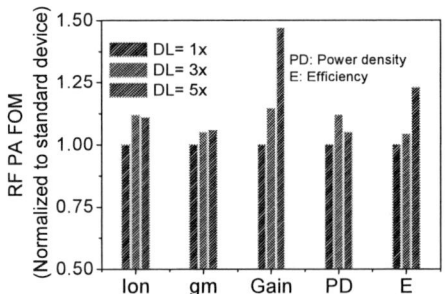

Fig. 11: RF PA FoM improvement after drain engineering. Figure depicts 40%-50% improvement in RF PA FoMs.

drain N+) and within the gate-to-Nwell overlap region (Fig.7b). Increased electric field results in significant mobility reduction across these regions (Fig.7c). As the charge modulation takes place when mobile carriers exceed the background doping, reducing the current density in critical regions should mitigate space charge modulation. Increasing the drain N+, drain diffusion length (DL) successfully reduces current density and mitigates electric field peaking and electron mobility reduction (Fig. 8). This has resulted in an improved device with increased ON-current, lower on-resistance, increased trans-conductance and load line swing when compared to the standard device. The RF PA performance (Fig. 9-11), shows 45% improvement in RF gain and 25% improvement in PAE. Fig. 10b shows 40% reduction in the on-resistance of the DeNMOS device by drain engineering.

ESD Robustness

PA circuits are directly connected to the antenna and exposed to ESD threats. Thus, ESD robustness is an essential feature of PA driver stage. Standard DeNMOS device with smallest DL value is intrinsically ESD weak (Fig. 15a). The root cause is the early current filament formation (Fig. 16a) due to field peaking and mobility degradation. Avoiding the early space charge modulation underneath the drain diffusion, by increasing DL, mitigates an early filament formation (Fig. 16b), which improves the ESD robustness by 5× (Fig. 15a & 15b). This highest reported ESD robustness of DeNMOS device [8] allows the implementation of a self-protecting PA stage without need of an additional ESD protection.

978-1-4799-8002-4/14 $31.00 © 2014 IEEE   65

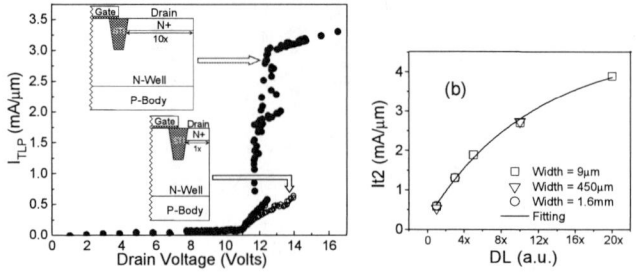

Fig. 15: (a) TLP characteristics of conventional and modified drain engineered device. Figure clearly depicts significant improvement in the device's ESD robustness by drain engineering. (b) Consistent improvement in the failure current (It2) by increasing drain diffusion length.

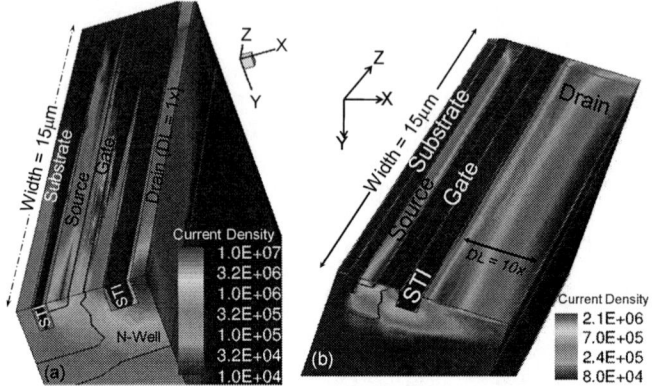

Fig. 16: (a) 3D TCAD pictures depicting current filamentation in the standard DeNMOS device at low currents (I_{TLP} = 0.5mA/μm). (b) Device survives filamentation after drain engineering, even at very high currents (I_{TLP} = 0.5mA/μm).

Self Heating and Technology Scaling

Increase in the die temperature due to high power consumption and limited efficiency of the circuit is a serious issue to the device reliability for power applications. In order to investigate temperature rise across the device, thermal mapping over the Si Die was performed using thermocouple probe tip. The Die temperature, at 24dBm of RF power level, was found to be 25°C above room temperature, which was scaled down to 18°C after device optimization (Fig. 17a). This is in agreement with an increased PAE after device optimization. A scaling of this temperature level to 1 Watt (30dBm) for SoC applications is therefore feasible. An area comparison shows that the DeNMOS based RF PA in 28nm node CMOS occupies 6× less area, for a given power level, compared to the 5 stack silicide blocked NMOS transistor based RF PA (Fig. 17b). As shown, the 5 stack silicide blocked RF PA requires separate RC network at each transistor gate in order to sustain high signal swing at the output node without exceeding gate oxide breakdown voltage. Another way to increase the output power is to stack more transistors. However, in 28nm and beyond, stacking of more than 5 transistors is not possible due to limited drain-to-pwell and nwell-to-pwell reverse bias breakdown voltages. To explore the on-chip integration of PA for advanced CMOS, the scaling trends are compared in Fig. 18. This work shows competitive FoM for RF PA in the most advanced CMOS technology node. Due to the simplicity and process

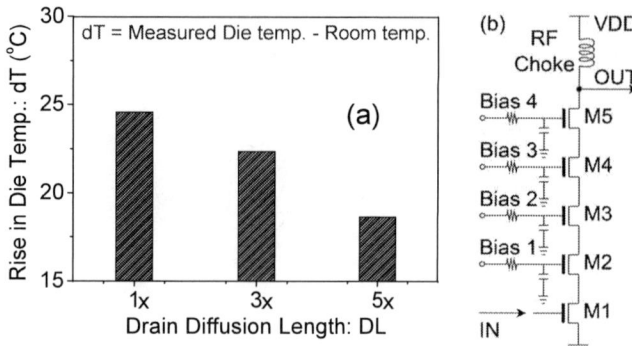

Fig. 17: (a) Rise in Die temperature as a function of drain diffusion length, measured at 24 dBm of RF power level. (b) 5 stack silicide blocked NMOS transistor based RF PA circuit. A stack of 5 transistors are required to achieve a maximum voltage swing in 28nm technology (10V). Note the additional area required by the bias network for individual transistor gates. (De-coupling cap and other details are skipped from this figure for simplicity.)

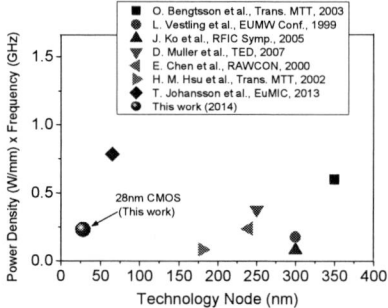

Fig. 18: Figure depicts power density vs. technology nodes of various DeMOS/LDMOS based RF PA implementations. A clear area advantage can be seen for an integrated RF PA in advanced CMOS nodes.

compatibility of DeNMOS device used in this work, it can be integrated into even further scaled technologies with only minor process cost penalty.

Conclusion

A power amplifier based on a STI DeMOS device has been realized in 28 nm CMOS. For a drain to source breakdown voltage of 10V, a high output power of 24dBm, power density of 0.16W/mm at 1GHz with efficiency of 50% has been reported. Thus, STI DeMOS can provide an option for integrating PAs with good intrinsic ESD robustness in advanced CMOS nodes for applications with limited frequency and performance requirements. To optimize the RF performance of the device the field distribution on the drain side has been discussed and an optimization strategy is provided.

References

[1] R. Kumar, et. al., *ISSCC*, 2014. pp. 328-329.

[2] C. H. Wu, et. al., *RFIC Symp.*, 2013. pp. 129-132.

[3] U. K. Mishra, et. al., *Proc. of IEEE*, 2008. pp. 287-305.

[4] T. Johansson, et. al., *Trans. on MTT*, 2014. pp. 111-124.

[5] D. Gruner, et. al., *Trans. on MTT*, 2010. pp. 4022-4030.

[6] A. Ferrara, et. al., *IEDM*, 2013. pp. 168-171.

[7] R. V. Langevelde, et. al., *IEDM*, 2000. pp. 807-810.

[8] M. Shrivastava, et. al., *IEEE TDMR*, 2012. pp. 615-625.

Heterogeneously integrated sub-40nm low-power epi-like Ge/Si monolithic 3D-IC with stacked SiGeC ambient light harvester

Chang-Hong Shen[1], Jia-Min Shieh[1,2*], Wen-Hsien Huang[1], Tsung-Ta Wu[1], Chien-Fu Chen[3], Ming-Hsuan Kao[4], Chih-Chao Yang[1], Chein-Din Lin[1], Hsing-Hsiang Wang[1], Tung-Ying Hsieh[1], Bo-Yuan Chen[1], Guo-Wei Huang[1], Meng-Fan Chang[3], and Fu-Liang Yang[5]

[1]National Nano Device Laboratories, No.26, Prosperity Road 1, Hsinchu 30078, Taiwan;

[2]Department of Photonics, National Chiao Tung University, Hsinchu 30010, Taiwan

[3]Department of Electrical Engineering, National Tsing Hua University, Hsinchu 30013, Taiwan;

[4]Institute of Electro-Optical Engineering, National Chiao Tung University, Hsinchu 30010, Taiwan

[5]Research Center for Applied Sciences, Academia Sinica, Taipei 11529, Taiwan.

[*]Tel:+886-3-5726100-7617, Fax:+886-3-5722715, E-mail: jmshieh@narlabs.org.tw

Abstract

For the first time, we report heterogeneously integrated sub-40nm epi-like Ge/Si monolithic 3D-IC with low-power logic/NVM circuits and efficient photovoltaic energy harvester. Threshold voltage engineering and driving current boosting technologies enable stackable Ge/Si UTB (<15nm) MOSFETs, CMOS inverter and SRAM (SNM=270mV@0.7V) achieve low operation voltage. Stackable 1-T NVM with high speed (100ns) and low driving-voltage operation provide power-off storage while SRAM serve as power-on working memory. 100% aperture ratio SiGeC ambient light energy harvester with maximum output power of $7mW/cm^2$ layered on the monolithic 3D-IC chip envisions a self-powered monolithic 3D-IC technology for advanced low-power wire-less sensor networks, wearable devices, and devices for Internet of Things.

1. Introduction:

Low power circuits/memory [1-2] and efficient energy-harvesters are key components for advanced low-power wire-less sensor networks, wearable devices, and devices for Internet of Things [3]. Recently, monolithic 3D-IC technology with advantages of high bandwidth, low power consumption, and cost-effective manufacturing is attractive since it can realize compact and energy-efficient products [4-5]. Furthermore, amorphous Si (a-Si) thin film photovoltaic (TFPV) as an ambient light energy harvester [6-7] with power management circuits provide an attractive way to power low power circuits for wire-less sensor networks [8-9]. However, low thermal budget fabrication technology and complex hetero-integration processes are still challenge for such an advanced monolithic 3D-IC technology. In this work, stackable, low-voltage, high on-currents, sub-40nm node Si UTB (<15nm) MOSFETs and heterogeneous dopant-free Ge junctionless (JL) FETs were developed for a self-powered monolithic 3D-IC technology by the invention of low

temperature (LT) super-CMP-planarized laser-crystallized epi-like Si/Ge pseudo-UTB [4], and self-aligned bottom-gate/embedded S/D structures. LT a-SiGeC TFPV ambient light-energy harvesters monolithically integrated on 3D-IC envisions advanced self-powered and low-power wire-less sensor networks and wearable devices **(Fig. 1)**.

2. Device Fabrication

In sequentially processed sub-40nm monolithic 3D-IC with self-powered modules, high performance stackable Si UTB MOSFETs, Ge JL FETs and NVMs were fabricated on LT epi-like Si/Ge UTB (<15nm) channels and with plasma-ALD gate dielectrics and microwave/laser-activation. Plasma-deposited ($<200^{\circ}C$) a-SiGeC TFPV ambient light-energy harvester was finally stacked on the roughened light-trapping surface formed from top metallization structures of chips **(Fig. 2)**.

3. Results and Discussion

(A) Heterogeneously integrated low-power Ge/Si UTB MOSFET, logic circuits and SRAM: Stackable Si UTB n/p MOSFETs with self-aligned embedded S/D current boosters (like raised source/drain) and back gate (BG) Vth adjusters **(Fig. 3)** and flat-band voltage-optimized gate structures [10] shows on-currents as high as 128/104 µA/µm ($|V_d|$=1V and $|V_g|$=1V) and steep subthreshold slope of 123/144 mV/dec **(Fig. 4(a) and (b))**. The typical voltage transfer characteristics of stackable Si UTB MOSFETs show quite symmetric and small V_{th} due to UTB and HfOx-based gate even without BG control, enabling low-voltage (1V) CMOS inverters **(Fig.5(a))**. With BG control **(Fig.4(c))**, inverters can operate at lower voltage of 0.8 V **(Fig.5(b))**. Moreover, the LT epi-like Ge shows ultra-high *p*-type Hall mobility (~800 cm^2/V-s) **(Fig. 6)**. Dopant-free Ge JL p-FETs **(Fig. 7)** reveals I_{on}/I_{off} >10^4 **(Fig. 8(a))**. A p-Ge/n-Si CMOS inverter was thus demonstrated with operation voltage below 2 V **(Fig. 8(b))**. The stackable 6T

978-1-4799-8002-4/14 $31.00 © 2014 IEEE

SRAM bit-cells with V_{th} engineering show a high SNM of 270 mV in the butterfly curve at operation voltage as low as 0.7V (**Fig. 9**). This low operation voltage, stackable Ge/Si UTB CMOS, logic units and SRAM ensure power-efficient monolithic 3D circuits.

(B) Embedded SiGeC photovoltaic ambient light energy harvester: The stackable a-SiGeC TFPV device as embedded ambient light energy harvesters for powering low-power circuits is a promising self-powered technology (**Fig. 10**). In order to gain high electricity from ambient light and reduce power-management complexity, several-types of multi-junction (single, tandem, and triple) a-SiGeC TFPV harvesters with various output voltage (0.6, 1.3 and 1.8V) can be adopted and all of them reveal similar power-output of ~7mW/cm^2 for outdoor condition (**Fig. 11**) and ~170 μW/cm^2 for indoor condition (**Fig. 12**). Angular dependent light-electricity reveals an attractive omni-directional property, only 40% electricity drop at 60 degree illumination (**Fig. 13**). Integrating UV-transparent a-SiC window layer alternatively enhances UV light-driven electricity, suitable for LED light environment (**Fig. 14**).

(C) Stackable low power-driving 1-T NVMs: Here, low temperature (<450°C) UTB (13 nm)-MONAOS NVMs with bandgap-engineered (BE) gate stacks consisted of plasma-ALD Al$_2$O$_3$ and high-density plasma-deposited Si-based dielectrics show lower subthreshold swing (155 mV/dec) and fast operation speed (100 ns) to lead to lower driving-voltage of ±9V, compared to planar-MONOS NVMs with high driving-voltage of ±16V [4]. The cross sectional TEM image and band-diagram are shown in **Fig. 15**. The program/erase speed of 100 ns, good retention with charge loss of 20% after extrapolation to ten years and good endurance of more than 10^5 cycles for LT UTB-MONAOS NVMs are demonstrated (**Fig. 16**). This stackable high speed and low operation voltage UTB-MONAOS NVMs can be monolithically integrated with SRAM to form a two-macro (SRAM+NVM) memory module. This enables a system to use NVM for program/data storage during power off to suppress standby current, while achieves high-performance computing with SRAM during power-on period (**Table1 and Fig. 1**).

4. Conclusion

Heterogeneously integrated low-power Ge/Si monolithic 3D-IC with stacked ambient light harvester was demonstrated. This 3D circuit-architecture provides low operation voltage logic circuits/NVMs/SRAMs and efficient energy harvesting function for self-powered, low-power electronic products.

Acknowledgements

The authors would like to thank the financial support from Ministry of Science and Technology (NSC 102-2218-E-492-001) and National Applied Research Laboratories (NARLabs) of the Republic of China.

References

[1] M. Wang. et al., "Nonvolatile SRAM cell," IEDM Tech. Dig., Dec. 2006, pp. 27-30.

[2] M. –F. et al., "Low Store Energy, Low VDDmin, 8T2R Nonvolatile Latch and SRAM With Vertical-Stacked Resistive Memory (Memristor) Devices for Low Power Mobile Applications," IEEE J. Solid-State Circuits, vol. 47, no. 6, pp. 1483-1496, Jun. 2012.

[3] R. Aitken et al., "Device and Technology Implications of the Internet of Things," VLSI Symp. Tech., p.1 (2014).

[4] C.-H. Shen et al., "Monolithic 3D Chip Integrated with 500ns NVM, 3ps Logic Circuits and SRAM," IEDM Tech. Dig., p. 232-235, 2013.

[5] P. Batude et al., "Advances, Challenges and Opportunities in 3D CMOS Sequential Integration," IEDM Tech. Dig., p. 151-154, 2011.

[6] J. Lu et al., "Above-CMOS a-Si and CIGS Solar Cells for Powering Autonomous Microsystems," IEDM Tech. Dig., p. 708-711, 2010.

[7] W. Rieutort-Louis et al., "Device Optimization for Integration of Thin-Film Power Electronics with Thin-Film Energy-harvesting Devices to Create Power-delivery Systems on Plastic Sheets," IEDM Tech. Dig., p. 283-286, 2011.

[8] K.W.R. Chew et al., "A 400nW Single-Inductor Dual-Input-Tri-Output DC-DC Converter with Maximum Power Point Tracking for Indoor Photovoltaic Energy Harvesting," ISSCC Dig. Tech. Papers, P.68-70, 2013.

[9] W. Jung et al., "23.3 A 3nW fully integrated energy harvester based on self-oscillating switched-capacitor DC-DC converter," ISSCC Dig. Tech. Papers, P.398-400, 2014.

[10] C.-C. Yang et al., "Record-high 121/62 μA/μm on-currents 3D stacked epi-like Si FETs with and without metal back gate," IEDM Tech. Dig., p. 731-734, 2013.

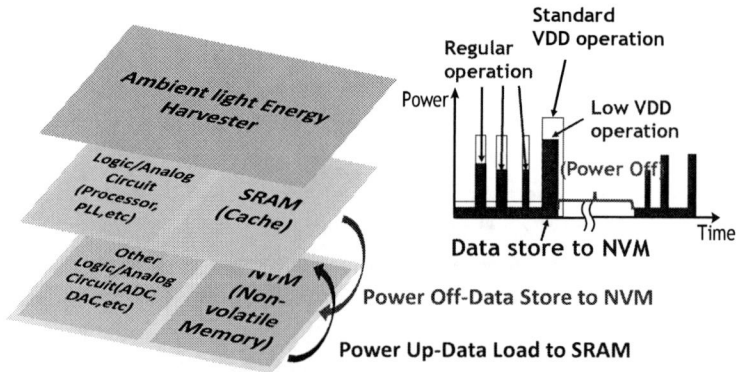

Fig 1. Schematic illustration of monolithic 3D-IC with multiple processors, analog circuits, volatile memory (SRAM), nonvolatile memories (NVMs), and ambient light energy harvester.

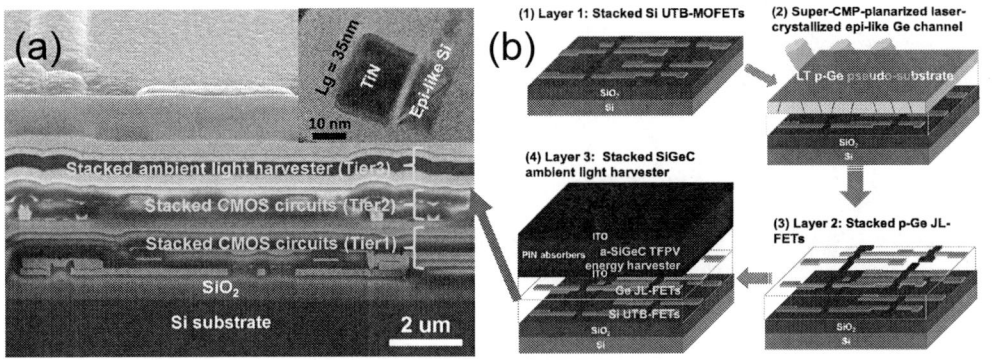

Fig 2. (a) The cross-sectional SEM image of a-SiGeC TFPV ambient light harvester stacked on monolithic 3D-IC chip. Inset shows a 35nm stackable UTB MOSFET. (b) Schematic diagram of process flow of 3D Ge/Si sequential integration Chip with stacked a-SiGeC TFPV ambient light harvester.

Figure 3. Cross-sectional SEM and schematic illustration of UTB MOSFETs with embedded S/D and back gate Vth adjusters.

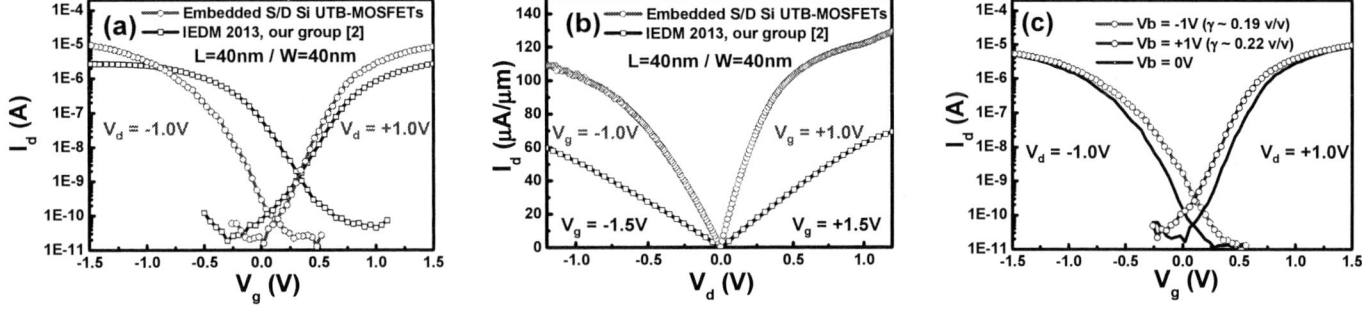

Fig 4. (a) Id-Vg and (b) Id-Vd characterization of Si UTB n/p MOSFETs with embedded S/D current boosting structure. (c) V_{th} adjusting capability in embedded S/D Si UTB n/p MOSFETs by back-gate control.

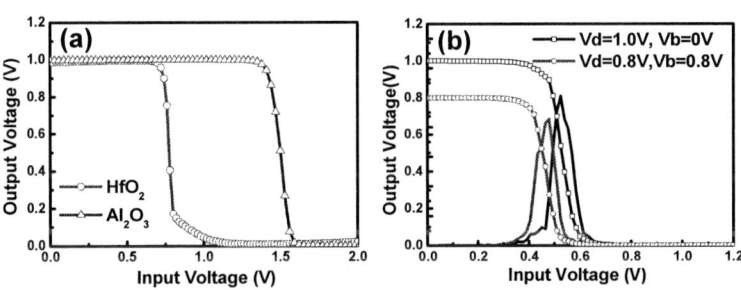

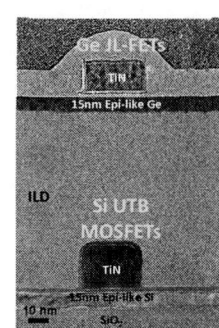

Fig 5. Voltage transfer characteristics of inverters with stackable UTB CMOS with threshold voltage adjusting functionality enabled by (a) flatband voltage control in gate structures and (b) back gate biasing.

Fig 6. (a) Hall mobility of laser-crystallized epi-like Ge as function of thickness.

Figure 7. The cross-sectional TEM image of double-layered Ge/Si monolithic 3D ICs.

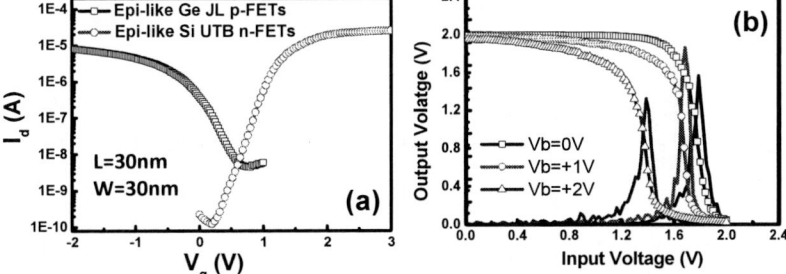

Fig 8. (a) Id-Vg characterization of p-type Ge JL-FETs and n-type Si UTB MOSFETs. (b) Voltage transfer characteristics (VTC) of p-Ge/n-Si CMOS inverters under adjusting V_{th} in n-type Si UTB MOSFETs by BG.

$SNM = min(SNM_1, SNM_2).$

Experimental Results :
⇒ SNM: 280mV @2V (2013 IEDM, our group [2])
⇒ SNM: 390mV @1V (this work)
⇒ SNM: 270mV @0.7V (this work)

Fig 9. Butterfly curve of the best stackable 6T SRAM bit-cells with high SNM of 270 mV at low operation voltage 0.7V.

978-1-4799-8002-4/14 $31.00 © 2014 IEEE   69

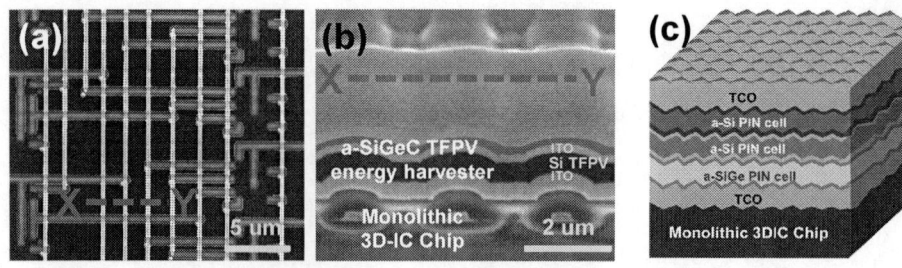
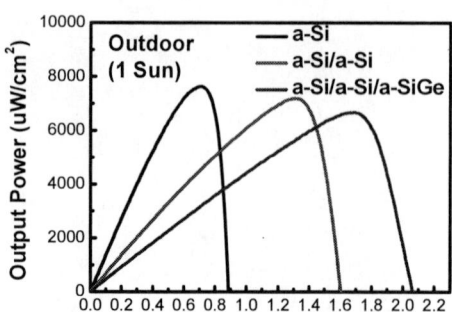

Figure 10. (a) The top-view and (b) cross-sectional SEM images of a-SiGeC TFPV light-energy harvester with inherent light-trapping textured structure formed from top metallization structure in monolithic 3D-IC chip. (c) Schematic illustration of triple junction a-SiGeC TFPV energy harvester on monolithic 3D-IC chip.

Figure 11. Output power of single, tandem, and triple junction a-SiGeC TFPV light-energy harvester for outdoor condition (1Sun).

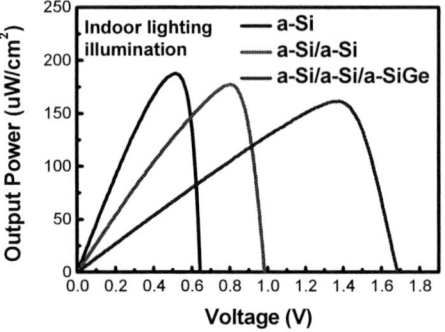
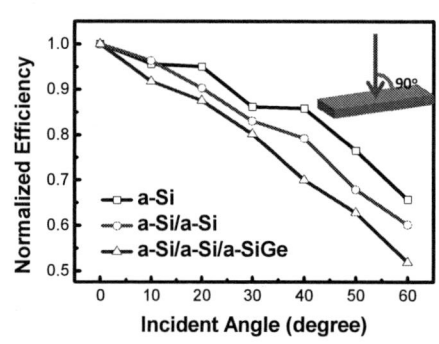
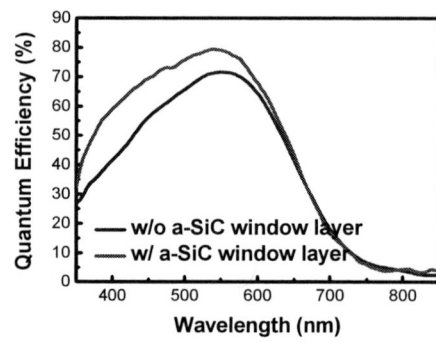

Figure 12. Output power of multi-junction a-SiGeC TFPV light-energy harvester for indoor condition (>2000 lux).

Figure 13. Angular dependent output power of multi-junction a-SiGeC TFPV light-energy harvesters for outdoor condition (1Sun).

Figure 14. Quantum efficiency of UV-enhanced a-Si TFPV light-energy harvester with high UV-transparent a-SiC window layer.

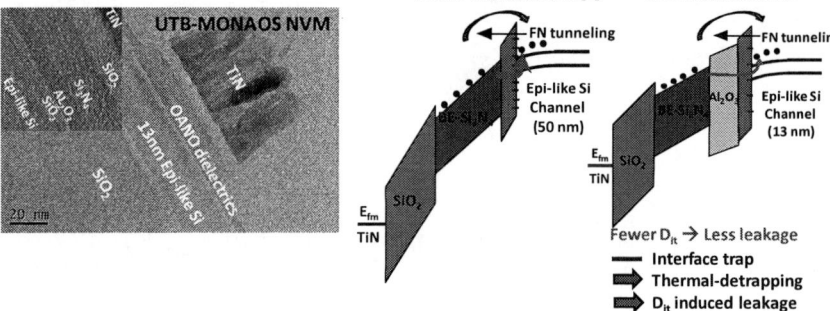

	Single Operation (power consumption)	Drive Capability (Indoor – 100uW/cm², 1 Sun – 7mW/cm²)
Inverter	0.6 nW / switch	150,000 inverters (Indoor)
SRAM	0.9 nW / write	100,000 SRAMs (Indoor)
NVM	2.1 uW / bit	2 kbits (1 Sun)

Fig 15. A TEM picture and band diagram of a UTB-MONAOS NVM with improved speed and retention by 13nm UTB and plasma-ALD Al_2O_3-enabled BE gate stack.

Table 1. Power consumption estimation of logic gates, SRAM and NVMs.

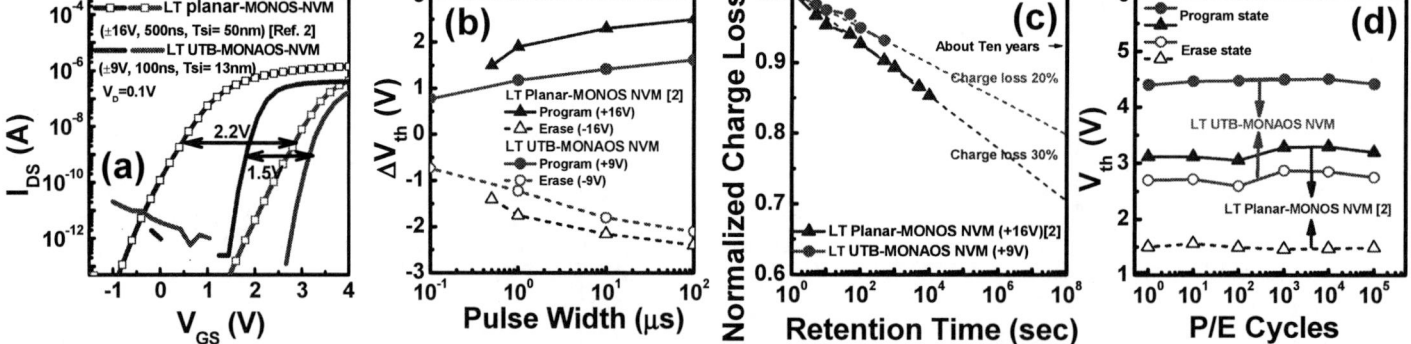

Fig 16. (a) Low temperature planar-MONOS and UTB-MONAOS NVMs made on Epi-like Si have Vt window of 2.2 V and 1.5V after sub-µs programming, respectively. (b) Program and erase speed characteristics of LT planar-MONOS and UTB-MONAOS NVMs under gate pulses of ±16 and ±9 V, respectively. (c) The improved retention characteristics in UTB-MONAOS NVMs by BE gate stacks and 13nm UTB with normalized charge loss (ΔVth (t)/ΔVth(0)) of 20% (30% for planar-MONOS NVMs) after extrapolation to 10 years. (c) Endurance of both NVMs with P/E cycles up to 10^5 and no degradation of memory window (1µs pulses applied for both retention and endurance testing).

A 14nm Logic Technology Featuring 2nd-Generation FinFET Transistors, Air-Gapped Interconnects, Self-Aligned Double Patterning and a 0.0588μm^2 SRAM cell size

ABSTRACT

INTRODUCTION

technology generations
σ
in 22nm. With
σ

RELIABILITY

INTERCONNECTS

KEY DESIGN RULES & TECHNOLOGY FEATURES

SRAM, PRODUCT AND YIELD

The 14nm
μ

CONCLUSIONS

TRANSISTOR PERFORMANCE AND VARIATION

We have shown a high
μ

REFERENCES

Symp. VLSI Tech. Dig.
IEDM Tech. Dig.
IEDM Tech. Dig.
Symp. VLSI Tech Dig.
IEDM Tech Dig.
Symp. VLSI Tech. Dig.

Layer	Pitch (nm)	Scale Factor to [1]
Fin	42	0.70
Contacted Gate Pitch	70	0.78
Metal 0	56	N/A
Metal 1	70	0.78
Metal 2	52	0.65

Table 1: Layer Pitches

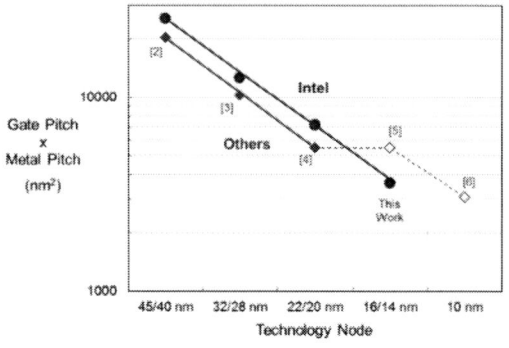

Figure 1: Multi-Generation Scaling Trend

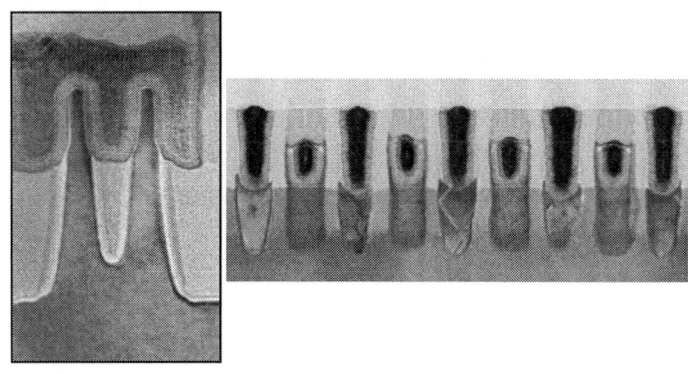

Figure 2: Transistor Fin and Gate-Cut Images

(not same scale)

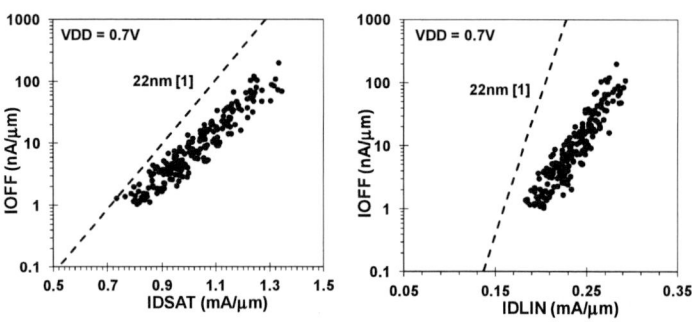

Figure 3: NMOS Idsat and Idlin curves

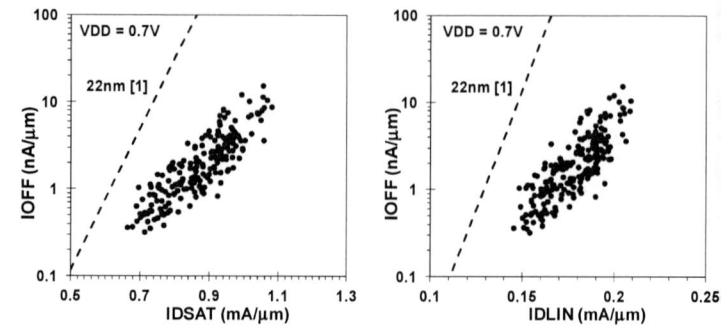

Figure 4: PMOS Idsat and Idlin curves

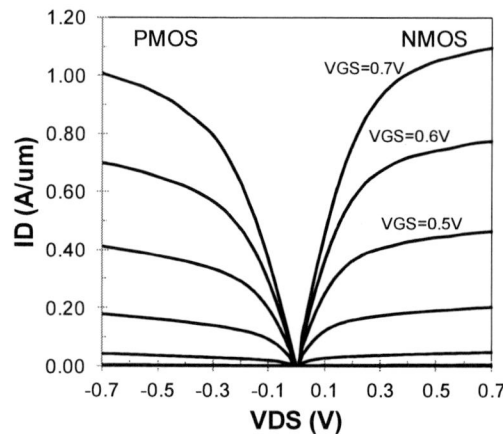

Figure 5: Transistor I-V Curves

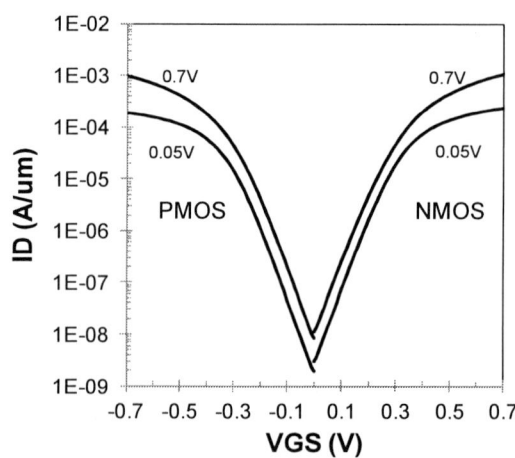

Figure 6: Subthreshold Curves

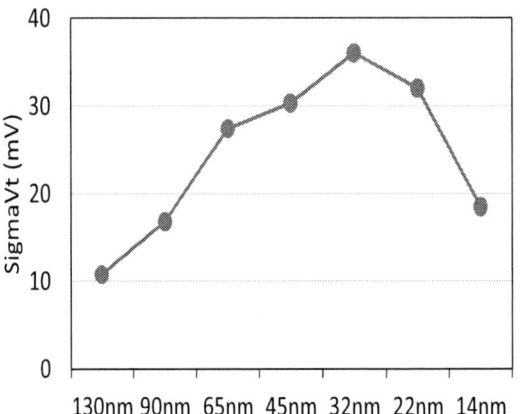

Figure 7: Random Varation Trend (σVt)

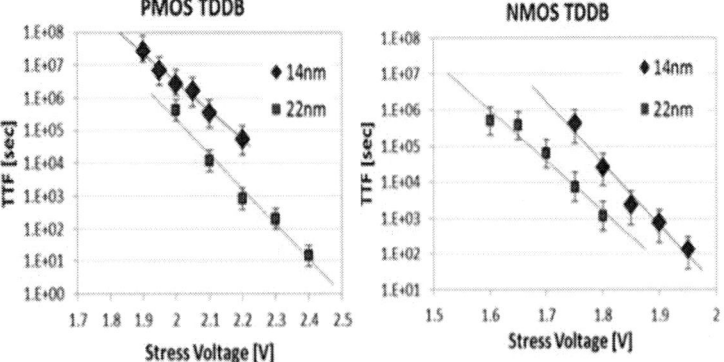

Figure 8: PMOS and NMOS TDDB

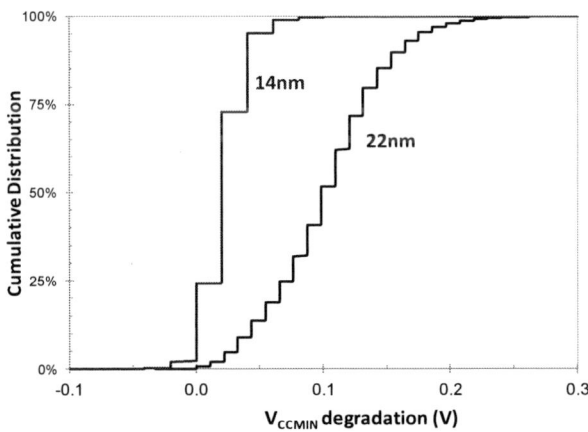

Figure 9: SRAM Aging Behavior

Figure 10: Interconnect Stack

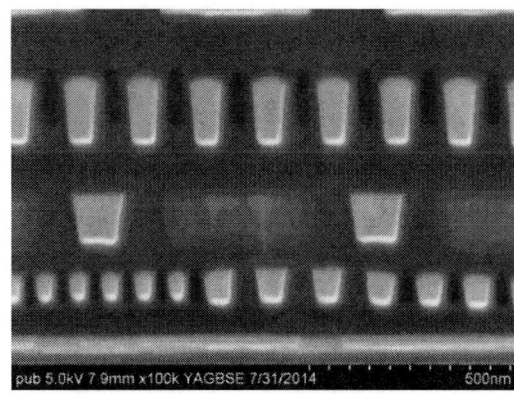

Figure 11: Air-Gapped Interconnects

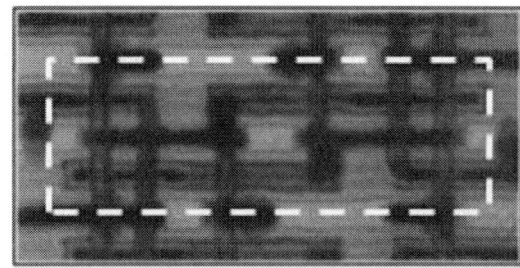

Figure 12: SRAM Bitcell

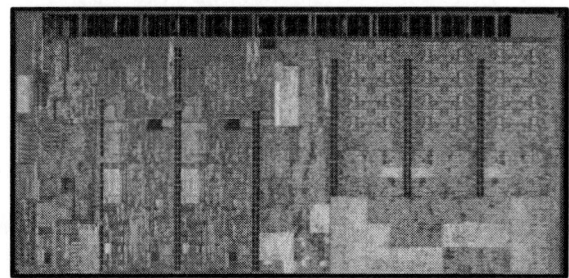

Figure 13: Broadwell Die Photo

High Performance 14nm SOI FinFET CMOS Technology with $0.0174\mu m^2$ embedded DRAM and 15 Levels of Cu Metallization

C-H. Lin, B. Greene, S. Narasimha, J. Cai, A. Bryant, C. Radens, V. Narayanan, B. Linder, H. Ho, A. Aiyar, E. Alptekin, J-J. An, M. Aquilino, R. Bao, V. Basker, N. Breil, M. Brodsky, W. Chang, L. Clevenger, D. Chidambarrao, C. Christiansen, D. Conklin, C. DeWan, H. Dong, L. Economikos, B. Engel, S. Fang, D. Ferrer, A. Friedman, A. Gabor, F. Guarin, X. Guan, M. Hasanuzzaman, J. Hong, D. Hoyos, B. Jagannathan, S. Jain, S-J. Jeng, J. Johnson, B. Kannan, Y. Ke, B. Khan, B. Kim, S. Koswatta, A. Kumar, T. Kwon, U. Kwon, L. Lanzerotti, H-K Lee, W-H. Lee, A. Levesque, W. Li, Z. Li, W. Liu, S. Mahajan, K. McStay, H. Nayfeh, W. Nicoll, G. Northrop, A. Ogino, C. Pei, S. Polvino, R. Ramachandran, Z. Ren, R. Robison, I. Saraf, V. Sardesai, S. Saudari, D. Schepis, C. Sheraw, S. Siddiqui, L. Song, K. Stein, C. Tran, H. Utomo, R. Vega, G. Wang, H. Wang, W. Wang, X. Wang, D. Wehelle-Gamage, E. Woodard, Y. Xu, Y. Yang, N. Zhan, K. Zhao, C. Zhu, K. Boyd, E. Engbrecht, K. Henson, E. Kaste, S. Krishnan, E. Maciejewski, H. Shang, N. Zamdmer, R. Divakaruni, J. Rice, S. Stiffler, P. Agnello

IBM *Semiconductor Research and Development Center* (SRDC), Hopewell Junction, NY 12533, USA, Contact: bgreene@us.ibm.com

Abstract

We present a fully integrated 14nm CMOS technology featuring finFET architecture on an SOI substrate for a diverse set of SoC applications including HP server microprocessors and LP ASICs. This SOI finFET architecture is integrated with a 4th generation deep trench embedded DRAM to provide an ultra-dense ($0.0174um^2$) memory solution for industry leading 'scale-out' processor design. A broad range of Vts is enabled on chip through a unique dual workfunction process applied to both NFETs and PFETs. This enables simultaneous optimization of both lowVt (HP) and HiVt (LP) devices without reliance on problematic approaches like heavy doping or Lgate modulation to create Vt differentiation. The SOI finFET's excellent subthreshold behavior allows gate length scaling to the sub 20nm regime and superior low Vdd operation. This leads to a substantial (>35%) performance gain for Vdd ~0.8V compared to the HP 22nm planar predecessor technology. At the same time, the exceptional FE/BE reliability enables high Vdd (>1.1V) operation essential to the high single thread performance for processors intended for 'scale-up' enterprise systems. A hierarchical BEOL with 15 levels of copper interconnect delivers both high performance wire-ability as well as effective power supply and clock distribution for very large >600mm^2 SoCs.

Technology Description

The critical dimensions for this 14nm technology are shown in Table1. The process flow, along with key device cross sections, is shown in Fig1. The 42nm fin pitch is achieved using sidewall image transfer. The SOI substrate provides multiple advantages for overall finFET integration. Use of an SOI substrate 1) minimizes the process complexity associated with both fin isolation and eDRAM integration, 2) minimizes the parasitic capacitance at the base of the fin, 3) simplifies patterning of the active fins, and 4) minimizes each component of fin structural variability (i.e. height, thickness, and profile).

It is important to underscore a key point concerning fin height selection in a HP finFET technology. To enable high frequency operation (near Vmax), it is imperative to minimize the back-end RC delay so that it represents a small fraction of the critical path's overall gate delay. In this paradigm, excessively tall fins lead to unfavorable power/performance trade-offs (Fig2). In this work, careful design/technology co-optimization has resulted in the selection of an ideal fin height that supports enterprise server class performance at the lowest power envelope.

The resulting SOI finFET device response is shown in Fig3. High G_m is achieved through a combination of gatestack, epi S/D, and contact interface optimization. The SOI finFET architecture enables a SCE that is well controlled down to the sub-20nm Lgate regime (Fig4). This device behavior, coupled to the optimized fin height

selection and parasitic capacitance optimization, has resulted in a performance improvement of >35% over the predecessor 22nm technology node [2] (Fig5). Furthermore, this performance gain stretches over a broad Vt range (from 100nA/um HP FETs to sub 1nA/um LP and SRAM array FETs which are featured in many ASICs). It is generally difficult to support such a wide Vt range in a finFET technology. Typically, Vt separation is achieved by doping, which carries with it many negative consequences to finFET response [3]. As a result, most industry standard finFET offerings restrict the allowable Vt range available to a designer. In this work, we apply an innovative dual WF process to generate widely spaced Vts (for both N/P) without reliance on doping to create the Vt separation [4]. The dopant removal enables 1) significant performance enhancement for HiVt (LP) devices due to the mobility gain (Fig6) and 2) significant Vt mismatch (Vmin) reduction for low leakage SRAM cells due to the RDF reduction (Fig7).

As mentioned earlier, one of the key advantages of the SOI substrate is the ability to co-integrate deep trench eDRAM with logic. In this work, the 14nm eDRAM unit cell has been scaled down to $0.0174um^2$, which provides a unique memory solution for cache starved processors (Fig8). This cell scaling and performance (access time) have largely been enabled by the finFET device architecture. The SCE improvement achieved in the pass gate has allowed for significant reduction of both Lgate and Vt, without compromising retention specifications. The unique challenge of DT integration with fins comes in the control of the strap resistance between the SOI crystalline fin and the highly doped poly-crystalline fin (Fig9). The epi S/D module has been engineered to minimize the resistance at this interface through growth optimization on the two different underlying materials. A top down of the interface after epi S/D is shown in Fig10. This overall finFET based cell design, together with optimization of the strap resistance, has resulted in a cell access time that is 0.7X of the previous best-in-class value achieved by our 22nm technology.

The TDDB and BTI results for this technology are shown in Fig11 and Fig12. Both pass the specifications required to support 10 year lifetime at Vmax >1.1V. This technology provides up to 15 levels of Cu metallization. The hierarchical BEOL architecture (Fig13) begins with 64nm pitch at M1/Mx and expands up to ultra thick levels required for efficient clock and power distribution across a >600mm^2 chip. Throughout the BE, metallization processes have been developed so that wire resistance and EM lifetime can be optimized simultaneously (Fig 14).

Conclusions

A HP 14nm SOI finFET technology has been developed featuring >35% performance improvement, an ultra-dense $0.0174um^2$ embedded DRAM memory cell, and dual WF gatestack enablement that achieves optimized HP and LP devices simultaneously on chip.

978-1-4799-8002-4/14 $31.00 © 2014 IEEE

References

[1] S-T. Chen et al., IITC/MAM 2011
[2] S. Narasimha et al., IEDM 2012
[3] C-H. Lin et al., VLSI 2012
[4] K. Seo et al., VLSI 2014

Table 1 Key technology scaling rules and attributes. Patterning details for the 64nm M1/Mx provided in Ref [1].

Level	Pitch
Fin	42nm
Gate (single pattern w/ cut)	80nm
Contact	80nm
M1 (Bi-directional)	64nm
Mx	64nm
BE Hierarchy: 1X,1.25X,2X,4X,8X,40X	
eDRAM cell area 0.0174 μm^2	

- Deep Trench
- SOI Fin Patterning
- Low-K Spacer
- Dual Epi S/D
- RMG
- Contact Patterning
- Ti Liner Silicide
- 15 BEOL Levels

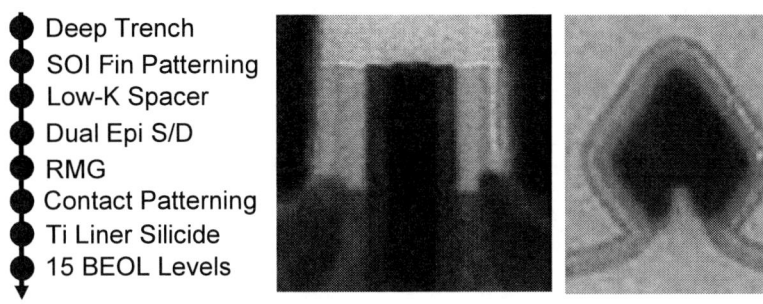

Fig. 1 Process flow and cross sections for 14nm SOI finFET technology.

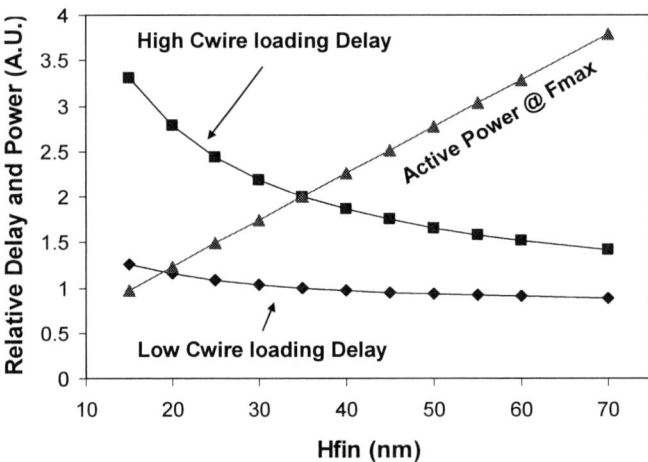

Fig. 2 Power performance optimization as a function of Hfin. High performance designs require low Cwire critical paths. These paths do not benefit from Hfin increase.

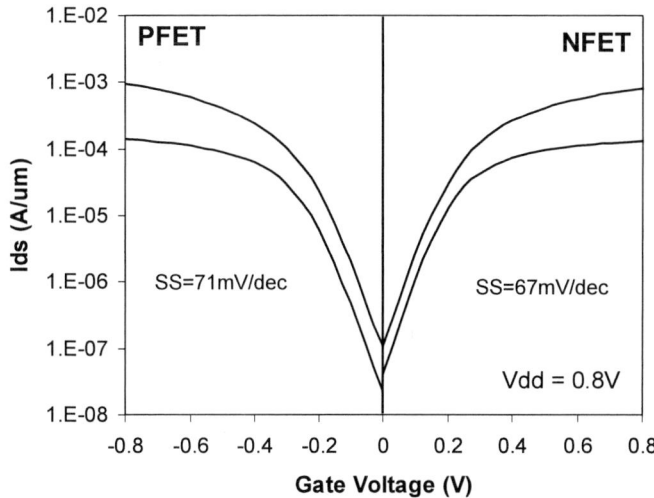

Fig. 3 Nominal DC Id-Vg response for HP devices at 0.8V. The curves reflect Idsat values of 808/935 $\mu A/um$ (N/P) for HP devices. These Id values are normalized by the true Si FIN perimeter (and not fin pitch). AC values (without self-heating) are 5% higher than the DC values shown here.

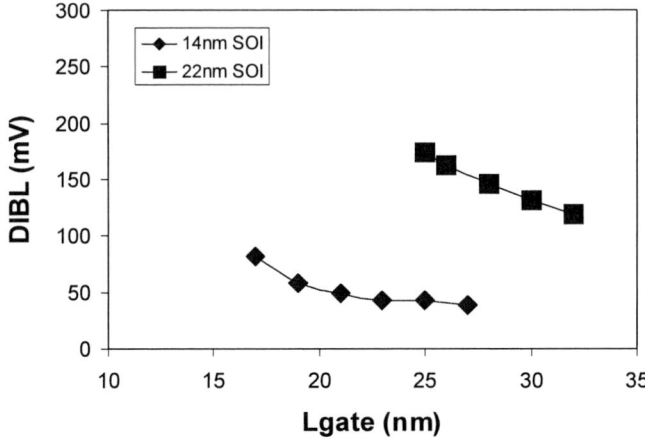

Fig. 4 DIBL response of 14nm SOI finFETs compared to 22nm SOI planar devices. SCE control is demonstrated down to the sub-20nm gate length regime.

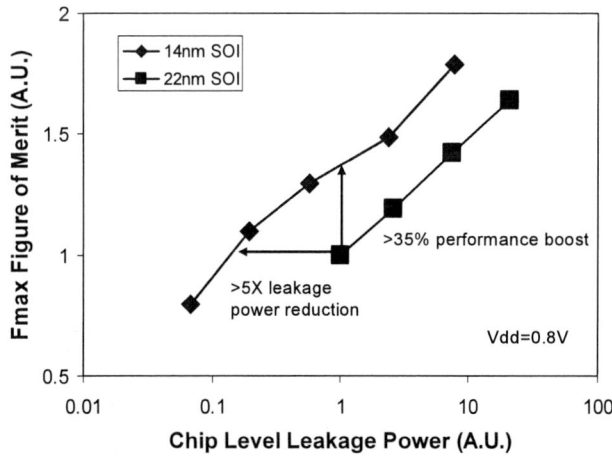

Fig. 5 14nm SOI finFET performance benchmarking at fixed leakage compared to our 22nm SOI HP technology. This 14nm gains exceed 35% (Vdd 0.8V).

978-1-4799-8002-4/14 $31.00 © 2014 IEEE

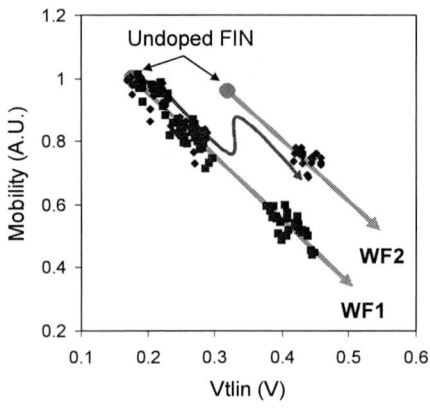

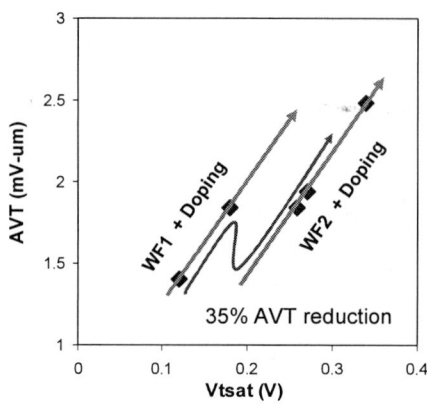

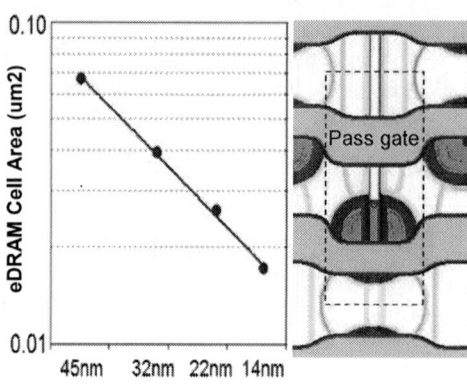

Fig. 6 Benefit of dual WF design for the mobility response of the hiVt (LP) device offerings. The grey arrows point in the direction of increasing doping.

Fig. 7 Benefit of dual WF design for optimization of SRAM AVt through Vt mismatch and RDF control.

Fig. 8 IBM eDRAM area scaling over the last 4 generations culminating in the 0.0174um^2 cell in 14nm SOI finFET technology.

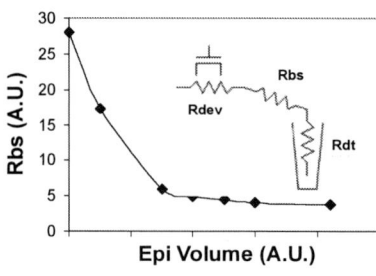

Fig. 9 Representative image after fin patterning. The interface between the crystalline (SOI) and polycrystalline (trench) segments defines the 'strap' region.

Fig. 10 Importance of the epi S/D process on the buried strap resistance. The TEM illustrates how the epi growth proceeds from both the SOI and poly seed regions

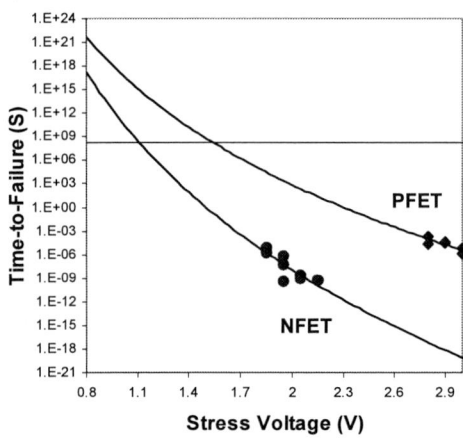

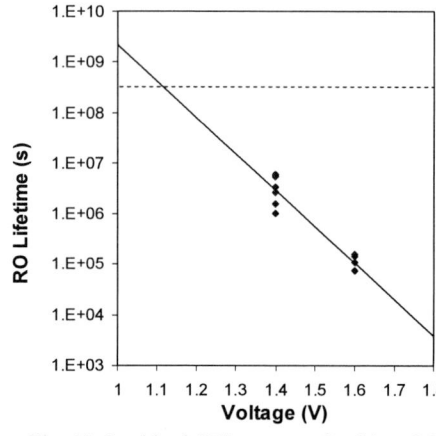

Fig. 11 TDDB response for 14nm SOI N/P finFETs. The yield projections support 10 year EOL specifications for Vmax of >1.1V.

Fig. 12 Combined BTI response for 14nm SOI N/P finFETs. The projections support 10year EOL specifications for Vmax of >1.1V.

Fig. 13 Hierarchical back end reflecting 1X, 1.25X, 2X, 4X, 8X, 40X layering.

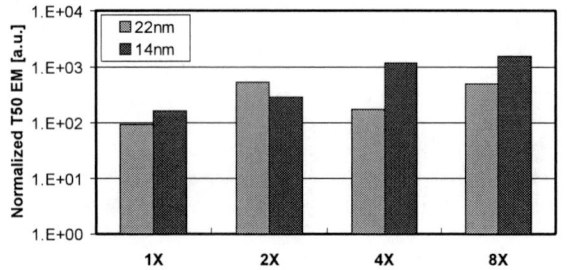

Fig. 14 EM optimization for the 1x, 4X, 8X BE levels in 14nm (compared to 22nm)

978-1-4799-8002-4/14 $31.00 © 2014 IEEE   76

A 55 nm Triple Gate Oxide 9 Metal Layers SiGe BiCMOS Technology Featuring 320 GHz f_T / 370 GHz f_{MAX} HBT and High-Q Millimeter-Wave Passives

P. Chevalier, G. Avenier, G. Ribes, A. Montagné, E. Canderle, D. Céli, N. Derrier, C. Deglise, C. Durand, T. Quémerais, M. Buczko, D. Gloria, O. Robin, S. Petitdidier, Y. Campidelli, F. Abbate, M. Gros-Jean, L. Berthier, J.D. Chapon, F. Leverd, C. Jenny, C. Richard, O. Gourhant, C. De-Buttet[*], R. Beneyton, P. Maury, S. Joblot, L. Favennec, M. Guillermet[*], P. Brun[*], K. Courouble, K. Haxaire, G. Imbert, E. Gourvest, J. Cossalter, O. Saxod, C. Tavernier, F. Foussadier, B. Ramadout, R. Bianchini, C. Julien, D. Ney, J. Rosa, S. Haendler, Y. Carminati, B. Borot

STMicroelectronics & CEA-LETI (*), 850 rue Jean Monnet, Crolles, France, Email: pascal.chevalier@st.com

Abstract

This paper presents the first 55 nm SiGe BiCMOS technology developed on a 300 mm wafer line in STMicroelectronics. The technology features Low Power (LP) and General Purpose (GP) CMOS devices and 0.45 µm² 6T-SRAM bit cell. High Speed (HS) HBT exhibits 320 GHz f_T and 370 GHz f_{MAX} associated with a CML ring oscillator gate delay τ_D of 2.34 ps. Transmission lines, capacitors, high-Q varactors and inductors dedicated to millimeter-wave applications are also available.

Introduction

High-speed Silicon Germanium (SiGe) BiCMOS technologies in volume production today for 77 GHz automotive radars and 100 Gb/s optical communications exhibit HBT with (f_T,f_{MAX}) ~ (200,300) GHz in average in 180 nm or 130 nm CMOS nodes for the most advanced ones [1-2]. In the past years lots of emphasis was put to increase the performance of SiGe HBT, especially in Europe [3] to reach (f_T,f_{MAX}) = (300,500) GHz [4]. However dense and fast CMOS transistors are increasingly important for digital signal processing and control circuits. A first 90 nm BiCMOS technology in 300 mm was published more than 10 years ago [5] but with an HBT featuring (f_T,f_{MAX}) = (130,100) GHz only. More recently a 90 nm BiCMOS technology featuring a faster HBT ((f_T,f_{MAX}) = (300,360) GHz) was announced in 200 mm [6]. In this paper we present the first 55 nm BiCMOS technology in 300 mm dedicated to millimeter-wave (mm-W) applications.

Technology features

BiCMOS055, also called B55, is derived from the 65 nm LP/GP Mix (triple gate oxide) CMOS platform (C065) presented in [7]. It benefits from a 10% linear shrink and the related process improvements done for an LP only version (C055). NMOS and PMOS high-V_T (HVT) and standard-V_T (SVT) flavors are available both for 1.2 V (LP) and 1.0 V (GP) transistors together with 2.5 V (IO) MOS transistors. A low-V_T (LVT) flavor option is proposed for both LP and GP transistors. High densities single-port and dual-port high speed and low power SRAM bit cells using HVT and/or SVT LP transistors are offered too. Minimum gate lengths achieved for GP NMOS and bit cell area for low power SRAM are 45 nm and 0.45 µm² respectively. SiGe HBTs feature a conventional double-polysilicon self-aligned (DPSA) architecture, for which emitter-base self-alignment is provided by Selective Epitaxial Growth (SEG) of the SiGe:C base [8]. Three collector flavors, leading to different f_T / BV_{CEO} trade-offs for the High Speed (HS), Medium Voltage (MV) and High Voltage (HV) HBTs are available. Collector of the HS HBT is formed by a standard 'buried layer + epitaxy + sinker / deep trenches / SIC' module. Emitters are scalable in width and length with a minimum area of 0.10×0.30 µm². A 3-µm thick copper layer + 1.5-µm thick via and a 5 fF/µm² MIM capacitor (option) have been added to C055 to end up in a back-end of line (BEOL) featuring 8 copper metal layers and 1 aluminum capping layer. This BEOL combines the advantages of being fully compatible with the existing 55 nm CMOS libraries and to provide enhanced

performance for millimeter-wave passives. All the natural resistors, capacitors and transistors coming from C065/C055 are proposed including a 6 kΩ/sq HIPO resistor option. Varactors, inductors and transmission lines have been specifically designed for millimeter-wave applications. Cross-sections of the technology are shown in Fig. 1, 2 and 3.

CMOS devices results

Providing CMOS devices with electrical characteristics unaltered compared to the mother technology is a challenge of the BiCMOS integration, especially with the number of CMOS devices offered in B55. I_{ON}-I_{OFF} plots and ring oscillator results (Fig. 4 to 8) for LP and GP devices show that it has been carried out successfully. This is also highlighted by the 5.25 Mb SRAM results presented in Fig. 9 and 10.

SiGe HBTs results

Performances of the 3 SiGe HBT flavors are presented in Table 1. Current gain transit frequency and maximum oscillation frequency (f_T,f_{MAX}) = (325,375) GHz, (180,385) GHz and (65,270) GHz are obtained for the HS, MV and HV transistors associated with BV_{CEO} values of 1.5 V, 1.9 V and 3.2 V respectively. HF characteristics are plotted in Fig. 11 to 13 with Fig. 13 emphasizing the reliability of the peak f_T and f_{MAX} extractions for the HS HBT. Gate delay measured on 23-stage CML ring oscillators using HS HBT with L_E = 1.5 µm, similar to those presented in [8], is 2.34 ps in average with a standard deviation below 1% only for a full wafer (Fig. 14), which illustrates the good process uniformity. DC characteristics of HS NPN are shown in Fig. 15 and 16 (with the Gummel plots of a yield structure). Peak of current gain (~1900) is reached at V_{BE} ~ 0.7 V. Fig. 17, presenting HS NPN results across process windows, shows that targeted HF performances are achieved with enough margin.

Millimeter-wave passives results

The offer of GO1 and GO2 P+/NWell accumulation MOS varactors, presented in Table 2, is dedicated to mm-W applications (frequency of resonance > 110 GHz). The evolution of the performance with the frequency of a GO2 varactor using GO1 minimum gate length is presented in Fig. 18. Q-factors larger than 10 are measured at mm-W frequencies. Geometries of mm-W inductors covering a wide inductance range are listed in Table 3 together with 3D view examples. Fig. 19 exhibits a Q-factor always larger than 10 too. Finally performances of the transmission lines are summarized in Table 4. An attenuation constant value as low as 0.5 dB/mm is measured at 60 GHz.

Conclusion

BiCMOS055 features a unique combination between CMOS node, SiGe HBT and mm-W passives performances. This is highlighted by the benchmark of Fig. 20, showing that digital density offered in B55 is 5 and 3 times larger compared to other existing BiCMOS technology in 130 nm and 90 nm respectively. It will serve demanding optical, wireless and high-performance analog applications for which high gain, low noise and low power consumption are required.

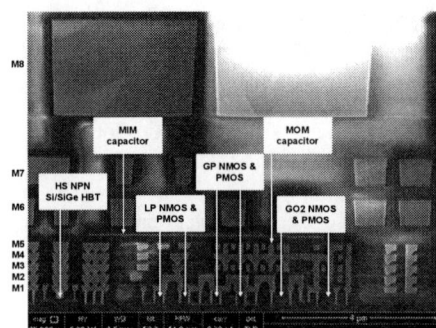

Fig. 1: BiCMOS055 SEM cross-section showing main devices and BEOL up to Metal 8

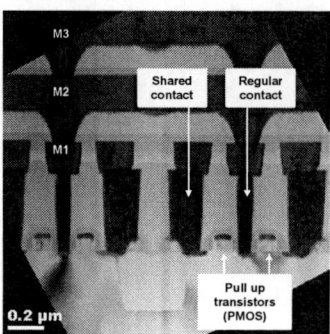

Fig. 2: TEM cross-section in a pull up row of a 0.45 µm² SRAM (5.25 Mb)

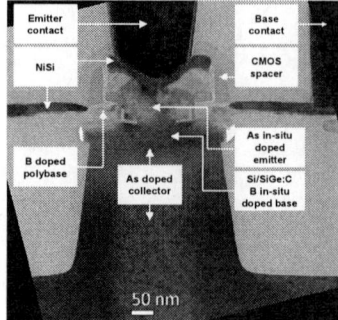

Fig. 3: TEM cross-section of a 0.1×4.9 µm² HS SiGe HBT (zoom in on Emitter and Base regions)

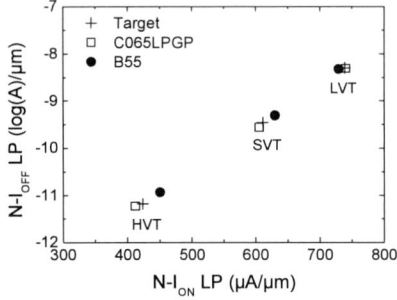

Fig. 4: I_{OFF} vs. I_{ON} plot for 0.06×10 µm² (drawn) LP NMOS transistors (B55 vs. C065)

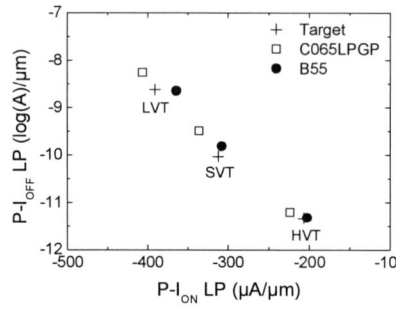

Fig. 5: I_{OFF} vs. I_{ON} plot for 0.06×10 µm² (drawn) LP PMOS transistors (B55 vs. C065)

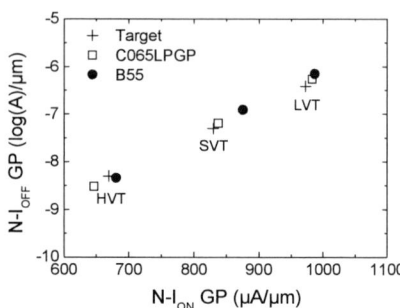

Fig. 6: : I_{OFF} vs. I_{ON} plot for 0.06×10 µm² (drawn) GP NMOS transistors (B55 vs. C065)

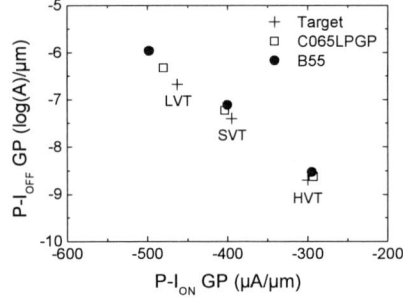

Fig. 7: I_{OFF} vs. I_{ON} plot for 0.06×10 µm² (drawn) GP PMOS transistors (B55 vs. C065)

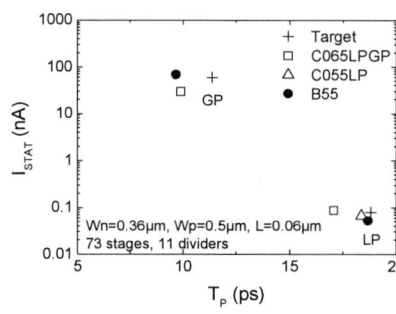

Fig. 8: SVT LP and GP ring oscillators static gate current leakage I_{STAT} vs. gate delay T_P (B55 vs. C055 & C065)

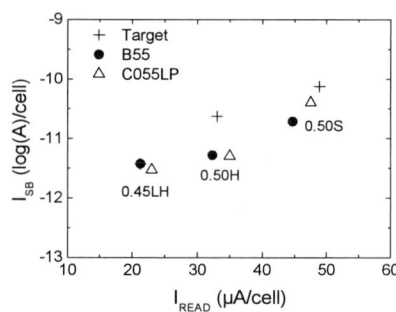

Fig. 9: SRAM bit cells leakage I_{SB} vs. read current I_{READ} (B55 vs C055)

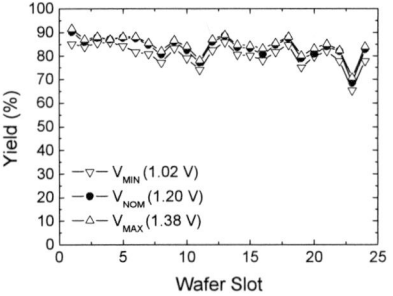

Fig. 10: 0.45 µm² SRAM yield vs. V_{DD} for a full B55 lot

	NPN HS	NPN MV	NPN HV
Max f_T (GHz)	326 (V_{CB} = 0.5V)	178 (V_{CB} = 1V)	64 (V_{CB} = 2.5V)
Max f_{MAX} (GHz)	376 (V_{CB} = 0.5V)	384 (V_{CB} = 1V)	269 (V_{CB} = 2.5V)
J_C (mA/µm²) at max f_T & f_{MAX}	19	6.1	1.9
BV_{CBO} (V)	5.4	7.3	14.4
BV_{CEO} (V)	1.5	1.9	3.2

Table 1: HS, MV and HV SiGe HBTs main characteristics (0.1×4.9 µm² CBEBC transistors)

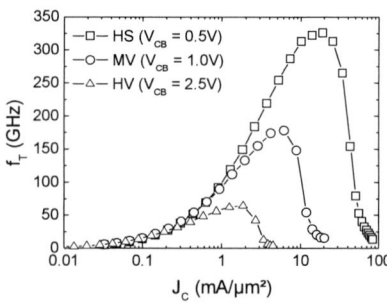

Fig. 11: Current gain transit frequency f_T vs. collector current density J_C for 0.10×4.9 µm² SiGe HBTs

978-1-4799-8002-4/14 $31.00 © 2014 IEEE

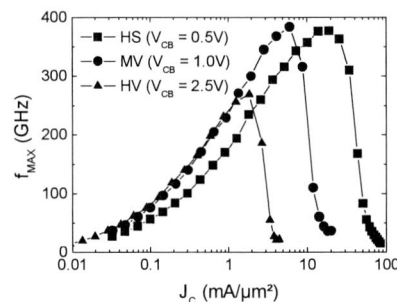

Fig. 12: Maximum oscillation frequency f_{MAX} vs. collector current density J_C for $0.10 \times 4.9 \ \mu m^2$ SiGe HBTs

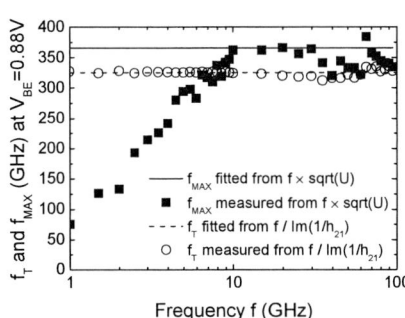

Fig. 13: Peak f_T and f_{MAX} extraction vs. frequency at $V_{BE} = 0.88 \ V$ ($V_{CB} = 0.5 \ V$) for a $0.10 \times 4.9 \ \mu m^2$ HS SiGe HBT

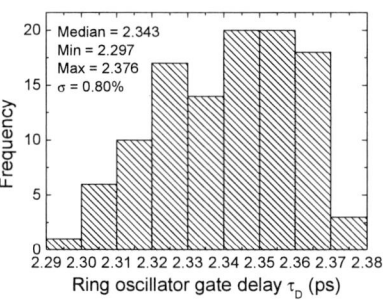

Fig. 14: 23-stage CML ring oscillator gate delay τ_D frequency histogram of a full B55 wafer (109 sites)

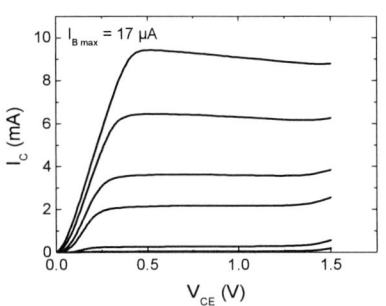

Fig. 15: I_C-V_{CE} characteristics for different I_B (Max value corresponds to peak f_T) of a $0.10 \times 4.9 \ \mu m^2$ HS SiGe HBT

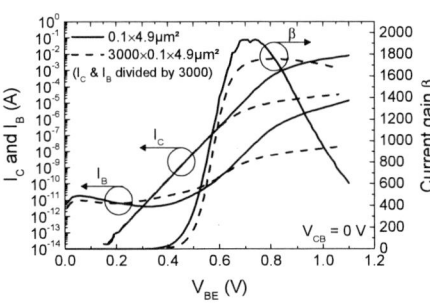

Fig. 16: Gummel characteristics of a single $0.10 \times 4.9 \ \mu m^2$ HS SiGe HBT and of a yield structure of 3000 transistors

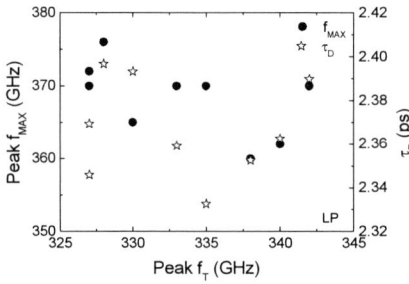

Fig. 17: Peak f_{MAX} & τ_D vs. peak f_T of HS HBT results coming from process windows on vertical profile only

	Varactor P+/NWell GO1 (1.2V)	Varactor P+/NWell GO2 (2.5V)
Capacitance range (Cmin / Cmax)	5 fF / 1 pF	5 fF / 1 pF
Typical tuning ratio at 25 GHz & C=100 fF	3 (max 5)	3 (max 5)
Max Q at 25 GHz, C=100 fF & 1.2 V (GO1) or 2.5 V (GO2)	20	30
Frequency of resonance at C=100 fF &1.2 V (GO1) or 2.5 V (GO2)	> 110 GHz	> 110 GHz

Table 2: Millimeter-wave single-ended GO1 and GO2 accumulation MOS varactors offer

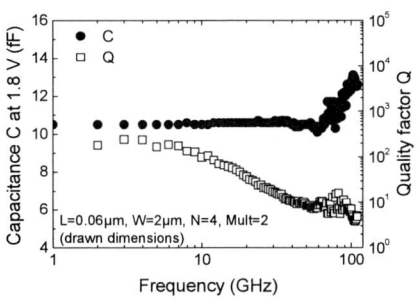

Fig. 18: Capacitance and Q-factor vs. frequency of a GO2 varactor featuring minimum GO1 gate length

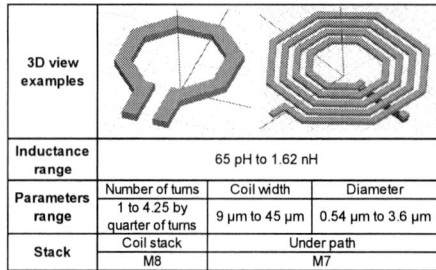

3D view examples		
Inductance range	65 pH to 1.62 nH	
Parameters range	Number of turns	Coil width	Diameter
	1 to 4.25 by quarter of turns	9 μm to 45 μm	0.54 μm to 3.6 μm
Stack	Coil stack	Under path
	M8	M7

Table 3: Millimeter-wave inductor geometries designed with the millimeter-wave specific BEOL

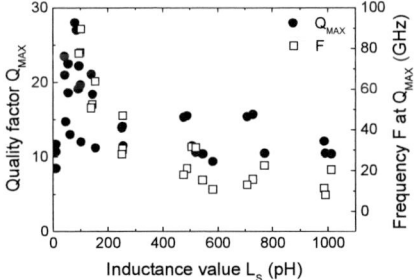

Fig. 19: Maximum Q-factor Q_{MAX} and frequency at Q_{MAX} vs. inductance values for different inductor geometries

T line configuration	Zc range (Line width range)	α at 60GHz	Stack
Single-ended	28 Ω to 87 Ω (0.54 μm to 16.74 μm)	0.51 dB/mm (50 Ω)	Line in M8 (Gnd plane in M1 or M4)
Differential	100 Ω diff. (2.00 μm if gnd M1)	0.82 dB/mm (100Ω)	Line in M8 (Gnd plane in M1 or M4)

Table 4: Summary of the transmission lines performance using the millimeter-wave specific BEOL

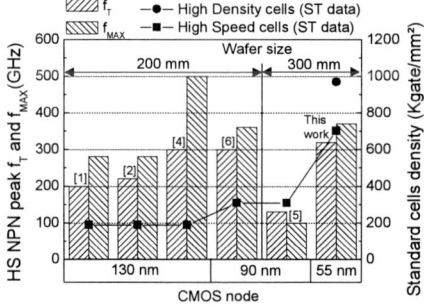

Fig. 20: B55 vs. BiCMOS platforms published by industrial companies [1][2][5][6] and research institutes [4]

Acknowledgement

This work was supported by the RF2THZ SiSoC project of the EUREKA program CATRENE in which the French partners are funded by the Ministry of Industry. The authors wish to acknowledge the staff of the 300 mm Si plant and the Silicon Technology Development teams in STMicroelectronics Crolles involved in the various aspects of this work. They also thank the management support of F. Boeuf, O. Noblanc, B. Sautreuil and R. Fournel.

References

[1] B.A. Orner et al, IEEE BCTM Proc., 2003, pp. 203–206
[2] G. Avenier et al, IEEE J. Solid State Circuits, 2009, pp. 2312–2321
[3] P. Chevalier et al, IEEE BCTM Proc., 2011, pp. 57–65
[4] H. Rücker et al, IEEE SiRF Proc., 2012
[5] K. Kuhn et al, IEEE IEDM Tech. Dig., 2002, pp. 73–76
[6] J.J. Pekarik et al, IEEE BCTM Proc., 2014, in press
[7] F. Arnaud et al, IEEE VLSI Tech. Dig., 2004, pp. 10–11
[8] P. Chevalier et al, IEEE CSICS Proc., 2012

MOS Capacitor Deep Trench Isolation for CMOS Image Sensors

N. Ahmed[1,2], F. Roy[1], G-N.Lu[2], B. Mamdy[1,2], J-P. Carrere[1], A. Tournier[1], N. Virollet[1], C. Perrot[1],
M. Rivoire[1,3], A. Seignard[3], D.Pellissier-Tanon[1], F. Leverd[1], B. Orlando[1]

[1] STMicroelectronics, 850, rue J. Monnet, BP. 16, 38921 Crolles, France
[2] INL-UMR5270, CNRS, Villeurbanne, F-69622, France
[3] CEA-LETI, 17 rue des Martyrs, 38054 Grenoble, Cedex 9, France
Tel : (+33) 4-3892-2741, E-mail: *nayera.ahmed@st.com*

Abstract

This paper proposes the integration of MOS Capacitor Deep Trench Isolation (CDTI) as a solution to boost image sensors' pixels performances. We have investigated CDTI and compared it to oxide-filled Deep Trench Isolation (DTI) configurations, on silicon samples, with a fabrication based on TCAD simulations. The experiment measurements evaluated on CDTI without Sidewall Implantation (SWI) exhibit very low dark current (~1aA at 60°C for a 1.4μm pixel), high full-well capacity (~12000e-), and it shows quantum efficiency improvement compared to DTI configuration. Pixels with optimized CDTI gate oxide thickness have demonstrated comparable angular response to oxide-filled DTI counterparts.

Introduction

With the continuous scaling down of pixel size of CMOS image sensors (CIS) [1], the crosstalk issue becomes critical. One solution is to introduce Deep Trench Isolation (DTI) between pixels, which suppresses electrical crosstalk resulting from diffusion of photo-generated carriers and reduces optical crosstalk thanks to the pixel Sidewall (SW) acting as light guide [2-5]. However, DTI integration in the pixel presents Si/SiO2 interface, which should be passivated by pinning the silicon surface Fermi level (E_f) close to either valence band or conduction band. To do so, Sidewall Implantation (SWI) is currently used [6]. However, it is difficult to obtain a desired doping profile for the DTI due to its very large aspect ratio. Moreover, SWI will also affect other pixel's performances such as Full-Well Capacity (FWC) and the size of the collecting region of photo-generated carriers.

We propose here an alternative by integrating MOS Capacitor Deep Trench Isolation (CDTI), which is also an efficient solution for crosstalk problem. It allows active SW interface passivation by biasing the CDTI in accumulation regime, without the need of SWI. Accordingly, better performances can be achieved.

Innovative solution and its integration

The proposed CDTI consists in filling trenches with doped poly-silicon. Its fabrication steps are inserted in front-end CIS process, as shown in Fig. 1. It should be mentioned that no extra mask is needed for the CDTI integration compared to the DTI one. Fig. 2 shows a fabricated CDTI 1.4-μm pixel structure. TCAD simulations have been performed on four pixel configurations for comparison: (a) DTI with SWI and additional shallow sidewall implantation (SSWI) around photodiode (PD); (b) CDTI with SWI and SSWI; (c) DTI with SWI only; (d) CDTI with neither SWI nor SSWI. It is noted that the additional SSWI serves to stop spreading of PD depletion region to SW interface.

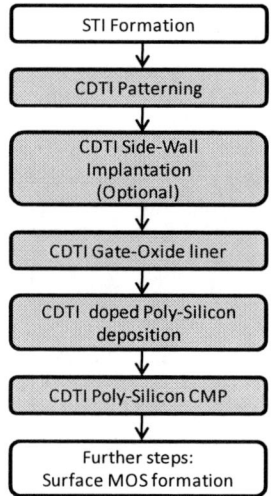

Fig.1: CDTI-integration process flow

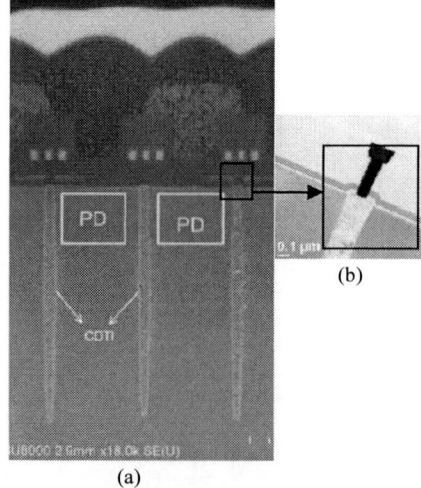

Fig.2: Cross-section of fabricated pixel with CDTI, a)Using Scanning Electron Microscope(SEM), b)Using Transmission Electron Microscopy(TEM)

Fig. 3 presents doping profile (in color change) of these four configurations. Fig. 4a draws simulated distributions of hole density near SW interface for these configurations. Thanks to MOS effect, CDTI (in accumulation) configurations provide a much higher hole density at SW interface than DTI. In case of CDTI without SWI, the hole population is concentrated at the interface, leaving a larger region for collection of photo-generated carriers. Fig. 4b shows the flat-band voltage (V_{FB}) at which holes start to be well accumulated at the Interface.

Pixels designs corresponding to these four configurations have then been fabricated and characterized. Fig. 5 shows that CDTI allows a clear improvement in the statistical distributions of pixel dark current (I_{dark}). In CDTI configurations, I_{dark} is monitored by the CDTI gate bias voltage (Fig. 5b). It is minimized to only 1aA at 60°C when MOS CDTI is in strong accumulation mode. These CDTI configurations, even without SW Implantations, have much lower I_{dark} than DTI counterparts. This may be explained by better active SW electrical passivation created by pinning the E_f due to much higher interface hole density.

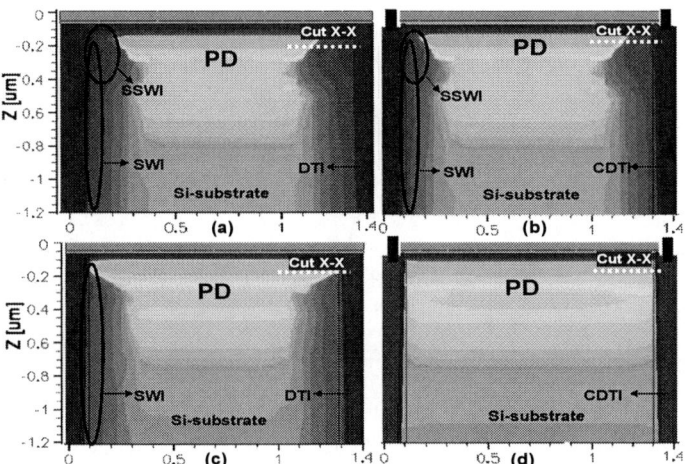

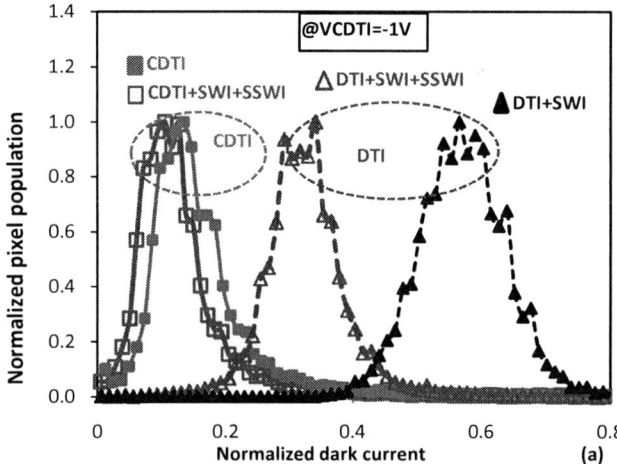

Fig.3: 3D-TCAD simulation, doping profiles are calibrated by secondary ion mass spectroscopy (SIMS), a) DTI+SWI+SSWI, b) CDTI+SWI+SSWI, c) DTI+SWI, d) CDTI

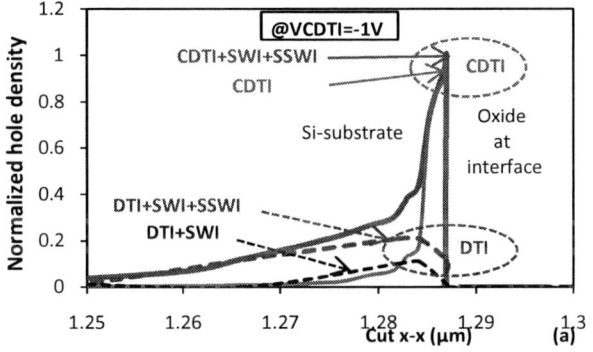

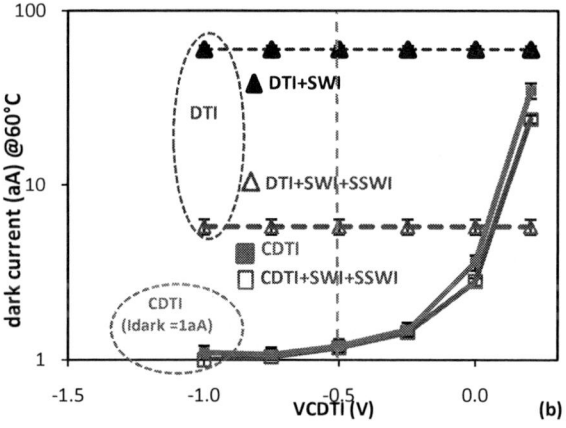

Fig. 5: Statistical distributions of pixel dark current: a) @ VCDTI=-1V, b) dark current *VS.* VCDTI @ 60°C

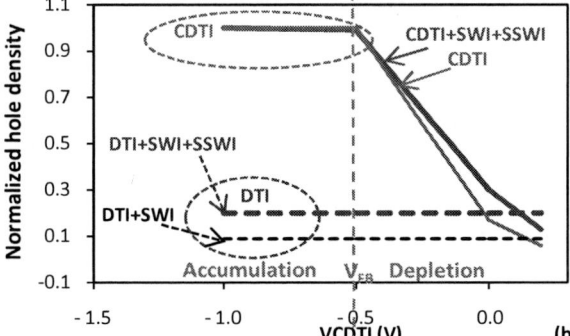

Fig.4: 3D –TCAD simulation of hole density near deep trench Interface Si/Sio$_2$@Z=-0.2μm: a) @VCDTI=-1V, b) In function of VCDTI

From Fig. 6, we can see that the DTI with only SWI configuration has activation energy (E_a) of 0.68eV, indicating dominant thermal generation mechanism of dark current due to a depleted interface.

This finding is consistent with Fig. 3c & 4a. It accounts for a much higher I_{dark} level as shown in Fig. 5. The three other configurations have an E_a of 1.1eV, which implies dominant diffusion mechanism of I_{dark}, and good interface holes accumulation.

978-1-4799-8002-4/14 $31.00 © 2014 IEEE   81

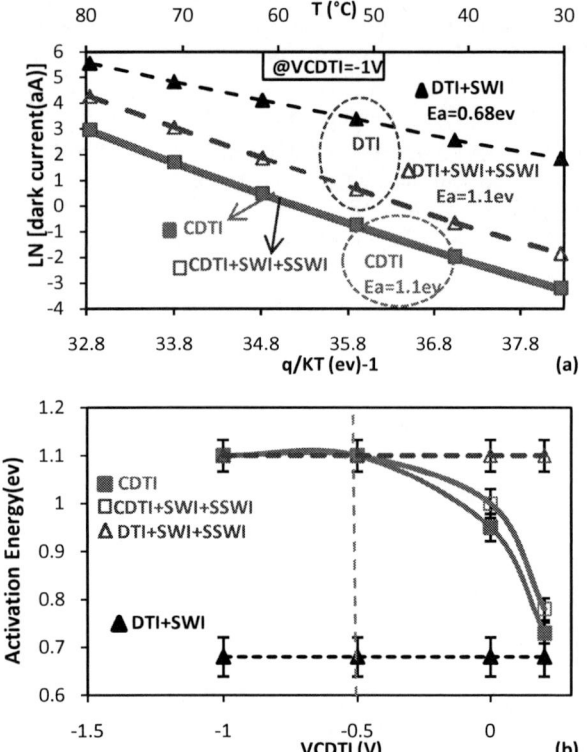

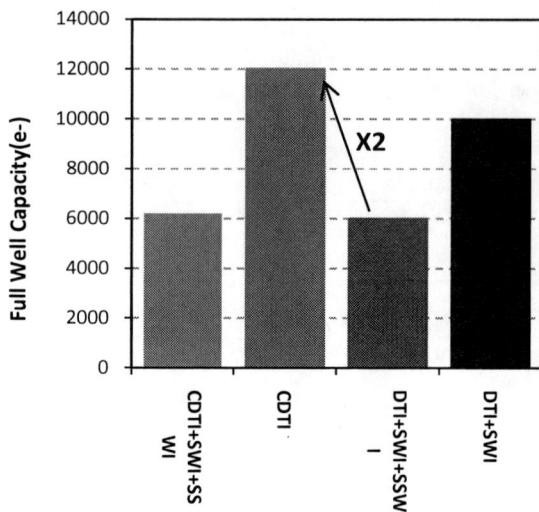

Fig.8: Full well capacity @VCDTI=-1V

Fig. 9 compares, in terms of quantum efficiency (QE), two selective configurations: DTI with SWI and SSWI versus CDTI alone. QE for CDTI is 6% better for blue and 3% for green, than the DTI with SWI and SSWI. This difference may be explained by SWI effect on the collection efficiency of photo-generated carriers.

Fig. 6: Activation energy (E$_a$) measurements, with temperatures ranging from 30 to 80°C: a) @VCDTI=-1V, b) E$_a$ VS. VCDTI

The evolution of E$_a$ versus VCDTI confirms that the transition from depleted interface towards a hole accumulated SW occurs around VCDTI=V$_{FB}$ (Fig. 6b).

Fig. 7 exhibits the transfer characteristic of the four configurations. Both CDTI and DTI without SSWI have doubled FWC compared to the two other configurations: ~12000e- vs ~6000e- (Fig. 8). This means that SSWI degrades FWC of the PD by restricting its extension region.

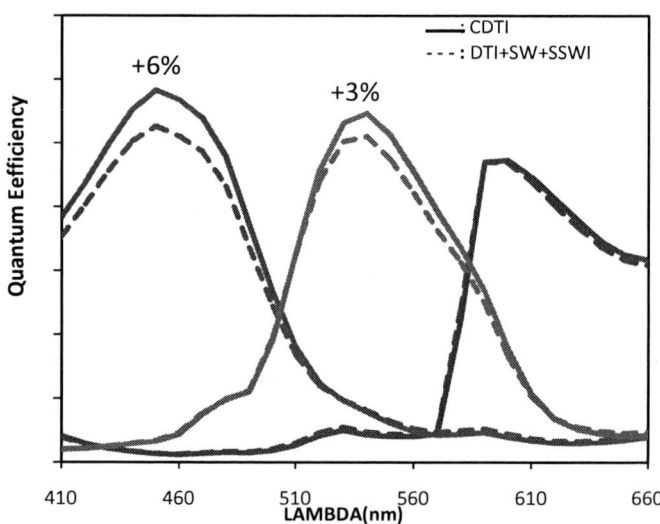

Fig. 9: QE curves comparison between with CDTI and with DTI+SWI+SSWI. Clear gain without crosstalk problem is observed

Process optimization targeting optical crosstalk through frustrated reflection has also been carried out. Fig. 10b shows angular response of the CDTI configuration with two oxide thicknesses, in comparison with that of DTI+SWI+SSWI configuration.

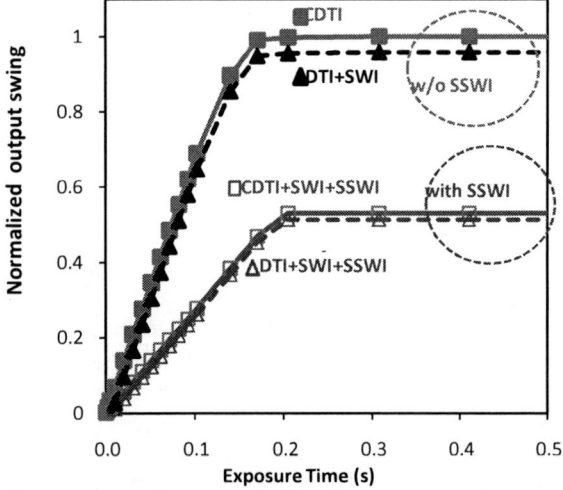

Fig.7: Transfer characteristic @VCDTI=-1V

978-1-4799-8002-4/14 $31.00 © 2014 IEEE

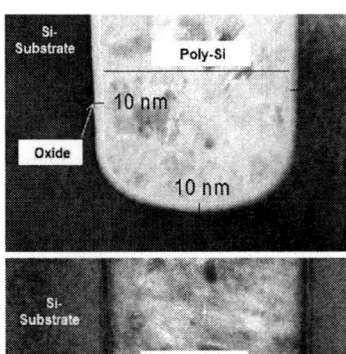

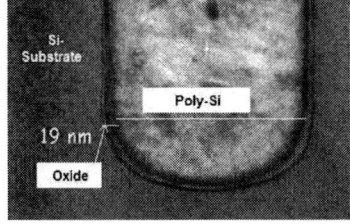

(a)

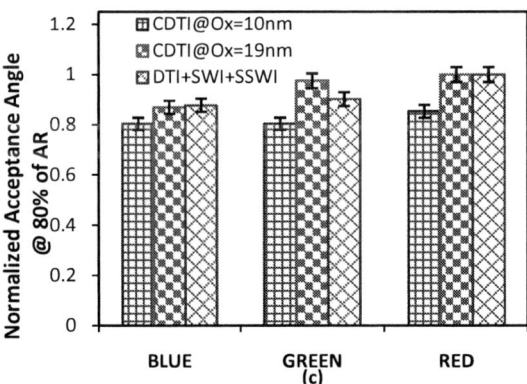

(b)

(c)

Fig. 10: a) Cross-section of fabricated CDTI with the two oxide thicknesses, b) Acceptance Angle curves for GREEN pixels, c) Normalized acceptance angle calculated at 80% of angular response

As in Fig. 10b&10c, optimized CDTI with 19nm gate oxide has comparable angular response with DTI 200nm oxide filling. It should be mentioned that results previously shown on the CDTI configuration correspond to the optimized 19 nm oxide case. Fig. 11 compares global performances between CDTI and DTI configurations.

The CDTI outperforms the DTI in all the aspects, with significant improvements in I_{dark} and FWC.

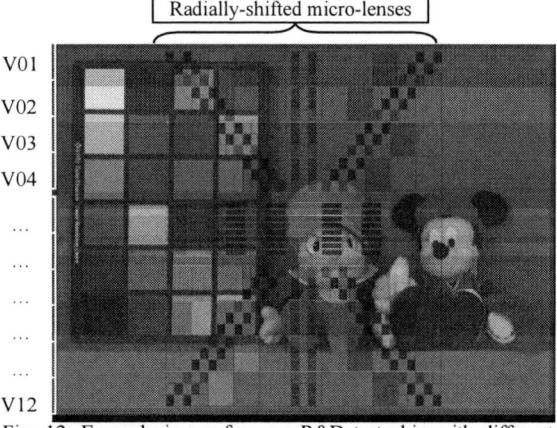

Fig. 11: Overall normalized pixel performances

Fig. 12 presents a picture taken from a CDTI CIS, which integrates, for the testing, different parametric variations, including off-axis variations of optical components.

Fig. 12: Example image from an R&D test chip with different parametrical variations, including off-axis variations of optical components

Conclusion

In this paper the successful integration of CDTI has been reported, featuring accumulation-biased active SW interface passivation which allows much lower dark current with improved full-well capacity and quantum efficiency, compared to oxide-filled DTI. CDTI with optimized gate oxide thickness is as efficient as DTI to deal with crosstalk. This technology is promising to boost the performance of different pixel types and sizes, especially below 1μm pixel pitch.

References

[1] J.C. Ahn, et al., "Advanced Image Sensor Technology for Pixel Scaling Down Toward 1.0μm,"IEDM, pp.275-278, 2008.
[2] A. Tournier et al., "Pixel-to-Pixel Isolation by Deep Trench Technology: Application to CMOS Image Sensor," IISW, 2011.
[3]Koen De Munck et al., "Backside Illuminated Hybrid FPA achieving Low Cross-Talk combined with High QE," IISW, 2011.
[4] Y. Kitamura et al., "Suppression of Crosstalk by Using Backside Deep Trench Isolation for 1.12μm Backside Illumination CMOS Image Sensor," IEDM,pp.24.2.1-24.2.4, 2012.
[5]J. Ahn, et al., "A 1/4-inch 8Mpixel CMOS Image Sensor with 3DBackside-Illuminated 1.12im Pixel with Front-Side Deep-Trench Isolation and Vertical Transfer Gate, "ISSCC, pp.124-125, 2014.
[6] H.In Kwon et al., "The analysis of dark signal sin the CMOS APS imagers from the characterization of test structures," IEDM, pp.178-184,2004.

Three-Dimensional Integrated CMOS Image Sensors with Pixel-Parallel A/D Converters Fabricated by Direct Bonding of SOI Layers

Masahide Goto[1], Kei Hagiwara[1], Yoshinori Iguchi[1], Hiroshi Ohtake[1], Takuya Saraya[2],

Masaharu Kobayashi[2], Eiji Higurashi[2], Hiroshi Toshiyoshi[2], and Toshiro Hiramoto[2]

[1] NHK Science and Technology Research Laboratories, 1-10-11 Kinuta, Setagaya-ku,

Tokyo 157-8510, Japan.   Email: goto.m-fk@nhk.or.jp

[2] The University of Tokyo, Tokyo, Japan

Abstract

We report the first demonstration of three-dimensional (3D) integrated CMOS image sensors with pixel-parallel A/D converters (ADCs). Photodiode (PD) and inverter layers were directly bonded with the damascened Au electrodes to provide each pixel with in-pixel A/D conversion. We designed ADC with a pulse frequency output and fabricated a prototype sensor with 64 pixels. The developed sensor successfully captured video images and confirmed excellent linearity with a wide dynamic range of more than 80 dB, which showed feasibility of pixel-level 3D integration for high-performance CMOS image sensors.

Introduction

The resolution and the frame rate of CMOS image sensors have increased to meet the escalating demand toward high-reality video systems [1]. Conventional CMOS image sensors have photodetectors and signal processors on the same substrate plane for column-parallel signal processing [2-4]; many pixels in each column time-share a signal processor, thereby making it difficult to improve the signal processing speed. Pixel-parallel signal processing is the best solution to this problem [5-7], in which every pixel has its own signal processor. Although a pixel-parallel sensor was reported with a high frame rate of 10,000 frame/s [7], the spatial resolution was limited because of the large area of signal processor to be implemented in a plane.

To overcome these problems, we have proposed the 3D integrated CMOS image sensor as illustrated in Fig. 1 [8-11]. The sensor has photodetectors and signal processors implemented in different layers that have been vertically stacked. The signal in each pixel is vertically transferred and processed for the pixel-parallel signal processing. Since the

spatial resolution is not degraded by 3D stacking, both high resolution and high frame rate are achieved at the same time.

Several different types of 3D stacked image sensors have recently been reported [12-20] by using through-silicon vias (TSVs) and microbump technology. However, a TSV or microbump was shared by multiple pixels, and pixel-parallel signal processing was not realized in such sensors, because the diameter of the TSVs or bumps was larger than the pixel size of a few micrometers or less.

In this paper, we demonstrate for the first time pixel-parallel 3D integrated CMOS image sensors. Our direct bonding method with damascened Au electrodes offers pixel-wise electrical interconnection. The developed sensor shows excellent linearity with a wide dynamic range thanks to the in-pixel ADC with a pulse output, thereby indicating its suitability for next-generation high-performance image sensing.

Structure, Design, and Fabrication

Fig. 2 shows a schematic diagram of the developed CMOS image sensor pixel. The PD and inverters formed on separate SOI layers are bonded with Au electrodes. In-pixel ADC with a pulse frequency output was designed to include a PD, inverters, and a reset transistor as shown in Fig. 3. When the PD potential (V_{PD}) reaches the threshold voltage of the inverter by illumination, a pulse is output and the PD is simultaneously reset. By repeating this operation, pulses whose number corresponds to the illumination intensity are obtained as shown by the SPICE simulation results in Fig. 4. The pixel can directly convert V_{PD} to a digital signal as the pulses to ensure high noise tolerance. In addition, the sensor can handle a wide range of illumination because the ADC has no saturation limit of the PD capacity.

978-1-4799-8002-4/14 $31.00 © 2014 IEEE

The fabrication process for the 3D integrated sensor is shown in Fig. 5. (a) FETs and aluminum electrodes are formed on the FDSOI wafers by the 0.2-μm-rule 1poly-3metal SOI-CMOS process. The thicknesses of the SOI and the gate oxide are 50 nm and 4.4 nm, respectively. (b) An intermediate SiO_2 layer was patterned to form via holes on aluminum electrodes. After a Ti/Au seed layer was sputtered, a Au layer was formed by electroplating. (c) Damascene process was used with the chemical mechanical polishing (CMP) to form a flat surface of Au/SiO_2. (d) After dicing into a 20 mm square, the chip is surface-activated by Ar plasma at 200 W for 30 s and O_2 plasma at 250 W for 30 s before the bonding process. (e) The inverter chip and an identically processed PD chip were aligned using an infrared microscope and then directly bonded at a force of 2000 N (a pressure of 5 MPa) and a temperature of 200 °C for 60 min. (f) The handle layer of the PD chip was thinned to 35 μm by grinding, and finally removed by XeF_2. Unlike typical TSVs, this direct bonding process allows Au interconnect electrodes of 1 μm or less. Stacking more than three layers is also possible by repeating the bonding and handle-layer removal processes. Therefore, the developed process is suitable for pixel-parallel image sensors.

Fig. 6 shows photographs of the Au and SiO_2 surfaces after CMP. (a) The diameter of the Au electrodes was set to 10 μm. (b) Average roughness (Ra) of SiO_2 was found to be 0.15 nm. Photograph of the pixel array before bonding is shown in Fig. 7. The sensor consists of 64 pixels, and each pixel size is 80 μm square with four Au electrodes. Figs. 8 and 9 show photographs of the bonded chips and its cross-sectional image. The final thickness of the upper PD chip was 6.5 μm. No voids were observed at the bonded interface. Photograph of a stacked pixel array is shown in Fig. 10, where PD with 60 μm square is stacked on inverters.

Measurement results

PD current generated by illumination was measured as shown in Fig. 11. The photocurrent increased in proportional to the illumination intensity. The developed image sensor was evaluated on the measurement setup shown in Fig. 12, where output pulses were counted by pulse counters on an FPGA circuit board. Fig. 13 shows the measured output of the sensor, which confirmed that the pulses corresponded to the illumination intensity as designed. The input-output

characteristics of the sensor are shown in Fig. 14. We obtained a linear output with a wide dynamic range of more than 80 dB, which exceeded the dynamic range of 60 dB of the conventional sensor. The sensor could be made more sensitive to lower illumination and to enhance its dynamic range to more than 100 dB by applying a thicker SOI layer for the PD. Fig. 15 shows examples of video images captured by the developed sensor, where alphabet letters printed in black on a transparent film were used as target objects. We thus successfully confirmed the operation of the first pixel-parallel 3D integrated CMOS image sensor.

Conclusion

3D integrated CMOS image sensors with pixel-parallel ADCs were demonstrated by stacking PD and inverter layers. The developed sensor successfully operated to capture video images. Excellent linearity with a wide dynamic range of more than 80 dB was confirmed, indicating that the in-pixel ADC improved the sensor performance. The sensor is promising to develop high-performance image sensors for the next generation.

References

[1] T. Watabe et al., ISSCC, p. 388, 2012.

[2] E. R. Fossum, IEEE T-ED, Vol. 44, p. 1689, 1997.

[3] J. Park et al., IEEE T-ED, Vol. 56, p. 2414, 2009.

[4] M. Shin et al., IEEE T-ED, Vol. 59, p. 1693, 2012.

[5] D. X. D. Yang et al., IEEE J-SSC, Vol. 34, p. 1821, 1999.

[6] F. Andoh et al., IEEE T-ED, Vol. 47, p. 2123, 2000.

[7] S. Kleinfelder et al., IEEE J-SSC, Vol. 36, p. 2049, 2001.

[8] M. Goto et al., IEEE S3S, 11.2, 2013.

[9] K. Hagiwara et al., Waferbond, p. 125, 2013.

[10] M. Goto et al., IEEJ TEEE, Vol. 9, p. 329, 2014.

[11] M. Goto et al., IEEE T-ED, Vol. 61, p. 2886, 2014.

[12] J. Aoki et al., ISSCC, p. 482, 2013.

[13] S. Sukegawa et al., ISSCC, p. 484, 2013.

[14] J.C. Liu et al., VLSI Tech, 21.3, 2014.

[15] K. Lee et al., IEEE T-ED, Vol. 60, p. 3842, 2013.

[16] K. Kiyoyama et al., 3DIC, 5-1, 2011.

[17] X. Zhang et al., IEEE Sensors, p. 1933, 2011.

[18] A. Xhakoni et al., IISW, P27, 2011.

[19] T. Asano et al., ECTC, p. 40, 2009.

[20] V. Suntharalingam et al., ISSCC, p. 38, 2009.

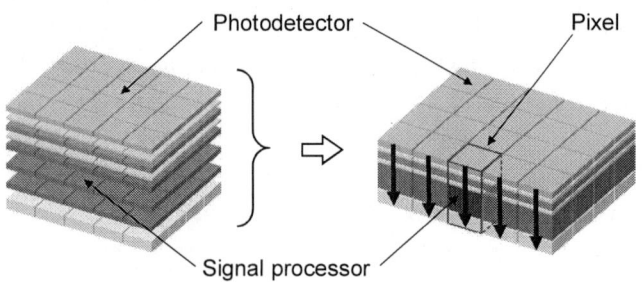

Fig. 1 – Schematic diagram of 3D integrated CMOS image sensor. The photodetectors and signal processors in the different layers are vertically stacked. The signal in each pixel is vertically transferred and processed, enabling pixel-parallel signal processing.

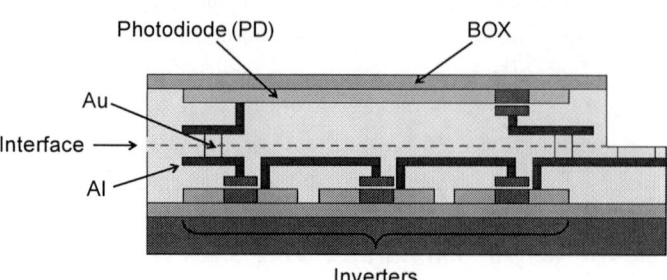

Fig. 2 – Conceptual diagram of developed CMOS image sensor pixel. The photodiode (PD) and inverters formed on separate SOI layers are bonded with Au electrodes.

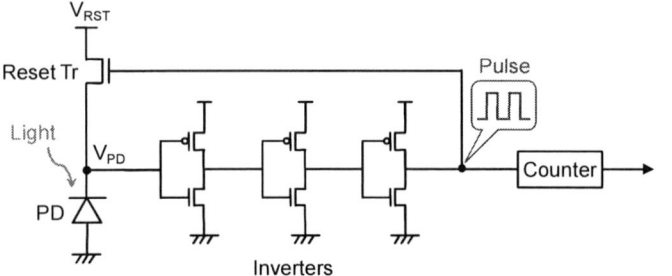

Fig. 3 – Circuit diagram of in-pixel A/D converter (ADC) with a pulse frequency output, which consists of a PD, inverters, and a reset transistor. Since the ADC has no saturation limit of the PD capacity, the sensor can handle a wide range of illumination.

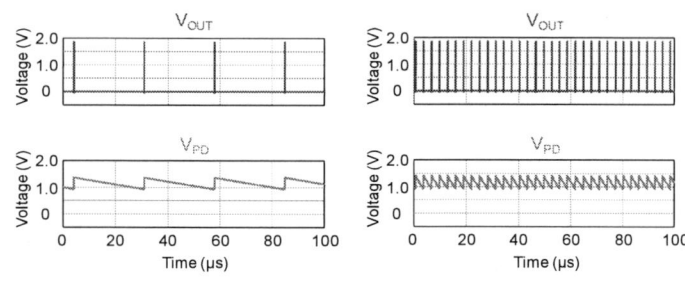

(a)   (b)

Fig. 4 – SPICE simulation results for designed ADC.
(a) Lower illumination. (b) Higher illumination.
A larger number of pulses is generated for a higher illumination.

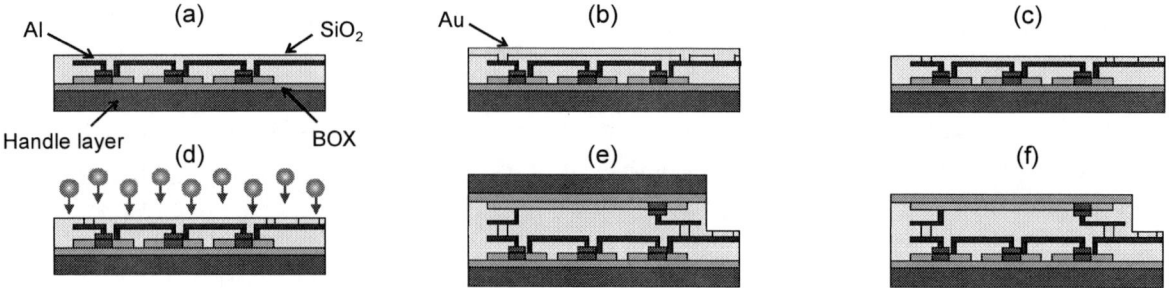

Fig. 5 – Fabrication process of 3D integrated CMOS image sensor.
(a) FDSOI FET forming. (b) Via hole forming and Au electroplating. (c) Damascene process by CMP. (d) Plasma treatment.
(e) Direct bonding. (f) Handle layer removing.

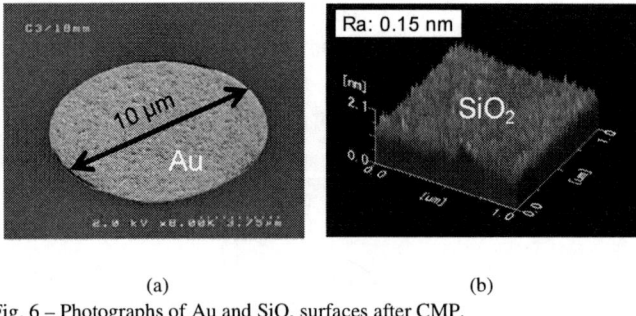

(a)   (b)

Fig. 6 – Photographs of Au and SiO$_2$ surfaces after CMP.
(a) SEM image of Au electrode with a diameter of 10 μm.   (b) Atomic force microscope (AFM) image of SiO$_2$ surface.

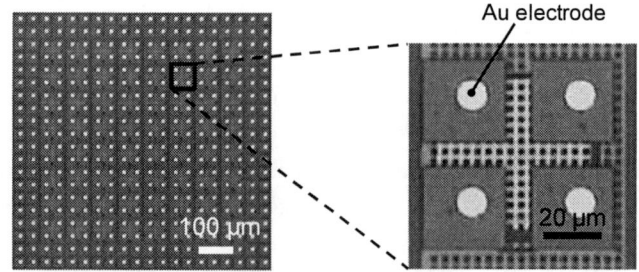

Fig. 7 – Photograph of developed CMOS image sensor pixel array before bonding. The sensor consists of 64 pixels, and each pixel size is 80 μm square with four Au electrodes.

978-1-4799-8002-4/14 $31.00 © 2014 IEEE   86

(a)　　　　　　　　　(b)　　　　　　　　　(c)

Fig. 8 – Photographs of bonded chips.
(a) After bonding. (b) After grinding. (c) After XeF$_2$ etching.
Handle layer was thinned to 35 μm by grinding, and removed by XeF$_2$.

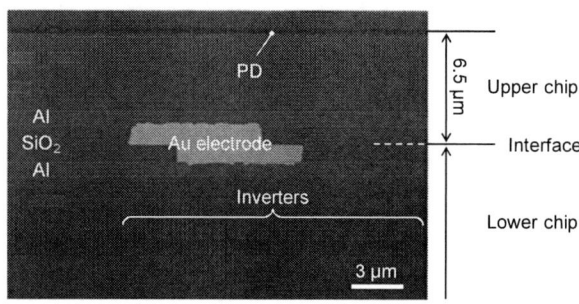

Fig. 9 – Cross sectional SEM image of bonded CMOS image sensor pixel. The upper chip was thinned to 6.5 μm. No voids were observed at the bonded interface.

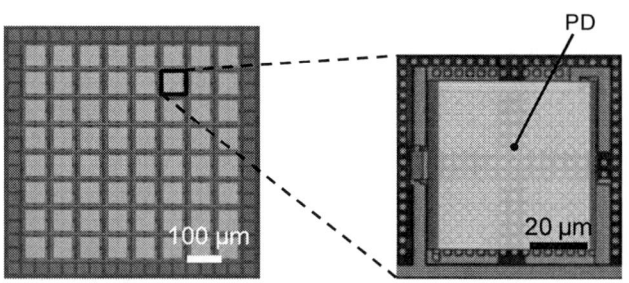

Fig. 10 – Photograph of bonded CMOS image sensor pixel array. PD with 60 μm square is stacked on inverters.

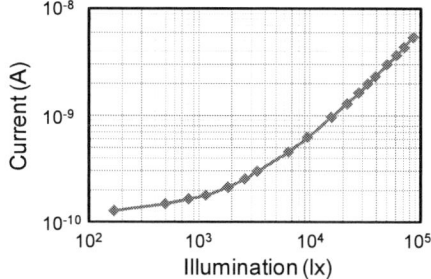

Fig. 11 – PD output current as a function of illumination. The photocurrent increased in proportional to the illumination intensity.

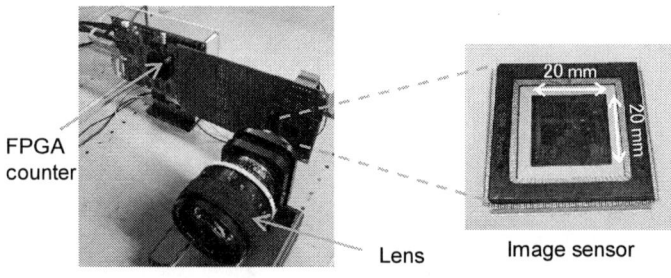

Fig. 12 – Photograph of measurement setup for developed sensor. Bonded chips were packaged to form an image sensor. Output pulses were counted by pulse counters on an FPGA circuit board.

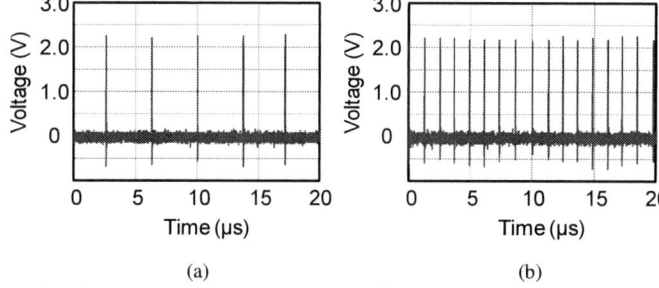

(a)　　　　　　　　　(b)

Fig. 13 – Measured output of 3D integrated CMOS image sensor.
(a) 3×10^5 lx. (b) 1×10^6 lx. The output of pulses whose frequency corresponded to the illumination intensity was confirmed.

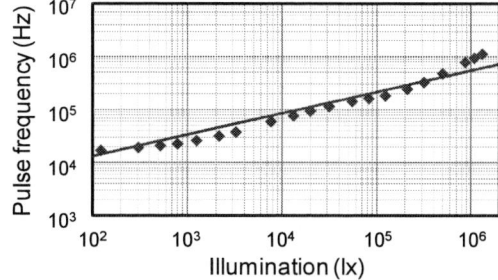

Fig. 14 – Input-output characteristics of 3D integrated CMOS image sensor, where the number of counted pulses was converted to the output frequency. We obtained a linear output with a wide dynamic range of more than 80 dB, which exceeded the dynamic range of 60 dB of the conventional sensor.

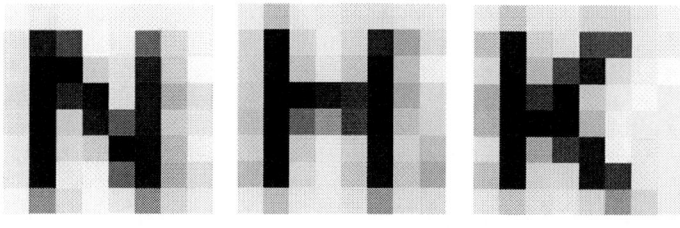

Fig. 15 – Examples of video images captured by the developed sensor, where alphabet letters printed in black on a transparent film were used as target objects. We successfully confirmed the operation of the first pixel-parallel 3D integrated CMOS image sensor.

High Sensitivity Image Sensor Overlaid with Thin-Film Crystalline-Selenium-based Heterojunction Photodiode

S. Imura[1*], K. Kikuchi[1], K. Miyakawa[1], H. Ohtake[1], M. Kubota[1], T. Nakada[2],
T. Okino[3], Y. Hirose[3], Y. Kato[3], and N. Teranishi[4]

[1]NHK Science and Technology Research Laboratories, 1-10-11 Kinuta, Setagaya-ku, Tokyo 157-8510, Japan
[2]Tokyo University of Science, SIC-2-206, 5-4-30 Nishihashimoto, Sagamihara, Kanagawa 252-0131, Japan
[3]Panasonic Corporation, 1 Kotari-yakemachi, Nagaokyo City, Kyoto 617-8520, Japan
[4]University of Hyogo, 1-1-2 Koto, Kamigori, Ako-gun, Hyogo 678-1205, Japan
[*]Tel: +81-3-5494-3214, Fax: +81-3-5494-3278, Email: imura.s-la@nhk.or.jp

Abstract

We have developed a stacked image sensor based on a thin-film crystalline selenium (c-Se) heterojunction photodiode. Uniform c-Se-based photodiodes laminated on complementary metal-oxide-semiconductor circuits were fabricated by tellurium-diffused crystallization. We herein present the first high-resolution images obtained with such a device. Furthermore, the dark current was significantly decreased by using an n-type wide-band-gap gallium oxide (Ga_2O_3) layer with a high hole-injection barrier.

Introduction

Because of the demand for ultrahigh-definition video cameras, high-resolution imaging devices are currently an area of rapid development. For example, the Japan Broadcasting Corporation, NHK, is currently researching Super Hi-Vision (SHV) [1], which has 8 K resolution (7680 × 4320 pixels), for next-generation broadcasting systems. The goal is to provide images so realistic that a physical presence is conveyed to the viewer. However, because of the reduced pixel size of complementary metal-oxide-semiconductor (CMOS) image sensors, the amount of light received per pixel has also significantly decreased. Therefore, a technology to markedly enhance the sensitivity of image sensors is highly desirable.

Avalanche multiplication by the impact ionization of photogenerated carriers in a photoconversion film is one of most promising approaches to achieve high sensitivity. Amorphous selenium (a-Se) has been used as a photoconversion layer in high-gain avalanche-rushing amorphous photoconductor (HARP) photodetectors [2], which are characterized by ultrahigh sensitivity and low noise. However, HARP photodetectors still suffer from poor long-wavelength spectral response and require high operation voltage for avalanche multiplication.

Owing to the excellent spectral response of crystalline-selenium-based (c-Se-based) photodiodes over the entire visible spectrum (see Fig. 1) [3], this material is an attractive alternative to a-Se for use in photoconversion layers in CMOS image sensors. Tellurium-diffused (Te-diffused) crystallization techniques [4] make it possible to produce uniform c-Se films, enabling the combination of c-Se films and CMOS circuits. Because the optical absorption coefficient of c-Se is over one order of magnitude greater than that of silicon (Si) throughout the visible spectrum, the thickness of c-Se can be reduced to less than 500 nm (see Fig. 2), which is advantageous for image sensors with a smaller pixel size.

In addition, a structure consisting of stacked CMOS image sensors overlaid with a thin-film photoconversion layer has several advantages, such as a high aperture ratio, low cross talk, and high sensitivity. Another significant advantage is that it has a high saturation level (or dynamic range) because the storage node of the photoconverted charge is separated from the photoconversion layers [5]. However, with the typical three-transistors pixel circuit configuration, they suffer from the reset noise due to incomplete charge transfer. Recently, it has been shown that the above noise can be suppressed down to commercial device level by employing the column feedback amplifier [6]. With this technology, it is now possible to apply various photoconversion materials to image sensors.

In the present study, we use c-Se to demonstrate visible-light CMOS image sensors overlaid with thin-film photoconversion layers. We also present the first high-resolution images acquired with such a device.

Device Fabrication

Figures 3(a) and 3(b) show a three-dimensional (3D) schematic and a schematic cross section of the photodiode structure fabricated on a CMOS circuit, respectively. We deposited an n-type gallium oxide (Ga_2O_3) layer over the pixel electrodes of CMOS circuits by radio-frequency magnetron sputtering. To prevent the Se film from peeling during annealing, an ultrathin Te layer was deposited by vacuum evaporation onto the Ga_2O_3 layer prior to Se deposition [4]. Vacuum evaporation was also used to deposit a 500-nm-thick a-Se film, followed by annealing at 200°C to completely convert a-Se into c-Se, as indicated in Fig. 4. Next a 30-nm-thick indium tin oxide (ITO) film was deposited by direct-current magnetron sputtering to serve as transparent electrodes. The reverse bias is defined as the pixel electrode being positive with respect to the ITO electrode. All fabrication was performed at a relatively low temperature (<200°C), which is advantageous for CMOS manufacturing. Figure 5 shows a photograph of the final stacked c-Se-based image sensor, which has 992 × 636 pixels, with each pixel being 3 μm × 3 μm over an active sensor area of 2.95 mm × 2.10 mm.

Fixed-Pattern Noise

Figures 6(a)–6(d) show surface images acquired by scanning electron microscopy (SEM) of c-Se films with thicknesses of 2 µm, 1 µm, 500 nm, and 200 nm, respectively, on glass substrates. These SEM images show that the c-Se films consist of polycrystalline particles whose size depends on the thickness of the c-Se film because the particle size gradually increases from the interface between Ga_2O_3 and c-Se during the crystallization.

Figures 7(a) and 7(b) show images captured by fabricated image sensors with 2-µm-thick and 500-nm-thick c-Se films, respectively. When some of the c-Se polycrystalline particles have a size comparable to the 3 µm × 3 µm pixel size, a fixed-pattern noise appears over the entire image [see Fig. 7(a)]. This comparison shows that to obtain a clear image, the polycrystalline particles that make up the c-Se film must be much smaller than the pixels of the image sensor.

Device Characteristics

A. Dark-Current Characteristics

Figure 8 shows the dark current as a function of integration time for image sensors overlaid with metal/c-Se Schottky diodes with 1- and 0.1-nm-thick Te layers at a reverse-bias voltage of 3 V at room temperature. Tellurium acts as a shallow interface acceptor [7], which contributes to the dark current of the photodiode. The results show that the dark-current characteristics strongly depend on the thickness of the Te layer and that the dark current decreases with decreasing Te-layer thickness. Therefore, the Te layer should be made as thin as possible. On the basis of these results, we reduced the thickness of the Te layer to 0.1 nm (within the precision allowed by our deposition process).

Figure 9 shows the dark current as a function of integration time for image sensors overlaid with a metal/c-Se Schottky photodiode and with an n-Ga_2O_3/p-c-Se pn photodiode at a reverse-bias voltage of 3 V at room temperature. Measurements reveal that the dark current of the Ga_2O_3/c-Se pn photodiode is significantly lower than that of the Schottky photodiode. This is attributed to the large valence-band offset between the metal electrode and Ga_2O_3, which acts as an effective barrier against hole injection from a metal electrode under reverse bias (see Fig. 10).

B. Dependence of Ga_2O_3 Layer Thickness on Signal Current

Figure 11 shows the sensitivity as a function of reverse bias for illuminated Ga_2O_3/c-Se pn photodiodes with 2- and 10-nm-thick Ga_2O_3 layers and a 0.1-nm-thick Te layer on a CMOS circuit. The depletion layer between Ga_2O_3 and c-Se mainly spreads into the Ga_2O_3 layer because of the smaller carrier concentration in the Ga_2O_3 layer than that in the c-Se layer; thus an external applied voltage is required for the depletion layer to spread into the c-Se layer. These results clearly indicate that, with decreasing Ga_2O_3-layer thickness, the sensitivity of the image sensor effectively increases at a lower applied voltage.

C. External Quantum Efficiency

Figure 12 shows the spectral external quantum efficiency (EQE) for an image sensor overlaid with a 10-nm-thick Ga_2O_3/500-nm-thick c-Se heterojunction photodiode at a reverse-bias voltage of 6 V. The EQE covers the entire visible spectrum, indicating the suitability of the device for use as a visible image sensor; however, the EQE is lower than that of the 10-nm-thick Ga_2O_3/500-nm-thick c-Se photodiode fabricated on a glass substrate (cf. Fig. 1). Further investigations of the crystallinity of c-Se films and the structure of the photodiodes are needed to optimize the EQE of c-Se-based photodiodes on CMOS circuits.

Conclusion

We reported the first high-resolution images obtained from a stacked image sensor based on a thin-film n-Ga_2O_3/p-c-Se heterojunction photodiode laminated on a CMOS circuit. By exploiting the structural advantages and excellent material properties of c-Se, including the capacity to generate avalanche multiplication [5], we demonstrated highly sensitive image sensors, opening the door to next-generation ultrahigh-definition imaging systems.

Acknowledgment

The authors would like to thank Y. Matsunaga at Panasonic Semiconductor Solutions Co., Ltd. for support and helpful discussions throughout this study.

References

[1] Recommendation ITU-R BT 2020, "Parameter values for ultra-high definition television systems for production and international programme exchange", 2012.

[2] K. Tanioka, J. Yamazaki, K. Shidara, K. Taketoshi, T. Kawamura, S. Ishioka, and Y. Takasaki, "An avalanche-mode amorphous selenium photoconductive layer for use as a camera tube target," *IEEE Electron. Device Lett.* **8**, 392-394, 1987.

[3] S. Imura, K. Kikuchi, K. Miyakawa, H. Ohtake, and M. Kubota, "Low-voltage-operation avalanche photodiode based on n-gallium oxide/p-crystalline selenium heterojunction" *Appl. Phys. Lett.* **104**, 242101, 2014.

[4] T. Nakada and A. Kunioka, "Efficient ITO/Se Heterojunction solar cells," *Jpn. J. Appl. Phys.* **23**, L587, 1984.

[5] M. Mori, Y. Hirose, M. Segawa, I. Miyanaga, R. Miyagawa, T. Ueda, et al., "Thin organic photoconductive film image sensors with extremely high saturation of 8500 electrons/µm²," *VLSI Tech. Symp.* p. 22, 2013.

[6] M. Ishii, S. Kasuga, K. Yazawa, Y. Sakata, T. Okino, Y. Sato, et al., "An ultra-low noise photoconductive film image sensor with a high-speed column feedback amplifier noise canceller," *VLSI Circuit*, p. 5, 2013.

[7] T. Kikuchi, Y. Ema, and T. Hayashi, "Rectification phenomena and photovoltaic effects in an amorphous $Se_{1-x1}Te_{x1}$-$Se_{1-x2}Te_{x2}$ heterostructure," *J. Appl. Phys.* **50**, 5043-5044, 1979.

978-1-4799-8002-4/14 $31.00 © 2014 IEEE

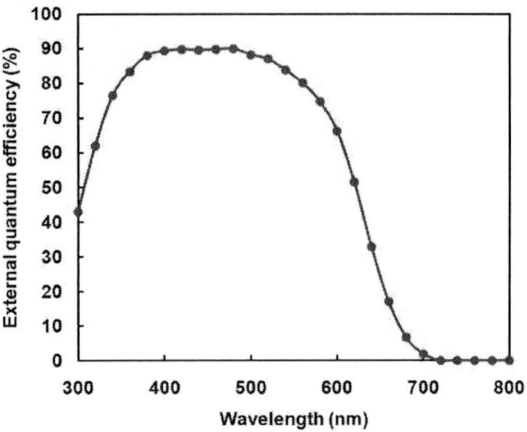

Fig. 1: Spectral external quantum efficiency for a 10-nm-thick-Ga$_2$O$_3$/500-nm-thick-c-Se heterojunction photodiode on a glass substrate at a reverse-bias voltage of 5 V. The EQE covers over entire visible region. The peak EQE for the photodiode is about 90%.

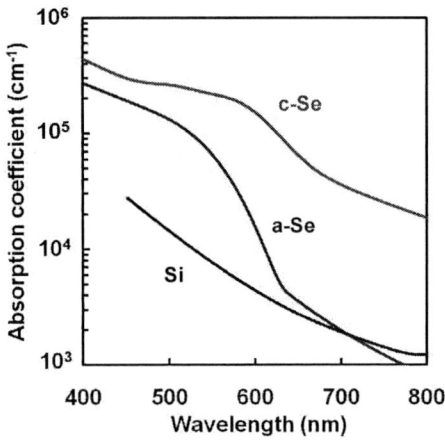

Fig. 2: Spectrum of absorption coefficient for c-Se, a-Se, and Si. The absorption coefficient of c-Se is much larger than that of a-Se and Si over the entire visible region.

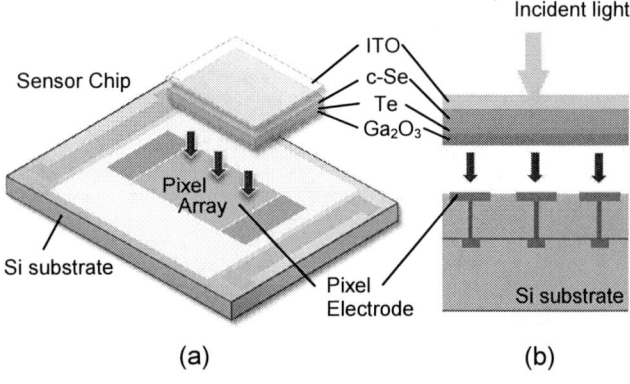

Fig. 3: (a) 3D schematic and (b) schematic cross section of stacked CMOS image sensor overlaid with an n-Ga$_2$O$_3$/p-c-Se heterojunction photodiode. The ITO transparent electrode is negatively biased with respect to the pixel electrode.

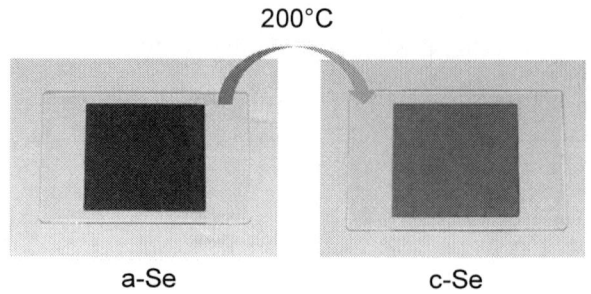

Fig. 4: Photograph of 500-nm-thick a-Se and c-Se. a-Se is converted into c-Se by annealing at 200°C. Because of the significant improvement in long-wavelength absorption, the color of the Se changes from red to dark gray noticeably during the crystallization process.

Fig. 5: Stacked CMOS image sensor consisting of 992 × 636 pixels with 3 μm × 3 μm pixels across an active sensor area of 2.95 mm × 2.10 mm.

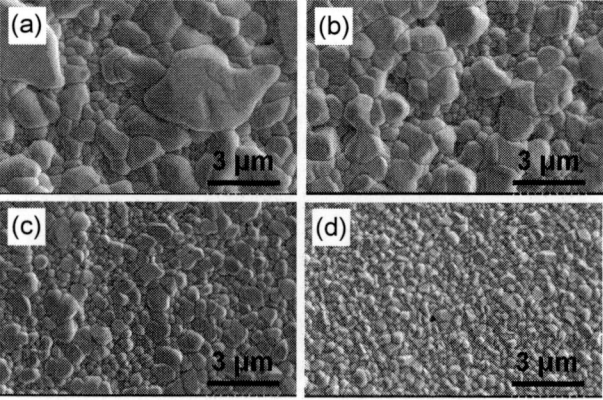

Fig. 6: SEM images of surface of c-Se film on glass substrate. Film thicknesses are (a) 2 μm, (b) 1 μm, (c) 500 nm, and (d) 200 nm. The size of polycrystalline particles decrease with decreasing c-Se–layer thickness.

978-1-4799-8002-4/14 $31.00 © 2014 IEEE

(a)　　　　　　　　(b)

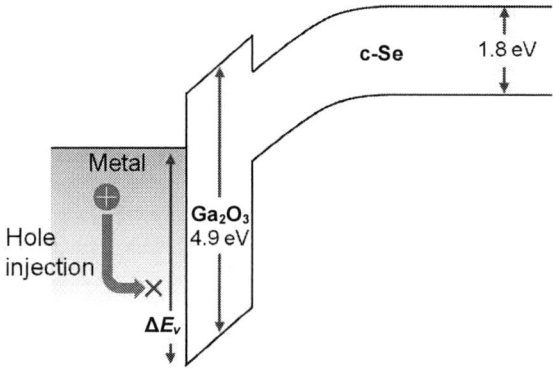

Fig. 7: Images captured by proposed image sensor based on c-Se films with thicknesses of (a) 2 μm and (b) 500 nm. Fixed-pattern noise appears when the size of c-Se polycrystalline particle is relatively larger than the pixel size of the CMOS circuit. Illumination: 330 lux, Aperture: F5.6.

Fig. 10: Schematic diagram of energy band of Ga_2O_3/c-Se pn photodiode under reverse bias. The large valence-band offset E_v between the metal electrode and Ga_2O_3, which serves as an effective barrier against hole injection, is attributed to the low-dark-current characteristics of the photodiode.

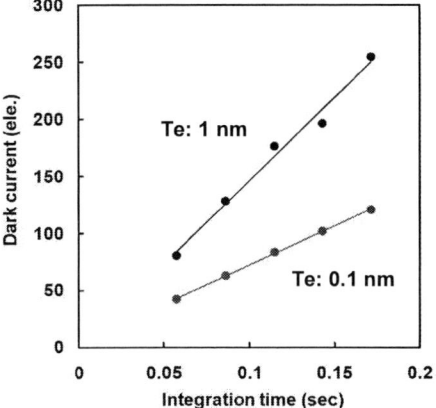

Fig. 8: Dark current as a function of integration time for image sensors overlaid with metal/c-Se Schottky photodiodes with 1- and 0.1-nm-thick Te layers at a reverse-bias voltage of 3V at room temperature. Dark current decreases with decreasing Te-layer thickness.

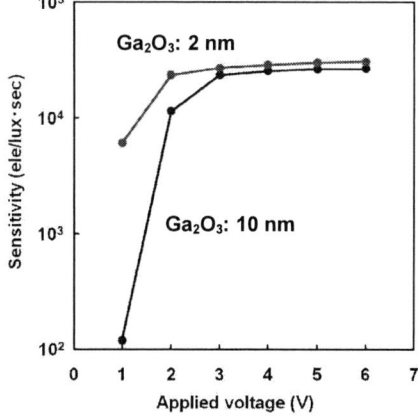

Fig. 11: Sensitivity as a function of reverse bias for image sensors overlaid with Ga_2O_3/c-Se photodiodes with 2- and 10-nm-thick Ga_2O_3 layers and 0.1-nm-thick Te layers. The sensitivity increases at lower applied voltage with decreasing Ga_2O_3-layer thickness.

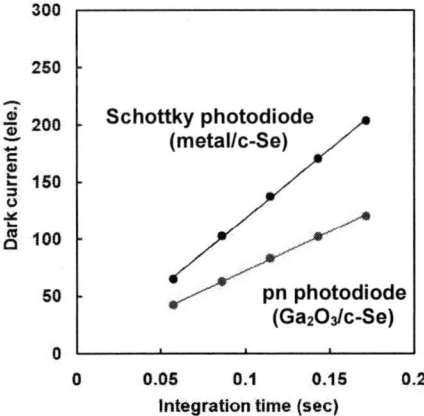

Fig. 9: Dark current as a function of integration time for image sensors overlaid with metal/c-Se Schottky photodiodes and Ga_2O_3/c-Se pn photodiodes with 0.1-nm-thick Te layer at a reverse-bias voltage of 3V at room temperature. Using a wide-band-gap Ga_2O_3 layer as a high hole barrier decreases the dark current.

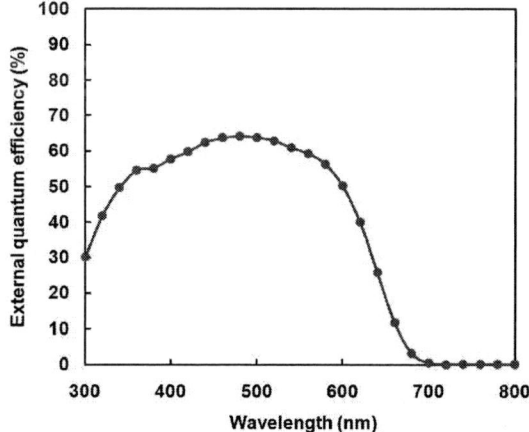

Fig. 12: Spectral external quantum efficiency for image sensors overlaid with 10-nm-thick-Ga_2O_3/500-nm-thick-c-Se pn photodiodes at a reverse-bias voltage of 6 V.

978-1-4799-8002-4/14 $31.00 © 2014 IEEE　　　　91

9.74-THz Electronic Far-Infrared Detection Using Schottky Barrier Diodes in CMOS

Zeshan Ahmad[1], Alvydas Lisauskas[2,3], Hartmut G. Roskos[2] and Kenneth K. O[1]

[1]Texas Analog Center of Excellence, The University of Texas at Dallas, Richardson TX, USA.
[2]Physikalisches Institut, JWG Universitat Frankfurt, D-60438 Frankfurt, Germany.
[3]Radiophysics Department, Vilnius University, Sauletekio av. 9, LT–10222 Vilnius, Lithuania.
Tel: (214) 202-5225, Fax: (972) 883-5550, Email: zeshan.ahmad@utdallas.edu

ABSTRACT

9.74-THz fundamental electronic detection of Far-Infrared (FIR) radiation is demonstrated. The detection along with that at 4.92THz was realized using Schottky-barrier diode detection structures formed without any process modifications in CMOS. Peak optical responsivity (R_v) of 383 and ~14V/W at 4.92 and 9.74THz have been measured. The R_v at 9.74THz is 14X of that for the previously reported highest frequency electronic detection. The shot noise limited NEP at 4.92 and 9.74THz is ~0.43 and ~2nW/√Hz, respectively.

I. INTRODUCTION

THz imaging in CMOS has gained great interest in recent years. This non-ionizing imaging modality has potential applications in material inspection, security, document authentication, medicine and others [1]-[3]. Both Schottky-barrier diode (SBD) and MOSFET based electronic detectors in CMOS operating between 0.3-1THz have been demonstrated [4]-[6]. Moving up in the FIR (Far Infrared) frequency range, NMOS fundamental electronic detection up to 4.25THz and 3rd order harmonic detection up to 9THz have been demonstrated. In [7], the feasibility of thermal detection at 28THz in 0.18μm CMOS is also investigated. This paper reports for the first time fundamental electronic detection at frequencies above 5THz. This and detection at 4.92THz were demonstrated using SBD's operating above their cut-off frequency (f_T) and fabricated without any process modifications in 130-nm CMOS [8].

The 14V/W optical R_v at 9.74THz is 14X higher than that of the previously reported highest frequency electronic detection (3rd order harmonic detection at 9 THz [6]). The measured NEP at 4.92THz with modulation frequency of 300Hz is 4.95nW/√Hz, and the shot noise limited NEP is ~0.43nW/√Hz. At 9.74THz, the shot-noise limited NEP of 10THz detection structure is ~2nW/√Hz. These suggest potential for an affordable electronic IR detector in foundry CMOS. Electronic detection mitigates the thermal isolation requirements of thermal detectors as well as providing a path for high density pixel arrays.

II. FAR INFRARED DETECTION TEST STRUCTURE

A. Direct-Antenna-Match Technique

Fig. 1 shows the test structures utilized to investigate FIR CMOS detection. The structures use direct-antenna matching with resistive match as opposed to conventional conjugate matching used at frequencies below 1THz [4], [5]. Above 2.5THz, conjugate matching suffers from higher matching loss and a non-realizable on-chip bypass capacitor necessary for extraction of rectified low-frequency signals. Radiation from transmission lines is the major loss mechanism for the former while self-resonance due to mutual-inductance between capacitor plates for the latter. Excluding the radiation loss in transmission-lines,

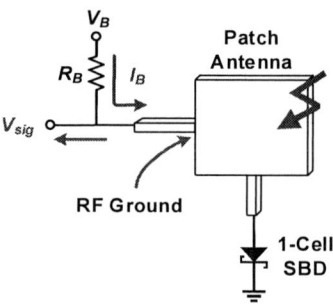

Fig. 1. Schematic of FIR detection structure.

the loss incurred in a matching network is comparable to the loss associated with using direct antenna matching. ANSYS HFSS simulations indicate that the best designed bypass capacitor with minimal mutual inductance is only operable up to 3THz. Use of a bypass capacitor is avoided by utilizing the inherent E-field symmetry of patch antenna shown in Fig 2. The E-field goes through zero at the center tap of the radiating edge creating a virtual RF ground [9]. This virtual ground can be utilized for dc-bias as well as rectified signal extraction with minimal impact on the antenna performance. The HFSS simulations indicate no degradation of antenna radiation efficiency (η_{rad}). Additionally, use of direct antenna match avoids low-frequency parasitics of read-out circuitry by removing

978-1-4799-8002-4/14 $31.00 © 2014 IEEE

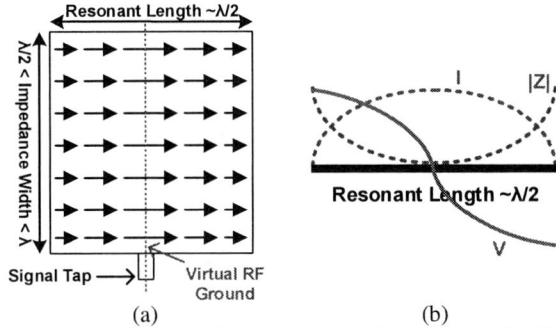

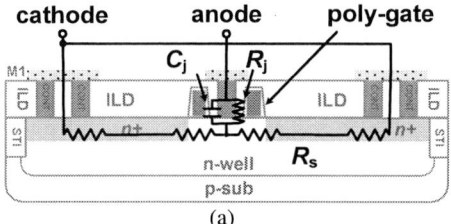

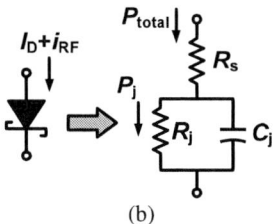

(a)   (b)

Fig. 2 (a) Current distribution on a Patch Antenna, (b) Voltage (V), Current (I) and Impedance (Z) distribution with respect to resonant length.

the bypass capacitor requirement which results in faster detector response and a smaller pixel area. Both aspects are critical for video-rate high-density pixel array implementation.

B. On-Chip Antenna Design

On-chip patch antennas (Fig. 3) tuned at 5 and 10THz were realized using metal levels 7 (M7) and 6 (M6), respectively with M1 & M2 layers shunted together to form a ground plane shielding the lossy silicon substrate below. Forming antennas using lower metal levels reduces the THz signal travel length to the detector by reducing the interconnect height (Fig. 3(b)). This lowers interconnect loss and parasitic capacitance leading to better sensitivity. The simulated antenna efficiencies, η_{rad} of 86% at 5THz and 88% at 10THz are achieved and mostly limited by the metal loss. Table 1 summarizes the

TABLE I

FIR ANTENNA PARAMETERS

Freq (THz)	†L x W (µm²)	†H (µm)	*Directivity (dBi)	*η_{rad} (%)
5	12 x 18	3.3	7.4	86
10	6.6 x 11	2.4	8.6	88

†Defined in Fig. 3(a), *Simulated

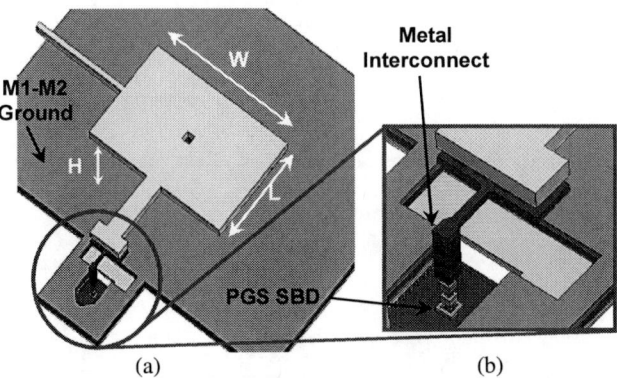

(a)   (b)

Fig. 3. (a) 3-D structure showing both on-chip patch antenna and interconnect to the Schottky-barrier diode, (b) Close-up of SBD interconnect.

Fig. 4 (a) Cross section of Poly-Gate Separated (PGS) Schottky diode, (b) Equivalent circuit.

design parameters and simulated performance of the on-chip patch antennas.

C. Diode Size

To minimize shunting of RF signal terminated at non-linear R_j of diode for better detection, the diode capacitance is minimized by using an 1-cell SBD shown in Fig. 4 with measured f_τ of ~2THz [8]. The 1-cell formed with the minimum-design rule 0.4x0.4µm² junction area exhibits zero-biased capacitance, C_{j0} of ~0.5fF.

III. MEASUREMENT RESULTS

A photomicrograph of the detectors fabricated in the UMC 130-nm CMOS process is shown in Fig. 5. The detectors were characterized using a tunable free-electron-laser (FEL) source with pulse lengths up to 30ps at the Helmholtz-Zentrum Dresden-Rossendorf, Germany. Fig. 6 shows the detector measurement setup including an FEL. The laser was mechanically chopped at 80 and 300Hz, and the response was measured using a high input impedance

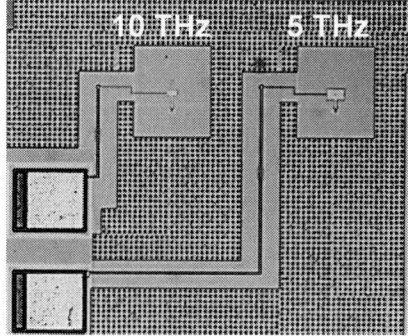

Fig. 5. Photomicrograph of FIR detectors.

978-1-4799-8002-4/14 $31.00 © 2014 IEEE

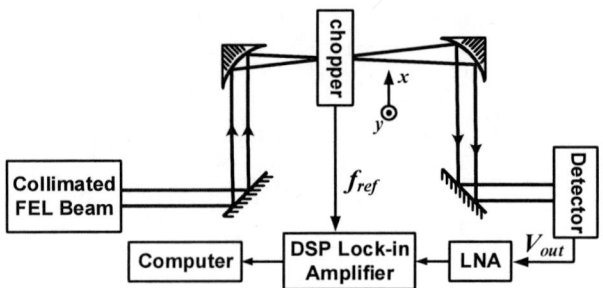

Fig. 6. Measurement setup including a free electron laser.

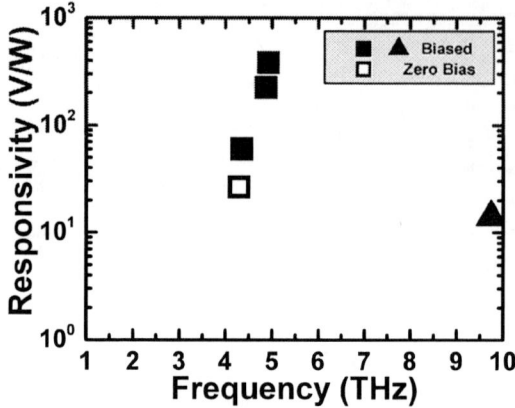

Fig. 7. Measured responsivity of 5 and 10-THz SBD detection structures.

(100MΩ, 25pF) LNA & lock-in amplifier (SR560/ SR5110). The beam-profile was imaged using an Ophir Pyrocam-III pyroelectric camera for spot-size calibration. A calibrated large-area Thomas-Keating (TK) absolute power meter was used for total beam-power (P_0) measurement. The power values were cross-checked with a FieldMaster power meter for accurate estimations. The peak beam intensity was computed using the Gaussian-beam relation of $2P_0/(\pi r^2)$.

The Pyrocam, FieldMaster, TK data, and simulated effective aperture area of patch antenna were used to estimate the received power. These results in ~4X larger received power or 4X lower responsivity compared to the values computed using the average beam intensity and physical antenna size. The 5-THz detectors were characterized with laser beams at 4.29, 4.35, 4.88, and 4.92THz. The beam attenuation was adjusted such that -20 to -30dBm power was incident to the antenna of detectors for small-signal detection. The responsivity data are shown in Fig. 7. The responsivity is strongly dependent on frequency near 5THz, which is expected from the antenna resonance behavior. The 383V/W optical R_v at 4.92THz is ~1.7X better than that of the 4.25THz detectors using a 90-nm MOSFET in [6]. A laser beam at 9.74THz with incident power to the antenna of approximately -28dBm was used to characterize the 10-THz detector. The optical R_v of 14V/W at 9.74THz which is ~14X higher than the highest R_v for electronic detection [6] is measured. The bias-dependence of R_v of FIR SBD detectors is shown in Fig. 8. The structures show improved response at a slightly forward biased condition due to the large zero-bias impedance of ~357MΩ that results in voltage division with the input impedance of the external LNA. At ~10nA, the dynamic resistance drops to 2.75MΩ with almost no degradation to the detected response. The measured optical R_v (Fig 8(a)) agrees well with the simulated peak R_v of 400V/W at 4.92THz. The faster R_v roll-off with I_{BIAS} is due to the parasitics of external read-out setup. There is significant discrepancy between the simulated and measured 9.74THz response (Fig. 8(b)). This may be due to operation close to the plasma frequency of cathode of SBD/n-well estimated to be ~12THz [10].

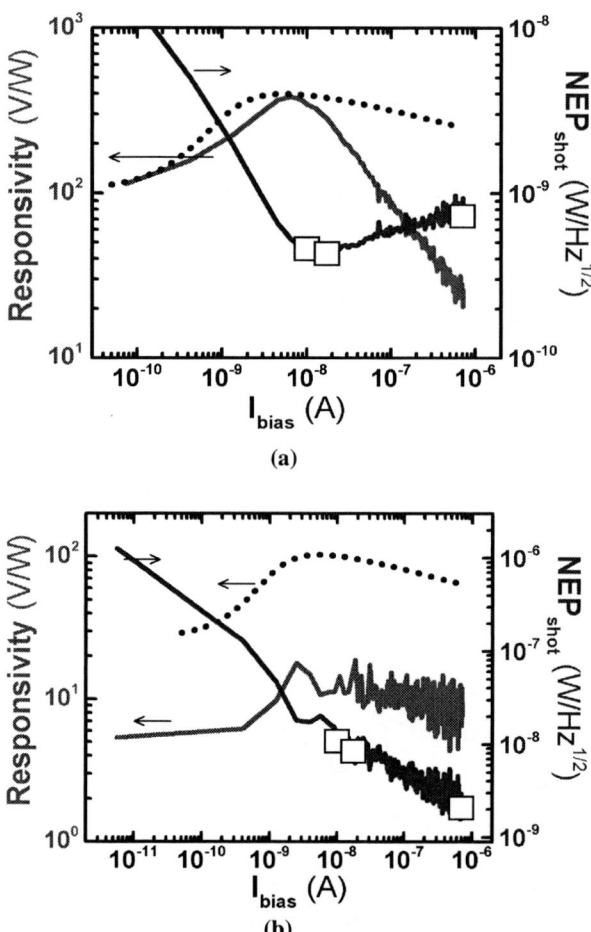

Fig. 8. (a) Measured (Simulated: dotted) bias-dependence of responsivity and shot noise limited (line: noise calculated, □: measured) NEP for 5THz SBD detector at 4.92THz, (b) Measured (Simulated: dotted) bias-dependence of responsivity and shot noise limited (line: noise calculated, □: measured) NEP for 10THz SBD detector at 9.74THz.

978-1-4799-8002-4/14 $31.00 © 2014 IEEE

The noise of SBD detection structures at 10, 17.3 and 700nA bias is measured up to 1MHz using an SR560 LNA and an Agilent 89441A Vector Signal Analyzer. The flicker-noise corner is ~30kHz at 17.3nA bias with shot noise of ~123nV/√Hz. Based on this, shot noise limited NEP of ~430pW/√Hz has been estimated at 4.92THz. At 0.7μA the shot-noise value of ~19.7nV/√Hz results in NEP of ~2nW/√Hz for the 10-THz detector. The bias-dependence of NEP for both 5 and 10-THz detectors is also shown in Fig. 8.

Fig. 9(a) shows the 10-THz detector scanned image of FIR emission from a dimmable compact fluorescent lamp (CFL). The FIR emission is only observed near the lowest brightness setting for the CFL. The imaging setup used is shown in Fig. 9(b) along with the THz detection board and CFL in Figs. 9(c) and 9(d), respectively. The almost isotropic emission from the CFL was confined to an area of ~5mm diameter using an aluminum foil wrapped funnel resulting in an unfocused spot size of ~4 x 4cm². Fig. 10 shows an FTIR (Bruker Vertex 80v) scanned spectrum of the CFL. The spectrum clearly shows peaks between 8-12 THz which lie within the detection band of 10-THz detector.

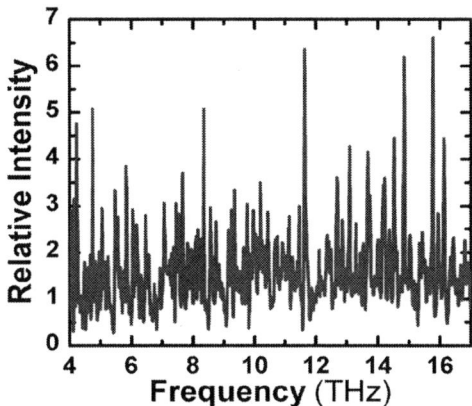

Fig. 10. Measured FTIR (Bruker Vertex 80v) spectrum of compact fluorescent-lamp (CFL) used for imaging experiment illustrated in Fig. 9.

setup, and HZDR Dresden Germany for FEL beam-time support.

IV. CONCLUSION

This work demonstrates that electronic detection up to 10THz is possible using ~2-THz f_T Schottky barrier diodes fabricated in CMOS. The SBD detector exhibits ~14X higher responsivity near 10THz than the highest previously reported R_v for electronic detection. This work also suggests electronic detection beyond 10 THz and potential for affordable imaging in the IR range using foundry CMOS.

ACKNOWLEDGEMENT

This work is supported by Texas Analog Center of Excellence (TxACE). The authors thank Martin Mittendorff and Dr. Stephan Winnerl for help with FEL

REFERENCES

[1] P. H. Siegel, IEEE MTT, vol. 50, no. 3, pp. 910-928, Mar. 2002.
[2] M. Tonouchi, THz Science & Tech., vol. 2, pp.90-101 2009.
[3] E. Seok et al., IEEE JSSC, vol. 45, no. 8, pp. 1554-1564, Aug. 2010.
[4] R. Han et al., IEEE JSSC, vol. 48, no. 10, pp. 2296-2308, Oct. 2013.
[5] D. Kim, S. Park, and R. Han, VLSI Circuits Symp., Jun. 2013, pp. 12-13.
[6] A. Lisauskas et al., J. Infrared, Millimeter & THz Waves, Jan. 2014.
[7] S.-H. Yang et al., IEEE MTT-S, May 2010, pp. 648–651.
[8] S. Sankaran et al., ISSCC, San Francisco, CA, USA, Feb. 2009, pp. 202-203.
[9] C. A. Balanis, Hoboken, NJ, USA: John Wiley & Sons, 2005.
[10] A.Van der Ziel, J. Appl. Phys., vol. 47, no. 5, 2059, 1976.

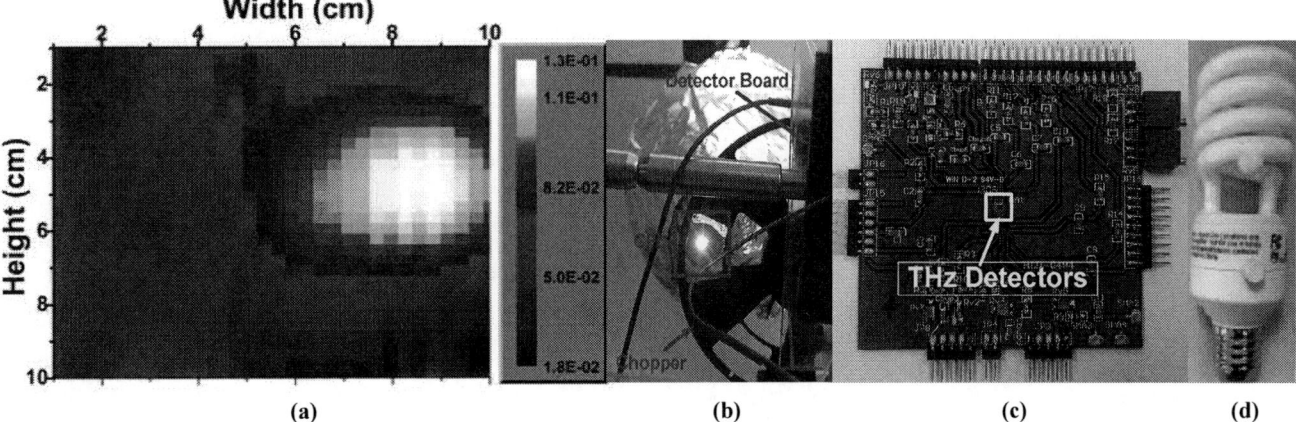

Fig. 9. (a) 2-D scanned image of compact fluorescent lamp (CFL), (b) CFL imaging setup, (c) THz detectors PCB, (d) CFL.

Experimental Demonstration of a Stacked SOI Multiband Charged-Coupled Device

Chu-En Chang[1], Julie D. Segal[2], Aaron J. Roodman[2], Christopher J. Kenney[2] and Roger T. Howe[1]

[1]Department of Electrical Engineering, Stanford University, Stanford, CA 94305, USA, email: cechang@stanford.edu
[2]SLAC National Accelerator Laboratory, Menlo Park, CA 94025, USA

Abstract

The multiband charge-coupled device (CCD) is a polychromatic image sensor fabricated on semiconductor-on-insulator substrates with multiple device layers. It can achieve a multifold improvement in efficiency compared to a conventional CCD, because photons are collected and resolved in the substrate without external optics. We present the first experimental demonstration of multiband light absorption and charge extraction for a single-pixel device. The ratio of the active layers' responsivities changes with wavelength, indicating that incident color information is captured in the multiple device layers.

Introduction

The multiband CCD is a polychromatic image sensor fabricated on semiconductor-on-insulator (SOI) substrates with multiple device layers (Fig. 1) (1). Because the stacked device layers in each pixel respond differently to photons of different wavelengths, each pixel encodes incident color information without the need for external optics, such as band-pass filters or prisms. Therefore, the multiband CCD can collect and resolve more photons per pixel or per unit area. The increased responsivity is multifold and approximately equals the number of stacked layers in the system (1). The multiband CCD is very attractive for applications that require efficient, low-light color imaging with a single image sensor, such as astronomical surveys (2) and X-ray detection. When compared to conventional color imaging techniques, the increased photon collection per pixel and the embedded color-resolving capability result in improvements in several performance metrics. When compared with prism-based systems (3), the system pixel count, form factor and sensitivity to focal plane misalignment can be reduced. Improved light sensitivity per unit area is possible compared with mosaic-filter-based systems (4). Compared with a system with a monochromatic image sensor and a filter exchange unit (5), the multiband imager has multifold higher throughput.

The concept of resolving colors in the substrate (6) was initially implemented in CMOS image sensors. In one implementation, each pixel was fabricated with an array of contacts that separately collected charge carriers from different depths (7). The color crosstalk depended on excess charge carrier density, because of electric field screening. Another implementation used nested triple p-n junctions in the pixel, with each junction collecting signals of one color (8). Both implementations require front-side metal contacts, which reduce the fill factor and degrade low-light performance. An SOI CCD implementation is attractive, because a) the fill factor can be 90% or more without complicated pixel design, and b) the SOI's buried insulating layers minimize electrical crosstalk between device layers during charge collection and transfer. In addition, the optical performance can be optimized by varying the SOI stack's layer thicknesses. The refractive index contrast between the insulator and the semiconductor can also be exploited in the optical design.

We present the first experimental demonstration of multiband light absorption and charge extraction. A single-pixel, front-illuminated, p-type multiband CCD on silicon SOI substrates with three active layers has been simulated, fabricated and tested with constant illumination and biasing. The ratio of the device layers' responsivities (A/W) is shown to change with wavelength, indicating that color information is captured in the pixel's output currents; this result is an important step toward the realization of a full-scale multiband SOI CCD.

Simulation

We modeled the single-pixel multiband CCD using commercial software (9) (10) (Fig. 2). It consists of the output transfer gate, the output summing well (OSW) and the sense nodes of a full multiband CCD (Fig. 1b). Unlike a typical CCD where serial gates and output circuitry are blocked from light, the OSW is open and serves as the photosensitive area. We have previously reported simulations results on the dynamic readout (1). In this work, the steady-state response is simulated and measured in wafer-level testing by applying constant biasing and constant optical excitations of different wavelengths. Currents at the sense nodes are recorded for each wavelength. The sense nodes are biased the same as for the dynamic readout. Because the sense nodes are biased and then left floating before charge measurement takes place in the dynamic readout, the DC experiment probes the device's charge extraction performance in the low-light limit.

Fig. 3 shows the simulated optical performance of the SOI stack, calculated by using the scattering-matrix method and published optical constants (11)(12). The absorptivity curves of different layers peak at different wavelengths, indicating the potential for multiband detection. Fig. 4 shows the

978-1-4799-8002-4/14 $31.00 © 2014 IEEE

electrical design for selective charge extraction. Each layer's electric potential distribution is plotted after they are biased and depleted, but before illumination. For each plot, the cross section is shown at the depth of the potential minimum. The device layers' doping concentrations and the positions, depths and biases of the channel-stop trenches and sense node contacts are designed such that each layer's potential monotonically increases from the sense node to the OSW. This allows different sense nodes to selectively extract charge from their respective layers with minimal crosstalk. Fig. 5 shows the simulated results for constant biasing and illumination. The quantum efficiency (QE) curves are obtained by dividing the charge carrier flux at the sense nodes by the incident photon rate. Background responses, obtained in a control simulation where the optical excitation is turned off, are removed from the data. Light absorption is simulated by increasing the carrier generation rate underneath the OSW. The ratio of total excess generation in each layer is the same as the calculated optical spectra. The generation profile is uniform in the first two layers, and spans uniformly for 4 μm at the top of the third layer; consequently charge collection at the OSW is assumed to be ideal. The QE curves follow the optical spectra closely, indicating effective charge extraction.

Fabrication and testing

Multilayer SOI substrates with three silicon layers were fabricated by direct wafer bonding and precision grinding. Vias to the buried device layers were made by deep reactive-ion etching in silicon and filled with doped polysilicon plugs. Fig. 6 and Fig. 7 show the images of the device halfway through and after completion of fabrication, respectively.

The test setup included three LEDs of different wavelengths as light sources. The wavelengths are close to the maxima of the optical spectra (Fig. 3). The light was coupled through a multimode fiber and a microscope objective to the device. The ambient light was kept below 0.1 nW/cm^2. Each LED was applied to the device separately. The same series of constant-current pulses was used to drive each LED. Each current pulse was accompanied by a 20-sec period of zero current to monitor the device's leakage. Each pulse was 30 seconds long, allowing effective measurement of the steady-state device response. The power outputs of the LEDs were separately measured. The incident power on the device was obtained by scaling the measured LED power with the ratio of the device's photosensitive area to the beam's spot size, assuming a uniform intensity profile (Fig. 8).

The test results demonstrate multiband light absorption and charge extraction. The current pulses from each layer show the same trend as the incident optical pulses (Fig. 9). Fig. 10 plots current versus incident power, with background signals removed. The response of each layer is linear to incident power for all wavelengths, suggesting linear charge generation and extraction. The ratio of the layers'

responsivities changes with wavelength, indicating that incident color information is incorporated in the output currents. Each layer is most responsive to a different wavelength, as desired. The bottom layer's signals are noisier because it was biased close to reverse breakdown voltage, which can be improved by including dielectric-wrapped sense nodes (1). Fig. 11 compares the measured and simulated quantum efficiency matrices. The fact that the measured QE matrix is invertible indicates that the wavelength of an unknown monochromatic source can be uniquely identified without external optics (1). Potential reasons for the discrepancy between the simulated and measured quantum efficiencies include (i) the 0.5-μm, within-wafer non-uniformity of silicon layers and (ii) variations in the doping concentration of the ultrahigh resistive substrate that can reduce the bottom channel's depletion depth and limit the infrared response.

Conclusion

We report the first demonstration of multiband light absorption and charge extraction on a single-pixel SOI multiband CCD having three device layers. The response of each layer is linear to incident power at all tested wavelengths. The ratio of the layers' responsivities changes with wavelength, indicating that color information is incorporated in the pixel without external optics. This proof of concept paves the way for the realization of a full-scale SOI multiband CCD.

Acknowledgement

We acknowledge the financial support by the U.S. Department of Energy through the SLAC National Accelerator Laboratory. The fabrication was performed in the Stanford Nanofabrication Facility supported by the National Nanotechnology Infrastructure Network.

References

(1) C.-E. Chang, J. D. Segal, A. J. Roodman, R. T. Howe, and C. J. Kenney, "Multiband charge-coupled device," *IEEE Nuclear Science Symp. and Medical Imaging Conf.*, pp. 743-746, 2012.
(2) Ž. Ivezić *et al.*, "LSST: from science drivers to reference design and anticipated data products," arXiv:0805.2366 [astro-ph], 2014.
(3) H. de Lang and G. Bouwhuis, "Optical system for a color television camera," U.S. Patent 3 202 039, 1965.
(4) B. E. Bayer, "Color imaging array," U.S. Patent 3 971 065, 1976.
(5) J. A. Tyson and the LSST Collaboration, "Large Synoptic Survey Telescope: overview," *Proc. SPIE*, vol. 4836, pp. 10-20, 2002.
(6) B. C. Burkey, R. S. VanHeyningen, R. A. Spaulding and E. L. Wolf, "Color responsive imaging device employing wavelength dependent semiconductor optical absorption," U.S. Patent 4 613 895, 1986.
(7) G. Langfelder, A. Longoni and F. Zaraga, "Further developments on a novel color sensitive CMOS detector," *Proc. SPIE*, vol. 7356, no. 73562A, 2009.
(8) R. B. Merrill, "Vertical color filter detector group and array," U.S. Patent 6 632 701, 2003.
(9) *Sentaurus Structure Editor User Guide*, Version G-2012.06, Synopsys, Inc., Mountain View, CA, 2012.

978-1-4799-8002-4/14 $31.00 © 2014 IEEE

(10) *Sentaurus Device User Guide*, Version G-2012.06, Synopsys, Inc., Mountain View, CA, 2012.

(11) D. M. Whittaker and I. S. Culshaw, "Scattering-matrix treatment of patterned multilayer photonic structures," *Phys. Rev. B*, vol. 60, pp. 2610-2618, 1999.

(12) E. D. Palik, *Handbook of Optical Constants of Solids*, New York: Academic Press, 1985.

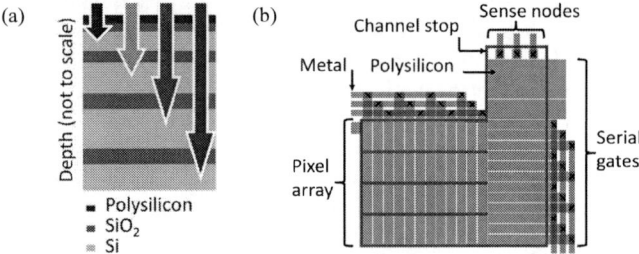

Fig. 1. SOI multiband CCD concept. (a) Each device layer in the SOI substrate interacts differently with different optical bands. (b) The layout is similar to that of conventional CCDs, except that a separate sense node is required to exclusively extract charge from each layer to avoid crosstalk. Charge collection and transfer for all layers are handled by the same set of gates that are on either the front or the back side.

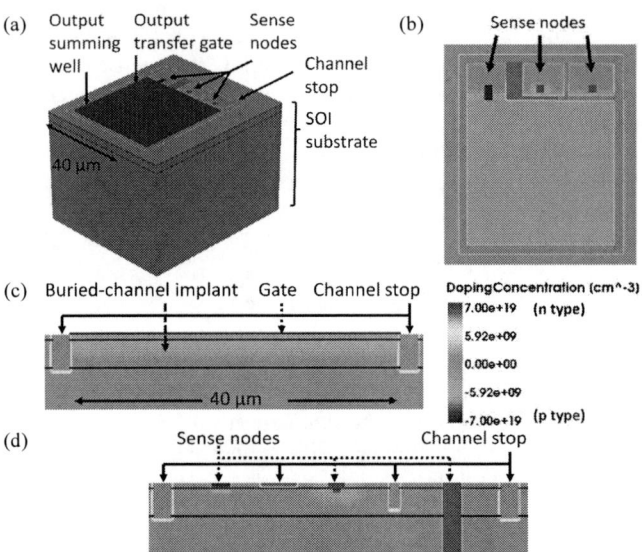

Fig. 2. The TCAD model of the single-pixel multiband CCD. It includes the last two gates and the sense nodes of a full device (Fig. 1b). (a) Perspective view. (b) Horizontal cross section at the depth center of the first layer. (c) Cross section across the output transfer gate. (d) Cross section along the sense nodes. The doping concentration color scale applies to all plots.

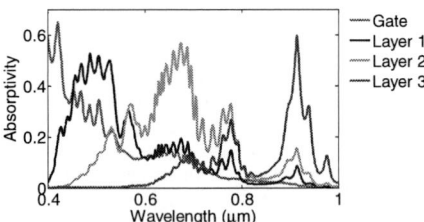

Fig. 3. The optical performance of the single-pixel device, calculated using the scattering-matrix method and published optical constants (11)(12). Note that the absorptivity curves peak at different wavelengths.

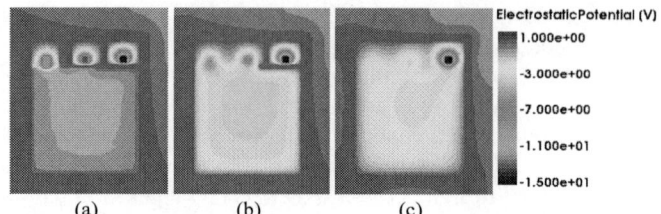

Fig. 4. The electrostatic potential of (a) the first, (b) the second and (c) the third device layer, after the device is biased and depleted. For each plot, the cross section is shown at the depth of the potential minimum. The layers' doping concentrations and the positions, depths and biases of the channel-stop trenches and the sense nodes are designed such that each layer's potential monotonically increases from the sense node to the OSW, which facilitates selective charge extraction and reduces crosstalk between layers.

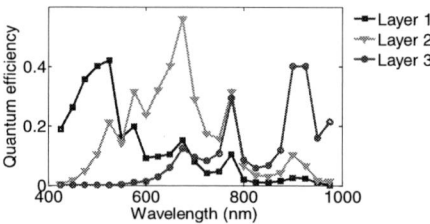

Fig. 5. DC measurements simulated using commercial software (10). The effects of multiband absorption and charge extraction are included. Charge collection at the OSW is assumed to be ideal. The currents at the sense nodes change with wavelength. Quantum efficiency is the carrier flux divided by the photon incident rate. The incident intensity is 0.5 mW/cm^2, which is approximately the maximum used in the experiments.

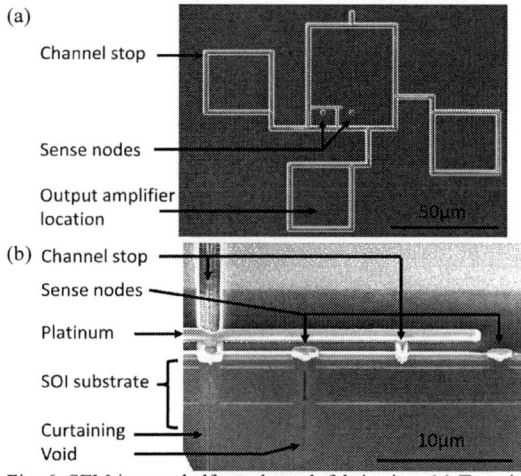

Fig. 6. SEM images halfway through fabrication. (a) Top view. (b) 52° cross section along the sense nodes. The platinum sacrificial layer and the curtaining artifact are the byproducts of focused ion beam sectioning.

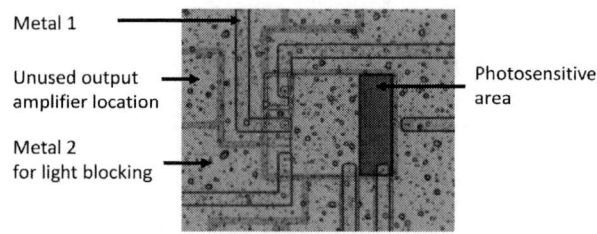

Fig. 7. Optical micrograph of the pixel after completion of fabrication.

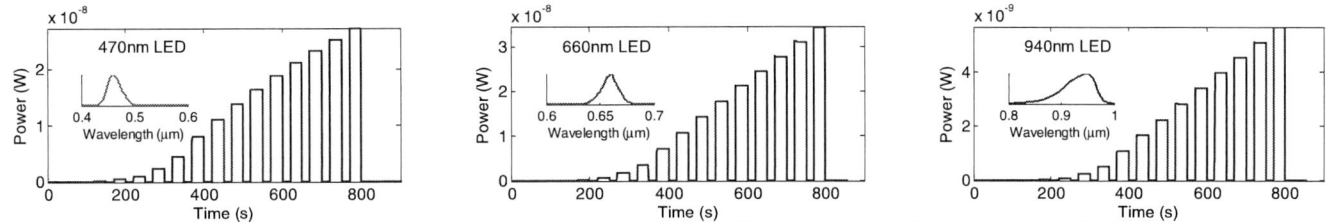

Fig. 8. Measurements of the LEDs used for testing. The LEDs are Thorlabs M470F1, M660F1 and M940F1. The insets show their power spectral densities, as provided by the vendor. The power is measured by a Newport 818-ST2 photodiode with a Newport 1835-C power meter. The vertical scales have been adjusted to account for the size difference between the photosensitive area of the device and the spot size of the beam.

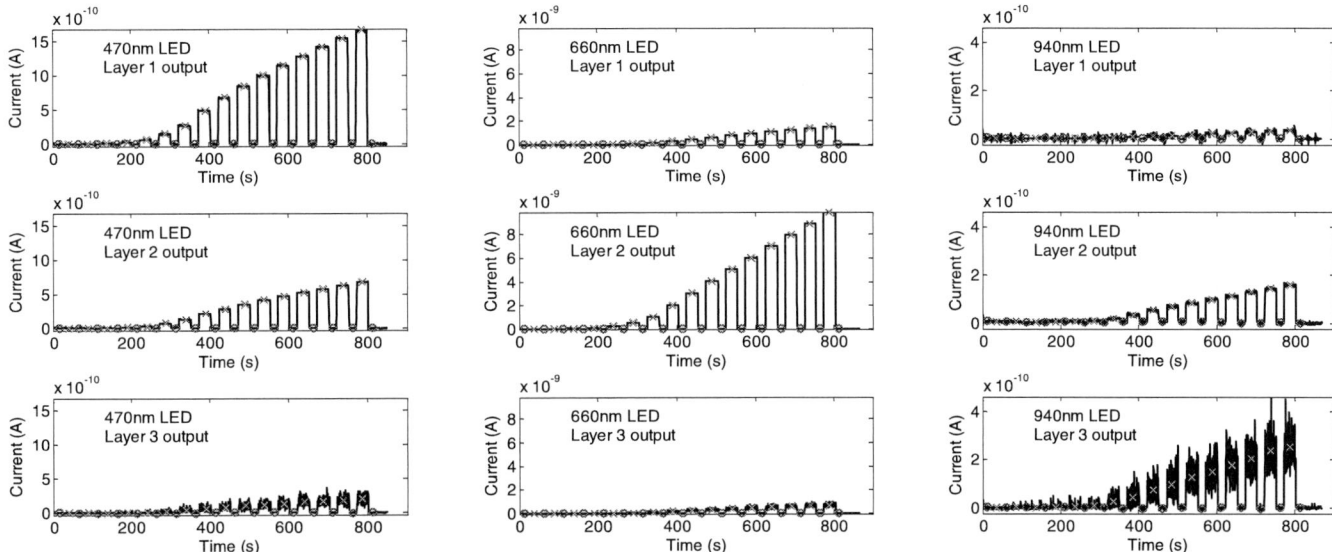

Fig. 9. Current measurements. The device was excited by each LED separately and the currents at the three sense nodes were recorded versus time. The curves are offset by the average leakage current for comparison purposes; however, no other scaling or filtering is used. Blue crosses indicate the average height of the current pulse. Red circles indicate the average value of the background.

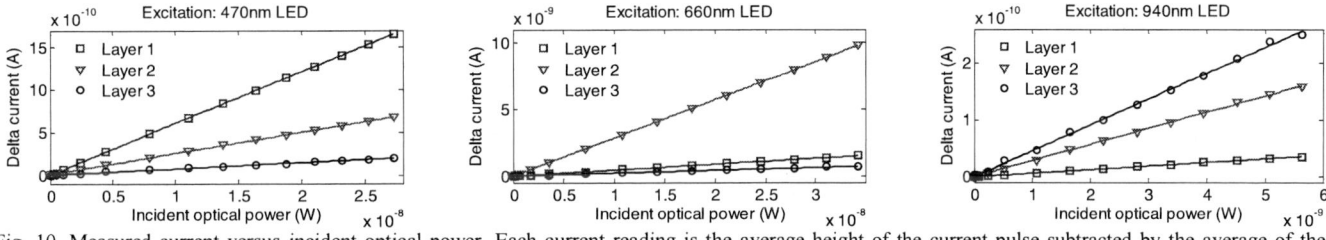

Fig. 10. Measured current versus incident optical power. Each current reading is the average height of the current pulse subtracted by the average of the adjacent background levels in Fig. 9. The incident power is obtained similarly from Fig. 8. Solid lines are linear fits (correlation coefficient r > 0.99 for all cases). The slopes are the responsivities (A/W). Each layer interacts differently with optical excitations of different wavelengths.

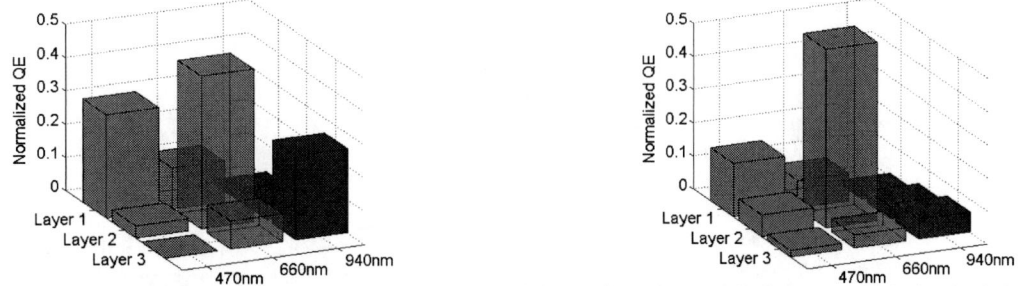

Fig. 11. Column plots of the simulated (left) and measured (right) QEs. Nominal device dimensions and doping were used in the simulation. The calculated optical spectra (Fig. 3) and the LEDs' power spectral densities were taken into account. The measured values are the responsivities in Fig. 10 multiplied by the photon energy and divided by the electron charge. The matrix of the measured QE values is invertible, indicating that the device can uniquely identify the wavelength of an unknown monochromatic source (1).

978-1-4799-8002-4/14 $31.00 © 2014 IEEE

Enhanced time delay integration imaging using embedded CCD in CMOS technology

Piet De Moor, Jo Robbelein, Luc Haspeslagh, Pierre Boulenc, Alper Ercan, Kyriaki Minoglou, Anne Lauwers, Koen De Munck and Maarten Rosmeulen

imec, Kapeldreef 75, B-3001 Leuven, Belgium
Tel: +32 16 28 8616, E-mail: Jo.Robbelein@imec.be

Abstract

This paper presents a new imager platform developed at imec enabling the monolithic integration of 130 nm CMOS image sensors (CMOS/CIS) with charge coupled devices (CCD). The process module was successfully developed and the potential of this embedded CCD in CMOS (eCCD) was demonstrated with the fabrication of a time delay integration (TDI) imager.

Introduction

A new imager platform based on the monolithic integration of the 130 nm CMOS image sensors (CMOS/CIS) platform and charge coupled devices (CCD) has been developed at imec. In this paper the successful development of several time delay and integration (TDI) imagers will be presented. The TDI imagers use a light sensitive CCD imaging part with a column based CMOS readout combining the best technologies for both integration of light induced charges and a fast readout. Several technology test structures have proven the good performance and yield of this process module.

Motivation

CMOS image sensors (CMOS/CIS) have been replacing traditional CCD based imagers for most consumer applications thanks to the integration of on-chip readout electronics. Still, CCD technology allows transfer and summation of signals in the charge domain in a nearly noiseless way, whereas CMOS is suffering from kTC noise in the voltage domain. However, CCD technology does not allow to integrate complex readout on-chip, resulting in a complex system. A combination of both technologies offers the best of two worlds: the charge domain operation of CCD's and the fast yet low power consumption readout of CMOS.

Technology development

Previous approaches [1] to realize an embedded CCD in CMOS technology tried to use standard CMOS design rules to enable CCD operation. Imec took a different approach (Fig. 1) by developing a dedicated new process module in addition to its 130 nm CMOS/CIS standard flow, containing dual gate thickness NMOS and PMOS transistors (1.2 and 3.3 V) and a dedicated pinned photodiode and transfer gate for a 4T CMOS pixel.

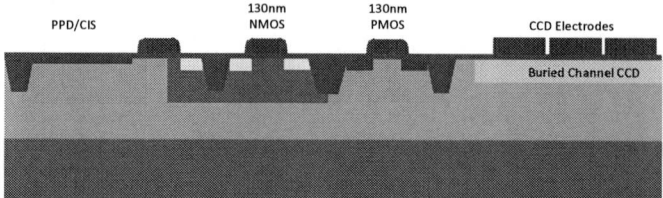

Figure 1: Schematic cross-section of the eCCD technology elements: pinned photodiodes and NMOS/PMOS in 130 nm CMOS and CCD in combination with CCD electrodes.

Advanced lithography tools and processes enabled a minimal poly gate-to-gate gap down to ~ 130 nm, assuring loss-less charge transfer (Fig. 2).

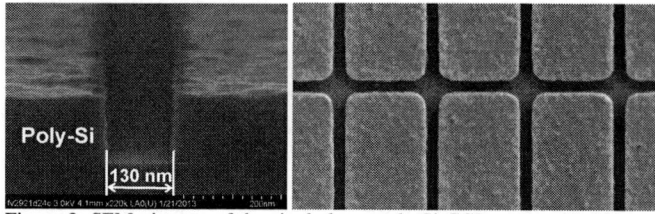

Figure 2: SEM pictures of the single layer poly-Si CCD gates with 130 nm spacing: cross-section (left) and top view (right).

Metal strapping using the CMOS Cu metals is used to reduce RC time constants. The eCCD process can be further extended with process modules enabling back side illumination. Figure 3 shows the excellent QE of the back illumination module demonstration using test diodes.

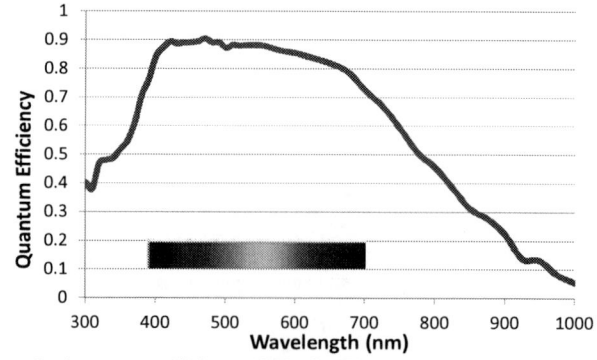

Figure 3: A quantum efficiency (QE) of > 70 % across the visible spectrum is obtained using backside illuminated test diode structures.

Technology evaluation

Very long interdigitated poly-Si yield evaluation structures were implemented and measured. Meander-fork structures with up to 10 meter length yielded non-shorted CCD gates with a poly gap of 130 nm (Fig. 4).

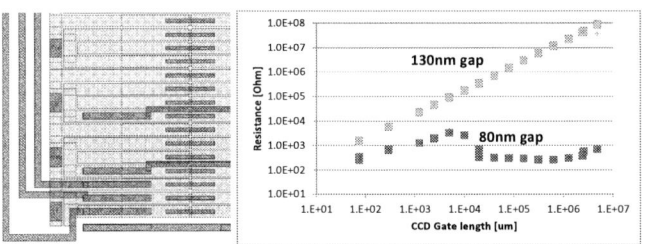

Figure 4: Large interdigitated poly-Si yield evaluation structure: layout (left); statistical data of meander resistance as a function of meander length for different gap widths, indication of very good yield for a gap of 130 nm (right).

A full set of CMOS (e.g.) transistor evaluation structures was used to capture the impact of the additional CCD module processing and no change in CMOS performance was observed (Fig. 5).

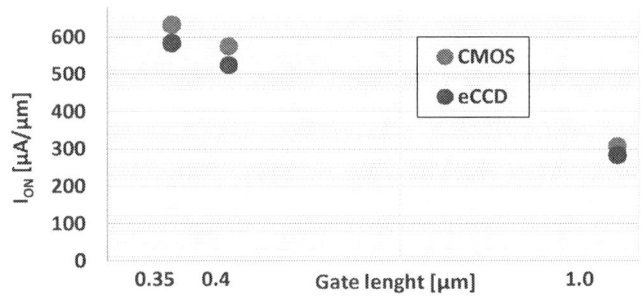

Figure 5: Comparison of CMOS transistor performance (I_{on} current density) with and without the extra CCD module implemented. No impact of the CCD module on the 130 nm CMOS characteristics was observed.

TDI imager design

In a TDI sensor a linear moving scene is imaged using oversampling of a pseudo-linear array (128x48 or 128x24 pixels in our case). Using the oversampling a signal to noise ratio (SNR) increase is obtained. CCD-based imagers are typically used for TDI as they combine collection and noiseless charge transfer synchronized with the scene movement. We designed a TDI imager in the new eCCD process with CCD pixels and a CMOS column based readout (Fig. 6) [2].

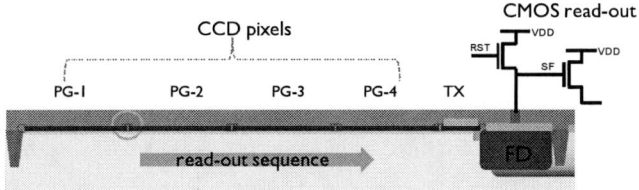

Figure 6: Conceptual cross-section of the TDI device in eCCD technology: The pixels are consisting of CCD gates while the (column) readout is realized using 130 nm CMOS.

Three different test imagers using a 4-phase CCD transfer and with a pixel pitch of 10 and 26 micron were designed (Fig. 7,8, Table 1).

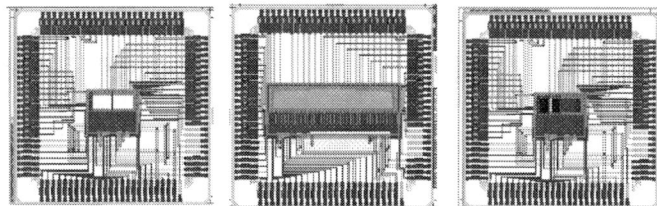

Figure 7: Top layout of 3 different TDI imagers: PAN, XS and test array. An identical readout and I/O configuration is used.

Table 1: Summary of parameters of TDI test vehicles.

parameter	PAN	XS	TEST
pixel pitch (um)	10	26	10
format (rows x columns)	48 x 128	24 x 128	48 x 128
number of selectable TDI stages	8	8	8

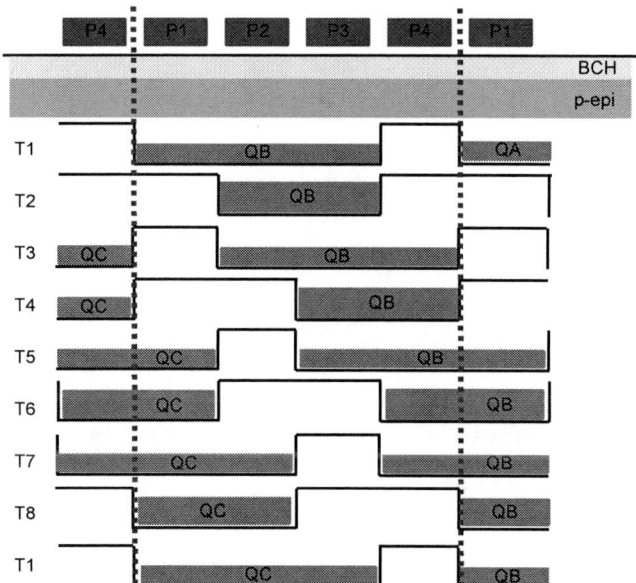

Figure 8: Charge transport in a 4 phase CCD array. 8 time cycles are required to transport a charge package over 1 pixel.

The imager consists of 8 selectable stages enabling optimization of SNR in different illumination conditions. The CCD operation was optimized with TCAD modeling using the standard available voltages of the 130 nm CMOS, i.e. maximum 3.3 V. Half of the array contains an anti-blooming feature based on a column-wise lateral overflow gate and drain (Fig. 9).

978-1-4799-8002-4/14 $31.00 © 2014 IEEE   101

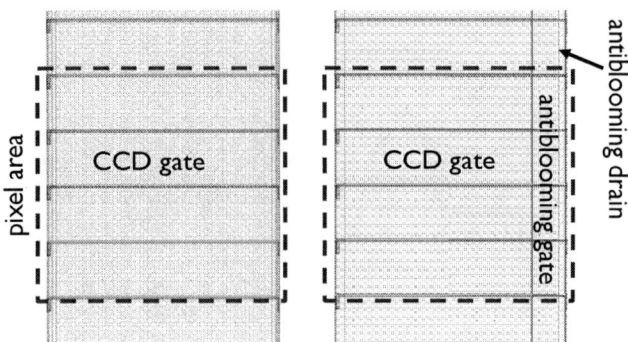

Figure 9: Pixel layout (for the XS sensor) without (left) and with (right) lateral anti-blooming channel. Metal straps are used to obtain low RC time constants.

The full well capacity is optimized for ~ 100 ke⁻ per pixel. At the top of the CCD pixel column dedicated charge injection gate structures were designed for testing purposes. At the bottom of the CCD pixel column extraction light-shielded CCD gate structures were designed to connect to the CMOS readout. The column readout consists of a floating diffusion node connected to a reset transistor and a source follower readout (Fig. 10). The total conversion gain is 10 uV/e-.

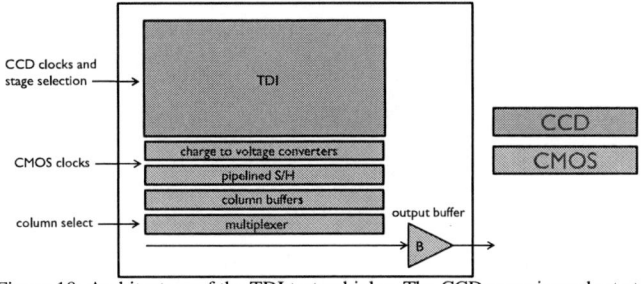

Figure 10: Architecture of the TDI test vehicles: The CCD array is readout at column level by CMOS electronics.

The sample and hold stage, address-decoded multiplexer and output buffer are all designed using the standard 130 nm CMOS (Fig. 11).

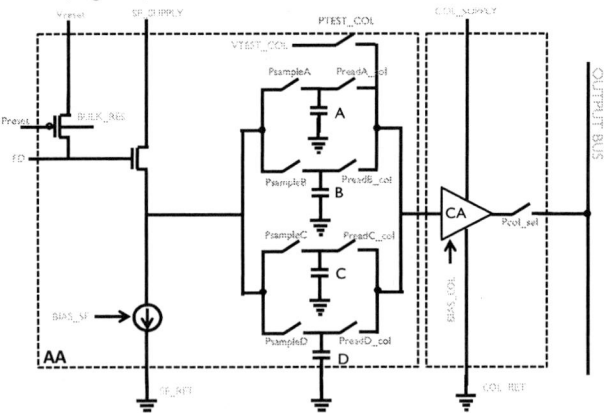

Figure 11: CMOS column readout structure featuring a source follower, sample and hold stages for both reset and signal allowing correlated double sampling and pipe-lined operation.

CCD driving electronics nor analog-to-digital converters (ADC's) were designed on-chip at this stage, but can be implemented easily in the future. A light shield covers the CMOS peripheral circuits.

TDI imager results

First of all, the CCD channel potential was measured using a dedicated test structure (Fig. 12).

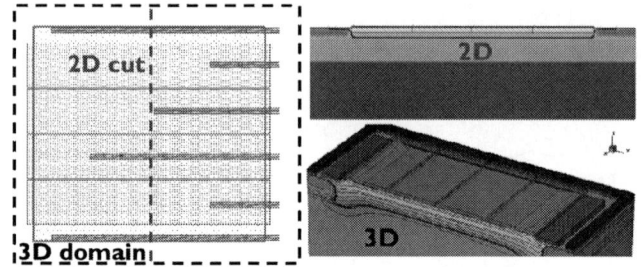

Figure 12: Channel potential test structure design (left). Corresponding TCAD 2D (right top) and 3D (right bottom) simulations used to optimize the implantation conditions.

This allowed to choose the best implantation conditions for the CCD channel. A TCAD study revealed that for small pixels a full 3D simulation of the electrostatic potential distribution underneath the gates is required as the simplified 2D TCAD predicts results which are incorrect (Fig. 13).

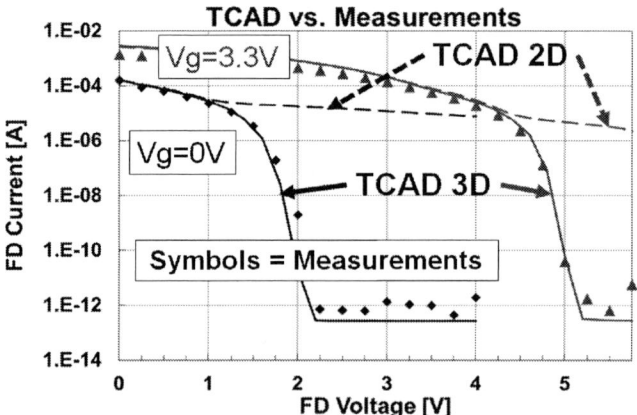

Figure 13: Channel potential measurements and simulations using 2D and 3D TCAD. 3D TCAD is required for reliable results.

Then the charge transfer efficiency (CTE) was measured using the fill-spill method, and a gate-to-gate CTE of > 99.9987 % was obtained on a chain of 1600 CCD gates, at a pixel line rate of 12.5 kHz (Fig. 14).

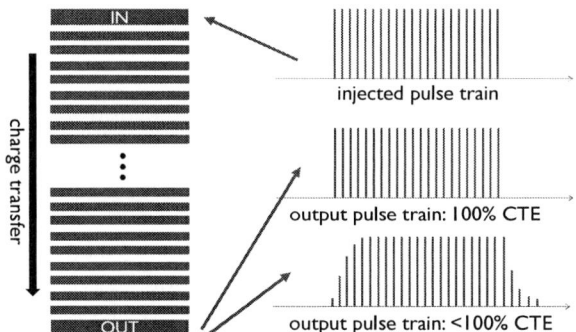

Figure 14: Measurement of the CCD charge transfer efficiency (CTE). A pulse train of charges is injected in a linear array of CCD gates and extracted at the other end. A CTE of >99.9987 % is calculated using the deformed output. [3]

An evaluation readout board with drivers allowing fine tuning of voltages, timing and slope of the CCD waveforms was used to readout the TDI imagers. (Fig. 15). The anti-blooming functionality was demonstrated by measuring with and without activating the anti-blooming gate, using long integration times (Fig. 16).

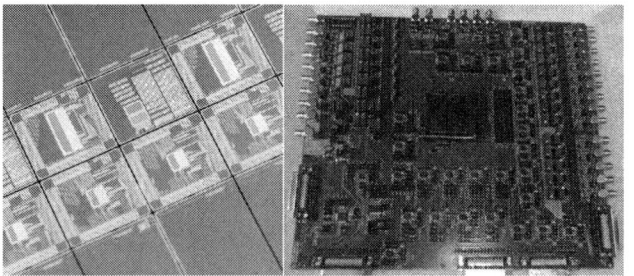

Figure 15: Wafer level picture of TDI imagers and large area yield structures (left). TDI imager readout board (right).

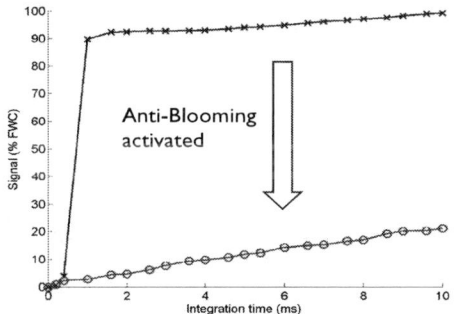

Figure 16: Measured effect of the controllable anti-blooming: For long integration times some pixels start to leak into neighboring pixels. When the anti-blooming gate is activated the effect of the blooming neighbor is totally avoided.

As a first imaging test, a small spot of light was projected onto the image sensor. Then the image was read out column by column. Figure 17 (b) shows a fiber through which light with a wavelength of 550nm is directed onto a part of the sensor (Fig. 17 (a)). To visualize the image, first a dark measurement is done to eliminate the average dark current, then the illuminated measurement is done. The correlated double sampling (CDS) values of the dark measurement are subtracted from the CDS values of the illuminated measurement. The light spot is clearly visible on Fig.17 (c).

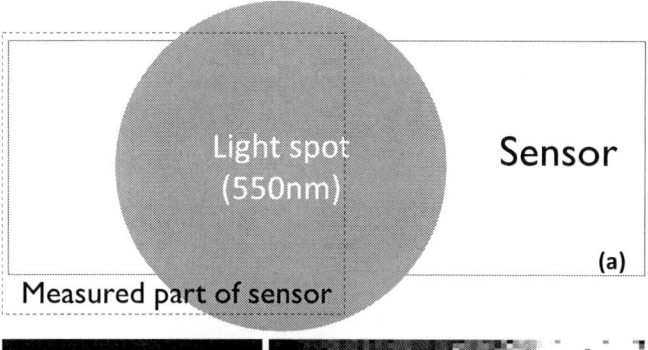

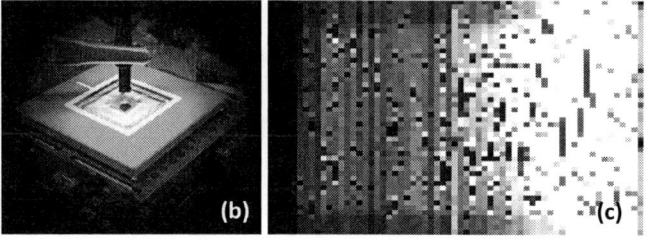

Figure 17: (a) Schematic of the illuminated part of the image sensor. (b) A glass fiber pointed towards the image sensor. (c) The calculated image of the light spot.

The non-uniformities in Figure 17(c) are caused by dark current non-uniformity. Clearly more work is needed in the future to decrease the dark current and to further improve the dark current non-uniformity.

Conclusion

We demonstrated the first TDI imager device in the newly developed embedded CCD in CMOS technology. In the near future this technology will be combined with the backside illuminated imager platform developed at imec, enabling beyond state-of-the-art TDI imaging for industrial inspection or earth observation applications.

Acknowledgements

The development of the TDI imager was supported by CNES (French Space Agency).

References

[1] J. Crooks, B. Marsha, R. Turchetta, K. Taylorb, W. Chanb, A. Lahavc, A. Fenigsteinc., "Kirana: a solid-state megapixel uCMOS image sensor for ultra-high speed imaging," Proc. SPIE 8659 (2013)

[2] A. Ercan, L. Haspeslagh, K. De Munck, K. Minoglou, A. Lauwers, P. De Moor, "Prototype TDI sensors in embedded CCD in CMOS technology," Proceedings IISW 2013 workshop

[3] R.W. Brodersen, D.D Buss, AL F. Tasch Jr., "Experimental Characterization of Transfer Efficiency in Charge-Coupled Devices," IEEE Transactions on Electron Devices Volume: 22 , Issue: 2, February 1975

A Solid State Thin Film Incandescent Light Emitting Device

Yue Kuo

Thin Film Nano & Microelectronics Research Laboratory, 235 J. E. Brown Eng. Bldg., Texas A&M University

College Station, TX, 77843-3122, USA, Email: yuekuo@tamu.edu

Abstract

A new type of solid state thin film white light emitting device has been studied. Light is emitted due to thermal excitation of the nano size conductive paths formed after the dielectric breakdown of the amorphous metal oxide thin film deposited on a silicon wafer. The mechanism of light emission, optical characteristics, reliability, driving methods, and efficiency are discussed. The complete device is made of the IC compatible materials and process.

Introduction

Lighting devices have been advanced from the bulky, low lifetime incandescent bulb to the environmental unfriendly, mercury discharged fluorescent tube to the expensive, single crystal semiconductor based light emitting device (LED), as shown in Fig. 1. For the broad band white light emission three LEDs of different band gap energies, e.g., for red, green and blue lights, or one UV or blue LED with a phosphor layer has to be used. The single crystal sapphire or compound semiconductor wafer and the MOCVD epitaxy process are required to fabricate LEDs.

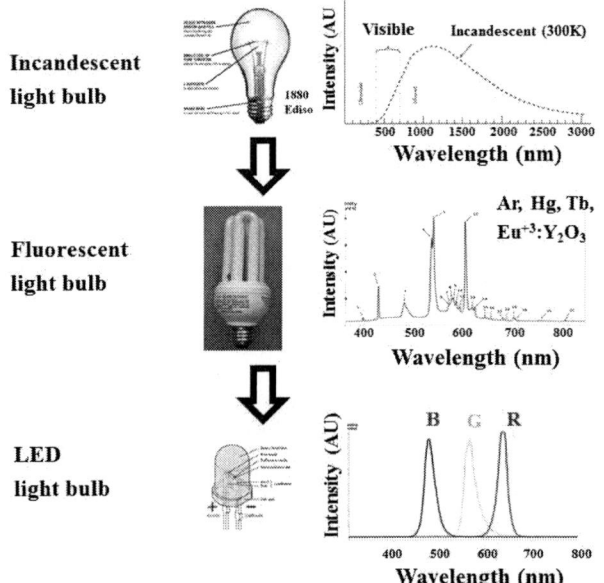

Fig. 1 Progress of light bulb and emission spectra.

Similar to the transformation of the vacuum-tube triode into the solid state transistor, as shown in Fig. 2, a new type of solid state incandescent light emitting device (SSI-LED) has been reported, recently (1-4). In this paper, authors discuss electrical and optical characteristics of this type of device.

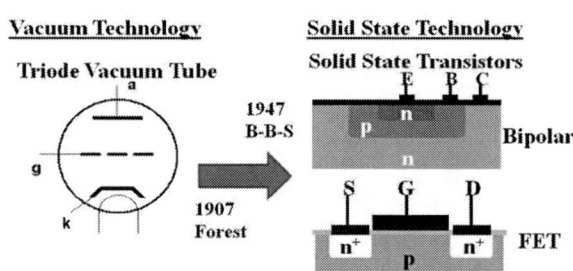

Fig. 2 Triode to solid state devices transformation.

Device Fabrication and Characterization

The SSI-LED has a MOS capacitor structure containing a metal oxide thin film gate dielectric with or without an embedded layer sputter deposited on a p-type Si wafer. After the post deposition annealing (PDA) using a rapid thermal annealing equipment, e.g., up to 1,000°C, an ITO layer was deposited and etched into the gate electrodes. Then, an aluminum layer was deposited at the back of the wafer followed by the final post metal annealing (PMA), e.g., 400°C in H_2/N_2 for 5 minutes. The device was measured for the *I-V*, *C-V*, and *I-t* (time) curves. The emitted light was measured with a spectrometer through an optical fiber located at approximately 2 mm from the ITO electrode.

Light Emission Phenomenon and Optical Characteristics

Fig. 3(a) shows bright light emissions from a single and a cluster of SSI-LEDs containing the HfO_2 dielectric film at the gate voltage V_g of -20V. The light emission phenomenon was only observed under the negative V_g not the positive V_g stress condition. Also, the brightness of the emitted light appears to increase with the increase of the magnitude of V_g. Fig. 3(b) shows several lighted patterns generated from an array of 300 μ diameter SSI-LEDs made of the ZrHfO dielectric stack driven through metal lines connected to contact pads located at the periphery of the chip.

(a)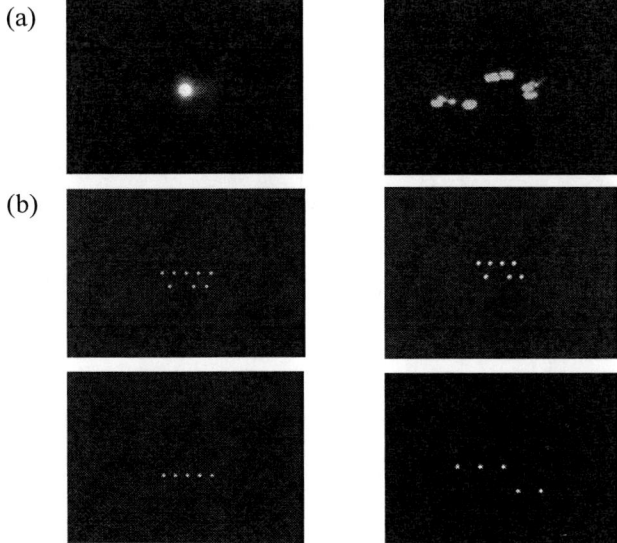

(b)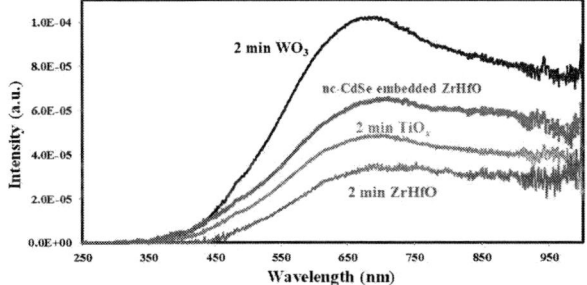

Fig. 3 (a) Light emission from a single and a cluster of SSI-LEDs of HfO$_2$ gate dielectric, and (b) lighted patterns from an array of SSI-LEDs.

Lights in the same visible and near IR wavelength range were emitted from SSI-LEDs containing different types of metal oxide dielectric thin films, as shown in Fig. 4. The spectrum shape, intensity, and the location of the peak wavelength (λ_{peak}) are dependent on the type of the metal oxide material. For all samples, the λ_{peak}'s are located in the visible wavelength range, which is similar to that of the sunlight. It is different from the light emitted from the tungsten incandescent lamp, which peaks in the IR region.

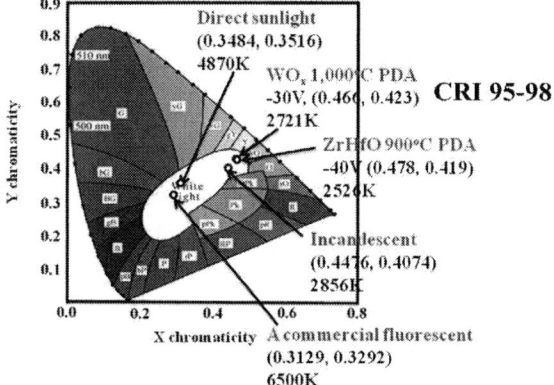

Fig. 4 Emission spectra of SSI-LEDs containing different metal oxide dielectrics. Light emission is influenced by the initial metal oxide material and structure.

Fig. 5 shows that lights from different SSI-LEDs are located in the warm white light region of the 1931 CIE diagram. The SSI-LEDs have high color rendering indices (CRI's) of 95-98. It is close to that of the black body emission of the incandescent bulb, e.g., 100, and higher than that of the commercial LED/YAG:Ce phosphor light, e.g., 79, or the standard cool white fluorescent lamp, e.g., 62 (5). The correlated color temperature (CCT) of the

SSI-LED is usually between 2,500K and 3,500K depending on the V_g, the metal oxide material, thickness, the PDA temperature, etc. (6).

Fig. 5 CIE coordinates, CRI's, and CCT's of SSI-LEDs, conventional incandescent bulb, fluorescent bulb, and sunlight.

Mechanism of Light Emission and SSI-LED Structure

Fig. 6 shows the typical J (leakage current density)-V_g curve of the SSI-LED. Before the dielectric breakdown, charges are transferred through the high-k dielectric following the Schottky emission mechanism at the very low $|V_g|$ and the Poole-Frenkel mechanism at the higher $|V_g|$ (7). However, after the dielectric breakdown, it follows the Ohm's law, i.e., the leakage current increases linearly with the increase of the voltage. The light emission occurs after the dielectric breakdown. The dielectric breakdown caused the formation of conductive paths in the dielectric film. The leakage current of the device after the dielectric breakdown is always higher than that before the dielectric breakdown, i.e., $|V_g|$ < $|V_{BD}|$ (3). The conductive paths are formed permanently.

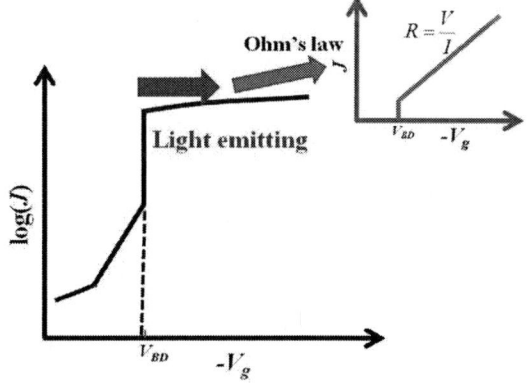

Fig. 6 J-V curve of the SSI-LED in regions before and after light emission. Light emission was only observed after dielectric breakdown under condition of hole-accumulation and -injection from the p-type Si.

The light emission phenomenon occurs only under the hole-injection condition, i.e., the negative V_g stress, not the electron-injection condition, i.e., the positive V_g stress. There is no obvious dielectric breakdown under the electron injection condition, i.e., the leakage current increases gradually without an abrupt change at a certain V_g.

The light emitted from the SSI-LED is composed of discrete tiny bright dots each of which has a sharp edge, as shown in Fig. 7(a) and (b). Therefore, each bright dot is contributed by a single conductive path. These dots are of different sizes but evenly distributed across the electrode surface, as shown in Fig. 7(c). This is different from the uniform light emission from a conventional semiconductor based LED.

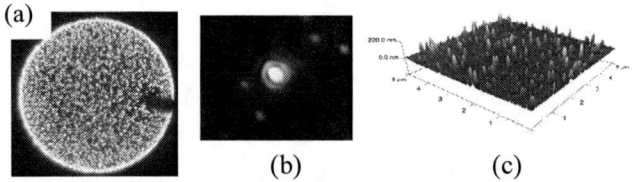

Fig. 7 (a) Enlarged view of a LED, (b) an individual bright dot, and (c) AFM topographic view of surface after light emission and ITO strip.

The conductive paths are physically formed from the breakage of chemical bonds in the original film. For example, Fig. 8 shows the Hf bond structures of a SSI-LED (a) before and (b) after light emission at V_g = -60V (6).

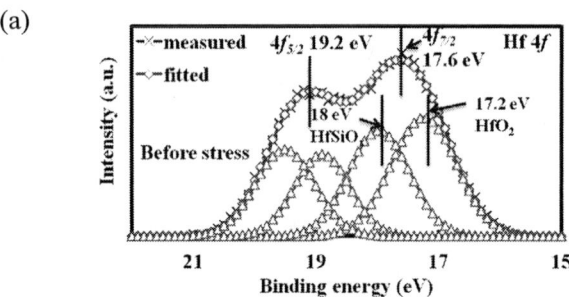

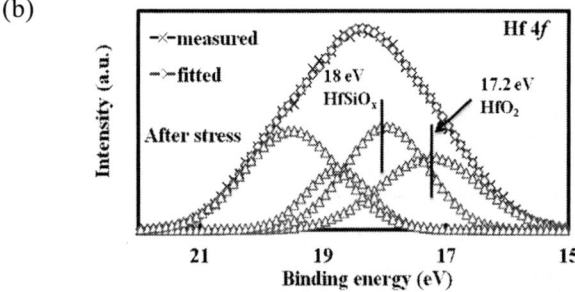

Fig. 8 XPS of Hf 4f spectra of a LED with 6 min deposited ZrHfO and 1,000°C PDA (a) before and (b) after 1 min V_g = -60V stress (6).

The original film contained stoichiometric HfO_2 in the bulk film and $HfSiO_x$ in the interface layer. After light emission, the HfO_2 peak ratio decreased due to the increase of the interface layer thickness. Separately, the interface layer became more SiO_2-rich after the light exposure process. Although the XPS spectrum does not provide the nanoscale information of the conductive path, the comparison of the spectra of the sample before and after the light emission helps the understanding of the bond change in the process.

The emission light intensity increases with the increase of the driving voltage and the λ_{peak} shifts to the red direction, as shown in Fig. 9 (4), which can be explained by the increase of the thermal energy in the conductive path.

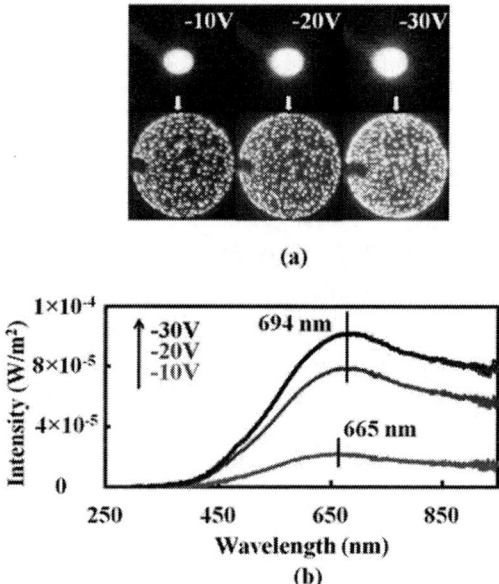

Fig. 9 (a) Low and high-magnification photos and (b) spectra of a WO_3 SSI-LED stressed at different V_g's (4).

In summary, the SSI-LED is a device composed of a large number of parallel-connected tiny incandescent light bulbs. The conductive filaments are embedded in the dielectric film instead of the vacuum, as shown in Fig. 10.

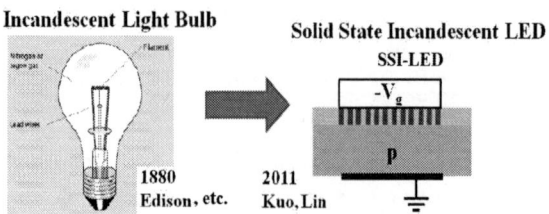

Fig. 10 Comparison of structures of a conventional incandescent lamp and the new SSI-LED.

Lifetime, Driving Method, and Efficiency

The SSI-LED light emission not only is repeatable but also can last for a long time. Fig. 11 shows that a nc-CdSe embedded ZrHfO SSI-LED can emit light continuously for more than 5,600 hours in air. The leakage current and light intensity reduce slightly and smoothly, which is different from the "droop" failure of the conventional single crystal based LED. The long lifetime is due to the unique structure of the dielectric material surrounded conductive paths.

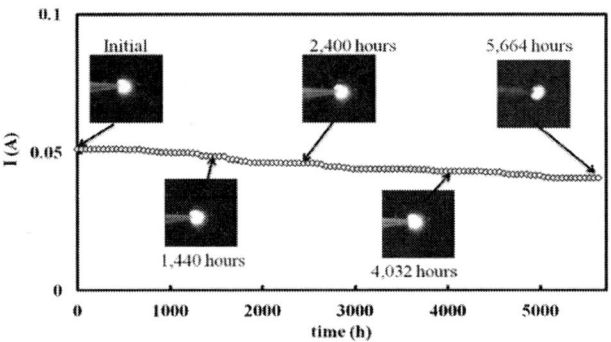

Fig. 11 Lifetime of a nc-CdSe embedded ZrHfO SSI-LED.

The light emission from the SSI-LED can be observed by stressing it with DC voltages or pulsed voltages at > 1,000 Hz frequency with various duty cycles (1,4). The device can also be driven with an AC voltage, as shown in Fig. 12.

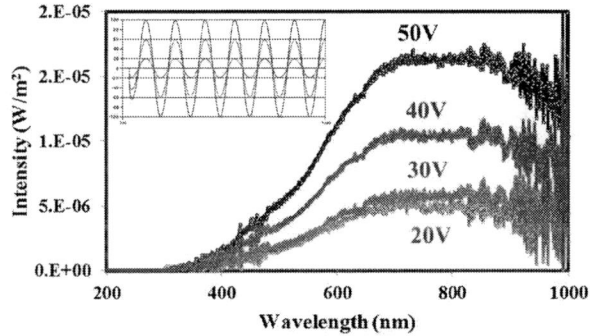

Fig. 12 Emission spectra of a SSI-LED at 50Hz AC.

Currently, due to the lack of the accurate characterization tool, the exact EQE of the SSI-LED is not available. The estimated EQE is < 1% based on the method of dividing the photon energy of the emission spectrum from 200 nm to 1,000 nm measured at 2 mm from the device with the input electric power. According to Forsyhe and Worthing (8), the conversion efficiency of the incandescent lamp is reversely proportional to the width and length of the conductive filament. Therefore, in principle, the SSI-LED should have a much higher EQE than that of the conventional incandescent bulb. The efficiency of the SSI-LED can be improved by many methods, such as the optimization of the dielectric material, thickness, and driving method. Fig. 13 shows that the emission light intensity could be increased by embedding a nanocrystal layer in the ZrHfO film to enhance the formation of conductive paths (1).

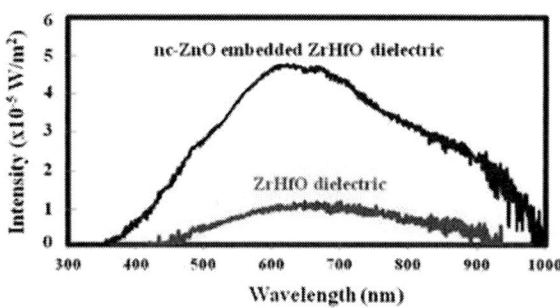

Fig. 13 Emission spectra of SSI-LEDs of ZrHfO and nc-ZnO embedded ZrHfO (1).

The author thanks Chi-Chou Lin, Xiaoning Zhang, and Shumao Zhang for device fabrication and characterizations.

References

(1) Y. Kuo and C.-C. Lin, "A light emitting device made from thin Zirconium-doped Hafnium Oxide high-*k* dielectric film with or without an embedded nanocrystal Layer," *Appl. Phys. Lett.*, Vol. 102, pp.031117, 2013.

(2) Y. Kuo and C.-C. Lin, "Electroluminescence from metal oxide thin films," *ECS Solid State Lett.*, Vol. 2, pp.Q59-61, 2013.

(3) Y. Kuo and C.-C. Lin, "Light emitting from sputter deposited ultra thin hafnium oxide films under electric bias conditions," *Solid State Electronics*, Vol. 89, pp.120-123, 2013.

(4) C.-C. Lin and Y. Kuo, "White light emission from ultra thin tungsten metal oxide film," *J. Vac. Sci. Tech. B*, Vol. 32, pp. 011208, 2014.

(5) http://www.lightbulbsdirect.com/page/001/ctgy/cri

(6) C.-C. Lin and Y. Kuo, "Factors affecting light emission from solid state incandescent light emitting devices with sputter deposited Zr-doped HfO_2 thin films," *ECS J. Solid State Sci. Technol.*, Vol. 3, pp. Q182-189, 2014.

(7) C.-H. Lin and Y. Kuo, "Ruthenium Modified Zr-Doped HfO_2 High-k Thin Films with Low Equivalent Oxide Thickness," *J. Electrochem. Soc.*, Vol. 158, pp. G162-G168, 2011.

(8) W. E. Forsythe and A. G. Worthing, "The properties of tungsten and the characteristics of tungsten lamps," *Astrophysical J.*, Vol. 61, pp.146-185 (1925).

Contact Resistance Reduction using Fermi Level De-pinning Layer for MoS$_2$ FETs

Woojin Park[1], Yonghun Kim[1], Sang Kyung Lee[1], Ukjin Jung[1], Jin Ho Yang[1], Chunhum Cho[2],
Yun Ji Kim[1], Sung Kwan Lim[2], In Seol Hwang[3], Han-Bo-Ram Lee[3], and Byoung Hun Lee[1,2,a)]

[1]School of Materials Science and Engineering, [2]Department of Nanobio Materials and Electronics,
Gwangju Institute of Science and Technology (GIST), Gwangju, South Korea,
[3]Department of Materials Science and Engineering, Incheon National University, Incheon, Korea,
Phone: +82-62-715-2308, E-mail: bhl@gist.ac.kr

Abstract

Achieving a low contact resistance for 2D materials is a critical challenge for device applications. In this work, the contact resistance of MoS$_2$ FETs has been drastically reduced by five times from the reference data using an optimized TiO$_2$ Fermi level de-pinning layer which reduced the effective Schottky barrier height to 0.1 eV. As a result, a very low contact resistance ~5.4 kΩ·μm was achieved without any doping technique.

I. Introduction

The contact resistance of MoS$_2$ FETs became a major challenge in the implementation of 2D device because it is difficult to introduce dopants into 2D materials [1]. Various approaches to reduce the contact resistance have been tried. Several kinds of doping methods have been experimentally studied, but their stability has not been confirmed [1,2]. Halogen doping was proposed as a theoretical possibility to reduce a depletion width [3]. Yet, the doping mechanism for MoS$_2$ has not been well understood yet. Also, the other approaches using a low work function contact metal to enhance the carrier injection efficiency were not effective because the Schottky barrier at metal/MoS$_2$ interface could not be easily modulated due to a Fermi level pinning [4]. Therefore, a stable and effective way to reduce the contact resistance is desperately required to fabricate practical MoS$_2$ FETs.

In our work, very thin Al$_2$O$_3$ and TiO$_2$ layers are inserted between contact metals and MoS$_2$ channel as a de-pinning layer, to achieve a lower contact resistance. After the optimization of interface dielectric layer, this approach indeed yielded a very low contact resistance even without doping in MoS$_2$ layer. The contact resistance of MoS$_2$ FETs was successfully reduced to ~5.4 kΩ·μm, which is five times lower than that of MoS$_2$ FETs without an interlayer. Drain current ~2.51 μA/μm has been obtained at EOT=90nm, L$_{CH}$=3μm, V$_D$=0.5V, V$_G$-V$_{TH}$=20V after improving the contact resistance. This is 10~20 times higher current drivability compared to the prior results [5,6].

II. Device Fabrication and Physical Characterization

Device fabrication flow for back-gate MoS$_2$ FETs are shown in Fig. 1. Mechanically exfoliated MoS$_2$ on Si/SiO$_2$(90nm) substrates is used as a starting material (Fig. 1(a)). IPA cleaning and RTA are performed to remove the residue of scotch tape. Then, an interface dielectric layer is deposited using ALD process followed by an RTA at 600°C for 5 min (Fig. 1(b)). Source and drain contact metals (Ti (50nm)/Au (5nm)) are deposited using PVD process through photolithography and liftoff process (Fig. 1(c)). Fig. 2 shows an SEM image of a MoS$_2$ FET. Fig. 3 shows Raman spectra for single- and multi-layer MoS$_2$ and their optical images. E$^1_{2g}$ - A$_{1g}$ modes are distributed within 18~20 cm^{-1} for single-layer MoS$_2$, while those of multi-layer are over 25 cm^{-1}. SEM photographs show that the coverage of ALD-deposited Al$_2$O$_3$ and TiO$_2$ layer on MoS$_2$ is improved as the thickness of interlayer increases to 2nm (Fig. 4). AFM topographic images and RMS roughness of ALD-deposited Al$_2$O$_3$ and TiO$_2$ film also indicate that the layer coverage is improved at 2nm compared to 1nm (Fig. 5). The surface with TiO$_2$ interlayer shows smoother than that with Al$_2$O$_3$ interlayer. The RMS roughness was ~0.33 nm for 2nm Al$_2$O$_3$ and 0.12 nm for 2nm TiO$_2$, respectively. Representative TEM images of a contact metal (Ti)/interlayer (3nm Al$_2$O$_3$) /MoS$_2$ stack are shown in Fig. 6. The lattice constant of MoS$_2$ is 0.65 nm. The MoS$_2$ has single crystalline with a direction of [0001]. Band diagram for Ti/MoS$_2$ stack and Ti/interlayer/MoS$_2$ stack is shown in Fig. 7, including an inset showing a band offset. The interlayer is expected to work as a Fermi level de-pinning layer which reduces the barrier height [7]. In addition to the de-pinning, dipole effect of interface layer is expected to partially contribute to the reduction of barrier height [8,9].

III. Electrical Characterization

Electrical measurements are conducted using a semiconductor parameter analyzer. Figure 8(a) and 8(b) are the transfer characteristics and the output characteristics of MoS$_2$ FETs with and without

978-1-4799-8002-4/14 $31.00 © 2014 IEEE

interlayers. MoS$_2$ FETs usually have negative threshold voltage showing depletion-mode behavior.

After adding the interlayer, the drain current drastically increased and total resistance decreased accordingly. The I$_{ON}$/I$_{OFF}$ ratio was over 10^3 for all devices. Average drain current values increased from 0.19 μA/μm to 0.93 μA/μm. Thus, average total resistance was reduced from 1080 kΩ·μm to 181~ 184 kΩ·μm with the inserting interlayer (Fig. 9, Fig. 10, and Table. 1). Interestingly, the average of hysteresis in I$_{DS}$-V$_{BG}$ curves also decreased from 16.8V to 9.6V with the interlayer (Fig. 11). The highest drain current ~2.51 μA/μm and the lowest total resistance of 37.4 kΩ·μm are obtained with TiO$_2$ 2nm (EOT=90nm, L$_{CH}$=3μm, V$_D$=0.5V, V$_G$-V$_{TH}$=20V). This drain current value is almost 10~20 times higher than that of prior works after a proper normalization considering EOT, L$_{CH}$, and V$_D$ [5,6].

The origin of this improvement is primarily attributed to the unpinning of Fermi level, which drastically reduced the contact resistance. Average barrier height values extracted using Temperature dependence of drain current was measured to extract the barrier height using Arrhenius plots as shown in Fig. 12. The barrier height is calculated using Richardson-Schottky equation shown below [10]

$$I = AA^*T^2 \exp\left[-\frac{\left(\phi_B - \sqrt{q^3 V/4\pi\epsilon_0\epsilon_r d}\right)}{k_B T}\right] \quad (1)$$

where A is the contact area, A^* is the effective Richardson constant, Φ_B is the Schottky barrier height, q is the electron charge, V is the applied forward bias, ε_0 and ε_r are the permittivities of the vacuum and MoS$_2$, respectively, d is the width of the interface barrier, k_B is the Boltzmann constant.

The barrier height of Ti/MoS$_2$ Schottky contact was 0.18 eV without the interlayer. This barrier height is reduced to 0.13 eV and 0.09 eV with 2nm Al$_2$O$_3$ and TiO$_2$ interlayer, respectively (Fig. 12). The difference between Al$_2$O$_3$ and TiO$_2$ is attributed to the difference in the interfacial dipole. Average barrier height reduction as a function of Al$_2$O$_3$ and TiO$_2$ interlayer thickness is shown in Fig. 13. Higher current with better dielectric coverage indicates that the de-pinning is strongly affected by the film coverage.

Series resistance (R$_{SD}$) is extracted using the ratio of two linear DC I$_{DS}$-V$_{BG}$ curves measured at different drain bias [12]. Fig. 14(a) shows a representative case for the devices with 2nm TiO$_2$ interlayer. Assuming R$_{SD}$ is not significantly changed by V$_D$, 3.5 kΩ of R$_{SD}$ is extracted from two I$_{DS}$-V$_{BG}$ curves. R$_{SD}$ values of devices with different interlayers are shown in Fig. 14 (b). The R$_{SD}$ is drastically reduced from ~125 kΩ to ~3.5 kΩ after adding the interfacial layer. The lowest

contact resistance is ~5.4 kΩ·μm. The contact resistance values of MoS$_2$ FETs reported in recent literature are summarized in Fig. 15. There are a few results reporting the contact resistance lower than ours. However, our result is obtained without using any doping in MoS$_2$ side. Thus, we expect further improvement with an addition of proper doping process in MoS$_2$ side [1].

Finally, there was a concern that the high-k dielectric used as an interlayer may cause a current fluctuation through a transient charge trapping. Thus, 1/f noise characteristics of drain current were analyzed as shown in Fig. 16. Interestingly, 1/f noise was reduced with an addition of better interlayer (Fig. 16 (a)). The linear correlation between 1/f noise versus resistance indicates that this result is primarily due to the current increase by the reduced contact resistance (Fig. 16 (b)).

IV. Conclusion

Drastic contact resistance decrease at contact metal and MoS$_2$ channel interface has been reported using TiO$_2$ de-pinning layer. The best contact resistance and drain current are ~5.4 kΩ·μm and ~2.51 μA/μm, respectively. This contact resistance reduction is attributed to the barrier height decrease and dipole effects. The interlayer technology combined with a proper doping can be a practical contact technology for 2D devices.

Acknowledgments

This research was supported by Global Frontier Program through the Global Frontier Hybrid Interface Materials(GFHIM) (2013M3A6B1078873) of the National Research Foundation of Korea(NRF) funded by the MOSIP, Korea and the Future Semiconductor Device Technology Development Program (10044842) funded By MOTIE and KSRC(Korea Semiconductor Research Consortium.

Reference
[1] L. Yang et al., Symp. VLSI Tech, p. 192. 2014.
[2] Y. Du et al., IEEE Electron. Device. Lett., vol. 34, p. 1328, 2013.
[3] Q.-Q. Sun et al., Appl. Phys. Lett, vol. 102, p. 093104, 2013.
[4] S. Das et al., Nano. Lett.. vol. 13, p. 100, 2013.
[5] S. Kim et al., Nat. Commun., vol. 3, p. 1, 2012.
[6] B. Radisvljevic et al., Nat. Nanotechnol.,vol. 6, p. 147, 2011.
[7] C. Gong et al., Nano. Lett., vol. 14 p. 1714, 2014.
[8] S.T. Zhang et al., Appl. Phys. Lett., vol. 89, p. 043502 102, 2006.
[9] B.E. Coss et al., IEEE Electron Device Lett., vol. 32, p. 862, 2011.
[10] J.W. Yoon et al., Small, 9, p.3295, 2013..
[11] J.P. Campbell et al., IEEE Electron Device Lett., vol. 32, p. 1047, 2011.
[12] H. Liu et al., ACS Nano., vol. 6, p. 8563, 2012.
[13] J. Lee et al., IEDM Tech. Dig., p. 19.2, 2013.

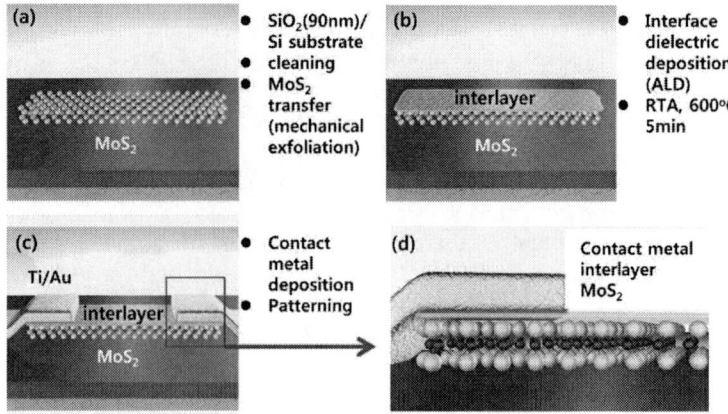

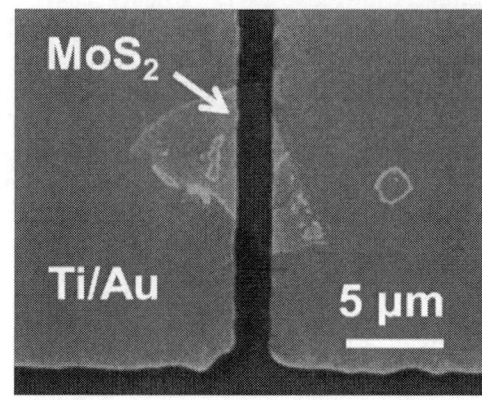

Fig. 1 Fabrication flow and schematic illustration of a MoS₂ back-gate FET with insulating interlayer in this work.

Fig. 2 SEM image of a MoS₂ FET (L$_{CH}$=3 μm, W=7 μm). Magnification is x 30000.

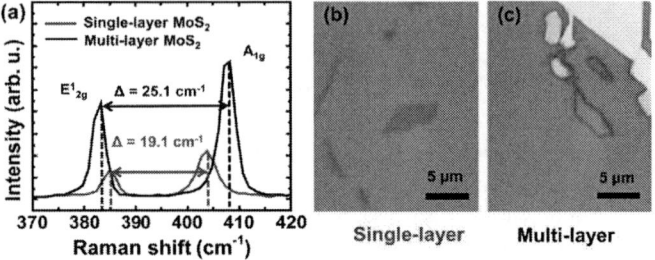

Fig. 3 (a) Raman spectra for single- and multilayer MoS₂. Optical images for (b) single- and (c) multilayer.

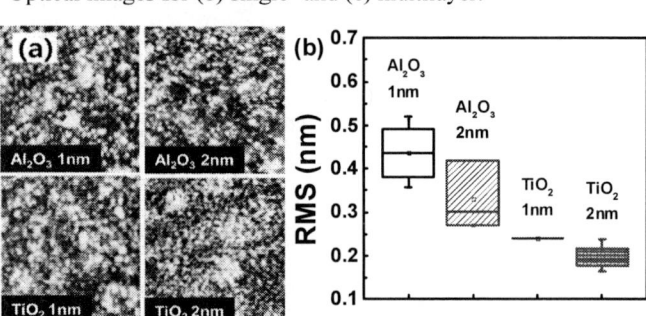

Fig. 5 (a) Non-contact AFM topographic images of TiO₂ and Al₂O₃ interlayer (1nm, 2nm). Scan area is 0.5 μm × 0.5 μm. (b) Average value of RMS roughness for different interlayer materials (1nm, 2nm)

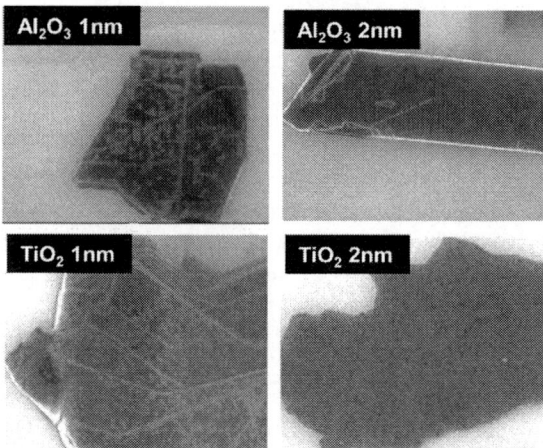

Fig. 4 SEM images of MoS₂ covered with Al₂O₃ and TiO₂. For 1nm Al₂O₃ and TiO₂ layers, film coverage on MoS₂ is poor with a large exposed area. Coverage is improved at 2nm.

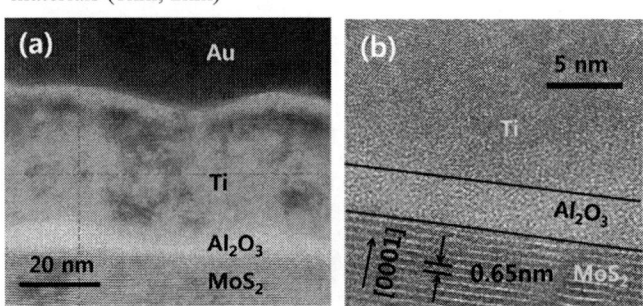

Fig. 6 (a) TEM image of contact metal (Ti/Au)/interlayer/MoS₂ structure. (b) HRTEM image. The lattice constant of MoS₂ is 0.65 nm. The MoS₂ is single crystalline with a direction of [0001].

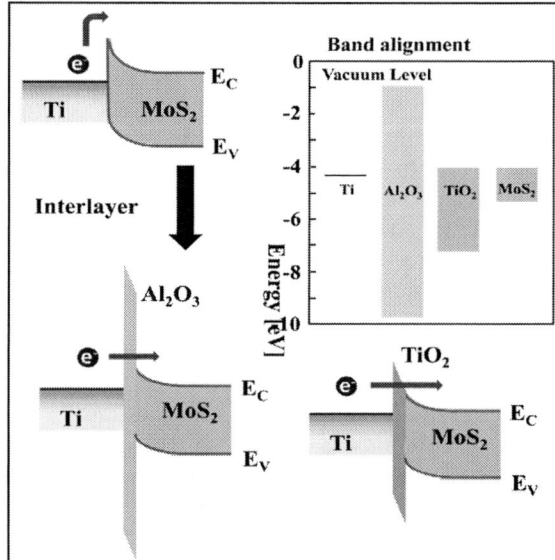

Fig. 7 Explanations of interlayer effects: 1) Interlayers can prevent Fermi level pinning. 2) Dipole effect reduce the barrier height.

978-1-4799-8002-4/14 $31.00 © 2014 IEEE   110

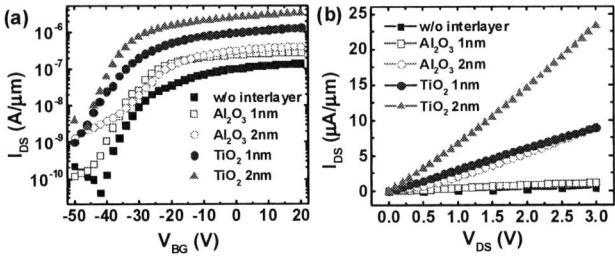

Fig. 8 (a) I_{DS}-V_{BG} and (b) I_{DS}-V_{DS} for with and without Al_2O_3 and TiO_2 interlayer. I_{DS}-V_{BG} is measured at V_D = 0.5V. I_{DS}-V_{DS} is measured at V_G = 0V. The highest drain current is 2.51 μA/μm and the lowest total resistance is 37.4

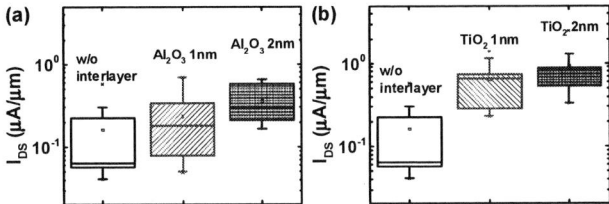

Fig. 9 Average values of drain current for MoS_2 FETs with and without (a) Al_2O_3 and (b) TiO_2 interlayer. Drain current is extracted at V_G-V_{TH} = 20V. (L_{CH} = 3μm, V_D = 0.5V)

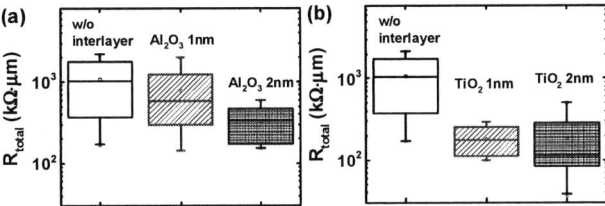

Fig. 10 Average values of total resistance for MoS_2 FETs with and without (a) Al_2O_3 and (b) TiO_2 interlayer. Total resistance is extracted at V_G-V_{TH} = 20V. (L_{CH} = 3μm, V_D = 0.5V)

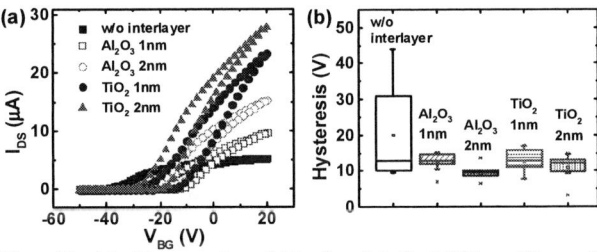

Fig. 11 (a) Hysteresis width for MoS_2 FETs with and without Al_2O_3 and TiO_2 interlayer. (b) Average value of hysteresis width. Average hysteresis width is reduced with interlayer. (L_{CH} = 3μm, V_D = 0.5V)

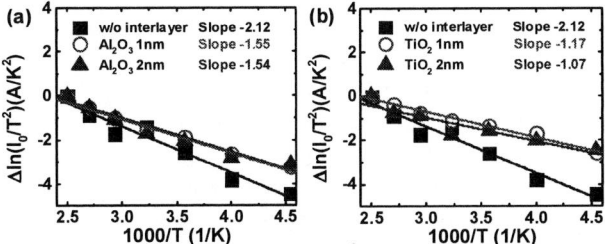

Fig. 12 Arrhenius plots [$\Delta\ln(I_0/T^2)$ versus 1000/T] for (a) with and without Al_2O_3 interlayer (b) with and without TiO_2 interlayer. Temperature dependency is reduced with interlayer. The lowest temperature dependency is observed for 1nm and 2nm TiO_2 interlayer cases.

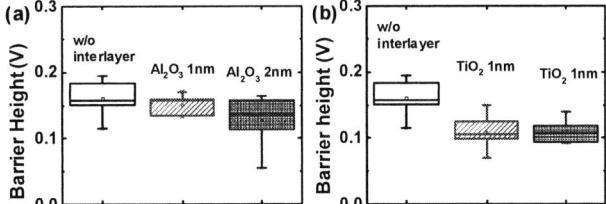

Fig. 13 Barrier height with various interlayer thickness for (a) Al_2O_3 and (b) TiO_2 interlayer.

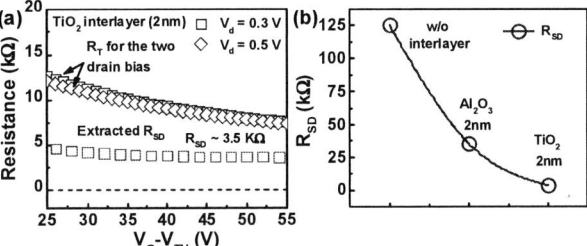

Fig. 14 (a) Total resistance (R_T = V_{DS}/I_{DS}) as a function of gate overdrive for V_D = 0.3 V (open squares) and V_D = 0.5 V (open diamonds) including the extracted R_{SD} for MoS_2 FET with TiO_2 (2nm) interlayer. (b) Extracted series resistance (R_{SD}) values for without interlayer, Al_2O_3 (2nm), and TiO_2 (2nm) cases.

	R_{total} (kΩ·μm)	I_{DS} (μA/μm)
w/o interlayer	1090	0.19
Al_2O_3 1nm	792	0.24
Al_2O_3 2nm	338	0.36
TiO_2 1nm	181	0.67
TiO_2 2nm	184	0.93

Table. 1 Average values of total resistance and drain current for MoS_2 FETs with Al_2O_3 and TiO_2 interlayer.

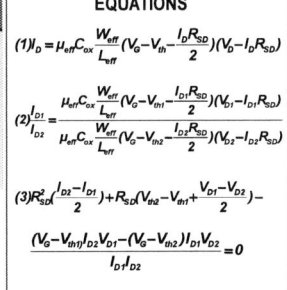

EQUATIONS

$$(1) I_D = \mu_{eff} C_{ox} \frac{W_{eff}}{L_{eff}} (V_G - V_{th} - \frac{I_D R_{SD}}{2})(V_D - I_D R_{SD})$$

$$(2) \frac{I_{D1}}{I_{D2}} = \frac{\mu_{eff} C_{ox} \frac{W_{eff}}{L_{eff}}(V_G - V_{th1} - \frac{I_{D1} R_{SD}}{2})(V_{D1} - I_{D1} R_{SD})}{\mu_{eff} C_{ox} \frac{W_{eff}}{L_{eff}}(V_G - V_{th2} - \frac{I_{D2} R_{SD}}{2})(V_{D2} - I_{D2} R_{SD})}$$

$$(3) R_{SD}^2(\frac{I_{D2} - I_{D1}}{2}) + R_{SD}(V_{D2} - V_{th1} + \frac{V_{D1} - V_{D2}}{2}) - \frac{(V_G - V_{th1})I_{D2}V_{D1} - (V_G - V_{th2})I_{D1}V_{D2}}{I_{D1}I_{D2}} = 0$$

Equations for R_{SD} extraction

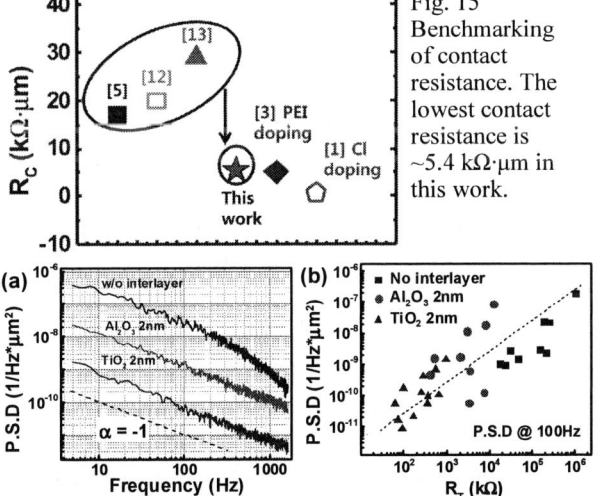

Fig. 15 Benchmarking of contact resistance. The lowest contact resistance is ~5.4 kΩ·μm in this work.

Fig. 16 (a) 1/f noise of MoS_2 FET with various interlayer conditions were measured. 1/f noise level was lowest for the TiO_2 interlayer (b) Channel noise and resistance have a linear correlation. The result is due to the increased current and the reduced resistance with interlayer.

Towards High-Performance Two-Dimensional Black Phosphorus Optoelectronic Devices: the Role of Metal Contacts

Yexin Deng[1,3*], Nathan J. Conrad[1,3], Zhe Luo[2,3], Han Liu[1,3], Xianfan Xu[2,3], Peide D. Ye[1,3#]

[1] School of Electrical and Computer Engineering, [2] School of Mechanical Engineering and [3] Birck Nanotechnology Center, Purdue University, West Lafayette, IN 47907, USA; Email: *deng58@purdue.edu, #yep@purdue.edu

Abstract

The metal contacts on 2D black phosphorus field-effect transistor and photodetectors are studied. The metal work functions can significantly impact the Schottky barrier at the metal-semiconductor contact in black phosphorus devices. Higher metal work functions lead to larger output hole currents in p-type transistors, while ambipolar characteristics can be observed with lower work function metals. Photodetectors with record high photoresponsivity (223 mA/W) are demonstrated on black phosphorus through contact-engineering.

I. Introduction

The discovery of graphene has driven the extensive research interests on 2D materials. However, a lack of bandgap limits its applications on electronic and optoelectronic devices [1-2]. This has led to the recent intensive research on other 2D layered materials with a bandgap, such as MoS_2. Recently, black phosphorus (BP) has been found to be an excellent candidate for 2D electronics and optoelectronics due to its high hole mobility (>10000 cm^2/Vs) and thickness-dependent direct bandgap [3-6]. BP is a stack of BP monolayers (termed 'phosphorene') with a puckered honeycomb structure, bound together by van der Waals interaction (Fig. 1). Unlike graphene, the bandgap (~0.3 eV or more) of few-layer black phosphorus enables an ON/OFF ratio of >10^5 in field-effect transistors (FET), and it has a hole mobility of up to ~1000 cm^2/Vs at room temperature [3-5]. The ultra-thin 2D nature makes it promising for use in aggressively scaled FETs, as well as thin film transistors for flexible electronics. Different from MoS_2 with a ~1.8 eV direct bandgap only in its monolayer form, BP shows a thickness-dependent direct bandgap, which exhibits ~0.3 eV in its bulk form and increases to >1 eV in its monolayer form -phosphorene [3-4]. Its relatively small direct bandgap makes it ideal for inferred optoelectronic applications. The first BP phototransistor shows a response to wavelength of up to 940 nm. However, it exhibits a relatively low photoresponsivity of 4.8 mA/W [6].

Metal contacts are one of the most important issues for 2D FETs and optoelectronic devices. A clear understanding of metal contacts on BP FETs and photodetectors is imperatively needed to improve the device performance. In this paper, we try to shed a light on the role of various metal contacts on BP FETs and photodetectors. The work function of metal plays an important role on the hole/electron conductions through the Schottky barriers at the metal-BP contacts. FETs with larger work functions metals as contacts exhibit larger hole drain currents, while ambipolar characteristics can be observed on devices with lower work function metals. A photodetector with a record high photoresponsivity (223 mA/V) is demonstrated on BP through contact engineering.

II. The Role of Metal Contacts on FETs Performance

a) **Device Fabrication**: Few-layer BP flakes were mechanically exfoliated from bulk material onto a p+ doped silicon substrate capped with a 90 nm SiO_2. Standard electron beam lithography was used to define the contact patterns. Metal contacts were formed by electron beam evaporation and lift-off process. Fig. 2 shows a schematic of a fabricated BP FET. The SiO_2 and p+ Si were used as a gate dielectric, and a back gate, respectively. To study the effects of various metals with different work functions on BP, two different metals were used as contact metals to form two FETs on the same flake (Fig. 3), negating the effect of flake-to-flake variability. Al (4.1 eV), Ti (4.3 eV) and Pd (5.1 eV) were used as metal contacts. The thicknesses of the flakes varied from 5 nm to 20 nm as determined by atomic force microscopy (AFM). The Raman-activated modes in Raman spectra are consistent with the previous work [4], confirming the nature of BP flakes (Fig. 4). All the measurements were performed at room temperature and in ambient atmosphere.

b) **Device Performances and Analyses**: First of all, the electrical characteristics of BP FETs are studied. As shown in the output curves of a typical BP FET using Pd as contacts (Fig. 5), a well-behaved current saturation is observed at a channel length of 1μm. Combined with its ultra-thin channel thickness, which is required for dimensions scaling by Moore's law, it shows the promise of BP for low-power high-speed FET applications. The device on the same flake using Ti as contacts shows a relatively lower on-state current under the same bias conditions. As the device structure is the same, this indicates a larger contact resistance, thus a larger Schottky barrier (SB) at the metal-semiconductor (MS) contact. This phenomenon is observed on all fabricated devices summarized in Fig. 6. Moreover, the devices with Pd as contacts always show larger current no matter how we change the order of metal deposition. The work function of metal is the determinant fact. From the transfer curves of this device (Fig. 7), a larger ON/OFF ratio can be observed on the devices with Ti as contacts. This underlines the nature of BP transistors as SB

978-1-4799-8002-4/14 $31.00 © 2014 IEEE

transistors in this case. As the contact resistance becomes relatively large in Ti device, the modulation of the SB width via back gating allows a larger ON/OFF ratio in Ti device [7]. The switch mechanism on BP FETs is controlled by two SBs as described explicitly in MoS_2 FETs [7]. This confirms that the SB height for hole transport in Ti devices is larger than that of Pd devices, and, unlike on MoS_2, metal contacts on BP are not strongly pinned. A small ambipolar characteristic can be observed on Ti devices (Fig. 7) and much clearly in Al devices (Fig. 8), as the SB height for electrons in Al is much smaller than that in Ti or in Pd. These results suggest the band diagram of devices with different metals on BP to be like in Fig. 9 (a), in which the SB height for holes at the contact increases as the work function of the metal decreases, and the SB height for electrons decreases as the work function of metal decreases. The strong ambipolar characteristics on Al devices can be illustrated by Fig. 9 (b) and (c). Under negative/positive gate bias, the band bending boosts the hole/electron injection at the MS contact at drain/source as the Al aligns close to the middle of bandgap, resulting in the ambipolar characteristics. Based on these understandings, a Schottky diode with asymmetric contact is demonstrated on the same flake with Ti/Pd as contacts (Fig. 10 and Fig. 11). A rectification ratio of ~100 is obtained and is expected to be larger if using metals with larger work function differences. The conduction mechanism can be understood by the band diagram in Fig. 12. As Pd can make a better p-contact on BP, the Ti SB is mainly used as a Schottky diode.

III. Contact-Engineered High-Performance Photodetectors

Based on its relatively small and thickness-dependent direct bandgap, BP photodetector is suitable for broadband photodetection. Here we fabricated the photodetectors using the same fabrication process as described above. The photocurrent measurement method is shown in Fig.13. A 633 nm He-Ne laser was focused on the active device region. Based on the study above, the band diagram of the devices with various metal contacts with and without illumination is shown in Fig. 14. Photoresponsivity (R) is defined as $R = I_{ph}/P_{laser}$ where I_{ph} is the photocurrent and P_{laser} is the laser power. With a larger SB height, the photo-generated holes are more easily trapped in the device. This constrains the R of the photodetector, as the photo-generated holes cannot be fully collected. To increase the relatively small R in the first reported BP photodetector [6], here we use Pd as metal contacts in order to reduce the SB height and improve the collection of photo-induced carriers. The device was under a periodic laser illumination and was biased at a small voltage (50 mV). The photoresponse is highly repeatable with long and short illumination period (Fig. 15, 16), demonstrating its capability for photodetection. The detected photocurrent increases as the incident laser power increases (Fig. 17). Moreover, the back gate (V_{gs}) can be used to modulate the band diagram, which results in different R with different V_{gs}

because the SB changes with V_{gs} or electrostatic doping of the channel (Fig. 18). As the back gate voltage increases, R decreases (Fig. 19). However, the ratio of $I_{illumination}/I_{dark}$, where $I_{illumination}$ and I_{dark} is I_d when the device is with and without illumination, increases as V_{gs} increases, suggests a better signal to noise ratio. These phenomena can be understood by the band diagram in Fig. 20. Under negative V_{gs}, the width of the SB near the contact becomes narrower, which makes it easier for holes to go through the barrier. Under positive V_{gs}, the width of SB becomes wider which limits the photo-generated electrons limits to reach metal contact and results a lower R. Moreover, the increase of electron density by increasing V_{gs} can also lead to a lower R due to a larger recombination rate in the channel. When V_{gs} is negative, the transistor is turned off, resulting in a larger $I_{illumination}/I_{dark}$ ratio. The maximum R in our devices is 223 mA/V at V_{gs}=-30 V and 76 mA/W at V_{gs}=0 (Fig. 19), which is 16 times larger than the value reported in [6] at the same V_{gs}=0 and with an even smaller V_{ds} in our case. R can be further improved by increasing the V_{ds} (Fig. 21). Finally, Table 1 summarizes the performance metrics of different 2D photodetectors reported in literature for benchmarking.

IV. Conclusion

1) Metals with larger work function as contacts can significantly reduce the SB height for hole injection on a BP FET to achieve a higher hole drain current. 2) Ambipolar characteristics are observed on the devices using lower work function metals as contacts. 3) High-performance photodetectors have been demonstrated using optimized metal contacts that can reduce the barrier height for photo-generated carrier collection and realize broadband photodetection, showing a record high photoresponsivity (223 mA/W) on this new 2D material.

Acknowledgement

This work is partly supported by NSF under Grants CMMI-1120577, ECCS-1449270 and ARO W911NF-14-1-0572 monitored by Dr. Joe X. Qiu.

Reference

[1] K. Novoselov et al., Nature, 438, p.197 (2005).
[2] Y. B. Zhang et al., Nature 438, p.201 (2005).
[3] L. Li et al., Nature Nanotechnol, 9, p.372 (2014).
[4] H. Liu et al., ACS Nano, 8, p.4033 (2014).
[5] F. Xia et al., Nature Commun. 5, p.4458 (2014).
[6] M. Buscema et al., Nano Letters,14, p.3347 (2014).
[7] H. Liu et al., ACS Nano, 8, p.1031 (2013).
[8] O. Lopez-Sanchez et al., Nature Nanotechnol, 8, p.497 (2013) .
[9] Z. Yin et al., ACS Nano, 6, p 74 (2012).
[10] N. Perea-López et al., Adv. Funct. Mater., 23, p.5511 (2013) .
[11] F. Liu et al., ACS Nano, 8, p.752 (2013).
[12] P. Hu et al., Nano Letters, 13, p.1649 (2013).
[13] P. Hu et al., ACS Nano, 6, p.5988 (2012)

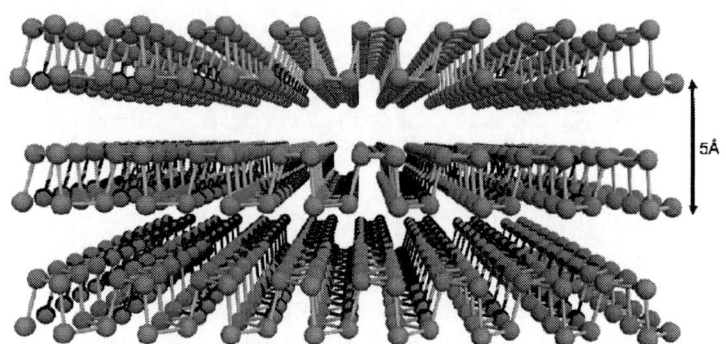

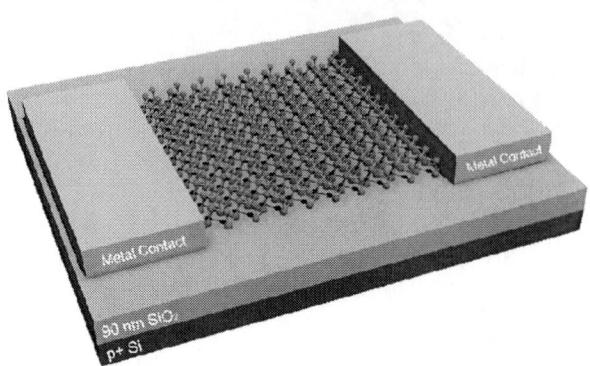

Fig. 1 Atomic structure of black phosphorus, which is a stack of black phosphorus monolayers, bound together by van der Waals interactions. In each of the layer, every phosphorus atom is covalently bonded with three adjacent phosphorus atoms to form a puckered honeycomb structure.

Fig. 2 Schematic of the black phosphorus field-effect transistor using 90 nm SiO_2 as gate dielectric and p+ doped silicon substrate as the back gate.

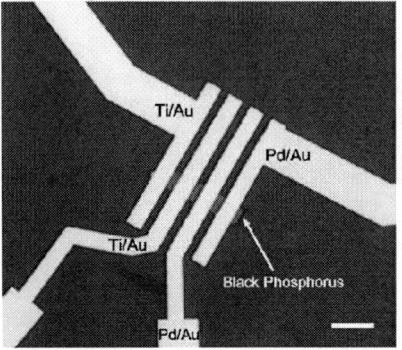

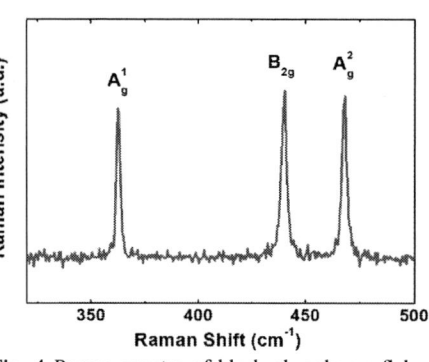

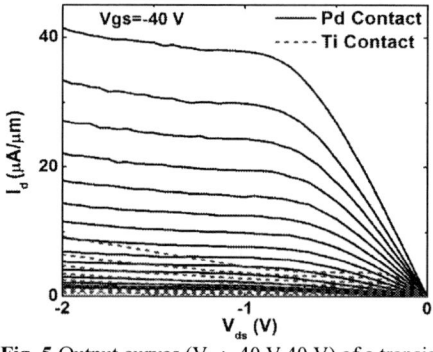

Fig. 3 Optical image of a fabricated device with different metal contacts on the same black phosphorus flake. Scale bar 5µm.

Fig. 4 Raman spectra of black phosphorus flake. The three Raman-activated modes corresponds to three different vibrational modes

Fig. 5 Output curves (V_{gs}: -40 V-40 V) of a transistors using Pd or Ti as source/drain contacts on the same flake. The Pd device shows larger output current.

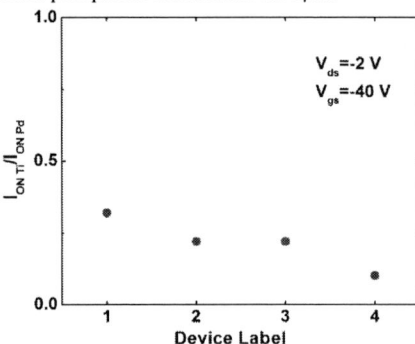

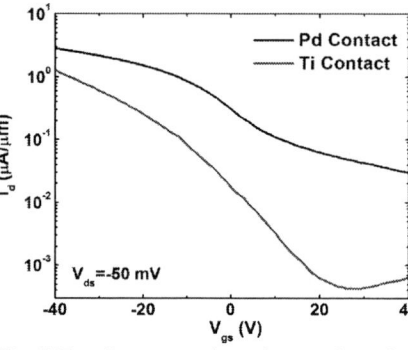

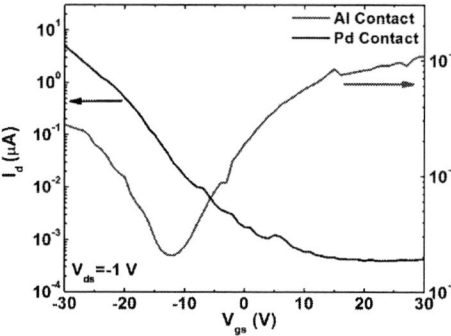

Fig. 6 The ratio of maximum output current of devices with Pd contacts over Ti contacts. The Pd devices show obviously larger current.

Fig. 7 Transfer curves of transistors using Pd or Ti as source/drain contacts. The Ti device shows lower current while maintaining larger ON/OFF ratio.

Fig. 8 Transfer curves of transistor with Pd or Al as source/drain contacts. The device with Al shows obvious ambipolar characteristics.

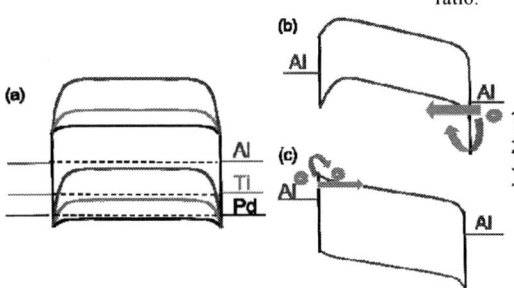

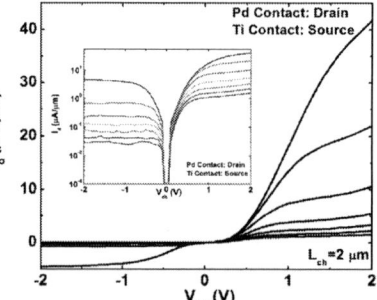

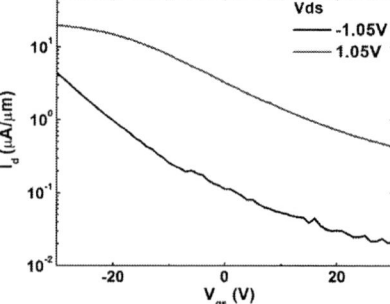

Fig. 9 (a) Band diagram of a transistor with different metal contacts under zero bias. Band diagram of a device with Al contacts with (b) negative and (c) positive back gate voltage.

Fig. 10 Output curves of the device with Ti/Pd as source/drain contact. The inset shows the curves in semi-log scale.

Fig. 11 Transfer curves of the device with Ti/Pd as source/drain contact with positive and negative V_{ds}.

978-1-4799-8002-4/14 $31.00 © 2014 IEEE

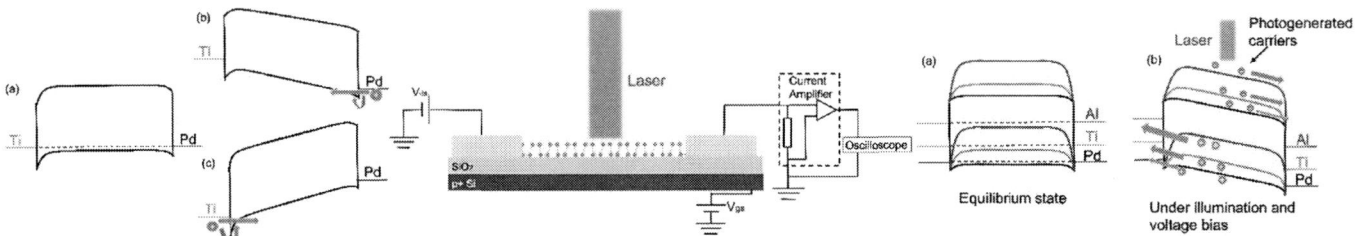

Fig. 12 Band diagram of the device with Ti/Pd as source/drain contact (a) equilibrium state, (b) $V_{ds}>0$ and (c) $V_{ds}<0$.

Fig. 13 Schematic of the setup of the photocurrent measurement under laser illumination. The current in the device is converted to voltage signal, and digitized by an oscilloscope.

Fig. 14 Band diagram of the device (a) at equilibrium state and (b) under illumination and a small voltage bias (V_{ds}) with $V_{gs}=0$..

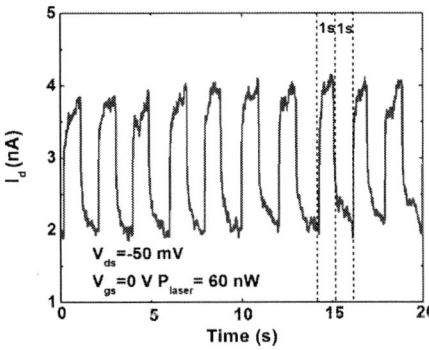

Fig. 15 I_d as a function of time when the laser is turned on/off with a period of 2 s.

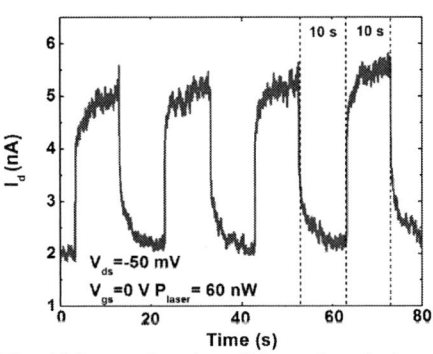

Fig. 16 I_d as a function of time when the laser is turned on/off with a period of 20 s

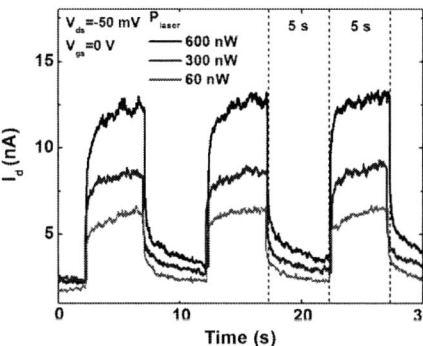

Fig. 17 I_d as a function of time when the laser is turned on/off with a period of 10 s under different incident laser power.

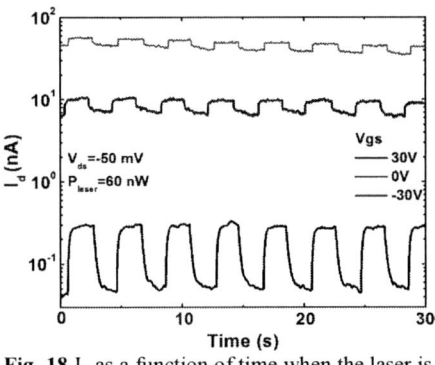

Fig. 18 I_d as a function of time when the laser is turned on/off with a period of 4 s under different back gate voltage.

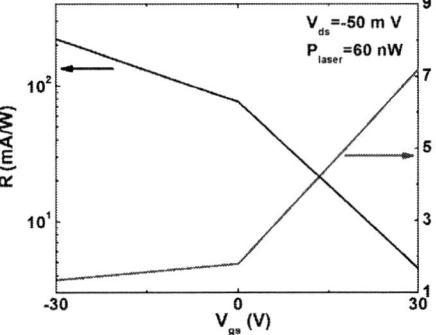

Fig. 19 The photoresponsivity (R) and the $I_{illumination}/I_{dark}$ as functions of back gate voltage. They show different trends when V_{gs} increases.

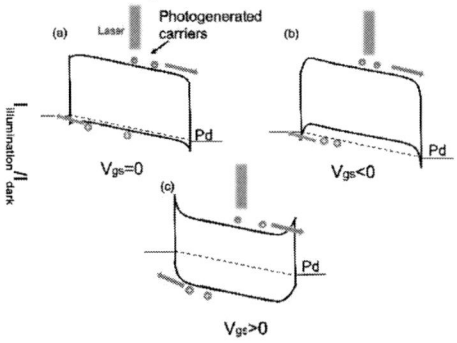

Fig. 20 Band diagram of the phototransistor under illumination with (a) $V_{gs}=0$, (b)$V_{gs}<0$ and $V_{gs}>0$.

Table 1 Summary of 2D Photodetectors

Materials	Measurement conditions	R(mA/W)	t_r (ms)	Spectral range	Ref
>1L BP	V_{ds} = 0.05 V, V_{gs} =0V, λ = 633 nm, P = 60 nW	76	100	Visible–IR	This work
>1L BP	V_{ds} = 0.2 V, V_{gs} =0V, λ = 640 nm, P = 10 nW	4.8	1	Visible–IR	5
1L MoS$_2$	V_{ds} =8V, V_{gs} = −70 V, λ = 561 nm, P = 150 pW	8.8×10^6	4000	Visible	6
1L MoS$_2$	V_{ds} =1V, V_{gs} =50 V, λ = 532 nm, P =80 µW	8	50	Visible	7
>1L WS$_2$	V_{ds} =30 V, V_{gs} = N.A., λ = 458 nm, P =2 mW	2.1×10^{-4}	5.3	Visible	8
>1L GaTe	V_{ds} =5V, V_{gs} =0 V, λ = 532 nm, P =3 × 10^{-5} mW/cm^2	10^7	6	Visible	9
>1L GaS	V_{ds} =2V, V_{gs} =0 V, λ = 254 nm, P = 0.256 mW/cm^2	4.2×10^3	30	UV-Visible	10
>1L GaSe	V_{ds} =5V, V_{gs} =0 V, λ = 254 nm, P =1 mW/cm^2	2.8×10^3	300	UV-Visible	11

Fig. 21 I_d as a function of time when the laser is turned on/off with a period of 4 s under different V_{ds}.

978-1-4799-8002-4/14 $31.00 © 2014 IEEE   115

Atomically Thin Graphene Plane Electrode for 3D RRAM

Joon Sohn*, Seunghyun Lee*, Zizhen Jiang, Hong-Yu Chen and H.-S. Philip Wong

Department of Electrical Engineering, Stanford University, 420 Via Palou Mall, Stanford, CA 94305, USA

Tel: +1 (214) 799-8020, Email: *joonsohn@stanford.edu, *seansl@stanford.edu

Abstract

3Å thick graphene edge was employed in the bit-cost scalable vertical RRAM structure to drastically reduce the total stack height to a single atomic layer. Two-layer 3D-stacked HfO_x RRAM with graphene planar electrode (G-RRAM) is demonstrated in a 3D cross-point architecture with the edge of the graphene plane electrode serving as the bottom electrode of the RRAM. Exceptional memory window (>80×), low reset current (~20 µA), and suitable set/reset voltages (2 to 4 V) were achieved. Large memory window and low SET compliance ensures low reset current and low power consumption. Resistance components were separately measured to verify the role of graphene/oxide interface and the graphene sheet resistance. This work is a significant step toward extreme vertical scaling of 3D vertical stacked memory structures.

1. Introduction

Metal oxide RRAM shows great promise as an emerging non-volatile memory due to its simple structure, scalability, and low power consumption [1]. In order to realize the high density RRAM, several groups have demonstrated bit-cost scalable 3D RRAM using the edge of the 3D stacked metal plane as one of the RRAM electrodes [2-4]. The total stack height (metal plane plus isolation dielectric) must be scaled down to enable a large number of 3D stacks due to limitations of the etch aspect ratio [3]. In this paper, we bring the plane electrode thickness scaling to its extreme by employing graphene as the plane electrode. This enables the vertical RRAM cell to have a metal plane electrode with atomic thickness of ~0.3 nm and yet maintain excellent electrical conductivity of the metal plane. In a previous demonstration of 3D RRAM using Pt as the plane electrode, 60 stacked layers in 3D RRAM were shown to be possible for a lithographic half-pitch (F) = 22 nm and etching aspect ratio of 30 [3]. Following the same approach, using 0.3 nm single-layer graphene and 6 nm thick isolation SiO_2, we project that one can achieve up to 200 stacks for 3D RRAM assuming the top electrode trench has a slope of 89 degrees. In order to compare the graphene electrode with the conventional metal-based electrode, two types of devices were fabricated: graphene based 2-layered 3D cross point G-RRAM and platinum electrode based 1-layered 3D cross point Pt-RRAM. 5 nm thick HfO_x was used as the resistive switching layer. We also illustrate the potential for high density 3D cross point G-RRAM array using SPICE simulations based on experimentally measured device and material properties.

2. Device Fabrication

Fig. 1 shows the fabrication flow of 3D G-RRAM. The schematic of the completely fabricated device of 2L G-RRAM is shown in Fig. 3. The TEM images of G-RRAM and Pt-RRAM are shown in Fig. 2 (a) – (c) and Fig. 2 (d), (e), respectively. In previous work, we mentioned the difficulty in obtaining a vertical side wall with Pt due to the difficulties of dry etching through metal [2]. In contrast, G-RRAM exhibits relatively steeper side wall due to the ease of etching through the thin graphene (compare Fig.2 (a) and (d)). The existence and the quality of graphene were verified with Raman spectrum after the whole process. D-peak to G-peak intensity ratio (I_D/I_G) from the Raman spectrum is a common indicator of disorders in graphene films. Fig. 4(b) shows the 2-dimensional map of I_D/I_G Raman peak ratio. The median value of I_D/I_G ratio is 0.11 over 30 µm × 60 µm area, confirming the low defect level of graphene [5] after the complete fabrication process. The typical single layer graphene Raman signature shown in Fig. 4(c) also verifies the monolayer thickness of graphene.

3. G-RRAM Device Performance

Fig. 5 shows the typical DC I-V curves of the top and the bottom cell of 2L G-RRAM sample. With TiN as the top electrode, the forming voltage is between 4 to 5 V (Fig. 11). A

978-1-4799-8002-4/14 $31.00 © 2014 IEEE

suggested switching mechanism of G-RRAM is shown (Fig. 6). The oxygen ions migrate from the HfO_x to the top electrode (TiN) under the presence of electric field, which would leave oxygen vacancy to form conducting filaments. This is consistent with the current understanding of the switching mechanism of HfOx RRAM with a TiN top electrode and a passive bottom electrode [6]. Fig. 7 shows the resistance distribution of G-RRAM and Pt-RRAM. From Fig. 7, we observe that the memory window of G-RRAM is significantly (>10×) larger than that of Pt-RRAM. The large memory window is a direct consequence of extremely scaled filaments formed from the edge of the very thin graphene electrode. This follows the general trend of RRAM where the HRS/LRS resistance ratio increases with device area down [1]. The high resistance of the HRS enables large memory window with a small SET compliance current and results in low reset current. We notice that both resistance values of the HRS and the LRS in G-RRAM are significantly higher than those of Pt-RRAM. To understand this, a breakdown of the resistance components of G-RRAM is shown in Fig. 8. There can be two contributing factors to the high HRS and LRS. First component is the R_{series} from the graphene electrode and the other is the R_{int} from the graphene edge/oxide interface. In order to confirm the role of R_{series}, we measured the sheet resistance (~6.7k $\Omega/\square$) and the contact resistance (~295 Ω) of graphene electrode (Fig. 9). Moreover, the resistance value for the device without the HfO_x is lower than 10k Ω (Fig. 10) confirming the negligible contribution of R_{series} in the total R. Both results support that high HRS and LRS is mostly attributed to the graphene/oxide interface (R_{int}) and not the series resistance (R_{series}). Fig. 12 shows the reset current distribution. Fig. 13 shows the DC I-V sweep cycling set and reset voltage distribution are similar for the first and second layer of G-RRAM cells. Room temperature retention of >300 min are demonstrated (Fig. 14). The variation of the set and reset voltages for 10 different G-RRAM cells are shown (Fig. 15). As shown in the figures, stacked G-RRAM exhibits consistent characteristics among the top and bottom layers.

4. Plane Electrode Resistance

As discussed in [2], [7], a low-resistance and thin plane electrode is required to minimize voltage drops and achieve large 3D RRAM arrays. Graphene exhibits superior resistance value per thickness (i.e. graphene is 20× thinner and 12× more resistive). From TLM measurements, Ti/Pt sheet (1nm/5nm thick) and graphene film (3Å thick) had sheet R of 558±24 $\Omega/\square$ and 6.7k $\Omega/\square$, respectively. Furthermore, metal films are known to exhibit exponential increase in sheet resistance as the thickness decrease under 10 nm. We expect the sheet resistance of metal to reach several $M\Omega/\square$ when thinned down to sub-1 nm [8]. The intrinsic sheet resistance of graphene is known to be ~6k $\Omega/\square$ [9] and this value can be lowered dramatically to 125-250 $\Omega/\square$ [9, 10] by charge transfer doping.

5. Analysis of 3D G-RRAM Array Size

We adopted the same resistance network and array simulation methodology for the worst-case selected cell of 3D RRAM as in ref. [2][7]. Using the criteria of >100 nA read margin and 75% write access voltage margin, an array size of >100G for a 200-stacked G-RRAM can meet both criteria, while an array of <100k of Pt-RRAM can only meet the read margin criterion. This is mainly because the smaller pillar height (6.3nm/stack) for G-RRAM and doped graphene RRAM (DG-RRAM) as compared to Pt-RRAM (11nm/stack) results in overall low pillar resistance in a 3D cross point architecture. An even larger array is achievable with the low sheet resistance of the doped graphene (100 $\Omega/\square$) from [11] (Fig. 16, Fig. 17). In Fig. 16, $\triangle I_{READ}$ is defined as the difference of the current flowing the read resistor (100kΩ) when RRAM cell is in the HRS or the LRS.

6. Conclusion

We demonstrated the first 3D RRAM with the world's thinnest plane electrode by using atomically thin (~3Å) graphene. 3D RRAM using *metal based* plane electrode with the same thickness will suffer severely from degradation of read/write margins due to high sheet resistance of the plane electrode. Graphene, on the other hand, exhibits superior sheet resistance per thickness compared to conventional metal and the additional capability to be doped to reach lower sheet resistance [9, 10]. Graphene is a key enabler for further vertical scaling of bit-cost scalable 3D RRAM arrays to more than 200 stacks while maintaining low total stack height for a low pillar resistance.

* J. S. and S. L. contributed equally to this work.

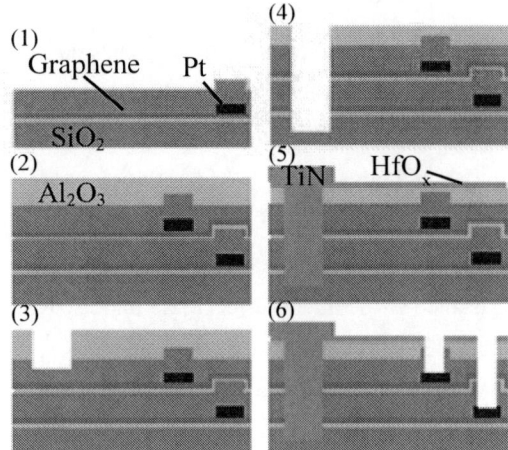

Figure 1. 3D G-RRAM process flow: (1) Graphene is transferred on to a dielectric surface with 5 nm Al_2O_3 and 100 nm of SiO_2. Ti/Pt (3 nm/30 nm) layers are deposited by evaporation and patterned by lift-off process. 60 nm of SiO_2 (LPCVD) is deposited. (2) These processes are repeated two times for two layers of single layer graphene; (3) 50 nm ALD Al_2O_3 is deposited on the top layer for etch hard mask. (4) A trench is etched down to the bottom SiO_2 layer. (5) 5 nm of HfO_x is conformally deposited as the active resistive switching layer and 200 nm of TiN is deposited by sputtering and patterned via lift-off. (6) The contacts are opened via dry etching.

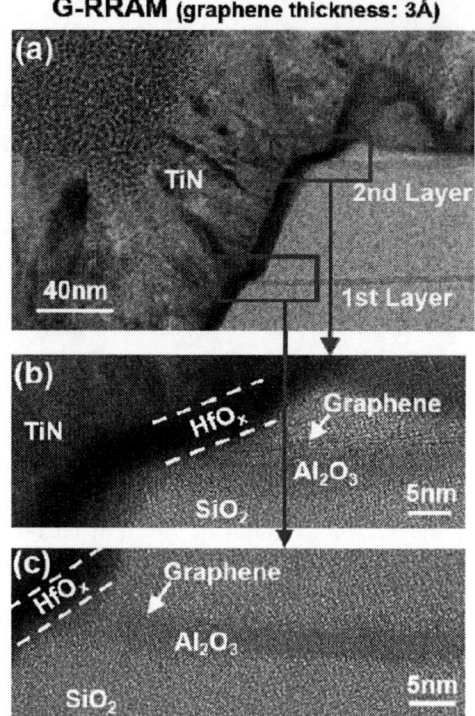

Figure 2. (a) TEM image of the double layer G-RRAM. (b) HR-TEM image of the 1st layer of G-RRAM. Graphene is barely visible since the thickness is close to the resolution limit. (c) HR-TEM image of the 2nd layer of G-RRAM. (d) TEM image of the Pt-RRAM of previous work [2]. (Pt thickness: 25nm) (e) TEM image of the Pt-RRAM of previous work [1]. (Pt thickness: 5nm)

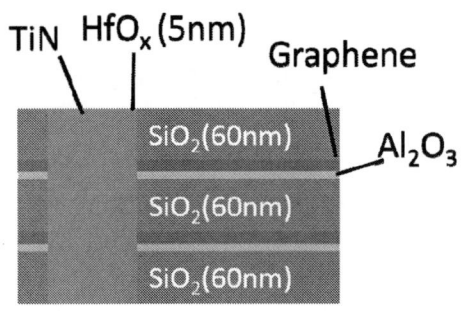

Figure 3. Schematic of completely fabricated 2L G-RRAM. A thin layer (5nm) of Al_2O_3 was deposited before the graphene transfer process to promote graphene adhesion to the surface.

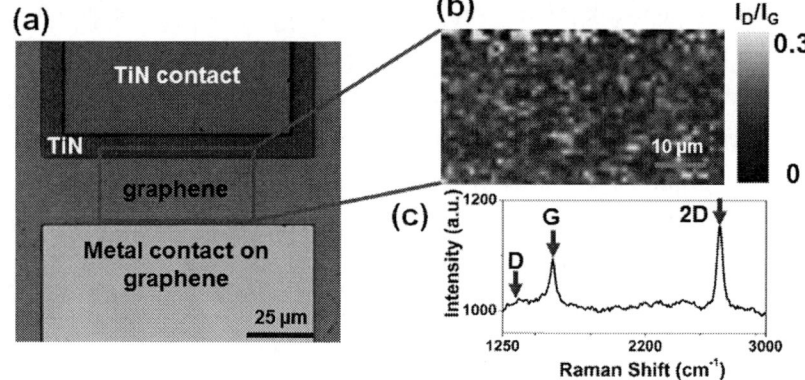

Figure 4. 2-dimensional Raman map of I_D/I_G ratio. The median value of I_D/I_G ratio is 0.11 over 30 μm × 60 μm area (shown as the blue rectangle) confirming the high quality of graphene after the complete fabrication process. The scale bar is 10 μm.

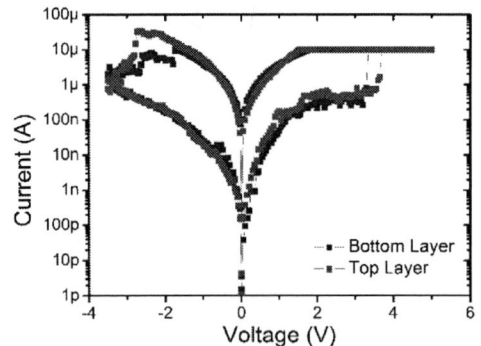

Figure 5. Typical DC I-V switching characteristics of bottom layer and top layer of G-RRAM.

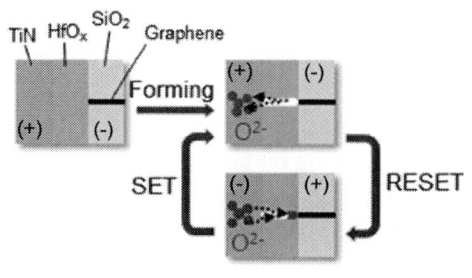

Figure 6. Switching mechanism of G-RRAM. Oxygen ions from the HfO_x are migrated to graphene layer to form a conductive filament. The conductive filament layer is extremely thin due to graphene's atomic thickness (~3Å).

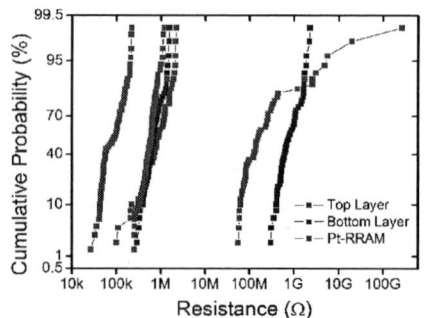

Figure 7. Resistance distribution of 50 cycles of DC sweep for bottom layer and top layer of G-RRAM and Pt-RRAM. Much larger memory windows are observed for G-RRAM compared to Pt-RRAM.

978-1-4799-8002-4/14 $31.00 © 2014 IEEE

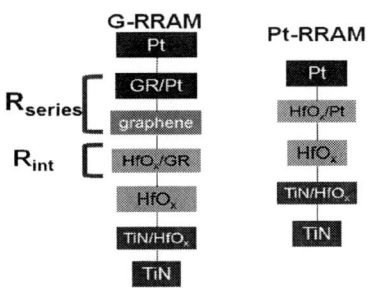

Figure 8. Resistance component breakdown of G-RRAM and Pt-RRAM. In comparison to Pt-RRAM, G-RRAM have two different resistive components: R_{int} from graphene/HfO$_x$ interface and R_{series} from graphene sheet resistance.

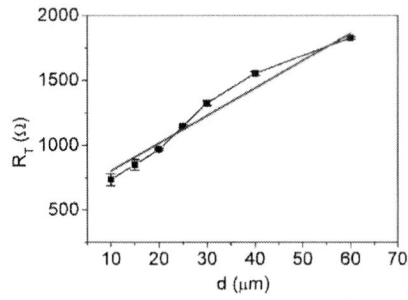

Figure 9. Sheet resistance extraction of graphene layer from circular transmission line test structures directly measured from the actual device wafer. The sheet resistance and the contact resistance are extracted to be 6.7kΩ/$\square$ and 295Ω, respectively.

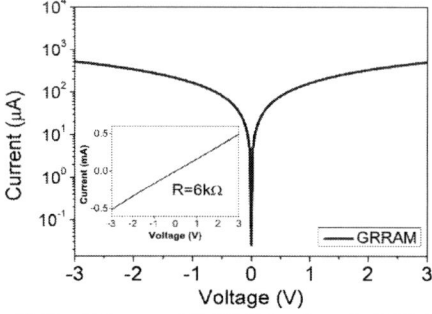

Figure 10. The I-V curve of G-RRAM *without* the HfO$_x$ layer (inset: linear scale). The resistance value is 6kΩ indicating that the contribution of graphene sheet R to the low HRS/LRS is negligible.

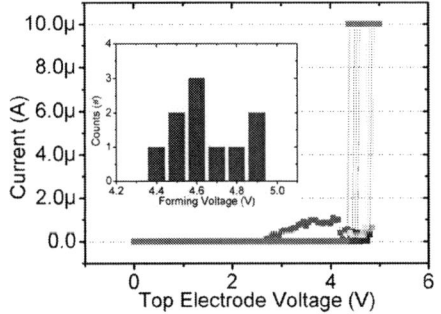

Figure 11. Forming curve collected from 10 cells with 10 μA compliance current. Inset: forming voltage distribution. Good uniformity of forming voltage is observed.

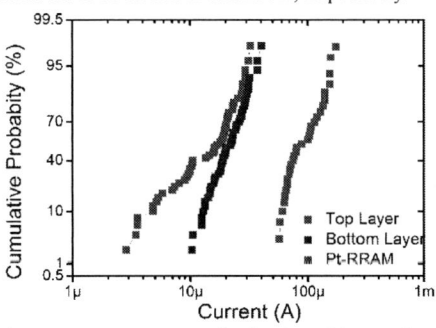

Figure 12. Reset current distribution of bottom layer, top layer, and Pt-RRAM. G-RRAM shows much lower reset current compared to Pt-RRAM.

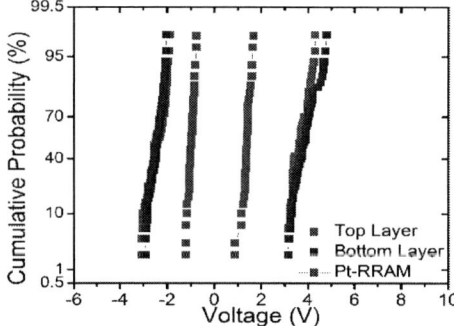

Figure 13. Reset and Set voltage distribution of bottom layer, top layer, and Pt-RRAM. Set and Reset Voltages of G-RRAM shows good uniformity as Set and Reset Voltages of Pt-RRAM.

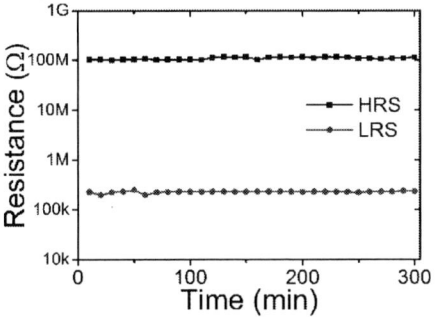

Figure 14. Retention test result for 1 G-RRAM cell : the device can maintain its resistance >300 min the room temperature.

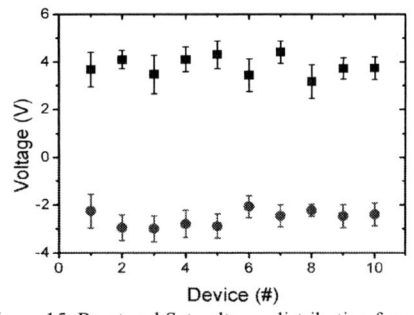

Figure 15. Reset and Set voltages distribution from 10 different cells. The uniformity of the voltage distribution is comparable to RRAM based on Pt[2].

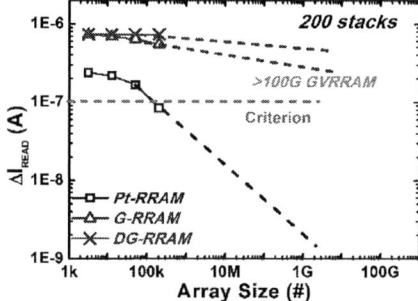

Figure 16. Read margin comparison between Pt-RRAM, G-RRAM and DG-RRAM for a 3D RRAM with 200 stacks of. For 100 nA read current margin (V$_{READ}$ = 1 V, R$_{IN-SERIES\ Read}$ =100kΩ, Worst case cell), an array size of >100G of G-RRAM and DG-RRAM is achievable, while maximum array size of Pt-RRAM is <100k.

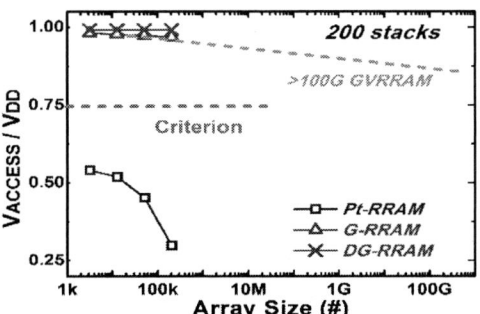

Figure 17. Write margin comparison between Pt-RRAM, G-RRAM and DG-RRAM for a 3D RRAM with 200 stacks For a 75% write access voltage margin (the criterion for minimum write voltage = 2/3 V$_{DD}$, Worst case cell), G-RRAM and DG-RRAM can attain > 80% write margin for an array size of >100G, while an array size of >1k for the Pt-RRAM cannot meet the criterion.

References

[1] H. −S. P. Wong, et al., Proc. IEEE, 100, 1951-1970, 2012.
[2] H. −Y. Chen, et al., IEDM, 497-500, 2012.
[3] H. −Y. Chen, et al., Nanotechnology, 24, 465201, 2013.
[4] E. Cha, et al., IEDM, 268-271, 2013.
[5] M. M. Lucchese, et al., Carbon, 48, 1592-1597, 2010.
[6] X. Guan et al, IEEE Trans. Electron Devices, 59, 1172- 1182, 2012.
[7] Y. Deng, et al., IEDM, 629-632, 2013.
[8] S. Yu, et al., Tech. Dig. Symp. VLSI Technol., 158-159, 2013.
[9] Y. Wang, et al., Advanced Materials, 23, 1514-1518, 2011.
[10] S. Bae, et al., Nature nanotechnology, 5, 574-578, 2010.
[11] F. Güneş, et al., ACS Nano, 4, 4595-4600, 2010.

Acknowledgement

This work is supported in part by the Office of the Director of National Intelligence (ODNI), Intelligence Advanced Research Projects Activity (IARPA) Trusted Integrated Circuits (TIC) Program, the member companies of the Stanford Non-Volatile Memory Technology Research Initiative (NMTRI) affiliate program, and Systems on Nanoscale Information Fabrics (SONIC) Center, one of six centers of Semiconductor Technology Advanced Research Network (STARnet), a Semiconductor Research Corporation (SRC) program sponsored by Microelectronics Advanced Research Corporation (MARCO) and Defense Advanced Research Projects Agency (DARPA). J. Sohn is additionally supported by the STX foundation scholarship for overseas studies. Z. Jiang is additionally supported by the M. Stanley Rundel Fellowship. Hong-Yu Chen is additionally supported by the Intel Ph. D Fellowship.

978-1-4799-8002-4/14 $31.00 © 2014 IEEE

Graphene Inductors for High-Frequency Applications –
Design, Fabrication, Characterization, and Study of Skin Effect

Xiang Li [1,2,*], Jiahao Kang [1,*], Xuejun Xie [1], Wei Liu [1], Deblina Sarkar [1], Junfa Mao [2] and Kaustav Banerjee [1,+]

[1] Department of Electrical and Computer Engineering, University of California Santa Barbara, Santa Barbara, CA 93106

[2] The Key Laboratory, Ministry of Education for Design and Electromagnetic Compatibility of High-Speed Electronic Systems, Shanghai Jiao Tong University, Shanghai, China 200240

* These authors contributed equally to this work.

+ Contact e-mail: kaustav@ece.ucsb.edu

Abstract

Graphene is very attractive for densely integrated and flexible high-frequency/RF IC applications due to its extraordinary electrical, thermal, and mechanical properties. This work presents the design, fabrication, and characterization of graphene on-chip inductors. The skin effect in graphene inductors is investigated experimentally for the first time based on a circuit model proposed and fitted from fabricated ¾-, 2-, and 3-turn spiral inductors. The operation frequencies are in 40-60 GHz range and Q-factors are around 3. Design and fabrication optimizations are performed to guide future studies.

I. Introduction

Graphene- composed of a single layer of carbon atoms arranged in a hexagonal lattice has extraordinary electrical and thermal conductivity that make it a promising candidate for a variety of high-frequency (HF) applications [1] including radio-frequency (RF) transistors [2] with unity current-gain cut-off frequency of up to 300 GHz [3],[4], electrostatic discharge protection [5], as well as low-loss interconnects and passives [6]-[8]. Besides, its atomically-thin 2-dimensional (2D) planar nature, superb mechanical strength and high flexibility (due to strong in-plane sp^2 bondings) as well as its band gap tunability (via lithographic control of its width) open up new opportunities in physically compliant electronics and systems such as flexible and wearable electronics [9],[10] and in the ultimate realization of *all-graphene* circuits [11],[12]. Fig.1 summarizes all of these properties and highlights future HF/RF applications of graphene on flexible substrates with a schematic proposal of an *all-graphene* analog circuit. As shown, on-chip inductors form an integral part of all such HF/RF circuits and hence, the design and fabrication of graphene on-chip inductors constitute a key step toward the realization of such flexible circuits and systems.

Theoretical works [7],[8] have identified that graphene can be very attractive for on-chip inductor application due to its large momentum relaxation time (leading to low-loss), and planar nature that allows it to be easily patterned into square helical structures without the need for corner metal contacts required in carbon nanotube bundle-based inductors [13]. Also, the low-loss property of graphene will lead to high quality factors (Q-factors) and hence, high energy-efficiency. Although theoretical studies have been reported, informative experimental studies on HF characteristics of graphene are rather sparse. Especially, the experimental study of practical graphene on-chip inductors have not been reported, neither have the design and fabrication optimizations. On the other hand, there is no experimental study

and evidence of the skin effect (SE) (an effect wherein HF current flow concentrates near the outer surfaces of conductors, as shown in **Fig.1**) in such graphene applications.

Therefore, in this work we perform the design, fabrication and characterization of graphene on-chip spiral inductors, as summarized in **Fig.2**. A circuit model is proposed and parameters are extracted from the fabricated inductors including ¾-, 2- and 3-turn inductors. For the first time, SE is investigated and is demonstrated to exist in each device. Fabrication optimization is also carried out to investigate the effects of isolation dielectric and contacts on the inductor performance. These results and findings provide necessary guidelines for future studies on graphene inductors as well as HF/RF applications of graphene and relevant 2D materials.

II. Design Optimization by Simulations

Q-factor of an inductor is an important metric in high-performance RF/mixed-signal circuits, which has to be optimized simultaneously w.r.t. its size (or design) and fabrication cost. To accomplish a superior design, it is very important to correctly understand the effect of each inductor parameter on the inductor performance.

Figs.3a and **3b** show the schematic views (cross-section and stratified view) of a graphene on-chip spiral inductor structure, where the parasitics of the structure are marked. The series inductance (L_G) and resistance (R_G) of graphene are obtained by employing the impedance extraction procedure developed in [7],[8]. **Fig.3c** shows the equivalent circuit model proposed in this work, which is developed from conventional equivalent circuit for inductors [14]. The dimensions of the inductors are defined in **Fig.3d**. The simulation is performed for ¾-turn inductors since they have small area and low inductance values that are suitable for ultra-high frequency operation [7],[8],[13]. **Fig.4** shows the effects of substrate resistance (R_{sub}), dielectric constant (ε_{ox}) and thickness (t_{ox}); and the total inductor length (L), inductor width (W), and inductor thickness (t) of graphene inductors on the Q-factor. It can be observed from **Fig.4a** that high Q-factor can be obtained by using low-loss (low-doping) substrate without shift of operation frequency (f_{op}) (the frequency at which the maximum Q-factor (Q_{max}) is achieved), due to the reduction of eddy current and energy loss. ε_{ox} and t_{ox} determine the dielectric capacitance C_{ox}, thereby affecting f_{op} and Q-factor, as shown in **Figs.4b** and **4c**. Q-factors first increase with L and W (**Figs.4d**, and **4e**) because of the increased inductance and reduced resistance, respectively, and then decrease with L and W due to the reduced magnetic coupling between the two ends and

978-1-4799-8002-4/14 $31.00 © 2014 IEEE

increased current proximity effect, respectively. However, as shown in **Fig.4f**, Q-factors increase with t due to increased number of conducting channels in graphene, which reduces the total resistance and energy loss. Hence, low-loss substrate, thick and low-permittivity substrate dielectric and thick graphene films should be chosen to obtain high Q-factors. W should be 2-4 µm, and L should be optimized to be as close to 400 µm as possible (albeit within the obtained graphene flake size).

III. Fabrication of Graphene Inductors

Multilayer graphene films are prepared by mechanical exfoliation of highly ordered pyrolytic graphite (HOPG) and transferred onto SiO_2 (300 nm)/Si (10 Ω.cm) substrate. Subsequently, graphene films are patterned into ribbon coils. For multi-turn inductors, an isolation dielectric layer (Al_2O_3) over graphene is grown and patterned, the thickness (50 nm) of which is optimized to eliminate the effects of overlap capacitance (C_{ol}, shown in **Fig.3a**). Metal contacts and pads (Ni/Au: 20 nm / 80 nm) are deposited and patterned, followed by an annealing process. The entire fabrication process is illustrated by **Figs. 5a-5e**. **Figs.5f-5i** show the micrographs and SEM images of some fabricated graphene inductors. The dimensions of all the devices are listed in **Table I**.

IV. Characterization

Measurement and De-embedding: The pad structures are designed as ground-signal-ground (GSG) coplanar waveguide (CPW) **[15]** with graphene test structure in the signal path. Subsequently, S-parameter measurements are performed in the frequency range of 100 MHz - 67 GHz using Agilent N 5227A Network Analyzer and a microwave probe station equipped with Cascade Infinity GSG-probes. As shown in **Fig.6**, to capture the intrinsic properties of the graphene inductors themselves, de-embedding procedures **[15],[16]** are performed to stepwise remove the parasitic effects of the CPW using the dummy structures shown in **Figs.6a-6d**.

Circuit Modeling: In order to simplify the circuit model (in **Fig.3c**), the de-embedded S-parameter matrix (**Figs.7a,7b**) of one test structure is converted to transmission (ABCD) matrix (**Figs.7c,7d**). According to the form of ABCD matrix, the circuit model can be simplified from **Fig.7e** to **Fig.7f**. Further simplifications are described in the **Fig.7** caption, which result in the simplified circuit shown in **Fig.7g**.

Skin Effect (SE): SE investigation is carried out through fitting the circuit parameters to the measured data. As shown in **Fig.8a**, large fitting error can be found in the the fitted curve without considering SE (keeping R_G and L_G as constants over the frequency range). Subsequently, an exponential model of R_G and L_G as function of frequency is employed to fit the impedance of the circuit model to the measured data of the DUT, as shown in **Fig.8a**, with the fitted resistance and inductance of all the fabricated devices plotted in **Figs. 8b, 8c** and listed in **Table I**. It is found that the proposed model gives very good fitting results for all the inductors, thereby demonstrating the existence of SE in graphene inductors, although having negligible SE is actually desirable.

Q-Factor Calculation: The Q-factors of all the six fabricated devices are shown in **Fig.9**, with Q_{max}, and corresponding f_{op} listed in **Table I**. It can be observed that f_{op} of the inductors are very high, especially for multi-turn inductors. This is because the inter-turn coupling capacitance (C_s) of these inductors is very small and cannot be appreciably increased due to the small thickness of the graphene films.

Fabrication Optimization: For multi-turn inductors, C_{ol}, and C_s (in **Fig.3a**) are investigated and optimized by tuning the dielectric. The fabrication optimization is performed and illustrated in **Figs.10a-10f**, on the same device. It can be observed in **Fig.10g** that C_s has significant influence on Q_{max}, and corresponding f_{op}, while C_{ol} only has capacitive influence on the Q-factor in the low-frequency range. C_{ol} becomes negligible when isolation dielectric thickness reaches 50 nm (as explained in **Fig.7** caption). Hence, Q_{max} and f_{op} can be optimized by dielectric engineering. **Fig.11** compares the Q-factors of inductor #3-1 before and after annealing. It can be clearly observed that the Q-factor increased by 2X after annealing, which is mainly due to the improvement of the quality of contacts (R_C reduced from 356 Ω to 318 Ω) **[17],[18]**. Hence, contact engineering is also a necessity for further improvement of inductor Q-factors.

V. Summary

In this work, various graphene on-chip inductors are designed and fabricated. A circuit model is proposed and parameters are extracted by characterizing the fabricated inductors. Based on that, the existence of skin effect in multilayer graphene ribbons is experimentally demonstrated for the first time. This work also established pathways for design and fabrication optimizations of graphene on-chip inductors for the first time, providing guidelines for future inductor design as well as any high-frequency application based on graphene, including the demonstration that the inductor performances (Q-factors and f_{op}) can be tuned by contact and dielectric engineering and doping techniques, aiming at different applications, as illustrated in **Fig.12**.

VI. Acknowledgements

This research was supported by the U.S. National Science Foundation under Grant CCF-1162633. All process steps for device fabrication were carried out using the Nanostructure Cleanroom Facility at the California NanoSystems Institute and the Nanofabrication Facilities at UCSB – part of the National Nanotechnology Infrastructure Network.

References

[1] H. Li, et al., *IEEE Trans. Elect. Dev.*, **56**, 9, 1799–1821 (2009).
[2] F. Schwieerz, *Nat. Nanotech.*, **5**, 487–496 (2010).
[3] L. Liao, et al., *Nature*, **467**, 7313, 305–8 (2010).
[4] Y. Wu, et al., *Nature*, **472**, 7341, 74–8 (2011).
[5] H. Li, et al., *IEEE Trans. Elect. Dev.*, **61**, 6, 1920–1928 (2014).
[6] C. Xu, et al., *IEEE Trans. Elect. Dev.*, **56**, 8, 1567–1578 (2009).
[7] D. Sarkar, et al., *IEEE Trans. Elect. Dev.*, **58**, 3, 843–852 (2011).
[8] D. Sarkar, et al., *IEEE Trans. Elect. Dev.*, **58**, 3, 853–859 (2011).
[9] M. Catrysse, et al., *Sens. Actuators A*, **114**, 302–311 (2004).
[10] D. J. Lipomi, et al., *Nat. Nanotech.*, **6**, 788–792 (2011).
[11] J. Kang, et al., *Appl. Phys. Lett.*, **103**, 8, 083113 (2013).
[12] S. Lee, et al., *Nat. Comm.*, **3**, 1018 (2012).
[13] H. Li, et al., *IEEE Trans. Elect. Dev.*, **60**, 9, 2862–2869 (2013).
[14] H. Li, et al., *IEEE Trans. Elect. Dev.*, **56**, 10, 2202–2214 (2009).
[15] J. J. Plombon, et al., *Appl. Phys. Lett.*, **90**, 063106 (2007).
[16] A. Tselev, et al., *Nano Lett.*, **8**, 2, 152-156 (2008).
[17] W. Liu, et al., *Nano Lett.*, **13**, 5, 1983–1990 (2013).
[18] W. Liu, et al., *IEEE Int. Elect. Dev. Meeting*, 499–502 (2013).

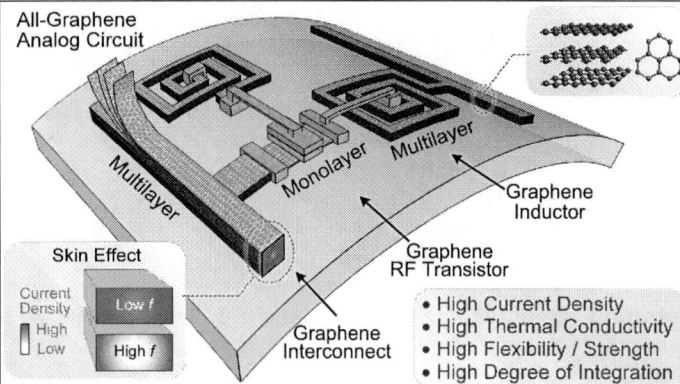

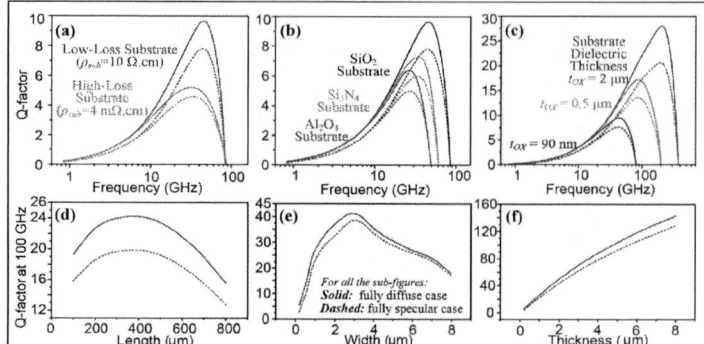

Fig. 4: The effects of **(a)** substrate resistivity, **(b)** substrate dielectric constant, **(c)** substrate dielectric thickness, **(d)** total inductor length, **(e)** inductor width, and **(f)** inductor thickness on the Q-factors of multilayer graphene-ribbon (GR) based ¾-turn inductors for ultrahigh frequency applications. Two extreme cases of fully diffuse (solid curves) and fully specular (dashed curves) are considered in our simulation for each parameter. The substrate is chosen as 500-μm-thick silicon. The default total length (=$3D_{out}$ for ¼- turn inductors), width, and thickness are 130 μm, 1 μm, and 1 μm, respectively. Low-loss substrate is used in (b-f). Substrate dielectric (SiO₂ in (a, c-f)) thicknesses are 90 nm for (a,b) and 0.5 μm for (d-f). According to these results, low-loss substrate, thick and low-permittivity substrate dielectric and thick graphene films should be chosen to obtain high Q-factors. W should be 2-4 μm, and L should be as close as possible to 400 μm.

Fig. 1: Schematic illustrating an all-graphene analog circuit (frequency mixer), in which all circuit components, including monolayer-graphene-based radio frequency (RF) transistor, multilayer-graphene-based inductors and interconnects are monolithically integrated on a single flexible substrate. The extraordinary physical properties of graphene, as listed on the lower right hand side, make it an ideal material for future flexible electronics. However, clear understanding of high-frequency effects (particularly skin effect) in graphene and design issues related to its passive applications (especially inductors) are currently missing. This is the main target of this work.

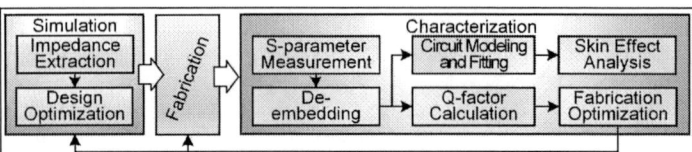

Fig. 2: The flowchart of this study. First, based on the *impedance extraction* results, *design optimizations* for inductors are performed using simulations (based on Fig.3c). Next, *fabrication* of test and de-embedding structures are performed. The *characterizations* are first performed by *S-parameter measurement*, followed by *de-embedding* procedures. By *circuit modeling and fitting*, the simplified circuit model is proposed and circuit parameters are extracted, based on which *skin effect analysis* is completed. On the other hand, using the de-embedded S-parameters, *Q-factor calculation* is performed, and subsequently *fabrication optimizations* (contact/dielectric effects) are carried out.

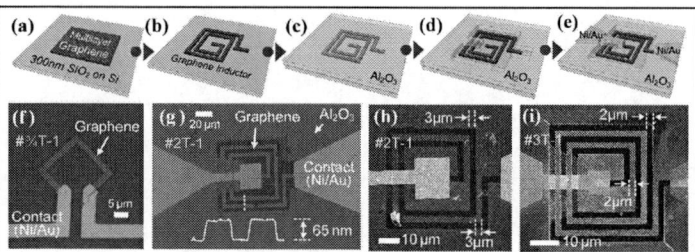

Fig. 5: **(a-e)** Fabrication processes of graphene on-chip inductor test structures: **(a)** Preparation of graphene films on SiO₂ (300 nm) /Si (low-loss) substrate by mechanical exfoliation and thermal transfer; **(b)** Patterning of GR-based inductors by electron beam lithography; **(c)** Growth of 60 nm Al₂O₃ as isolation dielectric by atomic layer deposition at 150 °C; **(d)** Patterning of Al₂O₃ by KOH wet etching; **(e)** Deposition and patterning of metal contacts (Ni/Au: 20nm/80nm). **(f, g)** Optical photos of fabricated **(f)** ¾-turn and **(g)** 2-turn inductors. **(h, i)** SEM images of fabricated **(h)** 2-turn and **(i)** 3-turn inductors. The yellow curve in (g) shows the thickness profile measured by atomic force microscope along the yellow dashed line. The dimensions and labels of all the fabricated inductors are listed in Table I.

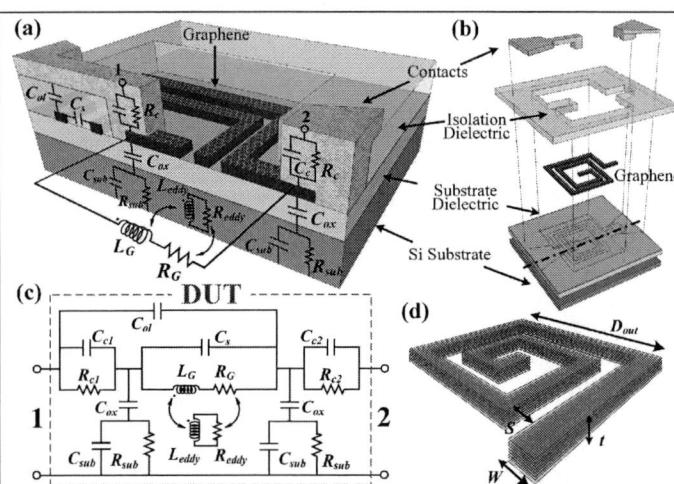

Fig. 3: **(a)** Schematic view of graphene on-chip spiral inductor with the equivalent circuit parameters. L_G and R_G are the series inductance and resistance of graphene, respectively. C_s, C_{ol}, C_{ox}, C_{sub}, and R_{sub} represent the inter-turn coupling capacitance, overlap capacitance, substrate dielectric capacitance, substrate capacitance and resistance, respectively. R_c and C_c represent the contact resistance and capacitance, respectively. Eddy inductance and resistance, L_{eddy} and R_{eddy}, capture the eddy current effects in the substrate. **(b)** Stratified view of (a). The dotted horizontal line indicates the plane where the cross-section in (a) is taken. **(c)** Equivalent circuit model for the device under test (DUT) in (a), which is a two-port network. **(d)** Schematic view of a graphene inductor coil, where D_{out}, W, t, and S are the outermost diameter, the conductor width, the conductor thickness, and the conductor spacing, respectively.

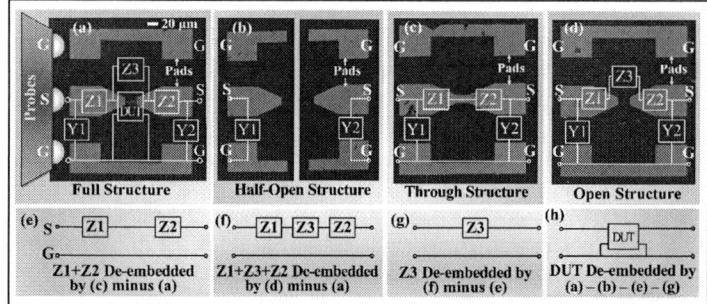

Fig. 6: **(a-d)** optical photos of de-embedding structures and parasitic impedance circuit topology (white): **(a)** *full structure* (DUT with coplanar waveguide), **(b)** *half-open*, **(c)** *thru*, and **(d)** *open* structure. The *open* structure is identical to the *full structure* except that no DUT is in the gap of the signal conductors (S). The *half-open* structure is exactly half-cut of the *open* structure. The *thru* structure has continuous signal conductor. **(e-h)** The stepwise de-embedding procedure adopted in this work. Parallel parasitics Y_1 and Y_2 are obtained from the S-parameters of the *half-open*, and series parasitics Z_1 and Z_2 are extracted from the *thru* structure by removing Y_1 and Y_2. Series parasitic Z_3 is then de-embedded from the measured S-parameter of the *open* structure by removing Z_1, Z_2, Y_1, and Y_2. Finally the parameters of the DUT are extracted by stepwise removal of parasitics in (b), (e) and (g) from (a).

978-1-4799-8002-4/14 $31.00 © 2014 IEEE

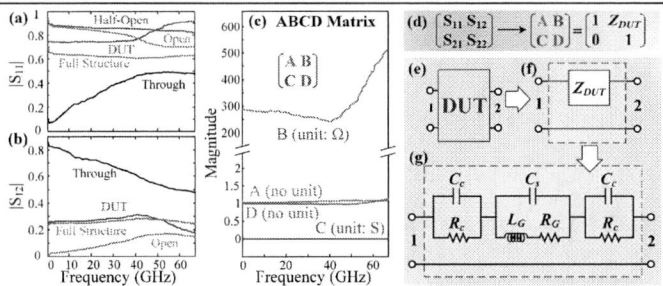

Fig. 7: Magnitude of **(a)** reflection (S_{11}) and **(b)** transmission (S_{12}) characteristics for *half-open*, *open*, *thru*, *full structure*, and *DUT* (taking device #2T-2 in Table I as an example). As shown in (a), because the input impedance of the graphene device is smaller (larger) than that of the *open/half-open* (*thru*) structure, there are smaller (larger) signal reflection (S_{11}) and larger (smaller) transmission (S_{12}) in the graphene device. **(c)** Plots of the four elements of the transmission (ABCD) matrix, which is directly converted from the de-embedded S-parameters. **(d)** The conversion of S-parameter matrix to ABCD matrix. **(e-g)** Simplification procedures of the equivalent circuit model. Since the ABCD matrix of all the fabricated samples possess the form of (d), where A and D are equal to 1 and C is 0, **(e)** *the DUT* can be reasonably equivalent to **(f)** *the series impedance topology*. Hence, C_{ox}, C_{sub}, and R_{sub} can be neglected, for the low-loss low-C_{OX} substrate chosen and the frequency range. R_{eddy} and L_{eddy} are also neglected because of the frequency range. Besides, C_{ol} can be neglected because its effect does not affect the parameter fitting when isolation dielectric thickness is 50 nm. Moreover, the contacts are designed to be symmetrical, and therefore R_{c1}, R_{c2} and C_{c1}, C_{c2} can be simplified to R_c and C_c, respectively. Hence, the circuit model of DUT (Fig. 3c) can be finally simplified to **(g)**.

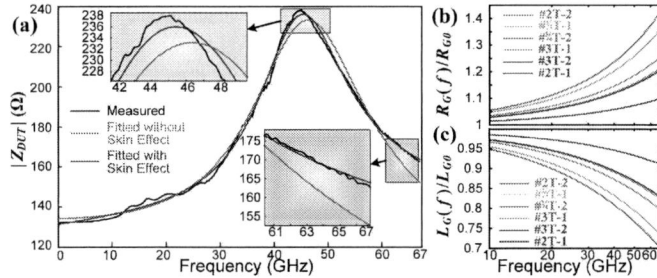

Fig. 8: **(a)** The magnitude of DUT impedance (Z_{DUT}) as a function of frequency, including the measured data (inductor #¾T-2) and the fitted data using the circuit model in Fig.7g. Insets zoom into the peak and high frequency region. Large fitting error can be found in the fitted curve without considering skin effect (keeping R_G and L_G as constants over frequency), especially near low frequency, high frequency and the peak. However, by modeling R_G and L_G as monotonical functions of frequency ($R_G(f)$ and $L_G(f)$), the circuit model can be perfectly fitted to the measured data. Several mathematic models (polynomials, square root, etc.) are utilized to model this SE, and the best fitting results are obtained from the exponential model, where $R_G(f)$ and $L_G(f)$ are modeled as exponential functions of frequency ($R_G(f)=R_{G0}\exp(Af)$, $L_G(f)=L_{G0}\exp(-Af)$, where A is a coefficient, and R_{G0} and L_{G0} are low frequency series resistance and inductance). This proves that SE exists in graphene. **(b)** Normalized resistance ($R_G(f)/R_{G0}$) and **(c)** normalized inductance ($L_G(f)/L_{G0}$) are fitted from the measurement data of all the six fabricated devices. The resistance increases with frequency while the inductance decreases with frequency for all the devices, which indicates SE exists in all of them. The curves are different for different samples, indicating that A is varying among samples because of the different device dimensions.

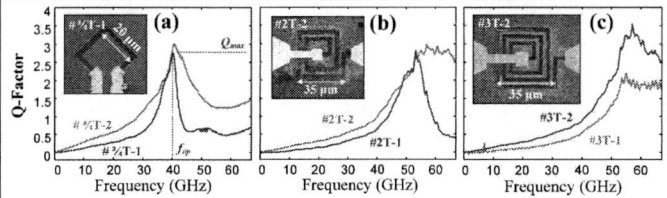

Fig. 9: **(a-c)** the Q-factors vs. frequency of all the six fabricated devices: **(a)** ¾-turn inductors; **(b)** 2-turn inductors and **(c)** 3-turn inductors. Insets are example SEM images for the three types. Q-factor is calculated by the equation $Q = -\text{imag}(Y_{in})/\text{real}(Y_{in})$, where Y_{in} is the input admittance at port 1 with port 2 shorted (Fig. 3c) calculated from de-embedded S-parameters. All of the calculated maximum Q-factor (Q_{max}) and corresponding operation frequency (f_{op}) are listed in Table I.

Table I: The sample labels, physical dimensions (D_{out}, W, t, and S), fitted circuit parameters (R_{G0}, L_{G0}, and C_s) using SE model, the SE model fitting coefficient (A), the maximum Q-factors (Q_{max}) and corresponding operation frequency (f_{op}) for all fabricated inductors. The first digit in the sample labels indicates the number of turns. It can be observed that f_{op} of the inductors are very high, especially for multi-turn inductors. This is because the inter-turn coupling capacitance, C_s, of these inductors is very small and cannot be appreciably increased due to the small thickness of graphene films.

Sample label	Dimensions				Fitted parameters				Q-factors	
	D_{out} (μm)	W (μm)	t (nm)	S (μm)	R_{G0} (Ω)	L_{G0} (nH)	C_s (fF)	A (THz⁻¹)	f_{op} (GHz)	Q_{max}
¾T-1	20	2	25	-	38.3	0.66	24.0	4.6	40.2	2.75
¾T-2	24	3	62	-	11.7	0.15	95.3	3.4	40.8	3.00
2T-1	50	3	64	3	20.4	0.27	26.3	1.4	53.3	2.80
2T-2	35	4	10	3	34.8	0.41	17.1	5.3	57.0	2.97
3T-1	40	2	30	2	30.7	0.33	21.6	2.9	57.0	3.52
3T-2	35	2	25	2	51.7	0.54	14.0	2.8	55.0	2.19

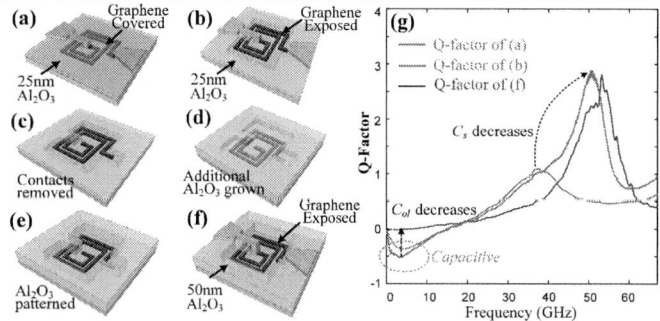

Fig. 10: **(a-f)** Schematics showing process flow of fabrication optimization of isolation dielectric on sample #2T-1: **(a)** device #2T-1 initially covered by 25-nm-thick Al₂O₃ isolation dielectric; **(b)** device #2T-1 with coils exposed in air, where most area of the isolation dielectric in (a) is etched off; **(c)** #2T-1 with contacts removed by HCl; **(d)** #2T-1 with additional 25 nm Al₂O₃ grown; **(e)** #2T-1 with the 50-nm-thick Al₂O₃ patterned; **(f)** #2T-1 with new contacts deposited/patterned on the 50-nm-thick Al₂O₃. **(g)** The Q-factors of (a), (b), and (f) extracted after de-embedding. As shown in (g), comparing (a) and (b), both Q_{max} and f_{op} increased, because after exposing the graphene coils in the air, the dielectric surrounding graphene is changed from high-k (Al₂O₃) to low-k (air) in (b), which decreases the inter-turn C_s. On the other hand, as shown in (g), the Q-factors of (a) and (b) are below zero in the low-frequency region, indicating the circuit is capacitive, which is due to the large C_{ol}. By increasing the thickness of the isolation dielectric, C_{ol} is decreased (in (f)), resulting in the returning of Q-factor from capacitive to inductive in the low-frequency range. Q_{max} and f_{op} are hardly affected by C_{ol}. As discussed earlier, C_{ol} in the equivalent circuit can be neglected for 50 nm Al₂O₃. However, the circuit model in Fig.7g cannot be fitted to devices with 25 nm Al₂O₃ isolation dielectric, indicating that C_{ol} cannot be neglected in that case.

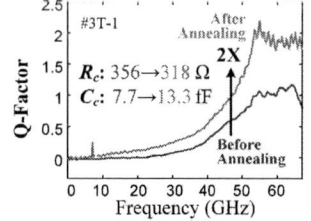

Fig. 11: The Q-factors of inductor #3T-1 before and after annealing process (at 400 K for 1 hour). It can be observed that by annealing process, the Q-factor is improved by 2X, which is mainly due to the improvement in the quality of metal contacts [17],[18]. The fitted R_c is changed from 356 Ω to 318 Ω, which serves as an evidence. The fitted C_c changes from 7.7 fF to 13.3 fF, one possible reason of which could be the improvement of interface adhesion and subsequent reduction in the gap between metal and graphene.

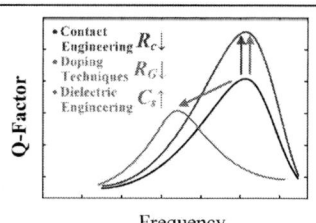

Fig. 12: Q-factor sketch illustrating three strategies that can be adopted to improve the performance of graphene on-chip inductors. R_c and R_G can be reduced by contact engineering and proper doping techniques, respectively, thereby improving the Q-factor of inductors. Though the small thickness of graphene films results in low inter-turn coupling capacitances that make the f_{op}'s of the fabricated inductors in range of 40-60 GHz, C_s can be tuned by dielectric engineering and then f_{op} can be tuned to the desired frequency.

Nanophotonics with two-dimensional atomic crystals

Thomas Mueller[*], Marco M. Furchi, Andreas Pospischil, and Dmitry K. Polyushkin

Vienna University of Technology, Institute of Photonics, Gußhausstraße 27-29, 1040 Vienna, Austria
[*]Email: thomas.mueller@tuwien.ac.at, Tel: +43-1-58801-38739, Fax: +43-1-58801-38799

Abstract

Electronic and photonic devices based on two-dimensional (2D) atomic crystals, such as graphene and layered transition-metal dichalcogenides (TMDCs), are perceived as potential candidates to complement, or even replace, conventional semiconductor devices in various applications. 2D crystals are of high material quality and stability, even so, they can be produced with large-area dimensions and at low cost. Moreover, the possibility of stacking different atomically-thin 2D layers on top of each other provides the opportunity of creating "artificial" designer materials, so-called van der Waals heterostructures. In this paper, optoelectronic devices based on 2D materials will be presented. We will discuss photodetection, light emission and photovoltaic energy conversion in 2D monolayers and van der Waals heterojunctions.

I. Photodetection

Despite being a semi-metal without band gap, graphene shows a surprisingly strong photoresponse. While photo-excitation in the bulk of the sheet is followed by fast carrier

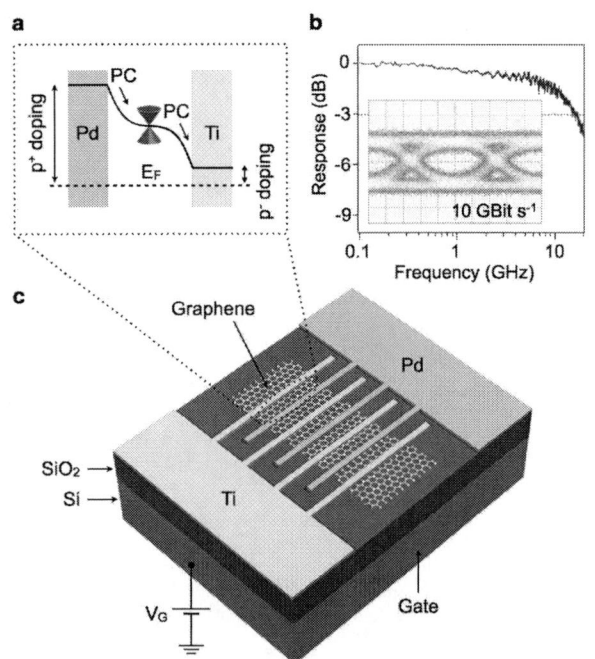

Fig. 1. (a) Band profile between two neighboring contact electrodes. (b) Photoresponse versus light intensity modulation frequency. A 3-dB bandwidth of 16 GHz is achieved. Inset: eye-diagram, showing an open eye at 10 Gbit/s data rate. (c) Schematic drawing of a MGM photodetector.

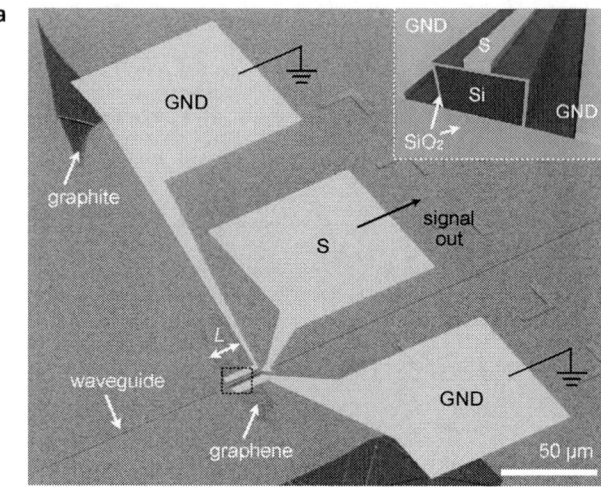

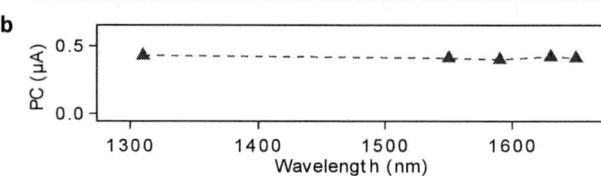

Fig. 2. (a) Scanning electron microscope (SEM) image of a waveguide-integrated graphene photodetector. Inset: Schematic of the device cross-section. (b) A spectrally flat photoresponse is obtained across all optical communication bands.

recombination, excitation near metal-graphene contacts or graphene p-n junctions leads to a photocurrent due to photoelectric and photo-thermoelectric effects (1). Fig. 1c shows a device that employs a lateral potential gradient near metal-graphene interfaces for charge separation (2). As the photocurrent is produced only in the vicinity of the metal contact, an interdigitated finger structure was used to increase the photoactive area. Neighboring contacts consist of two different metals, one with high and the other with low work function, allowing to break the mirror symmetry of the device (Fig. 1a) without having to apply a bias voltage. Zero-bias operation is essential in graphene-based photodetectors because of graphene's highly conductive behavior. Using ultrafast photocurrent measurements, a bandwidth of 16 GHz was determined (Fig. 1b). We also could demonstrate the faithful detection of an optical bit stream at a data rate of 10 GBit/s (inset in Fig. 1b), demonstrating the feasibility of using graphene photodetectors in optical communication systems.

The inherently weak optical absorption ($\approx$2.3 %) of the graphene monolayer, however, limits the achievable responsivities under normal incidence configuration (up to a few mA/W only). We therefore investigated several concepts to increase the interaction of light with graphene, one of which (3) relies on the integration of graphene with a silicon waveguide (Fig. 2a). In this device, the optical mode in the waveguide is absorbed as the light propagates in parallel to the graphene sheet. Although the overlap of the waveguide mode with graphene is small, 44% absorption was achieved in an only 24 μm long device. A careful design of the device geometry was performed in order to evaluate the tradeoff in photoresponsivity due to the optical absorption in the center metal electrode. A bandwidth of 18 GHz and a responsivity of up to 50 mA/W were demonstrated. The photoresponse is wavelength independent across all optical telecommunication bands (Fig. 2b), unlike the drastic decrease of the response beyond 1550 nm of germanium photodetectors, currently employed in silicon photonics.

Another example (4) is shown in Fig. 3a. This device benefits from the strong increase of the optical field amplitude inside a microcavity. The bottom mirror consists of a Bragg reflector, made of a quarter-wavelength-thick AlAs and $Al_{0.1}Ga_{0.9}As$ layers. The top mirror is made of a SiO_2 and Si_3N_4 layers. The absorption is 26-fold enhanced as compared to devices without cavity, reaching values of ≈ 60 %. A photoresponsivity of 21 mA/W is achieved. The device shows a response only at the design wavelength (≈ 860 nm), making it promising for wavelength division multiplexing (Fig. 3b). The concept of enhancing the light-matter interaction in graphene by use of an optical cavity is not limited to photodetectors alone; it can also be applied to electro-absorption modulators and other 2D crystal-based optoelectronic devices.

Although graphene offers the possibility of device operation over a wide wavelength range and with high electrical bandwidth, the responsivity is ultimately limited by the short carrier lifetime that results in a ≈ 10 % internal quantum efficiency. The carrier lifetimes in 2D semiconductors, such as MoS_2 and other TMDCs, are longer which results in higher responsivities (1). Moreover, application of a bias voltage may lead to photoconductive

gain. Fig. 4a shows a phototransistor (5) based on a MoS_2 monolayer. Fig. 4c shows the responsivity for bias voltages of +/-5 V. Our measurement were performed in N_2-atmosphere and at kHz-frequencies to exclude the influence of adsorbates at the MoS_2 surface. At low incident optical power, a responsivity of more than 6 A/W is achieved, corresponding to a photoconductive gain of ≈ 100. The gain results from valence band tail states (induced by disorder or structural defects; see Fig. 4b/left) into which photoexcited holes are trapped. Charge neutrality then requires the external circuit to provide additional electrons until recombination with the trapped hole occurs. We calculate the response by solving the rate equations for the model in Fig. 4b/right (electron trapping is neglected as we perform our analysis in the transistor ON-state) and obtain good agreement with the measurement (see solid line in Fig. 4c). The reduction of responsivity under strong illumination is due to filling of the hole traps. Electron trap states can be emptied by negative gate voltages that shift

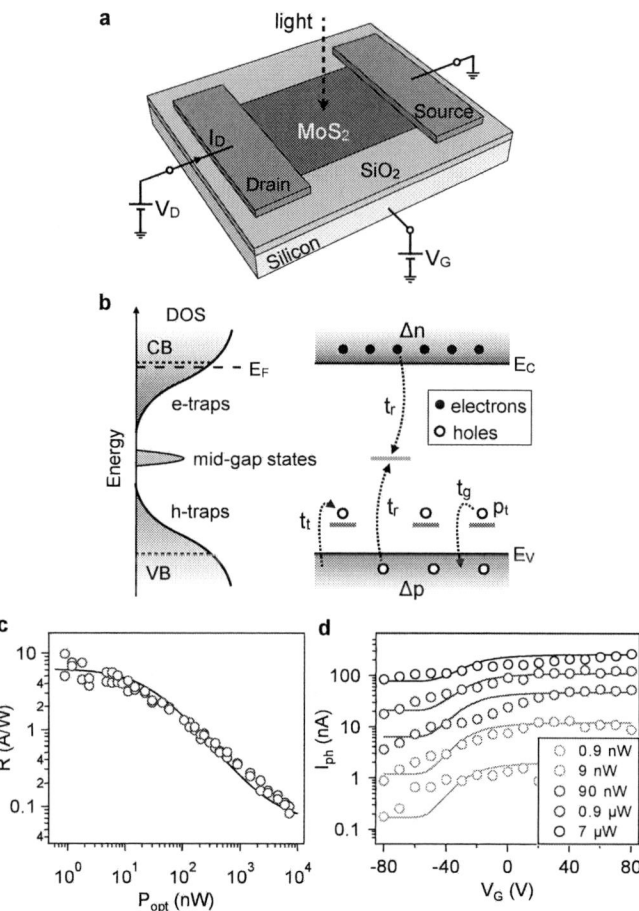

Fig. 4. (a) Schematic drawing of a MoS_2 phototransistor. (b) Left: Density of states (DOS) in monolayer MoS_2. Right: Simplified band diagram. The VB tail is approximated by a discrete distribution of hole traps. (c) Power dependence of photoresponsivity (symbols: measurement results for bias voltages of +5 V and -5 V; line: theoretical results). (d) Gate voltage dependence of the photocurrent (symbols: measurement; lines: theory).

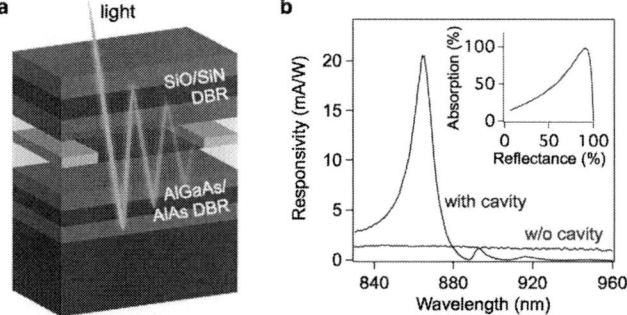

Fig. 3. (a) Schematic drawing of a microcavity-integrated graphene photodetector. (b) Spectral response of a cavity-integrated device (blue line) and a reference device without cavity (red line). The responsivity of the reference device is independent of wavelength, but more than an order of magnitude weaker than that of the cavity-integrated device. Inset: Theoretical dependence of the absorption in graphene on the reflectivity of the top mirror.

978-1-4799-8002-4/14 $31.00 © 2014 IEEE   125

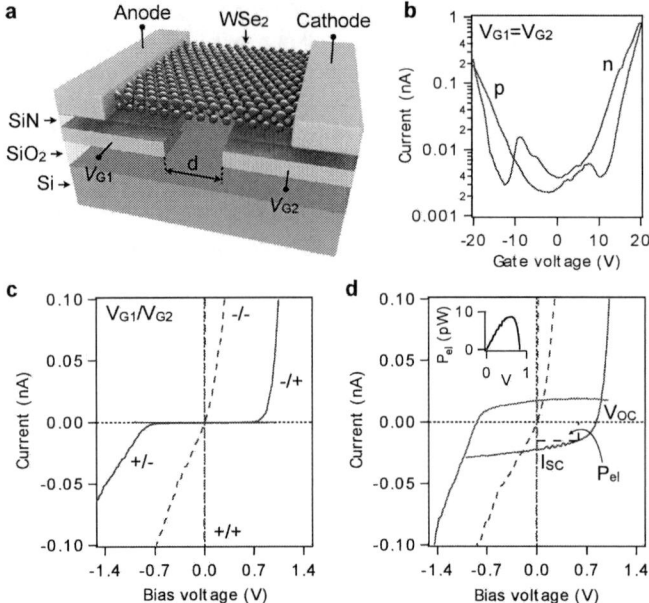

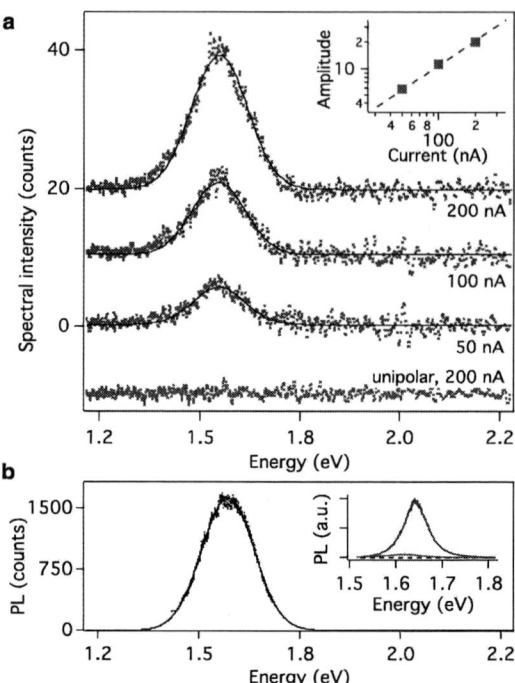

Fig. 5. (a) Schematic drawing of a monolayer WSe₂ device with split gate electrodes. (b) Gate characteristic of the device measured by interconnecting both electrodes ($V_G = V_{G1} = V_{G2}$). Both electrons (n) and holes (p) can be injected into the WSe₂ sheet. (c) IV-characteristics in the dark for different gate voltage polarities. (d) IV-characteristics under optical illumination with 1400 W/m². The bias voltages are the same as in (c). Inset: Electrical output power P_{el} versus bias voltage.

Fig. 6. (a) Electroluminescence emission spectra for different currents. Blue symbols: ambipolar device operation ($V_{G1} = -40$ V, $V_{G2} = 40$ V), black lines: Gaussian fits, red symbols: unipolar device operation ($V_{G1} = V_{G2} = 40$ V). Inset: Electroluminescence amplitude versus current. (b) PL of the WSe₂ sheet on the device. Inset: PL of monolayer (blue line), bilayer (red line) and multilayer (dashed black line) WSe₂.

the Fermi level towards mid-gap. Thus, the effective mobility is reduced which also reduces the photocurrent (Fig. 4d).

II. Electroluminescence

TMDCs are direct bandgap semiconductors when thinned to monolayer thickness, thus providing the opportunity to realize light emitters (see inset in Fig. 6b, where we show photoluminescence spectra of mono-, bi- and multi-layer flakes). Fig. 5a depicts a lateral p-n junction diode in which split gate electrodes, fabricated by electron-beam lithography and covered with a 100-nm-thick silicon nitride dielectric layer, couple to two different regions of a WSe₂ monolayer flake (6). As demonstrated in Fig. 5b, both electrons and holes can be injected into the channel by applying positive and, respectively, negative voltages to the gate electrodes. Applying voltages of opposite polarities to the gate electrodes leads to a rectification of the current, whereas by applying voltages with same polarity, our device operates as a resistor (Fig. 5c).

By driving a constant current through the device, light emission from the electrostatically defined p-n junction is achieved. The electroluminescence emission, depicted in Fig. 6a, peaks at 1.547 eV, in accordance with the photoluminescence (Fig. 6b). If, on the other hand, a unipolar current is driven through the flake (red line Fig. 6a), no light emission is obtained, demonstrating that the electroluminescence arises from the injection of an ambipolar current into the device. We estimate the electroluminescence

efficiency (ratio of emitted optical power and electrical input power) to be approximately 0.1 %, limited by resistive losses and non-radiative recombination in WSe₂. It may thus be increased by reducing the contact resistance or by using a crystalline substrate to reduce the density of recombination centers.

III. Photovoltaics

The IV-curve of the lateral p-n (n-p) junction shifts downwards (upwards) under optical illumination (Fig. 5d), meaning a current flow to an external load. The WSe₂ monolayer diode can therefore be utilized as a photovoltaic solar cell. We extract a maximum electrical power of $P_{el} = 9$ pW (inset in Fig. 5d), from which we derive a power conversion efficiency of ≈0.5 % at an incident illumination of 1400 W/m². This value appears to be small compared to conventional solar cells, but it needs to be assessed in light of the weak optical absorption (≈5 %) of the monolayer crystal.

It would be difficult, though, to scale the lateral p-n junction diode in Fig. 5a to macroscopic dimensions. We have therefore developed a vertical junction device (7), in which MoS₂ and WSe₂ monolayers are stacked on top of each other to realize a van der Waals heterostructure (Fig. 7a). We have chosen WSe₂ and MoS₂, as these materials have different electron affinities (WSe₂: 3.5–4.0 eV, MoS₂: 4.2–4.6 eV). As a result, a type-II band discontinuity is formed (Fig. 7b). The junction is electrically tunable by the back-gate electrode, and

978-1-4799-8002-4/14 $31.00 © 2014 IEEE

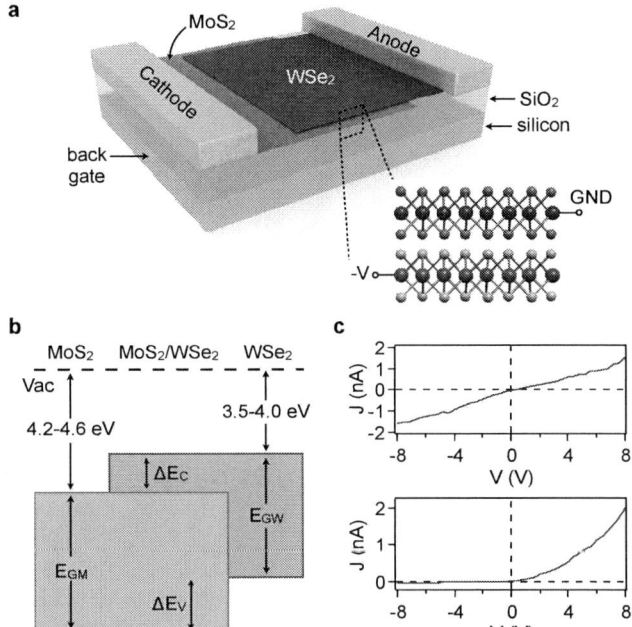

Fig. 7. (a) Schematic drawing of the device structure. (b) Energy band diagrams of WSe$_2$ (right; red) and MoS$_2$ (left; blue). The MoS$_2$/WSe$_2$ hetero-bilayer bands are a superposition of the MoS$_2$ and WSe$_2$ monolayer bands. (c) IV curves measured with V_G = 10 V (n–n regime; upper panel) and V_G = -59 V (p–n regime; lower panel).

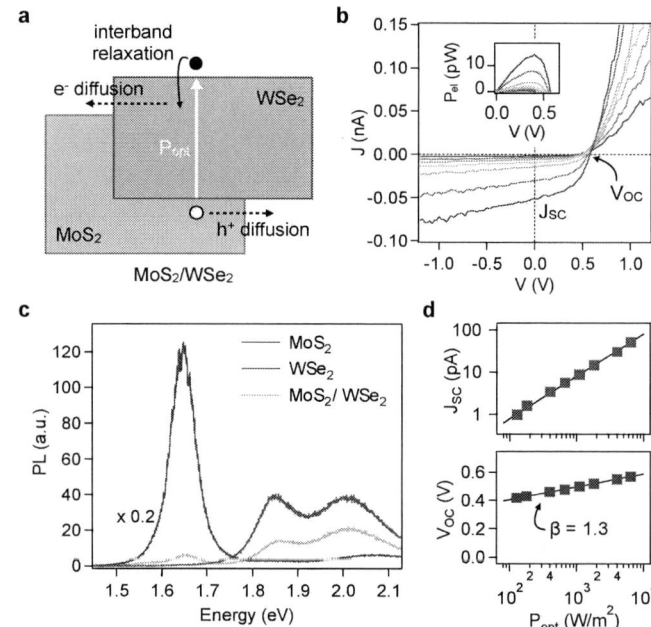

Fig. 8. (a) Illustration of the photovoltaic effect in a MoS$_2$/WSe$_2$ van der Waals heterojunction (photon absorbed in WSe$_2$; an equivalent diagram can be drawn for a photon absorbed in MoS$_2$). (b) IV curves under optical illumination (180–6400 W/m^2). Inset: Electrical output power P_{el} versus bias voltage. (c) PL emission from WSe$_2$ (red line), MoS$_2$ (blue line) and MoS$_2$/WSe$_2$ heterojunction (orange line). (d) Power dependence of short-circuit current (upper panel) and open circuit voltage (lower panel) as a function of incident optical power P_{opt}. Symbols: measurement, line: fit.

under appropriate gate voltage an atomically-thin vertical p-n junction is realized (Fig. 7c). Photoexcited electrons (holes) in the WSe$_2$ (MoS$_2$) sheet are transferred into the neighboring MoS$_2$ (WSe$_2$) layer and are extracted by the metal electrodes (Fig. 8a). Upon optical illumination, the device thus shows a photovoltaic effect (Fig. 8b) and we determine a power conversion efficiency of ≈0.2 %. This value is comparable to efficiencies reported for conventional bulk WSe$_2$ p-n junctions (0.1–0.6 %). Fig. 8c shows photoluminescence spectra measured at three different positions on the sample: WSe$_2$ (red), MoS$_2$ (blue), and MoS$_2$/WSe$_2$ heterojunction (orange). The photoluminescence from the junction is strongly quenched, evidencing efficient spatial electron–hole separation. Fig. 8d shows short-circuit current J$_{SC}$ (top) and open-circuit voltage V$_{OC}$ (bottom) as a function of illumination intensity. J$_{SC}$ scales linearly with P$_{opt}$ and shows no indication of saturation. V$_{OC}$ scales with ln(P$_{opt}$), as expected from p-n junction theory. From the slope of the curve, we extract β = 1.3 (β denotes the recombination order; β = 1 for monomolecular (Shockley Read Hall) recombination and β = 2 for bimolecular (Langevin) recombination), suggesting Schottky Read Hall interlayer recombination processes as limiting factor for device performance. The ≈90 % optical transparency of these devices could make them attractive for semi-transparent solar cells. Moreover, by choosing a proper 2D material thickness, the trade-off between optical transparency and efficiency may be adjusted according to the application requirements.

Acknowledgement

We acknowledge financial support by the Austrian Science Fund FWF (START Y-539) and the European Union Seventh Framework Programme (grant agreement no. 604391 Graphene Flagship).

References

(1) F.H.L. Koppens, T. Mueller, Ph. Avouris, A.C. Ferrari, M.S. Vitiello, and M. Polini, "Photodetectors based on graphene, related two-dimensional crystals, and hybrid systems," *Nature Nanotechnol.*, DOI: 10.1038/NNANO.2014.215, 2014.

(2) T. Mueller, F. Xia, and Ph. Avouris, "Graphene photodetectors for high-speed optical communications," *Nature Photon.*, vol. 4, pp. 297–301, 2010.

(3) A. Pospischil, M. Humer, M.M. Furchi, D. Bachmann, R. Guider, T. Fromherz, and T. Mueller, "CMOS-compatible graphene photodetector covering all optical communication bands," *Nature Photon.*, vol. 7, pp. 892–896, 2013.

(4) M. Furchi, A. Urich, A. Pospischil, G. Lilley, K. Unterrainer, H. Detz, P. Klang, A.M. Andrews, W. Schrenk, G. Strasser, and T. Mueller, "Microcavity-integrated graphene photodetector," *Nano Lett.*, vol. 12, pp. 2773–2777, 2012.

(5) M.M. Furchi, D.K. Polyushkin, A. Pospischil, and T. Mueller, "Mechanisms of photoconductivity in atomically thin MoS$_2$," ArXiv:1406.5640, 2014.

(6) A. Pospischil, M.M. Furchi, T. Mueller, "Solar-energy conversion and light emission in an atomic monolayer p-n diode," *Nature Nanotechnol.*, vol. 9, pp. 257–261, 2014.

(7) M.M. Furchi, A. Pospischil, F. Libisch, J. Burgdörfer, T. Mueller, "Photovoltaic effect in an electrically tunable van der Waals heterojunction," *Nano Lett.*, vol. 14, pp. 4785–4791, 2014.

Broadband 10Gb/s Graphene Electro-Absorption Modulator on Silicon for Chip-Level Optical Interconnects

Y. T. Hu[*], M. Pantouvaki[#], S. Brems[#], I. Asselberghs[#], C. Huyghebaert[#], M. Geisler[*], C. Alessandri[#], R. Baets[*], P. Absil[#], D. Van Thourhout[*] and J. Van Campenhout[#]

[#]IMEC, Kapeldreef 75, 3001 Leuven, Belgium
[*]Ghent University – IMEC, Photonics Research Group, Gent, Belgium
Tel: (32) 1628 7732, Email: Joris.VanCampenhout@imec.be

Abstract

We report the first silicon integrated graphene optical electro-absorption modulator capable of 10Gb/s modulation speed. We demonstrate low insertion loss and low drive voltage combined with broadband and athermal operation in a compact hybrid graphene-Si device, outperforming Si(Ge) optical modulators for future chip-level optical interconnect application.

Introduction

Integrated optical modulators with high modulation speed, small footprint and broadband, athermal operation are highly desired to realize high-density low-power optical interconnects tightly integrated with high-performance logic [1-2]. Silicon optical modulators have been strongly optimized over the past few years, but still suffer from large footprint [3], a narrow optical bandwidth or poor temperature tolerance [4]. Alternatively, compact broadband SiGe electro-absorption modulators (EAM) have been demonstrated, but with limited thermal stability [5]. Graphene, which exhibits 2.3% absorption over a wide spectral range from the visible to the infrared owing to its unique linear and gapless band dispersion, is a promising material for robust electro-absorption modulators [6-7]. In this paper, we demonstrate a broadband, athermal and compact graphene-Si EAM operating at 10Gb/s, challenging Si(Ge) modulators for thermally volatile and area sensitive applications such as chip-level optical interconnects.

Device Design and Fabrication

The graphene EAM consists of a graphene-oxide-silicon (GOS) capacitor structure implemented on top of a planarized silicon-on-insulator (SOI) rib waveguide, as shown in Fig. 1. The graphene sheet interacts strongly with the guided optical modes of the sub-micron Si waveguide, and the optical loss can be modulated by controlling the Fermi level in the graphene layer through metal contacts placed on the graphene and Si layer. A modified 130nm CMOS shallow-trench isolation (STI) module with two Si patterning steps was used to define the Si rib waveguides on 200mm SOI wafers. After planarization, three phosphorous ion implantation steps were carried out locally in the Si layer to reduce device resistance. Next, a 5nm thick gate oxide was thermally grown on top of the Si waveguide and the wafer was diced. Subsequently, a single-layer graphene sheet

grown by chemical vapor deposition (CVD) on a $Si/SiO_2/Cu$ substrate was transferred to the Si chips using an elastomer stamp [8]. The graphene was patterned by photolithography followed by an oxygen plasma etch. Two separate steps defined the graphene-metal contact (10nm Cr/50nm Au/20nm Ti/420nm Au) and the Si-metal contact (20nm Ti/480nm Au). SEM images of a fabricated device are shown in Fig. 1. Raman data (Fig. 2) indicates that the single layer graphene quality is preserved after processing. Grating couplers [9] were implemented for fiber-to-chip optical coupling.

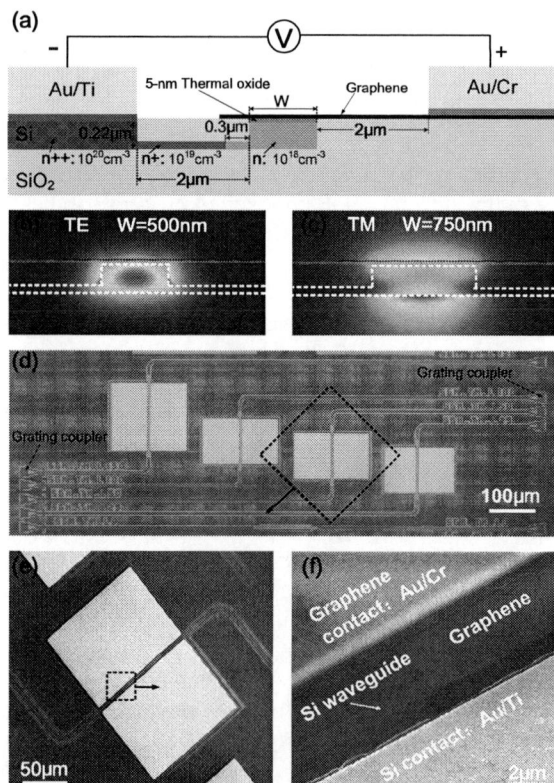

Fig. 1. Schematic cross section of the hybrid single-layer graphene-Si EAM showing the key structural parameters (a). Simulated optical mode profiles for hybrid graphene-Si waveguides designed for quasi TE (W = 500nm) (b) and TM (W = 750nm) (c) polarized guided modes. Optical microscope image showing the graphene EAM test site (d), and top-view (e) and tilted-view (f) SEM images of fabricated devices.

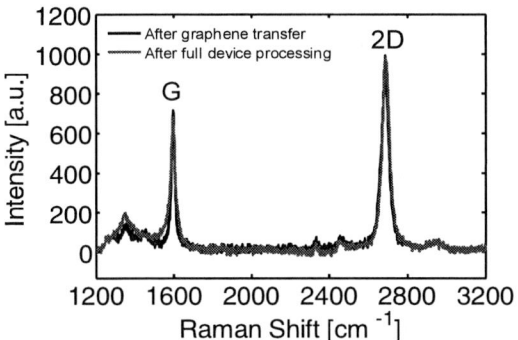

Fig. 2. Raman spectra of the single-layer graphene sheet after transfer onto the SOI chip and after full device processing. The intensity ratios of the D and G peak ($I_D/I_G \sim 0.3$), as well as the 2D and G peak ($I_{2D}/I_G \sim 1.4$) indicate that the graphene quality is maintained after processing.

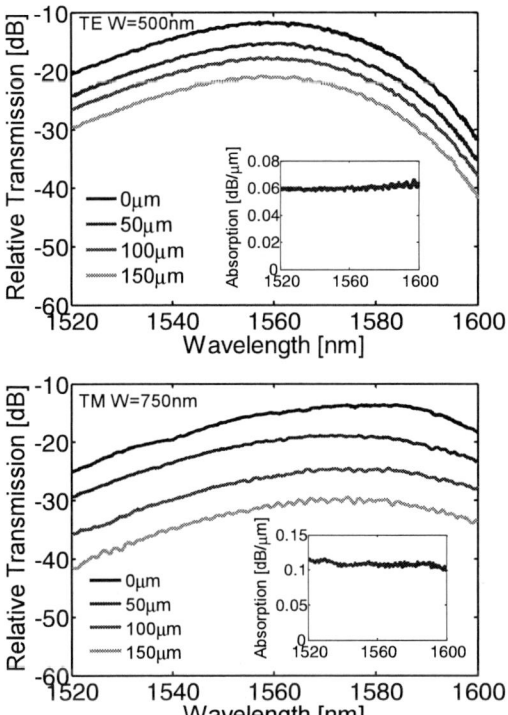

Fig. 3. Fiber-to-fiber transmission spectra for unbiased TE and TM graphene EAMs with graphene lengths of 50μm, 100μm, 150μm, as well as for reference waveguides without graphene. Graphene-induced waveguide losses are extracted to be 0.06dB/μm for the TE mode and 0.11dB/μm for the TM mode across the full wavelength range (insets). The Gaussian-like roll–off of the transmission at short and long wavelengths originates from the grating coupler pair used for fiber-chip coupling [9].

Static Performance

Transmission spectra of unbiased graphene EAM devices with lengths of 50μm, 100μm, 150μm and reference waveguides without graphene are plotted in Fig. 3, for quasi-transverse electric (TE) and quasi-transverse magnetic (TM) guided modes. Graphene-induced optical absorption losses of ~0.06dB/μm and ~0.11dB/μm are extracted for the TE and TM modes respectively over a wavelength range of 80nm (limited by the grating couplers) around 1560nm (Fig. 3). By

applying drive voltages V_D in the range -3V to 3V, an extinction ratio (ER) of ~5dB and on-state insertion loss (IL) below 4dB are obtained for a 50μm TM graphene EAM across the entire wavelength range (Fig. 4), owing to the strong interaction of the TM mode with the graphene sheet (Fig 1). Longer devices enable higher ER, exceeding 10dB for a 150μm device (Fig. 5), at the expense of higher IL. All tested graphene EAM devices exhibited strongest extinction at reverse drive voltages, likely caused by a Fermi-level shift in the graphene sheet due to doping. No significant changes in device response were observed for temperatures in the range 20°C to 49°C, implying robust athermal operation (Fig. 6). Fig. 7 shows the C-V and I-V curves of the graphene EAMs, illustrating GOS accumulation and depletion capacitance for the 50μm long device of ~220fF and ~100fF at V_D=2V and -2V respectively, and leakage currents in the nA range.

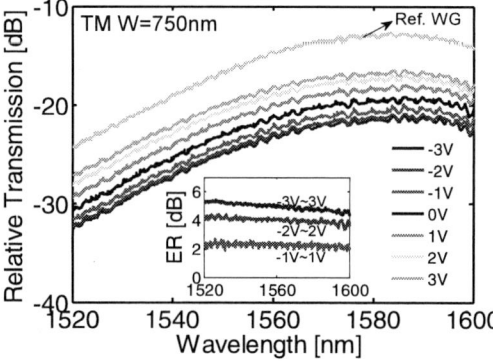

Fig. 4. Fiber-to-fiber transmission spectra of a 50μm TM graphene EAM at different drive voltages. An extinction ratio of ~5dB and insertion loss below 4dB is obtained across the full wavelength range for -3V to 3V voltage swing.

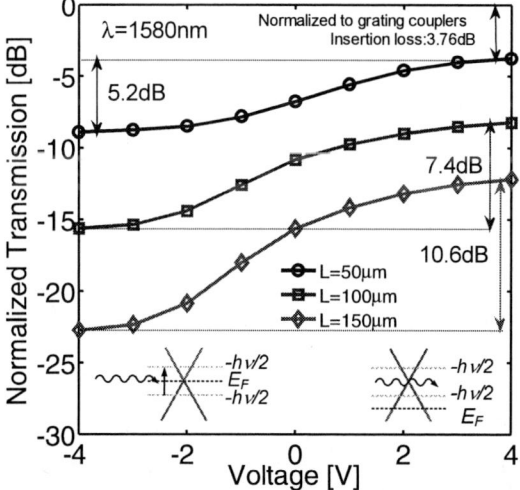

Fig. 5. Static switching curves versus applied voltage of TM graphene modulators with lengths of 50μm, 100μm and 150μm at 1585nm wavelength, normalized for the optical insertion loss of the grating couplers. All EAM devices exhibit maximum absorption at -4V, suggesting a graphene Fermi level close to the Dirac point, and strong absorption bleaching at 4V, suggesting strong hole accumulation ($E_F < -h\nu/2$, h = Planck's constant, ν = photon frequency).

High Speed Performance

Radio-frequency (RF) S11 measurements and fitting to an electrical circuit model suggest a series resistance of Rs~250 and accumulation (depletion) capacitance of CgSi~200fF (80fF) for the 50μm device (Fig. 8). The electro-optical S21 response shows RC-limited 3-dB bandwidths of 5.9GHz and 2.9GHz at reverse and forward bias respectively (Fig. 9), an up to 5 times improvement over earlier demonstrations [6-7], and in good agreement with the electrical model. Next, eye diagrams were generated for the 50μm graphene EAM with a drive voltage of 2.5Vpp and 1.75V forward bias, using the setup shown in Fig 10. Wide open eye diagrams with dynamic ER~2.5dB and low jitter are obtained at 6Gb/s and up to 10Gb/s (Fig. 11), across a 35nm wavelength range (Fig. 12). This is the first demonstration of high-quality optical modulation at high bit rates by a graphene EAM. The energy consumption is estimated to be ~350fJ/bit (E_{bit}~$CV^2/4$). Finally, the performance metrics are compared to state-of-the-art Si(Ge) modulators in Table 2, illustrating the thermal, bandwidth and footprint benefits of the graphene EAM. Based on the electrical model, design optimizations are expected to reduce R_s by a factor of 2-4, targeting 25Gb/s modulation speed and beyond.

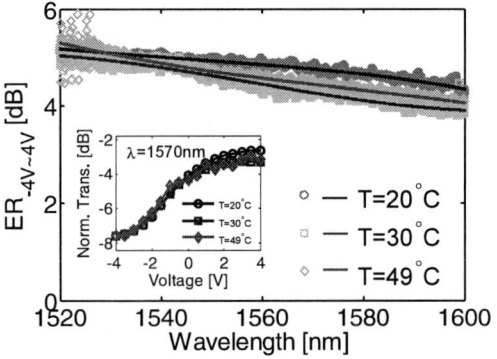

Fig. 6. Static extinction ratio versus wavelength of a 50μm TM graphene EAM measured for -4V to 4V voltage swing at chip temperatures of 20℃, 30℃ and 49℃. The inset shows the normalized static switching curves vs. applied voltage.

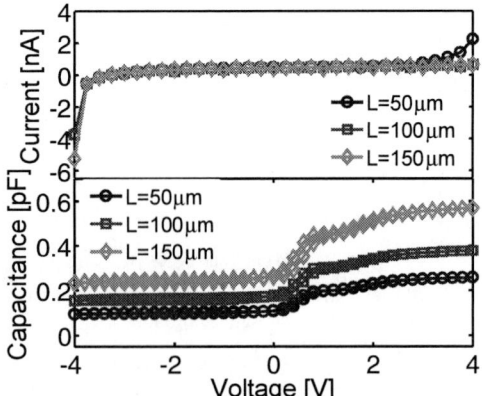

Fig. 7. Static I-V and quasi-static (100kHz) C-V curves for TM graphene EAMs with lengths of 50μm, 100μm and 150μm.

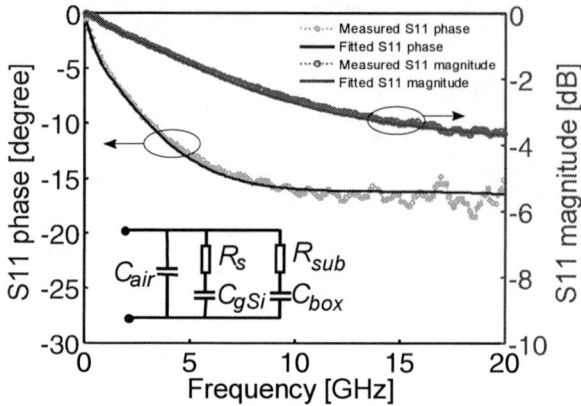

Fig. 8. Measured RF S11 response of the 50μm graphene EAM at 0V bias and the fitted curves using an equivalent electrical circuit model (inset). (C_{air} = the capacitance between the two electrodes through the air, R_s = the series resistance between the two electrodes, C_{gSi} = the capacitance of the GOS structure, R_{sub} = the resistance of the substrate, C_{box} = the capacitance of the buried oxide layer).

Table 1. S11 fitting parameters of the 50μm graphene EAM for bias voltages of 0, -1 and 2V, and electrical 3dB RF bandwidth estimated from the fit.

	$R_s(\Omega)$	C_{gSi}(fF)	$R_{sub}(\Omega)$	C_{box}(fF)	C_{air}(fF)	f_{3dB}(GHz)
0V	241	81.8	1140	114	10.2	6.1
-1V	269	78	1360	90	11.2	5.8
2V	246	194	1100	19	9.6	2.7

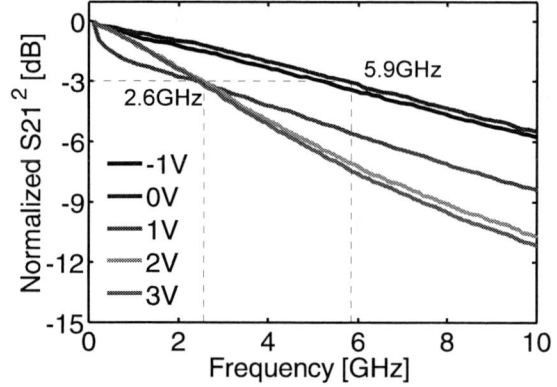

Fig. 9. Measured and normalized electro-optical S21 responses of the 50μm graphene EAM at different bias voltages showing electro-optical bandwidth of 5.9GHz in reverse bias and 2.6GHz in forward bias.

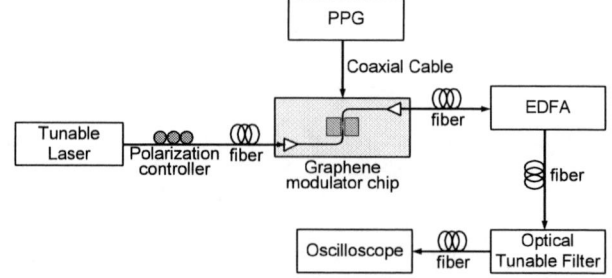

Fig. 10. Experimental setup used for high speed eye diagram measurements. (PPG = pseudo-random bit pattern generator, EDFA = erbium-doped fiber amplifier).

978-1-4799-8002-4/14 $31.00 © 2014 IEEE

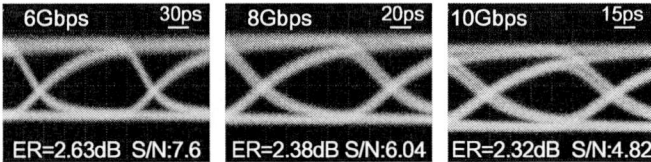

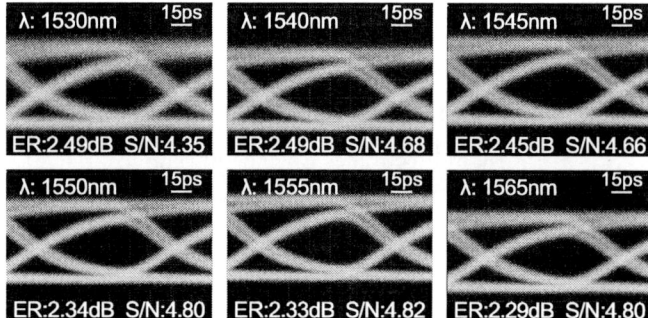

Fig. 11. Optical eye diagrams measured at 1560nm for the 50µm graphene EAM device at 6Gb/s, 8Gb/s and 10Gb/s modulation speed, using a drive voltage of 2.5V_pp swing and 1.75V forward bias delivered with a 50 terminated probe. Part of the noise originates from the EDFA used in the measurement setup.

Conclusion

We have reported the first compact graphene-Si electro-absorption modulator operating at 10Gb/s over a wide wavelength and temperature range, challenging best-in-class Si(Ge) modulators for future chip-level optical interconnects.

Fig. 12. Optical eye diagrams measured at 10Gb/s modulation speed for the 50µm graphene EAM device for wavelengths in the range 1530-1565nm, using a drive voltage of 2.5V_pp swing and 1.75V forward bias delivered with a 50 terminated probe. A dynamic extinction ratio of 2.3dB or better is obtained for all wavelengths.

Table 2. Benchmarking table comparing the performance of the presented graphene-Si EAM versus different types of state-of-the-art Si and SiGe optical modulators. The graphene EAM combines compact footprint with large optical bandwidth over a wide temperature range, with relatively low power consumption. Our work has substantially advanced the modulation speed of the graphene EAM over prior reports. Significant potential still exists to further improve the modulation speed of the graphene EAM closer to speeds obtained in highly optimized Si(Ge) modulators.

Modulator type	Ref.	Footprint	Wavelength	Drive voltage	Optical bandwidth	Temperature range	ER	Insertion Loss	Power		3dB frequency	Maximum Bit Rate
									Static	Dynamic		
		[µm²]	[nm]	[V]	[nm]	[°C]	[dB]	[dB]	[mW]	[fJ/bit]	[GHz]	[Gb/s]
Si Mach-Zehnder Modulator	[3]	~3000x500	1300	1.5	>80*	>80*	3.4	7.1	~20#	450	30	50
Si Ring Modulator	[4]	~10x10	1550	0.5	<0.1	<1	6.4	1.2	<0.01	~1	21	44
SiGe EAM	[5]	~55x10	1550	2.8	35 5§	<1 <40§	~5	~4	>2.2	60	>30	28
Graphene-Si EAM	[6]	~40x10	1500	3	>180	n/a	2.4	n/a	n/a	n/a	1.2	n/a
Graphene-Si EAM	This work	~50x10	1550	2.5	>80	>29	2.5	<4	<1x10⁻⁴	350	2.6-5.9	10

*Assuming balanced Mach-Zehnder modulator with broadband 3dB splitters, #Mach-Zehnder bias control, §Bandwidth can be traded for temperature tolerance.

Acknowledgements

This work was supported by imec's Core Partner Program. Prof B.J. Cho (KAIST, Korea) is greatly acknowledged for the generous supply of high-quality graphene.

References

[1]. G. T. Reed, G. Mashanovich, F. Y. Gardes and D. J. Thomson, "Silicon optical modulators," Nat. Photon. 4, 518–526 (2010).

[2]. I. A. Young, et al. Optical I/O Technology for Tera-Scale Computing. IEEE J. Solid-State Circuits 45.1, 235–248 (2010).

[3]. M. Streshinsky, et al, "Low power 50 Gb/s silicon traveling wave Mach-Zehnder modulator near 1300 nm," Opt. Express 21, 30350-30357 (2013).

[4]. E. Timurdogan, C. M. Sorace-Agaskar, J. Sun, E. S. Hosseini, A. Biberman and M. R. Watts, "An ultralow power athermal silicon modulator," Nat. Commun. 5, (2014).

[5]. D. Feng, et al., "High speed GeSi electro-absorption modulator at 1550 nm wavelength on SOI waveguide," Optics Express, Vol. 20, Issue 20, pp. 22224-22232 (2012).

[6]. M. Liu, et al., "A graphene-based broadband optical modulator," Nature 474, 64–67 (2011).

[7]. M. Liu, X. Yin and X. Zhang et al., "Double-layer graphene optical modulator," Nano Lett. 12(3) (2012).

[8]. J. Song, et al., "A general method for transferring graphene onto soft surfaces," Nature Nanotechnology 8, 356–362 (2013).

[9]. F. Van Laere, et al., "Compact Focusing Grating Couplers for Silicon-on-Insulator Integrated Circuits," IEEE Photon. Tech. Lett., 19(23), p.1919-1921 (2007).

Fast Visible-Light Phototransistor Using CVD-Synthesized Large-Area Bilayer WSe$_2$

Pang-Shiuan Liu[1], Chang-Hsiao Chen[2], Wei-Ting Hsu[3], Chih-Pin Lin[1], Tzu-Ping Lin[1], Li-Jen Chi[1], Chao-Yuan Chang[1], Shih-Chieh Wu[1], Wen-Hao Chang[3], Lain-Jong Li[2] and Tuo-Hung Hou[1*]

[1]Department of Electronics Engineering and Institute of Electronics, National Chiao Tung University, Hsinchu, Taiwan
[2]Institute of Atomic and Molecular Sciences, Academia Sinica, Taipei, 10617, Taiwan
[3]Department of Eletrophysics, National Chiao Tung University, Hsinchu, Taiwan
[*]Tel: +886-3-5712121 ext 54261; E-mail: thhou@mail.nctu.edu.tw

Abstract

P-channel transition metal dichalcogenide ultrathin-body phototransistor (UTB-PT) with a response time as fast as 100 μs has been demonstrated for the first time using the CVD-synthesized large-area bilayer WSe$_2$. Because of its excellent compatibility with mass production, the application of WSe$_2$ UTB-PT for high-speed proximity interactive display has been proposed.

Introduction

Since the discovery of graphene in 2004, a novel class of materials has emerged: 2D layered material. Various 2D layered materials, such as BN, MoS$_2$, WS$_2$, MoSe$_2$, WSe$_2$, NbSe$_2$, etc., were discovered soon after graphene [1]. In particular, transition metal dichalcogenides (TMDs) are suitable for electronic and optoelectronic applications because of their moderate bandgap values (Fig. 1). A TMD ultrathin body (UTB) FET [2] presents a more favorable opportunity for scaling than its Si counterpart because of its self-assembled molecular monolayer (<1 nm), high and symmetric electron and hole mobilities (>200 cm^2/Vs), and ideal surface property without dangling bonds. Therefore, TMD UTB-FET is a promising platform technology for the post-silicon era. Numerous new device functionalities, such as memories [3], phototransistors (PT) [4-8], and electro-luminescent devices [9] have been demonstrated using TMD UTB-FET structures. Furthermore, because of the weak van der Waals' interaction with the surroundings, 2D TMD layers can be easily transferred between substrates. This substrate-independent nature is ideal for flexible electronics [10], and developing p-channel TMD UTB-FET complements existing n-channel IGZO thin-film transistor (TFT) technology.

One particularly interesting application of TMD-based PT is the proximity interactive display with in-cell photosensors integrated in pixels. Developing PT that may be fabricated using a similar process as a pixel access transistor can significantly reduce integration cost and increase resolution (Fig. 2). Previous proximity interactive display was realized using IGZO TFT technology [11]; however, the slow response time because of persistent photocurrent (PC) complicates high-speed detection. Recently, n-channel MoS$_2$ UTB-PT has generated substantial interest because of its high responsivity [7] and wide spectral response [5, 6]. However, the slow response time ranging from 50 ms to tens of seconds [4-8] remained a challenge. In addition, TMD PTs using channel materials other than MoS$_2$ have yet to be investigated. In this study, we fabricate the first p-channel UTB-PT using CVD synthesized large-area bilayer WSe$_2$ [12]. In contrast to the prevalent mechanical exfoliation technique, the CVD synthesis method is mass-production compatible. WSe$_2$ has a narrower bandgap than MoS$_2$, thus offering a wider visible spectral response. In addition, WSe$_2$ UTB-PT shows a response time of 100 μs, the fastest among all reported TMD UTB-PTs, and promising for high-speed proximity interactive display.

Large-area Synthesis of Bilayer WSe$_2$

Bilayer WSe$_2$ was synthesized by the selenization of WO$_3$ powders in a hot-wall CVD furnace on a sapphire substrate as shown in Fig. 3a [12]. The WO$_3$ powders placed in the heating zone center of the furnace were heated to 925 $^\circ$C, while the Se powders placed at the upper stream side maintained at 270 $^\circ$C. The WO$_3$ and Se vapors were carried by an Ar/H$_2$ flow to the sapphire substrate placed at the down stream side at 850 $^\circ$C. Fig. 3b shows the optical microscope (OM) image of bilayer WSe$_2$ homogeneously grown on an 1 cm×2 cm sapphire substrate. Photoluminescence (PL) spectra (Fig. 4) shows an optical bandgap of 1.6 eV. Raman spectra (Fig. 5) shows a characteristic peak at 308 cm^{-1}, which is related to the interlayer interaction of bilayer WSe$_2$ [12].

Fabrication of Bilayer WSe$_2$ UTB-PT

Top-gate WSe$_2$ UTB-PT was fabricated using a four-mask process as shown in Fig. 6. A Pd (15 nm)/Ni (25 nm) bilayer was used as the source/drain contact, and a composite 2-nm HfO$_2$ (evaporation) /25-nm HfO$_2$ (ALD at 120 $^\circ$C) was used as the gate dielectric. Both opaque Ni and transparent ITO gates were fabricated. Fig. 7 shows the top-view OM images of WSe$_2$ UTB-PT with the ITO gate as compared with that with the Ni gate. The cross-sectional TEM and EDX analyses in Fig. 8 confirmed that the channel comprises highly crystalline bilayer WSe$_2$.

Device Characteristics

Fig. 9 shows typical I_D (dark)-V_G transfer curves of p-channel bilayer WSe$_2$ UTB-PT with no significant difference between the Ni and ITO gates. Pd is known to form p-type Schottky barrier (SB) with WSe$_2$ [13], and the dark current conduction is governed by the gate-controlled SB modulation (Fig. 10). PCs measured at various V_D, wavelengths, and incident light power (P_{light}), are shown in Figs. 11-14. Pronounced changes of transfer curves were found in the "OFF" state with a 10^2-fold increase in I_D because of the lower dark current, but the highest PC was obtained in the "ON" state. Significant PC was found across a wide range of visible spectra and increased with V_D and P_{light}. The linear dependence between PC and P_{light} indicates that PC is determined by the number of photo-generated carriers under illumination. The magnitude of PC is proportional to V_D, suggesting that the lateral electric field along the WSe$_2$ channel is critical for carrier separation. The calculated responsivity (PC/P_{light}) is comparable to several MoS$_2$ UTB-PTs reported [4-6], but shows a characteristic bell-shaped curve where a maximum exists at a small -V_G bias (Fig. 15). In addition, the device shows a negative threshold voltage (V_t) shift and hole trapping at a strong -V_G bias, and the trapped holes can be partially annihilated after light exposure at a zero gate bias (Fig. 16).

Fig. 17 illustrates the PC mechanism in bilayer WSe$_2$ UTB-PT. Photo-generated electron-hole pairs under illumination are separated by the lateral electric field, which depends strongly on both V_G and V_D through the SB modulation. A small -V_G bias provides more favorable separation field and thus higher PC. At a strong -V_G bias, hole trapping at the gate oxide results in significant V_t shift. Both the negative V_t shift and quenching of photo-generated holes may lead to additional PC reduction. At a zero gate bias, photo-generated electrons may recombine with trapped holes and partially recover the V_t shift. According to the model, reducing Pd/WSe$_2$ SB, for example by using additional NO$_2$ doping [13] or MO$_x$ contacts [14], and improving the quality of gate dielectric and its interface with WSe$_2$ would further enhance PC responsivity.

Time-dependent PC dynamics were measured by modulating the laser beam with a mechanical chopper. Continuous light ON/OFF cycles in Fig. 18 shows stable and fast PC response. Fig. 19 shows the PC response to a single light pulse at various V_G. Although enlarging V_G increased PC, it also induced hole traps and led to slow decay of PC with time, which should be avoided to obtain a stable and fast response. Finally, Fig. 20 shows the rising and trailing edges of the PC response. An 100-µs rising and 400-µs falling time is significantly faster than 50 ms to tens of seconds in previous MoS$_2$ UTB-PTs [4-8].

Conclusion

Proximity interactive display may revolutionize the way we interact with future information systems. The required high-speed, high-density, low-cost in-cell photosensor can be realized using the proposed WSe$_2$ UTB-PT. The device shows a wide visible spectral response and a fast response time of 100 µs. The PC mechanism has been discussed, and the effect of hole trapping deserves more attention in the future research.

Acknowledgments

This work was supported by National Science Council of Taiwan, Republic of China, under grant NSC 102-2119-M-001-005-MY3, MOST 103-2221-E-009-221-MY3, and NCTU-UCB I-RiCE program under grant NSC-102-2911-I-009-302. The authors would like to thank the Nano Facility Center at National Chiao Tung University and National Nano Device Laboratories, where the experiments in this paper were performed.

References

[1] K. S. Novoselov, D. Jiang, F. Schedin, T. J. Booth, V. V. Khotkevich, S. V. Morozov, and A. K. Geim, "Two- dimensional atomic crystals," *Proc. Nat. Acad. Sci.*, vol. 102, no. 30, pp. 10451–10453, Jul. 2005.

[2] B. Radisavljevic1, A. Radenovic, J. Brivio, V. Giacometti and A. Kis, "Single-layer MoS$_2$ transistors," *Nature Nanotech.*, vol. 6, pp. 147–150, Mar. 2011.

[3] M. S. Choi, G.-H. Lee, Y.-J. Yu, D.-Y. Lee, S. H. Lee, P. Kim, J. Hone and W. J. Yoo, "Controlled charge trapping by molybdenum disulphide and graphene in ultrathin heterostructured memory devices," *Nature Commun.*, vol. 4, 1642, Mar. 2013.

[4] Z. Yin, H. Li, H. Li, L. Jiang, Y. Shi, Y. Sun, et al., "Single-layer MoS$_2$ phototransistors," *ACS Nano*, vol. 6, no. 1 pp. 74–80, Dec. 2011.

[5] H. S. Lee, S.-W. Min, Y.-G. Chang, M. K. Park, T. Nam, H. Kim, J. H. Kim, S. Ryu, and S. Im, "MoS$_2$ nanosheet phototransistors with thickness-modulated optical energy gap," *Nano Lett.*, vol. 12, no. 7, pp. 3695–3700, Jun. 2012.

[6] W. Choi, M. Y. Cho, A. Konar, J. H. Lee, G.-B. Cha, S. C. Hong, S. Kim, J. Kim, D. Jena, J. Joo and S. Kim, "High- detectivity multilayer MoS$_2$ phototransistors with spectral response from ultraviolet to infrared, " *Adv. Mater.*, vol. 24, pp. 5832–5836, Nov. 2012.

[7] O. Lopez-Sanchez, D. Lembke, M. Kayci, A. Radenovic and A. Kis, "Ultrasensitive photodetectors based on monolayer MoS$_2$," *Nature Nanotech.*, vol. 8, no. 7, pp. 497–501, Jun. 2013.

[8] H.-M. Li, D.-Y. Lee, M.-S. Choi, D.-S. Qu, X.-C. Liu, C.-H. Ra, and W. J. Yoo, "Gate-controlled schottky barrier modulation for superior photoresponse of MoS$_2$ field effect transistor," in *IEDM Tech Dig.*, 2013, pp. 507–510.

[9] R. S. Sundaram, M. Engel, A. Lombardo, R. Krupke, A. C. Ferrari, Ph. Avouris, and M. Steiner, "Electroluminescence in single layer MoS$_2$," *Nano Lett.*, vol. 13, no. 4, pp. 1416–1421, Mar. 2013.

[10] J. Pu, Y. Yomogida, K.-K. Liu, L.-J. Li, Y. Iwasa, and T. Takenobu, "Highly flexible MoS$_2$ thin-film transistors with ion gel dielectrics," *Nano Lett.*, vol. 12, pp. 4013–4017, Jul. 2012.

[11] S. Jeon, S.-E. Ahn, I. Song, C. J. Kim, U.-I. Chung, E. Lee,I. Yoo, A. Nathan, S. Lee, J. Robertson and K. Kim. "Gated three-terminal device architecture to eliminate persistent photoconductivity in oxide semiconductor photosensor arrays," *Nature Mater.*, vol. 11, pp.301–305, Feb. 2012.

[12] J.-K. Huang, J. Pu, C.-L. Hsu, M.-H. Chiu, Z.-Y. Juang, Y.-H. Chang, W.-H. Chang, Y. Iwasa, T. Takenobu, and L.-J. Li, "Large-area synthesis of highly crystalline WSe$_2$ monolayers and device applications," *ACS Nano*, vol. 8, pp. 923–930, Dec. 2014.

[13] H. Fang, S. Chuang, T. C. Chang, K. Takei, T. Takahashi and A. Javey, "High-performance single layered WSe$_2$ p-FETs with chemically doped contacts," *Nano Lett.*, vol. 12, pp. 3788–3792, Jun. 2012.

[14] S. Chuang, C. Battaglia, A. Azcatl, S. McDonnell, J. S. Kang, X. Yin, M. Tosun, R. Kapadia, H. Fang, R. M. Wallace, and A. Javey, "MoS$_2$ P-type transistors and diodes enabled by high work function MoO$_x$ contacts," *Nano Lett.*, vol. 14, pp. 1337–1342, Feb. 2014.

[15] A. Kaula, "Two-dimensional layered materials: Structure, properties, and prospects for device applications," *J. Mater. Res.*, vol. 29, pp. 348–361, 2014.

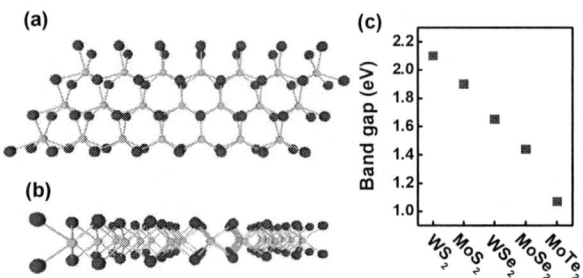

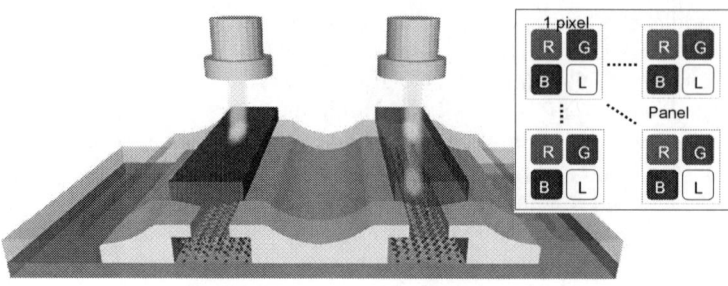

Fig. 1 (a) Top view and (b) cross-sectional view of the atomic structure of 2D TMD. (c) Bandgap values of various 2D TMD materials [15].

Fig. 2 Schematic of a monolithic 2D TMD-based light sensor using TMD UTB-FET (with an opaque top gate) as an access transistor and TMD UTB-PT (with a transparent top gate). Inset shows a schematic of proximity interactive display panel integrating red/green/blue (RGB) color cells and light sensor (L) in one pixel.

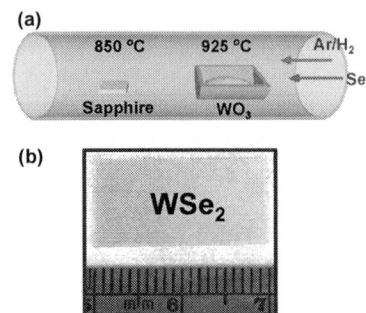

Fig. 3 (a) Schematic illustration for the CVD growth of WSe_2 thin film in a furnace. (b) Image of homogeneous WSe_2 grown on a 1 cm×2 cm double-side polished sapphire substrate.

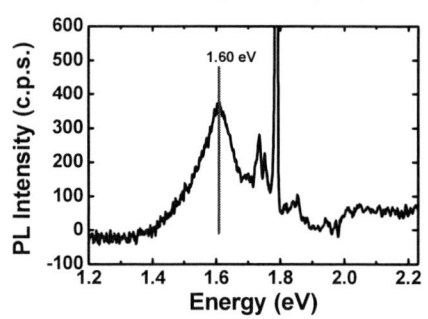

Fig. 4 Photoluminescence (PL) spectra of CVD bilayer WSe_2, showing an optical bandgap of 1.6 eV.

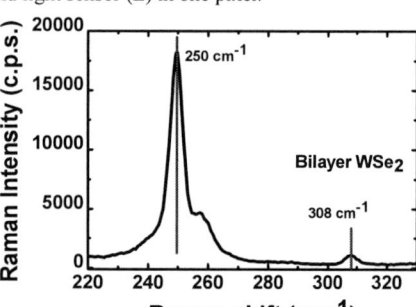

Fig. 5 Raman spectra of CVD bilayer WSe_2, showing two characteristic peaks at 250 cm^{-1} and at 308 cm^{-1}. The former is assigned to E^1_{2g} mode, while the later is due to the WSe_2 interlayer interaction.

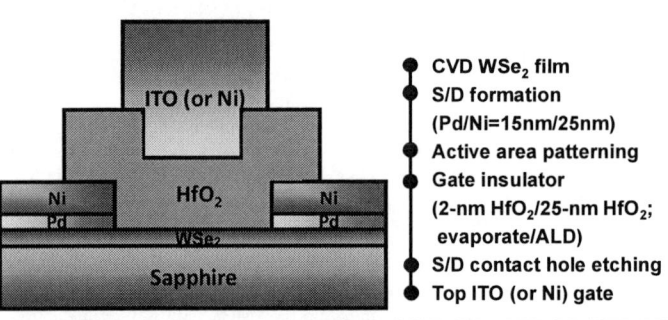

Fig. 6 Schematic diagram and process flow of CVD WSe_2 UTB-PT (UTB-FET) fabricated on a sapphire substrate.

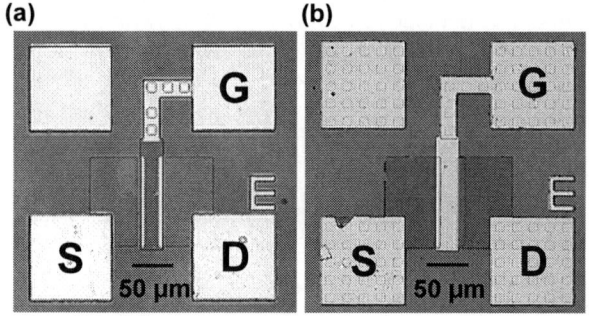

Fig. 7 Optical microscope images of the fabricated CVD WSe_2 (a) UTB-PT with a transparent ITO gate and (b) UTB-FET with an opaque Ni gate.

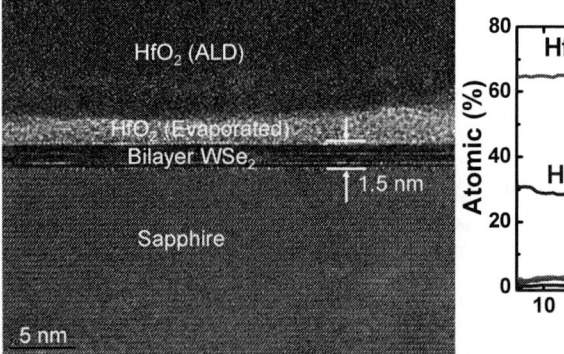

Fig. 8 (left) High-resolution cross-sectional TEM image and corresponding EDX analysis of the fabricated CVD WSe_2 UTB-PT confirm highly crystalline bilayer WSe_2 with a thickness of 1.5 nm and sharp interfaces with sapphire and HfO_2. (right) Corresponding EDX analysis showing the stoichiometric composition of WSe_2.

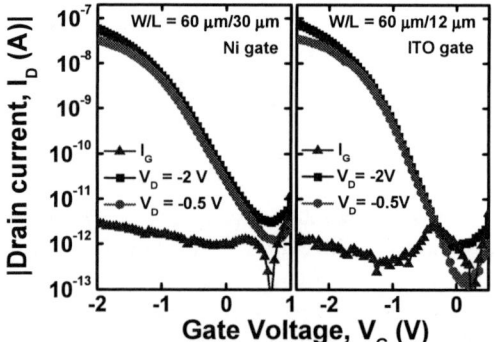

Fig. 9 Comparable I_D-V_G transfer characteristics of WSe_2 UTB-FET (Ni gate) and UTB-PT (ITO gate; w/o illumination).

978-1-4799-8002-4/14 $31.00 © 2014 IEEE

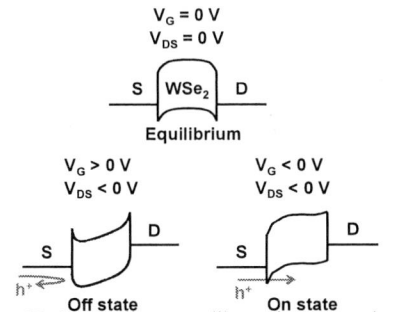

Fig. 10 Band diagrams illustrating operational mechanism of WSe$_2$ UTB-PT (dark current) by gate-controlled SB modulation.

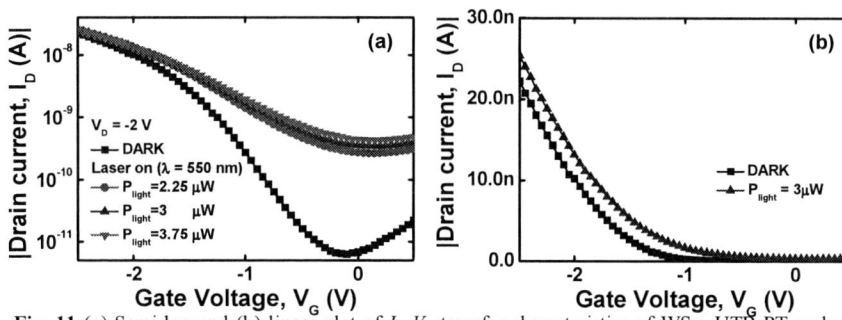

Fig. 11 (a) Semi-log and (b) linear plot of I_D-V_G transfer characteristics of WSe$_2$ UTB-PT under visible light illumination with varied incident laser power. Pronounced changes under illumination were found in the "OFF" state with a 10^2-fold increase in I_D because of the lower dark current, but the highest PC was obtained in the "ON" state, as clearly shown in the linear plot.

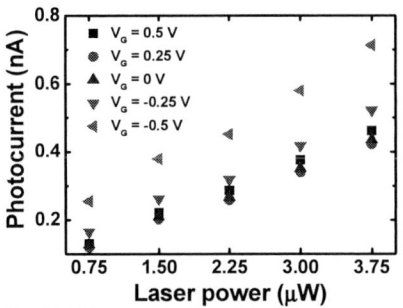

Fig. 12 PC as a function of incident laser power. Linear dependence across a wide range of V_G indicates that PC is determined by the number of photo-generated carriers under illumination.

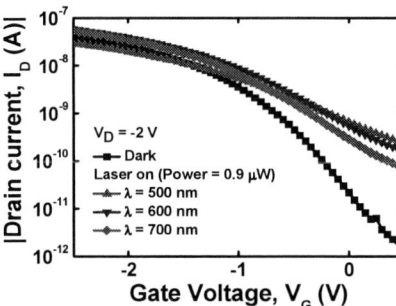

Fig. 13 I_D-V_G transfer characteristics of WSe$_2$ UTB-PT under visible light illumination at various wavelengths. Strong PC response was found at all wavelengths tested.

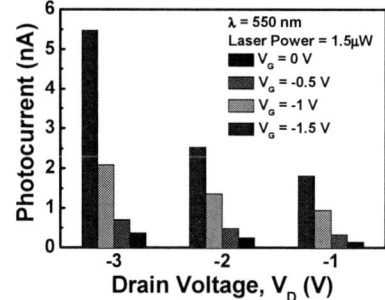

Fig. 14 PC of WSe$_2$ UTB-PT measured at various V_D and V_G. PC increases linearly with V_D because of the increased lateral electric field.

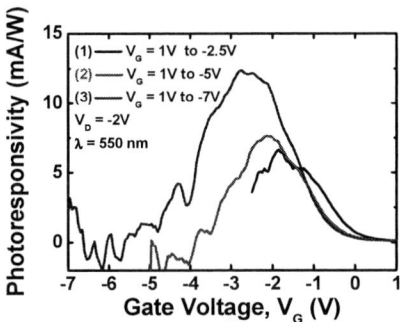

Fig. 15 Photoresponsivity (PC/P_{light}) as a function of V_G. The measuring sequence was labeled as (1) to (3). The responsivity peaks at a small -V_G bias, while hole trapping at a large -V_G bias leads to a negative shift of the peak.

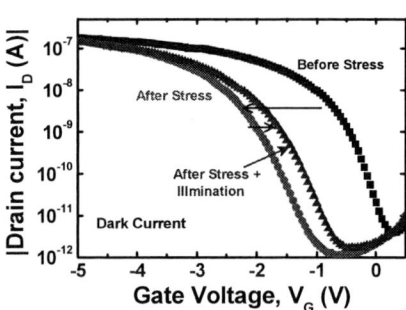

Fig. 16 Negative V_t shift after a large -V_G stress because of hole trapping in the oxide. After light illumination of 30 s at a zero bias, the V_t shift can be partially recovered by recombining trapped holes with photogenerated electrons.

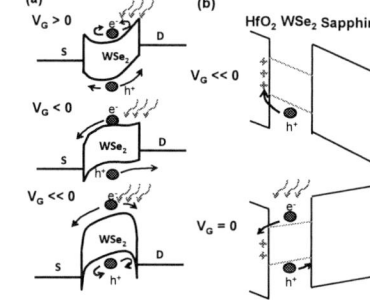

Fig. 17 Band diagrams of WSe$_2$ UTB-PT (a) (lateral direct.) illustrating the dependence of PC on V_G. (b) (vertical direct.) illustrating hole trapping in the oxide and recombination with photo-generated electrons under illumination.

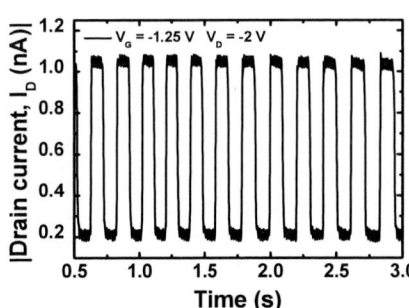

Fig. 18 Time-dependent PC dynamics measured by modulating the laser beam (550 nm) with a mechanical chopper. Reproducible PC response of light ON/OFF cycles can be achieved by choosing appropriate bias conditions.

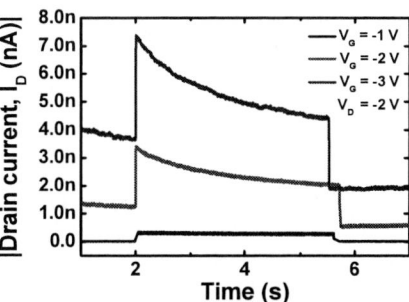

Fig. 19 Time-resolved photoresponse at various V_G. A large -V_G bias induces higher PC, but the pronounced hole trapping leads to significant and slow decay of PC with time.

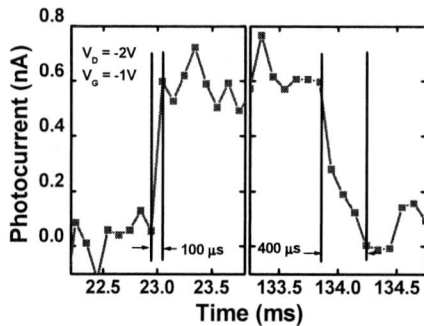

Fig. 20 Enlarged rising and falling edges of time-resolved PC. Rising time can be faster than 100 μs (setup limit) and falling time is 400 μs by using a 550-nm laser and P_{light} = 0.6 μW.

978-1-4799-8002-4/14 $31.00 © 2014 IEEE

Controlling oxygen vacancies in doped oxide based CBRAM for improved memory performances

G. Molas[1], E. Vianello[1], F. Dahmani[2], M. Barci[1], P. Blaise[1], J. Guy[1], A. Toffoli[1], M. Bernard[1], A. Roule[1], F. Pierre[1], C. Licitra[1], B. De Salvo[1], L. Perniola[1]

[1]CEA LETI Minatec Campus, 38054 Grenoble Cedex 9, France – gabriel.molas@cea.fr
[2]Altis Semiconductor, 224 Blvd. John Kennedy, 91105 Corbeil-Essonnes, France

Abstract

In this paper the concept hybrid CBRAM assisted by oxygen vacancies is presented for the 1st time. Doping the resistive layer of oxide/Cu based CBRAM with < dopant species and concentrations is proposed in order to improve the memory performances. By means of experimental characterizations, numerical model and atomistic calculations, we demonstrate that increasing the doping content ease the filament formation by facilitating the Cu injection in the resistive layer. The proper choice of the doping element and concentration allows to significantly reduce the forming voltage (up to a forming free behavior), or alternatively to increase the memory window of 3 decades, with no forming voltage increase and retention degradation (stable behavior at 200°C, 260°C soldering sustained).

Introduction

CBRAM is a promising technology for future nonvolatile memories due to its good scalability, fast read and write times < 10ns. It is compatible with CMOS technology and can be integrated in 3D architectures. Using an oxide as electrolyte instead of chalcogenide material, CBRAM insures good high temperature retention [1-5] and resistance to soldering, but at the price of higher forming voltages. In this paper we study the impact of doping the Metal Oxide resistive layer with Hf or Al in order to facilitate the Cu insertion during forming and improve the OFF resistance after RESET.

Technological details

The CBRAM technology studied in this work consists in a Metal Oxide (MO_x) resistive layer deposited on a metal plug, and combined with a Cu-based ion supply layer, deposited as a top electrode. The resistive layer was doped by co-sputtering with two types of dopants D (D being Hf or Al) with various concentrations, leading to $MO_x(Hf)$ and $MO_x(Al)$ alloys (fig.1.a). Note that $MO_x(Al)$ samples are thicker than $MO_x(Hf)$ samples. Hf and Al are chosen because their related oxides HfO_2 and Al_2O_3 are respectively more conductive (lower bandgap) and more insulating (higher bandgap) than the MO_x host material. Varying the doping content allows to modify the metal/oxygen (M+D)/O atomic ratio of the alloy (metal including the metal M of the resistive layer and the doping element D), from 0.5 to 1 (fig.1.b). Doping with Hf element reduces the bandgap while Al doping allows keeping a roughly constant value, due to a more conductive feature as (M+D)/O increases balanced by a higher intrinsic bandgap of Al_2O_3 with respect to MO_x (fig.2). The memory was integrated above the back-end of an advanced CMOS logic, in 1T1R configuration, and was tested in quasi static (QS: fig.3) and pulsed modes.

Results and discussion

A. Forming operation

Increasing the (M+Hf)/O ratio with Hf dopants leads to a lower resistance for the fresh state, hence a lower forming voltage (fig.4). With Al dopant, V_F shows a non monotonous behavior as Al concentration is raised. Indeed, two antagonist effects occur: V_F tends to increase as $V_F(Al_2O_3)>V_F(MO_x)$, while V_F tends to be reduced as the doping content increases due to a more metallic behavior of the $MO_x(Al)$ alloy (higher (M+Al)/O ratio). QS measurements illustrate that in doped samples, forming occurs at a lower voltage, but higher current (fig.5). Thus, increasing the doping concentration allows reaching forming at a lower applied voltage.

B. RESET operation

Ramped voltage-pulse RESET measurements were performed on SET devices, raising either the bit line V_{bl} or the word line voltage V_{wl} for 1µs pulse widths. In $MO_x(Hf)$, for V_{bl} lower than 1.3V, R_{OFF} remains independent of the doping concentration. On the other hand, for $V_{bl}>1.3V$ a higher memory window (resulting from a higher HRS resistance) is reached in the undoped samples (fig.6), while R_{OFF} saturates for doped samples. In other words, a larger memory window can be obtained in undoped samples for strong RESET conditions. Indeed, R_{OFF} is limited by the resistance level of the fresh state R_0, decreasing with the doping content. On the contrary, Al doping of MO_x enlarges the maximum R_{OFF} value: an optimum R_{OFF} is achieved for 10% Al doping (~3 decades increase: fig.7). Further increasing Al concentration starts to degrade R_{OFF} due to a too high (M+Al)/O ratio. Finally, V_{RESET} and I_{RESET} (defined as the voltage and current where RESET starts) extracted in QS mode are independent of the electrolyte doping (fig.8). This can be explained by the fact that RESET operation is mainly governed by R_{ON}.

978-1-4799-8002-4/14 $31.00 © 2014 IEEE

C. SET operation

In Hf-doped samples, V_{SET} decreases with the doping concentration (fig.9) due to a lower initial R_{OFF} (fig.6). Interestingly, the strongly doped sample is forming free, in the sense that the forming and SET voltages are equal. The Al-doped samples exhibit a higher V_{SET} than the undoped reference, again due to their higher initial R_{OFF}. Targeting the same initial R_{OFF} among all the samples, V_{SET} and I_{SET} (respectively defined as the bias and current at which SET occurs) are constant whatever the doping ratio, confirming that they are independent of the electrolyte doping and are governed by the HRS resistance level (fig.10).

D. Endurance and retention

Endurance was measured without correction algorithm on 64bits arrays. Both Hf- and Al-doped samples show a slight decrease of R_{OFF} after 1000 cycles. A memory window of > 1 decade is maintained after 40kcycles for Al-doped CBRAMs (fig.11). Resistance to 260°C 15min soldering was demonstrated in doped and undoped samples, making this technology suitable for embedded products (fig.12). Retention was then measured at 200°C for the various samples. For MO_x(Hf), stable memory window was maintained after 10^5s. For MO_x(Al), some degradation appears at 21% of doping content, both on LRS and HRS, while 13% exhibits good retention behavior (fig.13). To accelerate the retention loss [6], measurements were performed for ~30kOhm R_{ON} (fig.14). The faster information loss rate of Hf-doped samples was explained in our numerical model [4] by a higher Cu diffusion coefficient in the electrolyte. *Summarizing, Hf dopant can be used for reducing the forming voltage at the cost of the ON/OFF ratio; small concentration of the Al dopant can be used to improve the memory window without a huge change of the forming voltage. In both cases no major changes of switching parameters and data retention performances are observed.*

E. Discussion

First of all, we studied the impact of Cu inclusion in the electrolyte. By 1st principal calculations, we showed that gradual Cu incorporation in MO_x creates energy levels in the bandgap, leading to a conductive material for ~10% Cu concentration associated to LRS (fig.15). Then, we calculated the formation enthalpy to add a Cu atom in the resistive layer. Inserting Cu in interstitial sites in MO_x appears as the most favorable case (lowest formation enthalpy); however a high energy of 3.2eV is required. Assuming oxygen vacancies in the resistive layer, the energy cost to insert Cu in interstitial site is strongly reduced (1.3eV), indicating that the formation of a Cu filament should be easier (fig.16). Then we show that in MO_x(Hf), Cu insertion is easier than in MO_x and MO_x(Al), explaining the lower V_F. In MO_x(Hf) and MO_x(Al) doped samples, inserting oxygen vacancies strongly facilitates the

Cu inclusion, easing the filament formation (fig.17). To decorelate oxygen vacancy formation and Cu insertion during forming, structures integrating an oxygen vacancy source layer as bottom interface were investigated. Performing a pre-forming operation at negative bias and low current leads to an intermediate HRS level, due to the formation of a small V_O filament in the electrolyte like in an OXRAM device. Then, during the subsequent forming operation at $V_a>0$, the measured V_F decreased with the HRS intermediate resistance (fig.18), Cu being inserted more easily in the V_O sites. Based on the experimental and simulated results presented above, we propose the following phenomenological description (fig.19). In an oxide based CBRAM, the energy cost to insert Cu atoms in the electrolyte is elevated, leading to the requirement of a forming operation. During this phase, oxygen vacancies (V_O) may be created in the oxide, facilitating the Cu migration in the resistive layer. During RESET, Cu is removed but the remaining defects in the electrolyte prevent the recovery of the fresh resistance level. In doped electrolytes, Cu insertion in V_O is strongly facilitated, less V_O are required to form the filament. In MO_x(Hf), Cu insertion is facilitated, resulting in reduced V_F. In MO_x(Al), lower V_O generation during forming allows reaching a higher R_{OFF}, enhancing the window margin. *In conclusions, V_F is reduced in MO_x(Hf) due to dopant assisted forming (easier Cu insertion), and R_{HRS} is enhanced in MO_x(Al) due to lower V_O concentration after RESET.*

Conclusions

In this paper a phenomenological description of oxide based CBRAM is presented, elucidating the role of oxygen vacancies. Doping the electrolyte with Hf or Al respectively facilitates the Cu filament formation, reducing the forming voltage, or allows to take advantage of an alternative alloy with intrinsic improved window margin (3 decades of R_{OFF}/R_{ON} increase, stable 200°C retention) and keeping constant the operating voltages.

References

[1] J. R. Jameson, P. Blanchard, C. Cheng, J. Dinh et al., "Conductive-bridge memory (CBRAM) with excellent high-temperature retention," IEDM 2013 Tech. Dig., pp.30.1.1-30.1.4.
[2] A. Calderoni, S. Sills, N. Ramaswamy, "Performance Comparaison of oxide-ReRAM and Cu based ReRAM," Int. Mem. Workshop 2014, pp.5-8.
[3] K. Aratani, K. Ohba, T. Mizuguchi, S. Yasuda et al., "A Novel Resistance Memory with High Scalability and Nanosecond Switching", IEDM 2007 Tech. Dig., pp.783–786.
[4] J. Guy, G. Molas, E. Vianello, F. Longnos et al., "Investigation of the physical mechanisms governing data retention in down to 10nm Nano-trench Al2O3/CuTeGe Conductive Bridge RAM (CBRAM)," IEDM 2013 Tech. Dig., 30.2.1-30.2.4.
[5] L. Goux, K. Sankaran, G. Kar, N. Jossart et al., "Field-driven ultrafast sub-ns programming in W\Al2O3\Ti\CuTe-based 1T1R CBRAM system", VLSI 2012 Tech. Symp., pp.5-6.
[6] M. Barci, J. Guy, G. Molas, E. Vianello et al., "Impact of SET and RESET conditions on CBRAM high temperature data retention", proc. of IRPS 2014, 5E.3.1-5E.3.4.

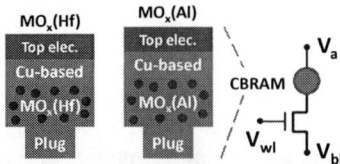

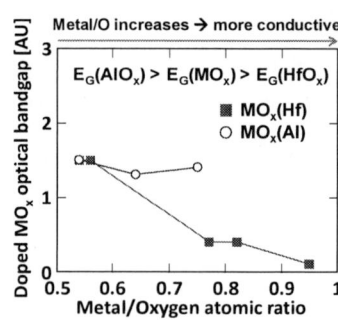

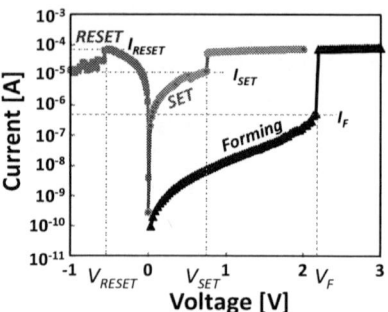

	MO_x	Hf Dopant MO_x(Hf)				Al Dopant MO_x(Al)	
Doping % (RBS)	0%	5%	15%	16%	20%	13%	21%
(M+Dopant)/O atomic ratio (RBS)	0,54	0,56	0,77	0,82	0,95	0,64	0,75

Fig.1 – (a) Schematics of the studied CBRAM. $MO_x(Al)$ samples are thicker than $MO_x(Hf)$ samples. (b) Metal/Oxygen atomic ratio measured by Rutherford Backscattering Spectrometry, "Metal" including the metal of the metal oxide M and the Hf or Al doping element.

Fig.2 – Measured optical bandgap of Hf and Al-doped metal oxides constituting the electrolyte.

Fig.3 – Typical IV transient characteristics during forming, SET and RESET operations indicating the forming, SET and RESET quasi static currents and voltages.

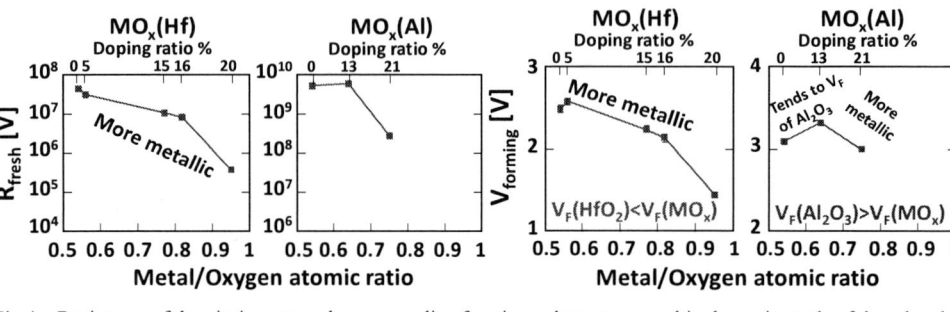

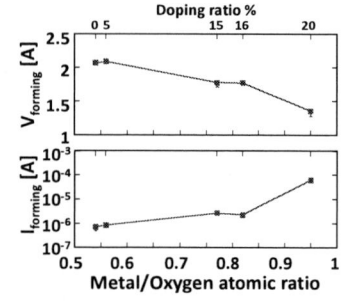

Fig.4 – Resistance of the virgin state and corresponding forming voltage measured in dynamic mode of doped and undoped metal oxide CBRAM. "Metal" in x-axis includes the metal of the metal oxide M and the Hf or Al doping element. Forming voltage is defined to form 50% of the cells for a forming pulse of 10µs.

Fig.5 – Forming voltage and current extracted from quasi static IV transient characteristics (see Fig.3) for $MO_x(Hf)$.

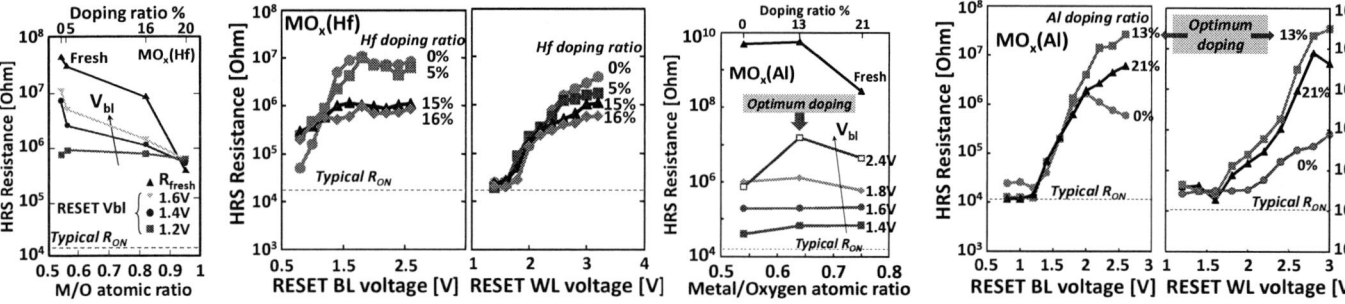

Fig.6 – Left: HRS resistance for various RESET Vbl, for Hf-doped oxide CBRAM (1µs pulse). Right: HRS resistance as a function of the Bit Line and Word Line RESET voltages for $MO_x(Hf)$ CBRAM (1µs pulse).

Fig.7 – Left: HRS resistance for various RESET Vbl, for $MO_x(Al)$ with various Al doping ratio. Right: HRS resistance as a function of the Bit Line and Word Line RESET voltages for various $MO_x(Al)$ CBRAM (1µs pulse).

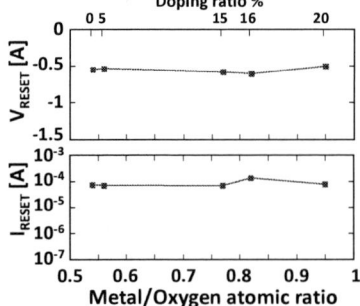

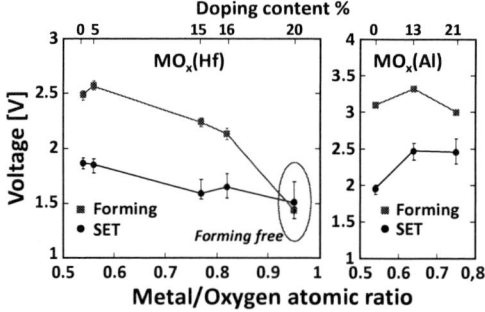

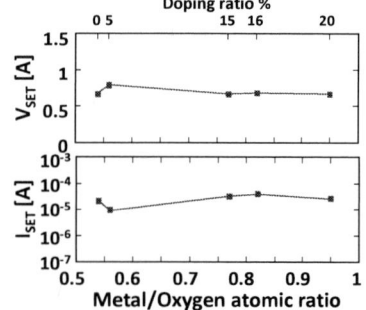

Fig.8 – Quasi static RESET voltage and current (see fig.3) for various $MO_x(Hf)$ CBRAM (same initial R_{ON}).

Fig.9 – SET and forming voltages for the various studied CBRAM devices. CBRAM with the highest doping concentration exhibit a forming free behavior.

Fig.10 – SET voltage and current (measured in quasi static, see fig.3) in $MO_x(Hf)$. No trend is observed (same initial R_{HRS}).

978-1-4799-8002-4/14 $31.00 © 2014 IEEE

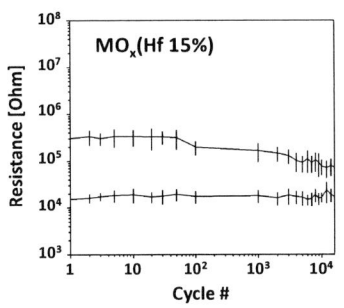
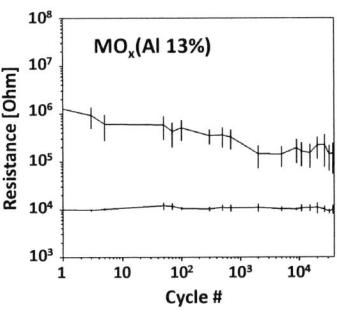
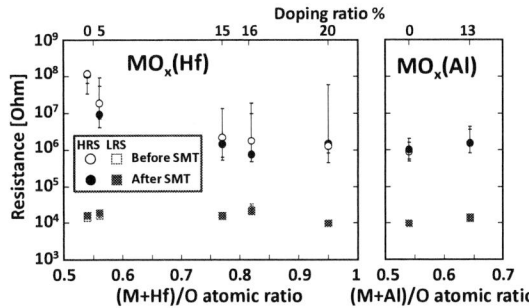

Fig.11 – Cycling characteristics for $MO_x(Hf)$ and $MO_x(Al)$ (mean value and standard deviation on 64 cells are shown). No correction algorithm is applied.

Fig.12 – Resistance for LRS and HRS before and after 15min at 260°C emulating typical soldering reflow.

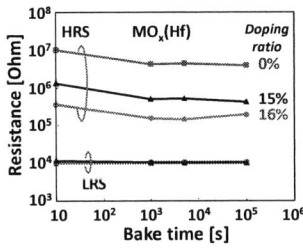
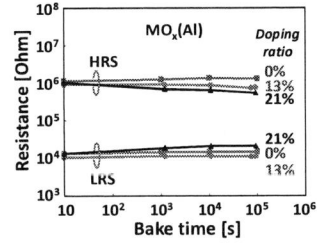
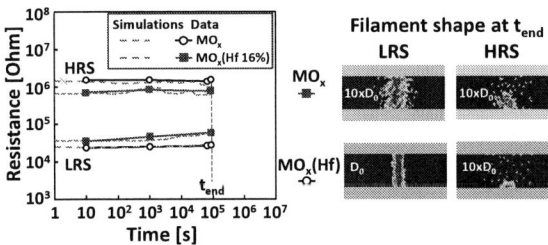

Fig.13 – 200°C measured retention characteristics for HRS and LRS of $MO_x(Hf)$ and $MO_x(Al)$ CBRAM. 10kOhm R_{ON} insures stable retention, except for $MO_x(Al$ 21%).

Fig.14 – 130°C measured and simulated [4] retention characteristics (mean value over ~100 cells) for MO_x and $MO_x(Hf)$. 30kOhm R_{ON} was used to accelerate LRS retention loss. 16% Hf-doping increases the Cu diffusion coefficient D_0 in the resistive layer (model from [4]).

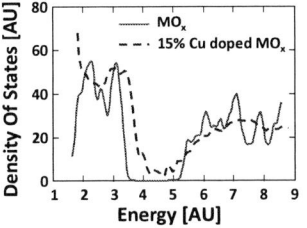
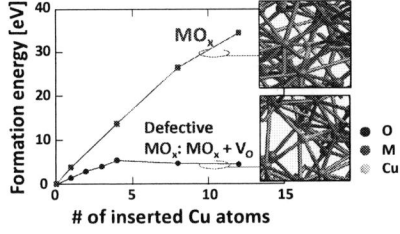

Fig.15 – Density of States calculated by Density Functional Theory for MO_x and MO_x with Cu inclusions. 15% Cu-doped MO_x is conductive.

Fig.16 – Calculated formation enthalpy energy to add Cu atoms in interstitial sites in (a) MO_x, and in (b) defective MO_x assuming the presence of V_O in the system. Cu insertion is strongly reduced when MO_x is O deficient.

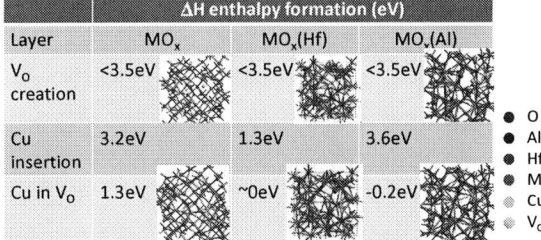

	ΔH enthalpy formation (eV)		
Layer	MO_x	$MO_x(Hf)$	$MO_x(Al)$
V_O creation	<3.5eV	<3.5eV	<3.5eV
Cu insertion	3.2eV	1.3eV	3.6eV
Cu in V_O	1.3eV	~0eV	-0.2eV

Fig.17 – Calculated formation enthalpy energy using 1st principal calculations (Density Functional Theory) for various structural modifications (V_O creation, Cu insertion and Cu insertion assuming V_O) in MO_x, $MO_x(Hf$ 14%) and $MO_x(Al$ 14%) resistive layers.

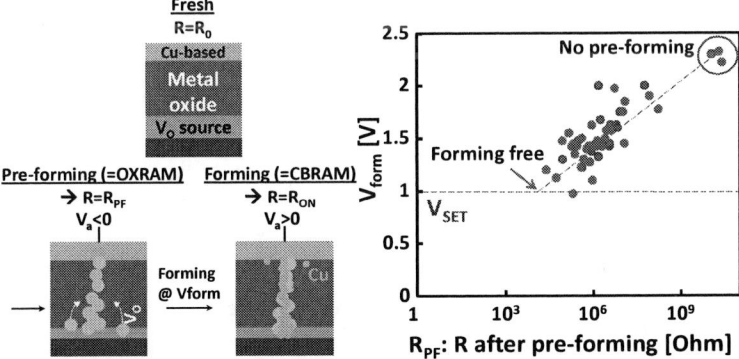

Fig.18 – Demonstration of a hybrid CBRAM/OXRAM memory, including an oxygen vacancy reservoir in the bottom of the stack. When a reverse pre-forming operation is performed at negative voltage, a thin filament composed of oxygen vacancies is created in the resistive layer (OXRAM) leading to an intermediate HRS resistance. This facilitates the CBRAM direct forming operation when a positive bias is applied (Cu injected in the vacancies), reducing the forming voltage. Right: corresponding forming voltage as a function of the cell resistance after reverse forming operation.

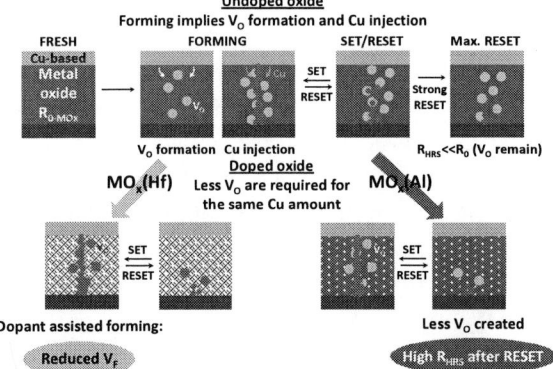

Fig.19 – Phenomenological description of oxide based CBRAM. During forming, oxygen vacancies are created in the resistive layer, facilitating the Cu insertion to form the filament. Doping the resistive layer, less V_O are required to create the filament. In $MO_x(Hf)$, Cu insertion is facilitated, resulting in reduced V_F. In $MO_x(Al)$, lower V_O generation during forming allows reaching a higher R_{OFF} after RESET, enhancing the window margin.

978-1-4799-8002-4/14 $31.00 © 2014 IEEE

Process Integration of a 27nm, 16Gb Cu ReRAM

John Zahurak[1], Koji Miyata[2], Mark Fischer[1], Murali Balakrishnan[1], Sameer Chhajed[1], David Wells[1], Hong Li[1], Alessandro Torsi[1], Jay Lim[1], Mark Korber[1], Keiichi Nakazawa[2], Satoru Mayuzumi[2], Motonari Honda[2], Scott Sills[1], Shuichiro Yasuda[2], Alessandro Calderoni[1], Beth Cook[1], Gowri Damarla[1], Hai Tran[1], Bei Wang[1], Chris Cardon[1], Kamal Karda[1], Jun Okuno[2], Adam Johnson[1], Takafumi Kunihiro[2], Jun Sumino[2], Masanori Tsukamoto[2], Katsuhisa Aratani[2], Nirmal Ramaswamy[1], Wataru Otsuka[2], Kirk Prall[1]

(1) Micron Technology Inc., Boise, ID, USA   (2) Sony Corporation, Atsugi, Kanagawa, Japan

TEL: (208)368-5174, FAX: (208)368-3002, Email: Jzahurak@micron.com

Introduction:

Resistive RAM (ReRAM) is a promising candidate for future non volatile memory applications. Despite an abundance of literature on resistive memory cells with numerous materials (1), demonstration of high density memories suitable for relevant commercial applications is absent. We have proposed the viability of a Cu-based ReRAM Cell capable of meeting storage class memory specifications (2, 3). This paper describes a 16Gb Non-volatile (NV) Cu-based ReRAM chip fabricated at the 27nm node. 180MB/s write and 900MB/s read performance is enabled. The array architecture is 6F2 and utilizes a highly modified BRAD (Buried Recessed Access Device) coupled with novel, self-aligned plugs contacting a Damascene Cu-ReRAM Cell. Line-SAC (Self Aligned Contact) Digit lines running over the array complete cell formation.

Cell Architecture:

ReRAM cell technology requires an access device to enable sufficiently large array blocks (4). Therefore we have leveraged 27nm 6F2 DRAM array technology as a starting point, which yields a $0.0044\mu m^2$ cell. Pitch multiplication is required for multiple levels to reach 27nm geometries. Bipolar ReRAM operation swings the cell plate bias for Program/Erase operations which requires patterning the cell plate into stripes called Common Source Plate (CSP), Fig 1.

Establishing planarity between the array and periphery circuitry is accomplished by building the arrays in blocks of raised epitaxial silicon, Fig 2. Silicon grown within confined regions typically suffers from high defect levels. We control defects within the epitaxial regions using a carefully engineered stress relief boundary accompanied by a small perimeter circuit "keep out" zone (or hedge). The thickness of Si epitaxy is chosen to align for planarization with the periphery CMOS gate stack.

DRAM access devices are engineered for extremely low leakage levels. However, the ReRAM cell places a much different set of requirements on access device performance; namely significantly higher drive with moderate I_{off}.

Access Device (A-FET):

While the A-FET for ReRAM superficially resembles a DRAM BRAD, significant modifications are required to deliver sufficient programming current. First, by relaxing the DRAM ultra-low I_{off} requirement, silicide can be introduced into the array to minimize pillar contact resistance. Silicide formation is challenging and must be carefully engineered to form at the bottom of the Exhumed Contact opening with limited reactive Si availability.

Modifications were also made to the A-FET Si trench and saddle profiles. Increasing the saddle height of the device while limiting the silicon trench depth serves to both minimize parasitic resistances and increase the effective fin ratio of the buried device. Optimizing the profile and ratio of these dimensions is a strong driver of device performance.

The gate oxide was developed to support elevated array voltages and suppress leakage current. Finally, implants and thermal budgets were re-designed to align with the new physical device structure and to maximize activation in the confined pillar. Completed A-FET images are shown in Fig 3.

978-1-4799-8002-4/14 $31.00 © 2014 IEEE

Typical A-FET electrical performance is shown in Fig 4a-b. Figure 4c illustrates performance improvements derived from the multiple optimizations discussed above which culminates in I_{on} of 40µA/device. As shown in the top curves of Fig 4c, test structure results are very well aligned with actual array curves. Finally, Figure 4d demonstrates current distributions taken from the array through 1M cycles under various bias conditions.

Array Contacts:

Contacting the S/D of the A-FET in the array using conventional techniques requires an orthogonal, double pitch multiplication process. Such an approach is burdened with poor registration margin and high process costs, so we have introduced a novel self-aligned contact. This contact is formed by the exhumation of a sacrificial layer which is introduced into the stack prior to STI formation. During STI and A-FET structure formation, the sacrificial layer is chopped to form an isolated plug which is then selectively removed. As described previously, contact silicide is critical for enhancing AFET drive capability. Our flow utilizes a single silicide process to form both array and periphery contacts simultaneously. Materials for the bottom electrode (BE) contacting the cell are selected to enhance cell performance. The BE is deposited on the silicide and then isolated using CMP prior to Cell Deposition, Fig 5.

Cell Formation:

ReRAM cells are sensitive to contamination and damage which can be introduced during a subtractive cell process; i.e. reactive ion etching. These process related defects manifest themselves as time zero fails, increased cell variability, and degraded reliability. Introduction of a Cu-Damascene cell module circumvents these issues by removing the switching regions of the cell from proximity to process related defects. The Damascene trench is filled with an electrolyte layer, a Cu-based ion reservoir, and a metal strapping layer. Nitride encapsulation of the cell metal strap and sidewalls is critical for both cell functionality and isolation of the cell from the subsequent bit line formation.

Set and reset distributions from the 16Gb array are shown in Fig 6 and BER in Table 2. We note that 16Gb array cell distributions are shown to be comparable to that of both an intrinsic cell (planar "edgeless" cell) as well as a "scaled cell" (damascene cell at relaxed pitch). Equivalent performance from the three different cell structures demonstrates a successful integration flow whereby the intrinsic cell performance is translated directly into dense array structures with no loss of capability.

Bit Line (BL) Formation:

Bit lines are formed using an industry-first Line-SAC etch process over the Damascene Cell, which eliminates the need for a separate plug contact layer (Fig 7). Use of an etch stop layer in the periphery allows the array BL metal to also serve as a local interconnect in the periphery. Metal fill is very challenging at the array geometries (AR > 10:1) and is enabled with an aggressive Cu Damascene process. Resistance meets design targets of 800mΩ/sq.

Chip Performance:

The chip design consists of an 8-bank interleaved architecture with DDR interface (2). Currently we demonstrate a throughput of 900MB/s for read and 180MB/s for write operations. These results are shown in Table 1. The on-chip programmable controller algorithm allows great flexibility in balancing cell functionality and reliability with chip performance. Ongoing cell and algorithm improvements are expected to enable throughput at peak design capability.

Summary:

A 27nm 16Gb Cu based NV Re-RAM chip has been demonstrated. Novel process introduction to enable this technology include a Damascene Cell, Line-SAC Digit Lines filled with Cu, exhumed-silicided array contacts, raised epitaxial arrays, and high-drive buried access devices.

References:

(1) P. Wong, H. Lee, S. Yu, Y. Chen, Y. Wu, P. Chen, B. Lee, F. Chen, M. Tsai, "*Metal-Oxide RRAM*", Proc. IEEE, Vol 100, No 6, June 2012

(2) R. Fackenthal, Makoto Kitagawa, et al, "*A 16Gb ReRAM with 200MB/s Write and 1GB/s Read in 27nm Technology*", ISSCC 2014, pp. 21-22

(3) S. Sills, S. Yasuda, J. Strand, A. Calderoni, K. Aratani, A. Johnson, N. Ramaswamy, "*A copper ReRAM Cell for Storage Class Memory applications*", VLSI 2014

(4) G. Burr, R. Shenoy, K. Virwani, P. Narayanan, A Padilla, B. Kurdi, H. Hwang, "*Access devices for 3D crosspoint memory*", J. Vac Sci. Technol. B 32(4), Jul/Aug 2014

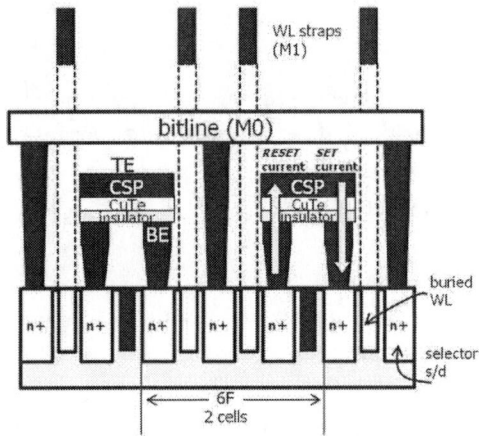

Figure 1: Schematic of Cu-ReRAM 6F2 Architecture (1)

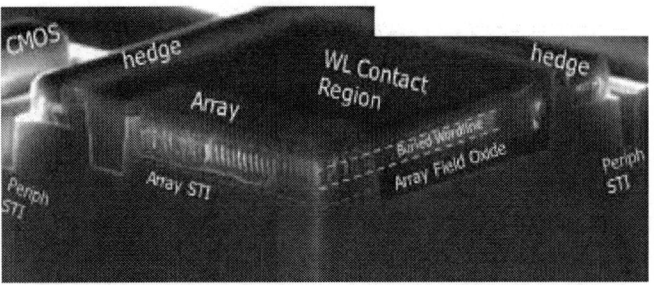

Figure 2: 3D Perspective showing Epi Array used for planarity

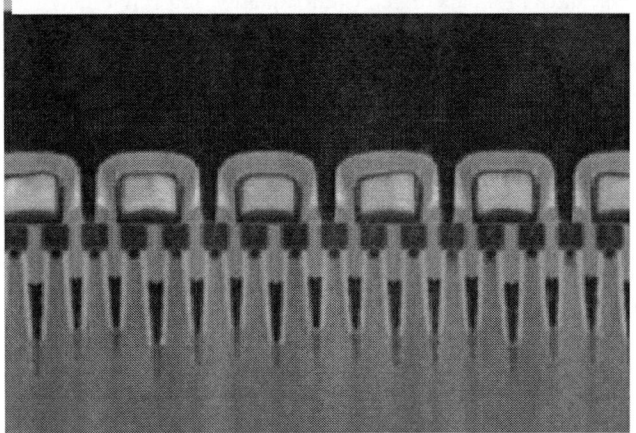

Figure 3: Completed 27nm array (x-direction) showing A-FET, silicide, metal plugs, Damascene cell, and SAC DL.

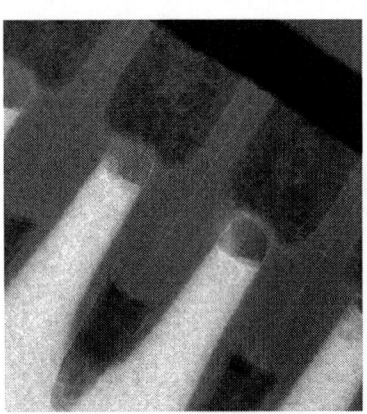

Figure 5: Exhumed Contact showing silicide and Metal Plug completed prior to Cell deposition.

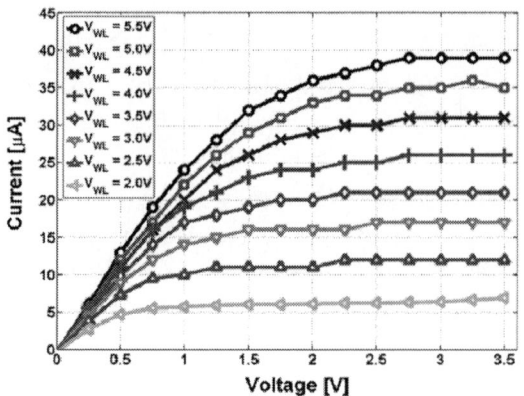

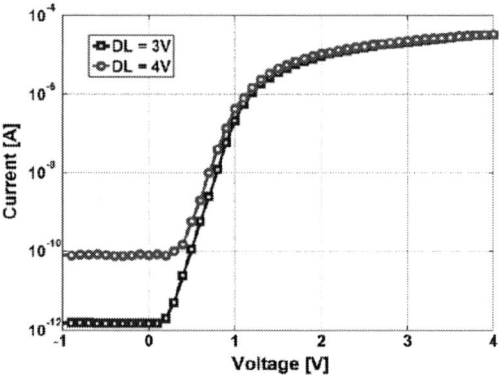

Figure 4a-b: A-FET IV curves.   SS=150mV/dec

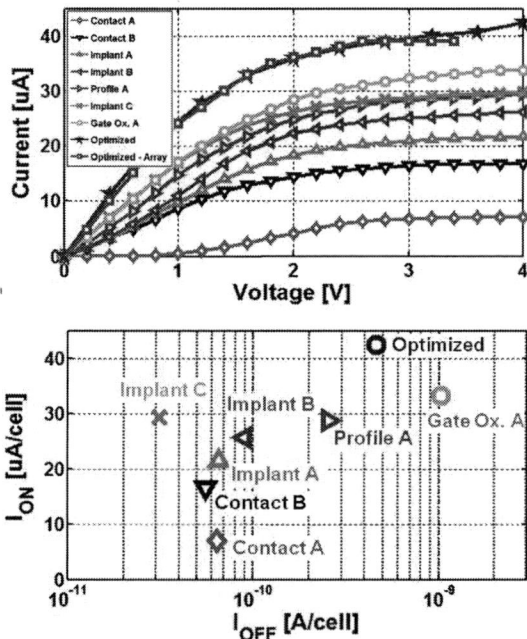

Figure 4c: A-FET performance improvements through contact, structure, implant, and gate oxide optimization. Note alignment of test structure and array results seen in top transistor curve overlay.

978-1-4799-8002-4/14 $31.00 © 2014 IEEE   142

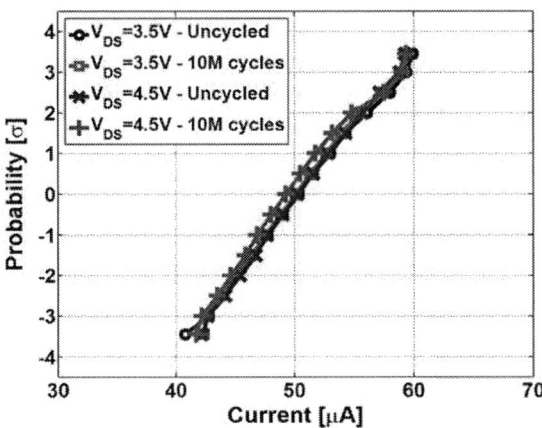

Figure 4d: A-FET array current distributions through 1M cycles. (V_{GS} = 5.5V)

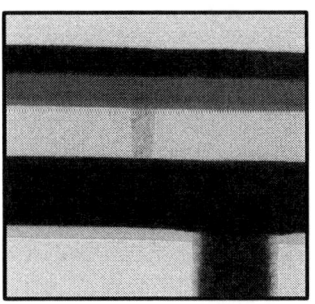

Figure 6: Cell Structures used in comparisons of cell performance: Intrinsic Edgeless Cell (left) vs. Scaled isolated Cell (right) vs. Full Integrated Cell shown in Figure 3.

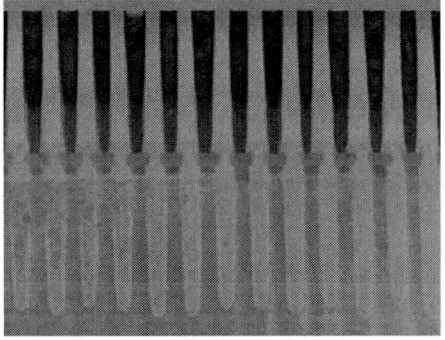

Figure 7: Cross-section of array in the y-direction showing DL runners and high aspect ratio Cu fill.

Table 1: Chip performance comparison of design target with silicon results to date.

Summary Table			
Feature		**Design Capability**	**Actual Demonstration**
Density		16 Gb	16 Gb
Tech node (nm)		27	27
Cell Size (nm²)		4374 (6F²)	4374 (6F2)
Die Size (mm²)		168	168
Selector		Buried WL MOS selector	Buried WL MOS selector
Read Performance	BW (MB/s)	1000	900
	Latency (uS)	2	2.3
Write Performance	BW (MB/s)	200	180
	Latency (uS)	10	11.7

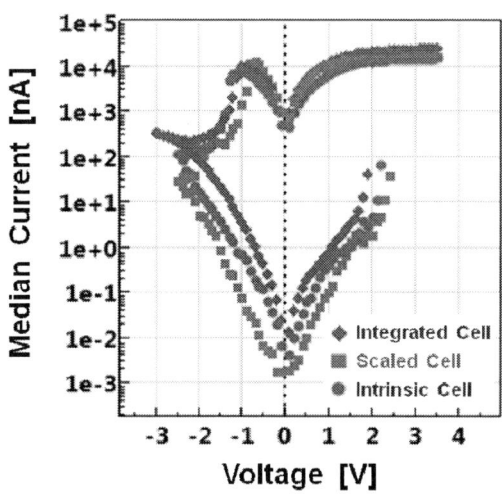

Figure 6a: DC IV curves comparing Intrinsic Cell (Planar edgeless cell), Scaled Cell (Damascene Isolated Cell), and the fully Integrated Cell at 27nm

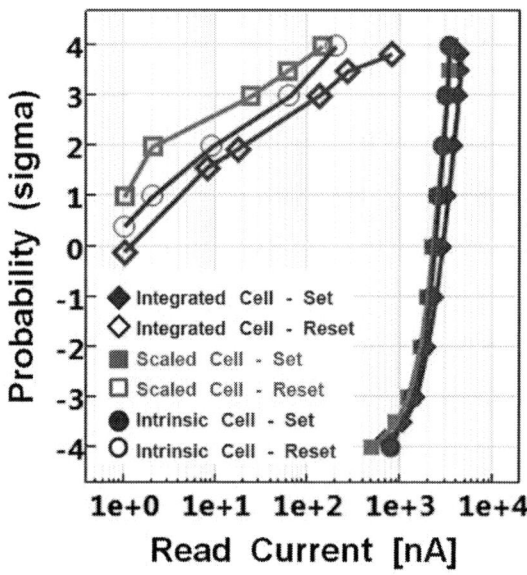

Figure 6b: Cell Current distributions after 1k cycles showing comparable behavior.

Table 2: Cell BER summary through cycling and retention bake.

Cycle #	BER Cycling	BER Cycle + Bake*
10^3	8.9E-05	1.4E-03
10^4	1.2E-04	8.9E-04
10^5	8.9E-05	2.2E-03

* 1yr 55°C Retention BER (E_A = 1.5eV)

978-1-4799-8002-4/14 $31.00 © 2014 IEEE   143

Resistive Memories for Ultra-Low-Power embedded computing design

E. Vianello, O. Thomas, G. Molas, O. Turkyilmaz, N. Jovanović, D. Garbin, G. Palma, M. Alayan, C. Nguyen, J. Coignus, B. Giraud, T. Benoist, M. Reyboz, A. Toffoli, C. Charpin, F. Clermidy, L. Perniola

CEA LETI Minatec Campus, 38054 Grenoble Cedex 9, France – elisa.vianello@cea.fr

Introduction

Recent announcements of 16 and 32 Gbits resistive memories (ReRAM) test chips [1-2] show that ReRAMs are emerging as leading candidates to replace conventional Flash memories thanks to their high density, good scalability and low voltage. Moreover ReRAM technology offers potential for new applications and markets, as new functionality coupled with logic circuits to save energy in mobile devices and Ultra Low Power (ULP) embedded systems. The introduction of the non-volatility enables advanced power management and freeze mode capabilities to achieve zero standby leakage. Depending on the application, strongly different specifications are required and it is possible that the choice of the ReRAM technology will be application and market segment specific.

Abstract

This paper addresses two technologies as an example of optimized devices for FPGA and fixed-logic IC design (as non volatile Flip-Flops).

In FPGAs, the replacement of the SRAM based configuration memory with ReRAM allows saving area and suppressing the standby power consumption. Although the non-volatility eliminates standby power consumption, the leakage current through the ReRAM during run time depends on the high resistive state (HRS) value. A Conductive Bridge RAM (CBRAM) cell with dual-layer electrolyte stack with a resistance ratio higher than 10^6 has been proposed to minimize the leakage current in the operating mode.

The introduction of ReRAM in fixed-logic IC circuits allows unifying non-volatility, zero standby leakage, and rapid power on/off operations. Endurance, low operating voltage and high speed are the main ReRAM specifications to enable the design of non-volatile Flip-Flop with fast switching. An HfO$_2$/Ti based OxRAM cell with a switching time lower than 10ns at 1V and a high endurance up to 10^8 cycles has been proposed for this application.

ReRAM technologies

The studied ReRAM devices are 1T-1R structures integrated in advanced CMOS technology. The CBRAM resistor is composed by a dual layer solid electrolyte, HfO$_2$ deposited by ALD and GeS$_2$ realized by RF-PVD, embedded between a Ag electrode and an W plug. Samples without the HfO$_2$ were processed for comparison [3-4]. The OxRAM active layer consists of ALD HfO$_2$, sandwiched between TiN and Ti/TiN electrodes [5] (Fig.1).

A. Interface Engineering of GeS$_2$/Ag-Based CBRAM for improved HRS reliability

Fig.2 shows the quasi-static CBRAM bipolar characteristic obtained in the samples with and without the HfO$_2$ layer. The

memory switches to the low resistance state (LRS) when a positive bias is applied on the Ag top electrode, forming a Ag-based conductive filament (CF) in the GeS$_2$ layer, while a negative voltage leads to the CF dissolution and thus to the HRS. In the dual layer CBRAM the SET voltage does not enable the migration of Ag$^+$ ions inside the HfO$_2$, because of the higher energy barriers for ions diffusion with respect to GeS$_2$. The role of the HfO$_2$ layer is to suppress the leakage current in the HRS (I$_{off}$) due to the residual conductive paths in the GeS$_2$ after the reset process. To corroborate this hypothesis an empirical model (Fig.3) was used to simulate the OFF/ON-state leakage current in the dual layer sample. The HRS is modeled as a metal (W)/insulator (HfO$_2$)/semiconductor (GeS$_2$) (MIS) capacitor; while in the LRS, the structure relies on a MIM structure, because of the metallic behavior of the CF in the GeS$_2$ layer. The increase of the HfO$_2$ barrier from 1 to 2 nm reduces I$_{off}$ of 1.5 decades, while its impact on the LRS current is negligible because of the increase of carrier concentration in the GeS$_2$ and of electric field in the HfO$_2$. Based on dynamic measurements, the SET time voltage dependence was extracted (Fig.4). An increase of the SET time appears in the dual layer sample; it is linked to the reduced electric field in the GeS$_2$ layer, for a given SET voltage, and it limits the thickness of the HfO$_2$ barrier layer. Fig.5 shows the RESET time as a function of the applied voltage. Low voltages have been studied to estimate the LRS voltage disturb performances. Projected 10 years voltages disturb immunities of 40 and 6mV have been extracted in the HfO$_2$/GeS$_2$ and GeS$_2$ samples respectively. Finally, a memory window of about 6 decades has been demonstrated on 10^3 cycles for the HfO$_2$/GeS$_2$ sample (Fig.6).

B. HfO$_2$/Ti-based OxRAM for low voltage/time switching

The basic working principle of OxRAM cells relies on the creation and rupture of a CF made of oxygen vacancies responsible for resistive switching. Figs.8, 9 and 11 respectively show the forming/SET/RESET time voltage dependence for the HfO$_2$/Ti OxRAM cells. The switching time has been extracted by dynamic measurements as shown in Fig. 7. SET time lower than 10ns has been achieved at 1V. Fig.12 compares the studied cell with a Ta$_2$O$_5$ based OxRAM technology; its set/reset write-pulse voltages are larger in the entire time range [6]. During the RESET process the current gradually decreases, therefore a current criterion has to be chosen to define the RESET time (black line in Fig.11). The RESET voltage impacts both the RESET time (Fig.12) and the HRS value (Fig. 13). Fig.14 demonstrates the extracted cell behavior up to more than 10^8 cycles without failure, demonstrating their potential high endurance. Fig.15 shows the impact of the SET/RESET pulse width on the resistance values: longer pulses improve both the LRS and HRS variability. The dependence of the endurance and of the LRS versus the compliance current during SET (I$_{comp}$) is shown in

Fig.16. In the low current operation region (I_{comp}<300µA), the endurance is strongly improved at the cost of higher LRS.

ReRAM technologies benchmark

Figs. 17-18-19 compare the two studied devices to different ReRAM technologies [6-12]. A tradeoff exists between memory window and endurance (Fig.17), the HfO_2/GeS_2 CBRAM and the HfO_2/Ti OxRAM cells are two representative cases of cells with optimized memory window and endurance, respectively. Moreover the higher HRS and memory window of the HfO_2/GeS_2 CBRAM and of the atom switch technology [7] are at the expense of a longer switching time (Fig.18). In all the samples the LRS is modulated by the compliance current (Fig.19).

ReRAM design

Improving energy efficiency is a major challenge for modern wireless systems. ReRAM co-integration with CMOS hold promise for area and energy saving for both FPGA and fixed-logic ICs. The introduction of the non-volatility opens the way for zero consumption in sleep mode by saving the system context while enabling a fast wake-up transition. ReRAM electrical characteristic requirements differ application-to-application as discussed next (Fig.20).

A. CBRAM-based FPGA

With increasing costs of advanced technologies, FPGAs are becoming a more suitable alternative for a large number of medium-volume applications than in the past due to their non-recurring engineering cost and ease of design. However for mobile and ULP embedded applications, the use of FPGAs still remains challenging due to chip area and energy efficiency. The FPGA configuration memory can consume up to the half of the total chip area [13] (Fig.21). Therefore FPGA performances are directly related to the memory technology used. SRAM-based FPGA are fast, but the SRAM volatility requires to reload configuration and routing data at power-up or to maintain a data retention voltage (DRV). Flash-based FPGA retains the data, but hybrid CMOS-Flash technology results in higher fabrication cost and slower performance. Shifting the configuration memory in the BEOL with ReRAM brings fast non-volatile re-configurability, energy efficiency and high density [14]. The 1T2R cell [3] offers the highest-density integration solution compared to a regular 6T SRAM cell, but in terms of programmability the 2T2R cell enables better performances and robustness (Fig.22) [15]. The resistance pair gives a severe requirement in HRS and HRS/LRS ratio to satisfy power consumption and retention time requirements. The record high HRS achieved with CBRAM GeS_2/HfO_2 stack (Fig.17) minimizes the leakage current of 2T2R cell less than 1.3nA at 1V, the lowest compared to 1T2R solutions reported so far [16-18], making it more competitive to SRAM cell leakage (Fig.24) [19]. The overall FPGA performance gains have been evaluated using the VPR flow described in [20] in 130nm CMOS. In comparison to SRAM-based FPGA, CBRAM-based FPGA reduces area by 32% leading to shorter and less capacitive wires. Improvements in critical path delay and power consumption are observed by 9% and 10%, accordingly. Compared to a SRAM-based FPGA supplied at 1.5V, the total

power consumption is reduced by 24% and more than 90% for applications having an activity of 50% and 1% respectively (Fig.23). The power gain for the CBRAM-based FPGA is achieved thanks to the 0V supply voltage applied in stand-by mode, enabling a significant higher power saving than an SRAM-based FPGA assuming a stand-by supply of 0.6V. Concerning data retention, the ReRAM devices in the 2T2R cell are oriented in order to avoid HRS disturb (transition from HRS to LRS) because of the higher energy required to reset compared to set. LRS disturb immunity higher than 10 years is ensured by the large memory window that minimizes the voltage drop across the ReRAM in LRS as low as 1mV at 1V of supply voltage (Fig.5). CBRAM endurance fulfills the FPGA programing cycle requirements ($\leq 10^3$).

B. OxRAM-based Non Volatile Flip-Flop

For saving and restoring the state of sequential fixed-logic ICs during the power-down mode balloon (or shadow) latches are commonly employed. Flip-flop (FF) data are retained by maintaining a minimum supply voltage of the balloon latch, reducing the system leakage current. Replacing balloon latch by ReRAM enables nullifying leakage currents. To minimize the energy consumption in sleep/wake-up transitions fast storing (ReRAM SET or RESET) and restoring of FF values are required, hence making OxRAM a good candidate. The non-volatile FF (NVFF) solution [21] shown in Fig.25 is designed to not degrade the performances and consumption in operating mode by decoupling the non-volatile part from the slave stage of the FF. Context saving is performed by storing the FF value in two ReRAMs before powering down the system which limits the number of programming operations and makes OxRAM endurance adequate. This approach successfully solves co-integration of ReRAM with advanced CMOS technology nodes, hampered by the difference between their operating voltages. For this purpose, a dual voltage scheme is employed, FF on CMOS operating voltage (VDD) and ReRAM part on the programming supply voltage (VDDH). The supply voltage difference is overcome by level shifting output FF data combined with the control signals. The simulations results (OxRAM, 28nm CMOS) shown in Fig.26 compare the sleep energy of the NVFF to a standard FF versus the time spent in sleep mode. The estimated sleep energy for NVFF is 28pJ and is independent of the sleep time. In FF case, it is evaluated for different supply voltages. Considering 1V and 0.5V FF supply in sleep mode, replacing it with NVFF solution reduces the power consumption for a period of inactivity longer than 10ms and 100ms, respectively.

Conclusions

This work demonstrates the benefits of ULP embedded computing design using optimized ReRAM technologies. Devices with high memory window, as the HfO_2/GeS_2-based CBRAM, enable high density and low-leakage cells for FPGA configuration memory. Cells with low voltage/time switching and high endurance, as the HfO_2/Ti based OxRAM, provide fast switching fixed-logic design. They open the way for fast freeze mode zero-leakage design in modern wireless systems.

Acknowledgements: the authors would like to thanks B. De Salvo, F. Dahamani, S. Jeannot and P. Candelier for helpful discussion and support.

RERAM technologies

[1] R. Fackenthal et al., 2014 ISSCC, pp.338-340. [2] Tz-Yi Liu et al., 2013, ISSCC, pp.210-212. [3] G. Palma et al., 2014 TED, vol.61 no.3 pp.793-800. [4] E. Vianello et al., 2013 IEDM, pp.741-744. [5] A. Benoist et al., 2014 IRPS, pp. 2E.6.1-5. [6] L. Goux et al.2014 VLSI [7] M. Tada et al., 2010, IEDM, pp.403-406. [8] Q. Liu et al., 2009 ESSDERC pp.221-224 [9] M. Tada et al., 2012, IEDM, pp.693-696. [10] K. Aratani et al., 2007 IEDM, pp.783-786. [11] W. Kim et.al., 2011 VLSI, pp. 22-23 [12] L. Goux et al., 2012, VLSI, pp.159-160. [13] M. Lin et al., 2007 CADICS, pp.216-229. [14] Y.Y. Liauw et al.,2012 ISSCC, pp. 406-408. [15] F. Clermidy et al., 2014, DATE, pp.1-6. [16] P.-E. Gaillardon et al., 2010 ICECS, pp.62-65.[17] S. Onkaraiah et al., 2013 ISCAS, pp.2440-2443. [18] S. Yasuda et al., 2011 IMW, pp.1-4. [19] M. Yamaoka et al., 2005 JSSC, vol.40, no.1, pp.186-194. [20] J. Luu et al., 2009, International Symp. on FPGAs. [21] N. Jovanovi et al. 2014, NEWCAS.

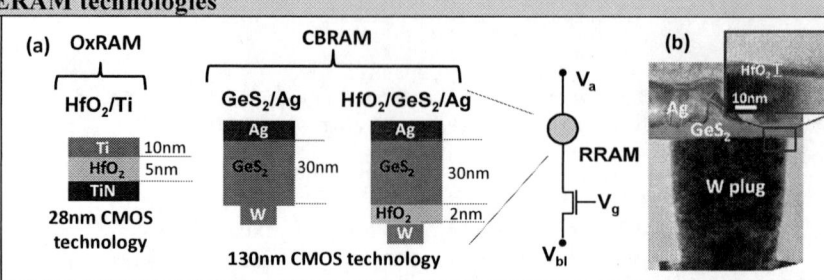

Fig.1: (a) Schematics of the ReRAM studied devices integrated in 1T1R structure ($V_{set}=V_a-V_{bl}$ and $V_{reset}= V_{bl} -V_a$) (b) TEM cross section of the CBRAM device with HfO_2/GeS_2 electrolyte.

Interface engineering of GeS₂/Ag based CBRAM for ideal switch with enlarged memory window and HRS

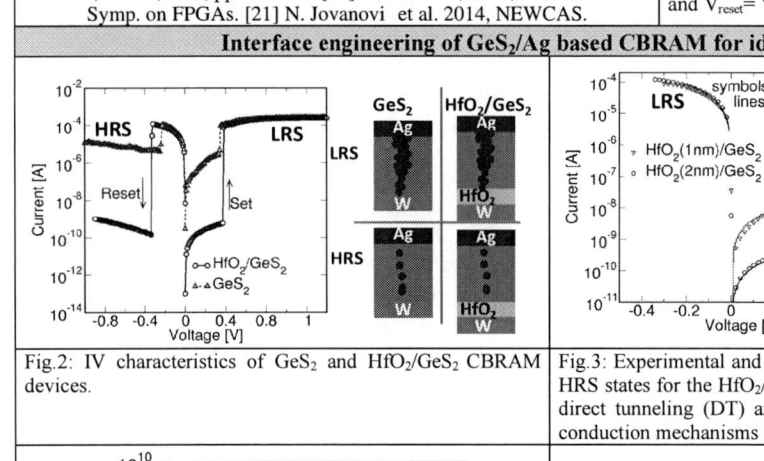

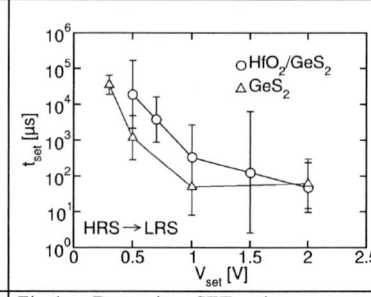

Fig.2: IV characteristics of GeS_2 and HfO_2/GeS_2 CBRAM devices.

Fig.3: Experimental and simulated I/V curves of the LRS and the HRS states for the HfO_2/GeS_2. The model takes into account both direct tunneling (DT) and the trap-assisted tunneling (TAT) as conduction mechanisms in the HfO_2 layer.

Fig.4: Dynamic SET time versus programming voltage for GeS_2 and HfO_2/GeS_2 CBRAM devices.

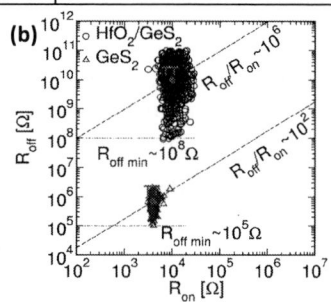

Fig.5: RESET time during low negative stress bias (HRS disturb immunity). Each point is the mean of 64 cells.

Fig.6: (a) Pulse cycling test for GeS_2 (top) and HfO_2/GeS_2 (bottom) devices. V_{set}=2.3V, V_{reset}=2V, I_{comp}~100µA, pulse width 10µs. No correction code has been used. (b) Corresponding corner plot analysis.

HfO₂/Ti based OxRAM for low voltage/time switching

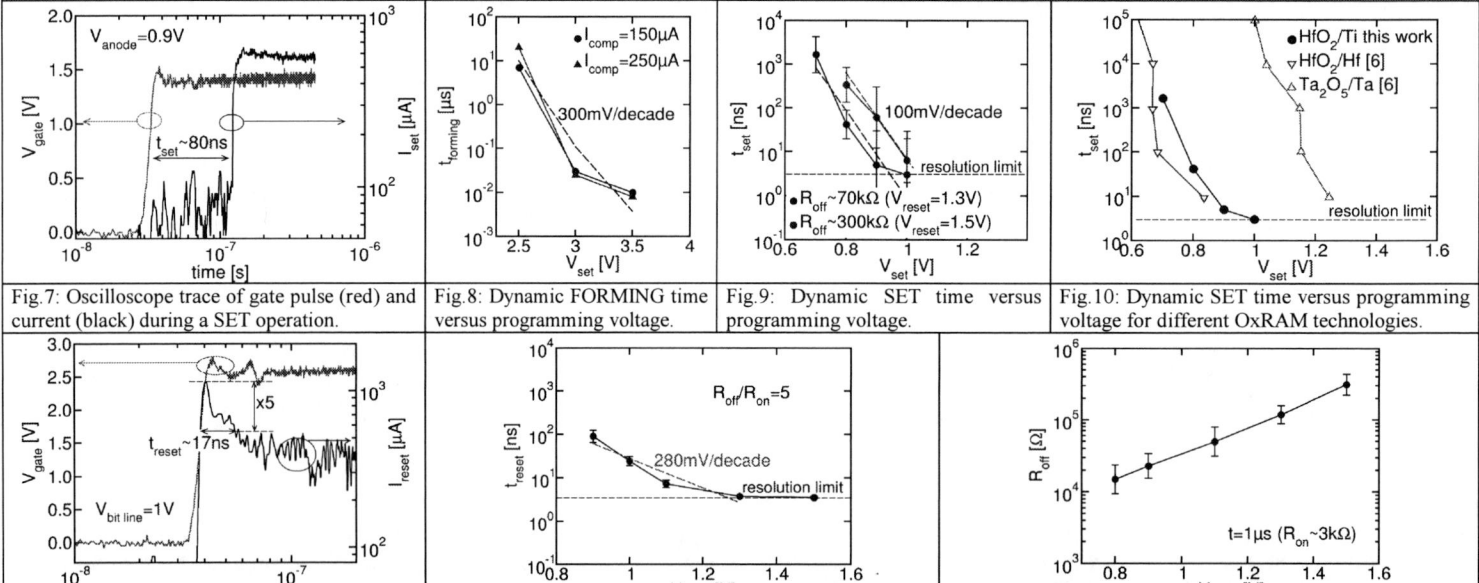

Fig.7: Oscilloscope trace of gate pulse (red) and current (black) during a SET operation.

Fig.8: Dynamic FORMING time versus programming voltage.

Fig.9: Dynamic SET time versus programming voltage.

Fig.10: Dynamic SET time versus programming voltage for different OxRAM technologies.

Fig.11: Oscilloscope trace of gate pulse (red) and current (black) during a RESET operation.

Fig.12: Dynamic RESET time versus programming voltage for an HfO_2 OxRAM cell for a R_{off}/R_{on} ratio of 5.

Fig.13: RESET resistance versus reset voltage for the HfO_2/Ti OxRAM cell.

978-1-4799-8002-4/14 $31.00 © 2014 IEEE

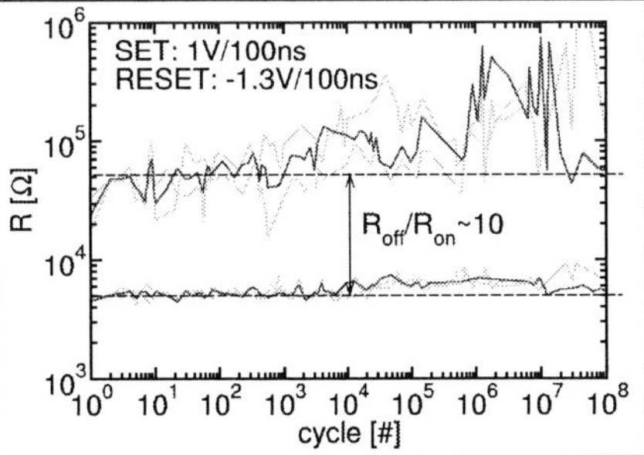

Fig.14: SET/RESET endurance evaluated up to 10^8 cycles with a switching time of 100ns.

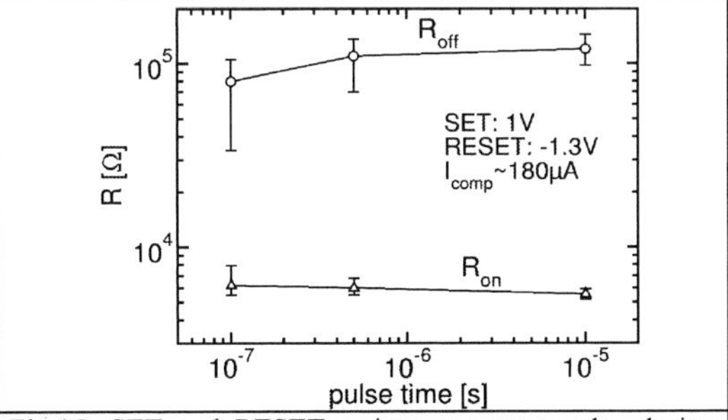

Fig.15: SET and RESET resistance versus pulse during endurance evaluated up to 10^8 cycles.

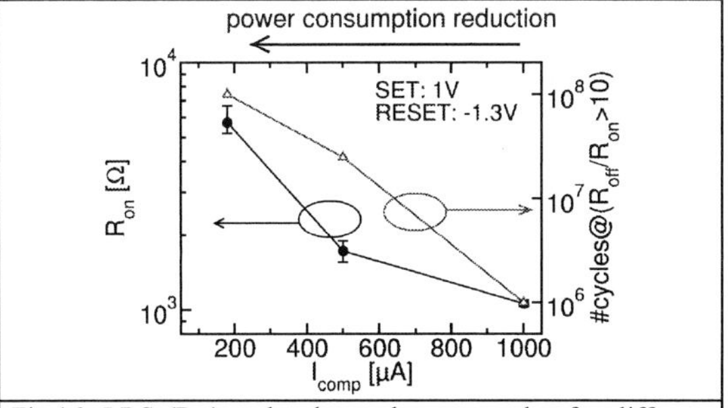

Fig.16: LRS (R_{on}) and pulse endurance cycles for different compliance current values.

RERAM technologies benchmark

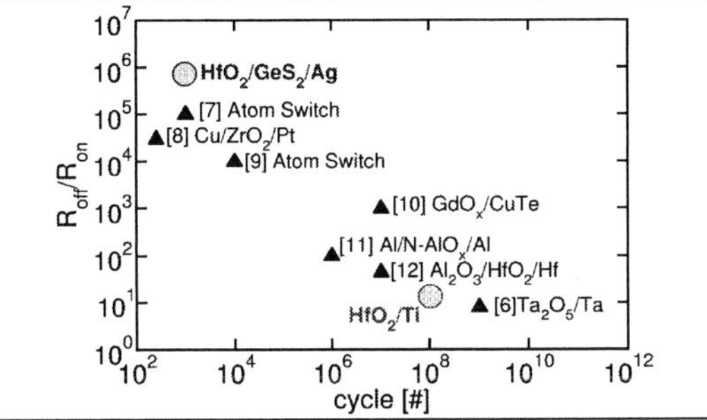

Fig.17: R_{off}/R_{on} vs endurance for the studied technologies. Black triangles are from the literature.

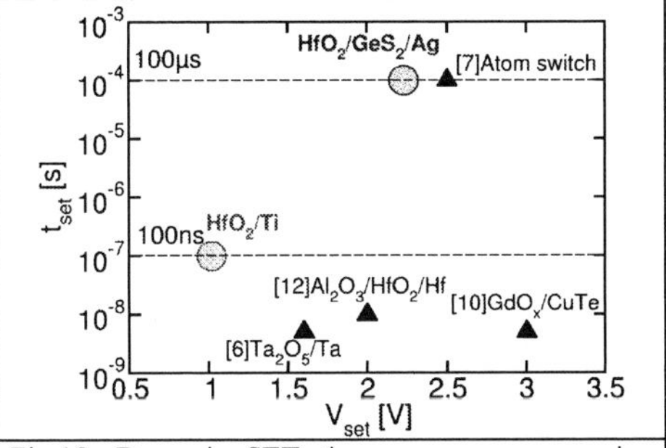

Fig.18: Dynamic SET time versus programming voltage. Black triangles are from the literature.

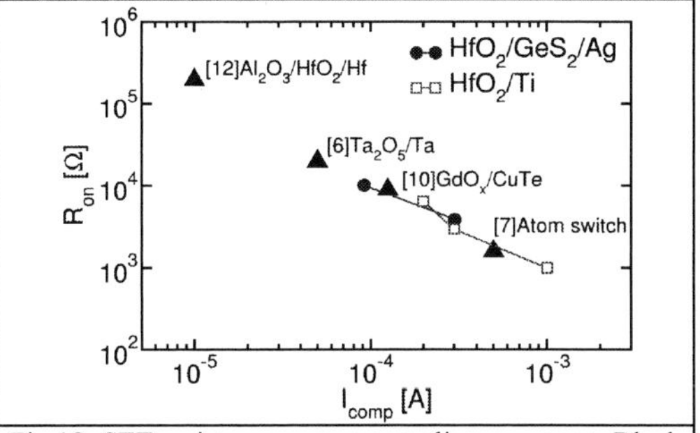

Fig.19: SET resistance versus compliance current. Black triangles are from the literature.

Novel RERAM hybrid integrated circuits

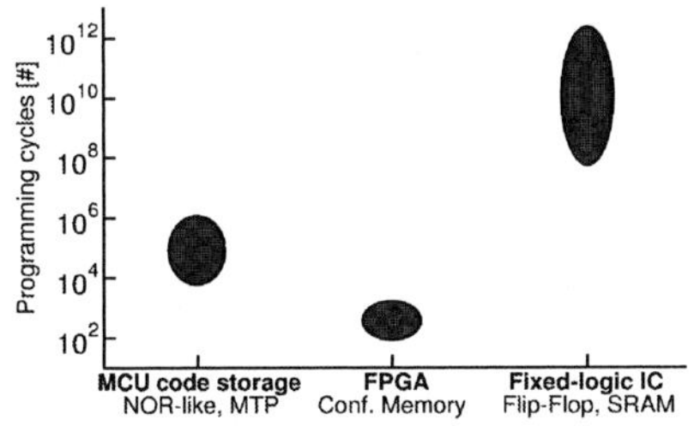

Fig.20: Required programing cycles versus embedded applications.

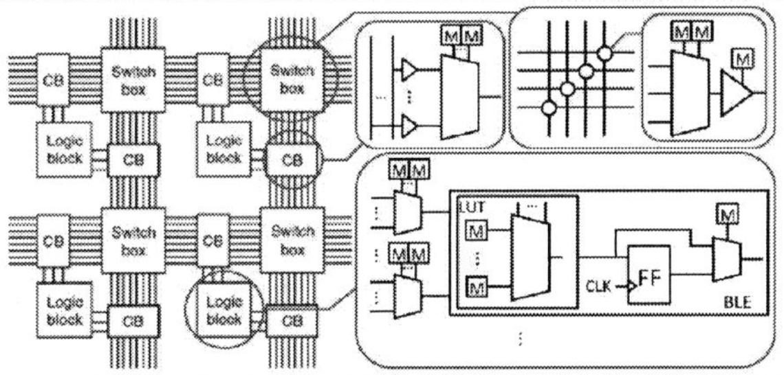

Fig.21: Island style FPGA. Logic Blocks (LB) perform the logical operations and the connection boxes (CB) with the switch boxes (SB) carry out the connection routing resources.

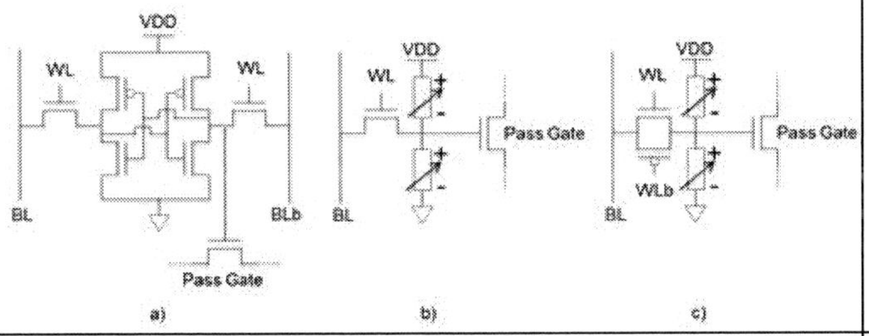

Fig.22: FPGA configuration memory cells: a) 6T-SRAM, b) 1T2R, c) 2T2R.

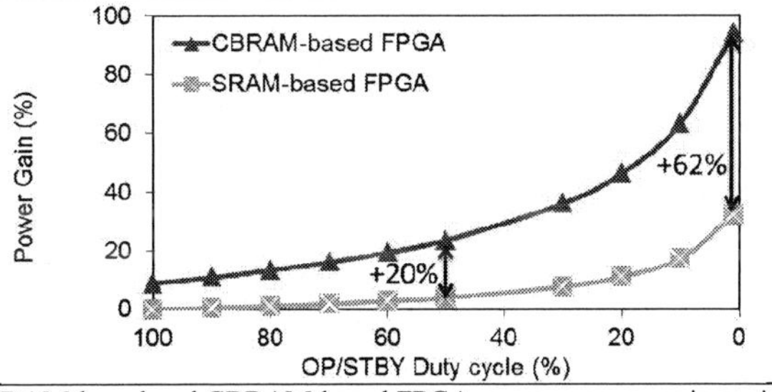

Fig.23: DRV SRAM-based and CBRAM-based FPGA average power gain saving compared to 1.5V fixed supply voltage SRAM-based FPGA versus activity ratio between operating mode and stand-by mode estimated through various representative applications listed in [15].

Ref.	Architecture	Technology	$I_{leak}@V_{dd}$ [pA]	V_{dd} [V]	Standby Cons.
[19]	6T-SRAM	-	25	1.2	yes
[16]	1T-2PCM	GST	5×10^7	1	no
[17]	1T-2CBRAM	W/GeS$_2$/Ag	10^7	1	no
[18]	1T-2CBRAM	Pt/ZnCdS/Ag	10^6	1	no
[This work]	2T-2CBRAM	W/HfO$_2$/GeS$_2$/Ag	1300	1	no

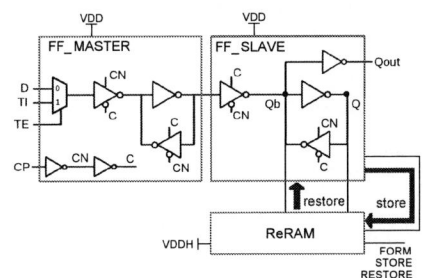

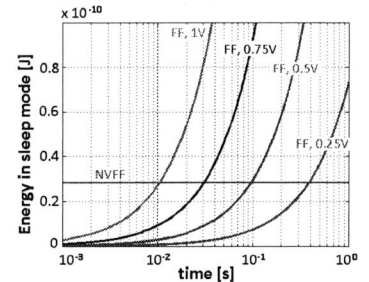

Fig.24: Comparison of the proposed 2T2CBRAM leakage to the 6T-SRAM and 1T2R cells designed for FPGA memory configuration with different ReRAM technologies.

Fig.25: Master-Slave NVFF block diagram: The non-volatile part stores/restores Q, Qb data from/to the slave latch.

Fig.26: Sleep mode energy for NVFF and FF supplied for different DRV. Constant NVFF energy corresponds to the data store and restore energy.

Point Twin-bit RRAM in 3D Interweaved Cross-point Array by Cu BEOL Process

Yung-Wen Chin[1], Shu-En Chen[1], Min-Che Hsieh[1], Tzong-Sheng Chang[2],
Chrong Jung Lin[1], and Ya-Chin King[1]

[1]Microelectronics Laboratory, Institute of Electronics Engineering, National Tsing Hua University, Hsinchu300,Taiwan
[2]Process Integration Division, Taiwan Semiconductor Manufacturing Company, Hsinchu 300, Taiwan
Phone/Fax:+886-3-5721804/886-3-5762186, E-mail: ycking@ee.nthu.edu.tw

Abstract

A self-rectifying twin-bit RRAM in a novel 3D interweaved cross-point array has been proposed and demonstrated in 28nm CMOS BEOL process. With TaO_x RRAMs on both sides of a single Via, the twin-bit RRAM cell is composed by Cu back-end layers only. Excellent selectivity by its asymmetric IV characteristic enables the twin-bit 1R cells to be efficiently stacked in 3D cross-point arrays.

Keywords: RRAM, self-rectifying, BEOL

Introduction

Resistive switching memory is one of the most promising candidates for future NVM due to its superior characteristics in size, speed and CMOS-process compatibility [1-3]. As the NAND Flash enters the 3D era [4], 3D RRAM arrays also emerge to maintain its competitiveness in data densities [5]. Unlike NAND array's gate control capability, cross-point RRAM arrays suffer from sneak currents during read operations, which limit its plausible array size [6]. Introducing a rectifying diode to each RRAM cell, i.e. 1D1R, [7] can suppress this problem at the cost of greater integration complexity. In our previous work on 3D Via-RRAM [8], a rectifier is stacked above a Via RRAM in a 3D cross-point 1D1R array. To further reduce lateral and vertical dimensions of a cell, a 3D-stackable self-rectifying twin-bit RRAM by Cu BOEL is proposed and demonstrated by pure CMOS back-end process, aimed for embedded NVM applications.

Cell Structure and Characteristics

To create a TaO_x based RRAM film in Cu back-end layers, a Via is placed at the center of two adjacent metal BLs. With careful spacing control, a TaO_x RRAM film is layered between Via and Metal. The TEM picture of the twin-bit RRAM with left and right bits at each side in Figure 1(a) indicates an expected point active RRAM region of less than 5nmx40nm, as illustrated in Figure 1(b). The EDX profile from a copper BL to its adjacent copper Via in Figure 2 reveals that a transition metal oxide film consists of TaO_x is well-defined between its BL edge and the Cu Via plug electrode by the dual-damascene process.

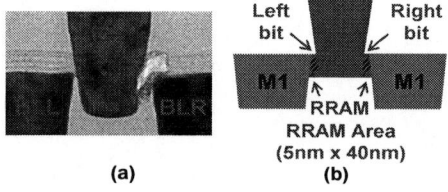

(a)   (b)

**RRAM Area
(5nm x 40nm)**

Figure 1 (a) TEM picture of a twin-bit RRAM. Via is center-placed between the right and left BLs, forming 2 bits, simultaneously. (b) Illustration of a twin-bit RRAM with overlap region of 5nmx40nm.

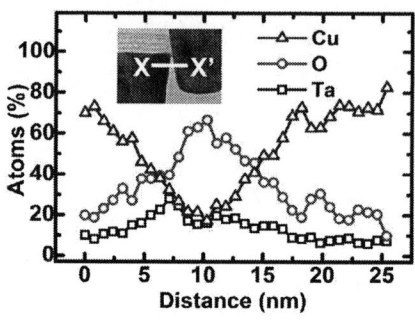

Figure 2 EDX profile along XX' cut line indicating TaO as its TMO layer.

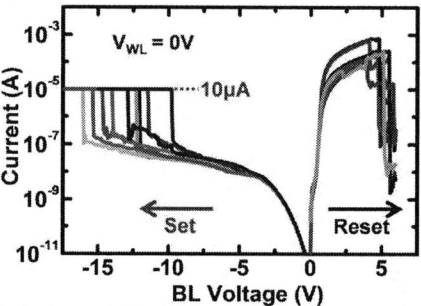

Figure 3 Set/reset I-V characteristics in bipolar DC sweeps. A compliance level of 10µA limits its reset current level.

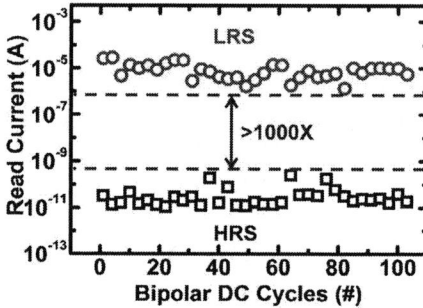

Figure 4 DC cycling characteristics of the twin-bit RRAM, read at BL=0.9V. 1000X read window is successfully demonstrated between cycles.

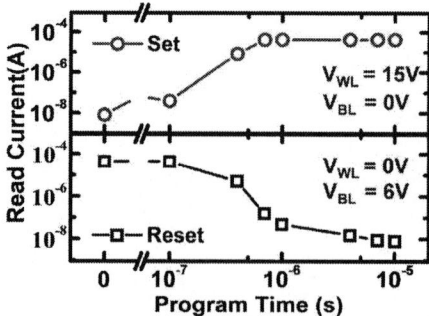

Figure 5 Time to program characteristics of the twin-bit RRAM. Both reverse set and forward reset operations can be completed in micro-second range.

978-1-4799-8002-4/14 $31.00 © 2014 IEEE   148

Figure 3 shows the bipolar set and reset characteristics of a typical twin-bit RRAM. In reverse set, current flows from Via to Metal, while in forward reset, current flows from Metal BL to Via. Through bipolar operations, over 1000X ON/OFF window is easily established, as shown in Figure 4. Figure 5 summarized the time to reset and set at a voltage of 6V and 15V, respectively. Data suggests that this point twin-bit RRAM can be operated within micro-second range. The cumulative plot in Figure 6 of the LRS/HRS states of a single cell between cycles shows sizable window. Initial states of the left and right bits summarized in Figure 7 reveal no significant difference or misalignment induced biases on either side. Figure 8 reveals that an initially symmetric device is transformed into a self-rectifying RRAM indifference to its forming directions. The cell shows prominent self-rectifying behaviors in its LRS, while the 1000X forward/reverse read ratio are kept even at a raised temperature of 125°C (Figure 9).

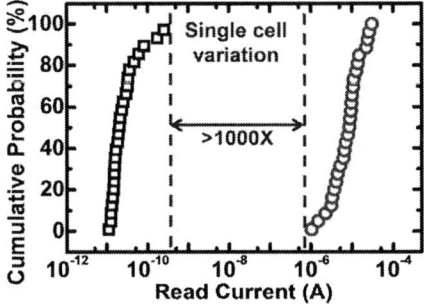

Figure 6 Cumulative plots of LRS/HRS of a single device between cycles. Data shows over 1000X read window remains in state alternating operations.

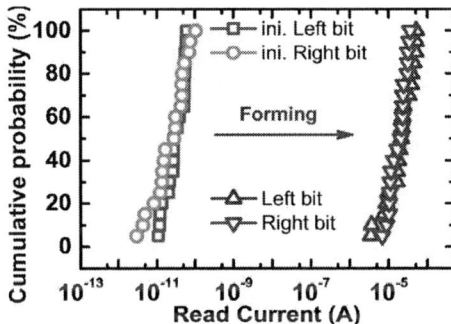

Figure 7 Read current distribution of the left and right bits in its initial state and after forming. No difference is observed between left and right bits.

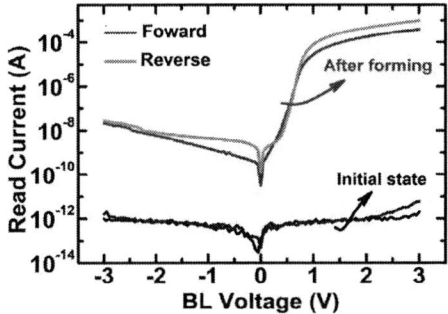

Figure 8 Initially symmetric IV characteristics switches into asymmetric IVs by forming operations, independent of its forming directions.

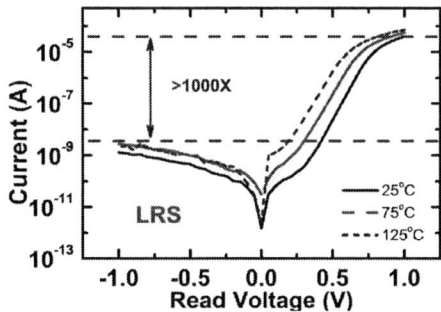

Figure 9 I-V characteristics with high selectivity in LRS of twin-bit RRAM are demonstrated under wide temperature range.

Array Architecture and Reliability Testing

Illustrated by its cross-sectional view in Figure 10 and top-view in Figure 11, a 3D interweaved cross-point array is proposed. A single bit only required half a Via, which greatly enhanced its horizontal packing density. The cell exhibits prominent self-rectifying behaviors in its LRS, the key for enabling a sizable 1R array to further stack with height reduction vertically, as reveal in the 3D illustration of the interweaved array in Figure 12. The SEM picture of a single layer in this 3D array is shown in Figure 13, while its operation conditions are summarized in Table 1.

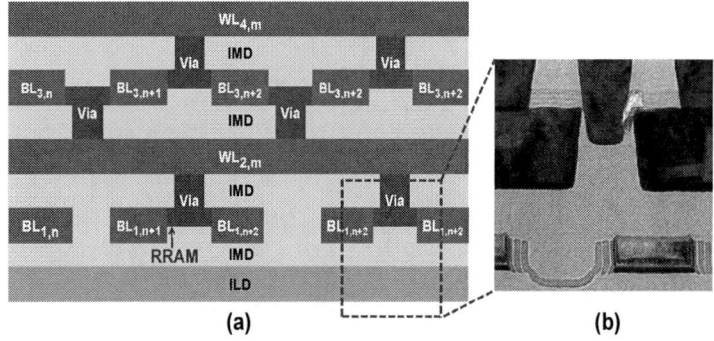

Figure 10 (a) 3D cross-section of twin-bit RRAM cross-point array. Each RRAM is selected by interweaved WLs and BLs. (b) TEM picture of twin-bit RRAMs in the bottom layer, suggesting that this 3D array can be placed on top of its peripheral circuit, achieving 100% array efficiency.

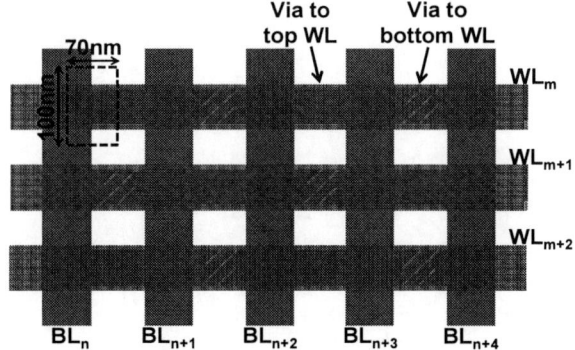

Figure 11 Top-view layout of a 3D interweaved cross-point array. The unit cell contains only half a Via, leading to a compact cell of 70nmX100nm.

978-1-4799-8002-4/14 $31.00 © 2014 IEEE   149

Long-term temperature and read stress tests in Figure 14 and Figure 15 indicate that the characteristics of the twin-bit RRAM cells remain stable of over 1000X LRS/HRS read window during and after stress. The program disturb test in Figure 16 shows that 1000X read window after 10^4 cycles is demonstrated. Unselected cells in both LRS and HRS suffer little to no set disturb. Figure 17 shows that as the number of stacked metal layers increases, the number of the stacked cells increases accordingly, which effectively enhance packing densities. The maximum size of a cross-point array is generally limited by the readout ratio between the on-current of the selected cell and off-state leakage currents from cells on the same selected WL. As suggested by data in Figure 18, the maximum read window occurs at a read voltage of 0.9V.

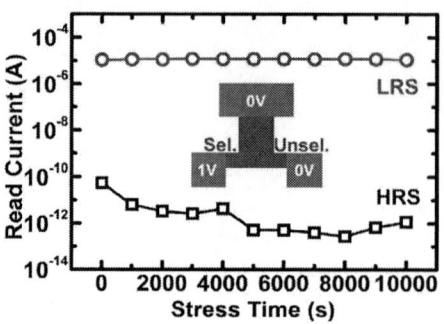

Figure 15 Read disturb of twin-bit RRAM. LRS and HRS are undisturbed without read window narrowing after 10k seconds of continuous read at 1V.

		BL,WL above BL₅	WL₄	BL₃	BL,WL below WL₂
Set	sel.		15	0	
	unsel.	7.5	7.5	7.5	7.5
Reset	sel.		0	6	
	unsel.	0	6	0	6
Read	sel.		0	0.9	
	unsel.	0	0.9	0	0.9

Table 1 Operation conditions of 3D array. Proper inhibit voltages are applied on unselected BLs and WLs.

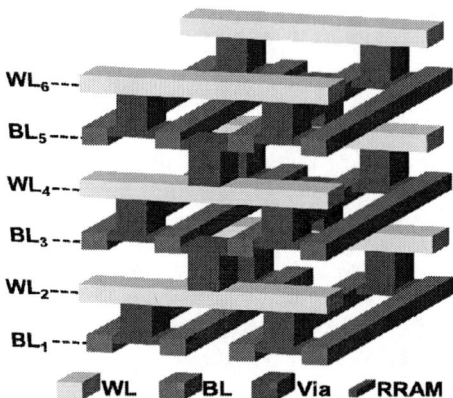

Figure 12 Illustration of the 3D interweaved cross-point array.

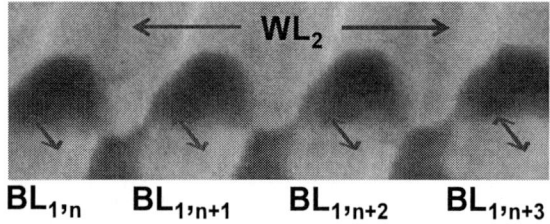

Figure 13 SEM picture of a single layer of the twin-bit RRAM. By repeat stacking of metal line with Vias, a 3D interweaved cross-point array can be constructed by Cu back-end layers.

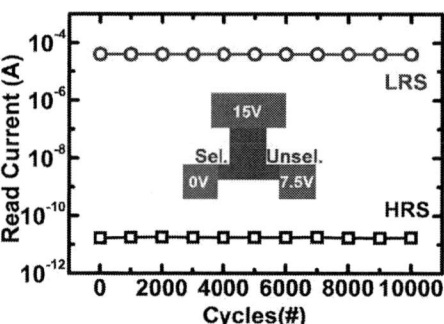

Figure 16 Set disturb tests for unselected cells at V_{WL}=15V and V_{BL}=7.5V. Excellent disturb immunity is demonstrated on unselected cell.

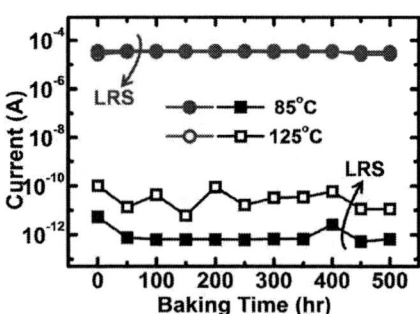

Figure 14 Retention of twin-bit RRAM. Read window remains stable after 85°C and 150°C 500hrs baking with read voltage of 0.9V.

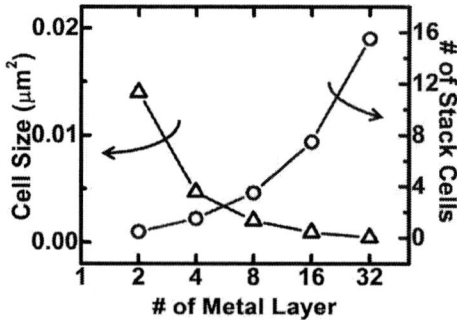

Figure 17 In interweaved cross-point array, the number of stacked cells increases with increasing metal layers, leading to increase packing density.

978-1-4799-8002-4/14 $31.00 © 2014 IEEE   150

Under this condition, a maximum 1Mbit Array with 1024 cells sharing a common WL is projected, see Figure 19. Figure 20 reveals that good uniformity in its HRS/LRS amongst cells. The cumulative probability of forward/reverse ratio in LRS of different RRAMs in Figure 21 shows enough selectivity remains under wide temperature range. Lastly, Figure 22 compares the scaling trend of 2D CRRAM with 3D-stackable twin-bit RRAM. Beside its extremely small half-Via cell, this new 3D stacking scheme results in high vertical scalability.

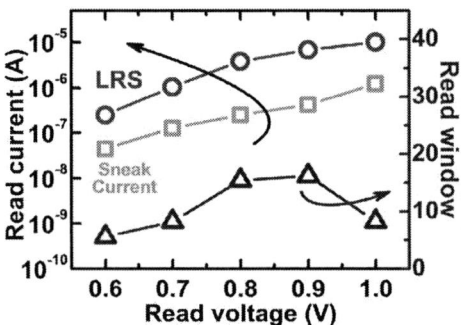

Figure 18 In a twin-bit RRAM array with 64 cell on a WL, the maximum read window occurs at a read BL voltage of 0.9V.

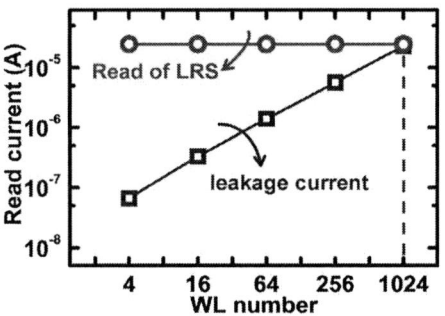

Figure 19 The self-rectifying characteristic of twin-bit RRAM, enables a maximum 1Mbit cross-point array without selectors.

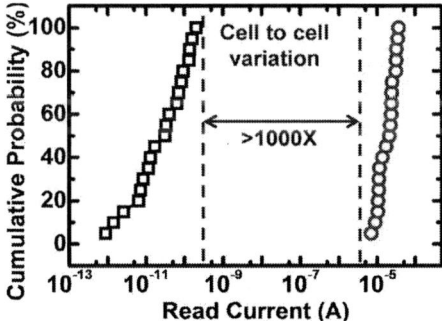

Figure 20 Cumulative plot of LRS/HRS of different twin-bit RRAM. Over 1000X read window remains in worse case conditions.

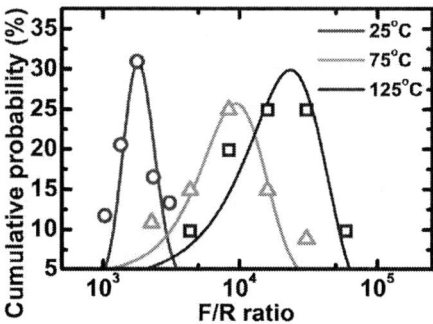

Figure 21 Distribution of F/R ratio under different temperature. Good uniformity of over 1000X F/R ratio in different RRAM is demonstrated.

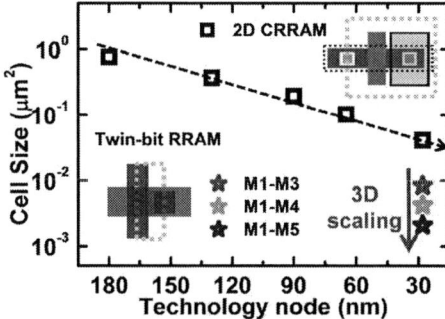

Figure 22 Scaling trend of 3D twin-bit RRAM. Great scalability is demonstrated both laterally and vertically.

Conclusion

In this work, a self-rectifying point twin-bit RRAM is proposed and successfully demonstrated in 28nm CMOS Cu BEOL process. In a novel interweaved cross-point array, the half-Via RRAM cell can be compacted stacked vertically, with competitive scalability in all dimensions.

Reference

[1] W.C.Shen, et al. "High-K Metal Gate Contact RRAM (CRRAM) in Pure 28nm CMOS Logic Process," *IEDM Tech. Dig*, p.745. 2012

[2] C.-H.Wang, et al. "Three-Dimensional 4F2 ReRAM Cell with CMOS Logic Compatible Process," *IEDM Tech. Dig*, p.664. 2010

[3] X.A. Tran, et al. "Self-Rectifying and Forming-Free Unipolar HfOx based-High Performance RRAM Built by Fab-Avaialbe Materials," *IEDM Tech. Dig*, p.713 2011

[4] J. Jang, et al. "Vertical Cell Array using TCAT(Terabit Cell Array Transistor) Technology for Ultra High Density NAND Flash Memory," *VLSI*, p.192. 2009

[5] H.-Y. Chen, et al. "HfOx Based Vertical Resistive Random Access Memory for Cost-Effective 3D Cross-Point Architecture without Cell Selector," *IEDM Tech. Dig*, p.497. 2012

[6] S. Yu, et al. "3D vertical RRAM-Scaling limit analysis and demonstration of 3D array operation," *VLSI*, p.158. 2013

[7] M.-J. Lee, et al. "2-stack ID-IR Cross-point Structure with Oxide Diodes as Switch Elements for High Density Resistance RAM Applications," *IEDM Tech. Dig*, p.771. 2007

[8] M.C. Hsieh, et al. "Ultra High Density 3D Via RRAM in Pure 28nm CMOS Process," *IEDM Tech. Dig*, p.260. 2013

Experimental and theoretical understanding of Forming, SET and RESET operations in Conductive Bridge RAM (CBRAM) for memory stack optimization

J. Guy, G. Molas, P. Blaise, C. Carabasse, M. Bernard, A. Roule, G. Le Carval, V. Sousa, H. Grampeix, V. Delaye, A. Toffoli, J. Cluzel, P. Brianceau, O. Pollet, V. Balan, S. Barraud, O. Cueto, G. Ghibaudo[*], F. Clermidy, B. De Salvo, L. Perniola

CEA, LETI, MINATEC Campus, 17 rue des Martyrs, 38054 GRENOBLE Cedex 9, France, jeremy.guy@cea.fr
[*]IMEP-LAHC, MINATEC/INPG, BP 257, 38016 Grenoble France

Abstract

In this paper, we deeply investigate for the 1st time at our knowledge the impact of the CBRAM memory stack on the Forming, SET and RESET operations. Kinetic Monte Carlo simulations, based on inputs from ab-initio calculations and taking into account ionic hopping and chemical reaction dynamics are used to analyse experimental results obtained on decananometric devices. We propose guidelines to optimize the CBRAM stack, targeting Forming voltage reduction, improved trade-off between SET speed and disturb immunity (time voltage dilemma) and window margin increase (RESET efficiency).

Introduction

Conductive Bridge RAM (CBRAM) are envisaged as a promising alternative to Flash memory due to their high speed, low voltage, low consumption and ease of integration in the back end of a logic process [1-2]. In order to improve the thermal stability (i.e. soldering and retention), oxides have been introduced as electrolyte in the CBRAM technology, but induce the necessity of a Forming operation. To respond to the electrical needs of this unique cycle a complex circuitry may be required. Moreover, a good control of the RESET mechanisms, governing the window margin and thermal stability [3] is essential for CBRAM industrialization but still has to be clearly understood. In this paper, we present Nano-trench CBRAM combining various material stacks and offering good scaling capability (down to 5x50nm effective area). The impact of electrical and thermal properties of the CBRAM layers on the Forming, SET and RESET operation is addressed and a stack optimization guideline is proposed.

Technological details

Nano-trench CBRAM devices (Fig.1) are composed by a bottom electrode (BE) defined by a TiN, Ta or WSi metal ring (the electrode length – 20nm to 5nm – being governed by the metal thickness [3]). A trench defined by e-beam lithography (down to 50nm) is then patterned in a SiN capping layer, allowing the reduction of the active top electrode length. Various oxides were used as resistive layer (RL: Al_2O_3, HfO_2, MO_x...) and capped by a PVD $CuTe_x$-based active top electrode (TE).

Results and discussion

1. Simulation framework

A Kinetic Monte Carlo (KMC) model was developed to describe the CBRAM forming and SET operations, computing the filament growth at the atomic level in the RL (Fig.2). For each Cu particle, we compute the events rates corresponding to the oxidation and reduction at the electrodes, and the field assisted transport in the RL (Fig.3). The events rates are calculated in the two $(\vec{x}, \vec{y})$ directions following these equations, (1) corresponding to the ionic diffusion events and (2) corresponding to the oxidation (+*) and reduction with a (−*):

$$\vec{\Gamma} = \nu \cdot \exp\left(-\frac{E_A - Q d_{Cu-Cu}\frac{\chi}{3}\vec{\varepsilon}}{k_B T}\right)(\vec{x}, \vec{y}) \qquad (1)$$

$$\vec{\Gamma} = \nu \cdot \exp\left(-\frac{E_A \pm^* \frac{1}{2}\left(Q d_{Cu-Cu}\frac{\chi}{3}\vec{\varepsilon} + \phi_{RL} - \phi_M\right)}{k_B T}\right)(\vec{x}, \vec{y}) \qquad (2)$$

where ν is the vibration frequency ($10^{13} s^{-1}$), E_A the Activation Energy ($E_{A\text{-oxidation}} = E_{A\text{-reduction}} = 3.1$ eV; $E_{A\text{-Volume-diffusion}} = 0.5$ eV; $E_{A\text{-Surface-diffusion}} = 0.6$ eV; $E_{A\text{-Surface-desorbtion}} = 0.8$ eV), Q the charges number , d_{Cu-Cu} the distance between 2 Cu sites, $\vec{\varepsilon}$ the electric field, χ the electric susceptibility [5], $\phi_{M/RL}$ the Metal and RL work function, k_B the Boltzmann Constant and T the Temperature. Each event is modeled as the transition between two states separated by an energy barrier E_A (Fig.4). From the even rates the events probabilities are calculated and for each iteration time, an event is randomly picked and the global system is updated (position, electric field and probabilities). The cell current is calculated through a percolation model [3] following the local Cu occupation, and allowing to compute I(time) forming characteristics (Fig.5). The forming time is thus calculated when compliance current is reached (Fig.6) depending on the operating voltage. Assuming a residual filament in the RL to describe High Resistive State (HRS), SET time can also be calculated.

2. Forming/SET Analysis

Our samples are formed by applying a positive voltage on the TE to create the Cu-based Conductive Filament (CF). Forming can either be considered in Quasi-Static mode to extract $V_{Forming}$ or in pulsed mode to plot the $t_{Forming}(V_{pulse})$ characteristics. We firstly studied the role of TE and BE work functions (Φ). For both electrodes, an increase of Φ leads to a reduction of $t_{Forming}$ and $V_{Forming}$ (Fig.7-10).

978-1-4799-8002-4/14 $31.00 © 2014 IEEE

Φ induces a flat band voltage shift $V_{FB}=\Phi_{BE}-\Phi_{TE}$ [4] but also affects the chemical reactions: increasing Φ_{TE} (resp. Φ_{BE}) reduces the effective oxidation (resp. reduction) energy barrier. While in the case of the TE the $V_{Forming}$ reduction is essentially due to the oxidation barrier decrease, for the BE, $V_{Forming}$ reduction essentially results from the V_{FB} variation (the reduction barrier change being negligible). In terms of $t_{Forming}(V_{pulse})$ curves (Fig.8, 10) the increase of both Φ_{BE} and Φ_{TE} shows a shift of the characteristics, favorable to low voltage operation but with a stronger effect for Φ_{TE} (3 $t_{Forming}$ decades per 0.3eV vs 1eV for the BE). Then, the impact of the RL thickness is shown in Fig.11. Thinner RL accelerates the forming operation due to a higher electric field in the system for a given voltage. In particular a shift of the $t_{Forming}(V_{pulse})$ characteristics modulated by a change of slope is observable in Fig.12; a reduction of 5Å resulting in a ~2 decades faster $t_{Forming}$ at 3.5V. Analyzing the $t_{SET}(V_{SET})$ simulations allows to address the time-voltage dilemma; a steeper slope indicating a better tradeoff between HRS disturb and SET speed. The role of the RL material is then exposed in Fig.13 including various physical parameters. First, a lower Cu energy level in the RL (Φ_{RL}) leads to a $V_{Forming}$ decrease by reducing the effective oxidation energy barrier (Fig.14). The RL density (tied to the deposition conditions) controls the Cu-Cu hopping distance. A long distance amplifies the electric field between hopping sites ($V_{Forming}$ reduction of 2V per 0.8Å). Moreover, increasing the permittivity from 9 (Al_2O_3) to 20 (HfO_2) leads to a $V_{Forming}$ decrease of ~2V, assuming the other materials properties unchanged. Clausius Mossetti relation [5] explains this observation trough an enhancement of the local electric field at high permittivity. Regarding $t_{Forming}(V_{pulse})$ characteristics (Fig.15), the hopping distance and permittivity increase result in a change of slope improving the disturb immunity, while a lower Φ_{RL} shifts the curve, leading to a faster SET. *Finally, the previous studied material properties can be divided in two groups depending of their impact on the $t_{Forming}(V_{Pulse})$ characteristics: the properties changing the energy levels (Φ_{RL}, Φ_{TE}, Φ_{BE}) only shift the curve while the RL features (thickness, permittivity, hopping distance) can enhance the impact of the applied voltage and improve the SET speed and disturb immunity.*

3. RESET Analysis

Our samples are erased by applying a negative bias on the TE to dissolve the Cu-based CF. Using WSi instead of TiN BE, the RESET voltage is reduced from about -0.2V to -0.5V, with a wider dispersion in the latter (Fig.16-17). Thermal simulations were performed to calculate the CF temperature elevation by Joule heating (T_{MAX}). A higher temperature for TiN BE is observed in the simulations due to the poor TiN thermal conductivity reducing the thermal dissipation (Fig. 18). Introducing the calculated temperature into oxidation/reduction Γ ratio Eq. (2), we identified for TiN and WSi BE the favorable RESET voltage regime (Fig.19), matching the previous V_{RESET} dispersion of Fig. 17. WSi

samples also showed a R_{OFF} improvement (1 decade) compared to TiN BE (Fig.20). Finite-elements simulations were used to compute the temperature along the CF during RESET (Fig.21). While T_{MAX} is localized in the center of the filament for a WSi BE, it gets closer to the BE for TiN. These locations correspond to the disrupting points of the CF during RESET (Fig.22). Once RESET starts, the CF is disconnected and the remaining part attached to the BE acts as a Cu source. The remaining CF tip is then progressively oxidized, which explains the progressive RESET behavior (Fig.16), and gradual R_{OFF} increase. Once the tip is completely dissolved, the maximal R_{OFF} is reached. *In conclusions, a higher BE thermal conductivity leads to a higher $|V_{RESET}|$ but a higher R_{OFF} after saturation.* Then, the CF morphology (tied to the programming conditions) was related to the RESET performances. Indeed, forming the cell with a low voltage and long pulse leads to a higher $|V_{RESET}|$ (Fig.23). Simulations of Forming at high or low voltages show two distinct CF morphologies: low $V_{Forming}$ enlarges the CF bottom radius, leading to a smaller heating part and a larger and taller dissipating area. Consequently, higher current (i.e. higher V_{RESET}) is required to reach a sufficient temperature so that RESET occurs. Finally we demonstrate that an optimized RL can improve RESET efficiency. MO_x RL shows faster forming but a lower R_{OFF} than HfO_2 (Fig.24). Adding a thin HfO_2 interface to the MO_x RL, we were able to increase R_{OFF} by 1 decade, with no $V_{Forming}$ degradation. This is due to the higher thermal conductivity of HfO_2 which pushes the T_{MAX} location of the CF back from the BE, allowing to dissolve a larger CF area during RESET.

Conclusions

In this paper, we deeply investigated the Forming, SET and RESET operations of CBRAM. We demonstrated experimentally and theoretically (based on Kinetic Monte Carlo simulations) that a careful optimization of the memory layer properties allows a forming voltage reduction (by electrode engineering), but also a better tradeoff between SET speed and disturb immunity (tuning the resistive layer thickness, dielectric constant, and density). We clarified the role of the bottom electrode thermal conductivity on the RESET efficiency, and proposed a bi-layer resistive layer combining low $V_{Forming}$ and high R_{OFF}.

References

[1] J.R. Jameson et al., "Conductive-bridge memory (CBRAM) with excellent high-temperature retention," IEDM 2013 Tech. dig., pp.30.1.1-30.1.4.

[2] A. Calderoni et al., "Performance Comparaison of oxide-ReRAM and Cu based ReRAM," Int. Mem. Workshop 2014, pp.5-8.

[3] J. Guy et al., "Investigation of the physical mechanisms governing data retentionin down to 10nm Nano-trench Al_2O_3/CuTeGe Conductive Bridge RAM (CBRAM)," IEDM 2013 Tech. dig., pp.30.2.1, 30.2.4.

[4] J.R. Jameson et al., "One-dimensional model of the programming kinetics of conductive-bridge memory cells," Appl. Phys., 99, 063506, 2011.

[5] J.W. McPherson et al., "Complementary model for intrinsic time-dependent dielectric breakdown in SiO_2 dielectrics," J. Appl. Phys. 88, 9, pp.5351-5359, 2000.

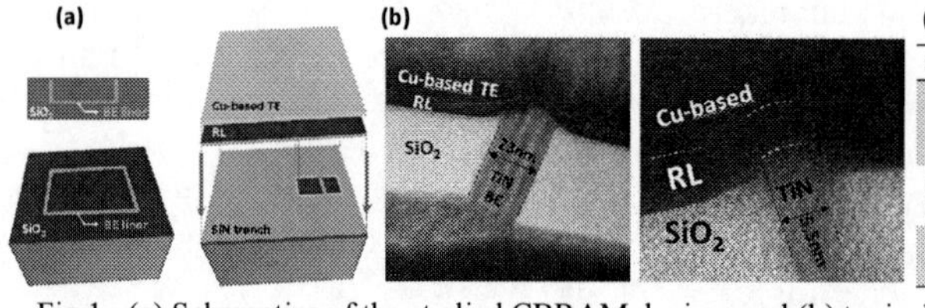

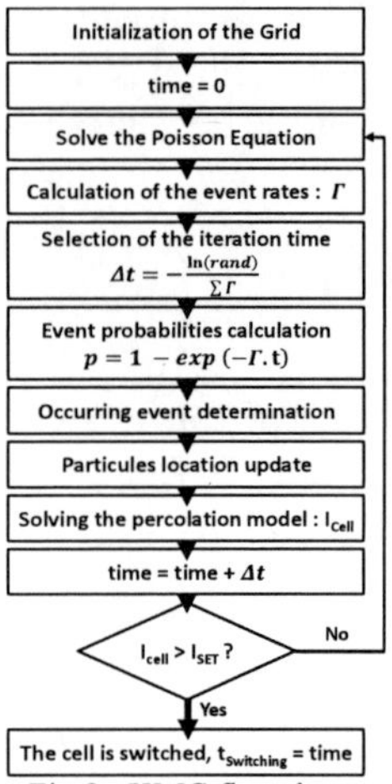

Initialization of the Grid

time = 0

Solve the Poisson Equation

Calculation of the event rates : Γ

Selection of the iteration time
$\Delta t = -\dfrac{\ln(rand)}{\Sigma \Gamma}$

Event probabilities calculation
$p = 1 - exp\,(-\Gamma.t)$

Occurring event determination

Particules location update

Solving the percolation model : I_{Cell}

time = time + Δt

$I_{cell} > I_{SET}$? — No

The cell is switched, $t_{Switching}$ = time

Fig 2. KMC flowchart.

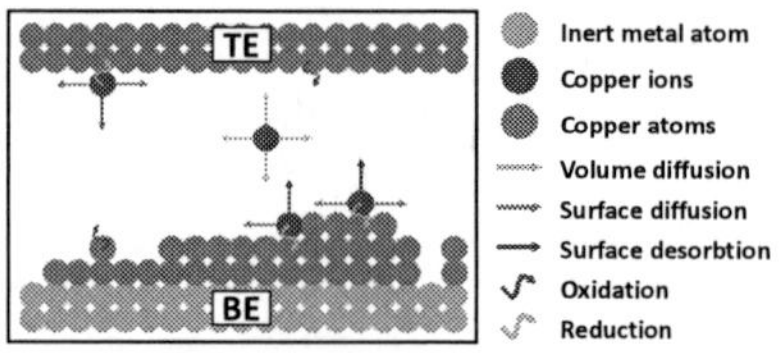

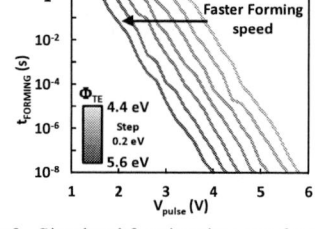

Fig 1. (a) Schematics of the studied CBRAM devices and (b) typical TEM cross sections of the 20nm studied devices. Nano-trench technology enables the bottom electrode (BE) scaling down to ~5nm. (c) Stack description of the studied samples.

Samples	S1	S2	S3	S4	S5	S6	S7	S8	S9
Bottom Electrode BE	TiN	TiN	WSi	TiN	TiN	WSi	Ta	Ta	Ta
Resistive Layer RL	Al_2O_3 3.5nm	Al_2O_3 3.5nm	Al_2O_3 3.5nm	Al_2O_3 5nm	MO_x 5nm	Al_2O_3 5nm	MO_x 5nm	HfO_2 5nm	HfO_2 1nm MO_x 5nm
Top Electrode TE	Cu	$CuTe_x$	$CuTe_x$	$CuTe_x$	$CuTe_x$	$CuTe_x$	$CuTe_x$	$CuTe_x$	$CuTe_x$

Fig 3. Physical mechanisms simulated in the KMC and common atoms/ions legend.

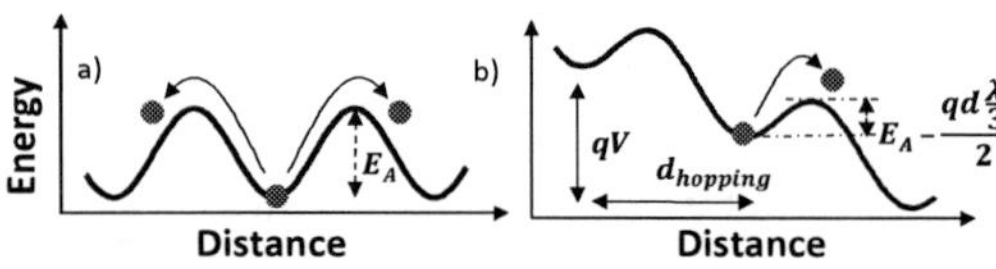

Inert metal atom
Copper ions
Copper atoms
Volume diffusion
Surface diffusion
Surface desorbtion
Oxidation
Reduction

Fig 4. Activation energy modified by electric field for Cu hopping and oxidation-reduction reactions.

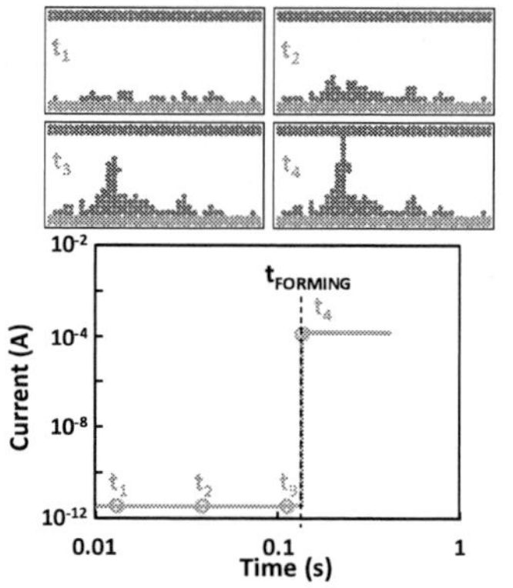

Fig 5. Illustration of simulated filament growth and corresponding $t_{Forming}$ extraction.

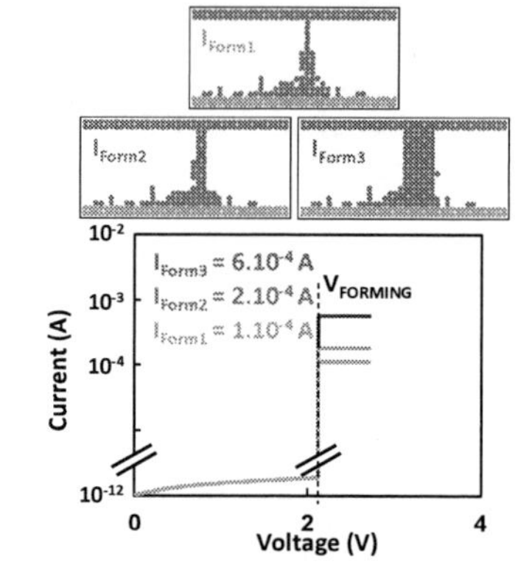

Fig 6. Simulated Quasi-Static IV for various Forming currents corresponding to the filament thickness.

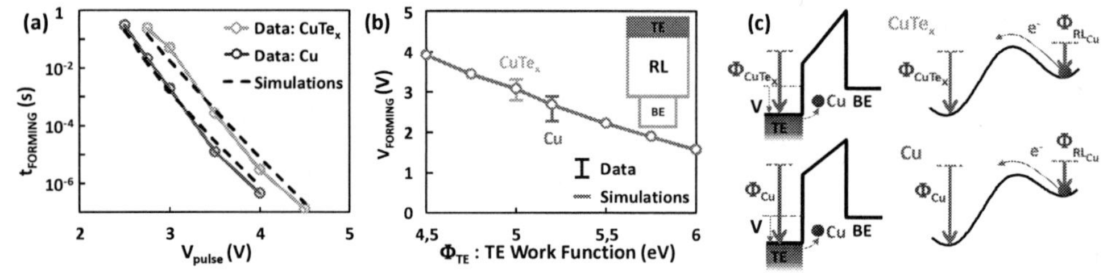

Fig 7. (a) Measured and simulated forming time as a function of the applied voltage for CBRAM integrating Cu (S1) or $CuTe_x$ (S2) as TE (ion supply layer). (b) Quasi static forming voltage dependence with TE work function. (c) Schematics illustrating the impact of Φ_{TE} on the electric field in the RL, and on the oxidation-reduction reaction.

Fig 8. Simulated forming time as a function of the applied pulse voltage for various TE work functions.

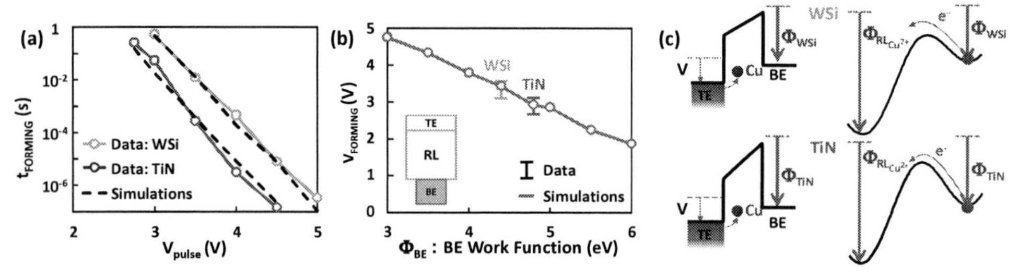

Fig 9. (a) Measured and simulated forming time as a function of the applied voltage for CBRAM integrating WSi (S3) or TiN (S2) as BE (inert electrode). (b) Quasi static forming voltage dependence with BE work function. (c) Schematics illustrating the impact of Φ_{BE} on the electric field in the RL, and on the oxidation-reduction reaction.

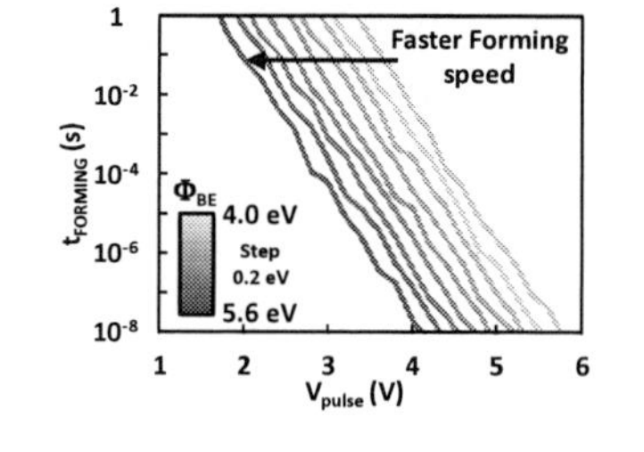

Fig 10. Simulated forming time as a function of the applied pulse voltage for various BE work functions.

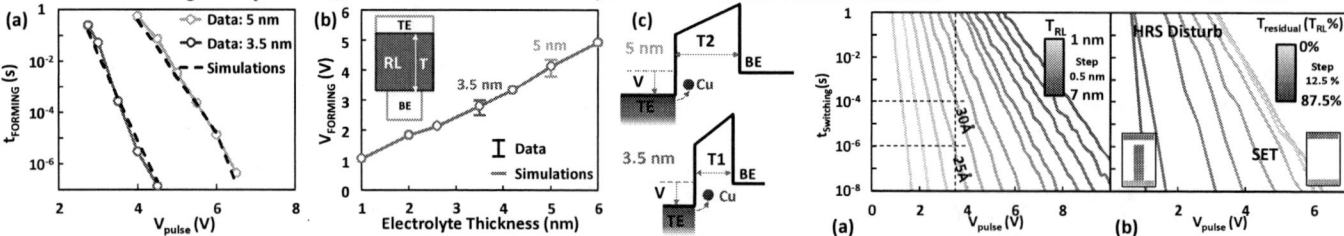

Fig 11. (a) Measured and simulated forming time as a function of the applied voltage for 3.5nm (S2) and 5nm (S4) Al_2O_3-based CBRAM. (b) Quasi-Static forming voltage dependence with RL thickness. (c) Schematics illustrating the impact of RL thickness on the electric field.

Fig 12. (a) Simulated $t_{Forming}$ as a function V_{pulse} for various RL thicknesses. (b) Simulated t_{SET} starting from various residual filament sizes (emulating various R_{OFF}) in a 3.5nm RL.

978-1-4799-8002-4/14 $31.00 © 2014 IEEE

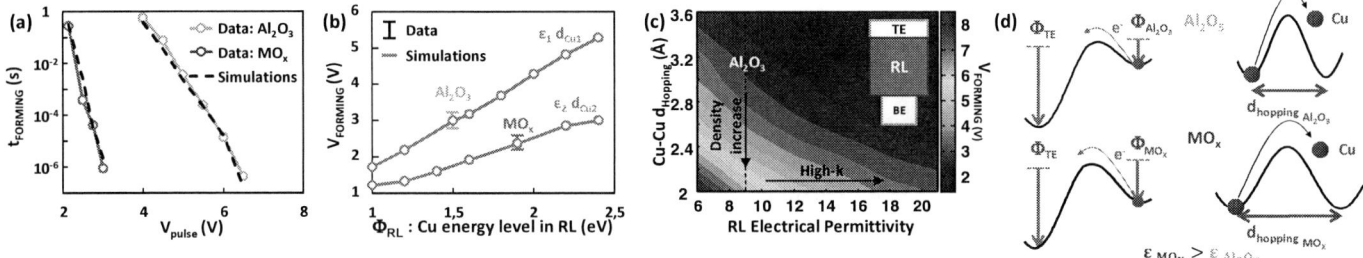

Fig 13. (a) Measured and simulated forming time as a function of the pulse voltage for Al₂O₃-based (S4) and MOₓ-based (S5) CBRAM. (b) Quasi-Static forming voltage dependence with Cu energy depth (from vacuum) in the RL. (c) Quasi-Static forming voltage dependence with Cu-Cu hopping distance in the RL, and RL permittivity. (d) Schematics illustrating the impact Φ_{RL} on the oxidation-reduction reaction.

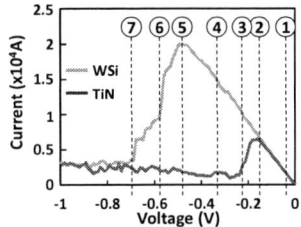

Fig 14. (a) Calculated DOS of γ-Al₂O₃ using G₀W₀ and corresponding energy levels for neutral and charged Cu (resp. Cu and Cu²⁺). (b) Corresponding simulated γ-Al₂O₃ structure.

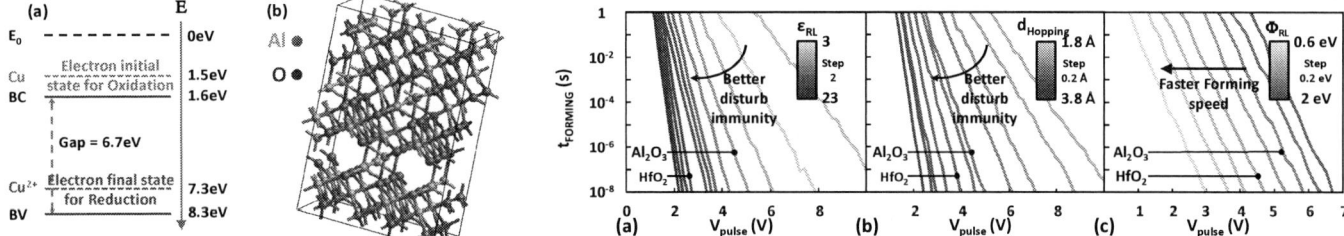

Fig 15. (a) Simulated forming time as a function of the applied voltage for various (a) RL permittivities, (b) Cu-Cu hopping distance in the RL, (c) Cu energy level in the RL.

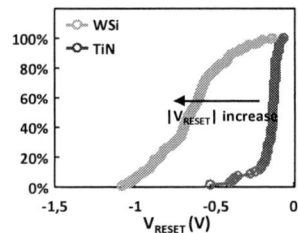

Fig 16. IV RESET characteristics indicating a change in V_{RESET} between WSi (S6) and TiN BE (S4), and illustrating a gradual RESET process.

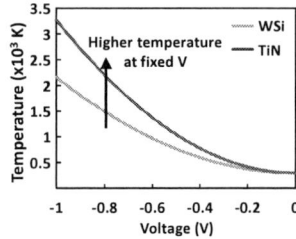

Fig 17. Measured |V_{RESET}| distribution for WSi (S6) or TiN BE(S4) (~10 cycles on 40 cells).

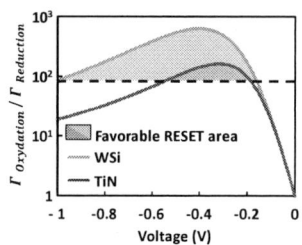

Fig 18. Calculated maximum temperature in the CF as a function of the RESET voltage for WSi and TiN BE.

Fig 19. Calculated oxidation/reduction ratio (see equation from Fig.4) for Cu in the filament as a function of the RESET voltage for WSi and TiN BE.

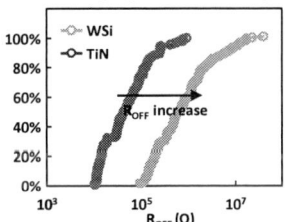

Fig 20. Measured R_{OFF} distribution after RESET for WSi (S6) and TiN (S4) BE (~10 cycles on 40 cells).

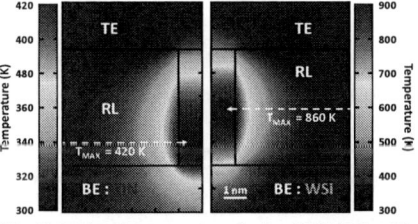

Fig 21. Calculated temperature in the CBRAM at V_{RESET} for WSi and TiN BE, showing the impact of the thermal dissipation in the BE on the localization of the maximum temperature (κ_{TE}=400W.m⁻¹.K⁻¹; $\kappa_{Filament}$=300W.m⁻¹.K⁻¹; κ_{RL}=15W.m⁻¹.K⁻¹; κ_{TiN}=30W.m⁻¹.K⁻¹; κ_{WSi}=170W.m⁻¹.K⁻¹).

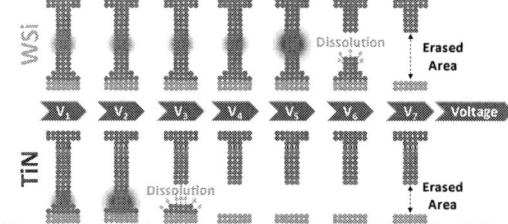

Fig 22. Schematics of the filament during RESET for WSi and TiN BE, for various voltages corresponding to Fig.17, illustrating the disrupting point (T_{MAX} location in the CF) and the residual filament after RESET.

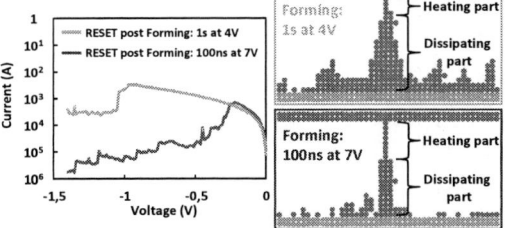

Fig 23. Measured IV RESET characteristics for TiN BE (S4) with Al₂O₃ RL and simulated filament morphologies after two extreme forming conditions: Cond.1: 1s at 4V, Cond.2: 100ns at 7V.

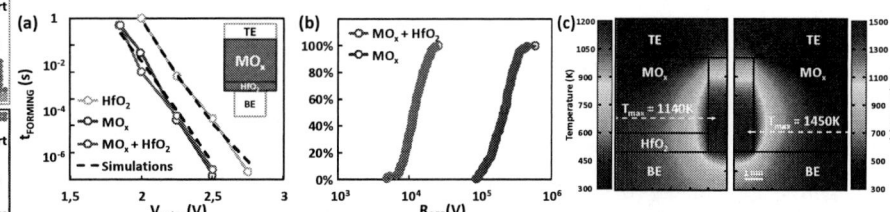

Fig 24. (a) Measured and simulated forming time as a function of the applied voltage for three different RL (S7, S8, S9). (b) Measured R_{OFF} after RESET for the three RL (~10 cycles on 40 cells). (c) Calculated temperature in the CBRAM during RESET for the MOₓ (S7) and bilayer (S9) RL; T_{Max} localisation is pushed towards the TE for the bilayer RL CBRAM (κ_{TE}=400W.m⁻¹.K⁻¹; $\kappa_{Filament}$=300W.m⁻¹.K⁻¹; κ_{MOx}=1W.m⁻¹.K⁻¹; κ_{HfO2}=20W.m⁻¹.K⁻¹; κ_{BE}=40W.m⁻¹.K⁻¹).

978-1-4799-8002-4/14 $31.00 © 2014 IEEE   155

Ultrathin (~2nm) HfO_x as the Fundamental Resistive Switching Element: Thickness Scaling Limit, Stack Engineering and 3D Integration

Liang Zhao*, Zizhen Jiang, Hong-Yu Chen, Joon Sohn, Kye Okabe, Blanka Magyari-Köpe, H.-S. Philip Wong, Yoshio Nishi

Department of Electrical Engineering, Stanford University, Stanford, CA 94305-4070, USA; *Email: lzhao10@stanford.edu

Abstract - This paper addresses the thickness scaling limit of HfO_x-based RRAM through a combination of theoretical calculations of electron transport and experimental demonstration of 2nm-HfO_x devices. The comparison of 2nm devices with thicker references and bilayer stacks confirm the switching thickness is less than 2nm. The 3D integration of 2nm-HfO_x devices enables much larger array size and leads to superior performances compared to planar devices.

I. Introduction

RRAM has emerged as a promising candidate for future high-density non-volatile memories [1]. Recently, horizontal scaling [2] and 3D vertical architectures [3-6] of RRAM have been extensively studied to improve scalability. However, the scaling limit in the material stack thickness has not been fully understood and pursued. Hf/3nm-HfO_x RRAM with a switching-stack thickness of 5nm has been demonstrated, with effective ON/OFF ratio of ~10 [7]. In addition, pulse-train studies of HfO_x RRAM indicated the effective switching thickness to be less than 2nm [8]. In this work, the theoretical scaling limit of HfO_x thickness was first estimated using non-equilibrium Green's function calculations based on density functional theory (DFT-based NEGF). Next, devices with 2nm HfO_x directly sandwiched between electrodes (TiN and Pt) were fabricated, and compared with thicker HfO_x devices. Bi-layer devices were further studied by stacking 2nm HfO_x with TiO_2 and Al_2O_3 layers, indicating the fundamental switching element to be potentially less than 2nm thick. Finally, the 2nm HfO_x device was integrated into 3D vertical geometry which demonstrates promising characteristics such as ON/OFF ratio (~100), switching speed (~20ns), endurance (10^8) and data retention (~10 years).

II. Theoretical Evaluation of Thickness Scaling Limit

The theoretical limit of thickness scaling is mainly determined by electron tunneling which limits the OFF-state resistance. To investigate the dependence of ON/OFF ratio on oxide thickness, TiN/HfO_x/TiN device structures were constructed with HfO_x thickness (t_{OX}) ranging from 1 to 3 nm (Fig. 1(a)). The conductive filament (CF) is modeled as a chain of oxygen vacancies (Fig. 1(b)) which creates a delocalized path of electron wavefunction (Fig. 2). Previous study has confirmed this assumption to be valid for order-of-magnitude estimation of ON/OFF-state resistances [9].

The atomic positions were first optimized using density functional theory with on-site Coulomb corrections (DFT+U), as implemented in VASP [10]. After the relaxed configuration was obtained, electron transport calculations were carried out using NEGF, as implemented in QuantumWise ATK [11]. Details of the calculation methods are summarized in Fig. 3.

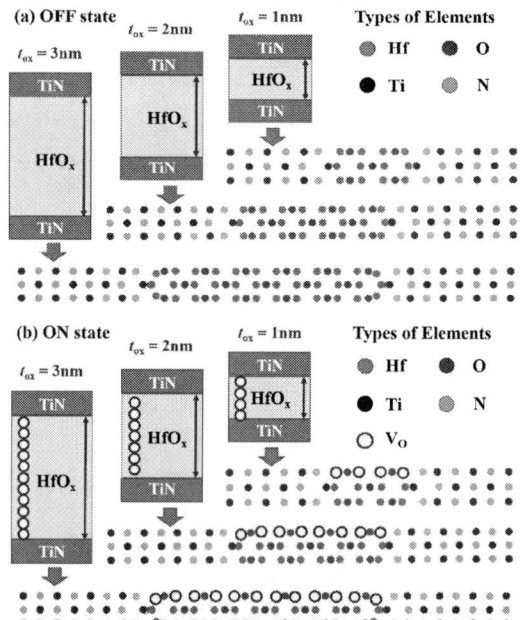

Fig. 1 Schematic diagrams and atomic configurations of the simulated TiN/HfO_x/TiN devices in (a) OFF state and (b) ON state. HfO_x thickness varies from 1 to 3 nm. The OFF state is represented by HfO_x layer without any oxygen vacancy, while the ON state is represented by a chain of oxygen vacancies across the HfO_x layer.

978-1-4799-8002-4/14 $31.00 © 2014 IEEE

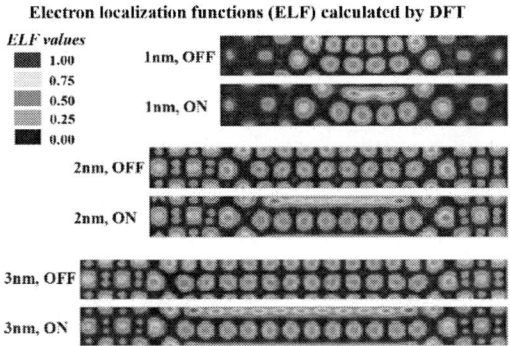

Electron localization functions (ELF) calculated by DFT

ELF values
1.00
0.75
0.50
0.25
0.00

1nm, OFF
1nm, ON
2nm, OFF
2nm, ON
3nm, OFF
3nm, ON

Fig. 2 The electron localization functions (ELF) of the simulated devices, obtained from DFT calculations. A delocalized conductive path connecting the two electrodes can identified in the ON-state devices.

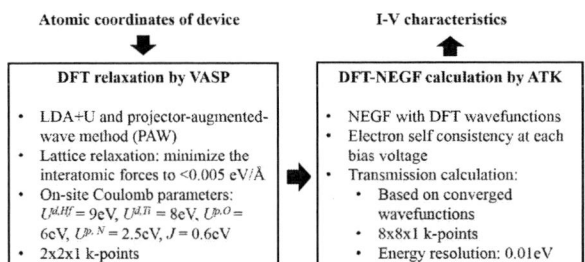

Fig. 3 Simulation flow to obtain I-V characteristics from initial atomic coordinates, through DFT relaxation and DFT-NEGF calculations.

The calculated I-V characteristics (Fig. 4(a-b)) suggest that ON-state devices exhibit Ohmic conduction behaviors, while OFF-state devices show non-Ohmic behaviors, which is typically seen in tunneling/hopping conduction [12]. Fig. 5(a) summarizes the calculated ON- and OFF-state resistances of the devices. The ON-state resistance is linearly dependent on t_{OX}, while the OFF-state resistance is exponentially dependent on t_{OX}. Assuming the filament size is independent of the cell area, the ON/OFF ratio is inversely proportional to the cell area which only affects the OFF-state resistance.

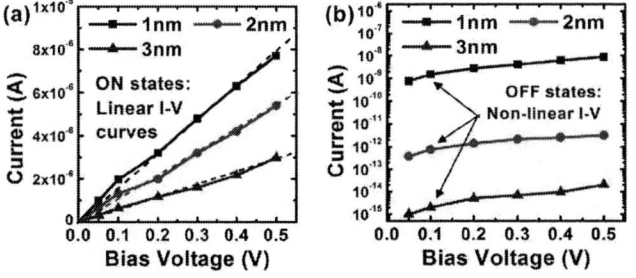

Fig. 4 Simulated I-V characteristics of the HfO$_x$ devices in (a) ON state and (b) OFF state. The linear I-V characteristics in ON state imply Ohmic conduction behaviors. The non-linear I-V characteristics in OFF state imply tunneling/hopping conduction.

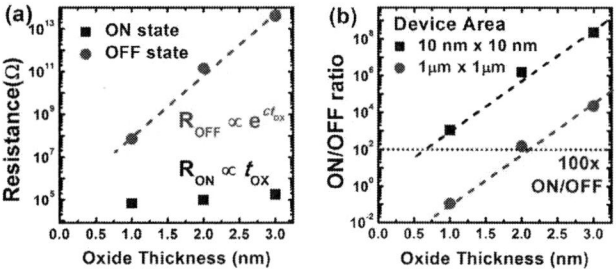

Fig. 5 (a) Cell resistances in both ON and OFF states, simulated by DFT-NEGF. (b) Estimation of the ON/OFF ratios of different device areas, assuming the CF diameter to be 10nm.

The ON/OFF ratios of different cell areas were calculated based on the approximated CF diameter of 10nm (Fig. 5(b)). These results suggest that 100x ON/OFF ratio is feasible using 2nm HfO$_x$, even for relatively large device area.

III. Experimental Results and Discussion

A. 2nm-HfO$_x$ planar device

TiN/ALD-HfO$_x$/Pt devices were fabricated with 1x1μm^2 cross-point area. The detailed fabrication processes can be found in [13]. Fig. 6(a) shows the cross-sectional TEM image of the fabricated RRAM stack. Fig. 6(b) shows the scaling down of forming voltages as a function of HfO$_x$ thickness.

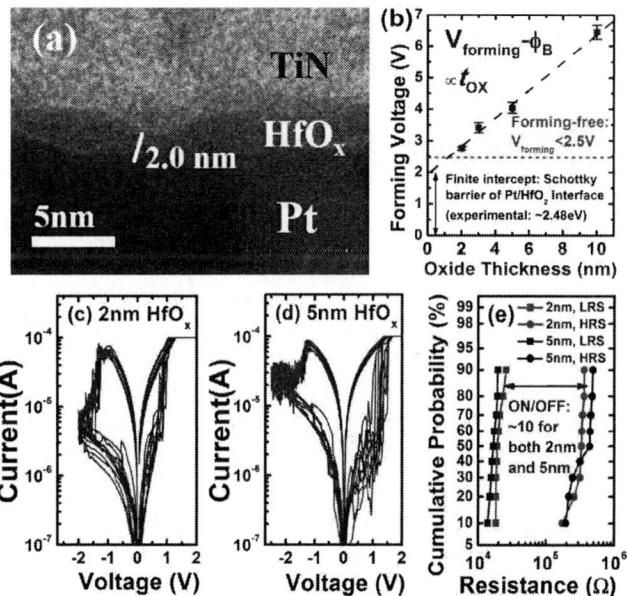

Fig. 6 (a) Cross-sectional TEM image of the planar 2nm-HfO$_x$ devices with 1x1μm^2 area. (b) Forming voltages of 1x1μm^2 HfO$_x$ RRAM, plotted as a function of oxide thickness. 2nm-HfO$_x$ devices are almost forming-free. (c-d) DC sweep characteristics of (c) 2nm- and (d) 5nm-HfO$_x$ planar devices. 10 DC cycles are plotted for each device and similar DC characteristics can be observed. (e) Resistance distributions of 2nm- and 5nm-HfOx devices.

978-1-4799-8002-4/14 $31.00 © 2014 IEEE

Fig. 6(c-d) compares the DC sweep characteristics of a 2nm- and a 5nm-HfO$_x$ device. Fig. 6(e) summarizes the resistance distributions of the two devices by DC sweep. Except for the forming voltages, 2nm- and 5nm-HfO$_x$ devices exhibit similar DC characteristics, which suggests the effective switching thickness to be potentially less than 2nm.

B. Bilayer stacks with 2nm HfO$_x$

To further confirm that 2nm of HfO$_x$ is sufficient for resistive switching, the excess HfO$_x$ thickness is replaced with 2nm of different ALD-based oxides (TiO$_2$ and Al$_2$O$_3$). Fig. 7(a-b) shows the DC sweep curves of the Al$_2$O$_3$- and TiO$_2$-stacked bi-layer devices. Fig. 7(c) compares their resistance distributions with the 2nm-HfO$_x$ device. The characteristics of Al$_2$O$_3$ stack are similar to that of 2nm HfO$_x$. TiO$_2$ stack has higher resistances in both ON and OFF states, but retains a similar ON/OFF ratio. These results confirm the 2nm-HfO$_x$ as a fundamental resistive switching element, while TiO$_2$ layer can be considered as a potential selector to suppress the sneak-path currents in high-density RRAM arrays [3].

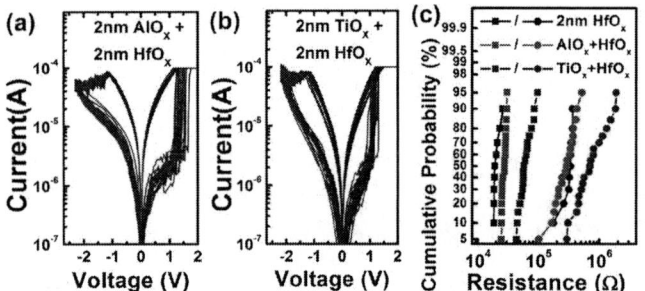

Fig. 7 (a-b) Comparison of DC sweep characteristics of (a) 2nm AlO$_x$ + 2nm HfO$_x$ and (b) 2nm TiO$_x$ + 2nm HfO$_x$ bilayer devices. 20 DC cycles are plotted for each device. (c) Comparison of the resistance distributions of the bilayer devices to that of the single-layer (2nm-HfO$_x$) device.

C. 3D vertical integration of 2nm HfO$_x$

Recently, the diameter of metal pillars has been identified as a limiting factor for the scaling of 3D vertical RRAM (VRRAM). A core-shell structure was proposed to enlarge array size [14]. The role of 2nm HfO$_x$ may provide an approach to further extend VRRAM's scaling limit. As shown in Fig. 8(a), the feature size of the vertical pillars is equal to the sum of metal-pillar diameter and twice the oxide thickness. At 15nm feature size, a reduction of HfO$_x$

thickness from 5nm to 2nm means an increase of metal-pillar diameter from 5nm to 11nm, which significantly lessens the impact of interconnect resistances. In order to evaluate the array scalability, SPICE circuits were built for 32-layer 3D VRRAM arrays to simulate the write access voltage (V_{DD} = 3V, [14]). The array size at which the access voltage drops below 1.8V is defined as the maximum array size. The simulation results suggest that the scaling of HfO$_x$ thickness greatly enlarges the total number of bits allowed in a VRRAM array (Fig. 8(b)).

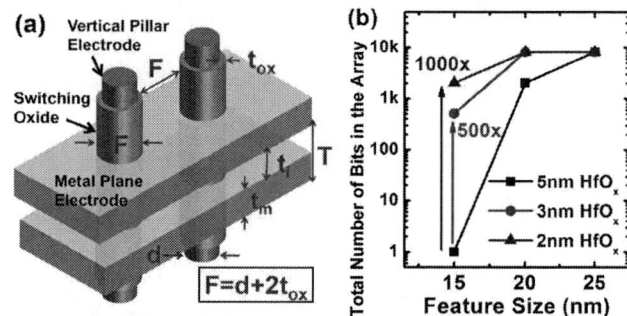

Fig. 8 (a) Schematic of the 3D structure and the dimension parameters of 3D VRRAM array. The feature size is affected by both pillar diameter and t_{OX}. (b) Comparison of the maximum array sizes achieved by using 5nm, 3nm or 2nm HfO$_x$, determined by the SPICE simulations [14]. As F scales down to 15 nm, the maximum array size can be increased up to 1000×/500× using 2nm/3nm HfO$_x$ instead of 5nm HfO$_x$.

The detailed fabrication process of 3D VRRAM can be found in [3], with the 5nm HfO$_x$ replaced by 2nm layer in this work. Fig. 9(a) shows the cross-sectional TEM image of the fabricated 2nm-HfO$_x$ VRRAM. Fig. 9(b) shows the I-V characteristics of 100 DC cycles of the as-fabricated device.

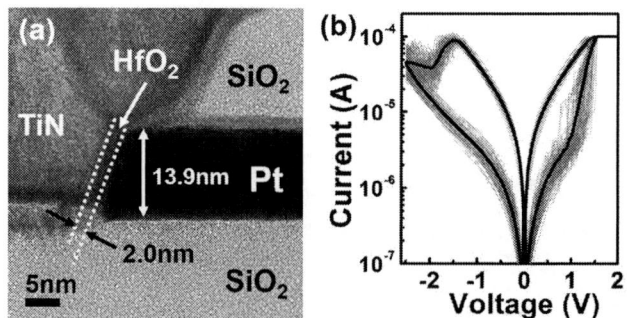

Fig. 9 (a) Cross-sectional TEM image of the fabricated 3D vertical RRAM (VRRAM) device with 2nm HfOX as the switching layer. (b) DC sweep characteristics of the 2nm-HfOx VRRAM. The black solid line represents the average values of 100 DC cycles.

Fig. 10(a) demonstrates 1000 cycles of pulse switching, using a pulse-verify scheme [7] to achieve the 100x ON/OFF ratio predicted by simulations (Fig. 10(b)).

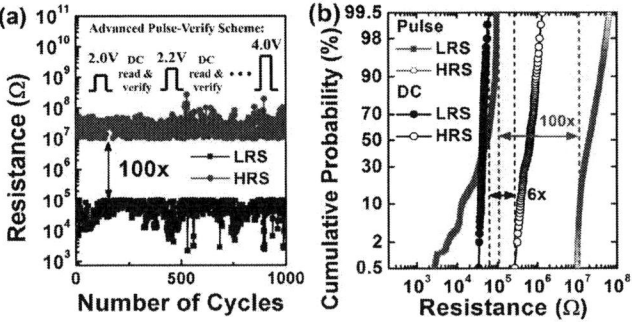

Fig. 10 (a) Pulse switching of the 2nm-HfO$_x$ device for 1000 cycles, achieved by the advanced pulse-verify scheme (inset, [7]). (b) Distributions of LRS and HRS resistances achieved by the two switching methods. Pulse switching helps to expand the ON/OFF ratio to 100.

Fig. 11(a) shows the endurance of ~10^8 cycles with an ON/OFF ratio of 20 measured in 2nm-HfO$_x$ VRRAM devices. Fig. 11(b) shows the retention characteristics, which can be extrapolated to almost 10 years at room temperature. Overall, the performances of 2nm-HfO$_x$ devices integrated in 3D vertical architecture are improved from planar devices.

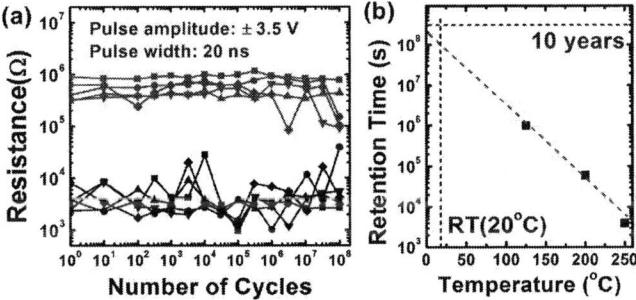

Fig. 11 (a) Endurance of five 2nm-HfO$_x$ VRRAM devices, measured up to 10^8 switching cycles. (b) Retention time at different temperatures, defined as the time at which ON/OFF<10.

The improved performance of ultrathin-HfO$_x$ VRRAM compared to planar devices can be partially attributed to the reduced surface roughness of the bottom electrode. Fig. 12(a) depicts the AFM images of the e-beam evaporated Pt bottom electrode, which shows the RMS roughness to be around 2nm. The rough surface causes fluctuation of the effective oxide thickness and increases cycle-to-cycle variations. In the case

of 3D VRRAM, the surface roughness can be neglected due to the small vertical dimension, thus leading to better electrical characteristics (Fig. 12(b)).

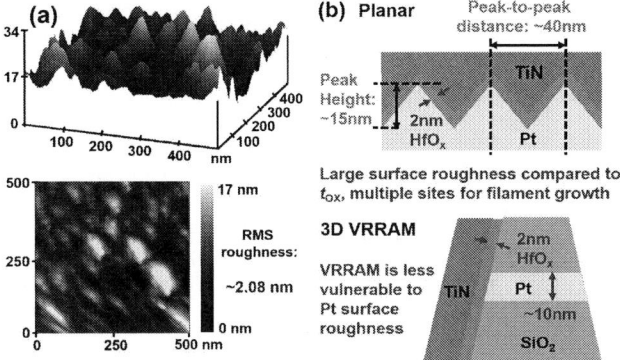

Fig. 12 (a) AFM images of the Pt electrode surface of planar devices, deposited by e-beam evaporation. (b) Schematic demonstration of the advantages of integrating thin HfO$_x$ in VRRAM architecture.

IV. Conclusion

Key achievements: (1) The thickness scaling limit of HfO$_x$ switching layer is theoretically estimated using DFT-based NEGF calculations. (2) The fabricated 2nm-HfO$_x$ devices exhibit similar switching characteristics to reference devices with thicker HfO$_x$, as well as structures stacked with TiO$_2$ or Al$_2$O$_3$ layer. 2nm HfO$_x$ is the fundamental switching element. (3) 2nm-HfO$_x$ devices were integrated into 3D vertical cell structure and demonstrated promising ON/OFF ratio (~100), switching speed (~20ns), endurance (10^8), and data retention (~10 years @ RT) for high-density vertical RRAM arrays.

References:
(1) H.-S. P. Wong et al., Proc. IEEE 6, 1951-1970(2012).
(2) B.Govoreanu et al., IEDM Tech. Dig., 729-732, 2011.
(3) H.-Y. Chen et al. IEDM Tech. Dig., 497-500, 2012.
(4) S. Yu et al., VLSI Symp., T158-T159, 2013.
(5) S. Yu, et al., ACS Nano. 7, 2320-2325(2013).
(6) H.-Y. Chen et al., VLSI Symp., 2014.
(7) B. Govoreanu et al., IMW, 48-51, 2013.
(8) L. Zhao et al., Nanoscale 6, 5698-5702(2014).
(9) O. Pirrotta et al., SISPAD 2014.
(10) G. Kresse et al., Phys. Rev. B 47, 558-561(1993).
(11) Atomistix ToolKit version 13.8, QuantumWise A/S.
(12) S. Yu et al., APL 99, 063507 (2011).
(13) S. Yu et al., VLSI-TSA, 2011.
(14) Y. Deng et al., IEDM Tech. Dig., 629-632, 2013.

3D-stackable Crossbar Resistive Memory based on Field Assisted Superlinear Threshold (FAST) Selector

Sung Hyun Jo, Tanmay Kumar, Sundar Narayanan, Wei D. Lu and Hagop Nazarian

Crossbar Inc. 3200 Patrick Henry Dr. Suite 110, Santa Clara, CA 95054
Tel: (408) 884-0281, Fax: (408) 884-0283, Email: sung-hyun.jo@crossbar-inc.com

Abstract

We report the integration of 3D-stackable 1S1R passive crossbar RRAM arrays utilizing a Field Assisted Superlinear Threshold (FAST) selector. The sneak path issue in crossbar memory integration has been solved using the highest reported selectivity of 10^{10}. Excellent selector performance is presented such as extremely sharp switching slope of < 5mV/dec., selectivity of 10^{10}, sub-50ns operations, > 100M endurance and processing temperature less than 300°C. Measurements on the 4Mb 1S1R crossbar array show that the sneak current is suppressed below 0.1nA, while maintaining 10^2 memory on/off ratio and > 10^6 selectivity during cycling, enabling high density memory applications.

Introduction

For ultra-high density nonvolatile memory applications (>1Tb), 3D-stackable 1TnR or 1S1R structures is needed. Suppressing the leakage (sneak) current in 1TnR or 1S1R crossbar arrays has been a main challenge for high density RRAM development. Various selector devices such as tunneling diode [1], bidirectional varistor [2], MIEC [3] and Ovonic threshold switch [4] have been proposed. Key requirements of selectors include high selectivity (ΔI @V_R, $1/2V_R$), steep turn on slope, high current density, fast turn on and recovery and high endurance. Previous reported selector devices showed selectivity of $150 \sim 10^5$ and turn on slope of $60 \sim 450$mV/Dec. In this work, we present a FAST selector with selectivity of $\sim 10^{10}$, turn on slope < 5mV/dec, fast turn on and recovery (<50ns). Furthermore, 4Mb 1S1R RRAM arrays are demonstrated with large memory on/off ratio and selectivity.

Selector Device

The FAST selector utilizes a superlinear threshold layer (STL) in which a conduction path is formed at the threshold

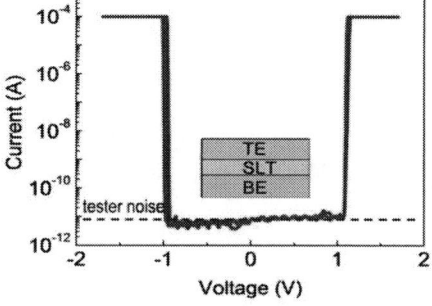

Fig. 1. I-V characteristics of FAST selectors (100nm x 100nm). The device shows bidirectional threshold switching with larger than 10^7 on/off ratio (test-limit).

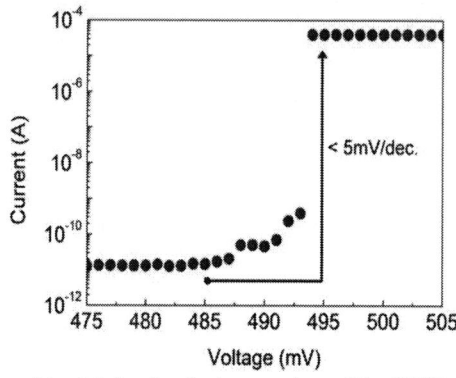

Fig. 2. Zoomed-in plot showing the turn on slope of the FAST selector. The device showed extremely sharp turn on slope of less than 5mV/dec.

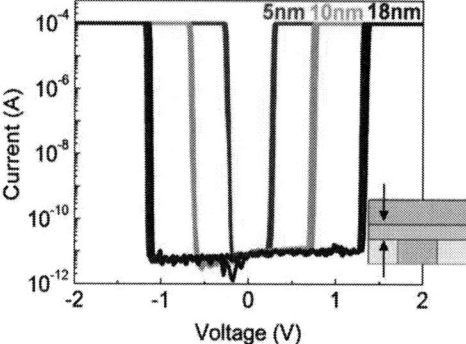

Fig. 3. The threshold voltages can be tuned by controlling the SLT layer thickness.

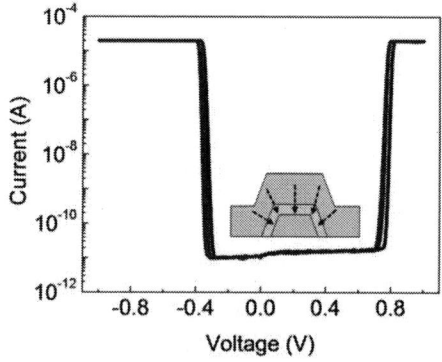

Fig. 4. Asymmetric threshold voltages can be achieved by modulating the device structure (by the modulation of electric field).

electric field. The device provides bidirectional volatile switching with large resistance ratio, high turn on current and steep turn on slope, as shown in Fig. 1. The measured selectivity for the 100nm x 100nm device is > 10^7, and is limited by the test setup. The switching slope is extremely sharp and is less than 5mV/dec. (Fig. 2), which is

978-1-4799-8002-4/14 $31.00 © 2014 IEEE 160

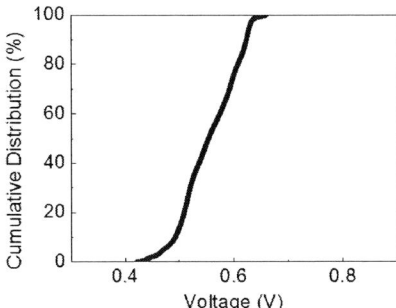

Fig. 5. Threshold voltage (V_{TH}) distribution of FAST selectors within a wafer. Target V_{TH}: 0.6V.

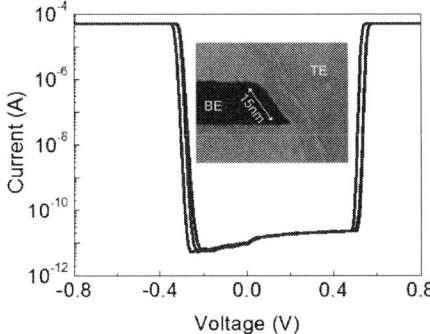

Fig. 6. Reliable switching was still maintained in 15nm x 100nm devices.

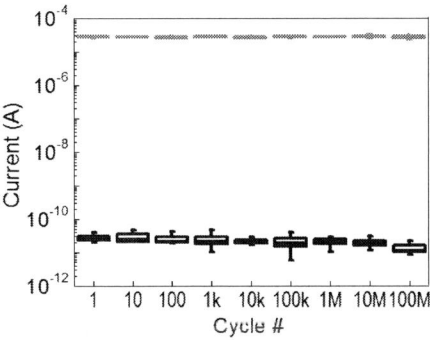

Fig. 7. Cycling test of FAST selectors. The selectors can be reliably cycled over 100M cycles while maintaining > 10^6 on/off ratio (test limited).

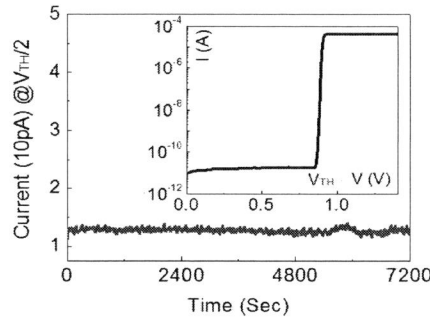

Fig. 8. DC stress test. The device did not turn-on at $0.5V_{TH}$ during two hour-long DC stress. Inset: DC switching characteristic.

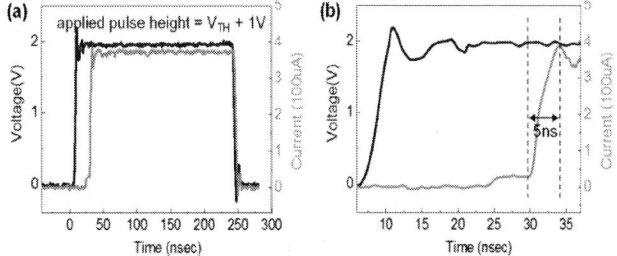

Fig. 9. Switching speed of a FAST selector. (a) The selector can be turned-on within 30ns in response to a 2V pulse. (b) Zoomed-in figure of (a). Off-to-on transition time is about 5ns (tester limited) for over 300uA of passing current through the selector.

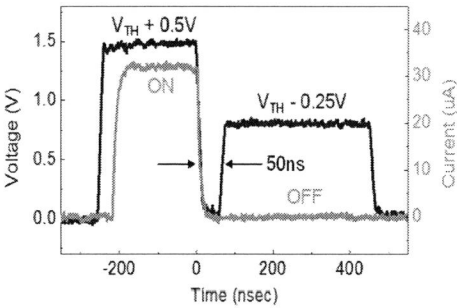

Fig. 10. The selector quickly relaxes to the off state in less than 50ns.

beneficial to array level operations (e.g. larger read voltage margin, faster read). The threshold voltages (V_{TH}) of the selector can be tuned by controlling either the STL thickness or device structure (Figs. 3-4). Fig. 5 shows good device uniformity and tight V_{TH} distribution. Reliable switching can be maintained in 15nm x 100nm devices (Fig. 6) with current density > $5 \times 10^6 A/cm^2$. The selector can reliably switch over 100M cycles (Fig. 7). To test disturb at half-selected bias, a constant DC stress at $0.5V_{TH}$ was applied for a two hour-long period, which did not turn-on the device (Fig. 8). For voltage above V_{TH}, the switching speed is faster than 50ns with the on-off transition time less than 5ns (Fig. 9). The device can quickly recover to the off state once the voltage is removed, with a recovery time < 50ns (Fig. 10). Once a device switches to the on-state, a much smaller hold voltage (V_H) is required to deliver a target current I_P (Fig. 11 (a)). V_H increases as I_P increases, but it was found to be independent of V_{TH} (Fig. 11 (b)-(d)). The small (< 0.3V for 200uA) V_H and very large off-state resistance minimize the voltage overhead when integrated with RRAM.

Array Integration

The FAST selectors have been integrated to a passive crossbar array (Fig. 12). Even for a 4Mb crossbar array, the sneak current has been suppressed below 0.1nA at both 25°C and 125°C, demonstrating very high device yield and low leakage current. We have also successfully integrated the FAST selectors with RRAM in passive crossbar 1S1R arrays.

978-1-4799-8002-4/14 $31.00 © 2014 IEEE

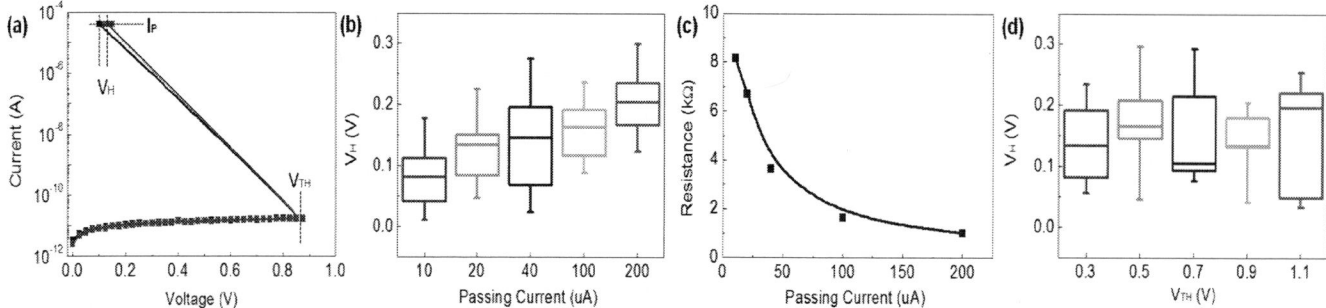

Fig. 11. Hold voltage (V_H) characteristics of FAST selectors. (a) Once a device switches on, a small V_H is required for passing a specific target current (passing current I_P). (b) I_P vs. V_H. (c) Median on resistances vs. I_P. (d) V_{TH} vs. V_H.

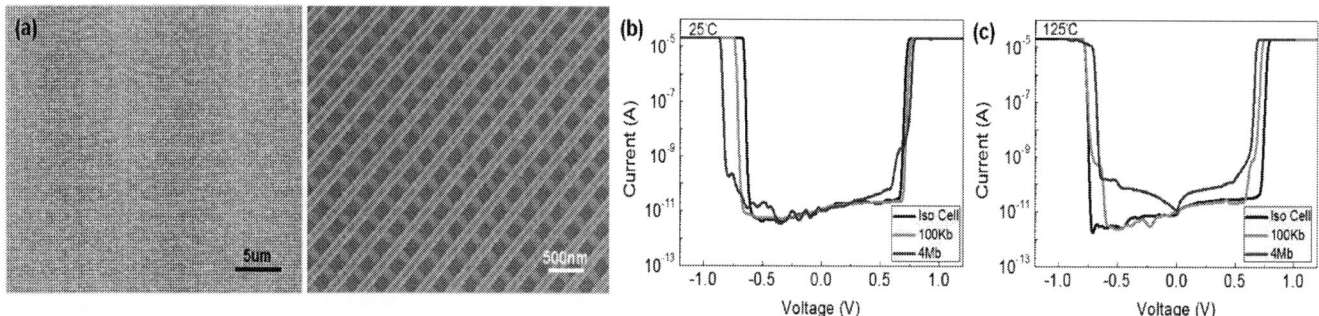

Fig. 12. Passive crossbar array integration of FAST selectors (single cell to 4Mb array). (a) Images of the fabricated selectors. (b) I-V characteristics of the integrated selectors (isolated single cell, cells in 100Kb and 4Mb arrays) at 25°C. (c) I-V at 125°C.

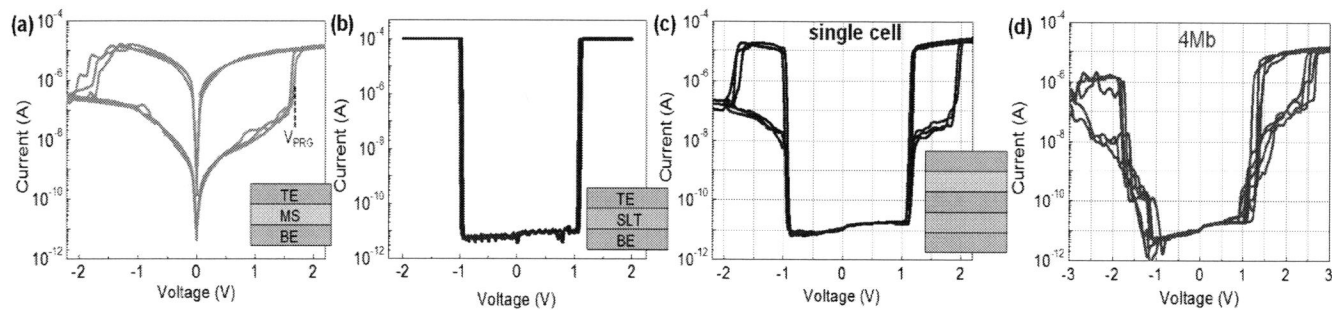

Fig. 13. Passive crossbar integration of RRAM devices with FAST selectors. I-V characteristics of a single cell level (a) RRAM, (b) selector, and (c) integrated 1S1R device. (d) I-V characteristics of a 4Mb passive crossbar array based on 1S1R.

For large array integration, we developed forming-free low current ($\leq$ 20uA) RRAM cells (Fig. 13 (a)) to minimize IR drop and power consumption, and we designed selectors of which V_{TH} is larger than $0.5V_{PRG}$ but smaller than V_{PRG} (Fig. 13 (b)) of the RRAM to suppress sneak currents during both the program and the read operations. The integrated 1S1R device shows > 10^2 memory on/off ratio and > 10^6 selectivity (Fig. 13 (c)). The device operations can be also maintained for the 4Mb 1S1R crossbar arrays (Fig. 13 (d)), which is the largest array size demonstrated to date in a passive crossbar structure. The integrated device can reliably switch more than

100K cycles while maintaining the large memory on/off and selectivity (Fig. 14). To extract the intrinsic leakage current of an individual selector, leakage current through an entire 40Kb selector array was measured (Fig. 15 (a)). The extracted selectivity is found to be 10^{10} (@100nm device). Fig. 15 (a) also shows that there is no single shorted selector device within the 40Kb array. By using the same test method, we calculated the selectivity for different device areas (Fig. 15 (b)). Circuit simulations showed that the selectivity larger than 10^5 is required to design Mb level passive crossbars (Fig. 16) and our FAST selectors surpass the requirement.

978-1-4799-8002-4/14 $31.00 © 2014 IEEE   162

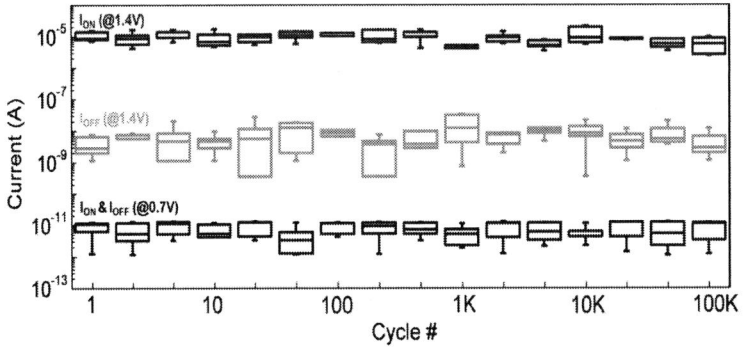
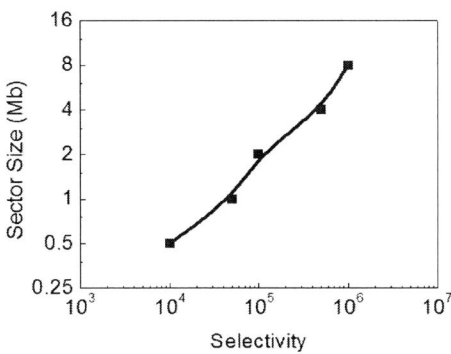

Fig. 14. Cycling demonstration of 1S1R devices. On, off states and half-selected currents are shown. The integrated 1S1R devices maintained $> 10^2$ memory on/off ratio and $> 10^6$ selectivity during the cycling.

Fig. 16. Practical passive crossbar sector size vs. selectivity. Assumed concurrent 2kb program with Icc max = 50mA.

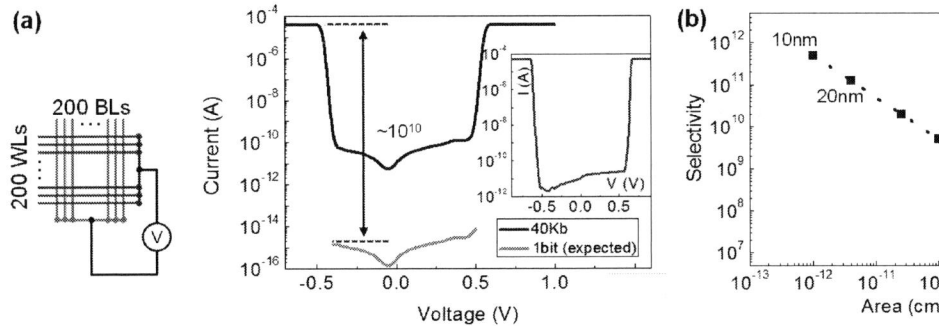

Fig. 15. Leakage current test of selectors and the projected selectivity. (a) Leakage current through entire 40Kb devices and projected 1bit (100nm x 100nm) leakage current. Inset: Typical I-V of a selector on the same wafer. (b) The selectivity vs. device area based on the leakage current measurement.

Table 1. FAST selector summary

Key Parameters	Performance
Selectivity (ΔI @V_R, 1/2V_R)	~10^{10} (@100nm) $\geq 10^{11}$ (@20nm, projected)
Endurance	$> 10^8$
Voltage overhead (Hold voltage)	$< 0.3V$ for passing 200uA
Switching slope	$< 5mV/dec.$
Max. current density	$> 5 \times 10^6$ A/cm^2
Processing Temp.	$< 300°C$

Conclusion

As summarized in Table 1, FAST selectors offer excellent performance metrics such as the largest reported selectivity (10^{10}) to date, steep slope and fast turn on/recovery for high density memory applications. Functional 4Mb passive crossbar RRAM arrays have been demonstrated based on the FAST selectors.

References

[1] A. Kawahara et al., IEEE ISSCC, 25.6, 2012

[2] W. Lee et al., VLSI Tech. Symp., 37-38, 2012

[3] K. Virwani et al., IEDM, 36-39, 2012

[4] S. Kim et al., VLSI Tech. Symp., T240-241, 2013

High-Drive Current ($>1MA/cm^2$) and Highly Nonlinear ($>10^3$) TiN/Amorphous-Silicon/TiN Scalable Bidirectional Selector with Excellent Reliability and Its Variability Impact on the 1S1R Array Performance

Leqi Zhang[*], Bogdan Govoreanu[#], Augusto Redolfi, Davide Crotti, Hubert Hody, Vasile Paraschiv[**],
Stefan Cosemans, Christoph Adelmann, Thomas Witters, Sergiu Clima, Yang-Yin Chen,
Paul Hendrickx, Dirk J. Wouters[*], Guido Groeseneken[*], Malgorzata Jurczak

imec, Kapeldreef 75, B3001, Leuven, Belgium; [*]also with Dept. of Electrical Engineering,
KU Leuven, Belgium; [**]also with SC Etch Tech Solutions, Iasi, Romania; [#]E-mail: govorean@imec.be

Abstract

An optimized TiN/amorphous-Silicon/TiN (MSM) two-terminal bidirectional selector is proposed for high density RRAM arrays. The devices show superior performance with high drive current exceeding $1MA/cm^2$ and half-bias nonlinearity of 1500. Excellent reliability is fully demonstrated on 40nm-size crossbar structures, with statistical ability to withstand bipolar cycling of over 10^6 cycles at drive current conditions and thermal stability of device operation exceeding 3hours at 125^0C. Furthermore, for the first time, we address the impact of selector variability in a 1S1R memory array, by including circuit simulations in a Monte Carlo loop and point out the importance of selector variability for the low resistive state and its implications on the read margin and power consumption.

Introduction

Although several reported resistive switching (RS) structures [1-3] showed excellent memory characteristics, implementation of high-density Resistive RAM (RRAM) arrays lags behind, mainly due to the sneak currents [4] issues. To suppress the leakage paths, a non-linear two-terminal selection device is required to connect serially with each RS element forming a 1S1R cell (Fig.1). In contrast to a conventional metal-silicon-metal (MSM) structure with doped-Si, acting as a back-to-back Schottky diode (Fig.2a), we recently proposed [5] an MSM RRAM selector using an ultrathin, undoped amorphous-Silicon (a-Si) layer, deposited by a PVD process, which behaves as a low bandgap insulator and provides bidirectional rectification by tunneling conduction (Fig.2b).

In this work, we propose an optimized MSM structure, performance improvement of which is achieved by both device and process engineering. In contrast to other selector concepts [6-9], this device has the advantages of using baseline CMOS process steps, manageable thermal budget and moreover it is scaling-friendly, projecting improved current drive and nonlinearity towards 10nm-size. Thanks to the amorphous nature of the Si film, the device performance is not affected by doping or poly-crystallinity induced fluctuations.

Furthermore, we address the impact of cell variability on 1S1R crosspoint arrays by employing circuit simulations. The impact of different variability sources is studied by using random array pattern generation with "variability injection", where specific numerical techniques are used to induce individual/multiple variability sources in a Monte Carlo array simulation loop.

Selector structure & Process

The MSM structure is implemented in a crossbar process [1], with device sizes ranging from $1um^2$ down to $40x40nm^2$. An in-situ PVD deposition of the a-Si/TiN top electrode stack eliminates formation of thin interfacial oxides. After passivation, featuring a thermal budget of $370^\circ C$, an additional N_2 annealing for up to $600^\circ C$, for 5min, is applied, which reduces the trap density in the Si layer without affecting its amorphous nature (Fig.3). Furthermore, careful a-Si thickness tuning is pursued to find the best balance between the voltage operation range and the current levels, for improving the current-voltage nonlinearity.

Electrical performance & Reliability

A. DC performance

Current-Voltage (IV) characteristics for devices with different a-Si thickness show that a 5min N_2 annealing for up to $600^\circ C$ improves the nonlinearity without affecting the maximum drive current (J_{max}), at any a-Si layer thickness (Fig.4). High current density of over $1MA/cm^2$ at 3.6V and half-bias nonlinearity ($NL_{1/2}$) of about 1500 are achieved for 15nm a-Si annealed devices. 8nm a-Si thickness annealed devices achieve maximum current drive ($>1MA/cm^2$) at 2.6V with a $NL_{1/2}$ of about 240. The low bias leakage improvement increases with the annealing temperature, with a tendency to saturate at about $500^\circ C$. This decreasing of low bias leakage current leads to NL improvement (Fig.5). Significant NL improvement and better J_{max}-NL tradeoff are achieved by combined annealing-thickness effect (Fig.6). 15nm a-Si devices display a larger voltage operation range, which further improves the half-bias NL, compared to 8nm devices. Moreover, larger voltage operation range improves the 1S1R full cell nonlinearity, especially for the write operation [10]. The J_{max}-$NL_{1/2}$ trade-off can be further adjusted by interface (barrier) engineering, e.g. using metal electrodes with a different workfunction. Selector performance is projected to improve by scaling. The breakdown (BD) voltage extrapolates towards >4V at 10nm-size, leaving a margin of about 0.4V for reaching a $1MA/cm^2$ drive current (which enables $1\mu A$ switching current for a 10nm-size RS structure). Alternatively, J_{max} extrapolates towards $10MA/cm^2$, for a 10nm-size selector (Fig.7), which may relax the requirement on the maximum switching current the selector is able to withstand.

Selector IV's show weak temperature dependence at high bias, consistent with a tunneling-based dominant conduction mechanism. Furthermore, temperature dependence is stronger at low bias (Fig.8) and slightly increases with a-Si thickness,

978-1-4799-8002-4/14 $31.00 © 2014 IEEE

indicating a different dominant conduction mechanism. A trap-assisted tunneling (TAT) model [11] is able to explain the low-bias IV characteristics for both annealed and unannealed samples **(Fig.9)**. The TAT model fits experimental data well, in the low bias region, with the same set of parameters, except for the extracted trap density, which decreased by a factor of ~5 by annealing. Furthermore, at higher biases, a defect-free tunneling model [12] explains the experimental data, for both annealed and unannealed samples.

B. Reliability

The devices show excellent stability under thermal stress **(Fig.10)**. Baking test at 125^0C for 3 hours shows no degradation of cell characteristics, indicating that temperature has a limited impact on the device performance. Negligible degradation is observed with AC stress after 10^6 cycles at $\pm3.6V/100ns$ **(Fig.11)**. Furthermore, to mimic practical use of selector in memory circuits, i.e. alternation of program and read cycles, a large read voltage (+3V/100us) is applied after each AC stress cycle, resulting in limited degradation across the whole voltage range **(Fig.12)**, after up to 10^6 cycles.

1S1R array performance & Variability impact

A. Variability-free analysis

Array-level simulations [10] **(Fig.13)** are carried out to assess the MSM selector performance in 1S1R crosspoint arrays. A generic model with fixed parameters is employed to describe the behavior of a RS element, in terms of resistance, i.e. Low Resistive State (LRS) or High Resistive State (HRS), voltage and current levels **(Fig.14)**. The full cell IV behavior is constructed by serially combining the characteristics of 1R (from the model) and 1S (from the experiment data on 15nm a-Si MSM of $40x40nm^2$ cell size) **(Fig.15)**. Calculated read and write array performance using the ideal (variability-free) cell characteristics **(Fig.15)** shows that the access write voltage (AWV) drop remains below 15% **(Fig.16)** and the Read Margin (RM) decreases to just about 10% **(Fig.17)** for an 1Mbit array size, when using a half-bias scheme for both read and write operations. RM during remains the more sensitive parameter to array size increase, compared to AWV drop during the write operation.

B. Impact of cell variability

The achievable array size is ultimately impacted by the cell variability. We introduce a variability-aware array performance assessment methodology, accounting for data pattern randomness [13], as well as for the selector and resistive element variability. Individual, as well as combined impact of cell variability on the array performance indicators (e.g. RM) is enabled by simulation methodology to identify the variability contributors in the array **(Fig.18)**.

1R data patterns have big impact on the read performance when considering 1R-only crosspoint arrays which causes wide spread of readout current [13]. However, the impact is limited for 1S1R arrays, since the resistance of the unselected cells is dominated by the selector, irrespective of the 1R state (i.e. LRS/HRS) in the unselected cells, therefore tight readout current distribution is observed, with variability-free 1S and 1R elements **(Fig.19)**.

Selector variability is injected into the full 1S1R cell, while forcing a variability-free 1R element **(Fig.20)**. Larger spread of LRS readout current indicates that the LRS is more sensitive to selector variability compared to HRS since a larger fraction of the voltage drops over the selector element at readout voltage. Randomly generated 1S1R cells with selector-only variability are distributed in arrays with a random data pattern in a Monte Carlo loop. This allows to calculate the current distribution. CDF of the current shows that the variability from the selected cell is dominant in the current spread **(Fig.21a)**. If the variability of the selected cell is disabled, tight distribution is observed **(Fig.21b)** indicating limited impact of the variation from the unselected cells. The extracted RM is affected by selector variability **(Fig.22)** and decreases by ~20% for the worst-case (i.e. tail-to-tail, inset Fig.22), compared to the ideal case **(Fig.17)**. Conversely, 1R-only variability is enabled while assuming an ideal selector **(Fig.23)**. The extracted RM affected by 1R-only variability results in a worst-case decrease of up to 18% **(Fig.24)** compared to the variability-free case. Finally, when all variability sources are activated, the net result is a larger distribution spread of the current in both LRS/HRS states **(Fig.25)**, where the lower LRS readout current tail results in a total penalty on the RM of about 30%, for the worst case. In contrast to RM degradation, the upper tail of the LRS readout current increases the power consumption in the worst case, however, the impact is limited **(Fig.26)**.

By injecting individual and multiple sources of variability in the 1S1R cell, the results show that the selector is an important array variability contributor and mainly affects the LRS readout current, which degrades the RM and increases the power consumption. Hence, an additional margin in selector NL is required in order to address this issue and enable robust, variability-aware 1S1R array design. Moreover, selector with low variation is important, to avoid NL improvement loss due to large cell variability.

Conclusion

We reported a TiN/a-Si/TiN MSM selector featuring an improved performance of >$1MA/cm^2$ current drive and a half-bias NL of ~1500 at 40nm-size, and with projected improvements towards 10nm-size. The fabrication process is simple, using a thin, undoped a-Si layer, thickness of which can be used to adjust the J_{max}-NL selector trade-off. The thermal budget does not exceed 600˚C. The devices show very good immunity to thermal stress and excellent reliability exceeding 10^6cy at $\pm3.6V/100ns$. We introduced a variability-aware array-sizing analysis methodology and identified selector variability as a key factor in 1S1R array RM degradation, with a penalty here of ~30%. Selectors with reduced variability and higher NL are required to enable robust 1S1R cross-point array design.

References

[1] B. Govoreanu, *et al, IEDM Tech. Dig.*, pp. 729, 2011.
[2] Y.B. Kim, *et al, VLSI Symp. Tech. Dig.*, pp.52, 2011.
[3] J. Yi *et al, VLSI Symp. Tech. Dig.*, pp.48, 2011.
[4] M.J. Lee *et al, IEDM Tech. Dig.*, pp.771, 2007.
[5] L. Zhang *et al, El. Dev. Lett.*, 35, pp.199, 2014.
[6] S. Kim *et al, VLSI Symp. Tech. Dig.*, pp.155, 2012.
[7] K. Virwani, *et al, IEDM Tech. Dig.*, pp. 2.7.1, 2012.
[8] W. Lee, *et al, VLSI Symp. Tech. Dig.*, pp.37, 2012.
[9] V.S.S. Srinvasan, *et al, El. Dev. Lett.*, 33, pp. 1396, 2012.
[10] L. Zhang et al, IMW, pp.34, 2014.
[11] B. Govoreanu et al, IEDM Tech. Dig., pp. 479, 2006.
[12] B. Govoreanu et al, Solid-St. Electron., 47, pp. 1045, 2003.
[13] H. Li et al, IMW, pp.30, 2014.

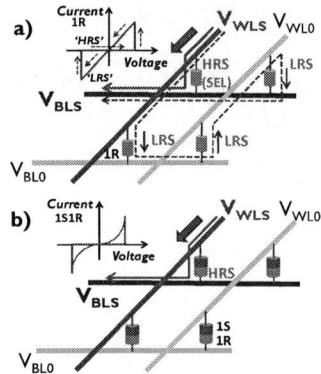
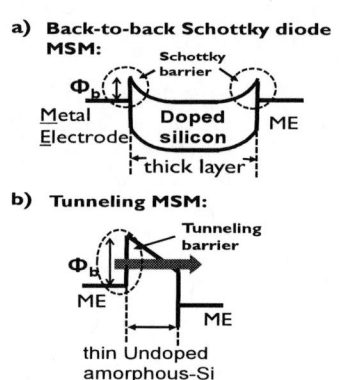
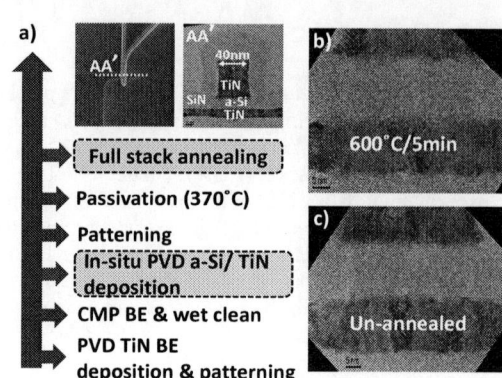

Fig.1: a) Sneak currents limit the functionality of crossbar arrays with (nearly) linear resistive switching elements (1R). b) Leakage can be suppressed by serially inserting a nonlinear selection element (1S), to form a 1S1R cell configuration.

Fig.2: a) Band diagram of the conventional MSM selector (at no bias), with doped Si, rectifying as a back-to-back Schottky diode. b) Band diagram of the new MSM selector (under bias), with ultrathin undoped a-Si, which behaves as a low-bandgap tunnel dielectric.

Fig.3: a) Process flow for crossbar MSM selectors, with device sizes ranging from 1um^2 down to 40x40nm^2 In-situ a-Si/TiN deposition process and anneal are the key process steps for performance improvement. b,c) TEM results for unannealed and 600°C-annealed structures. Microstructural analysis did not reveal any morphological difference in the deposited Si layer after annealing, which remained amorphous.

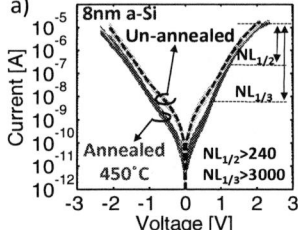
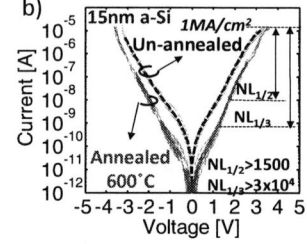
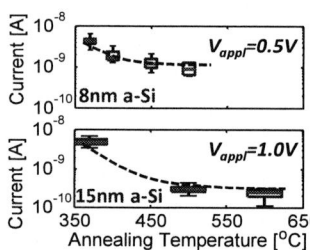
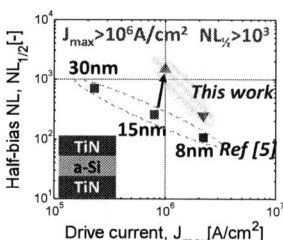

Fig.4: Current-Voltage (IV) plots for a) 8nm and b) 15nm a-Si thickness, MSM devices before and after annealing. Over 1MA/cm^2 current drive is maintained, while NL improves to >1.5*10^3 for 15nm a-Si devices after 600°C/5min/N$_2$ anneal. The anneal effect is clearly visible after a thermal budget of only 450°C for the thinner, 8nm a-Si devices. Data on 40nm-size devices.

Fig.5: Leakage current at low bias for 8nm (0.5V bias, top) and 15nm a-Si (1.0V bias, bottom) MSM selectors for different anneal temperatures (anneal time was 5min). Device size: 40nm.

Fig.6: A selector Figure-of-Merit NL$_{1/2}$-J$_{max}$ plot. Significant NL improvement and *better J-NL trade-off* are achieved by combined anneal-thickness effect. NL$_{1/2}$ is extracted at maximum current drive.

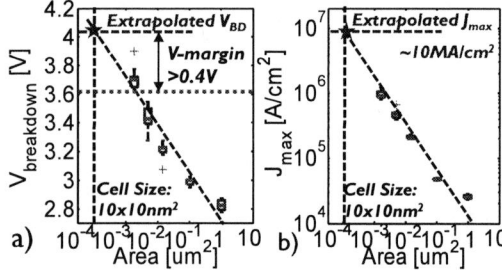
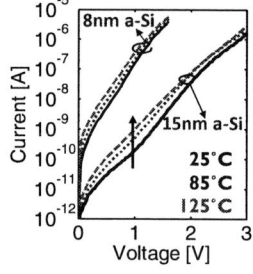
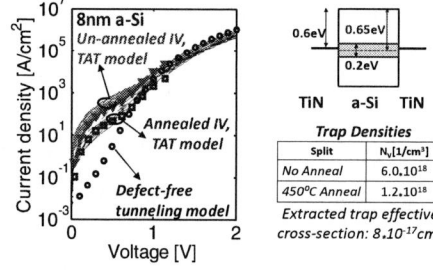

Fig.7: a) Extrapolated margin between the breakdown voltage and voltage for selector reaching the target current (assuming combing the selector with a resistive element switching at 1uA in 10x10nm^2) is larger than 0.4V. b) The estimated J$_{max}$ exceeds 10MA/cm^2 at 10x10nm^2, J$_{max}$ is limited by the breakdown voltage.

Fig.8: Temperature dependent IV characteristics for 8nm and 15nm a-Si devices, evidencing stronger I(T) variation in the low-bias range, for both splits.

Fig.9: (left): TAT (low-bias) [11] and defect-free tunneling (high bias) [12] conduction models explain experimental IV's for both 450°C-annealed and un-annealed samples. (right): The MSM band diagram, with a 0.65±0.1eV extracted narrow trap band and bulk trap densities corresponding to the 2 splits.

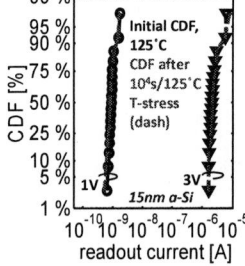
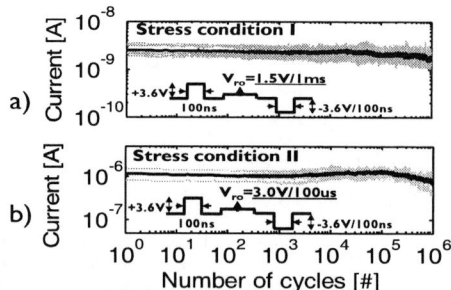
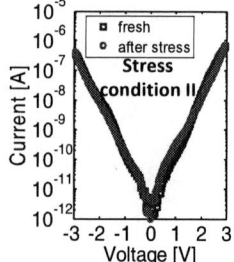

Fig.10: Cell current distribution before (circles) and after (dashed lines) application of a thermal stress of 125°C for 3hours, at different biases. All currents are read at 125°C. Data collected on 40nm devices.

Fig.11 a) Negligible degradation is observed with AC stress after 10^{6}cycles at ±3.6V (V at 1MA/cm^2). b) To mimic practical application operation (read/program), large V$_{read}$ (+3V, equivalent stress 100us*10^6~100s) is applied together with ±3.6V pulse stress.

Fig.12: Comparison of the IV characteristics before and after electrical stress, for the stress conditions II, as defined in Fig. 11.b), showing limited degradation at any bias point.

978-1-4799-8002-4/14 $31.00 © 2014 IEEE

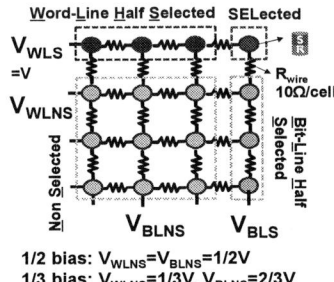

1/2 bias: $V_{WLNS}=V_{BLNS}=1/2V$
1/3 bias: $V_{WLNS}=1/3V$, $V_{BLNS}=2/3V$

Fig.13: Schematics of the crosspoint 1S1R array netlist for circuit performance assessment, used for an initial, worst-case scenario, [10] on a variability-free squared ($N_{BL}=N_{WL}$) array. The blocks show schematically the selected cell, the half-selected bit/word line cells as well as the nonselected cells. The line resistance was assumed 10Ω/cell.

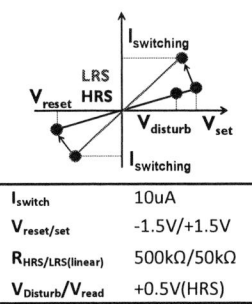

I_{switch}	10uA
$V_{reset/set}$	-1.5V/+1.5V
$R_{HRS/LRS(linear)}$	500kΩ/50kΩ
$V_{Disturb}/V_{read}$	+0.5V(HRS)

Fig.14: Generic model for the 1R element, with switching between Low- and High-Resistive States (LRS/HRS). Switching I/V and resistance levels are representative for HfO₂-based RRAM.

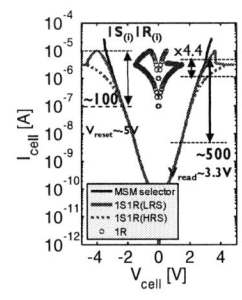

Fig.15: The variability-free IV characteristics of the full 1S1R cell is constructed by serially combining 1R (model-based) and 40nm-size experimental 1S data (15nm a-Si MSM selector).

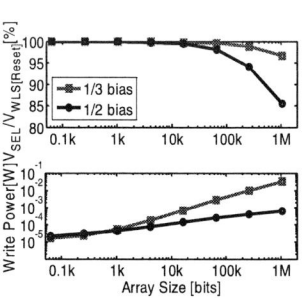

Fig.16: Calculated write voltage drop is less than 15% for 1Mbit array size. Even if this is larger for the ½ bias scheme, this is preferred over ⅓ bias, which has higher power consumption. All calculations assumed 1S1R characteristics as derived from Fig. 15.

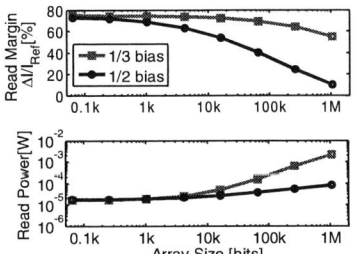

Fig.17: Read Margin (RM) and Read Power vs array size, showing that RM is the most critical parameter for a 1Mbit array, in a ½ bias scheme, which is more convenient for power consumption.

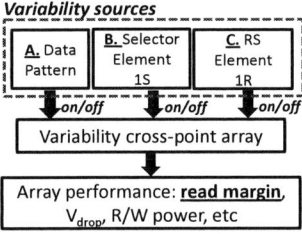

Fig.18: Variability-aware array level analysis methodology. The impact of different sources of variability on the read margin is studied using random array pattern generation and separate (off)/combined (on) 1S, 1R variability injection in a Monte Carlo array simulation loop.

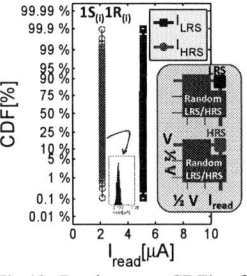

Fig.19: Read current CDF's of ideal $1S_{(i)}1R_{(i)}$ cells, with variability-free 1S and 1R elements, in an array with random data patterns (variability source A in Fig. 18) show marginal current spread.

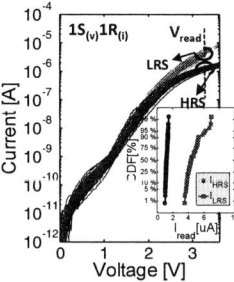

Fig.20: Constructed $1S_{(v)}1R_{(i)}$ cell IV characteristics for both LRS/HRS states, with experimental (variability-affected) selector IV's and ideal (variability-free) 1R characteristics. This enables injecting 1S-only variability into the 1S1R cell, extracted (inset) and further used in array simulations.

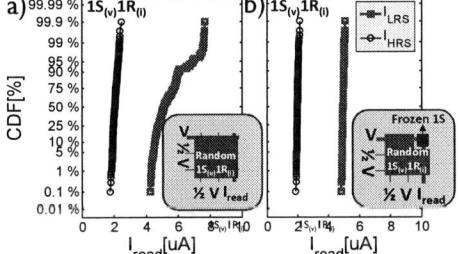

Fig.21: a) current CDF. Including selector variability for all the cells in the array. **b)** If the variability of the selected cell is disabled (right), tight distribution is observed. Data are on a 64x64bit array size. The results remain valid for arrays of up to at least 256x256bit (64kbit) size (see next figure). _**Note:**_ 64kbit is the max. array size used in simulations, due to computing resource limitations.

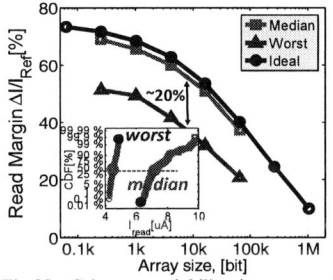

Fig.22: Selector variability impact on the RM, for worst-case (tail-to-tail) and median cell (inset, 64kbit) cases, show about 20% penalty compared to the variability-free, ideal case. Data pattern variability (see Fig.18, 19) is turned on.

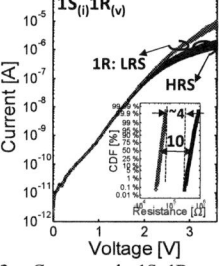

Fig.23: Constructed $1S_{(i)}1R_{(v)}$ cell IV characteristics for both LRS/HRS states, with ideal, variability-free selector IV's and variability-affected 1R characteristics. The 1R features: median 1R as in Fig.14 and tail-to-tail on/off~4.

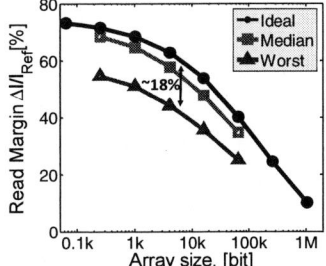

Fig.24: 1R variability impact on the RM, with up to 18% penalty for the worst-case, compared to the ideal case. The data pattern variability is turned on.

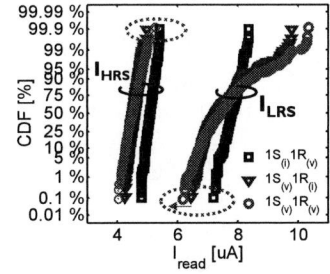

Fig.25: Readout current CDF's for 64kbit arrays enabling 1S-only, 1R-only and both 1S&1R variability sources. Data pattern variability is turned on.

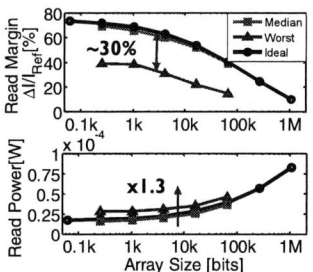

Fig.26: RM (top) and Read power (bottom) vs. array size when all variability sources are turned on. An overall penalty of ~30% for the worst-case variability (calculated at the current tails) is observed, while the impact on read power, although present, is lower.

978-1-4799-8002-4/14 $31.00 © 2014 IEEE   167

Impact of 3D integration on 7nm high mobility channel devices operating in the ballistic regime

W. Guo, M. Choi*, A. Rouhi, V. Moroz*, G. Eneman, J. Mitard, L. Witters, G. Van der Plas, N. Collaert, G. Beyer, P. Absil, A. Thean, E. Beyne

IMEC, Kapeldreef 75, Leuven, Belgium;
* Synopsys, Inc., 700 East Middlefield Road, Mountain View, CA 94043;
Ph.: +32 16 28 36 14 - Email: Wei.Guo@imec.be

Abstract

We report for the first time the impact of 3D IC process induced local thermo-mechanical stress effects on CMOS devices for 7nm technology node (N7) operating in the ballistic regime. We show that the ballistic current is less affected by the uniaxial stress than the drift current. The ballistic current ratio decreases for longer gate, it is 80%, 45% of the total current for respectively 14nm and 40nm gate lengths devices resulting in larger TSV proximity effects for longer devices. 4 point bending measurements for Ge channel p-FinFETs, confirm the simulated stress sensitivities used to provide a physical estimation of the Keep Out Zone (KOZ). For high mobility channel devices, whereas N7 p-FinFETs exhibits similar TSV stress sensitivities, the KOZ for n-FinFETs is a function of the channel material choice for the co-integration. Unstrained Ge n-FinFETs are largely affected by the TSV proximity, 40% more than Si, SiGe and strained Ge n-FinFETs. Materials like Ge and InGaAs have a reduced sensitivity to the vertical stress component whereas the impact is significant in the case of Si and SiGe n-type devices, with more than 10% change in drive current for 400MPa compressive vertical stress.

Introduction

We previously reported that TSV induced Keep Out Zone will be reduced for scaled silicon technologies [1]. However at N7, high-mobility channel materials like SiGe, Ge and III-V in narrow fin FinFET or nanowire gate all around devices are candidates to replace or complement Si devices and deliver the required electrostatic control and speed. These innovations might require different stress level limits for the 3D co-integration (Fig.1) and induce new types of layout dependent effects.

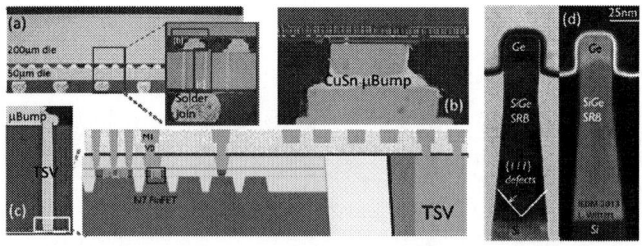

Fig. 1 : (a) 3D 2 tiers (50µm and 200µm thick Si dies) stacking X-section. (b) and (c) devices located at TSV and µbump proximity can be affected by thermal-mechanical stress. (d) for 3D N7 node integration, high mobility channel material FinFET are considered. X-section for Ge FinFET [2].

These new materials present key differences with respect to Si when considering their processing and material properties, especially mechanical stress sensitivities. Notably these new devices are also engineered to operate at much lower supply voltages (VDD =0.7 V) in order to counter balance the GIDL due to the narrower band gap inherent to these materials. Furthermore, the logic core devices with physical gate length Lg<20nm operate well into the ballistic regime.

Device assumptions and ballistic current ratios

For the 7nm node, both 6nm fin width FinFETs with (110)/<110> fin sidewall orientation and 5nm diameter nanowires with 14nm channel length are considered. Si, $Si_{0.5}Ge_{0.5}$ and Ge channels are introduced for n and p type devices and $In_{0.53}Ga_{0.47}As$ is considered for n type devices. The strain levels are shown in Fig.2, for Si, $Si_{0.5}Ge_{0.5}$, and Ge channels it is kept at 1.5 GPa to balance the leakage due to smaller band gap at larger strain levels. This uniaxial strain is applied along the channel direction. For InGaAs channel devices, the strain engineering is not considered.

	Si channel	SiGe channel	Ge channel
NMOS +1.5 GPa	Si / Si / 25% Ge	15% Ge / 50% Ge / 50% Ge	30% Ge / Ge / 75% Ge
PMOS -1.5 GPa	70% Ge / Si / 25% Ge	85% Ge / 50% Ge / 50% Ge	Ge / Ge / 75% Ge

Fig. 2 : To reach the 7nm node power / performance targets, stress engineering is required as a combination of SRB and S/D Epi (±1.5 GPa).

The ballistic current contribution to the total current is studied for both n and p-FinFETs for different channel materials. The ballistic ratio is calculated by using Monte Carlo simulations. Fig.3 shows that ballistic ratio is larger for shorter devices for both n and p-FinFETs. For 14nm gate length, the ballistic current is dominant, the ratio is around 80% for Si, SiGe and Ge. However in linear operation region the drain current is still dominated by the mobility limited current and exhibit larger stress sensitivity as demonstrated using stressed band structure in Sentaurus Band Structure [3].

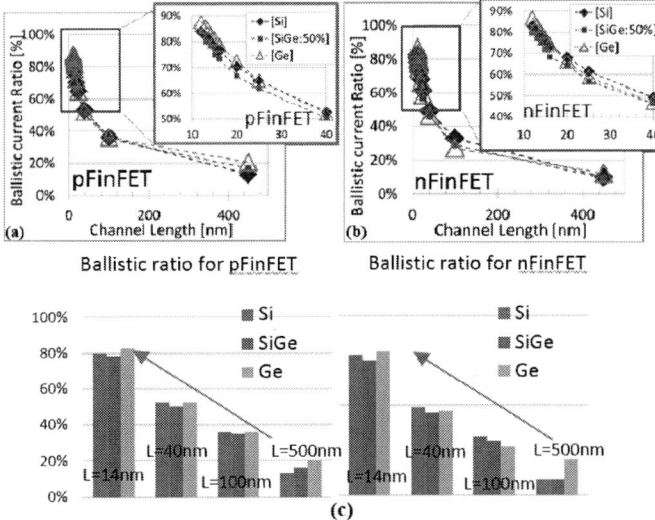

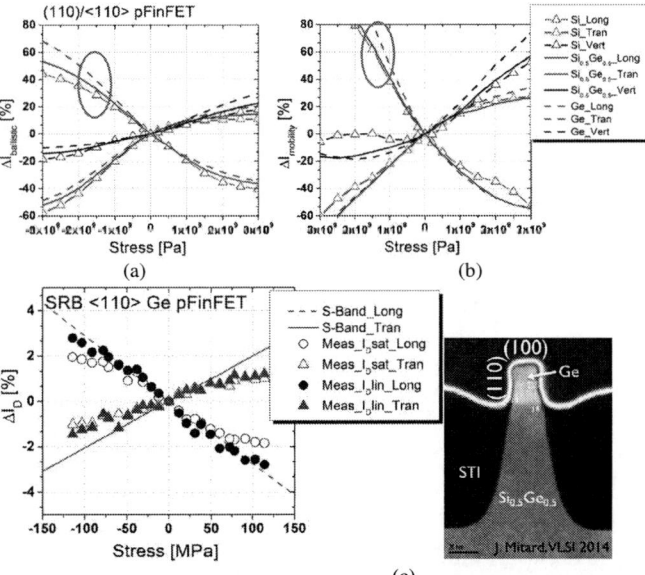

Fig. 3 : (a, b) the ballistic current contribution to total drive current increases when channel length decreases for n and p-type FinFET, W_{fin}=6nm. The ballistic current is calculated at fixed leakage I_{off}=1nA/μm for V_{DD}=0.7V. (c) Summary of ballistic ratio for n and p-FinFET with Si, $Si_{0.5}Ge_{0.5}$ and Ge channel.

Stress responses of 7nm Devices

To study the stress sensitivity on ballistic current and on mobility limited current, a multi-subband, Kubo-Greenwood mobility calculator based on the solution of the 1D/2D Schrödinger equation with non parabolicity effects and 6-band k•p band models was used for n and p type devices, respectively. Fig. 4 (a and b) illustrates the stress sensitivity of ballistic current and mobility limited current to uniaxial stress of p-FinFET with Si, $Si_{0.5}Ge_{0.5}$, and Ge channel materials.

Fig. 4 : S-band physical analysis of ballistic current stress sensitivity and mobility limited current stress sensitivity (a) and (b) respectively for p-FinFET with Si, SiGe and Ge channel material. (c) Good agreement of 4 point bending experiments and simulation in case of Ge channel p-FinFET, the processing detail can be found in [4].

The mobility limited current shows 2 times higher stress sensitivity compared to ballistic current. The blue circles highlight the typical strain engineering range along the channel

direction. The four-point bending experiment is used for verifying the physical analysis results. Fig.4c illustrates good agreement between the simulated stress sensitivity of Ge p-FinFET and experimental results. The processing detail of the devices can be found in [4].

Using this approach, the stress responses of ballistic current and drift current needed for the KOZ estimation are derived and shown in the Fig.5 (a,b,c) for Si, $Si_{0.5}Ge_{0.5}$ and Ge n-FinFET.

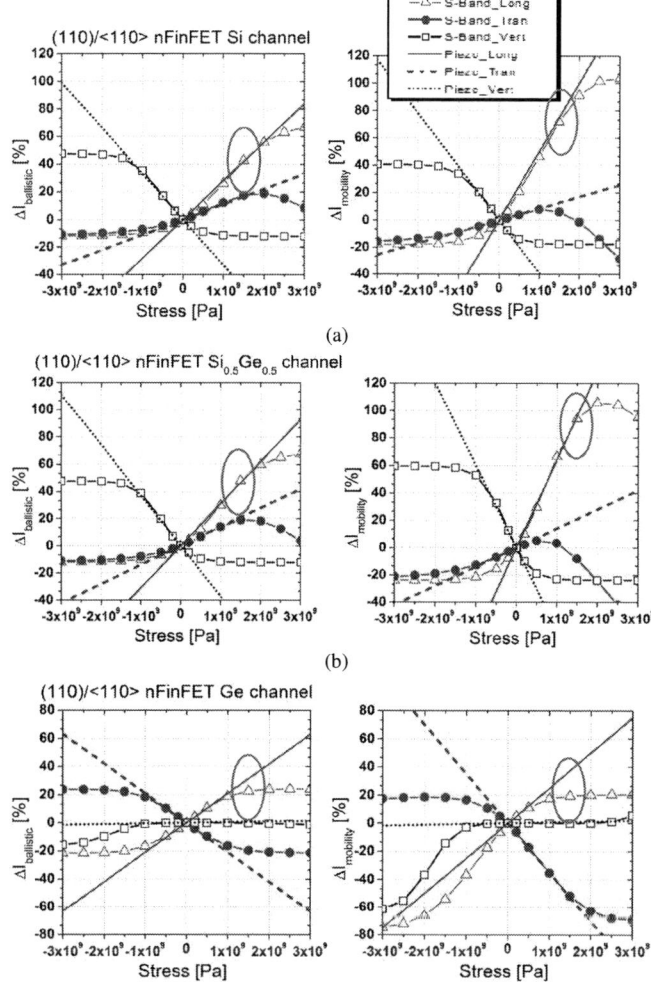

Fig. 5: S-band physical analysis of ballistic current stress sensitivity and of mobility limited current stress sensitivity (a) for Si , (b) for $Si_{0.5}Ge_{0.5}$ and (c) for Ge n-FinFET.

Unlike Ge, Si and $Si_{0.5}Ge_{0.5}$ n-FinFET show very similar stress sensitivity due to the same stress induced electron re-occupancy in X valley. Tab.1 comprehensively consolidates the impact of compressive and tensile stress on the valley's re-population affecting the effective mass and the density of states (DOS) and therefore the stress sensitivity of the drain current (I_{on}). For Ge n-FinFET, electrons occupy L valley. For tensile stress along channel direction Sxx, electrons move from L1,L3 valleys to L2,L4 valleys which have lower effective mass. It results into an increase of the current. For strain engineered Ge n-FinFET (>1GPa tensile stress Sxx), the stress sensitivity starts to saturate which is highlighted with blue circles in Fig. 5c. However strained Si and SiGe n-FinFET still show sensitivity to stress (Fig. 5a and b) which

978-1-4799-8002-4/14 $31.00 © 2014 IEEE

is because not only the electrons move to a lighter mass valley (from X1,X2 to X3 valley) but the effective mass m_{X3} also reduces with the increase of tensile stress.

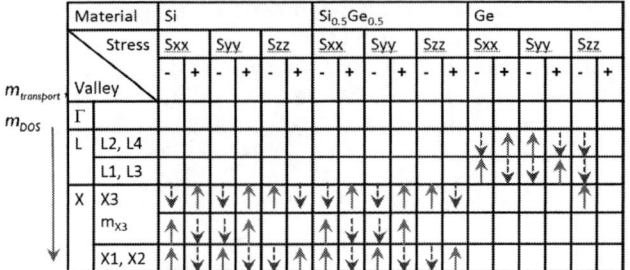

Material		Si						Si$_{0.5}$Ge$_{0.5}$						Ge					
	Stress	Sxx		Syy		Szz		Sxx		Syy		Szz		Sxx		Syy		Szz	
$m_{transport}$ / m_{DOS}	Valley	-	+	-	+	-	+	-	+	-	+	-	+	-	+	-	+	-	+
	Γ																		
L	L2, L4															↓	↑	↓	↓
	L1, L3															↑	↓	↑	↓
X	X3	↓	↑	↓	↑	↑	↓	↓	↑	↓	↑	↑	↓	↓	↑	↓	↑	↑	↓
	m_{X3}	↓	↑	↓	↑	↓	↑	↓	↑	↓	↑	↓	↑	↓	↑	↓	↑	↓	↑
	X1, X2	↑	↓	↑	↓	↓	↑	↑	↓	↑	↓	↓	↑	↑	↓	↑	↓	↓	↑

Tab. 1 : Electron valley repopulation table for Si, Si$_{0.5}$Ge$_{0.5}$ and Ge as a function of stress. Variation of the effective mass. Red continue arrow indicates the transport effective mass and DOS increasing.

In$_{0.53}$Ga$_{0.47}$As n-FinFETs exhibit no tensile stress sensitivity as shown in Fig.6. However it remains sensitive to large compressive stress. For compressive longitudinal and vertical stress (Sxx and Szz, respectively), the electrons repopulation to the heavier mass valleys (L1,L3 and X3, respectively) which results into a negative impact on the current. For the compressive transversal stress (Syy), electrons move from Γ valley to the L2,L4 valleys ($m < m_{L2,L4} < m_{L1,L3}$), but the larger DOS of the L-valley leads to the higher inversion change density and compensates for the decreasing injection velocity.

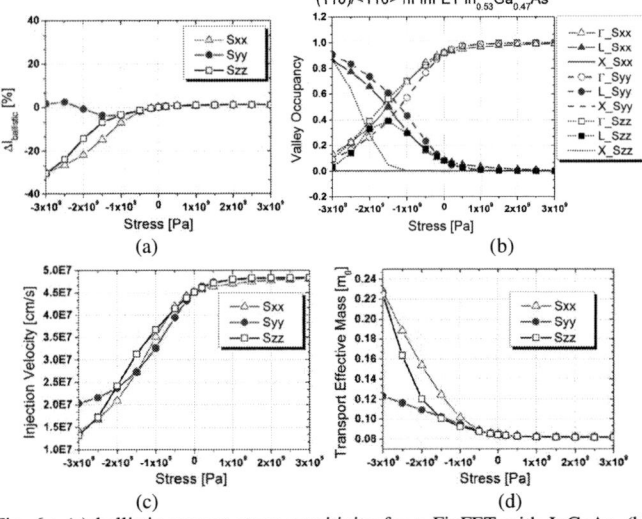

Fig. 6 : (a) ballistic current stress sensitivity for n-FinFET with InGaAs. (b) valley occupancy (c) injection velocity and (d) transport effective mass as function of stress for InGaAs n-FinFET.

Silicon based Nanowires (NW) is considered as an alternative option for N7 or beyond. Their stress sensitivity, shown in Fig.7, is explained by the electron repopulation and by the band warping for n and p Si NWFET respectively. For the horizontal NW (<110>), the trend of the ballistic current variation is similar to the Si FinFET, but the stronger quantum confinement of the NW makes the difference qualitatively. For the vertical NW (<100>), the tensile Sxx, makes electrons move to X2,X3 valleys as $m_{X2,X3}$ is smaller than m_{X1} and the ballistic current increasing. For the tensile and compressive stress Sxx, hole band structure warping leads to the lightest effective mass.

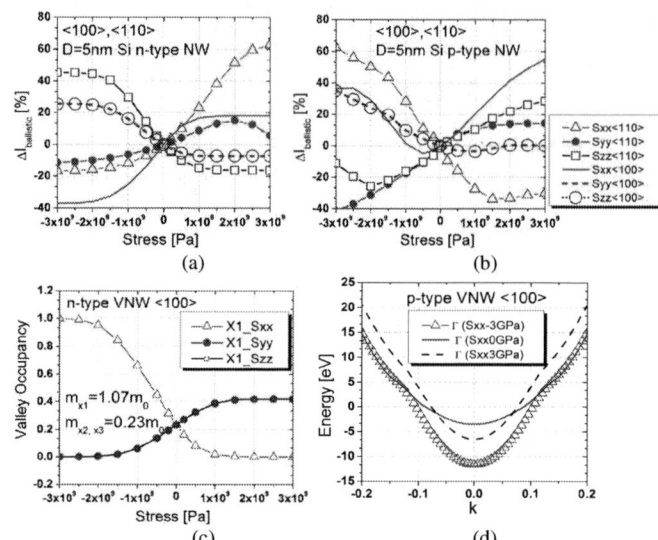

Fig. 7 : (a,b) Variation of the ballistic current for n and p type Si nanowire (NW) device. (c,d) Valley occupancy and energy dispersion for electrons in a Si nanowire of 5nm diameter.

Tab.2 and Fig.8 summarize the stress sensitivity for the different materials and device architectures considered for N7 that will be used for the KOZ estimation. As typically ±1.5 GPa stress is applied along the channel for the performance boosting, the stress sensitivity of Sxx to strain devices is extracted in the blue circle highlighted region of Fig. 4 and 5. For InGaAs n-FinFET and NW, the strain engineering is not considered in this work.

Stress sensitivity [%/100MPa]		Si		Si$_{0.5}$Ge$_{0.5}$		Ge		In$_{0.53}$Ga$_{0.47}$As *	Si HNW	Si VNW
		ΔI_b	ΔI_m	ΔI_b	ΔI_m	ΔI_b	ΔI_m	ΔI_b	ΔI_b	ΔI_b
Sxx	n	2.8	5	3.1	6.4	2.1	2.5	1.5	2	2.5
	p	-2	-3.9	-2.2	-4	-2.6	-5.5	x	-2.9	1.7
Syy	n	1.1	0.85	1.4	1.4	-2.1	-3.5	-0.5	0.8	-1.6
	p	1.5	2.1	1.6	2.3	1.4	1.5	x	2.2	-1
Szz	n	-3.2	-3.9	-3.7	-6	0.05	0.06	0.4	-2.7	-1.6
	p	0.7	1.5	0.6	1.6	1	2.5	x	2.2	-1
Sxx (@ 1.5 GPa strain Engineering)	n	2.8	5	3.1	6.4	0.5	0.25	*InGaAs only sensitive to compressive stress		
	p	-1.6	-3.9	-2.2	-4.5	-2.6	-6.2			

Tab.2: Summary of stress sensitivity for all channel material device studied

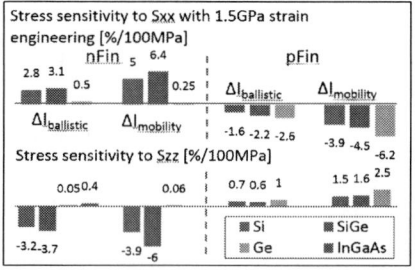

Fig. 8: Ge n-FinFET stress sensitivity to S$_{xx}$ becomes negligible after 1.5GPa stress engineering. Ge n-FinFET is insensitive to vertical stress

Strained Ge n-FinFET has up to 24 times lower sensitivity to Sxx compare to unstrained Ge. However no difference is estimated for Ge p-FinFET case. InGaAs shows sensitivity to vertical stress, around 0.4% impact per 100MPa is estimated. Ge p-FinFET shows up to 2.5%/100MPa vertical stress sensitivity.

Impact of TSV and μbump

TSV induces thermo-mechanical stress during 3DIC process integration. The performance of devices located at TSV proximity are affected. It depends on the location of the device, the device stress sensitivity and TSV technology [1]. The TSV induced influence on N7 CMOS performance is estimated in this work. The impact of 3x50μm TSV-middle on I_{on} is shown in Fig. 9 and was estimated for all channel materials and different channel length exhibiting ballistic current ratio from 80% to 30%. This estimation is for the worse-case scenario where the reduction of the device stress sensitivity due to the access resistances was not taken into account.

The results show for n-FinFET, TSV induced stress can affect more strongly on unstrained Ge device (±7%) compare to other device configurations. For strained Ge n-FinFET the impact of TSV on I_{on} is about ±4% (Fig. 9a). InGaAs n-FinFETs are insensitive to the TSV tensile stress, however they are sensitive to its compressive stress component that results in an electron re-population to the L and X heavier valleys responsible for a smaller injection velocity and larger effective mass (Fig.6 c, d). It results in around ±4% impact; The impact on p-FinFET is similar for Si, SiGe and Ge channel devices (Fig. 9c), due to the similar stress sensitivity (Fig. 4 a, b). Fig. 9b and Fig. 9d show the estimated KOZ figure of merit K [%·μm²] for different channel length.

n-FinFET KOZ is sensitive to the channel material choice whereas the p-FinFETs exhibits similar sensitivities (Fig.9a,c).

X_{KOZ}(3mm TSV) [mm]		W_{fin}=6nm, L=14nm/L=100nm					Si D=5nm	
		Si	$Si_{0.5}Ge_{0.5}$	Ge	sGe	$In_{0.53}Ga_{0.47}As$	HNW	VNW
Ion=5%	n	2/2.3	2.2/2.6	2.7/3	2/2.4	1.6	2.5	1.5
	p	2.6/3	2.7/3.1	2.9/3.5	2.9/3.5	x	2	3
Ion=0.5%	n	6.5/8	7/8.2	8.5/10	6.5/7	5.3	8.4	4.5
	p	8.6/9.5	8.9/9.7	9/10.2	9/10.2	x	6.8	9.2

Tab. 3: KOZ width is estimated for high mobility channel material and device architecture suitable for the 7nm technology node.

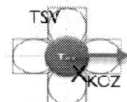

Due to the thermal expansion coefficient (CTE) mismatching between Si, CuSn based micro-bump (μbump) and polymer underfill (UF), thermo-mechanical stress can be induced after thermal compression bonding (TCB) process for achieving fine pitch die to die stacking. According [5], an arrange of few MPa vertical compressive stress can be expected. This stress level is highly dependent on the TCB process condition and selected UF material. Fig.10a shows the extracted stress components obtained from transistors based stress sensors located at the region on top of μbump in a dense array. One can see that the vertical stress is dominant (up to -400MPa).

For estimating the μbump impact on device drive current, we have used the extracted stress value from [5]. The impact is illustrated in Fig.10b. For n-FinFET, Ge channel devices are less sensitive than Si, SiGe and InGaAs devices. For Si and SiGe n-FinFET located at the top center of the μbump, an positive impact up to 14% can be expected; For p-FinFET, the effect is negative, around -3% can be expected for Si, SiGe and Ge devices.

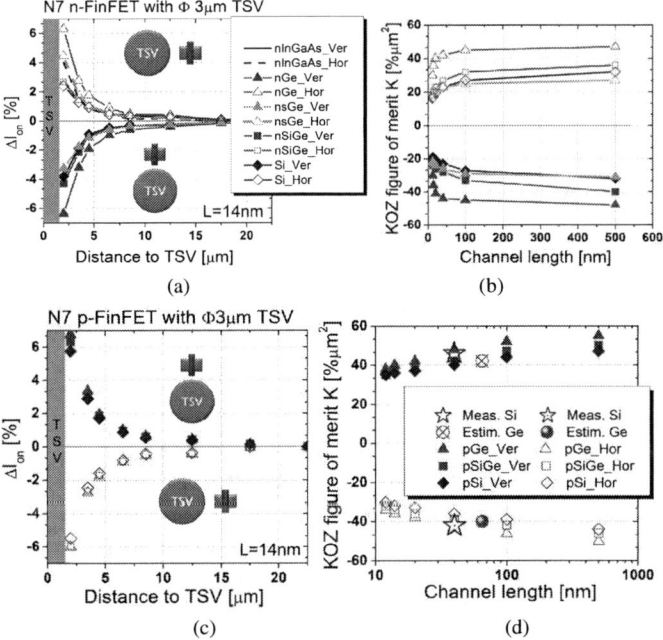

(a)

(b)

(c)

(d)

Fig. 9: (a, b) the impact of TSV on Ion current is estimated for Lg= 14nm n and p-FinFETs. (b)The KOZ figure of merit K estimated based on analytical model from [1]. In all cases considered the KOZ figure of merit K is lower when the channel length decreases.

The results for the KOZ, that is analytically modeled as shown in [1], are consolidated in Tab.3. This detailed analysis show that TSV induced KOZ area decreases for shorter devices due to higher ballistic contribution. Unstrained Ge n-FinFETs are largely affected by the TSV proximity, 40% more than Si, SiGe and strained Ge n-FinFETs showing similar KOZ (Fig.9 a,b and Tab.3). The sensitivity to Sxx stress along the channel is largely reduced when the Ge is strained (Tab.2). The

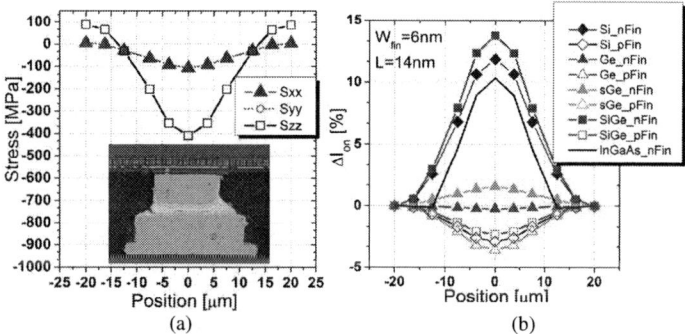

(a)

(b)

Fig. 10 : (a) measured μbump introduce large vertical stress component Szz [5]. (b) Impact on the devices (Lg=14nm) drive current. ΔIon is the current variation compared for the devices located below μbump and below UF (20μm away from the μbump center).

Conclusion

The impact of 3D IC process induced local thermo-mechanical stress effects on CMOS devices for 7nm technology node (N7) is smaller but has new dependencies with respect to the device operation, ballistic current ratio and the channel material considered.

References:

[1] W. Guo et al., IEEE-IEDM 2013, pp. 12.8.1-12.8.4.

[2] L. Witters et al., IEEE-IEDM 2013, pp.535-537.

[3] Sentaurus Band Structure, version H-2013.03.

[4] J. Mitard et al., IEEE-VLSI 2014, pp.1,2, 9-12 June 2014.

[5] V. Cherman et al., ECTC2014, pp.309,315, 27-30 May 2014.

A mobility enhancement strategy for sub-14nm power-efficient FDSOI technologies

B.DeSalvo[1], P.Morin[2], M. Pala[3], G.Ghibaudo[3], O.Rozeau[4], Q.Liu[2], A.Pofelski[5], S.Martini[4], M.Cassé[4], S.Pilorget[2], F.Allibert[6], F.Chafik[2], T.Poiroux[1], P.Scheer[2], R.G.Southwick[3], D.Chanemougame[2], L.Grenouillet[4], K.Cheng[7], F.Andrieu[4], S.Barraud[4], S.Maitrejean[1], E.Augendre[1], H.Kothari[2], N.Loubet[2], W.Kleemeier[2], M. Celik[2], O.Faynot[1], M.Vinet[4], R.Sampson[2], B.Doris[7]

[1]CEA-LETI (_barbara.desalvo@cea.fr_), [2]STMicroelectronics, [7]IBM, [6]SOITEC, 257 Fuller Road, Albany NanoTech, NY 12203, U.S.A.
[3]IMEP-LAHC, [4]CEA, LETI, MINATEC Campus, Grenoble, France [5]STMicroelectronics, Crolles, France

Abstract - Continuous CMOS improvement has been achieved in recent years through strain engineering for mobility enhancement. Nevertheless, as transistor pitch is scaled down, conventional strain elements (as embedded stressors, stress liners) are loosing their effectiveness [1]. The use of strained materials for the channel to boost performance is thus essential. In this paper, we present an original multi-level evaluation methodology for stress engineering design in next-generation power-efficient devices. Fully-Depleted-Silicon-On-Insulator (FDSOI) is chosen as the ideal test vehicle, as it offers the advantage of sustaining significant stress within the channel without plastic relaxation (the thin channel staying below the critical thickness [2]). Starting from 3D mechanical simulations and piezoresistive coefficient data, an original, simple, physically-based model for holes/electrons mobility enhancement in strained devices is developed. The model is calibrated on physical measurements and electrical data of state-of-the-art devices. Non-Equilibrium Greens Function (NEGF) quantum simulations of holes/electrons stress-enhanced mobility give physical insights into mobility behavior at large stress (~3GPa). Finally, the new strained-enhanced mobility model is introduced in an industrial compact model [3] to project evaluation at the circuit level.

Devices - Recently, a high-speed, power-efficient 14nm FDSOI technology with dual SOI/strained SiGeOI N/P channel has been demonstrated [4]. Further performance can be achieved by using a tensile channel (with strained-SOI substrate - sSOI) in NMOS and by enhancing PMOS channel compression (with more Ge, to compensate the sSOI tensile stress). Layout optimization reducing the transversal stress (detrimental for hole mobility [5]) should be explored. Despite scaling, embedded Raised-Source-Drain (RSD) efficiency can also be increased with more activated C and P (n-type) and Ge (p-type) dopants. Moreover, the use of removal gate (rather than gate-first) should be considered. To explore this path, we fabricate dedicated FD devices on SOI and sSOI. The cSiGe channel in PMOS devices is formed by SiGe epitaxy followed by a condensation process, with Ge content up to 35%. Fig.1 shows the NMOS/PMOS featuring a 6nm-thick channel and 20nm gate length.

Electrical Data - Fig.2 shows the I_{ON}-I_{OFF} characteristics of the studied devices, demonstrating the performance advantage of strained Si (SiGe) channel for NMOS (PMOS) devices over relaxed Si. Narrow cSiGe PMOS devices with uniaxial strain, the transverse strain being fully relaxed, show the best performance. To correlate I_{ON} gain to mobility enhancement, we extract the low field mobility (μ_0) in short channel devices using the Y-function [8] (which allows for series resistance effect suppression). Good agreement is obtained with the effective mobility (μ_{eff}) extracted by traditional split-CV method in long channel devices (Fig.3). To explore layout effects, the mobility is studied versus the gate length (L_G), width (W) and active area length (L_{act}). Fig.3 shows a 60% mobility gain measured on long, large sSOI transistors compared to SOI, due to tensile biaxial stress. A significant mobility enhancement is also achieved while increasing the Ge content in cSiGe PMOS, thanks to compressive biaxial strain. In both unstrained/strained devices, the mobility decreases for short gate lengths due to defect-enhanced scattering rates close to S and D region (as neutral and/or Coulombian centers) or to ballistic effect [9]. Nevertheless, it appears that the short channel mobility gain extracted in Fig.3(c) is higher than the one obtained in long channel devices, due to stronger effects of RSD stressors with short gate length. In NMOS, even low activated SiCP RSD grown on sSOI are tensile and generate more stress than in unstrained SOI. No variation of μ_0 is observed versus L_{act} and W in unstrained PMOS/NMOS (Fig.4). A strong layout effect appears in cSiGe PMOS, where the narrowing of the active width to 80nm leads to +60% μ_0 enhancement (in full agreement with the I_{ON} measurements, Fig.2). On the other side, when L_{act} decreases, a mobility gain degradation happens. A completely different behavior is obtained for sSOI NMOS, showing a weaker layout dependence (Fig.5).

3D Mechanical Simulations - 3D mechanical simulations of FDSOI structures are performed using finite elements and considering the elastic and anisotropic properties of the SiGe materials to study longitudinal (σ_x, along gate length) and transverse (σ_y, along gate width) stress profiles (Fig.6). Plastic relaxation effects were here neglected. The stress profiles show a significant decrease of the transverse stress as gate width is reduced, with an almost complete relaxation below 40nm. Conversely, the longitudinal stress is maintained to substantial values even for small W. Transmission Electron Microscopy (TEM) Nano-Beam Electron Diffraction and Electron Dispersive Spectroscopy are used to determine the lattice deformation and the Ge fraction, respectively. From this data, we calculate the longitudinal stress σ_x under the gate, with either 25% or 30% cSiGe. Reasonable matching is obtained with pure elastic simulations. To evaluate the effectiveness of embedded RSD stressors in scaled devices with decreased gate pitch, we calculate the channel stress induced both with cSiGe channel and eSiGe RSD as function of Ge fraction, and study the impact of a removal gate process compared to a gate first process (Fig.7). It appears that, the strained channel provides the largest stress contribution, while moderate stress comes from RSD and gate last.

Quantum Simulations (QS) - The transport properties of strained cSiGe FDSOI devices (with $t_{channel}$=6nm, t_{BOX}=25nm)

978-1-4799-8002-4/14 $31.00 © 2014 IEEE

are computed with a quantum approach based on the NEGF formalism and calibrated with experimental current-voltage curves. To simulate cSiGe PMOS, we employ a 6 bands k·p Hamiltonian accounting for the spin-orbit interaction and strain interaction matrix [10, 13], whereas NMOS are simulated with a 2-band k·p Hamiltonian [11]. Alloy scattering and acoustic and optical phonons are included within the self-consistent Born approximation. Effective mobility is extracted in long channel devices (L_G=500nm) according to the method proposed in [12]. Fig.8 shows the deformation of the highest valence subband of a cSiGe PMOS induced by different types of stresses: the compressive stress along [110] significantly reduces the hole effective transport mass, thus improving the transport properties. Fig.8 shows the mobility gain of cSiGe PMOS with Ge molar fraction ranging from 0.1 to 0.5. In order to emulate experiments, first the mobility gain of SiGe films w.r.t. the pure Si is computed for different stress levels. Then, the mobility gains induced by the SiGe intrinsic stress are plotted (red bullets). Considering the intrinsic SiGe strain curve, a 200% mobility improvement is achievable with 50% Ge fraction. Similarly, Fig.9 shows the I_{ON} gain induced by different Ge fractions for short devices with L_G=20nm, yielding a 75% I_{ON} improvement for 50% Ge (lower than mobility gain due to the high-field scattering processes impact). Fig.10 shows the electron mobility gain of NMOS induced by uniaxial/biaxial strain. It appears that, for an inversion density (N_{inv}) of 10^{13}cm^{-2}, the uniaxial component is dominant at high stress levels, in agreement with the literature [15, 16].

Analytical Model - Analytical model for stress-enhanced mobility is developed. For holes, the mobility gain is assumed exponentially dependent on the longitudinal/transverse stress (instead of being linearized). Ge%-dependent piezoresistive coefficients are used [14]. Saturation at high stress is forced in agreement with QS. For electrons, an empirical law is used inspired by piezoresistive coefficients [15] and quantum simulations [16]. The stress channel and SD profiles are taken from 3D mechanical simulations. Figs.3-5 show that model captures very well the stress effects, independently of device type and Ge%. Fig.12 quantitatively evaluates the PMOS improvement achievable with layout optimization, i.e. slicing the channel in narrow strips yielding transverse mechanical stress relaxation. The gain is amplified when Ge content increases. Fig.13 shows the variation of longitudinal and transverse stresses with the key stressors modules (both for n and p type devices). The longitudinal stress magnitude is increased, whereas the transverse stress is reduced. The corresponding mobility gain induced by each stressor, computed by means of mechanical simulations and mobility analytical model as presented above, is also shown. sSOI substrate provides the major NMOS gain (x2), whereas PMOS mobility is improved by a factor 8 both by Ge fractions increase in channel/RSD and by layout optimization (slicing) to reduce the transverse stress component.

Implementation in Circuit Compact Model - The new stress-enhanced mobility model has been included in the UTSOI2 SPICE model [3]. Pre-layout ring oscillators are simulated either with unstrained or strained P/NMOS devices at different V_{dd}. The oscillators consist in a loop of inverter stages with a fan-out=3 (i.e. each stage contains 3 inverters in parallel). A parasitic capacitance of 1.5fF is assumed on the output of each stage to emulate back-end effects. For each case, the threshold voltage has been adjusted to keep the static power constant. Fig.14 shows the oscillator frequency versus the dynamic power (at constant static power). When V_{dd} is reduced from 0.9V to 0.75V a dynamic power gain of 50% is achieved while maintaining frequency performance thanks to PMOS/NMOS mobility boosters.

Conclusions - In this paper, a simple but powerful physically-based model (validated by NEGF simulations) allowing stress engineering design in next-generation technologies was developed and implemented in a circuit simulator. Ring oscillator simulations showed that a dynamic power gain of 50% is achievable while maintaining circuit frequency performance by means of mobility boosters. By means of a multi-scale modelling and measurements, a clear scaling path to achieve high mobility, power-efficient sub-14nm FDSOI technologies was identified.

Acknowledgments - This work was carried out in the frame of the ST/IBM/LETI joint development program, the NANO2017 program supported by the French government and the French ANR (through the project ANR-13-NANO-0009 "NOODLES").

References - [1] A.Khakifirooz et al., "Hole Transport in Strained and Relaxed SiGe Channel Extremely Thin SOI MOSFETs", IEEE El. Dev. Lett., 34, 11, '13. [2] D.C. Houghton, "Strain relaxation kinetics in Si1−x Ge x /Si heterostructures", J. Appl. Phys. 70, 2136, '91. [3] T.Poiroux et al.,"UTSOI2: A complete physical compact model for UTBB and independent double gate MOSFETs", Techn. Dig. of IEDM'13. [4] O.Weber et al. , "14-nm FDSOI Platform Technology for High-Speed and Energy-Efficient Applications", VLSI Technology Symposium '14. [5] P.Packan et al., "High Performance Hi-K + Metal Gate Strain Enhanced Transistors on (110) Silicon", Techn. Dig. of IEDM'08. [6] L.Grenouillet et al., "UTBB FDSOI transistors with dual STI for a multi-Vt strategy at 20nm node and below", Techn. Dig. of IEDM'12. [7] K.Cheng et al., "High Performance Extremely Thin SOI (ETSOI) Hybrid CMOS with Si Channel NFET and Strained SiGe Channel PFET", Techn. Dig. of IEDM'12. [8] G.Ghibaudo, "New method for the extraction of MOSFET parameters", Electr. Letters, 24, 544, '88. [9] M.V. Fischetti, "Long-range Coulomb interactions in small Si devices. Part II. Effective electron mobility in thin-oxide structures", J. Appl. Phys., 89, 1232, '01. [10] M. Fischetti et al., "Band structure, deformation potentials, and carrier mobility in strained Si, Ge, and SiGe alloys", J. Appl. Phys., 80, 2234, '96. [11] J. C. Hensel et al., "Cyclotron Resonance in Uniaxially Stressed Silicon. II. Nature of the Covalent Bond", Phys. Rev., 138, A225, '65. [12] S. Poli et al, "Size Dependence of Surface-Roughness-Limited Mobility in Silicon-Nanowire FETs", IEEE Tr. On Electr. Dev., 55, 2968, '08. [13] D. Esseni et al., "Semi-classical transport modelling of CMOS transistors with arbitrary crystal orientations and strain engineering", Journal of Comp Elec., 8, 209, '09. [14] M.Cassé et al., «Study of piezoresistive properties of advanced CMOS transistors: Thin film SOI, SiGe/SOI, unstrained and strained Tri-Gate Nanowires", techn. Dig. of IEDM'12. [15] O.Weber et al., "Examination of Additive Mobility Enhancements for Uniaxial Stress Combined with Biaxially Strained Si, Biaxially Strained SiGe and Ge Channel MOSFETs", Techn. Dig. of IEDM '07. [16] K.Uchida et al., "Physical mechanisms of electron mobility enhancement in uniaxial stressed MOSFETs and impact of uniaxial stress engineering in ballistic regime", Techn. Dig. of IEDM 2005.

Devices and Electrical Data

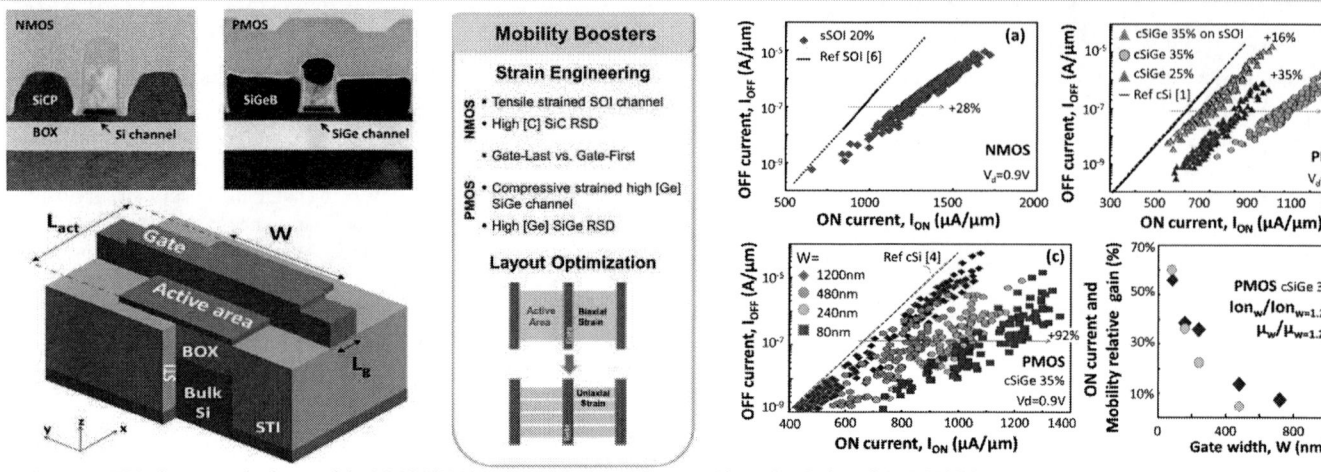

Figure 1. TEM images and schema of the NMOS/PMOS FDSOI devices. The mobility boosters explored in this paper are indicated on the rigth.

Figure 2. I_{on}/I_{off} of (a) SOI/sSOI NMOS, (b) cSi/cSiGe PMOS on SOI (sSOI), (c) 35% cSiGe PMOS with different W. (d) 35% cSiGe PMOS I_{on} (μ_0) gain versus W.

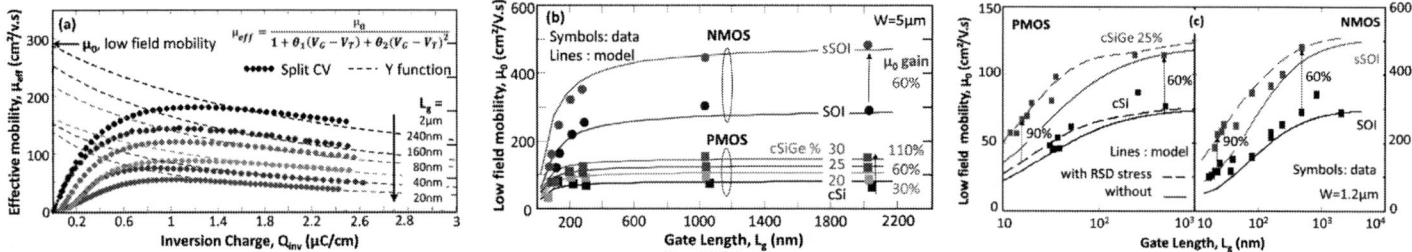

Figure 3. (a) Mobility extracted by Y function [8] and Split-CV methods. PMOS/NMOS mobility versus gate length L_g, (lin-lin plot (b) or lin-log plot (c)).

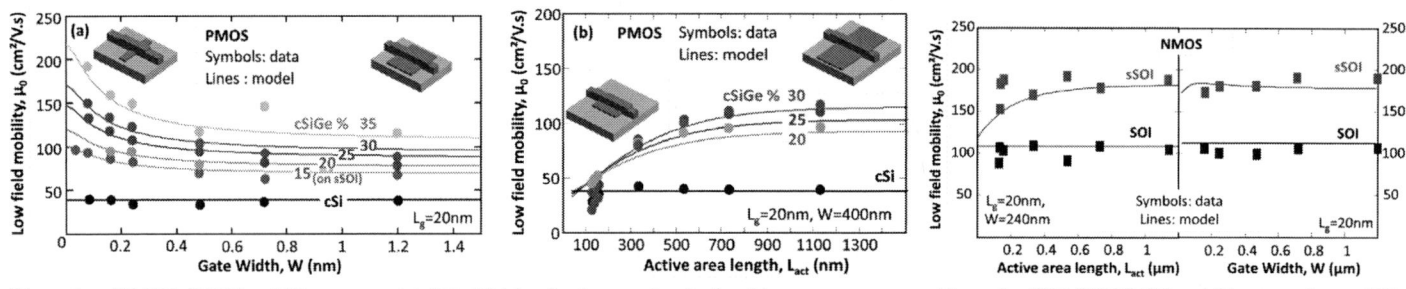

Figure 4. cSi/cSiGe PMOS mobility versus gate width, W (a) and active area length, L_{act} (b).

Figure 5. SOI/sSOI NMOS mobility versus L_{act} and W.

Mechanical Simulations

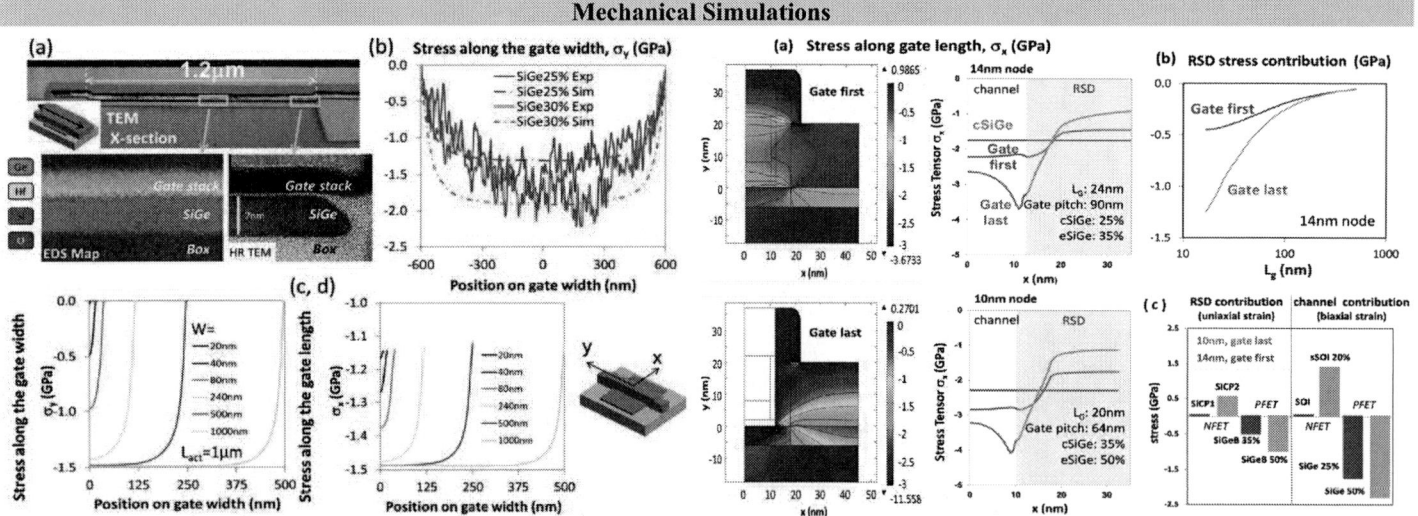

Figure 6. (a) TEM, EDS analyses on cSiGe PMOS along the transistor width (W=1.2µm). (b) Stress profiles measured by NBED and computed by 3D mechanical simulations. (c), (d) 3D mechanical simulations of relaxation in strained cSiGe PMOS. Stress profiles along gate length (σ_x) and along width (σ_y) with different W are shown (here L_{act}=1µm).

Figure 7. 2D FEM mechanical simulations: (a) contribution of cSiGe channel, channel and Raised Source Drain, with gate-first/last, to PMOS stress. (b) RSD stress contribution vs transistor gate length. (c) Summary of NMOS/PMOS stress contributions in nominal 10/14nm standard cells, considering the decreased pitch and transverse/ longitudinal strain values based on 3D simulations.

978-1-4799-8002-4/14 $31.00 © 2014 IEEE

Quantum Simulations

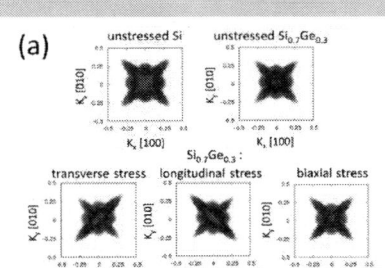

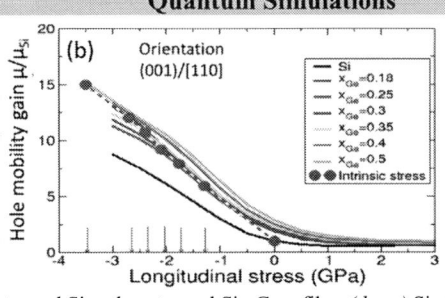

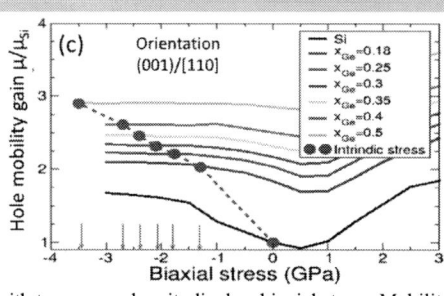

Figure 8. *(a)* Highest hole subband for pure *(up)* unstressed Si and unstressed Si$_{0.7}$Ge$_{0.3}$ film, *(down)* Si$_{0.7}$Ge$_{0.3}$ with transverse, longitudinal or biaxial stress. Mobility gain as a function of the longitudinal *(b)* and biaxial *(c)* stress for a FDSOI PMOS with a Si$_{1-x}$Ge$_x$ channel with different values of the Ge molar fraction x (μ_{Si} being the mobility of unstressed Si). Mobility gain values corresponding to intrinsic SiGe stress levels (for the various Ge fractions) are indicated by red bullets.

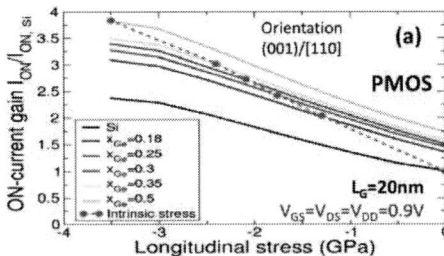

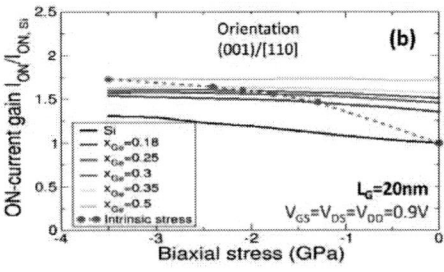

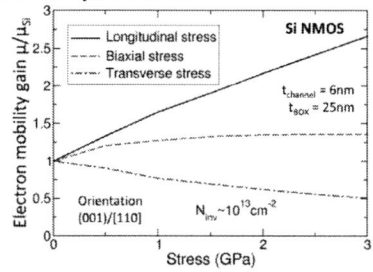

Figure 9. ON current gain versus *(a)* longitudinal and *(b)* biaxial stress for a FDSOI PMOS with Si$_{1-x}$Ge$_x$ channel with different values of the Ge molar fraction x ($I_{ON,Si}$ is the ON current of unstressed Si PMOS).

Figure 10. Mobility gain as function of longitudinal, biaxial and transverse stress for a cSi FDSOI NMOS.

Mobility Analytical Modeling

Basic equation of stress-enhanced mobility model:

PMOS mobility - Exponential law from piezoresistive coefficients:

$$\frac{\Delta\mu}{\mu} = -\pi(\%_{Ge}) \times \sigma \implies \ln(\mu) = -\int_0^\sigma \pi(\%_{Ge}) \times \sigma \cdot d\sigma \implies \frac{\mu_\sigma}{\mu_0} = \exp\left(-\int_0^\sigma \pi(\%_{Ge}) \times \sigma \cdot d\sigma\right)$$

$\pi_{L,T}(\%_{Ge})$, piezoresistive coefficients from [14].

NMOS mobility - Empirical law (from piezo-coeff [15] and QS [16]):

$$\mu_L(\sigma_x) = \frac{a + b \cdot \sigma_x}{a + \sigma_x} \quad \mu_t(\sigma_x) = \left[\left[1 - \exp\left(-\frac{\sigma_x}{c}\right)\right] \cdot d + 1\right]\frac{\sigma_x + a}{a + \sigma_x} \quad \mu_{glob}(\sigma_x, \sigma_y) \approx \mu_L(\sigma_x) \cdot \mu_t(\sigma_y)$$

with a/b/c/d (=4/3.4/0.7/0.65) from fitting of experimental data.

Mechanical Stress - From 3D simulations, the transverse stress:

$$\sigma_y(y, W, L_{act}) = \sigma_{maxy}(L_{act}, \%_{Ge})\left[1 - e^{\frac{y+W}{\lambda_y}}\right]\left[1 - e^{\frac{W-y}{\lambda_y}}\right]$$

with λ_y(=80nm), R_m(=0.24) from fitting of experimental data, the maximum stress vs Ge content being:

$$\sigma_{maxGe}(\%) = -\frac{7.45.\%}{100} + \left(\frac{1.35.\%}{100}\right)^2$$

The mobility degradation with **L$_G$ reduction** is modeled via an empirical formula:$1/\mu_{shortCh} = (1 + L_c/L_G)/\mu_{longCh}$, L$_c$ fitting parameter.

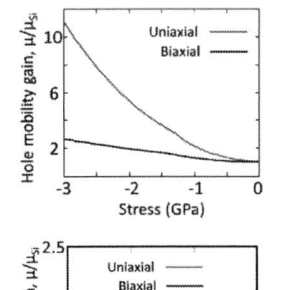

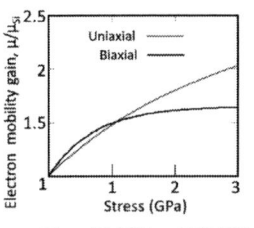

Figure 11. PMOS and NMOS mobility gain versus stress.

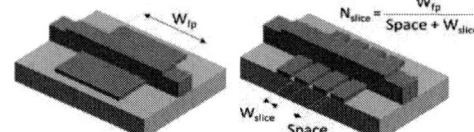

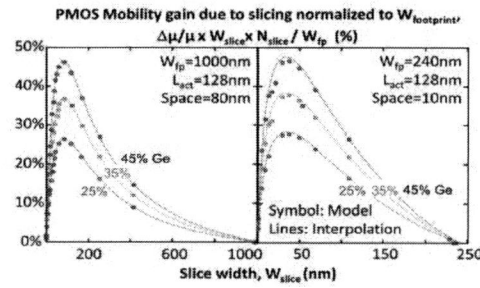

Figure 12. Quantification of slicing benefit (i.e. striped width) on PMOS mobility, based on analytical model.

Extrapolations and Compact Modeling Simulations

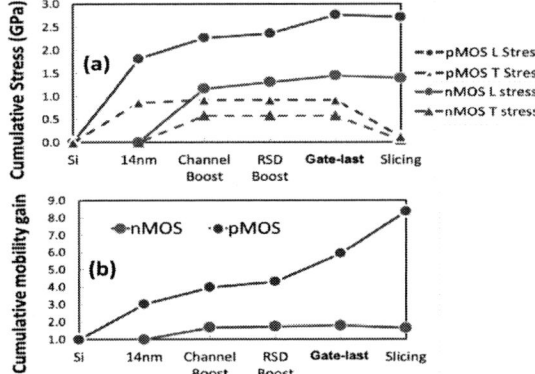

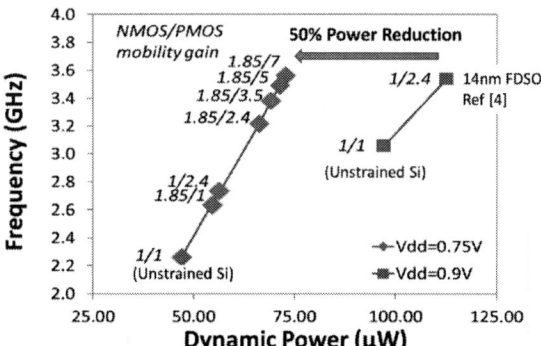

Figure 13. *(a)* Variation of transverse and longitudinal strain for NMOS/PMOS stressors (from 3D mechanical simulations of 14/10nm nominal cells). *(b)* Corresponding mobility gain, by the analytical model. *Note that carrier transport along the (001) Si planes presents a significant bonus for electrons versus holes, thus more hole mobility gain allows for n/p mobility balance.*

Figure 14. Oscillator circuit simulations (assuming same device geometries for PMOS and NMOS, FO=3 and 1.5fF parasitic capacitances) showing the reduction in dynamic power by reducing the V$_{dd}$ from 0.9V to 0.75V, while maintaining frequency performance through PMOS/NMOS mobility boosters.

Coupled Monte Carlo Simulation of Transient Electron-Phonon Transport in Small FETs

Y.Kamakura[1,2], I.N.Adisusilo[1], K.Kukita[1], G.Wakimura[1], S.Koba[3], H.Tsuchiya[2,3], and N.Mori[1,2]

[1]Division of Electrical, Electronic and Information Engineering, Osaka University, Suita, Osaka 565-0871, Japan
[2]Japan Science and Technology Agency (JST), CREST, Kawaguchi, Saitama 332-0012, Japan
[3]Department of Electrical and Electronic Engineering, Kobe University, Kobe 657-8501, Japan
Phone: +81-6-6879-4850, Fax: +81-6-6879-7791, E-mail: kamakura@si.eei.eng.osaka-u.ac.jp

Abstract

Using a coupled Monte Carlo technique for solving both electron and phonon Boltzmann transport equations, the transient electrothermal simulation of nanoscale FETs is performed. It is shown that the time constants for the electron and phonon transport are different in order of magnitude, and the self-heating has little impact on digital circuit delay, while it would affect the bias temperature instability because of the long decay time of the created hot spot. The effectiveness of introducing the lightly doped drain structure is also discussed to reduce the hot spot temperature.

Introduction

The heat conduction property is one of the main concerns for nanoscale FETs relating to reliability and performance. In particular, the non-planar structures surrounded by insulating films are considered to have worse thermal properties because there are only narrow paths to conduct the higher density power (> 10 TW/cm^3 (1)) away from the hot spot created mainly in the drain. However, the impacts of the self-heating are still controversy, because the time constant for self-heating in nanoscale devices is much longer than characteristic switching times of CMOS logic and memory, and hence the self-heating effect in digital circuits expected to be negligible (2). Furthermore, the effect of the localized drain self-heating on the AC negative bias temperature instability (NBTI) has been also discussed (3). Although the state of the art electrothermal simulations have been reported so far to analyze the steady-state conditions (e.g., (1, 4)), the time-domain simulations are also needed to assess the above problems. So in this study, we present a transient simulation of nanoscale FETs using a Monte Carlo (MC) method for solving the Boltzmann transport equation of electrons and phonons.

Simulation Method

Our motivation for using the phonon MC simulator is to rigorously calculate the heat conduction properties in nanoscale devices, where the quasi-ballistic transport effects would become significant. Fig. 1 shows an example (not for Si); the experimental data suggesting non-Fourier heat conduction (5) are well reproduced by the MC method. For Si, we have carefully calibrated the simulator to yield the correct thermal conductivities (6, 7) as Fig. 2 (8). Fig. 3 shows the comparison of the simulation results to the conventional heat diffusion analysis. Note that the Fourier's law overestimates the heat removal efficiency in case of small dimensions ($<\sim$ the phonon mean free path λ) due to the finite phonon group velocity (9). The transition from diffusive to ballistic transport is clearly confirmed in Fig. 4; although λ is significantly shortened in ultrathin films due to the frequent boundary scattering, the quasi-ballistic transport effect is not negligible in the regime relating to interests in nanoscale FET simulation.

The phonon MC simulator was coupled with the electron simulation to analyze the self-heating effect (Fig. 5). To simulate the electrons the standard MC model of (10) was employed, and the spatial distribution of the heat generation rate was monitored by counting the net number of emitted phonons from the electrons. We assumed that the contribution of the optical phonons to the heat conduction is negligible, and thus the thermal energy stored in the optical modes can only be dissipated through the conversion into the acoustic modes via phonon-phonon scattering. We have treated this process by a relaxation time approximation, and assumed two values for the optical phonon decay time ($\tau_{op} = 10$ or 2 ps) considering its ambiguity (1). According to the heat power density transferred from the hot electrons and optical phonons, the acoustic phonons were generated and their transport was simulated with the MC method. The concept of the simulation is basically the same as our previous report (11), but in this work several new features have been included. Firstly, the accuracy of the phonon transport models has been greatly improved by including, e.g., the realistic phonon dispersion relation and the phonon boundary scattering (the problems to use the approximate models were discussed in (8, 9)). Secondly, the double-gate structure as Fig. 6 was considered, whereas in (11) the n-i-n diode was simulated, whose temperature characteristics may be different from FETs. Fig. 7 shows the simulated I_D-V_G characteristics; the bias conditions where the FET exhibits a negative bias coefficient were used in this study.

Results and Discussion

Fig. 8 shows the transient drain current after turning on the gate obtained with the coupled MC simulation. The decrease of the drain current due to self-heating was observed at $t >\sim 10$ ps regardless of τ_{op}, which is much larger than the time constant of the electric delay (~ 0.2 ps). Fig. 9 indicates that the high-density heat is generated particularly inside the drain through the optical phonon emission, which results in the temperature increase as Fig. 10. In the early stage, T_{op} increases promptly, and then the heat energy is gradually transfered to the acoustic phonons. Note that, as Fig. 11, the temperature rise localized in the drain has little impact on the current (in FETs the saturated drain current is determined in principle by the source side injection), indicating that the current reduction is mainly caused by T_{ac} increase in the channel region. Since $T_{ac}(t > 10$ ps) is not strongly dependent on τ_{op} as Fig. 12, the current reduction observed in Fig. 8 is considered to be less sensitive to τ_{op}. The transient characteristics of $T_{op}(t)$ and $T_{ac}(t)$ are well understood with the help of the equivalent thermal circuit model suggested in Fig. 13. In the present simulation condition, the hot spot temperature is projected to reach near the melting point of Si at steady state (in this study, the device structure with aggressively scaled EOT and the low threshold voltage was assumed as Fig. 6 to enhance the drive current and the heat generation).

Figs. 14 and 15 shows another example of the coupled simulation. In this case, the self-heating showed no impact on the electric behavior, because the circuit time constant is less than that for the self-heating. However, the phonon temperatures remained to be raised for a long time even after the drain voltage transition has finished, which would affect the BTI reliability due to the high oxide field (@$V_G = V_{DD}$ and $V_D = 0$ V) along with the raised temperature.

On the other hand, in, e.g., analog applications, the self-heating effect is considered as a serious concern (12). Thus, in this work, a possible approach to engineer the hot spot temperature was explored. As shown in Fig. 16, we compared the three drain structures to modulate the potential profile (Fig. 17). The simulation results (Fig. 18) suggest that the lightly doped drain (LDD) structure is preferable to reduce the peak temperature. The gradual potential drop around the drain edge is effective to disperse the phonon emitting points and thus the resulting heat density was significantly suppressed.

Conclusion

Using a coupled MC technique for solving both electron and phonon Boltzmann transport equations, the transient electrothermal simulation of nanoscale FETs has been carried out. The time constants for the electron and phonon transport were different in order of magnitude, and the self-heating showed little impact on digital circuit delay, while it would affect the BTI reliability because of the long decay time of the created hot spot. Furthermore, it was shown that LDD structure is preferable to reduce the peak temperature of the hot spot.

Acknowledgements

The authors would like to acknowledge Prof. S. Uno, Dr. J. Hattori, Prof. T. Watanabe, and Mr. T. Zushi for many helpful discussions about fundamental physics of phonon transport in semiconductors.

References

(1) J.A. Rowlette and K.E. Goodson, "Fully Coupled Nonequilibrium Electron-Phonon Transport in Nanometer-Scale Silicon FETs," *IEEE Trans. Electron Devices*, vol. 55, pp. 220–232, January 2008.

(2) S. Lee et al., "Experimental analysis and modeling of self heating effect in dielectric isolated planar and fin devices," *Symp. VLSI Tech. Dig.*, pp. T248–T249, 2013.

(3) G.L. Rosa, S. Rauch, III, F. Guarin, and S. Boffoli, "Insights in the Physical Damage of $V_{GS} = V_{DS}$ High-k PMOSFET Degradation in AC Switching Conditions," *IEEE Trans. Device Mater. Rel.*, vol. 13, pp. 185–191, March 2013.

(4) R. Rhyner and M. Luisier, "Self-heating effects in ultra-scaled Si nanowire transistors," *Tech. Dig. IEDM*, pp. 790–793, 2013.

(5) A. Minnich and G. Chen, "Quasi-Ballistic Heat Transfer From Metal Nanostructures on Sapphire," *Proc. ASME/JSME 2011 8th Thermal Engineering Joint Conference*, pp. T30071-1–T30071-6, 2011.

(6) M.G. Holland, "Analysis of Lattice Thermal Conductivity," *Phys. Rev.*, vol. 132, pp. 2461–2471, December 1963.

(7) W. Liu and M. Asheghi, "Phonon-boundary scattering in ultrathin single-crystal silicon layers," *Appl. Phys. Lett.*, vol. 84, 3819–3821, April 2004.

(8) K. Kukita and Y. Kamakura, "Monte Carlo simulation of phonon transport in silicon including a realistic dispersion relation," *J. Appl. Phys.*, vol. 114, pp. 154312-1–154312-8, October 2013.

(9) K. Kukita and Y. Kamakura, "Monte Carlo simulation of diffusive-to-ballistic transition in phonon transport," *J. Comp. Electronics*, vol. 13, pp. 264–270, March 2014.

(10) C. Jacoboni and L. Reggiani, "The Monte Carlo method for the solution of charge transport in semiconductors with applications to covalent materials," *Rev. Mod. Phys.*, vol. 55, pp. 645–705, July 1983.

(11) Y. Kamakura, N. Mori, K. Taniguchi, T. Zushi, and T. Watanabe, "Coupled Monte Carlo Simulation of Transient Electron-Phonon Transport in Nanoscale Devices," *Proc. SISPAD*, pp. 89–92, 2010.

(12) T. Takahashi, N. Beppu, K. Chen, S. Oda, and K. Uchida, "Self-Heating Effects and Analog Performance Optimization of Fin-Type Field-Effect Transistors," *Jpn. J. Appl. Phys.*, vol. 52, pp. 04CC03-1–04CC03-6, February 2013.

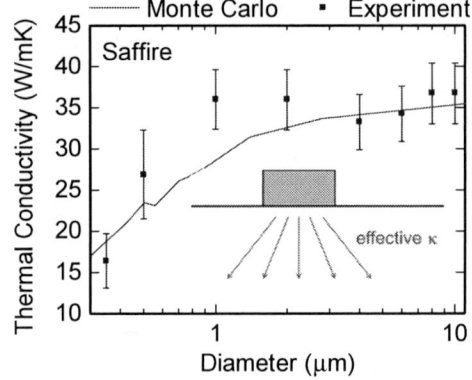

Fig. 1: Effective thermal conductivity (apparently reduced due to quasi ballistic phonon transport) of sapphire beneath the metallic dot with sub-micron diameters. The MC results are compared with the experimental data measured by using transient thermorefrectance (5). In this MC, the phonon mean free path was assumed to be 140 nm (constant).

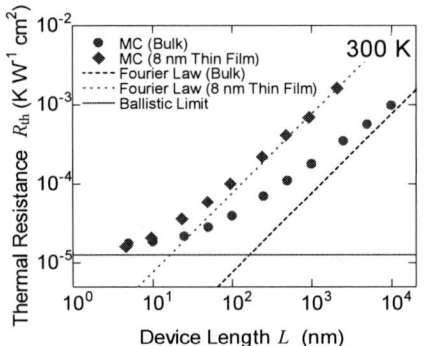

Fig. 4: Equivalent thermal resistance R_{th} extracted from the one-dimensional MC results for bulk Si (red dots) (9) and 8 nm Si film (green diamonds) simulated with various distances between the heat source and the sink. The characteristics expected from the Fourier law (dashed lines) and the ballistic transport theory (solid line) are also plotted.

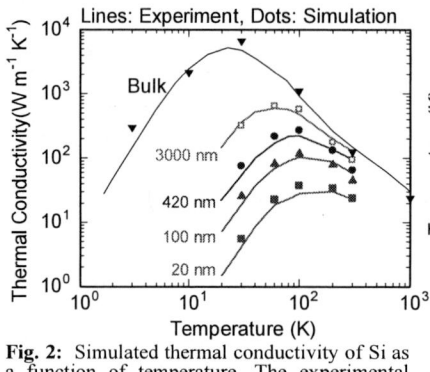

Fig. 2: Simulated thermal conductivity of Si as a function of temperature. The experimental data for bulk Si (6) and Si thinfilms (7) are also plotted for comparison.

(right) Fig. 3: Time evolution of the local temperature distributions simulated with the MC simulator (solid lines) and the Fourier based heat conduction equation (dashed lines) considering acoustic phonon transport in Si. One-dimensional structure with a heat source (yellow shaded area) and heat reservoirs at both ends (300 K) were assumed. The results for two sizes of (a) L = 500 nm and (b) 50 nm are compared.

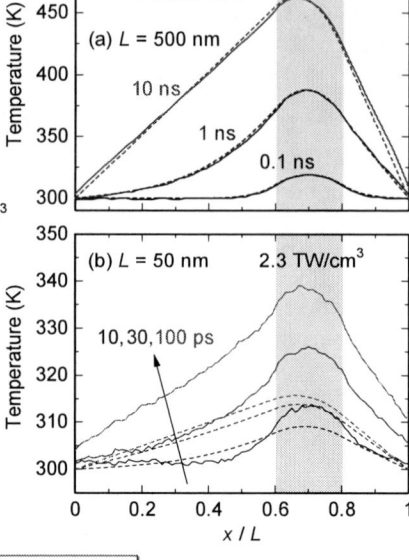

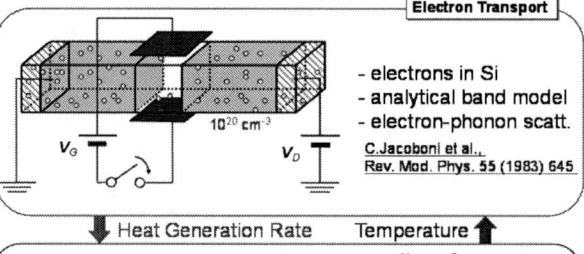

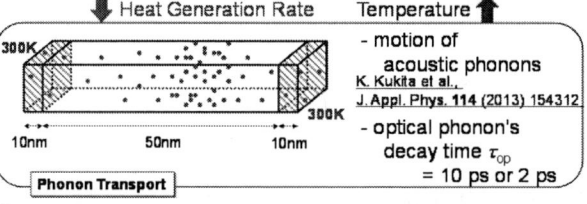

Fig. 5: Schematic view explaining the concept of the coupled MC simulation for electron and phonon transport in double gate FETs. The green shaded areas are reservoirs controlling the MC particle numbers.

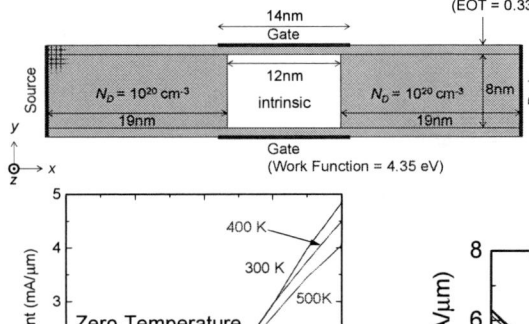

Fig. 6: Schematic view of the double gate FET structure simulated in this study.

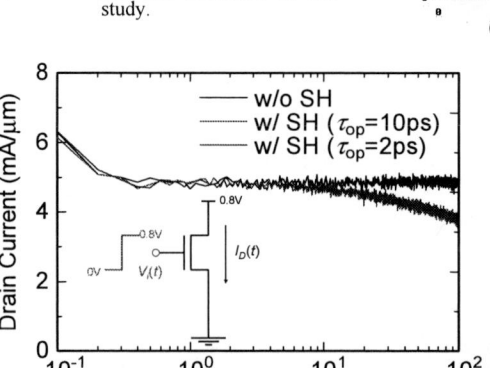

Fig. 7: Transfer characteristics simulated using the electron's MC simulator without self-heating effect. The device temperature was uniform throughout the device.

Fig. 9: The spatial distribution of heat generation rate simulated with the ensemble MC simulator for electron transport. Note that the heat generation is mainly caused by the optical phonon emission inside the drain.

Fig. 8: The transient drain current simulated by using the coupled MC simulator for electrons and phonons plotted as a function of time t after turning on the gate voltage. The results with and without taking account of self-heating effect are compared.

978-1-4799-8002-4/14 $31.00 © 2014 IEEE

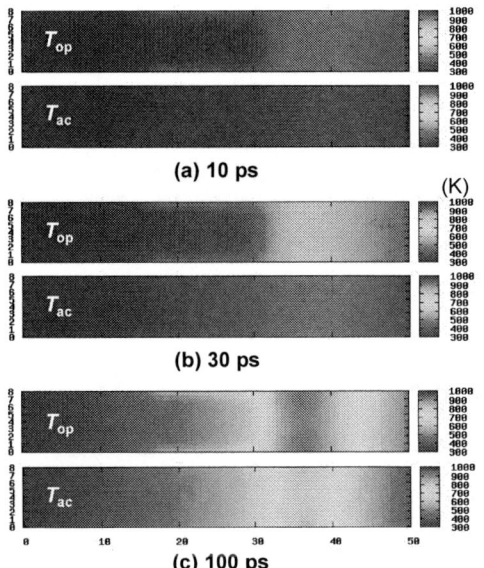

(a) 10 ps

(b) 30 ps

(c) 100 ps

Fig. 10: Time evolution of the local temperature distribution simulated with the ensemble MC simulator for phonon transport (τ_{op} = 10 ps). The temperatures of (a) optical phonons and (b) acoustic phonons are plotted as a parameter of time t after turning on the device.

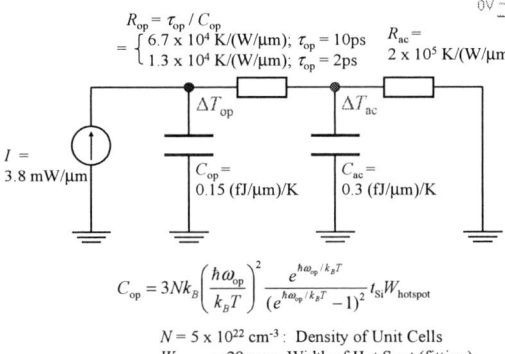

$$R_{op} = \tau_{op} / C_{op}$$
$$= \begin{cases} 6.7 \times 10^4 \text{ K/(W/}\mu\text{m)}; \ \tau_{op} = 10\text{ps} \\ 1.3 \times 10^4 \text{ K/(W/}\mu\text{m)}; \ \tau_{op} = 2\text{ps} \end{cases}$$

$$R_{ac} = 2 \times 10^5 \text{ K/(W/}\mu\text{m)}$$

$I = 3.8$ mW/μm

$C_{op} = 0.15$ (fJ/μm)/K

$C_{ac} = 0.3$ (fJ/μm)/K

$$C_{op} = 3Nk_B \left(\frac{\hbar\omega_{op}}{k_B T}\right)^2 \frac{e^{\hbar\omega_{op}/k_B T}}{(e^{\hbar\omega_{op}/k_B T} - 1)^2} t_{Si} W_{hotspot}$$

$N = 5 \times 10^{22}$ cm^{-3} : Density of Unit Cells
$W_{hotspot} = 28$ nm : Width of Hot Spot (fitting)

Fig. 13: Equivalent thermal circuit model for understanding the simulation results shown in Fig. 12. The heat generation source is modeled as a constant current source I, and the node voltages correspond to the temperatures $\Delta T_{op} = T_{op} - T_0$ and $\Delta T_{ac} = T_{ac} - T_0$, where T_0 (= 300 K) is the specified temperature at the phonon reservoirs. The value of I was obtained by integrating the heat density shown in Fig. 9 (a). C_{op} is calculated from the specific heat for optical phonons, and $R_{op} = \tau_{op} / C_{op}$. R_{ac} and C_{ac} are determined by the fitting procedure. All values are given per unit length of the device along the z direction.

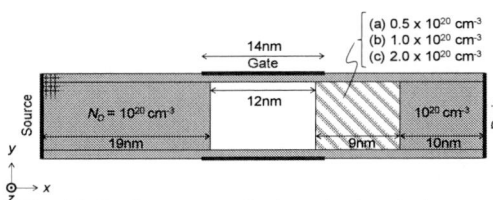

Fig. 16: Device structure for investigating the impact of drain potential profile on the hot spot temperature. The doping density at the drain edge (hatched area) was changed to (a) 0.5 x 10^{20}, (b) 1 x 10^{20}, and (c) 2 x 10^{20} cm^{-3}.

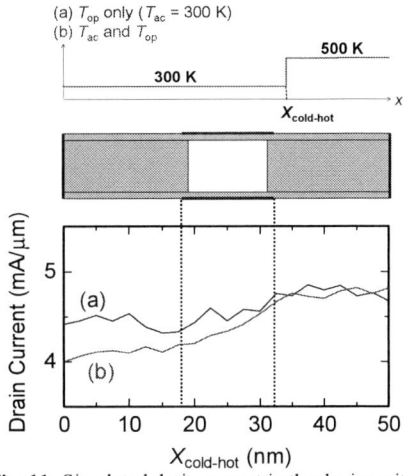

Fig. 11: Simulated drain current in the device with non-uniform temperature distribution. The device temperature ((a) T_{op} only, and (b) both T_{op} and T_{ac}) was raised to 500 K in $x > X_{cold-hot}$, while the other area was set to 300 K.

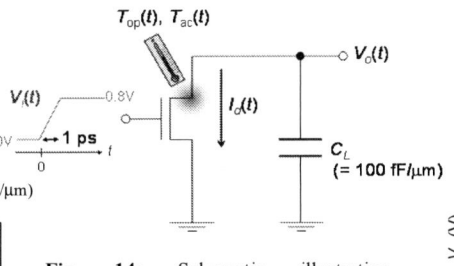

Fig. 14: Schematic illustration explaining the simulation mimicking the transistor switching in digital circuits. A relatively heavy load capacitance C_L (40x larger than the gate capacitance of the FET) was connected to the drain, and its discharge from $V_o = 0.8$ V to 0 V was simulated using the coupled MC simulator for electrons and phonons.

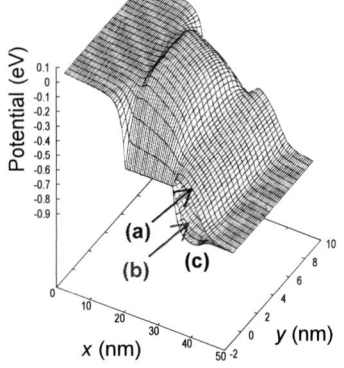

Fig. 17: Simulated potential profiles for the device structures depicted in Fig. 16.

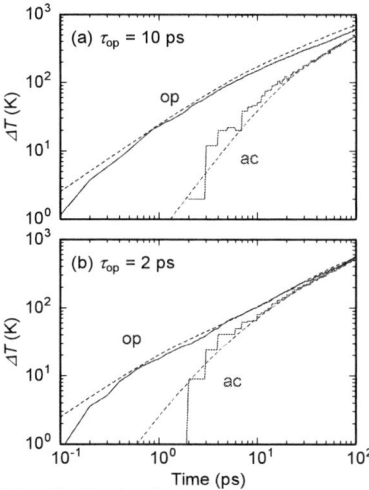

Fig. 12: The local phonon temperatures at the drain edge (x, y) = (35 nm, 0 nm) are plotted as a function of t. Solid lines: the results simulated with the MC method. Dashed lines: the results calculated with the equivalent thermal circuit model given in Fig. 13.

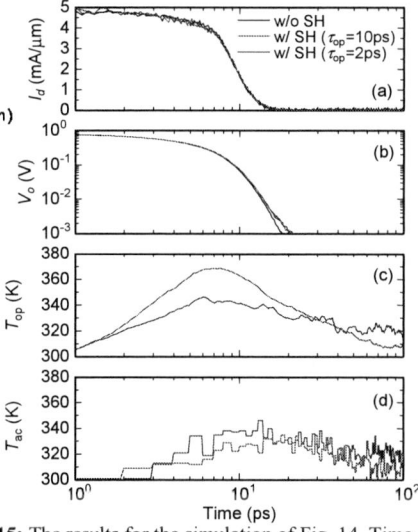

Fig. 15: The results for the simulation of Fig. 14. Time evolutions of (a) the drain current I_d, (b) the drain voltage V_o, (c) the optical phonon temperature T_{op}, and (d) the acoustic phonon temperature T_{ac} are plotted as a function of t after applying the gate voltage.

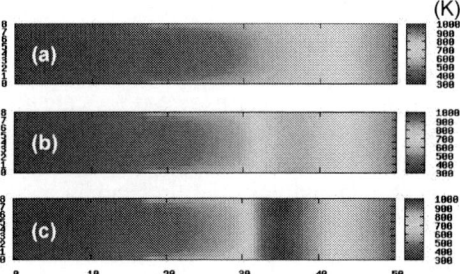

Fig. 18: Simulated optical phonon temperature distributions for the device structures depicted in Fig. 16. As the simulation in Fig. 8, the drain voltage of 0.8 V was applied, and the temperature distributions were observed 80 ps after turning on the gate voltage.

978-1-4799-8002-4/14 $31.00 © 2014 IEEE

Modeling and Optimization of Group IV and III-V FinFETs and Nano-Wires

Victor Moroz, Lee Smith, Joanne Huang, Munkang Choi, Terry Ma,
Jie Liu, Yunqiang Zhang, Xi-Wei Lin, Jamil Kawa, and Yves Saad

Synopsys, Inc., 700 East Middlefield Road, Mountain View, CA 94043 USA
(E-mail: victor.moroz@synopsys.com)

Introduction

Design of 7nm FinFETs and 5nm nanowire (NW) MOSFETs is exciting, as there are multiple unexplored options in terms of material choices, transistor architecture, and new approaches to building standard library cells. The expected design rules are determined by the requirement to double transistor density at subsequent technology nodes **(Fig. 1)**. The design rules are driving the choice of transistor architecture through electrostatics, leakage, patterning and mechanical integrity (i.e. manufacturability).

In this work, we present an overview of modeling methodologies that cover different aspects of designing 7nm and 5nm transistors and standard library cells using group IV and III-V materials.

Modeling methodology

The workhorse for Technology Computer Aided Design (TCAD) transistor analysis remains Drift-Diffusion model (DD) due to its computational efficiency, although for the expected channel lengths and fin/wire cross-section it is extended by engaging a variety of advanced models such as Density Functional Theory (DFT), Empirical Pseudopotential Method (EPM), Quantum Transport (QT), and Monte Carlo (MC) [1] **(Fig. 2)**.

Quantum confinement in narrow fins or NWs considerably affects the bandstructure, which in turn determines transistor behavior. The shape of the NW cross-section is determined by the manufacturing process [2] and can be described by minimizing surface energy using Wulff construction **(Fig. 3)**. The band structure of materials confined into specific shapes with tight dimensions is calculated with DFT and EPM models and then transferred into QT, MC, and DD tools for carrier transport analysis. Introducing ballistic mobility [3-6] enables to extend DD model to short channel transistors with ballistic transport contributing over 90%.

Mechanical integrity

The fin height is not expected to scale significantly whereas the fin width has to shrink in sync with the channel length to prevent excessive off-state leakage. This means higher fin aspect ratio and potential fin bending and collapse due to the forces acting on the fin during manufacturing, for example at wet cleaning process. Applying a force to the top of reference fin [7] leads to moderate deformation of 7nm **(Fig. 4)**. Same force applied to a 6nm wide fin shows excessive bending, but switching to shallower isolation with dual tapering or to SOI FinFET helps to bring bending back towards behavior of the reference fin that is known to be manufacturable.

Another important issue related to slow fin height scaling combined with aggressive fin pitch scaling is the merging of epitaxial SiGe source/drain that can introduce dislocations **(Fig. 5)**. One way to simultaneously address the fin bending and the epi merging issues is to replace fin with Gate All Around (GAA) NW.

Group IV and III-V channels for 7nm FinFETs

The confinement in a 6nm wide fin has moderate impact on the SiGe bandgap but significantly widens the InGaAs bandgap. For the N-FinFETs with 7nm design rules, only Si-Ge channels with <40% Ge **(Fig. 6)** and InGaAs channels with <20% In **(Fig. 7)** satisfy Low Power leakage spec. Notice that for III-V channel material with lower effective mass the bandgap has to be over 1.4 eV for LP spec due to the high Band-To-Band-Tunneling (BTBT) leakage

According to our MC analysis, about 64% of electrons experience ballistic transport in 14nm N-FinFETs **(Fig. 8)**. Contribution of ballisticity is expected to increase to 84% at 7nm node due to the shorter channel and due to > 1 GPa stress expected in N-FinFETs with Stress Relaxed Buffer (SRB) stress engineering [8]. In this section, we perform analysis of FinFET channel material engineering by using a subband-based Lundstrom top-of-the-barrier approach.

Comparative analysis of 7nm FinFETs with different channel materials shows that the best High Performance (HP) FinFETs are InGaAs with 30% In, with pure Ge channel only 8% behind **(Fig. 9a)**. For the Standard Process (SP) FinFETs, the best option by far is relaxed InGaAs channel with 30% In, with strained Si trailing by ~60% I_{dsat}. For the LP FinFETs, the best performer is strained Si channel.

This behavior is explained by inversion charge N_{inv} **(Fig. 9b)** and injection velocity V_{inj} **(Fig. 9c)**. SiGe with low Ge content has almost flat N_{inv} and V_{inj} that follows by N_{inv} drop at 85% Ge where the carriers move from delta valleys into lambda valleys with lower effective mass. This follows by N_{inv} recovery towards pure Ge accompanied by two-fold increase in V_{inj}. The InGaAs has systematically lower N_{inv} than SiGe, but considerably higher V_{inj}, especially towards InAs.

To summarize this section, at 7nm design rules III-V NMOS FinFETs can outperform Si channel by almost 2x for SP leakage spec, but trail Si for LP leakage spec. For the PMOS, the III-V material of choice is InGaSb with bandgap that is about half of the InGaAs, and therefore even worse BTBT leakage. The SiGe and Ge channels exhibit slight advantage over Si in certain conditions.

978-1-4799-8002-4/14 $31.00 © 2014 IEEE

2-Input NAND library cell

Considering mechanical integrity issues and transistor performance, let's compare 7nm strained Si bulk FinFETs with 7nm relaxed Ge SOI FinFETs and 5nm relaxed Si GAA NWs, all for the SP leakage spec of 1 nA/um at Vdd=0.7 V. Here, we use QT for modeling ballistic channel current and Lundstrom backscattering model with the low-field mobility due to phonon and surface roughness scattering computed with a Kubo-Greenwood approach. The DD model is calibrated to QT results in the channel and then DD is applied to the entire transistor, including contact resistances, source, and drain.

The 7nm process options have perfectly matched NMOS/PMOS, with relaxed Ge SOI exhibiting ~10% performance advantage **(Fig. 10)**. The 5nm NWs match NMOS FinFET performance, but lag ~20% in PMOS strength, mainly due to the heavier hole mass for <100> NWs. Nanowires provide surprisingly strong performance, with the on-state current of a NW matching that of the fin that has ~6x larger cross-section area.

Transition from a single GAA fin to a single GAA NW with a fixed fin pitch exhibits much stronger NW performance than would be suggested by naïve assumption of the current being proportional to the fin height **(Fig. 11)**. This is achieved by several simultaneous effects. First, as the fin height scales from 30nm down to 5nm, electrostatics improves and sub-threshold slope reduces from high 70's to low 60's. Second, there is a ~1.8x boost of N_{inv} due to gate-all-around NW geometry. Third, the NW exhibits ~1.6x higher carrier velocity due to the confinement-driven carrier repopulation into the two-fold delta valleys with lower effective mass. Some of the gain is set back by the the higher source access resistance, because of the scaled down contact area (conservatively assuming the same ρ=1.5e-8 $\Omega\,cm^2$) and because of the smaller cross-section of source extension. With everything accounted for, one NW strength is comparable to one fin strength. This enables building 5nm library cells with comparable number of lateral or vertical NWs to the number of fins in a 7nm FinFET 2-NAND cell.

The 10 track high 7nm 2-input NAND library cell is based on quadruple spacer fin patterning and quadruple LE-LELELE metal patterning with 5 NMOS and 5 PMOS fins **(Fig. 12)**. Parasitic Middle-Of-Line (MOL) capacitances are unusually high due to the tightly spaced metal 0 and gates **(Fig. 13)**. The 10 track high 5nm 2-NAND cell is based on EUV+DSA NW patterning with four NMOS and five PMOS vertical NWs **(Fig. 14)**. Parasitic capacitances are an order of magnitude lower due to the different way of building interconnects **(Fig. 15)**. Atomistic analysis is instrumental in optimizing the shape of epitaxially grown S/D on top of the vertical NWs to minimize access resistance **(Fig. 16)**.

Comparative analysis of the 11-stage ring oscillators built on different technology options of the 2-NAND library cell shows a huge spread in performance vs power consumption **(Fig. 17)**. The switching delay and the energy consumption per switch are shown per one ring oscillator stage, which

is a 2-NAND library cell for three different power supply voltages of 0.5 V, 0.6 V, and 0.7 V.

The 7nm Ge SOI FinFET has a slight advantage over the baseline strained Si FinFET process. Considering that the maximum gain that can be achieved by channel material engineering in previous section is no more than 2x, we evaluate a hypothetical channel material that doubles performance of both NMOS and PMOS FinFETs. At Vdd = 0.7 V, the double transistor strength option reduces the delay by 2.6x and simultaneously reduces power consumption by 24%.

Alternatively, replacing nitride spacers with oxide spacers achieves even better gains of 2.2x speedup with 2.4x power reduction. This happens due to significant reduction of parasitic MOL capacitances that have to be recharged at every switch cycle. however, further reduction of spacer permittivity from 3.9 for the oxide to 2.6 for a low-k material does not bring a noticeable performance gain, because MOL capacitances are already negligible with oxide spacers for the FinFETs with 7nm design rules.

Switching transistor architecture from the baseline Si FinFET to lateral GAA NW with identical layout achieves similar gains to the spacer material engineering, mainly due to reduction of the MOL capacitances for the NW being shorter than the fin. The best performance-energy trade-off is achieved by vertical Si NWs with 5nm design rules, with 3.6x gain in switching speed and simultaneous 5x power reduction.

Summary

We described simulation methodologies involving a variety of modeling techniques applied to design and optimization of several key aspects of the 7nm and 5nm transistors and standard library cells. Analysis of channel material engineering for the 7nm FinFETs points to different trade-offs for the HP, SP, and LP leakage specs. Mechanical stability of the fins with high aspect ratio is evaluated as a major factor determining fin shape engineering and transitions from bulk FinFET to SOI FinFET and then to NW. Comparative analysis of the 10 track high 2-input NAND library cells based on different channel materials, different spacer materials, and different transistor architectures suggests that the largest benefits of 3.6x speed gain with 5x reduction in power consumption is achieved by switching from 7nm Si baseline FinFET process to 5nm vertical Si NWs. Within lateral transistors at 7nm design rules, transition from fins to lateral NWs and replacing nitride spacers with oxide spacers offer significant speed/power advantage. The channel material engineering brings the weakest advantage on the library cell level.

References

[1] T. Ma et al., *IEDM Tech. Dig.*, p. 367, 2010.
[2] P. Hashemi et al., *IEDM Tech. Dig.*, p. 788, 2010.
[3] C. Jungemann, *ISDRS Proc.*, 2007.
[4] M. Lundstrom, *IEEE Trans. Electron Dev.*, p. 133, 2002.
[5] M. S. Shur, *Electron Dev. Lett.*, p. 511, 2002.
[6] M. Zilli et al., Electron Dev. Lett., p. 1036, 2007.
[7] C.-H. Jan et al., *IEDM Tech. Dig.*, p. 44, 2012.
[8] S. Gupta et al, *IEDM Tech. Dig.*, p. 641, 2013.

Technology	7nm	5nm
Gate pitch	44	31
L	15	11
Spacer	7	5
Fin width	6	5
Fin height	30	5
Fin pitch	24	16
Contact size	15	10
EOT	0.8	0.8

Fig. 1 Expected design rules for 7nm and 5nm nodes. All sizes are in nm.

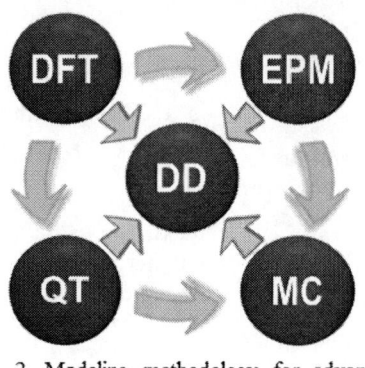

Fig. 2. Modeling methodology for advanced CMOS devices. DFT – Density Functional Theory, EPM – Empirical Pseudopotential Method, DD – Drift-Diffusion, QT – Quantum Transport, MC – Monte Carlo.

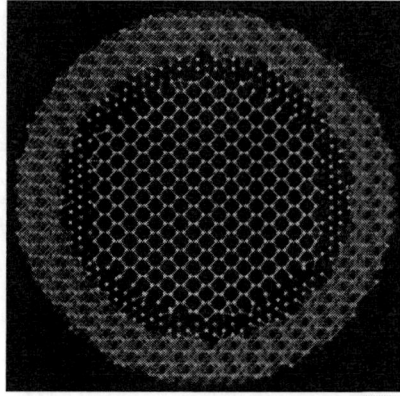

Fig. 3. Atomistic construction of 5nm Si nanowire with SiO/HfO dielectric

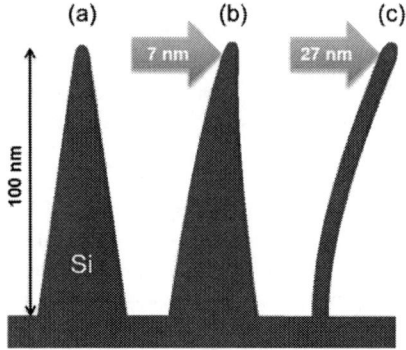

Fig. 4. Fin bending at wet clean process. (a) Reference fin from 22nm node, (b) Bending of reference fin, (c) 6nm wide fin, (d) 7nm node, (e) 7nm Ge SOI FinFET.

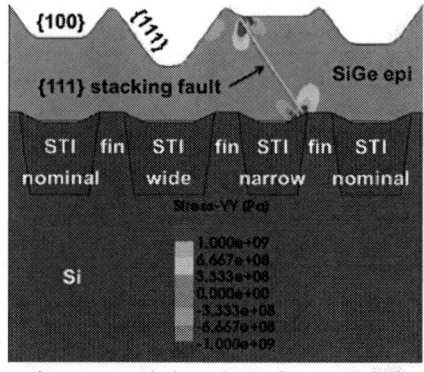

Fig. 5. SiGe source/drain epitaxy for parallel fins with uneven spacing due to patterning imperfections.

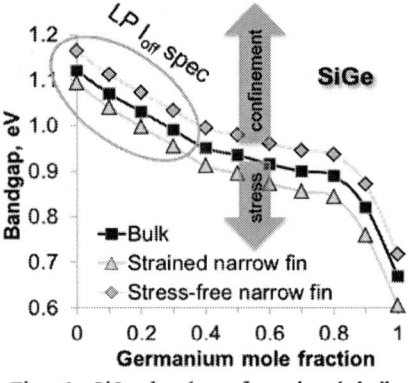

Fig. 6. SiGe bandgap for relaxed bulk, relaxed 6nm fin, and fin with +2 GPa stress.

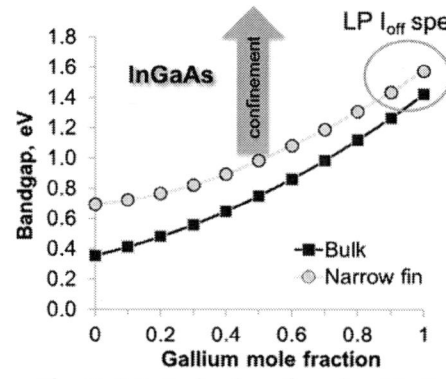

Fig. 7. InGaAs bandgap for relaxed bulk, relaxed 6nm fin, and fin with +2 GPa stress.

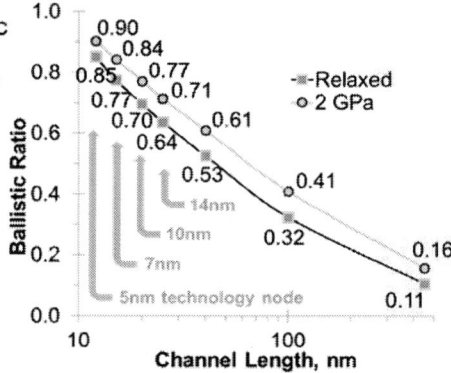

Fig. 8. Contribution of ballistic current to Idsat of Si N-FinFET with 6nm wide fin.

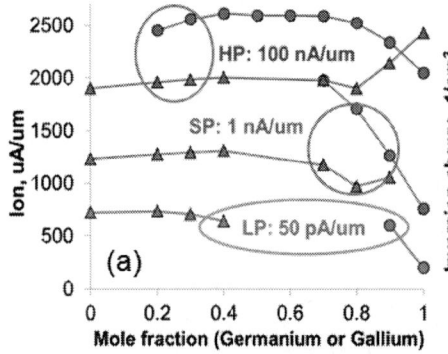

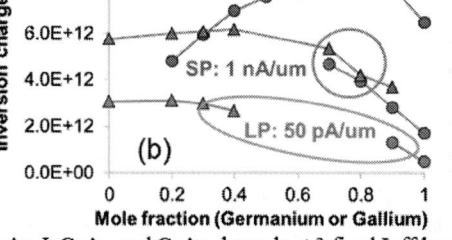

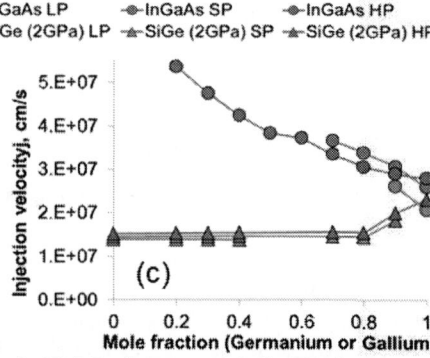

Fig. 9. The 7nm N-FinFETs with Si, SiGe, Ge, InAs, InGaAs, and GaAs channels at 3 fixed Ioff levels: High Performance, Standard Performance, and Low Power. (a) Idsat at Vdd=0.7 V, (b) Inversion charge, (c) Electron injection velocity. Triangles show SiGe and circles show InGaAs data.

978-1-4799-8002-4/14 $31.00 © 2014 IEEE

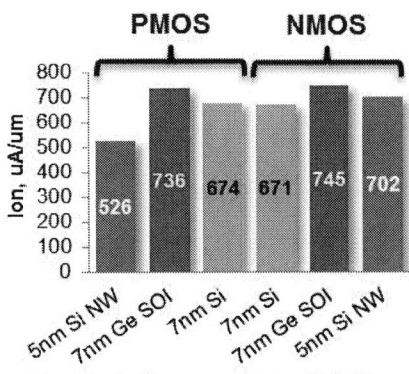

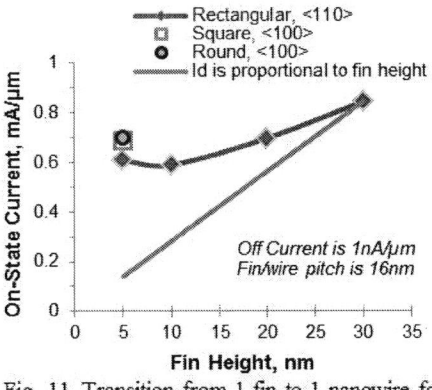

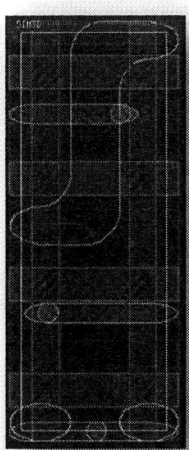

Fig. 10. Performance of 7nm FinFETs and 5nm nanowires at Vdd=0.7 V and SP Ioff.

Fig. 11. Transition from 1 fin to 1 nanowire for NMOS GAA Si channel with 5nm design rules.

Fig. 12. Layout of 10 track high 2-input NAND library cell implemented with 7nm strained Si FinFETs. Quadruple spacer transfer fin patterning. Double spacer transfer gate patterning. Quadruple LELELELE M0 and M1 patterning. Five parallel NMOS fins and five PMOS fins. Cell height is 360 nm and cell width is 132 nm.

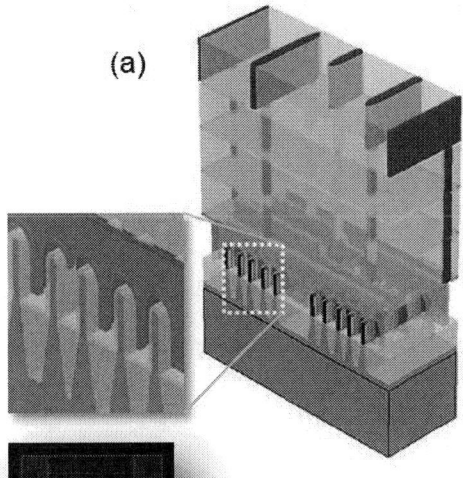

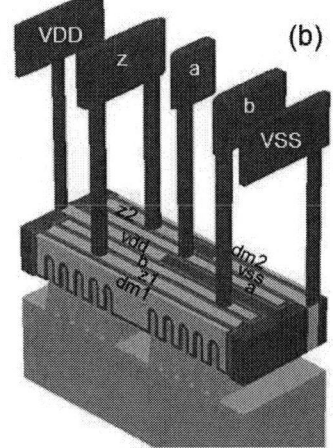

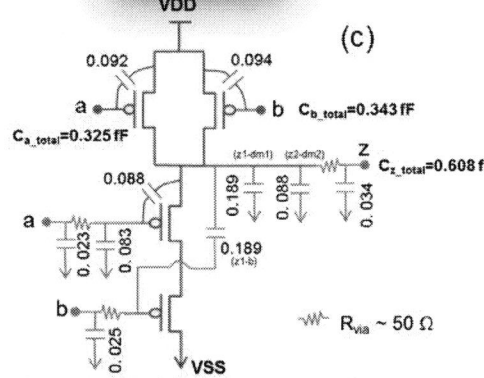

Fig. 13. 7nm 2-NAND library cell. (a) Overview, (b) MOL and BEOL, (c) Extracted parasitic capacitances.

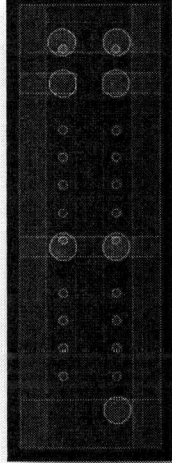

Fig. 14. Layout of 10 track high 2-input NAND library cell implemented with 5nm relaxed Si nanowires. EUV + DSA nanowire patterning. Barrier-less cobalt interconnects. Four parallel NMOS wires and five PMOS wires. Cell height is 240 nm and cell width is 62 nm.

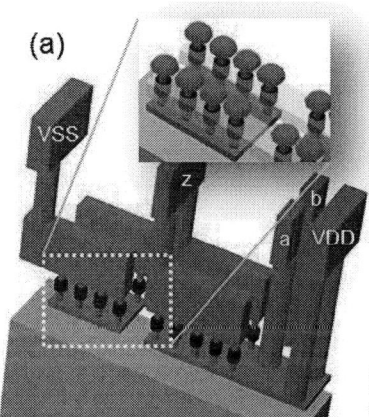

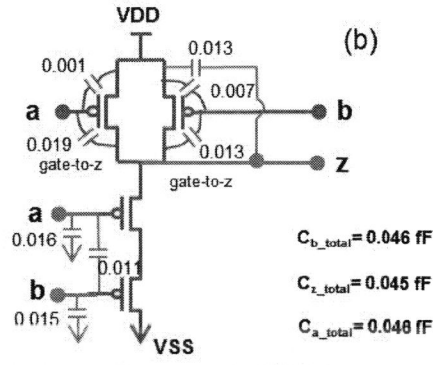

Fig. 15. 5nm vertical nanowire 2-NAND library cell. (a) Overview, (b) Extracted parasitic capacitances.

Fig. 16. Atomistic Lattice Kinetic Monte Carlo modeling of source/drain epitaxy evolution and merging with the neighbors for 5nm vertical nanowires.

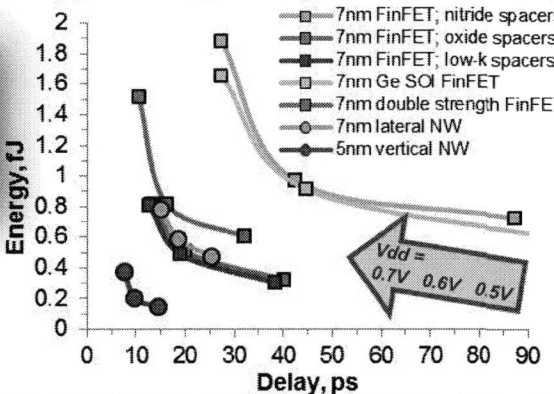

Fig. 17. Performance of 11-stage ring oscillators built with 7nm and 5nm 2-NAND library cells with 1 nA/um I_{off} @0.7V

978-1-4799-8002-4/14 $31.00 © 2014 IEEE

Performance Evaluation of InGaAs, Si, and Ge nFinFETs based on Coupled 3D Drift-Diffusion/Multisubband Boltzmann Transport Equations Solver

Seonghoon Jin, Anh-Tuan Pham, Woosung Choi, Yutaka Nishizawa[†],
Young-Tae Kim*, Keun-Ho Lee*, Youngkwan Park*, and Eun Seung Jung*

Samsung Semiconductor Inc., 3655 N First St., San Jose, CA 95134, USA, Tel: 408-544-5273, email: s.jin@samsung.com
[†]Samsung R&D Institute Japan *Semiconductor R&D center, Samsung Electronics, Korea

ABSTRACT

This paper presents a simulation study of InGaAs, Si, and Ge nFinFETs by solving the coupled drift-diffusion (DD) and the multisubband Boltzmann transport equation (MSBTE) in 3D domains. The effects of the quasi-ballistic transport, source/drain contact resistances, and band-to-band tunneling (BTBT) on the device performance are studied.

INTRODUCTION

As the traditional scaling down of the transistor size can no longer provide performance improvement, the advance of the current logic technology is mainly driven by innovations such as introducing new channel materials, strains, and novel 3D structures [1]. This inevitably increases the time and cost to develop a new technology node. Previously, TCAD device simulation was mainly used to optimize the device design for the given material and structure. Nowadays, it plays an important role to explore and narrow down various technology options at the early stage of the technology development.

In order to provide reasonable guidelines at the early stage where the experimental data are insufficient to calibrate the conventional macroscopic models, a physics-based microscopic model is preferable as the model parameters are mainly originated from intrinsic material properties and less dependent on the specific technology. In addition, the model should be able to simulate realistic 3D structures. Finally, the source/drain resistances need to be modeled accurately as their portions to the total resistance keep increasing due to the small contact area and the low active concentration [2], [3], [4].

The coupled 3D DD/MSBTE solver [5], [6] employed in this work fullfills these requirements as it accounts for both the detailed microscopic physics in the channel and the realistic source/drain design. Using the solver, we explore the performance of InGaAs, Si, and Ge nFinFETs. In addition, an improved density-gradient (DG) model and a nonlocal BTBT model with a quantum correction term is developed to estimate the leakage of the new channel materials.

MODEL DESCRIPTIONS

In the coupled DD/MSBTE solver, a 3D simulation domain is divided into the channel region and the source/drain regions. Then, the semi-classical MSBTE is deterministically solved in the channel region while the DD equation is solved in the source/drain regions in a fully coupled manner with appropriate boundary conditions [6].

In the MSBTE, the free streaming operator describes the ballistic transport along the quasi-1D k-space for each 2D quantized subband while the scattering operator provides the coupling between different momentums and subbands [5]. Therefore, the material dependent bandstructure and scattering parameters determine the carrier transport in the channel.

In order to obtain the quantized subband structure for the Γ, Λ, and Δ conduction band valleys for the given material, orientation, and stress conditions, the effective mass approximation (EMA) Schrödinger equation with nonparabolic correction [7] is solved at multiple 2D device cross-sections with calibrated band structure parameters.

Phonon (acoustic, non-polar, and polar optical modes), surface roughness (SR), and alloy scattering mechanisms are considered where the scattering models are extended to handle arbitrary 2D cross-sections [6]. The scattering parameters are determined from the measured low-field mobility curves [8].

In order to model the contact resistance, a distributed resistance is included at the semiconductor/metal interface.

The extension of the DG model and the BTBT model will be briefly explained in the next section.

RESULTS AND DISCUSSION

Fig. 1 shows the simulated SOI finFET structure. As for the channel material, InAs, $In_{0.53}Ga_{0.47}As$, Si, and Ge are considered. For Si and Ge, we consider relaxed and 2 GP tensile stressed conditions. In order to maximize the contact area, we assume that the contacts fully surround the source and drain.

Fig. 2 compares the electron density profiles at the 2D cross-sections of the different channel materials obtained from the self-consistent simulations. Fig. 3 compares the electron density, average injection velocity, and low-field mobility of the different channel materials. The Si channel provides the largest inversion charge, but it gives the lowest injection velocity. On the contrary, the InAs channel provides the highest injection velocity, but the inversion charge is the lowest due to the small density-of-states. The Ge channel provides the highest mobility with reasonable carrier density and injection velocity. Compared with the relaxed Si and Ge, the strained Si (sSi) and the strained Ge (sGe) provide significantly improved

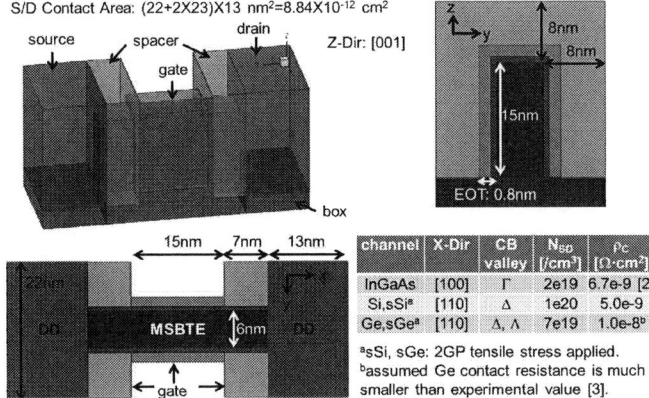

S/D Contact Area: (22+2X23)X13 nm²=8.84X10⁻¹² cm²

Z-Dir: [001]

EOT: 0.8nm

channel	X-Dir	CB valley	N_{SD} [/cm³]	ρ_c [Ω·cm²]
InGaAs	[100]	Γ	2e19	6.7e-9 [2]
Si,sSi[a]	[110]	Δ	1e20	5.0e-9
Ge,sGe[a]	[110]	Δ, Λ	7e19	1.0e-8[b]

[a]sSi, sGe: 2GP tensile stress applied.
[b]assumed Ge contact resistance is much smaller than experimental value [3].

Fig. 1. Simulated 3D SOI finFET structure with InAs, In₀.₅₃Ga₀.₄₇As, Si, and Ge channel materials. For Si and Ge, we consider relaxed and 2 GP tensile stressed conditions. The DD equation is solved in the source/drain while the MSBTE is solved in the channel.

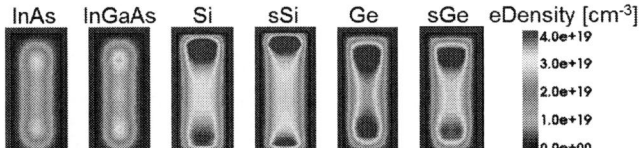

Fig. 2. Electron density profiles at the 2D cross-section of the channel center when $V_G = 0.6$ V and $V_D = 0$ V. The quantum confinement is more pronounced at the InAs and InGaAs channels due to small effective mass.

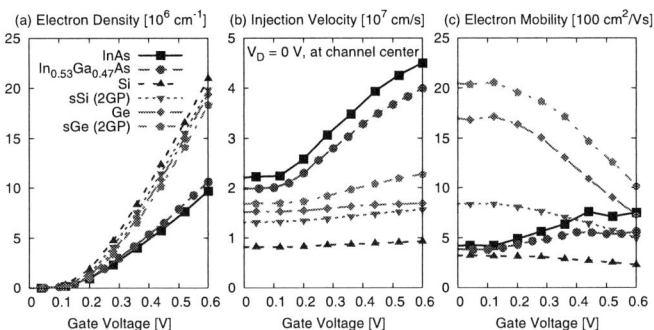

Fig. 3. (a) Integrated electron density, (b) average injection velocity, and (c) low-field electron mobility at the center of the channel as a function of V_G. The Si channel provides the largest inversion charge, but it gives the lowest injection velocity. The InAs channel provides the highest injection velocity, but the inversion charge is the lowest due to the small density-of-states. The Ge channel provides the highest mobility with reasonable carrier density and injection velocity. Compared with the relaxed Si and Ge, the strained Si (sSi) and the strained Ge (sGe) provide significantly improved injection velocity and mobility with only a minor reduction in the inversion charge.

injection velocity and mobility with only a minor reduction in the inversion charge.

Fig. 4 shows the obtained $I_D - V_G$ curves for the different channel materials. The sGe channel gives the largest on-current provided that $\rho_c = 10^{-8}$ Ωcm². However, the drain current is largely degraded when more realistic contact resistance is employed (see Fig. 5). Therefore, the contact resistance must be reduced to make the n-type sGe attractive. Apart from the sGe, our simulations show that the sSi channel still gives better performance than the InGaAs channel.

According to our simulations, the scattering processes in the InGaAs channel are still important (see Fig. 6 and Fig. 7). Among them, the SR scattering is most critical due to the strong geometric quantization along the width direction ($W_{fin} = 6$ nm). As the Γ valley of the In$_{1-x_{Ga}}$Ga$_{x_{Ga}}$As channel exhibits strong band nonparabolicity ($\alpha \approx 2 - x_{Ga}$), it is important to take into account the nonparabolic correction (NPC) also in the matrix element of the SR scattering (see (90) of [7]), which is usually overlooked. In Fig. 8, we compare the calculated low-field mobility of the single-gate ultrathin body (UTB) In₀.₅₃Ga₀.₄₇As channel as a function of body thickness with available simulation [9], [10] and experimental data [8], [11], [12]. The calculated mobility with the NPC in the SR scattering matches fairly well with the experimental data as well as our internal 8-band $\mathbf{k} \cdot \mathbf{p}$ Hamiltonian-based mobility

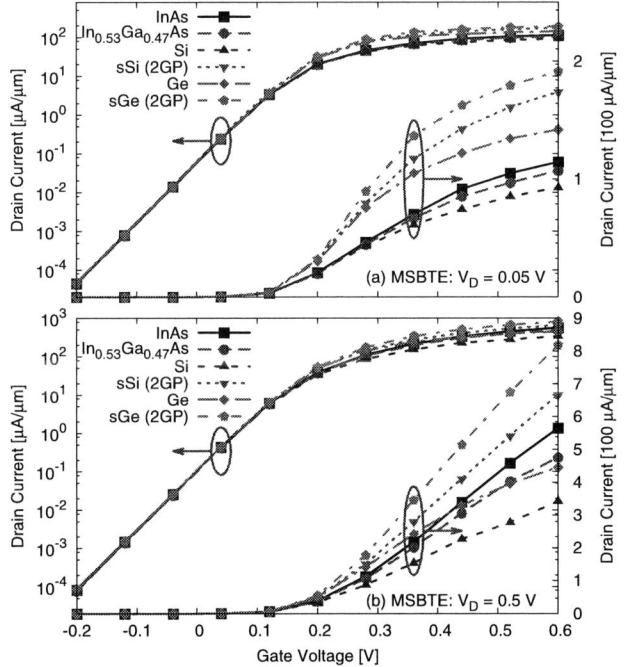

Fig. 4. Simulated $I_D - V_G$ curves at (a) $V_D = 0.05$ V and (b) $V_D = 0.5$ V. The gate workfunction is adjusted to obtain the same off-current (0.1 μA/μm) at $V_G = 0$ V and $V_D = 0.5$ V. The drain current is normalized by the gate overlapped fin perimeter (36 nm). The sGe channel gives the largest on-current provided that $\rho_c = 10^{-8}$ Ωcm². The sSi channel gives better performance than the InGaAs channel.

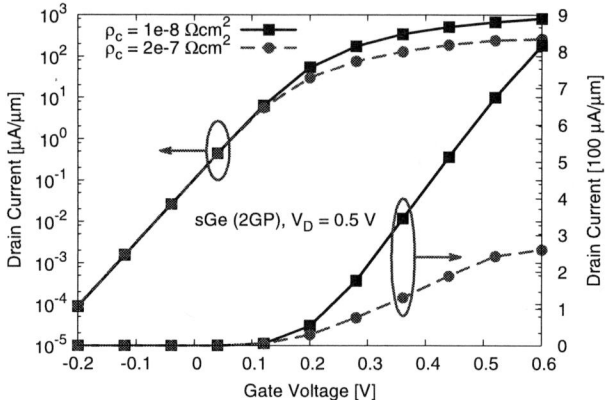

Fig. 5. Influence of the contact resistance on the drain current of the strained Ge channel finFET. The use of more realistic contact resistance (2×10^{-7} Ωcm^2) significantly degrades the performance.

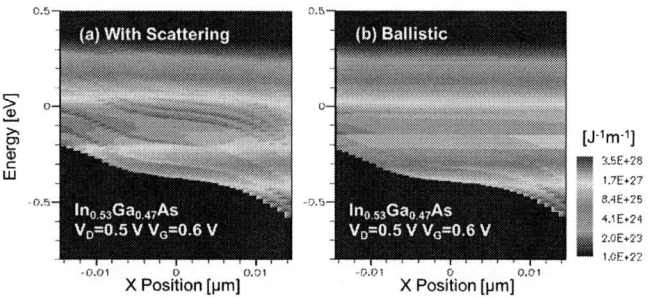

Fig. 6. Energy resolved electron density inside the $In_{0.53}Ga_{0.47}As$ channel (a) with and (b) without scattering. In the ballistic simulation, small amount of artificial inelastic phonon scattering needs to be added to avoid numerical instability.

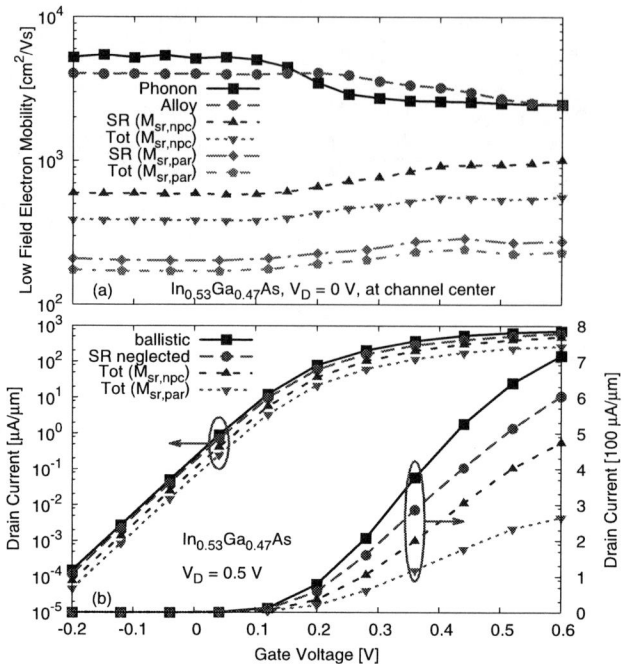

Fig. 7. (a) Component-wise mobility of the $In_{0.53}Ga_{0.47}As$ channel finFET and (b) the influence of the scattering mechanisms on the drain current. The scattering processes significantly degrade the performance. The SR matrix element from the parabolic EMA ($M_{sr,par}$) overestimates the scattering rate compared with that with the nonparabolic correction ($M_{sr,npc}$). When the fin width is small enough, $M_{sr,npc}^2 \approx M_{sr,par}^2 \left[1 + 4\alpha\left(E_\mu - U_\mu\right)\right]^{-1/2} \left[1 + 4\alpha\left(E_\nu - U_\nu\right)\right]^{-1/2}$ where α is the nonparabolicity factor, μ and ν are the subband indices, and E_μ and U_μ are the subband energy and the expectation value of the potential energy obtained from the EMA [7].

calculations [8] while the EMA strongly underestimates the mobility.

For V_{DD} of 0.5 V, the BTBT is not a serious problem for the Si and Ge channels. However, the BTBT can limit the off-current in the InAs channel due to its small bandgap (≈ 0.36 eV). To check the BTBT leakage in the InAs channel, we first calibrate the density-gradient (DG) model [13] to match the carrier density and the drain current (see Fig. 9). The conventional DG model fails to reproduce the inversion charge density of the InAs channel as it inherently assumes the parabolic band. A new correction term is added to the DG model to capture the nonparabolic effects as shown in Fig. 9.

Then, a dynamic nonlocal BTBT model is employed where the model accounts for the stress and orientation dependent BTBT from the valence band to the Γ, Λ, and Δ valleys. The anti-crossing of the valence band and the quantum correction of the conduction and valence bands (see Fig. 10) are properly taken into account. Fig. 11 shows the drain current with the BTBT turned on. Since the off-state channel barrier is modulated by the generated holes inside the channel (see Fig. 12), the off-current is larger than the BTBT generation

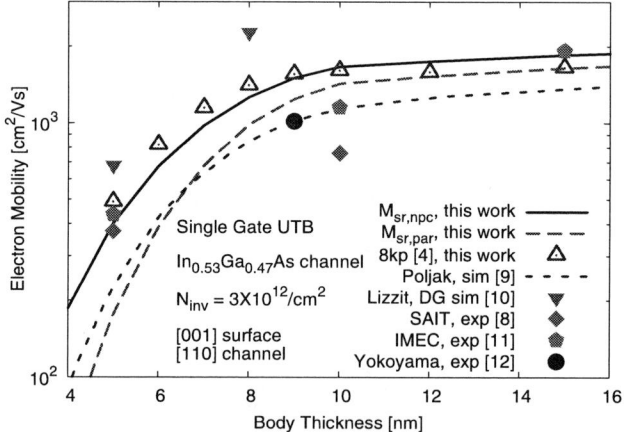

Fig. 8. Calculated low-field mobility of the single-gate UTB $In_{0.53}Ga_{0.47}As$ channel as a function of body thickness. Available simulation data [9], [10] and experimental data [8], [11], [12] are also shown for comparison. The calculated mobility with the NPC in the SR scattering matches fairly well with the experimental data as well as our internal 8-band $\mathbf{k} \cdot \mathbf{p}$ Hamiltonian-based mobility calculations [8] while the EMA strongly underestimates the mobility.

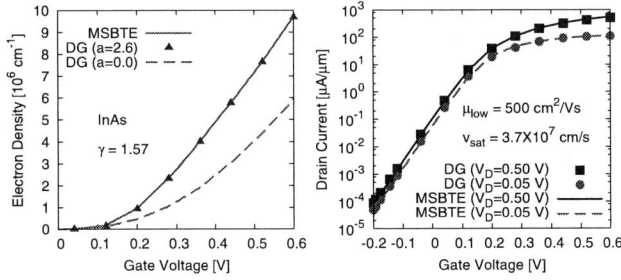

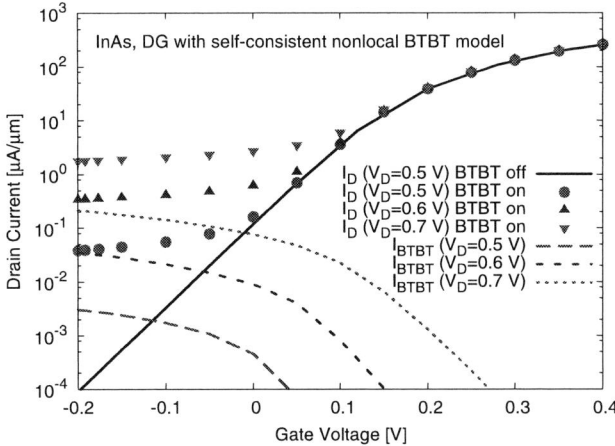

Fig. 9. The density-gradient (DG) calibration of the InAs channel finFET with an additional high density correction term. The new DG equation reads: $\Lambda = -\frac{\hbar^2 \gamma}{6m^*} \frac{\nabla^2 \sqrt{n}}{\sqrt{n}} - akT \ln\left[1 + \min\left(n/n_0, 10\right)\right]$ where n is the electron density and $n_0 = 10^{18}$ /cm^3. The correction term captures the enhancement of the inversion charge density due to the strong nonparabolicity of the Γ valley.

Fig. 11. $I_{\mathrm{D}} - V_{\mathrm{G}}$ curves obtained from the calibrated DG simulation with the nonlocal BTBT model for the InAs channel finFET. Since the off-state channel barrier is modulated by the generated holes (see Fig. 12), the off-current is larger than the BTBT generation current (BJT-like amplification). The full Jacobian (derivatives) needs to be considered in the nonlocal BTBT model for convergence.

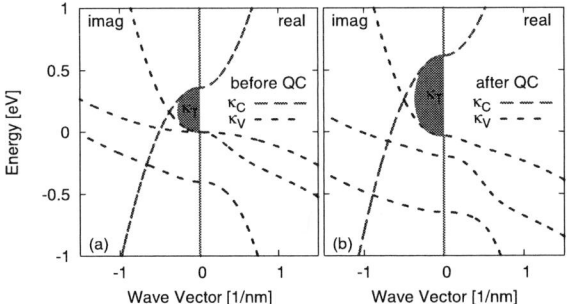

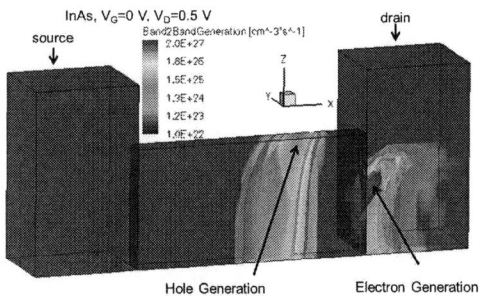

Fig. 12. Electron and hole band-to-band tunneling rate for the InAs finFET at $V_G = 0$ V and $V_D = 0.5$ V. Holes are generated inside the channel while electrons are generated at the drain side.

Fig. 10. The real and imaginary band dispersions along the [100] direction of the InAs channel (a) before and (b) after the quantum correction (QC). The quantum confinement reduces the BTBT rate as it reduces the tunneling energy window (the nonlocal overlap between E_C and E_V) and increases the WKB factor (proportional to the filled area in the uniform field limit). The present nonlocal BTBT model takes into account the valence band anti-crossing by adding the quantum correction to the 6-band $\mathbf{k}\cdot\mathbf{p}$ Hamiltonian similarly to the strain term and solving the inverse Schrödinger equation. For example, $H_{\mathrm{yy}} = L\pi^2/t^2$ and $H_{\mathrm{xx}} = H_{\mathrm{zz}} = M\pi^2/t^2$ are added when the quantization is along the y-direction with the body thickness t (L and M are the $\mathbf{k}\cdot\mathbf{p}$ parameters [14]). Due to the anti-crossing, the heavy hole band in the real k-space behaves like the split-off band in the imaginary k-space. The conduction band shift is modeled by: $\Delta E_{\mathrm{C}} = \left[-1 + \sqrt{1 + 2\alpha\hbar^2\pi^2/\left(m_{\mathrm{C}} t^2\right)}\right] / (2\alpha)$. The tunneling wavevector is obtained from the conduction and valence band imaginary dispersions as follows: $\kappa_{\mathrm{T}}^{-2} = \kappa_{\mathrm{C}}^{-2} + \kappa_{\mathrm{V}}^{-2}$.

current, which is similar to the BJT operation with the channel region as the base. Still, the BTBT induced leakage is marginal for $V_{\mathrm{D}} = 0.5$ V although it increases sharply with V_{D} and eventually limits the off-current (see Fig. 11).

Conclusion

In conclusion, we have demonstrated that the developed coupled DD/MSBTE solver can be an accurate and practical tool for studying the performance of the future node finFETs with new channel materials as it accounts for both the detailed microscopic physics in the channel and the realistic source/drain design. A few important improvements on the

modeling of the SR scattering, the DG calibration, and the BTBT have been presented. The simulation results show that the sGe channel can be an attractive option as long as the contact resistance can be reduced. Otherwise, the sSi channel will remain competitive as the InGaAs may not provide the expected ballistic performance due to the strong SR scattering.

References

[1] K. J. Kuhn, *IEEE Trans. on ElectronDevices*, 59, 1813 (2012).
[2] J. A. del Alamo, *et al.*, *IEDM*, 24 (2013).
[3] H. Miyoshi, *et al.*, *Jpn. J. Appl. Phys.*, 53, 04EA05 (2014).
[4] S. H. Park, *et al.*, *IEEE Trans. on ElectronDevices*, 59, 2107 (2012).
[5] S. Jin, *et al.*, *IEEE Trans. on Electron Devices*, 55, 2886 (2008).
[6] S. Jin, *et al.*, *SISPAD*, 348 (2013).
[7] S. Jin, *et al.*, *J. Appl. Phys.*, 102, 083715, (2007).
[8] A.-T. Pham, *et al.*, *ESSDERC*, accepted (2014).
[9] M. Poljak, *et al.*, *IEEE Trans. on Electron Devices*, 59, 1636 (2012).
[10] D. Lizzit, *et al.*, *IEEE Trans. on Electron Devices*, 61, 2027 (2014).
[11] A. Alian, *et al.*, *IEDM*, 437 (2013).
[12] M. Yokoyama, *et al.*, *Appl. Phys. Exp.*, 4, 054202 (2011).
[13] M. G. Ancona and H. F. Tiersten, *Phys. Rev. B*, 35, 7959 (1987).
[14] G. Dresselhaus, *et al.*, *Phys. Rev.*, 98, 368 (1955).

978-1-4799-8002-4/14 $31.00 © 2014 IEEE

Simulation analysis of III-V n-MOSFETs: channel materials, Fermi level pinning and biaxial strain

Enrico Caruso, Daniel Lizzit, Patrik Osgnach, David Esseni, Pierpaolo Palestri, Luca Selmi

DIEGM, University of Udine, Via delle Scienze 208, 33100 Udine, Italy, FAX:+39-0432-558251
email: caruso.enrico@spes.uniud.it

I. Abstract

In this work we employ a state-of-the-art Multi-Subband Monte Carlo simulator to investigate the performance of III-V n-MOSFETs with L_G = 11.7nm. We analyze GaSb versus InGaAs strained and unstrained channel materials and the implications of Fermi level pinning on electrostatic and transport. We found that InGaAs MOSFETs can outperform strained silicon for low V_{DD} applications. Advantages related to strained InGaAs are limited and mainly due to reduced Fermi Level Pinning.

II. Introduction

III-V compound semiconductors are being extensively studied as channel materials in n-type MOSFETs for high-performance, low V_{DD} applications [1]. However, low density-of-states (DoS) and Fermi-level pinning (FLP) due to interface states (N_{it}) result in a lower inversion density (N_{inv}) compared to silicon [2], that limits the improvements in on-current, (I_{ON}). To overcome these issues, I_{ON} boosters are being actively investigated, including strain [3] and the use of materials such as GaSb where L valleys with larger DoS with respect to the Γ valleys contribute to the current [4].

Several computational studies have appeared based on top-of-the-barrier (ToB) models [5], [6], on 3D semi-classical Monte Carlo [7], or on full quantum, but mainly ballistic, transport models [8], [9]; all these studies considered unstrained III-V compound materials and neglected FLP and N_{it} effects.

Recently, however, experimental data became available for mobility in n-type InGaAs MOSFETs [10], [11], [2], for the FLP impact [2], and strain induced mobility enhancements [3], [12], which make it compelling to extend previous performance analysis.

In this work we employ a state-of-the-art Multi-Subband Monte Carlo simulator (MSMC) to investigate the performance of nanoscale III-V n-MOSFETs with L_G = 11.7nm. Our model is systematically compared to or calibrated against a large set of experiments and previous theoretical results and we analyze: (a) GaSb versus InGaAs channel materials; (b) electrostatic and transport implications of FLP; (c) strain induced μ_{eff} and I_{ON} improvements and possible links to FLP.

III. Models and devices

Our MSMC employes a non-parabolic effective mass approximation (NP-EMA) energy model for both quantization and transport [13]. The model captures the changes in the transport effective mass in thin quantum wells and was previously validated by comparison with Tight-Binding (TB) results [14]. Here we complement this validation by showing in Fig.1 the energy vs. wave-vector in GaAs and InAs quantum wells comparing the NP-EMA model to quantized $\mathbf{k}\cdot\mathbf{p}$ results.

Fig.2 reports the effective mass m^* extracted from the curvature of the E($\mathbf{k}$) close to the subband minima: m^* increases when reducing the T_W and the NP-EMA model tracks the $\mathbf{k}\cdot\mathbf{p}$ results very well.

As a further check of the NP-EMA model, we have run MSMC simulations switching off the scattering and verified that the results are in quite good agreement with the ToB calculations employing a full-band TB description [15]. Sample results are reported in Figs.3 and 4 for $In_{0.53}Ga_{0.47}As$ and GaSb, respectively. NP-EMA parameters are summarized in Tab.I.

In the following, we simulate a double-gate MOSFET, sketched in Fig.5, designed according to the ITRS for III-V based MOSFETs [1]. Since the ITRS spec for well thickness (T_W=7nm) gives a DIBL>100 mV/V, we set T_W=6nm.

IV. $In_{0.53}Ga_{0.47}As$ and GaSb MOSFET.

Simulated I_{DS} for a $In_{0.53}Ga_{0.47}As$ channel is compared in Fig.6 with relaxed and strained Si channels. Scattering parameters for $In_{0.53}Ga_{0.47}As$ are the same as in [14]. Parameters for Si and sSi are from [16]. In Fig.6, we see that for V_{GS} larger than the threshold voltage (and up to ≈ 0.55 V) $In_{0.53}Ga_{0.47}As$ provides an I_{DS} improvement with respect to sSi, whereas for larger V_{GS} sSi outperforms $In_{0.53}Ga_{0.47}As$ because of the DoS bottleneck of III-Vs.

As for (111) GaSb MOSFETs, Fig.7 shows that, for T_W=6nm, GaSb provides no I_{ON} improvements compared to $In_{0.53}Ga_{0.47}As$ even in the ballistic limit, due to the large population of the three L valleys that in (111) inversion layers show large mass in the transport plane (3 ellipses spaced by 120°). In fact, possible advantages of (111) GaSb have been claimed for T_W <3 nm [4] where the L valley with large quantization mass (and low in-plane mass) is dominant. Fig.8 shows that also in our model GaSb becomes competitive for T_W =3 nm.

978-1-4799-8002-4/14 $31.00 © 2014 IEEE

V. Interface states and FLP

Interface states were added to the model as described in [17]. Validation and calibration of the model against FLP measurements is reported in Figs.9 and 10. Significantly high Dit values in the CB (see Fig.11) are needed to reproduce the experimental FLP.

In the presence of strain the FLP is mitigated due to the lowering of the CB: in fact, following [3] we have assumed that the Dit profile referred to the vacuum level does not change (i.e we shift rigidly the Dit profile with respect to the conduction band minimum, Fig.10) and obtained a fairly accurate estimate of the pinning in the presence of strain (see Fig.9).

The simulated I_{DS}-V_{GS} curves of the device in Fig.5 with $In_{0.53}Ga_{0.47}As$ channel, with and without interface states, are reported in Fig.12: Dit inducing pinning at inversion charges of approx $5 \cdot 10^{12} cm^{-2}$ has a detrimental effect on I_{ON}. Dit profiles inducing less pinning (i.e around $1 \cdot 10^{13} cm^{-2}$) are required to limit this effect.

In this analysis we set a low Dit in the gap (see Fig.10), much lower than experimentally reported in [2], [18], to assess the effect of FLP on the current, when the influence of Dit on the SS is essentially negligible, otherwise for matched I_{OFF} higher SS would result in lower I_{ON}.

VI. Strain in InGaAs MOSFETs

To investigate strain-induced mobility and then I_{ON} enhancements, we used the **k·p** method to calculate band edges, effective masses and non-parabolicity factors (α) in the presence of strain [19], and then fed these parameters to the MSMC transport model. Fig.13 shows that our **k·p** results for the bandgap (plot a) and for the effective mass (plot b) of unstrained $In_xGa_{1-x}As$ are in good agreement with both experimental data and previous calculations. Fig.14 compares our **k·p** results for the in-plane ($m_{in-plane}$, plot a) and out-of-plane ($m_{out-plane}$, plot b) masses in an $In_xGa_{1-x}As$ layer grown on InP: both masses are in good agreement with [20], while a different trend is observed for $m_{out-plane}$ with respect to [21].

Fig.15 reports experimental electron mobility enhancements for biaxial strain which appear to be smaller in $In_{0.53}Ga_{0.47}As$ than in Si MOSFETs.

Fig.16 compares simulations and experiments for mobility in unstrained and biaxially strained $In_{0.53}Ga_{0.47}As$ MOSFETs, including in the simulations, beside FLP, the effect of interface traps as Coulomb scattering centers: our simulation setup captures the main features of the experimental data. As suggested in [3], the main responsible for the mobility enhancement is the modified Dit (see profiles in Fig.10). In fact we verified that without interface traps the mobility with and without strain are very similar. As discussed in [17], the x-axis of Fig.16 is the sum of free inversion charge and trapped charge, thus assuming that interface traps in the CB are fast and respond to the CV measurements.

The $I_{DS}(V_{GS})$ for the device of Fig.5 considering strained and unstrained $In_{0.53}Ga_{0.47}As$ channel are reported in Fig.17. Without including traps, the impact of strain is negligible, whereas the weaker FLP associated with strain (see Fig.9) provides a small current improvement when interface traps are included in the simulation. Traps have no effect on SS since the Dit assumed in the gap is small.

VII. Conclusions

Accurate MSMC simulations including non-parabolic corrections, calibrated scattering rates and interface traps have shown that $In_{0.53}Ga_{0.47}As$ MOSFETs designed according to the ITRS can overperform strained silicon for low V_{DD} applications. Biaxial strain up to 0.5% gives limited advantages which are mainly due to mitigated FLP. GaSb (111) MOSFETs do not offer significant advantages in terms of I_{ON} compared to $In_{0.53}Ga_{0.47}As$ at the well thickness of interest for the L_G=11.7nm node.

VIII. Acknowledgments

The research leading to these results has received funding from the European Community's Seventh Framework Programme (FP7/2007-2013) under grant agreement III-V-MOS Project n619326).

Work supported in part by TSMC. We thank G.Doornbos (TSMC) for valuable discussions.

REFERENCES

[1] *ITRS 2013. http://www.itrs.net/Links/2013ITRS/Home2013.html.*
[2] N. Taoka *et al. Proc. IEDM*, p. 27.2.1, 2011.
[3] S. Kim *et al. Appl. Phys. Lett.*, vol. 100, no. 19, p. 193510, 2012.
[4] M. Rodwell *et al.* in *Device Research Confer.*, no. 100, p. 149, 2010.
[5] M. De Michielis *et al. IEEE TED*, vol. 54, no. 1, p. 115, 2007.
[6] C. H. Yu *et al. IEEE Tran. on Nanotech.*, vol. 11, no. 2, p. 287, 2012.
[7] K. Kalna *et al. IEEE TED*, vol. 55, no. 9, p. 2297, 2008.
[8] M. Luisier *et al. IEEE EDL*, vol. 32, no. 12, p. 1686, 2011.
[9] S. H. Park *et al. IEEE TED*, vol. 59, no. 8, p. 2107, 2012.
[10] Y. Xuan *et al.* in *Proc. IEDM*, p. 637, 2007.
[11] C. L. Hinkle *et al. IEEE EDL*, vol. 30, no. 4, p. 316, 2009.
[12] S. Kim *et al. Appl. Phys. Lett.*, vol. 104, no. 11, p. 113509, 2014.
[13] S. Jin *et al. Journal Applied Physics*, vol. 102, no. 8, p. 83715, 2007.
[14] D. Lizzit *et al. IEEE TED*, vol. 61, no. 6, p. 2027, 2014.
[15] R. Kim *et al. IEEE EDL*, vol. 32, no. 6, p. 746, 2011.
[16] N. Serra *et al. IEEE TED*, vol. 57, no. 2, p. 482, 2010.
[17] P. Osgnach *et al.* in *Proc. ULIS*, p. 21, 2014.
[18] G. Brammertz *et al. Appl. Phys. Lett.*, vol. 95, p. 202109, 2009.
[19] T. Bahder *Phys. Rev. B*, vol. 41, no. 17, p. 11992, 1990.
[20] M. Sugawara *et al. Phys. Rev. B*, vol. 48, no. 11, p. 8102, 1993.
[21] J. Kim *et al. Journ. Appl. Phys*, vol. 108, no. 1, p. 013710, 2010.
[22] I. Vurgaftman *et al. Journ. Appl. Phys*, vol. 89, no. 11, p. 5815, 2001.
[23] M. Fischetti *et al. IEEE TED*, vol. 38, no. 3, p. 634, 1991.
[24] R. G. Veprek *et al. Phys. Rev. B*, vol. 76, p. 165320, 2007.
[25] Y. S. Kim *et al. Phys. Rev. B*, vol. 82, no. 20, p. 205212, 2010.
[26] T. P. O'Regan *et al. J. Appl. Phys*, vol. 108, no. 10, p. 103705, 2010.
[27] C. Köpf *et al. Solid-State Electron.*, vol. 41, no. 8, p. 1139, 1997.
[28] S. H. Olsen *et al. Journ. Appl. Phys*, vol. 97, no. 11, p. 114504, 2005.
[29] K. Rim *et al.* in *Proc. IEDM*, p. 3.1.1, 2003.
[30] M. Currie *et al. J. Vac. Sci. Technol., B*, p. 2268, 2001.

978-1-4799-8002-4/14 $31.00 © 2014 IEEE

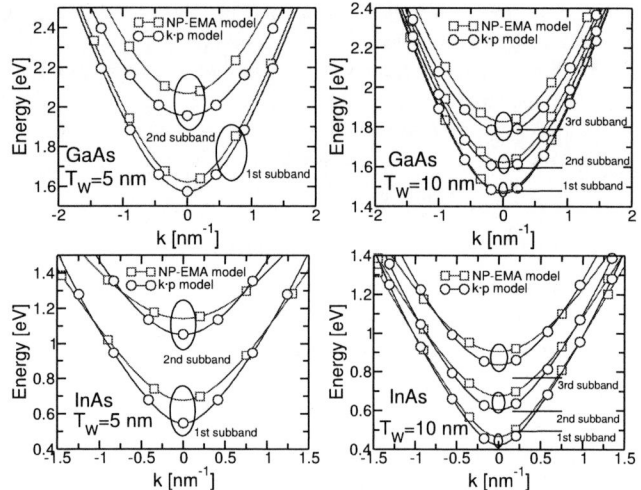

Material	m^* (Γ) [m_0]	α (Γ) [eV^{-1}]	E_Γ^L [meV]	m_l (Ł) [m_0]	m_t (Ł) [m_0]	α (Ł) [eV^{-1}]
InAs	0.026 [a]	2.5 [a]	716 [b]	0.64 [b]	0.05 [b]	0.45 [c]
GaAs	0.067 [d]	0.6 [d]	296 [b]	1.9 [b]	0.0754 [b]	0.4 [c]
GaSb	0.039 [a]	1.65 [a]	63 [b]	1.3 [b]	0.10 [b]	1.26 [e]
InGaAS	0.043 [d]	1.5 [d]	760 [f]	1.57 [f]	0.23 [f]	0.5 [f]

[a] Extracted using **k·p** [19], [22]; [b] From [22]; [c] From [23]; [d] Extracted using **k·p** [19], [22], [24]; [e] Extracted fitting E-K reported in [25]; [f] From [26].

TABLE I: NP-EMA parameters used in the simulations.

Fig. 1: Conduction band profile at the Γ point in GaAs and InAs quantum-wells with quantization in the [100] direction and T_W= 5 and 10 nm. Circles: quantized **k·p** results with the Hamiltonian in [19]; squares: NP-EMA using closed boundaries. In these plots the energy reference (E=0 eV) is taken at the maximum of the valence band of the bulk material.

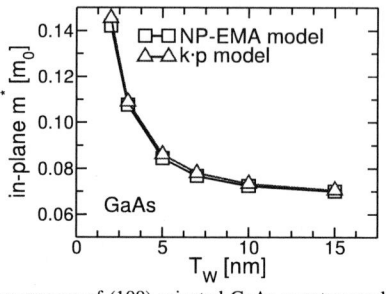

Fig. 2: In plane masses of (100) oriented GaAs quantum wells as a function of well thickness. Comparison between quantized **k·p** results [19] and the NP-EMA model [13].

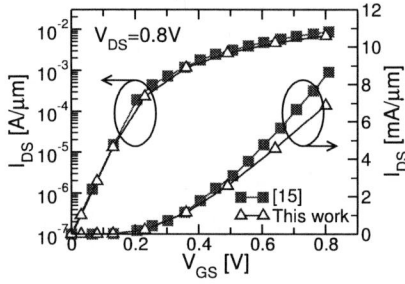

Fig. 3: Simulated ballistic current for a $(100)/[011]$ In$_{0.53}$Ga$_{0.47}$As MOSFET with EOT=0.5 nm. T_W=4 nm. Triangles: model of this work; Squares: ToB model from [15].

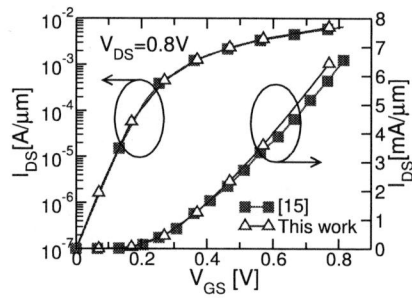

Fig. 4: Same as Fig.3 for a $(110)/[1\bar{1}0]$ GaSb MOSFET with EOT =1.0 nm.

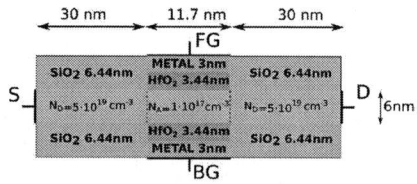

Fig. 5: Sketch of the double-gate MOSFET simulated in this work and designed according to the ITRS for III-V transistors [1]: L_G=11.7 nm, EOT=0.62 nm (3.44 nm of HfO$_2$), film thickness 6 nm, S/D doping 5·10^{19} cm^{-3}.

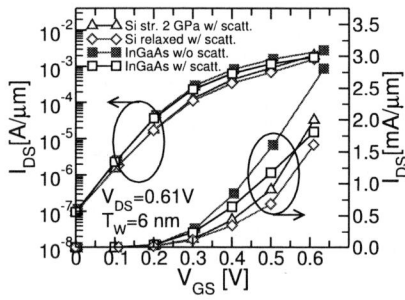

Fig. 6: Simulated I_{DS}-V_{GS} characteristics for the device in Fig.5 considering $(110)/[1\bar{1}0]$ In$_{0.53}$Ga$_{0.47}$As, $(110)/[1\bar{1}0]$ Si and 2GPa [110] uniaxial tensile strained silicon channel.

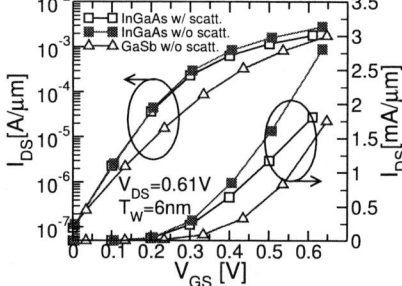

Fig. 7: Simulated I_{DS} vs V_{GS} characteristics for the device in Fig.5 considering $(111)/[1\bar{1}0]$ GaSb (triangles) and $(110)/[1\bar{1}0]$ In$_{0.53}$Ga$_{0.47}$As channel either with (blank square) or without (filled squares) scattering.

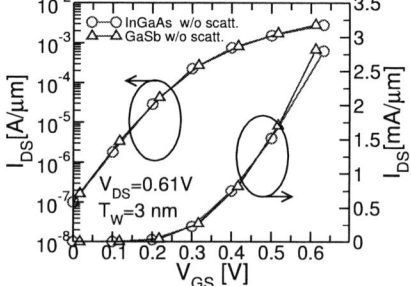

Fig. 8: Ballistic I_{DS} vs V_{GS} simulations for $(111)/[1\bar{1}0]$ GaSb (triangles) and $(110)/[1\bar{1}0]$ In$_{0.53}$Ga$_{0.47}$As (circles). Same device as in Fig.5 but with T_W=3 nm.

978-1-4799-8002-4/14 $31.00 © 2014 IEEE

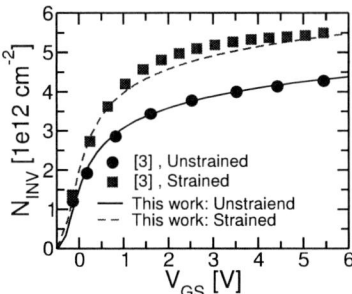
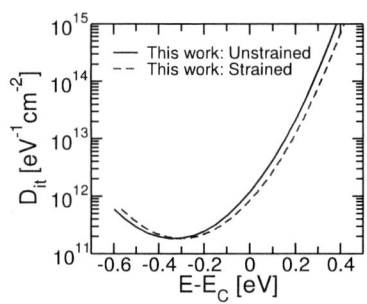
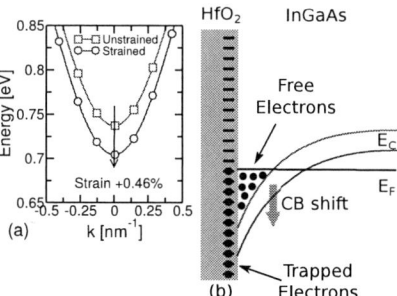

Fig. 9: Simulated inversion electron density compared with experimental data from [3] for a long channel device.

Fig. 10: Interface trap densities used to fit experimental N_{INV} in Fig.9. The blue-dashed line is the shift of the unstrained curve (black) by 33 mV that is the CB shift induced by strain (see Fig.11).

Fig. 11: (a) Bandstructure for strained and unstrained $In_{0.53}Ga_{0.47}As$ calculated using the **k·p** method. (b) Assuming the same Dit profile with respect to vacuum level for both strained and unstrained materials [3], the CB shift in strained InGaAs increase the free electrons.

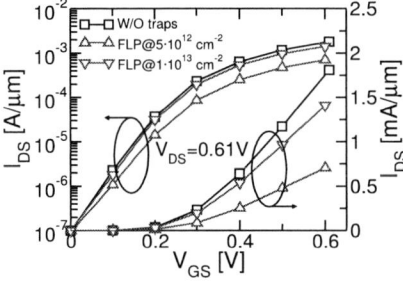
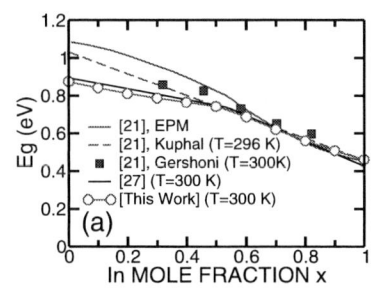
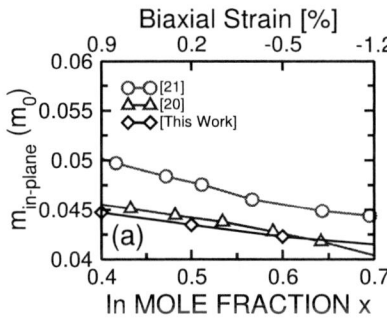

Fig. 12: I_{DS} vs. V_{GS} for the device in Fig.5 considering $In_{0.53}Ga_{0.47}As$ channel and Dit trap profile corresponding to FLP at different inversion densities (profile as in Fig.10 shifted to provide the wanted FLP). Traps are present at both interfaces of the double-gate structure.

Fig. 13: Energy Gap and effective mass for bulk $In_xGa_{1-x}As$ as a function of the In molar fraction.

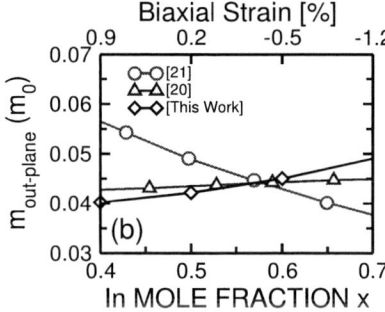
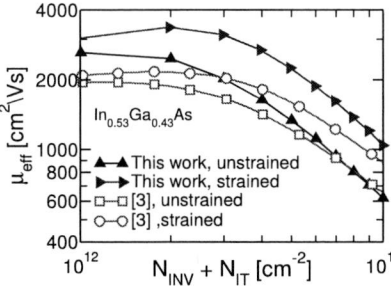

Fig. 14: Modulation of effective masses in $In_xGa_{1-x}As$ over InP versus the In molar fraction.

Fig. 15: Experimental mobility enhancement in biaxially strained Silicon and $In_{0.53}Ga_{0.47}As$ for a long channel MOSFET. Dashed line is a guide for the eyes.

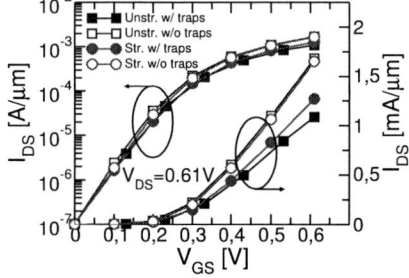

Fig. 16: Simulations and experiments for mobility in unstrained and biaxially strained (strain level 0.46%) long channel $In_{0.53}Ga_{0.47}As$ MOSFETs (structure as in [3]). Simulations include the effect of interface traps (profiles from Fig.10).

Fig. 17: I_{DS} vs. V_{GS} for the device in Fig.5 considering strained and unstrained (biaxial strain 0.46%) $In_{0.53}Ga_{0.47}As$ channel. Dit profiles are from Fig.10 and are applied to both interfaces of the double gate structure.

978-1-4799-8002-4/14 $31.00 © 2014 IEEE

Assessment of Hole Mobility in Strained InSb, GaSb and InGaSb Based Ultra-Thin Body pMOSFETs with Different Surface Orientations

Pengying Chang[1,2], Xiaoyan Liu[2,3]*, Gang Du[2] and Xing Zhang[2]

[1]School of Electronic and Computer Engineering, Peking University, Shenzhen, 518055, China.
[2]Institute of Microelectronics, Peking University, Beijing, 100871, China, E-mail: xyliu@ime.pku.edu.cn
[3]Innovation Center for MicroNanoelectronics and Integrated System, Beijing, 100871, China

Abstracts

This work presents a systematic assessment of hole mobility in InSb, GaSb and InGaSb based ultra-thin body (UTB) double-gate pMOSFETs employing a self-consistent method based on 8×8 $k \cdot p$ Schrödinger and Poisson equations and including important scattering mechanisms. Physical models are calibrated against experiments. The effect of body thickness, surface/channel orientation, biaxial and uniaxial strain, and heterostructure design on hole mobility in III-V materials has been systematically investigated in order to help in providing useful guidelines.

Introduction

Due to the development of high quality dielectrics on III-V surfaces, III-V materials have become as a credible alternative for high-speed and low-power logic [1]. This is driven by their excellent electron mobility. However, most III-V materials lack good hole mobility for p-type devices. Looking into future III-V CMOS technology, one challenge focus on enhancing the hole mobility in III-V materials. Recently, there have been great interests in InSb, GaSb and InGaSb based devices [2-4], which are considered to be promising candidates for PMOS application. Novel device structures, combined with strain engineering, have been used successfully in III-V based devices. Surface orientation optimization is needed to further booster hole mobility.

This work presents a systematic assessment of hole mobility in III-V based ultra-thin body (UTB) double-gate (DG) pMOSFETs employing a self-consistent method based on 8×8 $k \cdot p$ [5] Schrödinger and Poisson equations. Then hole mobility is calculated via Kubo-Greenwood formalism accounting for various scattering mechanisms [6]. The results are validated against experiments. The effect of body thickness, surface/channel orientation, biaxial and uniaxial strain, and heterostructure design on hole mobility have been investigated systematically and thoroughly. In this work, we compare and benchmark strained InSb, GaSb, and InGaSb p-channel DG devices in terms of hole mobility.

Simulation Method

The quantum confined and strained valence band structures are calculated by solving 8×8 $k \cdot p$ Schrödinger and Poisson equations self-consistently [7, 8]. Fig.1 shows the simulated device structure in this work. When dealing with different surface orientations of (001), (110) and (111), appropriate rotations of the k space must be performed [9]. Fig.2 plots the energy contours and band structure of InSb

with different orientations. According to [9], in the case of (001) and (111) orientations, mobility does not depend on channel direction; while in the (110) cases, mobility depends on channel direction due to its strong anisotropy, and both (110)/[$\bar{1}$10] and (110)/[001] channel directions are considered below.

Hole mobility of the 2DEG is calculated using the Kubo-Greenwood formula. The models include: nonpolar acoustic (AC) and optical phonon (OP), polar optical phonon (POP), surface roughness (SR), and alloy (AL) scattering (only considered in ternary materials like InGaSb) [10]. Simulation parameters are listed in Table I. Values for In$_x$Ga$_{1-x}$Sb are taken linearly interpolated from InSb and GaSb parameters [11, 12]. Physical models are calibrated and verified with experimental data in Fig.3 for InSb [2] and GaSb [4] based devices and in Fig.17 for In$_x$Ga$_{1-x}$Sb devices [4], which demonstrates an excellent agreement between simulations and experiments. In order to fairly compare the performance of InSb/GaSb/InAs/GaAs devices, we adopt the same SR parameters (Δ=1.6nm, Λ=2.5nm) for these binary materials in the simulation below.

Table I Scattering parameters for hole mobility calculation in InSb, GaSb and InGaSb in the simulation

Symbol	Unit	InSb	GaSb	In$_x$Ga$_{1-x}$Sb
D_{ac}	eV	7.02	6.88	
$(DK)_{op}$	10^8eV/cm	10.81	11.03	Linearly interpolated
$\hbar\omega_{op}$	meV	25	29.8	
$\hbar\omega_{pop}$	meV	24.3	28.89	
Δ	nm	1.6	1.6	1
Λ	nm	2.5	2.5	2.5
U_0	eV			0.3

Results and Discussion

A. Unstrained Mobility

Fig.5 (a-d) shows hole mobility (μ_h) vs. inversion density Ns in 5nm-thick unstrained InSb/GaSb/InAs/GaAs devices with different surface/channel orientations. For all these materials, mobility trend with surface orientations is (110)/[$\bar{1}$10]>(111)>(110)/[001]>(001) at a wide range of inversion density Ns. For III group materials, μ_h in In-based is higher than Ga-based, while more degradation with Ns increasing. For V group materials, μ_h in Sb-based is higher than As-based. As a results, μ_h in InAs and GaAs is comparable with universal Si [13], while μ_h in InSb and GaSb is higher than unstrained Ge [14] and following we focus on these two materials especially on the strain effect due to their high μ_h. Furthermore, the anisotropy of Ga-based mobility is stronger than In-based. The inversion carrier density Ns vs. gate voltage Vg with T_B=5nm in Fig.4 (a), reveals the Ns trend with materials is: InSb<GaSb<InAs<GaAs, and trend with surface orientations is (001)<(111)<(110) for all materials. And this can be understood that the confinement is strongest for (110) orientation, followed by (111) and (001) orientations as shown in Fig.4 (b). Fig.6 (a) depicts contributions of PH-limited and SR-limitsed mobility in GaSb. The large difference of PH-limited mobility between various orientation/channel directions mainly caused by the difference of average transport effective mass M_α^1 seen in Fig.6 (b), which is calculated as [15]. While the smaller quantization mass in the case of (001), leads to stronger SR scattering and thus smaller SR-limited mobility. The InSb's and GaSb's μ_h dependence on body thickness (T_B) ranging from 3nm to 20nm is shown in Fig.7. It is observed that μ_h is severely degraded for all the channel directions considered below 10nm in GaSb, but happens at smaller values of T_B in InSb. Furthermore, for extremely thin body thickness, mobility in non-(001) devices outperform that in the (001) case, especially in the (110)/[$\bar{1}$10] case. Contributions of different scattering mechanisms shown in Fig.8, reveals a dominant influence of SR scattering which exhibits a T_B^6 behavior due to thickness fluctuations [16]. The degradation can be understood by form factor for nonpolar PH, matrix element for SR and effective mass in Fig.9.

B. Biaxial Strain

μ_h vs. biaxial strain in 5nm-thick InSb (Fig.10) and GaSb (Fig.11) devices shows compressive strain can improve μ_h significantly, and orientation trend is similar with unstrained cases. In addition, the enhancement is larger than 2× under 2% biaxial compressive strain in both InSb and GaSb. Mobility enhancement factor vs. inversion density Ns under 1% biaxial compression is displayed in Fig.12. For both InSb/GaSb channel devices, mobility enhancement of (001) devices under biaxial compression gradually loses most of its initial gain with increasing Ns. While mobility enhancement of (110) and (111) retain most of the enhancement at higher Ns. Therefore, the enhancement factor trend with surface orientations at high Ns is also (110)/[$\bar{1}$10]>(111)> (110)/[001]>(001).

C. Uniaxial Strain

μ_h vs. uniaxial stress along channel direction in 5nm InSb (Fig.13) and GaSb (Fig.14) shows similar trend with biaxial strain but more significant enhancement of μ_h. In addition, the enhancement is approaching 5.1× in InSb under 2GPa uniaxial compressive strain along (110)/[$\bar{1}$10] direction. Fig.15 shows enhancement factor vs. Ns under 1GPa uniaxial compressive stress. Similarly, (110)/[$\bar{1}$10] case obtains the maximum mobility enhancement at high density. Mobility enhancement stems from changes in the effective mass and scattering rate both of which depend on the strain-altered valence band structures. Fig.16 shows mobility enhancement contributions from transport effective mass and scattering suppression in GaSb under (110)/[$\bar{1}$10]

uniaxial stress. It can be clearly seen that enhancement mainly results from the reduced effect mass.

D. QW Heterostructure Strained $In_xGa_{1-x}Sb$ pMOSFETs

The quantum well (QW) heterostructure FETs combine strain and quantum mechanical confinement to obtain desired transport properties with reduced off-state leakage. Fig.17 shows calculated μ_h validated against experiments [4] for $In_{0.2}Ga_{0.8}Sb$ with alloy scattering considered. The numerical value of alloy potential U_0 is extracted as a fitting parameter. Fig.18 shows device structures of strained InGaSb QW HFETs and corresponding hole mobility with T_B=5nm are presented in Fig.19. μ_h in both HFETs-on-Bulk and HFET-OIs can be improved by inserting a cap layer. Fig.20 shows μ_h in compressively strained $In_xGa_{1-x}Sb$ w/o and w/ cap layer of 1/2/3nm. When inversion becomes stronger, a fraction of holes in the low mobility cap region increases, which can degrade mobility. Therefore, the cap should be as thin as possible. Fig.21 shows μ_h in $In_xGa_{1-x}Sb$ QW vs. InSb fraction. Alloy scattering leads to an asymmetric 'U' shaped mobility behavior. Mobility enhancement is 2.4× with 2.55% biaxial compressive strain due to lattice mismatch at x=0.5

Conclusions

Based on a physical modeling, hole mobility dependency on III-V materials, surface orientations, strain effect, and different heterostructure design are systematically assessed and summarized in Table II. For all of these materials under both unstrained and strained cases, (110)/[$\bar{1}$10] is the optimal surface channel directions. Besides, the design of $In_xGa_{1-x}Sb$ QW HFET can be optimized by T_B, InSb fraction, and cap layer.

Acknowledgements

This work is supported by NKBRP 2011CBA00604, NSFC Grant No. 61306104, China Postdoctoral Science Foundation Grant No. 2013M540018.

References

[1] Jesús A. del Alamo et al., IEDM 2013, p. 24-27.

[2] M. Radosavljevic et al., IEDM 2008, p. 1-4.

[3] B. R. Bennett et al., Journal of Crystal Growth 311, 1 (2008).

[4] A. Nainani et al., IEDM 2010.

[5] T. B. Bahder, Physical Review B 41, 17 (1990).

[6] D. Esseni et al., Nanoscale MOS Transistors: Semi-Classical Transport and Applications (Cambridge Univ. Press, U.K., 2011), p. 233.

[7] Pengying Chang et al., SISPAD 2013, p. 388-391.

[8] Pengying Chang et al., IWCE 2014.

[9] Pengying Chang et al., ESSDERC 2014, in press.

[10] Pengying Chang et al., SSDM 2014, p. 660-661.

[11] I. Vurgaftman et al., Journal of Applied Physics 89, 11 (2001).

[12] http://www.ioffe.ru/SVA/NSM.

[13] S. Takagi et al., IEEE Transaction on Electron Devices 41, 12 (1994).

[14] T. Maeda et al., IEEE Electron Device Letters 26, 102 (2005).

[15] L. Donetti et al., Journal of Applied Physics 110, 063711 (2011).

[16] K. Uchida et al., IEDM 2002.

[17] K. Takei et al., Nano Letters 12, 4 (2012)

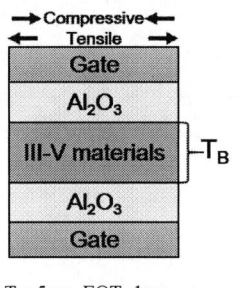

T_B=5nm, EOT=1nm

Fig.1 Simulated device structure of III-V materials (InSb/GaSb/InAs/GaAs) p-channel DG QW FET in this study.

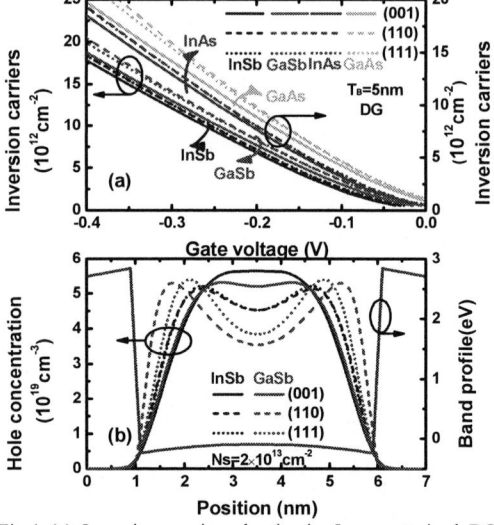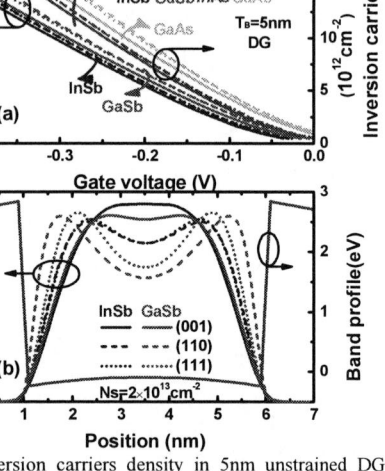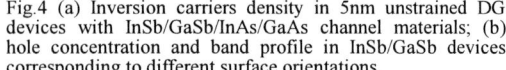

Fig.2 Energy contours for 5nm InSb devices with respect to (001), (110) and (111) orientations in (a,b,c). (001)/[110], (110)/[$\bar{1}$10], (110)/[001], (111)/[0$\bar{1}$1] surface/channel directions are considered and corresponding band structures are shown in (d,e,f,g).

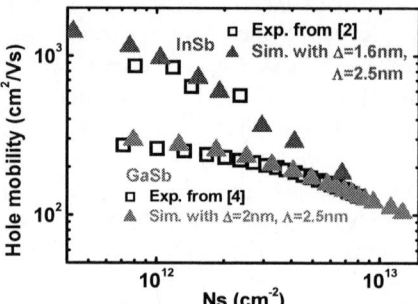

Fig.3 Comparison between the simulation and experimental hole mobility for InSb [2] and GaSb based [4] devices. Simulation reproduce well the experiment, which verify our physical models.

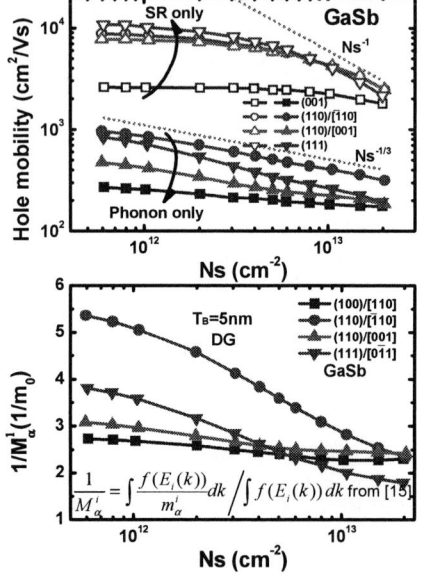

Fig.4 (a) Inversion carriers density in 5nm unstrained DG devices with InSb/GaSb/InAs/GaAs channel materials; (b) hole concentration and band profile in InSb/GaSb devices corresponding to different surface orientations.

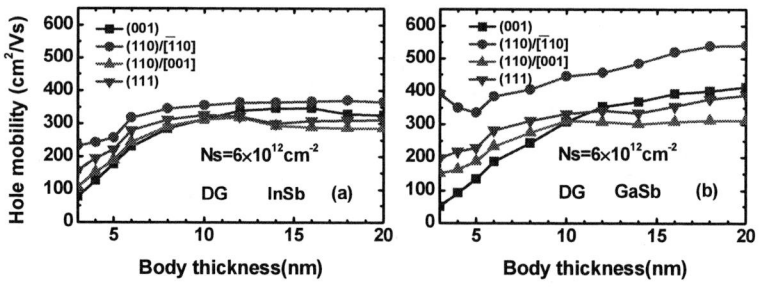

Fig.5 Hole mobility in 5nm-thick unstrained (a) InSb, (b) GaSb, (c) InAs, and (d) GaAs devices for (001), (110) and (111) orientations along various channel directions. **Overall**: Optimal directions of InSb and GaSb have ~2x higher hole mobility than unstrained Ge. **u_h trend with surface orientations for all materials**: (110)/[$\bar{1}$10]>(111)>(110)/[001]>(001).

Fig.6 (a) Contributions of phonon and SR scattering mechanisms to the total mobility for different surface and channel directions in 5nm DG GaSb devices as in Fig.5 (b). (b) Corresponding average transport effective mass M_α^1 vs. inversion density for the first subband. The difference between M_α^1 mostly contributes to the different PH-limited mobility behavior with orientations as mentioned in (a).

$$\frac{1}{M_\alpha^i} = \int \frac{f(E_i(k))}{m_\alpha^i} dk \Big/ \int f(E_i(k)) dk \text{ from [15]}$$

Fig.7 Dependency of hole mobility in unstrained (a) InSb and (b) GaSb DG devices on body thickness with different surface and channel directions at Ns=6×10^{12}cm^{-2}.

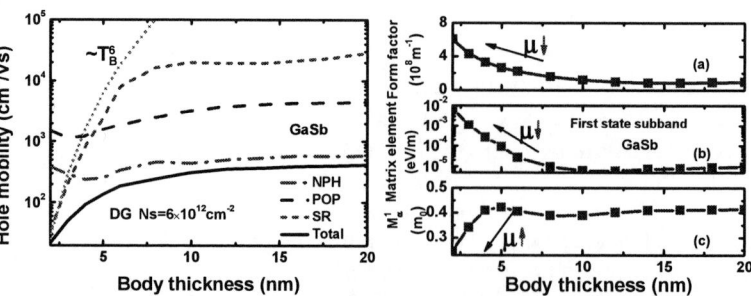

Fig.8 Contributions of different scattering mechanisms to the total mobility in (001) oriented GaSb DG devices. The T_B^6 trend is plotted for comparison with SR-limited mobility.

Fig.9 (a) Form factor for nonpolar phonon scattering, (b) matrix element for SR, and (c) average transport effective for the first subband in (001) oriented GaSb devices.

978-1-4799-8002-4/14 $31.00 © 2014 IEEE   194

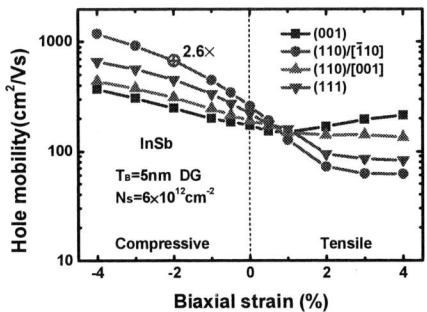

Fig.10 Hole mobility vs. biaxial strain in 5nm InSb DG devices. **Overall** compressive strain is better and tensile strain is worse. $\underline{\boldsymbol{u_h}}$ **enhancement**: 2.6× at 2% along (110)/[$\bar{1}$10].

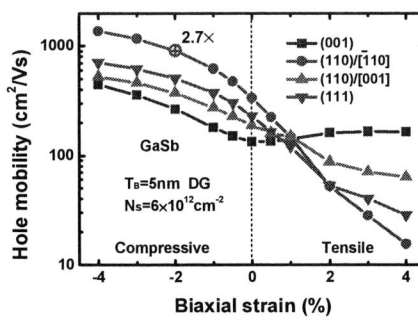

Fig.11 Hole mobility vs. biaxial strain in 5nm GaSb devices. **Overall** comp. strain is better and tensile strain is worse. $\underline{\boldsymbol{u_h}}$ **enhancement**: 2.7× at 2% comp. along (110)/[$\bar{1}$10].

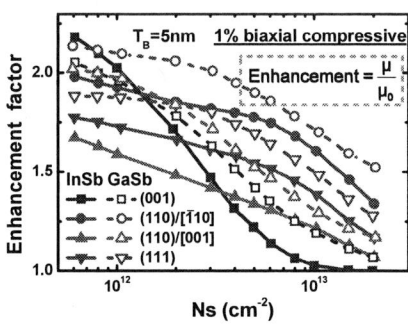

Fig.12 Hole mobility enhancement vs. inversion density in 5nm-thick InSb and GaSb DG devices under 1% biaxial comp. strain with various surface and channel directions.

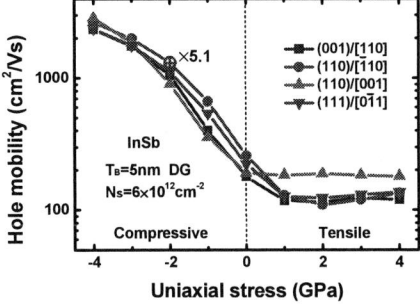

Fig.13 Hole mobility vs. uniaxial stress in 5nm InSb DG devices. **Overall** compressive strain is better and tensile strain is worse. $\underline{\boldsymbol{u_h}}$ **enhancement**: 5.1× at 2GPa uniaxial comp. along (110)/[110].

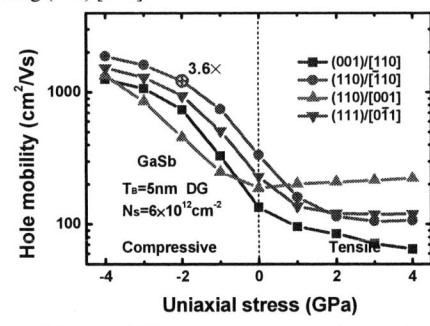

Fig.14 Hole mobility vs. uniaxial stress in 5nm GaSb DG devices. **Overall** compressive strain is better and tensile strain is worse. $\underline{\boldsymbol{u_h}}$ **enhancement**: 3.6× at 2GPa uniaxial comp. along (110)/[110].

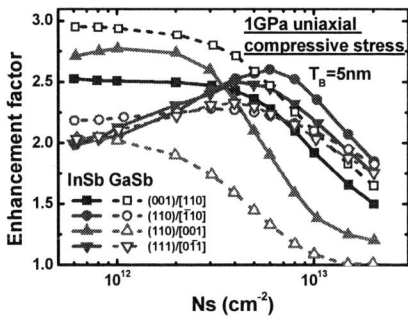

Fig.15 Hole mobility enhancement vs. inversion density in 5nm-thick InSb and GaSb DG devices under 1GPa uniaxial compressive stress.

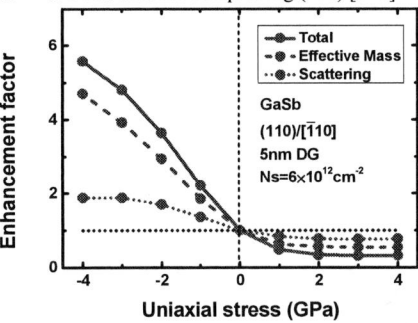

Fig.16 Mobility enhancement contributions in GaSb from transport effective mass and scattering suppression under uniaxial stress along channel direction (110)/[$\bar{1}$10] as show in Fig.14.

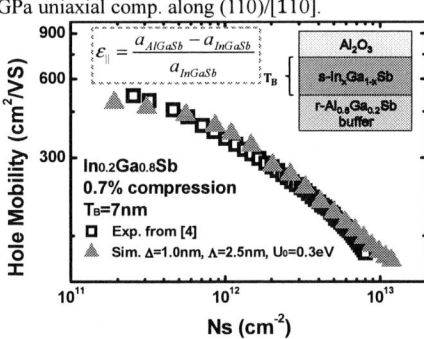

Fig.17 Comparison between calculated and experimental hole mobility for 7nm-thick $In_{0.2}Ga_{0.8}Sb$ [4] with comp. strain due to lattice mismatch between InGaSb and AlGaSb.

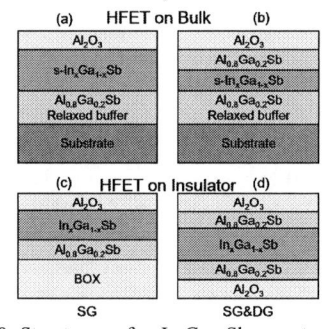

Fig.18 Structures of s-$In_xGa_{1-x}Sb$ quantum well (QW) heterostructure MOSFET (HFET) on bulk (a, b) and on insulator (c, d) in [4, 17].

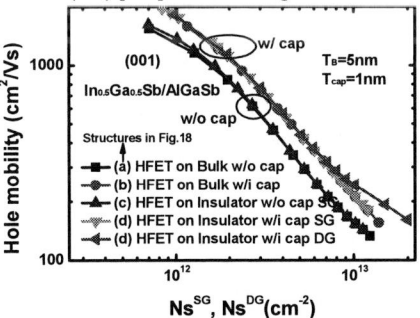

Fig.19 Calculated hole mobility vs inversion density Ns for different QW structures as shown in Fig.18 (a-d) with $In_{0.5}Ga_{0.5}Sb$ channel under 2.55% compressive biaxial strain due to lattice mismatch.

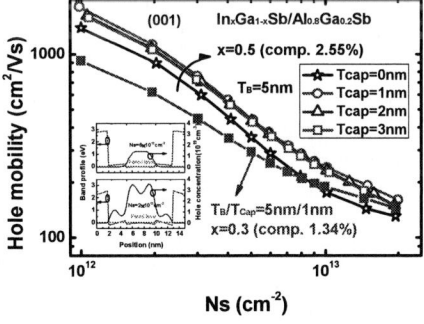

Fig.20 Hole mobility in $In_xGa_{1-x}Sb$ of structure in Fig.18(d) w/o and w/ cap layer under lattice mismatched biaxial strain. A thin WB cap can be help to enhance hole mobility.

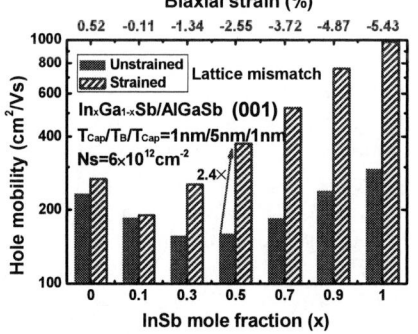

Fig.21 Hole mobility vs. InSb mole fraction x in 5nm relaxed and biaxial strain $In_xGa_{1-x}Sb$ due to lattice mismatch with DG HFET-OI (Fig.18(d)). $\underline{\boldsymbol{u_h}}$ **enhancement**: 2.4× at x=0.5.

Table II Summary of Hole Mobility in DGFET for Materials, Orientations and Strains

Unstrained Hole Mobility				Strain Effect		
Orientation/Channel Direction		**Materials at T_B=5nm**		T_B=5nm, N_s=6×10¹² cm⁻²	**Biaxial**	**Uniaxial**
Thick T_B	(110)/[$\bar{1}$10]>(001)> (111)>(110)/[001]	Sb-based have higher mobilities than As-based; Ga-based anisotropy is stronger than In-based		**Favorable Type**	Compressive	Compressive
				Orientation/Channel Direction	(110)/[$\bar{1}$10]>(111)> (110)/[001]>(001)	(110)/[$\bar{1}$10]>(111)/[0$\bar{1}$1]> (001)/[110]>(110)/[001]
Thin T_B	(110)>[$\bar{1}$10]>(111)> (110)/[001]>(001)	**Optimal (110)/[$\bar{1}$10]**	InSb	4.2×μ_h(Si), 2.6×μ_h(Ge),		
			GaSb	3×μ_h(Si), 1.9×μ_h(Ge)	**Materials Enhancement**	InSb<GaSb	InSb>GaSb

978-1-4799-8002-4/14 $31.00 © 2014 IEEE

Two-Dimensional Nanoelectromechanical Systems (2D NEMS) via Atomically-Thin Semiconducting Crystals Vibrating at Radio Frequencies

Philip X.-L. Feng[1*], Zenghui Wang[1], Jaesung Lee[1], Rui Yang[1], Xuqian Zheng[1], Keliang He[2], Jie Shan[2,3]

[1]Electrical Engineering & Computer Science, [2]Physics, Case Western Reserve University, Cleveland, OH 44106, USA
[3]Physics, Pennsylvania State University, University Park, PA 16802, USA
*Email: philip.feng@case.edu, Phone: (216) 368-5508, Fax: (216) 368-6888

Abstract

We report on the initial explorations of engineering atomically-thin semiconducting crystals into a new class of two-dimensional nanoelectromechanical systems (2D NEMS) that are attractive for realizing ultimately thin 2D transducers for embedding in both planar and curved systems. We describe the first resonant NEMS operating at radio frequencies (RF), based on MoS_2, a hallmark of 2D semiconducting crystals derived from layered materials in transition metal dichalcogenides (TMDCs). Through a series of careful measurements and analyses, we demonstrate a family of MoS_2 2D NEMS resonators possessing very high fundamental-mode frequencies ($f_0 \sim 120$MHz, in the VHF band), very broad dynamic range ($DR \sim 70-110$dB), rich nonlinear dynamics, and outstanding electrical tunability.

Introduction

Within the past 2 decades, fueled by advances in both top-down and bottom-up nanofabrication techniques, and enabled by breakthroughs in high-precision measurements, our capability of understanding and manipulating mechanical degrees of freedom in genuinely nanoscale structures (<100nm in at least 1 dimension) has experienced many exciting progresses. NEMS vibrating in their resonances at high frequencies, in particular, offer a versatile platform for innovating transducers at nanoscale. Resonant NEMS have already demonstrated a number of records in sensing toward the fundamental limits, such as displacement at fm (10^{-15}m), physical and chemical mass sensing at single-atom/ molecule (10^{-21}g to 10^{-24}g) levels, force detection at single-spin interaction limit (10^{-18} to 10^{-21}N), and energy (heat) measurement at attojoule (10^{-18}J) level. Today's mainstream NEMS resonators/transducers are primarily in 1D shapes of nanowires/tubes, beams/cantilevers, sculpted from materials such as Si, SiN, SiC, diamond, carbon nanotubes, and III-V compounds, *etc*. The advent of 2D materials creates unprecedented opportunities for realizing atomically-thin transducers that offer new advantages over the 1D NEMS. 2D crystals possess unique and outstanding mechanical properties, including ultimately low areal mass, high elastic modulus ($E_Y \sim 0.3-1$TPa), superb intrinsic strength ($\varepsilon_{int} \sim 25\%$) and flexibility, which make these crystals attractive for creating new 2D NEMS. Graphene, a herald of 2D crystals but a semi-metal, has recently been studied for enabling NEMS resonators toward sensing and ultralow-power RF signal processing [1-6]. Beyond graphene, MoS_2 and other 2D TMDCs have sizable bandgaps that depend on number of layers and can be tuned by strain continuously. These clearly open an avenue toward 2D semiconductor-based NEMS transducers, resonators, and 2D systems with tunability.

First MoS_2 2D NEMS Resonators at HF/VHF

Without any coherent external excitations but simply in thermal dynamic equilibrium with its environment, fluctuation-dissipation theorem (FDT) dictates any 2D NEMS to be in Brownian motions, which are manifested as thermomechanical modes of damped harmonic resonators. By carefully engineering ultrasensitive optoelectronic and purely electronic readout schemes at RF, we can detect such undriven thermomechanical resonances (Fig. 1a), which directly reveals some intrinsic characteristics of the devices.

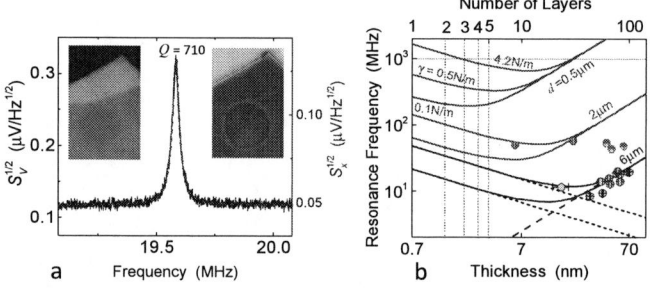

Figure 1: (a) Thermomechanical resonance of a multilayer MoS_2 resonator. *Insets*: Optical & SEM images. (b) Frequency scaling. Dashed lines show the ideal membrane and plate limits.

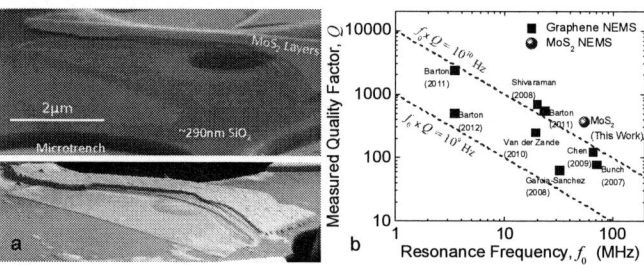

Figure 2: (a) SEM (*top*) and AFM (*bottom*) images of a multilayer MoS_2 nanomechanical resonator. (b) Comparison of the highest $f_0 \times Q$ values in recently reported graphene NEMS resonators with the MoS_2 NEMS demonstrated in our work.

We have prototyped several generations of MoS_2 2D resonators vibrating in the HF (3-30MHz) and VHF (30-300MHz) bands [7]. The achieved figure of merit ($f \times Q$ value) has surpassed those of graphene 2D NEMS (Fig. 2). Beyond thermomechanical resonances, we can also excite the devices' resonances by employing optical (Fig. 3) and electrostatic (Fig. 7) schemes.

Frequency Scaling Capability and Effects of Thickness in the Integer Numbers of Layers

Analyzing the interplay between built-in tension and thickness-dependent flexural rigidity, we gain insight in the elastic nature of the 2D NEMS and can quantitatively explain

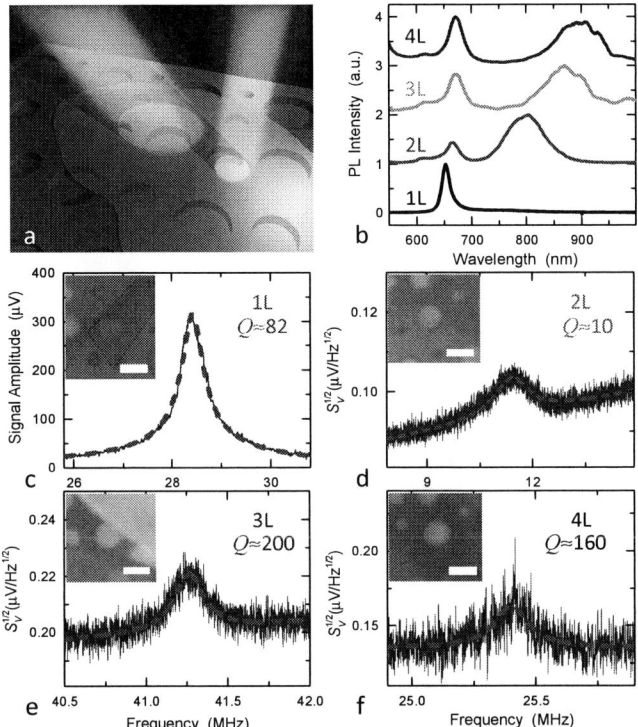

and predict their frequency scaling (Fig. 1b). Less than 10-layer-thick 2D devices generally operate in the membrane limit; thus controlling tension is very effective for tuning frequency. For the first time, we measure resonances of clearly defined 1-, 2-, 3-, and 4-layer MoS_2 drumheads (# of layers verified by photoluminescence, or PL, see Fig. 3b). Interestingly, the measured thermomechanical resonances show signatures of the discrete integer numbers of layers.

Surprisingly Broad Dynamic Range (DR)

Dynamic range (DR) is a key metric in evaluating a NEMS resonator and large DR translates into higher sensitivity (resolution, or limit of detection) and lower phase noise (in a closed-loop feedback oscillator). DR is defined as the range (ratio in dB) between the 'signal ceiling' (onset of nonlinearity) and the 'noise floor' (intrinsically, the Brownian motions). We first analyzed the DR of MoS_2 (Fig. 4) and other 2D NEMS [8], elucidating a fundamental feature that enables surprisingly high DR values in 2D NEMS as compared to conventional 1D nanowire/tube/beam NEMS. Here, we further show our first unambiguous measurements: in many mono- and few-layer devices, DR~70dB to ~110dB are attained (Fig. 5). These 2D NEMS also exhibit new air damping properties (Fig. 6), due to unique bulging and compressing-cavity effects [9].

Figure 3: 1-, 2-, 3- and 4-layer (1, 2, 3, 4L) MoS_2 resonators. (a) Illustration of optical excitation and detection scheme. (b) Photoluminescence (PL) data of 1L to 4L MoS_2 resonators. (c)–(f) Measured nanomechanical resonances from 1L to 4L MoS_2 resonators (scale bars: 2µm). Red dashed lines in resonance plots are fittings to a finite-Q harmonic resonator.

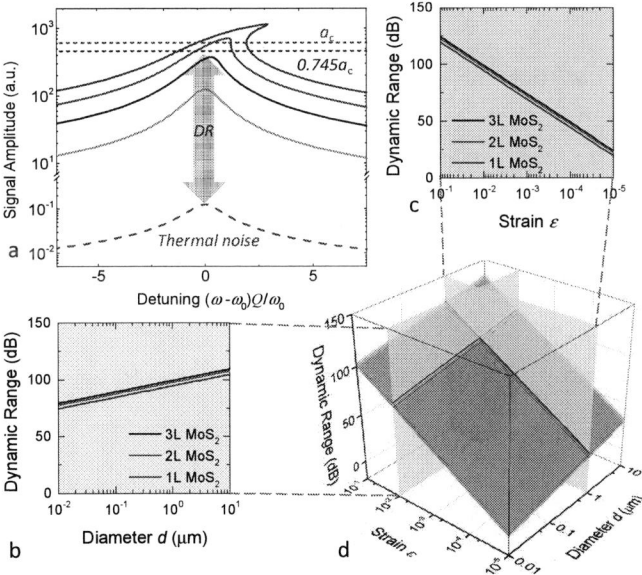

Figure 4: Intrinsic dynamic range (DR) of 1-, 2-, and 3-layer MoS_2 nanomechanical resonators. (a) Nonlinear response of a Duffing resonator. (b)-(d) DR of 1, 2, and 3 layer MoS_2 circular drumhead resonators of various diameters (d) and under different initial tension levels. The colored vertical planes represent different cross sections in the parameter space (b) 2D cross-section plot (DR vs. d) of the ε=0.01 plane in (d). (c) 2D cross-section plot (DR vs. ε) of the d=1µm plane in (d).

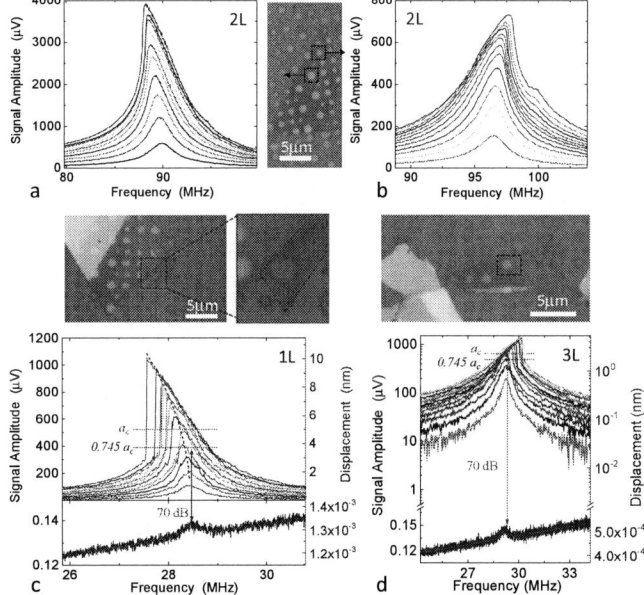

Figure 5: Direct measurement of dynamic range (DR) of 1-, 2-, and 3-layer MoS_2 nanomechanical resonators. (a)-(b) Duffing response of two 2L MoS_2 resonators (d=1.5µm and 1.25µm), showing softening and hardening, respectively. The optical image of both devices is shown in between. (c) Duffing response and thermomechanical noise spectrum measured from a 1L MoS_2 resonator (d=1.5µm, optical images of different magnifications shown on top). The *backbone* curve, the critical amplitude a_c, and the 1dB compression point $0.745a_c$ are highlighted. The dynamic range is determined by the ratio between the 1dB compression point and the on-resonance thermomechanical noise amplitude. (d) Measurement of a 3L MoS_2 resonator (d=1.5µm, optical image shown on top). The labels are the same as in (a), and the amplitude is plotted in logarithmic scale.

978-1-4799-8002-4/14 $31.00 © 2014 IEEE

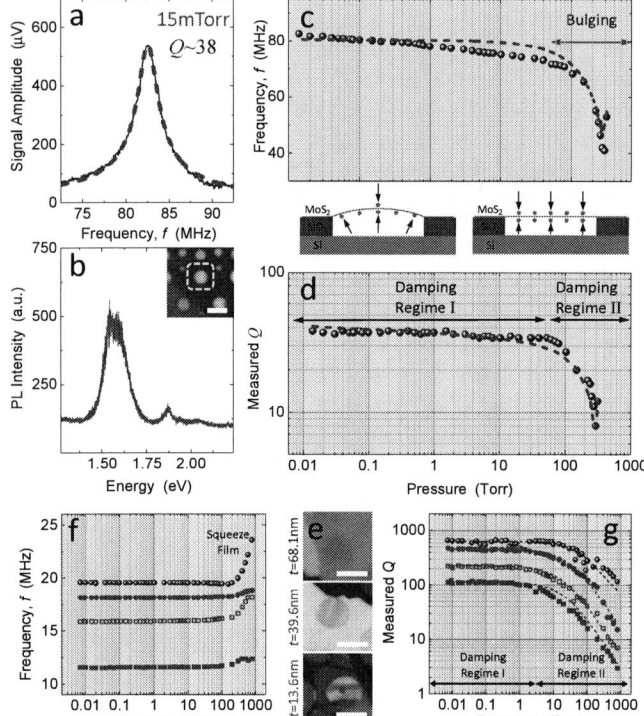

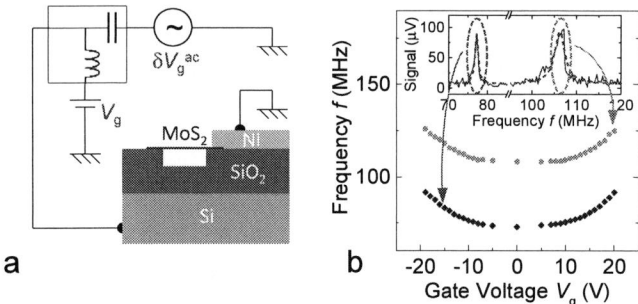

Figure 6: Air damping in MoS_2 resonators. (a) Resonance of a bilayer device at 15mTorr. Red dashed line is the fitting to a finite-Q harmonic resonator. (b) PL data confirming bilayer MoS_2. *Inset*: optical image of the sample with dashed box indicating the device (scale bar: 2μm). (c) The evolution of measured resonance frequency f under varying pressure p. Dashed line shows the fitting to our model. (d) The evolution of measured quality factor Q under varying pressure p. Dashed line shows the fitting to our model. Schematics illustrates device bulging. (e) Images of 3 multilayer devices (thickness labeled, scale bars: 5μm), and their (f) f vs. p and (g) Q vs. p.

Figure 7: Electrical driving and tuning of resonance. (a) Schematic of electrical tuning and driving. (b) Electrostatic frequency tuning of two resonance modes (inset) of a 4L MoS_2 resonator between −20 and 20V.

Electrically Tunable 2D NEMS Resonators

High tunability and large tuning ranges are desirable for NEMS oscillators and other signal processing functions. We demonstrate the first electrically tunable 2D NEMS in mono- and few-layer MoS_2. We explore gate electrical tuning of resonance in these MoS_2 VHF resonators by fabricating and measuring MoS_2 NEMS with engineered electrodes (Figs. 7a, 9a). By applying a dc+ac signal across the MoS_2 and its gate electrode (Fig. 7a), we electrically excite the mechanical

resonance and tune the resonance frequency (Figs. 7b, 9c). We show that many such devices have exceptional tunability (Fig. 8), with electrostatic tuning of $\Delta f/f \approx$ 84ppm per single electric charge. The tuning efficiency is comparable to the best reported in graphene NEMS (Fig. 8). We also study the electronic transport of doubly-clamped MoS_2 resonators (Fig. 9), and investigate the transistor behavior in both ambient and vacuum. This shall enable MoS_2 FET-NEMS integration.

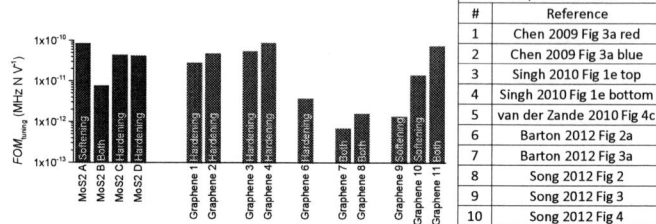

Graphene Devices	
#	Reference
1	Chen 2009 Fig 3a red
2	Chen 2009 Fig 3a blue
3	Singh 2010 Fig 1e top
4	Singh 2010 Fig 1e bottom
5	van der Zande 2010 Fig 4c
6	Barton 2012 Fig 2a
7	Barton 2012 Fig 3a
8	Song 2012 Fig 2
9	Song 2012 Fig 3
10	Song 2012 Fig 4

Figure 8: Electrical tuning and tuning *figure of merit* (*FOM*). By considering the trade-off between the resonance frequency and tuning range, we introduce the tuning $FOM = f_{res}\eta\varepsilon_0/E$ (η: tuning range in %, ε_0: strain, E: applied electrical field) and compare specifications across reported 2D NEMS resonators (*Blue*: MoS_2, *Red*: graphene) [2,10-13].

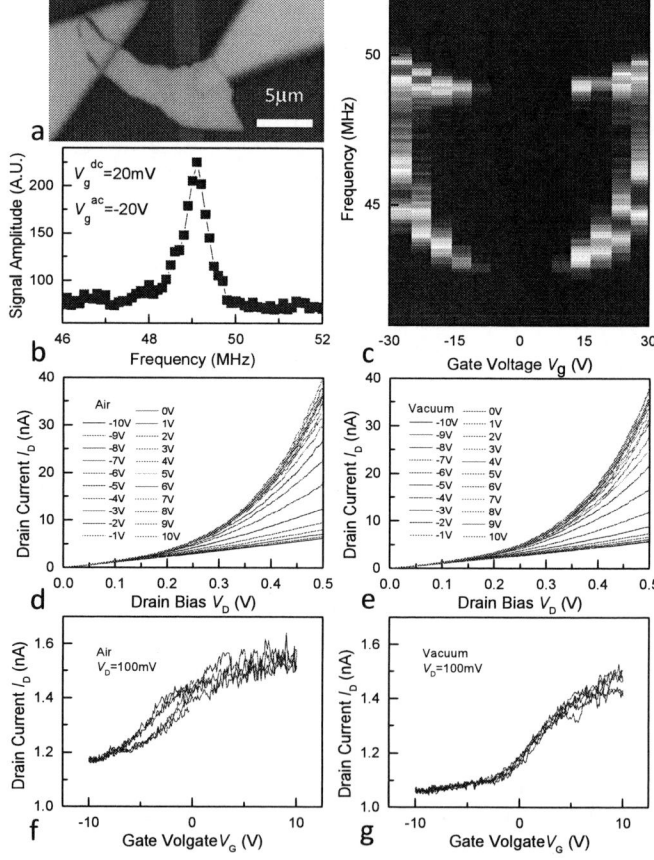

Figure 9: Measured data from a doubly-clamped MoS_2 resonator/transistor. (a) Optical image of the device. (b) Resonance measured at V_g^{dc}= −20V. (c) Gate tuning of the first two resonance modes, with the gate voltage sweep in the range of V_g^{dc}= −30V to +30V. (d)-(e) Measured transistor I_D-V_D data of the device in air and in vacuum. (f)-(g) Measured transfer characteristics of the device in air and in vacuum. Transistors are analyzed by following [14].

Benefits from Structural Nonidealities and Asymmetries

We have also investigated MoS_2 2D nanoelectromechanical resonators with structural nonidealities and asymmetries. We have found that compared with geometrically 'ideal' devices, the non-ideally structured can engender new and unique multimode resonance characteristics [15] (Fig. 10). Through experimental characterization and finite element modeling, we identify the correspondence between the asymmetry in the device geometry and the device's multimode resonance spectrum. Figure 10a shows the evolution of the resonance frequency and mode shapes of the first three flexural modes in a MoS_2 resonator with varying geometry. Figure 10b and 10c demonstrate that from the relative spacing between the frequencies of the different resonance modes, the geometry of the device can be unambiguously identified. These results demonstrates that geometry can be exploited and engineered for tuning the device performance, such as resonance frequencies, mode shapes, mode splitting, and frequency spacing, enabling features beyond regular or 'ideal' devices.

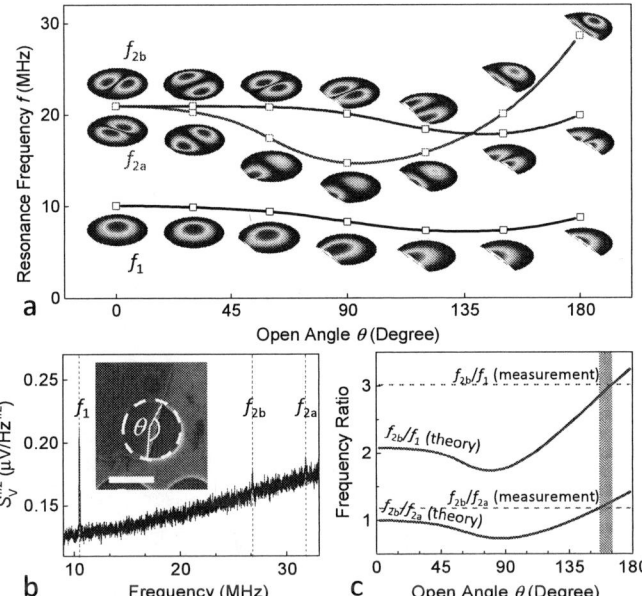

Figure 10: Multimode resonance frequency evolution with device nonideality and asymmetry. (a) Evolution of the first 3 modes with open angle θ. (b) Measured multimode resonance of a device with $\theta \approx 160°$. Red dashed lines are guides to the eyes, indicating the first 3 resonance modes. *Inset*: the SEM image of the actual device, with the open angle illustrated. (c) The measured frequency ratios uniquely identify $\theta \approx 160°$ for this asymmetric device, in excellent agreement with device image. Solid curves are the calculated frequency ratios as functions of the open angle θ. The horizontal dashed lines are the experimentally measured values. The vertical grey stripe indicates where the data intercept the theoretical prediction.

Conclusion and Perspective

These new ultimately thin semiconducting resonators will engender novel 2D transducers, integrated schemes for tuning the coupling effects among information carriers (electron, phonon, photon) on atomic sheets. New devices and functions [15,16] are emerging for sensing and information processing.

Acknowledgements

We thank the support from Case School of Engineering, National Academy of Engineering (NAE) Grainger Foundation Frontier of Engineering (FOE) Award (FOE2013-005), Case Western Reserve University (CWRU) Provost's ACES+ Advance Opportunity Award. Part of the device fabrication was performed at the Cornell Nanoscale Science and Technology Facility (CNF), a member of the National Nanotechnology Infrastructure Network (NNIN), supported by the National Science Foundation (Grant ECCS-0335765).

References

[1] J. S. Bunch, A. M. van der Zande, S. S. Verbridge, I. W. Frank, D. M. Tanenbaum, J. M. Parpia, H. G. Craighead, and P. L. McEuen, "Electromechanical resonators from graphene sheets", *Science* **315**, 490-493 (2007).

[2] C. Chen, S. Rosenblatt, K. I. Bolotin, W. Kalb, P. Kim, I. Kymissis, H. L. Stormer, T. F. Heinz, and J. Hone, "Performance of monolayer graphene nanomechanical resonators with electrical readout", *Nature Nanotechnology* **4**, 861-867 (2009).

[3] A. Eichler, J. Moser, J. Chaste, M. Zdrojek, I. Wilson-Rae, and A. Bachtold, "Nonlinear damping in mechanical resonators made from carbon nanotubes and graphene", *Nature Nanotechnology* **6**, 339-342 (2011).

[4] R. A. Barton, J. Parpia, and H. G. Craighead, "Fabrication and performance of graphene nanoelectromechanical systems", *Journal of Vacuum Science & Technology B* **29**, 050801 (2011).

[5] C. Chen, S. Lee, V. V. Deshpande, G.-H. Lee, M. Lekas, K. Shepard, and J. Hone, "Graphene mechanical oscillators with tunable frequency", *Nature Nanotechnology* **8**, 923-927 (2013).

[6] P. X.-L. Feng, "Tuning in to a graphene oscillator", *Nature Nanotechnology* **8**, 897-898 (2013).

[7] J. Lee, Z. Wang, K. He, J. Shan, and P. X.-L. Feng, "High frequency MoS_2 nanomechanical resonators", *ACS Nano* **7**, 6086-6091 (2013).

[8] Z. Wang and P. X.-L. Feng, "Dynamic range of atomically thin vibrating nanomechanical resonators", *Applied Physics Letters* **104**, 103109 (2014).

[9] J. Lee, Z. Wang, K. He, J. Shan, and P. X.-L. Feng, "Air damping of atomically thin MoS_2 nanomechanical resonators", *Applied Physics Letters* **105**, 023104 (2014).

[10] V. Singh, S. Sengupta, H. S. Solanki, R. Dhall, A. Allain, S. Dhara, P. Pant, and M .M. Deshmukh, "Probing thermal expansion of graphene and modal dispersion at low-temperature using graphene Nanoelectromechanical systems resonators", *Nanotechnology* **21**, 165204 (2010).

[11] A. M. van der Zande, R. A. Barton, J. S. Alden, C. S. Ruiz-Vargas, W. S. Whitney, P. H. Q. Pham, J. Park, J. M. Parpia, H. G. Craighead, and P. L. McEuen, "Large-scale arrays of single-layer graphene resonators", *Nano Letters* **10**, 4869-4873 (2010).

[12] R. A. Barton, I. R. Storch, V. P. Adiga, R. Sakakibara, B. R. Cipriany, B. Ilic, S. P. Wang, P. Ong, P. L. McEuen, J. M. Parpia, and H. G. Craighead, "Photothermal self-oscillation and laser cooling of graphene optomechanical systems", *Nano Letters* **12**, 4681-4686 (2012).

[13] X. Song, M. Oksanen, M. A. Sillanpää, H. G. Craighead, J. M. Parpia, and P. J. Hakonen "Stamp transferred suspended graphene mechanical resonators for radio frequency electrical readout", *Nano Letters* **12**, 198-202 (2012).

[14] R. Yang, Z. Wang, and P. X.-L. Feng, "Electrical breakdown of multilayer MoS_2 field-effect transistors with thickness-dependent mobility", *Nanoscale* **6**, DOI:10.1039/c4nr03472d (2014).

[15] Z. Wang, J. Lee, K. He, J. Shan, and P. X.-L. Feng, "Embracing structural nonidealities and asymmetries in two-dimensional nanomechanical resonators", *Scientific Reports* **4**, 3919 (2014).

[16] J. Lee and P. X.-L. Feng, "Atomically-thin MoS_2 resonators for pressure sensing", *Proc. IEEE Int. Freq. Contr. Symp. (IFCS)*, 282-285, Taipei, Taiwan, May 19-22 (2014).

Integrated On-Chip Energy Storage Using Porous-Silicon Electrochemical Capacitors

D. S. Gardner[1], C. W. Holzwarth III[1], Y. Liu[1], S. B. Clendenning[1], W. Jin[1], B. K. Moon[1], C. Pint[1,2],
Z. Chen[1], E. Hannah[1], R. Chen[1], C. P. Wang[3], C. Chen[3], E. Mäkilä[4], and J.L. Gustafson[1,5]

[1]Intel Labs, Intel Corp., 3065 Bowers Ave., Santa Clara, CA 95052,
[2]now at Vanderbilt University, [3]Florida Intl. University, [4]University of Turku, [5]now at Ceranovo
Tel: (408) 765-2025, Email: d.s.gardner@intel.com

Abstract

Integrated on-chip energy storage is increasingly important in the fields of *internet of things*, energy harvesting, and *wearables* with capacitors being ideal for devices requiring higher powers, low voltages, or many thousands of cycles. This work demonstrates electrochemical capacitors fabricated using porous Si nanostructures with very high surface-to-volume ratios and an electrolyte. Stability is achieved through ALD TiN or CVD carbon coatings. The use of Si processing methods creates the potential for on-chip energy storage.

Introduction

Capacitors can capture energy at high rates and low voltages as well as provide higher power, albeit at lower energy densities, and degrade minimally from charging and discharging. These properties follow because capacitors are electrostatic devices and do not rely on chemical reactions to store energy making them favored over batteries for certain applications or suitable to work in synergy with batteries.

In this work, electrochemical (EC) capacitors based on porous-silicon (P-Si) nanostructures with channel sizes 20 nm to 100 nm were synthesized and coated with atomic layer deposited (ALD) films or chemical vapor deposited (CVD) carbon. Measurements from the coated P-Si capacitors reveal that an areal capacitance of 3 to 6 mF/cm^2 can be achieved using 2 μm deep pores, over an order of magnitude higher than previous studies utilizing P-Si [1, 2]. Also, the energy density of the coated samples was significantly higher. The devices were fabricated using silicon process methods with the potential for on-chip integration. The pores can be formed in localized regions on the front side of a Si die or utilizing the backside bulk Si of integrated circuits for kinetic or solar energy harvesting systems such as silicon solar cells [3] or for localized low-energy storage such as for integrated sensors.

Experimental

To maximize the performance of P-Si electrochemical capacitors, it is important to optimize the pore structure and surface properties. Pore structure can be summarized with five interdependent parameters: pore size, surface area, porosity, depth, and morphology; likewise, the surface properties can be summarized by chemical stability, conductivity and wettability. These properties control the performance of the electrochemical capacitor: capacitance, operating voltage, cycle lifetime, and operating frequency.

A. Porous-Si Nanostructure

The total surface area needs to be high to obtain large capacitance which requires small pore diameters and deep etch depth resulting in high-aspect-ratio features. Contrariwise for high operating frequency, large pore diameters are preferred for fast ion transport. For optimal

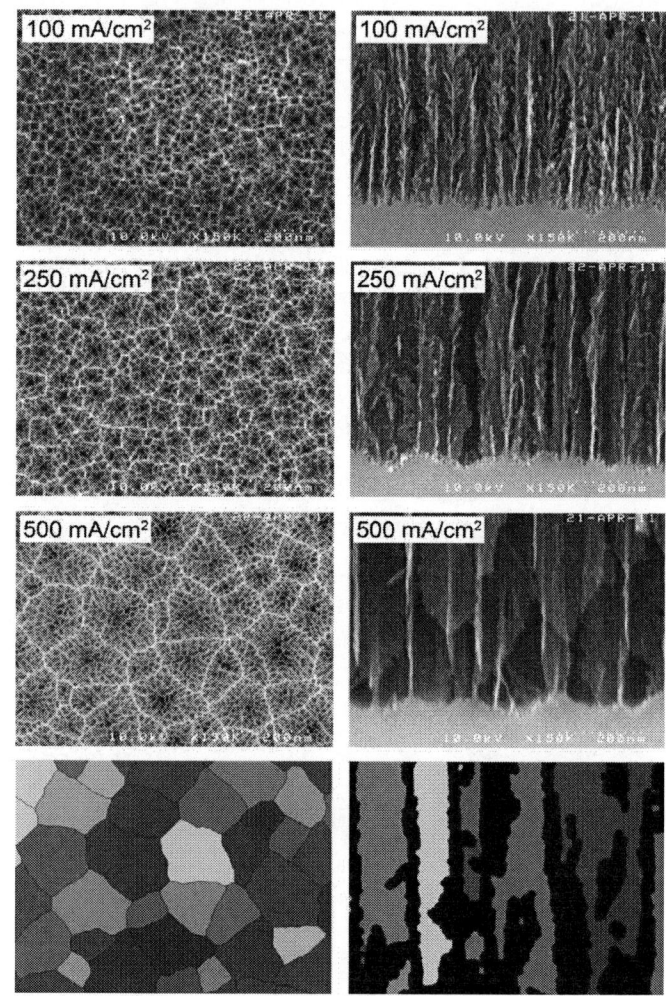

Fig. 1. SEM images demonstrate that the porous silicon structure can be varied by controlling etch conditions such as: electrochemical etching current density (shown), etchant, Si doping, etc. Image morphological analysis was used to quantify the P-Si structure (bottom).

performance, large and small pore diameters are needed.

Electrochemical etching using hydrofluoric acid and isopropyl alcohol (2:1 HF:IPA) was employed [4] to etch P-Si nanostructures up to 250 μm deep with a porosity of ~75%. The etching apparatus is a double tank electro-chemical cell developed by AMMT [5]. Highly doped (100) p-type silicon (0.01 to 0.001 Ω-cm) was chosen because it provides good electrode conductivity and depending on the current density used, wide smoother-walled pores and narrow highly-branched pores could be achieved (Fig. 1). A sacrificial layer process consisting of a high current density

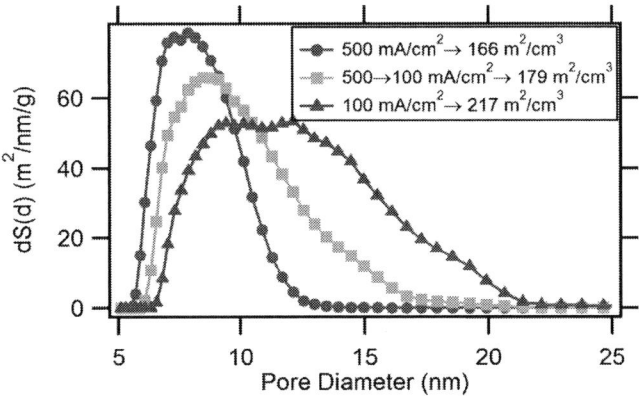

Fig. 2. Pore size distribution from BET absorption analysis of 3 different etching current densities demonstrating the inverse relationship between pore diameter and surface-area density.

etch and a KOH strip before the P-Si etch process is used to ensure fully open pores. This proved to be essential for ALD coating penetration and for device operation at high power densities.

Measurements using BET adsorption analysis shows that the pore size distribution demonstrates an inverse relationship between pore diameter and surface-area density (see Fig. 2). The high-aspect ratios can lead to high effective series resistance (ESR) because of the long path that ionic charge carriers have to traverse. In an effort to optimize these parameters, a tapered-pore nanostructure with main channel sizes tapering from 100 nm to 20 nm (Fig. 3) and extremely high surface-to-volume ratios was developed (see Table 1).

B. Surface Coatings

This P-Si nanostructure was combined with an ionic liquid to create an electrochemical capacitor. Bare silicon is not stable over a long period of time due to unwanted reactions with the electrolyte, resulting in decreasing capacitance with cycling. Conductive, passivating and/or electrochemically active surface coatings such as ALD titanium nitride (TiN) (see Fig. 4) [6], pseudocapacitive materials (e.g. VN, RuO_2), SiC [7], or carbon (see Fig. 5) [6, 8-10] were found necessary for long-term stability. Recent work by one of our coauthors

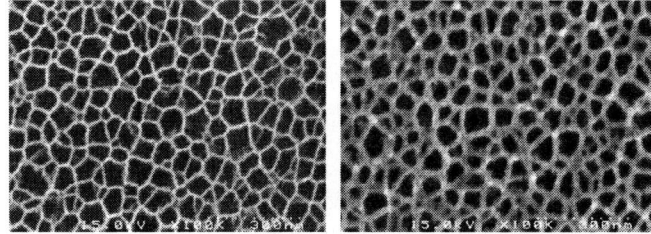

Fig. 4. SEM images of top of porous region (left) before and (right) after stop-flow ALD TiN deposition. The pore walls get thicker but the overall pore structure does not change.

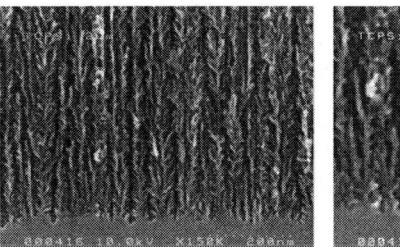 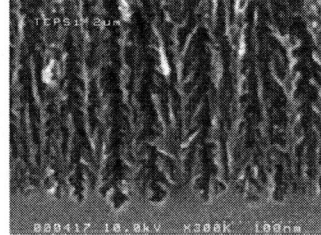

Fig. 5. Cross sectional images of carbonized P-Si deposited at 500°C, then 720°C using CVD with acetylene (N_2:C_2H_2) gas (see [8, 10] for details).

[11] showed that coating P-Si with graphene-like carbon deposited at high temperatures (>650°C) can reduce these unwanted reactions. The present study demonstrates that stability is also achievable at reduced deposition temperatures with ALD TiN enabling on-chip integration.

ALD of films in ultrahigh aspect ratio features (AR > 100:1) presents unique challenges. To obtain uniform coatings, efficient surface reactions are needed between high volatility, low molecular weight, small molecular diameter precursors without chemical vapor deposition side reactions [12, 13]. In this work, the $TiCl_4$ + NH_3 ALD TiN process met these criteria and resulted in passivating conductive films (see Fig. 4). ALD reactor design also plays a critical role. Whereas the majority of ALD is in a continuous viscous flow reactor, a pseudo stop-flow ALD reactor was used to obtain uniform coatings to a 12 μm pore depth (see Fig. 6). The substrate was allowed higher-pressure extended precursor soak times with intervening pump-purge cycles. Surface area measurements using BET analysis showed that the surface area is reduced with increasing TiN thickness (see Table 1).

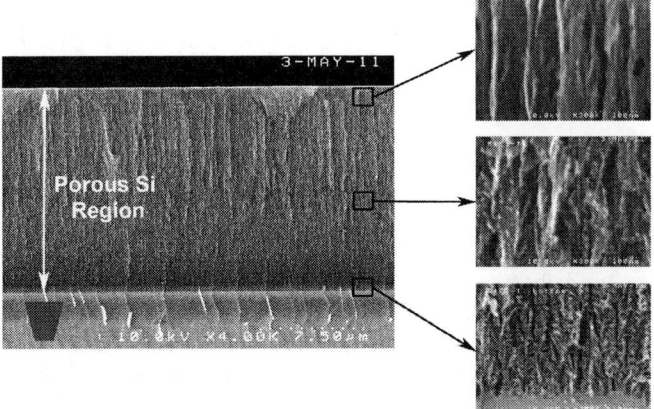

Fig. 3. Tapered porous silicon nanostructure 15 μm deep created by changing the current density during etching. The tapered profile can be optimized for desired device performance (e.g. faster speed or higher capacitance).

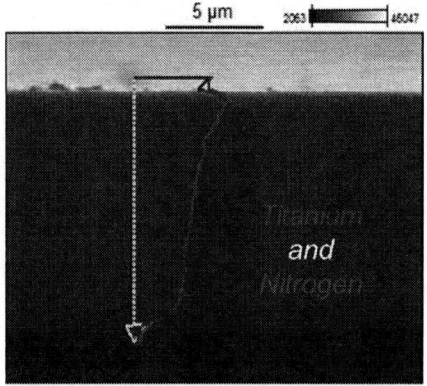

Fig. 6. SEM cross sectional image with an EDS profile superimposed on 12 μm deep pores of Ti and N concentrations showing Ti and N concentration as a function of depth.

978-1-4799-8002-4/14 $31.00 © 2014 IEEE

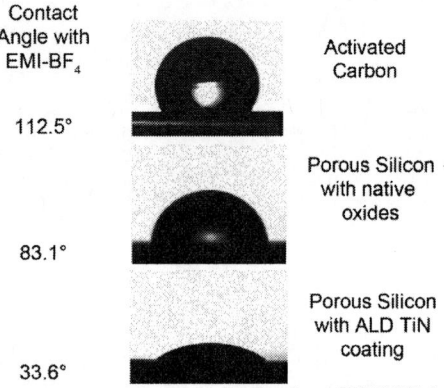

Contact Angle with EMI-BF$_4$

112.5° — Activated Carbon

83.1° — Porous Silicon with native oxides

33.6° — Porous Silicon with ALD TiN coating

Fig. 7. Surface energy measurements using droplets of EMI-BF$_4$ ionic liquid. The best wettability (smallest contact angle) is with TiN coatings.

Table 1. Porous Si surface area reduction from TiN coating.

Coating	Surface area m^2/cm^3	Percent reduction
Before ALD	241	—
2.5nm TiN	210	−13%
5 nm TiN	168	−30%
10 nm TiN	113	−53%

Material Characterization

Key characteristics of the bare and coated P-Si nanostructures (surface area, pore volume, and pore size distribution) and the effectiveness of the surface coatings were studied. SEM images were used to ensure fully open pores and to measure the tapered pore structures. ALD TiN films were first characterized as planar films for conductivity, wettability, and electrochemical stability. Penetration depth and chemical composition of the coatings were measured with energy-dispersive X-ray spectroscopy (see Fig. 6). The surface energy of ionic liquid was examined using liquid droplets (see Fig. 7) to enhance electrolyte penetration. Cyclic voltammetry and impedance spectroscopy were characterized using galvanostatic cycling with a two terminal testing cell (see Fig. 8). Coating quality was characterized by measuring the electrochemical window of fabricated EC capacitors.

Performance

Cyclic voltammetry and impedance spectroscopy was used to characterize P-Si capacitors that were uncoated or coated

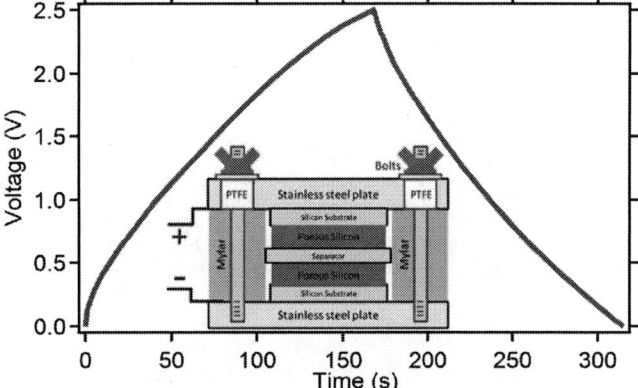

Fig. 8. Charge-discharge curve to 2.5 V using a constant 0.4 mA/cm^2 charging current of an ALD TiN coated 12 μm deep P-Si EC capacitor. Inset shows a two-terminal testing cell with two P-Si electrodes and a separator.

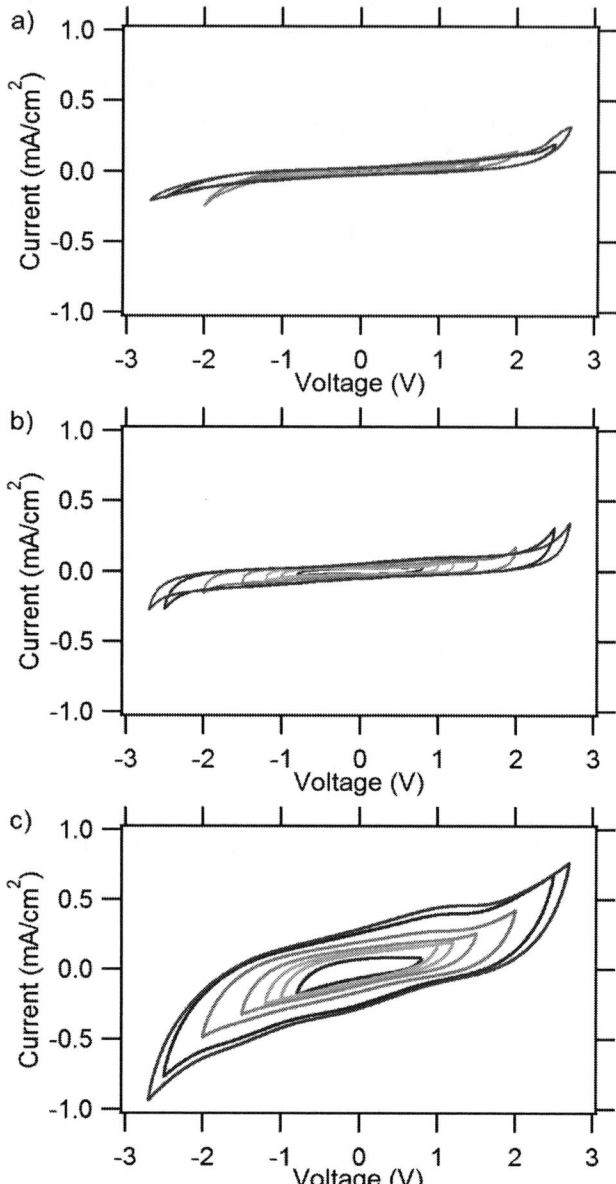

Fig. 9. Cyclic voltammetry measurements after 100 cycles at 50 mV/s of (a) uncoated P-Si silicon, (b) carbonized P-Si, and (c) P-Si with TiN coating.

with TiN or carbon. Devices were successfully prepared by vacuum impregnating EMI-BF$_4$ ionic liquid into the porous electrodes. A scan rate of 50 mV/s for 100 cycles was used (see Fig. 9) at increasing voltage windows to determine the electrochemical window for each device. The device coated with TiN or carbon can be stably cycled (with repeated overlapping curves) while the bare P-Si degrades. The TiN coated device curves are nearly rectangular indicating a capacitive behavior with no faradic reactions.

The capacitance, C = Q/ΔV where Q is obtained by integrating the discharged current of the cyclic voltammetry curves (see Fig. 10), dropped after the first cycle in bare P-Si from reactions while that of TiN coated P-Si increases during the first hundreds of cycles, which is known as *conditioning*.

978-1-4799-8002-4/14 $31.00 © 2014 IEEE

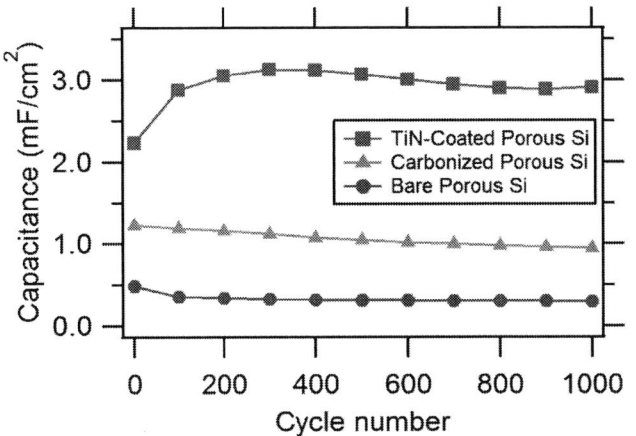

Fig. 10. The capacitance is stable after long term cycling at 50 mV/s to 0.8 V which is within the electrochemical window for all the devices.

After stabilization, the TiN coated P-Si has an aerial capacitance of 3.1 mF/cm^2 for 2 µm deep pores, 10× higher than bare P-Si and 3× more than the carbonized P-Si.

A Ragone chart showing the performance characteristics of the P-Si EC capacitors by plotting energy density (storage capacity) vs. power density (speed of charge and discharge) is shown in Fig. 11. The values are calculated by integrating the galvanostatic discharge curves for each device at various current densities and normalizing to the electrode's volume. Using TiN coatings, volumetric energy densities of up to 9×10^{-3} Wh/cm^3 were obtained which is >20× higher than bare P-Si at 1 V and comparable to commercial carbon-based EC capacitors. This converts to a specific energy of 6 Wh/kg using the P-Si and electrolyte masses (not accounting for surface coatings and packaging).

The improvement from coatings is also seen in impedance spectroscopy (Fig. 12) where the real capacitance C' =

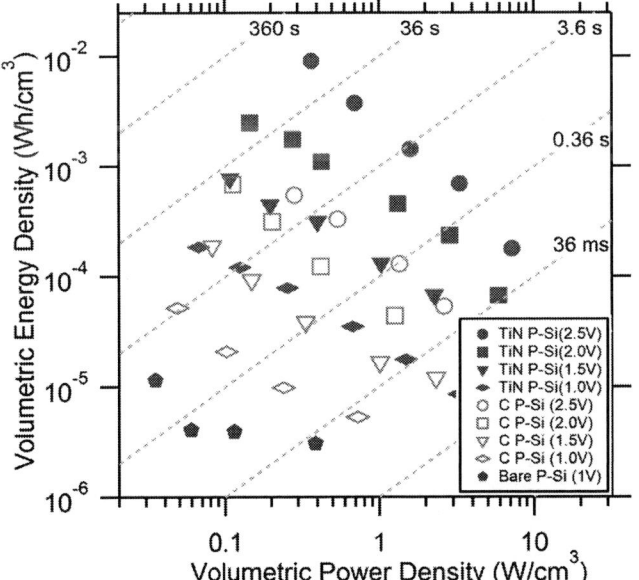

Fig. 11. Ragone plot for performance comparison of volumetric power density versus energy density of TiN coated P-Si, carbonized P-Si (C P-Si), and bare P-Si. The 100th cycle at each voltage is used.

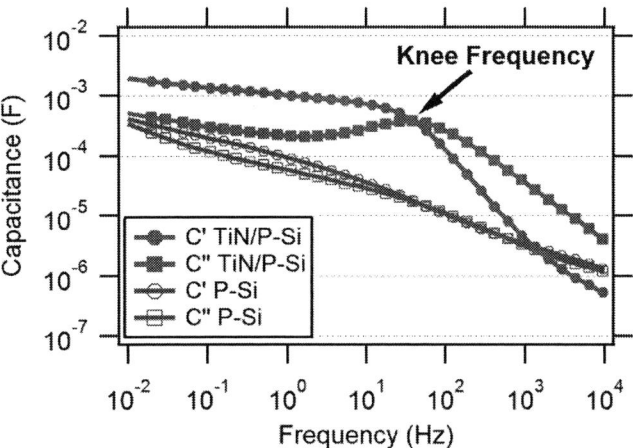

Fig. 12. The impedance was tested at open circuit potential. Real capacitance C' shows the storage capability whereas the imaginary capacitance C" corresponds to energy dissipation by an irreversible process.

$Z''/(\omega|Z|^2)$) and the imaginary capacitance $C''=Z'/(\omega|Z|^2)$ are calculated from Z' and Z" representing the real and imaginary part of the electrochemical impedance respectively and ω representing the frequency. At lower frequencies, the capacitance of TiN/P-Si is 3.0 mF/cm^2, similar to that from CV measurements, while bare P-Si is only 0.4 mF/cm^2. The time constant of the TiN coated P-Si EC capacitor is 17.6 ms while bare P-Si is too lossy to be measured.

Conclusion

Tapered porous-silicon nanostructures were characterized, optimized, passivated with TiN or carbon, and used with ionic liquid to form an electrochemical capacitor. Pseudo stop-flow low-temperature ALD TiN was successfully used to coat the high aspect ratio structures. The energy density is increased by coating with ALD TiN or carbon with ALD TiN producing the largest increase. The P-Si EC capacitors with carbon or TiN coatings both exhibited stable capacitance for over 1,000 cycles. Impedance testing showed that devices exhibited capacitive behavior (C' ≥ C") up to 60 Hz. P-Si based EC capacitors are demonstrated as having the potential to provide localized integrated on-chip energy storage.

References

[1] S.E. Rowlands, R.J. Latham, W.S. Schlindwein, Ionics 5, 144–149, 1999.
[2] S. Desplobain, G. Gautier, J. Semai, L. Ventura and M. Roy, Phys. Stat. Sol. (c), 4, 2180, 2007.
[3] D.S. Gardner and C.L. Pint, U.S. Patent #8,816,465, issued Aug. 2014.
[4] V. Lehmann, R. Stengl, and A. Luigart, Mater. Sci. Eng., B69-70, pp. 11-22, 2000.
[5] AMMT Adv. Micromachining Tools GmbH, www.ammt.com, Germany.
[6] D.S. Gardner, Z. Chen, W. Jin, S.B. Clendenning, E.C. Hannah, T.V. Aldridge, J.L.Gustafson, U.S.Patent App. 20130273261, filed Sept. 30, 2011.
[7] J. Alper, M. Vincent, C. Carraro, R. Maboudian, APL 100, 163901, 2012.
[8] E. Mäkilä, et. al., Langmuir, 28, pp. 14045-14054, 2012.
[9] J. Salonen, M. Bjorkqvist, E. Laine, L. Niinisto, Appl. Surf. Sci., 225, 389–394 (2004).
[10] D. S. Gardner, W. Jin, Z. Chen, C.W. Holzwarth, C.L. Pint, B.K. Moon, and J.L. Gustafson, U.S. Patent Appl. 20140078644, filed Sept. 17, 2013.
[11] L. Oakes, A. Westover, J.W. Mares, S. Chatterjee, W.R. Erwin, R. Bardhan, S.M. Weiss, C.L. Pint, Sci. Rep., 3, 3020, 2013.
[12] J. W. Elam, D. Routkevitch, P. P. Mardilovich, S. M. George, Chem. Mater. 15, 3507-3517 (2003).
[13] S. O. Kucheyev, J. Biener, T. F. Baumann, Y. M. Wang, A. V. Hamza, Z. Li, D. K. Lee, R. G. Gordon, Langmuir, 24, 943-948 (2008).

Comprehensive Analysis of Deformation of Interfacial Micro-Nano Structure by Applied Force in Triboelectric Energy Harvester

Myeong-Lok Seol[1], Jin-Woo Han[2], Jong-Ho Woo[1], Dong-Il Moon[1], Jee-Yeon Kim[1], and Yang-Kyu Choi[1*]

[1] Department of Electrical Engineering, KAIST, 291 Daehak-ro, Yuseong-gu, Daejeon 305-701, Republic of Korea

[2] Center for Nanotechnology, NASA Ames Research Center, Moffett Field, California 94035, USA

* Tel.: (+82) 42-350-3477, Fax: (+82) 42-350-8565, Email: ykchoi@ee.kaist.ac.kr

Abstract

The correlation between the deformation of an interfacial micro-nano structure and the applied pressure in a triboelectric energy harvester (TEH) is analyzed for the first time. The modeling, simulation, visualization experiment, and electrical measurements are conducted in order to clarify the effects of the structural deformation, which governs the triboelectric charge density. The results imply that a small-sized structure is advantageous in output power, while a large-sized structure is advantageous in the pressure sensing range.

Introduction

Triboelectricity has long been considered an unwanted phenomenon and is the cause of breakdown in electronics, fires, and human discomfort. However, strong triboelectricity can be an effective energy source when utilized in an energy harvester. A triboelectric energy harvester (TEH), i.e. a triboelectric nanogenerator, intentionally generates strong triboelectricity through contacting two different materials, and it uses the strong triboelectric charges to drive an iterative current via electrostatic induction (Fig. 1) [1, 2].

Micro-nano structures are often formed on triboelectric surfaces in order to increase the effective contact area at the contact interface. The enlarged contact area generates larger triboelectric charges, which results in increased induced currents. Various interfacial structures, e.g. well-ordered microstructures, nanoparticle array structures, and nature-replicated structures, have been introduced for performance improvements in TEHs [3-5].

For conventional TEHs without interface micro-nano structures, applied pressure is not considered to be an important variable because triboelectric charging occurs whether the contact pressure is strong or not. However, when micro-nano structures are formed on the triboelectric surface, the applied pressure could critically affect the performance of TEHs due to the deformation behaviors of the interfacial micro-nano structures (Fig. 2). In this study, the correlation between the deformation of the interfacial micro-nano structures and the applied pressure is comprehensively

analyzed using modeling, simulation, visualization experiments, and electrical measurements.

Experimental Setup

A rigid double-plate connected and supported by two springs was used as an experimental template in order to ensure iterative contact and separation operations (Fig. 3). An Ag-deposited silicon wafer was fixed to the top plate. The Ag layer had dual functions of being a triboelectric surface and a top electrode. A polymer layer (polydimethylsiloxane; PDMS), which was attached to the Au-deposited silicon wafer, was fixed at the bottom plate. The Au layer beneath the PDMS layer behaved as a bottom electrode and the polymer layer behaved as a counter triboelectric surface. According to the triboelectric tendency, PDMS is an excellent electron-accepting material while Ag is a good electron-donor material [5]. Therefore, PDMS can easily obtain electrons from Ag when they are brought into contact. The thickness of the PDMS layer was 400 μm.

A PDMS pyramid array was formed using the replica-molding process (Fig. 4) [6]. The master template of the replica-molding process was fabricated on a silicon wafer. A dot array was patterned on the wafer using photolithography, and the wafer was subsequently dipped into a KOH solution for the anisotropic etching of the silicon. Through splitting the sizes of the dot array, pyramid array structures with various sizes were fabricated (Fig. 5).

During the measurement, the shaker machine repeatedly applied pressure (P) to the top plate of the TEH (Fig. 6). The vibration frequency was 1 Hz and all electrical data were measured 3 min after the vibration began. The humidity, which is an important variable in TEH [5], was maintained at 55~65 % for equitable comparisons. The output open-circuit voltage and short-circuit current were measured using an electrometer (Keithley 6514).

Results and Discussion

Square-shaped open-circuit voltage and peak-shaped short-circuit currents were produced from the TEHs (Figs. 7 and 8). At the contact state, the distance between the top electrode and polymer surface was extremely close; therefore, positive charges accumulated at the top electrode in order to compensate the fixed negative triboelectric charges at the polymer surface. In contrast, in the separation state, the bottom electrode was closer to the polymer surface; thus, positive charges accumulated at the bottom electrode in order to compensate the fixed negative triboelectric charges at the polymer surface. The amount of positive charges flowing governs the output power of TEH (Fig. 7). Because the contact process caused by the instantaneous pressure occurs much faster than the separation process caused by the restoring force of the springs, the positive current peak at the contact process exhibited a higher intensity but a shorter flow time (Fig. 8). The total amounts of flowing charges during the contact and separation processes were the same. The strong instantaneous power of the TEHs was sufficient to directly light 25 serially connected LEDs without a storage component (Fig. 9).

For conceptual understanding of the pressure-deformation relationship, a finite-element simulation (COMSOL) as conducted in order to extract the displacement profiles (Fig. 10). Under a weak pressure, only the tip of the pyramid was affected by the top metal plate. In this 'partial contact' condition, only a small amount of triboelectric charge was generated. As the pressure increased, the entire polymer surface came into contact. The generated triboelectric charges became saturated in this 'full contact' condition. The expected pressure-deformation relationship was also experimentally visualized using a specifically designed experiment (Fig. 11). In this experiment, a transparent glass slide is used as the top plate instead of a metal-deposited silicon wafer. The top-view profiles of the deformed PDMS pyramid array placed under the glass slide can be observed using an optical microscope. From the result, the observed deformation profile was similar to the simulation.

For a deeper understanding of the pressure-deformation relationship, analytical modeling was conducted (Fig. 12). As the applied pressure increased, the increased deformation of the elastic pyramid produced a stronger restoring force in order to become force equilibrium. Therefore, the initial deformation in the early stage was sensitive to small pressure changes, but further deformation toward full contact required much stronger pressures. According to the modeling result, the open circuit voltage was proportional to $(\sigma T/\varepsilon)(TP/EL)^{2/3}$, where σ is the triboelectric charge density of the flat polymer surface, T is the thickness of the polymer layer, ε is the dielectric constant of the polymer, P is the vertical pressure, E is Young's modulus of the polymer, and L is the unit length of the pyramid structure.

From the electrical measurements, the open circuit voltage was sensitive to pressure changes in a small pressure range (i.e. partial contact conditions), but the increment ratio decreased in high pressure ranges (i.e. full contact conditions) (Fig. 13). The measured data for the small pressure range were well fitted to the partial contact model; however, the difference between the measured data and the model increased as the pressure increased (Fig. 14). The difference between the measured data and the partial contact model was severe for small-sized pyramid arrays, which could be understood because the partial contact range of the small-sized pyramid was smaller than that of the large-sized pyramid. That is, a TEH with a small-sized interface structure requires a small pressure to develop into the full contact condition, while the TEH with a large-sized structure requires a strong pressure to develop into the full contact condition. Based on these analyses, TEHs with small-sized nanostructures are advantageous for the output power, in particular for applications to harvest small forces. In contrast, large-sized microstructures are advantageous in the pressure sensing range when TEHs are used for pressure sensor applications (Fig. 15). In order to ensure the endurance of the experimented TEH, the iterative contact and separation operations were conducted continuously over two days with a vibration frequency of 2 Hz (Fig. 16). The output-induced current was not significantly degraded during and after the stress.

Conclusion

The results of this study confirmed that the triboelectricity of micro-nano structures in TEHs was strongly influenced by the applied pressure. The analyzed results could provide guidelines for designing multi-purpose triboelectric devices that could be used as energy harvesters and pressure sensors. Engineers should carefully design the interfacial micro-nano structures considering the target application of the TEH.

References

[1] F.-R. Fan, Z.-Q. Tian, and Z. L. Wang, *Nano energy* 1, **2012**.

[2] G. Zhu, C. Pan, W. Guo, C.-Y. Chen, Y. Zhou, R. Yu, and Z. L. Wang, *Nano Lett.* 12, **2012**.

[3] F.-R. Fan, L. Lin, G. Zhu, W. Wu, R. Zhang, and Z. L. Wang, *Nano Lett.* 12, **2012**.

[4] J. Chen, G. Zhu, W. Yang, Q. Jing, P. Bai, Y. Yang, T.-C. Hou, and Z. L. Wang, *Adv. Mater.* 25, **2013**.

[5] M.-L. Seol, J.-H. Woo, D.-I. Lee, H. Im, J. Hur, and Y.-K. Choi, *small* (in press).

[6] D. Qin, Y. Xia, G. M. Whitesides, *Nat. Protoc.* 5, **2010**.

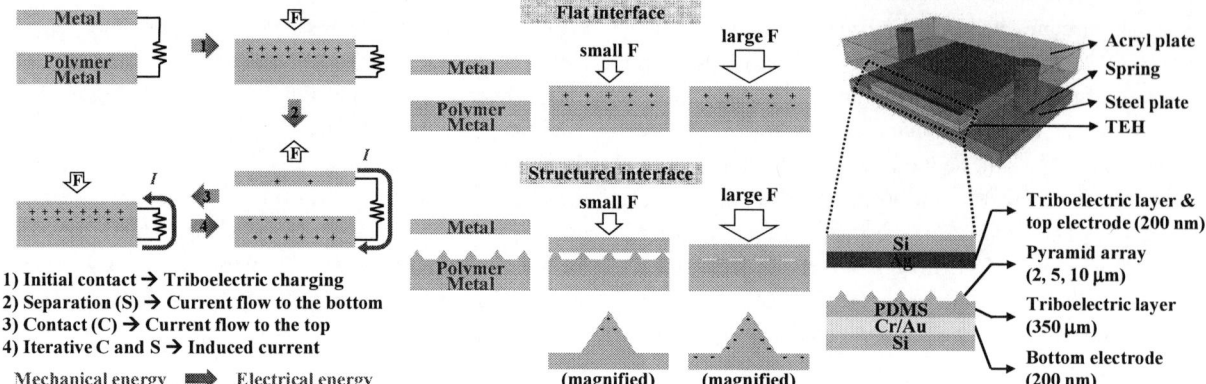

1) Initial contact → Triboelectric charging
2) Separation (S) → Current flow to the bottom
3) Contact (C) → Current flow to the top
4) Iterative C and S → Induced current

Mechanical energy ⟹ Electrical energy

Fig. 1. Operating principle of the TEH. Fixed triboelectric charges at the polymer surface attract counter charges at the top and bottom electrodes at the separation and contact states, respectively.

Fig. 2. Conceptual force dependency of TEH when the interface has a flat structure, and raised micro-nano structures. A greater degree of force dependency is anticipated when micro-nano structure exists at the interface.

Fig. 3. Schematic of experimental device. The platform device is composed of two rigid plates supported by two springs at the edges. Ag and PDMS layers are used as triboelectric layers. PDMS layer contains pyramid array structures.

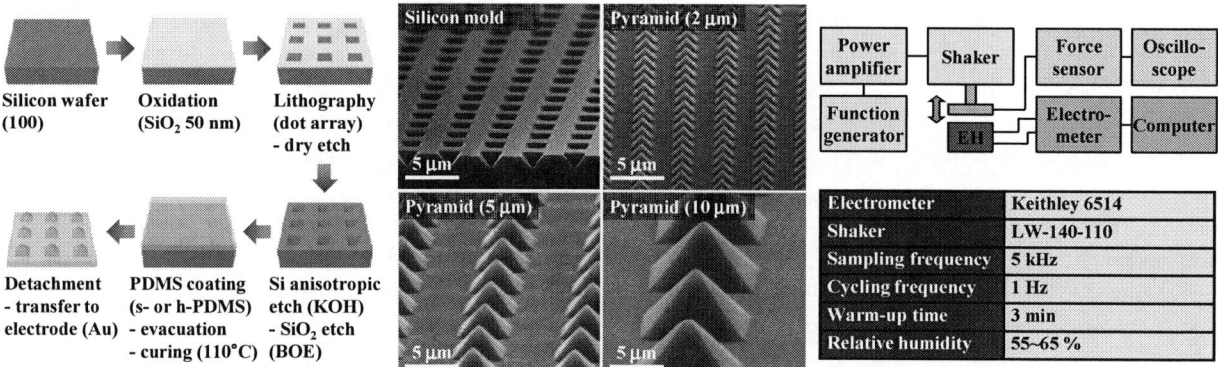

Fig. 4. Fabrication procedure of the PDMS layer with pyramid array structure using replica-molding process. A silicon mold template (an inverse pyramid array) is fabricated by natural anisotropic etch using KOH solution.

Fig. 5. SEM images of the silicon mold template and fabricated pyramid array structures with various sizes. Pyramid structures are selected due to their positive side-angle which is suitable for analyzing force dependency.

Electrometer	Keithley 6514
Shaker	LW-140-110
Sampling frequency	5 kHz
Cycling frequency	1 Hz
Warm-up time	3 min
Relative humidity	55~65 %

Fig. 6. Measurement set-up and experimental conditions. The shaker can apply tunable pressures to the triboelectric energy harvester, and output voltage and current are measured by electrometer.

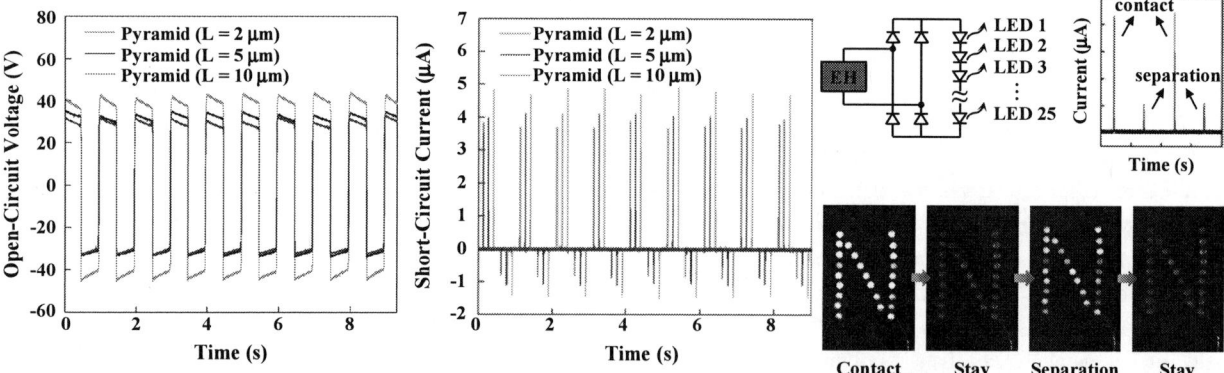

Fig. 7. Open-circuit voltage of TEH with pyramid array structures with various sizes. Voltage-decaying behavior at stationary state is caused by discharging of electrons through the parasitic resistance.

Fig. 8. Short-circuit current of TEH with pyramid array structures with various sizes. The instantaneous current is higher at the moment of contact because the contact process happens faster than the release process.

Fig. 9. Real-time operation of 25 LEDs directly connected to the TEH without any storage component. All LEDs are connected in series. Rectifying bridge circuit is inserted between the TEH and LEDs.

978-1-4799-8002-4/14 $31.00 © 2014 IEEE

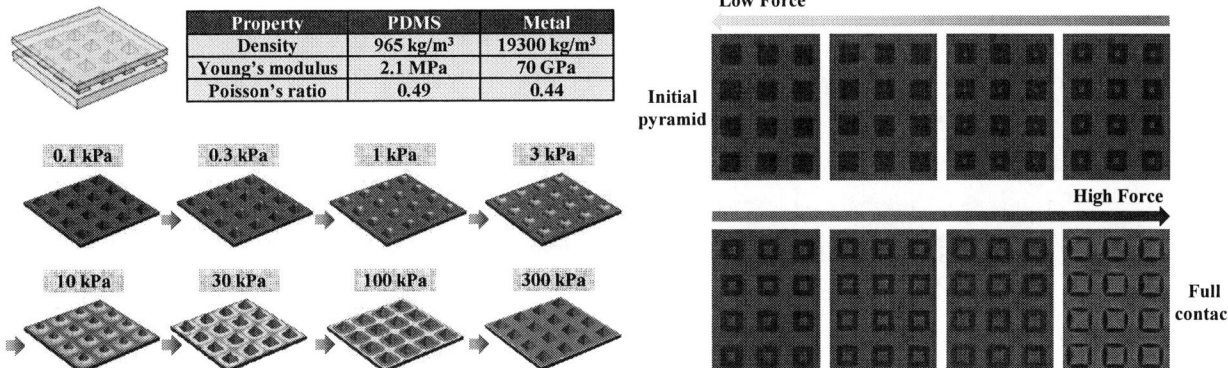

Fig. 10. Simulated displacement profile (COMSOL) with various pressures. The pressures are vertically applied to the top rigid metal plate, which delivers the force to the underlying pyramid array made of soft polymer (PDMS). The top plate is hidden in the color profiles for clearer visualization. Wider area are affected when stronger pressures are applied.

Fig. 11. Optical microscope images visualizing the deformation of the PDMS pyramid array. The images are taken while vertical forces are applied to a transparent glass slide on the pyramid array. The bright parts in the pyramids are the compressed regions where forces are focused, and the dark parts are uncompressed regions, which maintain initial shape.

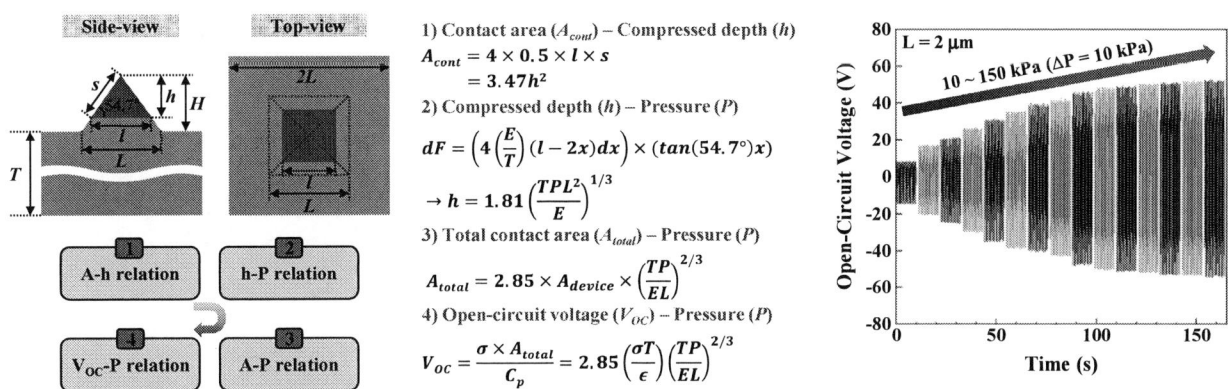

1) Contact area (A_{cont}) – Compressed depth (h)

$$A_{cont} = 4 \times 0.5 \times l \times s$$
$$= 3.47h^2$$

2) Compressed depth (h) – Pressure (P)

$$dF = \left(4\left(\frac{E}{T}\right)(l - 2x)dx\right) \times (tan(54.7°)x)$$
$$\rightarrow h = 1.81\left(\frac{TPL^2}{E}\right)^{1/3}$$

3) Total contact area (A_{total}) – Pressure (P)

$$A_{total} = 2.85 \times A_{device} \times \left(\frac{TP}{EL}\right)^{2/3}$$

4) Open-circuit voltage (V_{OC}) – Pressure (P)

$$V_{OC} = \frac{\sigma \times A_{total}}{C_p} = 2.85\left(\frac{\sigma T}{\epsilon}\right)\left(\frac{TP}{EL}\right)^{2/3}$$

Fig. 12. The analytical model of the relationship between open-circuit voltage (V_{OC}) and applied pressure (P). The model is only valid for a pyramid array structures made of soft polymer. The dark green part of the pyramid indicates a contacted part, *i.e.* a compressed part, and the bright green part indicates the other part without deformation. This model is only applicable for the partial contact condition, i.e., the compressed depth (h) is smaller than pyramid height (H).

Fig. 13. Measured open-circuit voltage of TEH with a pyramid array structure. As stronger pressure is applied, larger area is contacted generating larger triboelectricity, so open-circuit voltage is increased.

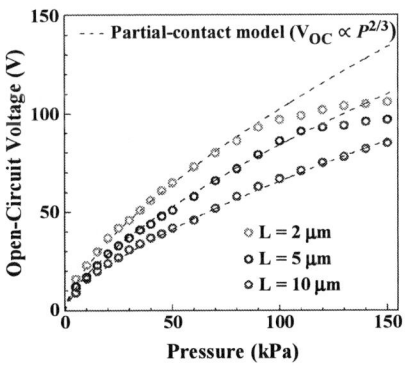

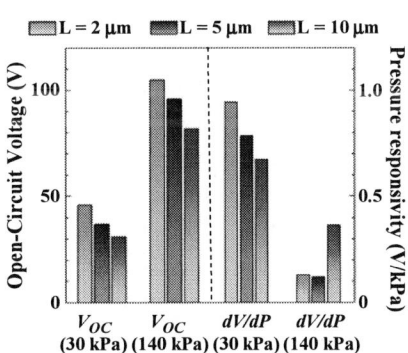

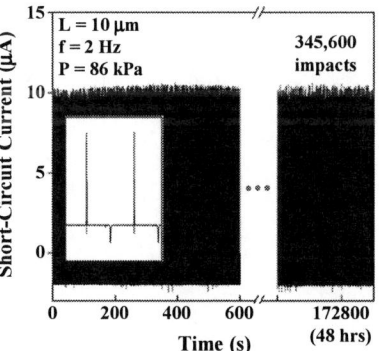

Fig. 14. Correlation of pressure and open-circuit voltage. Under strong pressure, the measured value from the small-sized pyramid is not fitted to the partial-contact model because small-sized pyramid easily get into full-contact mode.

Fig. 15. Open-circuit voltage (V_{OC}) and pressure sensitivity (dV/dP) under two different pressure conditions. The small-sized pyramid is advantageous on the output power, but disadvantageous on the pressure sensing range.

Fig. 16. Endurance characteristic of the experimented TEH. Output power is not significantly degraded for 2 days of stress. Applied pressure was 86 kPa, which is the strongest pressure used in this work.

978-1-4799-8002-4/14 $31.00 © 2014 IEEE

A high efficiency frequency pre-defined flow-driven Energy Harvester dominated by on-chip modified Helmholtz Resonating cavity

X. J. Mu, C. L. Sun, H.M Ji, L. Y. Siow, Q. X. Zhang, Y. Zhu, H.B. Yu, J.F. Tao, Y. D. Gu* and D. L. Kwong

Institute of Microelectronics, A*STAR (Agency for Science, Technology and Research)

11 Science Park Road, Singapore Science Park II, Singapore, 117685,

Email: guyd@ime.a-star.edu.sg; mux@ime.a-star.edu.sg; Tel: (65)-6770-5915, Fax: (65)-6774-5747

Abstract

We present a novel flow-driven energy harvester with its frequency dominated by on-chip modified Helmholtz Resonating Cavity (HRC). This device harvests pneumatic kinetic energy efficiently and demonstrates a power density of 117.6 $\mu W/cm^2$, and charging of a 1 μF capacitor in 200 ms.

Introduction

The emerging trend of self-powered electronic systems or extensive deployed wireless node sensors creates great demand for miniature energy harvesters (EHs). Current vibratory EH strategies, although showing some commercial traction, are inherently bulky, costly vacuum packaging dependent and inefficient for low frequency applications or constant fluctuation environment. These weaknesses limit the practical use of current EH [1-4].

We introduced a modified Helmholtz Resonating Cavity (HRC) to EH design (Fig.1.) that eliminates these drawbacks. In current status, the constant fluid flow is pressurized and regulated by EH modified HRC and in turn drives a high frequency (several~10s kHz) piezoelectric EH structure.

In this paper, we report on the successful realization of aluminum nitride (AlN) and P(VDF-TrFE) based MEMS EH platform. The most significant results are:

1. An AlN-based EH is fabricated with CMOS-compatible processes; converting constant air flow into a specific frequency motion for energy harvesting, delivering power density of 117.6 $\mu W/cm^2$.

2. An on-chip modified HRC of EH is the key structure for the maximum power output and determines the device's working frequency, which is independent of the input air flow.

3. With the same modified HRC and the P(VDF-TrFE) material substitution, the peak to peak voltage output is approaching 5 volts.

4. 500 mV has been achieved by using this modified HRC EHs to charge a 1 μF capacitor in 200 ms.

5. No vacuum packaging is required for this kind of EHs to maintain a maximum power output.

Design and Simulation

Traditional HRC (Fig.1 (a)) and whistle type HRC (Fig.1 (b)) are well known and extensively employed in the motorcycle and car exhausts systems, architectural acoustics and musical instrument design, which show an unique acoustic behavior at specific frequencies.

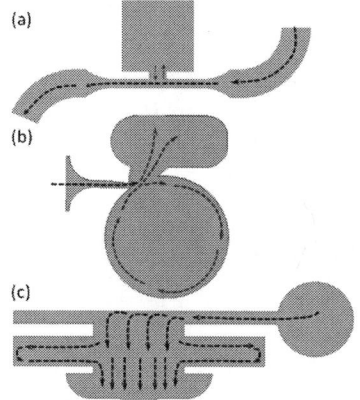

Fig. 1 Schematic of the flow trace in (a) Traditional Helmholtz Resonating Cavity (HRC); (b) Whistle type Helmholtz Resonating Cavity (HRC); (c) Modified Helmholtz Resonating Cavity (HRC)

The fluid motion behavior in aforementioned two types of cavities were simulated as shown in Fig.2 (a), and it is observed that the fluid flow trace is only confined in a narrow path area but not the major cavity area. In order to make full use of the large acoustic cavity and enhance the possibility of the physical interaction between fluid and the flexible functional structure, a modified HRC (Fig.1(c)) is proposed in this paper. In this construct, a functional piezoelectric micro-structure (e.g. micro-belts and micro-nets) (Fig.4) are placed in the specific designed acoustic cavities, in where they are forced into vibration by pressurized inlet fluid. The frequency of the vibration is solely related to the geometry dimensions of the cavity and the inlet narrow nozzles (Table. 1), and is independent of inlet fluid flow conditions including flow rate, pressure and temperature, which is as called Helmholtz resonating frequency. The fluid motion behavior of such modified HRC was also investigated by software fluent and shows a superior momentum transfer possibility between fluid flow and the piezoelectric micro-structure over the traditional HRCs (Fig.2 (b)).

978-1-4799-8002-4/14 $31.00 © 2014 IEEE

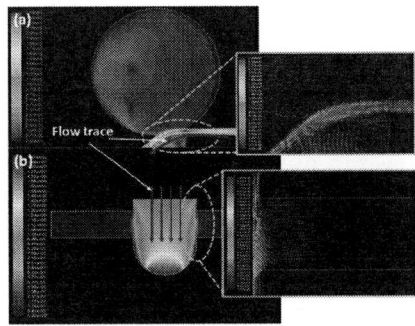

Fig. 2 Fluid motion behaviors of (a) traditional and whistle type Helmholtz Resonating Cavity (HRC) and (b) Modified Helmholtz Resonating Cavity (HRC)

Table 1. Equation of the resonance frequency of the Helmholtz Cavity

$$f = \frac{c}{2\pi} \sqrt{\frac{nA_0}{V(l_0 + 1.7r_0)}}$$

f: resonant frequency of the Helmholtz cavity	c: sound velocity
A_0: the cross section of the inlet nozzle	V: the volume of the Helmholtz cavity
L_0: the length of the inlet nozzle	r_0: nozzle end length correction

Integration and experiment

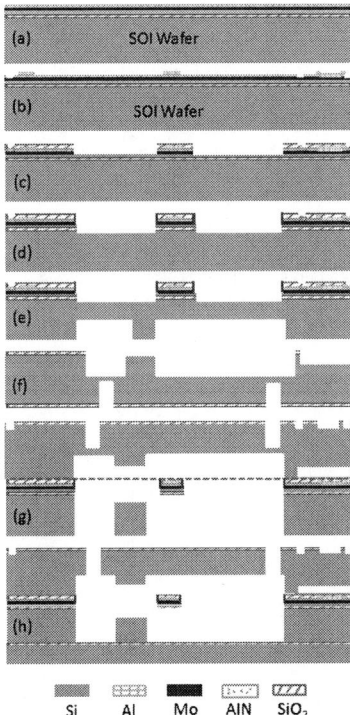

Fig. 3 Major process integration steps. (a) The piezoelectric 0.02 μm AlN /0.2 μm Mo/1.2 μm AlN stack were consecutively deposited on a SOI wafer; (b) Pattern the top and bottom Al electrodes; (c) Pattern the piezoelectric stack; (d) 20 μm Si+ 1 μm SiO₂+ 80 μm Si etch + Backside wafer Grinding to 550 μm; (e) Backside 250 μm partial Si etch; (f) Top cap wafer fabrication; (g) bonding top cap wafer with the device wafer and (h) Inlet and outlet hole forming by DRIE on top cap wafer, after that bonding bottom wafer to device wafer and final partial dicing to expose top electrodes.

In order to validate the proposed EH concept, a batch of different designed EH devices (first 4 designs in table 2) were fabricated with CMOS compatible process by using AlN/SOI platform. Key process integration steps are shown in Fig. 3(a)–(h). The process start with a functional AlN/Mo/AlN stacks (Fig. 3(a)) deposited on a SOI wafer; 0.7 μm thick Al was then deposited as electrical pad material. After that, two masks (Figs. 3(b)–(d)) were used for top Al, bottom Mo electrodes patterning and frontside release etching to define the device layer. Subsequently, backside of the device wafer was patterned and partial etched to define the cavity (Fig. 3(e)). At the same time, the top cap wafer was processed to define the top cavity and the inlet and outlet hole (Fig.3(f)). Then the device wafer was bonded to the top cap wafer and the backside of the device layer was released. Finally, a bare Si wafer was bonded to the backside of device wafer to seal the cavity, which also acts as the support wafer for the front side inlet and outlet hole release (Fig.3 (h)). To verify the universality of the modified HRC, P(VDF-TrFE) micro-belt (Devices 5 and 6 in table 2) was also employed to compare with AlN micro-belt. Fig. 4 shows the fabricated functional structure, including AlN micro-belt based modified HRC typed EH (Fig.4 (a)), AlN micro-structure based traditional HRC typed EHs (Figs.4 (b)-(c)), P(VDF-TrFE) micro-belt based modified HRC typed EH (Fig.4 (e)). The packaged silicon based EHs have a PDMS cap bonded on the top of the inlet hole (Fig. 4 (d)) so as to help coupling the air flow into the device inner cavity.

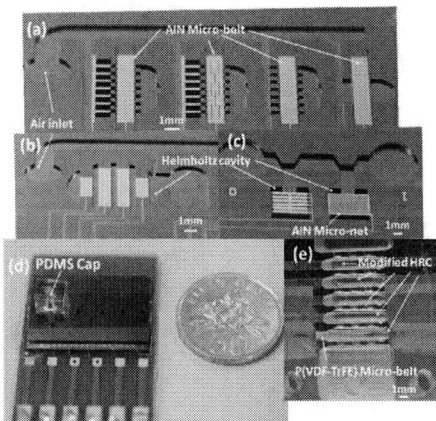

Fig. 4 Optical images of (a) Modified HRC typed EHs (Device 3 and Device 4); (b) Traditional HRC typed EHs (AlN micro-belt embedded) (Device 1); (c) Traditional HRC typed EHs (AlN micro-net embedded) (Device 2); (d) Packaged modified HRC based EHs (Device 3); (e) P(VDF-TrFE) micro-belt + modified HRC (Devices 5 and 6)

The schematic of the EH voltage and power measurement set-up is shown in Fig.5. The flow source is compressed air; a pressure controller (ALICAT) and a flow meter (Bronkhorst) were connected in series with the flow source and the EHs that are under tests. A Digital Phosphor Oscilloscope (Tektronix

DPO 7354) was employed to read out the voltage output of the EHs and its corresponding frequency. A external resistor is connected with EHs under tests in parallel to extract the power that the EHs are able to deliver to the load.

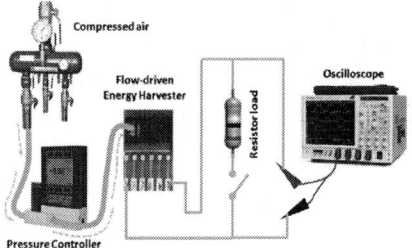

Fig.5 The schematic of the EH voltage, operating frequency, and power measurement set-up: compressed air is used to induce EH vibration; a 15kΩ resistor was used to extract the power

Table 2. Different EHs designs

Device 1	Design B	Whistle type HRC+ 10 mm ×1 mm ×20 μm AlN micro-belt
Device 2	Design A	HRC + 3 mm × 2 mm × 20 μm AlN micro-net
Device 3	Design C	Modified HRC +10 mm × 1 mm × 20 μm AlN micro-belt
Device 4	Design C	Modified HRC +15 mm × 2.5 mm × 20 μm AlN micro-belt
Device 5	Design C	Modified HRC+ 10 mm × 1 mm × 12 μm P(VDF-TrFE) micro-belt
Device 6	Design C	Modified HRC+ 30 mm × 1 mm × 12 μm P(VDF-TrFE) micro-belt

Results and discussion

The open circuit voltage spectra (V_{open}) vs. input air pressure (i.e. the flow rate) and corresponding frequencies of the first four different designs (Table. 2) are summarized in Fig. 6 and Fig.7. It is obvious that the modified HRC based EHs (devices 3 and 4) provide superior voltage output, although the traditional HRC EHs have a relative low start vibration pressure 0.3 psi vs. 2.5 psi (Fig.6). The dominating frequencies of the four devices vary from each other and are dependent of HRC shape and dimensions (Fig.7).

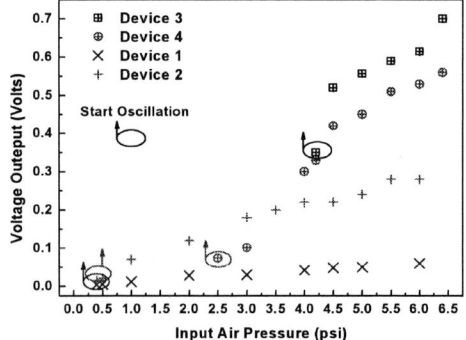

Fig. 6 Devices 1-4 performance comparison: Input air pressure vs. Voltage generation. The start oscillation points have been marked in the figure

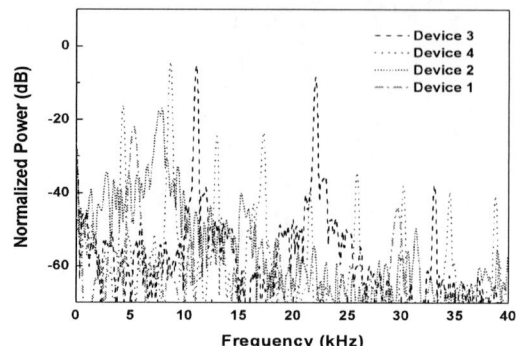

Fig. 7 The dominating frequencies of Devices

The open circuit voltage output under different input air pressures 4 psi, 5 psi and 6 psi and the corresponding frequencies are summarized in Fig.8. It is clear that higher input pressure only leads to higher V_{open} but does not change the frequency spectra. This makes the EH a promising candidate to capture energy from constant fluctuation environment, which greatly simplify the circuitry.

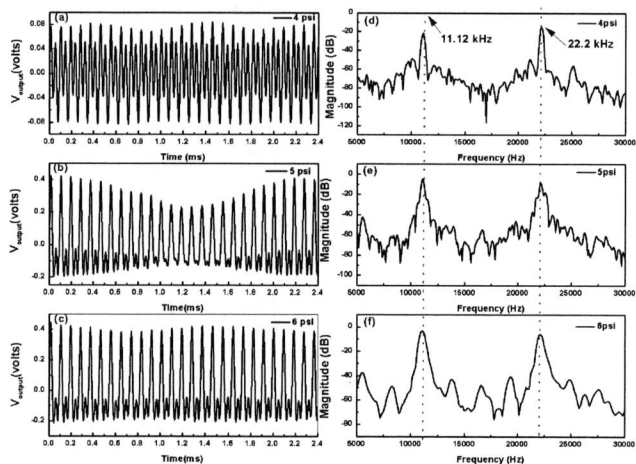

Fig. 8 Modified Helmholtz Resonating Cavity (Device 3): the output voltage (a)-(c) and corresponding frequency spectra of EHs (d)-(f) at input air pressure 4psi, 5psi and 6 psi respectively

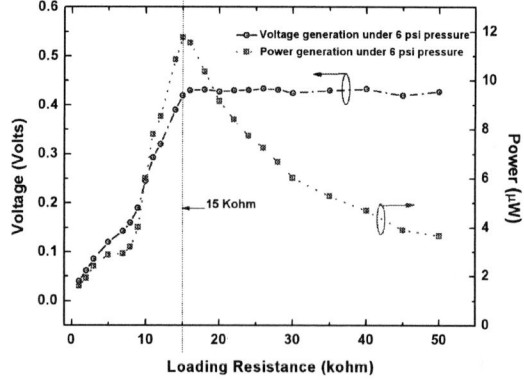

Fig. 9 the Resistance loading vs. Voltage generation and Power generation curves. The maximum power output is around 11.76 μW at 15 kΩ that respect to the power density of 117.6 μW/cm² (10 mm ×1 mm effective area for device 3)

A series of external resistance loading have been connected to the EHs in parallel and the optimized power output 11.76 μW is obtained with a loading resistance of 15 kΩ as shown in Fig.9.

Table 3. Material comparison

	AlN	P(VDF-TrFE)
d_{31}	-1.9 (pC/N)	6± 20% (pC/N)
d_{33}	5 (pC/N)	-16 ± 20% (pC/N)
E	329 GPa	0.95 GPa (machine direction), 1.5 GPa (transverse direction)
ε_{11}	8.0(10^{-11}F/m)	10.18(10^{-11}F/m)
ε_{33}	9.5(10^{-11}F/m)	10.18(10^{-11}F/m)
ρ	3260 kg/m^3	1.8 kg/m^3
v	0.24	0.18

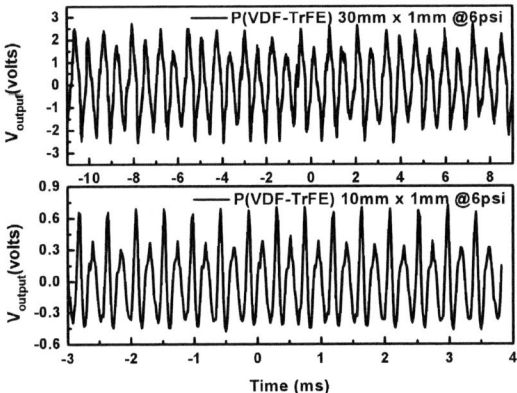

Fig. 10 (a) Voltage output of Device 5 and (b) Device 6 at 6psi pressure input

P(VDF-TrFE) is a type of soft piezoelectric material, which has a piezoelectric coefficient d_{31} three-folds that of AlN (Table. 3). In order to evaluate the versatility of the modified HRC, two different specs of P(VDF-TrFE) micro-belts were tailored and placed in a home-made modified HRC. This HRC consists of PMMA top cap and bottom cavity moulds, which were micro-milled separately and then assembled together as a whole structure to define a inner HRC. The two to seven-folds (up to 5 volts) of voltage outputs V_{pp} were approached by device 5 and device 6 respectively (Fig.10) compared with same size AlN micro-belt in the same designed modified HRC. Due to the smaller elastic module of the P(VDF-TrFE) (0.95 GPa) compared with that of Si (169 GPa), a larger deformation is achieved when the micro-belt vibrates at the nature frequency. This nature frequency of the P(VDF-TrFE) micro-belt is inevitable lower than that of AlN/Si of the same size (e.g. 10 mm × 1 mm).

A home-made power management circuit (Fig.11(a)) with 0.18 μm CMOS process technology is employed to verify the charging capability of the developed EHs. This circuit is composed of a negative voltage rectifier, a boost/Buck Boost

converter and a on-board capacitor. A 1 μF capacitor is successfully charged by the EHs. The capacitor is charged to 300 mV, 200 mV and 500 mV by Device 3, Device 5 and Device 6 respectively within 200-400 ms. This indicates the implementation potential of the prototype device for wireless sensor nodes powering or self-powered electronic systems.

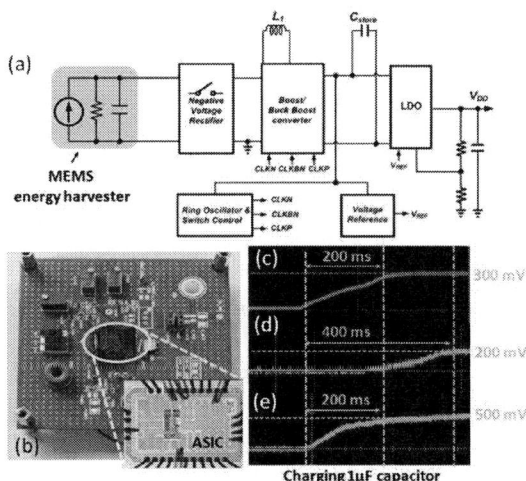

Fig. 11 (a) Schematic and (b) picture of the high efficiency power management circuit for MEMS EHs; Capacitor (1 μF) is charged by (c) AlN based EH (Device 3), (d) 10 mm × 1 mm × 12 μm P(VDF-TrFE) microbelt based EH (Device 5) and (e) 30 mm × 1 mm × 12 μm P(VDF-TrFE) micro-belt based EH (Device 6) through rectified by the 0.18 μm power management circuit

Conclusion

A novel frequency design solution for efficiently harvesting fluid kinetic energy is implemented in a MEMS format. The material AlN and P(VDF-TrFE) have demonstrated their impressing power generation capability through working together with modified HRC. A power density of 117.6 μW/cm^2 is tentatively achieved and 1 μF capacitor is successfully charged in 200 ms. This work points a new direction of building high performance, wideband, miniature EH for medical, automotive, and wireless applications.

Acknowledgement

This work was supported by the core project from Institute of Microelectronics, Agency for Science, Technology and Research (A*STAR), Singapore.

References

(1) R. Elfrink, et al., *IEDM Tech. Dig*, 2011, pp. 677-680.

(2) R. Andosca, et al., *Sens. Actuators, A*, 2012, pp. 76-87.

(3) C. Sun, et al., *Energy Environ. Sci.*, 2011, pp. 4508-4512.

(4) S.P. Matova et al., *J. Micromech. Microeng*, 2011, 104001.

Fabrication of Integrated Micrometer Platform for Thermoelectric Measurements

Maciej Haras*[†], V. Lacatena*[†], F. Morini*, J-F. Robillard*, S. Monfray[†], T. Skotnicki[†] and E. Dubois*

*IEMN UMR CNRS 8520, Avenue Poincaré, 59652 Villeneuve d'Ascq, France
[†]STMicroelectronics 850 rue Jean Monnet 38926 Crolles,France ;(✉)Maciej.Haras@isen.iemn.univ-lille1.fr

I. INTRODUCTION

Thermoelectricity is a promising energy harvesting method which converts heat into electricity in direct, silent, vibration-less and reliable way [1]. The lack of materials which are thermoelectrically efficient and economically attractive is confining thermoelectricity to niche and sophisticated applications. Materials with low thermal ($\kappa=\kappa_{ph}+\kappa_e$), high electrical ($\sigma$) conductivities and as high as possible thermopower (**S**) are required. To compare them, the dimensionless-figure-of-merit **zT** Eq.1 is used [2].

$$zT = \frac{S^2 \cdot \sigma}{\kappa} \cdot T = \frac{S^2 \cdot \sigma}{\kappa_e + \kappa_{ph}} \cdot T \qquad (1)$$

Thermal conductivity features two contributions, the dominant associated with phonons (κ_{ph}) while electrons and holes (κ_e) play a negligible role. Referring to Fig.1 thermo-generator efficiencies are around 1%-5% near room temperature, making this conversion unattractive.

Popularization of thermoelectricity may be achieved by conversion efficiency improvement and fabrication cost reduction. Silicon (**Si**) could be a good thermoelectric material offering: (i) industrial compatibility, (ii) direct integration on chip, (iii) preservation of environment and (iv) cost reduction, but it has too high bulk thermal conductivity. Recent studies suggested the possibility to reduce thermal conductivity in Silicon thin-film as shown in Fig.2, with moderate change on electrical conductivity and thermopower [3-5]. This reduction can make Silicon a competitive

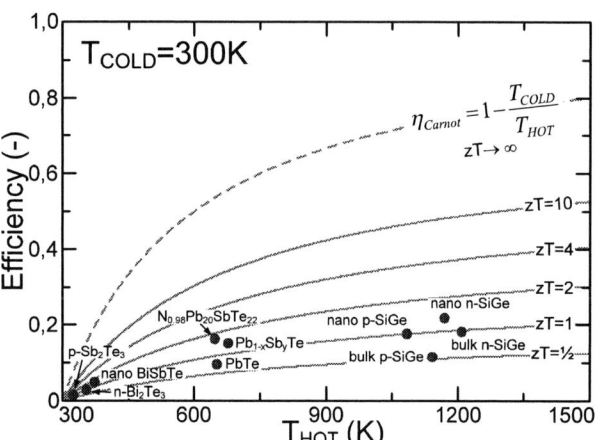

Fig.1 Maximu conversion efficiency versus temperature for different zT values. Dots show the highest efficiency for given material

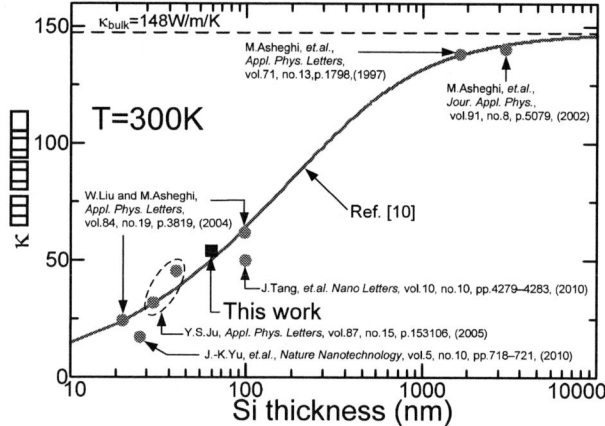

Fig.2 Thermal conductivity of silicon membranes, dots show measurement values, solid line presents the Sondheimer model [10]

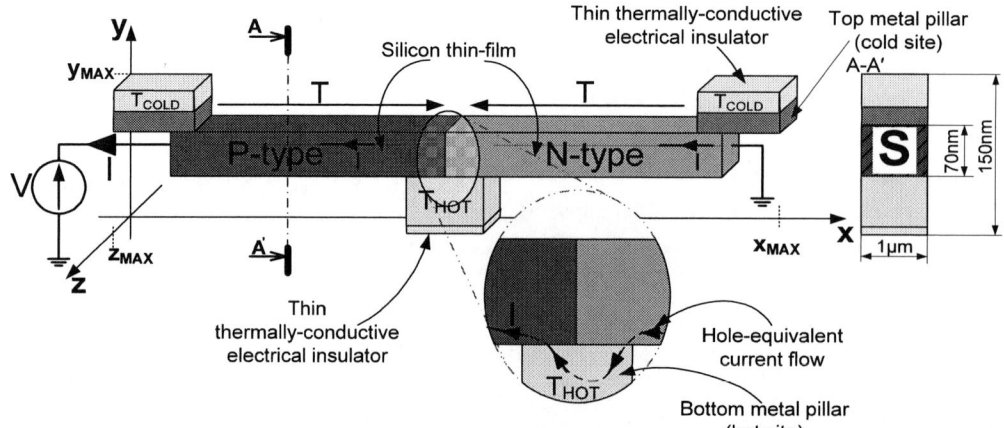

Fig.3 Simulated structure of Silicon lateral thermo-generator with x_{MAX}=10μm (5μm for each membrane) z_{MAX}=1μm

978-1-4799-8002-4/14 $31.00 © 2014 IEEE

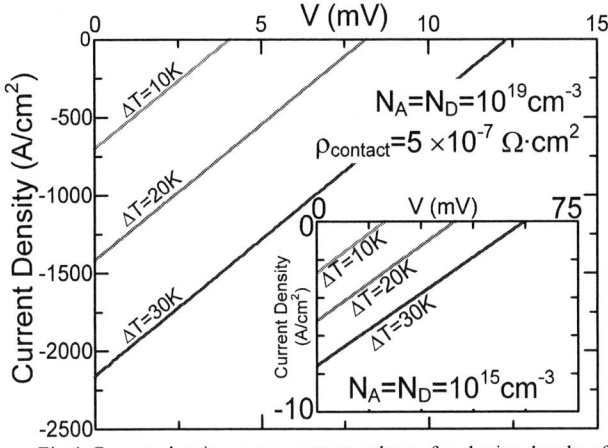

Fig.4 Current density versus output voltage for doping levels of $10^{19}cm^{-3}$ and $10^{15}cm^{-3}$ (*inset plot*) respectively. For doping level of $10^{19}cm^{-3}$ generator specific contact resistivity is 5×10^{-7} Ω.cm²

Fig.5 Harvested power density versus output voltage for generator with a doping level of $10^{19}cm^{-3}$ and specific contact resistivity of 5×10^{-7} Ω.cm²

phonon glass/electron crystal material [6] for which the dimensionless-figure-of-merit **zT** would converge towards $S^2/L \sim 4$ where $L = 2.45 \times 10^{-8}$ W·Ω/K² is the Lorenz factor. Such a Si-compatible material with **zT≈2-3** is a 'Holy Grail' [7] that would take off thermoelectricity.

II. MOTIVATIONS

To investigate harvesting capabilities of Silicon, a finite element simulation of lateral thermo-generator [8] as shown in Fig.3 was performed using non-isothermal drift-diffusion model [9] Eq. 2.

$$\begin{cases} \vec{j}_n(T) = -\sigma_n(T) \cdot \left[\vec{\nabla} \phi_{Fn}(T) - S_n(T) \cdot \vec{\nabla} T \right] \\ \vec{j}_p(T) = -\sigma_p(T) \cdot \left[\vec{\nabla} \phi_{Fp}(T) - S_p(T) \cdot \vec{\nabla} T \right] \end{cases} \quad (2)$$

The current densities in Eq.2 allow taking into account local lattice temperature impact on parameters such as electrical conductivities (σ_n, σ_p), carriers mobilities (μ_n, μ_p), carrier concentrations (**n**, **p**) or quasi-Fermi levels (ϕ_{Fn}, ϕ_{Fp}). The considered temperature is here the local lattice temperature (**T**).

The structure consists in thin-film Silicon membranes to take advantage of thermal conductivity reduction in low dimensionality 2D systems. Assuming a fixed temperature gradient, Fig.4 clearly shows that the output current and power density are governed by electrical conductivity. At high doping level ($10^{19}cm^{-3}$), a harvested power density close to 7W/cm² is obtained for ΔT=30K (Fig.5). The good electrical properties of Silicon result in a significant harvested power density that, in turn, demonstrates that Silicon is a good thermoelectric material provided that thermal conductivity is reduced. Moreover, use of CMOS technology enables to take advantage of low specific contact resistivity ($\rho_{contact}$) which has a parasitic impact on the generator performance.

III. DEVICE FABRICATION

Based on this encouraging result, an integrated micrometer platform for thermal conductivity measurements was designed and fabricated. The process flow is depicted in Fig.6. The departure point is a 70nm thick Silicon-On-Insulator (**SOI**) wafer upon which thermal oxide is grown

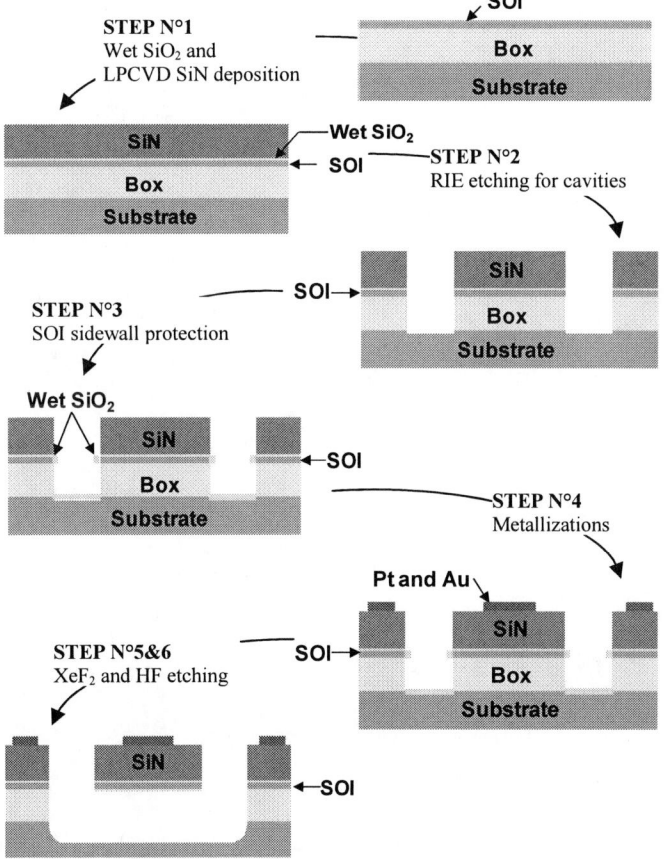

Fig.6 Process flow for micrometer platforms for thermoelectric measurements

followed by deposition of SiN (*STEP N°1*). Reactive ion etching (**RIE**) is subsequently used to etch side cavities that define the shape of the membrane to be suspended (*STEP N°2*). Afterwards the thermal oxide is grown on the **SOI** sidewalls to protect from etching in the following step (*STEP N°3*). Subsequently two metallization steps are realized. The first one uses a thin platinum layer structured in serpentines acting as both temperature sensing and heating elements. The second gold metallization step contributes to reduce probing resistances (*STEP N°4*). Finally, the membrane is released in two steps: first, the Silicon handler is under-etched using XeF$_2$ in gaseous phase (*STEP N°5*). Secondly, the buried oxide of the **SOI** stack is eliminated by vapor HF (*STEP N°6*). The final device is presented in Fig.7a while Fig.7-b shows a zoom of a 70nm thick Silicon membrane.

IV. CHARACTERIZATION

To better understand the measurement methodology, Fig.8 depicts the device cross-sectional view along A-A' cutline placed in Fig.7-b.A voltage applied to the heater produces a heat flow (**Q**) thanks to the Joule effect. Half of this heat (**Q/2**) flows in each branch of the thin-film membrane, causing a temperature rise in sensors. The measurement is realized in vacuum to eliminate convection losses, assuming that heat flow equals the electric power applied to heater (**P$_H$**). Thermal conductivity is calculated using Fourier's law assuming a constant temperature gradient along the membrane:

$$\kappa = \frac{1}{2} \frac{P_H}{T_H - T_S} \cdot \frac{L}{W \cdot t} \qquad (3)$$

where **T$_H$** and **T$_S$** are the heater and sensor temperatures. **L**, **W** and **t** denotes the membrane length, width and thickness respectively.

To determine the sensor and heater temperatures (**T$_S$** and **T$_H$**),

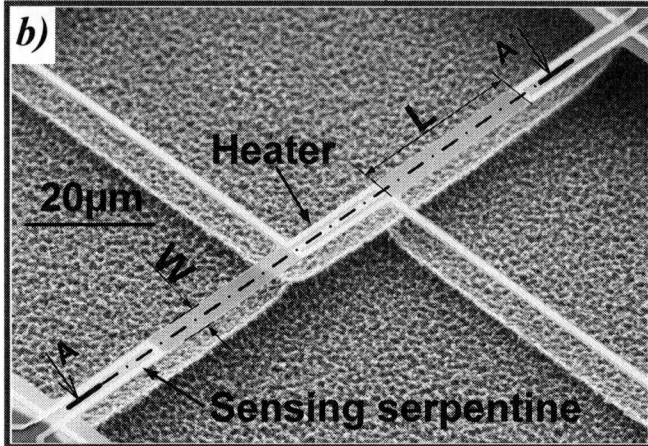

Fig.7 a) Complete view of realized device b) fully suspended membrane of 70nm thick crystalline Silicon equipped with heater and sensing serpentines

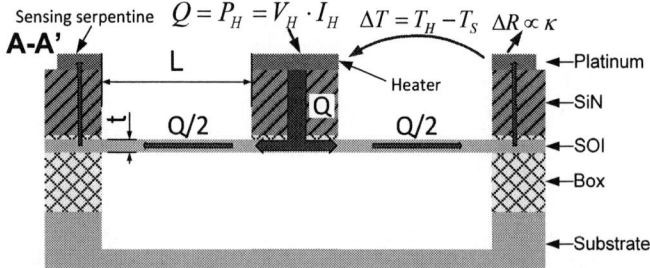

Fig.8 Thermal conductivity measurement methodology. Device's cross-sectional view following A-A' cutline Fig.7-b

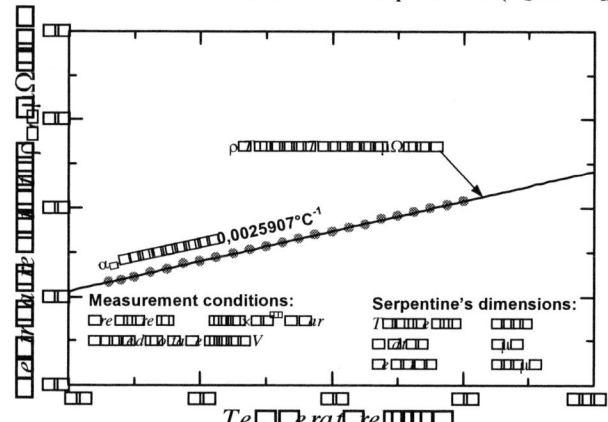

Fig.9 Measurement of heater's electrical resistivity ρ$_{Pt}$ versus temperature for constant applied voltage

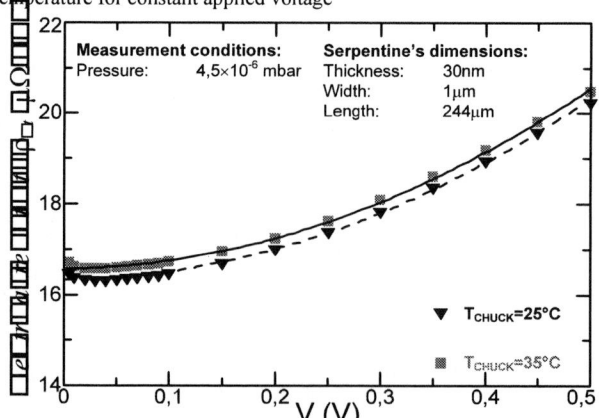

Fig.10 Measurement of heater's electrical resistivity ρ$_{Pt}$ versus voltage at constant chuck temperature

978-1-4799-8002-4/14 $31.00 © 2014 IEEE

the temperature coefficient of electrical resistivity (α_{Pt}) is first determined by measuring the electrical resistivity (ρ_{Pt}) at different chuck temperatures Fig.9.

According to Fig.11, the heater temperature T_H is subsequently derived from the measured electrical resistivity ρ_{Pt} under Joule heating (Fig.10) with the help of the previously calibrated α_{Pt} (Fig.9). Putting ρ_{Pt}, α_{Pt} and chuck temperature (T_{CHUCK}) into Eq.4 allows to calculate the heater temperature:

$$T_H(V) = T_{CHUCK} + \frac{\left(\dfrac{\rho_{Pt}(V)}{\rho_{Pt}(V=0)} - 1\right)}{\alpha_{Pt}} \quad (4)$$

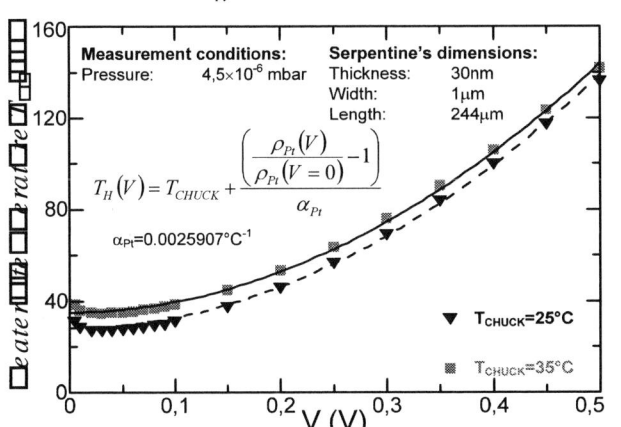

Fig.11 Heater temperature versus applied voltage at constant chuck temperature

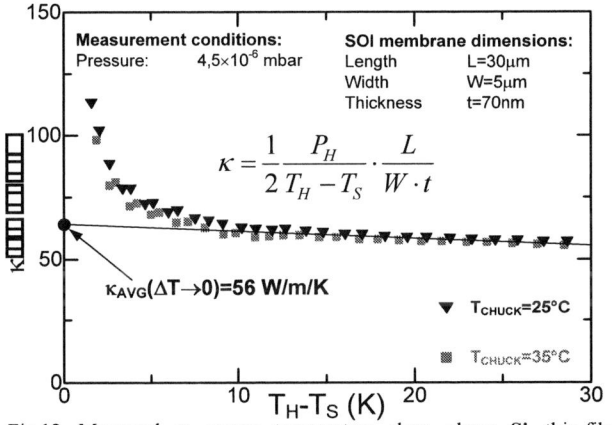

Fig.12 Measured κ versus temperature drop along Si thin-film membrane

V. RESULTS

Fig.12 presents the measured thermal conductivity κ against temperature drop along the Silicon membrane for two chuck temperatures. By its definition the thermal conductivity is a value for zero heat flux limit, thus the value of thermal conductivity is found using tangent line to measurement points. The measured mean value of the thermal conductivity for 70nm thick crystalline Silicon is 56±1.5W/m/K. The accuracy of this result is consistent with the theoretical prediction [10] reported in Fig.2.

Thermal conductivity is dependent on the material temperature, as it is visible on Fig.12. For T_{CHUCK}=25°C, measured values of thermal conductivity are slightly higher than at T_{CHUCK}=35°C. This reduction due to rise material's temperature is expected and linked to the increasing phonon-phonon interactions at higher temperature.

VI. CONCLUSIONS

Good electrical performance of Silicon enables high harvested power density allowing Silicon to compete with conventional thermoelectric materials such as Bi_2Te_3 or Sb_2Te_3. High value of thermal conductivity eliminating the use of Silicon in thermoelectricity is no longer an unbeatable drawback since the possibility of reduction has been established and confirmed. Thermal conductivity reduction by a factor 3 over bulk value is reported in 70nm thick Silicon. The Silicon thermal conductivity can be reduced even 100× over bulk when using dedicated patterning which can block the propagation of phonons responsible for heat transport [3]. This reduction will have significant influence in launching industrialization of CMOS compatible thermo-generators.

ACKNOWLEGMENT

The research leading to these results has received funding from the STMicroelectronics-IEMN common laboratory and the European Research Council (Grant Agreement no. 338179). This work was partly supported by the french RENATECH network.

REFERENCES:

[1] D. M. Rowe, *Thermoelectrics and its energy harvesting. Modules, sytems and applications in thermoelectrics.* CRC Press, 2012.

[2] D. M. Rowe, "Thermoelectric power generation," *Proc. IEE*, vol. 125, pp. 1113–1136, 1978.

[3] J.-K. Yu, S. Mitrovic, D. Tham, J. Varghese, and J. R. Heath, "Reduction of thermal conductivity in phononic nanomesh structures," *Nat Nano*, vol. 5, no. 10, pp. 718–721, Oct. 2010.

[4] P. E. Hopkins, C. M. Reinke, M. F. Su, R. H. Olsson, E. A. Shaner, Z. C. Leseman, J. R. Serrano, L. M. Phinney, and I. El-Kady, "Reduction in the Thermal Conductivity of Single Crystalline Silicon by Phononic Crystal Patterning," *Nano Letters*, vol. 11, no. 1, pp. 107–112, 2011.

[5] E. Chávez-Ángel, J. S. Reparaz, J. Gomis-Bresco, M. R. Wagner, J. Cuffe, B. Graczykowski, A. Shchepetov, H. Jiang, M. Prunnila, J. Ahopelto, F. Alzina, and C. M. Sotomayor Torres, "Reduction of the thermal conductivity in free-standing silicon nano-membranes investigated by non-invasive Raman thermometry," *APL Materials*, vol. 2, no. 1, p. 012113, Jan. 2014.

[6] D. Rowe, Ed., *CRC Handbook of Thermoelectrics.* CRC Press, 1995.

[7] T. M.Tritt, H. Böttner, and L. Chen, "Thermoelectrics: Direct solar thermal energy conversion," *MRS Bulletin*, vol. 33, no. 4, p. 366, 2008.

[8] M. Haras, V. Lacatena, S. Monfray, J.-F. Robillard, T. Skotnicki, and E. Dubois, "Unconventional Thin-Film Thermoelectric Converters: Structure, Simulation, and Comparative Study," *Journal of Elec Materi*, vol. 43, no. 6, pp. 2109–2114, Jun. 2014.

[9] G. K. Wachutka, "Rigorous thermodynamic treatment of heat generation and conduction in semiconductor device modeling," *Computer-Aided Design of Integrated Circuits and Systems, IEEE Transactions on*, vol. 9, no. 11, pp. 1141–1149, 1990.

[10] E. H. Sondheimer, "The mean free path of electrons in metals," *Advances in Physics*, vol. 1, no. 1, pp. 1–42, 1952.

Atomic Scale Engineering of Metal-Oxide-Semiconductor Photoelectrodes for Energy Harvesting Application Integrated with Graphene and Epitaxy SrTiO₃

Li Ji[1], Martin D. McDaniel[2], Li Tao[1], Xiaohan Li[1], Agham B. Posadas[3], Yao-Feng Chang[1], Alexander A. Demkov[3], John G. Ekerdt[2], Deji Akinwande[1], Rodney S. Ruoff[4], Jack C. Lee[1] and Edward T. Yu[1]

[1]Microelectronics Research Center, Department of Electrical and Computer Engineering, [2]Department of Chemical Engineering, [3]Department of Physics, [4]Department of Mechanical Engineering, The University of Texas at Austin, Austin, TX 78758
Tel: (512) 660-1326, Email: nmgjili@utexas.edu

Abstract

In this work, hydrogen production from water is demonstrated via a p-type silicon photocathode with a thin epitaxial strontium titanate, SrTiO3 (STO), as capping layer by molecular beam epitaxy. The advantages of using STO are the ideal conduction band alignment and perfect lattice match between single crystalline SrTiO3 and Si, so the photogenerated electrons can transport through the capping layer with a reduced recombination rate. The STO/p-Si photocathode exhibited a maximum photocurrent density and open circuit potential of 35 mA/cm2 and 450 mV, respectively. There was no observable decrease in performance after 10 hr operation in 0.5M H_2SO_4. We found the efficiency and performance were highly dependent on the size and spacing of the structured metal catalyst. Scaled down the metal catalysts feature size into nanometer region can greatly improve the efficiency. In addition, samples with graphene (Grahene/p-Si) as the lateral transport channel and capping layer shown an enhanced fill factor compared with that of STO/p-Si.

Introduction

A clean and efficient way to overcome the limited supply of fossil fuels and greenhouse effect is highly desired. Hydrogen production via solar water splitting through metal-oxide-semiconductor (MIS) photoelectrodes, where energy harvest and water electrolysis are combined, is a promising method for solving the energy issue. The metal serves as a cocatalyst and the oxide is used to protect the semiconductor from corrosion but allow photogenerated minority carriers to tunnel to the electrolyte and metal [1]. However, the conversion efficiencies of Si-based MIS photoelectrodes is still low, due to the unfavorable energy band alignment in MIS junctions and poor interface quality between silicon and oxide layers [2]. In this work, as shown in Fig. 1, single crystalline SrTiO₃ (STO) ultrathin film were grown on p-type Si (100) by Molecular Beam Epitaxy (MBE). STO provides a small conduction band offset with Si, facilitating tunneling, and the lattice mismatch between (111) STO and (100) Si is very small. The photo-generated electrons in the Si substrate will face a minimal barrier when tunneling through the STO layer hence a high current density could be obtained. In addition, we found that using structured metal cocatalyst, instead of

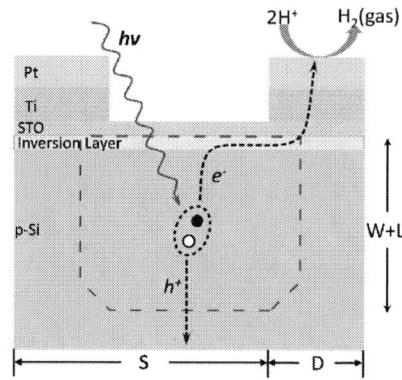

Fig. 1 Schematic of Metal/STO/p-Si photocathodes. The light is absorbed by silicon substrate. The photogenerated electrons will travel along the inversion layer formed at silicon surface and then transport through STO layer to Pt surface, where protons is reduced to hydrogen gas.

thin film coating, and using graphene will greatly increase the efficiency and fill factors.

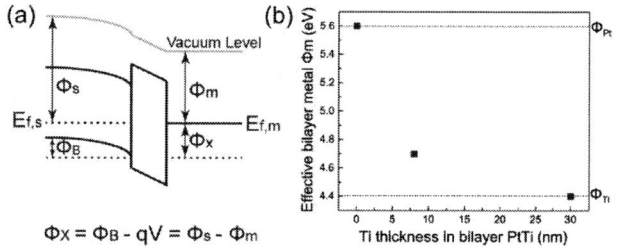

Fig. 2 (a) Band diagram of a MIS photocathode. The maximum photovoltage obtained depends on the difference of work function between semiconductor and metal. (b) The relationship between effective work function and Ti thickness in PtTi bilayer structure. The thickness of Pt is fixed at 20nm.

Engineering of Metal

Work Function Engineering As illustrated in Fig. 2 (a), the open-circuit-voltage (V_{oc}) depends on the difference between work function of metal and semiconductor ($_s$-$_m$) or on the height of Schottky barrier ($_B$). Early work pointed out that ultrathin oxide film will de-pin the Fermi level between metal and semiconductor, which is applicable to our device structures. Thought Pt exhibits excellent catalyst

978-1-4799-8002-4/14 $31.00 © 2014 IEEE

performance, its large work function (~5.5eV) reduced the maximum V_{oc} we can get. In addition, experimental results shown that the adhesion between Pt and STO were poor. After several minutes' testing in solution, Pt went off the STO surface. These problems could be solved by inserting a Ti layer. Fig. 2 (b) shows that the work function of PtTi bilayer metal will get close to that of Ti when the thickness of Ti layer is 30nm. And Ti is a well-known adhesion layer used for making metallic ohmic contact with silicon. So all devices discussed in this work has 30nmTi/20nmPt as metal cocatalyst.

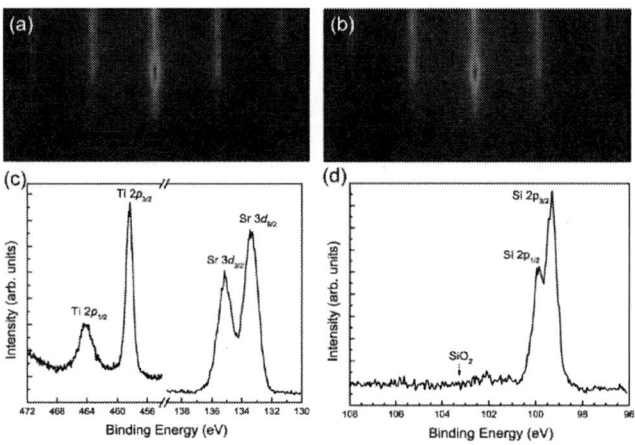

Fig. 3   RHEED images of STO/p-Si along (a) [110] and (b) [210]. (c) and (d) are *in-situ* XPS results of STO/p-Si, showing a SiO$_2$-free STO/Si interface.

Engineering of Oxide

Materials Analysis   In Fig. 3, the RHEED images are taken along the [110] and [210] azimuth of the cubic perovskite. The elongated streak patterns indicate a well-crystallized and atomically smooth film. Film composition and quality of the STO-Si interface were analyzed using *in situ* XPS. The absence of a peak at ~103.3 eV suggests that the interface is free of amorphous SiO$_2$.

Table 1. Lists of possible transport mechanism, expressions of their current density and normalized conductance in MOS structure.

Transport Mechanism	Current Density (J)	$G_n=$ (dI/dV)/(I/V)
Poole-Frenkel	$AV_{ox}exp(B\sqrt{V_{ox}}/kT)$	$1+B\sqrt{V_{ox}}/2$
Fowler-Nordheim	$AV_{ox}^2\exp(-B/Vox)$	$2+B/Vox$
Schottky	$AT^2exp(B\sqrt{V_{ox}}/kT)$	$B\sqrt{V_{ox}}/2$
Hopping	$AV_{ox}exp(BVox)$	$1+BVox$
Ohmic	$AV_{ox}exp(-B/kT)$	1

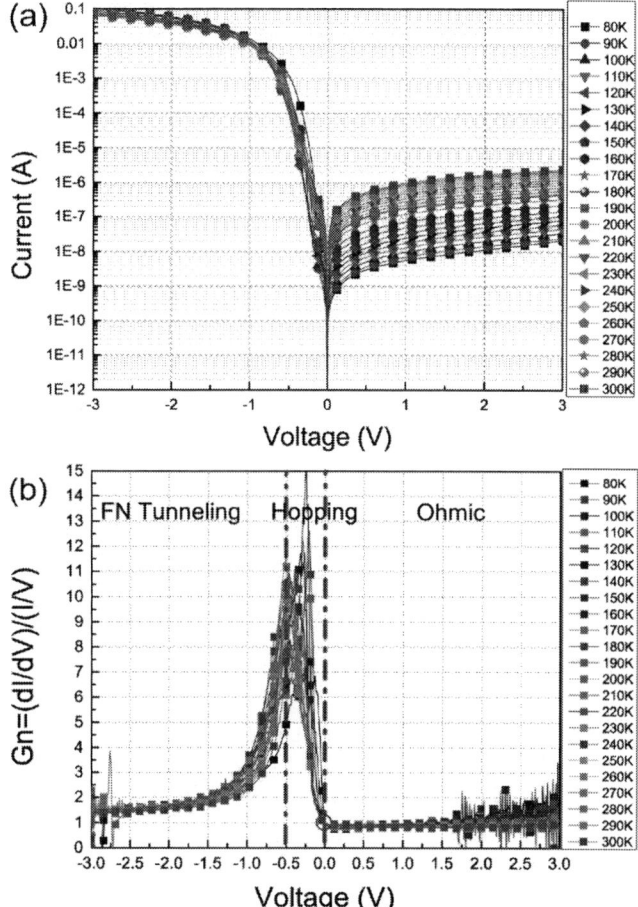

Fig. 4   I-V characteristic in rough vacuum (< 0.1 mTorr) and normalized conductance analysis.

Current Transport Mechanism   I-V characteristic under various temperatures in rough vacuum (< 0.1 mTorr) was performed to investigate the current transport mechanism. Normalized conductance G_n method is used. The possible transport mechanisms are listed in Table 1. As shown in Fig. 4, for V>0 region, it is ohmic conduction, confirming the zero CB band offset between STO and Si.

Engineering of Semiconductor

Metal Catalyst Size Effect   As shown in in Fig. 1, photogenerated electrons will travel in inversion layer to adjacent metal. By decreasing the metal feature size—if smaller than diffusion length or even further smaller than the depletion region width—the fill factor will be increased. Nanosphere lithography (NSL) is used to fabrication sub-100nm structures. The schematic and SEM results of NSL are shown in Fig. 5. Linear sweep voltammetry results of STO/p-Si samples are shown in Fig. 6 (a). As decreasing the feature size of metal catalysts, the fill factor is greatly increased.

978-1-4799-8002-4/14 $31.00 © 2014 IEEE   217

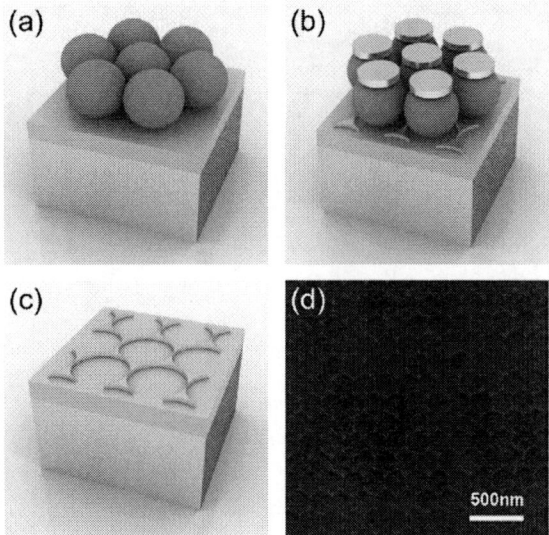

Fig. 5   Schematic of Nanosphere lithography (NSL): (a) nanosphere deposition, (b) bilayer metal deposition by e-beam evaporation and (c) nanosphere lift-off. The final devices is characterized by SEM as in (d).

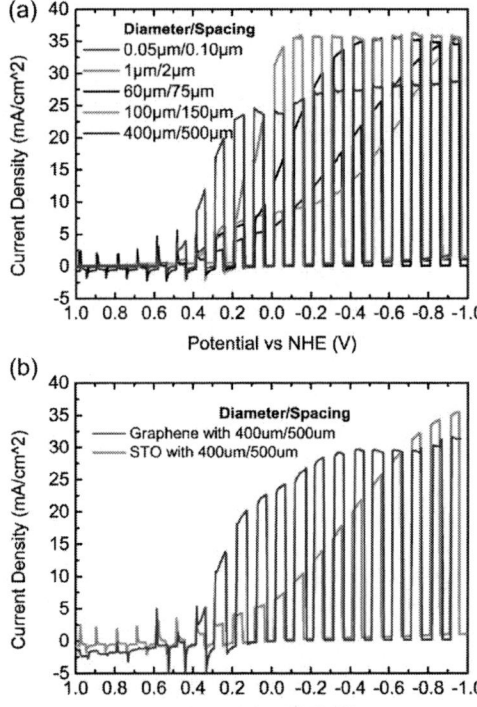

Fig. 6   Linear sweep voltammetry (LSV) in 0.5 M H_2SO_4 under 100 mW/cm² illumination of (a) STO samples with various diameters and spacings of metal catalysts and (b) comparison of STO/p-Si and Graphene/p-Si samples with same metal catalyst structures.

Graphene   Results of STO/p-Si confirmed the lateral transport of photogenerated electrons in inversion layer.

This can be further optimized by using graphene as the conducting channel due to its high mobility. With help of PMMA-assisted transfer process, large area (2 inch²) CVD grown graphene were transferred to p-type (100) Si wafer. As shown in Fig. 6 (b), compared with STO/p-Si sample with same metal catalysts structures, the fill factor of Graphene/p-Si is greatly improved. It is worth noting that Graphene/p-Si exhibits good stability and shows no dramatically degradation after several hours testing.

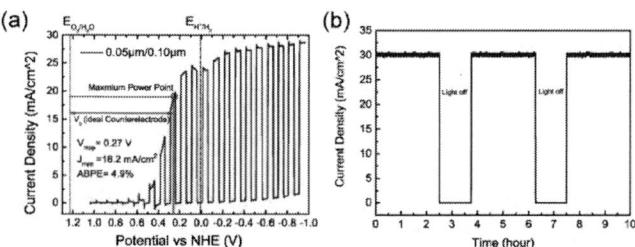

Fig. 7   (a) ABPE results of STO/p-Si sample with 50nm/100nm metals catalyst structures. (b) Stability testing in 0.5 M H_2SO_4 under 100 mW/cm² illumination. The light was turned off every 2.5 h.

Performance

Fig. 7 (a) shows the applied bias photo-to-current-conversion-efficiency (ABPE) for 50nm metal dots. The ABPE reaches 4.9%, highest value up to date for single junction silicon based photocathode. As shown in Fig. 7 (b), after 10 hours test, no degradation of current is observed, confirming an excellent stability.

Conclusions

We demonstrated that the epitaxial STO/Si heterojunction is an efficient and stable photocathode for water splitting. High photocurrent density (35 mA/cm²), onset potential shift (450mV), and long-time stability were achieved. In addition, we extended work on the relation between size of surface metal catalyst and efficiency. Results indicate that utilizing characteristic sizes smaller than the limiting factors -- diffusion length and depletion width -- would greatly increase the efficiency. Sub-100 nm nanostructures made by nanosphere lithography yielded the highest reported ABPE efficiency of 4.9%. Graphene/p-Si yield higher fill factor than STO/p-Si due to the enhanced current transport in lateral direction.

References

[1] Y. W. Chen *et al*, *Nature Materials*, **10** (2011) p539
[2] D. Esposito *et al*, *Nature Materials*, **12** (2013) p562

FDSOI CMOS Devices Featuring Dual Strained Channel and Thin BOX Extendable to the 10nm Node

Q. Liu[1], B. DeSalvo[2], P. Morin[1], N. Loubet[1], S. Pilorget[1], F. Chafik[1], S. Maitrejean[2], E. Augendre[2], D. Chanemougame[1], S. Guillaumet[1], H. Kothari[1], F. Allibert[4], B. Lherron[1], B. Liu[1], Y. Escarabajal[1], K. Cheng[3], J. Kuss[3], M. Wang[3], R. Jung[3], S. Teehan[3], T. Levin[3], M. Sankarapandian[3], R. Johnson[3], J. Kanyandekwe[1], H. He[3], R. Venigalla[3], T. Yamashita[3], B. Haran[3], L. Grenouillet[2], M. Vinet[2], O. Weber[5], E. Josse[6], F. Boeuf[6], M. Haond[6], J.-L. Bataillon[1], W. Kleemeier[1], T. Skotnicki[6], M. Khare[3], O. Faynot[2], B. Doris[3], M. Celik[1], R. Sampson[1]

[1]STMicroelectronics, [2]CEA-LETI, [3]IBM, [4]SOITEC, Albany NanoTech, NY 12203, U.S.A.; [5]CEA-LETI,
[6]STMicroelectronics, 850, rue Jean Monnet, 38920 Crolles, France.
Phone: 1-518-292-7218. Email: qliu@us.ibm.com (qing.liu@st.com)

Abstract

We report FDSOI devices with a 20nm gate length (L_G) and 5nm spacer, featuring a 20% tensile strained Silicon-on-Insulator (sSOI) channel NFET and 35% [Ge] partially compressive strained SiGe-on-Insulator (SGOI) channel PFET. This work represents the first demonstration of strain reversal of sSOI by SiGe in short channel devices. At V_{dd} of 0.75V, competitive effective current (I_{eff}) reaches 550/340 µA/µm for NFET, at I_{off} of 100/1 nA/µm, respectively. With a fully strained 30% SGOI channel on thin BOX (20nm) substrate and V_{dd} of 0.75V, PFET I_{eff} reaches 495/260 µA/µm, at I_{off} of 100/1 nA/µm, respectively. Competitive sub-threshold slope and DIBL are reported. With the demonstrated advanced strain techniques and short channel performance, FDSOI devices can be extended for both high performance and low power applications to the 10nm node.

Introduction

Planar FDSOI represents an important device architecture for continued CMOS scaling. [1-9] Its advantages include excellent short channel electrostatics, un-doped channels and effective back bias for performance boost and leakage lowering. Moreover, FDSOI is fabricated using a more conventional, lower cost process, enabling it as an attractive alternative to more complex FinFET architectures. In this work, we demonstrate the successful implementation of strained FDSOI devices with $L_{G,}$ spacer & BOX dimensions scaled to 10nm feature sizes.

Device Integration

A simplified FDSOI integration flow, featuring cSiGe PFET, gate first HK/MG, and a novel equal spacer scheme, is shown in Fig. 1. In previous work [6-8], the integrated process resulted in 2-3nm of spacer thickness difference between the NFET & PFET. The thicker spacer exaggerates the trade-off between external resistance and junction abruptness, which limits electrostatic performance at scaled L_G. The equal spacer integration is shown schematically in Fig. 2. After gate patterning, a thin layer of dielectric material is deposited and etched to form the first spacer, where the thickness determines the final junction design. A dual layer of sacrificial liners are then deposited. After opening NFET region, an isotropic and selective etch process is used to remove the top sacrificial layer, while the bottom layer acts as an etch stop. The bottom layer is subsequently removed prior to the in-situ doped SiCP RSD epitaxy. After another liner deposition to protect NFET region, a similar process is performed on the PFET side and in-situ doped SiGeB epitaxy grown, such that now both the NFET and PFET have exactly the same spacer thickness. The final device structure is shown in Fig. 3. The device channels were formed beginning with a 20% tensile strained sSOI wafer (NFET), followed by a patterned epitaxy and condensation to obtain a 35% [Ge] SGOI channel (PFET) with compressive strain. Condensation temperature was determined to strongly influence channel quality, evidenced in Fig. 4. A low temperature process was developed to form the defect free SiGe channel from the sSOI starting substrate.

Device Characteristics

- Devices on sSOI

Tensile strain is known to benefit NFET carrier mobility. As device dimensions scale, the strain induced from CESL or RSD diminishes, driving the need to identify new alternatives. Fig. 5 shows the NFET drive current (I_{on}) and effective current (I_{eff}) as a function of off current (I_{off}) at V_{dd}=0.9V. Here, the NFET is built using 20% tensile strained sSOI, where 20% refers to a Si layer with a lattice matching 20% [Ge] SiGe. At I_{off} of 100/1 nA/µm, the I_{on} reaches 1440/1120 µA/µm, while the I_{eff} reaches 850/590 µA/µm, respectively. Both values represent the highest reported drive currents for an FDSOI device. At V_{dd}=0.75V, The I_{on} is 1120/760 µA/µm, and I_{eff} is 610/360 µA/µm, at I_{off} of 100/1 nA/µm, respectively, as shown in Fig. 6. The tensile strain in the channel reduced the V_t by ~140mV vs. a non-strained Si channel, as shown in Fig. 7. The electron peak mobility improves by 60% with sSOI, as shown in Fig. 8. It was reported previously that compressively strained SGOI has significant mobility/performance improvement when the bi-axial strain is transformed into uni-axial strain. [6, 7] Narrower width devices leverage this behavior, as shown in Fig. 9. Additional improvement can be expected with further width scaling to 10nm dimensions. A comparison of reliability parameters, such as breakdown voltage (VBD) and positive bias temperature instability (PBTI), between sSOI and SOI channels, is shown in Fig. 10. The reliability of sSOI NFET is similar to that of SOI NFET, with both better than that of a comparable 20nm Bulk NFET.

For PFET, tensile strain is detrimental to hole mobility. Therefore, compensation and strain reversal is needed. The relaxation of sSOI by incorporating 20% [Ge] SiGe into the PFET channel was reported previously [11]. Here, the reversal to compressive channel strain through higher [Ge] SiGe, and the associated performance improvement are reported. Fig. 11 shows the comparison of PFET I_{on} and I_{eff} as a function of I_{off}. Four channel materials are compared: 35% [Ge] SiGe on sSOI, 20% [Ge] SiGe on sSOI, 18% [Ge] SiGe on SOI, and pure Si. When integrating 20% [Ge] SGOI on a 20% sSOI substrate, the channel becomes fully relaxed. The performance is similar to that of a pure Si channel, where the main difference is the V_t. By increasing [Ge] to 35%, the strain is reversed and becomes compressive. The performance is similar to 15% [Ge] SiGe on a non-strained starting SOI substrate and represents the first demonstration of strain reversal on short channel devices. The PFET V_t is largely determined by the Ge content and strain in the channel, while the performance is dominated more by the latter. A mobility difference is observed in the slope of the curves, with higher mobility having a flatter slope. Fig. 12 shows the sub-threshold performance of the CMOS devices, indicating a good short channel control.

- *Devices on thin BOX (20nm and 15nm)SOI*

When scaling to 10nm dimensions, BOX thickness (T_{BOX}) reduction improves short channel control and provides an extra pathway for the scaling of FDSOI devices. [3,8] At the same L_G and T_{Si}, a thinner T_{BOX} lowers DIBL, which enables a shorter L_G design, a corresponding larger contact area and a lower parasitic capacitance. Alternatively, it can be used to relax T_{Si} at a given L_G, which improves manufacturability. A thinner T_{BOX} also increases the body factor and therefore the ability to modulate V_t, which enables low V_{dd} applications. With T_{BOX} and L_G at 20nm, excellent electrostatics are obtained, shown in Fig. 13. The NFET I_{on} and I_{eff} as a function of I_{off} are shown in Fig. 14. At V_{dd} of 0.75V and I_{off} of 100/1 nA/µm, the NFET I_{on} and I_{eff} are 830/635 µA/µm and 495/300

µA/µm, respectively. Fig. 15 shows the PFET I_{on} and I_{eff} as a function of I_{off}. At V_{dd} of 0.75V and I_{off} of 100/1 nA/µm, the PFET I_{on} and I_{eff} reach 875/575 µA/µm and 495/260 µA/µm, respectively. Fig. 16 compares short channel Ron/DIBL with previously reported work [8], which featured a 25nm T_{BOX}. Both NFET and PFET DIBL are improved due to the combination of thinner BOX and optimized junctions. The PFET Ron improvement is derived from the higher mobility of the higher [Ge] channel. The I_d/V_G curves with various back bias (V_{bb}) at V_{dd} of 0.75V are shown in Figs. 17 and 18. For the NFET, the body factor increases from 70mV/V to 100mV/V, with T_{BOX} thinning from 20nm to 15nm. For the PFET, the body factor increases from 75mV/V to 95mV/V. It is also worth noting that for the NFET, the GIDL floor (lowest leakage point) becomes lower when applying a negative bias, due to a lower electric field at drain side [4]. However, the GIDL floor of the SiGe channel PFET increases when V_{bb} is >2V. It is more pronounced with a 15nm T_{BOX}, due to the higher electric field.

Table 1 shows a benchmark of device characteristics of this work and other previously reported state-of-the-art Bulk FinFET and FDSOI devices. Competitive electrostatic performance and drive current are achieved at a much smaller L_G, illustrating the capability to extend FDSOI devices to 10nm for both high performance and low power applications.

Further performance improvement

Further T_{inv} thinning on sSOI NFET and higher Ge SiGe PFET can lead to higher performance. In general, 1A T_{inv} scaling results in 5~7% I_{eff} increase. Higher mobility, from either higher strained channel materials, or layout optimization to transform bi-axial to uni-axial strain, is also critical for performance improvement. [7, 8] Relaxation of the sSOI PFET region before condensation, to more efficiently leverage the compressive SiGe channel strain, is being actively investigated. Furthermore, the introduction of a replacement metal gate (RMG) is expected to induce even higher mobility and performance, due to the channel stress enhancement. [12]

Conclusions

In this paper, we reported high performance & low power FDSOI devices with competitive drive current and electrostatic behavior. The 20nm L_G and 5nm spacers used are compatible with 10nm design rules. Two additional enabling elements for scaling FDSOI devices to 10nm were reported and discussed: advanced strain techniques for performance improvement, and reduced BOX thickness for better SCE & higher body factor.

978-1-4799-8002-4/14 $31.00 © 2014 IEEE

- cSiGe formation at PFET area
- STI RIE and liner deposition
- STI fill and CMP
- Ground plane (GP) implantation and annealing
- High-k / metal gate patterning
- **1st spacer deposition and etch**
- **Dual sacrificial liner deposition**
- **NFET top sacrificial liner selective removal**
- NFET ISPD SiC RSD EPI pre-clean and growth
- Hard mask deposition
- **PFET top sacrificial liner selective removal**
- PFET ISBD SiGe RSD EPI pre-clean and growth
- 2nd spacer formation
- Rapid thermal annealing (RTA) + laser annealing
- Salicide
- MOL and BEOL

Fig. 1 A simplified FDSOI integration flow, featuring cSiGe PFET, gate first high-k / metal gate, equal 1st spacer formation and dual in-situ doped raised source/drain epitaxy process. The starting material can be either strained SOI (sSOI) wafer or non-strained SOI wafer.

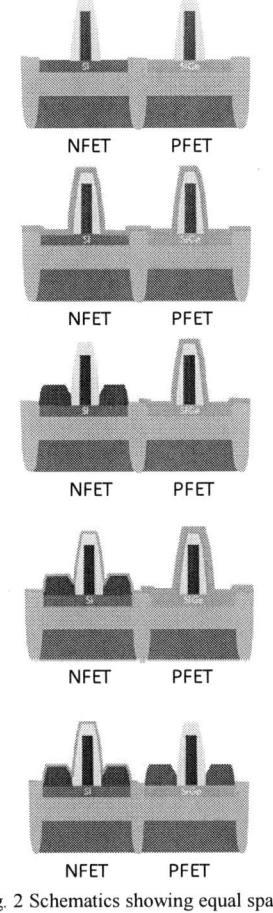

Fig. 2 Schematics showing equal spacer integration. Both NFET and PFET have the exact same spacer thickness after the RSD EPI.

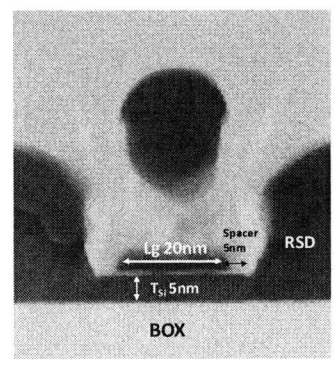

Fig. 3 TEM cross-section of final device structure.

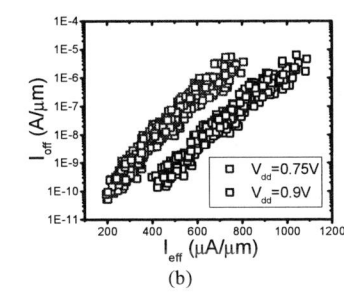

Fig. 4 Topdown SEMs showing post 35% SiGe condensation on sSOI wafer with (a) a high temperature process and (b) a low temperature process. Clearly, with optimized low temperature condensation, defect free SiGe channel is formed on sSOI wafer.

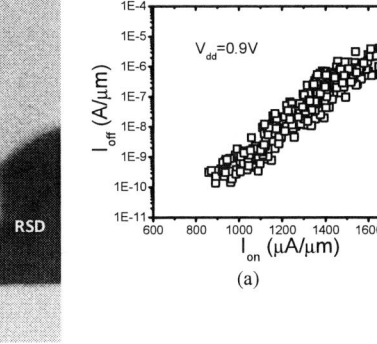

Fig. 5 At V_{dd}=0.9V, and an off current of 100nA/μm and 1nA/μm, the sSOI NFET (a) drive current reaches 1440μA/μm and 1120μA/μm, respectively; and (b) effective current reaches 850μA/μm and 590μA/μm, respectively. It is the best ever reported FDSOI NFET.

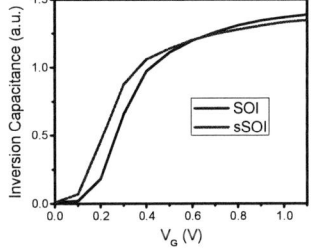

Fig. 6 sSOI NFET (a)I_{on} and (b) I_{eff} at V_{dd} of 0.9V and 0.75V. At V_{dd}=0.75V, and I_{off} of 100nA/μm and 1nA/μm, the I_{on} reaches 1120μA/μm and 760μA/μm, while I_{eff} reaches 610μA/μm and 360μA/μm, respectively.

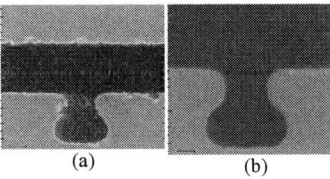

Fig. 7 NFET long channel C/V measurement showing a V_t decrease of 140mV from regular SOI to tensile strained SOI (sSOI).

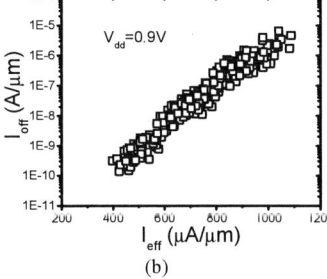

Fig. 8 NFET mobility plot showing that a 60% improvement of electron peak mobility is obtained at the same SOI thickness.

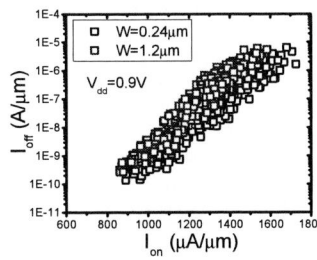

Fig. 9 I_{on}/I_{off} plot of sSOI NFET with active width at 0.24μm and 1.2μm, showing that with smaller width, the bi-axile strain transformed toward uni-axile strain, which benefits the mobility and performance

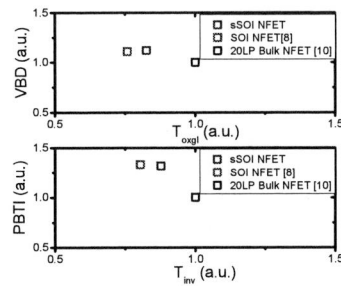

Fig. 10 VBD/PBTI comparison, showing sSOI NFET has similar reliability as SOI NFET, while both are superior to 20nm Bulk NFET.

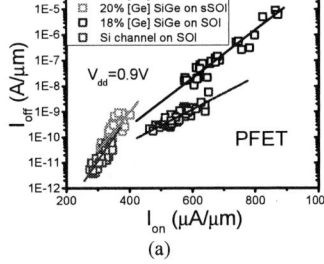

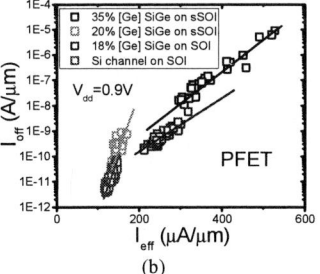

Fig. 11 PFET comparison between 35% [Ge] SiGe on sSOI, 20% [Ge] SiGe on sSOI, 18% [Ge] SiGe on SOI and Si channel. Both (a) I_{on}/I_{off} and (b) I_{eff}/I_{off} plots show the effect of strain compensation and reversal by the SiGe on sSOI. 20% [Ge] SiGe is able to compensate the tensile strain, while 35% [Ge] SiGe enables strain reversal to compressive and shows the performance resembling ~15% [Ge] SiGe on SOI substrate.

978-1-4799-8002-4/14 $31.00 © 2014 IEEE

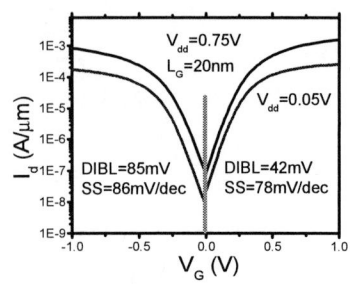

Fig. 12 I_d/V_G of N/PFET fabricated on sSOI substrate with a gate length at 20nm, showing good electrostatics.

Fig. 13 I_d/V_G of N/PFET with L_G at 20nm and T_{BOX} of 20nm. Excellent SCE is achieved. The large I_{off} on PFET is mainly from the incorporation of 30% [Ge] SiGe in channel, resulting in much lower V_t.

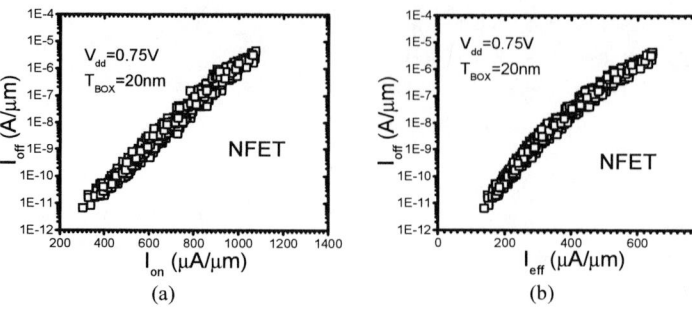

(a)   (b)

Fig. 14 At V_{dd} of 0.75V, NFET (a) I_{on} is 880μA/μm and 635μA/μm; and (b) I_{eff} is 495μA/μm and 300μA/μm, at I_{off} of 100nA/μm and 1nA/μm, respectively.

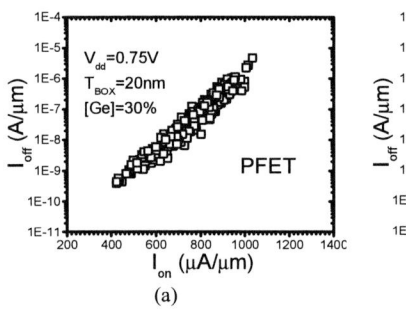
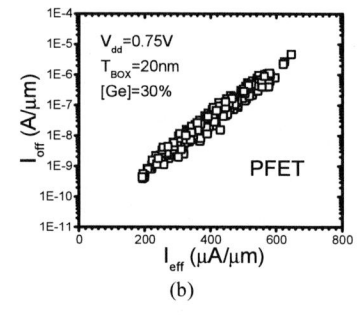

(a)   (b)

Fig. 15 At V_{dd} of 0.75V, PFET (a) I_{on} is 875μA/μm and 575μA/μm; and (b) I_{eff} is 495μA/μm and 260μA/μm, at I_{off} of 100nA/μm and 1nA/μm, respectively.

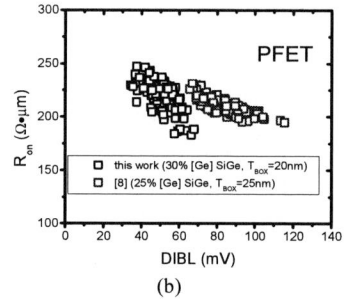

(a)   (b)

Fig. 16 The R_{on}/DIBL plots of (a)NFET and (b)PFET showing better DIBL with thinner T_{BOX} and optimized junction design, comparing with [8].

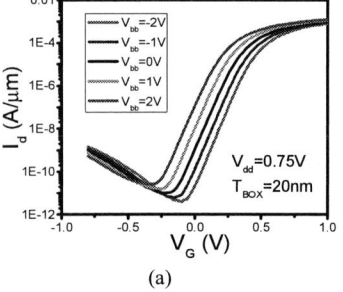
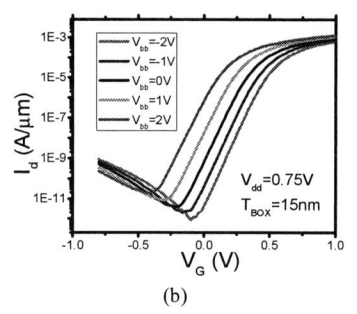

(a)   (b)

Fig. 17 I_d/V_G curves of NFET on (a)20nm BOX and (b)15nm BOX substrate with back bias from -2V to 2V, showing the larger body factor with thinner BOX.

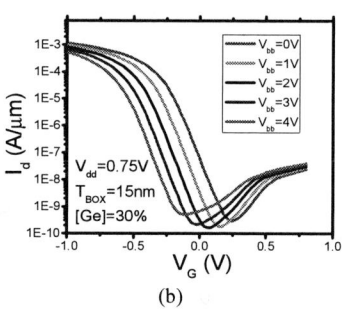

(a)   (b)

Fig. 18 I_d/V_G curves of PFET on (a) 20nm BOX and (b) 15nm BOX substrate with back bias from 0V to 4V. The GIDL floor starts to increase when large bias (V_{bb}>2V) is applied, which results in higher electric field and tunneling current.

Table 1 A comparison of devices in this work with State-of-Art Bulk FinFET and 14FDSOI devices.

	*Auth et al [13]	*Jan et al [14]	Wu et al [15]	Weber et al [9]	This work
V_{dd} (V)	0.8	0.75	0.75	0.8	0.75
N/P DIBL (mV/V)	46/50	30/35	52/42		56/51
N/P SS (mV/dec)	69/72	71/72	73/71		78/78
I_{off} (nA/μm)	100/1	100/1	100/1	100/1	100/1
L_G (nm)	30/34	30/34	30/34	20/34	20/20
NFET I_{on} (mA/μm)	1.26/0.88	1.08/0.71	0.95/0.76		1.12/0.76
NFET I_{eff} (mA/μm)	0.65/0.42			0.46/0.31	0.61/0.36
PFET I_{on} (mA/μm)	1.1/0.78	0.91/0.59	1.01/0.75		0.88/0.58
PFET I_{eff} (mA/μm)	0.56/0.38			0.39/0.25	0.5/0.26

*Drive/effective current is normalized to footprint, which is 20~30% higher than normalized to effective channel width.

Acknowledgement

We would like to thank J. Hartmann (STMicroelectronics) for managerial support. This work is performed by the research alliance teams at various IBM facilities.

Reference

[1] Q. Liu, et al, VLSI, p.61, 2010.
[2] O. Weber, et al, IEDM, p.58, 2010.
[3] O. Faynot, et al, IEDM, p.50, 2010.
[4] Q. Liu, et al, VLSI, p.160, 2011.
[5] N. Planes, et al, VLSI, p.133, 2012.
[6] A. Khakifirooz, et al, VLSI, p.117, 2012.
[7] K. Cheng, et al, IEDM, p.419, 2012
[8] Q. Liu, et al, IEDM, p.228, 2013.
[9] O. Weber, et al, VLSI, p.16, 2014
[10] H. Shang, et al, VLSI, p.129, 2012.
[11] A. Khakifirooz, et al, EDL, p.1358, 2013.
[12] S. Morvan, et al, IEDM, p.530, 2013.
[13] C. Auth, et al, VLSI, p.131, 2012
[14] C.-H. Jan, et al, IEDM, p.44, 2012
[15] S.-Y. Wu, et al, IEDM, p.224, 2013.

978-1-4799-8002-4/14 $31.00 © 2014 IEEE

Future Challenges and Opportunities for Heterogeneous Process Technology. Towards the Thin Films, Zero Intrinsic Variability Devices, Zero Power Era.

S. Deleonibus, O. Faynot, T. Ernst, M. Vinet, P. Batude, F. Andrieu, O. Weber, D. Cooper, F. Bertin, H. Moriceau, L. DiCioccio, T. Signamarcheix, M. Sanquer*, X. Jehl*, O. Cueto, H. Fanet, F. Martin, H. Okuno*, F. Nemouchi, G. Poupon, Y. Lamy, D. Gasparutto*, X. Baillin, L. Duraffourg, J. Arcamone, L. Perniola, B.de Salvo, E.Vianello, L. Hutin, C.Poulain, E. Beigne, R. Tiron, L. Pain, S. Tedesco, S. Barnola, N. Posseme, C. Le Royer, A. Villalon, R. Salot

Universités Grenoble Alpes, CEA, LETI MINATEC Campus, 38054 Grenoble, France, *Universités Grenoble Alpes, CEA/INAC
38000 Grenoble, France , Tel: +33 438785973, Fax: +33 438785183 Email: simon.deleonibus@cea.fr

Abstract

Linear scaling CMOS has encountered many hurdles which request new process modules, driven mainly by the maximization of energy efficiency. Fabrication at the sub 10nm node level will request Intrinsic Variability approaching to zero. The rapid growth of mobile, multifunctional and autonomous systems is hardly demanding to reach Zero Power consumption. The solutions to integrate Thin Film based devices, architectures and systems in order to face these challenges are described.

Looking desperately for an Energy Efficient Sustainable World?

By 2025, 25 % of the World Gross Domestic Product will depend on the development of Information and Communication Technologies(ICT) through strongly growing sectors such as health, communication, transport and energy. Today, data centers and the network consume more than 3% of the total electricity worldwide and generate at least 2% of CO_2[1]. In industrialized countries, the breakdown shows that ICT are responsible for more than 10% of electricity energy consumption. Scarcity of "spice metals"[2] and materials, such as rare earths and noble metals, will request sustainable recycling policies or challenge to thermodynamically viable alternative solutions. The amount of Internet protocols will reach the zettabyte(10^{21}) level in 2017 and will increase three folds in 5 years. Less greedy device, interconnect, computing technologies and architectures are essential to aim at x1000 less power consumption. In this context, it makes sense to aiming at global system level Zero Power consumption from a grid, while maximizing the Energy Efficiency for CMOS and Memories, which can be combined to contribute to the energy saving balance. Challenging tomorrow's exponentially growing electronic market, towards Internet of Things, Autonomous and Mobile systems for new societal needs, request a drastic reduction to Zero Power and Zero Intrinsic Variability, Heterogeneous and 3D integration at the device, functional and system levels. It is important to remind that instrisic variability is linked to the variation in objects constituents, while extrinsic variability is due to external interactions and uncertainty linked to the observation of measured physical quantities[3].

Silicon Thin Films towards Zero Intrinsic Variability, Zero Power

On one hand, SOI technology has reached a maturity level that makes it competitive and outreaches the most advanced bulk silicon platforms (Figs 1a-1h): Ultra Thin Body and Buried Oxide (UTBBOx) 300 mm wafers feature less than 10 nm Ultra Thin Silicon On Insulator(UTSOI) thickness which is controlled with a sub-nm dispersion, equivalent to bulk silicon state of the art surface roughness leading to record low variability, tunable biaxial and uniaxial strain [4-8]. Thanks to intentionally un-doped sSOI/SiGeOI channels, single midgap metal gate integration and dual channel, FDSOI has demonstrated low subthreshold slopes for High Energy Efficiency and tunability at reduced cost[4-8] (Fig 1) down to 10 nm gate lengths. For sub-8 nm transistors, nanowires architectures will be necessary to control device electrostatics [8-10](Fig.1g): Low Power to High Performance specifications can be reached, down to w=7nm, via boosters tuning. Improvement of power/speed figures can be obtained by using a dual gate last approach [11]. FDSOI performance has been demonstrated on 28nm high density circuits(Fig 2), thanks to body reduced parasitic capacitances and biasing techniques [12]. Access resistance minimization has been a major issue to integrate these devices by avoiding thin silicon agglomeration during selective dual stressors epi and salicidation[11,13]. Alternative contact and source and drain regions made by non-alloyed solutions [14-16] are still challenging.

Memory hierarchy revision, gives new opportunities to low voltage switching embedded Non Volatile Resistive Memories (RERAMs) [17,18] to be co-integrated with logic, beyond data storage purposes. They will enable new architecture based drastic power consumption reduction, latency and design reconfigurable, programmable or neuromorphic architecture circuits (Fig 3)[17,18] .

Removing the extrinsic doping from the CMOS transistors channel, leads to considering new device architectures for future systems on chip. Scaling at the sub-8 nm gate lengths will face the challenge of Zero Intrinsic Variability for Logic as well as for Memory devices. Scaling VDD and device length less than 5 nm might request new architectures that provide sufficient overdrive and static power minimization. Tunnel Field Effect Transistors (TFET) have shown high Ion/Ioff capabilities on SiGexOI [19] (Fig 4). Single to few electrons phenomena should be taken into account and be exploited in combination with CMOS for multivalued logic architectures[14].

Process technologies and variability for sub-8nm CMOSFETs will suffer from dopant stochastic diffusion or trap assisted tunneling[19](Fig.4) from source and drain. FETs are excellent sensors to detect an ionized Single Dopant atom by means of single electron effects (Fig 5) [20]. The unitary placement of several atoms(Fig. 5c) would enable more complex functions in a single device : coupling 2 atoms in one transistor to make an electron pump has been reported[21]. Deterministic doping techniques, promising to control Zero Variability by using the placement of down to 1 dopant atom et the time [22-24], are being developed (Fig 6 a,b,c). Single ion implantation[22], STM lithography combined with CVD [23]and chemical grafting[24] are among the most popular ones. Resistive memories will suffer from variability of conduction paths at the same scale: deterministic crystalline clusters placement [25] (fig 6d) would pave the way to near zero variability. The integration of 2D materials, such as Transition Metals Disulfides(TMD), could pave the way to new high level applications thanks to their optical and mechanical properties[26](Fig 7). Heterostructured source and drains without alloying(Fig 8a) [27] will reduce transistor access edge roughness-related assisted tunneling. Patterning solutions can bring Zero Variability: direct self-assembly (DSA) with high χ resists (Fig8b)[28] and Self-Limited Light Ion Implantation Etching (S2L2IE), combined to optical or Multibeam lithography are already under development. Post Cu interconnect can be addressed with C based materials: CNTs bundles [29] (Fig 8c) or graphene nanoribbons combined to DSA could generate monodisperse wires. Ultimately, metallized DNA templates could generate single atom wide metallic strings(Fig 8d)[30]. Top-down Nanowires and ribbons co-integrated with CMOS give access to increased sensing capabilities (Fig. 9a & b)[10,31] and drastic power reduction. Mechanical switches in an adiabatic architecture [32] could reduce power consumption by ca. a factor of 1000 as compared to CMOS [33] (Fig 9c&d).

More Moore and More than Moore meeting for 3D

Increasing CMOS density without dimensional scaling is a question, driven by the cost of optical lithography, which has drastically slowed down wavelength scaling. On one hand, Multibeam lithography [34] is a strong alternative to EUV. On the other hand, Monolithic 3D[35] (Fig 10) offers unique opportunities to partition CMOS design, integrate new materials at low temperature or memories [36], sensors, actuators, imagers,… and reduce power[36,37]. Based on a UTBBOx scalable architecture, Low Power and High Performance Nanoelectronics(More Moore) can be co integrated with Diversification devices(More than Moore) to

access new applications opportunities (Fig 11). To speed up development, Parallel 3D[38-41] has been proven to "cram more and more components in a package" (RF, MEMS, Passives,...), benefit to performance and power consumption. Added Power sources[40](Fig 12) could maximize Energy Efficiency to Zero Power consumption from a grid.

Conclusion

We have discussed the pathways for nanoelectronics research to maximize energy efficiency and aim at Zero Power and Zero Intrinsic Variability in the next decades. Thin films materials, integrated in 3D at the device, functional and system levels certainly represent major opportunities.

Acknowledgements

This work is supported by the Nano 2017 CEA,LETI/STMicroelectronics/IBM Alliance Program, CEA/SOITEC Joint Program, multiple Eureka and EU FP7 projects, CEA ZeroPOVA and A3DN Flagship Programs.

References

[1] Fettweis G., Zimmermann E., WPMC 2008, Dresden (2008).
[2] Reller A., Phys. Status Solidi RRL,1,(2011)/ DOI 10.1002/pssr.201105126
[3] Van Belle G.,(2008), Statistical Rules of Thumb, Wiley, p. 99
[4] Weber O. et al., IEDM 2008, 10.1109/IEDM.2008.4796663
[5] Faynot O. et al., IEDM Dig. pp50-53, San Francisco (CA), Dec 2010.
[6] Ventosa C. et al, *Electroch.and Solid-State Lett.,*,**12,**10,H373-H375,(2009)
[7] Weber O. et.al, VLSI Tech. Dig., p.16, Honolulu(HI), June 2014
[8] Barraud S. et al., VLSI Tech. Dig,, p230, Kyoto(Japan), June 2013
[9] Deleonibus S.,*Editor,* "Intelligent Integrated Systems",Pan Stanford Publishing Corp., Singapore, 2014
[10] Ernst T., IEDM 2008, 10.1109/IEDM.2008.4796804
[11] Coquand R. et al., VLSI 2012, 10.1109/VLSIT.2012.6242437
[12] Wilson R. et al. ISSCC 2014. 10.1109/ISSCC.2014.6757509
[13] Carron V. et al., IWJT 2014 Proc. , Shanghai (PRC),(2014)
[14] Vinet M. et al. IEDM 2013, 10.1109/IEDM.2013.6724697
[15] Hutin L. et al. IEDM 2009, 10.1109/IEDM.2009.5424425
[16] A. B. Fadjie Djomkam, V. etal, MRS 2013 Symposium **T2.10** (2013)
[17] Suri M. et al., IEDM2012, 10.1109/IEDM.2012.6479017
[18] E.Vianello et al., invited talk IEDM this issue
[19] Villalon A. et al., VLSI Tech. Dig.,p.84, Honolulu(HI), June 2014.
[20] Wacquez R. et al., VLSI 2010, 10.1109/VLSIT.2010.5556224
[21] Roche B. et al., *Nature Comm.*, 4, n° 1581, doi:10.1038/ncomms2544
[22] Shinada T., IEDM 2011, 10.1109/IEDM.2011.6131644
[23] Fueschle M. et al., Nat.Nanotech.,5, pp.502-505,(2010)
[24] Matthey L.et al, SSDM Proc.,Fukuoka(Japan), Sept. 2013
[25] Ghezzi G. et al, Applied Physics Letters 101, no. 23, 2012.
[26] Ma N., Jena D., Phys Rev. X, v4, p011043,2014
[27] Hutin L. et al, IWJT 2014 Proc. , Shanghai (PRC),(2014)
[28] Tiron R. et al., Proc. SPIE. 8680, 868012, 2013
[29] Dijon J. et al., IEDM 2010, 10.1109/IEDM.2010.5703470
[30] Clavé G. et al, Org. Biomol. Chem., 2014, 12, 2778
[31] Arcamone J. et al, IEDM 2011, 10.1109/IEDM.2011.6131637
[32] Milanina K.M. et al, J Appl Phys, 95, 183105,(2009)
[33] Houri S. et al, to be published
[34] Pain L. et al, EIPBN 2014 Proc., Washington DC, May 2014
[35] Batude P. et al. IEDM 2011, 10.1109/IEDM.2011.6131506
[36] Turkyilmaz O. et al, DATE 2014 , 10.7873/DATE2014.351
[37] Abe K. et al, ICICDT 2008, 10.1109/ICICDT.2008.4567279
[38] Poupon G., PanPacific Microel. Symp. 2014 Proc.,Hawaii(HI) , Feb 2014
[39] El Bouayadi O. et al, EuMW 2013 Proc,, Nuremberg(FRG) , Sept. 2013
[40] Phan V.P. et al, Adv. Funct. Mat., 22, n°12, p2580, (2012)
[41] Di Cioccio L. et al., Low Temperature Bonding Conference, Japan(2014)

978-1-4799-8002-4/14 $31.00 © 2014 IEEE

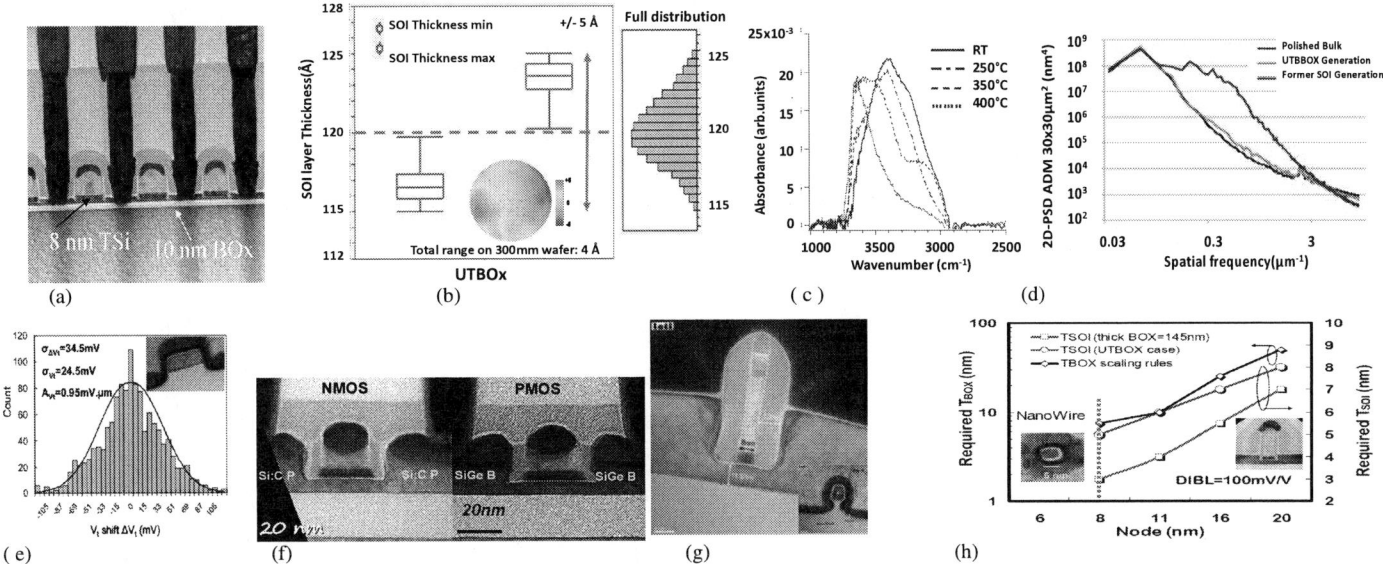

Figure 1 UTBB SOI features Low Power to High Performance devices: (a) 28 nm node 6TSRAM on Tsi=8 nm/TBOx=10nm (b) mapping of 12.0 nm thin SOI on 25 nm BOx/ average distribution on 1000 wafers; (c) IR spectrum reveals water desorption with increasing bonding temperature [4]; (d) roughness of UTSOI is comparable to commercial bulk silicon roughness; (e)VT distribution on 300 mm wafer: matching pairs VT parameter shows record values of 1mV.μm[4] ; (f) Dual selective S/D epi [7]; (g) Experimental Lg=8nm W=7nm nanowire CMOS SOI FETs [8]; (h) Scaling rules for UTSOI [5]

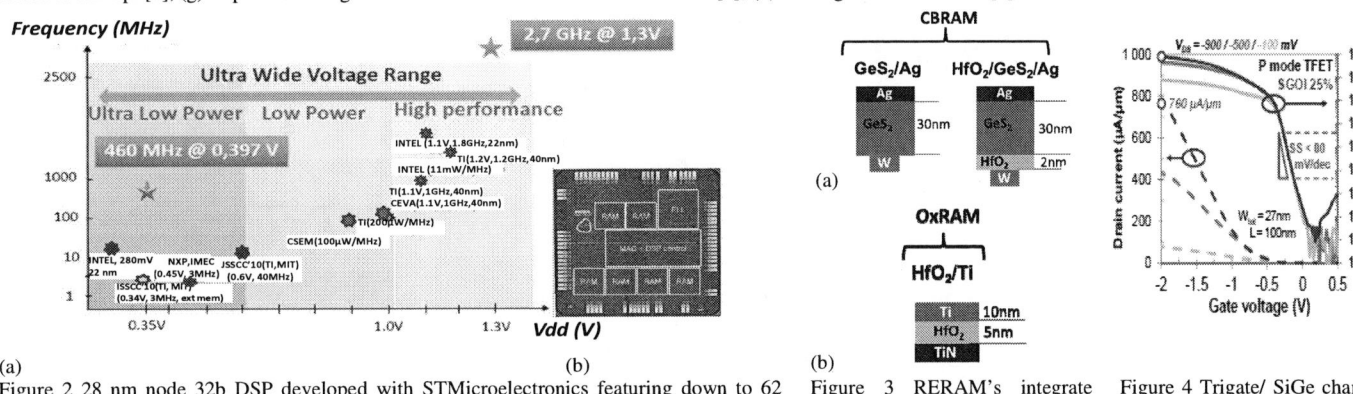

Figure 2 28 nm node 32b DSP developed with STMicroelectronics featuring down to 62 fJ/operation at 0.53V[12]: (a) using back biasing can be run at 460MHz/Vdd=0.397V and 2.7GHz/Vdd=1.3V (b) Chip micrograph : S=1mm²; 5 Million transistors.

Figure 3 RERAM's integrate 1T1R. Two types are described: (a) Conductive Bridge RAM; (b) Oxygen vacancy RAM [18]

Figure 4 Trigate/ SiGe channel Nanowire TFET features record saturation current, outperforms already published results [19]

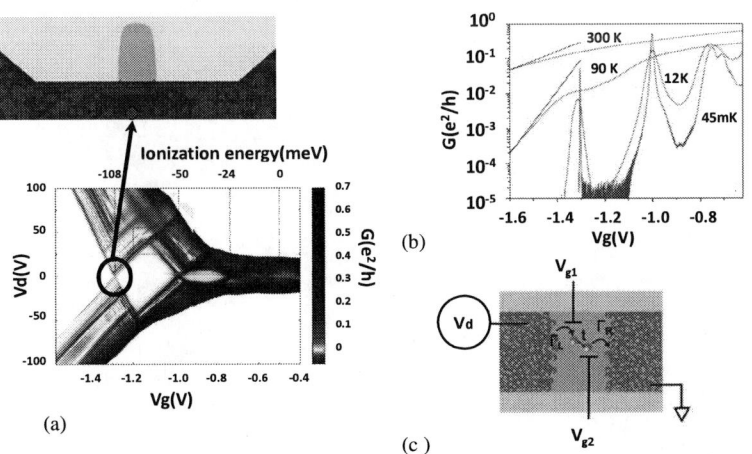

Figure 5 Single Atom Single Electron Transistor. Detection of single dopant in small MOSFETs[20] (a), featuring tunnel barriers demonstrating Coulomb blockade(b), appeals for Zero Intrinsic Variability; (c)"circuit in one device" : electron pump by association of 2 dopants in a MOSFET channel[21].

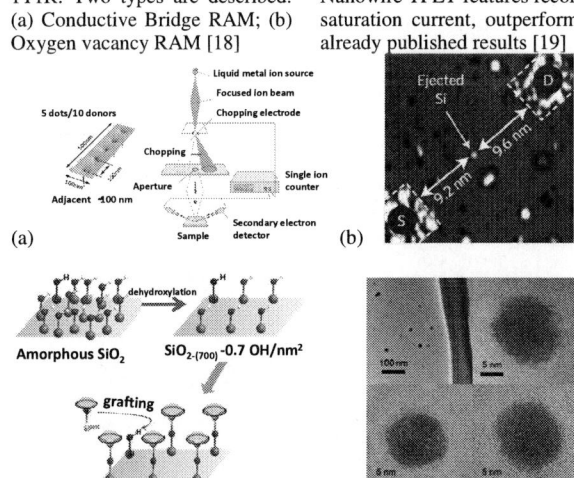

Fig 6 Techniques that should lead to near Zero Intrinsic Variability Deterministic doping by : (a) Single ion implantation [22]; (b) STM lithography and CVD[18]; (c) chemical grafting [24]; (d) Control of Nano-clusters placement in chalcogenides (GeSe)[25].

978-1-4799-8002-4/14 $31.00 © 2014 IEEE

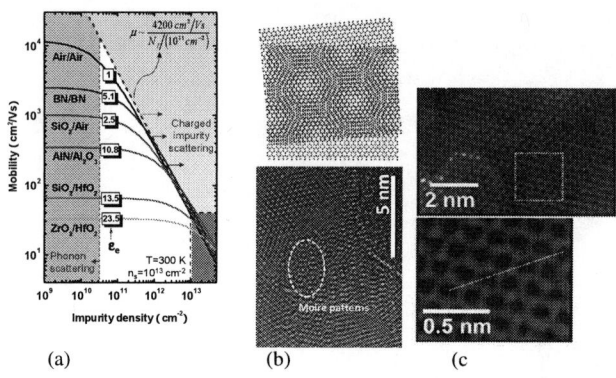

(a) (b) (c

Fig 7. 2D materials to ease Zero Intrinsic Variability: (a) The materials properties are very sensitive to substrate and top layer encapsulation. Case of MoS2[26]; (b) TEM of exfoliated MoS2 films and their Moiré patterns: prepared for mid-process to above IC integration on silicon chips;
(c) High Resolution Electron Holography: surface potential of 1 BN monolayer of. Few impurities observed on top left side (bright spots).

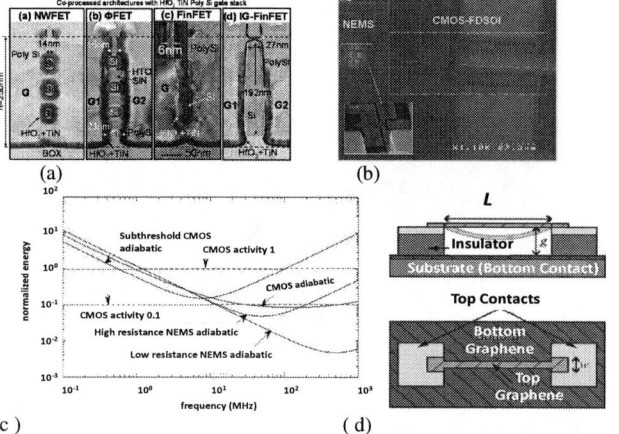

(a) (b)

(c) (d)

Fig 9 Nanowire by top down approach: (a) 3D stacking for flexible logic design with gate last process[7]; (b) co integration of CMOS readout circuit with NEMS for ultra-high sensitivity gas sensing[31]; (c)NEMS Adiabatic architectures reduce energy consumption by ca. 1000 vs. CMOS [33]; (d) Single sheet Graphene switches are candidates as nanoscale switches [32]

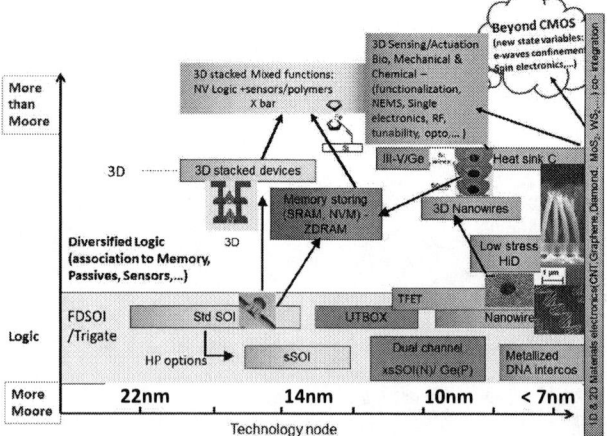

Fig 11 Advanced SOI based Roadmap for co-integration of More Moore and More than Moore devices

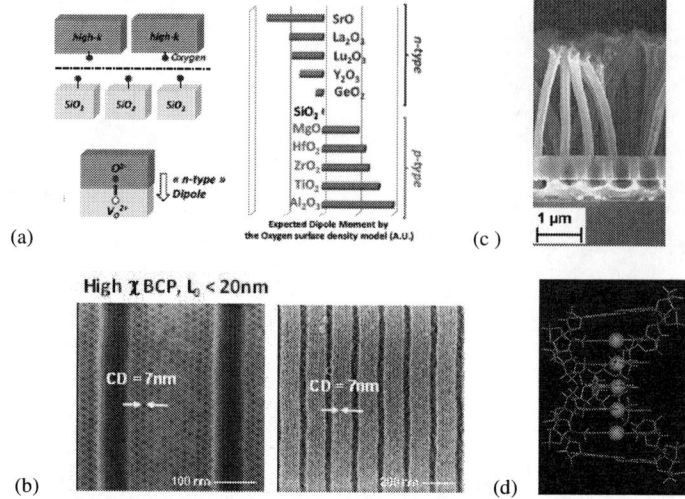

(a) (c)

(b) (d)

Fig 8. Process modules for Zero Instrinsic Variability: (a) Non alloyed contacts could help reduce variability in transistor access. Model including dipole brought by oxygen[27] ; (b) Block co-polymers DSA Patterning showing sub-10nm resolution monodisperse patterns[28]; (c) CNT bundles Interconnect [29] ; (d) DNA metallation by Cu5-[30].

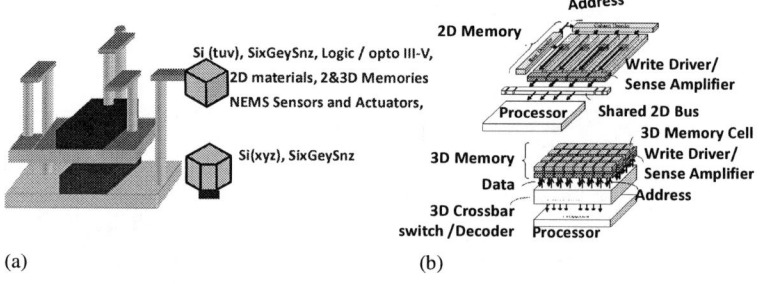

(a) (b)

Fig 10 (a)Monolithic 3D integration [35] makes possible Heterogeneous co-integration of materials and sensors thanks to cold end processing at mid process: 50% improved area and energy delay product can be obtained on partitioned Si CMOS FPGAs[36]; (b) High density or 3D memories (DRAM, RERAM)[37] embedded with digital circuit can deliver more bandwidth, reduce latency, introduce neuromimetic and programmable architectures

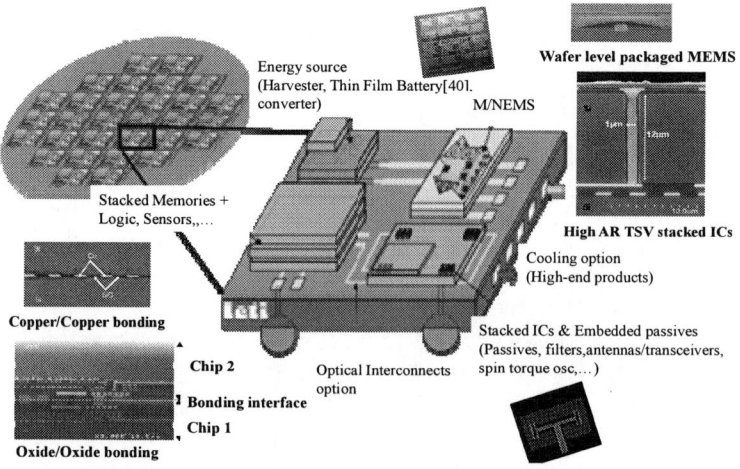

Fig 12 Parallel 3D integration, defined as a flexible Set of Tools[38-41], could profit by collective fabrication on a Silicon interposer, takes into account packaging at wafer level. The choice of organic, glass or silicon substrate depends on applications by challenging performance, thermal losses, compatibility to the environment, reliability and cost.

978-1-4799-8002-4/14 $31.00 © 2014 IEEE

First Experimental Demonstration of Ge CMOS Circuits

Heng Wu, Nathan Conrad, Wei Luo, and Peide D. Ye*

School of Electrical and Computer Engineering, Purdue University, West Lafayette, IN 47906, U.S.A.
*Tel: 1-765-494-7611, Fax: 1-765-496-6443, Email: yep@purdue.edu

Abstract

We report the first experimental demonstration of Ge CMOS circuits, based on a novel recessed channel and S/D technique. Aggressively scaled non-Si CMOS logic devices with channel lengths (L_{ch}) from 500 to 20 nm, channel thicknesses (T_{ch}) of 25 and 15 nm, EOTs of 4.5 and 3 nm and a small width ratio (W_n:W_p=1.2) are realized on a Ge-on-insulator (GeOI) substrate. The CMOS inverters have high voltage gain of up to 36 V/V, which is the best value among all of the non-Si CMOS results by the standard top-down approach. Scalability studies on Ge CMOS inverters down to 20 nm are carried out for the first time. NAND and NOR logic gates are also investigated.

Introduction

With the continuous device scaling and integration density increasing, Si CMOS technology is approaching its physical limit. High mobility channel materials such as Ge [1-5] and III-V [6-7] are intensively studied. However, most of works are limited to the single MOSFET level. There are only several reports discussing non-Si CMOS logic [8-17].

Recently, we reported a breakthrough in high-performance accumulation-mode Ge nFETs by the recessed channel and S/D [18]. Taking advantage of the doping density gradient along the depth axis, the recess process realizes both heavily doped S/D and lightly doped channel. Thanks to the fully-depleted (FD) ultra-thin-body (UTB) recessed channel, low resistivity recessed S/D contact, well-engineered threshold voltage (V_{TH}) and balanced electron and hole mobilities in Ge, nFETs and pFETs with near symmetrical performance and good voltage transition are achieved in the CMOS inverters with a wide range of supply voltage (V_{DD}), from 1.6 to 0.2 V.

Experiment

Fig. 1 shows the Ge CMOS inverter schematic, highlighting the recessed channel and S/D structures employed in the devices. Fig. 2 briefly summarizes the key fabrication processes. The experiment started with a GeOI wafer with 180 nm lightly n-doped (100) Ge and 400 nm SiO$_2$ on (100) Si from Soitec^{TM} as shown in Fig. 3(a). After a standard clean, the device isolation was carried out by SF$_6$ inductively coupled plasma (ICP) dry etching. After the mesa etching, the samples were selectively implanted with P (5×10^{15}/cm^2 at 30 keV) and BF$_2$ (4×10^{15}/cm^2 at 30 keV) for nFETs and pFETs, respectively, both of which were then activated by a *common* rapid thermal anneal (RTA) at 500 °C for 1 min in N$_2$ ambient. After that, an optimized *common* SF$_6$ ICP dry etching with a high aspect ratio was used to form the recessed channel, as shown in the testing structures in Fig. 3(b). Fig. 3(c-d) show the recessed channels in real fabricated devices with T_{ch} of 15 and 25 nm determined by different etching time and adopted in different samples. After smoothing the channel interface by a surface wet clean using cyclic 2% HF rinsing, 1 nm Al$_2$O$_3$ was first deposited by ALD at 250 °C and then a post-deposition oxidation (PO) was performed by RTA at 500 °C for 30 s in pure O$_2$ ambient to grow GeO$_x$ passivating the Al$_2$O$_3$/Ge interface. Next, the *common* ALD gate dielectric of 5 or 8 nm Al$_2$O$_3$ was deposited at 300 °C for different samples. The overall EOT is calculated to be 3 or 4.5 nm, considering both the Al$_2$O$_3$ and the GeO$_x$. After a post deposition anneal (PDA) at 500 °C for 1 min in forming gas ambient and etching away the oxide in the S/D area, an extra BCl$_3$/Ar ICP dry etching was used to remove the top Ge layer as the recessed S/D etching. Note that this is one of the key processes in this experiment and the recessed S/D dry etching is carefully calibrated to precisely control the etch rate and etched profile. The etching rate is around 15 nm/min. Fig. 4(a) shows the test recessed S/D structure, indicating that about 20 nm of top Ge was removed. 100 nm Ni was then deposited as the *common* S/D metal contacts, followed by a *common* ohmic anneal by RTA at 250 °C for 30 s in N$_2$ ambient. The metal gate was formed by 40/60 nm Ti/Au for pFETs and 40/60 nm Ni/Au for nFETs. Finally, devices were connected for CMOS logic gates.

The ratio of nFETs to pFETs gate width is carefully designed to be 1.2:1 (1 μm : 0.85 μm) for balanced performance. Three samples: A (T_{ch} = 15 nm, EOT = 4.5 nm), B (T_{ch} = 25 nm, EOT = 4.5 nm) and C (T_{ch} = 25 nm, EOT = 3 nm) are thoroughly investigated and presented.

Results and Discussion

Fig 4(b) shows the TLM structure under SEM and square Ni metal contacts are placed on isolated conductive Ge with different gaps between each other. Low resistivity Ohmic contacts are realized on both nFETs and pFETs by using common Ni recessed S/D, which greatly simplifies the process complexity compared to using multiple metal layers respectively for n- and p-type contacts. The contact resistances (R_c) are extracted to be 0.45 and 0.37 Ω·mm for Ge n- and p-contacts and 88 and 135 Ω/□ for the sheet resistance (R_{sh}), as shown in Fig. 4(c). The small standard deviations of measured results shown in the inserted figures verify the good uniformity of the recessed S/D contacts. The contact quality could be further improved by optimizing the etching depth of recessed S/D. Fig. 5 explains the basic principles in the recessed channel and S/D structures. Due to the near-Gaussian distribution profile of the doping ions in Ge [18-19], the ion concentration first increases then decreases rapidly along the depth axis into the body. The recess processes, combined with the doping density gradient, result in the realization of a heavily doped S/D region and a lightly doped channel region. Higher doping level in S/D region reduces Schottky barrier width at the metal-semiconductor interface, thus improves the contact resistance [17]. Meanwhile, lower doping level in channel region increases the maximum depletion width, thus enhances the gate control and realizes the enhancement-mode operation in the devices. It also reduces the Coulomb scattering generated by ionized dopants, thus improves the carrier mobilites for both electrons and holes.

978-1-4799-8002-4/14 $31.00 © 2014 IEEE

Fig. 6(a) depicts the top-down view of the smallest L_{ch} CMOS inverter under SEM and the dark region is SiO_2 in the area with top conductive Ge layer removed during the device isolation. The gate areas of the nFET and pFET in the same inverter shown in Fig. 6(a) are enlarged in Fig. 6(b-c) and the L_{ch} of both devices are 20 nm. Fig. 6(d) gives the bird's eye view of a CMOS inverter. Fig. 7 shows the transfer curves of the nFET and the pFET inside a 50 nm L_{ch} inverter in *sample A* at $|V_{ds}|$ = 0.05, 0.5 and 1 V. With a T_{ch} of 15 nm and an optimized gate stack, both of the two devices show good I_{ON}/I_{OFF} ratios > 1×10^5 and balanced threshold voltages ($|V_{TH}|$ ~ 0.5 V). For comparison, transfer curves of a longer channel (L_{ch} = 90 nm) device in *sample B* with a T_{ch} of 25 nm are given in Fig. 8. The short channel effects (SCEs) are greatly suppressed as proved by reduced DIBLs and further improved I_{ON}/I_{OFF} ratios. Fig. 9 presents the I_d-V_{ds} curves of the same two nFETs and pFETs in Fig. 7-8 with $|V_{gs}|$ from 0 V to 3 V in 0.2 V steps, showing near-symmetrical output characteristics.

Fig. 10 shows the voltage transfer curves of the same two inverters shown in Fig. 7-8 with a V_{DD} from 1.6 V to 0.2 V. Longer channel inverter shows a better voltage transition. Further increasing L_{ch} (400 nm), together with reduced EOT (3 nm) in *sample C* yields much steeper V_{OUT} versus V_{IN} curves as shown in Fig. 11. Fig. 12 compares the voltage gains of the same two inverters in Fig. 10 and the 90 nm L_{ch} inverter in sample B has larger voltage gain, proving that better gate electrostatics control leads to steeper voltage transition. Fig. 13 gives the voltage gains of the same long channel inverter in Fig. 11. A High peak voltage gain of 36 V/V is obtained at a V_{DD} of 1.2 V. Fig. 14 shows the peak voltage gain scaling metrics of the three samples at V_{DD} = 1 V. Thinner and longer channel results in an improved voltage gain, indicating a better gate electrostatic control.

Noise margin (NM) is the maximum departure from the ideal logical level that places the gate at a small-signal voltage gain of unity, quantifying the robustness of a gate with respect to the input signal interference. Fig. 15 compares the noise margins of the same two inverters in Fig. 10 in butterfly transfer curves at a V_{DD} of 1.2 V. Fig. 16 shows the NM of the same long channel inverter in Fig. 11 at V_{DD} = 1.2 V, showing a larger NM_H (NM for high input) of 0.5 V and NM_L (NM for low input) of 0.38 V. Both of the noise margins as a function of V_{DD} are depicted in Fig. 17. While the absolute values of NM increase with higher V_{DD}, the ratios to V_{DD} fluctuate around 40%. Scaling metrics of noise margin (NM_L+NM_H) at V_{DD} of 1 V for the three samples are given in Fig. 18. NMs decrease with smaller L_{ch}, due to stronger short channel effects, indicating worse immunity to noises in input signal. Larger NMs are obtained by enhancing the gate electrostatic control through thinning T_{ch} from 4.5 nm to 3 nm and reducing EOT from 4.5 nm to 3 nm.

Transition width (TW) as a function of V_{DD} for the same long channel inverter in Fig. 11 is shown in Fig. 19. TW is defined by the difference between low and high V_{IN} corresponding to a voltage gain of 1. It describes the range of V_{IN} needed to switch the inverter between "1" and "0" states and smaller TW means better voltage transition at certain V_{DD}. The absolute value of TW increases with higher V_{DD}, while the percentage value to V_{DD} decreases and tends to saturate. Fig. 20 provides the L_{ch} dependence of TW for the three samples at V_{DD} = 1 V. Similar to

the case of peak voltage gain and noise margin, smaller EOT, thinner and longer channels provide lower TW.

Fig. 21 compares the transient current (I_{DD}) of the three inverters in Fig. 10-11. I_{DD} is the current flowing through the inverter during switching from "1" to "0" states and it partially determines the speed of CMOS logic gates. Shorter channel device has significant large I_{DD} and the I_{DD} increases from 0.12 μA to 7.4 μA with V_{ds} rising from 0.2 V to 1.2 V. Fig. 22 summarizes the scaling trend of the switching current (max I_{DD}) of the three samples at V_{DD} = 1 V. The switching current increases with decreasing L_{ch}. Moreover, by employing larger T_{ch} and smaller EOT, the I_{DD} increases as expected.

Fig. 23 shows the output signals of a 100 nm L_{ch} CMOS inverter in response to input square-wave signals with different frequencies. The output signal still maintains good square-wave shape at 1 kHz. Fig. 24(a) and (b) show the top-down view of a fabricated NAND logic gate under SEM and its circuit diagram. Fig. 25 provides the output signal of a 100 nm L_{ch} NAND gate with two input signals at a supply voltage of 1.2 V. Four combinations of input states "1 1", "0 1", "1 0" and "0 0" are used and the output signal shows sharp transitions. Fig. 26(a) and (b) show the top-down view of a fabricated NOR logic gate under SEM and its circuit diagram. The output signal of a 100 nm L_{ch} NOR gate is provided in Fig. 27 and same testing conditions are applied as used in the NAND gate.

Table 1 compares all of the non-Si CMOS results reported in literature with this work. We have realized the smallest VLSI-related non-Si CMOS inverters fabricated by the top-down approach. A record high peak voltage gain at low V_{DD} (36 V/V at 1.2 V) is obtained on a 400 nm L_{ch} Ge CMOS inverter.

Conclusion

We experimentally demonstrate the first Ge CMOS circuits by a novel recessed channel and S/D technique. Inverters with high voltage gains up to 36 V/V and L_{ch} down to 20 nm are realized. The first scalability study on Ge CMOS inverters is carried out. NAND and NOR logic gates are also investigated in the time domain. This study provides strong evidences of Ge as a promising candidate to replace Si in future's low power and high speed CMOS logic applications.

Acknowledgement

The authors would like to thank J. J. Gu, L. Dong, M. Si, L. M. Yang, M. S. Lundstrom and K. K. Ng for the valuable discussions. This work is partly supported by the SRC GRC program.

Reference

[1] A. Toriumi, et al., *IEDM* 2011, p.28.4.1. [2] R. Pillarisetty, *Nature*, p. 324, 2011. [3] B. Duriez, et al., *IEDM* 2013, p.522. [4] J. Mitard, et al., *IEDM* 2008, p.876. [5] B. Liu, et al., *IEDM* 2013, p.657. [6] Y. Xuan, et al., *EDL*, p.294, 2008. [7] J. Alamo, *Nature*, p.317, 2011. [8] J. Nah, et al., *Nano Letters*, p.3592, 2012. [9] A. W. Dey, et al., *Nano Letters*, p.5593, 2012. [10] S. Nam, et al., *PNAS*, p.21035, 2009. [11] J. Feng, et al., *EDL*, p.911, 2006. [12] G. Jin, et al., *EDL*, p.1236, 2011. [13] L. Dong, et al., *VLSI* 2014, p.60. [14] T. Irisawa, et al., *VLSI* 2014, p.118. [15] L. Czornomaz, et al., *IEDM* 2013, p.2.8.1. [16] H. Sunamura, et al., *VLSI* 2014, p.180. [17] S. Takagi et al., *IEDM* 2012 p.23.1.1. [18] H. Wu, et al., *VLSI* 2014. p.96. [19] K. Suzuki, et al., *TED*, p.627, 2009.

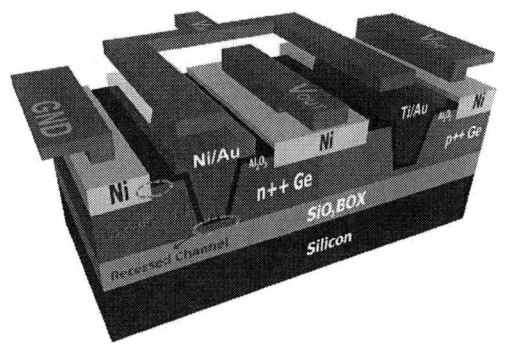

Fig. 1 Device schematic of a Ge recessed channel and S/D CMOS inverter. The recessed channel and S/D structures are highlighted for better illustration.

- ▣ Mesa Isolation (SF₆ ICP Dry Etch)
- ▣ N-Implantation (P 5×10¹⁵ 30 keV)
- ▣ P-Implantation (BF₂ 4×10¹⁵ 30 keV)
- ▣ **Common** Dopant Activation (500 °C 1min in N₂)
- ▣ **Common** Channel Formation (SF₆ ICP Dry Etch)
  - ⊕ Process I (T_ch=15nm) ⊕ Process II (T_ch=25nm)
- ▣ Surface Wet Clean (HF Cyclic Rinse)
- ▣ **Common** Gate Oxide Formation
  - ○ 1ˢᵗ ALD (250°C 1nm Al₂O₃)
  - ○ Oxidation (500°C 30s in O₂)
  - ○ 2ⁿᵈ ALD for Gate Dielectric
    - ♦ *Sample A* (300°C 8nm Al₂O₃) With T_ch=15nm
    - ▲ *Sample B* (300°C 8nm Al₂O₃) With T_ch=25nm
    - ★ *Sample C* (300°C 5nm Al₂O₃) With T_ch=25nm
  - ○ PDA (500°C 1min in N₂/H₂)
- ▣ **Common** S/D Contacts Formation
  - ○ Local Oxide Etch (BCl₃/Ar ICP Dry Etch)
  - ○ Top Ge Etch (BCl₃/Ar ICP Dry Etch)
  - ○ Metal Deposition (Ni)
  - ○ Ohmic Anneal (250°C 30s in N₂)
- ▣ Gate Metal Deposition
  - ○ pFETs (Ti/Au) ○ nFETs (Ni/Au)
- ▣ Device Interconnection

Fig. 2 Fabrication process flow of the Ge CMOS in this experiment. Three samples with different conditions are fabricated.

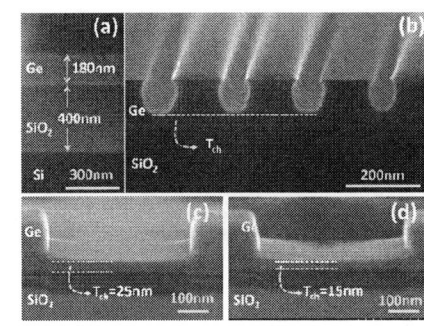

Fig. 3 (a) Cross section of a GeOI substrate. (b) Testing recessed channel structures. (c) The 25 nm T_ch channel in *sample B and C*. (d) The 15 nm T_ch channel in *sample A*.

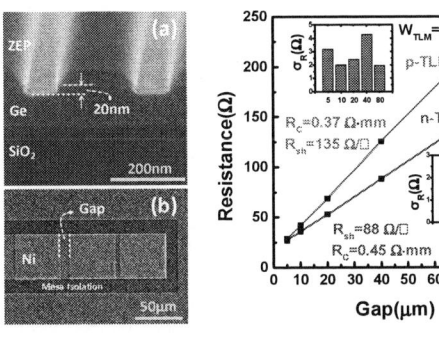

Fig. 4 (a) Testing recessed S/D structures with 20 nm top Ge layer removed. (b) TLM structure under SEM. (c) TLM data for both Ge n-contact and p-contact. Inserted figures show the standard deviation based on 10 measured devices.

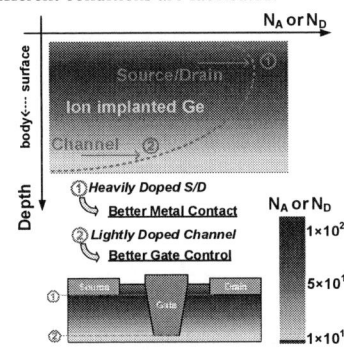

Fig. 5 Basic idea in the recessed S/D and channel. Doping density gradient in the implanted Ge helps to obtain heavily doped S/D and lightly doped channels.

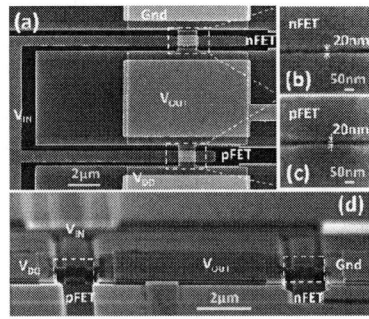

Fig. 6 (a) Top-down view of a fabricated Ge CMOS inverter with the smallest channel length. (b-c) zoom-in images of the gate area in the CMOS inverter in (a), the channel length is 20 nm. (d) Bird's eye view of a CMOS inverter under SEM.

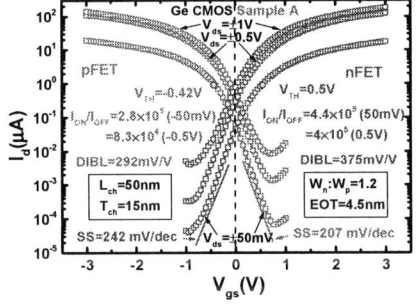

Fig. 7 Transfer curves of the nFET and pFET in a 50 nm L_ch CMOS inverter in *sample A*. With the ultra-thin channel, both of devices show good ON and OFF state.

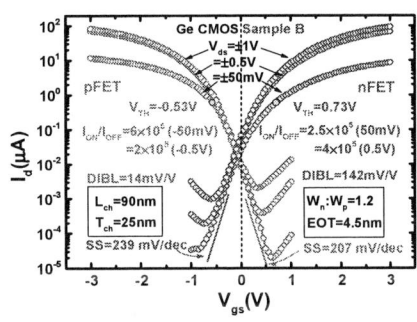

Fig. 8 Transfer curves of the nFET and pFET in a 90 nm L_ch CMOS inverter in *sample B*. Because of the longer channel, the SCE is further suppressed.

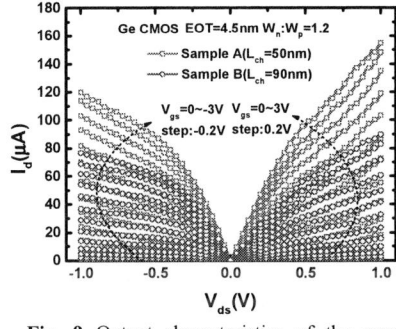

Fig. 9 Output characteristics of the same four devices shown in Fig. 7-8 with a V_gs sweeping from 0 V to 3 V for nFETs and 0 V to -3 V for pFETs.

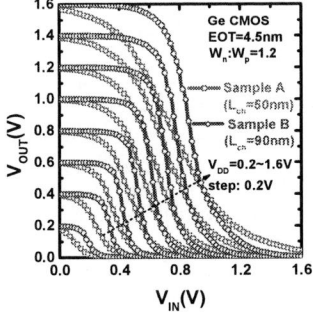

Fig. 10 V_OUT versus V_IN of the same two CMOS inverters shown in Fig. 7-8 in *sample A and B* with a V_DD from 0.2 V to 1.6 V in 0.2 V step.

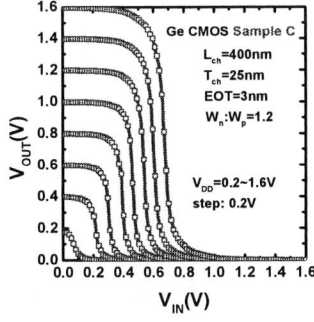

Fig. 11 V_OUT versus V_IN of a 400 nm long channel inverter in *sample C* with a V_DD from 0.2 V to 1.6 V in 0.2 V step.

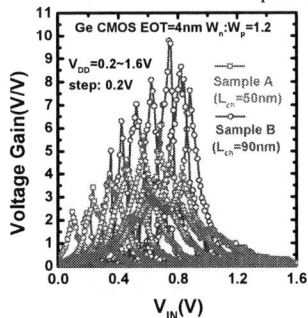

Fig. 12 Voltage gain versus V_IN of the same two inverters in *sample A and B* shown in Fig. 10 with a V_DD from 0.2 V to 1.6 V in 0.2 V step.

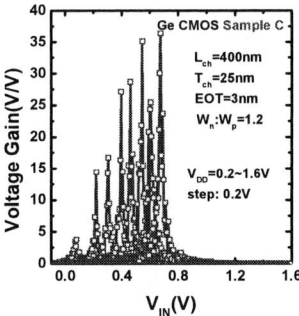

Fig. 13 Voltage gain versus V_IN of the same long channel inverter in *sample C* shown in Fig. 11 with a V_DD from 1.6 V to 0.2 V in 0.2 V step.

978-1-4799-8002-4/14 $31.00 © 2014 IEEE

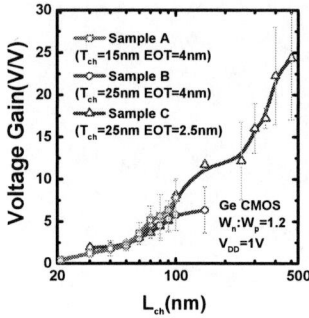
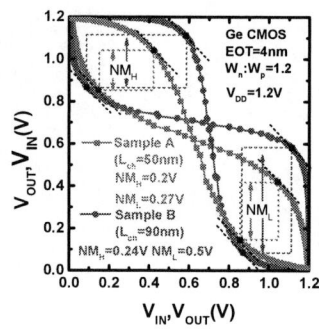
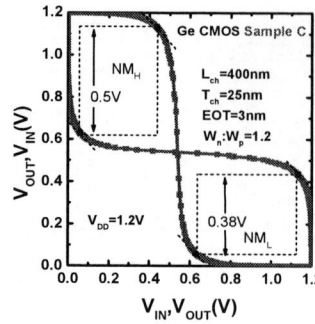
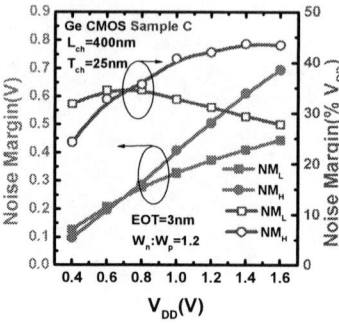

Fig. 14 Channel length dependence of the maximum voltage gain of CMOS inverters in *sample A, B and C* with a V_{DD} of 1 V.

Fig. 15 Noise margin of the same two CMOS inverters in *sample A and B* shown in Fig. 10 with a V_{DD} of 1.2V.

Fig. 16 Noise margin for the same long channel inverter in *sample C* shown in Fig. 11 with a V_{DD} of 1.2 V.

Fig. 17 Noise margin in absolute and percentage value to V_{DD} plotted against V_{DD} of the same long channel inverter in *sample C* shown in Fig. 11

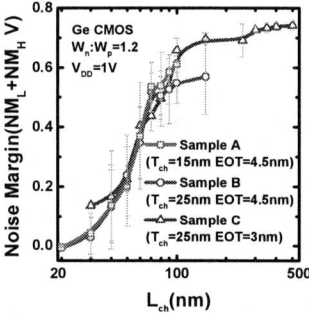
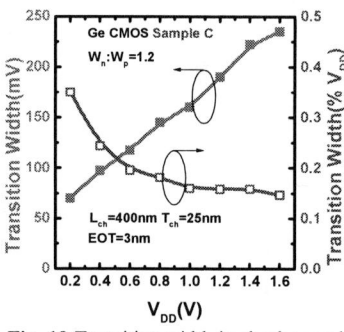
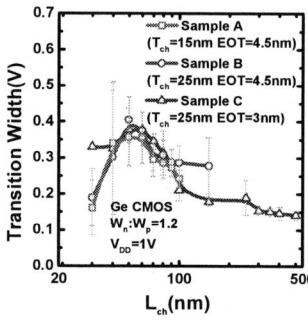
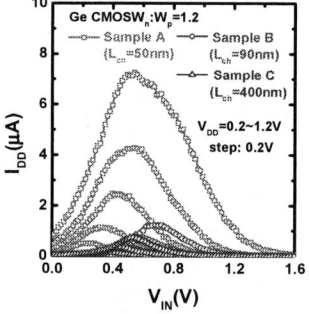

Fig. 18 Noise margin (NM_L+NM_H) scaling metrics of CMOS inverters in *sample A, B and C* with a V_{DD} of 1V.

Fig. 19 Transition width in absolute and percentage value to V_{DD} plotted against V_{DD} of the same long channel inverter in *sample C* shown in Fig. 11.

Fig. 20 Channel length dependence of transition width of CMOS inverters in *sample A, B and C* at V_{DD} of 1 V.

Fig. 21 Transient current versus V_{IN} curves of the same three CMOS inverters in *sample A, B and C* shown in Fig. 10-11 with a V_{DD} from 1.2 V to 0.2 V.

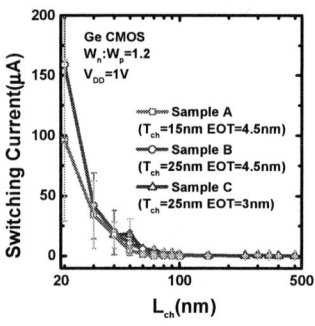
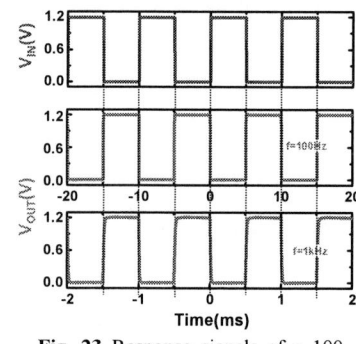
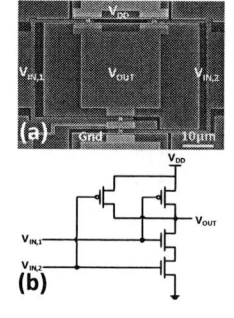
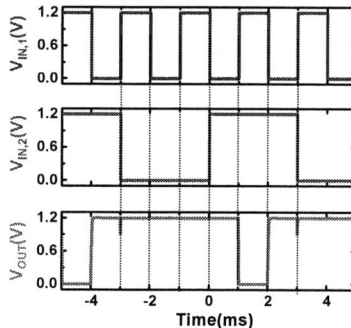

Fig. 22 Channel length dependence of switching current (maximum transient current) of CMOS inverters in *sample A, B and C* with a V_{DD} of 1 V.

Fig. 23 Response signals of a 100 nm L_{ch} CMOS inverter in *sample B* to square-wave input signals with different time periods.

Fig. 24 (a) Top-down SEM image of a NAND logic gate. (b) Circuit diagram of the NAND logic gate.

Fig. 25 Response signals of a 100 nm L_{ch} NAND logic gate in *sample B* to two square-wave input signals in the time domain.

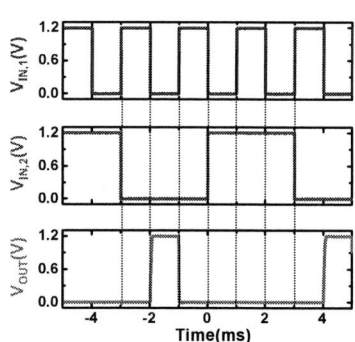

Fig. 26 (a) Top-down SEM image of a NOR logic gate. (b) Circuit diagram of the NOR logic gate.

Fig. 27 Response signals of a 100 nm L_{ch} NOR logic gate in *sample B* to two square-wave input signals in the time domain.

Table 1. Comparison of the Ge CMOS logic gates in this work with other non-silicon CMOS inverter results in literature.

Publication	Channel Material	Gate Dielectric	Fabrication Method	L_{ch}	Voltage Gain
NL 2012 UC Berkley[8]	InAs n-Channel InGaSb p-Channel	10 nm ZrO₂ SiO₂	PDMS transferred Nano-ribbon on SiO₂	2.85 μm (nFET) 2.6 um (pFET)	14V/V (V_{DD}=1V)
NL 2012 Lund[9]	InAs n-channel GaSb p-channel	4 nm Al₂O₃	Bottom-up nanowire on SiO₂	2.7 μm	10 V/V (V_{DD}=1V)
PNAS 2009 Harvard[10]	InAs n-channel SiGe p-channel	20 nm HfO₂	Bottom-up nanowire on SiO₂	1.5 μm	45V/V (V_{DD}=4V)
EDL 2006 Stanford[11]	Si n-channel Ge p-channel	SiO₂ (EOT=10nm, nFET) GeNₓ(EOT=20nm, pFET)	Top-down on Si substrate with RMG GeOI layer	1 μm (nFET) 1.5 μm (pFET)	4V/V (V_{DD}=5V)
EDL 2011 Samsung[12]	α-IGZO n-channel Poly Si p-channel	SiO₂ /SiNₓ	Top-down on Si substrate with multiple channel layer	40 μm (nFET) 10 μm (pFET)	18V/V (V_{DD}=7V)
VLSI 2014 Purdue[13]	GaAs (111)A n-channel GaAs (111)A p-channel	4 nm La₂O₃ / 4 nm Al₂O₃	Top-down on common GaAs(111)A substrate	1 μm	12V/V (V_{DD}=3V)
VLSI 2014 AIST[14]	InGaAs n-channel SiGe p-channel	4.5 nm HfO₂ (nFET) 7.8 nm Al₂O₃ (pFET)	Top-down on SGOI substrate with InGaAs layer by wafer-bonding	10 μm	26V/V (V_{DD}=1V)
IEDM 2013 IBM[15]	InGaAs n-channel SiGe p-channel	10 nm Al₂O₃	Top-down on SiGeOI substrate with InGaAs layer by wafer-bonding	500 nm	14V/V (V_{DD}=1V)
VLSI 2014 Renesas[16]	α-IGZO n-channel α-SnO p-channel	30 nm SiN / 20 nm SiO₂	Top-down on Wafer with IGZO and SnO	0.8 μm	12V/V (V_{DD}=5V)
THIS WORK	Ge n-channel Ge p-channel	8 nm Al₂O₃ (Sample A) 5 nm Al₂O₃ (Sample C)	Top-down on common GeOI substrate	50 nm -C 400 nm -A	5 V/V (V_{DD}=1.2V)-A 36 V/V (V_{DD}=1.2V)-C

978-1-4799-8002-4/14 $31.00 © 2014 IEEE

InAlP-Capped (100) Ge nFETs with 1.06 nm EOT: Achieving Record High Peak Mobility and First Integration on 300 mm Si Substrate

Xiao Gong,[1] Qian Zhou,[1] Man Hon Samuel Owen,[1] Xin Xu,[1] Dian Lei,[1]
Shu-Han Chen,[2] Gene Tsai,[2] Chao-Ching Cheng,[2] You-Ru Lin,[2] Cheng-Hsien Wu,[2] Chih-Hsin Ko,[2] and Yee-Chia Yeo.[2,*]

[1] Department of Electrical and Computer Engineering, National University of Singapore (NUS), Singapore 117576.
[2] Taiwan Semiconductor Manufacturing Company, Hsinchu, Taiwan 300, R. O. C.
*Phone: +886-3-5636688 ext. 7223015, Fax: +886-3-6687827, Email: yeo@ieee.org

I. Introduction

Ge is a promising alternative channel material for sub-10 nm CMOS technology due to its high electron and hole mobilities. Ge pFETs with excellent subthreshold characteristics and high drive current I_{ON} have been demonstrated [1]. While various surface passivation techniques have been reported for Ge nFETs [2]-[19], realizing Ge nFETs with small subthreshold swing S and high I_{ON} is still challenging, especially for enhancement mode Ge nFETs. In addition, Ge nFETs should preferably be formed on Si substrates.

In this work, InAlP-capped Ge nFETs were realized on 300 mm Si substrates using the buffer layer technique for the first time. With a gate stack comprising 2.3 nm of InAlP and 2 nm of HfO$_2$, an EOT of 1.06 nm was achieved. In addition, InAlP-capped Ge nFETs on bulk Ge substrates were also formed using sub-400 °C process modules. At a gate length L_G of 3 µm, we obtained a record high Ge (100) peak mobility of 1370 cm^2/V·s, the lowest reported hysteresis of 15 mV, and S of 103 mV/decade. By scaling L_G down to 500 nm, the highest reported I_{ON} of 127 µA/µm (at $V_{GS} - V_{TH} = 1$ V and $V_{DS} = 1$ V) and peak intrinsic transconductance $G_{m,int}$ of ~275 µS/µm (at $V_{DS} = 0.8$ V) were obtained for enhancement mode Ge (100) nFETs.

II. Key Highlights and Device Fabrication

Key highlights of this work are shown in Fig. 1 (a). First, an InAlP passivation layer with a thickness of 2.3 nm was inserted between Ge and the high-k gate dielectrics for mobility enhancement. Second, a process flow with low thermal budget of 400 °C was introduced to reduce the possible inter-diffusion among Ge, InAlP, and the gate dielectrics, and therefore maintains good interface quality. Third, scaling-down of EOT was realized by reducing the HfO$_2$ thickness or by direct deposition of HfO$_2$ on InAlP. Lastly, integration of Ge nFETs on 300 mm Si substrates was demonstrated for the first time.

The process flow for fabricating Ge nFETs is shown in Fig. 1 (b). A 2.3-nm thick InAlP layer was grown on the (100)-oriented Ge substrate by MOCVD. After pre-gate clean, 0.5 nm Al$_2$O$_3$ and 2 nm or 3 nm HfO$_2$ were deposited by ALD at a temperature of 250 °C, followed by the TaN deposition and patterning. After S/D phosphorus implant, forming gas annealing (H$_2$:N$_2$ = 1:9) was performed at 400 °C for 30 minutes. This step also

activated the implanted phosphorus, and is the highest temperature step in the entire process flow (excluding InAlP growth). NiGe contact formation was finally performed at a temperature of 350 °C to complete the transistor fabrication. For Ge nFETs integrated on the Si substrate, HfO$_2$ was directly deposited on InAlP without insertion of the Al$_2$O$_3$ layer. The Ge channel layer was grown on the 300 mm Si substrate using the buffer layer technique.

To illustrate the concept of inserting an InAlP passivation layer, Fig. 2 shows the energy band diagram of the InAlP-capped Ge nFETs at the strong inversion regime. There is a conduction band offset ΔE_C of 0.84 eV between InAlP and Ge. This ΔE_C value is large enough to confine the electrons in the Ge channel and separate them from the interface traps which may exist at the high-k/InAlP interface. This leads to reduced interface trap scattering, and therefore, higher drive current can be achieved.

III. InAlP-Capped Ge nFETs on the Ge Substrate

A. Long-Channel Ge nFETs

Fig. 3 shows the high resolution cross-sectional TEM image of the Ge/InAlP/Al$_2$O$_3$/HfO$_2$/TaN stack with InAlP of ~2.3 nm, Al$_2$O$_3$ of ~0.5 nm, and HfO$_2$ of ~2 nm. Excellent crystalline quality of the InAlP layer was observed. The Ge/InAlP and InAlP/Al$_2$O$_3$ interfaces are sharp and flat. Fig. 4 shows the EDX profile along the line A-A' in Fig. 3 and confirms the existence of In, Al, and P elements. The 'bumps' of the In and P EDX profiles also give the indication of the InAlP thickness of ~2.3 nm and is consistent with the one obtained from the TEM image in Fig. 3. RMS roughness of the InAlP-capped Ge surface is ~0.39 nm with a scan area of 3 µm × 3 µm, as indicated in the AFM image of Fig. 5.

I_{DS}-V_{GS} curves of a Ge nFET with L_G of 3 µm in Fig. 6 exhibits excellent transfer characteristics with S of 103 mV/decade and I_{ON}/I_{OFF} ratio close to 4 orders. The hysteresis is as small as 15 mV between forward and backward voltage sweeps. Reducing the HfO$_2$ from 3 to 2 nm leads to 8% enhancement of the drive current at V_{GS}-V_{TH} of 1 V and V_{DS} of 1 V, as shown in the I_{DS}-V_{DS} characteristics of the Ge nFETs with different HfO$_2$ thicknesses in Fig. 7. The cumulative statistical plot in Fig. 8 indicates that reducing the HfO$_2$ thickness from 3 to 2 nm also reduces the median S from 114 to 108 mV/decade.

978-1-4799-8002-4/14 $31.00 © 2014 IEEE

Fig. 9 shows the electron effective mobility μ_{eff} of Ge nFETs as a function of inversion carrier density N_{INV} measured by the split-CV method. The effect of S/D series resistance R_{SD} was taken out during the mobility extraction. For Ge nFETs with the InAlP capping, record high electron peak mobility of 1370 cm^2/V·s was achieved on the (100)-oriented Ge substrate.

B. Interface Study of the Ge/InAlP/Al$_2$O$_3$/HfO$_2$ Stack

To extract the D_{it} of the Ge/InAlP/Al$_2$O$_3$/HfO$_2$ stack, Ge capacitors were fabricated. A thicker (5.5 nm) HfO$_2$ was used to reduce gate leakage so that D_{it} can be accurately extracted. Fig. 10 (a) and (b) show the C-V curves measured at different frequencies ranging from 1 kHz to 1 MHz at 230 K and 77 K, respectively. Using the conductance method, D_{it} with values ranging from 6.2×10^{11} to 2.5×10^{12} cm^{-2}·eV^{-1} near the valence band edge and mid-gap was obtained, as shown in the plot of D_{it} as a function of energy in the Ge band-gap (Fig. 11). Surface potential fluctuations in the Ge/InAlP/Al$_2$O$_3$/HfO$_2$ layers were considered during the extraction.

C. Electrical Characteristics of Short-Channel Ge nFETs

Output characteristics of a Ge nFET with L_G of 500 nm are shown in Fig. 12. I_{ON} of 127 µA/µm was obtained at $V_{GS} - V_{TH}$ of 1 V and V_{DS} of 1 V. This is the highest drive current reported for enhancement mode Ge (100) nFETs so far. I_{ON} of our Ge nFETs is limited by the high and un-optimized R_{SD}, which was extracted to be ~ 8 kΩ·µm. This is due to the non-self-aligned NiGe formation with large gap between the edge of the gate and the S/D contact pads. After taking out the effect of R_{SD} and drain conductance G_D, peak intrinsic $G_{m,int}$ of 275 µS/µm was achieved at V_{DS} of 0.8 V, as shown in the peak $G_{m,int}$ vs. L_G plot in Fig. 13.

IV. InAlP-Capped Ge nFETs on 300 mm Si Substrate

A. Growth of Ge Channel Layer on 300 mm Si Substrate

The Ge layer was grown on the Si substrate using the buffer layer technique by MOCVD. The buffer layer thickness is ~1 µm, as shown in the cross-sectional TEM image of Fig. 14 (a). The RMS surface roughness of the substrate after the growth of InAlP is ~0.51 nm with a scan area of 3µm × 3µm (not shown here). High resolution TEM image in Fig. 14 (b) shows the excellent crystalline quality of the InAlP layer and the uniform and sharp Ge/InAlP interface.

B. Electrical Characteristics Ge nFETs on the Si Substrate

Fig. 15 shows the I_{DS}-V_{GS} curves of a Ge nFET integrated on the Si substrate with S of 131 mV/decade. Drive current of 45.6 µA/µm was obtained at V_{GS}-V_{TH} of 1 V and V_{DS} of 1 V (Fig. 16). Peak $G_{m,ext}$ at V_{DS} of 1.05 V as a function of L_G in Fig. 17 demonstrates good scalability of the InAlP-capped Ge nFETs. The C-V characteristics of the p-Ge/InAlP/HfO$_2$/TaN MOS capacitor in Fig. 18 shows that EOT of 1.06 nm was

achieved with direct deposition of HfO$_2$ grown on 2.3 nm InAlP.

For Ge nFETs with the same L_G of 5 µm, the drive current measured at the same gate overdrive of 1 V and V_{DS} of 1 V is 2.5 times higher than that reported in Ref. 17 despite a larger EOT, as illustrated in Fig. 19. The gate leakage current density J_G in Fig. 20 is less than 3×10^{-2} A/cm^2 in the gate voltage range of -1 to 1 V. At V_{FB} + 1 V, J_G is only slightly larger than the smallest value reported, as indicated in the plot of J_G as a function of EOT in Fig. 21.

V. Benchmarking

Fig. 22 benchmarks the peak electron μ_{eff} of (100)-oriented Ge nFETs with various passivation techniques. The InAlP-capped Ge nFETs using sub-400 °C process modules achieve the record high Ge (100) electron peak mobility of 1370 cm^2/V·s. In addition to the high μ_{eff}, InAlP-capped Ge nFETs also exhibit S comparable to the best reported values, as indicated in Fig. 23 where S values are plotted as a function of the EOT of long channel planar Ge nFETs. The benchmark of I_{ON} at $V_{GS} - V_{TH}$ of 1 V and V_{DS} of 1 V in Fig. 24 shows that a record high I_{ON} of 127 µA/µm for enhancement mode Ge nFETs was realized though the L_G is not the shortest.

VI. Conclusion

InAlP-capped Ge nFETs with sub-400 °C process modules were reported. Ge nFETs on Ge substrates with InAlP/Al$_2$O$_3$/HfO$_2$ as gate dielectrics demonstrate the highest reported Ge (100) peak μ_{eff} for inversion mode devices. In addition, the gate stack with HfO$_2$ directly deposited on the InAlP cap was implemented in Ge nFETs on 300 mm Si substrates for the first time. This leads to the realization of long-channel Ge nFETs with 1.06 nm EOT, high drive current, excellent S, and low gate leakage current. InAlP is a good passivation technique for Ge nFET gate stack formation, and could enable the use of Ge channel for both nFETs and pFETs in future high performance and low power logic applications.

Reference

[1] B. Duriez et al., IEDM 2013, p. 522.
[2] B. Liu et al., IEDM 2013, p. 657.
[3] C. H. Lee et al., VLSI Symp. 2013, p. T28.
[4] R. Zhang et al., VLSI Symp. 2013, p. T26.
[5] S. H. Hsu et al., IEDM 2012, p. 525.
[6] C. M. Lin et al., IEDM 2012, p. 509.
[7] C. T. Chung et al., IEDM 2012, p. 383.
[8] R. Zhang et al., IEDM 2012, p. 371.
[9] R. Zhang et al., IEDM 2011, p. 642.
[10] Y. C. Fu et al., IEDM 2010, p. 432.
[11] W. Chen et al., IEDM 2010, p. 420.
[12] C. H. Lee et al., IEDM 2010, p. 416.
[13] K. Morii et al., IEDM 2009, p. 681.
[14] C. H. Lee et al., IEDM 2009, p. 457.
[15] D. Kuzum et al., IEDM 2009, p. 453.
[16] G. Thareja et al., IEDM 2010, p. 245.
[17] R. Zhang et al., IEEE Trans. Elect. Dev. 60, p. 927, 2013.
[18] W. Bai et al., VLSI Symp. 2003, p. 121.
[19] H. Wu et al., VLSI Symp. 2014, p. 96.

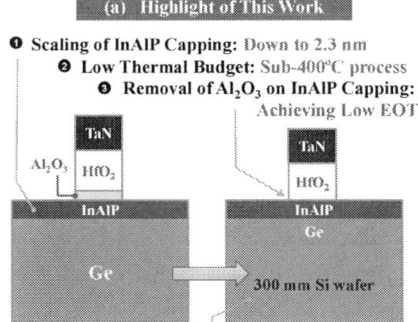

(a) Highlight of This Work

❶ Scaling of InAlP Capping: Down to 2.3 nm
  ❷ Low Thermal Budget: Sub-400°C process
    ❸ Removal of Al₂O₃ on InAlP Capping:
              Achieving Low EOT

❹ First Ge nFETs on 300 mm Si Substrate

(b) Process Flow for Device Fabrication

- MOCVD Growth of InAlP
- Pre-Gate Clean
- *In situ* TMA Surface Cleaning (250 °C, 50 cycles)
- High-κ Deposition (250 °C)
  - ❶ Al₂O₃ (0.5 nm)/HfO₂ (2 nm) ⎱ Ge nFET on
  - ❷ Al₂O₃ (0.5 nm)/HfO₂ (3 nm) ⎰ Ge Substrate
  - ❸ HfO₂ (2 nm) ⎱ Ge nFET on
  - ❹ HfO₂ (3 nm) ⎰ Si Substrate
- TaN Metal Deposition and Patterning
- Phosphorous S/D Implantation
- FGA and Dopant Activation
  (400 °C and 30 minutes)
- NiGe Contact Formation

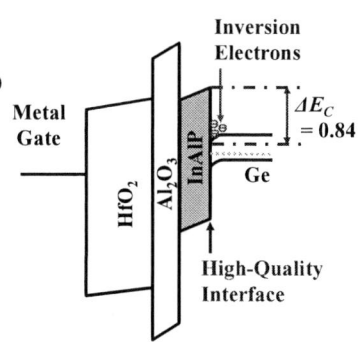

Fig. 1. (a) Key highlights of this work: An InAlP passivation layer with a thickness of 2.3 nm for mobility enhancement; Thermal budget is kept low to maintain a high-quality InAlP/Ge interface; EOT is scaled down by reducing HfO₂ thickness or by direct deposition of HfO₂ on InAlP; First integration of Ge nFETs on 300 mm Si substrate. (b) Key process steps for fabricating Ge nFETs. For Ge nFETs on Ge substrate, 0.5 nm of Al₂O₃ was used. For Ge nFETs integrated on Si substrate, no Al₂O₃ was used in the gate stack.

Fig. 2. Band diagram of an InAlP-capped Ge nFET at strong inversion regime. Electrons are confined in the Ge layer by the InAlP cap due to the large conduction band offset of 0.84 eV.

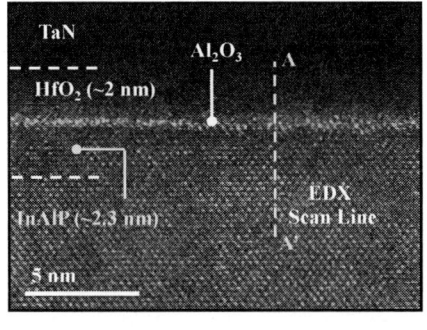

Fig. 3. High resolution cross-section TEM image of the Ge/InAlP/Al₂O₃/HfO₂/TaN stack. The InAlP and HfO₂ thicknesses are 2.3 and 2 nm, respectively. Excellent crystalline quality of InAlP, and sharp Ge/InAlP and InAlP/Al₂O₃ interfaces were observed.

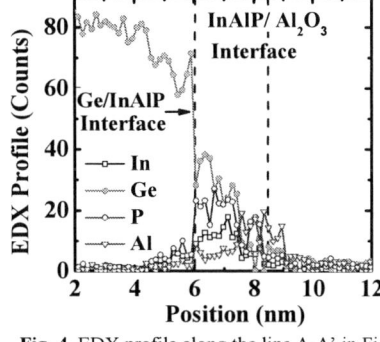

Fig. 4. EDX profile along the line A-A' in Fig. 3 confirms the InAlP grown on Ge. The InAlP thickness is ~2.3 nm and is consistent with the one obtained from the TEM image in Fig. 3.

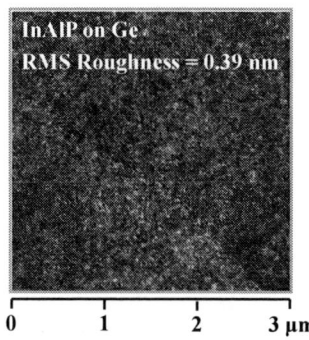

Fig. 5. RMS roughness of the InAlP-capped Ge surface is 0.39 nm with a scan area of 3μm×3μm.

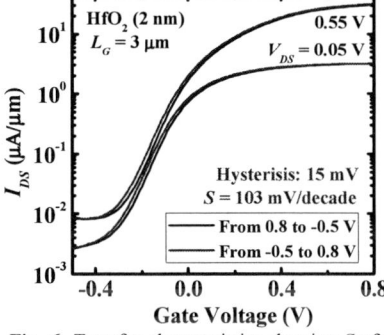

Fig. 6. Transfer characteristics showing S of 103 mV/decade, I_{ON}/I_{OFF} ratio close to 4 orders, and 15 mV hysteresis.

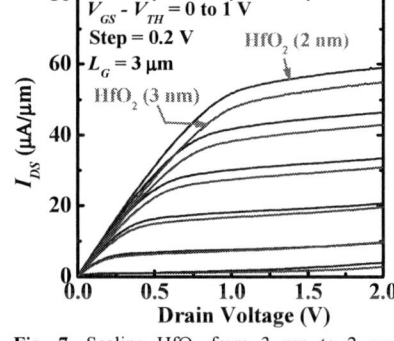

Fig. 7. Scaling HfO₂ from 3 nm to 2 nm enhances drive current by 8% at $V_{GS}-V_{TH}$ of 1 V and V_{DS} of 1 V.

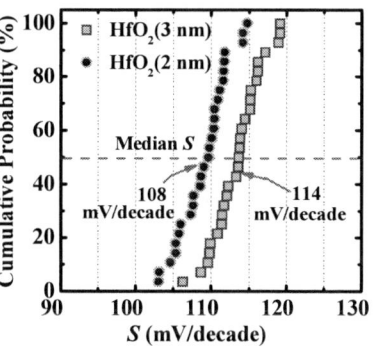

Fig. 8. Statistical plot shows tight distribution of S. Reducing the thickness of HfO₂ improves S.

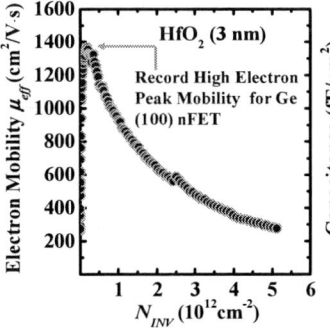
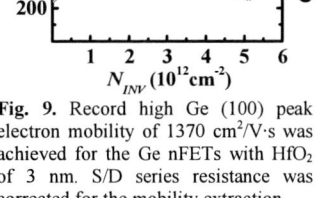

Fig. 9. Record high Ge (100) peak electron mobility of 1370 cm²/V·s was achieved for the Ge nFETs with HfO₂ of 3 nm. S/D series resistance was corrected for the mobility extraction.

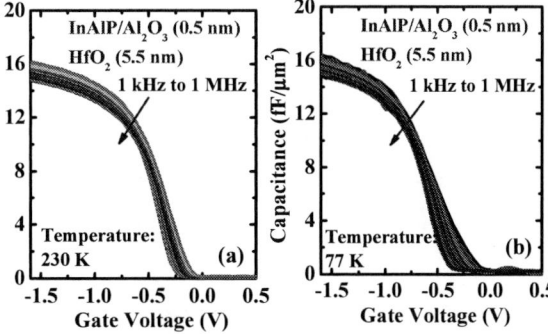

Fig. 10. *C-V* characteristics of the Ge MOS capacitors measured at (a) 230 K and (b) 77 K with different frequency values ranging from 1 kHz to 1 MHz.

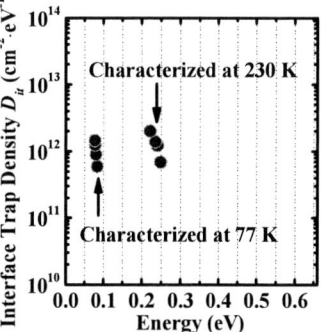

Fig. 11. D_{it} distribution near the valence band edge and mid-gap. D_{it} is in the range of $6.2×10^{11}$ to $2.5×10^{12}$ cm⁻²·eV⁻¹.

978-1-4799-8002-4/14 $31.00 © 2014 IEEE

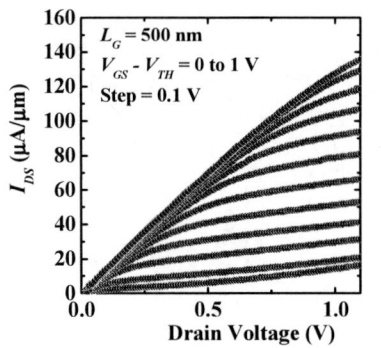

Fig. 12. I_{DS}-V_{DS} curves for a Ge nFET show I_{ON} of 127 µA/µm at $V_{GS} - V_{TH}$ of 1 V and V_{DS} of 1 V.

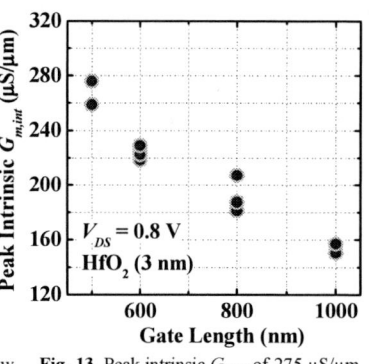

Fig. 13. Peak intrinsic $G_{m,int}$ of 275 µS/µm was achieved at V_{DS} of 0.8 V and L_G of 500 nm.

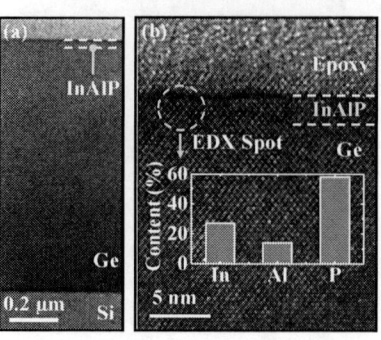

Fig. 14. (a) TEM image of an InAlP-capped Ge layer grown on 300 mm Si substrate. (b) High-resolution TEM showing the high-quality InAlP and the excellent InAlP/Ge interface.

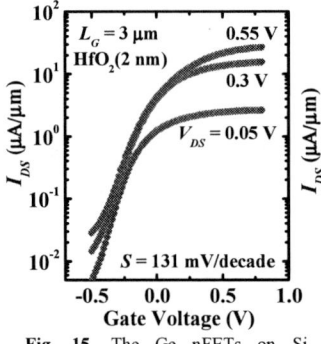

Fig. 15. The Ge nFETs on Si substrate show S of 131 mV/decade and I_{ON}/I_{OFF} ratio of more than 3 orders.

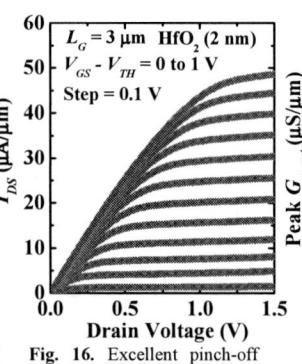

Fig. 16. Excellent pinch-off and saturation output characteristics were achieved for the same device in Fig. 15.

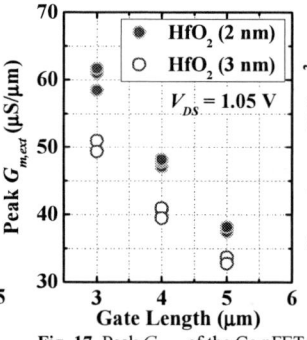

Fig. 17. Peak $G_{m,ext}$ of the Ge nFETs on Si substrate scale well with L_G. A thinner HfO_2 thickness improves peak $G_{m,ext}$.

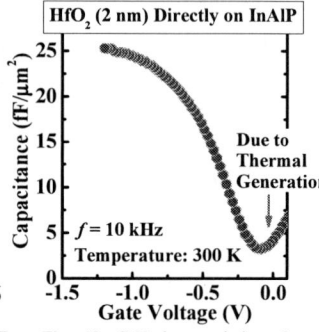

Fig. 18. C-V characteristics of the $TaN/HfO_2/InAlP/p$-Ge MOS capacitor. EOT of 1.06 nm was achieved.

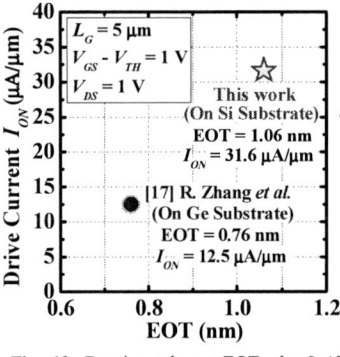

Fig. 19. Despite a larger EOT, the InAlP-capped Ge nFET on Si substrate show 2.5 times higher I_{ON} as compared with that in Ref. 17 at the same L_G of 5 µm and the same $V_{GS} - V_{TH}$ of 1 V and V_{DS} of 1 V.

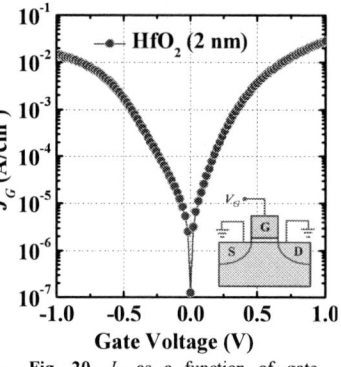

Fig. 20. J_G as a function of gate voltage for the Ge nFET on Si substrate with HfO_2 of 2 nm. The configuration for the measurement is also shown.

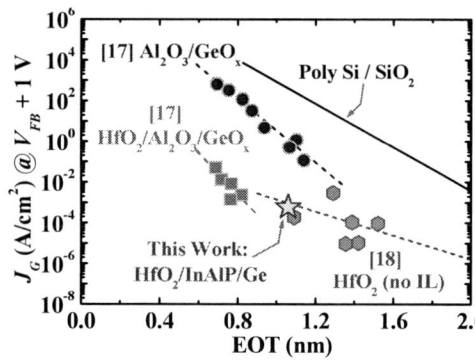

Fig. 21. J_G measured at $V_{FB} + 1$ V as a function of EOT. InAlP- capped Ge nFETs exhibit sufficiently suppressed gate current with EOT of 1.06 nm. This is attributable to the high-quality gate stack and maintaining of the physical thickness of gate dielectrics using HfO_2 with high permittivity.

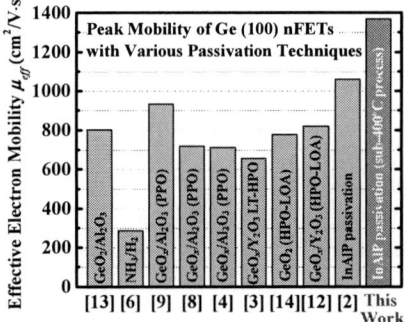

Fig. 22. Benchmark of peak μ_{eff} of Ge (100) nFETs using various passivation techniques in literature. Record high peak electron μ_{eff} of 1370 cm²/V·s was achieved for Ge (100) nFETs in this work. Note that μ_{eff} of Ge (111) is not included.

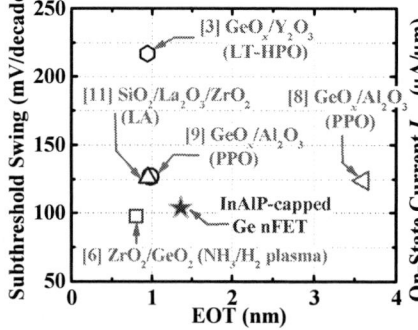

Fig. 23. S values as a function of EOT for long channel Ge nFETs with different surface passivation techniques. InAlP passivation leads to the realization of Ge nFETs with S comparable to the best reported values.

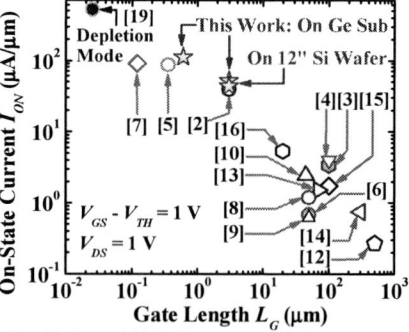

Fig. 24. Record high drive current of 127 µA/µm was achieved for enhancement mode Ge nFETs at $V_{GS} - V_{TH}$ of 1 V and V_{DS} of 1 V though the L_G is not the shortest. Note that Ref. 19 is a depletion-mode device.

978-1-4799-8002-4/14 $31.00 © 2014 IEEE

Ge n-channel FinFET with optimized gate stack and contacts

M.J.H. van Dal, B. Duriez, G. Vellianitis, G. Doornbos,
R. Oxland, M. Holland, A. Afzalian, Y.C. See, M. Passlack, C.H. Diaz[1]

TSMC, Kapeldreef 75, B-3001 Leuven, Belgium, [1]TSMC R&D Hsinchu 300-75 Taiwan, email: mark_van_dal@tsmc.com

Abstract

Whilst high performance p-channel Ge MOSFETs have been demonstrated [1-4], Ge n-channel MOSFET drive current has been lagging behind mainly hampered by high access resistance and poor gate stack passivation [5-9]. In this work, we address these issues on a module level and demonstrate Ge enhancement mode nMOS FinFETs fabricated on 300mm Si wafers implementing optimized gate stack ($D_{it} < 2 \times 10^{11}\,\mathrm{eV^{-1} \cdot cm^{-2}}$), n^+-doping ($N_d > 1 \times 10^{20}\,\mathrm{cm^{-3}}$) and metallization ($\rho_c = 1 \times 10^{-7}\,\Omega\mathrm{cm^2}$) modules. $L_G \sim 40$ nm devices achieved $I_{on} = 50\ \mu\mathrm{A}/\mu\mathrm{m}$ at $I_{off} = 100$ nA/um, $S \sim 124$ mV/dec, at $V_{DD} = 0.5$V. The same gate stack and contacts were deployed on planar devices for reference. Both FinFET and planar devices in this work achieved the highest reported g_m/S_{sat} at 0.5 V to date for Ge nMOS enhancement mode transistors to the best of our knowledge at shortest gate lengths.

Ge modules for NMOS

n-Ge doping - Using an n^+/p junction with P implantation, activation and diffusion was studied for different annealing techniques. Secondary Ion Mass Spectrometry (SIMS) data after activation are shown in Fig. 1. Activation of P in Ge can be achieved at low temperature [10] but we find active doping concentration N_{act} to not exceed 1×10^{19} cm^{-3} irrespective of annealing time. As suggested in [11], this may be related to implantation damage-induced acceptor formation resulting in lower P activation. At medium temperature, P-vacancy clusters become mobile leading to a typical box-like profile and N_{act} of mid-10^{19} cm^{-3} (Fig. 1). The optimized process achieves abrupt and high N_{act}

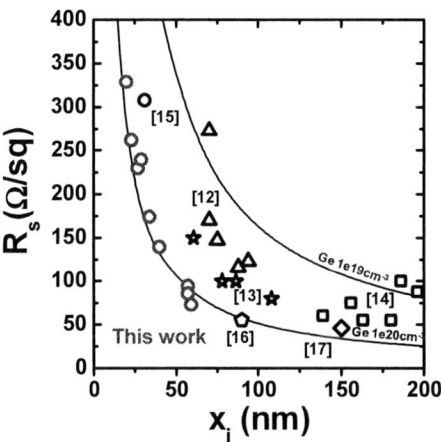

Figure 2: μ4PP R_s *vs.* x_j for optimized, scaled n-Ge junctions benchmarked with literature data [12-17]. Our data represent best R_s-x_j values to date following the $N_{act} = 1 \times 10^{20}$ cm^{-3} trend line down to 23 nm junction depth.

junctions yielding the best R_s-x_j envelope reported to date for Ge n^+/p junctions (Fig. 2) [12-17].

n-Ge metallization - As Me/Ge contacts show strong pinning close to the Ge valence band [18], high N_{act} at the metal/Ge interface is crucial to promote tunneling through the Schottky barrier and to achieve ohmic contacts. We investigated NiGe/n-Ge contacts by circular transmission line method (CTLM) structures. After implantation, anneal and patterning (248nm lithography), Ni was deposited and NiGe was formed. The unreacted Ni was removed by a selective wet etch. We developed a new CTLM contact resistivity (ρ_c) extraction model based on Reeves model [19] that captures the parasitic metal resistance R_{ME} contribution (*i.e.* NiGe sheet resistance). Fig. 3 shows extracted parameters. We found that for P-doped Ge, increased Ni thickness leads to

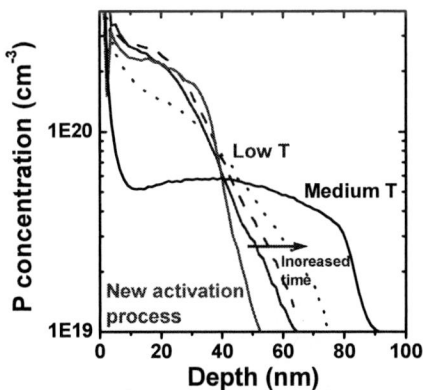

Figure 1: SIMS of phosphorus in Ge annealed using different annealing techniques. At low T we estimate 1×10^{19} cm^{-3} from R_s. The box-like profile at medium T is indicative of PV pair diffusion achieving N_{act} of mid-10^{19} cm^{-3}. Our new activation process provides high N_{act} and junction abruptness of ~ 4.5nm/dec.

CTLM: spacing 0.35 − 32 μm

Doping species	Ni	ρ_c ($\Omega \cdot \mathrm{cm^2}$)	R_{sc} (Ω/sq)	R_{me} (Ω/sq)
P	thin	5.0e-6	148	8.5
P	thick	8.8e-5	133	3.5
As	thin	1.1e-7	95	8.5
As	thick	1.9e-7	93	3.5
B	thin	5.4e-7	38	9

Figure 3: ρ_c and sheet resistance of n+ Ge, R_{sc} determined from CTLM using new extraction model (not shown). NiGe sheet resistance, R_{me}, used as input parameter, is based on μ4PP measurements. Results for P are in line with [20]. The lowest ρ_c achieved was $1.1 \times 10^{-7}\ \Omega\mathrm{cm^2}$.

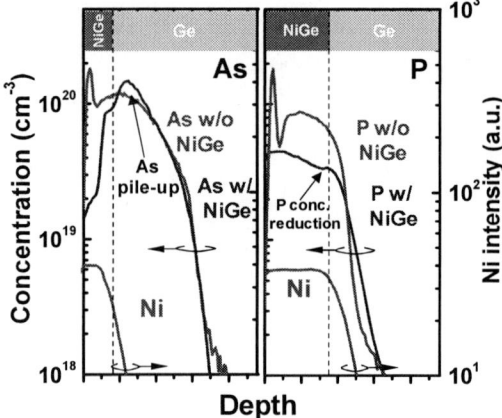

Figure 4: SIMS of As (left) and P (right) in Ge with or without NiGe formation. As piles up at the NiGe/Ge interface while the P concentration is reduced considerably.

higher ρ_c, while for As-doped Ge ρ_c remains low ($< 2\times10^{-7}$ Ωcm^2). The effect is largely due to higher dopant concentration at the NiGe/Ge interface as shown in Fig. 4. The lowest ρ_c achieved was 1.1×10^{-7} Ωcm^2 which compares well with lowest values reported [20-23].

Ge gate stack - Interfacial treatments, high-κ optimization and treatments of the gate stacks reported in [3] were deployed in this work aiming for a device design with sub 1nm EOT scaling capability. Multi-frequency C-V for a 100x100 μm² MOSCAP with EOT = 1.3 nm is shown in Fig. 5. The small frequency dispersion in depletion to accumulation region is indicative of a low density of electrically active traps at and close to the dielectric/Ge interface. Low-T C-V analysis was used to extract D_{it} across the Ge band-gap (Fig. 6); the Hi-Lo method gave similar D_{it}; there is no dependence on EOT (Fig. 7). $D_{it} \sim 2\times10^{11}$ $eV^{-1}cm^{-2}$ was found around mid-gap. The D_{it} levels achieved in this work are amongst the lowest reported on n-type Ge(100) and Ge(110) at scaled EOTs of 1-3 nm [24-25].

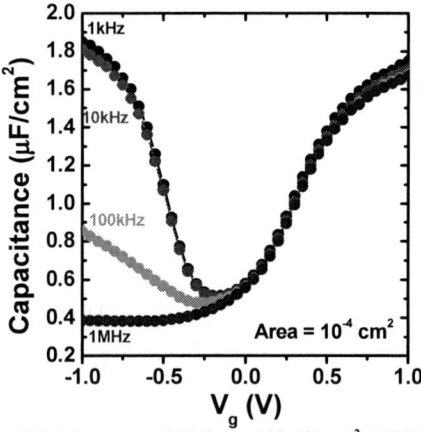

Figure 5: Multi-frequency C-V for a 100x100 μm² MOSCAP. The small frequency dispersion in depletion to accumulation region is indicative of a low density of electrically active traps at and close to the dielectric/Ge interface.

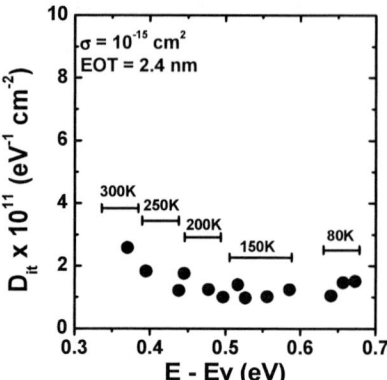

Figure 6: D_{it} *vs.* energy from conductance method at temperatures ranging from 80 to 300 K. D_{it} of 2×10^{11} $eV^{-1}cm^{-2}$ was found around midgap.

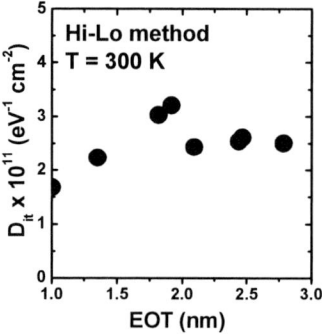

Figure 7: Midgap D_{it} dependence on EOT. D_{it} remains low at low EOT.

Ge enhancement mode NMOS Transistors

Enhancement mode n-channel Ge planar (001) FETs and (001)/⟨110⟩ FinFETs were fabricated using Ge-in-STI 300 mm Si wafers as in [1, 2] and the device process including aspect-ratio trapping as described in [1-3], respectively. Gate-to-channel capacitance (for several CETs) and C-V curves (for the same CETs) for the planar MOSFETs are shown in Fig. 8. Effect of annealing conditions for contacts on device performance becomes

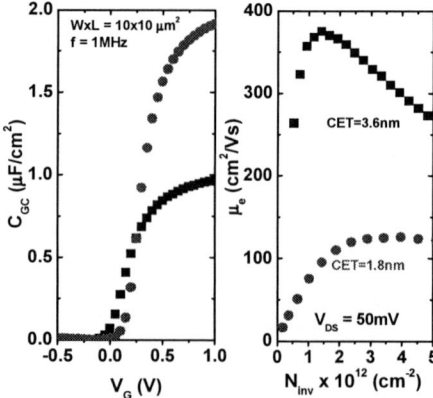

Figure 8: Gate-to-channel capacitance (for several CETs) and mobility curves (for the same CETs) for optimized gate dielectric (planar devices).

978-1-4799-8002-4/14 $31.00 © 2014 IEEE

apparent from Figs. 9-11. Significant improvement in g_m (Fig. 9) and short channel effect (SCE) control (Fig. 10) is observed when optimized contacts are used. R_{ext} was extracted as $500\,\Omega\cdot\mu m$ while for the reference devices $R_{ext} \sim 60\,k\Omega\mu m$ was estimated (Fig. 11). Further SCE immunity is achieved implementing the FinFET architecture (Fig. 10). Fig. 12 shows g_m and S as a function of L_g for the FinFETs. Sub-threshold swing of 76mV/dec at $V_{ds} = 50$ mV and $L_G = 60$ nm is achieved. Transfer and output characteristics of the best short-channel device are shown in Figs. 13-14. Our nGe planar and FinFETs are benchmarked in Fig. 15 and exhibit highest g_m/S_{sat} at shortest gate length and scaled EOT compared to published results for enhancement mode Ge nFETs on (100) or (111) surfaces [6-9].

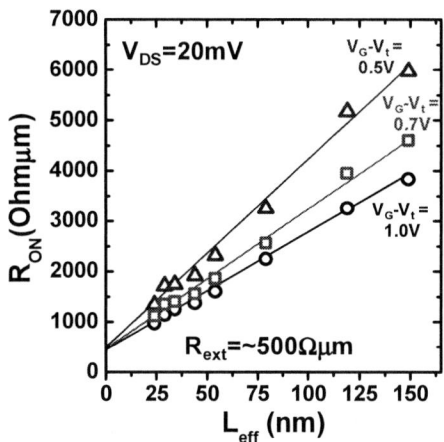

Figure 11: Average R_{on} ($\equiv V_{ds}/I_{ds}$ with $V_{gs} - V_t = 0.5$, 0.7, 1.0V and $V_{ds} = 20$mV) vs. effective gate length L_{eff} of Ge planar devices. R_{ext} was determined as $500\,\Omega\mu m$. For comparison R_{ext} of reference devices is $\sim 60k\Omega\mu m$.

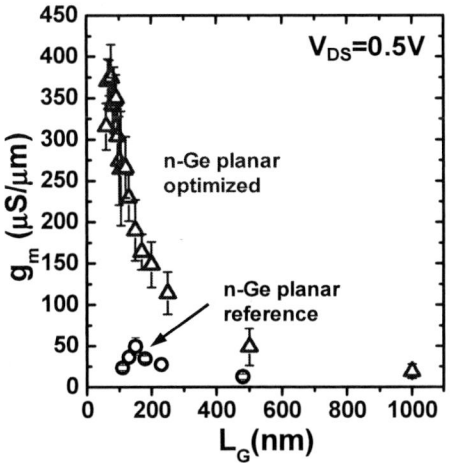

Figure 9: Peak-g_m at $V_{ds} = 0.5$ V as function of gate length for Ge n-channel planar FETs comparing reference and optimized devices.

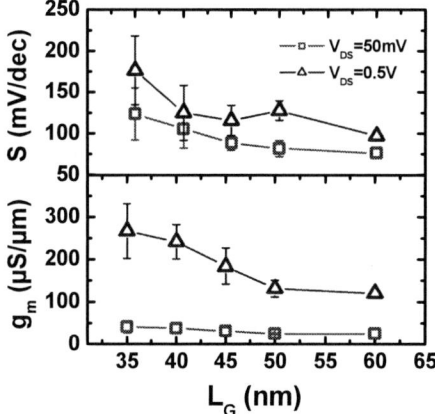

Figure 12: Sub-threshold swing and peak-g_m at $V_{ds} = 50$mV, 0.5 V as function of gate length for Ge n-channel FinFETs. At $L_G = 60$ nm, $S = 78$ mV/dec at $V_{ds} = 50$mV.

Conclusions

We report Ge n-channel FinFETs implementing optimized gate stack ($D_{it} < 2 \times 10^{11}$ eV^{-1}cm^{-2}), n$^+$-doping ($N_{act} > 1 \times 10^{20}$ cm^{-3}) and metallization ($\rho_c = 1 \times 10^{-7}\,\Omega cm^2$) modules. Both Ge nMOS FinFETs and planar transistors fabricated on (001) substrates in this work achieved the highest performance and g_m reported to date for Ge enhancement-mode NFETs with 0.5V V_{DD} and $I_{off} < 100$nA/um at the shortest gate-lengths (~ 40 nm on Fin-FETs, ~ 90nm on planar transistors) and lowest S_{sat} (~ 120mV/dec) to the best of our knowledge with sub 1nm EOT scaling capability.

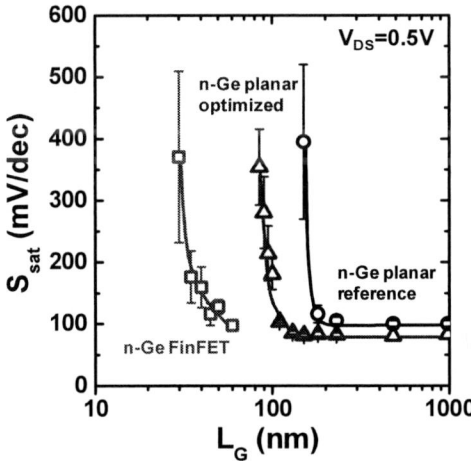

Fig. 10: Sub-threshold swing at $V_{ds} = 0.5$ V as a function of gate length for Ge n-channel planar FETs (reference and optimized) and FinFETs. Roll-up is improved when optimized contacts are implemented. FinFETs provide even better SCE control.

Acknowledgement - _The authors thank IMEC for processing capabilities and technical support, and Dr. Y. C. Sun for management support._

References

[1] M.J.H. van Dal *et al.*, *IEDM Tech. Dig.*, p. 521 (2012).
[2] M.J.H. van Dal *et al.*, *IEEE Trans. Electron Devices* 61(2) p. 430 (2014).
[3] B. Duriez *et al.* *IEDM Tech. Dig.*, p. 522 (2013).
[4] J. Mitard *et al.*, *VLSI Symp. Proc.*, p. 138 (2014).
[5] C.H. Lee *et al.*, *IEEE Trans. Electron Devices* 58(5) p. 1295 (2011).
[6] C.T. Chung *et al.*, *IEDM Tech. Dig*, p. 383 (2012).
[7] S.-H. Hsu *et al.*, *IEDM Tech. Dig*, p. 525 (2012).
[8] W.B. Chen *et al.*, *IEDM Tech. Dig*, p. 420 (2010).
[9] H. Wu *et al.*, *VLSI Symp. Proc.*, p. 96 (2014).
[10] R. Duffy *et al.*, *Appl. Phys. Lett.* 96, 231909 (2010).
[11] J. Kim *et al.*, *Appl. Phys. Lett.* 98, 082112 (2011).
[12] A. Satta *et al.*, *Appl. Phys. Lett.* 88, 162118 (2006).
[13] C.O. Chui *et al.*, *Appl. Phys. Lett.* 87 091909 (2005).
[14] M. Posselt *et al.*, *J. Vac. Sci. Technol. B* 26, 430 (2008).
[15] G. Hellings, Ph.D. thesis K. University of Leuven (2012).
[16] C. Wündisch *et al.*, *Appl. Phys. Lett.* 95, 252107 (2009).
[17] J. Kim *et al.*, *Appl. Phys. Lett.* 101, 112107 (2012).
[18] T. Nishimura *et al.*, *Appl. Phys. Lett.* 91 123123 (2007).
[19] G.K. Reeves, *Solid-State Electronics* 23, p 487 (1980).
[20] M. Shayesteh *et al.*, *IEEE Trans. Electron Devices* 60(7) p. 2178 (2013).
[21] G. Thareja *et al.*, *IEDM Tech. Dig*, p. 245, (2010).
[22] A. Firrincieli *et al.*, *Appl. Phys. Lett.* 99, 242104 (2011).
[23] P.P. Manik *et al.*, *Appl. Phys. Lett.* 101, 182105 (2012).
[24] C.H. Lee *et al.*, *VLSI Symp. Proc.*, p. 145 (2014).
[25] R. Zhang *et al.*, *IEEE Trans. Electron Devices* 60 p. 927 (2013).

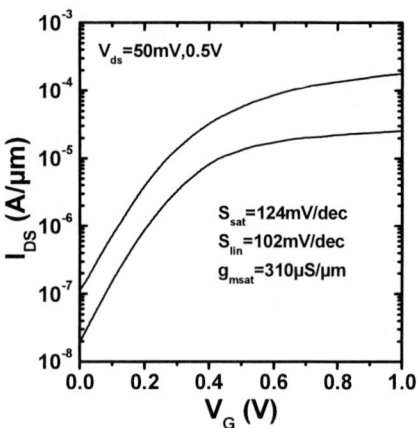

Figure 13: Linear and saturation (V_{ds} = 50 mV, 0.5 V) transfer characteristics of Ge n-channel FinFET (L_G = 40 nm). I_{on} = 50 μA/μm at I_{off} = 100 nA/μm (V_{DD} = 0.5 V)

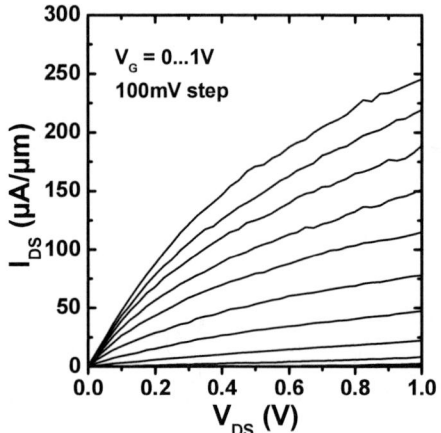

Figure 14: Output characteristics of Ge n-channel FinFET (L_G = 40 nm).

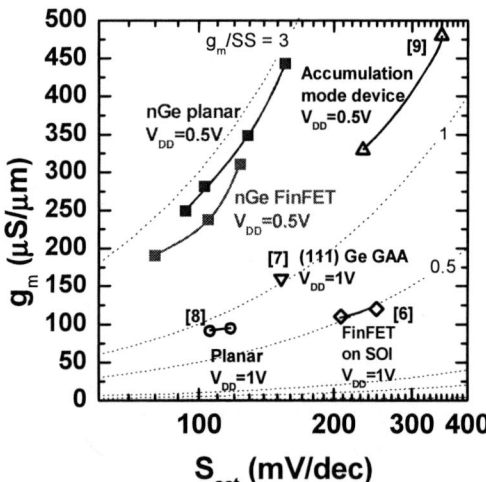

Figure 15: Extrinsic peak g_m *vs.* S_{sat} benchmark for Ge n-channel FETs [6-9], including our planar and FinFET data and accumulation mode nGe FET [9] at V_{DD}=0.5V and FinFET on SOI [6], Gate-All-Around (GAA) with (111) Ge side walls [7] and planar Ge nFET [8] all at V_{DD}=1V. Our data represent highest g_m/S for n-channel Ge FETs to date.

978-1-4799-8002-4/14 $31.00 © 2014 IEEE

In-situ Doped and Tensily Stained Ge Junctionless Gate-all-around nFETs on SOI Featuring I_{on} = 828 µA/µm, I_{on}/I_{off} ~ 1×10^5, DIBL= 16-54 mV/V, and 1.4X External Strain Enhancement

I-Hsieh Wong[1,2], Yen-Ting Chen[1], Shih-Hsien Huang[1], Wen-Hsien Tu[1], Yu-Sheng Chen[1], Tai-Cheng Shieh[1], Tzu-Yao Lin[1], Huang-Siang Lan[1], and C. W. Liu[1,2,*]

[1]Graduate Institute of Electronic Engineering and Department of Electrical Engineering,
National Taiwan University, Taipei, Taiwan,
[2]National Nano Device Laboratories, Hsinchu, Taiwan, *E-mail: chee@cc.ee.ntu.edu.tw

Abstract

In-situ CVD doping and laser anneal can reach [P] and tensile strain as high as $2x10^{20}$ cm^{-3} and 0.34%, respectively, in Ge on SOI with low defect density and high activation rate (nearly 100% near the surface), and enables high performance of the junctionless (JL) Ge gate-all-around (GAA) nFETs. The device with the W_{fin} of 13 nm, EOT of 10 nm, and nominal L_G of 280 nm has I_{on} = 350 µA/µm, I_{on}/I_{off} = $3×10^6$, SS = 185 mV/dec, and DIBL = 16 mV/V. The device with the W_{fin} of 9 nm and EOT of ~ 0.8 nm achieves the record high I_{on} of 828 µA/µm at V_{GS} - V_T = 1.5 V and V_{DS} = 2 V with DIBL = 54 mV/V, I_{on}/I_{off} = $1x10^5$ and SS = 150 mV/dec. Besides the epitaxial tensile strain (0.34%) generated by laser anneal due to the misfit of thermal expansion coefficients between Ge and Si, the enhanced tensile strain by the microbridge structure is also beneficial for I_{on}. The drain current enhancement of ~40% is achieved under the mechanical uniaxial tensile strain of ~0.25% due to sub-band splitting and carrier repopulation into the L4 valleys with the small conductive effective mass. The non-uniform shape of Ge channel with a minimum width at the center leads to enhanced I_{on} as compared to uniform channel. The extracted mobility of JL devices increases with increasing temperature, indicating the domination of impurity scattering. The threshold voltage of JL devices has the negative temperature coefficient and EOT scaling reduces the temperature dependence.

Introduction

Due to the conventional 6T SRAM design using nFET as the pass transistors, the higher I_{on} of nFET as compared to pFET is preferred. To reach manufacturability, the Ge nFETs are more favorable than III-V nFETs. The JL Ge nFETs is the solution that simplifies the S/D formation process [1-3] and improves the fast roll-off of mobility at high overdrive voltage. In this work, the JL-Ge GAAFETs were fabricated on *in-situ* doped epi-Ge layer on SOI with anisotropic etching to remove the defect near Ge/Si interface and to form the GAA structure [4]. The epitaxial and mechanical strains are also used to boost the performance.

Ge epitaxy and material analysis

The *in-situ* heavily-doped 105 nm Ge layer was grown on SOI by RTCVD with GeH$_4$ and PH$_3$. The laser anneal is used to remove the defects of epi-Ge and to generate tensile strain. The SiO$_2$ cap is deposited before laser annealing process to prevent dopant out-diffusion. Ge JL channel has intrinsic higher conductivity than Si [5, 6] (**Fig. 1**). The epi-Ge layer with laser anneal has doping concentration as high as $2x10^{20}$ cm^{-3}. Due to the channel controllability of the W_{fin} (down to 9 nm), the doping of $1.2x10^{19}$ cm^{-3} is used to turn off the channel. Note that, for the channel doping of $2x10^{20}$ cm^{-3}, the W_{fin} ~4 nm is required to turn off the channel, but it is beyond our process capability. The SIMS analysis shows the laser annealed epi-Ge (0.5 J/cm^2) has higher doping concentration than that of RTA and *in-situ* annealed samples with the abruptness of 2.6 nm/dec (**Fig. 2**). The high activation rate of ~100% near the surface is measured by the SIMS and SRP profiles (**Fig. 3**). The twins and dislocation defects of epi-Ge are significantly reduced after the laser annealing process (**Fig. 4**). The XRD of the epi-Ge on SOI after laser anneal shows FWHM of 0.22° and the tensile strain of 0.34% (**Fig. 5**). Note that our tensile strain much larger than the previous reported 0.196% [7] is due to the high temperature of laser anneal (melting temperature).

Ge Junctionless Gate-all-around nFETs
1. Device fabrication:
The process flow of the JL-Ge GAAFETs is shown in **Fig. 6**. After the epi growth on SOI, the Ge fins along [110] direction

978-1-4799-8002-4/14 $31.00 © 2014 IEEE   239

on (001) SOI are formed by the e-beam lithography and anisotropic etching. Due to the high etching rate at defective region near the Ge/Si interface, the Ge fin is separated from the SOI substrate to form the GAA structure (**Fig. 7**). The uniaxial tensile strain in the free standing Ge channel can be produced by the microbridge structure [8]. The SiO_2 layer of 400 nm is deposited by PECVD as field oxide and spacers. After the $Al/Al_2O_3/GeO_2$ gate stack formation by RTO and ALD, the contact and gate area are defined. For low EOT devices, the ZrO_2/Al_2O_3 are grown by ALD with *in-situ* TiN gate metal deposition to form the conformal gate stack around the GAA structure and is subsequently annealed in the forming gas. The cross-sectional TEM image viewed along channel direction is shown in **Fig. 8**. The devices with $Al/Al_2O_3/GeO_2$ and $TiN/ZrO_2/Al_2O_3$ gate stacks have the EOT of 10 nm and 0.8 nm, respectively, extracted from CV measurement of planar devices.

2. Electrical properties:

The I_{on}/I_{off} ratio of 3×10^6 and SS of ~ 185 mV/dec are obtained for the JL-Ge GAAFETs with W_{fin} of 13 nm, EOT = 10 nm and L_G = 280 nm (**Fig. 9**). The drain current at V_{GS} - V_T = V_{DS} = 2 V is 350 µA/µm (**Fig. 10**). **Fig. 11** shows the CV characteristic of $TiN/ZrO_2/Al_2O_3$ planar MISCAPs, which has the EOT of 0.8 nm and dispersion of 4%. For the W_{fin} = 9 nm devices, the improved SS of ~150 mV/dec, and I_{on}/I_{off} ratio of 1×10^5 are obtained with EOT = 0.8 nm (**Fig. 12**). The drain current at V_{GS} - V_T = 1.5 V and V_{DS} = 2 V is 828 µA/µm (**Fig. 13**). As compared to the previous work of Ge 3D and JL FETs, our device has higher drive current and better short channel control (**Table I**). The simulated I_d-V_g of the JL-Ge nGAAFETs with the non-uniform and uniform channel is shown in **Fig. 14**. Based on our device geometry (**Fig.7 (b)**), the non-uniform channel with W_{fin} of 9 nm at the center and 29 nm at the edge has 4.8X drain current of the uniform channel device (W_{fin} = 9 nm).

3. Mobility extraction and temperature dependence:

The effective mobility of JL Ge n- and p-GAAFETs are extracted by the channel resistance method and fitted using the Y-function method [12] (**Fig. 15**). To extract the mobility with GAA structure, the effective gate capacitance is obtained by planar device and simulation. The mobility shows less dependence on surface roughness at high overdrive region due to the nature of bulk conduction and low electric field. The electron mobility of JL devices increases with increasing temperature (**Fig. 16**) and is proportional to the 3/2 power of

temperature (**Fig. 17**), indicating that the mobility is dominated by impurity scattering. The temperature coefficient of V_T is ~-7 mV/K of device with 10 nm EOT is extracted (**Fig. 18**) and can be improved by scaling the EOT down to the 0.8 nm (-1 mV/K). The low EOT devices have smaller Q_{it}/C_{ox} leads to smaller V_T shift than 10 nm EOT devices.

4. Strain response:

The mechanical tensile strain is applied by wafer bending to further boost the current. Due to the suspended structure of GAAFETs and large size difference between channel and S/D, the strain is mainly generated in the channel, shown by simulation [13] (**Fig. 19**). The current have enhancement of 12% and 40% under uniaxial strain of 0.12% and 0.25 %, respectively, for W_{fin} =13 nm and EOT = 10 nm devices (**Fig. 20**). For the unstrained Ge with [110] channel direction, the carrier equally populated at L4 and L4' valleys (**Fig. 21**). As the uniaxial tensile strain increasing, the bandgap of L4 valleys decreases faster than that of L4' valleys (**Fig. 22**). The carriers then repopulate from L4' valley with m_c = 0.22 m_0 to L4 valley with smaller m_c = 0.08 m_0, and this leads to the current enhancement.

Conclusion

The superior Ge JL nFETs (8X of the INV devices) can be the CMOS solution using Ge channel for both n and pFETs. The increasing mobility with increasing temperature of Ge JL nGAAFETs can have higher drive current at higher operation temperature such as 85 ℃. The current of JL nFETs can be further boosted by applying proper strain.

Acknowledgments

This work is supported by Ministry of Science and Technology (No. 102-2622-E-002-014- and 102-2120-M-002-001-). The partial support of TSMC is also highly acknowledged.

References

[1] Che-Wei Chen et al., TED, 60.4, p.1334, 2013. [2] R. Rios et al., TED, 32, p.1170, 2011. [3] C.-W. Lee et al., APL, 94, 053511, 2009. [4] Shu-Han Hsu et al., Thin Solid Films 540, p.183, 2013. [5] D. B. Cuttriss et al., Bell Syst. Tech., p.509, 1961. [6] W. R. Thurber, Semiconductor measurement technology, 1981. [7] Jifeng Liu et al, Physical Review B, 70(15), 2004. [8] M. J. Süess et al., Nature Photonics, 7(6), p.466, 2013 [9] Cheng-Ting Chung et al., TED, 60.6, p.1878, 2013. [10] Shu-Han Hsu et al., IEDM, p.525, 2012. [11] Heng Wu et al., VLSI, 2014. [12] A. Cros et al., IEDM, 2006. [13] K.E. Moselund et al., TED, 57.4, p.866, 2010.

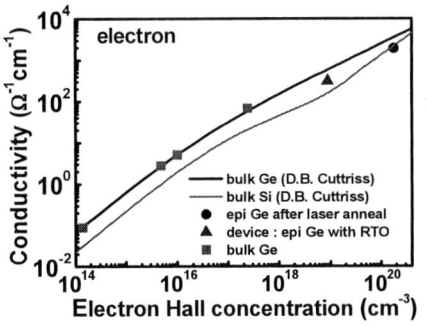

Fig. 1 Conductivity versus the Hall concentration. Ge has 3.7X conductivity of Si at concentration of 1.2×10^{19} cm^{-3}.

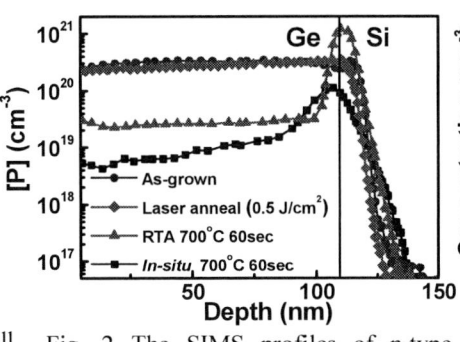

Fig. 2 The SIMS profiles of n-type epi-Ge layer on SOI. Laser anneal produces high [P] and a sharp profile.

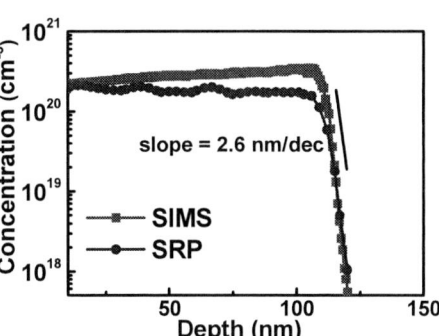

Fig. 3 The SIMS and SRP profiles of n-type epi-Ge layer after laser anneal. The activation rate reaches ~100 % near the surface and abruptness of phosphorus is 2.6 nm/dec.

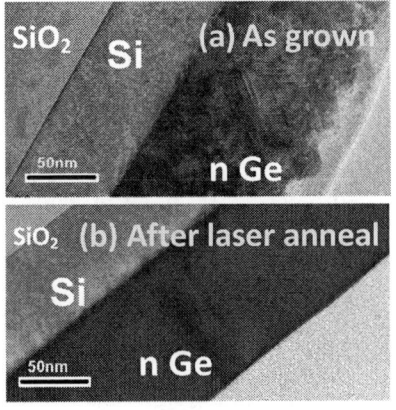

Fig. 4 TEM images of n-type epi-Ge layer (a) before and (b) after laser anneal. There are less defects in the laser annealed sample.

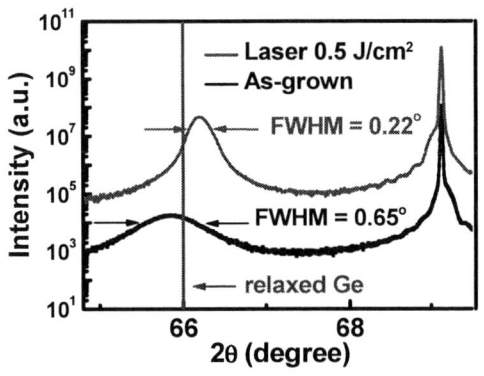

Fig. 5 The XRD profiles of epi-Ge layer. Laser anneal leads to small FWHM.

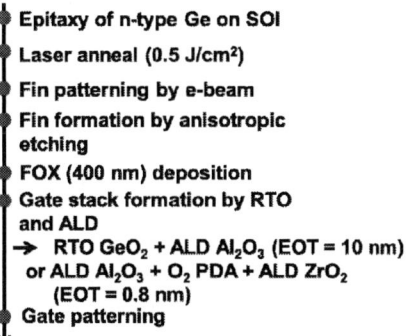

Fig. 6 The process flow of the JL-Ge nGAAFETs.

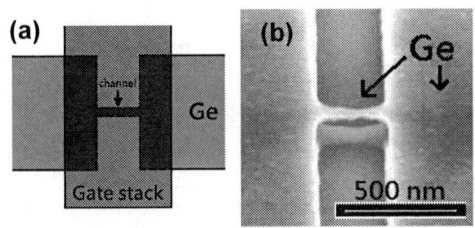

Fig. 7 (a) Top views and (b) tilted SEM image of the JL-Ge nGAAFETs. The GAA structure is formed by anisotropic selective etching.

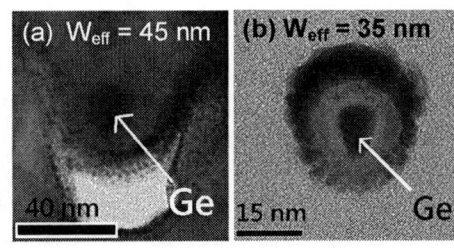

Fig. 8 Cross-sectional TEM images of the JL-Ge nGAAFETs viewed along width direction with (a) W$_{fin}$ = 13 nm/EOT = 10 nm and (b) W$_{fin}$ = 9 nm/EOT = 0.8 nm.

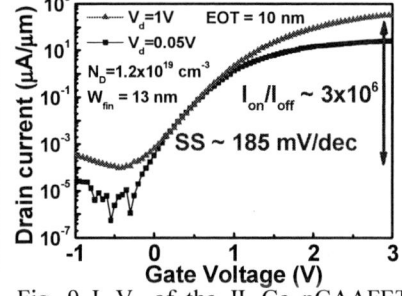

Fig. 9 I_d-V_g of the JL-Ge nGAAFET with I_{on}/I_{off} = 3×10^6 and SS = 185 mV/dec for the W$_{fin}$ = 13 nm/EOT = 10 nm device

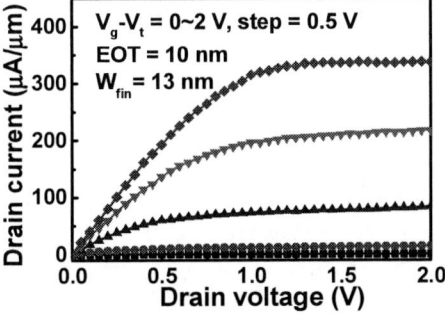

Fig. 10 Output characteristics of JL nGAAFET. I_{on} is ~350 µA/µm at V_{GS} - V_T = V_{DS} = 2 V with W$_{fin}$ = 13 nm and EOT = 10 nm.

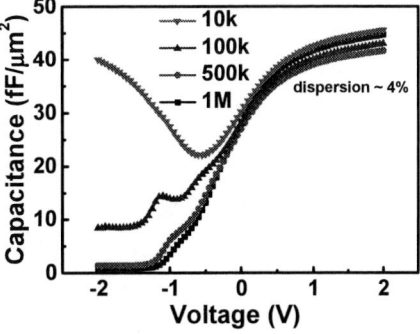

Fig. 11 C-V characteristics of the TiN/ZrO$_2$/Al$_2$O$_3$ MISCAPs with EOT ~ 0.8 nm.

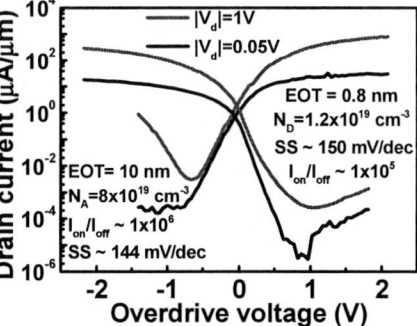

Fig. 12 I_d-V_g of the JL-Ge nGAAFET with W$_{fin}$ = 9 nm and EOT ~ 0.8 nm as well as the pFET for comparison. nFET has higher I_{on} than pFET

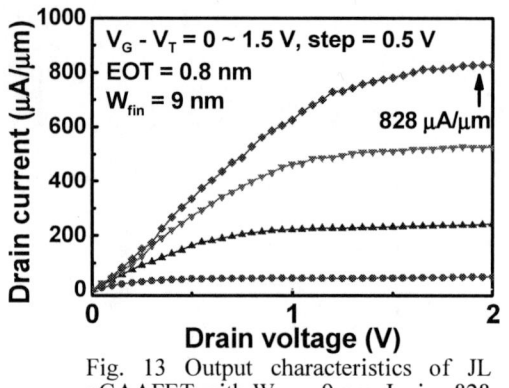

Fig. 13 Output characteristics of JL nGAAFET with W_{fin} = 9 nm. I_{on} is ~828 µA/µm at V_{GS} - V_T = 1.5V and V_{DS} = 2 V.

Table I Comparison of our JL-Ge nGAAFETs. (record data in red)

Structure	[1] JL nFinFET	[9] INV nFinFET	[10] INV nGAA	[11] JL n-V-groove	Our work JL pGAA	This work JL nGAA	
L_G (nm)	110	120	350	60	250	280	
EOT (nm)	4.9	3.6	5.5	5.2	10	10	0.8
SS (mV/dec)	110	144	94	220	144	185	150
DIBL (mV/V)	89		100	780	95	16	54
$g_{m.max}$@1V (µS/µm)		170	90	590	180	266	650
I_{on} (µA/µm)	102	120	110	714	271	350	828
@V_{ov}(V)	2	0.9	1.5	1.5	2	2	1.5
@V_D(V)	2	1	1	1.5	1	2	2

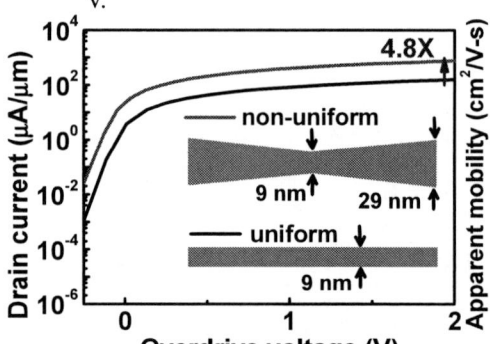

Fig. 14 Simulated I_d-V_g of the JL nGAAFET with the non-uniform (based on SEM image) and uniform channel. The I_{on} is enhanced by ~ 4.8X.

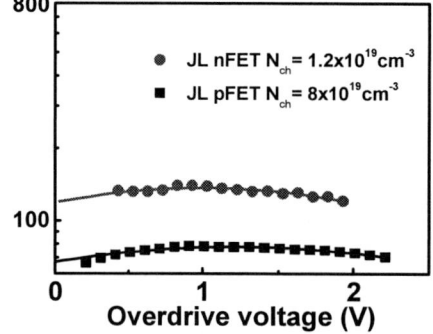

Fig. 15 Apparent mobility vs overdrive voltage (V_{ov}). The curves are fitted with the Y-function method.

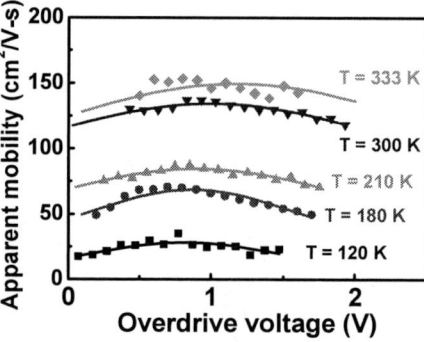

Fig. 16 Temperature dependence of apparent mobility of JL nGAAFETs. The mobility increases with the increasing temperature.

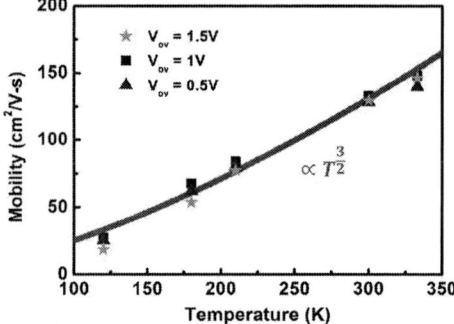

Fig. 17 Apparent mobility vs temperature. The mobility is proportional to the $T^{3/2}$, indicating impurity scattering is dominant.

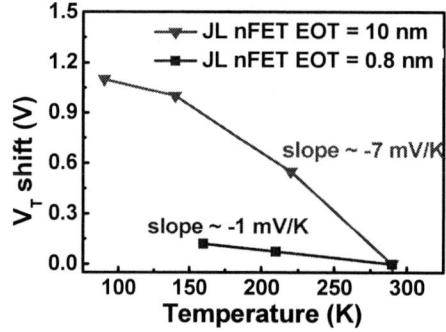

Fig. 18 V_T shift vs temperature. The V_T shift is reduced with the EOT scaling.

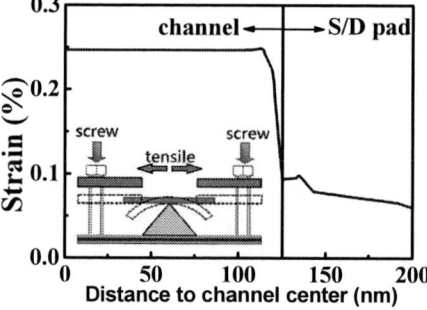

Fig. 19 Simulated strain profiles by wafer bending along the channel direction of GAA structure. The inset shows the stress mechanism.

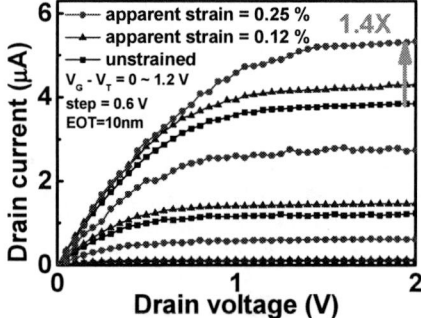

Fig. 20 Output characteristics of JL nGAAFET under mechanical uniaxial [110] tensile strain. 1.4X enhancement is achieved under 0.25% uniaxial tensile strain

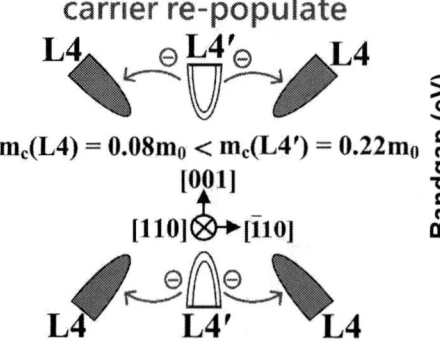

Fig. 21 The schematic valley distribution along with channel direction [110]. Carrier re-populates from L4' valleys to L4 valleys (small conductivity mass) under uniaxial tensile strain.

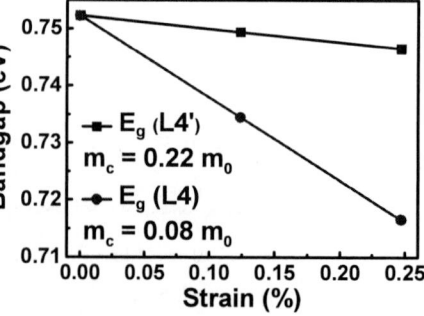

Fig. 22 The bandgap of L4 and L4' valleys under tensile strain. The L4 valleys with small conductivity mass have increasing electron population with increasing strain.

978-1-4799-8002-4/14 $31.00 © 2014 IEEE

Experimental Realization of Complementary p- and n- Tunnel FinFETs with Subthreshold Slopes of less than 60 mV/decade and Very Low (pA/μm) Off-Current on a Si CMOS Platform

Y. Morita, T. Mori, K. Fukuda, W. Mizubayashi, S. Migita, T. Matsukawa,
K. Endo, S. O'uchi, Y. Liu, M. Masahara, H. Ota

National Institute of Advanced Industrial Science and Technology (AIST)
Tsukuba, Ibaraki 305-8569, Japan.
E-mail: y.morita@aist.go.jp

Abstract

Complementary (p- and n-type) tunnel FinFETs operating with subthreshold slopes (SSs) of less than 60 mV/decade and very low off-currents (on the order of a few pA/μm) have been experimentally realized on the Si CMOS platform. Improvements in the SSs have been realized by optimizing epitaxial channel growth on heavily arsenic- and boron-doped source surfaces for purging interface defects at the epitaxial tunnel junctions. By improving the interface quality, SSs of 58 and 56 mV/decade and on/off current ratios (ON/OFF) of 2×10^6 and 3×10^4 (with $V_D = |0.2|$ V) were respectively obtained for p- and n- tunnel FETs (TFETs) simultaneously.

Introduction

In this study, we fabricate Si CMOS TFETs suitable for massive sensor-network applications. Because the TFET utilizes band-to-band tunneling (BTBT) of semiconductors as an operation principle, a steeper SS than that found in conventional metal oxide semiconductor FETs is possible. [1,2] In particular, for massive social devices, a steep SS value (less than 60 mV/decade) is a primary consideration, together with a low off-current and high cost effectiveness.

The most serious drawback of the Si TFET is that it has a smaller drain current than that of Ge or III-V TFETs. [3-7] However, a small drain current is advantageous for low off-current operation. Moreover, the cost effectiveness of Si is a well-known fact. Here, we choose a tunnel FinFET architecture; the scaled fin channel can effectively suppress the SS. In addition, we adopt an ultrathin epitaxial channel with a tri-gate configuration, which can enhance the ON current. [8,9] The optimization of the epitaxial growth condition is another key technology for the present tunnel FinFET, as it enables the purging of interface defects.

In this study, by adopting these technologies, SSs of 58 and 56 mV/decade and on/off current ratios (ON/OFF) of 2×10^6 and 3×10^4 (with $V_D = |0.2|$ V), respectively, are obtained

simultaneously for p- and n-TFETs with very low off-currents (on the order of pA/μm). Sub-60 mV/decade SS and large ON/OFF ratios make Si CMOS TFETs suitable for low operating-power and low cost sensor-network applications.

Device design

The tunnel FinFET in this study consists of an ultrathin undoped epitaxial channel, heavily-doped source layers, and a tri-gate configuration (Fig. 1b). Outcroppings of channel/source interface along the fin-sidewalls are covered by gate electrodes. Fig. 2 shows the TCAD-simulated cross-sectional potential distribution in the channel. [10,11] The combination of the side and top gates produces a stronger potential gradient at the channel edge. At the channel/source interface, the vertical BTBT is initiated by both the top and side gate electric fields. The electric fields are intensified by the cooperative top and side fields to effectively enhance the BTBT. This is the "synthetic electric field" (SE) effect that we proposed in a previous study. [9] Fig. 3 shows cross-sectional transition electron microscopy images of the typical tunnel FinFET with epitaxial channel. Secondary ion mass spectroscopy (SIMS) analysis shows very high dopant concentrations and steep dopant gradients (1 nm/decade) at the vertical tunnel junction for both the arsenic- and boron-doped source surfaces (Fig. 4). Fig. 5 shows the effect of an additional epitaxial channel on the performance of the TFET I_D–V_G. In comparison to the conventional 3-D tri-gate TFETs, the addition of the epitaxial channel drastically enhances the I_D value. In addition, the penalty of off-current increase is very limited, which is a promising result of the enhanced on-current (I_{ON}) and preserved off-current (I_{OFF}) in this experiment.

Process Optimization

Figs. 6 and 7 show the process flow of the present tunnel FinFET. [9,12] Surface cleaning prior to epitaxial growth is

978-1-4799-8002-4/14 $31.00 © 2014 IEEE 243

represented by "pre-epitaxial cleaning," which is a key step in the process, consisting of multiple cleaning sub-steps (Fig. 7h). To improve the interface quality for both n^{++} and p^{++} Si, chemical-oxidation using an H_2SO_4:H_2O_2 mixture and HF-etching cycle was repeatedly performed as a form of pre-epitaxial cleaning. The effects of the surface pre-cleaning were evaluated using X-ray photoelectron spectroscopy (XPS) analysis. On the heavily doped Si surface, the etchabilities of the Si oxide layers are modified by the implanted species (Fig. 8). [13] After surface cleaning with a 5-min dip in 1% HF solution, the n^{++} surface becomes hydrophobic, but the p^{++} surface remains hydrophilic (as is well-known). Generally, a hydrophobic surface is obtained for hydrogen-terminated Si without oxide. However, the oxide on the p^{++} surface is rather thinner than that on n^{++} surface, which is the complete opposite of the features of undoped surfaces. [14] With additional oxidation, the etchabilites of the oxide layers are improved. The oxide layer thicknesses decrease step-by-step with additional oxidation/etching cycles (Fig. 9). The cleaning cycle is directly responsible for improvements in the epitaxial quality (Fig. 10). By increasing the number of cleaning cycles to N = 5, high-quality epi-layers on the n^{++} and p^{++} Si surfaces can be successfully realized.

Discussion

Increasing the number of cleaning cycles improves the steepnesses and I_{ON}'s of the I_D–V_G curves of the p- and n-TFETs (Fig. 11). Even after five cycles of cleaning, SIMS analysis shows that there are still very high dopant concentrations and steep dopant gradients for arsenic and boron (Fig. 4). For five cleaning cycles, the SS_{MIN} value of the TFET decreases to 54 and 55 mV/decade for the p- and n-polarities, respectively (Fig. 11c). Increasing the number of cleaning cycles from N = 5 to 7, improves performance further (Fig. 11d). Fig. 12 shows a relationship between the ON/OFF ratio and cleaning cycle. 2×10^6 and 3×10^4 ON/OFF (V_D = |0.2| V) for p- and n-TFETs, respectively, are obtained for N = 7. In TEM measurements, interfaces seem to be clear after five cleaning cycles, but the TFET performances are further enhanced. This means that the BTBT performance itself can be regarded as an index of defects at the tunnel junction. Moreover, especially for the n-TFET, it is clear that the V_{TH} shifts toward a smaller V_G magnitude. It is a well-known fact that the wettability of the heavily B-doped Si surface after HF treatment is affected by B concentration. Fig. 13 shows the relationship between the approximate time for the Si surface to become hydrophobic through HF etching and the number of cleaning cycles. Generally, the hydrophobic nature of the Si surface reflects surface oxide removal. However, this trend disagrees with the measured oxide thickness As shown in Fig. 9, surface oxide layer for B-implanted surface is removed over three times of cleaning cycles. However, it takes longer to obtain hydrophobic surface by HF etching for three, four, and five times of N cases. It strongly suggests that surface charged layers or potential modulation exists on the heavily B-doped Si surface and initiates V_{TH} modulation, which could be a future issue for Si CMOS TFETs with epitaxial tunnel junctions. By realizing V_{TH} reduction on the n-TFET, symmetric CMOS operation with an SS less than 60 mV/decade could be expected.

Summary

Complementary (p- and n-) tunnel FinFETs operating with a subthreshold slope (SS) of sub-60 mV/decade and very low off-currents (on the order of pA/μm) have been experimentally demonstrated on the Si CMOS platform. Improvement of the SS and off-current has been realized by purging interface defects at the tunnel junctions for optimizing a epitaxial channel growth on heavily arsenic- and boron-doped source surfaces. By improving the interface quality, SSs of 58 and 56 mV/decade, and on/off current ratios of 2×10^6 and 3×10^4 (with V_D = |0.2| V) have been obtained for p- and n-TFETs, respectively. These performances were realized on a conventional Si CMOS platform, without special booster technologies. Very low off-current, Sub-60 mV/decade SS, and large ON/OFF ratios in the Si CMOS TFET make it suitable for low operating-power and low cost sensor-network applications.

Acknowledgement
The work was partly supported by a grant by JSPS through the First Program initiated by CSTP.

References
[1] A.C. Seabaugh, and Z. Qin, Proceedings of the IEEE 98 (2010) 2095.
[2] A.M. Ionescu, et al., IEDM Tech. Dig. (2011) 16.1.1.
[3] G. Dewey, et al., IEDM Tech. Dig. (2011) 785.
[4] G. Zhou, et al., IEDM Tech. Dig. (2012) 32.6.1.
[5] L. Knoll, et al., IEDM Tech. Dig. (2013) 4.4.1.
[6] T. Mori, et al., Symp. VLSI Tech. Dig. (2014) 86.
[7] A. Villalon, et al., Symp. VLSI Tech. Dig. (2014) 84.
[8] Y. Morita, et al., Electron Device Lett. 35 (2014) 792.
[9] Y. Morita, et al., Symp. VLSI Tech. Dig. (2013) T236.
[10] HyENEXSS ver. 5.5.
[11] K. Fukuda, et al., Proceedings of SISPAD (2012) 284.
[12] Y. Morita, et al., Jpn. J. Appl. Phys. 52 (2013) 04CC25.
[13] Y. Morita, et al., Proceedings of IEEE ESSDERC (2014).
[14] F. Yano, et al., J. Vac. Sci. Thechnol. B 14 (1996) 2707.

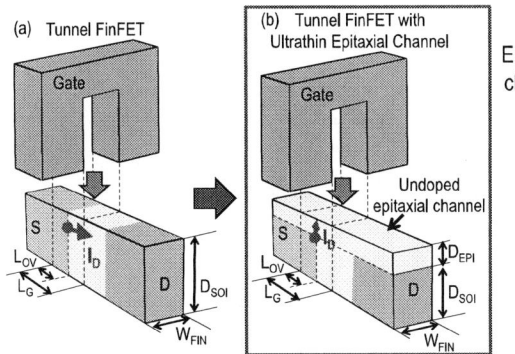

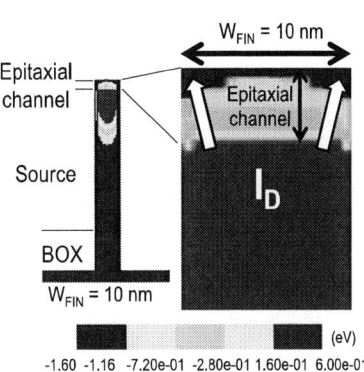

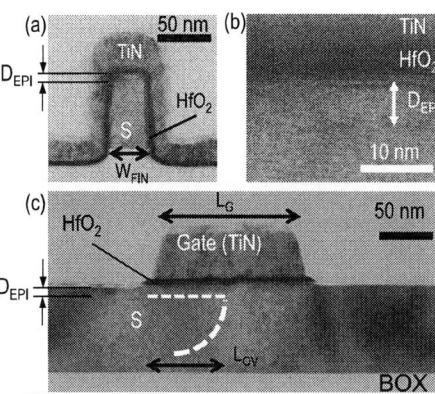

Fig. 1 Device structure of (a) conventional tunnel FinFET and (b) Tunnel FinFET with ultrathin undoped epitaxial channel used in this study. The epi-layers on the contact areas are removed before contact formation.

Fig. 2 TCAD simulation of potential distributions in the ultrathin epitaxial channel surrounded by the gate electrode. Distributions at the slice in the channel-cross sections are indicated. Strong potential modulation by an SE effect is initiated in the edge region.

Fig. 3 X-TEM images of the typical tunnel FinFET with ultrathin epitaxial channel. (a) Channel cross-section, (b) Magnified epitaxial channel interface, and (c) Gate cross-section.

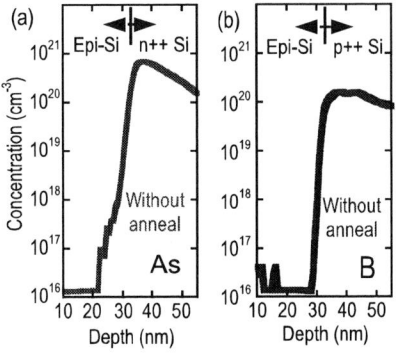

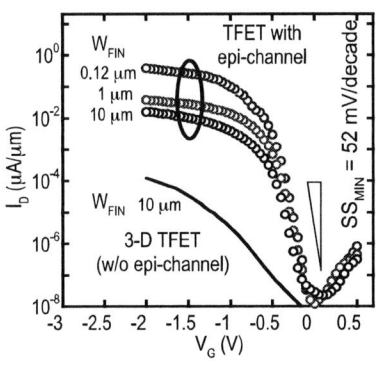

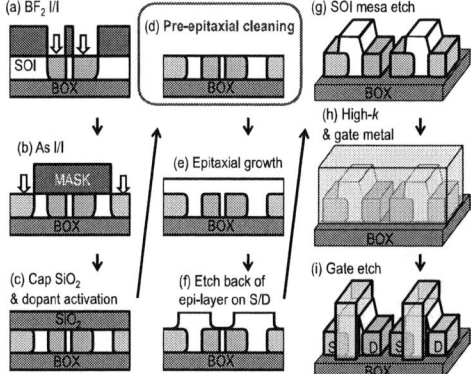

Fig. 4 Front-side SIMS analysis for undoped epi-Si/source interfaces. The number of cleaning cycles is N = 5. (See text) Very steep (1 nm/decade) gradients of the dopant concentration are realized.

Fig. 5 Measured I_D–V_G (V_D = −0.05 V) characteristics for conventional 3-D TFET without epitaxial channel (Fig. 1a) and TFET with epitaxial channel (Fig. 1b). Epitaxial channel thickness is 10 nm. (L_{OV} = 400 nm) It is clear that the fin width scaling enhances the I_D of TFETs with epitaxial channel.

Fig. 6 Process flow of CMOS-compatible tunnel FinFETs used in this study. The dopant concentration in the n++ (As) and p++ (B) source surface region is 2×10^{21} and 2×10^{20} cm^{-3}, respectively. Surface cleaning prior to the epitaxial growth is represented by "pre-epitaxial cleaning," which consists of multiple cleaning steps.

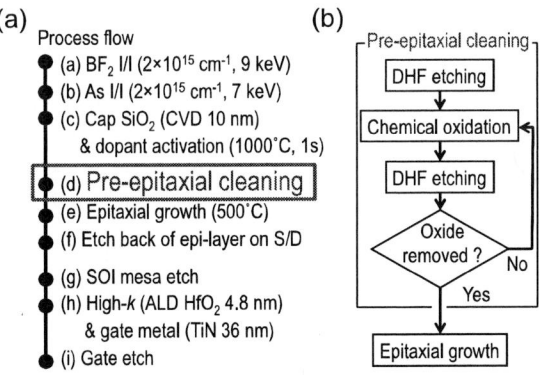

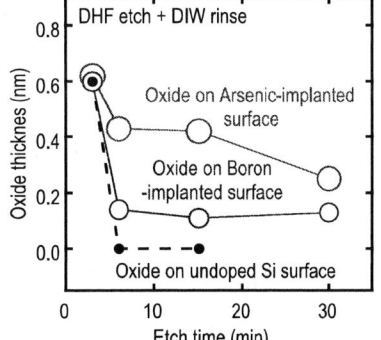

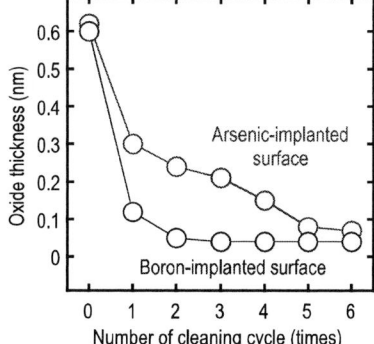

Fig. 7 (a) Process flow of tunnel FinFETs used in this study. (b) Process flow of the "pre-epitaxial cleaning," which consists of oxidation/etching sequential cleaning. N represents repeated numbers of additional oxidation/etching cycles.

Fig. 8 Oxide thickness vs. etching time by DHF. The oxide thicknesses were evaluated by XPS measurement.

Fig. 9 Oxide thickness vs. number of cleaning cycles. The oxide thicknesses were evaluated by XPS measurement.

978-1-4799-8002-4/14 $31.00 © 2014 IEEE

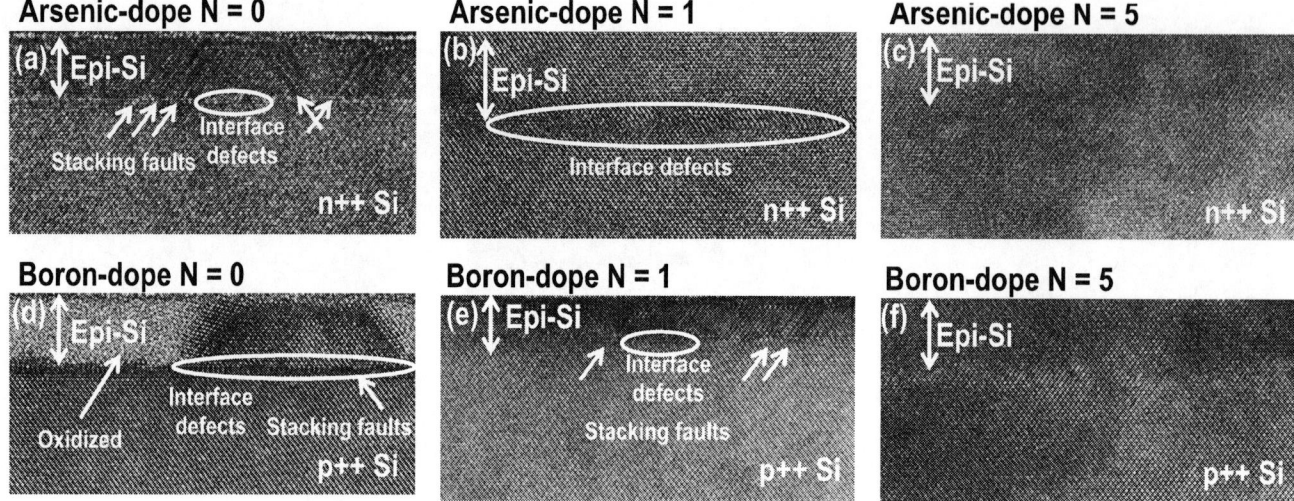

Fig. 10 X-TEM images (30×15 nm^2) of epi-Si interfaces prepared by the cyclic cleaning before epitaxial growth. In (a) and (d) (N = 0), a interface-quality is poor. The cleaning cycle is directly responsible for improvements in epitaxial quality. By adding a single cleaning cycle (N = 1), the epi-layer on n++ Si is drastically improved as compared to N = 0, but a small amount of defects still exist at the interface (b). As for the epi-layer on p++ Si (e), oxidation of the channel is suppressed. But a significant amount of stacking faults or defects still exists. By increasing the number of cleaning cycles (N = 5) (c and f), high-quality epi-layers both on the n++ and p++ Si can be successfully realized.

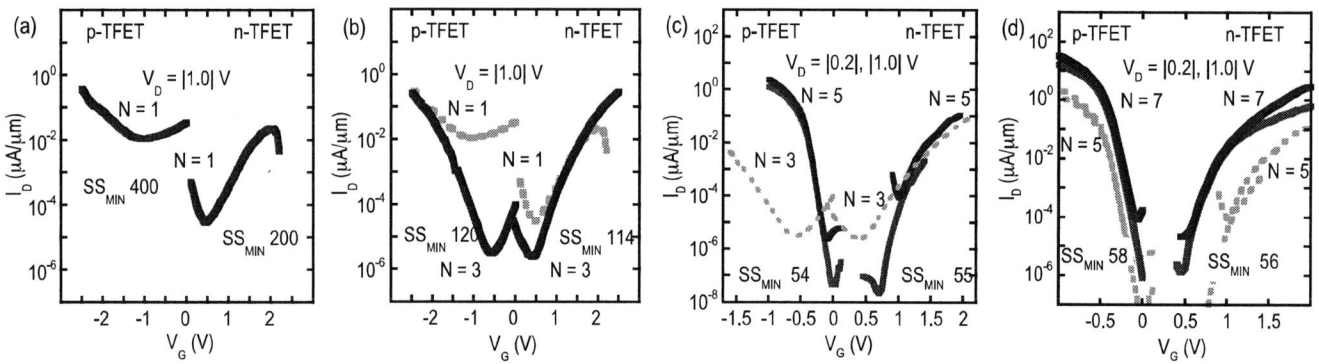

Fig. 11 I_D–V_G of p- and n-TFETs prepared by cyclic cleaning before epitaxial growth. Grey dashed curves represent I_D–V_G of fewer cleaning cycles case. L_G = 1000 nm for (a) and (b), and 100 nm for (c) and (d). W_{FIN} = 10 μm for (a) and (b), and 20-30 nm for (c) and (d). In (d), SS_{MIN} values are 58 and 56 for p- and n- polarity, respectively.

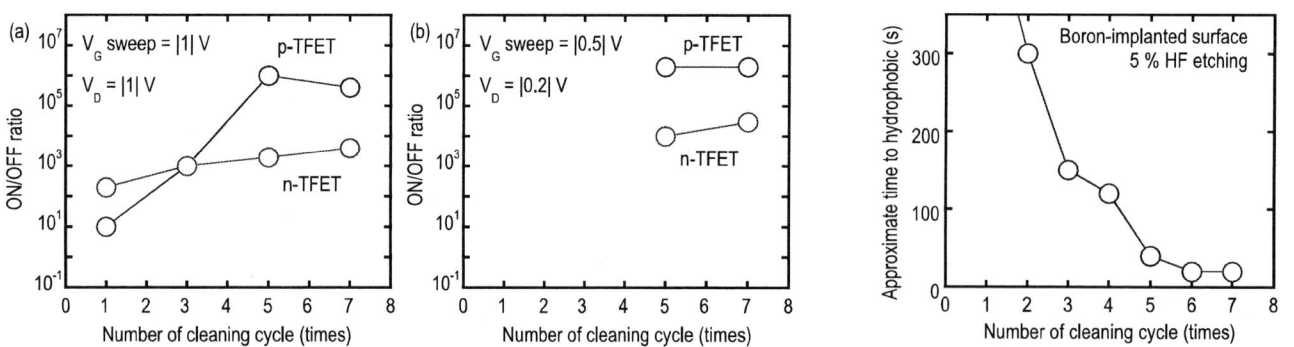

Fig. 12 ON/OFF current ratios of p- and n-TFETs prepared by cyclic cleaning before epitaxial growth. In (b), ON/OFF ratios of 2×10^6 and 3×10^4 (with $V_D = |0.2|$ V) for N = 7 have been obtained for p- and n-TFETs, respectively.

Fig. 13 Relationship between approximate time to obtain hydrophobic surface by HF treatment and number of cleaning cycles. Surface oxide layer for B-implanted surface is removed over three times of cleaning cycles. (Fig. 9) However, it takes longer to obtain hydrophobic surface by HF etching for three, four, and five times of N cases.

978-1-4799-8002-4/14 $31.00 © 2014 IEEE

Jot Devices and the Quanta Image Sensor

Jiaju Ma, Donald Hondongwa and Eric R. Fossum

Thayer School of Engineering at Dartmouth
14 Engineering Drive, Hanover, NH 03755 USA

Abstract

The Quanta Image Sensor (QIS) concept and recent work on its associated jot device are discussed. A bipolar jot and a pump-gate jot are described. Both have been modelled in TCAD. As simulated, the pump-gate jot has a full well of 200e- and conversion gain of 480uV/e-.

Introduction:

The Quanta Image Sensor (QIS) is a possible third generation solid-state image sensor concept that seeks to take advantage of shrinking pixel sizes. Pixel shrink, good for smaller cameras or larger format sensors, normally suffers from diminishing full well capacity and concomitant SNR and image quality deterioration. The QIS represents a different paradigm in image capture where small full well is overcome by digital integration with faster readout, and power dissipation offset by reduced ADC bit depth (1). The ultimate single-bit QIS has an effective full well of one electron and a single-bit ADC.

The commercial realization of the QIS requires overcoming many significant technical challenges. It is an oversampled image sensor in both space and time, using both sub-diffraction limit (SDL) pixel sizes (e.g. <900nm) and very high field rates (e.g. 1000fps). In addition to the SDL pitch, the photodetectors have small full-well capacity, typically less than 200e-, and high conversion gain (e.g., 1mV/e-) so that each photoelectron can be counted with low bit-error rate. In the single-bit QIS, each output is binary in nature and an output image pixel is composed from many such SDL photoelement outputs across bit planes from several (e.g., 16) fields as illustrated in Fig. 1. Thus photoelements are given a special name, "jot", meaning smallest thing. The QIS is intended to have of the order of 0.1-10.0 Gjots leading to output data rates of 0.1-10 Tb/s.

The potential advantages of the QIS over current CMOS image sensors include low-light sensitivity, adjustable tradeoff in resolution vs. sensitivity even after image capture, noiseless time-delay and integration (TDI) along an arbitrary track, also after image capture, and a film-like

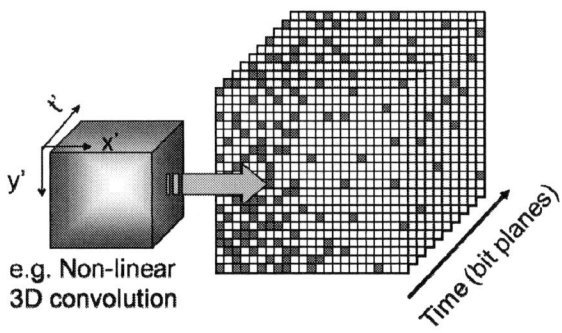

Figure 1. Conceptual schematic for creating pixels from QIS jots.

D-log(H) exposure characteristic that yields intrinsic overexposure latitude as shown in Fig. 2 (2). Signal, noise and exposure-referred SNR (SNRH) is shown in Fig. 3 for an example summation of a 16x16x16 (X,Y, t) "cubicle" of jots, used to create a single pixel. Combining fields with different shutter settings allows high dynamic range without sudden SNR dips. By moving the digital domain right up against photoelectron capture, the process of image formation is altered from having a physically defined pixel on the sensor, to permitting flexible, programmable pixel creation in the digital domain following signal capture.

The small jot size and the fast field-readout rate is more about what we term "flux capacity" than about spatial or

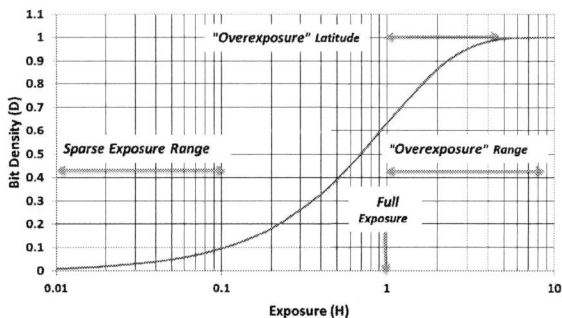

Figure 2. Bit density vs. exposure for QIS showing D-log(H) film-like characteristic. Bit density is effectively the probability that a jot has captured a photoelectron for a given quanta exposure H.

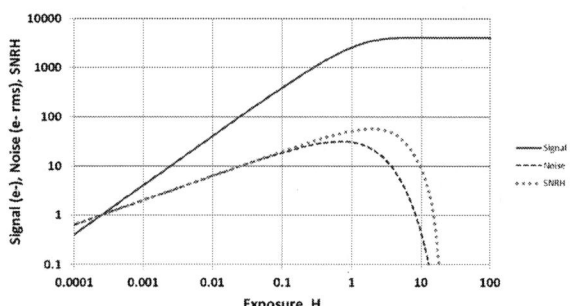

Figure 3. Signal, noise, and exposure-referred SNR as a function of exposure H for an (X,Y,t) cubicle of 16x16x16 = 4096 jots.

temporal resolution in the QIS, although higher resolution comes along as part of the nature of the device. The nominal "full well" photon flux ϕ_{wn} that can be handled by the QIS can be written as

$$\phi_{wn} = jf_r(2^n - 1)/\sigma\bar{\gamma} \qquad (1)$$

where j is the jot density, f_r is the field readout rate, n is the ADC bit depth ($n=1$ for single-bit QIS), σ is the electronic shutter duty cycle, and $\bar{\gamma}$ is the average conversion efficiency of incident photons to collected photoelectrons. For consumer applications, the flux capacity of the QIS should be similar to that of CMOS image sensors and this drives high values for jf_r. Multi-bit QIS devices are also being considered for increasing flux capacity, with $1 < n \lesssim 6$. Very bright light and flash photography remain possible issues. For other applications, such as low-light vision, the jot density can be relaxed.

Readout of the QIS is architecturally similar to that of conventional CMOS image sensors, with column-parallel readout consisting of sense amplifier and ADC. We have recently reported some readout circuit progress (3). We have since demonstrated about 5pJ/b for the sense amp and 1b ADC in a 1Mpix binary sensor with 3.6um pixel pitch operating at 1000fps (1Gb/s) in 0.18um CMOS that will be reported separately. Realization of a consumer gigajot sensor requires an order of magnitude reduction in energy/bit that will be accomplished by supply rail scaling, technology node scaling, and improved circuit design. It is expected that a gigajot sensor (e.g., 42k x 24k) would be best implemented in a sub-65nm process with $\lesssim$500nm jot pitch.

We have been making steady progress on several fronts – low-power bit-plane readout, the jot device itself, algorithms for output image creation, and understanding the general imaging characteristics of the QIS. In the remainder of this paper, we will concentrate on the electron device element at the heart of the QIS, the jot device.

Jot Devices

The jot device has many requirements in common with a modern CMOS image sensor (CIS) pixel. It must have high quantum efficiency (including fill factor), good photoelectron collection efficiency, and low dark current. In addition, it must have low readout noise and essentially no lag. Readout noise in state-of-the-art (SOA) CIS is of the order 100-200uV r.m.s. yielding input-referred read noise of a few electrons r.m.s. To get above the read noise for a single photoelectron, one needs at least 200uV/e- conversion gain, and for low bit-error rate (one photoelectron is one bit in the QIS) the read noise must be below about 0.15e- rms (2) suggesting a conversion gain of about 1mV/e- (0.16fF). This is perhaps 5x higher conversion gain than found in SOA CIS devices. However, compared to CIS devices, the full-well capacity (FWC) requirement is greatly relaxed (e.g., from at least 3000e- to less than 200e-). FWC is a major difficulty for CIS pixel shrink today and one of the motivating factors for exploring the QIS paradigm. Crosstalk between adjacent jots of the same color is not as critical as in SOA CIS because we are already below SDL pitch. Several jots may be under a single color filter and microlens.

In our work we consider only backside-illuminated (BSI) devices. Four basic approaches have been considered thus far. First is shrinking SPAD devices to SDL pitches. This seems a more distant horizon at this time. However, a low-density SPAD-based-jot QIS has been recently demonstrated with 77kjots with 8um pitch in 0.13um CMOS operating at 5000fps (4). Second is the use of a single-electron FET detector (5) that due to IP reasons we are not currently pursuing but remains interesting.

At Dartmouth, we have considered a bipolar-based jot and a pump-gate jot. Both start from structures similar to those used in modern CIS pixels. Adapting technology developed for CIS pixels is sensible in creating the QIS jots, speeding development and reducing barriers for adoption. A pump-gate jot has been designed in a modified 65nm CIS process and is being fabricated. The 1.4um pitch was determined by preexisting circuits and could be readily reduced in the future.

Pump-Gate Jot

The pump-gate jot is similar CIS pixels with several differences. First, the carriers are stored under the transfer gate (TG) to reduce jot area. Second, the floating diffusion (FD) is moved distally from TG to reduce overlap capacitance. Third, a tapered reset gate is proposed to further reduce overlap capacitance. Fourth, complete transfer of charge takes two steps: (a) a vertical transfer

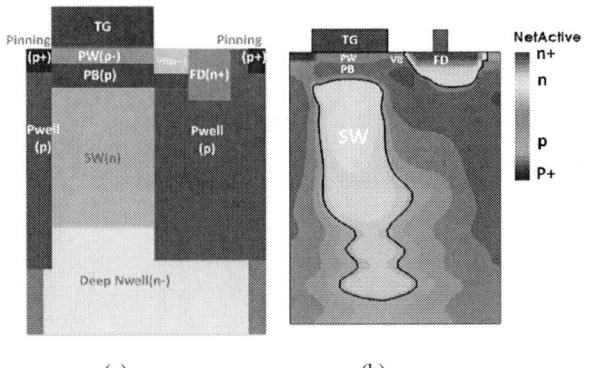

(a)   (b)

Figure 4. Schematic of (a) pump gate jot doping and (b) TCAD simulation.

upwards to surface under TG, followed by (b) a lateral transfer over the "virtual phase" barrier (VB) between TG and FD. The two steps constitute the "pump" action of the charge transfer – similar to a technique used in two-phase CCDs (6) and more recently in a global-shutter CIS (7).

Some recent reported CIS pixels have included storage under the transfer gate (8,9) but since high capacity is needed, implementation has been their major challenge. The small full-well capacity of a QIS jot makes it possible to shrink the size of the device compared to CIS pixels. In the proposed BSI device, a shallow buried-photodiode storage well (SW) is formed underneath the transfer gate, surrounded by a p-type doped well. As depicted in Fig. 4, an n-type SW photodiode well is formed below TG. On the top of SW, there are two p-type doped regions PB and PW, and PW is more lightly doped than PB, so that when TG is biased by flat band voltage, the potential of PW is higher than PB. Between PW and FD, is the p-type doped "virtual-phase" potential barrier (VB). It is more lightly doped than PW. Together they form a double-step potential profile. The p+ pinning layer on the side of TG helps quench surface-generated dark current.

The potential profile is shown in Fig. 5 for the cases of TG

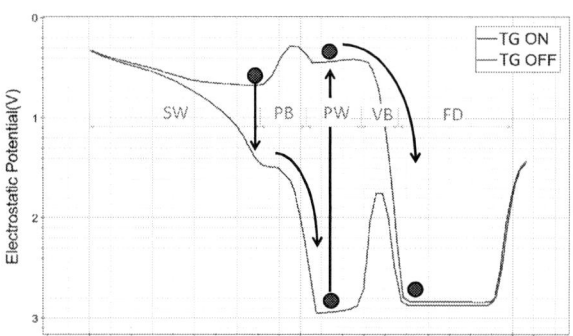

Figure 6. Potential along transferred photoelectron path.

On and Off. During integration TG is "Off" and photoelectrons are collected in SW. After integration, TG is positively biased (On), and carriers are transferred from SW to PW. This is the first step of the "pump" action charge transfer. VB is in depletion mode and forms a virtual barrier between PW and FD. When TG is turned off, the potential of PW returns to its initial level, and as soon as the potential of PW becomes lower than VB, charge in PW will flow over VB into the distal FD. Meanwhile, because of the doping concentration and the location, PB always has a lower potential than PW, which prevent charge from flowing back to SW. This is shown in Fig. 6. Thus, by double step "pump" action, complete charge transfer can be achieved with low TG overlap capacitance.

A complete 1.4um pump-gate jot design with distal FD and other layout improvements such as tapered reset gate is estimated to achieve 480uV/e- with 200e- full well. The distal FD design also reduces clock feedthrough to the column bus when TG is pulsed.

Additional jot-area savings comes from shared readout as is commonly used in CIS pixels. 4-way shared readout was implemented with a 1.0um jot pitch. With the distal FD, the additional TG gates will not increase FD capacitance due to overlap, but 4-way sharing has lower conversion gain of 250uV/e- due to FD size and the remote reset gate. Using

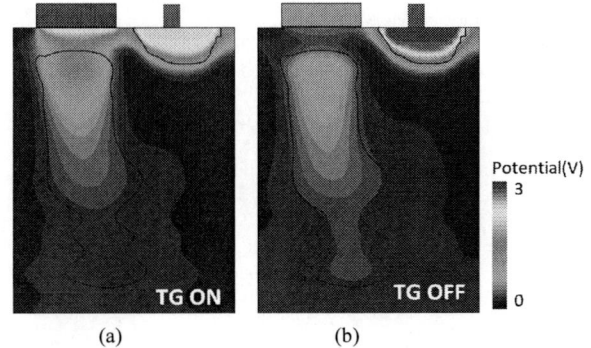

(a)   (b)

Figure 5. Potential in pump-gate jot for (a) TG ON and (b) TG OFF

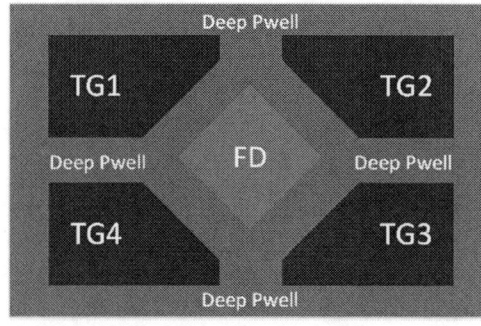

Figure 7. Four-way shared pump-gate jot floorplan, not including reset, selection and SF transistors.

978-1-4799-8002-4/14 $31.00 © 2014 IEEE   249

the same baseline process and simulation, a conventional 4-way shared CIS pixel has a conversion gain of 70uV/e-.

The four jots could be covered by a single color filter and micro lens. Such a microlens might be implemented with a center dimple to avoid guiding rays to the FD. With SDL jots, it is not critical which of the four adjacent jots photoelectrons randomly enter, provided that they are not lost to FD. A deep p-well will be used surrounding four adjacent jots for isolation and photoelectron deflection from FD. An example of a 4-way shared-readout jot floor plan is shown in Fig. 7.

Both the 1.4um jot with non-shared readout, and the 4-way shared-readout 1.0um jot, are currently being fabricated.

Bipolar Jot

A PNP bipolar jot device was also explored as shown in Fig. 8. As a starting point, the pinning layer of pinned photodiode (PPD) was configured as an emitter, and the storage well as the base, and the substrate and a collector common to all jots. The CIS pixel reset gate was used to reset the base so that all majority carriers are depleted.

The main idea is that a single photoelectron collected into the fully depleted base causes the base potential to shift, according to the capacitance of the base. By using an emitter-follower readout circuit configuration, the base-emitter capacitance contributes little to the base capacitance. The base voltage change is transferred to the column bus where it can be sensed. The conversion gain for the jot is determined by the base capacitance in emitter-follower configuration.

In fact, the emitter-collector hole current wants to flow through the base at point of minimal base potential, whereas the photoelectrons want to gather at a point of maximum base potential. Thus, a confinement of the photoelectron at

a point of minimum base width is needed to modulate hole current. A collar structure was created to implement the confinement, and some modulation of the base voltage was observed in simulation. It was also found that collecting photoelectrons modulated the effective base width. However, satisfactory operation of the bipolar jot was not obtained in simulation and we are focusing on the pump-gate jot.

Conclusions

The Quanta Image Sensor has been described and two approaches for jot implementation have been discussed. The pump-gate jot, implementing by modifying a conventional BSI CIS pixel, was simulated and implemented. We are awaiting experimental devices.

Acknowledgments

This work was sponsored by Rambus, Inc. The authors appreciate useful discussions with M. Guidash and with other members of our group at Dartmouth. The support of Synopsys for TCAD tools is also appreciated.

References

1. S. Chen, A. Ceballos, and E.R. Fossum, "Digital Integration Sensor," in Proceedings of the 2013 International Image Sensor Workshop, Snowbird, Utah USA June 12-16, 2013.
2. E.R. Fossum, "Modeling the performance of single-bit and multi-bit quanta image sensors," IEEE J. Electron Devices Society, vol.1(9) pp. 166-174 September 2013.
3. S. Masoodian, K. Odame and E.R. Fossum, "Low-power readout circuit for quanta image sensors," Electronics Letters, Vol. 50 No. 8 pp. 589–591 April 2014.
4. N.A.W. Dutton, et al., "320x240 Oversampled Digital Single Photon Counting Image Sensor," in 2014 Symp. On VLSI Tech. Dig. Of Tech. Papers.
5. E.R. Fossum, et al., "High sensitivity image sensors including a single electron field effect transistor and methods of operating the same," US Patent No. 8,546,901.
6. R.H. Krambeck, R.H. Walden and K.A. Pickar," Implanted Barrier Two-Phase Charge-Coupled Device," Applied Physics Letters, vol. 19, pp. 520-522, 1971.
7. S. Velichko, et al., "Low Noise High Efficiency 3.75μm and 2.8μm Global Shutter CMOS Pixel Arrays," in *Proc. 2013 Intl. Image Sensor Workshop (IISW)*, Snowbird, UT, USA, June, 2013.
8. J. Michelot, et al., "Back Illuminated Vertically Pinned Photodiode with in Depth Charge Storage," in Proc. 2011 Intl. Image Sensor Workshop (IISW), Hokkaido, Japan, June, 2011.
9. J.C. Ahn, et al., "A 1/4-inch 8Mpixel CMOS image sensor with 3D backside-illuminated 1.12μm pixel with front-side deep-trench isolation and vertical transfer gate," 2014 IEEE International Solid-State Circuits Conference Digest of Technical Papers (ISSCC), pp.124,125, 9-13 Feb. 2014.

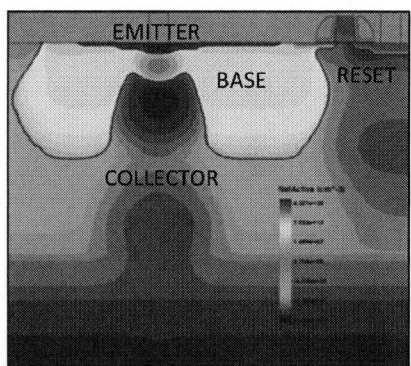

Figure 8. Bipolar jot doping

978-1-4799-8002-4/14 $31.00 © 2014 IEEE

SPAD based Image Sensors

Edoardo Charbon, Senior Member IEEE

Abstract[1]– The recent availability of miniaturized photon-counting pixels in standard CMOS processes has paved the way to the introduction of photon counting in low-cost time-of-flight cameras, robotics vision, mobile phones, and consumer electronics. In this paper we describe the technology at the core of this revolution: single-photon avalanche diodes (SPADs) and the architectures enabling SPAD based image sensors. We discuss tradeoffs and design trends, often referring to specific sensor chips and applications.

Single-Photon Avalanche Diodes

Miniaturized, solid-state photon counting is useful in applications, such as positron emission tomography (PET), LIDAR, fluorescence imaging, quantum key distribution, and wherever single-photon sensitivity and high precision time-of-arrival detection is critical. Photon counting can be achieved in CMOS since 2003, thanks to the creation of a Geiger mode APD or single-photon avalanche diode (SPAD) in a high voltage 0.8μm-process [1]. A SPAD is essentially a reverse-biased diode junction operating above breakdown, in so-called Geiger mode (Fig. 1).

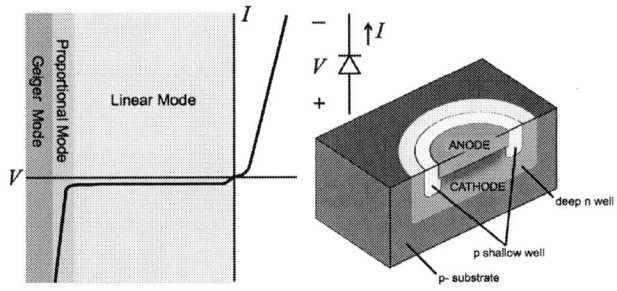

Fig. 1. APD biasing and operating regimes (left); physical structure (right).

In SPADs, a volume known as multiplication zone, where the electric field is above critical (3×10^5V/cm in silicon), is used to detect a photon and trigger an avalanche by impact ionization; the resulting current is sufficient to generate a current pulse of several μA.

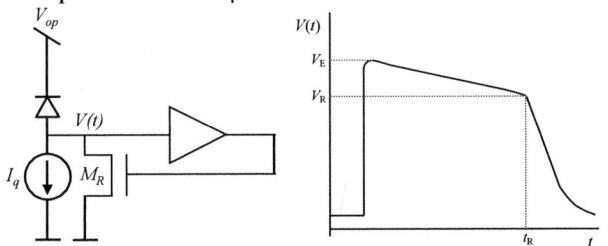

Fig. 2. Simplified schematics of active quenching and recharge (left); control of excess bias during detection cycle and dead time (t_R) (right).

Upon detection of the pulse, one must immediately quench the avalanche, so as to avoid the device destruction,

---

Edoardo Charbon is with Delft University of Technology, Delft, Netherlands (e-mail: E.Charbon@tudelft.nl).

and to recharge it to its pre-avalanche bias, which exceeds breakdown by a voltage known as excess bias (V_E). Fig. 2 shows a simplified feedback loop used to achieve quenching and recharge, so as to keep the detection cycle time ($\approx t_R$), or dead time, under strict control; this technique is known as active quenching and active recharge and it is advantageous to control afterpulsing and saturation [2],[3],[4].

Most pixels actually include some level of pulse shaping and processing, e.g. digital counting or time-to-digital conversion. This functionality is complex and CMOS compatibility of SPADs has been instrumental in achieving very high levels of integration necessary to implement it. Fig. 3 shows the cross-section of a pixel achieved in a standard 65nm CMOS technology [5]. Both N- and P-MOS transistors can be integrated in this technology adjacent to the SPAD.

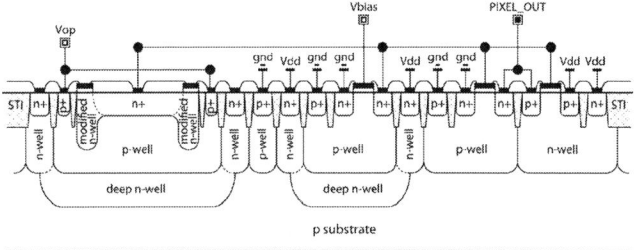

Fig. 3. Monolithic SPAD and CMOS compatible digital electronics.

In the near future, we expect to increase the level of integration even further but only for digital processing, leaving SPADs implemented in 'older' CMOS processes, where doping levels are more appropriate for high performance devices. This hybrid approach will be enabled by 3D integration, whereas a thin epitaxial layer will host the SPADs, possibly operating in backside illumination (BSI), and an electrically bound chip will perform electrical processing and communication at high speeds (see Fig. 4).

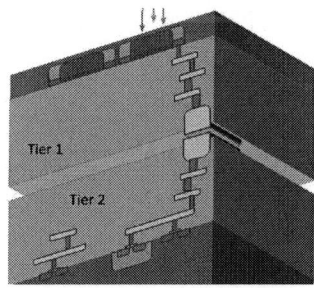

Fig. 4. 3D IC technology. BSI SPADs are integrated in a dedicated process (Tier 1); quenching/recharge circuit in high speed, standard CMOS (Tier 2).

SPAD based Imaging

SPAD image sensors are characterized in terms of noise, sensitivity, and timing resolution. Noise is measured as dark count rate (DCR), which is generally represented as cumulative distribution over a population of SPADs in the sensor, as shown in Fig. 5.

There are typically two knees in the plot, at 60-80% and at 96-98%, representing distinct populations of pixels known respectively as 'laughers' and 'screamers', whose activity is caused be tunneling and trap-assisted mechanisms [6].

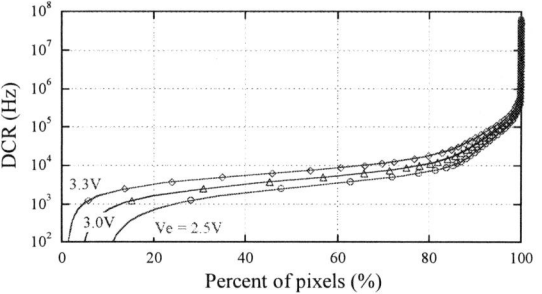

Fig. 5. Cumulative distribution of dark count rate in a typical CMOS SPAD image sensor. The first knee is indicative of higher probability of defects, generally present in larger SPADs ('laughers'). The second knee denotes the presence of much noisier pixels that are usually suppressed ('screamers').

Sensitivity is measured as photon detection efficiency (PDE), which is the product of quantum efficiency (QE), fill factor, and the probability of triggering an avalanche.

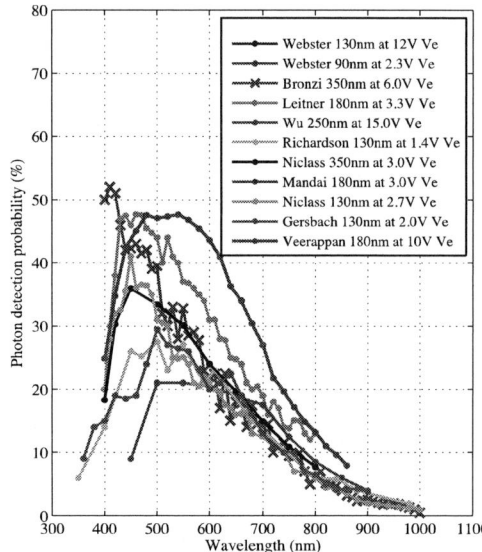

Fig. 6. State-of-the-art of SPAD photon detection probability. The references are [7],[8],[9],[10],[11],[12],[13],[14],[15], and [16], respectively.

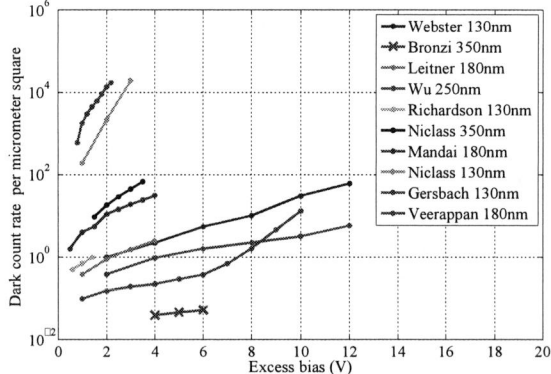

Fig. 7. Dark count rate vs. excess bias found in the same references.

When fill factor is excluded, PDE reduces to photon detection probability (PDP), which is a function of excess bias and wavelength; Fig. 6 shows a plot of PDP as a function of wavelength, as found in the literature [16]. Fig. 7 shows DCR as a function of excess bias.

Timing resolution is measured as the uncertainty of the optical response to a fast light source, generally a laser, in terms of sigma or FWHM of the Gaussian fit. The probability density function (PDF) of SPAD response is a combination of normal and exponential PDFs, due to carrier transport and multiplication mechanisms, as shown in Fig. 8. The figure also shows the impact of multiple photons seeding the avalanche, thereby improving response [17].

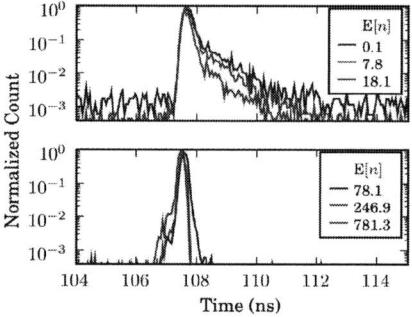

Fig. 8. Timing response as a function of the expected number of simultaneously impinging photons E[n], as measured in a SPAD by Fishburn [17].

SPAD based Image Sensor Architectures

Thanks to their digital nature, SPADs can be coupled directly, without amplification, to time-to-digital converters (TDCs) or other digital components. The first sensor to integrate SPADs and TDCs was LASP [18], where an event-driven mechanism was used to share 128 SPADs with a bank of 32 TDCs; an alternative method for resource redistribution was later used in [19]. Fig. 9 shows a micrograph of the chip that could achieve a timing resolution of 97ps (LSB), a fill factor of 5%, and a [0–100ns] range.

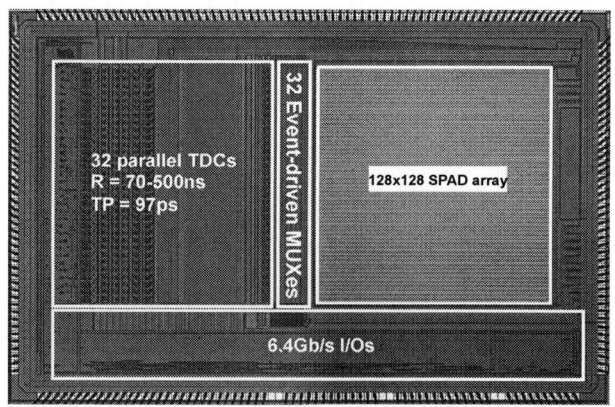

Fig. 9. Fully integrated event-driven SPAD-TDC image sensor LASP [18]. The chip measures 8x5mm².

Fig. 10 shows a face in 3D obtained via the time-of-flight of optical echos detected by LASP operating in TCSPC regime.

978-1-4799-8002-4/14 $31.00 © 2014 IEEE

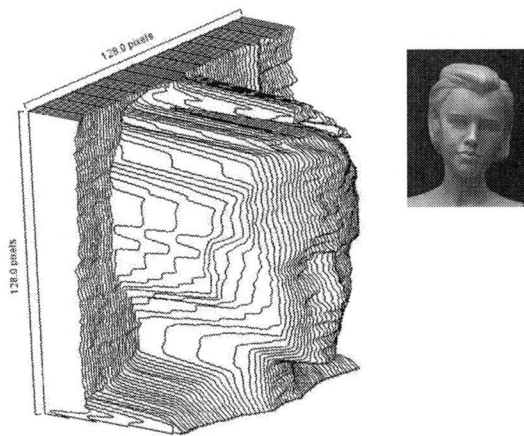

Fig. 10. Time-of-flight 3D image obtained with LASP [18].

Recently, TDC sharing has been pushed to higher limits in the EndoTOFPET-US sensor (Fig. 11), designed for endoscopic PET applications. Using 192 column-parallel TDCs, this chip can generate up to 32×10^6 17-bit timestamps per second, with a timing resolution of 44ps (LSB), and a fill factor of 57% [20].

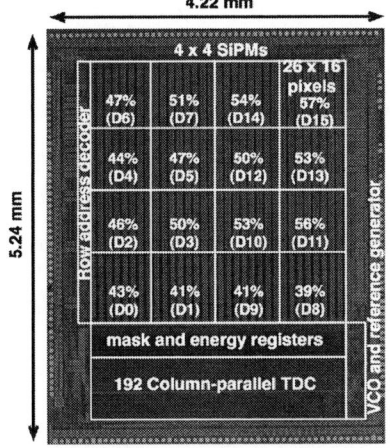

Fig. 11. Column-parallel TDCs in the EndoTOFPET-US chip [20].

Fig. 12 shows the time-domain tomography of photon bunches generated by 637nm laser (Advanced Laser Diode Systems GmBH, Germany) for various photon content. The EndoTOFPET-US sensor generates up to 192 timestamps for all the photon orders inside the bunch, thereby confirming that the probability density function (PDF) of photons is not i.i.d. but varies depending on the order [21]. This result is a confirmation of the theory based on order statistics introduced for SPADs and silicon photomultipliers (SiPMs) in [22].

With more advanced CMOS technologies, a continuous trend towards low pitch pixels and/or high functionality per pixel is emerging, fueled by the need for multi-megapixel photon counting cameras for scientific and biomedical applications [23]. The MEGAFRAME project [24],[25],[26],[27] addressed these very issues; the project yielded an image sensor where 160x128 SPAD-TDC pixels could generate over 20×10^9 10-bit timestamps per second

(one per pixel per µs) with a timing resolution of 52ps (LSB), but a fill factor of only 1% [28].

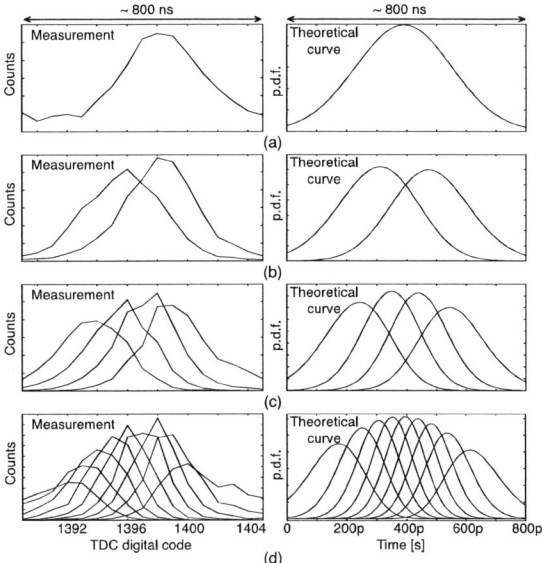

Fig. 12. Internal statistics in the photon bunch generated by a solid-state laser (ALDS GmBH, Berlin, Germany). Measured (left) and predicted (right) probability density functions for photon bunches, comprising 1 (a), 2 (b), 4 (c), and 9 photons (d) [21].

Fig. 13 shows a micrograph of the chip; low fill factor can be recovered, to some extent, in most applications (where light is relatively well collimated), by use of optical concentrators [29]; however, more research can optimize fill factor e.g. through more intelligent sharing and 3D integration.

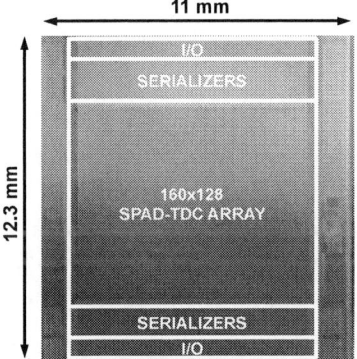

Fig. 13. The MEGAFRAME image sensor [28].

Fig. 14 shows a fluorescence lifetime image of a pollen grain obtained by MEGAFRAME in a FLIM setup.

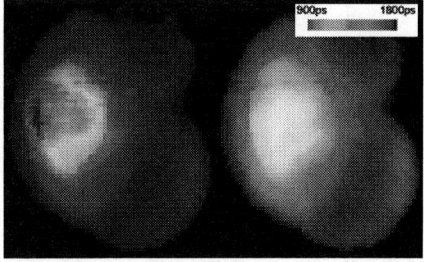

Fig. 14. Image of a Bisaccate Pine pollen grain (Carolina Biological Supply Company, NC, USA). FLIM (left); gray levels (right) [28].

978-1-4799-8002-4/14 $31.00 © 2014 IEEE

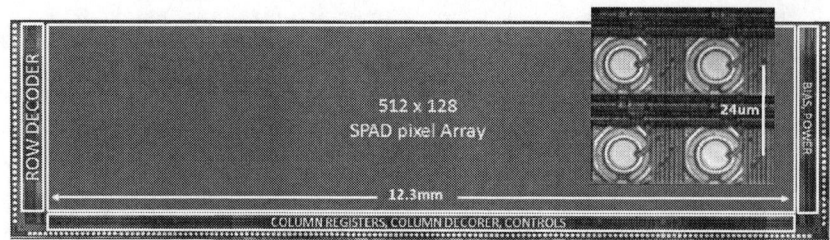

Fig. 15. Fully-parallel, gated SPAD image sensor SwissSPAD [30].

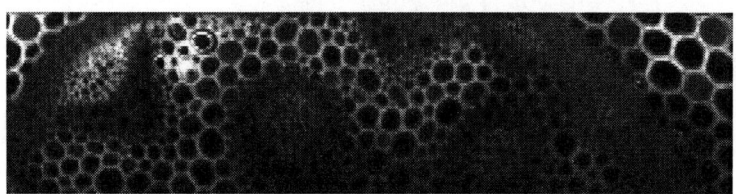 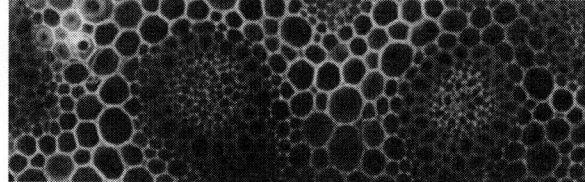

Fig. 16. Fluorescence intensity image obtained with SwissSPAD (left) and a EMCCD (right). For this experiment, polymer micro lenses were fabricated on SwissSPAD based on [29] (Courtesy: I.M. Antolovic).

Researchers have attempted to improve fill factor by using simplified digital [30],[31],[32] and analog [33],[34] counting *in situ*; electrical microlenses have also been proposed, requiring no post-processing steps [35]. In Fig. 15 an image sensor, SwissSPAD [30], is depicted, comprising 1/16 Mpixel, with a pixel pitch of 24μm and a 1-bit counter per pixel. The readout achieves 156kfps in continuous mode with gating capability (<5ns) and 35% fill factor (with microlenses). Fig. 17 shows the optical response of a pixel and the statistics of the its dynamic response across the array. A carefully designed clock and gate distribution network ensures 101ps (FWHM) uncertainty across the entirte chip. Further reduction in gate length [36],[37] and format [38] are in progress.

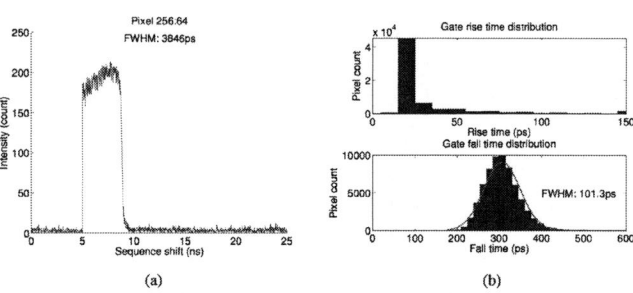

Fig. 17. Time response of SwissSPAD [30]. (a) optical response of a gated pixel; (b) statistics of rise and fall times over the entire array.

Fig. 16 shows a comparison between SwissSPAD and a commercial EMCCD (Andor, Northern Ireland) on a fluorecent sample imaged with a widefield microscope (Leica, Germany) under the same operating conditions.

Conclusions

CMOS compatibility has enabled significant growth in functionality and performance of SPAD pixels, paving the way to larger and more useful image sensors. Many applications that once required non solid-state sensors are now possible with SPAD based image sensors and the field is still expanding. Industry has begun integrating SPAD image sensors in mass products, further increasing the acceptance of SPAD imagers in consumer electronics. The trend will be accelerated when 3D ICs will be commonplace, enabling further growth of mass-produced SPAD based devices.

References

[1] A. Rochas *et al., Rev. Sci. Instrum.,* **74**(7), 3263–3270 (2003).
[2] E. Charbon, *Phil. Trans. R. Soc. A*, **372**, 20130100 (2014).
[3] S. Cova *et al., App. Opt.,* **35**(12), 1956-1976 (1996).
[4] A. Eisele *et al., International Image Sensor Workshop* (2011).
[5] E. Charbon *et al., IEEE IEDM* (2013).
[6] A. Spinelli and A.L. Lacaita, *IEEE Trans. Electron Devices*, **44**(11), 1931-1943 (1997).
[7] E.A.G. Webster *et al., IEEE El. Dev. Lett.,* **33**(11), 1589–1591 (2012).
[8] D. Bronzi *et al., SPIE* 8631, 86311B (2013).
[9] T. Leitner, *et al., IEEE Trans. Electron. Devices,* **60**(6), 1982–1988 (2013).
[10] J. Wu *et al., International Image Sensor Workshop* (2013).
[11] J. Richardson *et al., IEEE Photonics Technology Letters,* IEEE, **21**(14), 1020–1022 (2009).
[12] C. Niclass, *PhD Thesis*, Lausanne (2008).
[13] S. Mandai *et al., Opt. Express*, **20**(6), 5849–5857 (2012).
[14] C. Niclass *et al., IEEE J. Sel. Top. Quant. Electron.,* **13**(4), 863–869 (2007).
[15] M. Gersbach *et al., IEEE ESSDERC*, 270–273 (2008).
[16] C. Veerappan and E. Charbon, *IEEE J. Sel. Top. Quant. Electron.,* **20**(6), 1-7 (2014).
[17] M.W. Fishburn, *PhD Thesis*, Delft (2012).
[18] C. Niclass *et al., IEEE J. Solid-State Circuits,* **43**(12), 2977-2989 (2008).
[19] R.J. Walker *et al., IEEE ISSCC*, 410-411 (2011).
[20] S. Mandai, and E. Charbon, *JINST*, **8**, P05024 (2013).
[21] S. Mandai *et al., Opt. Lett.,* **39**(3), 552-554 (2014).
[22] M.W. Fishburn, *Trans. Nuclear Science,* **57**(5), 2549-2557 (2010).
[23] E.A.G. Webster *et al., El. Dev. Lett.,* **33**(5), 694-696 (2012).
[24] M. Gersbach *et al., IEEE ESSDERC*, 196-199 (2009).
[25] J. Richardson *et al., IEEE CICC*, 77-80 (2009).
[26] D. Stoppa *et al., IEEE ESSDERC*, 204-207 (2009).
[27] M. Gersbach *et al., IEEE J. Solid-State Circuits*, **47**(6), 1394-1407 (2012).
[28] C. Veerappan *et al., IEEE ISSCC*, 312-314 (2011).
[29] J. Mata Pavia *et al., Opt. Express* **22**, 4202-4213 (2014).
[30] S. Burri *et al., Opt. Express* **22** (2014).
[31] F. Guerrieri *et al., IEEE Photonics Journal*, **2**(5), 759-774 (2010).
[32] D. Bronzi *et al., IEEE J. Sel. Top. Quant. Electron.,* **20**(6), (2014).
[33] N.A.W. Dutton *et al., International Image Sensor Workshop* (2013).
[34] E. Panina *et al., IEEE Trans. Circuits Syst. II,* **61**(4), 214-218 (2014).
[35] C. Veerappan *et al., International Image Sensor Workshop* (2013).
[36] I. Nissinen *et al., IEEE ESSDERC*, 375-378 (2011).
[37] L. Pancheri *et al., International Image Sensors Workshop* (2011).
[38] Y. Maruyama *et al., IEEE J. Solid-State Circuits,* **49**(1), 1-11 (2014).

Toward 1Gfps: Evolution of Ultra-high-speed Image Sensors
-ISIS, BSI, Multi-Collection Gates, and 3D-stacking-

T. G. Etoh[*+], V. T. S. Dao[*], K. Shimonomura[*], E. Charbon, C. Zhang[**], Y. Kamakura and T. Matsuoka[***]

[*]Ritsumeikan University, Noji-Higashi, Kusatsu, 525-8577 Japan; [+]phone: 81-77-561-5052; [+]email: etoh@fc.ritsumei.ac.jp
[**]T U Delft, Delft, The Netherlands; [***]Osaka University, Suita, Osaka, Japan

Abstract

Evolution of ultra-high-speed image sensors is reviewed. Currently, the highest frame rate achieved by a solid-state image sensor is 16.7 Mfps, while the target of this project is 1 Gfps. To achieve this target, we propose an image sensing unit consisting of CCD pixels and a CMOS driver circuit, whereas the pixel has multiple collection gates, and the driver comprises a ring oscillator with XNOR circuits. The sensor and the driver are mounted on different IC chips by 3D stacking. Simulations confirm that the proposed technology achieves the target. In ultra-high-speed imaging, sensitivity is crucial. A technique to achieve very high sensitivity is also proposed.

Evolution Process

Since 1991, Etoh et al. have been updating the frame rate of high-speed video cameras : 4,500 fps in 1992 (1), 1 Mfps in 2002 (2), 16 Mfps in 2011(3), and 16.7 Mfps in 2013(4). We proposed a new sensor structure to achieve 1 Gfps (5). Table 1 summarizes the evolution process. Other important achievements in the field include the first practical in-situ storage image sensor by Kosonocky et al. (6), a CMOS ultra-high-speed image sensor with pixel-based storage in the periphery of the chip by Tochigi et al. (7), and an image sensor with in-situ CCD storage and CMOS readout by Crooks et al. (8).

Ultra-high frame rate is achieved by an in-situ storage image sensor, ISIS. Each pixel of the sensor is equipped with multiple memory elements to record a series of image signals simultaneously at all pixels. An ultra-high-speed video camera reproducing motion pictures requires more than one

hundred in-situ memories, while a multi-framing camera is equipped with four to sixteen in-situ memories.

In 1996, Kosonocky et al. developed an image sensor with a series-parallel-series (SPS) CCD in-situ memory, which "practically" achieved 0.5 M fps for 180x180 pixels (6). In 2001, an ISIS with more than one hundred in-situ memories achieved 1Mfps for 260x312 pixels (2). The in-situ memory was a linear CCD elongating in a direction slightly slanted to the photodiode grid. Curvilinear design is fully exploited in the layout to realize smoothly changing potential profiles. For example, the photodiode with the gradually widening n^+ doping made a linearly increasing potential profile which minimizes the travel time of the signal electrons in the photodiode (Fig. 1). However, the fill factor of the sensor was only about 15% due to a light shield covering the wide in-situ storage area.

Backside illumination increases the fill factor to practically 100%. However, the BSI ISIS had the following problems: (i) intrusion of remaining incident light after the absorption by the silicon layer to the in-situ storage on the front side, and (ii) migration of generated signal electrons to the storage. The thickness of the chip was increased to more than 30 um, which allows absorption of 99.9% of incident light with a wave length of 700 nm. The chip is processed in the p^-/n^- double epi-layers. A p-well is formed in the n^--epi layer, and the n^+ in-situ storage is made under the p-well, forming a pnpn cross-section structure (Fig. 2). The p-well prevents the signal electrons from migrating to the in-situ storage. The BSI structure also increased the frame rate with the additional metal wiring on the front side. The frame rate reached 16 Mfps (3). An EM CCD was also introduced in the readout HCCD for ultra-high sensitivity.

Table 1 Evolution of in-situ storage image sensors

Innovative Technologies	In-situ Storage (ISIS)	[+]Backside Illumination (BSI)		Multi-Collection-Gate (MCG)	3D Stacking	
Fundamental Inventions	Slanted linear CCD storage	pnpn cross-section structure		BSI MCG		
Supporting technologies	[*]Curvilinear CCD design	[+]EM-CCD (HCCD)	[#]Folded/Looped CCD	[**]Pipeline transfer	[++]Ana/Digi storage	[##]RO-XNOR driver
Frame rate	1 Mfps	16 Mfps	10 Mfps	50-100 Mfps	50-100 Mfps	1-10 Gfps
Advantages	[*]High transfer efficiency	[+]High sensitivity	[#]High SNR by in-pixel signal accumulation	[**]4-time faster than conventional ISIS	[++]Low noise	[##]Theoretically highest frame rate
Status	Developed in 2002	Developed in 2011	Test chip under evaluation	Test chip under design	Concept	Test chip under evaluation
Principle applications	←——————————— Ultra-high-speed imaging ———————————→					
	←——————————— Sensors for scientific measurement ———————————→					

The p-well in the BSI ISIS was designed to be thicker around the pixel boundary, thinner near the center, and none at the center (for example, see Fig. 6). A negative bias voltage is applied to the backside to deplete the travel route of the electrons. Guided by the potential profile created by the p-well and the backside voltage, a signal electron travels at a very high speed to the collection gate (Fig. 3).

An ISIS with a folded and looped CCD in-situ memory in each pixel was named Image Signal Accumulation Sensor, ISAS (Fig. 4). If the target event for imaging is reproducible, the S/N ratio is increased by the in-pixel accumulation of signals provided through repeating image captures, proportionally to the inverse of the square root of the number of the accumulations.

During the development of the BSI ISIS, we noticed that the travel time of an electron from the generation site to a collection gate can be less than 1 ns, which is much shorter than the transfer time of a collected signal charge packet from the collection gate to the attached storage area. Therefore, we proposed a backside-illuminated multi-collection-gate (BSI MCG) image sensor (5). The pixel has a group of collection gates surrounding the center (Fig. 5 and

Fig. 10). If image signals can be selectively collected by each of the multi-collection gates, the frame interval can be reduced to the electron travel time, 1 ns.

After collection of a signal charge packet by a collection gate, the charge packet can be transferred to the in-situ storage during collection of image signals by other collection gates. The pipeline transfer is useful for ultra-high-speed motion image capture, which requires more than one hundred consecutive frames (Fig. 10).

3D stacking is a promising technology to create intelligent image sensors. Ultra-high-speed imaging is one of the best application fields. In this paper, we are proposing two BSI MCG image sensors combined with 3D stacking technology: the first with a pixel-based driver circuit to increase frame rate, and the second with both analogue and digital in-situ memories to reduce noise.

The pixel based driver circuit consists of a ring oscillator (RO) made of inverters, each of which is equipped with an XNOR circuit. The sensing and recording circuits are placed on the first chip, and the RO-XNOR drivers are placed on the second one (Fig. 9). The sensor achieves a frame interval of 1ns as shown later.

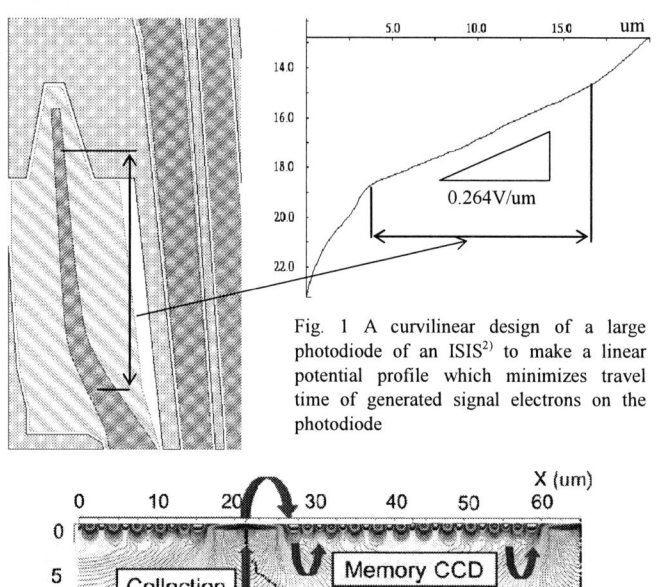

Fig. 1 A curvilinear design of a large photodiode of an ISIS[2] to make a linear potential profile which minimizes travel time of generated signal electrons on the photodiode

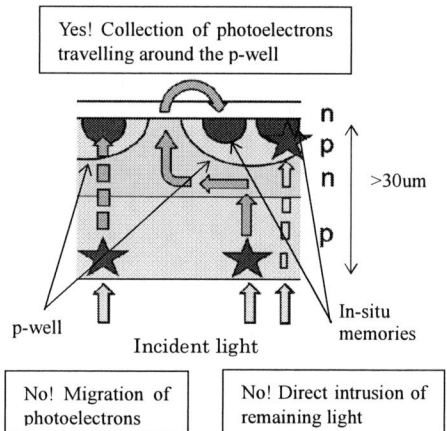

Fig. 2 The pnpn-structure of a BSI ISIS[3)4]

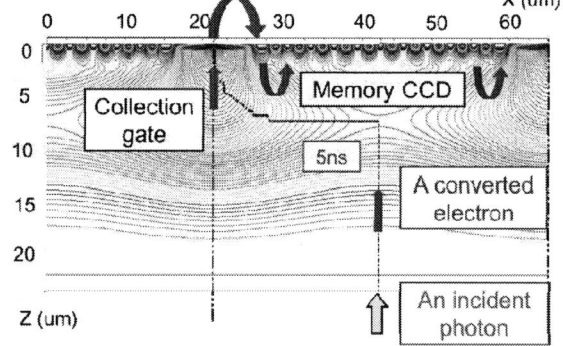

Fig. 3 A cross section view of a potential profile of an ISIS and a path of an electron generated at the pixel boundary near the backside: the pixel size is 43.2 um; the thickness of the chip is 32um.

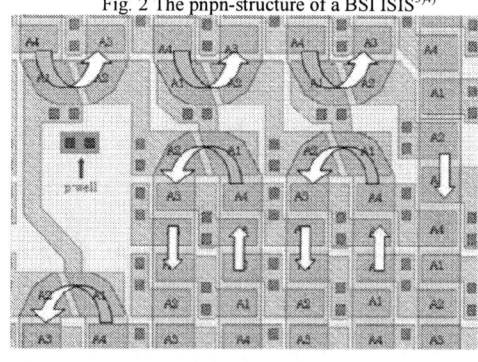

Fig. 4 A folded and looped CCD memory with Z-shaped electrodes that enables in-pixel signal accumulation for multiple image-capture operations of reproducible events with very low incident light

978-1-4799-8002-4/14 $31.00 © 2014 IEEE

An image sensor with analogue and digital in-situ memories digitizes only small signals recorded as low-bit digital signals on an attached chip right after the image capturing phase, leaving large signals in the analogue in-situ storage, which effectively reduces both dark current and readout noises.

Performance of a BSI-MCG image sensor

Fig. 5 shows a model of a BSI MCG image sensor. For the layout, the electron travel time is evaluated with Monte Carlo simulations (Fig. 7 to Fig. 8) (9). The temporal resolution of the electron travel time is defined as twice the standard distribution, which is less than 1ns. The performance of the RO-XNOR driver is evaluated by circuit simulations (Fig. 11). The temporal resolution is defined as the peak-to-peak distance, which is also about 1ns. Therefore, 1 Gfps can be achieved by the combination technologies. The selective collection of signal electrons is also confirmed.

Factors affecting the travel time distribution are analyzed. Major factors to delay the travel time are as follows: (i) semi-isotropic random motion of electrons in the backside p^{++} layer filled with holes for electron-hole recombination to reduce dark current; (ii) roundabout of the electrons generated near the pixel boundary travelling around the p-well (See Fig. 7 and Fig. 8).

Therefore, the temporal resolution of the sensor is further reduced by the following modifications: (i) thinning the backside p^{++} layer to much less than the absorption depth of incident light to the backside; (ii) focusing the light to the center of the pixel by using a light guide and a microlens. The former makes the photoelectron conversion mainly take place in the depletion layer under the p^{++} layer. The latter mitigates the roundabout of the electrons around the p-well. Then, the electrons are generated near the center of the backside and move straightforwardly to the front side. Now, the travel time is mainly distributed by scattering of electrons due to the collisions to atoms and vibration of a lattice. For the condition, the temporal resolution is 79.4 ps.

The temporal resolution of the proposed driver can be reduced by minimizing (i) the voltage swing of the driving pulses of the collection gates, and (ii) the capacitance of the collection gates by reducing the size with a finer process.

The test chip of the BSI MCG image sensor and the RO-XNOR driver is currently under evaluation (Fig. 12). The design also includes a built-in test structure based on a bank of time-to-digital convertors to assess the actual performance of the RO-XNOR.

An analogue/digital in-situ storage image sensor

Photon-counting sensitivity is another target of scientific image sensors. We propose a 3D-stacked ISIS with both analogue and digital in-situ memories to achieve both ultra-high speed and sensitivity. An in-pixel ADC and the in-situ digital memory are installed in an attached chip.

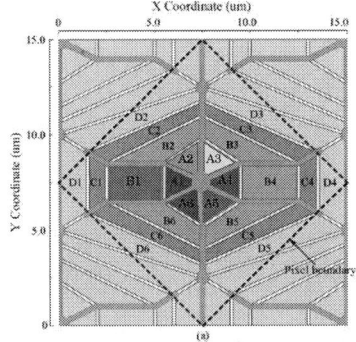

Fig. 5 Hexagonal BSI MCP image sensor: A1: Drain gate; A2-A6: Collection gates; B1: Drain; B2-B6: Storage gates; C: Barrier gates; D: Transfer gates

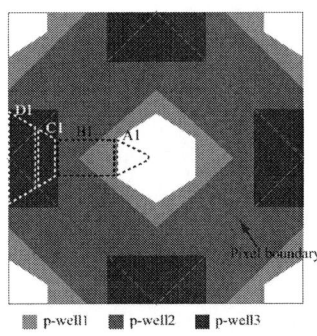

Fig. 6 Three p-well layers of the hexagonal BSI MCG image sensor shown in Fig. 5

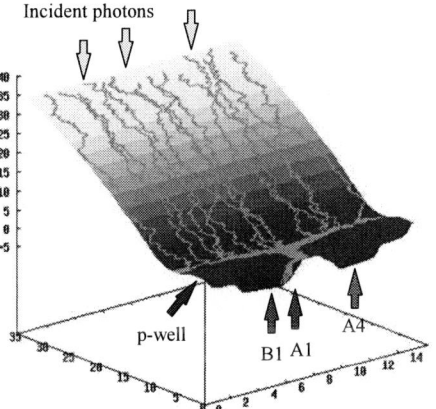

Fig. 8 A 2D potential profile and electron paths generated by Monte Carlo simulations for the hexagonal BSI MCP image sensor shown in Fig. 5: The electrons travel in the depleted layer created by the backside bias voltage, move around the p-well, go through the p-well hole (see Fig. 6), and are selectively collected only by the A1 gate, to which a higher voltage is applied.

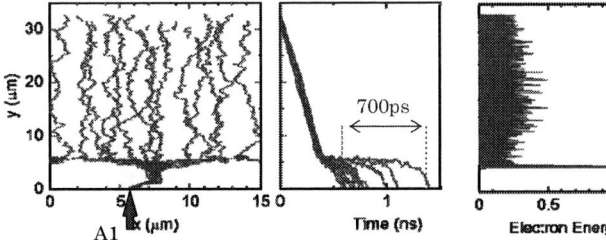

Fig. 7 Paths/Travel times/Energy of photoelectrons shown in Fig. 8

978-1-4799-8002-4/14 $31.00 © 2014 IEEE   257

Since the number of the analogue signals stored in each pixel is about one hundred, the readout time is very short. Even if the sampling rate is 100 kHz, it takes only 1ms. Therefore, the dark current noise is very low. The readout noise at 100 kHz from the analogue in-situ memory to the digital one is also almost negligible. By selecting a proper sampling rate, the ISIS with analogue and digital in-situ memories enjoys both ultra-high speed and sensitivity.

Introduction of selective digitization of low-bit signals further reduces the total noise to nearly photon-counting level with no impact ionization nor cooling.

References

(1) T. G. Etoh, High-speed video camera of 4,500 pps, *J. ITE*, 543-549, 1992 (in Japanese).

(2) T. G. Etoh D. Poggemann, A. Ruckelshausen, A. Theuwissen, G. Kreider, H.-O. Folkerts, et al., A CCD image sensor of 1Mframes/s for continuous image capturing of 103 frames, *ISSCC2002*, 46-47, 2002.

(3) T. G. Etoh, H. D. Nguyen, V. T. S. Dao, C. L. Vo, M. Tanaka, T. Okinaka et al., A 16 Mfps 165kpixel backside illuminated CCD, *ISSCC2011*, 406-407, 2011.

(4) T. Arai, J. Yonai, T. Hayashida, H. Ohtake, H. van Kuijk and T. G. Etoh, 252V/lux 16.7-Million per second 312-kpixel back-side-illuminated ultrahigh-speed Charge Coupled Device, *IEEE Trans. ED*, 60(10), 3450-3458, 2013.

(5) T. G. Etoh, V. T. S. Dao, T. Yamada and E. Charbon, Toward one Giga frames per second-Evolution of In-situ Image Sensors, *Sensors*, 13(4), 4640-4658, 2013.

(6) W. F. Kosonocky, G. Yang, C. Ye, R. K. Kabra, L. Xie, J. L. Lawrence, et al., 360x360-Element Very-High Frame-Rate Burst-Image Sensor, *ISSCC1996*, 182-183, 1996.

(7) Y. Tochigi, K. Hanzawa, Y. Kato, R. Kuroda, H. Mutoh, R Hirose, et.al., A global shutter CMOS image sensor with readout speed of 1 Tpixel/s burst and 780 Mpixel/s continuous, *ISSCC2012*, 382-383, 2012.

(8) J. Crooks, B. Marsh, R. Turchetta, K. Taylor, W. Chan, A Lahav, et al., Kirana, a solid-state megapixel uCMOS image sensor for ultrahigh speed imaging, *Proc. SPIE* 8659, 865903, 2013.

(9) V. T. S. Dao, K. Shimonomura, Y. Kamakura, and T. G. Etoh, Simulation analysis of a backside-illuminated multi-collection-gate image sensor, *ITE Trans. on MTA*, 2(2), 114-122, 2014.

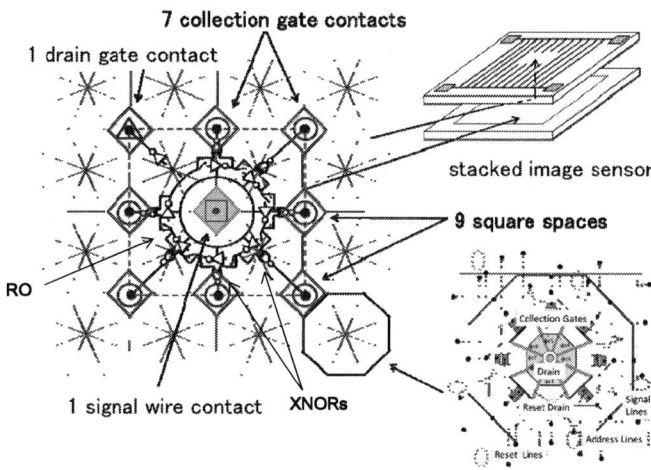

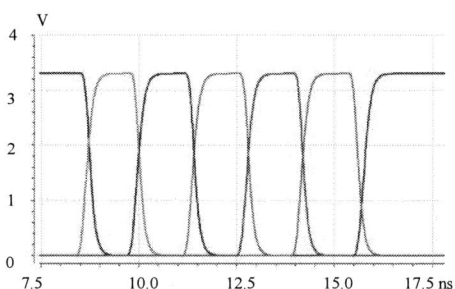

Fig. 9 A model of the pixels of a 3D-stacking MCG image sensor with octagonal MCG pixels and pixel-based RO-XNOR drivers

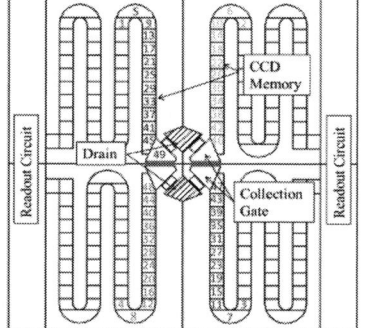

Fig. 10 Quadruple MCG image sensor with four collection gates and double drains: four CCD memory elements are operated with the same driving voltage pulses only by shifting the contact points to the electrodes from the metal inner buslines

Fig. 11 Outputs of a RO-XNOR driver for the hexagonal BSI MCG image sensor (Fig. 5)
dt = 1.43 ns for dV = 3.3 V; dt = 150 ps for dV = 1.2 V
[——— : Driving voltage of the drain collection gate A1;
 ═══ : Driving voltages of multi-collection gates]

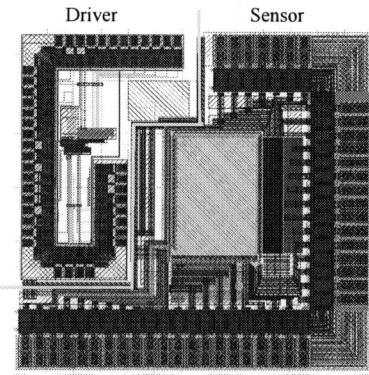

Fig. 12 The test chip layout of a BSI MCG image sensor and a RO-XNOR driver

Imaging with Organic and Hybrid Photodetectors

Sandro F. Tedde[1,*], Patric Büchele[1], Rene Fischer[1], Frank Steinbacher[1], Oliver Schmidt[1]

Abstract—**Organic semiconductors provide exiting new opportunities for the realization of flat panel image sensors as they can be processed from the solution phase on large areas at low cost. In particular the high charge separation efficiency obtained in a bulk heterojunction (BHJ) enables the realization of organic photodiodes (OPDs). The spectral sensitivity of OPDs can be tailored to cover wavelengths ranging from the visible to the near infrared region. These sensitivities match perfectly to a variety of X-ray scintillators enabling a further improvement in the sensitivity range. In combination with an amorphous silicon (a-Si) thin film transistor (TFT) backplane technology, visible, near infrared (NIR) and X-ray image sensors have been realized. Thin film OPDs have been used in combination with a cesium iodide (CsI) scintillator in a traditional stacked geometry, proofing state-of-the art performance. Even more, it is possible to blend X-ray absorbing particles directly into the organic semiconductor thereby enabling quasi-direct X-ray converters with the promise to achieve a modulation transfer function (MTF) that is as high as in direct converting materials such as amorphous Selenium.**

I. INTRODUCTION

ORGANIC semiconductors are attractive materials as active elements for a variety of optoelectronic devices, such as light emitting diodes, solar cells, and photodiodes. In particular the high charge separation efficiency obtained with an interpenetrating donor-acceptor heterojunction [1,2,3] was key for cost-efficient and large scale fabrication of organic solar cells and organic photodiodes (OPDs).

OPDs experienced in the last years a steady growth of interest which culminated in several recent publications and press releases by renowned institutions [4,5,6]. In particular the field of medical imaging seems to benefit mostly from OPD's potential. Most flat panel X-ray detectors today are realized by stacking an a-Si photo detector array and a scintillator which converts the incoming X-ray photons into visible light. Medical imaging requires large area X-ray detectors due to the limited ability to focus X-ray radiation and organic semiconductors provide the opportunity to economically process photosensitive elements on large areas.

Besides the potential cost-benefit of the OPD technology, it allows processing on flexible substrates (e.g. for mechanical

---

[1] Siemens AG, Corporate Technology, Erlangen, Germany
[*] email: sandro.tedde@siemens.com

This research was partially funded by the German Federal Ministry of Education and Research within the funding program Photonics Research Germany (Contract number 13N12377)

---

robustness). Ng et al. [7] have demonstrated an inkjet printed OPD matrix on an a-Si TFT backplane. IMEC did even replace the backplane technology by an OTFT array and demonstrated an all-organic photodetector array [6].

Here, we will discuss the performance of OPDs in combination with a-Si TFT backplanes to create a pixelated photodetector for visible, near infrared and X-ray imaging. The detector consists of 256x256 squared pixels with 98 μm pitch. On top of the backplane a photoactive bulk heterojunction (BHJ) layer is solution-processed consisting of a blend of a polymer as electron donor and fullerenes as electron acceptor. In our experience, the most interesting processing technique of OPDs is effortless spray–coating [8], which is a fast and high throughput fabrication method.

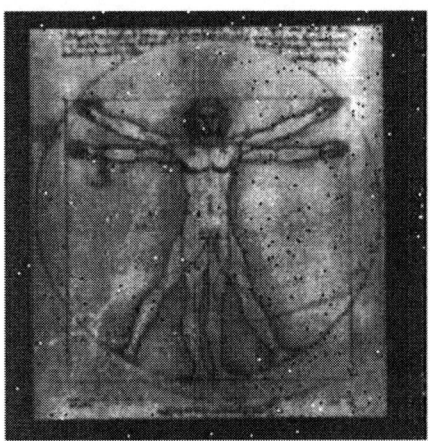

Fig. 1. Image recorded with a 256x256 pixel active matrix photodetector based on organic photodiodes under illumination of 10 μW/cm² at λ=530 nm and a pixel pitch of 98 μm.

Pixelated OPDs show high external quantum efficiency up to 55% in the visible spectrum, low dark currents densities of 5×10^{-6} mA/cm² at -5 V reverse bias, as well as high linearity and dynamic range. Figure 1 shows an image recorded with the spray coated organic detector array for the visible range. The illumination was performed with a pulsed light source at a wavelength of 532 nm and a light intensity of 10 μW/cm².

II. STACKED DETECTORS

The electron acceptor Phenyl-C61-butyric acid methyl ester (PCBM) and electron donor Poly(3-hexylthiophen-2,5-diyl) (P3HT) have been solved in a weight ratio of 0.75:1 in chlorobenzene. A ~600nm thick OPD was processed by spray-coating this solution onto an a-Si TFT backplane. The TFT array size was 256x256 pixels with a pixel pitch of 98μm and a geometrical fill factor of 70,2%, which is defined by a patterned indium-tin-oxide (ITO) electrode which serves as a bottom contact. A semitransparent Ca/Ag top contact was evaporated on top of the OPD and the device was encapsulated with a glass slide. A caesium-iodide scintillator (CsI) was attached to the device in order to convert the incoming x-ray radiation into visible light. Fig. 2 shows the layer stack of the resulting device as well as a microscope image of the individual pixels and a photograph of the processed panel (without scintillator).

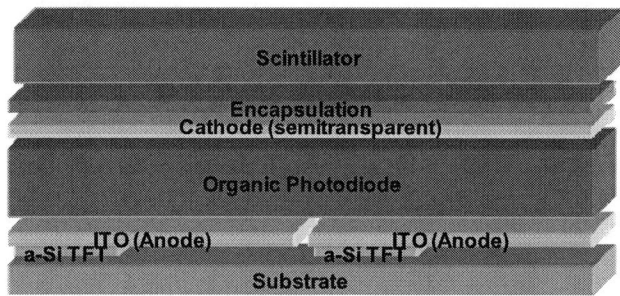

Fig. 2. Design of an organic photodetector matrix with stacked scintillator

III. Hybrid-Organic Detectors

The concept described above suffers from a limited image resolution as photons are emitted isotropically from the scintillator thereby-creating optical cross-talk between neighboring pixels, equally as in traditional a-Si/CsI flat panel X-ray detectors.

It has been shown that OPDs can be sensitized for the near-infrared (NIR) by the addition of QDs [9]. Similarly X-ray absorbers can be added to the BHJ in order to created quasi-direct X-ray detectors.

X-ray absorbers are mixed directly into the organic photo detector material, thereby minimizing the optical cross-talk and enabling x-ray detectors with high modulation transfer function (MTF). Layers with a thickness of up to 200μm are needed in order to achieve efficient x-ray absorption. At the same time thick layers might cause charge recombination and inefficient charge carrier extraction.

Here, we explore two different routes towards hybrid-organic photo detectors. In a direct-conversion approach lead sulfide (PbS) quantum dots have been processed in a matrix of a P3HT/PCBM bulk heterojunction (BHJ). The hybrid layer was deposited by spray-coating which enabled thick layers up to 50μm. X-ray conversion rates of 300 electrons/nGy/mm² could be observed, demonstrating the feasibility to extract charges from thick BHJ layers with high efficiency.

As a second route, scintillator particles were added to the organic photodiode matrix in order to realize a quasi-direct X-ray converter. The emitted light from an X-ray-excited gadolinium oxysulfide (GOS) scintillator is absorbed within 160nm from the particle in a P3HT/PBCM BHJ. This optical cross-talk is negligible compared to the gap between two neighboring pixels which is in the range of 10μm. Fig. 3 shows an X-ray image obtained with such a hybrid detector demonstrating the feasibility on system level and the excellent resolution. The accelerating voltage of the X-ray source was 70 kV while the incident dose 70 μGy.

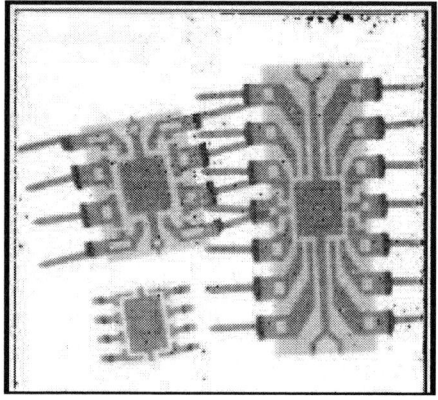

Fig. 3. X-Ray image of three ICs made with an hybrid detector were scintillating GOS particles have been embedded in the organic BHJ of P3HT:PCBM. Pixel pitch is 98 μm. The accelerating voltage of the X-Ray source is 70 kV while the incident dose is 70 μGy.

References

[1] G. Yu, J. Gao, J. C. Hummelen, F. Wudl, A. J. Heeger, *Science* 270, 1789 (1995).
[2] N. S. Sariciftci, L. Smilowitz, A. J. Heeger, F. Wudl, *Science* 258, 1474, (1992).
[3] A. J. Heeger, Bulk Heterojunction, *Adv. Mater.* 26, 10–28 (2014)
[4] E. Saracco, B. Bouthinon1, J.M. Verilhac, C. Celle, N. Chevalier, D. Mariolle, O. Dhez, J.P. Simonato, *Advanced Materials*, 25, 45, 6534–6538, (2013)
[5] M. Ihama, T. Mitsui, K. Nomura, Y. Maehara, H. Inomata, T. Gotou, Y. Takeuchi, *Proceedings IDW 2009*, INP1-4 (2009)
[6] X-ray detector on plastic delivers medical imaging performance [Online] http://www.holstcentre.com/en/NewsPress/PressList/PlasticXrayDetector.aspx (2014)
[7] T. N. Ng et al., *Appl. Phys. Lett.* 92, 213303 (2008)
[8] S. Tedde, J. Kern, J. Fürst, P. Lugli, O. Hayden, *Nano Letters* 9, 980-983 (2009).
[9] T. Rauch et al., Nature Photonics 3, 332 - 336 (2009)

A CMOS-compatible, integrated approach to hyper- and multispectral imaging

Andy Lambrechts, Pilar Gonzalez, Bert Geelen, Philippe Soussan, Klaas Tack and Murali Jayapala

Imec, Kapeldreef 75, B-3001 Leuven, Belgium. Tel: (+32) 16287847, Fax: (+32) 16281515 Email: lambreca@imec.be

Abstract

Imec has developed a unique hyperspectral sensor concept in which the spectral unit is monolithically integrated on top of a standard CMOS sensor at wafer level, hence enabling the design of compact, low cost and high speed spectral cameras with a high design flexibility. This paper presents the various demonstrated prototype sensors, with different filter arrangements and performance, linked to different usage modes and application domains. It also reviews the key aspects and challenges of imec's hyperspectral technology.

Introduction

Spectral imaging can reveal a lot of hidden details about the world around us (1), but is often still confined to laboratory environments due to the need for complicated, costly and bulky cameras. To bridge the gap between research and industry, imec has developed a unique hyperspectral sensor concept in which the spectral unit is monolithically integrated on top of a standard CMOS sensor at wafer level (Fig. 1), hence enabling an extremely compact form factor and mass-manufacturability. Different filter arrangements enable a variety of system level usage modes. The integrated filters can reach high transmission efficiencies and at the same time eliminate the need for camera level alignment of high-end lenses or gratings, as well as stray light generated by the unwanted reflections on discrete components or substrate layers (2).

This approach leverages on CMOS manufacturing technology, like deposition, high accuracy lithography and etching. This ensures the quality of the filter layers and enables pixel level filter definition, in terms of layout (covering different groups of pixels or depositing filters per pixel) and performance (i.e. number of bands, filter width), to match the requirements of specific applications. To prove the versatility of this technology, imec has successfully demonstrated three sensor types with different filter arrangements: a line-scan imager (3) and two non-scanning, frame-based spectral sensors, here called snapshot sensors (4, 5).

Context and Motivation

A rising number of hyper-/multi-spectral applications are on the verge of being deployed, but current camera prices and size/weight are often a blocking factor. Typical state of the art hyperspectral cameras are complex, bulky and expensive due to the optical components in their spectral unit. As a result, some emerging solutions try to overcome the limitations of RGB imaging for industrial, medical or even consumer systems, by making use of a limited number of discrete filters in a multi-sensor solution. This approach is

not scalable, and leads to additional problems with regard to sensor fusion. More challenging applications require more spectral bands to be captured. In imec's hyperspectral sensor, a large number of filters, e.g. 100, can be integrated and aligned with lithographic precision to very small pixels (5.5 micron pitch demonstrated). Thanks to this unique approach, compact, fast and cost-effective HSI cameras are available (Fig. 2), enabling the deployment of hyperspectral technology in the field.

Figure 1 200mm image sensor wafer onto which different hyperspectral filter designs have been processed. By varying the design, the number of bands, filter width, out of band blocking and the layout of the different filters over the pixel array, the hyperspectral imagers can be tuned to match the application requirements.

Figure 2 Prototype hyperspectral camera, with and without C-mount lens, integrating imec's hyperspectral mosaic sensor into a compact Ximea xiQ USB3 camera. A 2 euro coin and the 2 megapixel and 4 megapixel imagers are shown for scale.

978-1-4799-8002-4/14 $31.00 © 2014 IEEE

Sensor Design and fabrication

The core of a hyperspectral camera is the spectral unit, which is an optical component that implements the wavelength separation. The presented imagers use a set of Fabry-Pérot interferometers (6) with varying cavity lengths, integrated on top of the CMOS image sensor. Due to the thin-film filter technology, with a total thickness of ~1 micron, the mechanical dimensions of the imager are not modified, leading to extremely compact spectral cameras. Sensors with three different filter arrangements have been demonstrated (see Fig. 3-5):

- Line scan/wedge layout, where the filters are arranged in a staircase-like structure over the pixel array. This sensor is particularly suited for applications where the scene of interest has a natural translation movement (e.g. in a conveyor belt) and the hyperspectral imager will be used as a line-scanner. The acquired spectral images have 100 spectral bands and high spatial resolution, but require scanning.

- Snapshot/Tiled layout, in which filters are laid out in large squares on top of (groups of) pixels. Combined with an optical sub-system to replicate the scene onto each filter, this sensor enables real-time, low latency operation at video-rates, particularly suited for dynamic scenes. Acquired spectral images have 32 bands and lower spatial resolution, use an additional optical duplicator, but don't need scanning.

- Pixel-level mosaic layout, where the filters are arranged onto individual pixels, on a 5.5 micron pitch and in a 4x4 filter repeated configuration (Fig. 6), extending the traditional Bayer color imaging concept to multi- or hyperspectral imaging at video-rates without the need for dedicated fore-optics or linear scanning. Acquired spectral images have 16 bands, no additional optical duplicator, don't need scanning and can use demosaicing to improve spatial resolution.

Although the CMOS-compatible monolithic integration offers important advantages (i.e. cost-efficiency, improved performance and design flexibility), it also poses constraints: only processes and materials compatible with CMOS (in terms of thermal budget and contamination level) can be used. Since the performance of thin film optical filters depends on the optical parameters (refractive index and absorption coefficient) of the materials used to implement them, this limitation in material choice restricts the achievable spectral range, filter width and transmission efficiency. To overcome this limitation, imec has performed extensive material tuning, to enable the fabrication of sensors with a performance tailored to meet the requirements of very diverse applications. By varying the deposition conditions of a given material, its characteristics can be tuned to reach the optimal balance between its optical properties and the manufacturability. Combined with a design that takes into account different sources of variability during manufacturing, the approach can be ported to mass manufacturing.

(a)   (b)

Figure 3 (a) Conceptual representation of the hyperspectral line scan imager. Different spectral filters are arranged as a staircase, covering rows of pixels, where each spectral filter can cover one or more rows (both 8 and 16 rows per filter demonstrated, 100 spectral bands between 600 and 1000 nm). (b) A packaged hyperspectral line scan imager, in which the total filter stack thickness is less than 1 micron.

(a)   (b)

Figure 4 (a) Conceptual representation of the hyperspectral tiled snapshot imager. The 2 megapixel array is subdivided into 32 squares (4x8 arrangement), each 256x256 pixels tile is covered with a specific spectral filter. By using an optical duplicator element at camera level, the scene is images onto all 32 tiles in parallel. As such, 32 band hyperspectral video is possible, without the need for scanning, as all spectral information is captured in every frame. (b) Packaged sensor.

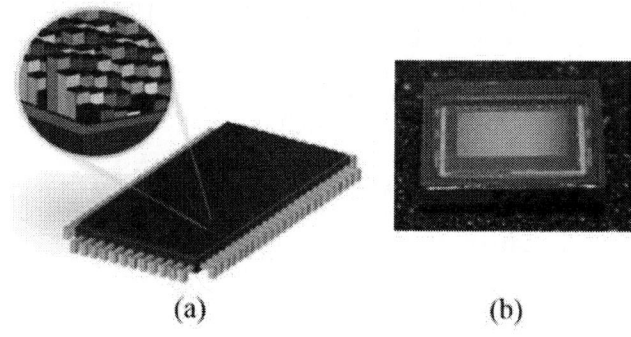

(a)   (b)

Figure 5 (a) Conceptual representation of the mosaic snapshot imager. Spectral filters are processed onto the pixel array in a 4 pixel x 4 pixel repeated pattern, each pixel measuring 5.5 x 5.5 micron. This enable the simultaneous acquisition of 16 spectral bands without the need for scanning or an optical duplicator. (b) Packaged mosaic sensor.

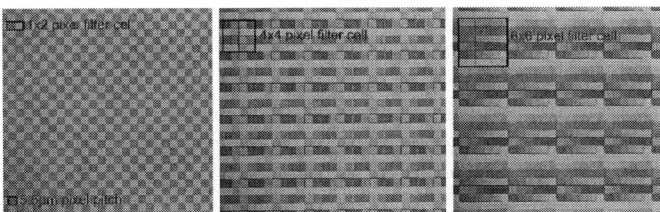

Figure 6 Microscope images of snapshot-mosaic imagers with different filter configurations. Each pixel (size 5.5 x 5.5 µm) within the cell is a different spectral filter. As can be seen, although our current mosaic prototype uses a 4x4 cell layout, other configurations (e.g. 1x2 cells or 6x6 cells) are possible.

Sensor evaluation

Table 1 summarizes the key specifications of our different prototypes. Fig. 7 shows the characterized spectral response of our wedge sensors, demonstrating a well-defined narrowband spectral response over a wide spectral range. These responses are obtained by illuminating the sensor with a collimated light and a monochromator setup (more information can be found in (3)).

Our per-pixel mosaic snapshot spectral sensors have been integrated into different commercially available cameras. Our prototype demonstrator cameras can acquire multispectral image cubes of 272x512 pixels over 16 bands, either in the VIS (470-620nm) or in the VNIR (600-1000nm). Fig. 8 shows the raw output image of the 4x4 VIS mosaic imager. This data was acquired without scanning, in an indoor office environment with standard lighting and camera settings (e.g integration time, gain). Fig. 9 provides a view of an image obtained using ambient outdoor illumination.

Fig. 10 illustrates the spectral response of the plant leaf, as measured using both our pixel-level mosaic snapshot sensor (Fig. 5) and our tiled snapshot sensor (Fig. 4). In order to interpret the reflectance results, a white reflectance tile was used to normalize for effects of illumination spectral power distribution, illumination non-uniformity, sensor quantum efficiency, vignetting and variations in filter bandwidth and peak transmission. The spectra of the tiled sensor have been plotted in green, while for the mosaic spectra a multitude of spectra from different cells is plotted in different colors, overlaid with red lines indicating the mean and ± 2 standard deviations. These spectra are consistent and both show the red edge of reflectance, which in leaves directly relates to the characteristics of chlorophyll (7). For certain applications, e.g. with wide field of views, the use of reflectance tiles may not be practical and additional efforts will have to be spent to calibrate the sensor output. However, since most of these (non-illumination related) effects are fixed, they may be addressed using calibration techniques similar to those used for flat-field correction and fixed pattern noise and dead pixel removal.

Table 1 Key specifications of our (a) line scan, (b) tiled snapshot and (c) mosaic snapshot imagers. For the line scan, a prototype on a CMOSIS CMV 4000 sensor (with 16 lines/band is also available. Although in all cases CMOSIS CMV imagers have been used, the spectral filters can be post-processed onto other imagers.

	(a)	(b)	(c)
Spectral range (nm)	600-1000	600-1000	470-620
Spectral bands/tiles	100	32	16
FWHM [a]	10 nm (using collimated light)		
Transmission efficiency	~85%		
Imager	CMOSIS CMV 2000 [b]		
Resolution	8 lines/band 2048 pixels/line	256x256 pixels/band	512x272 pixels/band
Scan/Frame rate	2880 lines/s	340 hypercubes/s	

[a] Full Width at Half Maximum
[b] For (a) sensors integrated on CMOSIS CMV4000 sensors are also available

Conclusion

Imec has developed a process for the monolithic integration of optical filters on top of the CMOS imager sensor, leading to compact, cost-efficient and faster hyperspectral cameras with improved performance. Fabricating the optical filters using solely techniques and tools derived from the IC industry, leads to a high degree of design flexibility. Customized filter layouts that match closely the application's requirements are therefore possible. To demonstrate the versatility of imec hyperspectral technology, prototype sensors with different filter arrangement and performance have been successfully fabricated: a 100-bands line-scan imager , a 32-bands tiled snapshot sensor and a 16-bands per-pixel mosaic snapshot imager.

References

(1) Y. Garini, I. T. Young and G. McNamara, "Spectral Imaging: Principles and Applications", Cytometry Part A 69A:735–747 (2006)

(2) J. Loesel and D. Laubier, Proc. of SPIE, 7100, 710013.1-710013.8 (2008).

(3) N. Tack, A. Lambrechts, S. Soussan and L. Haspeslagh, Proc. SPIE 8266, Silicon Photonics VII (2012).

(4) B. Geelen, N. Tack and A. Lambrechts, *Proc. SPIE* 8613, Advanced Fabrication Technologies for Micro/Nano Optics and Photonics VI (2013).

(5) B. Geelen, N. Tack and A. Lambrechts, *Proc. SPIE* 8974, Advanced Fabrication Technologies for Micro/Nano Optics and Photonics VI (2014).

(6) Dickman, "The Fabry-Pérot resonator", available at http://repairfaq.ece.drexel.edu/sam/MEOS/EXP03.pdf

(7) S. Seager, E.L. Turner, J. Schafer and E.B. Ford, Astrobiology, 5(3), 372–390 (2005).

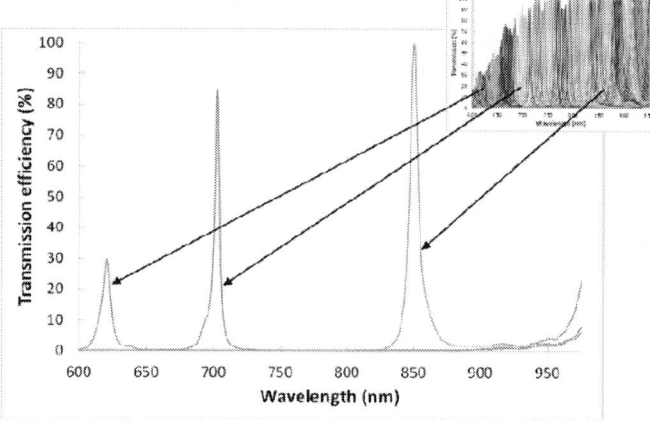

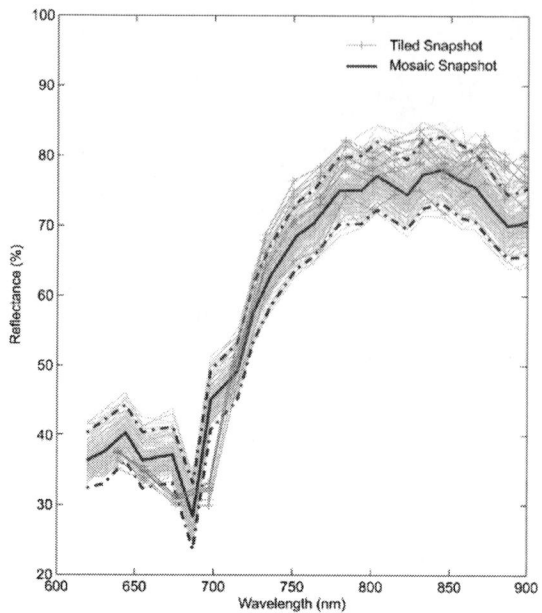

Figure 7 Three characterized filter responses for the 100 band, 600 to 1000nm linescan sensor, showing the filter transmission for the filters with peak responses at 620, 700 and 850nm. The full sensor response can be seen in the inset. Reported transmissions (which have been normalized) include the quantum efficiency of the sensor, without microlenses. The bands are equidistantly spaced within the range.

Figure 10 Measured reflectance spectra of a leaf, measured using tiled and mosaic snapshot sensors.

Figure 8 Raw output image of the imec mosaic imagers, combining 16 spectral bands onto a 4x4 grid of pixels.

Figure 9 Image of a rapidly moving outdoor traffic situation (reordered according to increasing wavelength) obtained with our 16-bands VIS mosaic sensor.

978-1-4799-8002-4/14 $31.00 © 2014 IEEE   264

Image Sensors for High-throughput, Massively-parallel DNA Sequencing: Requirements and Roadmap

Annette Grot, Pacific Biosciences, Menlo Park, CA 94025

Tel: (650) 521-8126, email: agrot@pacificbiosiences.com

Abstract: The cost of DNA sequencing has dropped significantly over the last decade, due in part to advances in high performance CCD and CMOS image sensors. Key performance specifications – such as resolution, sensitivity, and frame-rate, along with the performance improvements necessary for continued cost reduction – will be discussed.

Introduction: The first generation of sequencing instruments, developed in the 1970's, allowed scientists to read segments of DNA each roughly 1000 bases long[1]. Although this approach produced high fidelity reads, the throughput, or total number of bases read per day, was low because the process required a relatively large area on the sample plate.

With today's "next" sequencing instruments, sequencing occurs at specific locations by synthesis (a sequential addition of the complementary base to the single strand DNA template). The use of optical tags on the 4 different DNA base nucleotides together with high performance image sensor arrays has enabled dramatic increases in the number of parallel detection sites available on a single sample plate. Key parameters of these image sensors include resolution, sensitivity, and frame-rate. Advances in these parameters could help fuel the continued exponential decrease in DNA sequencing costs.

Resolution: In 2005, 454/Roche introduced the first "next generation" sequencer. Each sample plate is made up of millions of sites where clusters of DNA strands are immobilized [2]. Chemiluminescence is used to readout the sites where a nucleotide had been incorporated during each progressive wash-and-scan step. The plate is held in close proximity to the image sensor array both to maximize the number of sequencing sites and to reduce the complexity of the collection optics. As a result, the degree of parallelism is directly proportional to the resolution of the image sensor.

Other sequencing instruments, such as Applied Biosystems's (now ThermoFisher) SOLID, Illumina's HiSeq and Pacific Biosciences' RS II, use different color fluorescent tags on the nucleotides to read out the specific bases being sequentially incorporated along the strand [2]. A free-space optical system relays the fluorescence signals generated on the sample plate to the image sensor. Ideally, the resolution of the image sensor should match or exceed the optical resolution (space-bandwidth product) of the collection path. While very high NA microscope objectives have a space-bandwidth product around 4 million pixels, recent lower NA macrolens objectives can have even higher resolution. High resolution image sensors can also reduce the effects of optical cross-talk, but may in turn reduce the signal-to-noise ratio (SNR) and readout rate. This is an area where CMOS sensors are competitive with CCD arrays.

Sensitivity: Sensitivity refers to the minimum number of photons necessary to correctly identify a particular base in the sequence. High quantum efficiency at the pixel level is the main driver to reach high sensitivity. In addition, low light levels also require extremely low electronic read-noise and dark-currents. For those sequencing instruments which operate at relatively low frame-rates (<10fps), cooled CCD sensors are able to achieve some of the lowest dark current levels.

Frame-rate: Frame-rate is less critical in second generation sequencing instruments where each incorporation step is relatively long (many minutes). Because the sensor readout time is typically less than 1 sec, sequencing throughput can be increased by tiling several images together across the sample plate. During the "scan" step, the detection system scans across the plate taking 100s of images of different areas of the plate. On the other hand, the recently introduced PacBio RS II relies on the natural incorporation rate of the polymerase enzyme in order to achieve substantially longer read-lengths. The cameras operate at a much higher frame-rate in order to be able to correctly identify the incorporation events. Here CMOS image sensors offer a clear advantage over CCD arrays.

Future Improvements and Summary: The last decade has seen a tremendous increase in the throughput and read-lengths achievable in commercial DNA sequencing instruments. A large part of this technical advancement has been due to the availability of high resolution, low noise, high-speed image sensors. Subsequent generation sequencing instruments can only benefit from further improvements in performance as well as continued cost reduction.

978-1-4799-8002-4/14 $31.00 © 2014 IEEE

References:

1. Sanger F., Coulson A.R. *A rapid method for determining sequences in DNA by primed synthesis with DNA polymerase. J. Mol. Biol. 1975;94:441-448.*
2. Metzker M.L.*Sequencing technologies—the next generation. Nat. Rev. Genet. 2010;11:31-46.*

High Performance Silicon Imaging Arrays for Cosmology, Planetary Sciences, & Other Applications

Shouleh Nikzad, Michael E. Hoenk, John Hennessy, April D. Jewell, Alexander G. Carver, Todd J. Jones, Samuel L. Cheng, Timothy Goodsall, Charles Shapiro

Jet Propulsion Laboratory, California Institute of Technology
Pasadena, CA, 91109, USA
Phone: (818) 354-7496, Fax: (818) 393-4540, Email: shouleh.nikzad@jpl.nasa.gov

Abstract

High performance back-illuminated silicon arrays processed with precision atomic control of surface and interfaces are described. Precision atomic control of surface and interfaces parameters dictates quantum efficiency (QE), surface generated dark current, and cosmetic characteristics. Molecular Beam Epitaxy is used to grow delta layers and superlattice structures on the back surface of Si arrays. Photoelectron loss is greatly reduced and near 100% internal QE achieved. Photon losses are reduced by interface engineering using atomic layer deposition. Using these surface and interface treatments in different readout structures and formats creates a suite of devices with many applications. Si detectors spanning planetary science, astrophysics, medical diagnostics, machine vision, and commercial applications will be described.

Introduction

Silicon detectors have been widely used in scientific applications and commercial products since the invention of charge coupled devices (CCDs) in 1969 and their applications as imaging arrays [1,2]. NASA's Galileo mission marked the first use of CCDs in space, soon followed by Wide Field Planetary Camera (WF/PC) aboard Hubble Science Telescope (HST) [3]. Scientific discoveries and breathtaking images were obtained with this camera, which is modest in resolution and format by today's standards. Since the invention of Active Pixel Sensors (APS), the use of Si imaging arrays has exploded in commercial and consumer products based on large-scale production using complementary metal oxide semiconductor (CMOS) commercial foundries. In scientific applications, sensitivity or signal to noise ratio is the key parameter. In the same vein, QE is one of the most important metrics for scientific detectors.

Scientific ultraviolet (UV) observations often involve faint object imaging and spectroscopy. In addition to low flux sources, other challenges include high background from visible light, and in some astrophysics and planetary science observations, radiation-induced noise. This is in part why NASA UV instruments have used image-tube technologies with visible-blind photocathodes, such as microchannel plates (MCPs) and image intensified CCDs or CMOS arrays. Solid-state detectors offer many advantages over image-tube technologies, provided that challenges in UV sensitivity, out of

band rejection, and photon counting capability are successfully addressed.

We show in this paper that these challenges can be overcome by using precision atomic control of surface and interfaces. We briefly describe the physics of band structure engineering in back illuminated Si arrays, followed by a description of processes. We then discuss three applications. Our results demonstrate that back-illuminated Si detectors can replace image-tube based detectors, enabling higher efficiency and reliability.

Precision Atomic Control of Surfaces in Back-illuminated Detectors

The concept of back illumination was created to reduce the losses in the front circuitry of devices, effectively increasing the QE and expanding the spectral response into the UV range. Today back illumination has become commonplace even in commercial applications. Many of the processes needed for back illumination were initially developed for scientific detectors, with WF/PC-1 leading the way as the first implementation of the idea in space. Because of defects formed near the back surface of Si and near the Si-SiO$_2$ interface, the WF/PC-1 detectors exhibited QE hysteresis (QEH), which is characterized by low and unstable sensitivity, especially at the blue end of spectrum. QEH is a sign of poor passivation of the back surface, where trapping of photogenerated electrons and holes leads to time-variable surface charge. State-of-the-art ion-implanted CCDs used in space still exhibit QEH at a level of several percent, testament to the importance and difficulty of surface passivation [4].

Surface passivation of back-illuminated detectors is challenging because the standard processes would subject devices to excessively high temperatures, and because the passivated surface of the detector is necessarily exposed to the environment. Low temperature surface passivation technologies have been developed, but many suffer from lower QE or QE instability as a function of environment or illumination history. The precision control of surface band structure engineering and interface engineering afforded by superlattice doping and delta doping using molecular beam epitaxy shows high QE without any QEH [5–7].

Silicon detectors and imaging arrays are fabricated in CCD, CMOS or Avalanche Photodiode (APD) Array formats and readout structures in foundries. Fully fabricated arrays, complete with final metallization are received at JPL's Microdevices Laboratory (MDL). Different device structures and applications have slight variations in the process flow. Here we describe the nominal steps and processes to provide an overview of the process flow for producing high performance, back-illuminated Si arrays.

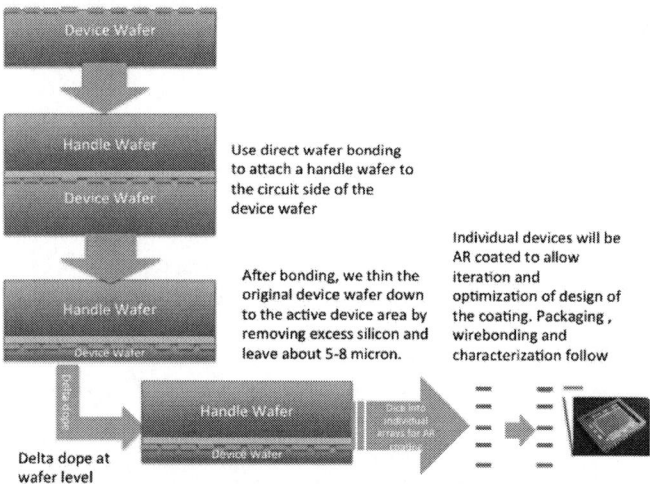

Fig. 1. Process flow of delta doped and superlattice doped arrays. Note that post delta doping, wafers can be coated (with antireflection layers or integrated filter) and then diced.

Fig. 2. Photos of with fully fabricated devices at different stages of post fabrication processing.

The fabrication starts with protecting the VLSI fabricated circuitry and pixel structure by using direct wafer bonding to attach a Si wafer to the device wafer's front surface. Because of the necessity of a 400 °C temperature in later processes, epoxy and organics are precluded. Following direct wafer bonding, the bulk of the detector substrate is removed by grinding and chemical mechanical polishing, reducing the device wafer thickness from ~800 microns to ~50 microns. The remainder of the substrate is removed by a selective thinning process, using an isotropic chemical etchant that removes the bulk P+ Si at a rate of >100:1 compared to the P-epilayer, essentially rendering the epilayer as an etch stop. Final polishing produces a smooth mirror finish surface that

will be subjected to a simple series of steps to prepare the surface for epitaxial growth.

The backside-thinned detector wafer is then passivated by molecular beam epitaxial (MBE) growth of 2D-doped Si layers, comprising single or multilayer stacks of delta-doped Si. MBE growth provides the precise band structure engineering of the Si surface to allow detection of higher energy photons (100 nm < λ < 400 nm) that are absorbed in shallow depths 4-10 nm with minimal to no loss of photoelectrons [6,8].

After passivation, wafers can be coated, diced and packaged. Atomic layer deposition (ALD) of AR coatings or filters can be performed on individual die, wafer sections with multiple die, or entire wafers. As described in the next section, ALD enables atomic layer control of structure and composition, including multilayer coatings with embedded metal for bandpass coatings with high in-band QE and excellent out-of-band rejection [9].

Atomic Layer Deposition of Coatings and Filters

Many of the concepts of antireflection coatings and visible rejection filters have been developed over the years, but could not be implemented with accuracy or fidelity. Atomic Layer Deposition (ALD) is a technique that is a close spinoff of chemical vapor deposition (CVD). ALD films are created by forming a single atomic layer at a time through a series of self-limiting chemical reactions with the substrate surface; sequential layers are separated by purging reaction byproducts after each layer is formed. This chemical-reaction-driven process allows for the precise control of stoichiometry, thickness, and uniformity, while producing highly dense and pinhole free films with sharp and well-defined interfaces in multilayer coatings. These characteristics make ALD nearly ideal for multilayer filters or antireflection coatings.

We have employed this technique to develop highly effective antireflection coatings in the challenging far UV spectral range [10,11]. Here, we describe their use in visible rejection, or more generally, out-of-band rejection filters that can be integrated onto the detector surface. Preliminary results of this work were recently reported based on die-level coating of APD detectors [6]. Fig. 3 shows QE measurements from superlattice-doped, APDs detectors, with both AR coatings and multilayer filters. MBE passivation and ALD coating were done at wafer scale. A comparison of dielectric-only coating with metal-dielectric multilayer stack demonstrates the concept. With a single embedded metal layer, a rejection ratio of an order of magnitude is achieved. With multiple embedded metal layers, rejection ratios from 10^4 to 10^7 are projected. This significant achievement now makes possible UV-sensitive Si detectors with tailorable visible- and solar-blindness.

Filtering technique is not limited to rejecting visible light but can be used to tailor the desired response of the detector over the required band while rejecting out-of-band photons. An example of that has been implemented for the far ultraviolet where important spectral lines of hydrogen and oxygen allow

978-1-4799-8002-4/14 $31.00 © 2014 IEEE

peering through the far ultraviolet window for planetary atmospheres in order to, for example, investigate the habitability of planets (Fig. 4).

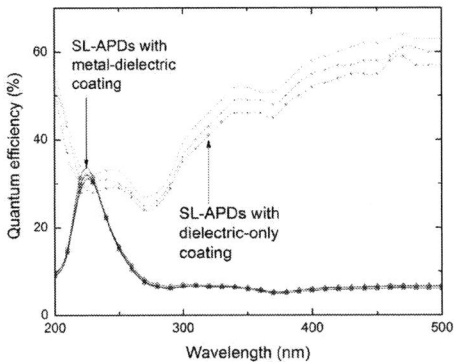

Fig. 3. ALD multilayer stacks of metal-dielectrics films. The results presented here represent the application of superlattice doping and AR coating technologies to APDs fabricated and characterized by RMD.

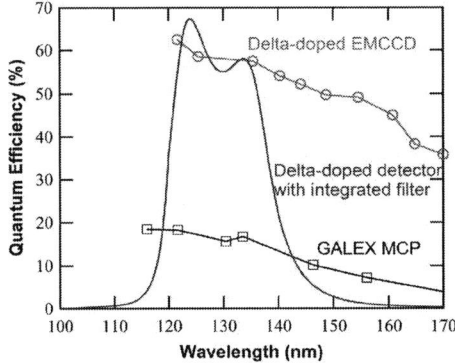

Fig. 4. Solar blind bandpass filter for planetary spectroscopy at far ultraviolet wavelengths. This multilayer metal-dielectric filter was designed to provide high in-band QE, and high out-of-band rejection (>10⁴).

A Brief Overview of Applications

The techniques described here can be applied to practically any Si imaging array. Depending on the application, readout structures, spectral range, out of band rejection, noise, QE, and more can be designed and customized.

In ground-based astronomy it is desirable to cover the wide spectral range in a single detector. A design that can cover the full spectral range offers system simplification and cost reduction. By extending typical Si response beyond the visible into the near ultraviolet and near infrared allows efficient detection in 320-1000 nm for ground-based astronomy [9]. By superlattice doping of Si CCDs or PIN diode arrays that have been especially designed and fabricated in thick, high purity Si [12–14], it will be possible to extend the response to blue and use the delta layer as an electrode for applying bias and fully deplete the 100-300 micron thickness of detectors.

Detecting extremely faint objects in cosmology and planetary science often requires photon counting ultraviolet imaging with narrow or medium bandpass. Photon counting detectors

typically use avalanche gain to amplify very faint signals, including ultraviolet sources in the intergalactic medium, traces of water in planetary atmospheres, and fluorescence signals from cancerous cells. The high electric field required for avalanche gain is achieved by increasing bias voltages (linear or Geiger mode). Avalanche photodiodes can be made into arrays and hybridized with CMOS readout arrays, avalanche mechanism can be included in CMOS pixel as in Single Photon Avalanche Photodiode (SPAD) [15], or avalanche gain can be can be introduced in a CCD structure by adding a an additional serial register [16,17].

We have implemented the post-fabrication processing described in this paper to avalanche photodiodes manufactured by Radiation Monitoring Devices (Fig. 3), and to electron-multiplying CCDs (EMCCDs) manufactured by e2v [6,10]. EMCCDs are two or four phase CCDs with an additional register, the so-called multiplication or gain register. With clock voltages of ~40-45 volts applied to the gain register, an electron gain of a thousand or more can be achieved. Note that the voltages required for EMCCD gain are lower by several orders of magnitude when compared to image tube detectors such as microchannel plates (MCPs).

It is important to evaluate the response of Si detectors for environments to study supernovae, planetary atmospheres, intergalactic medium and planetary surfaces. Radiation environments pose challenges that are mitigated by shielding. Fig. 5 demonstrates that a superlattice doped EMCCD introduced to increasing total dose exposures behaves essentially as expected.

Applications of terrestrial nature, such as machine vision and wafer and reticle inspection, also require detectors to be exposed to challenging radiation environments. The semiconductor industry requires back-illuminated CMOS imaging arrays to detect deep ultraviolet laser pulses with high QE and stable response. This is a challenging radiation environment, as detectors are blasted by many billions of deep UV excimer laser pulses over the instrument lifetime. DUV-induced damage poses severe problems of cost and reliability, as detectors passivated with ion implantation or oxide charging methods degrade rapidly under intense DUV illumination [18]. In collaboration with Alacron, Inc. and Applied Materials, we delivered superlattice-doped CMOS imaging arrays for integration and test of DUV cameras. Lifetime tests were performed by Applied Materials using pulsed DUV lasers at 193 nm and 263 nm. Superlattice-doped detectors exhibited QE greater than 60% at these wavelengths, with no measurable degradation in QE or dark current after exposure to billions of DUV laser pulses [5–7]. This stability is a unique capability enabled by the 2D doping methods developed at JPL [19].

Finally, as in astronomical observations, medical diagnostics requires detectors that are capable of measuring faint signals in the presence of out-of-band photons. Often this has to be achieved at high speed and in compact and low power instruments. High QE, high resolution imaging has been used

978-1-4799-8002-4/14 $31.00 © 2014 IEEE

for many years for medical diagnosis, mostly using x-rays [20], and more recently infrared imaging has been used for cancer detection [21]. Use of ultraviolet imaging for medical diagnostics is a relatively recent endeavor. Examples include tumor delineation aided by injected dyes that fluorescence in the near ultraviolet and skin cancer detection. The sensitivity and high resolution described here implemented in devices with fast readout rate can enable these challenging measurements for potentially less invasive diagnostic intraoperative tools.

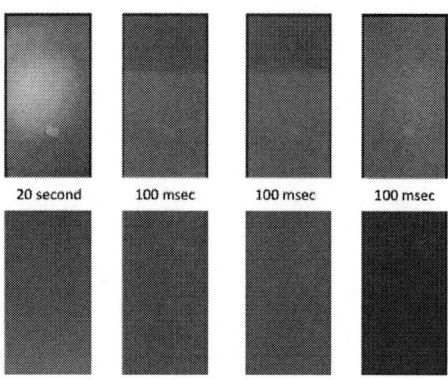

Fig. 5. Comparison of images of device before and after irradiation. The dark portion in top images are due to readout setup and not due to device.

Conclusion & Future Work

High QE back-illuminated Si detector arrays and their applications were described. Molecular beam epitaxy is used for surface band structure engineering to greatly reduce or remove traps for photoelectrons resulting near 100% internal QE. Atomic layer deposition is used to further tailor the response of the imaging array for given spectral range and applications. Scientific applications in cosmology, astrophysics and planetary science as well as machine vision and medical applications are summarily discussed. Radiation testing of delta-doped arrays showing normal performance were reported. In-band high QE is demonstrated and more significantly integrated filters that effectively reject out of band photons are demonstrated. This unprecedented results show that high performance Si imaging arrays have extensive use in scientific imaging and imaging spectroscopy in visible and ultraviolet parts of the spectrum.

Acknowledgements

The authors thank Mickel McClish and Richard Farrell of Radiation Monitoring Devices for QE measurements of superlattice-doped, AR-coated avalanche photodiodes. The authors thank Joseph Sgro of Alacron Inc. and Shraga Tsur of Applied Materials for their work on building and characterizing cameras with superlattice-doped CMOS detectors. This work was performed at Jet Propulsion Laboratory, California Institute of Technology, under a contract with NASA.

References

[1] S. Nikzad et al., "Digital Imaging for Planetary Exploration," in *Handbook of Digital Imaging*, M. Kriss, ed. (Wiley, 2015).

[2] F. Greer et al., "Atomically precise surface engineering of silicon CCDs for enhanced UV quantum efficiency," J. Vac. Sci. Technol. A **31**, 01A103 (2013).

[3] J. Trauger, "Sensors for the Hubble Space Telescope Wide Field and Planetary Cameras (1 and 2)," in *CCDs in Astronomy: Proceedings of a Conference Held in Tucson, Arizona, 6-8 September 1989* (Astronomical Society of the Pacific, 1990), pp. 217–230.

[4] N. R. Collins et al., "Wide Field Camera 3 CCD Quantum Efficiency Hysteresis: Characterization and Mitigation," in *Proc. SPIE 7439, Astronomical and Space Optical Systems*, P. G. Warren, C. J. Marshall, R. K. Tyson, M. Lloyd-Hart, J. B. Heaney, and E. T. Kvamme, eds. (2009), p. 74390B.

[5] M. E. Hoenk et al., "The DUV Stability of Superlattice-doped CMOS Detector Arrays," in *International Image Sensor Workshop* (2013).

[6] M. E. Hoenk et al., "Superlattice-doped silicon detectors: progress and prospects," in *Proc. SPIE 9154, High Energy, Optical, and Infrared Detectors for Astronomy VI*, A. D. Holland and J. Beletic, eds. (2014), p. 915413.

[7] M. E. Hoenk et al., "Superlattice-doped Imaging Detectors: Structure, Physics and Performance," in *Scientific Detectors Workshop* (2013).

[8] S. Nikzad et al., "High Throughput, High Yield Fabrication of High Quantum Efficiency Backilluminated Photon Counting, Far UV, UV, and Visible Detector Arrays," in *International Image Sensor Workshop* (2013).

[9] A. D. Jewell, J. Hennessy, M. E. Hoenk, and S. Nikzad, "Wide band antireflection coatings deposited by atomic layer deposition," in *Proc. SPIE 8820, Nanoepitaxy: Materials and Devices*, N. P. Kobayashi, A. A. Talin, A. V. Davydov, and M. S. Islam, eds. (2013), p. 88200Z–88200Z–9.

[10] S. Nikzad et al., "Delta doped Electron Multiplies CCD with Absolute Quantum Efficiency over 50% in the near to far Ultraviolet Range for Single Photon Counting Applications," Appl. Opt. **51**, 365–369 (2012).

[11] E. T. Hamden et al., "Ultraviolet antireflection coatings for use in silicon detector design," Appl. Opt. **50**, 4180 (2011).

[12] S. E. Holland et al., "Fabrication of back-illuminated, fully depleted charge-coupled devices," Nucl. Instruments Methods Phys. Res. Sect. A **579**, 653–657 (2007).

[13] Y. Bai et al., "Teledyne Imaging Sensors: silicon CMOS imaging technologies for x-ray, UV, visible, and near infrared," in *SPIE Proceedings*, D. A. Dorn and A. D. Holland, eds. (2008), pp. 702102–702102–16.

[14] Y. Bai et al., "Hybrid CMOS focal plane array with extended UV and NIR response for space applications," in *Proc. SPIE 5167, Focal Plane Arrays for Space Telescopes*, T. J. Grycewicz and C. R. McCreight, eds. (2004), pp. 83–93.

[15] E. Charbon et al., "SPAD-Based Sensors," in *TOF Range-Imaging Cameras*, F. Remondino and D. Stoppa, eds. (Springer, 2013), pp. 11–38.

[16] P. Jerram et al., "The LLCCD: low-light imaging without the need for an intensifier," in *Proc. SPIE 4306, Sensors and Camera Systems for Scientific, Industrial, and Digital Photography Applications II*, M. M. Blouke, J. Canosa, and N. Sampat, eds. (2001), pp. 178–186.

[17] J. Hynecek, "Impactron-a new solid state image intensifier," IEEE Trans. Electron Devices **48**, 2238–2241 (2001).

[18] F. Li and A. Nathan, *CCD Image Sensors in Deep-Ultraviolet: Degradation Behavior and Damage Mechanisms* (Springer, 2005), p. 231.

[19] M. E. Hoenk, "Surface Passivation by Quantum Exclusion Using Multiple Layers," U.S. patent 8,395,243 (2013).

[20] S. Nikzad, "High-performance silicon imagers and their applications in astrophysics, medicine and other fields," in *High Performance Silicon Imaging: Fundamentals and Applications of CMOS and CCD Image Sensors*, D. Durini, ed. (Elsevier, 2014), pp. 411–438.

[21] S. D. Gunapala et al., "Quantum Well Infrared Photodetector Technology and Applications," IEEE J. Sel. Top. Quantum Electron. **20**, 1–12 (2014).

Detecting elementary particles using Hybrid Pixel Detectors at the LHC and beyond

Michael Campbell, PH Department, CERN, 1211 Geneva, Switzerland

Abstract

At the CERN Large Hadron Collider bunches of high energy protons (and ions) collide at the heart of 4 massive experiments. Hybrid pixel detectors identify and tag individual particle tracks helping physicists to select the handful of interesting events from the millions created per second. Applications beyond LHC will also be described.

Introduction

On July 4th 2012 CERN announced the discovery of the Higgs Boson at the Large Hadron Collider. Englert and Higgs were awarded the Noble Prize for Physics in 2013 for postulating the existence of the boson along with Brout (now deceased) in 1964. The discovery was made possible by the combination of a machine capable of accelerating protons to unprecedented energies, and two huge detectors, called Atlas and CMS, able to record unambiguously the trajectories and energies of the particles produced by the proton collisions. Every 25 or 50ns bunches of protons are made to collide in the heart of the experiments and on average 20-30 proton interactions take place generating thousands of debris particles. In searching for the Higgs boson, the particles produced in a given interaction need all to be detected and tagged to a given bunch crossing (BC).

In the innermost regions of the detector the trajectory of charged particles must be recorded as precisely as possible. As charged particles produce around 80 electron-hole (e-h) pairs per μm in silicon one could imagine that detectors based on CCDs or regular CMOS sensors would be adequate. However, the needed time tagging precision imposes the use of high bandwidth detection circuitry and, with the very small signals available from such sensors at the time of the LHC construction, it was impossible to achieve a sufficiently high signal to noise ratio. Moreover, these devices are subjected to very high radiation doses throughout their lifetime and CCD transistors or CMOS diodes would not survive. This lead to the choice of hybrid pixel detectors to equip the inner layers of the experiments. These detectors offer 100% fill factor, and use high density CMOS circuitry to amplify the sensor signal and tag it to a given BC.

This paper will introduce hybrid pixel detectors and provide a brief overview of the large scale systems used at the LHC experiments. It will also describe how the same hybrid pixel detector approach is used in applications beyond high energy particle physics.

Hybrid pixel detectors at the LHC

The basic building block of a hybrid pixel detector is the combination of a monolithic matrix of reverse biased silicon diodes connected using high density bump bonding to a number of large area CMOS readout ASICs (1). Each individual sensor diode is connected using a solder bump to its own mixed-mode readout circuit on the ASIC side. An example of such a 'ladder' is shown in Fig 1. The sensor, which is made of high resistivity material (typically > 5kΩcm), is operated in full depletion allowing all of the

charge generated during one particle traversal of its ~200μm thickness to be collected promptly (within ~15 ns), see Fig 2. The readout circuit compensates any radiation-induced detector leakage current, integrates the charge pulse associated with the particle traversal, filters the resulting signal to minimize the electronic noise, determines if the signal is above a pre-defined threshold and, if so, associates the arrival time to a given BC. It might also record the amplitude of the collected charge. A typical pixel readout channel comprises a few hundred transistors. Probably the most important feature of hybrid pixel detectors is their ability to provide practically noise free 'images' of each BC. Analyzing each 'image' or 'frame' it is possible to disentangle the (on average) 20-30 simultaneously occurring events from each other. The front-end readout circuits typically have an input referred noise of around 100e- rms and are operated with a detection threshold of a few thousand electrons. As a typical particle deposits around 2.7fC (17 000 e-) spread over a few pixels one can be sure that all particle traversals are detected while essentially no noise hits contribute ambiguity to the 'image' of the BC. As these systems are in the heart of the experiments where the accumulated radiation dose is highest, the quantity of collected charge of each particle traversal degrades significantly with time, but, because of the high threshold to noise ratio of the electronics, an excellent detection efficiency is expected to be maintained throughout the lifetime of the experiment (~10 years). The ASICs used during the first LHC runs are designed in a 250nm CMOS process applying design techniques which guarantee a minimal radiation-induced degradation of performance throughout the expected lifetime (2).

Each of the four major LHC experiments uses a sub-detector system based on hybrid pixel detectors. In the case of Alice (3), Atlas (4), and CMS (5) these systems are deployed around the innermost part of the detector core, nearest to the particle interaction points. Given that their aim is to measure the vertices from which the particles emerge, these systems are referred to as vertex detectors. The pixel geometry, the number of pixels per chip, and the total area covered by the detectors vary between experiments and the main features are summarized in TABLE 1. An image of the Atlas pixel detector during construction is shown in Fig. 3 and some LHC interactions which have been reconstructed with the aid of data from the Atlas pixel detector are shown in Fig. 4.

One of the sub-detectors of the LHCb experiment aims to measure the momentum of each particle passing through a radiator which emits ultraviolet photons in a ring whose diameter is proportional to the momentum of the particle. Photons in the uv wavelength can generate a single e-h pair in Si or another suitable converter material. As hybrid pixel ASICs typically have a minimum detection threshold of many hundreds of electrons the direct detection of single electrons using a silicon sensor is impossible. The LHCb RICH photodetector, or pixel-HPD, contains a single hybrid pixel detector which acts as a photo-anode inside an evacuated phototube with a multialkali (S20) photocathode, see Fig. 5 (6). A single uv photon generates an electron by by photo-electric interaction with the photocathode. The

978-1-4799-8002-4/14 $31.00 © 2014 IEEE   271

electron is then accelerated and cross-focused onto the back of the Si sensor. In the LHCb pixel-HPD, the 5cm diameter entrance window is focused onto the back of a 16 mm x 16mm square Si diode array. An accelerating voltage of 20kV is applied and each electron then generates ~5000 e-h pairs in the high resistivity sensor. Around 500 of such tubes were installed in two systems at the experiment. Fig. 6 shows one event which has been produced from one plane (7). Once again the 'noise-free' operation of the detection system is evident.

Hybrid pixel detectors at the LHC are only one sub-component of the full detector which is composed of many different sub-systems. One of the greatest challenges at the LHC is to select from the 100's of millions of particle interactions being created per second, only those which are likely to contain new physics. This is done using a series of 'triggers' – electronic flags which identify potentially interesting BC's and are used to reduce the data volume being pushed off detector to a manageable level. The selection of a given BC (usually performed at either the pixel or pixel column level) is typically carried out using the 'first level trigger' which is generated by a combination of signals from the other detector sub-systems. As it takes a finite time to generate this trigger, it typically becomes available only ~100 BC's following a BC of interest. This reduces the number of 'images' to be read out by a factor of up to 1000. However, it also implies that each ASIC has to store each 'image' or 'frame' for up to ~100 BC's and then either send the data off-chip or dump it. While this functionality is essential in the LHC context it is rarely relevant to other applications where such triggers and often not available.

Hybrid pixel detectors in other applications:

It became clear fairly early on during the development of pixel detectors for LHC that such devices could be used in other applications. In particular, the idea of counting single X-ray photons was already being explored in the late 90's (8, 9, 10). These applications require a change in the ASIC readout architecture. Instead of selecting individual frames for readout, all hits above threshold in a given pixel are counted and then the entire accumulated image is sent off chip. This provided X-ray scientists with a new capability of imaging without the noise due to dark current which diminishes the performance of conventional integrating readout systems, especially in low flux environments (11), see Fig. 7.

X-rays of below ~10keV are fully absorbed and therefore detectable in standard high resistivity silicon. However, in medical diagnostics, where energies of 20keV-140keV are used (in order to have enough photons able to pass through the patient to form an image), higher stopping power sensors are needed. Counting chips, such as Medipix2, turned out to be extremely useful in characterization of the new materials leading to improvements in production processes which have helped in understanding how to make more uniform materials. A nice recent example of an X-ray image (12) taken using a GaAs sensor connected to the Medipix3 chip is shown in Fig. 8.

A whole spectrum of other applications were also addressed using such chips, most of which involved detecting charged particles (such as electrons or ions) directly. In other cases neutral particles produce charged particles in a converter material and these are subsequently detected. An example of the latter case is the detection of alpha or beta particles emitted during neutron absorption (13).

An alternative approach to single visible photon detection to the one developed by LHCb is shown in Fig. 9 (14). In this case a single photo-electron drifts under the influence on an electric field into the pore of a micro-channel plate (MCP) and, through collisions with the side of the pore under a very high electric field, the electron is multiplied many times producing a cloud of electrons at its base. A further drift field between the base of the MCP and the naked readout chip causes the charge cloud to be attracted to the input pad of the chip and this is detected by the readout circuit, just like a moving charge cloud in a sensor.

Ongoing and future developments of hybrid pixel detectors

The hybrid pixel ASICs developed for the first runs at LHC used a 250nm CMOS process but for the upgrades to the experiments deeper sub-micron processes are being explored. In the case of the upgrades for Atlas (15) and CMS (16) the aim is to use reduced chip and sensor thicknesses, at the same time modifying the readout ASIC architectures to be able to deal with the higher particle fluxes expected from the upgraded LHC accelerator. In the case of LHCb, a readout architecture is being implemented whereby all hit time stamps and amplitudes are sent off chip independently of an external trigger (17). The Timepix3 chip (18) which implements a data-push architecture already contains many of the features needed for the LHCb pixel readout chip. In the Alice experiment, the particle interaction rate and foreseen total accumulated radiation dose are much smaller than for Atlas and CMS. This allows them to work with reduced bandwidth detection circuitry and electronic noise. The aim is to use ultra-thin monolithic devices where the sensor and readout circuits are incorporated onto the same substrate (19), making use of special semiconductor processes which were developed for CMOS sensors.

In synchrotron science large, high–speed X-ray imaging systems have been and are being developed, often by groups and spin-off companies of groups who were originally involved in the design of the LHC pixel detectors. The Pilatus systems available form the company Dectris (whose first products were a spin off from the developments for CMS) have become the benchmark in this field (20).

Also in the synchrotron field, the development of free electron lasers (FEL's),themselves based on technologies and concepts developed for a future linear particle collider, has forced a move away from the photon counting readout approach. At FEL's the photons arrive in trains of closely spaced bunches and each bunch is some femtoseconds long. One wants to measure the number of photons impinging on a given pixel for each bunch. Typically a capacitor bank is integrated into every pixel and each capacitor contains a voltage (or charge) proportional to the charge detected in one bunch. The capacitors are read our between bunch trains. At the future European XFEL the ultimate aim would be to save 3000 frames at a rate of 5MHz and then read them out in the ~100ms inter bunch train gap (21). A number of developments are underway, but using present day CMOS only a fraction of frames can be recorded within a reasonable pixel size.

As most hybrid pixel systems detect particles one-by-one the hits can be binned according to their energy and this opens the way to spectroscopic X-ray imaging. In medical X-ray imaging there is a desire for better spectroscopic performance using high stopping power sensor materials. In these materials, as well as charge sharing between pixels due to carrier diffusion during charge drift, fluorescence photons can cause several hits to be produced for one impinging photon. Unless these events are properly treated the spectroscopic resolution is severely degraded. The Medipix3 architecture which uses simultaneous inter-pixel charge summing and hit allocation (22) has been shown to mitigate the degradation of energy resolution due to these effects, see Fig. 10. However, for applications such as spectroscopic computed tomography, the chips must be able to deal with extremely high hit rates. Such improvements may (finally) allow the introduction of spectroscopic X-ray imaging (with all of the exciting possibilities offered by new nano-particle based contrast agents) into clinical practice (23).

Engineers and physicists who develop hybrid pixel detectors try to make the most of progress in CMOS technology. Over the years pixel size has shrunk while the complexity of the circuit has grown. There is no end in sight to this trend at the time of writing. In high energy physics the push towards smaller pixel size is driven by a desire for better spatial resolution, but this only makes sense if the thickness of the materials used can be reduced. In fact, some groups who have worked on hybrid solutions in the past are now pursuing monolithic approaches. The challenge they face is to provide good signal to noise ratio at the required time tagging precision. Other groups are looking at reduced pixel pitch using newer CMOS processes while maintaining the hybrid approach (24). Other groups aim to produce sub-ns timing precision (25). In fields such as X-ray imaging one can try to use the advances in CMOS technology to develop circuitry which compensates for detector inhomogeneities. Through Si Via (TSV) technology offers the potential to tile a large area seamlessly.

References

(1) L. Rossi, P. Fischer, T. Rohe, N. Wermes, Pixel Detectors, Springer Publishing (2006) ISBN 978-3-540-28333-1.
(2) G. Anelli, M.Campbell, M. Delmastro et al, IEEE Trans. Nucl. Sci. Vol 46 pp 1690-1696 (1999)
(3) F. Meddi, Nucl. Instr. and Meth. A 465 (2001) pp 40-45.
(4) V. Vrba, Nucl. Instr. and Meth. A 465 (2001) pp 27-33..
(5) D. Bortello, Nucl. Instr. and Meth. A 579 (2007) pp 669-674.
(6) N. Styles, Nucl. Instr. and Meth. A 610 (2009) pp 57-60.
(7) D. L. Perego, Physics Procedia 37 (2012) pp. 606 – 612.
(8) M. Campbell, E.H.M.Heijne, G. Meddeler, E. Pernigotti, S. Snoeys., IEEE Nucl. Sci. Symp. (1997), pp 189-191
(9) P. Fischer, J. Hausmann, M. Overdick et al., Nucl. Instr. and Meth. A 405 (1998) pp 53-59.
(10) P. Delpierre, J.F.Berar, L. Blanquart, B. Caillot, J.C. Clemens, C. Mouget, , IEEE Nucl. Sci. Symp. (2000), pp 3/22-3/24 vol 1.
(11) J. Jakubek, Nucl. Instr. and Math. A 576 (2007) 223
(12) Image produced by S. Procz, FMF Freiburg, Germany
(13) J. Jakubek, C. Granja, T. Holy et al., Nucl. Instr. and Math. A 576 (2007) 223.
(14) J. Vallerga, J. McPhate, A. Tremsin et al.,. Nucl. Instr. and Math. A 546 (2005) 263-269.
(15) Clark, Elsing, Hessey, et al., ATLAS Phase II Letter of Intent: Backup Document. ATL-UPGRADE-PUB-2012-004. CERN, Geneva, Switzerland, 2012.
(16) H. Perrey, Proceedings of Science (Vertex2013) 014.
(17) M. van Beuzekom, Proceedings of Science (Vertex2013) 016.
(18) T Poikela et al 2014 JINST 9 C05013.
(19) M. Sitta, Proceedings of Science (Vertex2013) 018.
(20) Kraft, P. ; Bergamaschi, A. ; Bronnimann, C. ; Dinapoli, R. et al., IEEE Trans. Nucl. Sci. Vol 56 (2009), pp 758-764.

(21) A. Koch, M. Kuster, J. Sztuk-Dambietz and M. Turcato, 2013 J. Phys.: Conf. Ser. 425 062013.
(22) R. Ballabriga, G Blaj, M Campbell, M Fiederle et al., J. Instr. 2011 1748-0221 6 C01052
(23) M F Walsh, S J Nik, S Procz, M Pichotka, et al., J. Instr. 2013 1748-0221 8 P10012
(24) P. Valerio, J Alozy, S Arfaoui, R Ballabriga, et al., J. Instr. 2014 1748-0221 9 C01012
(25) M Noy, G Aglieri Rinella, M Fiorini, P Jarron, et al., J. Instr. 2011 1748-0221 6 C01086

TABLE 1. COMPARISON OF THE ALICE, ATLAS AND CMS HYBRID PIXEL DETECTOR SYSTEMS.

	Alice	Atlas	CMS
Readout chip name	Alice1LHCb	FEI3	PSI46
Pixel size (in μm)	50 x 425	50 x 400	100 x 150
Pixels/chip	256 x 32	160 x 18	80 x 52
Chips/ladder	5	16	16
Total # chips	1200	28 000	16 800
Total area (m²)	0.2	1.8	1

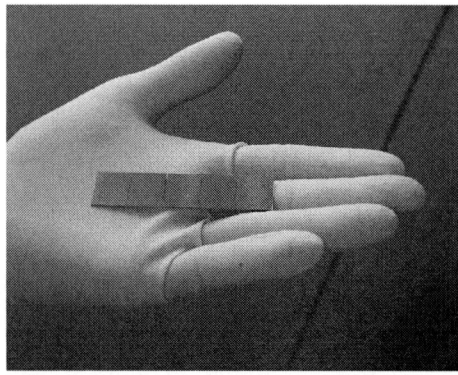

Figure 1. A hybrid pixel detector 'ladder' for the Alice experiment. A pixelated sensor is connected using micro-bumps to 5 ASIC readout chips. The rear side of the ASICs can be seen here. Each chip contains 8000 readout channels each measuring 50μm x 425μm.

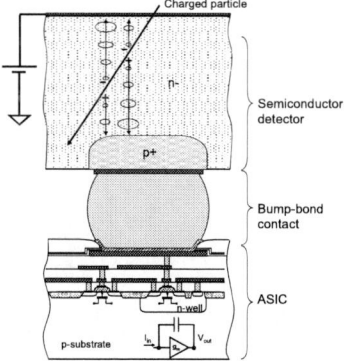

Figure 2. A cross sectional view of a single pixel of a hybrid pixel detector. A charged particle deposits e-h pairs within the fully depleted volume of the sensor. The charge drifts in the electric field creating a fast current signal on the diode electrode which is then transmitted to the readout circuit via the solder bump bond.

Figure 3. The Atlas pixel detector under construction.

Figure 4. Reconstruction of a multiple vertices in a single BC in the Atlas experiment. This image represents about 60cm in length.

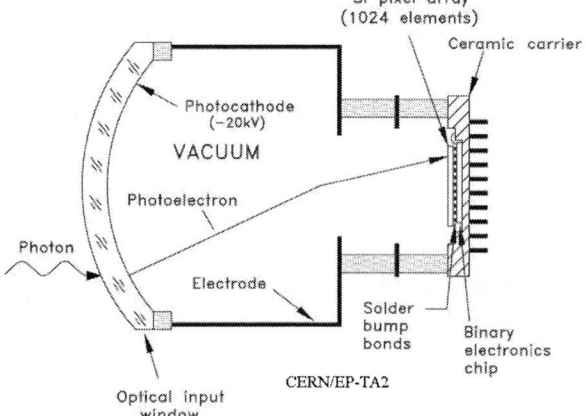

Figure 5. A schematic diagram of the LHCb pixel Hybrid Photon Detector (pixel-HPD). A photon impinging on the photocathode kicks out an electron which is accelerated and cross-focused onto the back of a hybrid pixel detector.

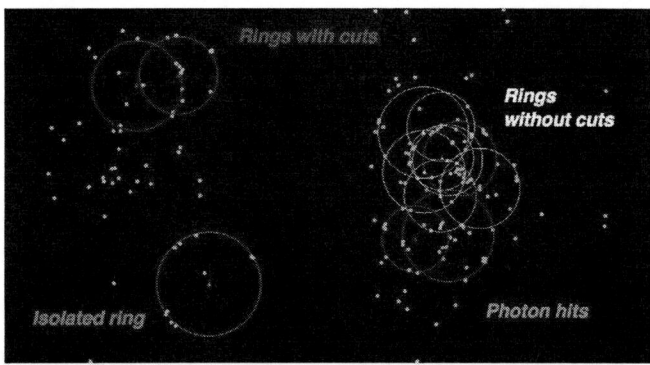

Figure 6. A screen shot illustrating the hit map from one layer of LHCb pixel-HPD tubes. The online reconstruction algorithm has been used to fit rings to the event. Additional rings are visible to the naked eye but have not been reconstructed.

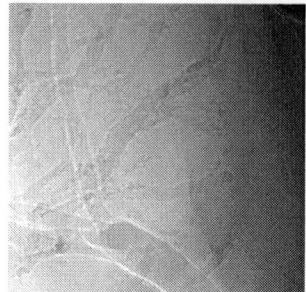

Figure 7. A close up view of the interior of the kidney of a mouse. This image was taken using a Medipix2 hybrid pixel detector and a micro-focus X-ray source. As the flux of such a source is very limited, and a high magnification is needed, this image took many hours to produce. It would be impossible to reproduce with a conventional integrating imaging panel.

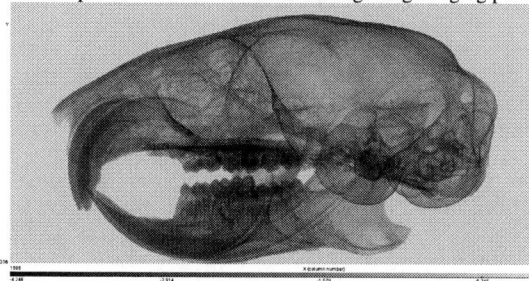

Figure 8 A tiled X-ray image of a mouse skull taken using the Medipix3 readout chip connected to a GaAs sensor.

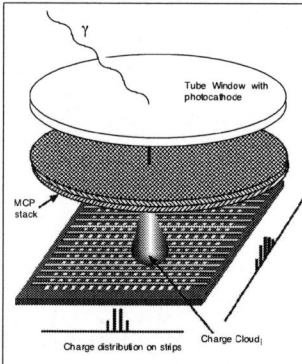

Figure 9. Schematic diagram of the micro channel plate based photon detection system. A single photo-electron falls into a pore of the micro channel plate and becomes amplified by multiple sequential collision with the plate walls and avalanches. The charge cloud at the base is drifted towards the bump bonding pad of the naked readout chip and detected.

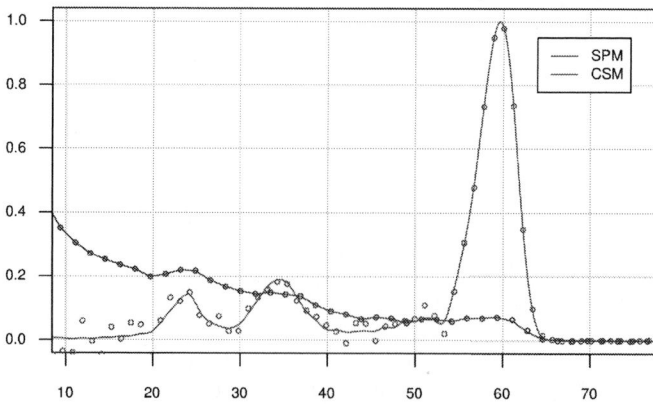

Figure 10. Measured spectra of a ^{241}Am source. In blue is shown the spectrum observed when the pixels work independently of each other. In red the Medipix3 charge summing and allocation feature has been enabled.

978-1-4799-8002-4/14 $31.00 © 2014 IEEE

Extremely low on-resistance Enhancement-mode GaN-based HFET using Ge-doped regrowth technique

Asamira Suzuki, Songbeak Choe, Yasuhiro Yamada, Shuichi Nagai, Miori Hiraiwa, Nobuyuki Otsuka, and Daisuke Ueda*

R&D Division   Device Solutions Center, Panasonic Corporation
*Department of Electronics, Kyoto Institute of Technology
3-1-1 Yagumo-naka-machi, Moriguchi City, Osaka 570-8501, Japan

Abstracts

In this paper, we present a normally-off GaN-based transistor with extremely low on-state resistance fabricated by using Ge-doped n^{++}GaN layer for ohmic contact. We developed a new GaN regrowth technique using Ge, which achieved extremely high doping level of 1×10^{20} cm^{-3}, and thereby the lowest specific contact resistance of 1.5×10^{-6} $\Omega \cdot cm^2$. Selectively deposited NiO gate using Atomic Layer Deposition (ALD) technique contributed to shorten the spacing between source and drain, making normally-off characteristics even with the 30% Al mole fraction of AlGaN. The fabricated device showed the record-breaking R_{on} of 0.95 $\Omega \cdot mm$ with maximum drain current ($I_{d,MAX}$) and transconductance (g_m) of 1.1 A/mm and 490 mS/mm, respectively. It is noted that the obtained V_{th} was 0.55 V. An on/off current ratio of 5×10^6 is also achieved.

Introduction

GaN-based heterojunction field effect transistors (HFETs) are very promising for power switching devices taking advantages of the superior material properties [1]. Recently, GaN-based transistors are expected to be used in the low voltage applications like DC-to-DC converters because GaN-based devices can keep the high blocking voltage even in the deep-submicron range keeping the high current handling capability. In this device, low on-resistance (R_{on}) and parasitic capacitances has been required for high frequency and low loss switching operation. From the viewpoint of safety, normally-off operation is strongly required. However, there arises the drawback between R_{on} and threshold voltage (V_{th}). This is because 2-dimensional electron-gas (2DEG) density has to be reduced to obtain the normally-off characteristics. Due to its relatively low 2DEG carrier density, normally-off GaN-based HFETs show higher R_{on}.

In order to reduce the R_{on}, we have developed low R_{on} enhancement-mode NiO-gate GaN-HFETs with regrown Ge-doped n^{++}GaN source/drain in direct contact with the 2DEG. The Ge-doped n^{++}GaN layer is grown by metal organic chemical vapor deposition (MOCVD) system. In this paper, we successfully fabricate short length NiO-gate GaN-HFETs and demonstrate extremely low R_{on} DC characteristics maintaining normally-off operation.

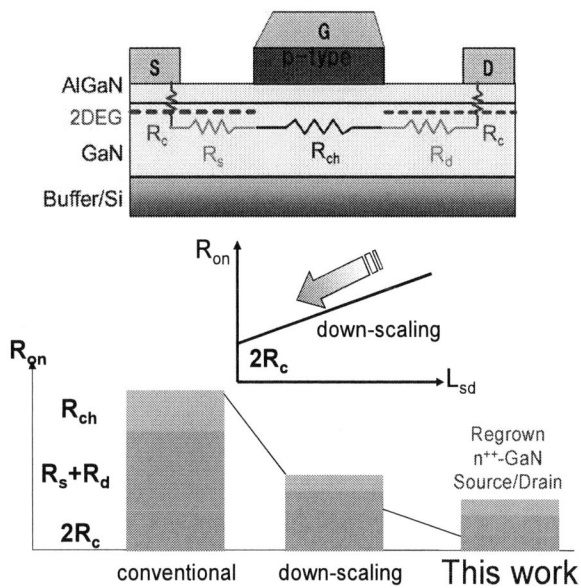

Fig. 1. R_{on} components in GaN-based HFET. As to reduce the device dimensions, the R_{on} decreases. But, the value of R_c becomes the dominant part of R_{on}.

Device Fabrication

As shown in fig. 1, R_{on} is composed of a series connection of R_c (contact resistance to source and drain areas), R_{ch} (channel resistance under the gate), and $R_{s/d}$ (access resistance to the channel). The reduction of R_c becomes more important as the gate-length (L_g) and source-drain spacing (L_{sd}) are reduced. In order to reduce R_c, we developed a new n-type doping technique using Ge. Fig. 2 shows the dependence of electron concentration of n^{++}GaN layer as a function of tetraethyl germanium (TEGe) supply, where carrier concentration over 1×10^{20} cm^{-3} is achieved by using Ge doping, while Si doping has a limit in MOCVD growth technique. The measured specific contact resistance as a function of TEGe supply is shown in fig. 3, where extremely low specific contact resistance of 1.5×10^{-6} $\Omega \cdot cm^2$ was achieved.

978-1-4799-8002-4/14 $31.00 © 2014 IEEE

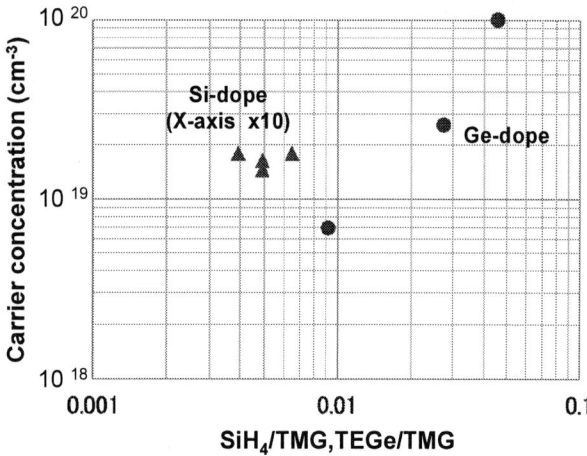

Fig. 2. Dependence of carrier concentration of MOCVD grown n^{++}GaN layer as a function of TEGe and SiH$_4$ supply. The highest electron concentration reaches 1 x 10^{20} cm^{-3} .

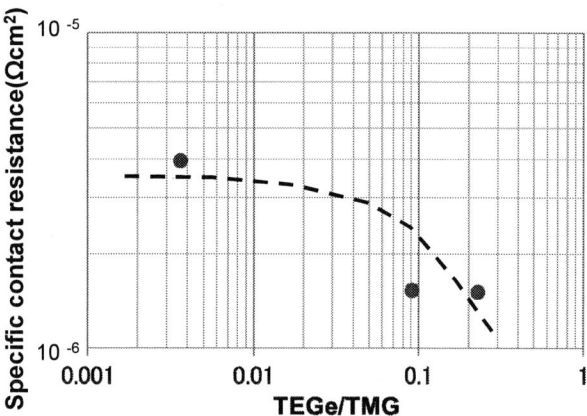

Fig. 3. TEGe supply dependence of specific contact resistance by transmission line measurements (TLM) on Ge-doped MOCVD grown n^{++}GaN layer.

Process flow

○ Selective regrowth of Ge-doped layer

over SiO2 mask after AlGaN/GaN growth on Si - - -(a)

○ SiN films formed over the source/drain electrodes

followed by gate delineation of 400nm - - - - - -(b)

○ Lift-off formation of the NiO film

only at the SiN groove- - - - - - -(c)

○ Gate metal deposition -(d)

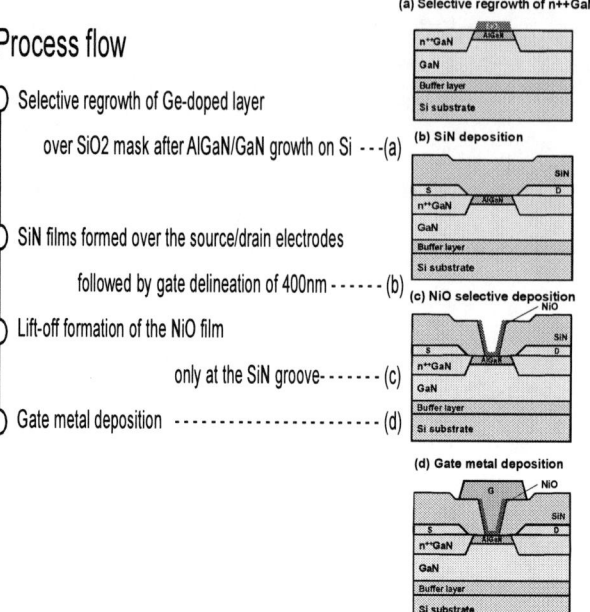

Fig.4 Processing steps of the device fabrication;
(a) selective regrowth of Ge-doped n^{++}GaN layer,
(b) SiN deposition on device surface,
(c) lift off formation of the NiO film using ALD,
(d) Gate metal deposition.

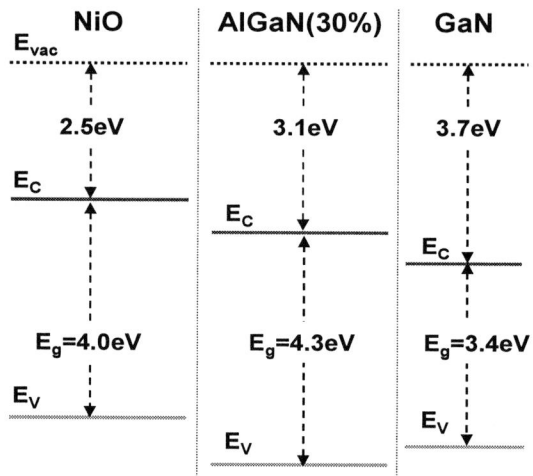

Fig. 5. Energy diagram compared with those of GaN-based materials.

The AlGaN/GaN hetero structure is grown on Si substrate by MOCVD system. The epitaxy layers included a 10-nm-thick undoped Al$_{0.3}$Ga$_{0.7}$N barrier layer, a 1-μm-thick unintentionally doped GaN layer, and a buffer layer on Si substrate.

Fig. 4 summarizes the processing steps of the device fabrication. Step 1: Selective regrowth of Ge-doped layer over SiO$_2$ mask after AlGaN/GaN growth on Si. Step 2: SiN films formed over the source/drain electrodes followed by gate delineation of 400nm. Step 3: NiO is formed only at the SiN groove using lift-off process. Step 4: Gate metal deposition.

We introduced the NiO gate to realize the normally-off operation [2]. The energy diagram of NiO and GaN-based materials is shown in fig. 5, where the smaller electron affinity and wider bandgap of p-type NiO material would lift the potential, which contributes to shift the threshold voltage to positive.

978-1-4799-8002-4/14 $31.00 © 2014 IEEE   276

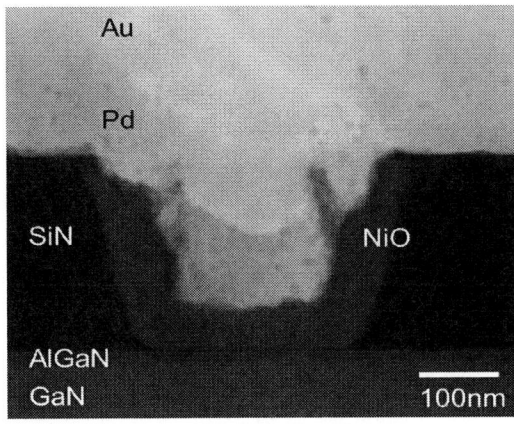

Fig. 6. Cross sectional TEM image of the selectively formed NiO gate structure fabricated on GaN HFET.

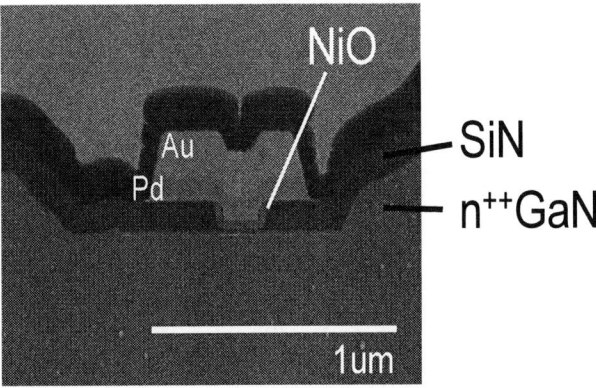

Fig. 7. Cross sectional SEM image of the fabricated NiO gate GaN HFET with n[++]GaN MOCVD regrowth ohmic layer.

The cross sectional transmission electron microscope (TEM) image of the ALD-grown NiO over the narrow opening of the dielectric films is shown in fig. 6. The cross sectional scanning electron microscope (SEM) image of the fabricated GaN transistor is shown in fig. 7.

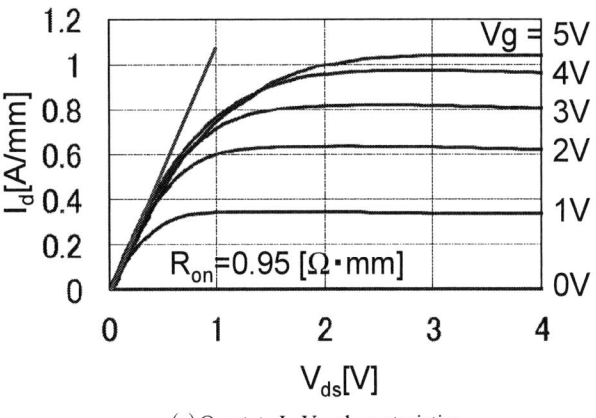

(a) On-state I_d-V_{ds} characteristics

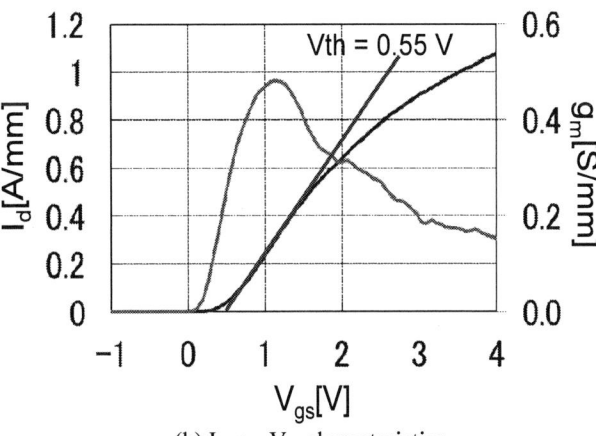

(b) I_d, g_m-V_{gs} characteristics

Fig. 8. DC current-voltage characteristics of the fabricated NiO gate GaN HFET. (a); the R_{on} of 0.95 $\Omega\cdot$mm and $I_{d,MAX}$ of 1.1 A/mm are shown in output characteristics. (b); a peak g_m of 490 mS/mm with the threshold voltage of 0.55 V is shown in transfer characteristics. Normally off characteristics with V_{th} of 0.55V is achieved.

Results and Discussion

The resultant DC current-voltage (I-V) characteristics of the fabricated NiO gate GaN HFET are shown in figs. 8 (a) and (b). Extremely small R_{on} of 0.95 $\Omega\cdot$mm is achieved at L_{SD}=1.0 μm.

A positive threshold voltage of 0.55 V, calculated as the x-intercept of a line tangent to the transfer curve, confirmed normally off operation. Off-state current of 2 x 10[-7] A/mm, $I_{d,MAX}$ of 1.1 A/mm, g_m of 490 mS/mm and on/off current ratio of 5 x 10[6] are achieved. The present device shows the R_{on} of 0.95 $\Omega\cdot$mm with the threshold voltage of 0.55 V, which is the best performance ever reported [3].

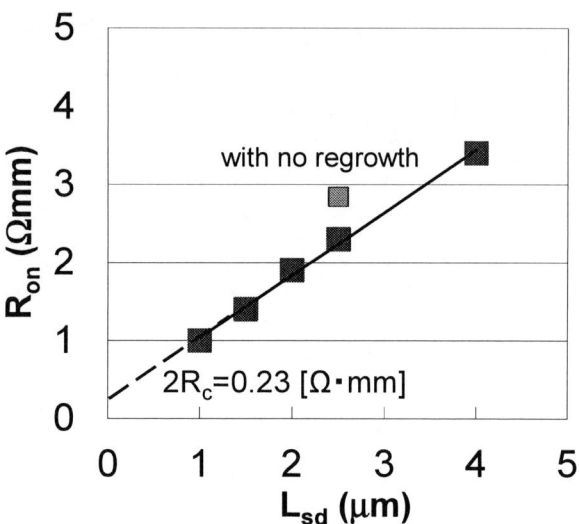

Fig. 9. Dependence of R_{on} on source-drain spacing (L_{sd}). The total contact resistance reaches down to as low as 0.23Ωmm. The R_{on} is successfully reduced due to introduction of Ge-doped n^{++}GaN ohmic regrowth layer to our device.

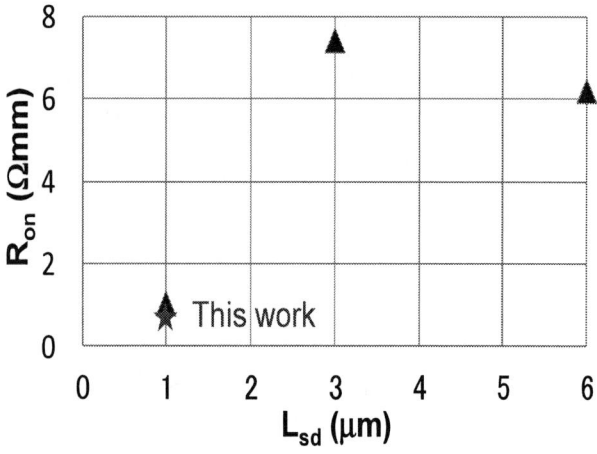

Fig. 10. Reported R_{on} of normally-off GaN-based HFETs in which the result using Ge-doped n^{++}GaN regrowth layer to our device is also shown.

As was extracted by the typical modeling technique, (in fig. 9), obtained $2 \times R_c$ reaches down to as low as 0.23 $\Omega \cdot$mm. Fig. 10 shows the present device comparing those reported R_{on} of normally-off GaN HFETs as a function of L_{sd}, where the present device achieved the lowest R_{on} ever reported.

We developed a new n-type doping technique using Ge, which is very effective to reduce R_c, which is dominant component of R_{on}. Furthermore, we introduce the NiO to gate material and successfully fabricate short length NiO gate by ALD method, which realize the normally-off characteristics without reduction of 2DEG density.

As a result of both n^{++}GaN layer and NiO gate, a normally-off GaN-based transistor with extremely low R_{on} was realized.

Conclusion

In conclusion, we successfully realized an extremely low R_{on} normally-off p-type NiO gate AlGaN/GaN HFETs with heavily Ge-doped MOCVD regrowth n^{++}GaN layer for the power switching applications. The NiO film is selectively formed by ALD at the narrow opening of the dielectric film. The fabricated HFET with the 400 nm NiO gate exhibits low R_{on} of 0.95 $\Omega \cdot$mm, $I_{d,MAX}$ of 1.1 A/mm, and a peak g_m of 490 mS/mm, with the threshold voltage of 0.55 V. An on/off current ratio of 5×10^6 is also achieved. The proposed processing realizes normally-off operation in conventional AlGaN/GaN HFET and also both the low R_{on} and the high $I_{d,MAX}$ without self-aligned process. We believe the proposed GaN-based transistor will contribute to the drastic increase of conversion efficiency in DC-to-DC converters represented by POL (Point Of Loads).

Acknowledgment

This work is partially supported by the New Energy and Industrial Technology Development Organization (NEDO), Japan, under the Strategic Development of Energy Conversion Technology Project.

References

[1] D. Ueda et al., IEDM Tech. Dig., p. 377, 2005.
[2] A. Suzuki et al., Extended Abstracts of the WOCSDICE, (2013) 77.
[3] A. L. Corrion et al., IEEE EDL, Vol. 31 p. 1116, 2010.
[4] H. Hilt et al., ISPSD Tech. Dig., p. 229, 2011.
[5] M. Kanamura et al., IEEE EDL, Vol. 31 p. 189, 2010.

Wide Temperature (10K-700K) and High Voltage (~1000V) Operation of C-H Diamond MOSFETs for Power Electronics Application

H. Kawarada[1], T. Yamada[1], D. Xu[1], H. Tsuboi[1], T. Saito[1], A. Hiraiwa[1]

[1] Faculty of Science and Engineering,
Waseda University, Tokyo 169-8555, Japan
Phone: +81-3-5286-3391 E-mail: kawarada@waseda.jp

Abstract

By forming a highly stable Al_2O_3 gate oxide on a C-H bonded channel of diamond, high-temperature and high-voltage metal-oxide-semiconductor field-effect transistor (MOSFET) has been realized. From -263°C (10K) to 400°C (673K), the variation of maximum drain-current is within 50% at a given gate bias. The maximum breakdown voltage ($V_{B,max}$) of the MOSFET without a field plate is 996V at a gate-drain distance (L_{GD}) of 9µm. We fabricated some MOSFETs satisfying $V_{B,max}/L_{GD} > 200V/µm$ (2MV/cm). This value is superior to those of lateral SiC or GaN FETs.

C-H Diamond MOSFET

Since diamond is a wide bandgap semiconductor (E_G =5.5eV), high-temperature operation and high-voltage switching are strongly desired. The advantage of p-channel FETs based on H-terminated (C-H) diamond surface (1-7) used in this study is its temperature-independent conductivity in contrast to that of boron-doped p-type diamond FET (8-11), where the acceptor level is as high as 0.37eV. Among diamond FETs, C-H diamond FETs (Fig.1) show much superior transistor performance such as high frequency operation in MESFETs (4,5) and MISFETs (6,7), because dense hole accumulation (~10^{13} cm^{-2}) near the surface is effectively modulated by applying gate bias from surface side. The formation mechanism of dense surface holes (2D hole gas, 2DHG) is still open question. The presence of C-H surface bonds is a necessary condition for the 2DHG. Several surface adsorbates or surface insulating films on H-terminated diamond have been proposed to explain the p-type conductivity based on surface acceptor without activation process (12) or surface negative charge attracted by spontaneous polarization of C-H surface (13). The reliability of C-H diamond FETs has been improved (14, 15) by the Al_2O_3 passivation of H-terminated diamond as shown in Fig.1 using atomic layer deposition (ALD). However, the FET process operation was limited around 200°C (15).

High Temperature and Wide Temperature Operation

Basic device structure C-H diamond MOSFET is shown in Fig.1(a). The substrate is high-pressure and high-temperature synthetic diamond where nitrogen atoms as deep donor were incorporated at the level of 10^{17} cm^{-3}. Nominally undoped diamond has been homoepitaxially grown by microwave plasma assisted CVD with thickness of 0.5µm. Source and drain regions have been formed firstly. Then, most of area except for source and drain metals has been H-terminated. After oxidation to form isolated region, a channel is limited only between source and drain as shown in Fig.2(b). Finally, gate metal was deposited and a gate was patterned by lift-off process.

Recently, we have reported a high-temperature ALD process for the deposition of Al_2O_3 at 450°C to reproduce 2DHG (16, 17) without adsorbates. Figure 2 shows the sheet resistivity and hole concentration of 2DHG at various Al_2O_3 thickness as a function of temperature (17). The diamond (100) homoepitaxial films passivated by Al_2O_3 with more than 20nm in thickness show temperature independent properties between room temperature and 500°C in the sheet resistivity and the hole density. It indicates, in principle, C-H diamond FET can operate up to 500°C without changing FET properties.

In Fig.3(a), we demonstrate FET operation up to 400°C in a MOSFET using a gate-insulating and passivating layer of ALD Al_2O_3 with thickness of 200nm. The pinch off and saturation behaviors in the I_{DS}-V_{DS} characteristics at 400°C are almost perfect. The property is very close to that of the room temperature operation shown at Fig.3(b). Although threshold voltage is as high as 16V, other FET characteristics such as drain current density, transconductance are better than recently reported values (18) of similar C-H diamond FETs with thinner gate oxide in the same lateral device size. It is an unexpected result from an ideal MOSFET theory. However, it is advantageous for high voltage application to use thick insulating layer as gate oxide.

The I_{DS}-V_{GS} characteristics at various temperatures are shown in Fig.4. The on-off ratio at room temperature is more than 10^8. However, it decreases as temperature increase and become 10^3 at 400°C. Since the activation of leakage current is about 1.5eV, it might be caused by the activation of deep donor (E_D=1.7eV) due to the nitrogen doped substrate and/or residual nitrogen in a homoepitaxial layer diamond. Other factor is imperfect device isolation by oxygen-termination (C-O) shown in Fig.1(b). The device isolation can be achieved by the difference of electron affinity χ between C-H diamond surface ($\chi = -1.3$ eV) and C-O surface ($\chi = +1.5$ eV). Only near the surface the potential difference keeps above 2.5 eV, but decreases deeper from the surface. To obtain high on-off ratio at high temperature, these leakage origins must be eliminated by new isolation and pure diamond without nitrogen.

Figure 5 shows the maximum drain current $I_{DS,max}$ at a given gate bias (V_{GS}=4V). As temperature raises from room temperature to 400°C, the increase of $I_{DS,max}$ is within 20% shown in Fig.5. In a lower temperature to 10K, $I_{DS,max}$ decreases by 30%. The low temperature behavior is expected from the conductivity measurement (19) by 4 point probe. In the present case, the increase of $I_{DS,max}$ is due to the increase of contact resistivity at low temperature, because the contact area between source metal (TiC) and H-terminated diamond is not face to face, but side by side as can be seen in Fig.1(b) and Schottky barrier height of TiC (~0.3eV) at C-H diamond is not negligible at lower temperature.

High Voltage Operation

We also have investigated the high voltage operation of the C-H diamond FETs as a function of gate-drain length L_{GD} as shown in Fig.6, 7, and 8. Figure 6 shows three breakdown characteristics of I_{DS} at different L_{GD}. The breakdown voltage of 996V is the highest in diamond FETs, but it does not mean electric field reached to breakdown field of diamond. Breakdown is not obvious in linear scale. Both drain current I_{DS} and gate leakage current I_{DGS} are shown at high drain voltage in a logarithmic I_{DS} and I_{DGS} in Fig.7. The I_{DGS} curve is a typical leakage characteristic of insulating film. It gradually increases at high voltage (high electric field), reaches the I_{DS} and finally starts to breakdown. The breakdown happened at gate oxide side.

On the other hand, an apparent breakdown at 363V in diamond side has been observed at a small L_{GD} of 1μm shown in Fig.6. The average electric field $V_{B,max}/L_{GD}$ is 360V/μm (3.6MV/cm) which exceeds the breakdown field of SiC and GaN. As shown in Fig.8, the breakdown field is not the high level (>2MV/cm) at long L_{GD}, but still keep >1MV/cm. $V_{B,max}/L_{GD}$ is often used as a criterion for high-voltage durability in planar FETs, with 100V/μm being a critical value for lateral power devices. Although the breakdown voltage of over 1000V obtained in lateral SiC MOSFETs (20,21), $V_{B,max}/L_{GD}$ is less than 100V/μm for SiC FETs, and slightly less than the value of AlGaN/GaN FET (~100V/μm). Recently an AlGaN/AlGaN FET with lateral breakdown field of 160V/μm has been reported (22), which is one of the highest values recorded for planar FETs. However, diamond has a potential to exceed 3MV/cm as indicated from Fig.8

Conclusion

By surface passivation of C-H bond diamond surface using ALD Al_2O_3 at high temperature, the following MOSFETs properties have been obtained.

1) Operation at a high temperature up to 400°C and a wide temperature range of 10K-673K (-263°C~ +400°C).

2) Nearly 1000V breakdown voltage without field plate and a maximum breakdown field of 3.6MV/cm.

These results show C-H diamond surface has a potential for next generation power MOSFETs.

Reference

(1) H. Kawarada, M. Aoki, I. Itoh, *Appl. Phys. Lett.*, **65**, 1563 (1994).

(2) H. Kawarada, *Surf. Sci. Rep.*, **26**, 205 (1996).

(3) P. Gluche, A. Aleksov, A. Vescan, W. Ebert, E. Kohn, *IEEE Electron Device Lett.* **18**, 547 (1997).

(4) H. Taniuchi, H. Umezawa, T. Arima, M. Tachiki, H. Kawarada, *IEEE Electron Device Lett.* **EDL-22**, 390 (2001).

(5) K. Ueda, M. Kasu, Y. Yamauchi, T. Makimoto, M. Schwitters, D. J. Twitchen, G. A. Scarsbrook, S. E. Coe, *IEEE Electron Device Lett.* **EDL-27**, 570 (2006).

(6) H. Matsudaira, S. Miyamoto, H. Ishizaka, H. Umezawa, H. Kawarada, *IEEE Electron Device Lett.* **EDL-25**, 480 (2004).

(7) K. Hirama, H. Kawarada et al. *IEEE IEDM* 4419088, p 873~876 (2007),

(8) A. Vescan, P. Gluche, W. Ebert, E. Kohn, *IEEE Electron Device Lett.*, **18**, 222 (1997).

(9) K. Ueda, M.Kasu, *Jpn. J. Appl. Pys.* **49**, 04DF16/1-4 (2010).

(10) T. Iwasaki, Y. Hoshino, K. Tsuzuki, H. Kato, T. Makino, M. Ogura, D. Takeuchi, H. Okushi, S. Yamasaki, M. Hatano, *IEEE Electron Device. Lett.*, **34**, 1175 (2013).

(11) T. Iwasaki, J. Yaita, H. Kato, T. Makino, M. Ogura, D. Takeuchi, H. Okushi, S. Yamasaki, M. Hatano, *IEEE Electron Device Lett.* **35**, 241, (2014).

(12) F. Maier, M. Riedel, B. Mantel, J. Ristein, L. Ley, *Phys. Rev. Lett.* **85**, 3472 (2000).

(13) K. Hirama, H. Takayanagi, S. Yamauchi, J. H. Yang, H. Kawarada, H. Umezawa, Appl. Phys. Lett., **92**, 112107 (2008).

(14) D. Kueck , A. Schmidt, A. Denisenko, E. Kohn, *Diam. Relat. Mater.* **19**, 166 (2010).

(15) M. Kasu, H. Sato, K. Hirama, *Appl. Phys. Express* **5**, 025701, (2012).

(16) A. Hiraiwa, A. Daicho, S. Kurihara, Y. Yokoyama, and H. Kawarada, *J. Appl. Phys.* **112** , 124504 (2012)

(17) A. Daicho, T. Saito, S. Kurihara, A. Hiraiwa, H. Kawarada, *J. Appl. Phys.* **115** , 223711 (2014).

(18) H. Kawarada, H. Tsuboi, T. Naruo, T. Yamada, D. Xu, A. Daicho, T. Saito, and A. Hiraiwa, *Appl. Phys. Lett.*,**105**, 013510 (2014).

(19) C.E. Nebel, C. Sauerer, F. Ertl, M. Stutzmann, C. F. O. Graeff, P. Bergonzo, O. A. Williams, R. Jackman, *Appl. Phys. Lett.* **79**, 4541 (2001).

(20) M. Noborio, J. Suda, and T. Kimoto, *IEEE Elec. Dev. Lett,* 30, 831 (2009).

(21) W.-S. Lee, C.-W. Lin, M.-H. Yang, C.-F. Huang, J. Gong, and Z. Feng, *IEEE Electron Device Lett,* 32, 360 (2011).

(22) T. Nanjo, A. Imai, Y. Suzuki, Y. Abe, T. Oishi, M. Suita, E.Yagyu, and Y.Tokuda, *IEEE Trans Electron Devices* **60**, 1046 (2013)

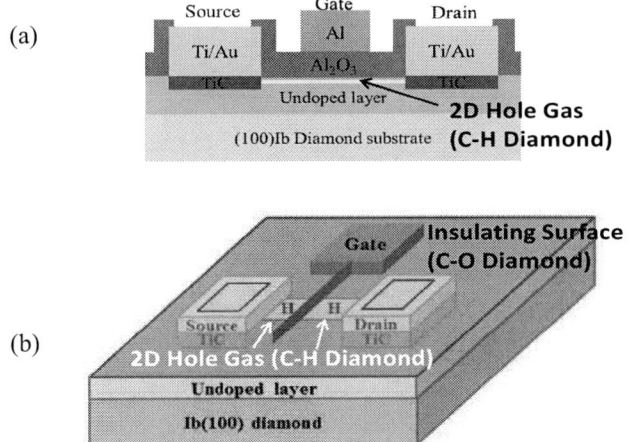

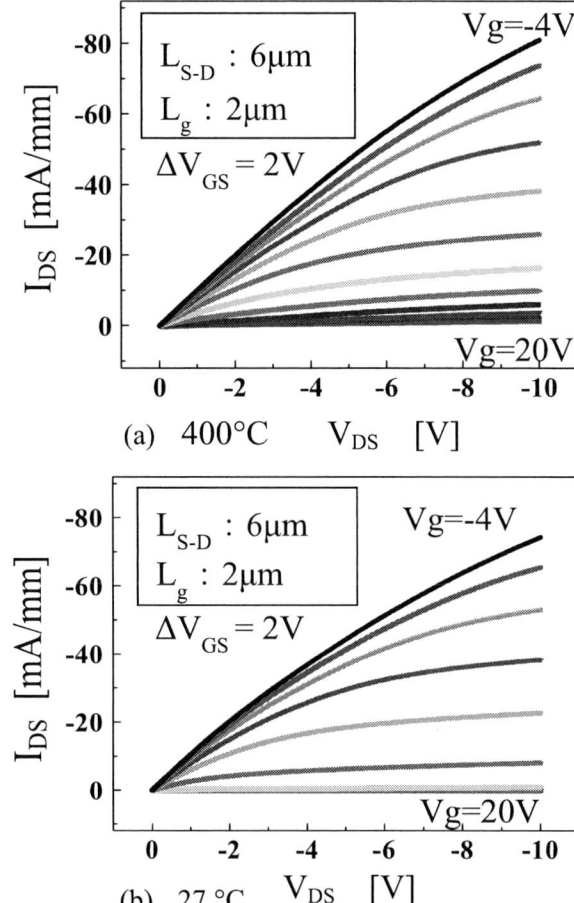

Fig.1 (a) Cross-sectional view of diamond MOS FET using H-terminated (C-H) surface channel. The C-H diamond surface is passivated by ALD Al_2O_3 at 450°C. The layer is also used for gate oxide (10-200nm in thickness). (b) 3D view of the C-H diamond MOSFET. Al_2O_3 covers almost whole area. O-terminated diamond surface is so insulating that the layer is used for device isolation. C-H diamond surfaces ("H" areas) exhibit 2D hole gas.

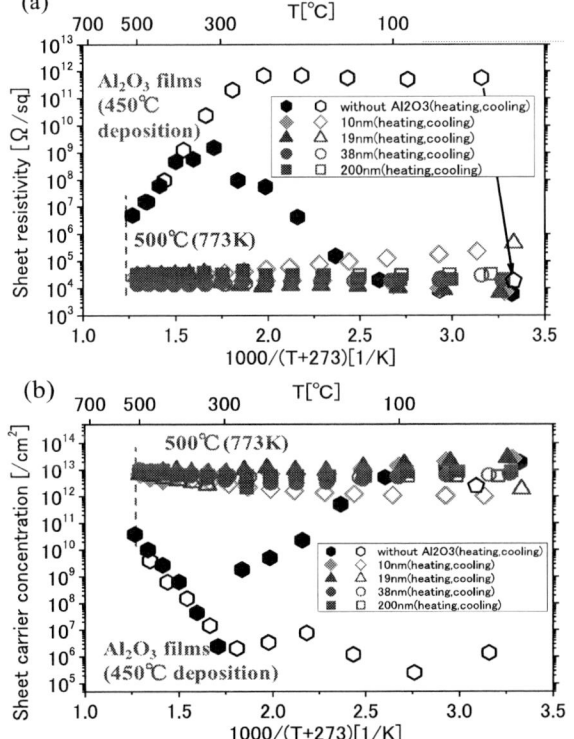

Fig.2 2D hole gas at Al_2O_3/C-H diamond interface. Temperature dependence of (a) sheet resistivity (Ω/sq) and (b) hole areal density (cm^{-2}). The Al_2O_3 thickness is varied from 10-200nm. All Al_2O_3 were formed by ALD at 450°C. Closed plots and open plots are heating and cooling process, respectively. (17)

Fig.3 I_{DS}-V_{DS} of characteristics of C-H diamond MOSFETs with 200nm thick Al_2O_3 gate oxide and undoped channel measured in vacuum (a) at 400°C and (b) at 27°C.

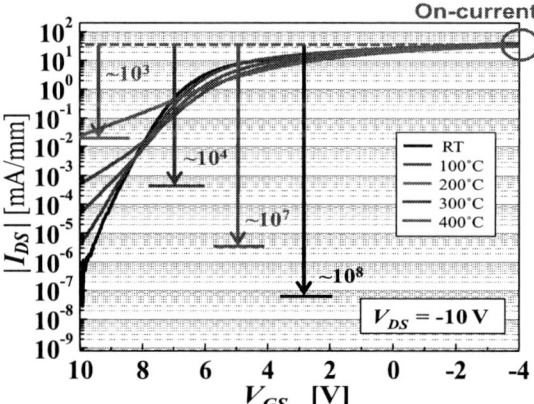

Fig.4 I_{DS}-V_{GS} characteristics measured from 400°C and to room temperature for the C-H diamond MOSFET The gate oxide thickness is 32 nm. The gate length and width are 2μm and 25μm, respectively.

978-1-4799-8002-4/14 $31.00 © 2014 IEEE   281

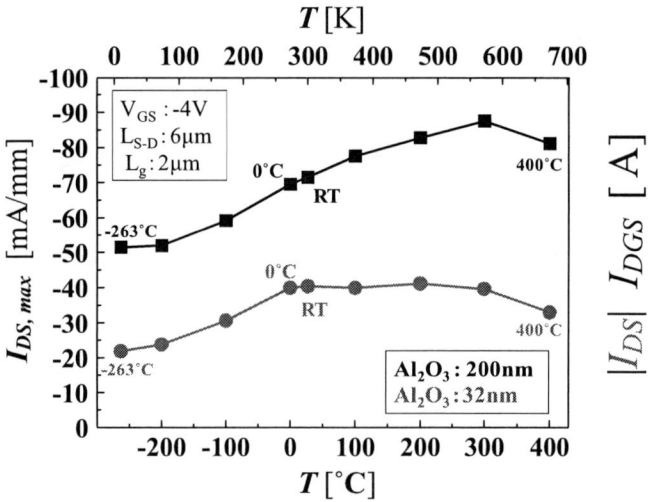

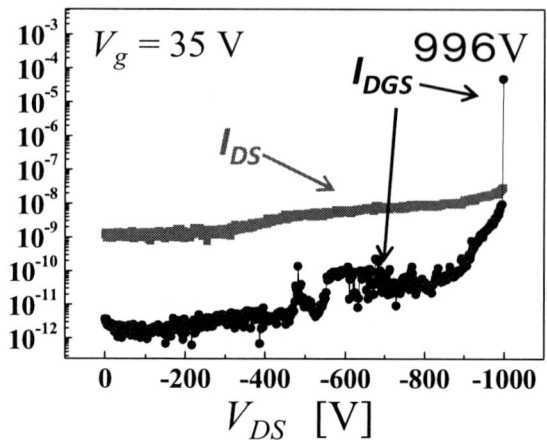

Fig.5 Temperature dependence of drain current maximum $I_{DS,max}$ at a fixed gate voltage $V_{GS} = -4V$. Two diamond MOSFETs with Al_2O_3 gate oxide thickness of 32nm and 200nm were shown. From 10K(-263°C) to 673K (400°C) the variation of drain currents are within 50%. The reduced drain currents at low temperature were due to high contact resistance by the small contact area due to TiC/C-H diamond interface.

Fig.7 Breakdown characteristics of a C-H diamond MOSFET with 200nm-thick gate oxide and passivation layer with the same gate width of 25μm. In addition to I_{DS}, a gate leakage current I_{DGS} was measured. Before I_{DS} start to breakdown, I_{DGS} increase rapidly and finally it reaches that of I_{DS}.

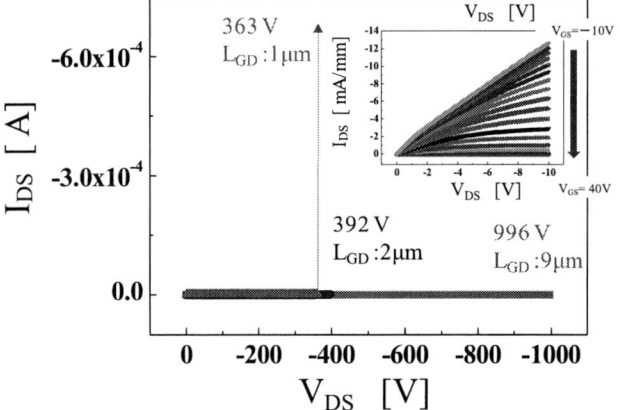

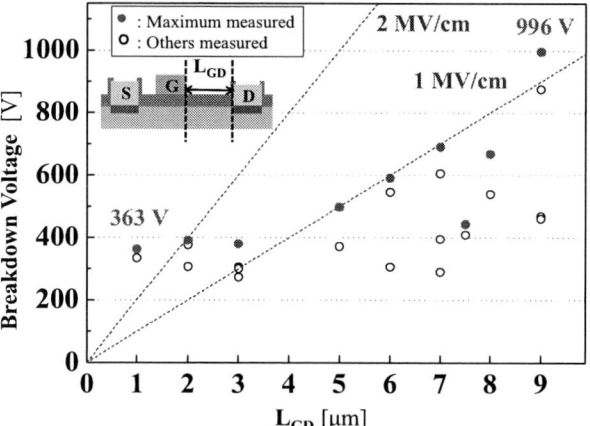

Fig.6 Breakdown characteristics of MOSFETs with 200nm-thick gate oxide and passivation layer with the same gate width of 25μm. In three MOSFETs with source-drain distance L_{SD} of 25 μm, the gate-drain distances L_{GD} are 1 μm , 2 μm and 9 μm, respectively. Breakdown voltages of the MOSFETs with L_{GD} of 1 μm, 2 μm, and 9 μm are 363V, 392V, and 996 respectively. The inset is the I_{DS}-V_{DS} characteristics of the MOSFET with the L_{GD} of 9 μm gate length showing the breakdown voltage of 996 V

Fig.8 Maximum (closed circles) and other (open circles) breakdown voltages as a function of gate to drain length L_{GD}. In $L_{GD} <$ 3μm, maximum breakdown voltages V_{Bmax} are around 400V. In $L_{GD} > 5μm$, V_{Bmax} are on the line of $V_{B\,max}/L_{GD} = $ 1MV/cm

978-1-4799-8002-4/14 $31.00 © 2014 IEEE   282

GaN-based Gate Injection Transistors for Power Switching Applications

Tetsuzo Ueda, Hiroyuki Handa, Yusuke Kinoshita, Hidekazu Umeda, Shinji Ujita, Ryo Kajitani, Masahiro Ogawa, Kenichiro Tanaka, Tatsuo Morita, Satoshi Tamura, Hidetoshi Ishida, and Masahiro Ishida

Green Innovation Development Center, Automotive & Industrial Systems Company, Panasonic Corporation
3-1-1 Yagumo-nakamachi, Moriguchi-shi, Osaka 570-8501, Japan

Abstract

GaN-based Gate Injection Transistors (GITs) with p-type gate over AlGaN/GaN heterojunction serve normally-off operations with low on-state resistances owing to the conductivity modulation by injection of holes. Established basic technologies on the GIT have shown promising features for switching applications. Further improvement of the performances would extend the applications and lead to the widespread use. In this paper, recent technologies on the GITs to improve the performances and extract the full potential are described. These include extension of the wafer diameter of Si up to 8 inch, InAlGaN quaternary alloy to reduce the series resistances, shortening the gate length to improve the device performances, integration of the gate driver and flip-chip assembly for faster switching.

Introduction

GaN-based transistors are very promising for power switching applications owing to the superior material properties. Recent progress on the epitaxial growth of GaN on cost-effective Si substrates has made the GaN devices emerged as a viable alternative replacing the currently available Si-based power devices [1]. The operation at higher frequencies together with higher efficiencies by GaN is expected to enable very compact switching systems. So far, a GIT has been proposed to achieve a normally-off operation by introducing p-type gate, of which the cross section and operating principle are shown in Fig.1 [2]. The demonstrated performances are superior to those by existing Si-based power devices; however, continuous improvement of the device including peripheral technologies would be required for extending the applications in large quantities.

In this paper, state-of-the-art technologies of GaN-based GITs are reviewed. These include the technologies for the epitaxial growth on a large-diameter substrate, device with high performances and its assembly that enable highly speed switching.

Epitaxial Growth Technologies

So far, novel epitaxial structures to overcome the lattice and thermal mismatches between GaN and Si have enabled high quality AlGaN/GaN hetero structure on large diameter Si by metal organic chemical vapor deposition (MOCVD). The wafer diameter is increased up to 8 inch as shown in Fig.2 [3]. The uniformity of the AlGaN thickness with the averaged value of 26nm measured by x-ray reflection is also plotted in Fig.3. The increase of the diameter would help the

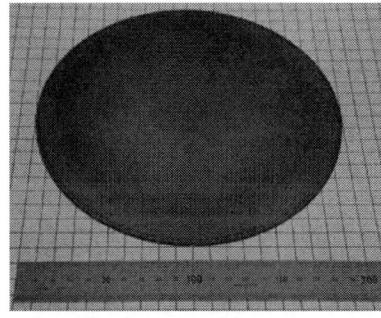

Fig.2 Photograph of an AlGaN/GaN hetero structure on an 8-inch Si substrate.

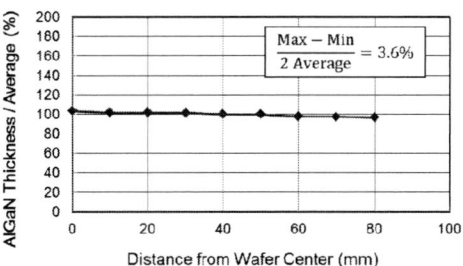

Fig.3 Variation of the thickness of AlGaN over 8-inch GaN buffer on Si.

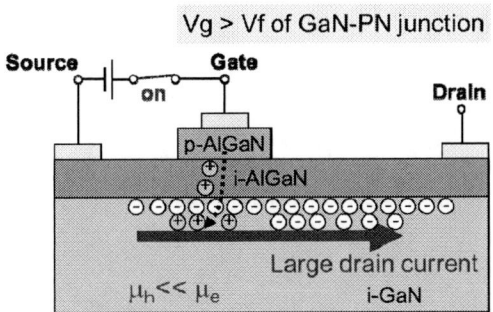

Fig.1 A schematic cross section and operating principle of GIT.

978-1-4799-8002-4/14 $31.00 © 2014 IEEE   283

reduction of the on-state resistance and the increase of the current by a single chip

High sheet carrier density at AlGaN/GaN is a very promising feature of the GaN transistors, which is caused by the material's unique polarization. InAlGaN quaternary alloy, of which the band gap energy and the lattice constant are plotted in Fig.4, can increase the magnitude of the polarization maintaining good lattice matching and thus more than doubles the sheet carrier density over GaN. The InAlGaN/GaN hetero structure is applied for p-AlGaN gate GIT of which the schematic cross section is shown in Fig.5 [4]. Note that the device employs an InAlGaN nearly lattice-matched to GaN with the sheet carrier concentration of 3 x

10^{13}cm^{-2}. The on-state resistance is reduced as well as the maximum drain current is increased maintaining a normally-off operation with the threshold voltage of 1.1V as summarized in Fig.6. The reduction of series resistance would greatly help to reduce the conduction loss in the switching operations.

Device and Processing Technologies

Reduction of the gate length in lateral GaN transistors, in general, reduces the gate capacitances and the channel resistances and thus enables faster switching. The GIT requires dry etching for the gate formation so that the processing damages and other processing issues had limited the shortening. The effect of the reduction of the gate lengths is examined with the improved processing [5]. Note that the GITs are fabricated by the self-aligned dry etching with the breakdown voltage of 30V. Shortening the gate lengths down to 0.5μm reduces the on-state resistance and increases the drain current as shown in Fig.7. Fig.8 is a cross sectional scanning electron microscope (SEM) image of the 0.5μm-gate GIT. The obtained $R_{on}Q_g$ (R_{on}: on-state resistance, Q_g: gate charge) as a figure-of-merit for high speed switching is

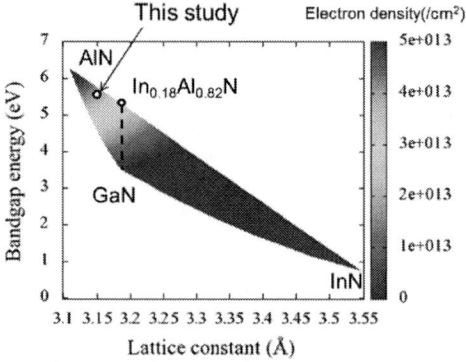

Fig.4 Bandgap energies of InAlGaN quaternary alloy system as a function of the lattice constants.

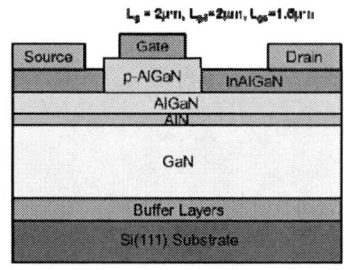

Fig.5 Schematic cross section of the fabricated p-gate GIT using InAlGaN/GaN hetero structure.

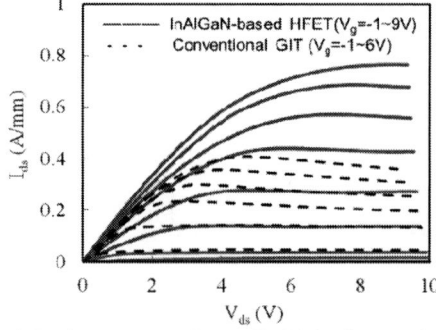

Fig.6 Drain current-voltage (I_{ds}-V_{ds}) characteristics of the GIT using InAlGaN/GaN in which those of the conventional GIT on AlGaN/GaN is also shown.

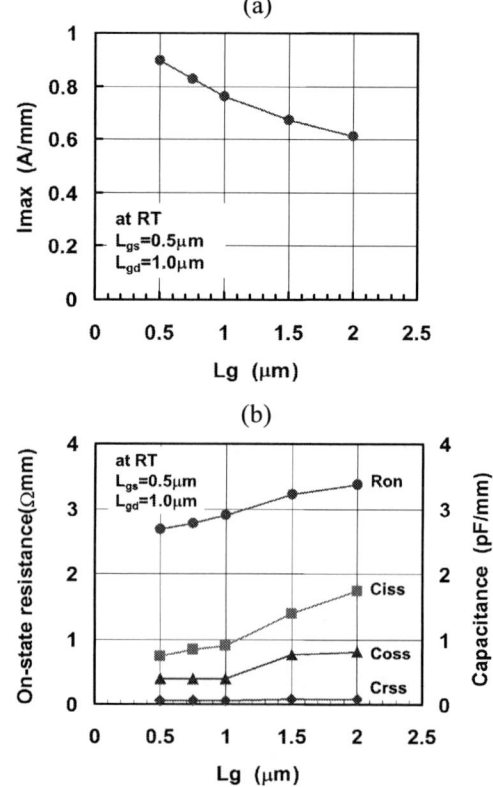

Fig.7 (a) I_{max}, (b) R_{on} and parasitic capacitances measured for the GITs with various gate lengths.

978-1-4799-8002-4/14 $31.00 © 2014 IEEE

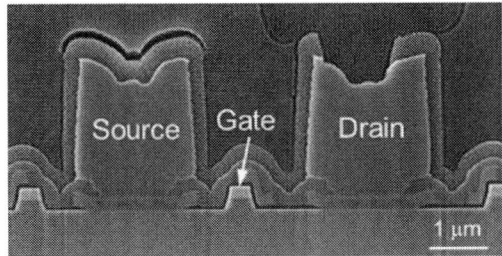

Fig.8 Cross sectional SEM image of the fabricated 0.5μm-gate GIT.

reduced down to 19mΩnC for the 30V device which is more than half from that of conventional Si MOSFETs. The short-gate GITs are applied to a low voltage point-of-load (POL), which successfully enable high frequency operation up to 5MHz with high efficiencies.

Possible integration of transistors is a very promising advantage of the lateral GaN devices. So far, a single chip inverter IC was demonstrated by using the device isolation using ion implantation [6]. Integration of the gate driver with GITs would enable faster switching by the reduction of the parasitic inductance and also reduce the size of the switching systems. In addition to the integration of the two GITs in a POL converter, gate drivers consisting of a DCFL (Direct Coupled FET Logic) with a buffer amplifier are integrated. The circuit diagram and the schematic cross section of the integrated devices are shown in Fig.9 and Fig.10, respectively [7]. Normally-on GaN HFETs (Hetero-junction Field Effect Transistors) are also fabricated by a part of the total processing for the GIT. The circuit using the buffer amplifier effectively reduces the power consumption in the driver. Fig.11 shows the photograph of the fabricated POL IC with high speed gate drivers. Successful conversions at 1-3MHz are confirmed, where the peak efficiency at 2MHz is as high as 88.2% for the 12V-1.8V conversion. The integration reducing the parasitic inductances would enable

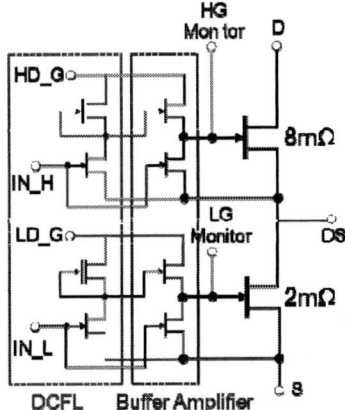

Fig.9 Circuit diagram of the DC-DC converter (POL) IC with gate drivers.

high frequency operations and help to reduce the size of switching systems.

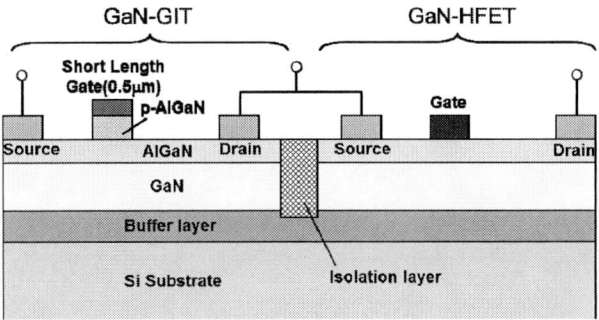

Fig.10 Schematic cross section of the devices employed in the POL IC. The normally-off GIT and the normally-off HFET fabricated by a single processing flow are integrated over a Si substrate.

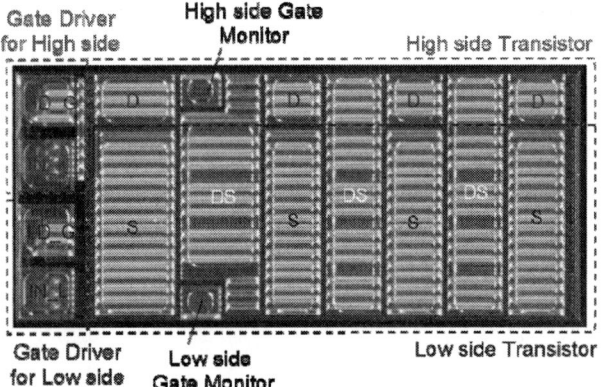

Fig.11 Photograph of the fabricated POL IC in which two GITs and two gate drivers are integrated.

Assembly Technologies

How the GITs are packaged is also a critical issue to enable faster switching. At present, the GIT is an only choice for a normally-off device by a single chip with 600V of the blocking voltage. The device is compatible with the flip-chip assembly as shown in Fig.12. The simulated parasitic inductance is dramatically reduced down to 2nH from 24nH in a conventional TO-220 package [8]. The switching waveforms are compared as shown in Fig.13, in which the switching time by the flip-chip is far smaller than that by TO-220. Note that the slew rate dV/dt is increased up to 170V/ns. The fast switching is applied for a DC-DC boost converter as a part of totem-pole bridgeless PFC (Power Factor Correction) in a power supply. The peak efficiency is 99.3% at 1kW while the maximum output power reaches 3kW as shown in Fig.14. The high efficiencies are originated from the low switching loss by the fast switching of the GIT with flip-chip bonding.

978-1-4799-8002-4/14 $31.00 © 2014 IEEE   285

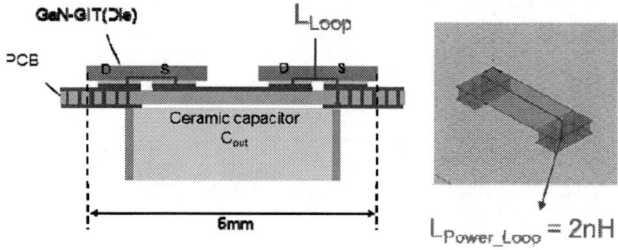

Fig.12 Flip-chip assembly of a GIT and simulation results on the parasitic inductance.

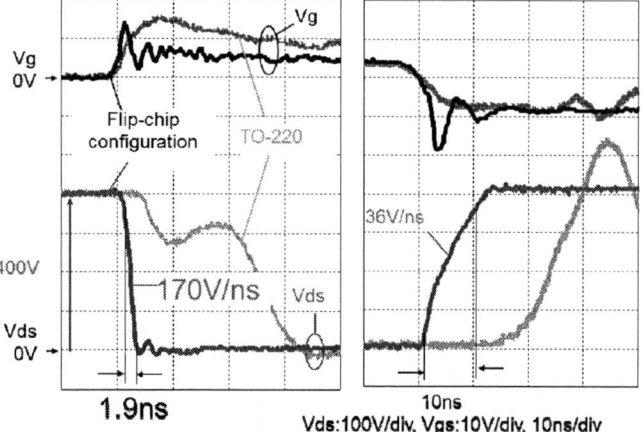

Fig.13 Switching waveform of GITs with flip-chip configuration as compared with that with TO-220 package.

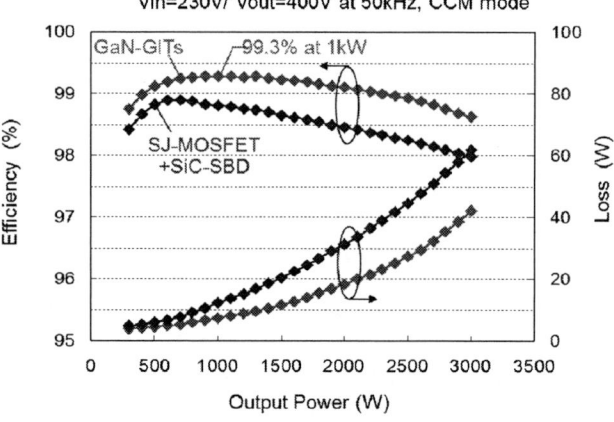

Fig.14 Operating efficiencies of DC-DC boost converter using 600V GITs with flip-chip bonding for various output powers.

Conclusions

Recent technologies on normally-off GaN-based GITs on Si are described. The demonstrated epitaxial growth, device, processing and assembly technologies extract the full potential of the devices enabling highly-efficient power switching systems. The GITs with these technologies are very promising for future energy-efficient power electronics.

Acknowledgments

The authors would like to gratefully acknowledge Dr. Minoru Kubo and Dr. Tsuyoshi Tanaka, Panasonic, for their sincere encouragement and technical advices throughout the work. They also would like to acknowledge Dr. Daisuke Ueda, Kyoto Institute of Technology, for his technical advices. The work is partly supported by the New Energy and Industrial Technology Development Organization (NEDO), Japan, under the Strategic Development of Energy Saving Innovative Technology Development Project.

References

(1) M. Hikita, M. Yanagihara, K. Nakazawa, H. Ueno, Y. Hirose, T. Ueda, Y. Uemoto, T. Tanaka, D. Ueda, and T. Egawa, "350V/150A AlGaN/GaN power HFET on Silicon substrate with source-via grounding (SVG) structure," *IEEE Trans. Electron Device*, vol.52, no.9, pp1963-1968, 2005.

(2) Y. Uemoto, M. Hikita, H. Ueno, H. Matsuo, H. Ishida, M. Yanagihara, T. Ueda, T. Tanaka, D. Ueda, "Gate Injection Transistor (GIT)—A Normally-Off AlGaN/GaN Power Transistor Using Conductivity Modulation," *IEEE Trans. Electron Device*, vol.54, no.12, pp3393-3399, 2007.

(3) M. Ishida, T. Ueda, T. Tanaka, and D. Ueda, "GaN on Si Technologies for Power Switching Devices," *IEEE Trans. Electron Device*, vol.60, no.10, pp3053-3059, 2013.

(4) R. Kajitani, K. Tanaka, M. Ogawa, H. Ishida, M. Ishida and T. Ueda, "A Novel High-current Density GaN-based Normally-off Transistor with Tensile Strained Quaternary InAlGaN Barrier," *Ext. Abst. of Int. Conf. Solid State Devices and Materials*, Tsukuba, Japan, September 2014.E-3-2.

(5) H. Umeda, Y. Kinoshita, S. Ujita, T. Morita, S. Tamura, M. Ishida, and T. Ueda, "Highly Efficient Low-Voltage DC-DC Converter at 2-5 MHz with High Operating Current Using GaN Gate Injection Transistors," *PCIM Europe*, Nuremberg, Germany, May 2014, PP45.

(6) Y. Uemoto, T. Morita, A. Ikoshi, H. Umeda, H. Matsuo, J. Shimizu, M. Hikita, M. Yanagihara, T. Ueda, T. Tanaka, and D. Ueda, "GaN Monolithic Inverter IC Using Normally-off Gate Injection Transistors with Planar Isolation on Si Substrate," *IEEE IEDM Tech. Dig.*, Baltimore, USA, December 2009, pp. 165-168.

(7) S. Ujita, Y.Kinoshita, H.Umeda, T.Morita, S.Tamura, M.Ishida and T.Ueda," A Compact GaN-based DC-DC Converter IC with High-Speed Gate Drivers Enabling High Efficiencies," *Int. Symp. on Power Semiconductor Devices and ICs (ISPSD) 2014*, Wikoloa, USA, June 2014, B1L-A-1

(8) T. Morita, H. Handa, S. Ujita, M. Ishida, and T. Ueda, "99.3% Efficiency of Boost-up Converter for Totem-pole Bridgeless PFC Using GaN Gate Injection Transistors," *PCIM Europe*, Nuremberg, Germany, May 2014.

978-1-4799-8002-4/14 $31.00 © 2014 IEEE   286

Schottky-on-Heterojunction Optoelectronic Functional Devices Realized on AlGaN/GaN-on-Si Platform

Baikui Li[1], Xi Tang[1], Qimeng Jiang[1], Yunyou Lu[1], Hanxing Wang[1], Jiannong Wang[2], and Kevin J. Chen[1]

[1]Department of Electronic and Computer Engineering, The Hong Kong University of Science and Technology, Hong Kong
[2]Department of Physics, The Hong Kong University of Science and Technology, Hong Kong
Phone: +852-23588969, Fax: +852-23581485, Email: eelibk@ust.hk, eekjchen@ust.hk

Abstract

We demonstrated that the metal-AlGaN/GaN Schottky diode is capable of producing GaN band-edge ultraviolet (UV) emission at 3.4 eV/364 nm under forward bias larger than ~2 V at room temperature. The underlying mechanism of the hole generation/injection and electroluminescence (EL) processes in this Schottky-on-heterojunction light-emitting diode (SoH-LED) was discussed based on the impact ionization of surface states presented in the (Al)GaN barrier layer. By replacing the conventional ohmic drain with a semitransparent Schottky drain, we demonstrated an AlGaN/GaN high-electron-mobility light-emitting transistor (HEM-LET) in which the drain current and EL emission are controlled simultaneously by gate voltage. Switching operation up to 120 MHz was obtained in SoH-LED to demonstrate its potential in providing high-speed on-chip light sources on the GaN electronic device platform.

Introduction

AlGaN/GaN heterostructure grown on silicon substrates (AlGaN/GaN-on-Si) is the dominant platform for the on-going intensive development of GaN-based power electronic devices. The *planar* nature of this platform offers the benefits of high-density integration that is being exploited in the form of GaN power integrated circuits [1, 2], peripheral sensing/protection circuits [3], and more recently, GaN-based pulse width modulation (PWM) circuit for integrated gate driver [4].

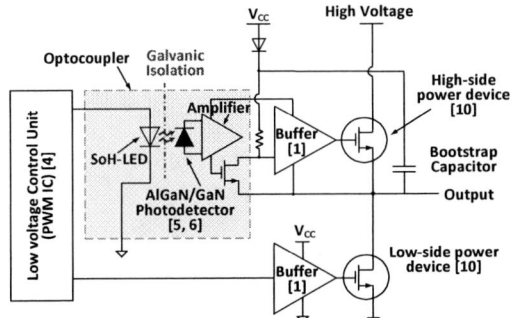

Fig. 1: Schematic diagram of a half-bridge power stage, using an opto-coupler for linking the low-side control circuit and the high-side drive circuit. An on-chip UV light source that can be seamlessly integrated with the AlGaN/GaN HEMT technology is highly desirable in the all-GaN power electronics solution.

For high-voltage switching applications (Fig. 1), opto-coupler is widely used to provide level-shifting between the low-side control circuit and high-side drive circuit with *galvanic* isolation. An opto-coupler that is material/process compatible and can be seamlessly integrated with the AlGaN/GaN-based power electronic devices/circuits is highly desirable for an *all-GaN* solution that promises reduced parasitics, compact size and enhanced reliability. Photodetectors based on AlGaN/GaN heterostructure [5, 6] have been widely reported with a cutoff wavelength at 365 nm (determined by the energy bandgap of GaN). However, UV light source was thought to be unprocurable from the unipolar AlGaN/GaN heterostructure. In this work, for the first time, we demonstrated that the metal-AlGaN/GaN Schottky diode is capable of producing GaN band-edge UV emission at 3.4 eV/364 nm under forward bias larger than ~2 V at room temperature. Using on-chip HEMT as a gate-controlled driver and the SoH-LED as a load, the SoH-LED was able to deliver switching operation up to 120 MHz, adequate for providing optical control signal for GaN high-side power switches.

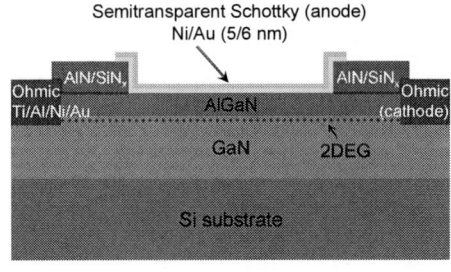

Fig. 2: Schematic cross-section of a SoH-LED on AlGaN/GaN-on-Si.

Fabrication of SoH-LED and HEM-LET

The AlGaN/GaN heterostructure consists of a 21 nm AlGaN barrier and a 3.8 μm GaN buffer/transition layer grown by MOCVD on a 4-in. Si(111) substrate. The 2DEG density is 10^{13} cm^{-2} and the carrier mobility is 2080 cm^{2}V^{-1}s^{-1} at room temperature. The cross-section of a SoH-LED was illustrated in Fig. 2. After formation of the ohmic electrode (cathode), an AlN/SiN$_x$ stack [7] by PEALD/PECVD was employed as the surface protection and passivation layer, followed by active device isolation using fluorine ion

978-1-4799-8002-4/14 $31.00 © 2014 IEEE

implantation. After defining the anode pattern and removing the dielectric stack, an remote plasma pretreatment (RPP) (using NH$_3$, Ar, N$_2$ plasma in sequence) was employed to remove native oxide and nitridize the surface [8]. Semitransparent Schottky anode was formed by Ni/Au (5/6 nm). AlGaN/GaN HEM-LETs with ohmic source and Shottky drain were fabricated simultaneously. A SiN$_x$/AlO$_x$ (15/8 nm) stack insulator was used as the gate dielectric in the HEM-LET to single out the Schottky drain. Finally, the devices were annealed at 400 °C for 10 min in N$_2$ ambience.

EL spectra and physical mechanism

A. EL emission image and spectra

Fig. 3(a) shows the EL emission images of a Ni/Au-AlGaN/GaN SoH-LED at different forward bias conditions taken by a CCD camera, with only the visible emission shown. The EL spectra (Fig. 3(b)) consist of not only a blue band and a yellow band, but also a **narrow GaN band-edge UV component**, similar to the photoluminescence (PL) spectra of the same AlGaN/GaN heterostructure excited by a 266 nm laser (Fig. 3(c)). Both the EL and PL spectra are from the GaN layer; no emission from the thin AlGaN barrier layer was detected. The relative intensity of the GaN band-edge UV emission increases with increasing drive bias/current and laser excitation intensity in the EL and PL spectra, respectively.

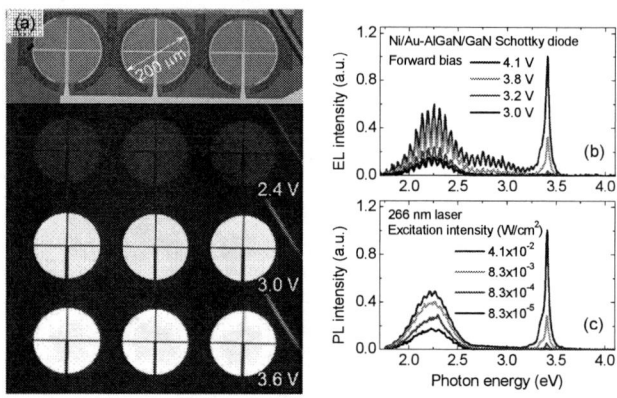

Fig. 3: (a) EL images of Ni/Au-AlGaN/GaN Schottky diodes with only visible light emission can be photographed. (b) EL spectra of the Schottky diode under different forward bias conditions. (c) PL spectra of the AlGaN/GaN heterostructure at various excitation intensities.

B. Current-voltage and EL-voltage characteristics

The current-voltage (*I-V*) characteristics, and the bias dependences of the full spectra integrated EL intensity and UV (285-390 nm) emission integrated intensity are plotted in Fig. 4. Aside from the current threshold at 1.1 V, an additional transition at ~2 V can be identified in the *I-V* curve. This additional transition voltage coincides with the threshold of the EL emission, indicating that holes start to be injected into the valence band of GaN layer at forward bias exceeding

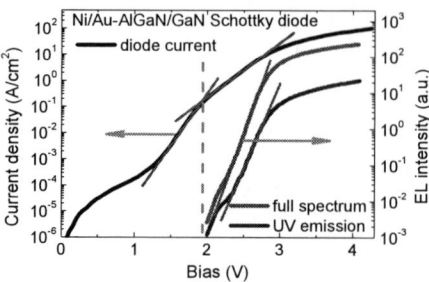

Fig. 4: *I-V* (black line), bias dependent full-spectra (green line) and UV-emission (blue line) integrated intensity characteristics of a Ni/Au-AlGaN/GaN SoH-LED at room temperature. The additional transition at ~2 V is indicated by the vertical dashed line. The light emission begins at this transition bias. Red lines are for visual guide.

2 V. The EL threshold voltage for a *pn* junction LED can be 100-200 mV lower than *hv/e* due to electron-hole thermal voltage, where *hv* is the photon energy [9]. Voltage even larger than *hv/e* is required for light emission from a conventional Schottky diode in order to achieve minority carrier injection condition [7]. Here, for the Ni/Au-AlGaN/GaN Schottky-on-heterojunction Schottky diode, the EL threshold voltage ~2 V is approximately 1.4 V lower than *hv/e* (GaN band-edge emission at 3.4 eV/364 nm), indicating an abnormal anti-Stokes light emission process.

C. Possible hole injection mechanisms

The band diagrams of Ni/Au-AlGaN/GaN Schottky diode under different bias conditions are schematically drawn in Fig. 5. The Schottky barrier height ~1.1 eV at Ni-AlGaN interface was extracted from the *I-V* characteristics in the bias range of 1.2 to 2 V shown in Fig. 4, indicating Fermi-level pinning effect. Under a forward bias slightly greater than 2 V, the metal Fermi level is above the GaN valence band, even in the extreme case when all of the external bias drops on the AlGaN layer, as shown in Fig. 5(b), ruling out the possibility of hole injection from metal to GaN through direct (or trap-assisted) tunneling or thermionic emission processes. The observed GaN band-edge 3.4 eV emission at a forward bias of ~ 2 V suggests a new hole injection process different from what exists in conventional *pn* junction LEDs.

The characteristic (Al)GaN surface-state distributions are illustrated in the band diagram shown in Fig. 5(c). The upper surface-band is responsible for the Fermi-level pinning at approximately 1 eV below the conduction band, whereas the lower surface-band has an energy distribution overlapping the valence band [11, 12]. Under forward bias, electrons are injected from the 2DEG and accelerated by the electric field in the AlGaN barrier. The electrons become 'hot' when they arrive at the Ni-AlGaN interface, and these 'hot' electrons can impact and ionize the surface-band states.

One possible model is that the metal Fermi level can be de-pinned at the Ni-AlGaN interface when the upper-surface band is fully ionized by the impact of hot electrons, as

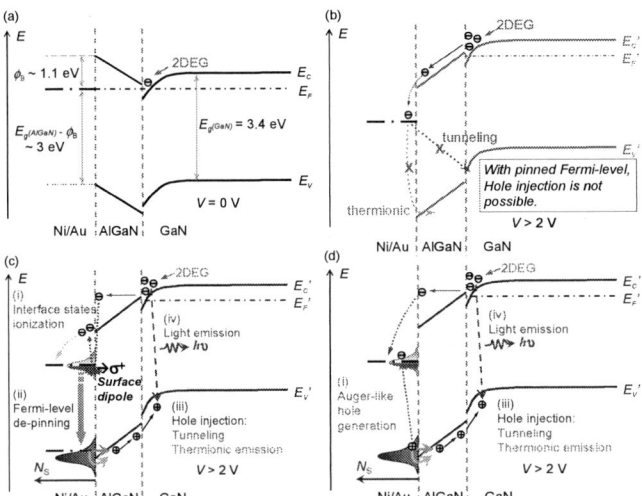

Fig. 5: Schematic band diagrams of the Ni/Au-AlGaN/GaN Schottky diode. (a) The energy quantities under zero bias are presented. (b) At a forward bias close to or larger than 2 V, the pinned metal Fermi level is far from aligned with the GaN valence band, eliminating the possibility of hole injection. Possible hole generation and injection mechanism, (c) Fermi-level de-pinning, (d) Auger process, at Ni-AlGaN interface under a forward bias close to or larger than 2 V, with the two distinct AlGaN surface bands illustrated schematically [11,12].

depicted in Fig. 5(c). As electrons in the upper surface-band are ionized, the metal Fermi-level is de-pinned simultaneously from its original position, subsequently moving downward and getting re-pinned at the lower surface-band. Then, holes can be injected from the re-pinned metal Fermi-level into the valence band of AlGaN through tunneling or thermionic emission, drift to the AlGaN/GaN interface region and recombine with 2DEG, generating the GaN band-edge UV emission. It should be noted that the AlGaN surface will get positively charged (compared with the condition of zero bias) as a result of the ionization of upper surface-band states, leading to the formation of a *surface dipole* pointing from surface into bulk. The field or potential of this *surface dipole* compensates the schematically drawn Fermi-levels' difference (in Fig. 5(c), which is apparently larger than the magnitude of applied bias) between metal and 2DEG at the state of Fermi-level de-pinning and re-pinning.

Another possibility is that the impact ionization happens at the lower surface-band. In this case, as one electron in the lower surface-band is ionized (emitted to the upper surface-band and drift into the metal), a hole will be generated simultaneously, as shown in Fig. 5(d). Since the energy distribution of lower surface-band overlaps with that of the valence band, the generated hole will be injected into the valence band of AlGaN through tunneling or thermionic emission, drift to the AlGaN/GaN interface region and recombine with 2DEG, generating the GaN band-edge UV emission. Such an hole generation and injection process is similar to an Auger process.

The hole injection processes assisted by tunneling and thermionic emission have been identified in low- temperature bias-dependent EL intensity characteristics [13]. Because both of the proposed hole generation/injection processes (i.e. Fermi-level de-pinning or Auger-process) are determined by the surface state distributions of AlGaN, the EL emission should be a 'universal' property of metal-AlGaN/GaN Schottky diodes. We have experimented with other Schottky metals, such as Pt/Au, Ni/Au and Ti/Au. Similar *I-V* and EL characteristics have also been observed.

In the above proposed models, one electron ionizes the upper (Fermi-level de-pinning) or lower (Auger process) surface-band state, inducing the generation/injection of one hole to recombine with another electron. This $2e\text{-}1h$ process maintains energy conservation and explains the anti-Stokes characteristic of EL emission.

DC characteristics of an HEM-LET

A light-emitting transistor (LET) can provide coherent electronic and photonic functionalities in a single device. The SoH-LED can be seamlessly integrated with a conventional MIS-HEMT, leading to the demonstration of an AlGaN/GaN HEM-LET, as shown in Fig. 6(a). Both the EL(UV) and drain current of the HEM-LET feature a turn-on voltage at ~2 V and can be controlled simultaneously by the gate voltage, as shown in Fig. 6(b).

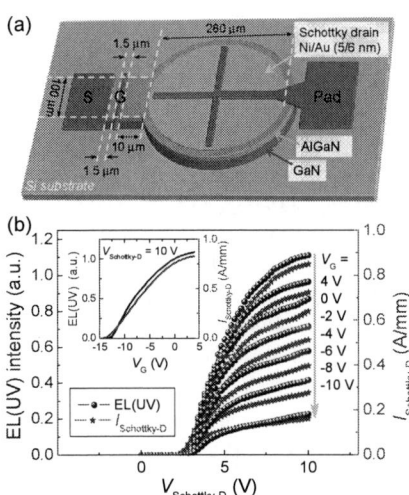

Fig. 6: (a) Schematic device structure of an AlGaN/GaN HEM-LET, with dimensions labelled. (b) $I_{\text{Schottky-D}}$ - $V_{\text{Schottky-D}}$ (red line) and EL (UV)-intensity vs. $V_{\text{Schottky-D}}$ (blue line) characteristics at various V_G of the HEM-LET. The inset provides the EL(UV)-intensity-V_G and $I_{\text{Schottky-D}}$-V_G characteristics of the HEM-LET at $V_{\text{Schottky-D}}$ = 10 V.

Switching operation of the SoH-LED

Figure 7(a) depicts the setup for measuring switched EL emission from the SoH-LED. Figure 7(b)-(d) shows the waveforms of drive biases and EL emissions at frequencies of

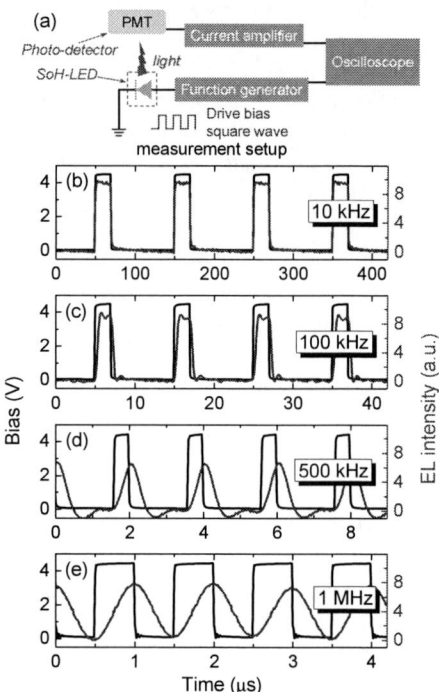

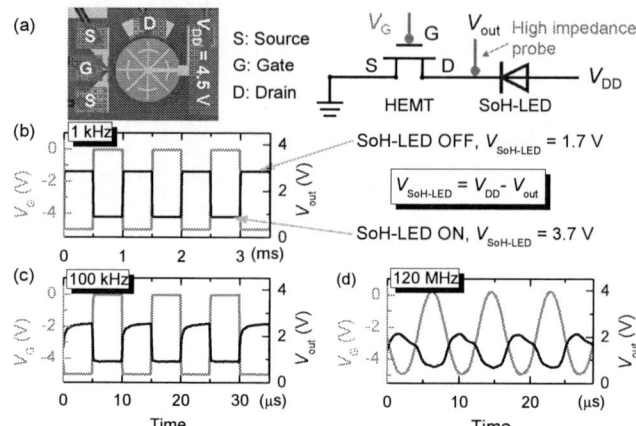

Fig. 7: (a) Measurement setup of pulsed operation of EL emission from a Ni/Au-AlGaN/GaN SoH-LED. (b)-(e) Waveforms of drive bias and EL emission at frequencies of 10 kHz, 100 kHz, 500 kHz and 1 MHz.

Fig. 8. Switching behaviors of a SoH-LED controlled by a HEMT. (a) Device photo and equivalent measurement circuit. A Ni/Au-AlGaN/GaN SoH-LED is in series with a HEMT and is set as the load during the switching measurement. Gate voltage V_G is controlled and the output voltage V_{out} at drain electrode (also the cathode of SoH-LED) is monitored. (b)-(d), Waveforms of V_G and corresponding V_{out} at frequencies of 1 kHz, 100 kHz, and 120 MHz. The voltage applied on Ni/Au-AlGaN/GaN Schottky junction equals to V_{dd}-V_{out}.

10 kHz, 100 kHz, 500 kHz and 1 MHz, respectively. The highest testing frequency is limited by the bandwidth of the current amplifier. The electrical switching operation of the SoH-LED was measured by integrating a SoH-LED (load) with a HEMT (driver), as shown in Fig. 8. Proper switching operation up to 120 MHz has been demonstrated.

Conclusion

Without any *p*-type doping layer, GaN band-edge EL(UV) emission has been generated from forward biased metal-AlGaN/GaN Schottky diode, facilitated by the creation of III-nitride surface with low surface state density, thus filling the void of UV light sources compatible with AlGaN/GaN-heterojunction-based photodetectors and power devices.

Acknowledgement

This work was supported by Hong Kong Research Grants Council under projects N_HKUST636/13 and 604410.

References

[1] W. Chen, K.-Y. Wong, and K. J. Chen, "Monolithic Integration of Lateral Field-Effect Rectifier with Normally-off HEMT for GaN-on-Si Switch-mode Power Supply Converters", in *IEDM Tech. Dig.*, pp. 141-144, 2008.

[2] E. Bahat-Treidel, O. Hilt, R. Zhytnytska, A. Wentzel, C. Meliani, J. Würfl, and G. Tränkle, "Fast-Switching GaN-Based Lateral Power Schottky Barrier Diodes with Low Onset Voltage and Strong Reverse Blocking", *IEEE Electron Device Lett.*, vol. 33, pp.357-359, 2011.

[3] A. M. H. Kwan, X. Liu, and K. J. Chen, "Integrated Gate-protected HEMTs and Mixed-Signal Functional Blocks for GaN Smart Power ICs", in *IEDM Tech. Dig.*, pp. 155-158, 2012.

[4] H. Wang, A. M. H. Kwan, Q. Jiang, and K. J. Chen, "A GaN Pulse Width Modulation Integrated Circuit". Proc. of the 26th *Int. Symp. on Power Semiconductor Devices & IC's*, pp.430-433, Hawaii, USA, June 15-19, 2014,.

[5] M. Martens, J. Schlegel, P. Vogt, F. Brunner, R. Lossy, J. Würfl, M. Weyers, and M. Kneissl, "High Gain Ultraviolet Photodetectors based on AlGaN/GaN Heterostructures for optical Switching", *Appl. Phys. Lett.*, vol. 98, 211114, 2011.

[6] T. Narita, A. Wakejima, and T. Egawa, "Ultraviolet Photodetectors using Transparent Gate AlGaN/GaN High Electron Mobility Transistor on Silicon Substrate", *Jpn. J. Appl. Phys.* vol. 52, 01AG06, 2013.

[7] A. J. Steckl, M. Garter, R. Birkhahn, and J. Scofield, "Green Electroluminescence from Er-doped GaN Schottky Barrier Diodes", *Appl. Phys. Lett.* vol. 73, pp. 2450-2452, 1998.

[8] S. Huang, Q. Jiang, S. Yang, C. Zhou, and K. J. Chen, "Effective Passivation of AlGaN/GaN HEMTs by ALD-grown AlN Thin Films", *IEEE Electron Device Lett.*, vol. 33, pp.516–518, 2012.

[9] E. F. Schubert, *Light Emitting Diodes* (Cambridge University Press, Cambridge, UK, 2006).

[10] Z. Tang, S. Huang, Q. Jiang, S. Liu, C. Liu, and K. J. Chen, "High-voltage (600V) low-leakage low-current-collapse AlGaN/GaN HEMTs with AlN/SiN$_x$ passivation," *IEEE Elec. Dev. Lett.*, vol. 34, no. 3, pp. 366-368, 2013.

[11] A. Rizzi and H. Lüth, "Electronic gap states on GaN (0001)-(1x1) surfaces studied by electron spectroscopies", *Nuovo Cim. D*, vol. 20, pp. 1039-1045, 1998.

[12] M. S. Miao, A. Janotti, and C. G. Van de Walle, "Reconstructions and Origin of Surface States on AlN Polar and Nonpolar Surfaces", *Phys. Rev. B*, vol. 80, 155319, 2009.

[13] B. K. Li, M. J. Wang, K. J. Chen, and J. N. Wang "Fermi-level Depinning and Hole Injection Induced Two-Dimensional Electron Related Radiative Emissions from a Forward Biased Ni/Au-AlGaN/GaN Schottky Diode", *Appl. Phys. Lett.*, vol. 95, 232111, 2009.

The Super-Lattice Castellated Field Effect Transistor (SLCFET):
A Novel High Performance Transistor Topology Ideal for RF Switching

Robert S. Howell[1], Eric J. Stewart[1], Ron Freitag[1], Justin Parke[1], Bettina Nechay[1], Harlan Cramer[1], Matthew King[1], Shalini Gupta[1],

Jeffrey Hartman[1], Megan Snook[1], Ishan Wathuthanthri[1], Parrish Ralston[1], Karen Renaldo, H. George Henry[1], R. Chris Clarke[2]

Northrop Grumman Electronic Systems[1], Linthicum MD, USA

Retired Northrop Grumman Electronic Systems[2]

Abstract

NGES reports the development of a novel transistor structure based on a GaN super-lattice channel with a 3D gate, named the SLCFET (Super-Lattice Castellated Field Effect Transistor). Transistor measurements provided median values of $I_{MAX} > 2.7$ A/mm, $V_{PINCH} = -8$V, with $R_{ON} = 0.41$ Ω-mm and $C_{OFF} = 0.19$ pF/mm, for an RF switch FOM of $F_{CO} = 2.1$ THz.

Introduction

The RF switch is a critical component for virtually all RF system architectures, because of its use in phase shifters, attenuators, true time delay, filters and T/R switches. This centrality of use within MMICs and RF components make the performance of the RF switch key to meeting the required system performance for many different system applications, ranging from advanced phased array radars and multi-function sensors, to wireless components of consumer electronics. In system architectures where these RF switch components are located in the front end of the systems, the overall system performance becomes closely tied with the RF switch performance, in particular the insertion loss and isolation as a function of frequency.

This high degree of sensitivity to the RF switch insertion loss and isolation performance has driven a great deal of interest towards investigation and development of low loss RF switching, particularly in the field of RF MEMS [1-3], which have exhibited a lower loss and higher isolation relative to semiconductor based RF switching techniques. However, the high actuation voltages, slower switching speeds, complex packaging requirements and problematic reliability of RF MEMS have continued to stimulate interest in solid state solutions, including semiconductor [4] and more recently chalcogenide phase change materials [5]. PiN diodes [6] offer extremely low losses and high isolation similar to RF MEMS, but their power consumption requirements and more complicated control bias networks limit their utility to those applications with relatively few switches. FET based switches offer low power consumption

and fast switching, as well as modest control voltages and excellent reliability, but have shown significantly higher insertion loss and lower isolation relative to RF MEMS and diode based switches. This has been true for FET based RF switches in a variety of different technologies, including CMOS [7], Silicon on Sapphire [8], GaAs pHEMTs [9], or InP [10], ABCS [11] and GaN HEMTs [12]. This relatively poor performance of the various FET technologies results from the close correlation between their respective ON resistances to their OFF capacitances. The utility of increasing gate periphery to decrease ON resistance is thus limited by the increase in the OFF capacitance in an approximately equal fashion. This relationship is generally described by F_{co}, the RF switch figure of merit [13]:

$$F_{co} = \frac{1}{2\pi R_{ON} C_{OFF}} \qquad (1)$$

Where R_{on} is the ON resistance of the switch and C_{off} is the OFF capacitance of the switch. Values for the RF switch figure of merit in FETs ranges from ~360 GHz for Si CMOS on SOI [14], to ~500 GHz in GaAs pHEMTs [15] and GaN HEMTs [12], to ~840 GHz for InP HEMTs [10]. This compares to an AlGaAs PiN diode F_{co} of ~2.4 THz [6], and an RF MEMS F_{co} of 3.8 THz [3]. The phase change RF switch, ideal for slow switching and reconfigurable RF applications, has demonstrated an F_{co} of 12.5 THz [5]. For a FET based RF switch to achieve a broadband insertion loss and isolation performance similar to RF MEMS and PiN diodes, an increase in the FET F_{co} to a similar THz level is required. Such an improvement would result in an optimal RF switch that combines the superior broadband loss and isolation performance of RF MEMS and PiN diode RF switch technology, with the attractive reliability, switching speed and low power consumption of FET RF switches.

Approach

In order to improve upon the basic FET structure for RF switching, it is important to understand the origins of the conventional device structure's limitations in RF switching.

978-1-4799-8002-4/14 $31.00 © 2014 IEEE

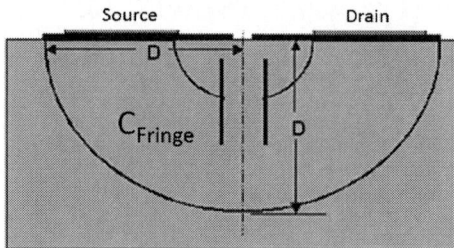

Figure 1: The fringing field capacitance of a FET is a three dimensional effect in contrast to the two dimensional current carrying sheet charge.

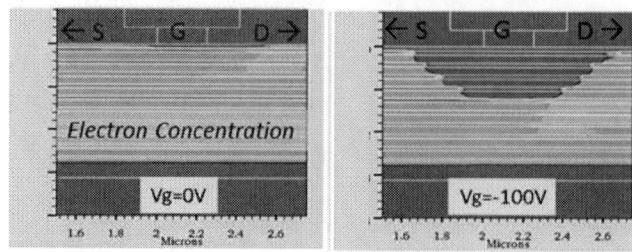

Figure 2: Simulation showing electron screening effect of the top electron layers in super-lattice that prevents a simple 'top' gate from depleting out the carriers in a super-lattice structure prior to applied fields that exceed the critical field strength of the super-lattice material.

Whereas an RF amplifier structure and resulting performance is strongly influenced by the gate to source and gate to drain capacitances, the RF switch performance is controlled by the OFF state capacitance, which is primarily composed of the source to drain capacitance. In a conventional FET, this source to drain capacitance, and thus the OFF capacitance, is determined in large part due to the fringing field capacitances between the source and drain, as depicted in Figure 1. This fringing field capacitance scales proportionally with gate periphery, just as ON resistance scales inversely with the gate periphery, which is why the switch figure of merit to first order is determined by the transistor type and material rather than its specific design. However, an important insight into this relationship is that in many FET structures, such as MOS and HEMT devices, the current carrying channel is a two dimensional sheet charge. Thus, if the current carrying channels are paralleled by stacking in the vertical dimension, the ON resistance can be scaled down without significantly altering the fringing field capacitance or increasing the OFF capacitance. This stacking of current channels can be accomplished with a super-lattice structure, such as with AlGaAs/GaAs or AlGaN/GaN heterojunctions.

Super-lattice structures have been examined for use in FETs in the past [16], and more recently to make optoelectronic semiconductor devices [17], micro-coolers [18] or as a means of reducing contact or access resistance to a HEMT [19, 20]. However, the use of a super-lattice as the channel for a FET has been limited by the inability to fully turn OFF the stacked paralleled current channels formed by the super-lattice [21, 22]. This is caused by the screening effect the top channels have on the bottom channels, preventing the gate field from depleting out the bottom channels. This limitation is illustrated in Figure 2, which shows a simulation of a super-lattice with several stacked two dimensional electron gas (2DEG) layers and a conventional gate. As the gate bias is increased, the electrons are depleted from the top most channel layers, but

the electric field from the gate cannot penetrate deeply into the super-lattice to deplete out the lower electron channel layers before catastrophic avalanche breakdown of the FET materials occurs.

The solution to this is to adopt a three dimensional, 'castellated' gate structure (so named for their resemblance to the crenellations on top of a medieval castle's walls), a gate structure first developed for improving the power handling capabilities of MESFETs [23]. This castellated gate allows the electric field applied by the gate to penetrate into each of the stacked 2DEG channels, by means of the sidewalls of the three dimensional gate structure. This permits the gate field action to deplete out the carriers of each of the channels in the super-lattice, and thereby control them. The combination of this super-lattice epitaxial structure in conjunction with the castellated, three dimensional gate is, to the best knowledge of the authors, a new device topology for FETs, that we have named the Super-Lattice Castellated Field Effect Transistor, or SLCFET. The super-lattice creates multiple current channels in parallel between the source and drain of the device, lowering ON resistance and the resulting RF switch insertion loss, while maintaining approximately the same OFF capacitance of a conventional HEMT, providing a dramatic improvement in RF switch performance.

Results

While the SLCFET structure can be applied to a number of material systems with various advantages and disadvantages, this reported SLCFET is based on a super-lattice of AlGaN/GaN heterojunctions, fabricated on a 100 mm diameter semi-insulating SiC wafer using an MOCVD process. The super-lattice of AlGaN/GaN dramatically lowers the sheet resistance of the epitaxy, as a single heterostructure is measured to have a sheet resistance of 300 $\Omega/\square$, while the super-lattice lowers the sheet resistance to 60 $\Omega/\square$. The castellated, three dimensional gate is formed using a series of trenches created using a dry etch process of the AlGaN/GaN super-lattice, on top of which the gate is overlaid, as shown pictorially in Figure 3. The SLCFET transistors fabrication process uses a combination of I-line stepper and electron beam lithography processes, resulting in a direct written 0.25 μm 'brick' gate. A representative SLCFET is shown Figure 4, with a close up view of the castellations and three dimensional gate above the super-lattice depicted in the SEM micrographs.

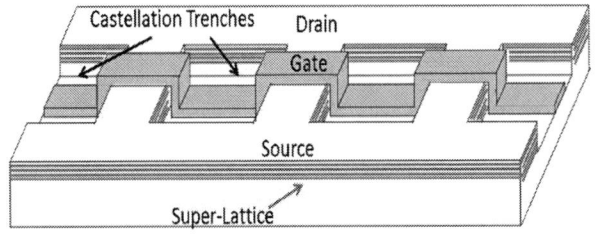

Figure 3: Illustration of the SLCFET device structure combining a super-lattice epitaxial channel with a three dimensional 'castellated' gate.

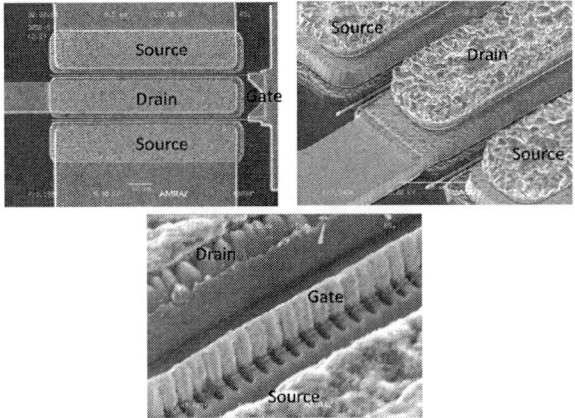

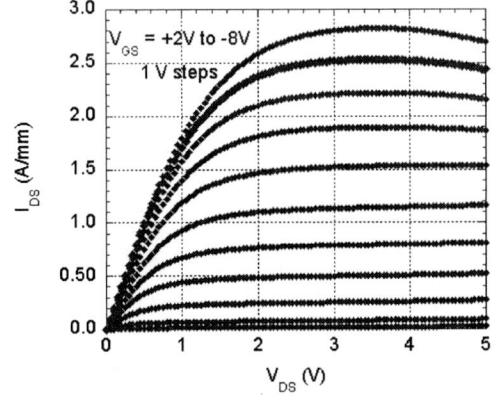

Figure 4: SEM micrographs depicting a representative 2x67 μm SLCFET device structure, with close up of castellated gate.

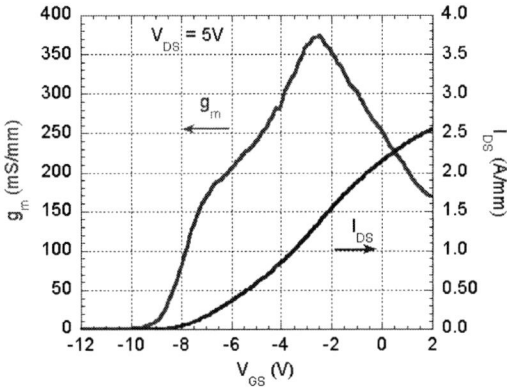

Figure 6: Measured I_{DS} and g_m vs V_{GS} of SLCFET, measured at V_{DS} = 5V

SLCFET transistors with I_{MAX}>2.7 A/mm, V_{PINCH} = -8V, g_{m_MAX}=375 mS/mm have been measured, shown in Figures 5 and 6. These transistors were measured to have a median R_{ON}=0.41 Ω-mm and C_{OFF}=0.19 pF/mm, for an RF switch Figure of Merit of F_{CO}=2.1 THz. This compares very favorably against the current state of art performance for conventional transistor topologies as reported in the literature, shown in Figure 7, with a FOM 3-10x greater than commonly measured for other FET RF switch technologies, and approaching the RF Switch FOM values reported for RF MEMS and PiN diodes.

While the SLCFET structure has not been designed or optimized for operation as an RF amplifier, the H21 and MAG performance versus frequency is shown in Figure 8. From this plot, values for f_T and f_{max} can be extrapolated using a -6 dB/octave relationship, which leads to an f_T of 52 GHz and f_{max} of 53 GHz.

Wideband single pole double throw (SPDT) RF Switch MMICs designed to operate at 1-18 GHz were built with the

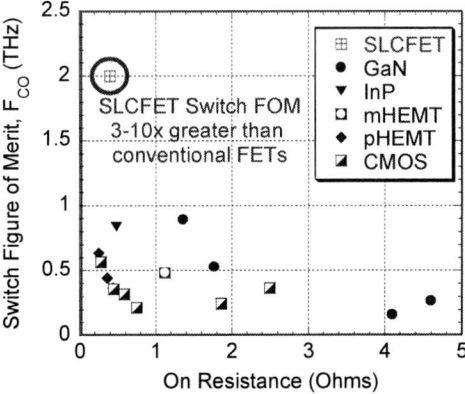

Figure 7: Comparison of reported F_{CO} values for various FET based RF switch technologies (F_{CO} = $1/2\pi R_{ON}C_{OFF}$).

Figure 5: Measured SLCFET I_{DS} vs V_{DS}, at V_{GS} = +2 to -8V in 1V steps.

978-1-4799-8002-4/14 $31.00 © 2014 IEEE

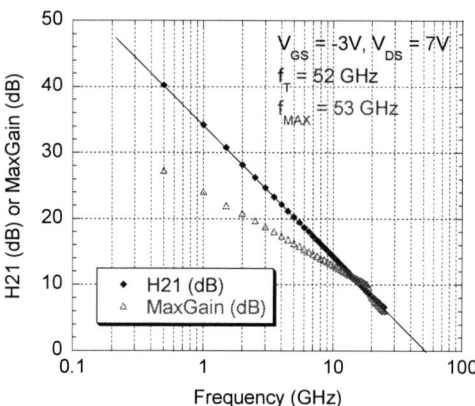

Figure 8: Measured H21 and MAG gain plots for SLCFET, which provides extrapolated values of f_T = 52 GHz and f_{max} = 53 GHz

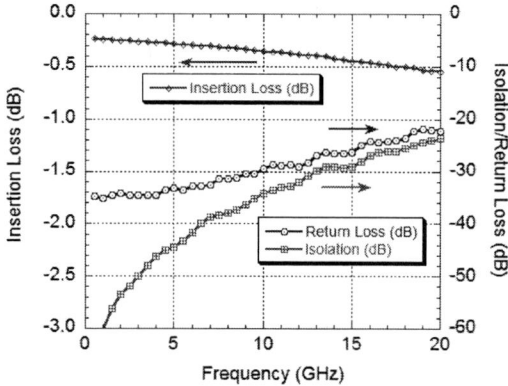

Figure 9: Representative SLCFET Series-Shunt 1-18 GHz SPDT measurement, demonstrating broadband, low loss and high isolation

SLCFET devices in a series-shunt circuit topology. Figure 9 shows the low insertion loss ($\leq$ 0.5 dB), high isolation ($\geq$ -26 dB) performance resulting from this technology, performance that compares favorably with that achieved by MEMS and diode based RF switches. By combining excellent RF switching performance with the numerous desirable aspects of FET based switching, the SLCFET is a key enabling technology for next generation RF systems.

References

[1] S. Lucyszyn and Suneat Pranonsatit, "RF MEMS for Antenna Applications", *7th European Conference on Antennas and Propagation EUCAP 2013*, pp. 1988-1992, 2013.

[2] G. Rebeiz, G-L Tan, J.S. Hayden, "RF MEMS phase shifters: design and applications", *Microwave Magazine* v. 3 no. 2, pp. 72-81, 2002.

[3] R. Stefanini, M. Chatras, P. Blondy, G. Rebeiz, "Miniature RF MEMS metal-contact switches for DC-20 GHz applications," *2011 IEEE MTT-S International Microwave Symposium Digest*, pp. 1-4, 2011.

[4] D. Gotch, "A Review of Technological Advances in Solid-state Switches-In-depth look at the technological advances in solid-state switches, from discrete and monolithic PIN diode switches to GaAs PIN and FET." *Microwave Journal* v. 50, no. 11, pp. 24-36, 2007.

[5] N. El-Hinnawy, et. al., "Improvements in GeTe-Based Inline Phase-Change Switch Technology for RF Switching Applications," *2014 CS MANTECH*, pp. 401-403, 2014.

[6] T. Boles, J. Brogle, D. Hoag, and Curcio, "AlGaAs PiN diode multi-octave, mmW switches," *COMCAS 2011*, pp. 1-5, 2011.

[7] M. Parlak and J.F. Buckwalter, "A 2.5 dB insertion loss, DC-60 GHz CMOS SPDT switch in 45-nm SOI," *2011 CSICS*, pp. 1-4, 2011.

[8] D. Kelly, C. Brindle, C. Kemerling, and M. Stuber, "The State of the Art of SOS CMOS RF Switches," *2005 CSICS*, pp. 200-203, 2005.

[9] Q. Xiao, G. Samiotes, T. Galluccio and B. Rizzi, "A high performance DC-20 GHz SPDT switch in a low cost plastic QFN package," *Proc. 39th European Microwave Conference*, pp. 1673-1676, 2009.

[10] H. Kamitsuna, Y. Yamane, M. Tokumitsu, H. Sugahara and M. Muraguchi, "Low-power InP-HEMT switch ICs Integrating 2x2 switches for 10 Gb/s systems," *IEEE J. Solid State Circuits*, v. 41, no. 2, pp. 452-460, 2006.

[11] C-T Yu, C-C.Shen, H-C. Ho, H-Y. Chang, J-S Fu and H-K Lin, "Low Voltage High-Speed Antimonide-Based Compound Semiconductor (ABCS) 2-μm InAs/AlSb HEMT MMIC Process and Its Broadband Switch Application," *Proceedings of APMC 2012*, pp. 196-198, 2012.

[12] C. Campbell and D. Dumka, "Wideband high power GaN on SiC SPDT switch MMICs," *2010 IEEE MTT-S International Microwave Symposium Digest*, pp. 145-148, 2011

[13] R. Gutmann and D. Fryklund, "Characterization of Linear and Nonlionear Properties of GaAs MESFETs for Broad Band Control Applications", *IEEE Trans. on Micr. Theory and Tech.*, Vol MTT-35 No. 5, May 1987, pp. 516-521.

[14] S. Parthasarathy, et. al., "RF SOI Switch FET Design and Modeling Tradeoffs for GSM Applications," *2010 23rd International Conference on VLSI Design*, pp. 194-199, 2010.

[15] Y. Tsukahara, et. al. "Millimeter-wave MMIC Switches with pHEMT Cells Reduced Parasitic Inductance," *2003 IEEE MTT-S Digest*, pp. 1295-1298, 2003.

[16] D.K. Arch, M. Shur, J. Abrokwah and R.R. Daniels, "Superlattice conduction in superlattice modulation-doped field-effect transistors," *J. Appl. Phys.*, v. 61, no. 4, pp. 1503-1509, 1987.

[17] K.Y. Cheng, "Molecular beam epitaxy technology of III-V compound semiconductors for optoelectronic applications," *Proc. IEEE*, v. 85, no. 11, pp.1694-1714, 1997.

[18] J. Zhang, N. Anderson and K.M. Lau, "AlGaAs superlattice microcoolers," *Appl. Phys. Lett.*, v. 83, no. 2, pp.374-376, 2003.

[19] S. Heikman, S. Keller, D. Green, S. DenBaars, U. Mishra, "High conductivity modulation doped AlGaN/GaN multiple channel heterostructures," *J. Appl. Phys.* v. 94, no. 8, pp. 5321-5325, 2003.

[20] T. Palacios, A. Chini, D. Buttari, S. Heikman, A. Chakraborty, S. Keller, S. DenBaars and U. Mishra, "Use of double-channel heterostructures to improve the access resistance and linearity in GaN based HEMTs," *IEEE Trans. Elec. Dev.*, v. 53, no. 3, pp. 562-565, 2006.

[21] N.H. Sheng, C.P. Lee, R.T. Chen, D.L. Miller and S.J. Lee, "Multiple-channel GaAs/AlGaAs High Electron Mobility Transistors," *IEEE Elec. Dev. Letters*, v. EDL-6, no. 6, pp. 307-310, 1985.

[22] D.K. Arch, M. Shur, J.K. Abrokwah and R.R. Daniels, "Superlattice conduction in superlattice modulation-doped field-effect transistors," *J. Appl. Phys.* v. 61, no. 4, pp. 1503-1509, 1987.

[23] R.C. Clarke, "A high-efficiency castellated gate power FET," *Proceedings IEEE/Cornell Conference on Advanced Concepts in High Speed Semiconductor Devices and Circuits*, August 1983, p.93-ill, IEEE Inc., New York, 1983.

MIT Virtual Source GaNFET–RF Compact Model for GaN HEMTs: From Device Physics to RF Frontend Circuit Design and Validation

Ujwal Radhakrishna, Pilsoon Choi, Sushmit Goswami,
Li-Shiuan Peh, Tomás Palacios, Dimitri Antoniadis

Microsystems Technology Laboratories, Massachusetts Institute of Technology, Cambridge, MA 02139, USA
{ujwal, pilsoon, sushmit, peh, tpalacios, daa}@mit.edu

Introduction

GaN HEMT based power amplifiers (PAs) are gaining foothold in high-power transceiver circuit design at microwave frequencies [1]. The high breakdown voltage, high-current capability together with low on-resistance and on-capacitance of GaN HEMTs enables improved efficiency and linearity at higher output-power PAs [2], [3]. To take advantage of the performance gains of these devices in RF circuit design, accurate non-linear, large signal device models suitable for high frequency and high power operation regimes are required. It is also desirable that these compact models be grounded on appropriate device physics in order to gain insight into the impact of the behavioral nuances of the GaN HEMTs on RF circuit performance, which is not the case with most of the available models such as EEHEMT, Curtice, and Angelov models [4]. This work is a first demonstration of a physics-based GaN HEMT compact model that is calibrated and verified all the way from device- to an RF- circuit-level.

The MVS-G-RF model captures static and dynamic device behavior through self-consistent current and charge expressions. In addition, access regions, which play an important role in device linearity [4] are modeled as *implicit-gate* transistors. The model includes the effect of self-heating, gate leakage, device parasitics and RF device-noise. The model requires a small number of parameters with straightforward physical meanings and is validated against DC–IV, S-parameter, load-pull, and noise figure measurements of fabricated devices. The model was then used to design and validate an RF circuit which in this case was an RF front-end for the IEEE 802.11p standard and fabricated on Cree's 0.25 μm GaN-on-SiC platform.

MVS-G-RF: Transport and charge model

The MVS-G-RF model sub-circuit along with the structure schematic of the device showing different regions of interest are given in Fig. 1.

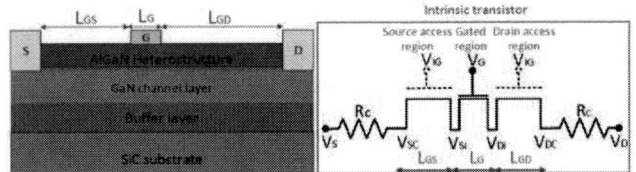

Figure 1. (a) Cross-sectional schematic of the commercial RF-AlGaN/GaN HEMTs on SiC. (b) The equivalent circuit for the device with gated-transistor, implicit-gate access region transistors, and contact resistances is shown.

$$I_D/W = Q_{inv,s}v_{sat}F_{sat} \quad F_{sat} = \frac{V_{DSi}/V_{DSAT}}{(1+(V_{DSi}/V_{DSAT})^{\beta})^{1/\beta}} \quad V_{DSAT} = \frac{L_g v_{sat}}{\mu} \quad (1)$$

$$Q_{inv,s(d)} = C_g n\phi_t ln\left(1 + exp\left(\frac{(V_{GSi(GDi)})-(V_T-\alpha\phi_t F_f)}{n\phi_t}\right)\right) \quad (2)$$

The gated-region transistor drain current in (1)-(2) has a similar form as in [5] and is based on the virtual-source top-of-barrier carrier transport theory. Drift-diffusion transport in access regions is captured by employing gradual channel approximation resulting in (3)-(4) which have dependency on access region drain-charge and source-charge (5a)-(5b). The transition from pinch-off to velocity saturation with access region length scaling is achieved using $F_{Vsatrs-rd}$ in (4) [6].

$$I_{DRS-RD}/W = \frac{(Q_{invsrs-srd}+Q_{invdrs-drd})}{2}v_{sat}F_{vsatrs-rd} \quad (3)$$

$$F_{Vsatrs-rd} = \frac{(Q_{invsrs-srd}-Q_{invdrs-drd})/C_{Ig}V_{DSATrs-rd}}{\left(1+((Q_{invsrs-srd}-Q_{invdrs-drd})/C_{Ig}V_{DSATrs-rd})^{\beta}\right)^{1/\beta}}$$

$$V_{DSATrs-rd} = \frac{L_{gs-gd}v_{sat}}{\mu} \quad ; \quad V_{IG}-V_{Trs-rd} = 1/(\mu R_{sh}C_{Ig}) \quad (4)$$

$$Q_{invsrs-rd} = 2C_{Ig}n\phi_t ln\left(1 + exp\left(\frac{(V_{IG}-V_{SC-Di})-(V_{Trs-rd}-\alpha\phi_t F_{f,srs-rd})}{2n\phi_t}\right)\right) \quad (5a)$$

$$Q_{invdrs-rd} = 2C_{Ig}n\phi_t ln\left(1 + exp\left(\frac{(V_{IG}-V_{Si-DC})-(V_{Trs-rd}-\alpha\phi_t F_{f,srs-rd})}{2n\phi_t}\right)\right) \quad (5b)$$

Self-heating, ΔT, is calculated through an RC thermal network and affects carrier mobility and velocity (6a)-(6b).

$$\mu = \mu_0\left(1+\frac{\Delta T}{T_0}\right)^{-\varepsilon}; \quad v_{sat} = v_{sat0}(1+Z\Delta T)^{-1} \quad (6a)\text{- }(6b)$$

The Schottky gate diode models in (7) capture low and high level injection effects in forward mode as well as recombination and gate-induced-drain-lowering (GIDL) in reverse mode.

$$I_{GSi,GDi} = WI_j e^{\frac{V_j}{\eta\phi_t}}\left(e^{\frac{V_{GSi,GDi}}{\eta\phi_t}}-1\right) + WI_{rec}e^{-\frac{V_j}{\eta_{rec}\phi_t}}\left(1-e^{\frac{V_{GSi,GDi}}{\eta_{rec}\phi_t}}\right) \quad (7a)$$

$$F_{fs,fd} = \frac{1}{1+exp\left(\frac{V_{GSi,GDi}-(V_{inj}-0.5\alpha\phi_t)}{\alpha\phi_t}\right)} \quad \frac{1}{\eta} = \frac{1}{\eta_1} + \left(\frac{1}{\eta_2}-\frac{1}{\eta_1}\right)F_{fs,fd} \quad (7b)$$

The drain and gate current model fits along with drain-current derivatives are given in Fig. 2 for different V_D and V_G ranges, using the parameter list of Table 1. The source and drain charges in (2) have similar form as in [5] and the channel charges and their derivatives, C_{gs} and C_{gd}, are self-consistently solved as in (8) using current continuity and charge partitioning.

$$Q_{S(D)} = \frac{2WL_g}{(Q_{inv,s}^2-Q_{inv,d}^2)^2}\left[+(-)\frac{Q_{inv,s}^5-Q_{inv,d}^5}{5} - (+)Q_{inv,d(s)}^2\frac{Q_{inv,s}^3-Q_{inv,d}^3}{3}\right] \quad (8)$$

978-1-4799-8002-4/14 $31.00 © 2014 IEEE   295

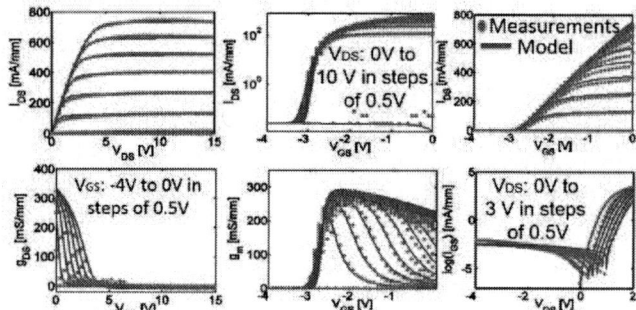

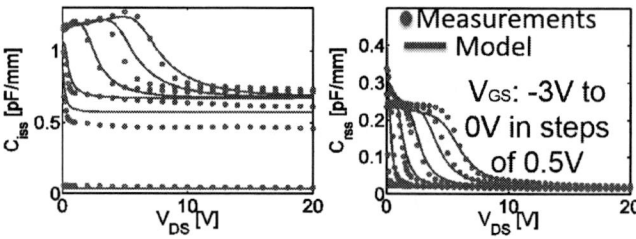

Figure 2. Output, transfer, output conductance, transconductance and gate current plots of L_g=0.25 µm RF device. Good agreement has been achieved between the model and measurement. The drop in g_m beyond V_t is caused in the model by the non-linear access region behavior and device self-heating.

The resulting input (C_{iss}) and reverse transfer capacitances (C_{rss}) are validated against S-parameter derived device measurements as shown in Fig. 3.

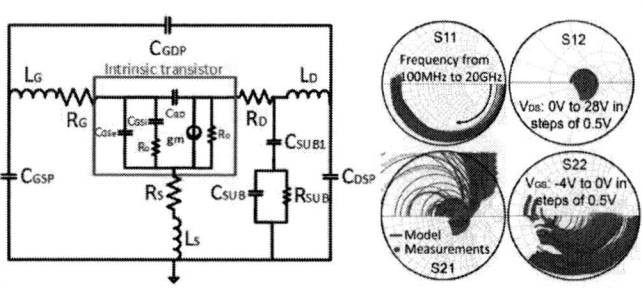

Figure 3. Charge model that is self-consistent with transport gives the input capacitance (C_{iss}) and reverse transfer capacitance (C_{rss}) fits which compare with the measured C-Vs obtained from S-parameter measurements, as shown above. Fringing capacitance extracted from measurements is the only additional parameter needed for the fits.

The small signal S-parameters are modeled by adding device-level parasitics associated with pads, leads, and substrate, as shown in the circuit of Fig. 4. The values of parasitic elements are extracted from S-parameter fits with the values listed in Table. 1, as described in Fig. 4.

Figure 4. Parasitics associated with contact pads, terminal leads, and substrate losses are added onto the core intrinsic model. The parameters are extracted from small signal S-parameter fits over the desired bias and frequency ranges required for RF circuit design. The input gate parasitics affect S11 parameter and output drain parasitics including substrate parasitics can be extracted from S22 parameters.

Device large signal modeling for PA design

The physical nature of MVS-G-RF model ensures that once small signal model parameters are accurately calibrated against measurements, the model can be used to predict large-signal device behavior without employing any additional parameters. The device level large signal metrics such as G_t, P_{out}, PAE, and I_{out} are measured using a wafer-level load pull measurement setup from Maury-Microwave. The measured power sweep data are compared against the model in Fig. 5a for a 0.25 µm gate length and 2×180 µm gate width Cree GaN-on-SiC device biased at class AB (V_{DS}=28 V and V_{GS}=V_t+0.25 V); this shows correct prediction of P_{out}=28 dBm and PAE=45% at 6 GHz. The calibrated model can also be used to make projections linking key device-level bottlenecks to RF device performance, as shown in Fig. 5b. The device-access-regions are responsible for gain compression at high input power and the model predicts that self-aligned devices could potentially boost PAE by 40% by increasing P_{out} in compression. Fabrication technologies that yield lower gate-parasitic-fringing capacitances are beneficial as well, with the limiting case of zero fringing capacitances yielding enhanced G_t= 33 dB and PAE=60%.

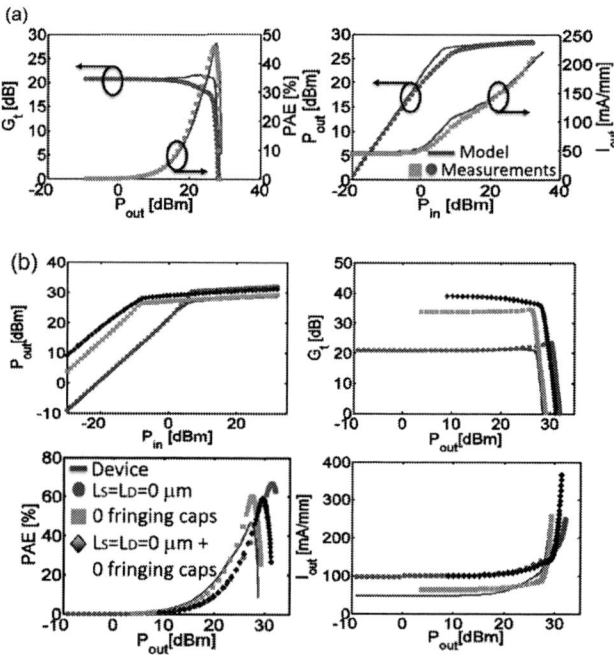

Figure 5. (a) Large signal metrics such as P_{out}, G_t, PAE and I_{out} of a 2×180 µm width commercial GaN HEMT compared against the model. The measurements were made using the on-wafer load pull setup. The transistor was biased in class AB mode which resulted in G_t=20 dB, PAE=50% (b) the benchmarked model was then used to study the effect of access regions and fringing capacitances on the large signal device performance. In class AB mode, self-aligned device would result in 40% higher PAE due to increased Pout in compression, while removing fringing capacitances would cause a significant increase in G_t by 50%. A self-aligned device with no parasitics would accrue both benefits.

978-1-4799-8002-4/14 $31.00 © 2014 IEEE   296

Device noise modeling for LNA design

The important white noise sources associated with GaN HEMTs are thermal noise associated with carriers flowing through the channel and device parasitic resistors. In addition, transport of carriers through the gate-Schottky barrier contributes to shot-noise. In MVS-G-RF, both of these noise sources are modeled self-consistently with the transport and require no additional parameters (9)-(10).

$$I^2_{S,th} = \frac{4kT}{R_S} \qquad I^2_{D,th} = \frac{4kT}{R_D} \qquad I^2_{G,th} = \frac{4kT}{R_G}$$

$$I^2_{GD,shot} = 2qI_{GD} \qquad I^2_{GS,shot} = 2qI_{GS} \qquad (9)$$

$$I^2_{ch,th} = 4kTg_m(q_s + q_d) \qquad q_s = \frac{Q_s}{Q_{inv,s}} \qquad q_d = \frac{Q_d}{Q_{inv,s}} \qquad (10)$$

Here k is Boltzmann constant, q is electron charge, g_m is small signal transconductance. The noise model is based on the previous work in [7]. Low frequency flicker and generation-recombination noise is not included in this work. The device level noise figure measurements are made on the 2×180 μm device using on-wafer Maury-Microwave noise measurement setup at 6 GHz. The noise elements employed in the model is shown in Fig. 6 along with the comparison of model fits to data. Aside from the bias-dependent device noise figure (NF), gain (G_{assc}) and noise resistance (R_n), the model is able to estimate the $NF_{min}= 2$ dB and $G_{assc}=15$ dB at $V_{DS}=12$ V and $V_{GS} \sim V_t + 0.15$ V (DC bias of LNA for 5.5 mA current). The discrepancy in device-off-state warrants further calibration of the noise setup.

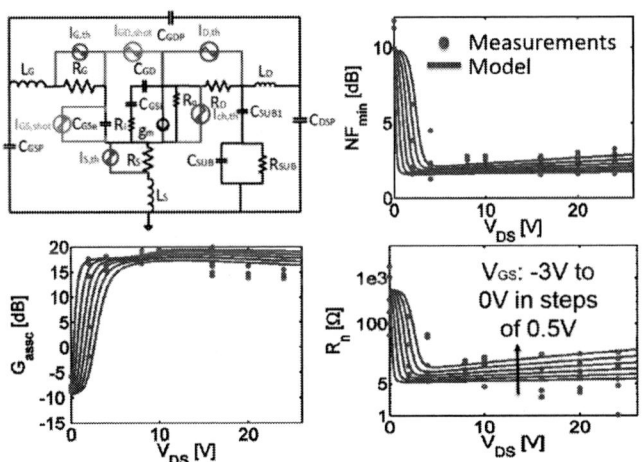

Figure 6. Adding thermal noise sources to parasitic and channel resistances and shot noise sources to the gate heterostructure diodes is sufficient to model RF device noise. The small signal sub-circuit along with device minimum noise figure, associated gain and noise resistance fits comparing the model results with measurements is shown. The noise model depends only on small signal gain terms and DC terminal currents and requires no additional parameters. At $V_{DS}=12$ V and $V_{GS}=V_t +0.15$ V, $NF_{min}= 2$ dB and $G_{assc}=15$ dB which is correctly predicted by the model.

Model validation against a GaNFET RF frontend circuit

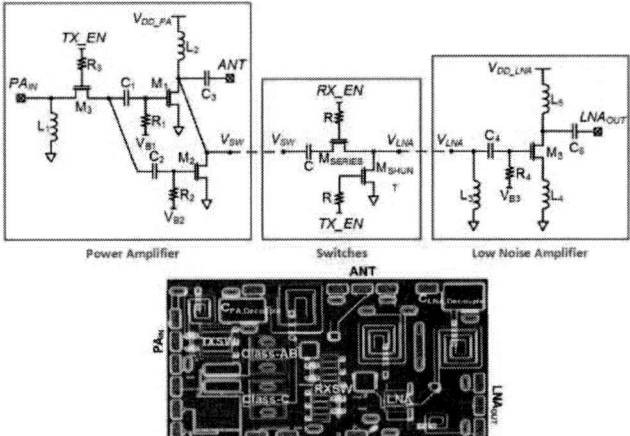

Figure 7. The circuit schematic for a high-efficiency high-power RF front-end design that employs dual-biased GaN HEMTs for T_X PA and a GaN HEMT for R_X-LNA - integrated with GaN HEMT switches is shown along with the die micrograph image of the circuit taped out using Cree 0.25 μm GaN-on-SiC process.

The model is used to simulate a fully integrated 0.25 μm GaN-on-SiC 5.9 GHz RF front-end for vehicle-to-vehicle communication [3] whose circuit schematic and die micrograph are shown in Fig. 7. The design consists of dual-biased PA in T_X mode and LNA in R_X mode with integrated switches to toggle between modes (refer to [3] for details).

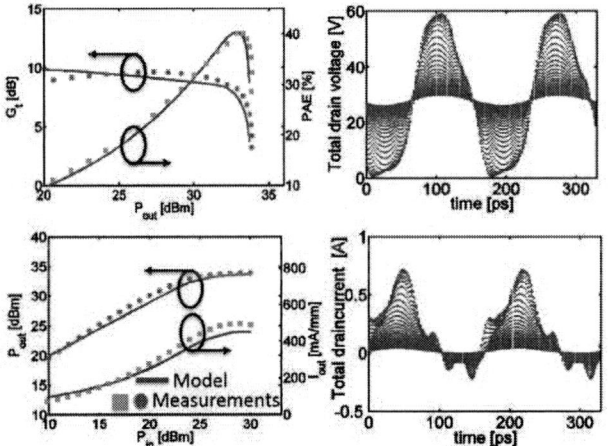

Figure 8. The T_X PA circuit power sweep measurements with the optimized class AB/C bias conditions with $V_{DS}=28$ V at 5.9 GHz are compared against the model. The T_X configuration achieves $P_{sat} =34$ dBm with 40 % PAE. The model matches these values along with a correct prediction of PAE =30 % and $G_T =10$ dB at back-off output power of 28 dBm as required in the IEEE 802.11p standard. The corresponding time domain drain voltage and current waveforms obtained by the MVS-G-RF model show that the current is maximum when the voltage is almost minimum and vice versa, for high efficiency, confirming class-AB/C operation reasonably.

978-1-4799-8002-4/14 $31.00 © 2014 IEEE

Figs. 8 and 9 show the model comparison with circuit measurements in T_X and R_X modes respectively. The calibrated device model is able to predict measured P_{sat} =34 dBm with 40% PAE along with a correct match to measured PAE =30% and G_t =10 dB at back-off P_{out} of 28 dBm in the post-layout simulation in ADS (to capture circuit-level parasitics) as shown in Fig. 8. In the R_X mode, good agreement is achieved with measured S11/S12, 3rd order intermodulation (IM3), and NF (3.7 dB at 5.9 GHz including switches).

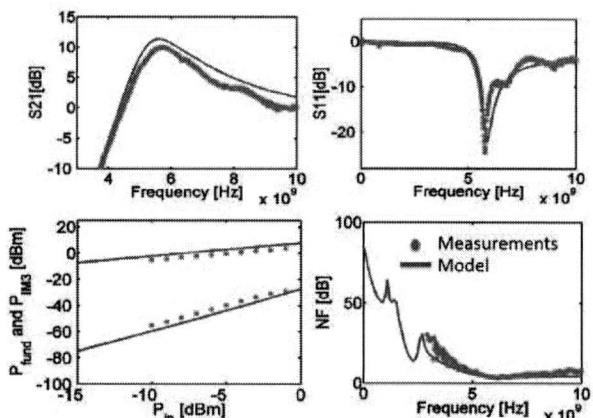

Figure 9. The small signal response of the R_X LNA is compared against the MVS-G-RF model. The small signal gain S21 is about 10 dB and S11 is less than -20 dB at 5.9 GHz, which meets the design requirements, with the model accurately predicting the S parameters. The non-linearity of LNA is tested with a two-tone signals (5.865 GHz + 5.885 GHz) and the resulting fundamental and third-order intermodulation signals are correctly predicted by the model. The resulting OIP3 is around 22 dBm. The NF measurements with calibrated noise source match the model simulation results closely with NF=3.7 dB at 5.9 GHz.

Conclusions

A physics-based compact transport and charge model for RF-GaN HEMTs has been developed, including device self-heating, non-linear access region behavior, noise, etc. The model is validated against measurements from device-level DC up to circuit-level. The model is implemented in Verilog-A that is a suitable base for circuit simulations.

Acknowledgements

The authors would like to thank Prof. Jesus Del Alamo for providing load pull measurement setup and Dr. James Fiorenza and Dr. Geoffrey Coram from Analog Devices Inc. for valuable suggestions. This work was supported by the MIT SMART-LEES program, the MIT GaN Energy Initiative and the DARPA NEXT program.

Table 1: Key parameters for transport and charge fitted to HEMTs

Parameters	Values	Notes
Extracted parameters		
L_g (μm)	0.25	Channel length
W (μm)	2×180	Number of fingers × Gate width
C_g (F/cm^2)	4.0e-7	Areal gate capacitance
R_{sh} $(ohm/sq.)$	150	Sheet resistance of access region
R_C $(ohm\text{-}mm)$	0.8	Contact resistance
μ (cm^2/Vs)	1500	Low-field mobility
V_{t0} (V)	-2.79	Threshold voltage for V_d~0V
$I_j(A/cm)$	4.9e-3	Reverse saturation gate leakage current
V_j (V)	1.0	Gate Schottky barrier potential
Electrostatic fitting parameters		
α	3.5	Sub-threshold transition parameter
δ (V/V)	0.004	Drain-induced-barrier-lowering
SS $[V/Dec]$	0.08	Sub-threshold swing
C_{Ig} (F/cm^2)	1.0e-7	Implicit-areal gate capacitance
Transport fitting parameters		
v_{sat} (cm/s)	1.3e7	Saturation velocity
β	1.5	Lateral-field saturation parameter
R_{th} (W/K)	30	Thermal resistance
C_{th} $(K.\ s/W)$	5e-7	Thermal Capacitance
ϵ	2.3	Mobility self-heating parameter
Device level parasitic parameters		
R_S, R_D (Ω)	0.1	Contact parasitic resistance
$R_G(\Omega)$	1	Contact parasitic resistance
L_S, L_D, L_G (pH)	10	Parasitic lead inductance
$C_{GSP}, C_{GDP}, C_{DSP}$ (fF)	80,20,40	Parasitic pad capacitance

References

[1] L. Dunleavy et. al, Microwave Mag.,vol. 11, No. 6, pp. 82-96, Oct. 10.

[2] P. Choi et al., IEEE MWCL, vol. 23, no. 8, pp. 433–435, Aug. 13.

[3] P. Choi et al., RFIC symposium, Jun. 2014.

[4] U. Radhakrishna, Massachusetts Inst. Of Tech., MS thesis, Jun. 2013.

[5] U. Radhakrishna et.al, IEDM, pp. 13.6.1-13.6.4, Dec. 2012.

[6] U. Radhakrishna et.al, PSS Wiley (c) vol. 11 (3-4), pp. 848-852, 2014.

[7] C. Sanabria, Ph.D. dissertation, UCSB, 2006.

978-1-4799-8002-4/14 $31.00 © 2014 IEEE

Scaling Breakthrough for Analog/Digital Circuits by Suppressing Variability and Low-Frequency Noise for FinFETs by Amorphous Metal Gate Technology

Takashi Matsukawa, Koichi Fukuda, Yongxun Liu, Junichi Tsukada, Hiromi Yamauchi,
Yuki Ishikawa, Kazuhiko Endo, Shin-ichi O'uchi, Shinji Migita, Wataru Mizubayashi,
Yukinori Morita, Hiroyuki Ota, and Meishoku Masahara

Nanoelectronics Research Institute, National Institute of Advanced Industrial Science and Technology (AIST),
1-1-1, Umezono, Tsukuba, Ibaraki 305-8568, Japan
Tel: +81(29) 861-5358, Fax: +81(29) 861-5170, Email: t-matsu@aist.go.jp

Abstracts

The effectiveness of amorphous metal gate (MG) in suppressing low-frequency noise (LFN) for FinFETs has been thoroughly investigated. It was demonstrated that the amorphous TaSiN MGs with various atomic compositions provide flexible tuning of threshold voltage (V_t) as well as small V_t variability, namely A_{Vt}. It was found that the TaSiN-MG FinFETs exhibit drastic reduction of LFN in comparison to the poly-crystalline TiN MG case. Modelling by 3D-TCAD reveals that work function variation (WFV) of the MG has a significant impact on LFN generation. Suppression of A_{Vt} and LFN is highly beneficial to conduct further scaling of analog/digital components in SoC.

Introduction

Variability of V_t which emerges with scaling is critical obstacle to proceed scaling of digital circuit components such as SRAM [1]. In addition to the V_t variability, LFN lies as serious obstacle for scaling analog circuits in SoC [2, 3], since flicker noise increases proportionally with inverse of transistor channel area. FinFETs [4] provide merit of suppressed V_t variability by elimination of random dopant fluctuation [5]. We have previously reported that introduction of an amorphous MG in the FinFETs instead of a poly-crystalline MG further suppresses the V_t variability achieving smallest A_{Vt} value [6] by elimination of WFV of the MG. As for the LFN issue, it has been reported that FinFET/nanowire structure has a potential to suppress the LFN in comparison to bulk planer MOSFETs [7, 8]. In this work, impacts of WFV elimination by the amorphous MG technology on the LFN of FinFETs are investigated comprehensively. TaSiN-based amorphous MGs with various compositions are used to realize both V_t tuning and small V_t variability. The LFN is characterized for the FinFETs with the amorphous MGs together with a poly-crystalline TiN MG as a reference. The impact of WFV on the LFN generation is discussed by using 3D-TCAD.

Sample Device Fabrication

Ternary alloys of TaSiN with sufficient thermal stability [9] and small V_t variability [6] are used for the MG of the FinFET. The TaSiN film was deposited by physical vapor deposition (PVD). In order to the control the atomic composition of TaSiN, N_2/Ar flow ratio and Ta/Si target composition were changed as shown in Table 1. The composition of the deposited films was analyzed by Rutherford backscattering spectroscopy (RBS) (Fig. 1) and the results are summarized in Table 1. The crystalline nature was analyzed by X-ray diffraction (XRD) after rapid thermal annealing (RTA) necessary for dopant activation (Fig. 2). Excepting for TaSi (label A) and poly-crystalline TiN as a reference, all the other films exhibit amorphous nature. The composition and the crystal structure for the TaSiN films are summarized in the ternary phase diagram (Fig. 3). The amorphous structure is obtained for broad variety of the composition. These amorphous films were introduced in the MG of FinFETs by the gate first process flow (Fig. 4). The TEM micrographs of the fabricated FinFETs show that the amorphous MGs are formed precisely on the fin channel as well as the TiN reference (Fig. 5).

Static Variability Characterization

I_d–V_g curves for the FinFETs with the various amorphous MGs vary reflecting the composition change (Fig. 6). Namely, increase in the N composition and increase in the Ta/Si ratio both contribute to increase V_t. The V_t roll-off remains unchanged regardless of the composition (Fig. 7). Pelgrom plot is obtained for V_t mismatch of the paired FinFETs with various gate length (L_g) to evaluate local variability of V_t (Fig. 8). Slope values of the Pelgrom plot, A_{Vt}, are summarized in Fig. 9. Regardless of the atomic composition difference, the amorphous MGs (B~D) exhibit significantly suppressed A_{Vt} in comparison to the poly-crystalline TiN MG, including smallest one (A_{Vt}=1.32 mVμm) for TaSiN (B). Namely, the V_t tunability by the composition control and the small variability are fully compatible. Interface trap density (N_{it}) was also evaluated for the MGs on the fin channels by charge pumping (CP) method [6, 10] (Fig. 10). The CP current amplitude is converted to the N_{it} values (Fig. 11). All the amorphous MGs together with the TiN reference exhibit comparable N_{it}. Namely statistical fluctuation in the interface trap charges, which also causes V_t fluctuation [11], is not responsible for the significant A_{Vt} difference. Thus, WFV suppression is the reason for the suppressed A_{Vt} of the amorphous MGs.

Noise Characterization

Power spectral density S_{Id} of LFN was obtained for the FinFETs with identically designed L_g of 90 nm. V_g was set approximately at averaged V_t ($<V_t>$) for each MG case. Because of the fluctuation in the LFN level, median of S_{Id} in 48 samples is used as reported in [12] and compared among the different MGs (Fig. 12). The TiN MG reference shows 1/f dependence due to flicker noise. Clearly seen, the amorphous MGs exhibit dramatic suppression of the LFN. The LFN waveforms of the median samples also show the significant suppression of the noise level for the amorphous MG (Fig. 13). The flicker noise level as the LFN is given by

$$\frac{S_{Id}}{I_d^2} = \frac{kT}{\gamma f W_g L_g}\left(\frac{1}{N} + \frac{\mu}{\mu_{C0}\sqrt{N}}\right)^2 N_t(E_F) \qquad (1)$$

where W_g is the channel width, N is the carrier density, $N_t(E_F)$ is the trap density responsible to the LFN, μ is the mobility, γ is the attenuation coefficient of the electron wave function, and μ_{C0} is the fitting parameter [2]. In order to discuss the reason for the different LFN levels, correlation with measured N_{it} as an index of $N_t(E_F)$ is examined. Median of S_{Id}/I_d^2 at 10 Hz is used in the following analysis. While the plots for the amorphous MGs obey the trend of $S_{Id}/I_d^2 \propto N_{it}$, the TiN MG shows discrepancy from the trend (Fig. 14). On the other hand, correlation with A_{Vt} (Fig. 15) clearly shows the trend that S_{Id}/I_d^2 reduces with the reduction of A_{Vt}. Namely, suppression of the WFV is responsible for the LFN reduction. Importantly, the significant WFV impact on the LFN is considered to hinder scaling for analog circuits.

The fluctuation in the LFN level among the 48 samples is also compared between the TiN and amorphous MGs (Fig. 16). Both the TiN and TaSiN (D) MG cases fit well with log-normal distribution as reported in [13]. The TiN MG exhibits significantly larger amount of fluctuation than the TaSiN (D) MG, which amounts to x8.4 for 3σ. The larger fluctuation in the LFN level needs increased design margins for analog circuits and thus would be additional hindrance to their scaling. The V_t fluctuation (δV_t) in the samples causes fluctuation in carrier density (δN) in the channel given by $\delta N \approx -\delta V_t C_{ox}/q$ at constant V_g, where C_{ox} is the gate capacitance, and N affects the LFN level according to (1). Thus, correlation between S_{Id}/I_d^2 and V_t fluctuation is examined as shown in Fig. 17. Negligible correlation between S_{Id}/I_d^2 and V_t shows that N variation is not the dominant reason for the S_{Id}/I_d^2 fluctuation. Namely, the WFV is also responsible for the LFN fluctuation.

Mechanism for the increased LFN is discussed by modelling WFV of the TiN MG having low/high work function (WF) grains [14, 15] using 3D-TCAD [16, 17] (Fig.

18). The current flow distribution in the fin channel at $V_g=V_t$ is simulated. The granular WFV causes narrowing of the current path and its centroid shift close to the channel/gate interface in comparison to the uniform WF case. As explained in Fig. 19, narrowing of the current path due to the WFV enhances the influence of the charged trap by the bottleneck effect [12]. The current centroid shift close to the charged trap also causes increased influence of the charge [7]. Both play significant roles for the LFN increase by the WFV.

Finally, the LFN level normalized by the channel size is compared with the state of art platform technologies [18,19] as benchmarking (Fig. 20). Impact of the A_{Vt} and LFN suppression on SoC scalability is summarized in Fig. 21. For the circuit elements such as SRAM, x0.65 reduction of A_{Vt} by replacing the poly-crystalline MG by the amorphous one for FinFETs enables x0.65 shrink of the die size. Importantly, for the circuit elements such as analog amplifier in which LFN limits the scaling as in (1), x0.21 reduction (Fig. 20) of the LFN level by introducing the amorphous MG enables scaling x0.21 in area and x0.45 in size.

Summary

The TaSiN amorphous MG provides drastic suppression of LFN for FinFETs in addition to the well-suppressed V_t variability. WFV plays significant role in the LFN generation for the MG FinFETs. Both the A_{Vt} and LFN suppression for the FinFETs by the amorphous MG technology enables further scaling of digital/analog SoC.

Acknowledgement

This work was supported in part by Grant-in-Aid for Scientific Research (No. 26289113) from JSPS, Japan.

References

[1] K.J. Kuhn *et al.*, *IEEE Trans. Electron Devices*, 58, p.2197 (2011).
[2] E.P. Vandamme *et al.*, *IEEE Trans. Electron Devices*, 47, p.2146 (2000).
[3] C.-H. Jan *et al.*, *IEDM Tech. Dig.*, 2010, p.604.
[4] C. Auth *et al.*, *VLSI Symp. Tech. Dig.*, 2012, p.131.
[5] T. Chiarella *et al.*, *Solid State Electron.*, 54, p.855 (2010).
[6] T. Matsukawa *et al.*, *IEDM Tech. Dig.*, 2012, p.175.
[7] W. Feng *et al.*, *IEDM Tech. Dig.*, 2011, p.630.
[8] T. Ohguro *et al.*, *VLSI Symp. Tech. Dig.*, 2012, p.149.
[9] Y.-S. Suh *et al.*, *VLSI Symp. Tech. Dig.*, 2001, p.47.
[10] G. Kapila *et al.*, *IEEE Electron Device Letters*, 28, p.232 (2007).
[11] X. Wang *et al.*, *IEDM Tech. Dig.*, 2011, p.103.
[12] M. Saitoh *et al.*, *VLSI Symp. Tech. Dig.*, 2013, p.T228.
[13] P. Srinivasan *et al.*, *IEDM Tech. Dig.*, 2012, p.458.
[14] A. Yagishita *et al.*, *IEEE Trans. Electron Devices*, 48, p.1604 (2001).
[15] K. Nakajima *et al.*, *VLSI Symp. Tech. Dig.*, 1999, p.95.
[16] HyENEXSS Ver.5.5.
[17] T. Matsukawa *et al.*, *VLSI Symp. Tech. Dig.*, 2014, p.142.
[18] D. Lopez *et al.*, *IEEE Trans. Electron Devices*, 58, p.2310 (2011).
[19] C.G. Theodorou *et al.*, *IEEE Trans. Electron Devices*, 61, p.1161 (2014).

Table 1 Deposition condition of Ta-Si-N films.

Label	Target material in PVD	N_2/Ar flow ratio [%]	Composition measured by RBS [at. %]		
			Ta	Si	N
A	$TaSi_2$	0	30	70	0
B	$TaSi_2$	4	14	36	50
C	$Ta_{67}Si_{33}$	0	79	21	0
D	$Ta_{67}Si_{33}$	2	37	16	47
E	$Ta_{67}Si_{33}$	4	26	16	58

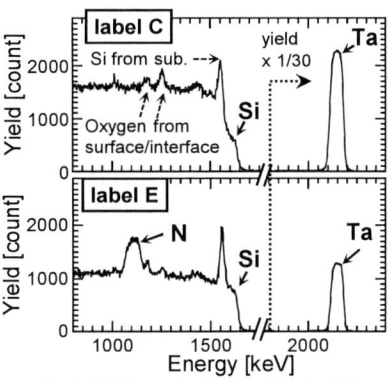

Fig.1 RBS spectra for Ta-Si (label C) and Ta-Si-N (label E) films for estimating atomic composition.

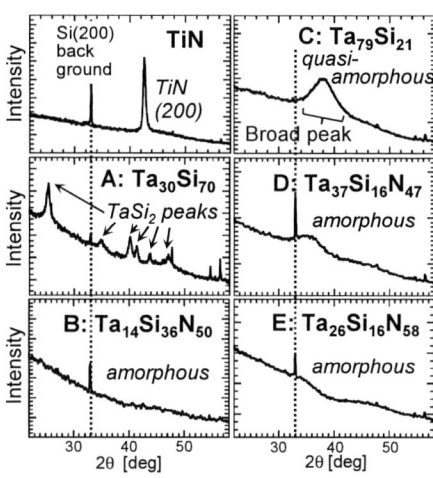

Fig.2 XRD for Ta-Si-N, Ta-Si and TiN films after RTA at 870°C equivalent to dopant activation. Samples excepting TiN and Ta-Si(A) exhibit amorphous nature.

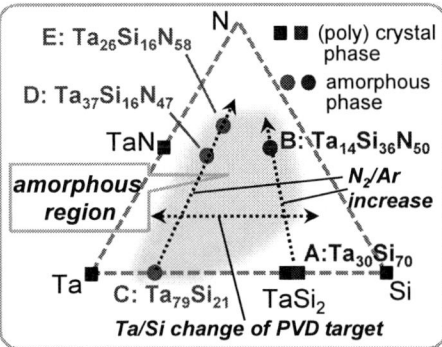

Fig.3 Ternary phase diagram for the Ta-Si-N system. Composition and crystallinity for the thin films (A-E) are also shown. Amorphous MG is realized for various composition.

- (100) SOI, T_{BOX}=400 nm
- (110) fin channel patterning
- Gate oxidation (T_{ox}~2 nm)
- Metal gate deposition (PVD)
  - **- Ta-Si, Ta-Si-N, TiN**
- n+ poly-Si HM depo.
- Gate patterning
- Extension I/I (As, BF_2 4×10^{14} cm⁻²)
- Side wall spacer formation
- S/D I/I (P, BF_2 10^{15} cm⁻²)
- Dopant activation RTA (~870°C, 2 s)
- NiSi formation on S/D
- Metallization

Fig.4 Process flow for the sample FinFET fabrication.

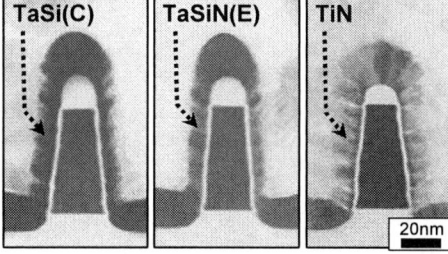

Fig.5 Features of the fabricated FinFETs with the TaSi(C), TaSiN(E) and TiN MGs. Fin height (H_{fin}) is set at 50 nm.

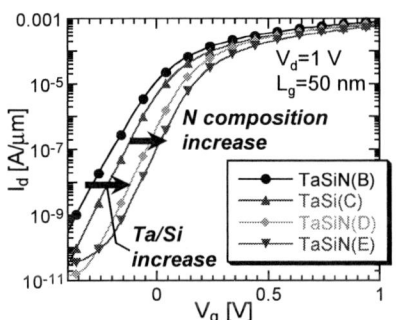

Fig.6 I_d-V_g curves of FinFETs with amorphous MGs having different composition.

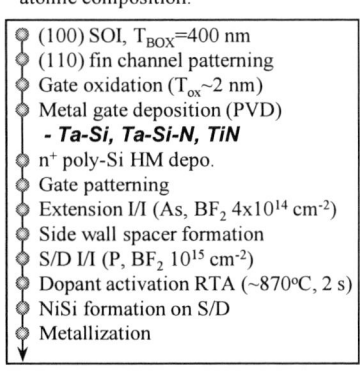

Fig.7 V_t roll-off of FinFETs with amorphous MGs having different composition.

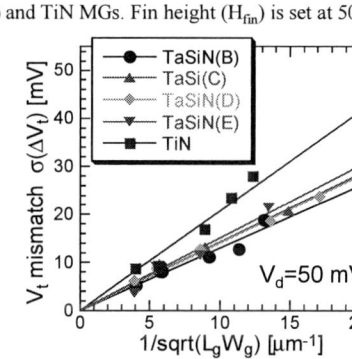

Fig.8 Pelgrom plots for FinFETs with different MGs. V_t mismatch (ΔV_t) of paired transistors was measured to obtain local variability.

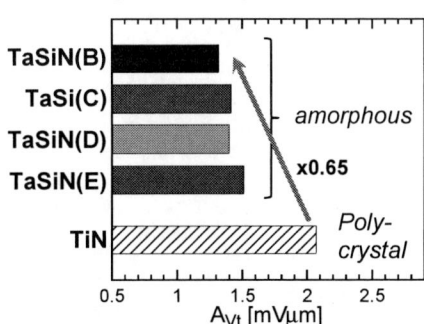

Fig.9 Comparison of A_{Vt} among different MGs for FinFETs. Amorphous MGs suppress V_t variability effectively regardless of atomic composition.

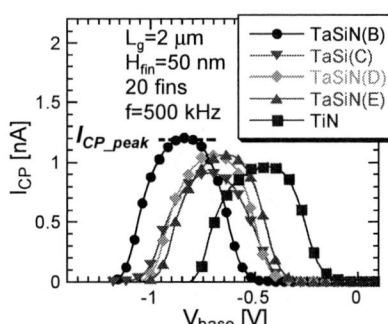

Fig.10 Charge pumping (CP) measurement for fin channels with the various MGs for evaluation of interface trap density (N_{it}).

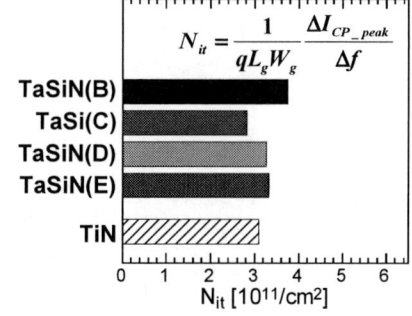

$$N_{it} = \frac{1}{qL_gW_g}\frac{\Delta I_{CP\_peak}}{\Delta f}$$

Fig.11 Summary of N_{it} obtained by CP measurement showing that N_{it} values are comparable for the amorphous TaSiN and poly-crystal TiN MGs.

978-1-4799-8002-4/14 $31.00 © 2014 IEEE

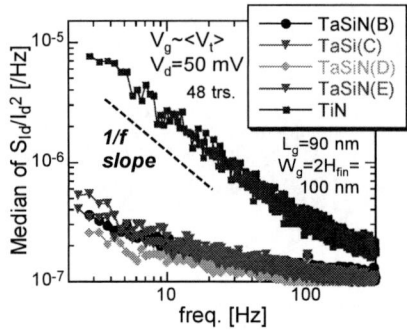

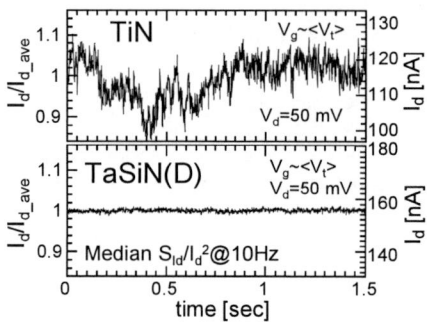

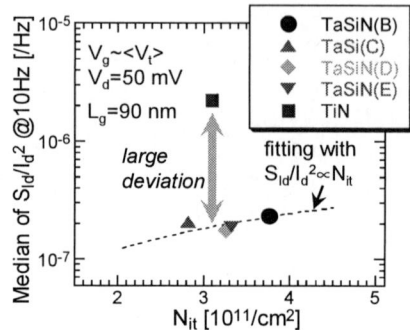

Fig.12 Power spectral density of I_d noise for median of 48 FinFETs with different MGs. Flicker noise with 1/f slope for the TiN MG is significantly suppressed for the amorphous MGs.

Fig.13 Comparison of the LFN waveform between the poly-crystalline TiN and amorphous TaSiN MGs. Sample FinFETs with median of S_{Id}/I_d^2 at 10 Hz are selected.

Fig.14 Correlation between LFN level S_{Id}/I_d^2 and N_{it} as an origin of flicker noise. While the amorphous MGs obey trend of $S_{Id}/I_d^2 \propto N_{it}$, the poly-crystalline TiN exhibits significant deviation from the trend.

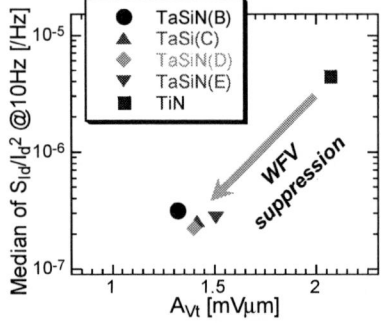

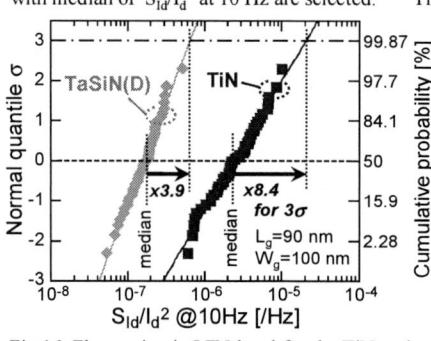

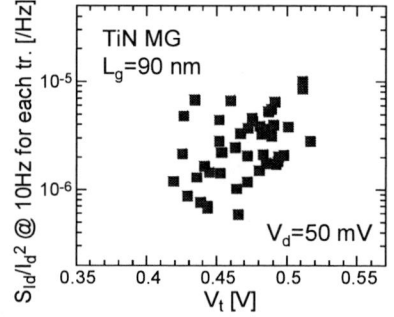

Fig.15 Correlation between LFN level and A_{Vt}. Suppression of WFV by the use of amorphous MGs is effective for S_{Id} suppression as well as the well-suppressed V_t variability.

Fig.16 Fluctuation in LFN level for the TiN and TaSiN MGs showing log-normal distribution. The TiN MG exhibits larger fluctuation in addition to the larger median of S_{Id}.

Fig.17 LFN level vs linear V_t fluctuation for the TiN-MG FinFETs, showing negligible correlation. Thus, fluctuation in carrier density due to V_t is not the reason for the S_{Id} fluctuation.

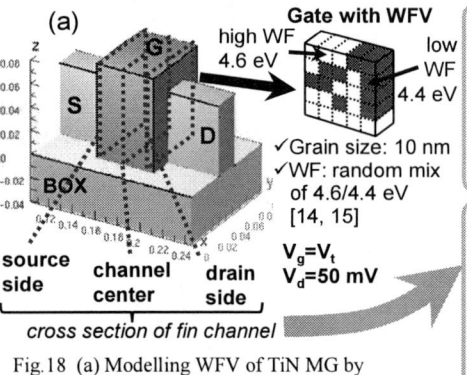

Fig.18 (a) Modelling WFV of TiN MG by 3D TCAD. Simulated current distribution (J_{nx}) in fin channel for (b) uniform gate WF and (c) granular WFV cases. Granular WFV causes narrowing of current path and its centroid shift to gate/channel interface.

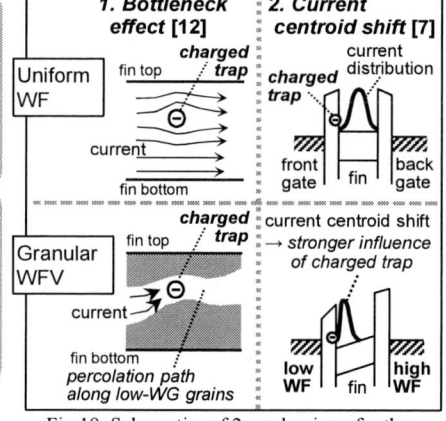

Fig.19 Schematics of 2 mechanisms for the increase in LFN level due to WFV.

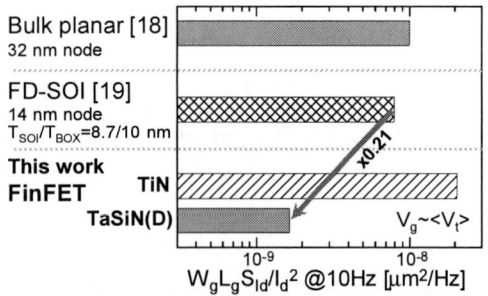

Fig.20 Benchmarking of the LFN level compared with those for bulk planar [18] and FD-SOI technologies [19].

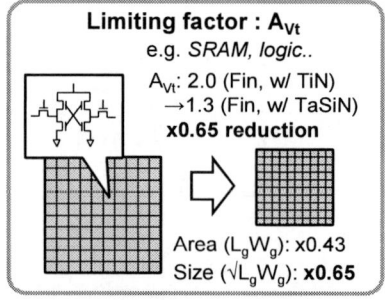

Fig.21 Impact of A_{Vt} and S_{Id} suppression on circuit scalability.

978-1-4799-8002-4/14 $31.00 © 2014 IEEE

302

A Circuit Level Variability Prediction of Basic Logic Gates in Advanced Trigate CMOS Technology

E. R. Hsieh[1], C. M. Hung[1], T. Y. Wang[1], Steve S. Chung[1], R. M. Huang[2], C. T. Tsai[2], and T. R. Yew[2]

[1]Department of Electronics Engineering and Institute of Electronics, National Chiao Tung University, Taiwan

[2]United Microelectronics Corporation (UMC), Hsinchu, Taiwan

Abstract- Variability has been one of the major scaling issues in advancing the CMOS technology. In this paper, a variation model from the device level to circuit level has been proposed and demonstrated on advanced trigate FinFETs. First, a simple and accurate transport model was developed to model variability at the device level. It was then implemented in Spice and the calculation of variation of basic logic gate building block was demonstrated with only W/L and the slopes, A_{vt}, A_{gm}, in the Pelgrom plot, as inputs. Finally, a unified simple analytic form was developed to predict the variability of various basic logic circuits regardless of the number of devices and the complexity of circuits.

1. Introduction

With the scaling of CMOS devices, the device or circuit faces various fluctuation sources of V_{th}, I_d, and noise-margin, which have been one of the major scaling issues. Among them, V_{th} and I_d are suffered mainly from random dopant fluctuation (RDF) [1] in the channel, random trap fluctuation (RTF) [2] in the gate dielectric, random telegraph noise(RTN) [3] of oxide trap, surface roughness of thin dielectric film [4], and metal grain fluctuation of metal gate [5], which have been extensively studied for over a decade. However, few studies have been focused on circuit-level variation [6] and the circuit variability in commercial tool, e.g., Spice, in not available.

In this work, a variation model from the device level to circuit level will be proposed and demonstrated on advanced trigate FinFETs. It will be constructed from the carrier transport of CMOS devices based on Virtual Source Model (VSM) [7], Fig.1, which describes carrier behavior with injection velocity, V_{inj}, being injected from the source to the drain, considering the series resistance, R_{sd}, effect. I_d can be modeled by VSM adequately. Then, the dominant variation factors of I_d will be analyzed. This leads to the development of a *variation model* which can be built-in Spice. A guideline for **circuit level variation** will be developed for basic logic gates such as NAND, NOR, CMOS inverter, and latch circuits.

2. Device Fabrication

The 28nm technology node of bulk trigate CMOS devices with poly-Si gate made on a pilot foundry platform, with EOT (SiON) = 20Å, various W/L sizes, were prepared. Also, bulk-planar devices on the same technology node were made for comparisons, Fig. 2.

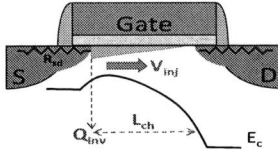

Fig. 1 The schematic to describe the carrier transport, in which V_{inj} is the velocity injecting from the peak of channel barrier, and Q_{inv} is the channel charge density at the same position.

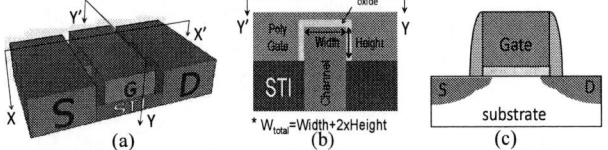

Fig. 2 The devices used in this work: (a) the trigate MOSFET, (b) the cross-section in the direction Y-Y' of (a), and (c) the planar MOSFET for comparisons.

3. Results and Discussion

A. The Device Transport Model

Figure 3 is the flow chart of a device model, improved from VSM (Virtual Source Model) [7]. I_d-V_{gs}, I_d-V_{ds}, and C-V are measured, and V_{th}, Q_{inv}, and R_{tot} can be extracted by following initial guesses of V_{inj} and μ, with iterations in just few loops. Finally, I_d-V_{gs} and I_d-V_{ds} are obtained from Eq. (1) in Table 1. Fig. 4 compares fitted I_d-V_{gs} results of trigate CMOS devices against the measured data, showing good matches. Fig. 5 shows the comparison between fitted I_d-V_{ds} of trigate CMOS devices and the measurement. Also, very good matches can be achieved. Fig. 6 shows the performance of trigate devices in terms of V_{inj}, which exhibits a higher V_{inj} obviously, compared to planar ones. On the improvement of the original VSM, the source-drain resistance effect, a gate bias dependent R_{sd} has been elaborated, Fig. 7, such that accurate mobility characteristics can be obtained, the blue curve in Fig. 8. This assures the accuracy of the device's level model.

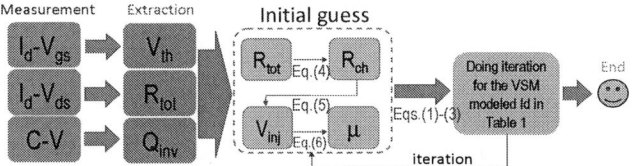

Fig. 3 The flow chart of Virtual Source Model (VSM). From the measured I_dV_{gs}, I_dV_{ds}, and C-V characteristics, V_{th}, total resistance(channel resistance-R_{ch} and source-to-drain resistance -R_{sd}), and inversion charge density can be extracted. Followed by initial guess from Eqs. (4)-(6) in Table 1, injection velocity, effective mobility, etc. parameters can be determined. Finally, the modeled I_d-V_{gs} and I_d-V_{ds} can be calculated, Eqs. (1)-(3) in Table 1.

$$I_d / W = Q_{inv} \cdot V_{inj} \cdot F_{sat} \quad \text{- Eq. (1)}$$

$$Q_{inv} = C_{inv} \cdot \frac{S.S.}{2.3} \ln(1 + e^{(V_{gs}' - V_{th} + \alpha \cdot \phi_t \cdot F_t)/n \cdot \phi_t}) \quad \text{- Eq. (2)}$$

$$F_t = \frac{1}{1 + e^{(V_{gs}' - V_{th} + \alpha \cdot \phi_t /2)/\alpha \cdot \phi_t}} ; F_{sat} = \frac{V_{ds}'/V_{dsat}}{[1 + (V_{ds}'/V_{dsat})^\beta]^{1/\beta}} \quad \text{- Eq.(3)}$$

$$R_{tot} = \partial I_d / \partial V_{gs} ; R_{sd} = R_{tot} - R_{ch} \quad \text{- Eq. (4);}$$

$$\frac{Q_{inj} \cdot V_{inj}}{V_{dsat}} = \frac{1}{W \cdot R_{ch}} \quad \text{Eq. (5);} \quad V_{dsat} = \frac{V_{inj} \cdot L_{ch}}{\mu} \quad \text{- Eq. (6)}$$

Table 1 The equations used in VSM. Eq. (1) models the drain current of MOSFETs, in which F_{sat} is the saturation function to determine the $V_{d,sat}$ in I_dV_{ds}; F_t is the transfer function to describe the sub-threshold characteristics.

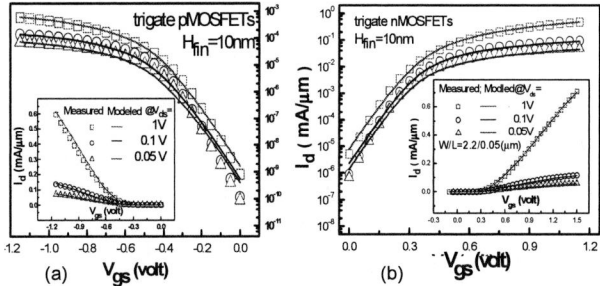

Fig. 4 Comparisons of the I_dV_{gs} curves for the measured data and Spice modeled data in trigate (a) nMOSFETs and (b) pMOSFETs, in logarithm scale and linear scale (the insert) with different V_{ds} values, which shows good matches.

978-1-4799-8002-4/14 $31.00 © 2014 IEEE   303

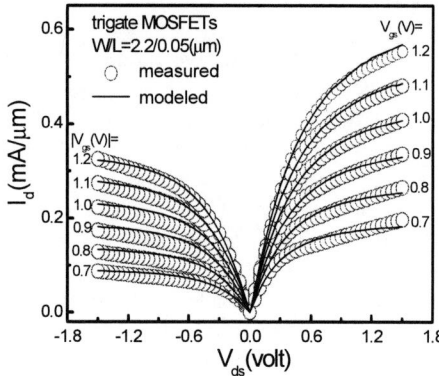

Fig. 5 Comparisons of the I_d-V_{ds} curves between the measured data and Spice modeled data in trigate CMOS devices in the linear scale with different V_{gs} values, which show good matches.

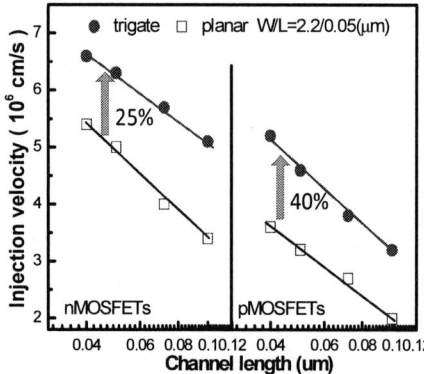

Fig. 6 The comparisons of V_{inj} between trigate and planar devices for (left) nMOSFETs and (b) pMOSFETs. It reveals that p-channel trigate devices show a much larger improvement since its sidewall has (110) orientation.

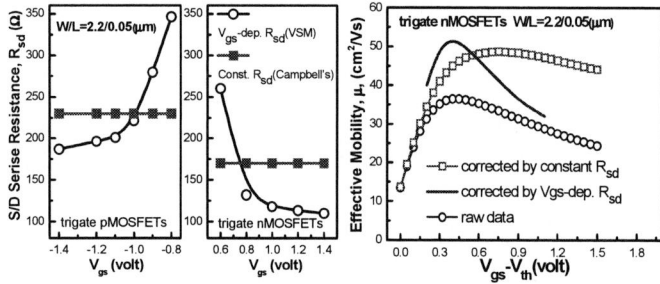

Fig. 7 Comparisons of the R_{sd} between Campbell's constant R_{sd} method and the current gate bias dependent method. The latter is more reliable for inclusion in the present VSM.

Fig. 8 Comparisons of the effective mobility extracted by split C-V and corrected by gate bias dependent R_{sd}. The blue curve in the present model is more reliable considering the R_{sd} effect.

B. The Variation Model-from Device to Circuit

To understand which parameters dominate I_d variation, Fig. 9 shows the method by decoupling I_d into V_{inj} and V_{th}. V_{th} can be further decoupled into *DIBL* and $V_{th,lin}$. From Fig. 10, σV_{inj} is closely related to $\sigma g_{m,max}$, whose scattering plot shows a strong dependency. Furthermore, multi-variant analysis (MVA) that we developed in [8] was adopted here to analyze the contribution of V_{th} and g_m variations to the I_d variation. Results in Figs. 11 and 12 show that both V_{th} and g_m variations exhibit strong dependency on I_d variation, which should be taken into account to construct the I_d variation model.

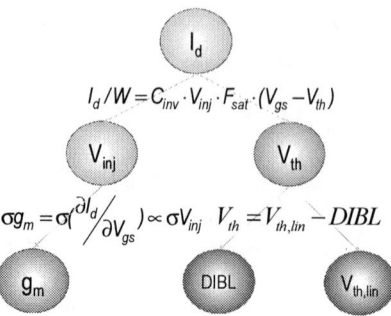

Fig. 9 The multi-variable analysis to show the variation sources of I_d equation based on VSM, in which the factors of V_{inj} and V_{th} can be extracted. The variation of V_{ing} is attributed to the g_m variation, while V_{th} variation is decoupled into DIBL and $V_{th,lin}$.

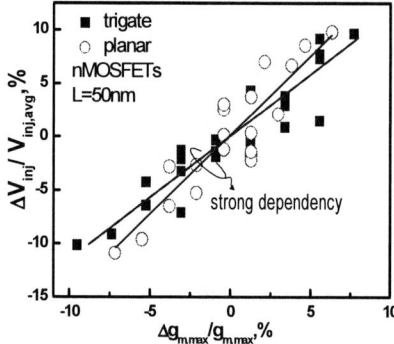

Fig. 10 The scattering plots of V_{inj} versus $g_{m,max}$ of trigate and planar nMOSFETs, showing strong dependency, i.e., the variation of V_{inj} comes from g_m variation.

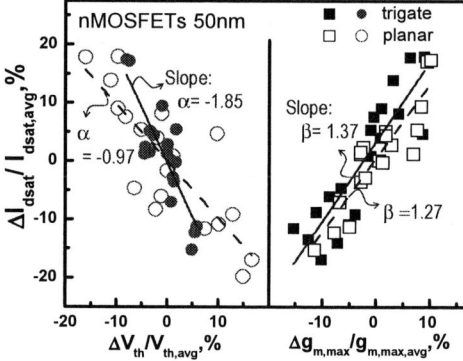

Fig. 11 I_d variation can be decoupled into the variations of V_{th} and g_m. This scattering plots show the strong dependency, and the slopes represent the strength of this dependency.

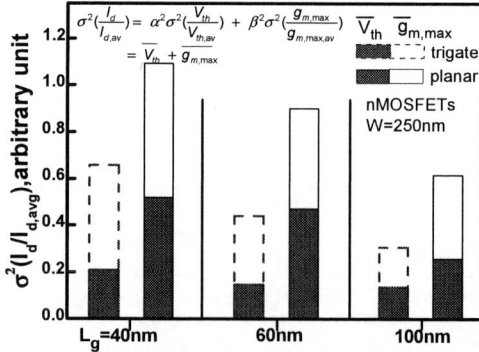

Fig. 12 By multi-variant analysis (MVA), the variance of normalized I_d ($\sigma^2 I_d$) is expressed as the contributions from the variances of normalized V_{th} and $g_{m,max}$. Both V_{th} and $g_{m,max}$ carry about the same weight to the variation of I_d.

Furthermore, in Fig. 9, we need to examine which factor dominates V_{th} variation, $V_{th,lin}$ or *DIBL*? Fig. 13 shows that, in trigate devices, $V_{th,lin}$ is dominant, but DIBL is a secondary factor with less contribution to the V_{th} variation. So, eventually, in the framework of Fig. 9, we may simply use V_{th} and g_m Pelgrom plots [9], Figs. 14 and 15, to model I_d variation for any device sizes. In short, by providing few experimental inputs, W/L, A_{vt}, and A_{gm}, we will be able to develop a built-in Spice model, as described in Fig. 16. For device-level verifications, Figs. 17 and 18 show I_d-V_{gs} and I_d-V_{ds} variations. Gaussian distributions of measured statistical data and the Spice-modeled function, showing good matches. Fig. 19 is the normalized I_d variation, which verified the accuracy of the modeled data in Figs. 17 and 18 against the experiment.

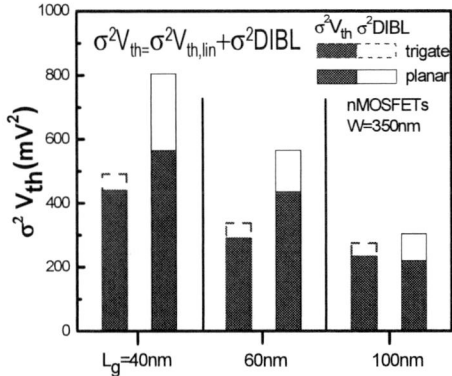

Fig. 13 By MVA, the variance of V_{th} can be further expressed as the summation of $V_{th,lin}$ and DIBL variations. Results show that $V_{th,lin}$ dominates in both trigate and planar devices.

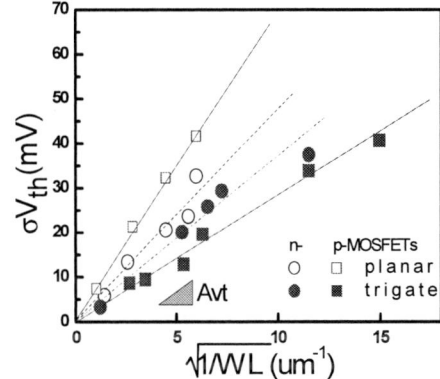

Fig. 14 In order to model I_d variation, the Pelgrom plot of V_{th} was first used to predict the V_{th} variation in terms of the device areas from the slope of the Pelgrom plot, A_{vt}.

Fig. 15 Because g_m variation is closely related to the I_d variation, a Pelgrom plot on g_m is utilized to model the g_m variation of any areas from the slope, A_{gm}, and the intercept with y axis is defined as $g_{m,max0}$.

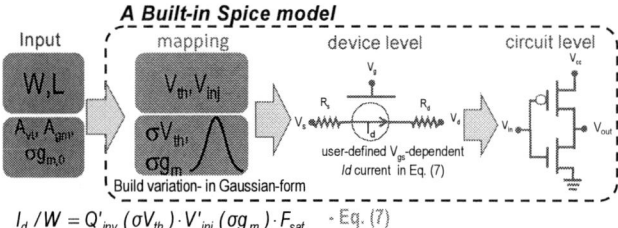

$$I_d / W = Q'_{inv} (\sigma V_{th}) \cdot V'_{inj} (\sigma g_m) \cdot F_{sat} \qquad \text{- Eq. (7)}$$

Fig. 16 The flow chart of the proposed Spice Variation Model which describes the variation of devices and circuits. Just few parameters, W, L, A_{vt}, and A_{gm}, as inputs, V_{th}, V_{inj}, σV_{th}, and σg_m can be directly calculated automatically in spice based on Table 1. Then, these variables are transformed into a user-defined element to simulate the CMOS devices at the device level. As a result, the variation of any complicated circuits can be modeled by building the block of devices, based on Eq. (7).

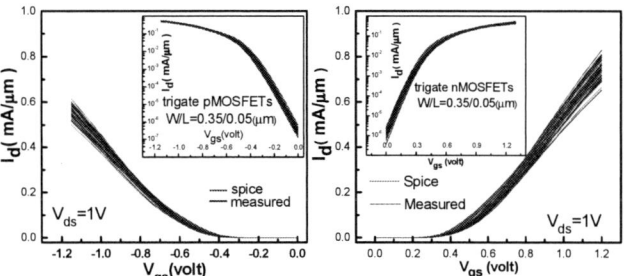

Fig. 17 Comparisons of measured and Spice modeled results of the variations of $I_d V_{gs}$ for (a) pMOSFETs and (b) nMOSFETs in the linear scale and logarithm scales, showing good matches.

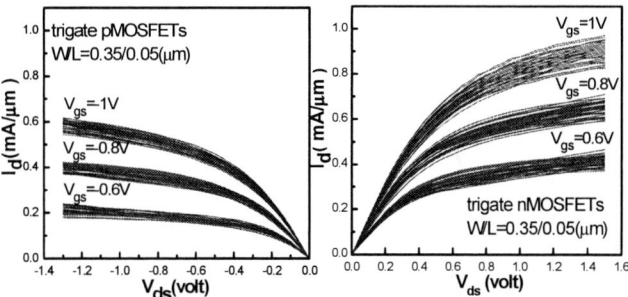

Fig. 18 Comparisons of measured and Spice modeled results of the variations of $I_d V_{ds}$ for (a) pMOSFETs and (b) nMOSFETs, showing good matches.

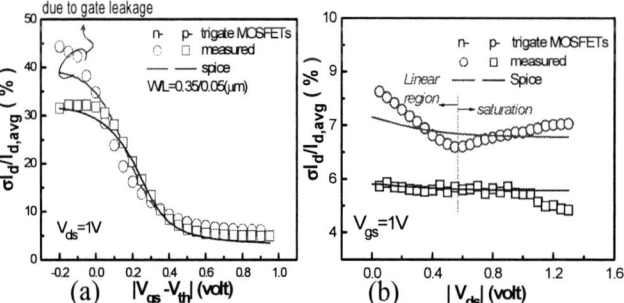

Fig. 19 The dependence of the I_d variation in various operating regions: (a) Validation of the Spice and measured I_d-V_{gs} variation data in Fig. 17. (b) Validation of the Spice and measured I_d-V_{ds} data in Fig. 18.

C. Circuit Variation- for Basic Logic Gates

For circuit level applications, three basic circuit connections are examined first, i.e., parallel, series, and two-port. Since parallel connection of devices is simply the summation of device area, which is naturally governed by the

Pelgrom plot, next, we focused on the other two connections. Fig. 20 compares series connection of two devices, whose behavior is similar to the single device. In other words, the variation of two series devices, exhibit a similar linear relationship in Pelgrom plot, Fig. 21. Again, Fig. 22 is the variation of CMOS inverters where its switching voltage, V_{sw} (Fig. 23), can be simply predicted by the A_{vt}. It was found that variation of switching voltages, such as, V_{th} or V_{sw}, of circuits can be described by a typical form of the square-root of the inverse of area, Eq. (8) in Table 2.

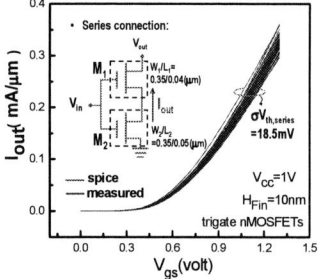

Fig. 20 The measured and Spice modeled characteristics of the variation for two series-connected trigate nMOSFETs. Both data show good matches. The insert is the schematic of this series circuit.

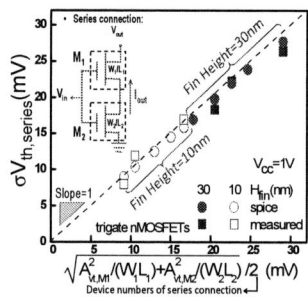

Fig. 21 $\sigma V_{th,series}$ is the V_{th} variation of the series-connected circuit, which is linearly proportional to the square-root of summation of each V_{th} variance divided by the device number, predicted by A_{vt}, i.e., this type of circuit can be directly calculated by A_{vt}.

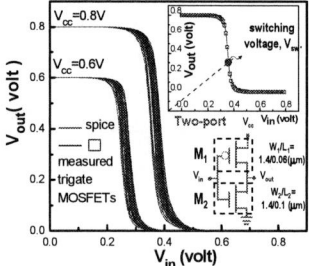

Fig. 22 CMOS inverter: This result compares the variation of inverters for spice and measured data, showing good matches. The variation parameter is defined as the switching voltage, V_{sw} (the insert).

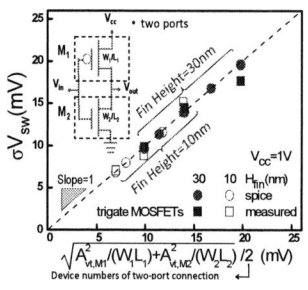

Fig. 23 CMOS inverter: σV_{sw} is linearly proportional to the square-root of the summation of each V_{th} variance divided by the device number, as predicted by A_{vt}, i.e., the variation of two-port CMOS connection can be directly predicted.

$$\sigma V_{sw} = \frac{\sqrt{\sum_{i=1}^{n}\left(\frac{A_{vt}^2}{LW}\right)_i}}{n} \quad \text{Eq.(8)}$$

Table 2 For three basic types of circuit connection, the variation of the switching voltage can be empirically described by a square-root of the inverse of area, divided by the number of connected devices.

Further, we demonstrated on two-input, NAND and NOR, CMOS circuits. Fig. 24 is the variation of the transfer curves for NAND and NOR, showing good matches between Spice and measured data. Again, Fig. 25 shows its variation of the butterfly curves that are the transfer curve of latch circuits, which serve as the storage component of 6T SRAM. *A summary was drawn* in Fig. 26 where σV_{sw} of NAND, NOR, and latch, also obey Eq. (8) in Table 2, in which V_{th} or V_{sw} variations can be easily calculated from the device size and the respective A_{vt} of Pelgrom plot. Finally, Fig. 27 is σV_{sw} of a

multi-stage inverter circuit versus the device numbers. Each curve belongs to a specific technology with different A_{vt}. It reveals that σV_{sw} decreases with the increasing number of stages, especially the trigate with shorter fin-height shows the lowest variation. The above results provide us a convenient way to predict the variations of any types of logic circuits regardless of the technology and the complexity of circuits.

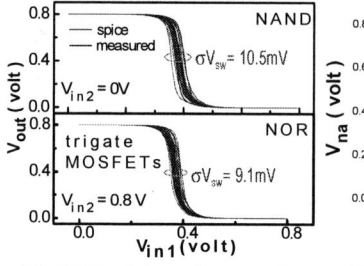

Fig. 24 The Spice and measured variation characteristics of (a) NAND and (b) NOR. The NAND and NOR circuit show almost the same variation, 10.5mV vs. 9.1mV.

Fig. 25 Comparisons of the variation of butterfly curve between Spice and measured data with different fin heights, showing that the taller fin height induces more fluctuations.

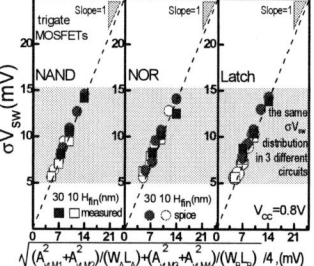

Fig. 26 The plot of σV_{sw} based on Eq. (8) for (a) NAND, (b) NOR, and (c) Latch circuits. It reveals a linear dependency in terms of the device area and Pelgrom plot coefficients for all the cases, which make it easier to estimate the circuit variations.

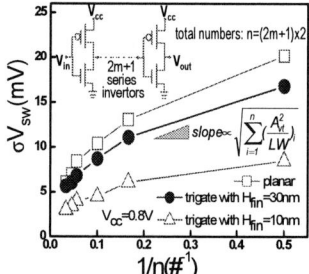

Fig. 27 The plot of σV_{sw} versus device numbers in a multi-stage CMOS inverters. The variations decrease with increasing number of stages.

In summary, *for the first time*, a *circuit level variation model* has been developed and demonstrated on an advanced trigate CMOS technology. The device model was improved from the Virtual Source Model (VSM) by including accurate R_{sd} and mobility calculations. Then, the model was implemented in Spice as a built-in model and for use in the circuit variation predictions. It was verified on several typical logic gates, NAND, NOR, inverter, latch etc. and the conclusion was reached with a ***unified form***, Eq. (8) in Table 2, which can be used to estimate the circuit variations with only few parameters, i.e., *W/L, A_{vt}, and A_{gm}*, required as inputs. This provides us a simple guideline to the variability design of more complex integrated circuits.

Acknowledgments This work was support in part by the Ministry of Science & Technology, Taiwan, under contract no. *NSC102-2221-E009-094* and ***NCTU-UC Berkeley*** I-RiCE program, under *MOST-103-2911-I-009-302*.

References: [1] E. R. Hsieh et al., *VLSI*, p.194 (2011). [2] H. M. Tsai et al., *VLSI*, p. 189 (2012). [3] E. R. Hsieh et al., *IEDM*, p. 454 (2012). [4] E. R. Hsieh et al., *IEDM*, p. 770 (2013). [5] X. Wang et al., *TED*, p. 2293 (2011). [6] Y. Cao et al., *TCAD*, p. 1866 (2010). [7] A. Khakifirooz et al., *TED*, p. 1674 (2009). [8] E. R. Hsieh et al., *VLSI-TSA*, p. T25 (2012). [9] M. J. M. Pelgrom et al., *IEEE J. Solid-State Circuits*, p. 1433 (1989).

Ultra-High-Q Air-Core Slab Inductors for On-Chip Power Conversion

Naigang Wang, David Goren, Eugene O'Sullivan, Xin Zhang, William J. Gallagher, Philipp Herget and Leland Chang

IBM Research Division, T. J. Watson Research Center, Yorktown Heights, NY 10598
E-mail: nwang@us.ibm.com ;   Phone: (914) 945-2663

Abstract

Air-core slab inductors with specially designed current return paths are proposed to achieve the ultra-high Q required for on-chip power delivery and management at >90% efficiency. Uniquely optimized for buck converter circuits, this CMOS-compatible structure avoids the challenges of thin-film magnetics. Q~25-30 at 200-300MHz is experimentally demonstrated.

Introduction

The power efficiency of VLSI products today strongly depends on the ability to perform on-chip power conversion at high efficiency. Such integrated voltage regulation circuits enable fast, fine-grain dynamic voltage scaling for efficient power management, high-voltage step-down conversion for efficient power delivery, and on-chip voltage generation to contain system supply proliferation. Microprocessor products today use linear regulators [1], which can be integrated on chip, but are fundamentally inefficient for large conversion ratios, or buck converters with package-integrated inductors [2], which achieve acceptable efficiency, but are not fully on-chip. For mainstream applications that may reach beyond 100W thermal design power (TDP), it is desirable to achieve >90% efficiency at ~A/mm^2 power densities in a fully integrated buck converter – a goal thus far limited by on-chip inductor Q.

With ~nH on-chip inductances, very high frequencies (VHF) in the ~100MHz range are needed for on-chip power conversion to balance output current ripple with FET losses [3]. Traditional air-core spiral inductors exhibit low Q at these frequencies due to on-chip metal thickness constraints (~10um), which has limited demonstrated efficiencies to ~70% [4]. In recent years, significant research has focused on improving inductor Q~ωL/R by integrating thin-film magnetics [5]; however, to date, converter efficiencies with such materials have still been <80% [6].

In this work, we clarify the power inductor requirements for practical on-chip voltage conversion circuits and highlight the challenges of thin-film magnetic-based inductors. We propose and demonstrate a new air-core slab structure with ultra-high Q, suitable for chip-scale buck converters at >90% efficiency.

Inductor Design for High-Efficiency Buck Converters

Fig. 1 shows that with an aggressive 90% target, MOSFET losses limit switching frequencies to <500MHz. On the other hand, >10MHz is desirable to balance efficient continuous conduction mode operation, ~A/mm^2 output current

density, ~ns transient response, and achievable on-chip inductance density. In this frequency regime with nH inductances, FET-related loss imposes an upper limit of ~7% inductor loss (an optimistic bound as interconnects and control circuits contribute additional losses).

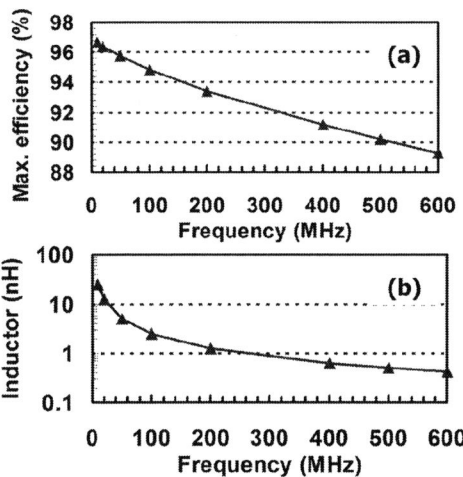

Figure 1. Simulated efficiency without inductor loss for a 2V-to-1V buck converter supplying 1A DC load current using 32nm CMOS technology. (a) maximum efficiency achieved by optimizing switch sizes; (b) required inductance to enable buck converter operation at the DCM/CCM boundary.

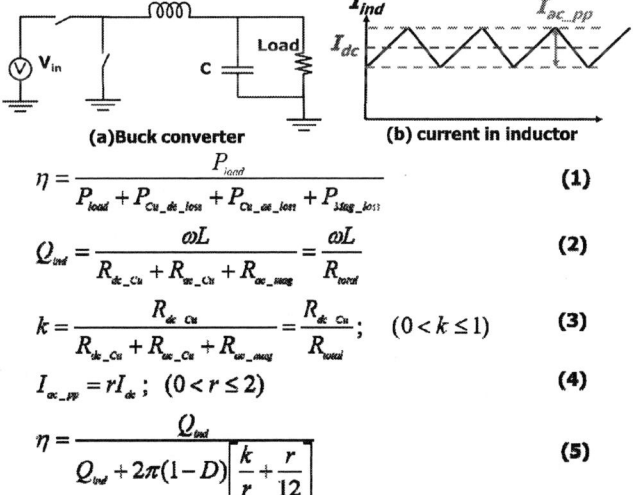

$$\eta = \frac{P_{load}}{P_{load} + P_{Cu\_dc\_loss} + P_{Cu\_ac\_loss} + P_{Mag\_loss}} \quad (1)$$

$$Q_{ind} = \frac{\omega L}{R_{dc\_Cu} + R_{ac\_Cu} + R_{ac\_mag}} = \frac{\omega L}{R_{total}} \quad (2)$$

$$k = \frac{R_{dc\_Cu}}{R_{dc\_Cu} + R_{ac\_Cu} + R_{ac\_mag}} = \frac{R_{dc\_Cu}}{R_{total}} ; \quad (0 < k \le 1) \quad (3)$$

$$I_{ac\_pp} = r I_{dc} ; \quad (0 < r \le 2) \quad (4)$$

$$\eta = \frac{Q_{ind}}{Q_{ind} + 2\pi(1-D)\left[\dfrac{k}{r} + \dfrac{r}{12}\right]} \quad (5)$$

Figure 2. (a) Buck converter and (b) current in inductor. Inductor efficiency is defined as the buck converter efficiency considering only inductor loss.

978-1-4799-8002-4/14 $31.00 © 2014 IEEE   307

In a buck converter, the inductor carries a DC-biased triangle wave current (Fig. 2). Inductor efficiency as expressed by Eq. (1) is comprised of DC wire resistance (R_{dc}), AC skin effect resistance (R_{ac}), and magnetic loss (R_{ac_mag}), which is zero for air-core inductors). Equation (5) shows that inductor efficiency strongly depends on r, the ratio between AC current ripple and DC current, and k, the fraction of total inductor resistance that is DC resistance. Buck converters achieve optimum efficiency at continuous vs. discontinuous conduction mode boundary [7], which implies r=2. The high required Q value is alleviated with smaller k (Fig. 3).

For air-core inductors, $R_{ac} \propto \omega^{0.5}$ enables continuous Q improvement up to self-resonance, which is generally beyond the frequency range of interest (Fig. 4a). For magnetic inductors, R_{ac_mag} often dominates, such that a peak Q exists due to the nonlinearity of magnetic eddy current loss ($\propto \omega^2$) and excess loss ($\propto \omega^2$ or $\omega^{1.5}$) [8]. For $R_{ac\_mag} \propto \omega^2$, Q_{peak} occurs when $R_{ac} = R_{dc}$ at k=0.5 (Fig. 4b). Normalization of Eq. (1) yields a Q_{peak} requirement for magnetic inductors of >17 to achieve ~5% inductor loss (Fig. 5). This is higher than results published to date even with new materials and complex lamination layers (Fig. 6). Significant advancements in magnetics and inductor design are thus needed.

$$Q_{required} \geq \frac{\eta}{1-\eta} 2\pi(1-D)\left[\frac{k}{r} + \frac{r}{12}\right] \quad (6)$$

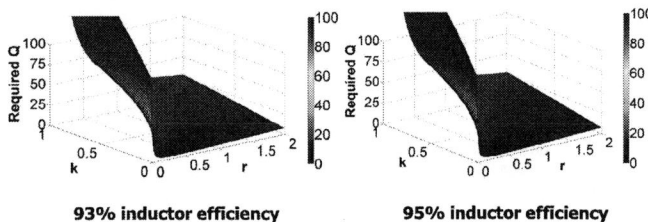

Figure 3. Required inductor Q for 93% and 95% inductor efficiency when D=0.5.

$$Q = \frac{\omega L}{R_{dc} + R_{ac}} \quad (7)$$

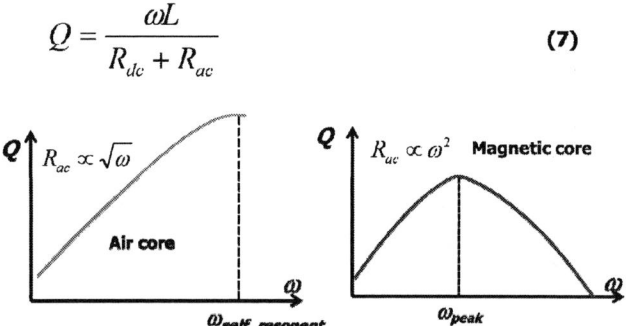

Figure 4. Inductor Q as function as frequency for (a) air core inductor, where Q increases until self-resonance; and (b) magnetic inductor where peak Q occurs when $R_{ac} = R_{dc}$, i.e. k=0.5.

$$Q_{ind} = \frac{\omega L}{R_{dc\_cu} + A \cdot \omega^2} = Q_p \frac{2\frac{\omega}{\omega_p}}{1+\left(\frac{\omega}{\omega_p}\right)^2} \quad (8)$$

$$\eta = \frac{Q_p \dfrac{2\dfrac{\omega}{\omega_p}}{1+\left(\dfrac{\omega}{\omega_p}\right)^2}}{Q_p \dfrac{2\dfrac{\omega}{\omega_p}}{1+\left(\dfrac{\omega}{\omega_p}\right)^2} + 2\pi(1-D)\left[\dfrac{\dfrac{1}{2}\dfrac{1}{1+\left(\dfrac{\omega}{\omega_p}\right)^2}}{r} + \dfrac{r}{12}\right]} \quad (9)$$

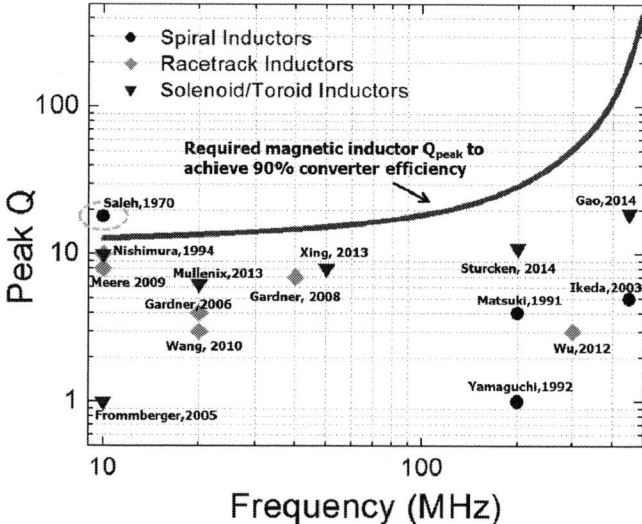

Figure 5. Inductor efficiency as function as Q_{peak} for magnetic inductors and required magnetic inductor Q_{peak} for 93% and 95% inductor efficiency

Figure 6. Comparison of required magnetic Q_{peak} (considering switch losses in Fig. 1) with reported magnetic inductors peak Q, excluding non-CMOS-compatible inductors (e.g. MEMS dimensions, high temperature processing). It should be noted that the circled point at ~10MHz uses a large number of wire turns to achieve ~uH inductance, which may be too high for fast transient response in on-chip applications [9-24].

Ultra-High-Q Air-Core Slab Inductors

Air-core inductors, however, can be designed for sufficient Q by minimizing R_{dc} via use of a wide (~200um), thick (~10 um) slab, and by specially designed, low-resistance inductive return paths on either side of the slab (Fig. 7). With a return path separation distance of 200-300um, ~nH inductance can be achieved. This structure maintains a similar inductance to traditional spiral inductors due to the return path loop, but R_{dc} is improved as load current need only flow through the slab (Fig. 8). To minimize impedance, DC return current will flow via the least resistive path (ground near the load), whereas AC return current will flow through the two designed return paths (Fig. 9). This structure thus achieves lower k and higher Q as compared to a spiral inductor.

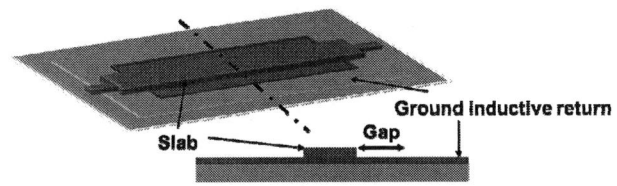

Figure 7. Basic slab inductor structure with two metal layers

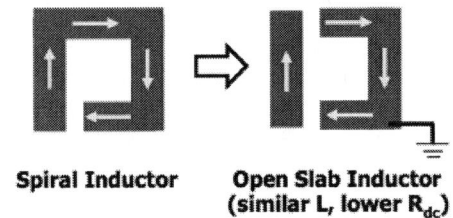

Spiral Inductor **Open Slab Inductor (similar L, lower R_{dc})**

Figure 8. Spiral inductor vs. slab inductor with inductive return

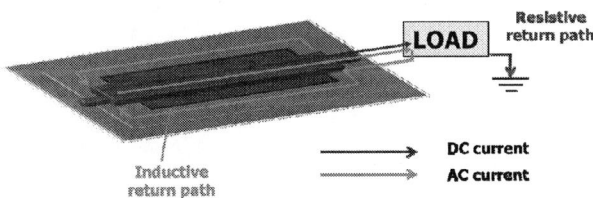

Figure 9. Schematic of DC and AC return paths

The slab inductor can be built in a CMOS back-end process (Fig. 10) by designing the structure in a thick metal layer (or multiple layers tied together). The return paths can reside in the same layer as the slab or in existing layers such as in the chip power grid. In the latter case, however, openings must be cut away from the power grid to ensure that return currents are sufficiently far away in the desired paths. These openings can take the form of small slots to minimize power grid disruption (Fig. 11).

- **FEOL+BEOL**
- **Deposit oxide**
- **Reactive Ion Etch to open via**
- **Through mask plate Cu slab**
- **Deposit dielectrics for passivation**

Figure 10. Fabrication flow for slab inductor

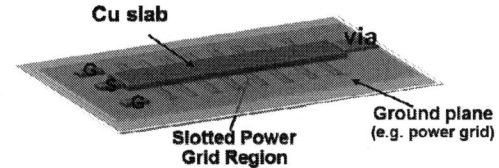

Figure 11. Schematics of slab inductor with slotted ground plane

To experimentally demonstrate the slab structure, a 10um through-mask electroplated Cu layer was fabricated above a 5um slotted power grid return. One end of the slab is shorted to ground with a via to imitate high capacitive loads for 1-port characterization. The inductors are 2mm long to enable high Q, low resistance measurements. Measured devices demonstrate nearly constant nH-range inductance and skin-effect-limited resistance to achieve Q as high as 25-35 at 200-300 MHz (Fig. 12). Due to high Q (25) and low k (0.25) at 200 MHz, a 96.6% inductor efficiency is achieved (5). Fig. 12 confirms that a ~200um gap is needed between the slab and return paths. As compared with prior work (Table 1), the proposed structure achieves ultra-high Q at an inductance density suitable for this application, and both inductance and Q can be further improved through coupling.

Using an analytic model of the slab structure calibrated both to experimental results and FEM simulation (Fig. 13), simulated buck converter designs in 32nm CMOS demonstrate >90% conversion efficiency for ~2:1 conversion ratios (Fig. 14).

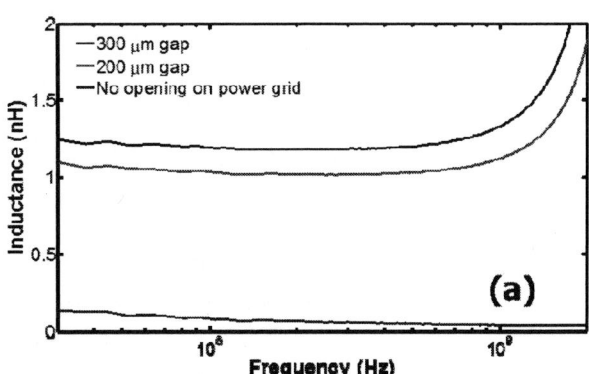

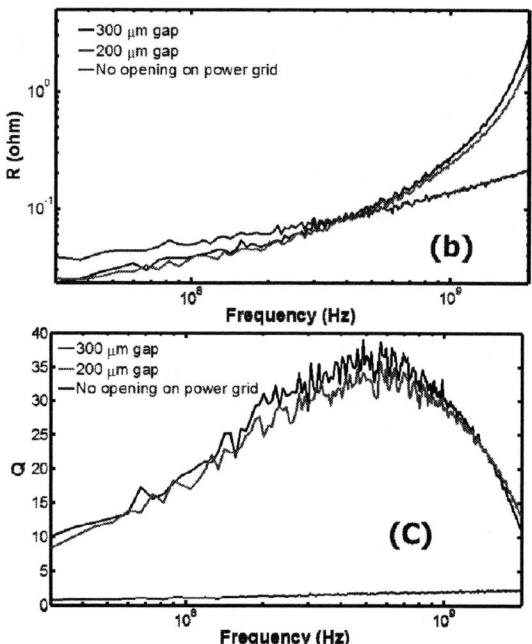

Figure 12. Measured (a) inductance, (b) resistance and (c) Q of slab inductors with 200 um and 300 um gaps between slab and return paths, respectively. As a reference, slab above a no-opening power grid is also shown.

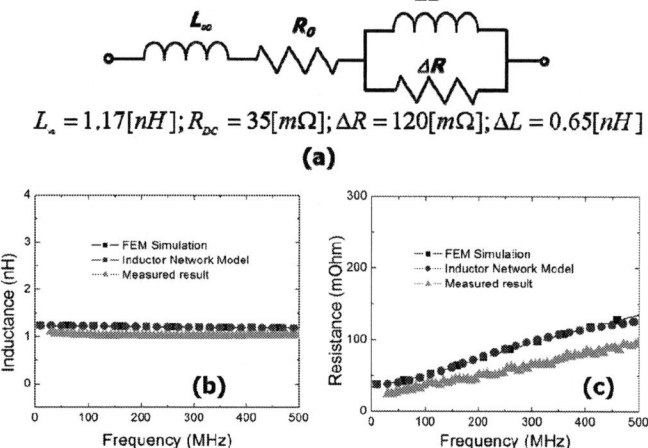

$$L_a = 1.17[nH]; R_{DC} = 35[m\Omega]; \Delta R = 120[m\Omega]; \Delta L = 0.65[nH]$$

(a)

Figure 13. (a) Inductor model; and FEM simulated, network modeled and measured (b) inductance and (c) resistance of a slab inductor.

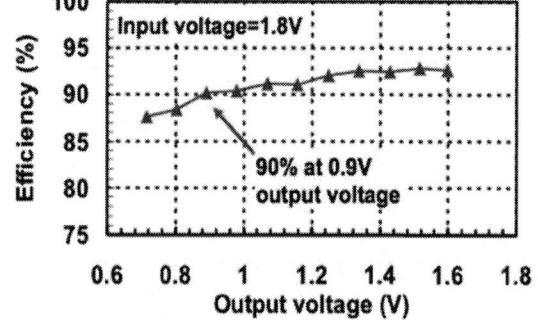

Figure 14. Simulated 32SOI buck converter design using a calibrated open slab inductor model demonstrating >90% efficiency for ~2:1 conversion.

Table 1. Compared with prior art, the slab inductor can achieve ultra-high Q for on-chip power conversion in a simple BEOL process.

	RF spiral [9]	Solenoid [10]	Stripeline [11]	Spiral [12]	Solenoid [13]	Slab [This work]
Q	1-3@200MHz	18@450MHz 15@200MHz	7@50MHz	10@10MHz	18@450MHz 15@200MHz	25@200MHz 30@300MHz
L (nH)	1-10	2-3	3-5	30-150	8-15	1.2
Area (mm²)	0.01-0.04	0.03-0.06	0.03-0.06	0.5-2.5	--	1.2
Metal Thickness (um)	3-5	4.5	5	<35	5	10
Structure	BEOL	BEOL	BEOL	BEOL	BEOL	BEOL
Magnetic materials	No	Yes	Yes	Yes	Yes	No

Conclusion

We have proposed and demonstrated an air-core slab structure to achieve ultra-high Q on-chip inductors to enable high-efficiency integrated power conversion. The device can be incorporated into a CMOS back-end process and avoids challenges associated with the integration of magnetic materials.

Acknowledgement

Device fabrication performed at the Microelectronics Research Laboratory in Yorktown Heights, NY.

References

[1] Z. Toprak-Deniz et al., *ISSCC* '14.

[2] E. Burton et al., *APEC* '14.

[3] G. Sizikov et al., *ICECS*, 2010.

[4] H. Krishnamurthy et al., Symp. VLSI Circ. '14.

[5] D. Gardner et al., *IEDM* '06.

[6] N. Strucken et al., *IEEE JSSC*, vol. 48, no. 1, pp. 244-254 , 2013.

[7] T. Karnik *ISSCC* 2012 Tutorials.

[8] G. Bertotti, *IEEE Trans. Magn.*, vol. 24, pp. 621-630, 1988.

[9] J. Burghartz and B. Rejaei, *IEEE Trans. Electron. Devices*, vol. 50, pp. 718-729, 2003.

[10] Y. Gao et al., *IEEE Trans. Electron. Devices*, vol. 61, pp. 1470-1475, 2014.

[11] D. Gardner et al, *J. Appl. Phys.*, vol. 103, pp. 07E927-1-6, 2008.

[12] R. Meere et al, *IEEE Trans. Magn.*, 45 (10), 2009.

[13] N. Strucken et al, *APEC*, 2014.

[14] N. Saleh and A. H. Qureshi, *Electron Lett.*, vol. 6, pp. 850-852, 1970.

[15] N. Wang et al., *J. Appl. Phys.*, vol. 111, pp. 07E732-1-6, 2012.

[16] G. Gardner et al., *IEEE Trans. Magn.*, vol. 45, pp. 4760-4766, 2009.

[17] X. Xing et al., *IEEE Trans. Magn.*, vol. 47, no. 10, pp. 3104-3107, 2011.

[18] H. Nishimura, et al., IEEE Translation Journal on Magnetics in Japan, vol. 9, pp. 76-84, 1994.

[19] H. Matsuki et al., IEEE Trans. Magn., vol. 27, no. 6, pp. 5438-5440, 1991.

[20] H. Wu et al., IEEE Trans. Magn., vol. 48, no. 11, pp. 4123-4126, 2012.

[21] M. Yamaguchi et al., IEEE Trans. Magn., vol. 28, no. 5, pp. 3015-3017, 1992.

[22] M. Frommberger et al., IEEE Trans. Microw. Theory Tech., vol. 53, no. 6, pp. 2096-2100, 2005.

[23] K. Ikeda et al., IEEE Trans. Magn., vol. 39, no. 5. Pp. 3057-3061, 2003.

[24] J. Mullenix, A. El-Ghazaly, and S. X. Wang, IEEE Trans. Magn., vol. 49, no. 7, pp. 4021-4027, 2013.

Efficient Wireless Power Transmission Technology Based on Above-CMOS Integrated (ACI) High Quality Inductors

Salahuddin Raju, Xing Li, Yan Lu, Chi-Ying Tsui, Ki Wing-Hung, Mansun Chan, and C. Patrick Yue

Dept. of Electronic and Computer Engineering, The Hong Kong University of Science and Technology, Kowloon, Hong Kong
Email: {rsalahuddin, eepatrick}@ust.hk

Abstract

Fully-integrated on-chip inductors with up to 200 nH/mm^2 inductance density and a peak qualify factor of 25 are demonstrated based on above-CMOS integration (ACI) post processing techniques. Utilizing a 380-nH ACI inductor, a 2.5×2.5 mm^2 wireless energy harvesting antenna was implemented. Measurement results show that it can receive 27 mW from a 250-mW transmitting power source at a distance of 5.3 mm. This represents a 7-fold increase in wireless power transfer efficiency compared to other reported technologies.

Introduction

Large on-chip inductance in the µH range is desirable for effective integration of power electronics [1] using standard IC processes. However, on-chip inductor performance is generally limited by thin metal thickness (typically about several µm) and low silicon substrate resistivity (~10–30 Ω-cm). To overcome these limitations of standard CMOS processes, an above-CMOS-integration (ACI) technology is proposed with a thick polymer dielectric, placed above the passivation layer of a standard CMOS IC, to serve as the base and a very thick metal layer to realize the spiral inductors.

Using the proposed ACI technology, a 4.5×4.5 mm^2 inductor can render 4.3 µH inductance (L) on a low resistivity substrate while achieving a peak quality factor (Q_{Peak}) of 26. To demonstrate the application of ACI inductor in miniaturized power electronics, a 2.5×2.5 mm^2 wireless power-receiving antenna was designed. It can wirelessly receive 27 mW from a 250-mW 27 MHz transmitting signal at a distance of 5.3 mm, and 1.1 mW at a distance of 21.3 mm. It is the best reported performance compared with published data.

Above CMOS Integration (ACI) Technology

To suppress substrate loss due to low resistivity CMOS IC substrate [2, 3], a thick low-k polymer dielectric (SU8, ε_r = 2.8) was placed on top of the passivation layer of the IC for isolation as shown in Fig. 1. A single lithography is used to pattern the SU8 to form deep trenches with very high aspect ratio (> 10) where the inductor is formed [4]. After thermal treatment at 150 °C, all deep trenches were filled with copper

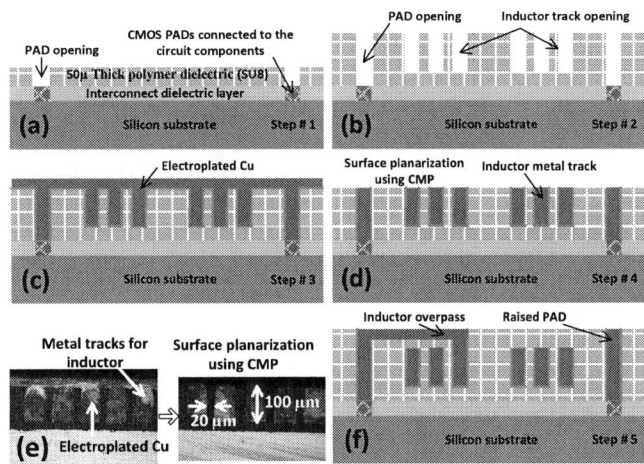

Fig. 1 The ACI fabrication process: (a) thick polymer dielectric deposition, (b) deep trench patterns for inductor, (c) seed layer deposition followed by Cu electroplating, (d) overburden Cu removal by CMP, (e) cross-section of the wafer before and after the CMP, and (e) inductor over pass to provide connectivity to the PAD.

using the plating process. Fig. 1(e) shows the void-free copper filled trenches that realized the spiral inductor traces. A micrograph of an ACI inductor and its cross-section are shown in Fig. 2(a). To demonstrate the CMOS integration capability, a CMOS IC with only dummy PADs was prepared (Fig. 2(b)). When the inductor was fabricated on top of the IC, and the dummy PADs on the CMOS IC were also raised by the ACI processing (Fig. 2(c)).

The improvement of self-resonance frequency and quality factor (Q) due to the thicker dielectric is experimentally verified in Fig. 3. The quality factor was improved more than 5 times. By changing the turn numbers (N), measured inductances with an area of 4.5×4.5 mm^2 range from 3.47 µH to 4.4 µH with a peak quality factor Q_{Peak} range from 27 to 22 at 10-20 MHz frequencies (Fig 4 & 5). The DC resistance (R_{DC}) was reduced by increasing the thickness of the metal layer (Fig. 5(b)). As a result, a high L to R_{DC} ratio up to 1752 nH/Ω and the inductor efficiency of more than 90% [5] is achieved. These are state-of-the-art FoM for integrated power electronic applications [5, 6].

In any inductor technology, higher L and Q are desirable in the per-unit area of the silicon footprint to lower the solution cost. Hence, these two factors should be the ultimate metric for fair comparison among different technologies [7]. Using

978-1-4799-8002-4/14 $31.00 © 2014 IEEE

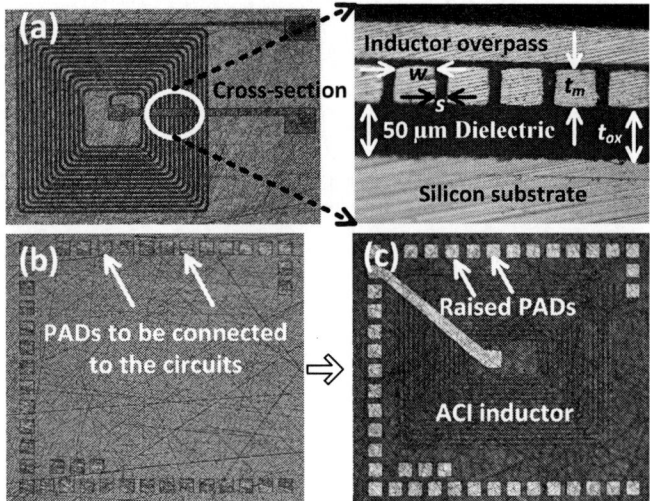

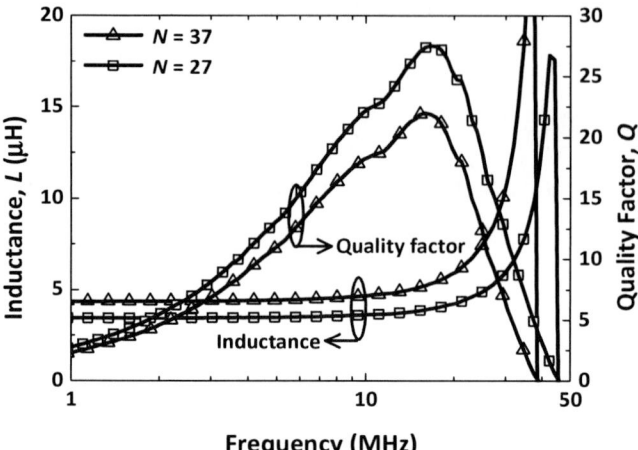

Fig. 2 (a) The micrograph of a test-inductor and its cross-section. (b) Dummy PADs to mimic a CMOS chip configuration. (c) The micrograph of an ACI inductor after the process.

Fig. 4 Frequency response of L and Q of 4.5×4.5 mm^2 ACI inductors on top of a 10 Ω-cm substrate. Design parameters: track width $w = 30$ μm, spacing $s = 15$ μm, metal thickness $t_m = 40$ μm, bottom dielectric thickness $t_{ox} = 50$ μm, and turn number $N = 37$ & 27.

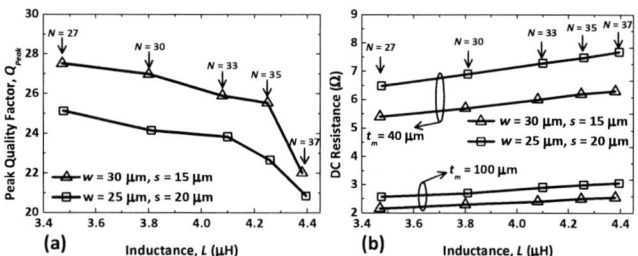

Fig. 5 (a) The effect of turn number (N) on Q_{Peak} and L. (b) DC resistance was reduced by increasing the thickness of the metal (t_m).

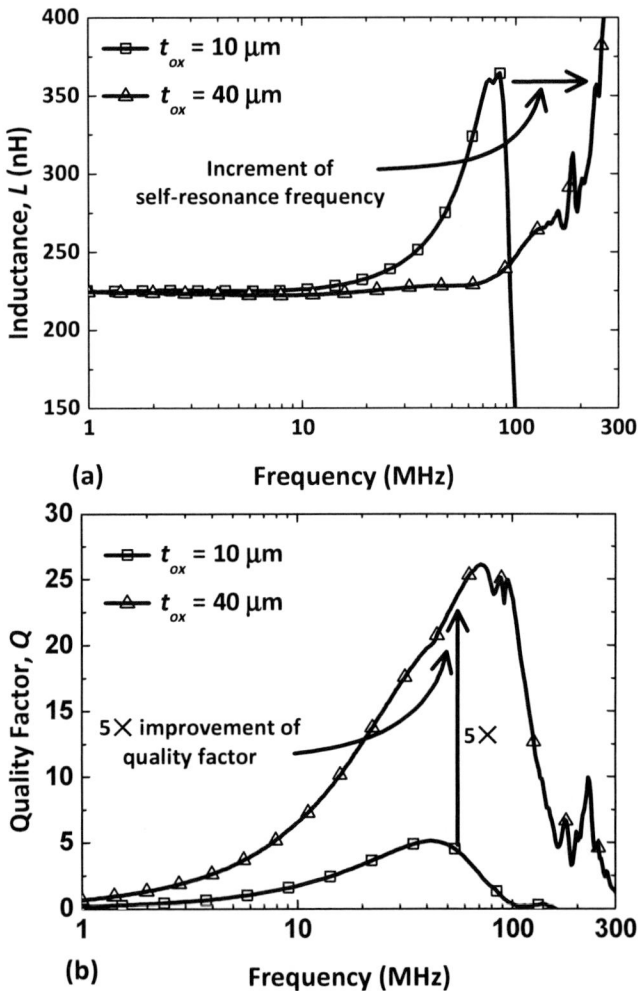

Fig. 3 Improvement of self-resonance frequency (a) and quality factor (b) due to higher dielectric thickness (t_{ox}).

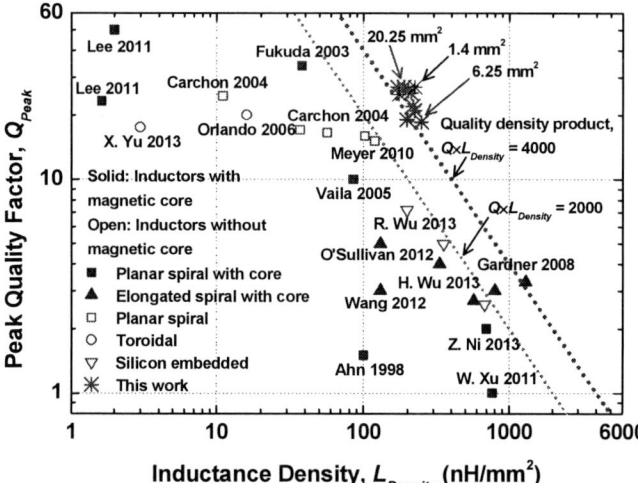

Fig. 6 Measured peak quality factor (Q) and inductance density ($L_{Density}$) of the leading technologies for inductor integration on silicon. Here, $L_{Density}$ is defined as inductance per mm^2.

the Q-$L_{Density}$ product as the performance comparison measure, the superiority of the ACI inductors can be observed. In particular, an inductance density of 200 nH/mm^2 with a Q more than 20 is the best performance, to date, in leading power inductor technologies, as shown in Fig. 6.

978-1-4799-8002-4/14 $31.00 © 2014 IEEE   312

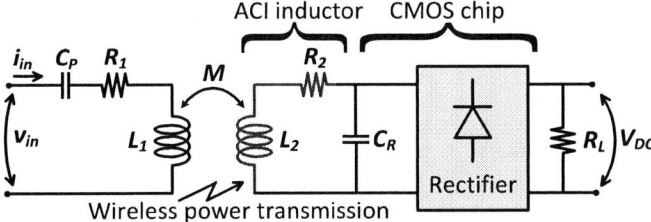

Fig. 7 Schematic diagram of wireless power transmission system.

TABLE I. DESIGN PARAMETERS OF WIRELESS POWER TRANSMISSION

Specifications	Power Transmitting Inductor (Tx)	Power Receiving Inductor-1 (Rx#1)	Power Receiving Inductor-2 (Rx#2)
Inductor Area	20×20 mm^2	2.5×2.5 mm^2	1.1×1.3 mm^2
Technology	PCB	ACI	ACI
Turn Number	3	13	15
Track Width	700 μm	30 μm	17.5 μm
Spacing	300 μm	30 μm	12.5 μm
Inductance	250 nH	380 nH	175 nH
Quality Factor	81 @ 27 MHz 95 @ 40 MHz	21 @ 27 MHz	23 @ 40 MHz

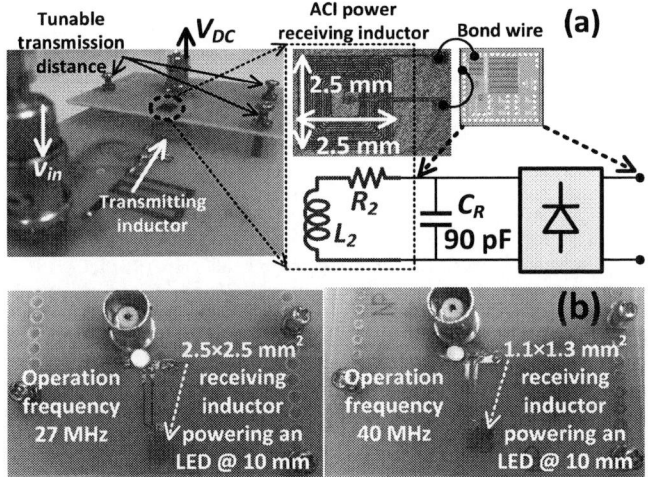

Fig. 8 (a) The experimental setup for measuring the characteristics of the wireless power transmission system. (b) The mm-size power receiving inductor is powering an LED at 10 mm distance.

Wireless Power Transmission to an IC Chip

The ACI inductor is used to construct a mm-size wireless power receiving antenna. The wireless power link schematic is given in Fig. 7. The ACI inductor replaced the traditional PCB or wire- wound inductors which usually take up a few cm. The design parameters are summarized in Table I.

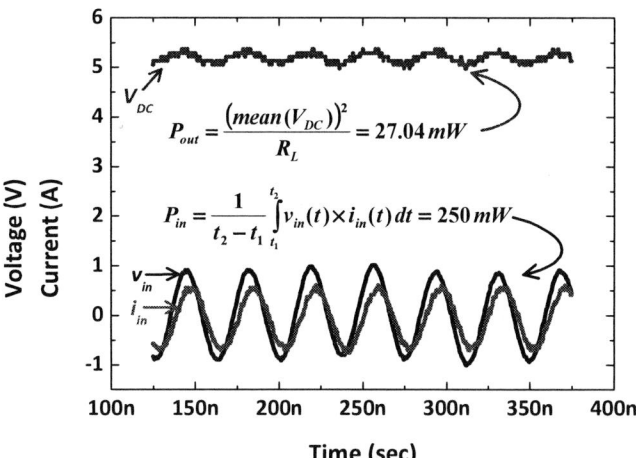

Fig. 9 Time domain measurement of transmitting signals v_{in} and i_{in}, and the received DC voltage V_{DC} across the 1 KΩ load resistance, at 5.3 mm distance. The efficiency was calculated as $\eta = P_{out} / P_{in} = 10.8$ %.

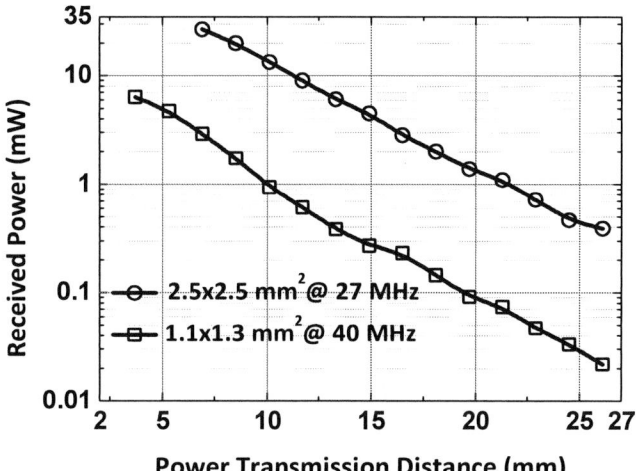

Fig. 10 Received power by the mm-size wireless power receiver at different transmission distances, provided that the transmitted power was 250 mW.

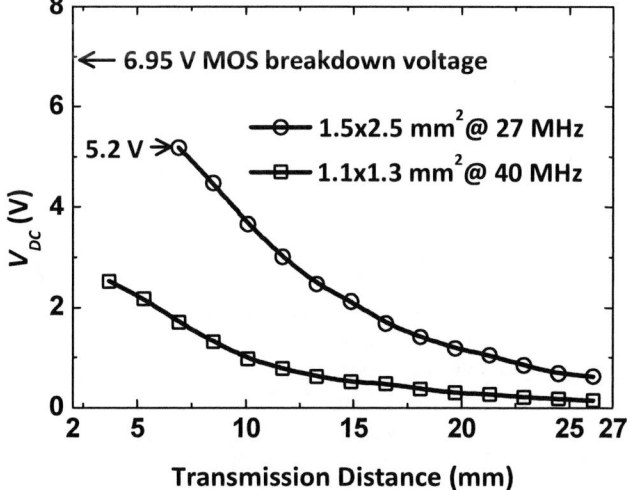

Fig. 11 Measured rectified DC voltage across the load resistance (R_L = 1 KΩ) at different transmission distances.

978-1-4799-8002-4/14 $31.00 © 2014 IEEE

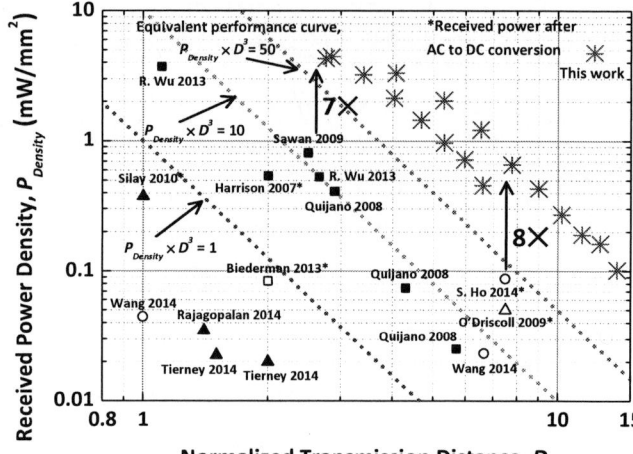

Fig. 12 Summary of reported wireless power transmission systems with different inductor integration schemes in the literature. **Legends:** *square* (■ □)- power receiving inductor on silicon, *triangular* (▲ △)- receiving inductor on PCB, *circle* (● ○)- wire wound receiving inductor, *open* (□ △ ○)- operation frequency is at GHz range, and *solid* (■ ▲ ●)- operation frequency is at MHz range. The normalize transmission distance, D, is defined as the ratio of transmission distance and the size of the power receiving inductor, and the received power density is defined as- $P_{Density} = P_{out}/Area = P_{in} \times \eta/Area$.

The ability of capturing electromagnetic energy by an inductor mainly depends on the number of turns (N) in a particular area [8] and the Q value. The ACI inductors offer significant advantages in terms of these metrics. The higher $L_{Density}$ lowers the required capacitance to the sub-100 pF range for operation frequency in the MHz range as shown in Fig. 8. As a result, the proposed ACI technology can effectively reduce the overall silicon footprint of a power electronic module.

The detailed measurement setup is shown in Fig. 8, and the results are described in Fig. 9, Fig. 10 & Fig. 11, respectively. The mm-size receiver can harvest 27 mW to 1.1 mW power in a transmission distance ranging from 5.3 to 21.3 mm from a 250-mW transmitter as shown in Fig. 10. More power could be harvested by the inductor by supplying more power from the transmitter. However, the rectifier IC has a breakdown voltage of about 7 V. To avoid breakdown, lower transmitted power was adopted to keep the DC voltage across the load to below 5 V as shown in Fig. 11.

A comparison is presented to evaluate the performance of the proposed wireless power transmission system with others reported in the literature. Due to the different inductor technologies and sizes used, the received power per unit area ($P_{Density}$) and the normalized transmission distance (D) are used as the performance metric for a fair comparison. The received $P_{Density}$ inversely varies with D^3 similar to that of mutual coupling between the transmitter and receiver [8]. Hence, the equivalent performance curve is defined as $P_{Density} \times D^3$. Fig. 12 shows that the proposed wireless power transmission system outperforms any other technologies

reported in the literature [9-12], transmitting mW levels of power even at a distance 10 times larger than the antenna size.

Conclusion

An area efficient, high performance and CMOS-compatible post-processing technology for μH-range inductor integration is presented. With this technology, we have demonstrated the most efficient mm-size wireless power harvesting devices reported to date.

Acknowledgment

This work was supported in part by a grant (T23-612/12-R) from the Research Grants Council of the Hong Kong Special Administrative Region Government under the Theme-Based Research Scheme, and by the Hong Kong's University Grant Committee via the Area of Excellence project AoE-P04-08.

References

[1] C. R. Sullivan, D. V. Harburg, J. Qiu, C. G. Levey, and D. Yao, "Integrating magnetics for on-chip power: a perspective," *IEEE Transaction on Power Electronics*, vol. 28, no. 9, pp. 4342-4353, September 2013.

[2] C. P. Yue, S. S. Wong, "Physical modeling of spiral inductors on silicon," *IEEE Transaction of Electron Devices*, Vol. 47, No.3, pp. 560-568, March 2000.

[3] J. N. Burghartz, "Progress in RF inductors on silicon - understanding substrate losses," *IEEE International Electron Devices Meeting (IEDM)*, pp. 523-526, 1998.

[4] A. D. Campo, and C. Greiner, "SU-8: a photoresist for high-aspect-ratio and 3D submicron lithography," *Journal of Micromechanics and Microengineering*, 17, pp. R81-R95, 2007.

[5] P. Herget, et al., "Limits to on-chip power conversion with thin film inductors," *IEEE Transaction on Magnetics*, vol. 49, no. 7, pp. 4137-4143, July 2013.

[6] C. Ó. Mathúna, N. Wang, S. Kulkarni, and S. Roy, "Review of integrated magnetics for power supply on chip (PwrSoC)," *IEEE Transaction on Power Electronics*, vol. 27, no. 11, pp. 4799-4816, November 2012.

[7] D. S. Gardner, et al., "Review of on-chip inductor structures with magnetic films," *IEEE Transaction on Magnetics*, vol. 45, no. 10, pp. 4760-4766, October 2009.

[8] S. Raju, R. Wu, M. Chan, and C. P. Yue, "Modeling of mutual coupling between planar inductors in wireless power applications," *IEEE Transaction on Power Electronics*, vol. 29, no. 1, pp. 481-490, January 2014.

[9] F. S. Quijano, J. G. Cantón, J. Sacristán, T. Osés, and A. Baldi, "Wireless powering of single-chip systems with integrated coil and external wire-loop resonator," *Applied Physics Letters*, 92, pp. 074102(1-3), 2008.

[10] M. Sawan *et al.*, "Multicoils-based inductive links dedicated to power up implantable medical devices: modeling, design and experimental results," *Biomedical Microdevices*, vol. 11, no. 5, pp. 1059-1070, June 2009.

[11] S. O'Driscoll, A. S. Y. Poon, and T. H. Meng, "A mm-sized implantable power receiver with adaptive link compensation," *IEEE International Solid-State Circuit Conference (ISSCC)*, pp. 294-295, 2009.

[12] J. S. Ho, et al., "Wireless power transfer to deep-tissue microimplants," *Proceedings of the National Academy of Sciences (PNAS)*, vol. 111, no. 22, pp. 7974-7979, June 2014.

A Magnetic Tunnel Junction Based True Random Number Generator with Conditional Perturb and Real-Time Output Probability Tracking

Won Ho Choi*, Yang Lv*, Jongyeon Kim, Abhishek Deshpande, Gyuseong Kang, Jian-Ping Wang, and Chris H. Kim

Dept. of ECE, University of Minnesota, 200 Union Street SE, Minneapolis, MN 55455, USA, Email: choi0444@umn.edu, *equal contribution

Abstract

This work experimentally demonstrates for the first time a True Random Number Generator (TRNG) based on the random switching probability of Magnetic Tunnel Junctions (MTJs). A conditional perturb and real-time output probability tracking scheme is proposed to enhance the reliability, speed, and power consumption while maintaining a 100% bit efficiency.

Introduction

True Random Number Generators (TRNG) are specialized circuits used in a wide variety of applications ranging from cryptography and hardware based security to statistical sampling and advanced simulation techniques. Traditional CMOS based TRNGs utilize physical noise present in CMOS circuits such as thermal noise, random telegraph noise, and oscillator jitter. However, existing CMOS based TRNGs require extensive post-processing to ensure a high level of randomness in the output bits, which incurs a significant performance, power, and area overhead [1]. The goal of this work is to develop a new class of TRNGs based on the random switching probability of Magnetic Tunnel Junctions (MTJs) for compact area, simpler design, high throughput, and reliable operation. In particular, we propose a conditional perturb and real-time output probability tracking scheme to achieve a 100% bit efficiency (or 100% useable bits) while improving the reliability, speed, and power. An additional benefit of MTJ-based TRNG is that it can be readily implemented using an existing STT-MRAM array with negligible circuit overhead.

The Spin Transfer Torque (STT) switching phenomenon in an MTJ (Fig. 1) is subject to random thermal fluctuation noise which gives rise to a switching probability contour map as shown in Fig. 2 [2]. By applying an optimal "perturb" pulse

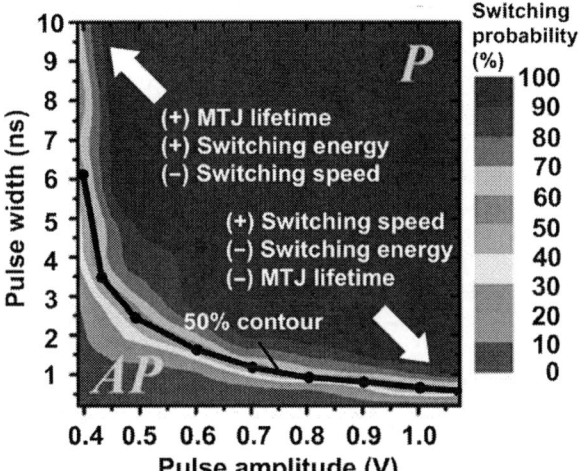

Fig. 2. MTJ switching probability as a function of pulse width and pulse amplitude [2] (AP→P switching direction).

whose width and amplitude correspond to the 50% switching probability contour, the final resolved state of the MTJ will depend solely on the random thermal noise, producing an unbiased random output bit. In this work, we address the two main considerations for a high quality TRNG design: (1) achieving the optimal trade-off between switching speed, power, and lifetime, and (2) ensuring a 50% switching probability under different PVT conditions.

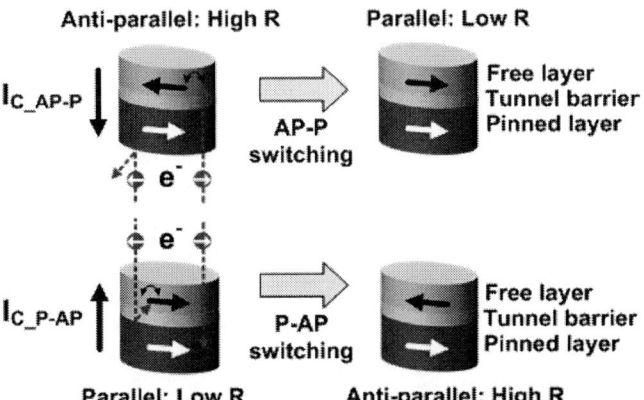

Fig. 1. Illustration of Spin Torque Transfer (STT) switching principle in Magnetic Tunnel Junction (MTJ).

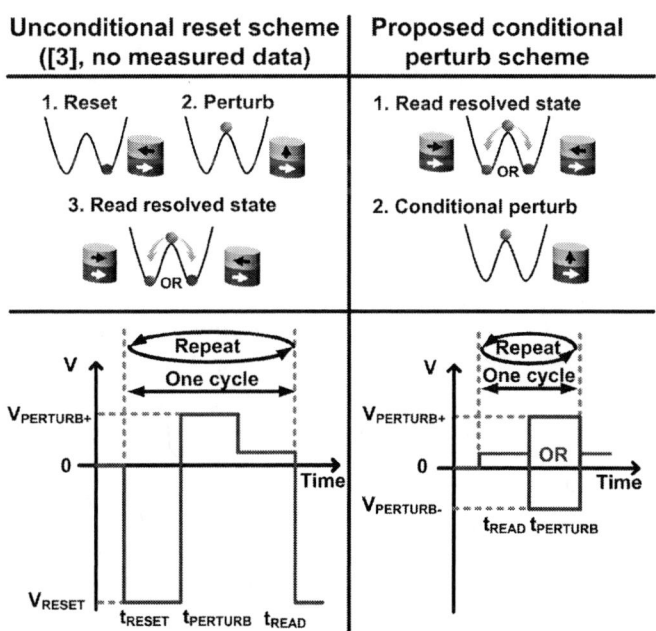

Fig. 3. Random number generation schemes: (left) unconditional reset scheme and (right) the proposed conditional perturb scheme.

978-1-4799-8002-4/14 $31.00 © 2014 IEEE

Conditional Perturb Scheme

The working principles of the conventional unconditional reset scheme [3] (concept only) and the proposed conditional perturb scheme are described in Fig. 3. The conventional technique applies an initial reset voltage (V_{RESET}) large enough to force the MTJ into a reset state (i.e., AP). Subsequently, a smaller perturbation voltage $V_{PERTURB+}$ in the opposite direction (i.e., AP to P) is applied to induce STT switching with a 50% probability. Finally, the resolved state is read out using a small read voltage (V_{READ}). The proposed scheme, on the other hand, perturbs the cell according to the previously sampled MTJ state, thereby eliminating the reset phase all together. Fig. 4 shows a high level comparison between the two schemes. The advantages of the proposed technique are threefold. First, the absence of a reset phase enhances the lifetime of the MTJ as illustrated in the time-to-breakdown measurements in Fig. 5 [4][5]. Unlike in STT-MRAM

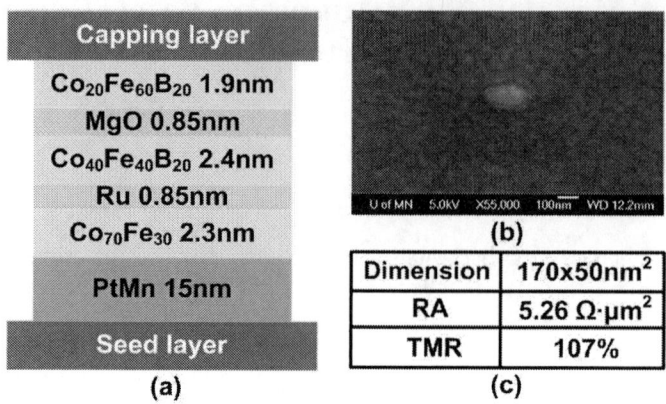

Fig. 6. (a) MTJ vertical stack structure (b) SEM image (c) key parameters of the fabricated MTJ device.

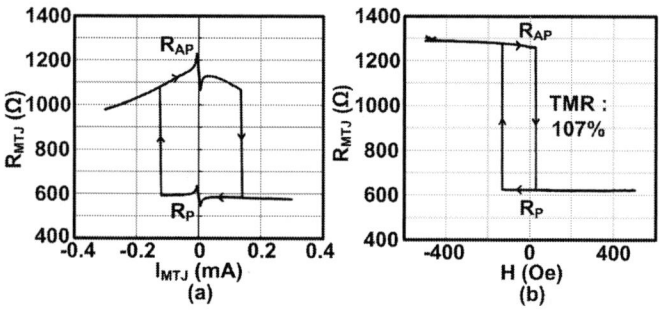

Fig. 7. Measured (a) R-I and (b) R-H hysteresis curves of the fabricated MTJ device. Data was collected while sweeping the MTJ current (a) and external field (b).

	Unconditional reset scheme ([3], no measured data)	Proposed conditional perturb scheme
Bit rate	1X (Slow) 1bit / (t_{RESET}+$t_{PERTURB}$+t_{READ})	1.67X (Fast) 1bit / (t_{READ}+$t_{PERTURB}$)
Switching energy	1X (High) E_{RESET}+$E_{PERTURB}$+E_{READ}	0.29X (Low) E_{READ}+$E_{PERTURB}$
MTJ lifetime	Short time-to-breakdown	Long time-to-breakdown
Design overhead	Strong reset driver	Polarity detection, symmetric AP→P and P→AP switching

Fig. 4. TRNG performance comparison between the unconditional reset and the proposed conditional perturb schemes.

application where the cell is accessed infrequently, MTJs for TRNGs need to be accessed continuously throughout the lifetime of the product (e.g., 10 years) making lifetime related issues a first rate concern. Second, random bits can be generated at a faster rate since no reset is required and the perturb and read operations can be made relatively fast. Finally, the energy dissipation is lower for the proposed conditional perturb scheme.

The MTJ stack structure, SEM image, and summary of measured MTJ parameters are given in Fig. 6. The measured R-I and R-H hysteresis curves are shown in Fig 7. The

Unconditional reset　Conditional perturb

Fig. 5. (a) Timing diagrams for MTJ Time-to-breakdown (t_{BD}) analysis (b) Lifetime comparison between the two TRNG schemes based on MTJ measurement data [4][5].

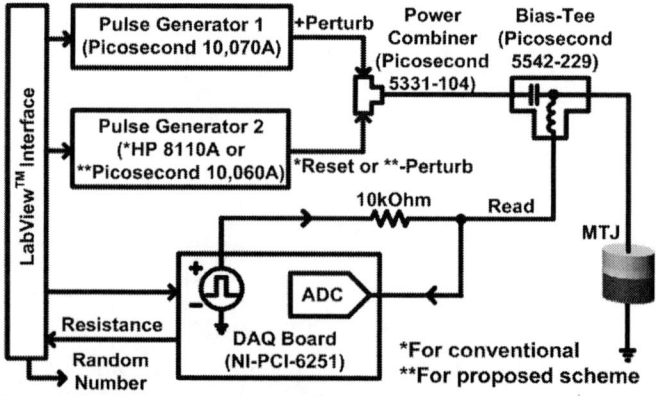

Fig. 8. Random number generator measurement setup with sub-50 picosecond pulse width resolution.

978-1-4799-8002-4/14 $31.00 © 2014 IEEE　　316

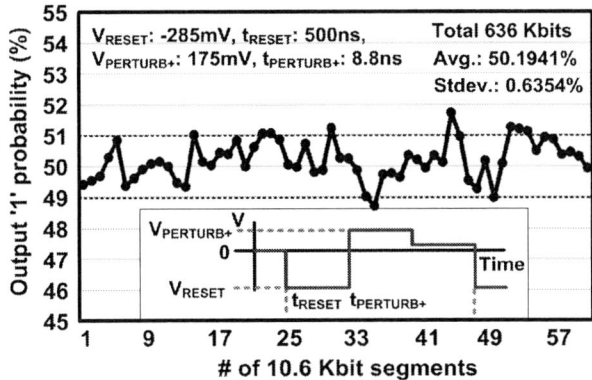

Fig. 9. Measured output '1' probability of each 10.6 Kbit segment for the unconditional reset scheme.

Unconditional reset scheme, # of segments: 55
Pass if P-value$_T$ (χ^2) > 0.0001 and Proportion > 0.9454

	Test	P-value$_T$ (χ^2)		Proportion		Pass/Fail	
1	Frequency	0.005358		0.9272		Fail	
2	Block frequency	0.637119		0.9818		Pass	
3	Cumulative Sums	0.080519 (Forward)	0.080519 (Reverse)	0.9090	0.9272	Fail	Fail
4	Runs	0.401199		1.0000		Pass	
5	Longest-Run-of-Ones	0.025193		1.0000		Pass	
6	Rank	0.266984		1.0000		Pass	
7	FFT	0.897763		1.0000		Pass	
8	Non-overlapping Template Matching	All sub-test: Pass					
9	Serial	0.224821 (P-value$_1$)	0.554420 (P-value$_2$)	0.9818	0.9636	Pass	Pass
10	Approximate Entropy	0.595549		1.0000		Pass	

Fig. 10. NIST randomness test result of 636 Kbits from the conventional unconditional reset scheme.

After Von Neumann Correction
Unconditional reset scheme, # of segments: 55
Pass if P-value$_T$ (χ^2) > 0.0001 and Proportion > 0.9454

	Test	P-value$_T$ (χ^2)		Proportion		Pass/Fail	
1	Frequency	0.181557		1.0000		Pass	
2	Block frequency	0.062821		1.0000		Pass	
3	Cumulative Sums	0.554420 (Forward)	0.055361 (Reverse)	1.0000	1.0000	Pass	Pass
4	Runs	0.514124		0.9818		Pass	
5	Longest-Run-of-Ones	0.145326		1.0000		Pass	
6	Rank	0.823537		1.0000		Pass	
7	FFT	0.000347		1.0000		Pass	
8	Non-overlapping Template Matching	All sub-test: Pass					
9	Serial	0.401199 (P-value$_1$)	0.366918 (P-value$_2$)	0.9818	0.9818	Pass	Pass
10	Approximate Entropy	0.924076		1.0000		Pass	

Fig. 11. NIST randomness test results for the unconditional reset scheme after applying the Von Neumann correction (bit efficiency: 25%).

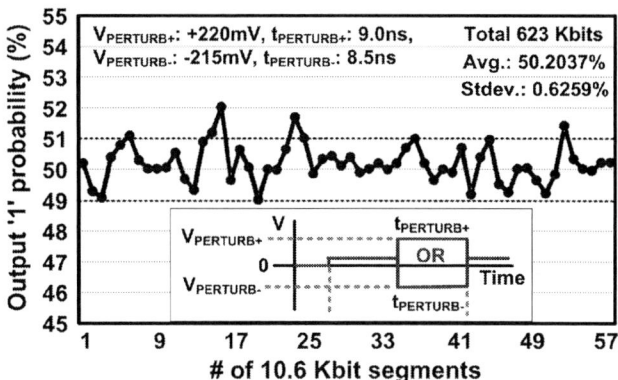

Fig. 12. Measured output '1' probability of each 10.6 Kbit segment for the proposed conditional perturb scheme.

Conditional perturb scheme, # of segments: 55
Pass if P-value$_T$ (χ^2) > 0.0001 and Proportion > 0.9454

	Test	P-value$_T$ (χ^2)		Proportion		Pass/Fail	
1	Frequency	0.000831		0.9272		Fail	
2	Block frequency	0.266918		0.9464		Pass	
3	Cumulative Sums	0.037566 (Forward)	0.021999 (Reverse)	0.9272	0.9272	Fail	Fail
4	Runs	0.437274		1.0000		Pass	
5	Longest-Run-of-Ones	0.474986		0.9818		Pass	
6	Rank	0.085953		1.0000		Pass	
7	FFT	0.437274		1.0000		Pass	
8	Non-overlapping Template Matching	All sub-test: Pass					
9	Serial	0.637119 (P-value$_1$)	0.202268 (P-value$_2$)	0.9818	0.9818	Pass	Pass
10	Approximate Entropy	0.115387		1.0000		Pass	

Fig. 13. NIST randomness test result of 623 Kbits from the proposed conditional perturb scheme.

After Von Neumann Correction
Conditional perturb scheme, # of segments: 55
Pass if P-value$_T$ (χ^2) > 0.0001 and Proportion > 0.9454

	Test	P-value$_T$ (χ^2)		Proportion		Pass/Fail	
1	Frequency	0.334538		1.0000		Pass	
2	Block frequency	0.637119		1.0000		Pass	
3	Cumulative Sums	0.249284 (Forward)	0.202268 (Reverse)	1.0000	1.0000	Pass	Pass
4	Runs	0.349121		0.9818		Pass	
5	Longest-Run-of-Ones	0.200936		1.0000		Pass	
6	Rank	0.597670		1.0000		Pass	
7	FFT	0.328827		0.9636		Pass	
8	Non-overlapping Template Matching	All sub-test: Pass					
9	Serial	0.798139 (P-value$_1$)	0.924076 (P-value$_2$)	1.0000	1.0000	Pass	Pass
10	Approximate Entropy	0.080519		0.9636		Pass	

Fig. 14. NIST randomness test results for the conditional perturb scheme after applying the Von Neumann correction (bit efficiency: 25%).

measurement setup in Fig. 8 consists of high speed pulse generators for providing V$_{PERTURB+}$, V$_{RESET}$ or V$_{PERTURB-}$ pulses, a data acquisition board for generating the read voltage pulse and sampling the MTJ state using the same signal line, a power combiner, and a bias tee. A software program controls the pulse generators and the data acquisition board generates the timing sequences described earlier. Fig. 9 shows the measured probability of each 10.6 Kbit segment for the conventional conditional reset scheme. Note that the output probability typically needs to stay within 50±1% to pass the NIST frequency test [6]. A small number of segments fail to

meet this criterion and consequently, the output data fails to pass the frequency and cumulative sums tests as shown in Fig. 10. Von Neumann's algorithm can be applied to remove skew in the TRNG output and pass all 10 NIST tests [1]. However, the bit efficiency (=fraction of useable bits) drops from 100% to 25%. The measurement data from the proposed scheme in Figs. 12, 13, and 14, indicate a similar level of randomness as compared to the conventional scheme.

978-1-4799-8002-4/14 $31.00 © 2014 IEEE

Real-Time Output Probability Tracking

To achieve good randomness without incurring any bit efficiency loss, a real-time output probability tracking scheme that actively unbiases the output bit stream is proposed. The circuit diagram is shown in Fig. 15 where two 10 bit counters are used to calculate the output probability of each consecutive 1 Kbit segment. $t_{PERTURB-}$ is adjusted according to the digital comparator outcome while all other parameters such as $V_{PERTURB+}$, $t_{PERTURB+}$ and $V_{PERTURB-}$ are kept constant for a simple single-parameter feedback control. Note that a segment size much shorter than 1 Kbit makes the output probability fluctuate while a segment size much longer than 1 Kbit increases the locking time unnecessarily. The real-time tracking scheme was implemented in software and the measurement setup in Fig. 8 was used to verify the concept using the fabricated MTJ device. The conditional perturb scheme was used for all measurements involving real-time tracking. Measured probability of each 1 Kbit segment and the corresponding $t_{PERTURB-}$ are illustrated in Fig. 16. The minimum $t_{PERTURB-}$ step was set as 0.05ns. After an initial locking period of 65 Kbits, the output data passed all NIST randomness tests while maintaining a 100% bit efficiency (Fig. 17). Finally, we show a conceptual diagram of a TRNG implemented using an existing STT-MRAM array (Fig. 18), which could potentially allow massive generation of random numbers with negligible circuit overhead.

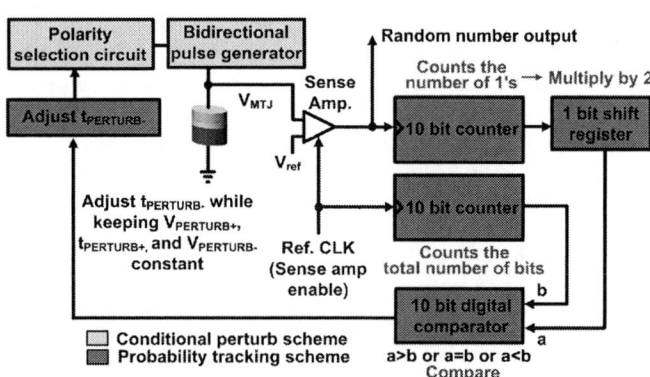

Fig. 15. Proposed MTJ-based TRNG with conditional perturb and real-time output probability tracking. The two techniques were implemented in software and experimentally verified using a real MTJ device.

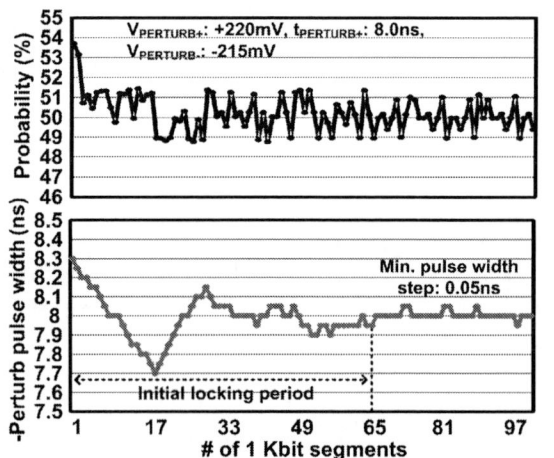

Fig. 16. Measured output '1' probability and -perturb pulse width for each 1 Kbit segment with the proposed real-time output probability tracking scheme.

Raw data after probability tracking
Conditional perturb scheme, # of segments: 55
Pass if P-value$_T$ (χ^2) > 0.0001 and Proportion > 0.9454

	Test	P-value$_T$ (χ^2)		Proportion		Pass/Fail	
1	Frequency	0.102947		1.0000		Pass	
2	Block frequency	0.019203		0.9636		Pass	
3	Cumulative Sums	0.012910 (Forward)	0.366928 (Reverse)	1.0000	1.0000	Pass	Pass
4	Runs	0.582910		1.0000		Pass	
5	Longest-Run-of-Ones	0.201928		1.0000		Pass	
6	Rank	0.693028		0.9818		Pass	
7	FFT	0.381291		1.0000		Pass	
8	Non-overlapping Template Matching	All sub-test: Pass					
9	Serial	0.283910 (P-value$_1$)	0.683921 (P-value$_2$)	0.9818	0.9636	Pass	Pass
10	Approximate Entropy	0.334538		1.0000		Pass	

Fig. 17. NIST randomness test results for the proposed MTJ-based TRNG with conditional perturb and real-time output probability tracking. Note that output bits after the initial locking period are used for the randomness test.

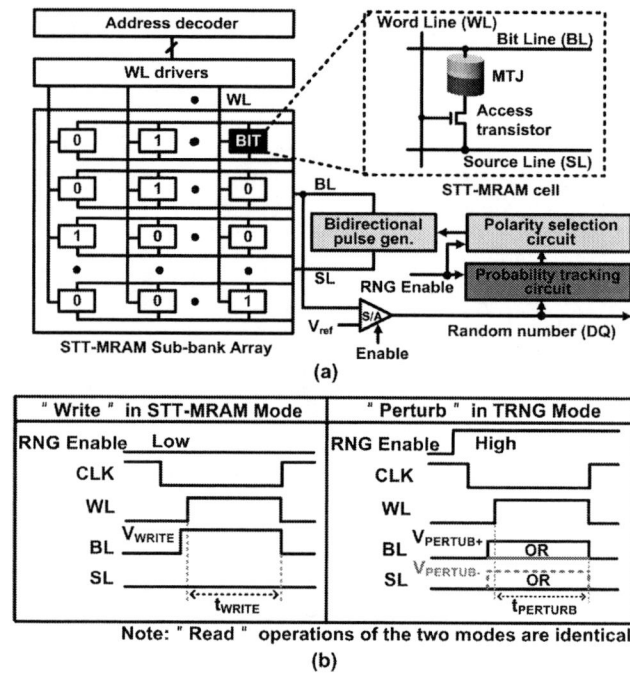

Fig. 18. (a) Conceptual diagram of a TRNG circuit implemented using an existing STT-MRAM array (b) "Write" and "Perturb" timing diagrams of STT-MRAM and TRNG modes.

Acknowledgements

This work was supported in part by C-SPIN, one of six centers of STARnet, a Semiconductor Research Corporation program, sponsored by MARCO and DARPA. The authors would also like to thank Vijay Reddy at TI for technical discussion.

References

[1] K. Yang, et al., ISSCC, pp. 280, 2014.
[2] H. Zhao, et al., JAP, pp. 07C720, 2011.
[3] S. Yuasa, et al., IEDM, pp. 3.1.1, 2013.
[4] C. Yoshida, et al., IRPS, pp. 139, 2009.
[5] W. R. Hunter, IRPS, pp. 72, 1999.
[6] A. Rukhin, et al., NIST Pub 800-22, 2010.

DTMOS Mode as an Effective Solution of RTN Suppression for Robust Device/Circuit Co-Design

Shaofeng Guo[1], Ru Huang[1,3,*], Peng Hao[1], Mulong Luo[1], Pengpeng Ren[1], Jianping Wang[2], Weihai Bu[2], Jingang Wu[2], Waisum Wong[2,3], Scott Yu[2,3], Hanming Wu[2,3], Shiuh-Wuu Lee[2,3], Runsheng Wang[1], Yangyuan Wang[1,3]

[1] Key Laboratory of Microelectronic Devices and Circuits (MOE), Institute of Microelectronics, Peking University, Beijing 100871, China
[2] Semiconductor Manufacturing International Corporation (SMIC), Shanghai 201203 and Beijing 100176, China
[3] Innovation Center for MicroNanoelectronics and Integrated System, Beijing 100871, China
*Email: ruhuang@pku.edu.cn

Abstract

In this paper, using DTMOS as an effective solution of RTN suppression without device/circuit performance penalty is proposed and demonstrated for the first time, with experimental verification and circuit analysis. The experiments show that RTN amplitude is greatly reduced in DTMOS mode, which is even better than the body-biasing technique of FBB, due to the efficient dynamic modulation mechanism. Circuit stability and performance degradation induced by RTN are much improved in the design using DTMOS. New characteristics of RTN physics in DTMOS are also observed and studied in detail. The results are helpful to the robust and reliable device/circuit co-design in future nano-CMOS technology.

Introduction

Recently, the random telegraph noise (RTN) has been drawing much attention as a significant source of dynamic variability [1-6]. Because it can lead to large V_{TH} (or I_D) degradation and cause transient probabilistic failure of digital circuits due to its stochastic nature as we have demonstrated [5-7], RTN will narrow the tight design margin in nano-CMOS technology. Several attempts through process adjustment are reported (e.g., buried channel), but with device performance degradation and/or ineffective for the newborn-trap induced RTN during circuit aging. Therefore, effective RTN suppression technique without performance penalty emerges as an urgent demand for device/circuit co-design.

In this paper, we propose, for the first time, to use the dynamic threshold MOS (DTMOS) operation mode as an effective method to suppress RTN without intrinsic device/circuit performance loss. The experimental results show that RTN amplitude is greatly reduced in DTMOS mode, and new characteristics of RTN time constants are found in DTMOS. The effectiveness of RTN suppression in DTMOS is also demonstrated in circuit analysis in terms of performance degradation and transient variation.

Strategy for RTN Suppression at Device/Circuit Level

Generally for variability reduction, apart from process adjustment and system redundancy, body-biasing technique (FBB/RBB) has been widely used at the device/circuit

level [8]. Recently, it is demonstrated that FBB is helpful for reducing flicker noise [9]. In this work we further examine whether FBB is also suitable to RTN suppression for the first time. Meanwhile, considering that DTMOS is more efficient in dynamic modulation, we propose to use it as a more effective way for device/circuit co-design against RTN. Since no study on the RTN characteristics of DTMOS has been reported, the new features of RTN physics will be investigated by experiments in this work as well.

Experimental Demonstration

A. Devices and RTN characterization. The devices used in this work are with 28nm high-k/metal-gate (HKMG) technology. Three types of device operation mode and the corresponding measured I_D-V_G curves are shown in Fig. 1, including MOS, FBB MOS and DTMOS modes. It is worth noting that, both the RTN amplitude (ΔV_{TH} or $\Delta I_D/I_D$) and time constants are crucial for accurate evaluating the impacts of RTN in circuits. The circuit static performance degradation depends on the RTN amplitude, while the time constants determine the dynamic variability and transient failure probability in digital circuits. Fig. 2 shows the typical experimental results of RTN in MOS, FBB and DTMOS modes, which have evident reduction in the aspect of amplitude. The deviation of time constants can also be found in Fig. 2(e)~(h).

B. RTN amplitudes reduction in DTMOS mode. Fig. 3 shows the extracted RTN amplitudes at different gate voltages in MOS mode, FBB mode and DTMOS mode. The RTN reduction in both FBB and DTMOS modes can be observed, but DTMOS mode has the largest amount of reduction. Fig. 4 shows that a maximum RTN amplitude reduction of 58% can be obtained in DTMOS mode in this work. The similar phenomenon have also been observed with the SiON gate dielectric devices (not shown here).

To get a clear understanding of the RTN reduction mechanisms in FBB and DTMOS modes, it is useful to plot $\Delta I_D/I_D$ with gate overdrive (V_G-V_{TH}), for normalizing the effect of body biasing. As shown in Fig. 5(a), the $\Delta I_D/I_D$ curves of FBB mode converge with the MOS curve, which indicates that the RTN reduction in FBB mode results from

978-1-4799-8002-4/14 $31.00 © 2014 IEEE

the body-bias induced V_{TH} shift only. However, the curve of DTMOS mode is still far below the others as expected, because of the different operation mechanism. For a given oxide trap, its RTN amplitude is mainly determined by the inversion charge (see equation 4), which is proportional to $(V_{GS}-V_{TH})$ and the gate capacitance. The equivalent gate capacitance in DTMOS is larger and still increases with increasing V_G after strong inversion [10], rather than keeping constant as in MOS and FBB modes. Large inversion charge in DTMOS leads to reduced RTN amplitude. Alternatively, it can also be interpreted as follows. Two boundary values of the V_{TH} shift are marked as ΔV_{TH1} and ΔV_{TH2} ($\Delta V_{TH1}<\Delta V_{TH2}$) in Fig. 5(b). It can be found that the left side of the V_{TH1}-shifted curve and the right side of the V_{TH2}-shifted curve are merged into the MOS curve [Fig. 5(c)], which indicates that the RTN reduction results from the dynamic V_{TH} modulation effect of DTMOS.

C. New observations on time constants in DTMOS mode.

Fig. 6 shows the extracted capture time constant (τ_C) and emission time constant (τ_E) of RTN, and reveal the same V_G dependence of τ_C in both MOS and DTMOS mode. However, a very different V_G dependence of τ_E is observed in DTMOS mode. As shown in Fig. 8&9, within the RTN test window, the opposite trend of the V_G dependence of surface potential (and thus $E_{C\text{-}eff}$) and E_T contribute to the non-monotonic trend of τ_E (see Eq. 1). Besides τ_C and τ_E themselves, it is even more important to focus on τ_C/τ_E, which reflects the trap spatial position (x_T/T_{OX}) by equation (2) and also indirectly determines the trap occupancy rate [5-7]. For DTMOS mode, τ_C/τ_E in Fig. 7(a)(b)(d) have non-monotonic V_G dependence, and the corner points exhibit a linear relationship with x_T/T_{OX} (Fig. 10). An effective surface Fermi level $E_{F\text{-}eff}$ is introduced here in DTMOS mode as shown in Fig. 8 due to the splitting of Fermi energy levels at the bulk and surface, and it decreases as V_G increasing in nMOS devices, with more rapid speed for larger V_G. Therefore, with the modified model in equation (3), the new observations of the V_G dependence of τ_C/τ_E in DTMOS can be well explained, and can also be predicted for the range out of the RTN test windows as shown in Fig. 11.

Fig. 12&13 shows the statistical V_G dependence of τ_C/τ_E and the relationship between the Lorentzian power spectral density (PSD) of single RTN and τ_C/τ_E. According to the results above, the distribution of Lorentzian PSD of multiple traps can be obtained as shown in Fig. 14, and it can be found that the Lorentzian PSD of most of the traps in DTMOS mode are smaller than MOS mode, which suggests DTMOS mode as an effective solution against RTN.

Effectiveness of RTN Suppression in Circuits Design

Based on the experimental results, transient circuit simulation of RTN is performed on our recently developed platform [7]. For the DTMOS mode, τ_C/τ_E are categorized into two types based on the comparison results between MOS and DTMOS mode (Fig. 15). Fig. 16 shows the ring oscillator (RO) adopted for simulation as the representative digital circuit and the corresponding eye diagram with the RTN effects. The V_G dependence of amplitude and time constants are included with the models in Table I.

Fig. 17&18 show the jitter and the mean shift of center frequency in a single RO operating at three modes, indicating the DTMOS mode as the best solution against RTN in the aspects of both variation and degradation. The jitters of RO under different RTN conditions are compared in Fig. 19, which suggests it important to consider the new properties of time constants in DTMOS mode. It shows that the suppression of type 2 RTN in DTMOS is not as good as type 1 (but still better than FBB). But fortunately, type 2 RTN is actually less than 20% in reality (see Fig. 12). As for the realistic applications, it is important to pay attention to the VDD dependence of the static degradation and the dynamic variability. Fig. 20 shows the jitters under different VDD in three modes, and the $\Delta f/f$ and σ(RO jitter) are extracted as show in Fig. 21, which suggest that for all these modes, the degradation and variability induced by single RTN increase with reduced VDD, and DTMOS is still the best mode. Therefore, as for low-power robust design against RTN, DTMOS is the best choice.

Summary

An effective method to suppress RTN without compromising intrinsic performance is proposed in this paper for the first time, which is demonstrated with the experiments and circuit analysis. It is found that using DTMOS mode can greatly suppress RTN impacts on device and circuits, and is even better than FBB technique. The results are useful for the robust and reliable design against RTN in the future.

Acknowledgement

This work was partly supported by the 973 Projects (2011CBA00601), NSFC (61106085), and National S&T Major Project (2009ZX02035-001).

References

[1] T. Matsumoto, et al., *IEDM*, 2012, pp.581-584. [2] H. Miki, et al., *IEDM*, 2012, pp.450-453. [3] N. Tega, et al., *VLSI*, 2009, pp.50-51. [4] S. Realov, et al., *IEDM*, 2010, pp.624-627. [5] J. Zou, et al., *VLSI*, 2012, pp.139-140. [6] J. Zou, et al., *VLSI*, 2013, pp.186-187. [7] R. Wang, et al., *IEDM*, 2013, pp.834-837. [8] S. Narendra, et al., *IEEE JSSC*, vol. 38, no. 5, pp. 696-701, 2003. [9] Arnoud P. van der Wel, et al., *IEEE JSSC*, vol. 42, no. 3, pp.540-550, 2007. [10] R. Huang, et al, *Solid-State Electronics*, vol. 47, no. 8, 2003.

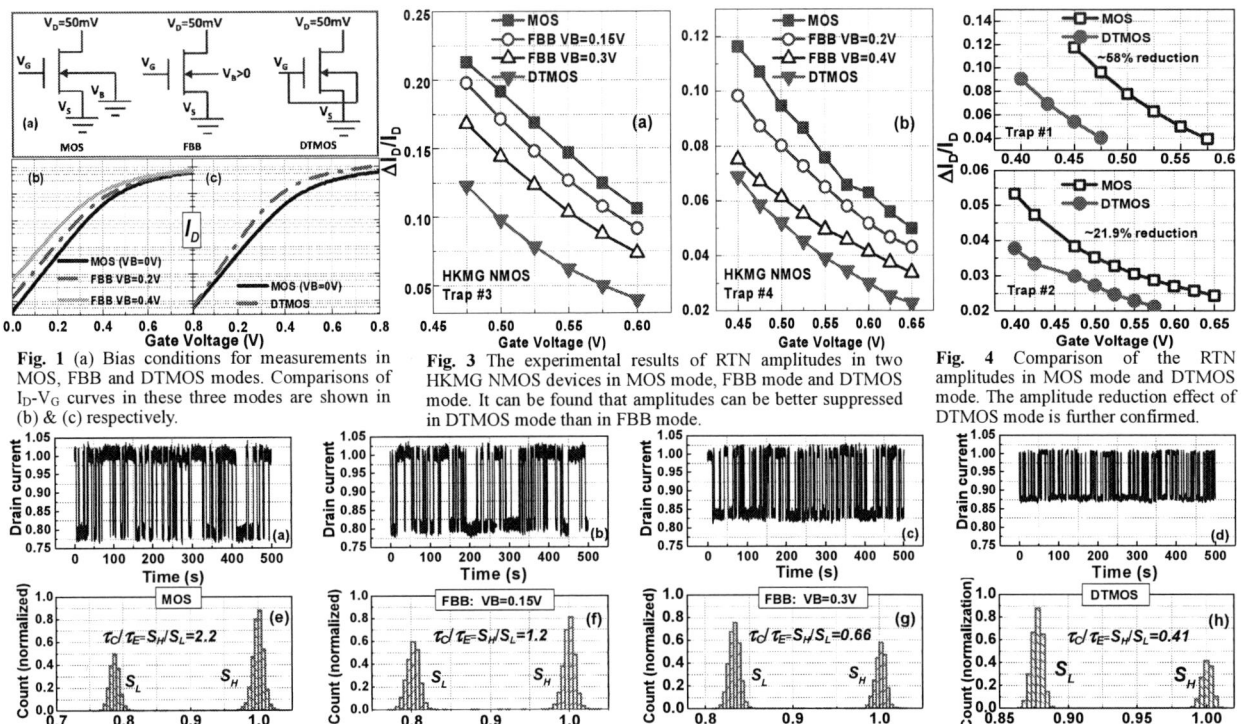

Fig. 1 (a) Bias conditions for measurements in MOS, FBB and DTMOS modes. Comparisons of I_D-V_G curves in these three modes are shown in (b) & (c) respectively.

Fig. 3 The experimental results of RTN amplitudes in two HKMG NMOS devices in MOS mode, FBB mode and DTMOS mode. It can be found that amplitudes can be better suppressed in DTMOS mode than in FBB mode.

Fig. 4 Comparison of the RTN amplitudes in MOS mode and DTMOS mode. The amplitude reduction effect of DTMOS mode is further confirmed.

Fig. 2 (a) (b) (c) (d) Raw data of the measured RTN induced by the same trap in MOS, FBB and DTMOS modes. Note: the drain currents are normalized to the high current states. (e) (f) (g) (h) Histogram of the measured drain currents. The ratio of the percentages in high and low states corresponds to the τ_c/τ_E as shown.

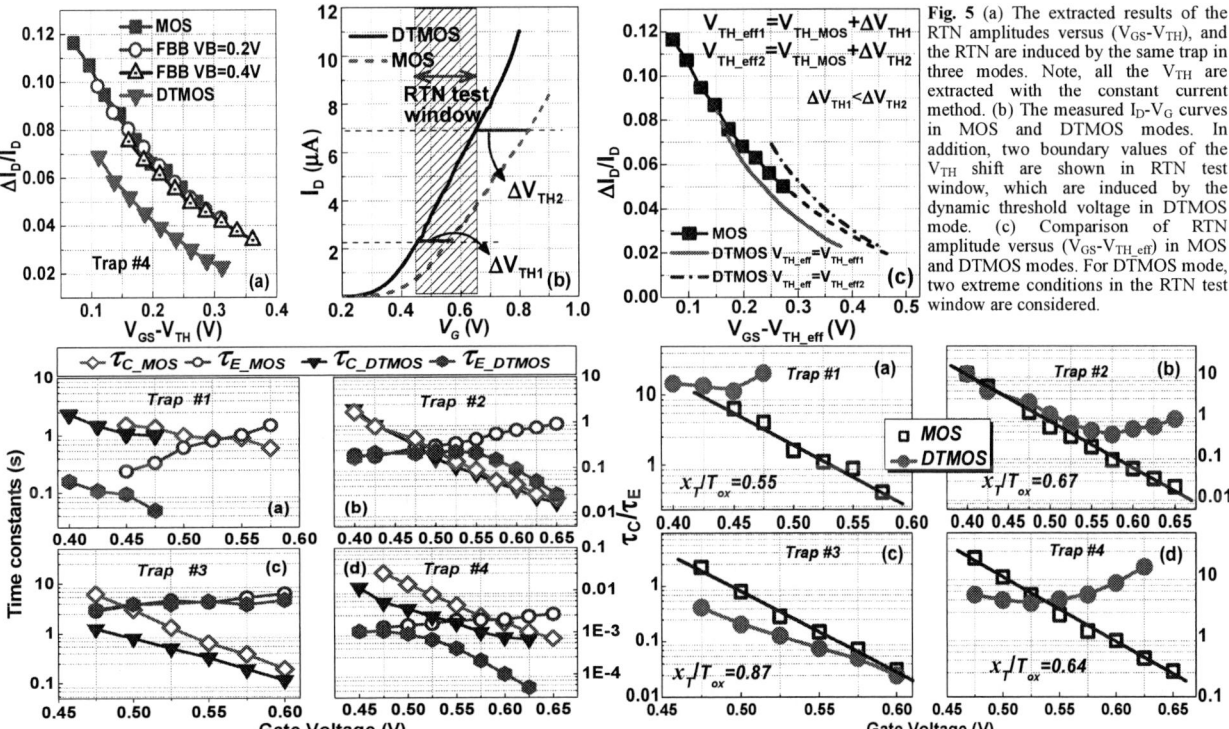

Fig. 5 (a) The extracted results of the RTN amplitudes versus (V_{GS}-V_{TH}), and the RTN are induced by the same trap in three modes. Note, all the V_{TH} are extracted with the constant current method. (b) The measured I_D-V_G curves in MOS and DTMOS modes. In addition, two boundary values of the V_{TH} shift are shown in RTN test window, which are induced by the dynamic threshold voltage in DTMOS mode. (c) Comparison of RTN amplitude versus (V_{GS}-V_{TH_eff}) in MOS and DTMOS modes. For DTMOS mode, two extreme conditions in the RTN test window are considered.

Fig. 6 The extracted time constants of RTN for different traps in MOS and DTMOS modes. An observed V_G dependence of τ_E in DTMOS mode is different from that in MOS mode.

Fig. 7 Comparison of the tendency of τ_c/τ_E changing with gate voltage between MOS and DTMOS modes. The positions of traps in the oxide with respect to the gate dielectric/channel interface are extracted according to the equation

Table I Some key equations used in analyzing the time constants and amplitude of RTN

(1) RTN capture & emission time constants :	(2) For MOS:

(1) RTN capture & emission time constants :

$$\tau_c = \frac{1}{\sigma v_{th} n} \qquad \tau_e = \frac{1}{\sigma v_{th} N_c \exp[(E_T - E_{C\_eff})/kT]}$$

$$\frac{\tau_c}{\tau_e} = \exp(\frac{E_T - E_{F\_eff}}{kT}) \qquad E_T = E_{T0} - q\frac{x_T}{T_{ox}}(V_{gs} - V_{FB} - \varphi_s)$$

$$ln\frac{\tau_c}{\tau_e} = -\frac{1}{kT}\begin{bmatrix}(E_{Cox} - E_{T0}) - (E_{C\_eff} - E_{F\_eff}) - \Phi_0 \\ + q\varphi_s + q\frac{x_T}{T_{ox}}(V_{gs} - V_{FB} - \varphi_s)\end{bmatrix}$$

(2) For MOS:
$$\frac{d(\ln \tau_c/\tau_e)}{dV_{gs}} = -\frac{q}{kT}\frac{x_T}{T_{ox}} \qquad when\ V_{gs} - V_{TH} > 0$$

(3) For DTMOS :
$$\frac{d(\ln \tau_c/\tau_e)}{dV_{gs}} = -\frac{1}{kT}(\frac{qx_T}{T_{ox}} - \frac{d(E_{C_{eff}} - E_{F_{eff}})}{dV_{gs}}) + q(1 - \frac{x_T}{T_{ox}})\frac{d\varphi_s}{dV_g}$$

(4) RTN amplitude dependence:
$$\frac{\Delta I_D}{I_D} = \frac{4r^2}{WL - 2r(L - 2r)} \qquad r = \sqrt{\frac{q}{Q}\frac{d_1}{d_1 + d_2}}$$
$$Q = C_{OX}(V_{GS} - V_{FB} - \varphi_s) - qN_b x_{dep}$$

978-1-4799-8002-4/14 $31.00 © 2014 IEEE

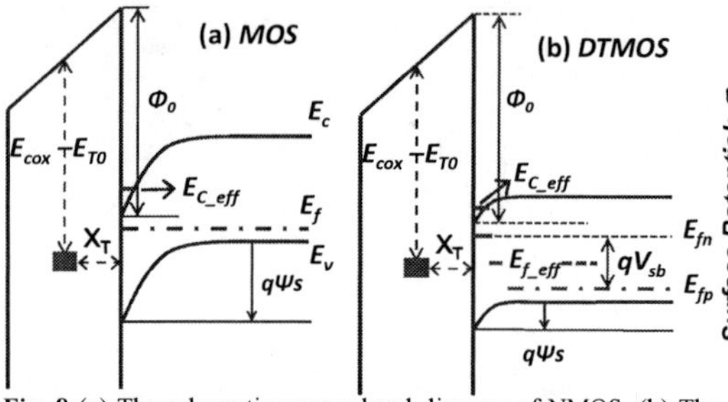

Fig. 8 (a) The schematic energy band diagram of NMOS. (b) The schematic energy band diagram of a n-type DTMOS. It is worth noting that E_{C_eff} is introduced due to the quantum confinement effect, and the effective Fermi energy level E_{f_eff} decreases with V_G increasing as a result of the increasing split of quasi-Fermi energy levels between the bulk and surface.

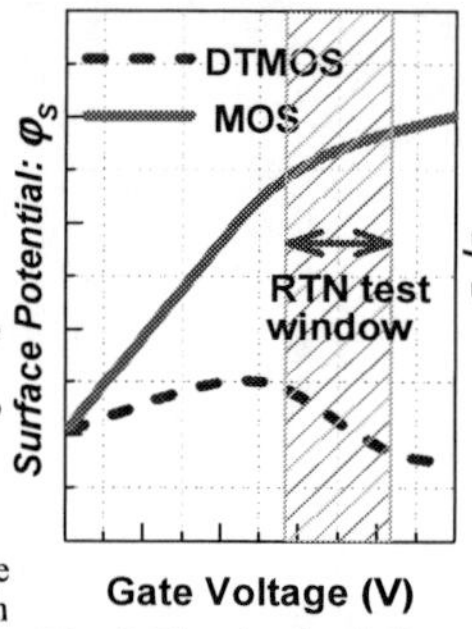

Fig. 9 The simulated V_G dependence of the surface potential in MOS mode and DTMOS

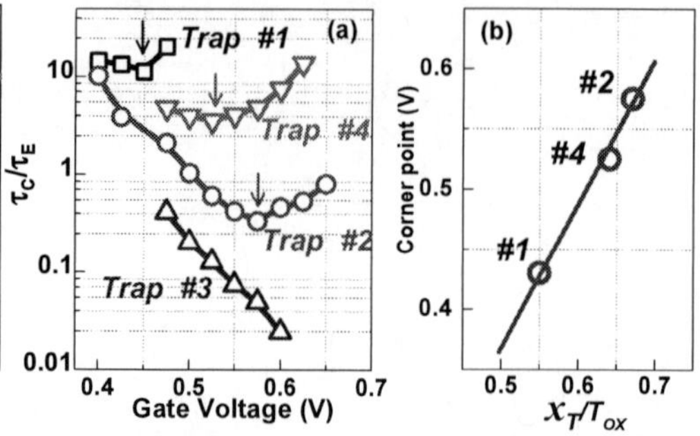

Fig. 10 (a) Comparison of the V_G dependence of τ_C/τ_E in each trap. Note that, the corner points are different, and it can be observed that the positions of the corner point have a linear relationship with the corresponding x_T/T_{OX}, as shown in (b).

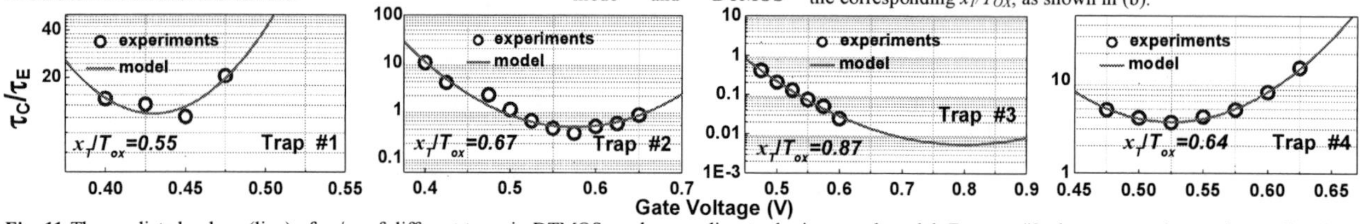

Fig. 11 The predicted values (line) of τ_C/τ_E of different traps in DTMOS mode according to the improved model. For trap #3, the corner point can be predicted at about 0.8V. With this method, more comparisons can be done for traps located at different positions in gate dielectric between MOS mode and DTMOS mode.

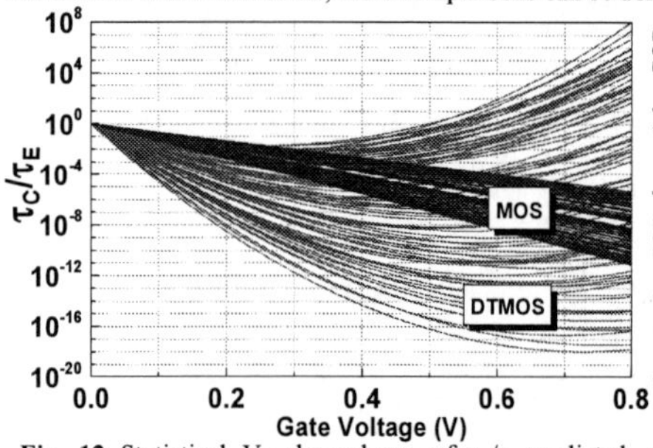

Fig. 12 Statistical V_G dependence of τ_C/τ_E predicted according to equation (2) & (3) in MOS mode and DTMOS modes. Note that, τ_C/τ_E has been normalized to 1 at $V_g=0$V.

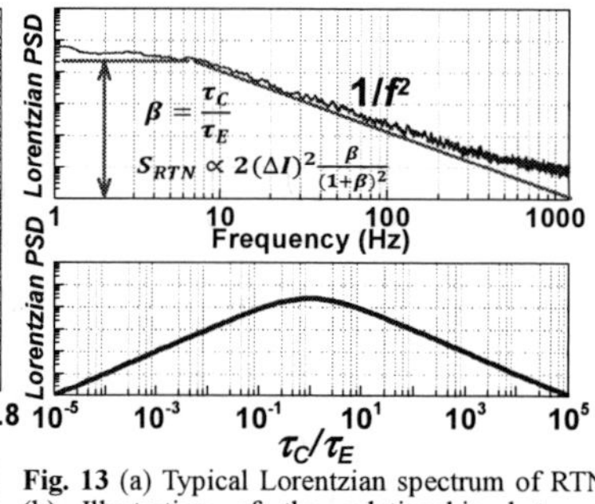

Fig. 13 (a) Typical Lorentzian spectrum of RTN (b) Illustration of the relationship between Lorentzian PSD and τ_C/τ_E.

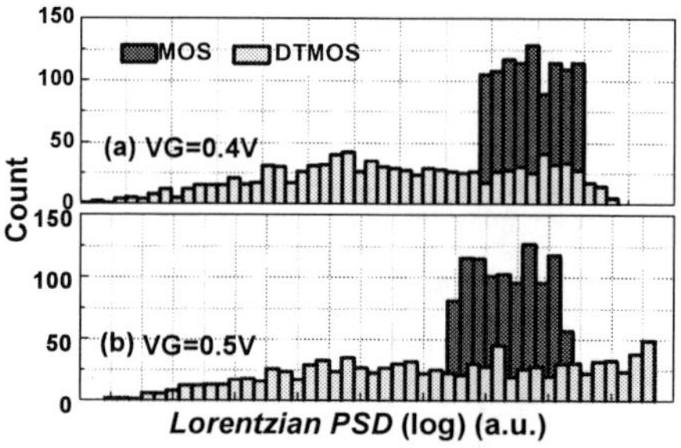

Fig. 14 Comparison of the distributions of Lorentzian PSD induced by different traps in MOS mode and DTMOS mode.

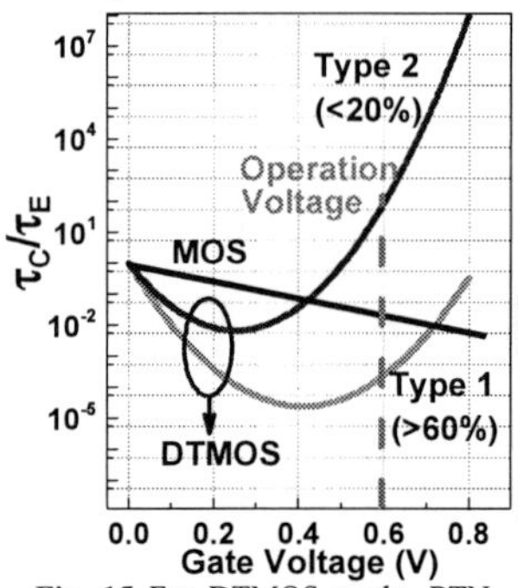

Fig. 15 For DTMOS mode, RTN are categorized into two types according to the trend of τ_C/τ_E changing with gate voltage.

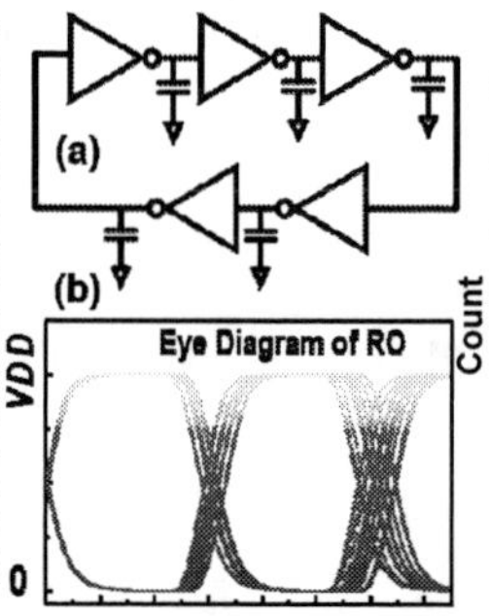

Fig. 16 (a) The 5-stage RO adopted for simulation in this work. (b) Typical eye diagram of RO with the effects of RTN.

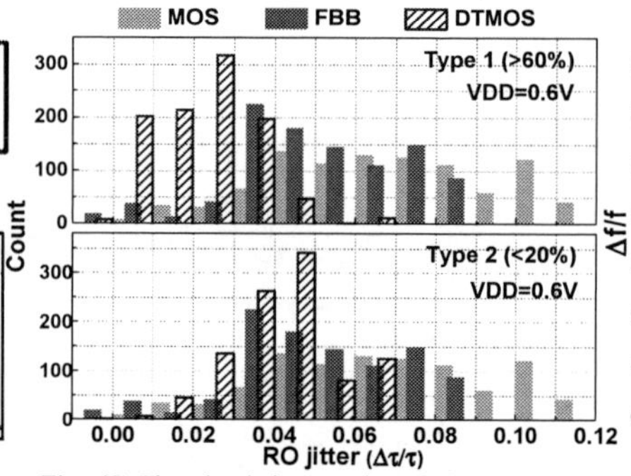

Fig. 17 The simulation results of the RO jitter distributions with the impacts of (a) type 1 and (b) type 2 traps in DTMOS mode. All of them are compared with the jitter distributions in the MOS mode and FBB mode.

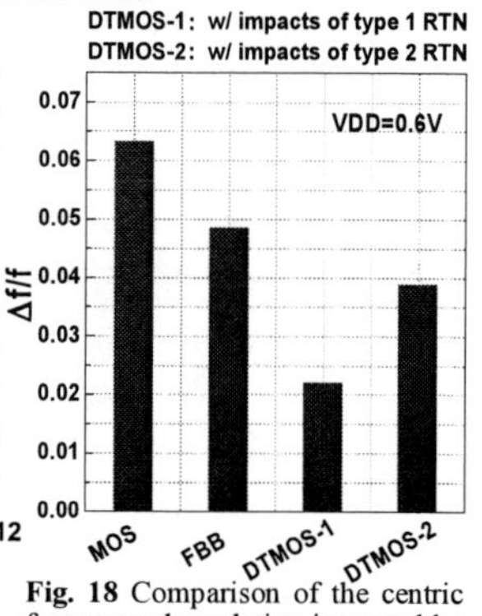

Fig. 18 Comparison of the centric frequency degradation impacted by RTN in different modes, including MOS mode, FBB mode and DTMOS mode.

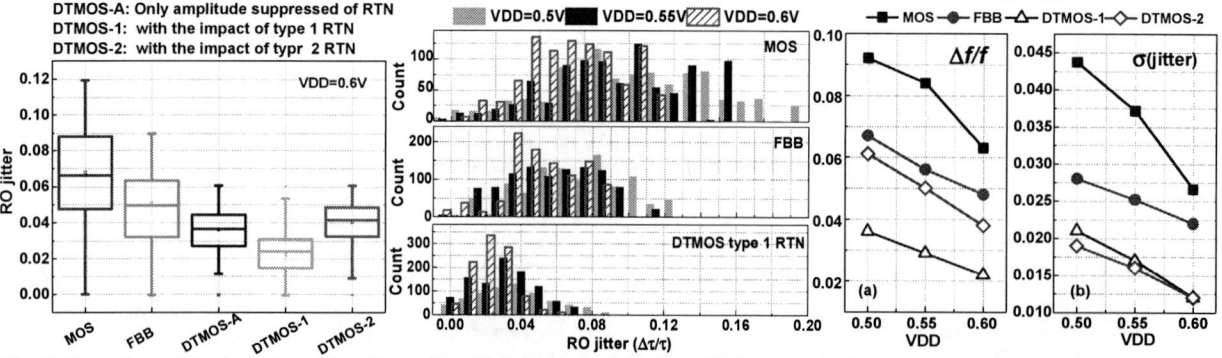

Fig. 19 Comparison of the simulation results of the jitter distributions in different modes.

Fig. 20 Comparison of the jitter distributions under different VDD in these three modes.

Fig. 21 The simulation results of the VDD dependence of $\Delta f/f$ and σ(jitter) in different modes.

978-1-4799-8002-4/14 $31.00 © 2014 IEEE

Poly Pitch and Standard Cell Co-Optimization below 28nm
Marlin Frederick, Jr.
ARM INC
3711 S Mopac Expressway, Bldg 1, Suite 400, Austin, TX, 78746 USA
Phone: +1-512-314-1017, Email: marlin.frederick@arm.com

Abstract

In sub-28nm technologies, the scaling of poly pitch while beneficial for area typically has a negative impact on device performance. The primary limitation is the non-scaling physical channel length and the device level parasitic impact on effective device performance. Therefore, it is important to scale the poly pitch in line with the maximum routable pin density supported by a process to achieve optimal power, performance, and cost (area).

Introduction

Process scaling is no longer following a path of traditional pitch or even device type scaling. With interconnect density and performance scaling differently from device density and performance and cost scaling as complex function of different processing options, co-optimizing a process technology and a standard cell platform is critical to achieve optimal power, performance, and area/cost. This is a complex multi-variable problem. To make matters worse, rarely during the optimization phase is it known in advance the specifics of SoC designs that will later prove to be the majority of the process volume. Given the limitations of space, this paper will focus only on poly pitch scaling impact on block level power, performance, and area for an ARM Cortex-A9 processor. Of course proper technology and standard cell co-optimization considers the impact of all layers for a wider range of logic types.

A quick survey of 28nm generation SoC designs reveals that standard cells typically consume 60-70% of the die area, are the primary or only component of the most timing critical paths, and typically consume more than 75% of the chip power. What is less well known is most yields are now limited by standard cell printability concerns. This implies that SOC scaling is dominated by standard cell scaling and needs to be the primary focus for the process definition moving forward.

Background

To better frame the problem, one must understand what limits the density of standard cell regions. There are three possibilities, cell area, interconnect density, or a combination/interaction of the two. Before one can make this determination, a few simplifying assumptions need to be made as the number of degrees of freedom are very large making the design space expensive to search by brute force.

The first assumption is moving to taller cell heights is not an option to try to relieve local routing congestion. Fig. 1 shows

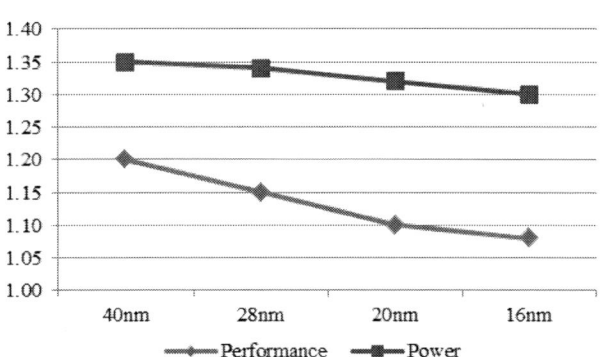

Figure 1: Relative power and performance of a 12 versus 9 track cell architecture from 40nm to 16nm for an ARM Cortex-A9.

the historical trend of power and performance as a function of cell height. Therefore, benchmarking will focus on a 9-track cell height as a baseline even though shorter cell heights are definitely possible and critical to consider.

The second assumption is there are enough metal layers of sufficient density to insure global routing congestion is not the dominate factor in determining block density. Why? In modern SoC design, this assumption is usually met because of the economics associated with adding an additional metal layer virtually always improves density more than the cost increases resulting in an overall lower per die cost.

Just because block density isn't being limited by global route congestion doesn't mean interconnect isn't the limiting factor. The third consideration is local route congestion is a potential limiting factor for density. While many forms of local route congestion are possible the dominate consideration is the limitation of the lower metal layers to access pins on cells. There is a pin density limit beyond which the lower levels of interconnect cannot support and therefore the design becomes local route congested. To truly understand local routing congestion requires extensive benchmarking of real designs for many different target objectives.

To help clearly define this issue a simple model is provided illustrate the problem. Local route congestion is dominated by many interacting factors. Let's assume everything not explicitly defined in the equation in Fig. 2 can be modeled as a maximum degree of M2 utilization, $M2_{UTIL}$. The main factors in $M2_{UTIL}$ are spacing of different net different layer vias, pin spreading both vertically and horizontally within a cell, and the ability of the router to optimize track assignments simultaneously on multiple layers. A good starting point for inexperienced standard cell designers is to

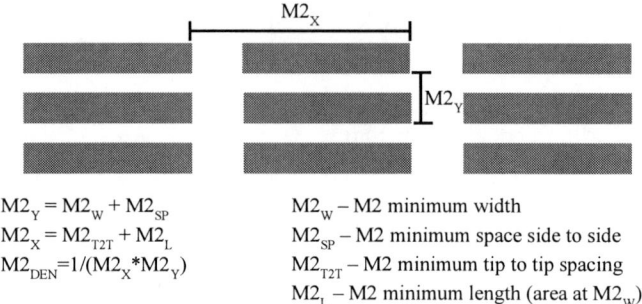

$M2_Y = M2_W + M2_{SP}$ $M2_W$ – M2 minimum width

$M2_X = M2_{T2T} + M2_L$ $M2_{SP}$ – M2 minimum space side to side

$M2_{DEN} = 1/(M2_X * M2_Y)$ $M2_{T2T}$ – M2 minimum tip to tip spacing

                                      $M2_L$ – M2 minimum length (area at $M2_W$)

Maximum Usable M2 Density = $M2_{UTIL} * M2_{DEN}$

Figure 2: Simple model for estimating maximum usable M2 density.

adjust $M2_{UTIL}$ values measured in previous generation designs by known relative differences in the next generation.

This equation can be compared to the weighted average pin density of cells to gain an understanding of whether cell area or local congestion is likely to drive block area. This is critical to narrow the scope because benchmarking requires development of multiple libraries and routing technology files, plus thousands of synthesis, place, and route trials per combination to understand the achievable density and specifically what is limiting for that combination.

Weighted average pin density is a strong function of cell drive strength, cell height, and topology. The histogram in Fig. 3 shows a typical cell drive strength distribution in sub-28nm technologies. The instance count is dominated by single fold cells. The histogram in Fig. 4 shows a typical topology distribution. The instance count is dominated by AOI21/OAI21, AOI22/OAI22, NAND2/NOR2, and INV topologies. Combined this indicates pin density is predominately determined by single fold cells where pin counts match the number of poly pitches the cells are wide. The flop is one key exception. This makes projection of the weighted average pin density relatively easy for comparison

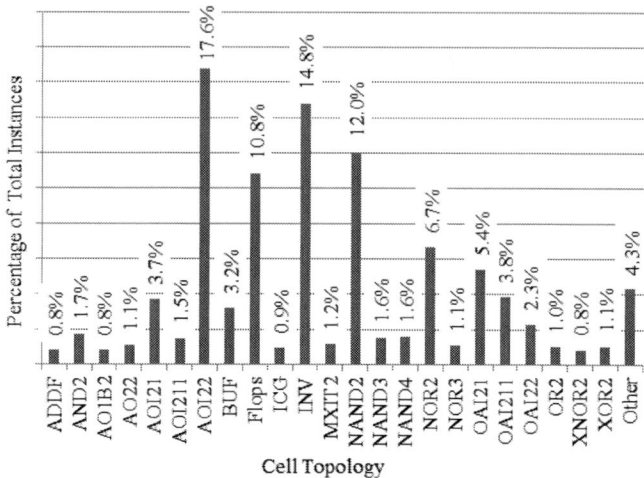

Figure 4: Histogram of typical topology distribution in sub-28nm. Sample taken from Cortex-A9 implemented at maximum achievable frequency.

to the usable M2 density to determine a starting point.

Cell area is always a candidate to be the limiting factor for block density. Cell area is a complex function of many factors and can only be determined by actual development of cells. Pin accessibility is a critical cell consideration and requires actual cell development to correctly model as well. In many cases, development of versions of cells with more poly pitches than the minimum possible to improve pin access is crucial. One saving grace is often cell layout for one poly pitch can be stretched to save time creating layout for another. If the cell in question wasn't already theoretical minimum area, re-optimization needs to be considered with increasing pitch. Pin optimization always needs to be considered when increasing poly pitch as there is often room for improvement when crossing over certain key thresholds.

Other than the potential to improve pin access, the main reason to consider a larger poly pitch is the potential to improve device power/performance characteristics. Fig. 5 shows the relative device performance possible with increasing poly pitch using ARM process models. Clearly with increasing pitch comes the option to increase the physical channel length and decrease the Vt while holding the I_{ON}/I_{OFF} ratios relatively constant for a given condition. In the

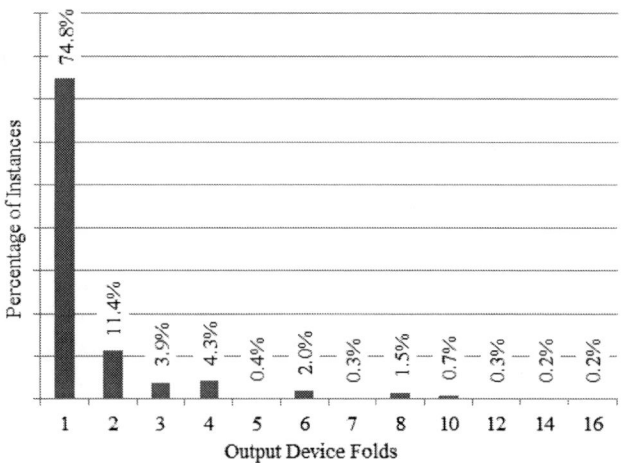

Figure 3: Histogram showing typical drive strength distributions sub-28nm. This sample was taken using Cortex-A9 implemented at maximum achievable frequency.

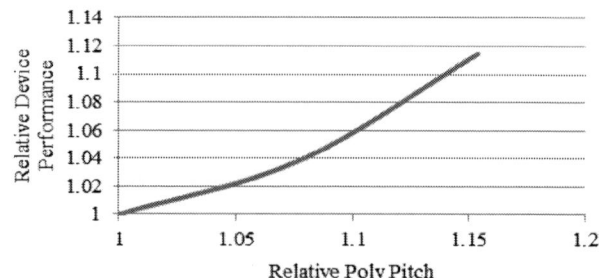

Figure 5: Device performance versus poly pitch using ARM device models assuming iso-leakage and iso-dynamic power for transistor plus weighted interconnect overhead to account for density loss.

results section, ARM took the liberty of adjusting these knobs to show some of the potential possible.

Method

Benchmarking of an ARM Cortex-A9 using industry standard synthesis, place and route flows was done to compare three different poly pitches relative to a fixed interconnect definition. To gain an appreciation for the multi-variable aspect of the co-optimization problem, a "bonus" fourth process variation is considered where the increased poly pitch enables a new cell level construct to be printable. This is a common result of the increased pitch and must be factored into the co-optimization effort along with any potential increase in mask costs or alternatively improvements in yield/process margin. These four ARM process definitions are based on existing industry capability, so they are representative of the real co-optimization potential possible.

For this investigation cell width was held constant in terms of number of poly pitches for all the process definitions except the fourth process. In the fourth process, the area of the low drive flops were reduced by two poly pitches to reflect enablement of a new layout construct made possible by the larger pitch. The two poly pitch reduction was applied only to the middle poly pitch to reduce benchmarking costs, but is equally valid and interesting for the largest poly pitch. Pin access was optimized uniquely for each poly pitch. The interconnect design rules were held constant for all four process variations. Table 1 summarizes the relative comparison of the four processes.

Table 1: Summary of Process Variations Benchmarked

Process Version	Relative Poly Pitch	Other Features
P1	1	-
P2	1.076	-
P3	1.153	-
P4	1.076	2 pitch smaller low drive flops

Results

Fig 6, 7, and 8 show the results of the benchmarking of all four processes. For completeness a full range of performance was investigated for the slowest and fastest devices modeled in each process variation. Increasing the spread of device performance is possible, but these extremes are representative of reasonable design points possible for each poly pitch.

Studying Fig 6, it is clear P1 produces cells with pin densities higher than the maximum density supported by the interconnect layers. This is shown by the larger P2 cells producing the same minimum area as the P1 cells at the lowest frequency levels. Given the inherent disadvantage in performance and power shown in Fig 6, 7, and 8, P1 makes little sense to pursue relative to P2 unless there is potential to improve the interconnect density. Let's assume the analysis showed higher interconnect density was either impossible to achieve or too costly. Remember there is an excess of routing resource as a whole. Only the lowest layers are overwhelmed, classic local route congestion, implying a tighter M2 is required to gain any relief.

Fig 6 shows the P3 designs are cell area limited. This can be observed by comparing the block area for the lowest frequencies of the P2 and P3 implementations. More interesting, is the fact the area increase of the P3 block over the P2 block is roughly the same as the pitch increase from P2 to P3, approximately 7 percent. This is strong evidence interconnect and cell area are near a balance point in the P2 process definition at the lower frequencies. The majority of standard cell area is implemented in this flat band. Don't overlook the higher performance potential possible in P3 relative to the P2. For some designs performance is the most important metric. Better yet, after a certain frequency the P3 definition can achieve a better performance and better area, but unfortunately at the cost of more power as shown if Fig 7 and 8. Consider that intermediate device types are possible and mixing device types in a single design may enable higher performance at iso-power and better area for the P3 process.

Studying Fig 6 more, it is clear P4 provides a useful density advantage over the P2. Don't forget P2 is near the balance of pin versus interconnect density, so how did the area improve? The larger poly pitch enabled cells that are not pin density limited to shrink in area (e.g. flops) resulting in an overall block area improvement. Of interest, but not shown here is the fact that P4 can enable some high pin density cells to shrink in area, but doing so actually hurt block area due to the placement engine being unable to spread the smaller high pin density cells optimally due to design rule limitations. This initially resulted in P4 and P2 areas being roughly equal until corrected. If absolute density is most critical, the P4 process is the best process option considered.

Conclusions

The smallest poly pitch may not produce the best overall process if the cell densities or pin structures enabled are not supported by the interconnect density. P4 shows a larger poly pitch can even save area if key cells can be reduced in pitch count. There is also a strong indication that enabling more than one poly pitch on a die may enable a process to support a wider range of performance while still supporting best case density on a block by block basis as needed.

Careful analysis of the benchmarking results is vital or a beneficial concept may be obscured by an unexpected interaction as was the case initially for P4. Often these situations provide an opportunity to feedback ideas to the process development which may be supportable, but initially not considered or deemed useful. Clearly co-optimizing can lead to a better result than pure aggressive pitch scaling.

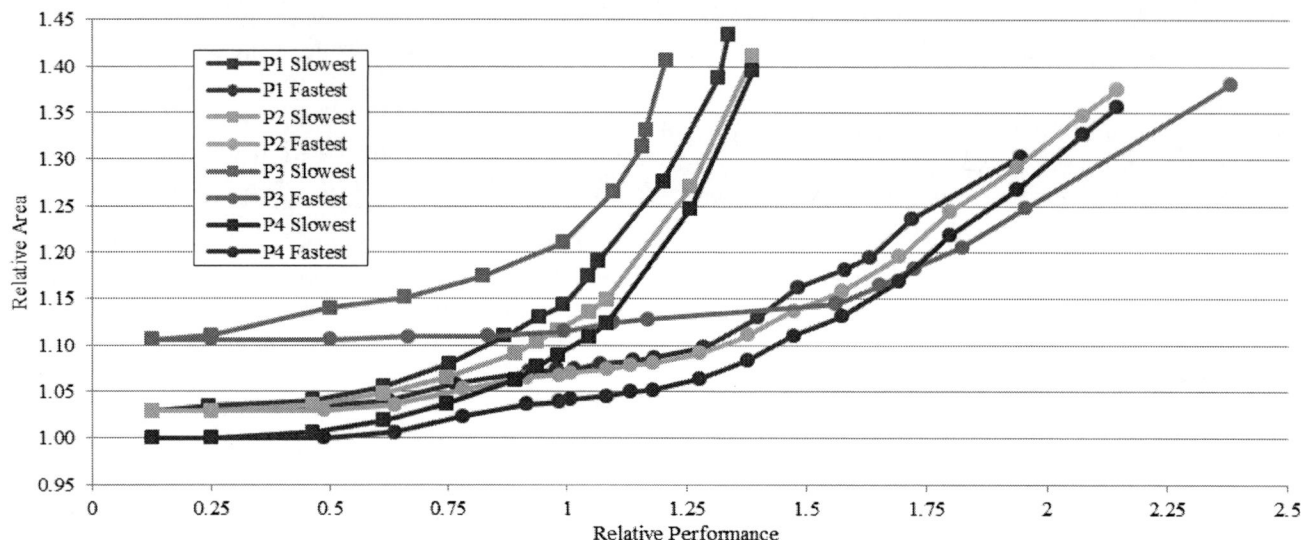

Figure 6: Block area versus performance results for all four process definitions showing the achieved spread possible using the slowest and fastest available devices for each process.

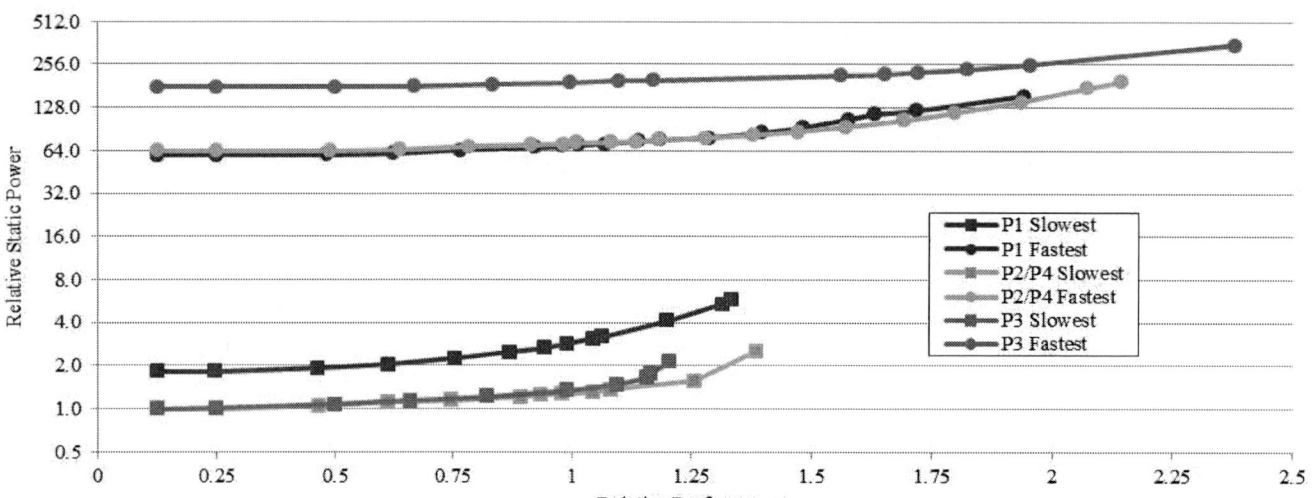

Figure 7: Block static power versus performance results for all four process definitions showing the achieved spread possible using the slowest (least leakage) and fastest (most leakage) available devices for each process. The difference between P2 and P4 is insignificant, so they are shown as combined lines.

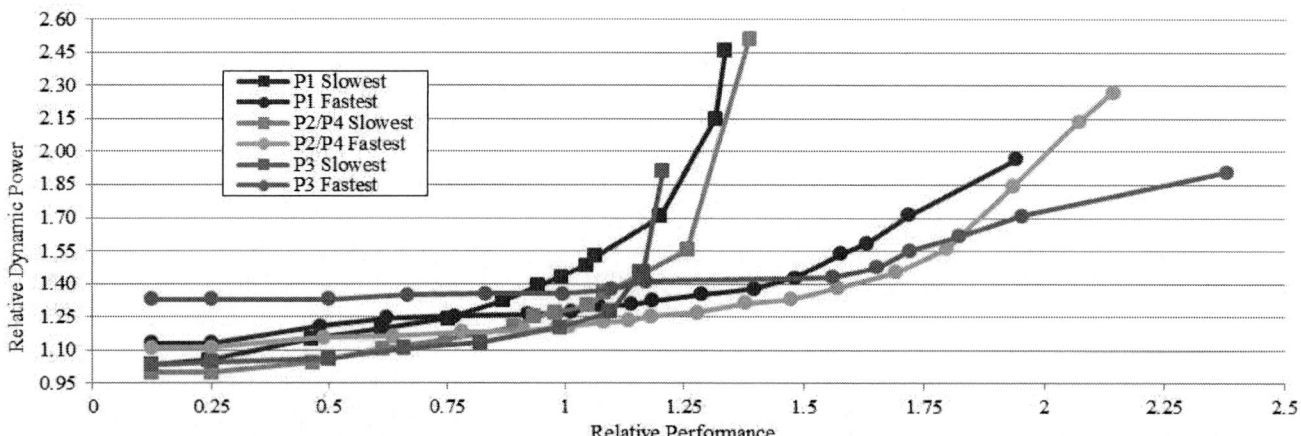

Figure 8: Block dynamic power versus performance results for all four process definitions showing the achieved spread possible using the slowest and fastest available devices for each process. The difference between P2 and P4 is insignificant, so they are shown as combined lines. Note the significant increase in short circuit current with faster devices.

978-1-4799-8002-4/14 $31.00 © 2014 IEEE

NEM Relay Design for Compact, Ultra-Low-Power Digital Logic Circuits

Tsu-Jae King Liu[1*], Nuo Xu, I-Ru Chen[1], Chuang Qian[1] and Jun Fujiki[2]

[1]Department of Electrical Engineering and Computer Sciences, University of California, Berkeley, CA 94720 USA
[2]Toshiba Corporation, Tokyo 105-8001, Japan
*Phone: +1-510-642-0253, Fax: +1-510-643-7846, E-mail: tking@eecs.berkeley.edu

Abstract

Since mechanical switches (relays) have zero off-state leakage and perfectly abrupt ON/OFF switching behavior, in principle they can be operated with a very small voltage swing and overcome the energy efficiency limit of CMOS technology. This paper discusses recent developments to address the remaining technical challenges for fully realizing the promise of ultra-low-power mechanical computing.

Introduction

The energy efficiency of any CMOS digital logic circuit is fundamentally limited by transistor off-state leakage (I_{OFF}) (1). A mechanical switch (relay) has zero I_{OFF} and perfectly abrupt ON/OFF switching behavior so that it can be operated with a very small voltage swing (2). Therefore, scaled relay technology potentially can overcome the energy efficient limit of CMOS technology. The endurance of micrometer-scale relays has been demonstrated to improve exponentially with decreasing operating voltage and is projected to exceed 10^{15} ON/OFF cycles at 1 Volt (3), which is more than adequate for applications such as wireless sensor networks which require ultra-low power consumption (4). In addition, relays can function well across a wide range of temperature (5) and can incorporate multiple input and output electrodes for greater functionality as compared with transistors, enabling reductions in device count (6). Remaining challenges for realizing the promise of ultra-low-power mechanical computing are contact adhesion, miniaturization to achieve a very small device footprint, and process-induced variability. This paper discusses approaches to overcoming each of these challenges.

Overcoming the Surface Adhesion Energy Limit

Fig. 1 illustrates the operation of a conventional three-terminal mechanical switch design. In the OFF state, an air gap separates the movable suspended electrode from the fixed contacting electrode so that no current can flow between them. To turn ON the switch, an electrostatic force (F_{elec}) is applied by applying a voltage between the fixed actuator electrode and the suspended electrode to bring it into contact with the contacting electrode. To turn OFF the switch, the applied voltage is lowered so that the spring restoring force (F_{spring}) of the suspended electrode causes it to come out of contact. In order for the switch to turn OFF properly, F_{spring} must be greater than the surface adhesive force (F_{adh}) between the contacting surfaces in the ON state. This in turn means that F_{elec} must be greater than F_{adh}, which limits the extent to which the actuation area and/or the actuation voltage can be reduced. Thus, contact adhesion sets a lower limit for the relay switching energy (7). Note that the actuation gap and area typically are significantly larger than the (OFF-state) contact gap and apparent contact area, respectively (i.e. $g_{act} > g_{cont}$ and $A_{act} > A_{cont}$), for reliable low-voltage operation (8).

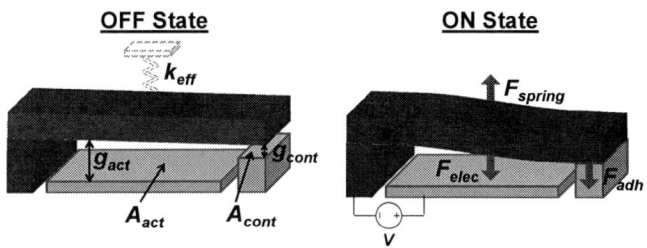

Fig. 1: Schematic isometric views illustrating the operation of a normally-OFF electrostatically actuated mechanical switch. $F_{elec} > F_{spring} > F_{adh}$.

To overcome this adhesion energy limit, a relay can be designed to be in the ON state as fabricated (Fig. 2) so that F_{spring} works in the same direction as F_{act} to overcome F_{adh}, and hence F_{act} can be smaller than F_{adh} (9). For the single-throw switch design, the as-fabricated contact gap must be very small (~1 nm) so that van der Waals force is sufficient to turn ON the relay. This challenging requirement can be mitigated by employing a single-pole/double-throw (SPDT) design as shown in Fig. 3, in which electrostatic force is used to switch between each of the two states. This 5-terminal switch can be engineered to be non-volatile (i.e. bi-stable) if $F_{spring} < F_{adh}$.

Compact BEOL Relay Design

The lateral dimensions of a relay should be proportionately scaled down to reduce its footprint. Lithographic limits make it difficult to scale down the width of the flexural beam and the gap sizes, however, which in turn limit the extent of beam-length scaling (10). One approach to overcome this challenge is to use a three-dimensional (3-D) structure that leverages an advanced CMOS back-end-of-line (BEOL) process with airgap interconnect structures (11). Fig. 4 illustrates a 5-terminal SPDT nano-electro-mechanical (NEM) switch implemented using 4 interconnect layers (3 for actuation, 1 for contact).

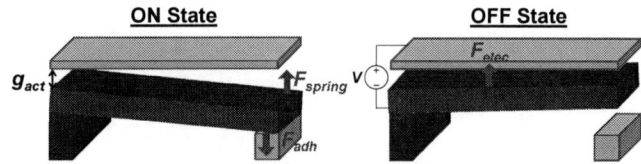

Fig. 2: Schematic isometric views illustrating the operation of a normally-ON electrostatically actuated mechanical switch. $F_{elec} + F_{spring} > F_{adh}$.

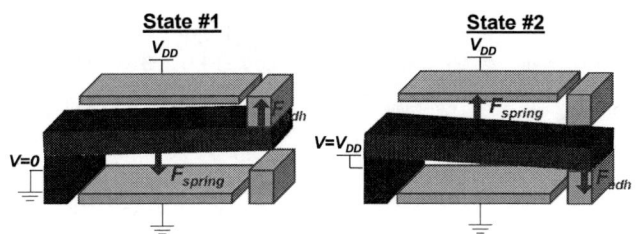

Fig. 3: Schematic isometric views illustrating the operation of a bi-stable electrostatically actuated mechanical switch. $F_{spring} < F_{adh}$.

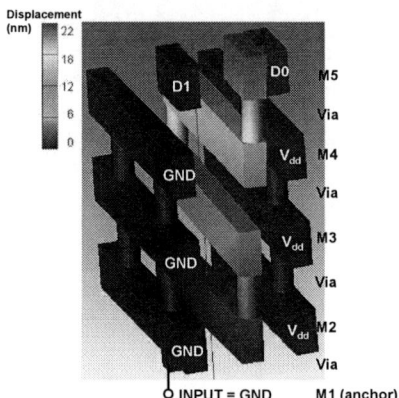

Fig. 4: Three-dimensional single-pole/double-throw (SPDT) relay design implemented with an advanced sub-20 nm CMOS BEOL process (with airgap interconnect structures). Depending on the input voltage applied to the beam, it will be in contact with either the D1 or D0 electrode. A large effective actuation area is achieved using 3 layers of metal, to provide for lower actuation voltage.

Note that the vias used to electrically connect the actuator portions of the structure also serve as torsional/flexural elements to reduce the effective spring constant of the structure for lower voltage operation. Fig. 5 plots the simulated pull-in voltage (minimum value of V_{DD}) and catastrophic pull-in voltage (maximum value of V_{DD}) as a function of layout area (*i.e.* for different actuator beam lengths), assuming sub-20 nm design rules (12). Sub-1 V operation is projected for a layout area less than 0.1 μm^2. Its relatively small footprint. Low switching delay and extremely low switching energy (Fig. 6) makes the bi-stable NEM relay attractive for non-volatile SRAM application (12).

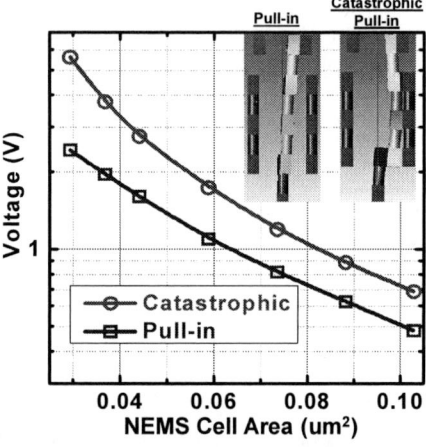

Fig. 5: Simulated pull-in voltage and catastrophic pull-in voltage (such that the beam contacts the actuation electrode) for the BEOL NEM switch design shown in Fig. 4, with varying device footprint (varying actuator/beam length). The inset illustrates the switch structure in the pull-in and catastrophic pull-in states (12).

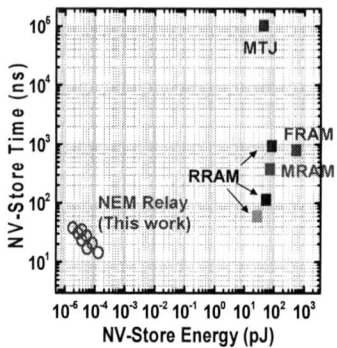

Fig. 6: Benchmarking of switching time *vs.* energy for bi-stable NEM relays of various areas (ref. Fig. 5) against other non-volatile memory devices (12).

Zero Crowbar Current Relay-Based Circuit Design

Because of their abrupt switching behavior, process-induced variations in switching voltage and/or switching time can temporarily result in a direct current path ("short circuit") between the power supply and ground, if a relay-based digital circuit is implemented using the conventional complementary logic circuit topology in which there is at least one "pull-up" device connected between the output node and power supply and at least one "pull-down" device connected between the output node and ground (13). The resultant "crowbar current" results in a component of dynamic power consumption that can be much larger than that for a CMOS implementation, since transistors have gradual switching behavior (14). To avoid this issue, a 6-terminal (6-T) SPDT relay design (Fig. 7), in which the input electrode is electrically isolated from the output electrode by incorporating an insulating dielectric material between the actuation and contact portions of the movable structure, should be used along with an appropriate circuit design methodology (15). The two actuation electrodes located on either side of the movable input electrode (IN) are biased at the supply voltage (V_{DD}) and ground (GND), respectively; the output electrode (OUT) is connected to one of two data electrodes (D_1 and D_2), depending on the input voltage. The circuit symbol and truth table (wherein "0" $\equiv$ GND and "1" $\equiv$ V_{DD}) for this switch are shown in Fig. 6; the actuation electrode biased at V_{DD} is denoted by "+" in the circuit symbol. Note that the MEM logic switch implements the multiplexer function: $OUT = \overline{IN} \cdot D_1 + IN \cdot D_2$.

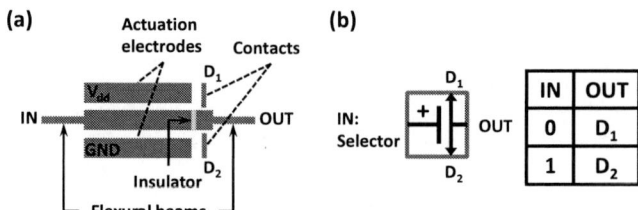

Fig. 7: Conceptual schematic illustrating a SPDT 6-terminal relay design (15). Note that it incorporates an insulator to electrically isolate the input from the output signal path. (a) schematic illustration showing the biasing scheme, (b) circuit symbol used in this work and truth table.

978-1-4799-8002-4/14 $31.00 © 2014 IEEE

Fig. 8 shows how the 6-T relay can be used to implement any basic logic function. One of the requirements for combinational logic is single-stage operation, *i.e.* the operation should be carried out within one clock cycle (one mechanical switching delay (16)). Examples are shown in Figs. 9(a) and 9(b) for multiple-input AND and OR functions, respectively, in which intermediate results are carried by OUT terminals each connected to a D_1 or D_2 (rather than IN) electrode of another switch. A 2N:1 multiplexer (MUX) is easily implemented using N(N+1)/2 switches, by feeding OUT of the previous stage into D_1 or D_2 of the next stage, as shown in Fig. 9(c). A N-bit decoder can be implemented using 2N+1-2 relays, by alternating the polarity of the actuation electrodes for a "0" vs. "1" input bit and then feeding OUT into D_1 or D_2 of the next lower bit, as shown in Fig. 9(d).

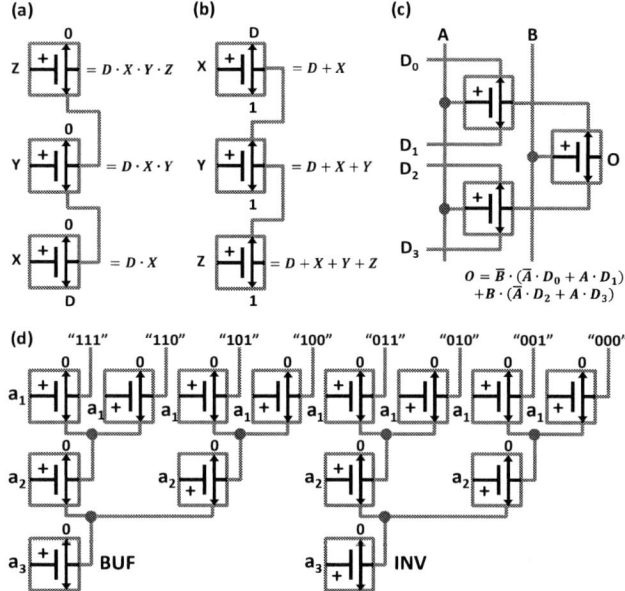

Fig. 9: MEM-based combinational logic gates: (a) 3-input AND, (b) 3-input OR, (c) 3-to-1 MUX, and (d) 3-bit decoder (15).

As shown in Fig. 10, a single-stage carry-generation and adder circuit suited for a carry-lookahead adder (15) can be realized with only 4 and 8 relays, respectively, by multiplexing AND, OR, XOR and XNOR outputs. It should be noted that, since each of these logic gates are single-stage, their mechanical switching delay does not depend on the number of inputs. Their electrical (capacitive charging, or "RC") delay increases with N, however. Due to the relatively large mechanical switching delay (> 1 ns), the insertion of buffers (BUF) to reduce the electrical delay is only beneficial if N is very large (> 100) (17), since an additional mechanical switching delay would be incurred.

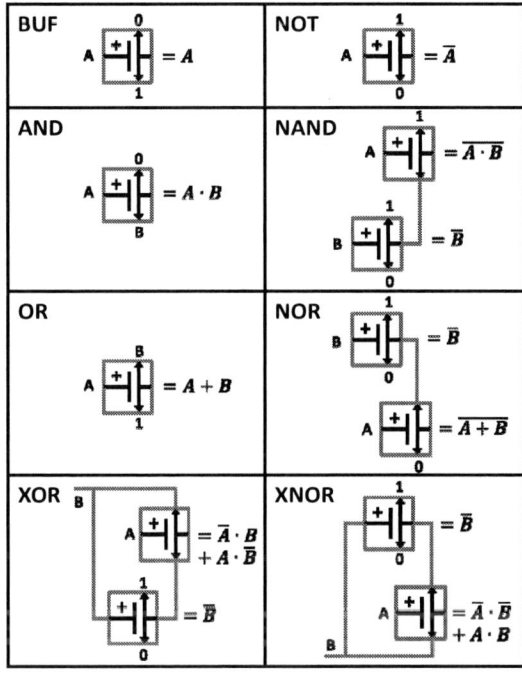

Fig. 8: Basic logic gates implemented with the 6-T relay design shown in Fig. 7 (15).

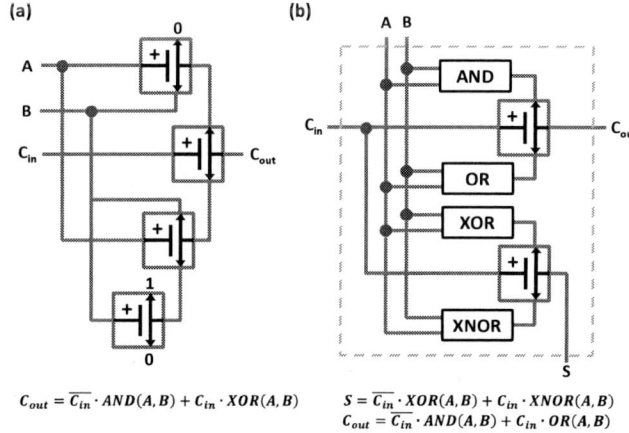

$$C_{out} = \overline{C_{in}} \cdot AND(A,B) + C_{in} \cdot XOR(A,B)$$

$$S = \overline{C_{in}} \cdot XOR(A,B) + C_{in} \cdot XNOR(A,B)$$
$$C_{out} = \overline{C_{in}} \cdot AND(A,B) + C_{in} \cdot OR(A,B)$$

Fig. 10: MEM-based (a) carry-generation, (b) full-adder circuit, with AND, OR, XOR and XNOR referring to Fig. 8 (15).

978-1-4799-8002-4/14 $31.00 © 2014 IEEE

Table I compares the number of switching devices used to implement various logic functions, for CMOS *vs.* relay technologies. It can be seen that the relay circuit design methodology described herein provides for lower device count and therefore potentially lower active power consumption. It also avoids the need to generate complementary signals, in contrast to the pass-gate logic design methodology used in (16).

Table I: Device Count Comparison (15)

FUNCTION	CMOS	6-T RELAY
BUF	4	1
NOT	2	1
NAND	4	2
XOR	6	2
MUX	8	1
Carry-gen.	8	4
Full-adder	24	8

Conclusion

The surface adhesion energy limit for a nanometer-scale relay can be overcome by designing it such that the spring restoring force counteracts the surface adhesive force to reduce the electrostatic force required for switching. An advanced CMOS BEOL process can be leveraged to fabricate 3-D relays with footprint less than 0.1 μm^2 and switching voltage below 1 V. A complementary relay design and circuit design methodology should be used to ensure zero crowbar current (in addition to zero standby current) despite process-induced variations in switching voltage, to guarantee ultra-low-power consumption.

Acknowledgement

This work is supported in part by the Center for Energy Efficient Electronics Science (NSF Award 0939514).

References

(1) B. H. Calhoun, A. Wang, and A. Chandrakasan, "Modeling and sizing for minimum energy operation in sub-threshold circuits," *IEEE Journal of Solid-State Circuits*, vol. 40, pp. 1778-1786, September 2005.

(2) U. Zaghloul and G. Piazza, "Sub-1-volt piezoelectric nanoelectromechanical relays with millivolt switching capability," *IEEE Electron Device Letters*, vol. 35, pp. 669-671, June 2014.

(3) H. Kam, E. Alon, and T.-J. K. Liu, "A predictive contact reliability model for MEM logic switches," *IEEE International Electron Devices Meeting Technical Digest*, pp. 399-402, 2010.

(4) T.-J. K. Liu, J. Jeon, R. Nathanael, H. Kam, V. Pott, and E. Alon, "Prospects for MEM-relay logic switch technology," *IEEE Int'l Electron Devices Meeting Technical Digest*, pp. 424-427, 2010.

(5) H. Kam, V. Pott, R. Nathanael, J. Jeon, E. Alon, and T.-J. K. Liu, "Design and reliability of a micro-relay technology for zero-standby-power digital logic applications," *IEEE International Electron Devices Meeting Technical Digest*, pp. 809-811, 2009.

(6) T.-J. K. Liu, L. Hutin, I-R. Chen, R. Nathanael, Y. Chen, and E. Alon, "Recent progress and challenges for relay logic switch technology," *Symposium on VLSI Technology Digest of Technical Papers*, pp. 43-43, 2012.

(7) C. Pawashe, K. Lin, and K. J. Kuhn, "Scaling limits of electrostatic nanorelays," *IEEE Transactions on Electron Devices*, vol. 60, pp. 2936-2942, September 2013.

(8) F. Chen H. Kam, D. Marković, T.-J. K. Liu, V. Stojanović, and E. Alon, "Integrated circuit design with NEM relays," in *Proceedings of the IEEE/ACM International Conference on Computer-Aided Design*, pp. 750-757, 2008.

(9) I-R. Chen, C. Qian, E. Yablonovitch, and T.-J. K. Liu, "Nanomechanical switch designs to overcome the surface adhesion energy limit, submitted to *IEEE Electron Device Letters*.

(10) D. T. Lee T. Osabe, and T.-J. K. Liu, "Scaling limitations for flexural beams used in electromechanical devices," *IEEE Transactions on Electron Devices*, vol. 56, pp. 688-691, April 2009.

(11) S. Nitta, D. Edelstein, S. Ponoth, L. Clevenger, X. Liu, and T. Standaert, "Performance and reliability of airgaps for advanced BEOL interconnects," in *Proceedings of the International Interconnect Technology Conference*, pp. 191-192, June 2008.

(12) N. Xu, J. Sun, I-R. Chen, L. Hutin, Y. Chen, J. Fujiki, C. Qian, and T.-J. K. Liu, "Hybrid CMOS/BEOL-NEMS technology for ultra-low-power IC applications," *Int'l Electron Devices Meeting Technical Digest*, 2014.

(13) R. Nathanael, V. Pott, H. Kam, J. Jeon, E. Alon, and T. King Liu, "Four-terminal-relay body-biasing schemes for complementary logic circuits," *IEEE Electron Device Letters*, vol. 31, pp. 890 -892, August 2010.

(14) H. J. M. Veendrick, "Short-circuit dissipation of static CMOS circuitry and its impact on the design of buffer circuits," *IEEE Journal of Solid - State Circuits*, vol. SC -19, pp. 468-473, August 1984.

(15) J. Fujiki, N. Xu, L. Hutin, I-R. Chen, C. Qian, and T.-J. K. Liu, "Microelectromechanical relay and logic circuit design for zero crowbar current," *IEEE Transactions on Electron Devices*, vol. 61, pp. 3296-3302, September 2014.

(16) M. Spencer, F. Chen, C. Wang, R. Nathanael, H. Fariborzi, A. Gupta, H. Kam, V. Pott, J. Jeon, T.-J. K. Liu, D. Marković, E. Alon, and V. Stojanović, "Demonstration of integrated micro-electro-mechanical relay circuits for VLSI applications," *IEEE Journal of Solid-State Circuits*, vol. 46, pp. 308-320, 2011.

(17) H. Fariborzi, F. Chen, R. Nathanael, J. Jeon, T.-J. K. Liu, and V. Stojanović, "Design and demonstration of micro-electro-mechanical relay multipliers," in *Proceedings of the IEEE Asian Solid-State Circuits Conference*, pp.117-120, 2011.

978-1-4799-8002-4/14 $31.00 © 2014 IEEE

High I_{on}/I_{off} Ge-source ultrathin body strained-SOI Tunnel FETs

- impact of channel strain, MOS interfaces and back gate on the electrical properties

Minsoo Kim[1,2], Yuki Wakabayashi[1], Ryosho Nakane[1], Masafumi Yokoyama[1,2],
Mitsuru Takenaka[1,2] and Shinichi Takagi[1,2]

[1]The University of Tokyo, 2-11-16 Yayoi, Bunkyo-ku, Tokyo 113-8656, Japan, [2]JST-CREST
Tel: +81-3-5841-6733, Email: minsoo@mosfet.t.u-tokyo.ac.jp

Abstract

High performance operation of Ge-source/strained-Si-channel hetero-junction tunnel FETs is demonstrated. It is found that tensile strain in Si-channels can enhance the tunneling current because of the reduced effective energy bandgap, $E_{g,eff}$. Nitrogen heat-treatment can improve the gate-to-channel MIS interface which causes SS improvement. The fabricated Ge/sSOI(1.1 %) tunnel FETs show high I_{on}/I_{off} ratio over 10^7 and steep minimum SS of 28 mV/dec. Back biasing effects are also investigated and the I_{on} and average SS are improved by positive back biasing.

Introduction

While tunnel field-effect transistors (TFETs) based on band-to-band tunneling are expected as ultra-low power devices, development of the optimum materials, structures and processing for TFET channel and source regions is still of paramount importance for realizing both low subthreshold swing (SS) of sub-60 mV/dec and high drain on-current/off-current ratio (I_{on}/I_{off}) at the same time. As for the TFET channels, strained-Si (sSi) is a realistic material, because of the smaller bandgap. Tensile strain TFETs with SiGe sources [1, 2] or nano-wire structures [3, 4] have already been reported. However, a systematic study of the impact of channel strain on TFET performance has not been performed yet. As for the TFET sources, pure Ge sources grown on Si are expected to provide higher tunneling current, because of the type-II staggered band alignment between Ge and Si [5]. While poly-Ge source TFETs on (100) Si [6] and Ge source TFETs on (100)/(110) Si [7] have already been reported, the examination of the Ge sources including the in-situ doping in TFETs is still limited. Also, operation of strained-Si TFETs combined with the Ge sources has not been demonstrated yet.

In this study, we realize Ge source strained-SOI nTFETs with the different amounts of strain by using in-situ doped Ge sources grown on strained-Si and optimizing annealing condition. The impact of tensile strain on the electrical properties of the TFETs is systematically examined. It is found that higher strain leads to better TFET performance. 1.1 % tensile strain SOI TFETs with the Ge source have exhibited I_{on}/I_{off} ratio higher than 2×10^7 and the minimum SS (SS_{min}) of 29 mV/dec at room temperature. It is also found that back bias (V_B) can effectively modulate the electrical properties of the present TFETs, providing another option to improve TFET performance.

Proposed device structure

Fig. 1 and 2 show the schematic structure of the fabricated TFETs and the band diagram of the Ge-source/sSi-channel junction, respectively. The smaller bandgap and higher E_v edge of the Ge-source, the smaller bandgap and lower E_c edge of tensily-strained Si and the resulting reduction in $E_{g,eff}$ at the type-II staggered hetero-junction can increase the tunneling probability with maintaining the relatively large E_g of sSi in the drain regions, leading to suppression of the ambipolar leakage current. Fig. 3 shows the band diagram for sSi used experimentally in this study. In addition, recent improvement of Ge MOS interface properties [8, 9] makes the Ge sources more promising than SiGe sources known to have higher D_{it}. Ultrathin body (UTB) sSOI allows us to reduce the short channel effects and junction leakage current. Also, the present structure can be fabricated by simple standard CMOS processes.

Fabrication Process

Fig. 4 shows the schematic process flow of the present Ge/sSi TFETs. Unstrained SOI and two types of sSOI substrates with 0.8 and 1.1 % biaxial tensile strain are used for studying the strain effect on the TFET performance. P ion implantation at 3 keV is carried out to form the drain regions, followed by RTA at 900 °C for activation. The Raman spectra (Fig. 5) indicate that strain is maintained during the processing. In-situ B-doped Ge layers are grown at 200 °C on the source regions of the substrates by MBE. The layer-by-layer growth high quality Ge films and the flat surfaces are confirmed by the streak RHEED patterns and the AFM images (Fig. 6). Fig. 7 shows the cross-sectional TEM images of the fabricated devices. It is found that higher strain channels lead to lower defect density of the Ge films, suggesting that channel strain can improve the Ge/sSi hetero-interface quality. The in-situ B doping concentration in the Ge layers is estimated to be as high as 1.3×10^{20} cm^{-3}. A 3-nm-thick $Al_2O_3/GeO_x/Ge$ gate stack with EOT of 2.5 nm is formed by ALD combined with ECR plasma post oxidation in order to realize the high quality MOS interfaces between Al_2O_3 and Ge [8, 9]. Ta is deposited as the gate metal, followed by Ni and Al deposition for the source contact and the contact pad, respectively. It is found that one of the key processes is post metallization annealing (PMA) in N_2 for 30 minutes. The PMA temperature is varied from 200 and 400 °C to optimize the TFET performance.

978-1-4799-8002-4/14 $31.00 © 2014 IEEE

Electrical properties of fabricated TFETs

Fig. 8 and 9 show the I_D-V_G and I_D-V_D characteristics, respectively, of fabricated un-strained SOI, 0.8 and 1.1 % strained SOI TFETs after PMA at 400 °C. It is found that an increase in strain leads to the increase in I_{on} and the decrease in I_G, resulting in high I_{on}/I_{off} ratio because of higher tunneling probability and higher barrier height between insulators and sSi [10]. Note that higher I_G than I_D of the Ge/SOI TFETs in low V_G region is attributed to the current flow from the gate to the source, while the influence of I_G on I_D is negligible. Thus, I_D is used as the TFET current instead of I_S. Fig. 10 shows the I_{on}/I_{off} ratio as a function of PMA temperature. Here, I_{off} is taken to be the minimum value of I_D. It is found that PMA temperature strongly affects the electrical properties of the fabricated TFETs. The I_{on}/I_{off} ratio is maximized after 400 °C PMA, which amounts to 4.4, 2.2 and 3.7×10^7 for the unstrained, 0.8 and 1.1 % strained SOI TFETs, respectively. The degradation of I_{on}/I_{off} with increasing V_D is ascribed to the increase of tunneling leakage current in the un- optimized drain junctions. Thus, the electrical characteristics in low V_D can be regarded as the inherent ones of TFETs with the optimized drain junctions. Fig. 11 shows SS extracted from the I_D-V_G curves in Fig. 8. Higher PMA temperature also significantly reduces SS, irrespective of the amount of the channel strain. Fig. 12 shows SS_{min} as a function of PMA temperature. The unstrained, 0.8 and 1.1 % strained SOI TFETs after 400 °C PMA yield SS_{min} of 55, 49 and 29 mV/dec at room temperature. In order to clarify the reason why PMA effectively enhances the TFET performance, Si and Ge MOS interface properties are evaluated. Fig 13 and 14 show the C-V curves of the Ge and Si MOS capacitors, respectively, with and without PMA. D_{it} at the Ge MOS interfaces (Fig. 15), extracted by the conductance method, does not significantly change with PMA temperature. On the other hand, D_{it} in the Si MOS capacitors (Fig. 16) is effectively improved with increasing PMA temperature. These results indicate that the better performance of TFETs with higher PMA temperature is attributed to more sensitive modulation of the surface potential of the Si channels with respect to V_G by reduction in D_{it} after PMA, as schematically shown in Fig. 17.

The measurement temperature dependence of the I_D-(V_G-V_{TH}) curves and SS-I_D curves is shown in Fig. 18 and 19, respectively. Here, V_{TH} is assumed as V_G at I_D of 1 nA/μm. It is found that I_D and the SS values are almost independent of temperature. This result means that tunneling current fully dominates I_D of the present TFET and that any defect-related

current such as TAT is effectively suppressed, in spite of the relaxed Ge sources. The dominance of the tunneling mechanism on I_D of the present TFETs are also confirmed by almost no L_G dependence of I_D of the present TFETs (Fig. 20). We have found that the electrical characteristics of the TFETs exhibit quite interesting substrate bias (V_B) dependence. Fig. 21 and 22 show the I_D-V_G and the extracted SS-I_D characteristics, respectively, of 0.8 % sSOI TFETs as a parameter of V_B from -20 to 20 V. The obtained V_B dependencies of SS_{min} and I_{on}/I_{off} are summarized in Fig. 23. While negative V_B introduces lower I_{off} and resulting lower SS_{min}, positive V_B yields significant increase in I_{on} and decrease in SS in a wide range of middle I_D region, in spite of the increase in I_{off} due to the increase in the ambipolar current. These strong V_B dependencies are different from previously reported ones [1, 2, 11]. The decrease in SS_{min} with negative V_B is attributed to the suppressed the leakage current in the drain region. On the other hand, the increase in I_{on} and decrease in SS in the middle I_D region under positive V_B can be explained by the modulation of the tunneling path in the vicinity of the source-channel junction. These V_B effects can be effectively utilized for improving the TFET performance. Fig. 24 and 25 show the benchmarks of the present TFET performance in terms of I_{on}/I_{off} ratio versus SS_{min} and average SS (SS_{avr}), respectively. Here, the SS_{avr} is defined as SS obtained over 3 orders of the magnitude of I_D from the minimum I_D (I_{off}). The high I_{on}/I_{off} ratio under low SS_{min} at V_D of 0.05 V has been demonstrated for the present TFETs. The I_{on}/I_{off} ratio and SS in higher target V_D can be improved by further optimizing the drain junction profiles.

Conclusion

Ge/strained-Si hetero-junction TFETs with in-situ B-doped Ge have been demonstrated. The increase in channel strain and optimization of PMA have successfully realized high performance of steep SS_{min} below 30 mV/dec and large I_{on}/I_{off} ratio over 10^7.

Acknowledgements

We would be grateful to SOITECH for providing strained SOI substrates.

References

[1] Q. T. Zhao et al., *EDL* **32** (2011) 1480 [2] Q. T. Zhao et al., *SSE* **74** (2012) 97 [3] L. Knoll et al., *EDL* **34** (2013) 813 [4] L. Knoll et al., *IEDM* (2013) 100 [5] O. M. Nayfeh et al., *EDL* **29** (2008) 1074 [6] S. H. Kim et al., *VLSI Symp.* (2009) 178 [7] G. Han et al., *APL* **98** (2011) 153502 [8] R. Zhang et al., *APL* **98** (2011) 112902 [9] R. Zhang et al., *TED* **59** (2012) 335 [10] T. Hoshii et al., *JJAP* **46** (2007) 2122 [11] A. Guo et al., *TED* **58** (2011) 3283 [12] M. Noguchi et al., *IEDM* (2013) 683 [13] T. Krishnamohan et al., *IEDM* (2008) 947 [14] Q. Huang et al., *IEDM* (2012) 187 [15] M.-H. Lee et al., *TED* **60** (2013) 2423 [16] W.-Y. Lo et al., *SSE* **65-66** (2011) 22 [17] G. Dewey et al., *IEDM* (2011)785 [18] W.-Y. Choi et al., *EDL* **28** (2007) 743

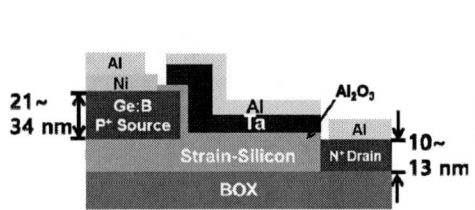

Fig. 1 Schematic structure of Ge/sSi hetero-junction TFET.

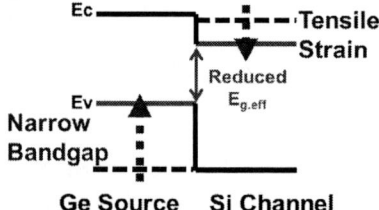

Fig. 2 Band diagram of Ge-source/sSi-channel hetero-junction. Combination of pure-Ge and tensile sSi can reduce the effective E_g ($E_{g.eff}$).

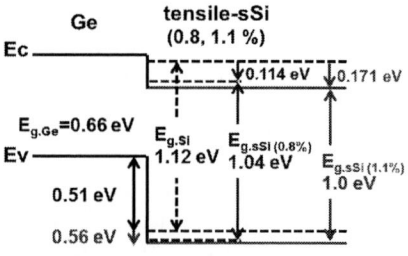

Fig. 3 Band alignment engineering using different tensile strain of Si-channel.

978-1-4799-8002-4/14 $31.00 © 2014 IEEE

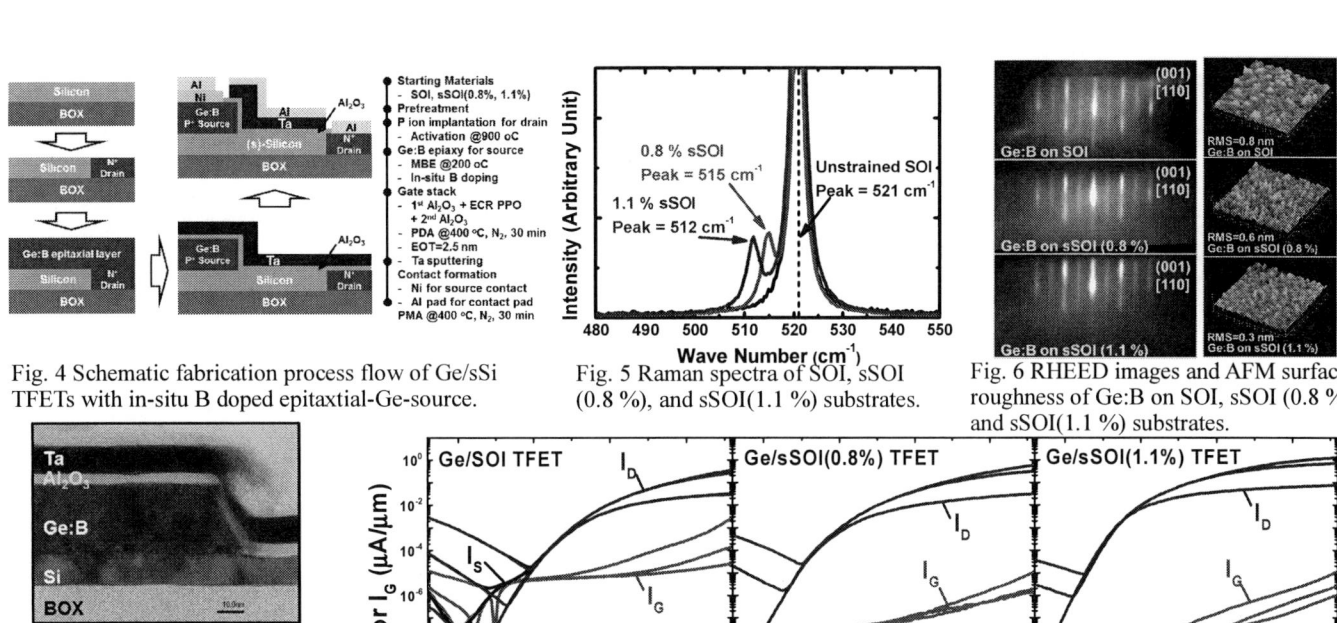

Fig. 4 Schematic fabrication process flow of Ge/sSi TFETs with in-situ B doped epitaxial-Ge-source.

Fig. 5 Raman spectra of SOI, sSOI (0.8 %), and sSOI(1.1 %) substrates.

Fig. 6 RHEED images and AFM surface roughness of Ge:B on SOI, sSOI (0.8 %), and sSOI (1.1 %) substrates.

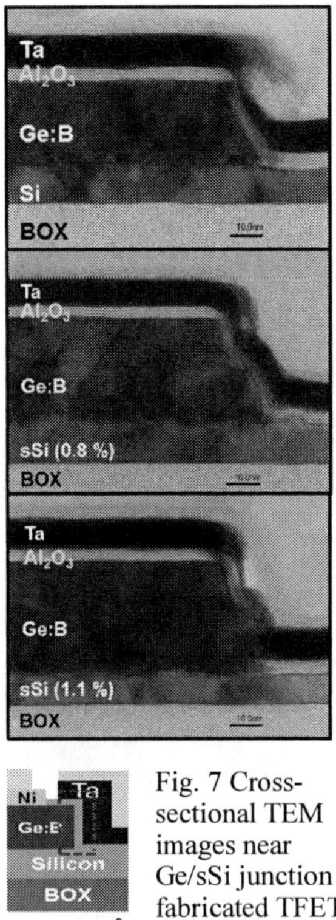

Fig. 7 Cross-sectional TEM images near Ge/sSi junction of fabricated TFETs.

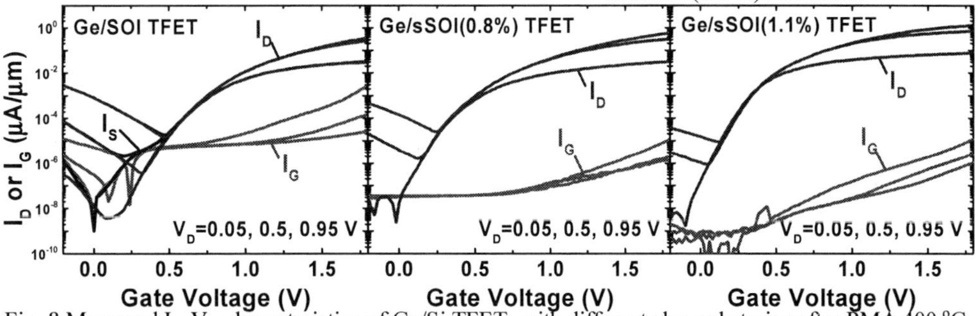

Fig. 8 Measured I_D-V_G characteristics of Ge/Si TFETs with different channel strains after PMA 400 °C. Strain devices show smaller gate-source leakage current.

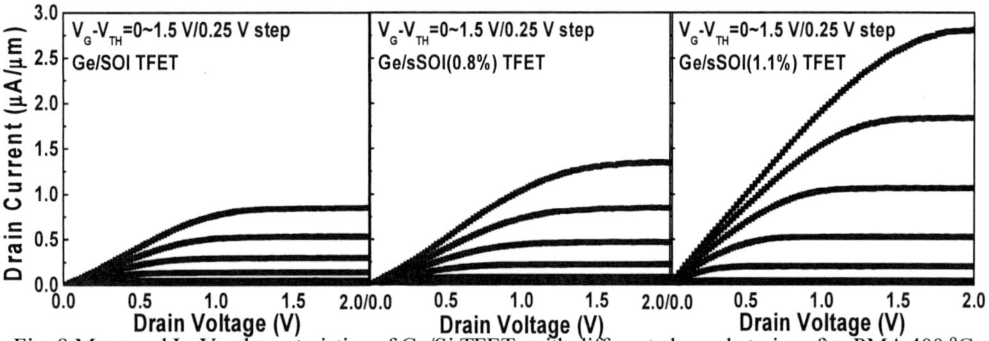

Fig. 9 Measured I_D-V_D characteristics of Ge/Si TFETs with different channel strains after PMA 400 °C. Here, V_{TH} is the gate voltage at $I_D=10^{-3}$ μA/μm.

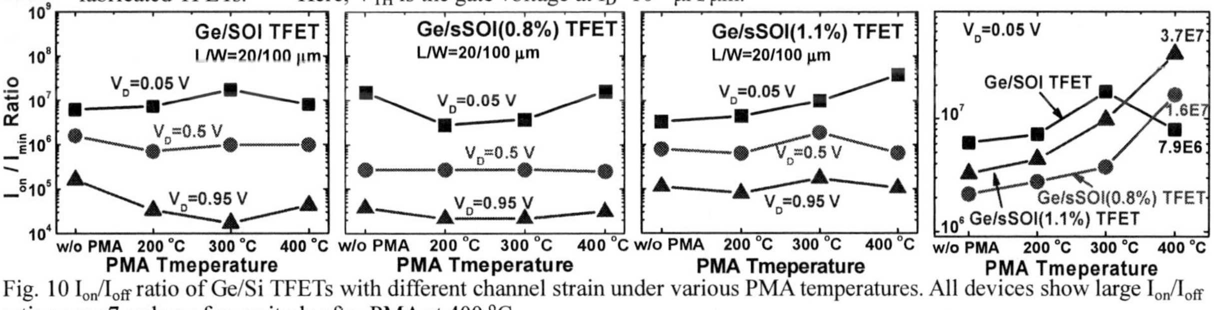

Fig. 10 I_{on}/I_{off} ratio of Ge/Si TFETs with different channel strain under various PMA temperatures. All devices show large I_{on}/I_{off} ratios over 7 orders of magnitude after PMA at 400 °C.

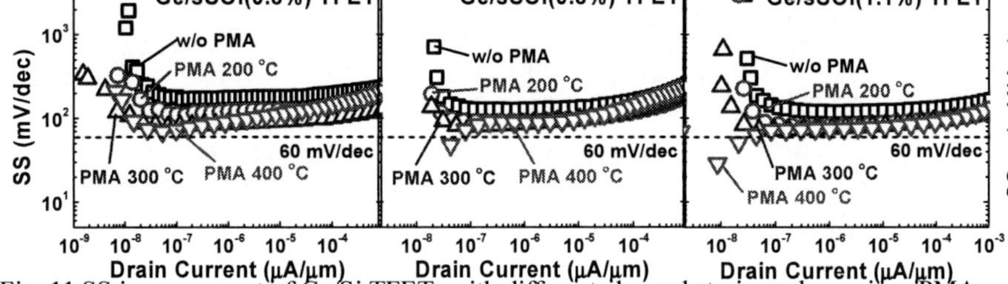

Fig. 11 SS improvement of Ge/Si TFETs with different channel strain under various PMA temperatures.

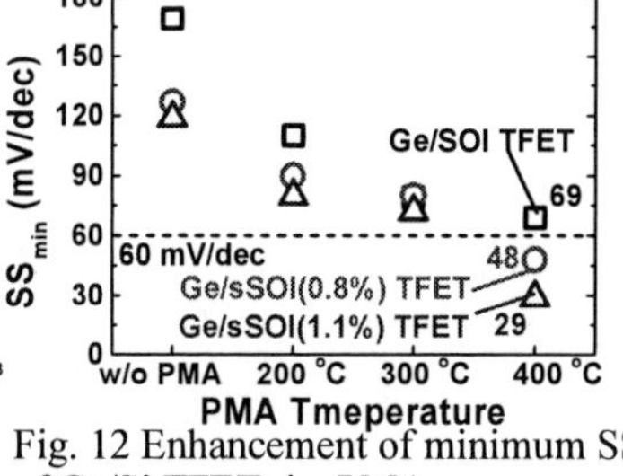

Fig. 12 Enhancement of minimum SS of Ge/Si TFETs by PMA.

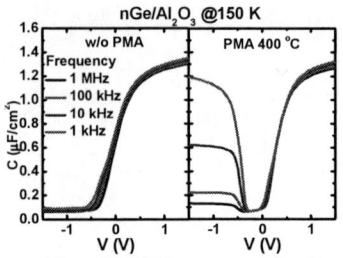

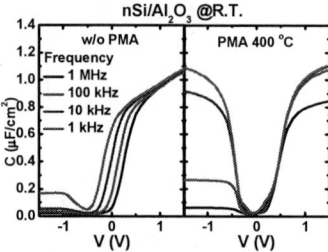

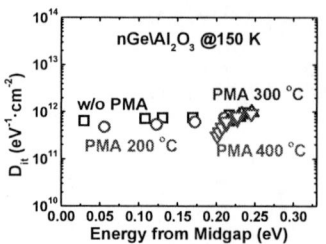

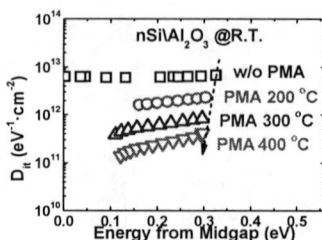

Fig. 13 C-V curves of nGe MOS capacitors before and after PMA.

Fig. 14 C-V curves of n-Si MOS capacitors before and after PMA.

Fig. 15 D_{it} of n-Ge MOS capacitors, measured by the conductance method at 150K, with various PMA temperatures.

Fig. 16 D_{it} of nSi MOS capacitors, measured by the conductance method at room temperature, with various PMA temperatures.

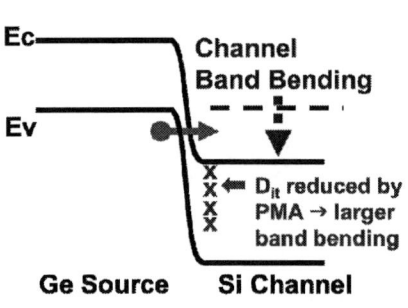

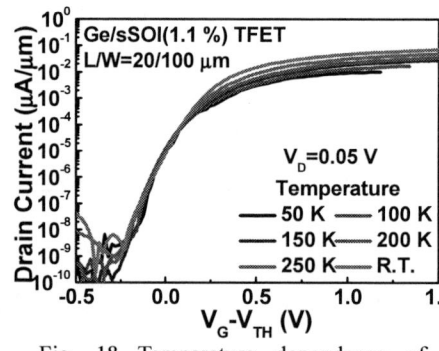

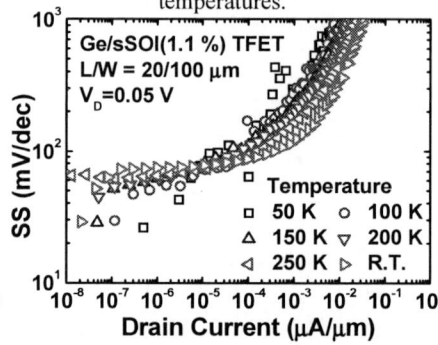

Fig. 17 Band diagram of Ge/Si hetero-junction with D_{it} at Si-channel/Al$_2$O$_3$ MOS interfaces.

Fig. 18 Temperature dependence of I_D-(V_G-V_{TH}) curves of sSOI (1.1%) TFET. Here, V_{TH} is defined as the gate voltage at $I_D=10^{-4}$ μA/μm.

Fig. 19 Temperature dependence of SS of sSOI (1.1%) TFET. Almost identical SS means the dominance of tunneling current.

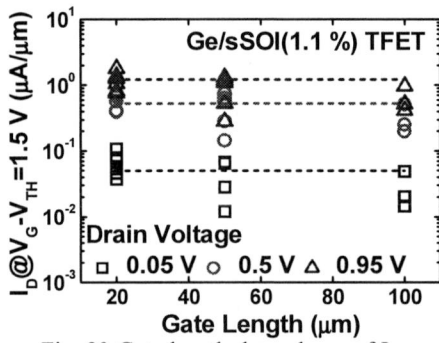

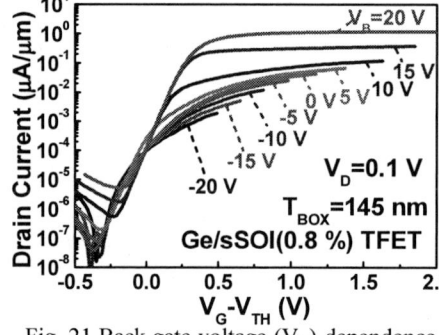

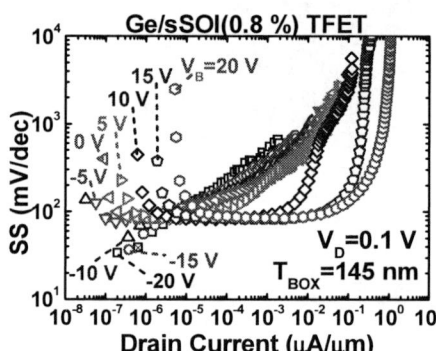

Fig. 20 Gate length dependence of I_D. Independence of I_D on L_G means that the dominant current is tunneling. Here, V_{TH} is defined as the gate voltage at $I_D=10^{-3}$ μA/μm.

Fig. 21 Back gate voltage (V_B) dependence of I_D-(V_G-V_{TH}) characteristics of sSOI (0.8%) TFETs. Here, V_{TH} is the gate voltage at $I_D=10^{-4}$ μA/μm.

Fig. 22 Back gate voltage (V_B) dependence of SS of sSOI (0.8%) TFETs. Negative V_B causes smaller minimum SS.

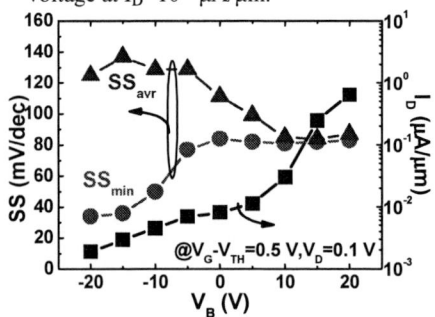

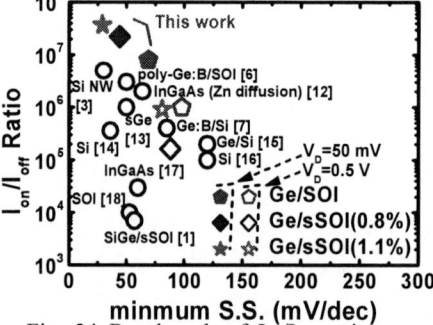

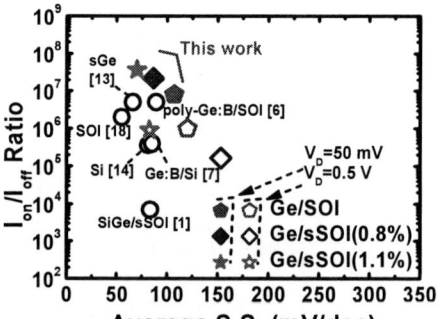

Fig. 23 Summary of back gate voltage (V_B) dependence of sSOI (0.8%) TFETs obtained from fig. 21 and 22. SS_{avr} is defined as SS averaged over 3 decades of I_D from the minimum SS.

Fig. 24 Benchmark of I_{on}/I_{off} ratio as a function of the minimum SS for the present TFETs and reported TFETs [1, 3, 6, 7, 12-18]

Fig. 25 Benchmark of I_{on}/I_{off} ratio as a function of the average SS for the present TFETs and reported TFETs [1, 6, 7, 13, 14, 18]. Here, the average SS is obtained over 3 orders of the magnitude of I_D from the minimum I_D (I_{off})

978-1-4799-8002-4/14 $31.00 © 2014 IEEE   334

Comprehensive Performance Re-assessment of TFETs with a Novel Design by Gate and Source Engineering from Device/Circuit Perspective

Qianqian Huang[1], Ru Huang[1,2*], Chunlei Wu[1], Hao Zhu[1], Cheng Chen[1], Jiaxin Wang[1], Lingyi Guo[1], Runsheng Wang[1], Le Ye[1] and Yangyuan Wang[1,2]

[1]Key Laboratory of Microelectronic Devices and Circuits, Institute of Microelectronics, Peking University, Beijing 100871, China
[2]Innovation Center for MicroNanoelectronics and Integrated System, Beijing 100871, China
Phone: 86-10-62757761, Fax: 86-10-62757761, *E-mail: ruhuang@pku.edu.cn

Abstract

In this paper, a novel TFET design, called Pocket-mSTFET (PMS-TFET), is proposed and experimentally demonstrated by evaluating the performance from device metrics to circuit implementation for low-power SoC applications. For the first time, from circuit design perspective, TFETs performance in terms of I_{ON}, I_{OFF}, subthreshold slope (SS), output behavior, capacitance, delay, noise and gain are experimentally benchmarked and also compared with MOSFET. By gate and source engineering without area penalty, the compatibly-fabricated PMS-TFET on SOI substrate shows superior performance with the minimum SS of 29mV/dec at 300K, high I_{ON} (~20µA/µm) and large I_{ON}/I_{OFF} ratio (~10^8) at 0.6V. Largely alleviated super-linear onset issue, reduced Miller capacitance and delay, and much lower noise level were also experimentally obtained, as well as high effective gain. Circuit-level implementation based on PMS-TFET also shows significant improvement on energy efficiency and power reduction at V_{DD} of 0.4V, which indicates great potential of this TFET design for low-power digital and analog applications.

Introduction

Tunnel FETs (TFETs) with band-to-band tunneling (BTBT) mechanism have emerged as one of the most promising candidates of MOSFET for low-voltage and low-power applications [1-2]. Due to its ultra-low I_{OFF} (~10^{-7}µA/µm) and steep subthreshold slope (SS) (<60mV/dec), Si TFETs can effectively reduce the supply voltage and also power consumption. However, low I_{ON} is the main challenge of Si TFETs due to its poor tunneling probability [3-4], and the resulting issues, including large intrinsic delay and low gain, may also limit the applications of Si TFETs. By introducing III-V materials with narrower bandgap, the I_{ON} can be efficiently boosted and many new III-V TFETs with high I_{ON} (>100µA/µm) have been reported [5-6], but with unsatisfactory I_{OFF} and SS values which may unfortunately increase the static power consumption. On the other hand, TFET devices, whether utilizing silicon or III-V materials, face many other challenges compared with MOSFET, such as large Miller capacitance [7], super-linear onset of output characteristics [8] and large noise level [9], which are detrimental to low-power SoC applications of TFETs. Therefore, merely improving I_{ON} and SS values is insufficient for TFET device design, and a fairly complete technology assessment of TFETs performance from the view of demands of circuits is intensively required.

In this paper, we experimentally evaluate the performance metrics of TFETs for both low-power digital and analog applications, and propose a novel TFET design by introducing comb-shaped gate and dopant segregated Schottky source. This new Pocket-mSTFET design can achieve significant performance improvement on I_{ON}, I_{ON}/I_{OFF} ratio and SS, as well as other above-mentioned performance requirements from low-power SoC application perspective. Compared with traditional TFET, Pocket-mSTFET-based inverter chain and full-differential operational amplifier show largely improved energy efficiency and gain bandwidth product.

I_{ON}, I_{OFF} and SS benchmark of Si, Ge and III-V TFETs

I_{ON}, I_{OFF} and minimum SS (SS_{min}) values are the three mostly-discussed device metrics of TFETs. Fig. 1 and Fig. 2 compare the experimental I_{OFF} vs. I_{ON} and SS_{min} vs. I_{ON} of the reported TFETs based on Si, SiGe, Ge and III-V homo- and heterojunctions. It can be seen that silicon-based TFETs show great potentials for both high I_{ON}/I_{OFF} ratio and ultra-steep SS_{min}, but a robust design is still needed to further enhance I_{ON}, as well as reduce average SS (SS_{avg}, SS value extracted within large drain current range) and improve other above-mentioned device metrics from the view of low-power SoC applications.

Pocket-mSTFET structure

A novel TFET design based on silicon is proposed to meet the performance requirement for both digital and analog applications by introducing comb-shaped gate and dopant segregated Schottky source.

A. Pocket-mTFET with comb-shaped gate

Based on our previously proposed Pocket-JTFET (PJ-TFET) structure [4] (Fig. 3b), a new comb-shaped gate configuration with multiple fingers and one comb-handle is introduced to the novel design, called Pocket-mTFET (PM-TFET), as shown in Fig. 3a. The multi-finger gate may induce larger tunneling area and thus larger I_{ON} than PJ-TFET, and the fully-depleted pocket at the source tunnel junction can increase the tunneling efficiency for TFET performance enhancement.

On the other hand, compared with PJ-TFET, the comb-shaped gate design in PM-TFET can also introduce two adaptively field enhancement mechanisms which are beneficial to the SS_{avg} improvement. For the side tunnel

978-1-4799-8002-4/14 $31.00 © 2014 IEEE

junctions at the gate finger region ("a" in Fig. 3a), the introduced junction depleted-modulation action proposed in [4] can effectively achieve more abrupt tunnel junction when the device turns on and thus larger electric field for steeper SS. For the tunnel junction at the gate handle region ("b" in Fig. 3a), the coupling effect between fingers induced by the multi-finger gate configuration may further increase the tunneling electric field with the decreased finger interval (Fig. 4). These combined two field enhancement mechanisms in PM-TFET may result in lower SS_{avg} within larger drain current range than PJ-TFET (Fig. 5) and traditional TFET.

Based on the Si CMOS-compatible process (Fig. 6), PM-TFET, PJ-TFET and traditional TFET of the same footprint are fabricated simultaneously on the bulk Si substrate, and the measured typical transfer characteristics are shown in Fig. 7. PM-TFET shows the steepest SS for more than 4 decades current (Fig. 8) and the highest I_{ON} while maintaining low I_{OFF} due to the comb-shaped gate design. In addition, compared with traditional TFET, the measured output characteristics of PM-TFET also show superior performance with better saturation and alleviated super-linear onset issue (Fig. 9). The positive I_{ON} dependence on temperature different from MOSFET further confirms the BTBT mechanism in PM-TFET (Fig. 10). However, compared with MOSFET, the I_{ON} of PM-TFET still needs further enhanced.

B. Pocket-mSTFET with gate and source engineering

Dopant segregated Schottky source is further experimentally introduced into the PM-TFET to boost the I_{ON}. As shown in Fig. 11, the new structure, called Pocket-mSTFET(PMS-TFET), has side BTBT junctions and additional Schottky junctions, which may result in steep switching dominated by BTBT like PM-TFET and also high I_{ON} from large Schottky current. Fig. 12 and 13 gives the measured I_D-V_{GS} and I_D-V_{DS} curves of fabricated PMS-TFET on the bulk Si substrate, showing MOSFET-comparable I_{ON} with large I_{ON}/I_{OFF} ratio ($>10^7$), steep SS_{min} and SS_{avg}, and also excellent output performance. The well fitted subthreshold characteristics with Kane's model [10] (Fig. 14) and negative I_{ON} dependence on temperature (Fig. 15) indicate two separated dominant current mechanisms in the subthreshold-state and on-state of PMS-TFET.

By fabricating PMS-TFET on SOI substrate, the device electrical characteristics can be further optimized (Fig. 16), and the minimum SS reaches 29mV/dec at 300K with I_{ON}/I_{OFF} ratio of $\sim10^8$. In order to compare the output performance of TFETs, a new V_{onset} is defined in Fig. 17 to assess the super-linear onset characteristics. Fig. 18 compares the measured SS_{min}, SS_{avg}, V_{onset} and I_{ON} of the fabricated different devices in this work. PMS-TFET on SOI substrate shows the superior performance and great potential as a promising TFET design. By further improving gate control capability of PMS-TFET by nanowires structure, even higher I_{ON} can be expected.

Performance assessment for low-power SoC applications

A. Device metrics evaluation for SoC applications

Other performance metrics (including C_{gd}, delay, noise and gain) for SoC applications of PMS-TFET, traditional TFET and MOSFET with the same footprint are also measured and compared to further evaluate this novel TFET design.

As shown in Fig. 19, compared with traditional TFET, the C_{gd} of PMS-TFET can be largely reduced, mainly due to the reduction of effective gate area induced by comb-shaped gate design. This may also lead to greatly-alleviated current overshoot and much smaller delay in PMS-TFET than traditional TFET (Fig. 20 and 21).

On the other hand, for analog applications, like traditional TFET [9], PMS-TFET samples show both flicker noise (Fig. 22) and random telegraph noise behavior (Fig. 23) with large device to device variation (Fig. 24a), which is mainly due to the limited turn-on area of Schottky junction. However, benefiting from the larger tunneling efficiency, PMS-TFET shows much lower noise level with lower normalized S_{id} than traditional TFET (Fig. 24b). In addition, the measured and calculated g_m, g_m/I_D, R_O and intrinsic gain are shown in Fig. 25 and 26. The results indicate that the PMS-TFET has the large capability of amplification and can even produce a higher gain at the same power level compared with traditional TFET.

Based on the above experimental demonstration of I_{ON}, SS_{min}, SS_{avg}, V_{onset}, C_{gd}, delay, S_{id}/I_D^2 and gain, Fig. 27 re-assesses the fabricated PMS-TFET, traditional TFET and MOSFET performances comprehensively. PMS-TFET can achieve high performance and low power simultaneously for SoC applications.

B. Evaluation of digital logic and analog circuit performance

By using verilog-A model calibrated with experimental data, a three-stage inverter chain with fanout of 4 and a full-differential operational amplifier are used for the PMS-TFET and TFET circuit analysis and comparison.

Compared with traditional TFET, the power consumption of PMS-TFET-based inverter chain largely decreases at V_{DD} of 0.4V (Fig. 28), and the energy efficiency is also significantly improved (Fig. 29) benefiting from its strongly reduced capacitance and SS_{avg}. For analog circuit performance, higher gain bandwidth product can be obtained in the PMS-TFET-based amplifier (Fig. 30), which makes PMS-TFET design suitable for low-power analog/mixed-signal SoC applications.

Summary

We experimentally and comprehensively evaluate the device metrics of TFETs for digital and analog applications, including I_{ON}, I_{OFF}, SS, output behavior, C_{gd}, delay, noise and gain. From device/circuit perspective, a novel TFET design with engineered gate and source is proposed and experimentally demonstrated and assessed, showing significant device and circuit performance improvement, which is very promising for high performance and low power SoC applications.

Acknowledgement

This work was partly supported by the 973 Projects (2011CBA00601), NSFC (60625403), and National Science & Technology Major Project 02 under Grant 2009ZX02035-001. The authors would like to thank the staff of National Micro/Nano Fabrication Laboratory of Peking University for their assistance in the device fabrication.

References: [1] A. M. Ionescu *et al.*, Nature, vol. 479, p. 329, 2011; [2] L. Knoll *et al.*, IEDM, p.100, 2013; [3] Q. Huang *et al.*, IEDM, p. 382, 2011; [4] Q. Huang *et al.*, IEDM, p. 187, 2012; [5] G. Dewey *et al.*, IEDM, p. 785, 2011; [6] G. Zhou *et al.*, IEDM, p.777, 2012; [7] Y. Yang *et al.*, IEEE EDL, vol. 31, p. 752, 2010; [8] E. Gnani *et al.*, ESSDERC, p. 105, 2012; [9] Q. Huang *et al.*, VLSI, p. 88, 2014; [10] Kane *et al.*, Tunneling Phenomena in Solids, p. 79, 1968.

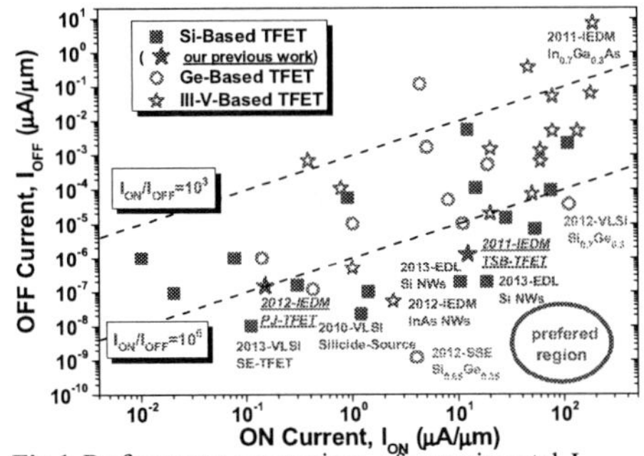

Fig.1 Performance comparison of experimental I_{OFF} vs. I_{ON} of reported TFETs based on Si, SiGe, Ge, and III-V homo- and heterojunctions.

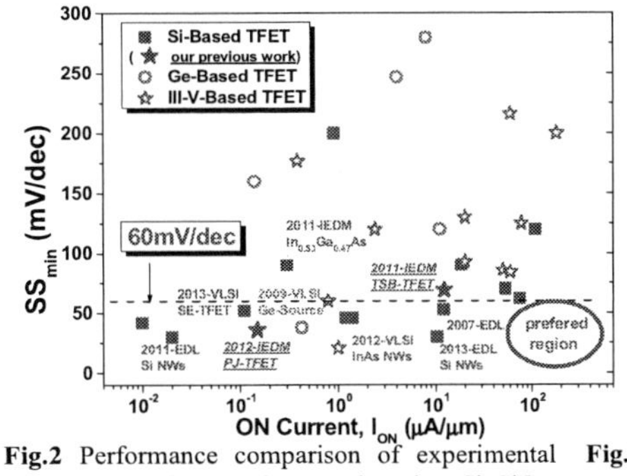

Fig.2 Performance comparison of experimental SS_{min} vs. I_{ON} of reported TFETs based on Si, SiGe, Ge, and III-V homo- and heterojunctions.

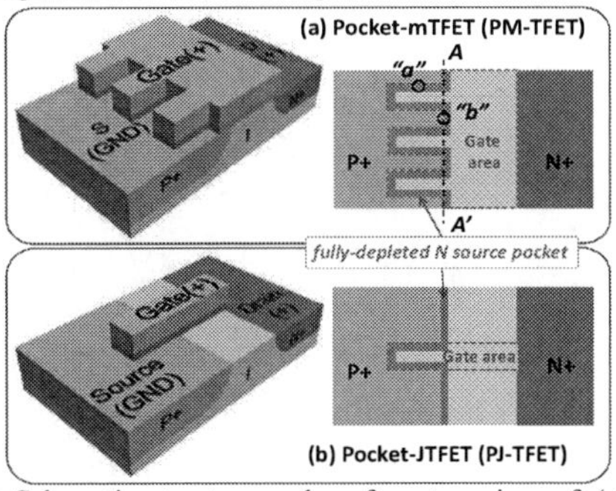

Fig.3 Schematic structure and surface top view of (a) proposed Pocket-mTFET with comb-shaped gate in this work and (b) our previously proposed Pocket-JTFET [4].

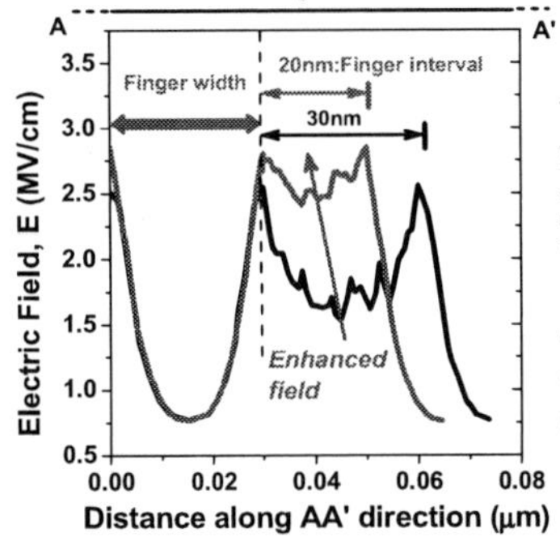

Fig.4 Simulated surface electric field of PM-TFET along AA' direction (shown in Fig.3a) with decreased finger interval.

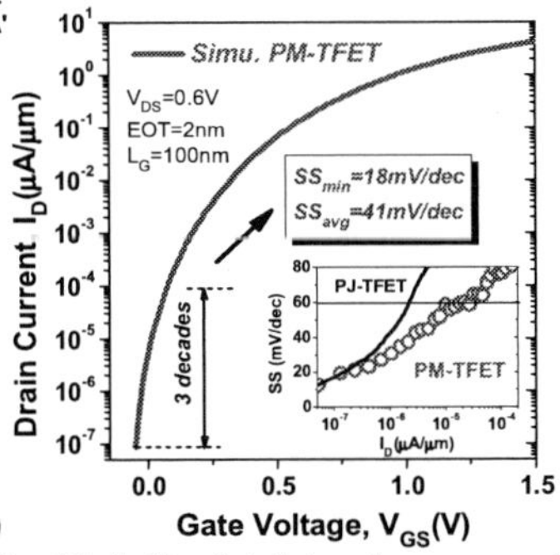

Fig.5 Simulated transfer curves of PM-TFET based on calibrated dynamic non-local BTBT model.

Fig.6 Experimental process flow of PM-TFET based on the bulk Si substrate.

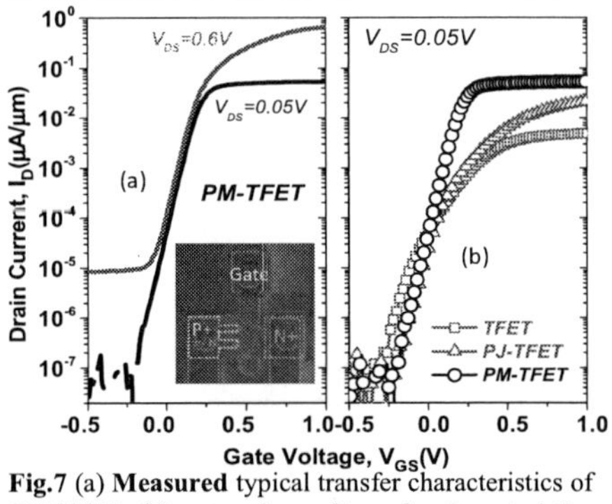

Fig.7 (a) **Measured** typical transfer characteristics of PM-TFET; (b) comparison of transfer characteristics between PM-TFET, PJ-TFET and traditional TFET.

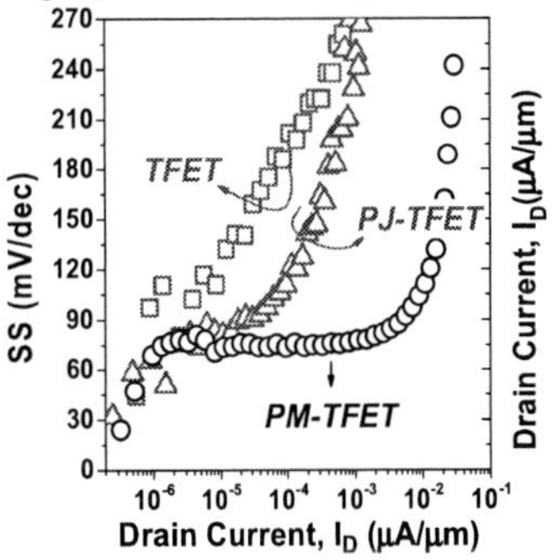

Fig.8 Subthreshold slopes of fabricated PM-TFET, PJ-TFET and traditional TFET as a function of I_D. PM-TFET shows the steepest SS within the largest current range.

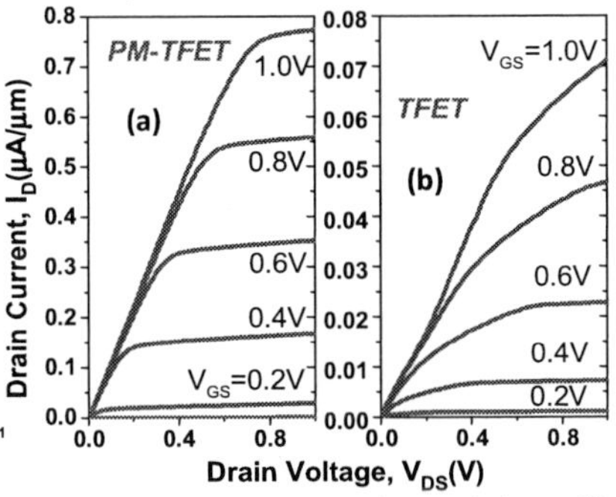

Fig.9 Measured output characteristics of fabricated (a) PM-TFET and (b) traditional TFET. PM-TFET shows alleviated super-linear onset issue due to the gate engineering.

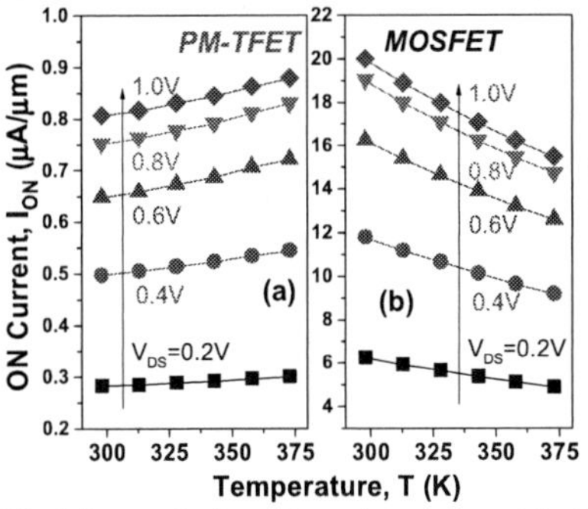

Fig.10 Measured I_{ON} dependence on temperature of PM-TFET and conventional MOSFET in this work; the positive I_{ON} dependence in PM-TFET indicates the dominant BTBT mechanism.

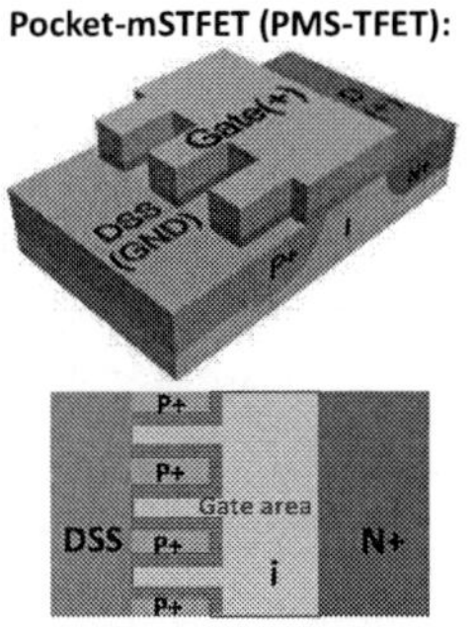

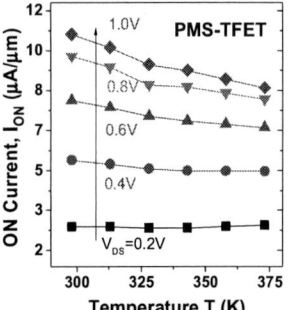

Fig.11 Schematic structure and surface top view of the proposed Pocket-mSTFET (PMS-TFET) with comb-shaped gate and dopant segregated Schottky (DSS) source.

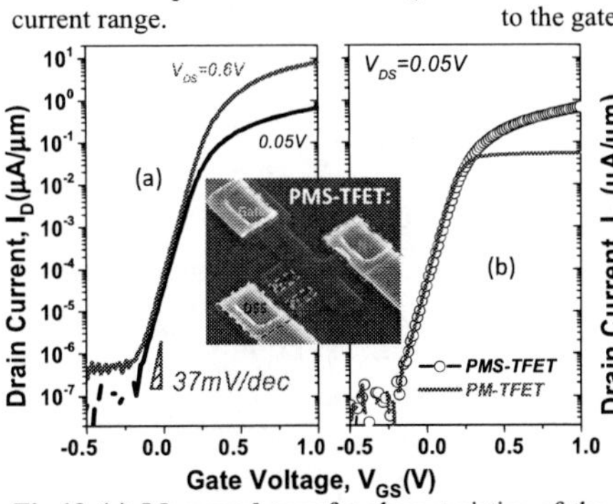

Fig.12 (a) **Measured** transfer characteristics of the fabricated PMS-TFET on the bulk Si substrate; (b) comparison of transfer characteristics of PMS-TFET and PM-TFET in the same die. PMS-TFET shows much higher I_{ON} due to the source engineering.

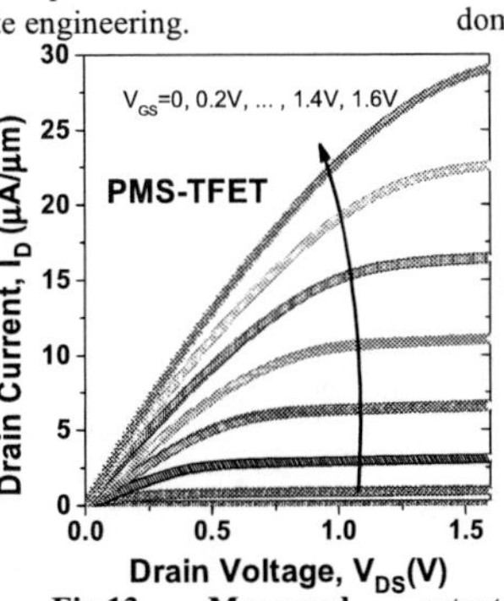

Fig.13 Measured output characteristics of the fabricated PMS-TFET on bulk Si substrate, showing good saturation behavior and low parasitic resistance.

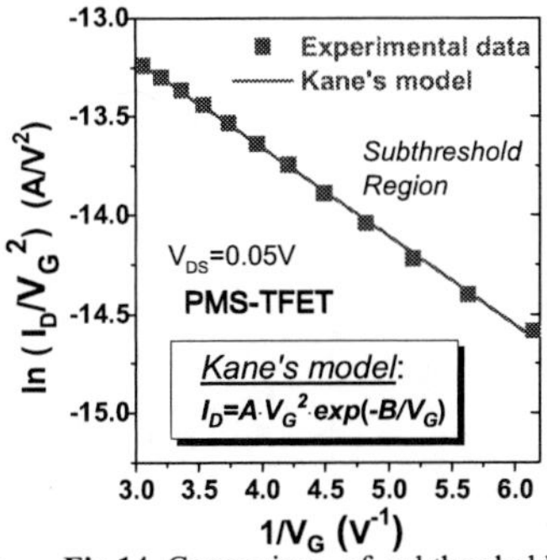

Fig.14 Comparison of subthreshold characteristics of PMS-TFET with Kane's model; perfect linear fitting demonstrates the dominant BTBT mechanism in subthreshold region.

Kane's model:
$$I_D = A \cdot V_G^2 \cdot exp(-B/V_G)$$

Fig.15 Measured ON-current dependence on temperature of PMS-TFET; negative dependence indicates the dominant Schottky current in the on-state.

978-1-4799-8002-4/14 $31.00 © 2014 IEEE

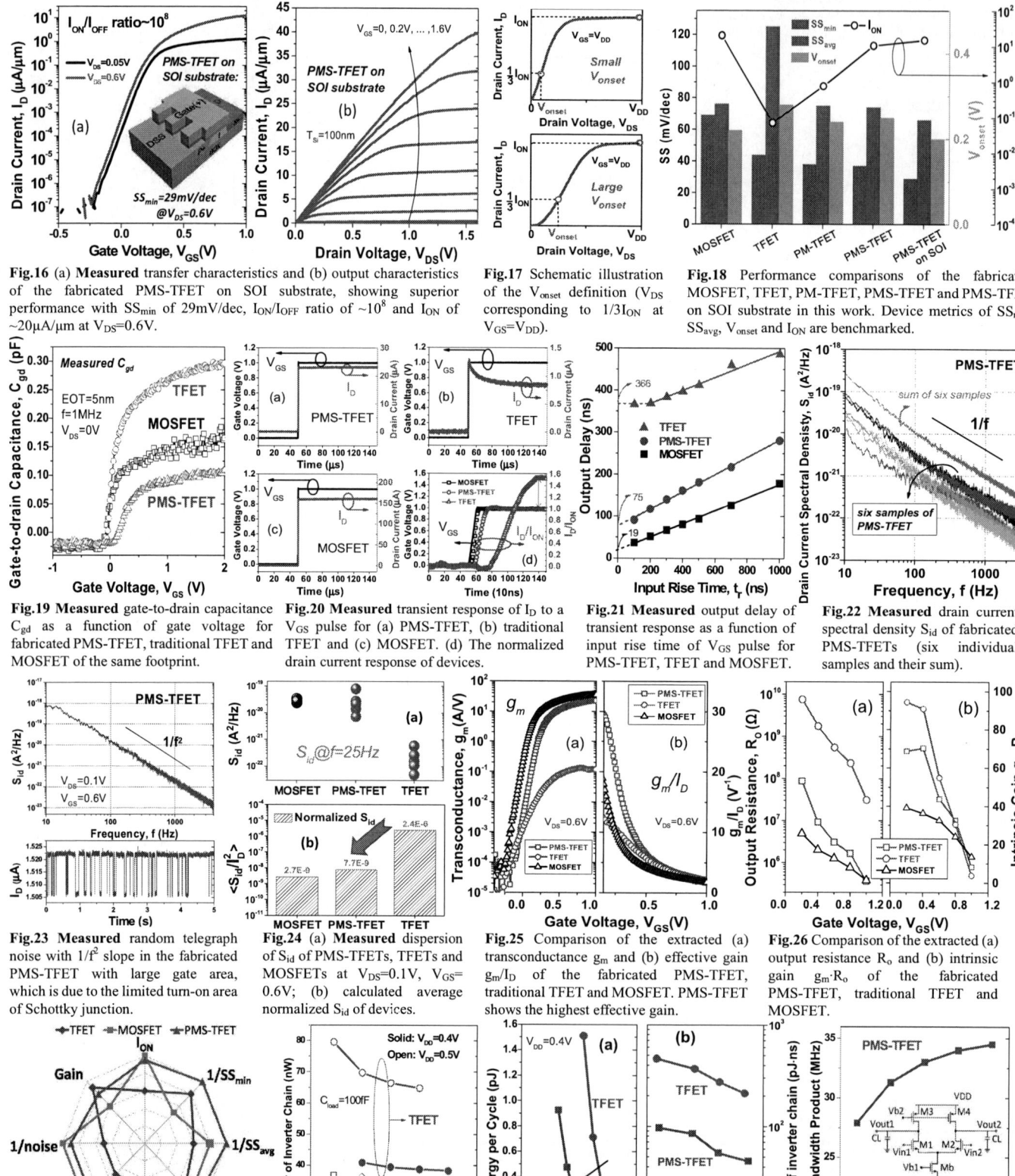

Fig.16 (a) **Measured** transfer characteristics and (b) output characteristics of the fabricated PMS-TFET on SOI substrate, showing superior performance with SS_{min} of 29mV/dec, I_{ON}/I_{OFF} ratio of ~10^8 and I_{ON} of ~20μA/μm at $V_{DS}=0.6V$.

Fig.17 Schematic illustration of the V_{onset} definition (V_{DS} corresponding to $1/3I_{ON}$ at $V_{GS}=V_{DD}$).

Fig.18 Performance comparisons of the fabricated MOSFET, TFET, PM-TFET, PMS-TFET and PMS-TFET on SOI substrate in this work. Device metrics of SS_{min}, SS_{avg}, V_{onset} and I_{ON} are benchmarked.

Fig.19 Measured gate-to-drain capacitance C_{gd} as a function of gate voltage for fabricated PMS-TFET, traditional TFET and MOSFET of the same footprint.

Fig.20 Measured transient response of I_D to a V_{GS} pulse for (a) PMS-TFET, (b) traditional TFET and (c) MOSFET. (d) The normalized drain current response of devices.

Fig.21 Measured output delay of transient response as a function of input rise time of V_{GS} pulse for PMS-TFET, TFET and MOSFET.

Fig.22 Measured drain current spectral density S_{id} of fabricated PMS-TFETs (six individual samples and their sum).

Fig.23 Measured random telegraph noise with $1/f^2$ slope in the fabricated PMS-TFET with large gate area, which is due to the limited turn-on area of Schottky junction.

Fig.24 (a) **Measured** dispersion of S_{id} of PMS-TFETs, TFETs and MOSFETs at $V_{DS}=0.1V$, $V_{GS}=0.6V$; (b) calculated average normalized S_{id} of devices.

Fig.25 Comparison of the extracted (a) transconductance g_m and (b) effective gain g_m/I_D of the fabricated PMS-TFET, traditional TFET and MOSFET. PMS-TFET shows the highest effective gain.

Fig.26 Comparison of the extracted (a) output resistance R_o and (b) intrinsic gain $g_m·R_o$ of the fabricated PMS-TFET, traditional TFET and MOSFET.

Fig.27 Radar chart for performance benchmark of the fabricated PMS-TFET, traditional TFET and MOSFET of the same footprint in this work.

Fig.28 Power of the inverter chain of PMS-TFET and traditional TFET as a function of input signal period at a V_{DD} of 0.4V and 0.5V.

Fig.29 (a) Energy per cycle of the inverter chain as a function of delay at V_{DD} of 0.4V; (b) Energy-delay product (EDP) as a function of static power consumption.

Fig.30 Gain bandwidth product at different bias current for PMS-TFET and traditional TFET. (Inset) schematic full-differential amplifier.

978-1-4799-8002-4/14 $31.00 © 2014 IEEE

A Schottky-Barrier Silicon FinFET with 6.0 mV/dec Subthreshold Slope over 5 Decades of Current

Jian Zhang, Michele De Marchi, Pierre-Emmanuel Gaillardon, Giovanni De Micheli

Integrated Systems Laboratory, EPFL, Lausanne, Switzerland
Tel: +41 21 6938164, E-mail: jian.zhang@epfl.ch

Abstract

In this paper, we demonstrate a steep *Subthreshold Slope* (SS) silicon FinFET with Schottky-barrier source/drain. The device shows a minimal SS of 3.4 mV/dec and an average SS of 6.0 mV/dec over 5 decades of current swing. Ultra-low leakage floor of 0.06 pA/μm is also achieved with high I_{on}/I_{off} ratio of 10^7.

Introduction

In low-power applications, small operation voltages and leakage currents are considered as the main technology means to reduce power consumption. However, the fundamental limitation of SS in conventional MOSFETs (~60 mV/dec at room temperature) becomes the bottleneck for continuously lowering the operation voltages. To break this limit, different types of devices have been proposed based on various mechanisms. *Tunnel FETs* (TFET) based on band-to-band tunneling demonstrate SS below 30 mV/dec [1]. Nevertheless, the sensitivity of SS to gate voltage leads to a worse average SS over the entire subthreshold region [1-3]. *Impact-ionization MOS* (IMOS) achieves sub-5 mV/dec SS based on avalanche breakdown [4], though requiring high V_{DS} and suffering from reliability issues. Based on an asymmetric structure similar to TFET and IMOS, feedback FET realizes steep SS with large hysteresis due to the charge trapping [5]. Recently, positive feedback based on weak impact ionization was proposed to achieve super-steep SS on UTBOX FDSOI substrate [6]. However, the steep transition only appears for 2 decades of current and rapidly degrades to ~100 mV/dec.

In this paper, we experimentally demonstrate a silicon FinFET with silicided source/drain, exploiting biased *Schottky Barriers* (SB). By combining a positive feedback induced by weak impact ionization with a dynamic modulation of the Schottky barriers, the device achieves a minimal subthreshold slope of 3.4 mV/dec and an average subthreshold slope of 6.0 mV/dec over 5 decades of current at room temperature. Ultra-low leakage current of 0.06 pA/μm and high I_{on}/I_{off} ratio of 10^7 are also obtained by effectively modulating the Schottky barriers.

Device Structure and Fabrication

The fin-based device structure is shown in Fig. 1. The *Schottky-Barrier Bias* (SBB) electrostatically modulates the Schottky barriers at S/D, while the gate controls the potential barrier in the channel to turn the device *on* or *off*.

Fig. 2a shows the fabrication steps of the device. Starting from a lightly *p*-type doped SOI substrate with 340 nm silicon device layer and 2 μm BOX, the channel is patterned into an 800 nm long and 50-70 nm wide fin shape with e-beam lithography (Fig. 2b). A layer of SiO_2 (~15 nm) is formed and followed by a conformal polysilicon deposition. Then, the SBB region is patterned. After a second oxidation (~15 nm) and polysilicon deposition, the gate is self-aligned to the SBB. The achieved gate length is 200 nm. After the gate process, silicon nitride spacers are formed to isolate the structures. A 20-nm nickel layer is deposited by sputtering and a specific annealing process is performed to produce NiSi at the S/D and gate contacts [7]. The device (Fig. 2c) has a channel length of ~600 nm and the final width of the fin is designed to be either 40 nm, 50 nm, or 60 nm. Although currently limited by our academic cleanroom facilities, the device can be scaled down easily thanks to the dopant-free process.

Working Principle

The working principle is shown in Fig. 3. For *n*-type behavior, i.e., when $V_{SBB} > 0$, electrons are selected to tunnel through the Schottky barrier into the channel. When V_G is at threshold voltage, a transition occurs in the device. Weak impact ionization generates electron/hole pairs (step 1). The generated electrons drift to the drain, and the holes accumulate in the potential well induced by the gate (step 2). This lowers the barrier, and provides more electrons for impact ionization. Then, more accumulated holes continue to lower the barrier and thus form a positive feedback [6]. In addition to the FinFET structure enhancing carrier multiplication [4], the second important contribution is from the dynamic modulation of the Schottky barrier. Parts of the generated holes are swept towards the source, increasing the hole density in SBB region (step 3). This helps to lower the energy band under SBB. Schottky barrier at source becomes thinner and more electrons tunnel through it. In the meantime, the potential well is kept until the final *on* state. This mechanism improves the I_{on}/I_{off} ratio, and is considered as a key to achieve steep transition for 5 decades of current. The operation of *p*-type is similar but with $V_{SBB} < 0$.

978-1-4799-8002-4/14 $31.00 © 2014 IEEE

Fig. 4 shows the TCAD simulation of the proposed device. Polarity of the device changes by adapting the polarity of V_{SBB}. Steep SS is obtained in both n-type and p-type (Fig.4a). A sudden change of the surface potential just before and after the *on* state is observed (Fig. 4b). The hole density in the device shows the accumulation of holes after the *on* state as predicted in the proposed mechanism (Fig. 4c).

Results and Discussions

All measurements are carried out at room temperature. Fig. 5 shows the characteristics of the steep SS FinFET. Minimum SS of 3.4 mV/dec is achieved. When decreasing V_{DS} (i.e., the lateral electric field), the impact ionization rate decreases, and the SS gradually degrades to 61 mV/dec at $V_{DS}=1V$. Compared to counterparts with significant hysteresis [5][8], double sweep confirms the negligible hysteresis in the characteristics of the proposed device. Fig. 5b illustrates the SS as a function of I_D. The average SS of 6.0 mV/dec is observed for 5 decades of current. V_{SBB} controls the tunneling of electrons from SB and thus modulates the SS and I_{on}/I_{off} ratio (Fig. 5c). n-type and p-type characteristics, obtained by changing the polarity of V_{SBB} in the same device, are demonstrated in Fig. 6. The SS of p-type is above 70 mV/dec, possibly due to a lower impact ionization rate of holes. Both n-/p-types show >10^6/10^7 I_{on}/I_{off} ratio. The steep SS in devices with different width of fin (W_{fin}) is shown in Fig. 7a. The SS is independent on W_{fin} within the range of 40-60 nm, while V_{th} increases in devices with a thinner fin. The statistics based on all the measured devices (~50) at lower voltages support this relation (Fig. 7b).

In order to show the effect of the SBB, we fabricated a FinFET with a single SBB region (Fig. 8). In this second device, the channel length and the impact ionization region are the same as in the proposed device. However, there is no potential well for accumulating holes under gate due to the absence of the SBB at source (Fig. 8b). As a result of the weak positive feedback, the SS is slightly below 60 mV/dec for less than 2 decades (Fig. 8d,e). This result verifies the importance of the potential well and the SBB at source. Fig. 9 shows statistical distribution of SS. At operation voltage of 5V, 63% of the measured devices demonstrate SS below 10 mV/dec. When reducing the operation voltage down to 4V, there are still 80% showing SS below 50 mV/dec. As observed in the output characteristics in Fig. 10, weak impact ionization occurs at low V_G and high V_{DS}. When increasing V_G, the lateral electric field decreases and impact ionization vanishes. Compared to IMOS, which is based on strong impact ionization and suffers from reliability problems, the weak impact ionization in the proposed device requires lower V_{DS} and only occurs during the transition. Thus, the device exhibits good reliability as shown in Fig. 11. In the test for hot carrier injection and bias temperature instability, no significant degradation is observed. Moreover, thanks to the FinFET structure, the device exhibits an excellent electrostatic control at low operation voltages (both V_{SBB} and V_{DS}) down to 0.5V (SS <70 mV/dec, and good control of DIBL effect) (Fig. 12). Table I compares the recently reported technologies with sub-60 mV/dec SS and the proposed device. Despite the thicker oxide and longer channel, the proposed device demonstrates ultra-low leakage current, high I_{on}/I_{off} ratio as well as good minimum and average SS at both high and low V_{DS}. Since the impact ionization rate strongly depends on the electric field, the performance at low V_{DS} can be further improved in scaled devices as suggested by the simulation shown in Fig. 4.

Conclusions

We experimentally demonstrated steep SS in a silicon FinFET, exploiting the electrostatic biasing of the S/D Schottky barriers. Minimal SS of 3.4 mV/dec and average SS of 6.0 mV/dec are achieved. Ultra-low leakage floor and high I_{on}/I_{off} ratio (0.06 pA/µm, 10^7 @$V_{DS}=5V$ and ~3 fA/µm, 10^8 @$V_{DS}=1V$) are suitable for low-power applications.

Acknowledgement

This work was supported by Grant ERC-2009-AdG-246810.

References

[1] K. Tomioka, M. Yoshinura, E. Nakai, F. Ishizaka, and T. Fukui, "Integration of III-V nanowires on Si: From high-performance vertical FET to steep-slope switch," *IEDM Tech. Dig. 2013*, pp. 88-91.

[2] Q. Huang, R. Huang, Z. Zhan, Y. Qiu, W. Jiang, C. Wu, and Y. Wang, "A novel Si tunnel FET with 36mV/dec subthreshold slope based on junction depleted-modulation through striped gate configuration," *IEDM Tech. Dig.* 2012, pp. 187-190.

[3] A. Villalon, *et al.*, "First demonstration of strained SiGe Nanowires TFETs with Ion beyond 700uA/um," *VLSI Tech Symp. 2014*.

[4] E.-H. Toh, *et al.*, "Impact ionization nanowire transistor with multiple-gates, silicon-germanium impact ionization region, and sub-5 mV/decade substheshold swing," *IEDM Tech. Dig. 2007*, pp. 195-198.

[5] A. Padilla, C. W. Yeung, C. Shin, C. Hu, and T.-J. K. Liu, "Feedback FET: A novel transistor exhibiting steep switching behavior at low bias voltages," *IEDM Tech. Dig. 2008*.

[6] Z. Lu, N. Collaert, M. Aoulaiche, B. De Wachter, A. De Keersgieter, J. G. Fossum, L. Altimime, and M. Jurczak, "Realizing super-steep subthreshold slope with conventional FDSOI CMOS at low-bias voltages," *IEDM Tech. Dig. 2010*, pp. 407-409.

[7] M. De Marchi, D. Sacchetto, S. Frache, J. Zhang, P.-E. Gaillardon, Y. Leblebici, and G. De Micheli, "Polarity control in double-gate, gate-all-around vertically stacked silicon nanowire FETs," *IEDM Tech. Dig. 2012*, pp. 183-186.

[8] C.-E. D. Chen, M. Matloubian, R. Sundaresan, B.-Y. Mao, C.C. Wei, and G. P. Pollack, "Single-transistor latch in SOI MOSFET's," *IEEE EDL*, vol. 9, no. 12, pp. 636-638, 1988.

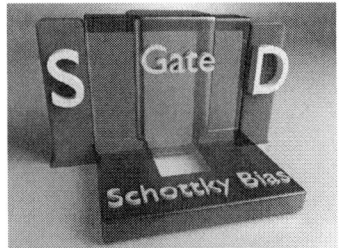

Fig.1. Sketch of the proposed FinFET. Violet: *Schottky-Barrier Bias* (SBB) region modulating the Schottky barriers. Red: Gate controlling the channel conduction.

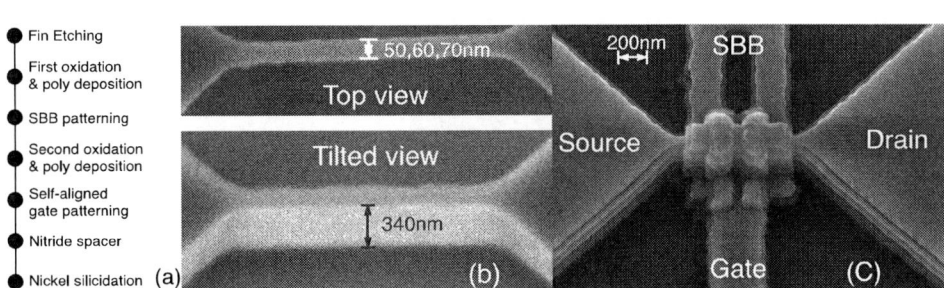

Fig.2. Fabrication of the proposed device: (a) Process flow. (b) SEM image of the fin, H_{fin}=340 nm, W_{fin} after fin etching are 50 nm, 60 nm and 70 nm. (c) SEM image of the final device, L_{gate}=200 nm, T_{ox}~15 nm. Total channel length ~600 nm, the final W_{fin} are 40 nm, 50 nm and 60 nm considering the silicon consumed during oxidation.

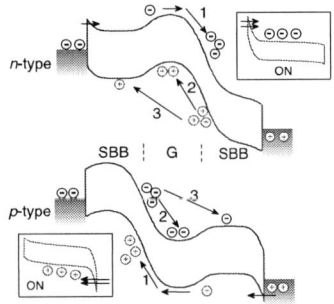

Fig.3. Band diagram of *n*-type (V_{SBB}>0) and *p*-type (V_{SBB}<0). Main switching mechanisms: 1: impact ionization, 2: generated carriers accumulating in the potential well under the gate, 3: carriers accumulating under the SBB region. When completely *on*, impact ionization and potential well vanish (inset diagrams).

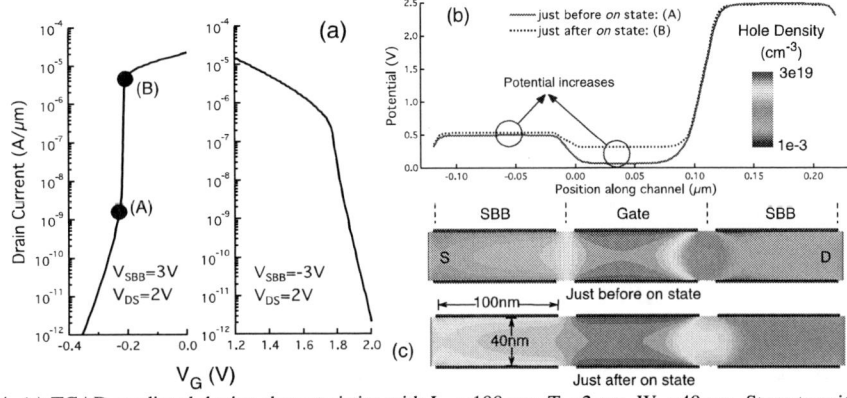

Fig.4. (a) TCAD-predicted device characteristics with L_{gate}=100 nm, T_{ox}=2 nm, W_{fin}=40 nm. Steep transition is observed in both *n*-type (~5 mV/dec) and *p*-type (~35 mV/dec). The device polarity changes by adapting the polarity of V_{SBB}. (b) Surface potential distribution just before and after *on* state (*n*-type). Potential increases under both Gate and SBB. (c) Hole density showing the accumulation of holes under Gate and SBB.

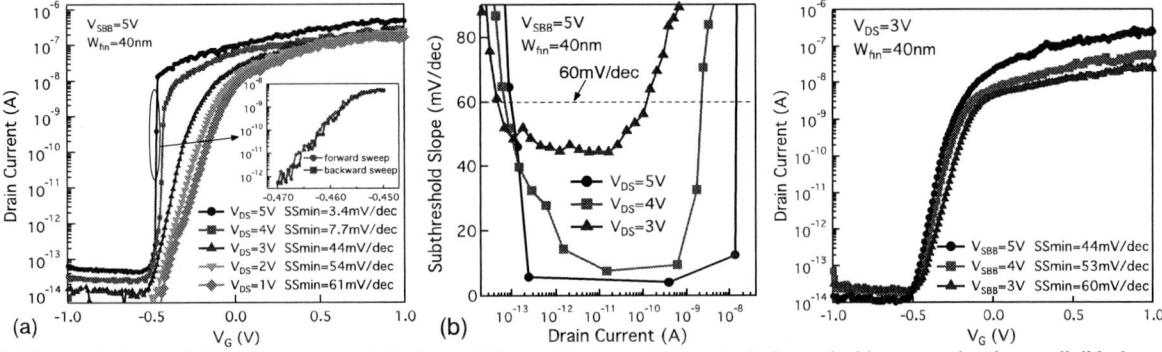

Fig.5. (a) Measured characteristics of the proposed device at different V_{DS} (room temperature). Inset: double sweep showing negligible hysteresis and minimal SS=3.4 mV/dec. GIDL is well suppressed. The applied V_{DS} are much smaller compared to IMOS [4]. (b) SS vs. I_D, showing average SS of 6.0 mV/dec for 5 decades of current. (c) I_D-V_G at different V_{SBB}. Higher V_{SBB} improves both SS and I_{on}/I_{off} ratio thanks to a better control of Schottky barriers.

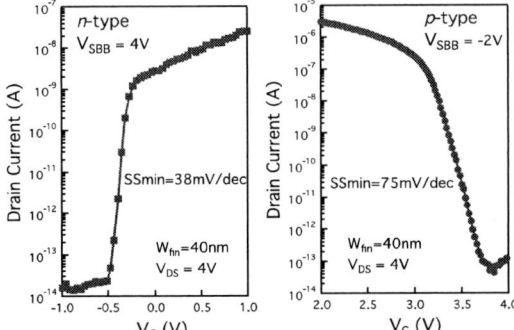

Fig.6. Measured *n*-type and *p*-type characteristics in the same device. The polarity of V_{SBB} determines the device polarity. *n*-type demonstrates 38 mV/dec SS, while the SS of *p*-type is above 70 mV/dec. Both show good I_{on}/I_{off} ratio and maintain good SS for over 5 decades of current.

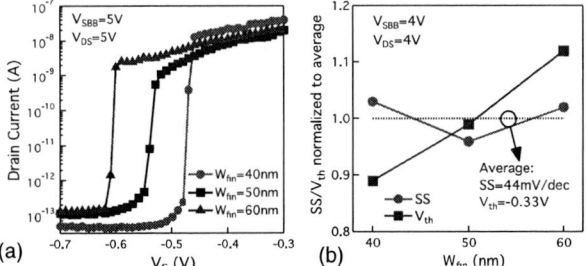

Fig.7. (a) Measured characteristics in devices with different W_{fin}. V_{SBB}=V_{DS}=5V. SS does not significantly depend on W_{fin}, while larger W_{fin} reduces the V_{th}. W_{fin}=40 nm shows better electrostatic control of the SB, thus enhancing I_{on}/I_{off}. (b) Statistics of SS and V_{th} based on all measured devices (~50) at lower voltage (V_{SBB}=V_{DS}=4V), confirming the dependence of V_{th} on W_{fin}. The values are normalized to the average value.

978-1-4799-8002-4/14 $31.00 © 2014 IEEE

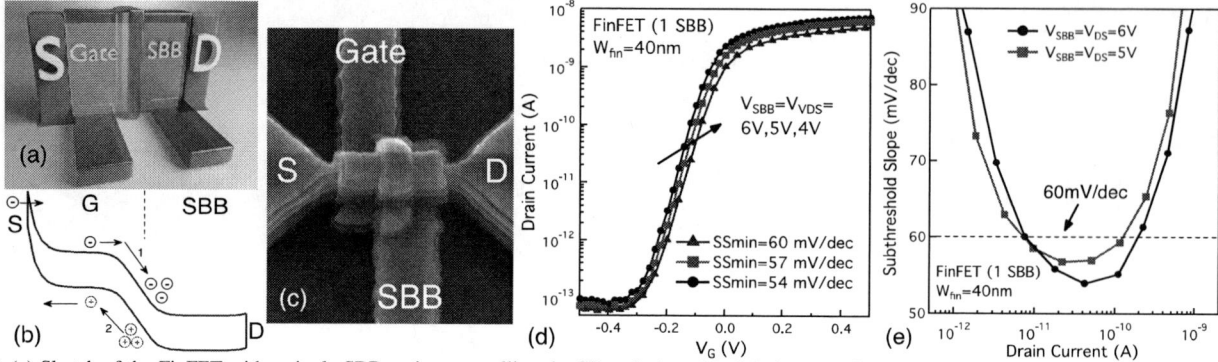

Fig.8. (a) Sketch of the FinFET with a single SBB region controlling the SB at drain to block holes tunneling. Gate controls the SB at source and electrons tunneling, thus turns the device *on* or *off*. (b) Band diagram shows no potential well in the channel, leading to a weak positive feedback. (c) SEM image. (d) Measured characteristics of the device. SSmin is slightly below the thermal limit. (e) SS vs. I_n. SS is below 60 mV/dec for less than 2 decades of current.

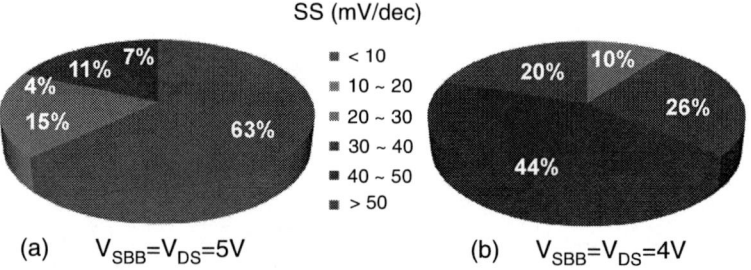

Fig.9. Statistical distribution of subthreshold slope of the proposed device illustrated in Fig. 1. (a) 63% of the measured devices are working with SS<10mV/dec at $V_{SBB}=V_{DS}=5V$. (b) 80% of the devices show SS below 50mV/dec at $V_{SBB}=V_{DS}=4V$.

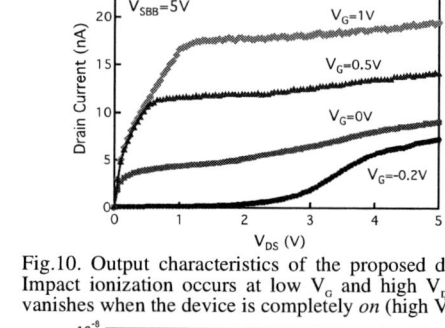

Fig.10. Output characteristics of the proposed device. Impact ionization occurs at low V_G and high V_{DS}, but vanishes when the device is completely *on* (high V_G).

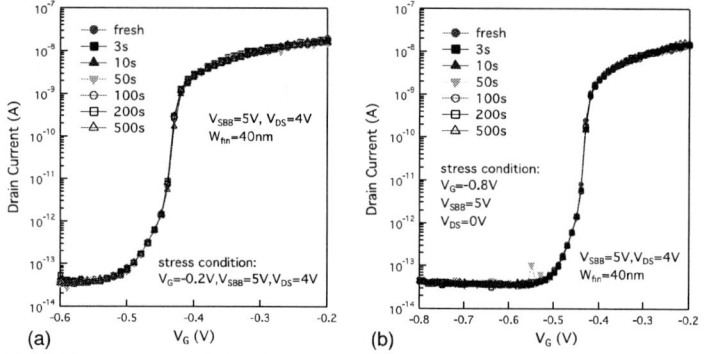

Fig.11. Reliability assessment during fast transition (SS<10 mV/dec): (a) hot carrier injection and (b) bias temperature instability with different stress period. No significant degradation is observed, proving the good reliability of the device.

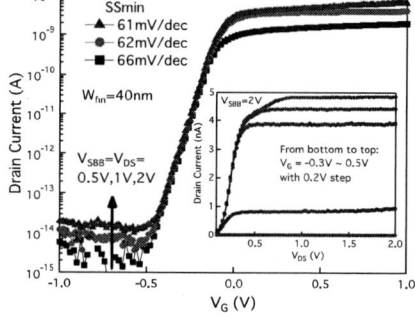

Fig.12. Characteristics under low operation voltages. Near-ideal SS and high I_{on}/I_{off} ratio show the excellent electrostatic control. Inset: Output characteristics show a good control of the DIBL effect.

Table I. Comparison between state-of-the-art technologies with sub-60 mV/dec SS

	[4] IEDM'07	[6] IEDM'10	[2] IEDM'12	[1] IEDM'13	This work	
Principle	Impact ionization	Positive feedback	Band-to-band tunneling	Band-to-band tunneling	Positive feedback + Schottky barrier modulation	
Technology	SiGe Nanowire	UTBOX FDSOI	Pocket + junction modulation in silicon TFET	InGaAs Nanowire/Si Heterojunction	Schottky-barrier silicon FinFET	
Symmetric	No	Yes	No	No	Yes	
EOT (nm)	3	2.5	5	1.91	15	
Channel length (nm)	90	55		150	600	
V_{DS}(V)	5.75	1.3	0.6	1	1	5
I_{on} (µA/µm)*	300	100	0.15	0.006	0.28	0.59
I_{off} (pA/µm)*	10000	1	0.42	0.06	0.003	0.06
I_{on}/I_{off}	10^4	10^8	10^5	10^5	10^8	10^7
SS min	<5	0.058	36	23	61	3.4
SS average (Decades of current)	<10 (4)	~70 (6)	81 (3)	~80 (4)	63 (6)	6.0 (5)

* In this work, the current is normalized to the effective width of the channel, i.e., $W_{eff}=W_{fin}+2\times H_{fin}$.

Can piezoelectricity lead to negative capacitance?

Justin C. Wong, Sayeef Salahuddin

Department of Electrical Engineering and Computer Sciences, University of California, Berkeley
Berkeley, CA 94720, USA

Abstract

A thermodynamic model was constructed to quantitatively analyze the negative capacitance effect in the presence of piezoelectricity, electrostriction, and ferroelectricity. The model shows that pure piezoelectricity and higher-order electromechanical coupling can provide a negative capacitance effect in principle, but are not strong enough in practice. Negative capacitance is predicted to occur due to ferroelectric polarization switching and not due to piezoelectricity.

Introduction

Negative capacitance has recently become a research area of great interest due to its potential ability to reduce subthreshold swing below the Boltzmann limit of 60 mV/decade (1). Much of the early research has focused on the intrinsic negative capacitance exhibited by ferroelectrics due to polarization switching. However, in a recent work, Then et al. presented compelling evidence of negative capacitance by adding a piezoelectric AlInN layer to the gate stack of a GaN MOS-HEMT (2). Jana et al. has also discussed the possibility of achieving negative capacitance using the piezoelectric property (3). This poses intriguing questions such as: is it possible to achieve negative capacitance purely from a piezoelectric effect? If yes, then what are the determining factors?

Here we examine these questions by constructing a thermodynamic model that allows for quantitative analysis of the negative capacitance effect in the presence of piezoelectricity, electrostriction, and ferroelectricity from a coherent platform. Our results show that (i) piezoelectricity and (ii) electrostriction could show a negative capacitance effect in principle, but for both of those mechanisms the electric fields needed are orders of magnitude larger than the breakdown fields of known piezoelectric oxides; and (iii) for a piezoelectric material which is also a ferroelectric, negative capacitance could arise at a more reasonable field but due to ferroelectricity and not due to piezoelectricity. To summarize, for all practical purposes, purely electromechanical coupling is found to be not strong enough to lead to negative capacitance.

Thermodynamic Model

Negative capacitance can be understood in terms of a positive feedback effect (1). Consequently, since piezoelectricity is a linear coupling between electric field and strain, then there

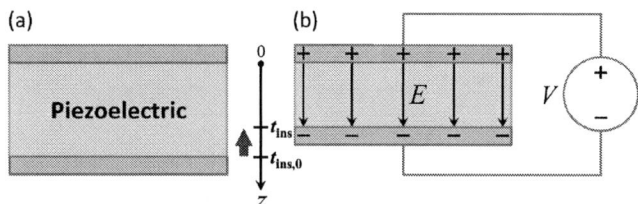

Fig. 1. (a) Schematic of a capacitor with a piezoelectric insulator of thickness $t_{ins,0}$. (b) Applying a voltage establishes an electric field that compresses the piezoelectric to a new thickness t_{ins} via the piezoelectric effect. This compression results in a higher electric field for the same voltage to further compress the piezoelectric. This process continuously repeats as a positive feedback effect resulting in negative capacitance.

must exist a positive feedback effect that exploits electromechanical coupling in order for piezoelectricity to provide negative capacitance. A simple scheme that achieves such a feedback effect is depicted in Fig. 1a. A piezoelectric is used as the insulator of a capacitor such that an applied voltage establishes an electric field that induces compressive strain in the piezoelectric along the capacitor axis (Fig. 1b). If the piezoelectric compresses, then the capacitance increases, resulting in additional charge stored and a stronger electric field for the same voltage. This higher electric field further compresses the piezoelectric, resulting in negative capacitance through a compressive positive feedback effect. To analyze this system, we developed a model using the thermodynamic framework shown in Fig. 2. The electric Gibbs free energy shown in Fig. 2c is minimized with respect to charge under constant voltage to obtain constitutive equations.

$$(a)\ \partial u = \underbrace{\sigma_{ij}\partial\varepsilon_{ij}}_{\substack{\text{elastic energy}\\\text{density}}} + \underbrace{E_i\partial D_i}_{\substack{\text{electric energy}\\\text{density}}}$$

$$(b)\ \partial U = \int_{\text{Volume}} \partial u\ dxdydz$$

$$(c)\ G = U - QV$$

Fig. 2. (a) Internal energy density differential for electroelastic materials. (b) Total internal energy differential (useful since our system couples energy density with volume). (c) Electric Gibbs free energy under isothermal conditions.

Results

The electric Gibbs free energy for a piezoelectric modeled after PZT is shown in Fig. 3. There is a clear thermodynamic instability corresponding to negative capacitance that can be

978-1-4799-8002-4/14 $31.00 © 2014 IEEE

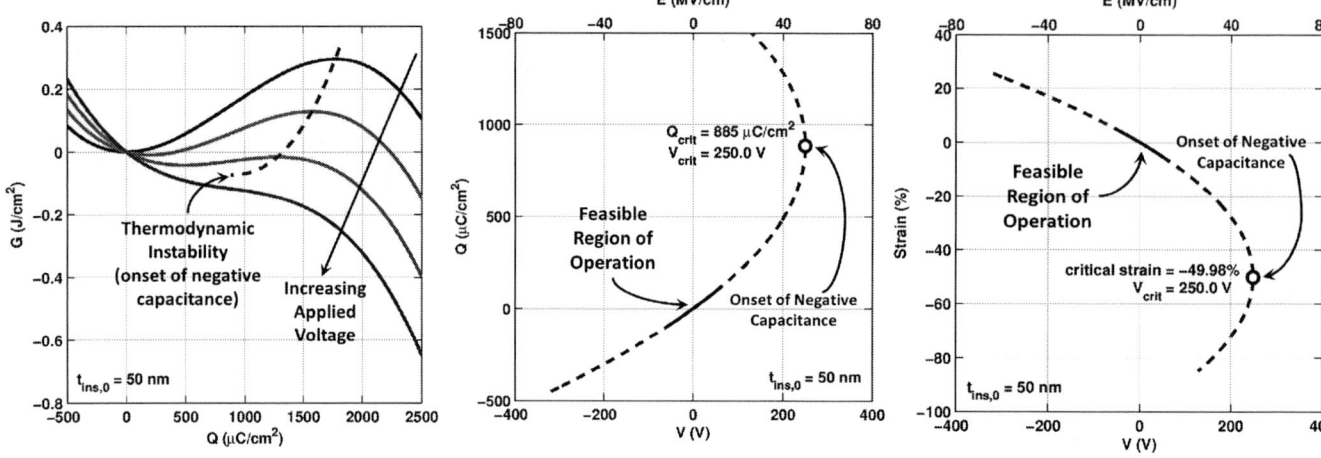

Fig. 3. Electric Gibbs free energy for a piezoelectric modeled after PZT. Notice the thermodynamic instability (negative capacitance) becomes accessible with sufficiently high voltage.

Fig. 4. Plot of charge versus voltage for the system analyzed in Fig. 3. The negative capacitance is present as predicted, but occurs at an unphysical amount of charge and electric field.

Fig. 5. Plot of strain versus voltage for the system analyzed in Fig. 3. Similarly to Fig. 4, the negative capacitance occurs at an unphysical amount of strain and electric field.

accessed in principle by applying sufficient voltage. Unfortunately, this instability is physically unachievable as shown in Fig. 4-5 due to the high amount of charge and electric field required. Piezoelectricity alone is not strong enough to achieve the targeted effect within a feasible operating range. Furthermore, PZT is considered a strong piezoelectric, which implies that piezoelectricity in weaker piezoelectrics—such as the nitrides used in Intel's MOS-HEMT—cannot by itself account for the negative capacitance observed.

The unphysical amounts of charge and electric field required prompted us to extend our model to include second-order effects such as electrostriction (the difference between piezoelectricity and electrostriction is depicted in Fig. 6). The corresponding electric Gibbs free energy for PZT with electrostriction included is shown in Fig. 7. The thermodynamic instability is no longer present due to PZT's positive electrostriction coefficient, which prevents the compressive positive feedback effect from occurring as shown in Fig. 9. Additionally, Fig. 8 shows that the relationship between charge and

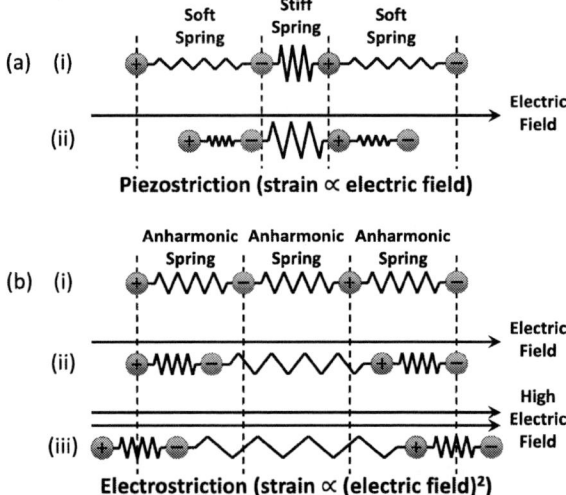

Fig. 6. (a) (i) Piezoelectricity can be modeled using springs with different spring constants (4). (ii) Applying an electric field results in net strain because each spring strains by a different amount depending on its spring constant. (b) (i) Electrostriction can be modeled by considering the anharmonicity of the spring model (4). (ii) Applying an electric field need not result in strain because each spring can strain equivalently to first order. (iii) Anharmonicity in this case allows for easier elongation but with tougher compression, resulting in a net tensile strain at high electric fields.

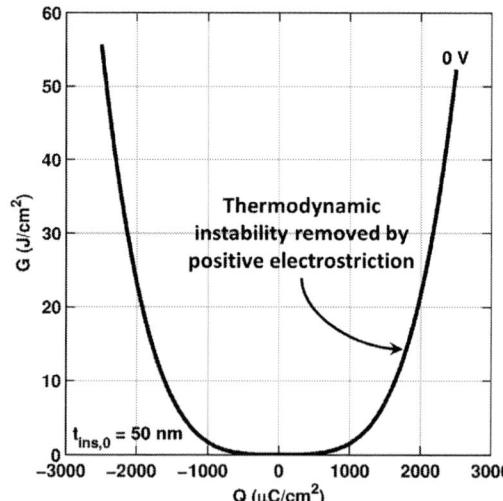

Fig. 7. Electric Gibbs free energy for PZT with electrostriction included. The thermodynamic instability has been removed by PZT's positive electrostriction.

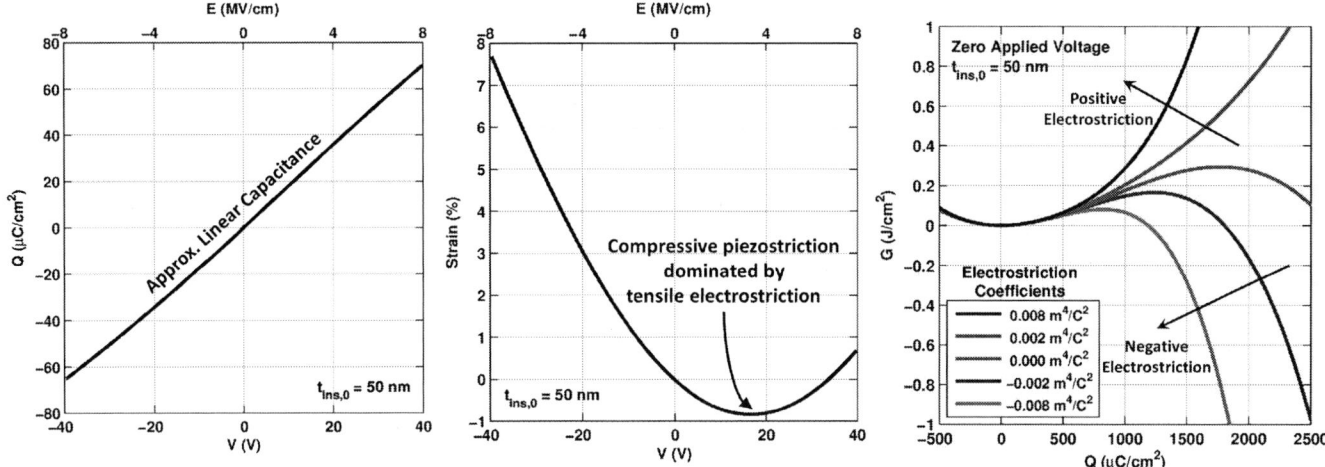

Fig. 8. The negative capacitance effect vanishes when electrostriction is included, and the overall capacitance becomes approximately linear.

Fig. 9. Tensile electrostriction prevents sufficient compression from occurring, which is needed to initiate the positive feedback effect.

Fig. 10. Electric Gibbs free energy with different electrostriction coefficients. Non-positive electrostriction is necessary for accessing the thermodynamic instability.

voltage is nearly linear. An extension to third-order effects was not performed because practical piezoelectric materials break down before third-order effects can occur.

These results prompted us to explore the consequences of different electrostriction coefficients. The electric Gibbs free energy under zero applied voltage is plotted for different electrostriction coefficients in Fig. 10. Positive electrostriction coefficients remove the thermodynamic instability, while non-positive electrostriction coefficients allow the instability to exist. Increasingly negative electrostriction coefficients make the thermodynamic instability more easily accessible. Fig. 11 shows charge and strain plotted against voltage for increasingly negative electrostriction coefficients. Although the required charge decreases for increasingly negative electrostriction coefficients, the required strains remain on the order of tens of percent, which are unphysical for piezoelec-

tric materials. Finally, Table I shows that most commonly used materials do not possess negative electrostriction coefficients (5).

TABLE I
ELECTROSTRICTION COEFFICIENTS OF COMMON MATERIALS

Material	Electrostriction Coefficient Q_{11} (m^4/C^2)
BaTiO$_3$	0.11
PbZr$_{1-x}$Ti$_x$O$_3$	0.04 – 0.09
LiNbO$_3$	0.016
LiTaO$_3$	0.021
PMN-0.28PT	0.055
PMN-0.32PT	0.056
PMN-0.37PT	0.056
CaF$_2$	-0.48
BaF$_2$	-0.33
SrF$_2$	-0.33

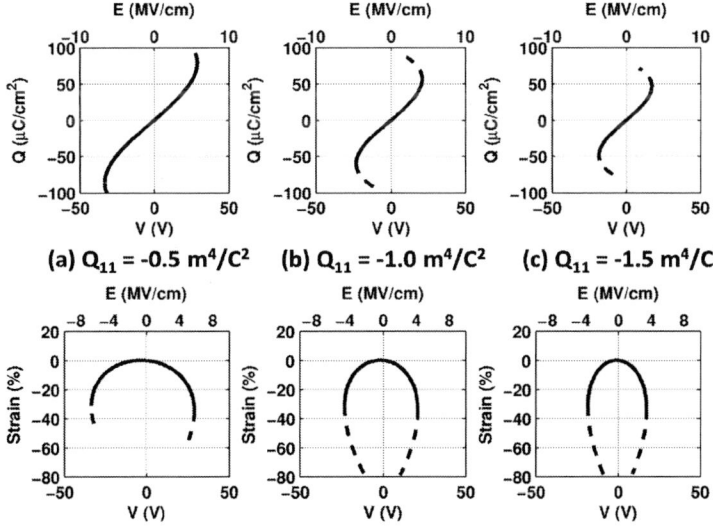

(a) Q_{11} = -0.5 m^4/C^2 (b) Q_{11} = -1.0 m^4/C^2 (c) Q_{11} = -1.5 m^4/C^2

Fig. 11. Charge and strain plotted against voltage for different negative electrostriction coefficients. Notice that the critical charge and voltage needed to enter the negative capacitance region reduces with increasingly negative electrostriction while the strains needed remain similar. ($t_{ins,0}$ = 50 nm)

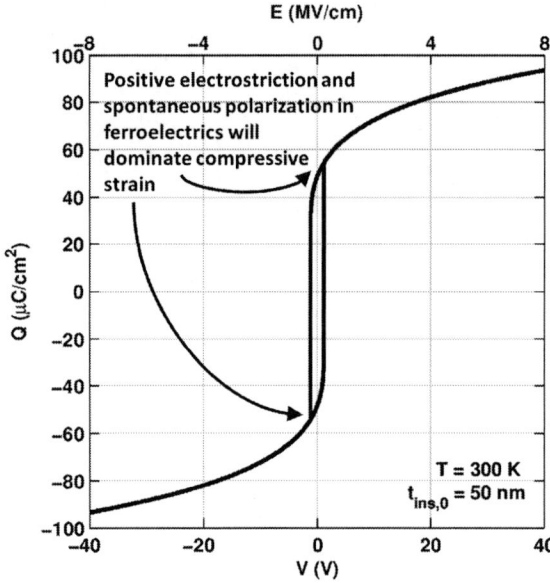

Fig. 12. Charge versus voltage plotted for PZT with electrostriction and ferroelectricity included. Positive electrostriction combined with spontaneous polarization tend to expand the material.

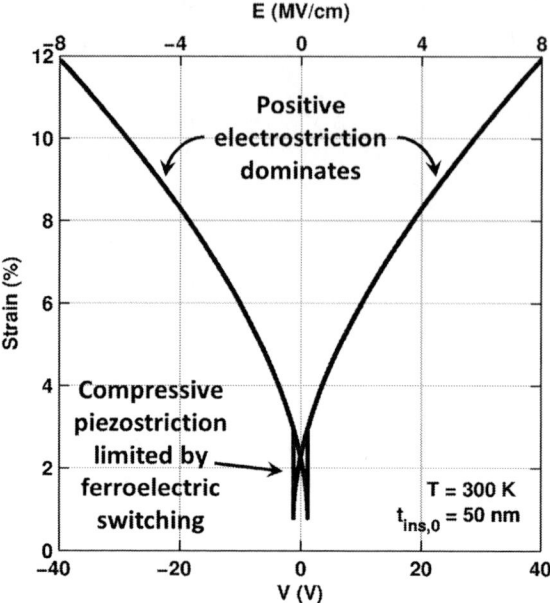

Fig. 13. Strain versus voltage for PZT with electrostriction and ferroelectricity included. Compressively straining the material below its spontaneous strain eventually switches the ferroelectric, resulting in expansion.

We further included the ferroelectric properties of PZT. The ferroelectric nature of PZT results in a spontaneous polarization that combines with electrostriction to produce a spontaneous strain. In order to compress PZT with an electric field via the converse piezoelectric effect, the electric field must be applied in the direction opposite to that of the spontaneous polarization. However, due to the ferroelectric nature of the material, applying too strong of an electric field switches the direction of the spontaneous polarization as shown in Fig. 12. Once this occurs, the material begins expanding in the direction of the applied electric field as shown in Fig. 13. Thus, a negative capacitance can be found but due to the ferroelectric polarization switching and not due to piezoelectricity.

Conclusion

To conclude, we find that piezoelectricity alone cannot provide a negative capacitance effect. This does not change even after including higher-order electromechanical coupling such as electrostriction. Piezoelectric strain might lead to dynamic change in threshold voltage by changing the mobili-

ty; however, strain is symmetric with electric field which means that the effect should arise in both branches of the hysteresis curve. Therefore, the effect observed by Then et al. is unlikely to have arisen from a purely electromechanical mechanism.

References

(1) S. Salahuddin and S. Datta, "Use of negative capacitance to provide voltage amplification for low power nanoscale devices," *Nano Lett.*, vol. 8, no. 2, pp. 405-410, Feb. 2008.

(2) H. W. Then et al., "Experimental observation and physics of 'negative' capacitance and steeper than 40mV/decade subthreshold swing in $Al_{0.83}In_{0.17}N/AlN/GaN$ MOS-HEMT on SiC substrate," in *2013 IEEE International Electron Devices Meeting*, 2013, vol. 117, no. 2004, pp. 28.3.1-28.3.4.

(3) R. K. Jana, G. L. Snider, and D. Jena, "On the possibility of sub-60 mV/decade subthreshold switching in piezoelectric gate barrier transistors," *Phys. Status Solidi*, vol. 10, no. 11, pp. 1469-1472, Nov. 2013.

(4) K. Uchino, "Piezoelectric and electrostrictive actuators," *Sixth International Symposium on Applications of Ferroelectrics*, 1986, pp. 610-618.

(5) F. Li, L. Jin, Z. Xu, and S. Zhang, "Electrostrictive effect in ferroelectrics: An alternative approach to improve piezoelectricity," *Appl. Phys. Rev.*, vol. 1, no. 1, p. 011103, Mar. 2014.

Sub-60 mV/decade Steep Transistors with Compliant Piezoelectric Gate Barriers

Raj K. Jana, Arvind Ajoy, Gregory Snider, and Debdeep Jena

Department of Electrical Engineering, University of Notre Dame, Notre Dame, IN, 46556 USA

Email: rjana1@nd.edu

Abstract

A novel mechanism is proposed for transistors that exploits the negative differential capacitance of piezoelectric gate barriers. Electric field induced electrostriction modulates the thickness of a piezoelectric barrier. Piezoelectricity and electrostriction in a compliant piezoelectric barrier combine to provide negative differential capacitance (NDC) with internal charge amplification. The effect of the NDC in the gate capacitor of a FET is to boost the on-current, and to provide an opportunity for switching steeper than the 60 mV/decade Boltzmann limit, both highly desirable.

Introduction

Steep transistors with sub-60 mV/decade subthreshold slope are being actively pursued for ultra-low voltage and low-power nanoelectronic applications [1, 2]. Transistors with steep switching can be potentially achieved either by replacing the conventional drift-diffusion transport mechanism by interband tunneling [1, 2], or by replacing conventional passive gate dielectrics with *active* gate barrier materials such as ferroelectrics [3] and piezoelectrics [4-7]. A ferroelectric gate insulator can take advantage of negative capacitance in the gate stack to internally amplify the surface potential ψ_s induced by the applied gate voltage V_{gs}, leading to an internal voltage gain $1/m = \partial\psi_s / \partial V_{gs} > 1$ [3]. Hence the subthreshold slope SS=$m\times$60 mV/decade can be lower than the Boltzmann limit [3]. Overcoming the subthreshold slope limit by NDC is possible because the net gate capacitance is effectively enhanced by adding a negative capacitance in series with a positive. Thus less voltage is necessary to induce the same channel charge as a conventional FET.

In this work, we first show that (a) negative differential capacitance (NDC) can potentially be achieved in a compliant *piezoelectric* gate barrier through the mechanism of electric-field-induced electrostriction [4, 5]. Then we investigate the behavior of a ballistic transistor, and show that (b) using a NDC gate stack enables higher on-currents. Finally, we discuss how one can potentially achieve sub-Boltzmann steep switching in a transistor using this NDC gate stack. We first study the behavior of an electromechanical capacitor, which is key to the problem.

Device Structure & Model Description

Fig. 1 (a) depicts the cross section of a parallel plate electromechanical capacitor. The 'dielectric' is a piezoelectric layer of dielectric constant ε_d and equilibrium thickness t_0 at V=0. What is the capacitance of the structure?

We first show that the geometric capacitance $C_0 = \varepsilon_d / t_0$ is actually the low-voltage limit of a richer dependence of the charge on voltage. Application of a voltage V exerts an electromechanical pressure on the compliant piezoelectric layer, leading to a compressive strain, which reduces the thickness. The strain creates more charge at the surfaces of the piezoelectric layer: this is an internal charge amplification. The pressure exerted is $P = \sigma_m^2 / \varepsilon_d$, where σ_m is the sheet charge density on the metal. Gauss' law and linear electromechanical coupling coefficients provide the relation between the metal charge σ_m and applied voltage V:

$$C_0 V = \sigma_m - \sigma_{sp} + \left(\frac{\sigma_{sp} - e_{33}}{\varepsilon_d C_{33}}\right)\sigma_m^2 - \frac{1}{\varepsilon_d C_{33}}\sigma_m^3 + \frac{e_{33}}{(\varepsilon_d C_{33})^2}\sigma_m^4, \quad (1)$$

where C_{33} is the elastic stiffness coefficient (unit: Pascal), e_{33} is the piezoelectric coefficient (unit: C/m^2), and σ_{sp} is the charge density (unit: C/m^2) due to spontaneous polarization, if present. This unconventional dependence of charge on voltage is interpreted later; for here we note that for a non-compliant ($C_{33} \to \infty$) dielectric with no piezoelectric and spontaneous polarization ($e_{33} = \sigma_{sp} = 0$), we recover $\sigma_m = C_0 V$. We next investigate the impact of this charge-voltage dependence on a ballistic FET [Fig. 1 (b)] in the spirit of Natori's device model [8] adding the quantum contact resistances of 0.026kΩ.μm at the source and drain ends.

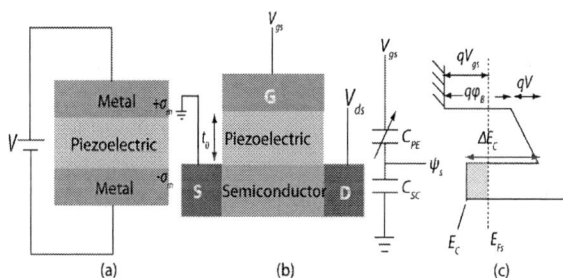

Fig. 1 – (a) A parallel-plate electromechanical capacitor with a piezoelectric (PE) barrier. (b) A piezoelectric gate transistor ("piezoFET") with a semiconductor channel and a piezoelectric gate barrier. The gate capacitance circuit is a series combination of the piezoelectric capacitance C_{PE} and the semiconductor capacitance C_{sc}. (c) The energy band diagram for the gate metal – piezoelectric - semiconductor channel stack of the transistor.

The carrier density in the semiconductor channel at the injection point induced by the gate $\sigma_m = qn_s$ is

$$qn_s = C_{SC} V_{th} \ln[1 + \exp[(E_{Fs} - E_C) / kT]], \quad (2)$$

where $C_{SC} = q^2 m^* g_s g_v /(2\pi\hbar^2)$ is the quantum capacitance, and $V_{th} = kT/q$ is the thermal voltage. From the energy band diagram in Fig 1(c), the relation between the applied V_{gs} and the voltage drop V across the piezoelectric barrier is $qV_{gs} = qV + (E_{Fs} - E_C)$. The gate-induced charge $\sigma_m = qn_s$ in the semiconductor channel is self-consistently calculated from Eqs 1 & 2. Finally, using this new dependence of charge on the voltages and the piezoelectric coefficients, the characteristic device parameters of piezoFETs are obtained from the ballistic transport model [8].

Results & Discussions

Using the model of Eq. 1, the characteristics of the electromechanical capacitor with different charge states and the resulting strain in the piezoelectric barrier are shown in Figs. 2, 3 & 4. The spontaneous polarization has been set to zero, and other parameters and indicated in these figures. At $V=0$ V, the piezoelectric layer is in equilibrium with no strain. The charge-voltage dependence then approaches the parallel-plate geometric capacitance $C_0 = \varepsilon_d / t_0$ shown in the green line in Fig 2. When $V > 0$, the strain in the layer increases and piezoelectricity induces higher charges as shown by the section of the $n_s - V$ curve labeled Q_2 in Fig. 2. At $V=V_{crit} <$ 1.1 V, the curve transitions into a *negative differential capacitance* branch ($Q_2 \rightarrow Q_3$ and $Q_1 \rightarrow Q_4$), where $C_{PE} = d\sigma_m / dV < 0$. This is highlighted in Figs 2 & 3, which show the flow of capacitance and strain with voltage. The resulting strain increases from ~ 4.6% at $V=V_{crit}$~1.2 V for the chosen parameters towards 100% by the time the $n_s - V$ curve hits $V=0$ beyond Q_3. This state is when the metal plates touch each other and short, meaning the compliant piezoelectric material has been 'squeezed' out, something that is not accessible in the solid state, but possible in gaseous plasmas. However, the NDC regime is accessible for lower strains and at lower voltages with high compliance piezoelectric barrier materials. The charge-voltage dependence has more curvature and the critical points can be accessed at smaller voltages for gate barriers with smaller stiffness coefficients (i.e., materials that are more compliant). For example, a material with stiffness coefficient C_{33}=1 GPa is compared to C_{33}=15 GPa in Figs. 2 and 4. For the dielectric constant and the piezoelectric coefficient, we use the material parameters of scandium aluminum nitride (ScAlN) barrier [9], but we let the elastic coefficient span a wider range to explore a wider range of the electromechanical phase space. The electromechanical effect is especially sensitive to C_{33}.

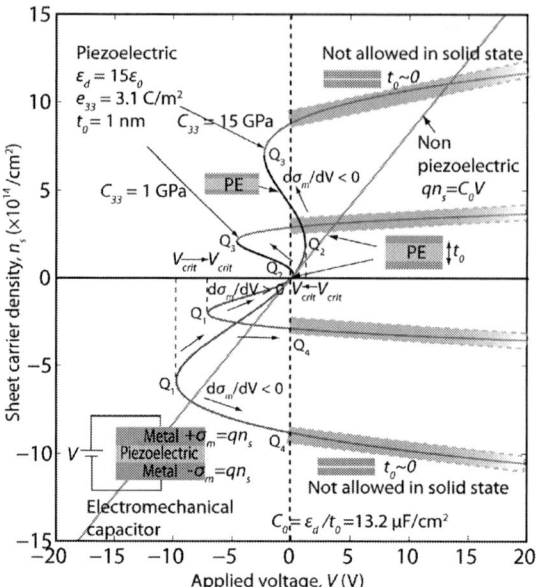

Fig. 2 – Charge-voltage (σ_m-V) characteristic of a piezoelectric (PE) electromechanical differential capacitor. It depicts the different charge states, such as positive capacitance branch $Q_1 \rightarrow Q_2$ where the slope $C_{PE} = d\sigma_m / dV > 0$ is positive, and negative capacitance branches branch $Q_2 \rightarrow Q_3$ and $Q_1 \rightarrow Q_4$ where $C_{PE} = d\sigma_m / dV < 0$. The green line is $qn_s = C_0 V$ for a passive capacitor.

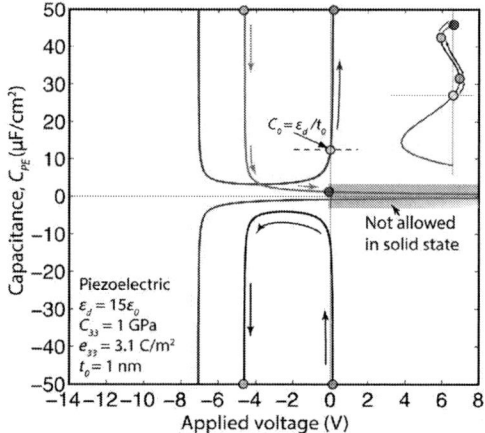

Fig. 3 – +ve & -ve differential capacitance branches of the electromechanical capacitor using a compliant piezoelectric barrier. At $V=0$ V, the electromechanical capacitance is $C_{PE} = C_0 = \varepsilon_d / t_0$ =13.2 μF/cm^2. The inset shows a section of the charge-voltage branch from Fig 2 with various points, with arrows showing the flow of charge with voltage, While Fig 3 shows the corresponding flow of capacitance with voltage. Note that the capacitance becomes infinite at points separating +ve and –ve differential capacitance branches.

978-1-4799-8002-4/14 $31.00 © 2014 IEEE   348

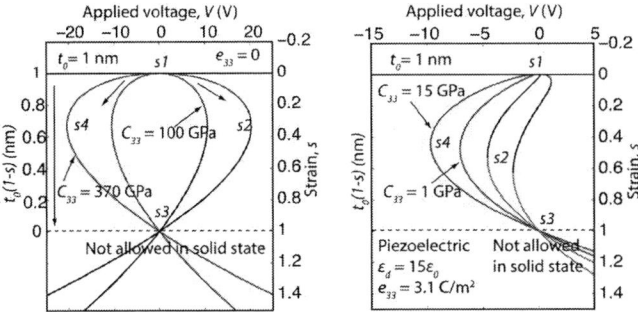

Fig. 4- Strain as a function of voltage in the piezoelectric layer. Under the applied voltage, electrostriction results in a strain $s = P / c_{33} \sim \sigma_m^2 / C_{33} \varepsilon_d$ [3, 4]. Physically accessible regime in solid state is $s_1 \rightarrow s_2 \rightarrow s_3 \rightarrow s_4$ where strain < 100%. Beyond this regime, the compressive strain along the thickness of the layer is > 100% [remaining layer thickness, $t_0 (1 - s) \sim 0$], which is not allowed in solid state materials.

Now by exploiting the non-linear charge-voltage characteristics of the compliant piezoelectric barrier in the gate stack of transistors, we calculate the gate capacitance C_g, the device characteristics (I-Vs), and the transconductance g_m of ballistic piezoFETs. The channel is chosen as GaN for Figs. 5-8, but the physics extends to any semiconductor channel material. An enhanced C_g due to the piezoelectric barrier compared to a passive dielectric in the on-state of the transistor is seen in Fig 5. This translates directly to a boost in the on-current (Figs 6, 7), which consequently improves the I_{on}/I_{off} ratio and provides a larger g_m (Fig 8). This is an attractive method to boost the on current of *any* transistor, including tunneling FETs that have low on-currents.

semiconductor gate stack. As a result of NDC using the compliant piezoelectric barriers the gate capacitance $C_g = C_{SC} C_{PE} / (C_{SC} + C_{PE})$ for a series piezoelectric/semiconductor stack is higher than $C_g = C_{SC} C_0 / (C_{SC} + C_0)$ for a series passive dielectric/semiconductor stack. Here C_{sc} is the density of states quantum capacitance, ~ 13.2 uF/cm^2 for GaN channel (electron effective mass, $m^* = 0.2m_o$).

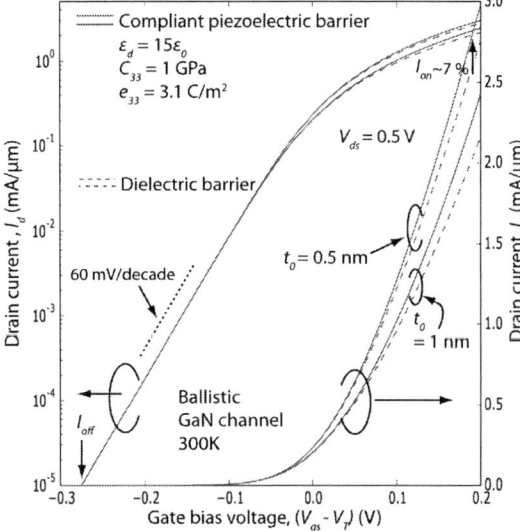

Fig. 6 – Transfer curves (I_d-V_{gs}) versus V_{gs} -V_T with logarithmic (left axis) and linear (right axis) scales at V_{ds}=0.5 V with t_0=0.5 nm, and 1 nm are shown. Higher I_{on} ~ 2.9 mA/μm at V_{gs} -V_T = 0.2 V & I_{on}/I_{off} ~ 10^5 are obtained for the transistor with piezoelectric barrier than the transistor (I_{on} ~ 2.7 mA/μm) with dielectric barrier of t_0=0.5 nm. We use electron effective mass m^*=0.2m_o, valley & spin degeneracy factors g_v = 1, g_s = 2 in GaN channel for calculation.

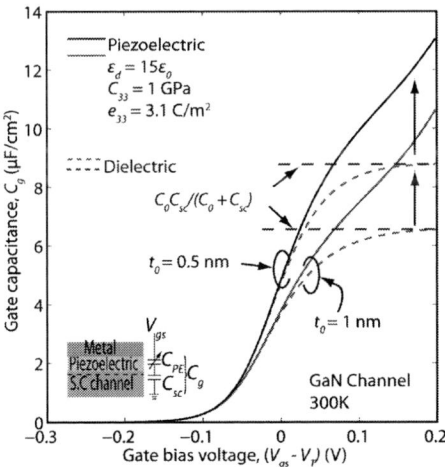

Fig. 5 – a) Gate capacitances $C_g = \partial(qn_s) / \partial V_{gs}$ versus V_{gs} –V_T plot for transistors with a compliant piezoelectric (solid line) and rigid dielectric (dashed line) barriers. The charge induced in the channel is obtained by self-consistent solution of Eqs. 1-2. V_T is the threshold voltage. Higher C_g (~ 47% increase in C_g at V_{gs} -V_T = 0.2 V) occurs due to internal charge amplification than the dielectric-

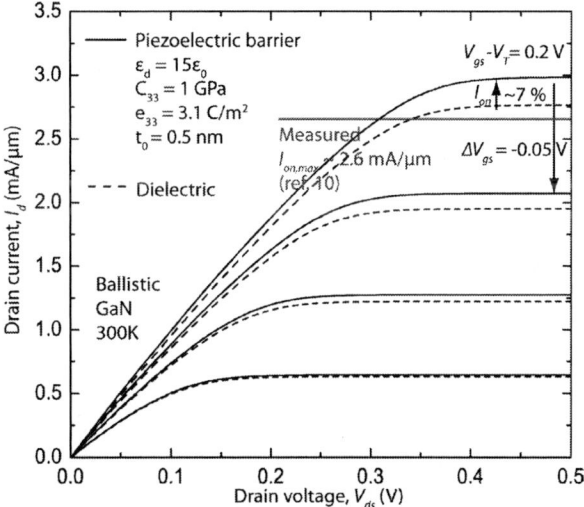

Fig. 7 – Output characteristics (I_d-V_{ds}) with different V_{gs}-V_T for GaN channel transistors are shown. Higher drain current (~ 7% increase in I_{on} at V_{gs} -V_T= 0.2 V) is resulted for the transistor with piezoelectric gate barrier (solid line) than the dielectric barrier (dashed line). Here, we indicate the measured maximum on-current $I_{on,max}$ ~ 2.6 mA/μm [10] for state-of-the art E-mode GaN HEMT device in the figure. Therefore, there is a possibility for improving drive on-current using piezoelectric barriers in transistors.

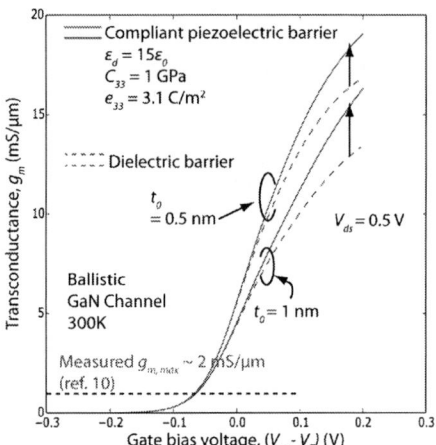

Fig. 8 – Transconductance, $g_m = \partial I_d / \partial V_{gs}$ versus gate bias voltage $V_{gs} - V_T$ at $V_{ds} = 0.5V$ is shown. Higher transconductance is resulted for the transistor with piezoelectric gate barrier (solid line) than the dielectric gate barrier (dashed line). Here, we indicate the measured maximum transconductance $g_{m,max} \sim 2$ mS/µm [10] for state-of-the art E-mode GaN HEMT device in the figure. Thus, transconductance can also be improved using piezoelectric barriers in transistors.

To show the possibility of accessing the NDC regime, we plot the load line characteristics (Eq. 2) of the charge versus voltage relation of the piezoelectric (Eq. 1) for different V_{gs} in Fig 9. The locus of intersection points spans the allowed operating points of the transistor. The operation of the piezoelectric/semiconductor system for $V_{gs} = 0.2$ V for example shows the internal amplification of charge at point (d) for the piezoelectric gate as compared to the passive gate charge at point (c). This is responsible for the boost in the on-state current. However, for the same material parameters, we obtain a subthreshold slope of 60 mV/decade. This is because of the inability to access the regime of NDC during the *off state operation* of transistor with the piezoelectric material properties used in the work. The steep switching behavior can be achieved if one can access the negative capacitance charge states during the off-state operation of the transistor, when $V_{gs} < V_T$. This may be possible by tuning the piezoelectric and compliant properties of the barrier material, and requires further study.

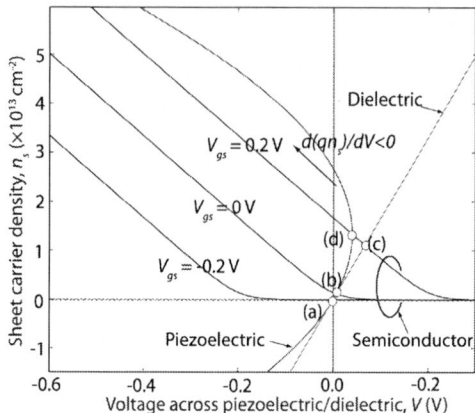

Fig. 9 – Load line analysis to obtain sheet carrier density n_s for different gate voltage, V_{gs}. Red(green) lines depict charge - voltage characteristics for piezoelectric(dielectric) capacitor. Blue curves show the semiconductor charge for different V_{gs}. Intersections of the above characteristic define the operating points of the system. At V_{gs} =0, -0.2 V, the operating points (a), (b) of the piezoelectric and dielectric barrier structures are identical, thereby giving similar currents and subthreshold slopes of 60 mV/decade. However, at V_{gs}=0.2 V, the operating point (d) lies in a regime of negative differential capacitance of the piezoelectric barrier structures. This provides an internal amplification of charge and a higher current (shown in Fig 6) as compared to point (c) of the dielectric barrier structure.

Conclusion

Transistors with compliant piezoelectric barriers based on NDC in the piezoelectric-semiconductor channel stack have been proposed. By using electrostriction and ballistic models, we show that compliant piezoelectric gate materials can boost the on current and the transconductance than conventional passive dielectric barriers. Steep switching of transistors may be possible if one can access negative capacitance regime in off-state operation of transistors by tuning the piezoelectric barrier properties. Therefore, the lower stiff, higher compliant piezoelectric barrier layers are favorable for realizing steep transistors with improved device performance.

Acknowledgment

This work is supported by the SRC/DARPA STARnet LEAST program.

References

[1] A. M. Ionescu , et al.,*Nature*, vol. 479, pp. 329-337, Nov. 2011.

[2] A.C. Seabaugh, et al., *Proceedings of the IEEE*, vol. 98, pp. 2095, Dec. 2010.

[3] S. Salahuddin , et al., *Nano Lett.*, vol. 8, pp. 405-410, 2008.

[4] R. K. Jana, et al., *Phys. Status Solidi (c)*, vol. 10, pp. 1469-1472, Oct. 2013.

[5] R. K. Jana, et al., *Energy Efficient Electronic Systems (E3S), Third Berkeley Symposium on*, Oct. 2013, DOI: 10.1109/E3S.2013.6705877.

[6] H. W. Then, et al., *Int. Electron Devices Meeting (IEDM) Tech. Dig.*, Dec. 2013, pp. 691–694.

[7] Z. Hu, et al., *Device Research Conference (DRC)*, 72nd Annual. IEEE. Jun. 2014, DOI: 10.1109/DRC.2014.6872283.

[8] K. Natori, et al., *J. Appl. Phys.*, vol. 76, pp. 4880, 1994.

[9] M. Akiyama, et al., *Appl. Phys. Lett.*, vol 102, pp. 021915, 2013.

[10] K. Shinohara, et al," *IEEE. Trans. Electron. Device.*, vol. 60, pp. 2982, 2013.

Progressive *vs.* Abrupt reset behavior in Conductive Bridging devices : a C-AFM tomography study.

U. Celano[1,2*], L. Goux[1], A. Belmonte[1,2], G. Giammaria[3], K. Opsomer[1], C. Detavernier[4], O. Richard[1], H. Bender[1], F. Irrera[3], M. Jurczak[1], W. Vandervorst[1,2]

[1]imec, Kapeldreef 75, 3001 Leuven – Belgium
[2] Katholieke Universiteit Leuven – Belgium
[3]Università "La Sapienza", Roma, Italy
[4]University of Gent - Belgium
*Umberto.celano@imec.be

Abstract

We investigated the physical origin of progressive and abrupt reset in conductive bridging memories. The conductive filaments for both types of reset are observed in 3D using C-AFM tomography, enabling the observation of broken and non-broken filaments respectively for abrupt and progressive reset.

Introduction

The Conductive bridging device (CBRAM) is considered as a valuable non-volatile storage technology, because it offers fast switching, high endurance and good scalability (1–4). CBRAM operation relies on the voltage-induced redox-based formation and rupture of a Cu- or Ag-based conductive filament (CF) in an insulating layer acting as a solid state electrolyte (3). Whereas the presence of a metallic CF is commonly accepted for the low resistive state (LRS) (1–5), the conduction nature of the high resistive state (HRS) is less clear. Generally, the formation of the HRS is described as an electrochemical-driven dissolution of the CF leading to a tunnel-barrier formation (5-7), a current constriction in a quantum-point-contact (QPC) (8), or conduction by trap-controlled mechanisms (9). In this paper, for the first time, we report on the observation of the HRS conduction mechanisms (tunneling in a broken filament, current constriction in a QPC) for 1T1R memory elements by mean of C-AFM tomography. Next, we relate the reset-behaviors experimentally observed, to specific configurations of the CF in the LRS.

Methodology

The memory device in this work is a crossbar structure (Fig.1) stacked on top of a selector transistor in a 1-Transistor/1-Resistor (1T1R) scheme, as shown in Fig. 2-3. Reset cycles of various devices are shown in Fig. 4 whereby a negative voltage is applied to the Cu electrode to break/dissolve the CF; the filament formation (not shown) is done with the opposite polarity. Though we use the same conditions for forming and cycling on similar (but different) devices, the I-V curves corresponding to their reset behavior are very different (Fig.4). In one case we observe an abrupt transition from the LRS to the HRS as evidenced by the steep drop in current when increasing V_{reset} (blue trace Fig.4). The

other type shows a progressive transition which also starts on V_{reset} and continuously decreases until the negative sweep is ramped (red trace Fig.4).

Results

In either case the I-V curves can be divided in two parts: (1) an initial part showing many fluctuations, which is considered characteristic for the creation of the final conduction path in the HRS and (2) a more stable part (termed as static) representative for the final state assumed by the CF in the HRS after the reset (Fig.5). In order to elucidate the nature of the CF in the abrupt and progressive HRSs, we form and cycle (with 100 µA current compliance) the CF in a device, leave it finally in HRS and then submit it to C-AFM tomography, termed scalpel C-AFM (10). Using scalpel C-AFM we collect C-AFM images of the conductive filament at different depths leading to a full 3D-characterization of the conductive volumes. The sectioning is induced by a controlled material removal through applying a strong pressure (GPa) between diamond-tip and sample during the contact-AFM scan. On each scan we collect topography as well as current in every pixel providing detailed maps of the spatial variation of local conductivity. Fig. 6a represents the results for an abrupt reset device, using traditional C-AFM whereas Fig. 6b shows some of the slices acquired at different depths within the switching layer (Al_2O_3) as sketched in Fig. 6c. The process leads to a uniform progressive removal (step-size below 1 nm) of the 5 nm Al_2O_3, as shown in Fig. 7. A 3D-tomogram of the conductivity variations within the sample is obtained by interpolation. The cross-section of the obtained 3D-tomogram is presented in Fig. 8. An interrupted (ruptured) CF is visible with a gap on the side of the inert-electrode. The filament gap shows a non-regular conical geometry with a gap-size ~ 1 nm (Fig. 8a-b). Returning to the last I-V (reset) cycle of this abrupt reset device before the scalpel C-AFM (Fig.9), one can observe that the static part of the IV can be fitted by means of the Fowler-Nordheim tunneling (FNT) mechanism (11). The FNT matches the experimental data well at high voltage but fails at low voltage (<0.5V Fig. 9). In this region a direct tunneling (DT) description is required. The observed non-symmetrical geometry of the broken filament (Fig. 8) suggests that a simple parallel-plate model cannot be used to fit the transport in the HRS. Based on our tomographic observation we model the remaining CF-shape (Fig. 10a) as a combination of three parallel plate capacitor-rings. We estimate the FNT-current as the sum of three contributions weighted for their effective area ($A_{1,2,3}$) and oxide thicknesses ($d_{1,2,3}$) (Fig. 10b). Due to the exponential dependence of the FNT-current with the barrier thickness, the overlapping planes are spaced by 0.1 nm each. Using our 3D tomogram we can exactly estimate the extent of these surfaces and use them for an optimized curve fitting as shown in Fig.11.

A. Tip induced reset

Whereas previous results relate to I-V's obtained from fully processed devices, similar results can be obtained when using C-AFM on blanket material (Fig.12). In this case the C-AFM tip is used to (locally) form and reset filaments on a Cu/Al_2O_3 (3nm)/tip structure (inset Fig.12a). The C-AFM switching cycle shows a leakage current during the filament rupture that follows a FNT dependence. Clear steps in leakage current which scale with an increasing barrier-thickness appear (Fig.12a). The same behavior is observed also for normal devices (Fig.12b). In essence these are a signature of steps in gap distance occurring during the tunnel-barrier (gap) formation. Using the parameters used in Fig. 9, we can extract the thickness of the subsequent gaps forming the tunnel-barrier formed after the CF rupture (Fig. 13a). Interestingly if we plot the histograms of these gaps (Fig.13b) we observe clear peaks at integers of 2.5 Å which is in the same order as the lattice parameter of the Cu metallic-phase. We repeated the C-AFM tomography on a device showing progressive reset (Fig. 14). Now the QPC model provides the only way to fit the static part of the IV. This model previously introduced to describe the conduction in OxRRAM (12) was recently successfully adapted to study the switching dynamic in CBRAM (8). The main assumption in QPC is that the filament is not broken and that the conduction is dominated by a constriction (inset Fig.14). Fig.15 shows the C-AFM tomography dataset related to the progressive reset device. The images indicate the presence of a conductive filament running through the oxide layer even though the device is in HRS. The local conductivity of the spot is low, suggesting that the filament shorting the two electrodes is not highly conductive as already observed in the case of LRS (10).

B. A model for the reset

Although the existence of broken and non-broken filaments for the HRS suggest different structural properties for the CFs, they might be viewed as the same process whereby the original formation of the CF controls the distinction between abrupt and progressive reset. Indeed previously we demonstrated that the LRS resistance is related to the volume of the CF (10) and therefore to the set process. This implies that different CFs may have a different volume. During reset,

electric-field driven and thermally activated hopping of Cu ions induces a structural rearrangement of the CF (enhanced at its constriction). This is limited in time and thus in the number of atoms which can be displaced. Starting from a weak/strong CF (low/high volume at the constriction) it is then not surprising that, under the assumption of displacing only a limited amount of Cu atoms, in one case the CF completely dissolves at the constriction (leading to the gap formation) whereas in the other case it merely reduces its dimension at the constriction. Hence although driven by the same phenomena, its impact on the final CF-shape creates a gap behavior or QPC behavior (Fig. 17). To support our suggestion, we show in Fig. 18 the LRS resistances of abrupt and progressive CFs and their correlation to the reset behavior. In line with our previous conclusion (10), Fig. 18 shows that weaker/stronger CF (i.e. higher/lower RLRS values), can be linked to abrupt or progressive reset behaviors.

Conclusion

We have successfully used C-AFM tomography for the (3D) observation of CF in CBRAM devices programmed in HRS. Different electrical reset behaviors have been linked to the existence of broken CF with tunnel-barrier formation for the case of abrupt reset, and unbroken lowly conductive filament associated to progressive reset. Both can be linked to the strength of the CF induced during the set stage.

Acknowledgements

Research funded by a Ph.D. grant of the Agency for Innovation by Science and Technology (IWT), we acknowledge the partial funding by IMEC's Industrial Affiliation program on RRAM.

References

1. R. Waser and M. Aono, *Nature materials*, 2007, *6*, 833–40.
2. I. Valov, R. Waser, J. R. Jameson, and M. N. Kozicki, *Nanotechnology*, 2011, *22*, 254003.
3. U. Russo, D. Kamalanathan, D. Ielmini, A. L. Lacaita, M. N. Kozicki, "Study of Multilevel Programming in Programmable", *IEEE Transactions on Electron Devices, 56*(5), 1040–1047, (2009).
4. L. Goux, K. Sankaran, G. Kar, N. Jossart, K. Opsomer, R. Degraeve, G. Pourtois, G. Rignanese, and C. Detavernier, in *VLSI Technology (VLSIT), 2012 Symposium on*, Honolulu, HI, pp. 69–70, (2012).
5. X. Guo, C. Schindler, S. Menzel, R. Waser, "Understanding the switching-off mechanism in Ag^+ migration based resistively switching model systems." *Applied Physics Letters, 91*(13), 133513. (2007).
6. J. Guy et al., "Investigation of the physical mechanisms governing data-retention in down to 10nm Nano-trench Al_2O_3 / CuTeGe Conductive Bridge RAM (CBRAM)", International Electron Devices Meeting, 2013, 742–745.
7. S. Menzel, B. Wolf, S. Tappertzhofen, U. Böttger, "Statistical modeling of electrochemical metallization memory cells", Memory Workshop (IMW), 2014 6th IEEE International, 2014.
8. A. Belmonte, R. Degraeve, A. Fantini, W. Kim, M. Houssa, M. Jurczak, L. Goux, " Origin of the deep reset and low variability of pulse-programmed $W/Al_2O_3/TiWCu$ CBRAM device", Memory Workshop (IMW), 2014 6th IEEE International, 2014.

9   D. Shang, Q. Wang, L. Chen, R. Dong, X. Li, W. Zhang, "Effect of carrier trapping on the hysteretic current-voltage characteristics in Ag/La0.7Ca0.3MnO3/Pt heterostructures", Physical Review B, 73(24), 245427, (2006).

10   U. Celano, L. Goux, A. Belmonte, K. Opsomer, A. Franquet, A. Schulze, C. Detavernier, O. Richard, H. Bender, M. Jurczak, W. Vandervorst, "Three-Dimensional Observation of the Conductive Filament in Nanoscaled Resistive Memory Devices." Nano Letters, 14(5), 2401–2406, (2014).

11   W. Frammelsberger, G. Benstetter, J. Kiely, R. Stamp, "Thickness determination of thin and ultra-thin SiO2 films by C-AFM IV-spectroscopy", Applied Surface Science, 252(6), 2375–2388, (2006).

12   R. Degraeve et al., "Generic learning of TDDB applied to RRAM for improved understanding of conduction and switching mechanism through multiple filaments", International Electron Devices Meeting, 2010, (1), 632–635.

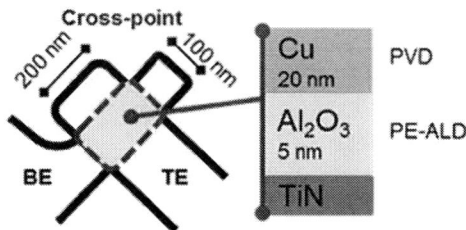

Fig. 1 Schematic of the cross-point memory element. Details on the cell stack are in the inset.

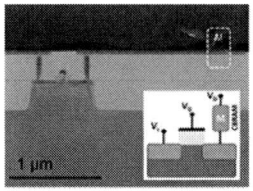 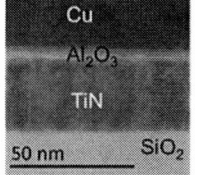

Fig. 2 Cross-section TEM image of our 1T1R test vehicle. In the inset, the schematic of the integrated configuration.

Fig. 3 Cross-sectional HRTEM of the memory element. The magnification of the region encircled in Fig.2 is shown.

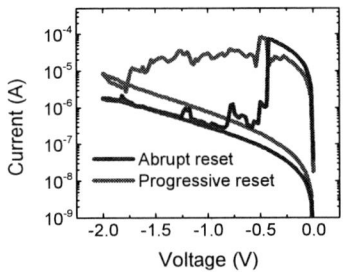 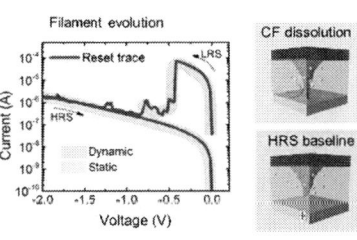

Fig. 4 Two devices operated under the same conditions can show very distinctive reset transitions. (1) sudden drop in conductance for the *abrupt reset* and (2) a continuous decrease for the *progressive reset*.

Fig. 5 A general reset sweep can be divided in two main components: first, the dynamic transition of the CF from LRS to HRS associated with current fluctuations. Second, the final (static) HRS leakage baseline after the CF has changed its conductance.

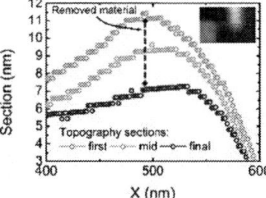

Fig. 7 The AFM topography is used during the tomography to control the removal of material in the active area. The Al2O3 is progressively removed as shown by the three scan lines in the same location at different stages of tomography.

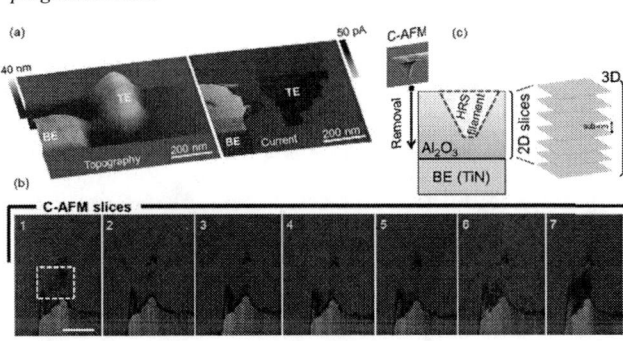

Fig. 6 (a) Traditional C-AFM on the memory element left in HRS, note the low intensity of the leakage in the case of HRS. We use the high sensitivity TUNA module as C-AFM sensor. (b) Set of 2D C-AFM images collected at different height within the oxide layer, representing the evolution of the CF in the HRS (bar 100 nm). (c) Schematic of the scalpel C-AFM concept, after removal of top electrode C-AFM images are collected at different height of the remaining CF.

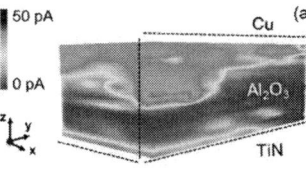 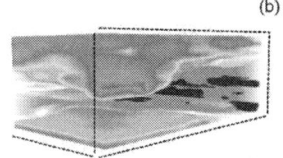

Fig. 8 (a) Reconstructed 3D sectional view of the conductive filament in the HRS for the abrupt reset. The images present the 5 nm-thick Al2O3 switching layer. The CF shows a rupture at the inert-electrode interface with a gap size between 0.5 – 1 nm. (b) We suppress the current level corresponding to Al2O3. Note, the gap-size is affected by the uncertainty induced by the non-zero leakage current detectable by C-AFM at the tunnel barrier.

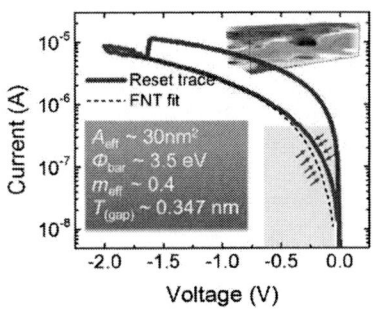

Fig. 9 The last cycle before the C-AFM tomography is reported for the case of abrupt reset. The static part of the IV can be fitted by FNT mechanism (using the fitting parameters shown in the inset).

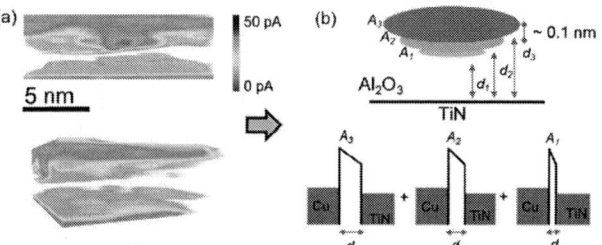

Fig. 10 Approximation for the tunneling junction. We estimate the FNT-current as the sum of three overlapping planes, separated by 0.1 nm. The emission area of each plane is the corona obtained by taking away the area of the front plane e.g. $A_{3eff} = (A_3 - A_2)$. $A_{1,2,3}$ are experimentally determined from the 3D tomogram.

978-1-4799-8002-4/14 $31.00 © 2014 IEEE   353

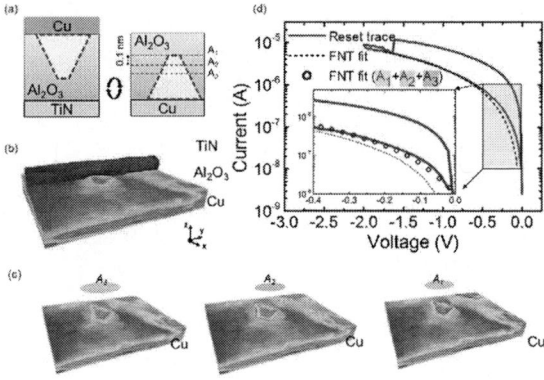

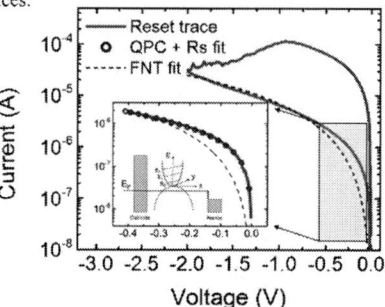

Fig. 12 (a) The C-AFM tip is used as a virtual electrode (inset). The bias is applied to the sample while the tip is grounded. In this configuration the rupture of the CF and the transition between LRS to HRS present a leakage current which follows a FNT dependence with increasing thickness of the tunnel barrier (t_{gap}). (b) The same effect can be observed also in devices.

Fig. 11 Optimization of the tunnel junction geometry. We inspect our 3D tomogram as in (a). We focus on the apex of the broken filament and using a contour-plot function we extract (b-c) the values for $A_{1,2,3}$. Finally, we use the sum of their three contribution for the curve fitting of the static state (d). The optimized geometry matches the low-bias current more accurately than the planar approximation.

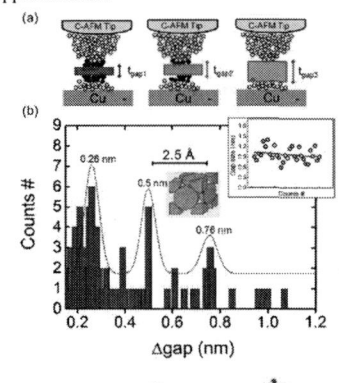

Fig. 13 (a) The increasing thickness of the tunnel barrier is representative of the growth of the tunnel gap after filament rupture in the memory element moving toward the HRS. (b) We introduce Δgap as the difference between two consecutive barrier thicknesses (t_n-t_{n-1}). Δgap can be interpreted as the size of the consecutive steps done by the CF during the formation of the tunneling gap after filament rupture. We plot all the steps into a histogram, and we clearly observe peaks appearing at integer of 2.5 Å. Those are fitted with a series of Gaussian distributions as a guide to the eye. In the inset we report the last value of t_{gap} extracted by the FNT fitting. The latter is representative for the average value of the tunnel gap formed at the end of the dynamic part of the IV.

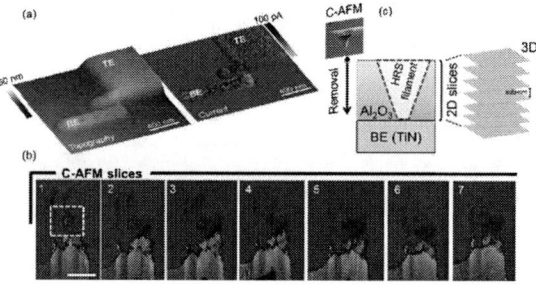

Fig. 14 The last cycle before the C-AFM tomography is reported in the case of progressive reset. In this case the FNT doesn't provide a good fit neither for low- or high-bias region. The inset shows the excellent fit provided by the QPC model for the static part of the IV. Note, in the previous case (Fig. 10) the situation was the opposite with ineffective fit of the QPC model. The inset shows the QPC constriction determined by parabolic potential well (ω_y) and barrier (ω_x) in the y and x directions respectively.

Fig. 16 Reconstructed 3D sectional view of the conductive filament in the HRS for progressive reset. The images present the 5 nm-thick Al_2O_3 switching layer. The CF appears as continuous path within the Al_2O_3. Note, the conductivity of the filament is order of magnitudes smaller than what observed for the LRS (10).

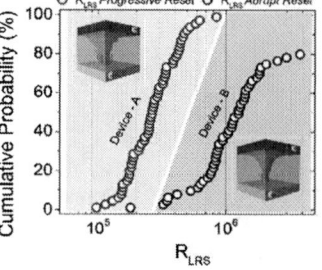

Fig. 15 (a) Traditional C-AFM on the memory element left in HRS showing progressive reset. (b) Set of 2D C-AFM images collected at different height within the oxide layer, representing the evolution of the CF in the HRS (bar 100 nm). (c) Schematic of the scalpel C-AFM concept, after removal of top electrode C-AFM images are collected at different height of the remaining CF.

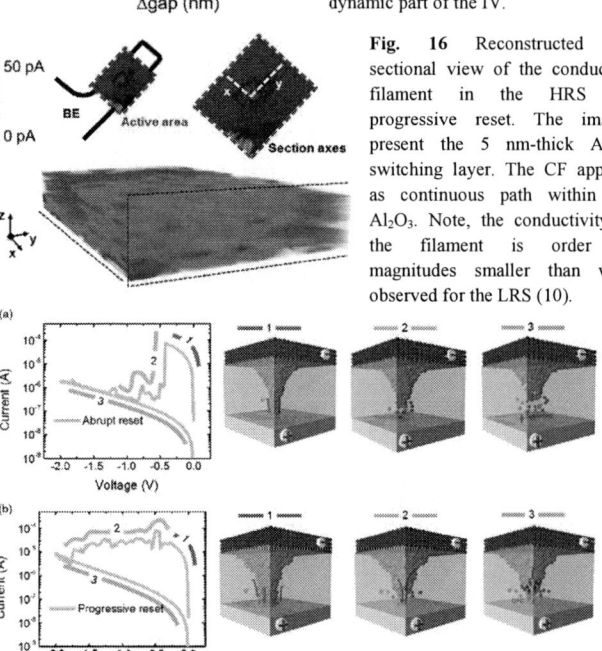

Fig. 17 Illustration of the reset process for both kind of reset transition. (a) In the case of *abrupt reset* (1) the CF is in the LRS when the reset sweep is ramped, negative polarity to Cu electrode. (2) The energy dissipated at the constriction of the CF induces the rupture of the CF. (3) The electric-field driven thermally activated hopping of Cu ions produce the formation of a tunnel-barrier at the CF/inert-electrode interface. (b) For *progressive reset* : (1) the CF is in the LRS when the reset sweep is ramped; (2) at the V_{reset} the IV shows fluctuations due to the rearrangement of particles in the CF; (3) due to its structural properties the CF does not undergoes a net rupture, nevertheless the particles within the CF rearrange creating a QPC constriction which leads to the HRS.

Fig. 18 Cumulative probability for the LRS resistances of devices showing both type of reset behaviors. The CFs of devices showing *progressive* reset show lower resistance as compared to the *abrupt* case. This supports the proposed model where the *abrupt* is related to weaker CFs and *progressive* to bigger (stronger) CFs.

978-1-4799-8002-4/14 $31.00 © 2014 IEEE

Understanding the Impact of Programming Pulses and Electrode Materials on the Endurance Properties of Scaled Ta₂O₅ RRAM Cells

C.Y. Chen[1,2], L. Goux[1]*, A. Fantini[1], A. Redolfi[1], S. Clima[1], R. Degraeve[1], Y.Y. Chen[1], G. Groeseneken[1], M. Jurczak[1]

[1] imec, Kapeldreef 75, B-3001 Leuven, Belgium; [2] KUL, Leuven, Belgium

*e-mail: gouxl@imec.be; phone: +32-16-28.87.47

Abstract

We demonstrate the strong impact of reset amplitude and duration on the endurance degradation of scaled TiN\Ta₂O₅\Ta cells, which from ab-initio and electrical switching simulation is attributed to O interaction with TiN. Clear improvements are obtained using (i) shorter write pulses, (ii) low O-affinity Ru bottom electrode, and or (iii) higher O-affinity HfO₂ dielectric.

Introduction

Today the resistive-switching RAM (RRAM) technology draws attention for high-performance application due to impressive progress in reliability, in particular regarding write endurance achievements both for HfO_2 [1-3] and Ta_2O_5 [4-6] based RRAM. However, although failure modes are analyzed in these works [1-7], there is still no clear understanding of the physical mechanisms causing endurance degradation. In this paper we systematically study the impacts of pulse-programming parameters on endurance characteristics of scaled TiN\Ta₂O₅\Ta cells. By means of hourglass (HG) electrical switching simulations and ab-initio calculations we analyze the physics leading to failure, drawing some guidelines towards improvements.

Test structures and Methodology

RRAM crossbar devices were integrated onto select transistor in a 1T1R circuit scheme (**Fig.1a,b**). Amorphous atomic-layer-deposited (ALD) Ta_2O_5 layers (6nm) were sandwiched between TiN (30nm) bottom electrode (BE) and Ta (10nm) top electrode (TE) realized by physical-vapor-deposition (PVD) (**Fig.1c,d**). The ref. device size is 100nm. 20nm-scaled devices having 3nm Ta_2O_5 and 5nm Ta TE and devices having Ru BE were also fabricated (**Fig.1e,f**). In this work, the switching current I_{cc} is fixed to 50uA, as modulated by the word-line voltage during forming and set. Positive (resp. negative) voltage is applied to the bit-line for set (resp. reset) switching. DC forming and switching characteristics of the TiN\Ta₂O₅\Ta cells are shown in **Fig.2**, while AC-pulse experiments confirmed the effective programming of the cells using pulse amplitudes V_{set} and V_{reset} lower than 1.5V for a pulse width (PW) in the range of few ns (**Fig.3**) [6]. In addition to these parameters, the leading (LE) and trailing edges (TE) parameters of the pulses were investigated (**Fig.4**). In all endurance tests, a statistical sample of more than 20 devices is used.

Fig.5 gives some information on the hourglass (HG) switching model used in this work to estimate the temperature of the conductive filament (CF) during switching [see more details in Ref. 8]. After determination of the number (n_c) of oxygen-vacancy (V_o) defects in the quantum-point-contact (QPC) constriction from I-V modeling [8], the CF temperature is estimated from the power per particle (**Fig.5**).

Endurance study

Fig.6 shows the impacts of varying PW_{reset} and V_{reset} from the standard test conditions ($PW_{set}=PW_{reset}=10ns$, $V_{set}=2.5V$, $V_{reset}=1.5V$, LE=TE=3ns). While the standard condition results in a lifetime of ~10^6 cycles, failing into a low resistance state (LRS) (failure mode A), the increase of PW_{reset} and/or V_{reset} results in a decreased lifetime and a more complex failure behavior (failure mode B). Firstly, the high resistance state (HRS) resistance is increased considerably from the start of the cycling (B1), then the cycling degradation is onset by a drift of LRS and HRS towards intermediate HRS levels, associated with failure to set (B2), and finally a reversed drift of both HRS and LRS towards LRS is observed with a definitive final stuck-LRS state (B3) (**Fig.6**). In **Fig.7a** we carried out reset-pulse cycling experiments (skipping the set pulse) in the same reset conditions as in **Fig.6b**, and the same drift of the HRS level was observed. Following this test, a regular set/reset cycling (**Fig.7b**) also showed failure to set exactly like in **Fig.6b**. These results indicate that the reset pulse is the main contributor to the failure B in **Fig.6b-e**. Interestingly, HG simulations evidenced that the CF temperature, although decreasing rapidly in the first tens of ns, remains >500K and fluctuates considerably during the whole reset pulse duration (**Fig.8a**). This large temperature is due to the large residual current density though the reset-induced narrower QPC constriction cross-section, while the temperature fluctuations are a signature of the thermal-induced competition between V_o fluctuations. Due to direct impact on power, the effect of V_{reset} on temperature is drastic (**Fig.8b**).

On the other hand, varying PW_{set} from the standard test conditions did not result in strong impact. As compared to

978-1-4799-8002-4/14 $31.00 © 2014 IEEE

Fig.6a, **Fig.9a** shows that the stuck-LRS failure is anticipated by only ~2 decades when increasing PW_{set} by 4 decades. Indeed, the CF temperature is by far lower during set than during reset (**Fig.9b**), which is due after set threshold to the snapback of the voltage dropping on the cell to the transition voltage, which is <1V (**Fig.9c**) [8].

Finally, varying LE and TE parameters did not results in large impact on endurance (**Fig.10**), as expected, because the CF is submitted to the maximum V_{reset} amplitude for a small portion only of the reset pulse.

Hence, *reducing PW_{reset} is most effective in improving endurance lifetime* (**Fig.11**), as summarized in **Fig.12**. Using thus shortest possible set and reset pulses allowed by the pulse-generator instrument (HP8110), we successfully reached 10^9 endurance lifetime for 20nm-scaled $TiN\backslash Ta_2O_5\backslash Ta$ cells (**Fig.13**).

From earlier works we know that the CF constriction is located close to the BE interface [9]. We thus considered the chemical interaction between the CF and the TiN layer. We investigated by ab-initio modeling the diffusion of Oxygen (O) within TaO_x as well as into various types of metals, and we calculated the energy barrier (ΔG) required for O to react with the metal (**Fig.14**). As far as an ideal TiN cristal is considered, a rather large ΔG is obtained, of similar magnitude as for the inert metal Ru (**Fig.14**). However, changing to pure Ti material results in drastic drop of ΔG. Experimentally the thermal instability of the stack $Ta_2O_5\backslash Ti$ was evidenced due to spontaneous soaking of O from the Ta_2O_5 layer by Ti [6]. In PVD-deposited TiN layers, any point defects such as Nitrogen vacancies (V_N) or Ti dangling bonds will have the same effect of locally decreasing ΔG and result in TiN oxidation. Regarding in-diffusion of O, calculations resulted in a significantly lower activation energy E_a for O diffusion into TiN than into Ru (**Fig.14**). These results suggest that O is expected not only to *react more* with TiN than with Ru, but also to penetrate deeper into TiN. Based on this, **Fig.15** shows a physical scenario accounting for both endurance degradation modes A and B (**Fig.6**). Regarding failure B, as induced by both the reset field and temperature, the reset efficiency of the constriction may be very high in the first cycles (B1); over the cycling test O might be trapped in TiN point defects, and O detrapping may be prevented due to gradual and extended oxydation of the TiN electrode, leading to the formation of an insulating TiO_xN_y material (**Fig.16**) resulting in the drift to intermediate HRS states and failure to set (B2). This reaction would induce a partial voltage redistribution over the TiO_xN_y material, preventing the switching. Final wear-out to stuck-LRS state (B3) might then be related to intermixing or gradual breakdown mechanism.

As expected from the above considerations, *the use of Ru BE materials experimentally suppressed the failure mode B*, as shown in **Fig.17a**. Even for large temperatures and fields associated to the large reset pulse V_{reset}=2.3V, no failure was observed before 10^9 write endurance cycles (**Fig.17b**), in agreement with the large ΔG and E_a parameters obtained for Ru BE (**Fig.14**). Interestingly, the stuck-LRS failure mode A observed on $TiN\backslash Ta_2O_5\backslash Ta$ cells cycled with *moderate reset stress* (**Figs.6a,9a,11**) is not observed either on $Ru\backslash Ta_2O_5\backslash Ta$ cells (**Fig.17**). This mode A may thus also be associated to the TiN BE. As a possible scenario, the more moderate field and temperature associated to short and low reset pulse may be insufficient for O oxydation (ΔG), however sufficient for O in-diffusion through TiN defects like grain boundaries or as interstitial, leading to O-depletion of the CF (**Fig.15a**). The moderate temperature associated to set-pulse induced degradation (**Fig.9a**) may result in a similar scenario in spite of the reversed direction of the field.

In **Fig.18** the reversed breakdown voltage of HRS state (V_{BD-HRS}) is characterized for all cells. Clearly, V_{BD-HRS} is increased from TiN to Ru BE, in agreement with the larger ΔG and E_a obtained in **Fig.14**. Note that V_{BD-HRS} is not cell-size dependent (**Fig.18**) and thus is a characteristic of the CF constriction.

Finally, following these guidelines the use of a different switching oxide having a lower enthalpy of metal oxidation should result in improved robustness to O-injection into the BE material. As expected, $TiN\backslash HfO_2\backslash Hf$ cells showed increased endurance lifetime in various cycling conditions (**Fig.19**).

Conclusion

In this work, we study the endurance degradation mechanism which is mostly governed by O interaction with TiN BE. Based on this, switching lifetime can be optimized by using short reset PW, more inert BE, or higher O-affinity dielectrics.

Reference:
[1] H.Y. Lee et al., IEDM 2010;
[2] A. Fantini et al., VLSI 2014;
[3] Y.Y. Chen et al., TED 2012;
[4] J.J. Yang et al., APL 97, 2010;
[5] M.J. Lee et al., Nature Materials 10, 2011;
[6] L. Goux et al., VLSI 2014;
[7] B. Chen et al., IEDM 2011;
[8] R. Degraeve et al., VLSI 2013;
[9] L. Goux et al., VLSI 2012, p159.

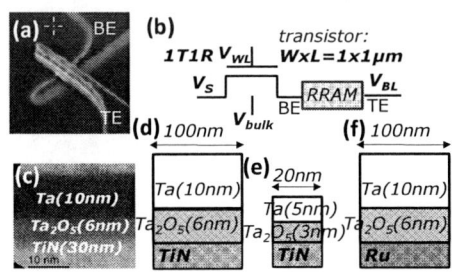

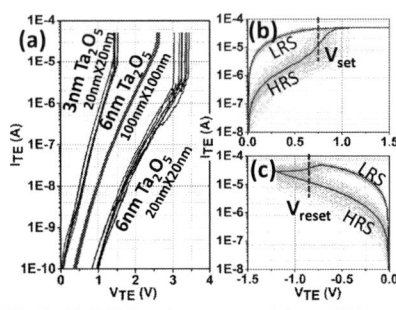

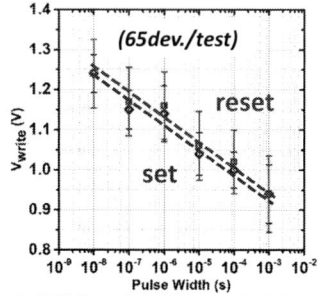

Fig.1: (a) RRAM cross-bar device, integrated in a 1T1R scheme (b), and (c) XTEM image of the ref. TiN\Ta₂O₅\Ta device; (d) schematic of ref. stack, (e) scaled device (20nm x 20nm), and (f) Ru\Ta₂O₅\Ta device

Fig.2: (a) DC forming traces of three different TiN\Ta₂O₅\Ta devices, showing size and oxide-thickness dependence of V_f; (b) DC set, and (c) reset characteristics of 20nm-scaled devices; I_{cc}=50µA

Fig.3: PW-dependent V_{write} required for reset and set of the ref. device, showing effective write at 1.5V for any PW down to few ns, serving as guideline for the following endurance study

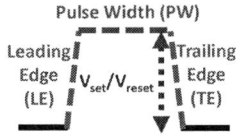

Fig.4: Write pulse parameters varied in the endurance characterization

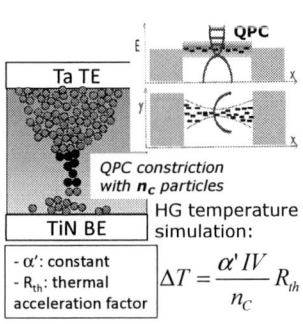

Fig.5: Schematic of hourglass switching model used in this work to estimate the temperature of the conductive filament during switching

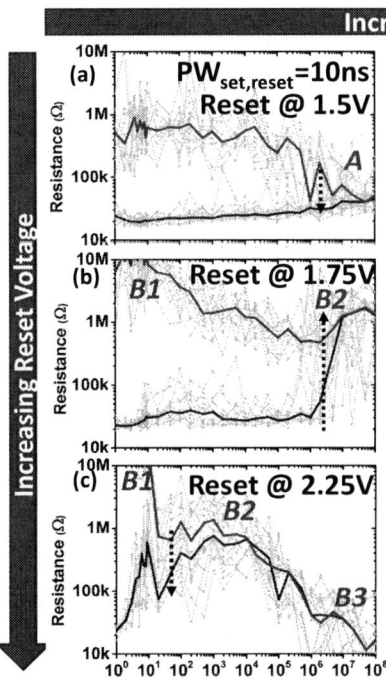

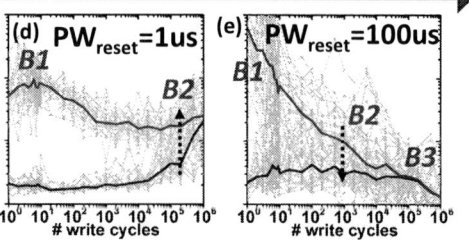

Fig.6: (a) AC endurance characteristics of the ref. device with standard test condition (pulse-width parameters PW_{set} = PW_{reset} = 10ns, edges LE = TE = 3ns, and amplitudes V_{set} = 2.5V and V_{reset} = 1.5V). It has a typical switching lifetime of 10^6 cycles with stuck-LRS failure mode (A). (b) Effect of increasing reset amplitude to 1.75V and (c) to 2.25V while keeping PW_{reset} = 10ns. In (d) and (e), PW_{reset} is increased while V_{reset} is held at 1.5V. Lifetime is dramatically decreased with the increase of V_{reset} and PW_{reset}, and the failure mode is changed to failure B, whereby a large initial HRS resistance (B1) drifts with the LRS towards intermediate HRS state failing to set (B2) and finally a reversed drift of both HRS and LRS is observed towards low-resistance levels with a definitive final stuck-LRS state (B3)

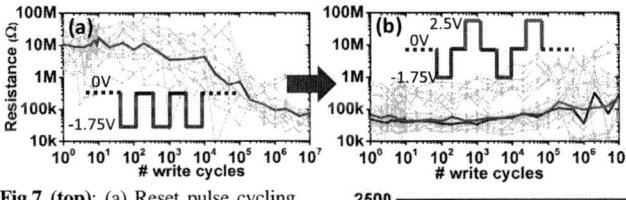

Fig.7 (top): (a) Reset pulse cycling without set pulse and (b) followed by regular set/reset cycling

Fig.8 (bottom): (a) Profiles of the CF temperature obtained during reset pulses of different PW_{reset} settings from 4ns to 100us and for amplitudes V_{reset} = 1.5V and (b) V_{reset} = 1.75V, as obtained from hourglass simulations; each plot contains 5 simulations. Temperature fluctuations are observed during the entire pulse duration, and temperature is highly increased by the amplitude V_{reset}. (c) Experimental DC reset data showing current fluctuation. Note the calculated temperature refers to power concentration.

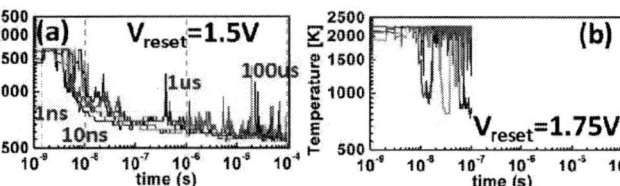

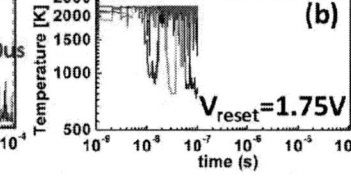

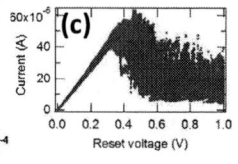

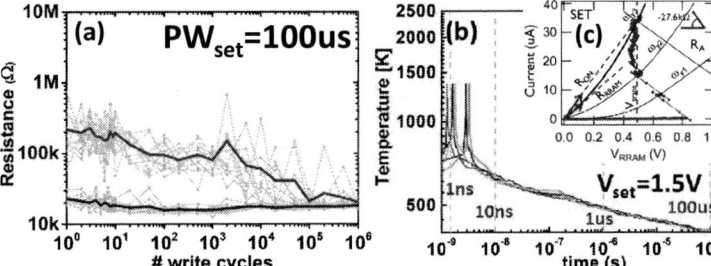

Fig.9 (left): (a) Endurance characteristics of the ref. device using standard condition except for PW_{set} = 100us, showing a ~2-decades decrease of the lifetime compared to the standard condition, which is clearly a different and lower impact than the effect of PW_{reset}; (b) Profiles of the CF temperature obtained during set pulses of different PW_{set} settings from 1ns to 100us, as obtained from hourglass simulations. Temperature is largely lower than during reset pulse, due to the voltage snapback on the device upon set transition, redistributing most of the applied voltage to the select transistor; (c) example intrinsic transient set trace

978-1-4799-8002-4/14 $31.00 © 2014 IEEE   357

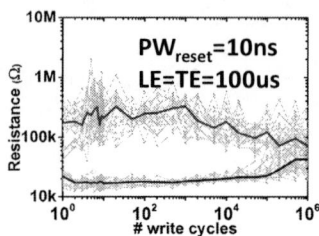
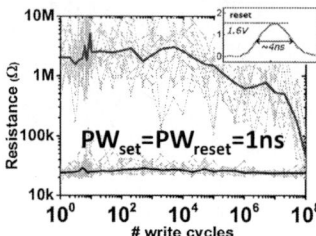
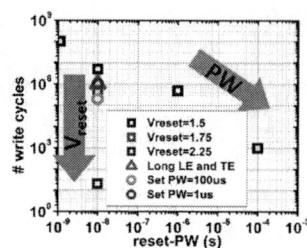
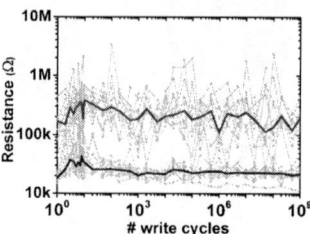

Fig.10: Endurance results for longer LE and TE, showing clearly lower impact than PW$_{reset}$ as the CF is submitted to large voltage for a small portion of the pulse duration

Fig.11: Endurance improvements obtained using ultra-short PW$_{reset}$ = PW$_{set}$ = 1ns (insert shows the reset pulse trace obtained from oscilloscope). Resistive window closure is seen only after 10^8 cycles

Fig.12: Summary of the endurance lifetime obtained on the ref. device tested in different conditions, as plotted vs critical parameter PW$_{reset}$

Fig.13: 10^9 unverified write cycles obtained using optimized ultra-short PW$_{reset}$ = PW$_{set}$= 1ns pulses on scaled devices (20nm x 20nm)

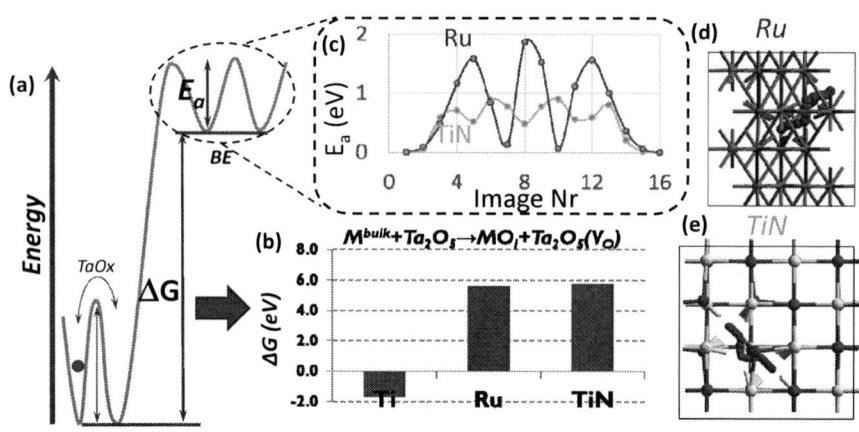

Fig.14: (a) ab-initio calculations of the energy barriers required for Oxygen to *(i)* diffuse within TaO$_x$, *(ii)* react with the BE material (ΔG), and *(iii)* diffuse in the BE material (E$_a$). TiN and Ru have similar ΔG (b), however the presence of defects in TiN such as N vacancies V$_N$ or Ti dangling bonds lead to a drop of ΔG, consistently with the result obtained for pure Ti BE (b); (c) ab-initio calculation of activation energy E$_a$ required for Oxygen to diffuse in crystalline metal along the lowest energy path in (d) Ru and (e) TiN, indicating that Oxygen is expected not only to react more with TiN than with Ru, but also to penetrate deeper in TiN. The crystal structures used in calculation for Ru is hexagonal closed packed and cubic for TiN

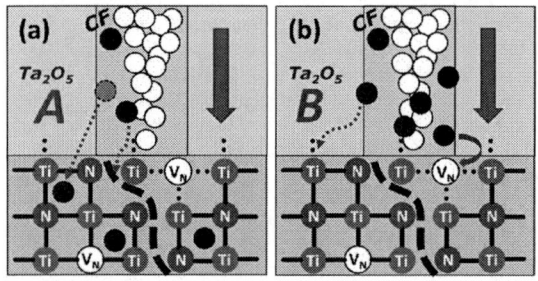

Fig.15: Schematics showing the interaction between the O anions (black) generated at forming and expected to penetrate (failure A) or react (a: failure B) with TiN, via defects resp. like grain boundary (a) and V$_N$ and Ti dangling bonds (b) in the conditions of high voltage and CF temperature during cumulated reset pulses

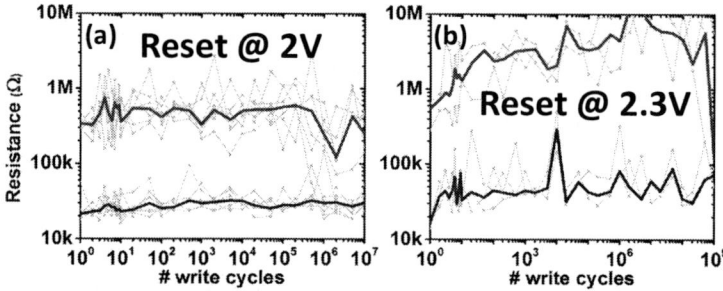

Fig.17: (a) Extended endurance lifetime obtained on Ru\Ta$_2$O$_5$\Ta devices tested in standard test condition except for V$_{reset}$ = 2V, and (b) 10^9 unverified cycles obtained in standard test condition except for V$_{reset}$ = 2.3V, confirming the suppression of the endurance degradation even in conditions of large CF temperature associated with large V$_{reset}$

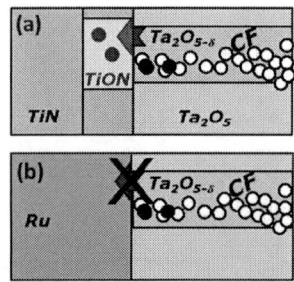
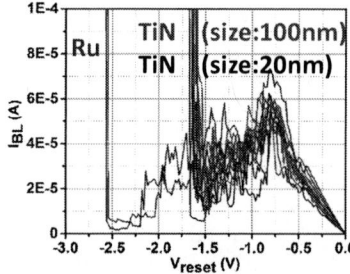
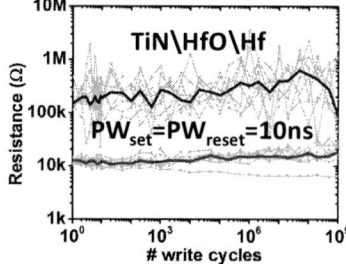
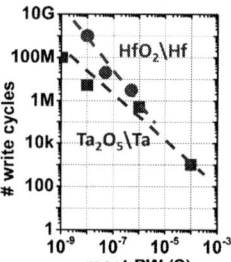

Fig.16: (a) Schematics showing the injection of O anions into TiN BE causing formation of parasitic TiO$_x$N$_y$ and accounting for failure mode B; (b) expected suppression of the reaction using Ru BE

Fig.18: Breakdown experiments carried out after DC reset of devices having either TiN or Ru BE material, showing the larger HRS breakdown voltage obtained for Ru BE, in agreement with the robustness of Ru\Ta$_2$O$_5$\Ta devices against HRS breakdown and stuck-LRS failure; Note the HRS breakdown voltage is independent of size

Fig.19: 10^9 unverified write cycle obtained with TiN\HfO$_2$\Hf cells. The cells are tested at V$_{reset}$ = 1.75V to accelerate the degradation.

Fig.20: Summary plot of the PW$_{reset}$ dependence of endurance lifetime obtained from two different dielectric stacks

978-1-4799-8002-4/14 $31.00 © 2014 IEEE

Pulsed cycling operation and endurance failure of metal-oxide resistive (RRAM)

S. Balatti[1], S. Ambrogio[1], Z.-Q. Wang[1], S. Sills[2], A. Calderoni[2], N. Ramaswamy[2] and D. Ielmini[1]

[1]DEIB, Politecnico di Milano and IU.NET, 20133 Milano, Italy, [2]Micron Technology Inc., Boise, USA

e-mail: simone.balatti@polimi.it

Abstract

Oxide-based resistive memory (RRAM) is under scrutiny for possible use for non-volatile storage and storage-class memory (SCM) complementing DRAM and SRAM. For SCM applications, set/reset times, variability and endurance are key concerns, which must be carefully understood to explore potential applications of RRAM. To that purpose we studied pulsed operation and endurance of oxide RRAM. We show that (i) resistance window (RW) is controlled by the negative voltage V_{stop} applied during reset, (ii) failure at high V_{stop} is due to negative set, causing filament overgrowth and RW collapse and (iii) endurance is independent of the pulse-width, which supports an Arrhenius model for endurance failure.

Introduction

Resistive switching memory (RRAM) is attracting a wide interest for possible future applications as mass storage and storage-class memory [1] thanks to optimal performances in terms of high speed [2], low power [3] and good scalability to small area [4]. However, many issues still need to be addressed, as *e.g.* set/reset times, variability and endurance. In particular, the dependence of the switching characteristics on the pulse-width/amplitude and the trade-off between time and amplitude must be carefully characterized and understood because these two parameters affect the set/reset variability and the endurance.

In this work we study the impact of the applied voltage during the reset pulse on the resistance window (RW), on the variability and on the endurance for different pulse widths. We show that the reset pulse voltage is the main parameter controlling the RW, the resistance variability and the endurance. For increasing reset voltage RW and variability improves while the

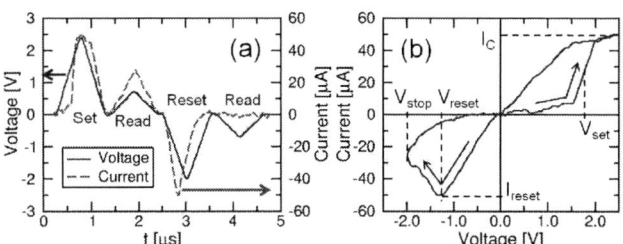

Figure 2: Measured TE voltage V and current I (a) and typical I-V characteristic (b) obtained from the measured V and I. Parameters are defined as follows: V_{set} is the set voltage marking the current crossing 25 μA, $I_C = 50$ μA is the compliance current, R_{LRS} is the resistance of the low resistance state (LRS), V_{reset} is the reset voltage marking the first decrease of R_{LRS}, I_{reset} is the corresponding reset current, R_{HRS} is the resistance of the high-resistance state (HRS).

endurance decreases, due to an increasing probability of set transition under negative voltage. From the time-dependence study we find that the pulse-width affects the set/reset voltages but not the endurance. The independence of the endurance on reset pulse-width can be explained by an Arrhenius model for degradation.

Experimental samples

We studied RRAM devices consisting of a TiN bottom electrode (BE), an amorphous HfSiO$_x$ switching layer and a Ti top electrode (TE) [5]. An on-chip integrated select transistor was used to limit the current during the set transition to a maximum compliance value I_C, as schematically shown in Fig. 1a. The setup for real-time monitoring of pulsed set/reset characteristics is reported in Fig. 1b: an arbitrary waveform generator applies a voltage pulse to the TE and to the gate of the transistor while an oscilloscope reads the voltage at the TE and the current flowing in the device [6]. Fig. 2a shows a typical waveform applied to the TE, consisting of a sequence of 4 triangular pulses, including (i) positive set, (ii) positive read, (iii) negative reset and (iv) negative read. Positive and negative voltages were used for the read pulses after set and reset, respectively, to avoid possible disturb. A pulse-width t_P of 1 μs was used in the figure for all program/read pulses. A relatively low gate voltage V_G was used during set to control the compliance current $I_C = 50$ μA, while a high V_G was used during read and reset pulses to minimize the transistor resistance. From the voltage/current in Fig. 2a, it is possible to obtain the I-V curve in Fig. 2b, indicating a RW of about 10x between low-resistance state (LRS)

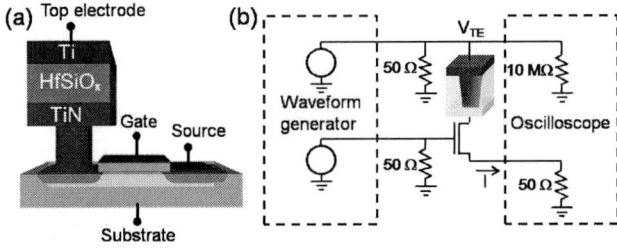

Figure 1: Schematic of the 1T1R cell (a) and of the experimental setup (b) used in this work. The RRAM stack includes a HfSiO$_x$ switching layer, a Ti cap layer and TiN BE.

978-1-4799-8002-4/14 $31.00 © 2014 IEEE   359

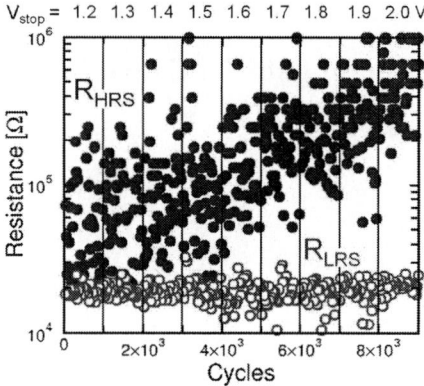

Figure 3: Measured R as a function of cycles. The stop voltage V_{stop} was increased every 10^3 cycles from 1.2 V to 2 V with step 0.1 V.

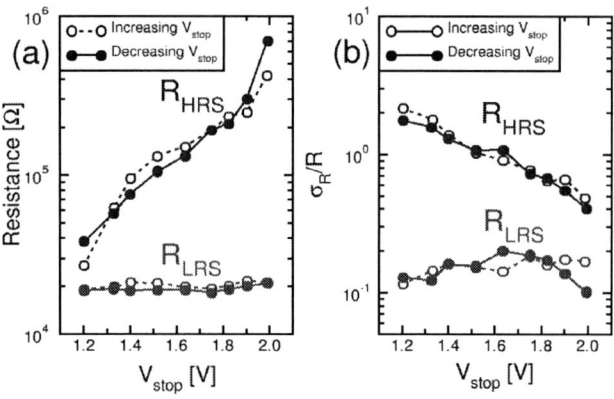

Figure 4: Median R_{HRS} and R_{LRS} (a) and relative standard deviation σ_R/R (b) as function of V_{stop}. Data are shown for both increasing V_{stop} (as in Fig. 3) and decreasing V_{stop}, showing consistent behavior and no degradation during cycling. Data indicate that increasing V_{stop} leads to a general improvement of HRS, where R increases and its cycle-to-cycle spread decreases.

and high-resistance state (HRS). The figure also shows the definition of the main parameters of the switching characteristics namely V_{set}, defined in correspondence of a current increase above 25 μA during the set transition, V_{reset}, defined in correspondence of the maximum current I_{reset} during the reset transition, and V_{stop}, namely the maximum negative voltage along the reset sweep.

Pulsed operation

Set and reset operations in Fig. 2 are controlled by I_C and V_{stop}, respectively. Limiting the current during the set transition allows to control the size of the conductive filament (CF) and the consequent I_{reset}, whereas the V_{stop} controls the HRS [7]. While I_C is kept as low as possible to limit the power consumption for high-density applications, V_{stop} can be tuned to optimize RW, switching variability and endurance. To study the impact of V_{stop} on switching, Fig. 3 shows the measured R_{LRS} and R_{HRS} across 9000 cycles, where V_{stop} was increased every 10^3 cycles from 1.2 to 2 V with a step of 0.1 V. The RW

increases with V_{stop} due to the increase of R_{HRS}. Application of a larger V_{stop} allows to open a wider gap with high resistivity within the CF, resulting in a larger R_{HRS} [8]. Fig. 4a shows the median R_{LRS} and R_{HRS} as a function of V_{stop}. Data for increasing and decreasing V_{stop} overlap with each other thus indicating no degradation during the experiment. The relative standard deviation of the resistance σ_R/R reported in Fig. 4b highlights that the variability of LRS is roughly independent on the V_{stop}. On the other hand, HRS variability decreases for increasing V_{stop}, indicating strong tightening of R_{HRS} distribution. These data demonstrate that increasing V_{stop} is beneficial not only for RW but also for switching variability.

Endurance failure

Cycling endurance was studied by monitoring the LRS and HRS as in Fig. 2 along several set/reset cycles. Fig. 5a shows R_{LRS} and R_{HRS} measured during a typical cycling experiment for $V_{stop} = 1.9$ V. Both R_{LRS} and R_{HRS} slightly decrease with cycling keeping a constant RW, then collapses around $N_{fail} = 1.7 \times 10^5$ cycles due to an increase of the R_{LRS} and a decrease of the R_{HRS}. The collapse of the RW is highlighted in Fig. 5b, showing a gradual decrease of RW across about 50 cycles. Note that failure mode differs from the usual high-resistance (open) or low resistance (short) failure modes which were previously reported [9, 10]. Fig. 5 clearly shows that the final state after the failure is neither HRS nor LRS but an intermediate state, with a resistance of about 30 kΩ.

We found that failure is always triggered by a negative set event as shown in Fig. 6a: reporting the I-V curve collected in correspondence of cycle A in Fig. 5b. After a typical positive set operation, the application of the negative pulse, causes a reset transition followed by an anomalous increase of the current.

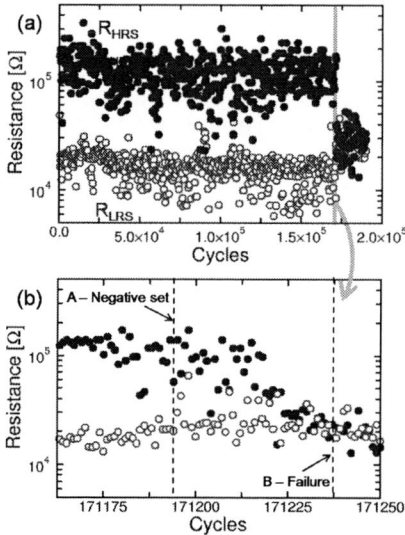

Figure 5: Measured R as a function of the number of cycles (a) and highlighted region in correspondence of failure around 1.7×10^5 (b). Two characteristic cycles are marked, corresponding to the negative-set event (A) and final failure due to RW collapse (B).

978-1-4799-8002-4/14 $31.00 © 2014 IEEE   360

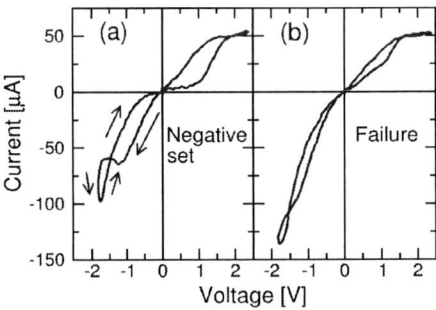

Figure 6: Measured I-V curves corresponding to the marked locations in Fig. 5, namely negative-set event (a) and failure (b). Negative set causes an excess current of about $2I_C$, initiating filament overgrowth which finally leads to the NS state with no RW in (b).

During the reset pulse the transistor gate is biased with a high voltage, to minimize the series resistance, therefore the unlimited current can freely rise to a high value, e.g., about $2I_C$ in Fig. 6a. We believe that the negative set transition consists of an injection of defects from the BE toward the TE, where the resulting filament largely overgrows due to the lack of a current compliance. As a result, R_{HRS} decreases due to the larger cross section of the depleted gap in the CF. The leaky HRS also inhibits the set operation: in fact, due to the relatively low R_{HRS}, there is no sufficient voltage drop across the RRAM to reach V_{set} and induce set transition. Fig. 6b shows the I-V curve corresponding to cycle B in Fig. 5b, after failure induced by the negative set transition. The I-V curve clearly shows no hysteresis, as a result of the lack of set transition in the leaky HRS. The state resulting from negative set will be therefore referred to as non-switching (NS) state in the following.

Cycling endurance was studied for different pulse amplitude V_{stop}. Fig. 7 shows the endurance N_{fail} as a function of V_{stop} for $t_P = 1\ \mu s$, showing a maximum N_{fail} around $V_{stop} = 1.7$ V. For V_{stop} decreasing below 1.7 V, the voltage is not enough to induce a reset transition, therefore the N_{fail}

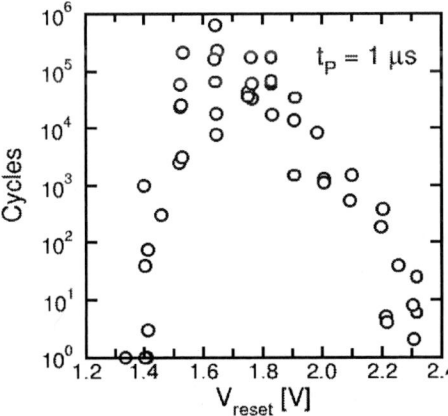

Figure 7: Measured N_{fail} as a function of V_{stop} for $t_P = 1\ \mu s$. Failure occurs by NS mode at high V_{stop}, while at low V_{stop} the device cannot reset properly.

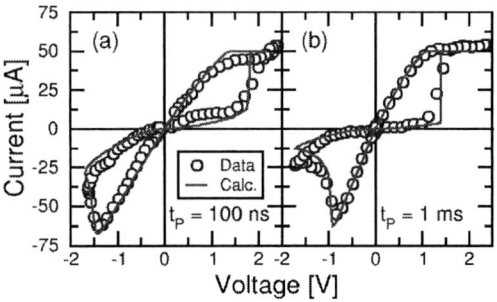

Figure 8: Measured and calculated I-V curves for $t_P = 100$ ns (a) and 1 ms (b). Note the decrease of set/reset voltages at increasing t_P, due to the voltage-accelerated nature of switching.

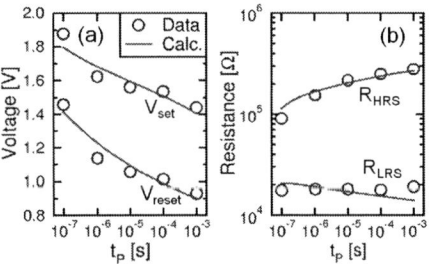

Figure 9: Measured and calculated V_{set}, V_{reset} (a), R_{HRS} and R_{LRS} (b) as a function of t_P.

sharply collapse to zero. Note that in this range of voltage ($V_{stop} < 1.7$ V), the failure is due to the insufficient voltage to reset the device, thus should not be viewed as a true failure event. For V_{stop} larger than 1.7 V, the endurance decreases due to an acceleration of the degradation with V_{stop}. As shown in Fig. 6, the failure is due to a negative set and the probability of negative set increases with V_{stop}. These results show that, although beneficial for RW and variability, he use of a large V_{stop} might be a concern from the viewpoint of endurance failure.

Endurance dependence on t_P

We studied the dependence of cycling endurance on pulse width t_P in the range from 100 ns to 1 ms. Fig. 8 shows the I-V curves for $t_P = 100$ ns (a) and 1 ms (c) at $V_{stop} = 1.7$ V. The figure also shows calculations by our analytical model [11]. Fig. 9 summarizes the measured and calculated switching parameters, namely V_{set}, V_{reset} (a), R_{LRS} and R_{HRS} (b) for $V_{stop} = 1.7$ V. Fig. 9a clearly shows the voltage-accelerated nature of the switching where both set and reset voltages decrease with t_P since less voltage is needed to trigger the set/reset transition at increasing time [12].

Fig. 10 shows the measured N_{fail} as a function of V_{stop} for increasing t_P. As the pulse-width increases, the N_{fail} curve shifts to lower V_{stop} following to the behavior of V_{reset} in Fig. 8a. Strikingly, however, the maximum $N_{fail} \approx 4 \times 10^5$ does not change with t_P, as summarized in Fig. 11a. To understand the behavior of N_{fail} as function of pulse width, we first considered an energy model where failure occurs as the energy E, dissipated over all set/reset cycles until failure and calculated

978-1-4799-8002-4/14 $31.00 © 2014 IEEE

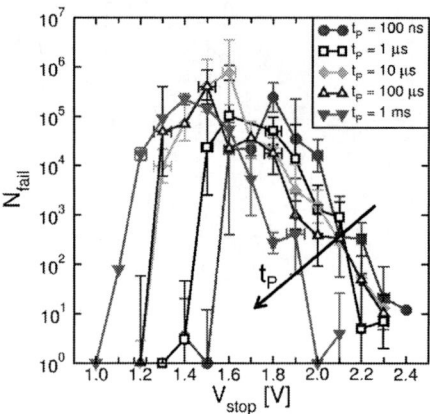

Figure 10: Measured N_{fail} as a function of V_{stop} for increasing t_P. The curves shift to lower V_{stop} at increasing t_P, in agreement with the t_P-dependence of switching voltages in Fig. 8a.

according to:

$$E = \int IV\,dt, \qquad (1)$$

reaches a threshold value E_{fail} representing the energy dose leading to destructive failure. Results of the energy model in Fig. 11a show that N_{fail} decreases as $N_{fail} \propto t_P^{-1}$, which cannot account for the t_P-dependence in the experimental data. To explain the independence of the endurance on t_P, we developed a second model based on an Arrhenius formula for triggering the negative set failure. The Arrhenius model predicts that failure occurs as

$$f = \int e^{-\frac{E_A}{kT}}\,dt, \qquad (2)$$

reaches a threshold, with T being the local temperature at the filament during set/reset processes. Calculations in Fig. 11b

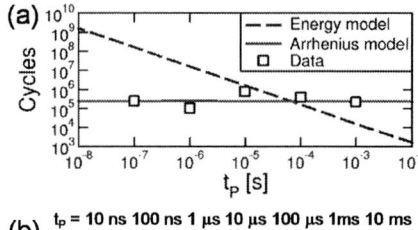

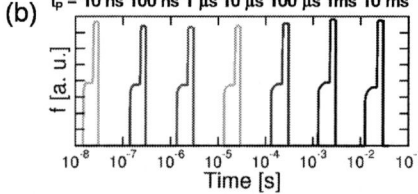

Figure 11: Optimum N_{fail} as a function of t_P (a), and calculated Arrhenius integral f in a set/reset cycle for increasing t_P (b). The Arrhenius model agrees with the t_P-independent N_{fail}, thus supporting T, combined with V_{stop}, as one of the main parameter controlling endurance failure.

show that f in one cycle is independent of t_P, since the decrease of the pulse-width is fully compensated by an increase of the voltages V_{set} and V_{reset}, as showed in Fig. 9a. These results support that local T induced by Joule heating, combined with the large electric field during reset, is one of the main driving forces of failure.

Conclusion

We studied pulsed operation of oxide RRAM, showing that V_{stop} controls RW and switching variability. Endurance failure at high V_{stop} is due to negative set, inducing a NS state with low R_{HRS} preventing set transition. Changing the set/reset pulse-width causes a negligible change of maximum endurance, which is explained by an Arrhenius model of failure.

References

[1] S. Sills, et al., "A copper ReRAM cell for storage class memory applications," *IEEE VLSI-Technology Tech. Dig.*, pp. 1–2, 2014.

[2] H. Y. Lee, et al., "Evidence and solution of over-RESET problem for HfO$_x$ based resistive memory with sub-ns switching speed and high endurance," *IEEE IEDM Tech. Dig.*, pp. 19.7.1–19.7.4, 2010.

[3] F. Nardi, et al., "Control of filament size and reduction of reset current below 10 μA in NiO resistance switching memories," *Solid-State Electronics*, vol. 58, pp. 42–47, 2011.

[4] B. Govoreanu, et al., "10x10 nm^2 Hf/HfO$_x$ crossbar resistive RAM with excellent performance, reliability and low-energy operation," *IEEE IEDM Tech. Dig.*, pp. 31–34, 2011.

[5] A. Calderoni, et al., "Performance comparison of O-based and Cu-based ReRAM for high-density applications," *IEEE IMW Tech. Dig.*, pp. 1–4, 2014.

[6] S. Koveshnikov, et al., "Real-time study of switching kinetics in integrated 1T/ HfO$_x$ 1R RRAM: Intrinsic tunability of set/reset voltage and trade-off with switching time," *IEEE IEDM Tech. Dig.*, pp. 20.4.1–20.4.3, 2012.

[7] S. Ambrogio, et al., "Understanding switching variability and random telegraph noise in resistive RAM," *IEEE IEDM Tech. Dig.*, pp. 31.5.1–31.5.4, 2013.

[8] F. Nardi, et al., "Resistive switching by voltage-driven ion migration in bipolar RRAM - Part I: Experimental study," *IEEE Trans. Electron Devices*, vol. 59, pp. 2461–2467, 2012.

[9] P. Huang, et al., "Analytic model of endurance degradation and its practical applications for operation scheme optimization in metal oxide based RRAM," *IEEE IEDM Tech. Dig.*, pp. 22.5.1–22.5.4, 2013.

[10] Y. Y. Chen, et al., "Balancing SET/RESET pulse for $> 10^{10}$ endurance in HfO$_2$/Hf 1T1R Bipolar RRAM," *IEEE Trans. Electron Devices*, vol. 59, pp. 3243–3249, 2012.

[11] S. Ambrogio, et al., "Analytical modeling of oxide-based bipolar resistive memories and complementary resistive switches," *IEEE Trans. Electron Devices*, vol. 61, pp. 2378–2386, 2014.

[12] C. Cagli, et al., "Modeling of set/reset operations in NiO-based resistive-switching memory RRAM devices," *IEEE Trans. Electron Devices*, vol. 56, pp. 1712–1720, 2009.

Impact of low-frequency noise on read distributions of resistive switching memory (RRAM)

S. Ambrogio[1], S. Balatti[1], V. McCaffrey[2], D. Wang[2] and D. Ielmini[1]

[1]DEIB, Politecnico di Milano and IU.NET, 20133 Milano, Italy, [2]Adesto Technologies, Sunnyvale, CA, USA
e-mail: stefano.ambrogio@polimi.it

Abstract

Resistive switching memory (RRAM) is one of the most promising emerging device technology for future storage and computing memories. As other emerging memories based on materials storage at the nanoscale, RRAM is affected by switching and read fluctuations. We addressed current fluctuation in RRAM at both cell and array levels. First, we present an analytical model for 1/f and random telegraph noise (RTN) in single (intrinsic) cells, allowing to predict time-dependent broadening of read current I_{read} distributions. Then we address tail cells with statistically-high noise in large arrays, revealing time-decaying random walk (RW) and intermittent RTN phenomena for the first time. A statistical noise model capable of explaining the current distribution broadening in RRAM arrays is finally developed and discussed.

Introduction

Resistive memories aiming at replacing Si-based Flash and DRAM devices rely on materials storage, where the device resistance is changed by manipulating the intimate structure of the active material. For instance, the resistive switching memory (RRAM) and the conductive-bridge memory (CBRAM) rely on the chemical composition within a conductive filament [1]. Similarly, the state of a phase change memory (PCM) is dictated by the atomic structure of a phase-change material [2] and the state of a magneto-resistive memory (MRAM) depends on the spin-polarization state within a magnetic tunnel junction [3]. As the device size is scaled down, resistance fluctuations arise due to atomic-level modifications in the storage material, such as the charging/discharging of a defect at the CF in RRAM [4], the bistable switching of a conductive defect in the amorphous chalcogenide [5], and the thermal hopping of magnetic domain walls in MRAM [6]. These noise sources must be carefully understood to understand and predict the scaling properties of each technology.

In this work we study the impact of low-frequency fluctuations on both single RRAM cells and arrays. First, we develop an analytical model for the current broadening distributions due to both 1/f and RTN noises. Then we extend the study to large RRAM arrays, aiming at understanding the behavior of tail bits with statistically-large noise. Tail bits reveal new noise behaviors, which we identify as time-dependent random walk (RW) and interrupted RTN. After discussing the possible physical descriptions of RW and RTN, we present a Monte-Carlo statistical model for predicting distribution broadening in RRAM arrays.

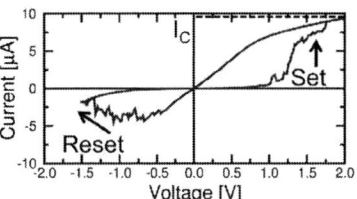

Figure 1: Measured I-V curve with a compliance current $I_C = 10$ μA.

Switching and noise characteristics

In this work a 1-Transistor 1-Resistor (1T1R) structure has been studied in which the resistive device was composed by an active metallic electrode, two different metal oxide layers, indicated by stack A and stack B, and an inert bottom electrode. All measurements have been performed on stack A, except where noticed. Fig. 1 shows a measured I-V curve in which the set transition is obtained through a positive voltage to the active electrode. The current increase is limited by the transistor in series with the device to a maximum value defined as compliance current I_C which is equal to 10 μA in Fig. 1. The set state achieved corresponds to approximately 100 kΩ. Reset transition is instead obtained through a negative voltage ramp.

Noise effects in intrinsic cells

Fig. 2a shows the measured read current I_{read} obtained by reading a 10 kΩ set state with a read voltage V_{read} equal to 10 mV. Current noise fluctuations dominate the overall behavior, leading to a broadening of the current distribution. The reason resides in the 1/f noise, which is evidenced by S_I, namely the power spectral density (PSD) in Fig. 2e, revealing a clear -1 slope. Fig. 2b shows the extracted relative broadening σ_I/I, namely the ratio between the standard deviation σ_I and the median value I of the read current samples from zero up to time t. The ratio is initially very low because of the reduced number of integrated current samples, then it increases with time due to the ongoing integration of the low frequency noise contributions. To predict the current broadening behaviour, we developed a simple analytical model for describing the evolution of the σ_I/I ratio. The variance σ_I^2 can be calculated through the integration of the S_I spectrum from a minimum frequency f_{min} up to a maximum frequency f_{max} [7],

$$\sigma_I^2 = \int_{f_{min}}^{f_{max}} S_I df, \qquad (1)$$

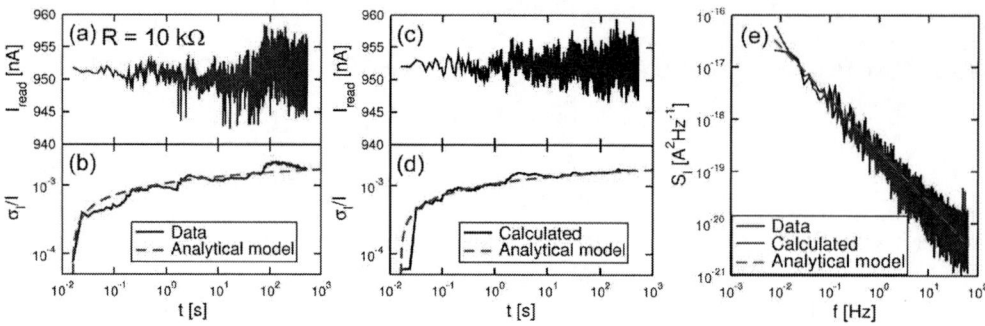

Figure 2: Measured I_{read} for a set state with $R = 10$ kΩ (a). Measured and calculated σ_I/I broadening (b). Calculated numerical 1/f noise (c) and σ_I/I broadening together with the analytical model (d). (e) shows the S_I of both measured and calculated noise, together with the analytical model, confirming the 1/f nature of noise.

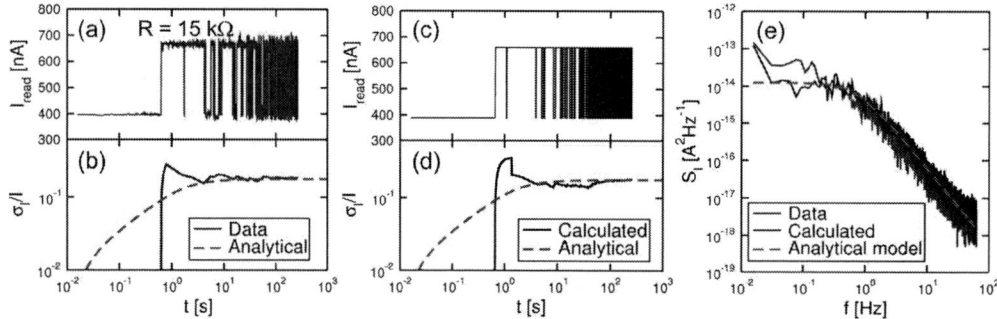

Figure 3: Measured I_{read} for a set state with $R = 15$ kΩ affected by RTN noise. The σ_I/I ratio is plotted in (b), together with the analytical calculation. (c) and (d) show the calculated numerical RTN noise. (e) shows the measured, calculated and analytical S_I, displaying a typical $1/f^2$ behaviour.

where $f_{max} = (2t_s)^{-1}$ and t_s is the sampling time, equal to 8 ms, and $f_{min} = t^{-1}$. Longer times lead to the integration of lower frequency noise contributions with a higher noise power, broadening the current distribution. Assuming a 1/f spectrum such as $S_I = A/f$, the parameter A can be evaluated from S_I at a frequency $f = 1$ Hz. Integrating S_I and dividing by I, the relative standard deviation σ_I/I can be analytically obtained by:

$$\frac{\sigma_I}{I} = \frac{\sqrt{A\ln(f_{max}t)}}{I}. \qquad (2)$$

The calculated results are reported in Fig. 2b, showing a good agreement with data, while Fig. 2e shows the measured S_I and calculated A/f. To further confirm Eq. 2, we simulated a numerical 1/f I_{read} noise and its corresponding time-dependent distribution broadening. Fig. 2c shows the calculated current as a function of time, while Fig. 2d shows the calculated σ_I/I and Fig. 2e shows the corresponding PSD. The numerical results agree with the experimental data and supports our analytical model in Eq. 2 as a useful tool for easily evaluating the broadening of a current level or, equivalently, of a resistance level based on the relationship $\sigma_I/I = \sigma_R/R$.

Similarly, we addressed random telegraph noise in single cells. Fig. 3a shows a measured I_{read} displaying RTN, obtained for a low resistance state with $R = 15$ kΩ with $V_{read} = 10$ mV. The extracted σ_I/I is reported in Fig. 3b. The relative standard deviation is approximately zero up to the first current step occurring at 0.7 s, due to the almost constant value of I_{read}, then

the ratio sharply increases due to the increased spread caused by the change in the current level. Finally, for longer times, it stabilizes at a constant value of 0.2. Differently from the 1/f case, the σ_I/I ratio does not continuously increase because only two current levels are integrated. An analytical model for describing the average σ_I/I evolution with time can be obtained, similar to the case of 1/f noise. For a random telegraph signal, the PSD S_I has a Lorentzian shape according to the formula [8]:

$$S_I = 4\Delta I^2 \frac{1}{\tau_{on} + \tau_{off}} \frac{\tau_P^2}{1 + (2\pi f)^2 \tau_P^2}, \qquad (3)$$

where τ_{on} and τ_{off} are the mean durations of the high-current phase and the low-current phase, respectively, $\tau_P = \frac{\tau_{on}\tau_{off}}{\tau_{on}+\tau_{off}}$ is the RTN time constant, and ΔI is the current amplitude of RTN. Following Eq. 1, the variance σ_I^2 is given by:

$$\sigma_I^2 = \frac{2\Delta I^2 \tau_P}{\pi(\tau_{on} + \tau_{off})}(atan(2\pi\tau_P f_{max}) - atan(\frac{2\pi\tau_P}{t})), \qquad (4)$$

where the minimum frequency in the integral was taken as $f_{min} = 1/t$. Fig. 3b shows σ_I/I obtained from Eq. 4. To support the analytical model in Eq. 4, an I_{read} affected by RTN was numerically generated, as shown in Fig. 3c. The corresponding σ_I/I is shown in Fig. 3d and the power spectral density is shown in Fig. 3e, all displaying a good agreement with data and supporting the analytical model. Note the smooth increase of σ_I/I by the analytical model, which is due to the convolu-

978-1-4799-8002-4/14 $31.00 © 2014 IEEE

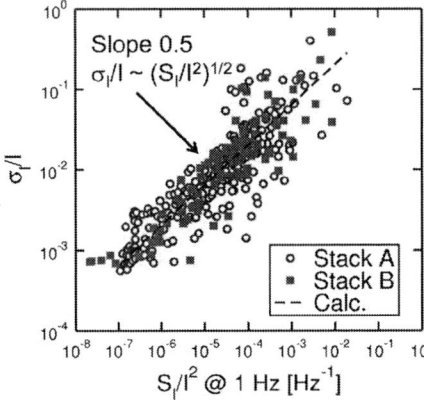

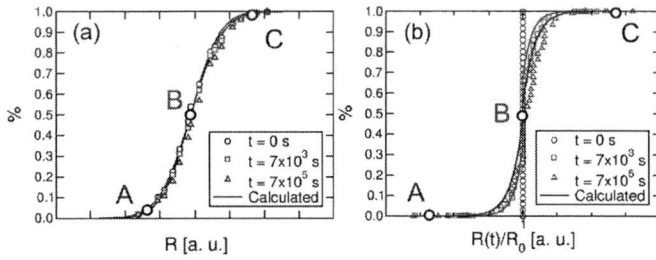

Figure 4: Scatter plot of σ_I/I at 520 s as a function of S_I/I^2 at 1 Hz. The 0.5 slope is consistent with the analytical model of Eq. 2.

Figure 5: Measured and calculated distributions of R (a) and of $R(t)/R_0$ at increasing times 0, $7\text{x}10^3$ and $7\text{x}10^5$ s. Fluctuations of R are best revealed by R/R_0 (b). Selected cells A, B and C are also shown in their final position within the distributions at $t = 7\text{x}10^5$ s.

tion of all possible step-change of σ_I/I seen in the experimental (Fig. 3b) or numerical (Fig. 3d) results.

Fig. 4 shows the scatter plot which correlates the measured broadening σ_I/I taken at 520 s and the corresponding S_I/I^2 taken at 1 Hz, for several programmed states. Data in the figure indicate a slope of 0.5, which can be explained by a dominant 1/f noise. In fact, σ_I/I is proportional to $\sqrt{A}$ in Eq. 2, while A is proportional to S_I, thus causing $\sigma_I/I \propto \sqrt{S_I/I^2}$. The calculated line is also shown in the figure, highlighting that 1/f noise constitutes the main source of broadening for resistive switching devices.

Noise effects in cell arrays

Reliability prediction in cell arrays requires the analysis of tail RRAM cells, as opposed to intrinsic cells considered in the previous section. To address noise effects in the distribution tail, we studied arrays of 1T1R devices. All array cells were programmed in the reset state since this state generally displays the highest noise [9]. Fig. 5a shows the resistance cumulative distribution function (CDF) of the entire array at different times $t = 0$, $7\text{x}10^3$ and $7\text{x}10^5$ s. The distribution shows a negligible dependence on time. However, resistance broadening is observed studying the CDF of the normalized resistance $R(t)/R_0$, as shown in Fig. 5b, where R_0 is the initial cell resistance. Three different cells A, B and C are reported in Fig. 6, showing random walk (RW) events at time t_{RW} and intermittent RTN active for a time t_{on}.

To study step changes of the resistance in Fig. 6, we defined $x = R_i/R_{i-1}$ as the ratio between resistances at times t_i and t_{i-1}. Fig. 7a shows the distribution $g(x)$ of the relative steps x among the entire set of cells for increasing times $t = 0$, $7\text{x}10^3$, $7\text{x}10^4$ and $7\text{x}10^5$ s. The distribution decreases at increasing time, meaning that RW and RTN steps become less and less probable with time. This cannot be explained by a stationary RW or RTN, as this would cause a time-independent $g(x)$.

Fig. 7b shows the distribution $g(t)$ defined as the number of cells with a relative resistance step larger than x at time t. The extracted $g(t)$ in the figure is shown for different $x > 2, 3, 4, 5$. Note that cells with $x > 2$ (or $x > 3$, etc.) also include cells with $x < 1/2$ (or $x < 1/3$, etc.), based on the symmetry of $g(x)$

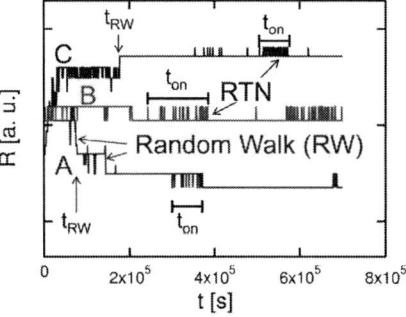

Figure 6: Measured R as a function of time for cells A, B and C from Fig. 5. RW occurs at t_{RW} and interrupted RTN lasts for time t_{on}.

with respect to $x = 1$ in Fig. 7a. All distributions show an initial decrease with a slope of -1 up to $5\text{x}10^4$ s, followed by a saturated region. The initial decay is attributed to RW steps reflecting CF stabilization after reset. After the initial decay, the number of steps stabilizes and the main contribution is due to intermittent RTN which is active only during times t_{on}.

Statistical model

We developed a statistical Monte-Carlo model to account for the resistance broadening, based on both RW and RTN contributions. RW steps were modelled as in Fig. 8 following an energy relaxation model. After the programming operation, the atomic structure stabilizes overcoming a sequence of increasing energy barriers E_{RW} (Fig. 8a). The corresponding times t_{RW} for RW steps are thermally activated:

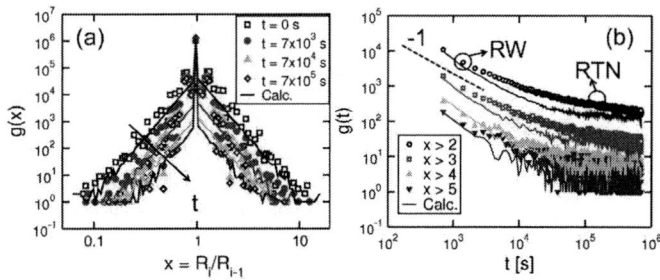

Figure 7: Measured and calculated distribution $g(x)$ of relative steps $x = R_i/R_{i-1}$ at increasing times, (a), and $g(t)$ for $x > 2, 3, 4, 5$ (b).

978-1-4799-8002-4/14 $31.00 © 2014 IEEE

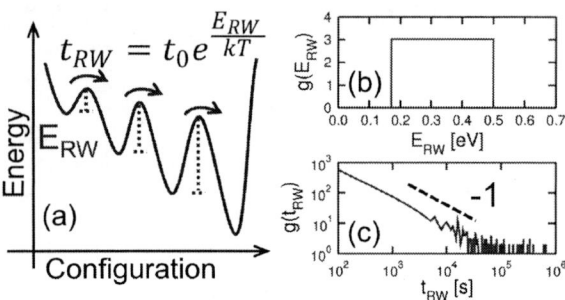

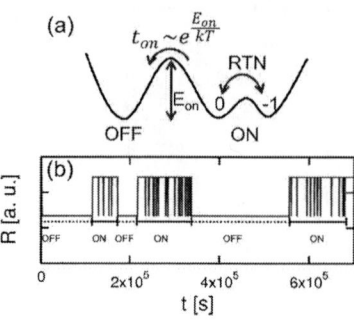

Figure 8: Energy relaxation model for RW based on distributed energy barriers E_{RW} (a), calculated distribution of E_{RW} (b) and relaxation times (c).

Figure 9: Energy profile for intermittent RTN with off (stable) and on (bistable) states (a), and experimental behavior of intermittent RTN (b).

$$t_{RW} = t_0 e^{\frac{E_{RW}}{kT}}. \tag{5}$$

Once the lowest barriers are overcome, the atomic structure faces higher energy barriers, hence longer times are needed to reach more stable states. This explains the dependence of RW from time and the initial high density of steps. Considering the relation $g(t_{RW})dt_{RW} = g(E_{RW})dE_{RW}$ where the density of time steps $g(t_{RW})$ is proportional to the density of energy states $g(E_{RW})$ and considering a constant energy distribution for RW $g(E_{RW}) = A$, Fig. 8b, the distribution $g(t_{RW})$ becomes

$$g(t_{RW}) = \frac{AkT}{t_{RW}}, \tag{6}$$

which explains the -1 slope, Fig. 8c.

Intermittent RTN can be explained as a consequence of a bistable defect which oscillates between an active or ON state and an inactive, OFF state, Fig. 9a. In the ON state, the defect oscillates between neutral and negative states, leading to RTN [9]. The measured intermittent RTN thus shows active and inactive states, as in Fig. 9b. In the model, RTN showed intermittent behaviour, leading to constant $g(t)$ for longer times. The calculated resistance and $R(t)/R_0$ CDF are plotted in Fig. 5, showing a good agreement with data and thus supporting the model as a reliable tool for statistical array broadening evaluation. The calculated $g(x)$ is plotted in Fig. 7a, highlighting the dependence of $g(x)$ from time due to RW, as also evident from the calculated $g(t)$ in Fig. 7b in which after an initial decay due to RW, a constant plateau corresponds to the RTN contribution.

Finally, different RRAM stacks were compared, namely stacks A and B. Figs. 10a-b report the $R(t)/R_0$ CDF for stack A (a) and B (b), showing negligible differences. Calculated results are also shown, evidencing a good agreement with data. Figs. 10c-d show the measured median value, $\pm 1\sigma$, $\pm 2\sigma$ and $\pm 3\sigma$ for stacks A (c) and B (d) as a function of time. The spreads are symmetrical and the median is constant, revealing no resistance drift. The calculated distributions show good accuracy with data. To sum up, resistance fluctuations cause distribution broadening that must be predicted and minimized to improve RRAM array distributions. Solutions to alleviate noise effects may include high-voltage sensing during read [10] and improvement of the resistance window by programming algorithms [9] and material engineering [11].

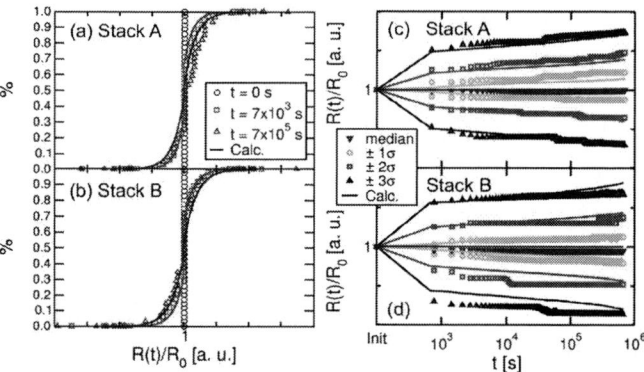

Figure 10: Measured and calculated $R(t)/R_0$ CDF (a-b) and spreads at increasing $\sigma = 0$ (median), $\pm 1\sigma$, $\pm 2\sigma$, $\pm 3\sigma$ (c-d) for stacks A (a-c) and B (b-d).

Conclusion

We studied low-frequency noise and distribution broadening in RRAM cells and arrays. A model is developed for the impact of 1/f and RTN on I_{read} distributions of single RRAM cells. Array distribution broadening is attributed to time-dependent RW and intermittent RTN phenomena. A Monte Carlo statistical model is developed for RW and RTN, allowing to predict distribution broadening at statistical level in RRAM arrays. The impact of current fluctuations on RRAM reliability can be alleviated through improved read/program schemes and novel active materials are needed.

References

[1] H.-S.P. Wong, et al., *Proc. IEEE* 100, 1951 (2012).

[2] S. Raoux, et al., *Chem. Rev.* 110, 240 (2010).

[3] C. Chappert, et al., *Nat. Mater.* 6, 813 (2007).

[4] D. Ielmini, et al., *Appl. Phys. Lett.* 96, 053503 (2010).

[5] D. Fugazza, et al., *IEDM Tech. Dig.* 723 (2009).

[6] L. Jiang, et al., *Phys. Rev. B* 69, 054407 (2004).

[7] P.R. Gray, et al., *Analysis and design of analog integrated circuits*, Wiley (2001).

[8] S. Machlup, *J. Appl. Phys.*, 25, 341 (1954).

[9] S. Ambrogio, et al., *IEDM Tech. Dig.* 782 (2013).

[10] S. Balatti, et al., *IRPS Tech. Dig.* 1 (2014).

[11] A. Calderoni, et al., *IMW Proc.* 1 (2014).

978-1-4799-8002-4/14 $31.00 © 2014 IEEE

A New Approach for Trap Analysis of Vertical NAND Flash Cell using RTN Characteristics

Daewoong Kang, Changsub Lee, Sunghoi Hur, Duheon Song, and Jeong-Hyuk Choi

Flash Process Architecture Team, Flash Development Center, Samsung Electronics Co.,
San #24, Nongseo-Dong, Giheung-Ku, Yongin-City, Gyenggi-Do, 449-711, South Korea
Tel) 82-31-209-1457, Fax) 82-31-209-0295, E-Mail) dw1461.kang@samsung.com

Abstract

We introduce new phenomena that show turn-on at back-side for Vertical NAND (V-NAND) with back-insulator and propose a new method to analyze the trap of back-interface related to the phenomena. Back-side traps have been analyzed with the back-gate structure [1]. However, V-NAND has no back-gate structure, so it's difficult to observe traps. With RTN method we proposed, it's possible for us to observe back-side traps.

Introduction

In order to satisfy the rapidly integrating high density NAND flash beyond 10 nm node generation, 3D NAND devices are considered as the most promising near-term solution [2]. The majority of the solutions presented nowadays use a deposited poly-silicon (poly-Si) channel [3-4] having SOI structure. Vertical NAND (V-NAND) relies on this poly-Si channel and consists of charge trapped devices (CTF) with a vertical cylindrical geometry. The channel is made of deposited poly-Si layer and the integration of these devices has been already proven [4]. Modeling of poly-Si channel has been studied thoroughly [5-6]. However, there has been no report on method which can analyze the interface trap as shown in Fig.1.

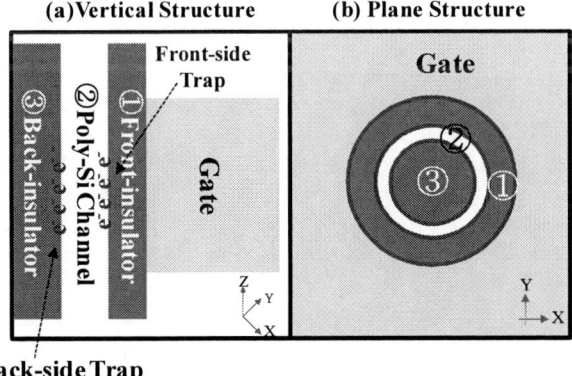

(a) Vertical Structure   **(b) Plane Structure**

Fig. 1 The schematic pictures for (a) vertical and (b) plane V-NAND structures.

In this paper, we introduce new phenomena related to the trap characteristics of back insulator. To study these phenomena, we analyzed it using TCAD simulator and evaluated Random Telegraph Noise (RTN) according to the cell Vth states in the V-NAND Flash test structure.

Phenomena and Measurement Method

Fig. 1(a) and (b) show the vertical and plane structures for V-NAND. In order to improve the back-side traps, the deposition method of the back-insulator was changed from process 'A' to 'B' which has almost the same heat budget as shown in Fig. 2.

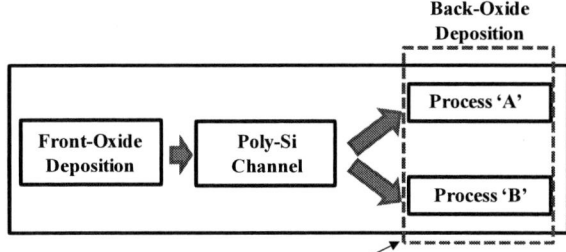

Process temperatures of 'A' and 'B' are almost the same

Fig. 2. The process flow chart about process 'A' and 'B'.

It was expected that their swings of initial Id-Vg were the same because only the back-side trap was changed. However the swing of process 'B' was improved as shown in Fig. 3. We simulated the current path to understand the phenomena. As a result, it was found that the current path of the program state was formed at the back-side. On the contrary, the current path of the erase state was not. The V-NAND consisting of a thin body, double gate and back-insulator gives rise to the current path a difference as shown in the simulation results of Fig. 4.

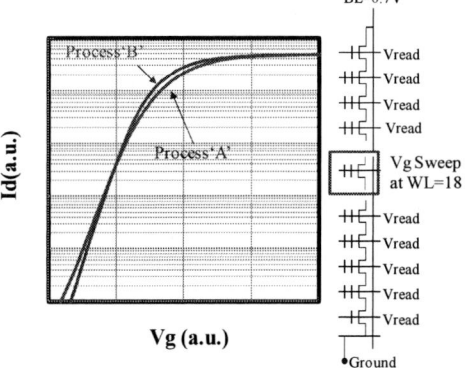

Fig. 3 Initial Id-Vg curves at WL=18 for Back-oxide deposition 'A' and Back-oxide deposition 'B' when all cells are initial state.

978-1-4799-8002-4/14 $31.00 © 2014 IEEE   367

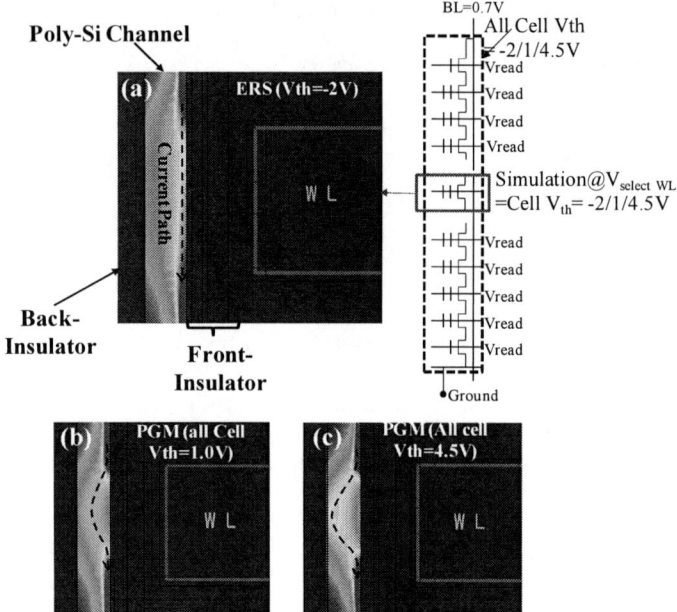

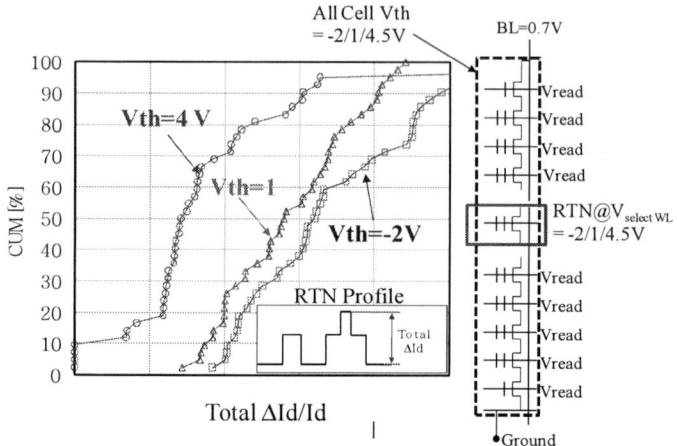

Fig. 4. The simulation results for current path of select WL in V-NAND for (a) $V_{select\ WL}$=Cell Vth=-2V (b) $V_{select\ WL}$= Cell Vth=1V (c) $V_{select\ WL}$=Cell Vth=4.5V, respectively.

From these observations, we have learned that it is very critical to characterize the back-side traps. Characterization of back-side traps has been analyzed in the back-gate structure [1]. However, the problem is that V-NAND has no back-gate structure, so it is very difficult to observe the traps. Therefore, we propose a new method which enables us to investigate the back-side traps of V-NAND using RTN measurement method depending on cell Vth states. RTN was measured at the cell test patterns of V-NAND. The current fluctuation can be expressed as the sum of ΔId/Id for front-insulator, poly-si and back insulator, respectively.

$$\frac{Total\Delta I_d}{I_d} \propto \sum \frac{\Delta I_d}{I_d} = \left[\frac{\Delta I_d}{I_d}\right]_{Front-Insulator} + \left[\frac{\Delta I_d}{I_d}\right]_{Poly-Si} + \left[\frac{\Delta I_d}{I_d}\right]_{Back-Insulator} \quad (1)$$

The magnitude of total ΔId/Id is proportional to the number of traps [7]. The front–insulator term should be the largest due to increased traps by lots of plasma damages. On the contrary, the poly-Si term is considered to be the smallest because RTN of the term is difficult to detect due to the average effect by many traps in Si-bulk. In Fig. 5, the RTN measured according to the cell Vth states shows the different distributions as expected in the simulation results of Fig. 4. As the cell Vth increases, total ΔId/Id decreases, which can be interpreted as follows. As the current path moves to back-side for the higher the cell Vth, the effect of front-insulator term of Eq.(1) should be smaller. From these reasons, the total current fluctuation decreases by increasing the cell Vth as shown in Fig. 5.

Fig. 5. The cumulative curves of current fluctuation(=total ▵ Id/Id) at the same current(Id=100 nA) according to cell states of 40 cells. Inset shows the schematic picture of RTN profile.

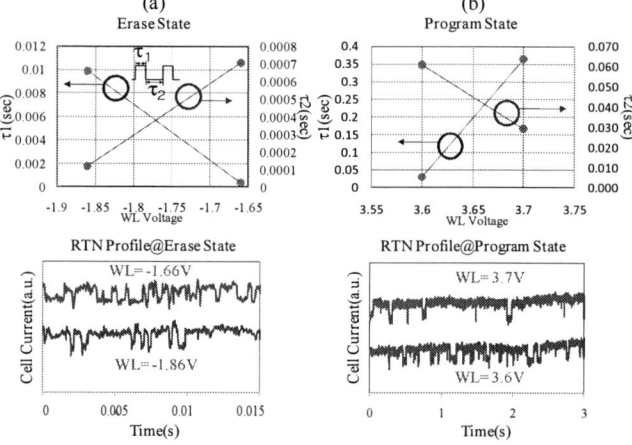

Fig. 6. The RTN profile and change capturing time(τ1) and emission time(τ2) as the increasing gate voltage for (a) Erase state and (b) PGM state.

The location of current path for the cell Vth states can be clearly proved through measuring capture/emission time of RTN as the gate bias increases. The capture time(=τ1) decreases for the erase state, to the contrary, the time increases for the program state as shown in Fig.6. It is because the capturing probability of traps was opposite at the front-insulator and the back-insulator as gate voltage increases as shown in Fig. 7.

Result and Discussion

Fig. 8 shows the RTN measurement results for the two different cell Vth states according to overdrive voltage for the back-side process 'A' and 'B' referred to Fig. 3. We observed the RTN of process 'B' is improved at only program state, which implies the number of back-side traps caused by the process 'B' was reduced and the front-side traps were not

978-1-4799-8002-4/14 $31.00 © 2014 IEEE

affected. This result explains well the changed swing for the process 'B' of Fig. 3 because the current flows at the backside. Fig. 9(a) and (b) show the RTN distributions for cell Vth=-2V/4.5V states before and after 10k P/E cycles. The RTN of program states is not changed according to P/E cycles. However, as the cell Vth states are lower, the RTN increased after P/E cycles, which means that generated traps after P/E cycles just affected the front-side interface as shown in Fig. 9(c).

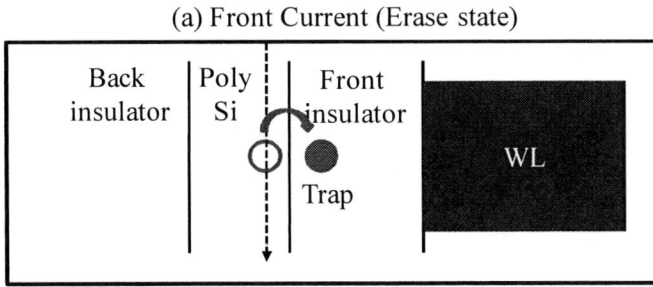

(a) Front Current (Erase state)

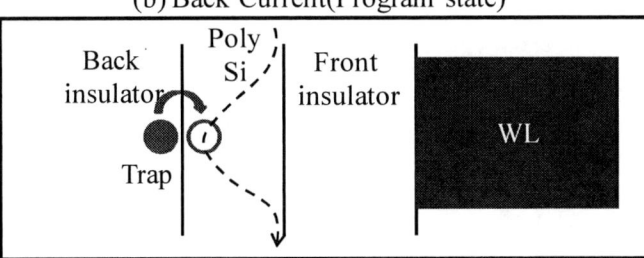

(b) Back Current(Program state)

Fig. 7. The schematic pictures for capturing/emission as increasing gate bias. (a) The capturing of electron decreases the current at front-insulator (b) The emission of electron increases the current at back-insulator.

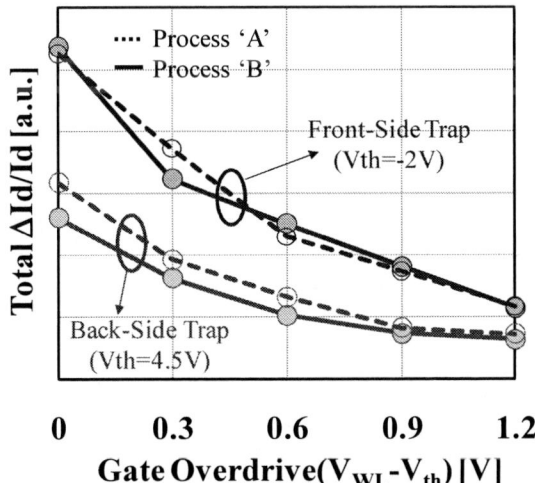

Fig. 8. The RTN measurement results according to gate overdrive voltage. The number of traps for process 'B' is reduced at only backside. Etch points represents the median values of 40 points.

These results show that the new method distinguishes well the back-side traps from the front-side traps. Fig. 10 shows the correlation between swing and total $\Delta Id/Id$ is weak. This is because that sub-threshold swing expressed the traps of all areas including the back-insulator. In Fig. 3, the change of sub-threshold swing for process 'B' was detected because the reduced traps were very large. In order to analyze the relation between main chip operation and back-insulator traps, the various processes were applied as shown in Fig. 11.

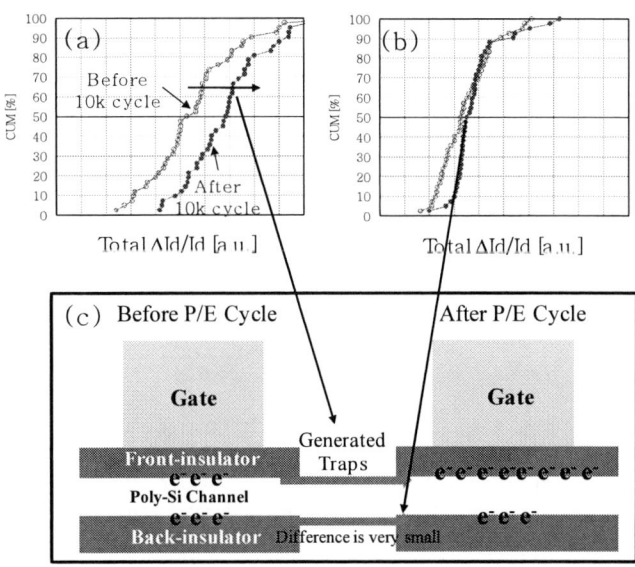

Fig. 9. The RTN distributions before and after 10 k P/E cycles at (a)Vth=-2V, (b) Vth=4.5V. (c) The location differences of traps before and after 10 k P/E cycle. Traps were generated between Poly-si and front –insulator.

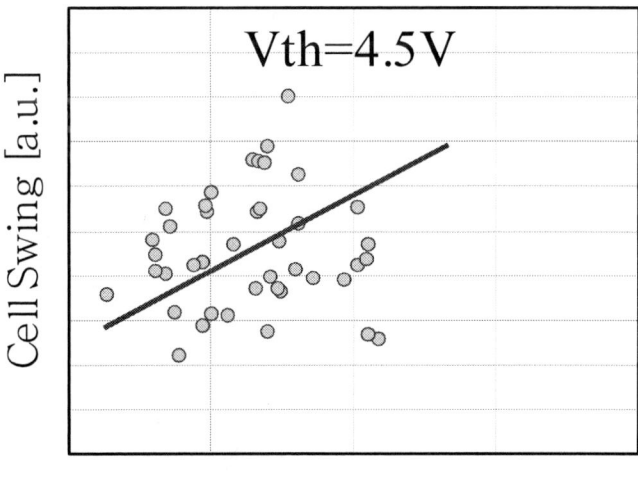

Fig. 10. The correlation between Total$\Delta Id/Id$ and cell swing at Cell Vth=4.5V.

978-1-4799-8002-4/14 $31.00 © 2014 IEEE

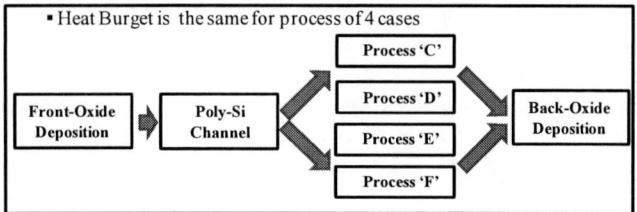

Fig. 11. The flow chart for various processes. Heat treatment is all the same.

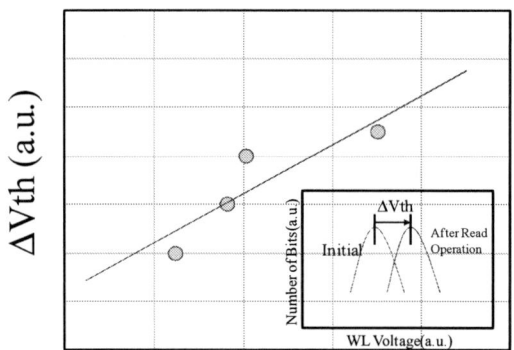

Fig. 12. The correlation of the Vth shift by read disturb and totalΔId/Id at Cell Vth=4.5V. Inset show the cell Vth shift by the stress situation related with read operation.

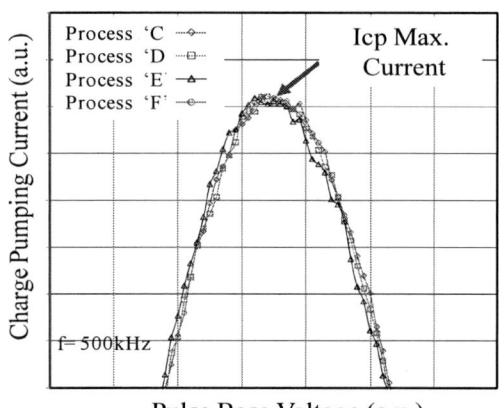

Fig. 13. The charge pump current for Process C,D,E,F. The maximum current of charge pump is almost same for processes, which means that process C,D,E,F do not affect the interface trap of front-insulator and poly-si.

It was observed that the traps of back-insulator affect the special read-disturb failure in main chip as shown in Fig. 12. The difference of front-insulator interface-trap is negligible because the charge pump current is almost the same as shown in Fig. 13. Fig 14 shows the sub-threshold swing and read disturb were not well correlated. This means that the ability of sub-threshold swing to detect the traps of back-side is worse.

The Vth shift by the read disturb before P/E cycle is almost the same as that of after P/E cycle as shown in Fig. 15. This shows that the traps generated at the front-side do not affect the read disturb characteristics.

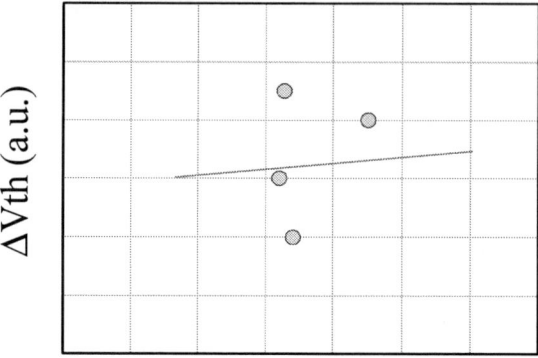

Fig. 14. The correlation of the Vth shift by Read disturb and cell sub-threshold swing at Cell Vth=4.5V.

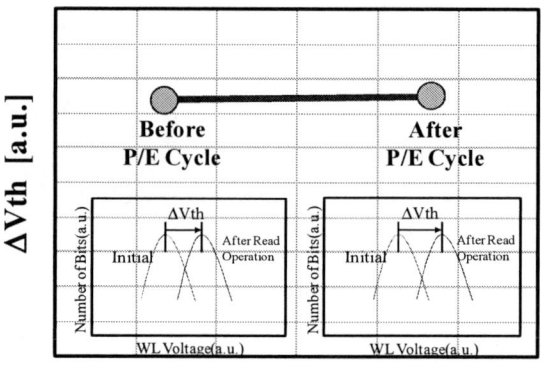

Fig. 15. The ΔVth by read disturb before and after P/E cycles respectively.

Conclusion

In this paper, we proposed the new method to analyze the back-side traps using the RTN measurement method for the various cell Vth states. We proved through the various experiments that it characterizes well the back-side traps in V-NAND with a thin poly-Si channel and a back-insulator. With RTN method that we have proposed, now it's possible for us to detect back-side traps for characterization and further analysis.

References

[1] M. Kimura et al, IEEE TED., vol. 57, no. 12, pp3426-3433, Dec. 2010.
[2] J. Choi et. al., VLSI Tech. Dig., pp178-179, 2011.
[3] R. Katsumata et al., VLSI Tech. Dig., pp126-137, 2009.
[4] J. Jang et al., VLSI Tech. Dig., pp.192-193, 2009.
[5] J. Kim et al., VLSi Tech. Dig., pp. 186-187, 2009.
[6] W. Kim et al., VLSI Tech. Dig., pp.188-189, 2009.
[7] D. Kang et al., IEEE VLSI Tech., Dig., pp.206-207 2

Through Silicon Via (TSV) effects on devices in close proximity– the role of mobile ion penetration - characterization and mitigation

C. Kothandaraman, S. Cohen, C. Parks, J.Golz, K.Tunga, S. Rosenblatt, J. Safran, C. Collins, W.Landers, J. Oakley, J.Liu, A.J. Martin, K. Petrarca, M. Farooq,T. L. Graves-Abe, N.Robson, and S.S.Iyer

IBM Microelectronics, Semiconductor R&D Center, Hopewell Junction, NY 12533,USA

Phone:+1-845-892-9851 email: raman1@us.ibm.com

ABSTRACT

A new interaction between TSV processes and devices in close proximity, different from mechanical stress, is identified, studied and mitigated. Detailed characterization via Triangular Voltage Sweep (TVS) and SIMS shows the role of mobile ion penetration from BEOL layers. An improved process is presented and confirmed in test structures and DRAM.

1. Introduction

Due to decreasing marginal improvements from channel length scaling there is significant interest in 3D integration. 3D integration allows for the stacking of chips, individually optimized for different applications such as memory or high-performance logic, thus increasing system level power-performance while providing smaller form factors. TSVs are an important enabling element for 3D integration as they form the conduit for both intra-chip and off-chip, power delivery and signal communication.

High aspect ratio TSVs are formed with deep silicon RIE followed by an insulator deposition and Cu metallization, followed by thinning, dicing and packaging (Fig.1). Typically the TSVs are introduced after one or more wiring / contact levels have been completed and the issues involved in the integration have been discussed (Fig 2). The mechanical stress created by the differential thermal expansion mis-match between Cu and Si is a subject of intense and careful study and is resolved using so called 'keep out zone' recommendations for the proximity of devices to TSV [1]. In this paper, we present data on device shifts in proximity to TSV that arise from a fundamentally different mode of interaction, an electrostatic effect rather than mechanical stress, and present a solution to the problem. This improved process is demonstrated via extensive characterisation in both test chips and product arrays.

2. Device Vt shifts in proximity to TSV

Fig 3a shows the Id-Vg characteristic of a PFET in a 3D enabled high performance SOI technology (process details are presented in [2]). The devices in close proximity to a TSV exhibit a threshold voltage (Vt) shift, of nearly 100 mV, absent in control devices on the same chip. These shifts are inconsistent with the mechanical stress effect from TSV as they shift the devices in the opposite direction predicted by TSV stress (fig 4); moreover, the NFETs with the same configuration are un-shifted (Fig 3b), and PFET shifts are reduced in other layouts with the same TSV proximity but the devices are present in the vicinity of other NFETs (Fig 5). Since these device Vt shifts suggest a change in the electrostatics of the gate, back-channel characterisations were performed on SOI devices utilizing the substrate as the gate and the buried oxide (BOX) as the gate insulator. These results (Fig 6) show that the back-channel Vt is also increased, again affecting only PFETs in proximity to TSV suggesting the presence of positive charges in the BOX as well as the gate oxide. Thermal annealing studies were conducted and showed increasing Vt shifts suggesting the presence of mobile charge (Fig 7). Similar indications of Vt shift are seen in the Id-Vg graphs in other publications of TSV to FET proximity evaluations but are not explicitly addressed [4];

3. Triangular Voltage Sweep characterisation (TVS)

In addition to biased temperature stresses, TVS was performed. TVS, by measuring the ionic drift current over and above the displacement current in the TSV / Silicon substrate capacitor, is effective in identifying both the presence of mobile ions as well as the species [5]. Capacitance hysteresis techniques rely on the flat-band shifts that could be due to interface charge states as well as mobile ions and lack the specificity to the ionic species. The charge concentration is estimated by measuring the area under the TVS spectrum. TVS data for the TSV showed sharp peaks, consistent with the presence of Na, with a concentration approaching 2E12 cm^2 (Fig 8). SIMS studies confirmed the presence of Na (fig 9) thus corroborating the TVS spectra; SIMS depth profiling showed increased concentration towards the BEOL confirming the source of the mobile ions. No evidence of Cu was found in TVS confirming the integrity of the TSV process (Fig 8).

4. Mitigation

The absence of shift in NFETs and in PFET layouts with nearby NFETs is consistent with earlier studies on mobile ion penetration in CMOS technologies, attributable to the gettering properties of As [6]. With improvements in the barrier layer properties along with improvements in BEOL chemistries the mobile ion penetration has been eliminated from advanced technology nodes. However, TSV processing must necessarily penetrate these barrier layers, to form the 'through silicon via', thus exposing the devices (Fig 10) and allowing the migration of the ions. Since, complete elimination of ionics in the BEOL involves significant alteration to existing logic technology, an improved post-etch cleaning process was developed and introduced, along with a modified TSV liner; this new process eliminated the movement and redistribution of ions preventing them from perturbing devices. PFETs in proximity to TSVs were characterised with this improved process and show no shifts relative to control devices. Subsequent thermal anneals confirmed the efficacy of this solution (Fig 11).

5. Product arrays

DRAM's are particularly sensitive to ionic penetration as the retention on the DRAM cell is very sensitive to mobile ions; earlier studies on TSV related Cu contamination were performed in DRAMs [7]. A 96 Mb embedded DRAM product array (Fig 12), with TSV utilizing the improved process was compared against an identical product without TSV. Fig 13 shows that the retention statistics are virtually identical confirming the efficacy of the new process.

6.Conclusion

A new interaction between TSV and devices is identified, studied and a process solution is demonstrated

Acknowledgements

The authors would like to thank all members of the 300mm semiconductor fabrication facility in East Fishkill, NY and the contributions of B. Himmel, R. Volant and R. Hannon.

References

1. W. Guo et al., in 'Cu TSV induced Keep Out Zone', IEDM (2013).
2. M.Farooq et al. in '3D Cu TSV integration, Testing and Reliability', IEDM (2011).
3. S.S. Iyer in ' The evolution of dense embedded memories in high performance logic technologies', IEDM (2012).
4. T.Kauerauf et al. in 'Effect of TSV presence on FEOL yield and reliability', IRPS, (2013).
5. D. K Schroeder in 'Semiconductor Material and Device Characterisation, 3'rd edition, pp 340, (2006)
6. C.Y. Wong et. al. in'Mobile ion gettering in passivated P+ gates', VLSI, (1990).
7. K.Lee in 'Impact of Cu contamination on Memory Retention in DRAM', Elec. Dev. Lett., Vol. 33., No. 9, (2012).

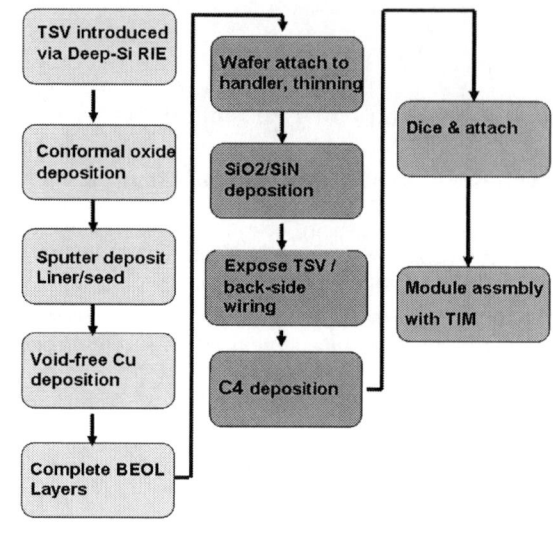

Fig1. A typical process flow for TSV integration in a high performance logic technology. Details can be seen in [1]. The process always involves etching through the SiN barrier layer that separates FEOL and BEOL and other barrier layers present between wiring levels.

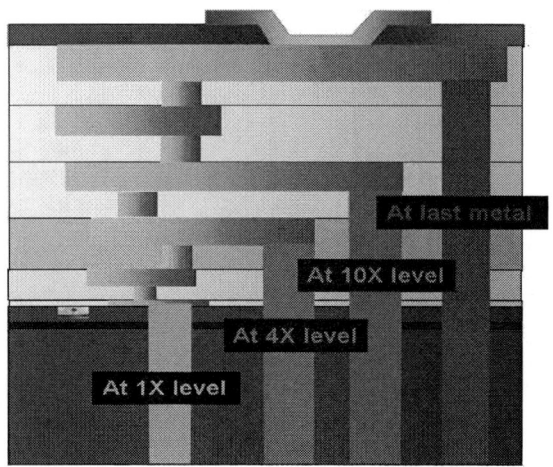

Fig2 TSV process options and compromises: TSVs are typically introduced after one or more contact and wiring levels are introduced; the choice is a compromise between process complexity of integrating a high aspect ratio via and the exclusion zones necessary for implementation [2, 3]

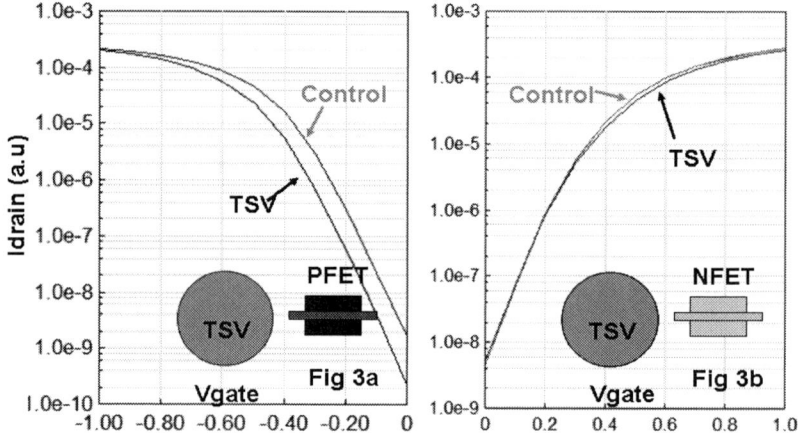

Fig. 3 a) PFETs in proximity to TSV show a significant change in Vt (~ 100mV) when compared to control devices while b) NFETs with identical proximity and orientation experience no shift; Control devices on the same chip and wafers without TSV process do not show any changes with comparable thermal budget.

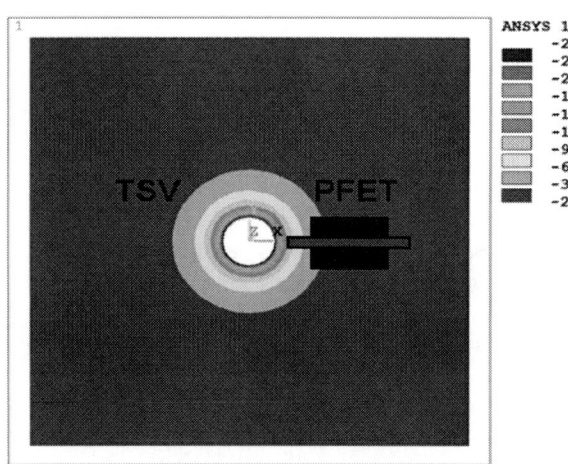

Fig 4: Finite element modeling of the TSV stress shows a tangential (channel orientation) compressive stress, which increases the carrier mobility; but the PFETs in proximity showed a decrease in current when compared to control devices on the same chip

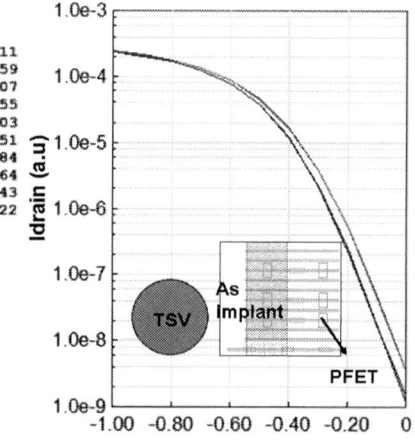

Fig 5: PFET's with the same channel orientation and proximity, but adjacent to other NFET's that receive an As implant, show reduced Vt shifts with reference to control devices.

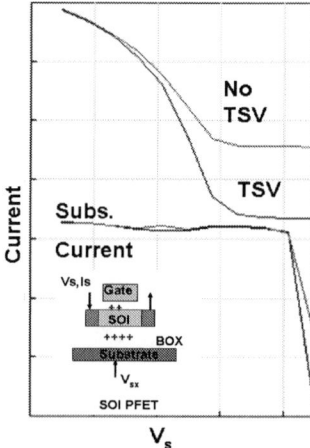

Fig 6: Is-Vsx of an SOI PFET, where the Vsx is swept captures the back-channel characteristics; again devices in proximity to TSV showed a decreased current when compared to control devices suggesting the presence of +ve charge in the BOX layer

978-1-4799-8002-4/14 $31.00 © 2014 IEEE   372

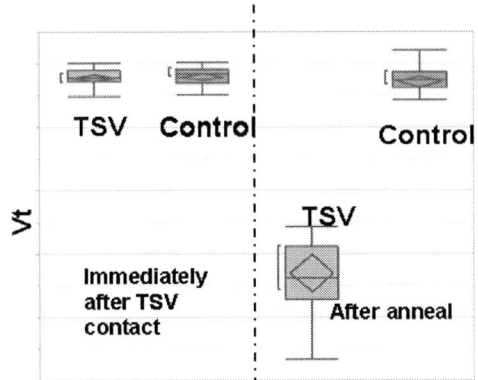

Fig. 7 Immediately after contact to the TSV the device shifts were not as pronounced as they are after the wafers go through further annealing (375C, 2 hours)

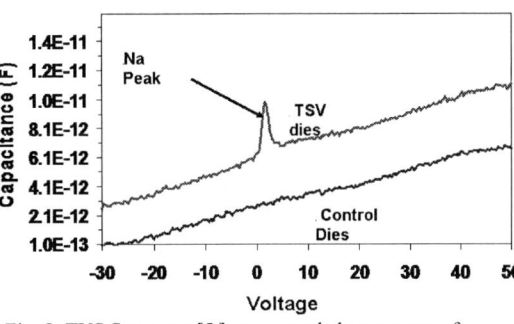

Fig. 8 TVS Spectrum [5] was recorded on a group of 10TSV at 250C; A calibrated ramp of 0.2V/sec was applied and the resulting current expressed as a capacitance (i/(dv/dt)); Sharp peaks represent fast moving ions such as Na; larger atomic number elements such as Cu show diffuse peaks at higher voltages (not present in these samples); the total charge is calculated as $Q = \int C\,dv$ (area under the peak) = 1E-11Coulombs representing a charge density of 2E12/cm^2

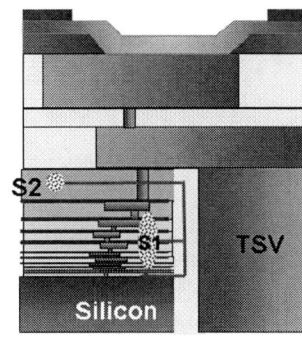

Fig. 10: TSVs intrinsically involve etching through the barrier layers that decouple FEOL and BEOL; This allows a pathway for potential mobile ions (S1,S2) to migrate, that otherwise would be contained by barriers; An improved post-etch cleaning process and modified liner processes eliminated their movement and redistribution

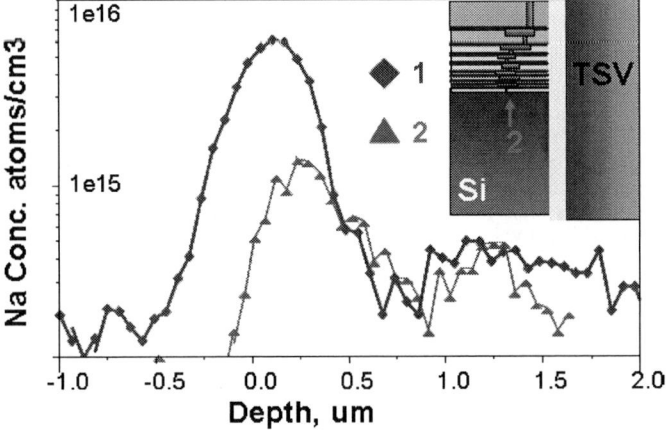

Fig. 9 SIMS profiling was conducted by careful delayering from the silicon side of the sample; the concentration decreased with increased distance from the TSV (2 vs 1); the concentration increased further as the ion beam approached BEOL (not shown). Concentration estimates ranged around E-12/cm^2, in agreement with TVS results.

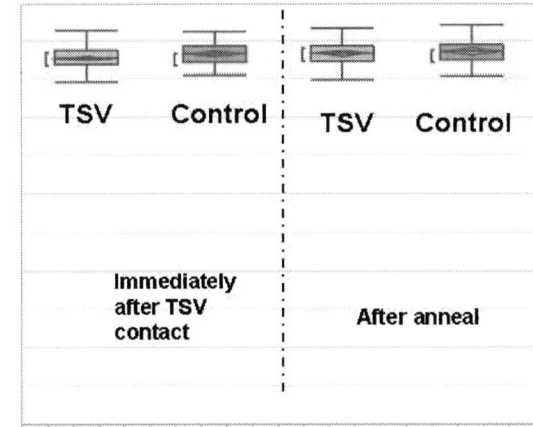

Fig. 11. The optimized process was characterized by comparing PFET in close proximity to control devices immediately after TSV contact as well as after annealing (375C 2 hours)

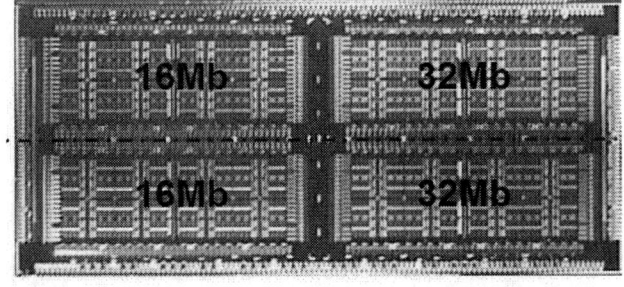

Fig. 12. A TSV enabled, 96Mb embedded DRAM product array was constructed in IBM's 32SOI high performance logic technology and the presence of TSV was characterized with the optimized TSV process

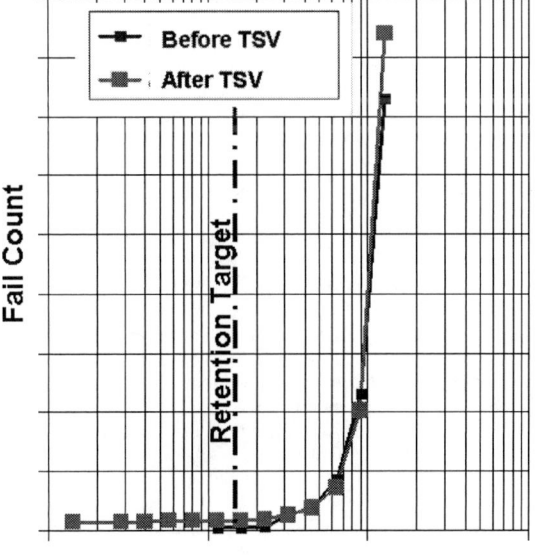

Fig. 13. Mobile ions when present have been shown to impact DRAM retention [7]; The relationship between fail count and retention time is a good metric to evaluate any degradations in retention. The retention statistics of the 96Mb eDRAM array was compared before and after the introduction of TSV. The curves showed no change due to TSV process

978-1-4799-8002-4/14 $31.00 © 2014 IEEE

Highly Beneficial Organic Liner with Extremely Low Thermal Stress for Fine Cu-TSV in 3D-Integration

M. Murugesan, T. Fukushima, J.C. Bea, Y. Sato, H. Hashimoto, K.W. Lee, and M. Koyanagi

GINTI, NICHe, Tohoku University, 6-6-01AzaAoba, Aramaki, Aoba-ku, Sendai, 980-8579, Japan.
Phone: +81-22-795-4119; Fax: +81-22-795-6907; E-mail: murugesh@bmi.niche.tohoku.ac.jp.

Abstract

The constructive role played by the Thermal-chemical vapor deposited (CVD) organic polyimide (PI) liner in the Cu-TSVs with diameter or width (ϕ) varying from 3 μm to 30 μm has been studied meticulously for its thermal stability, leakage current (LC), capacitance, TSV-chain resistance, stress absorbing ability, and the Si-lattice distortion arising from thermo-mechanical stress (TMS). The measured LC values for the CVD deposited PI liner is in the order of 10^{-13} to 10^{-15} A, which is on par with the value obtained for the conventional SiO_2 liner. The extremely low modulus value of PI liner helps not only to reduce the amount of Cu-extrusion, but also maintain an uniform Cu-extrusion. We were able to achieve a conformal deposition of PI liner even in ϕ = 3 μm via having the aspect ratio of 10 with the step coverage values of more than 0.8 (80%) at the TSV bottom corner. It was found that the d-space changing and thus the lattice stress is nearly five times smaller for the TSV with PI liner (~200 MPa) than for the TSV with SiO_2 liner (~1000 MPa). Nearly zero-degradation of PI liner was confirmed from C1s, O1s, and N1s core-level x-ray photoelectron spectra taken before and after annealing at 400 °C. We obtained the resistance value of as low as 18 mΩ per 10 μm-width TSV with 500 nm-thick PI liner fabricated on 12-inch wafer.

Introduction

In contrast to the conventional Moore approach in 2DLSI, in three-dimensional (3D) integration technique where several functional chips are vertically stacked and are interconnected by embedded TSV which runs across the thickness of the chip as shown in Fig. 1(a) and (b). Both Cu/W-TSVs and CuSn/InAu μ-bumps are widely used in 3D-LSIs for electrical interconnection. Out of several reliability issues in 3D-LSIs which were raised and extensively investigated (1-5), the formation of conformal low-k dielectric liner with step coverage close to 1 (100%) is highly preferred. Since this dielectric liner thickness decides the parasitic capacitance as shown in fig. 2, a relatively thick liner at the bottom corner of the TSV is a must one. The hardness of the liner material further aggravates the thermo-mechanical stress propagation into the TSV space region owing to its larger modulus value as shown in fig. 1(e) and (f). Again the harder liner material is responsible for the die-cracking and the bowed BEOL as shown in fig. 1 (g) and (h). Having learnt the required property of liner material, several authors have proposed organic material such as the epoxy liner (fig. 1(b) and BCB liner (fig. 1(d)). However, both have not met the complete requirements. Both are deprived of thermal stability and left with scaling challenges, since they were deposited by spin-on technique. Therefore it is required to find out a liner material as well as its deposition method that satisfies the following: (i) low modulus and low k, (ii) either low or equivalent thermal expansion co-efficient (CTE) to that of TSV metal, (iii) scalable to 12-inch level at low cost, (iv) conformal deposition with better step coverage, and (vi) last but not least free from scaling challenges.

In this study, we have investigated (i) the viability of physical vapor deposition of PI liner even in the 3 μm-width TSV on 12 inch wafer level by analyzing the uniformity, wettability and CTE, (ii) thermal-stability by x-ray photoelectron spectroscopy (XPS), (iii) the TMS induced by Cu-TSVs with PI liner before and after TC test by micro-Raman spectroscopy (μRS), (iv) Distortion in the Si lattice induced by TMS by using X-ray micro-diffraction (μ-XRD); (v)LC for PI liner and the TSV chain resistance by I-V measurements, and (vi)trapped charges in the PI liner by C-V measurement.

Reduced Cu-extrusion and enhanced step-coverage for PI liner in Cu-TSV

Fig. 3 reveals extremely better coverage for PI liner in various positions of the TSV sidewall as against the SiO_2 liner. The low modulus value (3-4 GPa) along with the CTE value of ~20 ppm/deg. (which is close the CTE of Cu, ~17 ppm/deg.) obtained for CVD PI liner greatly helps to reduced Cu-extrusion amount to more than one-half (fig. 4(a)) as against the popup amount observed for SiO_2 counterpart (fig. 4(b)). It is also important to note that with the CVD-deposited PI liner we have observed not only a low Cu-extrusion from the TSV, but also an uniform Cu popup height in the array of Cu-TSVs. fig. 5 (a & b). On the other, it is highly inhomogeneous for SiO_2 liner (fig. 5 (c & d)) that causes problems shown in fig 1(h). Therefore, it is proven that the softer, low modulus CVD PI liner partly accommodates the Cu expansion during heating, and thus the suppressed and uniform Cu extrusion for PI liner. Whereas the much harder (hardness 9500 MPa), high-modulus (Young Modulus ~80 GPa) SiO_2 liner cannot accommodate the Cu expansion, and thus enhanced Cu extrusion with non-uniformity. Fig. 6 and 7 comprehensively reveals that an overall increase in surface roughness for the PI liner upon heat treatment. Due to the soft nature of PI liner, there appears brindle pattern on the surface of PI liner along with a fourfold increase in the surface roughness after annealing at 400 °C. Considering various reliability issues associated with 3D-LSI, a highly scalable, with maximum aspect-ratio coverage, and low temperature formation of the proposed PI liner will play an important role in high-density 3D-LSI involving TSV with smaller size.

Leakage current test for PI liner in Cu-TSV _ I-V and C-V

The most important function of liner material is to what extent they are fool-proof to electrical loss with minimum parasitic capacitance that decides whether the liner is suitable for TSV applications. We have measured the parasitic capacitance by recording the C_p-V_G curve for CVD formed 0.5 μm-thick PI liner for various bias-voltages, at various frequencies, and the resultant hysteresis, and are respectively shown in fig. 8(a), (b) and (c). It is worth to note that the observed C-V curves are quite similar to the typical C-V curves obtained for SiO_2 liner (fig. 9(a) and (b)), but with the positive shift in the flat-band voltage, Vfb. However, the frequency dependent data revealed there is no fluctuation in the Vfb. Hence, the observed Vfb shift is not due to the trapped charges, but due to negative polarization with the bias voltage. The presence of hysteresis and the positive shift in Vfb in the C-V curve for the PI liner revealing the presence of negatively charged species, and obviously it is a reliability issue. Fig. 10(a) and (b) reveals that the leakage current for both the liners is comparable. The I-V data revealed that the leakage current(LC) for the CVD deposited PI liner is in the order of 10^{-13} to 10^{-15} A, which is on par with the value obtained for the conventional SiO_2 liner.

978-1-4799-8002-4/14 $31.00 © 2014 IEEE

Minimized TMS in Si induced by Cu-TSV with PI liner_ μ-RS

By now it is known that the TMS is an unavoidable reliability issue as long as TSV is used in for signal lines. Therefore, it is must for the 3D-research community to find a way how to reduce this TMS in the active Si caused by TSV metal. This can be solved either by choosing the TSV metal (3), or by deploying the stress absorbing material. Since the PI liner has low modulus value, it is possible for the PI to partly accommodate the expanded Cu. This is confirmed in fig. 12 that the presence of PI along the TSV side wall not only reduces TMS to more than 50 percent as compared to its SiO_2 counterpart. The quantity and the propagation of TMS for conventional SiO_2 liner, PI liner upon SiO_2 liner, and exclusive PI liner are respectively >500MPa, <300MPa, and <200MPa stress. Irrespective of the TSV size, two dimensional μ-Raman spectroscopic stress analysis revealed the existence of less than 200 MPa of stress in the Si around the TSV with PI liner which is in close agreement with the micro-diffraction results. It is noticed that the TMS produced by the Cu-TSVs were overlapped both orthogonally and ortho-diagonally for harder SiO_2 liner, whereas orthogonal only for low modulus PI liner upon SiO_2 liner and zero-stress overlapping for PI liner. Fig. 11 also confirms PI liner highly acting as stress absorber in both 5 and 20 μm TSV. This suppressed stress value in Si adjacent to TSV containing PI liner also supports the minimized Cu-extrusion as observed in the fig. 4.

Suppressed lattice-tilt and distortion in Cu-TSV with PI liner _ μ-XRD

Micro-diffraction reveals more precisely the lattice distortion and stress than μ-RS due to its reduced probe size and sensitivity to the lattice parameter (5). Also, μ-XRD reveals the stress in the lattice along different direction. Fig. 13 depicts the reciprocal space lattice (RLM) image for 3 μm and 30 μm-width TSV with PI liner. The four RLM images A B C and D in both figs. corresponds to the positions in-between the TSV and adjacent to the TSV as shown in top and bottom of center schematic diagram. It is clear from RLM of stress free Si (fig. 13 (c)), the lattice tilt is relatively large for 30 μm-width TSV. The deduced lattice tilt and stress from the RLM data are shown as counter diagram in fig. 14 and fig. 15 respectively for 3 μm- and 30 μm-width TSV with PI liner. It revealed that the Si-lattice structure is heavily distorted in the TSV space region for TSV with SiO_2 liner (Fig.16) as compared to that of TSV with PI liner. The extracted the d-space changing (and hence the lattice stress) is nearly five times larger for TSV with SiO_2 liner (~1000 MPa) than for the TSV with PI liner (~200 MPa). The observed lattice tilt for 3 μm and 30 μm TSV with PI liner was respectively 0.001 deg. and 0.012 deg., whereas the lattice tilt obtained for TSV with SiO_2 liner is three order of magnitude larger, and it is nearly 3 deg. Therefore the observed lattice tilt values for 3 μm and 30 μm TSV with PI liner was respectively 0.001 deg. and 0.012 deg. reveal that the low modulus-PI liner tremendously reduces the distortion in Si lattice near TSV and the TSV space region. Thus both the μ-RS and μ -XRD data clearly confirms that PI liner does absorb the TMS form Cu-TSV, and reduces the residual stress in the active Si many-folds.

Thermal stability of PI liner and the daisy chain resistance

The thermal stability of PI liner is also very important, as it decides the bonding temperature. In general, the higher the bonding temperature, the better is the yield. Therefore it is highly important to have the liner which is intact at the higher temperature. In order to confirm the thermal stability, the C1s, and O1s, and N1s core-level XP spectra were recorded for 250 nm-thick PI liner, before ((a) of fig.17,18.,19) and after annealing at 400 C for 30 min ((b) of fig.17,18.,19). The peaks at binding energy value of ~284.8 ev, ~286 eV and ~288 eV in the C1s spectra corresponds to the C=C, (sp2 hybridized) C-N, and (keto) C=O bond. The peak positions and fractions before and after annealing reveal the intactness of the PI liner formed by CVD. The chain resistances of 10 μm-width TSV for three different pitch values 40 μ m, 60 μ m, and 80 μ m are respectively shown in (a), (b), and (c) of fig. 20. We have obtained a minimum resistance value of 18 mΩ/TSV for 10 μm-width TSV. This confirms the potential use of PI as a dielectric-liner for Cu-TSVs on 12-inch wafer level for 3D-integration.

Conclusions

A PI liner material with a low modulus value (3-4 GPa) and a CTE (~20 ppm/deg.) comparable to that of Cu is successfully deposited along the 3 μm-width TSV sidewall with the coverage ratio >0.8 at the bottom corner of the TSV on 12-inch wafer by physical vapor deposition technique. The C-V and I-V data revealed that the leakage current for PI liner is quite par with that of SiO_2 liner. In Cu-TSV with PI liner, not only the Cu extrusion was suppressed to more than one-half, but also it is highly uniform which facilitates the remaining integration process. Both the μ-RS and μ-XRD data doubly confirms the tremendous reduction of TMS in the active Si caused by Cu, where the sidewall Si surface of the TSV is laminated with PI liner. The quantity and the propagation of TMS for the Cu-TSV containing conventional SiO_2 liner, PI liner upon SiO_2 liner, and exclusive PI liner are respectively >500 MPa, <300 MPa, and <200 MPa stress. Further the stress propagates both orthogonal as well as orthodiagonal, only along orthogonal, and neither orthogonal nor orthodiagonal direction, respectively. The PI liner largely assists to minimize the amount of both the lattice tilt (nearly three order less) and the Si-lattice distortion (around five times less) as compared to SiO_2 liner. XPS analysis of C1s, O1s, and N1s of PI liner strongly supports the suitability of PI liner in the 3D-integration processes at up to 400 ℃. Finally, by using PI as a dielectric liner in the 10 μm-width Cu TSV, we have obtained the resistance value of as low as18 mΩ/TSV.

Acknowledgment

The authors thank Dr. Y. Imai and Dr. S. Kimura for extending their help in carrying out micro-diffraction experiments at BL13XU, Spring-8.

References

[1] M. Murugesan, J.C. Bea, H. Kino, Y. Ohara, T. Kojima, A. Noriki, K.W. Lee, K. Kiyoyama, T. Fukushima, H. Nohira, T. Hattori, E. Ikenaga, T. Tanaka, and M. Koyanagi, "Impact of remnant stress/strain and metal contamination in 3D-LSIs with through-Si vias fabricated by wafer thinning and bonding", *Digests 2009 Int. Electron Devices Meeting. USA*, p. 499, 2009.

[2] M. Murugesan, H. Kino, H. Nohira, J.C. Bea, A. Horibe, F. Yamada, C. Miyazaki, H. Kobayashi, T. Fukushima, T. Tanaka, and M. Koyanagi, "Wafer thinning, bonding, and interconnects induced local strain/stress in 3D-LSIs with fine-pitch high-density micro-bumps and through-Si vias", *Digests 2010 Int. Electron Devices Meeting. USA*, p. 30, 2010.

[3] M. Murugesan, H. Kino, H. Hashiguchi, C. Miyazaki, H. Shimamoto, H. Kobayashi, T. Fukushima, T. Tanaka, and M. Koyanagi, "High Density 3D-LSI Technology Using W/Cu Hybrid TSVs", *Digests 2011 Int. Electron Devices Meeting. USA*, p. 139, 2011.

[4] M. Murugesan, H. Kobayashi, H. Shimamoto, F. yamada, T. Fukushima, J.C. Bea,K.W. Lee, T. Tanaka, and M. Koyanagi, "Minimizing the local deformation induced around Cu-TSVs and CuSn/InAu micro-bumps in high-density 3D-LSIs", *Digests 2012 Int. Electron Devices Meeting. USA*, p. 657, 2012.

[5] M. Murugesan, T. Fukushima, J.C. Bea, K.W. Lee, T. Tanaka, and M. Koyanagi, "Revisiting the silicon-lattice in the high-density 3D-LSIs — In the perspective of device reliability", *Digests 2013 IEEE Int. Electron Devices Meeting. USA*, p. 172, 2013.

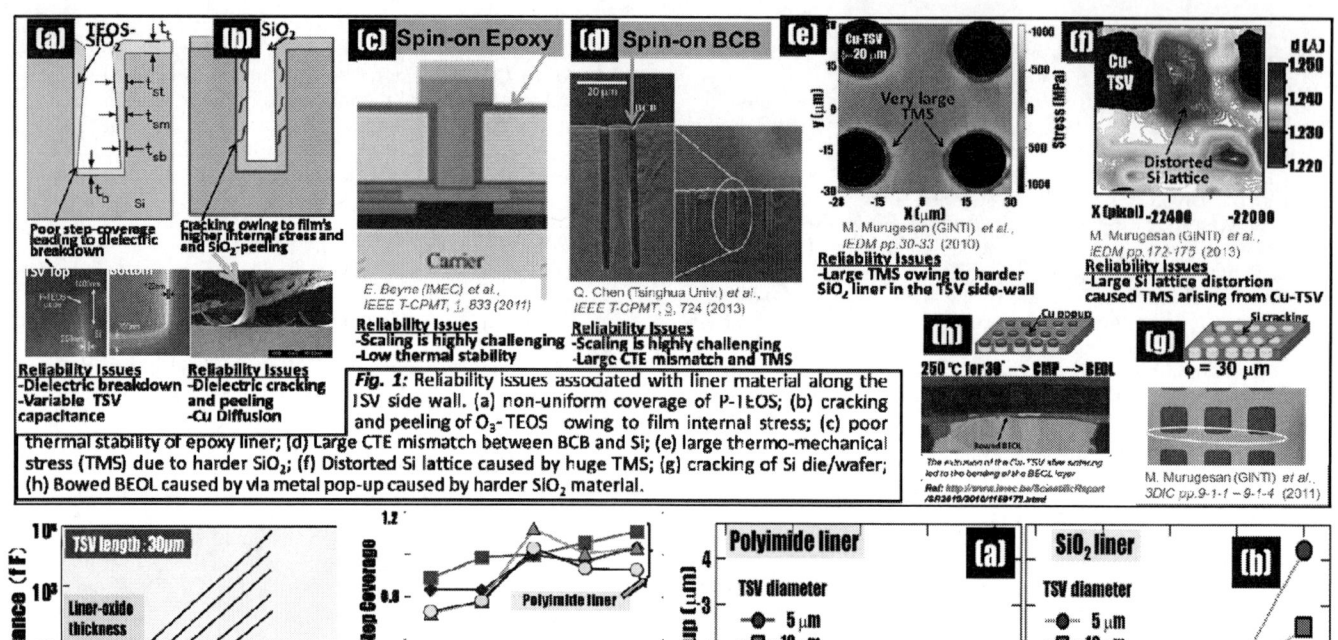

Fig. 1: Reliability issues associated with liner material along the TSV side wall. (a) non-uniform coverage of P-TEOS; (b) cracking and peeling of O_3-TEOS owing to film internal stress; (c) poor thermal stability of epoxy liner; (d) Large CTE mismatch between BCB and Si; (e) large thermo-mechanical stress (TMS) due to harder SiO_2; (f) Distorted Si lattice caused by huge TMS; (g) cracking of Si die/wafer; (h) Bowed BEOL caused by via metal pop-up caused by harder SiO_2 material.

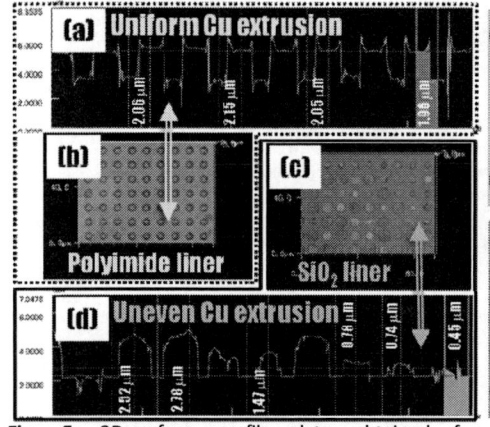

Fig. 2: Simulation data revealing the requirement of thicker liner material along the side wall of narrow TSV.

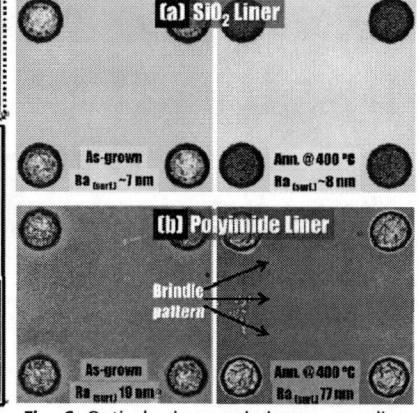

Fig. 3: Step coverage of polyimide liner for various TSV sizes. For comparison P-TEOS is included.

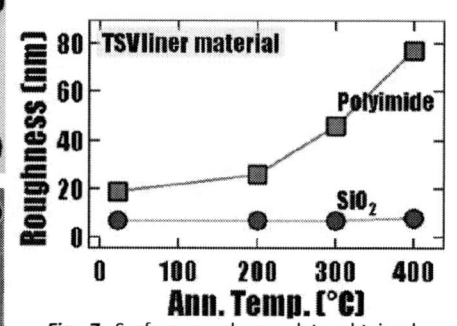

Fig. 4: Cu-extrusion data obtained from the white-light interference microscopic image for different TSV sizes: (a) TSV with polyimide liner, and (b) TSV with conventional SiO_2 liner

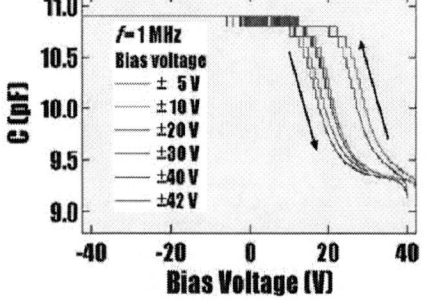

Fig. 5: 2D-surface profile data obtained for Cu-extrusion analysis revealing an uniform Cu popup for TSV with polyimide liner (a) & (b), and uneven Cu pop-up in the TSVs contained SiO_2 liner (c) and (d).

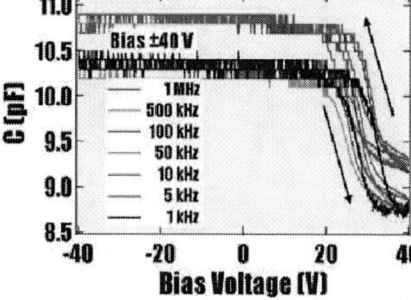

Fig. 6: Optical microscopic image revealing the uniform surface roughness for the SiO_2 liner (a), and the formation of brindle pattern for the organic liner (b).

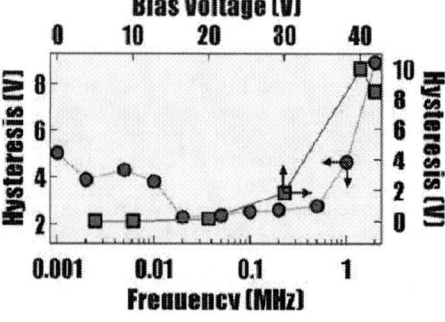

Fig. 7: Surface roughness data obtained from white light interference microscopic image taken for the samples post-heat treated at various annealing temperatures at room temperature, 200°C, 300°C and 400°C.

Fig. 8: Capacitance-voltage curves obtained for polyimide liner; (a) for various bias voltages at 1MHz, (b) at various frequencies for the bias voltage of +-40 V, and (c) the hysteresis value at different frequencies and various bias-voltages.

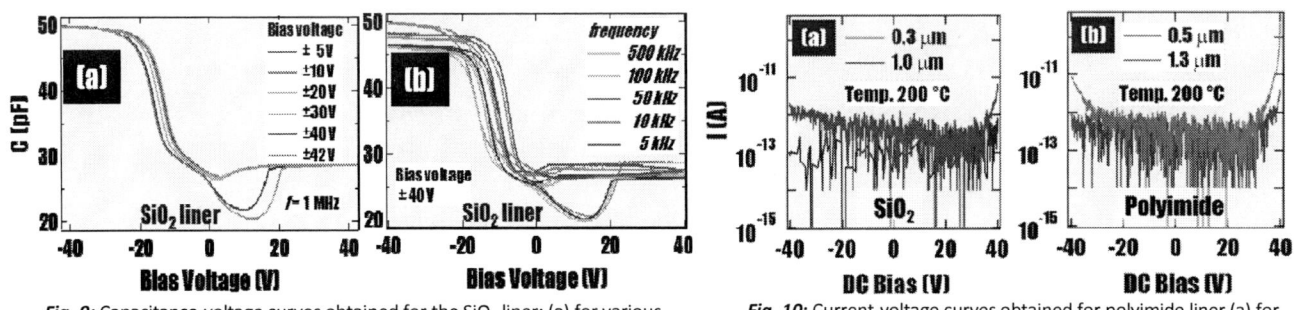

Fig. 9: Capacitance-voltage curves obtained for the SiO₂ liner; (a) for various bias voltages at 1MHz, (b) at various frequencies.

Fig. 10: Current-voltage curves obtained for polyimide liner (a) for two different thicknesses, and (b) for SiO₂ liner.

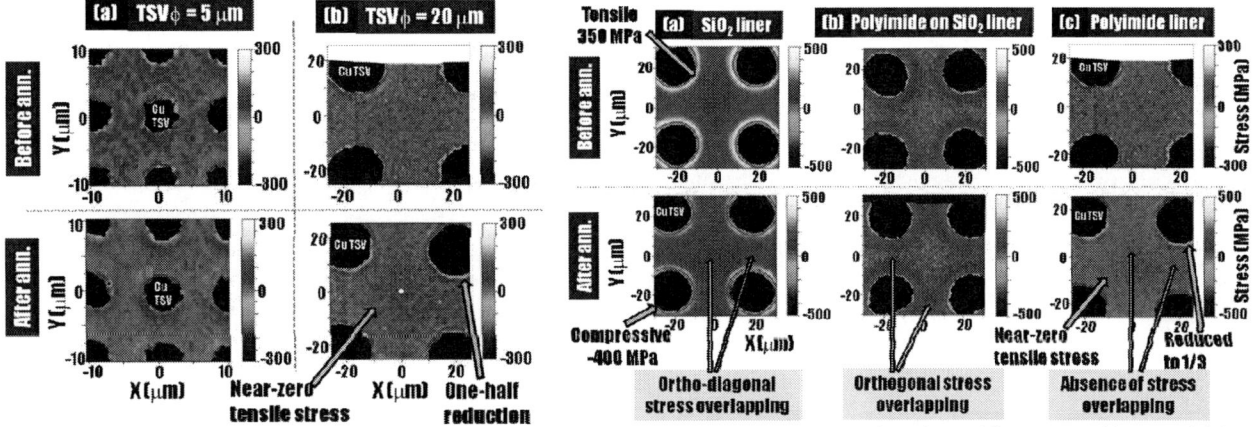

Fig. 11: 2D-μ-Raman stress mapping data obtained for 5 μm-width TSV (a), and 20 μm-width TSV with PI liner.

Fig. 12: 2D-μ-Raman stress mapping data obtained for 20 μm-width TSV with (a) SiO₂ liner, (b) PI liner on P-TEOS SiO₂ liner, and (c) PVD evaporated PI liner.

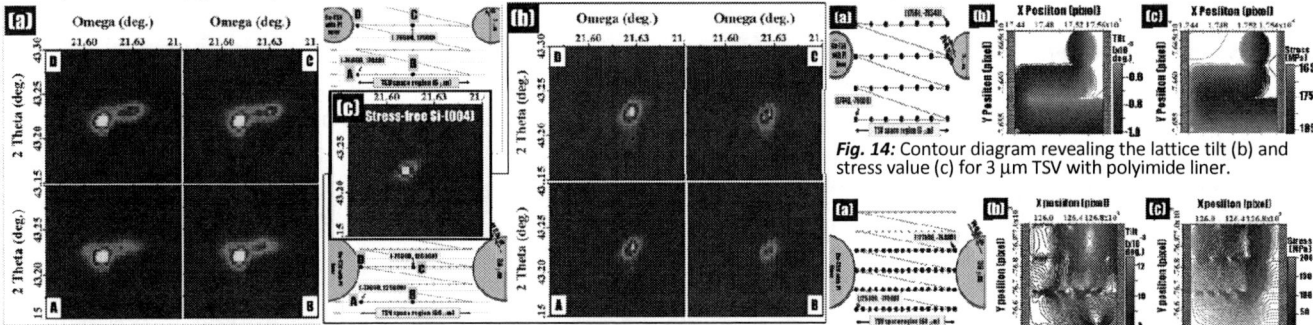

Fig. 13: Reciprocal lattice-mapping (RLM) image obtained for the 3 μm (a) and 30 μm(b) TSV with polyimide liner. Shown in the inset (c) is the RLM image obtained for the reference Si.

Fig. 14: Contour diagram revealing the lattice tilt (b) and stress value (c) for 3 μm TSV with polyimide liner.

Fig. 15: Contour diagram revealing the lattice tilt (b) and stress value (c) for 30 μm TSV with polyimide liner.

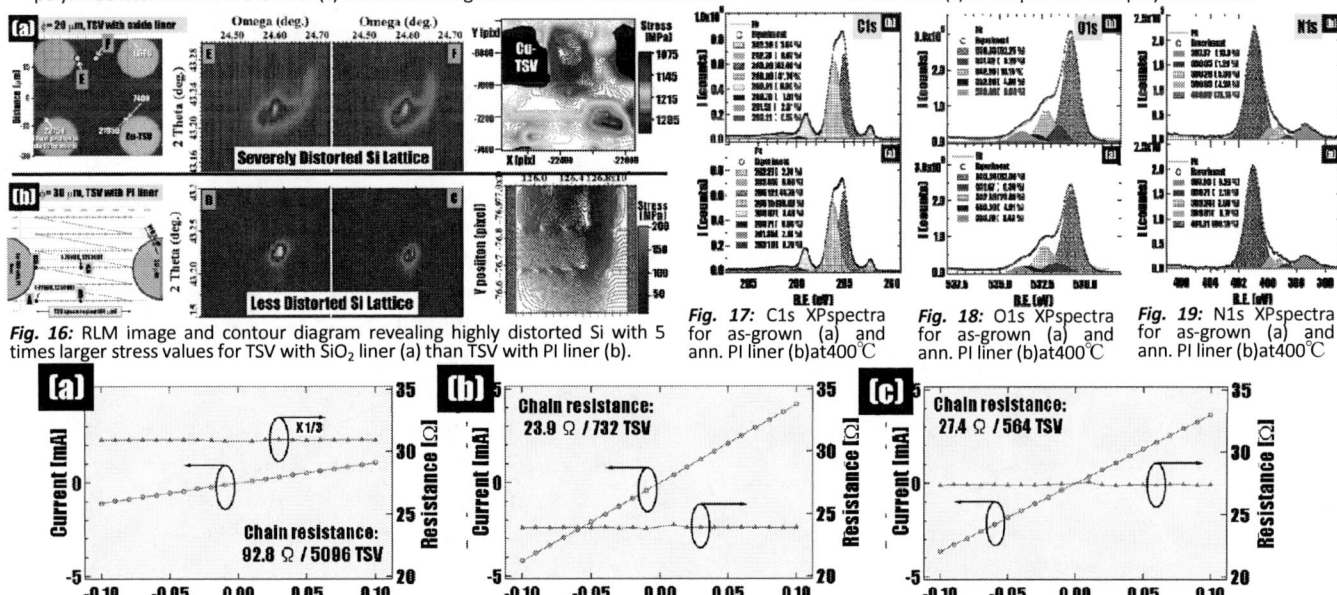

Fig. 16: RLM image and contour diagram revealing highly distorted Si with 5 times larger stress values for TSV with SiO₂ liner (a) than TSV with PI liner (b).

Fig. 17: C1s XPspectra for as-grown (a) and ann. PI liner (b)at400℃

Fig. 18: O1s XPspectra for as-grown (a) and ann. PI liner (b)at400℃

Fig. 19: N1s XPspectra for as-grown (a) and ann. PI liner (b)at400℃

Fig. 20: I-V and Resistance plots for TSV chain sample with three different pitch values, namely 40 μm (a), 60 μm (b), and 80 μm(c).

978-1-4799-8002-4/14 $31.00 © 2014 IEEE

An Ultra-Sensitive Resistive Pressure Sensor Based on the V-shaped Foam-like Structure of Laser-Scribed Graphene

He Tian,[1,2] Yi Shu,[1,2] Xue-Feng Wang,[1,2] Mohammad Ali Mohammad,[1,2] Cheng Li,[1,2] Yi Yang,[1,2] Tian-Ling Ren[1,2,*]

[1] Institute of Microelectronics, Tsinghua University, Beijing 100084, China

[2] Tsinghua National Laboratory for Information Science and Technology (TNList), Tsinghua University, Beijing 100084, China

[*]Email: RenTL@tsinghua.edu.cn

Abstract

We demonstrate a flexible, ultra-sensitive resistive pressure sensor based on the foam-like structure of laser-scribed graphene (LSG). Benefitting from the unique microstructure of the LSG, the sensitivity of the pressure sensor is as high as 0.96 kPa^{-1} in the low pressure range (0~50 kPa), which is the highest among all reported graphene-based pressure sensors. Moreover, the sensitivity in the high pressure range (50~113 kPa) is 0.005 kPa^{-1}. The response of the pressure sensor is highly stable up to 100 cycles with excellent performance. Our pressure sensor can meet the needs of specific applications, for example, high sensitivity for low-pressure applications and low sensitivity for high deformation applications. Moreover, the laser-scribing technology could enable large-scale production of the LSG pressure sensor with low cost in ~20 minutes. Our work indicates that laser scribed flexible graphene pressure sensors could be widely used for artificial electronic skin (e-skin), medical-sensing, bio-sensing and many other areas.

I. Introduction

Recently, flexible e-skin has attracted a lot of attention for its unique ability to sense low amounts of pressure, potentially initiating vast applications development in health monitoring. Recently, graphene [1-2] based resistive-type pressure sensors have been developed. Piezoelectric-type ZnO nanowire pressure sensors [3] and capacitive-type micro-structured PDMS pressure sensors [4] have also been reported. However, the sensitivity and the pressure sensing range in these devices are not enough for real-world applications. Moreover, the fabrication processes are complex. Here we develop an ultra-sensitive pressure sensor based on LSG, enabling pressure sensing in a large range. The laser-scribed fabrication process could enable a large-scale, low cost and time efficient production of such high performance pressure sensors.

II. Fabrication and Characterization

Fig. 1 and Fig. 2 show the device structure of the LSG pressure sensor, and the fabrication process, respectively. GO solution was synthesized from graphite powder using a common Hummers method. About 10 mL GO solution was drop-casted on the surface of a LightScribe DVD disc. The GO solution was left overnight to dry on the disc. The disc was then patterned by a Light-Scribe DVD Drive (HP Inc. 557S) with the GO undergoing laser-induced reduction to form LSG. The Nero Start Smart software was used to design and transfer arbitrary designs onto the DVD.

Our pattern consists of a simple grating structure. After the patterning process, two 1x1 cm^2 areas were cut out and assembled face-to-face. Both chips were oriented in such a way that the grating patterns were normal to each other. Finally each of the LSG films were wired-out using copper wire and packaged. The pressure sensing core element consists of two v-shaped LSG films in a face-to-face "crossbar" stack. The GO films are dense while the LSG (GO after laser irradiation) is composed of loosely stacked graphene layers (Fig. 3). The LSG is hence referred to as a foam-like structure. EDS mapping confirms the presence of carbon (Fig. 4). The darker areas showing a lack of oxygen (Fig. 5) demonstrate successful laser reduction. Raman spectrums (Fig. 6) further show the obvious increase of the 2D peak, which demonstrates the existence of stacked graphene layers. The 3D profile show the arrays of the LSG foam-like structure (Fig. 7). The height profile shows that the height and width of the LSG is 10.7 μm and 19.8 μm with a V-shape (Fig. 8).

III. Pressure Response

In order to test the response of our LSG pressure sensor under static and dynamic forces, a system containing a computer controlled stepping motor, a force sensor and electrical signal analyzer were used. In this system, static pressure up to 113 kPa and dynamic pressure up to 98 kPa could be loaded. The resistance change could also be simultaneously recorded.

As shown in Fig. 9, when a pressure ranging from 0 to 113 kPa is applied, the conductance increases significantly due to enhancement of contact between two LSG films. The sensitivity can be expressed as:

$$S = \delta(\Delta C / C_0) / \delta P \qquad (1)$$
$$\Delta C = C - C_0 \qquad (2)$$

where P is the applied pressure, C is the conductance when

978-1-4799-8002-4/14 $31.00 © 2014 IEEE

pressure is applied on the device, and C_0 is the conductance under base pressure. It is shown that the sensitivity is as high as 0.96 kPa^{-1} in the low pressure range (<50 kPa) while it lowers to 0.005 kPa^{-1} in the high pressure range. In the low pressure range, there is a significant change of contact-area between the two V-shaped foam-like structures. After the two V-shaped contacts become stable at high pressure, the change in contact resistance is minimal, causing a saturation in sensitivity (Fig. 10). Fig. 10 shows repeatable performance in the 1st, 50th and 100th cycles. An excellent operational stability of the LSG pressure sensor is demonstrated under 100 cycles with a 0~75 kPa force sweep. (Fig. 11). It is noted that a good signal-to-noise ratio has been obtained with negligible changes (Fig. 12). Fig. 13 shows the testing of 31 kPa pressure, 75 kPa pressure and the off state under 100 cycles. Fig. 14 shows the distribution of the conductance under 100 cycles. The 31 kPa and 75 kPa lines are quite uniform, while the off state has a larger fluctuation. This is because a small change in contact area of the two LSG films could induce a large change in conductivity.

In order to obtain the response time of the LSG pressure sensor, dynamic forces with 0.25 Hz and 0.5 Hz are applied. In the low pressure range, there is an obvious current change under different pressure due to its high sensitivity (Fig. 15). While in the high pressure range, there is a smaller difference of the current ratio (Fig. 16). The typical response time at low and high pressures are 72 ms and 0.4 ms, respectively (Fig. 17). Furthermore, the LSG pressure sensor can also be used to detect pressing (Fig. 18), bending (Fig. 19) and twisting (Fig. 20) forces. High signal-to-noise ratios are obtained in all three types of force measurements, further showing the high sensitivity of our LSG pressure sensor.

IV. Working Principle and Comparison

Our LSG pressure sensor is based on resistive change between two pieces of LSG films. The schematic in Fig. 21 shows the device structure and the principle behind current generation. The sensing mechanism could be explained by the force-dependent contact between y-direction LSG and x-direction LSG lines. The contact area between two LSG lines depend on the applied forces. When applying force on the device, a small compressive deformation could create more contact between two LSG lines, resulting in more electrical pathways. This can cause an increase in current since a fixed voltage bias is applied. After unloading, both LSG lines recover to their original shapes, reducing the contact area and therefore the current. The unique microstructure of the LSG is the core feature enabling ultra-sensitive pressure sensing due to a large change in contact area.

There are three main types of pressure sensors: piezoelectric, capacitive, and resistive. The sensitivity and pressure sensing range are the two key parameters to evaluate the performance of pressure sensors. As shown in table I, the sensitivity of our LSG pressure sensor is 0.96 kPa^{-1} which is higher than most recent reports showing $2.66 \times 10^{-5} \sim 0.55$ kPa^{-1} [1-4]. Moreover the pressure sensing range (113 kPa) is also the highest compared with other work.

V. Conclusion

A flexible, ultra-sensitive resistive pressure sensor based on LSG is demonstrated. The sensitivity of the LSG pressure sensor is as high as 0.96 kPa^{-1}, which is the highest among all the reported graphene based pressure sensors. The device has excellent performance and stability over 100 cycles. Our pressure sensor can meet the simultaneous requirements of high sensitivity for low-pressure applications and low sensitivity for high deformation applications. Our work shows that the LSG pressure sensors hold great promise for e-skin and other sensing applications.

Acknowledgement

This work was supported by the National Natural Science Foundation of China (61434001, 61025021, 61020106006) and the National Key Project of Science and Technology (2011ZX02403-002), and the Special Fund for Agro-scientific Research in the Public Interest (201303107). We are thankful for receiving support through the State Key Laboratory of Automotive Safety and Energy, Tsinghua University. M.A.M is thankful for receiving support from the postdoctoral fellowship program of the Natural Sciences and Engineering Research Council of Canada.

References

[1] A. D. Smith, et al., "Electromechanical piezoresistive sensing in suspended graphene membranes," *Nano Letters*, vol. 13, pp. 3237-3242, 2013.

[2] H.-B. Yao, et al., "A flexible and highly pressure-sensitive graphene–polyurethane sponge based on fractured microstructure design," *Advanced Materials*, vol. 25, pp. 6692-6698, 2013.

[3] W. Wu, X. Wen , Z. L. Wang, "Taxel-addressable matrix of vertical-nanowire piezotronic transistors for active and adaptive tactile imaging, " *Science*, vol. 340, pp. 952-957, 2013.

[4] C. B. Mannsfeld, et al., "Highly sensitive flexible pressure sensors with microstructured rubber dielectric layers," *Nature Materials*, vol. 9, pp. 859-864, 2010.

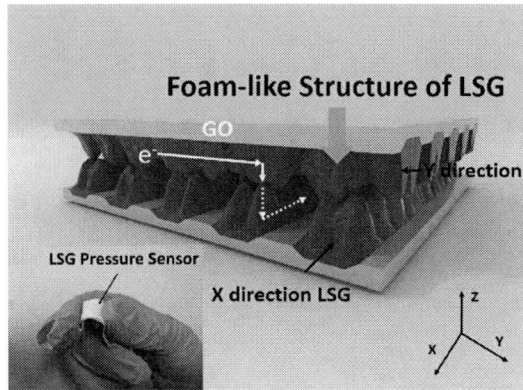

Fig. 1 Device structure of the pressure sensor based on foam-like LSG films in crossbar orientation. Inset showing flexible LSG pressure sensor in hand.

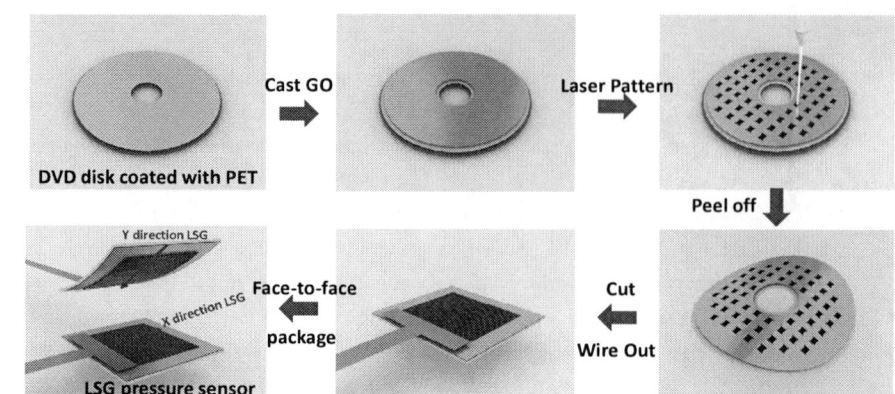

Fig. 2 The main fabrication processing steps of the LSG pressure sensor. A DVD burner with a laser-scribing function is used to convert GO into LSG. The upper LSG lines are in Y direction and lower LSG lines are in X direction to form a cross-bar structure. Two pieces of LSG are packaged face-to-face.

Fig. 3 Top view SEM image of the LSG surface in false color.

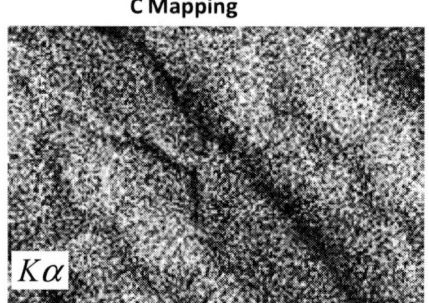

Fig. 4 EDS mapping of carbon reproducing the surface topography of the LSG and GO.

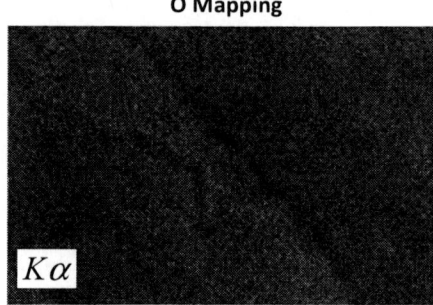

Fig. 5 EDS mapping of oxygen showing the reduction of GO.

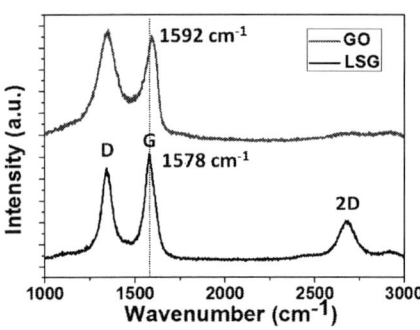

Fig. 6 The Raman Spectrum of LSG and GO.

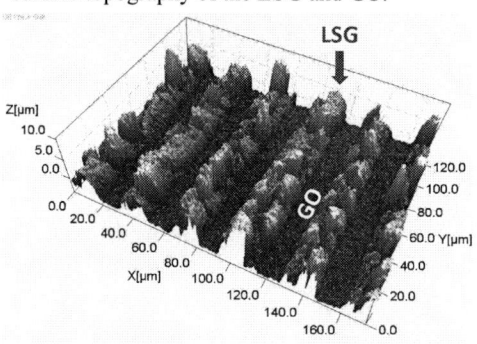

Fig. 7 A 3D profile of the LSG morphology captured by a white light interference microscope.

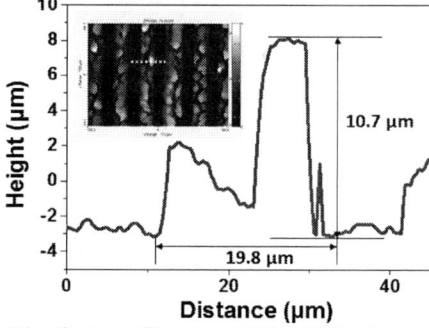

Fig. 8 A profile scan of the LSG shows a 10.7 μm height and 19.8 μm width.

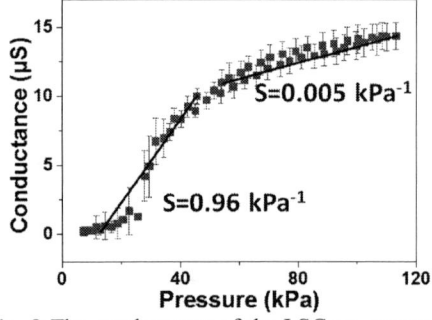

Fig. 9 The conductance of the LSG vs. pressure. At low pressures (<50 kPa), the sensitivity is 0.96 kPa^{-1}. At high pressures (50~113 kPa), the sensitivity is 0.005 kPa^{-1}.

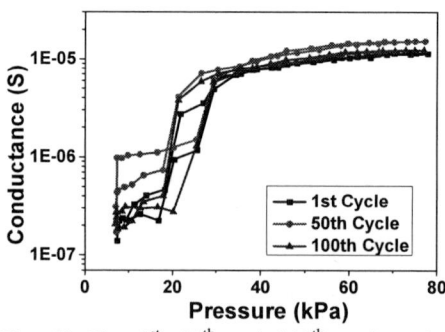

Fig. 10 The 1st, 50th and 100th cycles of the pressure response showing the repeatability of the performance.

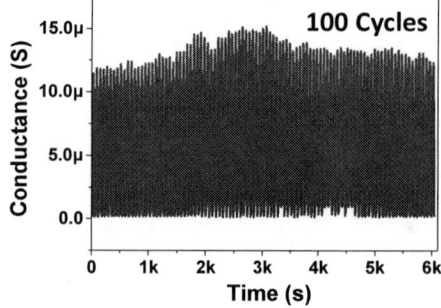

Fig. 11 The pressure sensor durability test (100 cycles). In each cycle the pressure is swept from 0~75 kPa.

978-1-4799-8002-4/14 $31.00 © 2014 IEEE   380

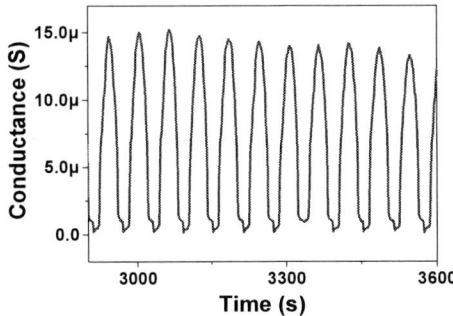

Fig. 12 A zoomed-in view of the curves in Fig. 13 after 50 cycles.

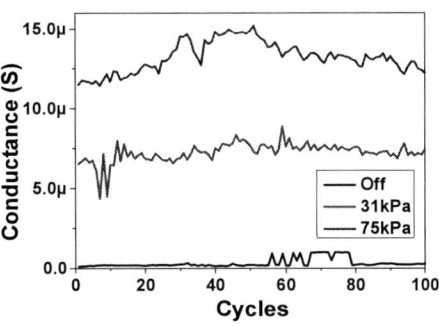

Fig. 13 The pressure sensor conductance over 100 cycles shown for three different pressures.

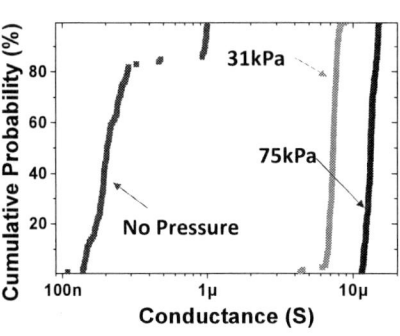

Fig. 14 Distribution of the conductance by applied pressure.

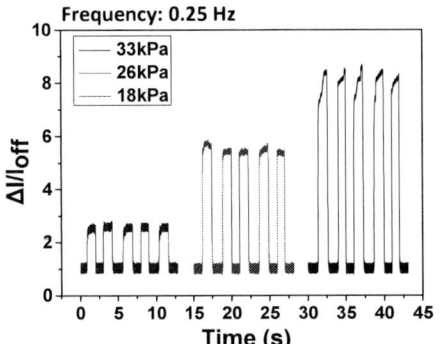

Fig. 15 Dynamic pressure response at low pressures.

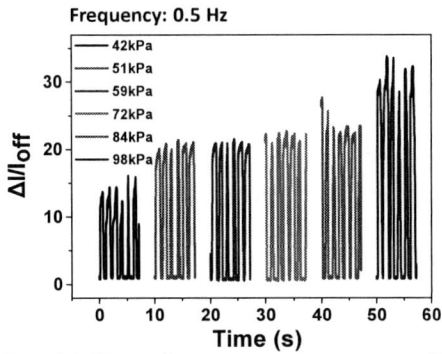

Fig. 16 Dynamic pressure response at high pressures.

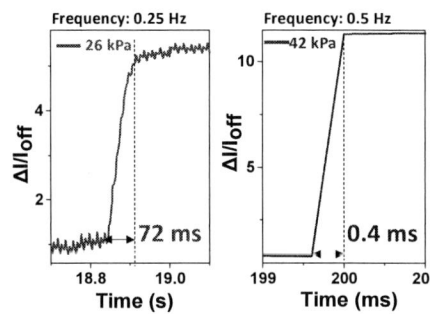

Fig. 17. The response time under low and high pressure.

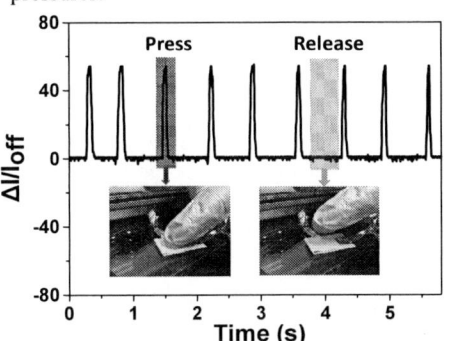

Fig. 18 The plot of current response to dynamic pressing and releasing cycles.

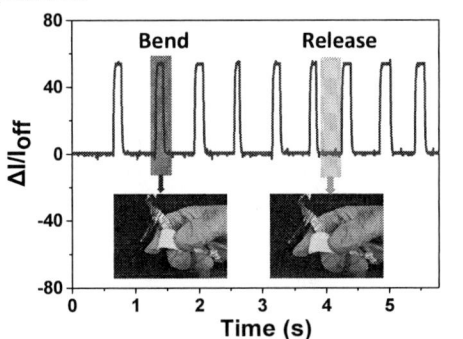

Fig. 19 The plot of current response to dynamic bending and releasing cycles.

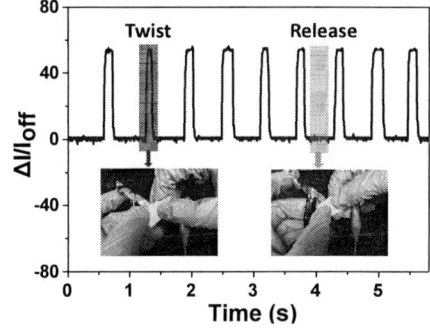

Fig. 20 The plot of current response to dynamic twisting and releasing cycles.

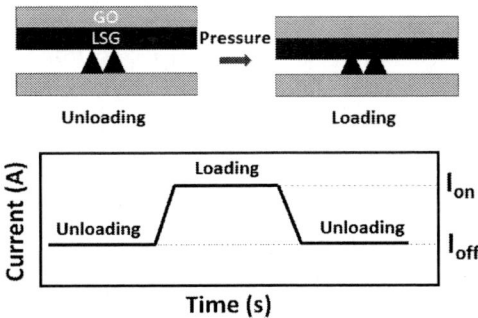

Fig. 21 Schematic illustration of the sensing mechanism and current changes in response to loading and unloading (I_{off}: unloading, I_{on}: loading).

Table I Performance comparison of recently reported pressure sensors. Our LSG pressure sensor with a foam-like structure has the highest sensitivity and pressure range.

Types of devices	Structure	Material	Sensitivity (kPa^{-1})	Detection limit (kPa)	References
Resistive	Suspended	CVD Graphene	2.66×10^{-5}	100	Nano Letters 2013 [1]
Resistive	Sponge	Graphene–Polyurethane	0.26	2	Advanced Materials 2013 [2]
Piezoelectric	Nanowire	ZnO	0.131	3.5	Science 2013 [3]
Capacitive	Pyramid	PDMS	0.55	7	Nature Materials 2010 [4]
Resistive	Foam	Laser-Scribed Graphene	0.96	113	This Work

Large-Scale Fabrication of Graphene-based Electronic and MEMS Devices

Debin Wang[1,†], He Tian[2,†], Iñigo Martin-Fernandez[1], Yi Yang[2], Tian-Ling Ren[2,*], and Yuegang Zhang[1, 2, 3,*]

[1]Lawrence Berkeley National Laboratory, Berkeley, CA 94720, USA

[2]Institute of Microelectronics & Tsinghua National Laboratory for Information Science and Technology (TNList), Tsinghua University, Beijing 100084, China

[3]Suzhou Institute of Nano-Tech and Nano-Bionics, Chinese Academy of Sciences, Suzhou 215123, China.

[†]These authors contributed equally to this work.

*Email: ygzhang2012@sinano.ac.cn, RenTL@tsinghua.edu.cn

-Invited Paper-

Graphene has been demonstrated great potential in electronic and optoelectronic applications. However, the zero band gap of graphene leads to the low on/off ratio in field effect transistors (FETs) and low optical wavelength selectivity in photo detectors. Moreover, the commonly used wet-transfer process for chemical vapor deposited (CVD) graphene could introduce contamination and defects that degrade the graphene device's performance. These problems could be resolved if we could find a graphene ribbon fabrication method that could precisely control the width, edge structure, as well as registries (location, orientation) on the substrates. Here, we will showcase recent works on novel transfer-free and contaminant-free CVD methods for direct-growth of graphene nanoribbons and microribbons on dielectric substrates.

By growing graphene nanoribbons (GNRs) on well-designed nano-templates, we demonstrated the fabrication of die-scale GNR-FET array on SiO₂/Si substrate using simple photolithography tools (Fig.1) [1]. The carrier mobility (> 1,000 cm²/V·s) is higher than those previously reported graphene nanoribbons fabricated on SiO_2 substrates, thanks to our novel transfer-free and contaminant-free direct growth process (Fig. 2). Because the GNR width in our method is defined by the metal-catalyst film thickness, which is not limited by the lithography resolution, it has the potential to be scaled down to sub-nanometer level. Direct GNR synthesis also avoids the problems of grain boundaries in conventional wafer-scale graphene film synthesis. This technological process was applied to the fabrication of die-scale FET arrays hosting hundreds of devices [1]. This method based on the surface-selective growth of graphene for GNR growth enables the tunable width growth of GNRs since the nanotemplates are controlled by the thickness of the catalytic material layer. Therefore, if the layer was scaled down to an only few angstroms thickness, GNRs with band gaps larger than 0.5 eV or even 1 eV are expected, thus, enabling room temperature GNR-FET based applications. In addition, we

have shown that the morphology of these nanotemplates defines the length and the position of the GNRs. Although we have only validated the process by using Ni and Al_2O_3 as the catalytic and noncatalytic materials, respectively, the method can be extended to other catalytic/ noncatalytic material combination as long as their interfaces are stable during the processing.

For the case of graphene microribbons (GMRs), by using fast annealing at a low temperature (750°C for 2-5 minutes) and dewetting of Ni, continuous few-layer GMRs (2-10 μm in width, up to a few millimeters in length) grow directly on bare dielectric substrates through Ni assisted catalytic decomposition of hydrocarbon precursors (Fig. 3) [2]. The dewetting of Ni on the dielectric substrates during the annealing process helps to facilitate the carbon diffusion on to the substrate surface and formation of graphene micro ribbons along the periphery of pre-patterned Ni template films. The short annealing time and low annealing temperature are the key factors to form continuous GMRs near the Ni film edges, and no graphene films under the Ni film. The low annealing temperature is particularly useful to improve compatibility of graphene integration with microelectronic technologies and allows for energy efficiency. These high quality GMRs exhibit low sheet resistance of 700 Ω ~ 2100 Ω, high on/off current ratio of ~3, and high carrier mobility of ~655 cm²V⁻¹s⁻¹ at room temperature (Fig.4). Since the size, geometries, and locations of the GMRs are well defined by the pre-patterned Ni film, this present approach can be therefore scaled up to wafer size fabrication of graphene based devices.

To demonstrate the temperature sensitivity of the CVD graphene devices, we measured the temperature dependence of the Raman G shift (Fig. 5) as well as the electrical resistance (Fig. 6). Increasing temperature led to a red shift of the G mode peak of CVD graphene. A temperature coefficient of -0.025 cm⁻¹/°C can be extracted from the slope of Fig. 5, which is consistent with previous reports for single layer graphene flakes. The thermal

978-1-4799-8002-4/14 $31.00 © 2014 IEEE

conductivity can be therefore determined in the range of 3.100 x 10^3 to 3.39 x 10^3 W/mK, which is in the mid-range of multi-walled and single-walled carbon nanotubes. The electrical resistance dependence on temperature in Fig. 6 shows good agreement with theoretical prediction. For $|V_G|>20V$ above Diract point, resistance decreases with temperature due to the increase of intrinsic carriers. In contrary, while $|V_G|>20V$ below Diract point, resistance increases with temperature due to increase of phonon scattering. This bimodal metallic-semiconducting behavior shows great potential for CVD graphene as an intriguing material for future thermal sensing platform.

CVD graphene can also be used in a transparent photodetector [3]. The fabrication process of a graphene PN junction device is shown in Fig. 7a. The device is highly transparent (Fig. 7b) and flexible (Fig. 7c). The transmittance in a broad wavelength range is ~90% (Fig. 7d). The photo response is tested with good performance under flat (Fig. 8a) and bend condition (Fig. 8b). This selectively doped CVD graphene photodetector showed a ~0.5 % modulation of conductance under global IR irradiation and a ~0.1 % under cold lamp (Fig. 9). From a comparison of a series of devices with various geometries, we identified that both the homogeneous and the PN junction regions contribute competitively to the photo response. The working principle of the graphene photodetector is shown in Fig. 10. Furthermore, two-terminal graphene photodetector can be fabricated on both transparent and flexible substrates without the need for complex fabrication processes used in electrically gated three-terminal devices. This represents the first demonstration of a highly transparent and flexible graphene-based IR photodetector that exhibits both good photo-responsivity and high bending capability.

Contrary to most other graphene-based IR photodetectors, the device described herein is derived through a selected-area chemical doping process. In addition to the broadband adsorption, our chemically doped CVD-grown graphene photodetector can be potentially fabricated on a large scale. Next, we deduced that the homogeneous and the junction regions contribute competitively to the photoresponse with

tunable dominating effects via device geometry design to achieve optimized sensitivity and speed. Because our described chemical doping process does not require complex fabrication process for metallic gate electrodes, it makes all-transparent and flexible IR photodetector possible.

Both of the GNR and GMR devices fabricated by the direct CVD growth methods have shown significant performance improvement over those made by conventional lithography methods. The new methods have overcome several limitations for the GNR and GMR patterning and are compatible with large-scale fabrication of graphene nano-devices. They can in principle fabricate graphene ribbons of any size and geometry, which is a feasible technology for the future integration with conventional semiconductor materials and the scalable production of graphene-based thermoelectric and optoelectronic devices.

Acknowledgement

This work was supported by the National Natural Science Foundation of China (61434001, 61025021, 61020106006) and the National Key Project of Science and Technology (2011ZX02403-002), and the Special Fund for Agro-scientific Research in the Public Interest (201303107). We are thankful for receiving support through the State Key Laboratory of Automotive Safety and Energy, Tsinghua University. The work performed at Lawrence Berkeley National Laboratory was supported by the Office of Science, Office of Basic Energy Sciences of the U.S. Department of Energy (No. DE-AC02-05CH11231).

References

1. Martin-Fernandez, I.; Wang, D.; Zhang, Y., Direct Growth of Graphene Nanoribbons for Large-Scale Device Fabrication. *Nano Lett.* 2012, 12, 6175-6179.
2. Wang, D. B.; Tian, H.; Yang, Y.; Xie, D.; Ren, T. L.; Zhang, Y. G, Scalable and Direct Growth of Graphene Micro Ribbons on Dielectric Substrates. *Sci Rep*, 2013, 3, 1348.
3. Liu, N.; Tian, H.; Schwartz, G; Tok, J. B. H.; Ren, T.-L.; Bao, Z., Large-Area, Transparent and Flexible Infrared Photodetector Fabricated Using P-N Junctions Formed by N-doping CVD-Grown Graphene. *Nano Lett.* 2014, 14, 3702-3708.

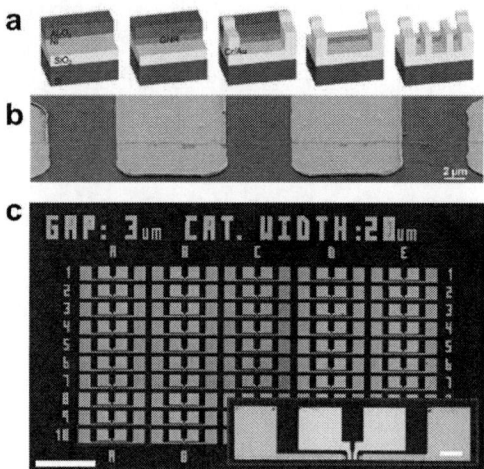

Figure 1. Direct growth of graphene nanoribbons (GNRs). (a) Schematic of the growth process showing the steps of GNR growth and the test structure fabrication. (b) SEM image of a GNR after the two sets of metal electrodes had been patterned. (c) Optical images of a GNR-FET array consisting of 50 device. Scale bars: 500 μm and 50 μm, respectively.

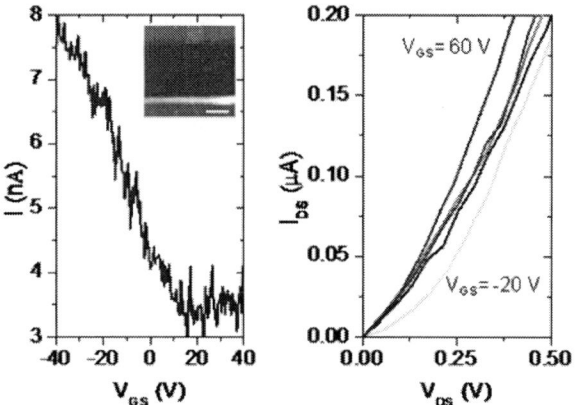

Figure 2. I−V_{GS} (V_{GNR} = 20 mV, left panel) and I_{DS}−V_{DS} (V_{GS} keeps at −20 V, 20 V, 40 V, and 60 V, respectively, right panel).

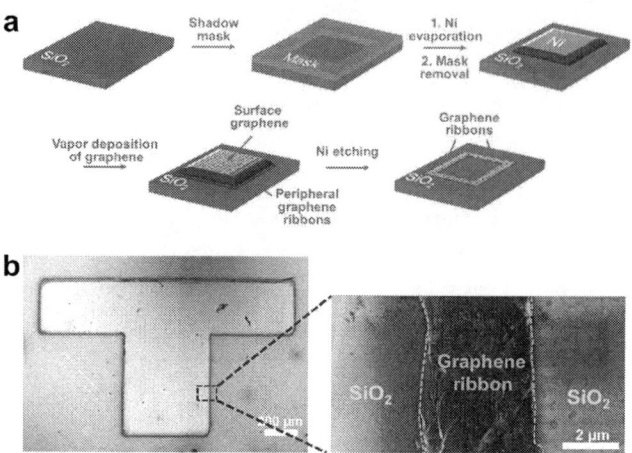

Figure 3. Scalable and direct growth of graphene microribbons (GMRs). (a) Schematic of the direct growth of GMRs along the periphery of pre-patterned Ni on a SiO_2 dielectric substrate. (b) The SEM micrograph reveals the formation of GMRs along the periphery of a "T" shape Ni film patterned by a shadow mask on a SiO_2 substrate.

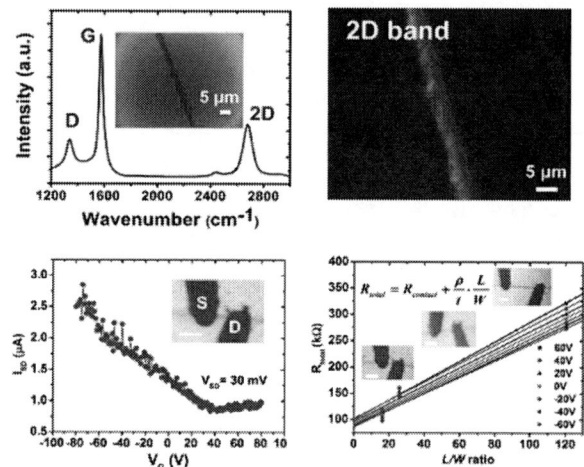

Figure 4. Raman characterization and electrical transport tests of the CVD-grown, transfer-free GMRs.

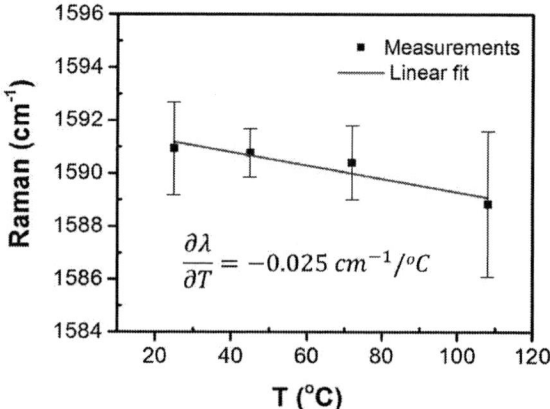

Figure 5. Raman G peak shifts of the CVD graphene as a function of temperature. The spectra are measured under a 532 nm excitation at ambient conditions.

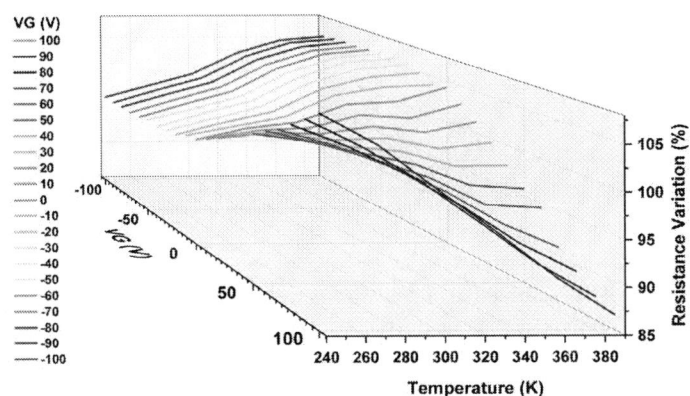

Figure 6. Temperature dependence of CVD graphene's electrical resistance at different gating voltages.

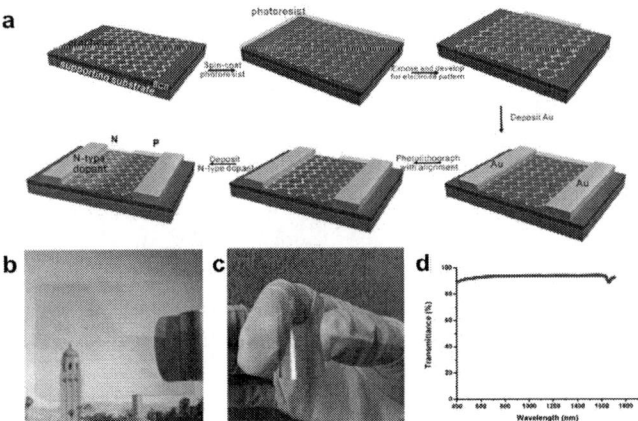

Figure 7. (a) Fabrication process of graphene photodetector. (b) The transparency and (c) flexibility of the device. (d) The testing showing ~ 90% transmittance in a broad wavelength range.

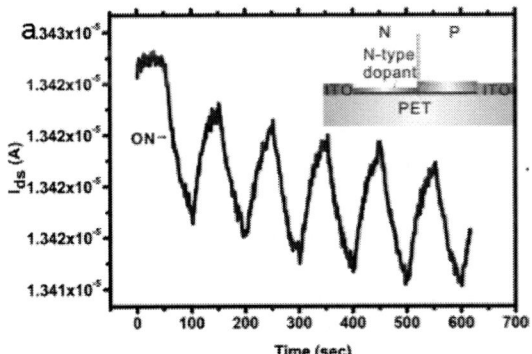

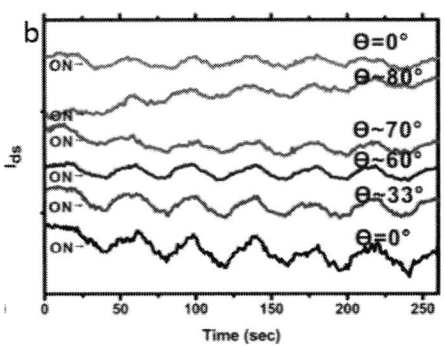

Figure 8. (a) Photo response of the graphene-based all transparent photodetector. (b) Photo responses of the Au-based photodetector on PET substrate at different bending angles (33°, 60°, 70° and 80°). Bottom and top curves are corresponding with pre- and post-bending photo response, respectively.

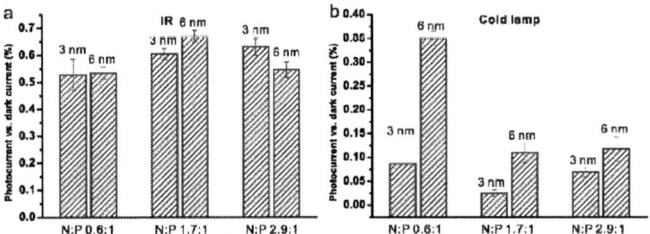

Figure 9. Comparison of the photoresponse of graphene PN junction with different thicknesses of n-type doper and area ratios of P- and N- region illuminated by (a) IR and (b) cold lamp.

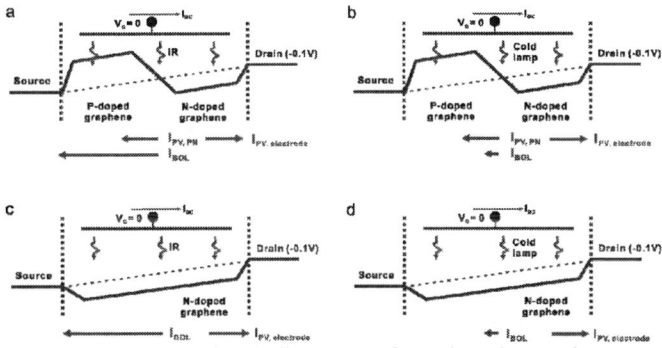

Figure 10. Energy band diagrams of PN junction and N-type graphene ($V_{ds} = 0.1$ V) under IR (>780 nm) and fluorescent cold lamp. Photoresponse under IR is dominated by the bolometric effect while the photovoltaic effect is more obvious under fluorescent cold lamp. I_(PV, PN): Photocurrent generated by photovoltaic effect in the PN junction. I_(PV, electrode): Photocurrent generated by photovoltaic effect in the N-doped graphene-metal junction. I_BOL: Photocurrent generated by the bolometric effect.

Flexible, Transparent Single-Layer Graphene Earphone

He Tian,[1,2] Yi Yang,[1,2] Cheng Li,[1,2] Mohammad Ali Mohammad,[1,2] Tian-Ling Ren[1,2,*]

[1] Institute of Microelectronics, Tsinghua University, Beijing 100084, China

[2] Tsinghua National Laboratory for Information Science and Technology (TNList), Tsinghua University, Beijing 100084, China

*Email: RenTL@tsinghua.edu.cn

Abstract

We demonstrate a novel flexible and transparent earphone based on single-layer graphene (SLG) for the first time. The SLG earphone operates in the frequency range of 20 Hz to 200 kHz and has a highest sound pressure level (SPL) of 70 dB with a 1 W input power. The SPL emitted from one to six layers of stacked SLG are compared. It is observed that the SPL decreases with an increasing number of stacked layers. The SLG earphone is packaged into a commercial earphone casing and can play music. Compared with a conventional earphone, the SLG earphone has a broader frequency response and a lower fluctuation. Testing results in both time- and frequency-domains show a frequency doubling effect, which indicates that the working principle is based on the electro-thermoacoustic (ETA) effect. As the SLG earphone operates in both the audible and ultrasonic frequency range, it can be used for a wide variety of applications, including for interspecies communication.

I. Introduction

Graphene research has been heavily focused on applications in electronics and photonics; however, little research has been done on acoustics [1]. Here we demonstrate a SLG earphone, operating according to the electro-thermoacoustic (ETA) effect. Exploiting such a working principle for the first time, the SLG earphone is fundamentally different from conventional ones. We also systematically study the sound generation from one to six stacked layers of SLG. The sound spectrum of the SLG earphone is significantly wider than a commercial earphone, covering both the audio and ultrasound frequencies up to 200 kHz. The earphone core is flexible, transparent and ultrathin. As a demonstration, the SLG earphone is packaged into a commercial earphone casing and tested successfully.

II. Stacked SLG Fabrication and Characterization

Stacked SLG (from one to six layers) were made and tested to understand the layer-dependence on sound performance. Large areas of SLG (5×10 cm^2) can be transferred on PET substrates with high transparency and flexibility (Fig. 1). Furthermore n (2~6) such layers can be transferred to the same PET substrate. This can be done by repeated PMMA wet transfer of SLG grown on copper by CVD. It is observed that the transparency of the stacked layers decreases with an increasing number of layers (Fig. 2). The sheet resistance of the SLG is ~90 $\Omega/\square$ (Fig. 3), showing a higher quality as compared to roll-to-roll SLG, whose resistance is reported to be 125 $\Omega/\square$ [2]. Stacking SLG layers decreases the sheet resistance down to 26 $\Omega/\square$ in a six-layer stack. The stacked SLGs were also transferred to SiO$_2$/Si substrates for optical and Raman analysis. As the stacked SLGs were not perfectly aligned, the layer number could be identified through the edges of the films (Fig.4). Raman spectra show that the SLG features yield stronger 2D peaks as compared with G peaks (Fig. 5). There are almost no D-peaks indicating a C-C lattice structure with very little defects. Due to the random stacking of SLG layers, unlike graphite, the intensities of the G and 2D peaks increase together. Statistical results show that 2D/G ratios are larger than 1 for stacked SLG layers (Fig. 6).

III. Layer Dependence on Sound Performance

The acoustic platform for testing the SLG earphones consisted of a standard microphone and a dynamic signal analyzer. A 1/4 inch standard microphone (Earthworks M50) was used to measure the sound pressure level of the loudspeakers. This microphone has a very flat frequency response reaching up to 50 kHz and a 31 mV/Pa high sensitivity. A signal analyzer (Agilent 35670A) was used to generate sine signals to drive the earphones, perform fast Fourier transform analysis and record the value of the sound pressure level. Our testing was performed in a soundproof box measuring $1.0 \times 0.5 \times 0.5$ m^3. In order to avoid echo, the box was filled with sound-absorbing sponges.

The sound performance of the stacked SLGs are tested by applying 5 V AC and 5 V DC signals. The frequency is swept from 20 Hz to 50 kHz. From layers 1 to 5, the sound pressure level (SPL) decreased (Fig. 7) which could be explained by an increased heat capacity per unit area (HCPUA). It is concluded that the SLG has the highest SPL due to its lowest HCPUA (Fig. 8). It is also noticed that there is an anomaly for a six-layer SLG stack (Fig. 8). Since six layers have a larger HCPUA, it is expected that

978-1-4799-8002-4/14 $31.00 © 2014 IEEE

the SPL would be lower. However, another SPL influencing factor is the thermal leakage from the substrate. As six layers have a larger total gap, the thermal coupling with the substrate is weaker, which in-turn could enhance the sound performance. In Table 1, the performance of our SLG earphone is compared with previous efforts [3-5].

IV. Fabrication of SLG Earphone

Since a single layer of graphene has the highest SPL, it was used to make an earphone prototype. The fabrication process of the SLG earphone has eight steps (Fig. 9). Briefly, the SLG is grown on copper by CVD at 1000 °C. The domain size is ~20 μm. Wet transfer is used to transfer the SLG on a PET or PI substrate. Subsequently, silver electrodes are applied to the SLG and wired out (Fig. 10a). Finally, the device is packaged into a commercial earphone casing (Fig. 10b, 10c). A drive circuit is also made to supply a DC bias so that the SLG earphone can directly connect to laptop for playing music (Fig. 10d).

V. Sound Performance of the SLG Earphone

The time domain response of the SLG earphone is investigated and compared to a commercial earphone. An input sine signal is swept from 0 Hz to 20 kHz in 20 s. It is observed that the SLG earphone reaches 20 kHz in 10 s (Fig. 11), indicating that the output frequency has been doubled. In order to reproduce the same frequency as the input signal, a DC bias is added and both single-frequency and double-frequency tones are generated (Fig. 12). It is also noted that the conventional earphone only has a single-frequency tone (Fig. 13). The SLG earphone also has a wider frequency response than a commercial earphone, especially in the ultrasound range (Fig. 14). The sound spectrums of the SLG earphone under different input power levels are also tested (Fig. 15) and a linear relationship is observed (Fig. 16). The frequency-doubling effect in the SLG earphone is also found in time-domain (Fig. 17). A sine-pulse response testing shows that the delay time is 20 μs (Fig. 18), which is similar to a commercial earphone. As shown in Fig. 19, this is the first experimental demonstration that the graphene earphone can emit sounds up to 200 kHz.

VI. Theoretical analysis of the SLG Earphone

A schematic shows the sound generation process in a SLG earphone (Fig. 20) through the ETA effect, which is radically different from conventional mechanical damping. A model was created based on the work by M. Daschewski et al. [6] and the simulation results show a good agreement with the experimental results (Fig. 21). The sounds generated by the SLG earphone are expected to show low directivity at low frequencies (Fig. 22). At higher frequencies, the signals are expected to concentrate with a high directivity.

VII. Conclusion

A transparent, flexible SLG earphone has been demonstrated for the first time operating based on the ETA effect. The earphone operates in the range of 20 Hz to 200 kHz, with a highest SPL of 70 dB. These metrics indicate that the SLG earphone has a performance exceeding other recent state-of-the-art sound generation devices. Furthermore, the SLG earphone has a broader frequency response and a lower fluctuation as compared to a commercial earphone. A systematic study of the sound generation performance of one to six stacked SLG layers was conducted. Further performance enhancement can be achieved by driving the SLG earphone using pulse density modulation. The SLG earphone technology is expected to bring transparent flexible earphone in the field of acoustics.

Acknowledgement

This work was supported by the National Natural Science Foundation of China (61434001, 61025021, 61020106006) and the National Key Project of Science and Technology (2011ZX02403-002), and the Special Fund for Agro-scientific Research in the Public Interest (201303107). We are thankful for receiving support through the State Key Laboratory of Automotive Safety and Energy, Tsinghua University. M.A.M is thankful for receiving support from the postdoctoral fellowship program of the Natural Sciences and Engineering Research Council of Canada.

References

[1] H. Tian, et al., "Graphene-on-paper sound source devices," *ACS Nano*, vol. 5, pp. 4878-4885, 2011.

[2] S. Bae, et al., "Roll-to-roll production of 30-inch graphene films for transparent electrodes," *Nature Nanotechnology*, vol. **5**, pp. 574-578, 2010.

[3] J. W. Suk, K. Kirk, Y. Hao, N. A. Hall, and R. S. Ruoff, "Thermoacoustic sound generation from monolayer graphene for transparent and flexible sound sources," *Advanced Materials*, vol. 47, pp. 6342-6347, 2012.

[4] Y. Wei, X. Lin, K. Jiang, P. Liu, Q. Li and Fan S., "Thermoacoustic chips with carbon nanotube thin yarn arrays," *Nano Letters*, vol. 13, pp. 4795-4801, 2013.

[5] H. Tian, et al, "Graphene earphones: entertainment for both humans and animals," ACS Nano, vol. 8, pp. 5883-5890, 2014.

[6] M. Daschewski, R. Boehm, J. Prager, M. Kreutzbruck, and A. Harrer, "Physics of thermo-acoustic sound generation," Journal of Applied Physics, vol. 114, pp. 114903, 2013.

 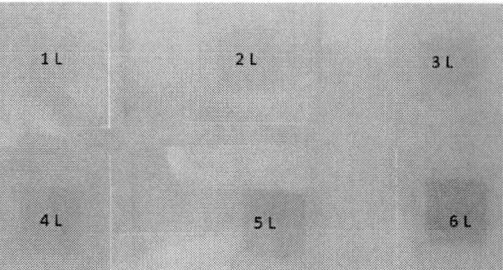 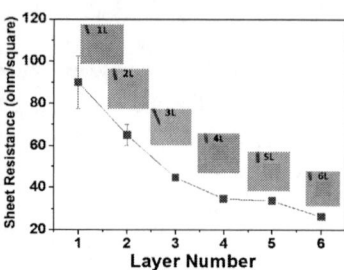

Fig. 1 A highly transparent 5 cm x 10 cm single graphene layer on PET. The dashed lines show the edges of the graphene film.

Fig. 2 Optical images of 1 to 6 layers of stacked single-layer graphene (SLG) films on PET substrates by wet transfer. The SLG dimensions are 1 cm × 1 cm.

Fig. 3 Sheet resistance vs. layer number of SLG films. The quality of SLG (<100 Ω/□) is even better than well-known roll-to-roll SLG [2].

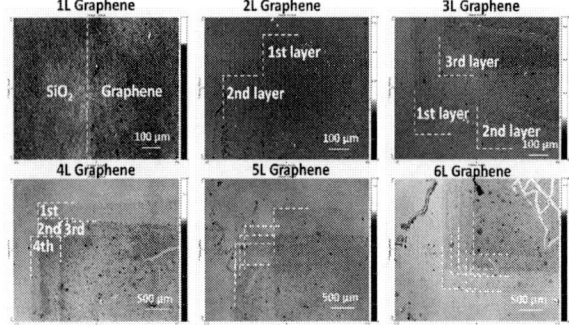 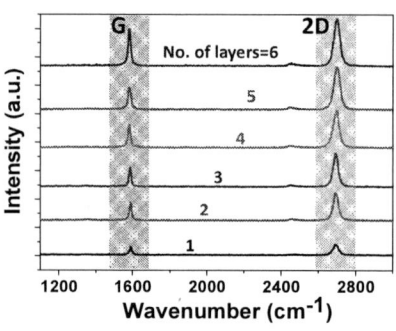 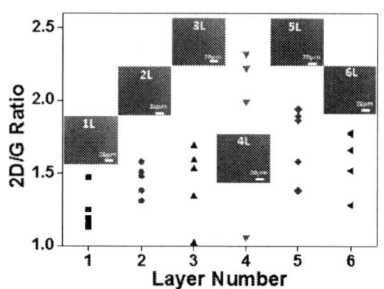

Fig. 4 Images of graphene films with 1 to 6 stacked layers on a SiO₂/Si substrate captured by a white light interference microscope. Each graphene layer can be identified by the color contrast and corner shape.

Fig. 5 Raman spectra of graphene films with different numbers of stacked layers. Due to the random stack of graphene layers, unlike graphite, the intensities of the G and 2D peaks increase together.

Fig. 6 2D/G ratio of graphene films with different numbers of stacked layers. The 2D/G ratios are larger than unity, indicating the presence of SLG.

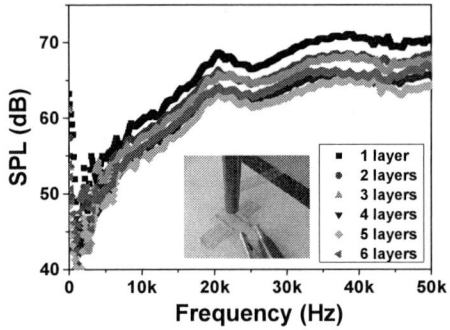 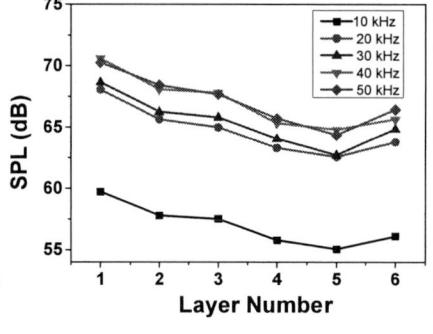

Fig. 7 The SPL vs. frequency for different graphene layers. The single-layer graphene has the highest SPL value.

Fig. 8 The SPL vs. layer number for different acoustic frequencies.

Table I A performance comparison of recent acoustic devices. Our SLG has the highest SPL and widest frequency range.

Materials	SPL (dB)	f Limit (kHz)	References
SLG	46	20	Advanced Materials 2012 [3]
CNT	60	20	Nano Letters 2013 [4]
rGO	35	50	ACS Nano 2014 [5]
SLG	70	200	This Work

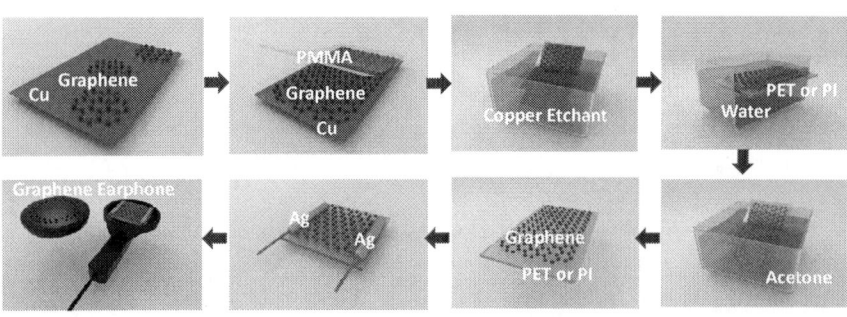 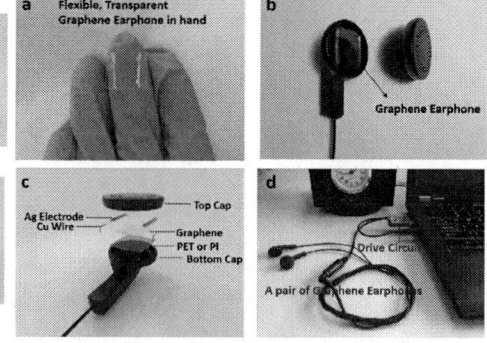

Fig. 9 Process flow of the SLG earphone fabrication. Since SLG has the highest SPL value, it was employed to make an earphone prototype. A SLG grown on a copper foil was coated with PMMA. The copper foil was etched and the SLG/PMMA stack was subsequently transferred to a PET or PI substrate. Finally, the SLG was wired out using copper wire and silver paste and packaged in a conventional earphone cap.

Fig. 10 SLG earphone images. (a) A SLG earphone in hand. (b) View of the SLG earphone in a commerical earphone casing. (c) Exploded view of the packaged SLG earphone. (d) A Pair of SLG earphones connected to a laptop through a drive circuit.

978-1-4799-8002-4/14 $31.00 © 2014 IEEE

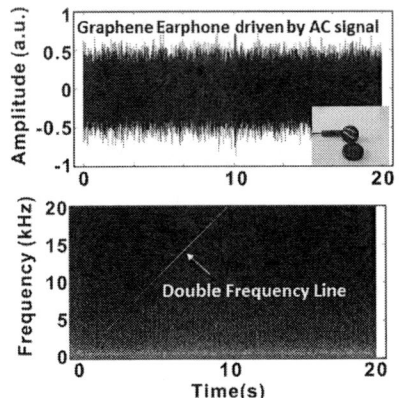

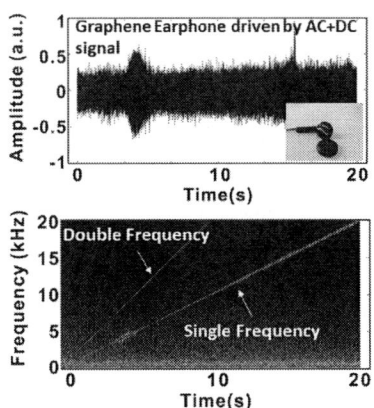

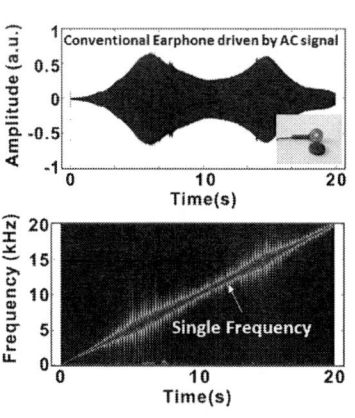

Fig. 11 The amplitude spectrum and its Fourier transform of the sound generated by the SLG earphone driven by an AC signal.

Fig. 12 The amplitude spectrum and its Fourier transform of the sound generated by the SLG earphone driven by an AC+DC signal.

Fig. 13 The amplitude spectrum and its Fourier transform of the sound generated by a conventional magnetic coil earphone.

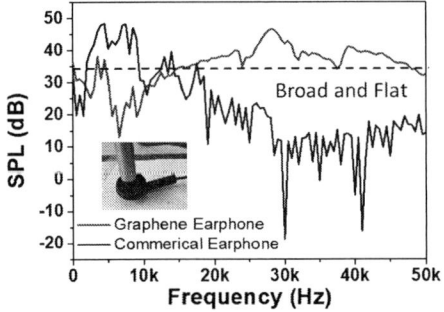

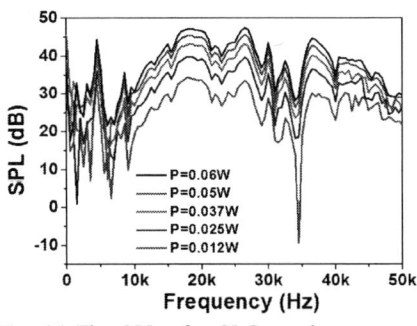

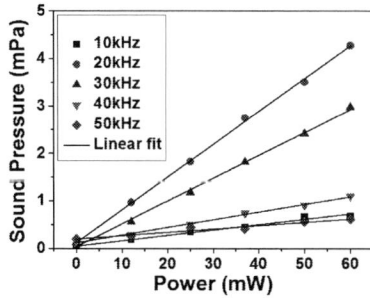

Fig. 14 SPL curves of a SLG earphone compared with a commercial earphone.

Fig. 15 The SPL of a SLG earphone operated under different input power levels.

Fig. 16 Power-dependent sound pressure measured under different frequencies.

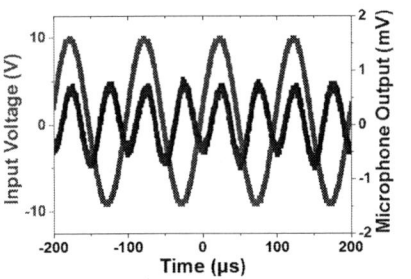

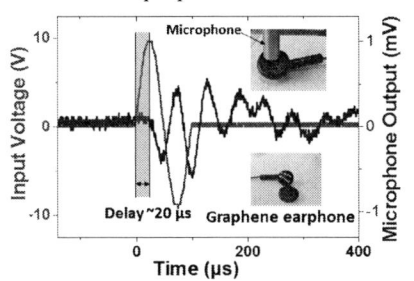

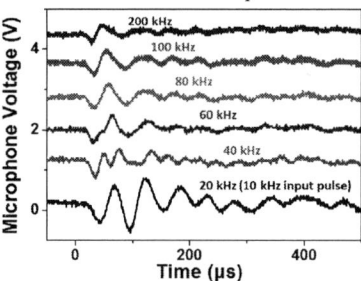

Fig. 17 A 10 kHz input signal fed to the SLG earphone and the corresponding sound signal collected by the microphone, showing the frequency-doubling effect.

Fig. 18 The measured acoustic response for a single 10 kHz sine-pulse input, showing a time delay of 20 μs.

Fig. 19 The measured acoustic response from different input AC sine-pulse frequencies. A 100 kHz input signal corresponds to a 200 kHz output sound frequency.

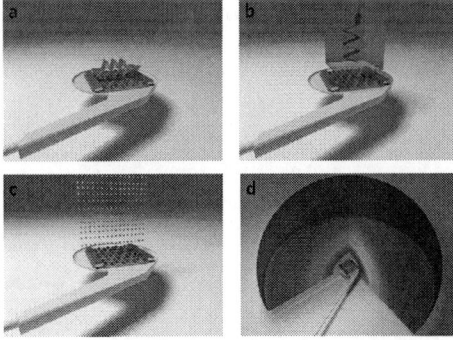

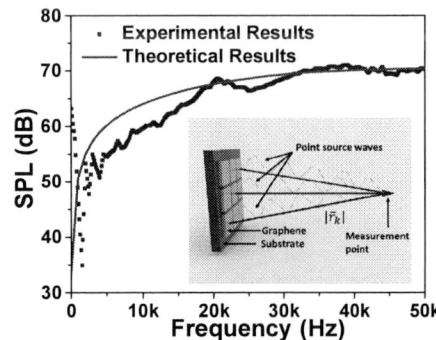

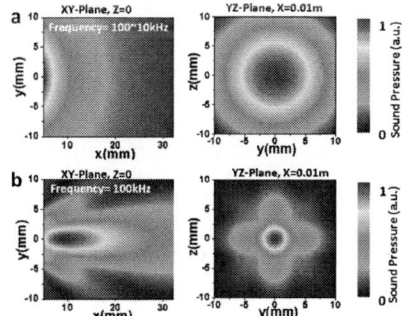

Fig. 20 Sound generation in a graphene earphone. An applied (a) signal causes (b) Joule heating, inducing (c) longitudinal, and (d) spherical sound waves in the near- and far-fields respectively.

Fig. 21 SPL vs. frequency showing that the model agrees well with the experiment. The inset illustrates the model of sound generation in a SLG earphone.

Fig. 22 Simulation of the SLG earphone performance at various frequencies, (a) ranging from 100 Hz~10 kHz, and at (b) 100 kHz.

978-1-4799-8002-4/14 $31.00 © 2014 IEEE   389

A Semiconductor Bio-electrical Platform with Addressable Thermal Control Circuits for Accelerated Bioassay Development

T. -T. Chen[1], C. -H. Wen[1], J. -C. Huang[1], Y. -C. Peng[1], S. Liu[1], S. -H. Su[1], L. -H. Cheng[1], H. -C. Lai[1], T. -C. Liao[1], F. -L. Lai[1], C. -W. Cheng[1], C. -K. Yang[1], J. -H. Yang[1], Y. -J. Hsieh[1], E. Salm[2], B. Reddy[2], F. Tsui[1], Y. -S. Liu[1], R. Bashir[2], M. Chen[1]

[1]Taiwan Semiconductor Manufacturing Company, Hsinchu, Taiwan; [2]University of Illinois at Urbana-Champaign, Champaign, IL

rbashir@illinois.edu; mark_chen@tsmc.com

Abstract

A 0.18μm SOI-CMOS bioelectrical sensing technology is introduced. An SOC chip integrates biosensor pixel arrays, controllers and amplifiers is used to demonstrate the performance of this technology. The pixel size in the pixel arrays is 10μm x 10μm, including biosensor, temperature sensor and heaters. The chip demonstrates detections of hydrogen ion concentration, enzymatic reactions and DNA hybridization with PCR. Experimental results show close to Nernst limit of 59mV/pH in ion concentration detection, sub-millimolar resolutions with 99.9% linearity in urea, and 400mV surface potential change in DNA hybridization. The SOC chip has an addressable temperature control for each pixel with embedded thermal sensors. A thermal time constant of 35msec/K and sub-degree localized temperature control are achieved. Order of magnitude improvements over previously reported are seen in both detectable minimum sample liquid volume and thermal time constant for PCR. This is a good demonstration of semiconductor technology for multi-biomarkers detection for medical applications.

Introduction

Diagnostic techniques play critical roles in modern health-cares as their outcomes are main factors in setting the courses of patient treatments. However, many bio-techniques used today yield very specialized data and, in general, come with high complexities and costs. People believe that semiconductor technology can enable possibilities to benefit health-cares because of 1) Micro bioelectrical devices exhibit higher sensitivity, smaller sample volume and faster reactions response. 2) High-throughput fabrication allows seamless mass production. 3) Robust and repeatable processing results in quantitative and reliable diagnosis. In this work, we introduce a bioelectrical pixel design and a 128-by-128 array with pixel-by-pixel integrated temperature controls on an SOC.

Bioelectrical Devices

Figure 1 shows the cross-section view of bio-sensor pixel with its surrounding heaters. The elements in the dashed box are made by post-processes for bio-sensing, and the others are made by 0.18μm SOI-CMOS technology.

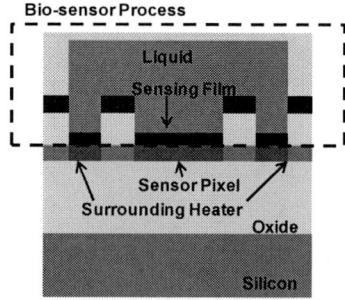

Figure 1: Cross-section view of sensor pixel with surrounding heaters.

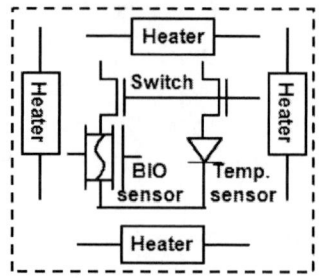

Figure 2: Schematic view of sensor pixel, which contains bio-sensor, temperature sensor, switches and surrounding heaters.

978-1-4799-8002-4/14 $31.00 © 2014 IEEE

Figure 2 shows the Schematic view of the pixel. Two switches are put in series with bio-sensor and temperature sensor. The bio-sensor, formed by a sub-micron FET with a high-k dielectric sensing film, is quite sensitive to ion-induced surface potential and the charges immobilized on the surface (Figure 3). It has a near-Nernst sensitivity of 56-59 mV/pH as an ion-sensitive transistor, and the sensitivity could be further improved using dual-gated structure [1].

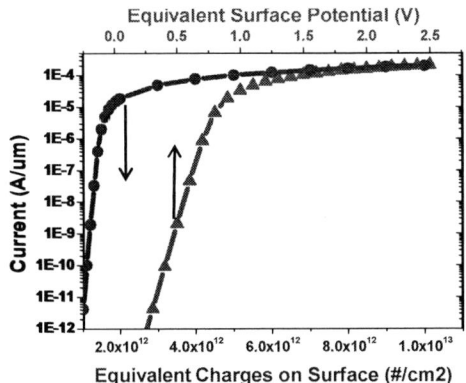

Figure 3: Bio-sensor sense the ion-induced surface potential and the charges immobilized on the surface.

Temperature Controls

In bio applications, high precision and quick response times in temperature controls are important. For example, a millimeter real-time PCR (Polymerase chain reaction) chip using bulk-silicon technology has achieved 40-thermal-cycle within half-hour [2]. The thermal time constant τ is governed by equation (1), where ρ, C_p and A are density, specific heat and surface area of sample body. ΔT in (1) is temperature difference when the heating power is applied. We reduced τ in our design by 1) minimizing sample volume-V (using shrunk components), and 2) enhancing power density q'' (using high sheet-resistance heaters and having good thermal isolation from SOI).

$$\tau = \frac{\rho c_p V \Delta T}{q'' A} \qquad (1)$$

In our work, a temperature accuracy of 0.64°C is attained by 2-point calibration from 25°C to 100°C as Figure 4.

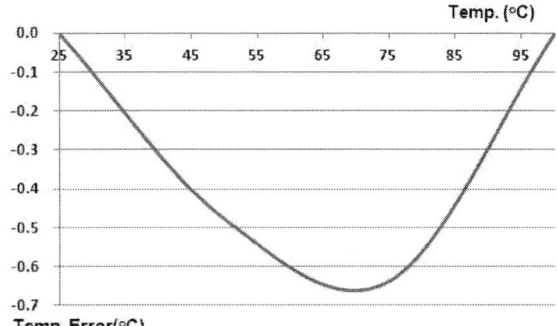

Figure 4: The nearby temperature sensor achieves sub-degree accuracy.

Figure 5 shows three areas A, B (20μm away from A) and C (420μm away) that we used to monitor the heating and thermal isolation among pixels. Figure 6 shows that we can heat up the area A to 100°C with 20mW of power while keeping thermal coupling to area C within 0.5%. Area B sees only 60% of the temperature increase of area A. The heating and cooling response times achieved are about 35msec/K. These measurements match well with our simulation results. The rapid thermal cycles and high accuracy temperature control are critical in multiplex assays.

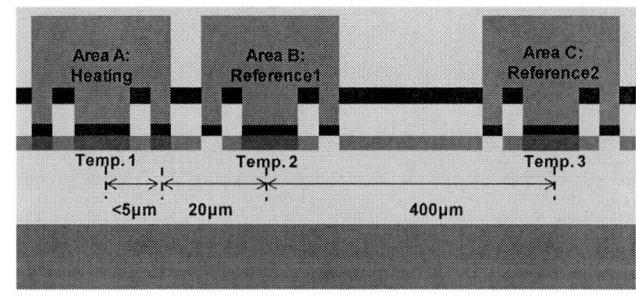

Figure 5: Three areas A, B (20μm away from A) and C (420μm away) that we used to monitor the heating and thermal isolation among pixels.

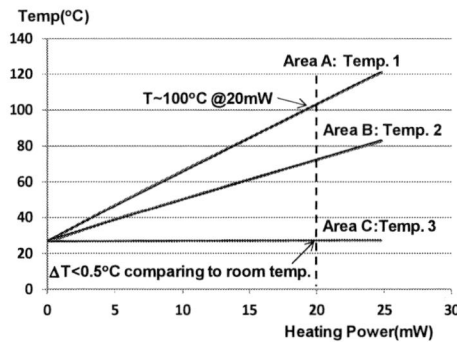

Figure 6: Area A is heated to 100°C by 20mW, ΔT of area B has only 60% ΔT of A and area C is isolated within 0.5% thermal coupling.

Bio Applications by 128x128 Micro-Arrays

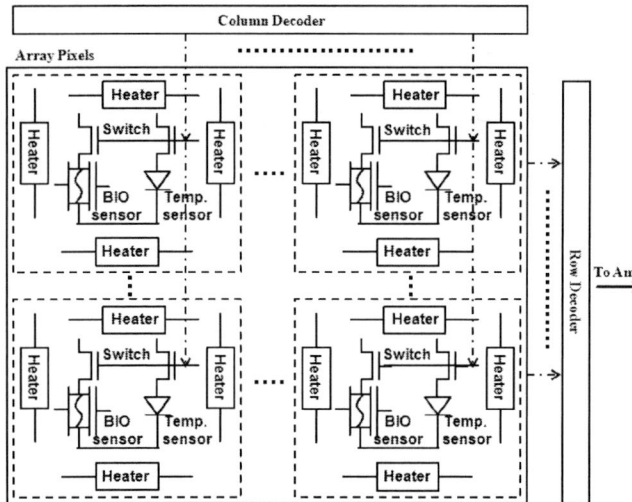

Figure 7: 128x128-pixel arrays with decoders.

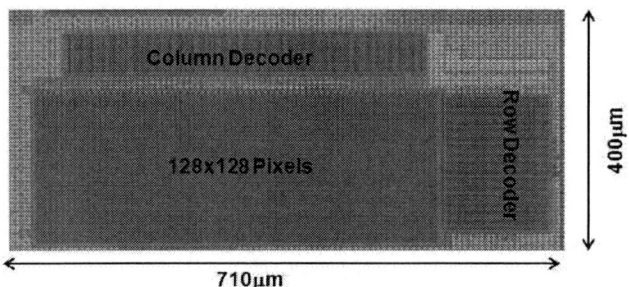

Figure 8: Photo of 128x128-pixel arrays with decoders.

Figure 7 (schematic) and Figure 8 (photo) show a 128-by-128 array of bio sensing pixels and associated decoders on a testchip. Collecting data in a large volume minimizes impacts of individual inaccuracy coming from process variation or environmental drifting. Figure 9 shows pixel-by-pixel pH sensing results measured from the entire array when we change the pH value from 8.2 to 6.45. Figure 10 shows that the mean and variance of pH sensing are -1.39μA/pH and 0.32μA, respectively. The testchip has been used to detect urine levels based on the enzyme (urease) -catalyzed reaction of equation (2):

$$CO(NH_2)_2 + 3H_2O \xrightarrow{urease} CO_2 + 2NH_4^+ + 2OH^- \quad (2)$$

As shown in Figure 11, we observe a highly linear response in the sub-millimolar range, using just 0.1μL of sample volume. This is about one order of magnitude improvement as compared to previously reported resolution and sample volume [3].

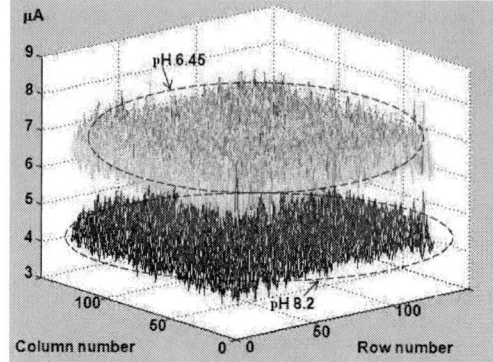

Figure 9: Current shifts of 128x128 array pixels due to pH changes.

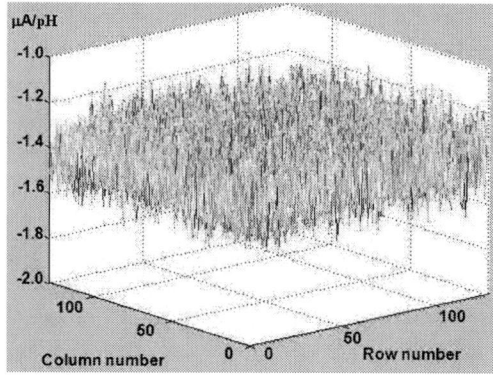

Figure 10: pH sensitivity map of 128x128 array pixels.

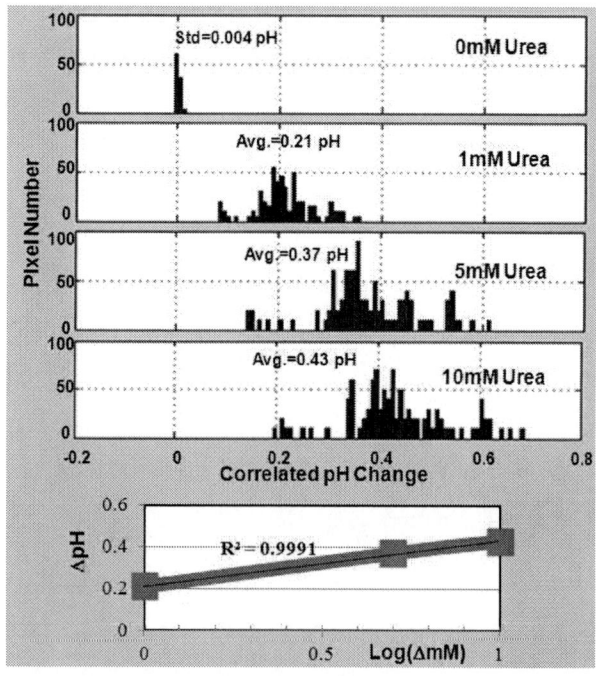

Figure 11: Urea detection with correlated pH change and 99.91%-linearity.

Deoxyribonucleic acid (DNA) detection of Hepatitis-B-Virus (HBV) was also conducted with this array. We see a 400mV equivalent surface potential change between 1μM matched and mismatched DNA. This is better than previous reported 25mV gate potential change [4]. In addition, we found that surface potential changes caused by immobilized DNA molecules have a good loading density (Figure 12-14).

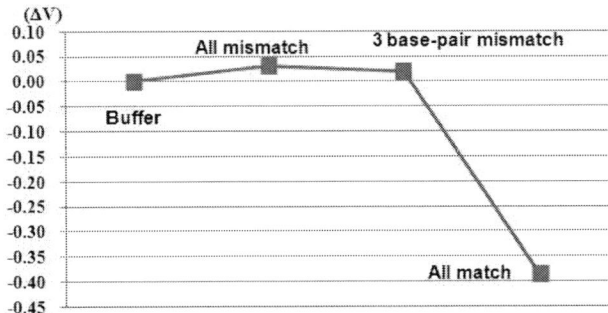

Figure 12: Equivalent surface potential change versus different DNA immobilized on the sensing film surface.

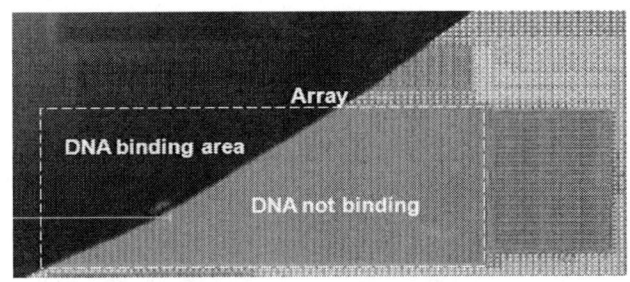

Figure 13: DNA selectively bound on the arrays.

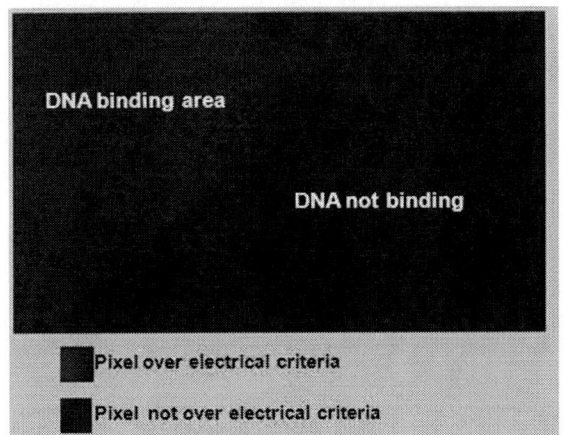

Figure 14: 2D image illustrates with a good loading density.

Table 1: The proposed highly integrated devices perform 1-2 order faster thermal response with compact devices and smaller volume samples.

	[2]	This design
Process	0.35μm CMOS	0.18μm SOI
Sample liquid volume	2μL	~0.1μL
Bio sensor size	104μm*34μm	0.66μm*0.32μm
Pixel size with heaters	~	10μm*10μm
Heater area with 100mW capacity	1mm*1mm (estimated)	100μm*100μm
Water thermal capacity	~8.4x10⁻³ JK⁻¹	0.42x10⁻³JK⁻¹
Temp. accuracy	Sub-degree	Sub-degree
Thermal time constant (Heater power=50mW)	0.35sec/K (estimated)	35x10⁻³sec/K
Ramp rate	3K/Sec	30K/Sec
Thermal cycle (ΔT=160K)	53sec	5.3sec (open loop)
40 thermal-cycle	35min	3.5min (open loop)
Thermal coupling (420μm distance)	Much larger	< 0.5%
Single-chip multi PCR	Hard to do	Available

Conclusion

This work presents an array device consisting of multiple cutting-edge semiconductor components to assist the development of electrical bioassays for medical applications. As a proof of concept, we demonstrated the capabilities of the device in terms of detections of enzymatic reaction and immobilization of bio-entities. The proposed highly integrated devices have the potential to largely expand its applications to all the heat-mediated bioassays, particularly with 1-2 order faster thermal response and smaller volume samples (Table 1).

Reference

[1] C. Duarte-Guevara, et al., "Enhanced Biosensing Resolution with Foundry Fabricated Individually Addressable Dual-Gated ISFETs," Anal. Chem., 2014, 86 (16), pp 8359–8367.

[2] C. Toumazou, et al., "Simultaneous DNA amplification and detection using a pH-sensing semiconductor system," Nat. Methods 10, 641–646 (2013).

[3] C. E. Lue, et al., "Optimization of Urea-EnFET Based on Ta2O5 Layer with Post Annealing," Sensors, vol. 11, pp. 4562-4571, May 2011.

[4] S. Kuga, et al., "Precise Detection Of Single Mismatched DNA With Functionalized Diamond Electrolyte Solution Gate FET," Tech. Dig. IEEE IEDM, p483 (2008).

Label-Free Optical Biochemical Sensor Realized by a Novel Low-Cost Bulk-Silicon based CMOS-Compatible 3-Dimensional Optoelectronic IC (OEIC) Platform

Junfeng Song,[1,2] Xianshu Luo,[1] Jack Sheng Kee,[1] Chao Li[1] and Guo-Qiang Lo[1]

[1]Institute of Microelectronics, A*STAR (Agency for Science, Technology and Research)
11 Science Park Road, Singapore Science Park II, Singapore 117685
[2]State Key Laboratory on Integrated opto-electronics, College of Electronic Science and Engineering,
Jilin University, Changchun, People's Republic of China, 130012
Email: songjf@ime.a-star.edu.sg; Tel: 65-6770-5756, Fax: +65 6773 1914

Abstract

We proposed and demonstrated label-free optical biochemical sensor realized on a novel bulk silicon-based platform with three-dimensional (3D) monolithic optoelectronic integration circuit (OEIC). The Ge-photodetector (Ge-PD) is integrated on bulk-silicon. Optical functional components are built with CMOS back-end-of-line (BEOL). Via this platform, the label-free optical biochemical sensor can get rid of the expensive SOI wafer. The costly tunable laser-source can be replaced by low cost wide-band light source. Optical signal is converted into electrical signal and read out by on-chip Ge-PD directly. This platform facilitates the development of cost-effective portable point-of-care (POC) diagnostic tool. Meanwhile, this also opens up new ways of electronic and photonic devices monolithically integration in 3D.

Introduction

Label-free optical biochemical sensor is testing refractive index changing by optical method (1). It is essential in the applications of medical diagnosis, healthcare and environmental monitoring. This is due to its merits of non-physical contact, high stability, high-speed detection, and high resolution, etc. Previously, we have developed a new electrical-tracing assisted optical biochemical sensing structure, which is monolithically integrated with multiple OEIC components on a single SOI chip (2-4). Such as grating coupler, thermal optical controlled tunable filter, label-free optical biochemical sensor and Ge-PD are integrated together. It has greatly reduced testing cost and seems suitable with POC request. However, it still has some of issues. For instance, SOI wafer is 10× costly than bulk-silicon. Silicon microring is with small size but the sensitivity not high. The buried oxide of SOI impedes heat-dissipation, which will made bad impact for electrical devices.

To overcome those issues, in this work, we propose a bulk-silicon based 3D-OEIC platform. Electrical devices can be fabricated on bulk-silicon. All optical devices use deposited materials and fabricated by BEOL process. This platform has, in addition to the aforementioned advantages of wafer cost and BOX issues, some other advantages: 1) Optical devices can be made with low-index contrasting material which has higher optical sensivity; 2) Different wavelength range can be selected, unlike SOI in which optical wavelength must be longer than 1.1 μm; and 3) Combining the advantages of existing electrical IC- and optical IC- technology.

Label-Free Optical Biochemical Sensor by Bulk-Si based 3D-OEIC Platform

A. Bulk-Si Based 3D-OEIC PD Structure

The basic concept of general SOI-based evanescent-coupled PD structure and the proposed bulk-Si based 3D-OEIC PD structure are illustrated in Fig. 1. The conventional silicon waveguide is replaced by PECVD SiN waveguide (core ~400 ×1000 nm^2). A scattering grating sits at the end of waveguide for vertical light coupling into the underneath Ge-PD. An Al mirror is designed on top of the scattering grating as a reflector to enhance the Ge-PD to collect light more efficiently. Detailed fabrication process is as described in (5). The Ge-PD is simply formed as P-i-N junction. For characterization purpose, the bulk-Si based 3D-OEIC Ge-PD is incorporated with a Y-branch, as shown in Fig. 2(a). Fig. 2(b) shows the zoom-in top view of the PD. Al mirror and electrodes are with same first metal layer. Fig. 2(c) shows the X-SEM for region along the scattering grating. Fig. 2(d) shows the SEM structure of the scattering grating of SiN. Ge-PD I-V characteristics with different optical input power are shown in Fig. 3(a). We can get the responsivity is ~0.2 A/W at 1550 nm. Fig. 3(b) shows the extracted responsivity at different reverse biased voltages and wavelengths. Shorter wavelength is with higher responsivity. Fig. 3(c) shows the dynamic response of PD at reverse bias of 5 V. 3-dB bandwidth is only 2 GHz. However, by means of Trans-Impedance Amplifier (TIA) and higher reverse bias (10 V), we can achieve 20 Gb/s data rate (see inset).

B. Bulk-Si Based 3D-OEIC Label-Free Optical Biochemical Sensor Structure.

In this study, we utilize this new platform for the application of label-free optical biochemical sensor. For cost saving benefit, we replace the SOI by bulk-Si wafer, and thus the typical Si waveguide is replaced by SiN based waveguide which has higher sensitivity as well. The schematics of device structure are shown in Fig. 4(a) and 4(b) for top and cross-section views, respectively. It consists of the basic building elements of coupler, tunable filter, sensor and PD. Coupler is made of 1D grating and a underneath metal mirror. The use of metal mirror is to enhance the light coupling efficiency. Tunable filter is made of an add/drop micro-ring resonator (MRR) with a heater above it. Using SiN waveguide along with thermo-optical effect, heater can modify the oscillation wavelength. Then simply, the sensor is just an exposed add/drop MRR. Lastly, the PD is 3D-OEIC Ge-PD described in Section (A).

C. Bulk-Si Based 3D-OEIC Label-Free Optical Biochemical Sensor Fabrication Process

The fabrication started with 8-inch bulk-silicon wafer. First, the wafer is pre-cleaned, followed with the epitaxial of 1 μm blank Ge deposition. To increase the electrical field in the intrinsic Ge region to increase the PD response speed, we etch away the P- and N- type regions with 400 nm step height. The implantations are performed with 45° tilting angle and 90° rotation angles, to ensure the doping species being well implanted into the waveguide sidewall. Following annealing of 500°C for 30s, 400 nm SiO$_2$ layer is deposited. With the contact holes formation, a metal stack of 25 nm TaN/750 nm AlSiCu/50 nm TaN is deposited and patterned as the first metal layer. Subsequently, another 4.5 μm oxide layer is deposited to clad between the Ge layer and metal layer, followed by chemical-mechanical planarization (CMP) to planarize the surface. SiN of 450nm was then deposited by low-temperature PECVD (400°C), followed by CMP again. SiN~400nm is remained after the planarization. SiN waveguide is then patterned, along with 2 μm SiO$_2$ deposition on top. Then 5 nm Ti/150 nm TiN are deposited as the heating element. After heater patterning, 600 nm SiO$_2$ is deposited and CMP planarization is repeated. After opening via holes to contact to heater and to the first metal layer, 2 μm AlSiCu is deposited as the top metal. With top metal patterning, a thick SiO$_2$ is deposited and planarized with CMP. The following steps are bond pad opening, sensing windows opening and local thermal trench patterning. The subsequent steps are dicing, micro-fluidic packaging to complete the sample preparation before testing.

D. Demonstration of Bulk-Si Based 3D-OEIC Label-Free Biochemical Sensor

As comparisons with Fig. 4(a) and (b), Fig. 4(c)-4(k) describe the top and cross-section of SEM pictures for each building element. The cross-section position is showed by red dished line in top SEM pictures. Fig. 5(a)-5(d) show the fabricated bio-chip, the chip-PCB wire-bonded system with microfluidic packaging, the testing setup and final prototype. Fig. 6 shows the bottom Al mirror reduced ~1-4dB of loss of SiN grating couplers. Fig. 7 shows the spectrum of tunable filter (with tracing ring) and it's FSR in top and bottom, respectively. FSR determines the incident light useful spectrum range. Fig. 8 shows incident light spectrum and tracing ring through port spectrum. The tracing ring regulation ability, sensing ring

sensitivity and system resolution determine total system detection limit. Fig. 9 shows tracing ring regulation ability is ~16.5 nm/W. Adopt the undercut technique, the regulation ability can be improve 20-ford (3,4). Fig. 10 shows sensing ring with sensitivity of 166.5 nm/RIU, which suggests ~2-3× higher than that of using Si-microring (2). Fig. 11 shows the resolution of $R=3\sigma=57.4$ μW. Fig. 12's top figure shows the sensing performance of the proposed 3D-OIEC biochemical sensor by scanning the heater voltage and recorded photo current response. Different powers are applied to thermally tune the microring in order to trace the resonance shift induced due to the solution refractive index change, which is summarized in bottom of Fig. 12. The bulk resolution sensing sensitivity is $S\approx7.9$ W/RIU. Therefore, the detection limit is estimated $DL=R/S\approx7.3$ μRIU. Fig. 13 shows the PD response for different PSS/PAH bi-layer periods. The operation processing is as same as (2). The fitting results show 24.5 mW heater powers for a bi-layer period. Therefore, the surface mass sensitive is $S\approx12.3$ mW/ng·mm^{-2} and detection limit is $DL\approx4.7$ pg·mm^{-2}. This system also can apply into continuously monitoring biochemical reactions. Fig. 14 shows the dynamic behavior of PSS/PAH deposition on the sensing ring surface.

Conclusion

We proposed a 3D monolithic OEIC integration platform on bulk-silicon, and applied for biochemical sensor as a case study. The bulk solution detection limit of ~7.3 μRIU and surface mass detection limit of 4.7 pg·mm^{-2} are achieved. Such bulk-Si based 3D-OEIC platform technology can also apply into on-chip optical interconnection.

Acknowledgments

This work was supported by the Science and Engineering Research Council of A*STAR (Agency for Science, Technology and Research), Singapore. First author is also partially supported by National Natural Science Foundation of China (NSFC, Grant No. 61177090, 61377048).

References

(1) X. Fan, I. M. White, S. I. Shopova, H. Zhu, J. D. Suter and Y. Sun, "Sensitive optical biosensors for unlabeled targets: A review," *Anal. Chim. Aata*, 620(1), 2008, 8-26.

(2) J. Song, X. Luo, X. Tu, M. K. Park, J. S. Kee, H. Zhang, M. Yu, G. Q. Lo and D. L. Kwong, "Electrical tracing-assisted dual-microring label-free optical bio/chemical sensors," *Opt. Express*, 20(4), 4189-4197, 2012

(3) J. Song, X. Luo, J. S. Kee, K. Han, C. Li, M. K. Park, X. Tu, H. Zhang, Q. Fang and L. Jia, "Silicon-based optoelectronic integrated circuit for label-free bio/chemical sensor," *Opt. Express*, 21(15), 2013, 17931-17940.

(4) J. F. Song, X. S. Luo, J. S. Kee, Q. Liu, K. W. Kim, Y. Shin, M. K. Park, K. W. Ang, and G. Q. Lo, "A Novel Optical Multiplexed, Label-Free Bio-Photonic-Sensor Realized on CMOS-Compatible OEIC Platform," in *IEDM Tech. Dig.*, 2013, 381-384.

(5) J. F. Song, X. Luo, X. Tu, L. Jia, Q. Fang, T.-Y. Liow, M. Yu and G. Q. Lo, "Three-dimensional (3d) monolithically integrated photodetector and WDM receiver based on bulk silicon wafer," *Opt. Express*, 22(16), 2014, 19546-19554.

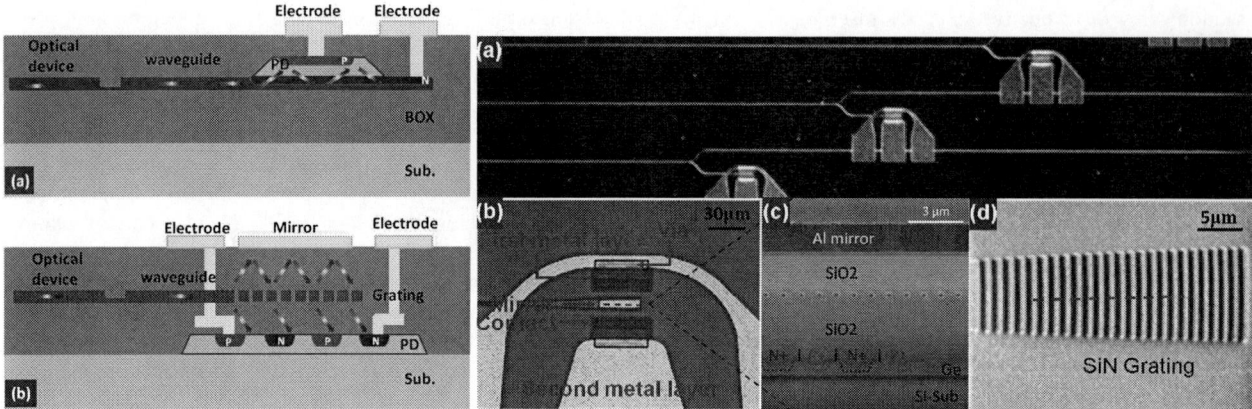

Fig. 1 Schematic illustrations. (a) The SOI-based PD. Optical and electrical devices are in same silicon film. (b) The proposed bulk Si-based 3D PD. Optical devices and electrical devices are separate up and down. A metal mirror is enhanced light collection for PD.

Fig. 2 (a) is optical microscope picture of the Y-branch integrated with a 3D PD. (b) is zoom-in of top view of 3D PD. (c) is the cross-sectional SEM image of the Ge PD region. Bottom is Ge film with interleaved P-i-N junction. The thickness of SiO₂ between SiN grating and Ge is 4 μm. Cladding SiO₂ is ~3μm. A 2μm Al mirror is on the top of SiN scattering grating. (d) is the top view SEM picture of the SiN scattering grating. Grating is with period of 1.1μm.

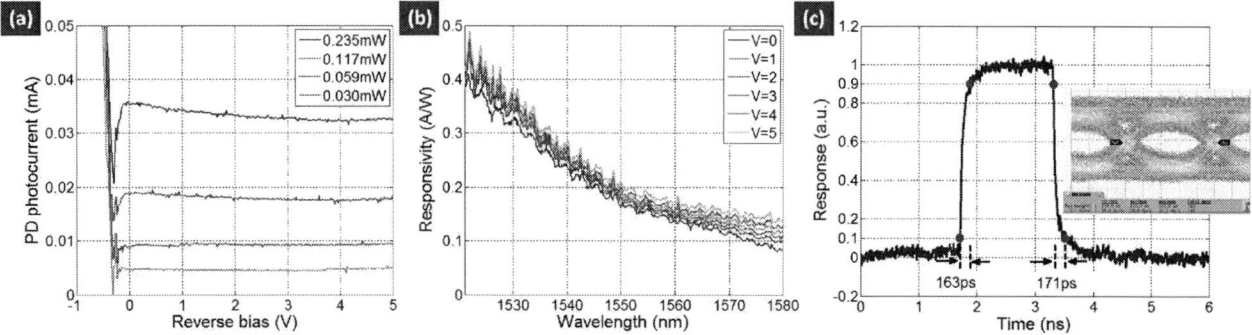

Fig. 3. Characteristics of 3D-PD. (a) The PD photocurrent with different optical input power. Laser wavelength is 1550nm. PD length is 48μm. (b) The PD responsivity vs. wavelength for different reverse bias. (c) PD dynamic response under optical square wave with 5V reverses bias. Rise time and fall time is 163ps and 171ps respectively. A 20Gb/s eye diagram with 10V reverse bias is inserted.

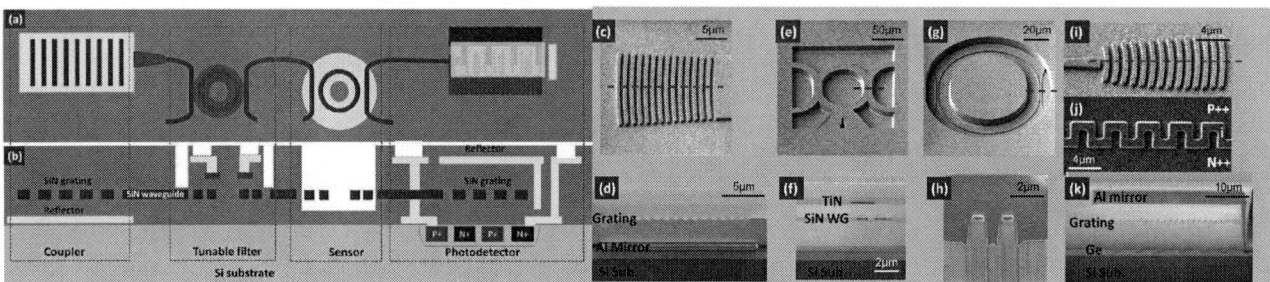

Fig. 4. Schematic illustrate map of 3D bulk-Si OEIC biochemical sensor system. (a) is top view and (b) is cross section view. (c) SEM image of vertical grating coupler. Period is 1.1μm and duty ratio is 50%. (d) Cross section of vertical grating coupler with an Al mirror underneath. The distance of grating and Al mirror is ~ 4μm. (e) Thermal tunable filter (tracing ring) with heat isolation trench. (f) Cross section SEM image of heater (TiN), microring resonator and waveguide coupler. The distance of TiN and SiN waveguide are 2μm. The thickness of TiN is 120nm. Gap of waveguide is 700nm. (g) Sensing microring resonator with opening window. The sensing microring is racetrack type. Radius is 40 μm and the length of straight waveguide coupler is 20μm. (h) Cross section of sensing ring waveguide with coupling waveguide. (i) SiN scattering grating. (j) Ge interleaved P-i-N junction. The width of intrinsic Ge is 1μm. (k) Cross section of photodetector, SiN grating, Ge interleaver P-i-N junction and Al mirror. The distance between grating and Ge film is ~5μm, Al mirror and grating is 2.5μm.

978-1-4799-8002-4/14 $31.00 © 2014 IEEE   396

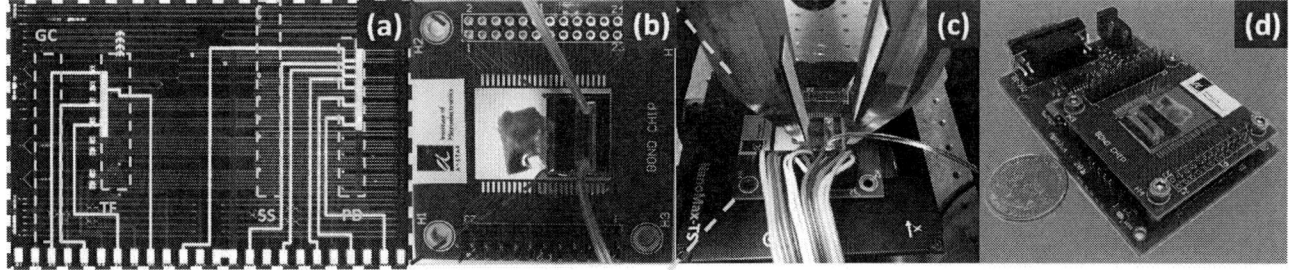

Fig. 5. (a) Microscope description of sensor chip. Yellow dished box denotes the grating coupler (GC), tunable filter (TF), sensor microring (SS) and PD. (b) Microscope of after wire bonding and Microfluidic packaging. (c) Testing setup. (d) Fully packaged Prototype in reference with a quarter coins. Overall size is < 6 cm^2 ×10 cm^2.

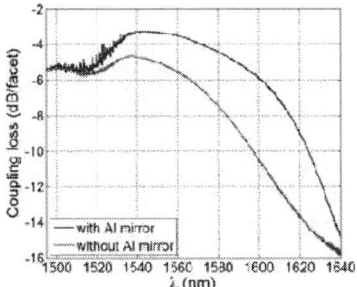

Fig. 6. Vertical grating coupler loss vs. wavelength. The blue color curve is with Al mirror underneath, and red color curve is without Al mirror.

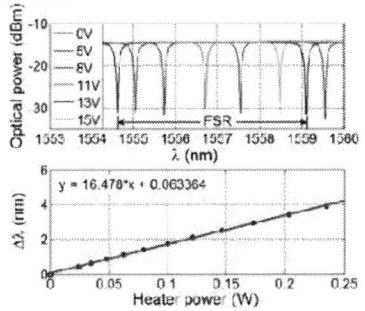

Fig. 9. Top figure shows the spectrum of tunable filter ring (tracing ring) with different heater bias. Bottom shows the resonance wavelength shift with heater power.

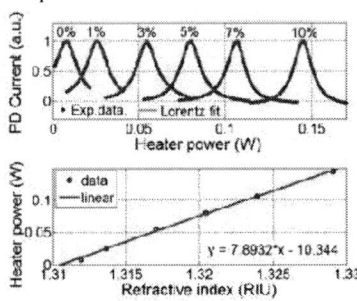

Fig. 12. Top shows PD current in response to heater power for different NaCl concentration. Dots are experimental results. Red color curves are Lorentz fitted. Bottom figure shows the optical reflect index of solution vs. heater power. Blue circles denote peak electrical power, and red color line is a linear fitted curve.

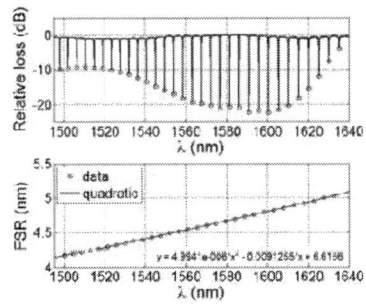

Fig. 7. Top shows spectrum of tracing ring. The transmission loss is balanced by reference waveguide. Bottom figure is FSR with wavelength.

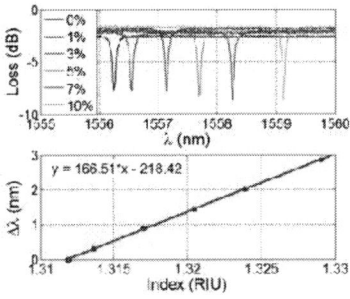

Fig. 10. Top figure shows spectrum of sensing ring with different concentration of NaCl solution. Bottom figure shows the resonance wavelength shift with refractive index of NaCl

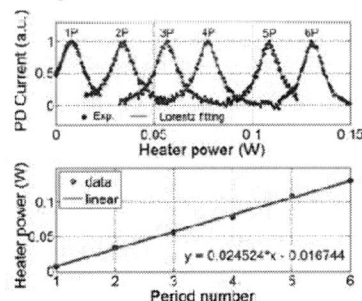

Fig. 13. Top figure shows the PD response for different number of PAH/PSS bi-layer periods. Blue dots are experimental data and red curves are Lorentz fitting results. Bottom figure is linear fitting results.

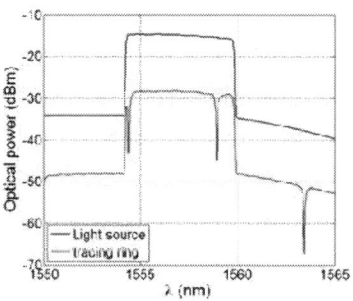

Fig. 8. According to FSR of Fig. 7. The input light spectrum width is set as ~5.1 nm that showed by blue curve. As comparison, tracing ring spectrum is shown as red color curve.

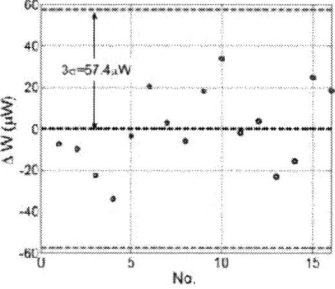

Fig. 11. Fitted electrical power centre shift of each measurement. σ is standard deviation and 3σ is resolution.

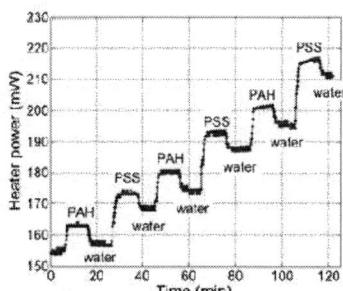

Fig. 14. Dynamic response of PAH and PSS deposition process. A single data point takes ~9 sec. Every step duration is ~10 minutes.

978-1-4799-8002-4/14 $31.00 © 2014 IEEE   397

An Integrated Tunable Laser using Nano-Silicon-Photonic Circuits

M. Ren[1,2], H. Cai[2], J. F. Tao[2], Y. D. Gu[2], K. Radhakrishnan[1], Z. C. Yang[3], D. L. Kwong[2] and A. Q. Liu[1,†]

[1]School of Electrical and Electronic Engineering, Nanyang Technological University, Singapore 639798
[2]Institute of Microelectronics, A*STAR (Agency for Science, Technology and Research), Singapore 117685
[3]Institute of Microelectronics, Peking University, Beijing 100871, China
([†]eaqliu@ntu.edu.sg)

Abstract

This paper reports an integrated tunable laser using nano-silicon-photonic circuits. In particular, all the necessary optical functions in an external cavity laser system, including the beam transmission, coupling, filtering and reflection, are realized using waveguide-based circuits. The high light-confinement capability of the nano-silicon waveguides avoids high optical loss in free-space as suffered in the conventional Microelectromechanical Systems (MEMS) tunable lasers, and thus guarantees superior performance. The proposed laser demonstrates large tuning range (45 nm), pure single-mode properties (45 dB side-mode-suppression ratio (SMSR)), and relatively high output power (~1 mW).

Introduction

Significant efforts have been made towards the miniaturization of the integrated wavelength tunable lasers. MEMS technology, which integrates mechanical structures into a single silicon chip using deep reactive ion etching (RIE) processes, is one of the most promising approaches (1-2). Most of the MEMS tunable lasers utilize mirror structures as the wavelength reflector in the free-space external cavity (1-3). To achieve higher coupling efficiency, the external cavity is required to be no more than 10-μm considering the poor focusing property of the mirrors (4). However, it is desirable to have longer cavity length while maintaining high coupling efficiency, in order to obtain a large wavelength tuning range. Although several attempts using 3D or 2D microlens have demonstrated for enhancing light confinement (4-5), the MEMS tunable lasers intrinsically suffer from either low output power or limited tuning range due to the large optical loss resulted from the beam divergence in the free-space optical MEMS.

The nano-silicon photonic integrated circuits are promising for external cavity tunable lasers (6-7), since the low optical loss is expected, thanks to the high light confinement capability of the micro/nano size waveguides. This property provides the possibility for broadband wavelength tuning with relative high output power and narrow linewidth. The photonic circuits are synergism with external cavity tunable laser. For example, various functional photonic devices, such as micro- ring resonator, Bragg grating, Mach-Zehnder interferometer, single/multimode mode coupler and photonic loop, have been developed (6-7). These demonstrated devices provide all the required functionalities in a tunable laser system, like optical transmission, filtering, coupling, splitting/combining and reflection. Additionally, the effective index of the photonic waveguide can be flexibly tuned through thermal-optic, electrical-optic or mechanical-optical schemes, which offers great potential for the laser's wavelength tuning. The photonic integrated tunable laser using InGaAsP material have been demonstrated (8-9), but the III-V material is of relatively high waveguide loss (10 dB/cm) and high cost. Silicon material is potentially a prior choice benefited from the relatively low transmission loss (1.5 dB/cm or even lower) of the nano-silicon waveguide.

In this paper, an integrated tunable laser using nano-silicon-photonic integrated circuits is presented with theoretical analysis, fabrication, packaging and experiments.

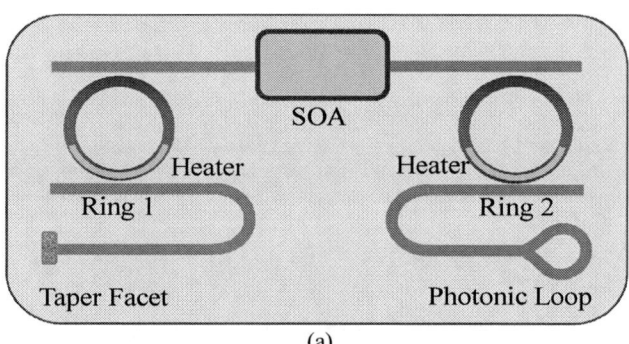

(a)

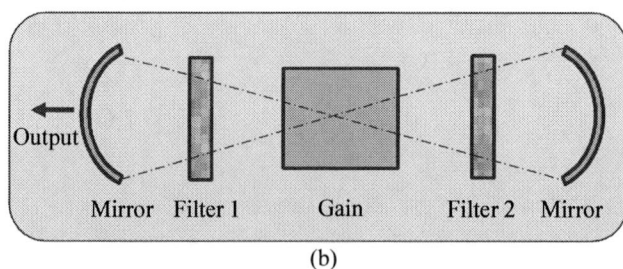

(b)

Figure 1: Illustration of: (a) the schematic of tunable laser using nano-silicon photonic circuits. The SOA provides gain, the silicon waveguides forms the external cavity, and two ring resonators function as wavelength filters. (b) The equivalent cavity model of the external cavity tunable laser.

978-1-4799-8002-4/14 $31.00 © 2014 IEEE

Design and Theoretical Analysis

The schematic of the proposed tunable laser is illustrated in Fig. 1(a). It integrates a semiconductor optical amplifier (SOA) chip, a photonic mirror loop, a taper waveguide and two separated ring resonators onto a silicon-on-isolator (SOI) platform. The two facets of the SOA chip are both anti-reflection coated and have a reflectivity of < 0.1%, in order to avoid the facet reflection. The external cavity is formed with nano-silicon photonic circuits. The silicon waveguide has a cross-dimension of 450 nm × 220 nm, which supports the single-mode transmission with TE polarization. The photonic loop works as a fully-reflective mirror, and the facet of the taper waveguide works as a partially-reflective mirror, and both mirrors are wavelength-independent. The two ring resonators, with their radii slightly different, play as wavelength selective elements. The wavelength tuning is achieved through electrically controlling the thermal-voltage applied on the micro-heaters over the ring resonators. The final output wavelength of the laser is collected with an optical fiber from the output taper. The equivalent cavity model of the integrated tunable laser is shown in Fig. 1(b).

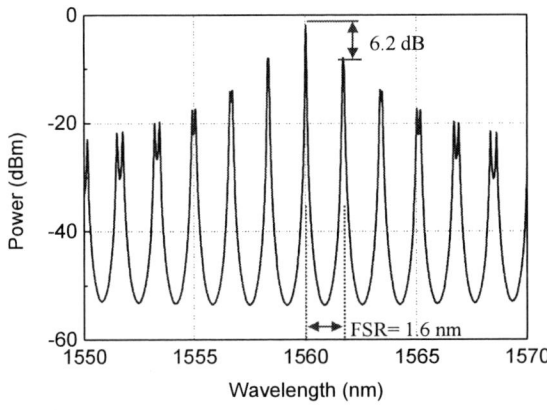

Figure 2: Transmission spectrum of the wavelength filter function results from the Vernier effect of the two ring resonators with slight radii difference. The transmission power difference between the main mode and the side mode is 6.2 dB.

The wavelength selection principle is implemented through two ring resonator filters, as demonstrated in Fig. 2. The radii of the two ring resonators are 62 and 60 μm, respectively, and the corresponding free spectrum range of the resonance is 1.55 nm and 1.6 nm, respectively. Due to the Vernier effect, the single-frequency lasing is obtained within the gain spectrum, when the resonant wavelength of the two filters align to each other, as shown in Fig. 2. As indicated in the transmission spectrum of the two ring resonators, the transmission power difference between the main mode and the side mode is 6.2 dB. This large transmission power difference ensures the over 40-dB SMSR on the lasing spectrum.

Fabrication Processes and Packaging

(a)

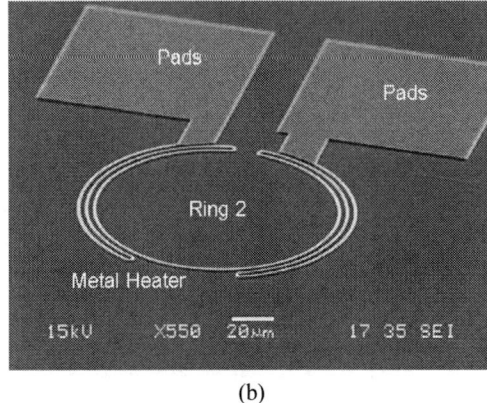

(b)

Figure 3: The SEM images of the fabricated tunable laser on the SOI-wafer: (a) the integrated tunable laser with the flip-chip bonded SOA on the SOI platform; (b) Ring 2 with oxide cladding and micro-heaters.

The tunable laser is achieved using both silicon photonics and MEMS processes. The main fabrication processes include photo-lithography, reactive ion etching, thin-film deposition, metal evaporation and lift off. The SEM images of the fabrication results are shown in Fig. 3. The SOA chip is bonded on the SOI platform with flip-chip bonding technology. Fig. 3(a) presents the fabricated tunable laser with the flip-chip bonded SOA chip. The silicon waveguides with silicon oxide cladding are highlighted. The coupling gap between the III-V and the silicon waveguides is within 5 μm. Fig. 3(b) presents the micro-heater, isolated from the Si ring waveguide by the 1.5-μm thick silicon dioxide cladding. The metal lines have a width of 1.8 μm and the line-space is 2 μm.

The tunable laser chip is packaged using the low-cost passive alignment approach. After the SOA chip is soldered using the flip-chip bonder, it is assembled with the heat sink and the substrate holder. The semi-assembly is loaded on a butterfly box. Afterwards, the SOA chip and metal electrodes on the SOI chip are electrically connected using golden wire

978-1-4799-8002-4/14 $31.00 © 2014 IEEE   399

bonding. Then the optical fiber is aligned and fixed using precision positioning machine. Finally, the wire-bonding is processed for the connection between metal electrodes and the cover attachment. The photo image for the packaged laser is shown in Fig. 4.

Figure 4: Photograph of butterfly-packaged tunable laser system with wire bonding, heat sink, substrate, and output fiber.

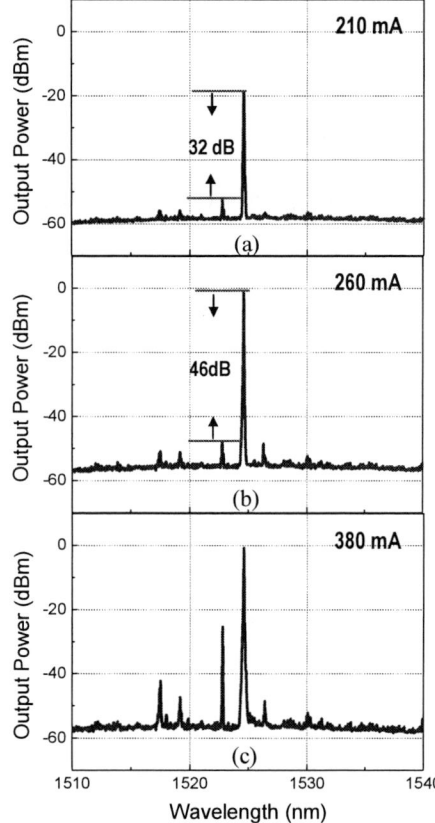

Figure 5: Experimental results of the output spectra when the SOA is operated at different current levels: (a) 240 mA, the laser is stimulated without saturation; (b) 300 mA, the laser works at the saturation status with 46 dB SMSR; (c) 380 mA, the laser works at the saturation status and multimode wavelengths.

Experimental Results and Discussions

Fig. 5 shows the experimental results of the output spectra of the laser, when the SOA is operated at different current levels. For example, the laser is stimulated at 210 mA, with the single wavelength at 1524.6 nm. It presents comparatively weak optical power (-20 dBm) and high SMSR of 32 dB. At 260 mA, the laser is stimulated and works at single mode status, ~ 0 dBm output power and 46 dB SMSR are obtained. At 380 mA, the laser works at its saturation status and the multimode lasing are observed. Although the output power at the main mode is above 0 dBm, the SMSR performance is decreased to 23 dB and the spectrum purity is degraded.

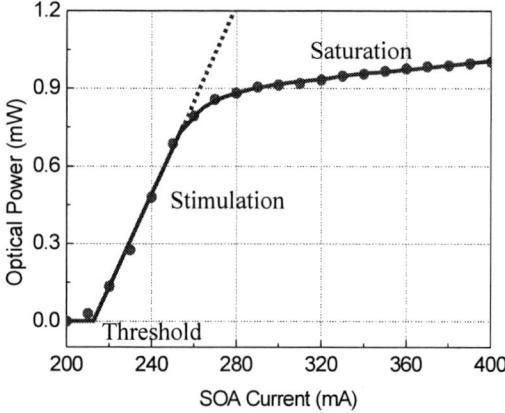

Figure 6: Comparison of the output power versus SOA current curves for the integrated tunable laser.

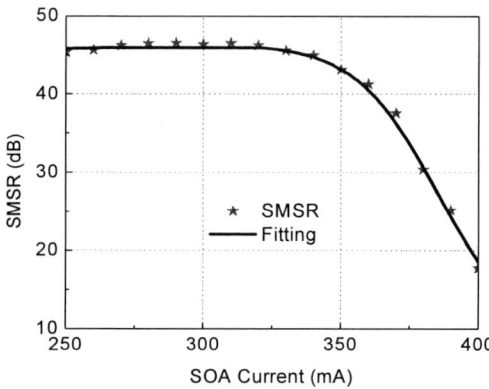

Figure 7: Output power and SMSR verse SOA current. Accordingly, the working status of the CCW light is characterized into 4 periods.

The lasing wavelength is kept at 1524.6 nm when tuning the driving current of the SOA is at the level of 200 ~ 400 mA. The curve of the output power versus current is shown in Fig. 6. The laser performance has three stages. When the SOA current is below 208 mA, the laser is below threshold. When the SOA current is increased from 209 mA to 260 mA, the laser is linearly stimulated with a slop efficiency of 21.7 W/A, and the output power increases from 0 to 0.87 mW. When the SOA current is above 260 mA, the laser works at the saturation stage, and the maximal output power is above 1 mW, as shown in Fig. 7. The SMSR is kept at above 45 dB with the SOA current varying over the range of 250 ~ 320 mA. The SMSR is degraded obviously when the SOA current is above 320 mA due to the side-mode lasing. Consequently, the optimized SOA operation range is 260 ~ 320 mA, considering both performances of the output power and the SMSR.

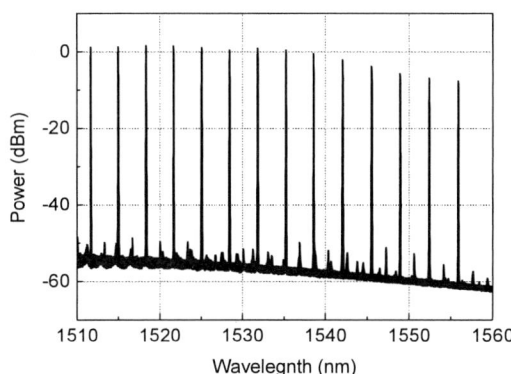

Figure 9: Experiential results of single-mode wavelength tuning for the integrated tunable laser using nano-silicon photonic circuits.

Summary

In this work, an external cavity tunable laser using nano-silicon photonic integrated circuits has been demonstrated. The experimental results have demonstrated large wavelength tuning range (~ 45.1 nm), excellent spectrum purity (SMSR > 45 dB) and relatively high optical power (~ 1mW). The specifications of the integrated tunable laser are of high competition, which enables its potential applications in the field of optical communication and remote sensing. Meanwhile, it opens up the possibility for large-scale integrated photonic circuits for the optical computing.

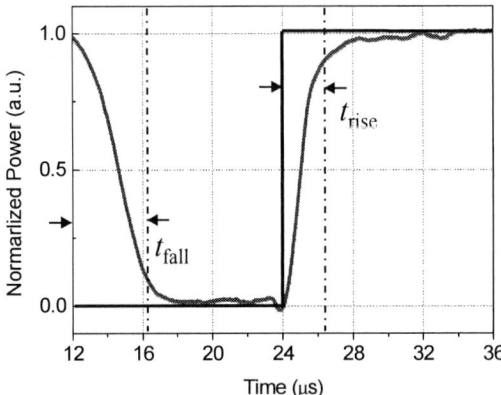

Figure 8: The switching speed of the thermal-optic wavelength tuning. The rise and fall time is 2.4 μs and 4.2 μs, respectively.

The lasing wavelength is tuned through the thermal-optic effect of silicon material, achieved by applying the thermal-voltage on the micro-heater of Ring 2. The tuning speed of thermal-optic wavelength tuning is measured, as incidated in Fig. 8. The rising time and the falling time of the optical single response are 2.4 μs and 4.2 μs, respectively. The falling time is longer than the rising one since the thermal dissipation of the nano-waveguide takes longer time as compared to the process of heating up waveguide.

Fig. 9 shows series optical output spectra of single wavelength output, when the SOA current is set to 250 mA. By increasing the thermal voltage applied to Ring 2 from 0.74 to 1.55 V, the lasing wavelength of the integrated laser is tuned from 1511.8 to 1556.9 nm, corresponding to the maximal achievable tuning range of 45.1 nm. Benefited from the strong light confinement capability of the nano-silicon waveguides, the SMSR of above 45 dB and the output power of 1 mW are achieved within the above obtained wavelength tuning range.

References

(1) A. Q. Liu, X. M. Zhang and V. M. Murukeshan, "A Novel device level micromachined tunable laser using polysilicon 3D mirror," *IEEE Photon. Tech. Lett.*, vol. 13, pp. 427-429, May 2001.

(2) J. D. Berger and D. Anthon, "Tunable MEMS devices for optical networks," *Opt. Photon. News*, vol. 14, pp. 42-49, 2003.

(3) X. M. Zhang, A. Q. Liu, C. Lu and D. Y. Tang, "Continuous wavelength tuning in micromachined Littrow external-cavity lasers," *IEEE J. Quantum Electron.* vol. 41, pp. 187-197, February 2005.

(4) A. Q. Liu, X. M. Zhang, "A review of MEMS external-cavity tunable lasers," *J. Micromech. Microeng*, vol. 17, pp. R1-R13, January 2007.

(5) H. Cai, J. F. Tao, Y. D. Gu, D. L. Kwong and A. Q. Liu, "Demonstration of a single-chip integrated MEMS tunable laser with a large wavelength tuning range," *IEEE International Electron Devices Meeting (IEDM)*, pp. 18.6.1-18.6.4, December 2013.

(6) T. L. Koch and U. Koren, "Semiconductor photonic integrated circuits ," *IEEE J. Quant. Electron.*, vol. 27, pp. 641-653, Mar 1991.

(7) J. Ahn, M. Fiorentino, R. G. Beausoleil, et.al, "Devices and architectures for photonic chip-scale integration," *Appl. Phys. Lett.*, vol. 95, pp. 9891-997, Jun, 2009.

(8) B. Liu, A. Shakouri and J. E. Bowers, "Passive mircoring-resonator-coupled lasers," *Appl. Phys. Lett.*, vol. 79, pp. 3561-3563, Nov, 2001.

(9) S. Matsuo and Toru Segawa, "Mircoring-resonator-based widely tunable lasers," *IEEE J. Sel. Top. Quant. Electron.*, vol. 15, pp. 545-554, 2009.

First Demonstration of High-Ge-Content Strained-$Si_{1-x}Ge_x$ (x=0.5) on Insulator PMOS FinFETs with High Hole Mobility and Aggressively Scaled Fin Dimensions and Gate Lengths for High-Performance Applications

Pouya Hashemi, Karthik Balakrishnan, Sebastian U. Engelmann, John A. Ott, Ali Khakifirooz, Ashish Baraskar*, Marinus Hopstaken, Joseph S. Newbury, Kevin K. Chan, Effendi Leobandung, Renee T. Mo and Dae-Gyu Park

IBM Research, GLOBALFOUNDRIES*, T. J. Watson Research Center, Yorktown Heights, NY 10598, USA; phone: 914-945-1958, email: hashemi@us.ibm.com

Abstract

For the first time, we report fabrication and characterization of high-performance s-$Si_{1-x}Ge_x$-OI (x~0.5) pMOS FinFETs with aggressively scaled dimensions. We demonstrate realization of s-SiGe fins with W_{FIN} =3.3nm and devices with L_G=16nm, in a CMOS compatible process. Using a Si-cap-free passivation, we report SS=68mV/dec and μ_{eff}=390±12 cm^2/Vs at N_{inv}=$10^{13}cm^{-2}$, outperforming the state-of-the-art relaxed Ge FinFETs. We also report the highest performance reported to date among sub-20nm-L_G pMOS FinFETs at V_{DD}=0.5V. In addition, hole transport as well as electrostatics, performance and leakage characteristics of SGOI FinFETs for various dimensions are comprehensively studied in this work.

Introduction

Strained $Si_{1-x}Ge_x$ FinFETs are promising pFET channel candidates for 10nm technology node and beyond, due to their excellent electrostatics and built-in uniaxial compression (1-12). While there are some outstanding work on FinFET devices with moderate Ge content (x=0.2-0.35) (2-5, 8, 12), very few data exists on the FinFETs with high-Ge content with x≥0.5, among them mostly focused on relaxed or strained pure Ge (6,7, 9-11). For Ge content around 50%, planar QW devices have shown to offer superior mobility benefit over Si, and decent sub-threshold slope (SS) using a Si cap surface passivation (13-17), and significant performance boost for narrow-width devices (16). On the other hand, poor SS have been reported for non-planar SiGe fins with x=0.48 (0.5V/dec) (1) and SiGe-cladded fins with x=0.45 (80mV/dec) (17), using a Si cap-free process. In addition, there is no outstanding report for transport in high-Ge content $Si_{1-x}Ge_x$ with x~0.5 at relevant fin and gate dimensions, suitable for future CMOS generations. In this work, for the first time, we demonstrate the fabrication and characterization of s-$Si_{1-x}Ge_x$-OI (x~0.5) FinFETs with aggressively scaled gate length (L_G=16nm) and fin dimensions (sub-10nm widths) using an improved Si-cap free HK/MG stack and ion-implant-free RS/D processes. Hole transport as well as short channel (SC) electrostatics, performance and leakage characteristics for various dimensions are comprehensively studied, for the first time.

Process Integration

Starting s-$Si_{1-x}Ge_x$-on-Insulator (SGOI, x~0.5) wafers were fabricated by a two-step Ge condensation process (**Fig. 1**). Compared to bulk SiGe fins which suffer from Ge out-diffusion into the substrate, SOI SiGe fins have better immunity to high-temperature (HT) processes and have potential of co-integration with HT/high-bandgap materials for other applications such as nFET or I/O devices (**Fig. 2**). SIMS analysis (**Fig. 3**) demonstrates very uniform Ge profile process with 50% Ge content and low background doping levels. The Ge content was confirmed by XRD, and ~1.7% biaxial compression was extracted (**Fig. 4**). **Fig. 5** shows the process flow developed to fabricate 50% s-SiGe pMOS FinFETs featuring gate first HK/MG, and implant-free RS/D processes. Fins were patterned using direct lithography and RIE, and no intentional trim was applied prior to the gate stack formation. Compared to the process reported in (12), a modified interface passivation and high-selectivity spacer RIE processes were developed. In addition, a relatively low-temperature RS/D epitaxy (Ge content ~0.4) was utilized to avoid fin agglomeration. Due to the retarded diffusion of the boron in high-Ge content SiGe, relatively HT RTA (1000°C) was used for S/D activation/drive-in purpose. **Fig. 6** shows an XTEM image of an s-SiGe (x~0.5) pMOS FinFET with physical L_G~16nm. **Fig. 7** illustrates the fin profile for various fin widths (W_{FIN}). We demonstrate, for the first time, the successful realization of sub-16nm-wide 50% SiGe fins with vertical sidewalls and W_{FIN} as small as 3.3nm (aspect ratio of 5.4). Fin height, H_{FIN}, is ~18nm for all different W_{FIN}s used in this study. Raman measurements (**Fig. 8**) show x~0.51 with ~1.8±0.1 % (~2.7GPa) biaxial/uniaxial compression for blanket film and fins, respectively.

Device Characterization

Split C-V measurements on array of fins (**Fig. 9**) reveal hysteresis-free, stable and excellent frequency response, indicating excellent dielectric passivation for 50% SiGe using an improved Si-cap-free passivation process (B). Compared to a standard Si-cap-free process (A) where V_{th} shift is beyond what expected by SiGe/HK/MG WF, the

978-1-4799-8002-4/14 $31.00 © 2014 IEEE

improved process B negatively shifts V_{th} due to the better interface trap passivation (**Fig. 9 (c)**). **Fig. 10** shows the extracted CET_{inv} for a wide range of device/fin widths. $CET_{inv} \sim 1.55nm$ and less than 1.4nm was extracted for planar devices and target fin widths, at $N_{inv} = 10^{13}$ cm^{-2}. Long channel (LC) transfer characteristics (**Fig. 11**) represents excellent SS=68mV/dec and 6 order of magnitude on-to-off ratio, using an improved process B. **Fig. 12-13** show as-extracted split-CV effective hole mobility of x~0.5 SGOI planar and FinFETs with various widths down to $W_{FIN} \sim 10nm$. Hole mobility monotonically increases (by 2.2X) with decreasing W_{FIN}, due to transformation of strain from biaxial to uniaxial, as shown in **Fig. 14**. This proves the strain is maintained in the SGOI fins, even with very high fabrication thermal budget (1000°C). As a result, impressive $\mu_{eff} = 390 \pm 12 cm^2/Vs$ at $N_{inv} = 10^{13}$ cm^{-2} is measured for the SGOI x~0.5 fins, which is 5X/1.6X over (100)/(110) Si hole universal and ~1.8X over relaxed Ge FinFET (10). **Fig. 15** shows the impact of the improved passivation (process B) to enhance mobility by 27% at $N_{inv} = 10^{13}$ cm^{-2}. Overlaying normalized g_D and integrated charge from split C-V reveals the presence of immobile charge (Q_{im}), leading to an under-estimation of mobility (**Fig. 16(a)**). After correction for Q_{im}, peak mobilities as high as 500cm^2/Vs is achieved for s-SiGe (x~0.5) FinFETs (**Fig. 16 (b)**). Compared to SGOI fins with x~0.3 (12), ~1.3X g_m benefit is observed at $L_G \sim 90nm$ (**Fig. 17**). **Fig. 18** shows transfer and output characteristics of a FinFET, with $L_G = 25nm$, representing $SS_{lin}/SS_{sat} = 73/95 mV/dec$ and impressive $I_{on} = 1.27 mA/\mu m$ at target HP $I_{off} = 100 nA/\mu m$ at $V_{DD} = 1V$ (after V_{th} shift by 0.295V). The GIDL-induced degraded SS_{sat} near 100nA/μm, degraded the performance at $V_{DD} = 1V$ at the target I_{off}. $V_{th, lin}$ vs. L_G for SGOI FinFETs (x~0.5) and those reported in (8) for (x=0, 0.27) are compared in **Fig. 19**. $V_{th, lin}$ shifts over (8) are consistent with SiGe band structure, considering similar MG WF. Moreover, the passivation-related V_{th} shift (B vs. A) is consistent with the trend observed in C-V. **Fig. 20** shows average DIBL (measured at $V_{DD} = 1V$) vs. L_G for various W_{FIN}, indicating excellent electrostatics integrity for devices of this work. **Fig. 21** shows (a) typical drain leakage characteristics of a SC FinFET (x~0.5, $L_G \sim 16nm$) for various V_{DS} between -0.5V and -1.0V and (b) plot of GIDL-dominated $I_{D, min}$ vs. V_{DD}. Considering the corner data, one can find out that even low V_{DD} values are a concern for low-power applications. **Fig. 22** shows the peak intrinsic g_m vs. L_G at $V_{DD} = 0.5V$, demonstrating devices with peak $g_{m, int}$ as high as ~2.6mS/μm. Finally, **Fig. 23** and **Fig. 24** show the transfer, output and g_m characteristics of the most-aggressively scaled high-Ge content (x~0.5) s-SiGe FinFET reported to date, with $L_G \sim 16nm$, SS_{lin}/SS_{sat} =79/85mV/dec, and peak $g_m/g_{m, int}$ =1.9/2.55 mS/μm at $V_{DD} = 0.5V$.

Summary and Conclusions

In summary, for the first time, we have demonstrated high-Ge-content SGOI (x~0.5) pMOS FinFETs, with aggressively scaled fins and gate dimensions (down to $L_G = 16nm$) using an improved Si-cap-free HK/MG and implant-free RS/D processes. LC devices represent impressive SS=68mV/dec and $\mu_{eff} = 390 \pm 12 cm^2/Vs$ at $N_{inv} = 10^{13}$ cm^{-2}, which is ~1.8X over state-of-the-art relaxed Ge FinFETs (10) and ~1.5X over s-SiGe FinFETs (x~0.3) (12). We also demonstrate SC devices with excellent electrostatics and HP, with $I_{on} = 1.27 mA/\mu m$ at equivalent HP I_{off} at $V_{DD} = 1V$. Our data shows that GIDL-limited $I_{D, min}$ at low V_{DD} is a concern for scaled s-SiGe FinFETs for LP applications. Finally, as shown in **Table 1**, we report the highest $I_{on} = 0.42 mA/\mu m$ reported to date among pMOS FinFETs with sub-20nm L_G, at the target HP $I_{off} = 100 nA/\mu m$ and $V_{DD} = 0.5V$.

Ref.	Device Structure	L_G	SS_{sat}	Drain-side I_{off} (V_{GS} shift by)	I_{on} at equivalent $I_{off} = 100 nA/\mu m$
(10)	bulk relaxed-Ge FinFET	20 nm	109 mV/dec	~700nA/μm (+0.3V)	not applicable*
(12)	s-Si$_{1-x}$Ge$_x$OI FinFET (x~0.25)	15 nm	82 mV/dec	100nA/μm (-0.03V)	0.38 mA/μm*
This Work	**s-Si$_{1-x}$Ge$_x$OI FinFET (x~0.5)**	**16 nm**	**85 mV/dec**	**100nA/μm (+0.325V)**	**0.42 mA/μm**

Table 1. Comparison of this work with non-Si channel pMOS FinFETs in the literature with $L_G \leq 20nm$ at $V_{DD} = 0.5V$. We report the highest $I_{on} = 0.42 mA/\mu m$ reported to date among pMOS FinFETs with sub-20nm L_G, at the target HP $I_{off} = 100 nA/\mu m$ and $V_{DD} = 0.5V$. *We acknowledge Ref. (10) and (12) reported devices with higher performance for devices with $L_G > 20nm$, due to the other factors such as better electrostatics.

Acknowledgements

This work was performed by the research alliance teams at various IBM Research and Development facilities. We thank T. Hook, M.-H. Na, T. Ando, G.G. Shahidi and T.C. Chen for stimulating discussions and management support.

References:

(1) T. Tezuka et al., *IEDM Tech. Dig.*, p. 887, 2007.
(2) T. Irisawa et al., *IEDM Tech. Dig.*, p. 709, 2005.
(3) C.E. Smith et al., *IEDM Tech. Dig.*, p. 309, 2009.
(4) I. Ok et al., *IEDM Tech. Dig.*, p. 777, 2010.
(5) L. Hutin et al., *VLSI Symp. Tech. Dig.*, p. 37, 2010.
(6) K. Ikeda et al., *VLSI Symp. Tech. Dig.*, p. 165, 2012.
(7) M.J.H. van Dal et al., *IEDM Tech. Dig.*, p. 521, 2012.
(8) P. Hashemi et al., *VLSI Symp. Tech. Dig.*, p. 18, Symp. 2013.
(9) K. Ikeda et al., Symp. *VLSI Symp. Tech. Dig.*, p. 30, 2013.
(10) B. Duriez et al., *IEDM Tech. Dig.*, p. 522, 2013.
(11) J. Mitard et al., *VLSI Symp. Tech. Dig.*, p. 138, 2014.
(12) P. Hashemi et al., *VLSI Symp. Tech. Dig.*, p. 18, 2014.
(13) I. Åberg, C. Ni Chleirigh, O.O. Olubuyide, X. Duan, and J.L. Hoyt, *IEDM Tech. Dig.*, p. 173, 2004.
(14) L. Gomez, P. Hashemi, and J.L. Hoyt, *IEEE Trans. Electron Dev.*, vol. 56, no. 11, p. 2644, 2009.
(15) L. Witters et al., *IEDM Tech. Dig.*, p. 654, 2011.
(16) S. Yamaguchi et al., *IEDM Tech. Dig.*, p. 829, 2011.
(17) H. Martens et al., *VLSI Symp. Tech. Dig.*, p. 58, 2014.

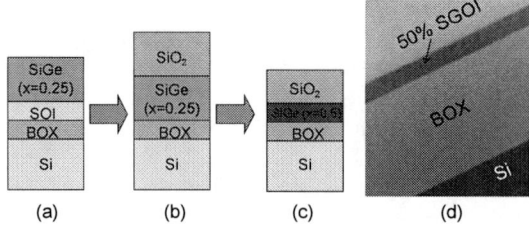

(a)   (b)   (c)   (d)

Fig. 1 Process flow schematic to fabricate strained $Si_{1-x}Ge_x$ on insulator substrates with x~0.5, after (a) s-$Si_{1-x}Ge_x$ (x~0.25) epitaxy on SOI, (b) the first Ge condensation at a temperature of T_1 and (c) the second Ge condensation at $T_2<T_1$ and (d) corresponding cross-section TEM after oxide removal.

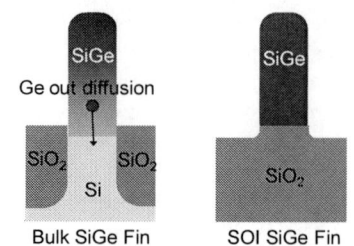

Bulk SiGe Fin   SOI SiGe Fin

Fig. 2 Comparison of bulk SiGe fin and SOI SiGe fin, in terms of immunity to high-temperature process. While bulk SiGe fin suffers from Ge out-diffusion into the substrate, SOI SiGe fin has better immunity to high-temperature processes, and has potential of co-integration with high-temperature materials for other applications, such as higher bandgap Si for nFET or I/O devices, etc.

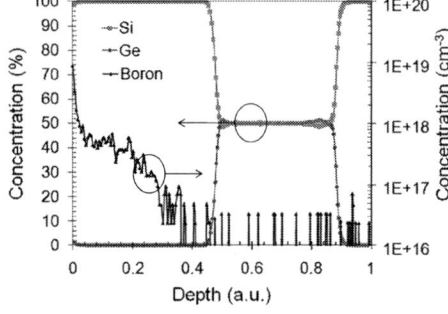

Fig. 3 Typical SIMS profile of SGOI substrates, after a two-step condensation process, showing very uniform Ge profile with near 50% Ge content and very low boron background doping level.

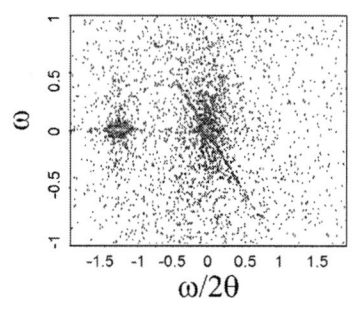

Fig. 4 Typical (004) XRD reciprocal space maps of the fabricated SGOI wafers. Based on the data for (004) and (113) scans at φ=±90°, Ge concentration of ~51% with ~1.7% biaxial compression was extracted.

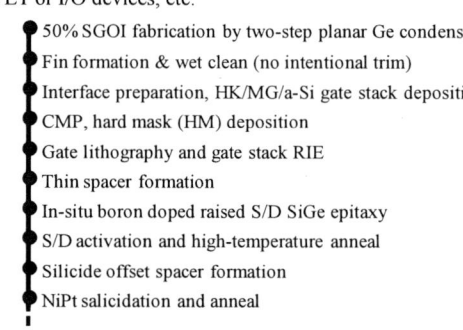

- 50% SGOI fabrication by two-step planar Ge condensation
- Fin formation & wet clean (no intentional trim)
- Interface preparation, HK/MG/a-Si gate stack deposition
- CMP, hard mask (HM) deposition
- Gate lithography and gate stack RIE
- Thin spacer formation
- In-situ boron doped raised S/D SiGe epitaxy
- S/D activation and high-temperature anneal
- Silicide offset spacer formation
- NiPt salicidation and anneal

Fig. 5 Process flow to fabricate s-SiGe (x~0.5) pMOS FinFETs featuring gate first HK/MG, ion-implant-free, and raised S/D processes. This process is immune to high temperature RTA as high as 1000°C.

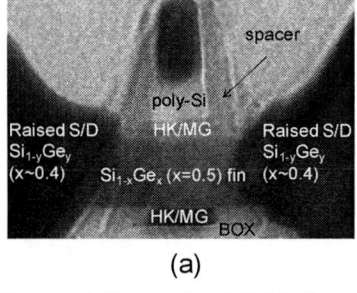

(a)

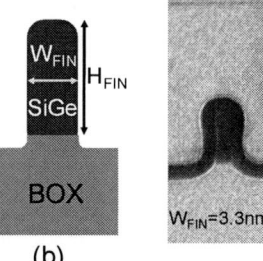

(b)

Fig. 6 (a) Cross-sectional TEM of a s-SiGe pMOS FinFET with physical L_G~16nm. The fin width, W_{FIN}, and fin height, H_{FIN}, are schematically defined in (b).

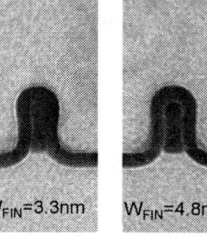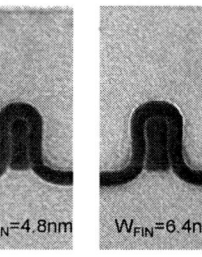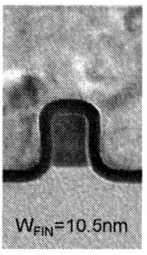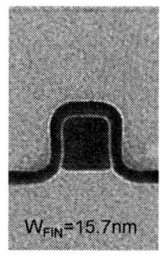

W_{FIN}=3.3nm W_{FIN}=4.8nm W_{FIN}=6.4nm W_{FIN}=10.5nm W_{FIN}=15.7nm

Fig. 7 Cross-sectional TEM images of the 50% SGOI Fins, perpendicular to the fin direction. Fins are defined by direct lithography and RIE and no intentional trim has been performed prior to gate stack formation. We demonstrate realization of sub 16nm fin widths with W_{FIN} *as small as 3.3nm* with aspect ratio of 5.4. H_{FIN} is about 18nm for all different widths.

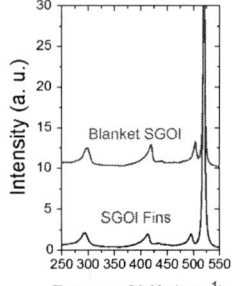

Fig. 8 Micro Raman spectra of SGOI substrate (blanket) and array of SiGe fins, with W_{FIN}~10nm, using a 532nm laser source. A Ge fraction of ~51% with ~1.8% (~2.7GPa) biaxial and uniaxial compression was extracted, for the blanket SGOI and fins, respectively.

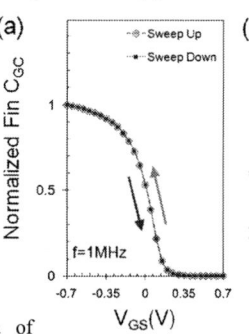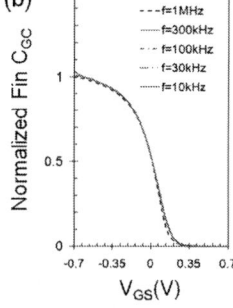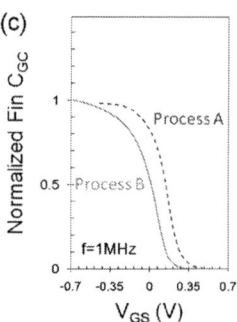

Fig. 9 (a) Hysteresis and (b) frequency dispersion characteristics of the s-$Si_{1-x}Ge_x$ (x~0.5) fins, from split C-V measurements of fins with L_G=8.5μm and W_{FIN}~10nm, using an improved passivation (process B). Hysteresis-free, stable and excellent frequency response, indicate excellent dielectric passivation for s-SiGe (x~0.5) using an improved Si-cap-free passivation (process B). (c) Compared to process A, improved passivation (process B) negatively shifts V_{th} due to the better interface trap passivation.

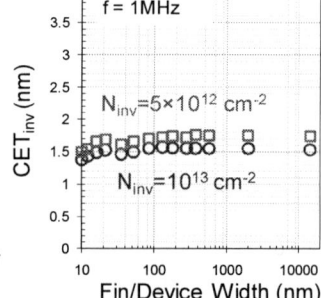

Fig. 10 Inversion Capacitance Equivalent Thickness (CET_{inv}) defined at N_{inv}=5×10^12 and 10^13 cm^-2 from split C-V measurements, for a wide range of s-SiGe active widths, including planar and narrow devices, down to ~10nm-wide fins.

978-1-4799-8002-4/14 $31.00 © 2014 IEEE   404

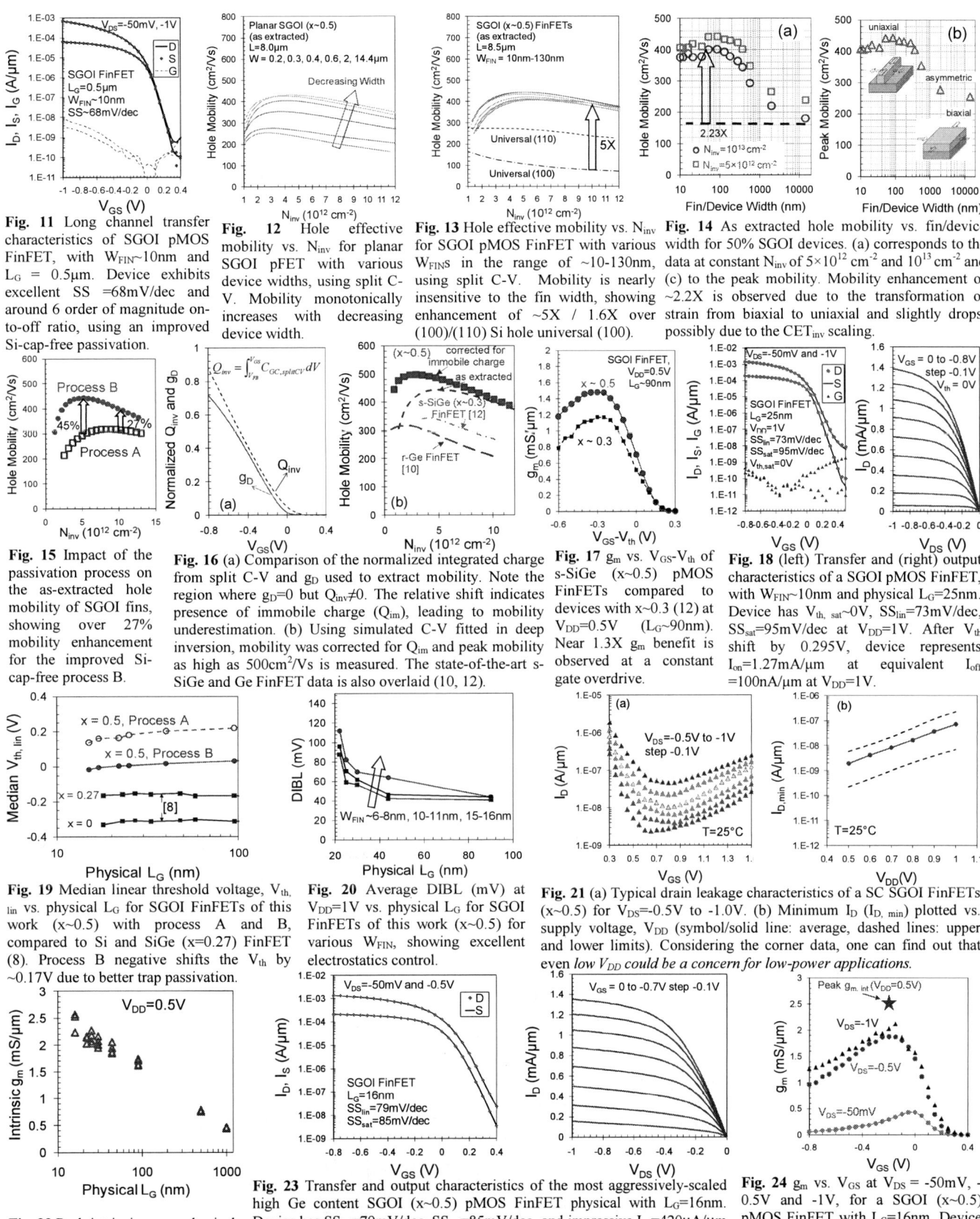

Fig. 11 Long channel transfer characteristics of SGOI pMOS FinFET, with W_{FIN}~10nm and L_G = 0.5µm. Device exhibits excellent SS =68mV/dec and around 6 order of magnitude on-to-off ratio, using an improved Si-cap-free passivation.

Fig. 12 Hole effective mobility vs. N_{inv} for planar SGOI pFET with various device widths, using split C-V. Mobility monotonically increases with decreasing device width.

Fig. 13 Hole effective mobility vs. N_{inv} for SGOI pMOS FinFET with various W_{FIN}s in the range of ~10-130nm, using split C-V. Mobility is nearly insensitive to the fin width, showing enhancement of ~5X / 1.6X over (100)/(110) Si hole universal (100).

Fig. 14 As extracted hole mobility vs. fin/device width for 50% SGOI devices. (a) corresponds to the data at constant N_{inv} of 5×10^{12} cm^{-2} and 10^{13} cm^{-2} and (c) to the peak mobility. Mobility enhancement of ~2.2X is observed due to the transformation of strain from biaxial to uniaxial and slightly drops, possibly due to the CET$_{inv}$ scaling.

Fig. 15 Impact of the passivation process on the as-extracted hole mobility of SGOI fins, showing over 27% mobility enhancement for the improved Si-cap-free process B.

Fig. 16 (a) Comparison of the normalized integrated charge from split C-V and g_D used to extract mobility. Note the region where g_D=0 but $Q_{inv}\neq0$. The relative shift indicates presence of immobile charge (Q_{im}), leading to mobility underestimation. (b) Using simulated C-V fitted in deep inversion, mobility was corrected for Q_{im} and peak mobility as high as 500cm^2/Vs is measured. The state-of-the-art s-SiGe and Ge FinFET data is also overlaid (10, 12).

Fig. 17 g_m vs. V_{GS}-V_{th} of s-SiGe (x~0.5) pMOS FinFETs compared to devices with x~0.3 (12) at V_{DD}=0.5V (L_G~90nm). Near 1.3X g_m benefit is observed at a constant gate overdrive.

Fig. 18 (left) Transfer and (right) output characteristics of a SGOI pMOS FinFET, with W_{FIN}~10nm and physical L_G=25nm. Device has $V_{th, sat}$~0V, SS_{lin}=73mV/dec, SS_{sat}=95mV/dec at V_{DS}=1V. After V_{th} shift by 0.295V, device represents I_{on}=1.27mA/µm at equivalent I_{off} =100nA/µm at V_{DD}=1V.

Fig. 19 Median linear threshold voltage, $V_{th, lin}$ vs. physical L_G for SGOI FinFETs of this work (x~0.5) with process A and B, compared to Si and SiGe (x=0.27) FinFET (8). Process B negative shifts the V_{th} by ~0.17V due to better trap passivation.

Fig. 20 Average DIBL (mV) at V_{DD}=1V vs. physical L_G for SGOI FinFETs of this work (x~0.5) for various W_{FIN}, showing excellent electrostatics control.

Fig. 21 (a) Typical drain leakage characteristics of a SC SGOI FinFETs (x~0.5) for V_{DS}=-0.5V to -1.0V. (b) Minimum I_D ($I_{D, min}$) plotted vs. supply voltage, V_{DD} (symbol/solid line: average, dashed lines: upper and lower limits). Considering the corner data, one can find out that even *low V_{DD} could be a concern for low-power applications.*

Fig. 22 Peak intrinsic g_m vs. physical L_G for SGOI (x~0.5) FinFETs at V_{DD}=0.5V.

Fig. 23 Transfer and output characteristics of the most aggressively-scaled high Ge content SGOI (x~0.5) pMOS FinFET physical with L_G=16nm. Device has SS_{lin}=79mV/dec, SS_{sat}=85mV/dec, and impressive I_{on}=420µA/µm at equivalent I_{off}=100nA/µm (V_{GS} shifted by 0.325V) at V_{DD}=0.5V, which is the highest on current reported to date for *sub 20nm L_G FinFETs*.

Fig. 24 g_m vs. V_{GS} at V_{DS} = -50mV, -0.5V and -1V, for a SGOI (x~0.5) pMOS FinFET with L_G=16nm. Device represents peak extrinsic/ intrinsic g_m=1.9/2.55 mS/µm at V_{DD}=0.5V.

Dual-Channel CMOS Co-Integration with Si NFET and Strained-SiGe PFET in Nanowire Device Architecture Featuring Sub-15nm Gate Length

P. Nguyen[1,2], S. Barraud[1], C. Tabone[1], L. Gaben[1,3], M. Cassé[1], F. Glowacki[1], J.-M. Hartmann[1], M.-P. Samson[2], V. Maffini-Alvaro[1], C. Vizioz[1], N. Bernier[1], C. Guedj[1], C. Mounet[1], O. Rozeau[1], A. Toffoli[1], F. Alain[1], D. Delprat[2], B.-Y. Nguyen[2], C. Mazuré[2], O. Faynot[1], M. Vinet[1]

[1] CEA-LETI, Minatec campus, 17 rue des Martyrs, 38054 Grenoble Cedex 9, France ; E-mail: sylvain.barraud@cea.fr
[2] SOITEC, Parc Technologique des Fontaines, 38926 Bernin, France
[3] STMicroelectronics, 850 rue J. Monnet, 38926 Crolles, France

Abstract

We have fabricated hybrid channel Ω-gate CMOS nanowires (NWs) with strained SiGe-channel (cSiGe) p-FETs and Si-channel n-FETs. An optimized process flow based on the Ge enrichment technique results in a +135% hole mobility enhancement at long gate lengths compared to Si. Effectiveness of cSiGe channel is also evidenced for ultra-scaled p-FET NWs (gate length L_G=15 nm) with +90% I_{ON} current improvement.

Introduction

NW transistors are today widely recognized as a promising solution to pursue the Moore's law beyond FinFET and Fully-Depleted Silicon-On-Insulator (SOI) CMOS technologies. During the last years, aggressively scaled NW transistors have already been demonstrated [1-3]. We previously showed high-performance uniaxial tensily-strained Ω-gate sSOI n-FET NWs down to 10nm gate length with an excellent electrostatic control [4]. We also recently fabricated and characterized high-performance Ω-gate p-FETs on compressively-strained-SiGeOI substrates obtained by the Ge enrichment technique [5]. Uniaxial compressive strain definitely improved hole transport in those SiGe NWs [5-7]. In this paper, for the first time, we report the successful co-integration of hybrid Si and SiGe channels in high AC performances NW CMOS devices that outperform state-of-the-art SOI nanowires. The strain is measured by using precession electron diffraction with a 1nm spatial resolution. We show that hybrid integration reduces the delay of CMOS ring oscillators (Fan-Out=3) by 50% at V_{DD}=0.9 V. We also demonstrate the most aggressively scaled hybrid CMOS NWs reported to date with NW width and gate length down to 7 nm and 11 nm, while maintaining high drive current (687 µA/µm for p-FET and 647 µA/µm for n-FET) with low leakage current and excellent short-channel-control (DIBL<50 mV/V). Finally, [110]-oriented NW is shown to be the best candidate to improve drive current under compressive strain.

Si/SiGe Hybrid Channel Patterning

Figs. 1-2 depict the integration scheme used for the fabrication of hybrid channel CMOS ($Si_{0.7}Ge_{0.3}$ for p-FET and Si for n-FET). An optimized Ge enrichment process [5] is used in p-FET regions (n-FET regions are protected by a hard mask during the oxidation enrichment process) for the formation of localized $Si_{1-x}Ge_xOI$ layers. Once the uniform Si and SiGe layers were formed in n-FET and p-FET regions, respectively, nanowires were etched with channels oriented along the [110] direction. NW widths (W_{top}) as small as 7nm were achieved.

The robustness of hybrid CMOS integration is shown in Fig. 3 with good pattern reproducibility in SRAM arrays with Si n-FETs and SiGe p-FETs. The channel thickness (NW height: H_{NW}) under the HfSiON/TiN gate is ~12 nm. The 18 nm-thick Si raised Sources and Drains (S/D) were grown to thicken the access regions of both Si and SiGe devices.

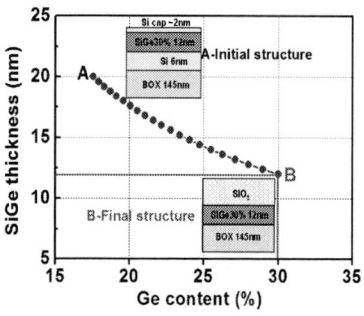

Fig.1: Ge enrichment process: A $Si_{1-x}Ge_x$ layer is first epitaxially grown on a thin SOI substrate and capped with a Si layer (structure A). The SiGe thickness is plotted as a function of the Ge content during the Ge enrichment process. Once the enriched SiGe film is 12 nm thick, the oxidation is stopped (structure B).

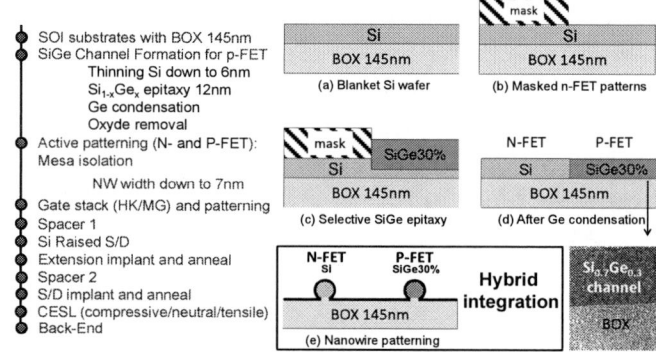

Fig.2: Key process steps of Ω-Gate CMOS nanowire transistors with hybrid Si and SiGe channels, gate-first HK/MG, and Si Raised-SD. Ge enrichment is used for the fabrication of hybrid channel CMOS (SiGe P-FET and Si N-FET omega-shaped gate nanowires).

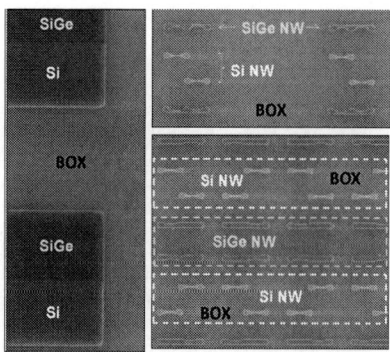

Fig.3: Top-view Scanning Electron Microscopy (SEM) images of 6-T Static Random Access Memory (SRAM) arrays after the formation of SiGe in p-FET regions (n-FET regions are made of Si).

978-1-4799-8002-4/14 $31.00 © 2014 IEEE

A second offset spacer was formed prior to S/D ion implantations, activation spike anneal and silicidation (NiPtSi silicide), in order to obtain low contact resistances. Fig. 4 shows TEM images of Ω-Gate n-FET (Si-channel) and p-FET (SiGe-channel) NW transistors with L_G down to 15nm. The EDX maps show the Ge content which is estimated to be 30%.

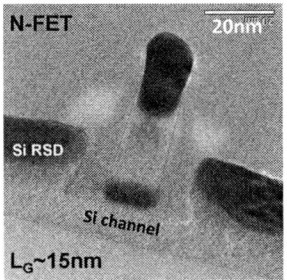

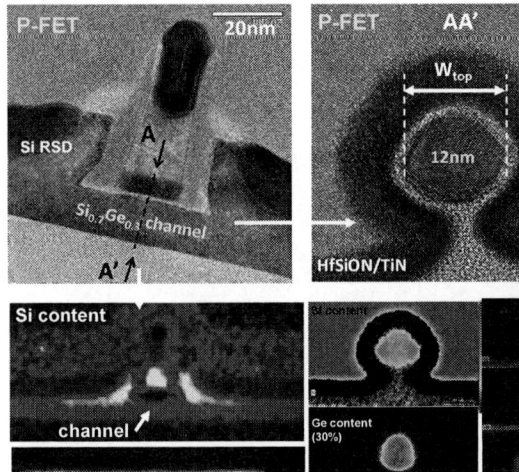

Fig.4: TEM images and EDX maps of Ω-Gate CMOS nanowire transistors (Si channel for N-FET and SiGe channel for P-FET) with diameter ∅=12 nm and L_G=15 nm. The Ge content of the SiGe channel is estimated to be 30% (by using semi-quantitative Energy Dispersive X-ray (EDX) spectroscopy). Uniform depositions of HfSiON and TiN metal gate were also confirmed by EDX analysis.

Low-field carrier mobility

Long channel carrier mobility has been extracted as a function of the inversion carrier density N_{inv} for different channel widths (Fig. 5). For the narrowest Si-NWs the electron mobility decreases, while hole mobility is improved [4,7]. No additional electron mobility degradation is observed on NMOS with a hybrid channel integration (Fig. 6a). However, a significant hole mobility improvement is evidenced for SiGe channels: +135% (resp. 60%) enhancement for NWs (resp. wide FETs) due to uniaxial (resp. biaxial) compressive strain (Fig. 6b). The interest of SiGe channel for p-FET NWs is thus well substantiated.

Device characteristics

Fig. 7a shows the threshold voltage V_{TH} versus W_{top} for Si and hybrid CMOS integration; the use of a SiGe channel is beneficial in terms of V_{TH} adjustment down to sub-20 nm L_G, with a 250 mV V_{TH}-shift for $Si_{0.7}Ge_{0.3}$ and W_{top}=10 nm (Fig. 7b). Fig. 7c shows the drain induced barrier lowering (DIBL) as a function of L_G for W_{top}=10 nm demonstrating excellent short channel effects (SCE) down to 10 nm gate length.

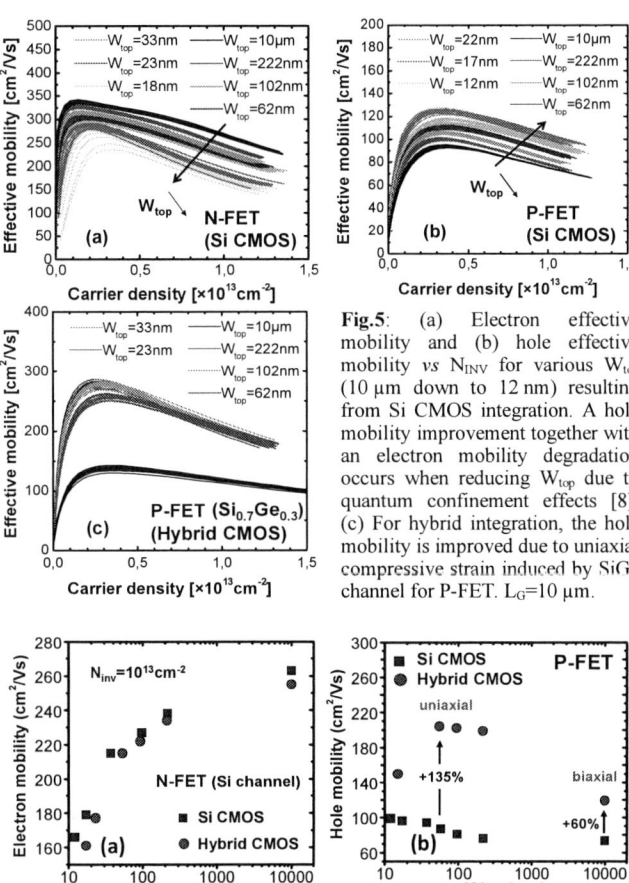

Fig.5: (a) Electron effective mobility and (b) hole effective mobility vs N_{INV} for various W_{top} (10 μm down to 12 nm) resulting from Si CMOS integration. A hole mobility improvement together with an electron mobility degradation occurs when reducing W_{top} due to quantum confinement effects [8]. (c) For hybrid integration, the hole mobility is improved due to uniaxial compressive strain induced by SiGe channel for P-FET. L_G=10 μm.

Fig.6: (a) μ_e vs W_{top} showing no additional mobility degradation in N-FET NW by hybrid CMOS integration. : (b) μ_h vs W_{top} for Si and hybrid integration. +135% μ_h improvement is observed due to uniaxial compressive strain induced by the SiGe channel (stronger improvement in NWs as compared to wide FETs). L_G=10 μm.

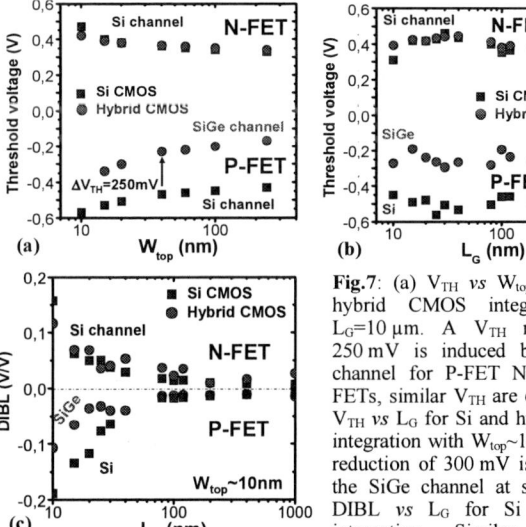

Fig.7: (a) V_{TH} vs W_{top} for Si and hybrid CMOS integration with L_G=10 μm. A V_{TH} reduction of 250 mV is induced by the SiGe channel for P-FET NWs. For N-FETs, similar V_{TH} are observed. (b) V_{TH} vs L_G for Si and hybrid CMOS integration with W_{top}~10 nm. A V_{TH} reduction of 300 mV is induced by the SiGe channel at short L_G. (c) DIBL vs L_G for Si and hybrid integration. Similar SCE are observed for N-FET whereas improved DIBL is achieved for the SiGe channel P-FET NW.

978-1-4799-8002-4/14 $31.00 © 2014 IEEE

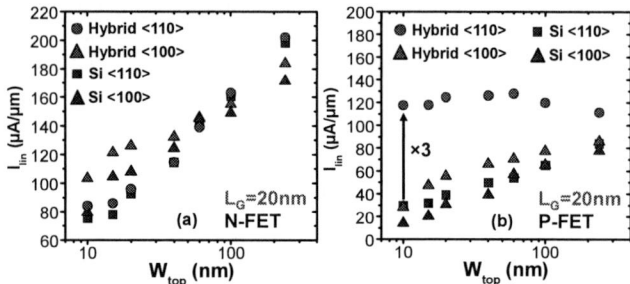

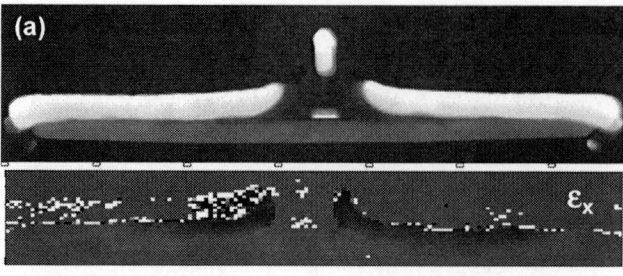

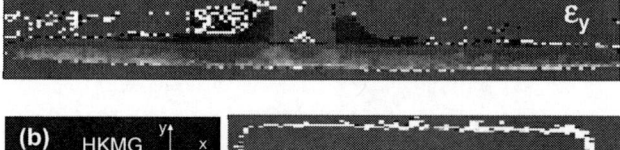

Fig.8: I_{lin} *vs* W_{top} for Si and hybrid integration along the [110] and [100] direction. (a) For N-FETs, the [100] Si NWs show the best linear current, closely followed by the [110] Si NWs. Few differences are observed between Si and hybrid CMOS integration. (b) For P-FETs, [110] direction yields the best I_{lin} enhancement under uniaxial compressive strain ($\times$3).

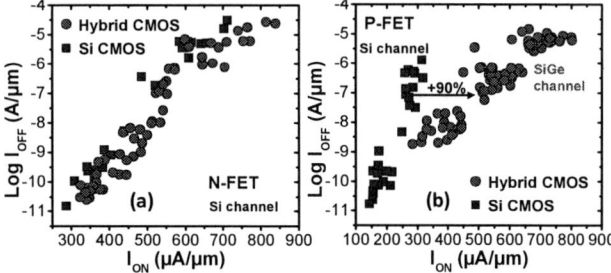

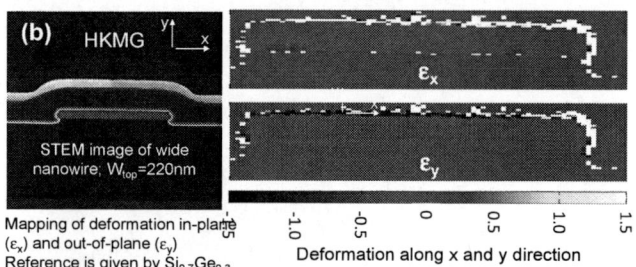

Mapping of deformation in-plane (ε_x) and out-of-plane (ε_y) Reference is given by $Si_{0.7}Ge_{0.3}$

Fig.9: N-FET (a) and P-FET (b) I_{OFF}-I_{ON} plots showing no N-FET performance degradation. I_{ON}=520 µA/µm at I_{OFF}=100 nA/µm and V_{DD}=0.9 V. P-FET performance is enhanced due to uniaxial compressive strain induced by the SiGe channel $\rightarrow$ +90% I_{ON} improvement for hybrid CMOS Ω-Gate P-FET NWs. Here, the gate length is in the range 10 nm<L_G<40 nm with $W_{top}\sim$25 nm.

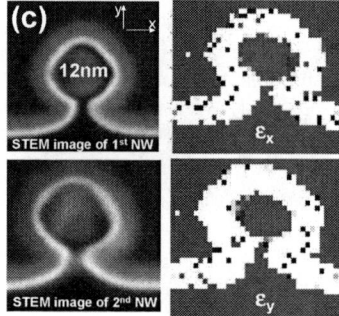

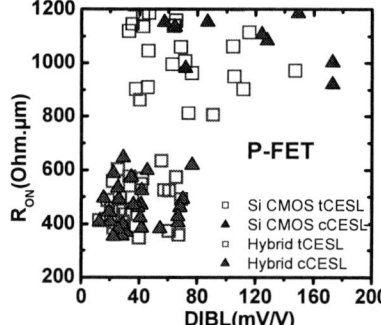

Fig.10: Deformation (ε) mapping in Ω-Gate NW along the S/D direction (a) and in the NW cross-section with W_{top}=220 nm (b) and W_{top}=12 nm (c). The precession electron diffraction technique evidences the compressive strain in NW down to 12 nm diameter. The deformation is calculated from relaxed $Si_{0.7}Ge_{0.3}$. The deformation is plotted along the growth direction (ε_y) and in-plane direction (ε_x).

The linear drain current I_{LIN} of [100] and [110] Si NW transistors (n- and p-FET) is plotted as a function of W_{top} in Fig. 8, for Si and hybrid channels. I_{LIN} is extracted at a source to drain voltage $|V_{DS}|$=0.05 V (L_G=20 nm). For electrons, a slight I_{LIN} enhancement is observed for the narrowest NWs along [100]. No n-FET performance degradation is observed when a hybrid channel integration is used. For SiGe channel p-FETs, uniaxial compressive strain can improve I_{LIN} by a factor of almost $\times$3 in [110]-NWs with respect to [100] SiGe NWs (or Si-NW). Compressive strain strengthens the light holes of the highest valence subbands of [110] oriented NWs by pushing heavy holes subbands down [8]. In Fig. 9 we show that uniaxial compressive strain in p-FET SiGe-channel NWs results in a +90% I_{ON} current improvement compared to Si devices (V_{DD}=0.9 V). No degradation of short-channel n-FET performance is evidenced. For a better understanding of strain-induced performance enhancement in NWs, precession electron diffraction (PED) with 1 nm spatial resolution has been used to measure strain. In Fig. 10, we plotted the deformation along the growth direction (ε_y) and in-plane direction (ε_x). Strain distribution with high spatial resolution evidences compressive strain in very narrow NWs (width down to 12 nm). This corroborates the p-FET performance increase evidenced in hybrid channel CMOS due to compressive strain. SiGe channel p-FET results in a strong R_{ON} reduction (R_{ON} is extracted at $|V_{GS}-V_{TH}|$=0.65 V and V_{DS}=-0.05 V) (Fig. 11). This reduction is due to both lower S/D resistivity and improved short channel mobility (Fig. 12).

Fig.11: R_{ON}-DIBL plot for Si and hybrid CMOS integration with tensile and compressive CESL. R_{ON} is measured at $|V_{GS}-V_{TH}|$=0.65 V and V_{DS}=-0.05 V. A strong R_{ON} reduction is observed and can be explained by lower S/D resistivity and improved short channel mobility.

As expected, a tensile Contact Etch Stop Layer (CESL) is the best option to improve n-FET performance. For short-channel p-FET NWs, uniaxial compressive strain definitely improve the low field hole mobility, consistently with the I_{LIN} enhancement shown in Fig. 8. Fig. 13 shows I_{DS}-V_{GS} of the most aggressively scaled hybrid CMOS NW reported to date (with W_{top}=7 nm for p-FET and W_{top}=11 nm for n-FET and L_G down to 11 nm) with excellent electrostatic control.

978-1-4799-8002-4/14 $31.00 © 2014 IEEE

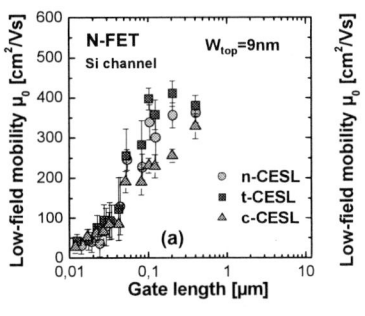
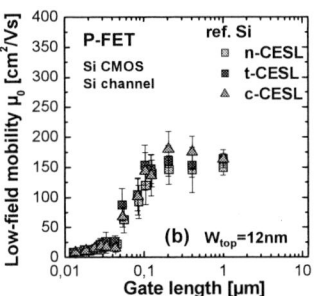
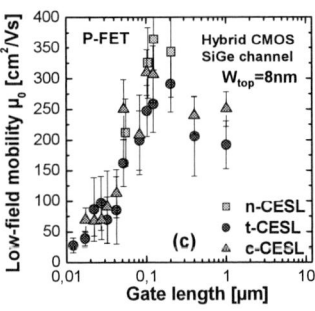

Fig.12: Low-field carrier mobility μ_0 vs L_G extracted by the Y-function method (a) in 9 nm n-FET NW. A better mobility is observed under tensile CESL. (b, c) For p-FET, hole mobility is enhanced for hybrid CMOS integration due to SiGe channel for gate length and NW width down to 15 nm and 8 nm, respectively. Hole mobility enhancement at short L_G in NW transistors is well correlated with the I_{ON} improvement previously observed on p-FET (Fig. 9).

AC performances

Ring oscillator (RO) operation is shown in Fig.14. ROs were designed with 80 stages. Faster ROs have higher dynamic ring oscillator current. As expected, RO delay decreases as the supply voltage V_{DD} increases. ROs with hybrid CMOS are faster; this is primarily due to lower series resistance and the higher mobility in p-FET NWs.

Conclusion

Hybrid channel CMOS nanowires with SiGe p-FETs and Si n-FETs have been successfully fabricated with diameter down to 7 nm and 11 nm gate length. In-depth electrical characterization evidenced higher p-FET performance (+90%) due to compressive SiGe channel in good agreement with strain measurements. For the first time, hybrid channel NW ring oscillators with 80 stages were demonstrated. A 50% reduction in RO delay due to SiGe-channel p-FET was evidenced. This leads us to conclude that hybrid channel CMOS integration can be easily used with high efficiency for NW-based technology.

Acknowledgements

This work was partially carried out in the frame of ST/IBM/LETI joint program. It was also partly funded by the French Public Authorities through NANO 2017 program. The authors thank SOITEC for providing SOI substrates.

References

[1] S.-D. Suk et al., IEDM, p.891 (2007)
[2] S. Bangsaruntip, VLSI, p.21 (2010)
[3] S. Barraud et al., IEEE EDL, **33**, p.1526 (2012)
[4] S. Barraud et al., VLSI, p.230 (2013)
[5] P. Nguyen et al., S3S conference (2014)
[6] P. Hashemi et al., VLSI, p.18 (2013)
[7] S. Barraud et al., IEEE EDL, **34**, p.1103, (2013)
[8] Y.-M. Niquet et al., Nanoletter, **12**, p.3545 (2012)
[9] M. Cassé et al., ECS Trans. 53, p.125 (2013)

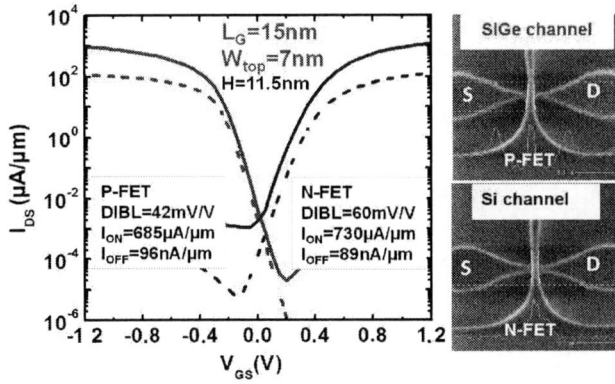

Fig.13: Transfer characteristics of the aggressively scaled Si N-FET and SiGe P-FET with physical gate length L_G=15 nm and NW width=7 nm.

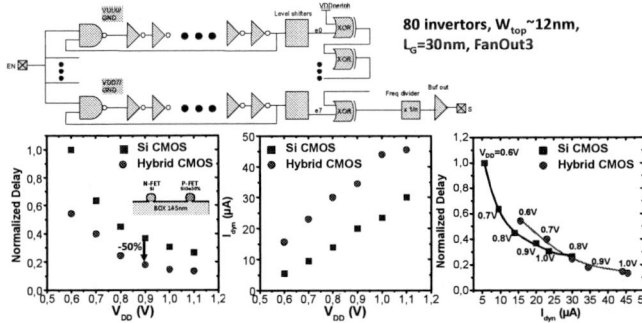

Fig.14: Normalized RO delay and dynamic ring oscillator (RO) current I_{dyn} vs supply voltage V_{DD}. Nanowire diameter $\varnothing$=12 nm and gate length L_G=30 nm. The ring oscillators are composed of 80 invertors with a fanout (FO) of 3. Hybrid CMOS integration shows a reduction of RO delay due to the SiGe channel (higher I_{ON} current). The dynamic current I_{dyn} increases when the supply voltage V_{DD} increases.

978-1-4799-8002-4/14 $31.00 © 2014 IEEE

V_{th} adjustable self-aligned embedded source/drain Si/Ge nanowire FETs and dopant-free NVMs for 3D sequentially integrated circuit

Chih-Chao Yang[1], Jia-Min Shieh[1*], Tung-Ying Hsieh [1], Wen-Hsien Huang[1], Hsing-Hsiang Wang[1], Chang-Hong Shen[1], Tsung-Ta Wu[1], Chun-Yuan Chen[2], Kuei-Shu Chang-Liao[2], Jung-Hau Shiu[3], Meng-Chyi Wu[3], and Fu-Liang Yang[4]

[1]National Nano Device Laboratories, No.26, Prosperity Road 1, Hsinchu 30078, Taiwan;
[2] Department of Engineering and System Science, National Tsing Hua University, Hsinchu, 30013, Taiwan;
[3] Department of Electrical Engineering, National Tsing Hua University, Hsinchu, 30013, Taiwan;
[4] Research Center for Applied Sciences, Academia Sinica, Taipei 11529, Taiwan.
[*]Tel:+886-3-5726100-7617, Fax:+886-3-5722715, E-mail: jmshieh@narlabs.org.tw

Abstract

3D stackable high-performance Si nanowire field-effect transistors (NWFETs) and dopant-free Ge junctionless nanowire non-volatile memories (JL-NWNVMs) with self-aligned embedded source/drain (S/D) current boosters and independent back gate (BG) V_{th} adjusters for 3D sequential integrated circuit are realized by low thermal budget process ($<450^\circ$C). The fabricated Si NWFETs exhibit low subthreshold swings (96 and 125 mV/dec.), high on-currents (232 and 110 μA/μm), and large γ value (>0.05) for V_{th} adjustment. The high-κ capped blocking dielectric bandgap engineered dopant-free Ge JL-NWNVM exhibits high I_{on}/I_{off} ratio ($>10^5$), large memory window (>4V), and low charge loss ($<40\%$, 10yrs). Thanks to the quantum confinement effect, such V_{th} adjustable nanowire devices perform well at higher temperatures, which give a wide design window for 3D sequential integrated circuit.

I. Introduction:

As scaling becomes more difficult, 3DIC fabrication has become the most important technology in recent years. 3D sequential integration (3DSI) approach [1], which is nearly compatible with nowadays semiconductor process, is one of the most promising ways to achieve high-performance, rich functions, and energy efficient integrated circuits. The most successful product is 3D NAND Flash [2][3], which uses multi-stacked poly-Si as channel material. However, the challenge is to fabricate high performance logic units using back-end compatible processes. Several groups are working on it, such as Batude et al. in CEA-Leti, Rajendran et al. and Park et al. in Sanford University, and J. Derakhshandeh et al. in Delft University of Technology [4]-[7].

In this article, we propose V_{th} adjustable embedded S/D Si NWFET and dopant-free Ge JL-NWNVM, which can be sequentially fabricated on a commercial bulk FinFET substrate (**Fig. 1**). The highly crystallized Si/Ge nanowire with self-aligned embedded S/D guarantees the device performance and back gate control widen the flexibility for circuit design (**Fig. 2**). Such heterogeneous integrated logic/NVM unit endorses future high-performance multi-layered 3DSI electronics (**Fig. 3**).

II. Device Fabrication:

The process flow of the self-aligned embedded S/D Si nanowire MOSFET and the dopant-free Ge JL-NWNVM with back gate are illustrated in **Fig. 4**. After back gate formation, the source/drain regions are self-aligned to the back gate structure during device fabrication. The core technology is the large-grained laser crystallized channel, followed by a novel super-CMP- planarization process (**Fig. 5**). The film properties are identified by Raman and XRD diffraction pattern.

III. Results and Discussion:

High contact resistance reduces drive current (I_{on}) of device as feature size keeps scaling. The embedded S/D, self-aligned to the back gate, increases the S/D thickness to reduce contact resistance. The drive currents have improvements of 20 to 30% (**Fig. 7,8**), compared to the device without embedded S/D. With nitridized HfO$_2$ gate dielectric, which has low interface defects (**Fig. 6**), the fabricated Si NWFETs exhibit low subthreshold swings (96 and 125 mV/dec.), high drive currents (232 and 110 μA/μm), and $I_{on}/I_{off} > 10^5$. The benchmark results are listed in **Table I** [8]-[10].

Feature size scaling of tri-gate devices enhances the front gate controllability. Therefore, the V_{th} are hardly shifted by body biasing in bulk Si or SOI FinFETs. With independent back gate and thin back gate oxide, we can more easily adjust the V_{th} in positive or negative direction (**Fig. 9**). As revealed, the V_{th} control factors (γ), extracted from the curves, can remain high (>0.05) even though the W_{Fin} of nanowire is shrunk to 20nm. Moreover, the device with independent back gate also can be optionally operated in quadruple gate control mode, which can further improve electrical characteristics. With multi-layered V_{th} adjustable Si NWFETs, the typical voltage transfer characteristics (VTC) of the vertically stacked inverter show good transfer characteristics (**Fig. 10**). The nanowire channel also brings another advantage due to quantum confinement effect. The device can be less sensitive

978-1-4799-8002-4/14 $31.00 © 2014 IEEE

to temperature (**Fig. 12**). Such V_{th} adjustable embedded S/D NWFETs with large γ value and low temperature sensitivity provide excellent current controllability to achieve low I_{off} and high I_{on} in 3D sequential integrated circuit.

Fabricating the highly crystallized Si or Ge channel, which usually need high laser power, on thin ILD layer without damaging the bottom-layer devices/circuits is the key step to realize the high performance 3D stackable devices. During the crystallization process, the laser light may damage the bottom devices due to the heat from crystallized Si/Ge film or the light penetrated through ILD layer, as illustrated in **Fig. 11**. Carefully controlling the laser power, α-Si/Ge film thickness, the ILD thickness can be as thin as 200nm, quite reducing the delay and power loss from interconnects.

Heterogeneous integration with 3D stackable Ge NVMs is a good approach to expand the high-density storage capability for hybrid chip application. An inherent dopant-free Ge channel (p-type, $N_d{\sim}2{\times}10^{18}$ cm^{-3}) using the plasma-deposition/laser-crystallization technology not only reduces implantation processes/thermal impact (caused by post-thermal or laser treatment) but enables JL FETs. Here, we have demonstrated a high-κ capped blocking dielectric bandgap engineered Ge JL-NWNVM (**Fig. 13**). The transfer characteristics of the device exhibit a high I_{on}/I_{off} ($>10^4$). After optimizing the gate stacked structure, the NVM performs a large memory window (>4V), program-erase speed (<1 ms), low charge loss (<40%, 10yrs), and good endurance ($>10^5$ cycles). Such dopant-free Ge JLNWNVM speeds up the development of heterogeneous Si/Ge integration owing to the lower thermal impacts.

IV. Conclusion:

V_{th} adjustable embedded S/D Si NWFETs and Ge JL-NWNVMs are demonstrated in this article. The cutting edge low thermal-budget (<450°C) devices enable the possibilities of Si/Ge heterogeneous integration and logic/NVM chip design for compact and energy-efficient mobile products.

Acknowledgements

The authors would like to thank the financial support from Ministry of Science and Technology (NSC 102-2218-E-492-001) and National Applied Research Laboratories (NARlabs) of the Republic of China.

References:

[1] C. C. Yang, S. H. Chen, J. M. Shieh, W. H. Huang, T. Y. Hsieh, C. H. Shen, T. T. Wu, H. H. Wang, Y. J. Lee, F. J. Hou, C. L. Pan, K. S. Chang-Liao, Chenming Hu and F. L. Yang, "Record-high 121/62 μA/μm on-currents 3D stacked epi-like Si FETs with and without metal back gate", International Electron Devices Meeting (IEDM) Tech. Dig., p. 731, 2013.

[2] J. G. Lisoni, A. Arreghini, G. Congedo, M. Toledano-Luque, I. Toqué-Tresonne, K. Huet, E. Capogreco, L. Liu, C. L. Tan, R. Degraeve, G. Van den bosch and J. Van Houdt, "Laser

Thermal Anneal of polysilicon channel to boost 3D memory performance", Symposium on VLSI Technology, pp. 20, 2014.

[3] H. T. Lue, P. Y. Du, W. C. Chen, T, H. Yeh, K. P. Chang, Y. H. Hsiao, Y. H. Shih, C. H. Hung and C. Y. Lu, "A novel dual-channel 3D NAND flash featuring both n-Channel and p-Channel NAND characteristics for bit-alterable flash memory and a new opportunity in sensing the stored charge in the WL spac", International Electron Devices Meeting (IEDM) Tech. Dig., pp. 80, 2013.

[4] P. Batude, M. Vinet, B. Previtali, C. Tabone, C. Xu, J. Mazurier, O. Weber, F. Andrieu, L. Tosti, L.Brevard, B. Sklenard, P. Coudrain, S. Bobba, H. Ben Jamaa, P-E. Gaillardon, A. Pouydebasque, O. Thomas, C. Le Royer, J.-M. Hartmann, L. Sanchez, L. Baud, V. Carron, L. Clavelier, G. De Micheli, S. Deleonibus, O. Faynot and T. Poiroux. "Advances, Challenges and Opportunities in 3D CMOS Sequential Integration", International Electron Devices Meeting (IEDM) Tech. Dig., p. 7.3, 2011.

[5] B. Rajendran, R. S. Shenoy, D. J. Witte, N. S.Chokshi, R. L. De Leon, G. S. Tompa and R. Fabian, "Low Thermal Budget Processing for Sequential 3-D IC Fabrication", IEEE Transactions on Electron Devices, vol. 54, no. 4, pp. 707, 2007.

[6] J. H. Park, M. Tada, D. Kuzum, P. Kapur, H. Y. Yu and H. S. P. Wong, K. C. Saraswat, "Low Temperature (≤ 380°C) and High Performance Ge CMOS Technology with Novel Source/Drain by Metal-Induced Dopants Activation and High-K/Metal Gate Stack for Monolithic 3D Integration", International Electron Devices Meeting (IEDM) Tech. Dig., p. 389, 2008.

[7] J. Derakhshandeh, N. Golshani, R. Ishihara, M. R.Tajari Mofrad, M. Robertson, T. Morrison and C. I. M. Beenakker, "Monolithic 3-D Integration of SRAM and Image Sensor Using Two Layers of Single-Grain Silicon", IEEE Transactions on Electron Devices, vol. 58, no. 11, pp. 3954, 2011.

[8] Y. H. Lu, P. Y. Kuo, Y. H. Wu, Y. H. Chen and T. S. Chao, "Novel GAA raised source / drain sub-10-nm poly-Si NW channel TFTs with self-aligned corked gate structure for 3-D IC applications" , VLSI, pp. 142, 2011.

[9] S. J. Choi, J. W. Han, S. Kim, D. I. Moon, M. Jang and Y. K. Choi, "A novel TFT with a laterally engineered bandgap for of 3D logic and flash memory", Symposium on VLSI Technology, pp. 111, 2010.

[10] H. B. Chen, Y. C. Wu, C. Y. Chang, M. H. Han, N. H. Lu and Y. C. Cheng, "Performance of GAA poly-Si Nanosheet (2nm) channel of Junctionless Transistors with ideal Subthreshold Slope", Symposium on VLSI Technology, pp. 232, 2013.

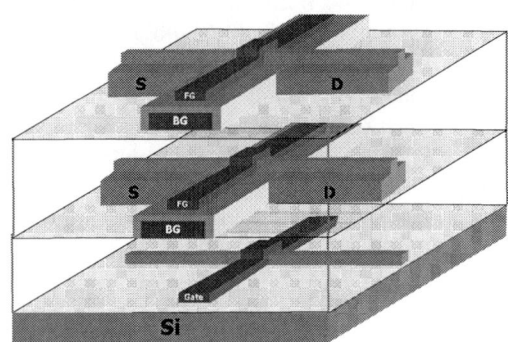

Fig 1. Schematic illustration of a 3D sequential integrated embedded S/D Si NWFET and Ge JL-NWNVM on a bulk Si FinFET.

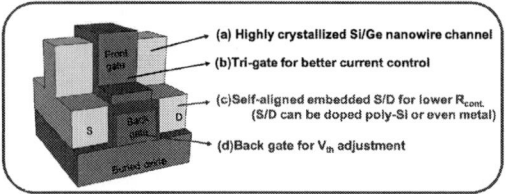

Fig 2. Advantages of the V_{th} adjustable self-aligned embedded S/D Si/Ge nanowire FET and dopant-free NVM.

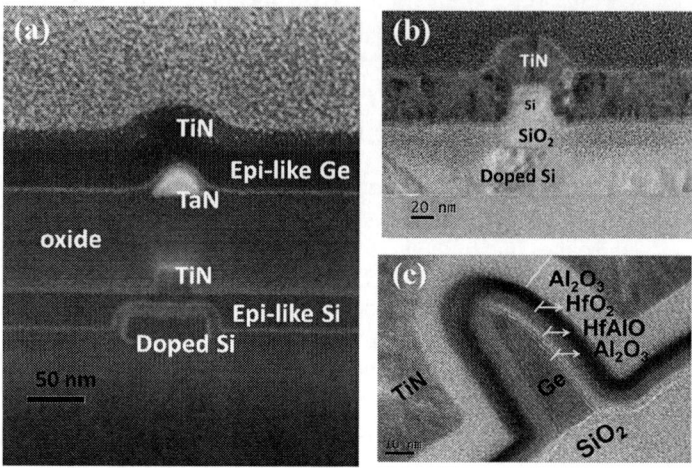

Fig 3. (a) Gate cross-section view (FIB) of a vertically stacked Ge JL-NWNVM on a Si NWFET with embedded S/D and back gate structures. TEM channel views of (b) a Si NWFET and (c) a high-κ capped blocking dielectric bandgap engineered Ge JL-NWNVM.

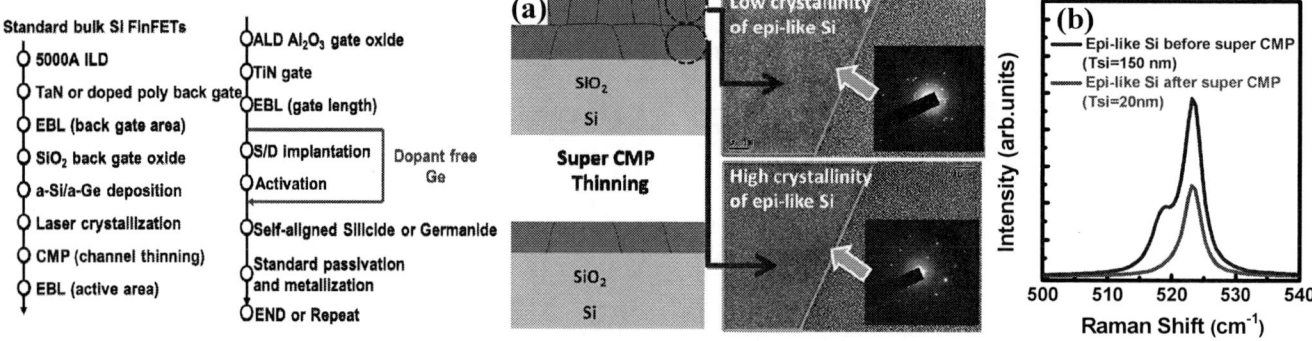

Fig 4. Process flow of the multi-layered high performance Si NWFET and dopant-free Ge JL-NWNVM for logic/memory hybrid 3D sequentially integrated circuit.

Fig 5. (a) The fabrication of highly crystallized epi-like Si channel by green nanosecond laser crystallization technology, followed by super-CMP-thinning process. (b) the Raman spectrums of the epi-like Si channel before and after the thinning process.

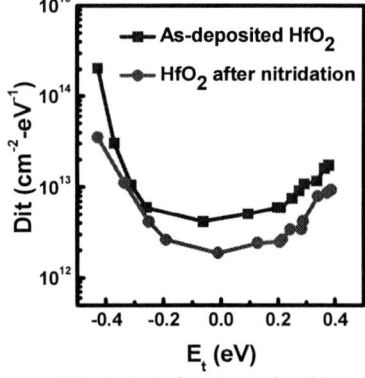

Fig 6. Interface trap densities of ALD-HfO2 gate dielectric before and after plasma-nitridation.

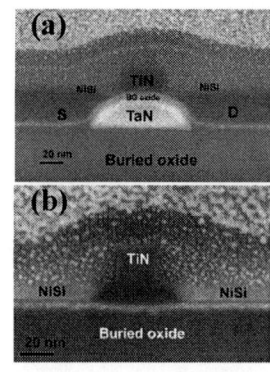

Fig. 7. Nickel-silicide (NiSi) formation on (a) Si NWFET with embedded S/D and TaN back gate (b) Si NWFET.

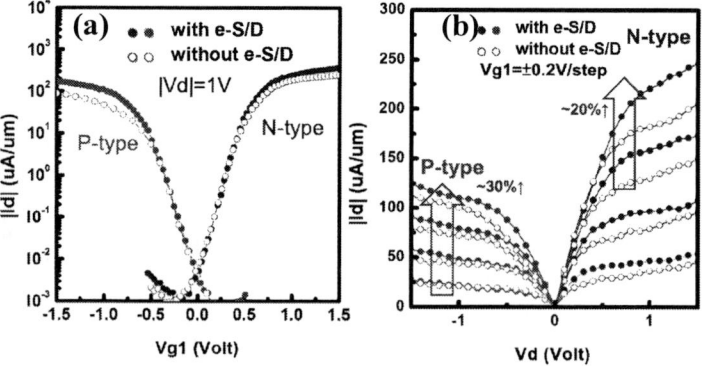

Fig. 8. The (a) Id-Vg1 (b) Id-Vd curves of the Si NWFETs demonstrate the drive current improvements from embedded S/D structure.

978-1-4799-8002-4/14 $31.00 © 2014 IEEE   412

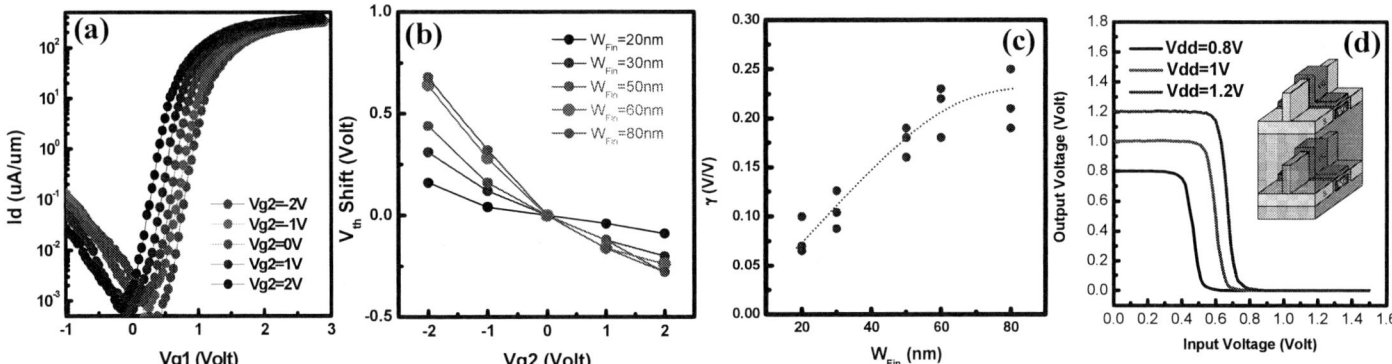

Fig. 9. The (a) Id-Vg1 curves with various back gate bias, (b) the normalized V_{th} shifts from the Id-Vg1 curves, and (c) the related γ values with different W_{Fin}.

Fig. 10. The typical voltage transfer characteristics (VTC) of a vertically stacked inverter.

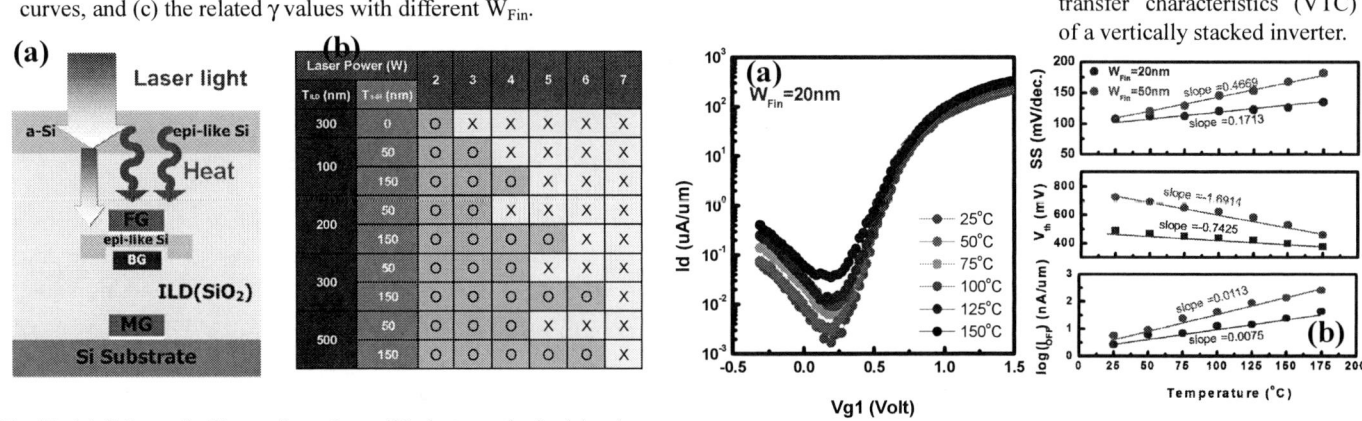

Fig 11. (a) Schematic illustration of possible bottom device/circuit damages from the laser light penetration and heat transfer, and (b) the test matrix with various laser power, a-Si and ILD thicknesses.

Fig 12. (a) The Id-Vg1 curves of a W_{Fin}=20nm Si NWFET at various operating temperature, and (b) the corresponding characteristics of SS, V_{th} and I_{off}, compared to a W_{Fin}=50nm Si NWFET.

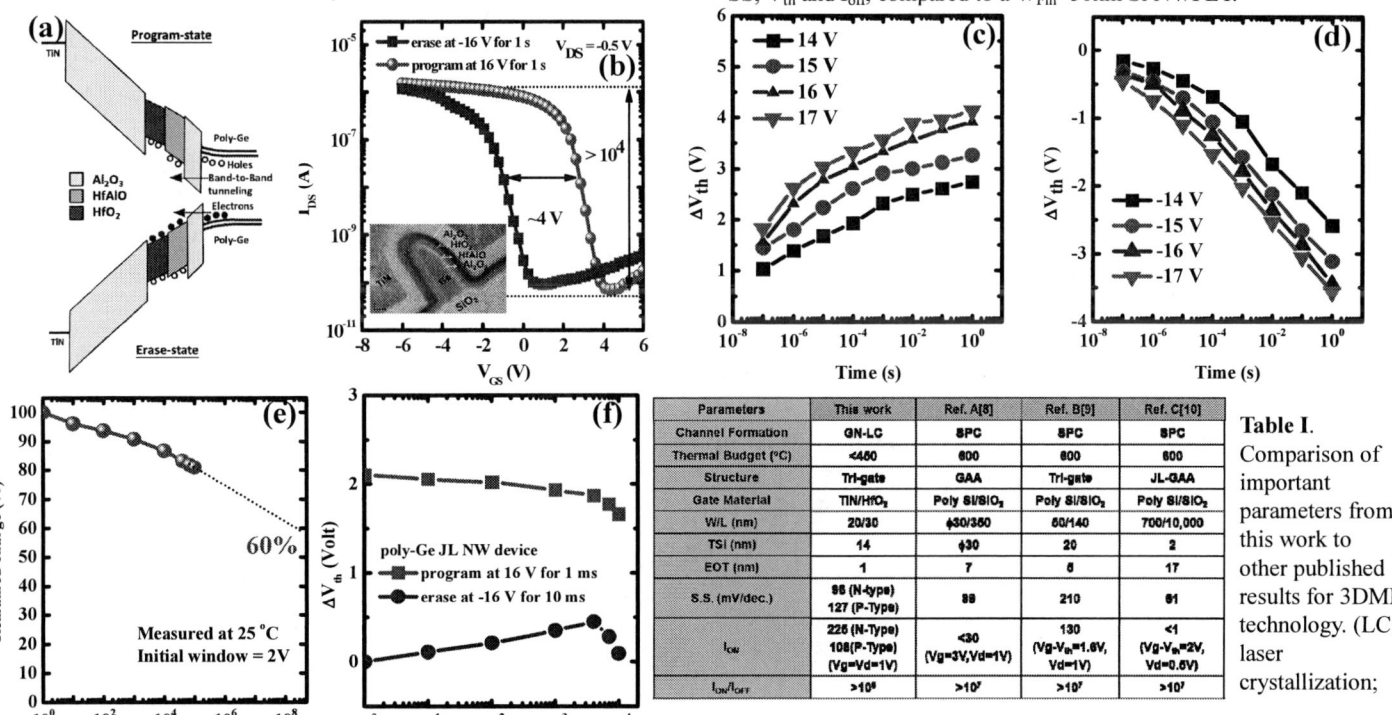

Table I. Comparison of important parameters from this work to other published results for 3DMI technology. (LC: laser crystallization;

Fig 13. The characteristics of Ge JL NW-NVM: (a) Band diagrams of Ge JL NW-NVM at program and erase states; (b) I_{DS}-V_{GS} curves; (c) Program-speed for various Vg; (d) Erase-speed for various Vg; (e) 60% remained charge after extrapolation to 10 years; (f) Endurance with P/E cycles up to 10^5 and slight degradation of memory window (1ms pulses applied for both retention and endurance testing)

Enhancement Mode Strained (1.3%) Germanium Quantum Well FinFET (W_{Fin}=20nm) with High Mobility (μ_{Hole}=700 cm^2/Vs), Low EOT (~0.7nm) on Bulk Silicon Substrate

A. Agrawal[1], M. Barth[1], G. B. Rayner Jr.[2], Arun V. T.[1], C. Eichfeld[1], G. Lavallee[1], S-Y. Yu[1], X. Sang[3], S. Brookes[3], Y. Zheng[1], Y-J. Lee[4], Y-R. Lin[4], C-H. Wu[4], C-H. Ko[4], J. LeBeau[3], R. Engel-Herbert[1], S. E. Mohney[1], Y-C. Yeo[4] and S. Datta[1]

[1]The Pennsylvania State University, University Park, PA, USA;
[2]Kurt J. Lesker Company, Pittsburgh, PA, USA;
[3]North Carolina State University, Raleigh, NC, USA
[4]Taiwan Semiconductor Manufacturing Company, Hsinchu, Taiwan
Email: ashish@psu.edu

Abstract

Compressively strained Ge (s-Ge) quantum well (QW) FinFETs with $Si_{0.3}Ge_{0.7}$ buffer are fabricated on 300mm bulk Si substrate with 20nm W_{Fin} and 80nm fin pitch using sidewall image transfer (SIT) patterning process. We demonstrate (a) in-situ process flow for a tri-layer high-κ dielectric $HfO_2/Al_2O_3/GeO_x$ gate stack achieving ultrathin EOT of 0.7nm with low D_{IT} and low gate leakage; (b) 1.3% s-Ge FinFETs with Phosphorus doped $Si_{0.3}Ge_{0.7}$ buffer on bulk Si substrate exhibiting peak μ_h=700 cm^2/Vs, μ_h=220 cm^2/Vs at 10^{13} /cm^2 hole density. The s-Ge FinFETs achieve the highest $\mu*C_{max}$ of 3.1×10^{-4} F/Vs resulting in 5x higher I_{ON} over unstrained Ge FinFETs.

Introduction

Ge pMOSFETs on bulk silicon substrate are a promising solution for improving the p-channel FET performance [1]. It is imperative to optimize the relaxed SiGe buffer counter doping with Phosphorus to achieve reliable isolation, maximize the uniaxial compressive strain in extremely scaled s-Ge fin to enhance μ_h, incorporate a gate stack with low EOT and D_{IT}, and mitigate the sidewall roughness to prevent μ_h reduction. In this work, we investigate the optimum depth of location of the buffer Phos doping, tri-layer optimization of an ultrathin EOT gate stack without Si interlayer (IL) (Fig. 1), residual uniaxial strain retention in scaled fins of a s-Ge QW heterostructure (Fig. 2), and temperature dependent μ_h characterization in s-Ge fins to quantify the effective mobility degradation. We experimentally evaluate enhancement mode, scaled s-Ge QW FinFETs with W_{Fin}=20nm and 80nm fin pitch to demonstrate record $\mu*C_{max}$ product of 3.1×10^{-4} F/Vs which is 2X higher than best reported till date.

Tri-layer Gate Stack

The tri-layer gate stack design and fabrication is shown in Fig. 3. GeOx is used to ensure low D_{IT} at high-κ/Ge interface, Al_2O_3 cap layer mitigates HfO_2-GeOx intermixing while HfO_2 alleviates gate leakage. The oxide layer thicknesses have been systematically optimized using in-situ spectroscopic ellipsometry (Fig. 4) to enable ultimate EOT scaling whilst preserving functionality of each layer as confirmed using TEM and EDX (Fig. 5). Direct correlation of the Al_2O_3 and HfO_2 thicknesses was established with the MOS capacitor C-V characteristics (Fig. 6). Excellent C-V response with ultrathin EOT of 0.72nm exhibiting the lowest gate leakage (Fig. 7, 8) was obtained with 17A HfO_2/5A Al_2O_3/6A GeOx/p-Ge gate stack. The D_{IT} analysis of these gate stacks (Fig. 9, 10) exhibited 4X better interface quality with thicker Al_2O_3 indicating more stable GeOx. Further, bulk trap density was reduced with Al_2O_3 thickness preventing HfO_2-GeOx intermixing (Fig. 11).

s-Ge FinFETs

Fig. 12 shows the schematic of s-Ge QW heterostructure on 300mm bulk Si substrate along with the fabrication process for scaled FinFETs with scaled and tight fin pitch using SIT process. Optimization of chlorine-based dry etch resulted in vertical fin sidewall profile with W_{Fin}=20nm and fin pitch of 80nm as seen under high resolution cross section TEM (Fig. 13)

Phos Doped Buffer Design

s-Ge QW MOSFETs with Phos doping placed at 150nm and 250nm depth in the $Si_{0.3}Ge_{0.7}$ buffer were characterized to identify the optimum depth of buffer counter doping. Four

978-1-4799-8002-4/14 $31.00 © 2014 IEEE

orders of magnitude lower I_{OFF} was obtained for 250nm deep Phos doping compared to no Phos in buffer, with no degradation in I_{ON} (Fig. 14). This indicated effective counter doping of acceptor defects in relaxed $Si_{0.3}Ge_{0.7}$ buffer that otherwise result in parallel conduction and affect device isolation.

Si cap vs. GeO$_x$ passivation

The incorporation of an ultrathin Si interlayer to passivate high-κ/Ge interface results in high EOT and is further incompatible with 3D FinFET manufacturing process flow. Hence, the tri-layer gate stack with GeO$_x$ passivation after Si cap removal was deposited on s-Ge QW MOSFET and FinFET (Fig. 15). 2X higher I_{ON} was obtained with tri-layer gate stack at V_{DS}=-0.5V, L_G=5μm compared to with Si cap on s-Ge QW MOSFET. This can be attributed to 2.3X higher capacitance as measured using split-CV along with excellent modulation from accumulation to depletion. Hence, Si cap removal and tri-layer dielectric with GeOx passivation was key in realizing low EOT with low D_{IT} on s-Ge QW.

E-Mode s-Ge QW FinFET

Excellent transfer characteristics with I_{ON}/I_{OFF}=2x10^4 were observed on s-Ge QW FinFETs with W_{Fin}=70nm, 45nm and 20nm fabricated with SIT process after Si cap removal and deposition of tri-layer gate stack (Fig. 16(a)) showing advantage of Phos doping in buffer in addition to ultrathin EOT gate stack on high mobility s-Ge channel. The combined effect of confinement due to quantization from QW and fin patterning resulted in enhancement mode operation for W_{Fin}=45nm and 20nm s-Ge QW FinFETs (Fig. 16(b)). Experimental effective hole mobility ($μ_{eff}$) for s-Ge QW MOSFETs and FinFETs extracted from split-CV is summarized in Fig. 17. Highest peak $μ_{eff}$ of 700 cm^2/Vs obtained for s-Ge FinFET with W_{Fin}=20nm shows 2.6X improvement compared to unstrained Ge [2] which is attributed to the residual asymmetric uniaxial strain in the fin as a result of patterning. In addition, 2X degradation in $μ_{eff}$ at N_s=10^{13} /cm^2 from planar to W_{fin}=20nm indicated increased scattering in FinFET compared to planar which was investigated using temperature dependent measurements.

Sidewall Scattering

Temperature dependent transfer characteristics for s-Ge QW FinFET with W_{Fin}=20nm showed higher modulation with temperature in the subthreshold region compared to s-Ge QW MOSFET for the same gate stack (Fig. 17, 18). Low subthreshold slope of 96 mV/dec for planar MOSFET indicates low D_{IT} at the high-κ/Ge top surface. In contrast, a degraded subthreshold slope of 150 mV/dec for s-Ge FinFET

is indicative of higher D_{IT} response from the high-κ/Ge interface at the sidewall due to higher density of dangling bonds as a result of fin etch. Temperature dependent hole mobility as a function of hole density for planar MOSFET (Fig. 19) revealed Ns^{-1} dependence at 300K and 77K, which is characteristic of phonon scattering limited mobility. For s-Ge QW FinFETs, a much stronger Ns^{-2} dependence and temperature independent mobility with varying fin width (Fig. 20) revealed sidewall roughness as the dominant scattering mechanism. Optimization of fin etch to reduce the sidewall roughness is key to achieving even higher hole mobility in s-Ge QW FinFET.

Benchmarking and Conclusions

In conclusion, optimized tri-layer high-κ gate stack exhibiting ultrathin EOT=0.72nm and low gate leakage on Ge was developed. Uniaxially s-Ge QW FinFETs with W_{Fin}=20nm was demonstrated with high $μ_{Peak}$=700 cm^2/Vs and 220 cm^2/Vs at 10^{13} /cm^2 with ultrathin EOT (Fig. 21). The high mobility s-Ge channel FinFET in conjunction with scaled gate stack resulted in the highest $μ*C_{max}$ product of 3.1x10^{-4} F/Vs (Table I) among high performance Ge FinFETs. The aforementioned enhancements in transport and gate stack resulted in 5X higher I_{ON} (Fig. 23) for s-Ge QW FinFET indicating promise for future alternate channel p-FinFET device technology.

Acknowledgement

This work was supported by the Pennsylvania State University Materials Research Institute Materials Characterization and Nanofabrication Laboratories and the National Science Foundation Cooperative Agreement No. ECS-0335765.

References

[1] S. Takagi et. al., IEDM 2003

[2] R. Zhang et. al., IEDM 2013

[3] R. Zhang et. al., IEDM 2012

[4] C. Choi et. al., EDL 2004

[5] R. Zhang et. al., IEDM 2011

[6] W. Bai et. al., VLSI 2003

[7] R. Xie et. al., IEDM 2008

[8] P. Hashemi et. al., EDL 2012

[9] K. Ikeda et. al., VLSI 2013

[10] Y. Kamata et. al., VLSI 2009

[11] B. Liu et. al., IEDM 2012

[12] P. Zimmerman et. al., IEDM 2006

[13] J. Mitard et. al., IEDM 2008

[14] R. Xie et. al., IEDM 2008

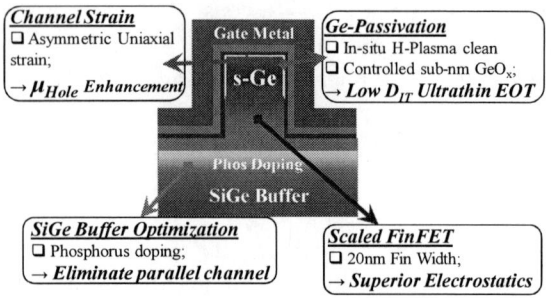

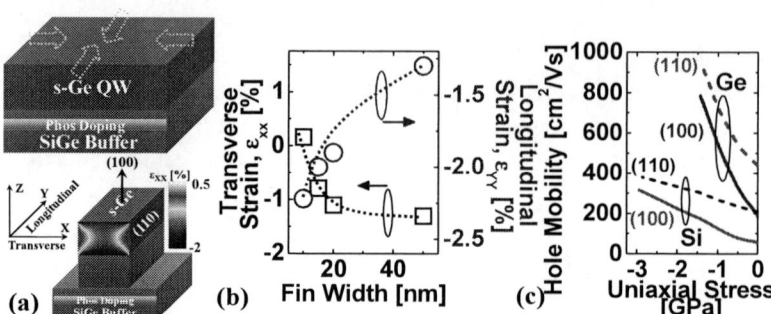

Fig. 1: (a) Schematic showing device parameters critically optimized and enhanced for high performance p-channel 1.3% compressively strained Ge QW FinFET grown on $Si_{0.3}Ge_{0.7}$ buffer on 300mm bulk Si substrate.

Fig. 2: Simulated transverse strain (ε_{xx}) profile for W_{Fin}=20nm Ge QW FinFET; color scale is in % strain, (b) Simulated transverse and longitudinal strain in channel as a function of fin width, (c) simulated hole effective mobility as function of uniaxial stress in channel for (100) and (110) orientated Ge and Si substrate.

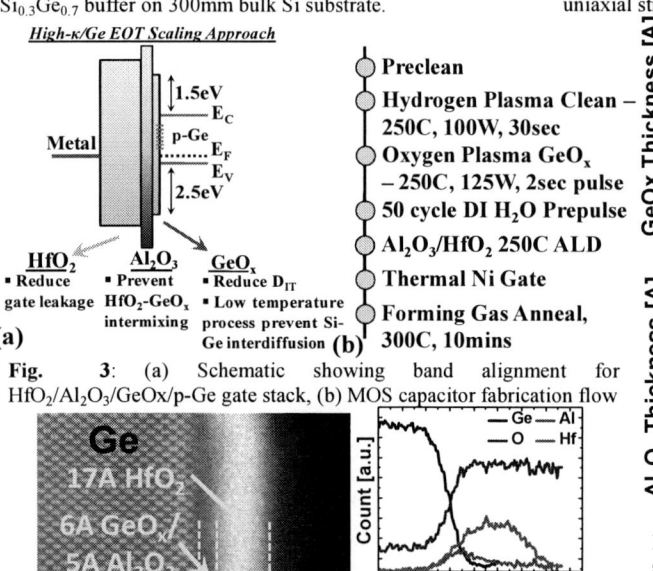

Fig. 3: (a) Schematic showing band alignment for $HfO_2/Al_2O_3/GeOx/p$-Ge gate stack, (b) MOS capacitor fabrication flow

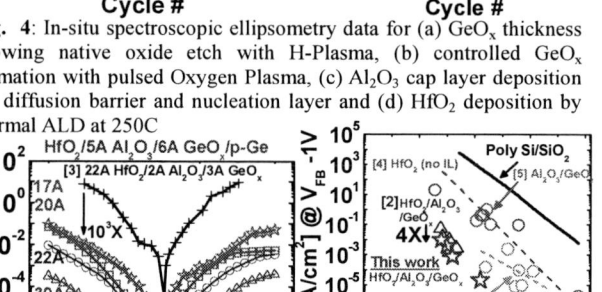

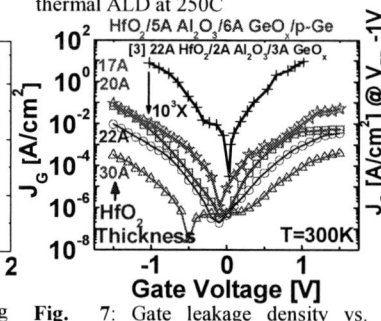

Fig. 4: In-situ spectroscopic ellipsometry data for (a) GeO_x thickness showing native oxide etch with H-Plasma, (b) controlled GeO_x formation with pulsed Oxygen Plasma, (c) Al_2O_3 cap layer deposition for diffusion barrier and nucleation layer and (d) HfO_2 deposition by thermal ALD at 250C

Fig. 5: (a) High resolution cross-section TEM of $HfO_2/Al_2O_3/GeO_x/Ge$ gate stack, (b) EDX line scan across the Ge gate stack

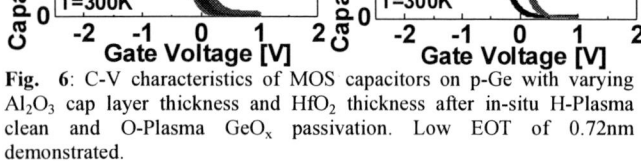

Fig. 6: C-V characteristics of MOS capacitors on p-Ge with varying Al_2O_3 cap layer thickness and HfO_2 thickness after in-situ H-Plasma clean and O-Plasma GeO_x passivation. Low EOT of 0.72nm demonstrated.

Fig. 7: Gate leakage density vs. voltage for $HfO_2/Al_2O_3/GeO_x/p$-Ge MOS capacitors with varying HfO_2 thickness indicating 10^3X reduction

Fig. 8: Gate leakage vs EOT benchmarking of $HfO_2/Al_2O_3/GeO_x/p$-Ge gate stack fabricated after H-Plasma clean with literature.

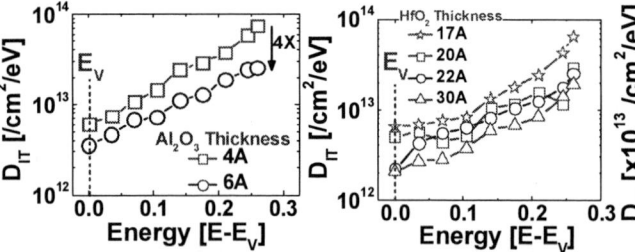

Fig. 9: Extracted density of interface states using equivalent circuit method as function of energy in the bandgap for varying Al_2O_3 and HfO_2 thickness.

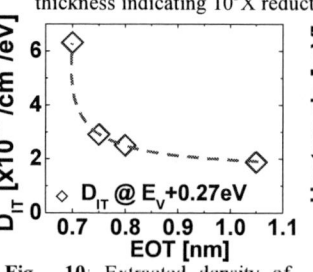

Fig. 10: Extracted density of interface states using equivalent circuit method vs. EOT at midgap

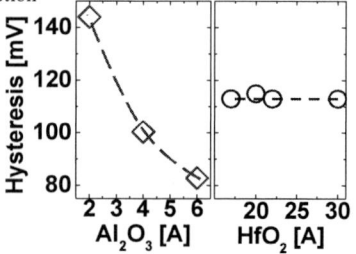

Fig. 11: Hysteresis vs Al_2O_3 and HfO_2 thickness with H-Plasma clean and Oxygen plasma oxidation indicating strong dependence on Al_2O_3 cap layer.

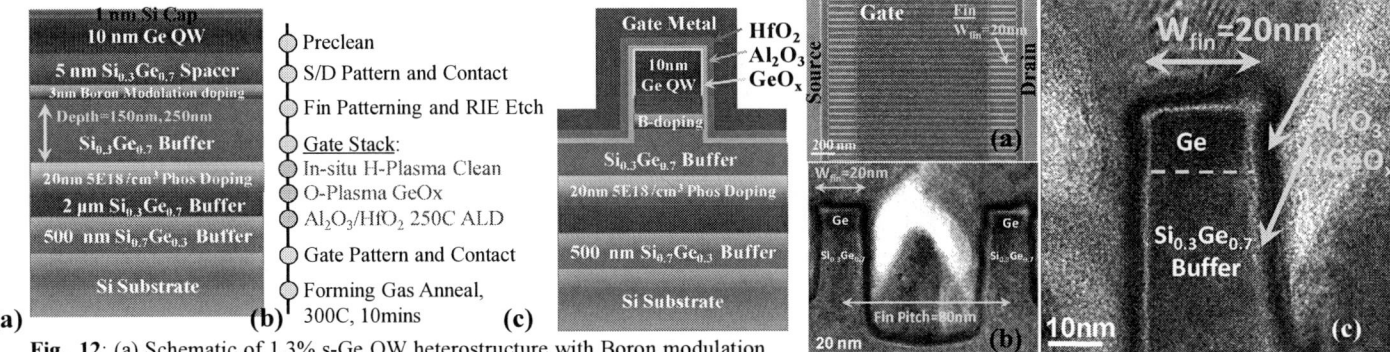

Fig. 12: (a) Schematic of 1.3% s-Ge QW heterostructure with Boron modulation doped SiGe buffer in addition to Phos doping to reduce parallel conduction in the buffer, (b) fabrication process flow for s-Ge QW MOSFET and FinFET with varying fin width, (c) schematic of s-Ge QW FinFET with in-situ H-Plasma clean, oxygen plasma GeO_x formation prior to high-κ deposition.

Fig. 13: (a) SEM showing long channel multi-fin device with W_{Fin}=20nm and fin pitch of 60nm, (b) cross section TEM showing s-Ge QW FinFET with 60nm fin pitch(c) HR-TEM showing vertical fin profile for 20nm W_{Fin} device

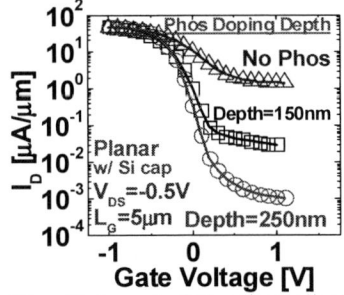

Fig. 14: Impact of Phosphorus doping depth in SiGe buffer on Id-Vg characteristics of s-Ge QW MOSFETs showing reducing I_{OFF} with increasing Phos doping depth.

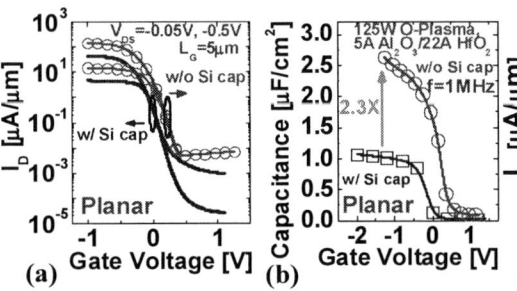

Fig. 15: (a) I_d-V_g characteristic showing impact of Si cap removal prior to high-κ deposition indicating 2X increase in I_{ON} (b) Split CV measured at 1MHz on planar QW MOSFET showing 2.3X higher C_{max} with Si cap removal and GeO_x passivation.

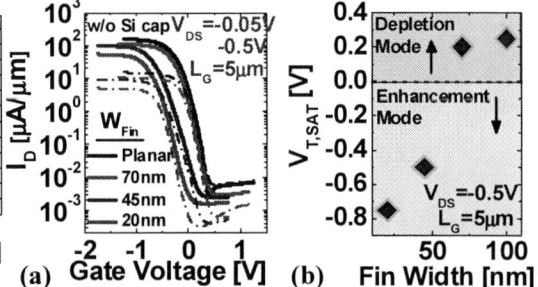

Fig. 16: (a) I_d-V_g characteristics w/o Si cap for high-κ/s-Ge QW MOSFET and FinFET at V_{DS}=-0.05V, -0.5V showing high I_{ON} and excellent I_{ON}/I_{OFF}=2x10⁴, (b) $V_{T,SAT}$ vs W_{Fin} indicating enhancement mode operation due to confinement for QW FinFETs. A conservative W_{Eff}=No. of fins *(W_{Fin}+2T_{QW}) was used

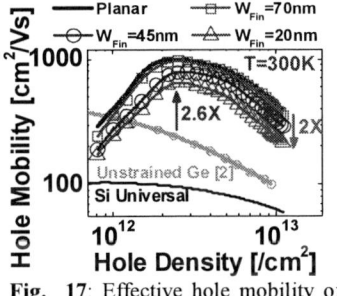

Fig. 17: Effective hole mobility of s-Ge QW MOSFET and FinFET with $HfO_2/Al_2O_3/GeO_x$ gate stack showing 2.6X improvement compared to unstrained Ge

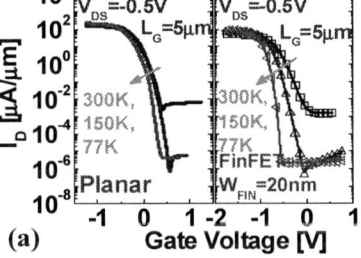

Fig. 18: (a) Id-Vg characteristics at V_{DS}=-0.5V, L_G=5um for high-κ/Ge QW MOSFET and FinFET with reducing temperature, (b) Subthreshold swing vs. temperature for planar MOSFET and FinFET

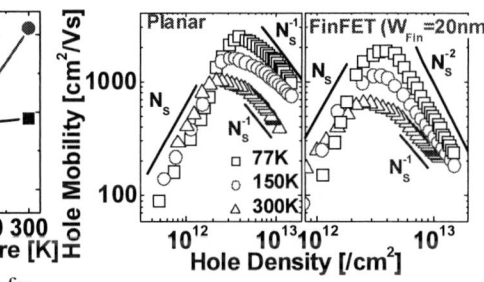

Fig. 19: Temperature dependent hole effective mobility vs hole sheet density for planar QW MOSFET and FinFET indicating Ns dependence.

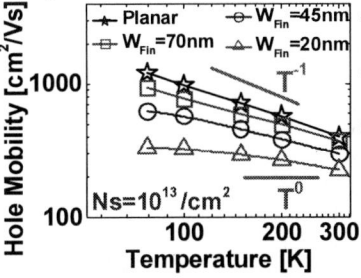

Fig. 20: Hole effective mobility vs. temperature for high-κ/s-Ge QW MOSFET and FinFET at Ns=10¹³ /cm².

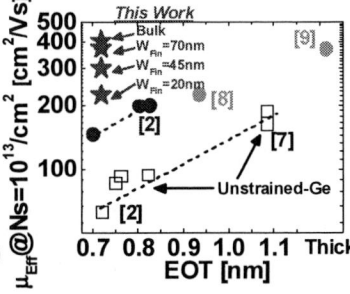

Fig. 21: High field mobility of high-κ/s-Ge QW MOSFET and FinFET in this work vs. EOT, compared with s-Ge pMOSFETs w/o Si cap to date

	W_{fin} (nm)	μ_{Peak} (cm²/Vs)	ε_{xx} (%)	EOT (nm)	μ_{Peak}*C_{ox} (x10⁻⁴)
This work	20	700	1.3	0.72	3.1
W. Chern IEDM,2012	18	850	2.5	1.5	1.9
R. Zhang IEDM,2013	800	380	2.5	0.8	1.6
J. Mitard VLSI,2014	15	600	1.3	1.7	1.6

Table I: Benchmarking of key device parameters demonstrated in this work with other high performance s-Ge FinFETs till date.

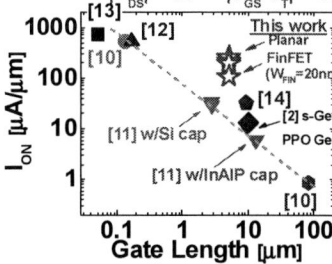

Fig. 23: Benchmarking of I_{ON} of the s-Ge MOSFET and FinFETs with in-situ oxidation achieved in this work with other devices from literature.

978-1-4799-8002-4/14 $31.00 © 2014 IEEE   417

First Demonstration of 15nm-W_{FIN} Inversion-Mode Relaxed-Germanium n-FinFETs with Si-cap Free RMG and NiSiGe Source/Drain

J. Mitard, L. Witters, H. Arimura, Y. Sasaki, A.P. Milenin, R. Loo, A. Hikavyy, G. Eneman, P. Lagrain, H. Mertens, S. Sioncke, C. Vrancken, H. Bender, K. Barla, N. Horiguchi, A. Mocuta, N. Collaert, A.V-Y. Thean

Imec, Kapeldreef 75, B-3001 Leuven, Belgium. E-mail address: Jerome.Mitard@imec.be

Abstract

This work demonstrates the feasibility of an inversion-mode relaxed Ge n-FinFET scaled down to 15-nm fin width and sub-40-nm gate length. CMOS-compatible processing steps such as STI formation, replacement metal gate (RMG), in-situ Phosphorus-doped raised-Source/Drain and a Ni-based contact scheme have been successfully implemented. This first industry-compatible Ge n-FinFET has a $G_{M,SAT,EXT}$ / SS_{SAT} of 250 $\mu S.\mu m^{-1}$ / 130 $mV.dec^{-1}$ (at the targeted V_{DS}=0.5V) which is on par with accumulation-mode junction-less Ge n-FETs.

Introduction

Recently, high-mobility Germanium-based p-FinFETs have been demonstrated to be a viable candidate for active power consumption restricted platforms thanks to its excellent drive current performance at significantly reduced operating voltage ($\pm$ 0.5V V_{DD}) [1-3]. The Ge n-FinFET counterpart promises similar high performance [4] enabling a CMOS solution for future devices but numerous key challenges still remain: the L_G and W_G scalability, gate stack and junction/contact engineering.

This work reports the successful development on a bulk FinFET-based test vehicle compatible with the current industry standards allowing a more in-depth study of all the aforementioned concerns. Limited thermal budget (<600°C) to allow better strain engineering of the Ge p-FinFET counterpart was also considered in this work. Leveraging the device learning from planar devices especially in the gate stack and junctions modules, functional inversion-mode Ge n-FinFETs at aggressively-scaled device dimensions are reported for the first time.

15nm-W_{FIN} & sub-40nm-L_G Device Demonstration

Key fabrication steps are summarized in Fig. 1. An heterogeneous Germanium growth on a 300-mm (100) Si wafer was carried out followed by an anti-punch through implant, fin etch and low temperature STI filled oxide. This module also called STI-Last has also been used to demonstrate high-performance and low-defectivity strained-Ge p-FinFETs [3]. After STI-recess for the fin definition, a low temperature SiO_2 deposition and dummy gate patterning were performed. Source/Drain were designed combining standard Phosphorus and Boron implants for extensions and halos respectively and a sacrificial $SiGe_{45\%}$ was grown to better control the germanidation step subsequently. Dopant activation at a temperature below 600°C was performed. After dummy gate removal, the gate stack was deposited, consisting of:

(1) an Al_2O_3/HfO_2 ALD-deposition with a 300°C oxygen plasma to allow the GeO_2 formation at the Ge channel interface

(2) a *single* replacement metal gate (RMG) ALD-TiN and W, enabling a good gate resistivity at 7nm node- L_G target [5-6].

Finally, Nickel-based Germano-silicidation through contact trenches and standard back-end of line (BEOL) processing completed the device fabrication.

Fig. 2 shows a high-resolution TEM image of the 15-nm wide fin after full device processing. Good conformality of the gate dielectric as well as the metal gate layer are obtained. Sub-40nm gate length scaling is demonstrated in Fig. 3. The I-V characteristics in saturation of a relaxed Ge n-FinFET at this aggressively scaled dimension is shown in Fig. 4 demonstrating excellent I_{ON}/I_{OFF} gain (>5decades), good sub-threshold swing and controlled OFF-state leakage.

Ge MOS Gate Stack Transfer to FinFET

Germanium channel passivation has triggered a lot of research over the past years [7-9]. A thin Si cap layer for Ge p-FET is one of the most promising schemes since it combines an outstanding NBTI robustness [10] with a good high field mobility in the 1.5nm-EOT range [3]. Additionally, near ideal 60mV/dec sub-threshold swing (SS_{LIN}) can be reached (Fig. 5) leading to a key advantage over the other passivation routes in terms of mid-gap interface traps density (D_{IT}) and therefore faster switching. However, the biggest concern is that this scheme does not provide decent electron mobility because of the Fermi-level pinning near the E_C(Ge) [11] or/and low electron mobility in a tensily-strained Si cap layer ($E_{C,Si}$ ~ $E_{C,Ge}$). GeO_2 as an interface layer between Ge and the high-κ offers the opportunity for a common gate dielectric for both Ge p- and n-FETs as shown in Fig. 6.

GeO_2-based channel passivation was therefore investigated on (110) Ge n-FinFETs. Multi-frequencies CV curves are shown in Fig. 7. A deep depletion region at low frequency (i.e. Mid-gap D_{IT}) is observed for both planar-like Ge n-FETs and narrow Ge n-FinFETs. This result is in line with the good SS_{LIN} extracted from these

978-1-4799-8002-4/14 $31.00 © 2014 IEEE

Si-cap free transistors (Fig. 8) and confirms the oxygen plasma passivation as an efficient driver to obtain low D_{ITs} also in 3D-gate Ge devices.

N+/P junction formation in Ge n-FETs

Low reverse diode current (I_R) is necessary to minimize I_{OFF} in transistors. Fig. 9 shows a IV characteristic of N+/P area-intensive diode and an I_R value as low as 0.2 A/cm^2 at $V_R = -1V$ is reported. This value is similar to that of our standard P+/N junctions (Fig. 10) owing a Ge pre-amorphization step which boosts the dopant activation in Ge ($\sim$1E20/cm^3 as classically reported [13]). Therefore, good N+/P diodes can be obtained and do not appear as a key concern for *planar* Ge n-channel transistors.

Ge n-FinFETs Transistor Scaling

While short channel effects in planar Ge n-FETs are found to be difficult to control, aggressive L_G scaling can be obtained thanks to the 3D-gate architecture. Fig. 11 shows no significant change in the SS_{SAT} parameter down to the shortest transistors. The extrinsic $G_{M,SAT}$–SS_{SAT} characteristic of Ge n-FinFET (Fig. 12) saturates

at large SS_{SAT} (short L_G) highlighting the presence of a large access resistance. $\sim$2.5kΩ.μm R_{EXT} is indeed found (Fig. 13) despite the use of junction conditions optimized for planar Ge nFETs. This demonstrates that further optimization is required in Ge n-FinFETs most likely in contact resistivity. Linearly-scaled I_D-V_G curve is shown in Fig. 14 and minor hysteresis is found indicating a reduced number of active defects at high V_{DS} and at short dimensions. However, the stochastic behavior of these defects remains to be studied in scaled Ge n-FinFETs. I_D-V_D curves are finally shown in Fig. 15 and demonstrates that similar performance than junction-less accumulation-mode Ge-OI n-FinFETs can be reported.

Conclusions

A bulk FinFET test vehicle has been successfully developed aiming at studying the key challenges faced by Ge n-FinFET. Functional 15-nm fin width and sub-40-nm gate length Ge n-FinFET was reported for the first time. Through an initial gate stack and junction development, the promising electrical performance of this industry-compatible devices was shown.

- GP/Well implants and anneal
- Low Temp. STI + recess (H_{FIN}=20nm)
- Dummy gate definition
- Phosphorus implantation + B Halos
- Nitride Spacer formation
- S/D SiGe 45% + Phosphorus implant (HDD)
- Junction Anneal (< 600°C)
- Inter Layer Dielectric deposition and CMP
- Dummy gate removal and high-k/MG/W fill + CMP
- Germanidation through Contact Hole (NiSiGe)
- W-plug + M1 (Cu)
- SiC passivation

Fig.1 : Process flow description of the Ge n-FinFET developed in this work.

Fig.2 : A high resolution TEM picture of the ~15-nm W_{FIN} transistor is shown (after fully processing). This device owns a GeO$_2$, Al$_2$O$_3$ and HfO$_2$ gate stack (Si cap free) together with a single TiN metal gate. Excellent conformity is obtained over the fin thanks the Atomic Layer Deposition (ALD) process.

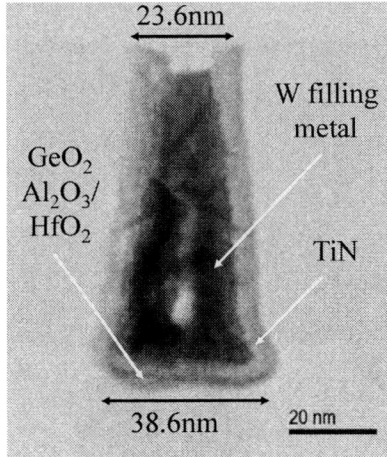

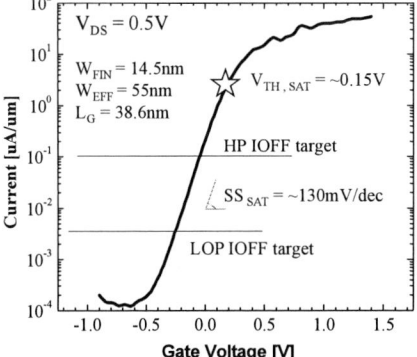

Fig.3-Left: High resolution TEM picture of the short-L_G device (after full processing) : ~24nm-L_G and ~40nm-L_G are measured respectively on top and bottom of the gate.

Fig.4-Right : Source current vs Gate voltage in saturation regime (V_{DS} = 0.5V) demonstrating that the 14.5nm-W_{FIN} and ~40nm-L_G Ge-nFinFET is fully functional. The single TiN/W metal gate together with GeO-based scheme provide a V_{TH_SAT} value close to the nMOS target (~0.2V). Good sub-threshold swing is obtained with controlled OFF-state leakage.

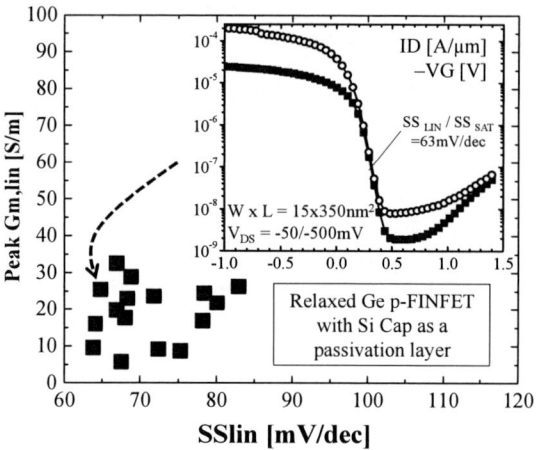

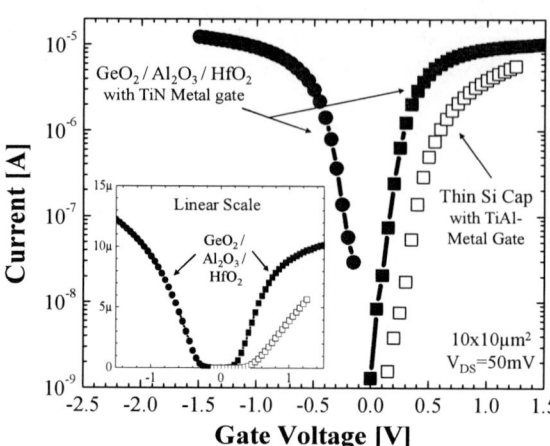

Fig.5 : Peak $G_{M,LIN}$ extraction as a function of SS_{LIN} for (110)-oriented relaxed Ge p-FinFETs using a thin Si cap as a passivation layer. Near ideal SS_{LIN} was found (~63mV/dec) using this scheme [inset].

Fig.6 : Full IV curves (log and linear scale) screening different Ge passivation options. The Si cap passivation scheme does not provide decent electron mobility while the GeO$_2$-Al$_2$O$_3$-HfO$_2$ option offers the opportunity to have common gate dielectrics for p- and n-Ge FETs with similar performance (inset).

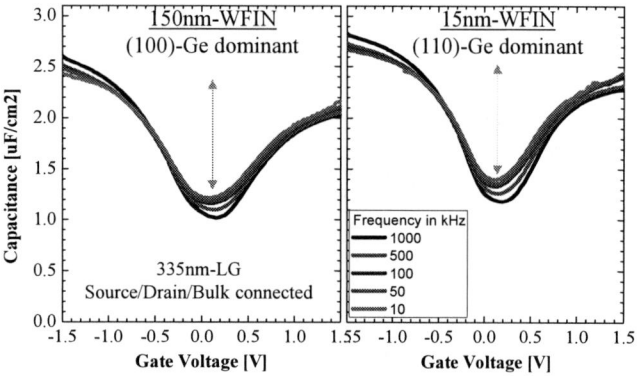

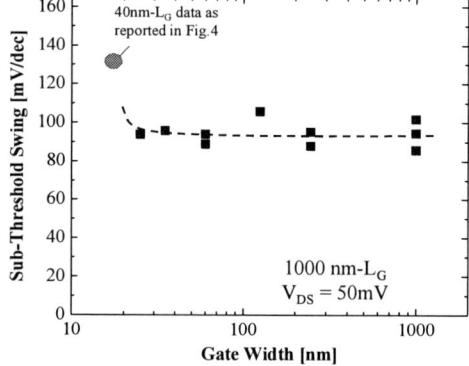

Fig.7 : Multi-frequencies CV curves from planar-like (Left side) and from the narrowest (Right side) transistors. The GeO$_2$-Al$_2$O$_3$-HfO$_2$ option is used in both characterization. Referring to the absolute C_{MIN} values and frequency dispersion in CVs, slight D_{IT} increase is seen at 15-nm W_{FIN}

Fig.8 : Linear sub-threshold swing as a function gate width. Down to 25nm-W_{FIN} dimensions, sub-100mV/dec is obtained. At 15-nm W_{FIN}, SS_{LIN} slightly increases in line with CV characterization (Fig.7)

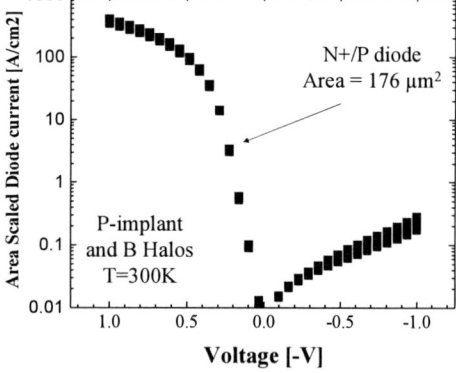

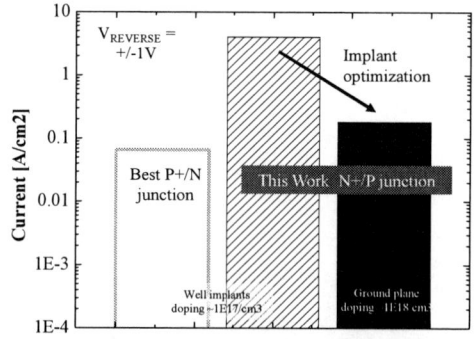

Fig.9 : IV characteristic of an n+ / p diode of Ge n-FETs. SiGe45% was grown on Ge followed by extension, halos and HDD implants. The SiGe 45% was fully silicided during the NiGe process Implanted well was used (N_D~1E18/cm^3).

Fig.10 : Reverse current benchmark (V_R=+/-1V). After junction optimization done on relaxed Ge planar devices, the reverse current of N+/P diode is reduced significantly while the ground plane doping has increased to better control the short channel effects. Best values are close to our reference P+/N diode having high Boron activation (Ge PAI and 500°C).

978-1-4799-8002-4/14 $31.00 © 2014 IEEE

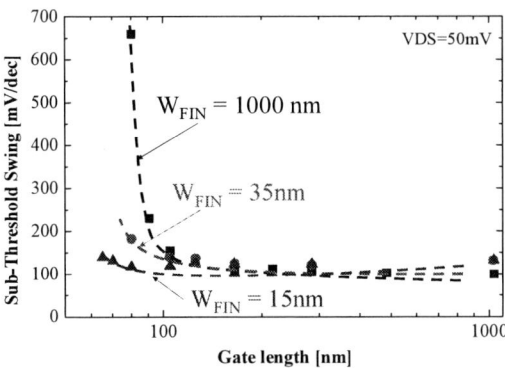

Fig.11 : Sub-Threshold swing as a function of gate length for different fin width. The electrostatics was drastically improved from planar to Ge n-FinFETs.

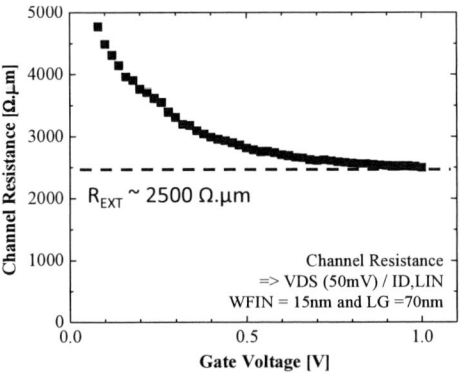

Fig.13 : External resistance extraction from a 15-nm W_{FIN} and short channel relaxed Ge n-FinFET. Despite the junction optimization carried out on planar Ge n-FETs (leading to an excellent reverse current ~0.15 A/cm^2 at N_D~1E18/cm3), further optimization is need in Ge n-FinFET.

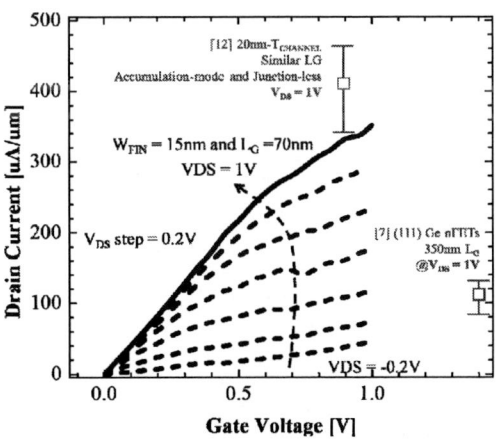

Fig.15 : I_D-V_D at different gate voltage. Despite the R_{EXT} penalty, the performance of the Ge n-FinFET from this work is on par with the best reported non-standard Ge n-FinFETs.

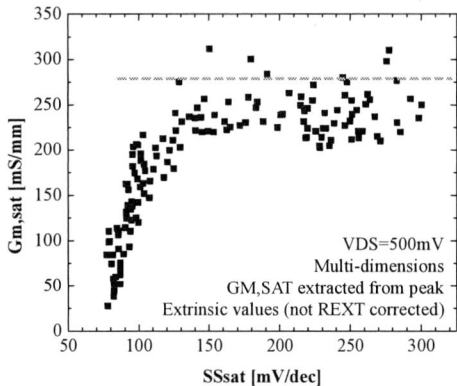

Fig.12 : Extrinsic peak $G_{M,SAT}$ as a function of sub-threshold swing in saturation. Classical $G_{M, SAT}$ increase with L_G scaling (SS_{SAT} increase) is first observed. Then, $G_{M,SAT}$ is kept fixed at the shortest dimensions.

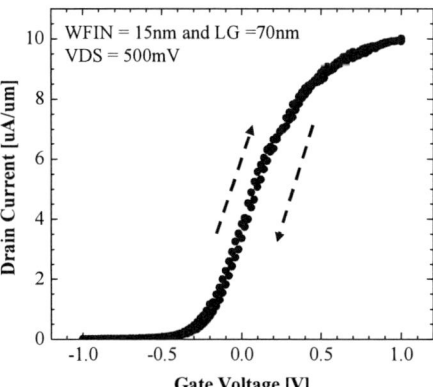

Fig.14 : I_D-V_G curve sweeping over 2V gate voltage and showing limited hysteresis in our GeO$_2$/Al$_2$O$_3$/HfO$_2$ gate dielectric. This result shows that few defects are active at short channel and operating drain voltage.

REFERENCE

[1] P. Hashemi, et al., VLSI proceedings, p. 18, 2014. [2] B. Duriez, et al., IEDM proceedings, p. 522, 2013. [3] J. Mitard, et al., VLSI proceedings, p. 138, 2014. [4] G. Eneman, et al., IEDM proceedings, p. 134, 2012 [5] L-Å Ragnarsson, et al., VLSI proceedings, p. 56, 2014. [6] M. Garcia Bardon, VLSI proceedings, p. 110, 2014. [7] S.H. Hsu et al., IEDM proceedings, p. 525, 2012. [8] B. Liu et al., IEDM proceedings, p. 657, 2013. [9] R. Zhang et al., IEDM proceedings, p. 371, 2012. [10]. J. Franco et al. IEDM proceedings, p. 397, 2013. [11] A. Dimoulas, et al., Appl. Phys. Lett., 89, 252110, 2006. [12] H. Wu, et al., VLSI proceedings, p. 96, 2014. [13] G. Hellings, et al., ECST, 12(12), H417, 2009.

ACKNOWLEDGEMENT

The imec's (sub-)10 nm CMOS partners involved in the logic program, ASM, European commission and local authorities, the imec pilot line and amsimec (test lab) are acknowledged for their support.

High-Performance Tri-Gate Poly-Ge Junction-Less P- and N-MOSFETs Fabricated by Flash Lamp Annealing Process

K.Usuda, Y.Kamata, Y.Kamimuta, T.Mori, M.Koike, and T.Tezuka

Collaborative Research Team Green Nanoelectronics Center (GNC), AIST,
Onogawa 16-1, Tsukuba, Ibaraki 305-8569, Japan
(Present address) Tel: +81-44-549-2312, Fax: +81-44+520-1257, e-mail: koji.usuda@toshiba.co.jp

Abstract

In order to realize high-performance stacking CMOS on insulator layers for 3D-LSIs, a technique for fabricating high-quality poly-crystalline Ge (poly-Ge) MOSFETs is mandatory. We employed the flash lamp annealing (FLA) method to grow high-quality poly-Ge by crystallization after melting an amorphous Ge layer. The Hall effect mobility of the p- and n-type poly-Ge film attained by the method was as high as 200 and 140 cm^2/Vs. Tri-gate junction-less (JL) p- and nMOSFETs were successfully fabricated on the poly-Ge channels without channel doping. The p- and nMOSFET exhibited high drive currents of up to 311 and 119 $\mu A/\mu m$ which were comparable to that of $L_g = 60$ nm node c-Si pMOSFET, and the record value of poly-Si nMOSFET, respectively.

Introduction

Sequential stacking of CMOS devices on inter-layers [1] is an attractive way to form 3D-LSIs with far denser interconnection density between upper and lower CMOS layers than those obtained by TSV technologies. Although amorphous Si [2], poly-crystalline Si [3] and poly-crystalline Ge (poly-Ge) transistors on insulator layers [4-5] have been investigated, their performances have been lower than that of crystalline-Si (c-Si). Therefore, we employed the flash lamp annealing (FLA) method [6] to form high-quality poly-Ge film on an insulator layer to overcome the performance issues of conventional poly- or c-Si MOSFETs.

In this study, we successfully obtained high Hall hole and electron mobilities of 200 and 140 cm^2/Vs for the FLA-grown poly-Ge film, which are larger than the drift hole and electron mobilities in c-Si. Using the n-type FLA poly-Ge layers, greatly improved ($\times 10$) drive currents of junctionless (JL) poly-Ge nMOSFETs were attained. Furthermore, high current drivability of the JL poly-Ge pMOSFETs comparable with or higher than $L_g = 60$ nm node c-Si pMOSFETs thanks to an effective hole mobility 22% higher than that of c-Si/SiO_2 inversion hole mobility is demonstrated.

Device structure and fabrication

Figures 1 and 2 are schematic diagrams of the target device structure of the stacked CMOS devices and the procedure of fabricating the JL poly-Ge p- and nMOSFETs with the FLA process. First, a 72-nm-thick amorphous Ge (a-Ge) layer for nMOSFET and 90 nm a-Ge layer for pMOSFET were deposited by sputtering on

a 200-nm-thick SiO_2 layer formed by thermal oxidation of Si (001) substrate. The substrate was located on a susceptor held at 400°C in nitrogen atmosphere. Then, a flash lamp pulse with the power of 80 J/cm^2 and duration of 10 ms was irradiated on the substrate to crystallize the layer. Since the poly-Ge channels were naturally p-type [5], the channels were undoped for pMOSFETs. On the other hand, the newly-employed two-step FLA process was carried out to form n-type poly-Ge layer with phosphorous (P) ion-implantation (I/I) and activation by the FLA after the first FLA process.

Figure 3 shows a schematic diagram of tri-gate FLA poly-Ge JL MOSFETs. Poly-Ge fins with a fin gate width (W_{fin}) of 15–57 nm were formed on the substrate by electron beam lithography and reactive ion etching (RIE). A 5-nm-thick HfAlO gate dielectric layer with an equivalent oxide thickness of 1.8 nm was deposited by atomic layer deposition at 250°C. A TaN gate electrode and a SiO_2 hard mask layer were deposited and patterned by electron-beam lithography and RIE, respectively. To form source/drain (SD) areas, a Ni germanidation process with low contact resistivity (ρC) was employed for both n- and pMOSFETs followed by a BEOL process at a temperature of 350°C or less. Two types of Ni germanidation process (Type I/II) were performed to form the SD contact for nMOSFETs. The Type I process was performed by using the conventional self-aligned Ni germanidation process, whereas the Type II process was selectively applied for the contact hole area on the n-type FLA poly-Ge layer by employing the two-step P ion-implantation [7].

To evaluate the poly-Ge crystal quality and orientation, Raman spectroscopy, cross-sectional transmission electron microscopy (TEM), and electron backscatter diffraction (EBSD) analyses were performed.

Results and discussion

(A) Physical properties of the FLA poly-Ge:

Figure 4 shows the EBSD images for the 30- and 72-nm-thick poly-Ge layers grown by the two-step FLA method. The grain size of the n-type FLA poly-Ge layers was found to be typically 3 $\mu m \times 300$ nm from the EBSD images, which showed larger size for the thicker poly-Ge layer. Figure 5 shows Raman spectra of the poly-Ge grown by the two-step FLA method as a function of the a-Ge thickness. Sharp poly-Ge spectra were obtained, which indicated fine crystalline nature both for the thicker and thinner Ge layers after the implanted P activation of

978-1-4799-8002-4/14 $31.00 © 2014 IEEE

the FLA. The full width at half maxima (FWHM) of the Raman peaks (Fig. 6) suggested better crystalline quality for the thicker poly-Ge layer because of the narrower FWHM irrespective of the P-dose concentration.

Figure 7 shows lower hole concentration and higher Hall hole mobility for thicker p-type poly-Ge layers before I/I, which is consistent with Fig. 6. Hall hole mobility of up to 200 cm^2/Vs was attained for the 100-nm-thick FLA poly-Ge layer. On the other hand, Fig. 8 shows the P dose dependence of Hall electron mobility and the concentration of the 72-nm-thick n-type poly-Ge layer grown by the two-step FLA method. The active carrier density is saturated to ~1×10^{19} cm^{-3} at the dose levels examined in this work, whereas the electron mobility increased with dose level up to 140 cm^2/Vs. Comparison with the mobility of the bulk-Si is also important. Note that the high Hall hole mobility of n-type poly-Ge was obtained at the P concentration of 5×10^{15} (cm^{-2}) at the ion energy of 10 keV, and that the implantation process for growth of the n-type poly-Ge for fabricating the nMOSFET channel layer was carried out under the condition of the P implantation. As shown in Fig. 9, the obtained Hall hole and electron mobility values overcome the drift mobility of holes and electrons in c-Si [8].

Figure 10 shows a cross-sectional TEM image of the n-type poly-Ge fin channel. The widths of the poly-Ge fin at the middle and the top were about 10 nm and 17 nm, respectively. Judging from the image, the fin was considered to be quasi-single crystalline, since the cross section showed no grain boundary.

(B) JL p-/nMOSFETs fabricated on the FLA poly-Ge:

Figure 11 shows I_d–V_g characteristics of the FLA poly-Ge tri-gate JL pMOSFETs (left) and the two-step FLA poly-Ge tri-gate JL nMOSFETs (Type I) (right) for the gate length (L_g) of 45 nm as a function of fin width. Here, the drain current (I_d) shown in this paper was normalized by the peripheral length of the fin (2 × H_{fin} + W_{fin}). It was confirmed that better cut-off characteristics were exhibited for channel widths narrower than 37 nm for pMOSFET and 22 nm for nMOSFET. Figure 12 shows reasonable I_d–V_g curves of the pMOSFETs (left) and the nMOSFETs (right).

Figure 13 shows effective hole and electron mobilities for the p- and nMOSFET obtained from split C-V measurements. The maximum hole mobility for the poly-Ge pMOSFET was 115 cm^2/Vs, which was 22% higher than that of a c-Si/SiO_2 pMOSFET in the high electrical field region of 1×10^{13} cm^{-2}. On the contrary, the peak electron mobility was 38 cm^2/Vs for the nMOSFET. The deviation of the Hall mobility suggests an additional degradation mechanism, which could be surface scattering mechanisms on the fin sidewalls, because the Hall factor for electrons in Ge is reported to be close to unity [9].

Figure 14 shows I_d-V_d characteristics of the FLA poly-Ge p- and nMOSFETs. A high drain current I_d of 311 µA/µm, which was comparable to that of c-Si pMOSFET with L_g = 60 nm node [10], was attained for the

pMOSFET and a high drain current I_d of 119 µA/µm was also attained for nMOSFETs. Figure 15 shows the effect of SD contact formation. A significant improvement of up to 10^5 in the Type-II devices was observed due to the larger offset between the NiGe-edge and the gate-edge. This result indicates that there is room to optimize the cut-off characteristics by adjusting the NiGe contacts. The benchmarking of current drivability (Table 1) indicates comparable and much greater values for poly-Ge p- and nMOSFETs, respectively, compared with poly-Si p-/nMOSFETs.

Conclusion

High Hall mobilities of up to 200 and 140 cm^2/Vs for p- and n-type poly-Ge layers on insulator layers were obtained by the FLA process. The mobility values were both higher than the drift mobility for p- and n-type c-Si with the same carrier concentration. Tri-gate JL p- and nMOSFETs were fabricated on the FLA poly-Ge layers. The pMOSFET exhibited a high drive current of up to 311 µA/µm at $V_g - V_{th} = -1$ V and $V_d = -1$ V which was comparable to that of the L_g = 60 nm node c-Si pMOSFET, due to the 22% higher effective hole mobility than that of a c-Si/SiO_2 pMOSFET. The nMOSFET also attained a high drive current of 119 µA/µm, which was almost the same as the record value [3] of poly-Si nMOSFETs. The high current drivability for p- and nMOSFET suggests that the FLA poly-Ge CMOS has great potential for future 3D LSIs.

Acknowledgement

This research was granted by the Japan Society for the Promotion of Science (JSPS) through the "Funding Program for World-Leading Innovative R&D on Science and Technology (FIRST Program)," initiated by the Council for Science and Technology Policy (CSTP).

References

[1] P.Batude et al., IEDM Tech. Dig., 151 (2011).

[2] T. Naito et al., VLSI Symp. Tech., 219 (2010).

[3] C.C.Yang et al., IEDM Tech. Dig., 731 (2013).

[4] Y. Kamimuta et al., VLSI-TSA, 109 (2013).

[5] Y. Kamata et al., VLSI Symp. Tech., 131 (2013).

[6] K.Usuda et al., APEX 7, 056501 (2014).

[7] M.Koike et al., APEX 7, 051302 (2014).

[8] C.Jacoboni et al., Solid State Elect. 20 p77 (1977).

[9] V. M Babich et al., Ukrain Fiz. Zhumal 14, 418 (1969).

[10] S. Thompson et al., IEDM Tech. Dig., 61 (2001).

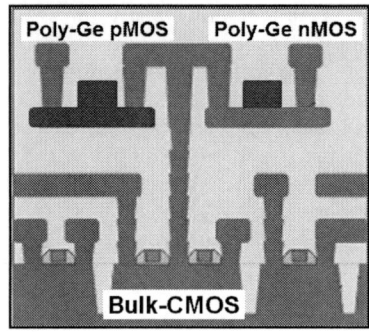

Fig. 1 Schematic of 3D-CMOS device. 3D-LSI is formed with sequential 3D stacking of CMOS on inter-layers.

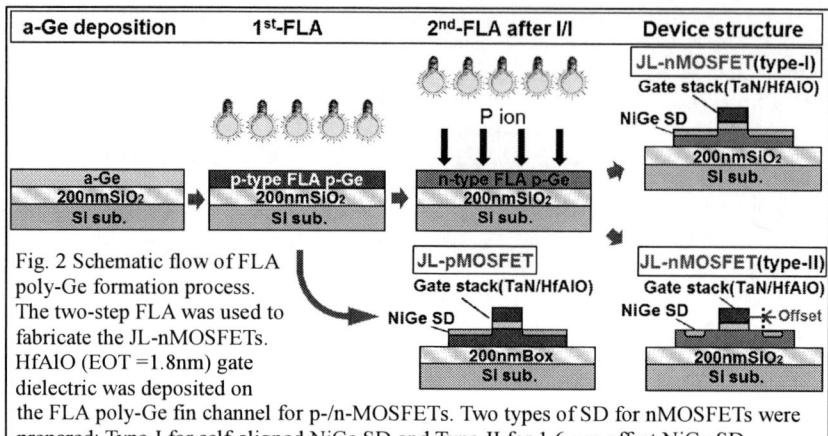

Fig. 2 Schematic flow of FLA poly-Ge formation process. The two-step FLA was used to fabricate the JL-nMOSFETs. HfAlO (EOT =1.8nm) gate dielectric was deposited on the FLA poly-Ge fin channel for p-/n-MOSFETs. Two types of SD for nMOSFETs were prepared: Type-I for self-aligned NiGe SD and Type-II for 1.6 μm offset NiGe SD.

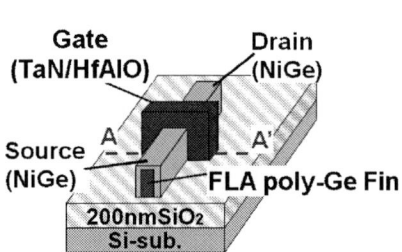

Fig. 3 Schematic structure of tri-gate FLA poly-Ge junction-less nMOSFET (Type-I). The cross-sectional TEM image shown in Fig. 10 was obtained along dotted line A-A'. Note the difference of SD structure with Ni germanide between Type-I and Type-II as shown in Fig. 2.

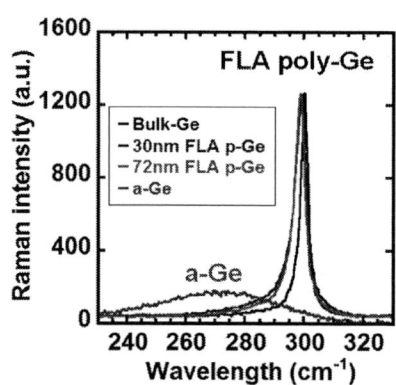

Fig. 5 Raman spectra of amorphous Ge, 30nm and 72nm poly-Ge grown by the two-step FLA method, and bulk-Ge. Sharp poly-Ge spectra were obtained after the two-step FLA process.

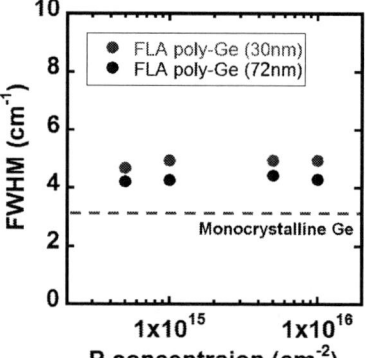

Fig. 6 P concentration dependence of FWHM of Raman spectra for poly-Ge grown by the two-step FLA method. The red dotted line denotes the FWHM of mono-crystalline Ge. The narrower FWHM suggests good crystallinity within the FLA poly-Ge.

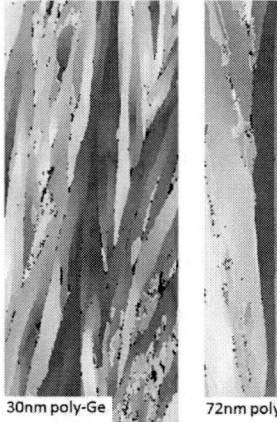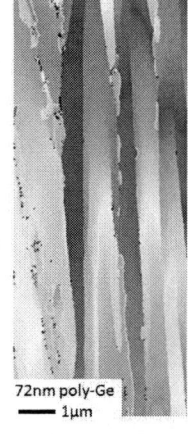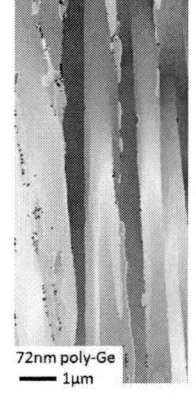

Fig. 4 EBSD images of 30nm (left) and 72nm (right) P doped (5x10^15 cm^-2) n-type poly-Ge layer grown by the two-step FLA method. The defects which were observed in the left figure as black contrast were significantly decreased for the 72-nm-thick poly-Ge layer.

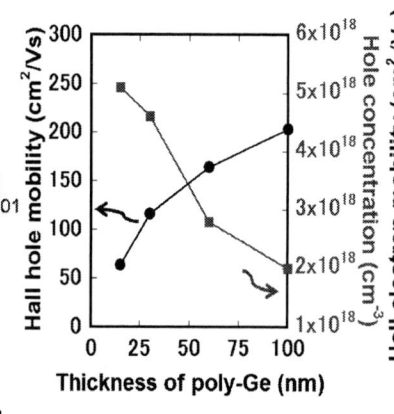

Fig. 7 Hole mobility and carrier concentration as a function of FLA poly-Ge thickness.

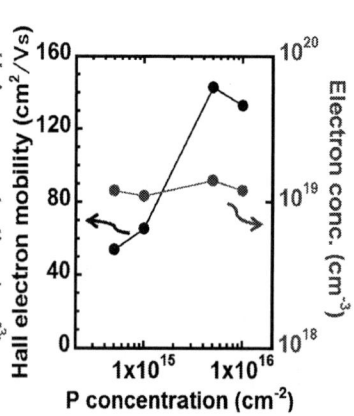

Fig. 8 P dose dependence of Hall electron mobility and the concentration of 72-nm-thick poly-Ge layer grown by the two-step FLA method.

978-1-4799-8002-4/14 $31.00 © 2014 IEEE

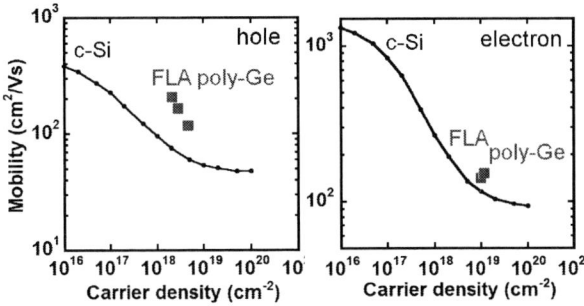

Fig.9 Comparison between Hall hole (left) and Hall electron (right) mobility for the FLA poly-Ge (red squares) and reported drift mobility for c-Si [8].

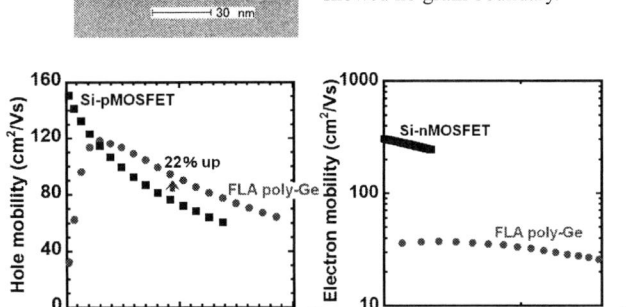

Fig. 10 Cross-sectional TEM image of a typical n-type two-step FLA poly-Ge fin. The widths of the poly-Ge fin at the middle and the top were about 10 nm and 17 nm, respectively. Note that the cross section showed no grain boundary.

Fig. 13 Mobility characteristics for FLA poly-Ge channel JL- p- (left) and nMOSFETs (right) obtained by the split C-V method (red circles). Carrier density is converted to surface carrier density values assuming surface channels, although the carriers should be distributed in the central region in the fin-channels for depletion-type JL devices in reality.

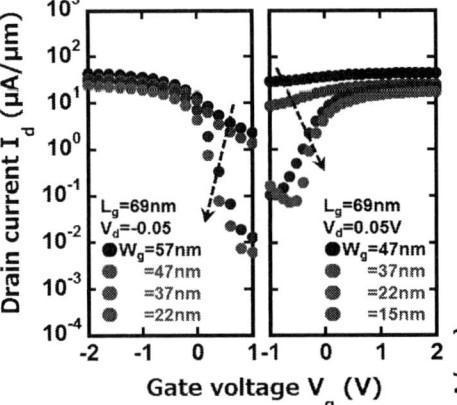

Fig. 11 I_d-V_g characteristics of the FLA poly-Ge tri-gate JL p-MOSFETs (left) and two-step FLA poly-Ge tri-gate JL nMOSFETs (Type-I) (right) as a function of fin width.

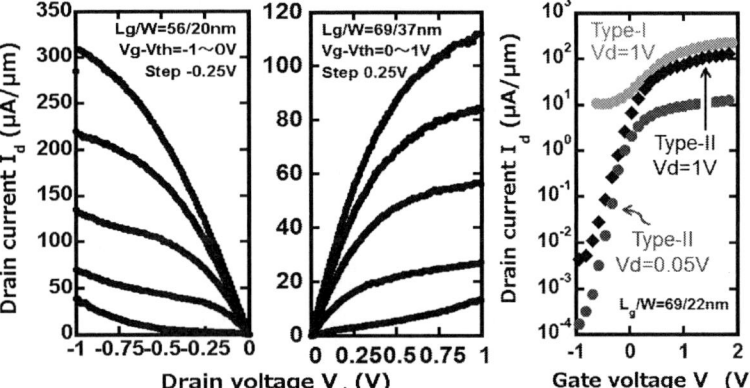

Fig. 14 I_d-V_d characteristics of the Type-I FLA poly-Ge JL-pMOSFETs (L_g/W_{fin}= 56/20 nm) (left) and two-step FLA poly-Ge JL-nMOSFETs (L_g/W_{fin}=69/37 nm) (right).

Fig. 15 I_d-V_g characteristics of the poly-Ge JL- nMOSFETs with two types of NiGe SD (Type-I/II).

Fig. 12 I_d-V_g characteristics of FLA poly-Ge tri-gate JL p-MOSFETs (left) and two-step FLA poly-Ge tri-gate JL nMOSFETs (Type-I) (right) (Lg = 69 nm) as a function of drain voltage.

	VLSI2013[5]	IEDM2013[3]	ISDRS2013[6]	This work
Channel	Ge(SPC)	Si(LC)	Ge(FLA)	Ge(FLA)
I_{on}(n)(μA/μm)	−	121	12	119
I_{on}(p)(μA/μm)	~103	62	280	311
Structure (n/p)	− /JL	Inversion	Inversion/JL	JL/JL
L_g/W_{fin} (nm)	(p)40/7	(n,p)52/50	(n,p)80/50	(n)69/37,(p)56/20

SPC: solid phase crystallization, LC: laser crystallization

Table 1 Electrical properties of poly Ge and Si channel MOSFETs. The tri-gate JL pMOSFET exhibited a drive current of up to 311 μA/μm (V_g-V_{th}=−1V, V_d=−1V) which was comparable to that of L_g= 60 nm node c-Si pMOSFET [10] and the nMOSFET attained a high drive current of 119 μA/μm, which was almost the same as the record value [3] of poly-Si nMOSFETs.

978-1-4799-8002-4/14 $31.00 © 2014 IEEE

Deep Sub-100 nm Ge CMOS Devices on Si with the Recessed S/D and Channel

Heng Wu, Wei Luo, Mengwei Si, Jingyun Zhang, Hong Zhou and Peide D. Ye*

School of Electrical and Computer Engineering, Purdue University, West Lafayette, IN 47906, U.S.A.

*Tel: 1-765-494-7611, Fax: 1-765-496-6443, Email: yep@purdue.edu

Abstract

We report on comprehensive studies of Ge CMOS devices with the recessed channel and S/D fabricated on a Ge-on-insulator (GeOI) substrate. Both nFETs and pFETs with channel lengths (L_{ch}) from 500 to 20 nm, channel thicknesses (T_{ch}) from 90 to 15 nm, EOTs from 5 to 3 nm, and gate stacks with and without the post oxidation (PO) are investigated. Benefiting from the fully depleted ultra-thin body (FD-UTB) channel with a reasonable interface, a low sub-threshold slope (SS) of 95 mV/dec is obtained in a 60 nm L_{ch} nFET and a record high I_{ON}/I_{OFF} ratio of 10^6 is realized in a 300 nm L_{ch} nFET. The recessed contact strongly dependents on the recessed depth and optimized recessed depth significantly improves the Ge contacts.

Introduction

Ge is considered as a promising channel material in the post Si CMOS era, due to its higher and near symmetrical carrier mobilities for both electrons and holes, large density of states, and Si-compatible low temperature process. Recently, many important progresses have been achieved [1-7]. In our previous report [8], a novel recessed channel and S/D technique is used to improve the Ge n-contact and the gate electrostatic control. In this paper, we carried out a comprehensive study of the device performance dependence on L_{ch}, T_{ch}, EOT and the interface passivation of both nFETs and pFETs with the recessed channel and S/D. The post oxidation process improves the $Al_2O_3/GeO_x/Ge$ interface. Determined by ion implantation profiles, the characteristics of nFETs are much more sensitive to T_{ch} than those of pFETs. To better understand the temperature-dependent device OFF- and ON-state characteristics, the OFF-state band-to-band-tunneling (BTBT) generation rate and the ON-state injection velocity are simulated by TCAD.

Experiment

Fig. 1 shows the experiment process flows for both Ge nFETs and pFETs. The device cross section at each main step is given in Fig. 2 and a brief summary on experimental splits is also included. The near-Gaussian distribution of the implanted ions in the Ge is plotted as the color-map shown in the cross section.

The starting material is a GeOI wafer with 180 nm lightly Sb-doped (100) Ge and 400 nm SiO_2 on (100) Si from Soitec (**step 1** in Fig. 2). An over-etched testing recess channel structure is given in Fig. 3, showing the layers of the substrate. The pFETs and nFETs were fabricated in parallel for better comparison. After cleaning and mesa isolation using SF_6 based inductively coupled plasma (ICP) dry etching, ion implantation (**step 2** in Fig. 2) was carried out for nFETs and pFETs using P and BF_2 respectively, which were then separately activated by rapid thermal anneal (RTA) in different conditions. Then, the channel was formed by a SF_6 ICP dry etching (**step 3** in Fig. 2). For the recessed channels, depending on the etching time, T_{ch} of 15, 25, 60 and 90 nm are realized, as shown in Fig. 3 (b-d). Next, the samples were

cyclically rinsed in 2% HF for 3 times as the surface wet clean. For the gate dielectrics, three conditions: 5 nm Al_2O_3 gate dielectric with the Ge post oxidation (PO) (*condition I*, EOT = 3 nm), 8 nm Al_2O_3 gate dielectric with the Ge PO (*condition II*, EOT = 4.5 nm) and 8 nm Al_2O_3 gate dielectric without the Ge PO (*condition III*, EOT = 5 nm), are studied. After the gate dielectric formation, the oxide in the S/D was first etched away and then a BCl_3/Ar ICP dry etching was conducted to partially remove the top doped Ge layer as the recessed S/D etch. The etch rate is calibrated to be 15 nm/min. Determined by etching time, different recess depths were studied to optimize the quality of metal contacts on Ge, as shown in Fig. 4(a-b). Next, 100 nm Ni was deposited as the metal contacts for both nFETs and pFETs (**step 4** in Fig. 2), followed by an Ohmic annealing by RTA. Finally, the gate metal was defined by 40/60 nm Ti/Au for the pFETs and 40/60 nm Ni/Au for the nFETs (**step 5** in Fig. 2). All of the split conditions are applied to both nFETs and pFETs.

Results and Discussion

Fig 4(c) shows the n-Ge TLM results with varying recess etching time (recess depths). The inserted figure depicts the contact resistance (R_c) versus the etching time, demonstrating that R_c first decreases and then increases quickly with the etching time (recess depth). This is resulted from the near-Gaussian distribution profile of doping ions: with extended etching time, the doping concentration at the newly formed surface first increases and then decreases rapidly, hence, there is an optimized recess etching time or depth for low-resistivity Ohmic contacts on ion-implanted Ge. Fig. 5(a) shows the best TLM results, where the low R_c of 0.32 and 0.15 $\Omega\cdot$mm and sheet resistance (R_{sh}) of 80 and 140 $\Omega/\square$ are achieved on n- and p-type Ge contacts, respectively. The difference of R_{sh} between n- and p-type contacts is mainly attributed to the difference of mobility between holes and electrons in Ge. Fig. 5(b-c) provide the ON-resistance (R_{ON}) versus L_{ch} of pFETs and nFETs in *condition III* at $|V_{gs}-V_{TH}| = 2$ V, $|V_{ds}| = 0.05$ V at different temperatures. The source/drain series resistance (R_{sd}) is extracted to be 0.7 and 1.1 $\Omega\cdot$mm for nFETs and pFETs. Although the R_c of p-type Ge contact is smaller than that of n-type Ge contact, R_{sd} of pFETs is larger than that of nFETs because of a factor of 2 larger R_{sh}. The R_{sd} shows negligible dependence on temperature, indicating the dominance of the tunneling current in the Ohmic contacts.

Fig. 6(a) provides the titled SEM image of a fabricated device. The device gate area is enlarged in Fig. 6(b). The recessed channel can be clearly seen in the testing device without the gate metal in Fig. 6(c). Fig. 7(a) shows the transfer curves of a 400 nm L_{ch} pFET in *condition III* at V_{gs} from 1 V to -3V and V_{ds} = -0.05, -0.5 and -1 V. The device has a reasonable SS of 151 mV/dec and a high I_{ON}/I_{OFF} ratio of 5×10^5, both at $V_{ds} = -0.5$ V. The g_m-V_{gs} curves of the same device are given in Fig. 7(b). For comparison, a 300 nm L_{ch} nFET in the same condition is given in Fig. 8, showing a decent SS of 139 mV/dec and a record high I_{ON}/I_{OFF} ratio of 10^6,

978-1-4799-8002-4/14 $31.00 © 2014 IEEE

both at $V_{ds} = 0.5$ V. By further improving the MOS interface using the PO technique, SS as low as 95 mV/dec is achieved in a 60 nm L_{ch} nFET in *condition II* at $V_{ds} = 0.05$ V as shown in Fig. 9(a). SS could to be further reduced by EOT scaling and interface optimization. The g_m-V_{gs} curves of the same device are given in Fig. 9(b). Fig. 9(c) provides the output characteristics of the same device with V_{ds} from 0 V to 1.5 V and V_{gs} from -1 V to 1.5 V in 0.2 V steps.

Fig. 10 provides the V_{TH} scaling metrics of nFETs with T_{ch} of 15/25 nm in *condition I* and T_{ch} of 25 nm in *condition III*. The V_{TH} roll-off resulted from the short channel effects (SCEs) is suppressed in thinner channel devices due to better gate electrostatic control of the channels. Compared with devices in *condition III* with a larger EOT and no PO, devices in *condition I* show slightly better SCE immunity. Figs. 11-12 provide the drain current (I_d) and maximum trans-conductance (g_{max}) scaling metrics of the same set of nFETs in Fig. 10 at $V_{ds} = 0.5$ and 1 V. Better interface after the PO process and smaller EOT provide better ON-state performance. However, smaller T_{ch} greatly reduces I_d and g_{max}, mainly due to the degraded electron mobility [9] and reduced cross-sectional area of current conduction. Fig. 13 shows the SS scaling metrics of the same set of nFETs in Fig. 10. SS increases with decreasing L_{ch}. Smaller T_{ch} and EOT and better interface offer better SS. The I_{ON}/I_{OFF} ratio scaling metrics at $V_{ds} = 0.5$ V in Fig. 14 show the similar trend as SS due to the SCEs.

Fig. 15(a-b) give the transfer curves of pFETs and nFETs with the same L_{ch} of 500 nm but different T_{ch} of 25, 60 and 90 nm, respectively. The nFETs are much more sensitive to T_{ch} than pFETs, as proved by the fast degradation in I_{ON}/I_{OFF} ratio with increasing T_{ch}. This is could be due to that the doping profile is different for n and p Ge ion implantation: the P ion is much more diffusive than BF_2 in Ge [10] and it has a much larger density gradient than that of BF_2 ions [11] in the channel region. The concentration of P ions decreases much faster than BF_2 ions in Ge, resulting in a worse gate control in nFETs than pFETs when T_{ch} increases. Figs. 16-17 show the g_m-V_{gs} and output curves of the same set of devices in Fig. 15. On the condition of same T_{ch} and L_{ch}, nFETs shows better ON-state performance because of a larger electron mobility than hole mobility. For pFETs, the 90 nm T_{ch} device has the largest g_{max} of 100 mS/mm at $V_{ds} = -1$ V. However, for nFETs, the g_{max} of the 90 nm T_{ch} is smaller than that of the 60 nm T_{ch} device which has a g_{max} of 230 mS/mm at $V_{ds} = 1$ V. This can be partly explained by 1) larger OFF-state leakage in thicker channel 2) the fact that the P ion concentration in the upper part of the 90 nm T_{ch} channel is much higher than that of the 60 nm T_{ch} channel and more Coulomb scattering lowers electron mobility.

Fig. 18 provides the V_{TH} scaling metrics of pFETs with T_{ch} of 25, 60 and 90 nm in *condition I* and T_{ch} of 25 nm in *condition III*, showing a similar trend as nFETs. The more negative V_{TH} of the pFETs in *condition III* could be due to more fixed charge in the Ge-oxide without the PO process. Fig. 19-20 show the I_d and g_{max} scaling metrics of the same set of pFETs in Fig. 18 at $V_{ds} = -1$ V. The I_d and g_{max} increase with decreasing L_{ch} as expected. Moreover, their relationship with T_{ch} show similar behaviors as in the nFETs shown in Figs. 11-12 because of the dependence of mobility on T_{ch}. Better interface quality by the post oxidation is confirmed by the improved I_d and g_{max} of the devices in *condition I* compared with the devices in *condition III* with the same T_{ch} of 25 nm. Fig.

21 shows the SS versus L_{ch} relationships of the same set of pFETs in Fig. 18. SS increases with lower L_{ch} and better immunity from SCEs are obtained with reduced T_{ch} and EOT, and higher interface quality. The I_{ON}/I_{OFF} ratio scaling metrics of the same set of pFETs are given in Fig. 22, showing similar trend like that of SS. The I_{ON} increase accounts for the ratio increase when $L_{ch} > 150$ nm and the I_{OFF} increase accounts for the ratio decrease when $L_{ch} < 150$ nm.

Figs. 23-24 show the transfer curves of a 50 nm L_{ch} pFET and a 60 nm L_{ch} nFET in *condition III* at temperatures from 300 K to 400 K at $|V_{ds}| = 0.5$ and 1 V, respectively. The I_{OFF} and SS increases at higher temperatures because of the temperature dependence of the diffusion current in OFF-state and sub-threshold region. Since the change of contact resistance at different temperatures is negligible as proved in Fig. 5, the reduced I_{ON} with the increase of temperature indicates the phonon scattering dominance in carrier motilies at room temperature.

All the error bars in the scaling metrics plots are based on the measurements over average 20 devices at each data point, indicating good device uniformity.

To better understand the device behavior, TCAD simulation is carried out on the recessed channel Ge nFET by Sentaurus. Fig. 25(a-b) shows the simulated electron density in a 50 nm L_{ch} and 15 nm T_{ch} nFET in the OFF and ON-state at $V_{ds} = 0.5$ V, respectively. The black lines in Fig. 25(a) mark the boundary of the depletion region with a full-depleted channel. In the ON-state, a layer with high density accumulation carriers is formed close to the interface between Ge and oxide, as shown in the inserted figure in Fig. 25(b). Fig. 26(a-c) show the BTBT generation rate in the same simulated device with different gate biases at 300 K, explaining the GIDL in the OFF-state. With increasing negative gate bias, the electrical field from drain to channel raises, increasing the band-bending, thus, the BTBT current becomes larger. Fig. 26(d) depicts simulated band-diagram of Ge along the BTBT current path marked by the dashed arrow in Fig. 26(c). Fig. 26(e-f) describe the electron velocity near the source side in device ON-state with different temperatures of 200, 300 and 400 K, showing much larger injection velocity at lower temperatures. Fig. 26(g) provides the simulated I_d-V_{gs} at different temperatures, in consistent with the results in Figs. 23-24.

Conclusion

We present comprehensive studies of the Ge nFETs and pFETs with the recessed channel and S/D on GeOI, fabricated for Ge CMOS applications. With further EOT scaling and interface optimization, the device process developed here is promising for ultimate low-power high-speed CMOS beyond Si.

Acknowledgement

The authors would like to thank J. J. Gu, L. Dong, M. S. Lundstrom and K.K. Ng for the valuable discussions. This work is supported by the SRC GRC program.

Reference

[1] R. Zhang, et al., *VLSI* 2011, p.56. [2] C. Lee, et al., *IEDM* 2009, p. 324. [3] J. Mitard, et al., *IEDM* 2008, p.876. [4] J. Park, et al., *IEDM* 2008, p.389. [5] B. Duriez, et al., *IEDM* 2013, p.522. [6] B. Liu, et al., *EDL*, p.1336, 2012. [7] J. Mitard, et al., *VLSI* 2014, p.138. [8] H. Wu, et al., *VLSI* 2014, p.96. [9] S. Kim, et al., IEDM 2013, p.429. [10] G. Thareja, et al., IEDM 2010, p.245. [11] K. Suzuki, et al., *TED*, p.627, 2009.

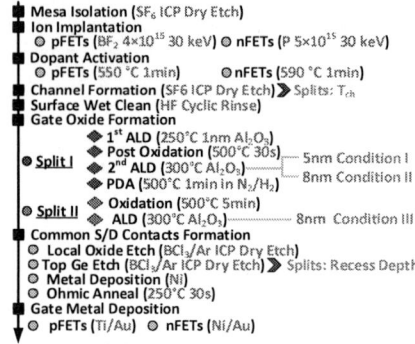

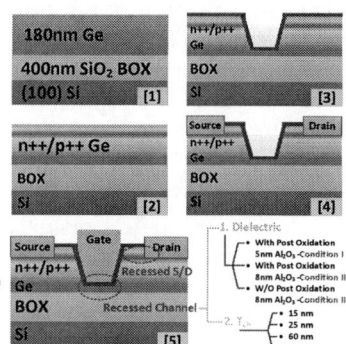

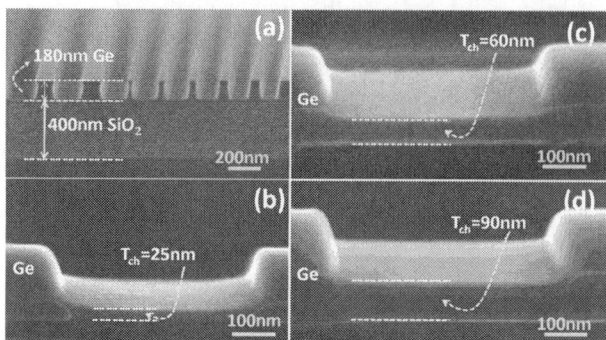

Fig. 1 Fabrication process flow of the Ge recessed channel and S/D MOSFETs. Different gate stacks / T_{ch} / recess depths are studied.

Fig. 2 Device cross sections at different fabrication steps. Experiment splits of gate dielectric and T_{ch} are listed.

Fig. 3 (a) Cross section of testing recessed channel. A substrate with 180 nm Ge on 400 nm SiO$_2$ was used. (b-d) Cross sections of recessed channels with T_{ch} of 25, 60 and 90 nm, respectively.

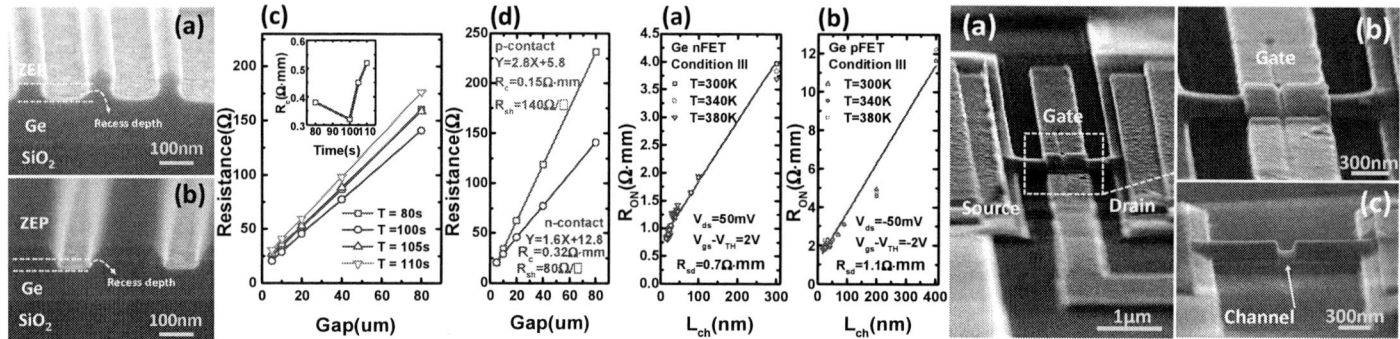

Fig. 4 (a-b) Testing recessed S/D structures with different recess depths. (c) TLM results of recessed contacts on n-Ge with different recess depths (etching time). The inserted figure shows the dependence of R_c on the etching time. (d) Best TLM results on n and p type contacts.

Fig. 5 On-resistance dependence on L_{ch} of nFETs (a) and pFETs (b) in *condition III* at different temperatures. The R_{sd} is extracted.

Fig. 6 (a) Tilted view of a fabricated device. (b) Zoom-in image of the gate area of the device in (a). (c) Tilted view of a testing device without the gate metal. The recessed channel can be clearly observed.

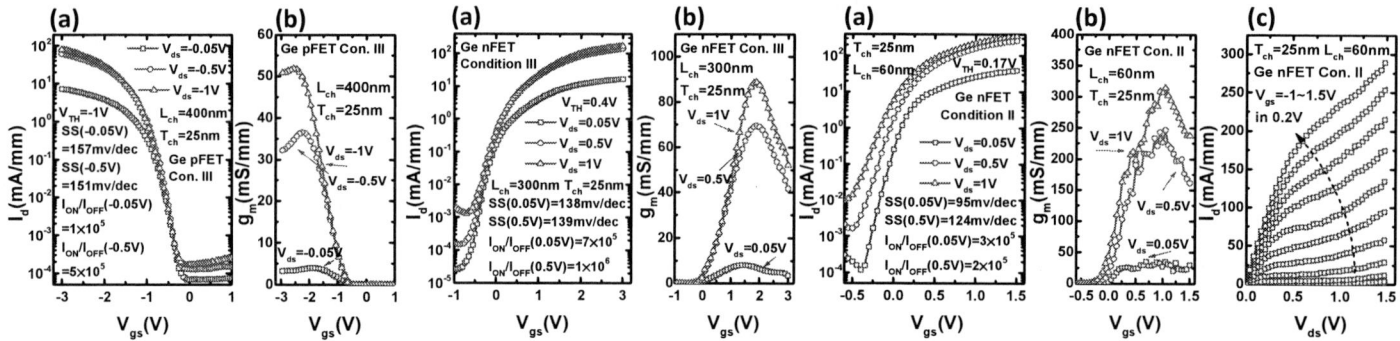

Fig. 7 I_d-V_{gs} curves of a 400 nm L_{ch} pFET in *condition III* with a high I_{ON}/I_{OFF}. (b) The g_m-V_{gs} curves of the device in (a).

Fig. 8 I_d-V_{gs} curves of a 300 nm L_{ch} nFET in *condition III* with a record high I_{ON}/I_{OFF}. (b) The g_m-V_{gs} curves of the device in (a).

Fig. 9 (a) I_d-V_{gs} curves of a 60 nm L_{ch} nFETs in *condition II*. A low SS of 95 mV/dec is obtained. (b) The g_m vs. V_{gs} relationship of the same device in (a). (c) Output curves of the same device in (a).

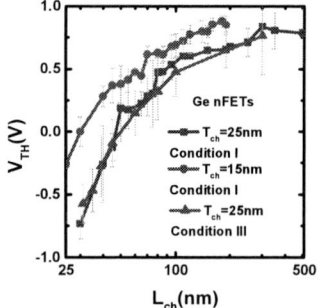

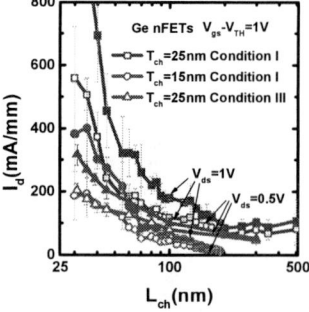

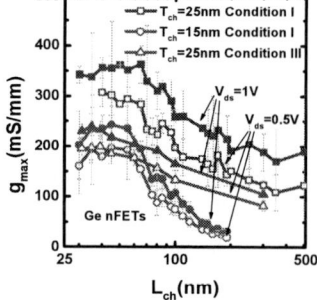

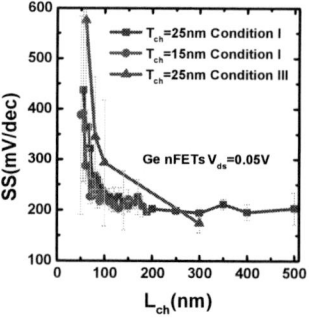

Fig. 10 V_{TH} scaling metrics of nFETs with T_{ch} of 15/25 nm in *condition I* and T_{ch} of 25 nm in *condition III*.

Fig. 11 I_d scaling metrics of nFETs with T_{ch} of 15/25 nm in *condition I* and T_{ch} of 25 nm in *condition III* at V_{gs}-V_{TH} = 1 V and V_{ds} of 0.5 and 1 V.

Fig. 12 g_{max} scaling metrics of nFETs with T_{ch} of 15/25 nm in *condition I* and T_{ch} of 25 nm in *condition III* at V_{ds} of 0.5 and 1 V.

Fig. 13 SS scaling metrics of nFETs with T_{ch} of 15/25 nm in *condition I* and T_{ch} of 25 nm in *condition III* at V_{ds} of 0.05 V.

978-1-4799-8002-4/14 $31.00 © 2014 IEEE

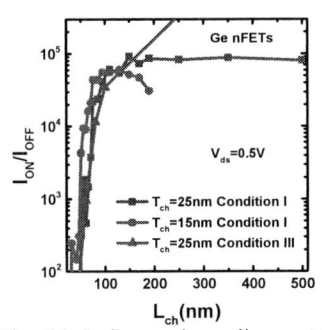

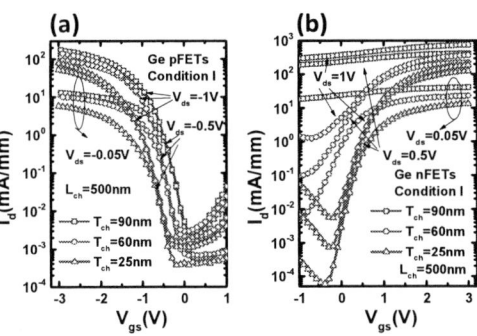

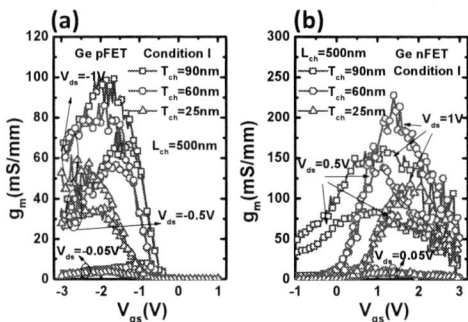

Fig. 14 I_{ON}/I_{OFF} ratio scaling metrics of nFETs with T_{ch} of 15/25 nm in *condition I* and T_{ch} of 25 nm in *condition III* at V_{ds} of 0.5 V.

Fig. 15 (a) Transfer curves of pFETs with T_{ch} of 25, 60 and 90 nm in *condition I*. (b) Transfer curves of nFETs with T_{ch} of 25, 60 and 90 nm in *condition I*.

Fig. 16 (a) g_m v.s. V_{gs} curves of pFETs with T_{ch} of 25, 60 and 90 nm in *condition I*. (b) g_m v.s. V_{gs} curves of nFETs with T_{ch} of 25, 60 and 90 nm in *condition I*.

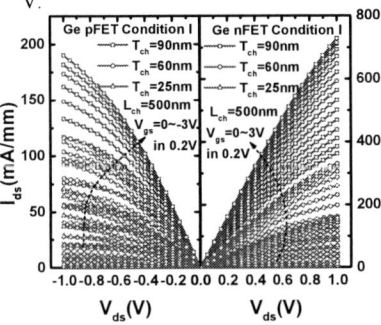

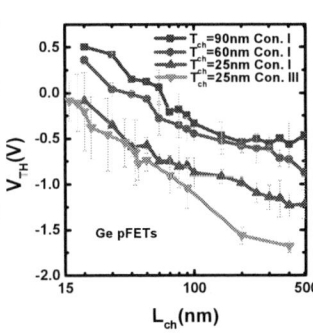

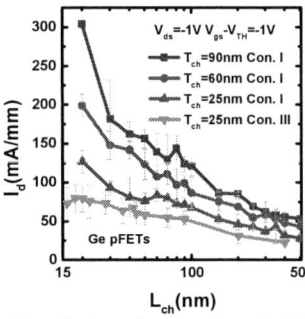

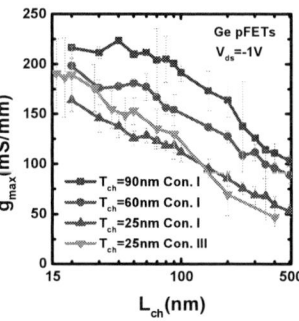

Fig. 17 left: Output curves of the same 3 pFETs in Fig. 15(a). Right: Output curves of the same 3 nFETs in Fig. 15(b).

Fig. 18 V_{TH} scaling metrics of pFETs with T_{ch} of 25/60/90 nm in *condition I* and T_{ch} of 25 nm in *condition III*.

Fig. 19 I_d scaling metrics of pFETs with T_{ch} of 25/60/90 nm in *condition I* and T_{ch} of 25 nm in *condition III* at V_{gs} -V_{TH} = -1 V and V_{ds} of -1 V.

Fig. 20 g_{max} scaling metrics of pFETs with T_{ch} of 25/60/90 nm in *condition I* and T_{ch} of 25 nm in *condition III* at V_{ds} of -1 V.

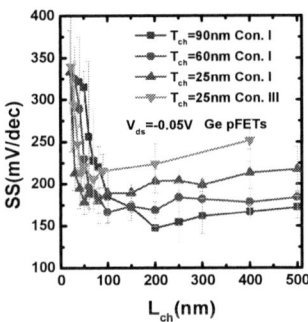

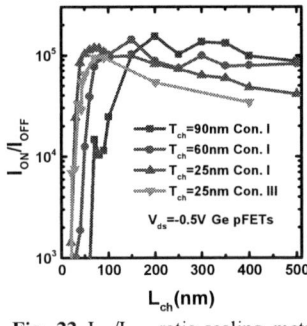

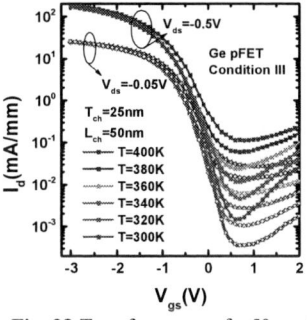

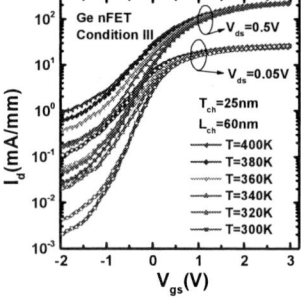

Fig. 21 SS scaling metrics of pFETs with T_{ch} of 25/60/90 nm in *condition I* and T_{ch} of 25 nm in *condition III* at V_{ds} of -0.05 V.

Fig. 22 I_{ON}/I_{OFF} ratio scaling metrics of pFETs with T_{ch} of 25/60/90 nm in *condition I* and T_{ch} of 25 nm in *condition III* at V_{ds} of -0.5 V.

Fig. 23 Transfer curves of a 50 nm L_{ch} pFET in *condition III* at different temperatures from 300 to 400 K.

Fig. 24 Transfer curves of a 60 nm L_{ch} nFET in *condition III* at different temperatures from 300 to 400 K.

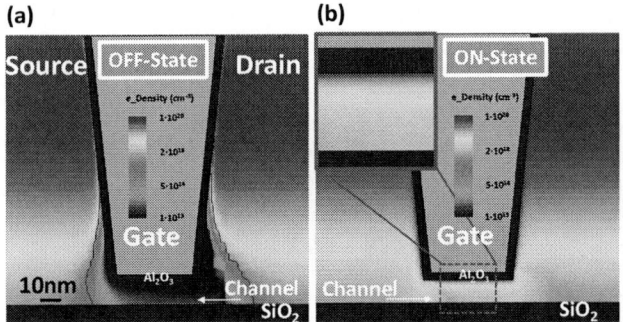

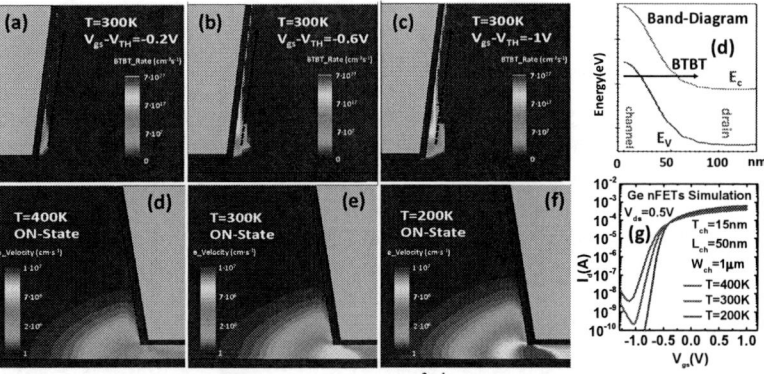

Fig. 25 (a) OFF-state simulation result of electron distribution of a 50nm L_{ch} recessed channel nFET by TCAD. The black lines mark the boundary of the depletion region. (b) ON-state simulation result of electron density of the device in (a). Inserted figure shows the enlarged channel area.

Fig. 26. (a-c) OFF-state BTBT generation rate ($cm^{-3}s^{-1}$) contours of the device in Fig. 25 with V_{gs}-V_{TH} = -0.2, -0.6 and -1 V at 300 K. (d) Simulated band-Structure along the tunneling path marked by the dashed arrow in (c). (d-f) ON-state electron velocity ($cm·s^{-1}$) contours of the same device at T of 400, 300 and 200 K. (f) Simulated transfer curves at different temperatures.

978-1-4799-8002-4/14 $31.00 © 2014 IEEE

The dynamics of surface donor traps in AlGaN/GaN MISFETs using transient measurements and TCAD modelling

Giorgia Longobardi[1], Florin Udrea[1], Stephen Sque[2], Jeroen Croon[2], Fred Hurkx[2] and Jan Šonský[2]

[1]Department of Engineering, Cambridge University, Cambridge, CB2 1PZ, U.K. **E-mail: gl315@cam.ac.uk**
[2]NXP Semiconductors, High Tech Campus 46, 5656 AE Eindhoven, The Netherlands

Abstract

This paper presents a detailed and correlated (i) Id-Vg, (ii) Cgg-Vg, and (iii) transient analysis of donor traps in a SiN/GaN/AlGaN/GaN Metal-Insulator-Semiconductor Field-Effect Transistor (MISFET) fabricated on a silicon substrate. We explain for the first time that the long-time constants are due to the close coupling between the emission/capture processes on one hand and the transient transport of electrons across the GaN/AlGaN barrier on the other. Emission and capture time constants were extracted for several bias conditions and temperatures. Moreover, we have developed a TCAD model that consistently gives a good match to DC, AC, and transient experimental results.

Introduction

Surface donor traps at the passivation/semiconductor interface are known to be of primary importance in the performance and reliability of AlGaN/GaN high-voltage transistors (1). Current collapse, off-state instability, and surface-induced breakdown are some of the negative effects associated with such defects (2, 3). In spite of several attempts in the literature to characterize these traps (4-7), there is a lack of in-depth study to explain their dynamic behavior. This paper analyses the trapping and de-trapping mechanisms into and from surface donor traps through measurements and TCAD simulations of a MISFET. A TCAD model was built in Sentaurus (software by Synopsys) to test the hypotheses made and further explain the measurement outcomes.

Description of the method

A dedicated MISFET with a large gate area is used for this study (Fig. 1). Forward (from $V_{gs} = -30$ to 20 V) and reverse sweeps (from $V_{gs} = 20$ to -30 V) were performed on the structure to obtain the transfer characteristics. The gate capacitance was measured at 1 kHz for the same range of gate biases. Long-time transient drain current (I_d) measurements were performed: the gate potential was stepped from 0 to -9 V with a step size of -1 V and from -9 to 0 V with a step size of 1 V. The response of the drain current to these steps was measured for a period of 10 ks at each bias condition.

The drain bias was kept constant at 0.1 V and the source and substrate potentials at zero bias for all the measurements. The same set of transient measurements was performed for temperatures of 25, 35, 45, and 55° C. For the extrapolation of the emission time constant, the measurement data was first fitted with a smoothing spline and then the derivative was calculated on the fitted curve (Fig. 2). The time constant τ is determined as the time at which the derivative is maximal.

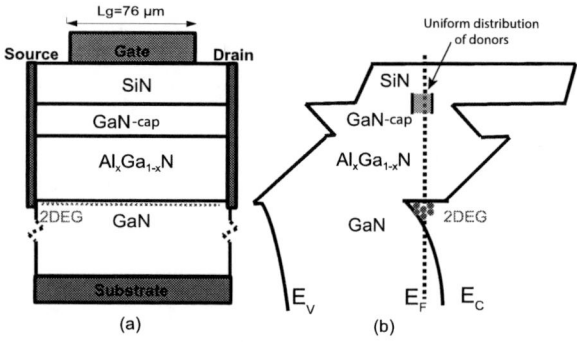

Fig. 1 (a) Cross-section of a large gate area MISFET; (b) band diagrams at zero bias.

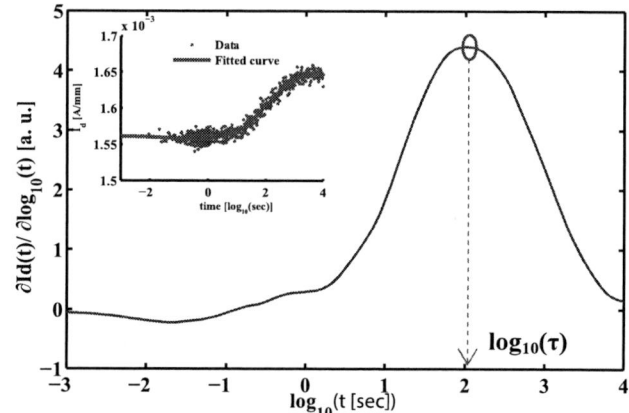

Fig. 2 Step response of the drain current to a gate-bias step from -8V to -9V (inset) and its derivative.

Id-Vg and Cgg-Vg

Fig. 3 shows the forward and reverse sweeps of the Id-Vg performed on the structure. The hysteresis observed has been associated to the presence of surface donor traps (8, 9). The study in (8) shows the presence of several distinct regions in

978-1-4799-8002-4/14 $31.00 © 2014 IEEE   430

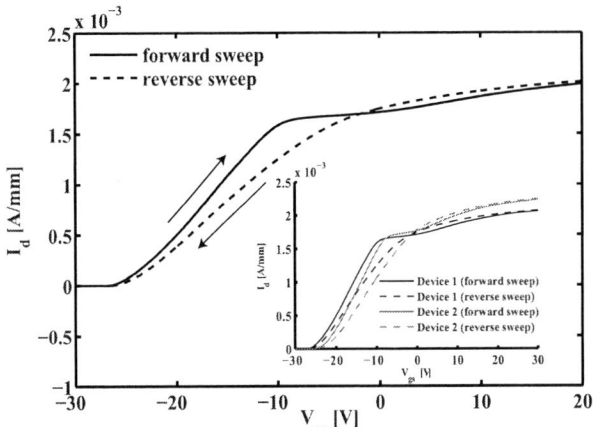

Fig. 3 Id-Vg measurements for the forward and reverse sweeps. Inset: all MISFETs on the same wafer displayed similar behaviour – two are shown here as an example.

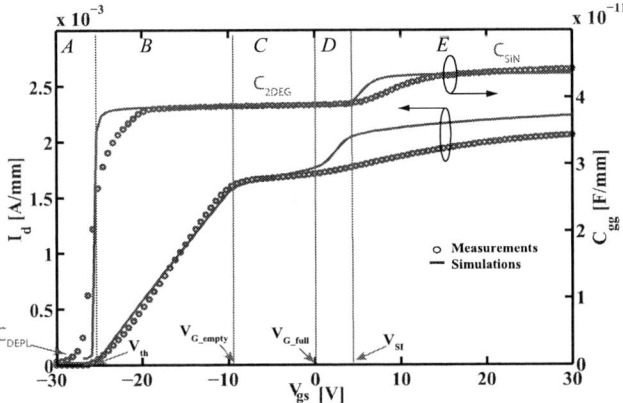

Fig. 4 Measurements and TCAD simulations of the Cgg-Vg (red) at 1 kHz and Id-Vg (blue) forward-sweep characteristics of the MISFET.

the transfer and Cgg-Vg MISFET characteristics (Fig. 4), in which the action of the donor traps as well as that of a surface electron layer are identified, as explained in Fig. 5 via TCAD simulations. A fixed charge density of 5.9×10^{12} cm^{-2} and a donor-like trap density of 1.5×10^{13} cm^{-2}eV^{-1} within an energy range of 0.32 to 0.52 eV below the conduction band in the TCAD input deck gave the best match to the measured characteristics (Fig. 4). The 'kink' (V_{G_empty}) during the forward sweep is due to the transition of the donor traps from being fully ionised (i.e., empty) to partially ionised (i.e., partially occupied) – a capture process. At this kink voltage, the quasi-Fermi level at the SiN/GaN interface is pinned by the donors. As a consequence, the gate potential no longer modulates the charge in the 2DEG (two-dimensional electron gas) layer, hence the drain current saturates, but it assists the de-ionisation of traps until all of them are filled (V_{G_full}). At higher gate voltages (V_{SI}), the conduction band at the SiN/GaN interface reaches the quasi-Fermi level and a surface electron layer forms, which screens the 2DEG from the gate and hence causes a second saturation in the current. As this electron layer is closer to the gate than the 2DEG is, a jump in the gate capacitance (C_{gg}) is observed (Fig. 4), which correlates with the beginning of the second saturation in

current in the Id-Vg (8). Interestingly, the Cgg-Vg sweep cannot capture the movement of donor charge, as the donors are too slow to react to the AC signal (1 kHz). During the reverse sweep in the Id-Vg, the ionisation process (i.e., emission) is considerably slower than the step applied at each bias, hence the amount of ionised donors is less than in the forward sweep for the same V_{gs} and the drain current is thus lower, creating the hysteresis.

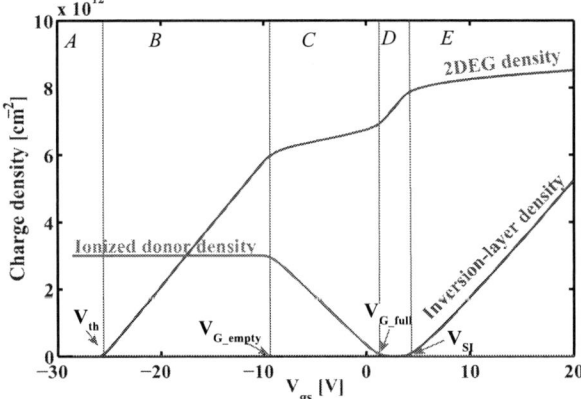

Fig. 5 Charge densities of the 2DEG, ionised donors, and the surface inversion (electron) layer as a function of gate bias in TCAD simulations.

Transient analysis

A. Id-Vg extrapolated from long step responses

Fig. 6 shows the Id-Vg extrapolated from the long step responses compared to the standard 'DC', which uses relatively short steps (1 s for every 200 mV). The higher transconductance and the movement of the kink towards more negative gate voltages in the long-time curve indicates that some donors are very slow to get ionized and consequently they give their contribution to the drain current after a very long time.

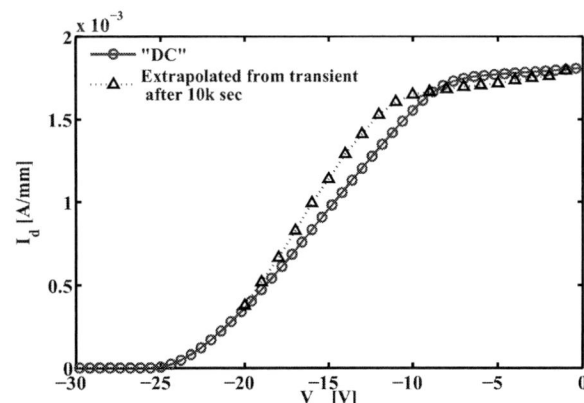

Fig. 6 Id-Vg extrapolated from long step responses (after 10 ks for every volt) compared to the standard DC (1 s for every 200 mV).

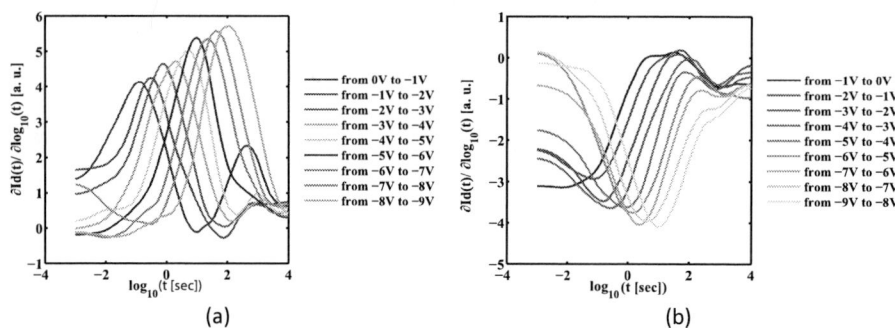

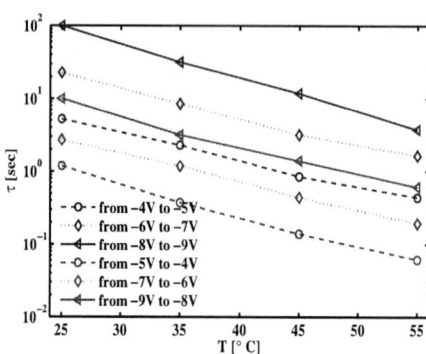

Fig. 7 First derivative of the measured transient current during (a) emission and (b) capture for different gate voltages. The gate voltages are stepped by 1 V and the drain current is monitored for 10 ks.

Fig. 8 Time constants associated with the emission (blue curves) from donor states to the 2DEG and capture (red curves) into donor states as functions of temperature.

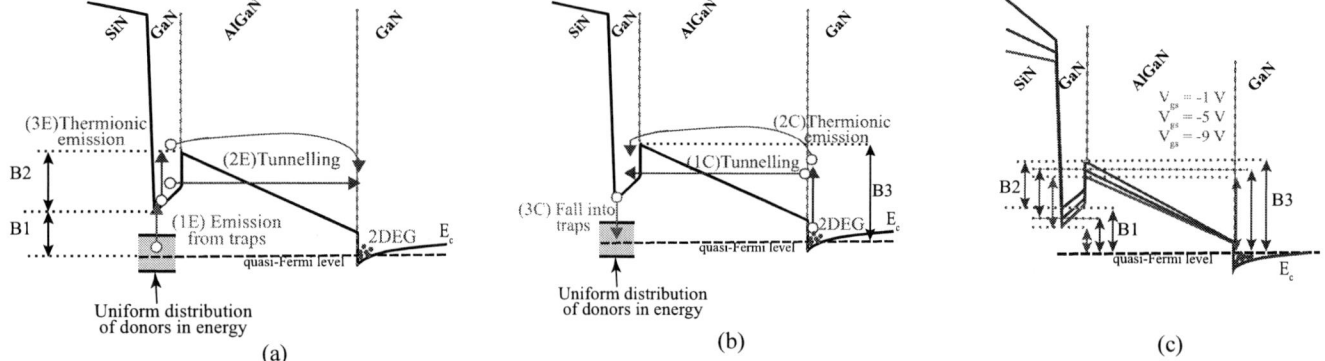

Fig. 9 Schematic representation of the energy bands and physical mechanisms involved in the transport of electrons (a) from the traps to the 2DEG (emission to 2DEG process); (b) from the 2DEG into the traps (capture into traps process); (c) Energy levels and barriers involved in the emission and capture processes for Vg = -1, -5, -9 V.

B. Gate bias and temperature dependence

In Fig. 7, the first derivative of the transient drain currents at T = 25° C is plotted for V_{gs} varying from 0 to −9 V (Fig. 7(a)) and from −9 to 0 V (Fig. 7(b)). It is very interesting to note that the time constants increase exponentially with more negative gate voltages. Furthermore, the emission process is slower than that of capture. Several other peaks are present in Fig. 7(a), which may correspond to different energy levels (or energy bands) of the donor traps, the first being dominant and shallower. Increasing the temperature speeds up the emission and capture exponentially as shown in Fig. 8. Here the extrapolated time constants are plotted as a function of temperature and at different gate biases.

It is important to highlight the fact that by looking at the variation of the drain current with time, the processes monitored are the emission from traps into the 2DEG and the capture of electrons from the 2DEG into the traps.

C. TCAD model and explanation of the physical mechanisms

In this section, the physical mechanisms of the (a) emission from donor states to the 2DEG and (b) capture from the 2DEG into donor states are explained in details.

The emission to the 2DEG comprises two coupled phases: (i) emission of electrons from the trap levels above the quasi-Fermi level to the conduction band and (ii) transport across the reverse[*] GaN-cap/AlGaN barrier to the 2DEG. The capture into traps is faster and comprises (i) transport from the 2DEG across the forward[*] AlGaN/GaN-cap barrier and (ii) capture of electrons from the conduction band into the traps. In Fig. 9, a schematic representation of these is shown. In particular, in Fig. 9(a) the following mechanisms are included: (1E) emission of electrons from the donor states, (2E) tunnelling of electrons through the reverse GaN-cap/AlGaN barrier B2, (3E) thermionic emission across the reverse barrier. The mechanism (1E) depends on the so-called trap barrier B1 while (2E) and (3E) depend on the reverse barrier barrier B2.

Fig. 9(b) includes the mechanisms relating to the capture process: (1C) tunnelling through the forward AlGaN barrier B3, (2C) thermionic emission across the forward AlGaN barrier B3, and (3C) capture into trap states.

---

[*] The names forward and reverse are associated with the forward and reverse sweeps in the Id-Vg.

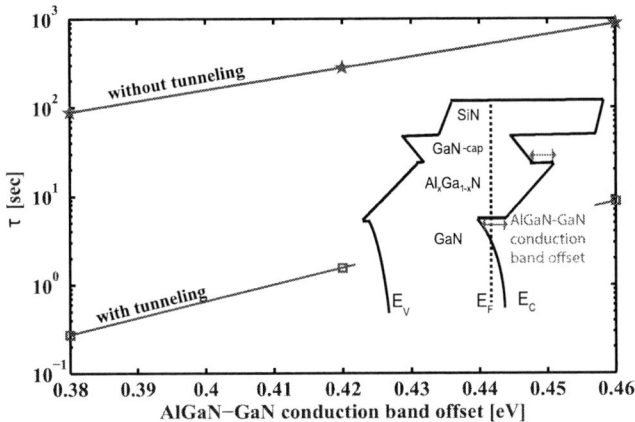

Fig. 10 Emission time constants versus AlGaN-GaN conduction band offset extracted from TCAD simulations. In the inset is shown the AlGaN-GaN conduction band offset modified in order to simulate a change in AlGaN barrier.

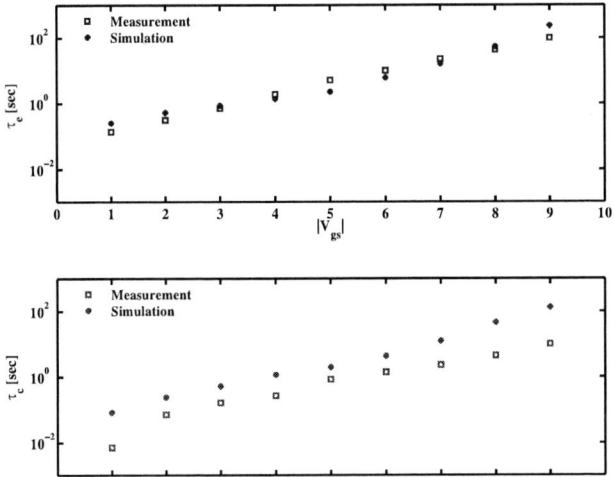

Fig. 11 Time constants of the emission (top/blue) and capture (bottom/red) processes as functions of the final gate bias.

Mechanisms (1C) and (2C) depend on the height and width of the forward barrier B3. The TCAD simulations in Fig. 10 show that the emission of electrons into the channel is exponentially dependent on the barrier height. This is because both the thermionic and the tunnelling processes are exponentially dependent on the negative value of the AlGaN barrier. The AlGaN barrier can be modified by changing the AlGaN-GaN conduction band offset as shown in the inset of the graph. Fig. 10 also highlights the fact that the tunnelling is the dominant transport process at room temperature.

By considering donors uniformly distributed in a larger band between 0.18 and 0.62 eV and by including the tunnelling model, the TCAD deck developed is able to predict well the dependence of both the emission and capture time constants as functions of the gate voltage (Fig. 11), thus demonstrating the validity of this model.

It is now possible to better understand how the emission and capture time constants decrease with the temperature (Fig. 8). The thermionic emission is in fact enhanced by an increase in temperature leading to faster transport of electrons across the barrier. Moreover, as the temperature increases, due to increased thermionic excitation, more electrons are available to tunnel through the barrier where it is narrower. In other words, both thermionic (directly) and tunnelling (indirectly) processes are enhanced substantially by an increase in temperature. As a result, the time constants are significantly shorter.

Conclusions

A detailed dynamic analysis of the donor traps at the surface of a GaN/AlGaN/GaN MISFET has been carried out by means of transient measurements and correlated to the Id-Vg and Cgg-Vg characteristics. It has been shown that the time constants associated with the emission and capture processes from and to donor states are not purely trap-related. The time taken by carriers to move across the forward or reverse GaN/AlGaN barrier between the surface and the 2DEG via tunnelling and thermionic emission is indeed embedded in the total extracted time constants. Both the tunnelling and thermionic processes are exponentially dependent on the AlGaN barrier height, the former being the dominant transport mechanism at room temperature. The emission times from the donor states to the 2DEG are longer than the capture times from the 2DEG into the donor states and both are exponentially dependent on temperature.

References

(1) B. Jogai, "Free electron distribution in AlGaN/GaN heterojunction field-effect transistors." Journal of applied physics 91.6, 2002, pp 3721-3729.

(2) J. M. Tirado, J. L. Sánchez-Rojas, and J. I. Izpura. "Trapping effects in the transient response of AlGaN/GaN HEMT devices." Electron Devices, IEEE Transactions on 54.3, 2007, pp 410-417.

(3) B. Liu, Ph.D. thesis, Massachusetts Institute of Technology (MIT), June 2013.

(4) S. Yang, Z. Tang, K. Y. Wong, Y. S. Lin, and Y. Lu. "Mapping of interface traps in high-performance Al2O3/AlGaN/GaN MIS-heterostructures using frequency-and temperature-dependent CV techniques." IEDM Tech. Dig , 2013, pp 152-155.

(5) G. Longobardi, et al.. "Impact of Donor Traps on the 2DEG and Electrical Behavior of AlGaN/GaN MISFETs", 2014,: 1-1.

(6) C. Liu, E. F. Chor, and L. S. Tan. "Investigations of HfO 2/AlGaN/GaN metal-oxide-semiconductor high electron mobility transistors."Applied physics letters 88, no. 17, 2006, pp 173504-173504.

(7) M. Fagerlind, et al.. "Investigation of the interface between silicon nitride passivations and AlGaN/AlN/GaN heterostructures by C (V) characterization of metal-insulator-semiconductor-heterostructure capacitors." Journal of Applied Physics 108, no. 1, 2010, pp 014508.

(8) G. Longobardi, F. Udrea, S. Sque, G. AM Hurkx, J. Croon, and J. Šonský., " The effect of the surface fixed charge and donor traps on the C(V) and transfer characteristics of a GaN MISFET – Experiment and TCAD simulations", ESSDERC 2014.

(9) G. Longobardi, F. Udrea, S. Sque, G. AM Hurkx, J. Croon, and J. Šonský, "Analysis of surface donor traps and fixed charges in GaN/AlGaN/GaN high-voltage transistors via the transfer characteristics of a MISFET", ISPS 2014.

Thermally Induced Threshold Voltage Instability of III-Nitride MIS-HEMTs and MOSC-HEMTs: Underlying Mechanisms and Optimization Schemes

Shu Yang, Shenghou Liu, Cheng Liu, Zhikai Tang, Yunyou Lu, and Kevin J. Chen

Department of Electronic and Computer Engineering, The Hong Kong University of Science and Technology, Hong Kong
Tel: +852-34692261, Fax : +852-23581485, E-mail: eesyang@connect.ust.hk

Abstract

The mechanisms of divergent V_{TH}-thermal-stabilities of III-nitride (III-N) MIS-HEMT and MOS-Channel-HEMT are revealed in this work. The more significant V_{TH}-thermal-instability of MIS-HEMT is attributed to the polarized III-N barrier layer that spatially separates the critical gate-dielectric/III-N interface from the channel and allows "deeper" interface trap levels emerging above the Fermi level at pinch-off. We also reveal the influences of the barrier layer's thickness and the fixed charges (e.g. F⁻) in the barrier layer on V_{TH}-thermal-stability and attempt to provide guidelines for the optimization of insulated-gate III-N power switching devices. A tailor-made normally-off MIS-HEMT with optimal tradeoff between performance and stability is thereby demonstrated, by conjunctively utilizing partially recessed gate and fluorine plasma implantation techniques.

Introduction

Wide bandgap III-N based power switching transistors, with the merits of delivering high conversion efficiency and high switching speed even at elevated junction temperatures (T_j), are destined to operate over broader temperature ranges. Hence, their temperature (T)-dependent behavior and stability are of critical importance to the III-N power circuits [1]. III-N MIS-HEMT and recessed-gate MOS-Channel-HEMT (MOSC-HEMT) (depicted in Fig. 1), as two predominant types of III-N power devices featuring insulated gate structures, are encountered with challenges of V_{TH} instability originating from the interface traps at the gate-dielectric/III-N interface [2]–[4]. With thermally sensitive electron emission processes of the interface traps, V_{TH}-thermal-stability of MIS-HEMTs [5], [6] and MOSC-HEMTs is of particular interest, yet has not been adequately investigated.

In this work, we revealed that thermally-induced V_{TH} shift is manifest in MIS-HEMT, but negligible in MOSC-HEMT, even with the same high-quality Al_2O_3(AlN interfacial-layer)/GaN interface [7], [8]. The distinction in V_{TH}-thermal-stability is attributed to the fact that the MIS-HEMT is a *buried-channel* device with the critical gate-dielectric/III-N interface spatially separated from the channel by a polarized III-N barrier layer. This polarized spacer raises the energy band at the critical interface, and allows interface trap levels over a broader energy range moving across the Fermi level more freely and changing their charge states via electron capture/emission processes,

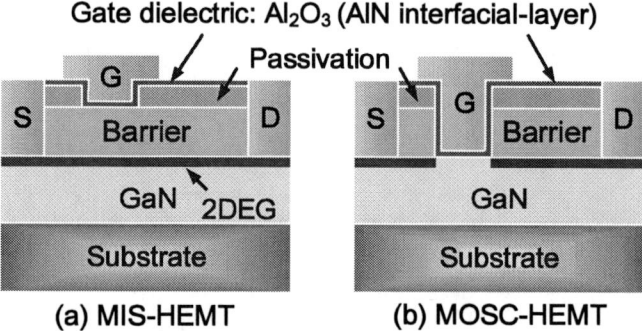

Fig. 1. Schematic cross sections of (a) gate-dielectric/GaN/AlGaN/GaN MIS-HEMT and (b) gate-dielectric/GaN MOS-Channel-HEMT (MOSC-HEMT). Gate dielectric: 25-nm ALD-Al_2O_3 with monocrystal-like AlN interfacial-layer [7], [8].

resulting in considerable thermally-induced V_{TH} shift. On the other hand, in MOSC-HEMT with the source/drain terminals laterally connected to the channel through the 2DEG access regions, deep interface trap levels along the channel are always kept below the Fermi level and maintain fixed charge states, leading to thermally stable V_{TH}. To overcome the relatively low channel mobility in MOSC-HEMT while realizing desirable normally-off operation with T-independent V_{TH}, a MIS-HEMT with optimally-engineered barrier is thereby developed by combining partially recessed gate and fluorine plasma implantation techniques.

Mechanisms of Thermally Induced V_{TH} Instability

The Al_2O_3(AlN)/GaN/AlGaN/GaN MIS-HEMT and Al_2O_3(AlN)/GaN MOSC-HEMT with the same gate dielectric stack process (Fig. 1) both deliver well-behaved DC performance at 25 °C and 200 °C, respectively (Fig. 2). In contrast to MIS-HEMT featuring negative V_{TH} shift with elevated T, MOSC-HEMT exhibits more stable V_{TH} with a tiny shift of -0.16 V from 25 °C to 200 °C (Fig. 3 and Fig. 4). In consistency with the T-dependent I_D-V_{GS} characteristics, C-V characteristics of MIS diode show larger T-dispersion around the 2DEG channel pinch-off voltage (i.e. "stretch-out" of slope 1 in Fig. 5(a)), compared with those of MOSC diode (Fig. 5(b)).

Interface trap density (D_{it}) in MOSC diode is in the range of ~8×10^{11}–6×10^{12} cm^{-2}eV^{-1} with activation energy (E_C–E_T) from 0.71 to 0.30 eV, which is determined by multi-T AC-conductance method (Fig. 6) [9]. Meanwhile, as the

978-1-4799-8002-4/14 $31.00 © 2014 IEEE
434

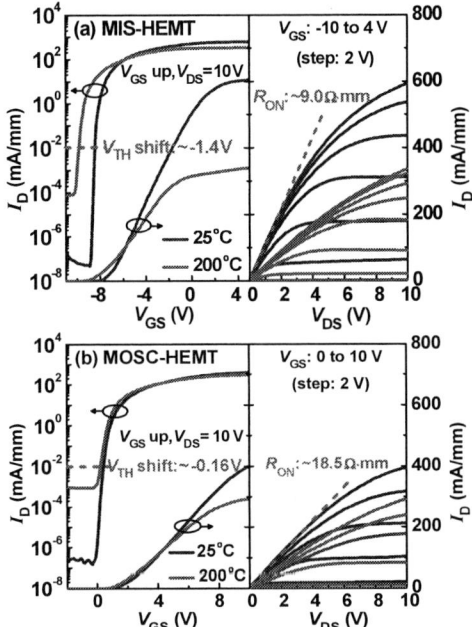

Fig. 2. DC transfer (left) and output (right) characteristics of (a) Al$_2$O$_3$(AlN)/GaN/AlGaN/GaN MIS-HEMT and (b) Al$_2$O$_3$(AlN)/GaN MOSC-HEMT at 25 °C and 200 °C, respectively. Device dimensions: L_G/L_{GD} = 2/10 μm. V_{TH} is defined at I_D = 10 μA/mm.

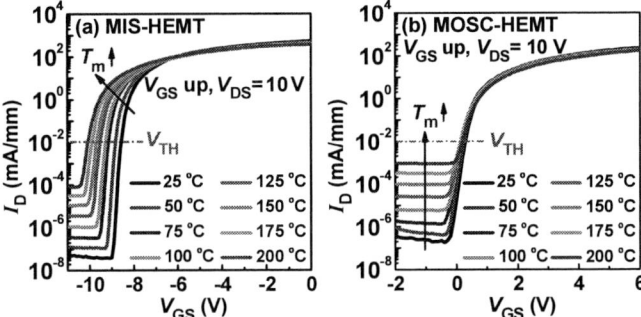

Fig. 3. (a) Multi-T DC transfer characteristics of (a) Al$_2$O$_3$(AlN)/GaN/AlGaN/GaN MIS-HEMT and (b) Al$_2$O$_3$(AlN)/GaN MOSC-HEMT with measurement temperature (T_m) increasing from 25 to 200 °C.

conventional conductance method could cause underestimation/uncertainty when applied to "buried-channel" MIS-heterostructures with two spatially separated interfaces, multi-T/f AC-CV techniques focusing on the second slope (slope 2 in Fig. 5(a)) have hereby been used to map the interface traps in MIS-heterostructures (blue symbols in Fig. 7(d)) [7]. In addition, pulsed I_D-V_{GS} characterization with long pulse width (W_{pulse}) of 2 s (equipment's upper limit) and elevated T is employed to extract deeper interface traps with E_C–E_T up to 1.1 eV in MIS-HEMT (Fig. 7(a)-(c)), as V_{TH}-thermal-stability of MIS-HEMT is susceptible to the dynamic charging/discharging of deeper interface traps (to be illustrated later). With E_C–E_T from 1.1 to 0.24 eV, D_{it} of MIS-heterostructure is in the range of ~1×10^{12}–6×10^{12} cm^{-2}eV^{-1} (Fig. 7(d)).

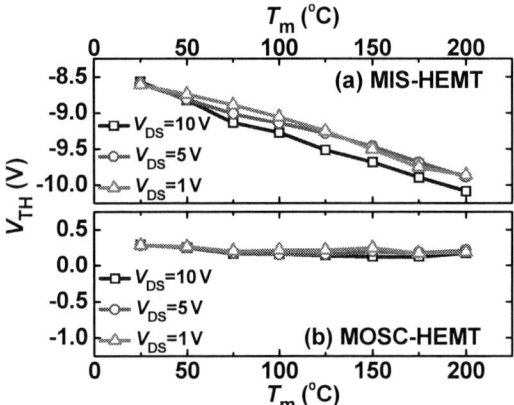

Fig. 4. T-dependence of V_{TH} in (a) MIS-HEMT and (b) MOSC-HEMT at V_{DS} = 1, 5, and 10 V, respectively. In MIS-HEMT, V_{TH} is dependent on V_{DS} when T_m > 50 °C, as larger V_{DS} enhances field-assisted de-trapping and leads to more negative V_{TH}.

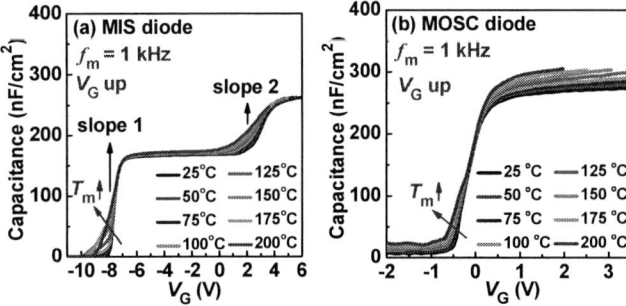

Fig. 5. Multi-T C-V characteristics of (a) Al$_2$O$_3$(AlN)/GaN/AlGaN/GaN MIS diode and (b) Al$_2$O$_3$(AlN)/GaN MOSC diode with AC measurement frequency (f_m) of 1 kHz and T_m increasing from 25 to 200 °C. Larger V_G span in (a) than in (b) is to show two rising slopes in the C-V characteristics of MIS diode.

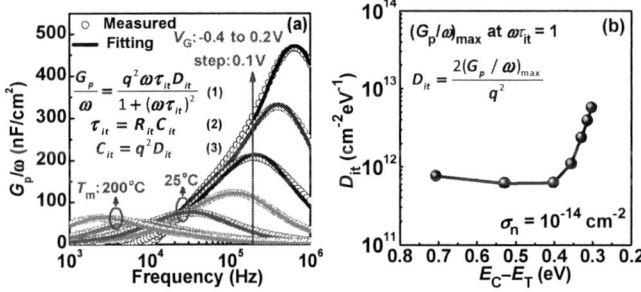

Fig. 6. (a) Measured G_p/ω-f characteristics of Al$_2$O$_3$(AlN) MOSC diode at 25 °C and 200 °C, respectively. (b) D_{it}-E_T mapping in MOSC diode using AC-conductance method.

Similar interface trap distributions in MIS-HEMT and MOSC-HEMT (Fig. 6(b) and Fig. 7(d)) rule out the possibility of different D_{it} responsible for the distinct V_{TH}-thermal-stabilities.

Examining the band diagrams at pinch-off as shown in Fig. 8, E_F relative to E_C at the dielectric/III-N interface in MIS-HEMT is deeper than that in MOSC-HEMT (i.e. $\Delta E_{MIS} > \Delta E_{MOSC}$, where $\Delta E_{MIS} = E_{C\_MIS} - E_{F\_MIS}$ and

978-1-4799-8002-4/14 $31.00 © 2014 IEEE

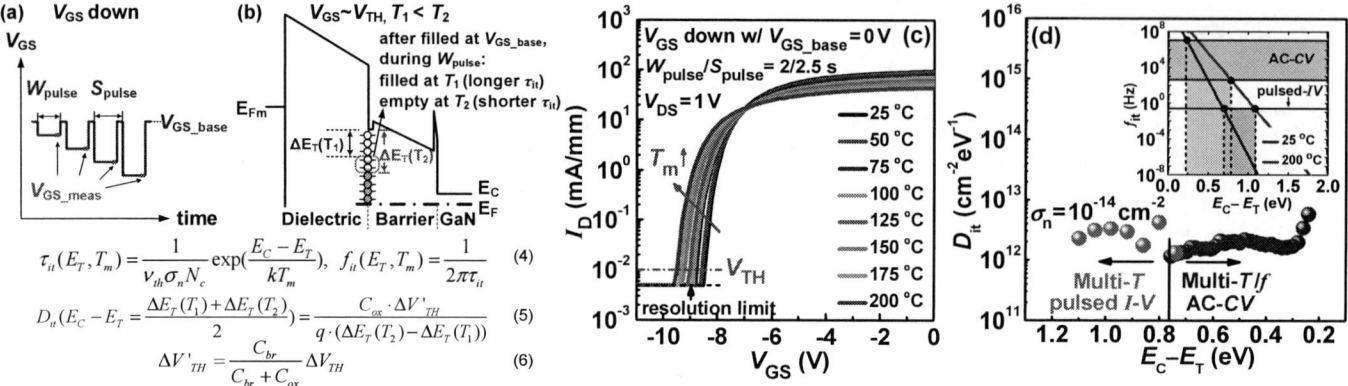

$$\tau_{it}(E_T, T_m) = \frac{1}{v_{th}\sigma_n N_c}\exp(\frac{E_C-E_T}{kT_m}), \quad f_{it}(E_T,T_m)=\frac{1}{2\pi\tau_{it}} \quad (4)$$

$$D_{it}(E_C-E_T) = \frac{\Delta E_T(T_1)+\Delta E_T(T_2)}{2} = \frac{C_{ox}\cdot\Delta V'_{TH}}{q\cdot(\Delta E_T(T_2)-\Delta E_T(T_1))} \quad (5)$$

$$\Delta V'_{TH} = \frac{C_{br}}{C_{br}+C_{ox}}\Delta V_{TH} \quad (6)$$

Fig. 7. D_{it}-E_T mapping in Al$_2$O$_3$(AlN) MIS-heterostructure: (a) Waveform of down-sweeping V_{GS} in pulsed transfer characterization. Pulse width/separation (W_{pulse}/S_{pulse}) = 2/2.5 s. (b) Schematic energy band diagram of MIS-HEMT at pinch-off, illustrating the principle of D_{it}-E_T mapping using multi-T pulse-mode transfer characterization. (c) Measured pulse-mode transfer characteristics of Al$_2$O$_3$(AlN) MIS-HEMT at V_{DS} = 1 V with T_m from 25 to 200 °C. Low V_{DS} is to maintain nearly uniform potential distribution, and thus, nearly uniform occupancy of interface traps in the gate region. (d) D_{it}-E_T mapping in Al$_2$O$_3$(AlN) MIS-heterostructures, by conjunctively using multi-T pulsed I-V and multi-T/f AC-CV techniques. Inset: relationship between detectable energy range (E_C–E_T) and characteristic frequency (f_{it}) of interface traps.

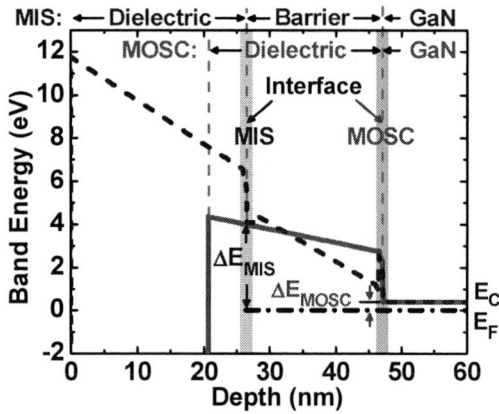

Fig. 8. Simulated energy band diagrams of the MIS-HEMT (blue line) and MOSC-HEMT (red line) at pinch-off under quasi-steady state.

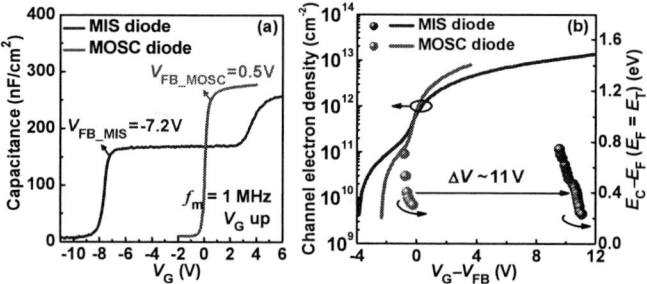

Fig. 9. (a) Measured C-V characteristics of Al$_2$O$_3$(AlN) MIS diode (blue line) and MOSC diode (red line) with f_m = 1 MHz at 25 °C. (b) Left vertical axis: V_G-dependence of channel electron sheet densities (n_s) in MIS diode (blue line) and MOSC diode (red line), extracted from C-V characteristics in (a). The lateral axis is offset by flat-band voltage (V_{FB}) for a fair comparison. Right vertical axis: V_G-driven Fermi level movements at the dielectric/III-N interface in MIS diode (blue symbols) and MOSC diode (red symbols).

$\Delta E_{MOSC} = E_{C\_MOSC}-E_{F\_MOSC}$), as a result of the buried-channel structure and strong polarization field in the AlGaN barrier layer. Accordingly, among the interface trap levels above the Fermi level (i.e. with E_C–E_T < ΔE_{MIS}), the relatively deep ("slow") ones act as "frozen states" at lower T but emit electrons and cause more negative V_{TH} at higher T, leading to thermally-induced V_{TH} shift in MIS-HEMT. In contrast, MOSC-HEMT exhibits enhanced V_{TH}-thermal-stability, as the deep interface trap levels cannot emerge above the Fermi level regardless of T. Even at a large reverse V_G at OFF-state (as in the case of the drain-side gate edge at higher V_{DS}), the negative V_{GD} will induce depletion region extension toward the drain in MOSC-HEMT, instead of further raising the energy band in the gate region, which is in agreement with V_{TH}'s independence on V_{DS} in Fig. 4(b). This mechanism is further experimentally validated by divergent V_G-driven Fermi level movements at the dielectric/III-N interface in MIS-HEMT and MOSC-HEMT (Fig. 9).

In interface trap admittance measurements, interface traps can respond to the AC measurement signal by contributing capacitance (as slope 2 in Fig. 5(a)) [7] or causing energy loss (represented by G_p/ω as in Fig. 6(a)) only when the characteristic frequency of interface traps (f_{it}) at the Fermi level (determined by V_G) matches f_m (i.e. f_{it} (E_T = E_F) = f_m), whereby the Fermi level position at the dielectric/III-N interface can be experimentally correlated with V_G. The positive lateral shift (~11 V) of (E_C–E_T) vs. (V_G–V_{FB}) in MIS diode suggests larger V_G–V_{FB} is required in MIS diode than in MOSC diode to make the Fermi level sweep through the interface traps with identical E_C–E_T. In other words, at pinch-off, E_F at the dielectric/III-N interface in MIS diode is substantially deeper than that in MOSC diode, which agrees well with the simulated results (ΔE_{MIS} > ΔE_{MOSC}) in Fig. 8. We can also find that when Fermi level sweeps through the interface traps with identical E_C–E_T, n_s is higher in MIS diode than that in MOSC diode.

978-1-4799-8002-4/14 $31.00 © 2014 IEEE

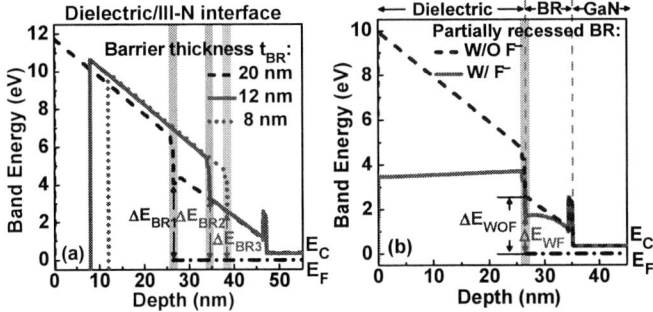

Fig. 10. Simulated energy band diagrams of the MIS-HEMTs at pinch-off under quasi-steady state: (a) with varied barrier thickness (t_{BR}) of 20, 12, and 8 nm, respectively; (b) with and without F^- when AlGaN barrier is partially recessed to 8 nm. The negatively charged F^- can reduce the upward band-bending in AlGaN barrier, leading to $\Delta E_{WF} < \Delta E_{WOF}$.

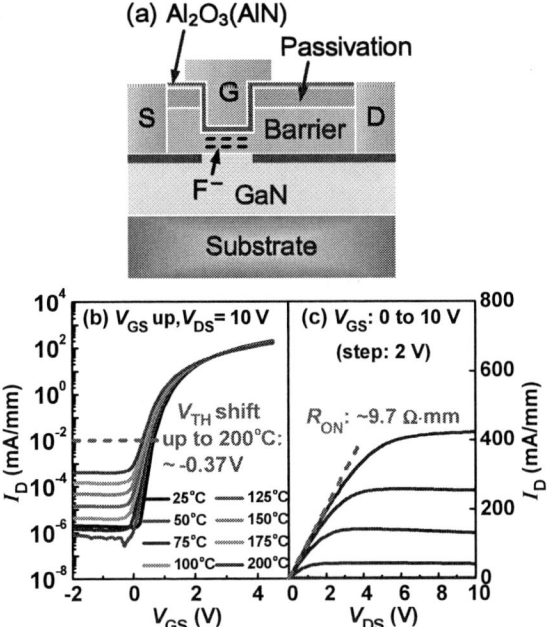

Fig. 11. (a) Schematic cross section of normally-off Al_2O_3(AlN) MIS-HEMT with optimally-engineered barrier structure, by jointly using partially recessed gate (t_{BR} = 8 nm) and fluorine plasma implantation. (b) Multi-T DC transfer characteristics of the optimally-engineered MIS-HEMT with T_m increasing from 25 to 200 °C. (c) DC output characteristics of the optimally-engineered MIS-HEMT at 25 °C.

Optimally-Engineered MIS-HEMTs

Despite the superior V_{TH}-thermal-stability, MOSC-HEMT exhibits higher ON-resistance (R_{ON}) (Fig. 2) due to lower channel electron mobility. To mitigate thermally induced V_{TH} instability, we need to narrow down the energy range of the interface trap levels above Fermi level at pinch-off. This target requirement can be fulfilled by thinning (or recessing) the III-N barrier layer (Fig. 10(a)) or incorporating negative charges (e.g. F^-) into the barrier layer, as F^- can reduce the upward band-bending (Fig. 10(b)). Therefore, MIS-HEMTs with optimally-engineered barrier structure are demonstrated by conjunctively using partially

recessed gate and fluorine plasma implantation techniques, which are capable of delivering normally-off operation, low R_{ON} comparable with conventional MIS-HEMTs, and enhanced V_{TH}-thermal-stability comparable with MOSC-HEMTs (Fig. 11).

Conclusion

We have studied the mechanisms of distinct V_{TH}-thermal-stabilities in III-N MIS-HEMTs and MOSC-HEMTs. In MIS-HEMTs, the polarized barrier layer spatially separates the critical gate-dielectric/III-N interface from the channel and allows "deeper" interface trap levels emerging above Fermi level more freely at pinch-off, leading to manifest V_{TH}-thermal-instability. Based on the effects of the thickness and incorporated fixed charges (e.g. F^-) of the III-N barrier layer on V_{TH}-thermal-stability, we demonstrate a normally-off MIS-HEMT with optimal tradeoff between performance and stability by combining partially recessed gate and fluorine plasma implantation techniques. Both the mechanisms responsible for V_{TH}-thermal-instability and the optimization schemes are valuable to develop high-performance and high-reliability III-N power devices for high-speed, high-power and high-temperature applications.

Acknowledgement

This work is supported in part by NSFC/RGC joint research project N_HKUST636/13 and ITF project ITS/192/14FP.

References

[1] P. G. Neudeck, R. S. Okojie, and L. Y. Chen, "High-temperature electronics - A role for wide bandgap semiconductors?" *Proc. IEEE*, vol. 90, no. 6, pp. 1065–1076, June 2002.

[2] S. Huang, S. Yang, J. Roberts, and K. J. Chen, "Threshold voltage instability in Al_2O_3/GaN/AlGaN/GaN metal-insulator-semiconductor high-electron mobility transistors," *Jpn. J. Appl. Phys.*, vol. 50, no. 11, pp. 110202-1–110202-3, Nov. 2011.

[3] J. H. Bae, I. Hwang, J. M. Shin, H. I. Kwon, C. H. Park, J. Ha, and J. Lee, "Characterization of traps and trap-related effects in recessed-gate normally-off AlGaN/GaN-based MOSHEMT," in *IEDM Tech. Dig.*, Dec. 2012, pp. 13.2.1–13.2.4.

[4] P. Lagger, C. Ostermaier, G. Pobegen, and D. Pogany, "Towards understanding the origin of threshold voltage instability of AlGaN/GaN MIS-HEMTs," in *IEDM Tech. Dig.*, Dec. 2012, pp. 13.1.1–13.1.4.

[5] R. Chu, D. Brown, D. Zehnder, X. Chen, A. Williams, R. Li, M. Chen, S. Newell, and K. Boutros, "Normally-off GaN-on-Si metal-insulator-semiconductor field-effect transistor with 600-V blocking capability at 200 °C," in *ISPSD Dig.*, June 2012, pp. 237–239.

[6] M. Miczek, C. Mizue, T. Hashizume, and B. Adamowicz, "Effects of interface states and temperature on the *C-V* behavior of metal/insulator/AlGaN/GaN heterostructure capacitors," *Jpn. J. Appl. Phys.*, vol. 103, no. 10, pp. 104510-1–104510-11, May 2008.

[7] S. Yang, Z. Tang, K.-Y. Wong, Y.-S. Lin, Y. Lu, S. Huang, and K. J. Chen, "Mapping of interface traps in high-performance Al_2O_3/AlGaN/GaN MIS-heterostructures using frequency- and temperature-dependent *C-V* techniques," in *IEDM Tech. Dig.*, Dec. 2013, pp. 6.3.1–6.3.4.

[8] S. Liu, S. Yang, Z. Tang, Q. Jiang, C. Liu, M. Wang, and K. J. Chen, "Al_2O_3/AlN/GaN MOS-channel HEMTs with an AlN interfacial layer," *IEEE Electron Device Lett.*, vol. 35, no. 7, pp. 723–725, July 2014.

[9] E. H. Nicollian and J. R. Brews, *MOS (metal oxide semiconductor) physics and technology*: Wiley, 1982.

Impacts of Fluorine-treatment on E-mode AlGaN/GaN MOS-HEMTs

X. Sun[1*], Y. Zhang[2], K. S. Chang-Liao[3] T. Palacios[2], and T. P. Ma[1]

[1]Yale Univ., CT, USA [2] Massachusetts Institute of Technology (MIT), MA, USA [3]National Tsing Hua Univ., Taiwan
Phone: +1 203-980-4789, [*]E-mail: xiao.sun@yale.edu

Abstract

The impact of fluorine treatment on AlGaN/GaN MOS-HEMTs has been investigated. Fluorine was found to suppress pre-existing traps in MOS-HEMT, which improves the off-state at high temperatures. Fluorine doping and associated etching, however, also generates slow border traps and fast interface states that degrade the MOS-HEMT performance. Multi-faceted mechanisms for drain current degradation due to F-doping and gate-recess-etch have been investigated in enhancement-mode MOS-HEMTs.

Introduction

GaN based MOS-HEMTs are emerging devices for high power and RF applications. Gate-recess etch and fluorine (F) doping are approaches commonly taken to achieve enhancement-mode (E-mode) operation [1, 2]. These technologies, however, tend to result in V_{th} instability and drain current (I_d) degradation in HEMTs [3-7], which could be attributed to F-induced bulk and interfacial traps. In an MOS-HEMT, the related scenario is more complex, because the inserted gate dielectric could weaken the screening effect of gate meatal and introduce traps in bulk oxides or at the oxide/barrier-layer interface [8]. In this work, we investigated multifaceted impacts of F-doping and associated etching on E-mode MOS-HEMTs over a wide range of fluorine doses.

Experimental

HEMT structures were fabricated on a commercial GaN-on-silicon wafer (Fig.1). To obtain E-mode behavior, the AlGaN (Al: 26%) barrier in the gated region was treated by CF_4 plasma at an ECR/RF power of 150W/20W for varying intervals of time (TEM in Fig.1). More experimental details can be found in [9]. SIMS study shows the peak of F-concentration at the top AlGaN surface (Fig.2). MOS-HEMT devices were also fabricated by depositing ALD-Al_2O_3 immediately after F-treatment, during which F was driven into Al_2O_3 from the AlGaN underneath. F treatment can effectively increase V_{th} up to ~ 4 V for CF_4 treatments of ~120 s, as shown in Fig. 3. In Fig. 4, the gate stack thickness decreases with CF_4 time, indicating that F also etches AlGaN barrier layer during the CF_4 plasma treatment. The resulting thinning of AlGaN, i.e., gate-recess-etch, along with negative charges induced by F ions can effectively increase V_{th}. Additional CF_4 plasma, however, caused serious degradation of on- and off- currents (I_{on}, I_{off}) as shown in Fig. 3 and elsewhere [5, 6].

F-passivation of traps in MOS-HEMT

In MOS-HEMTs, traps exist at the oxide/AlGaN interface and in the gate oxide [8-11]. In this work we have found that these traps appear in the MOS-HEMT with Al_2O_3 as gate

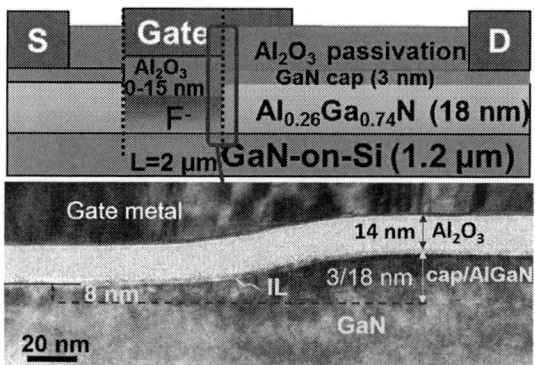

Fig. 1 Schematic and TEM views of MOS-HEMT device studied in this work; The TEM shows a device with 8 nm of AlGaN and 14nm Al_2O_3 on top. treated by CF_4 for 120s.

oxide, and are detrimental to the off-state at high temperatures. Fig.5 shows that, although the hysteresis and I_{off} are small for depletion-mode MOS-HEMTs at 300K, at higher temperatures they increase drastically. The Subthreshold Swing (SS) at different temperatures reveals that while more traps can respond to dc-V_g sweep at higher temperatures, they are positively charged above E_F,

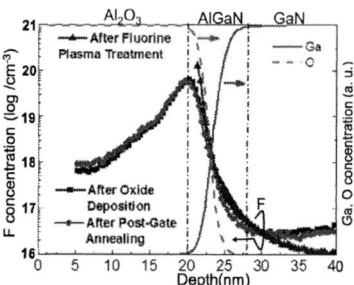

Fig.2 SIMS indicates that F concentration peaks at the AlGaN upper surface (Al_2O_3/AlGaN). F diffuses into the Al_2O_3 during the ALD process (after [9]).

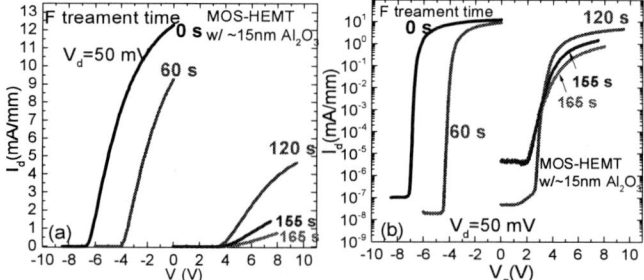

Fig. 3 The linear (a) and logarithmic (b) I_d-V_g of MOS-HEMTs with AlGaN treated by CF_4 for time intervals from 0s to 165s. I_{on} and SS degradation is observed after high F doses.

978-1-4799-8002-4/14 $31.00 © 2014 IEEE   438

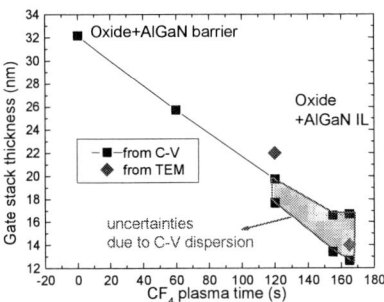

Fig.4 Total gate stack thickness (including Al_2O_3 and AlGaN layers) after CF_4 plasma treatment of varying intervals of time, as extracted from 1MHz C-V ($\varepsilon \sim 10$) and TEM.

resulting in a negative shift of V_{th} and thus a high I_{off}, i.e., difficult turn-off. It is observed that some deep-level traps—"fixed" positive charges commonly found in GaN MOS-HEMT [10, 11]—could also emit/capture electrons in the V_g operation range under high temperatures. For HEMT devices, that has no gate oxides, such degradation does not exist, as shown in the inset of Fig. 5b, proving that the traps are introduced by the Al_2O_3, including those in Al_2O_3 bulk or at Al_2O_3/AlGaN interface. Notably, the F-plasma treatment, even with a light dose that is not sufficient to achieve E-mode, is effective in passivating these traps. In Fig. 5b, the MOS-HEMT with 60 s of CF_4-treated AlGaN shows little hysteresis and well-behaved I_{off} even at 540 K, in strong contrast to the untreated AlGaN. Given that F plasma treatment has been reported to be stable in AlGaN at least up to 800 °C [12], fluorine doping could be a beneficial process for good high-temperature characteristics in GaN MOS-HEMT.

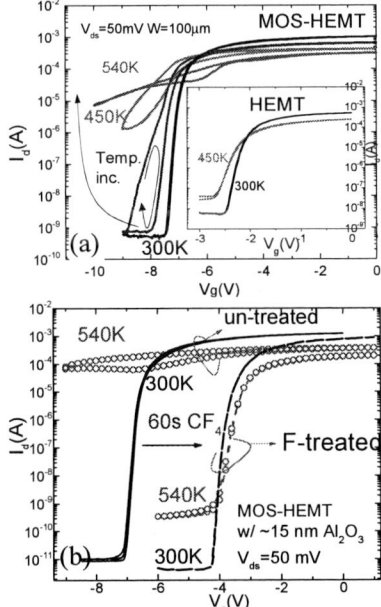

Fig.5 (a) degradation in the log I_d-V_g curves (including hystereses) with increasing temperatures from 300K to 540K in the untreated Al_2O_3-MOS-HEMT. The inset shows the untreated HEMT device without Al_2O_3. (b) log I_d-V_g curves (including hystereses) for the untreated- and the 60s-CF_4 treated Al_2O_3-MOS-HEMTs at 300 K and 540 K.

Slow border traps in AlGaN

We adapted the ac-transconductance (g_m) method to examine possible slow traps in AlGaN generated by F-treatment. In the ac-g_m method, under small ac-V_g, the occupation of traps in gate stacks fluctuates due to capture and emission of carriers. By modulating the ac-V_g frequencies, ac-g_m is able to reflect traps with varying time constant (τ). Unlike conventional gate admittance methods involving ac gate currents (such as C-V and G-V), the ac-g_m method probes much larger drain currents with much higher signal-to-noise ratios at low frequencies (down to 10 mHz), and hence is able to probe traps with longer τ ($100\mu s - 100s$). More details of the method can be found in [13, 14].

To focus on the traps in F-treated AlGaN, AlGaN/GaN HEMTs without gate oxides were tested. The results show that slow traps are introduced by F-treatment in AlGaN. In Fig. 6, for both the untreated and F-treated devices, ac-g_m shows little frequency dispersion at 300 K. But the F-treated one shows a large *positive* frequency dispersion at 540 K, suggesting F-induced traps capture channel electrons through *inelastic* tunneling. For untreated device, the frequency dispersion is negligible at 540 K (inset of Fig. 6b). Fig. 7 shows the normalized ac-g_m vs. frequency at different temperatures. The corresponding Arrhenius plots show an activation energy $E_a = E_T - E_F$ of 0.3 eV for the as-deposited, and $E_a = 0.75$ eV after 400°C annealing (Fig. 8), consistent with theories that annealing is critical for fluorine to form stable trapping centers [15-17]. Because the traps have an E_a larger than the AlGaN/GaN conduction band offset, they must not be at the interface, but instead should be in AlGaN as border traps, as in the band diagram shown in Fig. 9.

F-induced slow traps in AlGaN also result in trap-assisted-tunneling (TAT) in HEMT devices, as evidenced by the ac-g_m measurements. In Fig. 7c, a *negative* frequency dispersion appears in addition to the positive one, as the traps exchange electrons with the gate in addition to the channel [8] (Fig. 9), resulting in TAT in the gate stack of HEMTs. In MOS-HEMTs, however, the gate oxide suppresses TAT, so this ac-g_m feature is missing. As such, F-induced traps deep in AlGaN, with higher densities (Fig. 2), result in Fermi-level (FL) pinning by trapping electrons without detrapping them to the gate. For example, in Fig. 10, the dc-I_d degradation after 10 s of additional F-plasma in MOS-HEMTs can be simulated well by considering the FL pinning induced by a trap band (the inset of Fig. 10a).

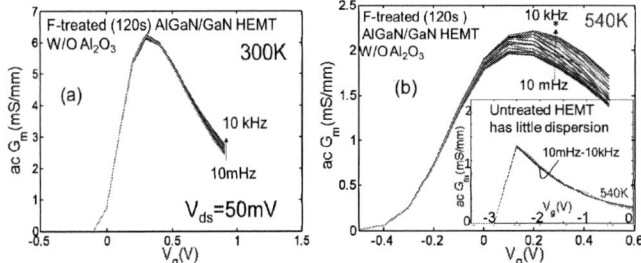

Fig. 6 AC-g_m vs. V_g at ac-v_g frequencies from 10 mHz to 10k Hz at (a) 300K and (b) 540 K for AlGaN/GaN HEMT with 120s CF_4 treatment. The inset in (b) shows the ac-g_m for the sample without F-treatement. * In Fig. 6(b), the frequency dependence is not completley monotonic. (see Fig. 7c)

978-1-4799-8002-4/14 $31.00 © 2014 IEEE

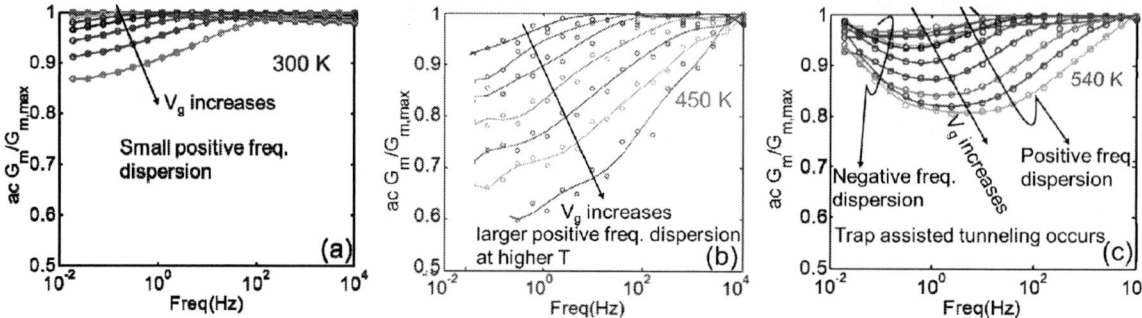

Fig. 7 Normalized ac-g_m vs. ac-v_g frequency for the AlGaN/GaN HEMT devices after 400° C annealing at (a) 300 K, (b) 450 K, and (c) 540 K. Fig.7(c) shows that a negative frequency dispersion (the recovery of the positive frequency dispersion) appears in the 540 K-annealed sample below ~1Hz.

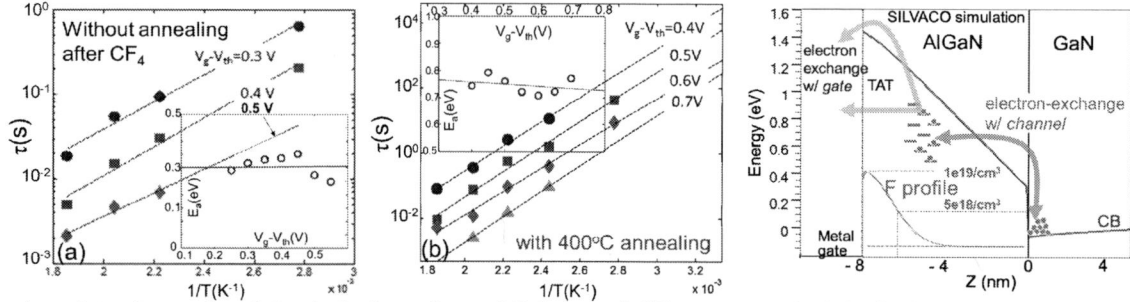

Fig. 8 Ahrrenious plots of ac-g_m degradation in the form of τ vs. 1/T for several different gate overdrives, for the (a) as-deposited and (b) 400°C annealed AlGaN/GaN HEMT with 120s F treatment. In the y axis, $\tau=1/f$, and f is the frequency where ac-g_m decreases by half of the maximum measured degradation in the frequency scope. Each inset shows E_a vs. dc gate overdrive V_g-V_{th}.

Fig. 9 Conduction band of the AlGaN/GaN gate stack in a HEMT simulated by SILVACO, resulting from inputing a F doping profile (all ionized) shown in the inset.

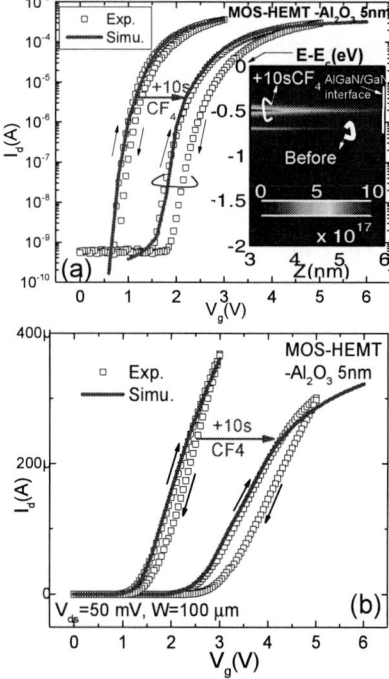

Fig. 10 Simulated and experimental I_d-V_g (with hystereses) in (a) linear and (b) log scale for two MOS-HEMTs (5nm Al$_2$O$_3$) having 10s difference in CF$_4$ treatment time interval. The inset of (a) shows the trap profile input to SIVACO: Trap density (contour) vs. Energy (E-E$_c$) and Distance (Z) from the Al$_2$O$_3$/AlGaN(~6nm) interface

Fast interface states at GaN Interface

In additional to slow traps, F plasma also induces fast interface states on GaN in the presence of an AlGaN blocking layer. C-V and G-V can probe such fast interface states in the subthreshold region when the trap time constants are longer with lower carrier density [18]. In Fig. 11, a large C-V dispersion appears after the MOS-HEMT is turned into E-mode by fluorine, when AlGaN is still relatively thick, e.g., 8 nm after 120 s of CF$_4$, as shown by TEM in Fig. 1. In Fig. 12, the frequency of the G$_p$/ω peak from the G-V measurements reflects FL movement under V_g, and the value thereof reflects the interface state density (D_{it}). A 60 s F plasma treatment can reduce D_{it} from 10^{11} to 10^{10}/cm^2/eV, indicating that light-dose F could passivate the pre-existing traps at/near the GaN interface. However, further treatment to 120 s increases D_{it} to 10^{11}/cm^2/eV, possibly due to fluorine accumulation at the AlGaN/GaN interface. Finally, 165 s increases D_{it} to 10^{12}/cm^2/eV, consistent with the D_{it} level of the Al$_2$O$_3$/GaN interface in GaN MOSFETs [19]. Indeed, TEM reveals that at 165 s, F treatment has etched all the AlGaN and reached the GaN, yielding an Al$_2$O$_3$/GaN interface (Fig. 13). Plasma over-etch can also damage the GaN surface [20].

I_d degradation mechanism in F-treated MOS-HEMTs

I_d degradation in F-treated MOS-HEMTs (Fig. 3) results from both mobility (μ) drop and FL pinning. In Fig. 14, I_d can be mostly recovered by 50-ns pulsed I_d-V_g measurement for medium F-dose (120 s), but can only be partly mitigated for heavy F-dose (165 s), partially because the time constants of generated interfaces states in the later are much shorter than 50 ns.

978-1-4799-8002-4/14 $31.00 © 2014 IEEE   440

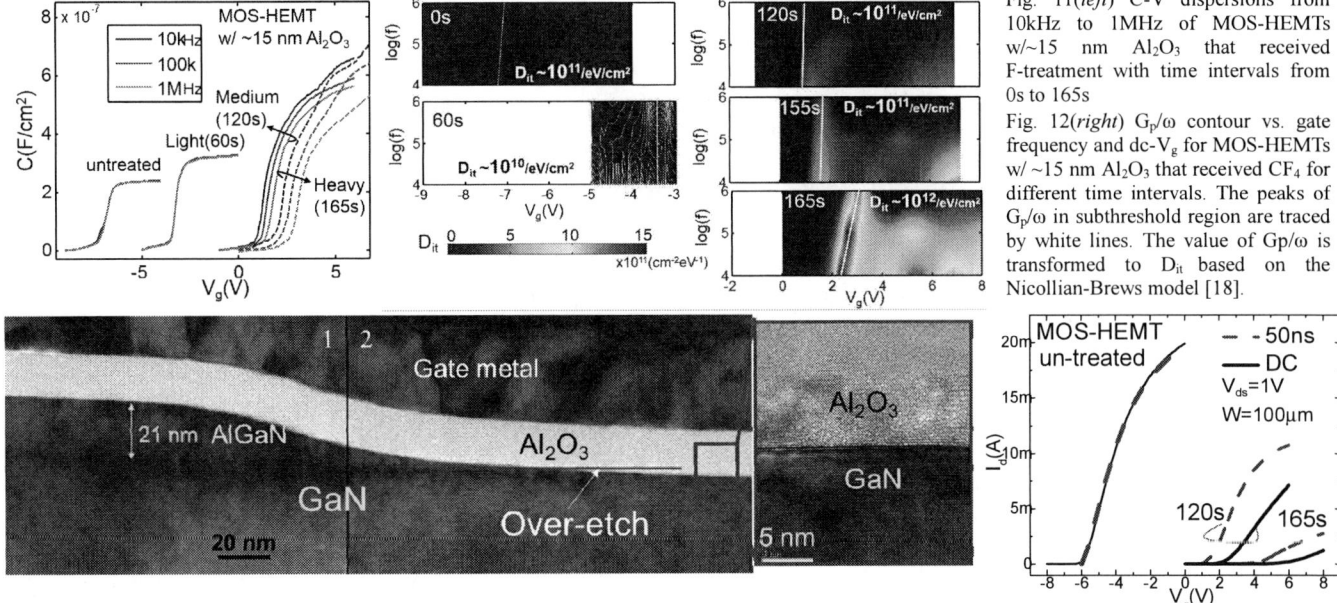

Fig. 11(*left*) C-V dispersions from 10kHz to 1MHz of MOS-HEMTs w/~15 nm Al_2O_3 that received F-treatment with time intervals from 0s to 165s

Fig. 12(*right*) G_p/ω contour vs. gate frequency and dc-V_g for MOS-HEMTs w/ ~15 nm Al_2O_3 that received CF_4 for different time intervals. The peaks of G_p/ω in subthreshold region are traced by white lines. The value of Gp/ω is transformed to D_{it} based on the Nicollian-Brews model [18].

Fig. 13 A cross sectional view of 165s-CF_4 treated MOS-HEMT by TEM. The frame on the right shows an enlarged view of the resulting Al_2O_3/GaN interface by F-plasma over-etch.

Fig. 14 DC and pulsed I_d-V_g (50 ns) for MOS-HEMTs w/~15 nm Al_2O_3 that received different CF_4 treatment time intervals

On the other hand, Fig. 15 shows the measured μ and V_{th} as functions of F plasma time for HEMTs and MOS-HEMTs. Unlike HEMTs, μ degrades dramatically in MOS-HEMTs with ~15 nm Al_2O_3, especially for the E-mode. The mechanisms of I_d degradation after F treatment are summarized below, in the order from low-dose to high-dose: (1) the thinning of AlGaN by F etching reduces 2DEG density and μ; (2) F generates border traps in AlGaN, resulting in FL pinning; (3) with a few nm of AlGaN left, F can generate interface states at AlGaN/GaN, resulting in stronger FL pinning and μ drop; (4) after AlGaN being completely etched out, plasma damage and a high-concentration of F in the GaN channel could degrade μ significantly by scattering (Fig. 15). Compared to MOS-HEMTs, HEMTs show little degradation in μ after 120 s F plasma treatment, due to TAT and gate screening of trapped charges, but at the cost of low V_{th}. A tradeoff therefore exists between V_{th} and I_{on} for these devices that use F implantation and/or gate-recess to achieve E-mode.

Summary

In summary, fluorine treatment enhanced SS and I_{off} of MOS-HEMT at high temperatures, by passivating traps contained in gate stacks of GaN MOS-HEMT devices. More Fluorine treatment, however, results in border traps in AlGaN with E_a~0.75eV, and interface states at AlGaN/GaN, degrading I_{on} through pinning FL and reducing mobility. Extremely low mobility can occur due to over-etch-induced GaN surface damage and high fluorine impurity concentrations within the channel.

Acknowledgement

This work is partially sponsored by ONR DEFINE MURI, and DTRA under HDTRA 1-10-1-0042.

Reference

[1] Y. Cai et al IEEE Tran. Elec. Dev., 53, p.2207, 2006
[2] T.Oka et al IEEE Elec. Dev. Lett., 29, p.668, 2008
[3] C. –Y. Ma, et al IEDM 2010, p. 476
[4] C.-W. Yi, et al IEDM 2007, p. 389
[5] A. Basu et al J. Vacuum Sci. Tech. *B* 25, p. 2607, 2007
[6] B. K. Li, et al Appl. Phys. Lett., 92, 082105, 2008.
[7] D. Bisi et al ESSDERC 2013, p.61
[8] X. Sun et al, Appl. Phys. Lett., 102, 103504, 2013
[9] Y. Zhang et al Appl. Phys. Lett. 103, 033524 (2013)
[10] J. P. Ibbetson et al, Appl. Phys. Lett. 77, 250, 2000
[11] S. Ganguly, et al Appl. Phys. Lett. 99, 193504, 2011
[12] M. J. Wang, et al J. Appl. Phys. 105, 083519, 2009
[13] X. Sun IEEE Elec. Dev. Lett. 33, p.438, 2012
[14] X. Sun et al IEDM 2012, 19.4.1
[15] K. Chen et al, IEDM 2011, p. 465
[16] A. Basua et al, J. Appl. Phys. 105, 033705, 2009
[17] M. J. Wang et al, Appl. Phys. Lett. 94, 061910, 2009
[[18] E. H. Nicollian and J. R. Brews, MOS Physics and Technology, Wiley,New York, 1982
[19] Y. Q. Wu et al, Appl. Phys. Lett., 90, 143504, 2007
[20] T. J. Anderson et al, J. Elec. Material., 39, p. 478, 2010

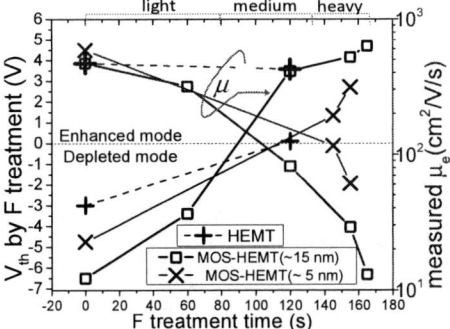

Fig. 15 V_{th} and measured peak mobility vs. CF_4 treatment time interval for HEMT and MOS-HEMTs with ~5 nm and ~15 nm Al_2O_3.

978-1-4799-8002-4/14 $31.00 © 2014 IEEE

High-Temperature Low-Damage Gate Recess Technique and Ozone-Assisted ALD-grown Al₂O₃ Gate Dielectric for High-Performance Normally-Off GaN MIS-HEMTs

S. Huang[1], Q. Jiang[2], K. Wei[1], G. Liu[1], J. Zhang[1], X. Wang[1], Y. Zheng[1], B. Sun[1], C. Zhao[1], H. Liu[1], Z. Jin[1], X. Liu[1], H. Wang[2], S. Liu[2], Y. Lu[2], C. Liu[2], S. Yang[2], Z. Tang[2], J. Zhang[3], Y. Hao[3], and K. J. Chen[2]

[1]Institute of Microelectronics, Chinese Academy of Sciences, Beijing, China
[2]Dept. of Electronic and Computer Engineering, Hong Kong University of Science and Technology, Hong Kong, China
[3]School of Microelectronics, Xidian University, Xi'an, China
Email: huangsen@ime.ac.cn, Fax: +86-10-62021601

Abstract

A high-temperature (180 °C) gate recess technique featuring low damage and *in-situ* self-clean capability, in combination with O₃-assisted atomic-layer-deposition (ALD) of Al₂O₃ gate dielectric, is developed for fabrication of high performance normally-off AlGaN/GaN metal-insulator-semiconductor high-electron-mobility transistors (MIS-HEMTs), which exhibit a threshold voltage of +1.6 V, a pulsed drive current of 1.1 A/mm, and low dynamic ON-resistance under hard-switching operation. Chlorine-based dry-etching residues (e.g. AlCl₃ and GaCl₃) are significantly reduced by increasing the wafer temperature during the gate recess to their characteristic desorption temperature, while defective bonds like Al-O-H and positive fixed charges in ALD-Al₂O₃ are significantly suppressed by substitution of H₂O with O₃ precursor.

Introduction

III-nitride (III-N) normally-off metal-insulator(oxide)-semiconductor high-electron-mobility transistors (MIS/MOS-HEMTs) have attracted much attention in power switching applications, for their capability of delivering lower gate leakage current, higher threshold voltage (V_{TH}) as well as larger gate swing than the Schottky-gate HEMTs [1-4]. Various structures have been proposed for normally off operation [1], among which the gate-recess scheme with thinned barrier layer is under intense investigation [3, 5]. The gate recess is usually performed with chlorine-based (e.g. BCl₃, Cl₂) inductively-coupled-plasma (ICP) dry etching processes that tend to induce physical damages, which consequently lead to high interface trap density and lower channel mobility. In addition, the gate-dielectric/GaN interface could be further deteriorated by two major etching byproducts AlCl₃ and GaCl₃, owing to their low desorption rate at temperature below 180 °C [6]. Therefore, a low-damage gate recess technique with *in-situ* self-clean capability by high-temperature desorption is highly desirable.

For the gate dielectrics, insulating Al₂O₃ grown by atomic layer deposition (ALD) is an attractive option [7] besides SiNₓ for its larger band-gap and higher breakdown strength. However, significant amount of defective bonds, such as Al-Al and Al-O-H, are commonly observed in ALD-Al₂O₃ films when H₂O is used as the oxygen precursor, resulting in high-density positive fixed charges and acceptor-like border/interface traps [7, 8]. The presence of positive charges hinder the formation of normally-off channel, and moreover, the charging and discharging of border/interface traps will induce significant V_{TH} instability [7], jeopardizing the safety of power switching devices. Therefore, it is necessary to use a hydrogen-free precursor to replace H₂O so that the formation of Al dangling bonds and H-related bonds can be suppressed.

In this work, a high-temperature (180 °C) chlorine-based low-damage gate recess technique is developed for the fabrication of GaN-based normally-off power MIS-HEMTs. By using O₃-sourced ALD-Al₂O₃ as the gate dielectric,

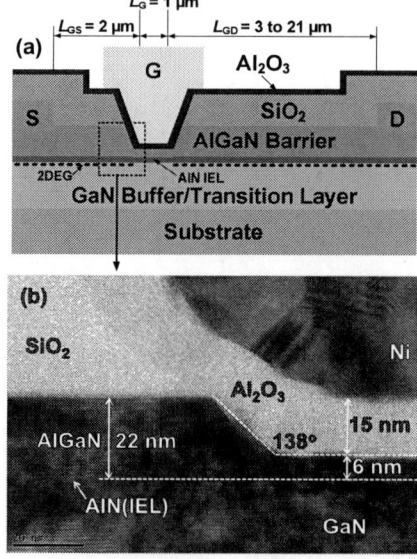

Fig. 1. (a) Schematic device structure of the normally-off Al₂O₃/AlGaN/AlN/GaN MIS-HEMT in this work. (b) TEM cross-sectional view of the device's gate edge.

positive fixed charges in Al_2O_3 and the dielectric defects are both significantly reduced, contributing to normally-off operation of MIS-HEMTs with high threshold voltage, low specific ON-resistance (R_{ON}), and high drive current.

Fabrication of Normally-off GaN MIS-HEMTs

The schematic cross section of the normally-off AlGaN/GaN MIS-HEMTs is depicted in Fig. 1. Device fabrication was performed on AlGaN/GaN heterojunction grown by metal-organic chemical vapor deposition (MOCVD) on sapphire substrate. The AlGaN barrier consists of a ~21-nm $Al_{0.25}Ga_{0.75}N$ layer and ~1-nm AlN interface enhancement layer (IEL). The source/drain ohmic contacts were first formed by Ti/Al/Ni/Au metallization annealed at 870 °C for 50 s in N_2 ambient with a measured contact resistance of ~1.2 $\Omega\cdot$mm. Then a 100-nm SiO_2 passivation layer was grown by plasma-enhanced chemical vapor deposition (PECVD) at 300 °C, which also serves as the *high-temperature ICP etching mask*. After mesa isolation and gate window opening, a 16-nm $Al_{0.25}Ga_{0.75}N$ layer was etched away by ICP, using low-RF-power mixed Cl_2/BCl_3 plasma at a substrate temperature of 180 °C. The power of ICP source and RF generator are set to be 50 and 15 W respectively, producing a slow etching rate of ~0.13 nm/s. Subsequently, a 15-nm Al_2O_3 gate dielectric was grown by thermal-mode ALD at 300 °C, with the first ~2-nm using trimethylaluminum (TMA) and H_2O as precursors, while the remaining 13-nm Al_2O_3 employing O_3 instead of H_2O as the oxygen precursor. Post-dielectric annealing at 500 °C in N_2 ambient for 1 min was then performed. Finally, Ni/Au bilayer was used as the gate metal. Fig. 1(b) presents the cross-sectional TEM of the gate region of the fabricated device. An etching angle of 138° is obtained and high-crystal-quality Al_2O_3 gate dielectric can be grown by using O_3 as the precursor.

High-temperature low-damage gate recess and ozone-assisted ALD-Al_2O_3 gate dielectric

By virtue of the increased recessing temperature, chlorine-based etching residues ($AlCl_3$ and $GaCl_3$) are effectively reduced in the gate window, as confirmed by X-ray photoelectron spectroscopy (XPS) of Cl 2p core-level shown in Fig. 2(a). The profile of the recessed gate window is remarkably improved, which is confirmed by the atomic force microscopy (AFM), as shown in Fig. 2(b) and (c). It is observed that, compared to the low-temperature gate recess, the surface roughness RMS (root mean square) in the gate window by high-temperature recess is reduced from 2.01 to 0.91 nm. XPS is also used to evaluate the properties of the gate dielectric. The substitution of O_3 for H_2O precursor in ALD effectively suppresses the formation of both Al-Al and Al-O-H bonds in Al_2O_3, as shown by Al 2p core-level spectra in Fig. 3(a). The relative composition of Al-O-H bonds, featuring a binding energy of ~74.2 eV, is reduced from 48.6% to 0.5%. This is because O_3 is able to supply ample radical O atoms without O-H groups, which ensures complete reaction with TMA [9]. Moreover, the energy bandgap (E_G)

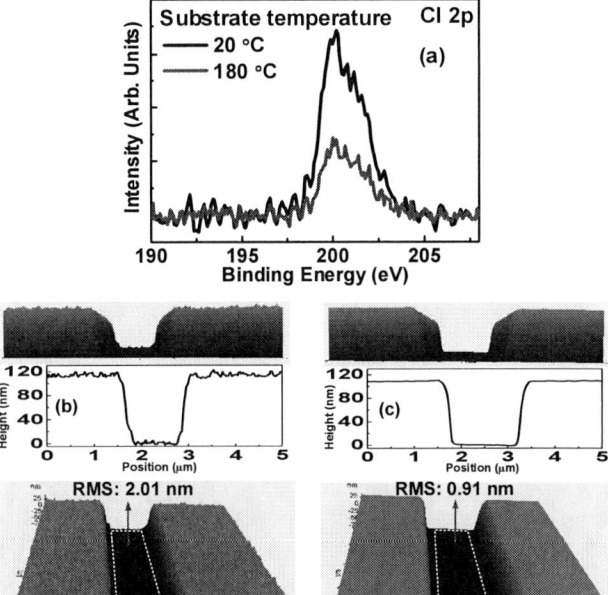

Fig. 2. (a) Cl 2p core-level spectra in the recessed gate region by XPS. During the recess etching, the substrate temperature is set at 20 °C or 180 °C, respectively. (b) and (c) show the trench profile of gate window recessed at 20 and 180 °C, respectively.

of the ALD-Al_2O_3 is increased from 6.9 to 7.1 eV by replacing H_2O with O_3, which is verified by the energy loss spectra of O 1s (shown in Fig. 3(b)).

The O_3-sourced ALD-Al_2O_3 exhibits excellent insulating property, verified by the current-voltage (*I-V*) characteristics of a MIS diode fabricated along with the MIS-HEMTs (Fig. 4(a)). The gate leakage of the MIS diodes at $V_G < +7$ V is lower than 2.5×10^{-8} A/cm^2, and the breakdown *E*-field of the dielectric is determined to be ~8.5 MV/cm.

Capacitance-voltage (*C-V*) characteristics of a Ni/Al_2O_3/AlGaN/AlN/GaN MIS diode are used to characterize the quality of the Al_2O_3/(Al)GaN interfaces, as shown in Fig. 4(b). Small frequency dispersion is observed from 1 kHz to 100 kHz in high-frequency *C-V* (HFCV) characterizations, while humps appear between 0 and 2 V in the quasi-static *C-V* (QSCV) measurements. Owing to the well-suppressed gate leakage, QSCV characterization is valid

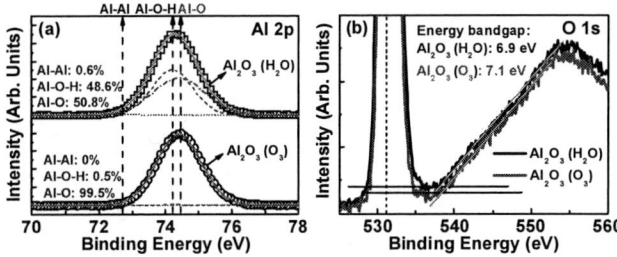

Fig. 3. Al 2p (a) and O 1s (b) core-level spectra of 15-nm ALD-Al_2O_3 grown using H_2O and O_3 as precursors for O, respectively.

978-1-4799-8002-4/14 $31.00 © 2014 IEEE

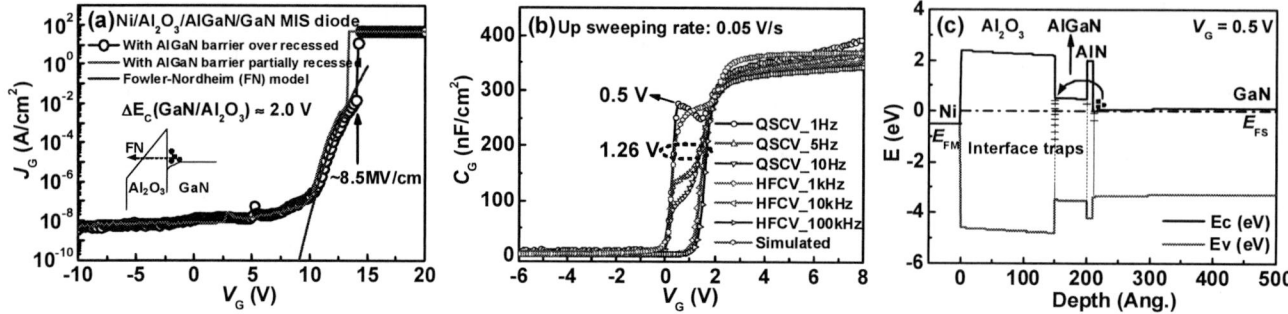

Fig. 4. (a) J_G-V_G characteristics of Ni/Al$_2$O$_3$/AlGaN/AlN/GaN MIS diode with AlGaN barrier layer (including AlN IEL) over recessed (open circle) and with AlGaN barrier partially recessed (red line). (b) High-frequency and quasi-static C-V characterization and simulation of the Ni/Al$_2$O$_3$/AlGaN/AlN/GaN MIS diodes. The barrier height of Ni/Al$_2$O$_3$ is 2.9 V as referred to Ref. 10. (c) Simulated energy band diagram of the MIS diode at $V_G = +0.5$ V.

and the electron-charging of deep traps at/or close to the ALD-Al$_2$O$_3$/AlGaN interface could be captured [7]. The hump grows as frequency decreases from 10 to 1 Hz. It indicates that the electron-filling (trapping) of some deep traps is captured by the up-swept QSCV, leading to a ~1.26-V positive shift of V_{TH}. Their trapping time constants are less than the reciprocal of the measured angular frequency, e.g., 1 s for 1 Hz.

Ideal (without traps) C-V curve and the corresponding energy band diagram are simulated according to the 1-Hz QSCV curve, as shown in Fig. 4(b) and (c). Two capacitance plateaus appear in the simulated C-V curve, implying the electron-accumulation at 2DEG channel and Al$_2$O$_3$/AlGaN interface, respectively. By comparing 1-Hz QSCV and 1-kHz HFCV, the density of the captured traps, with emission time constant longer than 1 ms, is about 2.91×10^{12} cm^{-2}. These acceptor-like traps are probably originated from the oxidation of the recessed AlGaN surface during ALD growth [7, 11]. Note that a net positive fixed charge density of $+0.9 \times 10^{12}$ cm^{-2} is introduced at the AlN(IEL)/GaN interface in the C-V simulation, in view of the polarization of the 6-nm un-recessed AlGaN barrier. The corresponding negative polarization charges at the upper surface of the AlGaN barrier are expected to be compensated by positive charges at ALD-Al$_2$O$_3$/AlGaN interface [8]. In this sense, the positive bulk/border charges in the ALD-Al$_2$O$_3$ should be less than $+0.9 \times 10^{12}$ cm^{-2}, a value smaller than that reported for ALD-Al$_2$O$_3$ deposited with H$_2$O (instead of O$_3$ in this work) source.

Device characterization

Fig. 5 plots the dc characteristics of the normally-off Al$_2$O$_3$/AlGaN/AlN/GaN MIS-HEMTs fabricated with the high-temperature gate recess (180 °C). The threshold voltage V_{TH}, maximum drain current density, and specific ON-resistance $R_{ON, sp}$ are +1.6 V, 615 mA/mm and 0.49 mΩ·cm^2, respectively, in MIS-HEMTs with a gate-to-drain distance of 3 μm. The specific ON-resistance is calculated by $R_{on, sp} = R_{ON} \cdot W_G \cdot (L_{SD} + 2 \times 1.5 \text{ μm})$, including a 1.5 μm extension for source/drain ohmic contacts. A small V_{TH} hysteresis (~0.11 V at $V_{DS} = 1$ V) and a high peak extrinsic

transconductance G_M (~134 mS/mm at $V_{DS} = 10$ V) are measured in the dc transfer characteristics. Moreover, good uniformity in V_{TH} can be obtained with the high-temperature gate recess technique, as shown in Fig. 5(d).

The OFF-state breakdown characteristics of the devices are plotted in Fig. 6(a). The breakdown voltage is capable of scaling up to 1420 V as L_{GD} is increased to 21 μm, dedicating lateral breakdown strength of 0.72 MV/cm. The inclined gate-trench sidewall and over-hang field plate effectively alleviate the high E-field on the drain-side gate edge, leading to enhanced breakdown voltage and suppressed gate electron injection into the access region, as shown in Fig. 6(c).

The normally-off MIS-HEMTs also exhibit good dynamic performance in both pulse I-V and hard-switching characterizations, as shown in Fig. 7. A high pulsed drive current of 1.1 A/mm is achieved at a pulse period of 10 μs with a 2% duty cycle (Fig. 7(a)), although they show larger current collapse at high OFF-state stress. In the switching characterization, an off-chip load resistor was mounted on Printed circuit board (PCB) and connected to the device

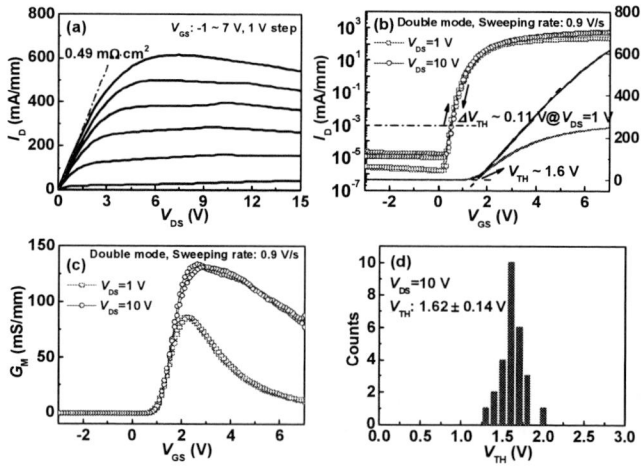

Fig. 5. dc I-V characteristics of normally-off Al$_2$O$_3$/AlN/AlN/GaN MIS-HEMTs. (a) dc output characteristics. (b)&(c) dc transfer characteristics measured at $V_{DS} = 1$ and 10 V. (d) Threshold voltage uniformity characterization.

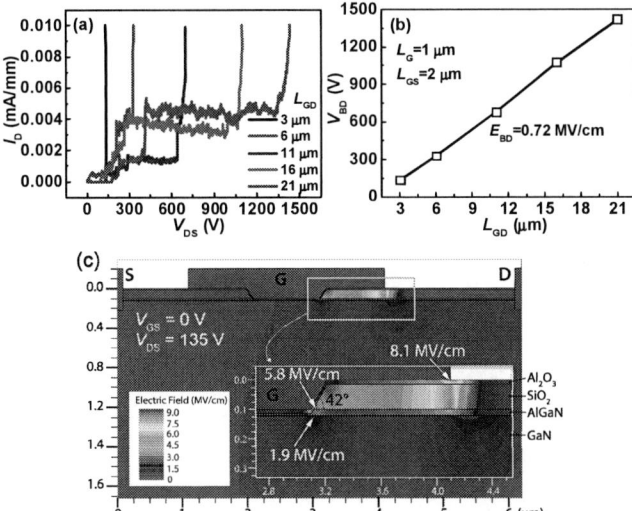

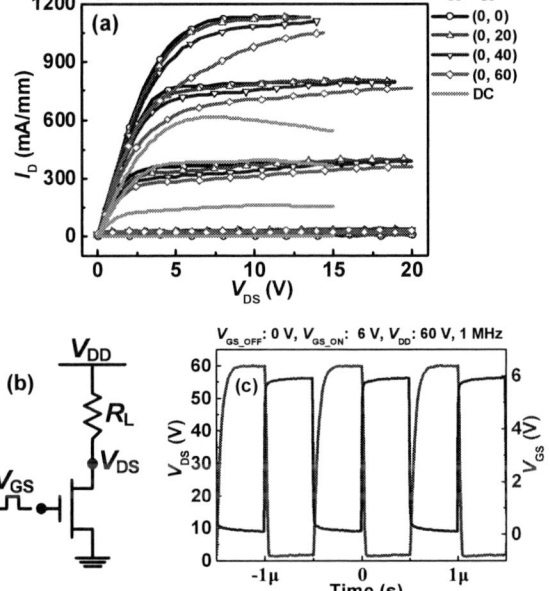

Fig. 6. (a)&(b) OFF-state breakdown characteristics of normally-off $Al_2O_3/AlGaN/AlN/GaN$ MIS-HEMTs. (c) Simulated E-field distribution of the MIS-HEMTs under an OFF-state bias point of $(V_{GS}, V_{DS}) = (0\ V, 135\ V)$.

Fig. 7. (a) Pulsed I_D-V_{DS} characteristics of normally-off $Al_2O_3/AlGaN/AlN/GaN$ MIS-HEMTs from various quiescent bias point (V_{GSQ}, V_{DSQ}). (b)&(c) ON-wafer hard-switching characterization of the MIS-HEMTs. The device is switched from V_{GS_OFF} of 0 V to V_{GS_ON} of 6 V at frequency of 1 MHz with a 50% duty cycle. V_{DD} is kept at 60 V.

through an RF probe to avoid gate oscillations and overshoots, as shown in Fig. 7(b). The dynamic ON-resistance is increased by 5% at $V_{DS} = 60$ V, owing to the suppression of electron injection at $SiO_2/AlGaN$ interface in the gate-drain access region (Fig. 6(c)). It is expected that the passivation performance can be improved by utilizing a charge-polarized AlN interfacial layer, or a high-quality SiN_x layer between AlGaN barrier and SiO_2 passivation layer.

Conclusion

A high-temperature gate recess technique featuring low damage and *in-situ* self-clean capability, in combination with O_3-assisted ALD growth of Al_2O_3 gate dielectric, is developed for fabrication of GaN-based normally-off power devices. The high-temperature gate recess technique leads to improved surface morphology and profile in the recessed gate window, while the O_3-assisted ALD-Al_2O_3 exhibit significantly suppressed positive fixed charge and lower leakage.

Acknowledgement

This work is supported by the Hong Kong Research Grant Council under grant ITS/192/14FP, the National Natural Science Foundation of China under Grant 60890191 and Grant 61474138, and the National Key Basic Research Program under Grant 2010CB327503.

References

[1] K. J. Chen and C. Zhou, "Enhancement-mode AlGaN/GaN HEMT and MIS-HEMT technology," *Phys. Status Solidi (a)*, vol. 208, no. 2, pp. 434–438, Feb. 2011.

[2] R. Wang, Y. Cai, C.-W. Tang, K. M. Lau, and K. J. Chen, "Enhancement-Mode Si_3N_4/AlGaN/GaN MISHFETs," *IEEE Electron Device Lett.*, vol. 27, no. 10, pp. 793–795, Oct. 2006.

[3] M. Kanamura, T. Ohki, T. Kikkawa, K. Imanishi, T. Imada, A. Yamada, and N. Hara, "Enhancement-Mode GaN MIS-HEMTs With n-GaN/i-AlN/n-GaN Triple Cap Layer and High-k Gate Dielectrics," *IEEE Electron Device Lett.*, vol. 31, no. 3, pp. 189–191, Mar. 2010.

[4] Z. Tang, Q. Jiang, Y. Lu, S. Huang, S. Yang, X. Tang, and K. J. Chen, "600-V normally-off SiN_x/AlGaN/GaN MIS-HEMT with large gate swing and low current collapse," *IEEE Electron Device Lett.*, vol. 34, no. 11, pp. 1373–1375, Nov. 2013.

[5] W. Saito, Y. Takada, M. Kuraguchi, K. Tsuda, and I. Omura, "Recessed-gate structure approach toward normally off high-Voltage AlGaN/GaN HEMT for power electronics applications," *IEEE Trans. Electron Devices*, vol. 53, no. 2, pp. 356–362, Feb. 2006.

[6] S. J. Pearton, J. C. Zolper, R. J. Shul, and F. Ren, "GaN: Processing, defects, and devices," *J. Appl. Phys.*, vol. 86, no. 1, p. 1, Jul. 1999.

[7] S. Huang, S. Yang, J. Roberts, and K. J. Chen, "Threshold Voltage Instability in Al_2O_3/GaN/AlGaN/GaN Metal–Insulator–Semiconductor High-Electron Mobility Transistors," *Jpn. J. Appl. Phys.*, vol. 50, no. 11, p. 110202, Oct. 2011.

[8] B. Shin, J. R. Weber, R. D. Long, P. K. Hurley, C. G. Van de Walle, and P. C. McIntyre, "Origin and passivation of fixed charge in atomic layer deposited aluminum oxide gate insulators on chemically treated InGaAs substrates," *Appl. Phys. Lett.*, vol. 96, no. 15, p. 152908, Apr. 2010.

[9] J. B. Kim, D. R. Kwon, K. Chakrabarti, C. Lee, K. Y. Oh, and J. H. Lee, "Improvement in Al_2O_3 dielectric behavior by using ozone as an oxidant for the atomic layer deposition technique," *J. Appl. Phys.*, vol. 92, no. 11, p. 6739, Nov. 2002.

[10] Z. Zhang, C. M. Jackson, A. R. Arehart, B. McSkimming, J. S. Speck, and S. A. Ringel, "Direct Determination of Energy Band Alignments of Ni/Al_2O_3/GaN MOS Structures Using Internal Photoemission Spectroscopy," *J. Electron. Mater.*, vol. 43, no. 4, pp. 828–832, Dec. 2013.

[11] S. Yang, Z. Tang, K.-Y. Wong, Y.-S. Lin, Y. Lu, S. Huang, and K. J. Chen, "Mapping of interface traps in high-performance Al_2O_3/AlGaN/GaN MIS-heterostructures using frequency- and temperature-dependent C-V techniques," in *2013 IEEE International Electron Devices Meeting*, 2013, pp. 6.3.1–6.3.4.

Trapping and High Field Related Issues in GaN Power HEMTs

Gaudenzio Meneghesso[1], Matteo Meneghini[1], Alessandro Chini[2], Giovanni Verzellesi[3], Enrico Zanoni[1]

[1]Department of Information Engineering, University of Padova, Italy, Via Gradenigo 6B, 35131 Padova - gauss@dei.unipd.it
[2]Dep, of Engineering "Enzo Ferrari", Univ. of Modena and Reggio Emilia, Italy
[3]Dipartimento di Scienze e Metodi dell'Ingegneria, Univ. of Modena and Reggio Emilia, Italy

Abstract

Gallium Nitride HEMTs grown on Si substrates are the most promising solution for the future technologies in the power electronics industry. Compensation of unintentional GaN n-type conductivity is specifically mandatory in the buffer for an optimum device blocking function. Carbon (C) or Iron (Fe) doping are the most common solutions that however are responsible also for the introduction of traps in the buffer, that induce large charge trapping and current collapse when devices are biased at high voltages as well as affect breakdown behavior of these devices. This paper reviews the main high field related issues recently reported in GaN-on-Si devices for power applications.

Introduction

Gallium Nitride HEMTs grown on Si substrates are the most promising solution for the future technologies in the power electronics industry [1]. Compensation of unintentional n-type conductivity is specifically mandatory in the buffer for an optimum blocking function. In this context, Carbon (C) or Iron (Fe) doping are the most common solutions that however are responsible also for the introduction of traps in the buffer, that induce large charge trapping and current collapse effects when devices are biased at high voltages. Also breakdown behavior of these devices is strongly affected by the compensated buffer. For this reason, present, state-of-the-art devices still suffer from parasitic and reliability issues that hamper their wide market penetration; more specifically, the most important issues are the following: a) charge trapping leading to current collapse and dynamic-Ron (dynamic on-resistance) increase [2,3]; b) premature lateral and vertical breakdown [4]; c) performance degradation observed when devices are submitted to high voltages both under open and closed channel conditions [5]. **The aim of this paper** is to review the main high field related issues (leading to charge trapping, breakdown mechanisms and device degradation) recently reported in GaN-on-Si power devices with some original results recently obtained in our labs.

Charge trapping and dynamic R_{on}

The most popular method to study the presence of charge trapping and current collapse in GaN HEMTs is the so called double-pulse I_D–V_D characterization. The method relies on the acquisition of pulsed I_D-V_{DS} and/or I_D-V_{GS} curves obtained starting from different quiescent bias points $[V_{GS,Q}, V_{DS,Q}]$, with the aim of evaluating the effects of a negative gate bias and of a high drain–gate voltage on the trapping process (**Fig.1**). **Figures 2** and **3** show representative examples of measurements carried out on two devices that show a different signature of the current collapse, i.e. i) threshold voltage shift without significant $g_{m,peak}$ reduction (**Fig.2**), and ii) transconductance peak ($g_{m,peak}$) reduction without significant threshold voltage change (**Fig.3**). Effects observed in **Fig.2** are attributed to trap charge in the device volume under the gate (see **Fig.4, left**), while those observed in **Fig.3** are attributed to trap charge taking place in the device volume of the access regions (see **Fig.4 right**). A more accurate description of the properties of the traps relevant for GaN-based transistors operation can be obtained by studying the current transients (in the on-state) induced by exposure to a trapping voltage in the off-state (typically a negative gate bias and a high drain-gate voltage), through current-mode DLTS (I-DLTS), see [6] and references therein.

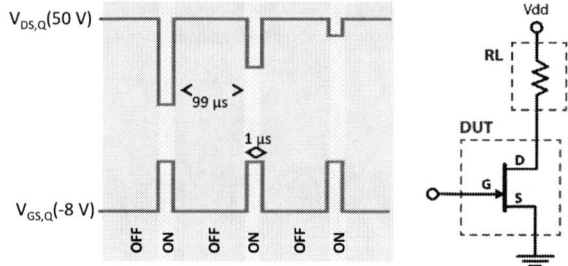

Figure 1. Schematic waveforms and circuit of gate and drain voltages during the double pulse measurements.

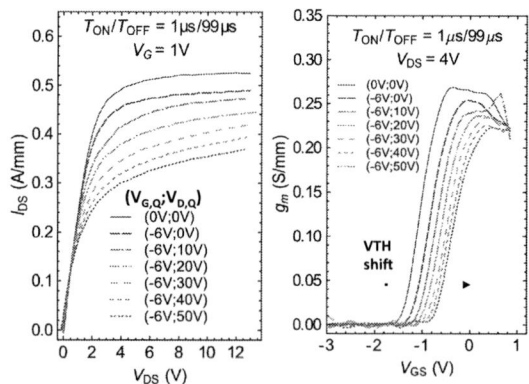

Figure 2. (a) I_{DS}-V_{DS} and (b) g_m-V_{GS} double pulsed characterization. The dynamic current collapse is mainly caused by the positive V_{TH} shift. Employed pulse width/period ratio is 1μs/99μs.

978-1-4799-8002-4/14 $31.00 © 2014 IEEE   446

Through the use of proper trap filling voltages, it is possible to achieve information about the location of the traps, as shown in **Fig.5**. By repeating the same measurements at different temperatures (see **Fig.6** for one particular filling

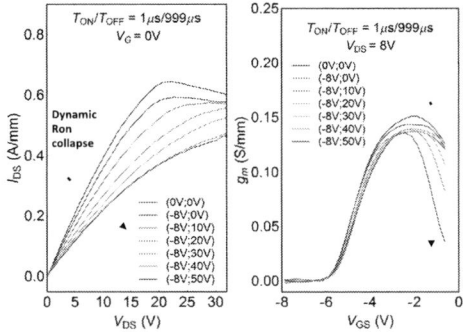

Figure 3. (a) I_{DS}-V_{DS} and (b) g_m-V_{GS} double pulsed characterization. The dynamic current collapse is mainly caused by Ron increase. Employed pulse width/period ratio is 1µs/999µs.

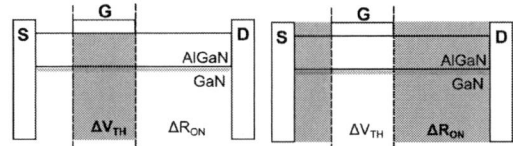

Figure 4. Sketch of traps localization leading to the pulsed measurements shown in Fig. 2 (left) and Fig. 3 (right).

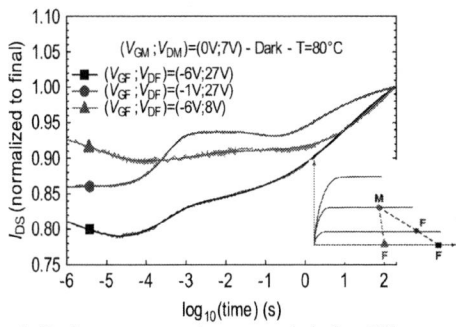

Figure 5. Drain current transients recorded after different trapping conditions i.e. filling trap (F): $(V_{GF}=-6V, V_{DF}=27V)$, $(V_{GF}=-1V, V_{DF}=27V)$, $(V_{GF}=-6V, V_{DF}=8V)$, see also inset. The measurement point (M) is always at $V_{GM}=0V$, $V_{DM}=7V$.

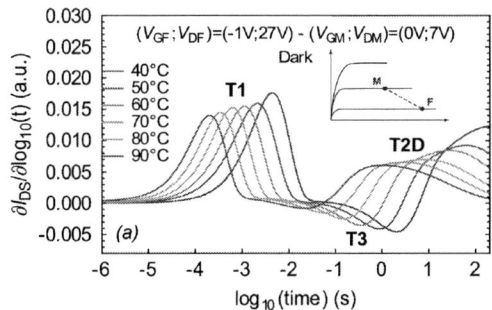

Figure 6. Derivative of the current transient with the filling pulse set at $(V_{GF};V_{DF})=(-1V, 27V)$ (see blue filled circle in Fig. 5, and the inset) carried out at different temperatures. Three traps are highlighted: T1, T3 and T2D. The calculated activation energies are reported in Fig.7.

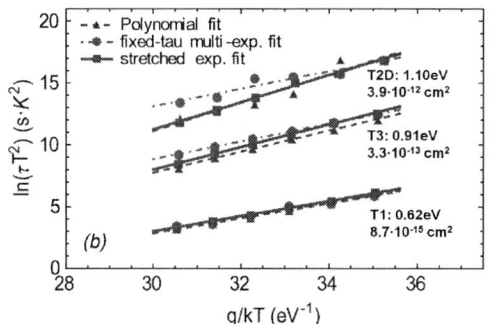

Figure 7. Arrhenius plot (of data in Fig. 6) with apparent activation energies (E_A) and capture cross-sections (σ_c) extrapolated by stretched exponential fit of T1- T2D- and T3-related de-trapping processes. Comparison with other fitting methods is also reported.

Table I: properties of the most frequent deep levels that affect the dynamic performance of GaN-based HEMTs [6]

E_A (eV)	Cross section cm^2	Origin
0.28	1x10^{-17}	Carbon in Gallium-substitutionals
0.37	1x10^{-16}	(open-core) dislocations, Nitrogen-vacancies
0.63	1x10^{-14}	GaN point-defects (clustering along dislocations), Promoted by Iron-doping
0.62	1x10^{-16}	GaN point-defects
0.85	4x10^{-14}	Carbon interstitials, (full-core) dislocations
1.10	1x10^{-12}	Gallium-antisites or interstitials
0.96	5x10^{-15}	Radiation-induced defects

condition), it is possible to extract the activation energy of the traps related to the observed transients, see **Fig.7**. T1 bas been attributed to point defects, T2D to the presence of highly localized defect states in the GaN buffer, and finally T3 to V_{Ga}-related deep acceptor (see [6] for more detailed analysis). **Table I** reports some of the most frequent deep levels recently reported in GaN HEMTs [6]. **Figure 8(a)** shows the transient analysis carried out at different temperatures. Two very important pieces of information can here be obtained: i) charge trapping requires long times (in the seconds regime) and ii) charge trapping increases with temperature (opposite with the common thinking that high temperature reduce the trapped charge). This indicates that there is a more complex trapping process in GaN devices, that can be explained by assuming there is a barrier for trapping (possibly related to the fact that electron must be injected from the substrate). To probe the effect of traps present in the buffer, backgating bias stresses have also been carried out, see **Fig. 8(b)**. In this case, source, gate and drain are grounded, and the bulk is biased at − 25 V, at different T.

In devices with non-optimum C-compensated buffer, large threshold voltage instabilities can also be observed [7], as shown in **Fig. 9** reporting double pulsed I_D–V_{GS} characteristics measured by using different baselines: 1) $(V_{GS-BL},V_{DS-BL})=(0V,0V)$: this is the initial baseline adopted as a reference for subsequent measurements; 2) $(V_{GS-BL},V_{DS-BL})=(-10V,400V)$: a significant negative V_{TH} shift is observed; 3) $(V_{GS-BL},V_{DS-BL})=(0V,0V)$: taken immediately after the previous measurement, this measurement leads to a large a positive V_{TH} shift (larger than the initial value). In [7] we have attributed the observed (recoverable) threshold-voltage shifts to the presence of channel/buffer trap modulation during device switching. Device simulations suggest that the observed behavior can be explained by the presence of electron traps in the GaN:UID channel and hole traps located in the GaN:C buffer. On the contrary, optimized C doping can result in negligible trapping effects [8]. Finally these "trapping" characterization must be carried out also at large V_{DS}, in fact current collapse and dynamic Ron increase can be negligible at relatively low voltages (say V_{DS}=100V) but can become very severe at higher voltages, as demonstrated in **Fig.10**.

Breakdown in power switching GaN HEMTs

Breakdown (BD) represents an important aspect to be fully understood in high power/high voltage GaN devices. What power circuit designers want is a device with sustainable breakdown that allows the device to withstand high current in breakdown even for a very short time (ns). Unfortunately, today, GaN devices cannot meet this requirement (sustainable breakdown has been reported either in vertical and lateral devices but at very low current levels, see **Fig.11**); for this reason it is necessary to overdesign the device breakdown for a certain voltage application (i.e. for 600 V device operation it is required 800V breakdown or higher). For this reason, over the past years several groups have investigated the physical origin of BD [9,10], with the aim of developing models to explain this phenomenon, and of proposing technological improvements to increase the robustness of the devices. In **Fig.11** we report the results of current-controlled breakdown measurements, carried out on devices with L_{GD}=4 μm and a pinch-off voltage of V_{TH}=-2 V.

This figure shows that the breakdown current may originate from gate-drain leakage or from drain-source leakage, depending on the applied gate-bias level: in fact, when the

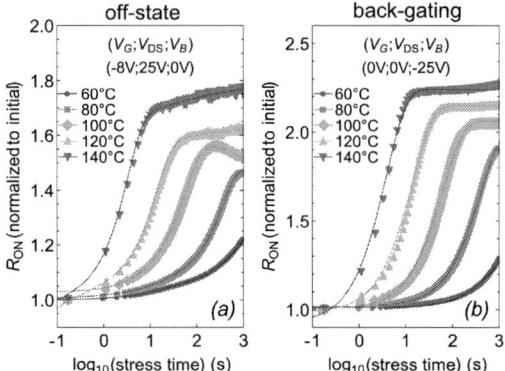

Figure 8. Kinetics of R_{ON}- increase at multiple temperatures in (a) off-state, and (b) back-gating bias conditions.

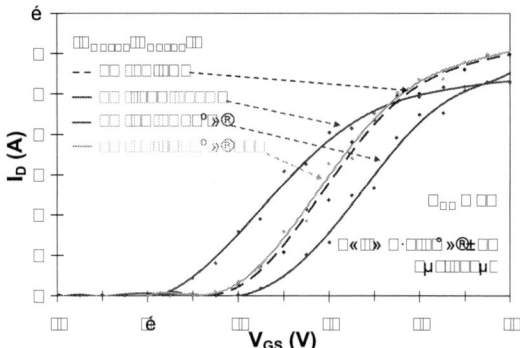

Figure 9. g_m-V_{GS} double pulsed characterization at V_{DS}=2 V obtained by applying the baseline sequence #1 to #4.

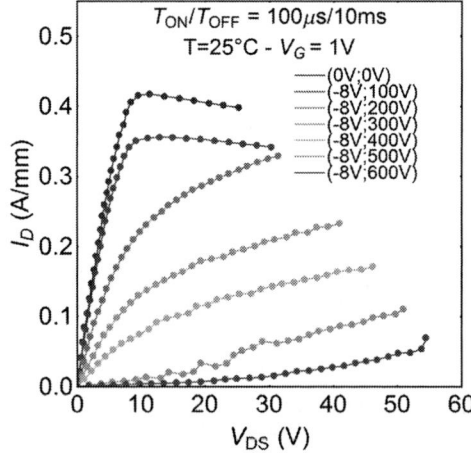

Figure 10. (a) I_{DS}-V_{DS} double pulsed measurements with baselines up to 600 V. Employed pulse width/period ratio is 100μs/10ms.

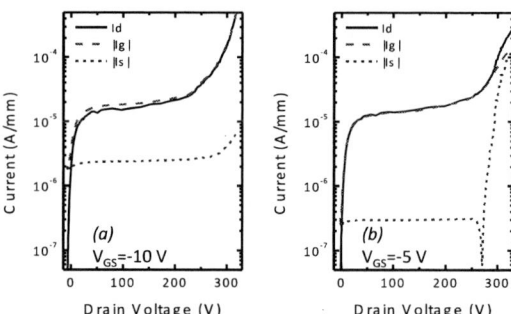

Figure 11. Breakdown measurements carried out in current-controlled mode on one of the analyzed samples. Measurements were taken with a gate voltage of (a) V_{GS}=-10 V, and (b) V_{GS}=-5 V.

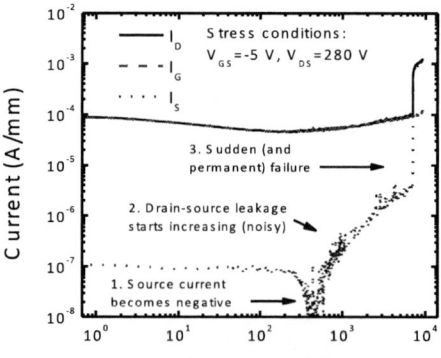

Figure 12. Constant voltage stress experiment in devices of Fig. 11 Stress conditions are V_{GS}=-5 V, V_{DS}=280 V. (L_{GD} = 4 um)

978-1-4799-8002-4/14 $31.00 © 2014 IEEE

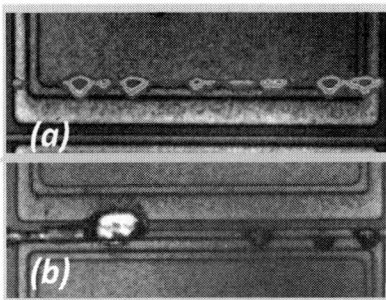

Figure 13. Spatially-resolved electroluminescence pattern measured (a) before the onset of sudden failure (end of phase 2 in Figure 12) and (b) after the permanent failure of one of the analyzed samples (stress conditions are V_{GS} = -5 V, V_{DS}=280 V).

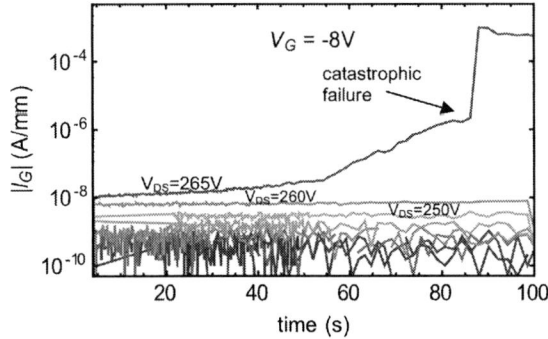

Figure 14. Variation of gate current measured during each of the final stages of the step-stress experiment carried out on a MIS-HEMT with L_{GD}=2μm. Failure occurs during the last stage and can be recognized as a sudden and permanent increase in gate leakage current.

breakdown curves are measured with a V_{GS} significantly lower than the pinch-off voltage, see **Fig.11 (a)**, breakdown (drain) current almost completely originates from gate-drain leakage, while the source current is an order of magnitude lower. On the other hand, when the breakdown curves are measured with V_{GS} closer to V_{TH}, drain-source leakage becomes comparable to gate-drain leakage when the drain voltage approaches BD, see **Fig.11 (b)**. This effect is due to a sub-threshold conduction mechanism or punch-through [4]. Fe or C compensation, or the use of Double Heterostructure devices have been proposed to tackle this problem [9]. Close to the breakdown conditions, GaN HEMTs can show time-dependent breakdown mechanisms: an example is shown in **Figs.12** and **13**, for a device biased near the breakdown condition i.e. with V_{GS}=-5 V, and a drain voltage of 280 V. As can be seen, by increasing stress times (around 400 s), the source current changes its sign from positive to negative, meaning that current is starting to flow out of the source. For longer stress times (here between 400 s and 7000 s) source current shows a further, and gradual, increase; the noise superimposed to source current increases after the change of sign. Finally a sudden increase in source current is detected for longer stress times (here after 7000 s); this leads to a strong increase in drain current, while the gate leakage remains almost stable. Damages of the gate are in correspondence of the major EL peaks, see **Fig.13** [12].

A time- and voltage-dependent degradation process has been detected also on MIS-HEMT devices (see **Fig. 14**); off-state stress may increase the probability of trapping between gate and drain [11]; the catastrophic failure can be related to a discharge in proximity of the edges of the drain contact.

Conclusions

The main charge trapping high filed related instability have been reviewed and briefly discussed. Breakdown mechanism has also been presented with some reliability related findings that will be largely discussed in the final paper.

Acknowledgements

The authors thank F. Rampazzo, I. Rossetto, D. Bisi, A. Stocco, R. Silvestri, C. de Santi, S. Dalcanale, for the fruitful discussion. This work is partially funded by the UE (ENIAC Joint Undertaking) in the E2CoGaN project (ENIAC-324280), by the project FP7-ICT-2011-7 (Project No. 287602) HiPoSwitch,and by the Progetto di Ateneo 2010 ''Development of normally off gallium nitride power devices for future green power applications''.

References

[1] B. J. Baliga, "Gallium nitride devices for power electronic applications," Semicond. Sci. Technol., vol. 28, p. 074011, 2013.

[2] O. Hilt, E. B.-Treidel, E. Cho, S. Singwald and J. Würfl, "Impact of Buffer Composition on the Dynamic On-State Resistance of High-Voltage AlGaN/GaN HFETs", 24th IEEE ISPSD 2012.

[3] D. Jin and J. A. del Alamo, "Mechanisms responsible for dynamic ON-resistance in GaN high-voltage HEMTs", 24th IEEE ISPSD 2012.

[4] E. Bahat-Treidel, O. Hilt, F. Brunner, J. Würfl, and G. Tränkle, IEEE Trans. on Electr. Dev. 55 (2008) 3354.

[5] M. Meneghini, G. Cibin, M. Bertin, G. A. M. Hurkx, P. Ivo, J. Sonsky, J. A. Croon, G. Meneghesso, and E. Zanoni, "OFF-State Degradation of AlGaN/GaN Power HEMTs: Experimental Demonstration of Time-Dependent Drain-Source Breakdown", IEEE Trans. on Electr. Dev. 61, 1987 (2014),

[6] D. Bisi, M. Meneghini, C. de Santi, A. Chini, M. Damman P. Brueckner, M. Mikulla, G. Meneghesso, and E. Zanoni, "Deep-Level Characterization in GaN HEMTs-Part I: Advantages and Limitations of Drain Current Transient Measurements", IEEE Trans. on Electr. Dev. 60 (10), pp. 3166-3175, 2013,

[7] G. Meneghesso, R. Silvestri, M. Meneghini, A. Cester, E. Zanoni, G. Verzellesi, G. Pozzovivo, S. Lavanga, T. Detzel, O. Häberlen, G. Curatola, "Threshold Voltage Instabilities in D-Mode GaN HEMTs for Power Switching Applications", IEEE IRPS, Kona HI, June 1-5, 2014.

[8] G. Verzellesi, L. Morassi, G. Meneghesso, M. Meneghini, E. Zanoni, G. Pozzovivo, S. Lavanga, T. Detzel, O. Haberlen, G. Curatola, "Influence of Buffer Carbon Doping on Pulse and AC Behavior of Insulated-Gate Field-Plated Power AlGaN/GaN HEMTs", IEEE Electr. Dev. Lett., Vol. 35, No. 4, pp. 443-445, Apr. 2014.

[9] J.Wuerfl, E.Bahat-Treidel, F.Brunner, E.Cho,O.Hilt, P.Ivo, A.Knauer, P.Kurpas, R. Lossy, M. Schulz, S. Singwald, M. Weyers, R. Zhytnytska, Microelectronics Reliability 51, 1710 (2011)

[10] M. Meneghini, D. Bisi, D. Marcon, S. Van Hove, T.-L. Wu, S. Decoutere, G. Meneghesso, and E. Zanoni, IEEE Trans. on Power Electronics, , 29, 2199 (2014)

[11] M. Meneghini, A. Zanandrea, F. Rampazzo, A. Stocco, M. Bertin, G. Cibin, D. Pogany, E. Zanoni, and G. Meneghesso, JJAP 52, 08JN17 (2013)

[12] M. Meneghini, A. Stocco, N. Ronchi, F. Rossi, G. Salviati, G. Meneghesso, and E. Zanoni, "Extensive analysis of the luminescence properties of AlGaN/GaN high electron mobility transistors", Appl. Phys. Lett, 97, 063508 (2010)

978-1-4799-8002-4/14 $31.00 © 2014 IEEE

CMOS-Compatible GaN-on-Si Field-Effect Transistors for High Voltage Power Applications

Man Ho Kwan, K.-Y. Wong, Y. S. Lin, F.W. Yao, M. W. Tsai, Y.-C. Chang, P. C. Chen, R.Y. Su,
C.-H. Wu, J. L. Yu, F. J. Yang, G. P. Lansbergen, H.-Y. Wu, M.-C. Lin, C. B. Wu, Y.-A. Lai,
C.-W. Hsiung, P.-C. Liu, H.-C. Chiu, C.-M. Chen, C. Y. Yu, H. S. Lin, M.-H. Chang, S.-P. Wang,
L. C. Chen, J. L. Tsai, H.C. Tuan, and Alex Kalnitsky

Power IC Program, Analog / RF & Specialty Technology Division, TSMC, Hsin-Chu, Taiwan.
Tel: (+886)-3-5636688 ext.708-4939, Fax :(+886)-3-5662051, Email: mhkwan@tsmc.com

Abstract

CMOS-compatible 100/650 V enhancement-mode FETs and 650 V depletion-mode MISFETs are fabricated on 6-inch AlGaN/GaN-on-Si wafers. They show high breakdown voltage and low specific on-resistance with good wafer uniformity. The importance of epitaxial quality is figured out in a key industrial item: high-temperature-reverse-bias-stress-induced on-state drain curent degradation. Optimization of epitaxial layers shows significant improvement of device reliability.

Introduction

High breakdown voltage, low on-resistance, fast switching, and high temperature operation of GaN power transistors have been demonstrated in high-efficient power conversion systems. In this paper, cost-effective CMOS-compatible AlGaN/GaN 100/650 V enhancement-mode field-effect transistors (E-FET) and 650 V depletion-mode MISFET (D-MISFET) are fabricated on 6-inch GaN-on-Si wafers. A low specific contact resistance R_c (0.35 Ω-mm) is achieved by Au-free process. The 650 V power devices show a high off-state breakdown voltage (BV) of > 820 V, and a low specific on-resistance $R_{on,sp}$ with good wafer uniformity. For device reliability, we figure out the importance of epitaxial quality in a key industrial item: high temperature reverse bias (HTRB) [1] stress-induced on-state drain current degradation. A major breakthrough on optimization of GaN epitaxial layers shows significant reduction of the degradation.

CMOS-Compatible GaN-on-Si Technology

E-FETs for 100 and 650 V ratings and 650 V D-MISFETs are fabricated on 6-inch GaN-on-Si wafers with CMOS-compatible process. The sheet resistance of Au-free gate metal is 0.17 Ω/square Specific ohmic contact resistance ($R_{c,sp}$) achieves 4×10^{-6} Ω-cm^2 with good wafer uniformity [2] and is comparable to conventional Au process (3×10^{-6} Ω-cm^2) [3]. The 100 V E-FET shows a threshold voltage (V_{th}) of 1.5 V and a specific on-resistance ($R_{on,sp}$) of 0.25 mΩ-cm^2, and the 650 V E-FET exhibits a V_{th} of 1.4 V and a $R_{on,sp}$ of 2.7 mΩ-cm^2. The gate of E-FET has small on-state gate leakage (on-I_g) (Fig. 1a) and sustains a high gate-source voltage (V_{GS}) up to

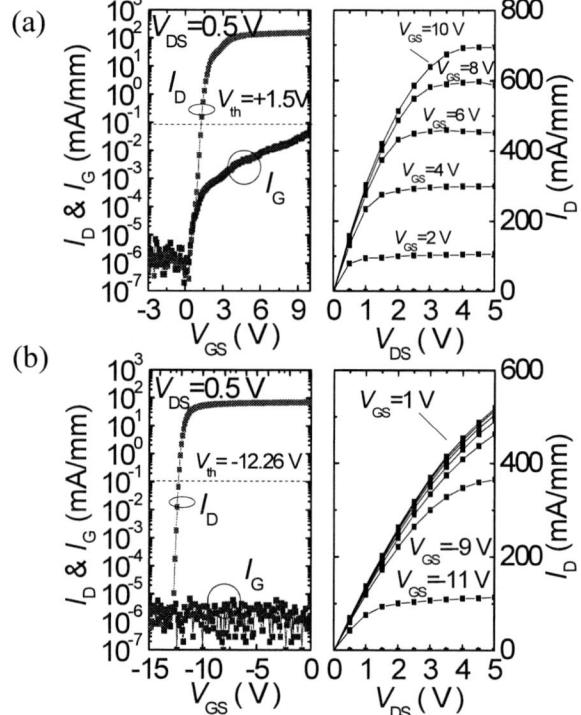

Fig. 1: DC output and transfer characteristics of (a) a 100 V E-FET, and (b) a 650 V D-MISFET. For an E-FET, the gate can sustain a high gate-source voltage (V_{GS}) up to 12 V without damaging the device.

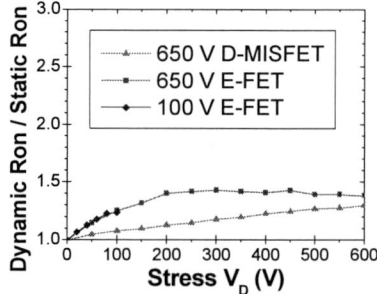

Fig. 2: Comparison of dynamic-to-static on-resistance ratio for various devices (switching frequency: 10 kHz, on-duty: 10%).

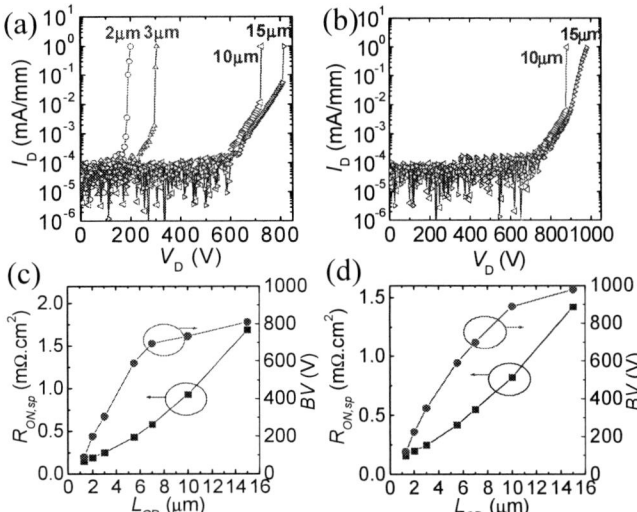

Fig. 3: Off-state characteristics at (a) $V_G = 0V$ for a E-FET and (b) $V_G = -20V$ for a D-MISFET with various gate-to-drain distances L_{GD}. Leakage current for both devices of $L_{GD} = 15\mu m$ is $1\mu A/mm$ at 700 V (Ion/Ioff ratio > 10^5). $R_{on,sp}$ and BV are plotted against of L_{GD} for (c) a E-FET and (d) a D-MISFET respectively.

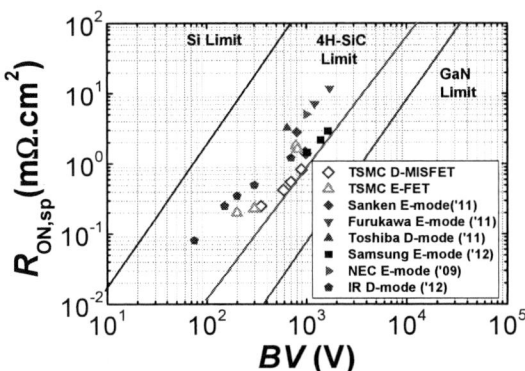

Fig. 4: Specific on-resistances $R_{ON,sp}$ versus BV for reported GaN-based transistors.

12 V in static mode without damaging the device, which enables a wide gate driving range. For D-MISFET ($V_{th} \sim -12.3$ V, $R_{on,sp} \sim 1.45$ mΩ-cm²), on-I_g is negligible (Fig. 1b).

The 650 V D-MISFET shows a dynamic-to-static on-resistance (R_{on}) ratio of around 1.2 at 520 V, while the 650 V E-FET exhibits the ratio at around 1.4 (Fig. 2). Switching figure-of-merit (FOM = $R_{on} \times Q_g$, where Q_g is the gate charge) for 650 V E-FET and D-MISFET are around 0.196 and 0.313 Ω-nC respectively, an order smaller than published Si and GaN devices [4-6]. The FOM for 100 V E-FET is around 41 Ω-nC, better than published GaN HEMTs [7]. The OFF-state drain leakage current of 650 V devices is 1 μA/mm at 700 V (Fig. 3a, b), with an on/off current ratio greater than 10^5.

With substrate grounded, 100 and 650 V E-FETs exhibit BV of 120 and 820 V respectively (Fig. 3c), while the 650 V D-MISFET shows a large BV of 950 V (Fig. 3d). They show

TABLE I
A LIST OF DISTRIBUTION OF DEVICE CHARACTERISTICS FOR 100/650 V E-FETs, AND 650 V D-MISFETs ON 6-INCH GaN-ON-SI WAFERS. DATA ARE FROM 14 WAFERS IN 4 LOTS.

Device Type	V_{th} (V)		$R_{on,sp}$ (mΩ-cm²)		BV (V)	$R_{c,sp}$ (Ω-cm²)	
	Median	Standard Deviation	Median	Standard Deviation		Median	Standard Deviation
100 V E-FET	+1.5	0.12	0.28	0.023	> 150		
650 V E-FET	+1.4	0.12	2.7	0.25	> 800	4×10^{-6}	1.36×10^{-6}
650 V D-MISFET	-12.3	0.50	1.8	0.20	> 800		

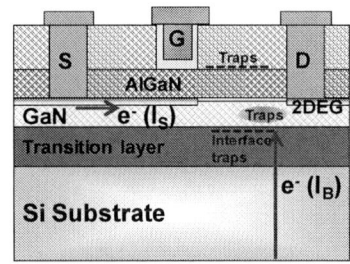

Fig. 5: A schematic diagram of a high voltage device illustrates possible electron trapping that partially depletes 2DEG in the channel under high voltage off-state stress and causes on-state I_D degradation

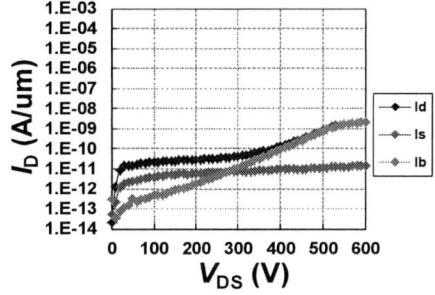

Fig. 6: Contributions of off-state substrate bulk (I_B) and source leakage and (I_S) of a 650 V device to the overall drain leakage I_D (150°C).

excellent $R_{on,sp}$ against BV close to 4H-SiC limit (Fig. 4) and exhibit uniform device characteristics on the wafers (Table I).

Device Reliability

Several reports have studied the reliability of AlGaN/GaN FETs [8-11], and try to identify some possible models leading to degradation of on-state performance (Fig. 5). Most of them focus on failures associated with the off-state source leakage (I_s), and/or off-state gate leakage (I_G). Nevertheless, the contribution of off-state substrate bulk leakage (I_B) to charge trapping during the stress cannot be ignored. For 100 V E-FET, off-state I_s dominates the drain leakage I_D, while for 650 V devices, off-state I_B beyond a drain-source voltage (V_{DS}) of 400 V dominates off-state I_D (Fig. 6). Initial off-state I_B of 650V D-MISFET is found related to the degradation of on-state I_D after HTRB stress [1] (Fig 7a). Under HTRB stress, V_{GS} of devices is biased below V_{th} at an ambient temperature of 150°C, while V_{DS} is biased at 80% of the specified ratings. When the device is soaked at high

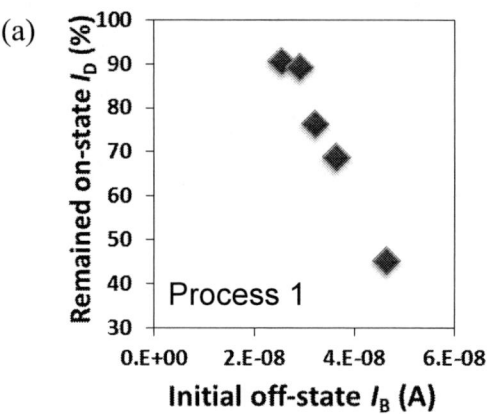

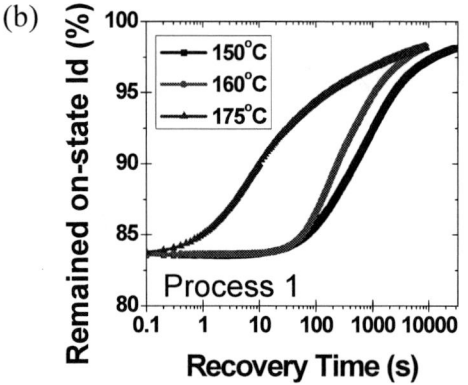

Fig. 7: (a) Initial off-state I_B closely relates on-state I_D degradation of a 650 V D-MISFET (with Process 1) after HTRB stress. Remained on-state I_D is defined as a ratio of $[I_D$ (current) $- I_D$ (before stress)] to I_D (before stress). I_B was measured at a V_D of 520 V and an ambient temperature of 150°C. (b) Recovery test of a 650 V D-MISFET was carried out after 12-hour HTRB stress. Under the test, the samples are biased at the on-state ($V_D = 0.1$ V, $V_G = 0$ V, $V_S = V_B = 0$ V). At higher ambient temperature, recovery rate is faster.

temperature after stress, on-state drain current recovers (Fig. 7b). Extraction of time constants [12] of the on-state I_D recovery curves against temperature indicates a deep trap state at ~1.2 eV (Fig. 8).

To further understand if the filling of traps is mainly contributed by off-state I_B of 650 V devices, off-state I_B at a V_{DS} of 520 V is plotted against temperature, and the extracted activation energy E_a is also ~1.2 eV (Fig. 9a), which matches the state of deep traps in the recovery test. In other words, on-state I_D degradation of 650 V devices after HTRB stress is mainly due to charge trapping associated with the substrate leakage during an off-state stress. A reduction of off-state I_B during HTRB stress further indicates the trap filling process (Fig. 9b). Nevertheless, location of traps along the path of off-state I_B can be various: GaN layer [10, 13], and/or the interface between the GaN layer and the transition layer (Fig. 5). During charge trapping, the energy band of the heterostructure shifts up and partially depletes the 2DEG channel. This is a cause leading to the on-state I_D degradation after stress (Fig. 10). After improving the quality of AlGaN/GaN-on-Si epitaxial structure with Process 2, those

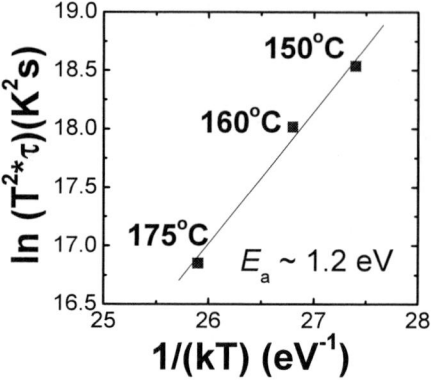

Fig. 8: Time constants τ at the rising portion of the curves (shown in Fig 8b) are extracted, and are plotted against $1/(kT)$, where $k = 8.62 \times 10^{-5}$ eV·K^{-1} is the Boltzmann constant and T is the ambient temperature. The extracted activation energy E_a is ~ 1.2 V.

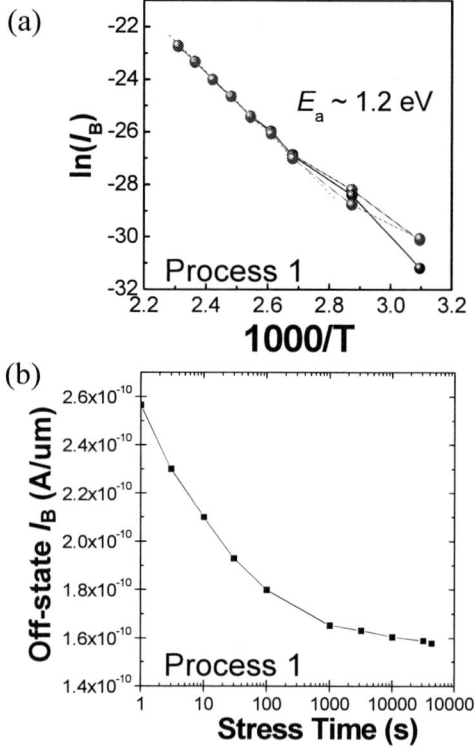

Fig. 9: (a) To understand the trap level associated to the off-state substrate leakage (I_B) of a 650 V device, its initial I_B was measured against temperature (T) to extract the activation energy (E_a). The E_a ~1.2 eV matches the level of traps performed in the time recovery test (Fig. 9). (b) During HTRB stress test, a reduction of off-state I_B against stress time indicates the trap filling process in the epitaxial structure.

traps are greatly reduced, and off-state I_B is also suppressed. Degradation of on-state I_D reduces significantly with Process 2 under HTRB stress (Fig. 11). The 650 V D-MISFET remains ~97% of on-state I_D after 500-hour stress, while the 100 V E-FET remains ~98% after 1000 hours.

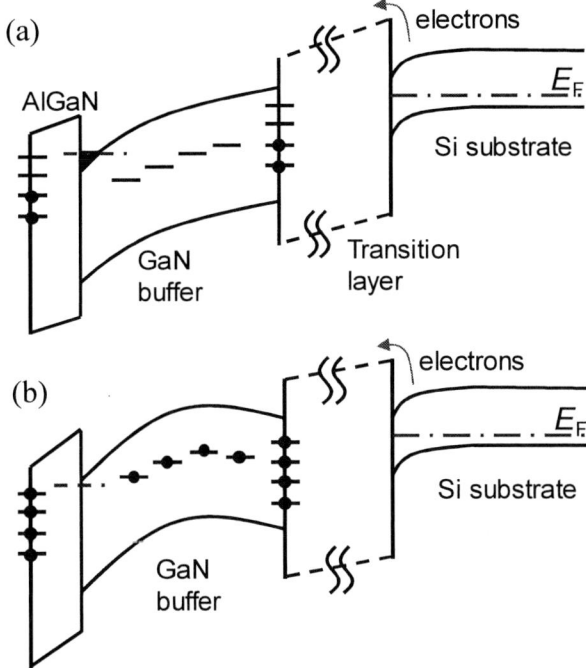

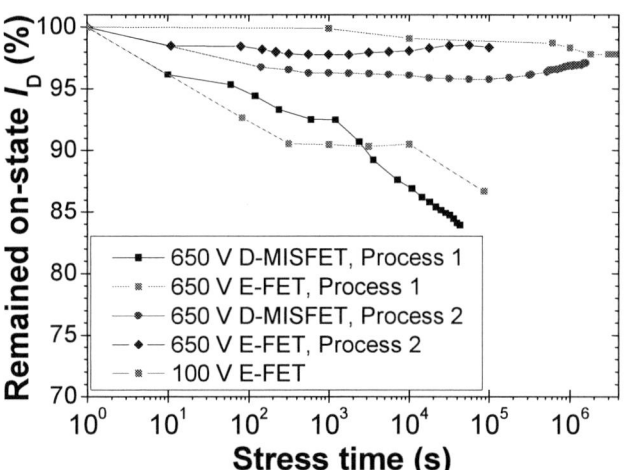

Fig. 11: Significant improvements of on-state I_D for 650 V D-MISFET. Reliability of on-state I_D of 100 V and 650 V E-FETs under HTRB stress is also added in the plot for reference. Drain biases of the devices are about 80% of the specified rating during stress.

Fig. 10: (a) The schematic band diagram shows the access region when the device is under high voltage off-state bias at the drain electrode. Electrons are injected from the substrate Si to GaN, and fill up the traps. (b) Filled traps shift up the band of the heterostructure, and partially deplete the 2DEG channel. This causes on-state I_D degradation after HTRB stress.

Conclusion

High performance of low cost CMOS-compatible GaN-on-Si 100/650 V E-FET and 650 V D-MISFET is demonstrated. The devices show high BV and low $R_{on,sp}$ close to 4H-SiC limit, as well as low FOM ($R_{on} \times Q_g$) which enables fast switching. Optimization of epitaxial layers is important to reduce stress-induced on-state I_D degradation in high-voltage power devices under a key industrial item: HTRB stress. A major breakthrough on the optimization of epitaxial structure reduces degradation of on-state I_D significantly. These devices show uniform distribution of performance on 6-inch GaN-on-Si wafers and are ready for manufacture.

REFERENCES

[1] JESD22-A108C standard, 2005.

[2] K.-Y. Wong, Y. S. Lin, C.W. Hsiung, G.P. Lansbergen, M.C. Lin, F.W. Yao, C.J. Yu, P.C. Chen, R.Y. Su, J.L. Yu, P.C. Liu, C.M. Chen, C.H. Chiang, H.C. Chiu, S.D. Liu, Y.A. Lai, C.Y. Yu, F.J. Yang, J. L. Tsai, C. S. Tsai, X. Chen, H.C. Tuan, and Alex Kalnitsky "AlGaN/GaN MIS-HFET with Improvement in High Temperature Gate Bias Stress-induced Reliability" *IEEE 26th International Symposium on Power Semiconductor Devices & IC's (ISPSD), Waikoloa, Hawaii, USA*, p. 55-58, June 2014.

[3] H. Lee, D.S. Lee, and T. Palacios, "AlGaN/GaN High-Electron-Mobility Transistors Fabricated Through a Au-Free Technology," *IEEE Electron Device Letters*, vol. 32, pp. 623-625, May 2011.

[4] L. Lorenz, G. Deboy, A. Knapp, and M. Marz, "COOLMOS™ – a New Milestone in High Voltage Power MOS," *IEEE 11th International Symposium on Power Semiconductor Devices & IC's (ISPSD), Toronto, Ont., USA*, p. 3-10, May 1999.

[5] Fairchild Semiconductor, Inc., *FGH40N65UFD 650 V, 40A Field Stop IGBT Datasheet, Rev C1*, pp. 1-9, November 2013.

[6] Transphorm, Inc., *TPH3002LD GaN Power Low-loss Switch Datasheet*, pp. 1-8, April 2014.

[7] D. Costinett, H. Nguyen, R. Zane, and D. Maksimovic, "GaN-FET Based Dual Active Bridge DC-DC Converter," *IEEE 26th Applied Power Electronics Conf. and Exposition (APEC), Fort Worth, Texas, USA*, p. 1425-1432, March 2011.

[8] G. Koley, V. Tilak, L. F. Eastman, and M. G. Spencer, "Slow Transients Observed in AlGaN/GaN HFETs: Effects of SiN$_x$ Passivation and UV Illumination," *IEEE Trans. Electron Devices*, vol. 50, pp. 886-893, April 2003.

[9] G. Meneghesso, G. Verzellesi, F. Danesin, F. Rampazzo, F. Zanon, A. Tazzoli, M. Meneghini, and E. Zanoni, "Reliability of GaN High-Electron-Mobility Transistors: State of the Art and Perspectives," *IEEE Trans. Device and Materials Reliability*, vol. 8, pp. 332-343, June 2008.

[10] M. Faqir, G. Verzellesi, G. Meneghesso, E. Zanoni, and F. Fantini, "Investigation of High-Electric-Field Degradation Effects in AlGaN/GaN HEMTs," *IEEE Trans. Electron Devices*, vol. 55, pp. 1592-1602, July 2008.

[11] J. A. del Alamo, and J. Joh, "GaN HEMT Reliability," *Microelectronics Reliability*, vol. 49, pp. 1200-1206, July 2009.

[12] J. Joh, and J. A. del Alamo, "A Current-Transient Methodology for Trap Analysis for GaN High Electron Mobility Transistors," *IEEE Trans. Electron Devices*, vol. 58, pp. 132-140, January 2011.

[13] Z.-Q. Fang, B. Claflin, D. C. Look, D. S. Green, and R. Vetury, "Deep traps in AlGaN/GaN heterostructures studied by deep level transient spectroscopy: Effect of carbon concentration in GaN buffer layers," *J. Appl. Phys*, vol. 108, pp. 063706, September 2010.

978-1-4799-8002-4/14 $31.00 © 2014 IEEE

Device Aware High-Speed Transceiver Design in Planar and FinFet Technologies

Ken Chang, Jafar Savoj, Parag Upadhyaya, Yohan Frans
Xilinx, Inc., San Jose, CA

Abstract

This paper studies the interaction between devices in advanced CMOS processes and the analog circuits used in high-speed transceivers. It describes the impact of variation on the design with both active and passive devices, and devises circuit techniques to mitigate their impact on the performance of transceivers. A case study of transceiver components in 20nm planar and 16nm FinFET processes are used to illustrate the interactions. With proper design techniques, the transceiver in 20nm operates up to 16.3Gb/s and achieves measured bit error rate lower than 10^{-15} over 28dB channels.

Introduction

High-speed wireline transceivers have enabled the explosion of the connectivity bandwidth for storage and communications in recent years. The advance of the process technology poses both advantages and challenges for transceiver design. Figure 1(a) shows the line rates only increase slightly over as process scales. This can be attributed primarily to clocking and equalization blocks, predominantly analog circuits, which are well known for not scaling well with process technologies. In contrast, Figure 1(b) shows how the "Figure of Merit", defined as the energy per bit normalized against the channel loss, scales across process nodes. A lot of this improvement is due to power scaling in digital CMOS circuits which constitute typically over 50% of all circuits in wireline transceivers.

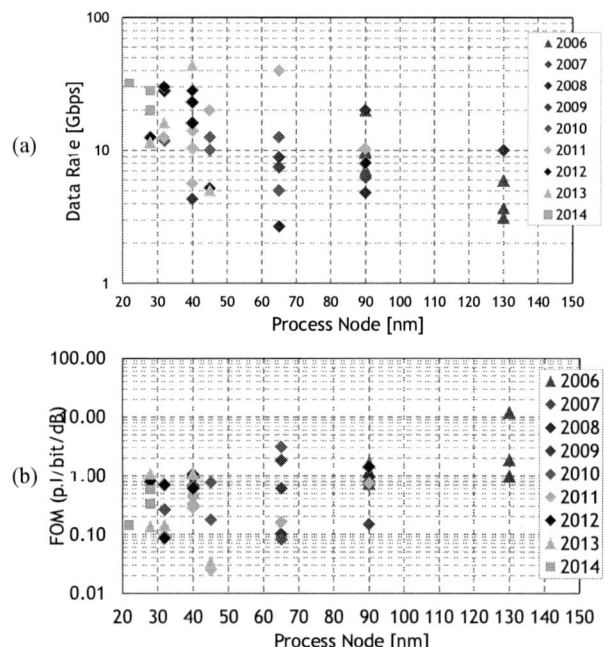

Figure 1: Wireline transceiver trend: (a) data rate (Gb/s) vs process nodes, showing only mild dependency (b) FOM (power/bit rate/channel loss in dB) vs process nodes, indicating much stronger improvement primarily due to power reduction benefit from advanced processes [data source: ISSCC / VLSI 2006-2014]

Clocking Circuits

Figure 2 shows the clock generation architecture for a quad transceiver in a 20nm FPGA [1]. To gain high performance, LC PLLs are used to provide high-precision, low-jitter clocks. Three LC VCOs are used to cover over a 2:1 frequency range to avoid any operating frequency gap. Two LC PLLs are used to offer two independent frequencies per quad transceiver. For lower line rates (< 6Gb/s), the ring PLL which provides a much wider operating range is used per channel.

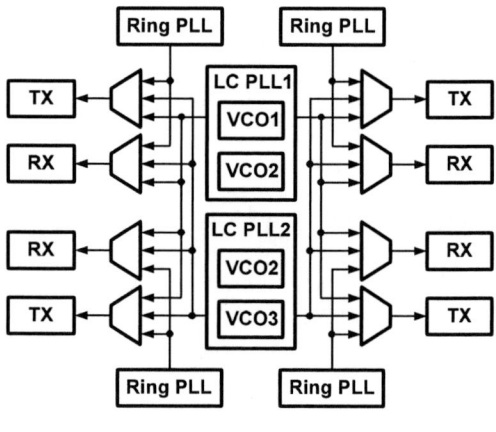

VCO1: 13-16.3GHz
VCO2: 9.8-13GHz
VCO3: 8-9.8GHz

Figure 2: Quad Transceiver Clocking Architecture. Combination of LC and ring PLLs provide full frequency ranges. High line rates use LC PLLs to provide low jitter, high precision clocks.

The frequency adjustment circuits of each LC VCO utilize varactors and switched metal capacitors. Figure 3(a) shows the LC VCO with its temperature compensation circuitry. Figure 3(b) compares capacitance versus voltage (CV) for the planar and FinFet accumulation mode NMOS varactors. The desired CV curve for low jitter PLL applications should provide a large linear range to maintain constant and moderately small CV gain over the full control voltage (*Vcnt*) dynamic range as shown in Figure 3(b). Higher CV gain translates to higher VCO gain and higher jitter in the LC VCO. In a FinFet process, Vt shift pushes the CV curve to higher voltages and sharper slope - higher CV gain. Both of them present challenges to wide range, low jitter LC PLL design. For a DC coupled NMOS LC VCO shown in Figure 3, the voltage across the varactor is defined by *Vgg=Voutp/n-Vcnt*. Higher threshold translates into lower tuning range for the varactor as

978-1-4799-8002-4/14 $31.00 © 2014 IEEE   454

only half of the CV curve can be utilized due to lower supply voltage. An alternative is to AC couple the varactor and set the bias voltage in the center to achieve higher tuning range. However, AC coupling capacitors are large and limit the tuning range of the VCO. The challenges for large tuning range of LC VCO design are made worse due to PVT (process, voltage and temperature) sensitivity, which results in frequency drift over the operating range.

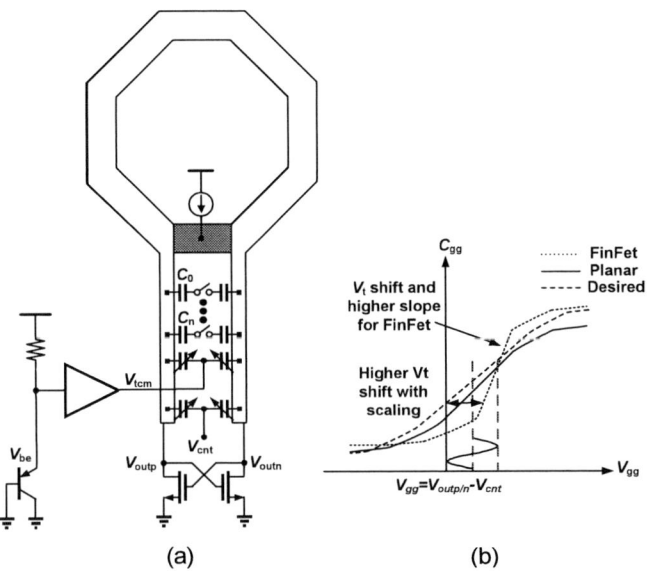

Figure 3: (a) LC VCO and its temperature compensation circuit (b) Varactor C-V curves in FinFet and Planar technology vs desired behavior. A low C-V slope is desired for low-jitter VCOs. FinFet varactors show higher threshold and also have a sharper C-V curve compared to planar varactors.

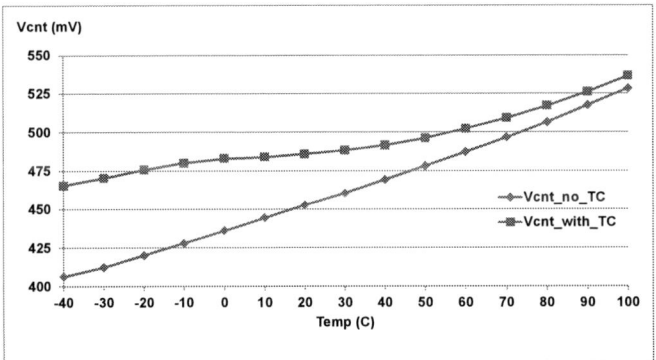

Figure 4: Measured Vcnt vs temperature variations with and without the temperature compensation circuits, indicating the circuit works as expected. The smaller dependency is desired to ensure maximum temperature tolerance within each frequency band (defined by the capacitor bank).

Typical LC VCOs are segmented into several small frequency bands. Once a frequency band is selected and PLL is locked, it must work over extreme temperature and voltage variations without resetting or switching the frequency band. Temperature stability or compensation circuit can be added to insure proper operation over such PVT conditions [2]. Figure 3(a) shows a simplified circuit for temperature compensation

used in the LC VCO. The circuit utilizes a bipolar transistor to sense temperature in the form of voltage which is amplified and used to control another set of varactors to counter the VCO frequency shift due to voltage and temperature shift. Figure 4 shows the measured results demonstrating the effect of the temperature compensation circuits. As expected, with the circuit enabled, the variation is reduced significantly.

Transceiver Circuits

Figure 5 shows the transceiver block diagram. The receiver, shown in Figure 5(a), includes the analog front-end, deserializer, and clock generation circuits. The analog front-end is composed of a continuous time linear equalizer (CTLE), slicers, slicer offset cancellation circuitry, and decision feedback equalizer (DFE). The receive data from analog front-end is then sent to the deserializer to parallelize the high-speed bit stream and is ultimately sent to Physical Coding Sublayer (PCS) for use by the FPGA fabric. The phase interpolator (PI) is used to provide the clocking for clock data recovery (CDR). Figure 5(b) shows the transmitter. It is composed of the serializer, a finite impulse response filter (FIR), output transmitter driver, and the clock generation circuitry. In the transmitter, the primary analog circuit is the output driver and the phase interpolator. As process node scales and line rate increases, several challenges must be addressed in transceiver design to deal with voltage supply scaling and reduced dynamic range, higher device variability and mismatch due to small geometry, and power density to name a few.

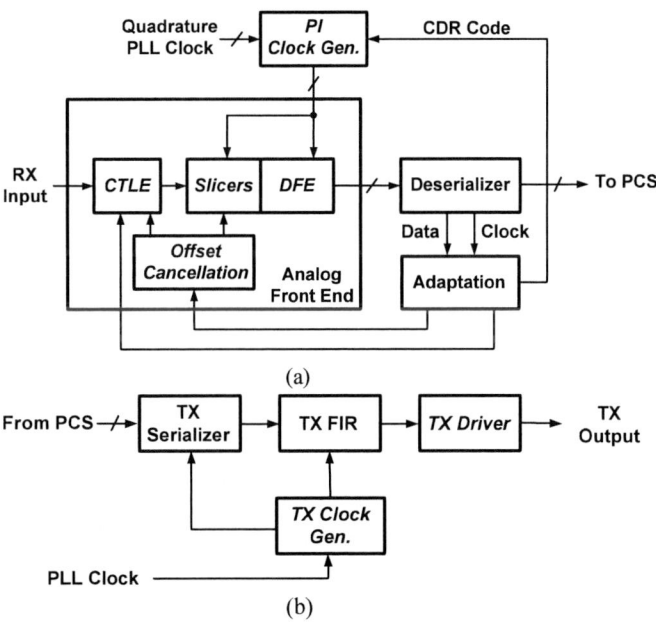

Figure 5: Transceiver block diagrams (a) receiver (b) transmitter. *Italic* are analog circuits. The rest are CMOS digital circuits.

Figure 6 illustrates the CTLE design and its bias circuit. The CTLE must operate linearly for DFE to achieve highest performance. Therefore, the TiN resistor is used for its load to achieve good linearity. While continuous adaptation loop is used to adjust the CTLE coefficients, the adjustment rate is so

978-1-4799-8002-4/14 $31.00 © 2014 IEEE   455

low that it is desirable to have the gain, the poles, and the zeros to be PVT independent. Unfortunately, the variation of TiN resistor can be as high as 20%. To maintain a constant transfer function over PVT, a constant Gm bias circuit [3] is used to provide the stable gain for CTLE. In addition to PVT variations, the device mismatches in the front-end, including CTLE and slicers, can impair the receiver input sensitivity – a critical performance parameter for the wireline receiver. To combat the random variations, offset cancellation circuit is added in each slicer, as shown in Figure 7(a) [4]. Figure 7(b) shows the measurement results over multiple lanes and parts. Without offset cancellation, the offset can be as large as 50mV (single-ended). After the offset cancellation, the receiver offset is reduced to 1 LSB (~3mW).

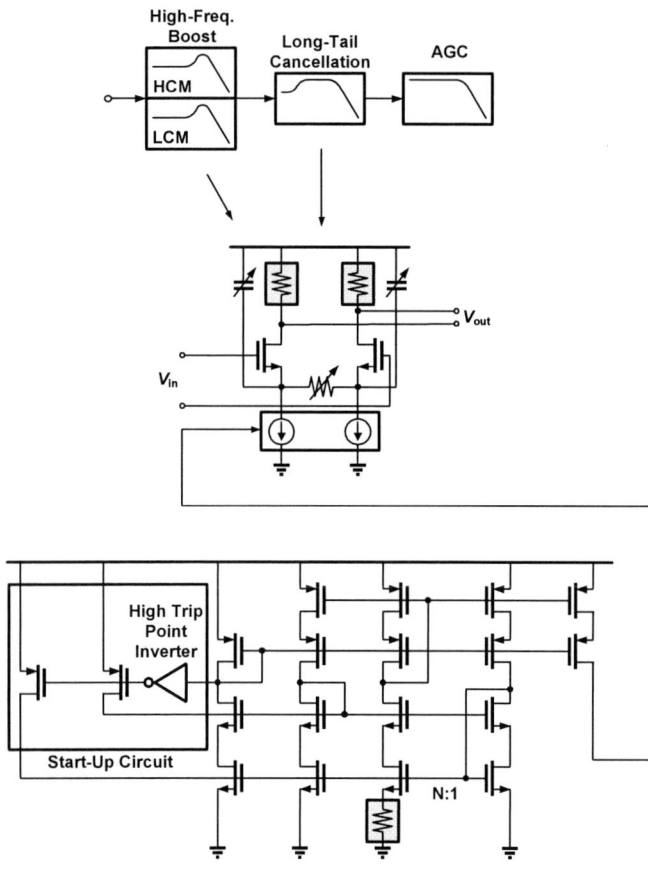

Figure 6: Receiver CTLE [1] and its constant Gm bias [4]. The CTLE must be in the linear mode for all PVT. Gm is ~ to 1/R. The overall gain is set by the ratio of two TiN resistors. The bias circuit has two stable operating points. The start-up circuit is to avoid the circuit get stuck in the all-zero state.

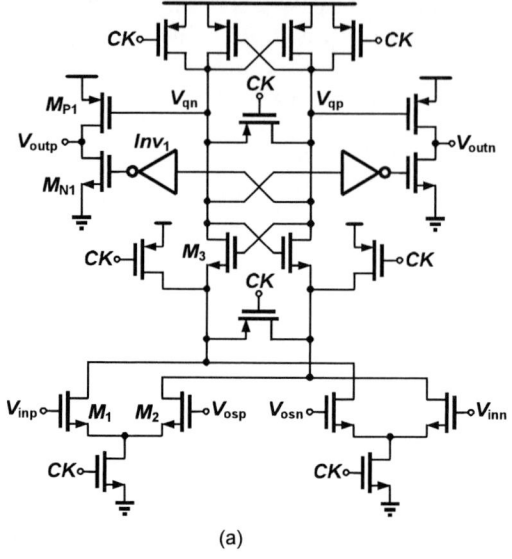

(a)

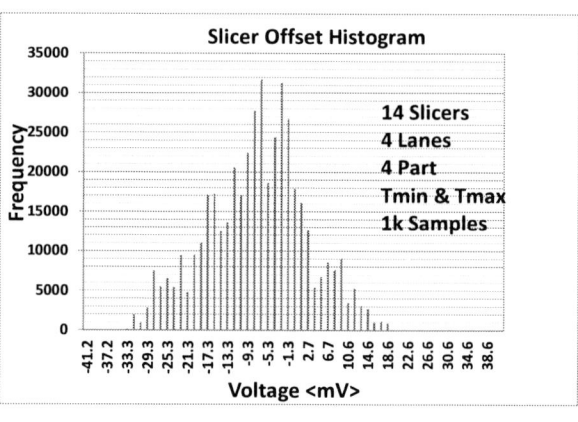

(b)

Figure 7: (a) Receiver slicer with offset cancellation. The slicer is based on StrongARM latch [4] but with added offset cancellation pair. (b) Measured offset before calibration is close to 50mV over 224 slicers over 4 parts. After cancellation, the offset is reduced to ~ 1LSB (3mV).

Figure 8 shows the transmitter circuitry and its biasing to account for load variability. The transmitter output needs to provide 100Ω differential load to avoid channel reflections. The load resistor is made of several parallel TiN resistors with PMOS switches to compensate PVT variations [5]. The impedance calibration employing a central replica circuit is used. The calibrated codes are then broadcast to every quad on the same chip. TiN resistors tend to be large in physical size to meet EM constraint and therefore its relative mismatch within the die is reasonably small. Figure 9 shows the measured resistance value across codes for few receiver termination schemes. The target 100Ω is close to the middle of the codes, providing the largest adjustment range. Figure 9(b) shows the measured resistance values across multiple lanes. It shows the variation is small, validating the calibration architecture.

978-1-4799-8002-4/14 $31.00 © 2014 IEEE

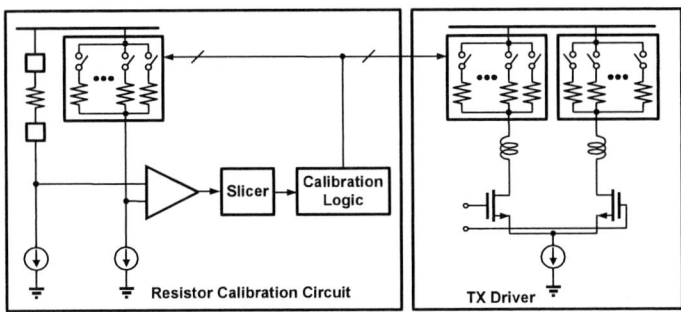

Figure 8: Transmitter and its centralized termination calibration circuit. The calibrated code is broadcast on the same die.

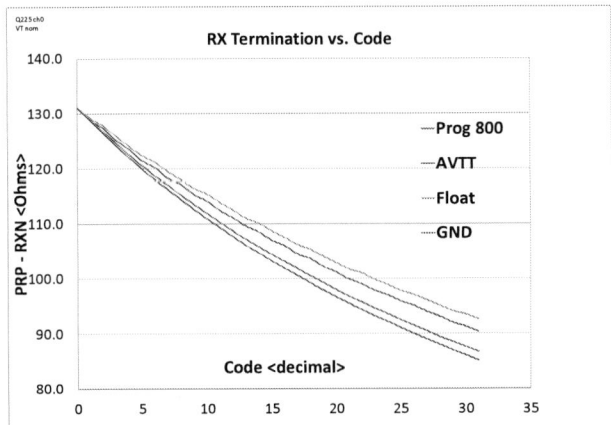

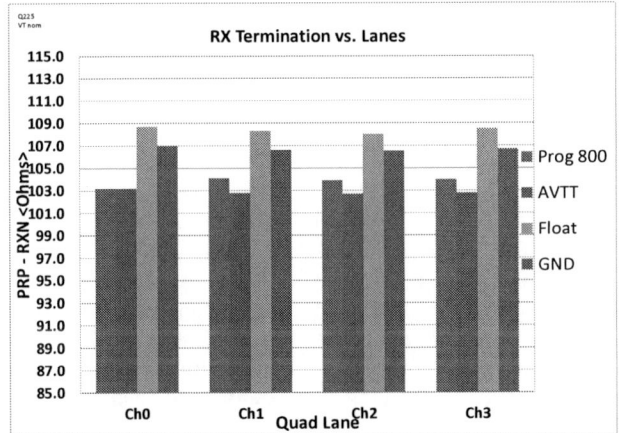

Figure 9: (a) Measured resistance across code for different RX termination voltages (b) Measure resistance across lanes, indicating small lane-to-lane variations, which is caused by within die variation on both TiN resistors and PFET transistors.

Summary

This paper studies the interaction between devices and analog circuits used in high-speed transceivers in both planar and FinFet processes. It describes the impact of variations on the design of both active and passive devices, and devises circuit techniques to mitigate their impact on the performance of transceivers. Case studies of transceiver components in 20nm planar and 16nm FinFET processes are used to illustrate the interactions. With proper design techniques, the transceiver in 20nm operates up to 16.3Gb/s. Figure 10(a) shows the 28dB

channel used for the testing for 20nm transceiver [1]. The resulting transmitter random jitter (RJ) is 313fs, shown in Figure 10(b). The LC PLL covers from 8Gb/s-16.3Gb/s without frequency dividers. Figure 10(c) shows the receiver eye opening via eye scan result and Figure 10(d) shows the measured bit error rate is lower than 10^{-15} over the 28dB channel. The robust performance validates all the circuit techniques used for the advanced process nodes.

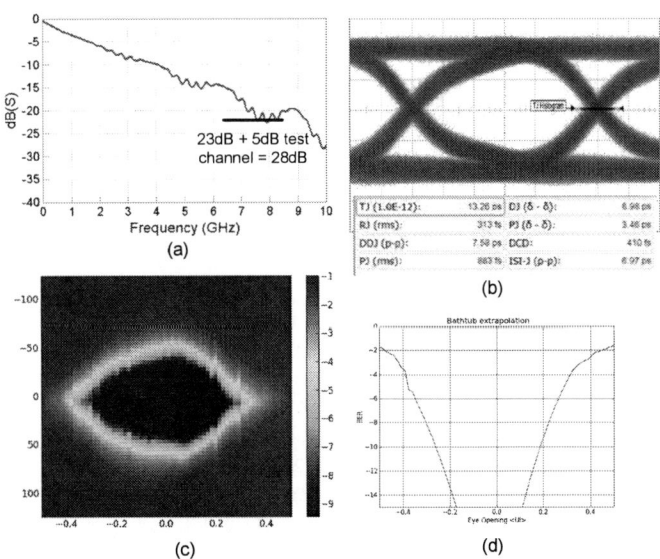

Figure 10: (a) channel loss (S21) for 28dB channel (including test cables) (b) TX output waveform (c) 16.3Gb/s Receiver Eyescan (d) 16.3Gb/s BER for 28dB channel [1]

Acknowledgments

The authors thank Santiago Asuncion, Jayesh Patil, Sai Lalith Chaitanya Ambatipudi for taking measurement data.

References

[1] J. Savoj, et.al., "Wideband Flexible-Reach Techniques for a 0.5-16.3Gb/s Fully-Adaptive Transceiver in 20nm CMOS," IEEE Custom Integrated Circuits Conference, Sept. 2014.

[2] M. S. McCorquodale, et.al., "A Monolithic and Self-Referenced RF LC Clock Generator Compliant with USB 2.0," IEEE Journal of Solid-states Circuits, Feb. 2007.

[3] T. Lee, "The Design of CMOS Radio-Frequency Integrated Circuits," Cambridge University Press, 2004.

[4] J. Montanaro, "A 160-MHz, 32-b, 0.5-W CMOS RISC Microprocessor," JSSC, Nov 1996.

[5] A. Garg, "A Quad-Channel 112-128Gb/s Coherent Transmitter in 40nm CMOS," IEEE Symposium on VLSI Circuits, June 2014.

Mismatch in High-K Metal Gate Process Analog Design

A. Woo, H. Eberhart, Y. Li, and A. Ito

Broadcom® Corporation

Irvine, CA USA

Abstract

This paper presents mismatch behaviors of high-K metal gate transistors when used in analog design applications. The data collected shows the sensitivity of mismatch in the high-K metal gate process to the overall layout environment, including top-metal routing placement, which was not reported before this work. We compared data from different fabrication sources, which indicated that mismatch can be improved through process as well as through layout strategy.

Introduction

In Systems-On-Chip (SoC) designs, analog and RF building blocks are often required to achieve certain high-precision functions. When technologies moved into the submicron range, there were concerns that the challenges introduced by the deep submicron technologies would eventually prohibit the integration of analog and RF functions into deep submicron SoC designs. This concern was promptly addressed by the introduction of design techniques and topologies that are more suitable for deep submicron technologies.

Among the successful techniques that enhance traditional analog and RF functions in deep submicron technologies are: SoC re-architecting [1], digitally assisted ADC [2], digitally assisted DAC [3], and others [4]. These techniques are enabled by the speed and integration level that the advanced deep submicron digital logic offered.

At the heart of all the success stories of analog/RF integration into deep submicron technologies are crucial analog circuit components that cannot be replaced by digital alternatives. It is, therefore, important to preserve a certain level of analog capability within a process to support the SoC integration trend of the industry. Transistor matching is one of the elements in analog designs that cannot be completely replaced by digital components [5]. In this work, we looked at MOS transistor matching for analog applications in the 28nm high-K metal gate (HKMG), gate last processes. It is essential for us to understand the mismatch behavior in order to optimize the analog performance and make architectural tradeoffs for any technology node.

This Work

This work focuses on the study of how transistor mismatch behavior in high-K metal gate MOSFET is influenced by the layout styles that analog/RF circuits employ. Our focus is on large-scale array-style structures of unit MOSFET. In HKMG MOSFET, new material high-K and metal gate reduce some of the transistor mismatch compared to poly-SiON MOSFET, but the new process steps, such as metal gate etch and CMP, introduce new sources for transistor mismatch. Therefore, gate size, density, and density gradient to neighboring blocks can all affect transistor mismatch.

Test Structures

Mismatch testlines: These are wafer probe-level testlines which consist of transistors of the same W and L placed in parallel with increasing number of fingers. As the number of fingers goes up, the local random variations would start canceling out. From the collective data, Avt [6] can be calculated by (1).

$$\sigma_{Total}^2 = \sigma_{Global}^2 + \sigma_{Local}^2 \qquad (1)$$

In this work, we compared the mismatch of different W and L. In analog applications, both small gate length and large gate length mismatches are important.

MOS arrays: Long gate length MOS arrays are common in precision circuits and were used in this experiment. Some arrays are ~400 um x 250 um in dimension, Fig. 1, and some arrays are ~200 um x 200 um with controlled neighborhood blocks of 100um in the surroundings, Fig. 2. Probes were placed such that measurements could be made at the array center and array edge locations. These arrays were varied to cover layout effects such as poly/active density and dummy fill environments.

Digital-to-analog converters (DAC): These DAC structures were based on a current steering architecture similar in concept to the DAC architecture found in [7]. Both NMOS and PMOS DACs were included in this study. Fig. 3 shows the layout of one of the DACs used in this study. The MOS current mirrors of the DACs are surrounded by the same number of dummy current mirrors that we use in DACs manufactured in poly-SiON gate technologies, and therefore, this experiment can be a fair comparison to previous technologies.

Fig. 4 shows a typical test setup used in this study. We chose to present MOS mismatch data collected from our DAC structures in terms of the differential non-linearity (DNL) measurement of the DACs. Differential non-linearity in this case measured the mismatch of two adjacent MOS units and reported the result in terms of the number of the least significant bit (LSB) in the DAC output. A DAC is commonly designed to have a DNL of +/-0.5 LSB over all codes. The x-axis in our DNL plots is arranged to represent the sequential location of the current mirrors from left-to-right in the MOS array. The current mirrors in each column over the x-position are summed together as one parallel MOS unit.

978-1-4799-8002-4/14 $31.00 © 2014 IEEE

Observations

Mismatch testlines: In poly-SiON MOS mismatch, longer gate length and increased number of MOS fingers both reduce the local mismatch variation of the devices. In our HKMG MOS mismatch data shown in Fig. 5, we found that short gate length MOSFETs follow the trend of local mismatch variation reduction with an increase of the number of MOS fingers. In the case of the HKMG long gate length devices, the P-type exhibited a different trend. The mismatch increased as we increased the number of fingers. A slight VT mean shift is also observed across the wafer when the number of fingers increases, as shown in Fig. 6.

MOS arrays: In the 400 um x 250 um arrays, the poly/active area density of the arrays varied from 79/75% to 40/40%. We compared the die-to-die variation at the lower left corner (L), center (M), and upper right (H) corner of each array. In one set of PMOS arrays, we saw the L and H location devices match well, while the M location devices demonstrated different patterns of VT distribution as a function of poly/active density percentage (Fig. 7). In the set of NMOS arrays shown in Fig. 8, we saw a slight variation of VT distribution in M locations, and poly/active density shows minor effect at the L locations.

We used the 200 um x 200 um MOS arrays to obtain the environment layout effect result shown in Fig. 9. The y-axis of Fig. 9 shows the edge-to-center mismatch percentage, normalized to the mean of all the samples of the same type. The x-axis of Fig. 9 shows the poly/active density and the environmental mix. The largest contributing factor in this case study in edge-to-center variation is the type of transistor of the arrays. All the other factors do not show strong correlations to the edge-to-center mismatch.

In some 200 um x 200 um arrays we overlaid the top metal along the edge dies as shown in Fig 10. The top metal overlaid introduced a layout effect that causes a current increase in PMOS and current decrease in NMOS. The effect is captured in Fig. 11 and Fig. 12, with devices under metal in circles. The data could be an indication of a tensile stress induced by the top metal layout, which could influence the NMOS and PMOS carrier mobilities in opposite directions.

Digital-to-analog converters (DAC): We have observed a much more prominent edge-to-center mismatch in the HKMG DACs. Fig 13 captures this edge-to-center mismatch pattern in the DNL measurement. When compared to a SPICE simulation, the simulated DNL, Fig 14, was below our silicon measurement, which agreed with our mismatch testline results.

In Fig 15, we are showing the DNL of one NMOS DAC and one PMOS DAC manufactured from one fab side-by-side with the DNL of an identical NMOS and PMOS DAC manufactured from a different fab. The NMOS DACs turn out to have less edge-to-center mismatch, and behave similarly between different fabrication sources. The PMOS DACs, on the other hand, show significantly more mismatch in one fab than the other.

One commonality found in the DAC DNL curves was a systematic pattern that is directly correlated to the presence of top metal over the layout patterns. Fig. 16 includes three different DACs with different top metal patterns and the resulting mismatch patterns. Fig 17 shows that this top metal layout effect manifests itself in different fabs with varying magnitude.

Conclusions

The high-K metal gate process introduced a new set of challenges for analog designs which require a high level of precision in device matching when using long gate length MOS transistors. The MOS mismatch is no longer dominated by traditional processing effects, like CD control, etc. A stronger mismatch vulnerability to metallization layout effect is also observed in HKMG process. The understanding and inclusion of these extra mismatch behaviors is important in the circuit architectural design phase. For transistor matching, the support from the process technology community in implementing mismatch counter balances would also be important to continue the SoC integration trend for our industry as we move into HKMG processes.

Acknowledgement

The authors would like to thank Chao-Yang Lu and Jay Shiau in supporting this work.

References

[1] M. Wakayama, "Nanometer CMOS from a Mixed Signal/RF Perspective", IEDM, p451-454, 2013

[2] J. Wu, et. al., "A 240-mW 2.1-GS/s 52-dB SNDR Pipeline ADC Using MDAC Equalization", JSSC, vol. 48, no. 8, p.1818-1828, 2013

[3] Y. Cong, et. al., "A 1.5-V 14-Bit 100-MS/s Self-Calibrated DAC", JSSC, vol. 38, no. 12, p2051-2060, 2003

[4] F. Lai, et. al., "Digitally-assisted analog designs for submicron CMOS technology", VLSI-DAT, p49-52, 2010

[5] K. Lakshmikumar, et. al., "Characterization and Modeling of Mismatch in MOS Transistors for Precision Analog Design", JSSC, vol, sc-21, no. 6, p.1057-1066, 1986

[6] M. Pelgrom, et. al., "Matching Properties of MOS Transisotrs", JSSC, vol. 24, no. 5, p1433-1440, 1989

[7] C. Lin, et. al., "A 10-b, 500-MS/s CMOS DAC in 0.6mm2", JSSC, vol. 33, no. 12, p1948-1958, 1998

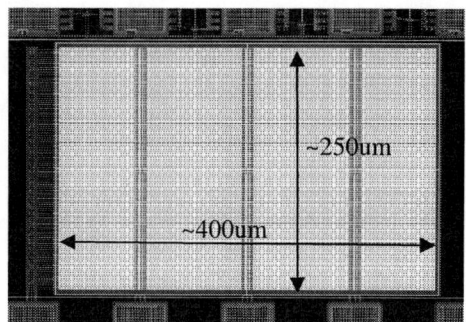

Figure 1 – Layout example of MOS array structure

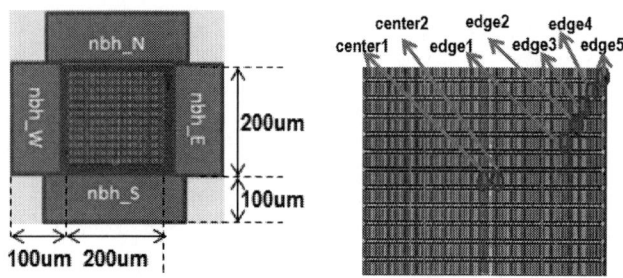

Figure 2 – Layout example of MOS array with controlled neighborhood. Probe locations are shown on the right.

Figure 3 – DAC layout example

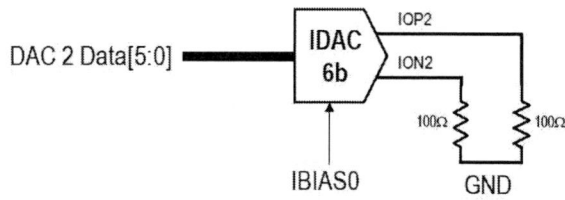

Figure 4 – A typical DAC test setup

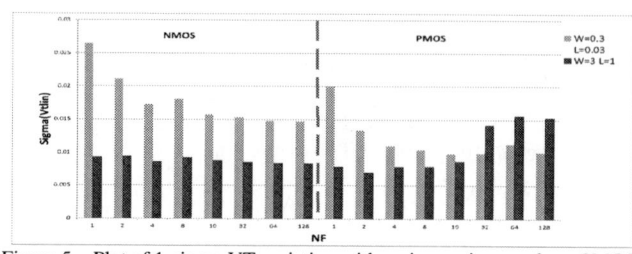

Figure 5 – Plot of 1-sigma VT variation with an increasing number of MOS fingers from left to right of the x-axis.

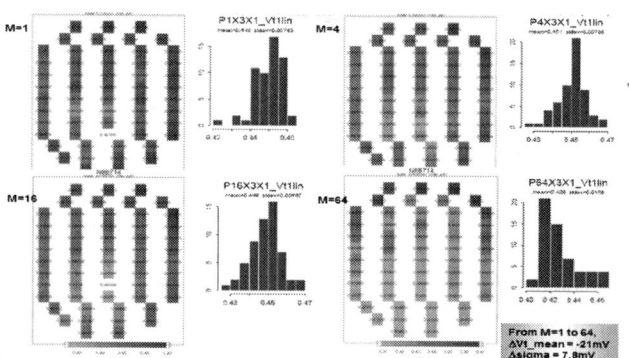

Figure 6 – A color-coded wafer mapping of long gate length Vt shift when the size of the MOS grouping increased, as indicated by the M factor at the upper left corner of the wafer maps.

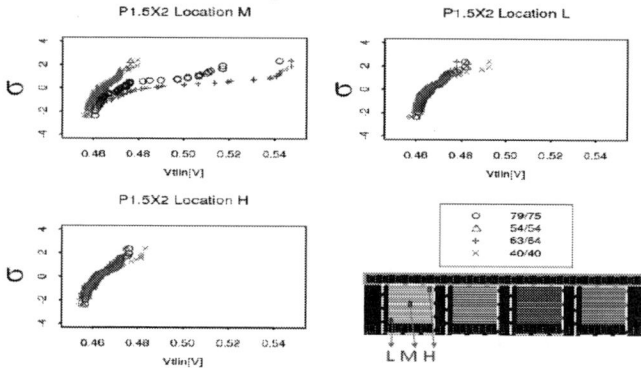

Figure 7 – PMOS array edge-to-center as a function of poly/active density.

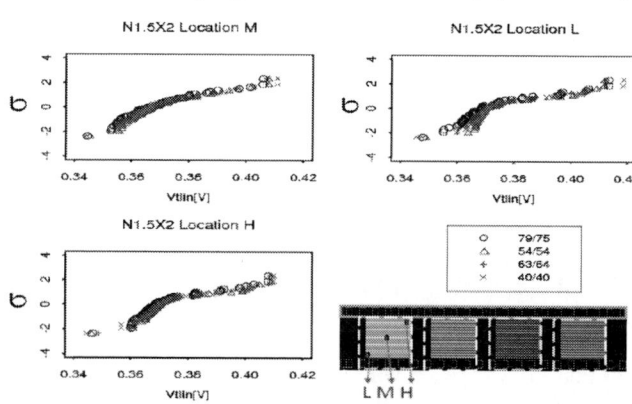

Figure 8 – NMOS array edge-to-center as a function of poly/active density.

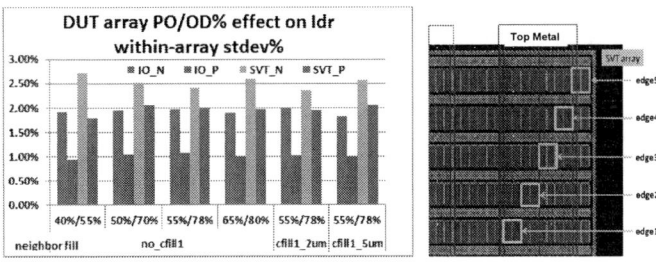

Figure 9 – Environment effects on 200 um x 200 um arrays

Figure 10 – Top metal placement

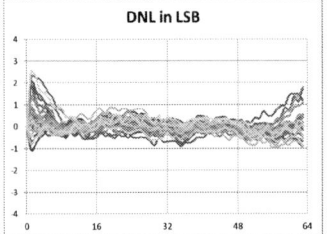

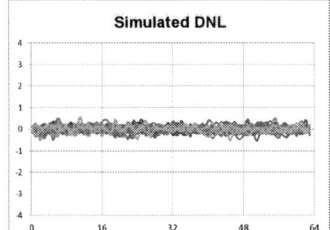

Figure 11 – Top metal effect on IO N and P mismatch

Figure 12 – Top metal effect on core N and P mismatch

Figure 13 – DAC DNL with edge effect in measurements

Figure 14 – DAC DNL SPICE simulation result

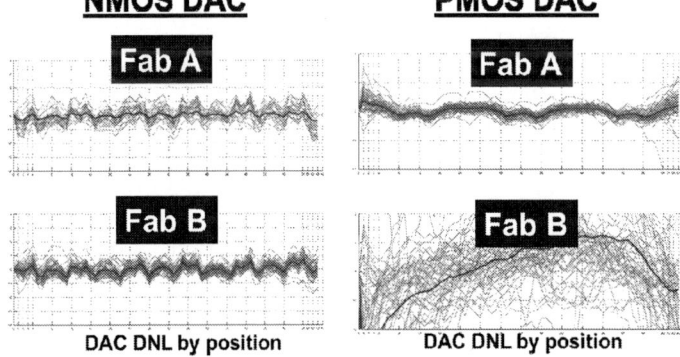

Figure 15 – DAC DNL measurements of identical DAC layout fabricated at different foundries.

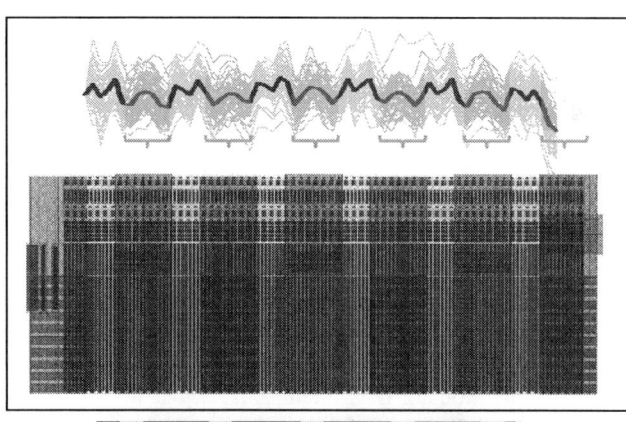

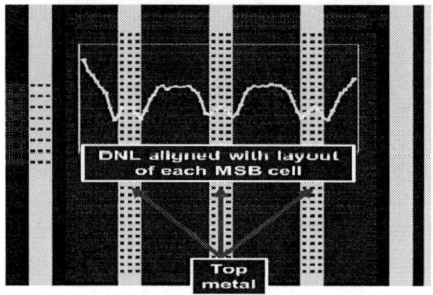

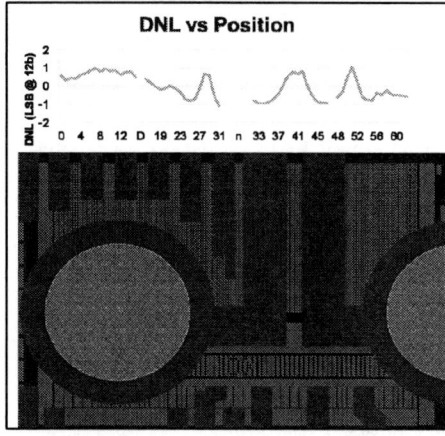

Figure 16 – Superposition of top metal location over measured DNL of three different DAC structures.

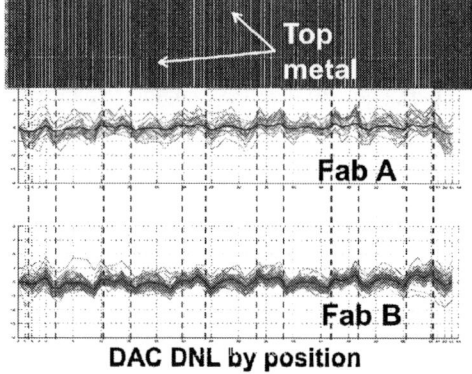

Figure 17 – Same top metal layout effect with different strengths at different foundries.

978-1-4799-8002-4/14 $31.00 © 2014 IEEE

Challenges of Analog and I/O Scaling in 10nm SoC Technology and Beyond

A.Wei, J. Singh, G. Bouche, M. Zaleski, R. Augur, B. Senapati, J. Stephens, I. Lin, M. Rashed, L. Yuan, J. Kye, Y. Woo, J. Zeng, H. Levinson, A. Wehbi, P. Hang, V. Ton-That, V. Kanagala, D. Yu, D. Blackwell, A. Beece, S. Gao, S. Thangaraju, R. Alapati, S. Samavedam

GLOBALFOUNDRIES, 400 Stonebreak Rd. Ext., Malta, NY, 12020, USA

Tel: 1-518-305-6409, Email: andy.wei@globalfoundries.com

Continuous process-level and system-level innovation has driven Moore's Law scaling for the last fifty years, and will continue to do so in the next decades. In the last two decades, there has been an acceleration of new materials and devices into semiconductor manufacturing, such as low-k, strained Si, high-k, and FinFET [1-4], in order to continue process and cost scaling. At the same time, ever increasing component integration on SoCs has further driven cost scaling, allowing the current mobile era to take shape [5]. In the next decade, the focus of SoC innovation will be on patterning and low-resistance materials on the process side, and multi-die package integration on the system side.

Contrary to the popular belief that Moore's Law ended with the introduction of double-patterning, innovation efforts have actually been accelerated due to the imminent non-arrival of EUV lithography technology. With the ever increasing cost of multi-pass 193i lithography, process innovation has focused on overscale, in order to maintain historical Moore's Law cost scaling. On the digital side of the SoC, process innovations based on 1-D patterning will allow for innovative layout techniques which overscale standard cells. Fig. 1 shows a layout study example of 193i-based ultra-high density 7-Track standard cells, targeted for low-variation and scalability to the next nodes. These standard cells are not only overscaled in cell height, but also in cell width. In addition, the layout techniques used actually enhance cell routability due to the resultant high pin count. As shown in Fig. 2, these 1-D layouts maintain high routing efficiency at short track height, compared to traditional 2-D layouts with limited pin access. On the memory side of the SoC, the slowdown in high-density SRAM scaling will likewise be reversed with process innovation, as shown in Fig. 3. The reversal started with the introduction of the FinFET, and will continue as new patterning methods allow bitcell scaling to follow the historic scaling trend.

The challenge will be on the analog I/O side of the SoC. As shown in Fig. 4, what was once a seemingly benign 5-15% of the SoC die area will rapidly increase to 30-40%, as the digital side overscales in an attempt to keep costs in line with Moore's Law. The traditional underscale of analog I/O severely limits and will quickly break cost scaling, regardless of how much overscale occurs on the digital side of the SoC.

The Fin Paradigm

The traditional analog I/O underscale arises from the fact that the I/Os typically do not scale node to node, i.e. scale factor ~1. Due to legacy standards, regardless of process technology, I/O FETs have to operate at a constant voltage and drive a constant load, while having enough sink capability to absorb the same ESD events. Thus, the width of the devices and passives cannot scale. Device performance improvements for I/O FETs have been historically limited because typically, the gate length cannot scale due to hot-carrier reliability. Device performance based on gate oxide scaling and uni-axial stressors have not applied to thick oxide long channel I/O FETs. The analog mixed/signal digital, analog, and analog passives has been able to scale combined, on the order of 20-40% per node. This is possible due to a heavy reliance on innovative mixed signal design, which is needed anyway to overcome scaling-induced analog device performance degradation [6]. Assuming that a larger portion of I/O and analog passives outweigh the scaling of the digital and analog FETs, analog and I/O, together, are generally lumped to be ~0.9x scale factor per node.

The FinFET enhances this scale factor somewhat due to fin pitch scaling. Fin pitch scaling allows the same amount of drive current to be produced from a smaller footprint, as shown schematically in Fig. 5 (a). Thus, if fin pitch scales 0.7x node to node, the footprint width of the output drivers and other FET-dominated circuits can also scale by ~0.7x node to node. How close to 0.7x is possible depends on the tradeoff of intrinsic per fin performance increase vs. increased parasitics such as BEOL resistance. Luckily for the fin, fin height increase is a large performance modulator which applies both to I/O FinFETs as well as core FinFETs.

The challenge in I/O FinFET scaling is RMG gap fill. The space between I/O fins decreases faster than core device fins, due to the extra I/O oxide thickness requirement. The reduced gap, show in Fig. 5 (b), results in different RMG gap fill behavior, and thus Vt may be outside of a useful range. The ability to tune the workfunction for each device type, as shown in Fig. 6, alleviates this limitation. It also allows for undoped FinFETs with superior Vt mismatch, which is very important for analog and I/O applications.

Analog Scaling

The FinFET reverses the planar trend of degrading gain and increasing variability, as shown in Fig. 7. With much improved electrostatics and reduced Vt mismatch, there could be a one-time shift of average gate length used in analog functions. In addition, fin pitch scaling reduces the degradation in signal to noise ratio that comes with scaling.

Again, the 0.7x fin pitch scale factor plays a role in reducing the SNR degradation due to the extra effective device width for a given footprint [7]. However, beyond these one-time or minor improvement in analog device characteristics, node to node analog scaling is still facing the same historical challenges of lower headroom, increased mismatch, reduced extrinsic F_T [5]. No matter how much the intrinsic device is improving, due to BEOL minimum pitch scaling, via resistances will continue to increase. Maximum via size cannot be significantly larger than the patterning process optimized minimum via size. As more restrictive 1-D patterning is required to scale to even finer pitches, more restrictions will be placed on maximum metal width as well. Thus, BEOL resistance will continue to degrade the overall analog FET performance in the coming nodes. While the obvious solution to further improve analog scaling is to move more functions over to the digital domain, there are limits in what can be done, as shown in Table 1. Thus, maintaining even a ~0.7x scale factor will become more and more challenging on the analog side.

3-D TSV Solution?

Even if a 0.7x scale factor can be achieved and maintained for analog I/O scaling, it still does not solve the analog I/O area explosion problem. As shown in Fig. 3, even with this optimistic scale factor, analog I/O will still be very large part of the SoC at 10nm and 7nm technology nodes. From a cost standpoint, it simply does not make sense to be using the most expensive leading-edge technology, for circuit blocks which inherently do not scale well, and are taking 30-50% of the die area. Taking as much of these analog I/O portions off the SoC, fabricating them in a mainstream low-cost technology, and then connecting these blocks back to the logic/SRAM through 3-D TSV, would seem to make a lot of sense. This is shown schematically in Fig. 8 (a). The logic/SRAM die would be fabricated in the leading edge technology with a simpler

process at reduced cost, complexity and die area, allowing higher yields. The analog I/O portion can be fabricated on a cheaper technology with provision for TSV area. This die would be added cost, along with extra test, assembly, and total stacked die yield loss. Adding together all these factors, a thorough cost analysis in Fig. 8 (b) shows that the cost crossover lies somewhere between 10nm and 7nm, and really shows major benefits at the 5nm node, based on current scaling projections. This cost analysis assumes a feature-neutral SoC profile based at 15% analog I/O area at the 16/14nm foundry nodes. As this SoC profile is scaled, the pitches and digital content have been overscaled, and the analog I/O component are underscaled as shown in Fig. 3. The analog I/O portion is assumed to be taken to a low-cost 28nm process with decreasing wafer cost as the base logic/SRAM technology scales to 10nm, 7nm, etc. Products with higher portion of analog I/O will benefit at earlier nodes.

While cost scaling can be enhanced with 3D-TSV, this will come at a total power tradeoff, as the total combined die area will increase compared to a monolithic die. The digital portions of the mixed signal circuits will increase substantially in size if moved up to the legacy technology process. Thus, innovative design techniques to partition will be required to minimize any power increase.

So while continued innovation and overscale will allow current SoCs to scale near the historic Moore's Law trend, the addition of 3-D TSV and proper partitioning will allow the opportunity to beat Moore's Law in the next technology nodes.

References: [1] M. Bohr, *et.al.* IEDM Tech. Dig., p. 448-450, 1995. [2] T. Gahni, *et.al.* IEDM Tech. Dig., p. 33-38, 2003. [3] K. Mistry *et. al.*, IEDM Tech. Dig., p. 253-266, 2007. [4] C. Auth, *et.al.* Symp. VLSI Tech., p. 218-219, 2012. [5] G. Yeap, *et.al.* IEDM Tech. Dig., p. 16-23, 2013. [6] M. Wakayama, *et.al.* IEDM Tech. Dig., p. 451-454, 2013, [7] G. Taylor, *et.al.* IEDM Tech. Dig., p. 441-443, 2013.

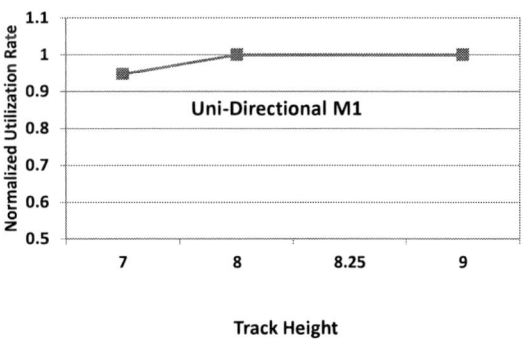

Fig. 2. Study of 1-D routing efficiency vs.track height for a standard cells using unidirectional M1. Utilization rate remains high down to 7-Track library, with DRC errors still manageable.

Fig. 1. Example of 7-Track high-density standard cells showing excellent pin accessibility.

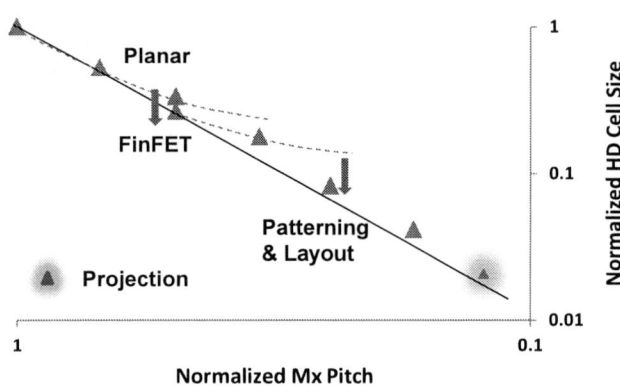

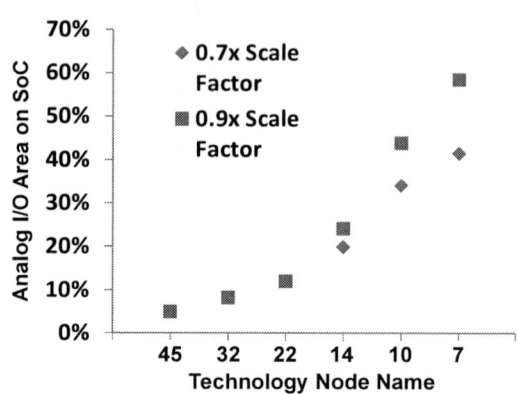

Fig. 3. High-density SRAM bitcell trends over multiple technologies. Multiple innovations have and are projected to maintain bitcell scaling at an 0.5x per generation trend.

Fig. 4. Example of analog I/O area as percentage of total SoC chip area by technology node, accounting for <0.5x logic/SRAM scale factor needed to maintain Moore's Law cost scaling at 14nm and beyond.

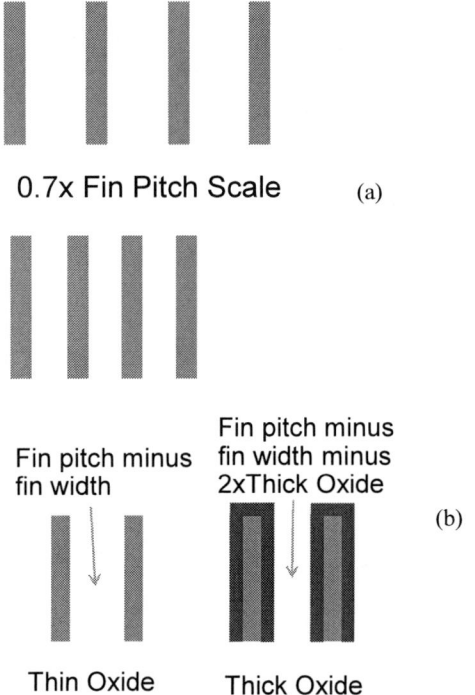

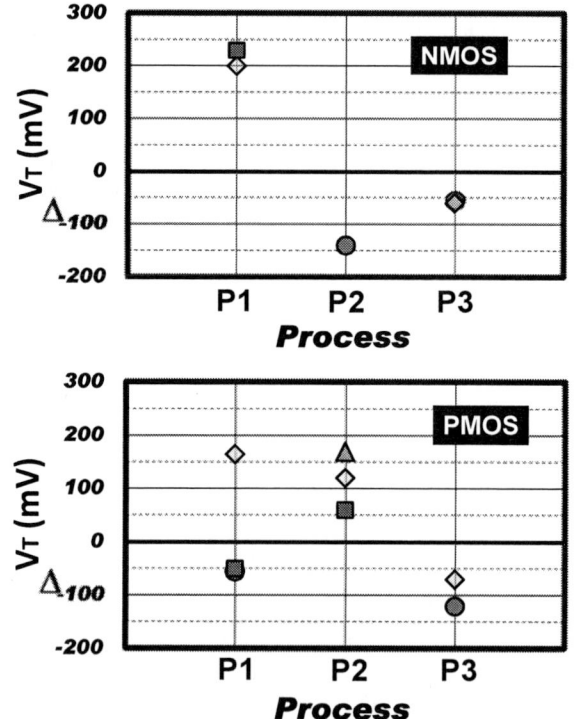

Fig. 5. Schematic diagram showing (a) fin pitch scaling maintaining drive current at 0.7x area, and (b) difference in thin oxide and thick oxide aspect ratio for RMG metal fill.

Fig. 6. Threshold voltage shift from RMG workfunction modification for various RMG processes on FinFETs.

978-1-4799-8002-4/14 $31.00 © 2014 IEEE   464

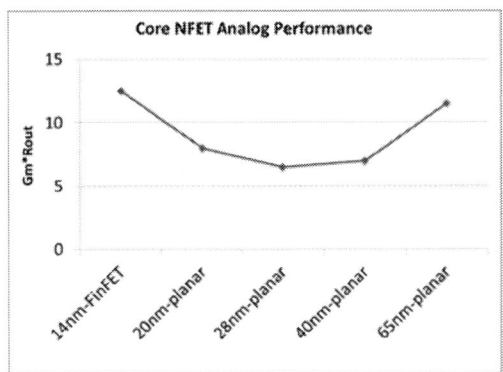

Fig. 7. Excellent analog performance of FinFETs allow a one-time gate length reduction for matched performance versus planar technologies.

Circuits	Can be Replaced By Digital?		Pros	Cons
PLL	—	can replace analog with digital PLL for some applications	Lower power, smaller area, faster acquisition time and easy to scale	Higher jitter, higher phase noise
Data converters	— —	There is no fully digital DAC/ADC. Sigma delta has the most digital components	NA	Analog blocks will not scale
Filters	—	For certain applications can be digital	Lower power, smaller area, linear phase accurate bandwidth	Lower bandwidth, more complex
GPIO	— —	Certain GPIOs can be replaced with core buffer. only for TSV application	NA	Lower speed, any high speed is pseudo open drain
LNA/Bandgap reference circuit	—	No digital solution so far	NA	NA
Temperature sensor circuits	—	Can be replaced by digital for	Lower power, smaller area	Lower accuracy

Table I. Analog circuit functionality can be replaced by digital/mixed signal circuits, but typical with performance penalty.

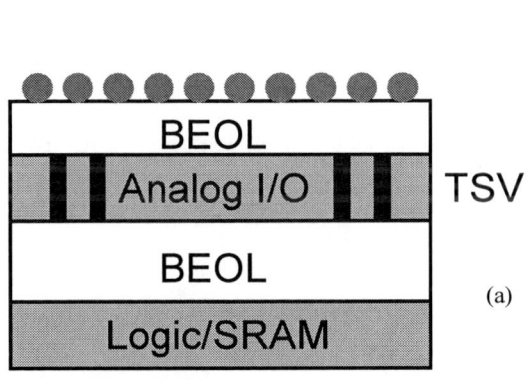

(a)

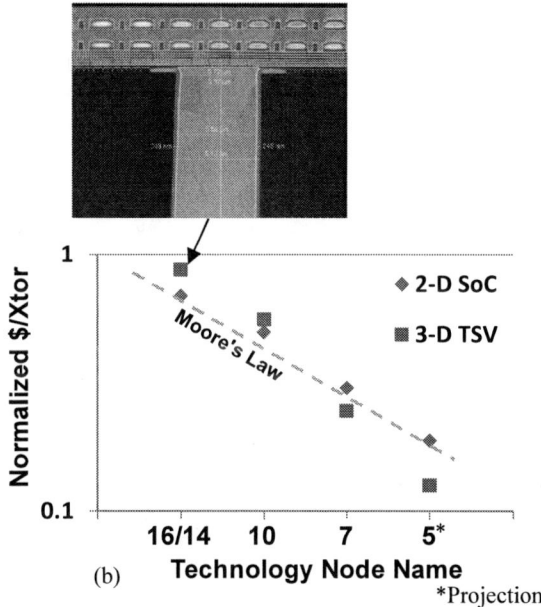

(b)

*Projection

Fig. 8. (a) Schematic of analog I/O die stack on logic/SRAM. (b) Cost per transistor with and without 3-D TSV. Anchor point is a 16/14nm SoC with 15% analog I/O area. Model accounts for TSV, test, assembly, yield deltas, wafer depreciation, and assumes matched die sizes to guarantee enough TSV area and bump density, so package costs are assumed to remain unchanged.

978-1-4799-8002-4/14 $31.00 © 2014 IEEE

Technology pathfinders for low cost and highly integrated RF Front End Modules

C. Raynaud (invited paper)

CEA-LETI, MINATEC Campus, 17 rue des Martyrs - 38054 GRENOBLE Cedex 9, France, christine.raynaud@cea.fr

Abstract

The focus of this paper is to highlight the challenges related to the increasing number of modes (GSM, WCDMA, LTE..) and frequency bands in mobile devices. It describes the technology pathfinders to get cheaper highly integrated multi-mode multi-band RF Front End modules.

Introduction

The RF Front End Module (FEM, Fig. 1&2) is located between the wireless transceiver (TCVR) and the antenna and aims to both transmit and receive the radio signal. It is composed of RF switches (antenna and mode switches), Power Amplifiers (PA). Up to now, System-in-Package approaches have been used with PA and switches on III-V substrate, packaged together with Integrated Passive Devices for harmonic filters, baluns, etc. But the increasing complexity to implement all bands combinations has created the need for lower cost integration.

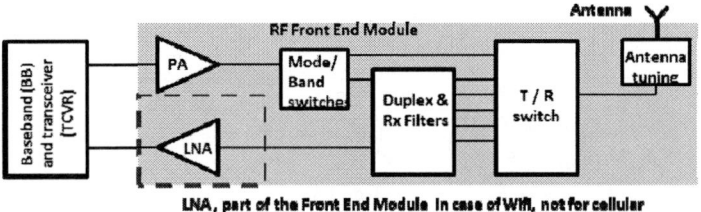

Fig. 1. Schematics of a RF Front End Module – Source: "3G/4G Multimode Cellular Front End Challenges" (RFMD white paper)

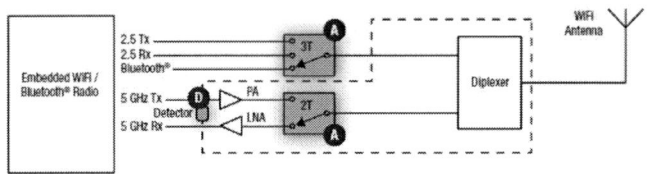

Fig. 2a. LNA, PA & Mode Switches in a dual band 2.5 GHz, 5GHz Wifi RF Front-End Module

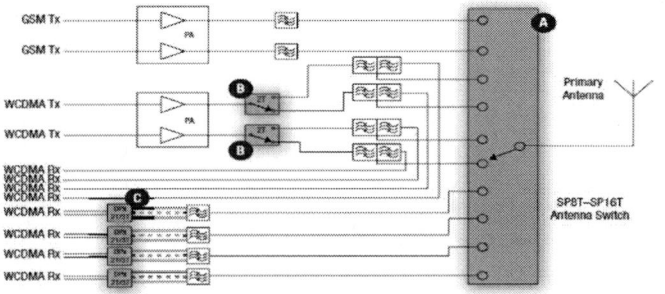

Fig. 2b PA & Switches (mode and antenna switches) in a cellular GSM & WCDMA RF Front-End Module
Source Fig2: http://www.skyworksinc.com/downloads/block_diagrams

RF switches

Previously, PIN diodes and GaAs pHEMT have been used as switches in RF Front End modules. SOS is also a good candidate ([1], [3]) but SOI MOSFETs (Fig. 3) have received widespread acceptance ([2], [4, 7-11]) to be used in

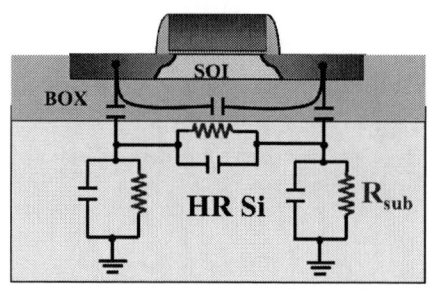

Fig. 3. Cross section of a SOI MOSFET for a RF switch

mobile devices as they offer, at a lower cost, the expected performance, for both insertion losses, isolation (Fig. 5a, 5b) and high linearity. SOI enables the integration on a same chip of switches with the CMOS controller. High voltage handling is obtained through 8 to 10 stacked SOI MOSFETs (Fig. 4, [2]), which have a better isolation than bulk switches, thanks mainly to a High Resistive (HR) substrate under the buried oxide (Fig. 3).

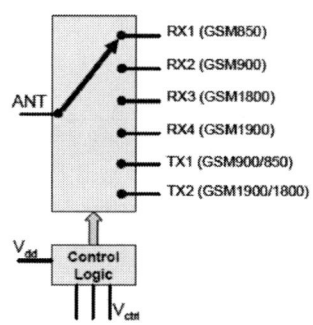

Fig. 4a. Single Pole, 6 Through (SP6T) RF switch [2]

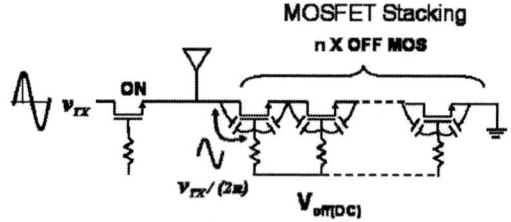

Fig. 4b. High voltage handling using SOI MOSFETs stacking in RF switches [2]

978-1-4799-8002-4/14 $31.00 © 2014 IEEE

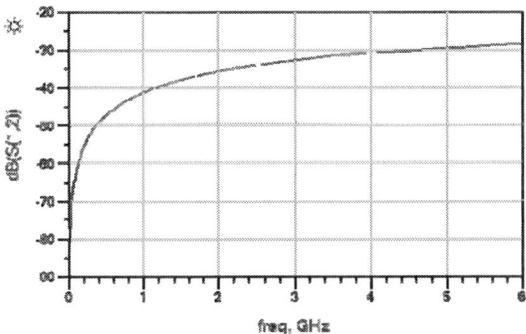

Fig. 5a. Isolation characteristics of a SP10T SOI RF switch fabricated by STMicroelectronics

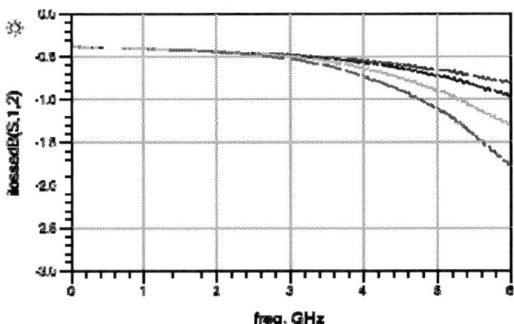

Fig. 5b. IL characteristics of the SP10T

This is more challenging to use pure HR bulk substrate because of latch-up and mainly much higher non-linear source-drain to substrate junction capacitances. SOS is good for linearity, which is measured through harmonics (Fig. 6).

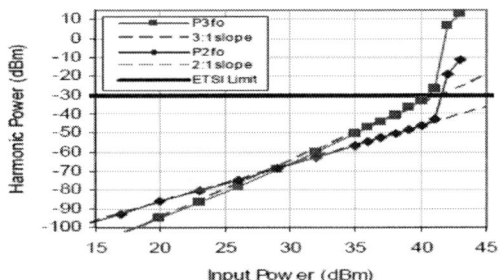

Fig. 6. SOS SP6T switches [3]

With a standard HR SOI substrate, it is challenging to reach the linearity specifications for 4G applications, due to remaining charge in the buried oxide, which induce a parasitic surface conductive layer under the buried oxide, degrading the performances. A solution has been proposed by adding specific trench and implant process steps [4] to contact the substrate below the buried oxide, like shown in Fig 7, but it adds process complexity and design tradeoffs. Another solution is to use a specific commercially available TR SOI substrate, where a Trap Rich layer has been added underneath the buried oxide [5, 6], as demonstrated by harmonics H2 & H3 measurements on a SP10T in Fig. 8, 9, showing a 25dB improvement on H2.

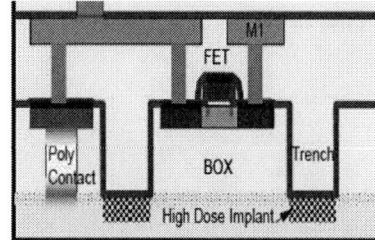

Fig. 7. SOI process with trench & implant [4]

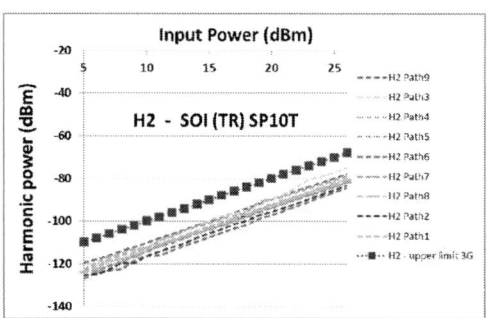

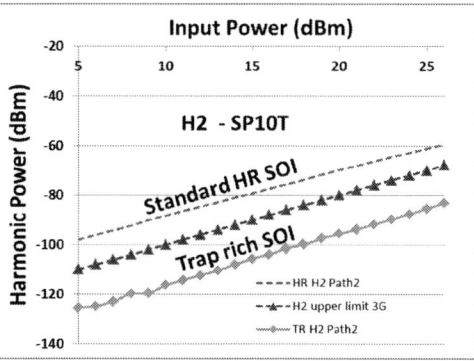

Fig. 8 – Harmonics H2 measured on a SP10T RF switch on Trap Rich SOI and standard HR SOI, fabricated by STMicroelectronics

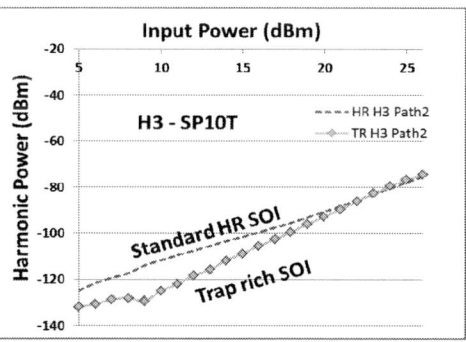

Fig. 9 (H3) – Harmonics measured on a SP10T RF switch on Trap Rich SOI and standard HR SOI, fabricated by STMicroelectronics

The Ron-Coff can be as low as 160 fs (Fig. 10) thanks to a process optimization. A trade-off in the device optimization (gate length reduction and source-drain extension doping profiles being the main key process parameters) can be found to keep a high RF breakdown voltage (> 3.5V) and a low Ron with short electrical gate length devices. Body contacted devices also help to reduce

978-1-4799-8002-4/14 $31.00 © 2014 IEEE   467

the Coff by applying a negative body bias, decreasing junction capacitances.

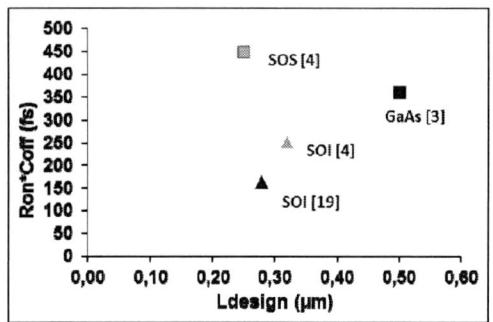

Fig. 10. Ron*Coff for RF switch devices

Power Amplifier (PA)

The PA is the most challenging circuit in the RF Front End Module. The Power Added Efficiency is one of the main criteria for this circuit, with the following definition: PAE = (Pout-Pin)/ Psupply To increase the amplifier ouput power and gain while keeping a high PAE, high f_T and High drain Voltage (HV) operation components are required. For MOSFETs in CMOS technologies, the gate length is generally reduced to increase gm and decrease the Miller gate to drain capacitance in order to reach higher f_T. But the Breakdown Voltage (BV) is degraded by using a thinner gate oxide, which is mandatory to enable the scaling. HV MOSFETs with BV > 12V and high f_T (> 20GHz) have been demonstrated in both bulk and SOI CMOS technologies (Fig. 11, 12) by shrinking the channel electrical length L_{ch} to increase gm and $L_{overlap}$ to reduce Cgd, with a trade-off between f_T, BV and reliability [12-18].

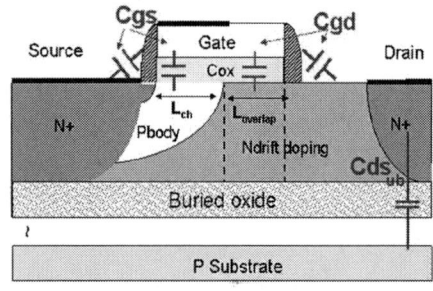

Fig. 11. Schematic cros section of an NLDMOS on a thin SOI film

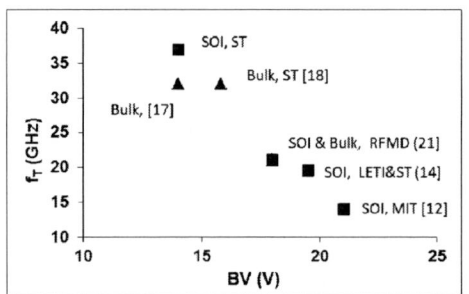

Fig. 12. Cut-off frequency as a function of the off breakdown voltage for HV devices (SOI and bulk)

For either bulk or SOI HV NLDMOS, only 2 masks are added to the CMOS Core process for a specific channel (Pbody) and drift (n- extension) doping. In case of a NLDMOS on a thin SOI film [13-16], a specific layout with body to source contacts is mandatory to suppress floating body effects (Fig. 13).

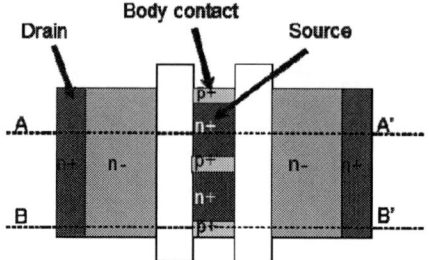

Fig. 13. Top view of a HV LDMOS with body contacted devices [13-16]

It is worth mentioning that a thin SOI film enables a lower drain to substrate capacitance as the source or drain junction reaches the buried oxide (Fig. 11). Progress on HV device integration and design on SOI [19-24] may accelerate the adoption of CMOS instead of InGaAs technology in PA Modules. With a NLDMOS PA on a thin HR SOI, up to 38dBm output power with a PAE > 60% has been measured @ 900 MHz (Fig. 14, [24]).

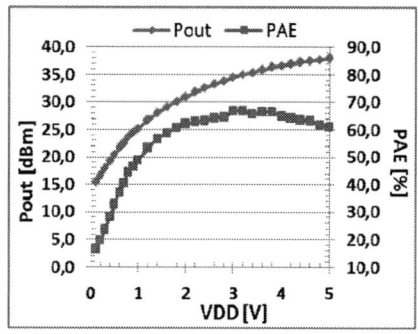

Fig. 14. 35dBm Pout and 68% PAE @ 900 MHz & 3.3V for a NLDMOS PA design on HR SOI [24]

A cascoded design, with a NLDMOS to sustain the high voltage stacked to a standard thick gate oxide MOSFET, measured @ 1.9 GHz, enables a high stability and low capacitances. A max PAE from 63% to 68% has been measured for a 28 dBm output power (Fig. 15, [19]).

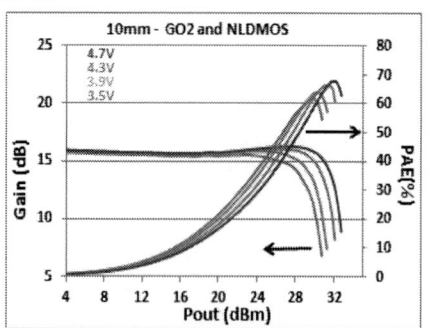

Fig. 15. 28dBm Pout and 68% max PAE @ 1.9 GHz for a cascoded NLDMOS & GO2 thick gate oxide MOS [19]

978-1-4799-8002-4/14 $31.00 © 2014 IEEE

Tunable filters in the future?
Thanks to a modified TCVR and FEM architecture in the future, tunable filters with lower Q factor than existing SAW filters could be the solution for higher integration. It can be made of high Q inductors with tunable capacitors. Different approaches have been proposed like Digitally Tunable Capacitor (RF switch plus integrated MIM capacitor), as well as Barium Strontium Titanate (BST) or MEMS. As summarized in Fig. 16 [25], MEMS are the best candidate. A new advanced concept, based on a stack of two piezo LNO substrate has also been proposed [26].

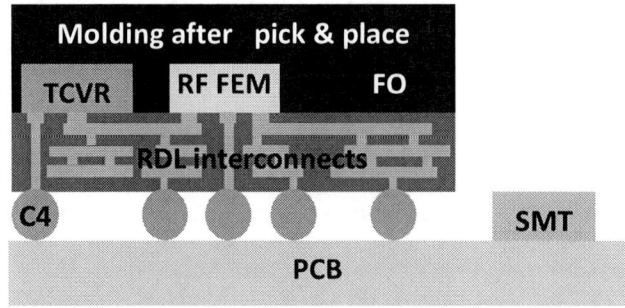

Fig. 18 e-WLP for multi-chip (TCVR & RF FEM) packaging

Strengths & Weaknesses for tunable RF technologies			
	SOS/SOI/GaAs Switches + Integrated MIM Capacitors	BST (Barium Strontium Titanate)	MEMS
Quality Factor (Q)	Low	Moderate	High
Intercept Point	Moderate	Moderate	High
Harmonics	Poor	Moderate	Excellent
Capacitance Ratio	High (for low Q) Low (for high Q)	Low	High
Voltage Handling	Low	Moderate	High
Switching Cycles	High	High	Moderate
Cost	Low	Moderate	Moderate

Fig. 16. Tunable capacitor benchmark with DTC, BST, MEMS [25]

High performances integrated passives and LNA
With a HR SOI substrate and thick metal back-end options, one can achieve inductance quality factor [27] competing with dedicated IPD technologies, to integrate the harmonics filters [28] or the duplexer [28] required in a dual band WiFi module. If considering WiFi dual band FEM which requires a 2.5 GHz and a 5 GHz LNA [29, Fig.17], one can then think about a low cost CMOS SOI WiFi FEM SOC.

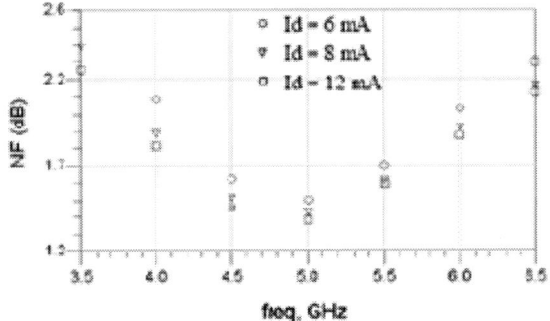

Fig. 17. FOM of a LNA on HR SOI [29]

System Packaging
e-WLB packaging technology could provide also a solution to integrate the RF modules with the transceiver (Fig. 18) in a low cost and thin package [30, 31]. It enables as well a smaller size system and integration of high performance low cost customized passives in the FO area [32]. The interest of a FO-WLP packaging approach is multi-chip packaging with interconnects independent of chip size and footprint.

Conclusion
Through an overview of the components required in the RF FEM, we have demonstrated that a higher integration is possible on a SOI substrate, with good performances for RF switches (IL, isolation, linearity) and with HV devices (f_T, BV) enabling efficient Power Amplifiers.

Acknowledgments
Thanks to A. Giry (CEA-LETI), B. Rauber, O. Bon, A. Monroy, G. Bertrand, F. Gianesello, V. Knopik, S. Gachon, R. Fournel (STMicroelectronics)

References
[1] D Nobbe, RWS 2008
[2a] C. Tinella, J.M Fournier, D. Belot, V Knopik, ESSIRC 2002
[2b] C. Tinella PhD, 2003, STMicroelectronics
[2c] C. Tinella et al, SiRF 2006
[3] D. Kelly, C. Brindle, C. Kemerling, M. Stuber, CSIC 2005
[4] A. Botula et al, SiRF 2009
[5] D. Lederer, JP Raskin, Solid State Electronics 2005
[6] E. Desbonnets, SOITEC White paper, Nov 2013
[7] M. Carroll et al, CSISC 2009
[8] T-Y Lee, S. Lee, IEEE RFIC 2010
[9] Q. Chaudry et al, IEEE RFIC 2012
[10] A. Joseph et al, IEEE RFIC 2013
[11] V. Blaschke, A. Unikovski, P. Hurwitz, S. Chaudry, IEEE RWS 2013
[12] J.G Fiorenza, D.A Antoniadis, J;A del Alamo, IEEE EDL 2001, vol. 22, N°3
[13] O. Bon et al, IEEE SOI Conference 2005
[14] O. Bon et al, IEEE SOI Conference2007
[15] O. Bon, O. Gonnard, F. Gianesello, C. Raynaud, F. Morancho, ISPSD 2007
[16] C. Raynaud et al, Workshop WSI, IEEE RFIC 2009
[17] Z. Lee et al, BCTM 2006
[18] B. Szelag et al, BCTM 2006
[19] V. Knopik, Workshop WSA, RFIC 2014
[20] J. Costa et al, IEEE-MTT-S, 2007
[21] S. Pornpromlikit et al, IEEE RFIC 2009
[22] L. Andia et al, APMC 2010
[23] A. Tombak, D.C Dening, M.S Carroll, J. Costa, E. Spears, TMTT, June 2012, vol. 60, N°6
[24] A. Giry, Workshop WSA, IEEE RFIC 2014
[25] V Steel, A Morris, Microwave Journal, Nov 2012
[26] A. Reinhardt et al, MTT-S, 2012
[27] C. Pastore et al., IEEE SOI Conference 2008
[28] F. Gianesello et al., IEEE RFIC 2010
[29] F. Gianesello, D. Gloria, C. Raynaud, S. Boret, ISIC 2007
[30] S. W Yoon et al, ECTC 2013
[31] G. Pares et al, EMPC 2013
[32] C. Durand et al, SiRF 2012, Session WE2D

978-1-4799-8002-4/14 $31.00 © 2014 IEEE

Digitally-Intensive RF Transceivers in Highly Scaled CMOS

Chih-Ming Hung
MediaTek Inc., Hsinchu, Taiwan

Abstract

Fast transistors and increasing device non-ideality in advanced nodes have motivated the development of digitally-intensive RF systems utilizing logic gates and algorithms to combat with analog impairments. In this paper, we will explore the power of such embedded intelligence. There are still constraints mandating device-level enhancement. Strong interactions between device and circuit communities are required to accomplish competitive products.

Introduction

Smartphones experienced explosive growth in the past decades, and accelerated the evolution of cellular standard and circuit technologies. The strong demand for high performance affordable handsets has spurred the requirement of using plain vanilla CMOS for RF. Fig. 1 shows a block diagram and a die photo of a fully-integrated SoC consisting of multiple radios, processor units, power management units, and a variety of peripherals. Since logic gates occupy most of the area, Moore's Law ripples through to analog circuits on the same die and any process mask adder only for RF is prohibited. However, there are growing analog impairments in advanced nodes greatly impacting circuit performance and cost. Fig. 2 shows a DAC size increases over straight process migration [1] due to worse device matching and tighter design rules on maximum poly density, minimum dummy fills, etc. Digital techniques have to be exploited to stay with or better than Moore's Law. Furthermore, along process technology scaling, the increase of f_t (Fig. 3), the decrease of voltage headroom (Fig. 4), and the reduction of transistor cost call for a paradigm shift from continuous-time (CT) to discrete-time (DT) operations to take advantages of digitally-intensive architectures where mixed-signal circuits are replaced by logic gates and algorithms natural in digital CMOS.

Discrete-Time Transceiver

Fig. 5 shows a generic diagram of a transceiver. The local oscillator (LO) generates precise RF frequencies that are synchronized with basestations. The transmitter sends RF signals to the air, and the receiver demodulates information from the antenna. Fig. 6 shows a discrete-time receiver (DRX) which brings power of DSP to the RF domain and uses only device components from standard digital CMOS [2]. The DT passive filter can easily be reconfigured with different orders and shapes, and can attain voltage gain by stacking capacitors [3]. Because there are only switches and capacitors, tradeoffs among voltage headroom, amplifier bandwidth, and linearity are mostly eliminated. High-speed transistors further enable high sampling rate in the switched capacitor filter, which in turn reduces noise folding and noise density $\propto kT/C/f_s$ where f_s is the sampling rate. Device accuracy over PVT is generally not required anymore because filter characteristic is now dominated by local device mismatch. Although the advancement of ADCs has triggered analog-light architectures which have fewer front-end amplifiers and filters [4], the benefits of DRX become even more critical in new wireless systems which require complex and flexible filter responses.

All-Digital Phase-Locked Loop (ADPLL)

An ADPLL and a conventional charge-pump PLL (CPPLL) are compared in Fig. 7. In the CPPLL, the PFD outputs timed pulses [5] whose duty cycles are proportional to the difference between FREF (reference frequency) and the divided VCO output. Although the pulses have digital waveforms, they don't represent any digital quantity and need to be converted to an analog VCO control voltage by a charge pump and a loop filter. A charge pump consists of switched current sources to deposit or remove charges in the loop filter. Active amplifiers may be included to enhance its performance. These analog circuits suffer from finite transistor output resistance, device mismatch, thermal noise and gate leakage current in highly scaled CMOS processes resulting in compromises in PLL in-band phase noise, loop response, phase locking accuracy, size and power consumption.

To eradicate the aforementioned deficiency, an ADPLL comprising only logic gates except the digitally controlled oscillator (DCO) to interface with RF world has emerged. The TDC which is analogous to an ADC converts phase errors (PE) into digital quantities. The PE comprehends key information internal to the PLL operations. It can assist built-in self calibration and can be further processed by a DSP to self-assess performance metrics such as phase modulation accuracy in real time without any expensive RF instrument. The digital loop filter sends bit streams to the DCO whose simplified schematic is shown in Fig. 8. Compared to a VCO, one unique feature is the switched varactors. A DCO does not require continuous frequency tuning but discrete frequency steps. Using low-cost IO NMOS transistors which share the same implants as those for core NMOS, and taking body effect into account, the V_t is near half V_{dd}, which is ideal for the switched varactors. It can be observed from Fig. 8 that under a large DCO voltage signal, the capacitance step (Δc between 0 and 1.4V gate bias) is relatively constant when the small-signal CV curve shifts slightly on the X axis. Therefore, the DCO frequency step has a negligible sensitivity to PVT.

To achieve low phase noise, a large voltage swing is required imposing reliability concerns. In particular, with the cross-coupled connections, there is >2.1V peak AC voltage across the 65nm transistors creating over stress for TDDB and NCS. It was discovered that the stress also caused noise degradation [6]. Voltage waveforms for the gm pair are shown in Fig. 9. Standard transistor reliability assessment based on modeling from DC stress is overly pessimistic by hundreds of times. However, the same modeling methodology can be applied to the product level under AC stress. When evaluating DCO phase noise drifts over various accelerated conditions and under real-life duty-cycled operations, 5 years product lifetime can be achieved. Although IO and DEMOS transistors could be options for the DCO cross-coupled pair, there is a penalty on frequency tuning range due to their large parasitic capacitance to maintain the same negative gm and phase noise performance.

978-1-4799-8002-4/14 $31.00 © 2014 IEEE

Inductors and capacitors occupy most of the space in a DCO. Even though MoM and gate capacitors shrink less than 30% per process node, which is shy from Moore's Law, since the inductors take more than half of a DCO real estate, the introduction of 1-2.8 μm AL and 0.8-3.4 μm top Cu layers helped DCO size to reach better than linear reduction. The improved Q also helps to lower the voltage swing. However, as noise and transistor reliability degrade over process scaling, and there has been no major inductor Q enhancement after 65nm, DCO size and voltage swing remain about constant in new processes. Fig. 10 shows measured phase noise (1/SNR). The -167 dBc/Hz at 20 MHz offset from 915 MHz RF carrier means a very high SNR that devices need to simultaneously sustain a large signal swing and exhibit low noise.

Digital Transmitter (DTX)

Both analog and digital transmitters are shown in Fig. 11. In the former, the DAC transforms digital quantities to an analog signal which is filtered to remove the sampling images. The reconstructed signal is then upconverted onto an RF carrier and transmitted by a PPA typically at <6 dBm. As new wireless standards demand wider signal bandwidth and higher linearity, it becomes very challenging to maintain amplifier gain, voltage headroom and filter characteristic in new process nodes. For PPA linearity, a larger backoff from its saturated output power is required to satisfy the increasing peak-to-average power ratio (Fig. 12). Although PPA supply voltage and impedance transformation ratio of the output matching network can be increased, there are tradeoffs between transistor reliability and power efficiency. On the other hand, it is possible to push the PPA operation toward the higher efficiency region in Fig. 12, then recover the system linearity by digital predistortion (DPD) [7]. As an extreme, the PPA always stays in the saturated condition and its output power is controlled by slicing the transistor to form an RF DAC. A simplified schematic of such digital PA (DPA) is shown in Fig. 13, which may require up to 17 bits resolution [8] depending on applications. Since the input signal is completely in digital format, digital signal processing can be deployed for the whole transmitter to form a DTX. The supply voltage can be adaptively reduced according to the instantaneous output power as linearity of individual transistors is no longer important. The overall power consumption can be further improved by using an efficient supply regulator such as a DCDC converter [9].

Unlike a linear PPA, the DPA AMAM curve in Fig. 12 has no linear region. Therefore, DPD is fundamentally required. A block diagram and the operating principle of the DPD are shown in Fig. 14. After measuring the DPA AMAM distortion (pink curve), a look-up table (LUT) is derived (green curve). Applying the LUT to the incoming data, the net DPA output follows the blue curve which becomes very linear. The same procedure is jointly applied for AMPM distortion. The resulting improvement on DPA output spectrum is shown in Fig. 14(b). The DPD technique can also be applied to a linear PPA in its high-power region to reduce the power backoff and to increase its power efficiency.

In addition to AMAM distortion which is inherent in a DPA, due to the high operating speed at RF, the mismatch in LO generation and distribution further produces AMPM distortion. Theoretically, DPD can completely recover both types of distortion. However, unlike a baseband DAC, the mismatch from each cell contributes differently to AMAM and AMPM distortion at various output power levels. The LUT in the DPD can become excessively large when the resolution requirement and order of modulation are increased. Fortunately, the switching operation greatly reduces the DPA sensitivity to transistor V_t and geometrical mismatch because in each RF period, DPA cells are duty-cycled and stay in deep triode and off states for most of the time. An auxiliary DAC can be used to further compensate random mismatch. In general, intrinsic device matching needs to be >10 bits even for small device sizes. Then, digital techniques are needed to reach a higher resolution.

DPA voltage waveforms are shown in Fig. 15. It has a low duty cycle and a high peak voltage at drain node. Even it has been designed to reduce the peak voltage, >2x V_{dd} was reached resulting in a significant reliability concern. The AC reliability involves NCS and drain TDDB. A product-level parametric shift evaluation is again needed to assess the lifetime which can reach 5 years with 65nm core transistors. If the DPA was implemented with redundant cells, each cell can be rotated in and out of operation by digital algorithms to reduce the accumulated stress or to allow devices to recover [10]. This makes economic sense thanks to the continuing scaling of transistor cost. The technique is not a remedy to transistor-level reliability limit but to increase the margin of product-level lifetime.

DPA performance is highly correlated with transistor speed. For a 5GHz DPA [7], the 25% duty cycle LO has only 50ps pulse width. The relatively large rise/fall time ~20ps in 28nm Poly/SiON process increases the distortion sensitivity to random mismatch in FEOL and BEOL. Hence, the DPD becomes very challenging in terms of resolution, calibration and design complexity. The output power which is proportional to $\sin(\omega_c t_p/2)$ is also impacted where ω_c is the carrier frequency and t_p is the effective duty cycle taking rise/fall time into account. Overall, a DPA would enjoy process scaling very well provided reliability and mismatch performance can be maintained.

Summary

Digitally-intensive architectures serve as great means to break the tradeoffs between device and circuit limitations. Using algorithms and switching circuits, ADPLL, DRX and DTX have wide spread. High-volume production of SoCs from multiple fabs without any circuit alteration proved that digital architectures can effectively reduce circuit sensitivity to most analog impairments and PVT variations. However, device characteristic such as reliability, noise and mismatch as well as BEOL variability still have major impacts to RF performance. Both device performance and circuit techniques have to be jointly optimized in future process generations.

[1] W.-H. Tseng et al., VLSI pp. 161-162 2014 [2] K. Muhammad et al., ISSCC pp. 268-269 2004 [3] M.-C. Lee et al., VLSI pp. 144-145 2009 [4] C.-M. Hung et al., VLSI pp. 12-14 2013 [5] F. Gardner, Wiley-Interscience Chap 12 2005 [6] V. Reddy et al., IEDM pp. 1-4 2009 [7] H. Wang et al., ISCAS pp. 501-504 2013 [8] Z. Boos et al., ISSCC pp. 376-377 2011 [9] G. Hanington et al, TMTT v47 pp. 1471–1476 1999 [10] T. Ong et al., TED v35 n7 pp. 978-984 1988.

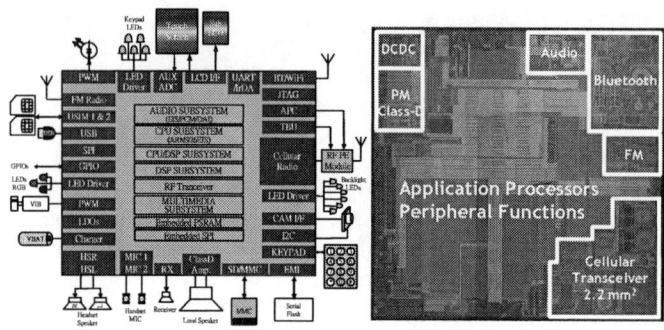

Fig. 1: A system diagram and a die photo of a fully-integrated radio SoC including three RF radios, modems, CPUs/GPUs, power management, and a variety of peripherals.

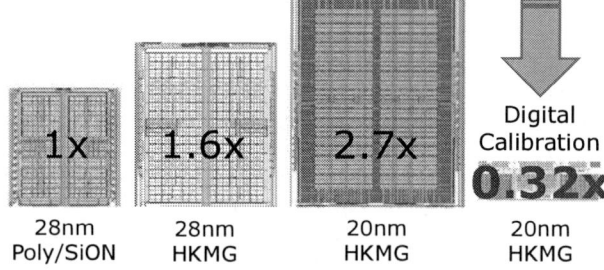

Fig. 2: Process migration for a DAC with the same design specification. Due to tighter design rules and worse device mismatch, the DAC size grows. Digital techniques helped to recover the deficiency and reach the goals.

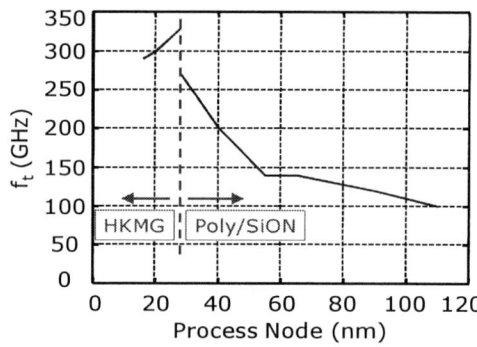

Fig. 3: f_t increases along node scaling for Poly/SiON processes. However, for HKMG, with higher extrinsic parasitic capacitance, f_t decreases in newer process nodes.

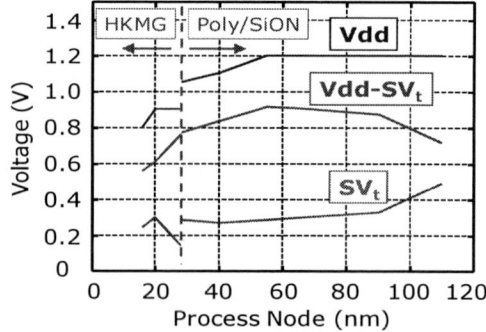

Fig. 4: Voltage headroom (V_{dd}-SV_t) generally decreases over process scaling after 90nm due to the fact that SV_t stays flat while V_{dd} decreases.

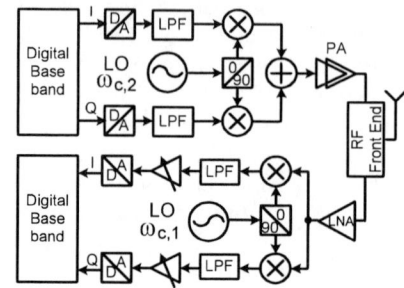

Fig. 5: Block diagrams of a generic wireless transceiver. The LO generates precise RF frequencies. The baseband data is converted to RF signals in a transmitter, and the receiver extracts the desired data from the antenna.

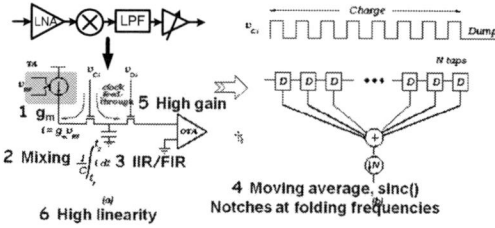

Fig. 6: A highly programmable discrete-time receiver which brings digital signal processing to the RF domain. It contains mostly switched circuits thus is far less sensitive to analog impairments than its conventional counterpart.

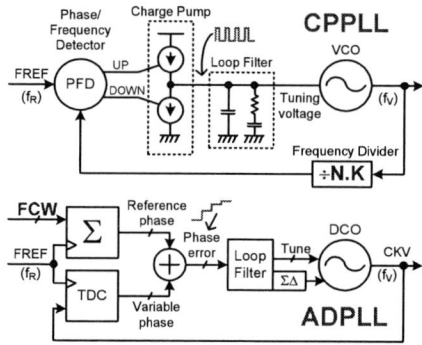

Fig. 7: Block diagrams of a CPPLL and an ADPLL. The ADPLL consists of only logic gates (except the DCO) to eliminate the analog tradeoffs in the CPPLL.

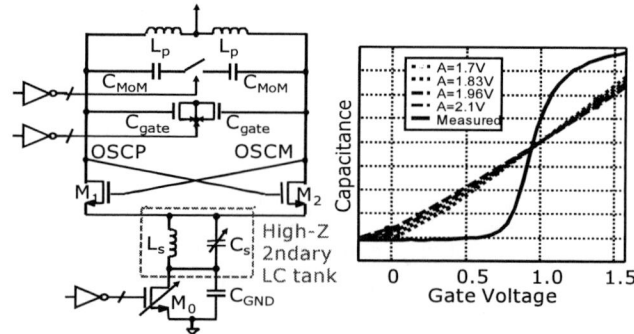

Fig. 8: A simplified DCO schematic. The spectral purity demands a large voltage swing on drain and gate terminals of M_1 and M_2 resulting in reliability concerns. The large-signal operation makes the capacitance difference between high and low gate voltage states nearly insensitive to V_t shifts.

978-1-4799-8002-4/14 $31.00 © 2014 IEEE

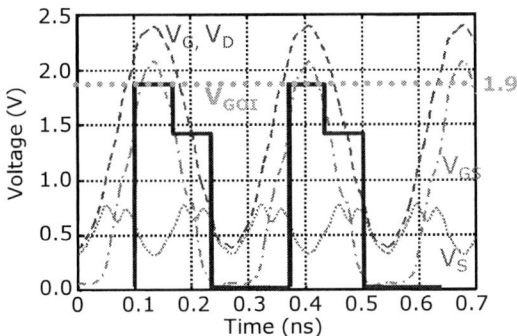

Fig. 9: Voltage waveforms of the cross-coupled transistors in the DCO. The effective gate oxide stress is ~1.9V for 25% duty cycle, 1.4V (V_{dd}) for another 25%. The peak voltage is near 2.5V. The peak V_{GS}/V_{DS} is ~ 2.1V.

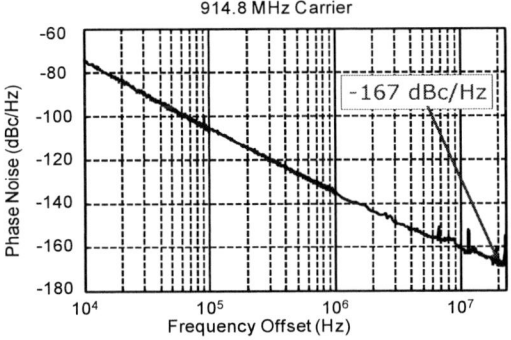

Fig. 10: Measured DCO phase noise. The large SNR (-167 dBc/Hz is 1/SNR/Hz) requires low noise and large voltage swing for both active and passive components.

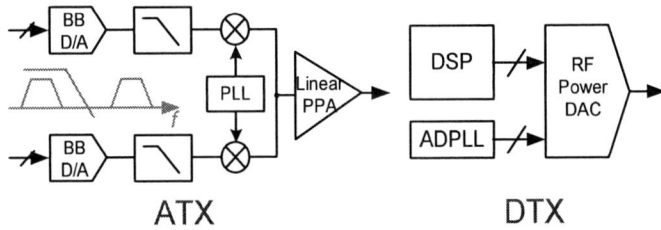

Fig. 11: Block diagrams of analog and digital transmitters. The DTX replaces all analog components in the ATX except the digital-to-RF power DAC which interfaces the RF world.

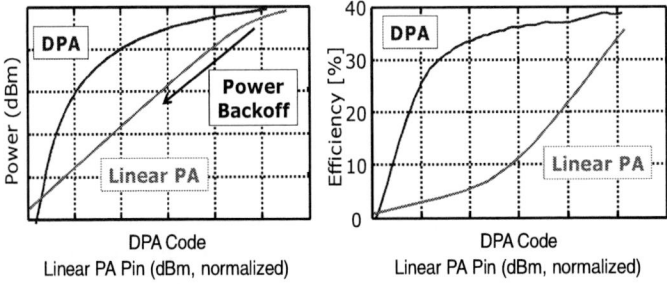

Fig. 12: Linearity and efficiency for a DPA and a linear PA. DPD is mandatory for DPA since each DPA cell operates in a switched condition and the AMAM curve is completely nonlinear. For linear PA, the operating point has to be backed off to maintain proper linearity resulting in degraded efficiency.

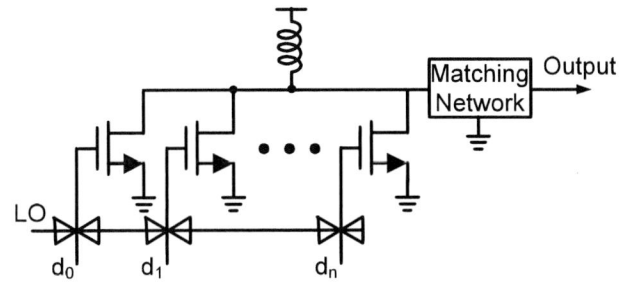

Fig. 13: A simplified schematic of a DPA. The transistors operate in switched conditions for better power efficiency. The output power is controlled by slicing the PA transistor to form an RF power DAC.

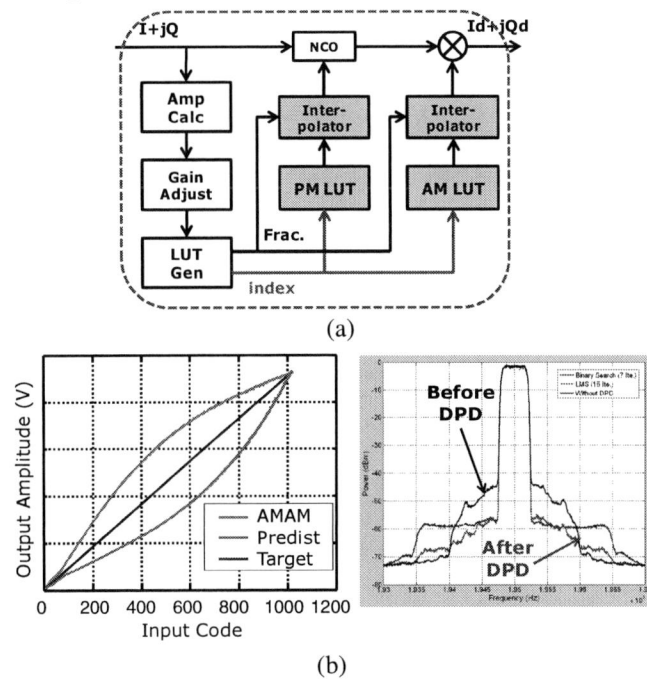

Fig. 14: A simplified block diagram of the DPD. I+jQ is the desired data which needs to be pre-distorted with the green curve in (b) to recover the nonlinear DPA behavior in pink. The improvement on RF spectrum is also shown in (b).

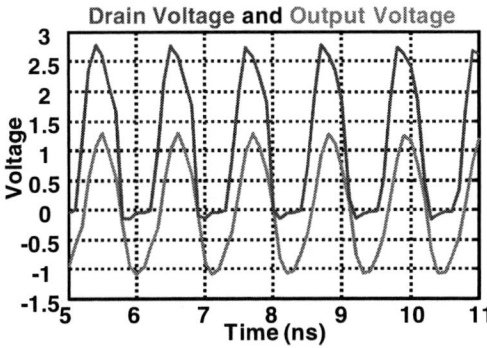

Fig. 15: The waveforms for drain and output nodes in Fig. 13. 65nm core transistors are chosen as good switches. The peak drain voltage is ~2x V_{dd} imposing reliability concerns. With an output power drift evaluation, the DPA can pass 5 years product life time.

978-1-4799-8002-4/14 $31.00 © 2014 IEEE

Circuit and Device Interactions for 3D Integration Using Inductive Coupling

Tadahiro Kuroda
Keio University, Yokohama Japan

Abstract

This paper presents a ThruChip Interface using inductive coupling and Highly Doped Silicon Via for power delivery. Design automation, manufacturability, applications, and scaling scenario are discussed.

I. Introduction

Device scaling enlarges the gap between chip performance and I/O bandwidth. Transistors in I/O periphery increase linearly with scaling, while those in chip area increase with the square. To cope with the scaling dilemma, Through-Silicon Via (TSV) has been investigated as an area interface. Although many die stacking approaches rely on TSV as the fundamental means for 3D integration, TSV is not generally available. It still has reliability/yield issues and its cost adder is limiting acceptance of 3D stacking. Replacing the mechanical approach, a low cost electrical solution is developed, namely ThruChip Interface, which uses inductive coupling.

II. ThruChip Interface (TCI) for Data Link
1) Near-field coupling

Near field is an electromagnetic field closer from a signal source than (wave length)/2π. Signal in the near field decay rapidly, in inverse proportion to the cube of distance, and can reach only short distances (Fig.1). In other words, just like there are invisible wires, no crosstalk occurs when multiple channels are formed.

Magnetic field can penetrate through IC chips but electric field cannot (Fig.2). Permeability of materials used in IC chips is all one, while permittivity differs, causing reflection and absorption of electric field. Furthermore, as degradation by eddy current in typical resistivity of Si substrate is negligibly small, magnetic field can penetrate it.

Typically distance of 1/3 of coil diameter can be reached (Fig.3). The shorter distance, the smaller coil size and the higher self-resonant frequency, yielding the higher bandwidth. For instance, when 3 chips of 10um thickness are stacked, a 60 μm coil can provide 90 Gbps bandwidth. When 9 chips of 10 μm thickness are stacked, a 240 μm coil can connect them by bus with 20 Gbps bandwidth.

2) ThruChip Interface (TCI)

TCI [1-37] is depicted in Fig.4. A coil is formed by using multi-layer standard wires and vias, so that digital wires can go across it. As the coil should not resonate, narrow wires and small vias can be used for low Q factor. Data communication uses baseband, not carrier, as the conventional digital on-chip communication. A transceiver is implemented by digital CMOS circuits of as small as 36 2NAND gates and it scales down by device miniaturization. TCI is a digital CMOS circuit solution, and hence eventually zero additional cost.

Comparison with TSV is summarized in Fig.5. TCI is much cheaper than TSV while bearing comparison in performance.

TCI performs high speed (30 Gb/s/ch)[32], low energy (0.01 pJ/b) [31], high integration (128-die stacking)[26], and low stacking height (bump less) and low thermal resistance (Fig.6). Data rate can be raised by increasing number of channels while keeping energy efficiency. The energy is lowered significantly, because an ESD protection device is eliminated. Chips can be stacked up without solder bumps.

3) Techniques to improve area and energy efficiency

To further improve area and energy efficiency circuit techniques have been developed. The coils formed by using multi-layer wires and vias allow overlapping. Crosstalk can be suppressed by phase division multiple access (Fig. 7)[37]. Using dual Tx coils eliminates PMOS, resulting in 0.55 V operation and 0.01 pJ/b (Fig.8)[31].

III. Reliability

TCI is as reliable as the conventional wireline as it uses near field coupling. Measured bit error rate is lower than 10^{-14} [01]. Measured jitter is smaller than 5% UI [01]. Crosstalk between channels is small enough if the channel pitch is the same as the coil diameter for a line arrangement and the channel pitch is twice as large for an array arrangement (Fig. 9) [03]. Measurement results show very small degradation by eddy current in chip substrate [05], power ground mesh [12], high-density bit lines and word lines (Fig. 10) [26], and also very small degradation by chip misalignment [15]. No interference from digital circuits to TCI [23] and no interference from TCI to SRAM [12] was observed. Influence from/to the environment (EMS, EMI) was also examined and no issue was found [22].

IV. Highly Doped Silicon Via (HDSV) for Power Delivery

Power can be delivered by conventional means (wire bond, TAB) or a new way with Highly Doped Silicon Via, HDSV (Fig. 11) [39]. A deeper than normal and more highly doped well is used to make a low resistance HDSV pathway directly through a thinned wafer using the silicon itself. Highly doped regions for power vias are first created by implants, followed by nominal process for transistors and wires, and metal caps are added on the HDSV. Wafer is then thinned to ~4 μm. The HDSV on one die and the electrodes on the next die are connected by pressure (solid intermetallic bonding by diffusion) using a room-temperature wafer bonding machine to create larger stacks. It is reported that DRAM chip substrate was thinned to 4 μm and yet no degradation of retention characteristics was found. TCAD simulations indicate the front-to-back resistance can be made lower than 3 mΩ when substrate thickness is below 5 μm and the HDSV net area is 0.7 mm^2 (Fig. 12), under conditions of 1×10^{16} cm^{-2} dose, 200 keV ion implantation, and 50 hour annealing at 1050°C. The HDSV can be divided into small regions and distributed. As large area is required to reduce resistance, the HDSV is not usable for high speed data, but TCI should be used instead. The HDSV should be low cost as it is made by implants.

V. Design Automation and Manufacturability

IC chips with TCI can be designed by the conventional design automation and manufactured with the conventional package and supply chain.

1) EDA

As the TCI is a digital CMOS circuit solution, conventional EDA tools and standard design flow can be applied (Fig. 13) [33].

2) Package

An Ultra-Thin Fan-Out Wafer Level package [35] enables package on package (Fig.14), which allows using the conventional supply chain. Power can be delivered by Thru Mold Via (TMV), and Re-Distribution Layer (RDL) which is fan out from the edge of a chip and connected to the center located pads for power and ground. Known-good-dies are placed to reform a wafer. The silicon can be thinned to 40μm to make the TCI area efficient. This wafer level package technology is available for mass production. Other conventional packages can be used as well.

VI. Application

TCI can be used for applications where TSV is expected for use. It can also be applied to contactless memory, contactless wafer testing, and bus proving through a package for debugging (Fig.15).

1) DRAM/SoC interface

A 352Gbps DRAM/SoC interface is developed (Fig.16) [35]. The overlapping coils with Quadrature Phase Division Multiplexing (QPDM) are employed. It outperforms WIO2 (TSV) in cost and LPDDR4 in power dissipation and latency (Fig.17).

2) NAND stacking for SSD

One-package SSD is made possible with TCI (Fig.18) [26]. I/O power dissipation is reduced to 1/6 (Fig.19). TCI consumes constant I/O energy and delay regardless of number of IOs that are connected, enabling low energy broadcasting (Fig.20)

3) 3D Processor

A 3D processor can adjust its performance by changing number of dies in the stacking [19]. Linux OS was installed, and several applications were performed and demonstration.

VII. Scaling Scenario and Future Directions
1) Constant Magnetic Field Scaling Scenario

A 3D scaling scenario will improve cost performance significantly. Just like the constant electric field scaling scenario in scaling a Field Effect Transistor, a constant magnetic field scaling scenario is found for scaling the inductive coupling channel (Fig. 21). Suppose device size and the supply voltage are scaled to half, and also chip thickness is scaled to half and coil turns are increased a little, electric field of FET as well as magnetic field of the inductive coupling channel are kept constant before and after the scaling. As a result, aggregated data bandwidth per area is increased by 8x, and energy per bit is reduced to 1/8. Thinning wafers is the future directions of competition.

2) Future perspectives

With the TCI and the HDSV, IO energy per bit will be reduced to 1/400 in NAND stacking (Fig. 22), and 1/10 in DRAM stacking (Fig. 23). Panel-level stacking as batch (wafer scale) process is a future scenario (Fig. 23).

VIII. Conclusion

The scaling dilemma of I/O bottleneck can be solved by moving from mechanical connection to electrical one using inductive coupling. One of the future directions of device and circuit interactions is found in thinning wafers and connecting them by inductive coupling.

Acknowledgement

The author is grateful to K. Uchida, T. Takahashi, D. Ditzel, S. Lee for discussion and TCAD simulation. This research was supported by JSPS KAKENHI S Grant Number 25220002.

References

[01] "A 1.2Gb/s/pin Wireless Superconnect Based on Inductive Inter-chip Signaling (IIS)," *ISSCC*, pp.142-143, 2004.

[02] "Analysis and Design of Inductive Coupling and Transceiver Circuit for Inductive Inter-Chip Wireless Superconnect," *Symp. VLSI Circuits*, pp. 246-249, 2004.

[03] "Cross Talk Countermeasures in Inductive Inter-Chip Wireless Superconnect," *CICC*, pp.99-102, 2004.

[04] "A 195Gb/s 1.2W 3D-Stacked Inductive Inter-Chip Wireless Superconnect with Transmit Power Control Scheme," *ISSCC*, pp.264-265, 2005.

[05] "Measurement of Inductive Coupling in Wireless Superconnect," *SSDM*, pp.670-671, 2005.

[06] "A 1Tb/s 3W Inductive-Coupling Transceiver for Inter-Chip Clock and Data Link," *ISSCC*, pp.424-425, 2006.

[07] "Perspective of Low-Power and High-Speed Wireless Inter-Chip Communications for SiP Integration," *ESSCIRC*, pp.3-6, 2006.

[08] "Constant Magnetic Field Scaling in Inductive-Coupling Data Link," *SSDM*, pp.606–607, 2006.

[09] "A 0.14pJ/b Inductive-Coupling Inter-Chip Data Transceiver with Digitally-Controlled Precise Pulse Shaping," *ISSCC*, pp.264-265, 2007.

[10] "CMOS Proximity Wireless Communications for SiP Integration (Invited)," *ISSCC*, 2007.

[11] "Low power technology for system LSI," *J. IEICE*, vol. 90, no. 11, pp. 977-981, Nov. 2007.

[12] "Interference from Power/Signal Lines and to SRAM Circuits in 65nm CMOS Inductive-Coupling Link," *A-SSCC*, pp.131-134, 2007.

[13] "An 11Gb/s Inductive-Coupling Link with Burst Transmission," *ISSCC*, pp.298-299, 2008.

[14] "Constant Magnetic Field Scaling in Inductive-Coupling Data Link," *IEICE T. Electronics, Vol. E91-C, pp. 200- 205, Feb. 2008.*

[15] "Misalignment Tolerance in Inductive-Coupling Inter-Chip Link for 3D System Integration," *SSDM*, pp.86-87, 2008.

[16] "A 2Gb/s 15pJ/b/chip Inductive-Coupling Programmable Bus for NAND Flash Memory Stacking," *ISSCC*, pp.244-245, 2009.

[17] "An Inductive-Coupling Link for 3D Integration of a 90nm CMOS Processor and a 65nm CMOS ...," *ISSCC*, pp.480-481, 2009.

[18] "3D System Integration of Processor and Multi-Stacked SRAMs by Using Inductive-Coupling Links," *Symp. VLSI Circuits*, pp. 256-257, 2009.

[19] "A Scalable 3D Processor by Homogeneous Chip Stacking with Inductive-Coupling Link," *Symp. VLSI Circuits*, pp. 94-95, 2009.

[20] "A 4.7Gb/s Inductive Coupling Interposer with Dual Mode Modem," *Symp. VLSI Circuits*, pp. 92-93, 2009.

[21] "47% Power Reduction and 91% Area Reduction in Inductive-Coupling Programmable Bus for NAND Flash Memory Stacking," *CICC*, pp.449-452, 2009.

[22] "Electromagnetic Interference and Susceptibility in Inductive-Coupling Link," *SSDM*, p.62-63, 2009.

[23] "An Extended XY Coil for Noise Reduction in Inductive-coupling Link," *A-SSCC*, pp.305-308, 2009.

[24] "A Wafer Test Method of Inductive-Coupling Link," *A-SSCC*, pp.301-304, 2009.

[25] "An 8Tb/s 1pJ/b 0.8mm2/Tb/s QDR Inductive-Coupling Interface Between 65nm CMOS and 0.1um DRAM," *ISSCC*, pp.436-437, 2010.

[26] "A 2Gb/s 1.8pJ/b/chip Inductive-Coupling Through-Chip Bus for 128-Die NAND-Flash Memory ...," *ISSCC*, pp.440-441, 2010.

[27] "Inductively Coupled ThruChip Interface," *ISSCC*, ES3, 2010.

[28] "A 0.7V 20fJ/bit Inductive-Coupling Data Link with Dual-Coil Transmission Scheme," *Symp. VLSI Circuits*, pp.201-202, 2010.

[29] "ThruChip Interface (TCI) for 3D Integration of Low-Power System (Invited)," *IEDM*, p.17.1.1, 2010.

[30] "A 2.7Gb/s/mm^2 0.9pJ/b/Chip 1Coil/Channel ThruChip Interface for NAND Flash Memory ...," *ISSCC*, pp.490-491, 2011.

[31] "A 0.55V 10fJ/bit Inductive-Coupling Data Link and 0.7V 135fJ/Cycle Clock Link with Dual-Coil Transmission Scheme", IEEE JSSC, pp. 965-973, April 2011.

[32] "A 30 Gb/s/Link 2.2 Tb/s/mm2 Inductively-Coupled Injection-Locking CDR for High-Speed DRAM Interface", JSSC, pp 2552-2559, November 2011.

[33] "A 5Gbps/ch ThruChip Interface and Automated P&R Design Methodology for 3-D Integration of 45nm CMOS Processors," *COOL Chips XV*, pp.1-3, 2012.

[34] "Dynamic power control with a heterogeneous multi-core system using a 3-D wireless inductive coupling interconnect," *ICFPT*, pp.293-296, 2012.

[35] "A Case for Wireless 3D NoCs for CMPs ," *ASP-DAC*, pp.23-28, 2013.

[36] "3D Clock Distribution Using Vertically/Horizontally Coupled Resonators," *ISSCC*, pp. 258-259, 2013.

[37] "A 352 Gb/s Inductive-Coupling DRAM/SoC Interfaces Using Overlapping Coils with Phase Division Multiplexing and Ultra-Thin Fan-Out Wafer Level Package," *Symp. VLSI Circuits*, pp.C44-45, 2014.

[38] "Ultra Thinning down to 4mm using 300-mm Wafer proven by 40-nm Node 2 Gb DRAM for 3D Multi-stack WOW Applications," *Symp. VLSI Technology*, pp.T22-23, 2014.

[39] "Low-Cost 3D Chip Stacking with ThruChip Wireless Connections," *Hot Chips*, 2014.

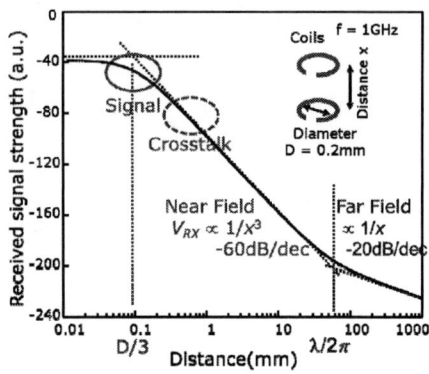

Fig. 1 Received signal rapidly decays at distance X > D/3 in the near field. Crosstalk is sufficiently suppressed.

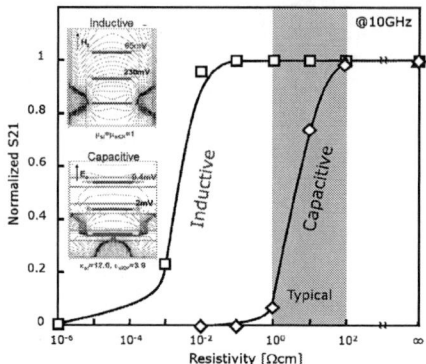

Fig. 2 Inductive vs. capacitive coupling. Magnetic field can penetrate through Si chips, while electric filed cannot.

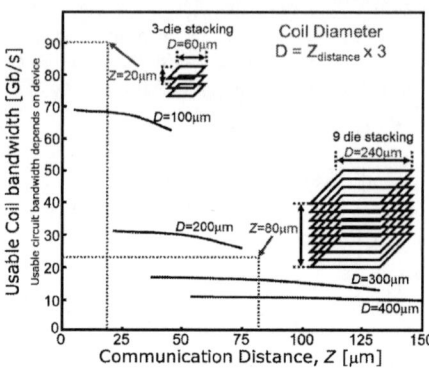

Fig. 3 Relation between coil size, communication distance and usable coil bandwidth.

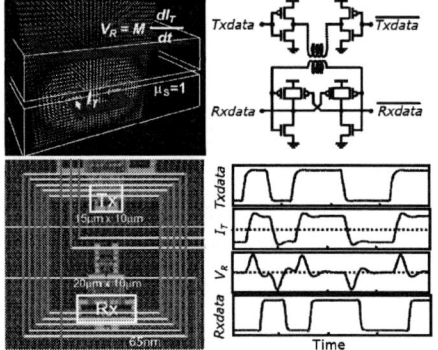

Fig. 4 ThruChip Interface (TCI). TCI is a digital CMOS circuit solution with eventually zero additional cost.

	TSV	TCI
Connection by	Wire	Wireless
Area/channel	TX/RX: 50x 2NAND TSV: 50μm x 50μm (area below/above TSV cannot be utilized for circuits)	TX/RX: 36x 2NAND Coil: 1μm x 250μm x 12 (area below/above TCI can be utilized for circuits)
# of channel	>250 ch	<50 ch
Data rate	<1Gb/s/ch	>5Gb/s/ch
Delay	40x 2NAND FO4	7x 2NAND FO4
Energy	40x 2NAND + 0.5pJ/b	80x 2NAND
Manufacturing	Additional Steps Req.	Standard CMOS Process
Additional Cost	>40%	<0.1%
Supply Chain	OSAT involved	Conv. business model

Fig. 5 TCI is much cheaper than TSV, while it bears comparison in performance. The conventional supply chain can be applied.

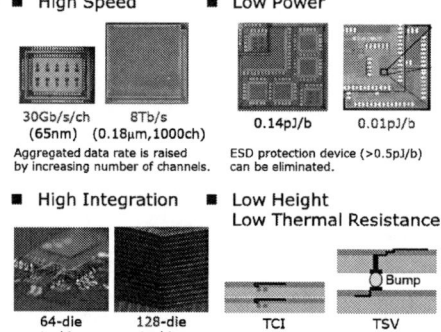

Fig. 6 TCI performs high speed (30 Gbps/ch), low energy (0.01 pJ/b), high integration (128-die stacking), low height (bump less).

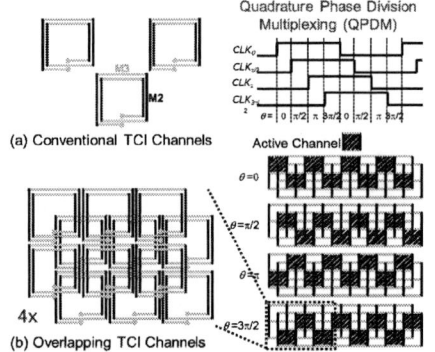

Fig. 7 Overlapping coils with phase division multiple access for improving area efficiency.

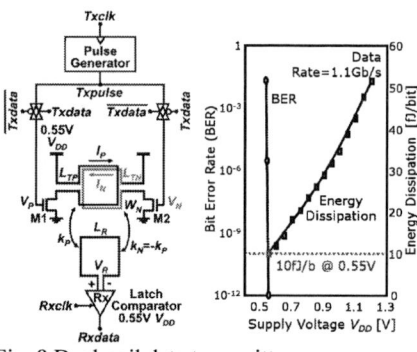

Fig. 8 Dual-coil data transmitter consumes world lowest energy dissipation (0.01 pJ/bit).

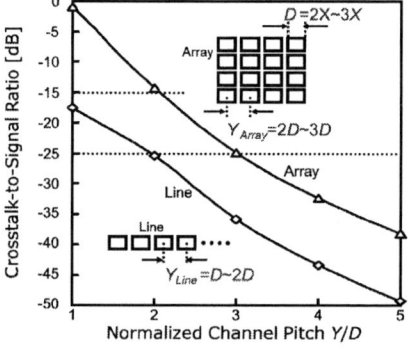

Fig. 9 Channel pitch vs. crosstalk. Crosstalk is suppressed at pitch of D~2D.

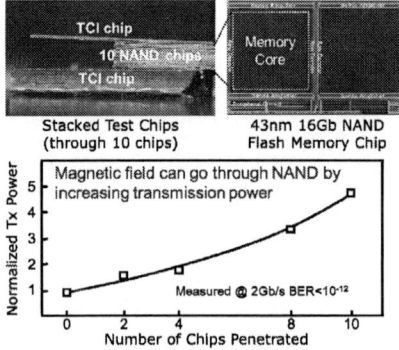

Fig. 10 Coupling degradation by eddy current in NAND chip stacking.

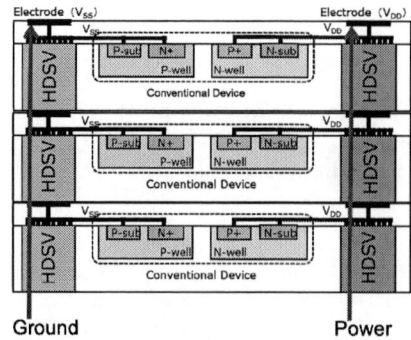

Fig. 11 Highly Doped Silicon Via (HDSV) by ion implant for power distribution.

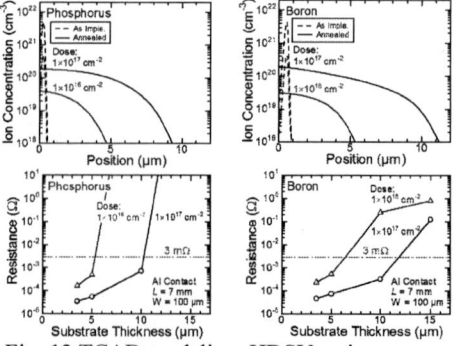

Fig. 12 TCAD modeling: HDSV resistance can be <3 mΩ when substrate is below 5 μm.

978-1-4799-8002-4/14 $31.00 © 2014 IEEE

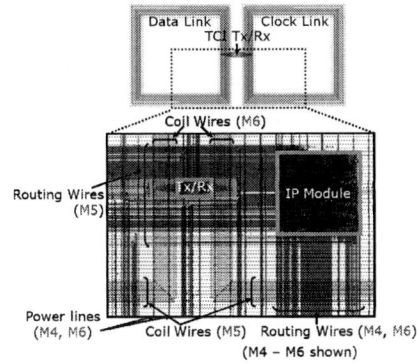

Fig. 13 Compatible with conventional EDA.

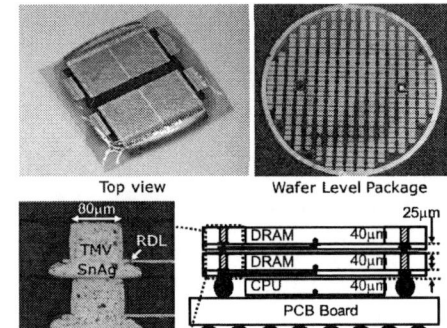

Fig. 14 Ultra-Thin Fan-Out Wafer Level Package for PoP.

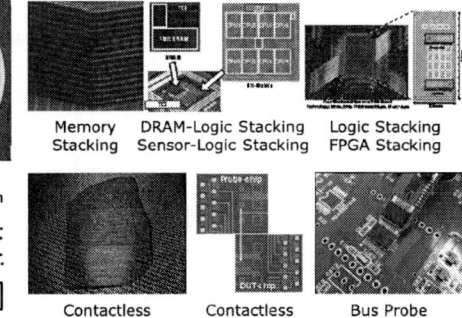

Fig. 15 Applications include homogeneous and heterogeneous chip stacking.

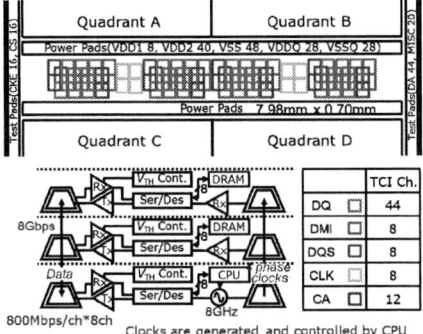

Fig. 16 352Gbps DRAM/SoC interface using overlapped coils and QPDM.

		TCI	TSV	LPDDR4
IO configuration		x44	x256	x64
Data rate		8 Gbps	1.1 Gbps	4.3 Gbps
Chip cost	Area SoC	-3%	+5%	0%
	DRAM	-0.6%	+25%	0%
	Process	0%	+25%	0%
	Total	-2%	+38%	0%
Power	IO	193 mW	212 mW	965 mW
Latency	IO	3 ns	2 ns	20 ns
	Total	43 ns	42 ns	60 ns

Assumption: 1) SoC die size = 8mm x 8.3mm, DRAM die size 7.2mm x 7.2mm. 2) LPDDR4 in SoC = 2.17mm². 3) SoC cost = $15, DRAM cost = $3

Fig. 17 TCI vs. TSV(WIO2) vs. LPDDR4.

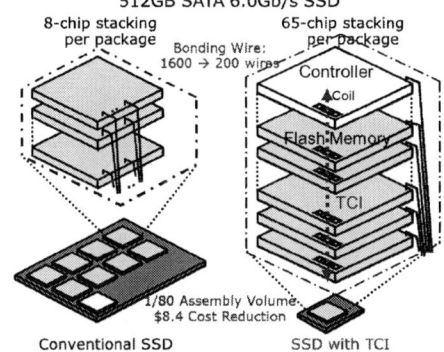

Fig. 18 One-Package SSD.

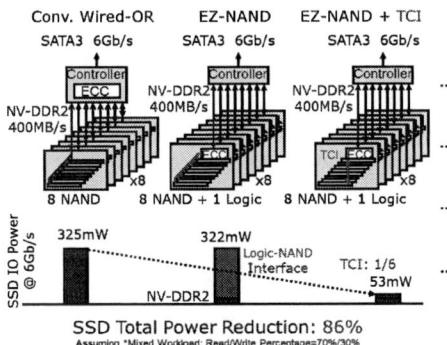

Fig. 19 SSD IO power reduction to 1/6.

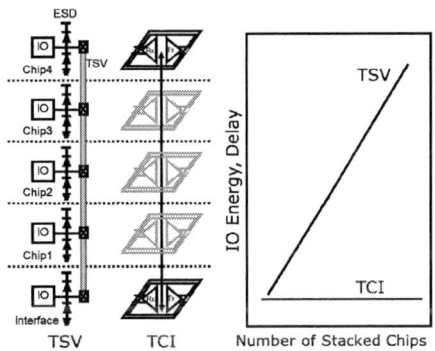

Fig. 20 Low energy broadcasting.

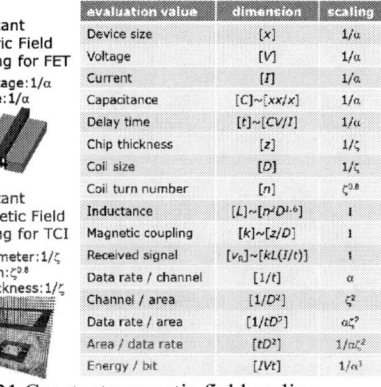

evaluation value	dimension	scaling
Device size	$[x]$	$1/\alpha$
Voltage	$[V]$	$1/\alpha$
Current	$[I]$	$1/\alpha$
Capacitance	$[C] \sim [xx/x]$	$1/\alpha$
Delay time	$[t] \sim [CV/I]$	$1/\alpha$
Chip thickness	$[z]$	$1/\zeta$
Coil size	$[D]$	$1/\zeta$
Coil turn number	$[n]$	$\zeta^{0.8}$
Inductance	$[L] \sim [n^2 D^{1.6}]$	1
Magnetic coupling	$[k] \sim [z/D]$	1
Received signal	$[v_R] \sim [kL(I/t)]$	1
Data rate / channel	$[1/t]$	α
Channel / area	$[1/D^2]$	ζ^2
Data rate / area	$[1/tD^2]$	$\alpha\zeta^2$
Area / data rate	$[tD^2]$	$1/\alpha\zeta^2$
Energy / bit	$[IVt]$	$1/\alpha^3$

Fig. 21 Constant magnetic field scaling. Area efficiency is improved by $\alpha\zeta^2$, and energy efficiency is improved by α^3.

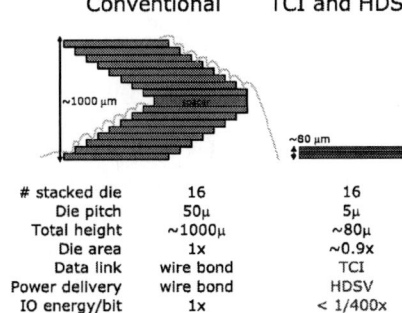

# stacked die	16	16
Die pitch	50μ	5μ
Total height	~1000μ	~80μ
Die area	1x	~0.9x
Data link	wire bond	TCI
Power delivery	wire bond	HDSV
IO energy/bit	1x	< 1/400x

Fig. 22 NAND stacking with TCI & HDSV.

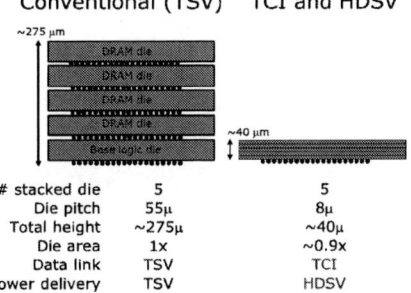

# stacked die	5	5
Die pitch	55μ	8μ
Total height	~275μ	~40μ
Die area	1x	~0.9x
Data link	TSV	TCI
Power delivery	TSV	HDSV
IO energy/bit	1x	< 1/10x

Fig. 23 DRAM stacking with TCI & HDSV.

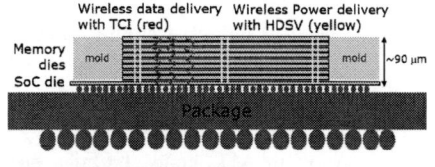

Panel-level stacking as batch (wafer scale) process
1) Known Good memory die placed face down on a support panel by the pitch of customer's chip size, mold is poured to the gap to form a memory panel by a memory vendor.
2) The memory panels are provided to an SoC vendor.
3) Known Good SoC die placed face down on a support panel by pitch of the SoC size by the SoC vendor.
4) The SoC panel is then thinned from the back.
5) The memory panel is placed on top of the SoC panel, face down, bonded by room temperature pressure bonding machine.
6) The panel thinned from the back.
7) Repeat the process to build up memory multi-layer tower on the SoC panel.

Fig. 24 Ultra-thin lowest cost 3D packaging.

978-1-4799-8002-4/14 $31.00 © 2014 IEEE

Co/Ni based p-MTJ stack for sub-20nm high density stand alone and high performance embedded memory application

G. S. Kar, W. Kim, T. Tahmasebi, J. Swerts, S. Mertens, N. Heylen, T. Min
imec, Kapeldreef 75, 3001 Leuven, Belgium, Phone:+32-16-28-8531 Fax:+32-16-28-1706
E-mail: kar@imec.be

Abstract

Excellent tunnel magneto resistance (TMR) values of 143% at resistance-area products (RA) of 4.7 $\Omega\mu m^2$ from 11nm thin Co/Ni based perpendicular magnetic tunnel junctions (p-MTJ) was achieved. Engineered wetting layer (WL), seed layer (SL) and the introduction of newly designed inner synthetic anti-ferromagnetic (iSAF) pinned layer in combination with ultra-smooth bottom electrode (roughness 0.5 Å) was yielded to vertically scaled 11nm thick Co/Ni p-MTJ stack with excellent magnetic properties. The introduction of iSAF layer demonstrates for the 1st time the free layer offset field controllability (< 100 Oe) of the spin-transfer-torque (STT) magnetic random access memory (MRAM) device down to 12 nm in diameter.

Introduction

Scalability of interface driven perpendicular magnetic anisotropy magnetic tunnel junctions (MTJs) has been improved down to 2X node[1-3], which verifies STT-MRAM for future high density standalone and high performance embedded memory applications[3-6]. Recent experimental result has demonstrated sufficiently low switching current densities with reasonable thermal stability factor, which can satisfy the requirements for technology node down to 2X [1,3,4,6] . To extend the scalability of STT-MRAM down to 1X node and beyond is still challenging. To maintain the high performance with scaling down, p-MTJ should keep high thermal stability (Δ) greater than 70 regardless of shrink of its volume. Moreover, the switching current density (Jc) should also be kept low simultaneously, which are in a trade-off relationship. It is also challenging to achieve a high tunneling magneto resistance (TMR) ratio at a low resistance area product (RA) product at the same time with p-MTJs. Vertical scaling of the stack brings advantages not only on the throughput of the MTJ stack, but also on the patterning of the MTJ stack which has become the major challenge for STT-MRAM commercialization. Moreover, thinner MTJ stack also allows to control free layer offset field more efficiently.

We demonstrate the thinnest p-MTJ stack with Co/Ni multilayer ever reported. It has thickness 10.6nm from the bottom electrode to the cap for the bottom pinned structure. It shows TMR value up to 143% at RA 4.7 $\Omega\mu m^2$. This stack has a novel iSAF pinned layer design that allows to reduce the free layer (FL) offset field down to 100 Oe for the STT-MRAM cells down to 10-nm diameter.

Experiment and result

Fig. 1 presents the detailed schematic and TEM of our Co/Ni based bottom pinned p-MTJ stack. Proper material selection and design of WL and SL to the bottom electrode is very important in order to maintain the well-developed (111) texture in Co/Ni super lattice, because strong PMA in Co/Ni supper lattice originates from the strong fcc (111) texture. M-H loops in figure 2 shows high Hk of around 12kOe and Hc about 3kOe were obtained from Co/Ni multilayer as shown in schematic cartoon. However, the super-smooth surface of the bottom electrode (BE) with 0.5Å roughness (polished TaN only) comes to have 2.7 Å roughness after those 6nm Hf and 2nm Ta (act as WL) together with 10nm NiCr alloy based seed layer. Furthermore, the roughness of the film surface increases up to 4.07 Å when Co/Ni based pinned layer is deposited on top of it (Fig. 3). Very high roughness at the CoFeB/MgO/CoFeB junction interface can be observed in the HRTEM image of the one on top of Co/Ni multilayer comparing to the other one on top of smooth BE, as shown in the lower panel of figure 3. Poor quality of MgO interface due to the high roughness causes the degradation of TMR at high temperature (>300 °C), as shown in figure 4. The, scaling down of WL and SL thickness will decrease the roughness and improve the thermal budget, as well as will result in the reduction of entire stack thickness. Fig. 5 shows the TMR and RA vs. annealing temperature for different SL thicknesses, as published [7]. A clear degradation of TMR is observed when the SL thickness reduced below 5nm for all annealing temperatures. Decrease in Hk was also observed when SL is smaller than 5 nm at all annealing temperatures (Fig. 6). Thus, it is required to develop a new design of SL and WL. *The thickness of the new design of SL was reduced down to 2nm keeping high Hk as shown in figure 7 of in-plane MH loops of the Co/Ni super-lattice. The CoFeB/MgO interface was also improved as shown in the HRTEM image (Fig. 8), which is accompanied by only slight degradation of TMR value at high (400 °C) annealing temperature (Fig. 8). The high* resolution X-ray diffraction (HRXRD) pattern shows the strong fcc (111) texture of the Co/Ni multilayer is still well-established with 2nm of thin seed layer (Fig 9). WL was scaled down to only a single 1nm Hf layer without compromising with the magnetic properties, where Hk is even improved from 8kOe (6nm Hf+2nm Ta) to 10kOe for 1nm Hf seed layer (Fig. 10). Finally, we observed the benefit in the MTJ annealing thermal budget, when we reduced the WL+SL thickness down to 3nm. A minor drop in TMR was observed at temperature 400 ℃, while the RA remains almost the same for the entire annealing temperature range (Fig. 11). This improvement is in agreement with the very sharp CoFeB/MgO interface (HRTEM, Fig. 11) and improved anisotropy field (Hk ~ 10 kOe). Strong PMA of the hard layer comes from the combined contribution of the improved WL, SL and Co/Ni super lattice in the MTJ stack. To confirm the strength of PMA of Co/Ni multilayer, the effect of the repetition number of bilayers on the PMA was studied. Figure 12, shows PMA correlation with the number of bilayers. TMR and RA vs. different number of Co/Ni bi-layers is shown in the figure 13. A clear degradation in TMR was recorded when number of bilayers were reduced below 6. Improvement in TMR was observed, while the total stack thickness was reduced (Fig. 14). This improvement is attributed to the improvement in the roughness at CoFeB/MgO interface, leading the improvement in the naturally grown MgO oxide quality.

We further developed our p-MTJ with Co/Ni HL into the inner SAF (iSAF) structure to compensate the offset field(Hoff) of FL, especially at sub-20nm technology nodes, keeping the advantage of scaling of the total stack thickness. Ru layer is introduced between Co/Ni multilayer and CoFeB reference layer (RL) in order to have their magnetizations coupled anti-ferromagnetically. Switching of RL and HL is clearly distinguished in the perpendicular M-H curve (Figure 15), comparing to that of the MTJ without SAF where only 2 steps were observed (Fig. 16). The iSAF structure is essential for MTJ to become a 1X node memory device considering the control of Hoff. As previously reported by IBM, iPMA between MgO and CoFeB layers on top of Ru layer cannot provide spontaneous p-magnetization in CoFeB layer [8]. However, our stack can provide adequate spin polarization (evidently by high TMR value) from the thin CoFeB reference layer while sufficiently strong anti-ferromagnetically coupled to the Co/Ni super-lattice to form an iSAF structure without losing any pinning strength. Micro-magnetic modeling on Hoff shows it can be controlled less than 100 Oe by minor tweaking of the repetition of Co/Ni multilayer in our iSAF MTJ with Co/Ni HL (Fig. 17) for MTJ dia. down to 10nm. However, in conventional SAF (cSAF) MTJs, Hoff is very hard to be controlled by changing thickness of the bottom HL, as shown by Fig. 18, due to the fact that the outer pin layer is less effective to compensate the inner pinned layer's offset field at very small MTJ diameter (<20nm). As an example, in MTJ with HLs of

Co/Pt multilayer, micro-magnetic modeling shows Hoff cannot be controlled under 100 Oe as MTJ shrinks below 40nm even if the Co/Pt repetition is increased up to 16 (Fig. 18). The stack parameters used in this micro-magnetic modelling is shown in figure 19.

Increase in PMA of Co/Ni super-lattice was also accompanied by the introduction of Ru layer, which enables further scaling in repetition of Co/Ni bi-layer even without sacrificing in TMR yielding (Fig. 20). Further optimization in MgO process and cap layer material optimization were required to achieve high TMR. The Rf-MgO still seems to significantly outperform the MgO by natural oxidation in our current process. Indeed, higher MR/RA and higher Ku.t (0.5 erg/cm^2) values for a similar CoFeB thickness are obtained . Introducing the improved RF-MgO process and new cap layer in the scaled Co/Ni iSAF stack, MR values as high as 143% at RA 4.7 has been obtained. The High resolution TEM of the ultra-thin stack is presented in figure 21. Smooth interface and high crystallinity was obtained for the entire stack.

Summary

Substantial reduction of the roughness and the thickness of wetting layer (WL) and seed layer (SL) are the key to the thin p-MTJ with high performance, since they strongly affect the quality of the interface between fixed layer and MgO barrier layer. Co/Ni based p-MTJ stack was scaled down to 10.6nm successfully without compromising with the magnetic properties of the stack. Excellent TMR of 143% at RA of 4.7 □ μm2 was achieved from this thin Co/Ni based p-MTJ stack The novel iSAF pinned layer design allows to reduce the free layer (FL) offset field down to 100 Oe for the STT-MRAM cells down to 10-nm diameter.

Acknowledgement

This work is supported by imec's Industrial Affiliation Program on STT-MRAM device.

References

1) H. Sato, et al., IEDM Tech. Digest. 3.2.1 (2013)
2) J. Kim, et al, Symposium on VLSI Technology Digest 76 (2014)
3) G. Jan, et. al., Appl. Phys. Express, 2012, 5, 093008.
4) H. Sato, M. Yamanouchi, S. Ikeda, et. al., Appl. Phys. Lett., 2012, 101, 022414.
5) S. Kang; Symposium on VLSI Technology Digest 44 (2014)
6) T. Min, et. al., IEEE Trans. On Magnetics, 2010, 46, 6, 2322.
7) G. Jan, et. al., US Patent, 2013, US8,508,006,B2.
8) D. C. Worledge, et. al., Appl. Phys. Lett., 2011, 98, 022501

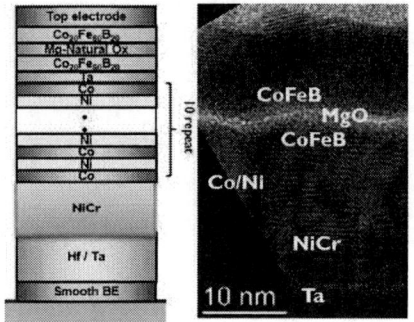

Figure 1. Figure 1: Schematic diagram of base line Co/Ni based bottom pinned p-MTJ stack and the HRTEM of that stack.

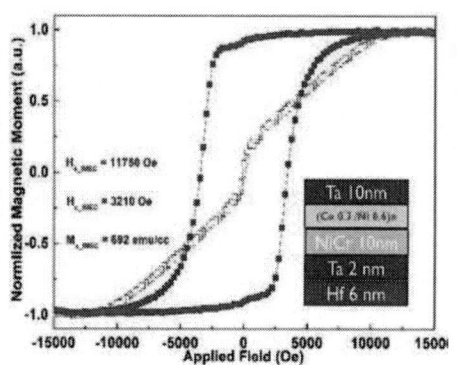

Figure 2. In-plane and perpendicular M-H loops of Co/Ni pinned layer deposited on Hf/Ta wetting layer and NiCr seed layer. Schematic of the stack is shown in the inset of the figure.

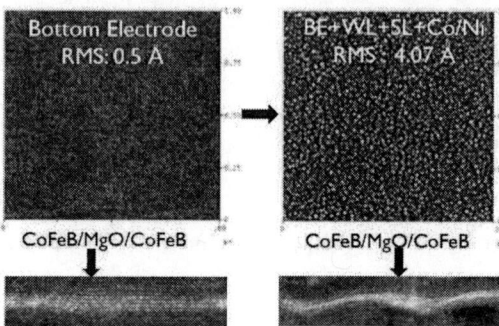

Figure 3. AFM images from super smooth (SS) BE and WL+SL+Co/Ni (10 repeat) deposited on the SS BE are in the upper panel. HRTEM images of CoFeB/MgO/CoFeB deposited on those respective surfaces are in the bottom panel.

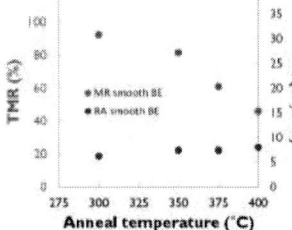

Figure 4. TMR and RA vs. Annealing Temperature of the base line Co/Ni based bottom pinned p-MTJ stack. TMR decreased at higher temp. (> 300 °C).

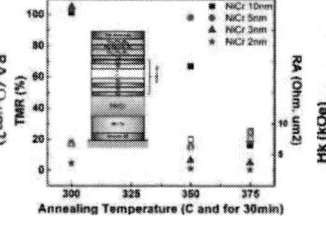

Figure 5. TMR and RA vs. annealing temperature for different NiCr seed layer thicknesses. Solid and open symbols are for TMR and RA, respectively.

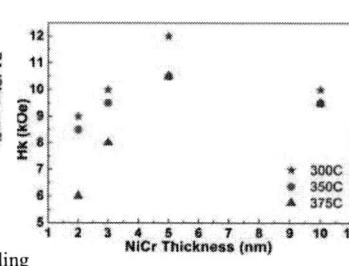

Figure 6. Hk Vs NiCr seed layer thickness at different annealing temperature.

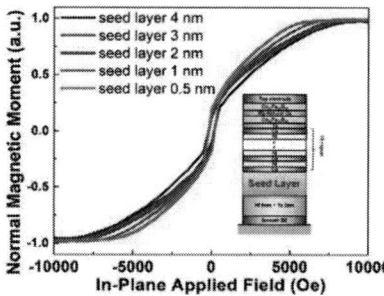

Figure 7. Normalized in-plan M-H loops for engineered seed layer with different thicknesses, thickness scaled down to 1 nm without sacrificing magnetic properties.

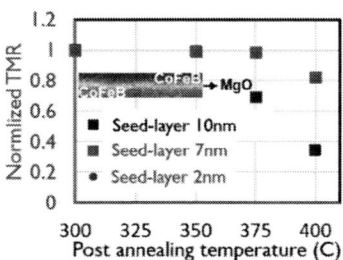

Figure 8. Normalized TMR vs. Annealing Temperature of the Co/Ni p-MTJ stack with different SL thicknesses. HRTEM of CoFeB/MgO/CoFeB is at the inset.

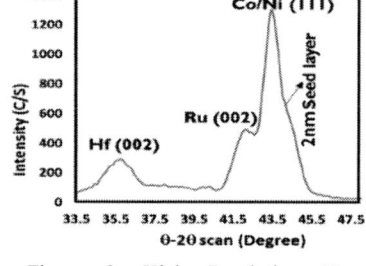

Figure 9. High Resolution X-ray diffraction pattern of Co/Ni multilayers deposited on Hf 6nm + Ta 2nm wetting layer followed by 2nm engineered seed layer .

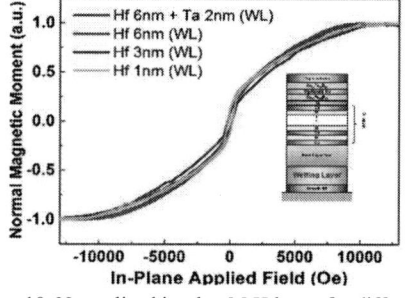

Figure 10. Normalized in-plan M-H loops for different wetting layers, thickness scaled down to 1 nm keeping the seed layer thickness constant at 2nm.

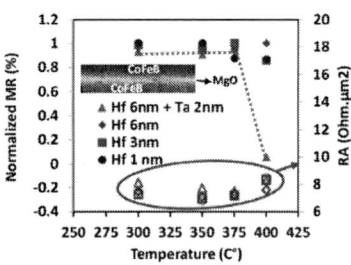

Figure 11. Normalized TMR & RA vs. Annealing Temperature of the Co/Ni p-MTJ stack with different WL thicknesses. HRTEM of CoFeB/MgO/CoFeB is at the inset. Solid and open symbols are for TMR and RA, respectively.

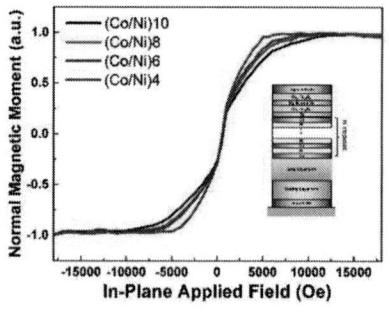

Figure 12. Normalized in-plan M-H loops for different number Co/Ni bi-layer repetition deposited on 3nm thick WL+SL.

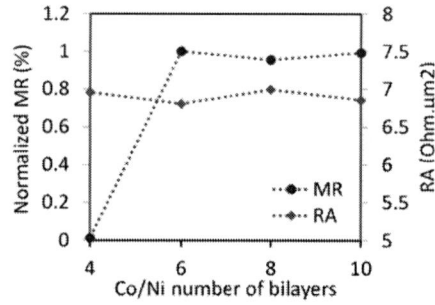

Figure 13. Normalized TMR & RA vs. Annealing Temperature of the Co/Ni p-MTJ stack with different Co/Ni by layer repetition.

978-1-4799-8002-4/14 $31.00 © 2014 IEEE

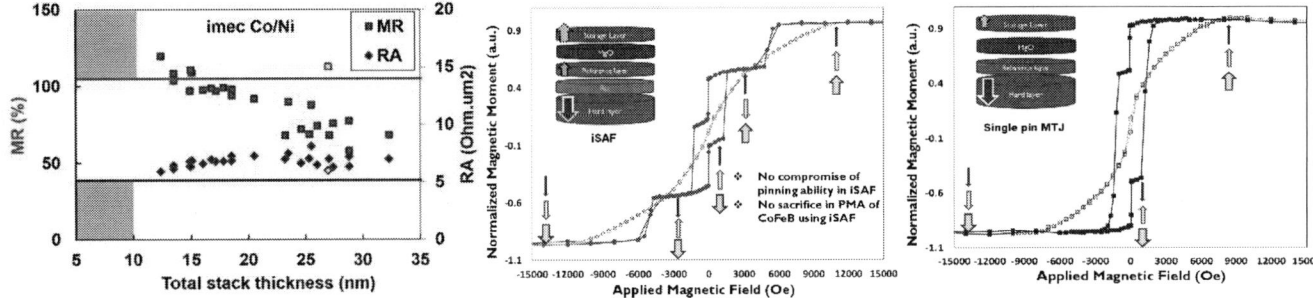

Figure 14. TMR and RA vs. total Co/Ni based bottom pinned p-MTJ stack thickness.

Figure 15. In-plane and perpendicular M-H loops of Co/Ni p-MTJ stack with *iSAF pinned* layer.

Figure 16. In-plane and perpendicular M-H loops of Co/Ni p-MTJ stack without cSAF pinned layer.

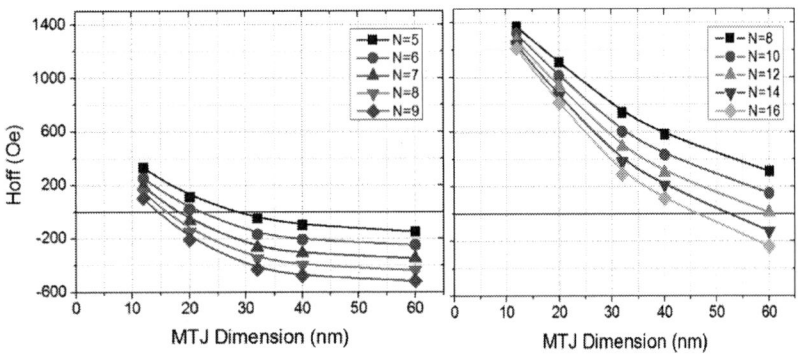

Figure 17. Calculated Offset field (Hoff) vs. MTJ dimension using micro-magnetic simulation from *iSAF* MTJ with Co/Ni HL with different repetition of Co/Ni multilayers.

Figure 18. Calculated Offset field (Hoff) vs. MTJ dimension using micro-magnetic simulation from cSAF MTJ with Co/Pt HL with different repetition of Co/Pt multilayers.

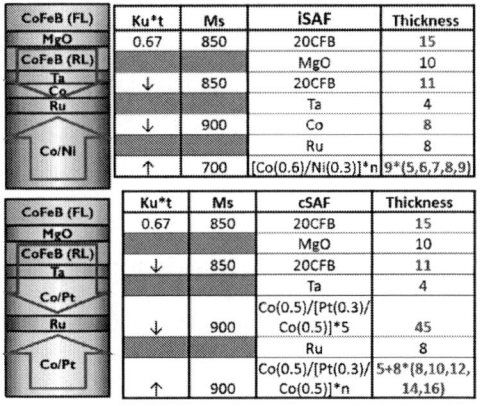

CoFeB (FL)	Ku*t	Ms	iSAF	Thickness
MgO	0.67	850	20CFB	15
CoFeB (RL)			MgO	10
Ta	↓	850	20CFB	11
Co			Ta	4
Ru	↓	900	Co	8
			Ru	8
Co/Ni	↑	700	[Co(0.6)/Ni(0.3)]*n	9*{5,6,7,8,9}

CoFeB (FL)	Ku*t	Ms	cSAF	Thickness
MgO	0.67	850	20CFB	15
CoFeB (RL)			MgO	10
Ta	↓	850	20CFB	11
Co/Pt			Ta	4
Ru	↓	900	Co(0.5)/[Pt(0.3)/Co(0.5)]*5	45
			Ru	8
Co/Pt	↑	900	Co(0.5)/[Pt(0.3)/Co(0.5)]*n	5+8*{8,10,12,14,16}

Figure 19. Schematic diagram explaining *iSAF* and cSAF in p-MTJ stack. Stack parameters used in the micro-magnetic modeling.

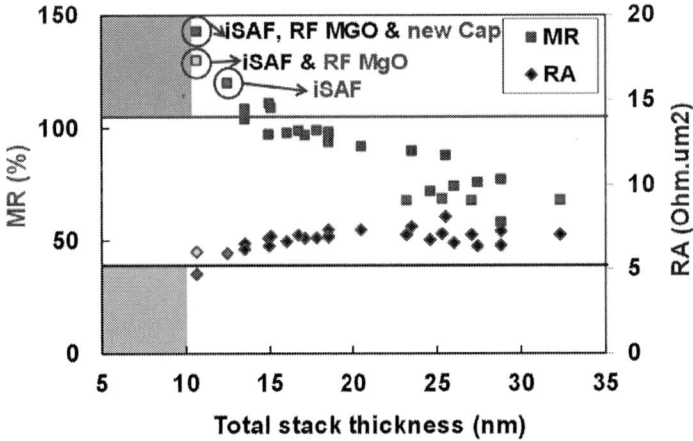

Figure 20. TMR and RA vs. total Co/Ni based bottom pinned p-MTJ stack thickness. Highest TMR obtained from thinnest Co/Ni p-MTJ stack

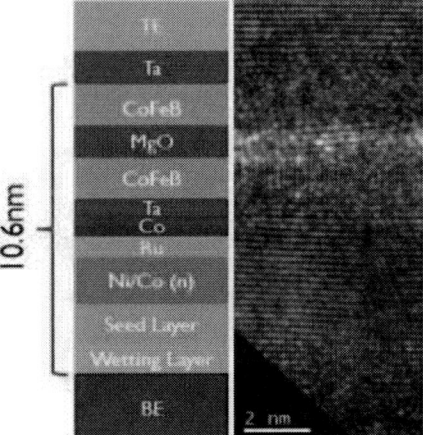

Figure 21. Schematic description of thinnest Co/Ni p-MTJ stack HRTEM of that shows strongly textured Co/Ni multilayer, high quality MgO and very sharp interface

Challenging Issues for Terra-bit-level Perpendicular STT-MRAM

J. G. Park[1], T. H. Shim[1], K. S. Chae[1,2], D. Y. Lee[1], Y. Takemura[1], S. E. Lee[1], M. S. Jeon[1], J. U. Baek[1], S. O. Park[2], and J. P. Hong[1]

[1]MRAM Center, Department of Electronics Engineering, Hanyang University, Seoul 133-791, Korea
[2]Samsung Electronics Co., Ltd., San #16 Banwol-dong, Hwasung, Gyeonggi-Do 445-701, Korea
Email: parkjgL@hanyang.ac.kr

Abstract

The current challenging issues for terra-bit-level perpendicular STT-MRAM cells have been reviewed in the view of four critical parameters such as TMR ratio, Δ, J_{ex}, and α. The TMR ratio of p-MTJ spin-valves are reaching to < 150% at the BEOL of >350°C. A single MgO based p- MTJ spin-valve could not satisfy Δ of > 74, proposing a double MgO based p-MTJ spin-valve. J_{ex} in SyAF layer adequately met > 0.7erg/cm^2 at BEOL of > 350°C. A $Co_2Fe_6B_2$ based p-MTJ spin-valve limits to α of 0.005, necessary to develop a low α material such as full Heusler half-metal. Thus, an essential challenge in the future is to satisfy four critical parameters simultaneously at > 350°C and 300-mm TiN electrode wafers.

Introduction

DRAM cells have been developed by innovatively modifying the structure of a selective transistor (n-MOSFET) and a capacitor as the design rule has been scaled-down; i.e., RCAT→SRCAT→S-FIN→VCAT for a selective transistor and a cylindrical structure with HfO_2 or Al_2O_3→ a pedestal structure with ZrO_2, $ZrO_2/Al_2O_3/ZrO_2$, or TiO_2/RuO_2 for a capacitor [1-9]. However, a capacitor of DRAM-cell should meet a physical limit as the design rule becomes less than 20 nm since a pedestal structural capacitors would collapse one another [10]. An alternative to overcome a physical limit of current DRAM cells would be p-STT MRAM because of its simple bi-stable magnetic resistance structure and a reasonable write/erase speed (several tens of ns). Dislike DRAM cells, p-STT-MRAM cells have nonvolatile memory characteristic, resulting in a lower power consumption than DRAM cells. For realizing terra-bit-level p-STT MRAM cells, perpendicular-magnetic tunnel junction (p-MTJ) spin-valves challenge to satisfy critical parameters such as a high tunneling magneto-resistance (TMR) ratio of >150%, thermal stability (Δ) of > 74 at 85°C, a low critical current density (J_C) of 13.4 MA/cm^2 [11], and anti-ferromagnetic-coupling-strength (J_{ex}) for a synthetic anti-ferromagnetic (SyAF) layer of > 0.7erg/cm^2 [12]. In particular, four critical parameters for p-MTJ spin-valves should be simultaneously performed at the back-end-line (BEOL) of > 350°C and 300-mm TiN electrode wafers.

TMR ratio at BEOL of > 350°C

In general, in p-STT-MRAM cells, p-MTJ spin-valves have been developed with the vertically stacking of amorphous CoFeB free layer/MgO tunneling barrier/CoFeB pinned layer with a $[Co/Pd]_n$-SyAF (synthetic anti-ferro-magnetic) layer[13]. The TMR of >150% in p-MTJ spin-valves should be achieved above the BEOL of >350°C, which is the first challenging issue. In particular, the TMR ratio would be practically much higher than 150% since the TMR ratio would decrease ~30 % after the p-MTJ spin-valve etching. To review what parameters determine mainly the TMR Ratio, the crystalline characteristic, chemical composition profile, and nanoscale free or pinned layer-thickness dependency, and annealing temperature dependency were investigated for Ta seed based $Co_2Fe_6B_2$/MgO p-MTJ spin-valves stacked with a $[Co/Pd]_n$-SyAF layer on 300-mm TiN electrode wafers, followed by an *ex-situ* annealing at 275°C under 3 tesla, as shown in Fig.1. The material structures of the bottom TiN electrode, Ta seed, $Co_2Fe_6B_2$ free layer, MgO tunneling barrier, $Co_2Fe_6B_2$ pinned layer, Ta capping layer, and $[Co/Pd]_n$-SyAF layer were NaCl crystalline, amorphous, amorphous, b.c.t. crystalline, amorphous, amorphous, and f.c.c. crystalline structure, as shown in Figs. 1(a) and (b). In particular, the b.c.t. crystalline MgO tunneling barrier with the lattice constant of 4.21Å was interfaced with both amorphous $Co_2Fe_6B_2$ free layer and pinned layers, as shown in Fig. 1(b). Thus, the interface perpendicular magnetic anisotropy (i-PMA) could be built up since it was originated from the orbital hybridization of O_{2p}-Fe_{3d} and O_{2p}-Co_{3d}, at the interfaces, as shown in Fig. 1(c). In addition, the TMR ratio is greatly determined by the achievement of the b.c.t. crystallinity of the MgO tunneling barrier, which strongly depends on the thickness of the $Co_2Fe_6B_2$ free layer (Fig. 1(d)), MgO tunneling barrier (Fig. 1(e)), and $Co_2Fe_6B_2$ pinned layers (Fig. 1(f)). Although the Ta seed based $Co_2Fe_6B_2$/MgO p-MTJ spin-valves stacked with a $[Co/Pd]_n$-SyAF layer was a surprising invention [13], they could not satisfy the requirement of the BEOL of > 350°C since the TMR ratio rapidly decreased when the *ex-situ* annealing temperature increased from 275 to 325°C, as shown in Fig. 1(e).

The decrease of the TMR ratio was associated with the crystallinity degradation of the b.c.c. MgO tunneling barrier caused by the Ta diffusion from a Ta seed into a MgO tunneling barrier (Fig. 2(a)) and by the Pd diffusion from a$[Co/Pd]_n$-SyAF layer into the MgO tunneling barrier (Fig. 2(b)). As a result, i-PMA characteristics of both $Co_2Fe_6B_2$ free (Figs. 2(c)-(e)) and pinned layer (Figs. 2(f)-(h)) degraded when the *ex-situ* annealing temperature (T_{ex}) increased from 275 to 350°C. In particular, the f.c.c. crystalline texturing from the $[Co/Pd]_n$-SyAF layer into the MgO tunneling barrier through the $Co_2Fe_6B_2$ pinned layer became much severe when

978-1-4799-8002-4/14 $31.00 © 2014 IEEE

T_{ex} increased from 275 to 350°C, which rapidly reduced the TMR ratio, as shown in Figs. 2(i)-(k). Up to now, the Ta seed based $Co_2Fe_6B_2$/MgO p-MTJ spin-valves stacked with a $[Co/Pd]_n$-SyAF layer could not achieve the TMR ratio of >150% at BEOL of > 350°C; i.e., 98.5 % at 240 °C [IBM]. Thus, a way to simultaneously avoid the diffusion of the Ta and Pd into the MgO tunneling barrier is essentially necessary.

Crystalline seed based $Co_2Fe_6B_2$/MgO p-MTJ spin-valves stacked with a $[Co/Pt]_n$-SyAF layer

In order to avoid the Ta diffusion into the MgO tunneling barrier from a Ta seed, a novel b.c.c. crystalline seed was introduced in the first. A b.c.c. seed layer with the lattice constant of 3.16Å could texture the $Co_2Fe_6B_2$ free layer from amorphous to b.c.c. crystalline structure after annealing at T_{ex}=400°C, as shown in Figs. 3(a), (b), and (c). As a result, the TMR ratio of ~137% could be achieved at T_{ex}=400°C due to no presence of the Ta diffusion into the MgO tunneling barrier and better Δ_1 coherent tunneling (Figs. 3(d), (e), and (f)). In addition, in order to avoid the Pd diffusion into the MgO tunneling barrier from a $[Co/Pd]_n$-SyAF layer, a $[Co/Pt]_n$-SyAF layer instead of a $[Co/Pd]_n$-SyAF layer for the b.c.c. seed based $Co_2Fe_6B_2$/MgO p-MTJ spin-valves was introduced, resulting in no degradation of the b.c.t. crystallinity of the MgO tunneling barrier due to no Pt diffusion from a $[Co/Pt]_n$-SyAF layer even at T_{ex}=400°C, as shown Figs. 4(a) with (b). Thus, The b.c.c. seed based $Co_2Fe_6B_2$/MgO p-MTJ spin-valves stacked with a $[Co/Pt]_n$-SyAF layer (Figs. 4(c) and (d)) could achieve the TMR ratio of ~160% satisfying the critical TMR ratio (150%) for terra-bit-level p-STT MRAM, as shown in Fig. 4(f), in the first.

Thermal stability (Δ) of > 74 at 85°C

Δ of > 74 at 85°C is essentially necessary to assure the data retention for 10 years, defined by $\Delta=K_uV/k_bT$. The Δ for the b.c.c. crystalline seed based $Co_2Fe_6B_2$/MgO p-MTJs at T_{ex}=400°C showed the Δ of ~24 for 20x20-nm in area, which was higher than that (Δ of ~19) for the Ta seed based $Co_2Fe_6B_2$/MgO PMA at T_{ex}=275°C. Note that Tohoku University [14] reported that Δ of ~ 43 was achieved at 40x40-nm in area and 300°C, which could be converted to Δ of ~ 11 at 20x20-nm in area. Thus, the achievement of Δ of > 74 at 85°C is another challenging issue. A solution to enhance Δ is an twice increase of the ferro-magnetic volume of the $Co_2Fe_6B_2$ free layer by using double MgO barrier(Fig. 6(a)), demonstrating twice magnetic moment of the $Co_2Fe_6B_2$ free layer (inset of Fig. 6(b)) and the TMR ratio of ~130% at 350°C (Fig. 6(b)). Thus, the double MgO based P-MTJ spin-valves could increase twice Δ, which could satisfy Δ of > 74. Note that Tohoku University [15] and AIST [16] presented the TMR of ~ 120% at 300°C and ~105 % at 350°C for a double MgO based p-MTJ spin-valves, respectively. However, the double MgO based p-MTJ spin-valves fundamentally reduces the TMR ratio compared to that of a single MgO based p-MTJ spin-valves since they consisted of two serial magnetic resistances. Remind that the TMR ratio for a single MgO

based p-MTJ spin-valve is defined by $(R_{AP}-R_P)/R_P$ while that for double MgO based p-MTJ spin-valve is defined by $\{(R_{AP}+R_B)-(R_P+R_B)\}/(R_P+R_B)$. Thus, in the case of a double MgO based p-MTJ spin-valve needs to achieve much higher TMR ratio, which is another challenging issue.

J_{ex} in SyAF layer of > 0.7erg/cm² at BEOL of > 350°C

A higher J_{ex} is essentially necessary to assure free cross-talks in p-MTJs-spin-valve arrays. The J_{ex} for the b.c.c. crystalline seed based $Co_2Fe_6B_2$/MgO p-MTJ spin-valve stacked with a $[Co/Pt]_n$-SyAF layer presented ~1.5erg/cm² at T_{ex}=400°C, satisfying a commercial memory-cell specification, as shown in Figs. 7(a) and (b). Note that there were no reports on J_{ex} in SyAF layer. In addition, J_{ex} in a $[Co/Pt]_n$-SyAF layer was higher than a $[Co/Pd]_n$-SyAF layer.

Critical current density (J_C) of 13.4 MA/cm²

J_{C0} is defined by $J_{C0}=\alpha\gamma eM_st(H_{ext}\pm H_{ani}\pm H_d/2)/u_Bg$, which could be obtained by a lower damping constant (α) such as full Heusler half-metal (Co_2FeAl). The damping constant for Co_2FeAl(0.001) was lower than that for $Co_2Fe_6B_2$(0.005) [17]. Thus, to develop the b.c.c. crystalline seed based Co_2FeAl or CoFeAlSi /MgO p-MTJ spin-valves stacked with a $[Co/Pt]_n$-SyAF layer would be another essential challenging engineering work [18-20].

Acknowledgement

This work was financially supported by the IT R&D program of the Ministry of Trade, Industry & Energy / Korean Evaluation Institute of Industrial Technology [No.10043398] and Basic Science Research Program through the National Research Foundation of Korea funded by the Ministry of Education [No.2014R1A2A1A01006474) & Brain Korea 21 PLUS Program in 2013.

References

[1] J.Y. Kim *et al.*, Symp. VLSI Tech. Dig., pp.11 (2003).
[2] J.Y. Kim *et al.*, Symp. VLSI Tech. Dig., pp.34 (2005).
[3] S.W. Chung *et al.*, Symp. VLSI Tech. Dig., pp.32 (2006).
[4] K.W. Song *et al.*, IEEE J. Solid-State Circuits, **45** 880 (2010).
[5] S.H. Oh *et al.*, Symp. VLSI Tech. Dig., pp.73 (2003).
[6] E. Gerritsen *et al.*, Solid State Elec., **49** 1767 (2005).
[7] D.S. Kil *et al.*, Symp. VLSI Tech. Dig., pp.38 (2006).
[8] A. Berthelot *et al.*, Proc. Solid-State Device Res. Conf., pp.343 (2006).
[9] S.K. Kim *et al.*, J. Mater. Res., **28** 313 (2012).
[10] D.H. Kim *et al.*, IEDM Tech. Dig., pp.69 (2004).
[11] K.C. Chun *et al.*, IEEE J. Solid-state circuits, **48**, 598 (2013).
[12] "SGMI Program". In http://www.samsung.com/mram (2013).
[13] G. Hu *et al.*, IEEE Mag. Lett. **4**, 3000104 (2013).
[14] S.Ikeda *et al.*, Nature Mater. **9**, 721 (2010).
[15] Hideo Sato *et al.*, IEEE Trans. Mag., **49**, 7, 4437 (2013).
[16] Kay Yakushiji *et al.*, Appl. Phys. Express **6**, 113006 (2013).
[17] S. Mizukami *et al.*, J. Appl. Phys. **105**, 07D306 (2009).
[18] Hiroaki Sukegawa *et al.*, Appl. Phys. Lett. **100**, 182403 (2012).
[19] Lichuan Jin *et al.*, IEEE Trans. Mag., **50**, 1, 1500104 (2014).
[20] Barman *et al.*, J. Appl. Phys. **101**, 09D102 (2007).

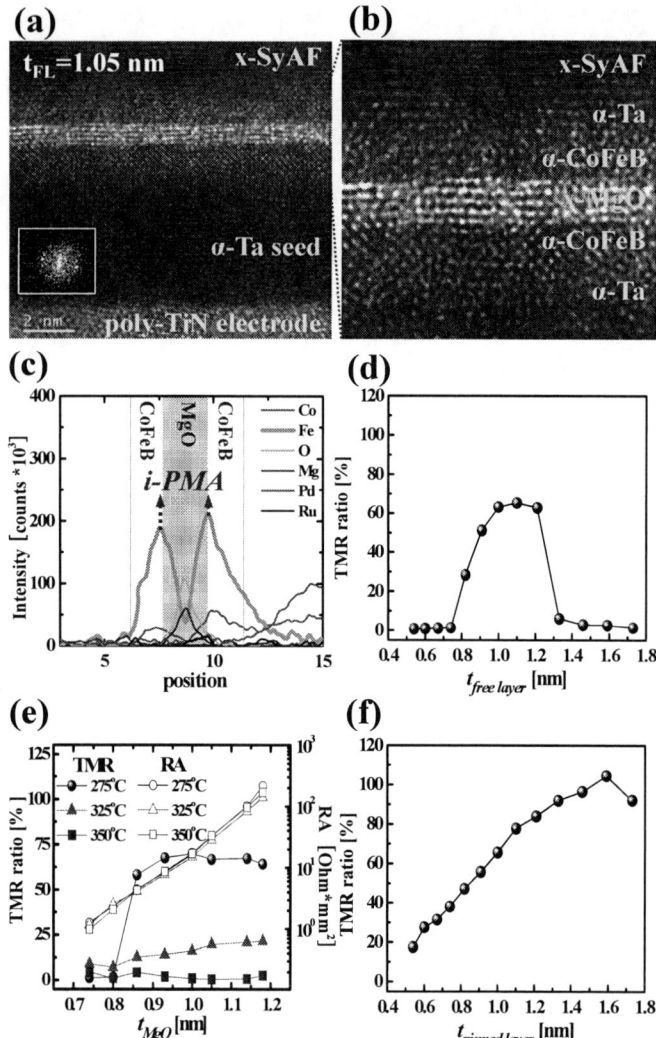

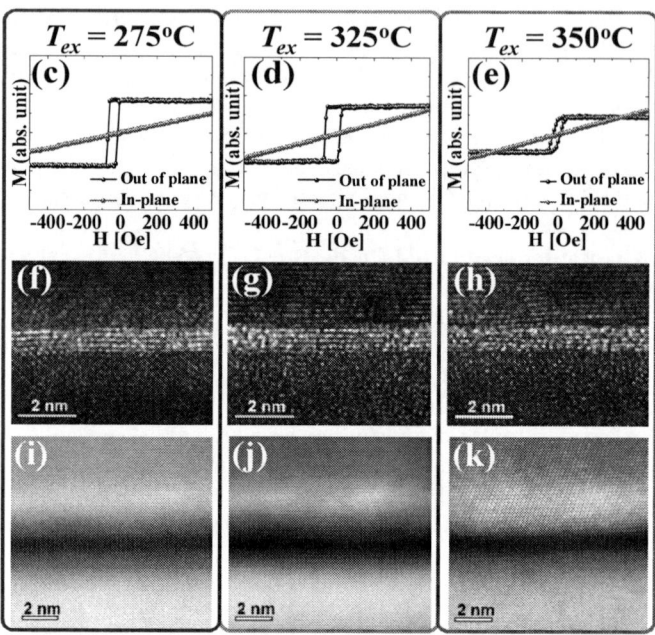

Fig. 2. Physical and chemical properties for Ta seed based $Co_2Fe_6B_2$/MgO p-MTJ spin-valves stacked with a $[Co/Pd]_n$-SyAF layer. (a) diffused Ta concentration profile by SIMS, (b) diffused Pd concentration profile, (c) M-H curve for the $Co_2Fe_6B_2$ free-layer at T_{ex}=275°C, (d) T_{ex}=325°C, (e) T_{ex}=350°C, (f) high-resolution x-TEM image at T_{ex}=275°C, (f) T_{ex}=325°C, (g) T_{ex}=350°C, high-resolution STEM image at T_{ex}=275°C, (h) T_{ex}=325°C, and (i) T_{ex}=350°C

Fig. 1. Ta seed based $Co_2Fe_6B_2$/MgO p-MTJ spin-valves stacked with a $[Co/Pd]_n$-SyAF layer. (a) x-TEM image of spin-valve, (b) high-resolution x-TEM image of p-MTJ, (c) chemical composition profile by EELS, (d) TMR ratio depending on the $Co_2Fe_6B_2$ free-layer thickness, (e) TMR ratio depending on the MgO tunneling-barrier thickness and T_{ex}, and (f) TMR ratio depending on the $Co_2Fe_6B_2$ pinned-layer thickness.

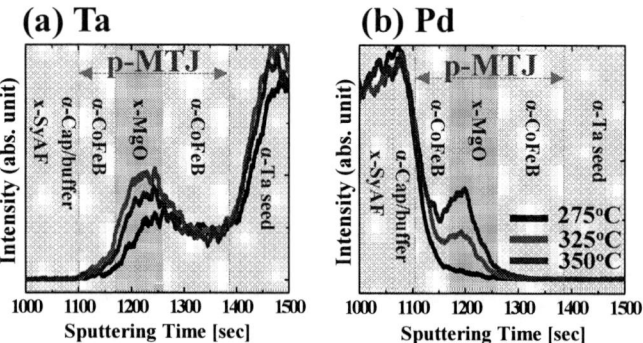

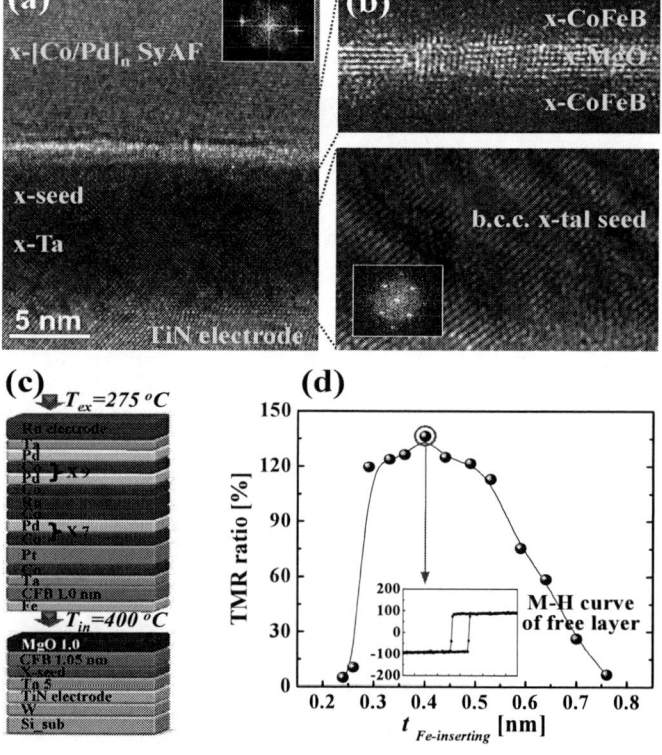

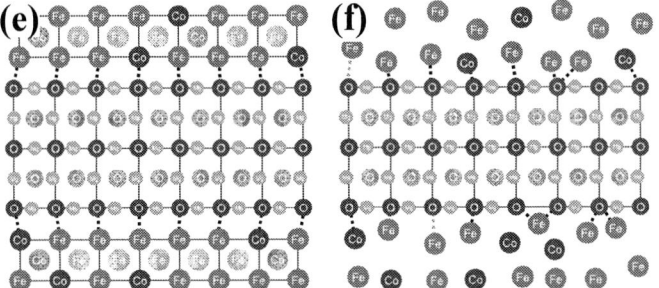

Fig. 3. b.c.c. seed based $Co_2Fe_6B_2$/MgO p-MTJ spin-valves stacked with a $[Co/Pd]_n$-SyAF layer. (a) x-TEM image of spin-valve, (b) high-resolution x-TEM image of p-MTJ, (c) schematic structure of spin-valve, (d) TMR ratio depending on the inserted Fe thickness, (e) atomic arrangement at the interface between b.c.c. seed based $Co_2Fe_6B_2$ and b.c.c. MgO tunneling barrier, and (f) at the interface between amorphous Ta seed based $Co_2Fe_6B_2$ and b.c.c. MgO tunneling barrier

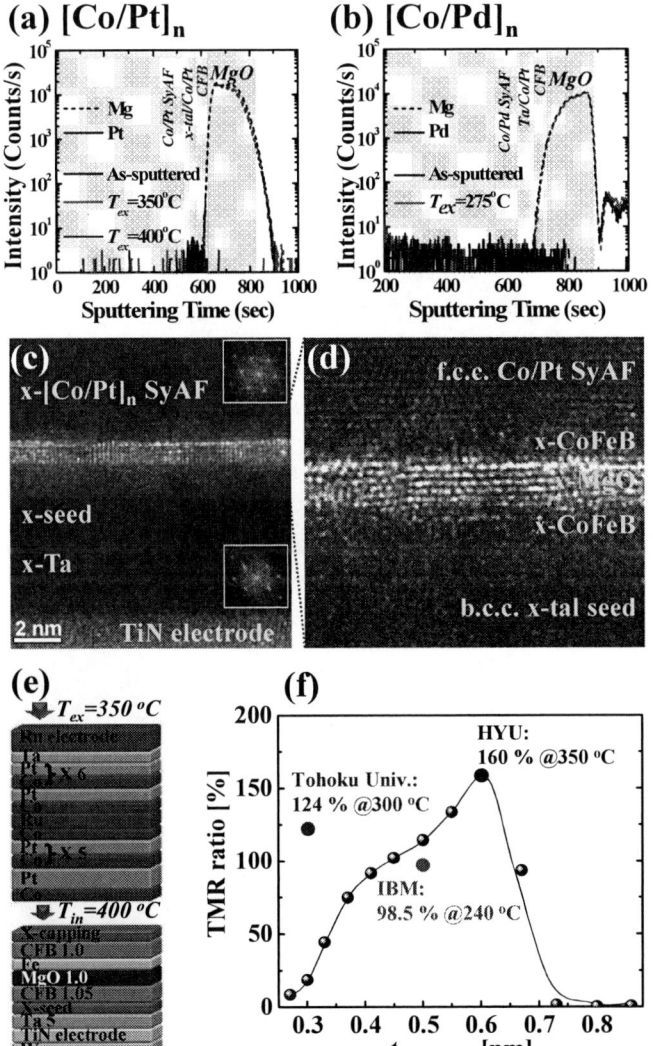

Fig. 4. b.c.c. seed based $Co_2Fe_6B_2$/MgO p-MTJ spin-valves stacked with a $[Co/Pt]_n$-SyAF layer. (a) diffused Ta concentration profile by SIMS, (b) diffused Pd concentration profile, (c) x-TEM image of spin-valve, (d) high-resolution x-TEM image of p-MTJ, and (f) TMR ratio depending on the x-capping layer thickness.

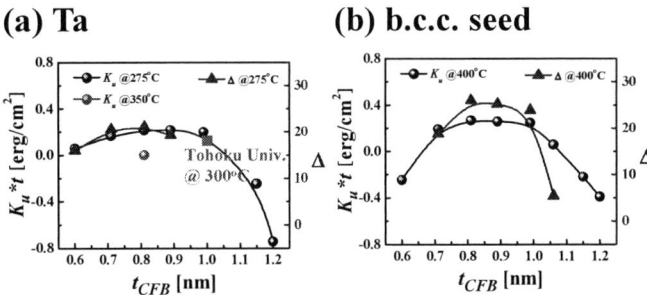

Fig. 5. Thermal stability (Δ). (a) Ta seed based $Co_2Fe_6B_2$/MgO p-MTJ and (b) b.c.c. seed based $Co_2Fe_6B_2$/MgO p-MTJ.

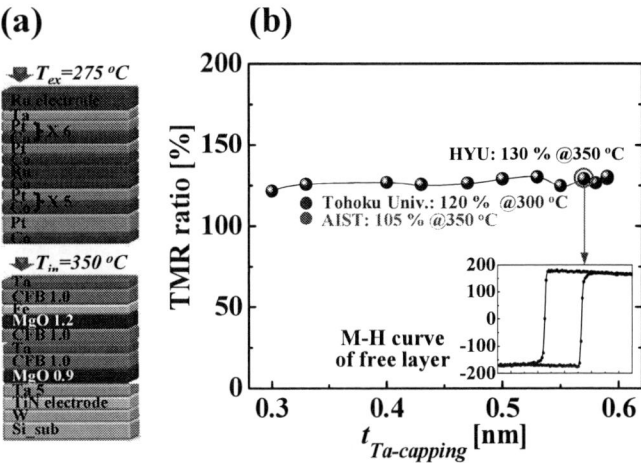

Fig. 6. b.c.c. seed based $Co_2Fe_6B_2$/MgO p-MTJ spin-valves stacked with a $[Co/Pt]_n$-SyAF layer. (a) schematic structure of spin-valve and (b) TMR ratio depending on the Ta capping layer thickness.

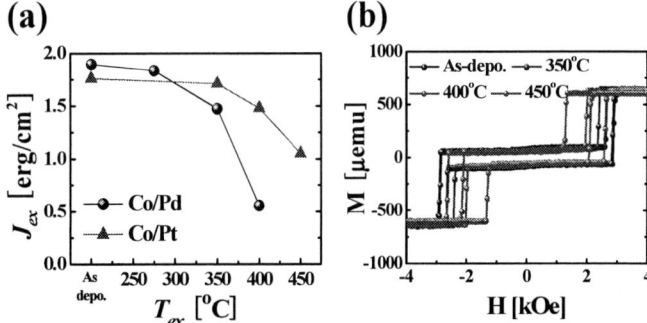

Fig. 7. (a) J_{ex} of the $[Co/Pd]_n$ and $[Co/Pt]_n$-SyAF layers (b) M-H curves for the $[Co/Pt]_n$-SyAF layers as a function of ex-situ annealing temperature.

978-1-4799-8002-4/14 $31.00 © 2014 IEEE

Area Dependence of Thermal Stability Factor in Perpendicular STT-MRAM Analyzed by Bi-directional Data Flipping Model

K. Tsunoda, M. Aoki, H. Noshiro, Y. Iba, S. Fukuda, C. Yoshida, Y. Yamazaki,
A. Takahashi, A. Hatada, M. Nakabayashi, Y. Tsuzaki, and T. Sugii

Low-power Electronics Association and Project (LEAP), 16-1 Onogawa, Tsukuba 305-8569, Japan
Phone: +81-29-879-8261, Fax: +81-29-856-2622, E-mail: tsunoda@leap.or.jp

Abstract

We report a statistical analysis of the thermal stability factor (Δ) for the top-pinned perpendicular magnetic tunnel junction (p-MTJ). By using a bi-directional data flipping model, the data retention characteristics of the "0" and "1" states can be fitted separately, including the saturation of failure probability. With the help of a resistance evaluation for the 16-kbit MTJ array, it became clear that the Δ of the "1" state increased as the device area increased, whereas the Δ of the "0" state remains constant regardless of the size. Moreover, we found that the p-MTJ exhibited a much smaller variation of Δ (9.6 ~ 14.3%) compared with the in-plane MTJ. Variations of Δ in both states decreased as the area increased. In combination with an intense magnetic measurement for the discrete monitor devices, the key parameter to increase the Δ and suppress its variation was investigated.

Introduction

Spin-transfer-torque MRAM (STT-MRAM) is one of the promising candidates as a scalable nonvolatile memory with its high-speed read/write and excellent cycling endurance. Since the switching current and thermal stability factor (Δ) are in a trade-off, accurate evaluation of Δ and its variation are important for the device design [1]. Interestingly, recent studies suggest that the Δ of a perpendicular magnetic tunnel junction (p-MTJ) does not depend on its area when the diameter is over 40 nm [2]. Therefore, we investigated the behavior of Δ for the top-pinned p-MTJ by focusing on the area dependence. To directly correlate the Δ with the device area, a 16-kbit test array that had digital address decoding and analog-mode resistance readout was designed.

Design and Fabrication of Top-pinned P-MTJ Array

Figure 1(a) shows a cross-sectional TEM image of the top-pinned p-MTJ. A counter bias magnetic field layer (CBF) was used to reduce the stray field from the pinned layer [3]. Though a double MgO interface is known to increase the Δ [4], we used a conventional Ta/CoFeB/MgO free layer to understand the basic features of Δ. A p-MTJ array was embedded in 65-nm-node CMOS Cu wires, as shown in Fig. 1(b). The typical STT switching of p-MTJs and corresponding magnetization directions are shown in Figs. 2(a) and 2(b). To investigate the relationship between the data flipping rate and device area including their variations, we used the 16-kbit MTJ array with analog-mode resistance readout. Figure 3 shows an example of resistance

distribution. Though the effective magnetoresistance (MR) ratio is decreased due to the large series resistance of the cell transistor and peripheral circuit, the data retention characteristics of the overlapped "0" and "1" states can be accurately evaluated with this method.

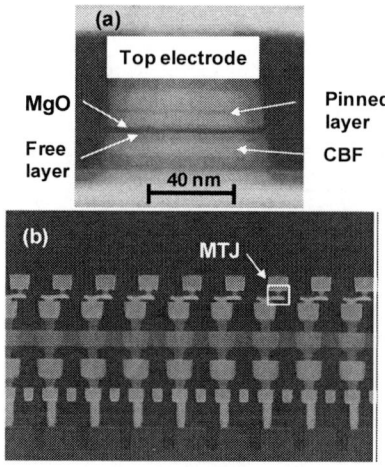

Fig. 1 (a) Cross-sectional TEM image of p-MTJ with top-pinned structure. CBF used to reduce stray field. (b) P-MTJ array embedded in 65-nm-node CMOS platform.

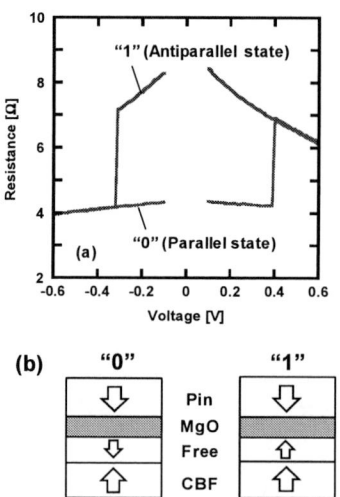

Fig. 2 (a) Typical STT switching of top-pinned p-MTJ after Cu metallization. External magnetic field is not required with CBF. (b) Schematic of "0" and "1" states. Arrows denote relative direction of magnetization.

978-1-4799-8002-4/14 $31.00 © 2014 IEEE

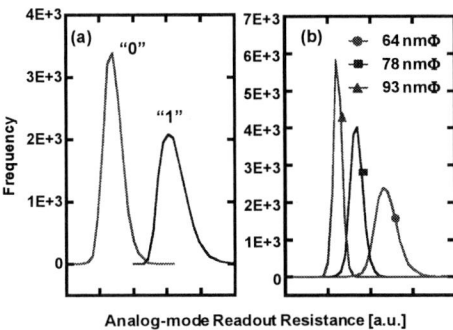

Fig. 3 Resistance distributions obtained from 16-kbit p-MTJ array. (a) Effective MR ratio decreased due to large series resistance of cell transistor and peripheral circuit, resulting in overlapped "0" and "1" resistance distributions. Even in this case, data retention characteristics can be evaluated in analog-mode readout.
(b) Resistance distributions of "0" state with different p-MTJ sizes.

Analysis by Bi-directional Data Flipping Model

A schematic of a data retention model and corresponding Neel-Brown (NB) equation are shown in Fig. 4. The data flipping probability F(t) can be expressed as a function of retention time and Δ. The F(t) of the 16-kbit p-MTJ array measured after all "1" writes is shown in Fig. 5. The baking temperature was 125°C. The number of flipping bits increased at a decreasing rate with the increase in time. This saturation in a Weibull plot cannot be fitted by using the NB model including the normal distribution of Δ [1].

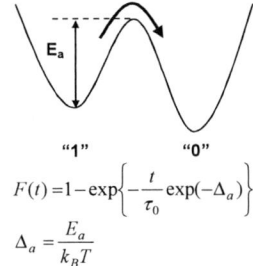

$$F(t) = 1 - \exp\left\{-\frac{t}{\tau_0}\exp(-\Delta_a)\right\}$$

$$\Delta_a = \frac{E_a}{k_B T}$$

Fig. 4 Schematic of data retention model and corresponding NB equation. Data flipping probability F(t) can be expressed as function of retention time (t) and thermal stability factor (Δ_a). τ_0 is inverse of attempt frequency.

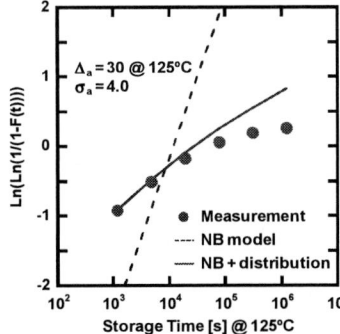

Fig. 5 F(t) of 16-kbit p-MTJ array measured after all "1" writes. In Weibull plot, F(t) increased at decreasing rate with increase in time, which cannot be fitted by NB model including normal distribution of Δ_a. σ_a is standard deviation of Δ_a.

Considering the variation in Δ, analysis of the retention behavior at the y-axis ~ 0 is very important. The discrepancy between theory and data is caused by bi-directional data flipping during the high temperature baking, as illustrated in Fig. 6. By taking both switching from "1" to "0" and switching back from "0" to "1" into consideration, F(t) can be expressed as in Fig. 6. Clear examples of bi-directional data flipping are shown in Fig. 7. We designed two MTJs that have the only difference in the stray fields. In device A (B), the stray field was positive (negative), and hence, $\Delta_a < \Delta_b$ ($\Delta_a > \Delta_b$). There was almost no data flipping of the "1" state in device B due to the large difference between Δ_a and Δ_b. This means that once switching from "1" to "0" occurred, switching back from "0" to "1" followed very quickly. F(t) in Fig. 5 was plotted again with the new model, as shown in Fig. 8. Measured data can be well fitted by using the NB model including the bi-directional data flipping and normal distribution of Δ.

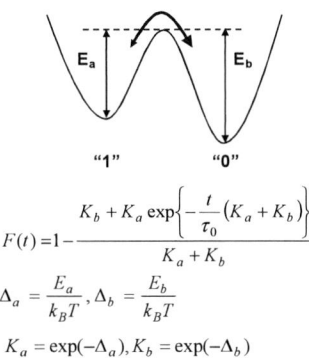

$$F(t) = 1 - \frac{K_b + K_a \exp\left\{-\frac{t}{\tau_0}(K_a + K_b)\right\}}{K_a + K_b}$$

$$\Delta_a = \frac{E_a}{k_B T}, \Delta_b = \frac{E_b}{k_B T}$$

$$K_a = \exp(-\Delta_a), K_b = \exp(-\Delta_b)$$

Fig. 6 Schematic of data retention model and corresponding data flipping probability F(t). Flipping from "0" to "0" and "0" to "1" taken into consideration.

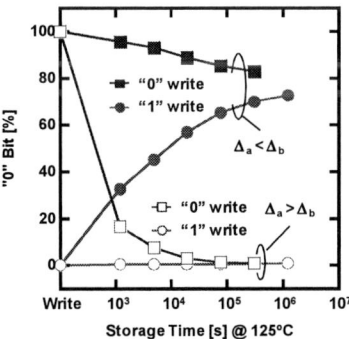

Fig. 7 Data retention characteristics of two 16-kbit p-MTJ arrays with different MTJ designs. Device A (filled circle/square): $\Delta_a < \Delta_b$. Device B (open circle/square): $\Delta_a > \Delta_b$. In device B, there was almost no fail in "1" state due to large difference between Δ_a and Δ_b.

978-1-4799-8002-4/14 $31.00 © 2014 IEEE

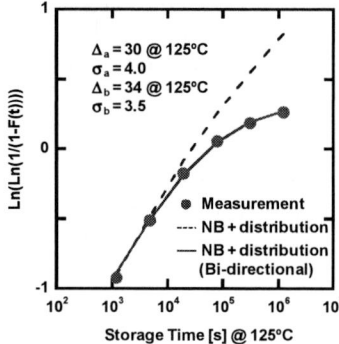

Fig. 8 F(t) of 16-kbit p-MTJ array measured after all "1" writes. Data can be fitted with NB model including bi-directional data flipping and normal distribution of Δ.

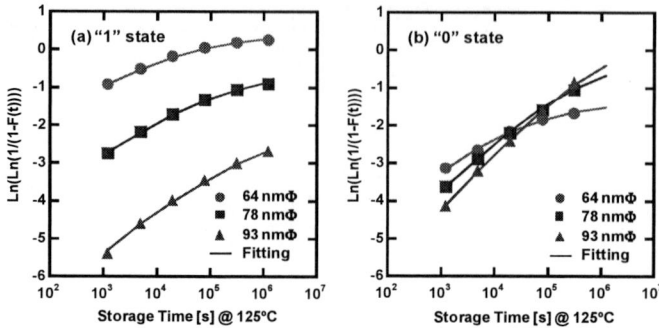

Fig. 10 (a) F(t) and fitting result of "1" state. (b) F(t) and fitting result of "0" state. It is worth noting that data in Fig. 10 (a) and (b) obtained from same 16-kbit array are fitted with same parameters (for example, same Δ_a, Δ_b, σ_a, σ_b for "1" and "0" states of 64-nmΦ MTJ array).

Area Dependence of Δ and Its Variation

Figure 9 shows the retention characteristics of three p-MTJ arrays with different MTJ diameters. With the new model, the F(t) of both the "1" and "0" states in Fig. 9 could be accurately fitted with the same parameters, as shown in Figs. 10(a) and 10(b). Various Δ obtained from different MTJ sizes are summarized in Fig. 11. The area of the MTJ was calculated from the average readout resistance of the 16-kbit array, assuming that the series resistance of the cell transistor and peripheral circuit is constant. As a result, the Δ of the "0" state were almost constant regardless of the size, which is consistent with the previous report [2]. However, the Δ of the "1" state had a clear dependence on the junction area. As for the standard deviations of Δ (1 σ/avg.), two states exhibited a similar decreasing trend as the area increased, as shown in Fig. 12. Interestingly, the Δ variation of the "1" state was much larger than that of the "0" state. Moreover, it is worth noting that in spite of the very small area of the p-MTJ, the variations of Δ (average of two states, 9.6 ~ 14.3%) were smaller than that of the in-plane MTJ (17%) [1].

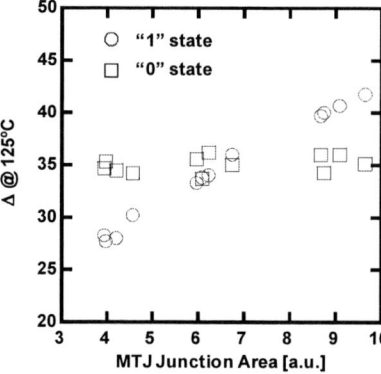

Fig. 11 Area dependence of Δ obtained from sixteen 16-kbit p-MTJ arrays with different MTJ sizes. Δ of "1" state have clear dependence on junction area, whereas that of "0" state almost constant regardless of size.

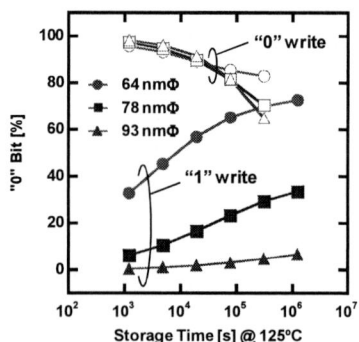

Fig. 9 Data retention characteristics of three 16-kbit p-MTJ arrays with different MTJ sizes

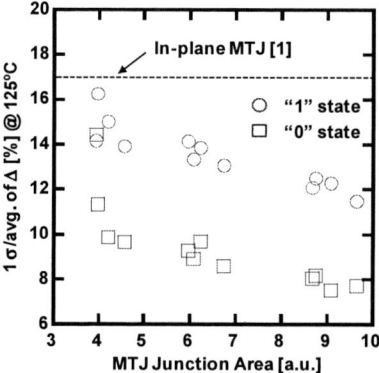

Fig. 12 Variation of Δ (1σ/avg.) obtained from sixteen 16-kbit p-MTJ arrays with different MTJ sizes. Variation of Δ decreased for both "0" and "1" states as junction area increased.

Figure 13 confirms that the increase of Δ in the "1" state as the junction area increased was reproducible even at 85°C. Moreover, the average Δ of the "1" and "0" states measured at 85°C and 125°C could be normalized to a single line of Δ at 25°C.

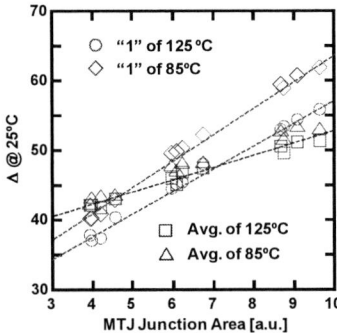

Fig. 13 Δ measured at 85°C and 125°C converted to Δ at 25°C. Average Δ of "1" and "0" states can be normalized to single line. Δ of only "1" state cannot be normalized well due to change of stray field depending on temperature. Lines are guides.

To investigate the reason for the asymmetric retention behavior in Fig. 11, the R-H loops of discrete monitor devices were analyzed as shown in Fig. 14. As a result, the asymmetry of Δ can be clearly explained by the asymmetric area dependence of two switching fields, H_{c0} and H_{c1}. As for the area dependence of variation in Fig. 12, the resistance variation relates to the Δ variation, as shown in Fig. 15(a). Strong correlation among the three parameters in Fig. 15(b) suggests that the suppression of variation in the junction area improved the variation of coercivity, which resulted in the decrease of Δ variation. By considering these results, the key parameter for controlling the Δ and its variation is the coercivity in the p-MTJ array.

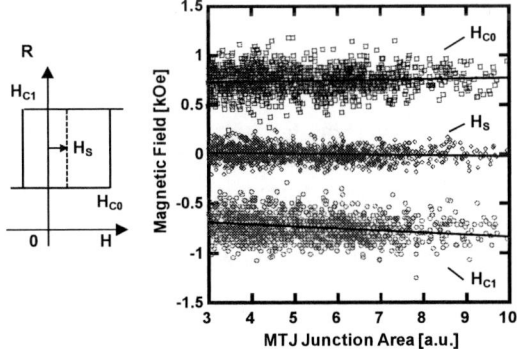

Fig. 14 Change of minor loop as function of junction area. Data obtained from 1460 discrete monitor devices. Lines are results of linear fitting. Switching field from "0" to "1" (H_{c0}) had almost no area dependence. However, switching field from "1" to "0" (H_{c1}) increased as junction area increased.

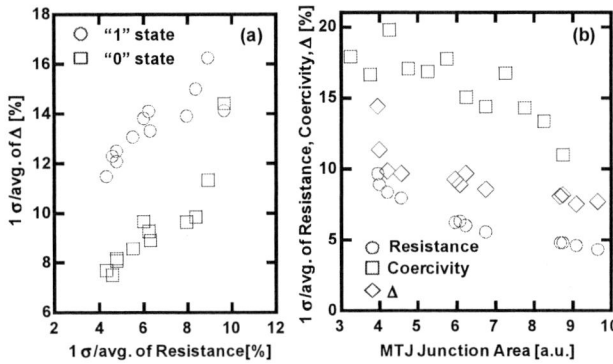

Fig. 15 (a) Change of Δ variation as function of resistance variation. (b) Variations of resistance, coercivity, and Δ of "0" state as function of junction area. Data of resistance and Δ obtained from same 16-kbit arrays, whereas that of coercivity obtained from discrete devices. Please note that Δ and coercivity of "0" state are almost constant regardless of junction area.

Conclusion

We investigated the area dependence of Δ in the 16-kbit p-MTJ array by using the bi-directional data flipping model. As a result, the Δ of the "1" state increased as the device area increased, whereas the Δ of the "0" state remained constant regardless of the size. The standard deviations of Δ for both the "0" and "1" states exhibited a similar decreasing trend as the area increased. From the analysis, the key parameter to control the Δ and its variation seemed to be the coercivity in the p-MTJ array. Device design not only to increase the coercivity but also to suppress its variation is important to improve the performance of perpendicular STT-MRAM.

Acknowledgements

This work was performed as "Ultra-Low Voltage Device Project" funded and supported by METI and NEDO. A part of the device processing was operated by AIST, Japan.

References

[1] K. Hofmann, K. Knobloch, C. Peters, and R. Allinger, "Comprehensive statistical investigation of STT-MRAM thermal stability," *VLSI Tech. Dig.*, p. 78, 2014.

[2] H. Sato, M. Yamanouchi, K. Miura, S. Ikeda, H. D. Gan, K. Mizunuma, R. Koizumi, F. Matsukura, and H. Ohno, "Junction size effect on switching current and thermal stability in CoFeB/MgO perpendicular magnetic tunnel junctions," *Appl. Phys. Lett.* 99, 042501, 2011.

[3] Y. Iba, C. Yoshida, A. Hatada, M. Nakabayashi, A. Takahashi, Y. Yamazaki, H. Noshiro, K. Tsunoda, T. Takenaga, M. Aoki, and T. Sugii, "Top-pinned perpendicular MTJ structure with a counter bias magnetic field layer for suppressing a stray-field in highly scalable STT-MRAM," *VLSI Tech. Dig.*, T136, 2013.

[4] J. -H. Park, Y. Kim, W. C. Lim, J. H. Kim, S. H. Park, J. H. Kim, W. Kim, K. W. Kim, J. H. Jeong, K. S. Kim, H. Kim, Y. J. Lee, S. C. Oh, J. E. Lee, S. O. Park, S. Watts, D. Apalkov, V. Nikitin, M. Krounbi, S. Jeong, S. Choi, H. K. Kang, and C. Chung, "Enhancement of data retention and write current scaling for sub-20nm STT-MRAM by utilizing dual interfaces for perpendicular magnetic anisotropy," *VLSI Tech. Dig.*, p. 57, 2012.

0.026μm² High Performance Embedded DRAM in 22nm Technology for Server and SOC Applications

C. Pei*, G. Wang, M. Aquilino, N. Arnold, B. Chandra, W. Chang, X. Chen, W. Davies, K. Hawkins, D. Jaeger, J. B. Johnson#, O.-J. Kwon, R. Krishnasamy#, W. Kong, J. Liu, X. Li, B. Messenger, E. Nelson##, K. Nummy, K. Onishi, D. Poindexter, S. Rombawa, C. Sheraw, T. Tzou, X. Wang, M. Yin, G. Freeman, T. Kirahata, E. Maciejewski, J. Norum, N. Robson, S. Narasimha, P. Parries, P. Agnello, R. Malik and S.S. Iyer

IBM Semiconductor Research and Development Center, Hopewell Junction, NY, 12533
#IBM Semiconductor Research and Development Center, Essex Junction, VT, 05403
##IBM Systems and Technology Group, Essex Junction, VT, 05403
*Phone: +1-845-8928458, Email: piec1@us.ibm.com

Abstract

This paper presents the industry's smallest Embedded Dynamic Random Access Memory (eDRAM) implemented in IBM's 22nm SOI technology. The bit cell area of 0.026μm² achieves ~60% scaling over the previous generation with deep trench (DT) capacitance optimized for performance and retention requirements. We report, for the first time, the asymmetric embedded stressor, cavity implant, through gate implant, and substrate n-band innovations to maintain aggressive cell scaling for the 22nm eDRAM technology.

Introduction

High density and low power on-chip eDRAM memory is critical to achieving high functional capability and power-performance goals in memory intensive systems such as microprocessors (1). With cell dimensions shrinking, critical issues for eDRAM in 22nm technology such as n-band resistance, junction butting, parasitic adjacent DT-induced leakage (PADIL), gate induced drain leakage (GIDL) and Vt variation have presented increased challenges. Innovations in cell design, device design, process architecture and integration were needed to address these issues. This paper presents the innovations employed for 22nm eDRAM technology which have successfully addressed these critical issues and achieved high density, performance and retention simultaneously.

Device architectural innovations

The 22nm eDRAM cell design has continued aggressive cell area scaling of 60% per logic generation since the 180nm node. Dimensions of transistor and DT, and space between each component have been scaled down as shown in Fig.1 compared to 45nm and 32nm nodes. In order to improve access FET performance, retention time and process yield, a unique asymmetric embedded stressor has been utilized for the access FET with a gate-first process

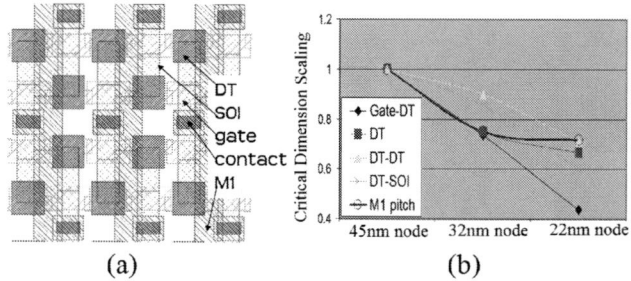

Fig.1 (a) 22nm eDRAM cell layout design and (b) aggressive scaling.

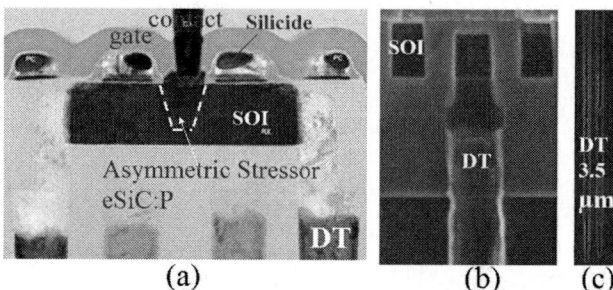

Fig. 2. 22nm eDRAM cell cross section (a) perpendicular to gate (b) parallel to gate; and (c) full DT image under BOX

flow as highlighted in Fig. 2 (a). By recessing silicon and regrowing highly doped epitaxial SiC:P on the bit line contact side only, we improve electron mobility for access transistor and lower contact resistance while maintaining integration simplicity due to commonality of process with all NFETs. Silicide is formed on bit line contact area and partially formed on gates to reduce resistance, but blocked from DT side to eliminate silicide encroachment effect on DT side for high yield. I-V curves measured from access FET are shown in Fig.3. Drive current is 48μA at bias condition of Vds=0.9V, Vgs=1.6V, with subthreshold slope 80mV/dec. FET leakage at Vg= -0.3V (word line low bias condition in operation) is as low as 6fA which is essential for high retention. The leakage floor increases 30X from room temperature to 105C° mainly coming from the expected Vt reduction and slope degradation at high temperature.

978-1-4799-8002-4/14 $31.00 © 2014 IEEE

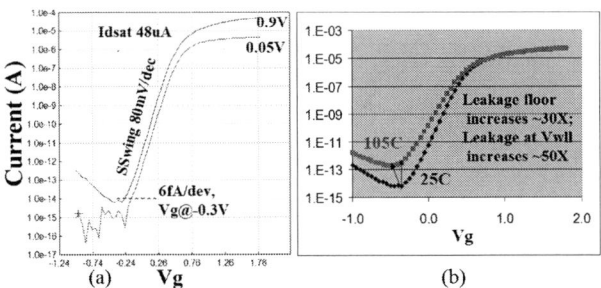

Fig. 3 (a) Array FET I-V curves; (b) and temperature dependency behavior

The asymmetric stressor also allows tuning the junction grading on the DT side which results in lower GIDL. 5X less GIDL was observed on the DT side compared to that measured on the bit line contact side, as shown in Fig.4. This provides significant writing benefit with higher body effect to lower Vt for performance improvement, while achieving longer retention time due to lower junction leakage and higher Vt in the data storage condition.

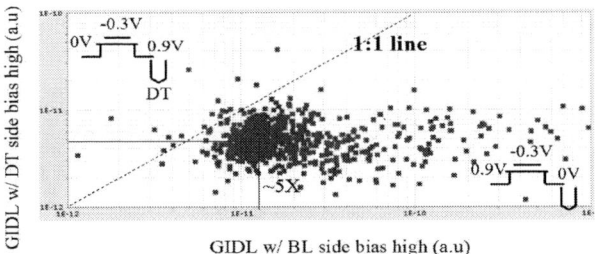

Fig. 4 Asymmetric stressor provides non-equal GIDL on bitline side vs DT side, benefiting writing speed and longer retention,

Cavity implant for junction butting

Under junction leakage in the bit line contact region allows charge flow between the two device floating bodies that share a bit line contact. This can lead to voltage and pattern sensitivity where leakage on one device influences the neighboring cell device characteristics. Junctions need to be strongly butted to buried oxide (BOX) to shut this leakage off completely. 22nm eDRAM takes advantage of the silicon recess on bit line side and utilizes low energy implantation into the cavity prior to epitaxial growth as shown in Fig.5 (a). Compared to conventional junction butting through source-drain implant alone, as shown in Fig.5 (b), the cavity

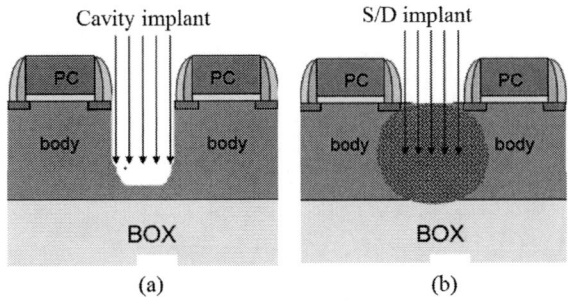

Fig. 5 (a) Cavity implant achieves strong junction butting and less SCE impact, superior to the junction butting (b) through deep S/D implant.

implant greatly allows effective junction butting with minimal impact on short channel effect (SCE) due to less lateral scattering. Without this implant, the silicon recess on bit line side may induce recess depth variation resulting in junction butting leakage sensitivity as Fig.6 (a) shows. 7nm recess change may introduce more than 100X junction butting variation. The cavity implant is able to tune and suppress this sensitivity completely as shown in Fig.6 (b).

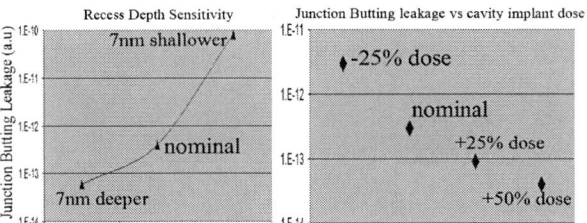

Fig. 6 (a) Junction butting leakage sensitivity to recess variation (b) Cavity implant able to suppress this sensitivity

The cavity implant also gives a lower electric field at the bottom body junction region as illustrated in the simulated results in Fig.7 compared to the case without cavity implant. This could be explained by the mechanism of lattice damage from cavity implant causes transient enhanced diffusion which better grades the junction thereby reducing electric field and thus improving junction leakage.

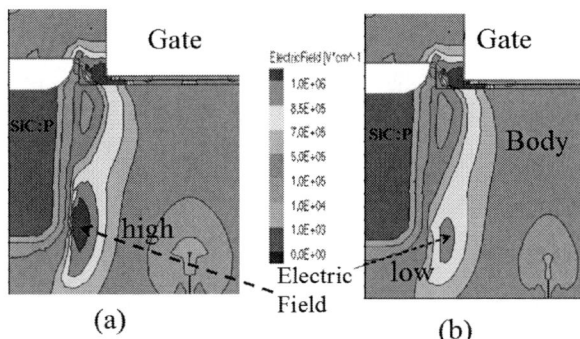

Fig. 7 Electric field reduces from (a) no cavity implant to (b) cavity implant

**Parasitic adjacent DT-induced leakage (PADIL)
and through gate implant (TGI)**

eDRAM cell scaling results in very close proximity between the access FET body and the adjacent DTs. PADIL results from the narrower DT-SOI space (4) where the bottom SOI corners and sidewall are electrically gated by adjacent DTs as shown in Fig.8(a). Conventional well implantation has a great difficulty to decouple the front FET Vt from back channel Vt of the parasitic FET since dopant diffusion causes the implant to raise back channel Vt to shut off back channel leakage to also increase front Vt which degrades eDRAM performance as shown in Fig. 8(b). The TGI scheme is superior to conventional pre-gate well implantation since it is put in after much of the gate related thermal processing and therefore is better able to create a retrograde doping profile with high doping concentration at

978-1-4799-8002-4/14 $31.00 © 2014 IEEE   491

SOI/BOX interface to control PADIL but not increasing front Vt that much. This approach also has the benefit of not adding significant counter doping in the source/drain diffusion areas as illustrated in Fig.9 (a)

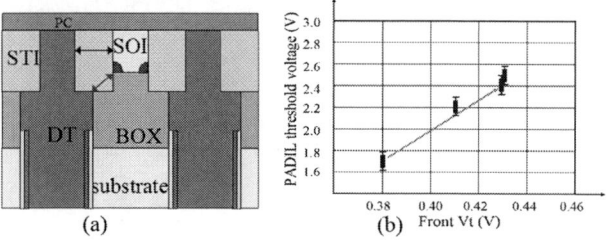

(a)

(b)

Fig. 8(a) PADIL issue is caused by DT-SOI space scaling; (b) PADIL Vt increases to shut off leakage also raises the front Vt.

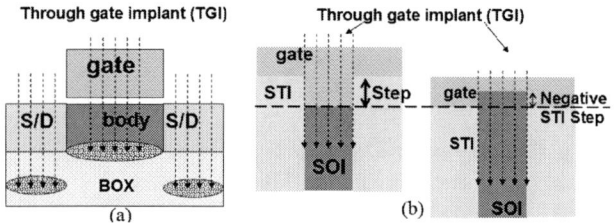

(a)

(b)

Fig.9 (a) TGI provides retrograded doping profile without counter doping S/D (b) TGI brings in an increased Vt sensitivity to STI Step.

The downside of TGI scheme is an increased Vt sensitivity to step height of STI (shallow trench isolation), as shown in Fig.9 (b), and to gate length. The gate polysilicon deposits conformally filling in the STI step which results in the implant physically going through thicker gate material and raises the implant doping closer to channel surface, causing higher front Vt; conversely a negative step causes lower Vt, thus resulting in Vt sensitivity, as shown in Fig.10.

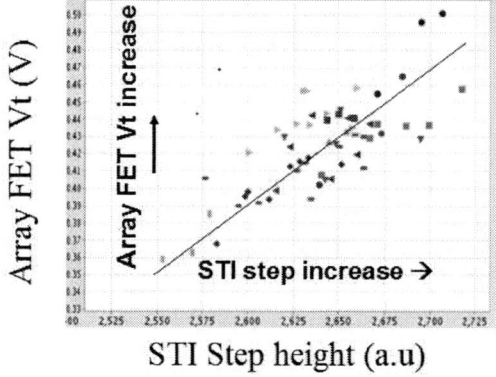

Fig. 10, Array FET Vt is proportional to STI Step height in TGI scheme.

The Vt variation has been significantly improved by TGI dose adjusting automatically, named i-APC, based on metrology data measured inline on upstream process variation, including both STI step height and gate length, which is fed forward to adjust the TGI implant dose. Vt variation of 25mV per sigma has thus been achieved, as shown in Fig.11.

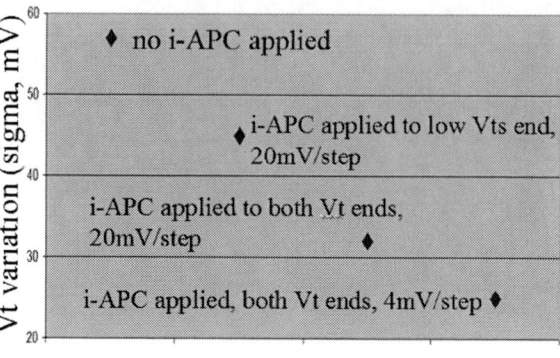

Fig. 11 Vt variation is significantly improved by i-APC scheme

Innovations of n+ epi plate and MOAT isolation

The 22nm process utilizes a very unique pre-doped n+ epi plate (2), as shown in Fig.12, which combines the n-band with the buried plate without process complexity and reduces the plate plus n-band resistance to only one fifth of resistance of the conventional n-band from the prior generation.

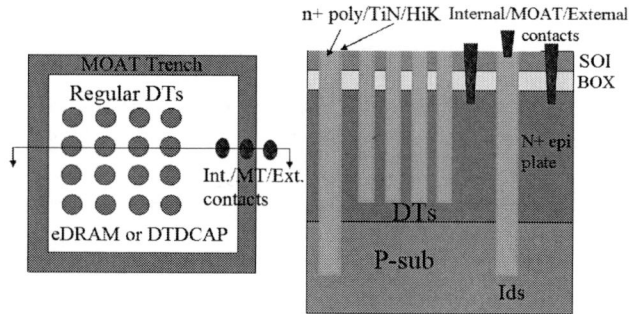

Fig. 12 n+ epi plate and MOAT Trench, top down and cross section views.

A MOAT trench provides isolation without extra lithographic steps or process complexity, also shown in Fig.12. MOAT is a ring trench feature extending through n+ epi layer into p-substrate with 3X wider trench dimension than regular DTs. Compared to 32nm node [5], superior frequency response for 22nm DT decoupling capacitors is shown in Fig.13.

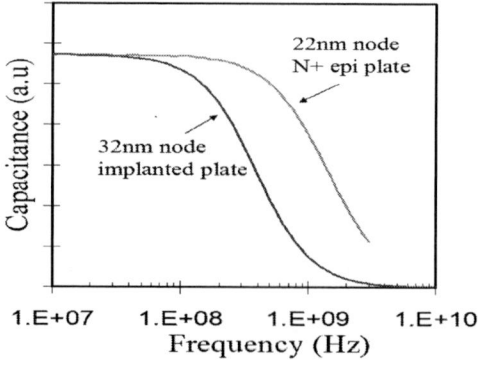

Fig.13 Superior frequency response of DTDCAP with n+ epi plate

978-1-4799-8002-4/14 $31.00 © 2014 IEEE   492

DT capacitance scaling

The n+ epi plate has significantly simplified the DT process compared to 32nm eDRAM by eliminating the need of a protective nitride spacer which protects the SOI device region from the heavy trench implant when forming the implanted plate. This allows smaller top DT diameter to achieve the required 3.5μm DT depth as shown in Fig.2 (c) which provides 12fF capacitance per DT to meet both retention and performance requirement. Overall cost and turn around time have been reduced. In-stead of DT bottling process [3, 4], 22nm eDRAM uses higher κ material HfSiOx to increase DT capacitance by 10% over HfO₂ to maintain DT scaling. Fig. 14 presents the DT node stack of 22nm node compared to 32nm node.

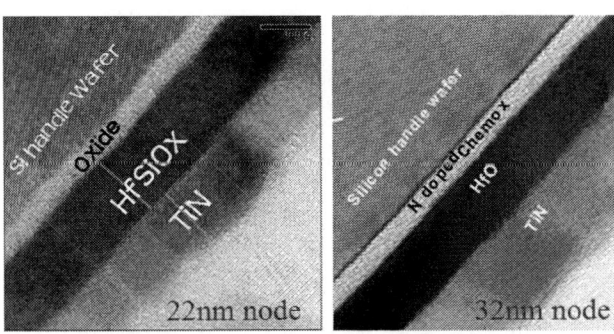

Fig.14 DT node stack, 22nm eDRAM uses HfSiOx with higher K value than that of HfO2 used by 32nm eDRAM.

Functional results

A representative functional bit fail map is shown in Fig. 15 (a), and perfect chip map of 20Mb eDRAM macros in Fig.15 (b) demonstrates excellent yield.

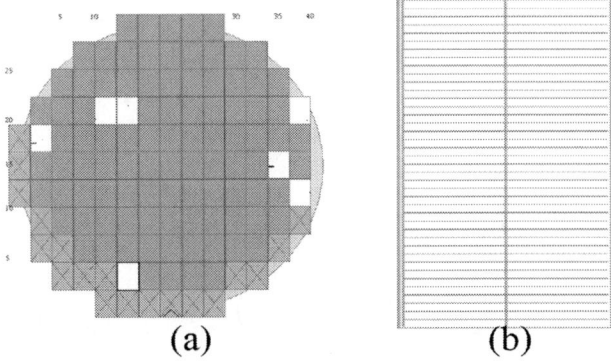

Fig.15 Bit fail maps (a) wafer map and (b) perfect chip map of 20Mb eDRAM macros.

Normalized fail count data suggests no fails up to 500μs retention, and retention time is as high as 2000μs at 4.5 sigma repairable at 85C, as shown in Fig. 16.

The 22nm eDRAM has various applications on IBM's server chips as on-chip embedded processor caches as well as associated applications. The fully integrated 256Mb product array has demonstrated capability of 1.4ns cycle time which is significantly faster than any other embedded DRAM. We have achieved latency of 700ps for certain cache instantiations,

which is faster than Static Random Access Memory (SRAM) for those sizes (1). 22nm eDRAM has been recently leveraged for IBM's 12-core 649mm² Server Processor POWER8™[6].

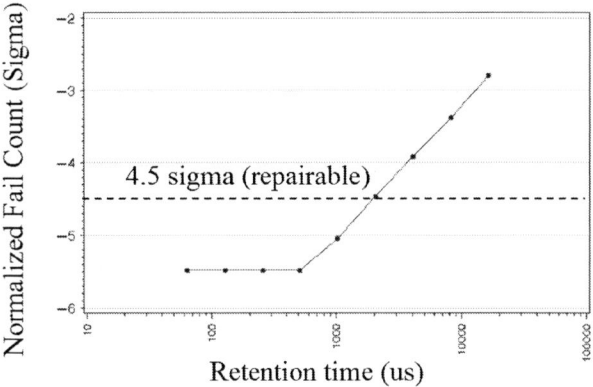

Fig. 16, Retention vs. fail count sigma. No fails up to 500 μs retention; retention as high as 2000 μs at 4.5 sigma at 85C.

Conclusion:

In this paper, we have described the critical issues for 22nm eDRAM technology including n-band resistance, junction butting, PADIL, GIDL and Vt variation with aggressive cell scaling. Innovations of pre-doped n+ epi plate, MOAT isolation, asymmetric stressor, cavity implant, through gate implant and DT node stack were introduced to address these issues. Excellent functional results have been presented.

Acknowledgements

The authors would like to acknowledge the 300mm fab of the Microelectronics Division of IBM at East Fishkill, NY and the members of patterning, unit process, device, integration, characterization teams for enabling this work.

References :

[1] S.S. Iyer et al, "45-nm silicon-on-insulator CMOS technology integrating embedded DRAM for high-performance server and ASIC applications", IBM J. Res. & Dev. Vol. 55 No. 3 pp. 5.1-5.14, 2011.

[2] S. Narasimha et al, "22nm High-Performance SOI Technology Featuring Dual-Embedded Stressors, Epi-Plate High-K Deep-Trench Embedded DRAM and Self-Aligned Via 15LM BEOL", IEDM Dig. Tech. Papers, pp. 52-55, 2012.

[3] G. Wang et al," Scaling Deep Trench Based eDRAM on SOI to 32nm and Beyond" IEDM Dig. Tech. Papers, 2009.

[4] N. Butt et al "A 0.039um2 High Performance eDRAM Cell based on 32nm High-K/Metal SOI Technology" IEDM Dig. Tech. Papers, 2010.

[5] B. Jayaraman et al, " Performance Analysis and Modeling of Deep Trench Decoupling Capacitor for 32 nm High-Performance SOI Processors and Beyond" ICICDT, pp 1-4, 2012

[6] E. Fluhr et al," POWER8TM: A 12-Core Server-Class Processor in 22nm SOI with 7.6Tb/s Off-Chip Bandwidth" , ISSCC, Dig Tech Papers pp. 96-97, 2014

A New Saw-Like Self-Recovery of Interface States in Nitride-Based Memory Cell

Yuh-Te Sung[1], Po-Yen Lin[1], Jim Chen[2], Tzong-Sheng Chang[2], Ya-Chin King[1] and Chrong Jung Lin[1]

[1] Microelectronics Laboratory, Institute of Electronics Engineering, National Tsing Hua University, Hsinchu 300, Taiwan
[2] Process Integration Division Fab12, Taiwan Semiconductor Manufacturing Company, Hsinchu 300, Taiwan

Phone/Fax: +886-3-5721804/886-3-5162182, E-mail: cjlin@ee.nthu.edu.tw

Abstract

A new saw-like self-recovery Self-Aligned Nitride (SAN) memory cell is proposed and fabricated in 28nm high-k metal gate (HKMG) CMOS process for high-density logic NVM applications. The cell is operated with Source-Side Injection (SSI) for programming and band-to-band hot holes (BBHH) for erasing. Two effective self-heating recovery mechanisms are proposed and performed to maintain a stable On/Off read window after cycling stresses. Besides, the characteristic and reliability comparison of the SAN cell in other technology nodes, 90nm/45nm/32nm, are characterized to further verify the saw-like self-detrapping and self-recovery operation. The new 28nm HKMG SAN memory cell with the self-detrapping recovery results excellent and superior endurance performance and can provide a very promising solution for logic NVM in advanced technologies.

Introduction

Logic nonvolatile memories have been widely discussed and built-in advanced CMOS logic technologies for its high density, high integrability, and low process complexity and cost. Some embedded nitride-based logic NVMs fabricated with a pure CMOS logic compatible processes have been demonstrated in our previous works[1-7]. The high scalability of Self-Aligned Nitride (SAN) cell has also been proved from 90nm CMOS logic process[1-4] to 45nm and 32nm with strain processes[5-7]. When the CMOS logic mainstream technology continuously evolves down to 28nm HKMG process, several challenges, such as soft breakdown, metal gate instead of poly gate, SILC problem, severely constrain further development of logic NVMs. In this study, a new nitride-based cell with saw-like self-recovery is proposed and successfully implemented in pure 28nm HKMG CMOS logic process for high-density logic NVM application in advanced nodes.

Cell Structure and Characteristics

The saw-like trap-recovery Self-Aligned Nitride (SAN) memory cell is completely fabricated by 28nm HKMG CMOS logic processes without extra masking or process step. Its schematic diagram and cross-sectional TEM picture are shown in Fig.1. Consisting of two serial NMOS, the SAN cell form a merged nitride spacer between the two metal gates (PG and SG). The merged nitride spacer is acting as charge storage node of this memory. The serial double gate memory cell can be arranged to a NOR-type

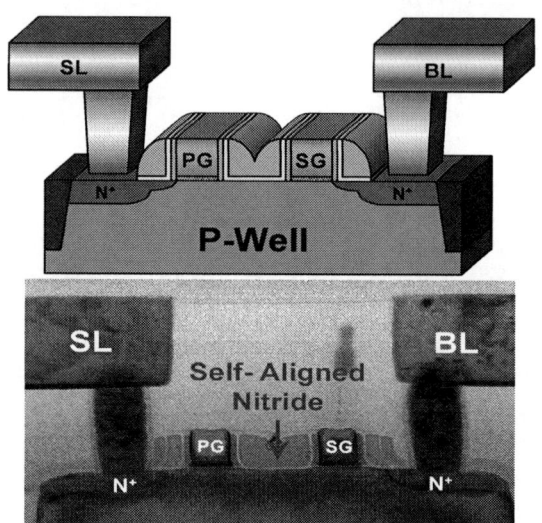

Fig.1. (a) Schematic illustration of SAN cell fabricated by 28nm HKMG CMOS process. (b) The cross-sectional TEM picture of SAN cell.

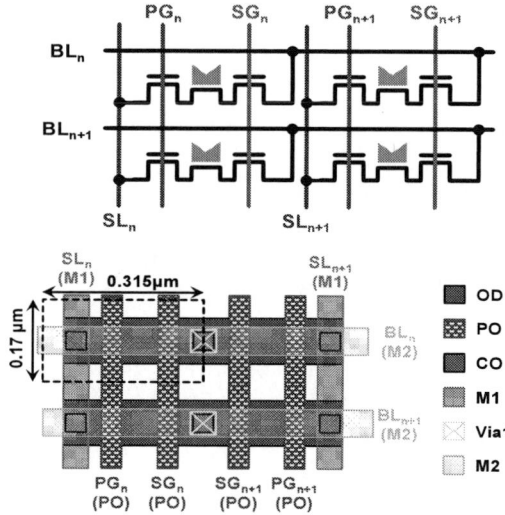

Fig.2. (a) 2x2 NOR-type array schematic with two WL (PG and SG), one BL and one SL. (b) 2x2 NOR-type array layout with shared BL contact and unit cell size of $0.0535\mu m^2$.

TABLE I
OPERATION CONDITIONS OF THE SAN CELL

		SL	PG	SG	BL
Read	*Selected*	0V	0.4V	1V	1V
	Unselected	0V	0V	0V	Float
Program	*Selected*	3V	3V	0.7V	0V
	Unselected	0V	0V	0V	Float
Erase	*Selected*	4.5V	-0.5V	0V	0V
	Unselected	0V	0V	0V	Float

978-1-4799-8002-4/14 $31.00 © 2014 IEEE

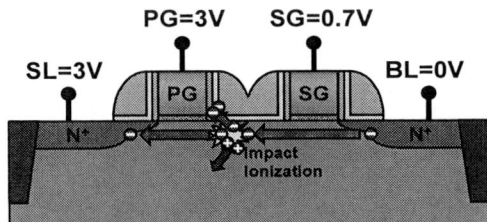

Fig.3. Source-side injection hot electrons near PG turn off the channel under SAN node for program.

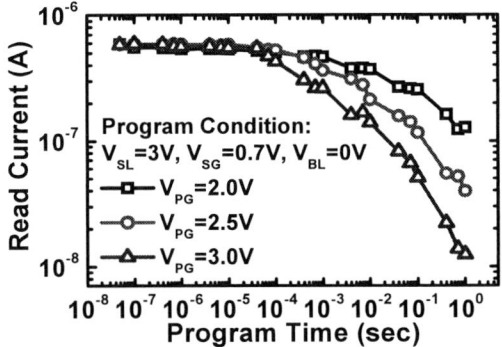

Fig.4. Time to program characteristics at V_{SL}=3V, V_{SG}=0.7V and V_{BL}=0V with different PG voltages.

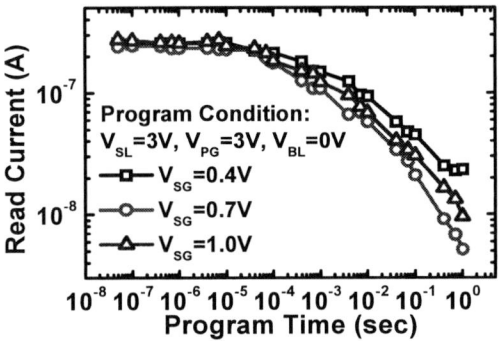

Fig.5. Time to program characteristics at V_{SL}=3V, V_{PG}=3V and V_{BL}=0V with different SG voltages.

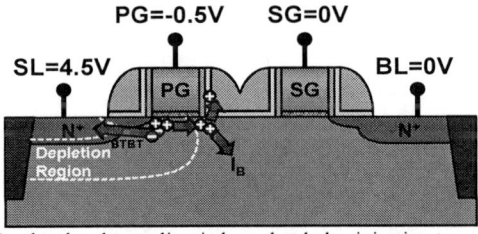

Fig.6. Band-to-band tunneling induces hot holes injection to neutralize the stored electrons in SAN.

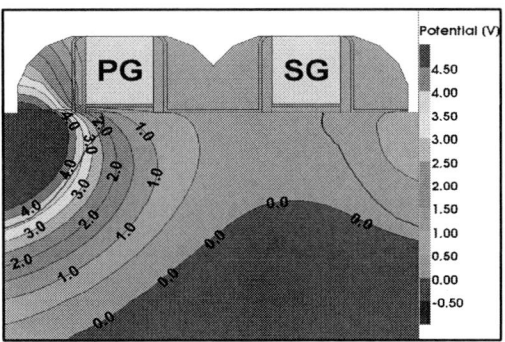

Fig.7. High voltage is applied to SL during erase operation, and extreme band bending enables band-to-band tunneling induced hot holes injection effectively.

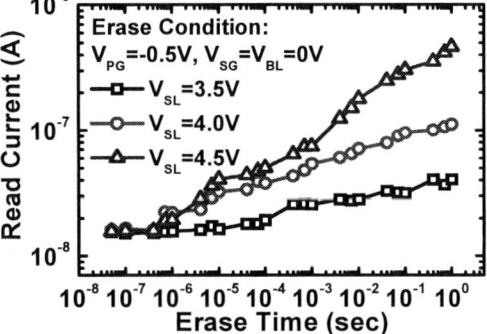

Fig.8. Time to erase characteristics at V_{PG}=-0.5V, V_{SG}=0V and V_{BL}=0V with different SL voltages.

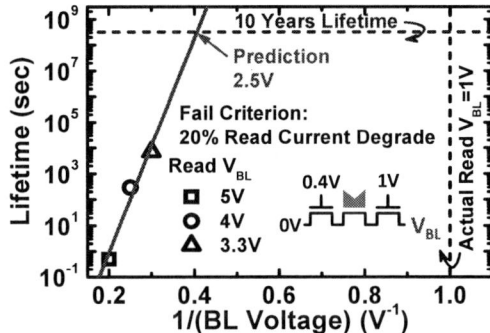

Fig.9. Read disturb characteristics project 10-years lifetime at BL voltage of 2.5V or below.

array as illustrated in Fig.2(a). Based on 28nm HKMG CMOS logic design rules, a very small cell size of 0.0535μm² can be achieved and fabricated as shown in Fig.2(b). The cell operation conditions in NOR-type array are summarized in Table.I. By floating the unselect BLs, the program and read disturb can be mostly suppressed by blocking the channel current. Fig.3 shows the direction and bias conditions of cell program operation. Time to program characteristics are shown in Fig.4 and Fig.5, the program data clearly shows higher PG voltage will have fast program speed by higher field. Fig.6 and Fig.7 depicts the bias conditions and potential contours of band-to-band tunneling

induced hot hole injection in erase operation. Fig.8 shows erase characteristics with different SL voltages for BBHH injection.

Reliability Characterization and Discussion

Long-term read stress is characterized and the BL voltage of ten-year lifetime is extrapolated at BL=2.5V or below as shown in Fig.9. In addition, the program disturb for unselect cells is shown in Fig.10. As unselect BLs are kept floating, the unselect cells can remain their state for 100k times of program disturb cycles. Fig.11 shows the result of data retention of 28nm HKMG SAN memory under different temperatures for 1000 hours bake. The margin of normalized I_{off}/I_{on} declines to 43%, 19%, and 12% of the original levels at 150°C, 85°C and 25°C for ten-year lifetime prediction. Fig.12 further shows 2-bits/cell operation results of the 28nm HKMG SAN memory by forward and reverse operations. However, due to the serious

978-1-4799-8002-4/14 $31.00 © 2014 IEEE

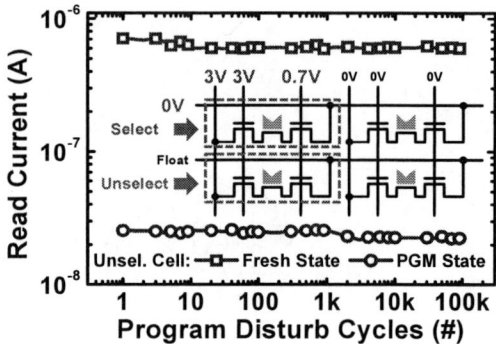

Fig.10. Program disturb characteristics for unselected cells at V_{SL}=3V, V_{PG}=3V, V_{SG}=0.7V for 100k cycles where BL is floating.

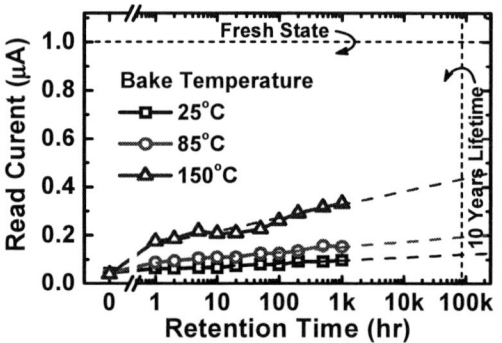

Fig.11. Retention study and 10-years lifetime prediction of SAN cell at three different temperatures for 1000 hours bake.

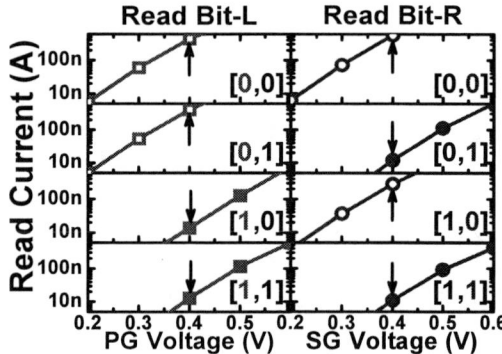

Fig.12. 2-bits per cell operation with distinctive four states by forward and reverse read operations.

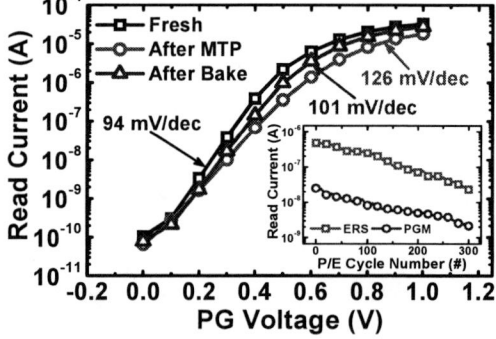

Fig.13. The degradation of read current and sub-threshold swing after hundred times of MTP operation stresses can be recovered and pulled-back through 150°C high temperature bake.

interface damage by SSI and BBHH injection, Fig.13 shows the 28nm SAN memory cell has a lot degradation after hundred times of cycling stresses. Moreover, the degraded

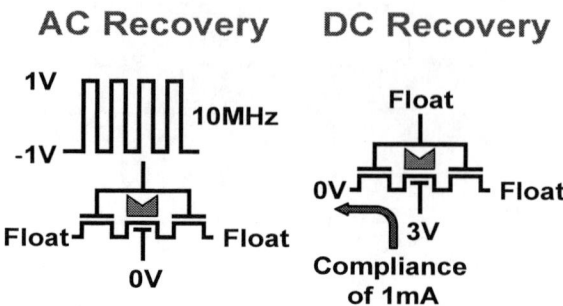

Fig.14. (a)10MHz high frequency voltage with amplitude of 1V is applied to PG and SG in AC recovery. (b)3V and 0V are applied to Bulk and SL with current compliance of 1mA in DC recovery.

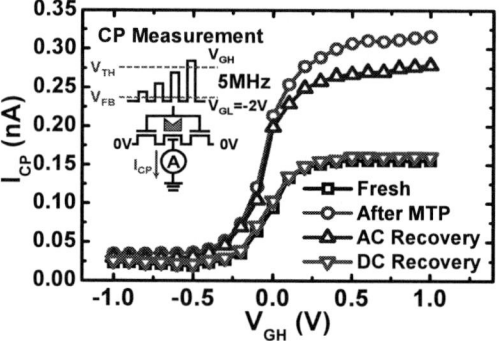

Fig.15. The decrease of charge-pumping current after performing AC recovery and DC recovery indicates the reduction of hot carrier injection induced interface traps.

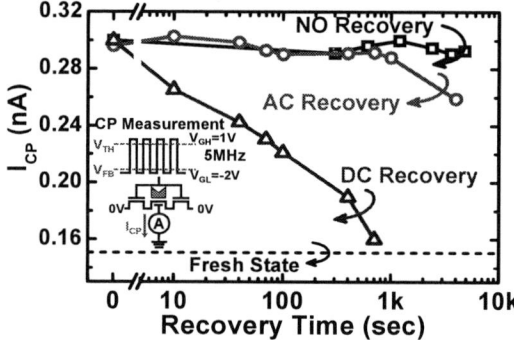

Fig.16. Time to charge-pumping current characteristics indicate that DC recovery results a lower pumping current and a better healing effect.

mobility and subthreshold swing can be pulled-back and recovered after 150°C bake only for few seconds. To perform a self-heating and on-chip recovery operation, two novel electrical recovery methods, AC and DC, are implemented to compensate cycling damages. In Fig.14(a), a series of 10MHz pulses with 1V amplitude on PG and SG is performed as an AC recovery method. By the back and forth current on PG and SG, the self-heating effect on gates can efficiently heal the interface traps, which is similar to do a high temperature bake. Another method to get self-heating effect is to apply 3V on p-substrate and 0V on n$^+$ SL as a DC recovery method. The forward current of n$^+$ junction suddenly heats up the cell and effectively recovers the cycling interface damage in the bottom oxide of the merged nitride spacer as depicted in Fig.14(b). For comparison, charge-pumping analysis for interface state is applied and summarized in Fig.15 and Fig.16. By performing the

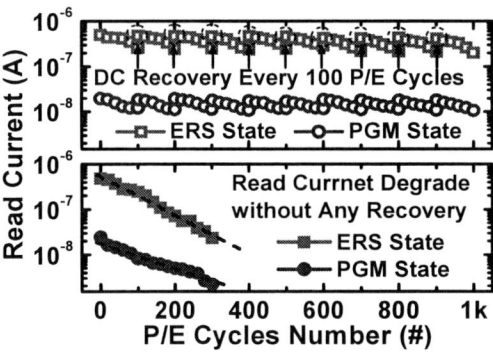

Fig.17. With DC recovery every 100 P/E cycles, SAN cell can easily achieve more than 1000 times MTP operation with 10X On/Off window.

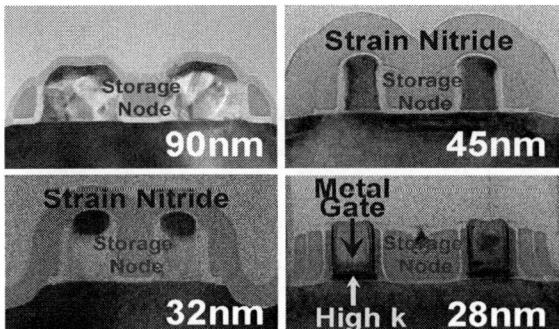

Fig.18. TEM pictures of SAN cells under different generations including 90nm, 45nm (strain nitride), 32nm (strain nitride) and 28nm (high-k metal gate).

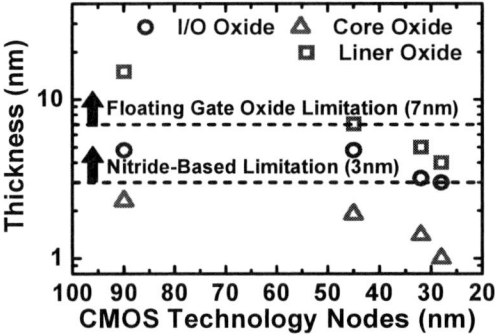

Fig.19. Oxide thickness comparison between different generations shows no SILC problem for SAN cell.

effective DC recovery per 100 cycles, the 28nm HKMG SAN memory cell can easily achieve more than 1,000 times of cycling stresses without current degradation as shown in Fig.17. Furthermore, the SAN cells in different CMOS logic technologies are summarized in Fig.18. As exhibited in Fig.19, the thickness of liner oxide, as a bottom oxide of the merged nitride spacer, is kept almost the same in different CMOS logic technologies, and their logic NVM solutions can remain superior reliability by adapting the new saw-like self-recovery function. The scaling trend of the critical dimensions of SAN cells is summarized in Fig.20. And the operation voltages for generations are also summarized and compared in Fig.21.

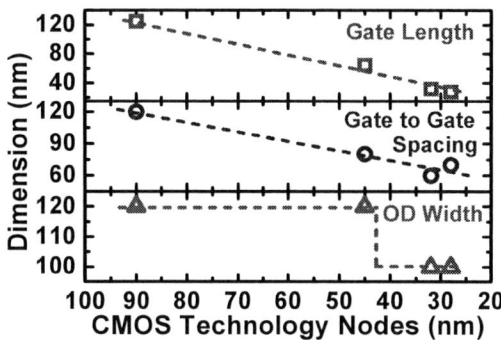

Fig.20. Scaling factors in the SAN cell over different generations of CMOS technologies.

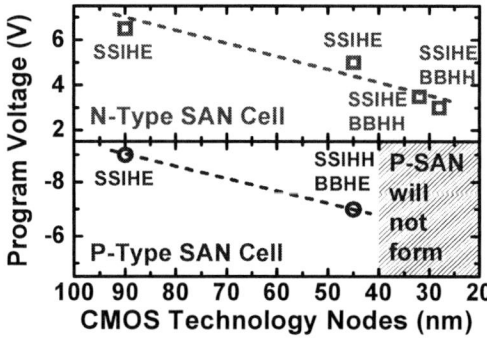

Fig.21. Required program voltages and carriers injection mechanism of the SAN cells in scaling CMOS logic technologies.

Conclusions

The new saw-like self-recovery SAN memory in fully compatible 28nm HKMG CMOS process is firstly proposed and demonstrated in this study. With a very small size, high scalability, high stability, high reliability with low power consumption, the new on-chip self-recovery 28nm HKMG SAN cell is a promising logic NVM solution in advanced CMOS logic platforms.

REFERENCE

(1) H.-C. Lai, K.-Y. Cheng, Y.-C. King, and C.-J. Lin, "A 0.26-μm^2 u-shaped nitride-based programming cell on pure 90-nm CMOS technology," *IEEE Electron Device Lett.*, vol. 28, no. 9, pp. 837-839, Sep. 2007.

(2) H.-C. Lai, C.-E. Huang, Y.-C. King, and C.-J. Lin, "Novel self-aligned nitride one time programming with 2-bit/cell based on pure 90-nm complementary metal- oxide-semiconductor logic technology," *Jpn. J. Appl. Phys.*, vol. 47, no. 11, p. 8369-8374, 2008.

(3) Y.-J. Chen, et al., "A Novel 2-Bit/Cell p-Channel Logic Programmable Cell With Pure 90-nm CMOS Technology," *IEEE Electron Device Lett.*, vol. 29, no. 8, pp. 938-940, Aug. 2008.

(4) C.-E. Huang, Y.-J. Chen, H. OuYang, C.-J. Lin and Y.-C. King, "Source side injection programmed p-channel self-aligned-nitride one-time programming cell for 90nm logic nonvolatile memory applications," *Jpn. J. Appl. Phys.*, vol. 49, no. 8, 04DD05, 2010.

(5) C.-E. Huang, et al., "A new self-aligned nitride MTP cell with 45nm CMOS fully compatible process," in *IEDM Tech. Dig.*, Dec. 2007, pp. 91–94.

(6) C.-E. Huang, Y.-J. Chen, H.-C. Lai, Y.-C. King and C. J. Lin, "A study of self-aligned nitride erasable OTP cell by 45nm CMOS fully compatible process," *IEEE Trans. Electron Device*, vol. 59, pp. 1228-1234, Jun. 2009.

(7) W. C. Shen, C.-E. Huang, H. OuYang, Y.-C. King, and C. J. Lin, "32nm strained nitride MTP cell by fully CMOS logic compatible process," in *VLSI-TSA*, Apr. 2012, pp. 987.

A Novel Double-Trapping BE-SONOS Charge-Trapping NAND Flash Device to Overcome the Erase Saturation without Using Curvature-Induced Field Enhancement Effect or High-K (HK)/Metal Gate (MG) Materials

Hang-Ting Lue, Roger Lo, Chih-Chang Hsieh, Pei-Ying Du, Chih-Ping Chen, Tzu-Hsuan Hsu, Kuo-Ping Chang, Yen-Hao Shih, and Chih-Yuan Lu

Macronix International Co. Ltd, Emerging Central Lab.
16, Li-Hsin Road, Hsinchu Science Park, Hsinchu, Taiwan.
E-mail: htlue@mxic.com.tw

Abstract - Erase saturation issue is a fundamental challenge for SONOS-type charge-trapping NAND Flash devices. Nowadays the most popular way to solve this issue is to pursue either curvature-induced field enhancement effect in the nano-wire SONOS device, or HK/MG to suppress the gate injection. However, both approaches have its drawback and reliability challenges. In this work, we propose a completely different approach that utilizes a double-trapping (or double storage) layer in a barrier engineered (BE) SONOS device to overcome the erase saturation ideally. A second nitride trapping layer (N3) is stacked on top of the first blocking oxide (O3) and 1st trapping layer (N2) of the original BE-SONOS device. Both theoretical model and experimental measured results indicate that when N3 stores sufficient electron charge it can greatly suppress gate injection, allowing continuous hole injection into N2 that gives a very deep erased Vt ~ -6V. A fully-integrated 3D Vertical Gate (VG) NAND Flash test chip using this novel device has been fabricated which demonstrates excellent MLC operation window and reliability. The flat and planar topology of this double-trapping BE-SONOS device enables minimal design rule of 3D NAND Flash array and possesses superb read disturb immunity.

I. Introduction

The fundamental challenge of SONOS device is that electron stored in nitride is hard-to-detrap, thus erase generally requires a very high electric field (>15MV/cm) [1]. To apply such high electric field for erase, not only tunnel oxide reliability is concerned, but also it induces strong gate injection that causes high erase saturation. To overcome erase saturation, people often pursue either curvature effect [2] or HK blocking oxide with a high work function MG [3] that reduces gate injection. A barrier engineered (BE) tunnel barrier was proposed to offer hole injection without direct-tunneling leakage at retention [4]. This avoids the difficulty in electron de-trapping. In principle, the BE-SONOS device can be combined with curvature or HK/MG method to eliminate gate injection. However, the field enhancement of curvature effect also causes program/read disturb, while the non-optimized high-K materials easily generates shallow traps (or dipole relaxation) that causes a fast initial charge loss, leading to program-verify (PV) offset [5, 6].

In this work, we propose a novel approach to suppress gate injection. We design a second trapping layer (N3) on top of the first blocking oxide/trapping layer (O3/N2) in BE-SONOS device. Table 1 briefly illustrates the concept and compares the three approaches.

II. Device Structure, Basic Characteristics, and Theoretical Model

Figures 1 (a) and (b) illustrate the TEM pictures of the double-trapping BE-SONOS device grown in a silicon substrate and 3DVG NAND Flash, respectively. The typical thicknesses for O1/N1/O2/N2/O3/N3/O4 are ~ 1/1.5/2/5.5/5.5/4/4 nm, respectively. Figure 2 shows the band diagram during –FN erasing. When N3 stores electrons, the top blocking oxide (O4) E field is greatly reduced, in turn it suppresses the gate injection. Meanwhile, the first trapping layer N2 can be continuously erased by the substrate hole injection. Figure 3(a) shows the ISPP and ISPE characteristic of a capacitor measured experimentally. Large programming window with ISPP ~0.9 is obtained. Figure 3(b) shows the erase transient characteristics. Low erase saturation of V_{FB} <-5V can be achieved. Compared with previous P+- gate BE-SONOS [4] without

double-trapping layer, the erase saturation is improved (lowered) by more than 3V. The much lower erase saturation also enables a larger erasing bias to shorten the erasing time. At -22V, 1msec erase can already produce V_{FB} < -4V.

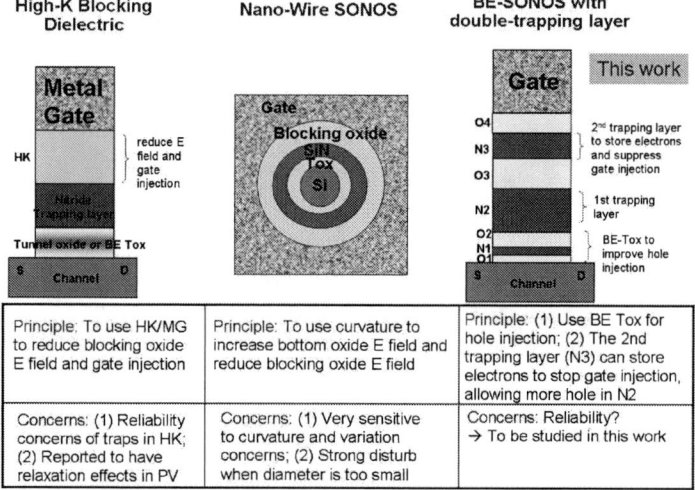

Table 1 Comparison of three approaches to overcome the erase saturation of SONOS-type charge-trapping NAND Flash devices.

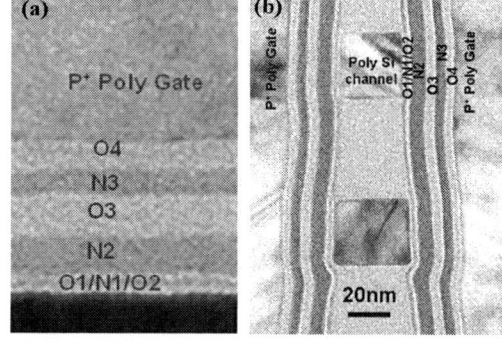

Figure 1 (a) The TEM picture of double-trapping BE-SONOS device with O1/N1/O2/N2/O3/N3/O4 multi layers grown on Si substrate. (b) The integrated double-trapping BE-SONOS in 3D Vertical Gate NAND Flash. Typical WL CD ~25nm, while BL CD ~30nm in our 3DVG device [9].

Theoretical WKB tunneling model [7] is used to analyze the device. In Fig. 4(a), the simulation shows good agreement with the experimental data in erase transient. The simulation can estimate the trapped charge in N2 and N3, as shown in Fig. 4(b). It is found that the stored charge areal density in N3 will exceed 5E12 cm^{-2} at long erasing time. Meanwhile, the first trapping layer (N2) will continue to be erased by substrate hole injection with hole density >1E13 cm^{-2}. Because N2 has larger weighting factor for Vt shift (it's closer to the channel) thus Vt can be continuously lowered even though N3 traps electrons at the same time. In other words, we separate the charge centroid of gate injection and channel injection, creating a larger erase window.

We also carry out the gate-sensing and channel sensing (GSCS) technique [8] to directly measure the stored charge distribution

978-1-4799-8002-4/14 $31.00 © 2014 IEEE

experimentally. The GSCS method compares two capacitors to extract the two variables (QN2 and QN3). The principle is shown in Fig. 5(a). Figure 5(b) and (c) show the first-time programming transient from fresh state. It is found that the +FN injected electrons are mostly stored in N2, while N3 has much fewer stored electrons. This indicates that N2 has a good capture efficiency and O3 can block most out tunneling from N2 toward N3. Figure 6 shows the subsequent 1st erasing and 2nd programming transient. For the first erasing, N3 traps electrons at longer erasing time, while N2 continues to trap hole allowing a deep erase for channel-sensing capacitor. The behavior of CS and GC capacitors are quite different due to the different V_{FB} weighting factor of N2 and N3. For the second programming, the stored electrons in N3 stay inert, but the trapped holes in N2 are recombined with electrons injected from the substrate. Therefore, the GSCS method directly confirms the charge storage dynamics of double-trapping BE-SONOS device, and is totally consistent with our theoretical model.

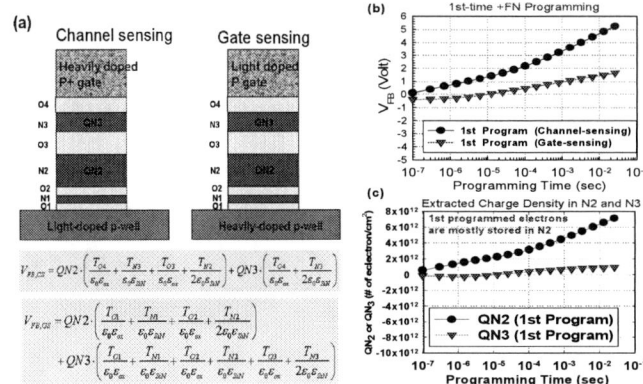

Figure 5 (a) Gate-sensing and channel-sensing (GSCS) method [8]. The equation of V_{FB} shift for GS and CS capacitors are shown. (b) The measured V_{FB} shift during 1st-time +FN (+20V) programming from fresh state. (c) The extracted QN2 and QN3 according to GSCS method.

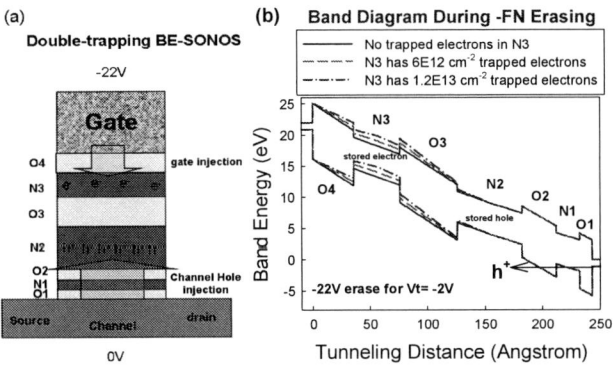

Figure 2 (a) Schematic plot of double-trapping BE-SONOS device during –FN erasing. Holes inject from substrate and are stored in N2. Meanwhile, gate injected electrons are trapped in N3. (b) The band diagram shows that when N3 trap electrons, the top blocking oxide O4's electric field is reduced, thus gate injection can be suppressed.

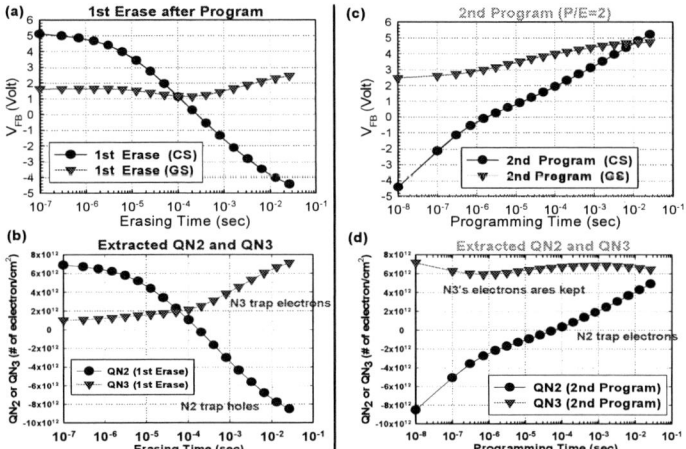

Figure 6 To continue Fig. 5 to perform 1st erasing and 2nd programming. (a) V_{FB} transient during 1st erasing from program. (b) Extracted QN2 and QN3 during 1st erase. It shows that N3 trap electrons, while N2 is erased by hole injection. (c) 2nd programming (P/E=2) transient. (d) Extracted QN2 and QN3. It shows that the previous injected electrons in N3 (by erase) is kept almost constant during 2nd +FN programming.

III. Performances of Double-Trapping BE-SONOS in 3DVG NAND

The typical PE memory window of the double-trapping BE-SONOS integrated in 3DVG NAND array [9] is shown in Fig. 7. The single cell possesses a large available memory window of 12V, with erase saturation Vt as low as -6V.

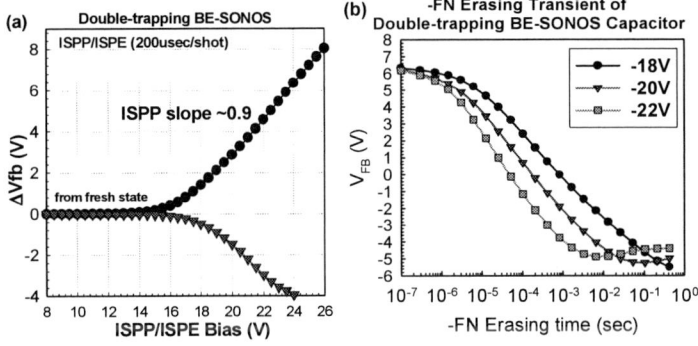

Figure 3 (a) Experimental data of ISPP/ISPE results of double-trapping BE-SONOS capacitor grown in bulk silicon (Fig. 1(a)). (b) The –FN erasing transient of double-trapping BE-SONOS capacitor. Low erase saturation of $V_{FB} < -5V$ is obtained. At -22V, it is possible to erase the device to $V_{FB} < -4V$ at 1msec.

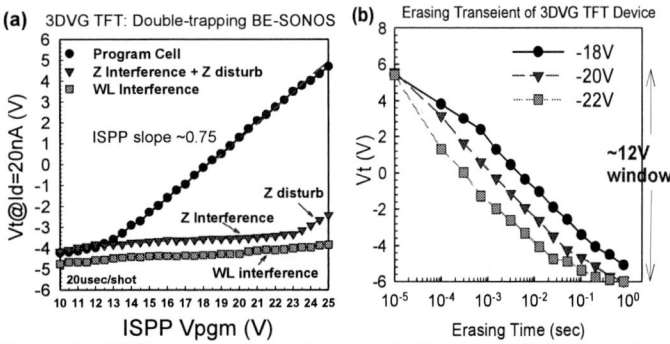

Figure 7 (a) ISPP programming and program-inhibit characteristics in a split-page 3DVG TFT NAND Flash using the double-trapping BE-SONOS device (Fig. 1(b)). (b) The –FN erasing transient. The 3DVG NAND erase is slower than capacitor, possibly due to the GIDL-induced erase that limits the channel hole generation speed. A P+ source dual-channel NAND [10] can boost the erase speed.

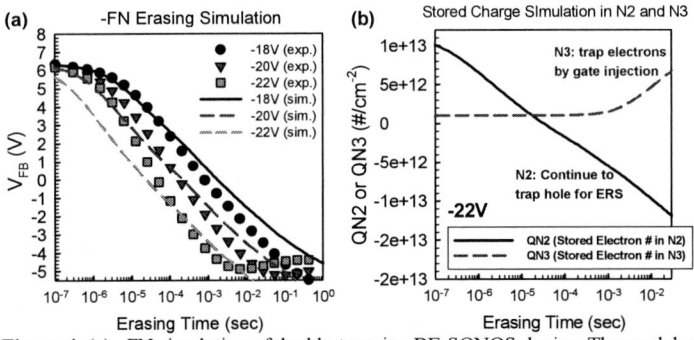

Figure 4 (a) –FN simulation of double-trapping BE-SONOS device. The model of BE-ONO barrier follows Ref. [7]. (b) The simulated trapped charge density in N2 and N3. At long erasing time, N3 starts to trap electrons, while N2 can be continuously erased by hole injection. Note that trapped electron density in N3 must exceed 5E12 cm^{-2} to stop gate injection.

The ISPP slope of 3DVG TFT device is ~0.75, which is smaller than the capacitor in Fig. 3(a). This is due to the fringe field effect in a small

3D transistor that modulates the tunnel oxide and blocking oxide E field and in-turn alters the FN tunneling ISPP behavior [11]. The Z-directional and WL interferences are minimized by optimizing the 3DVG topology.

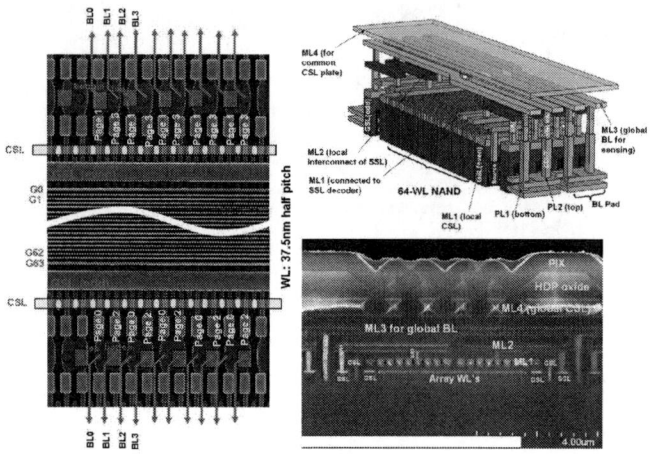

Figure 8 A full-integrated split-page 3DVG test-chip [9] is used to study the memory window of double-trapping BE-SONOS device.

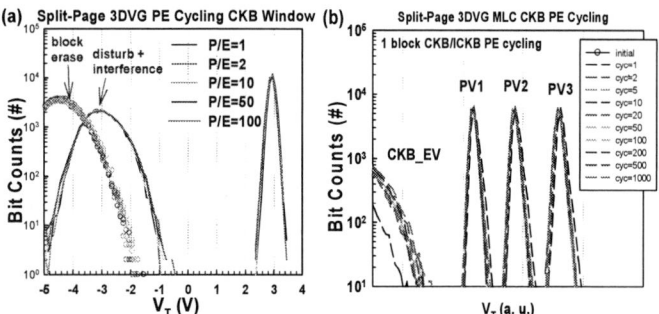

Figure 9 (a) A typical SCL CKB memory window result. Deep erased Vt distribution can be obtained by the help of double-trapping BE-SONOS device. As a result, an excellent SLC CKB programming window is obtained. Some shift of CKB-EV state from erased state is expected due to program disturb and interference. (b) 1000 PE Cycling of MLC CKB operation of one block. MLC operation suffers more disturb and interference modes than SLC Operation. The suitable memory window is still obtained.

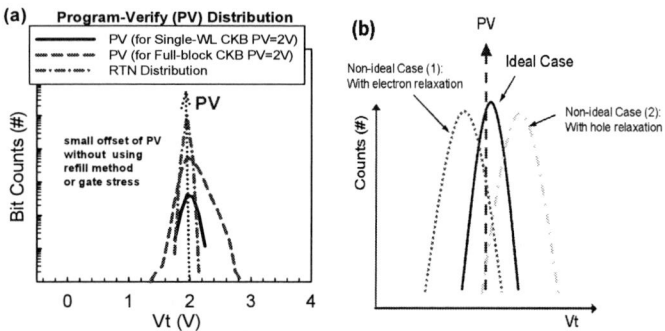

Figure 10 (a) Program-verify (PV) Vt distribution comparison. A single-WL PV shows tight distribution and small offset from the defined PV level, indicating small fast initial charge loss. (b) The schematic diagram to illustrate the effect of charge relaxation on the PV offset. PV state can be either shifted leftward (electron relaxation) or rightward (hole relaxation), if top blocking dielectric trap charges during +FN programming and is relaxed afterwards.

We utilize the 3DVG test chip [9] to study the memory window of double-trapping BE-SONOS, as shown in Fig. 8. Figure 9 shows the test-chip level SLC and MLC Vt distribution window when completing a full-block (64-WL) checkerboard (CKB) programming. Figure 9(a) shows that the erased Vt high bound can be lower than -2V after block erase. The interference and program disturb will shift the erase distribution

higher, but is still controlled much below Vt<0V, allowing a large design window. Figure 9(b) illustrates the MLC memory window. To create MLC operation there are more disturb and interferences that degrade the memory window, but the PV distribution can be kept reasonably tight.

An important characteristic is shown in Fig. 10(a). A single-WL CKB programmed state shows very tight distribution and small offset from the program-verify (PV) level defined in the sensing circuit. The tight distribution of PV is consistent with the RTN distribution. Figure 10(b) illustrates the impact of non-ideal PV distribution for comparison. If there is a fast electron loss after programming, then PV is shifted leftward, as commonly observed in most HK/MG devices [5, 6]. On the other hand, if the blocking dielectric trap holes after +FN programming (which is observed when nitride is in direct contact with poly gate in our other experiments), the hole is immediately relaxed after programming, leading to PV rightward offset.

For double-trapping BE-SONOS device, the PV offset (for one WL CKB) is minimal thus it suggests no fast charge relaxation effect. Thanks to the reliable top blocking oxide (O4) that block any unwanted charge relaxation toward gate.

It should be mentioned that the full-block CKB PV is still wider than a single-WL CKB PV, simply due to the many interference and back-pattern effects that inevitably broaden the distribution but not caused by the charge relaxation effect. The MLC PV tends to be rightward shifted due to the interferences but not caused by charge relaxation.

The PE cycling endurance is shown in Fig. 11. At high PE cycling number, the erased Vt has a higher Vt roll-up than programmed state. Detailed IdVg curves show that the S.S. is increased due to generation of interface traps. The merge point (corresponding to the charge neutrality level, [12]) of IdVg curves is different for erased and programmed states. We have clarified that the effect is due to the sidewall fringe field effect that moves the subthreshold current toward the sidewall at programmed state, as illustrated in Fig. 11(d). Thus the endurance degradation is not originated from the double-trapping layer, but simply from the O1/Si interface state generation. It can be further improved by strengthening the O1 post-stress immunity.

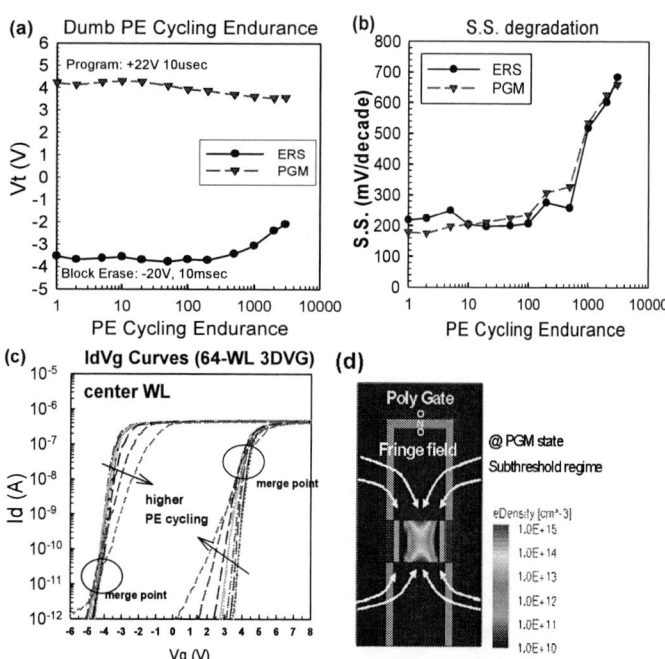

Figure 11 (a) Typical PE cycling endurance of the device under dumb-mode (1-shot) P/E test. Degradation is observed at high cycling counts. Erased state has more Vt shift than programmed state. (b) The measured subthreshold slope (S.S.) during cycling test. It shows dramatic increase at high cycling test. (c) The IdVg curves during cycling. The merge point of IdVg curves is different for erased and programmed state. (d) The 3D fringe field effect of PGM state. The inversion electrons tend to be started near sidewall at PGM state due to the fringe field effect, leading to less sensitivity of interface state trap (Dit) at Si/O1 interface and smaller Vt shift of PGM state after cycling.

IV. Reliability

High-temperature 150C baking results of the intrinsic CV capacitors are shown in Fig. 12. BE-SONOS retention is sensitive on the O2 quality and thickness. For the typical condition, BE-SONOS can retain excellent high-injection state (up to V_{FB} =+7V) even after 150C 1000hour baking.

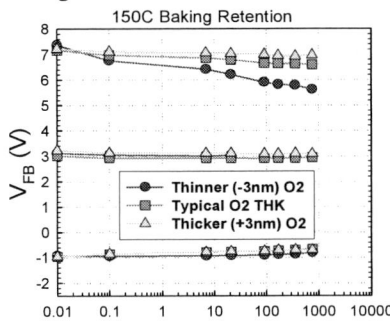

Figure 12 150C retention of double-trapping BE-SONOS CV capacitors. Thicker O2 can improve the retention significantly at high-injection states.

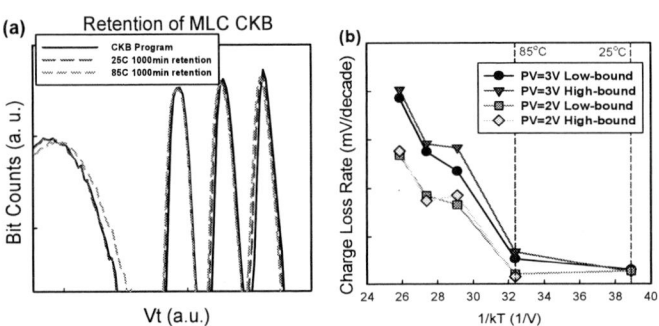

Figure 13 (a) Typical retention drift of MLC distribution at RT and 85C baking. The charge loss is group-behavior without tail distribution. (b) Charge loss rate of BE-SONOS at various baking temperature. The charge loss rate at <85C is below 30mV/decade, but it increases significantly at higher baking temperatures.

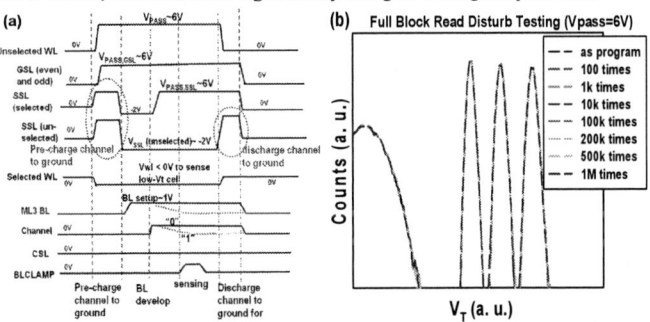

Figure 14 (a) optimized read waveform to suppress hot-carrier induced read disturb. During WL setup, the SSL is pre-turned-on to pre-charge channel to ground to avoid the channel boosting during WL setup. (b) Read disturb test of 3DVG BE-SONOS device. Optimized read waveform [9] is adopted to avoid hot-carrier injection. It shows small read disturb after 1M full-block read stress.

The retention results of the test chip are shown in Fig. 13(a). The fully-integrated device shows worse retention than CV capacitors because of much more complex integrated issues and some corner problems. On the other hand, the good news is that the charge loss is generally a group-behavior without tail distribution. The long-term charge loss summary of BE-SONOS test chip retention is shown in Fig. 13(b). At lower storage temperature (<85C), the charge loss rate can be smaller than 30mV/decade, providing enough sensing window after long-term storage. At higher temperatures, the charge loss rate significantly increases and does not follow a simple Arrhenius model. New retention qualification method is necessary for 3D charge-trapping devices [13].

The read disturb property of the 3DVG BE-SONOS device is shown in

Fig. 14. We adopt the optimized read waveform to suppress hot-carrier induced read disturb in 3D NAND [9]. Due to the flat topology without field enhancement effect, a high read endurance (>1M read stress for the full-block read) immunity is obtained. The device is very robust against gate stressing.

V. Summary and Advantages of this Novel Device

Table 2 summarizes the advantages of double-trapping BE-SONOS device. It can be designed with a nearly flat and planar topology which is extremely advantageous to be implemented in a minimal ~4F^2 design rule 3D NAND in order to maximize the memory density of 3D NAND Flash.

The barrier engineer tunnel dielectric solves the trade-off difficulty of erase and retention fundamentally. A practical device engineering strategy is to find the proper O1 and O2 that meets both erase speed and retention simultaneously.

The double-trapping engineering avoids the strong demand of very high-K materials together with very high-work function metal gate that complicates the integration challenges, and minimizes the reliability challenges of HK/MG as well.

Contrary to the curvature effect (which is to create asymmetrical E field of top and bottom oxide) of gate-all-around (GAA) devices, the double-trapping BE-SONOS instead creates asymmetrical "charge centroid" of gate injection (upper) and hole injection (lower) which also provides very effective erase saturation suppression. A substantial advantage is that the device is much more robust against a non-ideal etching angle of 3D NAND, while the GAA device would be highly sensitive to the curvature where the non-90 degree angle of channel hole introduces some device variations from top- to bottom-layer devices.

Lower gate work function or some poly gate doping variation are all tolerable because the device physics is to utilize the trapped electrons in N3 to stop gate injection but not to rely on work function or curvature effect. This greatly simplifies the process challenges in 3D NAND.

❏ The flat and planar topology without curvature enables minimal design rule for 3D NAND Flash that gives a cost efficient array, and also avoid read and program disturb by field enhancement effect 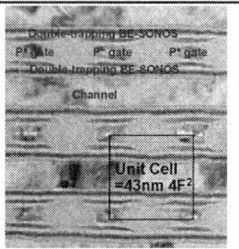
❏ Ready-to-use conventional materials (oxide, nitride, poly) without complex high-K/metal gate. Minimized learning time for reliability optimizations
❏ Solve erase and retention trade-off fundamentally → thinner O1 for faster erase, but thicker O2 for better retention. Thus it decouple the trade-off issue
❏ Robust against gate material work function change (such as poly gate doping variation)

Table 2 Summary of the advantages of double-trapping BE-SONOS.

References:
[1] S. C. Lai, et al, IEEE EDL, 2007, pp643-645. [2] R. Katsumata, et al, Symposium on VLSI Technology 2009, pp. 136-137, 2009. [3] C. H. Lee, et al, IEDM 2003, pp. 613-616. [4] H. T. Lue, et al, IEDM 2005, pp.555-558. [5] C. P. Chen, et al, IEDM 2010, pp. 118-121. [6] K. T. Park, et al, ISSCC 2014, pp. 334-335. [7] H. T. Lue, et al, IEEE TDMR 2010, pp. 222-232. [8] H. T. Lue, et al, IEEE-TED 2008, pp. 2218-2228. [9] C. H. Hung, et al, IEDM 2012, pp. 227-230. [10] H. T. Lue, et al, IEDM 2013, pp. 80-83. [11] H. T. Lue, et al, IEDM 2009, pp. 839-842. [12] W.C. Chen, et al, VLSI 2014. pp. .[13] E. S. Choi, et al, IEDM 2012, pp. 211-214.

Low-Frequency Noise and RTN on Near-Ballistic III-V GAA Nanowire MOSFETs

N. Conrad, M. Si*, S. H. Shin, J. J. Gu, J. Zhang, M. A. Alam, P. D. Ye**

School of Electrical and Computer Engineering, Purdue University, West Lafayette, IN 47906, U.S.A.

*Tel: 1-765-494-7611, Fax: 1-765-496-6443, *Email: msi@purdue.edu **Email: yep@purdue.edu*

Abstract

In this work, we report the first observation of RTN in highly scaled InGaAs GAA MOSFETs fabricated by a top-down approach. RTN and low frequency noise were systematically studied for devices with various gate dielectrics, channel lengths and nanowire diameters. *Mobility fluctuation* is confirmed to be the source of low-frequency noise, showing 1/f characteristics. Low-frequency noise was found to decrease as the channel length scaled down from 80 nm to 20 nm, indicating the near-ballistic transport in highly scaled InGaAs GAA MOSFET.

Introduction

InGaAs has been considered as one of the promising channel materials for CMOS logic circuit beyond 10 nm node because of its large electron injection velocity [1]. InGaAs gate-all-around (GAA) MOSFETs have been demonstrated to offer large drive current and excellent immunity to short channel effects down to deep sub-100 nm channel length [2]. On the other hand, classical theories suggest that low-frequency noise increases inversely with decreasing channel length. If it is true, this may negate some of the performance gain of short channel transistors [3-5]. Meanwhile, traditional oxide characterization methods, such as C-V and charge pumping, cannot be used for ultra-small devices without a body contact. Therefore, noise and RTN characterizations can be used as alternative probes to quantitatively analyze performance, variability and reliability of highly scaled devices [6-10]. Several groups have recently reported RTN of bottom-up synthesized *long-channel* InAs nanowire MOSFETs [11-13]. However, there is no report on RTN and low-frequency noise studies of highly scaled III-V GAA MOSFETs by top-down approach. In this work, we (i) report the first observation of RTN on top-down fabricated InGaAs GAA MOSFETs, (ii) examine the origin of low-frequency noise on highly scaled InGaAs GAA MOSFETs, (iii) systematically study the low-frequency noise and RTN characteristics of near-ballistic InGaAs GAA nanowire MOSFETs.

Experiments

Fig. 1(a) shows the schematic diagram and cross-sectional view of an InGaAs GAA MOSFET. The top-down fabrication process is shown in Fig. 1(b), which is the same as reported in Ref. [2]. The samples used for noise characterizations and device dimensions are summarized in Table 1. Samples A and B have a 0.5 nm Al_2O_3/4 nm $LaAlO_3$ stack (EOT = 1.2 nm), where Al_2O_3 was grown before $LaAlO_3$ for sample A and reverse order for sample B. Sample C has 3.5 nm Al_2O_3 as gate dielectric (EOT = 1.7 nm). The

RTN and low-frequency characterization setup is shown in Fig. 2. The gate voltage (V_{gs}) is supplied by a digital controllable voltage source. A Stanford SR570 battery-powered current amplifier is used as source voltage supply and monitor for the source current (I_s). I_s is used due to the relatively large junction leakage current in drain current (I_d). I_s shows more clearly the fundamental transport properties inside the nanowire. The amplifier output is directly connected to a Tektronix TDS5032B oscilloscope to record RTN signal and an Agilent 35670A dynamic signal analyzer to obtain the power spectrum density (PSD) of the noise of I_s. All noise measurements were performed at V_{ds} =50mV and at room temperature unless otherwise specified.

Results and Discussion

Figs. 3-4 show the well-behaved output and transfer characteristics of a GAA MOSFET with L_{ch}=W_{NW}=20 nm. PBTI measurement at V_{gs}=0.6, 0.8V on Sample A and B is shown in Fig. 5. PBTI measurement on Sample C can be found in [14]. V_T shifts less than 10 mV during noise measurements is ensured under maximum V_{gs}=0.4 V conditions. Fig. 6(a) and (b) show the RTN signals of an InGaAs GAA MOSFET, with L_{ch}=20 nm, W_{NW}=80 nm and 3.5 nm Al_2O_3 as gate dielectric, in time domain at V_{gs}=-0.025 V and V_{gs}=-0.075 V at 15°C. The histogram and lag plot at V_{gs}=-0.025 V of the same device are shown in Fig. 7(a) and (b). Two distinct current switching levels are observed, which clearly indicates the existence of a single active trap. Fig. 8 shows 'PSD of I_s' normalized by I_s^2 (i.e., S_{Is}/I_s^2) of the RTN signal shown in Fig. 6(a). A typical Lorentzian spectrum is shown in the noise spectrum with $1/f^2$ characteristics. Clear RTN signals are observed on about 1/3 of devices measured on Sample A, B and C, but only when V_{gs} is near threshold voltage (V_T). Fig. 9 shows the comparison of noise spectrum between a device with RTN signal and a device without RTN. The two devices share the same device dimension with L_{ch}=20 nm, W_{NW}=25 nm and 3.5 nm Al_2O_3 as gate dielectric. It is clear that the noise spectrum of the device without RTN shows 1/f characteristics, while the noise spectrum of the device with RTN is the superposition of 1/f noise spectrum and a Lorentzian spectrum. Fig. 10 shows (a) I_s histogram and (b), (c) RTN signals in time domain on a L_{ch}=20 nm, W_{NW}=25 nm device of Sample B, showing the superposition of RTN signal and mobility fluctuation (1/f) noise. This phenomenon suggests the fact that mobility fluctuation (rather than carrier number fluctuation) is the origin of low-frequency noise on devices without RTN signal. Fig. 11 shows S_{Is}/I_s^2 as a function of I_s on Sample A, B and C with L_{ch}=20nm and W_{NW}=20nm at f=10Hz which are weakly dependent on the different interfaces. All the three selected devices show 1/f spectrum

978-1-4799-8002-4/14 $31.00 © 2014 IEEE

without RTN. That S_{Is}/I_s^2 can be modulated by I_s indicates that the noise source is from channel rather than series resistance. In addition, S_{Is}/I_s^2 depends only weakly on the gate oxide, suggesting that oxide trapping and de-trapping induced carrier number fluctuation is not the source of low-frequency noise in this work. Fig. 12 shows the scaling metrics of S_{Is}/I_s^2 versus L_{ch} at f=10 Hz and W_{NW}=20 nm for Sample B. Normalized I_s noise is reduced as L_{ch} scaling down, which is opposite to the conventional noise L_{ch} scaling characteristic ($S_{Is}/I_s^2 \sim 1/L_{ch}$). This phenomenon leads to the main conclusion of this work: that near-ballistic transport of electrons in the channel is achieved, as determined through noise studies. As electrons from the source cannot equilibrate to lattice temperature immediately at drain contact, the conventional mobility fluctuation model, which assumes diffusive transport, can no longer be applied. In our near-ballistic InGaAs GAA MOSFETs, electrons encounter less scattering at smaller L_{ch} during transport from source to drain. Therefore, scattering induced mobility fluctuation decreases at small L_{ch} so that normalized I_s noise is reduced at small L_{ch}. This further confirms that mobility fluctuation is the origin of low-frequency noise for highly scaled InGaAs MOSFETs. Fig. 13 shows the relation between S_{Is}/I_s^2 and V_{ds} at V_{gs}=0V. Smaller normalized I_s noise is obtained at high V_{ds} because ballistic efficiency is higher at high V_{ds} than low V_{ds}. Fig. 14 shows the thermo-reflectance image of an InGaAs GAA MOSFET with L_{ch}=80 nm, W_{NW}=30 nm at V_{gs}=1 V. The drain side is heated at high V_{ds} by ballistic electrons, indicating that electrons travel substantial distance into the contact before reaching equilibrium with the lattice. This supports the conclusion of Fig. 13 that InGaAs GAA MOSFETs in this work are near-ballistic. Hot Carrier Injection measurement also confirms the near-ballistic transport in the devices as we reported in [15]. Fig. 15 shows the relation between normalized I_s noise with W_{NW}. S_{Is}/I_s^2 shows weak dependence on W_{NW}. Figs. 16-21 show the property of RTN in the same condition as in Fig. 6 from V_{gs}=-0.15V to 0V. Capture/emission time constants (τ_c/τ_e) are extracted and τ_c/τ_e distributions exactly follow Poisson distribution as predicted theoretically, as shown in Fig. 16. Fig. 17 shows the relation between τ_c, τ_e and V_{gs}. Fig. 17 studies the relation between τ_c/τ_e and V_{gs}. The positive correlation between τ_c/τ_e and V_{gs} indicates that electrons trapping and de-trapping occur between channel and gate oxide, as suggested in Ref. [10]. Fig. 19 shows the relation between time constants and the reciprocal of temperature (1000/T). τ_c and τ_e are extracted at 15°C, 30°C and 45°C and the activation energy (E_a) is also extracted. Fig. 21 shows the I_s histogram at different V_{gs}. A single peak at both low V_{gs} and high V_{gs} are obtained. Between -0.15V to 0.05V, double peaks are observed indicating the existence of RTN. At low V_{gs}, it is hard to observe the RTN signal because τ_e is longer than the measurement time. At high V_{gs}, RTN is negligible comparing with noise induced by mobility fluctuation because carrier number fluctuation induced noise drops faster than mobility induced noise with I_s increases. ΔI_d and $\Delta I_d/I_d$

relation with different V_{gs} of RTN signals is shown in Fig. 20, which confirms RTN signals are hard to be observed at high V_{gs}. Two-trap RTN signals are also observed on some of the devices, as shown in Fig. 22.

Conclusion

For highly scaled InGaAs GAA MOSFETs, mobility fluctuation is the source of low-frequency noise, showing 1/f characteristics. Low-frequency noise is suppressed at shorter channel length due to the near-ballistic transport at deep sub-100nm. RTN is for the first time observed on top-down InGaAs GAA MOSFETs only around threshold voltage because RTN is negligible compared to mobility fluctuation induced noise at high V_{gs}.

Acknowledgement

The authors would like to thank X. W. Wang, X. Lou, R. G. Gordon for the valuable discussions and technical supports. The work is partly supported by AFOSR Task project.

Reference

[1] J. A. del Alamo, "Nanometre-scale electronics with III-V compound semiconductors." *Nature* 479, 317-323, (2011).

[2] J. J. Gu *et. al.*, "20–80nm Channel length InGaAs gate-all-around nanowire MOSFETs with EOT= 1.2 nm and lowest SS= 63mV/dec", *IEDM Tech. Dig.* 633 (2012).

[3] P. Ren *et al.*, "New observations on complex RTN in scaled high-κ/metal-gate MOSFETs — The role of defect coupling under DC/AC condition", *IEDM Tech. Dig.* 778 (2013).

[4] H. Miki *et al.*, "Statistical measurement of random telegraph noise and its impact in scaled-down high-κ/metal-gate MOSFETs ", *IEDM Tech. Dig.* 450 (2012).

[5] N. Tega *et al.*, "Increasing threshold voltage variation due to random telegraph noise in FETs as gate lengths scale to 20 nm," *VLSI Tech. Dig.* 50 (2009).

[6] T. Grasser *et al.*, "The Paradigm Shift in Understanding the Bias Temperature Instability: From Reaction–Diffusion to Switching Oxide Traps," *Electron Devices, IEEE Transactions on*, 58, 3652 (2011).

[7] J. P. Campbell *et al.*, "Random telegraph noise in highly scaled nMOSFETs," *Reliability Physics Symposium, IEEE International*, 382 (2009)

[8] W. Feng *et al.*, "Fundamental origin of excellent low-noise property in 3D Si-MOSFETs ~ Impact of charge-centroid in the channel due to quantum effect on 1/f noise", *IEDM Tech. Dig.* 630 (2011).

[9] J. Chen *et al.*, "Experimental study of channel doping concentration impacts on random telegraph signal noise and successful noise suppression by strain induced mobility enhancement", *VLSI Tech. Dig.* 184 (2013).

[10] T. Nagumo *et al.*, "Statistical characterization of trap position, energy, amplitude and time constants by RTN measurement of multiple individual traps", *IEDM Tech. Dig.* 628 (2010).

[11] J. Salfi *et al.*, "Direct observation of single-charge-detection capability of nanowire field-effect transistors", *Nature Nanotechnology*, 5(10), 737-741 (2010).

[12] J. Salfi *et al.*, "Probing the gate–voltage-dependent surface potential of individual InAs nanowires using random telegraph signals", *ACS Nano*, 5(3), 2191-2199 (2010).

[13] G. Holloway *et al.*, "Trapped charge dynamics in InAs nanowires", *Journal of Applied Physics*, 113(2), 024511 (2013).

[14] S. H. Shin *et al.* "Impact of Nanowire Variability on Performance and Reliability of Gate-all-around III-V MOSFETs", *IEDM Tech. Dig.* 188 (2013).

[15] S. H. Shin *et al.*, "Origin and Implications of Hot Carrier Degradation of Gate-all-around nanowire III-V MOSFETs," *Reliability Physics Symposium, IEEE International*, 978 (2014).

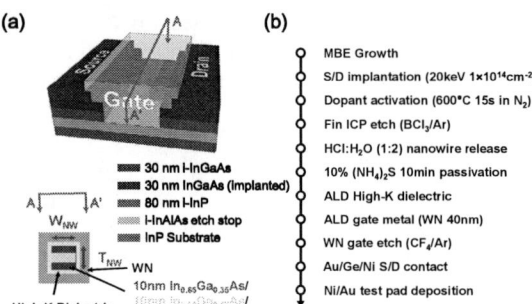

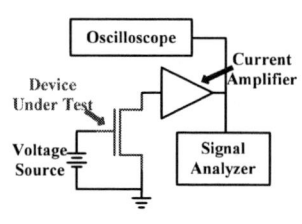

Table 1 Description of samples and device dimensions used in this work.

	Sample A (Al₂O₃ first)	Sample B (LaAlO₃ first)	Sample C (Al₂O₃ only)
Channel Material	10 nm In$_{0.65}$Ga$_{0.35}$As/ 10 nm In$_{0.53}$Ga$_{0.47}$As/ 10 nm In$_{0.65}$Ga$_{0.35}$As	10 nm In$_{0.65}$Ga$_{0.35}$As/ 10 nm In$_{0.53}$Ga$_{0.47}$As/ 10 nm In$_{0.65}$Ga$_{0.35}$As	10 nm In$_{0.65}$Ga$_{0.35}$As/ 10 nm In$_{0.53}$Ga$_{0.47}$As/ 10 nm In$_{0.65}$Ga$_{0.35}$As
L$_{ch}$ (nm)	20	20,30,50,80	20
W$_{NW}$ (nm)	20	20,25,30,35	20
T$_{NW}$ (nm)	30	30	30
Gate oxide	0.5nm Al₂O₃/ 4nm LaAlO₃	4nm LaAlO₃/ 0.5nm Al₂O₃	3.5nm Al₂O₃
EOT (nm)	1.2	1.2	1.7

Fig. 1 (a) Schematic diagram, cross-section view and (b) fabrication process of an InGaAs GAA MOSFET.

Fig. 2 Noise characterization setup diagram.

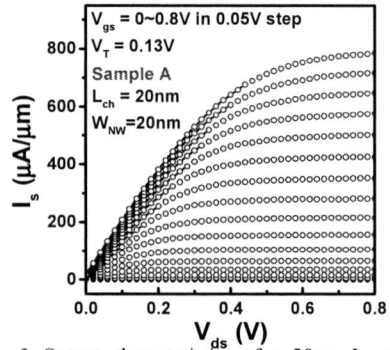

Fig. 3 Output characteristics of a 20nm L$_{ch}$ GAA MOSFET with Al₂O₃/LaAlO₃ gate dielectric (Sample A, EOT=1.2nm) and W$_{NW}$=20nm. I$_s$ is used due to relatively large junction leakage current in I$_d$.

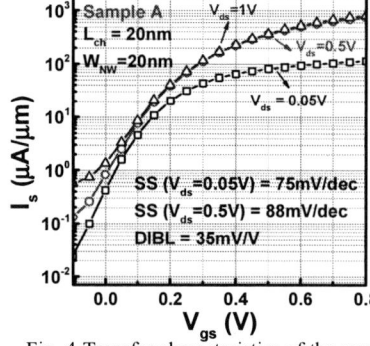

Fig. 4 Transfer characteristics of the same device shown in Fig. 3.

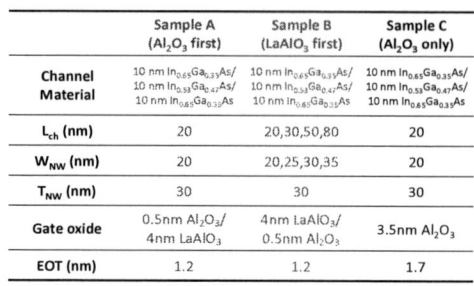

Fig. 5 Time evolution of ΔV$_T$ in PBTI (0-10⁴s) under stress of 0.6V and 0.8V for Sample A and Sample B devices with L$_{ch}$=80nm and W$_{NW}$=20nm.

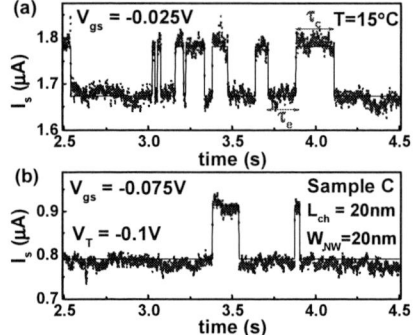

Fig. 6 I$_s$ fluctuation due to RTN in (a) V$_{gs}$=-0.025 V, (b) V$_{gs}$=-0.075 V on InGaAs GAA MOSFETs measured at 15°C.

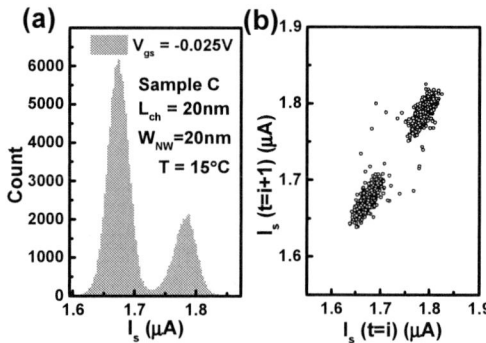

Fig. 7 (a) Histogram and (b) lag plot of a typical RTN signal shown in Fig. 6(a). The histogram and lag plot show two RTN levels.

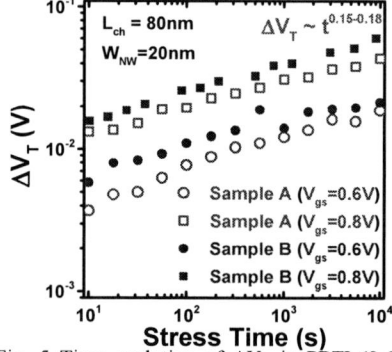

Fig. 8 Normalized I$_s$ noise of RTN signal shown in Fig. 6(a), showing 1/f² characteristics.

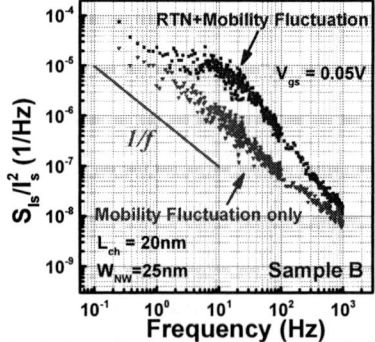

Fig. 9 Normalized I$_s$ noise of Sample B devices with RTN signal and without RTN signal. Noise spectrum of device without RTN is attributed to mobility fluctuation.

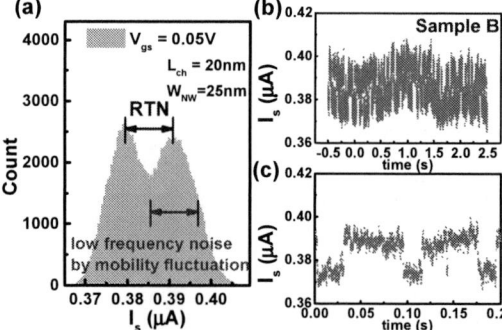

Fig. 10 (a) Histogram of a RTN signal of sample B with L$_{ch}$=20nm and W$_{NW}$=25nm. (b) and (c) RTN signals in time domain of the same signal as (a). (c) is a time segment inside (b).

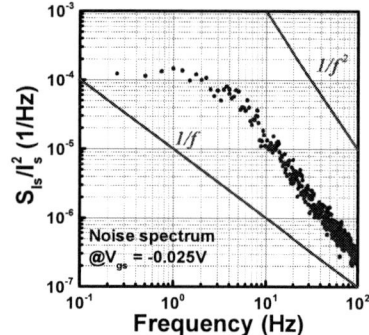

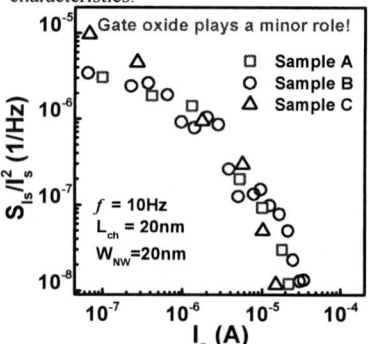

Fig. 11 Normalized I$_s$ noise at f=10 Hz for Sample A, B and C devices with L$_{ch}$=20 nm and W$_{NW}$=20 nm. Devices with different gate oxides exhibit similar noise level showing weakly dependent on interfaces.

978-1-4799-8002-4/14 $31.00 © 2014 IEEE

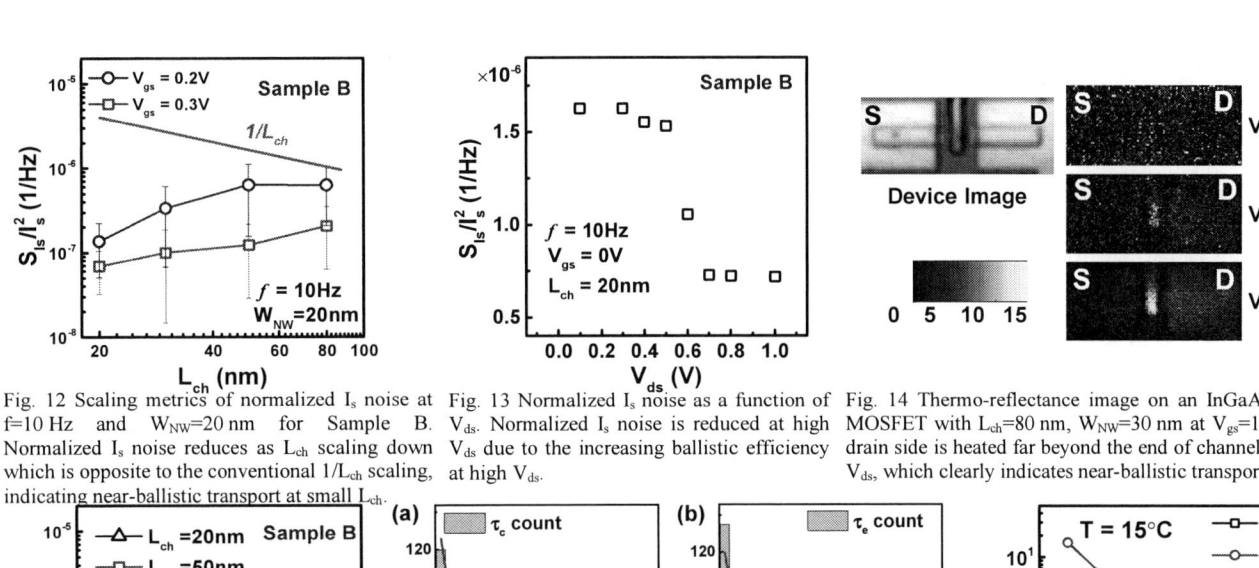

Fig. 12 Scaling metrics of normalized I_s noise at f=10 Hz and W_{NW}=20 nm for Sample B. Normalized I_s noise reduces as L_{ch} scaling down which is opposite to the conventional $1/L_{ch}$ scaling, indicating near-ballistic transport at small L_{ch}.

Fig. 13 Normalized I_s noise as a function of V_{ds}. Normalized I_s noise is reduced at high V_{ds} due to the increasing ballistic efficiency at high V_{ds}.

Fig. 14 Thermo-reflectance image on an InGaAs GAA MOSFET with L_{ch}=80 nm, W_{NW}=30 nm at V_{gs}=1 V. The drain side is heated far beyond the end of channel at high V_{ds}, which clearly indicates near-ballistic transport.

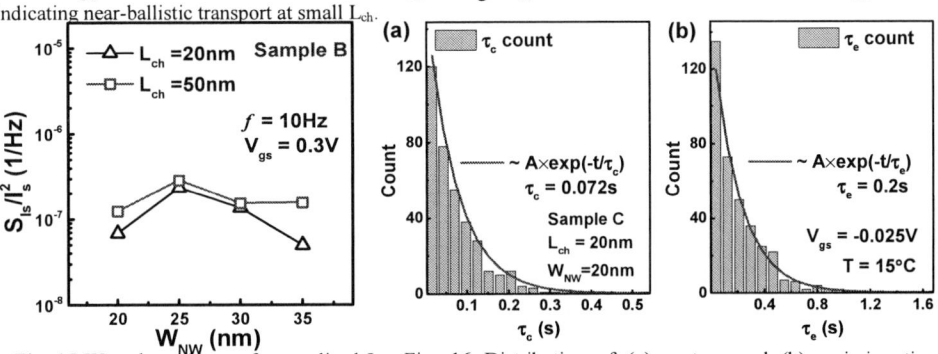

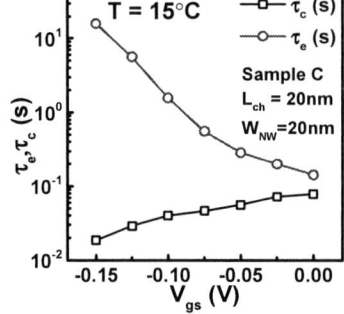

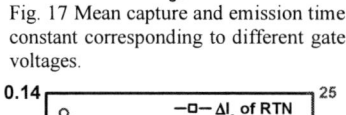

Fig. 15 W_{NW} dependence of normalized I_s noise at f=10 Hz and L_{ch}=20, 50 nm for Sample B. Normalized I_s noise shows weak W_{NW} dependence.

Fig. 16 Distribution of (a) capture and (b) emission time constant of RTN signal shown in Fig. 4.

Fig. 17 Mean capture and emission time constant corresponding to different gate voltages.

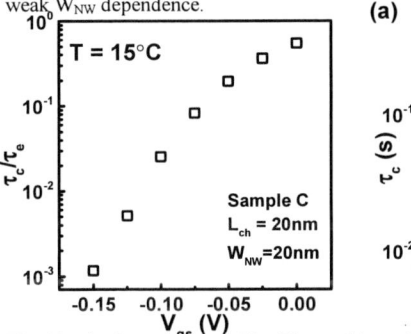

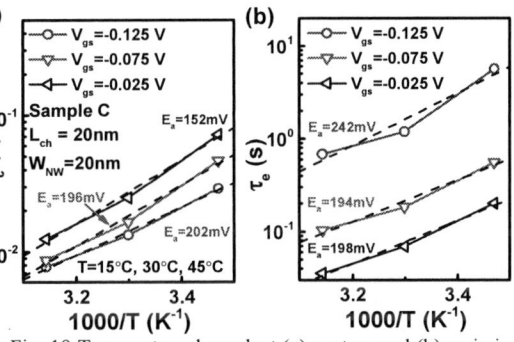

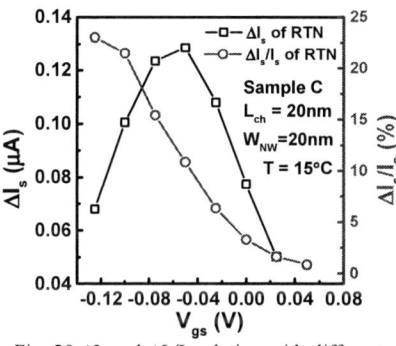

Fig. 18 τ_c/τ_e dependence on V_{gs}. The positive correlation indicates electron trapping happens between channel and gate oxide.

Fig. 19 Temperature dependent (a) capture and (b) emission time constant of RTN in device shown in Fig. 4.

Fig. 20 ΔI_s and $\Delta I_s/I_s$ relation with different V_{gs} of RTN signals shown in Fig. 6.

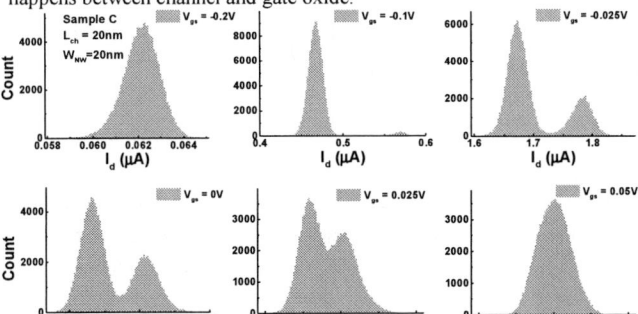

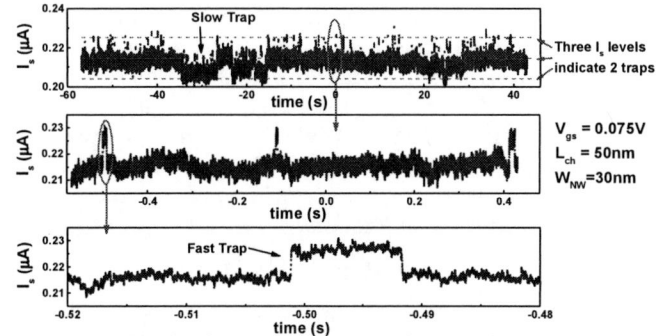

Fig. 21 Relation between I_s histogram and V_{gs} of RTN signals shown in Fig. 4. RTN signals are observed around V_T in most of devices of Samples A, B and C. This characteristic confirms that carrier number fluctuation induced I_s fluctuation is negligible under high V_{gs} and mobility fluctuation is the dominant noise source at on-state.

Fig. 22 RTN with 2 level trap response measured on Sample B at L_{ch}=50 nm, W_{NW}=30 nm and V_{gs}=0.075 V. (a) RTN signal in 100 s time length. A slow trap with τ_c~3 s can be observed. (b) and (c) show the RTN signals in 1s and 0.04 s time length. A fast trap with τ_c=10 ms is observed.

978-1-4799-8002-4/14 $31.00 © 2014 IEEE

RTN and PBTI-induced Time-Dependent Variability of Replacement Metal-Gate High-k InGaAs FinFETs

J. Franco[*], B. Kaczer, N. Waldron, Ph.J. Roussel, A. Alian, M. A. Pourghaderi, Z. Ji, T. Grasser[1], T. Kauerauf, S. Sioncke, N. Collaert, A. Thean, and G. Groeseneken[2]

imec, Kapeldreef 75, 3001 Leuven – Belgium

[1]Technische Universität Wien – Austria; [2]also with ESAT Dept., KU Leuven – Belgium; [*]Jacopo.Franco@imec.be

Abstract—We study RTN and PBTI in nanoscale InGaAs FinFETs fabricated on 300mm Si wafers. The average instability is found to be comparable to planar structures, but significantly larger when compared to Si devices. Although the novel devices follow the same time-dependent variability statistics and the corresponding area-scaling as their Si counterparts, a larger stochastic impact of single defects on the device characteristic is found to induce larger aging-related variance. We ascribe this to a more percolative channel conduction induced by still excessive interface and channel defectivity.

Introduction

High mobility channel FinFETs will be required for further CMOS performance enhancement in N7 and beyond [1]. Ge and III-V compounds are the frontrunners for *p*- and *n*-type channels respectively [2]. High intrinsic electron mobility ($\sim$3000cm^2/Vs [3,4]) has been recently demonstrated in InGaAs channel devices with "self-cleaning" [5] Al_2O_3–based dielectric stacks, with promising interface quality (D_{it} $\sim$3-5$\times$10^{11} cm^{-2}eV^{-1} [6]) and subthreshold swings ($\sim$75 mV/dec [3]). Moreover, imec has recently demonstrated the integration of *InGaAs FinFETs on 300mm Si wafers*, with replacement metal gate Al_2O_3/HfO_2 gate stack (Fig. 1) [7]. In this paper, we focus on the *PBTI reliability* of these InGaAs FinFETs. While the mean instability is found to be identical to previously studied planar structures [8] and significantly larger w.r.t. Si devices, *the nanoscale fins* with varying effective area (Table I) *enable microscopic insights into the instability via Random Telegraph Noise* (RTN) *and time-dependent variability* induced by capture/emission of channel electrons in/from oxide defects. We observe that *InGaAs FinFETs follow the same time-dependent variability statistics and the corresponding area-scaling as their Si counterparts* [9-12]. However, experimental data show *larger-than-Si device-to-device variability* in InGaAs devices: *we trace it to larger stochastic impact of single defects* on the device characteristics and *we discuss its possible origins*. We argue that while showcasing the best, i.e., tail, behavior demonstrates the potential of InGaAs, *correct understanding and description of the instability distributions in nanoscale non-planar devices (e.g., FinFETs and Nanowires) are essential for process optimization of the new technology.*

PBTI, Hysteresis and RTN

PBTI measurements in large area FinFETs reveal the *same reliability challenge* previously observed in planar InGaAs devices (Fig. 2, solid symbols). In particular, a detrimental weak voltage dependence of PBTI is consistently found

($\gamma\sim$1.5), ascribed to a wide distribution of defect levels in Al_2O_3 (Fig. 2b) [8]. *High pressure* H_2 *and* D_2 *anneals,* expected to passivate primarily dangling bonds, *did not yield any significant electron trapping reduction* (Fig. 2a, crosses), which therefore points to an intrinsic issue of this gate oxide. Along with excessive hysteresis induced by multiple charged defects with long emission times, frequent single electron (de-)trapping events inducing significant fluctuations (i.e., *noise*) in the I_D-V_G characteristic of nanoscale FinFETs have been observed (Fig. 3). Single defect charging with gigantic electrostatic impact (e.g., ΔV_{th} $\sim$50mV, Fig. 4) are not rare. Such large fluctuations in nanoscale FETs are already understood in terms of a potential perturbation of the underlying channel percolation configuration induced by single charged oxide defects (Fig. 5) [10,13]. By collecting the observed single-defect-induced ΔV_{th} in a Cumulative Distribution plot (Fig. 6a), an exponential distribution is seen, same as in Si devices [9]. The average impact per charge η (i.e., $\sim$ the inverse of the slope of the distribution) is found to scale inversely with area in InGaAs FinFET (Fig. 6b), also consistently with Si data [10,11]. However η is significantly larger (i.e. $\sim$4$\times$) w.r.t. the charge sheet approximation for the electrostatic impact of a single charge (η_0=q/C_{ox}, Fig. 6b dashed), and also w.r.t. Si FinFETs where we have previously observed η/$\eta_0$$\sim$2$\times$ (Fig. 6b dotted, based on experimental data in [10]). Since η is expected to scale proportionally with the oxide Capacitance Equivalent Thickness (CET) [14,15], InGaAs devices suffer a penalty due to the larger channel carrier centroid displacement from the interface (i.e., 'dark space') induced by quantum effects in a low DOS semiconductor [3,16]. The CET-dependence, however, cancels out when normalizing η by η_0 (cf. Table II)—the *larger (normalized) single defect impact* has to be therefore ascribed to more percolative channel conduction. We have previously shown that excessive N_{it} [17] and channel material defectivity [18] induce larger η (cf. Fig. 5). We argue these two factors are responsible for the larger normalized single defect impact in InGaAs devices.

Hysteresis distributions

Each nanoscale device is expected to include a Poisson-distributed number of charging oxide defects, each defect inducing the above discussed exponentially distributed ΔV_{th} [9]. Therefore, nominally identical devices can show very disparate hysteresis (Fig. 7). Note that *it is relatively common to find devices showing almost negligible hysteresis (Fig. 7a)—this is just a consequence of the stochastic nature of trapping in small area devices, and cannot be claimed as*

978-1-4799-8002-4/14 $31.00 © 2014 IEEE

improved reliability of some devices. Fig. 8a shows V_{th} distributions on the up- and down-sweep of I_D-V_G hysteresis measurements of multiple devices. Along with the median V_{th}-shift, an increased variance is found in the down-sweeps. This and the hysteresis ΔV_{th} distribution (Fig. 8b) are again both clear consequences of the statistics previously proposed based on Si planar data (Eq. 1, Table II). Note that the hysteresis of each device is uncorrelated with the device V_{th0} (Fig. 8c), confirming the stochastic nature of the phenomenon. Consistent with the statistics (Eq. 3) is the median hysteresis observed to be independent of the device area (Figs. 9,10), as well as larger variance observed in small area devices (Fig. 9, bars) [17]. Based on measured hysteresis variance for varying device areas it is possible to derive η independently again through Eq. 3. In agreement with the RTN step heights study, η is found to scale inversely with the device area, and to be ~4× larger than η_0 and ~2× larger than in Si devices (Fig. 11, cf. Fig. 6b). A good agreement of the values extracted with the two methods serves as a *proof of the correctness of our statistical framework in describing time-dependent variability also for InGaAs FinFETs.*

Device-to-device PBTI variability

Fig. 12 shows relaxation traces measured on a set of nominally identical devices after the same PBTI stress. Individual discharge events with distributed step heights are visible. The median PBTI shift shows a typical ~log(t) relaxation trend (Fig. 12b), and it is found to be independent of the device geometries (Fig. 13). This result implies that *gate stack optimization for improved reliability pursued on simple planar test vehicles is valid also for FinFETs.* By considering datasets pertaining to increasing relaxation times (see Fig. 12a), PBTI distributions with decreasing $\langle \Delta V_{th} \rangle$ and therefore decreasing variance (cf. Eq. 3) are obtained (Fig. 14, symbols). These PBTI-induced ΔV_{th} distributions are clearly non-Gaussian; the asymmetry and the distinct relation between median and variance are well described by Eq. 1 assuming η/η_0 =6 (Fig. 14, lines). To understand this larger η/η_0 value, hysteresis distributions were measured with increasing maximum V_G (Fig. 15): the median shift follows a weak power law of the gate voltage (γ~1.5) consistently with large area device data (cf. Fig. 2); a weak dependence of η on the maximum stress is also observed (Fig. 15b), and ascribed to *progressively worse percolative channel conduction due to increasing amounts of charged near-substrate oxide defects* inducing additional potential perturbations (see Fig. 5). Nevertheless, at low operating voltages of relevance for III-V logic, the ratio η/η_0 =4 observed above is confirmed.

Projections

Based on the experimental results summarized in Table III, Fig. 16 shows projected ΔV_{th} distributions at device end-of-life. Compared to Si n-FinFETs, ~10× *larger median shifts are expected* (Fig. 16b), *with significant additional variance to be addressed by additional design margins* (Fig. 16c). Such projections are valuable when optimizing new-technology

designs [19]—the larger ΔV_{th} variance might be, e.g., tackled by designing with multi-fin devices, at some expense of wafer area (Fig. 16a).

Conclusions

In nanoscale devices I_D-V_G hysteresis and PBTI-induced ΔV_{th} need to be studied in terms of their statistical distributions. *InGaAs FinFETs showed both larger median shifts and aging-induced variance w.r.t. Si counterparts.* The latter was traced to *a larger stochastic impact of single defects on the device characteristic*, ascribed to the CET-penalty of low DOS semiconductors (i.e., wider 'dark space') and chiefly to a more percolative channel conduction induced by still excessive interface and channel defectivity. *Suppressing defectivity* in channel, interface, and gate oxide appears *crucial for successful introduction of III-V devices.* This path is orthogonal to introducing suitable dielectrics with favorable defect level distribution to suppress excessive electron trapping at operating voltages [20].

References

(1) M. Bohr, "The evolution of scaling from the homogeneous era to the heterogeneous era", in *Proc.* IEDM, pp. 1.1.1-6, 2011;

(2) K. J. Kuhn, "Considerations for Ultimate CMOS Scaling", in TED 59(7), pp. 1813-1828,2012;

(3) A. Alian *et al.*, "Impact of the channel thickness on the performance of ultrathin InGaAs channel MOSFET devices", in *Proc.* IEDM, pp. 16.6.1-4, 2013;

(4) S. Takagi, M. Takenaka, "High mobility CMOS technologies using III-V/Ge channels on Si platform", in *Proc.* ULIS, pp. 1-4, 2012;

(5) M.L. Huang *et al.*, "Surface passivation of III–V compound semiconductors using atomic-layer-deposition-grown Al_2O_3", in APL(87), pp. 252104.1–3, 2005;

(6) T. Hoshii *et al.*, "Reduction in interface state density of Al_2O_3/InGaAs metal-oxide-semiconductor interfaces by InGaAs surface nitridation", in JAP 112(7), pp. 073702.1-8, 2012;

(7) N. Waldron *et al.*, "An InGaAs/InP quantum well finfet using the replacement fin process integrated in an RMG flow on 300mm Si substrates", in *Proc.* VLSI, pp. 1-2, 2014;

(8) J. Franco *et al.*, "Suitability of high-k gate oxides for III–V devices: A PBTI study in $In_{0.53}Ga_{0.47}As$ devices with Al_2O_3", in *Proc.* IRPS, pp. 1-6, 2014;

(9) B. Kaczer *et al.*, "Origin of NBTI variability in deeply scaled pFETs", in *Proc.* IRPS, pp. 26-32, 2010;

(10) J. Franco *et al.*, "Impact of single charged gate oxide defects on the performance and scaling of nanoscaled FETs", in *Proc.* IRPS, pp. 1-6, 2012;

(11) C. Prasad *et al.*, "Bias temperature instability variation on SiON/Poly, HK/MG and trigate architectures", in *Proc.* IRPS, pp. 6A.5.1-7, 2014;

(12) S.H. Shin *et al.*, "Impact of nanowire variability on performance and reliability of gate-all-around III-V MOSFETs", in *Proc.*IEDM, pp. 1-4, 2013;

(13) A. Asenov *et al.*, "RTS Amplitude in Decananometer MOSFETs: 3-D Simulation Study", in TED 50(3), pp. 839-845, 2003;

(14) A. Ghetti *et al.*, "Comprehensive Analysis of Random Telegraph Noise Instability and Its Scaling in Deca-Nanometer Flash Memories", in TED 56(8), pp. 1746-1752, 2009;

(15) J. Franco *et al.*, "Reduction of the BTI time-dependent variability in nanoscaled MOSFETs by body bias", in *Proc.* IRPS, pp. 2D.3.1-6, 2013;

(16) K. Kalna *et al.*, "Benchmarking of Scaled InGaAs Implant-Free NanoMOSFETs", in TED 55(9), pp. 2297-2306, 2008;

(17) M. Toledano-Luque *et al.*, "Degradation of time dependent variability due to interface state generation", in *Proc.* VLSI, pp. T190-191, 2013;

(18) M. Toledano-Luque *et al.*, "Quantitative and predictive model of reading current variability in deeply scaled vertical poly-Si channel for 3D memories", in *Proc.* IEDM, pp. 9.2.1-4, 2012;

(19) H. Kukner *et al.*, "Scaling of BTI reliability in presence of time-zero variability", in *Proc.* IRPS, pp. CA.5.1-7, 2014;

(20) J. Franco *et al.*, "Understanding the suppressed charge trapping in relaxed- and strained-Ge/SiO_2/HfO_2 pMOSFETs and implications for the screening of alternative high-mobility substrate/dielectric CMOS gate stacks", in *Proc.* IEDM, pp. 15.2.1-4, 2013.

978-1-4799-8002-4/14 $31.00 © 2014 IEEE

Acknowledgement: This work was performed as part of imec's Core Partner Program. It has been in part supported by the European Commission under the 7th Framework Programme (Collaborative project MORDRED, contract No. 261868). Discussions with Prof. M. Heyns, Drs. L-Å Ragnarsson, R. Degraeve, M. Toledano-Luque, P. Weckx, H. Kukner, A. Vais, D. Lin, T. Ivanov are acknowledged.

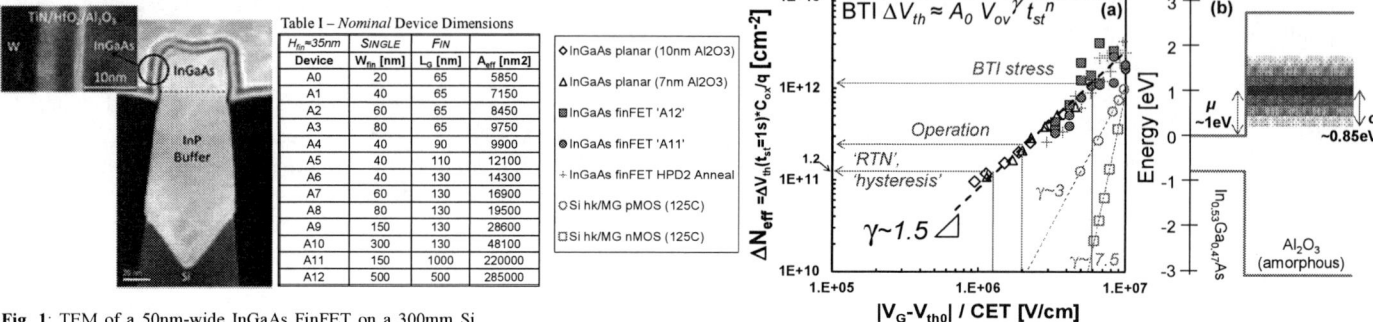

Table I – *Nominal* Device Dimensions

H_{fin}=35nm	*SINGLE*	*FIN*	
Device	W_{fin} [nm]	L_g [nm]	A_{eff} [nm2]
A0	20	65	5850
A1	40	65	7150
A2	60	65	8450
A3	80	65	9750
A4	40	90	9900
A5	40	110	12100
A6	40	130	14300
A7	60	130	16900
A8	80	130	19500
A9	150	130	28600
A10	300	130	48100
A11	150	1000	220000
A12	500	500	285000

◇ InGaAs planar (10nm Al2O3)
△ InGaAs planar (7nm Al2O3)
■ InGaAs finFET 'A12'
● InGaAs finFET 'A11'
+ InGaAs finFET HPD2 Anneal
○ Si hk/MG pMOS (125C)
□ Si hk/MG nMOS (125C)

Fig. 1: TEM of a 50nm-wide InGaAs FinFET on a 300mm Si wafer [7] used in this work. Inset shows the gate stack consisting of 2nm Al2O3/3nm HfO2 layers (expected EOT ~1.6nm). Table I reports *nominal* dimensions for the devices used in this work, later identified by the labels 'A0→A12'.

Fig. 2: **(a)** BTI shifts [$\Delta V_{th}(t_{stress}=1s)$] plotted in a CET-independent benchmark $\Delta N_{eff}(E_{ox})$ [$\Delta N_{eff}=\Delta V_{th}C_{ox}/q$, $E_{ox}\sim V_{ov}$/CET]. Similar shifts are observed for InGaAs planar [8] and FinFET devices without or with high pressure anneals (crosses). Note the large BTI shifts and the very weak voltage dependence ($\gamma\sim1.5$, formula inset) as compared to Si high-k/MG BTI data (green symbols), ascribed to **(b)** a wide energy distribution of oxide electron traps in Al2O3 [8].

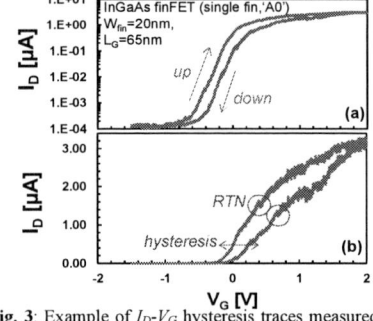

Fig. 3: Example of I_D-V_G hysteresis traces measured on nanoscale InGaAs FinFETs [**(a)** log-lin, **(b)** lin-lin scales]. Note the noise induced by (dis)charging of individual defects (RTN). Defects with τ_e>>τ_c remain charged after the V_G ramp-up, inducing hysteresis.

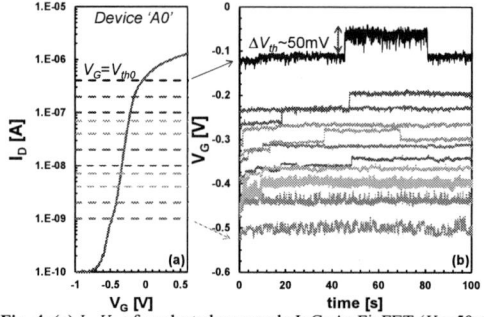

Fig. 4: **(a)** I_D-V_G of a selected nanoscale InGaAs FinFET (V_D=50mV). The biases corresponding to the indicated current levels (dashed lines) are maintained for 100 s to measure **(b)** RTN traces. An individual defect with gigantic impact on the device characteristic ($\Delta V_{th}\approx$50mV at $V_G\approx V_{th0}$ [10]) is consistently observed in this device.

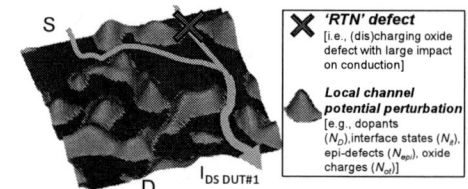

Fig. 5: A sketch of the current flow in a nanoscale channel, proceeding through percolation paths in a non-uniform potential profile [10,13,14]. An oxide defect located *on-top* of a percolation confinement point (×) can induce a large RTN fluctuation. Point charges of different nature (dopants [13-15], interface states [17], epitaxial-defects [18], *near-substrate* oxide charges [cf. Fig.15]) induce a more percolative conduction, thus enhancing the impact of individual defects.

'RTN' defect [i.e., (dis)charging oxide defect with large impact on conduction]

Local channel potential perturbation [e.g., dopants (N_D),interface states (N_{it}), epi-defects (N_{epi}), oxide charges (N_{ox})]

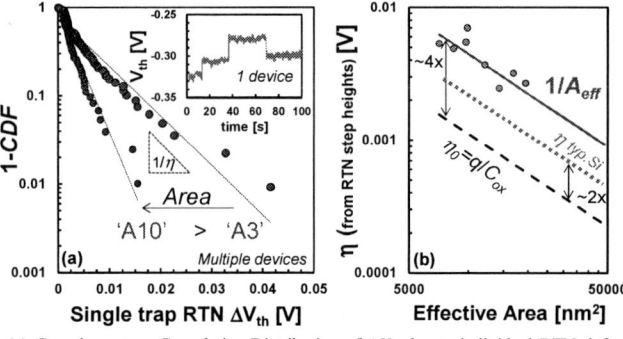

Fig. 6: **(a)** Complementary Cumulative Distribution of ΔV_{th} due to individual RTN defect (inset) observed in nanoscale InGaAs FinFETs. Note the wider exponential distribution for the smaller device area ('A3'<'A10', cf. Table I). **(b)** The average impact per defect η is inversely proportional to the device effective area, as for Si devices [10,11]. Note the large magnitude of η, ~4× larger than the charge sheet approximation for a single charge (dashed line η_0=q/C_{ox}) and ~2× larger w.r.t. Si finFETs (dotted line, from experimental data in [10]).

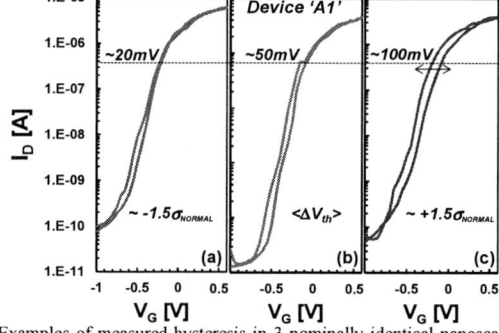

Fig. 7: Examples of measured hysteresis in 3 nominally identical nanoscale InGaAs finFETs (area 'A1', cf. Table I). Each device shows a different hysteresis magnitude, [**(a)**~20mV, **(b)**~50mV, **(c)**~100mV]. Despite the large median hysteresis, it is relatively easy to find devices showing reduced shifts. Note individual defect (dis)charging events inducing *noise* in the I_D-V_G traces.

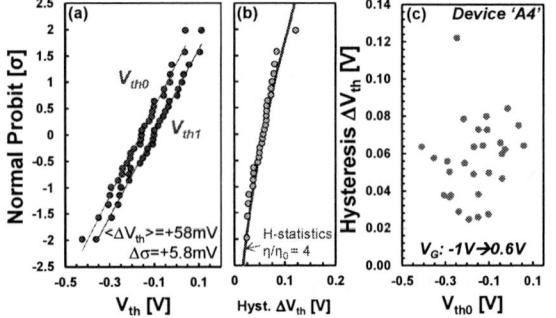

Table II: H-statistics (Ref. [9])
(Poisson-distributed number of charged defects with exponential-distributed impacts)

$$H_{\eta,\langle N_T\rangle}(\Delta V_{th}) = \sum_{n=0}^{\infty}\frac{e^{-\langle N_T\rangle}\langle N_T\rangle^n}{n!}\left[1-\frac{n}{n!}\Gamma(n,\Delta V_{th}/\eta)\right] \quad \text{Eq. (1)}$$

$$\text{Eq. (2)}\left\langle \Delta V_{th}(t)\right\rangle = \eta\left\langle N_T(t)\right\rangle, \quad \sigma^2_{\Delta V_{th}}(t) = 2\eta\left\langle \Delta V_{th}(t)\right\rangle \text{Eq. (3)}$$

Average single defect impact (η): dependences [14,15,17,18]

$$\eta \propto \frac{CET\sqrt{N_{pp}}}{A}, \quad \eta_0 = \frac{q}{C_{ox}} \propto \frac{CET}{A}, \quad \frac{\eta}{\eta_0} \propto \sqrt{N_D+N_{it}+N_{epidefects}+\cdots}$$

Fig. 8 (left): **(a)** Example of V_{th} distributions measured on the *up* (blue) and *down* (red) sweeps of hysteresis measurements (devices 'A4'), cf. Fig. 7. Hysteresis induces both a median V_{th}-shift (~58mV) and additional variance ($\Delta\sigma$~5.8mV). **(b)** The distribution of ΔV_{th} experienced by each device (symbols) is well described by the H-statistics (line; cf. Table II) [9]. **(c)** The ΔV_{th} measured in each device is uncorrelated with its initial V_{th}, confirming the stochastic nature of the phenomenon.

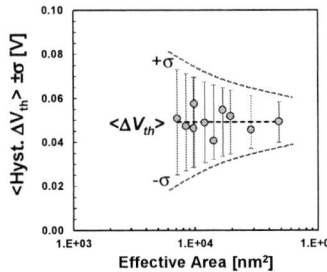

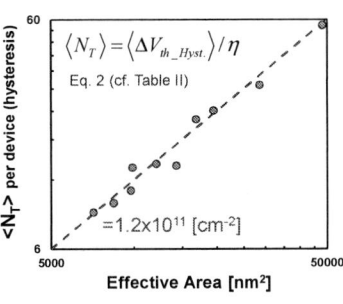

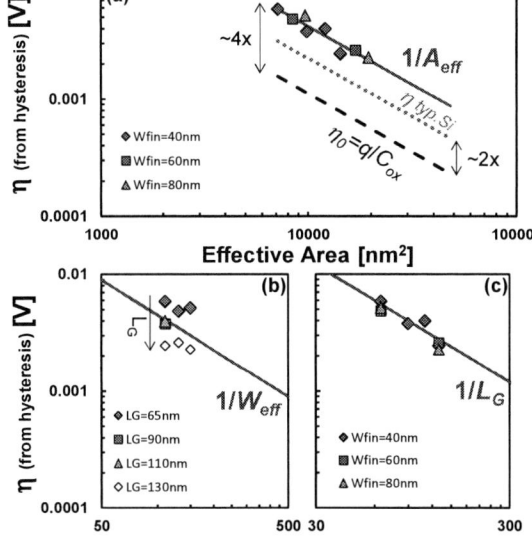

Fig. 9: Median hysteresis ΔV_{th} is shown to be independent of the device geometry [17]. This is a consequence of the average number of oxide defects per device $\langle N_T \rangle$ scaling proportionally with area (Fig. 10) and the average impact per defect η scaling inversely with area (cf. Table II, Eq. 2). However, larger variance is observed in smaller devices due to enhanced single defect impact η (cf. Eq. 3).

Fig. 10: The average number of defects contributing to the hysteresis increases proportionally with the device area [17]. Note: the same defect density in nanoscale devices as in large area planar and FinFET devices [cf. Fig. 2(a)] is found.

Fig. 11: Average single defect impact η extracted by modeling measured hysteresis distributions with H-statistics (cf. Eq. 3, Table II). η is observed to scale inversely with **(a)** the device effective area, and thus also with **(b)** the device effective width, and **(c)** the gate length. As for the extraction based on RTN traces (cf. Fig. 6), η values ~4× larger than η_0 and ~2× larger w.r.t. Si [10] are found. The consistency of the extracted η values confirms the validity of Eq.1 ('H-statistics') for describing the ΔV_{th} distributions also in InGaAs FinFETs.

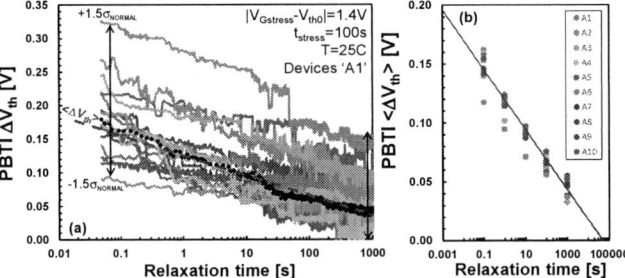

Fig. 12: **(a)** Example of PBTI relaxation traces measured in nanoscale InGaAs FinFETs (devices 'A1'). Individual defect discharge events are visible (cf. RTN traces in Fig. 4b) [9]. The median ΔV_{th} (dotted) is observed to follow a ~$\log(t)$ relaxation (dashed) from ~0.1 to 1000 s, **(b)** independently of the device geometry. As observed for planar structures [8], BTI recovery is faster in InGaAs devices as compared to Si (i.e., the ΔV_{th} recovers almost completely in a measurable period); hence, at short time scales of relevance for logic (i.e., ps), even larger than measured shifts are expected.

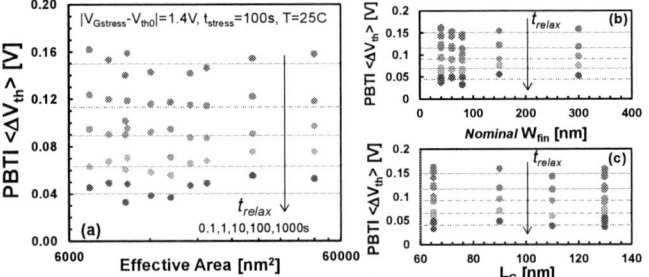

Fig. 13: **(a)** The median PBTI-induced ΔV_{th} is observed to be independent of the InGaAs FinFET effective area. No significant dependencies of the **(b)** fin width, nor **(c)** gate length are observed. Datasets for different relaxation times (i.e., for varying $\langle \Delta V_{th} \rangle$) are shown.

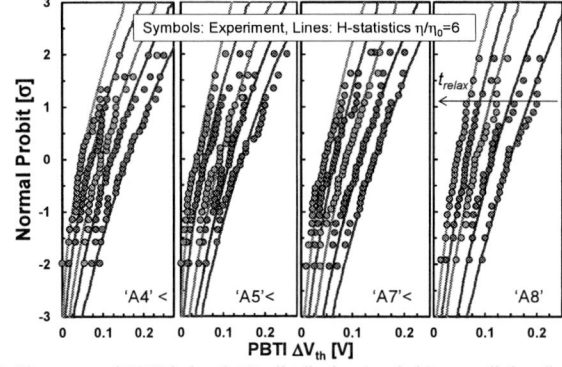

Fig. 14: The measured PBTI-induced ΔV_{th} distribution (symbols) are well described by the H-statistics with $\eta/\eta_0 = 6$ (lines). Data for 4 different nanoscale device areas ('A4'<'A5'<'A7'<'A8') are shown. Measured distributions for increasing relaxation times are shown for each area to highlight the relation between the median V_{th}-shift and the variance of the distribution, perfectly captured by the H-statistics (Table II).

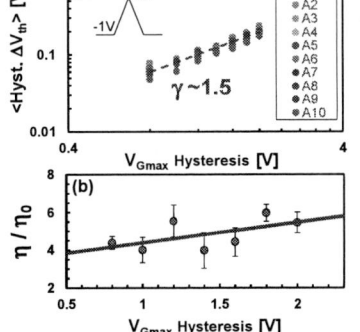

Fig. 15: **(a)** Independently of the device area, the median hysteresis shifts follow a power-law of the maximum applied V_G (inset), with exponent $\gamma \sim 1.5$, consistently with large area planar and finFET (cf. Fig. 2a). **(b)** A weak dependence of the extracted η/η_0 values is noted and ascribed to progressive worsening of the percolative conduction (see Fig. 5). Even at low voltages a large η/η_0 value (~4×, compared to ~2× for Si [10]) is observed.

Table III – Target Operation and Extracted BTI Parameters

N-finFET	InGaAs	Si
Target Operating V_{DD} [V]	0.5	1
Operating overdrive [V] (=2/3 * V_{DD})	0.33	0.66
Target CET: EOT+darkspace [nm]	1.8	1.4
Operating E_{ox} [MV/cm]	~2	~5
H_{fin} [nm]	35	35
Target W_{fin} [nm]	10	10
Target L_G [nm]	30	30
Effective Area (single fin) [nm²]	2400	2400
PBTI time exponent (n) [8,19]	0.1	0.16
η/η_0 (from Figs. 6,11) [10]	4	2
$\Delta N_{eff}(t_{st}=1s)$ @Op.E_{ox} (from Fig.2)[cm⁻²]	~2e11	~1e10
$\Delta V_{th}(t_{st}=10Y)$ @Op.E_{ox} [mV]	118	15

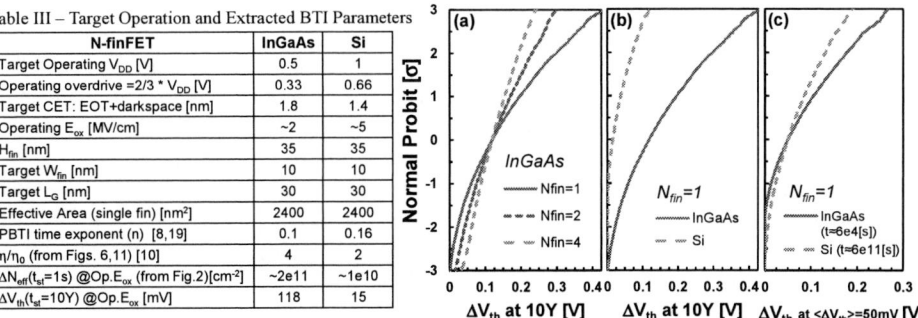

Fig. 16: Projections of device end-of-life ΔV_{th} distribution based on experimental data shown here and summarized in Table III. **(a)** The wide ΔV_{th} variance induced by the larger impact of individual defects in InGaAs FinFETs (i.e., $\eta/\eta_0 \sim$4×) could be tackled by increasing the effective device area by using multi-fin devices (projections shown for N_{fin}=1,2,4), at some expense of wafer area; **(b)** compared to Si FinFETs, ~10× larger $\langle \Delta V_{th} \rangle$ at 10 year is expected, together with an increased variance. **(c)** Projected distributions for a fixed $\langle \Delta V_{th} \rangle$=50mV in InGaAs and Si finFETs highlight the additional variance.

Direct Observation of Self-heating in III-V Gate-all-around Nanowire MOSFETs

S. H. Shin*, M. Masuduzzaman, M. A. Wahab, K. Maize, J. J. Gu, M. Si,
A. Shakouri, P. D. Ye, and M. A. Alam*

*E-mail: {shin136, alam}@purdue.edu, Phone: (765) 494-5988, Fax: (765)-494-2706
Department of ECE, Purdue University, West Lafayette, IN 47907, USA

Abstract

Gate-all-around MOSFETs use multiple nanowires to achieve target I_{ON}, along with excellent 3D electrostatic control of the channel. Although self-heating effect (SHE) has been a persistent concern, the existing characterization methods, based on indirect measure of mobility and specialized test structures, do not offer adequate spatio-temporal resolution. In this paper, we develop an *ultra-fast, high resolution* thermo-reflectance (TR) imaging technique to (*i*) directly observe the increase in local surface temperature of the GAA-FET with different number of nanowires (NWs), (*ii*) characterize/interpret the time constants of heating and cooling through high resolution transient measurements, (*iii*) identify critical paths for heat dissipation, and (*iv*) detect in-situ time-dependent breakdown of individual NW. Our approach also allows indirect imaging of quasi-ballistic transport and corresponding drain/source asymmetry of self-heating. Combined with the complementary approaches that probe the internal temperature of the NW, the TR-images offer a high resolution map of self-heating in the surround-gate devices with unprecedented precision, necessary for validation of electro-thermal models and optimization of devices and circuits.

Introduction

Multi-gate devices, such as, FinFET, Gate-all-around transistors (GAA-FET) improve 3D electrostatic control of the channel, but the corresponding increase in self-heating may compromise both performance and reliability. Although the self-heating effect of FinFET appears significant, but tolerable [1], the same may not be true for GAA geometry [2, 3], especially in quasi-ballistic regime where hot spots and non-classical heat-dissipation pathways may lead to localized heating and damage. The existing reports of the SHE on the SOI, FinFET or GAA-FET have so far relied either on indirect electrical measurements such as AC output conductance and gate resistance [4, 5] with inherent temporal delays, or on optical infra-red ($\lambda >$ 1.5um) imaging that cannot resolve deep sub-micron features so that it requires customized large structure. As a result, although the SHE has become a critical issue in GAA-FET, it has so far been impossible to fully resolve the spatio-temporal features of the SHE.

In this paper, we first develop an *ultra-fast, high resolution* thermo-reflectance (TR) imaging technique to directly observe the local time-dependent rise in the surface temperature, $\Delta T(x, y, t)$. A variety of devices with different number of nanowires (NW) and oxide thicknesses are explored,

and the effect of these parameters on the self-heating are interpreted and analyzed. For example, high resolution transient measurements allow us to characterize the time constants of heating and cooling of the channel, define surround-gate oxides as the primary heat conduction pathways for thermal dissipation, and interpret NW-specific self-heating and degradation of the transistor.

Experimental Setup

The devices used in this study are InGaAs GAA n-MOSFET (Fig. 1), with different oxides thicknesses (T_{ox}), channel lengths (L_{ch}), and # of NWs. The fabrication process is described in [6, 7] and the device dimensions are listed in Table 1. The steep subthreshold slope, reported experimentally in [6, 7], is reproduced by the 3D Sentaurus simulation (Fig. 2), confirming excellent electrostatic control on GAA-FET [8].

During the TR imaging [9, 10], a high-speed LED ($\lambda =$ 530nm) pulse illuminates the device, and a synchronized CCD camera captures the reflected image with $\sim$ 250nm spatial resolution (Fig. 3). Theoretically, this technique relies on the change of the complex refractive index of a material with differential increase in temperature(ΔT), so that the change in local reflectance of the device surface relates to ΔT as,

$$\frac{\Delta R}{R_0} = \frac{1}{R_0} \cdot \frac{dR}{dT}\bigg|_{T=T_0} \Delta T \equiv \boldsymbol{k} \cdot \Delta T$$

where $\boldsymbol{k}$ (K^{-1}) is the thermoreflectance coefficient [10]. The calibration of $\boldsymbol{k}$ allows a CCD image to be interpreted as a map of $\Delta T(x, y)$, with 50mK resolution. For the transient measurement of $\Delta T(x, y, t)$, the device is periodically turned on and off by V_{DS} pulse (Fig. 3b) allowing the channel to heat and cool, respectively. By controlling the delay of the LED pulse with respect to the beginning of the V_{DS} pulse, the TR image can capture different phases of the transient heating and cooling kinetics, with $\sim$ 50ns resolution. As a basic validation, Fig. 4 shows that $\Delta T \propto$ Power, as expected.

Characterization of Self-Heating

The high spatio-temporal resolution of TR imaging provides new insights into the transient heating/cooling of a GAA-FET as a function of #NW. Fig. 5 shows that during the ON (OFF) state of the V_{DS} pulse, the channel region heats (cools) at $\sim$ 200ns timescale. The steady state temperature (ΔT_{SS}) scales with the #NW, indicating significant thermal cross-talk among the NWs. Indeed, $\Delta T_{SS} \sim$ 50K at the gate metal surface for a 19-NWs transistor implies even higher self-heating inside the channel [2].

978-1-4799-8002-4/14 $31.00 © 2014 IEEE   510

Time Constants: To understand the dynamics of heating/cooling at the operating frequency, it is also important to characterize the time constants for heating and cooling carefully and precisely. To determine the time resolution needed to capture the transient temperature rise, we reduce LED pulse width (τ_{LED}) from 1.6us to 50ns, and check if the heating transients are fully resolved and independent of τ_{LED}. Fig. 6a shows the transient heating of the channel surface after V_{DS} pulse is turned ON and characterized with different τ_{LED}. The heating transients overlap for $\tau_{LED} \leq 400$ns, suggesting that $\tau_{LED} \sim 400$ns provides sufficient temporal resolution. A plot of the effective thermal time-constants, obtained by fitting the heating transients in Fig. 6a and summarized in Fig. 6b, confirms the assertion. Once the required τ_{LED} is determined, the transient heating and cooling for transistors with different number of NWs are measured (Fig. 7a). Fig. 7b shows that the increase in thermal cross-talk with the # of NWs increases saturated ΔT_{SS}. The time constants also increase with #NW indicating that the devices with larger geometry need more time to reach the maximum temperature. Physically, the increase in the time constants reflects the increase *effective* thermal resistance, confirmed by the increasing slope of the power dissipation vs. ΔT curves for transistors with different #NW (Fig. 8).

Optimization of Self-Heating and Reliability

Measurements

Identification of Heat Conduction Channel: The ON current can be improved by reducing SHE; this requires identification and subsequent optimization of the heat conducting channels. To find the primary heat conduction channel among the substrate, source-drain contacts, or the gate contacts (see Fig. 1), we measured self-heating for transistors with different T_{ox} (Fig. 9a). We find that the transistor wrapped with thicker oxide (Al_2O_3) shows reduced SHE, indicating dominant heat flow along S/D: the thicker oxide offers *lower* thermal resistance to S/D (Fig. 9b); the opposite would have been true if substrate or gate channels were dominant. A solution of the heat equation in the relevant geometry confirms this assertion that the heat flux escapes laterally to the S/D contacts along the oxide itself (Fig. 9c) (note that Al_2O_3 has higher thermal conductivity ($= 35W/(m \cdot K)$) than the immediate gate metal-WN). The premise is also supported by the observation of increased temperature on S/D contact metal in Fig. 10. In order to see clear ΔT_{SS} change depending on power dissipation, V_{DS} is varied from 0 to 4V under same $V_{GS} = 1$V. As expected, the asymmetry of heating near drain side (vs. source) reflects asymmetry in heat generation in quasi-ballistic sub-100nm GAA transistor. The spatial extent of heating in the drain pad (~1um), marked should not be misinterpreted as being due to direct heating by the ballistic electrons. Instead, the ballistic electrons are likely to act as a heat source at the drain-edge (~100nm) [8]; the subsequent diffusion of heat in the metal is observed by the TR-approach.

Reliability Measurements: Unlike the indirect methods used to date, the high spatio-temporal resolution of TR images can be used to detect the variability and degradation of individual NW (e.g. V_{th} shift, breakdown, which impacts the local ON current and consequently local temperature). As an illustrative example, following a gate stress for certain time (Fig. 11), the channel abruptly becomes very hot, reflecting dielectric BD and thermal cross-talk among the NWs. Soon thereafter, a few of the NWs are destroyed. With the broken NWs excluded and the cross-talk suppressed, $\Delta T(x, y)$ of the remaining NWs is restored to pre-BD levels (Fig. 11 (right)).

Conclusions

The high spatio-temporal resolution of the TR imaging offers unprecedented and fundamentally new insights into the mechanics and kinetics of self-heating (e.g., degree of self-heating, dominant heat conduction channel, dynamics of channel breakdown) of the emerging multi-gate technology. A nuanced use of this versatile technique will help calibrate quasi-ballistic electro-thermal modeling tools, assess the relative merits of different multi-gate topologies, and can eventually improve the cell/circuit layout to suppress SHE as a source of variability and reliability in modern hyper-scaled IC technology.

Acknowledgement

We acknowledge Birck Nanotechnology Center for the fabrication and characterization facilities. Prof. Ye thanks Xinwei Wang and Prof. Roy G. Gordon from Harvard University for the technical support in device fabrication.

References

[1] S. Ramey, A. Ashutosh, C. Auth, J. Clifford, M. Hattendorf et al., in IEEE International Reliability Physics Symposium (IRPS), 2013 pp. 4C.5.1-4C.5.5.

[2] S. H. Shin, M. Masuduzzaman, J. J. Gu, M. A. Wahab, N. Conrad, M. Si, P. D. Ye, M. A. Alam, in IEEE International Electron Device Meeting (IEDM), 2013 pp. 7.5.1-7.5.4.

[3] R. Wang, J. Zhuge, C. Liu, R. Huang, D. W. Kim, D. Park, Y. Wang, in IEEE International Electron Device Meeting (IEDM), 2008 pp. 753-756.

[4] W. Jin, W. Liu, Samuel K. H. Fung, Philip. C. H. Chan, Chenming Hu, IEEE Transactions on Electron Devices, vol.48, no.4, pp.730-736, April 2001.

[5] K. Jenkins, J. Sun, J. Gautier, IEEE Transactions on Electron Devices, vol.44, no.11, pp.1923-1930, Nov 1997.

[6] J. J. Gu, Y. Q. Liu, Y. Q. Wu, R. Colby, R. G. Gordon, and P. D. Ye, in IEEE International Electron Device Meeting (IEDM), 2011 pp. 769-772.

[7] J. J. Gu, X. W. Wang, H. Wu, J. Shao, A. T. Neal, M. J. Manfra, R. G. Gordon, P. D. Ye, in IEEE International Electron Device Meeting (IEDM), 2012 pp. 633-636.

[8] S. H. Shin, M. A. Wahab, M. Masuduzzaman, M. Si, J. J. Gu, P. D. Ye, M. A. Alam, in IEEE International Reliability Physics Symposium (IRPS), 2014 pp. 4A.3.1-4A.3.6.

[9] K. Maize, E. Heller, D. Dorsey, A. Shakouri, in IEEE International Reliability Physics Symposium (IRPS), 2013 pp. CD.2.1-CD.2.3.

[10] Zhang, Radiometric Temperature Measurements, Volume 43: II. Applications (Experimental Methods in the Physical Sciences), Academic Press, 2009.

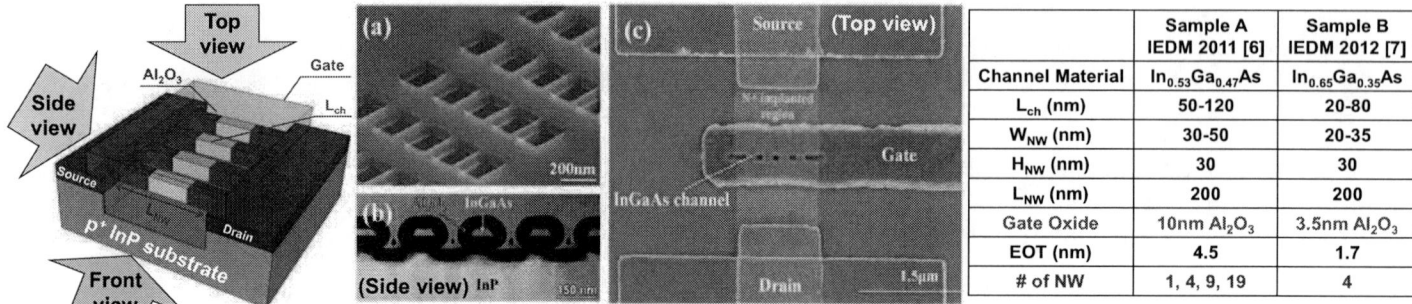

Fig. 1: (left) Schematic image of an InGaAs GAA NW n-channel MOSFET. (a) SEM image of parallel NWs. (b) STEM image of the cross section of the InGaAs NWs (Side view). (c) SEM image of the parallel InGaAs NWs (Top view). Images taken from Ref. [6].

Table. 1: The description of the two types of samples (different T_{ox} and different number of the NWs, $L_{ch} = 70\sim80nm$, and $W_{NW} = 30nm$) used in this study.

	Sample A IEDM 2011 [6]	Sample B IEDM 2012 [7]
Channel Material	$In_{0.53}Ga_{0.47}As$	$In_{0.65}Ga_{0.35}As$
L_{ch} (nm)	50-120	20-80
W_{NW} (nm)	30-50	20-35
H_{NW} (nm)	30	30
L_{NW} (nm)	200	200
Gate Oxide	10nm Al_2O_3	3.5nm Al_2O_3
EOT (nm)	4.5	1.7
# of NW	1, 4, 9, 19	4

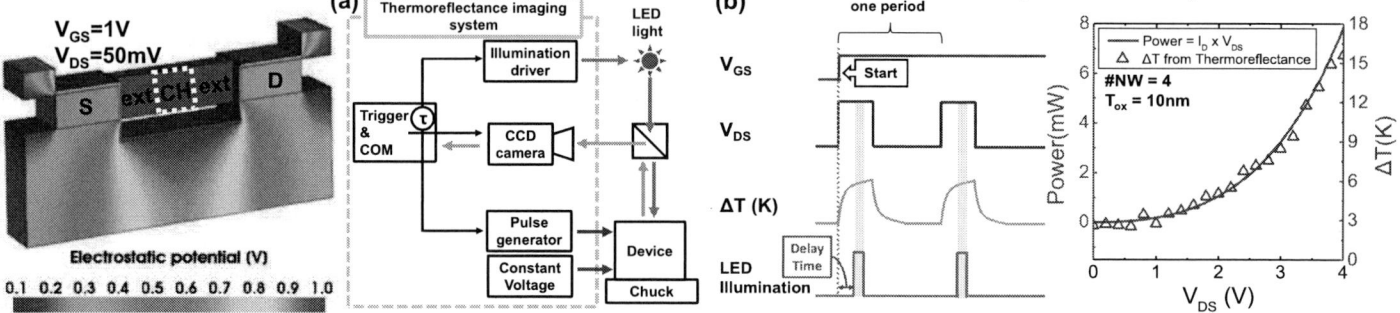

Fig. 2: Simulated potential profile of the GAA MOSFET (Sample A) for $V_{GS} = 1V$ and $V_{DS} = 50mV$. Strong gate controllability over the nanowires is confirmed [8].

Fig. 3: (a) Schematics of thermoreflectance (TR) imaging system. A pulse generator (V_{DS}) and a constant voltage source (V_{DS}) drive the transistor. A control computer triggers the illumination driver and the CCD camera for a given delay time with respect to V_{DS}. (b) The timing diagram for transient TR imaging with a given LED delay time.

Fig. 4: Both measured ΔT and power ($= V_{DS} \times I_D$) follow similar dependence with drain bias, indicating $\Delta T \propto Power$, as expected.

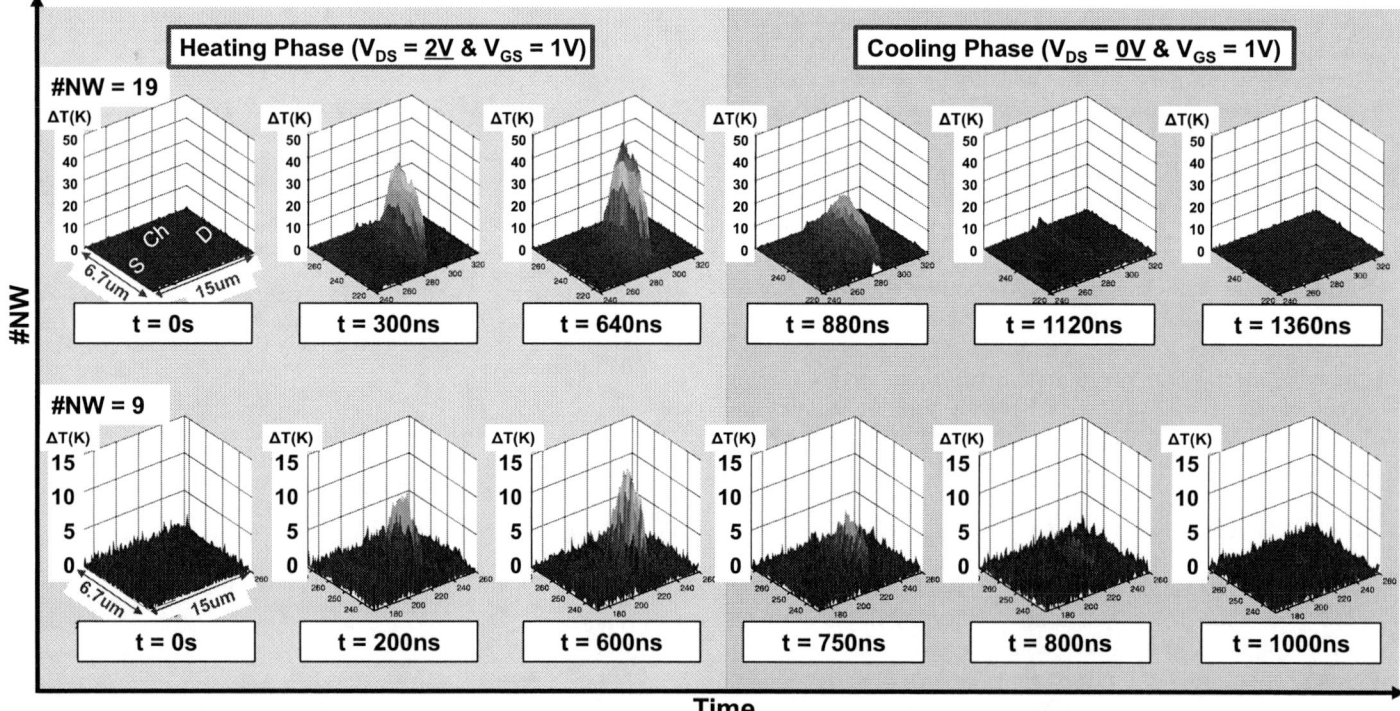

Fig. 5: Three dimensional TR images for heating ($V_{DS} = 2V$ & $V_{GS} = 1V$) and cooling ($V_{DS} = 0V$ & $V_{GS} = 1V$) phases. For clarity, only 3 images (out of more than 15) per cycle per device are shown. Also, images of 4 NWs are available, but not shown. The device with 19 NWs (top) shows higher saturation temperature compared to the device with 9 NWs (bottom). The heating and cooling time constants lie on the order of 100-500ns, depending on #NW, oxide thickness, etc.

978-1-4799-8002-4/14 $31.00 © 2014 IEEE

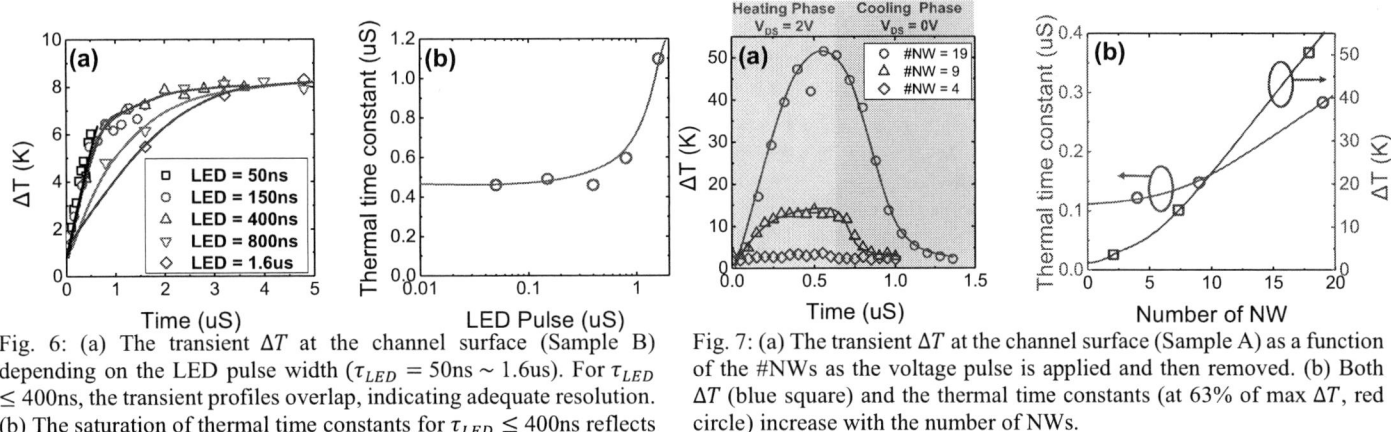

Fig. 6: (a) The transient ΔT at the channel surface (Sample B) depending on the LED pulse width (τ_{LED} = 50ns ~ 1.6us). For τ_{LED} ≤ 400ns, the transient profiles overlap, indicating adequate resolution. (b) The saturation of thermal time constants for τ_{LED} ≤ 400ns reflects the overlap of $\Delta T(t)$ in part (a).

Fig. 7: (a) The transient ΔT at the channel surface (Sample A) as a function of the #NWs as the voltage pulse is applied and then removed. (b) Both ΔT (blue square) and the thermal time constants (at 63% of max ΔT, red circle) increase with the number of NWs.

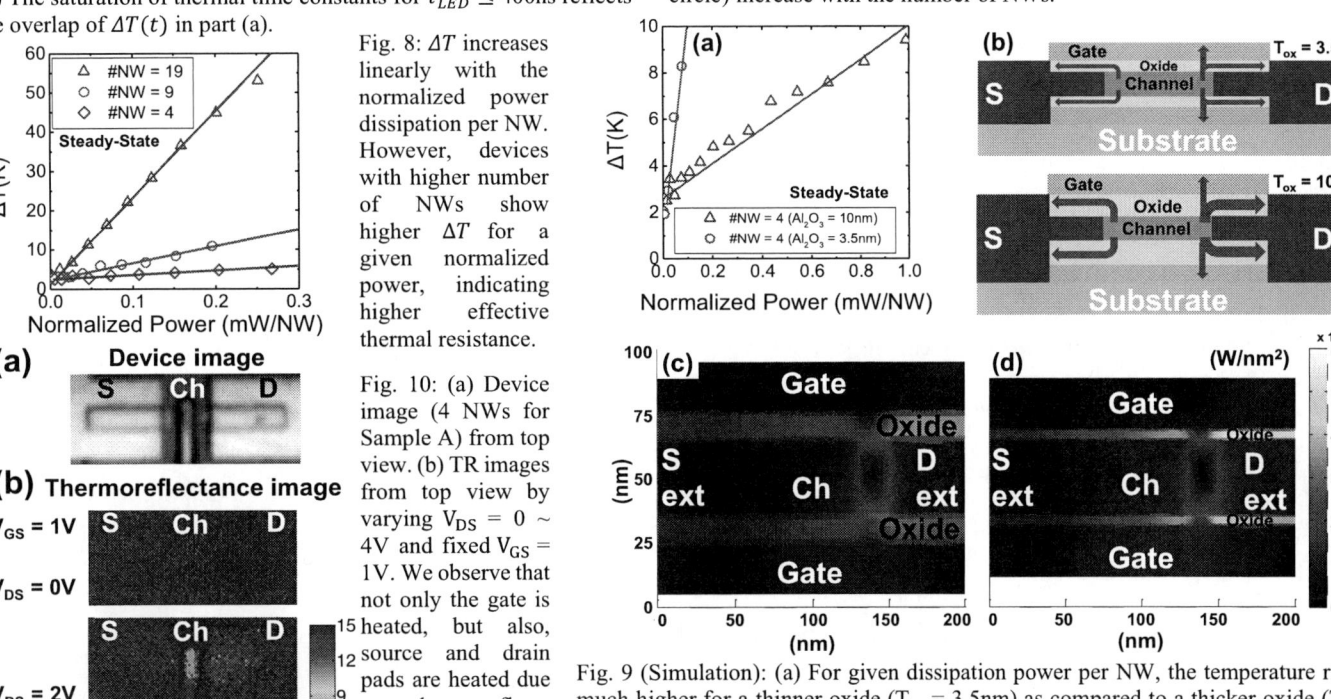

Fig. 8: ΔT increases linearly with the normalized power dissipation per NW. However, devices with higher number of NWs show higher ΔT for a given normalized power, indicating higher effective thermal resistance.

Fig. 10: (a) Device image (4 NWs for Sample A) from top view. (b) TR images from top view by varying V_{DS} = 0 ~ 4V and fixed V_{GS} = 1V. We observe that not only the gate is heated, but also, source and drain pads are heated due to heat flows through the oxide layer from heat source at drain-edge.

Fig. 9 (Simulation): (a) For given dissipation power per NW, the temperature rise is much higher for a thinner oxide (T_{ox} = 3.5nm) as compared to a thicker oxide (T_{ox} = 10nm). (b) For thicker oxide heat can flow more easily to the source/drain contacts, as shown schematically. (c, d) Simulation of heat flux (W/nm²) for T_{ox}=10nm (c) and 3.5nm (d) indicates higher heat flow through the oxide, due to the relatively higher thermal conductivity of Al₂O₃ compared to the first layer of gate metal (WN).

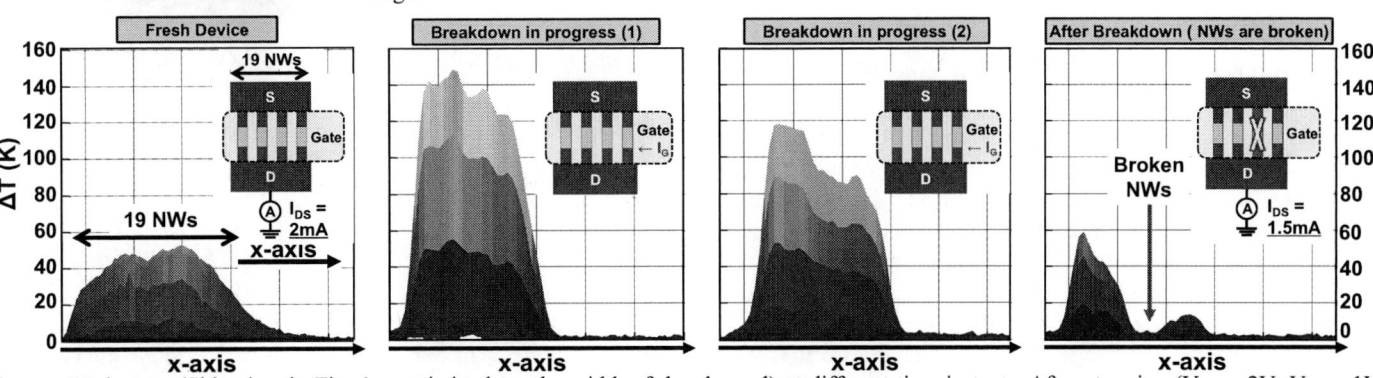

Fig. 11: TR images (Side view in Fig. 1, x-axis is along the width of the channel) at different time instants. After stressing (V_{DS} = 2V, V_{GS} = 1V) Sample A (with 19 NWs) for certain time, the channel region is suddenly heated due to the increased gate leakage (2nd image). Eventually, a fraction of the NWs are broken and the remaining NWs settle the pre-BD temperature (right image). Correspondingly, I_{ON} is decreased from 2mA at the beginning to 1.5mA at the end. This clearly indicates that about one-fourth of the NWs are no longer functioning. Schematic of breakdown of the NWs are shown in inset.

Gated and STI defined ESD diodes in advanced Bulk FinFET technologies

S.-H. Chen, D. Linten, J.-W. Lee[1], M. Scholz, G. Hellings, A. Sibaja-Hernandez, R. Boschke[2],
M.-H. Song[1], Y. See[1], Guido Groeseneken[2], and A. Thean

imec, B-3001 Leuven, Belgium, phone: +32 16 28 1124, fax: +32 16 28 1706, e-mail: Shih-Hung.Chen@imec.be

[1]TSMC, 300 Hsinchu, Taiwan. [2] also at ESAT Department, KU Leuven, B-3001 Leuven, Belgium.

Abstract

In CMOS scaling roadmap, bulk FinFET is the mainstream technology for sub-20nm nodes. However, newly introduced process options in advanced bulk FinFET technologies can result in significant impacts on intrinsic ESD performance. In this work, two types of ESD protection diodes are studied and the corresponding TCAD simulations bring an in-depth understanding on the failure mechanism of these ESD diodes.

Introduction

Bulk FinFET is now a mainstream technology in the CMOS roadmap [1-6]. ESD reliability has been investigated in both SOI and bulk FinFET [7-11]. A gate structure defined diode (gated diode) has been the proposed ESD diode solution [7, 8, 11]. Fin width (W_{fin}) is a crucial design parameter for ESD protection devices. Although the TLP failure current per layout width ($It2/W_{layout}$) is independent of W_{fin} for a gate diode [6, 7], a wider W_{fin} results in an improved clamping capability of the ESD diode because of the lower fin access resistance [8]. A wide W_{fin} may not be available in sub-20nm FinFET technology nodes due to the double patterning and epitaxial re-growth on source/drain (S/D) [12, 13]. Another process module is a local interconnect (LI) [14, 15] with LI-defined silicide last process [15]. Its impact on ESD device characteristics has not been reported so far.

The purpose of this work is to study the influence of these process options on ESD diode characteristics in a state-of-the-art bulk FinFET technology. Next to the gated diode, an STI defined ESD diode is introduced and characterized. 3D TCAD simulations are used to bring an in-depth understanding on the current conduction in these ESD protection diodes.

Process and Devices Description

A bulk FinFET technology platform with an EOT of ~1nm and a target W_{fin} of 10nm is used. The fin pitch is 45nm, using a self-aligned double patterning (SADP) lithography. The fin height is 30nm with S/D silicon epitaxial growth without merging fins. The advanced LI process is further integrated by two metallization layers which are LI1 as the bottom contact and LI2 on the top of LI1, as shown in Fig. 1. LI1 is used for contacting S/D area and LI2 is also used for gate contact. Before LI1 Tungsten metallization is filled in, a nickel silicide (NiSi) layer is formed at the bottom of LI trench. The minimum widths of LI1 and LI2 are 36 nm and 38 nm, respectively, with a pitch of 110 nm in both layers. Two types of diodes are investigated as ESD protection

devices: a gated diode and a STI diode for the first time, see Fig. 2. The design parameters, summarized in Table I, are: anode and cathode spacing (D) for the STI diode, and the gate length (L_G) for the gated diode, fin numbers (N_{fin}), and the numbers of LI strips (N_{LI}) on the anode/cathode side. The layout width (W_{layout}) which is determined by the N_{fin} and is proportional to total occupied layout area (A_{layout}) for the ESD device is usually used for failure current ($It2$) normalization. The diodes are measured in forward conduction mode by using a 100ns TLP setup with 2ns rise time. The device DC leakage current in reverse mode is monitored after each TLP stress step. A 10x increase in leakage current was considered as device failure.

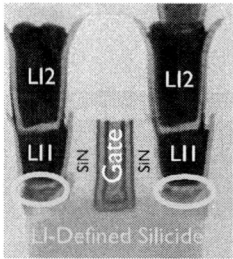

Fig. 1: TEM cross section (not to scale) of a core MOSFET in this technology platform. The gate oxide (GOX) is ~1nm HK dielectric. Two layers of local interconnects (LI) are used with 36 and 38 nm width. LI-defined silicidation is formed only at the bottom area of the LI trenches.

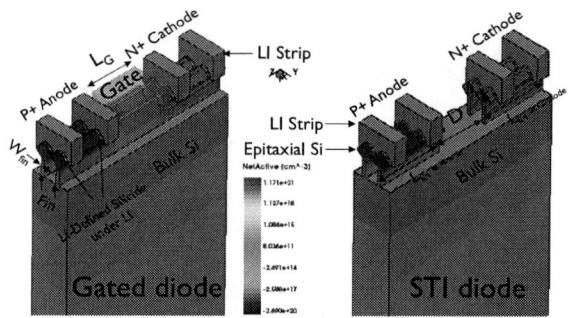

Fig. 2: 3D TCAD simulation structures of gated diodes (left) and STI diodes (right). Both diodes have two local interconnect (LI) strips in its anode and cathode sides. The gate length (L_G) in the gated and the anode to cathode spacing (D) in the STI diodes are 140nm, respectively.

Results and Discussion

A. Gated vs. STI Diodes

Fig. 3 shows the TLP IV characteristics of the gated diode and the STI diode. The STI diode shows a ~20% higher $It2$ compared to the gated diode. The STI diode however has a higher on resistance (Ron). The differences on the $It2$ and Ron are related to the different current distribution in these two

978-1-4799-8002-4/14 $31.00 © 2014 IEEE

diodes. The gated diode has a shallower and shorter anode to cathode current path through the Si fins, as shown in Figs. 4(a) and 4(b), compared to the STI diode, Figs. 4(c) and 4(d). Although this shorter current path can lead to a lower Ron, it also results in a current crowding effect that deteriorates the $It2$ level.

Table I: Three split parameters of the ESD diodes in this FinFET technology platform. The gate length (L_G) in the gated diode is used to define the anode and cathode spacing (D). With the fixed fin width and pitch, the layout width (W_{layout}) is exactly defined by the fin numbers (N_{fin}). Also, the fixed LI width and pitch directly give the corresponding anode and cathode lengths by the definition numbers of LI strips (N_{LI}).

Gated/STI Diodes	Splits
Anode to Cathode Spacing: (L_G or D)	140 to 580 nm
Layout width - W_{layout}	20 to 60 μm
Fin numbers – N_{fin}	440 to 1320
LI strip numbers - N_{LI}	2 to 6
Corresponding fin length in anode/cathode side	190 to 629.50 nm

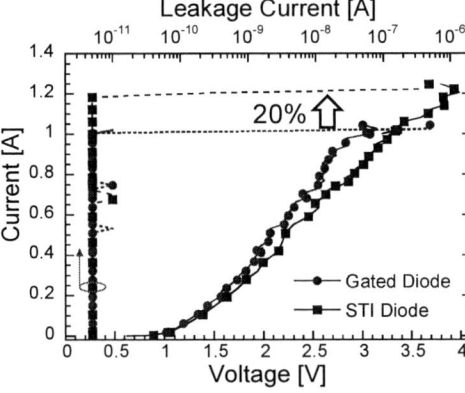

Fig. 3: Measured TLP IV curves of the gated diode and the STI diode. The two diodes have same layout parameters: the W_{layout} of 40μm with N_{fin} of 880, 4 LI strips (N_{LI} = 4) in its anode and cathode sides, and the anode to cathode spacing of 140nm in these two diodes (L_G or D).

B. Design Parameters

The impact of N_{LI} on the normalized $It2$ levels are shown in Fig. 5. The $It2$ normalized by W_{layout} increases whereas the $It2$ normalized by A_{layout} decreases with increasing the N_{LI}. The small N_{LI} is a more area-efficient design for the ESD protection diodes. Fig. 6 shows the normalized $It2$ and the Ron as a function of L_G (or D) for both diode types. The normalized $It2$ slightly increases with increasing L_G for the gate diodes due to the improved current uniformity for longer L_G [16]. However, the $It2$ levels are almost independent of the D in the STI diodes. This is attributed to the different current distributions between these two diode types in Figs. 4. The Ron of the gated diodes are ~0.3 to 0.5Ω lower than that of the STI diodes, and they both increase with increasing L_G (or D), see Fig. 6. The W_{layout} is the other key parameter impacting the $It2$ and Ron. Fig. 7 shows the normalized $It2$

and Ron versus W_{layout}. The normalized $It2$ is independent of W_{layout} whereas the Ron strongly decreases with increasing the W_{layout}. Capacitive load of an ESD device is an essential design parameter. Fig. 8 shows the parasitic capacitance versus L_G or D of the gated and STI diodes. The ~50% reduction of parasitic capacitance of the STI diode compared to the gated diode, makes the STI diode a preferred diode type for high frequency applications. The higher Ron of the STI diode can be reduced by an alternative layout style.

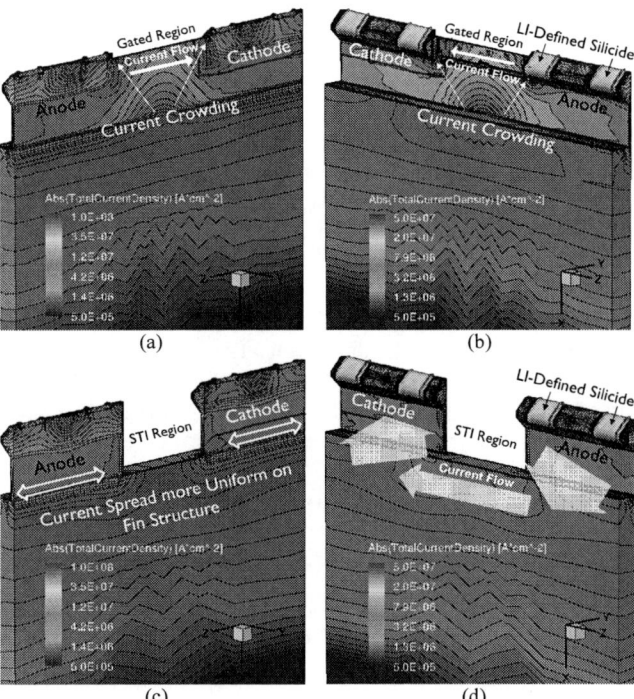

Fig. 4: 3D TCAD simulated results of current density distribution for the gated diode in (a) and (b), and for the STI diode in (c) and (d) under the same TLP stress level of 5V pre-charge voltage. A half fin architecture is used for the 3D TCAD device simulations. (a) and (c) are the 3D views at the middle of a fin structure while (b) and (d) are the 3D views at the middle of the fin spacing. These diode structures have the same N_{LI} of 2 in each anode and cathode side. The L_G in the gated diode and the D in the STI diode are 140nm.

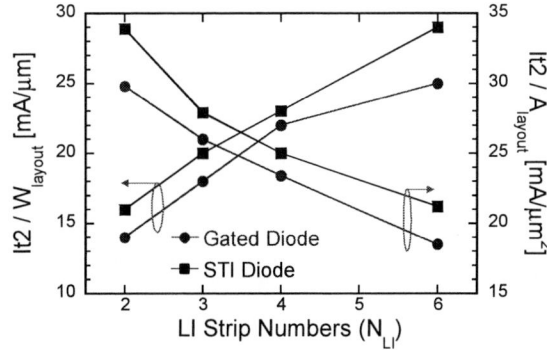

Fig. 5: Measured TLP IV curves of the gated diode and the STI diode. The two diodes have same layout parameters: the W_{layout} of 40μm with N_{fin} of 880, 4 LI strips (N_{LI}= 4) in its anode and cathode sides, and the anode to cathode spacing of 140nm in these two diodes (L_G or D).

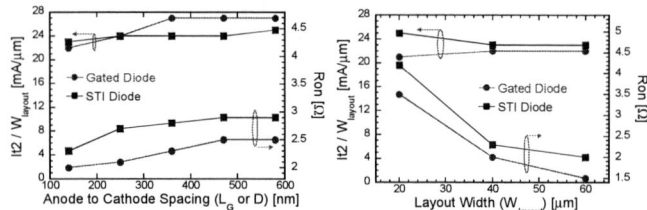

Fig. 6 (Left): Measured results of the normalized *It2* by W_{layout} and *Ron* versus the anode to cathode spacing (L_G or D) of the gated diodes and STI diodes. The splits of D (or L_G) are 140nm, 250nm, 360nm, 470nm, and 580nm. The normalized W_{layout} is 40μm (N_{fin}= 880) in all of the diodes.

Fig. 7 (Right): Measured results of the normalized *It2* by W_{layout} and *Ron* versus the W_{layout} of the gated diodes and the STI diodes. The splits of W_{layout} are 20μm (N_{fin}= 440), 40μm (N_{fin}= 880) and 60μm (N_{fin}= 1320). The D (or L_G) is 140nm in all of the diodes.

C. Rotated Diode

Instead of separating the anode and cathode of a diode along the same fin with an STI cut (of length D), a rotated STI diode structure is proposed. The anode and cathode are separated by 1 or more fin pitches. Fig. 9(a) and 9(b) show a simplified 3D and top view TCAD simulated device structure. Fig. 10 compares the TLP IV characteristics of the rotated and non-rotated STI diodes. The rotated one has a lower *Ron* and slightly higher *It2* level, because of its larger effective perimeter area under the same layout footprint. Table II shows that the rotated STI diode has 20% lower *Ron*, 8% higher *It2*, parasitic capacitance (*C*), and leakage current.

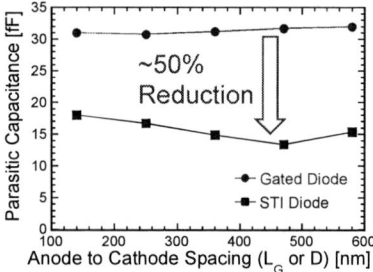

Fig. 8: Measured results of the parasitic capacitance versus the anode to cathode spacing (L_G or D) of the gated diodes and the STI diodes. The splits of D (or L_G) are 140nm, 250nm, 360nm, 470nm, and 580nm. The normalized W_{layout} is 40 μm in all of the diodes.

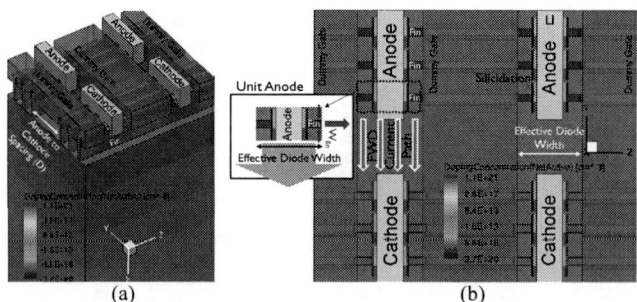

Fig. 9: 3D TCAD simulated structure of the proposed rotated diode architecture based on a commercial FinFET technology platform. The anode to cathode spacing was defined by more than one fin pitch. (b) The top view of the TCAD simulated structures of the proposed rotated diode structure. From the 2D top view, the forward (FWD) current path (white arrows) becomes a vertical direction from the anode to the cathode by the rotated

diode architecture. Also, the effective diode width is extended from the W_{fin} of 10nm to the extended LI width of >50nm for a unit anode, as the zoomed-in view in (b). However, there is also a certain layout penalty in this rotated diode structure, for example, the acceptable anode to cathode spacing will be much more than the standard fin pitch of 45nm.

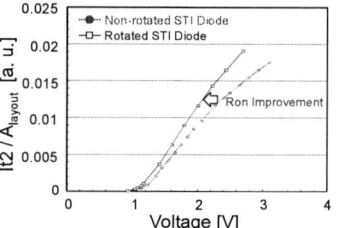

Table II Comparison of the key ESD characteristics: the on resistance (*Ron*), failure current (*It2*), parasitic capacitance (*C*), and reverse DC leakage current between the diodes with and without rotated orientation.

	Ron	It2	C	Leakage
Non-Rotated	100%	100%	100%	100%
Rotated	80%	108%	108%	108%

Fig. 10: Measured TLP IV characteristics of the diodes with different orientations. Compared with the non-rotated STI diode, the rotated one shows slightly higher *It2* level, but obviously improved *Ron*.

D. Planar vs. FinFET

Planar processed diodes with the same LI and BEOL are available to access the impact of the fin architecture on the TLP IV characteristics, see Figs. 11(a) and 11(b). In the FinFET gated diodes, only less than 30% surface silicon (anode/cathode) area is connected to bulk silicon. The *It2* decrease with less than 15% and 10 % in the gated diodes and the STI diodes, respectively, when moving from planar to FinFET device architecture with the same layout area (Figs. 11(a) and 11(b)). The *It2* levels do not decrease proportionally with the surface silicon area. The reason is partly related to the fin structure which forces the TLP current to spread more into the depth along the full fin height, compared to the planar architecture [9]. But, this current path confines the local hot spot inside the fin structure. In order to prevent this, deep implanted diodes have their junction fully below the fins and have an enlarged junction area in the bulk region under the same layout footprint, see Fig. 12. However, their *It2* levels do not improved by the deep junction (Figs. 13(a) and 13(b)). This indicates that the *It2* level is limited by LI and BEOL process.

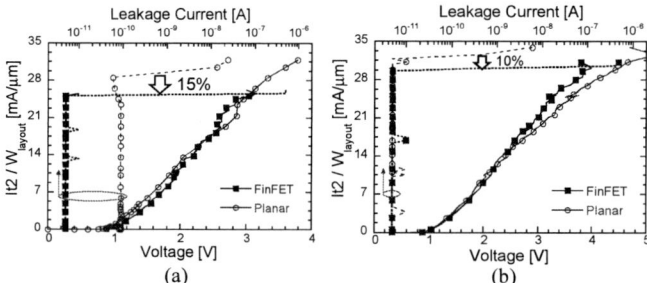

Fig. 11: Measured TLP IV characteristics of (a) gated diodes and (b) STI diodes in the different process wafers with planar and fin architectures, respectively. These diodes have the same the N_{LI} of 4 in their anode and cathode sides. The anode to cathode spacing (L_G or D) are 140nm in the gated and STI diodes, respectively.

E. Impact of LI and BEOL

With a wider LI and hence a wider silicide, the *It2* of the STI diode increases 80 %, compared to the one with the standard

978-1-4799-8002-4/14 $31.00 © 2014 IEEE   516

LI width in Fig. 14. The *It2* level of the standalone standard LI metallization is over 100mA/μm. Therefore, the root cause of failure is the LI-defined localized silicidation area (Figs. 1 and 2). The small silicide area defined by LI induces a higher spreading resistance and higher current non-uniformity, which generate a local damage under high current stress. Fig. 15 presents the simulated current ratio distribution along the fin length of the STI diodes with wide and LI-defined silicidation. The current ratio is more uniform in the one with the wide silicide architecture. Fig. 16 indicates that the STI diode with the wide silicide has higher critical diode current before the local temperature drastically increases. With the more uniform current distribution and the higher critical diode current, the wide silicide architecture enhances the *It2* of the ESD protection diodes.

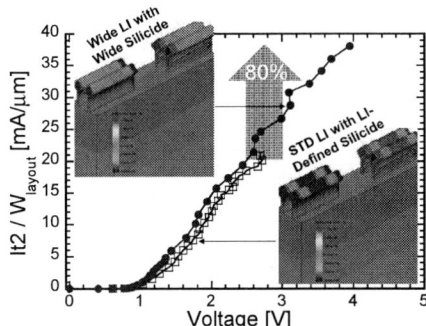

Fig. 14: Measured TLP IV characteristics of the diodes with different local interconnect (LI) formation area. The wider LI is an extended whole piece with the width of 368nm. The narrow LI is several LI strips with each width of 36 nm. The D of the diodes is 140nm.

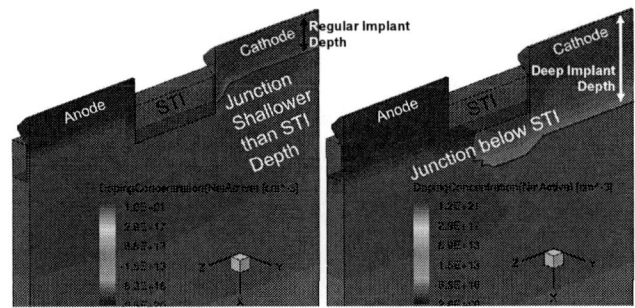

Fig. 12: 3D TCAD simulated structures of the STI diodes with a regular implant depth (left) and deep implant depth (right). The junction of the deep implanted diode is below the STI and fully under the fin architecture for preventing hot spot inside the fin. It has an enlarged junction area in the bulk region, compared to the regular implanted diode under the same layout footprint.

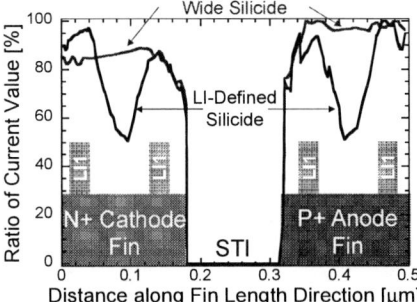

Fig. 15: Simulated results of the current ratio (to max. current value in each case) along the fin length direction. Most of the current ratio are above 70 % on the anode and cathode sides of the diode with wide LI. However, in the one with standard LI, the current ratio drops to 50% in the middle of the anode and cathode sides where the silicidation was not formed.

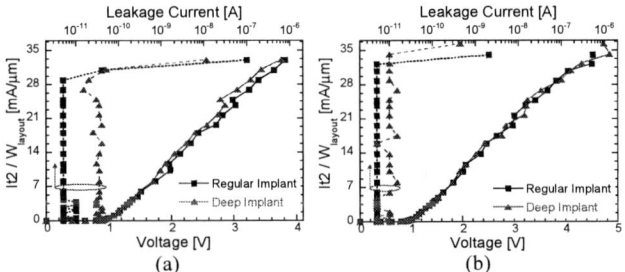

Fig. 13: Measured TLP IV characteristics of (a) gated diodes and (b) STI diodes in the different processed wafers with a deep implantation architecture, respectively. These diodes have the N_{LI} of 4 in each anode and cathode side. The anode to cathode spacing (L_G or D) are 140nm in the gated and STI diodes, respectively.

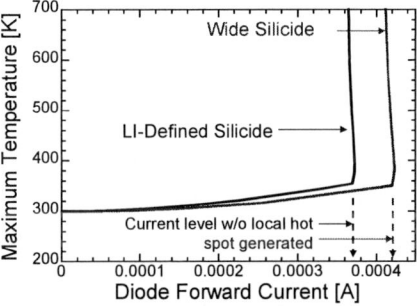

Fig. 16: Simulated temperature versus FWD diode current for the ESD diodes with different silicidation construction. The wide silicidation can tolerant a higher critical current before the diode induces the drastic local temperature increase.

Conclusion

The device characteristics of two different ESD protection diodes, a gated diode and an STI diode, are presented in a bulk FinFET technology platform. The STI diode is the better ESD protection device in the advanced bulk FinFET technology due to the excellent *It2/C* ratio, compared to the gated diode. Its higher *Ron* can be partially compensated by enlarging diode width or by rotating the diode orientation. Finally, the proposed wide LI architecture can boost the *It2* with 80% for the ESD protection diodes.

References. [1] C. Hu, *Device Research Conference*, 2001, pp.3-4. [2] L. Risch, *ESSDERC*, 2005, pp. 63- 68. [3] K. Okano, et al., *IEDM*, 2005, pp.721-724. [4] J.-M. Park, et al., *IEDM*, 2006, pp.1-4. [5] T.-S. Park, et al., *IEEE Trans. on Electron Devices*, vol.53, no.3, pp. 481-487, 2006. [6] C. Auth, et al, *VLSI Technology*, 2012, pp. 131-132. [7] A. Griffoni, et al., *EOS/ESD Symp.*, 2009, pp. 59-68. [8] S. Thijs, et al., *EOS/ESD Symp.*, 2011, pp. 27-34. [9] Y.-K. Choi, et al., *IEDM*, 2001, pp.19.1.1-19.1.4. [10] S.-H. Chen, et al., *EOS/ESD Symp.*, 2013, pp. 1-8. [11] D. Linten, et al., *IRPS*, 2013, pp. 2B.5.1-8. [12] H. Kawasaki, et al., *IEDM*, 2009, pp.1-4. [13] H. Shang, et al., *VLSI Technology*, 2006, pp. 54-55. [14] P. Packan, et al, *IEDM*, 2009, pp. 28.4.1-4. [15] P. Schuddinck, et al., *IEDM*, 2012, pp. 25.3.1-4. [16] D. Linten, et al., *EOS/ESD Symp.*, 2013, pp. 22-29.

Study of the piezoresistive properties of NMOS and PMOS Ω-Gate SOI Nanowire transistors: scalability effects and high stress level

J. Pelloux-Prayer[1], M. Cassé[1], S. Barraud[1], P. Nguyen[1,2], M. Koyama[1,4], Y.-M. Niquet[3], F. Triozon[1], I. Duchemin[3], A. Abisset[3], A. Idrissi-Eloudrhiri[1], S. Martinie[1], J.-L. Rouvière[3], H. Iwai[4], and G. Reimbold[1]

[1]CEA-Leti, MINATEC Campus, 17 rue des martyrs 38054 Grenoble Cedex 9, France (johan.pelloux-prayer@cea.fr); [2]SOITEC, Parc Technologiques des Fontaines, 38926 Bernin, France ; [3]CEA DSM/INAC 38054 Grenoble, France ; [4]Tokyo Institute of Technology, 226-8502 Yokohama, Japan

Abstract

We hereby present a comprehensive study of piezoresistive properties of aggressively scaled MOSFET devices. For the first time, the evolution of the piezoresistive coefficients with scaled dimensions is presented (gate length down to 20nm and channel width down to 8nm), and from the low to high stress regime (above 1GPa). We have shown that the downscaling of geometrical parameters doesn't allow the use of the conventional definition of piezoresistivity tensor elements. The obtained results give a comprehensive insight on strain engineering ability in aggressively scaled CMOS technology.

Introduction

The ever going scale reduction in CMOS industry pushes us toward new technological blocks like strain engineering *via* strained channel materials (SiGe, sSOI), or technological stressors like CESL, SiGe S/D... since recent technological nodes [1]. This technological step, based on the piezoresistive effect, is an extremely efficient tool to improve CMOS electrical performances [1-3]. In transistors this efficiency is described by the piezoresistive coefficients (PR), which linearly relate the strain in the transistor channel to the relative variation of the carrier mobility, and therefore of the drain current [4-9]. Also, the 3D architecture of multi-gate devices adds some complexity to strain engineering due to confinement and geometrical effects [10-17]. In that context the optimization of strain in NWs requires a comprehensive approach as a function of scaled dimensions and in a large range of applied stress.

We present in this paper a detailed study of the piezoresistive coefficients measured in different type of Ω-Gate devices (**Fig.1**) as a function of the channel width, gate length, and from low to high stress levels in the GPa range obtained by strained channel materials (compressive SiGe and tensile sSOI).

MOSFET type	Channel orientation	Technological stress	Channel material
N & P	[100]	–	Si
N & P	[110]	–	Si
P	[110]	S/D Si$_{0.7}$Ge$_{0.3}$ compressive	Si
N	[110]	sSOI tensile	Si
P	[110]	channel material compressive	Si$_{0.8}$Ge$_{0.2}$

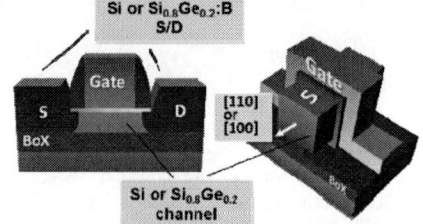

Fig. 1: Description of all the NW technological splits studied here (table), and schematics of the Ω-Gate NW transistors.

Device fabrication and experimental set-up

All the devices studied here (**Fig.1**) were fabricated using a top-down approach from 300mm SOI or sSOI (~1.4GPa biaxial tensile strain) wafers with a BOX thickness of 145nm [10]. SiGe channel with 20% Ge (~1.4GPa biaxial compressive strain) have been processed using Ge enrichment technique, with SiGe:B S/D [18]. The height of the channel (H$_{NW}$=11nm) is defined by the Si film thickness (**Fig.2**) and is the same for all our devices. The high-k/metal gate stack is composed of HfSiON/TiN (EOT=11-13 Å). For Si PMOS, both devices with Si or Si$_{0.7}$Ge$_{0.3}$:B S/D [10] were fabricated. Devices with varying gate length L$_G$ (10µm to 18nm), channel width W$_{top}$ (10µm to 8nm), and channel orientation ([110] *vs.* [100]) were measured.

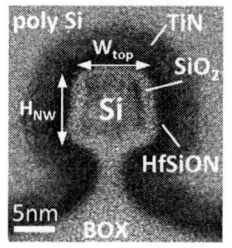

Fig. 2: TEM images of the NW along the channel direction (a), and in the cross-section (b) defining the dimensions of our NWs: L$_G$ (varying from 10µm down to 18nm), W$_{top}$ (varying from 10µm down to 8nm) and H$_{NW}$=11nm (gate stack: HfSiON/TiN). (c) Corresponding Z-contrast HR-STEM image used to perform the strain analysis of Fig.3.

The PR coefficients were extracted using a 4-point bending system allowing us to apply a uniaxial tensile stress calibrated from 0 to 125MPa [19], in both longitudinal and transversal configurations (**Fig.3**).

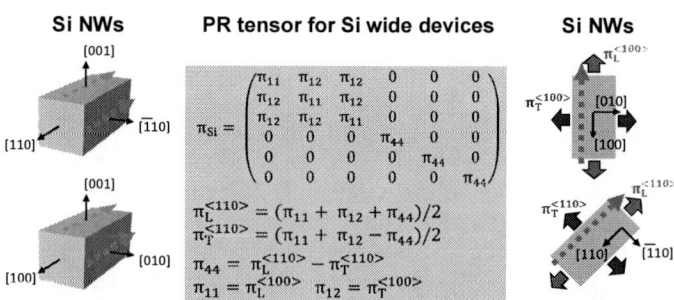

Fig. 3: Schematics of surface orientations in [110]- and [100]- oriented NWs. Equations of the piezoresistive theory for planar devices [8,9]. π_{11}, π_{12} and π_{44} can be extracted from the measured π_L and π_T values in [100]- and [110]-oriented transistors.

For strained devices, either by sSOI or SiGe material, the initial biaxial stress in wide devices reduce to a uniaxial stress in narrow transistors, typically W_{top}<100nm, as demonstrated by STEM measurements [20] performed in the cross-section of NW transistor (W_{top}=13nm), and by TCAD simulations (**Fig.4**).

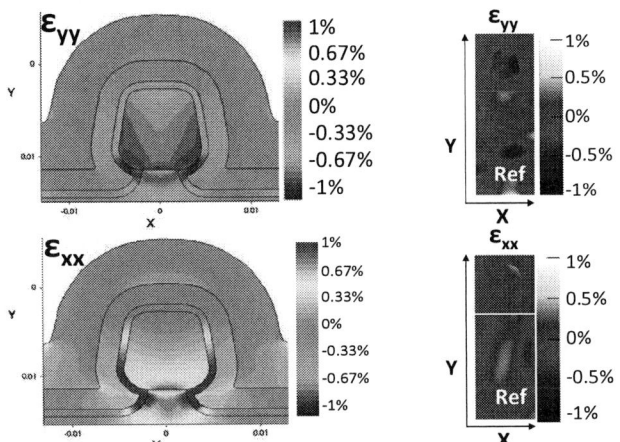

Fig. 4: Comparison of strain measurements by STEM [20] (right) and 2D mechanical simulation (left) on sSOI NW clearly shows the relaxation of biaxial strain toward uniaxial strain in the direction of the channel (*i.e.* $\varepsilon_{yy} \bullet$ - 0.5%; $\varepsilon_{xx} \approx 0\%$).

Effect of geometrical spatial confinement: From planar to NW transistor

Fig.5 demonstrates the change of the PR coefficients π_L and π_T with W_{top} below a critical value W_{crit}~100nm, for both NMOS and PMOS, in agreement with earlier measurements [19,21]. In particular, π_T decreases to 0 in NWs, whereas π_L strongly increases (~+50%) for NMOS [110]-NWs and decreases for PMOS [110]-NWs (~ -50%). For [100]-NW, sidewall and top surfaces are both (100) (schematics in **Fig.4**); all conduction surfaces remain (100) whatever W_{top}, and we thus expect π_L to be independent of W_{top}. **Fig.5** clearly shows that it is not the case, indicating specific confinement effects below W_{crit}. This also rules out the simple decomposition using the ratio of the different surface orientations to describe the PR properties of NWs (see Eqs. in **Fig.4**) [22].

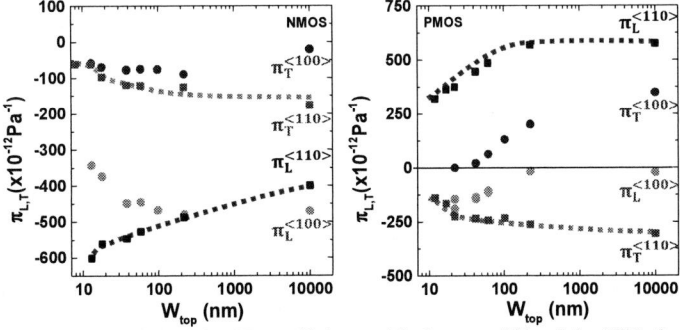

Fig. 5: Evolution of the PR coefficients with the top width of the NWs for NMOS (left) and PMOS (right) (L_G=10µm, N_{inv}=0.8×10^{13}cm^{-2}). Below a critical value of W_{top} (~100nm) both π_L and π_T strongly differ from the planar values (dashed lines are only guides for the eyes), due to the confinement effect induced by NWs.

The decomposition of the measured coefficients into fundamentals tensor components π_{11}, π_{12} and π_{44} is no more valid as indicated by the calculation of the sum $\pi_L+\pi_T$ for [100]- and [110]- oriented devices (**Fig.6**). For wide transistors this sum is equal to $\pi_{11}+\pi_{12}$ in both orientations in agreement with the PR theory (**Fig.4**), whereas for NWs this simple theory fails. To assess the effect of strain on mobility in confined transistors, the specific bandstructure of NWs has to be calculated.

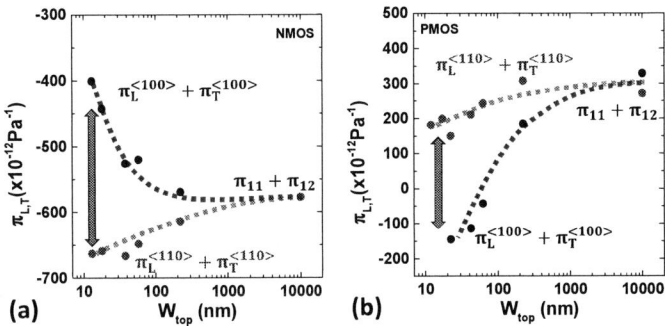

Fig. 6: Comparison of the sum $\pi_L+\pi_T$ obtained from [100]- and [110]-oriented devices as a function of W_{top} for NMOS (a) and PMOS (b). For wide transistors both values are equal to $\pi_{11}+\pi_{12}$, in agreement with the theory of the piezoresistivity tensor (see Eqs in **Fig.3**). For NWs (W_{top}<100nm), the *bulk* theory fails.

Fig.7a,b shows the effect of a 120MPa tensile strain on the VB structure calculated using 3 band k.p model [23,24] for NWs *vs.* wide planar. The phonon-limited mobility μ_{ph} calculated as a function of strain shows a reversed sign of π_L with change of [110] to [100]-NWs (**Fig.7c**), in agreement with experimental data of **Fig.5**. For NMOS, a tensile strain along [110]- and [100]-NW lowers the light valleys at Γ with respect to the heavier off-center valleys leading to higher electron mobility [23] and so a negative π_L coefficient. Calculated μ_{ph} is in qualitative agreements with the measured π_L data in both directions (**Fig.7c**).

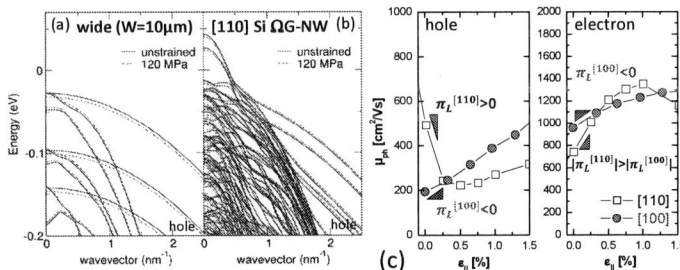

Fig. 7: Valence band structure of (a) wide planar, and (b) NWs (W_{top}=11nm×H_{NW}=11nm) calculated by 3band k.p method for a 0MPa (unstrained) and a 120MPa tensile strain along the [110] channel direction (N_{inv}=0.8×10^{13}cm^{-2}). (c) Phonon-limited mobility μ_{ph} calculated for a [110] and [100] Si-NW as a function of strain for electron and hole, explaining π_L in Fig.4 [23] (d_{NW}=8nm).

From low to high stress (SiGe and sSOI)

cSiGe and sSi channels are very effective to respectively increase the hole and electron mobility in planar devices as well as in NW transistors (**Fig.8a**). Measuring the PR coefficients on these materials allow exploring the effect of a high level of stress (above 1GPa) on mobility.

978-1-4799-8002-4/14 $31.00 © 2014 IEEE   519

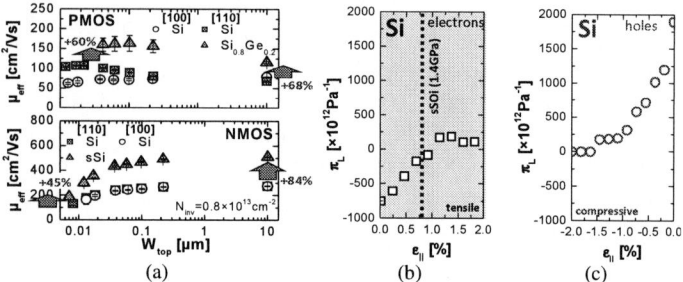

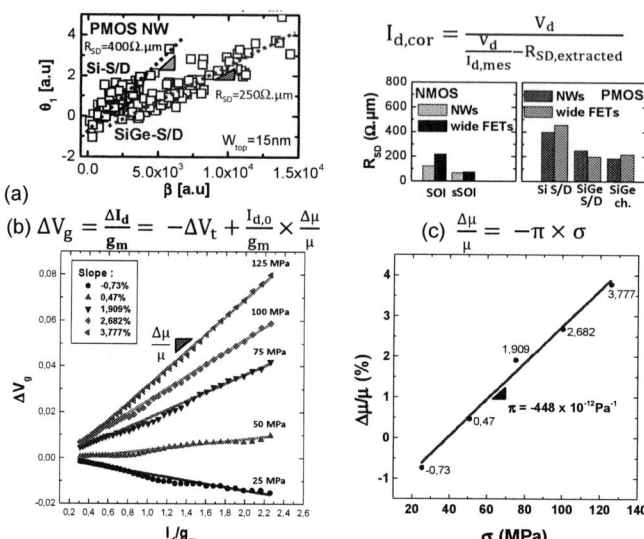

(a)

(b) $\Delta V_g = \dfrac{\Delta I_d}{g_m} = -\Delta V_t + \dfrac{I_{d,0}}{g_m} \times \dfrac{\Delta \mu}{\mu}$

$I_{d,cor} = \dfrac{V_d}{\dfrac{V_d}{I_{d,mes}} - R_{SD,extracted}}$

(c) $\dfrac{\Delta \mu}{\mu} = -\pi \times \sigma$

Fig. 8: (a) μ_{eff} measured by CV-split on long devices (L_G=10μm) as a function of W_{top} for all the technological splits NMOS & PMOS. Mobility improvement is obtained for sSi NMOS and cSiGe PMOS NWs. (b,c) π_L calculated from μ_{ph} [23] for a NW as a function of the uniaxial stress applied along the channel from 0 to ±2% for electrons (b) and holes (c).

NMOS sSOI tensile channels.– The addition of a tensile strain (sSi) on NMOS reduces the PR coefficients (**Fig.8b,9**), as the lighter subbands are already fully populated in sSi [23,10]. For π_L, this reduction is more severe for narrower width of the channel. In wide devices the biaxial contribution ($\pi_L + \pi_T$) is reduced to 0 in sSi and so $\pi_L = -\pi_T = \pi_{44}$. As shown before, for narrow devices this definition of the piezoresistive tensor isn't valid anymore so π_L and π_T are no more linked and π_T values remain ~0 in sSi NWs.

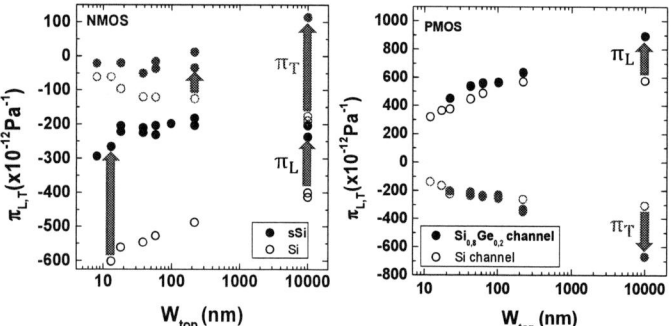

Fig. 9: PR coefficients π_L and π_T measured as a function of W_{top} for Si vs. sSi channel NMOS. (L_G=10μm, N_{inv}=0.8×10^{13}cm^{-2})

Fig. 10: PR coefficients π_L and π_T measured as a function of W_{top} for Si vs. cSiGe channel PMOS. (L_G=10μm, N_{inv}=0.8×10^{13}cm^{-2})

PMOS SiGe compressive channels.– SiGe channel can improve the piezoresistive characteristics of a PMOS by a factor 2 w.r.t. Si channel [19]. **Fig.10** shows that these improvements are reduced for narrow channels, as the PR coefficients for Si and SiGe NW are equals for W_{top}<100nm. The values for narrower NW (π_L=300×10^{-12}Pa^{-1}) still ensures anyway a strong effect of a uniaxial compressive stress up to at least 1.4GPa.

From long to short channel

The extraction method for long channel devices is straightforward, since the access resistance R_{SD} is negligible as compared to the channel resistance. For short channel devices, the effect of R_{SD} and the variation of mobility μ_{eff} with L_G complicate the applicability of the previous extraction method. Therefore, we used a first order derivation of the drain current method (**Fig.11**) which takes into account both of these parameters [25].

Fig. 11: Three steps extraction method of piezoresistive coefficient for a short L device (L_G=38nm, W_{top}=38nm). (a) R_{SD} correction of I_d using R_{SD} extracted by the Y-function technique for NMOS and PMOS devices. (b) First order derivation of drain current for small variation of I_d with stress. Resulting curves are plotted. (c) The slopes determined from Fig.12b vs. the related stress σ are used to extract the piezoresistive coefficient.

Effect of L in wide devices.– Both PR coefficients slightly decrease for **wide PMOS** with decreasing L_G down to 23nm (**Fig.12a**). Nonetheless, they still allow large improvement of mobility for aggressively scaled devices (**Fig. 17**) through compressive stress. For **wide NMOS** however, both PR coefficients strongly decrease with L_G, especially below L_G=100nm (**Fig.12b**).

Effect of L in NWs devices.– For **PMOS NWs** devices the tendency is reversed and both coefficients are increased with downscaled L_G. In particular a record value π_L~1000×10^{-12}Pa^{-1} is measured for the shortest L_G highlighting the strong potential of a compressive uniaxial stress in PMOS NWs (**Fig.12a**). For **NMOS NWs** the reduction of the PR coefficients with L_G is less severe than for wide devices (**Fig.12b**), leading to a strong effect of a longitudinal tensile strain (π_L=-300×10^{-12}Pa^{-1} for W_{top}=23nm).

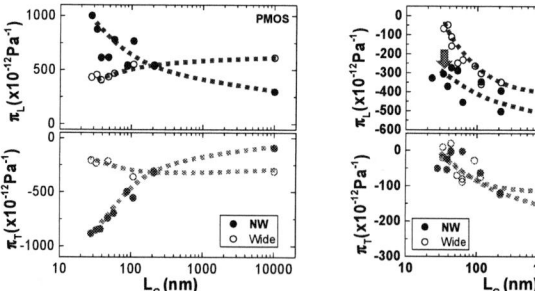

Fig. 12a: PR coefficients π_L and π_T as a function of L_G for wide (W_{top}=10μm) and NW (W_{top}=12nm) Si PMOS transistors.

Fig. 12b: PR coefficients π_L and π_T as a function of L_G for wide and NW (W_{top}=38nm) Si NMOS transistors.

SiGe vs. Si channel.– For wide devices, SiGe channel induces a larger PR effect (π_L and π_T) by comparison with Si PR, from long to short L_G (**Fig.13a**). Although this advantage disappears on

978-1-4799-8002-4/14 $31.00 © 2014 IEEE

NWs (π_L=1000×10^{-12}Pa^{-1} for both SiGe and Si NWs, **Fig.13b**), an additional compressive strain will however result in enhanced hole mobility for SiGe NWs compared to Si NWs due to higher hole mobility in SiGe [13,18].

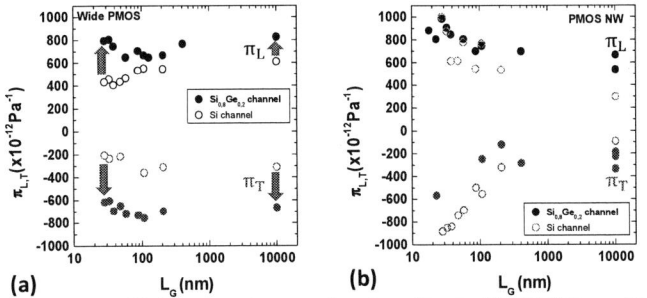

(a)   (b)

Fig. 13: PR coefficients π_L and π_T as a function of L_G for PMOS SiGe *vs.* Si channel: planar (a), NW (W_{top}=42nm for SiGe and W_{top}=12nm for Si) (b).

SiGe S/D.– SiGe S/D result in an increasing compressive stress for decreasing channel length L_G below ~500nm, up to the GPa range for shortest L_G [1,26]. As a consequence, the carrier mobility μ_0 strongly increases as well with decreasing L_G especially in NWs (**Fig.14**), in agreement with higher π_L coefficient in short NWs with Si S/D (**Fig.13b**).

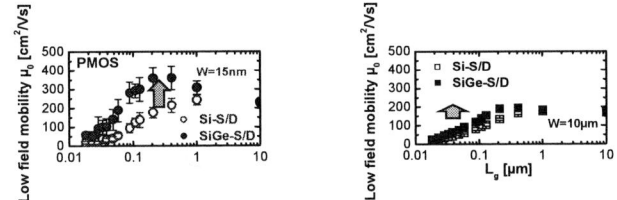

Fig. 14: Low field mobility μ_0 extracted by Y-function *vs.* L_G for NW and planar PMOS showing the effect of the compressive strain induced by SiGe S/D at L_G<500nm.

This additional stress does not change the PR properties of **wide PMOS** whatever the channel length (**Fig.15**). For **NWs**, π_L and π_T are smaller in strained channel (SiGe S/D) than in unstrained one (Si S/D), in agreement with theoretical calculation for high stress level (**Fig.8c**). The difference seen between high compressive stress in SiGe channel and in Si channel with SiGe S/D suggests that the PR properties of cSiGe channel are rather governed by the Ge composition (alloy scattering, bandstructure…) than by the stress itself.

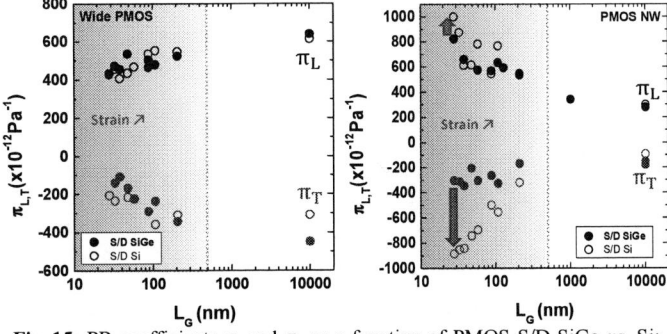

Fig. 15: PR coefficients π_L and π_T as a function of PMOS S/D SiGe *vs.* Si: wide (left), NW (right, W_{top}=12nm).Green area shows the SiGe S/D effect (cf. Fig.15).

Summary of results and conclusion

Fig.16 summarizes our results for scaled W and L. Extremely scaled PMOS NWs shows high enhancement effect with compressive stress, up to high stress level. For NMOS, the effect of a tensile strain along [110] is slightly reduced in scaled NWs, and decreases for high tensile strain. However the non-null values of π_L indicates that the mobility loss inherent to NW confinement can be compensated by a tensile strain, even in nanoscaled transistors.

MOSFET Type	Channel orientation	Beneficial stress	Relative mobility gain from our results per 100MPa of applied stress			
			Long L Large W	Long L Narrow W	Short L Large W	Short L Narrow W
N	[100]	tensile	4.7%	3.4%	N/A	N/A
P		tensile	0.2%	1.7%	0.9%	N/A
N	[110]	tensile	4.0%	6.0%	0.7%	3.3%
P		compressive	5.8%	3.5%	4.3%	10%
P S/D Si$_{0.7}$Ge$_{0.3}$	[110]	compressive	5.8%	2.8%	4.3%	8.2%
N sSOI	[110]	tensile	2.4%	2.0%	N/A	N/A
Si$_{0.8}$Ge$_{0.2}$ channel	[110]	compressive	8.9%	4.5%	8.0%	9.9%

Fig. 16: Summary of the effect of L shortening and W narrowing per 100MPa of additional uniaxial stress along the channel (from π_L data).

Acknowledgements: *This work was performed within the frame of the IBM-ST-CEA/LETI Development Alliance and in the DYNAMICULP, PLACES2BE and ANR NOODLES projects. The calculations were run using allocation from PRACE.*

References:

[1] S. Thompson et al., IEEE TED, p.1790 (2004)

[2] C. Ortolland et al., VLSI Tech. Symp., p.78 (2006)

[3] C. Auth et al., VLSI Tech. Symp., p.128 (2008)

[4] O. Weber et al., IEDM, p.719 (2007)

[5] A. Bradley et al., IEEE TED, p.2009 (2001)

[6] S. Thompson et al., IEDM Tech. Dig., p.221 (2004)

[7] K. Uchida et al., IEDM Tech. Dig., p.229 (2004)

[8] C. S. Smith, Physical Review, p.42 (1954)

[9] Y. Sun et al., "Strain effect in semiconductors" (2010)

[10] S. Barraud et al., VLSI Tech. Symp., T230 (2013)

[11] M. Saitoh et al., VLSI Tech. Symp., p.18 (2008)

[12] S. Bangsaruntip et al., IEDM Tech. Dig., p.526 (2013)

[13] P. Hashemi et al., VLSI Tech. Symp., T18 (2013)

[14] M. Garcia Bardon et al., VLSI Tech. Symp., T114 (2013)

[15] J. Mitard et al., VLSI Tech. Symp., T20 (2013)

[16] F. Driussi et al., Proc. ESSDERC, p.468 (2009)

[17] S. Gupta et al., IEDM Tech. Dig., p.641 (2013)

[18] P. Nguyen et al., S3S conference, to be published (2014)

[19] M.Cassé et al., IEDM Tech. Dig., p.637 (2012)

[20] D. Cooper et al., Appl. Phys. Lett., 233121 (2012)

[21] Y. Jeong et al., IEDM Tech. Dig., p.761 (2008)

[22] R. Coquand et al., VLSI Tech. Symp., p.13 (2012)

[23] Y.-M. Niquet et al., Nanolett., p.3545 (2012)

[24] V. Nguyen et al., IEEE TED, p.1506 (2013)

[25] O. Roux-dit-Buisson et al., IEE Circ.Dev.Syst., p.123 (1993)

[26] L. Smith et al., IEEE EDL, p.652

978-1-4799-8002-4/14 $31.00 © 2014 IEEE

Reliability Challenges for the 10nm Node and Beyond

James H. Stathis[1,a], M. Wang[1,b], R.G. Southwick[1,b], E.Y. Wu[2,c], B.P Linder[1,a],
E.G. Liniger[1,a], G. Bonilla[1,a], H. Kothari[3,b]

[1]IBM Research, [2]IBM Microelectronics, & [3]STMicroelectronics
[a]Yorktown Heights, NY, [b]Albany Nanotechnology Center Albany, NY, and [c]Essex Junction, VT

Abstract

Technology elements for the 10nm node and beyond include FINFETS on bulk or SOI, replacement gate process, multi-workfunction gate stacks, self-aligned contacts, and alternative channel materials. This paper describes current trends and how improved physics understanding and models can enable us to anticipate the effects of scaling on reliability even in early stages of development.

Introduction

New device structures and materials, coupled with aggressive gate and metallization pitch scaling for integration density, give rise to significant new challenges in reliability characterization and modeling. Technology elements for the 10nm node and beyond include FINFETs formed on either bulk or SOI [1], replacement gate process flow, multi-workfunction (WF) gate stacks, self-aligned contacts (SAC), and potentially alternative channel materials such as III-V and SiGe [2,3].

Voltage and Tinv Scaling

Electrical thickness (T_{inv}) scaling for improving performance will require reduced operation voltage in order to maintain equivalent reliability. Using the observed voltage and thickness dependence, Fig. 1 shows the voltage reduction requirement for interfacial layer (IL) scaling [4]. The 2013 ITRS roadmap anticipates sufficient voltage reduction to contain the expected impact on nFet T_{inv} scaling, but NBTI

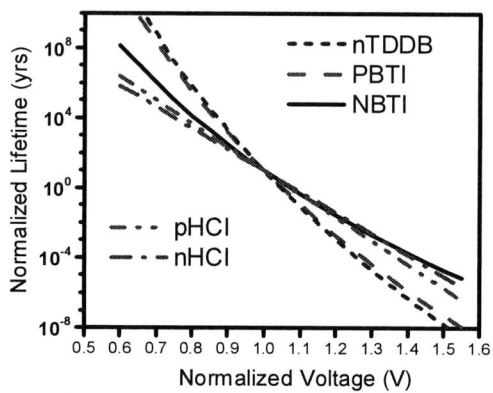

Fig. 2. Voltage dependent normalized lifetime for transistor FEOL reliability modes.

will limit pFet scaling unless new materials such SiGe channel devices are adopted [5]. The voltage dependence of hot carrier injection (HCI) lifetime for mid-Vg stress condition (i.e, Vg=½Vd+Vt) is weaker than for other mechanisms (TDDB or BTI) as shown in Fig. 2. Thus, as Vdd is reduced, the lifetime associated with HCI will increase by a lesser amount and HCI could become a larger contribution to the total degradation at end of life.

Pitch Scaling

Density is now one of the main drivers of scaling, and the spacing between the gate and the self-aligned S/D contact

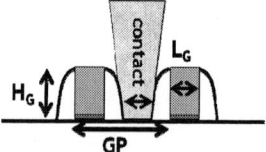

Fig. 3. Schematic of gate and contact geometry scaling.

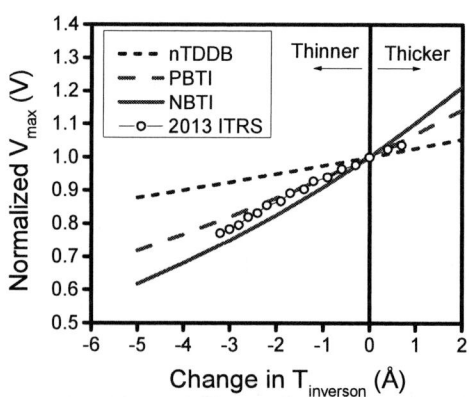

Fig. 1. Trend of maximum operation voltage V_{max} for constant reliability for the three major gate dielectric failure modes, and ITRS roadmap trend for Vdd and EOT scaling from 10nm node.

TABLE I
GATE AND CONTACT GEOMETRY SCALING

Node	Gate Pitch	Gate Length	Contact	Best case Spacer
32	120-130	30-35	30-40	25-30
22-14	80-100	25-30	20	18-25
10	65-75	20-23	20	12-16
7	45-55	12-18	15	9-12
5	35-45	9-16	12	7-9
3?	25-35	7-12	<12?	3-6?

All dimensions in nm. Best case spacer dimension (GP-LG-Contact)/2 reflects ideal case without contact taper.

978-1-4799-8002-4/14 $31.00 © 2014 IEEE

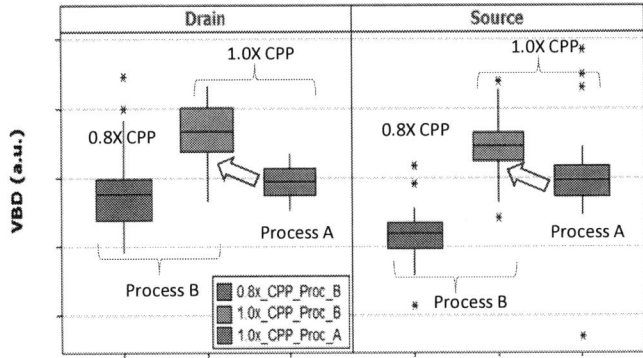

Fig. 4. Breakdown voltage between gate metal and self-aligned S/D contact with scaled gate pitch at L_{gate}=20nm.

will become less than 10nm (Fig. 3 and Table I) which is comparable to the gate dielectric thickness in early CMOS technology. Thus, the intrinsic as well as extrinsic reliability of the gate sidewall spacer will become increasingly important. Fig. 4 shows that good gate-to-contact breakdown voltage can be achieved in scaled pitch structures. Adoption of low-k (compared to Si_3N_4) spacer material is needed to maintain low contact capacitance. Voltage acceleration [6] and Weibull slope (Fig. 5) of low-k spacer material are comparable to Si_3N_4. Intrinsic max Vdd exceeds 1V for

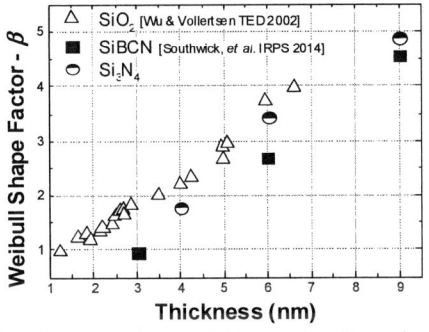

Fig. 5. Weibull slope as a function of thickness for SiO_2 and various sidewall spacer materials.

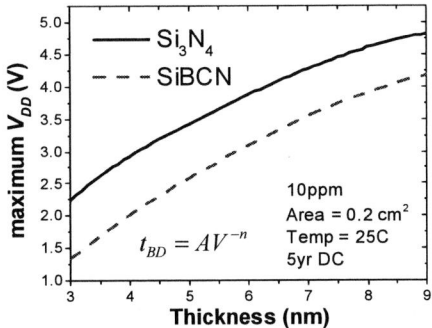

Fig. 6. Projected intrinsic Vdd_max for sidewall materials, using power law voltage acceleration.

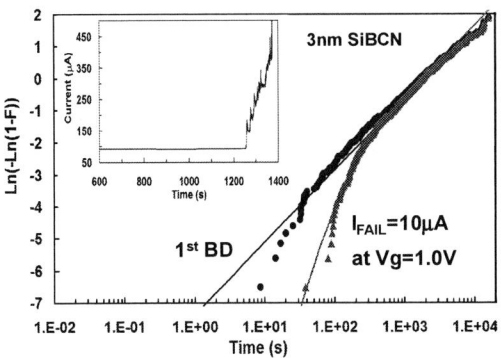

Fig. 7. First breakdown distribution and failure distribution (I_{Fail}=10µA) including progressive breakdown in SiBCN. Inset: Example of progressive BD in SiBCN.

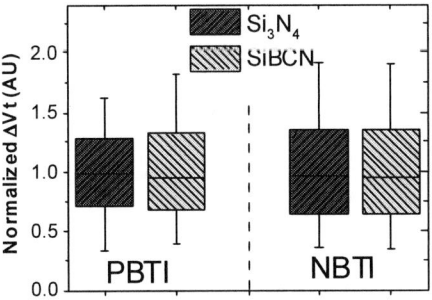

Fig. 8. Measured NBTI and PBTI are unaffected by spacer material in RMG FinFET.

anticipated thickness, and using power law voltage model gives more margin compared to $\sqrt{E}$ model (Fig. 6). As shown in Fig. 7, additional margin is obtained from progressive breakdown similar to gate dielectrics. Fig. 8 shows that the change from Si_3N_4 to SiBCN spacer material does not impact transistor BTI reliability.

Fin Geometry

The use of fully-depleted FinFETs has given rise to several new reliability considerations, including [110] interface orientation, increased self-heating during accelerated stress [7], and new solutions for Vt tuning and junction optimization [1]. At matched Vt (by metal workfunction tuning), undoped fins offer significant PBTI advantage over doped planar FET, as shown in Fig. 9 [8]. Multi-Vt devices with equivalent reliability can be obtained by process optimization as shown in Fig. 10 [1]. The sidewall (110) orientation reduces the FinFET advantage for pFET NBTI, and also for HCI damage. Reducing the fin thickness also increases HCI due to increased carrier capture cross section [7].

Without channel doping and with reduced thermal budget, control of the junction profile becomes more difficult and

978-1-4799-8002-4/14 $31.00 © 2014 IEEE

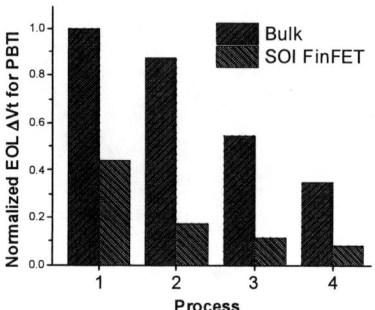

Fig. 9. Measured PBTI is up to 4x lower in SOI FinFET compared to bulk device at constant overdrive (Vg- Vt) for various dielectric stacks. After Wang *et al.*, EDL 2013.

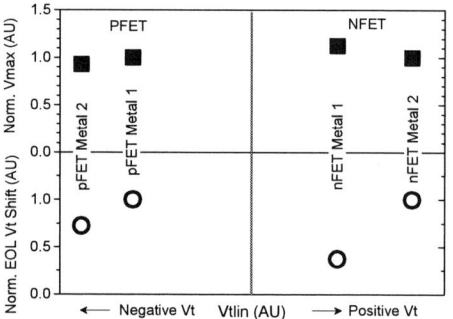

Fig. 10. Comparable TDDB (top) and BTI (bottom) reliability is achieved with multi-Vt stacks.

underlapped devices become more sensitive to HCI degradation (Fig. 11a), with HCI degradation more localized at the drain for underlapped devices (Fig. 11b). As shown in Fig. 12, the Id degradation sensed at high Vg overdrive with linear drain bias, i.e. ΔIdlin with Vg=Vdd and Vd=0.05V (and corresponding to ΔVt_idlin) is more sensitive to resistance or mobility degradation, including Rsd increase from defects generated in the spacer region, compared to Vt shift sensed at Vg near threshold (ΔVtlin). As gate stress voltage increases, from just above Vt toward Vd, the

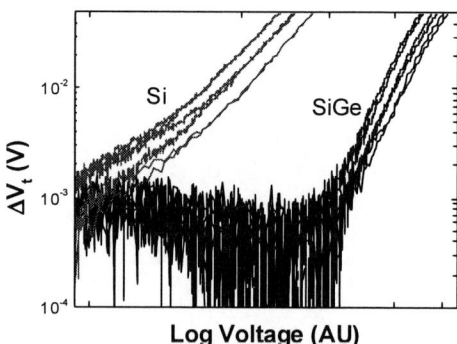

Fig. 13. Ramp BTI data for SiGe pFET compared to Si pFET.

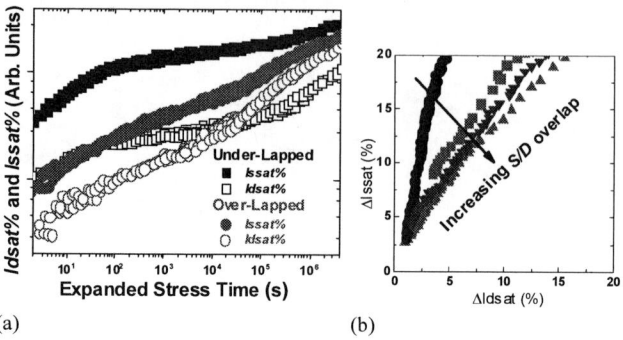

(a) (b)

Fig. 11. (a) Forward (Id) and reverse (Is) current degradation by HCI. (b) Underlapped devices show larger ratio of reverse/forward degradation implying more localized damage close to drain.

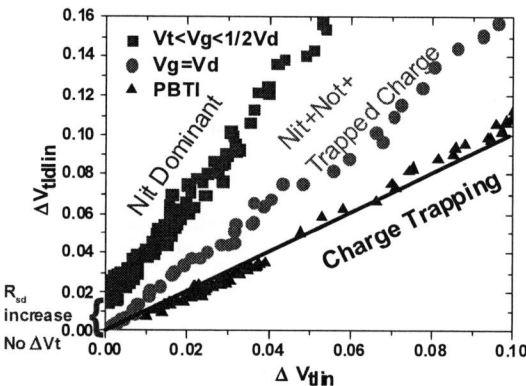

Fig. 12. ΔVt_idlin is the shift extracted by constant current method from Idlin measured near Vg=Vdd with Vd=0.05V. For charge trapping (no Rsd increase or mobility decrease), ΔVt_idlin = ΔVtlin. Dit (mobility degradation) leads to increased ΔVt_idlin. Rsd increase alone causes ΔVt_idlin without Vtlin change.

degradation mode changes from Nit dominant (mobility degradation) to trapped charge dominant (Vt shift).

New substrate materials

The search for higher performance in addition to density scaling is driving exploratory work on Si replacements. SiGe pFET channel was first introduced as a Vt adjustment element and was found to have greatly improved NBTI [5], as demonstrated in Fig. 13. Higher Ge content and/or III-V materials may be used in future for higher performance, with IL optimization key to realizing the full potential of the alternate channel material.

Variability

As dimension scaling and process complexity increase, variability inevitably becomes worse. Vt and BTI variability are well known issues, but the impact of variability on breakdown is not as well appreciated. Process non-uniformity

978-1-4799-8002-4/14 $31.00 © 2014 IEEE

makes the measured breakdown distributions broader, leading to overly conservative lifetime projection, and fundamentally invalidates the assumption of Poisson statistics that underlies the usual area-scaling relation for failure rate, $F_2=1-(1-F_1)^{(A2/A1)}$. Local variation can be separated from cross-wafer variation by extensive cross-wafer and within die measurements [9], and non-Poisson area scaling can be efficiently modeled using a defect clustering model [10] as shown in Fig. 14. With this approach, the low percentile failure times can be obtained even in the presence of multiple unknown sources of random variation.

Interconnect

The introduction of ultra-low k materials to reduce capacitance, and the selection of barrier materials, metal alloys, and metal capping layers required to meet resistance targets at finer pitches (<50nm) in scaled interconnect structures entail a significant increase in integration complexity and reliability challenges, due to degraded thermo-mechanical properties and lower dielectric breakdown strength. This drives a need to evaluate failure mechanisms in these heterogeneous structures at lower voltages, closer to operating conditions. Long-term studies (Fig. 15) show that the canonical √E acceleration model is overly conservative for predicting TDDB lifetime [11]. Dense wiring in SRAM cells is a potential vulnerable location for BEOL breakdown. Typical cell layouts have wordlines (WL) passing over Vdd, closely spaced metal lines between bitlines (BL) and ground, and sometimes bitlines crossing over the internal node connection. Breakdowns at these locations (Fig. 16) affect the cell differently than transistor gate breakdown within the cell. Sensitivity to breakdown-induced leakage between WL and Vdd, or BL and Gnd, will depend on sense amp design. Careful attention to voltage acceleration and progressive breakdown is required to model the impact of BEOL breakdown.

Acknowledgements: We thank Terry Hook, Geordie

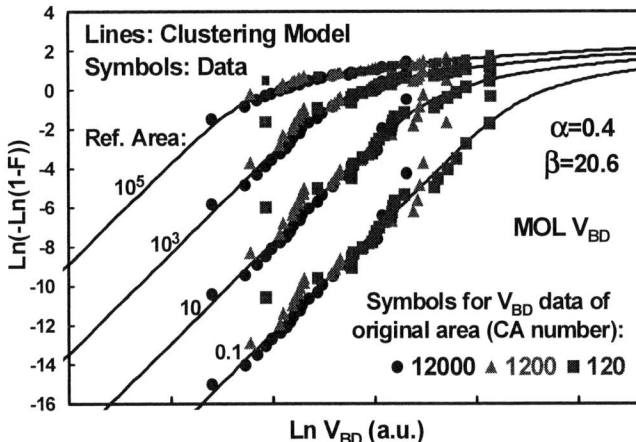

Fig. 14. Breakdown voltage data for gate-to-contact (PC-CA) showing non-Poisson area scaling and excellent agreement with clustering model at various reference areas. After E.Wu *et al.*, IRPS 2014.

Braceras, and Ed Cartier for insights. This work was performed by the Research Alliance Teams at various IBM Research and Development Facilities.

References

[1] K.-I. Seo *et al.*, VLSI-Technology 2014.
[2] P. Hashemi *et al.*, VLSI-Technology 2013, p. T18-T19.
[3] Y. Sun *et al.*, IEDM 2013, p. 48; R.T.P. Lee *et al.*, IEDM 2013, p. 44.
[4] B.P. Linder *et al.*, VLSI-DAT 2013.
[5] J. Franco *et al.*, IEDM 2011, p.445; S. Krishnan *et al.*, IEDM 2011, p. 634.
[6] R.G. Southwick *et al.*, IRPS 2014.
[7] S. Ramey *et al.*, IRPS 2013, 4C.5.1-4C.5.5.
[8] M.Wang *et al.*, Electron Dev. Lett. 34, p. 837-839, 2013.
[9] F. Chen *et al.*, IRPS 2014.
[10] E. Wu *et al.*, Appl. Phys. Lett 103, 152907, 2013, IEDM 2013, p. 401; R. Achanta *et al.*, IITC 2014; E.Wu *et al.*, IRPS 2014; E.Wu *et al.*, IEDM 2014.
[11] E. Liniger *et al.*, IRPS 2014.

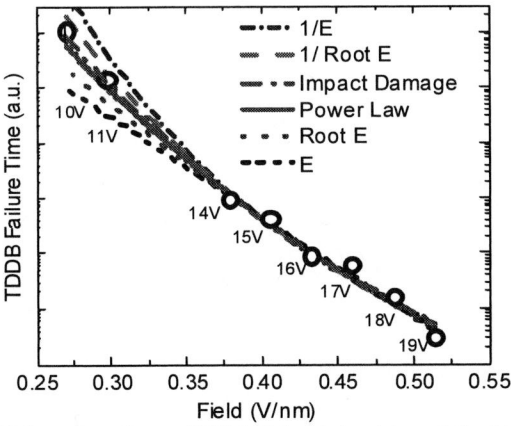

Fig. 16. Voltage dependence of BEOL dielectric breakdown (after Liniger IRPS 2014). Data lie between E and 1/E models.

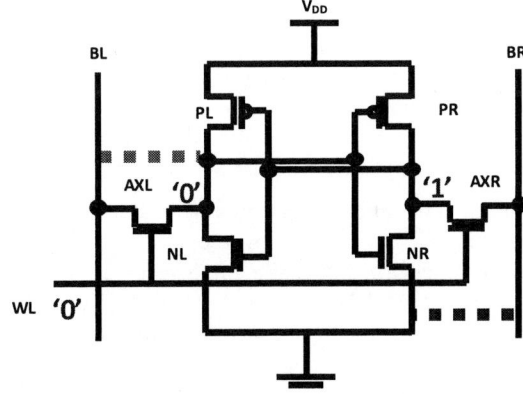

Fig. 16. Inter- and intra-metal level vulnerabilities in typical SRAM cell layout include bitline to node or to gnd, and Vdd to wordline (not shown).

Will Reliability Limit Moore's Law?

Anthony S. Oates

TSMC Ltd., 168 Park Ave. 2, Hsinchu Science Park, Hsinchu, Taiwan 30075; aoates@tsmc.com

Abstract– **Up to the present time reliability has not limited the rapid evolution of Si process technologies. However, the near future will bring a continual stream of innovations in transistor architecture and gate dielectric and interconnect materials. Maintaining historical high levels of reliability in this environment will be challenging. In this paper we discuss the reliability issues that have the potential to limit the future pace of technology progress.**

I. INTRODUCTION

The rapid pace of progress experienced by the semiconductor industry over the past 40 years would not have been possible without a concomitant improvement in long – term reliability of circuits [1]. By the mid-1990s, circuit failure rates had saturated and were subsequently held constant with technology progression by continuous improvements in the reliability of individual circuit elements (transistors and interconnects). This progress was the result of strong interdisciplinary collaboration in reliability engineering, failure analysis, materials and process engineering, and circuit design. Consequently circuit reliability has been maintained at a sufficiently high level to prevent it from constraining the rate of technology scaling. The need to improve reliability with each successive technology generation remains; however, the difficulty in meeting this challenge increases with each successive node since failure times of the most common IC degradation mechanisms decrease with technology progression. Looking forward, circuit performance and density improvements necessitate almost continual innovation in architectures and materials of transistors and interconnect. The question then naturally arises whether reliability can continue to be contained at historical levels, or will it degrade as technology progresses and pose a constraint on the future rate of progress inherent in the pursuit of Moore's Law. In this paper we will examine the reliability issues that have the potential to impact the pace of technology progression. This review is not intended to be exhaustive in scope, but will focus only on those degradation mechanisms that can most obviously be viewed as imposing potential technology limitations in the near future.

II. TRANSISTORS

Industry roadmaps indicate a continuous evolution of transistor architecture and materials to overcome increasingly severe limitations to performance and manufacturability as technology progresses and feature sizes are reduced and integration increases. The industry has successfully transitioned the most advanced process nodes to gate dielectrics that incorporate high-k / metal stacks. The reliability complications arising from the use of high-k layers, where charge trapping is significantly higher than in thermally grown SiO_2 [2] have been

minimal. PBTI in NMOS has been added as a new reliability issue to be monitored [3], while other transistor degradation mechanisms have been readily optimized. Now, the transition to the FinFET transistor architecture is underway, with all indications pointing to successful implementation. The consequence of the introduction of FinFETs to reliability also appears to be relatively benign since the structures do not exhibit new mechanisms of degradation [4]. However, self-heating in the confined fin geometry is exacerbated compared to the planar configuration [5], which complicates assessments of circuit reliability as well interpretation of accelerated test results.

Replacements for the FinFET architecture are already being contemplated. The Gate-All-Around (GAA) structure, where the gate entirely surrounds the channel is a primary candidate to replace FinFETs [6]. Reliability data that are available for GAA structures indicate more pronounced NMOS hot carrier degradation compared to PBTI, while the degradation characteristics are consistent with accepted models [7]. More data for BTI, and the impact of the circular gate geometry on gate dielectric reliability are required to provide a comprehensive reliability assessment, but these initial data are encouraging, and suggest the reliability of the GAA structure may not be a constraint for future development.

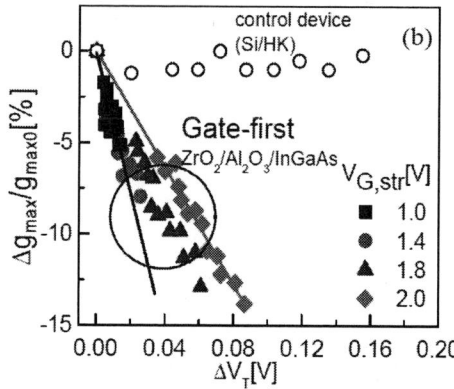

Fig. 1: PBTI stressing of InGaAs NMOS with Al_2O_3/ZrO_2 gate dielectric results in degradation of threshold voltage, V_t, and transconductance, g_{max} showing that both charge trapping in the dielectric, as well as interface state generation occurs during stress.

High mobility channel materials are being pursued to replace the Si channel. Ge is favored for PMOS, and when implemented with a SiO_2/HfO_2 gate dielectric, has been reported to significantly improve NBTI characteristics compared to a Si channel [8]. This results from the increased valence band offset, which reduces the coupling between channel carriers and defects in the bulk of the gate dielectric. InGaAs ternary III-V compounds are favored for NMOS. Unfortunately, InGaAs channels exhibit more serious PBTI degradation than Si channel devices, as shown in fig.1, resulting from charge trapping in the high-k gate dielectric stacks that are required (e.g. Al_2O_3 interfacial layers topped with a thicker ZrO_2 layer) [9]. Moreover, unlike Si channels, InGaAs exhibits interface state generation during PBTI, suggesting interface engineering will be critical aspect for reliability assurance.

Finally, novel gate stacks with higher k values will be required, even in the absence of high mobility channels to limit gate current increases and improve performance with scaling. The intrinsic TDDB capability of Al_2O_3 is sufficient to replace SiO_2 as the interfacial layer for the Si channel (fig.3) [10], but demonstrations of sufficient intrinsic capability for other high-k dielectrics for high mobility channels are so far lacking. However, it is clear that new high-k gate materials that are being considered for the interfacial layer all exhibit increased charge trapping compared to SiO_2 and can be expected to exacerbate BTI degradation.

We infer that the primary challenge for future technology development may not be in transitions to new transistor architectures or channel materials *per se*, but in the development of high quality, process compatible gate dielectric stacks and interfaces that minimize charge trapping and defect generation.

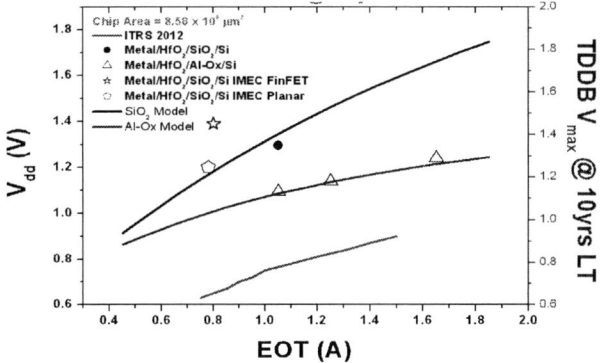

Fig. 3: Maximum gate bias for 10 years operation at 0.1% probability, 125 °C for SiO_2 and Al_2O= IL compared to the ITRS 2012 road map (Vdd).

III. INTERCONNECTS

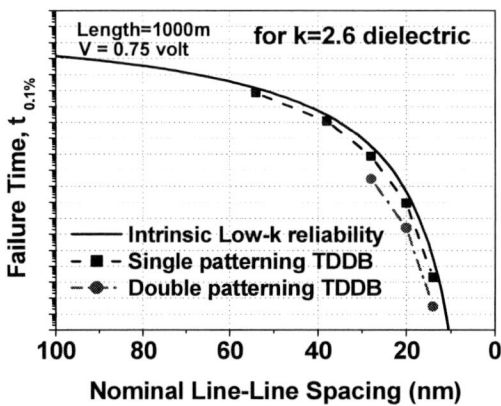

Fig. 4: Use condition low-k failure time prediction for L=1000m interconnect line length at V=0.75Volt.

Potential technology limitations associated with the reliability of interconnects are imminent. Most pressing is the rapid reduction of reliability of low-k porous dielectrics due to time dependent dielectric breakdown (TDDB). Porous materials exhibit failure times that decrease rapidly with technology scaling as an inherent materials characteristic [11]. The presence of pores reduces that length of the breakdown path within the dielectric and, when coupled with the decreasing thickness between metal lines and increasing electric fields at operating conditions of circuits, this results in a rapid decrease in failure time as the dielectric thickness is reduced, as shown in fig. 4 for a dielectric with k=2.6. At present, k~2.5 is a practical materials limit for scaling [12].

The dielectric reliability is also sensitive to local geometry variation [13], which is exacerbated as feature sizes are reduced. These variations consist of line edge roughness (LER) arising from the difficulty of patterning narrow Cu trenches, and from mis-alignments between features within and between metal levels. As a result the lithography technique used to define interconnects impacts failure times causing reductions below the intrinsic capability [12]. Single patterning can reduce failure times by an order of magnitude. The degradation associated with double patterning techniques is larger, reducing failure times by up to 2 orders of magnitude compared to the intrinsic capability of the material because of the increased variability within metal levels.

The electric field dependence of the failure time is critical in determining the continued viability of industry standard porous dielectrics at 10 nm and below. There are a multitude of models that have been proposed to allow extrapolation of failure times from accelerated test conditions to operating conditions. However, as fig.5 shows, the evidence to support models that provide more aggressive lifetime extrapolation than the standard √E model [14] is weak. Therefore, progress in lowering k, or even maintaining it at present levels, with continued technology progression requires integration of (novel) non-porous materials or changes in interconnect architecture to include air-gaps.

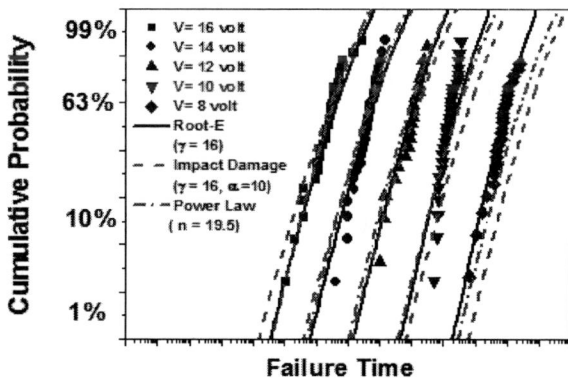

Fig. 5: Field dependence of low-k dielectric TDDB. Shown are fits to the data using several field dependence models; the root-E model provides the best fit to the complete data set.

Much attention has been paid recently to the diminished electromigration current carrying capability with scaling (fig. 6) [15,16]. Compounding the difficulty posed by this reliability

978-1-4799-8002-4/14 $31.00 © 2014 IEEE

degradation is the circuit performance driven requirement of increasing current density, J_{max} with technology progression. In response, to limit the reliability degradation, process solutions have been proposed to incorporate Cu alloys to replace pure Cu, and to use metal capping layers for Cu trenches to replace the standard dielectric cap. These process modifications are designed to reduce the rate of electromigration transport along Cu grain boundaries and the top Cu / dielectric interface in trenches respectively. However, at the most advanced technology nodes metal cap layers are ineffective [15] since electromigration occurs predominantly along grain boundaries. Cu alloys may be needed to provide sufficient electromigration performance at 10 nm and beyond. However, the scalability of these solutions to future nodes is not straightforward since they tend to increase Cu resistance. The reduction of interconnect RC is an important aspect in enhancing circuit performance. Since the potential to reduce the k (i.e. C) of porous low-k dielectrics is limited by reliability considerations discussed above, the focus is on development of lower metal resistance. The direction of this effort is the development of processes with novel ultra-thin barriers, which again poses challenges for electromigration reliability [16].

The difficulties of identifying scalable interconnect process solutions points to the necessity of identifying circuit design solutions to the electromigration issue. Improved current-carrying capabilities have been introduced with "short-length" EM design rules. The latter take advantage of the mechanical stress-induced backflow that occurs as a result of electromigration over relatively small length scales (10 µm) to significantly increase current carrying capability. In the past several years short-length electromigration design rules have moved from being IP specific to general purpose. However, as the circuit level J_{max} requirement increases, the short-length benefit is restricted to increasingly lower lengths. Going forward, more sophisticated short-length rules can be envisioned to take advantage of our understanding of electromigration reservoir effects in realistic circuit interconnect configurations, which offer the possibility of further improvements in current carrying capability [17].

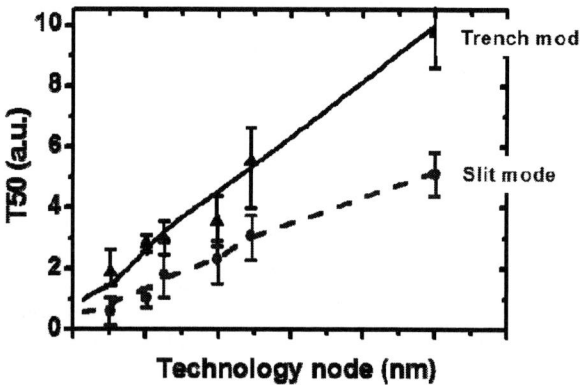

Fig.6: The reduction of electromigration failure times with technology progression from the 65 to the 20 nm node. Electromigration lifetimes are limited by slit-voids that form directly under vias

IV. SOFT-ERRORS

Soft-errors resulting from the interaction of alpha particles emitted from package and bump materials and cosmic ray neutrons with the silicon substrate have been of increasing concern as high levels of integration have substantially increased the amount of memory (SRAM, DRAM) and logic storage elements (flip-flops, FF) on circuits. Fig. 7 shows measured SER trends for memory and flip-flops as a function of technology progression over the past 15 years (combined α and neutron events.) The rate of soft errors (SER) at the device level has been declining with technology progression as SRAM and FF cell area has reduced more quickly than reduction in the critical charge required to be collected at sensitive nodes to cause a change of logic state. The introduction of the FinFET has resulted in a further reduction in device level SER by about an order of magnitude compared to an equivalent planar structure due to a reduction in charge collection associated with the restricted fin geometry [18]. This SER reduction is particularly striking for low energy particles such as package alpha particles where the reduction is so large (approximately a factor of 50) they can now be considered negligible compared to cosmic ray fast neutron – induced errors. However, at the system level increased integration overwhelms this positive device level trend, and SER continues to be a serious technology reliability issue.

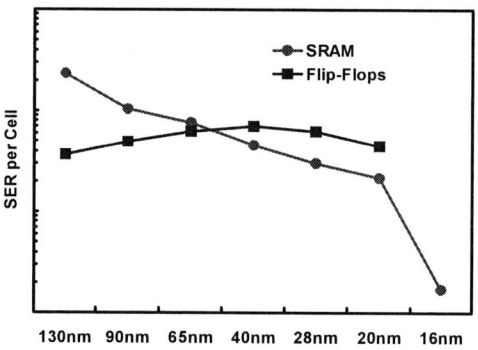

Fig. 7: Experimental soft-error rate trends for combined alpha particle and fast neutron strikes on SRAM and flip-flops (FF). Error rates have continuously declined for SRAM while FF error rates have decreased from the 40nm onward. Error rates for FinFETs (16 nm node) display almost an order of magnitude decrease compared to planar geometry.

Looking forward, there are several major concerns related to SER for future technology development: first, the potential for increased sensitivity of circuits resulting from either new upsets mechanisms (thermal neutrons and muons), or increased relative contribution from combination logic; and second, the impact of changes in transistor architecture and channel materials. Although upsets due to thermal neutron capture by naturally occurring B^{10} nuclei in the vicinity of transistors have been exacerbated by recent trends in interconnect processing techniques, the relative contribution of this mechanism decreases rapidly with technology progression and should not be a concern below the 20 nm node [19]. More worrisome is the potential for energetic muons to act as a new source of upsets.

978-1-4799-8002-4/14 $31.00 © 2014 IEEE

Muons are the most abundant energetic particles in the terrestrial environment, and as the charge required to upset a node in an SRAM has reduced with technology progression, the relative contribution to SER from terrestrial muons has increased. However, recent calculations of the impact of muon SER on SRAM suggest it will not be a dominant mechanism going forward (fig.8) [20]. Given the novelty of this upset mechanism more studies are required to unambiguously to determine the significance to circuit reliability of muon-induced upsets and the mechanisms involved. Upsets can also occur in combinational (random) logic circuits with increasing severity as operating frequency increases. Recent work suggests the potential for SER associated with combinational logic to be at least as large as that for unprotected FF for GHz operational frequencies [21].

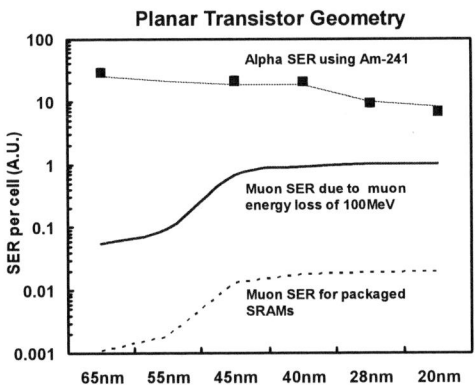

Fig. 8: Estimates of muon induced SER for SRAM compared to package induced alpha particles. An energy loss of 100 MeV represents a worst case estimate, from the perspective of SER of local environmental impact on muon energies.

As fig. 7 shows, changes in transistor architecture that limit the volume of Si available to collect charge associated with incident energetic particles has a positive impact on SER. The relative susceptibility GAA structures to SER, therefore, might be expected to be similar to that of FinFETs, if not smaller. However, the introduction of high mobility channels presents concerns because lower band gap energies coupled with higher carrier mobility may exacerbate SER despite the charge collection benefits inherent in 3D architectures. Fig. 9 shows increased TCAD derived SER estimates for Ge channel SRAM compared to Si channel for the FinFET architecture. The Ge channel exhibits a stronger dependence on the details of the geometry of the FinFET. The sensitivity of SER to the voltage in the range 0.7 to 0.9V is larger for the Ge channel. Simulations of SRAM and FF with III-V NMOS structures also indicate an increased SER compared to the Si channel above 0.5V [22].

V. CONCLUSIONS

Reliability will be a key enabler of the architectural and materials innovations required for future Si process technologies. The major challenges to be overcome lie the development of: high quality semiconductor/dielectric interfaces and gate di-

electric stacks for novel transistor structures; back-end dielectrics that eliminate the inherent vulnerability of porous materials; and the incorporation of more sophisticated backend reliability – design interactions into design kits.

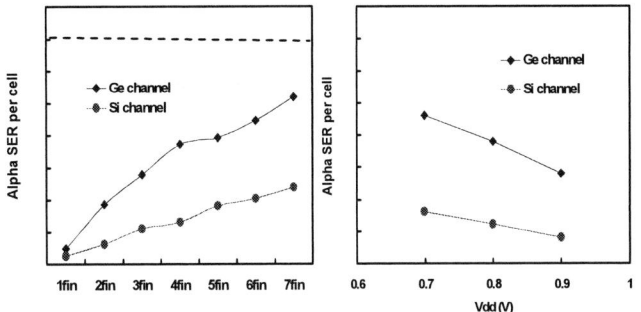

Fig. 9: TCAD-derived estimates of relative SRAM SER due to package–related alpha particles for FinFETs for Si and Ge channels. The charge collection is significantly increased with the Ge channel due to the lower band gap energy and high carrier mobility. The Ge channel also leads to a stronger voltage depe dence of SER.

ACKNOWLEDGMENTS

The author thanks Y. C. Sun for his support and encouragement of this work.

REFERENCES

[1] A.S.Oates in "Reliability Characterisation of Electrical and Electronic Systems, Swingler, to be published March 2015", Elsevier Publishing; [2] A.S. Oates, IEDM Tech. Dig. 2003, pp.38.2.1; [3] S. Zafar, A. Gallegari, E. Gusev, M. V. Fishetti, IEDM Tech. Digest 2002,pp. 517; [4] S. Ramey, A. Ashutosh, C. Auth, J. Clifford, J. Hicks, R. James, A. Rahman, V. Sharma, A. St Amour, C. Wiegand, Int. Rel. Phys. Symp., 2013, pp. 4C.5.1; [5] C. Prasad, L. Jiang, D. Singh, M. Agostinelli, C. Auth, et al, Int. Rel. Phys. Symp. 2013, 5D.1.1; [6] J.P. Colinge, M.H. Gao, A. Romano-Rodriguez, H. Maes, and C. Claeys, IEDM Tech. Dig., 1990, pp. 25.4.1; [7] S.H. Shin, M.A. Wahab, M.Masuduzzaman, M.W. Si, J.J. Gu, P. D. Ye, M.A. Alam, Int. Rel. Phys. Symp., 2014, pp. 4A.3.1; [8] J. Franco, B. Kaczer, J. Mitard, M.Toledano-Luque, P.J. Roussel, L. Witters, T. Grasser, G. Groeseneken, IEEE Trans. Dev. Mat. Rel. 13, 497, 2013; [9] S. Deora, G. Bersuker, W.-Y. Loh, D. Veksler, K. Matthews, T. W. Kim, R. T. P. Lee, R. J. W. Hill, D.-H. Kim, W.-E. Wang, C. Hobbs, and P. D. Kirsch, IEEE Trans. Dev. Mat. Rel. 13, 507, 2013; [10] K.C.Sahoo, A.S. Oates, IEEE Trans. Dev. Mat. Rel., 14, 327, 2014; [11] S.C. Lee, A.S. Oates, K.M Chang, Int. Rel. Phys. Symp. 2009, pp. 481; [12] S.C. Lee, A.S. Oates, Int. Rel. Phys. Symp, 2014, pp. 3A3.1; [13] S.C. Lee, A.S. Oates, K.M. Chang, IEEE Trans. Dev. Mat. Rel., 10, 307, 2010; [14] F. Chen et al, Int. Rel. Phys. Symp. 2005, pp. 464; [15] A.S. Oates, M.H. Lin, Int. Rel. Phys. Symp., 2012, pp. 6B2.1; [16] A.S. Oates, M.H. Lin, Int. Rel. Phys. Symp., 2013, pp. 3F.1.1; [17] A.S. Oates, M.H. Lin,. Int. Rel. Phys. Symp., 2014, 5A.2.1; [18] Y.P.Fang, A.S. Oates, IEEE Trans. Dev. Mat. Rel. 11, 551, 2011; [19] Y.P. Fang, A.S. Oates, IEEE Trans. Dev. Mat. Rel., 14, 583, 2014; [20] Y.P. Fang, A.S. Oates, IEEE Trans. Dev. Mat. In press. [21] N.N. Mahatme, N.J. Gaspard, T.Assis, S. Jagganathan, I. Chatterjee, T.D. Loveless, B.L. Bhuva, L.W. Massengill, S.J. Wen, R. Wong, Int. Rel. Phys. Symp, 2014, pp.5F.2.1; [22] H. Liu, M. Cotter, S. Datta, V. Narayanan, IEEE Trans. Dev. Mat. Rel., 14, 732, 2014

On the Microscopic Structure of Hole Traps in pMOSFETS

T. Grasser[◇], W. Goes[◇], Y. Wimmer[◇], F. Schanovsky[◇], G. Rzepa[◇], M. Waltl[◇], K. Rott[•], H. Reisinger[•],
V. V. Afanas'ev[‡], A. Stesmans[‡], A.-M. El-Sayed[†], and A.L. Shluger[†]

[◇]TU Wien, Vienna, Austria   [•]Infineon Munich, Germany   [‡]KU Leuven, Belgium   [†]UCL, London, UK

Abstract

Hole trapping in the gate insulator of pMOS transistors has been linked to a wide range of detrimental phenomena, including random telegraph noise (RTN), $1/f$ noise, negative bias temperature instability (NBTI), stress-induced leakage currents (SILC) and hot carrier degradation. Since the dynamics of hole trapping appear similar in various oxides such as pure SiO_2, SiON, and high-k, the responsible defects should have a related microscopic structure. While a number of defects have been suspected to be responsible for these phenomena, such as oxygen vacancies/E' centers, K centers, hydrogen bridges or hydrogen-related defects in general, the chemical nature of the dominant charge trap remains controversial. Based on extended time-dependent defect spectroscopy (TDDS) data, we investigate the statistical properties of a number of defect candidates using density functional theory (DFT) calculations. Our results suggest *hydrogen bridges* and *hydroxyl E' centers* to be very likely candidates.

Introduction

Frequently studied defects in silica are the oxygen-vacancy-related defects observed in irradiation studies, which have been investigated theoretically [1–4] as well as experimentally [5, 6]. In addition, hydrogen-related defects have been also widely studied [2, 4, 7] and linked to SILC data [2]. On the other hand, the defects contributing to RTN and NBTI have not yet been unanimously identified [8, 9].

Recent TDDS experiments on the defects responsible for RTN and NBTI [10, 11] have revealed crucial features to aid defect identification: Most importantly, defects can show metastability in both the neutral and the positive charge state. Pertinently, some defects were found to behave like switching oxide traps [12], while others have bias-independent emission time constants. Secondly, some defects were found to be volatile, meaning that they can become electrically inactive for random amounts of time, a feature previously observed for RTN [13, 14]. Another intriguing observation is the widely distributed defect properties, consistent with the structural disorder of the amorphous oxides employed in Si technologies. Thus, for a comparison of theory with experiment, it is mandatory to evaluate the statistical distributions of the defect parameters in amorphous materials.

So far, however, only few DFT calculations have been done on amorphous SiO_2 [3, 15–17]. In the following we will investigate the distributed defect parameters obtained from TDDS and compare them with DFT calculations for three defect candidates.

Experimental

As the defect properties vary over a wide range, we have identified and analyzed 35 defects using TDDS in six pMOSFETs ($W \times L = 150\,nm \times 100\,nm$, 2.2 nm SiON [10]). Capture (τ_c) and emission (τ_e) times of one extracted defect, A1, are shown in Fig. 1, where our recently suggested [10, 18] four-state non-radiative multiphonon (NMP) model is used to fit the data. The model requires the specification of a configuration coordinate (CC) diagram such as the one shown in Fig. 2 to describe

the dynamics of the transitions between, in general, four states. Various simpler cases occur as well, e.g. state 2 could lie higher in energy than state 2', resulting in a 2-state model, or $\varepsilon_{T1'}$ could be too high for state 1' to be reachable, giving a 3-state model.

Depending on the number of accessible states, 5, 7, or 11 fitting parameters are required, see Figs. 3 and 4. However, except for the position of the trap in the oxide, x, and the prefactor k_0 [18], all parameters could in principle be obtained from DFT calculations, *provided the microstructure of the defect was known*. Conversely, DFT parameters can be compared to experiment to help identify possible defect candidates. Due to the large differences in the CC diagrams obtained in amorphous structures, any match has to be validated at the statistical level, which will be attempted here for the first time in this context.

The extracted defect positions and energy levels are shown in a band diagram in Fig. 5. Note that the switching traps have a second energy level, $E_{1'} = \varepsilon_{T1'} - E_T$, which determines if the defect discharges via the pathway $2 \to 1' \to 1$ rather than via $2 \to 2' \to 1$ [19]. Next, the correlation between the capture and emission times, τ_c and τ_e, vs. position is shown in Fig. 6 and Fig. 7. While this weak correlation is dominated by the WKB coefficient, it is important to realize that there is no 1:1 correlation between τ_c and τ_e as assumed in simple SRH-like models [19, 20]. Also, as suggested previously [21], τ_c and τ_e are correlated ($\rho \approx 0.7$), see Fig. 8. Using our NMP model, we observe that defects measured at say 125–175 °C can have very large τ_c and τ_e ($\geq 300\,years$) when extrapolated to room temperature. Similarly, the effective activation energies range from 0.4 to 1.4 eV, see Fig. 9, demonstrating that the large capture/emission times are due to a thermally activated process.

Ab-initio Calculations

While our samples have an SiON dielectric, we performed our calculations for the simplest case, SiO_2, in order to minimize the enormous number of possible defect configurations. However, hydrogen as the most abundant element in Si processing was added, to include frequently suggested H-related defects [2, 7, 9, 14, 22, 23]. This choice is based on the rationale that pMOS/NBTI is very similar in SiO_2, SiON, and HK gate stacks, suggesting common defects to be responsible. Previous calculations [2, 24] have already shown, however, that the natural candidate, the E' center, has a very deep trap-level ($E_V(Si) - E_T \approx -3.5\,eV$), making it incompatible with NBTI/RTN.

For our DFT calculations we created large a-SiO_2 structures containing 216 atoms using ReaxFF [25]. To obtain more realistic bandgaps essential for aligning the oxide defect energy levels with the valence band in Si, a non-local PBE0_TC_LRC hybrid functional as implemented in the CP2K code [26] was used, yielding $E_g(SiO_2) = 8.1\,eV$ (cf. experiment: 8.9 eV).

As a first candidate, we consider the hydrogen bridge, see Fig. 10, which has previously been linked to SILC [2] and NBTI [24] based on calculations in c-SiO_2. In our amorphous structures

978-1-4799-8002-4/14 $31.00 © 2014 IEEE

all O atoms were each replaced with a single H atom to create 144 defect configurations. Each such defect was checked for having the required 4 states employing a similar criterion as suggested in [15]. For 12 of those defects, the full CC diagram was calculated using the nudged elastic band method [27], requiring on average 2×10^5 CPU hours for each defect.

Next, we checked the reactions of neutral hydrogen with defect-free a-SiO$_2$. Our calculations suggest that for Si-O bonds longer than 1.65 Å, energy can be released by reactions and reconfigurations induced by hydrogen. This confirms previous speculations [7, 28] that under certain circumstances this configuration can also be stable in the *neutral* charge state. As a result, a 3-fold coordinated Si facing a hydroxyl group is obtained, see Fig. 11, which has the 4 required states and is termed *hydroxyl E' center*. For 13 such defects the full CC diagram was calculated in a similar manner as for the hydrogen bridges.

Discussion

The crucial parameter of any defect is the thermodynamic energy level E_T as shown in Fig. 12. In our experimental data, only defects between -1 and $0\,\text{eV}$ below $E_V(\text{Si})$ are accessible. Unfortunately, DFT energy levels contain some uncertainty, making precise statements difficult. In our particular case, one could relate the defect levels to $E_V(\text{Si})$ calculated using the same hybrid functional, which would place 60%/75% of our hydrogen bridges/hydroxyl E' centers above $E_V(\text{Si})$ and thus render them permanently positive under NBTI conditions. To retain a larger fraction of our defect population (58%/50%) and thus improve our statistics, we allow for an energy correction of $-0.4\,\text{eV}$, corresponding to $\approx 50\%$ our SiO$_2$ bandgap error (0.8 eV).

Some selected CC diagrams of the calculated defects are shown in Fig. 13. While the E' center shows the required 4 states, due to its energetic position it would remain neutral under typical NBTI conditions. The hydrogen-bridge, on the other hand, is in good agreement with the experimentally observed behavior. This is also visible in Fig. 14 and Fig. 15, which show the calculated emission times and activation energies. Like the hydrogen-bridge, the hydroxyl E' center has CC diagrams compatible with experimental data, see Fig. 16. As such, it also has active defects within the experimental window, both from a time constant perspective (1 μs – 1 ks) as well as from an activation energy perspective (0.5 eV – 1.3 eV), see Fig. 17 and Fig. 18. Experimental data suggest that the defect distributions are much wider than our TDDS window [21, 29], so it is essential to compare the distributions of the parameters rather than single selected defects. The calculated distributions for both defect candidates are in general well in line with the experimental distributions, see Fig. 19. The most significant deviation is observed for $\varepsilon_{22'}$, which determines the emission time constant, but is notably smaller than the experimental values, on average by 0.5 eV. Whether this is an artifact of our bulk amorphous oxide structure or evidence for a different microscopic nature of the defect remains to be clarified.

Experimental data suggest that defects such as hydrogen bridges and hydroxyl E' centers are introduced during processing. As to hydrogen bridges, it has been shown that oxygen vacancies can trap hydrogen, which is available in abundance, thereby forming a hydrogen bridge. A possible mechanism for the creation of hydroxyl E' centers, on the other hand, would be through the reaction of H$_2$ with defect-free a-SiO$_2$ during high-temperature process steps [30], which can result in a fully passivated defect with two trapped H. Exposure to atomic H could form an active defect in an exothermic reaction with a small barrier, followed by the release of H$_2$. The calculated barriers for these reactions are consistent with these considerations, see Fig. 20. Subsequent H/H$_2$ reactions could then explain the observed volatility.

At a first glance, the involvement of hydrogen-related defects in the charge-trapping component of NBTI appears at odds with the experimental observation that the recoverable NBTI component is independent of the hydrogen concentration of the sample [31, 32]. However, even conservative estimations of the hydrogen concentration seem to suggest that hydrogen is available in abundance even in the driest oxides [33]. In such a case the defect concentration would be limited by the availability of oxygen vacancies and stretched Si-O bonds rather than the (very high) hydrogen concentration. If in these previous studies [31, 32] the hydrogen concentration could only be controlled in a limited manner without resulting in very low hydrogen concentrations, then these experiments would not be able to reveal a hydrogen concentration dependence of the recoverable component of NBTI.

Conclusions

In an extensive study we have compared the distributed parameters of possible hole trap candidates against experimental data. Our results suggest *hydrogen bridges* and *hydroxyl E' centers* to be very likely candidates consistent with various observations and allow understanding of the widely distributed defect properties, which is *essential for accurate reliability extrapolations*.

Acknowledgments

The research leading to these results has received funding from the Austrian Science Fund (FWF) project n°23390-M24, the European Community's FP7 project n°261868 (MORDRED), and the Intel Sponsored Research Project n°2013111914. The computational results presented have been achieved in part using the Vienna Scientific Cluster (VSC) and the UK's national high-performance computing service HECToR and Archer via the Materials Chemistry Consortium (EPSRC EP/F067496). Valuable discussions with P. Lenahan and J. Campbell are gratefully acknowledged.

References

[1] F. Feigl *et al.*, Solid State Comm. **14**, 225 (1974).
[2] P. Blöchl, PRB **62**, 6158 (2000).
[3] C. Nicklaw *et al.*, T-NS **49**, 2667 (2002).
[4] D. Fleetwood *et al.*, T-ED **49**, 2674 (2002).
[5] D. Griscom, PRB **22**, 4192 (1980).
[6] P. Lenahan *et al.*, T-NS **48**, 2101 (2001).
[7] J. de Nijs *et al.*, APL **65**, 2428 (1994).
[8] J. Campbell *et al.*, T-DMR **7**, 540 (2007).
[9] M. Houssa *et al.*, APL **90**, 043505 (2007).
[10] T. Grasser *et al.*, in *IRPS* (2010), pp. 16–25.
[11] T. Grasser *et al.*, in *IRPS* (2014), pp. 4A.5.1–4A.5.7.
[12] J. Conley Jr. *et al.*, T-NS **42**, 1744 (1995).
[13] M. Uren *et al.*, PRB **37**, 8346 (1988).
[14] T. Grasser *et al.*, in *IEDM* (2013).
[15] Z.-Y. Lu *et al.*, Phys. Rev. Lett. **89**, 285505 (2002).
[16] A. Alkauskas *et al.*, Physica B **401-402**, 546 (2007).
[17] A. Kimmel *et al.*, in , R. Sah *et al.* (Ed.) (ECS T, 2009), Vol. 19, pp. 2–17.
[18] W. Goes *et al.*, in *BTI*, T. Grasser (Ed.) (Springer, 2014), pp. 409–446.
[19] T. Grasser, MR **52**, 39 (2012).
[20] A. McWhorter, S.Surf.Phys 207 (1957).
[21] T. Grasser *et al.*, in *IEDM* (2011), pp. 27.4.1–27.4.4.
[22] R. Stahlbush *et al.*, T-NS **41**, 1844 (1994).
[23] V. Afanas'ev *et al.*, JAP **78**, 6481 (1995).
[24] F. Schanovsky *et al.*, in *SISPAD* (2013), pp. 1–4.
[25] A. van Duin *et al.*, J.Phys.Chem.A **105**, 9396 (2001).
[26] M. Guidon *et al.*, J.Chem.Theor.Comp. **5**, 3010 (2009).
[27] G. Henkelman *et al.*, J.Chem.Phys. **113**, 9901 (2000).
[28] V. Afanas'ev *et al.*, PRL **80**, 5176 (1998).
[29] G. Pobegen *et al.*, T-ED **60**, 2148 (2013).
[30] V. Afanas'ev *et al.*, J.Electrochem.Soc. **148**, 279 (2001).
[31] V. Huard, in *IRPS* (2010), pp. 33–42.
[32] T. Aichinger *et al.*, in *IRPS* (2010), pp. 1063–1068.
[33] J. Krauser *et al.*, J.Non-Cryst.Solids **187**, 264 (1995).
[34] K. Huang *et al.*, Proc.R.Soc.A **204**, 406 (1950).

978-1-4799-8002-4/14 $31.00 © 2014 IEEE

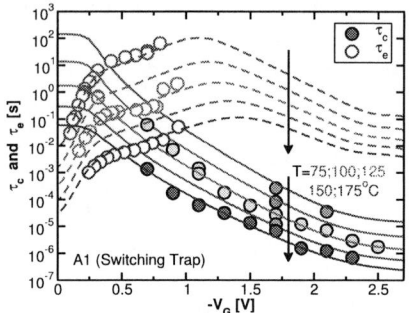

Fig. 1: The capture and emission times (symbols) of trap A1 extracted over a wide temperature and bias range. Our four-state non-radiative multiphonon model [10, 18] can describe the data very well (lines). Note the strong temperature activation as well as the strong bias-dependence of the emission time below $V_{th} \approx 0.5\,V$ and the saturation of the capture time at high biases.

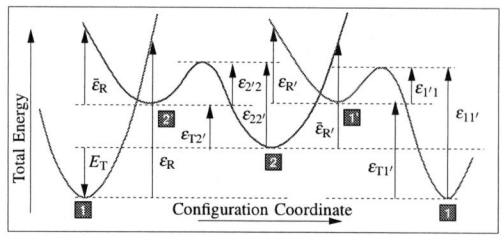

Fig. 2: Following the Born-Oppenheimer approximation we use potential energy surfaces along 1D configuration coordinates (CC) in the various states to describe the transitions using non-radiative multiphonon [19, 34] theory (1 ↔ 2 as 2 ↔ 1') and transition-state theory (2' ↔ 2 and 1' ↔ 1). The CC diagram of a typical four-state defect is specified by eleven fitting parameters but describes the bias-, temperature- and frequency dependence of the capture and emission times. In addition, the position of the trap into the oxide (x), the capture cross section k_0, and an attempt frequency ($\nu = 10^{13}\,s^{-1}$) need to be specified to fully define the transitions [18, 19].

Par	A1	Min	Mean	Max	Unit	
E_T	-0.08	-0.7	-0.36	0.07	eV*	2-State
ε_R	3.22	0.37	1.43	5.34	eV	
$\bar{\varepsilon}_R$	2.95	0.31	2.14	5.05	eV	
$\varepsilon_{T2'}$	0.25	0.16	0.37	0.88	eV	3-State
$\varepsilon_{2'2}$	0.56	0.01	0.63	1	eV	
$(\varepsilon_{22'})$	0.81	0.19	0.99	1.3	eV	
$\varepsilon_{T1'}$	0.31	0.24	0.68	1.82	eV	Full 4-State
$\varepsilon_{R'}$	0.38	0.09	0.82	1.48	eV	
$\bar{\varepsilon}_{R'}$	0.38	0.15	1.09	2.59	eV	
$\varepsilon_{11'}$	1.05	0.94	1.29	1.88	eV	
$(\varepsilon_{1'1})$	0.75	0.06	0.61	0.9	eV	
x	-0.61	-0.04	-0.68	-1.04	nm	
k_0	10^{-9}	10^{-14}	10^{-7}	10^{-6}	eV m³/s	

▨ Could be calculated by DFT *relative to E_V
▨ Remaining fit parameter ( ) overdetermined

Fig. 3: From the 11 parameters required for each defect, 9 can be in principle obtained from DFT, provided the microstructure and it variations are known. Shown are the parameters of defect A1 together with their distribution. Not all states may be accessible, reducing the general 4-state model to a 3- or even 2-state model.

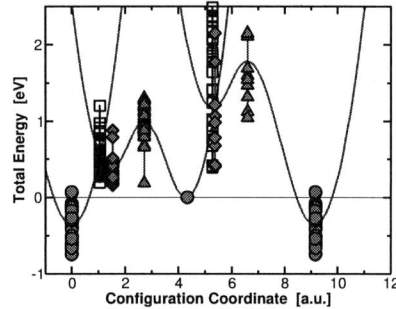

Fig. 4: The average CC diagram as shown in Fig. 2 extracted from 35 defects together with the distribution of the individual barriers with the energy measured relative to E_V(Si). Many distributions are rather wide (about 1 eV).

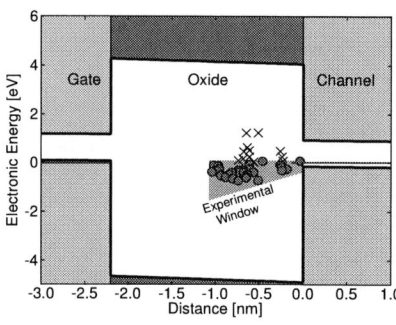

Fig. 5: The trap-levels E_T (circles) and $E_{1'} = \varepsilon_{T1'} - E_T$ (crosses) versus x coordinate of the 35 extracted TDDS defects in a band diagram. The energy is measured relative to E_V(Si). Only traps within the experimental window defined by V_{stress}^{max} are accessible.

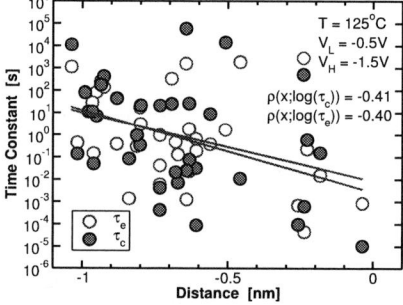

Fig. 6: A weak correlation between the extracted time constants and the position of the trap is observed. While this correlation is dominated by the WKB coefficient, no 1:1 correlation between τ_c and τ_e is obtained.

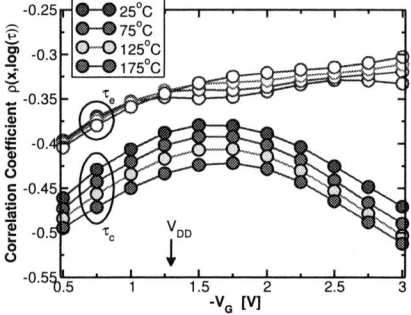

Fig. 7: The correlation coefficients $\rho(x, \log(\tau_e))$ and $\rho(x, \log(\tau_c))$ are small and depend weakly on bias.

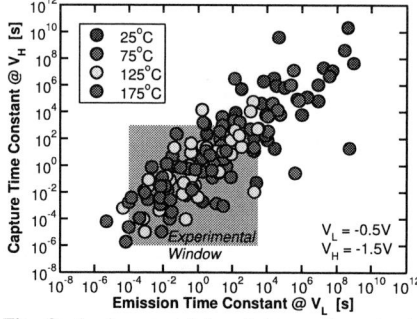

Fig. 8: A clear correlation between the extracted capture and emission times is observed ($\rho \approx 0.7$).

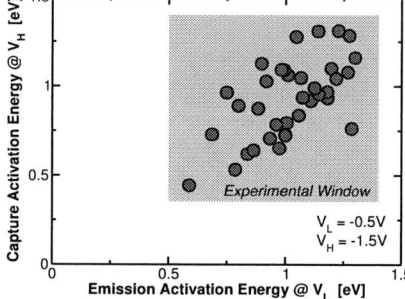

Fig. 9: The effective capture and emission activation energies show a similar correlation and distribution. The experimental window is shown for later reference, cf. Fig. 15 and Fig. 18.

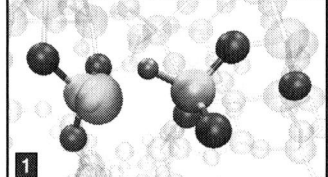
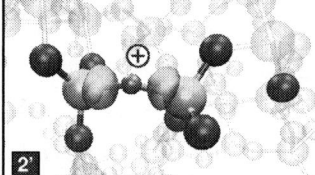
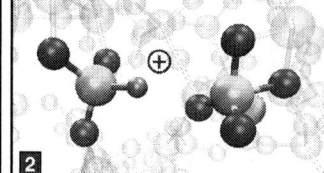

Fig. 10: The four states of the hydrogen bridge: In the initial configuration 1, H (silver) sits between two Si atoms (yellow) which themselves are surrounded by three O atoms (red). The electron density of the localized Kohn-Sham-eigenstate is shown as turquoise 'bubbles'. *Upon hole capture* the defect can go into the metastable state 2', where the Si atoms move closer together. Depending on the gate bias, the defect either goes back to state 1 or, *eventually into the stable positive state 2*, where the right Si has moved through the plane of its three O neighbors, forming a puckered configuration by bonding to the O in the far right. In the metastable state 1', the defect is neutralized but remains in the puckered configuration.

978-1-4799-8002-4/14 $31.00 © 2014 IEEE

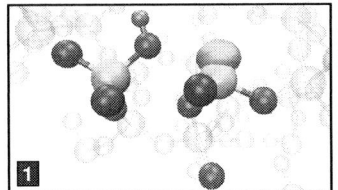 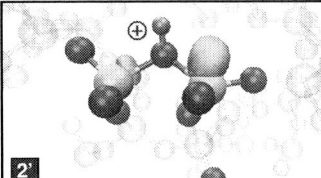 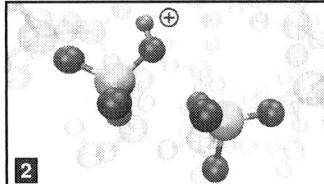 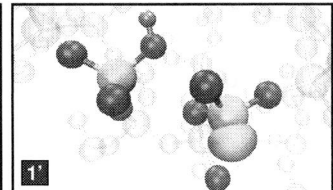

Fig. 11: The four states of the hydroxyl E' center: In the neutral configuration 1, a hydroxyl group sits at the left Si while the other carries a dangling bond. After hole capture, in state 2', the dangling bond has lost its electron and reforms the Si-O-Si bridge, resulting in the typical proton sitting on a bridging O. In state 2, the right Si moves through the plane of its O neighbors, forming a bond with the O in its back. In state 1', the dangling bond is restored but points into the other direction.

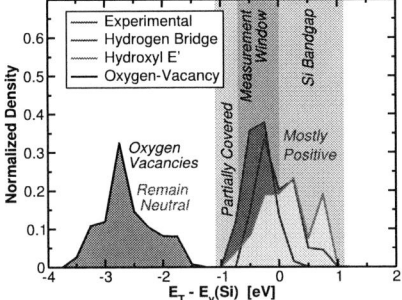

Fig. 12: The fundamental parameter which decides on which trap can be charged for a certain stress (V_G^H) and recovery (V_G^L) combination is the energy level E_T. Clearly, the oxygen vacancy/E' center is too low in energy, while both the hydrogen bridge and the hydroxyl E' center are in good agreement with the data inside the experimental window. Recall the uncertainty in DFT energy-levels and the $-0.4\,$eV energy shift used.

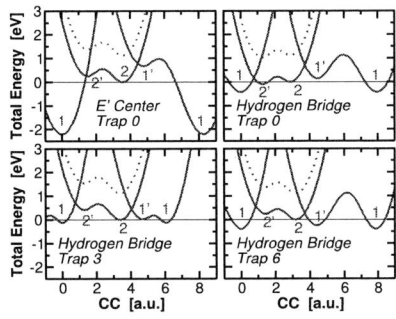

Fig. 13: Some selected DFT CC diagrams for the hydrogen bridge in a-SiO$_2$, all of which being different, reflecting the amorphous nature of the oxide. The CC diagram of the E' center is also shown at the top-left for comparison, for which state 1 is too low, resulting in a permanently neutral trap.

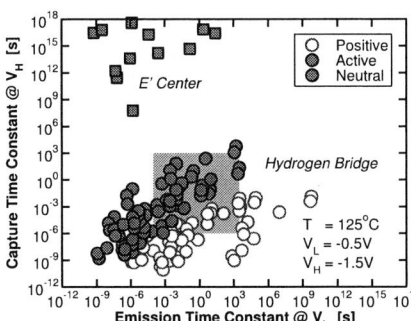

Fig. 14: The theoretical capture and emission times at stress and recovery bias for the defects in Fig. 13. Each DFT defect is considered at 10 random locations in the oxide between 0 and $-t_{ox}/2$. Some defects are energetically too low and remain neutral (blue circles), some are in the active region and can charge and discharge (red circles), while some are too high in energy and are always positive (empty circles). The oxygen vacancy cannot become charged during typical stress conditions and thus remains neutral.

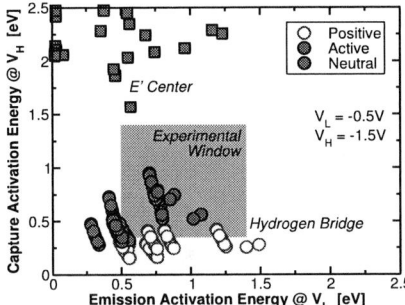

Fig. 15: The theoretical effective capture and emission activation energies for the hydrogen bridge and the regular E' center. While the E' centers would remain neutral under typical NBTI conditions, the hydrogen bridge would be visible in the experimental window.

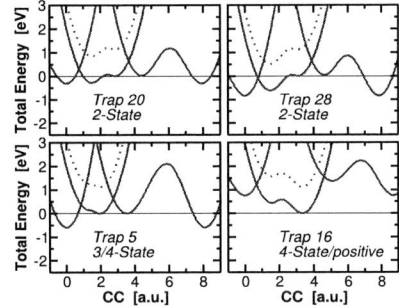

Fig. 16: The DFT CC diagrams of four selected hydroxyl group E' centers. Clearly, marked differences exist. From 13 defects, 3 are found to have all 4 required states in addition to sensible barriers separating them.

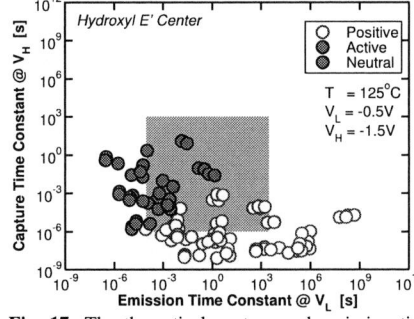

Fig. 17: The theoretical capture and emission time map of the hydroxyl group E' center. Just like the hydrogen bridge, the hydroxyl E' center would be visible in the experimental window.

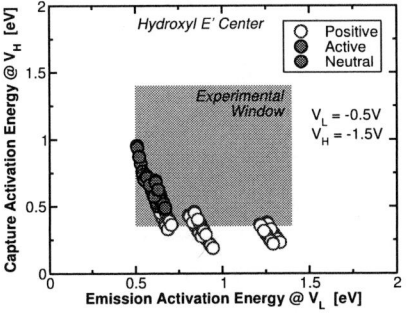

Fig. 18: The theoretical effective capture and emission activation energies for the hydroxyl E' center. Just like the hydrogen bridge, the hydroxyl E' center would be inside the experimental window.

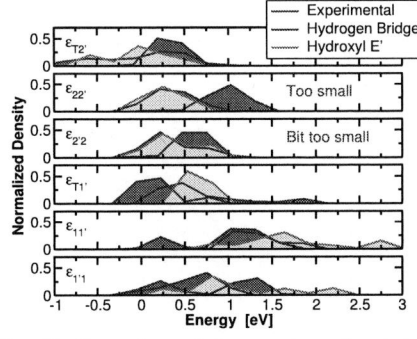

Fig. 19: Comparison of the experimental and theoretical barriers. While overall good agreement is obtained, the theoretical barrier $\varepsilon_{22'}$ is too small in general. Note that the defects with negative $\varepsilon_{T2'}$ are 2-state defects at the border of our experimental window.

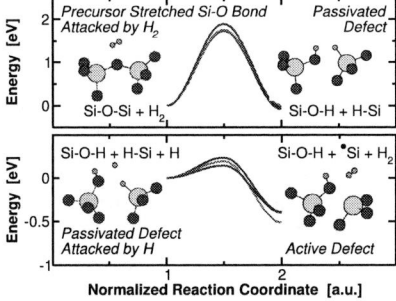

Fig. 20: Possible creation scenario of the hydroxyl E' center during high-temperature processing via the attack of a stretched Si-O-Si bond by H$_2$, which is later attacked by H to eventually form the active defect. Creation is limited by the precursor-density, not the H concentration [31, 32].

978-1-4799-8002-4/14 \$31.00 © 2014 IEEE

Analytical Formulation of SiO$_2$-IL scavenging in HfO$_2$/SiO$_2$/Si gate stacks
- A key is the SiO$_2$/Si interface reaction -

Xiuyan Li, Takeaki Yajima, Tomonori Nishimura, Kosuke Nagashio, and Akira Toriumi

Department of Materials Engineering, The University of Tokyo, Tokyo 113-8656, Japan

Phone : +81-3-5841-7161, e-mail : xiuyan@adam.t.u-tokyo.ac.jp

Abstract

The scavenging kinetics of ultra-thin-SiO$_2$ interface layer (IL) in HfO$_2$/SiO$_2$/Si stacks is investigated by focusing on SiO$_2$/Si interface reaction in addition to both O and Si atom kinetics. SiO$_2$/Si interface serves as a stage that the oxygen vacancy (V_O) is converted to Si release from SiO$_2$ with the help of Si substrate. Based on both diffusion kinetics and possible reaction, an analytical model for two-stage SiO$_2$-IL scavenging in high-k gate stack is proposed.

Introduction

SiO$_2$-IL scavenging phenomenon in high-k gate stack has been reported by many researchers since the early stage of high-k research [1, 2]. Recently it was carried out by introducing a scavenging metal to the gate electrode to achieve the equivalent-oxide-thickness (EOT) scaling [3, 4]. Although the scavenging is phenomenologically with the opposite direction against the oxidation, its mechanism has not been fully characterized yet, which is critical for interface control as well as for further EOT scaling. Thus, the objective of this paper is to understand the SiO$_2$-IL scavenging kinetics and to formulate its physics-based analytical model.

Experiments Characterizing Scavenging Kinetics

All of scavenging experiments in this work have been carried out in ultra-high-vacuum (UHV) instead of metal gates [5], because in UHV, it is not necessary to take account of side effects possibly caused by metal gates, and it is much simpler to characterize the scavenging process. We first discuss experimental results from following four viewpoints. (1) V_O kinetics, (2) Si kinetics, (3) substrate contribution, and (4) initially deposited HfO$_2$ effect.

(1) V_O diffusion kinetics

What we have so far clarified on the V_O diffusion are summarized as follows.

(i) SiO$_2$-IL scavenging occurs just below the silicidation temperature, and the upper HfO$_2$ is mandatory for scavenging (**Fig. 1**) [5].

(ii) O-atom in SiO$_2$-IL diffuses into HfO$_2$ layer, which was revealed by ^{18}O isotope tracing experiment [6].

(iii) Preexisting V_O in HfO$_2$ plays a significant contribution to the scavenging [7].

(2) Si diffusion kinetics

Naively thinking, Si in SiO$_2$ may epitaxially grow on Si substrate, or diffuse out from the gate stack in the scavenging. However, no direct evidence has been reported so far. This work presents *three kinds of new experimental results to clarify this issue*.

(i) Si-related desorption was not detected previously by quadrupole mass spectrometry due to the fact that SiO was overlapped with CO$_2$ in m/z=44. In the present experiment, ^{29}SiO$_2$-IL was grown for enhancing the detection sensitivity. The thermal desorption experiments in **Fig. 2** shows a clear ^{29}SiO desorption from HfO$_2$/^{29}SiO$_2$/Si stack, and no ^{29}SiO from ^{29}SiO$_2$/Si stack. This is the first and direct evidence of Si desorption in SiO$_2$-IL scavenging through HfO$_2$.

(ii) Si desorption was further confirmed by inspecting the adsorbed species on TiN/Si cap, which was set apart from HfO$_2$ surface. With the scavenging, SiO$_x$ peak appears

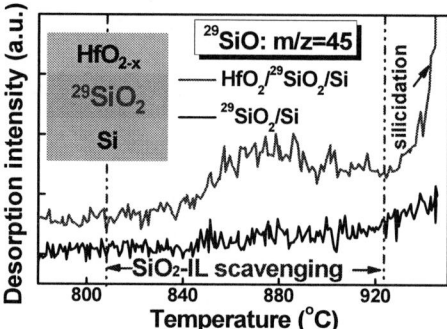

Fig. 2 TDS results of HfO$_2$/^{29}SiO$_2$/Si and ^{29}SiO$_2$/Si stacks. A ^{29}SiO peak is observed in scavenging of HfO$_2$/^{29}SiO$_2$/Si stack. No peak from ^{29}SiO$_2$/Si stack indicates SiO$_2$-IL scavenging causes ^{29}SiO desorption.

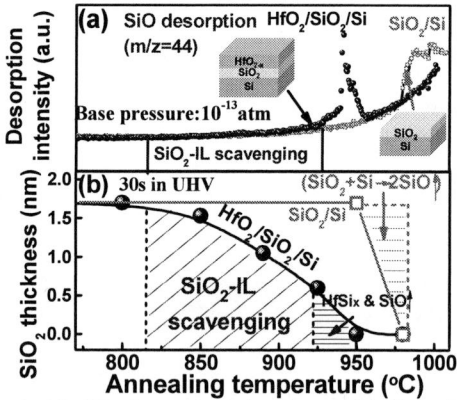

Fig. 1 **(a)** SiO desorption (m/z=44) and **(b)** SiO$_2$ change in HfO$_2$/SiO$_2$/Si and SiO$_2$/Si stacks in UHV-PDA. SiO$_2$-IL scavenging occurs just before silicidation followed by SiO desorption in HfO$_2$/SiO$_2$/Si stack and HfO$_2$ is mandatory for this process.

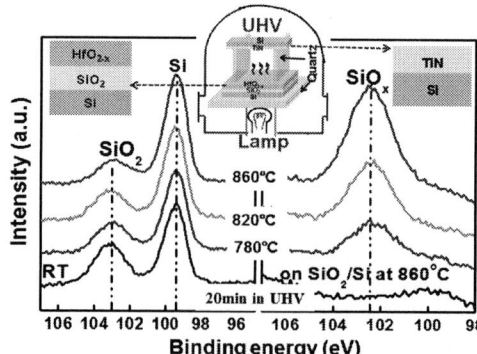

Fig. 3 XPS (Si2p) results of HfO$_2$/SiO$_2$/Si stack and TiN/Si cap separated from HfO$_2$ surface after UHV-PDA, including the as-deposited sample and TiN layer set on SiO$_2$/Si stack. With decrease of SiO$_2$-IL peak in HfO$_2$/SiO$_2$/Si, SiO$_x$ peak appears and increases on TiN, while there is no Si peak detected on TiN set on SiO$_2$/Si stack.

978-1-4799-8002-4/14 $31.00 © 2014 IEEE   534

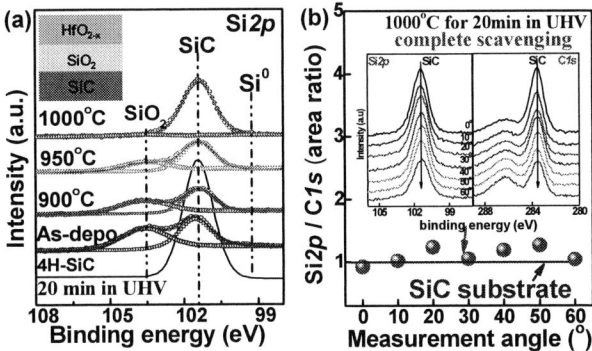

Fig. 4(a) XPS results of HfO₂/SiO₂/SiC stack after UHV-PDA (Si2p peak was normalized by C1s peak from SiC). With decrease of SiO₂-IL peak, even with complete scavenging (at 1000°C), no grown Si⁰ peak appears. **(b)** Angle resolved XPS for completely scavenged stack (inset) shows the value of Si2p/C1s (area ratio) does not change with increasing measurement angle. This value is comparable with that from SiC substrate, suggesting there is no Si regrowth on substrate after scavenging.

and increases on TiN layer (**Fig. 3**). This fact also assures that Si is really desorbed through HfO₂ during SiO₂-IL scavenging.

(iii) SiC in place of Si substrate was used to detect possible Si growth on the substrate, because it was difficult to distinguish grown-Si from Si substrate. **Fig. 4(a)** shows that no Si⁰ peak appears even after SiO₂-IL completely disappears. The angle-resolved XPS for the completely scavenged sample further shows that there is no indication of newly grown Si on SiC within our experimental resolution (**Fig. 4(b)**).

(3) Substrate effect on scavenging

We have already reported a significant difference of scavenging effects among on Si, sapphire and SiC [6]. In this paper, the temperature dependence of scavenging on Si and SiC is discussed. Note that although dielectric film stack was prepared as exactly the same, the activation energy of the scavenging on Si and SiC was different from each other (**Fig. 5(a)**). On the other hand, the apparent relationship among activation energies in scavenging on Si and SiC are same with that in oxidation of them, respectively (**Fig. 5(b)**). This fact suggests that the oxidation of Si in substrate may be involved in the scavenging. It is possible that substrate-Si is *oxidized* by SiO₂ in UHV-PDA, which may enhance the scavenging

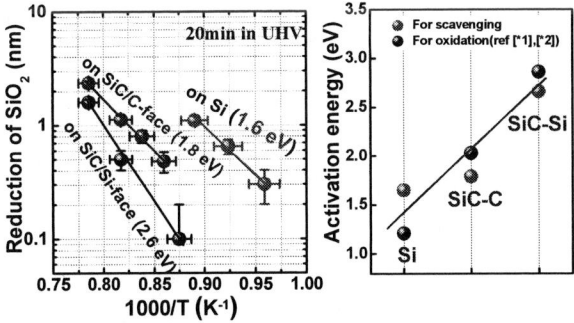

Fig. 5 (a) Activation energies of scavenging on Si, C-face and Si-face SiC substrates were estimated to be 1.6, 1.8 and 2.6 eV respectively. **(b)** The activation energies of scavenging on Si and SiC are similar to that of oxidation of them. It is demonstrated that easily oxidized substrate enhance scavenging. ([*1] B. E. Deal and A. S. Grove, *J. Appl. Phys.* **36**, 1965, 3770. [*2] M. Mukherjee, *Silicon Carbide-Materials, Processing and Applications in Electronic Devices,* (2011) 225).

reaction at SiO₂/Si interface

(4) Effect of initially deposited HfO₂

We found that scavenging occurred very fast within a few minutes and then became slower on both Si and SiC substrates, as shown in **Fig. 6(a)**. We have reported that preexisting V$_O$ in HfO₂ contributed to scavenging [7]. For studying the effect of preexisting Vo on the scavenging speed, some samples were pre-annealed in O₂ ambient for filling the preexisting V$_O$. SiC substrate was chosen in this experiment, because SiO₂-growth was ignored in O₂-annealing at such a low temperature. **Fig 6(b)** shows that in case of pre-O₂-annealed sample, the scavenging rate keeps slow from the initial stage and same amount of SiO₂ is scavenged in a longer time. 600°C/5min and 550°C/30s pre-O₂-annealled samples do not show any difference of the scavenging rate, indicating preexisting V$_O$'s in HfO₂ are filled mostly. These results suggest two stages with different V$_O$ source should be considered in the scavenging. Namely, **Stage-1:** the initial stage of scavenging attributable to pre-existing V$_O$ in HfO₂.
Stage-2: the slow scavenging attributable to V$_O$ created by UHV.

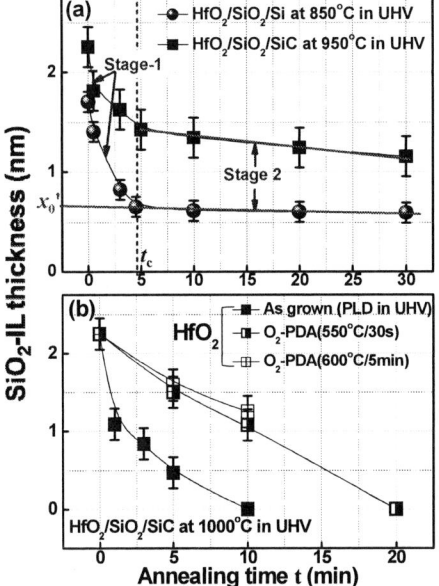

Fig. 6(a) SiO₂-IL change as a function of time in UHV-PDA in HfO₂/SiO₂/Si (850°C) and HfO₂/SiO₂/SiC (950°C) stacks. In both cases, scavenging occurs quickly at early stage and become much slower after that. **(b)** SiO₂-IL change as a function of time in UHV-PDA in HfO₂/SiO₂/SiC (1000°C) stack w/ and w/o pre-O₂-annealing. This shows that preexisting V$_O$ in HfO₂ contributes quick scavenging in initial stage.

Analytical Model Formulation

Based on the above experimental results newly found and those previously obtained, we propose a model for scavenging as follows (schematically depicted in **Fig. 7**).

i) V$_O$ in HfO₂ is dependent on HfO₂ quality. It changes with the time in stage-1 and reaches equilibrium state with UHV in stage-2, as shown in **Fig. 8**.

ii) V$_O$ in HfO₂ diffuses to SiO₂/Si interface through SiO₂-IL.

iii) V$_O$ reacts with SiO₂ at SiO₂/Si interface with the help of substrate Si, as the net equation 2Vo+SiO₂→Si.

iv) Generated Si diffuses to HfO₂ through SiO₂-IL, and is desorbed out to UHV through HfO₂.

978-1-4799-8002-4/14 $31.00 © 2014 IEEE

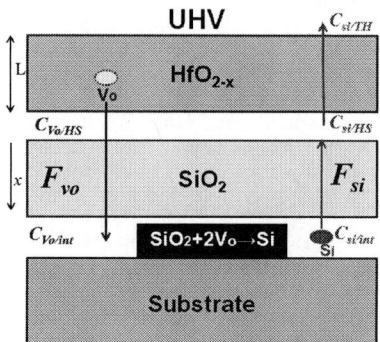

Fig. 7 Schematics of scavenging kinetics in $HfO_2/SiO_2/Si$ stack. V_O from HfO_2 diffuses through SiO_2, and reacts with SiO_2 at SiO_2/Si interface with help of Si from substrate. Generated Si diffuses through SiO_2 and HfO_2 and then is desorbed to UHV. It is regarded that V_O diffusion is converted to Si diffusion through the reaction at SiO_2/Si interface.

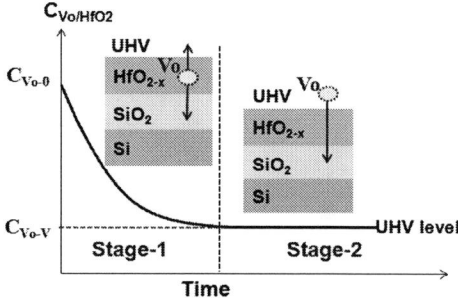

Fig. 8 $C_{Vo/HfO2}$ change with time. In stage-1, Preexisting V_O in HfO_2 diffusion into both SiO_2 and HfO_2, C_{V_O} decreases with time. In stage-2, V_O in HfO2 reach an equilibrium state with UHV, C_{V_O} become a constant decided by P_{O2} in UHV.

(1) Stage-2 in scavenging

Since stage-2 in Fig. 6(a) is more common and simpler, we firstly discuss this stage. The diffusion-reaction process is formulated by assuming that:

a) the flux approaches the steady state after stage-1,

b) the scavenging reaction is in the equilibrium state (the scavenging rate is relatively slow).

c) V_O distribution in HfO_2 is homogeneous (V_O diffusion in HfO_2 are extremely quick [8]).

Table 1 Definition of parameters used in formulation.

parameter	Definition	
x	SiO_2 thickness	$f(t)$
L	HfO_2 thickness	constant
$F_{A/B}$	Flux of species A in oxide B	$f(t)$
$D_{A/B}$	Diffusion coefficient of A in B	constant
$C_{A/CD}$	Concentration of A at C/D interface	$f(t)$
K	Reaction constant for $SiO_2+2Vo\leftrightarrow Si$	constant
$K_{A/VH}$	Diffusion constant of A from HfO_2 into UHV	constant
$F_{A/VH}$	Flux of A from HfO_2 into UHV	
$C_{Vo/HfO2}$	Concentration of Vo in HfO_2	$f(t)$
x_0	Initial SiO_2 thickness	constant
C_{Vo-0}	Concentration of preexisting Vo in HfO_2	constant
$C_{Vo-V(M)}$	Vo Concentration generated by UHV(metal)	$f(P_{O2})$

The parameters used in the formulation are defined in **Table 1**. Each step is described as follows.

i) V_O in HfO_2

In this stage, V_O in HfO_2 reaches the equilibrium state with UHV, $C_{Vo/HfO2}$ become a constant which is decided by P_{O2} in UHV as (Fig. 8),

$$C_{Vo/HS} = C_{Vo/HfO_2} = C_{Vo-V} \qquad (1)$$

ii) V_O diffusion to SiO_2/Si interface

$$F_{Vo/SiO2} = \frac{D_{Vo/SiO2}(C_{Vo/HS} - C_{Vo/int})}{x} = F_{in} \qquad (2)$$

iii) Reaction at SiO_2/Si interface reaction

$$\kappa = \frac{C_{Si/int}}{C_{SiO2} \cdot C_{Vo/int}^{2}}, \qquad C_{Si/int} = \kappa C_{Vo/int}^{2}, \qquad (3)$$

iv) Si diffusion from SiO_2/Si interface into UHV

$$F_{Si/SiO2} = -\frac{D_{Si/SiO2}(C_{Si/HS} - C_{Si/int})}{x}, \qquad (4)$$

$$F_{Si/HfO2} = -\frac{D_{Si/HfO2}(C_{Si/VH} - C_{Si/HS})}{L}, \qquad (5)$$

$$F_{Si/VH} = K_{Si/VH}C_{Si/VH} = F_{out}. \qquad (6)$$

By combining Eq. (2)~Eq. (6), $F_{Si/VH}$ is given by

$$F_{Si/VH} = K_{Si/VH}\left(\sqrt{\frac{C_{Vo/HS}}{Ax} + \frac{B+Cx}{4A^2x^2}} - \frac{\sqrt{B+Cx}}{2Ax}\right)^2, \qquad (7)$$

$$A = \frac{2K_{Si/VH}}{D_{Vo/SiO2}}, \quad B = \frac{1}{\kappa}\left(\frac{K_{Si/VH}}{D_{Si/HfO2}}L + 1\right), \quad C = \frac{1}{\kappa}\frac{K_{Si/VH}}{D_{Si/SifO2}}.$$

If N_{Si} and N_{Vo} are the numbers of removed Si from and incorporated V_O into a unit of SiO_2 layer during scavenging. Thus, $N_{Vo}=2N_{Si}$. The scavenging rate is described by

$$-\frac{dx}{dt} = \frac{F_{Si/VH}}{N_{Si}} = \frac{F_{Si/HfO2}}{N_{Si}} = \frac{F_{Si/SiO2}}{N_{Si}} = \frac{F_{Vo/SiO2}}{N_{Vo}} = \frac{F_{Vo/SiO2}}{2N_{Si}}. \qquad (8)$$

By putting Eq. (1) and (7) into Eq (8), the time-dependent scavenging is described as follows.

$$-\frac{K_{Si/VH}C_{Vo-V}}{N_{Si}}(t+\tau) = \frac{1}{2}\left(C_{Vo-v}A + \frac{C}{2}\right)x^2 + \frac{B}{2}x + \frac{1}{2}\sqrt{C\left(C_{Vo-v}A + \frac{C}{4}\right)} \cdot F(x) \qquad (9)$$

$$\alpha = C\left(C_{Vo/HS}A + \frac{C}{4}\right), \qquad \beta = B\left(C_{Vo/HS}A + \frac{C}{4}\right) + \frac{BC}{4}, \qquad \gamma = \frac{B^2}{4},$$

$$F(x) = \left(x + \frac{\beta}{2\alpha}\right)\sqrt{\left(x + \frac{\beta}{2\alpha}\right)^2 + \left(\gamma - \frac{\beta^2}{4\alpha^2}\right)} + \left(\gamma - \frac{\beta^2}{4\alpha^2}\right)\ln\left(\left(x + \frac{\beta}{2\alpha}\right) + \sqrt{\left(x + \frac{\beta}{2\alpha}\right)^2 + \left(\gamma - \frac{\beta^2}{4\alpha^2}\right)}\right)$$

Here, τ is a constant. This is the exact form of the scavenging in the stage-2 in our model. Since it does not provide us a clear view, we go back to Eq. (7) and obtain approximated formula by assuming that Si diffusion process predominantly may limit the total scavenging rate ($B+Cx \gg C_{Vo/SH}Ax$). Thus, the equation is simplified to be

$$\frac{1}{2}(C + 2C_{Vo-v}A)(x_0^2 - x^2) + B(x_0 - x) = \frac{1}{N_{Si}}K_{Si/VH}C_{Vo/HS}t. \qquad (10)$$

Note that it looks like the Deal-Grove model in Si oxidation (linear-parabolic formula). It is quite reasonable, because the net formula is determined by diffusion and reaction. But, in the present, coefficients of the linear-parabolic formula are not independent but related to each other.

Since the scavenging rate is extremely slow in the stage-2, resulting in $x \sim x_0'$ (x_0' is defined in Fig. 6(a)). There is,

$$x \approx x_0' - \frac{K_{Si/VH}C_{Vo-V}}{N_{Si}[x_0'(C + 2C_{Vo-v}A) + B]}t. \qquad (11)$$

Let us compare the model with experimental results. Eq.

(11) shows SiO$_2$-IL thickness x is proportional to t. Experimental results in Fig. 6(a) clearly demonstrate the linear relationship between x and t in this stage.

(2) Stage-1 in scavenging

In stage-1, pre-existing V$_O$ in HfO$_2$ diffuses into both UHV and SiO$_2$, $C_{Vo/HfO2}$ decrease with time and gradually approach the steady state with transition to the stage-2.

The time evolution of V$_O$ in HfO$_2$ can be described by the Gaussian type with considering HfO$_2$ layer as a slab of V$_O$ source. Here we use an effective diffusion constant D$_{SiO2}^*$ for V$_O$ diffusion into SiO$_2$ by taking account of the diffusion to UHV (**Fig. 9**). Namely,

$$C_{Vo}(x,t) = \frac{C_{Vo-0} - C_{Vo-V}}{2\sqrt{\pi D_{SiO_2}^* t}} \exp\left(-\frac{x^2}{4 D_{SiO_2}^* t}\right) \quad (12)$$

When $x=0$, the following equation describes the time dependent $C_{Vo/HS}$.

$$C_{Vo/HS}(t) = \frac{C_{Vo-0} - C_{Vo-V}}{2\sqrt{\pi D_{SiO_2}^* t}} \quad (13)$$

Since the time scale for V$_O$ diffusion in HfO$_2$ is much shorter than that for C_{Vo} change in SiO$_2$, we assume the V$_O$ flux into SiO$_2$/Si interface can be approximated as that in the steady state (from t$_2$ in Fig.9). Then we can formulate the stage-1 from Eq. (2) –Eq. (8) and (13). Again, we discuss the case where Si diffusion predominantly limits the total scavenging rate. The equation is simplified to be

$$\frac{1}{2}C\left(x_0^2 - x^2\right) + B(x_0 - x) = \frac{K_{Si/VH}(C_{Vo-0} - C_{Vo-V})}{N_{Si}\sqrt{\pi D_{Vo/SiO2}}}\sqrt{t} \quad (14)$$

In Fig. 6(a), Δx is mainly caused by scavenging in the stage-1 in the present work. Therefore we compare this model with the experiments results of the scavenging dependence on x_0, L in addition to t (t<t_c; t_c is defined in Fig. 6(a)).

Since $x_0+x=2x_0$ at the very initial stage, Eq. (14) is simplified to be

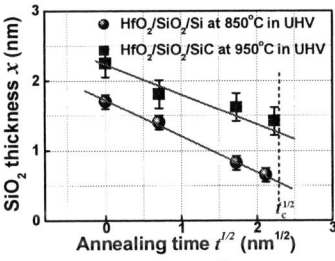

Fig. 9 SiO$_2$ thickness x as function of $t^{1/2}$ (t<t$_c$). The linear relationship fits well with equation (15).

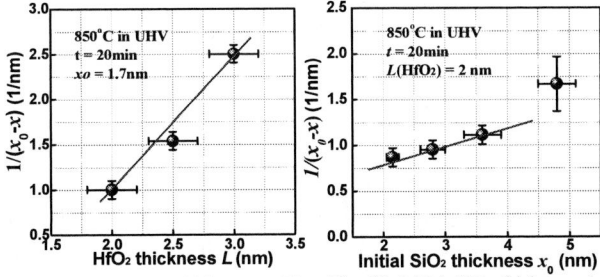

Fig. 10 HfO$_2$ thickness (L) dependence of scavenging above 2nm. The linear relationship between $1/(x_0-x)$ and L fits well with equation (15). Above 4 nm of HfO$_2$, scavenging is totally suppressed.

Fig. 11 Initial SiO$_2$ thickness (x_0) dependence of scavenging above 2nm. The linear relationship between $1/(x_0-x)$ and x_0 fits well with equation (15). Above 4 nm of SiO$_2$, scavenging is suppressed.

$$x \approx x_0 - \frac{K_{Si/VH}(C_{Vo-0} - C_{Vo-V})}{N_{Si}\sqrt{\pi D_{Vo/SiO2}}(Cx_0 + B)}\sqrt{t} \quad (15)$$

Eq. (15) shows that SiO$_2$-IL thickness x is in proportion to $t^{1/2}$, while $1/(x_0-x)$ is in proportion to both L and x_0 (L:HfO$_2$ thickness, x_0; SiO$_2$-IL thickness). **Fig. 9-11** clearly show those dependences.

(3) Metal gate effect in scavenging

In case of metal scavenging, the difference from UHV case lies on C_{Vo-V} and C_{Vo-0}. The "vacuum" created by metal might be much higher than UHV, thus the scavenging approaches the stage-2 much faster and the scavenging rate becomes faster. **Fig. 12** schematically shows the general view of the scavenging. Although there is difference among each case, the kinetic part of the model should be the same.

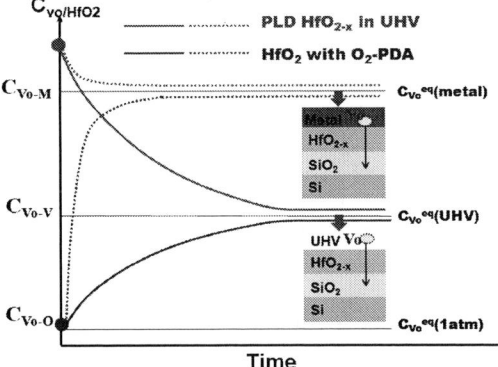

Fig. 12 Schematics of scavenging with different HfO$_2$ and "vaccum" layer. With metal, the "vacuum" level might be higher. V$_O$ in HfO$_2$ reaches the equilibrium state earlier and scavenging rate becomes faster.

Conclusion

We have experimentally demonstrated that V$_O$ reaction with SiO$_2$/Si interface is significantly important to understand the scavenging. We have found the two-stage scavenging depending on HfO$_2$ quality as well, and then formulated it analytically. The formula is like the Deal-Grove model in Si oxidation, but each coefficient is related to others. This formula will be extendable to metal scavenging by changing the V$_O$ concentration.

Acknowledgement

This work was partly performed in collaboration with STARC.

References

[1] V. Misra, G. P. Beuss, and H. Zhong, "Use of metal–oxide–semiconductor capacitors," *Appl. Phys. Lett.*, vol. 78, pp. 4166-4168, 2001.
[2] M. P. Agustin, H. Alshareef, M. A. Q-Lopez, and S. Stemmer, "Influence of AlN layers," *Appl. Phys. Lett.*, vol. 89, pp. 041906.1-041906.3,2006 .
[3] T. Ando," Ultimate Scaling of High-*k* Gate Dielectrics:" *Materials*, vol. 5 pp. 478-500, 2012.
[4] A. Nichau, A. Schafer, L. Knoll, S. Wirths, T. Schram, L. Ragnarsson, and *et al*, "Reduction of silicon dioxide interfacial layer," *Microelectronic Engineering*, pp. 109 -112, 2013.
[5] X. Li, T. Yajima, T. Nishimura, K. Nagashio, A. Toriumi, "HfO$_2$-assited interfacial SiO$_2$ reduction," *Thin Solid Films*, vol. 557, pp. 272-275, 2014.
[6] X. Li, T. Yajima, T. Nishimura, K. Nagashio, A. Toriumi, "Kinetic Model for Scavenging", June 2014, [*IEEE Si-Nanoelectronics Workshop, Hawaii*, S2-5]
[7] X. Li, T. Yajima, T. Nishimura, K. Nagashio, A. Toriumi, "Study of the interfacial SiO$_2$ scavenging," September 2013, [*45th international conference on Solid State Device and Material, Fukuoka*, PS1-4.)
[8] S. Ferrari and M. Fanciulli, "Diffusion reaction of oxygen" *J. Phys. Chem. B*, vol. 110, pp.14905-14910, 2006.

First Principles Study of SiC/SiO$_2$ Interfaces towards Future Power Devices

K. Shiraishi[1,2], K. Chokawa[2], H. Shirakawa[2], K. Endo[1], M. Araidai[1], K. Kamiya[3], and H.Watanabe[4]

[1]Graduate School of Engineering, Nagoya University, Nagoya, 464-8603, Japan,
[2]Graduate School of Pure and Applied Sciences, University of Tsukuba, Tsukuba 305-8571, Japan,
[3]Kanagawa Institute of Technology, Atsugi 243-0292, Japan.
[4]Graduate School of Engineering, Osaka University, Suita, 565-0871, Japan,
Phone and fax: +81-52-789-3715,   E-mail: shiraishi@cse.nagoya-u.ac.jp

Abstract

We clarify the intrinsic problems of SiC/SiO$_2$ interfaces by the first principles calculations. The unique nearly free electron like characteristics of SiC conduction band bottom causes unexpected formation of interface states near the conduction band bottoms by process induced strain. These results indicate that strain free process is necessary for fabricating high quality NMOSFET. Another proposal is developing PMOSFET instead of presently popular NMOSFET. Moreover, we also discuss the Vth instability caused by proton diffusion.

1. Introduction

Silicon Carbide (SiC) has for a long time been thought as one of the most promising power device material that substitutes silicon devices. It is well known that SiC/SiO$_2$ interfaces formed by thermal oxidation of SiC contain many interface trap sites [1]. As for trap sites at SiC/SiO$_2$ interfaces, it has been reported that there are a lot of trap states near conduction band bottom (CB) of SiC [2]. In this paper, we discuss physical origin of the oxidation and roughness induced interface states generation by the ab initio calculations, and proposed guiding principles for fabricating interface state free SiC-NMOSFET [3]. Moreover, we also propose the use of PMOSFET instead of NMOSFET.

2. Calculation Method

First principles DFT calculations were performed within GGA approximation [4,5]. We used ultra-soft pseudo-potentials [6]. We used a plane wave basis set with a cutoff energy of 36 Ry. We sampled 2×2×2 k-points for the Brillouin zone integration. After the optimizations, all the atomic forces were less than 10^{-3} Ht/Å

3. Nearly Free Electron State at SiC CBM

We first explain the very curious characteristics of SiC conduction band bottom wave functions. According to the recent first principles report, wave functions at SiC conduction bottoms are nearly free electron (NFE) like states as shown in Fig.1 [7]. The amplitude of SiC conduction band wave function is not located on the atoms, but distributed in the internal space like a free electron wave function. It is also noted that the shape of internal space sensitively depends on the SiC poly-types. In case of 3C-SiC, shape of internal space is straight and its length is infinite (Fig. 2(a)). Therefore, quantum confinement effect is essentially weak in 3C-SiC, leading to the relatively lower energy level of the NFE state and a small band gap.   In case of 4H-SiC, 6H-SiC or 2H-SiC, however, shape of internal space is zigzag-like (Fig. 2(a)). Quantum confinement effect for NFE states located inside the zigzag-shaped internal space should become very large which raises the energy level of NFE states. As a result, 3C-SiC band gap is smaller than that of 4H-SiC by about 1eV. This is the reason why band gaps of SiC sensitively depend on the poly-types as schematically illustrated in Fig. 2(b) and Fig.3. It is naturally expected that the above discussed SiC conduction band states which have NFE characteristics are sensitively affected by the modification of the shape of internal space. It means that interface defect, strain and roughness can affect NFE states resulting in the local change of conduction band energies, which degrades the NMOSFET properties such as mobility.

4. Oxidation Induced Defects at SiC/SiO$_2$ Interfaces

4.1 Interface States Caused by C-C Bonds Generation

We first discuss the oxidation induced C-C bond formation. As seen in Fig. 4. O incorporation causes large bond rearrangement. Two Si-O bonds and one C-C bond are formed by breaking two Si-C bond (Fig.4(a,b)). A C-C bond causes local strain around it and it modifies the shape of internal space of NFE states. This results in the lowering of conduction band by about 100 meV (Fig.4(c,d)), leading to the formation of interface states near conduction band bottoms (Fig.8).

978-1-4799-8002-4/14 $31.00 © 2014 IEEE

4.2 Interface States Caused by Si-Si Interaction

Next, we discuss the interface defects caused by 2nd nearest neighbor Si-Si interactions. By inserting O between Si-C bonds, 2nd nearest neighbor Si-Si distance changes from 3.1Å to 2.9 Å (Fig.5). As a result, conduction band bottom lowers by about 100 meV in case of 4H-SiC (Fig.6(a,b)). In case of 3C-SiC (Fig. 6(c,d)), however, no conduction band change is observed. This poly-type dependent conduction band change can be explained by the unique band structure of SiC (Fig.7). The bonding states related to the 2nd nearest neighbor Si-Si are generated and it appears below conduction band bottoms in large band gap 4H-SiC(Fig.7(a)). However, in case of small band gap 3C-SiC (Fig.7(b)), defect states do not appear. Above discussed states also lead to the formation of interface states near conduction band bottoms (Fig.8). These considerations indicate that strain free process is inevitable for fabricating SiC-NMOSFET. This implies that the use of PMOSFET instead of NMOSFET can be another solution (Fig.9).

5. Vth Instability Caused by Proton Mobile Ion

Finally, we comment on the Vth instability of SiC-MOSFET caused by proton. It has been reported that C impurity in SiO_2 can form CO_3^- like ions [8]. If proton forms complex with CO_3^- in SiO2 (Fig.10), hopping from one CO_3^- ion to another CO_3^- ion can cause Vth instability since proton act as positively charged mobile ions. This consideration coincides with recent experimental finding which indicate the existence of mobile ions in SiC-MOSFET [9]

6. Conclusions

We have considered the physical origin of SiC/SiO₂ interface states near the conduction band bottoms. We clarify the intrinsic problems of SiC/SiO₂ interfaces by the ab initio calculations. The unique nearly free electron like characteristics of SiC conduction band bottom causes unexpected interface states generation by process induced strain. We have proposed two recipes. One is the strain free processes. The other is developing PMOSFET instead of NMOSFET. Moreover, we have proposed a Vth instability mechanism caused by proton mobile ions.

References

[1] H.Yano, et al. IEEE Trans. Electron Devices 3 (1999) 504.
[2] G. Y. Chung et al., Appl. Phys. Lett. 76 (2000) 1713.
[3] K. Chokawa et al., Mat. Res. Forum. 740-742 (2013) 469.
[4] W. Kohn, and L. T. Sham, Phys. Rev. 140, A1133 (1965).
[5] J. P. Perdew, et al., Phys. Rev. Lett. 77, 3865 (1996).
[6] D. Vanderbilt, Phys. Rev. B 41, 7892 (1990).
[7] Y. Matsushita, et al., Phys. Rev. Lett. 108 (2012) 246404.
[8] Y. Ebihara et al. Appl. Phys. Lett., 100 , (2012) 212110.
[9] A. Chanthaphan et al. Appl. Phys. Lett. 102, (2013) 093510.

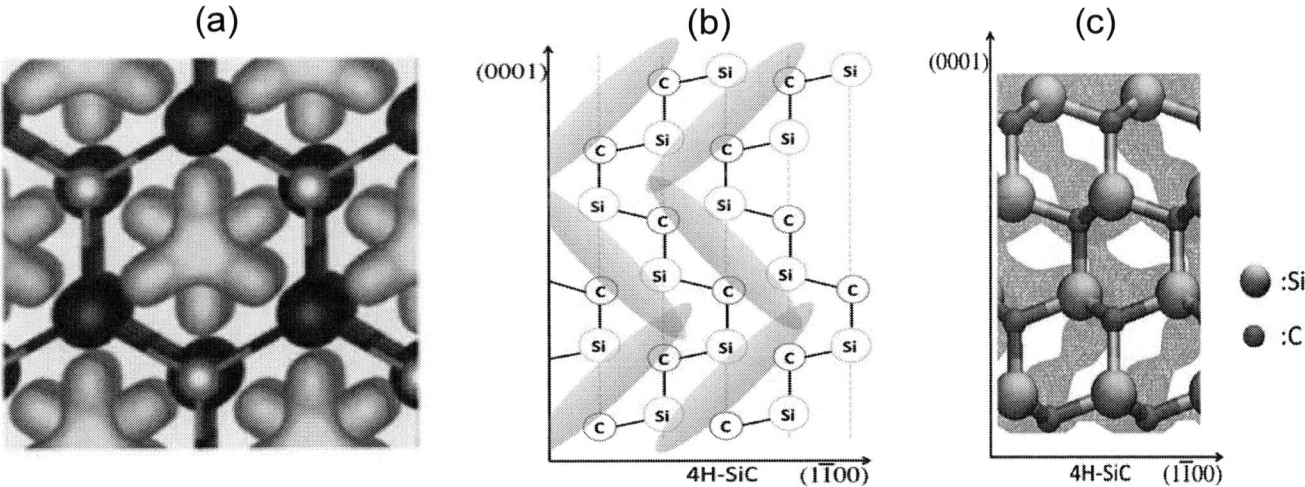

Figure 1. (a) Wave function of SiC conduction band bottom. [Ref. 1] (b) Schematic illustration of 4H-SiC atomic structure and shape of internal space. (c) Wave function of 4H-SiC. Wave function distributes in the internal space as drawn in Fig. 1. (c).

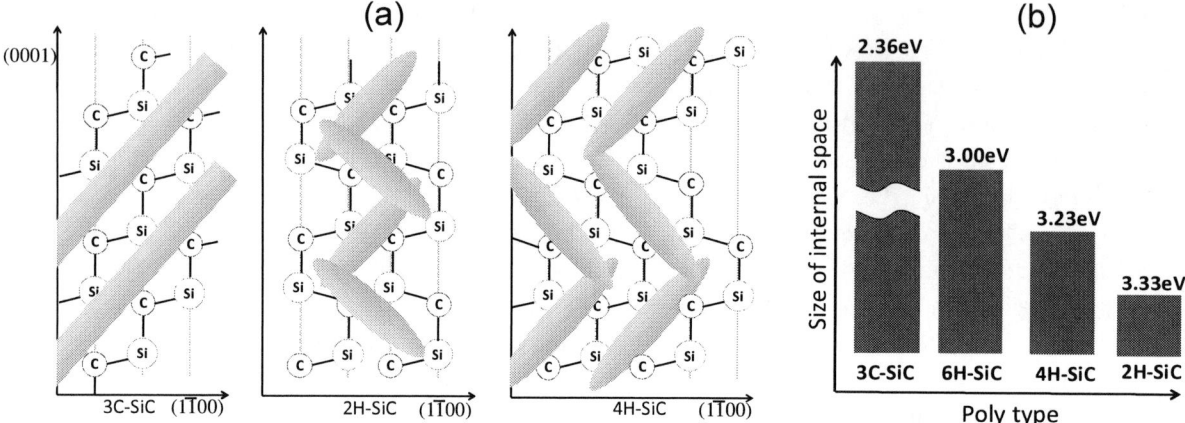

Figure 2. (a) Schematic illustrations of the shape of internal space of 3C-SiC, 2H-SiC and 6H-SiC. In case of 3C-SiC, shape of internal space is straight and its length is infinite. However, in case of 4H, 2H, and 6H–SiC, shape of internal space is zigzag-like. (b) Schematic illustration of poly-type sensitive band gap of SiC.

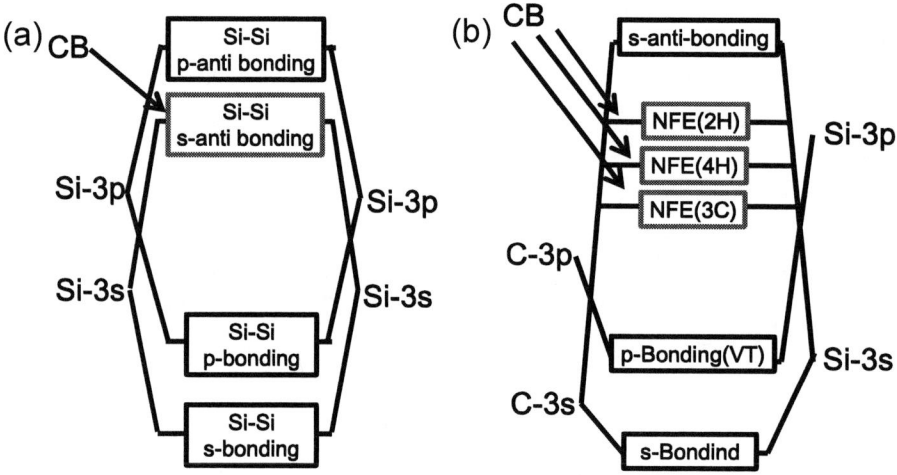

Figure 3. Schematic illustrations of band diagram of (a) Si (Typical of covalent bond semiconductor) and (b) SiC (NFE state decides band gap).

.

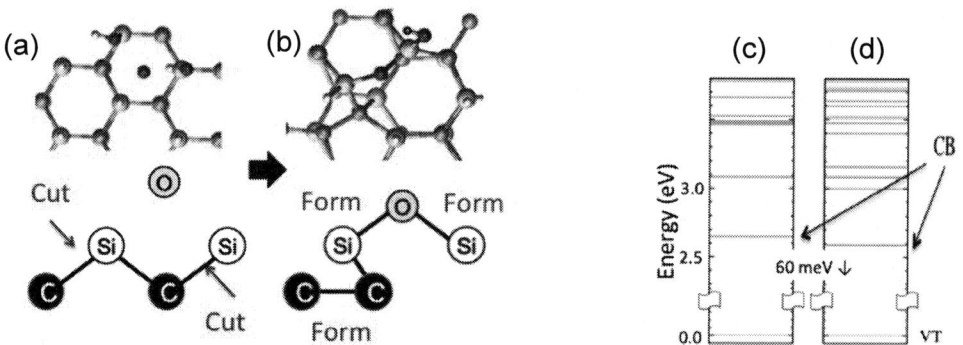

Figure 4. (a) Initial and (b) oxidized structure. O atom incorporation induces large bond rearrangement and forms C-C defects with large energy gain. Band gap values of (c) initial and (d) oxidized model. CB level is decreased by about 60meV by forming C-C defect. (Ref.3)

978-1-4799-8002-4/14 $31.00 © 2014 IEEE

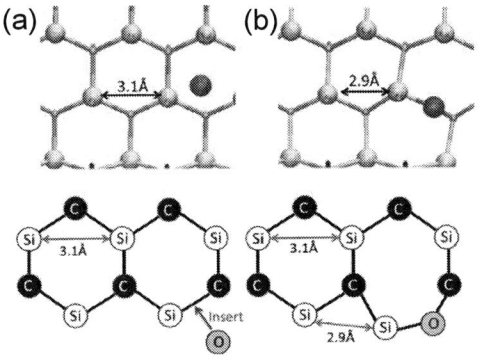

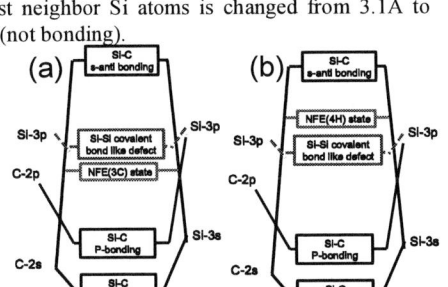

Figure 5. (a) Initial and (b) oxidized structure. O atom insertion induces strain. Distance between 2nd nearest neighbor Si atoms is changed from 3.1Å to 2.9Å (not bonding).

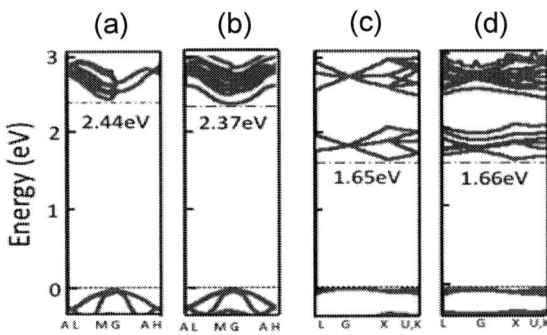

Figure 6. Band structure of (a) before, and (b) after oxidation of 4H-SiC. Band structures of (c) before and (d) after oxidation of 3C-SiC. In case of 4H-SiC, band gap decreases. However, in case of 3C-SiC, band gap does not change.

Figure 7. Schematic illustration of the appearance of interface states in case of (a) 3C-SiC and (b) 4H-SiC. In case of 3C-SiC, Si-Si level is higher than 3C-NFE level. In case of 4H-SiC, however, Si-Si level appears below the 4H-NFE level.

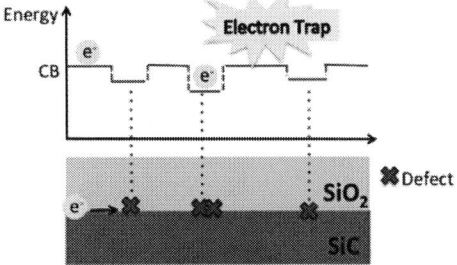

Figure 8. Schematic illustration of SiC MOSFET with defect structure such as C-C and Si-Si defect which cause local band gap modulation. Modulation of band gap induces interface states which causes electron trap near the conduction bottom.

Figure 9. (a) Intrinsic problem of SiC conduction band bottom induced by NFE characteristics. (b) Schematic illustration of p-type MOSFET and its simple band structure. In case of p-type MOSFET, carrier conduction is governed by the valence band. Thus, we need not consider the oxidation induced interface states originated from NFE characteristics.

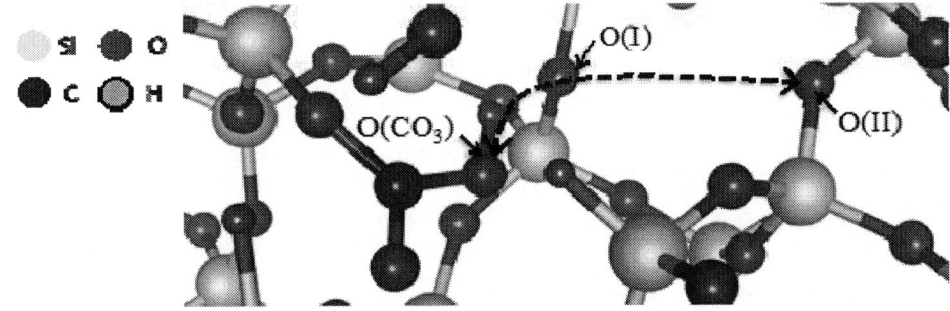

Figure 10. Atomistic structures of proton and CO_3^- ion complex and proton diffusion path.

978-1-4799-8002-4/14 $31.00 © 2014 IEEE

New Framework for the Random Charging/Discharging of Oxide Traps in HfO₂ Gate Dielectric: *ab-initio* Simulation and Experimental Evidence

Jingwei Ji [1], Yingxin Qiu [1], Shaofeng Guo [1], Runsheng Wang [1,*], Pengpeng Ren [1], Peng Hao [1], Ru Huang [1,2,*]

[1] Key Laboratory of Microelectronic Devices and Circuits (MOE), Institute of Microelectronics, Peking University, Beijing 100871, China
[2] Innovation Center for MicroNanoelectronics and Integrated System, Beijing 100871, China
*Email: ruhuang@pku.edu.cn; r.wang@pku.edu.cn

Abstract

A new framework for first-principle simulation on random charging/discharging of individual oxide traps is established and adopted for detailed studies on HfO₂ high-k gate dielectrics for the first time. The proposed framework provides an effective solution to the challenges in conventional multi-phonon simulation methodology, and successfully explains various experimental results in HfO₂ devices. 1-DOV defect, instead of traditionally assumed SOV, is found to be the crime oxide trap in HfO₂. And the anomalous RTN observations strongly support the high-order four-state model, which can be well explained by the two metastable states found in the 2-DOV defect. The framework is helpful for the fundamental understanding of RTN and NBTI reliability.

Introduction

With CMOS devices continuously scaling down, the random charging/discharging (C/D) of individual gate oxide traps has been paid great concern, due to its direct contribution to both random telegraph noise (RTN) [1-3] and the recoverable component of NBTI reliability [4-6] (Fig. 1). However, the microscopic origin of this phenomenon remains unclear, due to the extreme difficulty in time-dependent in-situ characterization of defects. On the other hand, first principle study has been carried out on the C/D of oxygen-vacancy trap in SiO₂ [7-9], which indicates that the current simulation framework faces great challenges such as unexpected high energy barrier calculated during charge transition [9]. Thus the framework for C/D of oxide traps needs to be extensively studied and further calibrated to serve as the powerful tool for RTN and NBTI optimization.

In this paper, a systematic and improved framework for precise first-principle simulation on C/D of oxide trap is established, and adopted for detailed studies on HfO₂ for the first time, as it is the mainstream high-k gate dielectric material. The new framework provides an effective solution to the key challenges in conventional methodology, and successfully explains various experimental results in devices with HfO₂ gate dielectrics. Thus the proposed framework is helpful for the fundamental understanding of RTN and NBTI reliability.

Challenges and the New Framework

The currently most accepted model for C/D of oxide traps is based on multi-phonon (MP) theory [10]. The basic MP theory is rather a qualitative model (as shown in Fig. 2a). Quantifying its model parameters from first principle calculation faces several challenges in the conventional MP simulation framework. One is the frequently calculated excessively high energy barriers for C/D [9], which deviates from realistic experiment data, indicating something missing in the base of the framework. With careful rethinking on the origins of trapping/detrapping, we refine a proper physics-based energy scale with the help of chemical potential adjustment, as in the simulation methodology below. The proposed framework is shown in Fig. 3.

The second challenge is that you need to find the right trap for C/D in HfO₂ devices, since there are many kinds of oxide defects. Traditionally single oxygen vacancy (SOV) is assumed as the criminal trap in gate oxides. But unreasonable discrepancy between ab-initio calculation and experimental data has been found in SOV in SiO₂ recently [9]. Similarly, as for HfO₂, the legitimacy of SOV should also be examined. Therefore, different types of defects in HfO₂ are studied based on our new framework. As will be shown later, with the strong support of experiments, a new type of defect, double oxygen vacancy (DOV), has been proved to be the dominating oxide trap contributing to random C/D in HfO₂.

The third challenge comes from the observations of anomalous RTN [11] and temporary RTN in NBTI recovery [12] reported in previous work, which indicates a multi-state MP model [10], thus the high-order description of C/D beyond the conventional harmonic approximation simulation is required. To address this, detailed reaction paths are drawn to find the metastable states of HfO₂ defects in our new framework. And one type of DOV shows great agreement with our recently observed another anomalous RTN, as will be shown later.

Simulation Methodology

To obtain reasonable energy barriers in ab-initio simulation, the physical picture behind needs to be re-investigated. Consider the defect as a system of two neighboring charge states q_n and q_{n+1} with Gibbs free energy G_n and G_{n+1}, respectively. Compared with q_{n+1} system, q_n system contains one more electron whose energy is included in G_n. The defect system can swap electron with outside, and the energy of an electron, inside or in the proximity of defect system, is numerically equal to the chemical potential μ of defect system. Due to thermal activation, the atomistic configuration of the defect keeps changing therefore G_{n+1} can be higher than G_{n+1min} under the most stable circumstance. If G_{n+1} plus the

978-1-4799-8002-4/14 $31.00 © 2014 IEEE

energy of one electron in the proximity reaches the threshold energy, or cross-section energy, then the defect could capture an electron and alter to qn state. The physical picture is similar for electron emission or hole capture.

Knowing that the cross-section energy is the energy sum of qn+1 system and an proximate electron, and μ remains a constant during charge transition, the minimum of energy curve of qn+1 state should be scaled into Gn+1+|μ|, while that of qn state remains Gn since it already contains the energy of one excess electron (as shown in Fig. 2b). Based on the discussion above, relative minimum energy level of different charge states could be settled.

In this study, the atomic configurations of SOV and three DOVs with three ways of oxygen proximity (Fig. 4) are built from the perfect supercell with 96 atoms of cubic-HfO2. Then the structure relaxation is done for various charge states: SOV for neutral/+1/+2 and DOV for neutral/+1/+2/+3/+4. To settle the relative minimum energy level, $(GDFT,n)min$ and μ, are calculated by Vienna ab-initio simulation package (VASP) [13]. The c-NEB reaction path search [14] is done between neighboring charge states to simulate the charging/discharging behavior. Then with the calculated energy points, reasonable and accurate energy curve is obtained.

New Defect Candidates for Oxide Traps in HfO2

Structure relaxation and defect charge density calculation gives out the atomistic images of various types of oxide traps (Fig. 5). As an ionic lattice, HfO_2 does not show apparent distortion when one or two oxygen are missing. In 1- and 2-DOV, the electron cloud of two vacancies shows clear overlap, thus these two types should be treated as a whole electron trap instead of two separate ones; while in 3-DOV electron cloud resembles two SOVs with no overlap. Energy curve of random charging/discharging could then be obtained in our new framework (Fig. 6&7). When external electric field is applied, the total energy would change with different V_G, thus the relative minimum energy level of two neighboring charge states would shift correspondingly, causing the change of activation energy and thus time constants (Fig. 8).

By comparison of the energy barrier from one stable configuration to the other one in neighboring charge state, barriers for SOV is much higher than DOV (Fig. 9), which indicates the DOV to be the criminal trap most frequently observed in HfO_2. Another attribution for different types of defects is their density of trap states (DOT) distribution (Fig. 10). In experimental extraction of oxide trap energy distributions (Fig. 11), large area MOSFETs with HfO_2 gate dielectric are measured using the method in [15], in which the dominant traps are counted. The measurement results greatly agree with the simulated DOT of 1-DOV in +1 charge state (Fig. 12), proving that 1-DOV is very likely to be the dominating type of oxide traps in HfO_2.

Exploring Metastable Defect States

If considering the existence of metastable states in random charging/discharging, at least three patterns of state transition are involved [10] (Fig. 13). The first pattern (two-state model) is common, and the second pattern (three-state model) is also found in previous work [11-12]. Recently, we have observed an anomalous RTN coincides with the third pattern (Fig. 14&15). The waveform can be divided into two zones: A with τ_e larger than τ_c, and B with inverse time constants (Fig. 15). Statistical data show same RTN amplitude and trap location in two zones (Fig. 16), so it is certain that zone A and zone B indicate the same trap yet with different metastable states. The three-state simulation of measured time constants shows huge difference from the real anomalous RTN waveform and typical three-state result (Fig. 17), thus the most likely pattern should be four-state model. The trap should have at least two stable and two metastable states. And from the waveform, the transition between two zones occurs rather slower than random charging/discharging process.

In our work, the atomistic origin of this trap type is found by ab-initio simulation based on our new framework. The c-NEB reaction path search on 2-DOV between +3 and +4 charge states has found metastable states in both directions (Fig. 18). On the other hand, the activation energy required for +3 to alter to +2 is higher than +3 to +3' or to +4'(Fig. 19), therefore the four-state transition could be observed on long time scale. Activation energy calculated on our framework shows great agreement with anomalous waveform observed in experiment: the activation energy required for charge transition (+3⇔+4', +4⇔+3') is always lower, so will the related time constants, than those for transition between stable and metastable states (+3⇔+3', +4⇔+4'), with or without V_G (Fig. 20&21).

Summary

A new physics-based framework for first-principle study on random C/D of gate oxide traps is built and applied to HfO_2 gate dielectric for the first time. The framework offers a solution to the problem of unreasonably high thermal barriers calculated in conventional studies. Further investigation based on this framework shows that DOV is evidently to be the criminal defect causing random C/D in HfO_2, with solid experimental support. The framework also successfully explains the anomalous RTN observed in experiments. The proposed framework is helpful to ab-initio simulations on various types of defects for random C/D behaviors.

Acknowledgment: This work was partly supported by the 973 Projects (2011CBA00601), NSFC (61106085), National S&T Major Project (2009ZX02035-001). The authors would like to thank Tibor Grasser and Yuchen Shen for the helpful discussions.

References: [1] P. Ren, et al, *IEDM 2013*, p.778. [2] J. Zou, et al., *VLSI 2012*, p.139. [3] J. Zou, et al., *VLSI 2013*, p.186. [4] T. Grasser, et al., *IEDM 2010*, p.85. [5] C. Liu, et al., *IEDM 2011*, p.571. [6] C. Liu, et al., *IEDM 2012*, p.466. [7] F. Schanovsky, et al., *J. Comp. Elect.*, p.135, 2010. [8] F. Schanovsky, et al., *JVST-B* 29(1), 01A201, 2011. [9] F. Schanovsky, et al., *SISPAD 2013*, p.1. [10] T. Grasser, et al., *Microelec. Reliab.*, p.39, 2011. [11] M.J. Uren, et al., *PRB*, 37, p.8346, 1988. [12] T. Grasser et al., *IRPS 2010*, p.16. [13] G. Kresse, et al., *PRB* 59, 1758, 1999. [14] G. Henkelman, et al., *J. Chem. Phys.* 113, 9901, 2000. [15] Z. Ji, et al., *TED*, p. 228, 2010.

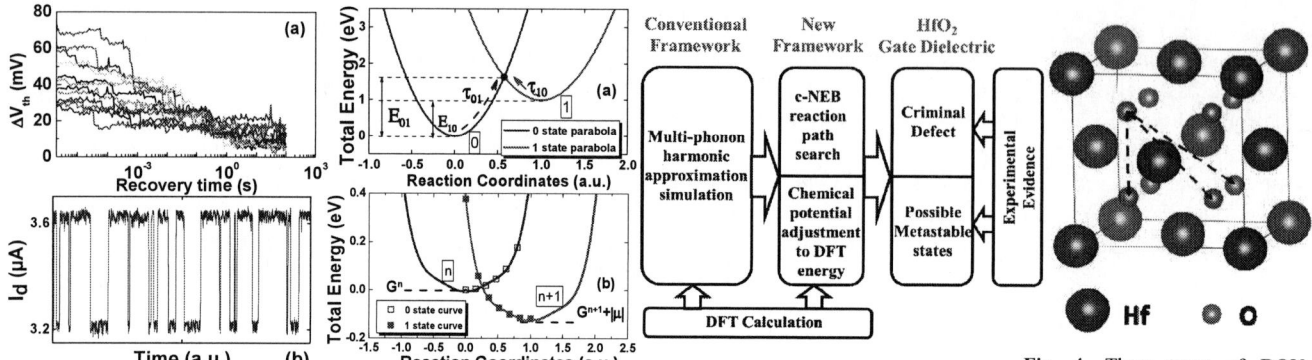

Fig. 1. Typical (a) NBTI recovery and (b) RTN experimental results in nanoscale devices. τ_c and τ_e are capture and emission time constants, respectively.

Fig. 2. (a) Conventional MP framework with harmonic approximation. (b) New framework with detailed reaction path and reasonable energy level. Reaction Coordinates are normalized.

Fig. 3. New ab-initio simulation framework used in this paper. The new framework is adopted for oxide traps in HfO₂, with comparison to experimental observed data.

Fig. 4. Three types of DOV. Dashed line indicates three ways of oxygen proximity. In our work, the vacancy of the nearest, 2nd nearest, 3rd nearest oxygen are named 1-, 2-, 3-DOV, respectively.

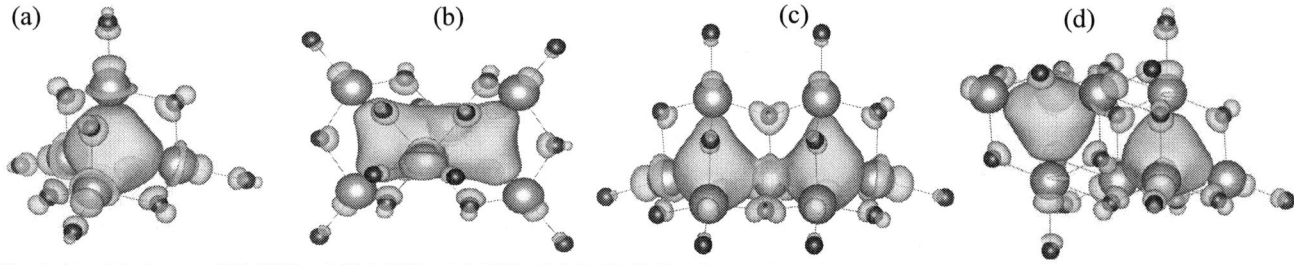

Fig. 5. Atomistic images of (a) SOV and (b) 1-DOV, (c) 2-DOV, (d) 3-DOV. Gold and Red spheres for Hf and O respectively, yellow part for electron cloud. Only atoms and bonds near the defect are shown.

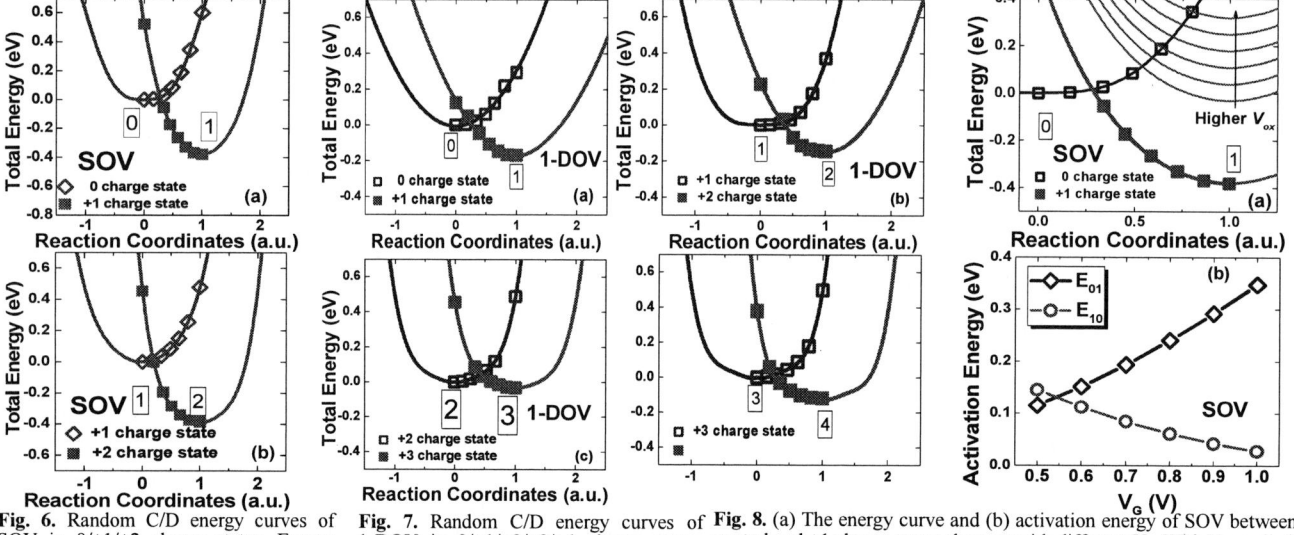

Fig. 6. Random C/D energy curves of SOV in 0/+1/+2 charge states. Energy curves are higher-order fitted than harmonic approximation.

Fig. 7. Random C/D energy curves of 1-DOV in 0/+1/+2/+3/+4 charge states. Energy curves are higher-order fitted than harmonic approximation.

Fig. 8. (a) The energy curve and (b) activation energy of SOV between neutral and +1 charge states changes with different V_g. With V_g applied, the free energy of defect will change by $\Delta G = V_{ox} \ast x / T_{ox}$, where x for the depth of the defect into the oxide, V_{ox} for voltage drop in oxide.

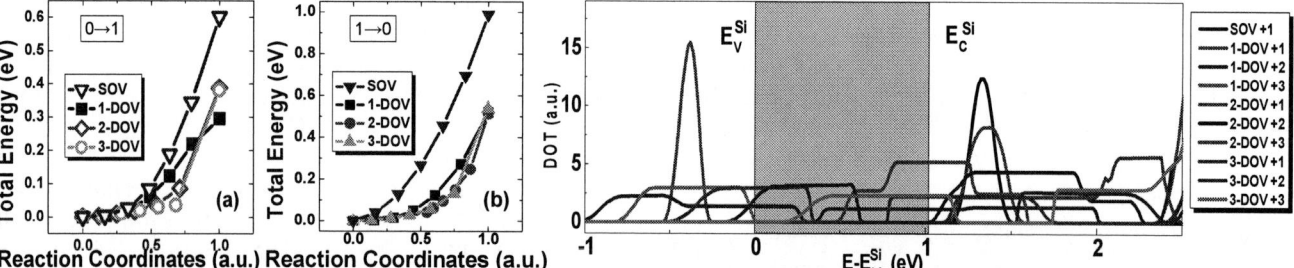

Fig. 9. Reaction paths between neighboring charge states of SOV and 1,2,3-DOV for (a) neutral to +1 state, and (b) +1 to neutral state.

Fig. 10. Ab-initio simulated density of trap states (DOT) distribution of various types of oxide traps in different charge states. Since neutral state traps will not be measured in our experiment, the DOT of neutral states are not shown here.

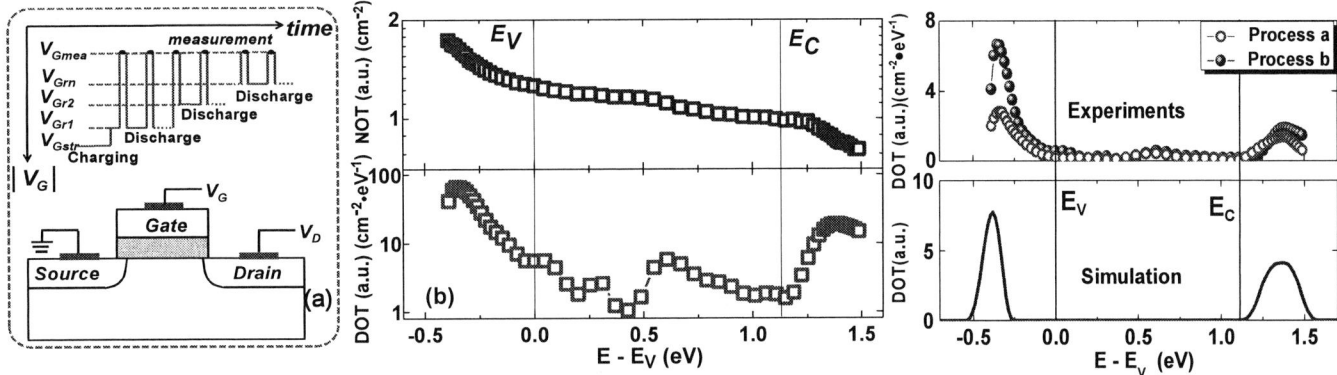

Fig. 11. (a) Measurement schematics of extracting the energy distributions of oxide traps. After the charging phase, discharging occurs, during which the recovery of ΔVth is monitored until the discharging completes. (b) Measured results of number of traps (NOT) per cm² and energy density of traps (DOT) per cm².

Fig. 12. Comparison of experimental DOT data of two HfO₂ processes and simulated DOT of 1-DOV in +1 state. The simulated relative height of two highest peaks and energy position agree well with experimental data.

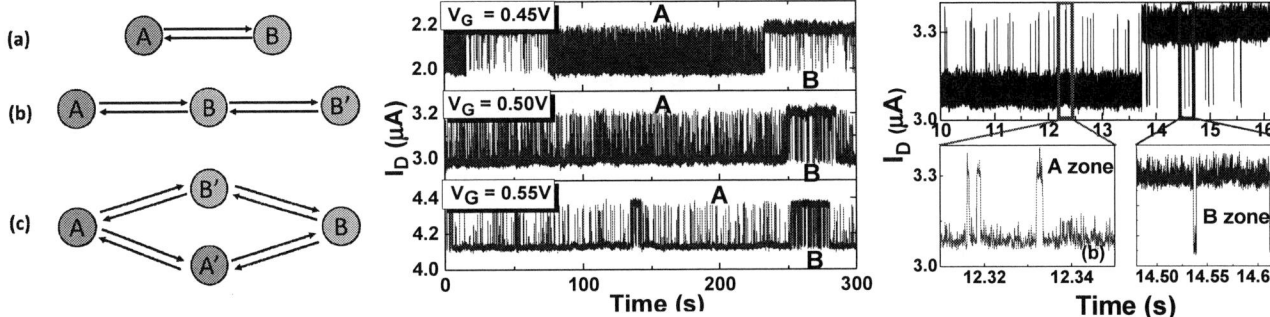

Fig. 13. Three typical patterns of state transition of charging/discharging (a) Common situation without metastable states. (b) Three-state with only one metastable state. (c) Four-state with two metastable states in different charge state.

Fig. 14. Anomalous RTN measurement result with different Vg. The waveform can be divided into A and B zones with different charge transition time constant. The transition time between A and B is of the order of hundreds of seconds.

Fig. 15. A closer view of the anomalous RTN waveform with $V_G = 0.50V$ in A and B zones separately. The emission time constant τ_e in A is longer than capture time constant τ_c, while $\tau_c > \tau_e$ in B.

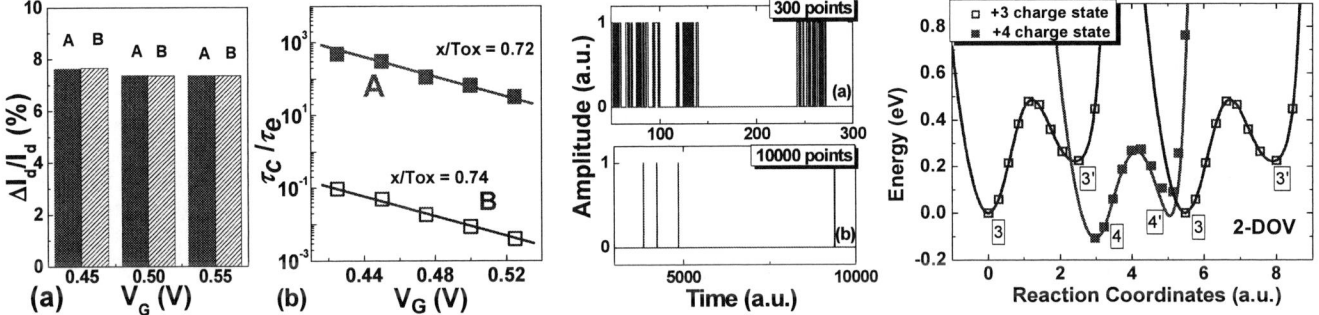

Fig. 16. The comparison of (a) RTN amplitude and (b) extracted trap location of A and B zones. The great similarity indicates that A and B zones represent the same defect in oxide. Trap location x/T_{ox} is extracted by method in [10].

Fig. 17. (a) The typical simulation result of three-state charge transition. (b) The simulation result of three-state charge transition using time constant extracted from anomalous RTN measurement. The comparison denies the possibility of anomalous RTN to be three-state pattern.

Fig. 18. Random C/D energy curve of 2-DOV between +3/+4 charge states. Two metastable states are labeled with 3' and 4'. To show the transition between 3 and 4', the curve of +3 charge state is duplicated.

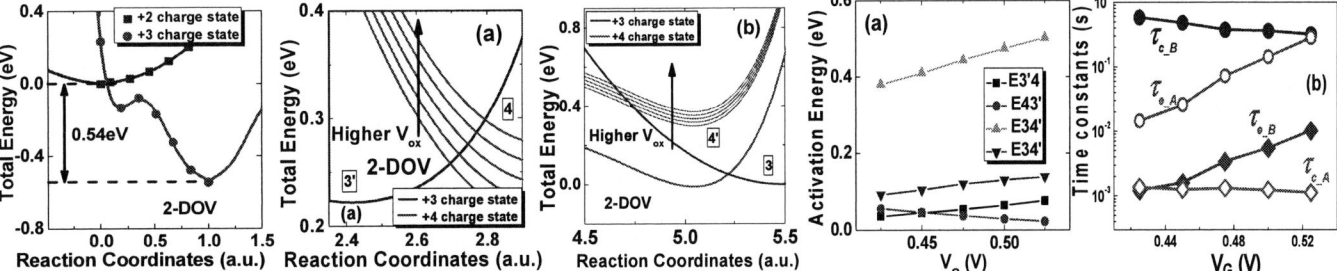

Fig. 19. Random charging/discharging energy curve of 2-DOV between +2/+3. The energy barrier shows that it is very hard for +3 state to transit into +2 state.

Fig. 20. The energy curve of 2-DOV between (a) 3' and 4, (b) 4' and 3 charge states changes with different V_{ox}. ΔG is calculated using $x/T_{ox} = 0.73$.

Fig. 21. (a) Activation energy between different states changes with V_G. (b) Extracted time constants in anomalous RTN.

978-1-4799-8002-4/14 $31.00 © 2014 IEEE   545

Microscopic understanding of the low resistance state retention in *HfO₂* and *HfAlO* based *RRAM*

B. Traoré[1], P. Blaise[1], E. Vianello[1], H. Grampeix[1], A. Bonnevialle[2], E. Jalaguier[1], G. Molas[1], S. Jeannot[2], L. Perniola[1], B. DeSalvo[1], Y. Nishi[3]

[1]*CEA, LETI, MINATEC Campus, 17 rue des Martyrs, 38054 Grenoble Cedex 9, France,*[2]*STMicroelecteronics, 850 rue Jean Monnet38920 Crolles, France.* [3]*Department of Electrical Engineering, Stanford University, California, USA.* E-mail: boubacar.traore@cea.fr

Abstract: *We study in detail the impact of alloying HfO₂ with Al (Hf₁₋ₓAl₂ₓO₂₊ₓ) on the device characteristics through materials characterization, electrical measurements and atomistic simulation. Indeed, movements of individual oxygen atoms inside the dielectric are at the heart of RRAM operations. Therefore, we performed diffusion barrier calculations relative to the oxygen vacancy (V_O) movement involved in R_on data retention. Calculations are performed at the best level using ab initio techniques. Our study provides an insight on the improved R_on stability of Hf₁₋ₓAl₂ₓO₂₊ₓ RRAM, via a simple explanation based on its higher atomic density (atoms/cm³) associated with shorter bond lengths between cations and anions in the presence of Al.*

Introduction:

Oxide based RRAM may revolutionize the future computing systems and are serious candidates for replacing FLASH technology[1-2] due to their low power operation, high switching speed [3] and capability to be used in neuromorphic applications [4]. Various companies [5] and research laboratories [6] have adopted the *HfO₂/Ti* solution as the active layer that casts the memory effect. *Ref* [7] used *HfO₂* incorporated with *Al* which resulted in the improvement of the low resistance (R_{on}) retention. However, a clear microscopic understanding of the devices is still missing as well as the impact of doping/alloying. For this purpose, we study in detail the impact of alloying HfO₂ with Al (Hf₁₋ₓAl₂ₓO₂₊ₓ) on the device performance though a systematic analysis that combines materials characterization, electrical measurements and atomistic simulation.

Samples description/Process:

5nm of *HfO₂* and *Hf₁₋ₓAl₂ₓO₂₊ₓ* (9:1) were deposited by *ALD* at 300°C where Hf₁₋ₓAl₂ₓO₂₊ₓ (9:1) stands for 9 cycles of *HfO₂* followed by 1 cycle of *Al₂O₃*. *HfAlO* shall refer to *Hf₁₋ₓAl₂ₓO₂₊ₓ*. Dielectric layers sandwiched between *PVD* deposited bottom electrode (*BE*) *TiN* (35nm) and top electrode (*TE*) *Ti* (10nm) were fabricated in a mesa process as schematically represented in **Fig.1a**. The devices were integrated in 65-nm *CMOS* technology. **Fig.1b** shows the typical bipolar switching characteristics which is associated with the creation and rupture of a conductive filament (*CF*), sufficiently enriched in oxygen vacancies *(V_o)*, allowing for resistive switching [8].

Results and discussion:
Material properties:

Table 1 highlights the material properties of both sample types as well as their available counterparts from ab initio. An annealing at 400°C for 1 hour in N₂ ambient has been performed to simulate the back end temperature (*T*) seen by the devices used for electrical characterization.The ab initio data were calculated using the structures shown in **Fig.2a** and **Fig.3a**. Densities obtained from ab initio are slightly higher than the experimental ones because of the more crystalline structure of ab initio models (as-deposited (as-dep.) samples have lower density compared to annealed ones due to crystallization process). With the same volume, *HfAlO* have a higher atomic concentration compared to

HfO₂ resulting in shorter *Hf-O* and *Al-O* bond lengths. We shall see the implications of this bond shrinking later on the diffusion process. The atomistic model of *HfAlO* (**Fig.3a**) was constructed based on the *ALD* cycle of *HfAlO* (9:1) with a thin *AlO* layer, *Al* atoms being colored green (dashed ellipse). The atomic proportion of each species (*Hf*, *Al*, and *O*) was calculated based on the *RBS* (Rutherford Backscattering) data of Table1 to reflect the correct *Hf/Al* ratio respecting *Hf₁₋ₓAl₂ₓO₂₊ₓ* stoichiometry. The following chemical equation was used to calculate the atomic proportions:

$$(1-x)HfO_2 + xAl_2O_3 \rightarrow Hf_{1-x}Al_{2x}O_{2+x}$$

x being 0.1 corresponding to the proportion of Al₂O₃ deposition in ALD cycle. The colored atoms numbered 1 to 4 in **Fig.2b** and **Fig.3b** represent the V_o diffusion paths that are investigated later in the paper.

Switching characteristics:

Fig.4a compares the switching voltages of *HfO₂* and *HfAlO*. The main difference between the two sample types is the forming voltage (V_F) which is higher for *HfAlO* as is further evidenced in **Fig.4b**. The Low V_F for *HfO₂* may be related to its lower band gap (5.6 eV for *HfO₂* vs. 5.7 eV for *HfAlO* favouring more electron injection. The band gap was measured by ellipsometry) and by the increase of its local electric field due to its higher dielectric constant (Table1) enhancing *HfO* bond breakage [9]. R_{on} and R_{off} distributions are compared in **Fig.5** and no significant difference is noted. Fig.6 shows the 10⁶ cycling endurance for both samples (**Fig.6**).

Data retention:

Fig.7 shows R_{on} retention evolution of *HfO₂* with time at 200°C with the devices programmed at 100μA. R_{on} fails towards high resistance values as the retention time increases which we associate to the diffusion of V_o at the tip (constriction) of the conductive filament (*CF*) as shown in the pictorial model in **Fig.7**. **Fig.8** compares the mean R_{on} of *HfO₂* at different temperatures (*T*) and we can see that as *T* increases R_{on} fails faster towards R_{off} which indicates that R_{on} retention is a highly activated process as *T* is raised. Similarly, **Fig.9** and **Fig.10** respectively show R_{on} retention of *HfAlO* at 200°C and the comparison of mean R_{on} at different *T*. As in *HfO₂*, R_{on} is a *T* activated process and fails faster towards high resistance level as *T* is raised. **Fig.11** compares mean R_{on} retention for *HfO₂* and *HfAlO*. It can be clearly seen that *HfAlO* has a better R_{on} stability at all the investigated *T*. The extracted failure times based on the failure criteria of **Fig.11** are plotted in an Arrhenius plot (**Fig.12**) using the Arrhenius equation($\tau=\tau_o exp(-E_a/k_bT)$). The linear trend of the data of **Fig.12** shows that the retention failure process follows an Arrhenius type of law. Higher E_a is extracted for *HfAlO* consistent with the previous figures and more than 10 years retention of at least 125°C is extrapolated for both samples.

Ab initio calculations:

All simulations were based on density functional theory (*DFT*) [10] using *SIESTA* code [11], with the GGA-PBE functional to

978-1-4799-8002-4/14 $31.00 © 2014 IEEE

describe the exchange-correlation term, and Troullier-Martins pseudopotentials were used for each atomic species to account for the core electrons [12]. Polarized Double-Zeta (DZP) basis set with an energy shift of 50 meV and a Mesh cut-off of 300 Rydberg were used for the calculations. A 2x2x2 monoclinic HfO_2 supercell with 96 atoms was set up for V_o formation energy and diffusion barriers calculations. For $HfAlO$ supercell optimization, all the atomic coordinates and cell geometry were relaxed until the maximum residual forces were less than 0.02 eV/A. The activation energies (E_a) for V_o diffusion were calculated using the Nudged Elastic Band (NEB) technique [13] as implemented in the Atomistic Simulation Environment (ASE) [14]. For the NEB calculations, at least 5 intermediate images were used and a maximum force of 0.05 eV/A was used as the convergence criteria. E_a corresponds to the height of the barrier when a diffusing species migrates from an initial state (IS) to a final state (FS) through a point of high potential energy called Transition State (TS). E_a is an important quantity in harmonic Transition State Theory (TST) which states that the reaction rate (τ) from IS to FS is given by $\tau= \tau_o exp(-E_a/k_bT)$ following the Arrhenius equation [15]. Therefore, since the fail time shown in **Fig.12** also follows Arrhenius expression, V_o diffusion barriers (**Fig.13**) that are calculated using ab initio can in principle, be related to E_a extracted from R_{on} retention experiment.

V_o as dominant diffusing species in $TiN/HfO_2/Ti$ or $TiN/HfAlO/Ti$ R_{on} retention:

In order to understand which $V_o^{+/-q}$, $+/-q$ being the charge state, may be dominant during R_{on} retention when CF (V_o rich) is created, we calculated the formation energy of V_o at different charge states and the most favourable cases are shown in **Fig.14** with respect to the Fermi level (charge injection level). We can see that neutral V_o becomes thermodynamically more favourable from close to mid-gap. **Fig.15a,b** respectively show V_o energy level and the band diagram of HfO_2 when no bias is applied which corresponds to the zero bias case of data retention test with neutral V_o level being slightly above mid-gap. **Figs.14, 15** provide a strong indication that neutral V_o are essentially the dominant diffusing species during R_{on} retention from thermodynamic perspective.

V_o diffusion process:

As an attempt to have a deeper microscopic understanding of R_{on} retention which we associated to V_o diffusion process (**Fig13**), we calculated the diffusion barriers of V_o by considering different paths in $HfAlO$ and HfO_2 as previously shown in **Fig.2b** and **Fig.3b**. The diffusion paths are numbered from 1 to 4. **Fig.16** shows the formation energy of the initial and final V_o states that are considered during the diffusion and are circled using color codes that are followed for the corresponding diffusion barrier calculation. We see that V_o formation energy is slightly lowered by Al incorporation when O is removed from AlO bonds in agreement with [16, 17]. **Fig.17** shows the diffusion profile of V_o in HfO_2 and an E_a of 2.16 eV is calculated consistent with the 2.19eV of [18]. Ref. [18] further states that V_o diffusion in HfO_2 is isotropic with an average activation energy of 2.4 eV. The atomistic models of Fig.17 show V_o migration corresponding to different states. In order to compare this E_a to that of V_o diffusion in $HfAlO$, 3 different paths were considered and are shown in **Figs.18, 19 and 20**. Depending on the initial state, the calculated E_a varies from 2.23eV (**Fig.20**) to 2.69 eV (**Fig.18**). Since diffusion is a statistical process with different paths involved, the activation energy is determined by the energy of the highest transition state during long range diffusion. Comparing V_o

diffusion barrier in HfO_2 to those obtained in $HfAlO$, we can see that Al incorporation increased V_o diffusion barrier which is consistent with the high E_a for $HfAlO$ extracted from experiment. However, experimental values are lower compared with ab initio based calculated E_a. We investigated this effect by inducing $4V_o$ in HfO_2 as pre-defects and re-calculated the diffusion of V_o (path 1) as shown in **Fig.21**. We notice that $4V_o$ as pre-defects resulted in E_a lowering by ~0.4 eV which becomes close to the experimental data. Indeed, HfO_2 becomes slightly defective when Ti is deposited on top of it due to the reactivity of Ti with O_i[19] and the fact that CF creation during forming also affects HfO_2 surrounding it.

Why high E_a for HfAlO?:

In order to have an insight on why V_o diffusion barrier increased with Al incorporation, we calculated HfO bond lengths in both HfO_2 and $HfAlO$ supercells as well as that of AlO as shown in **Fig.22**. Al incorporation induced the shrinkage of the different bonds (1.93 Å for AlO; 2.16Å for HfO) with some relative dispersion (slightly amorphised structure) while HfO bonds in HfO_2 are very peaked around 2.17 Å (Table2). This bond shrinkage in $HfAlO$ as well as its higher atomic concentration (Table1) reduces atomic mobility due to probable higher coulomb interaction resulting in increased V_o diffusion barrier. Therefore, alloying HfO_2 with a proper material (Al in this case) rendering the structure more amorphous is an efficient way of enhancing devices' thermal stability.

Conclusions:

A comprehensive experimental and simulation study of HfO_2 and $HfAlO$ RRAM devices has been presented. The study points out the impact of the atomic composition and structure on the resistive switching layer's thermal stability. We used ab initio based V_o diffusion barrier calculations to explain the microscopic behaviour of R_{on} retention. Al incorporation in HfO_2 results in better R_{on} retention of the devices due to their shorter bond lengths associated with their higher atomic concentration at the cost of a higher forming voltage. Thus, alloying HfO_2 with a proper dopant (Al in this study) is an efficient methodology to improve devices' thermal stability.

Acknowledment:

This work is financially supported by the Nanosciences Foundation within the framework of Nishi's Chair of Excellence in Grenoble, France.

References:

[1] H. Akinaga and H. Shima, Proc. IEEE, vol. 98, p. 2237, 2010.
[2] V. Sriraman et al., IPCSIT, vol. 32, p.101, 2012
[3] B. Govoreanu et al., IEDM Tech. Dig., pp. 729-732, 2011
[4] S. Park et al., Nanotech. Vol. 24, No.38, 2013
[5] Ch. Walczyk et al., J. Vac. Sci. Technol. B 29, 01AD02, 2011
[6] Y.S.Chen et al., IEDM Tech. Dig,pp. 31.3.1 - 31.3.4,2011
[7] A. Fantini et al., IMW,pp.18-21,2014.
[8] G. Bersuker et al., JAP 110, p. 124518, 2011
[9] J. McPherson et al., A.P.L 82, 2121,2003
[10] W. Kohn and L. J. Sham, Phys. Rev., vol. 140, p. A1133, 1965
[11] J. Soler et al., J. Phys., Condens. Matter14, 2745, 2002
[12] N. Troullier and J. L. Martins. Phys. Rev. B, vol. 43, pp. 1993-2006 (1991)
[13] G. Henkelman et al., J. Chem. Phys. 113, 9901, 2001
[14] S. R. Bahn, K. W. Jacobsen, Comput. Sci. Eng., Vol. **4**, 56-66, 2002
[15] D.S. Sholl, J.A. Steckel, Wiley-Blackwell, ISBN-13: 978-0470373170, 2009, Print
[16] L. Zhao et al., VLSIT, T106-T107,2013
[17] Z. Yuanyang et al.,Journ. Semi., 35, 4,2014
[18] N. Capron et al., APL 91, 192905,2007
[19] B. Traoré et al, IRPS, pp. 5E.2.1 - 5E.2.6, 2013

Device structure/Introduction

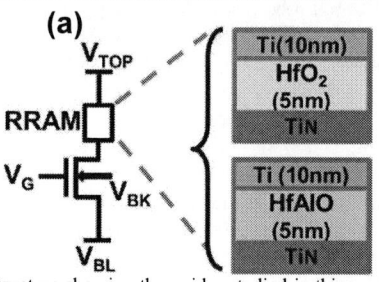

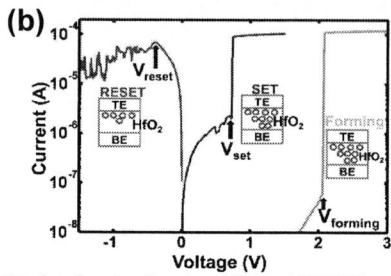

Figure 1. (a) 1T1R structure showing the oxides studied in this work (b) Typical IV plot showing forming, SET and RESET operations of the *RRAM*. The pictorial models show the conductive filament (*CF*) region rich of oxygen vacancies (*Vo*).

Material Properties/ Ab-initio models

	HfO₂			HfAlO (9:1)		
	Experiment		Ab initio (DFT)	Experiment		Ab initio (DFT)
	as-dep.	Annealed		as-dep.	Annealed	
Density (g/cm³) [XRR]	9.56	9.63	9.78	8.76	8.79	8.83
Ratio Hf/(Hf+Al) (%) [RBS]	100	-	100	86	-	82
Dielectric constant	22.2	-	-	18.9	-	-
Atomic density (atoms/cm³)	-	-	8.39x10²²	-	-	8.53x10²²

Table 1. Table showing the extracted physical parameters of the two sample types from physical characterization and ab-initio calculations. Red names in square bracket correspond to the technique used for the physical characterization

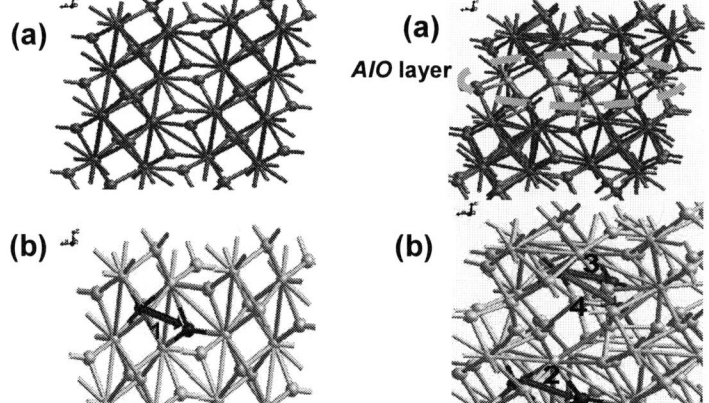

Figure 2. (a) *HfO₂* monoclinic structure used for ab-initio calculations. Blue spheres for *Hf* atoms and Red for *O*. (b) Atomic position, numbered **1**, is considered for *Vo* diffusion in *HfO₂*.

Figure 3. (a) *HfAlO* structure used and constructed based on RBS data of Table 1 for the different atomic concentrations. Blue spheres for *Hf*, Red for *O* and Green for *Al* (b) Atomic positions, numbered **2, 3 and 4**, are considered for *Vo* diffusion in *HfAlO*.

Device switching characteristics/Performance

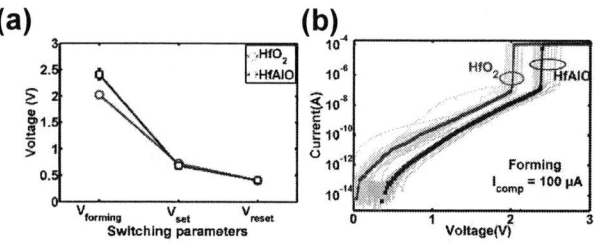

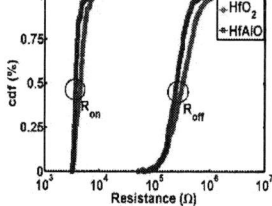

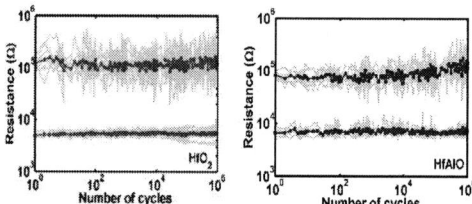

Figure 4. (a) Comparison of switching parameters for both sample types. The main difference is the higher forming voltage (*VF*) for *HfAlO*. (b) Forming IV for both samples. Thick curves represent median of over 25 devices for each sample type. The lower *VF* for *HfO₂* is related to its lower band gap (more injection) and the increase of its local electric field due to its higher dielectric constant following Lorentz-Mossoti relation[9].

Figure 5. Cdf plot of *Ron* and *Roff* for both samples. No significant difference is noted.

Figure 6. Cycling endurance for both samples. As in Fig.5, no significant difference is noted. Both samples exhibhit good cycling endurence.

Data retention

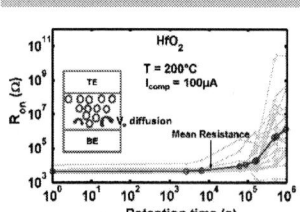

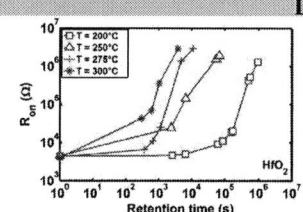

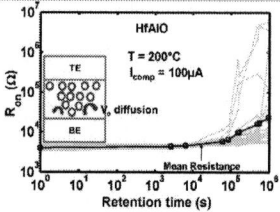

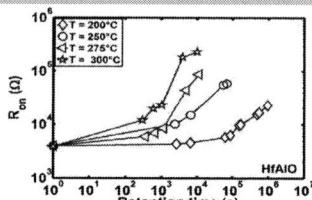

Figure 7. *Ron* retention behavior of *HfO₂* at 200°C. The red curve shows the mean trend of the devices. The pictorial model attributes the *Ron* failure to *Vo* diffusion at the *CF* tip (constriction) that is investigated later.

Figure 8. Mean *Ron* retention of *HfO₂* at 4 different temperatures *T*. As expected, *Ron* failure is accelerated as *T* is raised. This is attributed to increased *Vo* diffusion at high *T*.

Figure 9. *Ron* retention behavior of *HAfO* at 200°C. the blue curve shows the mean trend of the devices. The pictorial model attributes the *Ron* failure to *Vo* diffusion that is further investigated later.

Figure 10. Mean *Ron* retention of *HfAlO* at 4 different temperatures T. As for *HfO₂*, *Ron* failure is accelerated as T is raised.

978-1-4799-8002-4/14 $31.00 © 2014 IEEE

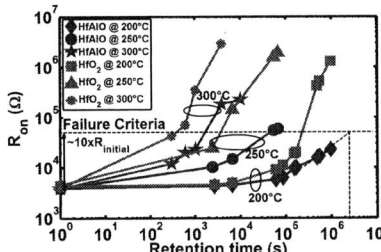

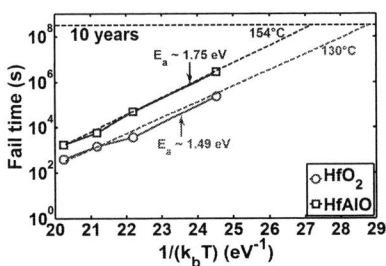

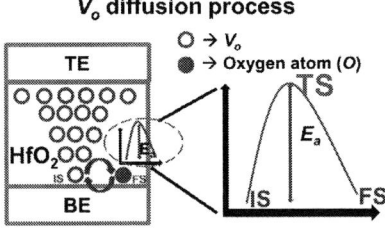

Figure 11. Comparing R_{on} retention behavior for HfO_2 and $HfAlO$. $HfAlO$ has a better R_{on} stability. HfO_2 shows a more abrupt failure while it is more gradual for $HfAlO$.

Figure 12. Extracted failure time based on the criteria shown in Fig.11. An higher activation energy (E_a) is extracted for $HfAlO$(using Arrhenius). Both samples present more than 125°C retention of 10 years.

Figure 13. Model showing V_o diffusion process during R_{on} retention where the CF ruptures at the tip (constriction). The diffusion process is associated with an E_a that is calculated with the NEB technique.

Ab-initio : V_o energy level and formation energies

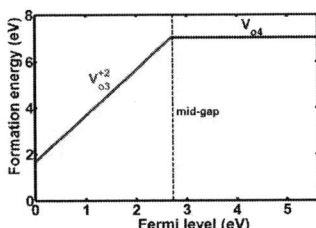

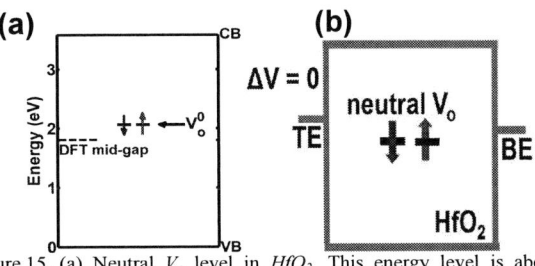

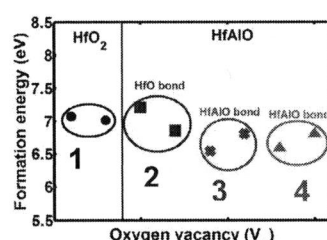

Figure 14. V_o formation energy in HfO_2 with respect to Fermi level. Neutral V_o becomes thermodynamically favorable from experimental mid-gap condition.

Figure 15. (a) Neutral V_o level in HfO_2. This energy level is above midgap from DFT calculations. (b) Band diagram showing neutral V_o level in HfO_2 at mid-gap. This corresponds to the case when no bias is applied which is the scenario of data retention.

Figure 16. Formation energy (ΔH^{Vo}) of V_o in HfO_2 and $HfAlO$ from the models shown in Fig.2. $\Delta H^{Vo}$is slighly lower for $HfAlO$. The 4 different ellipses with different colors correspond to V_o diffusion paths that are investigated and shown in Fig.2b and Fig.3b.

Ab-initio : V_o diffusion barriers calculation

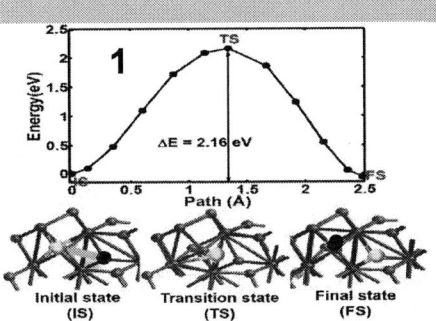

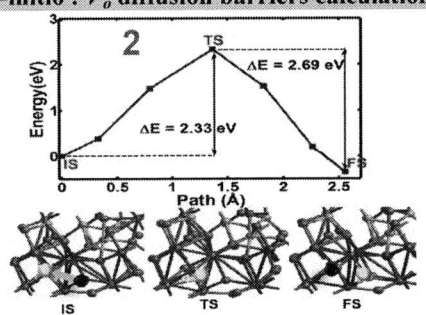

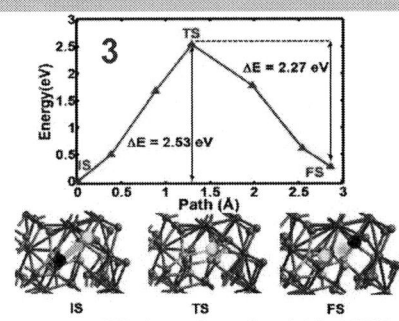

Figure 17. V_o diffusion in HfO_2 numbered **1** in Fig.2. An E_a of 2.16 eV is calculated in agreement with [18]. [18] gives an average E_a of 2.4 eV with different long range V_o diffusion paths investigated. Black sphere represents the initial V_o position while yellow sphere represents the moving O.

Figure 18. V_o diffusion number **2** in $HfAlO$. Depending on the initial point, 2.33 eV and 2.69 eV are calculated. 2.69 eV is considered since diffuison is dominated by the highest barrier. This E_a is higher than in HfO_2 in agreement with the experimental trend of Fig.11.

Figure 19. V_o diffusion path numbered **3** in $HfAlO$. E_a of 2.27 eV and 2.53 eV are calculated which are still higher than in HfO_2.

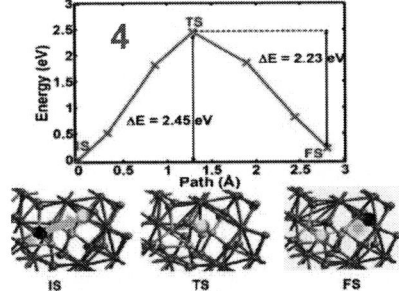

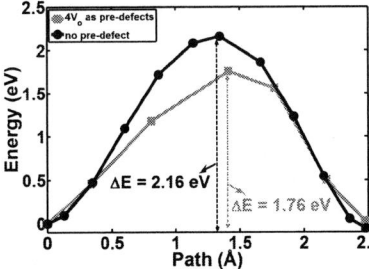

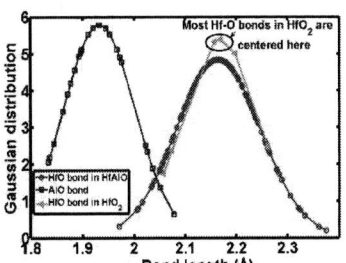

Figure 20. V_o diffusion path numbered **4** in $HfAlO$. E_a of 2.45 eV and 2.23 eV are calculated which are still higher than in HfO_2.

Figure 21. Decrease of E_a by ~0.4 eV with $4V_o$ as pre-defects in HfO_2 compared to the case with no-predefect (Fig17). This may explain why the experimental E_a are lower than the calculated ones with ab-initio where the structures are crystalline while the devices are polycristallin/amorphous with defects.

Figure 22. Comparing bond lengths in HfO_2 and $HfAlO$. AlO has the shortest bond length. HfO bond length in $HfAlO$ also reduced with Al incorporation. This bond shrinkage associated with higher atomic concentration may be the main effects related to a higher V_o diffusion barrier in $HfAlO$.

Table2. Summary of bond lengths in HfO_2 and $HfAlO$.

bond	Mean length (Å)
AlO	1.93
HfO in HfAlO	2.16
HfO in HfO2	2.17

978-1-4799-8002-4/14 $31.00 © 2014 IEEE

Nanosystems monolithically integrated with CMOS: emerging applications and technologies

J. Arcamone, J. Philippe, G. Arndt, C. Dupré, M. Savoye, S. Hentz, T. Ernst,
E. Colinet, L. Duraffourg and E. Ollier

Univ. Grenoble Alpes, F-38000 Grenoble, France
CEA, LETI, MINATEC Campus, 17 rue des Martyrs, F-38054 Grenoble, France
Email: julien.arcamone@cea.fr

ABSTRACT

This paper reviews the last major realizations in the field of monolithic integration of NEMS with CMOS. This integration scheme not only drastically improves the efficiency of the electrical detection of the NEMS motion. Our analysis is that it also represents a compulsory milestone to practically implement breakthrough applications of NEMS, such as mass spectrometry, that require large capture cross section (VLSI-arrayed NEMS) and individual addressing (co-integration of NEMS arrays with CMOS for closed-loop operation).

INTRODUCTION

NanoElectroMechanical Systems (NEMS) are essentially beam-shaped mechanical resonators with two out of their three main dimensions below 1µm. They constitute a very promising field both in terms of scientific interest, for instance for the observation of quantum mechanics phenomena, and from the practical applications point of view. NEMS should enable novel breakthrough applications, mainly in the field of chemical analysis (1), life science (2) and computing (3). In those first two fields (gas sensing and mass spectrometry – *MS* – respectively), they appear as excellent candidates since they exhibit unprecedented mass resolution, thanks to their inherent low mass, high resonance frequency f_0, and good frequency stability. These works have not only confirmed their intrinsic exceptional sensing potential, but also illustrated their increasing maturity as they were successfully integrated into complex systems (4). Nevertheless, these new applications tend to involve high-density NEMS arrays with a large number of electrical contacts per device, making their individual addressing complex. Moreover, the electromechanical transduction scheme of devices with such ultimate size and frequency range (10-100MHz) is critical. In this context, monolithic NEMS-CMOS integration seems the most adequate way to overcome those challenges.

IMPROVEMENT OF ELECTRICAL RESPONSE

Whatever the transduction scheme, scaling down the mechanical structure from MEMS to NEMS induces some electrical detection issues when using a direct (homodyne) measurement scheme (without heterodyning), such as (i) high connection losses due to pad and cable capacitances (the NEMS resonance frequency, generally ranging from 1 up to 100MHz, being far higher than the cutoff frequency of the output parasitic low-pass filter), and (ii) an increase of feedthrough parasitic coupling between the resonator input and output. These issues result in both degrading the signal-to

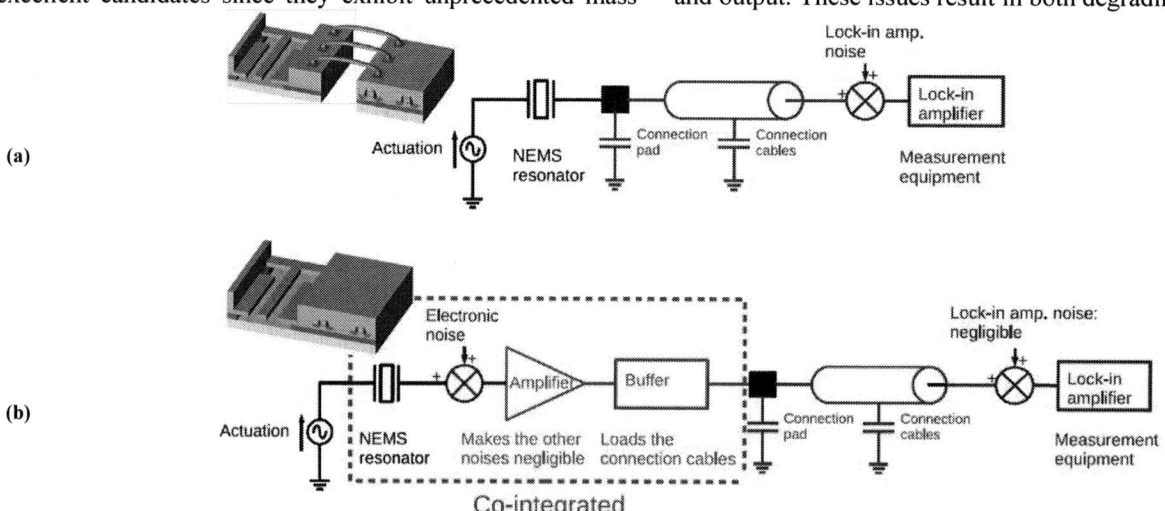

(a)

(b)

Figure 1 (inspired from (5)). Equivalent electrical schematics (a) of a stand-alone NEMS resonator with off-chip readout electronics. Parasitic capacitances at its output generate large losses of the NEMS signal; (b) of a NEMS resonator co-integrated with on-chip CMOS readout electronics that provides impedance matching for enhanced electrical detection.

978-1-4799-8002-4/14 $31.00 © 2014 IEEE

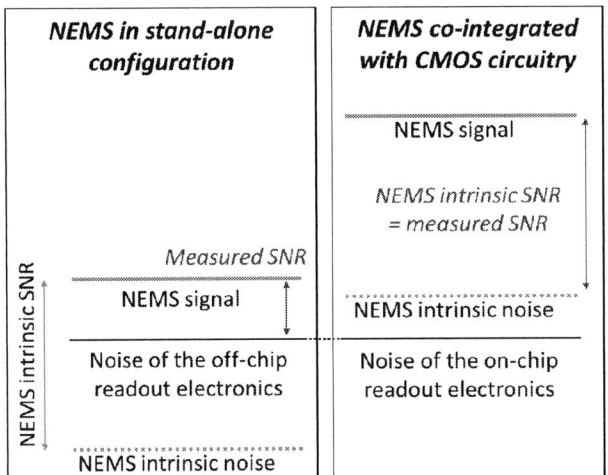

Figure 2. Schematic representation of SNR improvement provided by monolithic integration. In this case, the system can get the maximum SNR, fundamentally limited by the NEMS intrinsic noise.

-background ratio (SBR), and the signal-to-noise ratio (SNR). This ultimately induces a larger frequency noise of the whole system S_{f_SYSTEM}, as stated by the Robins formula (6):

$$S_{f\_SYSTEM} = \frac{\partial f}{\partial \varphi} S_{\varphi\_SYSTEM} = \frac{\partial f}{\partial \varphi} \frac{1}{SNR_{SYSTEM}} \quad (1)$$

S_φ being the system's phase noise. The left underlined term is linked to the SBR (in the base case, $\partial f/\partial \varphi = f_{0\text{-}NEMS}/2Q_{NEMS}$), the right one to the SNR (in the best case $SNR_{SYSTEM} = SNR_{NEMS}$, see Fig.2). In this context, co-integrating NEMS resonators with a dedicated adjacent CMOS readout and conditioning electronics (Fig.1b) is a very efficient way to overcome these difficulties. NEMS-CMOS devices indeed benefit from better SNR and immunity to external parasitic coupling (i.e. better SBR): see Fig.2 for the theoretical basis and Fig.3 for the experimental demonstration.

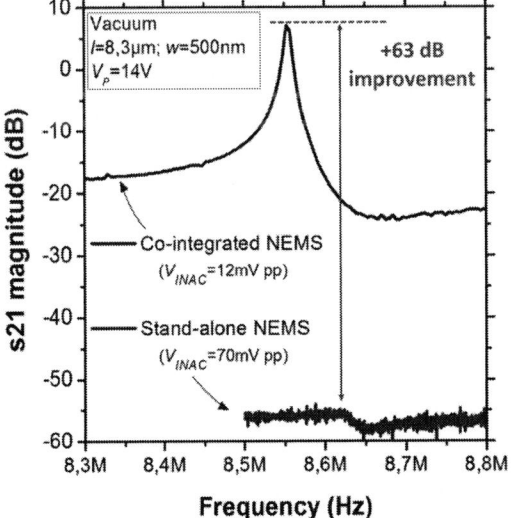

Figure 3 (from (5)). Experimental demonstration of SBR and SNR improvement through the comparison of the electrical response of two identical resonators (with same operating points), one in stand-alone configuration (cf Fig.1a), the other co-integrated with CMOS.

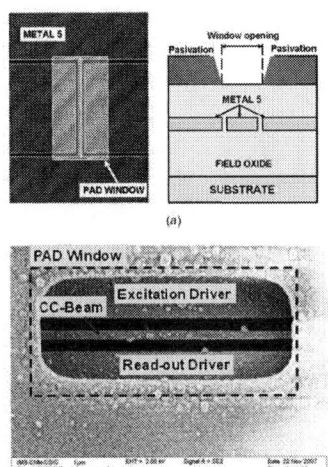

Figure 4 (from (8)). Post-CMOS integration of NEMS in CMOS back-end. Capacitive NEMS are made in the last metal level.

STATE-OF-THE-ART IN TERMS OF NEMS MATERIALS & PROCESSES

Monolithic integration of NEMS resonators with CMOS has been demonstrated in the last decade, mainly through three approaches. One consists in using one of the metal layers of back-end levels as NEMS structural layer (7,8) (Fig.4). Some MEMS-CMOS devices use part of, or the whole back-end stack (9-11), or above-IC electroplated metal (12). Another one relies on the middle-end polySi layers (13-15), commonly used as capacitors. In the third one, NEMS are integrated at the front-end, using the top single-crystal Si layer of SOI wafers (5,16,17) (Fig.5&6). In this category too, (18,19) report on an unreleased device using the whole MOS stack. Most of these examples do not induce any modification of the core CMOS process. The trend is on continuously miniaturizing the NEMS (fig.7), and unlike MEMS-CMOS devices, the novelty is that NEMS are converging towards advanced CMOS in terms of processes and dimensions: the smallest monolithic NEMS (up to our knowledge) (16,17) have been integrated at the front-end in the same 20nm thick layer as transistors channels of an FDSOI CMOS technology.

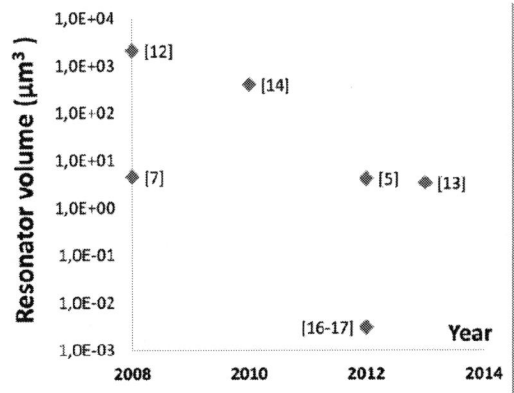

Figure 7. Benchmark of recent NEMS-CMOS realizations in terms of NEMS volume (of the vibrating part).

978-1-4799-8002-4/14 $31.00 © 2014 IEEE 551

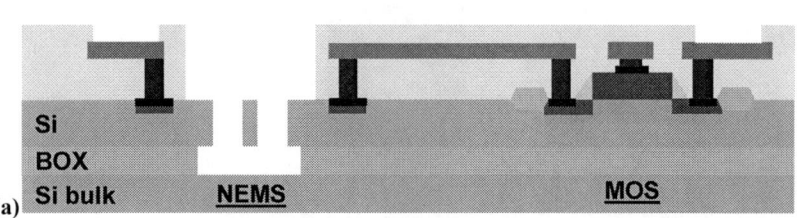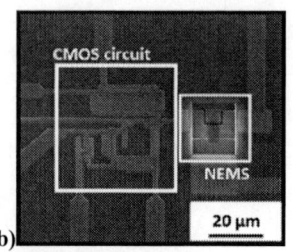

(a) Si / BOX / Si bulk / **NEMS** / **MOS** **(b)** CMOS circuit / NEMS / 20 μm

Figure 5 (from (5)). Example of pre-CMOS front-end integration. NEMS release is performed after CMOS completion (0.35μm CMOS process). Capacitive NEMS are defined in the 1μm thick top single-crystal Si layer of SOI wafers.

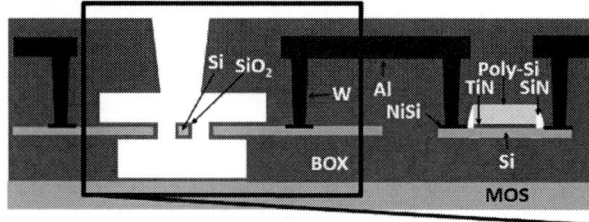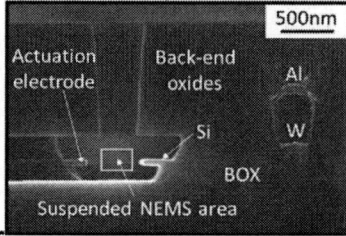

Figure 6 (from (16,17)). Example of pre-CMOS front-end integration. NEMS release is performed after CMOS completion (FD-SOI CMOS process). Piezoresistive NEMS are defined in the 20nm thick top single-crystal Si layer of SOI wafers (same layer as transistor channels).

NEMS-CMOS CLOSED-LOOP OSCILLATORS

Tracking in real time the resonance frequency of thousands of NEMS deployed in arrays will require an individual addressing that will probably be enabled only by monolithic integration. In practice, such arrays might consist of thousands of pixels, each one containing a NEMS-CMOS closed-loop oscillator, under the form of a PLL or a self-oscillator. In the latter case, the NEMS and its sustaining amplifier have to fulfill the Barkhausen criteria. A few pioneer works relying on this approach have been reported. Ref.(13) (Fig.8) reports on an 11MHz Pierce-type oscillator whose resonant element is a double-ended tuning fork NEMS resonator. The CMOS sustaining circuit (0.35μm AMS process) is a differential voltage (cascode) amplifier, buffered

by a source-follower output stage for impedance matching. The capacitive detection is so efficient, thanks to a 40nm gap, that it is possible to generate self-oscillations with a NEMS dc polarization smaller than the CMOS supply voltage (3.3V). Ref.(20) (Fig.9) describes a very compact, capacitive 8MHz NEMS-CMOS self-oscillator: the circuit (fabricated with a 0.35μm STMicro process) only contains 7 transistors and the NEMS-CMOS cell is as small as 50x70μm² (Fig.5b).

MAIN TRENDS FOR THE FUTURE

Regarding transduction schemes, one trend is the emergence of alternatives to capacitive detection, that seems to provide too low SNR for f_0 larger than ~20MHz. Piezoresistive (PZR) detection seems suitable for ultra-thin (sub-200nm), high f_0

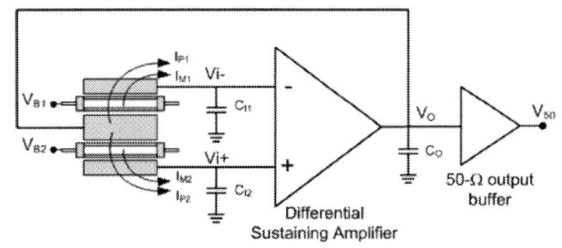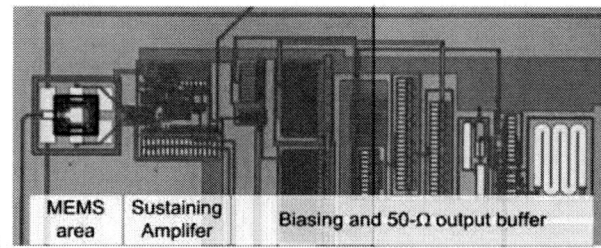

Figure 8 (from (13)). (a) Equivalent electrical schematic of this NEMS-CMOS self-oscillator; (b) Optical image of the corresponding device

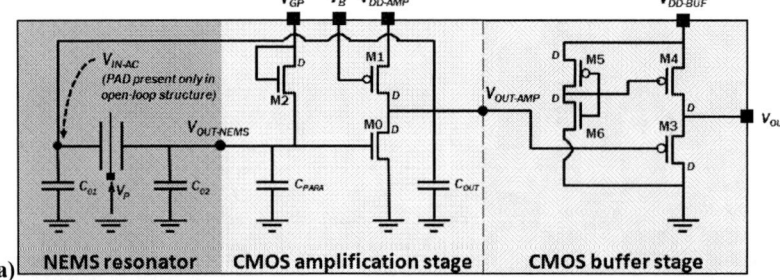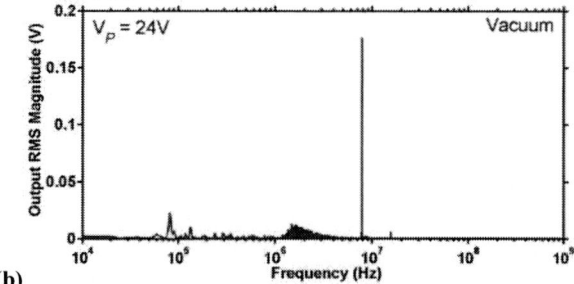

Figure 9 (from (20)). (a) Equivalent electrical schematic of this NEMS-CMOS self-oscillator; (b) Fast-Fourier transform of output electrical response

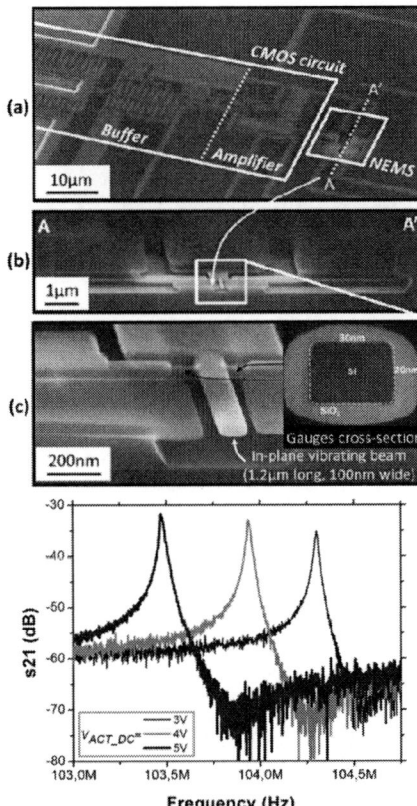

Figure 10 (17). Top: Tilted SEM images of ultimate single-crystal Si PZR 100MHz NEMS-CMOS devices (1.2μm long, 100nm wide and 20nm thick NEMS beam); bottom: electrical response in vacuum with 30dB SBR (using a 2.25mV rms ac actuation voltage).

(>20MHz) NEMS. PZR Si nanogauges indeed benefit from favorable scaling laws: the cross-sectional area on which the force is applied being smaller, the induced stress is magnified and the resulting relative resistance variation as well. Ref. (16,17) report on the monolithic integration of PZR NEMS (Fig.10); although such NEMS beams have a ultimate size (1.2μm long, 100nm wide, 20nm thick), their direct (homo-dyne) electrical response provides SBR as high as 30dB.

Detection schemes based on transistor architectures are another alternative. Ref.(18,19) report on unreleased 11GHz RF acoustic resonators implemented as resonant body transistors (RBT) with capacitive actuation (Fig.11). Their electromechanical transduction is based on a PZR modulation of the drain current of an n-type field-effect transistor, the PZR effect inducing a modulation of carriers mobility. Ref.(21) relies on a similar principle: this stand-alone device also has the potential for a monolithic front-end integration.

Besides their potential for computing and as RF components, NEMS-CMOS applications might be driven by gas and bio-sensing. In the latter case, the practical implementation of NEMS-based *MS* (2) and sensors for genomics (22) will probably require high-density NEMS-CMOS arrays for real-time monitoring and increased capture area. Besides TSV integration, one promising way to achieve this may be the so-called 3D sequential integration (23). This promising above-

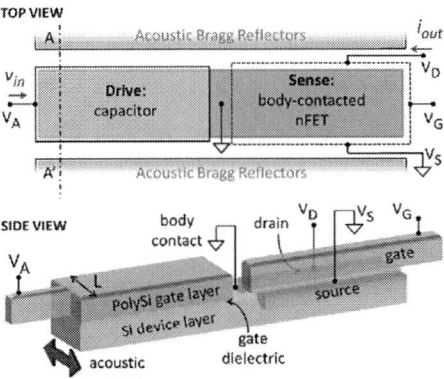

Figure 11 (19). Schematic representation of a resonant-body transistor, made at the front-end of an IBM SOI CMOS technology.

IC technology is based on the molecular bonding of CMOS back-end with an inverted SOI wafer; on which NEMS resonators are subsequently fabricated. In this configuration, the density of NEMS resonators could be unprecedented.

REFERENCES

(1) J. Arcamone et al., "VLSI silicon multi-gas analyzer coupling gas chromatography and NEMS detectors", *IEEE IEDM*, 2011, pp. 669–672

(2) M.S. Hanay et al., "Single-protein nanomechanical mass spectrometry in real time", *Nature Nanotechnology* 7, pp. 602–608, 2012

(3) O.Y. Loh and H.D. Espinosa, "Nanoelectromechanical contact switches", *Nature Nanotechnology* 7, pp. 283–295, 2012

(4) see for example www.apixtechnology.com

(5) J. Arcamone et al., "VLSI platform for the monolithic integration of single-crystal Si capacitive resonators with low-cost CMOS", *IEEE IEDM*, 2012

(6) W.P. Robins, "Phase Noise in Signal Sources", *Peter Peregrinus*, 1982

(7) J. Verd et al., "Monolithic CMOS MEMS Oscillator Circuit for Sensing in the Attogram Range", *IEEE Electron Device Letters* 29, pp. 146–148, 2008

(8) J.L. Lopez et al., "Integration of RF-MEMS resonators on submicrometric commercial CMOS technologies", *J. Microm. Microeng.* 19, 015002, 2009

(9) C.H. Chin et al., "Fabrication and characterization of a charge-biased CMOS-MEMS [...] field effect transistor", *J. Microm. Microeng.* 24, 2014

(10) M.-H. Tsai et al., "A Three-Axis CMOS-MEMS Accelerometer [...] Integrated Fully Differential Sensing Electrodes", *IEEE JMEMS* 21, 2012

(11) S-S Li, "Advances of CMOS-MEMS Technology for Resonator Applications", *IEEE NEMS*, 2013, pp. 520-523

(12) W.-L. Huang et al., "Fully monolithic CMOS Nickel micromechanical resonator oscillator", *IEEE MEMS*, 2008, pp. 10-13

(13) J. Verd et al., "A 3V CMOS-MEMS oscillator in 0.35μm CMOS technology", *IEEE Transducers*, 2013, pp. 806-809

(14) M.K. Zalalutdinov et al, "CMOS-Integrated RF MEMS Resonators", *IEEE JMEMS* 19, pp. 807-815, 2010

(15) J. Arcamone et al., "Full-wafer fabrication [...] mechanical mass sensors monolithically integrated with CMOS", *Nanotechnology* 19, 305302, 2008

(16) E. Ollier et al., "Ultra-scaled high-frequency single-crystal Si NEMS [...] for high sensitivity applications", *IEEE MEMS*, 2012, pp. 1368-1371

(17) J. Arcamone et al, "VHF NEMS-CMOS piezoresistive resonators for advanced sensing applications", *Nanotechnology*, 2014

(18) R. Marathe et al., "Si-based unreleased hybrid MEMS-CMOS resonators in 32nm technology", *IEEE MEMS*, 2012, pp. 729-732

(19) R. Marathe et al., "Resonant Body Transistors in IBM's 32 nm SOI CMOS Technology", *IEEE JMEMS* 23, pp. 636-650, 2014

(20) J. Philippe et al., "Fully monolithic and ultra-compact NEMS-CMOS self-oscillator [...] CMOS circuitry", *IEEE MEMS*, 2014, pp. 1171-1174

(21) S.T. Bartsch et al., "Nanomechanical Silicon Resonators [...] Tunable Gain and Sub-nW Power Consumption", *ACS Nano* 6, pp. 256-264, 2012

(22) J. Zhang et al., "Rapid and label-free nanomechanical detection of biomarker transcripts [...]", *Nature Nanotechnology* 1, pp. 214-220, 2006

(23) P. Batude et al., "3-D Sequential Integration: A Key Enabling Technology [...] of New Function With CMOS" *IEEE JESTCAS* 2, pp. 714-722, 2012

A Self-sustained Nanomechanical Thermal-piezoresistive Oscillator with Ultra-Low Power Consumption

Kuan-Hsien Li, Cheng-Chi Chen, Ming-Huang Li, and Sheng-Shian Li

Institute of NanoEngineering and MicroSystems
National Tsing Hua University, Hsinchu, Taiwan

ABSTRACT

This work reports wing-type thermal-piezoresistive oscillators operating at around 840 kHz in vacuum with ultra-low power consumption of only 70 µW for the first time. The combination of N-type heavily doped silicon with negative piezoresistive coefficients and sub-micron cross-sectional dimensions of the thermally actuated beams is key to ensuring self-sustained oscillation under sub-100µW level. In particular, shrinking thermal beam cross-sections in 2D is a novel and effective approach to achieve better figure of merit (e.g., FOM defined by the ratio of transconductance g_m to dc power consumption P_{DC}) of the self-sustained oscillators. By using proper control of silicon etching (ICP) recipe, the sub-micron cross-sectional dimensions of the thermally actuated beams can be easily and reproducibly fabricated in one process step. The phase noise of the proposed wing-type oscillators is also investigated in this work with -93.41 dBc/Hz at 1-kHz offset and -97.95 dBc/Hz at 100-kHz offset in air, and -95.9 dBc/Hz at 1-kHz offset and -95.7 dBc/Hz at 100-kHz offset in vacuum, respectively.

I. INTRODUCTION

Owing to zero amplifier circuit and simple low-cost processes, the thermal-piezoresistive self-sustained oscillators have recently attracted significant attention [1][2]. The thermal-mechanical-electrical coupling mechanism enables an internal positive feedback loop when a stable dc current and ac noise passing through the structures fabricated by heavily N-doped silicon wafer (i.e., offering negative piezoresistive coefficient, π_l). Upon the negative motional impedance ($1/g_m$) which provides sufficient energy to compensate resonant loss, the devices would introduce self-sustained oscillation.

However, the large dc power consumption (P_{DC}) places a bottleneck for the thermal-piezoresistive oscillators as compared to the piezoelectric and capacitive MEMS oscillators. By scaling down the thermal beams into a sub-micron dimension of width (1D shrinking) using E-beam lithography [3] or specifically lattice alignment processes [4], the power consumption of the thermal-piezoresistive self-sustained oscillators can approach mW level with higher resonance frequency. Nevertheless, high frequency design is less amenable to

achieve a sufficient transconductance (g_m), given by

$$g_m \propto QE\pi_l AR_{th} \frac{I_{dc}^2}{C_{th}\omega_n^2} \tag{1}$$

where Q, E, A are quality factor, Young's modulus, and thermal beam cross-sectional area, respectively; R_{th} and C_{th} are thermal resistance and thermal capacitance, respectively, of the thermal beam actuators; I_{dc} is the applied dc current and ω_n is the angular resonance frequency. As indicated in (1), it needs more dc current to satisfy the self-sustained oscillating condition once higher resonance frequency is designed. Therefore, the combination of the tiny cross-sectional area (sub-micron size) of the thermal beams and low resonance frequency design would lead to low-power self-sustained oscillation.

In this work, the wing-type self-sustained thermal-piezoresistive oscillators are fabricated using a conventional SOI-MEMS process. By a particular silicon DRIE (ICP) recipe, the cross-sections of the thermally actuated beams can be dramatically reduced to a submicron dimension in a 2D configuration while not affecting the resonance frequency of the resonators. This feature successfully decouples the design of desirable thermal beams and the determination of the resonance frequency, thus resulting in maximized FOM. The regular power consumptions of the proposed oscillators are all below 1 mW; and the record-low dc power of the oscillators reaches 70 µW with phase noise of -95.9 dBc/Hz at 1-kHz offset and -95.7 dBc/Hz at 100-kHz offset, respectively, showing great potential of the self-sustained thermal-piezoresistive oscillators in ultra-low power applications.

II. STRUCTURE AND DESING CONCEPT

Fig. 1 presents the wing-type MEMS oscillator structure and its out-of-plane vibrating mode shape. The piezoresistive signal can be detected under a simple setup depicted in Fig. 1 without additional electronic circuits for oscillation. In order to attain low driving power and high motional transconductance (g_m), the device design necessitates small thermal capacitance on its thermally actuated beams as indicated in (1). At the

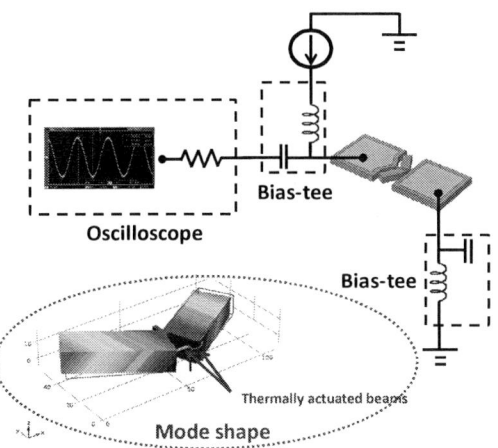

Fig. 1: Schematic of the proposed thermal-piezoresistive oscillator system and finite element simulated mode shape of the wing-type resonator at 843 kHz with out-of-plane motion. The device electrical resistance is 1.97 kΩ.

same time, the submicron-sized thermal beams, as shown in Fig. 4 and its inset, has very little effect on the resonance frequency while maximizing the drive/sense efficiency. The main parameters to determine the resonance frequency of the resonator is its rotational springs located in the pins (i.e., pivots) of the wings and the proof-masses. In addition, the negative π_l is an essential parameter to offer negative g_m and enable the resonant system to possess an internal positive feedback loop. Therefore, the fabricated wafer should be N-type heavily doped silicon with negative piezoresistive coefficients, and the thermal beams are aligned to the [100] crystal orientation to ensure the absolute longitudinal piezoresistive coefficient value (π_l) is maximum [5].

Fig. 2 presents an equivalent electrical circuit model for the thermal-piezoresistive oscillator. In order to satisfy the Barkhausen criterion, the energy loss from the electrical domain (e.g., resistive loss) and mechanical domain (damping) should be compensated through an inner energy pumping mechanism induced by the negative resistance ($1/g_m$).

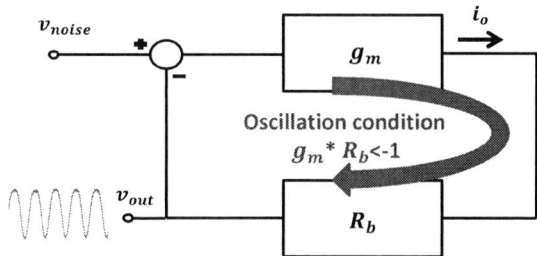

Fig. 2: Overall equivalent electrical circuit model of the oscillator. To satisfy the Barkhausen criterion, $|g_m * R_b|$ needs to be greater than 1 and $\angle g_m * R_b$ should be -180° to form a positive feedback loop. g_m is the motional transconductance and R_b is the electric resistance of thermally actuated beams

III. FABICATION PROCESS

The wing-type thermal-piezoresistive oscillators are fabricated using an SOI-MEMS process with only two lithography steps. The device and buried oxide layers of the SOI wafer are 15 μm and 2 μm thick, respectively. Fig. 3(a) shows that the pads and interconnects of Cr significantly reduce the resistive loss. In contrast to the vertical sidewall of conventional thermal beams [3], the inclined sidewall realized by an ICP scalloping effect and etching bias in Fig. 3(b) is key to contributing tiny cross-sections of thermal beams, as shown in Fig. 4, which is difficult to be realized in a conventional one-step photolithography. This etching technique offers a low cost and efficient way to shrink the dimensions of thermal beams into sub-micron level in a 2D configuration, thus creating near 100 times thickness reduction as compared to the proof-masses and mechanical springs of the resonator. Fig. 3(c) presents the final release step utilizing 49% hydrofluoric acid (HF) after the ICP silicon etching.

IV. EXPERIMENTAL RESULTS

The open-loop setup to characterize the resonators is depicted in Fig. 5. A KEITHLEY power supply provides a stable dc current from 0.5 mA to 0.7 mA while an ac power of -15 dBm from the network analyzer is applied into the device.

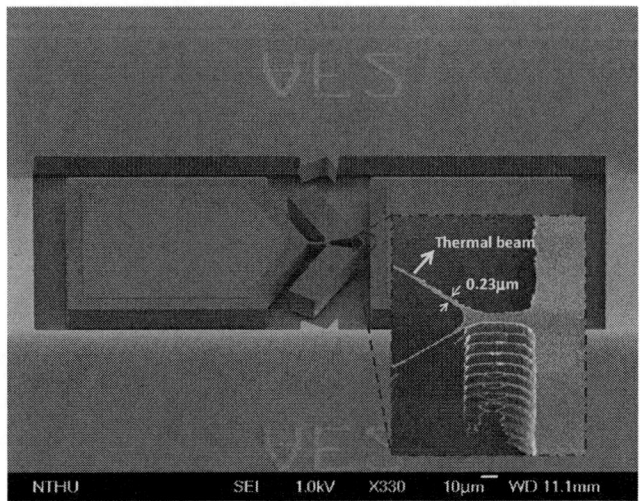

Fig. 4: SEM of a fabricated wing-type oscillator with a global view and its zoomed thermal beam where 0.23-μm beam width is clearly seen.

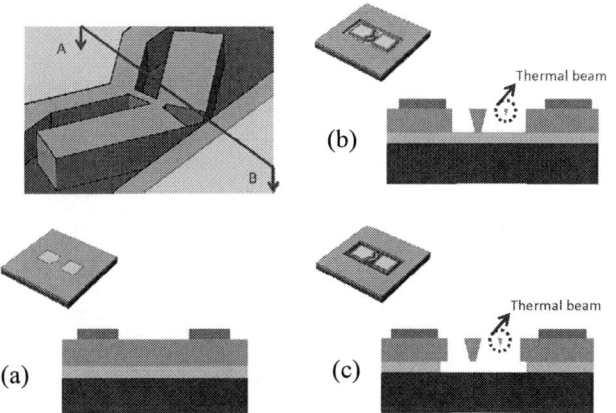

Fig. 3: Standard SOI-MEMS process flow. (a) PR is used to pattern the Cr pads (b) the structure is patterned by ICP; (c) final etch of the sacrificial oxide is performed using 49% HF solution to release the structure.

978-1-4799-8002-4/14 $31.00 © 2014 IEEE

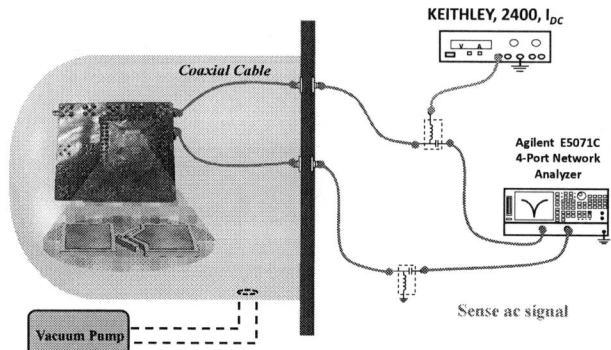

Fig. 5: Open-loop measurement setup in a vacuum chamber.

As the wing-type device is operated using the setup of Fig. 5, the frequency response and phase can be measured as shown in Fig. 6. It is evident that the resonant dips become stronger as the dc current increases with a turnover point at a critical current level. In addition, the phase would also suddenly change at this turnover point, indicating the oscillation condition is fulfilled. Before oscillation, the phase at the resonance frequency is close to 180° due to the positive resistance ($1/g_m$ // R_b) of the total system where R_b is the

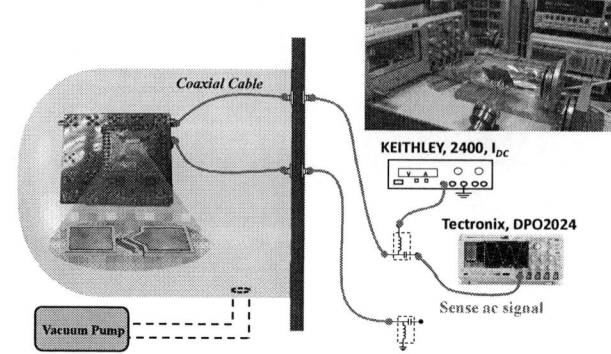

Fig. 7: Experimental setup of closed-loop measurement in vacuum (about 10^{-3} Torr).

resistance of the thermal beams. As oscillation ensues at the dc current of 0.65 mA, the network analyzer detects negative system resistance due to a sufficiently high transconductance (g_m) so as to shift the phase to -180° at resonance. After the de-embedding and post-data processing is performed, the frequency responses of the pure motional signal can be extracted as shown in Fig. 6(b).

The closed-loop measurement setup of the thermal-piezoresistive oscillator is depicted in Fig. 7. Fig. 8 shows the phase noise performance of the oscillator in air (top) and vacuum (bottom) under the lowest (threshold) dc driving powers. The vacuum case shows the record-low power consumption among published self-sustained thermal-piezoresistive oscillators [1][2], and is comparable to the state-of-the-art monolithic MEMS oscillators [6]. The peak-to-peak output voltage signals of the measured oscillator are 60.8mV in air and 22.4mV in vacuum, respectively, as shown in the insets of Fig. 8. Furthermore, Fig. 9 presents the phase noise performance under various dc driving powers. In a lower dc driving power at the critical oscillating condition, the phase noise slope shows a $1/f^4$ dependency, leading to a worse close-in phase noise. In contrast, the close-in phase noise can

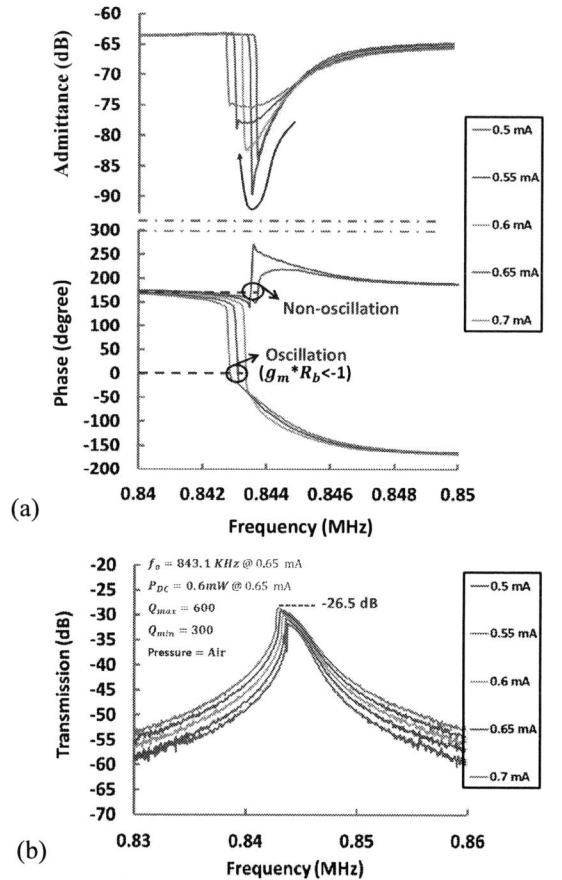

Fig. 6: (a) Open-loop measured frequency responses and phase in air. The resonant point phase can distinguish the operating condition. The phase has drastic change once oscillation condition is fulfilled. (b) The device only consumes 0.6 mW to launch oscillation at 843 kHz.

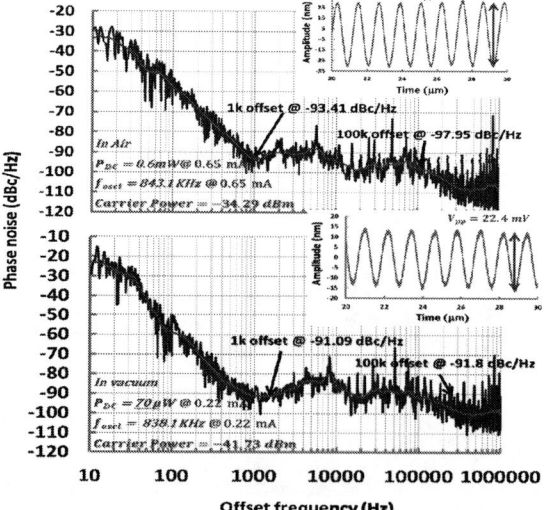

Fig. 8: Measured phase noise in air (top) and vacuum (bottom) of the 840-kHz wing-type oscillator.

978-1-4799-8002-4/14 $31.00 © 2014 IEEE

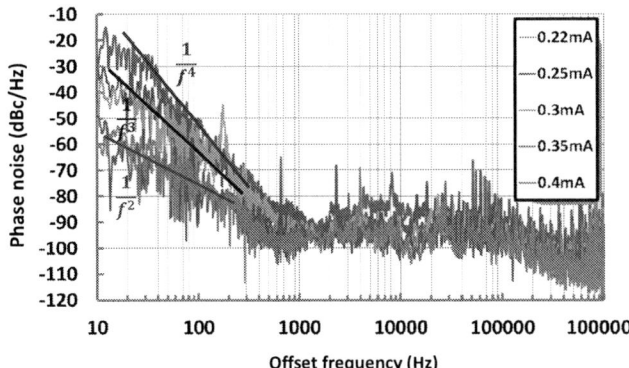

Fig. 9: Phase noise variations with increasing dc driving power of the 840-kHz wing-type oscillator in vacuum. At critical oscillating condition, the phase noise approximately follows $1/f^4$. As the device operates using higher dc driving power, the close-in phase noise performance is greatly improved.

be significantly reduced using larger dc driving powers; and the best performance in vacuum exhibits -58.48 dBc/Hz at 10-Hz offset and -95.94 dBc/Hz at 1-kHz offset, respectively, with a 24.8-mV peak-to-peak voltage output under dc power consumption of 254 µW.

The physical mode shape can be also verified using a LDV system (Laser Doppler velocimetry, cf. Fig. 10(a)) and the vibration displacements are investigated as shown in Fig. 10(b) and (c) to monitor the motion behaviors (the laser head aims at the wings) of the self-sustained oscillator. It is evident that the oscillator vibrates in an out-of-plane mode and the carrier frequency is about 700 kHz, which is in good agreement with the finite-element simulated results. The phase transition of 180° around the resonance frequency is also observed, corresponding to a pure mechanical vibration response. In the time domain plots of Fig. 10, each end of the wing (i.e., proof-masses) possesses 2 to 4-nm displacement, thus enabling sufficient stretching/contraction in the thermally actuated beams.

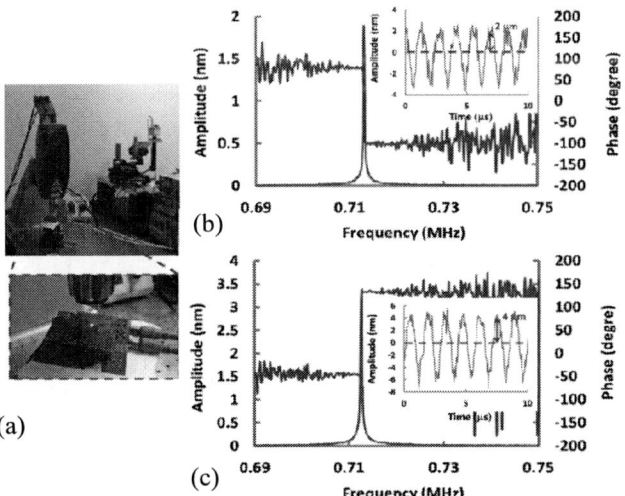

Fig. 10: (a) Measurement setup of a LDV system (Laser Doppler Velocimetry). (b) Frequency and time responses of the left-ended proof-mass. (c) Frequency and time responses of the right-ended proof-mass.

Table 1: Comparison and Summery of State-of-the-art Oscillators

	This work		Piezo-thermal oscillators		MEMS + Circuit	Circuit only	
	(1)	(2)	Ref [1] Nature physics' 11	Ref [2] IEDM' 10	Ref [6] IEDM' 13	Ref [7] TCS'13	Ref [8] TCS'13
P_{DC}	0.6 mW	70 µW	1.19 mW	4.44 mW	1.6 mW	0.86 µW	10 mW
Active Circuit	NO		NO	NO	Yes	Yes	Yes
Fabrication	Batch		Batch	Batch	0.35 2P4M	0.18 CMOS	65 nm CMOS
f_{osci}	843.1 kHz	838.1 kHz	1.26 MHz	5.01 MHz	1.2 MHz	1.1 MHz	645 MHz
Q factor	483	6,265*	13,300	48,500*	2,400	N/A	N/A
Output Signal	60.8 mV	22.4 mV	54 mV	30 mV	320 mV	N/A	237 mV
$|g_m|/P_{DC}$	1.14 A/v·w	0.863 A/v·w*	N/A	0.1 A/v·w*	N/A	N/A	N/A
PN (dBc/Hz) @ 1 KHz	-93.41	-91.09	N/A	N/A	-103	-38.92	N/A
@ 100 KHz	-97.95	-91.85			-107	-79.48	-88.11
PN FOM (dBc/Hz) @ 1 KHz	-154.15	-161.1	N/A	N/A	-162.54	-130.4	N/A
@ 100 KHz	-118.69	-122.32			-126.54	-130.96	-154.3
Pressure	Air	Vacuum (1e-4 torr)	Vacuum (7.5e-3 torr)	Vacuum	Vacuum	N/A	N/A

* Data from similar device

$$\text{PN FOM} = (f_m) - 20log\left(\frac{f_0}{f_m}\right) + 10log(P_{DC})$$

V. CONCLUSION

An ultra-low power thermal-piezoresistive oscillator has been successfully demonstrated with dc power consumption of only 70 µW. Compared to other state-of-the-art oscillators shown in Table 1, the wing-type oscillators exhibit 10 times higher FOM (defined in the bottom of the table) than other thermal-piezoresistive oscillators. The power consumption in this work outperforms the MEMS-circuit integrated oscillators, and is also on par with pure electrical oscillators (with poor phase noise) in micro-watt range.

Acknowledgements. This research was sponsored by the MOST of Taiwan (MOST-101-2628-E-007-008-MY2) and the Toward World-Class University Project. The authors are also grateful to the CNMM of National Tsing Hua University for the use of fabrication and measurement facilities.

REFERENCE

[1] P. G. Steeneken, *et al.*, "Piezoresistive heat engine and refrigerator," *Nature Physic*, vol. 7, pp. 354-359, 2011.

[2] A. Rahafrooz and S. Pourkamali, "Fully micromechanical piezo-thermal oscillators," *in IEEE Tech. Dig. IEDM*, pp. 7.2.1-7.2.4, 2010.

[3] A. Rahafrooz, *et al.*, "Thermal actuation, a suitable mechanism for high frequency electromechanical resonators," *in 23rd IEEE MEMS*, pp. 200-203, 2010.

[4] A. Rahafrooz and S. Pourkamali, "Controlled batch fabrication of crystalline silicon nanobeam-based resonant structures," *in 24th IEEE MEMS*, pp. 1345-1348, 2011.

[5] K. Nakamura, *et al.*, "Simulation of piezoresistivity in n-type single-crystal silicon on the basis of the first-principles band structure," *The American Physical Society, Phys. Rev.* B 80, 045205, 2009.

[6] M.-H. Li, *et al.*, "Foundry-CMOS integrated oscillator circuits based on ultra-low power ovenized CMOS-MEMS resonators," *in IEEE Tech. Dig. IEDM*, pp. 18.4.1- 18.4.4, 2013.

[7] Y. H. Chiang, *et al.*, "A submicrowatt 1.1-MHz CMOS relaxation oscillator with temperature compensation" *TCSII. Syst.*, vol. 60, pp. 831-841, 2013.

[8] J. M. Kim, *et al.*, "A low-noise four-stage voltage-controlled ring oscillator in deep-submicrometer CMOS technology," *TCSII. Syst.*, vol. 60, pp.71-75, 2013.

Optimizing the Close-to-Carrier Phase Noise of Monolithic CMOS-MEMS Oscillators Using Bias-dependent Nonlinearity

Ming-Huang Li, Chao-Yu Chen, Chi-Hang Chin, Cheng-Syun Li, and Sheng-Shian Li

Institute of NanoEngineering and MicroSystems, National Tsing Hua University, Hsinchu 30013, Taiwan
Phone: +886-3-571-5131 ext.80631, Fax: +886-3-574-5454, Email: s100035807@m100.nthu.edu.tw

Abstract

A fully monolithic 1.12-MHz CMOS-MEMS nonlinear oscillator comprising a double-ended tuning fork (DETF) resonator and a transimpedance sustaining amplifier has been proposed to enable significant close-to-carrier phase noise (PN) reduction while maximizing its output power for far-from-carrier phase noise improvement. The best-case PN of -77 dBc/Hz at 10-Hz offset, -97 dBc/Hz at 100-Hz offset, and -113 dBc/Hz at 1-kHz offset is realized in a monolithic CMOS-MEMS flexural-mode resonator oscillator for the first time, which is on par with bulk-mode MEMS oscillators using resonator $Q > 100,000$.

I. Introduction

Miniaturized micromechanical resonators recently attracts significant attention for timing reference and spectral processing applications due to their small size, low power consumption, and batch-fabrication cost effectiveness. In particular, foundry CMOS integrated resonator circuits [1]-[3] offer a very promising solution for frequency-domain signal processing with a preferable circuit integration feature. After the temperature compensation scheme is developed for CMOS-MEMS oscillators [1], one remaining puzzle that needs to be addressed is their phase noise (PN) performance. Since the PN of an oscillator is inversely proportional with the carrier power, the output driving voltage (V_{OUT}) has to be maximized. However, as the driving power increases, the close-to-carrier PN becomes worse due to the increase of a nonlinear amplitude-to-phase modulation (AM-to-PM) noise conversion [4]. To reduce the close-to-carrier PN, two general approaches were explored in past years: (1) implement high-power handling resonator designs, such as high-stiffness driving [5] and resonator array [6], to enhance resonator linearity, (2) implement automatic gain control (AGC) circuit [7][8] for output power regulation that ensures the resonator stays in its linear regime.

Counter to the conventional wisdom, significant close-to-carrier PN reduction is recently achieved by taking advantage of nonlinear oscillator operation [9][10]. In this work, we also explore this technique for CMOS-MEMS oscillator PN optimization. By operating the resonator in its anharmonic regime using bias-dependent nonlinearity, we have experimentally demonstrated the PN improvement more

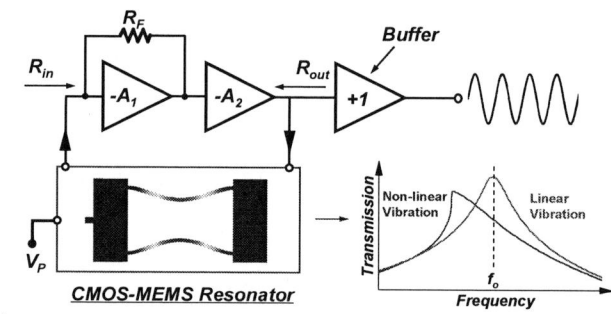

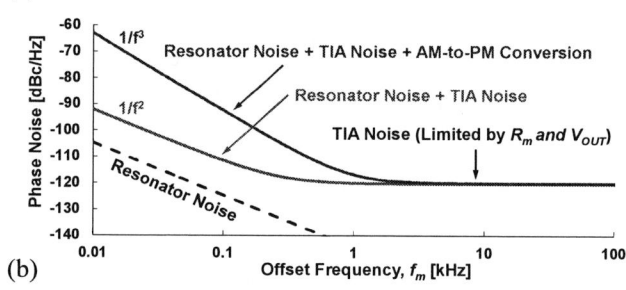

Fig. 1: (a) Top-level circuit schematic and mode shape of a CMOS-MEMS double-ended tuning fork (DETF) resonator oscillator. (b) Typical phase noise plot of a nonlinear CMOS-MEMS oscillator. The $1/f^3$ noise component comes from the nonlinear AM-to-PM noise conversion.

than 27 dB in the close-to-carrier frequency offsets as compared to our previous DETF oscillator [1].

II. Phase Noise of Nonlinear Oscillators

Fig. 1(a) depicts the schematic of a CMOS-MEMS oscillator which consists of a transimpedance amplifier (TIA) and a DETF MEMS resonator. Typically, the electrostatic nonlinearity in a capacitive MEMS resonator is significant, which strongly depends on the dc-bias voltage V_P. For a nonlinear oscillator operated at frequency f_o, the PN model encompassing nonlinearity can be expressed as [4],

$$S_{PN} \approx \frac{1}{2V_{OUT}^2} \left[4kTR_m + \frac{\overline{i_n^2}}{\Delta f} R_m^2 \right] \left[1 + \left(\frac{f_c}{f_m} \right)^2 \right] + \frac{S_{fn}(f_m)}{f_m^2} \quad (1)$$

where R_m is the motional impedance of the resonator, $i_n^2/\Delta f$ is the TIA input-referred current noise, f_c is the cutoff frequency ($f_c = f_o/2Q_L$ for a *linear* oscillator and Q_L is the loaded quality factor), f_m is the offset frequency, and S_{fn} is the frequency

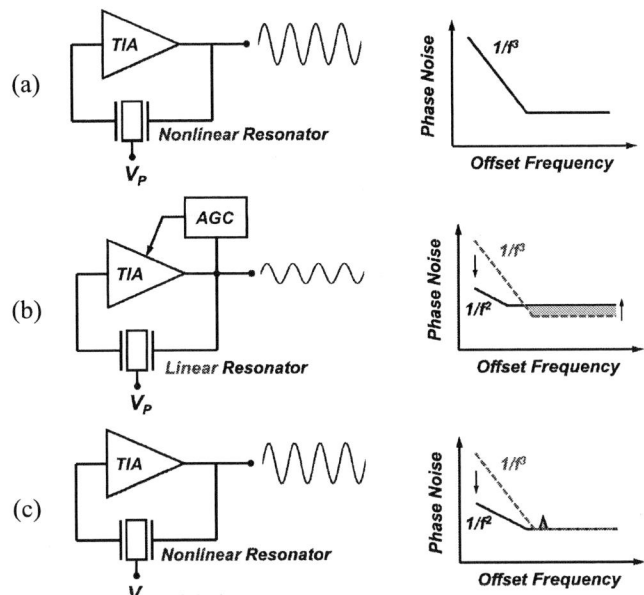

(a)

(b)

(c)

Fig. 2: Comparison of capacitive MEMS oscillator architectures. (a) Nonlinear MEMS oscillator without amplitude control and phase noise optimization. (b) Linear MEMS oscillator using AGC for close-to-carrier PN reduction. (c) Nonlinear MEMS oscillator with phase noise optimization by cancelling AM-to-PM noise conversion process via a specific feedback phase.

noise spectrum. For a nonlinear oscillator, f_c and S_{fn} can be further expressed by the following equations,

$$f_c = \left[\frac{\partial \phi}{\partial f}\right]^{-1} \approx \frac{d\Omega}{d\Delta}, \quad S_{fn} \approx \left|\frac{d\Omega}{da}\right|^2 \left|\frac{da}{dV_{OUT}}\right|^2 S_{An} \quad (2)$$

where Ω is the resonance frequency offsets, Δ is the phase delay, a is the resonator vibration amplitude, and S_{An} is the amplitude noise spectrum. Note that $(d\Omega/da)^2$ delineates the coupling between amplitude noise and phase noise, which is usually considered to be zero for a *linear* oscillator. Fig. 1(b) shows a typical PN plot for a nonlinear CMOS-MEMS oscillator where the electronic noise dominates the PN floor. For the close-to-carrier PN, the capacitive nonlinearity causes the AM-to-PM noise conversion, hence worsening the close-to-carrier PN and generating a $1/f^3$ PN component [4].

Fig. 2(a) illustrates the typical PN plot for a MEMS oscillator without amplitude control, showing a $1/f^3$ PN component. To reduce the AM-to-PM conversion, automatic gain control (AGC) circuit is widely adopted [7][8] to reduce the resonator nonlinearity by lowering the oscillator output power (cf. Fig 2(b)). However, this sacrifices the PN floor due to the output power reduction. In this work, we explore an alternative solution to reduce the close-to-carrier PN without affecting the PN floor by operating the resonator in its deep nonlinear regime (cf. Fig. 2(c)). As the oscillator is operated at a specific feedback point where the frequency-phase slope ($d\Omega/d\Delta$) and frequency-amplitude slope ($d\Omega/da$) approach zero, eq. (2) vanishes and the AM-to-PM noise conversion is effectively evaded [9][10].

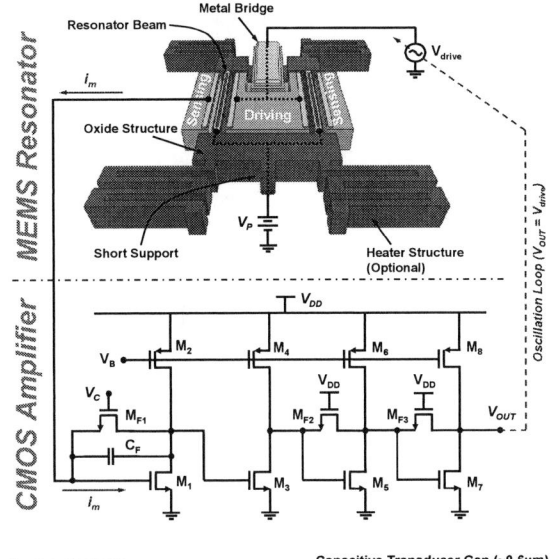

(a)

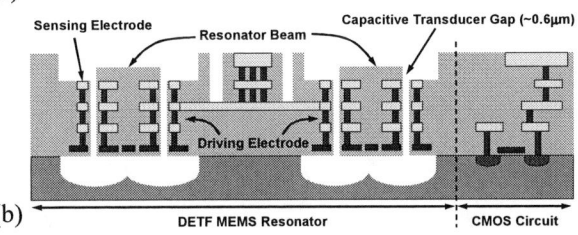

(b)

Fig. 3: (a) Detail schematic of the CMOS-MEMS resonator and sustaining TIA circuit. Control voltage V_C is used for TIA gain tuning. (b) Cross-sectional view of the resonator after post-CMOS fabrication process.

Since changing the dc-bias voltage V_P simultaneously modifies the cubic nonlinear spring constant [6] of the resonator and the oscillation frequency, electrostatic (i.e., bias-dependent) nonlinearity is expected to harness the overall feedback phase, thus optimizing the PN.

III. Design of CMOS-MEMS Oscillators for PN Tuning

The CMOS-MEMS oscillator was designed in a TSMC 0.35μm 2P4M standard CMOS process. Fig. 3(a) presents the perspective view of the CMOS-MEMS resonator and transistor-level circuit schematic. Under a typical operation, the oscillator circuit is biased with V_{DD} of 2.5V and consumes dc-power of 1.3 mW. The four-stage TIA designed in this work ensures high gain and low noise, which is suitable for oscillator implementation. Moreover, the gain-adjusting voltage, V_C, can be used to perform phase noise optimization. In this implementation, the bandwidth of TIA is determined by the feedback transistor M_{F1} (equivalent resistance R_{F1}) and capacitor C_F ($\omega_{3dB} \sim 1/R_{F1}C_F$). Therefore, the change of V_C not only modifies the TIA gain, but also moves the location of the dominant pole. With this voltage-controlled phase shifter and dc-bias V_P, we can select a proper operating point on the frequency-phase curve for phase noise reduction. Fig. 3(b) provides the cross-sectional view of the CMOS-MEMS resonator after a maskless metal wet etching process, where the detailed process flow can be found in [1]. The fabricated chip photo and SEM picture are shown in Fig. 4.

978-1-4799-8002-4/14 $31.00 © 2014 IEEE

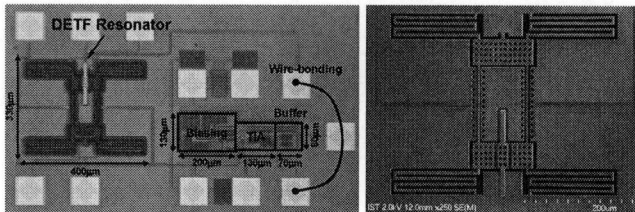

Fig. 4: Optical view and SEM picture of the monolithic CMOS-MEMS oscillator using a DETF MEMS resonator.

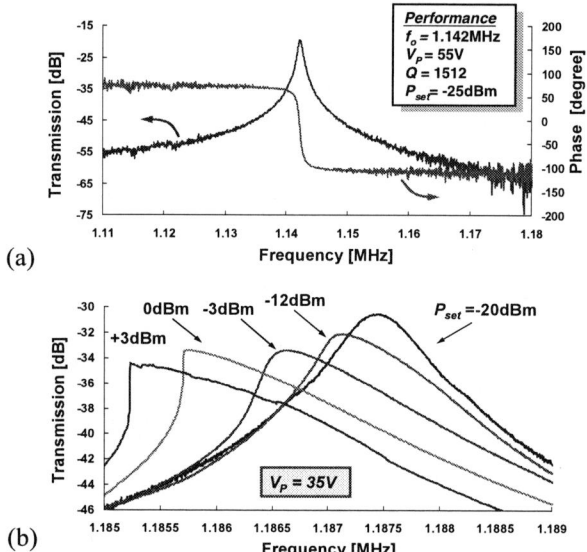

Fig. 5: Open-loop testing of the CMOS-MEMS resonator circuit. (a) Typical transmission response of the resonator with $Q > 1,500$ and stop-band rejection > 30dB. (b) Nonlinear behavior testing of the resonator under medium bias voltage ($V_P = 35$V).

IV. Experimental Results

The open-loop resonator circuit is firstly characterized in vacuum. Fig. 5(a) shows a decent transmission spectrum of the resonator circuit with $Q > 1,500$ and stopband rejection > 30 dB. The nonlinear behavior of the resonator is measured under different level of driving power (cf. Fig. 5(b)). For $V_P = 35$V, the bifurcation point (onset of nonlinearity) is measured at driving power (P_{set}) of 0 dBm.

To optimize the phase noise, the close-loop oscillator circuit is performed under several biasing conditions. Fig. 6 plots the measured PN at different offsets (f_m) with respect to V_P. In Fig. 6(a), the close-to-carrier PN ($f_m = 10$ and 100-Hz) becomes worse as V_P is greater than 35V, which is related to the increase in AM-to-PM conversion. However, the close-to-carrier PN suddenly decreases at $V_P = 68$V. It is an evident that the AM-to-PM conversion is greatly alleviated since a desirable feedback phase at a specific V_P is met. Fig. 6(b) shows the PN characteristics between 50V and 75V with fine resolution, where combination of $V_{DD} = 2.5$V and $V_C = 1.75$V is the best circuit biasing condition we found to achieve the lowest PN. Under this condition, the best-case PN is found at $V_P = 70.6$V with a notable PN performance of -77 dBc/Hz at 10-Hz offset, -97 dBc/Hz at 100-Hz offset, and -113 dBc/Hz

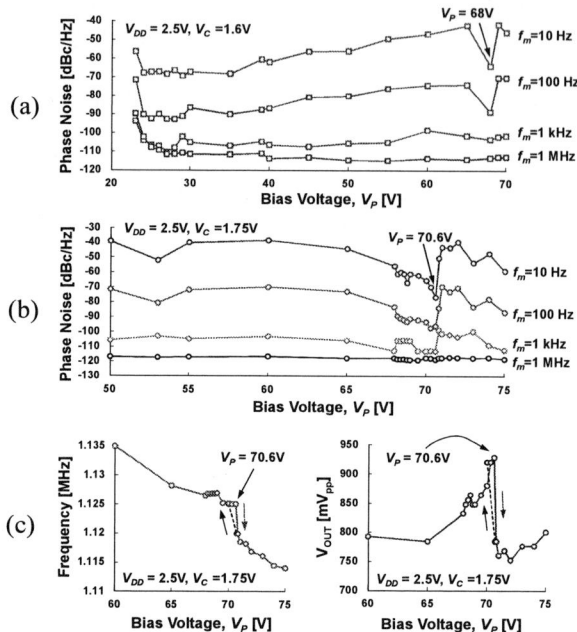

Fig. 6: Characteristic curves of the nonlinear oscillator. (a) Typical PN-V_P dependency of the nonlinear oscillator. (b) PN-V_P dependency for high-V_P. The optimized PN occurs at $V_P = 70.6$V. (c) Frequency and V_{OUT} tuning characteristics versus V_P.

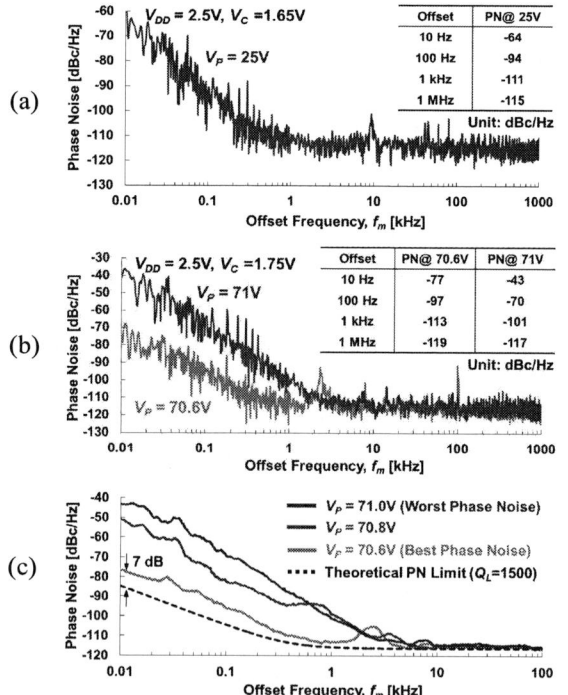

Fig. 7: Phase noise comparison of the nonlinear oscillator. (a) Phase noise performance at $V_P = 25$V (Medium-V_P). (b) Comparison of best-case and worst-case phase noise in highly nonlinear oscillator system (High-V_P). (c) Comparison with theoretical phase noise limit of a linear oscillator.

at 1-kHz offset, respectively. Note that there is a sharp transition in oscillation amplitude and frequency as $V_P > 70.6$V (cf. Fig. 6(c)). This indicates that the nonlinear behavior dynamically switches to another steady state, but

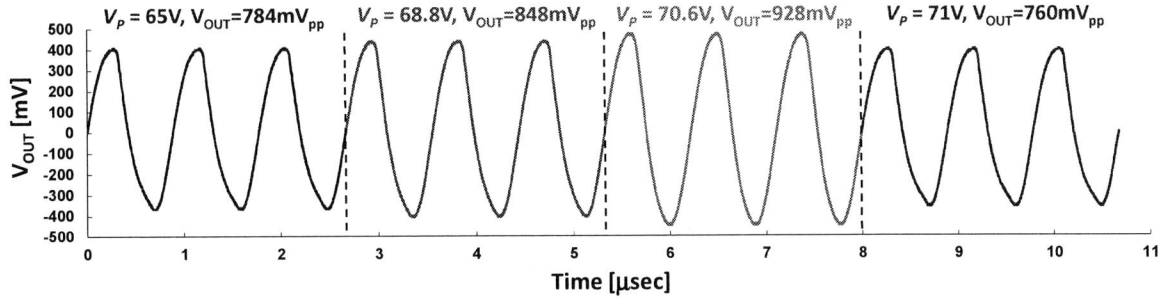

Fig. 8: Steady-state output voltage waveforms of the nonlinear oscillator using V_{DD} = 2.5V and V_C = 1.75V.

Table I: Comparison table with published state-of-the-art capacitive MEMS oscillators.

	ASFLM1 [11]	EMK23H2H [11]	JSSC'07 [7]	JSSC'04 [8]	Eurosens. XXIII [3]	IEDM'13 [1]	This Work
	Commercial Oscillator Products		Two-Chip MEMS Oscillators (w/ AGC)		Monolithic CMOS-MEMS Resonator Oscillators		
Circuit Integration	Wire-bond	Wire-bond	Wire-bond	Wire-bond	Monolithic	Monolithic	Monolithic
Resonator Material	Poly-Silicon	SCS	SCS	Poly-Silicon	Poly-Silicon	Oxide/Metal	Oxide/Metal
Resonator Type	Beam	Beam	*IBAR	Wide CC-Beam	DETF	DETF	DETF
f_o [MHz]	14.3	25	5.5	9.1	11.4	1.2	1.125
†Q	-	-	~100,000	1,036	-	1,700	1,500
P_{DC} [mW]	-	-	**1.8	0.78	4.0	1.6	1.3
PN @10Hz [dBc/Hz]	-40	-13	-66	-50	-25	-50	-77
PN @100Hz [dBc/Hz]	-78	-47	-92	-75	-80	-80	-97
PN @1kHz [dBc/Hz]	-88	-76	-112	-95	-100	-103	-113
PN @1MHz [dBc/Hz]	-116	-114	-135	-110	-107	-110	-119
FOM1 (10Hz) [dB]	-163.1	-140.9	-180.8	-169.1	-146.1	-151.5	-178.0
FOM2 (10Hz) [dB]	-	-	-178.2	-170.2	-140.1	-149.5	-176.9
FOM2 (100Hz) [dB]	-	-	-184.2	-175.2	-175.1	-159.5	-176.9

$FOM1(f_m) = L(f_m) + 20\log(f_m/f_o)$ [dB] $FOM2(f_m) = L(f_m) + 20\log(f_m/f_o) + 10\log(P_{DC}/1\text{mW})$ [dB]

*Bulk-mode resonator. **Power consumption including temperature compensation circuit. †Q-factor is extracted from linear resonators.

with poor PN performance. The oscillator performance can be recovered by reducing V_P back to 70V, and a small hysteresis zone can also be observed (ΔV_P = 0.7V, depicted as black dash lines in Fig. 6(c)). Fig. 7 compares the nonlinear oscillator performance under different biasing conditions. Compared to the medium-V_P oscillator in Fig. 7(a), biasing the resonator in its deep nonlinear regime (i.e., V_P = 70.6V, cf. Fig. 7(b)) leads to better performance at both close-to- and far-from-carrier offsets. Compared to the worst-case PN (V_P = 71V), the optimized nonlinear oscillator achieves 30 dB PN improvement at f_m = 10-Hz. The optimized PN in this work is only 7 dB away from the theoretical limit predicted by linear oscillator model (cf. Fig. 7(c)); further circuit and MEMS resonator design co-optimization is helpful for approaching this fundamental limit in future. Fig. 8 provides the steady-state oscillation waveforms for the nonlinear MEMS oscillator where V_P = 70.6V shows the most symmetric waveform with the largest V_{OUT} of 928 mV$_{pp}$.

Finally, Table I summarizes the performance of MEMS oscillators to date. To perform a fair comparison, this table is mainly based on flexural-mode capacitive resonator oscillators (except [7] is bulk-mode). The nonlinear CMOS-MEMS oscillator in this work shows a very competitive figure-of-merit (*FOM2*) at f_m = 10 and 100-Hz, which is better than that of other flexural-mode oscillators and comparable with the bulk-mode oscillator with the resonator Q > 100,000 [7].

V. Conclusions

In this work, we have experimentally demonstrated the PN tuning characteristics in a 1.12-MHz, 1.3-mW monolithic CMOS-MEMS oscillator. The optimized close-to-carrier PN of -77 dBc/Hz at 10-Hz offset and -97 dBc/Hz at 100-Hz offset is now comparable to high-Q bulk-mode oscillators, yielding a *FOM* of -176.9 dB. The PN optimization technique can be easily extended to general CMOS-MEMS oscillators serving as timing and spectral processing building blocks.

Acknowledgements. This research was sponsored by the MOST of Taiwan (MOST-103-2221-E-007-113-MY3). The chip fabrication is supported by CIC and TSMC, Taiwan.

Reference

[1] M.-H. Li *et al.*, *Tech. Dig. IEDM'13*, pp. 18.4.1-18.4.4, Dec. 2013.

[2] R. Marathe *et al.*, *J. Microelectromech. Syst.*, vol. 23, no. 3, pp. 636-650, Jun. 2014.

[3] J. L. Lopez *et al.*, *Procedia Chemistry*, vol. 1, pp. 614-617, 2009.

[4] H. K. Lee *et al.*, *Tech. Dig. Transducers'11*, pp. 510-513, Jun. 2011.

[5] L.-J. Hou *et al.*, *Proc. MEMS'12*, pp. 700-703, Jan. 2012.

[6] M.-H. Li *et al.*, *IEEE Trans. Ultrason., Ferroelect., Freq. Contr.*, vol. 59, no. 3, pp. 346-357, Mar. 2012.

[7] K. Sundaresan *et al.*, *IEEE J. Solid-State Circuits*, vol. 42, no. 6, pp. 1425-1434, Jun. 2007.

[8] Y.-W. Lin *et al.*, *IEEE J. Solid-State Circuits*, vol. 39, no. 12, pp. 2477-2491, Dec. 2004.

[9] E. Kenig *et al.*, *Phys. Rev. E.* **86**, 056207, Nov. 2012.

[10] L. G. Villanueva *et al.*, *Phys. Rev. Lett.* **110**, 177208, Apr. 2013.

[11] R. Henry and D. Kenny, *Proc. IFCS'08*, pp. 396-401, May 2008.

978-1-4799-8002-4/14 $31.00 © 2014 IEEE

High performance polysilicon nanowire NEMS for CMOS embedded nanosensors

I. Ouerghi, J. Philippe, L. Duraffourg, L. Laurent, A. Testini, K. Benedetto, A. M. Charvet, V. Delaye, L. Masarotto, P. Scheiblin, C. Reita, K. Yckache; C. Ladner, W. Ludurczak and T. Ernst.

CEA, LETI, MINATEC Campus, Université Grenoble Alpes, 17 rue des Martyrs - 38054 GRENOBLE Cedex 9, France, thomas.ernst@cea.fr

Abstract

We present for the first time sub-100nm poly-Silicon nanowire (poly-Si NW) based NEMS resonators for low cost co-integrated mass sensors on CMOS featuring excellent performance when compared to crystalline silicon. In particular, comparable quality factors (130 in the air, 3900 in vacuum) and frequency stabilities are demonstrated. The minimum measured Allan deviation of 7×10^{-7} leads to a mass resolution detection down to 100 zg (100×10^{-21} g). Moreover a novel method for in-line NW gauges factor (GF) extraction is proposed and used.

Introduction

NWs made from poly-Si have been studied recently for many applications such as: logic transistors (1), NAND flash memories (2), and biosensors (ISFET) (3). The common motivation is to provide low cost solutions for new embedded functionalities. Embedded poly-Si based NEMS can be envisioned in different levels of the CMOS process (Fig.1), from the use in the front end (4) or above IC (with specific annealing or SiGe) (5).

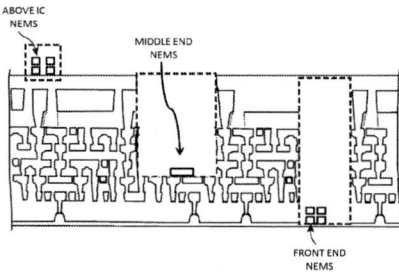

Fig.1 Schematic of the different possible levels for NEMS integration in a CMOS circuit.

However, if poly-Si is well known for MEMS accelerometers, resonators, no study exists on the behavior of such material at sub-100nm size for ultimate mass resolution. Early studies with poly-Si show good results but for piezoresistive detection, it is the first time such performances are obtained with sub 100 nm detection gauge (see Table 1). Especially, the quality factor, noise level, resistance value, gauge factor (for piezoresistive transduction), frequency stability may depend on the crystalline structure and doping of the Si NEMS. For ultimate resolution of mass detection (mass spectroscopy, gas detection) (6-7), sub- 100 nm dimension is mandatory for obtaining resolutions as fine as 1 zg with mono-crystalline silicon (c-Si) (8-9).

We propose in this paper a deep study of nano-scaled resonators by evaluating their performances for ultimate mass sensing.

Devices structure and fabrication

The crossbeam NEMS structure is constituted of a suspended P-doped piezoresistive NWs connected to a resonating lever arm in a symmetric bridge configuration (Fig.2) (10).

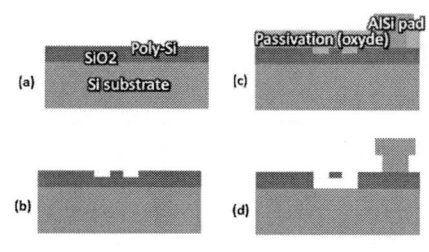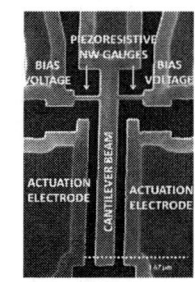

Fig.2. NEMS fabrication sequences: (a) 400nm thermal oxidation, 100nm LPCVD poly-Si deposition. (b) Etching of the silicon top layer to define the NEMS (c) Passivation, metal deposition and etching to define the metallic pads and (d) Release of the structure with a HF vapor. SEM micrograph (right) of a poly-Si NEMS made by a hybrid e-beam/DUV lithography.

Table 1. Benchmarking of poly-Si M/NEMS.

Ref.	Mat.	Transduction	Critical dimension (nm)	Thickness (nm)	F0 (MHz)	Max Q	Mass resolution	GF	df/f
(13)	Poly-Si	Capacitive	200	600	1.49	8000	n/a	n/a	n/a
(4)	Poly-Si	Capacitive	60	100	232	300	n/a	n/a	n/a
(14)	Poly-Si	Capacitive	840	5000	0.703	21	65 ag	n/a	5.35×10^{-7}
(15)	Poly-Si	Capacitive	20000	320	47.9	7980	1.5fg	n/a	1.6×10^{-5}
(17)	Poly-Si	Capacitive	350	282	22	4400	n/a	n/a	n/a
(16)	Poly-SiGe	Piezoresistive	25000	400	n/a	n/a	n/a	14	n/a
This work	**Poly-Si**	**Piezoresistive**	**80**	**100**	**40**	**3900**	**100 zg**	**33**	**<10⁻⁶**

978-1-4799-8002-4/14 $31.00 © 2014 IEEE

The differential bridge architecture provides intrinsic signal amplification and the background signal suppression. We have already demonstrated that the detection through c-Si gauges with this configuration has a large signal to noise ratio at room temperature (dynamic range of 100 dB) (10). The fabrication process is summed up in Fig.2. The lever arm and the lateral gauges are structured in a 100nm LPCVD poly-Si layer above a 400nm thermal oxide buffer. The silicon layer is implanted with Boron dopant atoms with 10^{19} cm^{-3} and 10^{20} cm^{-3} dopant atoms concentrations. Different poly-Si layer and doping level are compared: Those layers were characterized by XRD (Fig.3) to extract the preferential orientations and the average grain sizes, and a TEM cross section image is presented. The different samples are: a LPCVD columnar poly-Si ((220) weak texture) called PolyA, a (111) oriented LPCVD poly-Si called PolyB, (111) and (311) oriented LPCVD poly-Si called PolyC, and mono crystal silicon (SOI wafer) called c-Si. Average grain size is different for each orientation, but never exceeds 80nm. Table 2 presents the process conditions and doping concentrations as well as the deposition temperature and pressure. Electrical wires and pads are made with an aluminum silicide (AlSi) metallic layer. The structure is then released by vapor hydrofluoric acid (HF) etching.

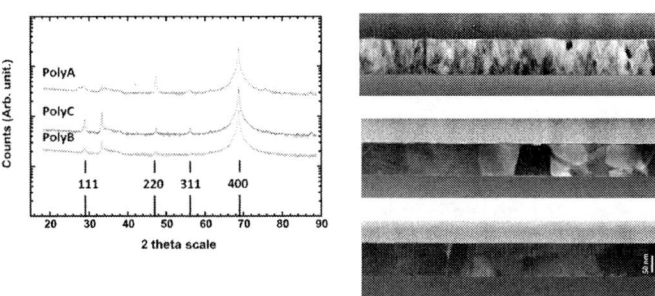

Fig.3. XRD characterization (left) of several poly-Si depositions. The 69° peak is the bulk orientation, some deposition type exhibit mostly a preferential (111), (220) or (311) orientation. TEM micrograph cross-sections (right) of 100 nm poly-Si layers above thermal oxide. Abrupt lines show the grain boundaries regions. From the top to the bottom: PolyA, PolyB and PolyC layer.

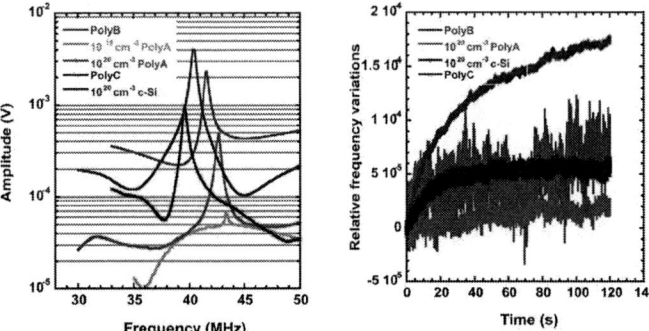

Fig. 4. Resonance peak in air (left) for different poly-Si and c-Si NEMS. Only NEMS with high dopant concentration show a high quality factor. Resonance frequency variation as a function of the time (right). Lower frequency stability is observed for poly-Si, but the c-Si exhibit a larger drift for long times.

Electrical performances

We first performed a resonance analysis of the samples. We measured the electromechanical response at high frequency (HF) of every device (poly-Si and c-Si) in an open loop scheme with a downmixing method (Fig.4) (11). The highest dopant concentration (10^{20} cm^{-3}) exhibited the best electromechanical performances in terms of amplitude at resonance. Although the gauge factor becomes low at high dopant concentration (well known for c-Si), the signal amplitude decrease may be compensated by the low gauge resistance. Indeed the system input impedance becomes comparable to the NWs impedance. When we compare the resonance frequency peaks between c-Si and poly-Si NEMS, we find that despite a slight reduction of the amplitude, the signal to background ratio (SBR) is conserved i.e. about 20 dB for PolyA and c-Si. We observe the same for the quality factors with a value of 130 both for PolyA and c-Si. To follow their resonance frequency variation, the nano-systems are embedded in a phase-locked loop (PLL). Fig.4 shows the resonance frequency variation as a function of the time. The curve widths illustrate the frequency stability and allow estimating the limit of detection in mass. The c-Si NEMS curve is the thinnest among all samples. However, a large drift appears through long times (seconds) more particularly for c-Si sample. It might be attributed to a temperature variation in NWs (c-Si NWs have the lowest resistance), but it does not perturb frequency stability measurement. Indeed the electrical power to thermal conductance ratio leads to a maximum self-heating effect of 50° C (Temperature computed by Finite Element Method with Comsol Multiphysics® for a bias voltage of 2V).

Table 2. Main process conditions for the poly-Si layers.

	Doping level	Deposition temperature (°C)	Deposition pressure (Torr)
PolyA	P-doped 10^{19}cm^{-3} and 10^{20}cm^{-3}	620	0.2
PolyB	n-doped 2.9 ×10^{20}cm^{-3}	580	0.175
PolyC	n-doped 7 ×10^{19}cm^{-3}	580	0.375

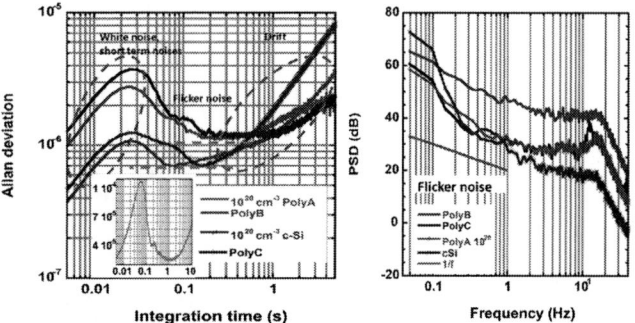

Fig.5. Allan deviation (left) as a function of integration time for different poly-Si and c-Si NEMS. At 100 ms, best poly-Si NEMS and c-Si are almost similar. At low integration time, the white noise is mostly due to the Johnson noise. Inset shows the result for a PolyA NEMS with lower doping level. PSD (right) as a function of frequency for different poly-Si and c-Si NEMS. 1/f noise is observable between 0.1 and 10 Hz.

978-1-4799-8002-4/14 $31.00 © 2014 IEEE

Moreover we focus on lower integration time (100 ms in the case of gas sensor application). Fig.5 reports the results of Allan deviation (12) and power spectral density measurement. We found an Allan deviation minimum below 10^{-6} for some poly-Si NEMS and c-Si at 100 ms. The Allan deviation is computed from the frequency data (shown in Fig.4). The higher values of Allan deviations within low integration times are attributed to short term noises or white noises such as the Johnson noise. This bandwidth does not concern also applications as mass spectrometry or gas sensor for example. Inset of Fig.5 (left) shows the 10^{19} cm^{-3} PolyA. The Allan deviation is one order of magnitude higher than other poly-Si NEMS. That can be explained by a low ohmic contact between pad and active layer and also a signal loss due to the input impedance (much lower than NW impedance for 10^{19} cm^{-3} PolyA). At high integration time, we measure a higher drift for c-Si NEMS, Allan deviation is better for poly-Si NEMS. The global degradation of Allan deviation for poly-Si is due to a higher 1/f noise when compared to c-Si NEMS. 1/f noise (represented by a constant value in Allan deviation figure or a 1/f trend in PSD figure) is quite similar for each poly-Si layer excepted for the low doping level. This result gives us supplementary information about flicker noise origins. As showed in TEM photos and XRD analysis, the poly-Si structures are different (texture and grain size) but the impact of quality layer on Flicker noise seems to be small. We studied the gauge temperature dependence (TCR) as shown in Fig.6. We found opposite values of TCR in nanowire gauge measurement, -1147 and 895 for PolyA and c-Si respectively.

This positive TCR of c-Si could be an explanation of higher frequency drift compared to poly-Si. To conclude the comparison p-/c-Si, thanks to the frequency stability study and resonance peak measurement, we demonstrate a quite equivalent SBR and stability (20 dB and below 10^{-6} respectively). To complete the study, resonance (Fig.7) and frequency stability measurement (Fig.8) in vacuum was achieved (useful for mass spectrometry). A quality factor value of 3900 and 4500 for poly- and c-Si respectively was found. About the Allan deviation, the vacuum allows to get rid of white noise more particularly at low integration time. Flicker noise remains even in vacuum, poly-Si exhibits equivalent minimum Allan deviation compared to in air.

In line NW gauge factor method and results

In order to understand why the amplitude at resonance is different between poly-Si NEMS and c-Si, we developed a new method for GF extraction. Unlike some conventional methods as the four-point bending measurement, this new method enables to measure directly the GF value of a NEMS. To extract the GF, the resistance variation of the gauges is measured (Fig.9).

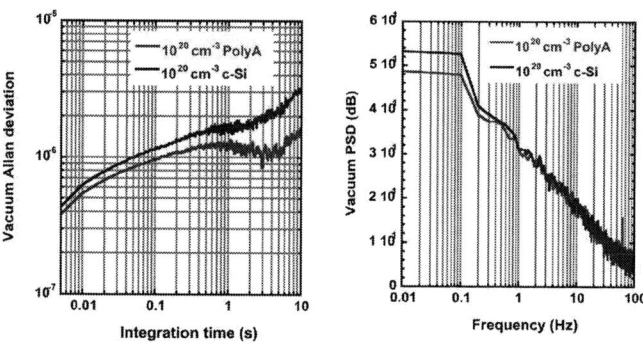

Fig. 8. Electromechanical response deviation as a function of integration time in vacuum (left). PSD as a function of frequency in vacuum (right).

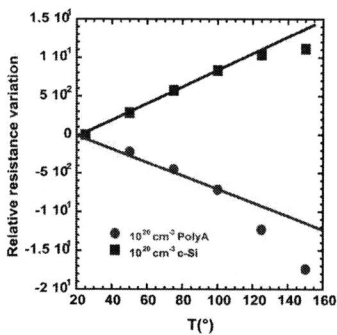

Fig.6. Relative resistance variation as a function of temperature fitted in linear regime. The trends are opposite. The extracted TCR are,-1147 and 895 for PolyA and c-Si respectively.

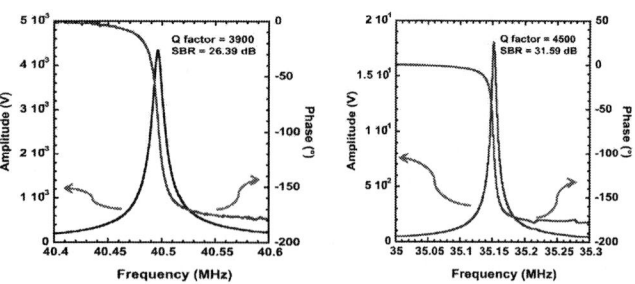

Fig.7. Vacuum resonance peak of a PolyA NEMS (left). Vacuum resonance peak of a c-Si NEMS (right).

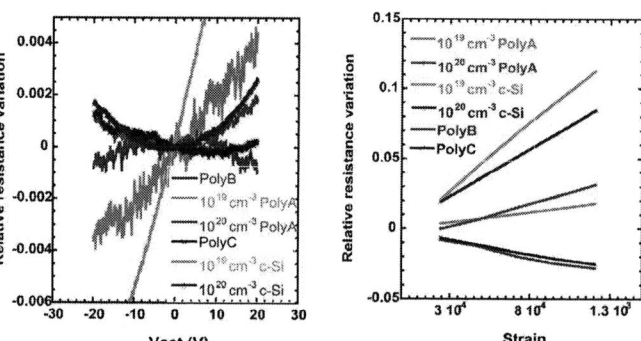

Fig.9. New GF extraction method (left). The resistance variation as a function of the actuation voltage, a quadratic relationship is expected. Only one NW gauge is polarized at 0.25 V. A voltage shift appears due to a field effect between NW and actuation. Four point bending measurement (right). Increase lines coresspond to a p doped layer, decrease lines to a n doped layer.

Then thanks to a finite element simulation (Fig.10), the induced stress in NW for any actuation voltage is known. The comparison between quadratic coefficients from experimental and simulation data enables to extract the GFs. The results are then compared with the conventional method as shown inFig.9. Table 3 summarizes every extracted GF and shows that the two methods provide compatible results. For polyA, it must be noted that the higher is the doping level, better is the GF. This variation is opposite to the well known trend of c-Si where GF decreases with the doping level. We could make the most of this result to raise the doping level, including an increase in the amplitude signal, a diminution of the noise and still keep a correct value of the GF.

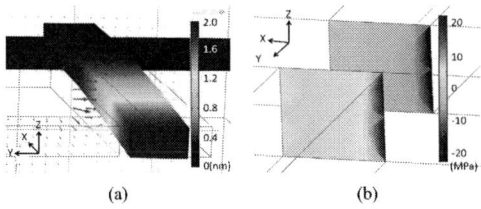

(a) (b)

Fig.10. (Color) Electromechanical simulation of the actuated NEMS for an actuation voltage of 25 V. (a) The color legend and red arrows represent the beam deflection and the electric field lines respectively. (b) Simulated gauge stress on the surfaces between the gauges and the beam. Negative and positive values represent a tensile and compressive stress respectively.

Table 3. Comparison between extracted GFs with the conventional method and the new method. Results are in good agreement, we found an increase of GF as a function of the doping level for PolyA (opposite trend of c-Si).

	PolyB	PolyA		PolyC	Single crystal silicon	
Doping level	$2.9\ 10^{20}cm^{-3}$	10^{19} cm^{-3}	10^{20} cm^{-3}	$7\ 10^{19}cm^{-3}$	10^{19} cm^{-3}	10^{20} cm^{-3}
Nanowire Resistivity [mOhm.cm]	3.53	1580	29,8	27.9	43.2	5.65
GF from conventional method	22	15	33	19	95	68
GF from the new method	18	13	32	15	98	53

Conclusion

In the high importance issue of low cost NEMS-CMOS co-integration, we studied several poly-Si deposition types for NEMS applications dedicated to mass sensing. For every high doped layer based poly-Si NEMS, the devices exhibit a good quality factor and frequency stability, those features are as competitive as c-Si NEMS ones. We also found that poly-Si does not suffer from frequency drift, this might be due to its higher resistance and negative TCR. At a high doping level, the poly-Si GF does not decrease with the doping level (and the Johnson noise is reduced). For all these reasons, because of performances close to c-Si, the poly-Si appears as an excellent low cost candidate for transduction. Finally mechanical properties have been studied thanks to a new GF extraction method, which is directly coming from NEMS output. This allows a promising direct measurement of the GF at the wafer level which would be suitable for the systematic characterization of NEMS devices integrated at large scale (LSI).

Acknowledgements

This work was supported by the European Research Council, Grant No. 240382 – DELPHINS project

(1) S. D. Suk, M. Li, Y. Y. Yeoh, K H. Yeo, "Characteristics of sub 5nm tri-gate nanowire MOSFETs with single and poly Si channels in SOI structure", *Symposium on VLSI Technology*, pp. 142-143, June 2009.

(2) S. H. Chen, H. T. Lue, Y.-H. Shih, C. F. Chen, T. H. Hsu, Y. R. Chen *et al.*, "A highly scalable 8-layer Vertical Gate 3D NAND with split-page bit line layout and efficient binary-sum MiLC staircase contacts", *IEEE International IEDM 2012*, pp. 231-234, December 2012.

(3) C. H. Lin, C. H. Hung, C. Y. Hsiao, H. C. Lin, F. H. Ko, Y. S. Yang, "Poly-silicon nanowire field-effect transistor for ultrasensitive and label-free detection of pathogenic avian influenza DNA", *Biosensors and Bioelectronics*, vol. 24, pp. 3019-302, June 2009.

(4) J. L. Munoz-Gamarra, P. Alcaine, E. Marigo, J. Giner, A Uranga, J. Esteve *et al.*, "Integration of NEMS resonators in a 65 nm CMOS technology", *Microelec. Eng.*, vol. 110, pp. 246-249, October 2013.

(5) P. Gonzalez, L. Haspeslagha, K. de Meyer, A. Witvrouw, "Evaluation of the piezoresistive and electrical properties of polycrystalline silicon-germanium for MEMS sensor applications", *2010 IEEE 23rd International Conference on MEMS*, pp. 580-583, January 2010.

(6) J. Arcamone, M. Savoye, G. Arndt, J. Philippe, C. Marcoux; E. Colinet *et al.*, "VLSI platform for the monolithic integration of single-crystal Si NEMS capacitive resonators with low-cost CMOS", *IEEE International IEDM 2012*, pp. 15.4.1-15.4.4, December 2012.

(7) E. Ollier, C. Dupré, G. Arndt, J. Arcamone, C. Vizioz, L. Duraffourg, *et al.*, "Ultra-scaled high-frequency single-crystal Si NEMS resonators and their front-end co-integration with CMOS for high sensitivity applications", *2012 IEEE MEMS*, pp. 1368-1371, January 2012.

(8) M.S. Hanay, S. Kelber, A.K. Naik, D. Chi, S. Hentz, E.C. Bullard *et al.*, "Single-protein nanomechanical mass spectrometry in real time", *Nature Nanotechnology*, vol.7, pp. 602-608, September 2012.

(9) T. Ernst, L. Duraffourg, C. Dupré, E. Bernard *et al.*, "Novel Si-based nanowire devices: Will they serve ultimate MOSFETs scaling or ultimate hybrid integration?", *IEDM*, pp. 1-4, December 2008.

(10) E Mile, G Jourdan, I Bargatin, S Labarthe, C. Marcoux, *et al.*, "In-plane nanoelectromechanical resonators based on silicon nanowire piezoresistive detection", *Nanotec.*, vol. 21 165504, March 2010.

(11) I. Bargatin, E.B. Myers, J. Arlett, B. Gudlewski, M.L. Roukes, "Sensitive detection of nanomechanical motion using piezoresistive signal downmixing", *Applied Physics Letters*, vol 86, 13, March 2005.

(12) E. Sage, O. Martin, C. Dupre, T. Ernst, G Billiot, L. Duraffourg *et al.*, "Frequency-addressed NEMS arrays for mass and gas sensing applications", *Transducers & Eurosensors*, pp. 665-668, June 2013.

(13) J. Arcamone, J. M.A.F. van den Boogaart, F. Serra-Graells, S. Hansen, J. Brugger, F. Torres *et al.*," Full wafer integration of NEMS on CMOS by nanostencil lithography", *IEDM '06*, pp. 1-4, December 2006.

(14) J. Verd, Bellaterra, G.Abadal, J. Teva; M.V. Gaudo, A. Uranga, X. Borrise, "Design, fabrication, and characterization of a submicroelectromechanical resonator with monolithically integrated CMOS readout circuit", *JMEMS*, vol 14 pp. 508-519, June 2005.

(15) M.K. Zalalutdinov, J.D. Cross, J.W. Baldwin, B.R. Ilic, W. Zhou, B.H. Houston, "CMOS-Integrated RF MEMS Resonators" *JMEMS*, vol 19, pp. 807-815, August 2010.

(16) P. Gonzalez, P, L. Haspeslagha, S. Severi, *et al.*,"Piezoresistivity and electrical properties of poly-SiGe deposited at CMOS-compatible temperatures", *ESSDERC*, pp. 476–479, September 2010.

(17) J.L. Lopez, J. Verd, J. Giner, A Uranga, G. Murillo, E. Marigo *et al.*, "High Q CMOS-MEMS resonators and its applications as RF tunable band-pass filters", *International Solid-State Sensors, Actuators and Microsystems Conference. Transducers 2009*, pp. 557-560, June 2009.

Integration of RF MEMS resonators and phononic crystals for high frequency applications with frequency-selective heat management and efficient power handling

Humberto Campanella*[1], Nan Wang[1], Margarita Narducci[1], Jeffrey Bo Woon Soon[1], Chong Pei Ho[2], Chengkuo Lee[2], Alex Gu[1]

[1] Institute of Microelectronics, A*STAR (Agency for Science, Technology and Research), 11 Science Park Road, Singapore 117685
[2] Department of Electrical and Computer Engineering, National University of Singapore, 4 Engineering Drive 3, Singapore 117576
*Email: campanellaph@ime.a-star.edu.sg, Tel: +65 6770 5616, Fax: +65 6773 1914

Abstract

We report a radio frequency micro electromechanical system (RFMEMS) device integrated with phononic crystals (PnC) that provide a Lamb-wave resonator with frequency-selective heat management, power handling capability, and more efficient electromechanical coupling at ultra high frequency (UHF) and low microwave bands. The integrated device is fabricated in a silicon-on-insulator (SOI) aluminum nitride (AlN) platform and boosts thermal performance by 40%, power handling by 3 dB, and coupling coefficient by three times. Design approach is scalable to higher frequencies.

Introduction

Existing technologies for mobile wireless communications deal with power handling restrictions to realize filters and duplexers for ultra high frequency (UHF) and low microwave frequencies. It is assumed filters should withstand power levels of +30 dBm (or 1 W) for radio frequency RF filters based on acoustic-wave, thin-film MEMS resonators. As building cell of RFMEMS applications, their thin-film structure limits the power handling and restricts the heating dissipation capability.

To tackle this issue, designers have played with the stack design, electrode interconnection scheme of filter resonators, and splitting of filter-composing resonators, among other strategies [1]. Process modifications can also address the issue by reducing the acoustic impedance of electrodes [2], and by controlling the dispersion of acoustic waves to reduce vibrations at resonance [3]. Most of these techniques are used in commercial filters and duplexers, and are combined with others like cascading and anti-parallel connection to cancel harmonic waves [4-5]. Acoustic reflectors are another feature that contributes to improvement of the quality factor (Q) of acoustic wave resonators [6]. Bulk acoustic wave resonators (BAW) of the Solidly Mounted Resonator type (SMR) stack Bragg reflectors just below the resonator area [7]. In surface acoustic wave resonators (SAW), designers employ metallic reflectors to concentrate the lateral waves so as to improve the Q factor [8].

On the other hand, phononic crystals (PnC) are structures with acoustic wave band gaps, local resonance, negative refraction, sound focusing and thermal cloaking capabilities [9]. The integration of PnC and thin-film piezoelectric resonators has proved to enhance the quality (Q) factor while reducing die area [10], and to reduce the anchor and support loss for quartz resonators [11].

The main motivation of this work is thus to deliver RFMEMS resonators that implement a practical power and thermal management strategy for miniaturized RF front-end modules suitable to microwave frequencies. Our approach combines the benefits of MEMS and PnCs to deliver a composite device that has increased electromechanical coupling coefficient by three times. As a consequence of this approach, our RFMEMS resonator features thermal management and power handling that are 40% and 3 dB better than the stand-alone MEMS without PnC. Another major design contribution is that the MEMS resonator has an efficient footprint design that requires no extra area for the PnCs.

Device Concept and Fabrication

Fig. 1 shows the device concept, wherein the active layer of the acoustic resonator and the PnC array are built using a common material. Outbound acoustic waves in the plane of the plate are confined by PnC-enhanced reflection back to the resonator, thus increasing the electromechanical conversion efficiency.

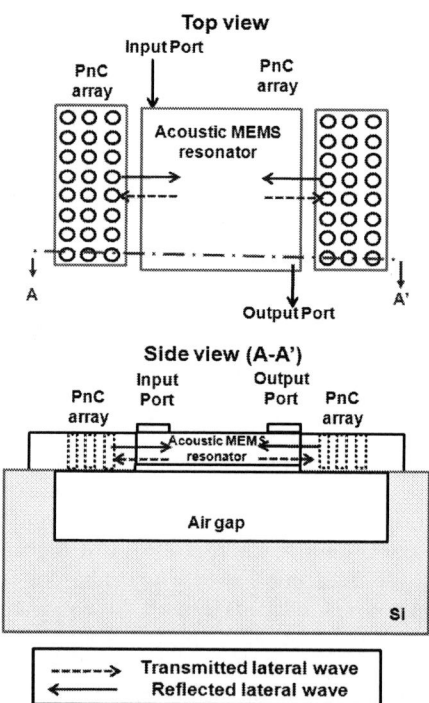

Figure 1: Working principle of the composite MEMS-plus-PnC resonator device, wherein PnC arrays co-located to the resonator act as acoustic reflectors of lateral Lamb waves.

The resulting device exhibits acoustic cloaking capability that confines the Lamb acoustic wave while dissipating heat at the target resonance frequency: As a phononic bandgap structure, the PnC array works as a band-rejection filter [12] for the outbound lateral acoustic waves which are reflected back to the resonator core, thus boosting the effective electromechanical coupling and power handling figures. At the same time, the PnC array allows more efficient heat management since heat is released at the operating frequency [13]. The design re-uses the release window area to co-locate the PnC array, allowing release etching at the same time along with PnC patterning. The design is modular and compatible with existing MEMS resonator designs, only integration of customized PnC to the resonator's frequency being required.

The integrated device is fabricated in a silicon-on-insulator (SOI) aluminum nitride (AlN) platform, according to the process flow shown in **Fig. 2**. The resonator uses AlN as structural layer with molydbdenum (Mo) checkered electrodes [14]. Same slab of AlN is used to create the PnC array that surrounds the resonator area.

Starting from SOI substrate with SiO$_2$ deposition, process flow is as follows: (a) AlN seed layer and bottom Mo electrode layer and bottom Mo electrode patterning with planarization of overfilled SiO$_2$; (b) AlN patterning with planarization of overfilled SiO$_2$; (c) Top Mo electrode patterning; (d) SiO$_2$ passivation with contact via defined and metallization; and (e) Release holes definition and release in XeF$_2$. Patterning of AlN on the PnC region creates a suspended membrane with thru-holes whose dimensions are designed to provide a bandgap centered at the resonance frequency of the fundamental symmetric Lamb wave mode (S0) of the resonator. A realization of the composite device features the PnC arrays co-located to the MEMS release windows, so as to improve the electrical and thermal performance of the resonator, like the device show in the right-hand optical microscope picture of **Fig. 3**. The impact in footprint is negligible, as we compare the composite with a similar stand-alone resonator without PnC arrays (left image in **Fig. 3**).

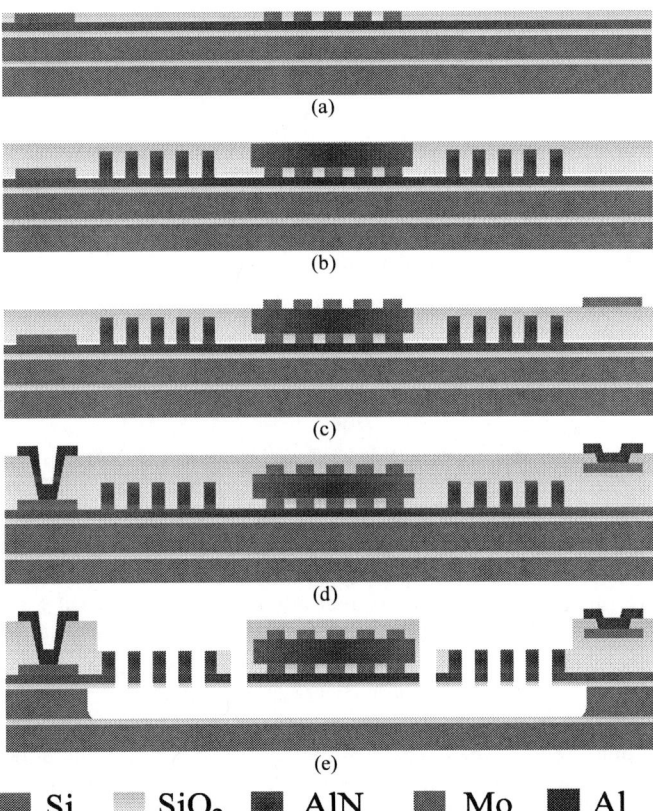

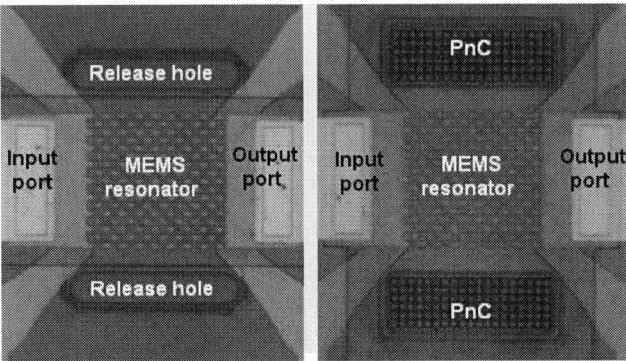

Figure 3. Implementation of integrated AlN MEMS Lamb-wave resonator and Phononic Crystal arrays: Non-PnC implementation (left) and composite device with co-located PnC arrays (PnCs are located at the same release holes area)

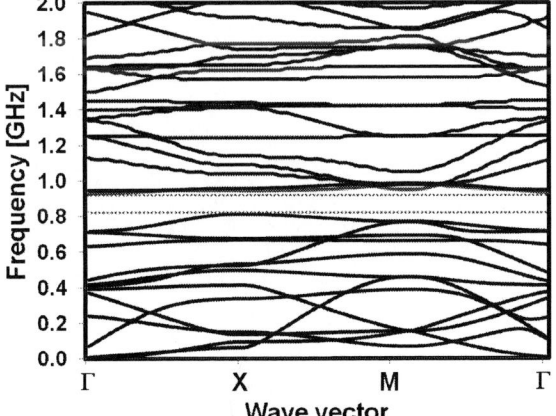

Figure 4: Acoustic cloaking design of PnC arrays for 850 MHz resonators: Brillouin diagram shows a 100-MHz band gap between 0.8 and 1.0 GHz. This indicates no acoustic modes can exist in the PnC within the frequency range of the band gap. Therefore, acoustic waves transmitted through the PnC in this frequency range will be suppressed and reflected back to the RFMEMS resonator.

■ Si   ■ SiO$_2$ ■ AlN   ■ Mo   ■ Al

Figure 2. Fabrication process flow of the integrated MEMS resonator and phononic crystal array (PnC): (a) Starting from SOI substrate with SiO$_2$, AlN seed layer and bottom Mo electrode layer and bottom Mo electrode patterning with planarization of overfilled SiO$_2$. (b) AlN patterning with planarization of overfilled SiO$_2$. (c) Top Mo electrode patterning. (d) SiO$_2$ passivation with contact via defined and metallization. (e) Release holes definition and release in XeF$_2$.

978-1-4799-8002-4/14 $31.00 © 2014 IEEE

The Brillouin diagram in **Fig. 4** shows the PnC array is designed to feature acoustic cloaking for 850-MHz resonators. The diagram shows a 100-MHz band gap between 0.8 and 1.0 GHz. This indicates no acoustic modes can exist in the PnC within the frequency range of the band gap. Therefore, acoustic waves transmitted through the PnC in this frequency range will be suppressed and reflected back to the RFMEMS resonator. The consequence of having more mechanical energy reflected to the resonator is the electromechanical coupling coefficient (k^2) is improved.

Thermal characterization

Fourier transform infrared (FTIR) spectroscopy was used to study reflectance of near IR spectra of the composite. To do so, we first scanned large areas of the device and identified the spectral components of the composite that are sensitive to heating, and more specifically those device regions where stack comprises Mo/AlN/Mo. We also carried out temperature calibration to extract the frequency sensitivity to temperature of the composite, by placing sample devices with and without PnCs in a temperature-controlled chamber inside the FTIR tool and sweeping the temperature to extract the slope of wavelength shifting over temperature [nm/°C]. Calibration will give us a reference of the total temperature change later on when we carry out power testing, so we can extract the power-induced heating of devices. **Fig. 5** shows an IR reflectance topography contour plot at a specific wavelength of 3.1 μm, which we found to be sensitive to heating for the Mo/AlN/Mo device stack. Picture shows device area (in blue), Mo interconnect lines (yellow to red), and oxide on substrate (cyan to green). **Fig. 6** shows a typical example of the comparative thermal performance between PnC composite vs. non-PnC devices after sweeping power from +20 to +35 dBm. The composite PnC+MEMS device happens to have 40% better thermal management or, in other words, to present lesser IR peak shifting within the observed range.

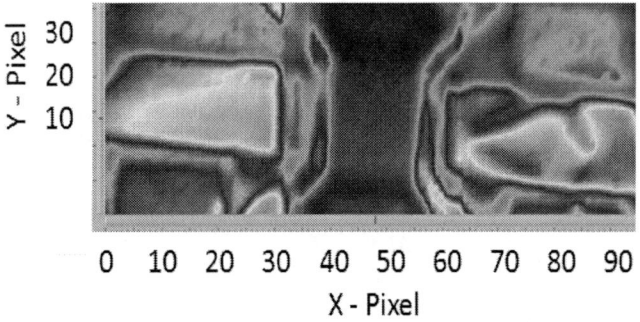

Figure 5: Thermal characterization of PnC vs. Non-PNC devices using near infrared (IR) reflectoscopy of PnC and non-PNc devices: IR reflection intensity topography is shown at the peak wavelength of 3.1 μm.

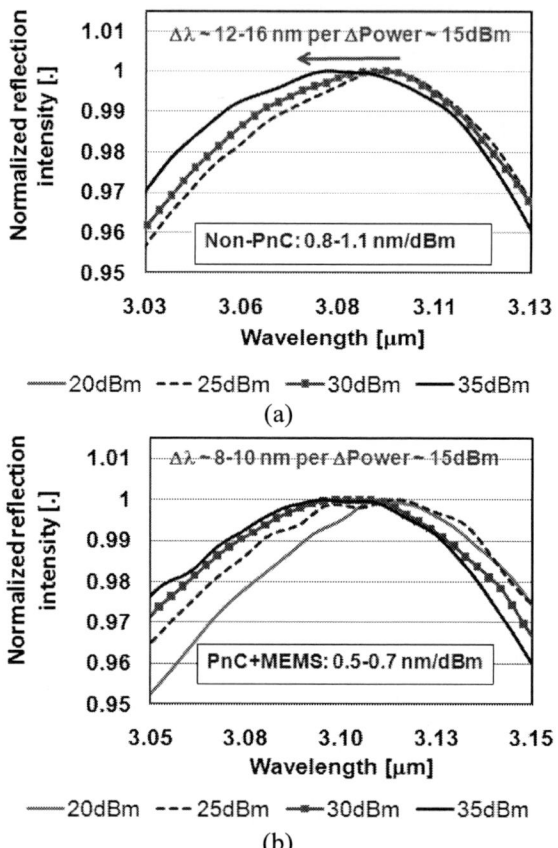

(a)

(b)

Figure 6: Thermal performance of PnC vs. Non-PNC devices using near infrared (IR) reflectoscopy: (a) IR spectral profile of non-PnC device at the MEMS resonator area (spectrum detailed around the 3.1 μm peak); (b) IR profile of co-located PnC device at same resonator point and wavelength showing reduced heating-driven shifting for the PnC device. This demonstrates the PnC-integrated MEMS resonator has > 40% better thermal management (IR shift ~ 0.5-0.7 nm/dBm).

Electrical Characterization

Power testing involved analysis of gain compression point (P1dB) and intermodulation intercept points of the third and second order (IIP3 and IIP2, respectively). We carried out analyses for both the composite and the stand-alone resonator. First, we swept the input power at one of the resonator ports from low to high power near +30 dBm to observe the power level when devices started compressing the signal, so as to find the P1dB value. To do so, we used a signal generator and a low noise amplifier to deliver the exciting signal to the resonator, and a spectrum analyzer connected to the resonator's output port to extract the actual power level at that port. **Fig. 7** showed two plots comparing the results of this test. PnC+MEMS showed typical compression points around +27 dBm, which resulted 3dB better than for non-PnC devices.

978-1-4799-8002-4/14 $31.00 © 2014 IEEE   568

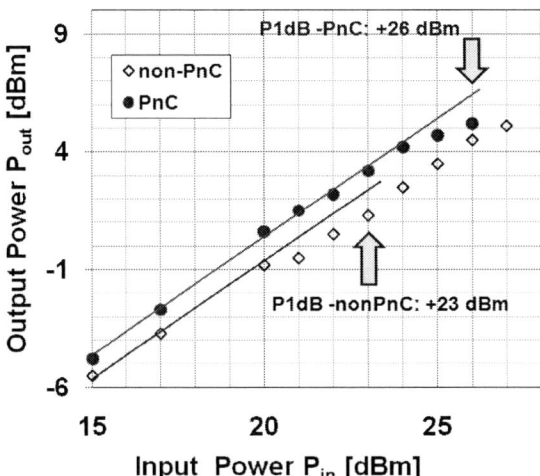

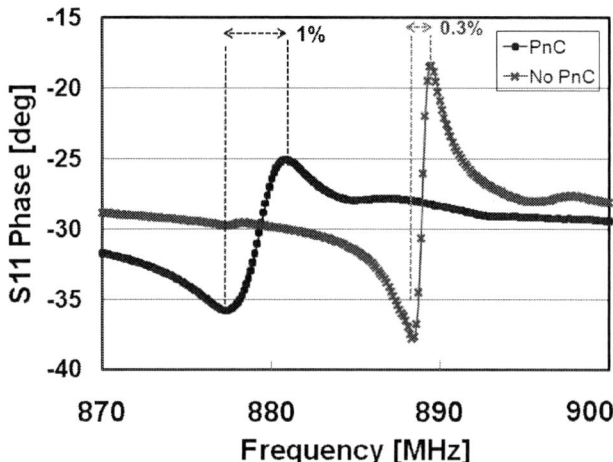

Figure 7: Comparative power linearity testing plots to extract the gain compression point values (P1dB) of devices with and without PnCs. A minimum gain of +3 dB in the P1dB is attained with PnCs

Figure 8: Electromechanical frequency response: PnC and non-PNc devices

Conclusion and Future Work

Integration of RF MEMS resonators and PnCs enable more efficient thermal management and power handling. Design approach is scalable, which put RFMEMS resonators in an advantageous position to push the next generation of high power, high frequency radio frequency (RF) applications.

Future developments of this work will include MEMS geometry optimization towards power handling maximization, based on thermal profiles vs. geometry mapping.

A second setup was employed to analyze the intermodulation products, wherein the signals at the output ports of two signal generators are combined and filtered and delivered to the resonator's input port. The frequency separation between the two signals is 100 kHz, one above and the second one below the resonance frequency of the device. **Table I** shows the analyses results for two lots of devices. PnC+MEMS composites exhibit IIP3 points that are at least +4 dB better than control devices without PnCs. IIP2 values are consistently better +2 dB than non-PnC control devices. As composite devices keep cooler at high power, compared to stand-alone MEMS devices, intermodulation products start to show up at higher power levels, thus causing the intercept points to be higher.

Last but not least, electromechanical testing showed the effective coupling coefficient of the resonators is also boosted up to three times for composite devices due to the acoustic wave trapping feature of the PnC arrays surrounding the MEMS device, boosting values from 0.3% to 1% (**Fig. 8**).

References

(1) R. S. Ketcham, "Optimized piezoelectric resonator-based networks", US patent 5,231,327, 1993.
(2) Ylilammi, "Thin-film bulk acoustic resonator with enhanced power handling capability," US patent 6,515,558, 2003.
(3) H. Heinze, P. Tikka, and E. Schmidhammer, "Bulk acoustic wave resonator filter," US patent 7,961,066, 2011.
(4) R. Aigner and M. Handtmann, "BAW apparatus," US patent 7,365,619, 2008.
(5) T. Jamneala and P. Bradley, "BAW filter having reduced second harmonic and method of reducing second harmonic generation in a BAW filter," US patent 7,548,140, 2009.
(6) R. H. Olsson, J. G. Fleming, and M. R. Tuck, "Contour mode resonators with acoustic reflectors," US patent 7,385,334, 2008.
(7) K. Umeda, "Piezoelectric resonator including an acoustic reflector portion," US patent 7,868,519, 2011.
(8) C. K. Campbell, "Applications of surface acoustic and shallow bulk acoustic wave devices," *Proceedings of the IEEE* 77, 1453-1484, 1989.
(9) M. Maldovan, "Sound and heat revolutions in phononics," *Nature* **503**, 209-17, 2013.
(10) C.-Y. Huang, J.-H. Sun, and T.-T. Wu, "A two-port ZnO/silicon Lamb wave resonator using phononic crystals," *Applied Physics Letters* **97**, 031913, 2010.
(11) C. H. Hung et al., "Design and fabrication of an AT-cut quartz phononic Lamb wave resonator," *J. Micromech. Microeng.* **23**, 065025, 2013.
(12) N. Wang et al., "Experimental Investigation of a Cavity-Mode Resonator Using a Micromachined Two-Dimensional Silicon Phononic Crystal in a Square Lattice," *IEEE Electron Dev. Lett.* **32**, pp. 821-3, 2011.
(13) M. Maldovan, "Narrow Low-Frequency Spectrum and Heat Management by Thermocrystals," *Phys. Rev. Lett.* **110**, 025902, 2013
(14) H. Campanella, L. Khine, J.M. Tsai, "Aluminum nitride Lamb-wave resonators for high power high frequency applications," *IEEE Electron Dev. Lett.* **34**, pp. 316-318, 2013.

TABLE I: Comparative intermodulation products of the third and second order of devices with and without PnCs (two lots analyzed). The PnC-improved devices have IIP3 points at least +4 dB better than control devices without PnCs. IIP2 values are consistently better +2 dB than non-PnC control devices

Lot	Device	IIP3 [dBm]	IIP2 [dBm]
Lot1	**PnC**	**+51**	**+82**
	Non-PnC	+47	+80
Lot2	**PnC**	**+46**	**+72**
	Non-PnC	+41	+70

A Monolithic 9 Degree of Freedom (DOF) Capacitive Inertial MEMS Platform

Ilker E. Ocak*, Daw D. Cheam, Sanchitha N. Fernando, Angel T.H. Lin, Pushpapraj Singh,
Jaibir Sharma, Geng L. Chua, Bangtao Chen, Alex Y.D. Gu, Navab Singh and Dim-Lee Kwong

Institute of Microelectronics, A*STAR (Agency for Science, Technology and Research)
11 Science Park Road, Singapore Science Park II, Singapore 117685
*Email: ocakie@ime.a-star.edu.sg, Tel: +65 6770 5480, Fax: +65 6773 1914

Abstract

A monolithic 9 degree of freedom capacitive inertial MEMS platform is presented in this paper. This platform for the first time integrates 3 axis gyroscopes, accelerometers, and Lorentz Force magnetometers together on the same chip without using any magnetic materials. This reduces the assembly cost, and fully eliminates the need of magnetic material processing and axis misalignment calibration. The fabricated sensors, vacuum packaged (vacuum ~100mTorr) at wafer level with epi-polysilicon through silicon interposer (TSI) wafer using eutectic bonding, performed within 10% of the simulation results.

Introduction

Inertial Measurement Unit's (IMU) are widely used for many applications in our daily lives to detect the location, orientation and motion of objects or platforms. Most of the IMU's currently available in the market are integrating three axis accelerometers and gyroscopes on the same platform. 3axis magnetometers are attached to these combo sensors as a separate die to build a 9DOF IMU. This method of integration increases the packaging cost and introduces axis misalignment error to the system which needs to be cancelled with further calibration and digital compensation methods. To reduce these additional costs most manufacturers are looking for a method to integrate 9DOF inertial sensors monolithically (1). The novel technology platform presented in this paper achieves this target successfully.

Design of the Platform

The most important challenge of building a 9DOF inertial MEMS combo is monolithic integration of magnetometers with other sensors. Current magnetometers available in the market are mostly using magnetic materials and these materials require dedicated tools to process. Being non-CMOS compatible is another drawback of the magnetic materials. Lorentz force type capacitive magnetometers are preferred for not requiring any magnetic material, being foundry friendly (CMOS compatible) and can achieve similar noise and dynamic range performance with other technologies (2). Langfelder et. al. mentioned that a Lorentz Force magnetometer having ten multi-paths can surpass in performance all of the existing technologies currently available (3). Fig. 1 shows the features of 9DOF capacitive inertial MEMS platform in which the multipath approach is applied to magnetometers by fabricating 3 layers of copper lines connected with via on the cavity SOI wafer (cSOI).

Using these copper lines, coils having multiple numbers of windings are implemented for both vertical and lateral axis magnetometers to increase the magnetic sensitivity.

TSI wafer with getter is used to create and maintain the vacuum level of the platform. Accelerometers does not generally need vacuum since they require over-damped behavior to achieve wider bandwidths. On the other hand capacitive type gyroscopes and magnetometers require very high vacuum levels to achieve better sensitivities. In order to achieve both vacuum levels at the same time, TSI - MEMS packaging is performed at a high vacuum level including getters for gyroscopes and magnetometers, but ventilation channels are placed on the sealing rings and/or cavity layers on the MEMS side for the accelerometers. These channels help to fill the ambient pressure into accelerometer cavities after the dicing of the wafer is done.

TSI wafer also hosts top electrode for the devices which has vertical capacitors. Gap between TSI and MEMS wafers is very important in determining the performance of vertical axis inertial sensors and it is precisely controlled by the oxide spacers (stand-off) deposited below the bonding material on the TSI wafers. The presented platform allows 2.5D integration of fabricated sensors with the help of the top redistribution lines (RDL). These top RDL's can be interfaced to any other sensor, PCB or ASIC via flip chip or wire bonding techniques.

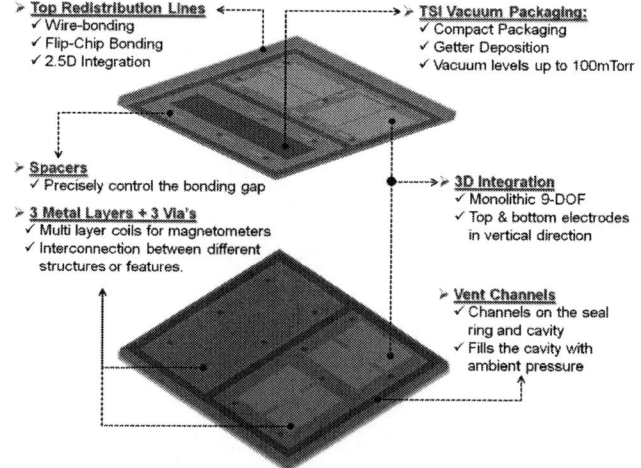

Figure 1: 9DOF Capacitive Inertial MEMS Platform overview

Fabrication

Fig. 2a- 2c shows the MEMS wafer which starts with 20µm deep cavity etching on the handle layer. It is followed with buried oxide (BOX) growth, fusion bonding and thinning the device layer of the cavity SOI wafer to 20μm's. Metal and via layers are formed using via last dual damascene process. Primary reason to fabricate these metal and via layers are to build the coils for magnetometers, but they can also be utilized for building more complex gyroscope and accelerometer architectures or interconnections between different structures and features.

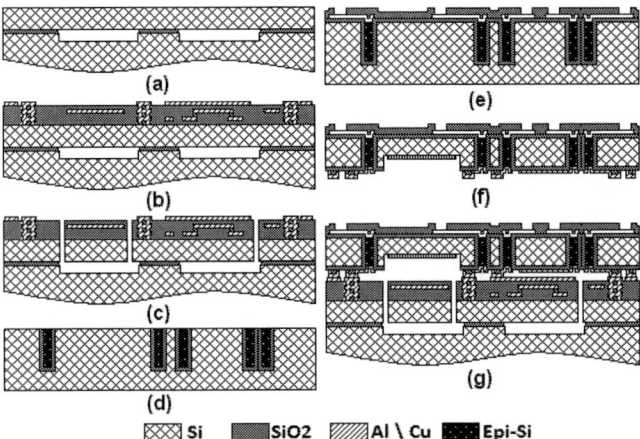

Si   SiO2   Al \ Cu   Epi-Si

Figure 2. MEMS wafer (a) cavity wafer (b) 3 metal and via layer deposition (c) structural etch (d) TSI etch and filling (e) Front side RDL and pad etch (f) grinding, stand-off, getter & back RDL (g) Bonded TSI & MEMS wafer bonding.

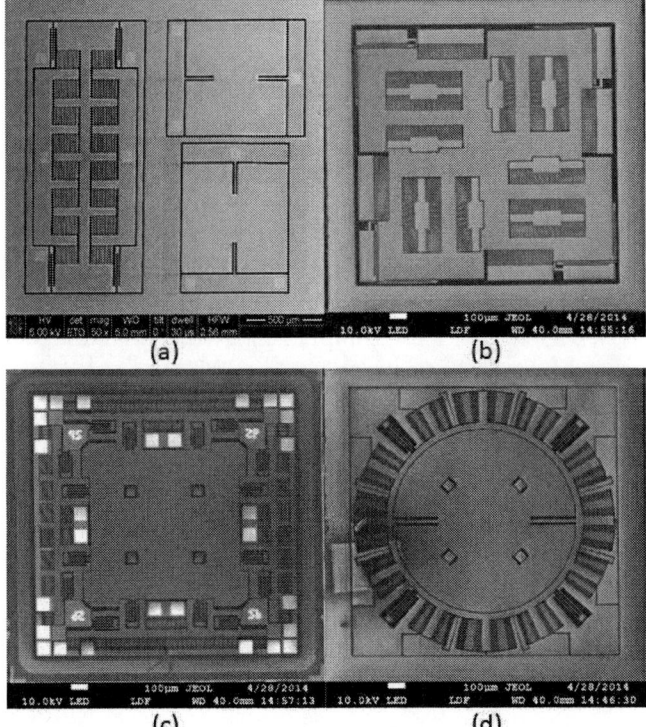

Figure 3: SEM images of fabricated sensors. (a) Three axis Lorentz Force magnetometer (b) Dual axis accelerometer (c) Z - axis tuning fork gyroscope (d) X & Y axis gyroscope

Approximately 2μm thick dielectric hosting the Cu lines is then patterned and used as a hard mask to etch the 20μm

thick device layer underneath. This final DRIE step suspends all the inertial sensors with the help of the cavity underneath them and completes the fabrication of the MEMS wafer.

TSI wafer fabrication shown in fig. 2d- 2g starts with via etching, liner oxide growth on the inner walls, and filling with highly doped n++ epi-poly silicon. It is followed by grinding of extra poly-silicon and then front side Al redistribution layer (RDL) fabrication. The wafer was then thinned down to 300 μm from backside to reveal TSVs and then followed by deposition of the spacers, RDL and getter on the backside of the wafer. We have used Titanium getter, capped with Nickel to achieve and stabilize at the required vacuum level (4). Process was completed by bonding MEMS to TSI wafer using Al\Ge bonding (5). Fig. 3 shows SEM images of inertial sensors before TSI vacuum packaging, fabricated using the 9DOF inertial MEMS platform.

Optimization of the Platform

Due to large mass requirement of inertial sensors large cavity areas are employed in the designs. Vacuum inside cavities of cSOI wafers was found to be significantly bending the large diaphragm like areas of the cavity wafer. Fig. 4 shows the excess amounts of remaining copper in the floppy regions after CMP of the damascene process. To overcome this issue, supporting pillars shown in fig. 4c are implemented in the design. These pillars support the device layer during the process, but they are detached from the structure during the final DRIE step for the release. The pressure inside the cavities of the SOI wafer is also optimized. It was found that using 400mTorr pressure instead of the original value of 40mTorr reduced the bending of the device layer significantly during the process. After all these measures, still remaining sags are filled with oxide and planarized up to silicon surface using CMP to remove all the roughness on the surface.

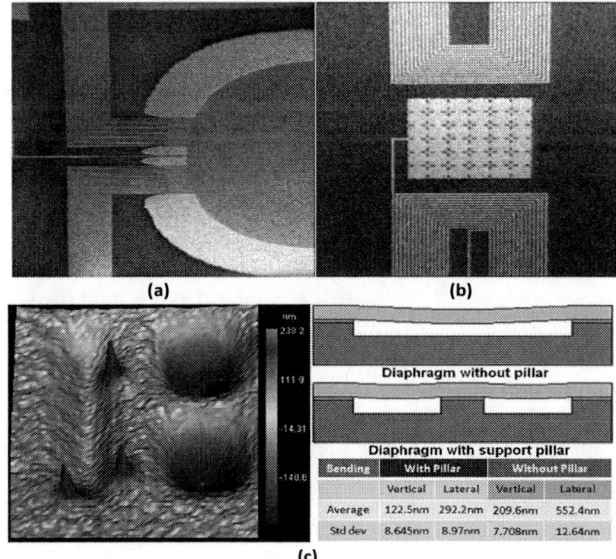

Figure 4: (a) Remaining Cu at the center of the cavities after CMP (b) Successful Cu CMP after surface planarization (c) Surface topology scan and average bending values for diaphragms with and without pillars

Another issue observed during the fabrication was the bending of the devices caused by the stress hysteresis of dielectric layers in which the metal coils are formed (6, 7). This problem is solved by tuning the stress levels in various dielectric layers. Optimization is done by preparing different test SOI wafers with different dielectric layer compositions and heating them up to the eutectic bonding temperature as the final step to simulate the exact processing conditions. Fig. 5 shows the variation of the total stack stress after each deposition and the final annealing step. According to these measurement results Tensile-Tensile-Compressive stack was found to have the least overall dielectric layer stress (25MPa) among the tested dielectric stack compositions. For the solution of this problem some other improvements are also employed from the sensor design point of view like reducing the sensor area, using circular structures instead of rectangular ones or placing trenches on the dielectric layers to further relieve the stress levels.

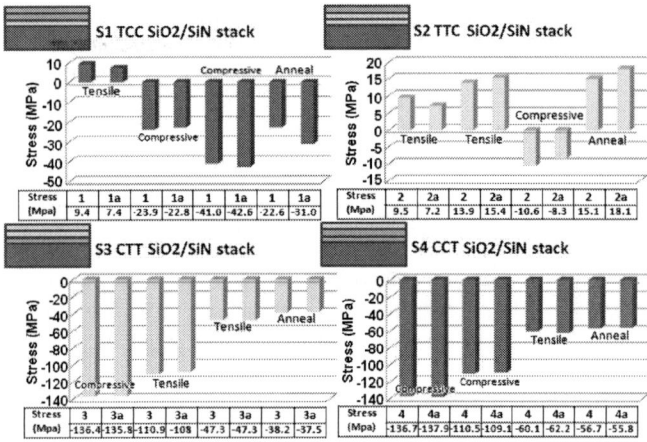

Figure 5: Short Loop results to optimize the film stress before and after 430C° Al/Ge bonding temperature.

Characterization

As described in the previous sections three axis accelerometer, gyroscope and magnetometer sensors are fabricated monolithically using 9DOF inertial MEMS platform. In order to verify that these sensors could be successfully fabricated using the claimed platform, mechanical transfer function of each device is measured and compared with the simulation results obtained using Coventorware. Fig. 6 below shows the comparison of measured and simulated resonance frequencies of magnetometer, gyroscope and accelerometers. The results show that the frequency mismatch is within 10% range of the simulated results. The reason to find slightly lower resonance frequency than the simulation results is the spring softening effect due to small undercuts during DRIE and application of DC bias voltage during the characterization which creates a negative spring constant effect for the resonating structures. Vacuum level of the cavities after the TSI – MEMS wafer level packaging is also very important in determining the performance of the inertial sensors. In order to monitor the

vacuum pressure after the packaging, a capacitive type resonator device is fabricated within the platform (8). Measurement results showed that the resonator structure had a Q factor of 4453. Using the theory for Q-factor estimation for rarified squeeze film air damping vacuum level is calculated to be around 100mTorr range for the designed resonator (9). Fig. 7 shows the measured resonance frequency and calculated Q-factor of the resonator, which was designed for vacuum level monitoring.

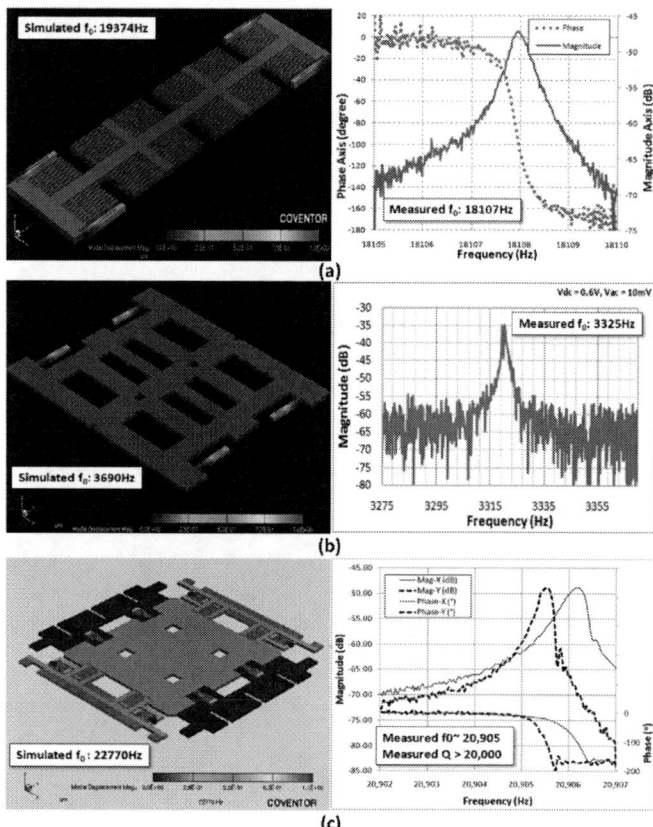

Figure 6: Comparison of measured and simulated resonance frequencies. (a) Magnetometer measured 18.1kHz and simulated 19.3kHz. (b) Accelerometer measured 3.3kHz and simulated 3.6 kHz. (c) Gyroscope measured 20.9kHz and simulated 22.7kHz.

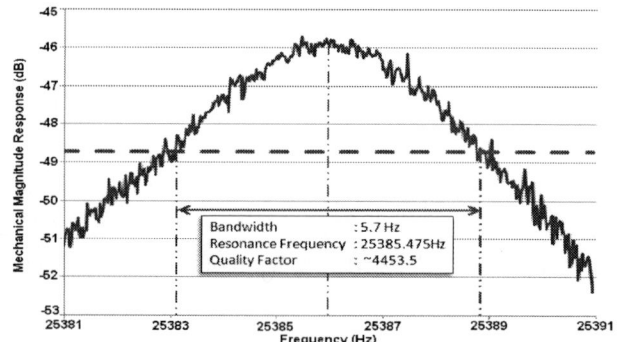

Figure 7: Q-factor of the TSI capped resonator. Estimated vacuum is around 100mTorr.

Capacitance vs. voltage behavior of the fabricated magnetometers is also measured to compare them with the simulation results obtained from Architect. Fig. 8(a) shows the complete model of the platform including all the parasitic capacitances and resistive paths through the structure. The parasitic capacitances from the liner oxide, top and bottom RDL capacitances are very important in determining the performance of the inertial sensors. High resistance (~5kΩ·cm) wafers are used as the TSI base wafer to reduce the effects of TSI parasitic capacitances on system performance. Design of the sensors should also be done in a way to reduce these additional parasitic effects. C-V simulations after adding all the parasitic effects like DRIE undercut, spring thinning, TSV and RDL capacitances shows ±5% matching between the simulation and test results as shown in fig. 8(b). Fig. 9 shows the images of the fabricated chips before and after TSI vacuum packaging.

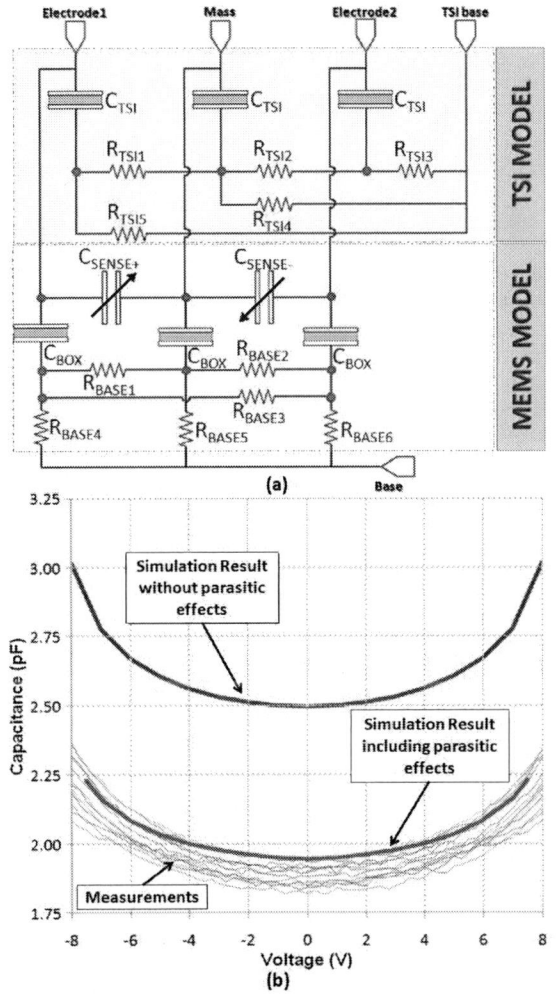

Figure 8: (a) Equivalent circuit model of the platform (b) Comparison of C-V measurement results with the simulations.

Conclusion and Future Work

A fully CMOS compatible 9DOF inertial MEMS platform, hosting 3 axis gyroscope, accelerometer and magnetometer, with wafer level packaging through epi-poly silicon TSI, is

successfully demonstrated for the first time without using any magnetic materials. This platform can reduce assembly and calibration costs and die size of sensors which will allow widespread usage in wearable technologies, Internet of Things (IOT) and many other similar applications. With its 2µm vertical and 1µm lateral capacitive gaps and 20µm thick structure layer, this platform is compatible with and can easily accommodate all the capacitive type inertial sensor designs currently available. By reducing the min. feature size for anchors and Al/Ge bonding pads more design space will be available to create advanced device architectures and shrink the die sizes. Number of metal lines and dielectric layers can also be reduced to simplify the process flow and decrease dielectric stress levels on the released structures.

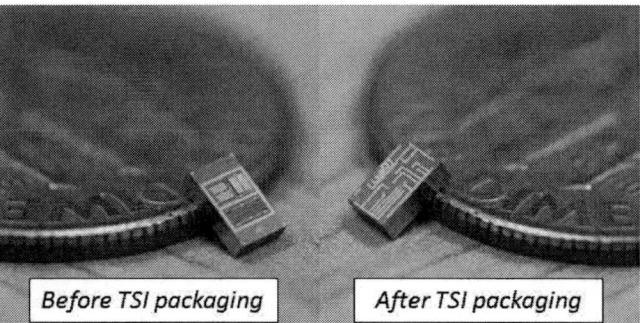

Figure 9: Fabricated magnetometer chips before and after TSI vacuum packaging.

Acknowledgement

Authors would like to thank Okmetic and Soitec for their helps with the fabrication of the cavity wafers and Global Foundries for their help with the fabrication of TSI wafers. This work has been partially funded by IME \ MEMS Consortium 2 and all the participating companies.

References

(1) L. Robin, and E. Mounier, "Inertial sensor market moves to combo sensors and sensor hubs," MEMS' Trends magazine, Oct. 2013, no.16, pp.16-18.
(2) AL. Herrera-May, LA. Aguilera-Cortés, PJ. García-Ramírez, and E. Manjarrez, "Resonant Magnetic Field Sensors Based On MEMS Technology," Sensors 2009, vol. 9, pp.7785-7813.
(3) G. Langfelder, and A. Tocchio, "Operation of Lorentz-Force MEMS Magnetometers with a Frequency Offset Between Driving Current and Mechanical Resonance," IEEE Transactions on Magnetics 2014, vol.50, no.1.
(4) V. Chidambaram, L. Xie, and B. Chen, "Titanium-Based Getter Solution for Wafer-Level MEMS Vacuum Packaging," Jour. of Elec. Mat., 2013, Volume 42-3, pp 485-491.
(5) B. Vu, and PM. Zavracky, "Patterned eutectic bonding with Al/Ge thin films for Micro Electromechanical Systems," Jour. Vac. Sci. Tech. B, 1996, vol.14, no.4, pp.2588-2594.
(6) G. Smolinsky, and TPHF. Wendling, "Measurements of Temperature Dependent Stress of Silicon Oxide Films Prepared by a Variety of CVD Methods," Jour. Electrochem. Soc., 1985, vol.132, no.4, pp.950-954.
(7) C. Zhiqiang, and X. Zhang, "Density change and viscous flow during structural relaxation of plasma-enhanced chemical-vapor-deposited silicon oxide films," Jour. of Appl. Phys., vol. 96, no. 8, pp. 4273-4280.
(8) J. Xu, and JM. Tsai, "A process-induced-frequency-drift resilient 32 kHz MEMS resonator," Jour. of Micromech. and Microeng., vol. 22-10, 105029.
(9) M. Bao, and H. Yang, "Squeeze film air damping in MEMS," Sensors and Actuators 2007, vol.136 pp.3-27.

Novel Intrinsic and Extrinsic Engineering for High-Performance High-Density Self-Aligned InGaAs MOSFETs: Precise Channel Thickness Control and Sub-40-nm Metal Contacts

Jianqiang Lin, Dimitri A. Antoniadis, and Jesús A. del Alamo

Microsystems Technology Laboratories, MIT, 60 Vassar St., 39-619, Cambridge, MA 02139, USA.

E-mail: linjq@mit.edu. Phone: 1-617-253-0714. Fax: 1-617-324-5341.

Abstract

We have fabricated self-aligned tight-pitch InGaAs Quantum-well MOSFETs (QW-MOSFETs) with scaled channel thickness (t_c) and metal contact length (L_c) by a novel fabrication process that features precise dimensional control. Impact of t_c scaling on transport, resistance and short channel effects (SCE) has been studied. A thick channel is favorable for transport, and a mobility of 8800 $cm^2/V \cdot s$ is obtained with $t_c=11$ nm at $N_s=2.6 \times 10^{12}$ cm^{-2}. Also, a record $g_{m,max}$ of 3.1 mS/μm and R_{on} of 190 $\Omega \cdot \mu$m are obtained in MOSFETs with $t_c=9$ nm and gate length $L_g=80$ nm. In contrast, a thin channel is beneficial for SCE control. In a device with $t_c=4$ nm and $L_g=80$ nm, S is 111 mV/dec at $V_{ds}= 0.5$ V. For the first time, working front-end device structures with 40 nm long contacts and gate-to-gate pitch of 150 nm are demonstrated. A new method to study the resistance properties of nanoscale contacts is proposed. We derive a specific contact resistivity between the Mo contact metal and the n^+ InGaAs cap of $\rho=(8 \pm 2) \times 10^{-9}$ $\Omega \cdot cm^2$. We also infer a metal-to-channel resistance of 70 $\Omega \cdot \mu$m for 40 nm long contacts.

Introduction

InGaAs is a promising channel candidate for CMOS applications [1,2]. However, the potential of InGaAs MOSFETs at realistic footprints has yet to be demonstrated. Advancing towards this goal requires precise lateral and vertical dimensional control in the intrinsic and extrinsic regions of the device. This work takes advantage of recent technological developments to study the impact of channel thickness and contact length on the performance of nanoscale InGaAs MOSFETs.

Recent prototype InGaAs QW-MOSFETs have attained great performance with $g_{m,max} \geq 2.7$ mS/μm devices demonstrated by several groups [3-6]. In this work, we explore further performance gains by exploiting recently developed techniques to precisely thin-down the intrinsic portion of the channel of self-aligned InGaAs QW-MOSFETs. This has allowed us to carry out a systematic study of the impact of channel thickness on the electrical characteristics of transistors with minimum resistance parasitics. The devices obtained through this effort outperform previous demonstrations.

Realistic CMOS devices require scalable pitch and ohmic contact size. Very few MOSFET structures reported to date feature a tight contact pitch. The tightest devices still exhibit a rather large pitch length (L_p) such as $L_p=500$ nm in [8]. In this work, we fabricate self-aligned MOSFET test arrays with a sub-150 nm pitch size. This is a significant leap in an effort to meet the requirements of the ITRS 2013 roadmap for III-V CMOS logic [2]. In these test structures, we study the resistance properties of contacts down to 40 nm in length. In spite of the excellent results that have been obtained, this effort has revealed the need for further progress on nanoscale contacts.

Device fabrication

This work builds on our prior research on contact-first, self-aligned InGaAs QW-MOSFETs [6]. The device structure used in this research is sketched in Fig. 1a and a TEM of a $L_g=40$ nm transistor is shown in Fig. 1b.

The starting heterostructure (Fig. 2a) includes a 10 nm thick composite channel that consists (from top to bottom) of $In_{0.7}Ga_{0.3}As$ (3 nm), InAs (2 nm) and $In_{0.7}Ga_{0.3}As$ (5 nm). Above this there is a 3 nm InP barrier. Following a similar process as in [6], the heterostructure is etched with 1 nm precision in the intrinsic portion of the device through a combination of Cl_2-based RIE and digital etch (DE) (Fig. 2b). RIE stops a few nm above the cap. Then DE is used to finely thin down the channel to its final thickness [7]. As discussed in [7], achieving 1 nm channel thickness resolution does not require precise control of the final etching point of the RIE process.

Using this technique, we have fabricated devices in a process run that includes different splits with between 2 and 14 cycles of DE. The correlation between the number of DE cycles and the resulting t_c is shown in Fig. 2a. Here t_c denotes the total thickness of the channel heterostructure which includes all layers between the InAlAs buffer and the gate oxide including the InP barrier if present. Thus, a buried-channel device is obtained when $t_c > 10$ nm, otherwise a surface-channel device results. Our process yielded identical transistors except for the value of t_c which varied from 12 nm down to 3 nm. TEMs of final devices for two etch depths are shown in Fig. 2c for devices with t_c of 4 nm and 8 nm, confirming our calibrations.

The rest of the fabrication process is similar to that of [6]. The gate oxide is 2.5 nm HfO_2 deposited by ALD. Device L_g spans from 40 nm to 5 μm. The length of the access region measured from the edge of the gate to the edge of the contact between the ohmic metal and the n^+ cap is $L_{access}=15$ nm (Fig. 1). The contacts have a length $L_c=10$ μm.

To examine the lateral scalability of our process and to study the resistance of nanoscale contacts, we simultaneously fabricated MOSFET arrays on the same chip. These consist of multiple series-connected transistor cells with scaled gates and contacts. The devices have L_g between 30-130 nm, L_c between 40-800 nm and $L_{access}=15$ nm. More details are given below.

High-Performance InGaAs MOSFETs

Our results show that channel thickness scaling exerts a very significant influence on the MOSFET electrical characteristics. From the ON-current point of view, $t_c=9$ nm is most favorable. A $L_g= 80$ nm device with $t_c=9$ nm exhibits a peak transconductance of $g_{m,max}=3.1$ mS/μm at $V_{ds}=0.5$ V (Fig. 3). We believe this is the highest g_m III-V transistor ever made exceeding the most advanced InGaAs MOSFETs and HEMTs [3-6, 10]. The output characteristics of this device, shown in Fig. 4, indicate an unprecedented R_{on} of 190 $\Omega \cdot \mu$m. A minimum subthreshold swing (S_{min}) of 159 mV/dec at $V_{ds}=0.5$ V and DIBL=310 mV/V are also obtained in the same device (Fig. 5a). Benchmarking of this device against recently published InGaAs MOSFETs (Fig. 6) indicates an outstanding balance between transport and subthreshold characteristics.

Thin channel devices exhibit superior subthreshold behavior as seen in Fig. 5b for a $t_c=4$ nm transistor where S=111 mV/dec at $V_{ds}=0.5$ V and DIBL=126 mV/V. This comes at the expense of g_m and ON-current that are both seriously compromised.

978-1-4799-8002-4/14 $31.00 © 2014 IEEE

In a systematic study of the impact of t_c on key figures of merit, Fig. 7 shows $g_{m,max}$ vs. t_c for devices with L_g from 80 nm to 5 μm. $g_{m,max}$ always peaks at t_c=9 nm. For $t_c \geq 10$ nm, the gate dielectric sits on the InP barrier (buried channel). Thus capacitive coupling to the channel is reduced. For $t_c \leq 8$ nm, the transport and resistance characteristics degrade prominently. This is also shown in Fig. 8 that graphs R_{sd} (=R_s+R_d) vs. t_c. R_{sd} is obtained from measurements of R_{on} vs. L_g at V_{gs}-V_t=0.5 V. The rise of R_{sd} for very thin t_c highlights the role of the spreading resistance associated with the link region at the gate edge of the channel.

In contrast with transport figures of merit, SCE control is improved by reducing t_c, as shown in Figs. 9, 10 and 11. Thinner t_c leads to improved S, DIBL and $V_{t,sat}$ vs. L_g characteristics. For $t_c \leq 4$ nm, the improvements tend to saturate.

To further understand charge control and transport in these devices, we have carried-out split-C-V measurements on long-channel MOSFETs (L_g=5 μm) at 1 MHz (Fig. 12). The ON-state capacitance increases monotonically as the channel is thinned down. Two factors contribute to this. First, as t_c is reduced, the centroid of charge moves closer to the gate. Second, for thin enough t_c, the InAs core is eventually removed and the electron effective mass increases. This yields an improved saturation behavior of the C-V characteristics for t_c=3 nm. In all cases, very low gate leakage current flows, as shown in the inset of Fig. 12 for t_c=9 nm.

The effective mobility (μ_{eff}) was extracted by split-C-V method on the L_g=5 μm devices after correcting for R_{sd} (Fig. 13). We find significant mobility degradation as the channel is thinned down due to increased interface scattering and the eventual extinction of the InAs core. This is consistent with [11]. In an MBE calibration heterostructure (same channel configuration but with a thin cap) we obtained a Hall mobility of $\mu_{H,max}$= 11,000 cm^2/V·s at N_s=2.6x10^{12} cm^{-2}. This confirms the excellent intrinsic transport properties of our as-grown material. For t_c=11 nm, μ_{eff} at N_s=2.6x10^{12} cm^{-2} is 8800 cm^2/V·s, about 80% of $\mu_{H,max}$. To the authors' knowledge, this is the highest μ_{eff} [1,12] in an InGaAs/InAs channel MOSFET. The degradation of μ_{eff} with respective to $\mu_{H,max}$ could be due to interface roughness scattering that is exacerbated by the 1 nm InP barrier (vs. 3 nm in the calibration heterostructure) that is present in the t_c =11 nm device.

High-Density InGaAs MOSFET Arrays

To study the characteristics of InGaAs MOSFETs with scaled pitch, we have fabricated MOSFET arrays and gate-less arrays as sketched in Figs. 14 a and b. The gate-less array has a similar structure as the MOSFET array except that the n$^+$ cap is not recessed and a gate is not fabricated. In the MOSFETs array, one cell (one pitch size) consists of one gate length, one contact length and two access lengths. Arrays with different numbers of cells and various gate and contact dimensions have been fabricated. These structures allow us to extract all the relevant resistance components of a tight-pitch MOSFET.

We have fabricated arrays with 1 to 4 transistor cells, as illustrated in Fig. 14. Pitch sizes varied between L_p=100 nm and ~1 μm. The channel thickness was 9 nm. Cross sectional TEM views of two MOSFET arrays are shown in Fig. 15. Fig. 15a shows a 1-cell array with L_p=200 nm, L_c=40 nm, and L_g=130 nm. Fig. 15b shows a 2-cell array with L_p=150 nm, L_c=80 nm, and L_g=40 nm. Gates in each cell are connected together and biased at V_g, while the inner contacts are floating. The two ends of the arrays are biased at V_d and V_s respectively. In all these arrays, we measure the total resistance vs. number of cells. In addition,

for the MOSFETs array the gate is biased at V_{gs}-$V_t \gg V_{ds}$. The unit-cell resistance (R_{cell}) is determined from the slope of a graph of total resistance vs. number of cells (Fig. 16). Figs. 17 and 18 show unit-cell resistance vs. L_c for both types of arrays.

A resistor network model is developed to analyze these results. Cross-section schematics and equivalent circuit models of both array types are shown in Fig. 19. For small enough V_{ds}, the channel in both arrays can be modeled as a simple resistor. In the MOSFET array, the contact can be modeled by a 2D resistor network composed of three coupled lateral conducting layers (Fig. 19b). The two vertical resistive couplings are characterized by two unknown contact resistivities, ρ_{12} and ρ_{23}. In the gate-less array, a single contact resistivity ρ_{12} couples two lateral conducting layers. ρ_{12} and ρ_{23} can then be extracted by fitting the model to the experimental results of Figs. 17 and 18 after the rest of the model parameters are obtained from independent TLM or Hall measurements (Table I).

We obtain ρ_{12} =(8±2)x10^{-9} Ω·cm^2 for the contact resistivity between the Mo contact and the n$^+$ cap from the gate-less array (Fig. 17), and ρ_{23} =(2±0.8)x10^{-8} Ω·cm^2 for the contact resistivity between the n$^+$ cap and the channel from the MOSFET array (Fig. 18). The value of ρ_{12} is consistent with independent measurements carried out in nano-TLMs [13] and other Mo/InGaAs contact experiments [14] and is much lower than other metal to InGaAs contact technologies [9].

We use these extracted values to estimate the vertical contact resistance, R_c, of scaled MOSFETs fabricated by our technology. R_c here refers to the resistance from the metal contact to the edge of the access region, or points C and A in Fig. 14b. The result for different values of L_c is plotted in Fig. 20. We infer a metal-to-channel resistance of 70 Ω·μm for 40 nm long contacts. These are encouraging results but more research is required to attain the required R_c in nanometer-scale contacts.

Conclusions

We have studied the role of body thickness on the electrical characteristics of self-aligned InGaAs QW-MOSFETs. In an optimized design with thick channel thickness, we have obtained a record $g_{m,max}$ of 3.1 mS/μm and a mobility of 8800 cm^2/V·s at N_s=2.6x10^{12} cm^{-2}. We have also fabricated front-end MOSFET test structures with 40 nm contact size and scaled contact pitch. A model is developed to study the characteristics of nano-scale contacts. A low contact resistivity of ρ=(8 ± 2)x10^{-9} Ω·cm^2 between Mo and n$^+$ InGaAs is derived from this study.

Acknowledgements

This work is sponsored by Donner Chair at MIT, SMA and SMART/LEES programs. Device fabrication was performed at MIT's Microsystems Technology Laboratories and SEBL.

References

[1] J. A. del Alamo, D. A. Antoniadis, *et al.*, *IEDM Tech. Dig.*, 2013, p.24.
[2] "III-V/Ge HP Logic Technology Requirements" in ITRS 2013 Edition.
[3] S. W. Chang, *et al.*, *IEDM Tech. Dig.*, 2013, p.417.
[4] S. Lee, C-Y Huang, *et al.*, *Appl. Phys. Lett.*, vol. 103, p. 233503, 2013.
[5] S. Lee, C-Y Huang, *et al.*, *IEEE Electron Dev. Lett.*, vol.35, p. 621, 2014.
[6] J. Lin, X. Zhao, *et al.*, *IEDM Tech. Dig.*, 2013, p.421.
[7] J. Lin, X. Zhao, *et al.*, *IEEE Electron Dev. Lett.*, vol. 35, p. 440, 2014.
[8] L. Czornomaz, N. Daix, *et al.*, *IEDM Tech. Dig.*, 2011, p.517.
[9] M. Egard, S Johansson, *et al.*, *IEDM Tech. Dig.*, 2011, p.303.
[10] H. Matsuzaki, T. Maruyama, *et al.*, *Electronic Lett.*, 42 (2006) p. 883.
[11] A. Alian, M.A. Pourghaderi, *et al.*, *IEDM Tech. Dig.*, 2013, p.437.
[12] J. Lin, D. A. Antoniadis, J. A. del Alamo, *IEDM Tech. Dig.*, 2012,p.757.
[13] W. Lu, A. Guo, *et al.*, *IEEE Electron Dev. Lett.*, vol. 35, p. 178, 2014.
[14] A. K. Baraskara, M. A. Wistey, *et al.*, *J. Vac. Sci. Technol. B*, vol. 27, p.2036, 2009.

978-1-4799-8002-4/14 $31.00 © 2014 IEEE

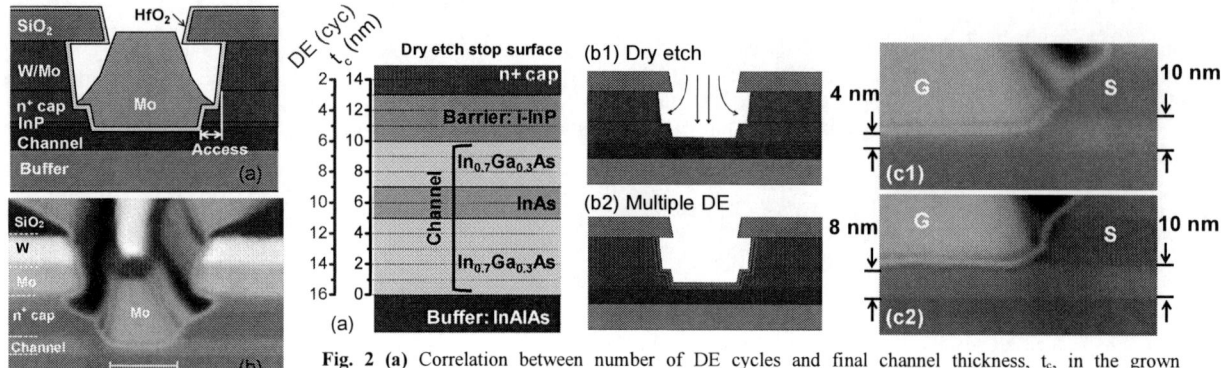

Fig. 1 (a) InGaAs MOSFET cross section schematic, and **(b)** TEM image of a complete device with L_g=40 nm and 2.5 nm HfO$_2$ gate dielectric.

Fig. 2 (a) Correlation between number of DE cycles and final channel thickness, t_c, in the grown heterostructure. **(b)** Precise etching process of the intrinsic region of the device is performed by RIE (b1) and DE (b2). The RIE stops a few nm above the channel surface. The final channel thickness is controlled by DE with 1 nm precision. **(c)** TEM images of finished devices with two t_c thicknesses. In the extrinsic portion, the as-grown total channel thickness is 10 nm. In the intrinsic portion, the resultant t_c are 4 nm (c1) and 8 nm (c2).

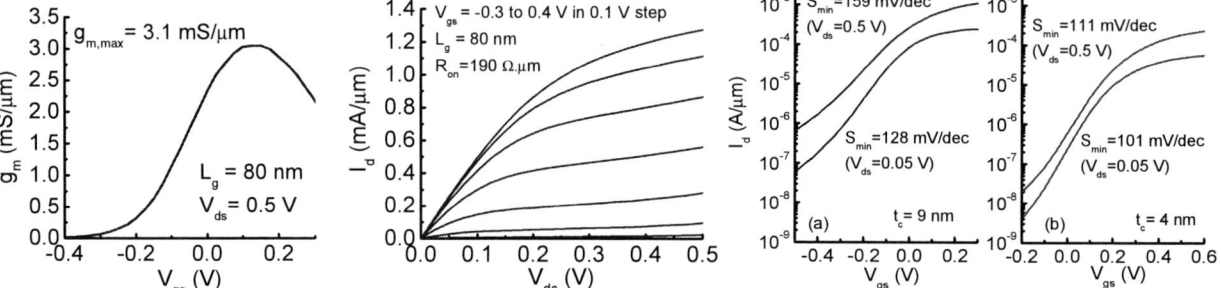

Fig. 3 Transconductance of InGaAs MOSFET with t_c =9 nm and L_g=80 nm.

Fig. 4 Output characteristics of device of Fig. 3.

Fig. 5 Subthreshold characteristics of two L_g= 80 nm MOSFET with **(a)** t_c=9 nm (same device of Fig. 3, optimized for ON-current) and **(b)** t_c=4 nm (optimized for SCE).

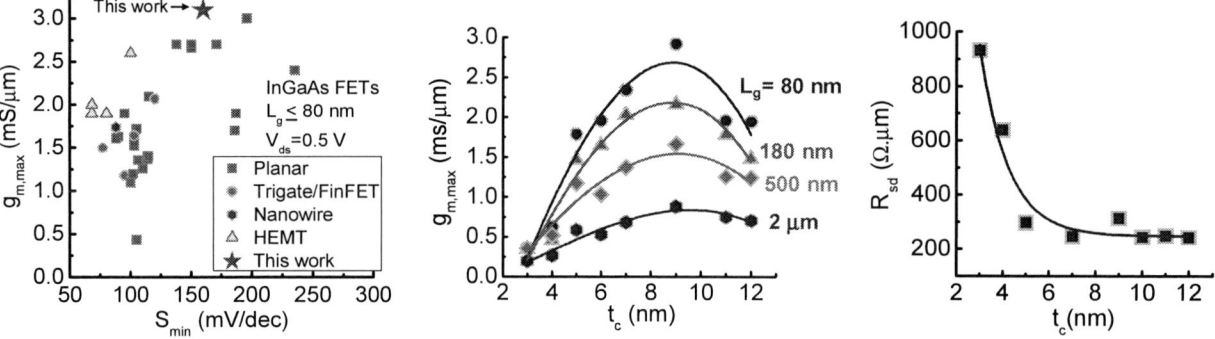

Fig. 6 Benchmarking of $g_{m,max}$ vs. S at 0.5 V for III-V FETs with L_g≤80 nm showing record device obtained in this work.

Fig. 7 Peak transconductance vs. t_c for transistors with L_g from 80 nm to 2 μm.

Fig. 8 R_{sd} vs. t_c. R_{sd} is extracted from measurements of R_{on} vs L_g.

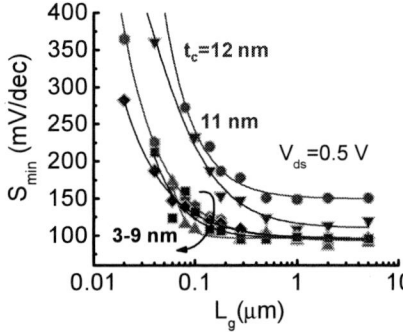

Fig. 9 Subthreshold swing at V_{ds}=0.5 V vs. L_g for MOSFETs with t_c from 3 nm to 12 nm.

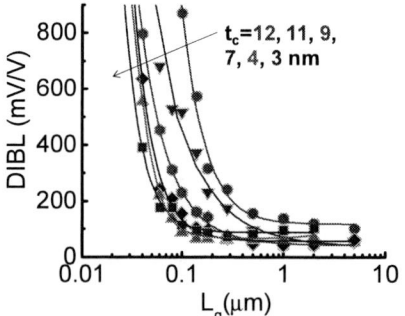

Fig. 10 DIBL vs. L_g for MOSFETs with t_c from 3 nm to 12 nm.

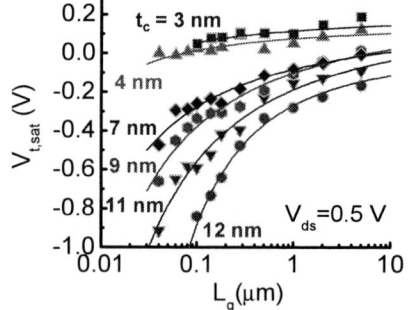

Fig. 11 $V_{t,sat}$ vs. L_g for MOSFETs with different t_c.

978-1-4799-8002-4/14 $31.00 © 2014 IEEE

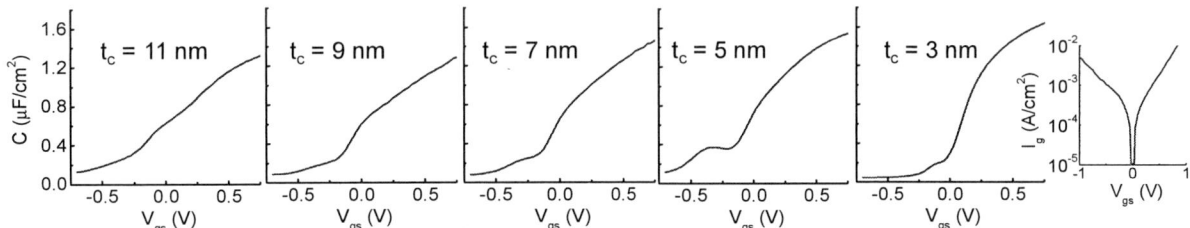

Fig. 12 C-V characteristics measured on devices with L_g=5 μm for different t_c at 1 MHz. Inset gives the typical gate leakage current density, in this case for t_c=9 nm.

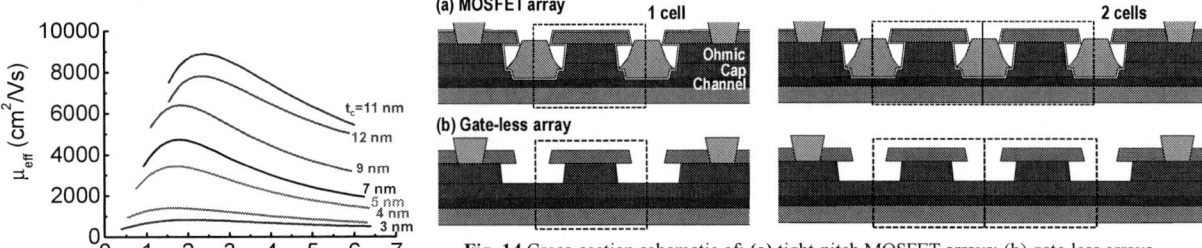

Fig. 13 Mobility *vs.* sheet carrier density for t_c from 12 nm to 3 nm.

Fig. 14 Cross-section schematic of: **(a)** tight-pitch MOSFET arrays; **(b)** gate-less arrays. Schematics show examples with 1 cell (left) and 2 cells (right).

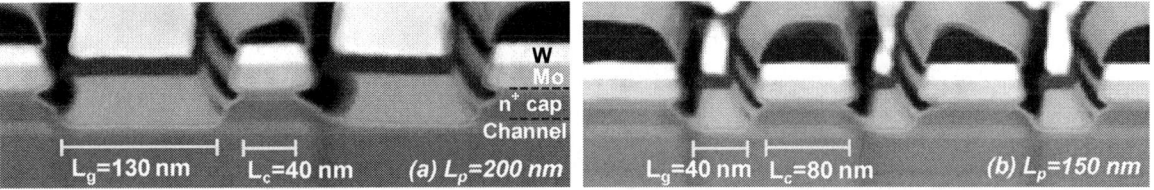

Fig. 15 TEM of tight-pitch MOSFET array with different pitch size (L_p), contact length (L_c) and gate length (L_g): **(a)** One-cell array with L_p=200 nm, L_c=40 nm, and L_g=130 nm; **(b)** Two-cell array with L_p=150 nm, L_c=80 nm, and L_g=40 nm.

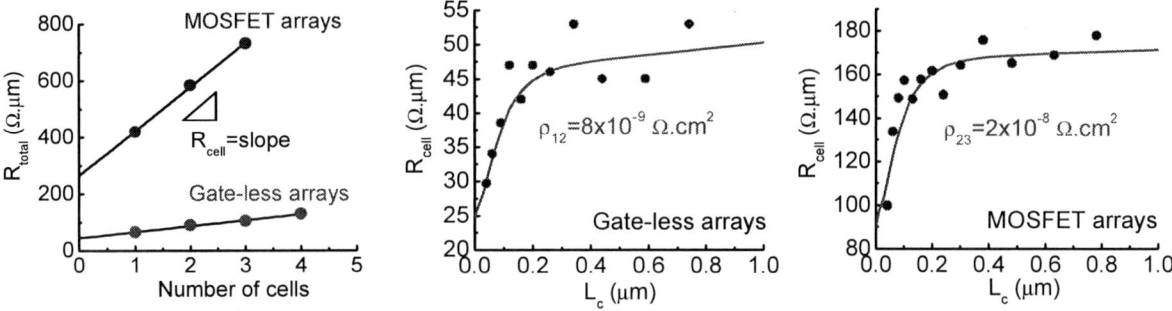

Fig. 16 Extraction of unit-cell resistance (R_{cell}) from slope of total resistance *vs.* number of cells in gate-less arrays and MOSFET arrays.

Fig. 17 Unit-cell resistance *vs.* contact length in gate-less arrays (Fig. 14b). ρ_{12} is extracted.

Fig. 18 Unit-cell resistance *vs.* contact length in MOSFET arrays (Fig. 14a). ρ_{23} is extracted.

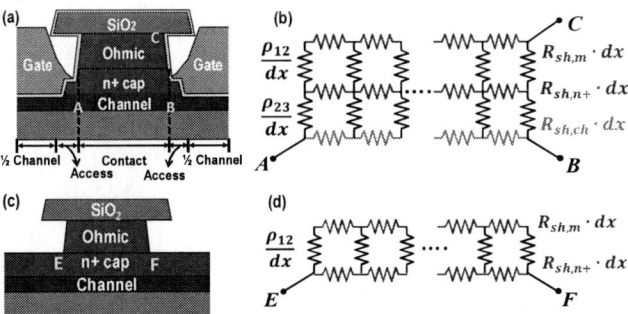

Fig. 19 (a) Cross-section schematic of unit cell of MOSFET array and **(b)** equivalent circuit model. **(c)** Cross-sectional schematic of unit cell of gate-less array and **(d)** equivalent circuit model.

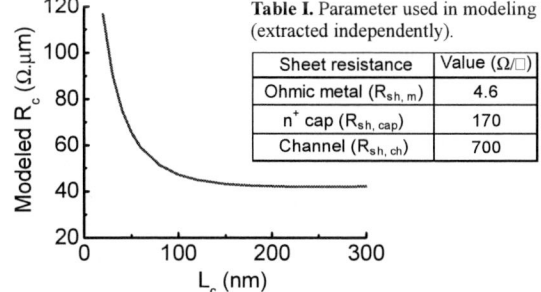

Table I. Parameter used in modeling (extracted independently).

Sheet resistance	Value (Ω/□)
Ohmic metal ($R_{sh, m}$)	4.6
n^+ cap ($R_{sh, cap}$)	170
Channel ($R_{sh, ch}$)	700

Fig. 20 Modeled vertical contact resistance *vs.* contact length of InGaAs QW-MOSFETs based on the extracted parameters that characterize our contact system. This is the estimated resistance between nodes A and C with B floating in Fig. 19a.

978-1-4799-8002-4/14 $31.00 © 2014 IEEE

(Invited) High-Performance III-V devices for future logic applications

D.-H. Kim[1,2], T.-W. Kim[1], RH. Baek[1], P. D. Kirsch[1], W. Maszara[2], J. A. del Alamo[3], D. A. Antoniadis[3], M. Urteaga[4], B. Brar[4], HM. Kwon[5], C.-S. Shin[5], W.-K. Park[5], Y.-D. Cho[6], SH. Shin[6], DH. Ko[6] and K.-S. Seo[7]

[1]SEMATECH, [2]GLOBALFOUNDRIES, [3]MIT, [4]Teledyne Scientific, [5]KANC, [6]Yonsei University and [7]Seoul National University
E-mail: Dae-Hyun.Kim@sematech.org

Abstract: High-mobility III-V transistors are poised to take the lead on future high performance logic operation. If this happens, indium-rich $In_xGa_{1-x}As$ is the most promising n-channel material. Indeed, remarkable progress has been made, including III-V gate-stacks with ALD-grown gate dielectrics. This paper reviews the evolution of high-performance III-V devices for future logic applications and discuss a possible path forward to further improve their logic figure-of-merits.

Introduction: In early 2000s, indium-rich $In_xGa_{1-x}As$ (x>0.53) has recently emerged as the most promising non-Si n-channel material for post Si CMOS logic applications [1]. This is thanks to the outstanding electron transport characteristics, excellent interfacial quality of the high-k/InGaAs gate stack by ALD and co-integration of InGaAs-based heterostructures with Si [1-2]. Recently, significant progress has been made on a variety of GaAs and InGaAs MOSFETs by many different groups. This paper reviews high-performance III-V devices for future logic applications, covers recent advances in some of the key enabling technology of InGaAs MOSFETs, and finally discusses options to further improve the performance of InGaAs MOSFETs.

How good are III-V's for future logic applications?: As a way to assess the prospects for a future III-V MOSFET technology with gate lengths in the sub-10 nm range, we started our research in 2005 on state-of-the-art III-V High-Electron-Mobility-Transistors (HEMTs). The HEMT in itself a device with near THz capabilities, was an excellent prototype Field-Effect-Transistor (FET) for future logic. The HEMT is a FET with a "medium-k" gate barrier and outstanding carrier transport properties. From 2005 to 2009, we investigated the logic characteristics of scaled-down InGaAs HEMTs [3-5]. From this time, **Figs. 1** and **2** show cross-sectional schematic and typical subthreshold/transfer characteristics of $L_g = 30$ nm $In_{0.7}Ga_{0.3}As$ HEMTs with $t_{ins} = 4$ nm at $V_{DS} = 0.5$ V [4-5]. DIBL, S and g_{m_max} are 120 mV/V, 90 mV/dec. and > 2.5 mS/•m at $V_{DS} = 0.5$ V, respectively, which outperform today's state-of-the-art Si nMOSFETs with equivalent gate length [6]. An I_{ON}/I_{OFF} ratio in excess of 10^4 in well-designed devices, even with supply voltage of 0.5 V is obtained. The device exhibits $I_{ON} = \sim 0.62$ mA/μm at an $I_{Leak} = 100$ nA/μm. This is the highest I_{ON} in any III-V FET on any material system, to date. This is about 20% higher I_{ON} than state-of-the-art high-performance 22 nm nMOSFET with comparable physical gate length and I_{Leak} at $V_{DD} = 0.5$ V [6]. It was data like these that showed that InGaAs-channel HEMTs prototyped in a university environment could outperformance state-of-the-art Si nMOSFETs of similar gate length at $V_{DD} = 0.5$ V that strongly highlighted the potential of InGaAs MOSFETs for logic.

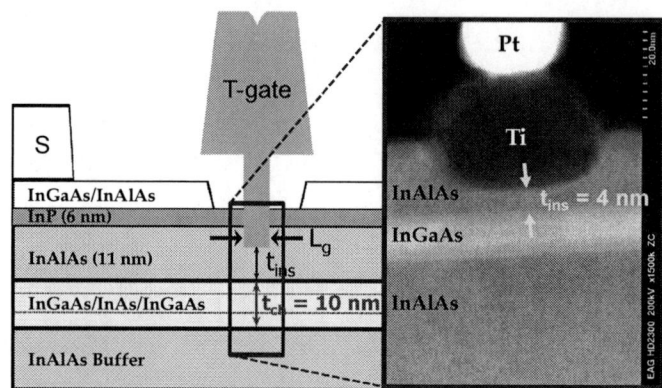

Fig. 1 Cross-sectional schematic of state-of-the-art InGaAs HEMT [4].

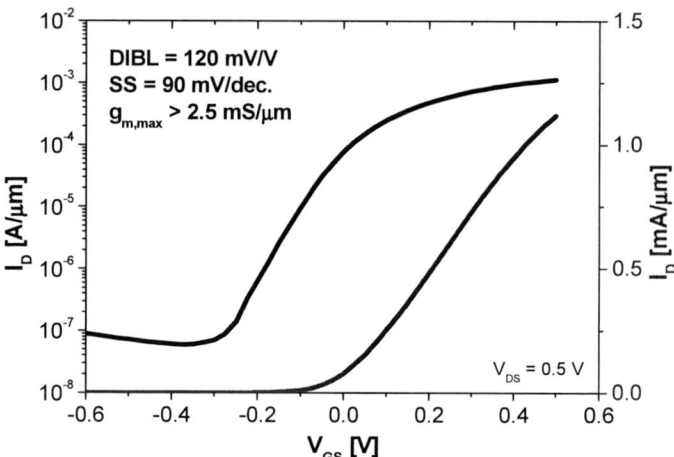

Fig. 2 Subthreshold and transfer characteristics of $L_g = 30$ nm InGaAs HEMT with $t_{ins} = 4$ nm [4-5].

Evaluation of III-V Gate Stack: The key enabling technology for InGaAs MOSFETs is a high-quality oxide/semiconductor interface by ALD. During ALD, a kind of 'self-cleaning effect' or 'clean-up effect' takes places such that III-V surface oxides are effectively removed [7]. Many different groups have demonstrated excellent Al_2O_3/InGaAs interfaces and surface-channel III-V MOSFETs. However, this is insufficient for future scaled CMOS. In particular, a dielectric with higher k is required. Also, in a surface-channel design, interface roughness scattering severely degrades the mobility (**Fig. 3**). The pressing need for the 7-nm technology node and/or beyond is for an ultra-scaled surface-channel design with total EOT well below 1 nm while maintaining excellent transport properties. We have been investigating a composite Al_2O_3/HfO_2 gate stack, where a thin Al_2O_3 interfacial layer serves as passivation. This leads to a far better mobility-EOT trade-off, while maintaining lower D_{it} on InGaAs which is confirmed by our device results (**Fig. 4**) [8].

978-1-4799-8002-4/14 $31.00 © 2014 IEEE

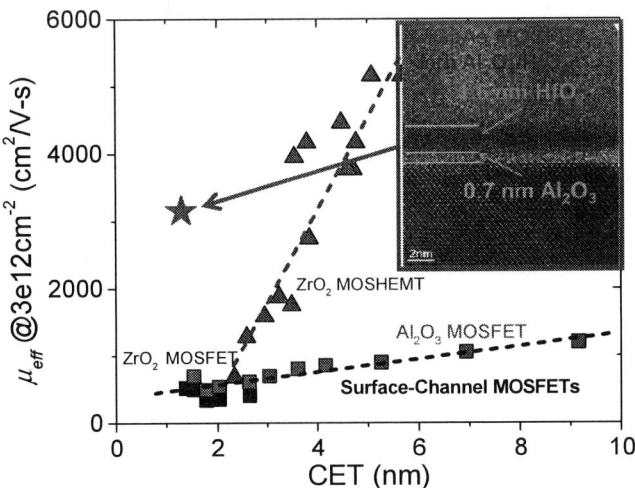

Fig. 3 μ_{eff} at $n_s = 3\times10^{12}\text{cm}^{-2}$ vs. CET for planar InGaAs QW MOSFET with Al_2O_3/HfO_2 [8] and other reports on various types of III-V MOSFETs. Inset is the cross-sectional TEM image for Al_2O_3/HfO_2 gate stack on InGaAs [8].

Fig. 4 Subthreshold characteristics of long-channel InGaAs

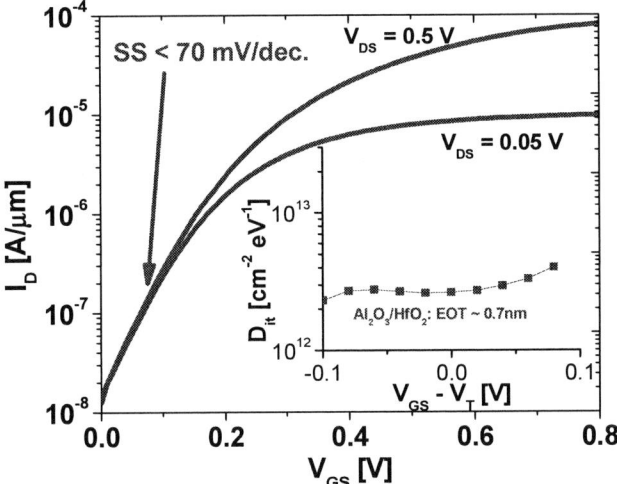

MOSFET with Al_2O_3/HfO_2 at $V_{DS} = 0.05/0.5$ V. Inset is extracted D_{it} for Al_2O_3/HfO_2 on InGaAs channel. EOT < 1 nm, SS < 70 mV/dec., and $D_{it} \sim 2 \times 10^{12}/\text{cm}^2\text{eV}$ [8].

Progress and Prospects of InGaAs MOSFETs: Starting from III-V MOSFETs with state-of-the-art characteristics, it is very useful to benchmark InGaAs MOSFETs against InGaAs HEMTs [9]. **Figs. 5 (a), (b)** and **(c)** show the evolution of transconducntace (g_m), on-resistance (R_{on}) and the current-gain cut-off frequency (f_T) of InGaAs MOSFETs as well as InGaAs HEMTs for the last three decades. InGaAs MOSFETs have now matched the g_m of HEMTs and surpassed the R_{on} of HEMTs. The high g_m behavior in InGaAs MOSFETs stems from the ability to scale the effective oxide thickness in the MOS gate stack as well as the excellent interface-state density (D_{it}, shown in the inset of **Fig. 4**). These are also responsible for the very low R_{on}. The excellent parasitic resistance characteristics in InGaAs MOSFETs arise from the absence of an energy barrier under the S/D contracts, which is the bottleneck in InGaAs HEMTs. Recently, several groups have successfully demonstrated sub-100 nm InGaAs MOSFETs with record R_{on} and g_m behavior, approaching $g_{m\_max} = 3$ mS/μm at $V_{DS} = 0.5$ V [10-12].

Figs. 6 and **7** exhibit microwave and high frequency noise-figure characteristics of an $L_g = 35$ nm $In_{0.7}Ga_{0.3}As$ MOSFET with Al_2O_3/HfO_2 bilayer gate stack. Excellent high-frequency characteristics can be observed in this device, such as current-gain cutoff frequency (f_T) = 440 GHz, maximum oscillation frequency (f_{max}) = 305 GHz and minimum noise-figure (NF_{min}) < 0.5 dB at 26 GHz. The results are record values for f_T and f_{max}, and the lowest NF_{min} at 26 GHz of any III-V MOSFET. It is interesting to see that today's InGaAs HEMTs still exhibit a far better f_T than InGaAs MOSFETs. This is mostly due to low-parasitic capacitance design employed in InGaAs HEMTs, such as T-gate. With the introduction of low parasitic capacitance MOSFET designs, it is expected that InGaAs MOSFETs that match or exceed the high-frequency characteristics of InGaAs HEMTs will be developed.

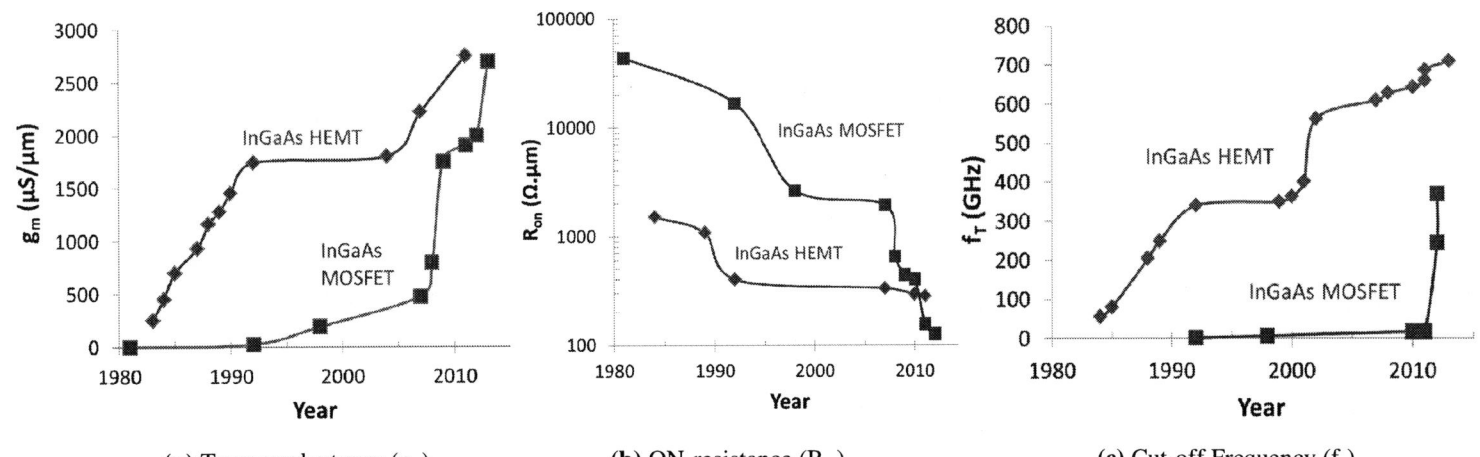

(a) Transconductance (g_m) **(b)** ON-resistance (R_{on}) **(c)** Cut-off Frequency (f_T)

Fig. 5 Benchmarking of Inversion-type InGaAs MOSFETs against InGaAs HEMTs, as a function of year [9].

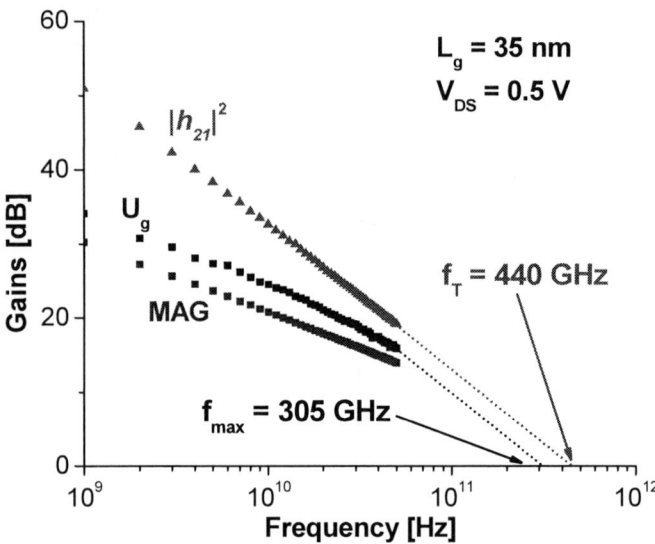

Fig. 6 RF gains (h_{21}, MAG and unilateral gain) against frequency for L_g = 35 nm $In_{0.7}Ga_{0.3}As$ MOSFETs with Al_2O_3/HfO_2 at V_{DS} = 0.5 V. The device shows a record value of f_T = 440 GHz.

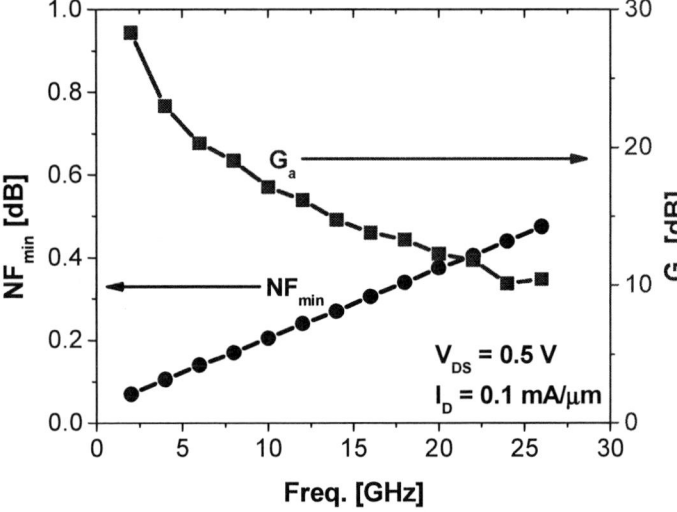

Fig. 7 High-frequency minimum noise-figure (NF_{min}) and associated gain (G_a) against frequency for L_g = 35 nm $In_{0.7}Ga_{0.3}As$ MOSFETs with Al_2O_3/HfO_2 at V_{DS} = 0.5 V.

Alternative MOSFET designs are being pursued. Epitaxial selectively-regrown S/D contacts (**Fig. 8**) have recently yielded record R_{on} and effectively mobility ($\mu_{n,eff}$) behavior in InGaAs MOSFETs [13-14]. This allows the introduction of tensile strain to boost performance [15] and enables a gate-last integration scheme that prevents degradation of the III-V gate stack [14]. Sub-10 nm MOSFETs will require 3D device architectures, such as Tri-gate FETs [16-17] or Gate-all-around FETs [18]. In Tri-gate FETs, how to define an InGaAs fin on Si is a major concern. For this, we have been investigating RIE in combination with post wet treatment. **Fig. 9** exhibits subthreshold characteristics of the state-of-the-art Tri-gate InGaAs MOSFET with $L_g/W_{fin}/H_{fin}$ = 50/30/20 nm [17]. A concern in Tri-gate MOSFETs is the sidewall MOS interface. From the subthreshold

characteristics in our previous report [17], coupled with TCAD simulation, we preliminary extract $D_{it} \sim 4 \times 10^{12}$ eV^{-1}cm^{-2} at the sidewall gate stack interface between Al_2O_3 and etched InGaAs fin, which is about 2 times larger than planar gate stack.

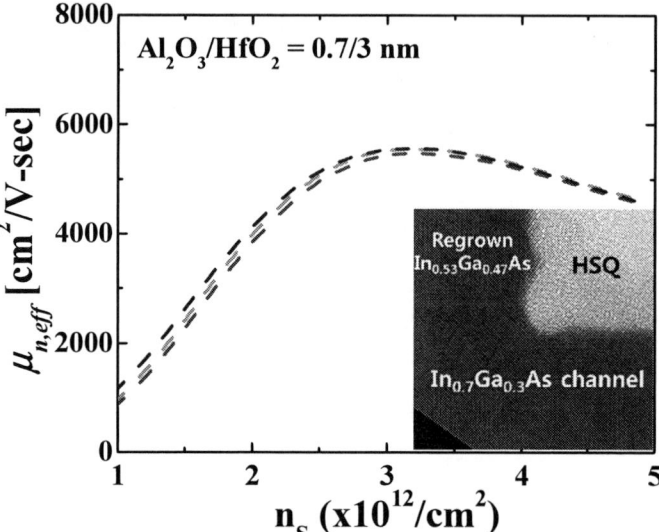

Fig. 8 Effective mobility ($\mu_{n,eff}$) as a function of carrier density (ns) for InGaAs MOSFET with epitaxial selectively-regrown S/D contact. This allows a gate-last integration scheme that prevents degradation of III-V gate-stack. This leads to a record $\mu_{n,eff}$ behavior, in excess of 5,500 cm^2/V-s at 300K [14].

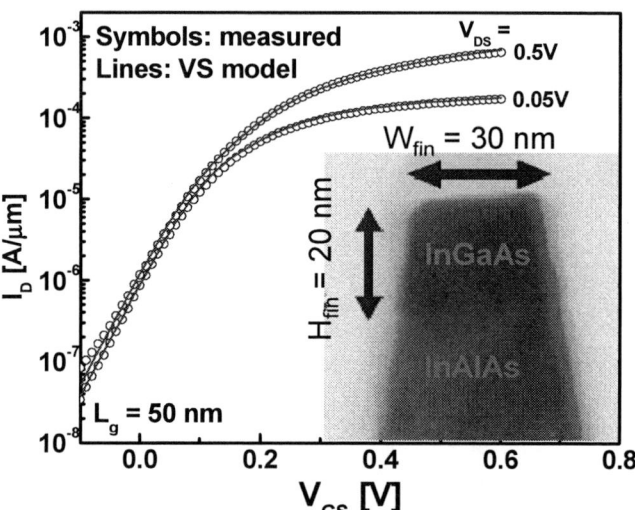

Fig. 9 Subthreshold characteristics of L_g = 50 nm InGaAs MOSFET with W_{fin}/H_{fin} = 30/20 nm. At V_{DS} = 0.5 V, the device exhibits SS < 80 mV/dec, DIBL < 20 mV/V, g_{m_max} = 1.5 mS/μm and I_{ON} = 380 mA/μm [8].

Fig. 10 highlights TCAD simulation result on the impact of D_{it} onto subthreshold-swing in double-gate (DG) InGaAs MOSFETs with L_g = 100 nm and EOT = 1 nm. Appropriate surface treatment and/or post-etch annealing process should be explored to further improve the sidewall MOS gate stack behavior. Finally, **Fig. 11** plots benchmarking of I_{ON} against L_g for the state-of-the-art InGaAs HEMT [5] and recently published InGaAs MOSFETs. Well-designed InGaAs MOSFETs exhibit a peak g_m of 2.7 mS/μm and a record I_{ON} ~ 0.4 mA/μm at I_{OFF} = 100 nA/mm and V_{DD} = 0.5 V.

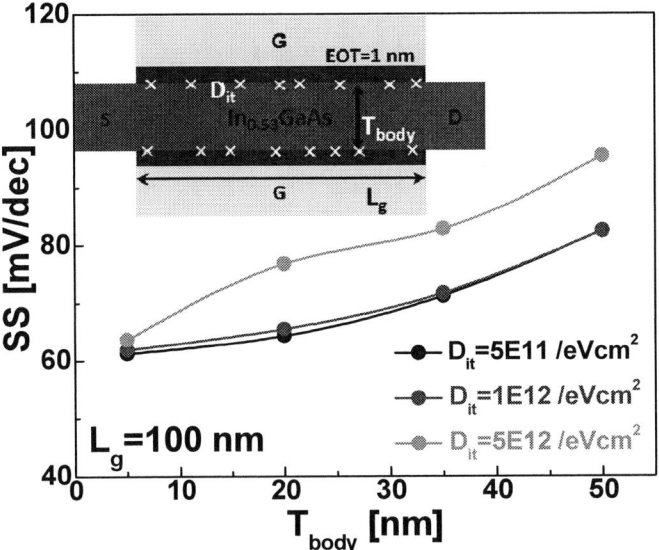

Fig. 10 Impact of D_{it} onto subthreshold-swing (SS) behavior in double-gate (DG) InGaAs MOSFETs with L_g = 100 nm and EOT = 1 nm, as a function of body thickness (T_{body}). T_{body} corresponds to fin width (W_{fin}) in tri-gate configuration.

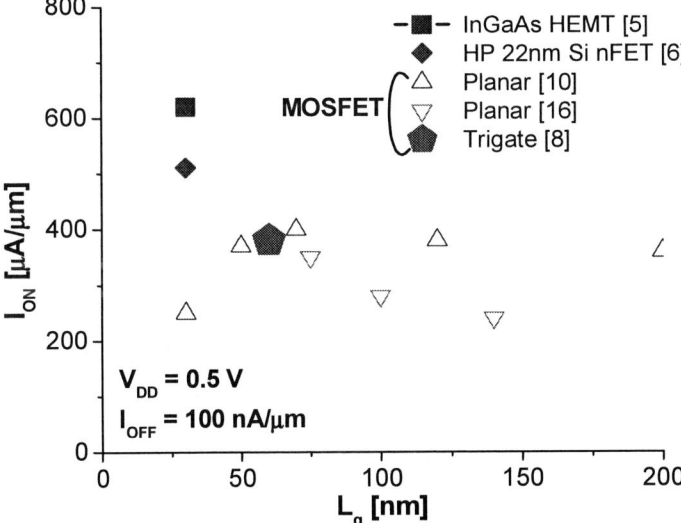

Fig. 11 Benchmarking of I_{ON} against L_g for state-of-the-art InGaAs HEMTs [5] and recent InGaAs MOSFETs with planar [10, 16] and non-planar architecture [8]. For all devices, I_{OFF} = 100 nA/mm and V_{DD} = 0.5 V.

Summary: CMOS might be about to take the disruptive step of replacing Si in the channel. If this happens, InGaAs is the most promising new channel material for n-MOSFETs. Remarkable progress has taken place, but many challenges

still remain. This paper reviews high-performance III-V transistor technology for future logic applications.

Ackonwledgement: The portion of KANC's research was supported by a grant from the R&D Program for Industrial Core Technology funded by the Ministry of Trade, Industry and Energy, Republic of Korea (Grant No. 10045216), and internal R/D program at KANC

Reference:
[1] G. Dewey *et al.*, VLSI Tech. Dig., p. 45 (2012).
[2] N. Waldron *et al.*, ECS Trans., Vol. 45, p. 115 (2012).
[3] R.S. Chau *et al.*, SSDM Ext. Abst., p. 68 (2003).
[4] D.-H. Kim *et al.*, IEDM Tech. Dig., p. 133 (2007).
[5] D.-H. Kim *et al.*, IEDM Tech. Dig., p. 765 (2011).
[6] C.-H. Jan *et al.*, IEDM Tech. Dig., p. 44 (2012).
[7] J. A. del Alamo, Nature, p. 317 (2011).
[8] T.-W. Kim et al., APEX, p. 074201 (2014).
[9] J. A. del Alamo *et al.* IEDM Tech. Dig., p. 24 (2013).
[10] J. Lin *et al.*, IEDM Tech. Dig., p. 421 (2013).
[11] SH. Lee *et al.*, APL, p. 233503 (2013).
[12] SW. Chang *et al.* IEDM Tech. Dig., p. 417 (2013).
[13] G. Zhou et al., IEDM Tech. Dig., p. 773 (2012).
[14] C.-S. Shin *et al.*, VLSI Tech. Dig., p. 36 (2014).
[15] H. C. Chin et al., IEEE EDL, pp. 805 (2009).
[16] M. Radosavljevic *et al.*, IEDM Tech. Dig., p. 765 (2011).
[17] T.-W. Kim *et al.*, IEDM Tech. Dig., p. 425 (2013).
[18] J. J. Gu *et al.*, IEDM Tech. Dig., p. 633 (2012).

High-Performance CMOS-Compatible Self-Aligned $In_{0.53}Ga_{0.47}As$ MOSFETs with G_{MSAT} over 2200 $\mu S/\mu m$ at V_{DD} = 0.5 V

Y. Sun, A. Majumdar, C.-W. Cheng, R. M. Martin, R. L. Bruce, J.-B. Yau, D. B. Farmer, Y. Zhu, M. Hopstaken, M. M. Frank, T. Ando, K.-T. Lee, J. Rozen, A. Basu, K.-T. Shiu, P. Kerber, D.-G. Park, V. Narayanan, R. T. Mo, D. K. Sadana, and E. Leobandung

IBM Research Division, T. J. Watson Research Center, Yorktown Heights, NY 10598
Phone: 914-945-3083; Email: yansun@us.ibm.com

ABSTRACT

We demonstrate high-performance self-aligned $In_{0.53}Ga_{0.47}As$-channel MOSFETs with effective channel length L_{EFF} down to 20 nm, peak transconductance G_{MSAT} over 2200 $\mu S/\mu m$ at L_{EFF} = 30 nm and supply voltage V_{DD} = 0.5 V, thin inversion oxide thickness T_{INV} = 1.8 nm, and low series resistance R_{EXT} = 270 $\Omega.\mu m$. These MOSFETs operate within 20% of the ballistic limit for $L_{EFF} \leq$ 30 nm and are among the best $In_{0.53}Ga_{0.47}As$ FETs in literature. We investigate the effects of channel/barrier doping on FET performance and show that increase in mobility beyond ~ 500 cm^2/Vs has progressively smaller impact as L_{EFF} is scaled down. Our self-aligned MOSFETs were fabricated using a CMOS-compatible process flow that includes gate and spacer formation using RIE, source/drain extension (SDE) implantation, and in-situ-doped raised source/drain (RSD) epitaxy. This process flow is manufacturable and easily extendable to non-planar architectures.

I. INTRODUCTION

Many III-V materials have significantly better electron transport properties than Si, and therefore, have been actively investigated as the NFET solution for high-performance CMOS applications [1]. While most III-V research has been focused on HEMTs [2]-[18] or MOS-HEMTs [19]-[32], very few works have focused on MOSFETs [33]-[39]. MOSFETs are required for high performance and device density scaling because MOS-HEMTs have high overlap capacitance while HEMTs have large foot-print and high gate leakage [36]. Therefore, we have focused on self-aligned MOSFETs and processes that are not only manufacturable but also compatible with CMOS applications [36]. In this work, we demonstrate high-performance self-aligned $In_{0.53}Ga_{0.47}As$ MOSFETs with L_{EFF} down to 20 nm and peak G_{MSAT} over 2200 $\mu S/\mu m$ at V_{DD} = 0.5 V. This work presents 2× higher performance compared to [36], and our devices are among the best $In_{0.53}Ga_{0.47}As$ FETs in literature.

II. DEVICE FABRICATION

The self-aligned MOSFET structure and our process flow are shown in Figs. 1 and 2, respectively. Details of device fabrication can be found in [36]. A TEM image of an entire device after processing to M1 level is shown in Fig. 3.

III. ELECTRICAL RESULTS

We first present data from $In_{0.53}Ga_{0.47}As$ channel MOSFETs with channel thickness T_{CH} = 20 nm and low

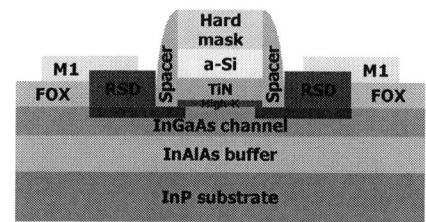

FIG. 1. Schematic of a self-aligned III-V MOSFET with implanted Si extensions and raised source/drain (RSD) grown by selective epitaxy. The devices are isolated using field oxide (FOX). The gate structure consists of high-κ gate dielectric, TiN metal gate, a-Si cap, and hard mask.

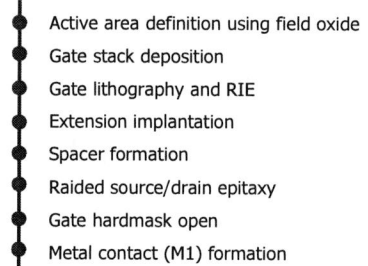

- Active area definition using field oxide
- Gate stack deposition
- Gate lithography and RIE
- Extension implantation
- Spacer formation
- Raided source/drain epitaxy
- Gate hardmask open
- Metal contact (M1) formation

FIG. 2. Process flow for fabricating self-aligned III-V MOSFETs. Our process flow is compatible with CMOS manufacturing, and includes multiple RIE steps, self-aligned SDEs, and self-aligned RSD.

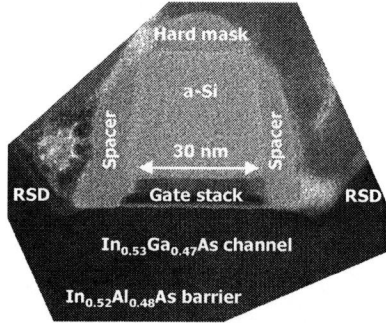

FIG. 3. TEM image of a self-aligned 20-nm-thick $In_{0.53}Ga_{0.47}As$ channel MOSFET after processing to M1 level. The gate length L_G = 30 nm. The gate stack consists of high-κ gate dielectrics and TiN metal gate.

channel/barrier doping N_A ~ 1 × 10^{17} cm^{-3}. We plot peak G_{MSAT} at drain bias V_{DS} = 0.5 V versus L_{EFF} in Fig. 4. We observe that peak G_{MSAT} > 2000 $\mu S/\mu m$ for $L_{EFF} \leq$ 60 nm. Our best devices have peak G_{MSAT} > 2200 $\mu S/\mu m$ at L_{EFF} = 30 nm. Also shown in Fig. 4 is our quasi-ballistic model [40], [41] fit to the data. Our model leads to excellent fits to the data and

978-1-4799-8002-4/14 $31.00 © 2014 IEEE

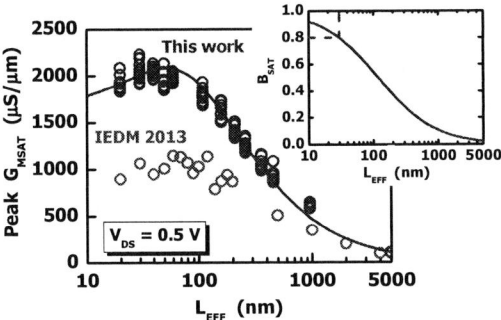

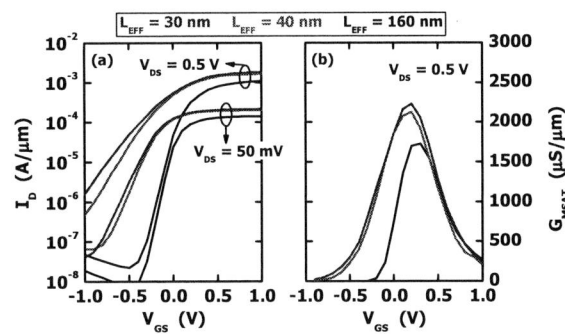

FIG. 4. Peak transconductance G_{MSAT} vs. effective channel length L_{EFF} of self-aligned 20-nm-thick In$_{0.53}$Ga$_{0.47}$As channel MOSFETs with channel/barrier doping $N_A \sim 1 \times 10^{17}$ cm^{-3} at drain bias $V_{DS} = 0.5$ V. The symbols are data (top 20 percentile) while the line is model fit. Inset: Intrinsic ballistic factor B_{SAT} vs. L_{EFF} at $V_{DS} = 0.5$ V. B_{SAT} is obtained from the model fit. Our best devices have peak $G_{MSAT} > 2230$ μS/μm at $L_{EFF} = 30$ nm while devices with $L_{EFF} \le 30$ nm operate within 20% of the ballistic limit.

FIG. 5. (a) Drain current I_D vs. gate voltage V_{GS} characteristics and (b) saturation transconductance G_{MSAT} vs. V_{GS} characteristics of self-aligned 20-nm-thick In$_{0.53}$Ga$_{0.47}$As channel MOSFETs with channel/barrier doping $N_A \sim 1 \times 10^{17}$ cm^{-3}, and effective channel length $L_{EFF} = 30$, 40, and 160 nm. The drain bias $V_{DS} = 50$ mV and 0.5 V in (a) and 0.5 V in (b). Our best device has peak $G_{MSAT} = 2230$ μS/μm at $L_{EFF} = 30$ nm and $V_{DS} = 0.5$ V.

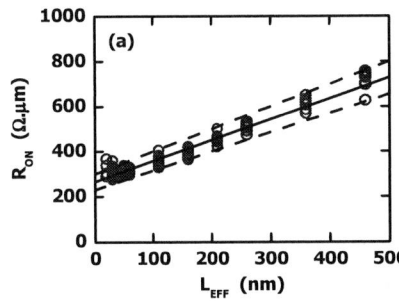

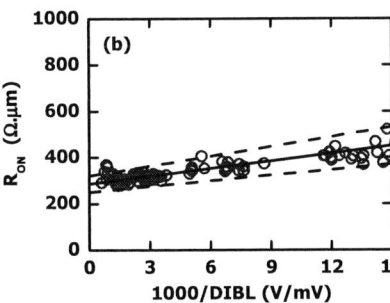

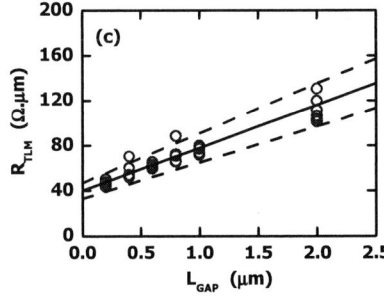

FIG. 6. (a), (b) On-resistance R_{ON} of self-aligned 20-nm-thick In$_{0.53}$Ga$_{0.47}$As channel MOSFETs with channel/barrier doping $N_A = 1 \times 10^{17}$ cm^{-3} plotted versus effective channel length L_{EFF} in (a) and versus 1/DIBL in (b). R_{ON} is extracted at constant gate overdrive $V_{GS} - V_{TLIN} = 0.7$ V and drain bias $V_{DS} = 50$ mV. The R_{ON} data is shown for the same set of devices shown in Fig. 4. In (b), 1/DIBL is used as a measure of L_{EFF}. We obtain series resistance $R_{EXT} = 265 \pm 35$ Ω.μm from (a), $R_{EXT} = 275 \pm 35$ Ω.μm from (b), and therefore, average $R_{EXT} = 270 \pm 40$ Ω.μm. The values of R_{EXT} obtained from (a) and (b) are within 4% of each other, which confirms that our estimate of L_{EFF} is fairly accurate. (c) Resistance of on-chip TLM structures R_{TLM} versus contact gap L_{GAP}. The $L_{GAP} = 0$ intercept gives total contact resistance $2R_C W = 40 \pm 7$ Ω.μm. The slope leads to sheet resistance of the raised source/drain $R_{S,RSD} = 38 \pm 6$ Ω/square.

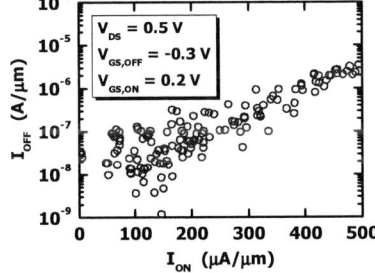

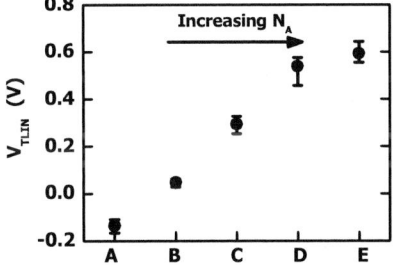

FIG. 7. Off-current I_{OFF} vs. on-current I_{ON} of self-aligned 20-nm-thick In$_{0.53}$Ga$_{0.47}$As channel MOSFETs with channel/barrier doping $N_A = 1 \times 10^{17}$ cm^{-3} at drain bias $V_{DS} = 0.5$ V. I_{OFF} is extracted at gate voltage $V_{GS,OFF} = -0.3$ V and I_{ON} at $V_{GS,ON} = V_{GS,OFF} + V_{DD} = 0.2$ V.

FIG. 8. Long-channel linear threshold voltage V_{TLIN} of five self-aligned 20-nm-thick In$_{0.53}$Ga$_{0.47}$As channel MOSFET wafers A-E with increasing channel/barrier doping N_A. V_{TLIN} is extracted at drain bias $V_{DS} = 50$ mV. As expected, V_{TLIN} increases with increasing N_A.

enables us to infer intrinsic ballistic factor B_{SAT} at $V_{DS} = 0.5$ V. As shown in the inset of Fig. 4, our short-channel devices with $L_{EFF} \le 30$ nm have $B_{SAT} \ge 0.8$, which implies that these FETs operate within 20% of the ballistic limit. Plots of drain current I_D and G_{MSAT} versus gate voltage V_{GS} characteristics of FETs with $L_{EFF} = 30$, 40, and 160 nm are shown in Fig. 5.

We plot on-resistance R_{ON} data in Figs. 6(a) and 6(b), and extract average $R_{EXT} = 270$ Ω.μm, which is 200 Ω.μm lower than that in [36]. This lower R_{EXT} is due to lower contact

resistance and lower RSD sheet resistance, as shown in Fig. 6(c). One should note that R_{EXT} extracted from the $L_{EFF} = 0$ intercept of R_{ON} versus L_{EFF} data includes the ballistic contact resistance R_B [41], [42]. Our simulations show that $R_B \sim 50$ Ω.μm, which is $\sim 20\%$ of the total R_{EXT} of our devices [42].

From Fig. 6(a), we obtain channel resistance $dR/dL = 935$ Ω/sq, which translates to effective mobility $\mu_{EFF} = 1000$ cm^2/Vs at sheet density $N_S = 6.7 \times 10^{12}$ cm^{-3}. This value of μ_{EFF} is $\sim 1.5\times$ higher than those reported in [36], and is due to

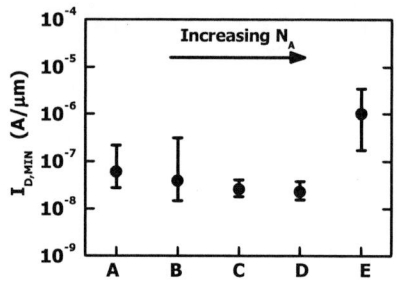 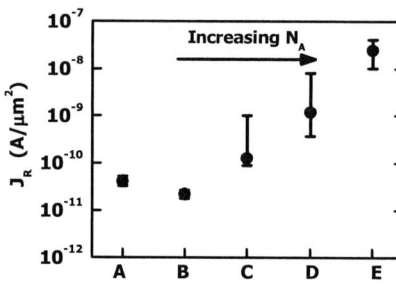 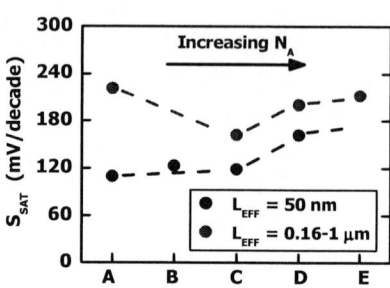

FIG. 9. (a) Minimum long-channel drain current $I_{D,MIN}$ of five self-aligned 20-nm-thick $In_{0.53}Ga_{0.47}As$ channel MOSFET wafers A-E with increasing channel/barrier doping N_A. $I_{D,MIN}$ is extracted at drain bias $V_{DS} = 0.5$ V. $I_{D,MIN}$ has a minimum as a function of N_A. (b) Reverse-bias leakage current density J_R of on-chip n^+ raised source/drain to p-type channel/barrier/substrate diodes from wafers A-E. J_R is extracted at reverse bias $|V_R| = 1$ V. At high N_A, J_R increases with increasing N_A due to reduction of tunneling barrier width. (c) Sub-threshold slope S_{SAT} ($V_{DS} = 0.5$ V) of short-channel (effective channel length $L_{EFF} = 50$ nm) and long-channel ($L_{EFF} = 0.16$ to 1 μm) MOSFETs from wafers A-E. The lines are guides to the eye. Short-channel S_{SAT} exhibits a minimum as a function of N_A due to an interplay of lower thermionic leakage and higher tunneling leakage at higher N_A.

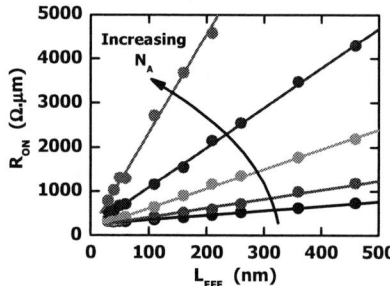

FIG. 10. On-resistance R_{ON} vs. effective channel length L_{EFF} of self-aligned 20-nm-thick $In_{0.53}Ga_{0.47}As$ channel MOSFETs with channel/barrier doping N_A in the range from $\sim 1 \times 10^{17}$ to $\sim 1 \times 10^{19}$ cm^{-3}. R_{ON} is extracted at constant gate overdrive $V_{GS} - V_{TLIN} = 0.7$ V and drain bias $V_{DS} = 50$ mV.

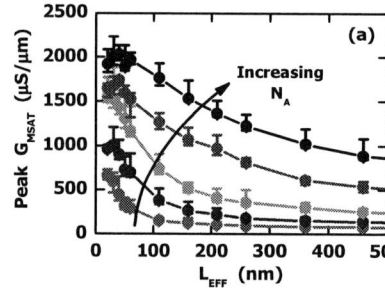 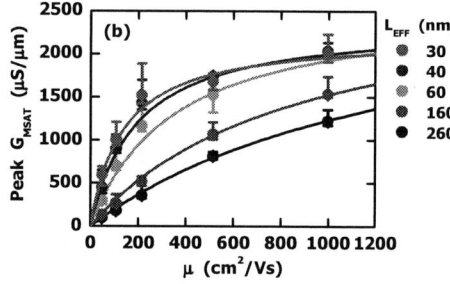

FIG. 11. (a) Peak saturation transconductance G_{MSAT} (drain bias $V_{DS} = 0.5$ V) vs. effective channel length L_{EFF} of self-aligned 20-nm-thick $In_{0.53}Ga_{0.47}As$ channel MOSFETs with channel/barrier doping N_A in the range from $\sim 1 \times 10^{17}$ to $\sim 1 \times 10^{19}$ cm^{-3}. (b) Peak G_{MSAT} vs. effective mobility μ_{EFF} for L_{EFF} is in the 30 to 260 nm range obtained for the data shown in (a) with μ_{EFF} being estimated from the slope dR/dL of the on-resistance R_{ON} vs. L_{EFF} data shown in Fig. 10. In (a) and (b), the symbols are 90 percentile data, while the error bars represent 80 and 100 percentile data. In (b), the lines are model fits to the data.

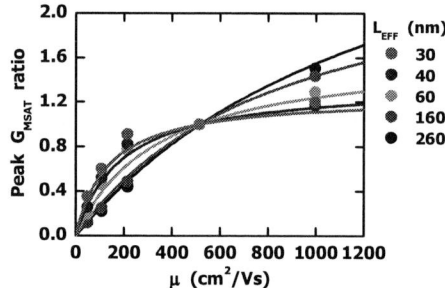

FIG. 12. Peak saturation transconductance G_{MSAT} ratio = peak G_{MSAT} / peak G_{MSAT} ($\mu_{EFF} = 515$ cm^2/Vs) vs. effective mobility μ_{EFF} of self-aligned 20-nm-thick $In_{0.53}Ga_{0.47}As$ channel MOSFETs with channel/barrier doping N_A in the range from $\sim 1 \times 10^{17}$ to $\sim 1 \times 10^{19}$ cm^{-3} and effective channel length $L_{EFF} = 30-260$. The symbols are data from Fig. 11(b) while the lines are model fits. We observe that an increase in μ_{EFF} beyond 515 cm^2/Vs has progressively smaller impact on G_{MSAT} as L_{EFF} is scaled down to below 40 nm.

lower N_A and improved gate dielectric process. Thus, the excellent performance shown in Figs. 4 and 5 is due to low R_{EXT}, thin T_{INV}, and high μ_{EFF}.

We plot off-current I_{OFF} versus on-current I_{ON} in Fig. 7 for $V_{DD} = 0.5$ V. Figs. 5 and 7 show that the shortest L_{EFF} devices have $I_{OFF} > 100$ nA/μm. In order to lower I_{OFF} and improve short-channel control, we fabricated a set of 20-nm-thick $In_{0.53}Ga_{0.47}As$ channel MOSFETs with N_A in the $\sim 10^{17}$ to $\sim$

10^{19} cm^{-3} range. As expected and shown in Fig. 8, higher N_A leads to higher long-channel threshold voltage.

As shown in Fig. 9(a), the minimum FET drain current $I_{D,MIN}$ has a minimum versus N_A. This is due to an interplay of lower classical thermionic leakage current (larger source-to-channel barrier) and higher quantum tunneling leakage current (thinner tunneling barrier) at higher N_A. The effect of higher N_A on tunneling current is clearly observed in reverse-bias leakage current density J_R of on-chip n^+-RSD to p-type channel/barrier/substrate diodes shown in Fig. 9(b). As shown in Fig. 9(c), long-channel sub-threshold slope S_{SAT} at $V_{DS} = 0.5$ V increases with increasing N_A, as expected, due to thinner depletion width. However, short-channel S_{SAT} exhibits a minimum versus N_A, which is again due to the interplay of lower thermionic leakage and higher tunneling leakage at higher N_A.

Higher N_A, however, leads to lower mobility, which is quite clear from the slope dR/dL of the R_{ON}-L_{EFF} data shown in Fig. 10. In Fig. 11(a), we show peak G_{MSAT} versus L_{EFF} for different values of N_A. We use dR/dL from Fig. 10 and $T_{INV} = 1.8$ nm to estimate effective mobility μ_{EFF}. We plot peak G_{MSAT} versus μ_{EFF} for $L_{EFF} = 30-260$ nm in Fig. 11(b) and replot this data as peak G_{MSAT} ratio versus μ_{EFF} in Fig. 12. In both Figs. 11(b) and 12, the lines are model fits to the data. We observe that an increase in μ_{EFF} beyond 515 cm^2/Vs has

978-1-4799-8002-4/14 $31.00 © 2014 IEEE

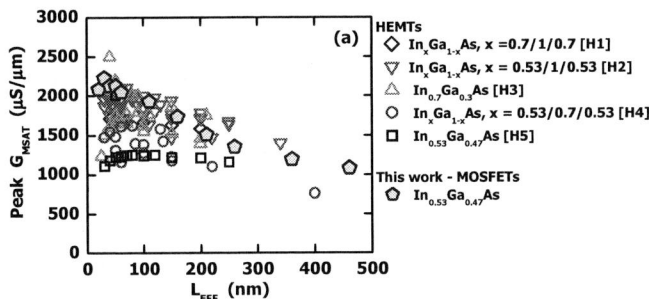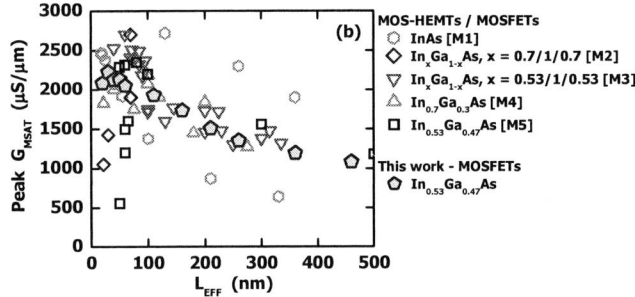

FIG. 13. Benchmarking of peak saturation transconductance G_{MSAT} (drain bias $V_{DS} = 0.5$ V) of our $In_{0.53}Ga_{0.47}As$ channel MOSFETs against best (a) III-V HEMTs, and (b) III-V MOS-HEMTs and MOSFETs. Our $In_{0.53}Ga_{0.47}As$ MOSFETs are among the best $In_{0.53}Ga_{0.47}As$ FETs [H5], [M5].

Benchmarking references: [H1] = [14]; [H2] = [3], [9], [11], [14]; [H3] = [4], [5], [15]-[18]; [H4] = [6]-[8], [10], [12]; [H5] = [2], [13]; [M1] = [30], [37]; [M2] = [35], [38]; [M3] = [22], [24], [25], [27]; [M4] = [20], [23], [29], [31]; [M5] = [21], [26], [28], [32], [39].

progressively smaller impact on G_{MSAT} as L_{EFF} is scaled down to below 40 nm. This conclusion is consistent with quasi-ballistic FET theory: for a given channel material (same electron effective mass m^*), as L_{EFF} is scaled down, there is less increase in G_{MSAT} due to an increase in effective mobility $\mu_{EFF} = q\tau/m^*$ (q = charge of an electron) via an increase in momentum relaxation time τ and mean free path λ [43], [44].

Finally, we benchmark the performance of our MOSFETs against the best III-V HEMTs, MOS-HEMTs, and MOSFETs in published literature. As shown in Fig. 13, our $In_{0.53}Ga_{0.47}As$ channel MOSFETs are among the best $In_{0.53}Ga_{0.47}As$ channel FETs.

V. CONCLUSIONS

Using a CMOS-compatible process flow that can be easily extended to non-planar architectures, we have demonstrated high-performance self-aligned $In_{0.53}Ga_{0.47}As$-channel MOSFETs with L_{EFF} down to 20 nm, peak $G_{MSAT} > 2200$ µS/µm at $L_{EFF} = 30$ nm and $V_{DD} = 0.5$ V, thin $T_{INV} = 1.8$ nm, and low $R_{EXT} = 270$ Ω.µm. Our devices with $L_{EFF} \leq 30$ nm operate within 20% of the ballistic limit and are among the best $In_{0.53}Ga_{0.47}As$ FETs in literature. In spite of the use of multiple RIE steps and SDE implantation, we obtain mobility $\mu = 1000$ cm²/Vs at high N_S for low channel/barrier N_A. We have also investigated the effects of channel/barrier doping on FET performance and shown that increase in μ beyond ~ 500 cm²/Vs has progressively smaller impact as L_{EFF} is scaled down to below 40 nm. Finally, we note that in order to further scale L_{EFF} and improve short-channel electrostatic integrity, future work will address the use of thinner $In_{0.53}Ga_{0.47}As$ channels.

ACKNOWLEDGEMENTS

We thank G. Shahidi and T. C. Chen for encouragement and management support. We thank the staff of IBM Yorktown Microelectronics Research Laboratory, J. J. Bucchignano, C. Cabral, and J. P. Silverman for help with device fabrication. We also thank D. A. Antoniadis, M. Chudzik, W. Haensch, H. Jagannathan, C.-H. Lin, and J. W. Sleight for stimulating discussions and helpful suggestions.

REFERENCES

[1] J. del Alamo, *Nature*, vol. 479, pp. 317-323, 2011. [2] S. Suemitsu *et al.*, *IEDM Tech. Dig.*, pp. 223-226, 1998. [3] D. Xu *et al.*, *IEEE Trans. Electron Devices*, vol. 47, pp. 33-43, 2000. [4] Y. Yamashita *et al.*, *IEEE Electron Device Lett.*, vol. 23, pp. 573-575, 2002. [5] K. Shinohara *et al.*, *IEEE Electron Device Lett.*, vol. 25, pp. 241-243, 2004. [6] D. H. Kim *et al.*, *IEDM Tech. Dig.*, pp. 1-4, 2005. [7] D. H. Kim and J. del Alamo, *IEDM Tech. Dig.*, pp. 1-4, 2006. [8] D. H. Kim *et al.*, *IEEE Trans. Electron Devices*, vol. 54, pp. 2606-2613, 2007. [9] D. H. Kim and J. del Alamo, *IEDM Tech. Dig.*, pp. 629-632, 2007. [10] D. H. Kim and J. del Alamo, *IEEE Trans. Electron Devices*, vol. 55, pp. 2546-2553, 2008. [11] D. H. Kim and J. del Alamo, *IPRM Tech. Dig.*, pp. 132-135, 2009. [12] T. W. Kim *et al.*, *IEDM Tech. Dig.*, p. 483-486, 2009. [13] G. Dewey *et al.*, *IEDM Tech. Dig.*, pp. 487-490, 2009. [14] T. W. Kim *et al.*, *IPRM Tech. Dig.*, pp. 1-4, 2010. [15] D. H. Kim *et al.*, *IEDM Tech. Dig.*, pp. 692-695, 2010. [16] T. W. Kim *et al.*, *IEDM Tech. Dig.*, pp. 696-699, 2010. [17] T. W. Kim *et al.*, *Electron. Lett.*, vol. 47, pp. 406-407, 2011. [18] D. H. Kim *et al.*, *IEDM Tech. Dig.*, pp. 319-322, 2011. [19] Y. Sun *et al.*, *IEDM Tech. Dig.*, pp. 1-4, 2008. [20] M. Radosavljevic *et al.*, *IEDM Tech. Dig.*, pp. 319-322, 2009. [21] M. Radosavljevic *et al.*, *IEDM Tech. Dig.*, pp. 765-768, 2011. [22] T. W. Kim *et al.*, *VLSI Tech. Dig.*, pp. 179-180, 2012. [23] D. H. Kim *et al.*, *IEDM Tech. Dig.*, pp. 761-764, 2012. [24] D. H. Kim *et al.*, *IEEE Electron Device Lett.*, vol. 34, pp. 196-198, 2013. [25] S. Lee *et al.*, *VLSI Tech. Dig.*, pp. T246-T247, 2013. [26] S. Lee *et al.*, *IPRM Tech. Dig.*, pp. 1-2, 2013. [27] S. Lee *et al.*, *App. Phys. Lett.*, vol. 103, art. 233503, 2013. [28] T. W. Kim *et al.*, *IEDM Tech. Dig.*, pp. 425-428, 2013. [29] C.-S. Shin *et al.*, *VLSI Tech. Dig.*, pp. 30-31, 2014. [30] S. Lee *et al.*, *VLSI Tech. Dig.*, pp. 54-55, 2014. [31] V. T. Arun *et al.*, *VLSI Tech. Dig.*, pp. 72-73, 2014. [32] X. Zhou *et al.*, *VLSI Tech. Dig.*, pp. 166-167, 2014. [33] Y. Q. Wu *et al.*, *IEEE Electron Device Lett.*, vol. 30, pp. 700-702, 2009. [34] U. Singisetti *et al.*, *IEEE Electron Device Lett.*, vol. 30, pp. 1128-1130, 2009. [35] J. Lin *et al.*, *IEDM Tech. Dig.*, pp. 757-760, 2012. [36] Y. Sun *et al.*, *IEDM Tech. Dig.*, pp. 48-51, 2013. [37] S. W. Chang *et al.*, *IEDM Tech. Dig.*, pp. 417-420, 2013. [38] J. Lin *et al.*, *IEDM Tech. Dig.*, pp. 421-424, 2013. [39] N. Waldron *et al.*, *VLSI Tech. Dig.*, pp. 26-27, 2014. [40] A. Majumdar and D. A. Antoniadis, *IEEE Trans. Electron Devices*, vol. 61, pp. 351-358, 2014. [41] A. Majumdar *et al.*, to appear in *IEEE Trans. Electron Devices*, 2014. [42] M. Rodwell *et al.*, *IPRM Tech. Dig.*, pp. 1-6, 2010. [43] M. S. Lundstrom, *IEEE Electron Device Lett.*, vol. 18, pp. 361-363, 1997. [44] M. S. Lundstrom *IEEE Electron Device Lett.*, vol. 22, pp. 293-295, 2001.

978-1-4799-8002-4/14 $31.00 © 2014 IEEE

Low Power III-V InGaAs MOSFETs Featuring InP Recessed Source/Drain Spacers with I_{on}=120 µA/µm at I_{off}=1 nA/µm and V_{DS}=0.5 V

C. Y. Huang[1], S. Lee[1], V. Chobpattana[2], S. Stemmer[2], A. C. Gossard[1,2], B. Thibeault[1], W. Mitchell[1], M. Rodwell[1],

[1]ECE Department, [2]Materials Department, University of California, Santa Barbara, CA 93117, USA

Phone:1-805-886-2630, Email: cyhuang@ece.ucsb.edu

Abstract

We report InGaAs-channel MOSFETs using recessed InP spacer layers in the regrown source and drain. By replacing narrow band-gap InGaAs with wide band-gap InP within the high-field region near the drain end of the channel, band-to-band tunneling (BTBT) leakage is significantly reduced. A 30 nm gate length device using InP spacers shows a minimum I_{off}~60 pA/µm, approximately 100:1 smaller than a similar device using InGaAs source/drain spacers. A FET using InP spacers, with 45 nm gate length, and with a 3 nm ZrO_2 gate oxide shows I_{on}=150 µA/µm at I_{off}=1 nA/µm and V_{DS}=0.5 V. The low off-state leakage current observed with InP source/drain spacers makes InGaAs MOS technology viable for low-power logic.

Introduction

Planar ultra-thin-channel InAs III-V MOSFETs have recently shown performance (0.5 mA/µm I_{off} at V_{DS}=0.5 V) comparable to Si 22 nm FinFETs for ITRS high performance (HP) applications [1] where the off-state leakage current I_{off} is set at 100 nA/µm. Though In(Ga)As MOSFETs have high on-state transconductance (g_m), the small band-gap of the In(Ga)As channel gives rise to high I_{off} at short gate lengths (L_g) through band-to-band tunneling (BTBT) [2], rendering prior reported III-V MOSFETs unsuitable for low-power (LP; I_{off}=30 pA/µm) and standard-performance (SP; I_{off}=1 nA/µm) applications [1-10]. Recently, we reported that either InP channel cap layers or recessed InP source/drain (S/D) spacers can further reduce I_{off} because InP (E_g~1.35 eV) is less prone to BTBT [11]. Similarly, Mo et al. reported reduced leakage current using an asymmetric InP drain electrode [12]. In [11], placing InP in the regions of highest electric field reduced I_{off} to 1 nA/µm, but thick InP spacers also increased the parasitic source/drain resistance and degraded I_{on}. Here we report low-leakage InGaAs MOSFETs with gate lengths as small as 22~30 nm. Thinning the InGaAs channel from 4.5 nm to 3 nm allows low I_{off} (1 nA/µm at V_{DS}=0.5 V), but degrades the transconductance and I_{on}. Using instead a 4.5 nm InGaAs channel with a partially recessed, doping-graded InP spacer, the minimum I_{off} can be reduced to 60 pA/µm at 30 nm L_g. With a recessed InP spacer and thin oxide, a high 150 µA/µm on-state current I_{on} at I_{off}=1 nA/µm can be obtained for 45 nm-L_g devices. This high on-off ratio ($1.5 \cdot 10^5$) in a short L_g device demonstrates the potential for III-V MOSFETs in low-power logic.

Fig. 1 shows a progression of III-V MOSFETs designs for reduced I_{off}. P-doped InAlAs or wider-bandgap AlAsSb back barriers reduce back barrier leakage currents [13]. Particularly for InAs, with its small band gap, BTBT leakage arises primarily in the high field region, either at the drain

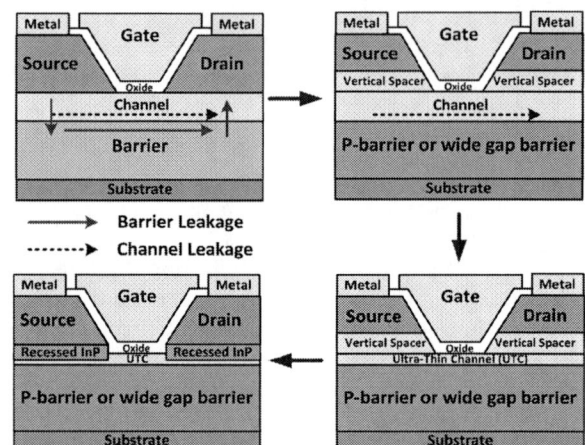

Fig. 1: Progression of III-V MOSFET designs for reduced off-state leakage current.

end of the channel next to the gate edge or at the junction between the channel and the regrown heavily-doped drain [14]. Thinning the channel increases the quantized band-gap, reducing BTBT, and improves electrostatics, improving the subthreshold swing (S.S.). Drain vertical field spacers reduce the gate-drain field, reducing BTBT, and also improve electrostatics [1]. Unfortunately, in thin channels, electron mobility is decreased and electron effective mass increased. Here, we show that using an InP recessed spacer provides an alternative means to reduce the I_{off} leakage floor without extreme thinning of the channel.

Device Fabrication

Epitaxial layers were first grown with 6 nm InGaAs channels and P-doped InAlAs barriers (**Fig. 2**). To form dummy gates, 30 nm hydrogen silsesquioxane (HSQ) was spun and patterned by e-beam lithography. The vertical spacer and N+ source/drain contact layers, either InGaAs or InP, were selectively regrown by MOCVD. Device mesas were isolated and the dummy gates removed in buffered oxide etch. For all samples, in the gate region, the channel surface was removed (2 cycles for Sample B; 1 cycle for the others) by digital etching to remove damage and adjust the channel thickness [15]. Sample A is the control sample, with a 4.5 nm thick channel, and was reported in [11]. Sample B has a 3 nm thick channel and an InGaAs spacer, while the other samples (sample C, D, and E) have a 4.5 nm thick InGaAs channel with an InP source/drain spacer recessed ~1.5 nm into the channel. The samples were then immediately transferred into

978-1-4799-8002-4/14 $31.00 © 2014 IEEE

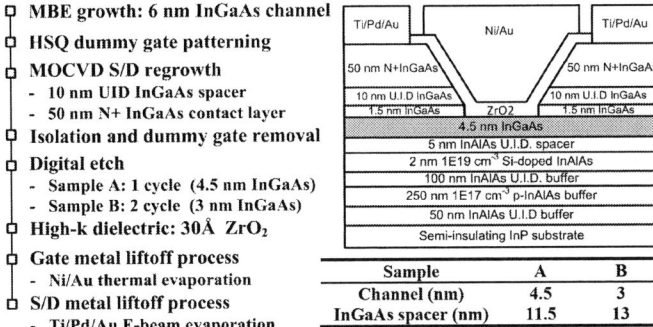

- ☐ MBE growth: 6 nm InGaAs channel
- ☐ HSQ dummy gate patterning
- ☐ MOCVD S/D regrowth
  - 10 nm UID InGaAs spacer
  - 50 nm N+ InGaAs contact layer
- ☐ Isolation and dummy gate removal
- ☐ Digital etch
  - Sample A: 1 cycle (4.5 nm InGaAs)
  - Sample B: 2 cycle (3 nm InGaAs)
- ☐ High-k dielectric: 30Å ZrO₂
- ☐ Gate metal liftoff process
  - Ni/Au thermal evaporation
- ☐ S/D metal liftoff process
  - Ti/Pd/Au E-beam evaporation

Sample	A	B
Channel (nm)	4.5	3
InGaAs spacer (nm)	11.5	13

Fig. 2: Gate-last process flow and device structures of samples A and B.

the ALD reactor, and pre-cleaned and passivated by cycles of N_2 plasma and TMA [16]. ZrO_2 gate dielectric was then deposited (38 Å for sample E, 30±1 Å for all others). The samples were then annealed at 400 °C in forming gas for 15 minutes. Ni/Au was thermally deposited as the gate electrode, and Ti/Pd/Au source/drain contacts were deposited.

Results and Discussion

Figs. 3 and **4** show the transfer and output characteristics of samples A and B. Thinning the channel from 4.5 nm to 3 nm reduces the minimum I_{off} to 3.5 nA/μm but degrades I_{on} and the transconductance. **Fig. 5** compares the gate I_G and drain I_D leakage for sample B over a larger bias range; the minimum off-state leakage is still dominated by BTBT rather than by gate leakage. **Fig. 6** shows the computed band structures of samples A and B. Thinning the channel increases the quantized band gap, thus reducing BTBT. However, even for a 3 nm InGaAs channel, the minimum BTBT leakage floor is still ~1 nA/μm at V_{DS}=0.5 V. For lower-power logic applications, we seek to further reduce BTBT leakage while maintaining high on-state performance.

Fig. 7 shows the structures of samples C, D and E. All have a 4.5 nm InGaAs channel, and have InP source/drain spacers, with the regrowth interface recessed 1.5 nm below the oxide/III-V interface. Sample C has a 5 nm undoped InP spacer. Samples D and E have a 5 nm undoped spacer, and above it an 8 nm linearly doping-graded InP spacer. Samples D and E have 30 Å and 38 Å ZrO_2, respectively.

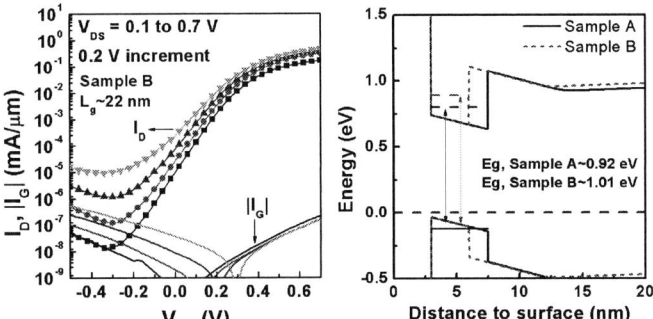

Fig. 5: I_G-V_{GS} and I_D-V_{GS} characteristics of L_g=22 nm devices on sample B.

Fig. 6: Energy band diagram in the channel for samples A and B.

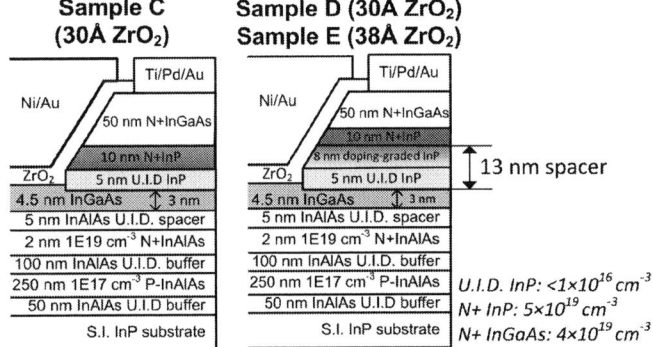

Fig. 7: Device structures of samples C, D, and E with a recessed InP source/drain spacer. All the samples have the symmetric source/drain and only the drain side is shown here.

Fig. 8 and **Fig. 9** compare the transfer and output characteristics of samples A and C at L_g=60 nm. The off-state leakage floor of sample C is reduced 10:1 at V_{DS}=0.7 V and 3:1 at V_{DS}=0.5 V. For sample C, the minimum I_{off} is limited by gate leakage I_g, while in sample A the minimum I_{off} is limited by BTBT. All samples have a ~1 μm overlap between the gate and the source and drain, hence I_g would be 10:1-50:1 smaller in a self-aligned device. As compared to FETs

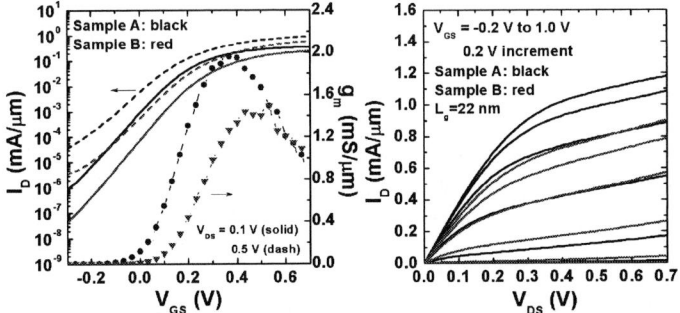

Fig. 3: Transfer characteristics of sample A and sample B.

Fig. 4: Output characteristics of sample A and sample B.

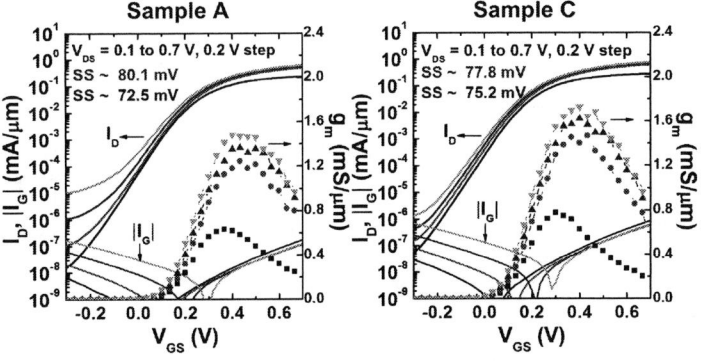

Fig. 8: Transfer characteristics of sample A and C for 60 nm L_g. The minimum off-state leakage floor is dominated by BTBT for sample A and gate leakage for sample C.

978-1-4799-8002-4/14 $31.00 © 2014 IEEE

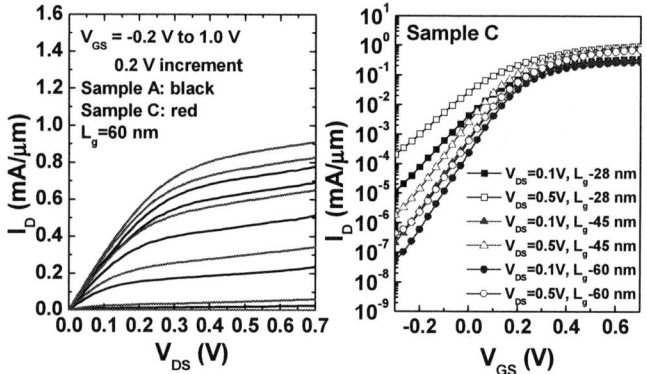

Fig. 9: Output characteristics of samples A and C at L_g=60 nm.

Fig. 10: I_D-V_{GS} characteristics vs. L_g for sample C.

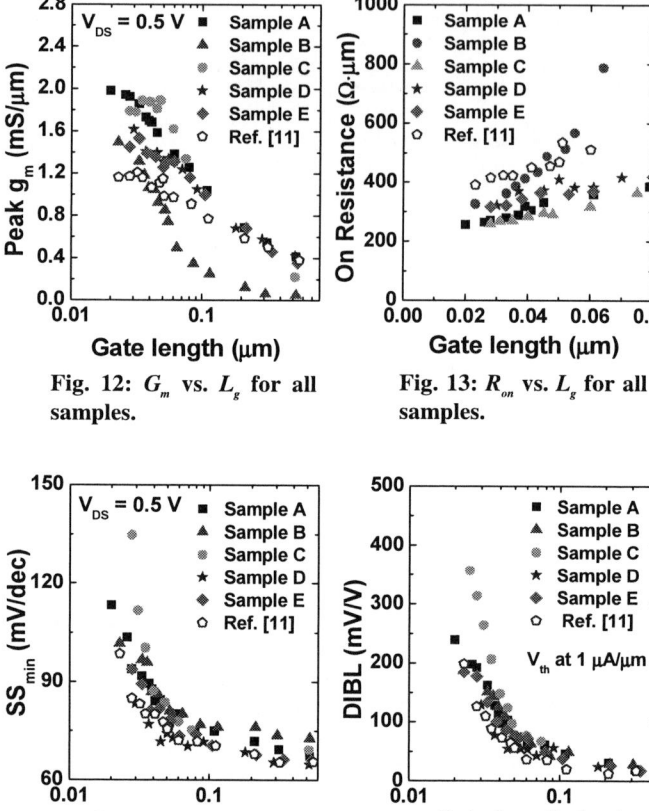

Fig. 12: G_m vs. L_g for all samples.

Fig. 13: R_{on} vs. L_g for all samples.

Fig. 14: SS vs. L_g for all samples.

Fig. 15: *DIBL* vs. L_g for all samples.

using 13 nm InP spacers [11], sample C, with 5 nm InP spacers, has smaller source/drain resistance. Higher added resistance is observed with InP than InGaAs spacers; we attribute this to the 0.2 eV InP-InGaAs band offset and the consequently smaller surface inversion sheet charge density of InP spacers. Although thinner InP spacers reduce R_{on} and increase I_{on}, FETs with thinner InP spacers have poorer electrostatics, hence increased *S.S.*, at short gate lengths (**Fig. 10**). To maintain good electrostatics, hence low *S.S.*, the spacer must have some minimum thickness, yet for low BTBT leakage only a fraction of this at the high-field region need be InP; thick, fully-depleted InP spacers reduce the on-state transconductance. For simultaneous high I_{on} and low I_{off}, this suggests the use of spacers alloy-graded from InP to InGaAs. Alternatively, a doping-graded InP spacer would be lightly depleted in the source, minimizing access resistance, yet heavily depleted in the drain, minimizing BTBT and *S.S.*.

Fig. 11 shows the transfer characteristics of samples D and E. By using a 5 nm undoped recessed InP spacer in combination with an 8 nm doping-graded InP spacer, the drain electric field is further smoothed, and low BTBT leakage and high I_{on} are achieved. The L_g=30 nm devices show 300 pA/μm minimum off-state leakage at V_{DS}=0.5 V, being limited by gate leakage. The peak transconductance is 1.6 mS/μm.

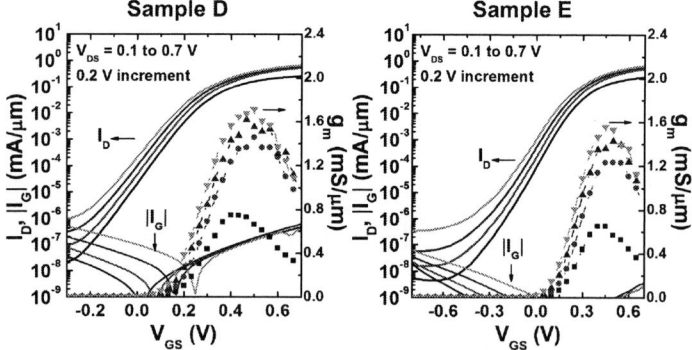

Fig. 11: I_G-V_{GS} and I_D-V_{GS} characteristics of L_g=30 nm devices for samples D (30 Å ZrO₂) and E(38 Å ZrO₂).

Further increasing the oxide thickness from 30 Å (sample D) to 38 Å(sample E) decreases the gate leakage, and the minimum I_{off} is reduced to 60 pA/μm, with a 100:1 smaller BTBT leakage floor than obtained using InGaAs source/drain spacers (sample A, **Fig. 3**). To our knowledge, this is the lowest leakage current observed in an InGaAs MOSFET at a VLSI-relevant gate length.

Fig. 12 shows the variation of g_m with L_g. Compared to [11], the on-state performance is greatly improved by thinning the InP spacer from 13 nm to 5 nm (Sample C). Using a doping-graded InP spacer, high g_m and low I_{off} can be obtained, with g_m reduced 10~20% compared to that with InGaAs spacers at small L_g. **Fig. 13** shows R_{on} vs. L_g; reducing the InP spacer thickness reduces R_{on}. The R_{on} extrapolated to zero L_g is ~207 Ω·μm for an 11.5 nm InGaAs spacer (Sample A), ~199 Ω·μm for a 5 nm InP spacer (Sample C), and 260~280 Ω·μm for a 13 nm doping-graded spacer (Samples D, E). Compared to [11], linearly doping-graded InP spacers improve the on-resistance from 364 Ω·μm to ~270 Ω·μm, indicating significantly reduced parasitic source/drain resistance.

Fig. 14 and **Fig. 15** show the variation of *S.S.* and DIBL with L_g. Increasing the spacer thickness improves DIBL through improved electrostatics, and improves *S.S.* through both

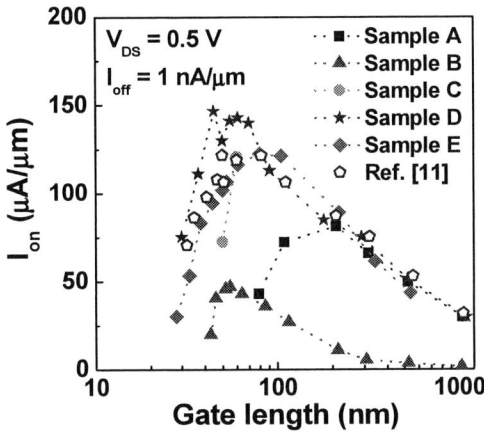

Fig. 16: I_{on} vs. L_g at I_{off}=1 nA/μm and V_{DS}=0.5 V.

improved electrostatics and reduced BTBT. Sample C shows high *S.S.* and large DIBL at small L_g because of poor electrostatics from the thin spacer (**Fig. 10**). In contrast, with a 13 nm doping-graded InP spacer (Samples D, E, also [11]), SS_{min} is ~90 mV/dec at L_g=30 nm.

Fig. 16 shows I_{on} vs. L_g at I_{off}=1 nA/μm and V_{DS}=0.5 V. Samples A and B shows low I_{on} at smaller L_g because the high BTBT leakage not only degrades the *S.S.*, but also increases I_{off} above 1 nA/μm at small L_g. Instead of further thinning the channel, samples C, D and E use recessed InP spacers, and show significantly improved I_{on} at small L_g due to reduced BTBT leakage. Sample D shows the maximum peak I_{on}=150 μA/μm at L_g= 45 nm. Sample E shows slightly reduced I_{on}, due to the thicker gate dielectric, which decreases g_m and increases *S.S.* For samples C, D, and E, I_{on} decreases rapidly as L_g is reduced below 40 nm. This is a consequence of poor electrostatics, hence large *S.S.*, at these gate lengths.

Fig. 17 shows I_{on} vs. L_g at V_{DS}=0.5 V, but at a larger I_{off}=100 nA/μm, benchmarking recent III-V MOSFETs. The FETs reported here use $In_{0.53}Ga_{0.47}As$ channels and show performance comparable to leading III-V FETs. Given an I_{off}=100 nA/μm metric, the 2.7-nm-thick InAs channel MOSFETs of [1] show highest I_{on}, as a consequence of larger gate capacitance and good electrostatics. In contrast, for low-power applications, a wider band-gap $In_{0.53}Ga_{0.47}As$ channel more readily provides low leakage current. To further improve I_{on} at small L_g, a tri-gate or nanowire structure would provide improved electrostatics and hence improved *S.S.* Combined with recessed InP source/drain spacers for low BTBT leakage, InGaAs MOSFETs would then be suitable for low-power logic.

Conclusions

We have demonstrated that recessed InP source/drain spacers can reduce band-to-band leakage current in InGaAs-channel MOSFETs while maintaining high transconductance. The minimum I_{off} for a 30 nm gate length device was 60 pA/μm,

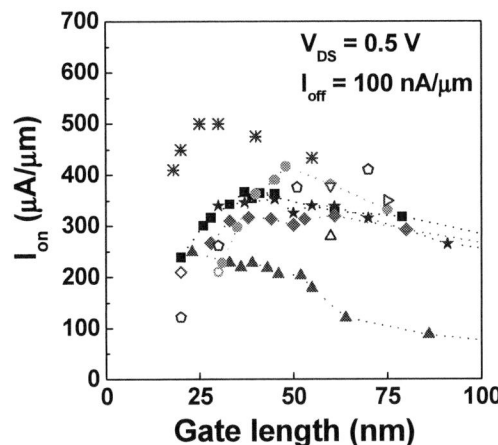

Fig. 17: I_{on} vs. L_g at I_{off}=100 nA/μm and V_{DS}=0.5 V and the benchmark with recent III-V FETs.

100:1 smaller than the leakage of a FET using InGaAs S/D spacers. The devices with thin oxide show the on-state current I_{on}=150 μA/μm at I_{off}=1 nA/μm and V_{DS}=0.5 V. These recessed InP source/drain spacers greatly reduces the leakage floor, making III-V InGaAs MOSFETs feasible for low-power logic.

Acknowledgement

This work was supported by the SRC Non-classical CMOS Research Center (Task 1437.009). A portion of this work was done in the UCSB Nanofabrication facility, part of the NSF funded NNIN network and MRL Central Facilities supported by the MRSEC Program of the NSF under award No. DMR 1121053.

References

[1] S. Lee et al., VLSI Symp. Tech. Dig., 2014, pp. 54-55.
[2] J. Lin et al., IEDM 2013, pp. 421-424.
[3] M. Radosavljevic et al. IEDM 2009, pp. 319-322.
[4] M. Radosavljevic et al. IEDM 2011, pp. 765-768.
[5] J. J. Gu et al., IEDM 2012, pp. 633-636.
[6] M. Egard et al., IEDM 2011, pp. 303-306.
[7] S. W. Chang et al., IEDM 2013, pp. 417-420.
[8] K. H. Goh et al., IEDM 2013, pp. 433-436.
[9] T. W. Kim et al., IEDM 2013, pp. 425-428.
[10] D. H. Kim et al., IEDM 2012, pp. 761-764.
[11] C. Y. Huang et al., Proceedings of Lester Eastman Conference 2014, in press.
[12] J. Mo et al., Appl. Phys. Lett. 105, 033516 (2014).
[13] C. Y. Huang et al., App. Phys. Lett. 103, 203502 (2013).
[14] R. Chu et al., Electron Dev. Lett. 29, pp. 974-976 (2008).
[15] S. Lee et al., IEEE IPRM 2013, 1.
[16] V. Chobpattana et al., Appl. Phys. Lett. 104, 182912 (2014).

978-1-4799-8002-4/14 $31.00 © 2014 IEEE

InGaAs/InAs Heterojunction Vertical Nanowire Tunnel FETs Fabricated by a Top-down Approach

Xin Zhao, Alon Vardi and Jesús A. del Alamo

Microsystems Technology Laboratories, Massachusetts Institute of Technology, Cambridge, MA 02139, USA

Email: xinzhao@mit.edu, Phone: 1-857-756-6001

Abstract

We demonstrate for the first time InGaAs/InAs heterojunction single nanowire (NW) vertical tunnel FETs fabricated by a top-down approach. Using a novel III-V dry etch process and gate-source isolation method, we have fabricated 50 nm diameter NW TFETs with a channel length of 60 nm and EOT=1.2 nm. Thanks to the insertion of an InAs notch, high source doping, high-aspect ratio nanowire geometry and scaled gate oxide, an average subthreshold swing (S) of 79 mV/dec at V_{ds}= 0.3 V is obtained over 2 decades of current. On the same device, I_{on}= 0.27 μA/μm is extracted at V_{dd}= 0.3 V with a fixed I_{off}= 100 pA/μm. This is the highest ON current demonstrated at this OFF current level in NW TFETs containing III-V materials.

Introduction

In light of the increased emphasis on energy efficiency in electronics, the tunnel FET (TFET) has become attractive due to its potential for low voltage operation [1]. In TFETs, InGaAs-based heterojunctions promise a combination of steep slope, high ON-current due to the reduced tunnel barrier height [2], and a well passivated surface. To enable continued scaling, a nanowire (NW) transistor geometry with wrapped-around gate is highly favorable due to the strong charge control and its robustness to short-channel effects [3]. To date, vertical NW TFETs with III-V materials have only been demonstrated through bottom-up techniques with complex manufacturing issues [4-8]. In this work, for the first time, we demonstrate InGaAs/InAs heterojunction vertical NW TFETs fabricated via a more manufacturing relevant top-down approach. Devices with a diameter (D) of 50 nm and EOT= 1.2 nm exhibit I_{on}= 0.27 μA/μm at a fixed I_{off}= 100 pA/μm, and V_{dd}= 0.3 V. This is the highest ON current demonstrated at this OFF current among NW TFETs based on III-V materials.

Fabrication Process

Fig. 1(a) shows a schematic view of the transistor fabricated in this work. The starting heterostructure, grown by MBE on an InP wafer, is similar to that in [2]. The tunneling junction consists of a p^+-i $In_{0.53}Ga_{0.47}As$ heterostructure in which a 2 nm i-InAs/8 nm i-$In_{0.7}Ga_{0.3}As$ "notch" has been inserted to reduce the tunnel barrier height and yield steeper subthreshold characteristics and high ON current [2]. The p^+ source and n^+ drain have a nominal 10^{20} cm^{-3} C and 6×10^{19} cm^{-3} Si doping, respectively. In Fig. 1(b), the energy band diagram along the nanowire is simulated with *Nextnano3* by solving Schrodinger-Poisson self-consistently in a double-gate geometry

(V_{ds}=0 V, V_{gs}=0.5 V). The efficacy of the InAs/$In_{0.7}Ga_{0.3}As$ notch to lower the tunnel barrier is evident [2].

The process flow is described in Fig. 2. We leverage process and etching technologies used in our previous work [9]. One major change is an improved RIE technology [10] to define the NW. Fig. 3 shows a comparison of D= 60 nm InGaAs NW etched in [9] and in this work. By increasing the substrate temperature during etch and optimizing the etching conditions, we are able to realize more vertical and smooth sidewalls. This improvement ensures tighter control of the NW diameter and yields better scalability. An exponent of the capabilities of this technology is shown in Fig. 4 which features a D=15 nm scale InGaAs NW with an aspect ratio greater than 15.

After NW formation, we introduce a spin-on glass (SOG) planarization and etch back step. This forms a 50 nm SOG layer that covers the bottom p^+ source on which the gate stack is deposited. In this way, we reduce the source-to-gate leakage current that affected our previous NW-MOSFETs [9], and are able to scale the gate dielectric thickness. Our process proceeds with a digital etch [11] to smooth the NW sidewalls and reduce its diameter to 50 nm. 2.5 nm ALD Al_2O_3 (EOT=1.2 nm) and W metal are deposited as gate stack. A 30 min 350°C forming gas anneal is performed. Mo is sputtered as contact metal to both p^+ source and n^+ drain. All devices have a final diameter in the intrinsic region of 50 nm and a channel length of 60 nm given by the undoped InAs/InGaAs layer thickness.

Results and Discussion

Fig. 5 shows subthreshold characteristics of one of the best performing single-NW TFETs. A subthreshold swing of 75 mV/dec averaged over I_d from 10^{-9} to 10^{-7} A/μm is obtained at V_{ds}= 0.3 V. ON current of 0.27 μA/μm is extracted with I_{off}= 100 pA/μm and V_{dd}= 0.3 V (V_{ds}=0.3 V, ΔV_{gs}=0.3 V). Interestingly, S slightly improves at higher V_{ds} [4]. The ON/OFF current ratio exceeds 10^5 in this device. The gate leakage current is below 10^{-12} A/μm in the subthreshold regime. The drain current fluctuations are attributed to the single NW nature of the device [4]. As a result, we report average values for the figures of merit. Hysteretical behavior is observed in the subthreshold characteristics. Measurements in a narrower V_{gs} range from -0.2 to 0 V yield S=79 mV/dec when averaging both sweeping directions.

The transfer characteristics (g_m obtained after smoothing due to current fluctuations) at V_{ds}= 0.3 V are shown in Fig. 6. A peak g_m value of 8 μS/μm is obtained. The output characteristics of

978-1-4799-8002-4/14 $31.00 © 2014 IEEE

the same device are shown in Fig. 7. Triode-like characteristics are observed [5] with a typical low V_{ds} super-linear behavior characteristics of TFETs [12]. Fig. 8 shows the output characteristics in a semilog scale including the reverse regime. Clear negative differential resistance (NDR) is observed for $V_{ds}<0$ and high V_{gs}, confirming the tunneling nature of the device operation in the ON regime. At $V_{gs}=0.8$ V, we observe a peak-to-valley ratio in I_d of 6.2. This is the highest value reported at room temperature in III-V NW TFETs.

Across the sample a spread of device characteristics is observed. Fig. 9 presents subthreshold characteristics of three different single-NW TFETs including the one shown in Figs. 5-8 (black curve) at $V_{ds} = 0.3$ V. The device with the most positive V_t shows the steepest subthreshold regime. Output characteristics of these devices in an identical scale are presented in Fig. 10. Devices with positive V_t show triode-like characteristics but devices with negative V_t exhibit saturating behavior and significantly more current. While low-V_{ds} super-linear behavior is observed at all V_{gs} values in devices with the most positive V_t, devices with more negative V_t only show super-linear onset at high V_{gs}. At low V_{gs}, MOSFET-like turn-on with V_{ds} is observed. The reasons for the wide distribution of device characteristics are not clear at this moment, but the high sensitivity of TFET characteristics to small geometrical variations is likely a contributing factor.

To further understand the physics of device operation, temperature (T) dependent measurements on another device (with relatively negative V_t) are performed. Fig. 11 shows the I_d-V_{ds} characteristics at $V_{gs}= 0.5$ V and different T in a semilog scale. NDR is clearly seen at all T in the reverse regime. I_d in the low $V_{ds}<0$ region shows little T dependence, confirming direct band-to-band tunneling (BTBT) current conduction. In the positive V_{ds} region, I_d increases slightly with T, which can be attributed to bandgap reduction with increasing T [13].

In Fig. 12, subthreshold characteristics at different T and V_{ds} =0.05 V are shown. A T-independent leakage current floor below pA/um range is reached, a unique feature of the NW geometry not seen in planar TFETs [13]. Another observation in Fig. 12 is the sharp saturation of I_d at high V_{gs}, which could be due to high interface states (D_{it}) density inside InGaAs conduction band or the depletion of p+ source in the overlapping region with the gate. The subthreshold current is very sensitive to T, which is not expected from a pure BTBT conduction mechanism. This suggests that a thermal process is involved. The average S at $V_{ds}=0.05$ V is shown as a function of T in Fig. 13, indicating that ideal behavior is never reached. The Arrhenius plot [ln $(I_d/T^{3/2})$ vs. $(1/kT)$] at several V_{gs} in the subthreshold regime is shown in Fig. 14. As observed in Fig. 15, the extracted thermal barrier height ($q\Phi_B$) from the Arrhenius plot is linearly dependent on V_{gs}. T-dependent gate efficiency due to interface states (D_{it}) alone is unlikely to be the root cause as a different T signature due to D_{it} is observed on MOSFETs on

similarly etched InGaAs surfaces [14, 15]. A Poole-Frenkel (PF) mechanism described in [13] involving field-enhanced thermal excitation of carriers from trapped states in the bandgap seems inconsistent with the linear dependence of $q\Phi_B$ on $V_{gs.}$. A possible explanation is that the subthreshold current is bottlenecked by a thermal-assisted tunneling process of electrons from the valence band of the p+ source into the lowest states available in the conduction band at the InAs notch. Simulations of the energy band diagram at different V_{gs} in Fig. 16 allow us to estimate the thermal energy ΔE between the bottom of the InAs notch and the Fermi level at the source $E_{F,S}$ that is required for this process. The inset confirms a linear relationship between V_{gs} and ΔE with values that are broadly consistent with measurements. Further analysis is needed to fully identify the responsible mechanism.

Fig. 17 benchmarks I_{on} vs. I_{off} among published vertical NW TFETs based on III-V materials at $V_{dd}= 0.3$ V ($V_{ds}=0.3$ V, $\Delta V_{gs}= 0.3$ V). Compared to other vertical III-V NW TFETs, our devices exhibit an excellent combination of steep slope and ON current, delivering high I_{on} at low I_{off}. This is testimony to the increased flexibility and precision heterostructure growth that is afforded by a top-down fabrication approach.

Conclusions

InGaAs/InAs heterojunction NW TFETs have been fabricated via a top-down approach for the first time. With improved InGaAs RIE technology and a thin SOG layer isolating source from gate, we have been able to obtain near vertical-sidewall NWs and scale the EOT to 1.2 nm. An average S of 79 mV/dec at $V_{ds}=0.3$ V is obtained over 2 decades of current in our best performing devices. $I_{on}= 0.27$ µA/µm is extracted at $V_{dd}=0.3$ V and a fixed $I_{off}=100$ pA/µm. This demonstrates an excellent combination of steep slope and ON current compared to other NW TFETs with III-V materials.

Acknowledgement

Research funded in part by NSF E3S STC (Award #0939514). Devices fabricated at MIT's Microsystems Technology Laboratories and Scanning Electron Beam Laboratory. The authors thank T. Yu, J. Lin, W. Lu and W. Chern for valuable discussions.

References
[1] Ionescu, A.M. et al., *Nature*, 2011, 479 (7373): p. 329-337.
[2] Dey, G. et al., in *Proc. IEDM*, 2011, p. 785-788.
[3] Avci, U.E. et al., in *Proc. IEDM*, 2013. p. 96-99.
[4] Schmid, H. et al., in *Proc. DRC*, 2011, p. 181-182.
[5] Riel, H. et al., in *Proc. IEDM*, 2012, p. 391-394.
[6] Tomioka, K. et al., *APL*, 2011, 98, 083114.
[7] Tomioka, K. et al., in *Proc. IEDM*, 2013, p. 88-91.
[8] Dey, A.W. et al., *EDL*, 2013, 34 (2): p. 211-213.
[9] Zhao, X. et al., in *Proc. IEDM*, 2013, p. 695-698.
[10] Zhao, X. et al., *EDL*, 2014, 35 (5): p. 521-523.
[11] Lin, J. et al., *EDL*, 2014, 35 (4): p. 440-442.
[12] De Michielis, L. et al., *EDL*, 2013, 33(11): p. 1523-1525.
[13] Mookerjea, M. et al., *EDL*, 2010, 31 (6): p. 564-566.
[14] Vardi, A. et al., in *Proc. DRC*, 2014, p. 219-220.
[15] Lin, J. et al., in *Proc. IEDM*, 2013, p. 421-424.

978-1-4799-8002-4/14 $31.00 © 2014 IEEE

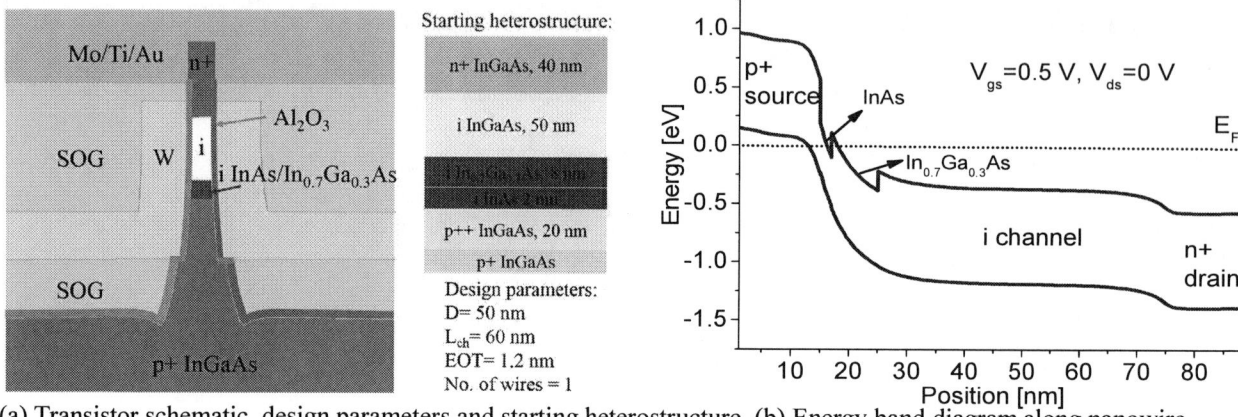

Fig. 1 (a) Transistor schematic, design parameters and starting heterostructure. (b) Energy band diagram along nanowire.

- ○ Nanowire formation by dry etch
- ○ ALD 4 nm Al$_2$O$_3$ as protection
- ○ Planarization and etch back to isolate gate metal from p$^+$ source
- ○ Digital etch
- ○ ALD 2.5 nm Al$_2$O$_3$ gate dielectric
- ○ Sputter 40 nm W gate metal
- ○ 30 min forming gas anneal at 350°C
- ○ Gate/top electrode isolation by planarization and etch back (2 times)
- ○ Contact via opening by dry etch
- ○ Sputter Mo as contact metal and S/G/D pad formation

Fig. 2 Process flow for InGaAs/InAs heterojunction vertical nanowire (NW) tunnel TFETs.

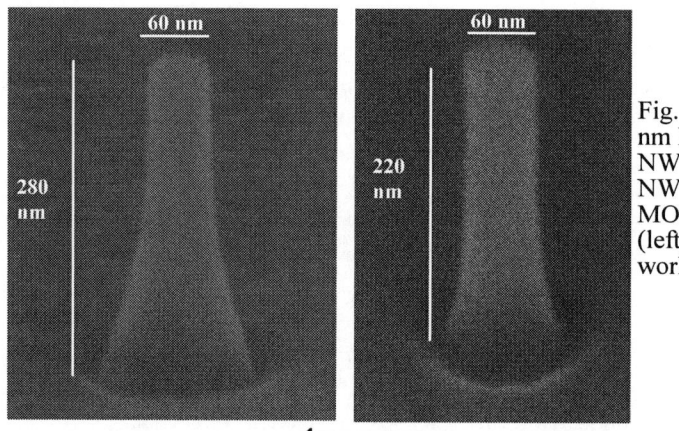

Fig. 3 D= 60 nm InGaAs NW used in NW MOSFETs [9] (left) and this work (right).

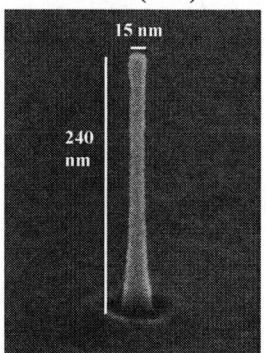

Fig. 4 D=15 nm InGaAs nanowire defined by optimized RIE technique.

Fig. 5 Subthreshold characteristics of a D= 50 nm single-NW TFET.

Fig. 6 Transfer characteristics at V$_{ds}$=0.3 V of the device shown in Fig. 5.

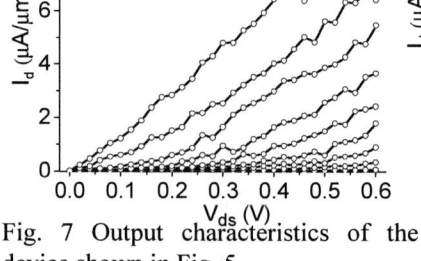

Fig. 7 Output characteristics of the device shown in Fig. 5.

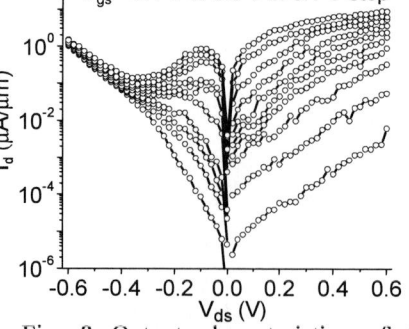

Fig. 8 Output characteristics of the device shown in Fig. 5 in semilog scale for positive and negative V$_{ds}$.

Fig. 9 Subthreshold characteristics of three single-NW TFETs at V$_{ds}$=0.3 V indicating device-to-device variability.

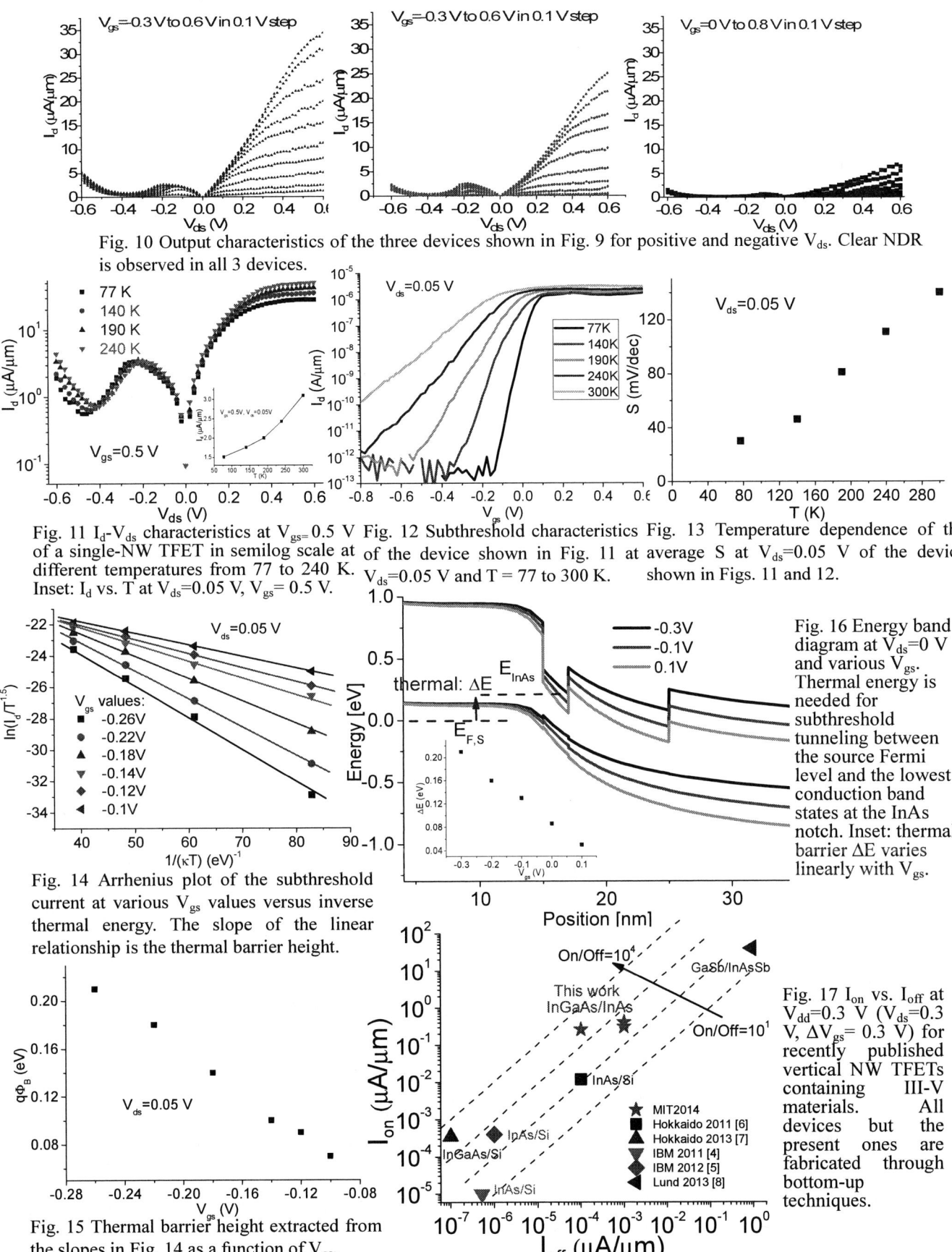

Fig. 10 Output characteristics of the three devices shown in Fig. 9 for positive and negative V_{ds}. Clear NDR is observed in all 3 devices.

Fig. 11 I_d-V_{ds} characteristics at V_{gs}= 0.5 V of a single-NW TFET in semilog scale at different temperatures from 77 to 240 K. Inset: I_d vs. T at V_{ds}=0.05 V, V_{gs}= 0.5 V.

Fig. 12 Subthreshold characteristics of the device shown in Fig. 11 at V_{ds}=0.05 V and T = 77 to 300 K.

Fig. 13 Temperature dependence of the average S at V_{ds}=0.05 V of the device shown in Figs. 11 and 12.

Fig. 14 Arrhenius plot of the subthreshold current at various V_{gs} values versus inverse thermal energy. The slope of the linear relationship is the thermal barrier height.

Fig. 16 Energy band diagram at V_{ds}=0 V and various V_{gs}. Thermal energy is needed for subthreshold tunneling between the source Fermi level and the lowest conduction band states at the InAs notch. Inset: thermal barrier ΔE varies linearly with V_{gs}.

Fig. 15 Thermal barrier height extracted from the slopes in Fig. 14 as a function of V_{gs}.

Fig. 17 I_{on} vs. I_{off} at V_{dd}=0.3 V (V_{ds}=0.3 V, ΔV_{gs}= 0.3 V) for recently published vertical NW TFETs containing III-V materials. All devices but the present ones are fabricated through bottom-up techniques.

978-1-4799-8002-4/14 $31.00 © 2014 IEEE

In$_{0.17}$Al$_{0.83}$N/AlN/GaN Triple T-shape Fin-HEMTs with g_m=646 mS/mm, I_{ON}=1.03 A/mm, I_{OFF}=1.13 μA/mm, SS=82 mV/dec and DIBL=28 mV/V at V_D=0.5 V

S. Arulkumaran[†], G. I. Ng[†‡], C. M. Manojkumar[†], K. Ranjan[†], K. L. Teo[†], O. F. Shoron[+],
S. Rajan[+], S. B. Dolmanan[*] and S. Tripathy[*]

[†]Temasek Laboratories@NTU, Nanyang Technological University, 50 Nanyang Drive, Singapore 637553.
[‡]School of EEE, Nanyang Technological University, 50 Nanyang Avenue, Singapore 639798
[+]Electrical and Computer Engineering Department, The Ohio State University, Columbus, OH43210, USA
[*]Institute of Materials Research and Engineering, A*STAR (Agency of Science, Technology, and Research), Singapore 117602.
Tel: +65-6592-7792; Fax: +65-6790-0215; E-mail: SArulkumaran@pmail.ntu.edu.sg

Abstract

We report the first 3D Triple T-gate InAlN/GaN nano-channel (NC) Fin-HEMTs on Si substrate with record high device performances at V_D as low as 0.5 V. Utilizing a T-gate approach on NC Fin-HEMT with stress engineered techniques, enhanced device transport properties with g_m=646 mS/mm, I_{on}=1.03 A/mm, I_{OFF}=1.13 μA/mm, I_{ON}/I_{OFF}~10^6, SS=82 mV/dec at V_D=0.5V were achieved. In addition, the Fin-HEMT also exhibited 3.2 times lower DIBL of 28mV/V. The dramatic improvement of device performance is due to the tensile stress induced by SiN passivation in the NC Fin-HEMT.

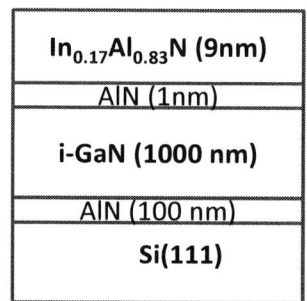

Figure 1. Schematic cross-section and key process steps for the fabrication of Fin-HEMT and conventional InAlN/GaN HEMTs on Si

Introduction

To have superior gate control, sub-nm fins have been formed on Si to realize the next generation high speed digital ICs with sub-threshold swing (SS) of ~60 mV/dec [1]. For example Intel has announced the manufacturability of fully depleted Si FinFETs [2]. To increase the switching speed in the logic circuit, high electron mobility III-V materials (GaAs, AlAs, InAs, InP) and related alloys have also been explored [3]. However, no known reports have achieved high I_{ON}, g_m, and low SS at a V_D=0.5 V with small Drain Induced Barrier Lowering (DIBL) using GaN transistor technology. In this work, we introduce a novel device structure which incorporates stress engineering technique to achieve simultaneously high performance in on-current (I_{ON}), off-current (I_{OFF}), extrinsic transconductance (g_m), SS and DIBL for high speed device applications. The 3D Triple 170-nm T-gate InAlN/GaN 200-nm nano-channel (NC) Fin-HEMTs achieved record high performances of I_{ON}=1.03 A/mm, g_m=645 mS/mm, I_{OFF}=1.13 μA/mm, SS=82 mV/dec and DIBL=28 mV/V at V_D=0.5 V. Our device uses more relaxed device geometries e.g. 170 nm T-shape gate versus other reported devices with 70 to 80-nm I-shape gate on 88-nm fins [4]. This is the first report of T-shaped gate approach on GaN based NC Fin-HEMTs with significantly improved device performance.

Device Fabrication and Characterization

In$_{0.17}$Al$_{0.83}$N/AlN/GaN HEMT structure was grown on Si(111) by MOCVD system (Fig 1). The lattice-matched HEMT structure exhibited 2DEG charge density of 2.74×10^{13} cm^{-2}, 2DEG mobility of 760 cm^2/V·s, and sheet resistance of 302 Ω/sq. Fig 1 shows the process flow for the fabrication of 3D Triple T-gate InAlN/GaN NC Fin-HEMTs. The height of the fin (H_{Fin}) is 12 nm. The effective fin width is calculated using the formula, W_{eff}= [number of fins ×{W_{Fin}+(2×H_{Fin})}]. Figure 2 shows (a) 3D

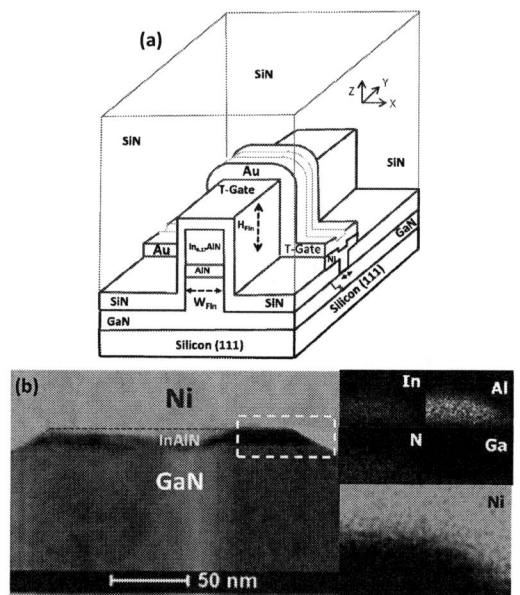

Figure 2. (a) 3D schematic view of a fabricated Triple T-shape gate InAlN/GaN NC Fin-HEMT, (b) Cross-sectional HR-TEM image {x-axis cut view of Figure 2(a)} and EDX mapping of formed single InAlN/GaN (176-nm) Fin (W_{eff}=200nm).

schematic view and (b) 2D HR-TEM image {x-axis cut of Fig 2(a)} and EDX mapping of a single fin width W_{Fin}=176 nm. Fig. 3(a) shows the cross-sectional HR-TEM image {y-axis cut of Fig 2(a)} of 170-nm gate-length InAlN/GaN HEMTs. For comparison, typical 170-nm-gate conventional T-gate [5] InAlN/GaN HEMTs were also fabricated on the same wafer with and without SiN passivation. The source-drain distance (L_{sd}) of 2.65 μm is maintained for both Fin-HEMTs and conv.HEMTs.

978-1-4799-8002-4/14 $31.00 © 2014 IEEE

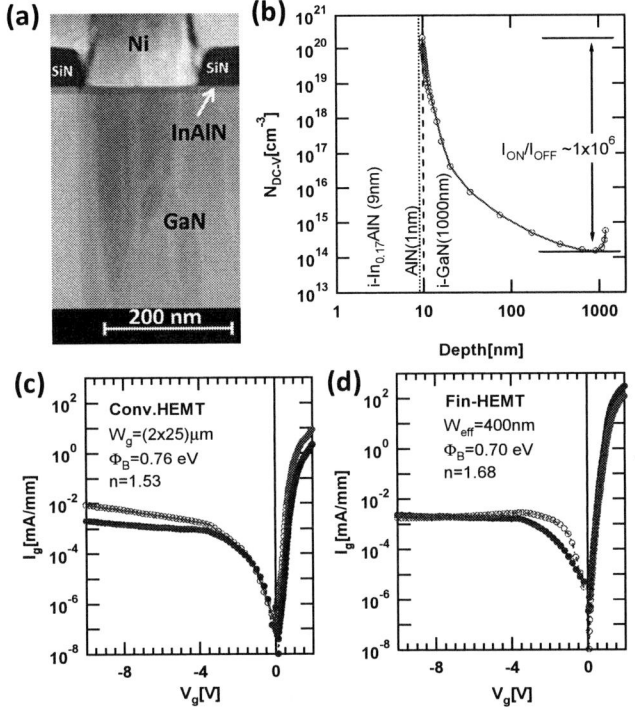

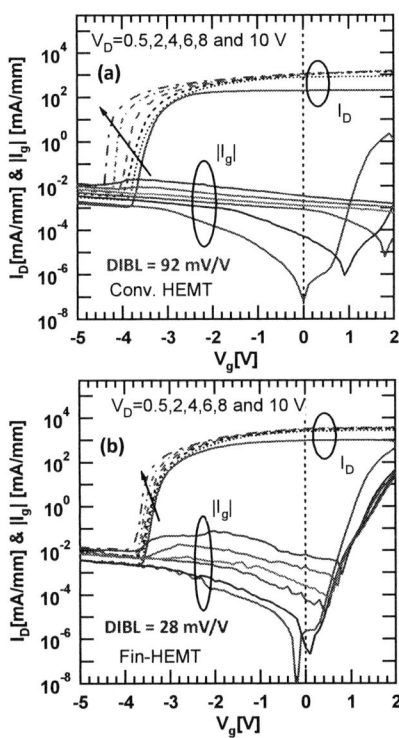

Figure 3. (a) Cross-sectional HR-TEM image of T-gate foot-print on InAlN/GaN HEMTs {y-axis cut of Fig 2(d)}, (b) 2DEG carrier profiling for InAlN/GaN HEMTs measured on Schottky diodes bv C–V measurements.. Two-terminal I₀-

Figure 5. $I_D(I_g)$-V_g characteristics of (a) Conv.HEMTs and (b) 3D TT-gate InAlN/GaN NC Fin-HEMT on Si for different drain voltages

Results and Discussions

The 2DEG carrier profiling of InAlN/AlN/GaN HEMT structure matches the grown structure on Si {See Fig 3(b)}. The gate-leakage current and Schottky barrier height values of NC Fin-HEMT are comparable with conv.HEMTs {See Fig 3(c) and (d)}. Figure 4 shows I_{DS}-V_{DS} characteristics of (a) 200-nm NC

Fin-HEMTs and (b) conv.HEMTs. Figure 3 shows (c) $I_D(I_g)$-V_g and (d) g_m-V_g characteristics measured at V_D=6V for conv. HEMT and NC Fin-HEMTs. The 200-nm NC Fin-HEMT exhibited I_{Dmax} of 3940 mA/mm at V_g=+1V and record high g_m of 1417 mS/mm at V_{DS}=6V. With reference to conv. HEMT, NC

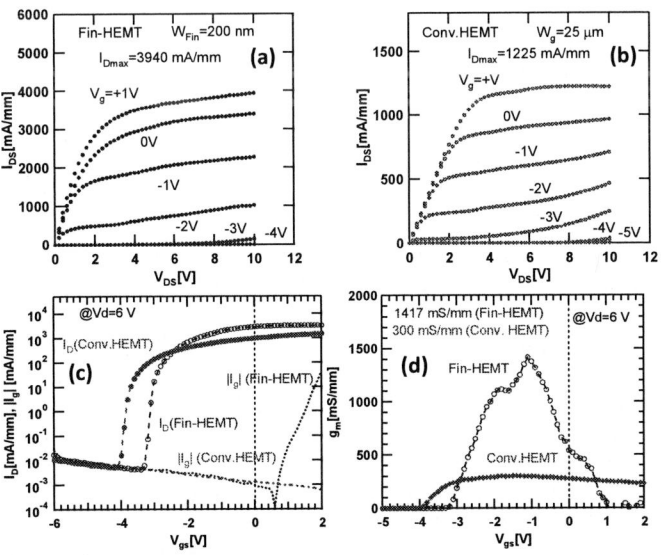

Figure 4. I_{DS}-V_{DS} characteristics of (a) Fin-HEMT and (b) conv. HEMT, (c) $I_D(I_g)$-V_g characteristics of Fin-HEMT and Conv. HEMTs at V_D=6V, (d) g_m-V_g characteristics of Fin-HEMT and Conv. HEMTs at V_D=6V.

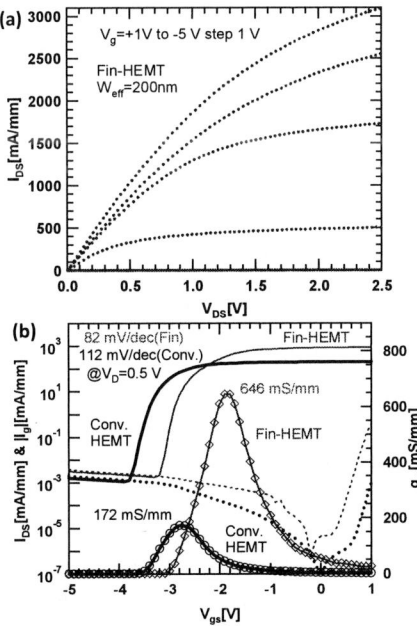

Figure 6. (a) I_{DS}-V_{DS} (0 to 2.5V) characteristics of 3D TT-gate InAlN/GaN 200-nm NC Fin-HEMT, (b) Log(I_D,|I_g|) (g_m)-V_g characteristics measured at V_D=0.5 V for InAlN/GaN Fin-HEMT and Conv. HEMTs.

978-1-4799-8002-4/14 $31.00 © 2014 IEEE

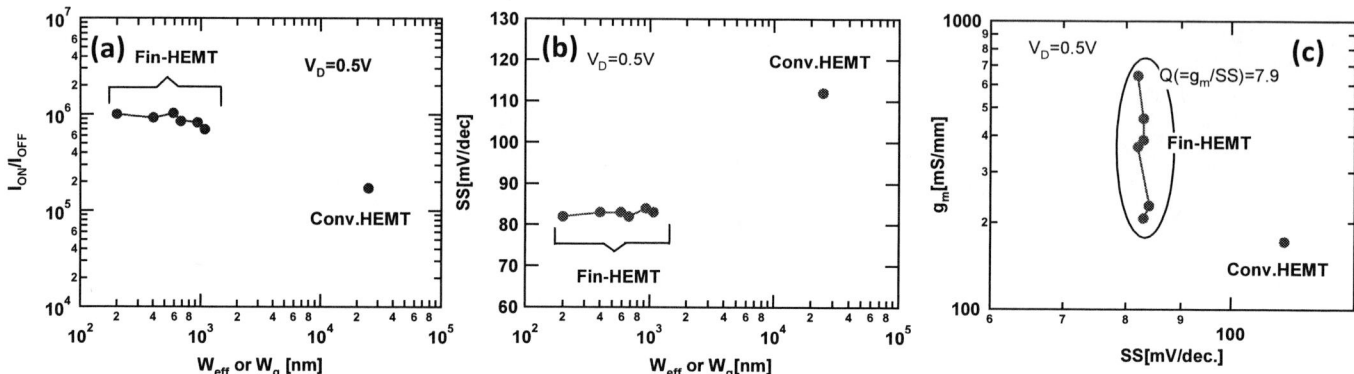

Figure 7 (a) I_{ON}/I_{OFF} ratio and (b) SS for different fin-width Fin-HEMTs and Conv. HEMTs, (c) g_m versus SS for different fin-width Fin-HEMTs and Conv. HEMTs at V_D=0.5 V. The Q-factor (=g_m/SS) is reached 7.9 for 200-nm NC Fin-HEMTs.

Table I. Benchmarking of NC Fin-HEMTs with Conv. InAlN/GaN HEMTs

Parameters	Conv. HEMT	NC Fin-HEMT	Remarks
I_{Dmax} [mA/mm] at V_g=+1 V	1225	3940	3.2× higher
g_{mmax} [mS/mm] at V_{DS}=6V	300	1417	4.7× higher
DIBL [mV/V]	92	28	3.3× lower
I_{ON} [mA/mm] at V_{DS}=0.5 V	193	1030	5.3× higher
SS [mV/dec.] at V_{DS}=0.5 V	112	82	1.4× lower
g_m [mS/mm] at V_{DS}=0.5 V	172	646	3.7× higher

Fin-HEMT exhibited 3.2× higher I_{Dmax}, 4.7× higher g_{mmax} and 3.3× lower DIBL (See Fig 5(a), (b) and Table I). It is worth to note that these performances are achieved using less stringent process techniques and more relaxed geometries. In contrast, Shinohara et al., achieved I_D of ~4000 mA/mm and g_m of ~1000 mS/mm but with a self-aligned 20-nm gate AlGaN/GaN HEMTS with regrown heavily doped n^+-GaN ohmic contacts by MBE on SiC [6].

At low driving voltage of V_D=0.5 V and V_g=+1V, the Fin-HEMT exhibited impressive device performances of I_{ON}=1030

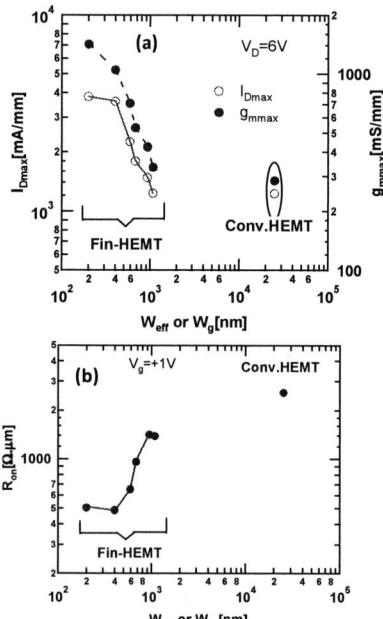

Figure 8 (a) I_{Dmax} and g_{mmax} for different Fin-width InAlN/GaN Fin-HEMT and Conv. HEMTs. (b) ON-resistance (R_{on}) at V_g=+1V for different fin-width Fin-HEMTs and Conv. HEMTs

mA/mm, g_m=646 mS/mm, SS=82 mV/dec, and I_{ON}/I_{OFF}=1×10^6 {See Fig 6(a) and (b)}. At V_D=0.5 V, the I_{ON}/I_{OFF} ratio and SS of different fin-width Fin-HEMTs exhibited values in the range of 8×10^5 to 1×10^6 and 84 to 82 mV/dec, respectively {See Fig 7(a) and (b)}. The observed I_{ON}/I_{OFF} ratio is in agreement with the peak to valley of 2DEG carrier profiling (See Fig 3(b)). The measured Q-factor (g_m/SS) of NC Fin-HEMT is 7.9 at V_D=0.5V. Similarly, the R_{on} at V_g=+1V increases with the increase of fin-width {See Fig 7(c)}. The I_{Dmax} and g_{mmax} of NC Fin-HEMT decreases with the increase of fin-width {See Fig 8(a)}. The ON-resistance (R_{on}) is also small for 3D TT gate InAln/GaN Fin-HEMTs on Si substrate {(See Fig 8(b)}. Because of these promising electrical properties, the proposed 3D TT-gate NC InAlN/GaN HEMTs on Si can be utilized for future ultra-high speed device applications. Previous investigations of Fin-FETs with I-shape gate suggest that the improvement in I_D and g_m could be due to an un-changed source resistance (R_s) even at high I_D conditions [4]. However, in this experiment, we do not expect any enhancement in I_D or g_m due to R_s. The R_s of the NC Fin-HEMT increases with the increase of drain current which is similar to the R_s behavior of conventional InAlN/GaN HEMTs (see Figure 9). To investigate the cause of the dramatic improvement in transport properties (I_D and g_m), we extracted electron velocity {($v_e(V_g)=I_D(V_g)/\{en_s×[V_g-(R_s×I_D(V_g))]\}$} for both NC Fin-HEMT and conv. HEMT using the measured transfer characteristics, (I_D-V_g) characteristics and capacitance-voltage measurements [7]. As shown in the figure, the velocity approaches v_e ~6×10^7 cm/s for the shortest fin widths, suggesting a significant enhancement of velocity in the strained FIN-HEMT structures. The cut-off frequency (f_T) of different fin-width Fin-HEMTs and conv. HEMTs are extracted using the equation (f_T= $v_e/(2\pi L_{eff})$). Based on the measured g_{mmax}=1417mS/mm, the extracted v_e ~6×10^7 cm/s and f_T~560 GHz are the highest ever reported in InAlN/GaN HEMTs on Si substrate {See Fig 10(a)}. Recently, Yue et al., reported f_T of 400 GHz from 30-nm gate conv. InAlN/GaN HEMTs (L_{sd}=270 nm) with re-grown n^+-GaN ohmic contact on SiC substrate [8]. The measured f_T of the conv. InAlN/GaN HEMT (70 GHz) is in good agreement with the extracted f_T values {See Fig 10(b)}. This dramatic improvement of transport properties in InAlN/GaN NC Fin-HEMTs are due to the influence of stress engineered InAlN/GaN Fin-HEMT with

978-1-4799-8002-4/14 $31.00 © 2014 IEEE

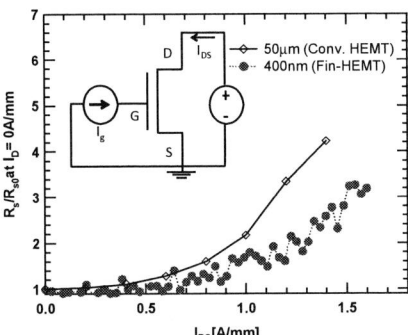

Figure 9. The ratio of measured source resistance (R_s) and R_{s0} measured at I_D=0 A/mm versus I_{DS}. R_s was measured at different I_{DS} using a set-up (inset)

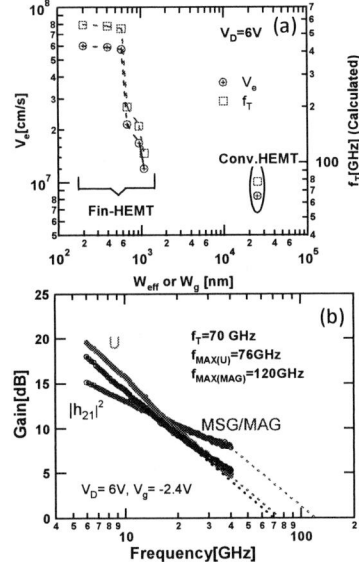

Figure 10 (a) Extracted electron velocity and calculated cut-off frequency (f_T= $v_e/(2\pi L_{eff})$) for different fin-width Fin-HEMT and conv. HEMTs (b) Small-signal characteristics for InAlN/GaN HEMTs ($W_g/L_g/L_{gd}$=(2×75)/0.17/1.7 µm)

three stressors {See Fig 11(a)}. The first stressor is a SiN layer which wraps over the formed fins. The second stressor is the Schottky gate material with T-shape that can provide additional localized stress to the channel by metal in the foot-print and the head by a combination of SiN and metal. The third stressor is another SiN passivation layer which covers the entire active region. An improvement of I_{Dmax} and g_{mmax} is observed in the SiN passivated conv.HEMTs {See Fig 11(b)}. Due to high parasitic resistance, un-passivated Fin-HEMT exhibited very low I_D and g_m (not shown here). As a result, the tensile stress induced in the formed NC/fin gives rise to enhanced device transport properties. The increase of tensile stress by SiN passivation has also been realized by optical methods (micro-Raman and micro-Photoluminescence). The devices show high knee-voltage {See Fig 6(a)} but this can be reduced by shrinking L_g, L_{sd} and W_{Fin} of the Fin-HEMTs.

Conclusions

The proposed 3D Triple T-gate InAlN/GaN nano-channel Fin-HEMTs were demonstrated for the first time on Si substrate.

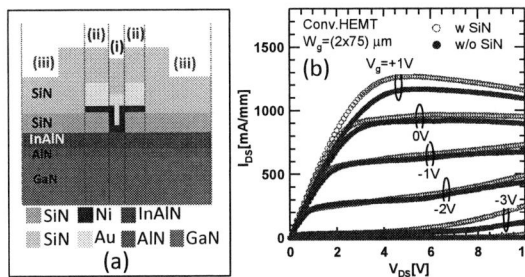

Figure 11 (a) Schematic cross-sectional diagram of T-shape gate configuration {y-axis cut of Fig 1(d)} with different stress regions {(i)- SiN/Au/Ni stress layers, (ii) SiN/Au/Ni/SiN stress layers and (iii) SiN/SiN stress layers} to InAlN/GaN NC. (b) I_{DS}-V_{DS} characteristics of conv.HEMTs with and without SiN passivation.

The Fin-HEMT exhibited 3.2× lower DIBL=28mV/V. At V_D=0.5 V, the promising electrical transport properties {g_m=646 mS/mm, I_{ON}=1.03 A/mm, I_{OFF}=1.13 µA/mm, I_{ON}/I_{OFF}~10^6, SS=82 mV/dec} of 3D TT-gate InAlN/AlN/GaN NC Fin-HEMTs on Si demonstrate their good potential for future ultra-high speed device applications.

Acknowledgements

The authors thank S. Vicknesh, K. S. Ang, N. Louis, S. C. Foo, M. Bryan, T. Lihuang and K. S. See for their help. Two of the authors S.R and O.F.S acknowledge funding from the Office of Naval Research DATE and EXEDE MURI programs (PM: Dr. Paul A. Maki).

References

1) B. Yu, L. Chang, S. Ahmed, H. Wang, S. Bell, C. Yang, C. Tabery, C. Ho, Q. Xiang, T. King Liu, J.Bokor, C. Hu, M. Lin, and D. Kyser, "FinFET scaling to 10nm gate length," IEDM Tech. Dig., 251-254, Dec. 2002.

2) B. Doyle, S. Datta, M. Doczy, S. Hareland, B. Jin, J. Kavalieros, T. Linton, A. Murthy, R. Rios and R.Chau., "High performance fully-depleted tri-gate CMOS transistors". IEEE Electron Device Lett. 24,263–265, Apr. 2003.

3) T.-W. Kim, D.-H. Kim, D.H. Koh, H.M. Kwon, R.H. Baek, D. Veksler, C. Huffman, K. Matthews, S. Oktyabrsky, A. Greene, Y. Ohsawa, A.Ko, H. Nakajima, M. Takahashi, T. Nishizuka, H. Ohtake, S. K. Banerjee, S.H. Shin, D.-H. Ko, C. Kang, D. Gilmer, R.J.W. Hill, W. Maszara, C. Hobbs and P.D. Kirsch, "Sub-100 nm InGaAs Quantum-Well Tri-Gate MOSFETs with Al₂O₃/HfO₂ (EOT < 1 nm) for Low-Power Logic Applications," IEDM Tech. Dig., 425-428, Dec. 2013.

4) D. S. Lee, H. Wang, A. Wu, M. Azize, O. Laboutin, Y. Cao, J. W. Johnson, E. Beam, A. Ketterson, M.L. Schuette, P. Saunier and T. Palacios, "Nanowire Channel InAlN/GaN HEMTs with High Linearity of g_m and f_T", IEEE Electron Device Lett., 34, 969-970, Aug.2013.

5) S. Arulkumaran, G. I. Ng and S. Vicknesh, "Enhanced Breakdown Voltage With High Johnson's Figure-of-Merit in 0.3µm T-gate AlGaN/GaN HEMTs on Silicon by (NH₄)₂Sₓ Treatment", IEEE Electron Device Lett., 34, 1364-1366, Nov. 2013.

6) K. Shinohara, D. Regan, A. Corrion, D. Brown, Y. Tang, J. Wong, G. Candia, A. Schmitz, H. Fung, S.Kim, and M. Micovic, "Self-Aligned-Gate GaN-HEMTs with Heavily-Doped n⁺-GaN Ohmic Contacts to 2DEG", IEDM Tech. Dig., p. 27.2.1, Dec. 2012.

7) T. Fang, R. Wang, H. Xing, S. Rajan and D. Jena, "Effect of optical phonon scattering on the performance of GaN transistors" IEEE Electron Device Letters, 33, 709-711, May. 2012.

8) Y. Yue, Z. Hu, J. Guo, B. S. Rodriguez, G. Li, R. Wang, F. Faria, B. Song, X. Gao, S. Guo, T. Kosel, G. Snider, P. Fay, D. Jena and H. Xing, "Ultrascaled InAlN/GaN High Electron Mobility Transistors with Cutoff Frequency of 400GHz", Jpn. J. Appl. Phys., 52, 08JN14, May. 2013.

Ultralow power transponder in thin film circuit technology on foil with sub - 1V operation voltage

Tung-Huei Ke[*,1], Kris Myny[1], Adrian Chasin[1,2], Robert Müller[1], Paul Heremans[1,2], and Soeren Steudel[1]

1) imec, Kapeldreef 75, 3001, Leuven, Belgium
2) ESAT, KU Leuven, Kasteelpark Arenberg 10, 3001, Leuven, Belgium
*e-mail: tunghuei@imec.be

Abstract

An ultra low power (ULP) transponder chip (XPDR) with sub 1V operation voltage is demonstrated by organic complementary (CMOS) thin film circuits (TFTs) on foil. The lowest operation voltage of the XPDR is down to 0.55V with a data rate of 35 bits/second. The power consumption (P_{tot}) of the XPDR is 2.5 µW at V_{dd} of 0.9 V and is 16 µW at V_{dd} of 2 V. A commercial AAA battery is employed to power the XPDR with an estimated run-time of more than 20 years.

Introduction

Ubiquitous sensing applications such as body area network (BAN) and environmental monitoring wireless sensor networks have driven the innovations of TFTs on foil in the past decade (1). Organic TFT technology is of interest for advantages like low process temperature and its potential in low cost flexible (2) and disposable (3) electronics. However, tens of volts are required in the previous publications to power advanced circuits such as line driver (4) and radio frequency identification tags (2). This increases the challenge to drive the circuits by battery or energy harvester e.g. via radio frequency interfaces (5). Here we demonstrate for the first time an ULP organic CMOS technology applying it to circuits with various complexity. Functional circuits including inverter, 19-stage ring oscillator (RO), and 8 bit XPDR with sub 1V supply voltage are demonstrated. The supply voltage (V_{dd}) of the XPDR is down to 0.55V with data rate up to 35 bits/sec. The total power consumption (P_{tot}) of the XPDR is between 2.5 µW to 16 µW when V_{dd} is between 0.9 V to 2 V. A commercial AAA battery is employed to power the XPDR with an estimated run-time of more than 20 years.

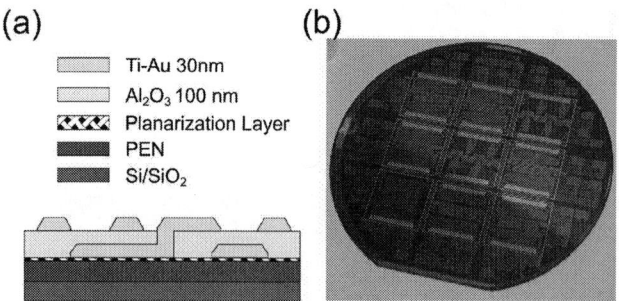

Figure 1: (a) Schematic structure of the foil on carrier (FOC) substrate with coplanar device structure (b) Photograph of the 6-inch FOC substrate

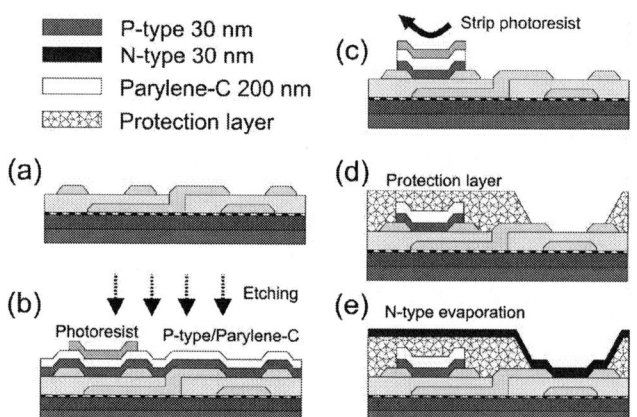

Figure 2: Schematic process flow to integrate organic p-type and n-type semiconductors side-by-side with high density: (a) substrate cleaning (b) P-type/Parylene-C deposition, patterning P-type/Parylene-C with photoresist and reactive ion etch (c) strip photoresist (d) protection layer deposition and patterning (e) n-type organic semiconductor evaporation

Sample fabrication

The schematic and photograph of the 6-inch polyethylene naphthalate (PEN) foil on carrier (FOC) substrates are shown in Fig. 1a-b. Atomic layer deposition (ALD) grown Al_2O_3 is employed as dielectric layer (ε=8) and bilayer Ti/Au (2 nm/30 nm) metallization are applied as source, drain and gate electrodes. The schematic integration flow for organic semiconductors are shown in Fig. 2. The FOC substrates are first treated by pentafluorobenzenethiol (PFBT) for carrier injection enhancement. Next, a layer of poly(α-methylstyrene) (PαMS) is spin-coated for dielectric passivation. Afterwards, 30 nm of p-type material, 3,9-diphenyl-peri-xanthenoxanthene (Ph-PXX), is thermally evaporated. We further deposit 200 nm of parylene-C for p-type semiconductor passivation. The p-type/parylene-C stack is then patterned by photolithography with the orthogonal resist technology (6). A protection layer is deposited afterward to protect the patterned p-type TFTs from the degradations due to the n-type TFTs fabrication. With photolithography, we etch through the protection layer to open the n-type area. A n-tetradecylphosphonic acid (C14-PA) solution is then employed as surface treatment of the Al_2O_3 in the n-type area for the n-type semiconductor, N3004 (7). Finally, 30 nm of N3004 are thermally evaporated as n-type semiconductor.

978-1-4799-8002-4/14 $31.00 © 2014 IEEE

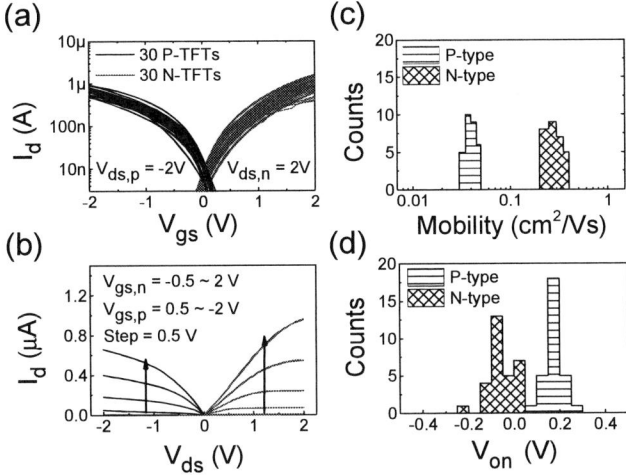

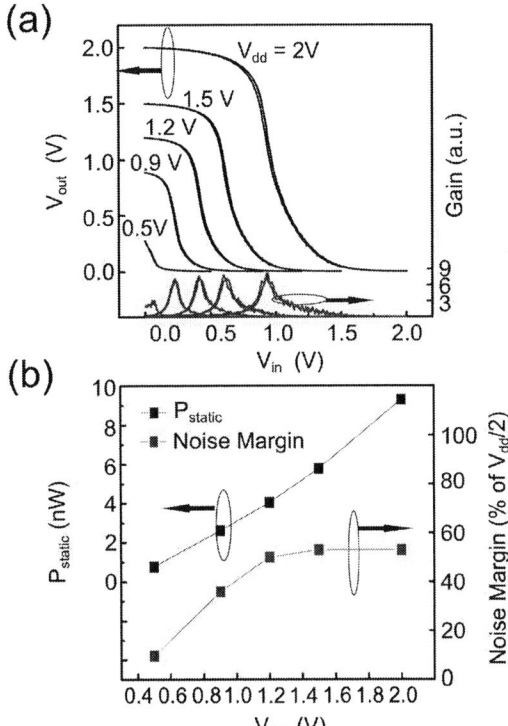

Figure 3: (a) Transfer characteristics of 30 integrated p-type and 30 integrated n-type TFTs. The W/L of the p-type and n-type TFTs are 560/5 μm/μm and 140/5 μm/μm, respectively. (b) representative output characteristics of the p-type and n-type TFT. (c) mobility and (d) turn on voltage (V_{on}) distributions in the 30 p-type and 30 n-type TFTs.

Characteristics of integrated TFTs

The transfer and output characteristics of the integrated p-type and n-type TFTs are shown in Fig. 3a-b. Thirty transistors for both p-type and n-type TFTs are characterized to study the parameters distribution. The average value of the mobilities and V_{on} of the p-type TFTs are 0.04 cm²/Vs and 0.33 V, respectively, while the average value of the mobilities and V_{on} of the n-type TFTs are 0.29 cm²/Vs and -0.006 V. The parameter distribution of the mobilities and V_{on} of the p-type and n-type TFTs are shown in Fig. 3c-d. The high uniformity of the parameters is a pre-requirement to design advanced integrated circuits at a low supply voltage. The output characteristics of the p-type and n-type transistors are shown in Fig 3b. We observed some effects of contact resistance at low V_{ds} in the n-type output characteristics. This effect can be reduced by increasing the channel length which would compromise the speed of the circuits.

Characteristics of the inverter.

The schematic and microscope image of the organic CMOS inverter are shown in Fig. 4. The voltage transfer curves and

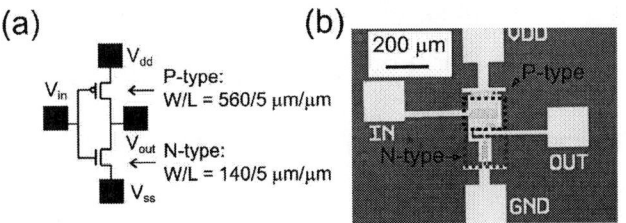

Figure 4: (a) The schematic and (b) the microscope image of the organic complementary inverter

Figure 5: (a) The voltage transfer curves and the gain values of the inverter versus V_{dd}. (b) The static power consumption (P_{static}) and noise margin of the inverter versus V_{dd}

the corresponding characteristics of the inverter are shown in Fig 5. The inverter starts to operate at V_{dd} of 0.5V with a gain of 3 and a switching threshold (V_{th}) of 0.05 V. The low peak to peak voltage (V_{p-p}) of 0.27 V and low V_{th} are due to the imbalanced pTFT and nTFT current at V_{dd} of 0.5 V. The noise margin (NM) and the static power consumption (P_{static}) of the inverter versus V_{dd} are shown in Fig. 5b. The P_{static} of the inverter is 1 nW at V_{dd} of 0.5V and is 9 nW at V_{dd} of 2 V. The low P_{static} can be attributed to the well-designed CMOS inverter. With increasing V_{dd}, the V_{th} moves toward half V_{dd}, thus the NM increases accordingly. The inverter shows rail to rail voltage swing from V_{dd} of 0.9 V and the gain remains at the level of 8 at higher V_{dd}. The noise margin is 53% of $V_{dd}/2$ at V_{dd} of 2V. It is worth to notice that the NM is still 9 % of $V_{dd}/2$ even with V_{dd} of 0.5V. The large NM at low operation voltage is essential to realize low voltage advanced circuits.

Characteristics of 19-stage RO

The schematic and microscope image of the RO are shown in Fig. 6a-b. Forty TFTs are integrated together to realize the circuit. The output signal of the RO at V_{dd} of 0.5 V and total power consumption (P_{tot}) versus V_{dd} are shown in Fig.7a-b, respectively. The RO start to operate at V_{dd} of 0.5V with operation frequency up to 82 Hz and with a P_{tot} of 0.17 μW.

978-1-4799-8002-4/14 $31.00 © 2014 IEEE

(a)

19-stage ring oscillator

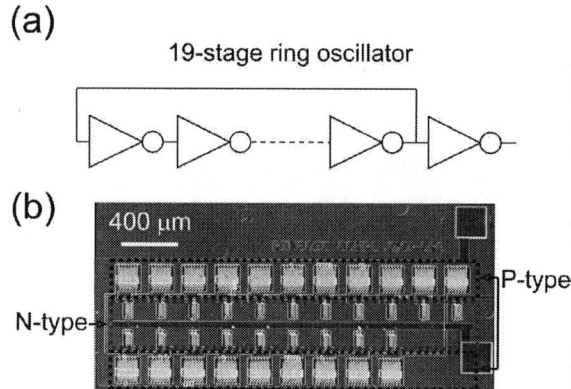

Figure 6: (a) The schematic and (b) microscope image of the organic complementary 19 stage ring oscillator (RO).

The corresponding V_{p-p} of 55 mV is consistent with the inverter characteristics. The P_{tot} of the RO is 0.17 μW at V_{dd} of 0.5 V and is 3 μW at V_{dd} of 2 V. The stage delay versus V_{dd} of the RO and the results from literature are shown in Fig. 8. The stage delay of the RO at $V_{dd} < 2$ V is comparable to the best results report in the literature with much shorter channel length (L = 1 μm) at the same V_{dd} (8).

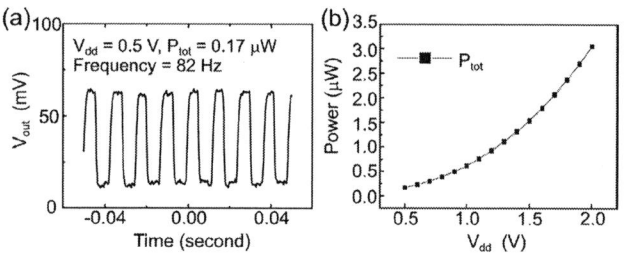

Figure 7: (a) The output signal of the 19 stage RO at $V_{dd} = 0.5$ V (b) total power consumption of the RO versus V_{dd}

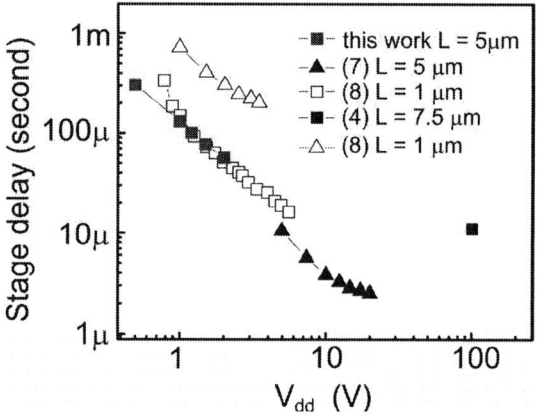

Figure 8: Stage delay of different organic complementary technology versus V_{dd}. (L = channel length)

Characteristics of 8 bit XPDR

The block diagram, photograph and the microscope image of the 8 bit XPDR are shown in Fig. 9. The output signal of the XPDR at V_{dd} of 0.55V are shown in Fig. 10. The 8 bit XPDR starts to render the correct embedded code from $V_{dd} = 0.55$V with data rate of 35 bits/second. This is the lowest operation voltage ever reported for transponder chip fabricated by thin film technology. We further measured the circuit up to 2V and the measured data rate is up to 582 bits/sec. The data rate and the P_{tot} of the transponder chip versus V_{dd} are shown in Fig. 11. The P_{tot} of the transponder chip is 2.5 μW at V_{dd} of 0.9 V and is 16 μW at V_{dd} of 2 V. Furthermore we power the XPDR by a commercial AAA battery with 1.6V output voltage. The photograph of the battery and output signal of the XPDR is shown in Fig 12a-b. Battery powered XPDR with data rate of 465 bits/sec are demonstrated. Assuming the charges of the battery is 1200 mAh, the battery could ideally (assuming no self-discharge) drive the XPDR for 21.9 years. We can therefore envision to combine this technology with printed thin film batteries for ubiquitous sensing applications.

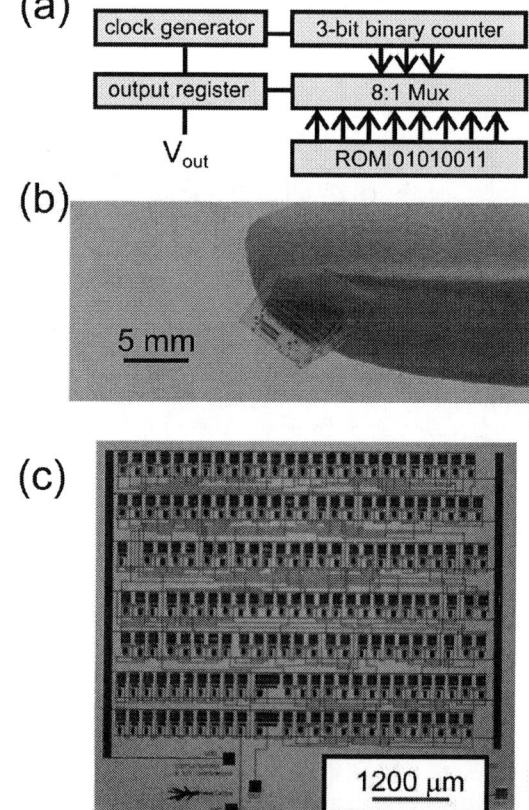

Figure 9: (a) Block diagram of the 8 bit organic complementary transponder chip (XPDR) (b) Photograph of the 8 bit organic complementary XPDR on top of a fingertip (c) Microscope image of the 8 bit XPDR

Conclusion

We demonstrate an organic complementary 8 bit XPDR with V_{dd} down to 0.55V. Ultralow power consumption of 2.5 μW of the XPDR is realized. The XPDR could be powered by commercial battery with ideal run-time of more than 20 years. The ultralow power thin film circuits on foil could open up new opportunities in ubiquitous sensing applications such as body area networks and environmental monitoring wireless sensor networks.

Acknowledgement

The authors would like to acknowledge financial support through the EU project COSMIC (FP7 ICT- 247681). This work was performed in a collaboration between IMEC and TNO in the frame of the HOLST Centre.

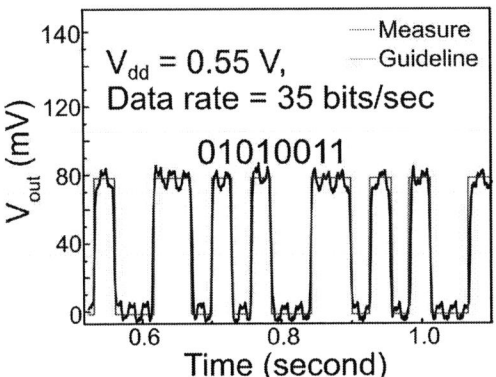

Figure 10: Output signal of the 8 bit transponder chip at V_{dd} of 0.55 V with data rate up to 35 bits/second

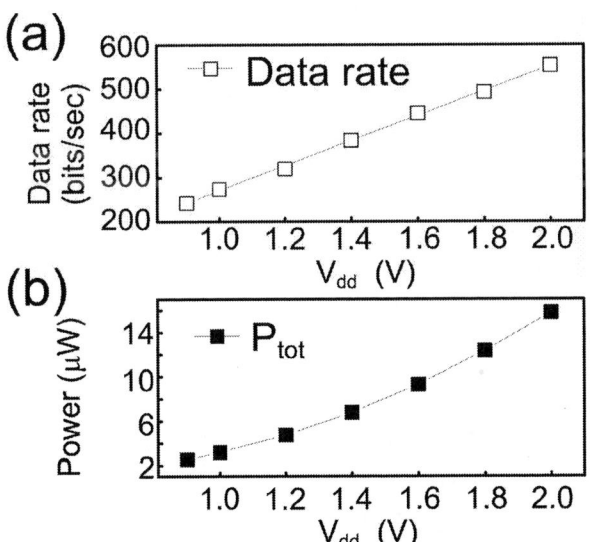

Figure 11: (a) The data rate of the 8 bit transponder chip versus V_{dd} (b) The power consumption of the transponder chip versus V_{dd}

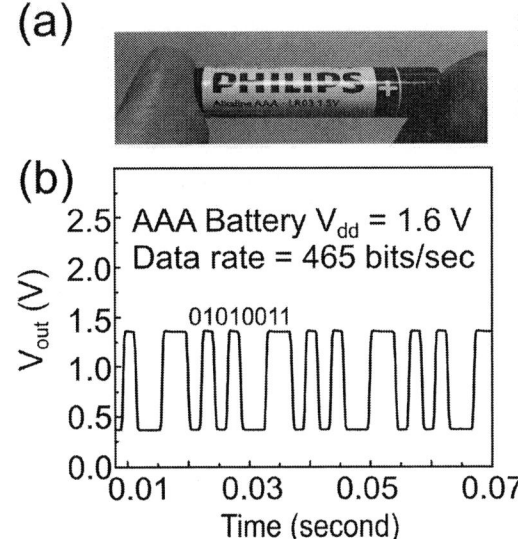

Figure 12: (a) Commercial AAA battery with measured voltage of 1.6V to power the XPDR (b) Output signal of the XPDR with a commercial AAA battery

Reference

(1) A. Nathan, A. Ahnood, M. T. Cole, Y. Suzuki, P. Hiralal, F. Bonaccorso, *et al.*, "Flexible Electronics: The Next Ubiquitous Platform," *Proc. IEEE*, vol. 100, Special Centennial Issue, pp. 1486–1517, May 2012.

(2) K. Myny, M. J. Beenhakkers, N. A. J. M. Van Aerle, G. H. Gelinck, J. Genoe, W. Dehaene, *et al.*, "Unipolar Organic Transistor Circuits Made Robust by Dual-Gate Technology," *IEEE J. Solid-State Circuits*, vol. 46, no. 5, pp. 1223–1230, 2011.

(3) H. Fuketa, K. Yoshioka, T. Yokota, W. Yukita, M. Koizumi, M. Sekino, *et al.*, "Organic-Transistor-Based 2kV ESD-Tolerant Flexible Wet Sensor Sheet for Biomedical Applications with Wireless Power and Data Transmission Using 13.56MHz Magnetic Resonance," in *isscc digest of* technical papers, 2014, vol. 3, pp. 490–492.

(4) B. Crone, A. Dodabalapur, Y. Lin, R. Filas, Z. Bao, A. LaDuca, et al, "Large-scale complementary integrated circuits based on organic transistors," Nature, vol. 403, no. 6769, pp. 521–3, Mar. 2000.

(5) A. Chasin, V. Volskiy, M. Libois, M. Ameys, M. Nag, M. Rockele, *et al.*, "Integrated UHF a-IGZO energy harvester for passive RFID tags," *2013 IEEE Int. Electron Devices Meet.*, no. 2, pp. 11.3.1–11.3.4, Dec. 2013.

(6) A. Zakhidov, J.-K. Lee, J. A. DeFranco, H. H. Fong, P. G. Taylor, M. Chatzichristidi, *et al.*, "Orthogonal processing: A new strategy for organic electronics," *Chem. Sci.*, vol. 2, no. 6, pp. 1178–1182, 2011.

(7) T.-H. Ke, R. Müller, B. Kam, M. Rockele, A. Chasin, K. Myny, *et al.*, "Scaling down of organic complementary logic gates for compact logic on foil," *Org. Electron.*, vol. 15, no. 6, pp. 1229–1234, Jun. 2014.

(8) F. Ante, F. Letzkus, J. Butschke, U. Zschieschang, K. Kern, J. N. Burghartz, and H. Klauk, "Submicron low-voltage organic transistors and circuits enabled by high-resolution silicon stencil masks," *2010 Int. Electron Devices Meet.*, pp. 21.6.1–21.6.4, Dec. 2010.

Thin-Film Heterojunction Field-Effect Transistors for Ultimate Voltage Scaling and Low-Temperature Large-Area Fabrication of Active-Matrix Backplanes

Bahman Hekmatshoar and Ali Afzali-Ardakani

IBM T. J. Watson Research Center, Yorktown Heights, NY
Email: hekmat@us.ibm.com

Abstract

Thin-Film heterojunction field-effect transistor (HJFET) devices with crystalline Si (*c*-Si) channels and gate regions comprised of hydrogenated amorphous silicon (*a*-Si:H) or organic materials are demonstrated. The HJFET devices are processed at 200°C and room temperature, respectively; and exhibit operation voltages below 1V, subthreshold slopes in the range of 70-100mV/dec and off-currents as low as 25 fA/μm. The HJFET devices are proposed for use in active matrix backplanes comprised of low-temperature poly-Si (LTPS) as the *c*-Si substrate. Compared to conventional LTPS devices which require process temperatures up to 600°C and complex fabrication steps, the HJFET devices offer lower process temperature, simpler fabrication steps and lower operation voltages without compromising leakage or stability.

Introduction

With recent advancements in quality and uniformity, low-temperature poly-Si (LTPS) is emerging as a promising material for manufacturing the TFT backplane of large-area organic light emitting diode (OLED) displays. However, the TFT process conventionally used for LTPS (which is similar to transistor fabrication on silicon-on-Insulator (SOI) substrates in micro-electronics) involves high-temperature gate dielectric deposition and implantation/activation of the source/drain and lightly-doped drain (LDD) regions, which may be challenging and/or expensive to scale to large areas. In addition, reducing the operation voltage and/or subthreshold slope of the TFT requires reducing the thickness of the gate dielectric which may result in excessive gate leakage and/or compromise the gate dielectric reliability in large-area devices. Moreover, even though LTPS can be prepared on low-cost flexible plastic substrates using appropriate buffer layers, the TFT process temperature is too high for such substrates.

In this paper, we show that the above-mentioned issues can be addressed by the recently proposed heterojunction field-effect transistor (HJFET) structure [1, 2]. Particularly, (i) the gate dielectric is replaced with a stack of hydrogenated amorphous silicon (*a*-Si:H), (ii) the implanted source/drain regions are replaced with hydrogenated crystalline silicon (*c*-Si:H), and (iii) the LDD regions are omitted. The *a*-Si:H and *c*-Si:H layers are grown using the same plasma-enhanced chemical vapor deposition (PECVD) reactor at temperatures close to 200°C and may be as thin as ~25nm and ~10nm, respectively. As a result, the thin-film transistor (TFT) fabrication process temperature is reduced from ~600°C to ~200°C and essentially the same equipment used for the fabrication of the conventional *a*-Si:H TFTs for large-area active-matrix liquid-crystal displays (AMLCD) can be used for the fabrication of HJFET devices on LTPS substrates as well. In addition, the operation voltage and subthreshold slope of the TFTs are substantially reduced by eliminating the gate dielectric without increasing the gate leakage or compromising the TFT reliability. Furthermore, we show for the first time, that the *a*-Si:H gate stack of the proposed HJFET devices can be replaced with organic materials to fabricate the TFTs essentially at room-temperature. Given the maturity of the *a*-Si:H technology, the *a*-Si:H/*c*-Si HJFET offers a lower barrier for insertion in manufacturing, while the organic/*c*-Si HJFET offers the opportunity of further cost reduction in the long term. We benchmark both types of HJFET devices against conventional TFTs and demonstrate significant advantage in terms of voltage scaling for the HJFET devices particularly in the regime of interest for OLED displays. The demonstrated devices are fabricated on standard SOI substrates; however the device structures are readily applicable to LTPS. For a fair comparison, the benchmarking has been also performed against conventional TFTs fabricated on standard SOI substrates.

HJFET with a-Si:H Gate Stack

A. Device Fabrication

The schematic cross-sections of a conventional TFT and an *a*-Si:H/*c*-Si HJFET are shown in Fig. 1(a) and (b), respectively.

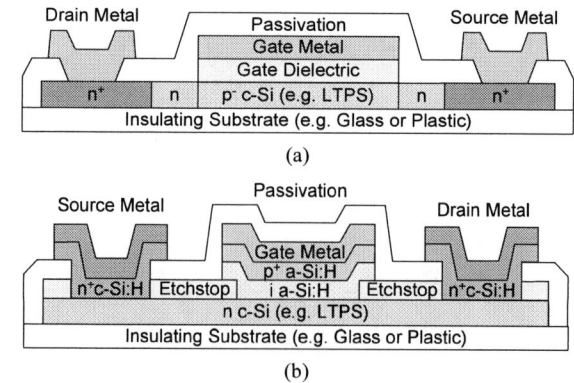

Fig. 1. Schematic cross-section of (a) a conventional thin-film transistor (TFT) and (b) a heterojunction field-effect transistor (HJFET) with a hydrogenated amorphous Si (*a*-Si:H) gate stack on crystalline Si (*c*-Si). The *c*-Si substrate can be comprised of low-temperature poly-Si (LTPS).

978-1-4799-8002-4/14 $31.00 © 2014 IEEE

The conventional TFT requires process temperatures close to 600°C for the deposition of the gate dielectric and implantation/activation of the source/drain and LDD regions, whereas the *a*-Si:H and *c*-Si:H regions of the HJFET are grown by PECVD at temperatures close to 200°C. Since the parasitic bipolar junction transistor inherent to the conventional TFT does not exist in the HJFET structure, the LDD regions are omitted. The optional etch-stop layer used in the HJFET can be deposited by PECVD or other low-temperature techniques.

B. Device Characteristics

The HJFET device structure provides the opportunity for reducing the TFT operation voltage without compromising the gate leakage or TFT stability. In VLSI devices, thin high-*k* dielectrics are used to reduce the operation voltage of the MOS devices (Fig. 2(a)). Since the gate leakage is proportional to the gate area, the small gate areas of the VLSI devices result in small gate leakage currents in spite of the large leakage current density through the thin high-*k* dielectrics. However, this is not the case for display devices due to their larger gate areas. As shown in Table I, the gate leakage of the exemplary VLSI MOS device with an ~8nm-thick HfO_2 gate dielectric (equivalent to ~1.2nm-thick SiO_2) is ~10pA at ~1V, which is sufficient for digital applications. However, the gate leakage of an MOS display device having the same W/L ratio but 10X larger W and L dimensions (i.e. 100X larger gate area) is ~1nA at ~1V using the same gate dielectric. This gate leakage is ~1000X larger than desired for display applications. In contrast, the leakage current through the *a*-Si:H/*c*-Si heterojunction is dominated by electron injection from the gate metal (Fig. 2(b)) and can be maintained well below 10^{-5} A/cm². The stability of the *a*-Si:H/*c*-Si heterojunction is also well-established from the field studies of commercial Si heterojunction solar cells [3]. Therefore, the *a*-Si:H/*c*-Si HJFET structure allows low operation voltages without compromising the gate leakage or TFT stability.

The schematic cross-section of a "normally-ON" HJFET demonstrated on a Si-on-insulator (SOI) substrate and the

Example: Gate Leakage per unit Gate Area = 10^{-2}A/cm²		
VLSI Device	W/L=1μm/100nm	Gate Leakage = 10pA
Display Device	W/L=10μm/1μm	Gate Leakage = 1nA

Table I. Example of the dependence of the gate leakage on the gate area.

experimental I-V characteristics of the device are given in Fig. 3(a) and 3(b); respectively. The HJFET device shows an operation voltage below 1V, subthreshold slope of ~70mV/dec and low leakage (and off-current) of the order of 1pA despite the large gate area of 10μm×100μm. The HJFET device can be used as a normally-ON TFT for both switching and driving applications provided that the TFT is operated at negative or small positive gate voltages to avoid forward- biasing the gate heterojunction [1]. However, if desired, a normally-OFF TFT can be realized by including a blocking stack in the gate of the HJFET [2] (Fig. 4(a)). The blocking stack forms an n-i-p diode in series with the gate heterojunction. When a positive voltage is applied to the gate, the n-i-p diode is reverse-biased, preventing excessive current flow through the gate. A blocking stack with a thin (<10nm) intrinsic (i) *a*-Si:H layer can substantially reduce the gate leakage at positive voltages with only a small penalty in subthreshold slope and off-current, thus allowing normally-OFF TFT operation (Fig. 4(b)).

HJFET with Organic Gate Stack

A. Device Fabrication

The schematic cross-section of the fabricated HJFET devices is illustrated in Fig. 5. In the *a*-Si:H/*c*-Si heterojunction, the surface passivation provided by the hydrogen supplied from *a*-Si:H results in surface recombination currents as low as 100 fA/cm² [5]. As a result, the gate leakage of the HJFET is not dominated by recombination at the *a*-Si:H/*c*-Si interface. In order to enable the use of organic materials as the gate contact of the HJFET devices, it is pertinent to provide adequate

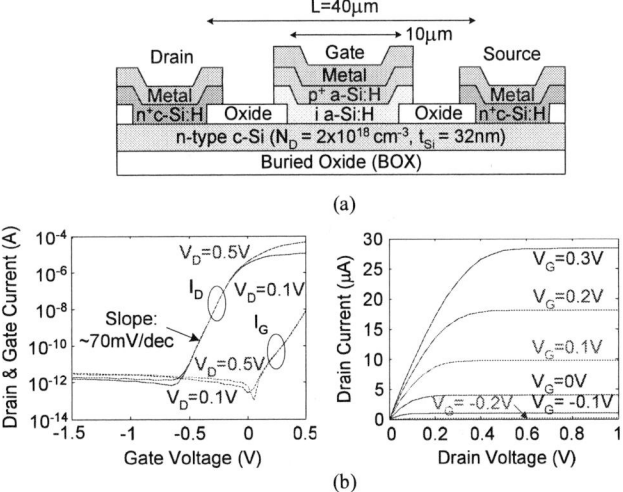

(a)

(b)

Fig.3. (a) Schematic cross-section and (b) experimental transfer and output characteristics of a "normally-ON" *a*-Si:H/*c*-Si HJFET fabricated on an SOI substrate. (W=100μm).

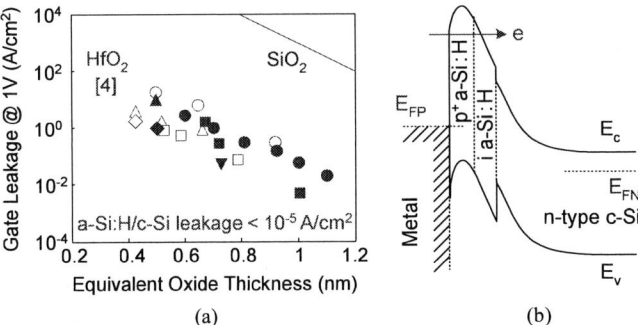

(a)   (b)

Fig. 2. (a) Gate dielectric leakage per unit gate area of conventional MOS transistors with SiO_2 or HfO_2 gate dielectrics vs. the equivalent oxide thickness (EOT) of the gate dielectric at an operation voltage of 1V [4], and (b) schematic energy band diagram illustrating the gate leakage of an *a*-Si:H/*c*-Si HJFET dominated by electron injection from the gate metal.

978-1-4799-8002-4/14 $31.00 © 2014 IEEE   603

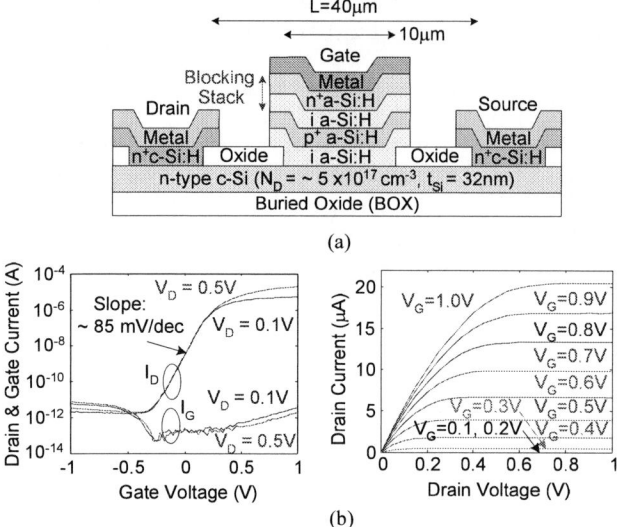

(a)

(b)

Fig.4. (a) Schematic cross-section and (b) experimental transfer and output characteristics of a "normally-OFF" a-Si:H/c-Si HJFET fabricated on an SOI substrate. (W=100μm).

surface passivation at the c-Si/organic interface. This is achieved by dipping the c-Si substrate in dilute hydrofluoric acid to remove the native oxide, followed by immersing the c-Si substrate in a long-chain alcohol such as 1-dodecanol to form an organic monolayer that saturates the dangling bonds on the surface of c-Si (Fig. 6). The passivating property of the organic monolayer was verified by photo-conductance decay measurements (Fig. 7(a)). The measured minority carrier lifetime corresponds to a surface recombination current of ~ 2pA/cm². As a result, the HJFET gate leakage is rather dominated by electron injection from the metal electrode which is enhanced by the presence of surface states as well as Schottky barrier lowering due to image force (Fig. 7(b)).

B. Device Characteristics

In the HJFET devices with thin (~5nm) pentacene layers, the off-current is dominated by the gate leakage due to excessive electron injection from the metal electrode. In contrast, thicker pentacene layers (~15nm) are more effective in blocking the electrons from the metal contact, thus sufficiently reducing the gate leakage. Thicker pentacene layers however compromise the electrostatics by reducing the gate control over the channel (Fig. 8). The transfer and output characteristics of the HJFET

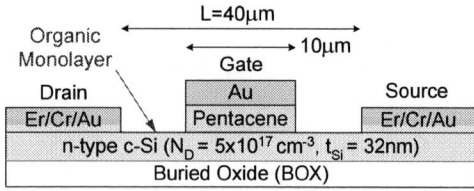

Fig. 5. The schematic cross-section of a HJFET device with an organic monolayer passivation and a pentacene gate contact fabricated on an SOI substrate. (W=100μm).

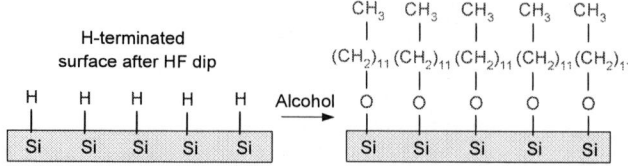

Fig. 6. Schematic illustration of the surface passivation of the c-Si substrate by dipping in dilute HF acid to remove the native oxide, followed by immersion in a long-chain alcohol (1-dodecanol).

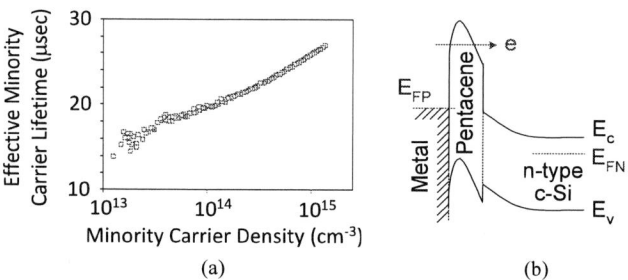

(a)   (b)

Fig. 7. (a) The effective minority carrier lifetime as a function of minority carrier density measured by photo-conductance decay on a test c-Si wafer (n-type, ~1 Ω.cm, ~300μm-thick) passivated with the organic monolayer on both sides, and (b) the schematic energy band diagram of the gate heterojunction of a c-Si/pentacene HJFET illustrating the injection of electrons from the gate metal into the c-Si substrate.

device with a ~15nm-thick pentacene layer are given in Fig. 9. Operation voltages below 1V, subthreshold slope of ~100 mV/dec and gate leakage of the order of ~10pA is achieved despite the large gate area of 10μm×100μm.

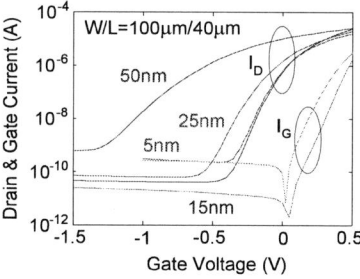

Fig. 8. The experimental drain and gate currents of the c-Si/pentacene HJFET devices fabricated with different pentacene thicknesses.

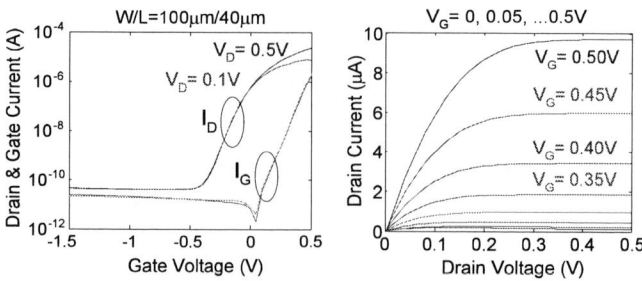

Fig. 9. The experimental output and transfer characteristics of a c-Si/pentacene HJFET device with a pentacene thickness of ~15nm.

Benchmarking vs. Conventional TFT

In a field-effect transistor, the drive current depends on the channel sheet resistance (R_{sh}). Therefore R_{sh} can be used as a metric to compare the drive capability of the HJFET and the conventional TFT. In a conventional TFT, R_{sh} is given by the well-known relationship $R_{sh} = 1/[\mu_n C_{ox}(V_{GS}-V_T)]$, where μ_n is the electron mobility, C_{ox} is the gate dielectric capacitance and V_T is the threshold voltage. In a HJFET, $R_{sh} = \rho_s/(t_{Si}-W_D)$, where ρ_s is the resistivity of c-Si ($\rho_s = 1/[qN_D\mu_n]$), t_{Si} is the c-Si thickness and W_D the depletion region width under the gate can be expressed as $W_D = [t_{Si}^2 - 2\varepsilon_s(V_{GS}-V_p)/(qN_D)]^{1/2}$, where ε_s is the dielectric constant of Si, q is the electron charge, N_D is the concentration of donors and V_p is the pinch-off voltage.

Plotting the gate voltage (V_{GS}) or overdrive voltage ($V_{GS}-V_T$ or $V_{GS}-V_P$) versus R_{sh} for the HJFET devices and conventional TFTs fabricated on SOI substrates [6,7] reveals the significant advantage of the HJFET devices particularly in the regime of interest for active-matrix OLED displays, where the typical drive currents are in the range of 1-5µA per pixel at maximum pixel brightness (Fig. 10). The normally-OFF a-Si:H/c-Si HJFET requires comparable overdrive voltages but lower gate voltages compared to the conventional TFT thanks to the small V_p and subthreshold slope, while the normally-ON a-Si:H/c-Si HJFET requires even lower overdrive voltages. Comparable overdrive voltages are required for the pentacene/c-Si HJFET.

Note, for simplicity, Fig. 10 is plotted under the assumption that R_{sh} is constant across the channel, which is the case in the linear regime at small drain voltages. However, the above conclusions are applicable to the saturation regime as well.

Summary and Conclusion

In summary, thin-film heterojunction field-effect transistors with c-Si channels and gate regions comprised of a-Si:H or organic materials were demonstrated. As summarized in Table

TFT Process Steps/Temperature			
Process:	**Conventional TFT**	**HJFET**	
		a-Si:H Gate	**Organic Gate**
Gate Stack	Dielectric CVD @ 600°C	a-Si:H PECVD @ 200°C	Organic Evaporation @ Room Temp.
Source/Drain	Implantation & Activation @ 600°C	c-Si:H PECVD @ 200°C	Metal Evaporation @ Room Temp.
LDD	Implantation & Activation @ 600°C	Not Needed	Not Needed
TFT Device Characteristics			
Operation Voltage	> 1-2 V	< 1V	< 1V
Subthreshold Slope	100-150 mV/dec	70-90 mV/dec	~ 100mV/dec

Table II. Summary of the main differences in the TFT fabrication process and device characteristics of the HJFET devices with a-Si:H and organic gate stacks presented in this work versus the conventional TFTs. The HJFET devices reduce the process temperature from 600°C to 200°C and room-temperature with a-Si:H and organic gate stacks, respectively. They also reduce the subthreshold slope and the operation voltage particularly in the regime of interest for AMOLED displays.

II, significant reduction in the process temperature and large-area low-cost processing capability is enabled by these devices. In addition, smaller operation voltages and subthreshold slopes are achieved compared to conventional TFTs. The demonstrated HJFET devices are of particular interest to TFT backplanes for active matrix OLED displays on low-temperature poly-Si (LTPS) substrates.

Acknowledgements

The authors gratefully acknowledge Dr. Ghavam G. Shahidi and Dr. T-C. Chen of IBM Research for technical discussion and managerial support. The authors are also grateful to Prof. Sigurd Wagner of Princeton University for allowing the usage of his PECVD facility for this work.

References

[1] B. Hekmatshoar, "Thin-Film Heterojunction Field-Effect Transistors with Crystalline Si Channels and Low-Temperature PECVD Contacts", *IEEE Electron Device Letters*, vol. 35, no. 1, pp. 81-83, Jan. 2014

[2] B. Hekmatshoar, "Normally-Off Thin-Film Silicon Heterojunction Field-Effect Transistors and Application to Complementary Circuits", *IEEE Electron Device Letters*, vol. 35, no. 5, pp. 545-547, May. 2014

[3] G. Graditi, *et al.* "Are heterojunction modules with intrinsic thin layer also reliable as they are good?", in *Proc. 24th Eur. Photovolt. Solar Energy Conf.*, Sep. 2009, pp. 3363–3366.

[4] T. Ando, "Ultimate Scaling of High-κ Gate Dielectrics: Higher-κ or Interfacial Layer Scavenging?", *Materials*, vol. 5, no. 3, pp. 478-500, March 2012

[5] M. Taguchi, et. al., "24.7% record efficiency HIT solar cell on thin silicon wafer", *IEEE J. Photovolt.* vol. 4, no. 1, pp. 96-99, Dec. 2013

[6] R. G. Manley, et. al., "Demonstration of High Performance TFTs on Silicon-on-Glass (SiOG) Substrate", *SID Tech. Dig.* pp. 287-289, 2007

[7] R. G. Manley, et. al. "Development of integrated electronics on silicon-on-glass (SiOG) substrate", *ECS Transactions*, vol. 16, no. 9, pp. 371-380, 2008

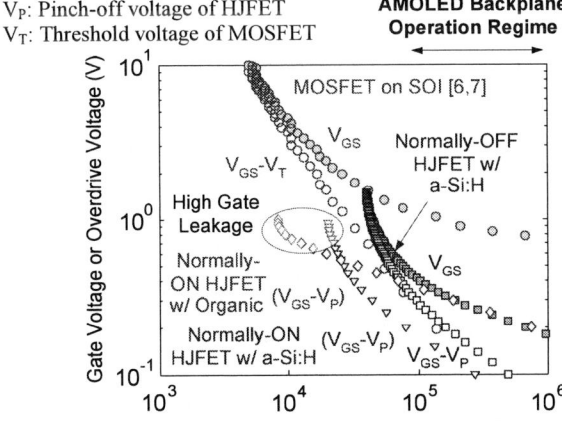

Fig. 10. Gate voltage and/or over-drive voltage vs. channel sheet resistance ($V_D / I_D \times$ W/L at $V_D = 0.1$V) of HJFET devices fabricated in this work benchmarked vs. conventional devices fabricated on SOI substrates [6,7].

High Performance Metal Oxide TFT and its Applications for Thin Film Electronics

Gang Yu, Chan-Long Shieh, Juergen Musolf, Fatt Foong, Tian Xiao, Guangming Wang, Kristoffer Ottosson
CBRITE Inc., 421 Pine Ave., Goleta, CA93117, USA; E-mail: gangyu@cbriteinc.com

Abstract

Recent progress on metal-oxide TFT with mobility and stability as good as LTPS-TFT and with uniformity and off current as good as pristine a-Si TFT are presented. Their applications for high pixel density displays and image arrays are discussed with emphasis on pixel and peripheral circuits with analog functions.

Introduction

Advanced display and sensor arrays call for transistors with high carrier mobility and low "OFF" current with low manufacture cost. Transistor and circuit based on crystalline silicon wafer are too costly for such applications. Thin film transistors (TFT) on glass or plastic substrates are promising in term of process cost, but they are limited by transistor and circuit performance. For examples, next generation flat-panel displays in 4Kx2K and 8Kx4K formats with frame time up to 480Hz and with gray shades of 10-12 bits call for TFT with mobility larger than 50 cm^2/Vsec, with switch ratio larger than 10^8 and with "OFF" current $\ll$ 1pA. Displays with LED emitter as display element, or with integrated gate and/or data driver call for high current operation stability and high performance uniformity over a large area. Large size image arrays with gain amplifiers at each image pixel set up even more rigorous specifications on carrier mobility, "OFF" current, switch ratio and device uniformity over the entire array. Amorphous silicon based transistors cannot meet such requirement due to low mobility and short operation lifetime. Low temperature poly-silicon based transistors cannot meet such requirements due to its high "OFF" current under Vds bias and performance inhomogeneity over long range (often called MURA defect).

CBRITE has developed MOTFT with high carrier mobility and superb operation stability competitive to high quality ELA-LTPSTFT was demonstrated [1-4]. These MOTFT also possess "OFF" current at <1 fA that a switch ratio of $>10^{14}$ was achieved without special design/process. Such high performance MOTFT can also be made onto flexible substrates with low process temperature [3]. Examples of using such high mobility and stability MOTFT for bottom emission OLED TV was demonstrated with an AMOLED with 85 ppi full-color pixel pitch, such pixel pitch can be used to construct 26" FHD, 52" 4Kx2K, or 104" 8Kx4K

OLED TVs [3]. Mobility over 50 cm^2/Vsec and stable operation over 10^2 coulomb at 60 °C validate such MOTFT for 480Hz, 8Kx4K OLED TV for more than 10 years operation lifetime [1,2]. Such high mobility and stability MOTFT is also a perfect candidate for large size, high frame rate, advanced LCD TV in 4K2K and 8K4K formats.

In addition to TV applications, such high performance MOTFT is also attractive for high pixel density, portable or wearable display applications. In this area, a question of general interest is whether such high performance can be retained in TFT with geometric dimensions down to design rules of existing TFT manufacture lines. Another question is what kind of products one could address for such small size MOTFT. We have developed MOTFT in bottom gate, top source/drain configuration with dimension down to design rules of existing display production lines for portable displays. 2T1C pixel driving circuit can be arranged in 29 μmx29μm. 880 ppi FHD white or colored monochrome AMOLED can be achieved in 2.5" diagonal with power budget competitive to all existing display technologies. The same backpanel can also be used for 2.5" qFHD and 5" FHD True-FC AMOLED Displays with 440 ppi. Each full-color display pixel takes 58μmx58μm space and comprises four sub-pixels arranged in square shape with different color coordinates. Such high pixel density backpanel can be fabricated with 5-6 mask steps. We thus provide a low cost AMOLED backpanel competitive to backpanel for inplane-switching (IPS) LCD which typically needs two more mask steps.

Methods and Results

We adopted bottom-gate, top source/drain TFT configuration with either an etch-stop (ES) layer on top of channel layer, or with back-channel-etching (BCE) structures. Both can be achieved with the same mask counts. For the case with ES, a self-aligned process can be adopted which guarantee alignment between ES and gate, which enables ES type MOTFT with channel length down to 5-6 μm. We have also been able to achieve performance similar to that in ES TFT than that in TFT made with so-called four-mask BCE process. A cross-section structure is sketched in Fig. 1. The gate and source/drain layers in these MOTFT can be made with conventional metals available in a-Si TFT manufacture lines. The gate dielectric can be made with common gate-insulators processed by PECVD, or with oxide insulator processed with non-PECVD methods such as atomic layer deposition, reactive sputter or

978-1-4799-8002-4/14 $31.00 © 2014 IEEE

annodization. The metal-oxide channel layer was deposited by means of sputter at room temperature. The highest process temperature was below 350 °C. We have been able to retain the TFT performance for device with dimension down to design rules for TFT manufacture lines dedicating for portable display products. An example I_d-V_{gs} data set in MOTFT with In-O channel and with W/L=3μm/3μm are shown in Fig. 2.

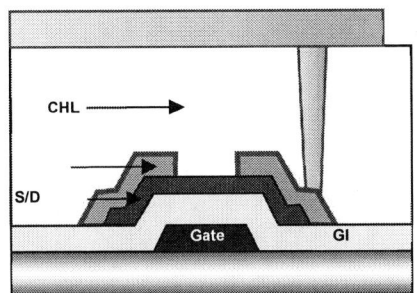

Fig.1 Cross-section view of the MOTFT backpanel.

MOTFT with such small dimensions show not only good Id-Vgs characteristics, but also superb output performance. Fig.3 shows an Id-Vds set from a device of W/L=5μm/6μm. The output impedance reaches ~250MΩ at 1 μA level. Such high impedance reduces brightness sensitivity related with voltage drop on Vdd line, enables narrow Vdd linewidth and top emission, AMOLED pixel driver with 2T1C design in 29μmx29μm area.

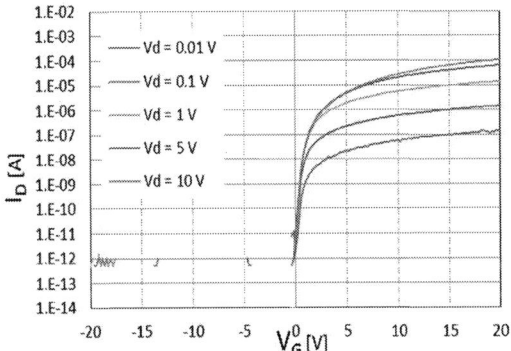

Fig. 2 Id-Vgs of a BCE MOTFT with W/L=3μm/3μm.

Such MOTFT shows superb stability bias-temperature stresses (DC NBTS and PBTS). The NBTS and PBTS were -60mV and 50mV respectively under Vg=+/-20V and Vds=0.1V for 3600 seconds at 60 °C [4].

High stability was also observed under accelerated current stress condition. Fig. 4 shows a data set taken under Vgs=Vds=5V (Id~10.1μA) at 60 °C for 60 hours. It was taken from a TFT with W/L=8μm/6μm which is close to the parameters used for the driving TFT in the pixel driver. The drain current follows a simple exponential decay within the test period with an index of 2.1×10^6 hours. Assuming the operation stability follow the same decay mechanism in longer operation period, one anticipates only ~6% current drop after 10^4 hours continuing operation at this accelerated driving condition. The shift of Vgs at 1nA level is approximately -50 mV over 60 hour test period, less than gray shade change in 8 bit data driver (1/256 of maximum pixel current). Such stability data are rarely seen in BCE type TFTs, including LTPS based devices.

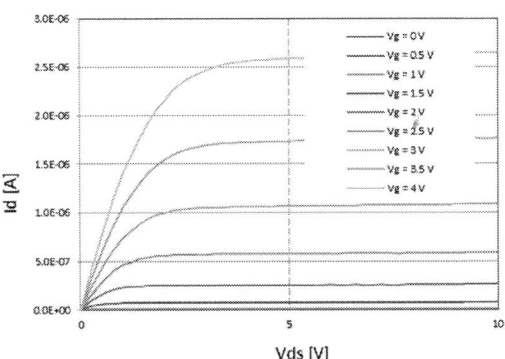

Fig. 3 Id-Vds of a BCE MOTFT with W/L=5μm/6μm.

Such high performance, small dimension MOTFT enables high pixel density AMOLED backpanel with simple 2T1C design without compensation. Data in Fig. 4 suggest that the current change over operation lifetime of a mobile device at room temperature is negligible. Taking mobile phone as an example, 3 years effective usage translates into ~10^4 hours at mean pixel current ~0.1μA at room temperature.

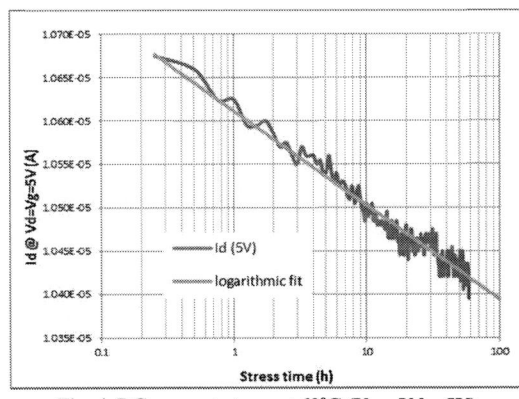

Fig. 4 DC current stress at 60°C (Vgs=Vds=5V).

Scaling effect was observed in MOTFT with large channel width, e.g. W=200um and L=3 or 6 μm. Drain current at Vgs=10V is substantially higher than 2mA. This result validates such MOTFT for peripheral driver circuits on glass substrate. In portable display application, a desire is to integrate the gate driver onto glass that one could design a portable products (such as smart phones) with narrow edge on both sides. The high current driving capability for TFT with such small geometric dimensions enables portable displays with narrow edges only achieved with LTPS-TFT.

978-1-4799-8002-4/14 $31.00 © 2014 IEEE

Operation stabilities for the integrated peripheral drivers have been tested, and confirmed to meeting the specifications for commercial applications.

With such TFT performance, top-emission, AMOLED backpanels were developed with single color sub-pixel in 29μmx29μm. Aperture ratio reaches 60-70% for 29 μm. Performance of the TFT in pixel driver is similar to that shown in Figs. 2-4. Uniformity test to TFTs in PCM zones at 4 corners of the panel revealed threshold voltage variation ~50 mV. Conventional 2T1C pixel driver can thus be used for the needed display quality. The cross-section of the TFT in backpanel is the same as shown in Fig. 1. Such a top emission AMOLED pixel circuit can be constructed with two more mask steps over S/D: planarization/passivation and pixel electrode. The high pixel density, top emission AMOLED backpanel can thus be fabricated with 6 mask steps (GE/GI-via/CHL/SD/PV-PLN/Pixel-electrode), two masks less than the backplane for AMLCD with similar pixel density. Introducing multi-tone photoresist process or other advanced designs or processes such as self-aligned process, the entire backpanel process can be reduced down to 5 or less mask steps. It is worth to mention that such backpanal can be made with existing manufacture lines originally set up for a-Si TFT, for color-filters and/or for touch-panels. For example, we have demonstrated bottom emission AMOLED backpanel with Gen-2.5 size (370mmx470mm) color filter lines [2].

Aiming at smart watch applications, a 2.5in diagonal FHD (1920x1080) white AMOLED was demonstrated with 29 μm pixel size [4, 5]. With a WOLED with 30 cd/A and 5V at 300 nits, a smart AMOLED watch display can be operated at approximate 30mW (assuming 30% pixel on at full brightness). Such a power budget is promising for battery powered smart watch and other wearable electronic devices.

The same AMOLED pixel design can also be used for a 2.5" QFHD (960x540), 5" FHD, 10" 4Kx2K True-FC AMOLED Displays with 440 ppi. Each FC pixel comprises 4 sub-pixels providing the color primaries achieved from the un-patterned OLED and an un-filtered OLED with broad-band emission without optical loss. These designs are thus promising generally for smart-watches, smartphones, tablets and flexible/wearable displays.

To overcome the energy loss issue in color-from-white design, CBRITE has developed a full-color OLED approach based on a sky-blue OLED with emission covering blue and green zones [6]. Fig. 5 shows an OLED data set from a sky-blue OLED developed at New Vision Opto-Electronic Technology. This OLED turns on at voltage above 3V, reaches 1050 nits at 4V (and 2.5mA/cm^2). The corresponding current efficiency is 42 cd/A and the corresponding power efficiency is 33 lm/Watt. Its emission profile is shown in Fig. 6A, covering blue and green primary colors. The emission

profile does not change over 2 orders of intensity range. This sky-blue OLED is thus perfect candidate for the OLED array without pixel level patterning process.

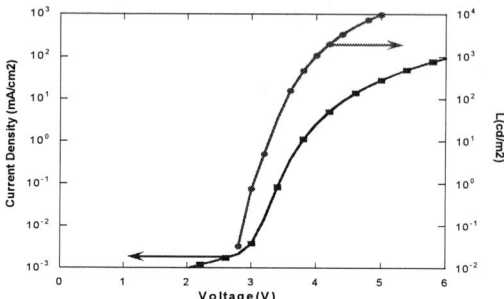

Fig. 5 V-I-L data set of the skyblue OLED.

Integrating such an OLED array onto the MOTFT backpanel with 29mm sub-pixel pitch, a True-FC AMOLED can be achieved with four sub-pixels in each FC pixel area. The blue pixel is formed by overlaying a magenta filter over the sky-blue OLED sub-pixel, the green pixel is formed by overlaying a long-wavelength pass yellow filter over the sky-blue OLED sub-pixel. The color filters only cut-off the light in "unwanted" band while keeping ~100% transmission without optical loss in the designated color zone. More than 50% of the emission photons from the skyblue emitter can be used effectively for green and blue pixels. The red emitter is constructed with a red luminescent filter. For example, CdSe/ZnS core/shall quantum dot filter (Aldrish 694606). >70% quantum efficiency can be achieved. The corresponding transmission/emission profile of the filters are shown in Fig. 6B and the resulting emission profile of the red, green, blue and sky-blue sub-pixels are shown in Fig. 6C.

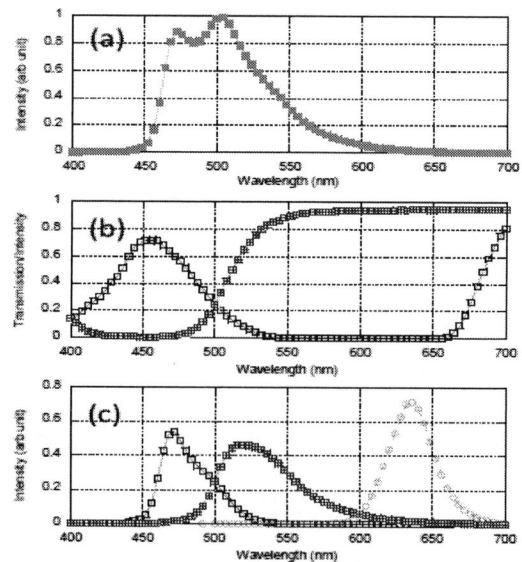

Fig.6 RGB color pixels from a skyblue OLED array.

The color gamut is ~72% NTSC 1931, a competitive number for portable display products. Simulation results reveals that the FC display can be operated with Vdd-Vcom=6-8V. With operation current at 80 mA for brightness of 300 nits (30% pixels on). Tiling 4 pieces 2.5" qFHD panels together, once could achieve FHD (1920x1080) AMOLED smart phone with 320 mW for brightness of 300 nits (30 % pixel on). Such a display represents one of portable displays with best power efficiency.

The high uniformity and stability MOTFT presented here can be operated in analog range as shown in the 2T1C AMOLED circuit. The output of the driving transistor regulating the OLED current based on the analog voltage sending onto the storage capacitor. In addition to display applications, another promising application for such MOTFT is pixel readout circuit in large size image array with amplification circuit in each pixel. Such pixel design has been called "active pixel sensor", APS. Again, the high uniformity and high mobility and operation stability in such MOTFT is perfect for the gain amplifier. The large switch ratio and low off current is also perfect for the reset and switch transistors in the APS circuit. such high performance. A X-ray radiation test was carried out with the transistors shown in Fig. 1. No change in Id-Vgs performance was observed after X-ray radiation with DOS equivalent to10^7 of standard chest X-ray test.

Discussion and Remarks

A current barrier for a white OLED to be used for portable AMOLED display is the substantial energy loss (and thus the high power consumption and shorter OLED lifetime) at the RGB color filters. The color-from-skyblue approach presented in this work breaks this barrier from several perspectives. First the RGBsB FC pixel design based on a monochrome sky-blue OLED emitter array reduces the optical energy loss at the filter layer to less than 50%. Secondly, white color which has high usage for all window based applications can be constructed with the unfiltered skyblue sub-pixel along with the red emitter. A portable display can also be designed with multiple operation modes, such as monochrome white color or skyblue color with lowest power consumption, and with full-color modes optimized for video image contents. We also point out the FC OLED array can be achieved with larger aperture ratio than those made with fine masks. The larger aperture ratio improves the power efficiency of the FC display to level practical for portable applications.

In addition to the fine-mask-free approach above, we have also developed a FC OLED approach based on a single fine mask [7]. This approach improves power efficiency further, eliminates color cross-talks related with fine mask alignment and/or deformation, and improve process bottleneck and cost considerably.

Extended from our early MOTFT work for OLED TV [4], we demonstrated that the high performance MOTFT can be made with dimension down to design rules in existing display production line. They are the best solution for portable/wearable displays. In addition to AMOLED this type of MOTFT can also be used for portable FC AMLCD with high ppi and with high aperture ratio. The switch TFT can be made on top of gate metal line in AMLCD backpanel with <<10 μm gate linewidth.

Summary

In this presentation, we demonstrate a new type of thin-film transistor with channel layer made of metal-oxide semiconductors. The performance of such MOTFT is as good as high-end LTPS-TFT, the uniformity and the off current are as good or even better than pristine a-Si TFT. Example of using such MOTFT for analog pixel circuits for AMOLED and for APS X-ray image array are discussed. Top emission, AMOLED backpanel has been realized with 29 um subpixel pitch. Such design can be used for 880 ppi FHD monochrome AMOLED in a 2.5" diagonal [4]. The same backpanel can also be used for 2.5" qFHD smart watch, 5" FHD and 10" 4Kx2K full-color AMOLED displays with 440 ppi pixel pitch. The same transistor can also be used for next generation AMLCD with pixel density beyond 400ppi.

The large switch ratio of these MOTFT enable X-ray imager with more pixel elements, with improved detectivity and with large dynamic range. Fabricating such image array onto curved or flexible substrates enables large size image sensor with variable focal plane. The superb stability under X-ray exposure makes it a great candidate for CT applications.

References

[1] G. Yu, C.-L. Shieh, F. Foong, G. Wang, A. Kuo, K. Yang, J. Wang, F. Chang, J. Peng and B. Nilsson, SID Symposium Digest, Vol.42, p.483 (2011).

[2] G. Yu, C.-L. Shieh, F. Foong, G. Wang, T. Xiao, J. Musolf, K. Ottosson, B. Berkoff and B. Nilsson, SID Symposium Digest, Vol.43, p.1123 (2012).

[3] C.-L. Shieh, G. Wang, J. Musolf, F. Foong, T. Xiao and G. Yu, SID Symposium Digest, Vol.44, p.717 (2013).

[4] G. Yu, C.-L. Shieh, J. Musolf, F. Foong, T. Xiao, G.-M. Wang, K. Ottosson, Z. Chen, F. Chang, C. Yu and J.-W. Park, Invited Presentation, Session 21-2, SID 2014.

[5] G. Yu, C.-L. Shieh, J. Musolf, F. Foong, T. Xiao, G.-M. Wang, K. Ottosson, Z. Chen, F. Chang, C. Yu, J.-H. Zou and L. Wang, Invited talk, IMID14, Daegu, Korea, August 25-29, 2014.

[6] G. Yu and C.L. Shieh, CBRITE Inc., US7,977,868

[7] G. Yu and C.L. Shieh, CBRITE Inc., US Publication 2011030938.

Integration of Solution-Processed (7,5) SWCNTs with Sputtered and Spray-Coated Metal Oxides for Flexible Complementary Inverters

L. Petti[1], F. Bottacchi[2], N. Münzenrieder[1], H. Faber[2], G. Cantarella[1], C. Vogt[1], L. Büthe[1], I. Namal[3], F. Späth[3], T. Hertel[3], T. D. Anthopoulos[2], and G. Tröster[1].

[1]Electronics Laboratory, Swiss Federal Institute of Technology, Gloriastrasse 35, Zürich, 8092, Switzerland,
TEL: +41-446320513, FAX: +41-446321210, email: luisa.petti@ife.ee.ethz.ch
[2]Department of Physics and Center for Plastic Electronics, Blackett Laboratory, Imperial College, London SW7 2BW, UK
[3]Institute of Physical and Theoretical Chemistry, University of Würzburg, 97074 Würzburg, Germany

Abstract

We report the integration of solution-processed high-purity semiconducting (7,5) single walled carbon nanotubes (SWCNTs) with metal oxides for the fabrication of high-performance CMOS inverters on free-standing plastic foils. Flexible inverters based on spin-coated SWCNTs and sputtered amorphous InGaZnO (IGZO) exhibit gains up to 85 V/V, even while bent to a tensile radius of 1 cm. To our knowledge, this is the highest gain ever reported for flexible and strained hybrid inverters, supplied at $V_{DD} \leq 10$ V. We also realize flexible inverters based on fully solution-deposited SWCNTs and InO_x semiconductors.

Introduction

Complementary MOS technology, combining n- and p-type transistors, enables simplification of circuit design whilst lowering power consumption, and minimizing noise. The realization of a thin-film equivalent to silicon CMOS electronics suffers from the lack of complementary organic or inorganic semiconductors with similar performance. To date, several groups have reported logic circuits based on n-type metal oxides and organic [1] or inorganic [2] p-type semiconductors. While n-type metal oxides, especially amorphous IGZO [3], possess high mobility >10 $cm^2V^{-1}s^{-1}$ and ambient stability, organic and inorganic p-type materials with matching performance are difficult to achieve. Recently, solution-based SWCNTs have drawn a considerable attention, owing to their good electrical and mechanical properties. Using inkjet-printing methods, in combination with mixed semiconducting SWCNTs and ZTO, complementary circuits have now been demonstrated [4]. In this paper, we advance the current state-of-the-art, by integrating spin-coated highly-selected semiconducting (7,5) SWCNTs [5] with sputtered and spray-coated n-type metal oxides, offering a route towards flexible and strained complementary inverters with high gains and excellent noise margins.

Fabrication

Hybrid complementary inverters were fabricated on 4" Si/SiO_2 wafers and on 50 μm-thick polyimide foils (7.6×7.6 cm²), using a bottom-gate coplanar TFT geometry. Fig. 1 shows: (a) the fabrication process flow and (b) the schematic cross-section. Cr gate contacts, Al_2O_3 gate isolator, and Ti/Au electrodes were realized using standard thin-film deposition methods and UV lithography [6]. Subsequently, the substrates

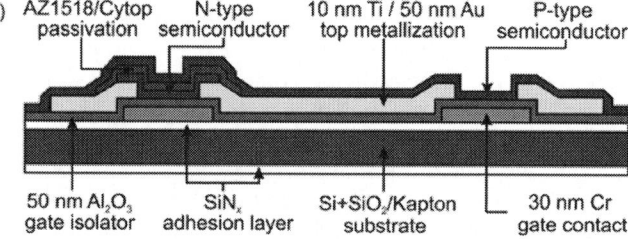

Fig. 1: Hybrid complementary inverter based on solution-processed (7,5) SWCNT TFTs and metal oxide TFTs: (a) process flow and (b) schematic cross-section. Fabrication process was optimized concerning low temperature fabrication, thickness of brittle materials, and adhesion between the different device layers aiming at good device performance and bendability.

were diced into chips of 1.5×1.5 cm². In the case of vacuum-processed metal oxide, a 15 nm-thick IGZO [3] film was deposited at room temperature, using RF magnetron sputtering and a ceramic $InGaZnO_4$ target. The IGZO was then patterned into islands by wet etching and passivated with photoresist (AZ1518®). For the solution-processed metal oxide, a 10 nm-thick InO_x film was deposited in air via ultrasonic spray pyrolysis, using a solution of 30 mg/ml indium nitrate hydrate in deionized water. During the automated spray process, the substrate was heated to 250°C. The InO_x was patterned using a shadow mask, and then passivated with a Cytop® layer. To complement the n-type IGZO and InO_x, a semiconducting (7,5) SWCNT solution, behaving as p-type material, was used. Single-chirality was achieved by means of polymer-wrapping [5]. The SWCNT solution was spin-coated and annealed at 90°C in nitrogen (N_2), resulting in a 10 nm-thick layer, which was left unstructured and unpassivated. Fig. 2 shows the surface topography of the three semiconductors obtained with atomic force microscopy (AFM). Both metal oxides present a homogeneous structure without any grain boundaries. Fig. 2c shows the randomly oriented SWCNT network, exhibiting continuous current percolation paths. Fig. 2d displays the absorption spectrum of the SWCNT solution. The peak at λ= 1050 nm corresponds to the first excitonic transition of the (7,5) diameter, proving that our SWCNTs are single-chirality.

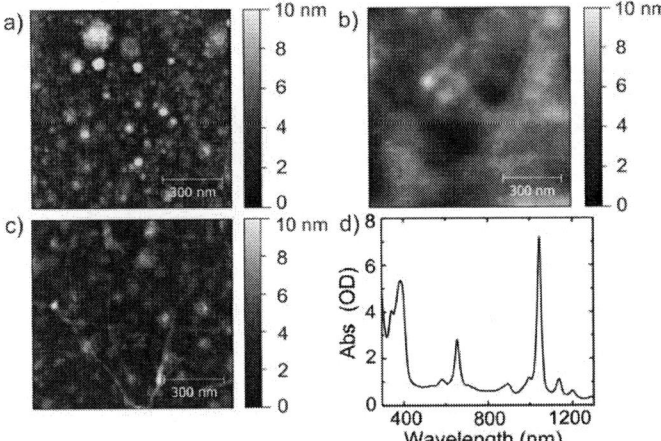

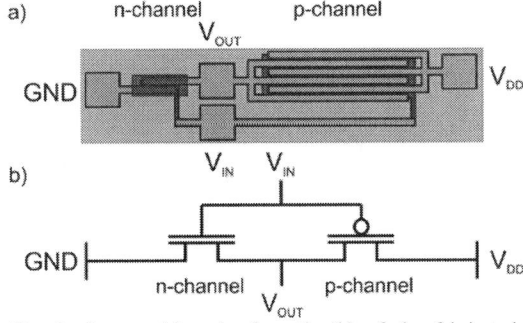

Fig. 2: AFM images showing surface topography of: (a) IGZO (15 nm), (b) InO_x (10 nm), and (c) (7,5) SWCNT (10 nm) films on $Si/SiO_2/SiN_x/Al_2O_3$ surface. (d) Absorption spectrum of the (7,5) SWCNT solution.

Fig. 4: Circuit diagram (a) and schematic (b) of the fabricated hybrid complementary inverter based on metal oxide TFTs and solution-processed (7,5) SWCNT TFTs. All TFTs have a channel length L of 25 μm, with an interdigitated channel, as visible in (a).

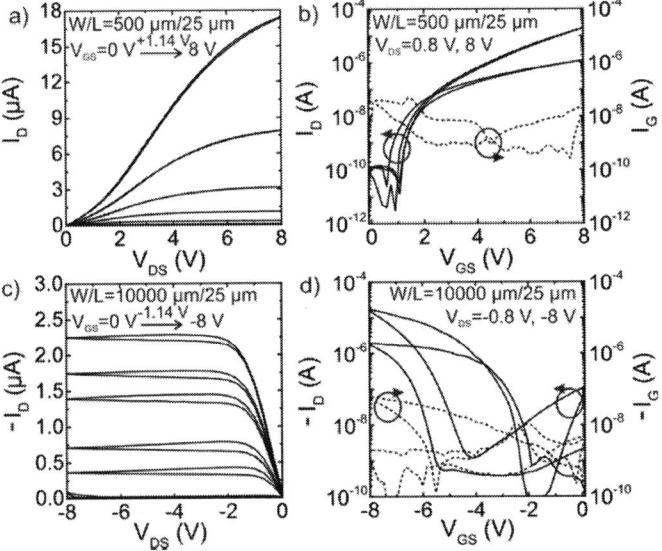

Fig. 3: Output and transfer characteristics of sputtered IGZO [(a) and (b)] and spin-coated (7,5) SWCNT [(c) and (d)] TFTs on Si/SiO_2 substrate. TFT performance parameters are: $\mu_{FE,n}$= 0.7 $cm^2V^{-1}s^{-1}$, $V_{TH,n}$= 4 V, SS_n= 0.12 V/dec, $\mu_{FE,p}$= 0.02 $cm^2V^{-1}s^{-1}$, $V_{TH,p}$= -2.8 V, and SS_p= 0.42 V/dec. Electrical measurements were carried out in nitrogen (N_2).

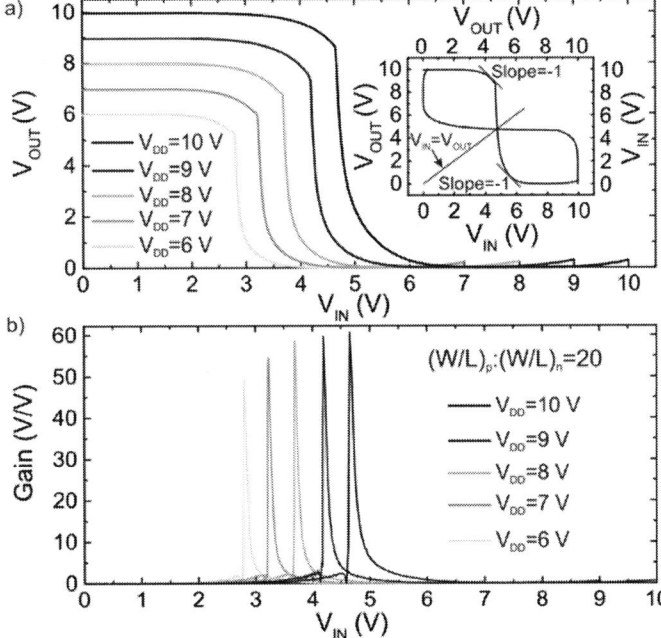

Fig. 5: Voltage transfer characteristic (VTC) (a) and static gain (b) of the complementary inverter based on sputtered IGZO and spin-coated (7,5) SWCNT TFTs on Si/SiO_2 substrate, measured at V_{DD}= 6, 7, 8, 9, 10 V. The inset shows the bi-stable hysteresis VTC at V_{DD}= 10 V. For V_{DD}= 10 V, V_M= 4.75 V, G= 60.6 V/V, NM_H= 4.33 V, and NM_L= 4 V.

IGZO/SWCNT TFTs and Inverters

Fig. 3 shows the I_D-V_{DS} and I_D-V_{GS} characteristics of the IGZO and SWCNT TFTs on Si/SiO_2 substrate, measured in nitrogen (N_2). TFT performance parameters were extracted using standard MOSFET equations. The IGZO TFT exhibits a saturation field-effect mobility $\mu_{FE,n}$= 0.7 $cm^2V^{-1}s^{-1}$, a threshold voltage $V_{TH,n}$= 4 V, and a sub-threshold swing SS_n= 0.12 V/dec. The low effective mobility compared to [6] is attributed to the non-annealed IGZO film, as well as to the coplanar TFT structure with Au contacts. The SWCNT TFT yields $\mu_{FE,p}$= 0.02 $cm^2V^{-1}s^{-1}$, $V_{TH,p}$= -2.8 V, and SS_p= 0.42 V/dec. The hysteresis in the drain current is caused by the polymeric presence in the SWCNT solution, whereas the slightly ambipolar behavior is due to the intrinsic properties

of the SWCNTs. The overall process yield of IGZO and SWCNT devices fabricated in multiple runs on 1.5×1.5 cm^2 chips, each containing 20 TFTs, is ≈ 70%. Process variability results in the following parameter distribution: $V_{TH,n}$= (3.2 ± 0.8) V, $\mu_{FE,n}$= (0.7 ± 0.5) $cm^2V^{-1}s^{-1}$, $V_{TH,p}$= (-2.5 ± 0.4) V, and $\mu_{FE,p}$= (0.026 ± 0.01) $cm^2V^{-1}s^{-1}$. The complementary inverter was realized by connecting SWCNT and IGZO TFTs as shown in Fig. 4. An extensive TFT characterization was used to design inverters with centered midpoint voltages V_M≈ $V_{DD}/2$. Fig. 5 displays: (a) the voltage transfer characteristics (VTC) and (b) the static gain G, measured at different supply voltages. At V_{DD}= 10 V, the inverter exhibits a V_M= 4.75 V, a G= 60.6 V/V, a rail-to-rail swing (V_{OH}= 9.95 V, V_{OL}= 0.28 V), and excellent noise margins (NM_H= 4.33 V, NM_L= 4 V). When V_{IN} is swept from low to high voltage bias, V_M shifts by +20 mV for V_{DD}= 5 V, and by +1.9 V for V_{DD}= 10 V. This

978-1-4799-8002-4/14 $31.00 © 2014 IEEE

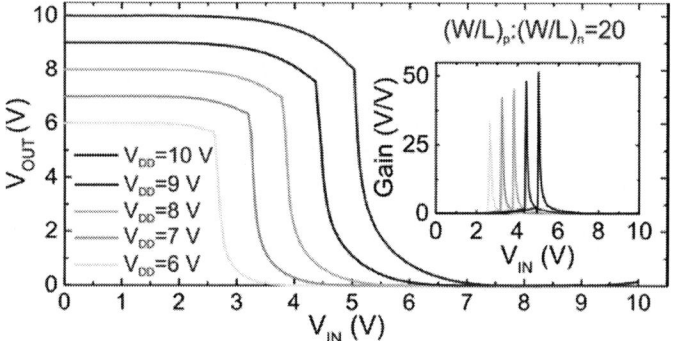

Fig. 6: VTC and gain (inset) of IGZO/SWCNT complementary inverter on Si/SiO$_2$ substrate, measured at V$_{DD}$= 6, 7, 8, 9, 10 V. For V$_{DD}$= 10 V, V$_M$= 5.1 V and G= 51.6 V/V. Measurements were carried out in ambient conditions.

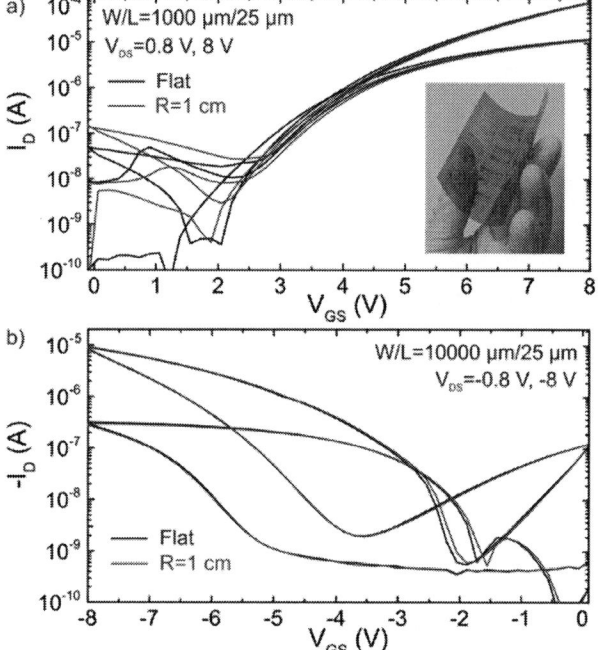

Fig. 7: Transfer characteristics of flexible IGZO (a) and SWCNT (b) TFTs on plastic foil, measured while flat and bent to a radius of 1 cm, which corresponds to a tensile strain ε≈ 0.29% parallel to the TFT channels. Bending changes V$_{TH,n}$ by +30 mV, V$_{TH,p}$ by +20 mV, µ$_n$ by -0.8%, µ$_p$ by +2%, SS$_n$ by 3%, and SS$_p$ by -3.7%. The inset shows a photograph of the processed flexible substrate. Electrical measurements were carried out in N$_2$.

is attributed to the gate-induced current hysteresis of the SWCNT TFT, which can be reduced by decreasing the polymeric content in the solution. Furthermore, IGZO and SWCNT TFTs are electrically stable. A positive (negative) bias stress up to 100 MV/m for a period of 300 s induces a positive (negative) ΔV$_{TH}$ up to 0.55 (-0.6) V in IGZO (SWCNT) TFTs, due to charge carrier injection in the gate isolator. The static power dissipation (V$_{DD}$= 10 V) is less than 5 µW for V$_{IN}$ HIGH or LOW, and reaches a maximum of 17 µW at V$_{IN}$= V$_M$. Measurements were carried out in N$_2$. Fig. 6 shows the VTC and the static gain of the inverter measured under ambient conditions. After being exposed to ambient air, the inverter is operational, with V$_M$= 5.1 V and G= 51.6 V/V.

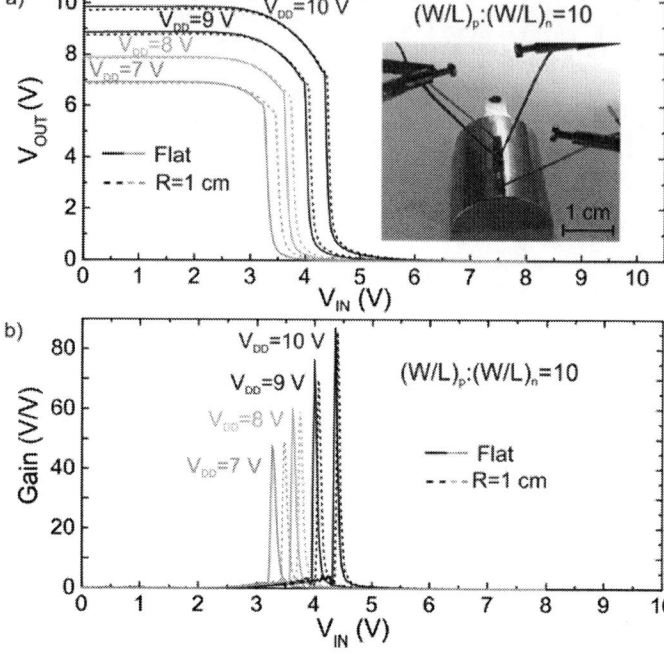

Fig. 8: VTC (a) and static gain (b) of flexible complementary inverter based on IGZO and SWCNT TFTs on plastic foil, measured at V$_{DD}$= 7, 8, 9, 10 V while flat and bent to a tensile radius of 1 cm (ε≈ 0.29% parallel to all TFT channels). For V$_{DD}$= 10 V, V$_M$= 4.39 (4.43) V and G= 87.1 (85.7) V/V for the flat (bent) circuit. The inset shows a photograph of the contacted flexible complementary inverter, bent to a tensile radius R=1 cm. The changes in V$_M$ (+40 mV) and G (-1.6%) are explained by the variations in V$_{TH}$, µ$_{FE}$, and SS of the TFTs. Electrical measurements were carried out in N$_2$.

These variations are due to the changes in the performance of the unpassivated SWCNT TFT, and can be strongly reduced by encapsulating the SWCNT devices. Fig. 7 shows the transfer characteristics of the IGZO and SWCNT TFTs on free-standing plastic foil, measured while flat and wound around a rod of 1 cm radius, which induces a tensile strain of ≈ 0.29% parallel to the channels (Fig. 7). The TFTs stay fully functional while bent and show only minor variations in the device performance (V$_{TH,n}$ +30 mV, V$_{TH,p}$ +20 mV, µ$_{FE,n}$ -0.8%, µ$_{FE,p}$ +2%, SS$_n$ +3%, and SS$_p$ -3.7%). Compared to electrical stress, mechanical stress results in a smaller effect on the TFT performance. Fig. 8 shows the VTC and the static gain of the mechanically flexible complementary inverter, measured while the circuit is flat and bent. A gain of 87 V/V and a midpoint voltage of 4.39 V were obtained. The changes in V$_M$ (+40 mV) and G (-1.6%) are explained by the variations in the V$_{TH}$, µ$_{FE}$, and SS of the TFTs.

INO$_x$/SWCNT TFTS AND INVERTERS

Fig. 9 shows the electrical characteristics of InO$_x$ and SWCNT TFTs on Si/SiO$_2$ substrate, measured in N$_2$. N-type InO$_x$ TFT parameters are: µ$_{FE,n}$= 0.02 cm^{2}V^{-1}s^{-1}, V$_{TH,n}$= 1.2 V, and SS$_n$= 1.5 V/dec. High off current hint at large carrier conductivity in the InO$_x$ layer, caused by the generation of intrinsic defects during processing. The slightly different characteristics of the SWCNT TFT in Fig. 9, compared to Fig. 3, is due to process variability. Fig. 10 and 11 display the

978-1-4799-8002-4/14 $31.00 © 2014 IEEE   612

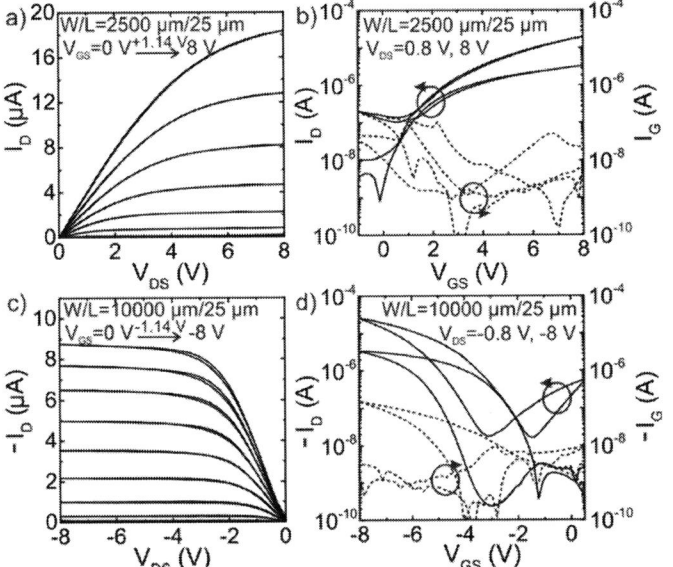

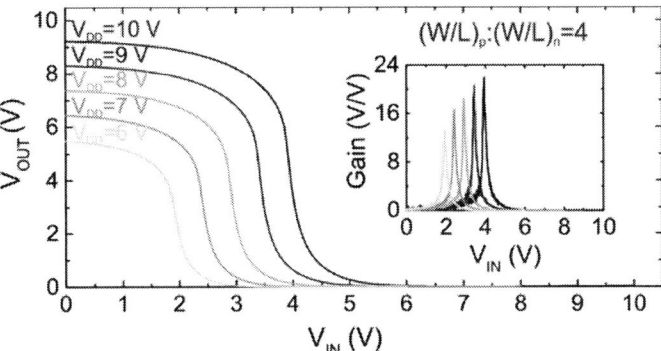

Fig. 11: VTC and static gain (inset) of flexible inverter with fully solution-processed semiconductors, based on spray-coated InO_x and spin-coated SWCNT TFTs on plastic foil, measured at V_{DD}= 6, 7, 8, 9, 10 V. For V_{DD}= 10 V, V_M= 3.97 V and G= 22 V/V. Measurements were carried out in N_2.

InO_x to realize mechanically flexible inverters with fully solution-deposited semiconductors. These results pave the way toward large-area solution-processed flexible electronics.

Fig. 9: Output and transfer characteristics of spray-coated InO_x [(a) and (b)] and spin-coated (7,5) SWCNT [(c) and (d)] TFTs on Si/SiO$_2$ substrate. TFT performance parameters are: $\mu_{FE,n}$= 0.02 cm^{2}V^{-1}s^{-1}, $V_{TH,n}$= 1.2 V, SS$_n$= 1.5 V/dec, $\mu_{FE,p}$= 0.01 cm^{2}V^{-1}s^{-1}, $V_{TH,p}$= -2.4 V, and SS$_p$= 1.15 V/dec. Electrical measurements were carried out in N_2.

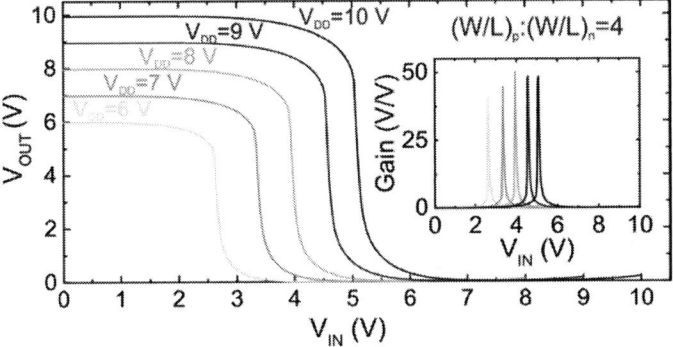

Fig. 10: VTC and static gain (inset) of rigid complementary inverter with fully solution-processed semiconductors, based on spray-coated InO_x and spin-coated SWCNT TFTs on Si/SiO$_2$ substrate, measured at V_{DD}= 6, 7, 8, 9, 10 V. For V_{DD}= 10 V, V_M= 5.1 V and G= 48.4 V/V. Electrical measurements were carried out in N_2.

VTCs and static gains of InO_x/SWCNT inverters on Si/SiO$_2$ and polyimide substrates. For V_{DD}= 10 V, the rigid (flexible) inverter yields V_M= 5.1 (3.97) V, G= 48.4 (22) V/V, and V_{OH}= 9.6 (9.2) V. These differences are attributed to the comparatively poor performance of the flexible InO_x TFT, caused by the increased roughness and the lower thermal conductivity of the polyimide when compared to Si/SiO$_2$.

Conclusion

Solution-based highly-selected semiconducting (7,5) SWCNTs and metal oxides enabled the fabrication of high-performance CMOS inverters on flexible plastic foils. Flexible complementary inverters based on spin-coated SWCNT and sputtered IGZO TFTs exhibited gains up to 85 V/V (V_{DD}= 10 V), even while bent to a tensile radius of 1 cm. Furthermore, we also integrated SWCNTs with spray-coated

Acknowledgement

The authors would like to acknowledge S. Rossbauer from Imperial College London for his support during electrical measurements. This work has been supported by the European Commission through two Seventh Framework Projects (FP7): Flexible multifunctional bendable integrated light-weight ultra-thin systems (FLEXIBILITY, grant agreement: FP7-287568) and Polymer-Carbon Nanotubes Active Systems for Photovoltaics (POCAONTAS, grant agreement: FP7-316633).

References

(1) M. Rockelé, M. Nag, T. H. Ke, S. Botnaras, D. Weber, D-V. Pham, J. Steiger, S. Steudel, K. Myny, S. Schols, B. van der Putten, J. Genoe, and P. Heremans, "Solution-Processed and Low-Temperature ZnO-Based N-Channel TFTs on Polyethylene Naphthalate Foil, Suited for Hybrid Complementary Circuitry", Proc. IDW, pp. 299-302, 2012.

(2) R. Martins, A. Nathan, R.Barros, L. Pereira, P. Barquinha, N. Correia, R. Costa, A. Ahnood, I. Ferreira, and E. Fortunato, , Advanced Materials, "Complementary Metal Oxide Semiconductor Technology With and On Paper", Advanced Materials, vol. 23, pp. 4491-4496, 2011.

(3) K. Nomura, H. Ohta, A. Takagi, T. Kamiya, M. Hirano, and H. Hosono, "Room-temperature fabrication of transparent flexible thin-film transistors using amorphous oxide semiconductors", Nature, vol. 432, pp. 488-492, 2004.

(4) B. Kim, S. Jang, M. L. Geiger, P. L. Prabhumirashi, M. C. Hersam, and A. Dodabalapur, "High-Speed, Inkjet-Printed Carbon Nanotube/Zinc Tin Oxide Hybrid Complementary Ring Oscillators", Nano Letters, vol. 14, pp. 3683-3686, 2014.

(5) A. Nish, J. H. Hwang, J. Doig, and R. J. Nicholas, "Highly selective dispersion of single-walled carbon nanotubes using aromatic polymers", Nature Nanotechnology, vol. 2, pp. 640-646, 2007.

(6) N. Munzenrieder, L. Petti, C. Zysset, G. A. Salvatore, T., Kinkeldei, C. Perumal, C. Carta, F. Ellinger, and G. Tröster, "Flexible a-IGZO TFT amplifier fabricated on a free standing polyimide foil operating at 1.2 MHz while bent to a radius of 5 mm", Proc. IEDM, pp. 96-99, 2012.

Solution-processed Poly-Si TFTs Fabricated at a Maximum Temperature of 150°C

M. Trifunovic, J. Zhang, M. van der Zwan, and R. Ishihara

Delft University of Technology, Delft Institute of Microsystems and Nanoelectronics Technology (DIMES)
Feldmannweg 17, 2628CT Delft, the Netherlands
Tel: (31)152781474 Email: m.trifunovic@tudelft.nl

Abstract

Printing liquid silicon devices using cyclopentasilane as the precursor have led to solution processed devices with high mobilities. The fabrication process, however, required a relatively high process temperature of 350°C, incompatible to inexpensive plastics. A novel processing method is presented that decreases the maximum processing temperature of poly-Si TFTs to 150°C, compatible to low-cost plastics and paper, by using a XeCl excimer laser treatment that would directly transform a solution with polysilane chains into solid polycrystalline silicon. Mobilities as high as 23.5 and 21.0cm²/Vs were obtained for the PMOS and NMOS devices respectively.

Introduction

Printing flexible transistors in roll-to-roll processing has gained a lot of attention in recent years, and have for a bigger part been linked to organic and metal-oxide semiconductor materials [1][2]. These materials have low processing temperatures and can be printed as a solution for TFT fabrication. However, the TFTs lack in their field-effect mobilities (in the order of 1 and 10cm²/Vs for organic and metal-oxide devices respectively)[1][2] as well as reliability compared to silicon. Poly-Si and single-grain Si TFTs fabricated through a solution-process, using a silicon compound commonly referred to as liquid silicon, [3] have several advantages such as high mobilities of a several hundred cm²/Vs[4] for both electrons and holes and high device reliability against moisture and electrical bias. The lowest processing temperature so far for the solution process of Si is 350°C and this obstructs the use of low-cost plastic and paper substrates [3][4][5]. The relatively high temperature is required for a thermal annealing process for the formation of an a-Si:H precursor from polysilane.

In this work, we present for the first time the fabrication of poly-Si TFTs using liquid silicon at a maximum processing temperature of 150°C by directly transforming polysilane into solid poly-Si using a XeCl excimer laser. This would allow solution processed poly-Si TFT fabrication on substrates with a low cost and low thermal budget such as PET, PEN, or even paper.

Experimental

Pure cyclopentasilane (Si_5H_{10}) has been coated on top of a substrate at an elevated substrate temperature of 80°C using a doctor blade in an oxygen free environment. This layer is polymerized into a polysilane solution under UV light for 30 minutes at a temperature of 100°C. The substrate could then directly be crystallized with a single shot of XeCl Excimer Laser (308nm, 28ns) (Fig. 1). This has also been observed with various laser energy densities as well as numbers of pulses. This made it possible for the thermal annealing step, for the a-Si formation, to be omitted. Depending on the energy density and the number of irradiated pulses, the polysilane directly transforms into a-Si, μc-Si or poly-Si. Multiple low energy density irradiation led to the formation of an a-Si:H film, whereas higher energy densities lead to crystallization of the polysilane film, as can be seen from the Raman spectra in Fig. 2 and the TEM images and EDS results in Fig. 3.

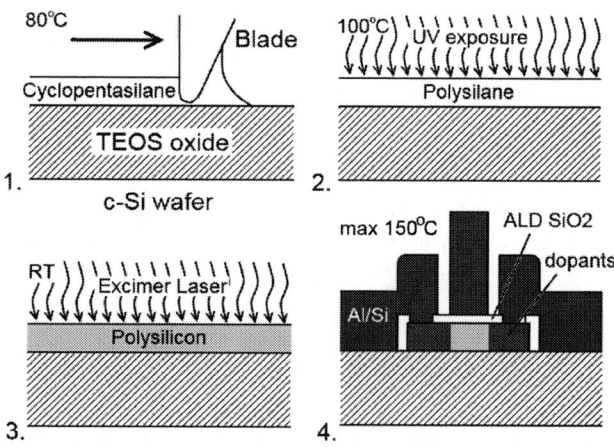

Figure 1: Solution processed poly-Si TFT fabrication process flow

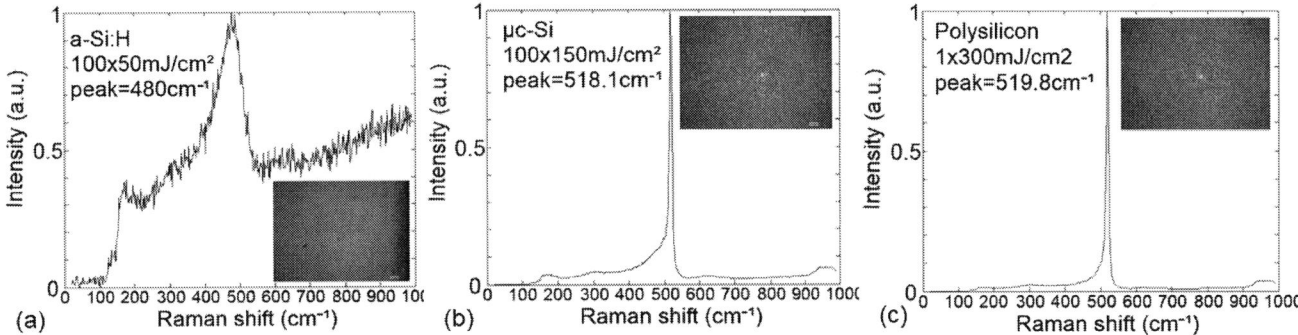

Figure 2: Raman spectra of Si films irradiated with various excimer laser conditions. 100 shots of 50mJ/cm² resulted in a-Si:H (a), 100 shots of 150mJ/cm² resulted in μc-Si(b), and 1 shot of 300mJ/cm² lead to poly-Si(c).

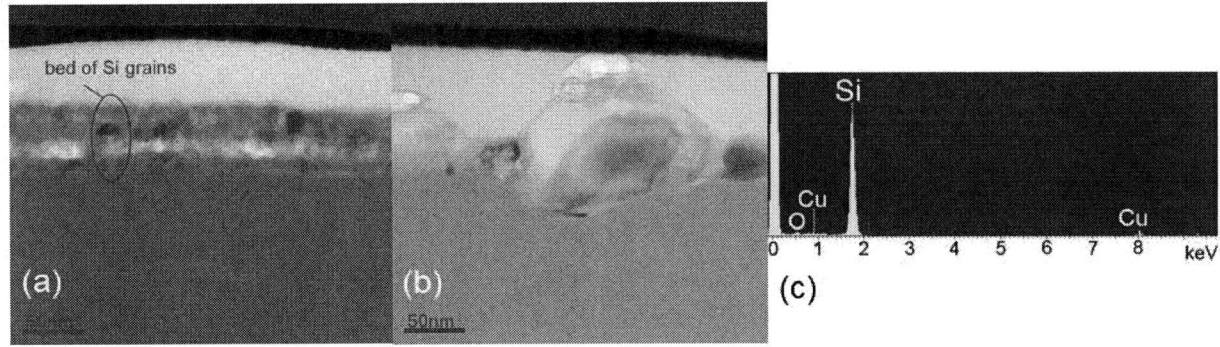

Figure 3: TEM images of Si grain formation when irradiating at polysilane with multiple shots (25 shots of 25mJ/cm²) (a), and a single shot (300mJ/cm²)(b). Energy Dispersive Spectroscopy results of the Si grains observed in the TEM images(c)

The EDS measurements were taken from inside the grain shapes on the TEM images. The result is a clear Si peak and very low levels of other contaminants, showing that the grains have not been oxidized even after having the sample left in open air for over a week. The TEM images show that a higher energy single shot could lead to larger crystal grain sizes, while multiple shots creates a bundle of smaller grains. Although the larger grains would generally lead to higher mobility TFTs, it is evident that the roughness is much lower in the case of the film irradiated with multiple shots. In both cases, the laser irradiation would only treat the deposited film without significantly harming the substrate which is an important difference when comparing to a thermal annealing treatment that was formerly used. This allowed us to even create a poly-Si film on top of a sheet of paper. (Fig. 4)

The physics behind the transformation from polysilane to poly-Si is yet to be understood. One possibility is that this process directly transforms polysilane into polysilicon in a photonic reaction. Another hypothesis is more related to a

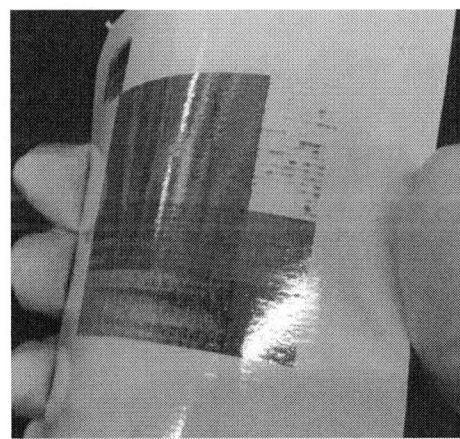

Figure 4: Picture of poly-Si (brown part) on a sheet of paper

thermal process, in which the film reaches 350°C for a transformation into amorphous silicon and subsequently is crystallized by the same laser pulse. The analysis of this phenomenon is being investigated by means of a model for which different physical constants are necessary.

978-1-4799-8002-4/14 $31.00 © 2014 IEEE

TFT Fabrication

A crystalline silicon substrate has been used with 800nm thick PECVD TEOS SiO_2 layer on top. A silicon film is formed on top of this using the previously described process with CPS. A silicon island pattern is plasma-etched on the film, and a gate dielectric SiO_2 of 50nm was deposited by ALD at 100°C using $H_2Si[N(C_2H_5)_2]_2$ as the Si precursor gas. A gate from Al/Si (99%/1%) is sputtered on top and dry-etched. Self-aligned ion implantation of P and B ions is used for the source and drain doping of the NMOS and PMOS devices respectively. The dopants are annealed using again the excimer laser. Finally the contacts are made by sputtering Al and dry-etching, without employing a passivation layer. The maximum TFT process temperature was 150°C, which was needed for baking photoresist during the lithography process. The final device schematic is shown in Fig.1.

Polysilicon Thin-Film Transistor Characteristics

Fig. 5 shows the transfer curves of the single and multiple excimer laser irradiated shot PMOS and NMOS poly-Si TFTs. Both single shot devices were crystallized with only 1 shot with a fluence of 300mJ/cm². The devices exposed to multiple shots did not exceed an energy density of 200mJ/cm². The highest mobilities obtained came from the multi-shot devices with 23.5 and 21.0cm²/Vs for the PMOS and NMOS respectively, which were higher than the highest single shot device mobilities. Although more grain boundaries were formed in the multiple shot devices, this does not seem to be the dominating factor concerning the mobility. The devices had a lower roughness of the film as was shown in the TEM images in Fig 3. The I_{on}/I_{off} ratio of the single shot devices was on the other hand much higher. Statistically the multiple shot devices gave a higher spread in mobilities compared to the single shot devices. A summary of the device parameters from Fig. 5 are listed in Table 1.

The output characteristics of the multiple shot PMOS and NMOS devices are shown in Fig. 7. For both, current crowding near the origin was observed due to high parasitic resistance of the S/D region which limits the mobility of the TFTs

Overall, the I_{on}/I_{off} ratios were reasonable, although the V_{th} is quite high. Processing conditions are yet to be optimized, not only in the liquid silicon film fabrication but also other steps such as doping of the liquid silicon film, and the low temperature ALD gate oxide.

Conclusion

This paper has shown that TFTs using liquid silicon can be fabricated at temperatures compatible to inexpensive plastics and paper by using the excimer laser instead of a thermal annealing treatment of polysilane. Mobilities of 23.5 and 21.0cm²/Vs were obtained for the PMOS and NMOS respectively with the maximum processing temperatures of 150°C.

References:

[1] J. Li et al., *Scientific Reports,* 2:754, DOI: 10.1038/srep00754 (2012)

[2] C.-L. Chen et al., *SID 2013 DIGEST,* pp. 760-762 (2013)

[3] T. Shimoda et al., Nature 440, 783-786, DOI: 10.1038/nature04613 (2006)

[4] J. Zhang et al., *Applied Physics Letters,* 102, 243502, (2013)

[5] T. Shimoda et al., *Jpn. J. Appl. Phys.* 53, 02BA01 (2014)

Table 1: Summary of single and multishot PMOS and NMOS device characteristics

	Single shot PMOS	Single shot NMOS	Multishot PMOS	Multishot NMOS
μ_{FE} (cm^2/Vs)	3.9	5.4	23.5	21.0
V_{th}(V)	-21	30	-11.2	-1.5
S(mV/dec)	341	363	1017	1968
I_{on}/I_{off}	10^4	10^4	10^2	10^3

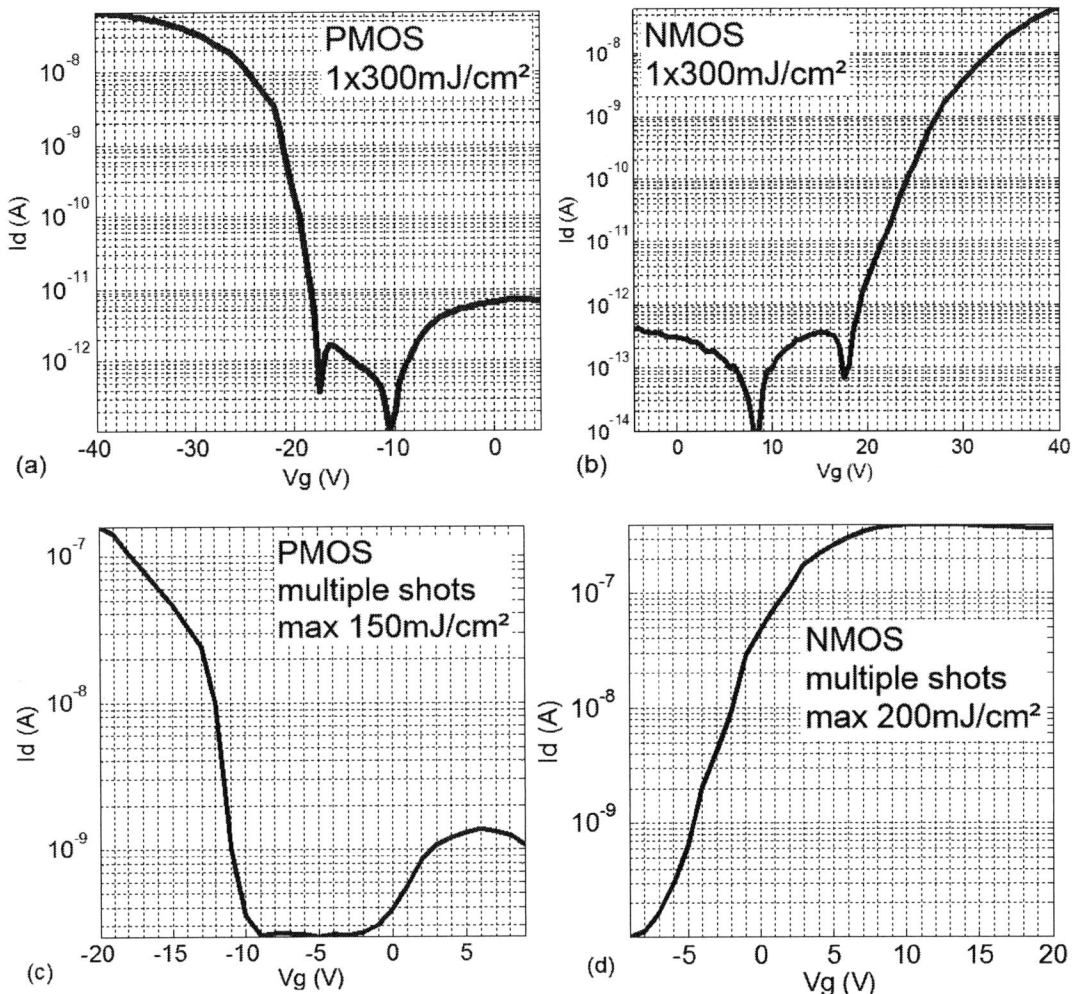

Figure 5: Transfer characteristics of the PMOS(a) and NMOS(b) poly-Si TFTs both crystallized with a single excimer laser shot of 300mJ/cm², and of the PMOS(c) and NMOS (d) of multiple laser irradiated TFTs:1 shot of 25, 50, 75, 100 and 20 shots of 150mJ/cm² for the PMOS, and 20 shots of 25, 50, 75, 100, and 200 for the NMOS. All measurements were conducted with a drain voltage of 20mV. All devices had a W/L ratio of 2. For the single shot devices, gate voltage has been swept in the off direction prior to measurement.

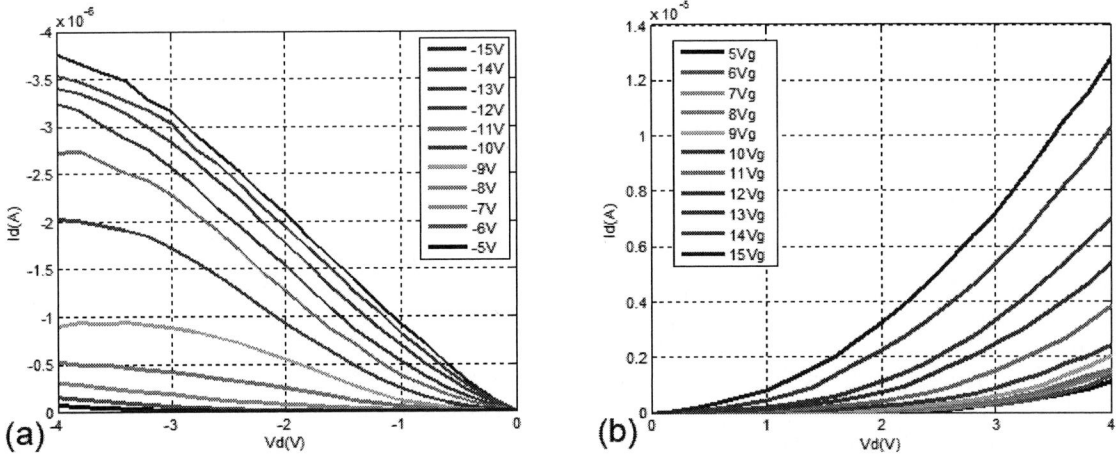

Figure 6: Output characteristics of the PMOS(a) and NMOS(b) devices. Both exposed with multiple laser shots. For the NMOS, a short-channel effect at higher Vd is observed.

978-1-4799-8002-4/14 $31.00 © 2014 IEEE

High Performance Ultra-Thin Body (2.4nm) Poly-Si Junctionless Thin Film Transistors with a Trench Structure

Mu-Shih Yeh, Yung-Chun Wu*, Min-Hsin Wu, Yi-Ruei Jhan, Ming-Hsien Chung, and Min-Feng Hung

Department of Engineering and System Science, National Tsing Hua University, Hsinchu 30013, Taiwan

TEL: +886-3-5715131 ext. 34287, FAX: +866-3-5720724, E-mail: ycwu@ess.nthu.edu.tw

Abstract

The novel trench junctionless poly-Si thin-film transistor (trench JL-TFT) with ultra-thin body (2.4 nm) is utilized to simple dry etching process. This novel devices show excellent performance in terms of steep SS (99 mV/dec.) and high I_{ON}/I_{OFF} (>10^7). The I_{ON} current of the ultra-thin body (UTB) JL-TFT is increased by quantum confined effect.

Introduction

The dry etching process is utilized in the fabrication of trench JL-TFT is used to form a trench and define the channel thickness (T_{CH}) and the gate length (L_G). The JL-TFT is fabricated simply by heavily doping the channel and source/drain (S/D) regions simultaneously. Moreover, the short channel effect (SCE) can be minimized while achieving a high drive current at an extremely-scaled down L_G of the JL-TFT[1]. Additionally, SCE can be minimized while a high drive current is reached with an extremely reduced L_G of the JL-TFT. The device exhibits the following advantages over the conventional inversion mode (IM) MOSTFT: (1) lack of need for an ultra shallow S/D junction, which simplifies the fabrication process; (2) low thermal budget owing to the elimination of the need for implant activation annealing following formation of the gate stack; and (3) concentration of the current of the JL device on the bulk of the semiconductor, which reduces the adverse effects of imperfect interfaces between the semiconductor and the insulator. The trench JL-TFT is required to ensure T_{CH} uniformity to suppress variations in device performance for ultra-thin devices T_{CH} is very thin[2-4]. The trench JL-TFT has great potential for use in advanced AMLCD, AMOLED and 3-D stacked ICs applications[5].

Device Fabrication

The proposed devices were fabricated by firstly growing a 400 nm thermal silicon dioxide layer on 6 inch silicon wafers with a (100) orientation as substrates. A 50 nm undoped amorphous silicon (a-Si) layer was then deposited by low-pressure chemical vapor deposition (LPCVD) at 550 °C. Then, the a-Si layer was solid-phase recrystallized (SPC) at 600 °C for 24 hrs in a nitrogen ambient atmosphere, forming large grains. The SPC layer was implanted by 16-keV with phosphorous ions at a dose of 2×10^{14} cm^{-2}, followed by furnace annealing at 600 °C for 4 hrs. The active NWs of the device were patterned by electron beam (e-beam) direct writing and transferred by RIE. The trench structure with 15 nm poly-Si channels were then defined by e-beam lithography and anisotropic etched by time-controlled RIE. Next, a 10 nm-thick thermal oxide layer, consuming 10 nm thickness poly-Si to form

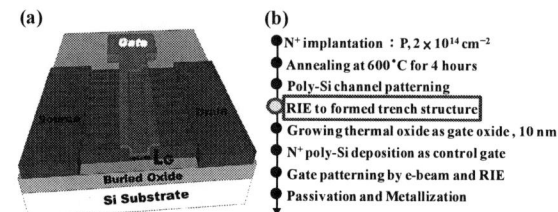

Fig. 1 (a) Trench JL-TFT device with ten NWs. (b) Details of the process flowchart of the fabricated devices are provided.

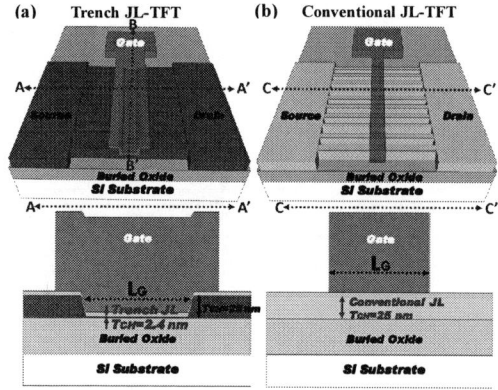

Fig. 2 (a)-(b) Cross-section of the channel structures of trench JL-TFT and conventional JL-TFT. Process flowcharts of the fabrication of devices are provided. A and A' (C and C') indicate cross-section form source to drain, and B and B' indicate region of channel.

5 nm channels, was grown as the gate oxide layer. In thermal oxidation process, growing a thermal 10 nm-thick SiO$_2$ consumes 10 nm-thick poly-Si (1:1). The value of L_G was determined by the trench structure. Additionally, 150 nm in-situ doped n$^+$ poly-silicon deposition as a thickness of to form a gate electrode, and patterned by e-beam lithography and RIE. A 200 nm SiO$_2$ passivation layer was then deposited. Finally, 300 nm Al-Si-Cu metallization was performed and sintered.

Results and Discussion

Fig. 1 (a)-(b) shows the key process flows of fabricating the trench JL-TFT with ten NWs. Fig.2 (a)-(b) are presents the structures of the trench JL-TFT and the conventional JL-TFT. As shown in the image, the thickness of the raised S/D is 25 nm; the thickness of the trench JL-TFT channel is 2.4 nm and the thickness of the conventional JL-TFT channel is 25 nm. The channel thickness of the conventional JL-TFT is thicker than that of the trench JL-TFT, so the series resistance of the conventional JL-TFT is lower than that of the trench JL-TFT.

978-1-4799-8002-4/14 $31.00 © 2014 IEEE

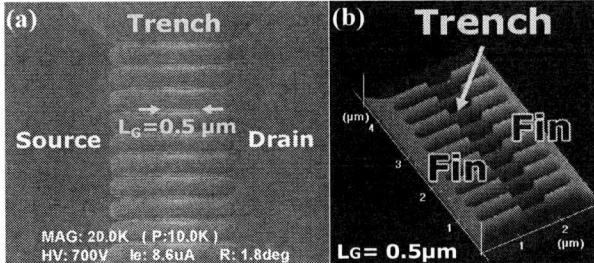

Fig. 3 (a) Top-view SEM image of active region of trench JL-TFT with $L_G = 0.5$ μm. (b) AFM image of channel region of trench JL-TFT with ten NWs.

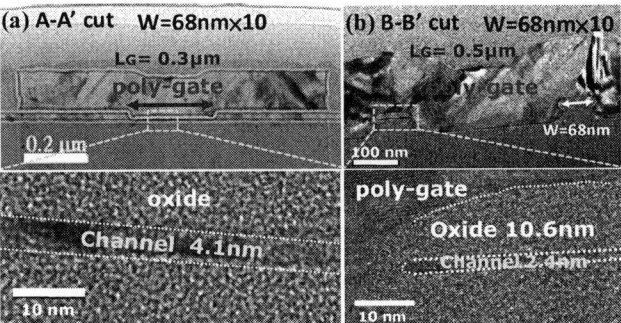

Fig. 4 (a) TEM images of trench JL-TFT with L_G=0.3 μm and T_{CH}=4.1 nm. (b) TEM images of trench JL-TFT with L_G=0.5 μm and T_{CH}=2.4 nm. The effective channel width (W_{eff})=[(68 nm+2.4 nm×2)×10].

Fig. 3(a) displays a top-view scanning electron microscopic (SEM) image of the active region of the trench JL-TFT with $L_G =$ 0.5 μm. Fig. 3(b) shows atomic force microscopic (AFM) images of the channel region of the trench JL-TFT with ten NWs. Fig. 4(a) presents transmission electron microscopic (TEM) images of the L_G is 0.3 μm and the T_{CH} is 4.1 nm. Fig. 4(b) presents TEM images of trench JL-TFT with $L_G = 0.5$ μm and $T_{CH} = 2.4$ nm. The effective channel width (W_{eff}) = [(68 nm+2.4 nm×2) ×10]. Fig. 5 plots the I_D-V_G characteristics of the trench JL-TFT with different L_G from 0.2 μm to 0.5 μm. The SS values of all devices are about 103 mV/dec. and high I_{ON}/I_{OFF} current ratio. Owing to the ultra-thin T_{CH} can provide good gate control ability. Fig. 6 shows the I_D-V_G characteristics of the NWs and the planar trench JL-TFT with $L_G = 0.5$ μm. The NWs device has a steeper SS (99 mV/dec.), a lower DIBL (0 mV/V) and an apparent threshold voltage (V_{TH}) (0.09 V) than those of the planar device. Where V_{TH} refers to the gate voltage at $I_D = 10^{-9}$A. The channel layer must be sufficiently thin to allow full depletion of the carriers in the channel by the gate, so that the trench JL-TFT can be effectively turned off.

Fig. 7 shows the I_D-V_G characteristics of the trench and the conventional JL-TFTs with $L_G = 0.5$ μm. The trench JL-TFT has a steep SS (103 mV/dec.) and an apparent V_{TH} (0.07 V). The g_m curves exhibit a different trend between the trench JL-TFT and the conventional JL-TFT devices. Fig. 8(a)-(d) show the mobility, SS, V_{TH} and mean DIBLs at various L_G from 0.2 μm to 0.5 μm for the trench JL-TFT devices and the conventional JL-TFT devices. The field-effect mobility was calculated from

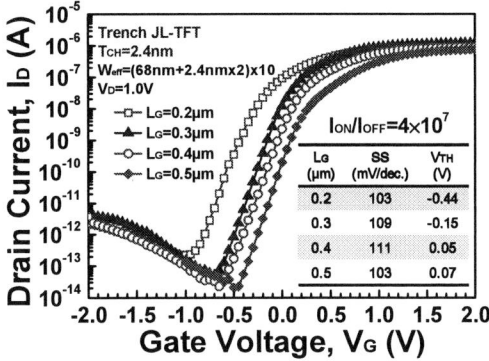

Fig. 5 Comparison of the I_D-V_G curves trench JL-TFT devices. The nanoscale device with L_G=0.2 μm to L_G=0.5 μm.

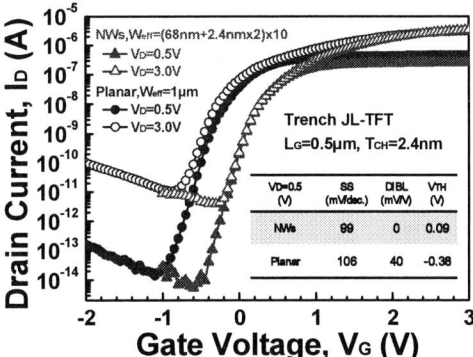

Fig. 6 Comparison of the I_D-V_G curves of NWs trench JL-TFT and planar trench JL-TFT devices with L_G=0.5 μm.

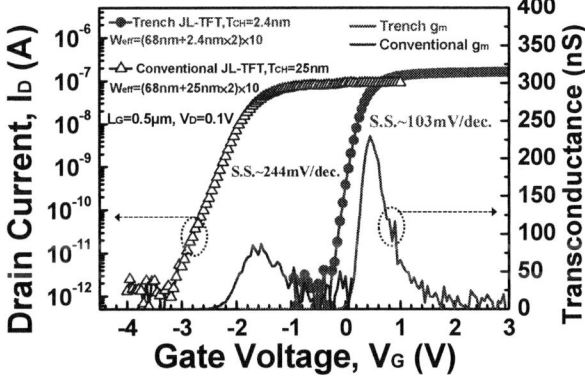

Fig. 7 Comparison of the I_D-V_G and g_m curves of trench JL-TFT and conventional JL-TFT devices with $L_G = 0.5$ μm. Trench JL-TFT device has a better SS than conventional JL-TFT device.

the maximum transconductance at $|V_{DS}|$=0.1 V by $\mu=(L_G \times g_m)/(W_{eff} \times C_{OX} \times V_{DS})^6$. The mean value is obtained from results for five devices with each L_G. The trench JL-TFT devices have lower DIBLs (~0 mV/V) and SS (~100 mV/dec.) than that of the conventional JL-TFT devices.

Fig. 9 compares the I_D-V_D output characteristics of the trench and the conventional JL-TFTs for $L_G = 0.5$ μm, and trench JL-TFT has a higher saturation current than the conventional JL-TFT. Most of electrons in the ultra-thin body

978-1-4799-8002-4/14 $31.00 © 2014 IEEE

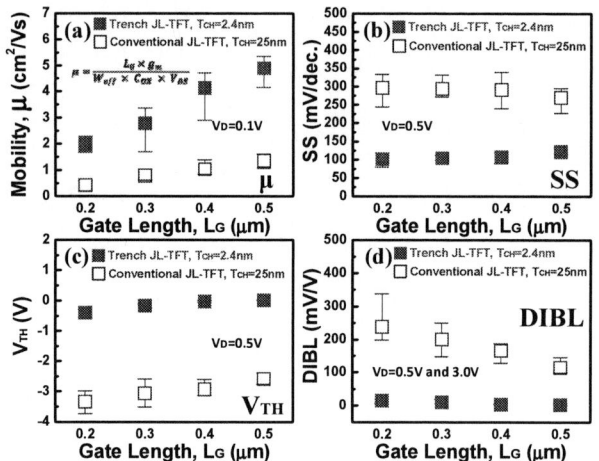

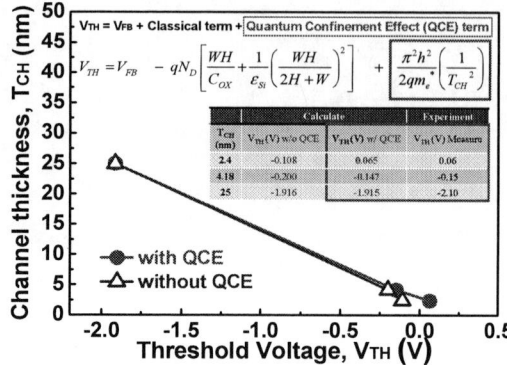

Fig. 8 (a)-(d) Comparison of the mobility, SS, V_{TH} and DIBLs at various L_G from 0.2 μm to 0.5 μm for the trench JL-TFT devices and the conventional JL-TFT devices. The mean value is obtained from results of five devices

Fig. 10 Calculation with quantum confinement effect (QCE) is more reasonable to the electrical measurement.

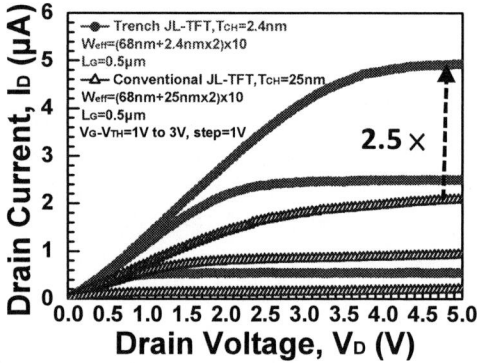

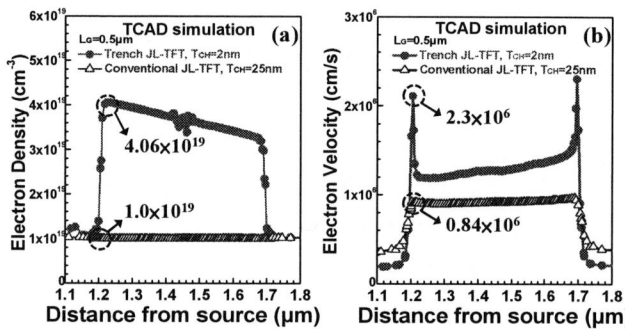

Fig. 11 TCAD simulations reveal (a) electron density and the (b) electron velocity of trench JL-TFT and conventional JL-TFT.

Fig. 9 I_D-V_D curves of trench JL-TFT and conventional JL-TFT with $L_G = 0.5$ μm. The ultra-thin channel JL-TFT has a higher I_{ON} than that conventional JL-TFT.

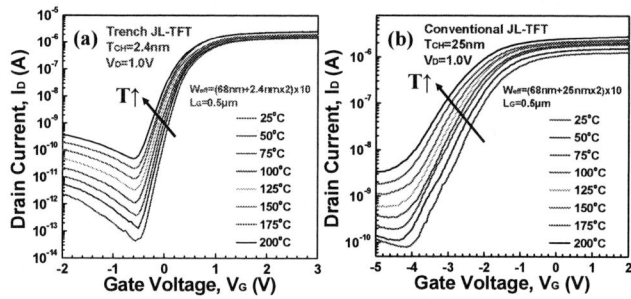

Fig. 12 Temperature dependence (25 °C to 200 °C) on I_D-V_G characteristics of trench JL-TFT and conventional JL-TFT. The I_{ON} and V_{TH} exhibit a positive and negative variation with increasing temperature.

(2.4 nm) remain in the 2-fold valleys which has a higher mobility and therefore a higher velocity[7]. Another reason is the reduction of interface scattering. For the JL-TFT, most of electrons pass through the middle area of the ultrathin-channel, and the current of the trench JL-TFT does not pass through the surface of the ultrathin-channel (body current). Therefore, interface scattering is reduced and the mobility is thereby improved. Based on the above results, the ultra-thin body of 2.4 nm has high electron density and fast electron velocity. Fig. 10 shows the calculated V_{TH} with quantum confinement effect (QCE) is more reasonable to the electrical measurement. By technology computer-aided design (TCAD) simulation[8], fig. 11(a)-(b) show the electron's density (n) and the electron's velocity (v) of different T_{CH} (2.4 nm and 25 nm). I_D is given by, $I_D = q \int \int n(x,y) \times v(x,y) \, dxdy$. Therefore, the proposed trench JL-TFT with the very thin T_{CH} has a higher saturation current than the conventional JL-TFT as presented in Fig. 9.

Fig. 12(a)-(b) show temperature dependence on I_D-V_G curves of trench JL-TFTs and conventional JL-TFTs. Fig. 13(a)-(d) show its impact on I_{ON}, I_{OFF}, SS and V_{TH}. The V_{TH} for

trench JL-TFTs are less sensitive to temperature than conventional JL-TFTs, that maybe due to the single crystal-like and QCE of the ultra thin channel. Table I. shows comparison of parameters of trench JL-TFT with other research.

Fig. 14(a) compares breakdown voltages (V_{BD}) of trench JL-TFT in this work, gate all around (GAA) JL-TFT and IM TFT with similar size in our previous work. The trench JL-TFT shows great V_{BD}, which seems likely that there are no doping concentration gradients and P-N junctions caused less literal electric field in the channel. Fig. 14(b), the cumulative distribution of V_{BD} for trench JL-TFT and IM-TFT is similar, indicating that the film of trench structure is uniform.

978-1-4799-8002-4/14 $31.00 © 2014 IEEE

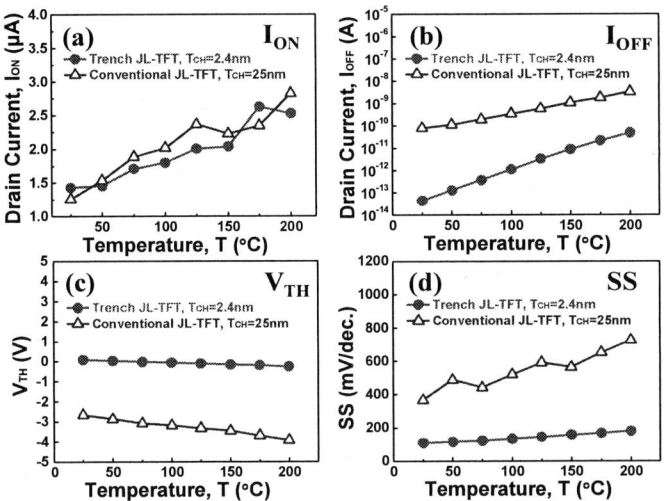

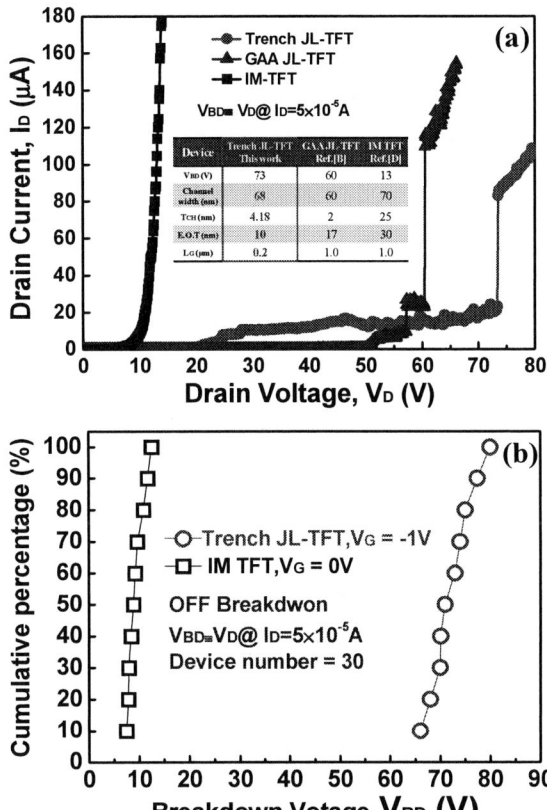

Fig. 13 (a)-(d) Impact of temperature dependence on the ON currents, OFF currents, V_{TH}, and SS for trench JL-TFT [W_{eff}=(68 nm+2.4 nm×2)×10, L_G=0.5 µm] and conventional JL-TFT [W_{eff}=(68 nm+25 nm×2)×10, L_G=0.5 µm]. The V_{TH} and SS for trench JL-TFTs are less sensitive to temperature than conventional JL-TFTs.

Table I

Lists important parameters in comparison with the other works

	This work	Ref.[A]	Ref.[B]	Ref.[C]
Channel structure	N poly-Si JL Tri-gate	N poly-Si JL-Planar	N poly-Si JL-GAA	N poly-Si JL-Planar
W/L (µm/µm)	0.07×10/0.5	2/4	0.07×10/1	10/0.4
EOT (nm)	10	8	17	8
T_{CH}(nm)	2.4	2	2	10
V_{TH}(V)	0.09	1	0.25	-0.15
SS(mV/ dec.)	99	101	61	250
µ(cm²/Vs)	4.56	0.45	n/a	n/a
I_{on}(A)	10^{-6}	10^{-5}	10^{-7}	10^{-5}
I_{on}/I_{off} (V_g:V_d)	>10^7 (2V:1.0V)	>10^8 (3V:0.1V)	>10^7 (3V:0.5V)	>10^7 (3V:0.1V)

Ref. .[A] J. K. Park et al., VLSI., pp. T115,2014. [B] H. B. Chen et al., VLSI., pp. T232,2013. Ref. [C] H. C. Lin et al., TED., pp. 1142, 2013. Ref [D] M. F. Hung et al.,APL, vol.98,pp.162108, 2012.

Conclusions

This work demonstrated the fabrication process and the excellent performances of the trench JL-TFTs with ultra-thin body (2.4 nm). The trench structure was successfully and easily integrated into the JL-TFT device. Additionally, the trench JL-TFT with NWs has excellent electrical characteristics, including low I_{OFF} and SS, negligible DIBL and high I_{ON}/I_{OFF} ratio. In addition to these improvements, our process is also compatible with the existing CMOS technologies. Such an ultrathin-channel trench JL-TFT with NWs is promising for use in 3-D stacked IC applications.

References

(1) J.P. Colinge, C.W. Lee, A. Afzalian, N. D. Akhavan, R. Yan, I. Ferain, et al., "Nanowire transistors without junctions," *Natural Nanotechnology*, vol. 5, pp. 225-229, Mar. 2010.

Fig. 14 (a) OFF breakdown (V_{BD}) comparisons of trench JL-TFT, GAA JL-TFT and IM-TFT. (b) The cumulative distribution of V_{BD} for trench JL-TFTs (V_{BD}=73V) and IM-TFTs (V_{BD}=13V).

(2) H.B. Chen, Y.C. Wu, C.Y. Chang, M.H. Han, N.H. Lu, and Y.C. Cheng, "Performance of GAA poly-Si nanosheet (2nm) channel of junctionless transistors with ideal subthreshold slope," *VLSI Tech. Dig.*, pp. T232-T233, Jun. 2013.

(3) S. Migita, Y. Morita, M. Masahara, and H. Ota, "Electrical performances of junctionless-FETs at the scaling limit (L_{CH} = 3 nm)," *IEDM Proc.*, pp. 8.6.1-8.6.4, Dec. 2012.

(4) K.H. Goh, Y. Guo, X. Gong, G.C. Liang, Y.C. Yeo, "Near ballistic sub-7 nm Junctionless FET featuring 1 nm extremely-thin channel and raised S/D structure," IEDM Proc., pp. 16.5.1-16.5.4, Dec. 2013.

(5) R.J. Lyu, H.C. Lin, M.H. Wu, B.S. Shie, H.T. Hung and T.Y. Huang, "Film profile engineering (FPE): A new concept for manufacturing of short-channel metal oxide TFTs," *IEDM Proc.*, pp. 11.2.1-11.2.4, Dec. 2013.

(6) J.K. Park, S.Y. Kim, K.H. Lee, S.H. Pyi, S.H. Lee, and B.J. Cho, "Surface-controlled Ultrathin (2 nm) Poly-Si Channel Junctionless FET towards 3D NAND Flash Memory Applications," *VLSI Tech. Dig.*, pp. 98-99, Jun. 2014.

(7) S.i. Takagi, J. Koga and A. Toriumi, "Subband structure engineering for performance enhancement of Si MOSFETs," *IEDM Proc.*, pp. 219-222, Dec. 1997.

(8) TCAD Sentaurus Device, Ver.E-2010.12, Synopsys, 2011.

978-1-4799-8002-4/14 $31.00 © 2014 IEEE

Performance Enhancement of a Novel P-type Junctionless Transistor Using a Hybrid Poly-Si Fin Channel

Ya-Chi Cheng[1], Hung-Bin Chen[1,2], Chi-Shen Shao[2], Jun-Ji Su[1], *Yung-Chun Wu[1], #Chun-Yen Chang[2,3] and Ting-Chang Chang[4]

[1]Department of Engineering and System Science, National Tsing Hua University, Taiwan;
[2]Department of Electronics Engineering and Institute of Electronics, National Chiao Tung University, Taiwan;
[3]Research Center for Applied Sciences, Academia Sinica, Taiwan; [4]Department of Physics, National Sun Yat-Sen University, Taiwan
E-mail: *ycwu@ess.nthu.edu.tw; #cyc3562@gmail.com;

Abstract

The hybrid poly-Si fin channel junctionless (JL) field-effect transistors (FET) are fabricated first. This novel devices show stable temperature/reliability characteristics, and excellent electrical performances in terms of a steep SS (64mV/dec), a high I_{on}/I_{off} current ratio (>10^7) and a small DIBL (3mV/V) by reducing the effective channel thickness that is caused by the hybrid P^+ channel and n-type substrate (hybrid P/N) junction. In addition, the novel P/N JL-TFT shows smaller series resistance and less current crowding than convectional JL-TFT with ultra-thin channel. Furthermore, our device can be supported by simulated results using technology computer-aided design (TCAD) simulation. Hence, the proposed hybrid P/N JL-TFTs are highly promising for future further scaling.

Introduction

Recently, the concept of the bulk and silicon-on-insulator (SOI) junctionless (JL) transistor, which contains a heavily, uniformly, and homogeneously doping species in the channel and source/drain (S/D), has been proposed and researched [1]. The advantage of these devices include their (1) avoidance of the use of an ultra shallow S/D junction, which greatly simplifies the process flow; (2) low thermal budgets owing to implant activation annealing after gate stack formation is eliminated, and (3) the current transport is in the bulk of the semiconductor, which reduces the impact of imperfect semiconductor/insulator interfaces. Such a feature has also been demonstrated with poly-Si thin film transistors (TFTs) [2-4] which are suitable for monolithic 3D vertically stacked ICs and to continue the applicability of Moore's law [5]. In addition, JL channel must be small enough to achieve outstanding turn-off characteristics. The small-dimensional channel makes process control difficult and possibly leads to an increase of the series resistance in the S/D accompanying a decrease of drain current in bulk and SOI JL FETs. To overcome these problems, the novel concept of JL hybrid P/N channel, which consists of p+ channel with n-type substrate on the buried oxide, has been implemented. This work successfully demonstrates a novel hybrid fin channel for the first time with excellent device performances. The electrical, temperature and reliability characteristics are characterized. The experimental performances could be supported by device simulation performances using technology computer-aided design (TCAD) simulation [6].

Device Fabrication

Fig. 1 shows the detailed process flows of the fabrication in the hybrid P/N fin channel junctionless (JL) TFTs. The device is fabricated by initially growing a 400nm thermal silicon dioxide layer on 6 inch silicon wafers. A 45 nm undoped amorphous silicon (a-Si) layer was deposited by low-pressure chemical vapor deposition (LPCVD) at 550°C. Then, the a-Si layer is solid-phase recrystallized (SPC) and formed large grain size at 600°C for 24 hours. The SPC layer is implanted with 16-keV phosphorous ions at a dose of 2×10^{14} cm^{-2} serving as n-type substrate layer, followed by furnace annealing at 600°C for 4 hours. Subsequently using the same method for the production of 45 nm SPC layer serves as p-type channel, and the SPC layer is implanted with 30-keV boron difluoride ions at a dose of 2×10^{14} cm^{-2}. Then, the channel layer is trimmed down to 35 nm by dry oxide. The active layers, composed of p+ channel and n-type substrate, are defined by e-beam lithography and then anisotropic etched by time-controlled reactive-ion etching. Next, a 9-nm-thick dry oxide is deposited as the gate oxide layer, consuming around 9-nm-thick poly-Si to form 24-nm-thick channels. Finally, gate formation, passivation and metallization are performed.

Results and Discussions

Fig. 2(a) schematically presents the proposed device structure of the JL-TFT with hybrid P/N fin channel. Fig. 2(b) shows scanning electron microscope (SEM) image of the active region for the hybrid P/N JL-TFTs with ten nanowires (NWs) at L_g=1μm. Fig. 2(c) and 2(d) shows the cross-sectional transmission electron microscopic (TEM) images of a single NW along the AA' direction. The channel dimensions of each NW are around 24 nm high × 28 nm wide and the n-type substrate dimensions are 35 nm high × 30 nm wide. A control sample, conventional NW JL-TFT device with single p-type channel layer of 12 nm channel thickness (T_{ch}), is also formed as the above processes except the processes to form the n-type substrate and different oxide trimming time. To analyze the doping profile within hybrid P/N poly-Si fin channel, the secondary ion mass spectroscopy (SIMS) measurement in Fig. 3 is completed on an un-patterned Si wafer that is performed through the identical processes except the patterning processes. The channel region is doped with average boron concentration of 5×10^{19} cm^{-3} and the underlying n-type region is doped with phosphorus concentration of around 4×10^{19} cm^{-3}. This doping profile indicates the formation of an abrupt p-n junction to achieve junction isolation and approaches the TEM profile in Fig. 2(d). Fig. 4(a) shows plots of I_d–V_g characteristics of hybrid P/N (T_{ch} = 24nm) and conventional JL-TFTs (T_{ch} = 12 nm) with L_g=1μm. The on-current (I_{on}) is defined as the drain current at V_g=-6V for hybrid P/N and conventional JL-TFTs. The off-current (I_{off}) is defined as the lowest drain current. As compared to conventional JL-TFTs, the hybrid P/N JL-TFTs has a low subthreshold slope (SS) value of 64 mV/dec, high I_{on}/I_{off} ratio of 2×10^7, small drain induced barrier lowering (DIBL) of 3mV/V. The linear field-effect mobilities (μ_{eff}) in

978-1-4799-8002-4/14 $31.00 © 2014 IEEE

the inset of Fig. 4(a) are 4 and 0.1 cm^2/V-s for hybrid P/N and conventional JL-TFTs, respectively. The high μ_{eff} in our novel structure is attributed to thicker channel thickness and less surface mobility scattering. In Fig. 4(b) exhibits the DIBL values with different L_g ranging from 1μm to 0.06 μm. The V_{th} is defined as the gate voltage at $I_d = 10^{-9}$ A. The n-type substrate affects channel/substrate junction in the hybrid P/N JL-TFT and produces an additional depletion region; the effective channel thickness is reduced to strengthen the controllability of the gate over that in the conventional JL-TFT. Hence, using a hybrid P/N channel structure can achieve a superior short-channel effect (SCE) control, and meanwhile, its V_{th} can also be easily tuned by controlling the n-type substrate doping concentration to fulfill the multi-V_{th} circuit design. The cumulative distribution of V_{th} and SS in two types of devices are shown in Fig. 5; the median values of V_{th} and SS (hybrid P/N: V_{th} =−0.9V, SS =93mV/dec; conventional: V_{th} =−1.2V, SS =113mV/dec) are extracted from cumulative distributions at 50%. The statistical device-to-device variations of electrical parameters for hybrid P/N JL-TFTs are slightly larger than conventional JL-TFTs. It is suggested that the epitaxial growth technology with in-situ doping would be preferred to further improve variations in the future. Fig. 6(a) compares the I_d–V_d output characteristics of the hybrid P/N and conventional JL-TFTs. The hybrid P/N JL-TFT has around 9 times saturation current than that of the conventional JL-TFT at V_g-V_{th} =−5V. Total resistance (R_{total}) of the hybrid P/N and conventional JL-TFT as a function of gate voltage at V_d =−0.4V are shown in Fig. 6(b). The relationship between the S/D series resistance (R_{SD}) and the R_{total} is described in the inset of Fig. 6(b). The R_{ch} represents the channel resistance, W_{eff} is the effective channel width, μ_{eff} is the effective mobility, and C_{ox} is the oxide capacitance. The R_{SD} value of hybrid P/N JL-TFT is around 48 times reduction as compared to that of the conventional JL-TFT devices, which is consistent to the trend of channel thickness. To examine thoroughly the phenomena that are evident in hybrid P/N JL-TFT devices, the simulated hole density and electrical field (E-field) distributions in the center of the channel region are determined at off-state (V_g=−1V) in Fig. 7(a) and 7(b), respectively. The doping concentrations in the source/drain/channel are set to 5 × 10^{19} cm^{-3}. The substrate doping is opposite type with 4 × 10^{19} cm^{-3}. The simulated hybrid P/N devices are calibrated with experimental data and all structures have the same V_{th}. The SOI JL-FET without P/N structure (F_h=24nm, F_{sub}=0nm) shows high leakage in the middle of the channel. The P/N JL-FET has thicker p-channel (F_h =24nm, F_{sub}=35nm) with opposite type substrate, which demonstrate a lower leakage value and records a better performance than the ultra thin channel (F_h=12nm, F_{sub}=0nm) SOI JL-FET. Hence, thicker film with opposite-type substrate has advantages of better film uniformity, lower leakage and smaller series S/D resistances. Additionally, in Fig. 7(b), the electrical field occuring in P/N interface of hybrid P/N JL-FET (F_h=24nm) performs the reduction in the effective channel thickness by the channel/substrate junction, which helps to reduce the leakage

current, and improve SCE in comparison to SOI JL-FET (F_h = 24nm and F_h=12nm). The simulated results agree with experimental data in Fig. 4(a). Table I. shows comparison of key parameters of JL-TFTs to other research. Our proposed novel structure shows excellent SS and I_{on}/I_{off} current ratio. Fig. 8 shows temperature (T) dependence from 77K to 475K on I_d–V_g curves for both devices and its impact on V_{th}, I_{on}, as shown in Fig. 9. The high T (300~475K) has larger V_{th} sensitivity than low T (77~300K). The V_{th} of hybrid channel is less sensitive to single channel, which could be attributed less phonon scattering to current transport near center of channel [7]. The I_{on} variation of conventional JL-TFT is larger due to the T dependence of R_{SD}. Table II shows various bias conditions at different stress mode. In Fig. 10, P/N fin channel exhibits better immunity against negative bias stress (NBS) stress at V_g=−6V, V_d=V_s=0V. In Fig. 11, the simulated E-field at the surface could be relaxed by adding the channel/substrate junction in hybrid P/N structure, which could alleviate NBS effect. Fig. 12 depicts hot carrier injection (HCI) stress at V_g=−2V, V_d=−6V, V_s=0V. The conventional JL-TFT is resistant to HCI than hybrid fin channel JL-TFT, probably due to larger R_{SD} effect. Fig. 13 shows drain avalanche hot electron injection (DAHE) stress [8] at V_g=0V, V_d=−3V, V_s=float. The conventional JL-TFTs suffer serious degradation on V_{th} and SS. It is suggested that hybrid channel could balance E-field distribution at drain-side when drain is at reverse bias, alleviating DAHE stress.

Conclusion

In this study, the hybrid P/N JL-TFTs with omega-gate are successfully fabricated and characterized. The electrical and physical performances, temperature characteristics and reliability of hybrid P/N fin channel and conventional devices were explored. Since the channel/substrate junction produces an additional depletion region, the effective channel thickness is reduced, the hybrid P/N JL-TFTs provide a favorable SCE control, which results in a better electrical performance (such as SS, I_{on}/I_{off}, DIBL) and reliability (such as NBS, DAHE stress). Hence, the proposed hybrid P/N fin channel JL-TFT is highly promising for use in advanced system-on-chip and 3D stacked ICs applications.

Acknowledgements

The authors would like to thank the National Nano Device Laboratories of Taiwan for the technical support.

Reference:
[1] J. P. Colinge et al., Nature Nanotech., vol. 5, pp. 225, 2010.
[2] H. C. Lin et al., EDL, pp. 53, 2012.
[3] C. J. Su et al., EDL, pp. 521, 2011.
[4] H. B. Chen et al., VLSI, pp. 232, 2013.
[5] S. J. Choi et al., VLSI, pp. 111 , 2010.
[6] User's Manual for Synopsys Sentaurus Device.
[7] C. W. Lee et al., ED, pp. 620, 2010.
[8] S. S. Chung et al., SSDM, p.612, 2002.

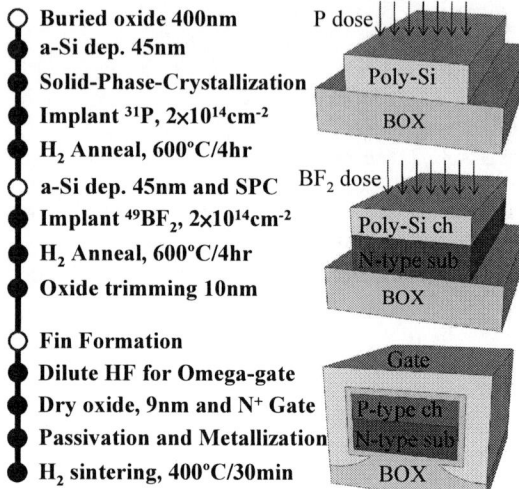

- ○ Buried oxide 400nm
- ● a-Si dep. 45nm
- ● Solid-Phase-Crystallization
- ● Implant ^{31}P, $2\times10^{14}cm^{-2}$
- ● H_2 Anneal, 600°C/4hr
- ○ a-Si dep. 45nm and SPC
- ● Implant $^{49}BF_2$, $2\times10^{14}cm^{-2}$
- ● H_2 Anneal, 600°C/4hr
- ● Oxide trimming 10nm
- ○ Fin Formation
- ● Dilute HF for Omega-gate
- ● Dry oxide, 9nm and N^+ Gate
- ● Passivation and Metallization
- ● H_2 sintering, 400°C/30min

Fig. 1. The detailed process flows of the fabrication n the hybrid P/N fin channel junctionless (JL) TFT

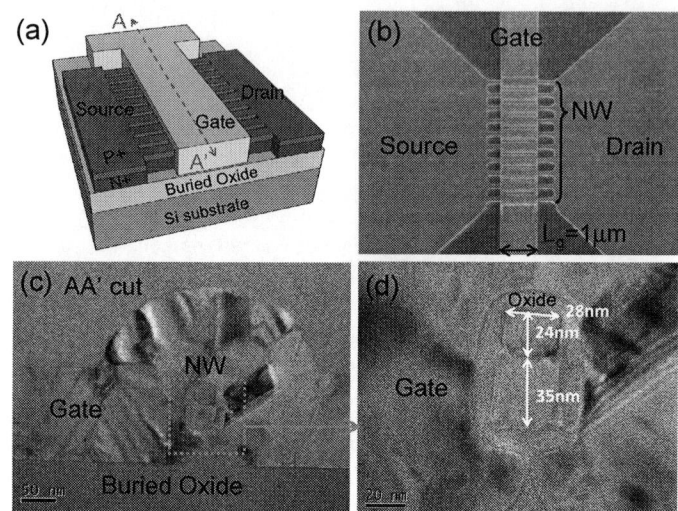

Fig. 2. (a) Schematic diagram of the proposed hybrid P/N JL-TFT devices with ten NWs. (b) Top view SEM image of the active region of the device with $L_g = 1\mu m$. (c) Cross-sectional TEM images of hybrid P/N channel along AA' direction with omega-gate structure. (d) The enlarged TEM images in Fig. 2(c) with $T_{ch} = 24$ nm, fin width = 28 nm.

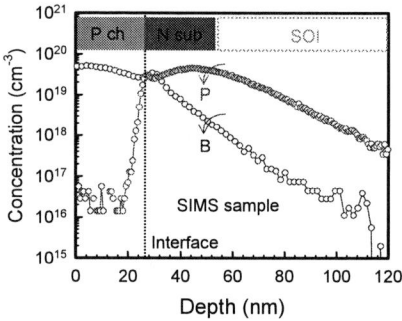

Fig.3. SIMS depth profile of boron and phosphorus dopant measured on an unpatterned wafer.

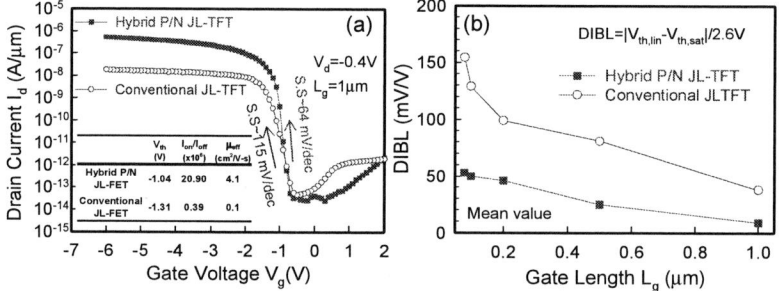

Fig. 4. (a) The transfer I_d-V_g characteristics, and (b) DIBL, in hybrid P/N and conventional JL-TFTs with $L_g = 1\mu m$ at $V_d = -0.4V$. The hybrid fin channel shows a steep SS of 64mV/dec and a small DIBL value from $L_g=1\mu m$ to 60nm.

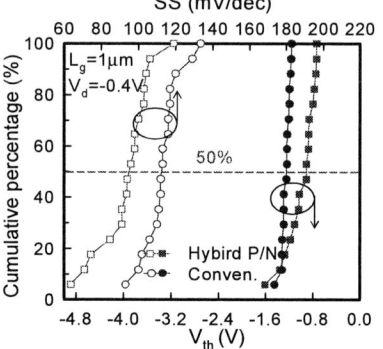

Fig 5. The cumulative distribution of V_{th} and SS for hybrid P/N (V_{th}=-1V, SS=95mV/dec) and conventional JL-TFTs (V_{th}=-1.2V, SS=115mV/dec).

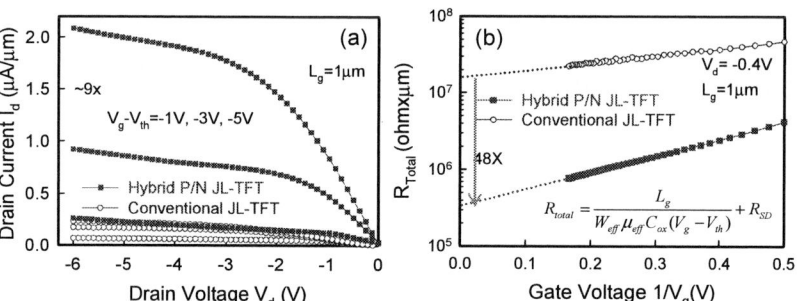

Fig. 6. The (a) I_d-V_d curve and (b) R_{total} as a function of gate voltage in hybrid P/N and conventional JL-TFTs with $L_g = 1\mu m$ at $V_d = -0.4V$.

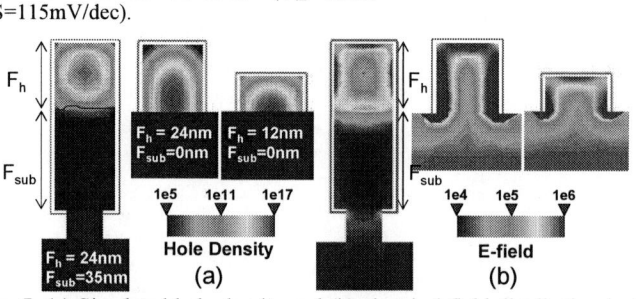

Fig. 7. (a) Simulated hole density and (b) electrical field distribution in the middle of the channel at off-state (at $V_g = -1V$) in hybrid P/N channel and single p-channel JL devices with different channel thickness and structure.

Junctionless TFT	This Work	Ref. [2]	Ref. [3]	Ref. [4]
Cross-section	Hybrid Ch. Rectangular	Flat Rectangular	Rough Rectangular	Nano Sheet
Channel Structure	P-SPC JL Pi-Gate	N-SPC JL Planar	N-SPC JL GAA	N-SPC JL GAA
W_{eff}/L (μm/μm)	0.076x10/1	10/5	0.07x2/1	0.07x10/1
V_{th} (V)	-1	~ -0.2	-0.3	0.25
EOT (nm)	9	8	15	17
T_{ch} (nm)	24	10	12	2
S.S. (mV/dec.)	64	240	199	61
I_{on}/I_{off} (V_g:V_d)	>10^7 (-6V;-0.4V)	>10^7 (3V;0.1V)	>10^6 (5V;0.5V)	>10^7 (3V;0.5V)

Table I. key parameters comparison in junctionless TFTs.

978-1-4799-8002-4/14 $31.00 © 2014 IEEE   624

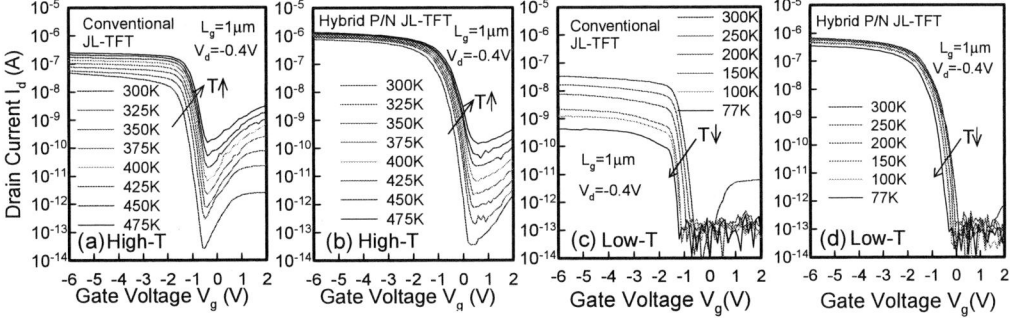

Fig. 8. (a)-(d) Temperature dependence (77K to 475K) on I_d–V_g characteristics of hybrid P/N JL-TFTs and conventional JL-TFTs with L_g=1μm. The on-current (I_{on}) and $|V_{th}|$ exhibit a positive and negative variation with increasing temperature.

Fig. 9(a). The impact of high temperature dependence on the V_{th} for hybrid P/N and conventional JL-TFTs.

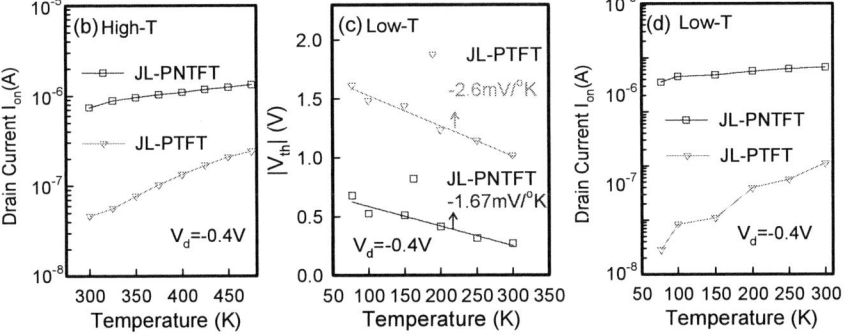

Fig. 9. (b)-(d) The impact of temperature dependence on the V_{th}, I_{on} for hybrid P/N and conventional JL-TFTs. The V_{th} and I_{on} for hybrid P/N fin structure are less sensitive to temperature than conventional JL-TFTs.

JL-TFT	Hybrid P/N			Conventional		
Stress condition	V_g	V_d	V_s	V_g	V_d	V_s
NBS (V)	-6	0	0	-6	0	0
HCI (V)	-2	-6	0	-2	-6	0
DAHE (V)	0	-3	Float	0	-3	Float
Read (V)	X	-0.4	0	X	-0.4	0

Table II. The various stress mode at different bias condition for hybrid PN and conventional JL-TFTs.

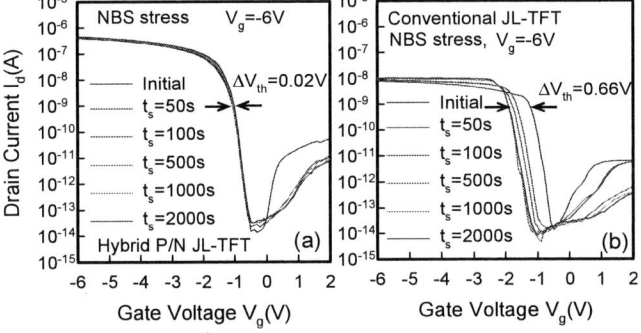

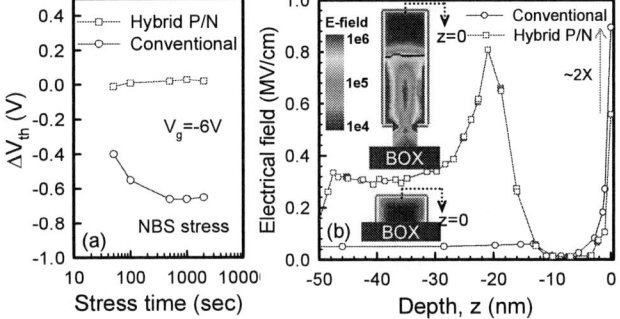

Fig. 10. The negative bias stress (NBS) at V_g=-6V with same electric field for (a) hybrid P/N and (b) conventional JL-TFTs. The hybrid P/N fin channel structure exhibits better immunity against stress than single fin channel structure.

Fig. 11. (a) the ΔV_{th} of negative bias stress. (b) Simulated electrical field distributions on the on-state from the fin top to fin bottom. The E-field of single fin structure is 2 times than one of hybrid P/N fin channel structure.

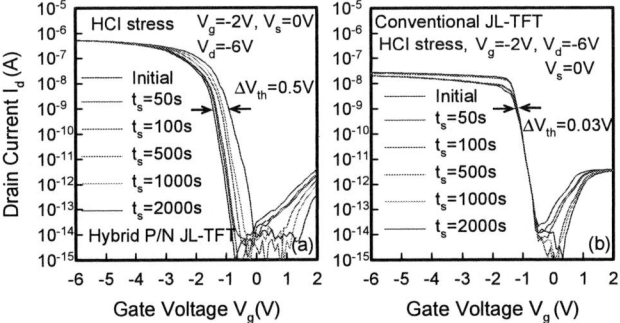

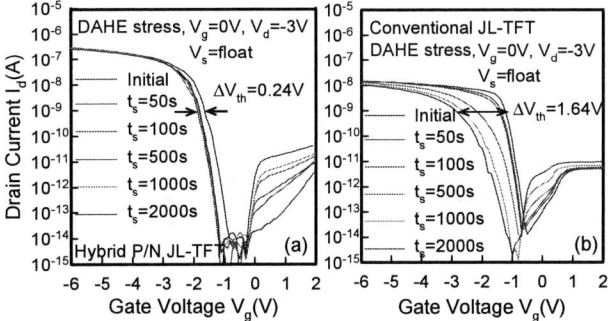

Fig. 12. The hot carrier injection (HCI) at V_g=-2V, V_d=-6V for (a) hybrid P/N and (b) conventional JL-TFTs. The conventional JL-TFTs exhibits better immunity against HCI stress, probably due to the series resistance.

Fig. 13. The drain avalanche hot electron injection (DAHE) at V_g=0V, V_d=-3V for (a) hybrid P/N and (b) conventional JL-TFTs. The conventional JL-TFTs exhibit serious degradation on V_{th} variation and SS.

978-1-4799-8002-4/14 $31.00 © 2014 IEEE

Wafer Level System Integration for SiP

Douglas C.H. Yu

Taiwan Semiconductor Manufacturing Company R&D, 168 Park Ave. 2, Science Based Industrial Park, Hsinchu, Taiwan, R.O.C.

Abstract

A family of novel wafer-level-system-integration technologies (WLSI) was proposed. This paper reviews WLSI feasibility work first. Further results on the reliability, the compatibility of the integration with both more advanced node Logic and DRAM devices, and the higher-level system integration of the WLSI technologies are then presented. Foundry has established a comprehensive system integration technology portfolio in wafer form to fulfill the needs from mobile to cloud computing for the future growth of the Si-based nano-electronics industry.

Introduction

In contrast to conventional organic substrate package-level or board-level system integration, wafer level system integration (WLSI) enables not only tight-pitch system interconnect in both horizontal (< 1μm) and vertical (40μm) directions, as well as chip-partition, but also a seamless integrated design and production flow from chip to system and sub-system [1-3]. Heterogeneous system integration of not only Logics and DRAM [4], but also more-than-Moore functionality, such as sensors, passives, analog, and other memory devices has been realized [4-10]. 3DIC with a CoWoS (chip-on-wafer-on-substrate) flow has evolved and matured becoming a manufacturing technology platform in recent years [4, 8]. Furthermore, a cost effective integrated fan-out (InFO) wafer level packaging has been shown to provide small form factor, low system leakage and efficient power, high memory bandwidth with much simplified process flow [3]. This paper presents a wide spectrum of products and applications that have been integrated by the two flexible technology platforms, CoWoS and InFO.

Advanced Nodes Integration with 3DIC TSV

CoWoS flow with TSV-middle scheme has been adopted to implement 3DIC in Logic device at 28nm nodes and beyond. Fig 1 shows up to 10-tiers stacking with excellent interconnection continuity. For 3D stacking purpose, FinFET wafer had been thinned to 20~50μm. FinFET device characteristics show no degradation after wafer thinning to even 20μm. The result is comparable to planar device in 30μm-thinned wafer. Cumulative plots in Fig. 2 reveal that ΔI_{dsat} and ΔV_{tl} are all less than 2% after thinning for planar and FinFET devices [4]. Chip stacking and bonding is another source of mechanical stress, and Chip-on-Wafer (CoW) bonding shows little or no impact to the transistors either, as seen in Fig 3.

Fig 1 10-tier stacked without I-V performance degradation (insert) has been demonstrated.

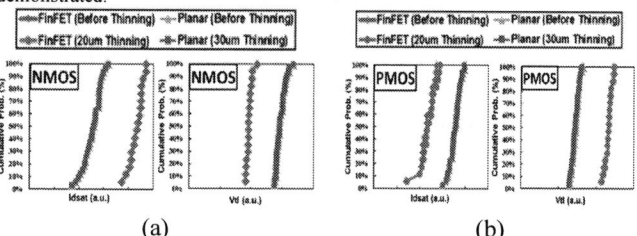

(a)   (b)

Fig 2 Cumulative plots of I_{dsat} and V_{tl} of (a) NMOS and (b) PMOS before and after thinning process for planars and FinFETs.

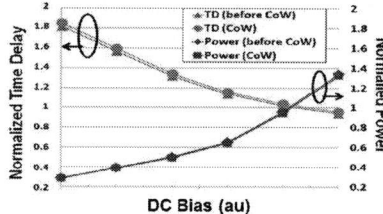

Fig 3 Chip-on-Wafer (CoW) stacking has little or no effect on ring oscillator performance, as shown are the power and time delay comparison before and after CoW.

At circuit level, 128M SRAM cell I_{cell}/I_{sb} and I_{cell}/V_{ts} also shows no yield degradation after TSV process integration, as shown in Fig 4. On multi-tier stacking, Fig 5 shows excellent and identical key transistor performance (I_d-V_g) at bottom and top tier after 5+1 tier stacking.

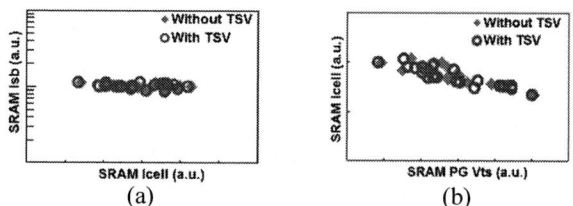

(a)   (b)

Fig 4 No impact on (a) I_{cell}/I_{sb} and (b) I_{cell}/V_{ts} in 128M SRAM after TSV process integration.

978-1-4799-8002-4/14 $31.00 © 2014 IEEE   626

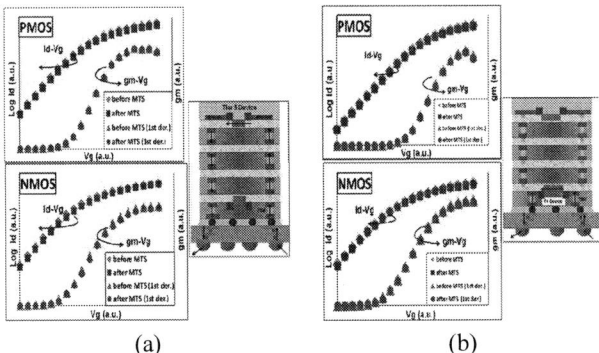

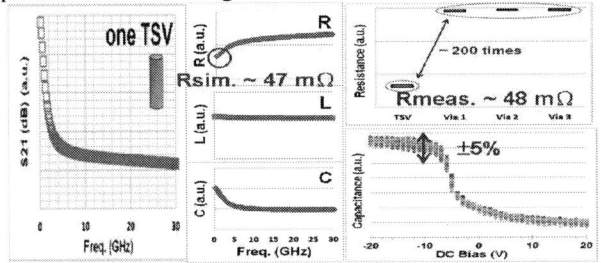

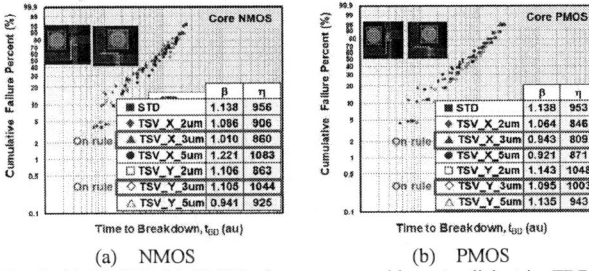

(a)   (b)

Fig 5 Key transistor performance with 5-tier stacks on (a) top and (b) bottom tiers. Both show no impact on N/P MOS I_d-V_g, g_m-V_g, and I_{off} performance.

For design readiness, TSV electrical model is verified with Si process as shown in Fig 6.

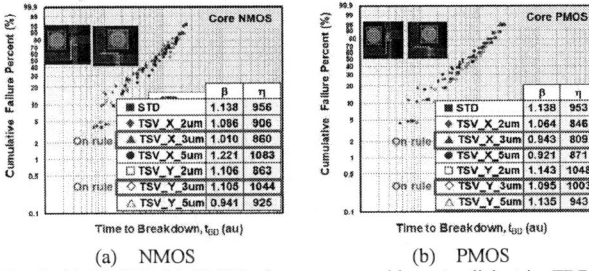

Fig 6 Extracted single TSV's R, L, and C from simulation data and compared to measurement.

Transistor gate oxide TDDB with TSV keep out zone (KOZ) as small as 2 μm, is proven without reliability performance degradation, Fig 7.

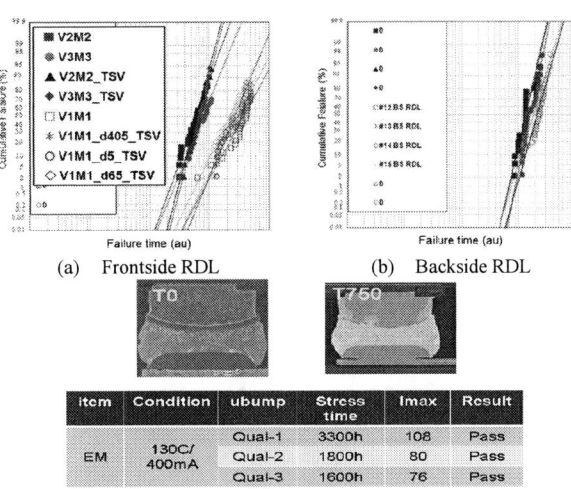

(a)   NMOS   (b)   PMOS

Fig 7 (a) NMOS (b) PMOS shows comparable gate dielectric TDDB performance for without TSV compared to with TSV at 2-5 μm keep out zone. 2μm KOZ shows no reliability degradation.

On the interconnection reliability evaluation, front-side back-end-of-line (BEOL), backside redistribution layer (RDL), and μbump joints all have demonstrated with excellent eletromigration resistance, Fig 8. Fig 9 summarized advanced node 3DIC integration reliability evaluation, which includes frontside front-end-of-line (FEOL) transistor HCI, BTI, V_{BD}, Gox TDDB, and PID, frontside BEOL Cu EM, SM, and ultra-low-k TDDB, and TSV/μbump/RDL EM.

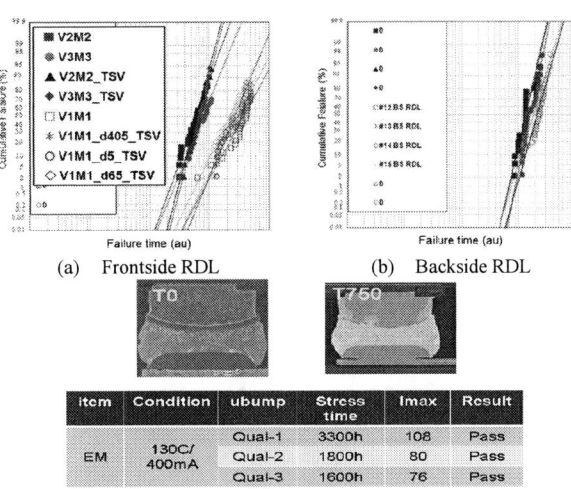

(a)   Frontside RDL   (b)   Backside RDL

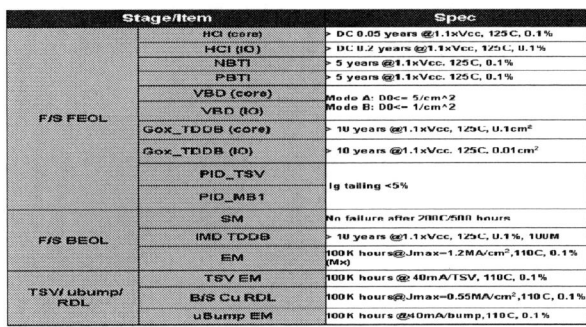

item	Condition	ubump	Stress time	Imax	Result
EM	130C/ 400mA	Qual-1	3300h	108	Pass
		Qual-2	1800h	80	Pass
		Qual-3	1600h	76	Pass

(c)   μbump

Fig 8 Electromigration reliability evaluation on (a) front-side BEOL, (b) baside RDL, and (c) solder μbump joints show excellent EM life performance.

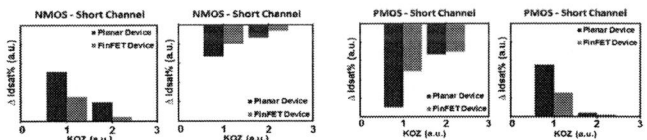

	Stage/Item	Spec
F/S FEOL	HCI (core)	DC 0.05 years @1.1xVcc, 125C, 0.1%
	HCI (IO)	DC 0.2 years @1.1xVcc, 125C, 0.1%
	NBTI	5 years @1.1xVcc, 125C, 0.1%
	PBTI	5 years @1.1xVcc, 125C, 0.1%
	VBD (core)	Mode A: D0<~ 5/cm^2 Mode B: D0<~ 1/cm^2
	VBD (IO)	
	Gox_TDDB (core)	10 years @1.1xVcc, 125C, 0.1cm²
	Gox_TDDB (IO)	10 years @1.1xVcc, 125C, 0.01cm²
	PID_TSV	Ig tailing <5%
	PID_MB1	
F/S BEOL	SM	No failure after 200C/500 hours
	IMD TDDB	10 years @1.1xVcc, 125C, 0.1%, 100UM
	EM	100K hours @Jmax~1.2MA/cm², 110C, 0.1% (M×)
TSV/ ubump/ RDL	TSV EM	100K hours @40mA/TSV, 110C, 0.1%
	B/S Cu RDL	100K hours @Jmax~0.55MA/cm², 110C, 0.1%
	uBump EM	100K hours @40mA/bump, 110C, 0.1%

Fig 9 28nm HPM 3DIC process reliability evaluation including transistor FEOL, BEOL, TSV, μbump, RDL have passed stringent multiple wafer lots reliability evaluation.

Good integration compatibility of 3D-IC and 3D transistor (FinFET) has also been verified. Fig 10 shows that FinFET is less sensitive, as much as 50%, to TSV mechanical stress than planar CMOS devices. This gives higher vertical interconnect density and better design margin as the industry moves into all-3D era of integrating 3D-transistor (eg. FinFET) and 3DIC.

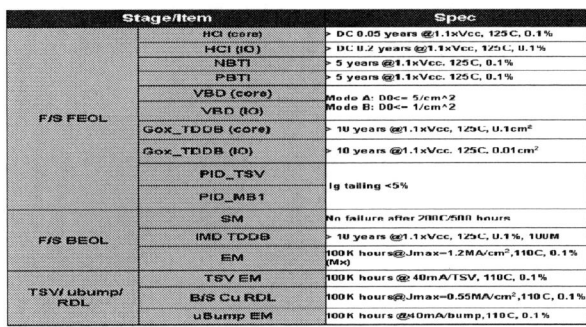

(a) Impact of single TSV positioned in horizontal (left) and perpendicular (right) directions to NMOS short channel device.

(b) Impact of single TSV positioned in horizontal (left) and perpendicular (right) directions to PMOS short channel device.

Fig 10. Benchmarking TSV impact on short channel planar and FinFET MOSFET. FinFET is less sensitive to TSV mechanical stress, as much as 50%, than planar transistors.

Advanced 16nm FinFET integrated with 3DIC TSV is also in reliability evaluation. Fig 11 shows multi-layer BEOL metal-via daisy chain test structures have exceeded stringent electromigration reliability requirement.

978-1-4799-8002-4/14 $31.00 © 2014 IEEE

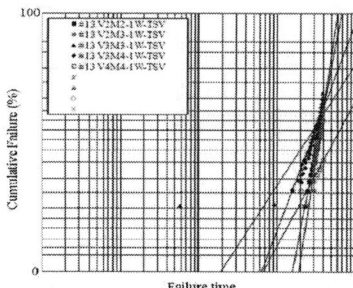

Fig 11 16nm FinFET BEOL integration with TSV. Electromigration evaluation shows excellent life distribution.

CoWoS and Ultra-Large Si Interposer

Contemporary high-end computing and storage markets require the integration of sub-systems with high performance. For both homogeneous FPGA or other system partition and heterogeneous logic, memory, and SoC integration, this demands the stacking of more chips with bigger size on an extra-large size interposer, CoWoS-XL. Figs 12-14 show a "stitching" technique, which has been developed to meet the CoWoS-XL needs. Fig 12 shows metal RDL and via daisy chain resistance across stitching without resistance variation and with good yield across the wafers. Metal line electromigration, stress-migration, are both robust, Fig 13. And inter-metal dielectric TDDB are also proven with good reliability, Fig 14.

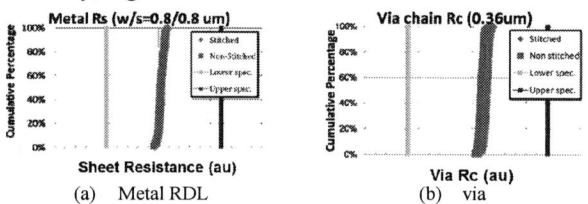

Fig 12 Comparable redistribution metallization (RDL) and via resistance and process reliability stress results for normal and stitched ultra-large test structures.

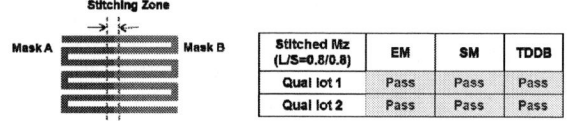

Fig 13 Stitched metallization process reliability evaluation results.

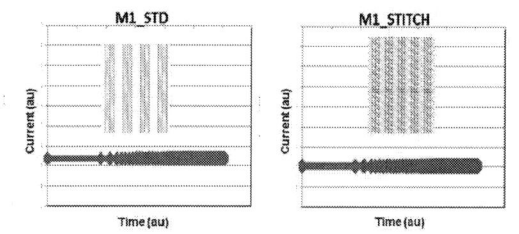

Fig 14 Stitched metallization dielectrics TDDB stressing at 150V, 125°C shows no degradation (T0.1%) after prolonged stress.

Figs 15 show the package and cross section structures. Good component warpage and co-planarity control are achieved,

which is essential for reliability and yield of surface mount on board for such enormous device and package sizes.

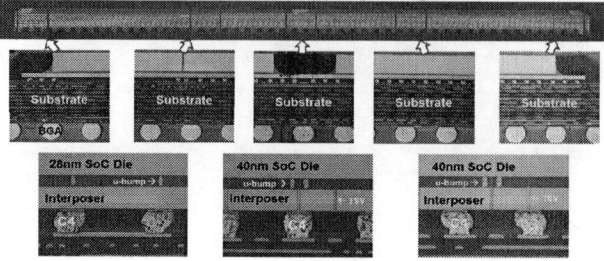

Fig 15 Optical and SEM cross section on CoWoS-XL assembly. Warpage and co-planarity control is a fundamental challenge for such an enormous package assembly process.

High density MiM capacitor has been successfully integrated in CoWoS to increase functionality as shown in Fig 16. A 3D IC HK (K=13.6) interposer MIM capacitor with thin dielectric (EOT=20Å), high Cap. density (17 fF/μm^2), low I_{LK} (< 1 fAmp/μm^2 at 125°C), large MiM area size (> 1056 cm^2) and intrinsic TDDB lifetime of 322 years at 1.8V (for 100 mm^2) has been demonstrated [11].

Fig 16 High density MiM capacitor has been successfully integrated in CoWoS Interposer. SEM cross section from left to right show progressively higher magnification of CoWoS TSV, μbumps, RDL, and MiM structures.

A high performance sub-system that integrates third party high bandwidth memory (HBM) with 28nm SoC logic chip, is shown. Fig 17 shows CoW in wafer form two SoC chips stacked side-by-side with 3 HBM cubes, each with 4-tier DRAM stack. Fig 18 shows CoWoS with final substrate and package assembly. 3D X-ray tomography reveals details of interconnection including μbumps, TSV's, and C4 bumps, as shown in Fig 19.

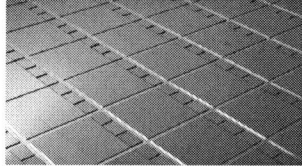

Fig 17 CoW two 28nm SoC and three 4-high HBM stacks.

Fig 18 Final CoWoS package of 2 SoC and 3 HBM stack completed with substrate

978-1-4799-8002-4/14 $31.00 © 2014 IEEE   628

Fig 19 3D X-ray reconstruction of local HBM stack on interposer, showing 4 tier µbump and TSV within HBM stack plus µbump and TSV in interposer. C4 at the bottom of interposer is also observed.

Integrated Fan-Out (InFO) and Package-on-Package (PoP)

Integrated fan-out (InFO) wafer level package (WLP) has been introduced [3] as a cost effective system integration solution. It can integrate chips of various functions with design and manufacture flexibility. Figs 20 and 21 show the largest package envelope of InFO for high-performance applications with high pin-counts, small form factor and simple structure. Different versions of high performance, cost competitive InFO schemes for single and multiple die integration with or without passives are being developed to meet a large varieties of mobile market needs.

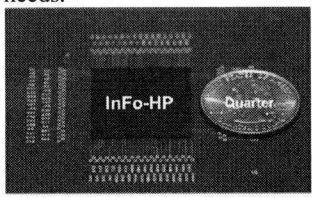

Fig 20 High performance InFO, InFO-HP, developed for high bandwidth cost competitive multi-die integration with fan-out ratio at 1.2-2.5.

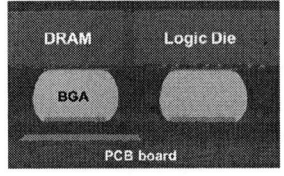
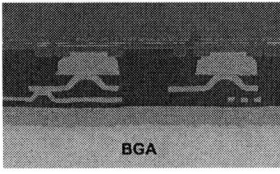

(a) Die edge   (b) RDL interconnection

Fig 21 Cross section SEM on InFO-HP (a) die edge and (b) close up at RDL interconnections.

Fig 22(a) shows structure of built-in 2D and 3D inductors. 2D inductor is smaller in footprint, while 3D is larger but has better electromagnetic isolation. Fig 22(b) is simulated Q-factors using Fig 22(a) structure. 3D inductor can have better Q-factor than 2D when using lower resistance vias and wider Cu lines. Fig 22(c) shows various 2D and 3D inductors that have been demonstrated with high values of Q-factor, and a wide range of inductance, small footprints, and good isolation.

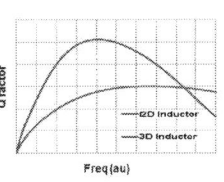
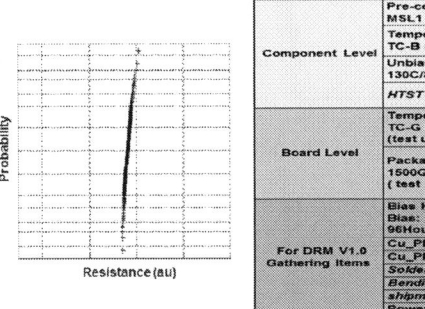

InFO inductor	2D	3D
Q_{fac}	30~64	34~50
Inductance	5~1.5 nH	5~2 nH
Area	0.2~0.09 mm²	0.28~0.14 mm²
Isolation	30~55 dB	50~65 dB
Heat dissipation	-	Thermal via

(a)   (b)   (c)

Fig 22 InFO is integrated with off-chip on-package 2D and 3D inductors. (a) On rule schematic 2D and 3D InFO inductor design. (b) Typical Q-factor of 2D (blue) vs 3D (red) inductor. (c) Best known in house InFO inductor characteristics benchmarking, 2D vs 3D. Both 2D, 3D and mixture of both are available to meet different design requirement.

We also grow from 2D (InFO) to 3D (InFO_PoP), to integrate DRAM Package on InFO Package (PoP). Fig 23 shows good yield and reliability of the 3D PoP package.

	Pre-condition test MSL1 at 260C x 3 reflow
Component Level	Temperature cycling TC-B (-55~125C) 500/1000 cycle
	Unbiased HAST 130C/85%, 33.3 psi, 96/168 Hour
	HTST 1000 Hour (150C)
Board Level	Temperature cycling TC-G (-40~125C) (test up to 1000 cycle)
	Package drop 1500G/0.5ms (test up to 150 drop)
For DRM V1.0 Gathering items	Bias HAST test Bias: TBD, 130C/85%RH/33.3psi 96Hour
	Cu_PPI EM
	Cu_PPI/Via EM
	Solder Ball EM
	Bending test
	shipment test -drop/vibration test
	Power Cycling

(a)   (b)

Fig 23 InFO-PoP package with (a) excellent joint yield, 100%. (b) Process and package reliability evaluation has passed both component level and board level evaluation.

Summary

Wafer level system integration (WLSI) for system in package with a variety of technology options is presented in this paper. The new WLSI SiP is innovated to meet a wide range of future market challenges from mobile to high performance, and from smart mobile device to Internet of Things (IoT). The WLSI technologies, as highlighted in this study, not only leverage existing wafer technology, capacity, design IP and infrastructures, but also leverage among WLSIs each other for manufacturability and cost-competitiveness.

References

[1] Douglas C.H. Yu, 2011 IMAPS Device Packaging Plenary Speech.
[2] B. Banijamali et al., IEEE ECTC, 35-40, 2013.
[3] Doug C.H. Yu, invited paper, to be published in Sep 2014. IEEE CICC, 2014.
[4] W.S. Liao, et. al., IEEE VLSI 2013, C18-C19.
[5] M. Koyanagi, IEEE IEDM, C1.2.1-1.2.8, 2013.
[6] A.V. Samoilov, et. al., IEEE Electron Devices Meeting (IEDM), 25.2.1-25.2.4, 2013.
[7] D.H. Kim, et. al., IEEE, ICCAD 2009. IEEE/ACM International Conference, 674 – 680, 2009.
[8] D. Ibbotson, et. al., IEEE VLSI-T, T38-T39, 2013.
[9] Douglas C.H. Yu, Plenary speech, 2013 RTI ASiP Symposium.
[10] S. M. Chen, et. al., IEEE VLSI-T, 2013, T46-T47.
[11] W.S. Liao, et al., IEEE IEDM 2014, this conference.

High-precision wafer-level Cu-Cu bonding for 3DICs

Masashi Okada[1], Isao Sugaya[1], Hajime Mitsuishi[1], Hidehiro Maeda[1], Toshimasa Shimoda[1],
Shigeto Izumi[1], Hosei Nakahira[1], and Kazuya Okamoto[1,2]

[1]Nikon Corporation, Nagaodai-cho 471, Sakae-ku, Yokohama, Kanagawa 244-8533, Japan
[2]Osaka University, Yamadaoka 2-1, Suita, Osaka 565-0871, Japan
Phone: +81-45-853-8510, FAX: +81-45-853-8518, E-Mail: Masashi.Okada1@nikon.com

Abstract

A high-precision Cu-Cu bonding system for three-dimensional integrated circuits (3DICs) fabrication adopting a new precision alignment methodology is proposed. A new pressure profile control system is applied in the thermocompression bonding process. Experimental results show that the alignment capability is 250 nm or better, with similar overlay accuracy (|average| + 3σ) for permanent bonding. These developments are expected to contribute to the fabrication of future 3DICs.

Introduction

The increasingly high speed and performance of CMOS devices, the basic component in current semiconductor devices, have been realized through hyper-miniaturization technologies. Beyond the 1x nm node range, however, the signal delay in global wiring used to couple individual IPs within a chip will become a crucial issue. The conventional countermeasures taken so far against signal delay will no longer be sufficiently effective and it will be necessary to use additional compensating circuits (repeaters). The repeaters will inevitably increase the chip size and also the power consumption. At the same time, to keep pace with the increasing popularity of mobile devices, even smaller ICs with more diverse functions will be required. The use of three-dimensional integrated circuits (3DICs) employing TSV technology, formed by vertically stacking chips with two-dimensional circuit patterns, is becoming increasingly common as a solution to meet these requirements [1-2]. 3D chips can be formed through stacking by adopting any of the following three modes: die-to-die (D2D), die-to-wafer (D2W), and wafer-to-wafer (W2W). Because of the reduced cost of chip fabrication and future 3DIC formation, the use of the W2W mode is expected to be inevitable [3]. According to the 2013 ITRS, a bonding overlay accuracy of 500-1000 nm will be required for the years 2015-2018 [4]. The precision wafer bonding of Cu-Cu interconnects is expected to be the key to 3DICs fabrication, although many challenges must be overcome to achieve this such as the compensation of distorted wafers, low-force wafer contact, and precision alignment capability. Recently an overlay alignment accuracy of 570 nm (3σ) was demonstrated for 300 mm W2W Cu-Cu thermocompression bonding [5]. However, this result only meets the ITRS requirement until 2018. On the basis of our long experience of CMOS lithography tools, we propose a new alignment and bonding methodology with a precision that cannot be achieved by using conventional assembly tools.

Bonding System

A. Compensation of wafer distortion

A wafer has a low thickness relative to its surface area, thus does not have high mechanical strength and is prone to brittle failure. In addition, a CMOS wafer itself has local and global distortion including bow and warpage. Therefore, to handle a wafer without damaging it and to achieve higher alignment accuracy, the wafer should be fixed to flat wafer holders (WHs) using an electrostatic chuck (ESC). Figs. 1(a) and (b) show upper and lower WHs, respectively. The bonding process can be easily achieved via the WHs by holding the wafers between them.

It is essential to prevent damage caused by the temperature difference between the inside and outside of the WHs due to the rapid increase in temperature in the thermocompression bonding process. To prevent damaging the WHs, hinge structures are formed on the metal frame as shown in Figs. 1 as above (a) and (b). Fig. 2 shows the result of stress analysis in the case of an increase in temperature from 23°C to 450°C in 15 min. The maximum principal stress was a tensile stress of 37 MPa, while the allowable tensile stress of AlN is 120 MPa. Thus, there is no danger of damaging the WHs during the bonding process.

B. Precision alignment procedure

To obtain higher alignment accuracy, we adopted enhanced wafer global alignment (EGA). The basic EGA procedure for bonding wafers is based on the standard coordinate system, determined by fiducial marks. Here we consider the bonding of two wafers, wafer-1 and wafer-2. As described above, actual diffused wafers are distorted; thus, to precisely align the two wafers, the optimization requires EGA with multiple alignment marks. For instance, when the alignment mark positions of wafer-2 in the fiducial coordinate system are at (A_{xi}, A_{yi}), where i is the mark number, after translation and rotation using the determined values, the positions in the converted coordinate system (M_{xi}, M_{yi}) are

$$\begin{pmatrix} M_{xi} \\ M_{yi} \end{pmatrix} = \begin{pmatrix} \cos\theta & \sin\theta \\ -\sin\theta & \cos\theta \end{pmatrix} \begin{pmatrix} A_{xi} \\ A_{yi} \end{pmatrix} + \begin{pmatrix} T_x \\ T_y \end{pmatrix}. \tag{1}$$

978-1-4799-8002-4/14 $31.00 © 2014 IEEE

Next, denoting the coordinates of the alignment marks of wafer-1 in the fiducial coordinate system as (D_{xi}, D_{yi}), we can determine the optimized shift (T_x, T_y) and rotation (θ) by minimizing the function $F(\theta, T_x, T_y)$ given by

$$F(\theta, T_x, T_y) = \sum \{(D_{xi} - M_{xi})^2 + (D_{yi} - M_{yi})^2\}. \qquad (2)$$

Fig. 3 shows typical position- and force-control time axis data obtained using Cu bump wafers with a Cu size of micron order. After the EGA procedure, the wafers are in soft contact with a force of less than 20 N per load cell. During the application of load and the clamping of the wafers, no position shift exceeding 100 nm was detected by interferometers.

Fig. 4 shows five results in the case of continuous pre-bonding over 300 mm wafers. Deviations in the X and Y directions of less than 250 nm were confirmed.

C. Thermocompression bonding procedure

Pre-bonded wafers, which are aligned in an alignment unit, are transferred to a bonding unit for thermocompression bonding. In the bonding, the pressure distribution is an important factor and is highly dependent on the diffused wafer profile and the device production process conditions. To cope with this issue, we have developed a pressure profile control module (PPCM). A schematic of the lower part of the core engine of the bonding unit is shown in Fig. 5. In an actual bonding unit, the upper and lower structures are symmetrically placed, and the wafers and WHs are wedged between the upper and lower heat plates. The principle of the pressure profile control is as follows. When P = F/S, the PPCM plate and the pressure profile at the wafer are flat. When P > F/S or P < F/S, the PPCM plate and the pressure profile at the wafer become concave and convex, respectively. Fig. 6 shows the pressure profile at the wafer calculated using the finite element method (FEM), which indicates a variable pressure profile. Experimental results obtained by applying a loading pressure of 1 MPa are shown in Fig. 7, and the controllability of the pressure profile at the wafer is confirmed.

D. IR metrology tool

We also developed a tool for measuring the overlay alignment accuracy that employs an IR microscope. A schematic of the IR metrology tool is shown in Fig. 8. The temperature of the tool is controlled at 23 ± 0.1°C. The overlay accuracy is determined by measuring the deviation between the line and space marks patterned on the two wafers. Fig. 9 shows a representative IR reflection image of overlay accuracy measurement marks on a pair of 300 mm wafers. A measurement repeatability of better than 20 nm was achieved at five measurement points as shown in Fig. 10.

Experimental Results

An overlay deviation map of wafers bonded by thermocompression is shown in Fig. 11(a). The wafers were pressed at a pressure of 1 MPa for 60 min at 300°C using the PPCM. The overlay accuracy of the bonded wafers, given by |average| + 3σ at 140 measurement points, was 160 nm in the X direction and 260 nm in the Y direction. This result shows that the bonding error after alignment by thermocompression bonding in our bonding system is very small and meets the ITRS requirements for 2015-2018 and beyond [4]. Figs. 11(b), (c), and (d) show the linear error components of shift, rotation, and magnification, respectively. Fig. 11(e) shows the nonlinear component, which is derived by subtracting the linear error components from the result of overlay measurement. The value of 3σ for the nonlinear deviation was 99 nm in the X direction and 138 nm in the Y direction. The main linear component of the overlay error is the shift, which can be easily compensated for in the alignment unit. The analysis of error components is useful for identifying the causes of overlay error and to eliminate them.

Fig. 12 shows cross-sectional transmission electron microscope (TEM) images of the bonded Cu interface of a 1 μm via. Although the bonding interface still existed, no interfacial voids were observed.

Conclusions

A high-precision Cu-Cu bonding system using a new alignment methodology and a pressure profile control module was proposed. We experimentally obtained an alignment capability of 250 nm or better and similar overlay accuracy for permanent bonding. These developments are expected to contribute to the fabrication of future 3DICs such as DRAM and MPUs.

Acknowledgments

We would like to acknowledge the engineers in Nikon Precision Equipment Company and Core Technology Center.

References

[1] M. Koyanagi, H. Karino, Kang Wook Lee, K. Sakuma, N. Miyakawa, M.Itani, "Future system-on-silicon LSI chips," *IEEE Micro*, vol. 18, no. 4, pp. 17-22, July/August 1998.
[2] Y. Xie, G. Loh, B. Black, K. Bernstein, "Design space exploration for 3D architectures," *J. Emerging Technologies in Computing Systems*, vol. 2, no. 2, pp. 65-103, April 2006.
[3] K. Okamoto, "Importance of wafer bonding for the future hyper-miniaturized CMOS devices," *ECS Trans.*, vol. 16, no. 8, pp. 15–29, 2008.
[4] *International Technology Roadmap for Semiconductors 2013, Interconnect Section.*
Available: http://www.itrs.net/Links/2013ITRS/2013Chapters/2013Interconnect.pdf
[5] W.H. Teh, C. Debb, J. Burggraf, D. Arazi, R. Young, C. Senowitz, and A. Buxbaum, "Post-bond sub-500 nm alignment in 300 mm integrated face-to-face wafer-to-wafer Cu-Cu thermocompression, Si-Si fusion and oxide-oxide fusion bonding," *3DIC 2010*, pp. 1-6.

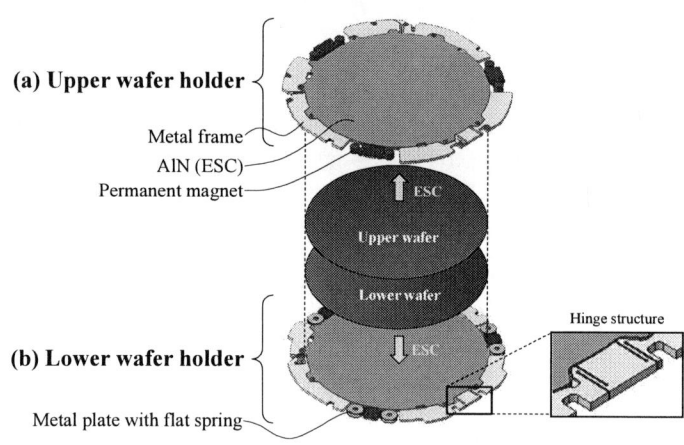

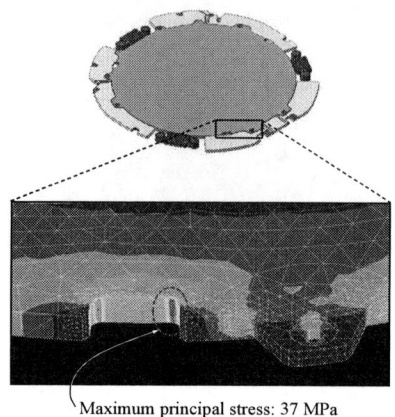

Fig. 1: Schematic of wafer holders. The thermal expansion of the ESC is absorbed by the elastic expansion of the hinge formed on the metal frame.

Fig. 2: Result of thermal stress analysis in the case of heating from 23°C to 450°C in 15 min.

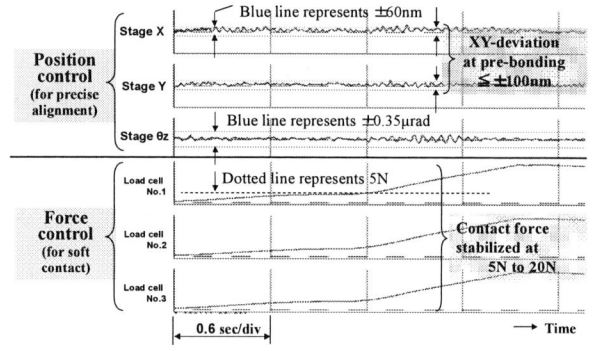

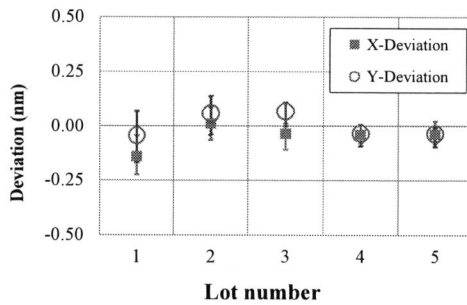

Fig. 3: Results of wafer bonding process using in situ position/load monitoring.

Fig. 4: XY deviation in the pre-bonding of 300 mm wafers in a measurement lot. Symbols are average values for nine measurement points and error bars represent the maximum and minimum deviation in each measurement lot.

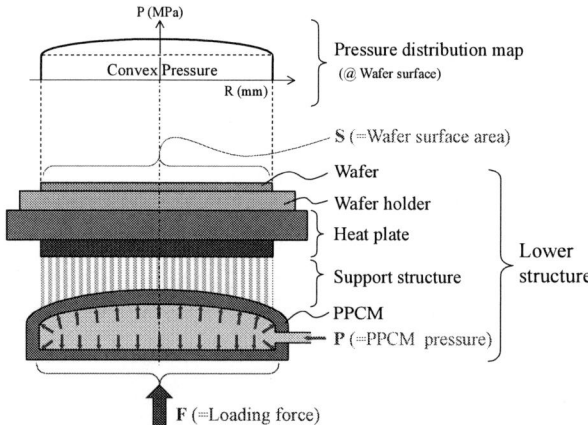

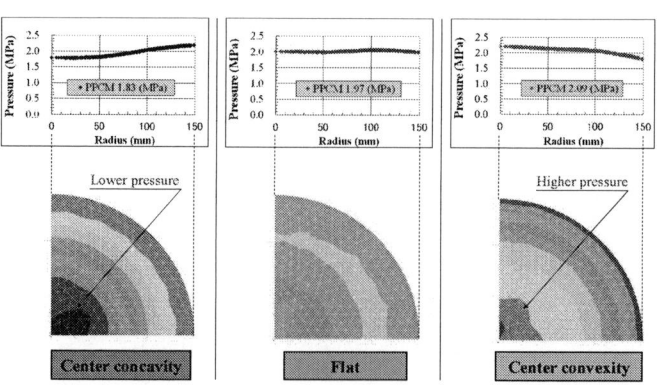

Fig. 5: Schematic of lower part of core engine of bonding unit. F: loading force for wafer, P: inner pressure of PPCM, S: area of wafer surface.

Fig. 6: FEM analysis result of pressure profile control using PPCM under a loading pressure (2.25 MPa).

978-1-4799-8002-4/14 $31.00 © 2014 IEEE

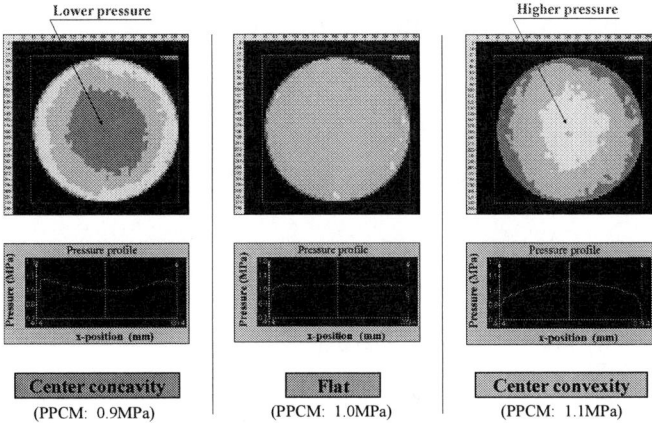

Fig. 7: Experimental results obtained using PPCM with a tactile pressure sensor film. The pressure profile is tunable under a constant loading pressure (1.0 MPa).

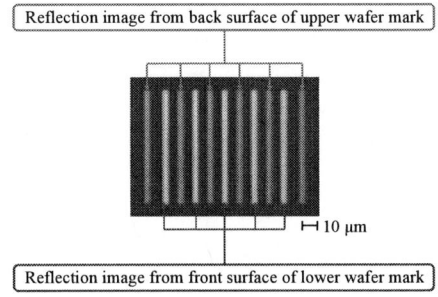

Fig. 9: Representative IR image of overlay measurement marks on a Cu bonded wafer pair.

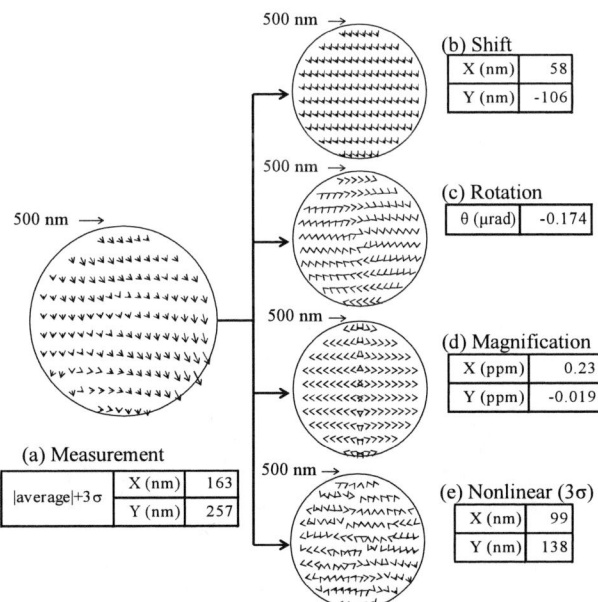

Fig. 11: Overlay deviation map of 140 measurement points on bonded 300 mm wafers measured by IR metrology tool.

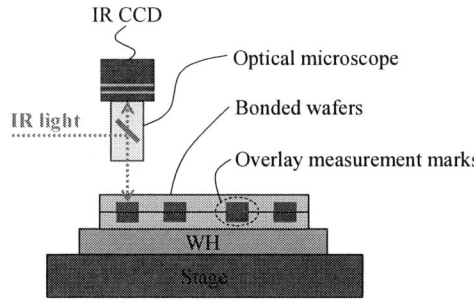

Fig. 8: Schematic of the IR metrology tool. The reflection images of the marks in the bonded wafers are measured.

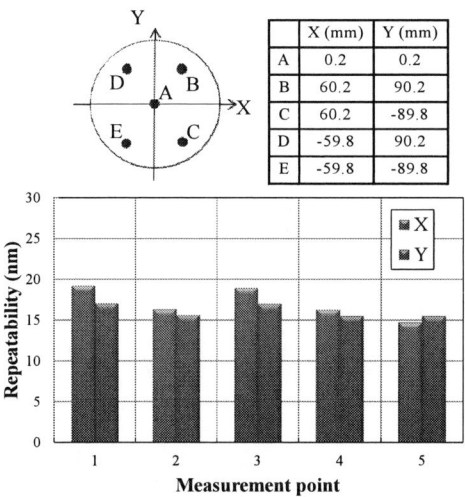

Fig. 10: Measurement repeatability results of IR metrology system. The repeatability is the value of 3σ for 11 measurements at each measurement point.

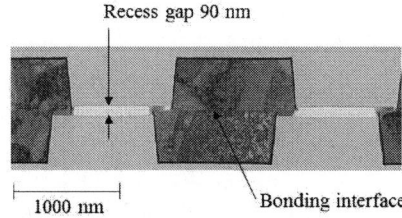

Fig. 12: TEM images of bonded interface of 1 μm via.

978-1-4799-8002-4/14 $31.00 © 2014 IEEE

A manufacturable interposer MIM decoupling capacitor with robust thin high-K dielectric for heterogeneous 3D IC CoWoS wafer level system integration

W.S. Liao, C.H. Chang, S.W. Huang, T.H. Liu, H.P. Hu, H.L. Lin, C.Y. Tsai, C.S. Tsai, H.C. Chu, C.Y. Pai, W.C. Chiang, S.Y. Hou, S.P. Jeng and Doug Yu

Research and Development, Taiwan Semiconductor Manufacturing Company, Ltd.,

No. 168, Park Ave. II, Hsinchu Science Park, Hsinchu, Taiwan 30075, R.O.C.,

Tel: 886-3-5636688 Ext. 722-5769, Fax: 886-3-6687827, Email: wsliaoc@tsmc.com

Abstract

A reliability proven high-K (HK) metal-insulator-metal (MiM) structure has been verified within the silicon interposer in a chip-on-wafer-on-substrate (CoWoS) packaging for heterogeneous system-level decoupling application. The HK dielectric has an equivalent oxide thickness (EOT) of 20Å, intrinsic TDDB lifetime of 322 years at an operation voltage (V_{cc}) of 1.8V, and a leakage current (I_{LK}) below 1 fA/μm^2 under +/-2V bias at 125°C. The measured unit area capacitance density for the single, 2- and 3-in-series Si-interposer HK-MiM combination is 17.2, 4.3 and 1.9 fF/μm^2, respectively, with their corresponding I_{LK} below 0.48, 0.19 and 0.09 fAmp/μm^2. Process reliability related defect density (D_0) of the interposer HK-MiM is as low as 0.095% cm^{-2} as judged by a 10 years lifetime breakdown voltage (V_{bd}) criterion at V_{cc}=3.2V. This low D_0 ensures the Si-interposer HK-MiM to be used in a large area over 1056 cm^2 within the Si interposer. Moreover, the V_{bd} tolerance of the HK-MiM can be drastically enhanced to be 9.75 and 14.25V, respectively, by 2- and 3-in-series HK-MiM configuration connection. At the package level during all steps of CoWoS processing, no distinguishable process induced damage (PID) and performance degradation (Cap., I_{LK} & V_{bd} tailing) were detected. Therefore, this high capacitance, low leakage, large area and reliability-proven Si-interposer decoupling capacitor (DeCAP) within CoWoS greatly enhances the merit of using Si-interposer HK-MiM capacitors for multi-chip system-level integration.

Introduction

3D IC chip-on-wafer-on-substrate (CoWoS) technology is attractive due to its heterogeneous system-level integration capability to integrate multiple dies with different functions on a common Si interposer for achieving drastic form factor shrinkage, power consumption reduction, operation speed & bandwidth enhancement, as well as system packaging yield improvements.

The insertion of the passive components or devices into Si interposer is highly desirable to replace the bulky passive components which were typically mounted on a package substrate (1~9). The metal-insulator-metal (MiM) capacitor also has been reported to integrate into silicon interposer and acts as a 3D IC package decoupling capacitor (DeCAP) with advantages of relatively high capacitance (Cap.) as well as low manufacturing cost (3,6). In this work, we will report our development of a wafer level system integration (WLSI) high-K (HK) MiM integrating into the 3D IC CoWoS silicon interposer in a 12 inch FAB as well as its electrical and reliability characterization results.

Fabrication

A HK-MiM with an equivalent dielectric constant (K) value of 13.6 and equivalent oxide thickness (EOT) of 20Å has been fabricated within the interconnect region of a 100μm-thick Si-interposer. Fig. 1 depicts the schematic diagram of such a structure, showing that the HK-MiM is inserted between metal-1 and metal-2 within the Si interposer. Continuously, Fig. 2 displays a cross-sectional SEM image after the process insertion of a HK-MiM.

Electrical Characterizations

A. HK-dielectric

Fig. 3 plots the normalized capacitance density versus voltage curve measured at a fixed frequency of 100 KHz, with its unit capacitance density of 17.0 fF/μm^2 at zero voltage for the HK-dielectric film. Fig. 4 dispays two normalized I-V curves for this HK-dielectric film measured at 25 and 125°C, respectively. It is clear that the measured leakage current (I_{LK}) at +/-2V bias is still below 1 fAmp/μm^2 even at a higher testing temperature of 125°C.

B. Si-interposer HK-MiM

Three different silicon interposer HK-MiM configuration types were designed and characterized. Fig. 5 shows that for all of the single, 2- and 3-in-series interposer HK-MiM combinations their measured unit capacitance density in average (at 50% cumulative distribution) is 17.15, 4.29 and 1.90 fF/μm^2, respectively. Fig. 6 exhibits that the Weibull plotting of breakdown voltage (V_{bd}) for the single MiM has

978-1-4799-8002-4/14 $31.00 © 2014 IEEE

the narrowest distribution with its V_{bd} at 50% cumulative distribution (V_{bd}@50%) being 4.75V. Only one premature V_{bd} failure point occurs over a large accumulative capacitor area of 1056 cm² between the V_{cc} of 1.0V and a pre-defined V_{bd} criterion of 3.2V, which is derived separately from the time dependent dielectric breakdown (TDDB) testing to represent the minimum V_{bd} required for 10 years lifetime operating at V_{cc}= 1.0V. Thus, the reliability related defect density (D_0) is 0.095% cm⁻² for a single HK-MiM to support a large area (1056 cm² for $D_0 \le 0.1\%$ criteria) system-level DeCAP application.

To enable the Si-interposer HK-MiM application at a higher V_{cc}, 2- and 3-in-series HK-MiM configuration structures are also employed and displayed in Fig. 6, with their V_{bd}@50% greatly improving to 9.75 and 14.25V, respectively. For these two multiple interposer HK-MiM connection types, their reliability related D_0 can be significantly reduced so that there is not a single premature V_{bd} failure observed over 1056 cm² area tested. This means that both 2- and 3-in-series HK-MiM capacitors are suitable for relatively higher V_{cc} requirement such as 1.8V, 3.3V,...., etc, or D_0 guaranteed ($D_0 \le 0.1\%$) relatively larger DeCAP areas for the WLSI 3D IC CoWoS packaging.

Subsequently, Fig. 7 shows the cumulative distribution curves of measured leakage current (I_{LK}) at 25°C is 0.48, 0.19 and 0.09 fAmp/μm² in average (I_{LK} at 50% cumulative distribution) for the single, 2- and 3-in-series HK-MiM combination, respectively, exhibiting robust Si-interposer HK-MiM capacitors with their I_{LK} well below 1 fAmp/μm². Moreover, it also indicates that a drastically reduced leakage current can be achieved once employing 2- or 3-in-series Si-interposer HK-MiM configuration connection for future WLSI CoWoS packaging designing.

Reliability Characterizations

A. TDDB

Fig. 8 displays the TDDB results that the maximum operation voltage to sustain 10 years lifetime at 125°C is 2.02V for a single HK-MiM with a fixed capacitor area of 100 mm², as predicted by the lower dashed line; moreover, at a little lower operation voltage of V_{cc}=1.8V, its predicted lifetime can be drastically increased to 322 years.

B. PID

For the process induced damage (PID) during chip on wafer (CoW) processing, Fig. 9 shows that there is no charge damage detection for all of the single Si-interposer HK-MiM combination configurations with various capacitor areas of 0.6, 0.8, 1.2 and 1.8 mm², respectively, under the same PID antenna ratio of 1.

Additionally, Fig. 10 shows that for the other PID antenna ratios of 2, 2.4 and 4 at the same single Si-interposer HK-

MiM configurations and areas, all of their measured I_{LK} cumulative curves are well below 15 pAmp which also indicates no sign of PID occurrence being detected.

Finally, for the PID characterizations in-between and after all steps of CoWoS processing, all of the confidence-level performance comparisons of Cap., I_{LK} and V_{bd} Weibull cumulative distribution curves and their tailings still exhibit no occurrence of process induced charge damage.

Conclusion

A 3D IC HK (K=13.6) silicon interposer MIM capacitor with thin dielectric (EOT=20Å), high unit capacitance density (17 fF/μm²), low leakage current ($I_{LK} < 1$ fAmp/μm² at 125°C), large MiM area size (> 1056 cm²) and intrinsic TDDB lifetime of 322 years at 1.8V (for 100 mm² area size) has been developed. Electrical (Cap., I_{LK}, V_{bd} tailing and D_0) and reliability (TDDB, PID) characterizations prove design and process flexibility for varied silicon interposer HK-MiM configurations with single, 2- or 3-in-series connection. Finally, no detection of a single PID failure during all processing steps of CoWoS manufacturing is also a demonstration of robust and manufacturable wafer level system integration (WLSI) of Si-interposer with a HK-MiM structure in a 12 inch FAB suitable for 3D IC heterogeneous packaging with system-level DeCAP capability.

References

(1) H. Jacquinot and D. Denis, "Characterization, Modeling and Optimization of 3D Embedded Trench Decoupling Capacitors in Si-RF Interposer," IEEE ECTC, 2013, pp. 1372~1378.

(2) E. Song, K. Koo, J. S. Pak, and J. Kim, "Through-Silicon-Via-Based Decoupling Capacitor Stacked Chip in 3-D-ICs," IEEE CPMT, 2013, v.3, n.9, pp. 1467~1480.

(3) S. Gandhi, et. al., "A Low-Cost Approach to High-k Thin Film Decoupling Capacitors on Silicon and Glass Interposers," IEEE ECTC, 2012, pp. 1356~1360.

(4) C. Auth, et. al., "A 22nm High Performance and Low-power CMOS Technology Featuring Fully-depleted Tri-gate Transistors, Self-aligned Contacts and High Density MIM Capacitors," IEEE VLSI, 2012, pp. 131~132.

(5) Z. Li, H. Shi, J. Xie, and A. Rahman, "Development of an optimized power delivery system for 3D IC integration with TSV silicon interposer," IEEE ECTC, 2012, pp. 678~682.

(6) A. Takano, et. al., "Development of Si Interposer with Low inductance Decoupling Capacitor," IEEE ECTC, 2011, pp. 849~854.

(7) K. Kikuchi, "Low-Impedance Evaluation of Power Distribution Network for Decoupling Capacitor Embedded Interposers of 3-D Integrated LSI System," IEEE ECTC, 2010, pp. 1455~1460.

(8) B. Dang, et. al., "Three-Dimensional Chip Stack With Integrated Decoupling Capacitors and Thru-Si Via Interconnects," IEEEE EDL, 2010, v. 31, n.12, pp. 1461~1463.

(9) C. T. Black, et. al., "High-Capacity, Self-Assembled Metal–Oxide–Semiconductor Decoupling Capacitors," IEEEE EDL, 2004, v.25, n.9, pp. 622~624.

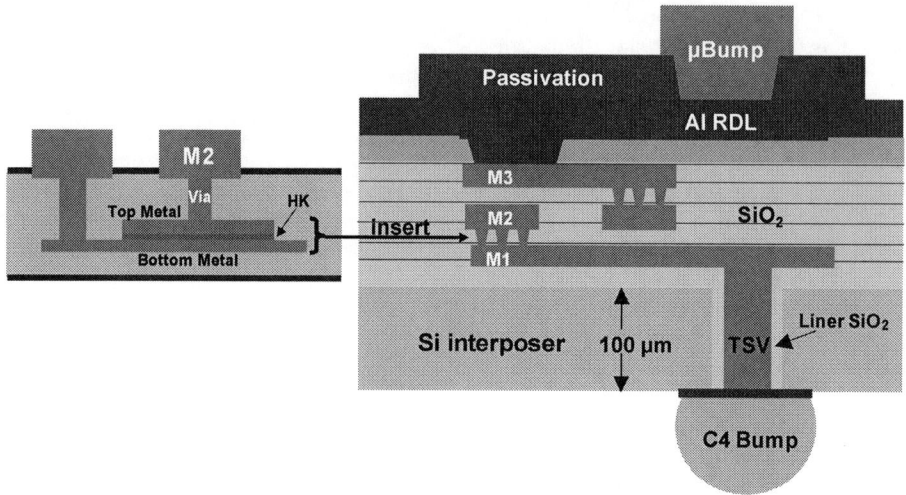

Fig. 1 – Schematic diagram showing the insertion of a high-K MiM structure in-between metal-1 and metal-2 of a 3D IC CoWoS silicon interposer interconnect region.

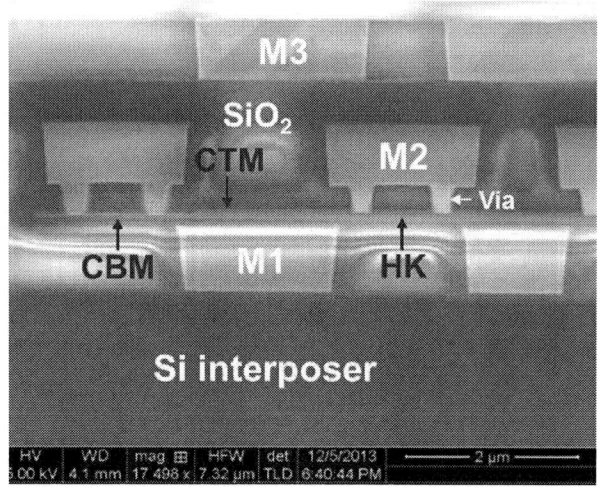

Fig. 2 – Cross-sectional SEM image showing the process integration of the high-K MiM into the CoWoS Si-interposer interconnect region in-between metal-1 and metal-2.

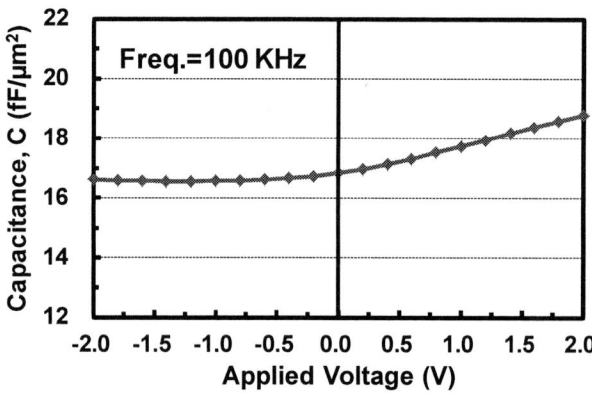

Fig. 3 – The normalized capacitance density vs. voltage curve of the high-K dielectric film measured at a fixed frequency of 100 KHz.

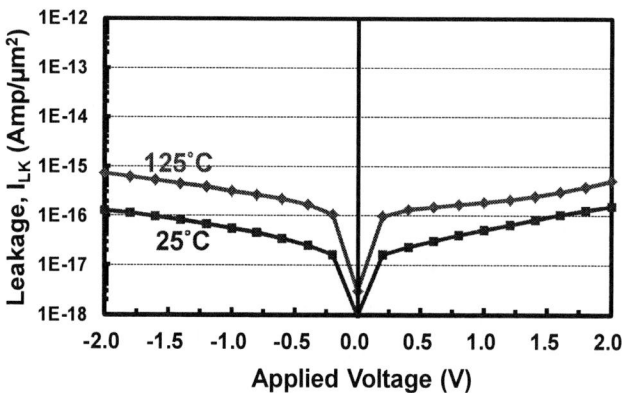

Fig. 4 – The normalized leakage current density vs. voltage curves of the high-K dielectric film measured at varied temperatures of 25 and 125°C, respectively.

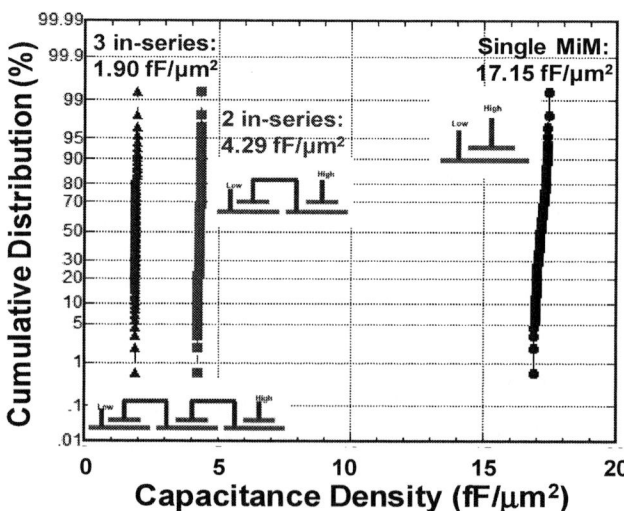

Fig. 5 – The measured capacitance density distribution curves of Si-interposer HK-MiM fabricated with single, 2- and 3-in-series HK-MiM configuration connection, respectively.

978-1-4799-8002-4/14 $31.00 © 2014 IEEE   636

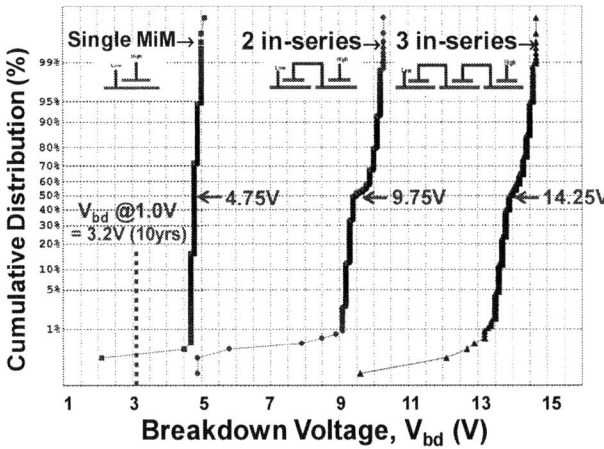

Fig. 6 – The measured breakdown voltage (V_{bd}) distribution curves of the Si-interposer HK-MiM fabricated with single, 2- and 3-in-series HK-MiM configuration, respectively.

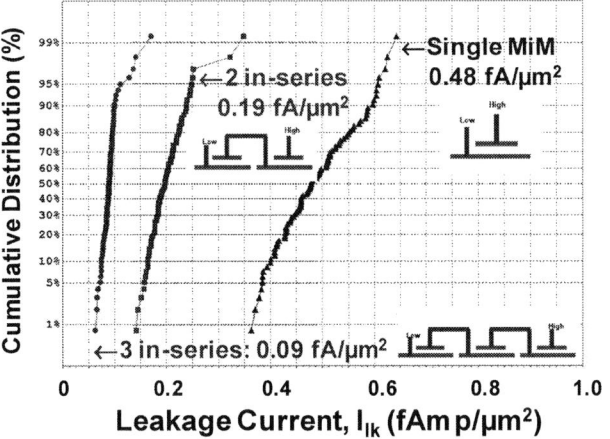

Fig. 7 – The measured leakage current (I_{LK}) density distribution curves of the Si-interposer HK-MiM fabricated with single, 2- and 3-in-series configuration at 25°C, respectively.

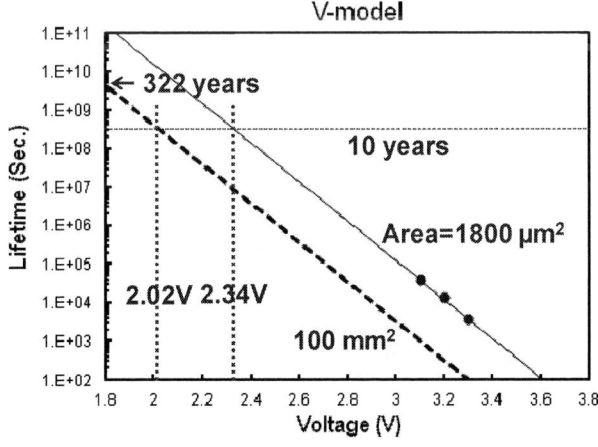

Fig. 8 – The TDDB lifetimes (solid line) of the high-K dielectric film with capacitor area of 1800 μm^2 after accelerating conditions of 3.1V, 3.2V, 3.3V and 125°C. The lower dashed line represents its predicted TDDB lifetimes for the normalized HK-MiM capacitor area of 100 mm^2.

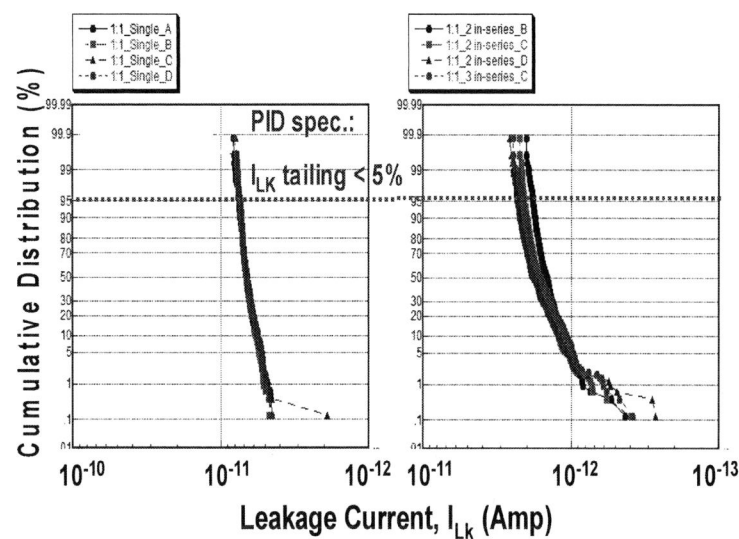

Fig. 9 – The measured leakage current (I_{LK}) distribution curves after CoW (Chip on Wafer) processing for the Si-interposer HK-MiM capacitors with a certain PID antenna ratio of 1, as well as for the single HK-MiM (left chart) and 2- & 3-in-series HK-MiM (right chart) with various capacitor areas of 0.6, 0.8, 1.2 and 1.8 mm^2, respectively.

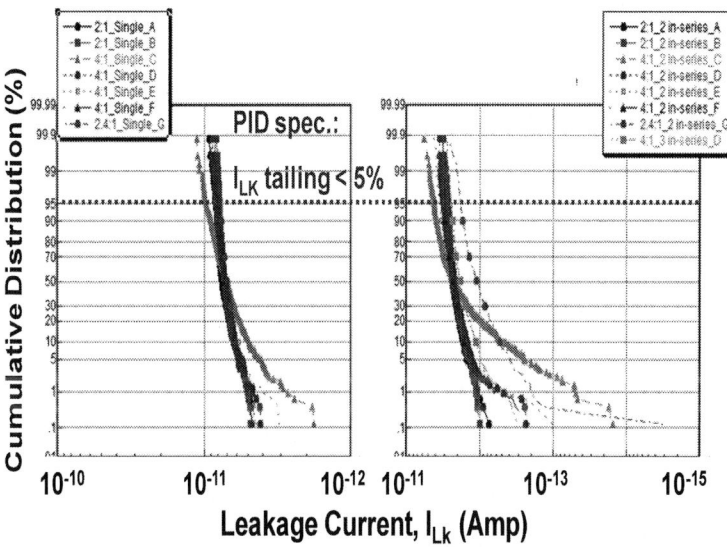

Fig. 10 – The measured leakage current (I_{LK}) distribution curves of the Si-interposer HK-MiM capacitors after CoW (Chip on Wafer) processing for the single HK-MiM (left chart) and 2- & 3-in-series HK-MiM (right chart) with different antenna ratios of 2, 2.4 and 4, as well as with varied HK-MiM capacitor areas of 0.6, 0.8, 1.2 and 1.8 mm^2, respectively.

978-1-4799-8002-4/14 $31.00 © 2014 IEEE

Monolithic 3D Integration of Logic and Memory:
Carbon Nanotube FETs, Resistive RAM, and Silicon FETs

Max M. Shulaker[1]*, Tony F. Wu[1], Asish Pal[1], Liang Zhao[1], Yoshio Nishi[1], Krishna Saraswat[1],
H.-S. Philip Wong[1], Subhasish Mitra[1,2]

Department of Electrical Engineering[1] and Computer Science[2]
Stanford University, Stanford, CA, 94305, U.S.A. *Email: maxms@stanford.edu

Abstract

We demonstrate monolithic 3D integration of logic and memory in arbitrary vertical stacking order with the ability to use conventional inter-layer vias to connect between any layers of the 3D IC. We experimentally show 4 vertically-stacked layers (logic layer followed by two memory layers followed by another logic layer), enabled by the integration of traditional silicon-FETs (on the bottom-most layer) with low-processing-temperature emerging nanotechnologies: metal-oxide resistive random-access memory (RRAM), and carbon nanotube-FETs (CNFETs). As a demonstration, we show a routing element of a switchbox for a field-programmable gate array (FPGA), with each component of the routing element (involving both logic and memory elements) on their own vertical layer.

Introduction

Monolithic Three-Dimensional Integration

Three-dimensional (3D) integration is a promising technology option for improving the performance, energy efficiency, and footprint of electronic systems [1]. Today's 2.5D and 3D integration are achieved through chip-stacking, with multiple vertical circuit layers connected using Through-Silicon Vias (TSVs). Monolithic 3D integration, whereby each circuit layer is thin and is fabricated directly over the previous circuit layers on the same substrate, can use conventional inter-layer vias (ILVs) to connect between various layers. The use of conventional vias rather than TSVs allows for massive vertical interconnect density, potentially maximizing the benefits of 3D integrated circuits (ICs) [1]. Moreover, monolithic 3D integration of logic *and* memory can enable new architectures and potentially alleviate the logic-memory communication bottleneck [2,3].

Overcoming Monolithic 3D Integration Obstacles

While monolithic 3D integration of logic and memory in arbitrary stacking order is an attractive technology, processing obstacles have prohibited its demonstration. Specifically, the processing temperatures for all upper layer circuitry must be low (<400^{0}C), so as to not damage or destroy the lower layers of logic, memory, or metal interconnects. Previous work achieved monolithic 3D integration of logic through the use of CNFETs due to their low processing temperature (<250^{0}C) [4], and with low-temperature wafer bonding of silicon SOI substrates [1]. Monolithic 3D integration of a layer of memory over a layer of logic has likewise been enabled by low-temperature RRAM processing [3]. To realize monolithic 3D integration of logic and memory in arbitrary stacking order, we employ both CNFETs and RRAM as the upper-layers of logic and memory, while performing all fabrication on a starting silicon-FET substrate, thereby showing that the entire process is compatible with existing silicon technologies [5].

In addition to enabling monolithic 3D integration, both CNFETs and RRAM are promising emerging nanotechnologies. CNFETs promise both improved performance and energy efficiency (~10x benefit in energy-delay product (EDP) compared to silicon-CMOS) [6], while RRAM potentially realizes a high-capacity storage and BEOL-compatible non-volatile memory [7]. Such a monolithically-integrated 3D IC, fabricated with 4 vertical layers of logic and memory elements, is illustrated in Fig. 1 (step 6), and Fig. 2.

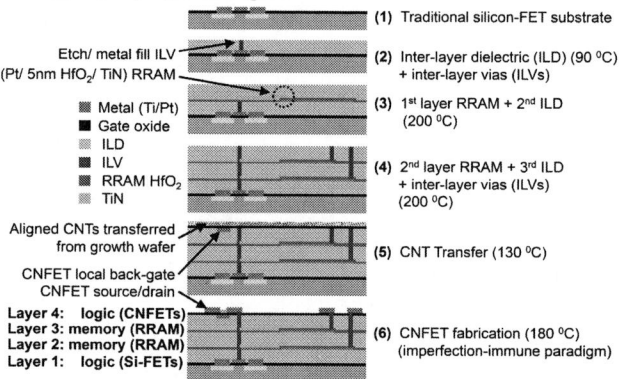

Figure 1: Process flow for monolithic 3D integration of silicon-FETs+RRAM+CNFETs. All post-silicon fabrication processing is <200 ^{0}C and VLSI compatible. The imperfection-immune paradigm is used to overcome mis-positioned and metallic CNTs for CNFETs on the 4th layer [11].

Fabrication Process

The key enabler for monolithic 3D integration of logic and memory elements is the use of emerging nanotechnologies with low processing temperatures over a starting silicon-FET substrate. The fabrication process is shown in Fig. 1. We begin with the fabrication of conventional silicon-FETs as the first layer of logic, which may be used as an access transistor for the upper layers of RRAM. The high-temperature dopant activation rapid thermal anneal (1050^{0}C) is performed during the silicon-FET fabrication and before the fabrication of any circuitry on upper layers. Following silicon-FET fabrication, a low-temperature (90^{0}C plasma-enhanced chemical vapor deposited (PECVD) 100 nm SiO$_X$) inter-layer dielectric (ILD) is deposited, and inter-layer vias (ILVs) are etched and filled with metal. The 2nd layer of the 3D IC (1st layer of RRAM cells) is deposited directly over this ILD. The RRAM uses a 10 nm Pt/ 5 nm HfO$_X$/ 3 nm TiN/ 10 nm Pt stack, which is fabricated with maximum processing temperature of 200^{0}C [8]. The 3rd layer of the 3D IC (2nd layer of RRAM cells) is fabricated in an identical manner, with an additional 100 nm ILD with ILVs connecting between the layers. Following the 3rd vertical layer, a 3rd 100 nm ILD is deposited, in preparation for the 4th layer of the 3D IC (2nd layer of logic using CNFETs). The CNTs are first grown on a crystalline quartz substrate, yielding >99.5% highly aligned CNTs [9]. Following growth, the CNTs are

978-1-4799-8002-4/14 $31.00 © 2014 IEEE

transferred from the quartz growth substrate onto the 3D IC, using a low-temperature (130^{0}C) transfer process that maintains both the alignment and density of the CNTs [9]. This low-temperature transfer is essential, as it decouples the high temperature growth (~900^{0}C) from the 3D IC, which would otherwise damage or destroy both the bottom layers of logic, memory, and metal interconnects.

To prepare the 3D IC for CNT transfer, the ILD undergoes chemical-mechanical polishing (CMP) followed by an argon sputter etch to planarize the surface. The local bottom gates for the CNFETs (1 nm Ti/ 10 nm Pt) are patterned [9], followed by depositing the CNFET high-κ gate dielectric (16 nm Al$_2$O$_X$) through atomic layer deposition (ALD) at 200^{0}C. The CNTs are transferred onto the CNFET gate dielectric, followed by the CNFET source and drain definition (2 nm Ti/ 12 nm Pt). We use the imperfection-immune paradigm to overcome the substantial imperfections inherent in CNTs: mis-positioned CNT-immune design renders the logic immune to mis-positioned CNTs, while VLSI-compatible Metallic CNT Removal (VMR) selectively removes >99.99% metallic CNTs from the circuit [11]. Finally, ILVs connecting between any remaining vertical layers are etched and filled. Importantly, the fabrication of every layer is performed on the same starting substrate above the previous layers of logic and memory, in both a silicon-CMOS and VLSI-compatible manner. Transmission-electron microscopy images of each layer of the 3D IC are shown in Fig. 2.

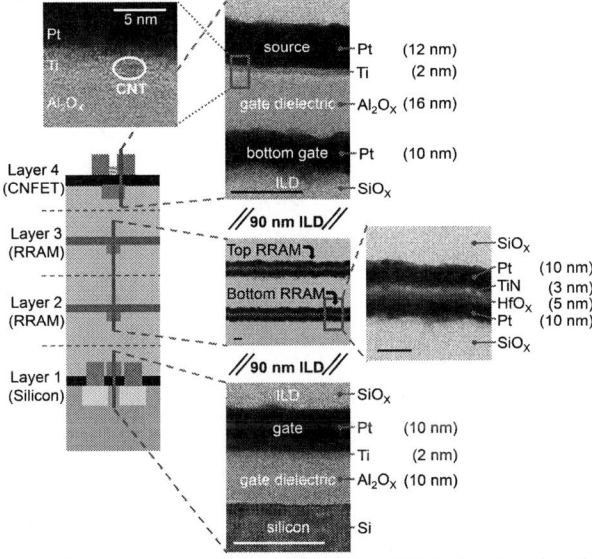

Figure 2: Transmission electron microscopy (TEM) images of each vertically stacked layer in the 3D IC. Silicon-FETs are on the bottom layer, followed by two layers of RRAM, followed by the top layer of CNFETs. Scale bar in all TEMs is 20 nm, unless otherwise noted.

Experimental Results

Arbitrary Connectivity Between Logic and Memory Layers
Fig. 3-4 demonstrate that the monolithic 3D processing does not destroy or damage the bottom layer circuitry. Fig. 3 shows negligible change in the I$_D$-V$_{GS}$ curve for typical silicon-FETs, measured both after the silicon-FET fabrication but before any 3D fabrication, and then after all 3D fabrication. Fig. 3 also shows negligible change in

performance for a typical RRAM on the 2nd layer of the 3D IC (bottom layer of RRAM), measured both before and after the rest of the 3D process is performed. The RRAMs across the 2nd and 3rd layers of the 3D IC also yield similar distributions for forming, set, and reset (Fig. 4).

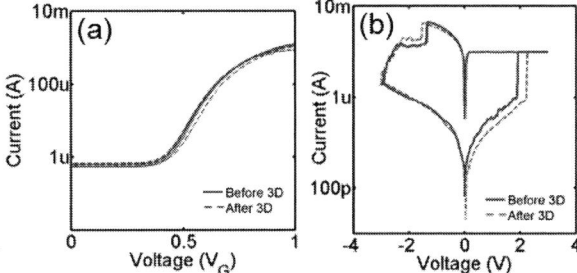

Figure 3: (a) I$_D$-V$_{GS}$ curve of typical silicon-FETs before and after monolithic 3D processing, showing silicon-FETs are unaffected by the monolithic 3D processing (10 different silicon-FETs are overlaid in the figure). (b) Set-reset curves for a typical RRAM both before and after subsequent monolithic 3D processing, showing RRAM is unaffected by the monolithic 3D processing.

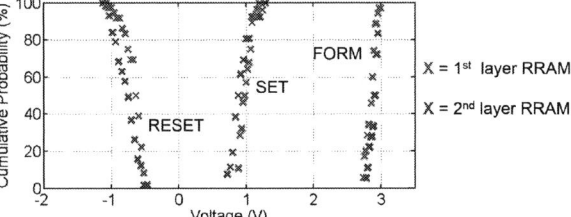

Figure 4: RRAM performance is invariant to its layer in the monolithic 3D stack. The form, set, and reset voltage distributions for the RRAM between the two layers is unchanged.

To demonstrate the ability to connect between any two layers of logic and memory, we fabricate 1 transistor-1 RRAM (1T1R) structures between every possible combination of logic and memory in our monolithic 3D stack, shown schematically in Fig. 5. Scanning electron microscopy (SEM) images of each of the 1T1R structures across the different circuit layers are shown in Fig. 6. Fig. 7 shows the 1T1R structures for the 1st layer of silicon-FETs controlling the 1st and the 2nd layers of RRAM above them. Since none of the circuit layers are affected by the 3D processing, the performance of both 1T1R structures are the same.

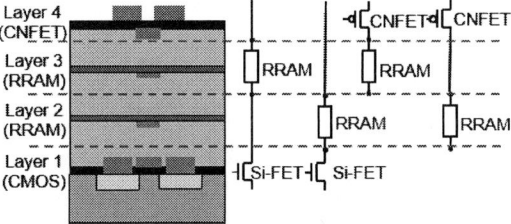

Figure 5: Arbitrary connections between logic and memory in 3D stack demonstrated with 1T1R structures, using each layer of logic and memory.

Fig. 8 shows the 1T1R structures for the CNFET top logic layer controlling the 1st and the 2nd layers of RRAM underneath it. Again, both 1T1R structures involving different layers of the RRAM exhibit similar forming, reset, and set characteristics. Importantly, both Fig. 7 and Fig. 8 confirm that the memory and logic performance is invariant

978-1-4799-8002-4/14 $31.00 © 2014 IEEE

to the fabrication order and placement in the 3D IC stack. The difference in the set and reset curves between the 1T1R structures including the silicon-FET versus the CNFET is attributed to the CNFET not exhibiting the same saturation in the I_D-V_{DS} characteristics as the silicon-FET (Fig. 7-8, these CNFETs saturate at higher V_{DS} due to un-optimized source and drain contact resistance, similar to previously reported I_D-V_{DS} CNFET characteristics [12,13]).

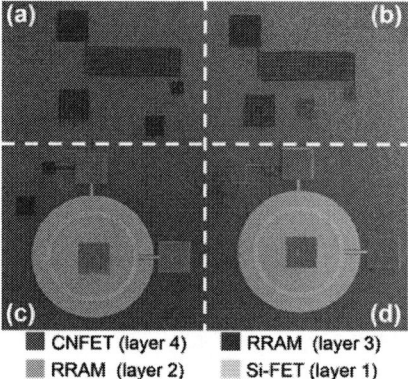

Figure 6: Scanning electron microscopy (SEM) images (false colored) of all 1T1R structures. (a) CNFET select transistor with RRAM on the 3rd layer. (b) CNFET select transistor with RRAM on the 2nd layer. (c) Si-FET select transistor with RRAM on the 3rd layer. (d) Si-FET select transistor with RRAM on the 2nd layer.

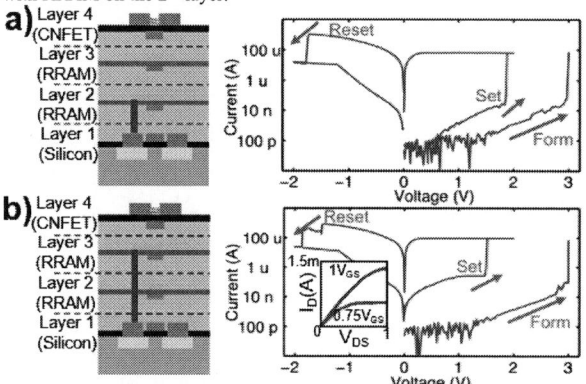

Figure 7: 1T1R results for silicon-FET select transistor underneath (a) 1st layer of RRAM, and (b), 2nd layer of RRAM. Inset shows silicon-FET I_D-V_{DS}.

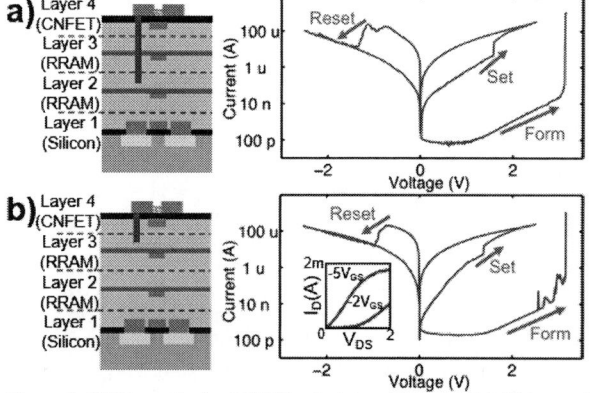

Figure 8: 1T1R results for CNFET select transistor over (a) 1st layer of RRAM, and (b), 2nd layer of RRAM. Inset shows CNFET I_D-V_{DS}.

While Fig. 8 shows proper 1T1R operation, the CNFET must be able to operate as a selector for the RRAM. As this is the first demonstration of CNFET and RRAM integration (in addition to monolithic 3D integration of CNFETs and RRAM), we further characterize the CNFET's ability to act as the selector for the RRAM. Fig. 9 shows additional characterization of the 1T1R using the CNFETs as the select transistor. The CNFET can successfully be used to both set the compliance current for the set operation of the RRAM, and can be used as a pass-gate to select whether or not to perform the reset operation of the RRAM. These operations can be only performed after VMR is used to selectively remove >99.99% metallic CNTs, resulting in >5,000 I_{ON}/I_{OFF} for the CNFET, while inadvertently removing <4% semiconducting CNTs [12] and therefore retaining enough I_{ON} to perform the RRAM reset.

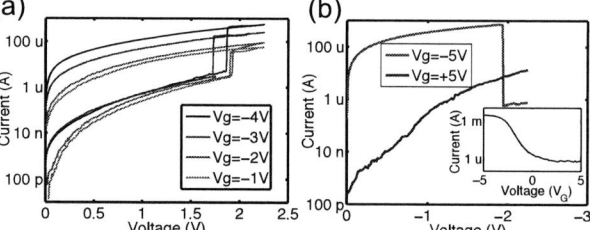

Figure 9: Characterization of CNFET as select transistor for RRAM in 1T1R structure. (a) The gate bias of the p-CNFET can be used to set the set compliance current for the RRAM. (b) The CNFET functions as a select transistor, and based on the value of the gate bias can either allow or prohibit the RRAM reset. The p-CNFET I_D-V_{GS} curve is shown bottom-right.

Demonstration

As an applied demonstration of monolithic 3D integration of logic and memory, we fabricate a 4-layer switching element of a switchbox for an FPGA. Switchboxes for an FPGA implemented with RRAM-based configuration memory have been shown to save 40% area and 28% in EDP, primarily attributed to the cell-size reduction due to the integration of RRAM over the silicon logic [3] instead of the use of a 6T SRAM cell in silicon. With the ability to fabricate memory and logic on any arbitrary layer, both the RRAM select transistor and the routing transistor can be integrated on top of each other, further reducing the cell footprint. Ideally, the routing element cell footprint can be further reduced to $4F^2$ [14]. Fig. 10 and Fig. 11 show the schematic and scanning electron microscopy image (respectively) of the fabricated 4-layer switching element containing 2 RRAM cells, a CNFET select transistor and silicon-NMOS routing transistor, with each element on its own layer. Fig. 12 shows proper operation of the switching element, with repeated switching between the on and off states. The I_{ON}/I_{OFF} is ~1000, which is attributed to the un-optimized silicon-FET fabrication with non-ideal threshold adjustment and inverse subthreshold slope (Fig. 3a). The fabricated switching element exhibits high switching endurance of >10^5 cycles with consistent programming voltages (Fig. 13-15), suitable for FPGAs.

978-1-4799-8002-4/14 $31.00 © 2014 IEEE

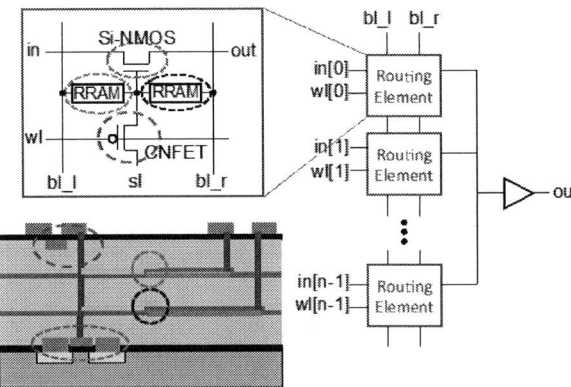

Figure 10: Schematic of the 4-layer switching element containing 2 RRAM cells, a CNFET select transistor and silicon routing transistor, with each element on its own layer.

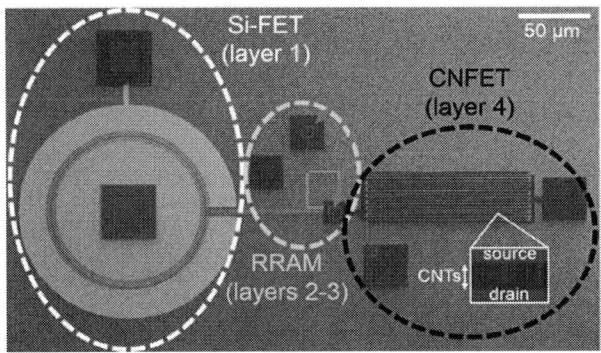

Figure 11: SEM (false colored) of switching element consisting of a CNFET access transistor for the 2 RRAM cells and the silicon NMOS pass transistor. All elements are on different layers.

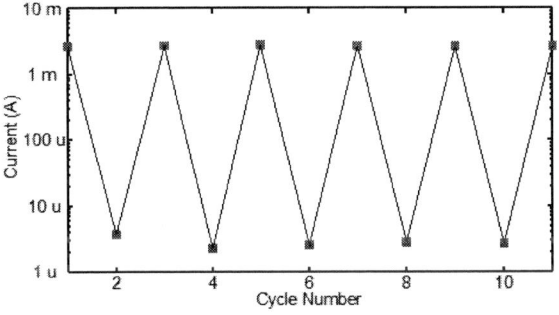

Figure 12: Results for the FPGA switching element, showing difference of ~1000x between I_{ON} and I_{OFF}.

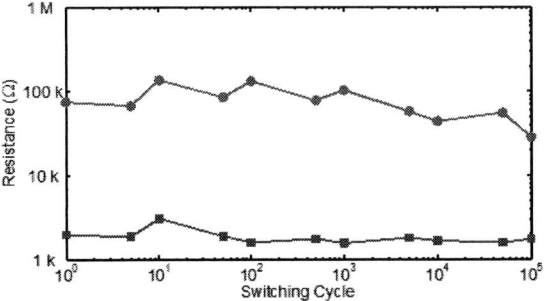

Figure 13: Characterization of a CNFET as a select transistor for RRAM in a 1T1R structure. Endurance testing of 10^5 cycles with 300 ns second pulsing with set/reset voltage of +/-2.3V.

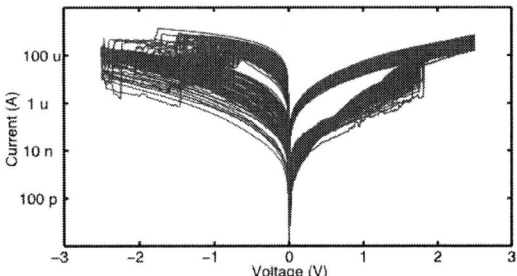

Figure 14: Characterization of a CNFET as a select transistor for RRAM in a 1T1R structure. DC cycling of set/reset for 100 cycles.

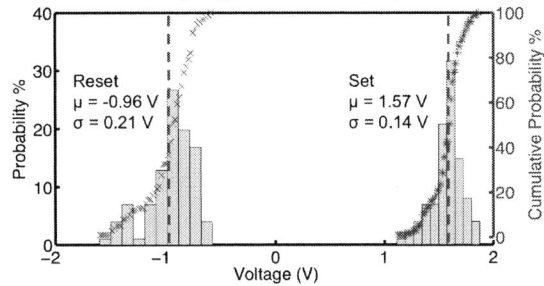

Figure 15: Characterization of a CNFET as a select transistor for RRAM in a 1T1R structure. Distribution of set/reset voltages from Figure 14.

Conclusion

We realize monolithic 3D integration of logic and memory in arbitrary vertical stacking order with the ability to use conventional inter-layer vias to connect between any layers of the 3D IC. This is enabled by the use of emerging nanotechnologies such as RRAM and CNFETs with low processing temperatures. This demonstration highlights the feasibility of monolithic 3D integration of logic and memory which promises major speed and energy benefits for digital systems [15].

Acknowledgements

This research was supported in part by STARnet SONIC, NSF, and Hertz/SGF for M.M.S.. We thank Prof. Eric Pop and Prof. Zhenan Bao of Stanford for fruitful discussions.

References

[1] P. Batude, et al., "Advances, challenges and opportunities in 3D CMOS sequential integration," *IEDM*, pp. 151-154, 2011.
[2] S. Wong, et al., "Monolithic 3D integrated circuits," *VLSI-TSA*, pp. 1-4, 2007.
[3] Y. Liauw, et al., "Nonvolatile 3D-FPGA with monolithically stacked RRAM-based configuration memory," *JSSCC*, pp. 406-407, 2012.
[4] H. Wei, et al., "Monolithic Three-Dimensional Integration of Carbon Nanotube FET Complementary Logic Circuits," *IEDM*, pp. 511-514, 2013.
[5] M. Shulaker, et al., "Monolithic three-dimensional integration of carbon nanotube FETs with silicon CMOS ," *VLSI Symp.*, pp. 214-215, 2014.
[6] L. Chang, "Short Course," *IEDM*, 2012.
[7] H.-S. P. Wong et al., "Metal-oxide RRAM," *Proc. IEEE*, vol. 100(6), pp. 1951-1970, 2012.
[8] L. Goux, et al., "On the gradual unipolar and bipolar resistive switching of TiN\HfO2\Pt memory systems," *Electrochem. Solid State Lett.*, vol. 13, pp. G54–G56, 2010.
[9] N. Patil, et al., "Wafer-scale growth and transfer of aligned single-walled carbon nanotubes," *Nanotechnology*, vol. 8(4), pp. 498-504, 2009.
[10] A. Franklin, et al., "Current scaling in aligned carbon nanotube array transistors with local bottom gating," *Electron Device Lett.*, vol. 31(7), pp. 644-646, 2010.
[11] J. Zhang, et al., "Robust digital VLSI using carbon nanotubes," *TCAD*, vol. 31(4), pp. 453-471, 2012.
[12] M. Shulaker, et al., "Sensor-to-digital interface built entirely with carbon nanotube FETs," *Journal of Solid-State Circuits*, vol. 49(1), pp. 190-201, 2014.
[13] M. Shulaker, et al., "Carbon Nanotube Computer," *Nature*, vol. 501(7468), pp. 526-530, 2013.
[14] X.P. Wang, et al., "Highly compact 1T-1R architecture (4F2 footprint) involving fully CMOS compatible vertical GAA nano-pillar transistors and oxide-based RRAM cells exhibiting excellent NVM properties and ultra-low power operation," *IEDM*, pp. 493-496, 2012.
[15] M. Ebrahimi, et al., "Monolithic 3D integration advances and challenges: from technology to system levels," *SOI-3D-Subthresh. Micro. Tech. Unified Conf.*, 2014.

978-1-4799-8002-4/14 $31.00 © 2014 IEEE

New insights on bottom layer thermal stability and laser annealing promises for high performance 3D VLSI

[1]C. Fenouillet-Beranger, [1]B.Mathieu, [1]B. Previtali, [2,1]M-P. Samson, [1]N.Rambal, [1]V. Benevent, [1]S. Kerdiles, [1]J-P. Barnes, [2]D. Barge, [1]P. Besson, [1]R. Kachtouli, [1]M. Cassé, [1]X. Garros, [1]A. Laurent, [1]F. Nemouchi, [3]K. Huet, [3]I. Toqué-Trésonne, [1]D.Lafond, [1]H. Dansas, [1]F. Aussenac, [2]G. Druais, [1,2]P. Perreau, [2]E. Richard, [2]S. Chhun, [2]E. Petitprez, [2]N. Guillot, [1]F. Deprat, [2,1]L. Pasini, [1]L. Brunet, [2,1]V. Lu, [1]C. Reita, [1]P. Batude, [1]M. Vinet

[1] CEA-LETI MINATEC Campus 17 rue des Martyrs 38054 Grenoble cedex 9, [2] STMicroelectronics, Crolles, France,
[3] LASSE (LAser System and Solutions of Europe), Gennevilliers, France; claire.fenouillet-beranger@cea.fr

Abstract

For the first time the maximum thermal budget of in-situ doped source/drain State Of The Art (SOTA) FDSOI bottom MOSFET transistors is quantified to ensure transistors stability in Sequential 3D (CoolCube™) integration. We highlight no degradation of Ion/Ioff trade-off up to 550°C. Thanks to both metal gate work-function stability especially on short devices and silicide stability improvement, the top MOSFET temperature could be relaxed up to 500°C. Laser anneal is then considered as a promising candidate for junctions activation. Based on in-depth morphological and electrical characterizations it demonstrates very promising results for high performance Sequential 3D integration.

Introduction

An alternative approach to conventional planar integration for future nodes is the CoolCube™ 3D or sequential 3D integration [1]. Compared to TSV-based 3D ICs, CoolCube™ integration offers the possibility to stack devices with a lithographic alignment precision (few nm) enabling via density > 100 million/mm² between transistors tiers. However this integration faces the challenge to realize a high performance transistor at the top level without degrading the electrical characteristics of the bottom one. One of the challenges consists in integrating transistors with low temperature process steps. Our previous works highlighted that silicide is the main responsible of bottom transistor thermal stability degradation beyond 500°C in "implanted junctions" SOTA FDSOI (Fully Depleted SOI) technology [2]. The maximum post process thermal budget acceptable for in-situ doped source/drain transistors has to be determined precisely and breakout analysis provides ways to push upwards the thermal limitations. Laser anneal thanks to its low in-depth thermal diffusion is a promising opportunity for top transistor dopant activation: it is expected to provide high activation together with surface confined heating [3].

In this paper, for the first time, the maximum thermal budget for the bottom MOSFETs is quantified on an advanced in-situ doped source/drain SOTA FDSOI technology. Thanks to in-depth 2D simulations study and electrical and morphological characterizations, the interest of laser anneal for CoolCube™ 3D is highlighted.

Bottom MOSFETs stability

The goal of the experiment is to determine the maximum allowable temperature for the top MOSFET: it depends on the temperature at which the bottom transistor performance degrades within a CoolCube™ integration.

The transistors integration scheme is described in Fig.1 and based on SOTA in-situ doped source/drain FDSOI technology [4].
The additional thermal anneals have been performed in two places in the route either after the PMD CMP step just before contacts processing or after contact filling with W (Fig. 1). Figs.2 & 3 highlight that the additional thermal budgets have no impact on the Ion at a fixed Ioff for both NMOS and PMOS devices.

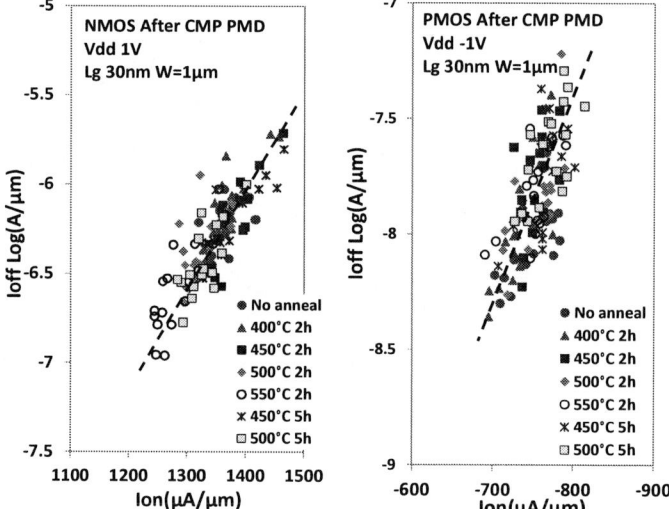

Figure 2: NMOS Ion/Ioff of bottom transistor versus thermal anneals. Lg 30nm W=1µm. Vdd 1V.

Figure 3: PMOS Ion/Ioff of bottom transistor versus thermal anneals. Lg 30nm W=1µm. Vdd -1V.

However we observe a slight Ion/Ioff shift above 500°C. It is well correlated with the linear $V_T(Lg)$ evolution of Figs.5 and 6 . As the thermal budget increases beyond 500°C a reverse short channel effect and an accentuated short channel effect is seen for NMOS and PMOS respectively (+30mV and -50mV shift of V_T between the reference and 550°C 2h anneal for NMOS and PMOS respectively).

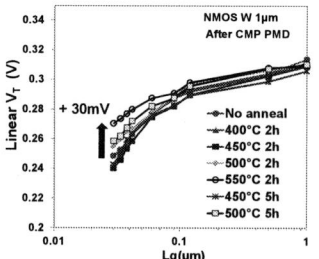 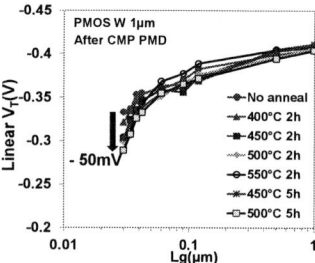

Figure 5: NMOS linear $V_T(Lg)$ of bottom transistor versus thermal anneals for variable Lg, W=1µm, Vdd 0.1V.

Figure 6: PMOS linear $V_T(Lg)$ of bottom transistor versus thermal anneals for variable Lg, W=1µm, Vdd -0.1V.

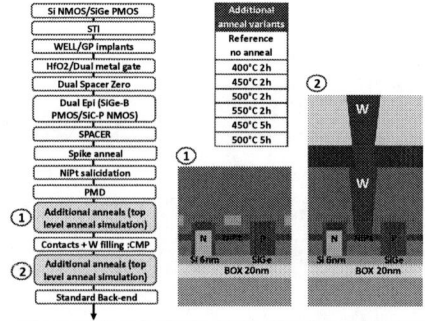

Figure 1: Process flow scheme [4]. Structure studied and table of variants.

978-1-4799-8002-4/14 $31.00 © 2014 IEEE 642

This effect is related to metal gate work-function instability for short devices (oxygen diffusion through the spacer) especially with HfO_2 as already described in [5].

However, whatever the thermal budget the DIBL is unchanged indicating that there is no change in the junction shapes reflecting no dopant diffusions (Figs.7 and 8). The unsalicided active resistance is also unaltered (Fig.9), reflecting no dopant deactivation phenomenon.

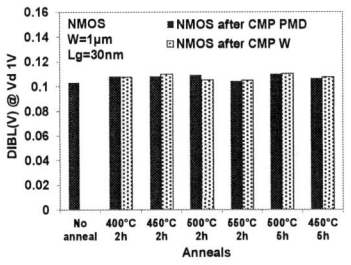

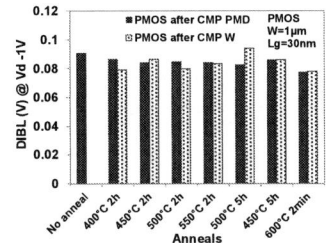

Figure 7: NMOS DIBL(Lg) of bottom transistor versus thermal anneals for Lg 30nm, W=1μm, Vdd 1V.

Figure 8: PMOS DIBL(Lg) of bottom transistor versus thermal anneals for Lg 30nm, W=1μm, Vdd -1V.

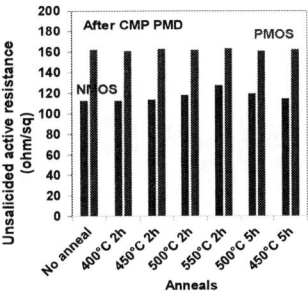

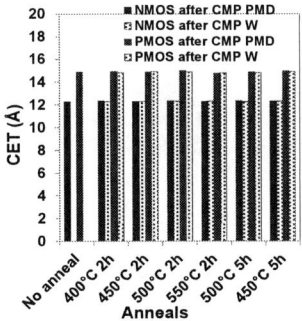

Figure 9: N+/P+ unsilicided active resistance versus additional thermal budgets.

Figure 10: NMOS and PMOS CET(Å) of bottom transistor versus thermal anneals for Lg 1μm W 1μm.

In addition, the CET is not modified (Fig.10) as well as the effective mobility versus the gate length, for example for the NMOS as shown on Fig.11, which is identical whatever the temperature. Only a slight modification of the access resistance is seen for temperature beyond 500°C 2h (Fig.12).

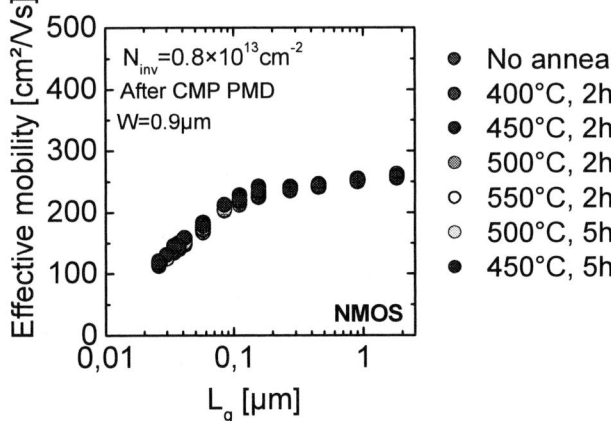

Figure 11: NMOS effective mobility μ_{eff} @ N_{inv}=0.8×10^{13}cm^{-2} versus Lg for several additional thermal anneals. W=0.9μm.

This global access resistance change is certainly linked to the silicide instability beyond 500°C as confirmed by the evolution of the salicided active resistance versus thermal anneals that follows an opposite trend for NMOS and PMOS on large active area (Fig.13).

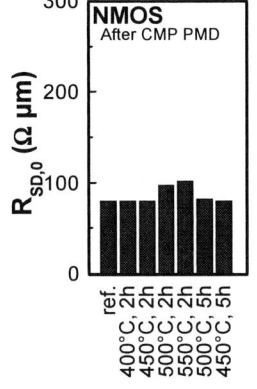

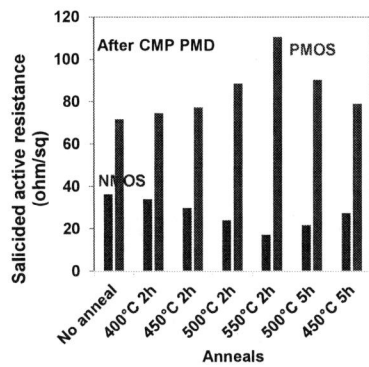

Figure 12: RSD extraction on NMOS versus several thermal budgets

Figure 13: N+/P+ salicided active resistance of bottom transistor versus thermal anneals.

Finally there is no influence on the conclusion whether the additional thermal budgets are performed after CMP or after W filling. No modifications of Ion/Ioff current up to 500°C (Fig.14 left) if the anneal is performed after W filling and only 5% after 550°C 2h (Fig.14 right).

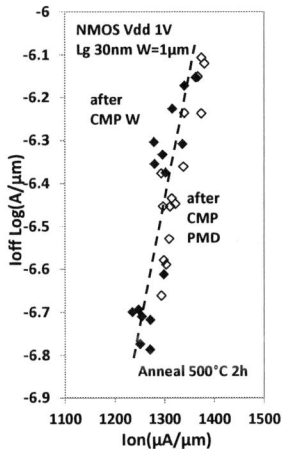

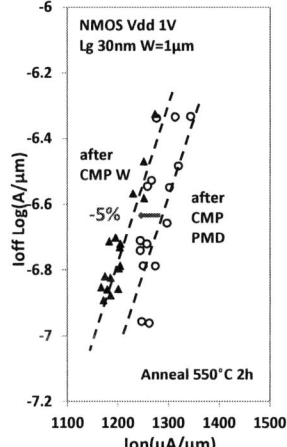

Figure 14: NMOS Ion/Ioff of bottom transistor for 500°C (left) and 550°C (right) 2h anneal. Comparison between anneal after PMD CMP or W CMP. Lg 30nm W=1μm. Vdd 1V.

As a summary, SOTA in situ doped FDSOI transistors exhibit a good thermal stability from junction (no dopant deactivation), mobility and EOT perspectives up to 550°C. To extend further the thermal stability in order to relax the thermal budget limitation beyond 500°C for the top MOSFETs, thermal stability of the metal gate work-function on small transistors and silicide should be improved [6].

Laser anneal

Laser activation for the top CMOS junctions is an opportunity to reduce in-depth thermal diffusion (for bottom MOSFET performance preservation) while keeping high activation level (for top MOSFET performance boost). The laser tool used is an Excimer (wavelength: 308nm, pulse duration: ~200ns).

TCAD simulations

2D TCAD simulations have been performed to determine the best process structure for the source and drain activation and recrystallization without degrading the gate (Fig.15).

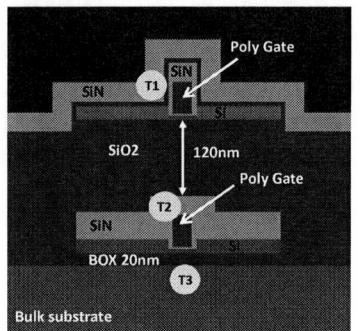

Figure 15: Optimized 3D simulated structure. SiN 30nm, lower BOX 20nm, interlevel oxide 120nm. Lg 30nm.

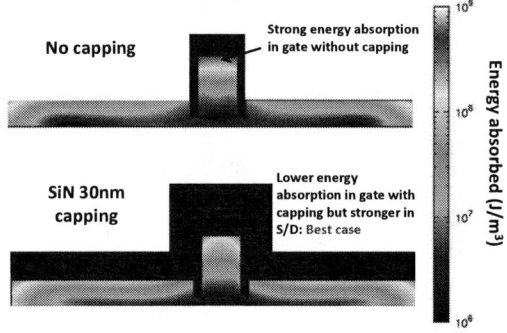

Our custom simulator solves the Maxwell equations with a Finite Difference Time Domain numerical method and provides a picture of the instantaneous laser electric field and of the absorbed power density in the 2D structure. Power distribution is then provided to a transient Poisson equation solver giving the temperature field as a function of time. With this tool, we determined that adding a 30nm SiN capping layer over the top transistors increases the amount and uniformity of the absorbed laser power compared to no capping case (Fig.15).

The best combination of oxide layers thicknesses of Fig.15 from a thermal point of view (with laser pulse tuned to heat the top transistor at 1200°C (melting point)), is the increase of the interlevel oxide thickness from 50nm up to 120nm and reducing the bottom BOX from 145nm down to 20nm (Fig.16). Thus, the lower gate temperature does not exceed 600°C during 200ns (best case) (Fig.16).

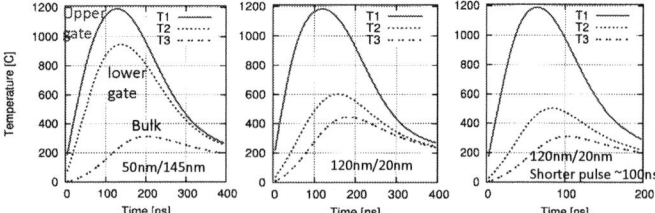

Figure 16: Temperature versus time for several interlevel oxide and bottom oxide layer. Best case 120nm upper/20nm lower with 308nm wavelength laser and ~200ns pulse duration. T1: T° in the gate stack of the upper layer, T2: T° in the gate stack of the lower layer, T3: T° at top of bulk silicon layer. Very best case with shorter pulse 100ns on the right.

In addition, to decrease further the temperature, laser anneal pulse duration could be reduced by a factor 2 (~100ns) to achieve the required 500°C at the bottom level (Fig.16) and/or by the additional introduction of low thermal conductivity materials or mirror layers.

Morphological and electrical results

Figs.17 to 20 show for both NMOS and PMOS planar FDSOI devices the as implanted and after laser anneal TEM pictures for the SiN 30nm capping layer case.

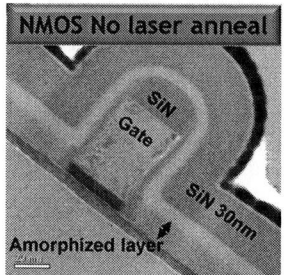 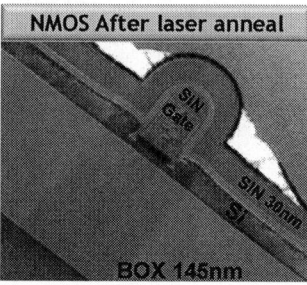

Figure 17: TEM picture as implanted for NMOS planar FDSOI with As LDD implant.

Figure 18: TEM picture after laser anneal for NMOS planar FDSOI with As LDD implant.

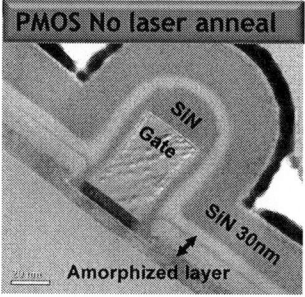 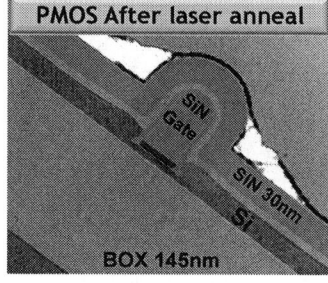

Figure 19: TEM picture as implanted for PMOS planar FDSOI with BF_2 LDD implant.

Figure 20: TEM picture after laser anneal for NMOS planar FDSOI with BF_2 LDD implant.

The whole source/drain silicon film is recrystallized for both NMOS and PMOS without gate morphological degradation. The residual defects observed in the NMOS source and drain can be reduced by using either a lower As implantation energy, Phosphorous, or even a hot implantation (crystalline seed layer increases). Sheet resistance measurements have been done on blanket SOI wafers with a silicon/BOX thickness of around 20nm/25nm respectively corresponding to the source and drain configuration. Several species (As, P, BF_2, B), doses and energies have been implanted and then covered by the 30nm SiN and annealed either by laser with various energy or with 1050°C spike RTA as reference.

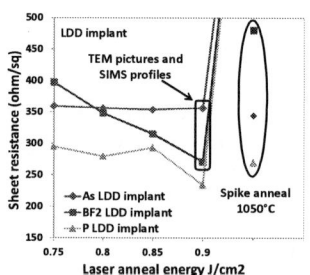 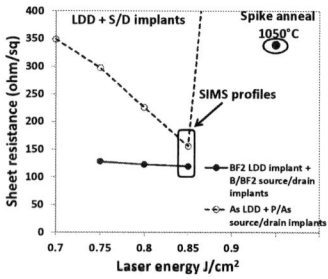

Figure 21: Sheet resistance measurements for As, BF2, P versus laser anneal energy. RTA spike

Figure 22: Sheet resistance measurements for As, BF2, P versus laser anneal energy. RTA spike sheet

sheet resistance is also plotted as resistance is also plotted as reference.
reference.

Fig.21 reveals that for a wide range of energy the sheet resistances for laser anneal are of the same order of magnitude as the RTA ones and even lower when the dose increases (Fig.22).
The TEM pictures for the energy of $0.9J/cm^2$ (Figs.23 and 24) confirm the results observed on FDSOI transistors (quasi no defects for Phosphorus or Boron implants during the crystallization phase and a few defects for As).

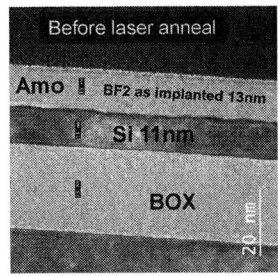 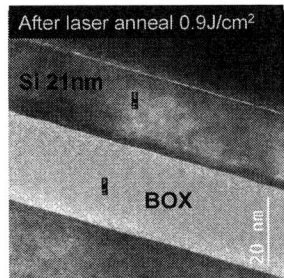

Figure 23: TEM picture as implanted and after laser anneal at an energy of $0.9J/cm^2$ for a wafer implanted with BF$_2$ LDD implant.

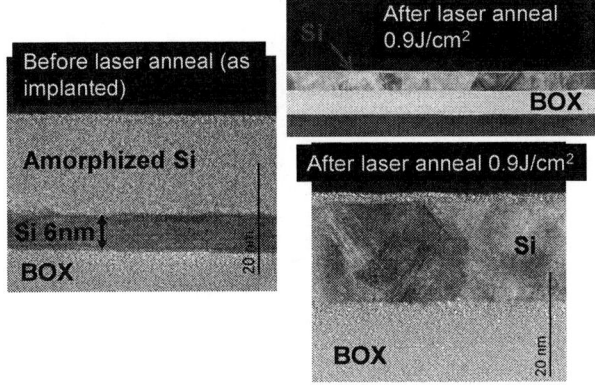

Figure 24: TEM picture as implanted and after laser anneal at an energy of $0.9J/cm^2$ for a wafer implanted with As LDD implant.

The corresponding SIMS profile for the best laser conditions determined from Figs.21 and 22 reveals a low dopant diffusions as compared to RTA anneal and lower dopant segregation at the silicon/BOX interface (Figs.25 to 29).

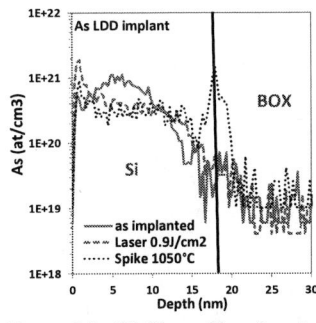 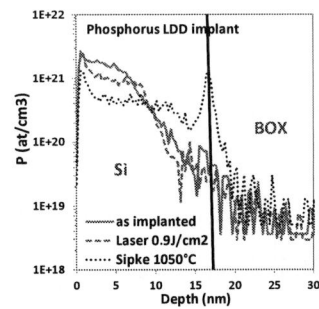

Figure 25: SIMS profiles for As LDD implant: as implanted, after laser anneal at $0.9J/cm^2$ and RTA spike 1050°C.

Figure 26: SIMS profiles for Phosphorus LDD implant: as implanted, after laser anneal at $0.9J/cm^2$ and RTA spike 1050°C.

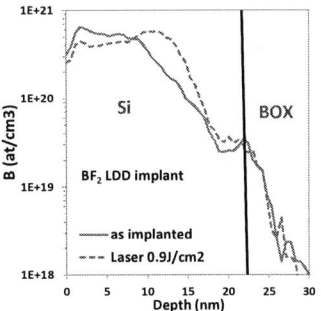 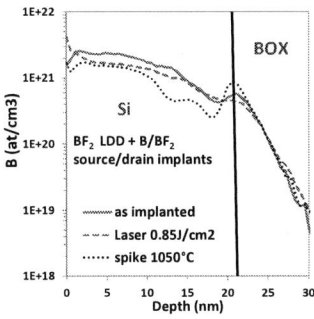

Figure 27: SIMS profiles for BF$_2$ LDD implants: as implanted, after laser anneal at $0.9J/cm^2$.

Figure 28: SIMS profiles for BF$_2$ LDD and B/BF$_2$ source/drain implants: as implanted, after laser anneal at $0.85J/cm^2$ and RTA spike 1050°C.

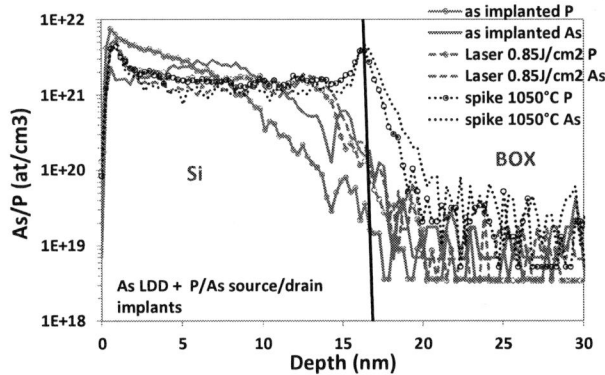

Figure 29: SIMS profiles for As LDD and P/As source/drain implants: as implanted, after laser anneal at $0.85J/cm^2$ and RTA spike 1050°C.

Conclusions

For the first time the maximum thermal budget of in-situ doped source/drain SOTA FDSOI bottom MOSFET transistors is quantified. SOTA in situ doped FDSOI transistors exhibit no Ion/Ioff trade-off degradation, a good thermal stability from junction (no dopant deactivation), mobility and EOT perspectives up to 550°C showing an improved thermal stability as compared to implanted technologies [2]. However the metal gate work-function to overcome the V_T variation on short MOSFETs and silicide stability should be improved in order to relax the top thermal budget limitation beyond 500°C. Laser anneal, by optimizing the 3D layers stack (material and thickness) is a promising solution to not exceed the 500°C limit at the bottom level while ensuring excellent top transistor performance.

References

[1] P. Batude et al, IEDM 2011, [2] C. Fenouillet-Beranger et al, ESSDERC, 2014, [3] A-K. Henning et al, S3S conference, 2013, [4] O. Weber et al, VLSI 2014, [5] L. Brunet et al, VLSI 2008, [6] F. Nemouchi et al, MAM 2014.

Acknowledgements: This work was partially carried out in the frame of the ST/IBM/LETI joint program. It was also partly funded by the French Public Authorities through NANO 2017 program.

Flexible High-performance Nonvolatile Memory by Transferring GAA Silicon Nanowire SONOS onto a Plastic Substrate

Ji-Min Choi[1], Jin-Woo Han[2], and Yang-Kyu Choi[1]

[1]Dept. of Electrical Engineering, KAIST, Korea, [2]Center for Nanotech., NASA, CA, USA,
Email: ykchoi@ee.kaist.ac.kr, Phone: +82-42-350-5477, Fax: +82-42-350-8565

Abstract

Flexible nonvolatile memory is demonstrated with excellent memory properties comparable to the traditional wafer-based rigid type of memory. This achievement is realized through the transfer of an ultrathin film consisting of single crystalline silicon nanowire (SiNW) gate-all-around (GAA) SONOS memory devices onto a plastic substrate from a host silicon wafer.

Introduction

Flexible electronics [1-5] are one of the most promising future technologies. Enabled by their flexibility, their advantages over traditional semiconductor devices on rigid substrates support their widespread application. Organic semiconductors [1,5] are typically utilized for low-cost applications despite their poor device performance and low density, while inorganic semiconductors [2-4] have recently been considered for high-performance flexible devices. Meanwhile, flexible types of memory [5,6] have been developed for fully functional flexible systems. However, the flexible memory devices reported thus far have large feature sizes with poor performance compared to conventional silicon wafer-based types of memory. This situation is caused by material limitations and the limited fabrication methods feasible for use with plastic substrates, such as high-temperature manufacturing processes and processes requiring highly sophisticated equipment. In this paper, we show, for the first time, flexible high-performance inorganic memory transistors that perform excellent memory properties with highly scaled-down dimensions similar to those of traditional wafer-based rigid memory devices.

Fabrication of the Flexible GAA SONOS FETs

The photographs shown in **Fig.1** are of GAA SONOS FETs on a SOI wafer (Fig. 1a), and these devices after being transferred onto a plastic substrate (Fig. 1b). The transferring sequence from a SOI wafer to flexible polyimide tape is schematically illustrated in **Fig. 2**. The process begins with coating a protection layer on the GAA SONOS devices, followed by attachment to a temporary handler. Afterwards, the silicon handle layer underneath the buried oxide (BOX) layer is chemically etched out until the BOX layer is exposed, where the BOX acts as an etch stop layer of Si. A 5% tetramethyl ammonium hydroxide solution at 90°C is used for the Si etching. The remaining ultrathin film layer consisting of the GAA SONOS devices is approximately 1 μm thick. Finally, the ultrathin film is transferred onto a flexible film with a thickness of 35 μm, followed by the removal of the temporary handler and the protection layer. **Fig. 3** depicts the details of the fabrication process used to create GAA SONOS FETs on a wafer [7] and associated TEM images. The starting wafer was a SOI wafer with a top silicon thickness of 110 nm. The top silicon was thinned down to 50 nm by sequential oxidation and wet etching processes. The top silicon layer was patterned to form a SiNW with 50 nm width. The width of the SiNW was further reduced to 30 nm by sacrificial oxidation and removal process. The Si NW was then suspended by wet etching the BOX underneath the SiNW with diluted HF solution. The large size source/drain probe pads served as the anchors, physically supporting the suspended SiNW. Afterwards, the charge trapping trilayer structure was sequentially formed with O/N/O layers having thicknesses of 3/7/12 nm, respectively. A n+ in-situ poly-Si layer was then deposited, and poly-Si/O/N/O layers were patterned simultaneously. The S/D was formed by implantation and activation processes. Then, a PECVD silicon dioxide layer was deposited and the via holes were patterned. Finally, aluminium was deposited and then patterned to form the metal pads for the probing.

Results and Discussions

All electrical characterizations of the GAA SONOS FETs were done with a gate length of 100 nm and a SiNW width and height of 30 nm each. **Fig. 4** shows a comparison of the transfer and output characteristics of the GAA SONOS FETs

978-1-4799-8002-4/14 $31.00 © 2014 IEEE

before and after the transfer process. The stability of the process during transferring steps is confirmed, since no device degradation is observed. **Fig. 5** shows the results of bending analysis of the fabricated flexible devices under different bending conditions. Mechanical stability is also confirmed by the negligible difference in performance when the flexible substrate is bent. The high-performance electrical characteristics of the fabricated flexible devices arise from the traditional CMOS fabrication processes utilized, and the excellent gate controllability due to the GAA structure despite the thick gate dielectric.

As with the preceding typical transistor characterizations, all characterizations of memory functionalities were carried out with GAA SONOS FETs under three different conditions: (i) with the initial devices fabricated on a SOI wafer, (ii) after the devices are transferred onto the flexible substrate, and (iii) with the flexible devices in a bent condition with a bending radius of 1 cm (corresponding to a strain of 0.1%, smaller than the strain at the fracture strength as shown in **Fig. 12b**). Program and erase operations are carried out by the FN tunneling mechanism.

Typical I_D-V_G hysteresis curves of the GAA SONOS for various program and erase voltages (V_{PGM}, V_{ERS}) are shown in **Fig. 6** and **Fig. 7**, respectively. Duration of the program and erase times (t_{PGM}, t_{ERS}) are 10 ms, and 100 ms, respectively. Acceptable NVM characteristics are achieved with the aid of O/N/O dielectric stacks as the charge storage node. A V_T window of about 4 V is achieved with a V_{PGM} value of 14 V and a V_{ERS} value of -14 V. **Fig. 8** and **Fig. 9** show the detailed program and erase transient behaviors, respectively. Moreover, the nonvolatile-memory (NVM) transient characteristics of the GAA SONOS are retained after the devices are transferred onto the polyimide from the host silicon wafer, and even under the bent condition.

To analyze the reliability of the NVM, typical values of V_{PGM} = 14 V, t_{PGM} = 10 ms, V_{ERS} = -14 V, t_{ERS} = 100 ms are used. A V_T window of more than 2.5 V is sustained for retention behavior longer than 10 years, as shown in **Fig. 10**. Endurance characteristics exceeding 10^4 program/erase cycles are obtained as well, as shown in **Fig. 11**. It should be noted that the NVM reliability is also maintained after the GAA SONOS devices were transferred onto the polyimide from the host silicon wafer, and even under the bent condition.

The present flexible NVM performance levels show not only the best performance among all flexible memory types to the best of our knowledge, but they also satisfy the traditional technical specifications of commercialized NVM. Moreover, there is no doubt that current commercially available products with extremely scaled-down devices can be turned into flexible products by adopting the presented transferring techniques.

The influence of plastic thickness on mechanical stability was analytically investigated, as shown in **Fig. 12**. The strain suffered by the devices varies according to the thickness of the flexible substrate. This supports that the optimization of the total film thickness can move ultra-flexible electronics with improved practicality toward realization. A film design in which the devices lie near the neutral mechanical plane would be the best option for the ultra-flexibility.

Conclusions

High-performance GAA SONOS nonvolatile memory was demonstrated on a flexible substrate through a transfer process, and its reliability was confirmed by a bending test. The proposed flexible nonvolatile memory has nanoscale dimensions, and shows the best nonvolatile memory performance to date on flexible substrates. Moreover, this device also satisfies the technical specifications of commercialized non-flexible nonvolatile memory.

References

[1] T. Sekitani et al., "Flexible organic transistors and circuits with extreme bending stability," *Nature Materials*, pp. 1015-1022, 2010.

[2] Y. Zhai et al., "High-Performance Flexible Thin-Film Transistors Exfoliated from Bulk Wafer," *Nano Lett.*, pp. 5609-5615, 2012.

[3] D. Shahrjerdi et al., "Advanced flexible CMOS integrated circuits on plastic enabled by controlled spalling technology," *IEDM Tech. Dig.*, pp.92-95, 2012.

[4] J. P. Rojas et al., "Transformational Silicon Electronics," *ACS Nano*, pp. 1468-1474, 2014.

[5] T. Sekitani et al., "Organic Nonvolatile Memory Transistors for Flexible Sensor Arrays," *Science*, pp. 1516-1519, 2009.

[6] Q.-D. Ling et al., "Polymer electronic memories: Materials, devices and mechanisms," *Prog. Polym. Sci.*, pp. 917-978, 2008.

[7] S.-W. Ryu et al., "One-transistor nonvolatile SRAM (ONSRAM) on silicon nanowire SONOS," *IEDM Tech. Dig.*, pp.633-636, 2009.

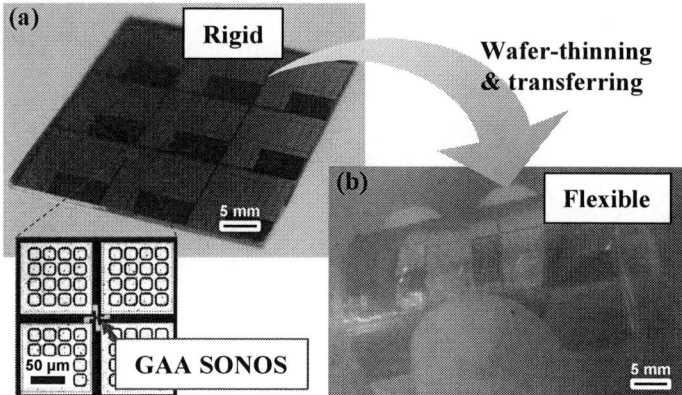

Fig. 1 (a) Photograph of the as-fabricated GAA SONOS FETs on a SOI wafer. **(b)** Photograph of the ultrathin GAA SONOS devices transferred onto a flexible substrate.

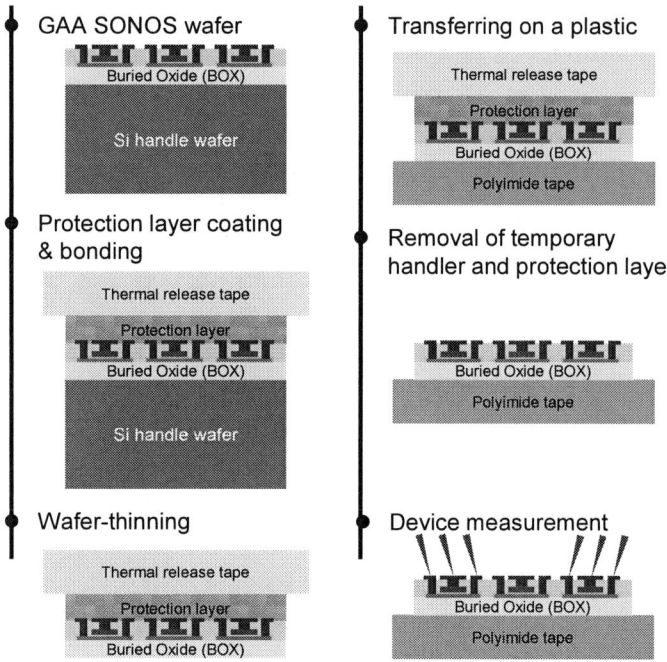

Fig. 2 Schematic illustration of the wafer thinning procedures and subsequent steps for transferring the high performance GAA SONOS onto a flexible substrate.

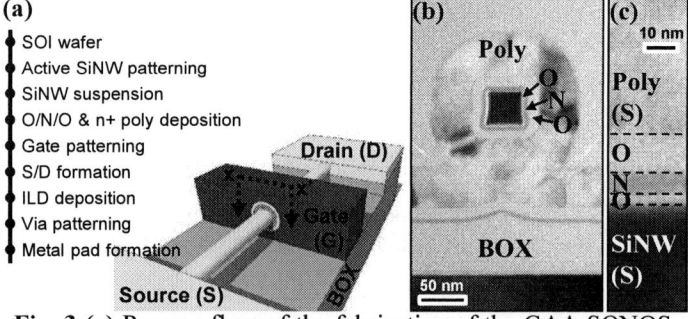

Fig. 3 (a) Process flow of the fabrication of the GAA SONOS and a 3-D schematic diagram of the GAA SONOS before ILD deposition. **(b)** A TEM image of the GAA SONOS along the x-x' direction of Fig. 3(a). **(c)** A TEM image of O/N/O (3/6/13 nm) layers for the nonvolatile memory operation.

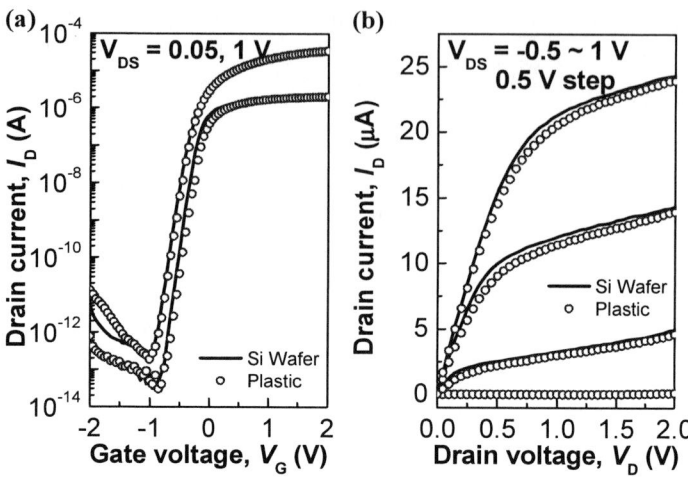

Fig. 4 Representative **(a)** Transfer and **(b)** Output characteristics of the GAA SONOS devices, before and after the ultrathin film transferring process. The changes in the I_D-V_G and I_D-V_D characteristics are negligible, proving the stability of the wafer-thinning and transfer processes.

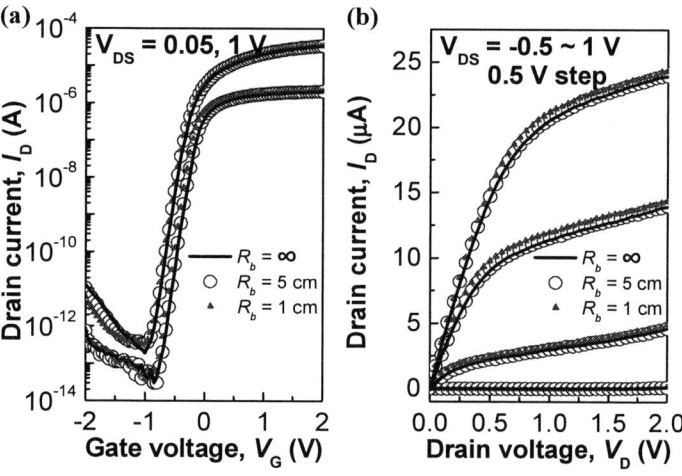

Fig. 5 (a) Transfer and **(b)** Output characteristics of the flexible GAA SONOS devices under different bending conditions. Mechanical stability is confirmed that nearly unchanged device characteristics are shown.

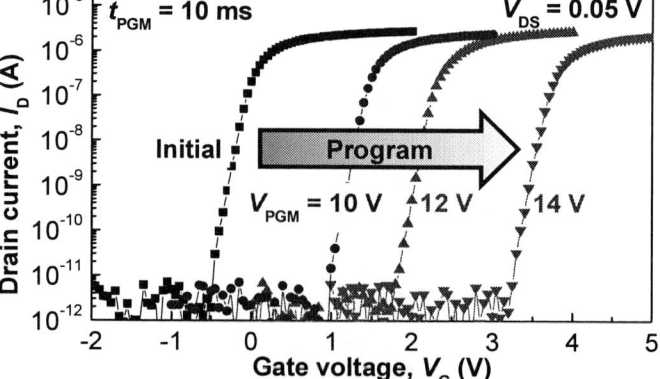

Fig. 6 Typical program characteristics. I_D-V_G curves with various program voltage (V_{PGM}) on the GAA SONOS device.

978-1-4799-8002-4/14 $31.00 © 2014 IEEE

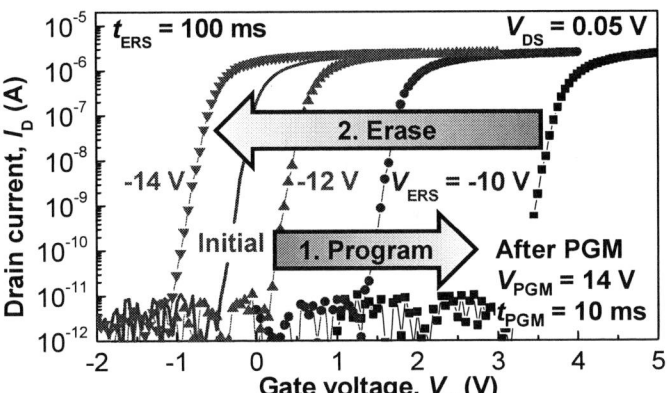

Fig. 7 Typical erase characteristics. I_D-V_G curves with various erase voltage (V_{ERS}) on the GAA SONOS device.

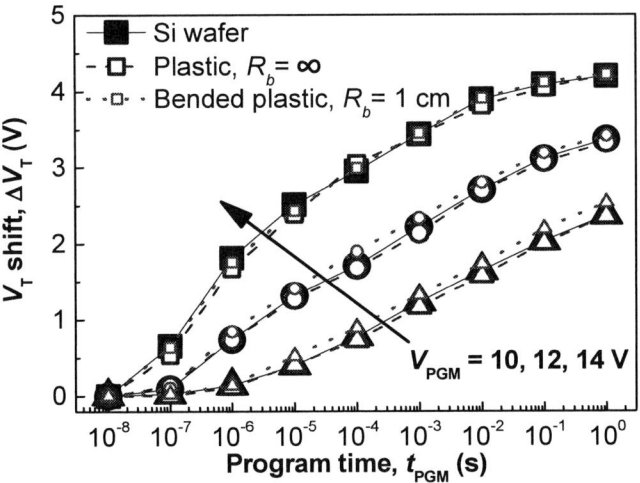

Fig. 8 Speed response of programming for nonvolatile memory operation with various V_{PGM} conditions. Program transient characteristics of the GAA SONOS is preserved after the devices are transferred onto a plastic, and even under bending.

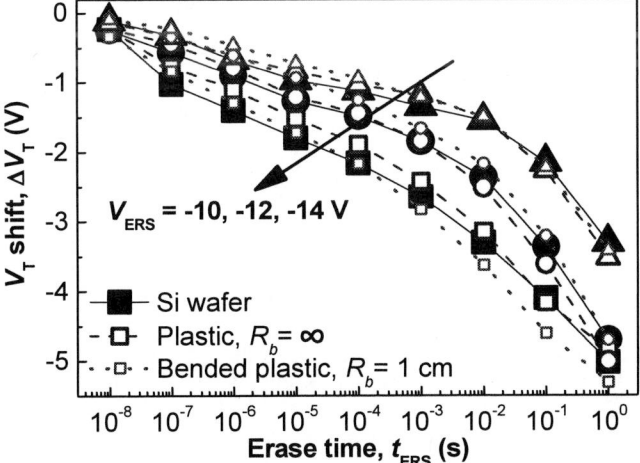

Fig. 9 Speed response of erasing for nonvolatile memory operation with various V_{ERS} conditions. Erase transient characteristics of the GAA SONOS is preserved after the devices are transferred onto a plastic, and even under bending.

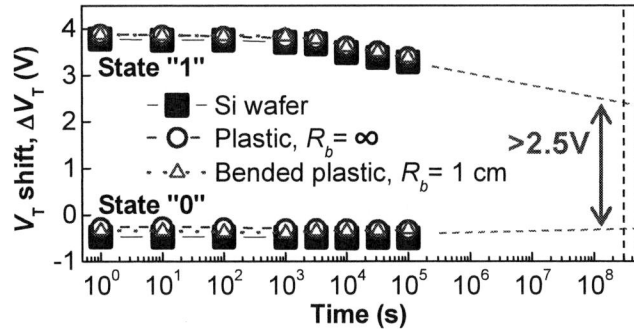

Fig. 10 Retention characteristics of the GAA SONOS. 10 years of retention is guaranteed with V_T window of more than 2.5 V. Retention characteristic of GAA SONOS is preserved after the devices are transferred onto a plastic, and even under bending.

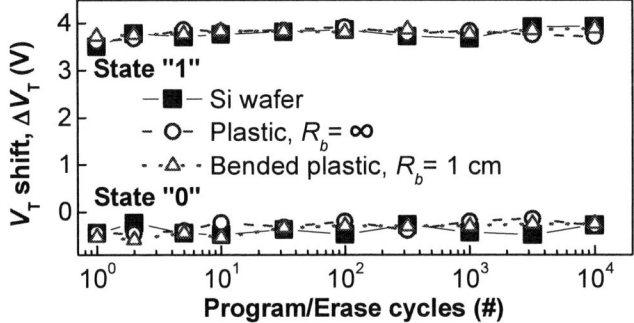

Fig. 11 Endurance characteristics of the GAA SONOS. The V_T window is maintained after 10^4 P/E cycles. Endurance characteristic of GAA SONOS is preserved after the devices are transferred onto a plastic, and even under bending.

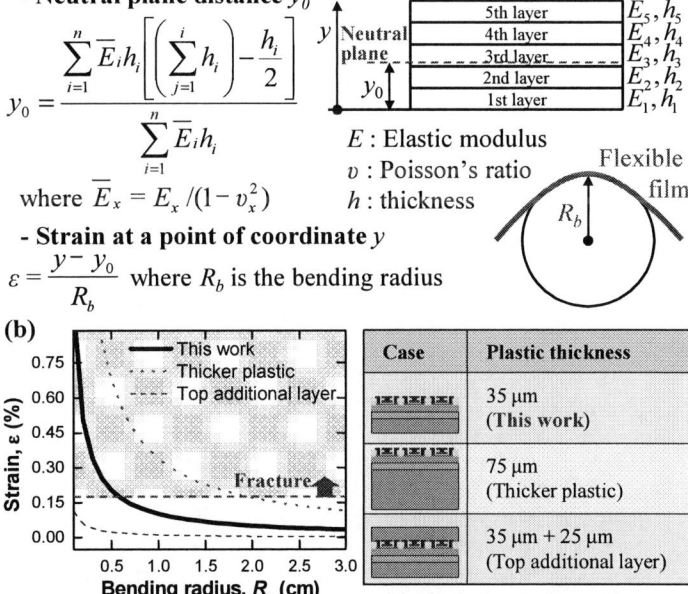

(a)

- **Neutral plane distance y_0**

$$y_0 = \frac{\sum_{i=1}^{n} \overline{E}_i h_i \left[\left(\sum_{j=1}^{i} h_i \right) - \frac{h_i}{2} \right]}{\sum_{i=1}^{n} \overline{E}_i h_i}$$

where $\overline{E}_x = E_x /(1 - v_x^2)$

E : Elastic modulus
v : Poisson's ratio
h : thickness

- **Strain at a point of coordinate y**

$$\varepsilon = \frac{y - y_0}{R_b}$$ where R_b is the bending radius

(b)

Case	Plastic thickness
	35 μm (This work)
	75 μm (Thicker plastic)
	35 μm + 25 μm (Top additional layer)

Fig. 12 (a) Simple modeling of the strain under bending. (b) The strain suffered by the devices versus bending radius (R_b). Mechanical endurance can be controlled by the film thickness engineering of the substrate.

Low power and high density STT-MRAM for embedded cache memory using advanced perpendicular MTJ integrations and asymmetric compensation techniques

K. Ikegami, H. Noguchi, C. Kamata, [†]M. Amano, K. Abe, [†]K. Kushida, [†]E. Kitagawa, [†]T. Ochiai,
N. Shimomura, S. Itai, D. Saida, C. Tanaka, [†]A. Kawasumi, [†]H. Hara, J. Ito, S. Fujita

Corporate R&D Center, Toshiba Corporation, Kawasaki Japan

† Center for Semiconductor Research & Development, Toshiba Corporation, Kawasaki, Japan

Email: kazutaka.ikegami@toshiba.co.jp

Introduction

Since it has been difficult to increase clock frequency of processors due to power budget, there is a trend toward increase in number of processor cores and cache capacities (Fig. 1) to improve the processor performance. According to this trend, there have been two serious issues on the cache memories. One issue is large leakage power of SRAM-based cache (Ex. About 80% of average processor power in a mobile usage case [1]). Another one is large memory area of SRAM especially for last level cache (LLC) like L4 cache. Recently, eDRAM is used to reduce memory area for LLC (Fig. 1). However, gate length of eDRAM is difficult to be reduced less than 40-50 nm, and its power is not small due to frequent refresh (retention time ~ 100µs.) . To reduce the cache power and decrease memory area further at the same time, advanced STT-MRAM based cache has been considered promising from theoretical analysis [2]. However, both low power and high density LLC have not been ever clarified based on a realistic MTJ (magnetic tunneling junction) integration and circuit design. This paper presents solutions for the power and memory density with more advanced STT-MRAM cell technologies by low-temperature process development and novel cache memory architecture based circuit design.

Advanced perpendicular STT-MRAM cell for low power and small memory cell

We analyzed STT-MRAM memory cell layout on an advanced CMOS technology. We assume that 2T-2MTJ cell [3] is used for L2 cache and 1T-1MTJ cell [4] is for LLC since the former operates faster and the latter has larger memory density. We confirmed that single bit memory cell area is determined *not* by "MTJ size" but by "write current (I_w)" for MTJ programming, since large size transistor is necessary for large I_w, as shown in Fig.2. Also, major part of power in MTJ is the write power and it is described as:

$$P = I_w \times t_w \times V_{dd} \ \dots \ (1) . \ (t_w: \text{write time})$$

Therefore, **reduction in I_w** is the most important point to decrease the power and memory area at the same time. To reduce I_w, it is vital to reduce the damping factor, α [4] since critical current (Ic) is proportional to α as shown in Fig. 3 (b). It has been reported that advanced perpendicular (p-) MTJ (30nmϕ) which has smallest α (~0.004) to date [4]. This paper presents more advanced p-MTJ (21nmϕ) having smaller α (~0.003) than that, as shown in Fig. 3 (a). As a result, I_w could be effectively reduced, as shown in Fig.

4 and Fig. 9 (b). It should be noted that I_w strongly depends on t_w. Figure 5 shows comparison of various I_w vs. t_w ever reported, indicating I_w and t_w of this paper have the lowest write power determined by (1).

By comparing memory cell of SRAM (140~200F^2, F is feature size for CMOS technology) and eDRAM (60F^2@22nm CMOS, 37F^2@28nm CMOS, where 37F^2 is area of advanced eDRAM converted in half-pitch 28 nm CMOS) [5]. To replace these for high capacity LLC [2], 37F^2 is a target for embedded STT-MRAM cell area. MRAM cell area on 28 nm logic CMOS technology was calculated, as shown in Fig. 6. Figure 4 shows that when t_w is 2ns, I_w is 50.7 uA. Figure 6 indicates that even 2T-2MTJ cell utilizing presented advanced p-MTJ can be smaller than that of eDRAM when t_w is 2ns. If we have to decrease memory cell area to minimum cell size (24F^2 for 2T-2MTJ cell and 12F^2 for 1T-1MTJ cell) to increase cache capacity further, I_w has to be smaller than 45 uA. In that case, presented advanced p-MTJ can satisfy write current requirements when t_w is 3 to 4ns. Previous processor simulation results [1] showed that such t_w is fast enough for cache access speed of mobile processor. Figure 7 is the relationship between write energy and unit cell area. Based on comparison with average energy of hp-SRAM including leakage current energy, the write power less than 0.3 pW is required to replace SRAM with embedded STT-MRAM [4]. In terms of cell size, required unit cell area of STT-MRAM cell to replace SRAM and eDRAM for LLC is 140F^2 and 37F^2. The presented advanced p-STT-MRAM cells only meet both energy and area requirements and can replace SRAM- and eDRAM-based cache memory.

LT-integration process for MTJ

For integration of advanced p-MTJs, we have developed MTJ-Last process. This process has advantage on low temperature (LT-) integration. It is widely known that magneto resistance (MR) ratio of MgO and CoFeB based MTJ decreases rapidly above 350 °C annealing (Fig. 8) which is lower than sintering temperature as reported in [6]. In other words, thermal budget issue will prohibit integrating high-performance CMOS and low-power MTJ. To resolve this issue, we investigated LSI manufactured by MTJ-Last process, in which MTJ is integrated at the end of metal layer fabrication to avoid adverse effect on MTJ by high-temperature annealing. Figure 9 (a) shows process flow and TEM micrograph of test chip fabricated by MTJ-Last process, and Ic of single MTJ @ 3ns is shown in Fig. 9 (b). Transistors and most of all metal layers are

978-1-4799-8002-4/14 $31.00 © 2014 IEEE

constructed in a 65nm CMOS logic process, where all high-temperature thermal process include sintering (~450° C) is completed. Then p-MTJ, top and pad metal are constructed with low temperature process. By integrating p-MTJ at the end of fabrication process, we can remove adverse effect on MTJ caused by CMOS process. To investigate performance overhead of MTJ-Last process, we evaluated delay increase due to parasitic resistance and capacitance by post-layout simulation (Fig. 10). Circuit structure for measuring delay increase is shown in the inset, which aims to evaluate operation speed of single 1T-1MTJ cell. It was confirmed that delay increase caused by parasitics is within 50 ps and negligible for LLC application.

Measurements of p-STT-MRAM cell switching characteristics

Test chips of 1Mb STT-MRAM fabricated by the MTJ-Last process were measured. Thermal stability factor (Δ) was evaluated from the distribution of H_c of 100 times switching. Δ is evaluated as 63 as shown in Fig. 11. To clarify necessary Δ for cache application, we analyzed access interval of cache data for various workloads. To precisely analyze cache data access interval, we used gem5 system simulator [8] and SPEC 2006 [9] workloads. Fig. 12 shows access interval of cache data for various workloads. As shown in figure, most of cache access interval is in several thousand cycles which means that required retention is around ~µs when processor frequency is around 1 – 10 GHz. In other words, retention of 1 second is long enough for cache memory. We evaluate necessary room temperature Δ considering chip retention failure rate [17]. We consider various memory capacity and chip operating temperature. Figure 13 shows the relationship between necessary Δ and required retention when we require chip error rate is smaller than 3×10^{-6}. If the required retention is 1 second, necessary room temperature Δ is 61 when memory capacity is 10MB and chip operating temperature is 80° C. In other words, presented p-MTJ has enough Δ for cache application. Fig. 14 is shmoo plot of MTJ fabricated by MTJ-Last process. 0.7 V, 4ns switching is demonstrated by MTJ integrated on high-performance CMOS. To evaluate performance improvement of CPU utilizing advanced p-MTJ, we evaluated performance when we apply it to L2 cache memory. We benchmarked with SPEC 2006 [9] workloads and used gem5 system simulator [8]. Fig. 15 (a) and (b) shows instruction per cycle (IPC) and energy per instruction (EPI) compared to conventional SRAM-based L2 cache. These results show that IPC and EPI is both degraded for previously reported STT-MRAM [19-20]. On the other hand, by utilizing advanced p-MTJ, we can reduce 60 % energy of cache memory with only 7% performance degradation compared to SRAM cache.

Further Reducing Power by Asymmetric Compensation

To further reduce write current, we propose asymmetric compensated MTJ (Fig. 16). We can make write current of AP to P switching much smaller than P to AP switching by asymmetric compensation. To take advantage of this, we implement circuit technique to make P to AP switching

more frequent than AP to P switching [7]. Write data to cache memory is divided into several data units. If number of 1 is larger than 0 in the data unit, we write as-is data with flag 0. If 0 is larger than 1, write data is flipped and set flag 1 (Fig. 17 (a)). When outputting data, original data is restored by taking XOR of every data bit and the flag (Fig. 17 (b)). Figure 18 is shmoo plot of asymmetric compensated MTJ fabricated by MTJ-last process. Write current of AP to P is lowered compared to Fig. 14. Figure 19 shows write energy of asymmetrically compensated MTJ. Write energy of AP to P is lowered by 13%. Figure 20 shows the relationship between data unit size and power. Write energy can be decreased more with a few additional bits. If we use 16 bit data unit size, we can reduce power by 4.6%. Figure 21 shows write energy of symmetric and asymmetric compensation when data unit is 16 bit. Energy reduction becomes larger as data bias become larger. When 0 consists 80 % in data unit, write energy is reduced by 9 %.

Conclusion

This paper presents novel solutions to realize high density memory cell integration and low power at the same time for STT-MRAM based cache memory. In terms of area and power reduction, it is most important to reduce write current of MTJ. We demonstrated lowest write current ever by decreasing damping constant of advanced p-MTJ and showed that both one bit area and write energy can be smaller than SRAM and eDRAM based cache memory. In terms integration, thermal process of CMOS fabrication is known to have adverse effect on MTJ. To integrate high performance CMOS and low power MTJ, we propose MTJ-Last process which can integrate MTJ by low temperature process. We demonstrated high speed and low power operation by the test chip of 1Mb advanced p-STT-MRAM fabricated by MTJ-Last process. Also, we precisely analyze cache access interval and chip-failure rate and showed that presented MTJ has long enough retention for cache memory application. To further reduce power of STT-MRAM based cache memory, we propose lowering write current by asymmetric compensation. With the circuit technique to increase smaller power switching direction, it can reduce cache memory power by 60% with only 7% performance degradation compared to SRAM cache.

Acknowledgement

This study was partly supported by NEDO.

References

[1] S. Fujita et al., ASPDAC, p.6, 2014.
[2] S. H. Kang et al. (Qualcomm) , Symp. on VLSI, 2014 , p.44.
[3] H. Noguchi et al., Symp. on VLSI circuit, P.108, 2013.
[4] E. Kitagawa et al., IEDM Tech. Dig., P.677, 2012.
[5] F. Hamzaoglu et al. (Intel), p.230, ISSCC 2014.
[6] K. Mizumura et al., J. Appl. Phys., 109, 07C711, 2011.
[7] S. Cho et al., MICRO, P. 347, 2009.
[8] The gem5 simulator: http://gem5.org/.
[9] SPEC CPU2006 benchmark suite: http://www.spec.org/.

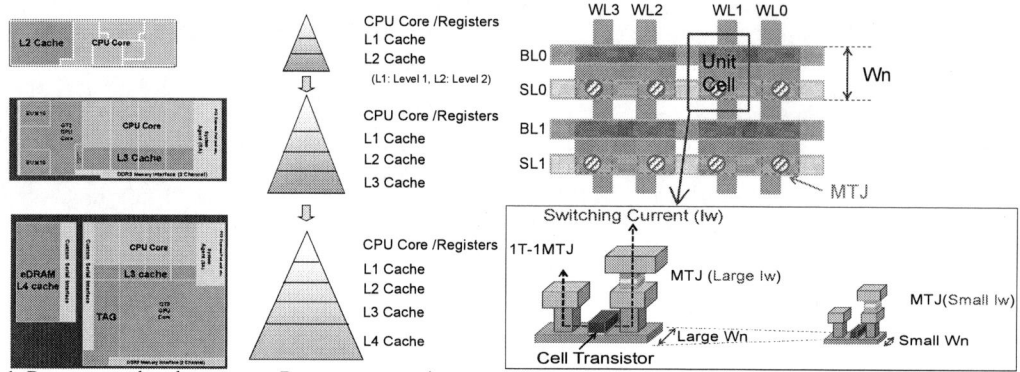

Fig. 1. Processor and cache memory. Recent processor increases memory hierarchy and cache capacity to improve performance.

Fig. 2. Layout of a single bit memory unit cell based on 1T-1MTJ and illustration for scalability by decreasing *Iw*. Note that the memory cell area is not dominated by MTJ size.

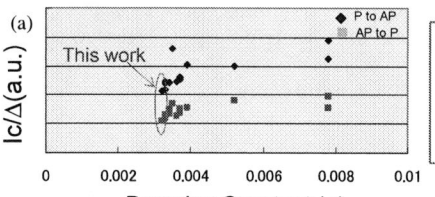

(b)

$$Ic \propto \frac{\alpha}{g(\theta)} \times \Delta \quad \dots (2)$$

Ic: Critical programming current ($I_w \propto I_c$)
g(θ): Spin polarizations efficiency
Δ : Thermal stability factor

Fig.3: Measured damping constant, α, for transition from P (Parallel state) to AP (Anti-parallel state) and the other transition of various p-MTJ structures. MTJs having the lowest α (~0.003) was obtained. As the damping constant is reduced, the programming current is decreased as shown in the equation (2).

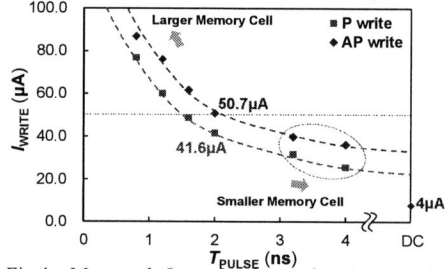

Fig.4. Measured Iw of advanced p-MTJ and its dependence on write time.

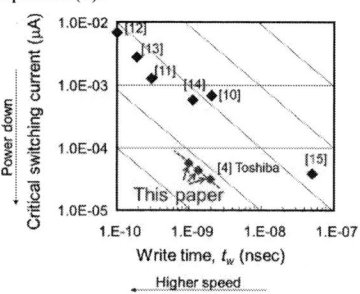

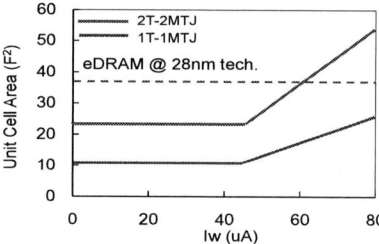

Fig. 7. Comparison of write energy for single bit MTJ reported. Based on comparison with average energy of hp-SRAM including leakage current energy, the write power less than 0.3 pW is required to replace SRAM with STT-MRAM[4]. Required unit cell area of STT-MRAM to replace SRAM and eDRAM for LLC is 140F² and 37F². The presented advanced p-STT-MRAMs only meet all of these requirements.

Fig. 5 Programming time and programming current of various MTJs. Advanced pMTJ is low power switching among various STT-MRAM.

Ref. for Fig. 5. [10] M. Hosomi et al., IEDM Tech. Dig., p.459, 2005. [11] H. Liu et al., APL 97, 242510 (2010). [12] O. J. Lee et al., APL 95, 012506 (2009). [13] H. Zhao et al., J. Phys. D: Appl. Phys. 45, 025001 (2012). [14] D. C. Worledge et al., Appl Phys Lett 98, 022501 (2011). [15] G. Jan et al., Applied Physics Express 5 093008 (2012)

Fig. 6. Iw dependence of unit cell area estimated for 28 nm process. MTJ diameter is assumed to be 30 nm.

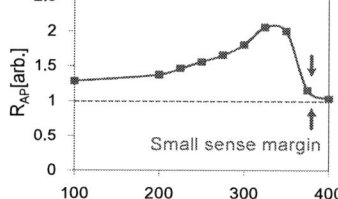

Fig. 8. R_P and R_AP of MgO/CoFeB based MTJ reported in [6]. Sense margin reduces rapidly above 350 ° C annealing.

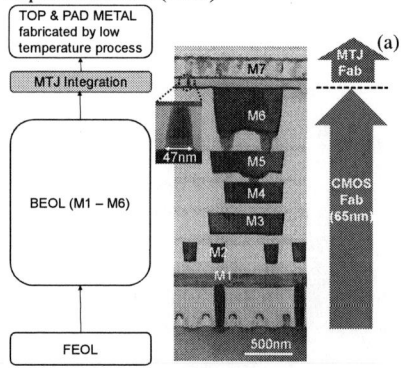

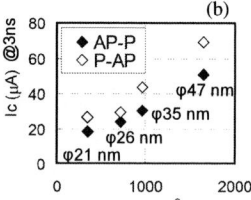

Fig. 9. (a) Process flow and TEM micrograph of test-chip fabricated by MTJ-Last process. MTJ is fabricated according to 65 nm process and can be scaled down with advanced technology.
(b) Ic of single MTJ in the test chip.

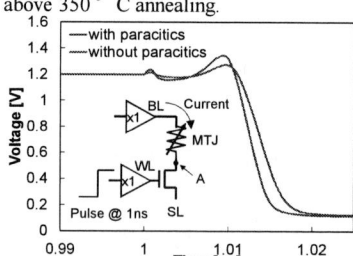

Fig. 10. Delay overhead evaluation of MTJ-Last process. Delay is evaluated by post-layout simulation with circuit shown inset. Delay increase is only 50ps.

978-1-4799-8002-4/14 $31.00 © 2014 IEEE

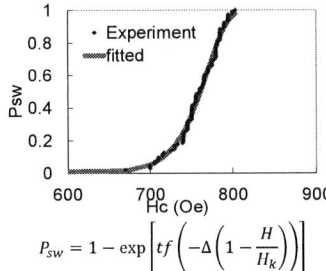

$$P_{sw} = 1 - \exp\left[tf\left(-\Delta\left(1 - \frac{H}{H_k}\right)\right)\right]$$

Fig. 11. Δ of MTJ fabricated by MTJ-Last process. [16]. From the distribution of H_c of 100 times switching, Δ is evaluated as 63. ([16] M. P. Sharrock., IEEE Trans. Mag. 35, 4414, 1999.)

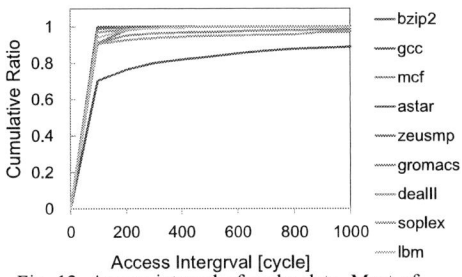

Fig. 12. Access interval of cache data. Most of data is accessed within 1000 cycles

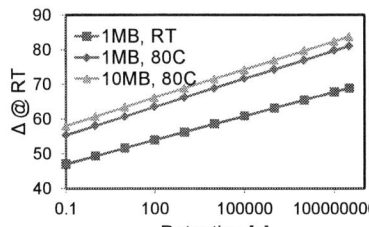

Fig. 13. Necessary room temperature Δ for chip failure rate 3×10^{-6} [17]. Δ=60 is sufficient for cache memory. ([17] R. Takemura et al., JSSC. 45, 849 (2010))

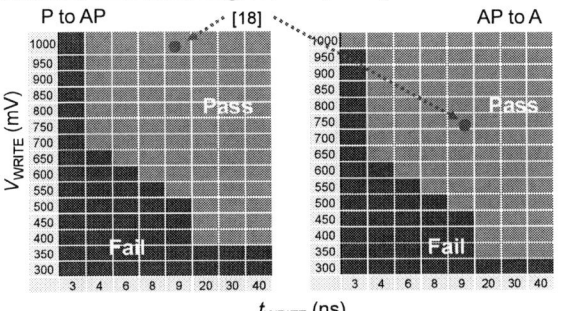

Fig. 14. Shmoo plots of MTJ fabricated by MTJ-Last process. ([18] Y.Lee et al., Symp. On VLSI TSA., p.1, 2013.)

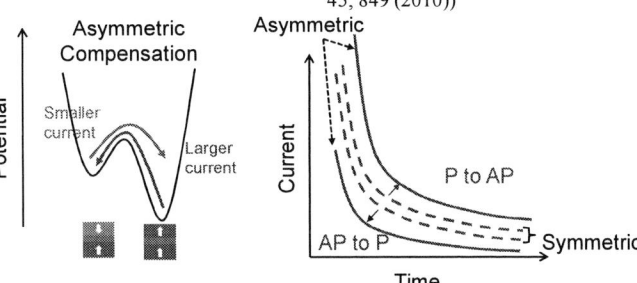

Fig. 16. Asymmetric compensation. By asymmetric compensation, write current of one direction is reduced whereas other direction increases. If we make cache data to switch smaller current direction more often, we can reduce total power.

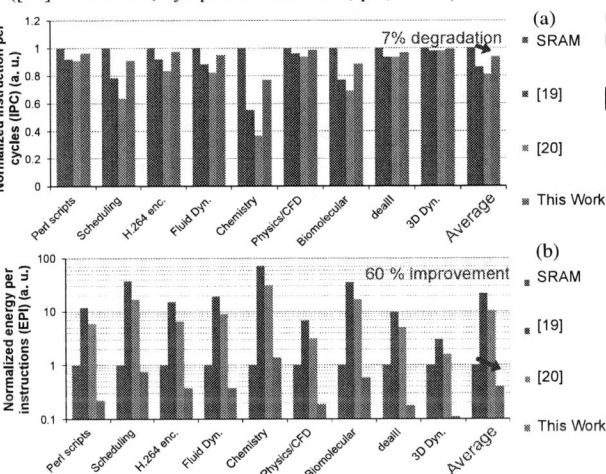

Fig. 15. CPU (a) instruction per cycle and (b) energy per instruction performance of asymmetric compensated MTJ.([19] R. Nebashi et al., ISSCC Dig. Tech Papers, pp.462-463, 2009. [20] K. Tsuchida et al., ISSCC Dig. Tech. Papers, pp.258-260, 2010.)

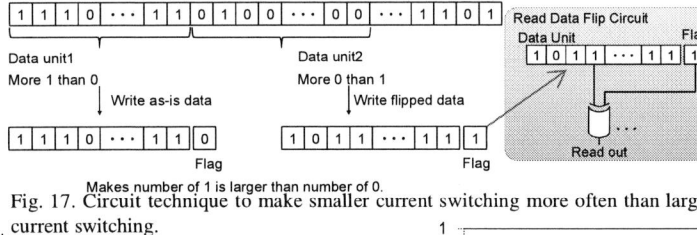

Fig. 17. Circuit technique to make smaller current switching more often than larger current switching.

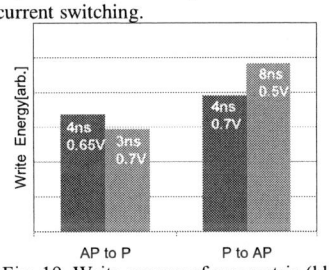

Fig. 19. Write energy of symmetric (blue) and asymmetric (green) compensation.

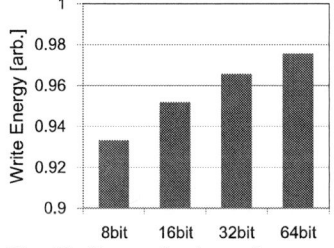

Fig. 20. Data unit size and energy. Energy is smaller when data unit is smaller.

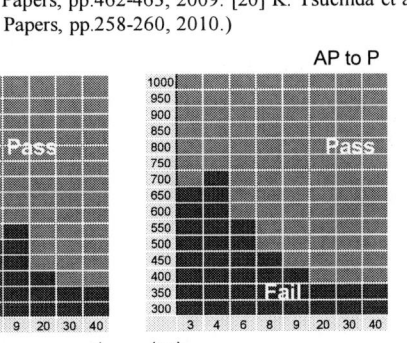

Fig. 18 Shmoo plots of asymmetrically compensated MTJ.

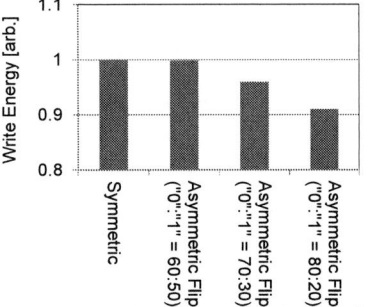

Fig. 21. Write energy of Symmetric and Asymmetric compensation when data unit is 16 bit. Energy reduction become larger as data bias become larger. When 0 consists 80 % in data unit, write energy is reduced 9 %

978-1-4799-8002-4/14 $31.00 © 2014 IEEE

Challenge of MOS/MTJ-Hybrid Nonvolatile Logic-in-Memory Architecture in Dark-Silicon Era

Takahiro Hanyu, Daisuke Suzuki, Akira Mochizuki, Masanori Natsui, Naoya Onizawa, Tadahiko Sugibayashi,

Shoji Ikeda, Tetsuo Endoh, and Hideo Ohno

1) Center for Spintronics Integrated Systems (CSIS), Tohoku University, JAPAN, 2) Center for Innovative Integrated Electronic Systems (CIES), Tohoku University, JAPAN, 3) Laboratory for Brainware Systems, RIEC, Tohoku University, JAPAN, 4) Laboratory for Nanoelectronics and Spintronics, RIEC, Tohoku University, JAPAN, 5) Dept. of Electrical Eng., Faculty of Eng., Tohoku University, JAPAN, 6) Frontier Research Institute for Interdisciplinary Sciences, Tohoku University, JAPAN, 7) NEC Corporation, JAPAN

1. Introduction

In this paper, we present a new architecture-level approach, called "nonvolatile logic-in-memory (NV-LIM) architecture," to solving performance-wall and power-wall problems due to the present CMOS-only-based logic-LSI processors [1]. Figure 1(a) shows a conventional logic LSI chip architecture, where global interconnections between logic and volatile memory modules dominates performance and power dissipation as well as leakage power continuously consumed by volatile memories. In contrast, since nonvolatile storage elements such as magnetic tunnel junction (MTJ) devices are easily distributed over a logic-circuit plane by using a 3D stack structure as shown in Figure 1(b), performance degradation due to intra-chip global wires can be drastically mitigated, which leads to a high- performance, ultra-low-power and highly reliable (or highly resilient) logic LSI.

One of the most useful methods to cut off the leakage power is to use power gating. Figure 2(a) shows a time chart of power dissipation in conventional logic LSI without power gating. If you apply the power gating in the conventional logic LSI, a part of standby power can be eliminated, but two additional operations, "back-up" and "boost-up" procedures, must be performed before and after applying the power gating, which may discourage to apply the power-gating technique as shown in Figure 2(b). In contrast, non-volatility is a good combination of applying the power gating, which ideally eliminates the wasted power dissipation as shown in Figure 2(c).

2. Configuration of nonvolatile logic LSIs:

Figure 3 shows nonvolatile VLSI processor architecture, where a high-density and high-speed MRAMs and nonvolatile flip-flops are used to simply realize a nonvolatile logic LSI. When you merge a part of nonvolatile on-chip memory into logic-circuit modules, it would be possible to improve the performance of the nonvolatile logic LSI. In the following description, some concrete design examples using MTJ-based NV-LIM architecture such as nonvolatile FPGA [2-6],

nonvolatile ternary content-addressable memory (TCAM) [7,8], and nonvolatile Micro Control Unit (MCU) [9] are demonstrated and their usefulness is discussed.

3. Design example of NV-LIM-based FPGA

Since multi-input nonvolatile MTJ-based lookup table (LUT) circuit requires series-connected MOS transistors and MTJs, their characteristic variations becomes critical. In order to mitigate the variation effect, we are inserted redundant MTJs to adjust the operating point of the LUT function. Figure 4 shows a block diagram and the fabricated chip of the proposed NV-LUT circuit. Figure 5 shows switch block of NV-FPGA. The use of asymmetric write-current characteristic of MTJs makes the effective area shrunk by sharing a single large-size MOS transistor. Figure 6 shows an MTJ-based NV-FPGA test chip with its chip features.

4. Design example of NV-LIM-based TCAM

Figure 7 shows chip photomicrograph of an MTJ-based nonvolatile TCAM test chip which is used as a high-speed index search engine. Robustness against soft error due to particle strike is one of the most important properties in the practical applications. Figure 8 shows a single-event transient (SET)-resilient NV-TCAM word circuit. Table 1 summarizes the performance comparison of TCAMs. The proposed NV-TCAM is implemented compactly with maintaining the variety of soft- error resilience.

5. Design example of NV-MCU

Figure 9 shows a chip photomicrograph of the proposed NV-MCU and its features, where all the data storages including on-chip RAM and flip-flops are nonvolatile and then it is possible to quickly cut off (or re-boot up) the power supply. Figure 10 shows the variety of the power dissipation in the proposed NV-MCU.

Acknowledgments

A part of this research was supported by JSPS FIRST Program, MEXT, and CIES.

References

[1] T. Hanyu, SPIN, vol. 3, no.4, 1340014, 2013.

[2] D. Suzuki, et al., J. Appl. Phys. vol. 111, 07E318, 2012.

[3] D. Suzuki, et al., J. Appl. Phys. vol. 115, 17B742, 2014.

[4] D. Suzuki, et al., IEEE Trans. Magn. 2014 (in press).

[5] D. Suzuki, et al., IEICE ELEX, 20130772, 2013.

[6] M. Natsui, et al. ISSCC, 194-195, 2013.

[7] S. Matsunaga, et al., Symp. VLSI-C, 106-107, 2013.

[8] N. Onizawa, et al., ASYNC, 1-8, 2014.

[9] N. Sakimura, et al., ISSCC, 184-185, 2014.

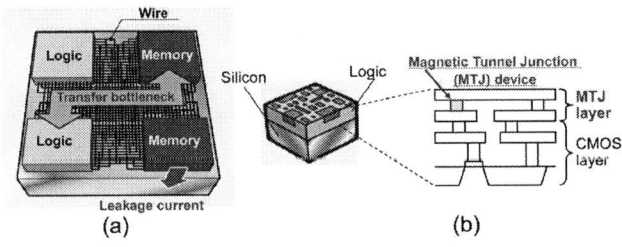

Fig.1: Comparison of logic-LSI architectures; (a) conventional, (b) nonvolatile logic-in-memory.

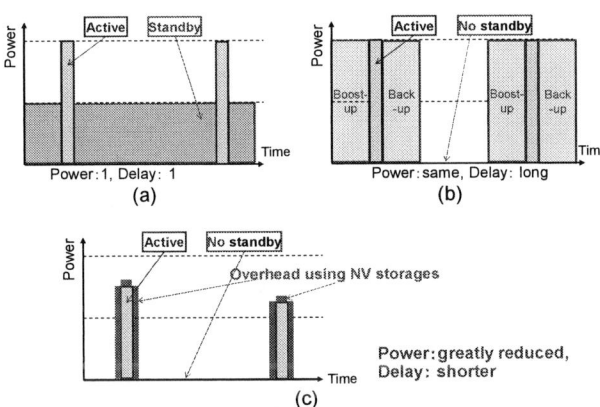

Fig.2: Combination of power-gating and nonvolatile logic techniques; (a) Conventional CPU without power gating, (b) Conventional CPU with power gating, (c) NV-LIM CPU with power gating.

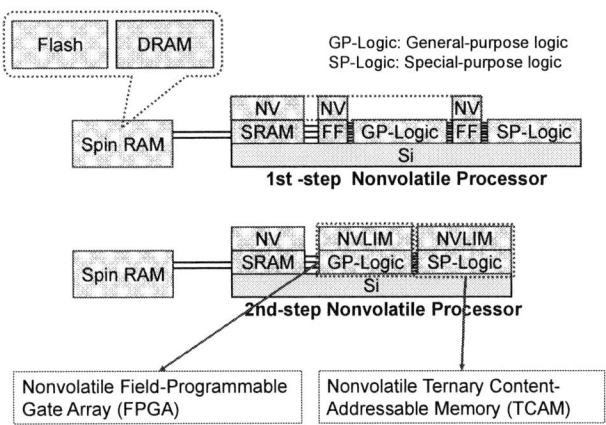

Fig. 3: Configuration of nonvolatile logic LSIs.

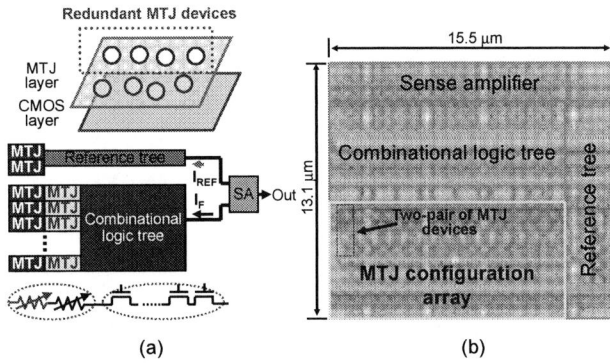

Fig. 4: Multi-input NV-LUT; (a) Schematic diagram, (b) Chip Photo of 4-input NV-LUT circuit.

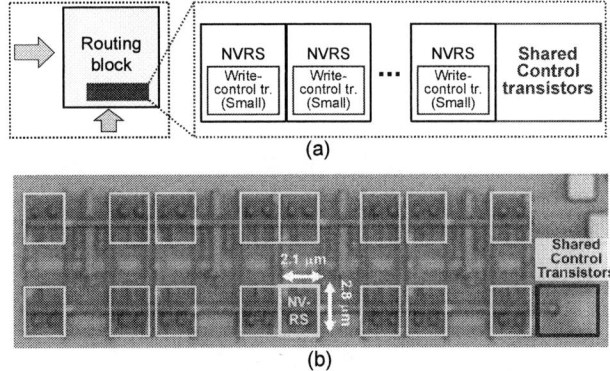

Fig. 5: Nonvolatile routing switch (NVRS) using shared control transistor structure; (a) Block diagram. (b) Chip photo of NVRS array.

978-1-4799-8002-4/14 $31.00 © 2014 IEEE

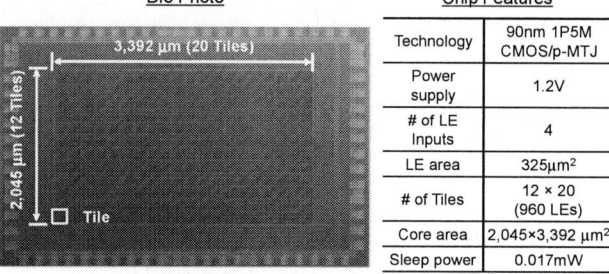

Chip Features	
Technology	90nm 1P5M CMOS/p-MTJ
Power supply	1.2V
# of LE Inputs	4
LE area	325μm²
# of Tiles	12 × 20 (960 LEs)
Core area	2,045×3,392 μm²
Sleep power	0.017mW

Fig. 6: Chip photo of NV-FPGA with 960 NV-LUTs.

Table 1: Performance comparison of TCAMs.

	Synchronous (CMOS)	Extension of ASYNC'13 (CMOS)	Proposed (CMOS/MTJ)
Cycle time [ns]	3.398	N/A	3.410
(search delay [ns])	1.699	N/A	2.330 (data)
(precharge delay [ns])	1.699	N/A	1.060 (spacer)
Energy metric [fJ/bit/search]	0.580	N/A	0.686
TCAM cell	24T	48T (2x 24T)	20T-4MTJ
SEU tolerant in cell	Yes	Yes	Yes
SEU free in cell	No	No	Yes (<10⁻⁵⁰)
Delay-variation tolerant	No	Yes	Yes
Soft-error detection	No	No	Yes

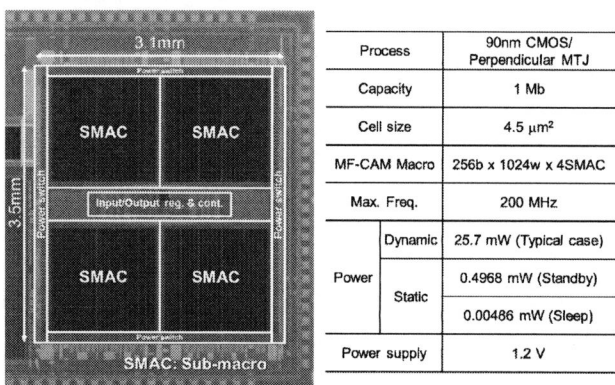

Process	90nm CMOS/ Perpendicular MTJ
Capacity	1 Mb
Cell size	4.5 μm²
MF-CAM Macro	256b x 1024w x 4SMAC
Max. Freq.	200 MHz
Power — Dynamic	25.7 mW (Typical case)
Power — Static	0.4968 mW (Standby)
	0.00486 mW (Sleep)
Power supply	1.2 V

SMAC: Sub-macro

Fig. 7: Chip photomicrograph and feature of the proposed NV-TCAM.

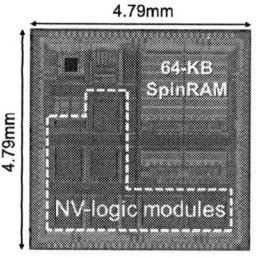

Process	90nm MVT CMOS/MTJ
Supply voltage	3.0V (DV_CC)
	1.0V (V_CORE)
Max clock freq.	20MHz
Dynamic power	145μW/MHz
Static power	1.6-117μW (PG mode)
	<0.1μW (Sleep mode)
Wakeup time	120ns

Fig. 9: Chip photomicrograph and feature of the proposed NV-MCU.

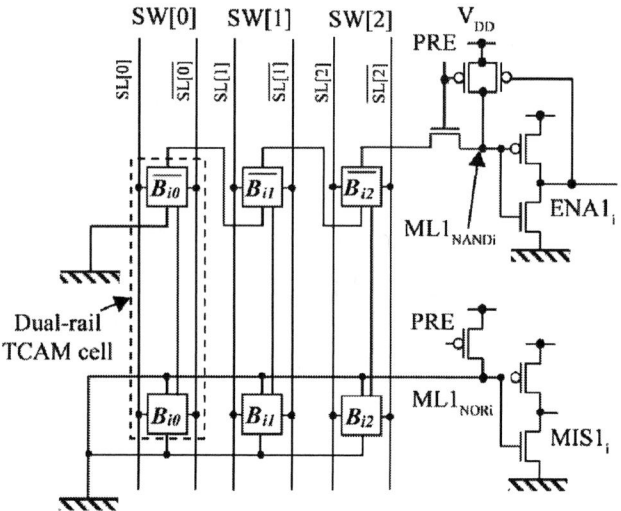

Fig. 8: Asynchronous Dual-Rail NV-TCAM Word Circuit.

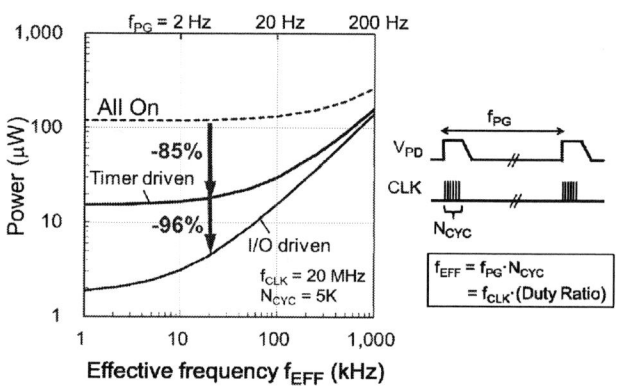

Fig. 10: Estimated power dissipation of the proposed NV-MCU.

Technology and Circuit Optimization of Resistive RAM for Low-Power, Reproducible Operation

D. C. Sekar, B. Bateman, U. Raghuram, S. Bowyer, Y. Bai[+], M. Calarrudo, P. Swab, J. Wu, S. Nguyen,
N. Mishra, R. Meyer, M. Kellam, B. Haukness, C. Chevallier, H. Wu[+], H. Qian[+], F. Kreupl[*], G. Bronner

Rambus, [+]Tsinghua University, [*]Technische Universität München

E-mail: dsekar@rambus.com, Phone: 650-804-8415

Abstract

Low-power, reproducible operation of Resistive RAM (RRAM) requires control of capacitive surge currents during write. We propose a fab-friendly TiN/conductive TaO_x/HfO_2/TiN RRAM with a built-in surge current reduction layer. It reduces worst case write current by 33% and fail bit count by 23x compared to conventional RRAM. A novel circuit to control surge current is demonstrated that improves write current by 40% and endurance by 63%. Switching, endurance and retention data for a 256kb chip with these concepts is presented.

Introduction

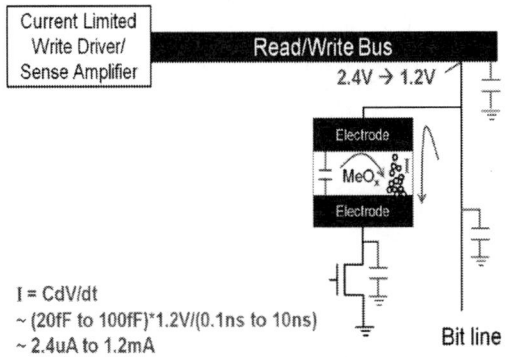

$I = CdV/dt$
~ (20fF to 100fF)*1.2V/(0.1ns to 10ns)
~ 2.4uA to 1.2mA

Figure 1: Capacitive surge currents can be high for RRAM chips.

RRAMs are actively pursued for embedded non-volatile memory due to their low power [1] and ease of integration into a logic process [2] compared to flash memory. An important issue for RRAM chips is control of capacitive surge currents during switching. If capacitive surge currents are high during FORM and SET operations (Fig. 1), filaments are known to become stronger and require higher RESET current to break [2][3]. Reproducibility of switching, which is crucial for building commercial products, is also sensitive to capacitive surge currents.

Figure 2: Die photo of 256kb 1T-1R RRAM chip.

In this work, we present results from a 256kb RRAM chip in in which memory devices, circuits and programming algorithms were developed to lower capacitive surge currents (Fig. 2).

TiN/Conductive TaO_x/HfO_2/TiN RRAM

Our proposed RRAM device uses *fab-friendly* materials such as TiN, HfO_2 and conductive TaO_x. Fig. 3 shows the device structure and its measured I-V curve. The conductive metal oxide electrode made of TaO_x plays an important role in device operation, as will be shown later in this paper.

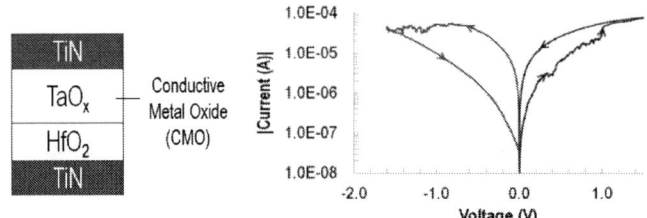

Figure 3: Our proposed device (left), and its measured I-V curve (right).

Fig. 4 shows a comparison of our proposed RRAM with well-studied TiN/Ti/HfO_2/TiN RRAM [4]. 1kb 1T-1R arrays were used for experiments. Median ON state resistance of both RRAM devices is similar in Fig. 4. However, for the same programming algorithm, the TiN/conductive TaO_x/HfO_2/TiN RRAM has $8.6k\Omega$ worst case ON resistance (R_{on}) compared to $5.8k\Omega$ for TiN/Ti/HfO_2/TiN RRAM - a 50% increase. This is important since RESET current is normally a function of R_{on} ($I_{reset} \sim V_{reset}/R_{on}$). Thus, the RRAM with the conductive metal oxide electrode has approximately $5.8k\Omega/8.6k\Omega = 33\%$ lower worst case RESET current. Cell size is a function of the worst case RESET current, so this is valuable. Fig. 4 shows that the TiN/Ti/HfO_2/TiN RRAM has ~88.5% bit yield, compared to ~99.5% bit yield for our proposed RRAM. This represents a 23x difference in fail bit count. A majority of the failed bits in Fig. 4 come from cells having $R_{on} < 15k\Omega$ that have higher RESET currents and see higher temperatures on their oxygen vacancy filaments. Voltages and data retention for the two RRAM devices are quite similar (Fig. 4).

The comparison between these two device types was kept as fair as possible – both device types had their structure and process flow optimized rigorously and the programming algorithm was kept the same, as previously noted. By changing the programming algorithm for TiN/Ti/HfO_2/TiN, improved bit yields were obtained. However, such algorithms degraded power or write time of TiN/Ti/HfO_2/TiN further compared to TiN/conductive TaO_x/HfO_2/TiN, or resulted in write times that were unrealistic for practical applications.

978-1-4799-8002-4/14 $31.00 © 2014 IEEE

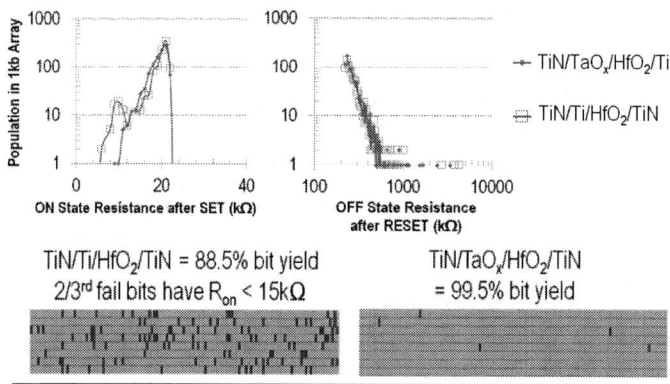

TiN/Ti/HfO₂/TiN = 88.5% bit yield
2/3ʳᵈ fail bits have R_{on} < 15kΩ

TiN/TaOₓ/HfO₂/TiN
= 99.5% bit yield

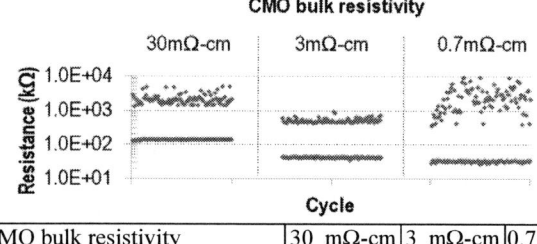

1kb arrays, 32 pulse write	Well-studied RRAM TiN/Ti/HfO₂/TiN [4]	Our proposal TiN/TaOₓ/HfO₂/TiN
FORM Voltage	4V	4V
SET Voltage	1.75V	1.75V
RESET Voltage	1.8V (median)	1.6V (median)
Retention	Similar	

Figure 4: Comparison of the proposed device to well-studied TiN/Ti/HfO₂/TiN [4] RRAM. 1kb 1T-1R arrays are used for experiments. Bit yield is defined as percentage of cells in the 1kb array that pass 5 cycles. In the bit maps, ■ = failed bits, ▨ = passed bits.

Understanding Switching Improvement for RRAMs with Conductive Metal Oxide Electrodes

We observed in Fig. 4 that for a typical programming algorithm, RRAMs with a conductive metal oxide (CMO) electrode provided ~33% lower worst case RESET current and 23x less fail bit count compared to conventional RRAM devices. The following discussion explains the reasons.

A. Reduction in Capacitive Surge Current

CMO bulk resistivity

CMO bulk resistivity	30 mΩ-cm	3 mΩ-cm	0.7 mΩ-cm
ON state resistance	166kΩ	44kΩ	35kΩ
Retention test (20min @ 110°C)	Failed	Passed	Passed

Figure 5: Higher Conductive Metal Oxide (CMO) resistance gives higher ON state resistance, indicating CMO resistance reduces surge current during SET.

Fig. 5 shows ON and OFF state resistances measured during cycling for three different CMO resistivity values. A higher resistivity CMO causes higher ON resistance. This indicates that a higher CMO resistance reduces current flowing through the RRAM during SET, leading to weaker filaments that can switch well but have degraded retention (Fig. 5). A device designer can therefore pick an optimal CMO resistivity that gives a good tradeoff between switching and retention properties. Simulations in Fig. 6 indicate that significant voltage can drop across the CMO during SET. This confirms that the CMO resistance and the IR drop across it play an

important role in device operation. Tighter ON state resistance distributions in Fig. 4 are also likely due to the CMO reducing capacitive surge currents during SET.

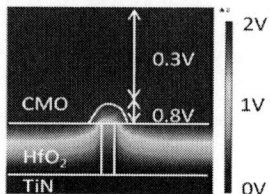

Figure 6: Simulations show current crowding at the CMO-filament interface impacts CMO voltage drop significantly.

B. Improved Thermal Efficiency

Fig. 6 shows an interesting property. A substantial amount of the CMO voltage drop occurs due to current crowding resistance as current spreads from the narrow filament to the wide top electrode. This current crowding causes Joule heating and temperature increases in the RRAM (Fig. 7). In addition, the higher thermal resistance of CMOs compared to metal electrodes prevents heat flow to the surroundings and leads to higher temperatures. RRAMs with conductive metal oxide electrodes can therefore provide several times higher temperature than conventional RRAM devices for the same write current, as shown in Fig. 7. This is advantageous, since ion motion and RRAM switching require high temperatures on filaments, and one can get these high temperatures with lower write currents when RRAMs have conductive metal oxide electrodes. This could be one reason why the RESET voltage in Fig. 4 does not increase much despite the IR drop one would expect from the CMO resistance.

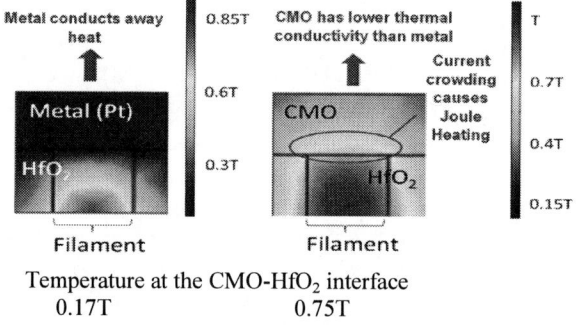

Temperature at the CMO-HfO₂ interface
0.17T   0.75T

Figure 7: Simulations show that a device with a CMO has higher temperature for the same write current, allowing easier ion motion.

To understand conduction mechanism, data from 128 TiN/TaOₓ/HfO₂/TiN RRAM cells was analyzed. Fits of the data to Schottky conduction models [5], Nearest Neighbor Hopping [5], Variable Range Hopping [6] and Frenkel Poole conduction [5] were attempted. Nearest Neighbor Hopping theory provided the best explanation for OFF state conduction while Variable Range Hopping theory provided the best explanation for ON state conduction, as shown in Fig. 8. The observation that hopping is the conduction mechanism indicates HfO₂ film optimization is crucial for improving our RRAM device.

978-1-4799-8002-4/14 $31.00 © 2014 IEEE

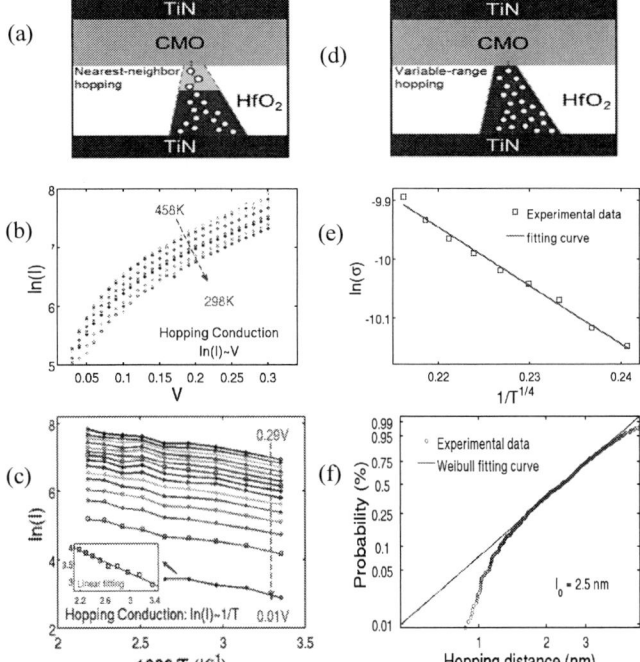

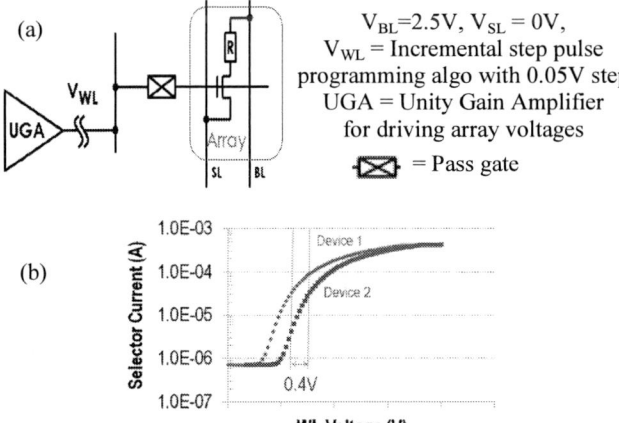

Figure 8: (a) Nearest Neighbor Hopping (NNH) model for OFF state.
(b) Measured OFF state I-V curves fit NNH model (c) Measured temperature trends fit NNH model (d) Variable Range Hopping (VRH) model for ON state (e) Measured temperature trends for ON state conductivity fit VRH model (f) Extracted hopping distance is 2.5nm.

Circuits to Reduce Capacitive Surge Currents in RRAMs
A common circuit technique to reduce capacitive surge current is shown in Fig. 9(a). Gate voltage of the select transistor is incremented so that the select transistor resistance limits surge current during SET. Fig. 9(b) reveals the challenges with this approach. To get a write current of ~100µA, the select transistor needs to be biased near its threshold voltage (V_t). V_t variation and exponential change of current near V_t make current control difficult.

Figure 9: (a) A common circuit approach to reduce capacitive surge currents during SET and FORM (b) Current of different transistors in an array can be substantially different near the threshold voltage.

To resolve this, a novel circuit was developed. One component of the circuit is shown in Fig. 10. A cell-level current mirror is used to limit current flow and provide immunity to temperature variations. Any die-to-die or wafer-to-wafer V_t variation impacts both transistors T1 and T2 in the current mirror, so their impact is minimized compared to Fig. 9(a). However, since T2 is a small-size memory FET, V_t mismatch between T1 and T2 could be a challenge.

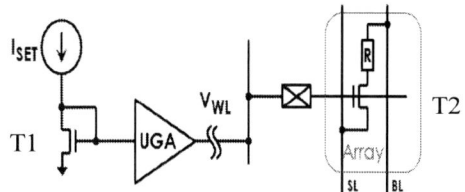

V_{BL}=2.5V, V_{SL} = 0V, I_{SET} = Incremental step pulse programming algorithm with 10µA step
Figure 10: Current mirror circuit to limit T2's current to I_{SET}.

To tackle mismatch between T1 and T2, transistors T3 and T4 are added to form the capacitive surge current reduction circuit in Fig. 11. The Source Line transistor T4 is shared between all cells on a Source Line (SL) and can be sized big to reduce mismatch effects. T2 reduces surge currents due to BL capacitance.

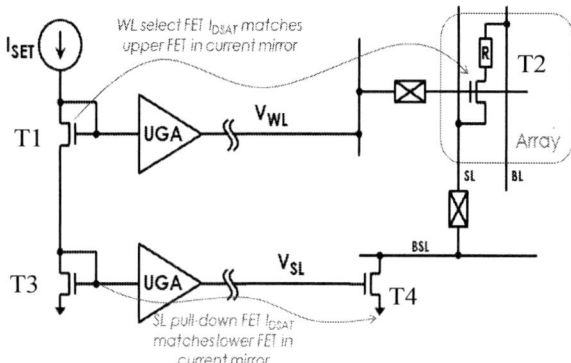

V_{BL}=2.5-3V, V_{SL} = 0V, I_{SET} = Incremental step pulse programming algorithm with 10µA step
Figure 11: Proposed capacitive surge current reduction circuit for RRAM.

Fig. 12 indicates a 40% reduction in RESET current is obtained with a circuit implementation of Fig. 11 for a 256kb array. In addition, reduction of capacitive surge currents improves control of oxygen vacancy filaments and gives a 63% increase in endurance for the 256kb array (Fig. 13).

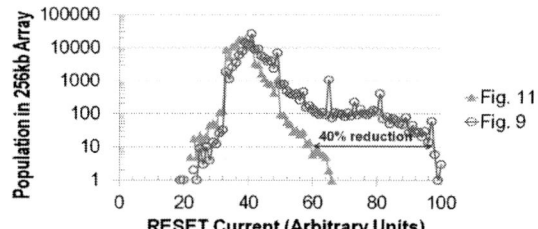

Figure 12: Approximate RESET current distributions for a 256kb array with two current compliance circuits.

978-1-4799-8002-4/14 $31.00 © 2014 IEEE   659

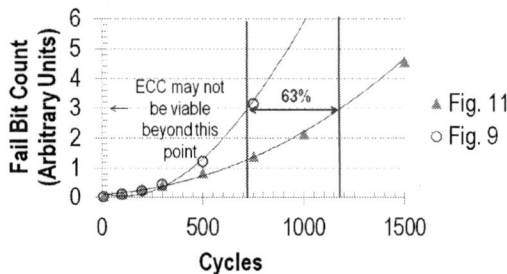

Figure 13: Endurance for a 256kb array with two current compliance circuits.

256kb Chip Results

Our 256kb 1T-1R RRAM chip (Fig. 2) has the memory device shown in Fig. 3 and the capacitive surge current reduction circuit shown in Fig. 11. In addition, the 256kb chip has state machines, sense amplifiers and various analog blocks. The chip's proper operation is demonstrated by the written bit patterns shown in Fig. 14.

Figure 14: Various bit patterns demonstrated in the 256kb chip, indicating it works well. ▓ ='1', ■ ='0'.

As Fig. 15(a) indicates, SET voltages are less than 2.25V, with bits in the 256kb array requiring an average of fourteen 40ns SET pulses. RESET voltages are less than 3V, with a large percentage of bits switching in the first 2V pulse, as shown in Fig. 15(b)

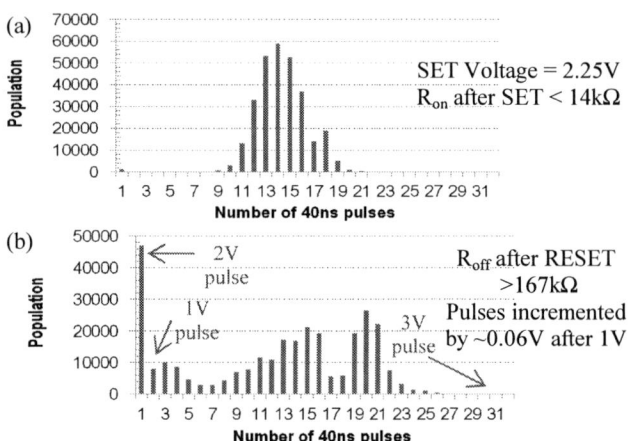

Figure 15: Number of pulses after which bits in the 256kb array finish (a) SET (b) RESET.

Endurance is greater than 1000 cycles, as shown in Fig. 16. This is sufficient for many applications such as code storage. Some degradation is occurring during switching since the number of pulses required for write increases with cycling. Fig. 16 also indicates tail bits in the 256kb array meet 70°C 10 year data retention specs.

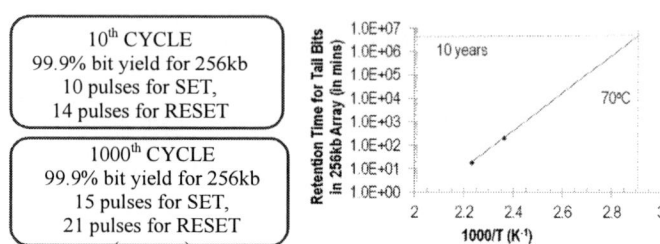

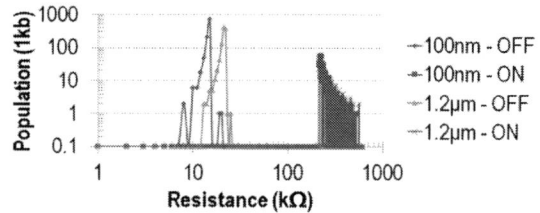

Figure 16: 256kb chip results as a function of Cycling (left) and Temperature of Retention Bake Test (right).

Scaling Trends

Fig. 17 shows scaling behavior of the TiN/conductive $TaO_x/HfO_2/TiN$ RRAM by analyzing two very different device sizes. The switching behavior is similar or improved at small dimensions.

1kb arrays	1.2 μm	~100 nm
RESET Voltage (median)	1.7V	1.75V
SET Voltage (median)	<3V	<3V
ON-OFF Ratio (worst case)	10x	15x
Retention	x	6x

Figure 17: Scaling trends for 1kb arrays.

Summary

To build large RRAM arrays suitable for practical applications, it is important to have reproducible RRAM switching and low write power. A fab-friendly TiN/conductive $TaO_x/HfO_2/TiN$ RRAM was proposed and demonstrated in this paper. For the same programming algorithm, it improved worst case write current by 33% and fail bit count by 23x compared to well-studied TiN/Ti/HfO_2/TiN RRAM. Our studies show this improvement is because the conductive metal oxide layer serves as a built-in capacitive surge current reduction layer and because it improves thermal efficiency. A circuit to reduce capacitive surge current was proposed, which improved write current by 40% and endurance by 63%. The proposed RRAM device and surge current reduction circuit were demonstrated on a 256kb chip. Results in this paper indicate technology and circuit optimizations can reduce capacitive surge currents in RRAM circuits and improve device operation quite significantly.

References

[1] R. Aitken, et al., Symp. on VLSI Technology, 2014.
[2] H. S. P. Wong, et al, Proceedings of the IEEE, 2012.
[3] Y. Sato, et al., IEEE Transactions on Electron Devices, vol.55, no.5, pp.1185,1191, May 2008
[4] H. Y. Lee, et al., Intl. Electron Device Meeting, 2008.
[5] F-C. Chiu, Advances in Materials Science and Engg., 2014.
[6] N. F. Mott, E. Davis, Oxford University Press, 1979.

Variability-tolerant Convolutional Neural Network for Pattern Recognition Applications based on OxRAM Synapses

D. Garbin[1,3], O. Bichler[2], E. Vianello[1], Q. Rafhay[3], C. Gamrat[2], L.Perniola[1], G. Ghibaudo[3], B. DeSalvo[1]

[1]CEA, LETI, Minatec Campus, F-38054 Grenoble, France. [2]CEA, LIST, F-91191 Gif-sur-Yvette, France.
[3]IMEP-LAHC, F-38016 Grenoble, France. Contact: daniele.garbin@cea.fr

Introduction

Software implementations of artificial Convolutional Neural Networks (CNNs), taking inspiration from biology, are at the state-of-the-art for Pattern Recognition (PR) applications and they are successfully used in commercial products [1]. However, they require power-hungry CPU/GPU to perform convolution operations based on computationally expensive sums of multiplications. This hinders their integration in portable devices. Some full CMOS-based hardware implementations of CNN have been suggested, but they still require the computation of multiplications [2]. *In this work, we present for the first time to our knowledge a spike-based hardware implementation of CNN using HfO₂ based OxRAM devices as binary synapses. OxRAM devices are chosen for their low switching energy [3] and promising endurance performance [4]. We perform an experimental and theoretical study of the impact of programming conditions at both device and system levels. A complex visual pattern recognition application is demonstrated with a spike-based hierarchical CNN, inspired from the mammalian visual cortex organization. A high accuracy (pattern recognition rate >94%) is obtained for all the tested programming conditions, even if the variability associated to weaker programming conditions is larger.*

Electrical characterization and physical modeling

We tested 1T-1R OxRAM devices, integrated in standard 65nm CMOS technology [5] (Fig.1). The OxRAM device is composed of a 5 nm thick HfO_x layer deposited by ALD embedded between a 10 nm thick Ti and a 35 nm TiN electrodes. OxRAM operating relies on formation/dissolution of an oxygen vacancy rich conductive filament (CF). Typical I-V characteristics are reported in Fig.1. Fig.2 shows the impact of the programming conditions on both the high resistance state (HRS) and the low resistance state (LRS) during pulsed cycling (T=100ns). The LRS is controlled by the compliance current during SET operation ($I_{C\,SET}$), while the HRS is tuned by means of the RESET voltage V_{BL}. The corresponding LRS and HRS Cumulative Distribution Function (CDF) are reported in Fig. 3(a) and (b), respectively. While a bending of LRS CDF is observed when decreasing $I_{C\,SET}$ [6], the width of HRS CDFs corresponding to different reset voltages remains constant [7]. The values corresponding to the mean μ_R, and standard deviation σ_R, are also defined in Fig.3. A 2D multi-Trap Assisted Tunneling model (Fig.4)

based on [8] is developed in order to interpret these results. The model is based on the hypothesis that during SET/RESET operations the CF is only partially disrupted/reformed in a L_{GAP} region (Fig.4) [9-10]. We model only this critical region, since the cell resistance value is mostly determined by its properties (gap length L_{GAP}, cross-section D and Oxygen Vacancy concentration $<V_O>$). We simulate the LRS values assuming that an $I_{C\,SET}$ increase corresponds to a higher $<V_O>$. The increase of the HRS for high V_{BL} values is modeled tuning L_{GAP}, while keeping $<V_O>$ constant (i.e. equal residual defect concentration given by the stack materials properties). The resistance variability is due to the random V_O positions in the L_{GAP} region. Fig.5c shows the experimental and simulated σ_R vs. μ_R obtained from different programming conditions (extracted from Figs. 5a-b). It is apparent that two different resistance regimes exist: the saturation of the σ_R is explained by a constant $<V_O>$. The OxRAM devices show a $>10^8$ endurance for a programming pulse width of 1μs, whatever the forming current. Switching energy can be reduced by shortening the programming pulse width to 100ns. As shown in Fig. 6, with 100ns pulse, a 10^8 endurance can be obtained only with high compliance current during Forming. We argue that this is due to increased V_O reservoir which delays the failure of the device.

OxRAM devices as synapses

Fig. 7 shows how OxRAM devices are used as synapses in the CNN. Multiple binary devices in parallel configuration are used to obtain multi-level synapses. A driver circuit is used to individually program OxRAM devices and propagate signals between neurons. System-level simulations of the CNN were performed with Xnet neural network simulator [11]. Fig. 8 illustrates how the model presented in Fig.4 allows implementing in Xnet the impact of the programming conditions on μ_R and σ_R. Note that when weak SET/RESET programming conditions (low voltage and current) are used the memory window is reduced and LRS σ_R increases.

CNN vs. FC NN

Fig. 9 shows a comparison between Fully Connected NN [12-14] and Convolutional NN topologies. In FC NNs each neuron is connected to every neuron of the upper layer by a large number of synapses. CNNs are inspired from visual cortex structure [15], where neurons are sensitive to small sub-regions of the input space, called receptive fields,

978-1-4799-8002-4/14 $31.00 © 2014 IEEE

exploiting the strong spatially local correlation present in images. In CNNs a small set of synapses (kernel) is shared among different neurons to connect layer N and $N+1$ through a convolutional operation. Fig. 10 illustrates the operation of convolution. The kernel corresponds to a feature that has to be localized in the input image. A peak in the convolution signal means that the feature is present in the input image, and the feature map indicates where the feature is present in the input field. The kernel feature must be learned in an initial phase (Section 5), and then the network can be used in read mode for visual pattern recognition.

CNN architecture: read-mode operation

Fig. 11 shows the CNN designed for the identification of handwritten patterns as digits. We use MNIST database [16] as a test bench. The architecture of the proposed CNN is shown in Fig.11, featuring two cascaded convolutional layers and a classification module (made of two fully connected layers) to associate the extracted feature maps to the 10 digit categories [17]. Fig. 12 shows how the input static images are encoded into Address Event Representation (AER) format, and the propagation of spiking activity through neuron layers. The neuron in the output layer with the highest activity gives the digit category. Fig. 13 shows the proposed implementation of a convolutional kernel with OxRAM synapses and illustrates the read-mode operation. Upon activation of an input neuron, an address decoder is used to dynamically map the kernel synapses to the output neurons that have the input neuron in their receptive field. A spike is then propagated through the synapses to the mapped output Integrate & Fire (IF) neurons, performing multiplication and accumulation operations in one single step, for each input event. Fig 14 shows the results of the network in read mode as a function of LRS/HRS mean value and associated device variability, for a wide range of programming conditions (results obtained in Fig.5). For strong programming conditions (Fig.8 b), a performance of 98.3% recognized digits is achieved, which gets close to the performance obtained with the equivalent static and ideal software implementation of the same CNN topology (98.5%). Even at weak programming conditions (Fig.8a) the network performance is still good (94% recognized digits) while switching energy is reduced from 60pJ to <10pJ.

STDP learning

The resistance of the kernel synapses can be defined in two ways: a) off-line supervised learning using backpropagation algorithm [17], where the LRS/HRS status of each OxRAM device is determined with computer simulations, then discretized and imported in the memory array with a one-time programming operation or b) on-line unsupervised learning, where the LRS/HRS status of the devices is learned in-situ with STDP algorithm presented in Fig. 15. Off-line learning is performed on a static, frame-driven, CNN, which is then mapped into an equivalent event-driven version [18]. Off-line

learning offers the best performance but requires a set of training patterns to perform the learning. STDP learning, on the other hand, allows learning kernel features in an unsupervised way and it is particularly useful when a training dataset is not available. Table 1 reports a comparison of network statistics between CNN and Fully Connected approach for STDP learning. For the same number of connections, the amount of programming events per device is up to 3 orders of magnitude higher for CNN, due to shared weights. Device endurance becomes therefore a critical factor for spike-based learning in CNN, with an estimated endurance requirement of $>10^6$ for a relatively small database like MNIST. On the other hand, it allows reducing the number of synapses (memory array size), resulting in smaller neuron fan-out and parasitic capacitance, thus easier hardware implementation.

Conclusions

We present for the first time a CNN spike-based architecture for pattern recognition using OxRAM devices as synapses for convolutional kernels. The impact of the programming conditions on the resistance levels and variability of OxRAM synapses on the CNN performances has been evaluated. Ultra-low power programming conditions (<10pJ per switching event) allow to achieve a good recognition rate (94%). The implementation of such system, thanks to synaptic sharing, allows a reduction of the memory array size at the expense of increased required switching events with respect to fully connected NN. Thanks to the promising endurance performances ($>10^8$ at $V_{SET}=1V$) and low power consumption OxRAM are good candidates for the CNN implementation.

Acknowledgements

The authors would like to thank STMicroelectronics for providing the OxRAM devices for this study. This work has been partially supported by the LabEx Minos ANR-10-LABX-55-01.

References

[1] LeCun et al., Proc. IEEE, vol.86, no.11, pp.2278,2324, 1998.
[2] Camuñas-Mesa et al.,IEEE J. Solid-State Circuit vol.47, no.2, pp.504,517, 2012.
[3] Goux et al., VLSIT, pp. 159,160, 2012.
[4] Kim Y.-B. et al., VLSIT, pp. 52,53, 2011.
[5] Diokh et al., IRPS, pp. 5E.4.1,5E.4.4, 2013.
[6] Fantini et al., IMW, pp.30,33, 2013.
[7] Benoist et al., IRPS, pp. 2E.6.1,2E.6.5, 2014.
[8] Guan et al., IEEE Trans. ED, vol.59, no.4, pp.1172,1182, 2012.
[9] Ielmini et al, IEEE Trans.ED, vol.58, no.10, pp.3246,3253, 2011.
[10] Wong et al., Proc. IEEE, vol.100, no.6, pp. 1951-1970, 2012.
[11] Bichler et al., NANOARCH, pp. 7,12, 2013.
[12] Suri et al., IEDM, pp. 10.3.1,10.3.4, 2012.
[13] Kuzum et al., IEDM, pp. 30.3.1,30.3.4, 2011.
[14] Suri et al., IEDM, pp. 4.4.1,4.4.4, 2011.
[15] Hubel and Wiesel, J. Physiology, 195 (1), pp. 215,243, 1968.
[16] LeCun et al., http://yann.lecun.com/exdb/mnist/.
[17] Simard et al., ICDAR, pp. 958,963, 2003.
[18] Perez-Carrasco et al., TPAMI, vol.35,no.11,pp.2706,2719, 2013

Electrical characterization and physical modeling

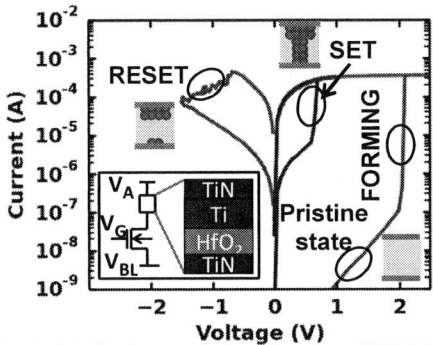

Figure 1 Typical I-V characteristics showing FORMING, SET and RESET operations. Inset: 1T-1R OxRAM device schematic.

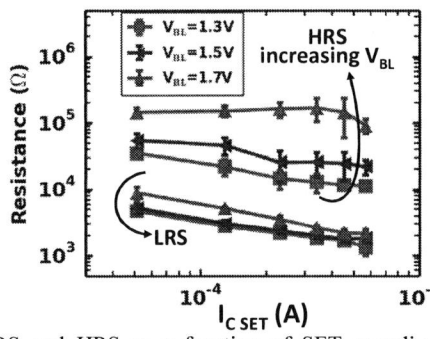

Figure 2 LRS and HRS as a function of SET compliance current ($I_{C\ SET}$) and different RESET voltages (V_{BL}).

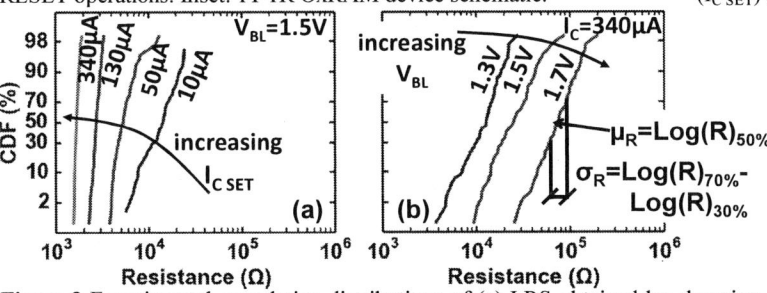

Figure 3 Experimental cumulative distributions of (a) LRS obtained by changing $I_{C\ SET}$ keeping V_{BL} constant and (b) HRS obtained by changing V_{BL} keeping $I_{C\ SET}$ constant. The extractions of the mean value (μ_R) and of the standard deviation (σ_R) are indicated.

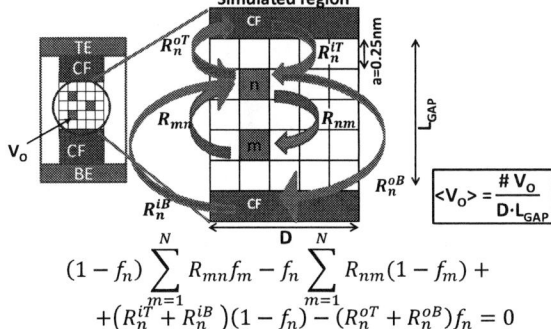

$$(1-f_n)\sum_{m=1}^{N} R_{mn}f_m - f_n\sum_{m=1}^{N} R_{nm}(1-f_m) + \\ +(R_n^{iT}+R_n^{iB})(1-f_n)-(R_n^{oT}+R_n^{oB})f_n = 0$$

Figure 4 Schematic view of the simulated oxide region and master equation used for 2D modeling of OxRAM variability.

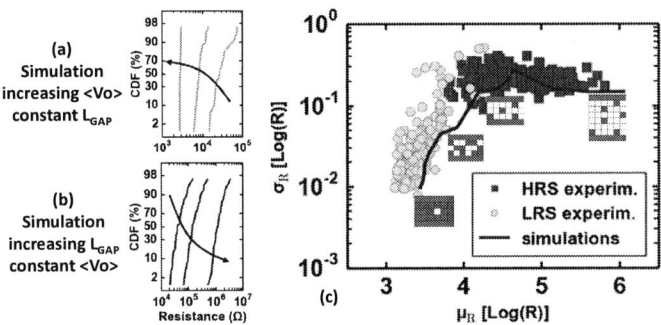

Figure 5 (a) Simulations for LRS obtained for constant gap size (L_{GAP}) and increasing $<V_O>$. (b) Simulations for HRS obtained by increasing gap size while keeping $<V_O>$ constant. (c) Experimental (symbols) and simulated (line) σ_R and μ_R extracted from Fig.3, and Fig. 5 (a-b) respectively.

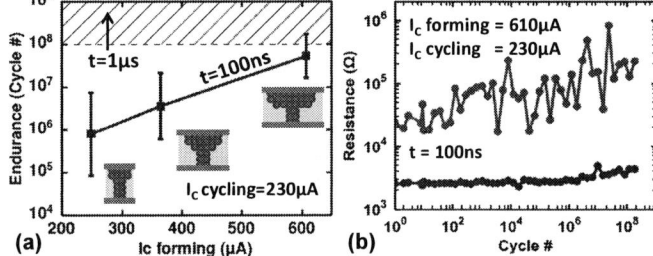

Figure 6 (a) Endurance as a function of compliance current during forming operation. Using programming pulse width of 1µs allows to achieve $>10^8$ cycles endurance whatever the I_C forming. When programming pulse width is decreased to 100 ns, 10^8 cycles endurance requires a higher compliance current during forming. (b) Endurance curve obtained at 100ns.

OxRAM devices as synapses

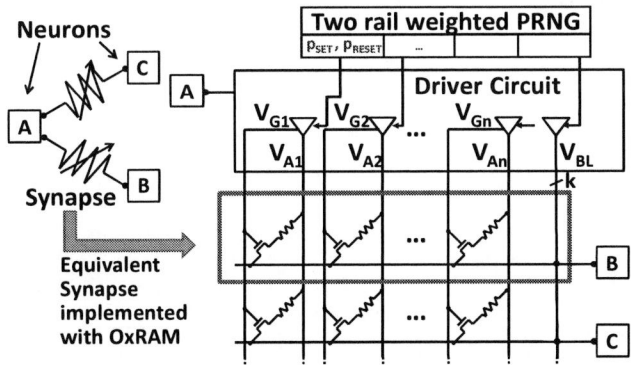

Figure 7 Schematic of OxRAM based synapses in a convolutional kernel. All the OxRAM devices on the same row build one equivalent synapse. Driver circuit is used to individually program OxRAM devices and propagate spikes to next neuron layer. The weighted PRNG is used for on-line learning, to implement the stochastic STDP programming presented in Fig.15.

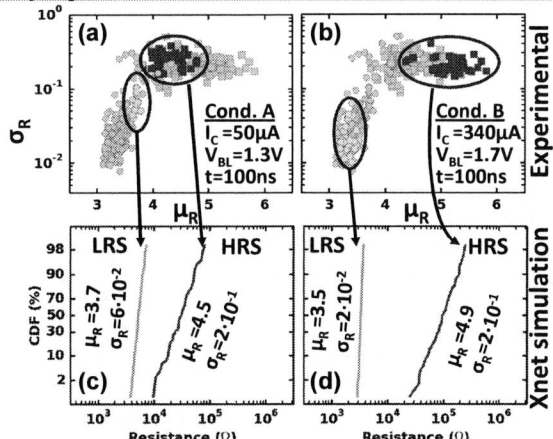

Figure 8 Experimental resistance levels and associated variability for (a) strong and (b) weak programming conditions. (c-d) Corresponding simulated synaptic distributions introduced in Xnet.

978-1-4799-8002-4/14 $31.00 © 2014 IEEE

CNN for visual pattern recognition

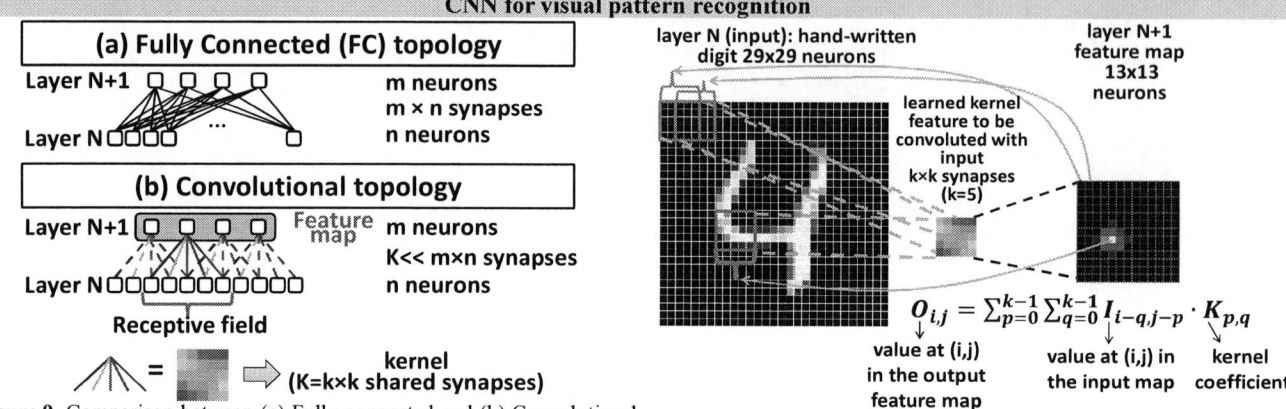

Figure 9. Comparison between (a) Fully connected and (b) Convolutional topologies. The use of synaptic kernels in CNNs (solid colored lines), shared between neurons, results in smaller number of required synapses.

$$O_{i,j} = \sum_{p=0}^{k-1} \sum_{q=0}^{k-1} I_{i-q,j-p} \cdot K_{p,q}$$

Figure 10. Schematic illustration of the convolution between an input image representing digit "4" and a learned kernel feature.

CNN architecture: read-mode operation

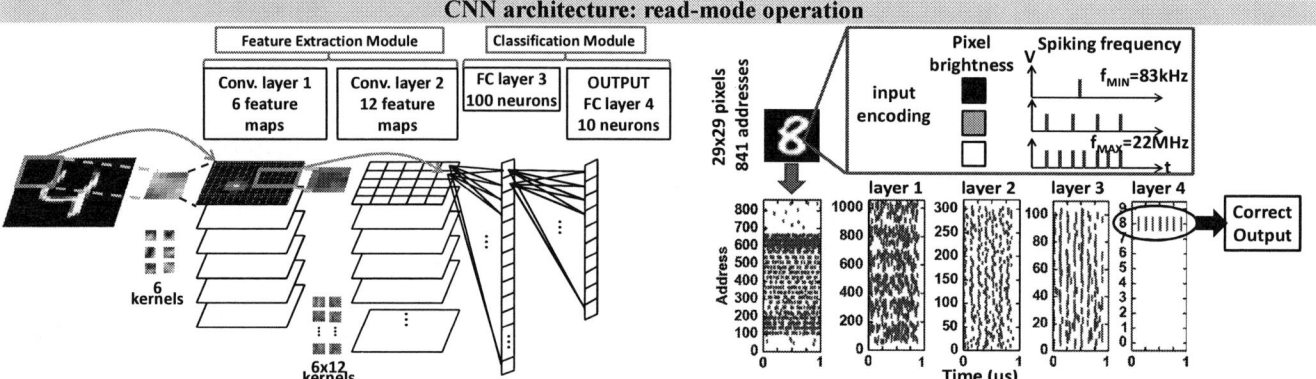

Figure 11. CNN architecture for handwritten digits recognition.

Figure 12. Propagation of spiking activity through neuron layers.

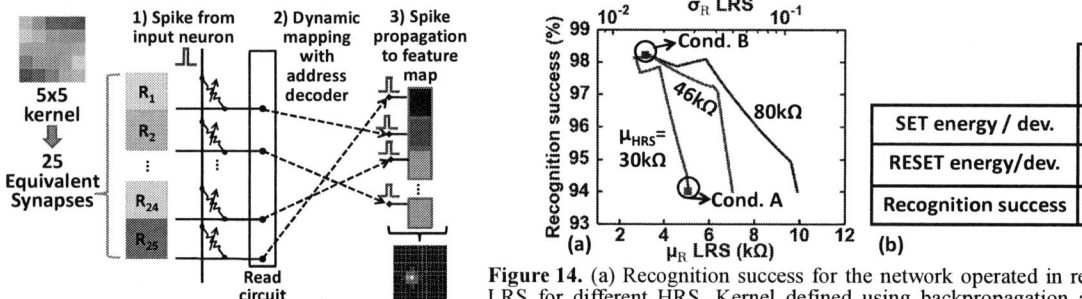

Figure 13. Proposed implementation of convolutional kernel. Each synapse is composed of 20 OxRAM devices as illustrated in Fig.7

	Cond. A Fig. 8a	Cond. B Fig. 8b
SET energy / dev.	5 pJ	34 pJ
RESET energy/dev.	9 pJ	58 pJ
Recognition success	94.0%	98.3%

Figure 14. (a) Recognition success for the network operated in read-mode as a function of LRS for different HRS. Kernel defined using backpropagation algorithm. Both LRS and HRS variability are taken into account. Highlighted points correspond to weak and strong programming conditions of Fig. 8a and 8b. (b) SET/RESET energy and recognition success associated to the selected programming conditions.

STDP learning

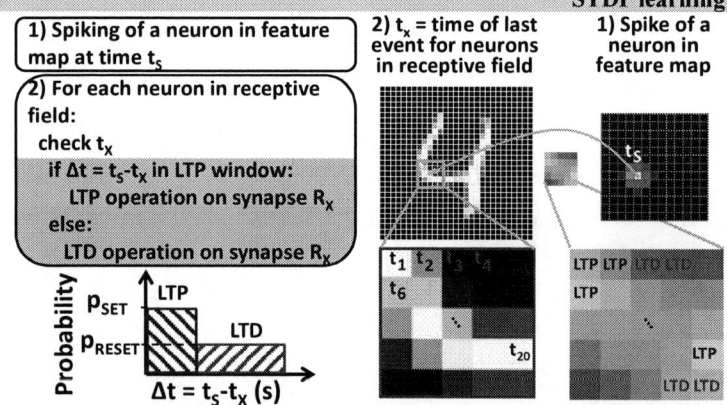

Figure 15. Proposed implementation of STDP for on-line, on-field learning. Synaptic weights are changed with SET/RESET pulses applied on OxRAM devices that constitute the equivalent synapses with STDP rule.

Learning phase duration: 60 ms 60 000 patterns of 1μs 10 digit categories	
Fully Connected NN	**Convolutional NN**
Nb. of connections: $9.5 \cdot 10^4$ Synapses: $9.5 \cdot 10^4$ Av. SETs/synapse: $2.4 \cdot 10^2$ Av. RESETs/synapse: $5.0 \cdot 10^2$	Nb. of connections: $8.6 \cdot 10^4$ <u>Shared</u> synapses: $8.8 \cdot 10^3$ Av. SETs/synapse: $4.2 \cdot 10^5$ Av. RESETs/synapse: $2.2 \cdot 10^5$

Table 1. Comparison of learning statistics for CNN and Fully Connected Network for STDP learning on full MNIST database, using same input encoding. Weight sharing in CNNs is more demanding in terms of number of SET/RESET operations, but it allows reducing the number OxRAM synapses.

3D Synaptic Architecture with Ultralow sub-10 fJ Energy per Spike for Neuromorphic Computation

I-Ting Wang, Yen-Chuan Lin, Yu-Fen Wang, Chung-Wei Hsu, and Tuo-Hung Hou[*]

Department of Electronics Engineering and Institute of Electronics, National Chiao Tung University, Hsinchu, Taiwan
[*]Tel: +886-3-5712121 ext 54261; E-mail: thhou@mail.nctu.edu.tw

Abstract

A high-density 3D synaptic architecture based on self-rectifying Ta/TaO$_x$/TiO$_2$/Ti RRAM is proposed as an energy- and cost-efficient neuromorphic computation hardware. The device shows excellent analog synaptic features that can be accurately described by the physical and compact models. Ultra-low energy consumption comparable to that of a biological synapse (<10 fJ/spike) has been demonstrated for the first time.

Introduction

Existing IT systems based on Si devices and von Neumann architecture suffer from the fundamental problems of device scaling and low energy efficiency, which may ultimately impede the realization of true human-level artificial intelligence. As a result, many believe that neuromorphic computation, similar to how our brains operate, is a promising direction to pursue for future artificial intelligence systems. However, present neuromorphic computations are implemented using complex Si-based CMOS circuits or software algorithms, and cannot overcome the limitations of low density and high energy consumption. To achieve true human-level artificial intelligence, a low-power two-terminal electronic synapse—a fundamental neuromorphic computation hardware that emulates the functions of a biological synapse—must be developed. RRAM, sometimes called memristor, is now being actively developed as a low-power electronic synaptic device [1-3]. It also inspired intriguing applications, such as pattern recognition and auditory processing [4-7]. RRAM-based synapses exhibit excellent scalability, compact 4F^2 cell size, full CMOS compatibility, and ultralow pJ energy consumption per spike [6], outperforming other electronic synapse candidates. However, two critical challenges remain. First, high-density 3D synaptic networks rather than 2D ones are essential to realize innumerable connections among neurons in the brain (Fig. 1). However, a cost-effective 3D architecture that overcomes the sneak current problem [8] has yet to be developed. Second, reducing the energy consumption to that of a biological synapse (~10 fJ per spike [3]) remains challenging. Recently, a self-rectifying Ta/TaO$_x$/TiO$_2$/Ti cell has shown great potential for implementing 3D vertical RRAM (V-RRAM) arrays over 10 Mb [9-11]. In this paper, we further demonstrate promising analog synaptic features in this device, and develop complete physical-based simulation and analytical compact model to facilitate future

neuromorphic system design. Furthermore, the self-rectifying characteristic enables the first high-density 3D synaptic architecture using easily fabricated V-RRAM. We also first demonstrate the ultralow energy consumption of less than 10 fJ per spike that represents 100x reduction as compared with the previous best synaptic device [3]. This extremely energy- and cost-efficient high-density 3D synaptic architecture is a breakthrough hardware for future neuromorphic computation.

Ta/TaO$_x$/TiO$_2$/Ti Cell

The device fabrication of the 2D Ta/TaO$_x$ (20 nm)/TiO$_2$ (60 nm)/Ti cell was similar to that described in [9]. The device area was 10^4 μm^2. Figs. 2(a) and (b) illustrate its SET-controlled and RESET-controlled multi-level-cell (MLC) operations, respectively. The gradual SET and RESET operations are crucial for analog synaptic devices where the synaptic weight (conductance) is affected by the input (training) strength. The conductance values can be read at -2 V, while very low current with no difference among different states was observed at a positive voltage bias.

Synaptic Functions

Fig. 3 illustrates alternating potentiating and depressing cycles in the Ta/TaO$_x$/TiO$_2$/Ti device by using consecutive 2000 training pulses. In contrast to the previous filamentary RRAM-based synapse [6], this device shows monotonically synaptic weight changes with little variation and robust cycling endurance. The potentiating and depressing ability in an identical device allows simultaneous implementation of both excitatory and inhibitory synapses. In addition, the asymmetric current readout at positive and negative biases provides a high self-rectifying ratio that suppresses sneak current through inactive synapses in the network. Furthermore, the synaptic weight function is tunable using the number of potentiating and depressing pulses (Fig. 4). Fig. 5 illustrates an extremely wide tunable range, showing no saturation after 800 pulses. A large number of analog synaptic weight states are known to improve capacity and robustness of neuromorphic systems [3]. Fig. 6 shows the measurement result of spike-timing-dependent plasticity (STDP). An action potential-like waveform (Fig. 7) was used to emulate STDP in biological synapses. STDP is a critical function for learning and memory in our brains based on the Hebbian learning rule where synaptic weight change depends on the relative timing of pre- and post-synaptic spikes. Fig. 8 illustrates the measurement result of paired-pulse facilitation (PPF). PPF is a

978-1-4799-8002-4/14 $31.00 © 2014 IEEE   665

short-term synaptic plasticity where reducing time interval between two sequential potentiating pulses enhances synaptic weight. The fitting time constants agree with that of biological synapses [12]. Fig. 9 illustrates memory retention characteristics. Increasing the number of training pulses (frequent rehearsal) facilitates the transition from short-term memory (STM) to long-term memory (LTM).

Physical Simulation and Compact Model

The switching and conduction mechanism of the $Ta/TaO_x/TiO_2/Ti$ cell has been explained elsewhere [9-11] by oxygen ion migration and homogeneous barrier modulation. The physical equations shown in Fig. 10 were used to simulate the SET-controlled (Fig. 11(a)) and RESET-controlled (Fig. 11(b)) MLC, alternating potentiating and depressing cycles (Fig. 12), and STDP (Fig. 13). Furthermore, a compact model of the proposed synaptic device must be developed for future large-scale neuromorphic system design. A set of analytical equations in Fig. 14 is derived from the physical-based simulation, and thus accurately describe potentiating (Fig. 15) and depressing (Fig. 16) characteristics with multiple pulse amplitudes and STDP (Fig. 6).

3D Synaptic Network

Cost-effective 3D V-RRAM was fabricated as 3D synaptic network using a simple process described in [10-11]. Fig. 17 shows the cross-sectional TEM image of double-layer V-RRAM with an effective device area of 0.2 μm^2. The 3D device shows stable potentiating and depressing characteristics (Fig. 18) and biologically equivalent STDP (Fig. 19), similar to those in the 2D device, but the cell conductance in the area-scaled 3D device is drastically reduced because of the homogeneous current conduction [10]. Reducing cell conductance is critical for implementing high-density crossbar array [10] and lowering energy consumption per training pulse. Fig. 20 shows that the energy consumption can be as low as 7 fJ per training pulse for depression. Although the high input voltage requires further improvement, this result first demonstrates that an electronic synapse can be as energy efficient as a biological one. Fig. 21 shows the schematic and fabricated 3D (double-layer) synaptic network prototype. Both 2D feature-size scaling and increasing the number of vertically stacked layers would increase integration density in the future. Commercial 3D vertical NAND technology using a similar fabrication concept is capable of integrating 10^{11} bits (connections) on a single chip [13].

Conclusion

Numerous promising synaptic features in a $Ta/TaO_x/TiO_2/Ti$ RRAM-based synaptic device have been successfully demonstrated. The analog synaptic plasticity can be precisely simulated using the physical and compact models.

Furthermore, the 3D synaptic network with self-rectifying characteristics for high-density integration and ultra-low training power comparable to that in biological synapses have been realized, suggesting promising potential of the 3D synaptic architecture for future neuromorphic computation.

Acknowledgments

This work was supported by National Science Council of Taiwan, Republic of China, under grant NSC 102-2221-E-009 -188-MY3 and NCTU-UCB I-RiCE program, under grant NSC-102-2911-I-009-302.

References

[1] S. H. Jo, T. Chang, I. Ebong, B. B. Bhadviya, P. Mazumder, and W. Lu, "Nanoscale memristor device as synapse in neuromorphic systems," *Nano Lett.*, vol. 10, pp. 1297–1301, 2010.

[2] T. Ohno, T. Hasegawa, T. Tsuruoka, K. Terabe, J. K. Gimzewski, and M. Aono, "Short-term plasticity and long-term potentiation mimicked in single inorganic synapses," *Nat. Mater.*, vol. 10, pp. 591–595, 2011.

[3] D. Kuzum, S. Yu, and H.-S. P. Wong, "Synaptic electronics: materials, devices and applications," *Nanotechnology*, vol. 24, 382001, 2013.

[4] S. Park, H. Kim, M. Choo, J. Noh, A. Sheri, S. Jung, K. Seo, J. Park, S. Kim, W. Lee, J. Shin, D. Lee, G. Choi, J. Woo, E. Cha, J. Jang, C. Park, M. Jeon, B. Lee, and H. Hwang, "RRAM-based synapse for neuromorphic system with pattern recognition function," in *IEDM Tech. Dig.*, 2012, pp. 231–234.

[5] M. Suri, O. Bichler, D. Querlioz, G. Palma, E. Vianello, D. Vuillaume, C. Gamrat, and B. DeSalvo, "CBRAM devices as binary synapses for low-power stochastic neuromorphic systems: Auditory (Cochlea) and visual (Retina) cognitive processing applications," in *IEDM Tech. Dig.*, 2012, pp. 235–238.

[6] S. Yu, B. Gao, Z. Fang, H. Yu, J. Kang, and H.-S. P. Wong, "A neuromorphic visual system using RRAM synaptic devices with sub-pJ energy and tolerance to variability: Experimental characterization and large-scale modeling," in *IEDM Tech. Dig.*, 2012, pp. 239–242.

[7] S. Park, A. Sheri, J. Kim, J. Noh, J. Jang, M. Jeon, B. Lee, B. R. Lee, B. H. Lee, and H. Hwang, "Neuromorphic speech systems using advanced ReRAM-based synapse," in *IEDM Tech. Dig.*, 2013, pp. 625–628.

[8] A. Flocke and T. G. Noll, "Fundamental analysis of resistive nanocrossbars for the use in hybrid nano/CMOS-memory," in *Proc. 33rd ESSCIRC*, 2007, pp. 328–331.

[9] C. W. Hsu, I. T. Wang, C. L. Lo, M. C. Chiang, W. Y. Jang, C. H. Lin, and T. H. Hou, "Self-rectifying bipolar TaO_x/TiO_2 RRAM with superior endurance over 10^{12} cycles for 3D high-density storage-class memory," in *VLSI Symp. Tech Dig.*, 2013, pp. 166–167.

[10] C. W. Hsu, I. T. Wang, C. C. Wan, I. T. Wang, M. C. Chen, C. L. Lo, Y. J. Lee, W. Y. Jang, C. H. Lin, and T. H. Hou, "3D vertical TaO_x/TiO_2 RRAM with over 10^3 self-rectifying ratio and sub-μA operating current," in *IEDM Tech. Dig.*, 2013, pp. 264-267.

[11] C. W. Hsu, Y. F. Wang, C. C. Wan, I. T. Wang, C. T. Chou, W. L. Lai, Y. J. Lee, and T. H. Hou, "Homogeneous barrier modulation of TaO_x/TiO_2 bilayers for ultra-high endurance three-dimensional storage-class memory," *Nanotechnology*, vol. 25, 165202, 2014.

[12] R. S. Zucker and W. G. Regehr, "Short-term synaptic plasticity," *Annu. Rev. Physiol.*, vol. 64, pp. 355–405, 2002.

[13] K.-T. Park, et al., " Three-dimensional 128Gb MLC vertical NAND Flash-memory with 24-WL stacked layers and 50MB/s high-speed programming," in *ISSCC Dig. Tech.*, 2014, pp. 334-335.

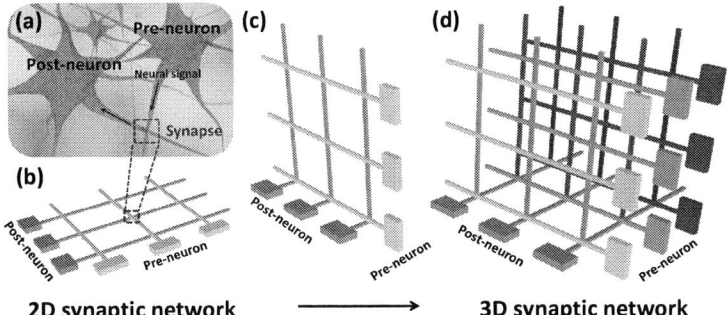

2D synaptic network ⟶ **3D synaptic network**

Fig. 1 (a) Intricate 3D synaptic network connects billions of neurons in human brains. (b) Integration density is low in 2D synaptic network of present neuromorphic systems. (c) 3D implementation of synaptic network in (b) significantly increases integration density. (d) High-density 3D synaptic network emulates that in biological systems.

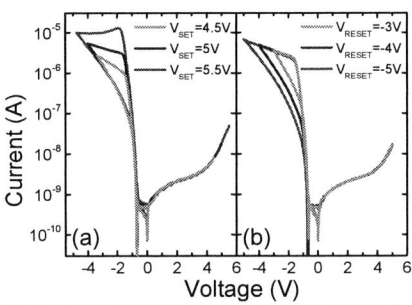

Fig. 2 (a) SET-controlled and (b) RESET- controlled MLC operations in the 2D Ta/TaO$_x$/TiO$_2$/Ti device. By varying V$_{SET}$ or V$_{RESET}$, multiple resistance states can be read out at -2 V, while very low current was observed at a positive voltage bias.

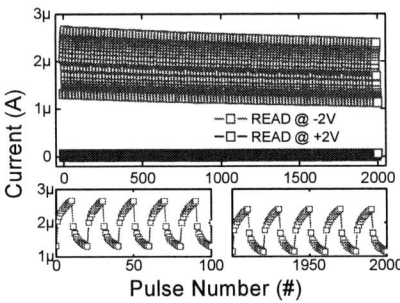

Fig. 3 Alternating potentiating (P) and depressing (D) cycles for 2000 training pulses (P: +9 V 50 µs/D: -8 V 50 µs) in the 2D Ta/TaO$_x$/TiO$_2$/Ti device. The response of the first and last 100 pulses are enlarged.

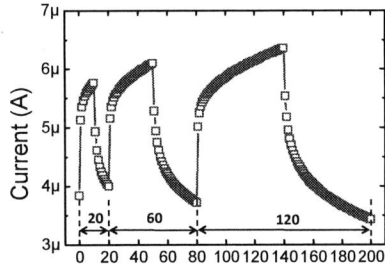

Fig. 4 Potentiating and depressing characteristics as a function of various numbers of training pulses. This phenomenon mimics biological memory enhancement by increasing stimuli.

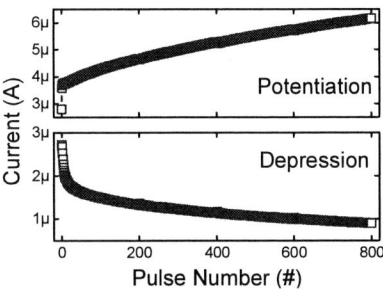

Fig. 5 Continuous potentiating and depressing characteristics over 800 training pulses, showing monotonically synaptic weight (conductance) changes without saturation.

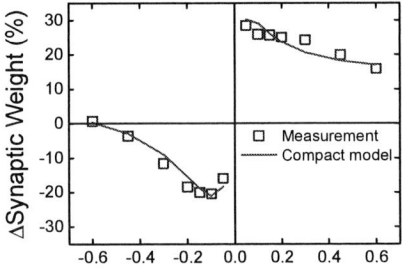

Fig. 6 STDP measurement shows a similar trend as that of a biological synapse, and is in good agreement with the compact model.

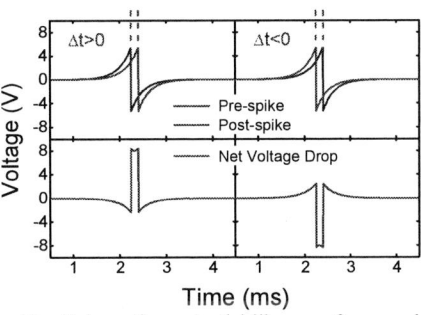

Fig. 7 An action potential-like waveform used to measure STDP (Fig. 6). The interval Δt and net voltage drop are the relative timing and voltage difference between pre- and post-spike.

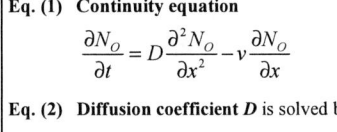

Eq. (1)	**Continuity equation**

$$\frac{\partial N_O}{\partial t} = D\frac{\partial^2 N_O}{\partial x^2} - v\frac{\partial N_O}{\partial x}$$

Eq. (2) **Diffusion coefficient** D is solved by

$$D = \frac{a^2 f}{2} \cdot \exp\left(-\frac{E_A}{kT}\right)$$

Eq. (3) **Drift velocity** v is solved by

$$v = af \cdot \exp\left(-\frac{E_A}{kT}\right) \cdot \sinh\left(-\frac{q\gamma aF}{kT}\right)$$

Eq. (4) **Poisson equation** to obtain F

$$\frac{\partial F}{\partial x} = \frac{2q}{\varepsilon_r \varepsilon_0}(-N_O + N_{V+})$$

Fig. 10 Equations used in the physical-based simulation [11]. Equation (1) is the continuity equation used to calculate time-dependent evolution of O^{2-} concentration N_O. In equation (2), a is the effective hopping distance, f is the attempt-to-escape frequency, E_A is the activation energy of O^{2-} migration, and kT is the thermal energy. In equation (3), γ is the fitting parameter for field dependence, q is the elementary charge, and F is the local electric field. In equation (4), N_{V+} is the oxygen vacancy concentration, ε_r is the dielectric constant of TaO$_x$ or TiO$_2$, and ε_0 is the permittivity of vacuum.

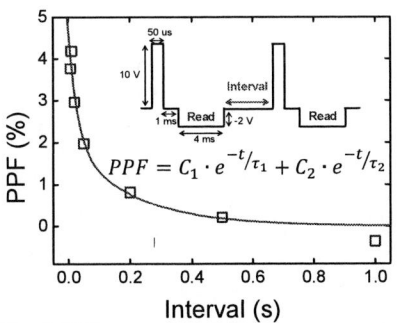

Fig. 8 PPF measurement (blue open square) and fitting (red line) using the listed formula (τ_1=10 ms, τ_2=175 ms). Inset shows the sequence of potentiating and read pulses.

$$PPF = C_1 \cdot e^{-t/\tau_1} + C_2 \cdot e^{-t/\tau_2}$$

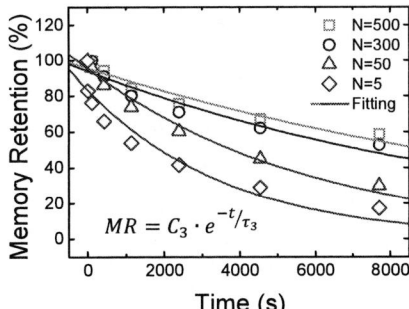

$$MR = C_3 \cdot e^{-t/\tau_3}$$

Fig. 9 Memory retention (MR) as a function of various numbers of training inputs (N). Increasing the number of training pulses facilitates the transition from STM to LTM.

978-1-4799-8002-4/14 $31.00 © 2014 IEEE

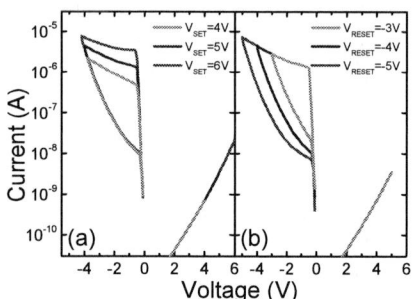

Fig. 11 Simulated *I-V* characteristics of (a) SET-controlled and (b) RESET-controlled MLC using physical equations in Fig. 10.

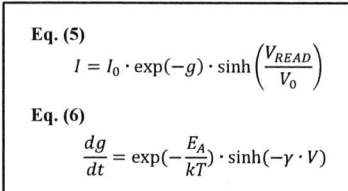

Eq. (5)
$$I = I_0 \cdot \exp(-g) \cdot \sinh\left(\frac{V_{READ}}{V_0}\right)$$

Eq. (6)
$$\frac{dg}{dt} = \exp\left(-\frac{E_A}{kT}\right) \cdot \sinh(-\gamma \cdot V)$$

Fig. 14 Analytical equations derived from the physical-based simulation. Equation 5 describes the relation between current and voltage where I_0 and V_0 are fitting parameters, g is the tunneling gap distance at the Ta/TaO$_x$ Schottky barrier, and V_{READ} is the read voltage. Equation 6 defines the time dependence of function g. V is the programming voltage, and E_A and γ are g-dependent fitting parameters.

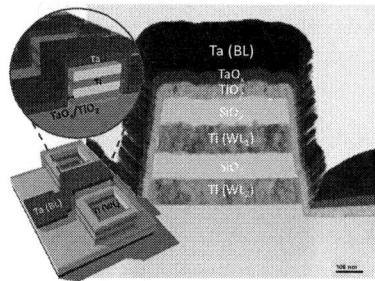

Fig. 17 Cross-sectional TEM image of 3D Ta/TaO$_x$/TiO$_2$/Ti double-layer V-RRAM. Inset illustrates the cutline where the TEM image was taken. The effective device area is 0.2 μm^2.

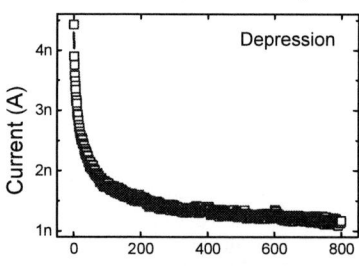

Fig. 20 Depressing characteristics (-18 V/10 ns) in the 3D device. The initial resistance is ~450 MΩ, and the energy consumption per spike is ~7 fJ. The required high voltage with a short pulse is in part caused by unoptimized RC delay in the device structure and measurement setup.

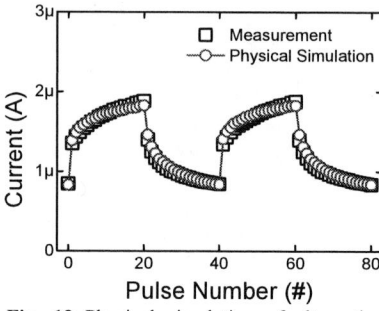

Fig. 12 Physical simulation of alternating potentiating and depressing characteristics, showing excellent agreement with the measurement.

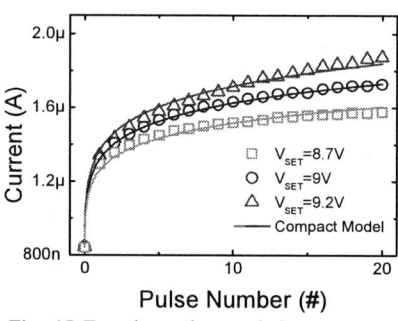

Fig. 15 Experimental potentiating characteristics (symbols) by using various voltage amplitudes. The initial resistance state is fixed approximately at 2.4 MΩ. Experiments and compact model calculation (color lines) are in good agreement.

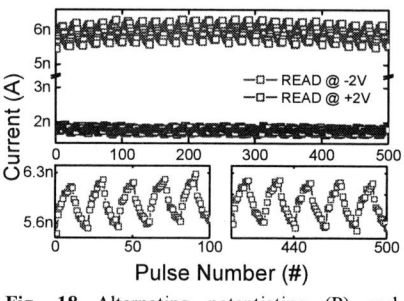

Fig. 18 Alternating potentiating (P) and depressing (D) cycles for 500 training pulses (P: +10 V 50 μs/D: -7 V 50 μs) in the 3D Ta/TaO$_x$/TiO$_2$/Ti device. The response of the first and last 100 pulses are enlarged.

(a)

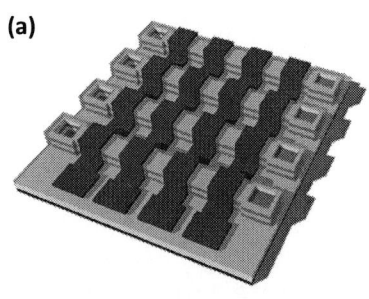

Fig. 13 Physical simulation of STDP shows a similar trend as that of a biological synapse.

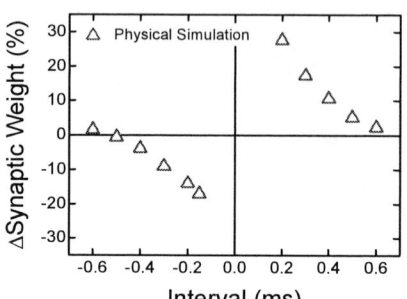

Fig. 16 Experimental depressing characteristics (symbols) by using various voltage amplitudes. The initial resistance state is fixed approximately at 1.2 MΩ. Experiments and compact model calculation (color lines) are in good agreement.

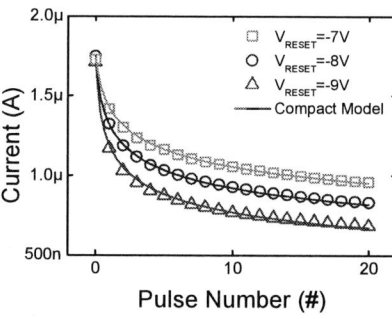

Fig. 19 STDP measurement result in the 3D Ta/TaO$_x$/TiO$_2$/Ti device. The result shows a similar trend as that of a biological synapse, and is in good agreement with the compact model. The measurement input waveform is shown in Fig. 7.

(b)

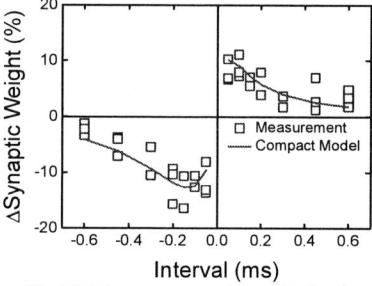

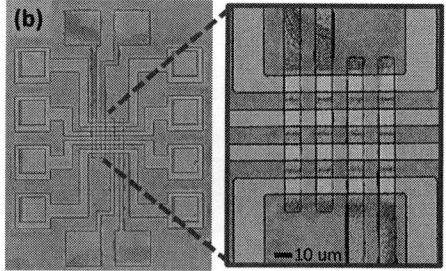

Fig. 21 (a) Schematic illustration and (b) top-view optical microscope image of the 3D double-layer V-RRAM array fabricated using a four-mask process. High-density 3D synaptic network in Fig. 1(d) can be realized by further increasing the number of vertical layers and 2D feature-size scaling.

978-1-4799-8002-4/14 $31.00 © 2014 IEEE   668

Highly Dependable 3-D Stacked Multicore Processor System Module Fabricated Using Reconfigured Multichip-on-Wafer 3-D Integration Technology

[1,2]K-W. Lee, [1]H. Hashimoto, [1,2]M. Onishi, [1]S. Konno, [1]Y. Sato, [1]C. Nagai, [1,2]J-C Bea,
[1,2]M. Murugesan, [1,2]T. Fukushima, [3]T. Tanaka, [1,2]M. Koyanagi

[1]New Industry Creation Hatchery Center (NICHe), [2]Global INTegration Initiative (GINTI),
Tohoku University, Sendai, Japan
[3]Department of Biomedical Engineering, Tohoku University, Sendai, Japan
Phone; +81-22-795-4119; Fax:+81-22-795-6907 ; E-mail: kriss@bmi.niche.tohoku.ac.jp

Abstract

A highly dependable 3-D stacked multicore processor module composed of 4-layer stacked 3-D multicore processor chip and 2-layer stacked 3-D cache memory chip is implemented using reconfigured multichip-on-wafer 3-D integration and backside TSV technologies for the first time. Tier boundary scan, self-repair circuits, and BIST circuits in the 4-layer stacked 3-D multicore processor chip and the basic read/write functions of memory circuits in the 2-layer stacked 3-D cache memory chip are successfully evaluated. High-density TSVs and micro-joining characteristics in the 3-D stacked chip were evaluated by a non-destructive method using high resolution X-ray CT scanning tool.

Introduction

Three-dimensional (3-D) hetero-integration technology allows the possibility of assembling various kinds of functional blocks such as processor, memory, sensors, logic, analog, photonic, and power ICs into one stacked chip [1–2]. Therefore it can create many potential applications beyond mobile and consumer products. One of the important potential applications is the intelligent vehicle electronic system [3]. Large numbers of and various kinds of LSIs and sensor devices such as radars, sensors, local area network, microprocessor, and electronic control unit are loaded in an automotive to prevent abnormal accidents and to assist an autonomous driving. To realize the intelligent vehicle system, we have proposed high-speed sensing, highly parallel processing image sensor system module as shown in Fig. 1, where image sensor system consists of two 3-D stacked image sensors with four layers of image sensor, analog, ADC, and interface chips (we have already developed [4-5]) for high speed signal sensing and eight 3-D stacked multicore processors with high dependability for highly parallel data processing. A highly dependable 3-D stacked multicore processor consists of four memory layers and eight processor layers including a supervisor processor layer to execute self-restoration function as shown in Fig. 2 to simultaneously satisfy high functionality, high availability, and high dependability. In this study, to evaluate the basic function of the 3-D stacked multicore processor system module with high dependability, the 4-layer 3-D stacked multicore processor chip and the 2-layer 3-D stacked cache memory chip are designed and fabricated using reconfigured multichip-on-wafer 3-D integration and backside TSV technologies, respectively.

Fabrication of 3-D stacked multicore processor

Fig. 3 shows the configuration (a) and the block diagram (b) of 3-D stacked multicore processor chip where four core processor layers are vertically stacked and electrically connected by vertical buses using TSVs [6]. To achieve high dependability, one of four core processor layers is used as a supervisor processor (SVP) layer to manipulate self-test and self-repair functions. The SVP also acts as a master processor to supervise the task control of whole system such as task scheduling among multicores. Fig. 4 shows the configuration (a) and the block diagram (b) of 3-D stacked cache memory chip, where cache memory layers are electrically connected by vertical memory bus using TSVs. Fig. 5 shows photos of a core processor chip with self-test and self-repair functions fabricated by 90-nm CMOS technology (a) and a cache memory chip fabricated by 130-nm CMOS technology, respectively. The processor chip has totally 1,920 TSVs and the cache memory has totally 1,200 TSVs, respectively. We have developed a novel reconfigured multichip-on-wafer 3-D integration technology [7]. We also developed both wafer-level and die-level backside TSV technologies [4] to construct TSVs. Fig. 6 shows the process flow for fabricating 3-D stacked chips by the reconfigured multichip-on-wafer 3-D integration and die-level backside TSV technologies. We use an SAE (self-assembly & electrostatic) wafer as a carrier wafer to fabricate 3-D stacked chips as shown in Fig. (a). Hydrophilic and hydrophobic areas are formed on the SAE wafer for chip self-assembly using the surface tension of liquid. Liquid droplets are simultaneously provided to hydrophilic areas in the self-assembly which act as bonding areas. Bipolar electrodes of metal lines are also formed on the SAE wafer surface to generate an electrostatic force for temporary bonding. Many KGDs (known-good-dies) of processor and memory with Cu/Sn bumps are simultaneously aligned with face-up by a self-assembly method using the surface tension of liquid and electrostatically bonded onto the SAE carrier wafer (b). Then, these KGDs with the SAE carrier wafer are glue-bonded with face-down to a support wafer and the SAE carrier wafer is de-bonded by applying a voltage with opposite polarity. Thus all KGDs are transferred from the SAE carrier wafer to the support wafer and we can obtain a reconfigured wafer with all known-good-dies (c). After that all KGDs are simultaneously thinned from the backside down to around 50-μm in thickness and a hard mask dielectric layer is deposited on the backside surface (d). After the backside TSV patterning, via holes of 10-μm diameter are etched from the chip backside until the first metal layer (M1) in each chip is exposed. Dielectric liner is deposited into via holes and then the liner oxide at the bottom of via holes is etched by dry etching to expose the M1 again (e). Metal barrier layer and Cu seed layer are deposited into via holes by sputtering. Via holes are completely filled with Cu by electroplating and then Cu/Sn

978-1-4799-8002-4/14 $31.00 © 2014 IEEE

bumps and Cu RDLs (re-distribution line) are formed (f). Each support wafer is diced into many support chips and consequently we obtain many KGDs with Cu TSVs and micro-bumps bonded onto support chips. These KGDs with support chips are bonded onto a Si interposer with high alignment accuracy (g). After open/short check and simple function test to confirm the joining quality between KGDs and Si interposer using the evaluation pads, support chips are simultaneously de-bonded by removing the glue layer (h). Next, another KGDs for the second layer are face-up-bonded onto KGDs of the first layer and after the test to confirm the joining qualities between top and bottom layer chips, support chips for the second-layer KGDs are de-bonded (i). By repeating these processes, we can fabricate a number of 3-D stacked chips with all KGDs (j). Highly reliable backside TSV formation is a key to fabricate such 3-D stacked chip with all KGDs. Specifically, high alignment accuracy of backside TSV patterning to M1 is strongly required for highly reliable backside TSV formation as shown in Fig. 7. In addition, the process optimization is essential for highly reliable backside TSV formation. Fig. 8 shows SEM cross-sectional images of Cu-TSV formed by the backside TSV technology before the process optimization. The liner oxide under-etch at the bottom of via (a), Si sidewall etching (notch) at the bottom interface (b) and Cu void in electroplating (c) are observed in the TSV formed by the backside TSV technology without process optimization. On the other hand, any liner oxide under-etch, Si notch and Cu-voids are not observed in the TSV formed by the backside TSV technology with process optimization as shown in Fig.9. As a result, we obtained the good electrical contact between Cu TSV and M1 as shown in Fig. 10. Backside TSV approach provides better reliability compared with via-middle TSV approach due to less impact of Cu TSV to BEOL and smaller probability of Cu diffusion. The impact of Cu TSV-induced stress on BEOL layers is smaller in the backside TSV approach, because the induced stress released to the backside of TSV not to BEOL as shown in Fig. 11 [8]. Furthermore the impact of Cu contamination from Cu TSV on device reliability is also smaller in the backside TSV approach, because the maximum process temperature remains as low as below 300°C. Higher processing temperature than 400°C might cause a serious Cu diffusion from Cu-TSVs and consequently a deterioration of minority carrier lifetime as shown in Fig.12 [9]. One of potential concerns in the backside TSV technology is a plasma-induced damage when the M1 layer is exposed during the via hole formation by plasma etching. MOS transistors connected to M1 might suffer from plasma-induced damages through the M1 when the M1 is exposed to the plasma. We evaluated the impact of plasma etching process for TSV formation on device reliability using the TEG pattern, which has various numbers of via hole pattern as shown in Fig. 13. Significant degradation was not observed in the I_d-V_g characteristics of MOS transistor which was connected to metal layer with via hole pattern even after 21 backside TSVs are formed as shown in Fig.14. Figures 15 and 16 shows the top views (a) and the SEM cross-sectional images (b) of the fabricated 4-layer stacked 3-D multicore processor chip and the fabricated 2-layer stacked 3-D cache memory chip, respectively. It is clearly observed in the figures that the backside Cu TSVs have good contact to the M1 layer with high alignment accuracy and without any notching and Cu voids.

Characterization of 3-D stacked multicore processor

Fig. 17 shows X-ray CT scanning images measured at TSV array area in the 3-D stacked multicore processor chip (a) and the 3-D stacked cache memory chip (b), respectively. A number of Cu TSVs, RDLs, metal micro-bumps, and BEOL interconnects are clearly seen. Thus X-ray CT scanning tool is very useful for a non-destructive 3-D failure analysis to characterize high-density

TSVs and metal micro-bump joining in 3-D stacked chip. We evaluated the basic functions of such 3-D stacked multicore processor chip. First of all, we confirmed that the processor core exhibited the performance of approximately 350 MIPS (Dhrystone 2.1) at 200MHz operation. In addition, we evaluated boundary scan circuits, self-repair circuits for TSVs and memory BIST circuits through TSVs in the 4-layer stacked 3-D multicore processor chip. Fig. 18 shows measured output waveforms of resister circuits at the first layer (a) and the second layer (b) in the 3-D stacked multicore processor chip, respectively. It was confirmed in the figure that the data were successfully read out from the register circuits of the first layer and the second layer through TSVs. Fig. 19 shows measured output waveforms of TSV boundary scan circuits in the peripheral areas which extend to the upper and lower layers to check the open/short of TSVs. The boundary scan test results showed a high TSV connectivity of 99.4% (171/172 TSV blocks, 4 TSVs/block) between the upper and lower layers. Fig. 20 shows measured output waveforms from the memory in the 4-layer stacked 3-D multicore processor chip obtained by functional test using memory BIST. Test data were written into the memory and the data read out from the memory were compared with the expected values by memory BIST. It was confirmed from the result of Fig.20 that both memory circuits and memory BIST circuits well worked without the degradation of the performance. Fig. 21 shows measured waveforms for cache-write and register-read in the 2-layer stacked 3-D cache memory chip. It was confirmed from the figure that data write/ read operations were successfully executed between the cache memory and registers through TSVs. In this work, we evaluated the feasibility of our reconfigured multichip-on-wafer 3-D integration technology by fabricated 4-layer 3-D stacked processor chip and 2-layer 3-D stacked cache memory chip, respectively. We successfully characterized basic functions both of 4-layer stacked 3-D processor chip and 2-layer stacked 3-D cache memory chip. In the future, we will stack the 4-layer 3-D stacked cache memory chip on the 8-layer 3-D stacked multicore processor chip to realize the high dependable 3-D stacked multicore processor module.

Conclusion

The 4-layer stacked 3-D multicore processor chip and the 2-layer stacked 3-D cache memory chip are fabricated using reconfigured multichip-on-wafer 3-D integration and backside TSV technologies. Boundary scan circuits, self-repair circuits, and memory BIST circuits in the 4-layer stacked 3-D multicore processor chip and memory and register circuits in the 2-layer stacked 3-D cache memory chip are successfully evaluated.

References

[1] M. Koyanagi, et al., *Symp. on Future Electron Devices.*, pp. 50, 1989
[2] T. Kunio, et al., *IEDM*, pp. 837, 1989
[3] K.W. Lee, et al., *IEEE T-ED*, vol. 58, no.3, pp. 748, 2011
[4] K.W. Lee, et al., *IEDM*, pp. 785, 2012
[5] K.W. Lee, et al., *IEEE T-ED*, vol.60, no.11, pp. 3842, 2013
[6] H. Hashimoto, et al., *IEEE 3D System Integration Conference*, pp. 978, 2013
[7] T. Fukushima, et al., *IEEE ECTC*, pp. 58, 2013
[8] D. Zhang, et al., *IEEE ECTC*, pp. 1407, 2013
[9] K.W. Lee, et al., *IEEE IRPS*, pp.3E.4.1, 2014

Acknowledgement

This research was supported by the Dependable VLSI Project of Core Research for Evolutional Science and Technology (CREST) of Japan Science and Technology Corporation (JST). We thank Y. Kitamura, Denso Ltd. for carrying out X-ray CT-scan analysis.

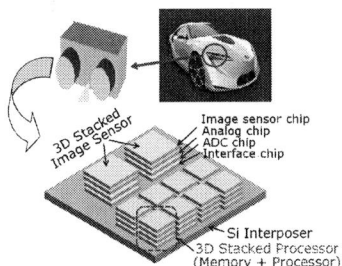

Fig.1. Concept of high-speed, highly parallel processing image sensor system module composed of 3-D stacked image sensors and 3-D stacked multicore processors for autonomous driving assist

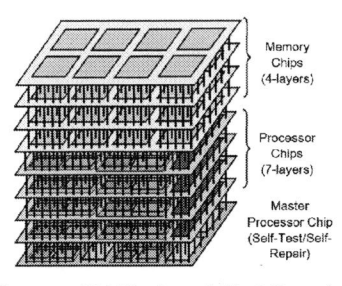

Fig.2. Concept of highly dependable 3-D stacked multicore processor module composed of 4-layer memories and 8-layer processors with self-restoration functions

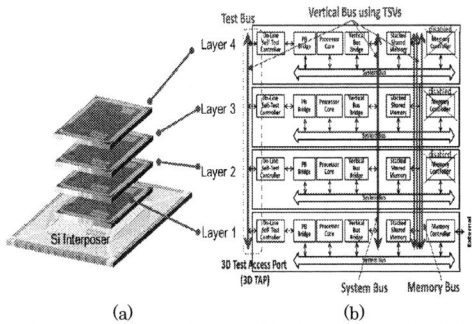

Fig.3. Configuration (a) and block diagram (b) of 3-D stacked multicore processor chip

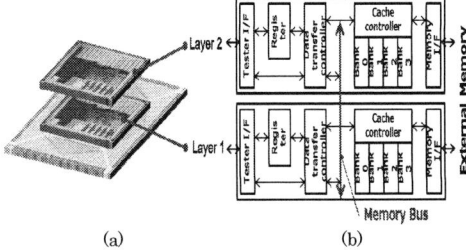

Fig.4. Configuration (a) and block diagram (b) of 3-D stacked cache memory chip

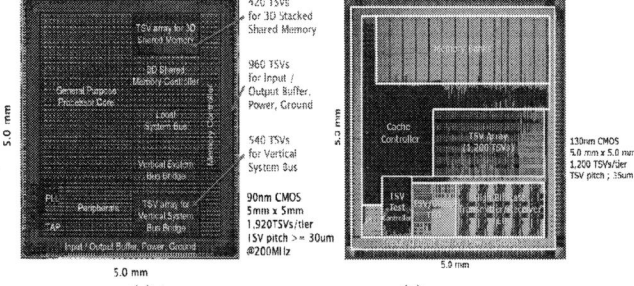

Fig.5. Photograph of core processor chip (a) and cache memory chip (b)

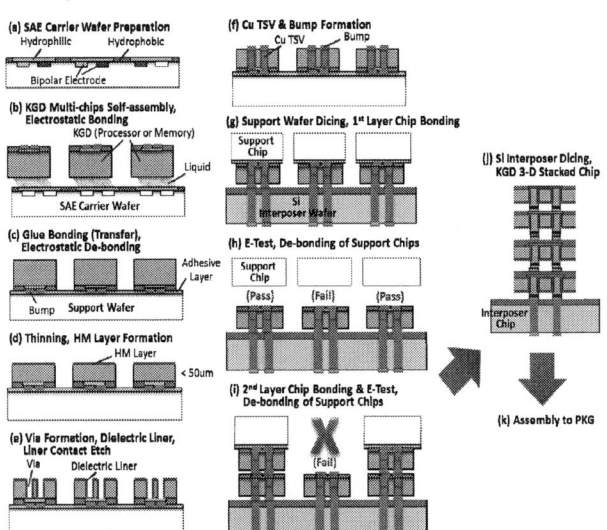

Fig.6. Process flow for KGD 3-D stacked chips using novel reconfigured multichip-on-wafer 3-D integration and reliable backside TSV technologies

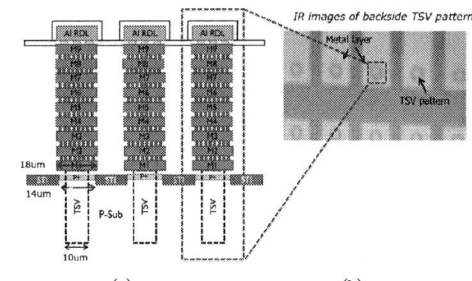

Fig.7. Design rule of backside TSV (a) and IR images after the backside TSV patterning to M1 layer (b)

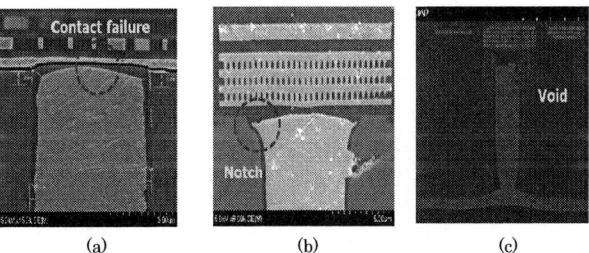

Fig.8. SEM cross-sectional images of backside Cu TSVs formed before the process optimization

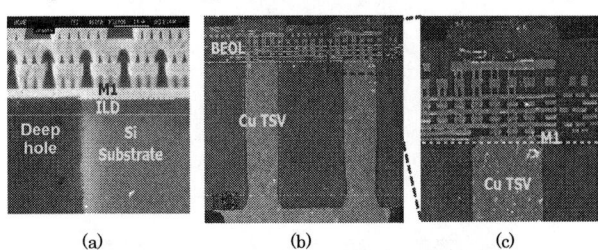

Fig.9. SEM cross-sectional images of backside Cu TSVs formed after the process optimization (a) after TSV etch, (b) and (c) after Cu TSV formation

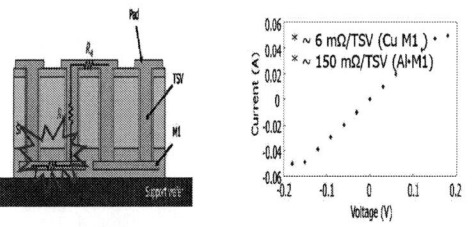

(a) TSV chain   (b) Contact resistance

Fig.10. Contact resistances between backside Cu TSV and M1 layer

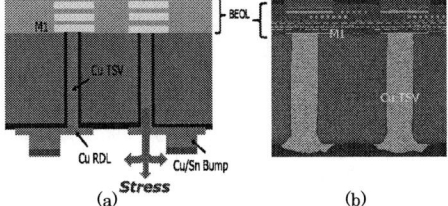

Fig.11. Cross-sectional structure (a) and SEM image (b) of backside Cu TSV with Cu RDL layer

978-1-4799-8002-4/14 $31.00 © 2014 IEEE

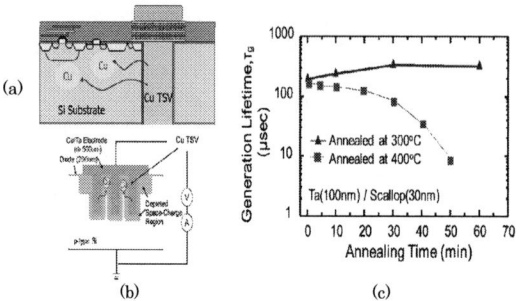

Fig.12. Conceptual image of Cu diffusion from Cu TSV (a) and configuration of trench MOS capacitor (b) and generation lifetime of minority carrier measured by C-t method using the trench capacitor as s function of annealing time at different temperature (c)

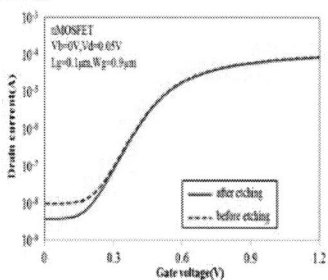

Fig.14. Id-Vg characteristics of MOSFET (Lg/Wg:0.1μm/0.9μm) before and after plasma etching for the via-hole fabrication using 21 via hole pattern

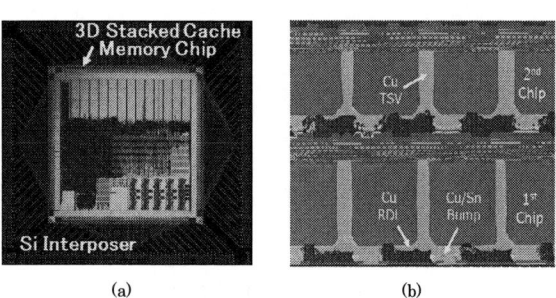

Fig.16. Top view (a) and SEM cross-sectional image (b) of the fabricated 2-layer stacked 3-D cache memory chip (Cu RDLs and Cu/Sn bumps are seen)

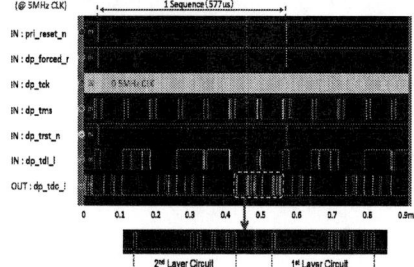

Fig.18. Measured output waveforms of register circuits at 1st layer and 2nd layer in the 4-layer stacked 3-D multicore processor chip

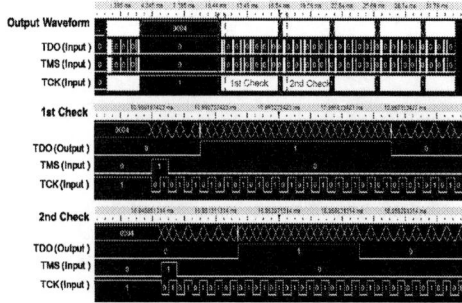

Fig.20. Measured output waveforms from memory in the 4-layer stacked 3-D multicore processor obtained by functional test using memory BIST

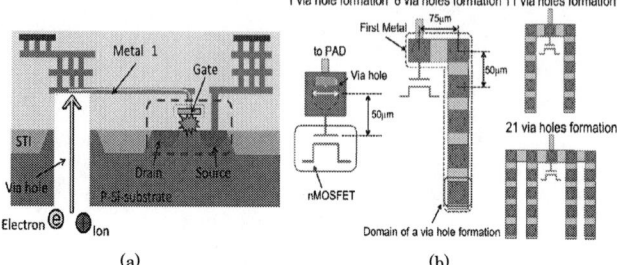

Fig.13. Configuration (a) and TEG pattern (b) for the evaluation of plasma-induced-damage effect on device reliability

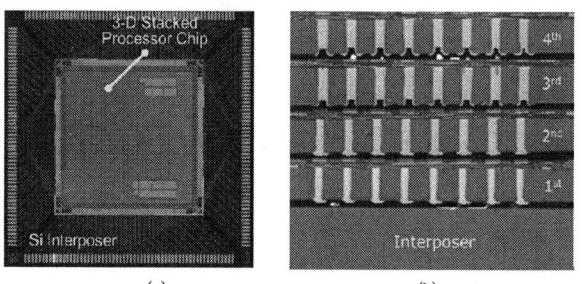

Fig.15. Top view (a) and SEM cross-sectional image (b) of the fabricated 4-layer stacked 3-D multicore processor chip, where Cu RDLs and bumps are not seen at the same line of Cu TSVs

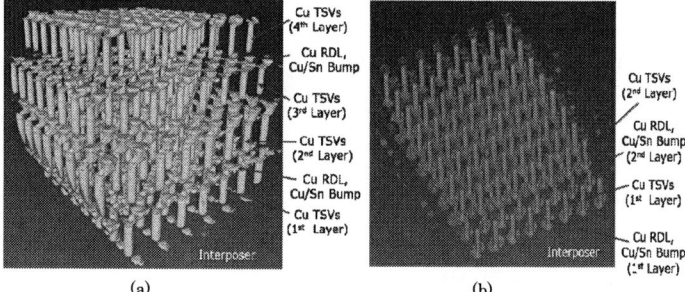

Fig.17. X-ray CT scanning image measured from TSV array areas in the 4-layer stacked 3-D multicore processor chip (a) and the 2-layer stacked 3-D cache memory chip (b), respectively

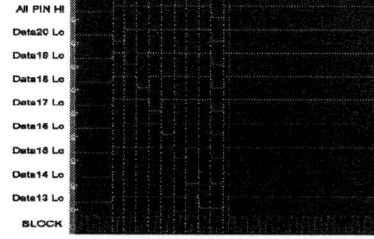

Fig.19. Measured output waveforms of TSV boundary scan in the peripheral area between the upper and lower layers in the 4-layer stacked 3-D multicore processor chip

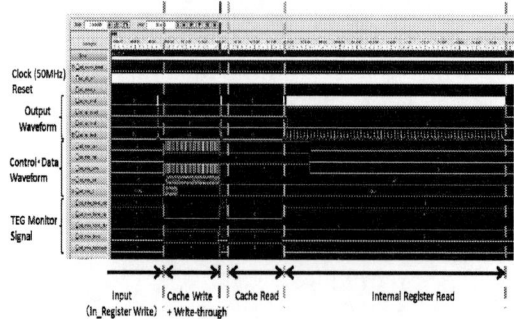

Fig.21. Measured waveforms for cache-write and register-read in the 2-layer stacked 3-D cache memory chip

978-1-4799-8002-4/14 $31.00 © 2014 IEEE

Pairwise Coupled Hybrid Vanadium Dioxide-MOSFET (HVFET) Oscillators for Non-Boolean Associative Computing

N. Shukla[1], A. Parihar[2], M. Cotter[1], M. Barth[1], X. Li[1], N. Chandramoorthy[1], H. Paik[3], D. G. Schlom[3], V. Narayanan[1], A. Raychowdhury[2] and S. Datta[1]

[1]The Pennsylvania State University, University Park, PA-16801, USA;
[2]Georgia Institute of Technology, Atlanta, Georgia, USA;
[3]Cornell University, Ithaca, NY, USA
Phone: (814) 777-8997; Fax: (814) 865-7065; Email: nss152@psu.edu

Abstract

Information processing applications related to associative computing like image / pattern recognition consume excessive computational resources in the Boolean processing framework. This motivates the exploration of a non-Boolean computing approach for such applications. In this work, we demonstrate, (i) novel hybrid set of pair-wise coupled oscillators comprising of vanadium dioxide (VO_2) metal-insulator-transition (MIT) system integrated with MOSFET; (ii) degree of synchronization between oscillators based on input analog voltage difference; (iii) implementation of hardware platform for fast and efficient evaluation of L_k fractional distance norm (k<1); (iv) improved quality of image processing and ~20X lower power consumption of the coupled oscillators over a CMOS accelerator.

Introduction

Complementary metal-oxide-semiconductor (CMOS) based Boolean logic is the cornerstone of the information technology industry. A class of problems, related to associative processing, requires massive computational resources in the Boolean framework [1]. While this has motivated the development of faster and more efficient algorithms, they are mapped back to the Boolean hardware which ultimately limits their performance. Fundamentally, this "Boolean bottleneck" arises from the requirement of a large number of power-intensive multiply–accumulate (MAC) operations, common to many associative-processing algorithms. This encourages the development of a non-Boolean approach utilizing coupled oscillatory systems (Fig. 1) [2-8]. Here, we experimentally demonstrate for the first time, various degrees of synchronization among hybrid VO_2-MOSEFT (HVFET) oscillators in response to an input analog signal difference. The synchronization dynamics enables fast and efficient calculation of a 'fractional distance norm' in the analog domain which is suitable for pattern matching in high dimensional space [9].

Single VO_2 Relaxation Oscillator

VO_2 is a correlated electron system that exhibits a metal-insulator transition (MIT) with up to 4 orders of abrupt resistivity change in 15nm thick epitaxial films (Fig.2b) (bulk films show up to 5 orders). The MIT can be triggered using various external stimuli including electrical triggering [10] (Fig. 2c). This abrupt resistivity change can be engineered into a current controlled negative differential resistance (NDR) regime using an optimum negative feedback to generate sustained charge oscillations [11]. Fig. 3a shows the single VO_2 oscillator circuit consisting of a two-terminal VO_2 device (Fig. 2a) with a MOSFET in series (HVFET-Oscillator). The details of the VO_2 film growth on the (001) TiO_2 substrate and the fabrication process are reported elsewhere [12]. All measurements are carried out at 273K. When the insulator-to-metal transition (E>E_2) is electrically induced, there is an abrupt change in the conductivity of VO_2 (Fig. 2c). The MOSFET (acting like a current source) provides a negative feedback to the VO_2 inducing an NDR such that the electric field across VO_2 drops below E_1 (Fig. 3b). This makes the metallic phase unstable causing VO_2 to return to insulating state resulting in a high field drop across VO_2 again. The process repeats inducing a stable oscillatory state [11] (Fig. 3c). The operating load line of the oscillator superimposed on the switching characteristics of the VO_2 device is illustrated in Fig. 3b. The oscillation amplitude superimposed on the output characteristics of the MOSFET (Fig. 3d) shows its operating region. Further, an input gate voltage modulates the MOSFET's operating point and enables the HVFET oscillator frequency to be programmed over a decade (Fig. 3e).

Pairwise Coupled Oscillators

To make such oscillators relevant to computing applications, we explore their synchronization dynamics. Fig. 4a shows the false-colored SEM of the fabricated VO_2 devices and illustrates the capacitively coupled oscillator configuration. The high-pass filtering configuration formed by the coupling capacitor C_C blocks DC interaction while allowing reactive power to couple. The synchronized time domain waveforms are shown in Fig. 4b along with the corresponding power spectrum (Fig. 4c). It is clear from Fig. 4d that the resonant frequency of the coupled oscillators can be tuned with an input gate voltage difference, ΔV_{GS} (= $V_{GS,2}$ - $V_{GS,1}$).

Computing with Coupled HVFET Oscillators

The nonlinear dynamics of the coupled oscillators are analyzed using the equivalent circuit in Fig. 5. Fig. 6 shows that the relative phase of the two oscillators can be tuned with the gate voltage inputs. The oscillators lock out of phase ($\sim 180°$) when ΔV_{GS} =0 V and the phase difference deviates from this value as ΔV_{GS} increases. The corresponding synchronized oscillator trajectories in phase space can be used to capture the difference between the inputs $V_{GS,1}$ and $V_{GS,2}$. An averaged exclusive-or (XOR) measure is used to analyze the oscillator output. The averaged XOR measure is defined as (i) thresholding the analog output to a binary stream, (ii) applying XOR operation on these binary values at every time instance (iii) averaging this XOR output for a finite time duration. The averaged XOR output is equivalent to the fraction of time the dynamical system spends in the grey region (Fig. 7a) where XOR = 1. The XOR output for various input V_{GS} values is shown in Fig. 7b. This output of the synchronized oscillators calculates a fractional distance norm, (L_k norm; k = 0.5), as seen by its close resemblance to the $(x_i^{0.5} - y_i^{0.5})^2$ distance map (Fig. 7c). The experimental and the simulated XOR outputs as a function of input ΔV_{GS} are shown in Fig. 7d, e, respectively.

Associative Processing Application

We investigate the application of these oscillators for visual saliency approximations (detecting parts of the image that visually standout) [13] (Fig. 8a). Oscillator-based edge detection is performed using an array of pairwise oscillators to approximate the degree of dissimilarity between a given image pixel and its immediate neighbors. Different edges, vertical, horizontal, diagonal, are detected based on the selection of neighboring pixels for comparison. As this concept is expanded to include the comparison of pixels within a larger neighborhood (pixels surrounding reference pixel; a 3x3 neighborhood is used here), the output approximates the visual saliency (Fig. 8b). It is evident that the oscillators show higher sensitivity to image contrast in comparison to a CMOS ASIC accelerator (Fig. 8a) that uses a linear $\sum |x_i - y_i|$ norm (Fig. 8b).

Power Consumption

The projected scaling of the input DC voltage and frequency with the VO_2 channel length is shown in Fig. 9a,b respectively. Fig. 10a,b shows the pareto chart of the power consumption for each component of the HVFET oscillator processor and the CMOS ASIC accelerator for visual saliency, respectively. All digital circuits were implemented with 11nm node transistor models projected from the 22nm node PTM model [14]. The coupled oscillators provide a power reduction of $\sim 20X$ over CMOS reflecting the advantage of 'let physics do the computing' approach [1] and removing the "Boolean bottleneck".

Conclusions

In this work, we provide (i) first experimental demonstration of coupled HVFET oscillators with input programmable synchronization; (ii) hardware platform capable of efficiently computing a fractional distance norm and its application in visual saliency; (iii) $\sim 20X$ reduction in power dissipation over CMOS.

Acknowledgement

This work was supported by Office of Naval Research through award N00014-11-1-0665. AP was partially funded by a gift from Intel Corporation.. SD and VN acknowledge funding, in part, from the National Science Foundation Expeditions in Computing Award-1317560.

References

[1] Shibata *et al.* CNNA, 2012.

[2] Levitan *et al.* CNNA, 2012.

[3] Narayanan *et al.* DATE, 2014.

[4] Hoppenstedt *et al.* PRL, 82,14, 1999.

[4] Datta *et al.* DAC 2014.

[5] Csaba *et al.* CNNA, 2012.

[6] Nikonov *et al.* arXiv: 1304.6125.

[6] Vassilieva *et al.*, IEEE Trans. Neural Netw. 22,84–95 (2011).

[7] Wang *et al.* IEEE Trans. Neural Netw. 6, 283–6 (1995).

[8] Izhikevich, IEEE Trans. Neural Netw.10, 508–26 (1999).

[9] C. Aggarwal *et al.* ICDT 200, 2001.

[10] Freeman *et al.*, APL, 103, 26, 2013.

[11] Shukla *et al.* Sci. Rep, 4, 4964, 2014.

[12] Tashman *et al.* APL, 104,063104 (2014).

[13] N Bruce PhD thesis, York University, Toronto, 2008.

[14] http://ptm.asu.edu/latest.html.

978-1-4799-8002-4/14 $31.00 © 2014 IEEE

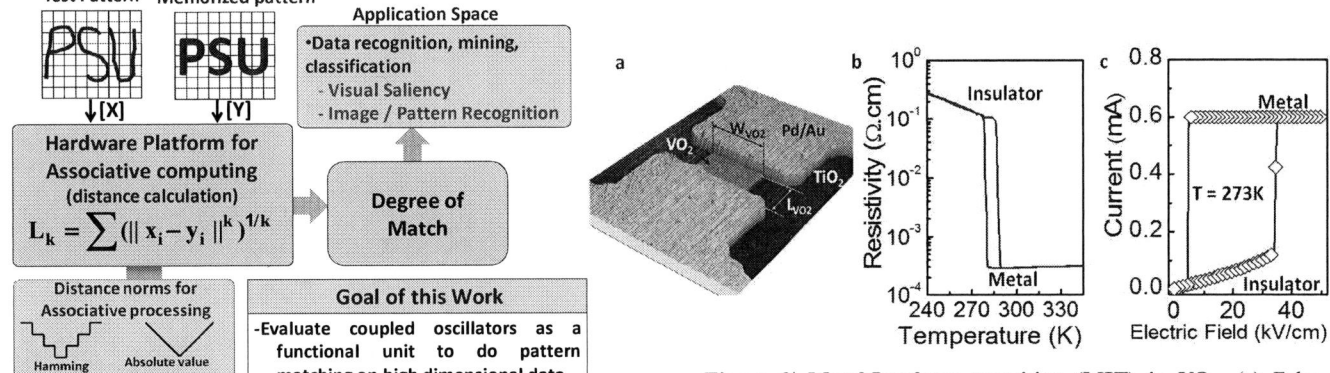

Figure 1| Motivation. We evaluate coupled oscillators as fundamental hardware block for non-Boolean associative computing. The power projection comparison of coupled oscillators with CMOS is also compared.

Figure 2| Metal-Insulator transition (MIT) in VO₂. (a) False colored AFM image of the two-terminal device with a strained VO$_2$ channel (thickness=15nm) on (001) TiO$_2$. (b) Typical resistivity of a compressively strained VO$_2$ film as a function of temperature. An abrupt change in resistivity (MIT) occurs near room temperature (c) Typical Current-Electric field characteristics of a two terminal VO$_2$ device showing the electrically induced MIT.

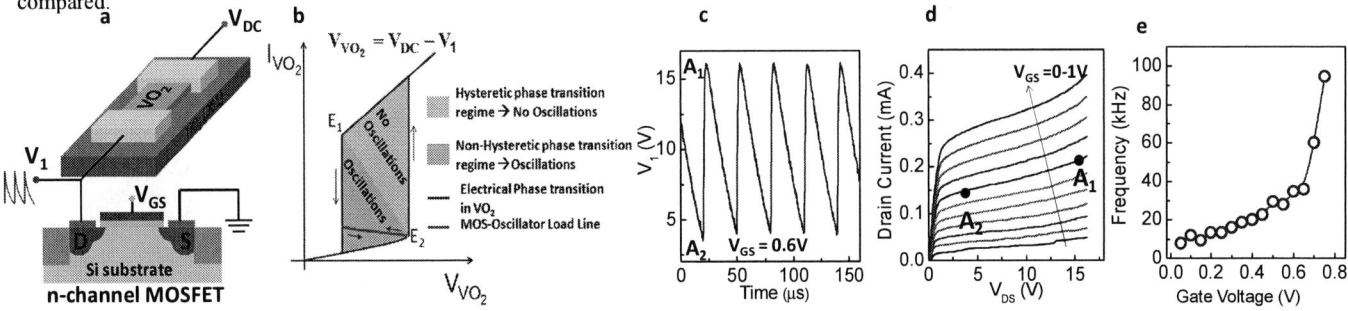

Figure 3| Hybrid VO₂-MOSFET oscillator (HVFET oscillator) with gate programmable oscillation frequency. (a) Schematic of a VO$_2$ relaxation oscillator consisting of a two terminal VO$_2$ device in series with the source-drain of a MOSFET. (L$_{VO2}$= 4 μm; W$_{VO2}$=6μm) (b) Schematic of the operating load line (red) of a HVFET oscillator superimposed on the two terminal I-V characteristics of the VO$_2$ device (blue). The MOSFET induces a non-hysteretic oscillatory regime in VO$_2$ through negative feedback. (c) Time-Domain waveform of the HVFET oscillator. These oscillations have been shown to be stable over 2.5x10^9 cycles. (d) Oscillation amplitude (A$_1$-A$_2$) superimposed on the output characteristics of the MSOFET.(e) Oscillation frequency as a function of gate voltage (V$_{GS}$) enabling VCO operation.

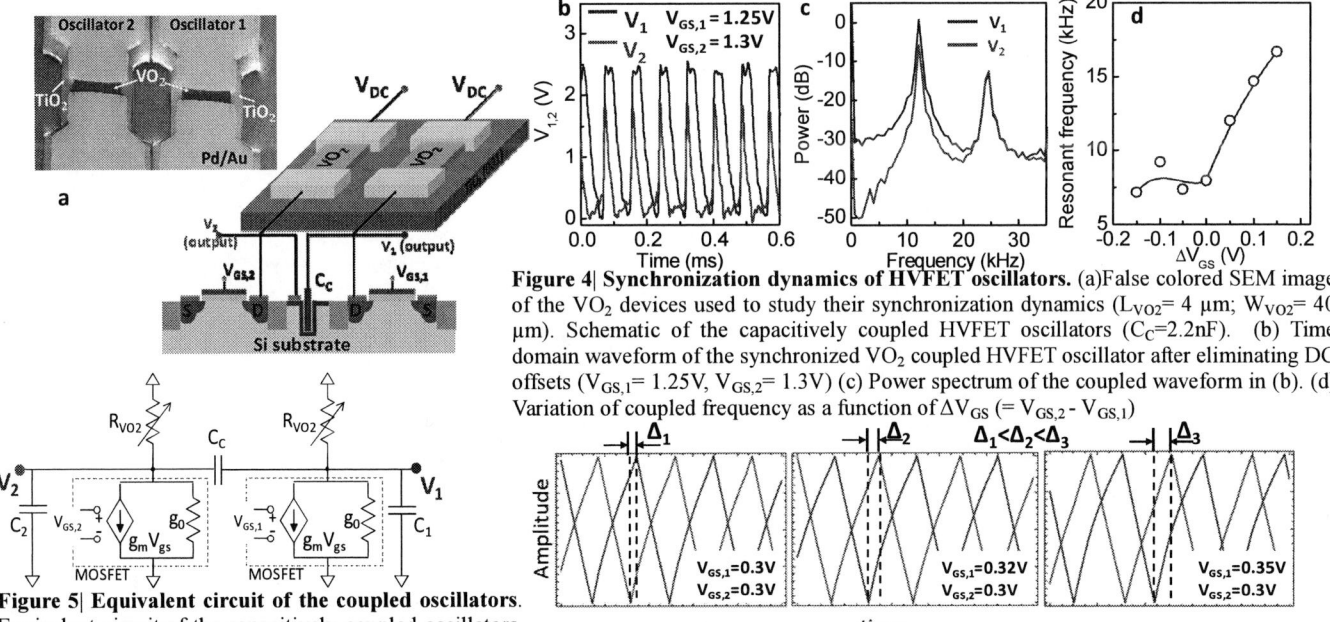

Figure 5| Equivalent circuit of the coupled oscillators. Equivalent circuit of the capacitively coupled oscillators. C$_1$ and C$_2$ are the equivalent capacitance of the VO$_2$ device and the output circuit of the MOSFET.

Figure 4| Synchronization dynamics of HVFET oscillators. (a)False colored SEM image of the VO$_2$ devices used to study their synchronization dynamics (L$_{VO2}$= 4 μm; W$_{VO2}$= 40 μm). Schematic of the capacitively coupled HVFET oscillators (C$_C$=2.2nF). (b) Time domain waveform of the synchronized VO$_2$ coupled HVFET oscillator after eliminating DC offsets (V$_{GS,1}$= 1.25V, V$_{GS,2}$= 1.3V) (c) Power spectrum of the coupled waveform in (b). (d) Variation of coupled frequency as a function of ΔV_{GS} (= V$_{GS,2}$ - V$_{GS,1}$)

Figure 6| Input gate voltage controlled synchronization characteristics of oscillators. The relative difference (indicated by black lines) between the synchronized waveforms can be tuned and increases with ΔV_{GS}.

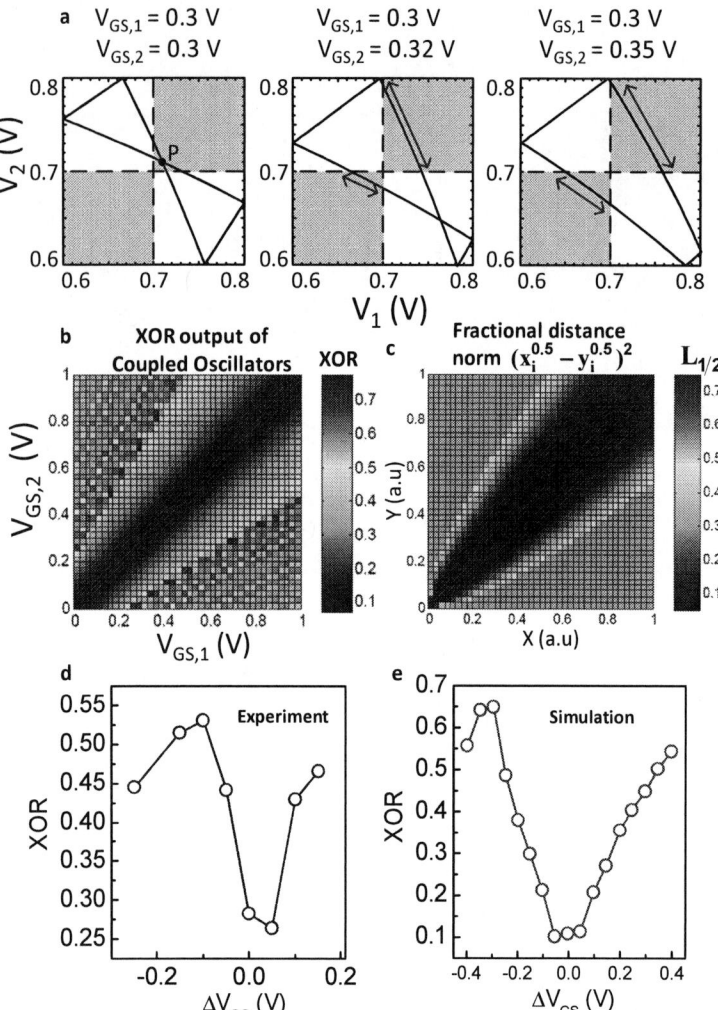

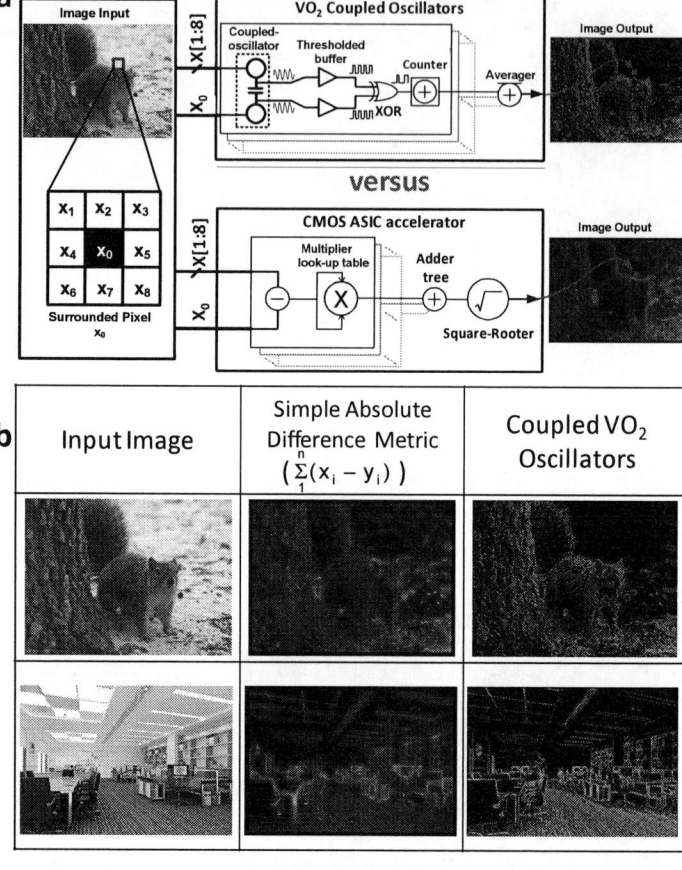

Figure 8| Visual Saliency processing using coupled HVFET oscillators and comparison with CMOS ASIC accelerator. (a) Schematic of the processing scheme for saliency detection with VO_2 coupled oscillators and CMOS ASIC accelerator. (b) Saliency detection outputs obtained with VO_2 coupled oscillators and using

the simple $\sum_{1}^{n}(x_i - y_i)$ metric.

Figure 7| XOR of oscillator output as a fractional distance norm and a computing metric for pair-wise coupled oscillators. (a) Plots showing the relation between averaged XOR output of the oscillators and the steady state periodic orbit of the coupled oscillator system. The averaged XOR output is the fraction of time spent in the gray region, which also captures the phase difference between $V_{GS,1}$ and $V_{GS,2}$. (b) Simulated XOR output as a function of $V_{GS,1}$ and $V_{GS,2}$. The XOR value is minimum when $V_{GS,1}=V_{GS,2}$. (c) Fractional distance norm $(x_i^{0.5} - y_i^{0.5})^2$. The XOR metric computes the Euclidean distance as seen by the similarity between the (b) and (c). (d) Experimental (e) Simulated XOR as a function of ΔV_{GS}.

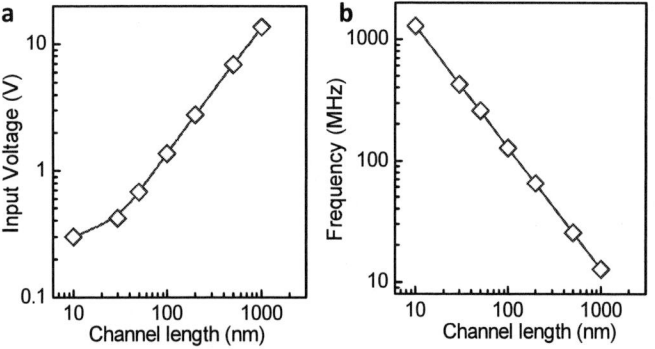

Figure 9| Projected VO_2 oscillator scaling. (a) Scaling of input DC voltage of the oscillator (b) Projected scaling of oscillator frequency with the VO_2 channel length.

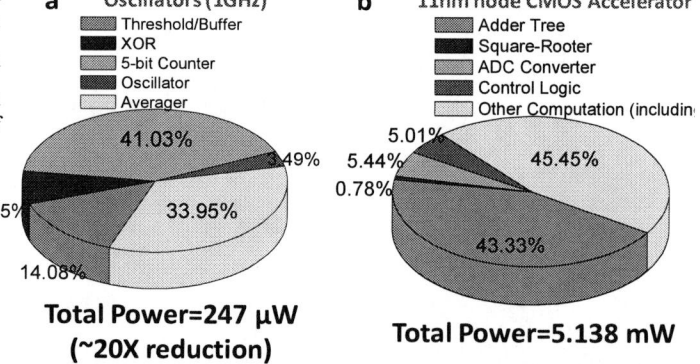

Figure 10| Power comparison. (a) Power dissipation break-down of individual components in the oscillatory processing scheme. (b) Power dissipation breakdown of individual components in a conventional synthesized CMOS ASIC processing accelerator. The Adder tree and the MAC instructions contribute to the bottle-neck in this Boolean processing scheme. All the digital circuits were implemented with 11nm node transistor models projected from the 22nm PMT model.

978-1-4799-8002-4/14 $31.00 © 2014 IEEE

Hybrid CMOS/BEOL-NEMS Technology for Ultra-Low-Power IC Applications

Nuo Xu[*], Jeff Sun, I-Ru Chen, Louis Hutin[§], Yenhao Chen, Jun Fujiki[†], Chuang Qian and Tsu-Jae King Liu

Department of Electrical Engineering and Computer Sciences, University of California, Berkeley, CA 94720 USA

currently with [§]CEA-LETI Minatec, 38054 Grenoble Cedex 9, France; [†]Toshiba Corporation, Tokyo 105-8001, Japan

Phone: +1-510-643-2639, Fax: +1-510-643-2636, [*]E-mail: nuoxu@eecs.berkeley.edu

Abstract

Three-dimensional (3-D) nano-electro-mechanical (NEM) switches (relays) are proposed to reduce the die area and power consumption of digital logic and memory circuits.

Introduction

Reductions in chip power consumption and area per function drive continued advancements in integrated circuit (IC) technology. Nanometer-scale electro-mechanical switches (NEMS) are potentially advantageous for ultra-low-power circuit applications because they have the ideal properties of zero OFF-state leakage and zero sub-threshold swing (SS) which in principle provide for switch operation with very small voltage swing [1-6]. However, logic NEMS demonstrated to date based on various technologies have relatively large operating voltages as well as large footprint [4, 5], and would require extra process steps to be monolithically integrated with (higher-performance) CMOS circuitry [5, 6]. We propose to leverage an advanced back-end-of-line (BEOL) process, i.e. with aggressively scaled metal wires and air-gap interconnect structures, to integrate three-dimensional (3-D) NEM switches with CMOS transistors for ultra-low-power, compact digital logic and memory circuits. NEMS performance and circuit operation are analyzed via 3-D mechanical simulations and compact modeling and circuit simulations, respectively.

Experimental

Fig. 1 shows a planar (2-D) prototype implementation of the single-pole/double-throw (SPDT) relay switch design investigated in this work, fabricated using either poly-SiGe or aluminum (Al). The switch comprises an Input node/electrode extending into a movable beam; electrostatic-actuation electrodes on either side of the movable beam and separated from the beam by actuation air-gaps; data electrodes $D_{\{0,1\}}$ on either side of the beam end and separated from the beam end by contact air-gaps smaller than the actuation air-gaps; and an Output node/electrode. Depending on the Input voltage, the beam end is actuated into contact with one of the data electrodes as well as the Output. The fabrication process flow is outlined in Fig. 1; photoresist overexposure was used to form sub-lithographic contact gaps to achieve relatively low pull-in voltage (V_{PI}) [7]. Fig. 2 shows measured terminal currents as a function of the voltage applied between the beam and one actuation electrode, for a poly-SiGe switch. V_{PI} can be reduced by scaling down the actuation gap size and beam width (Fig. 3), and by using Al as a softer beam material (Fig. 4).

BEOL NEM Relay Design

The International Technology Roadmap for Semiconductors (ITRS) projects that the metal wire width

and pitch will be scaled down very aggressively [8], which is advantageous for compact implementation of NEM switches. Recently developed air-gap interconnect technologies [9] enable a 3-D NEM switch to be fabricated without the need for special chip packaging. Fig. 5(a) illustrates how a SPDT NEM relay can be implemented using multiple metal layers in a state-of-the-art CMOS BEOL technology, leveraging air-gap interconnect structures. Metal layers 2-4 (M_{2-4}) are used to form the actuation portions of the movable structure while metal layer 5 (M_5) is used to form the contact portion. Note that a portion of the movable structure can be formed with an inter-metal dielectric (IMD) to electrically insulate the Input from the Output. Figs. 5(b) and 5(c) show the layouts for the actuator and contact layers, respectively. The contact gap is assumed to be minimally sized ($1F$, where F is the minimum feature size), while the as-fabricated actuation gaps are $1.5F$ to avoid catastrophic pull-in (i.e. beam contacts with an actuation electrode). Reductions in metal wire width and pitch following the industry roadmap [8] are beneficial for lowering V_{PI}, as well as for reducing device footprint. (Note that the device width is only $6F$.) The Coventor MEMS+ software tool [10] was used to simulate the operation of the BEOL NEM relay. Fig. 6 shows an optimized design with a serpentine movable structure comprising multiple layers of BEOL wires and vias, to minimize V_{PI}. As expected, contact first occurs in the M_5 layer as the actuation voltage is increased. Key process and design parameters are summarized in Table I. Fig. 7 shows how V_{PI} and the catastrophic pull-in voltage (V_{CA}) change with the device footprint which depends on actuator beam length. $V_{CA} - V_{PI}$ represents the maximum gate overdrive voltage and limits the device switching speed [7]. To estimate the switching delay, transient mechanical simulation was performed for the NEMS structure and a compact model (Fig. 8) similar to that in [1, 7] was used to evaluate the electrical (i.e. RC) delay, assuming that the beam is initially in contact with one of the data electrodes and is actuated by applying a voltage (in the range from V_{PI} to V_{CA}) between the beam and the actuation electrode on the opposite side. The simulated voltage waveforms and beam (contact portion) displacement vs. time shown in Fig. 9 indicate that the mechanical delay (~17 ns) limits the switching speed. Fig. 10 shows how the switching delay can be reduced by increasing the operating voltage (V_{DD}), for switches of different size. ~1 V operation with 20 ns switching delay is projected for a BEOL NEM relay with footprint < 0.1 μm^2.

For the NEM relay to switch states, the sum of the electrostatic force (F_{elec}, induced by an applied actuation voltage) and the spring restoring force of the beam (F_{spring}) must exceed the contact adhesive force ($F_{adhesion}$). For the relay to maintain its state with the actuation voltage removed (i.e. $V_{DD} = 0$), F_{spring} must be less than $F_{adhesion}$. In this case, it functions as a non-volatile (bi-stable) switch (Fig. 11).

978-1-4799-8002-4/14 $31.00 © 2014 IEEE

Compact Hybrid CMOS-NEMS Circuits

Digital Logic with Large Capacitive Loads

For large-scale data link applications, CMOS buffer circuits usually occupy a large die area and consume large amounts of power [11]. NEM relay switches integrated with CMOS (C+N) offer a potential solution to this issue. Fig. 12 compares CMOS *vs.* C+N implementations of various logic buffer circuits. The CMOS circuits are designed with the same logic depth (LD) so that they have comparable delay. For the C+N designs, NEM switches are used to isolate the output from the input and to reduce the CMOS load capacitances. For both implementations, the electrical effort and P/NMOS width ratio are adjusted according to the designs in [12] to achieve optimal layout area and power efficiency. Fig. 13 shows the equivalent circuit for 250 parallel buffer circuits, with total load capacitance of 100 fF for CMOS and only 50 fF for C+N due to the lower parasitic capacitance of NEM switches as compared with CMOS transistors (which have significant gate-overlap and fringe capacitances [8]). Figs. 14 and 15 compare the layout area and standby power of CMOS *vs.* C+N buffers based on 20nm low-standby power (LSTP) multi-gate CMOS technology [13], clearly showing the advantages of the C+N implementations. By adjusting the size of the transistors and V_{DD}, better energy efficiency can be achieved at the trade-off of increased delay, as shown in Fig. 16.

Non-Volatile SRAM and CAM

Power gating is effective for reducing the standby power consumption of inactive circuit blocks to maximize energy efficiency, but degrades system performance due to the need to store/restore state to/from memory blocks each time the logic block is powered off/on. The ability to remember the state of digital logic circuits and static random-access memory (SRAM) cells when they are powered down would eliminate this trade-off. Recently, researchers have investigated the 3-D integration of one or more non-volatile memory (NVM) element(s) within a single SRAM cell to achieve fast power-off/on speed [14, 15]. Various types of NVM devices such as magnetic memory (MRAM), ferroelectric memory (FRAM), and resistive memory (RRAM) have been used to implement NVSRAM cells [14-19]. However, each of these requires either high voltage or significant current to program, so that additional transistors are required to isolate the NVM device(s) during SRAM operation and to separately implement a NV-store function just prior to power-off. In other words, conventional NVM devices are not sufficiently energy efficient to shadow the content of an SRAM cell or a latch. A simplified NV NEM switch (with no Output electrode or insulating portion in the movable structure) is proposed to solve this issue. Fig. 17 shows the circuit schematic for a NVSRAM cell design comprising a standard 6T SRAM cell and a NV NEM switch with data electrodes driven through 2 additional access transistors. (To ensure proper write operation without disturbing another cell sharing the same bit lines, the cell data electrodes cannot be shared across a memory array.) The operating voltages for this 8T1N cell are summarized in Fig. 17 (bottom). Ordinarily this cell functions as a regular SRAM cell when the CMOS latch is powered (*i.e.* when EN is biased at V_{DD}); the Restore Word Line (RWL) is Low (ground) so that the voltage of the NEM switch beam corresponds to the data (Q) at the storage node (SN), causing it to be actuated into contact with the D_0 electrode if Q is Low, or alternatively with the D_1 electrode if Q is High, as ST is ordinarily biased High. To power down the cell, EN and ST are simply gated off, for zero power consumption in the sleep state. This is a distinct advantage of using a bi-stable NEM switch as compared with other NVM devices for this application. To restore the state of the cell (*i.e.* to write Q back into the SN), D_0 is driven to ground while D_1 is driven to V_{DD}; then RWL is enabled (biased at V_{DD}) to drive SN as the CMOS latch is powered on (*i.e.* as EN is ramped up to V_{DD}). The simulated voltage waveforms during a Restore operation are shown in Fig. 18. Fig. 19 benchmarks the bi-stable NEM switch technology against other NVM technologies [14-19] with respect to storage delay time and storage energy (calculated based on the formalism developed in [1, 7]). The far superior energy efficiency of the NEM switch makes it the only practical option for continual SRAM shadowing, which eliminates the need for performing an explicit NV-Store operation before powering down. It should be noted that the input capacitance of the NEM switch (~ 0.01 fF) is much lower than that of a storage node (> 0.1 fF) in a standard 6T SRAM cell, so that NVSRAM cell should operate with speed and energy comparable to that of a conventional 6T SRAM cell.

Content Addressable Memory (CAM) is used for high-speed searching applications and is used to classify and forward internet protocol (IP) packets [20]. A modified version of the 8T1N NVSRAM cell with the full 6-terminal NEM relay design (with the Output electrode portion of the beam electrically isolated from the Input electrode portion) can be used to implement a CAM cell, as shown in Fig. 20. The state of the NEM relay (*i.e.* the data electrode to which the Output terminal is connected) is determined by the content of the SRAM cell, and the data electrodes are connected to ground via access transistors gated by the Searching line (SL). The Matching Line (ML) is pre-charged high before the data in the SRAM cell is compared against the data on the Searching Line (SL); if the comparison yields a mismatch, then ML is pulled Low (connected to ground) through a NEM relay contact and an access transistor. The simulated waveforms in Fig. 21 demonstrate the matching operation of this NVCAM cell.

Conclusion

By leveraging advanced BEOL technology, low-voltage, small footprint and energy-efficient NEM relays can be realized. The potential applications of this technology include large-scale logic buffer circuits and non-volatile SRAM and CAM.

Acknowledgements

This work was supported in part by the Center for Energy Efficient Electronics Science (NSF Award 0939514). Nuo Xu would like to thank Professor Shimeng Yu from Arizona State University for helpful discussions.

References:

[1] M. Spencer, *JSSC*, 46, p.308, 2011. [2] H. Kam, *IEDM*, p.809, 2009. [3] D. Lee, *IEEE T-CADICS*, 32, p.653, 2013. [4] J. Jeon, *IEEE EDL*, 31, p.515, 2010. [5] T. He, *IEDM*, p.108, 2013. [6] N. Sinha, *IEDM*, p.813, 2009. [7] H. Kam, *IEEE T-ED*, 58, p.236, 2011. [8] *ITRS*, 2013 Edition. [9] B. Shieh, *IEEE EDL*, 19, p.16, 1998. [10] *MEMS+ User Guide*, 2013. [11] A. P. Chandrakasan, *Proc. IEEE*, 83, p.498, 1995. [12] R. Zimmermann, *JSSC*, 32, p.1, 1997. [13] *PTM Models for 20nm LSTP Technology*, 2012. [14] P.-F. Chiu, *VLSI Symp. Circ.*, p.229, 2010. [15] W. Wang, *IEDM*, 2006. [16] S. Yamamoto, *CICC*, p.531, 2009. [17] T. Miwa, *JSSC*, 36, p.522, 2001. [18] N. Sakimura, *JSSC*, 44, p.2244, 2009. [19] R. Nebashi, *ISSCC*, p.462, 2009. [20] K. Pagiamtzis, *JSSC*, 41, p.712, 2006.

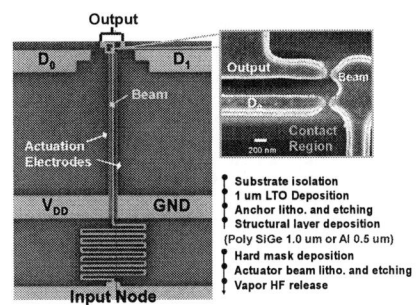

Fig. 1: Plan-view micrograph of a fabricated prototype relay, and the process flow.

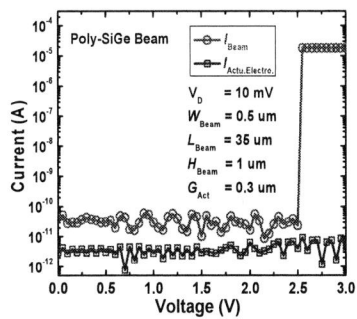

Fig. 2: Measured terminal currents *vs.* voltage for a poly-SiGe relay. The Input node is grounded while one of the actuation electrodes is used to pull in the beam.

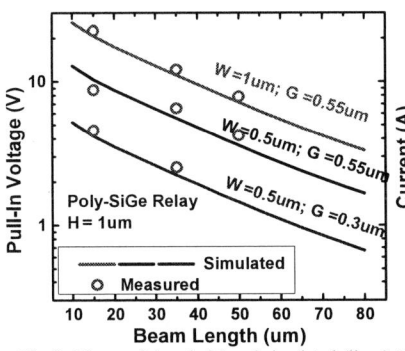

Fig. 3: Measured (symbols) and simulated (lines) V_{PI} from different relay designs. *Coventor MEMS+* Software [17] was used for the simulation.

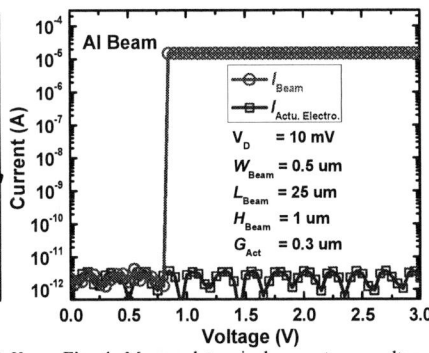

Fig. 4: Measured terminal currents *vs.* voltage for an Al relay, showing the possibility to reduce pull-in voltage (V_{PI}).

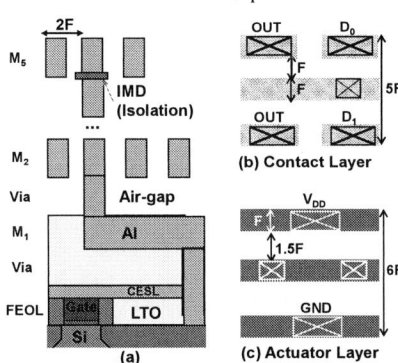

Fig. 5: (a) Illustration of the BEOL implementation of a NEM relay; and layout designs for (b) contact and (c) actuation beam layers of the NEMS.

Technology/Design	Properties
Metal Layer(s) used	$M_{2\text{-}4}$ Actuation M_5 Contact
Metal/Via Material	Al
Metal/Via Pitch (2F)	42 nm [8]
Metal/Via Width (F)	21 nm [8]
Metal/Via Aspect Ratio	1.9 [8]
IL Dielectric	Air-gap [9]

Tab. I: Key technology and design parameters for NEMS, based on advanced BEOL technologies [8, 9].

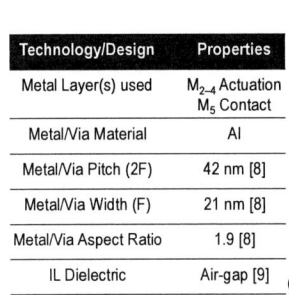

Fig. 6: Device structure of a BEOL NEMS using M_1 as the anchor, $M_{2\text{-}4}$ as the actuation layer, and M_5 as the contact layer.

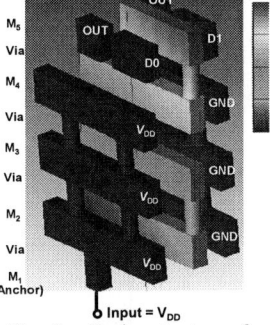
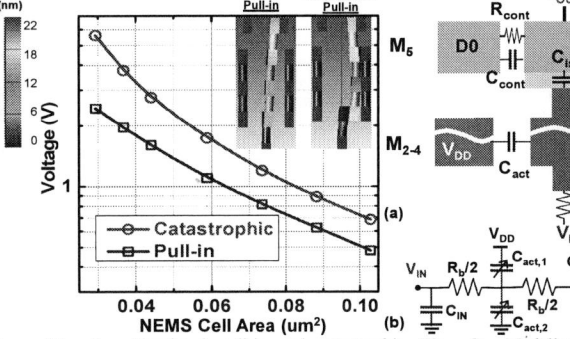

Fig. 7: Simulated pull-in and catastrophic voltages for the BEOL NEMS *vs.* footprint, adjusted by varying the actuator beam length. The inset shows the situations of normal pull-in and catastrophic pull-in of the beam.

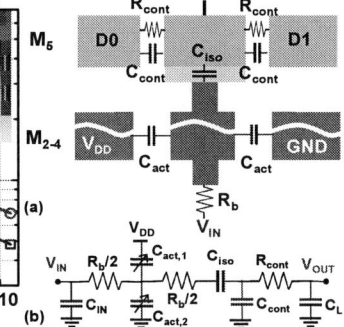

Fig. 8: Modeling of the electrical components in a BEOL NEM relay: (a) illustration of all relevant components and (b) the SPICE model.

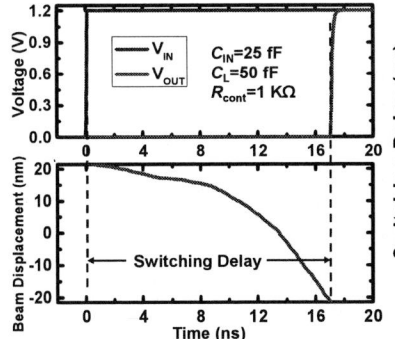

Fig. 9: Simulated (a) input and output voltage pulses; and (b) the displacement of the NEMS beam (contact portion) during switching.

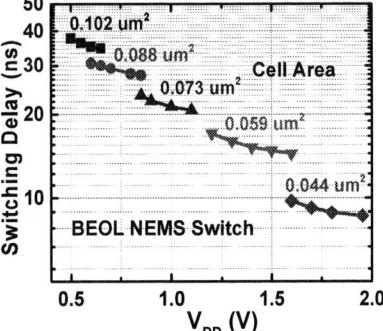

Fig. 10: Simulated switching delay *vs.* V_{DD} for BEOL NEMS with different cell areas.

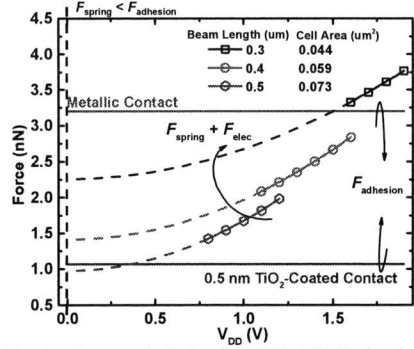

Fig. 11: Force analysis for the BEOL NEMS showing the condition to form a (non-)volatile switch. $F_{adhesion}$ is calculated based on the experimental results in [10].

978-1-4799-8002-4/14 $31.00 © 2014 IEEE

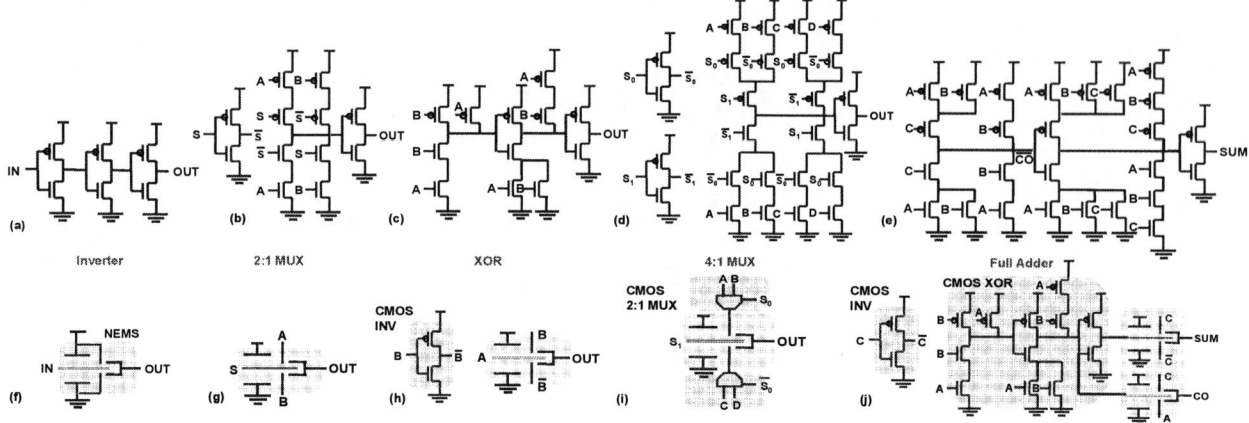

Fig. 12: Illustration of various types of buffer circuits: (a, f) inverter, (b, g) 2:1 MUX, (c, h) XOR, (d, i) 4:1 MUX, and (e, j) full adder, based on (a-e) CMOS and (f-j) CMOS+NEMS (C+N) designs. All the CMOS buffers are designed to have the same logic depth ($LD = 3$) and therefore comparable delay.

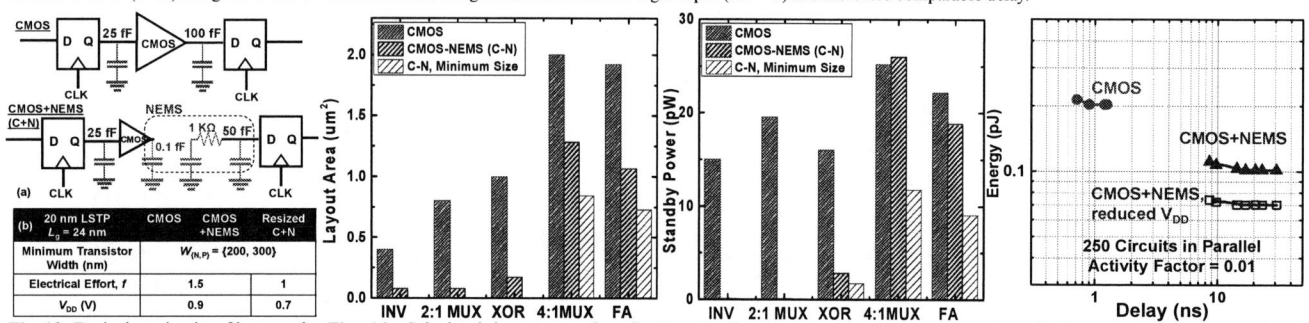

Fig. 13: Equivalent circuits of large-scale data buffers, based on CMOS and C+N. The table lists key design parameters.

Fig. 14: Calculated layout area for all logic buffers, based on CMOS and C+N; a resized C+N design is also included.

Fig. 15: Simulated standby power consumption of all logic buffers, based on CMOS and C+N; a resized C+N design is included. 20 nm LSTP technology was assumed [13].

Fig. 16: Energy vs. Delay for buffer circuits, for all technology variations.

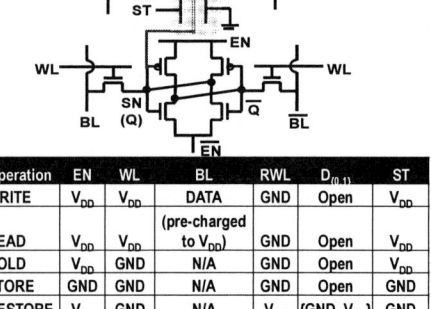

Fig. 17: (top) Proposed 8T1N non-volatile SRAM (NVSRAM) cell design, and (bottom) bias voltages for Write, Read, Hold, Store, and Restore operations.

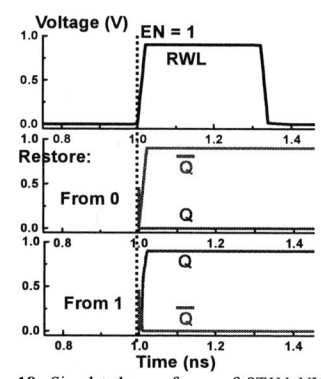

Fig. 18: Simulated waveforms of 8T1N NVSRAM cell during Restore operation, using 20 nm LSTP models [13] with $W_{N/P}$= 50/100 nm; and $C_{BL} = C_{RWL} = C_{WL}$ = 100 fF.

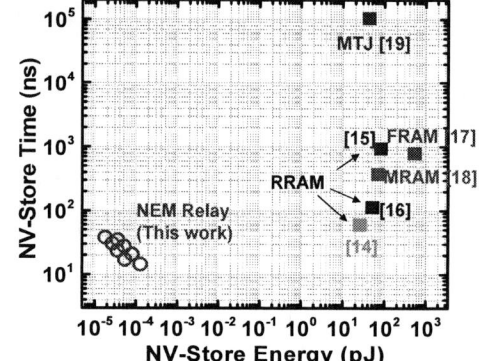

Fig. 19: Simulated store time vs. store energy of the NV NEM relay, compared with other NVM devices [14-19]. NEMS designs are from **Figs. 7 & 10.**

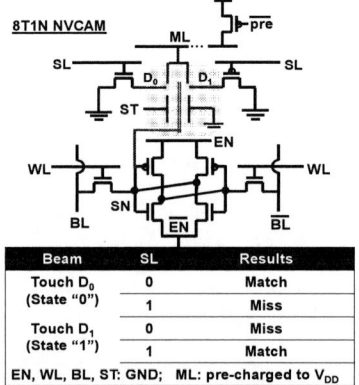

Fig. 20: (top) Proposed non-volatile CAM (NVCAM) cell design, and (bottom) operation summary.

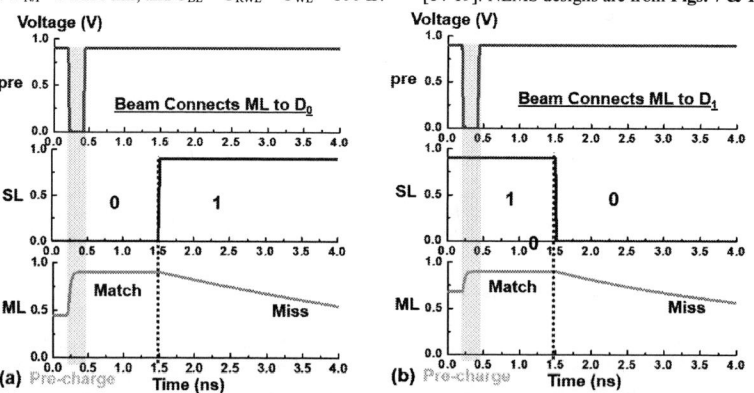

Fig. 21: Simulated waveforms of NVCAM cell during Matching operations, using 20 nm LSTP technology models [13] with W_N= 50 nm, W_P= 100 nm; and $C_{ML} = C_{SL}$ = 100 fF.

Optimization Metrics for Phase Change Memory (PCM) Cell Architectures

M. Boniardi, A. Redaelli, C. Cupeta, F. Pellizzer, L. Crespi[§], G. D'Arrigo†, A. L. Lacaita[§] and G. Servalli

Micron Semiconductor Italia s.r.l., R&D, Via C. Olivetti 2, 20864, Agrate Brianza, Italy, EU – Email: mboniardi@micron.com
[§]Politecnico di Milano, Dipartimento di Elettronica, Informazione e Bioingegneria (DEIB), P.zza L. da Vinci 32, Milano, Italy
†Consiglio Nazionale Ricerche - Istituto per la Microelettronica e Microsistemi (CNR-IMM), VIII Strada 5, Catania, Italy

Introduction

In Phase Change Memory (PCM) the storage element can be realized with different cell architectures [1-4], involving two main device approaches to heat up and program the cell; those can be gathered together in the *self-heating* and in the *heater-based* families. Here we report on different PCM cell architectures, based on standard $Ge_2Sb_2Te_5$ (GST) material, being fabricated with dedicated processes and studied in terms of their program/read efficiency and integration features. Despite the realized self-heating approaches showing slightly better efficiency, the heater-based *Wall* architecture is claimed to be the best one for process integration, still matching the electrical target. Further optimization guidelines for the *Wall* architecture, based on a simple model, are then provided, involving both electrical and thermal arguments.

Cell Architectures

We developed a variety of cell structures, exploring *Self-Heating (SH)* approaches with different aspect-ratios, both planar and vertical, to be compared with our conventional heater-based *Wall* architecture [1]. The electrical targets used for architecture comparison are related to product specifications, in particular the set resistance, $R_{SET} < 13k\Omega$, defines the read latency, while the programming current, $I_{PROG} < 200\mu A$, is linked to program throughput and power consumption [5]. Fig. 1a and b report a vertical *Self-Heating Wall (SHW)* cell architecture in which GST is deposited on a dielectric trench and then etched to realize a spacer. It has height L = 60 nm, GST thickness t = 9 nm and different widths, in the range W = 60 nm – 200 nm, obtained with a

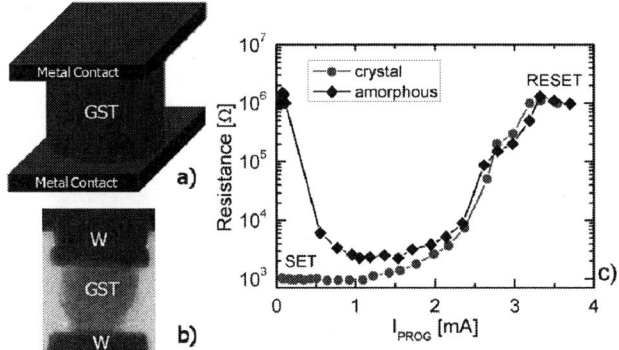

Figure 2. Vertical *Self-Heating Pillar* architecture (a), TEM y-section image of the integrated cell (b), R-I curves of a d≈50nm cell (c).

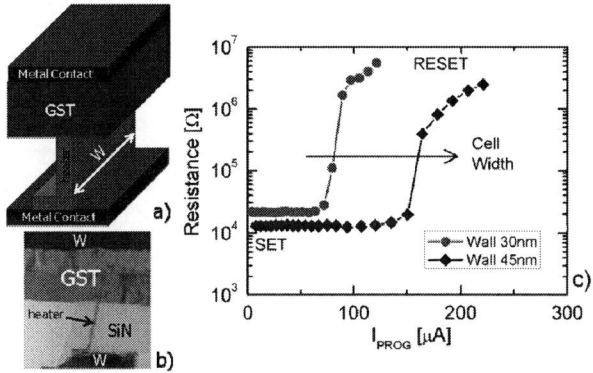

Figure 3. Vertical, heater-based, *Wall* architecture (a), TEM y-section image of the integrated cell (b), R-I curves of cells realized with two different widths, W (c).

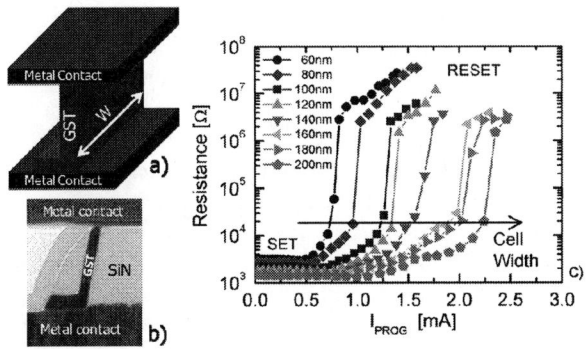

Figure 1. Vertical *Self-Heating Wall (SHW)* architecture (a), TEM y-section image of the integrated cell (b), R-I curves of cells with different widths, W (c).

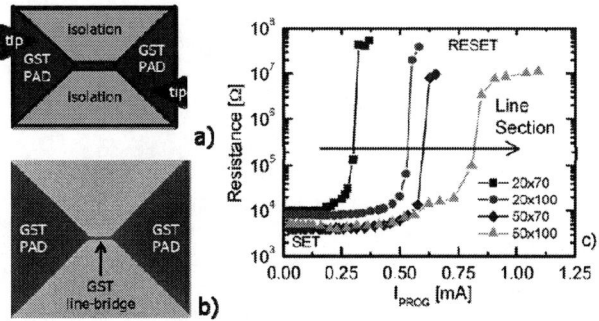

Figure 4. Planar *Self-Heating Line-Bridge* architecture concept (a), in plane SEM image of the realized cell (b), R-I curves of lines with different section, S=W·t (c).

978-1-4799-8002-4/14 $31.00 © 2014 IEEE

dedicated masking step (cell section is S=W·t). GST is in contact with two tungsten electrodes. As expected, R-I curves in Fig. 1c report increasing reset current I_{RESET} and decreasing set resistance R_{SET} as a function of cell W (or, equivalently, section S).

Figure 2 shows a self-heating cell in which GST forms a confined pillar, contacted by two tungsten electrodes. This cell is called in the following *SH-Pillar* and has a height L = 60 nm and a section diameter in the range of 50 nm. R-I curves in Fig. 2c show a quite low set state resistance.

The heater-based *Wall* architecture [1] is reported in Fig. 3, with a heater thickness t_H = 5 nm, here realized with two wall widths, W = 45 nm [1] and shrunk to W = 30 nm. Also here R-I curves in Fig. 3c show increasing I_{RESET} and decreasing R_{SET} with increasing the cell width W.

Finally, Fig. 4 reports a planar self-heating *Line-Bridge* cell architecture, obtained by e-Beam lithography. Line cells have fixed length, L = 400 nm, different widths ranging from W = 70 nm to 100 nm and different thickness t = 20 nm and 50 nm (cell section is S=W·t). R-I curves in Fig. 4c show the expected trend of R_{SET} and I_{RESET} as a function of geometry.

Efficiency Metrics and Architecture Comparison

In PCM the previously presented parameters, I_{PROG} and R_{SET}, are intrinsically linked together thanks to the *Joule effect* [6]: in general the higher R_{SET}, the lower I_{PROG} required to reach the reset state, but the more critical the read latency time. Similarly, the lower R_{SET}, the higher I_{PROG}, but the more critical is power consumption and program throughput. Figure 5 reports the I_{PROG}-R_{SET} correlation plot for the realized architectures. The best working area for PCM would be the left-bottom side of the plot, that is both low R_{SET} and I_{PROG}, or minimized R_{SET}·I_{PROG} product. As a case study we report in Fig. 6 the equations needed to define the R_{SET}·I_{PROG} metrics in the simple case of a *SH* pillar cell. The *Fourier's law* of heat transfer, the *Joule heating* equation, the cell thermal resistance (accounting for two parallel heat paths towards both electrodes, with the assumption of negligible lateral heat loss, according to data in Fig. 5) [7] and the on-

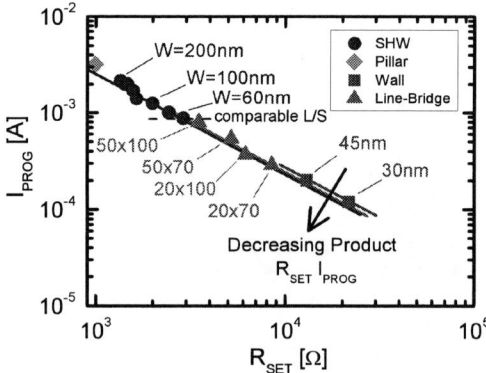

Figure 5. I_{PROG}-R_{SET} diagram. A slope -0.7 was expected (anisotropic scaling) [7]; instead, found slope -1 may be ascribed to a negligible lateral heat loss dependent on technology node and/or on additional lateral thermal resistances resulting from process.

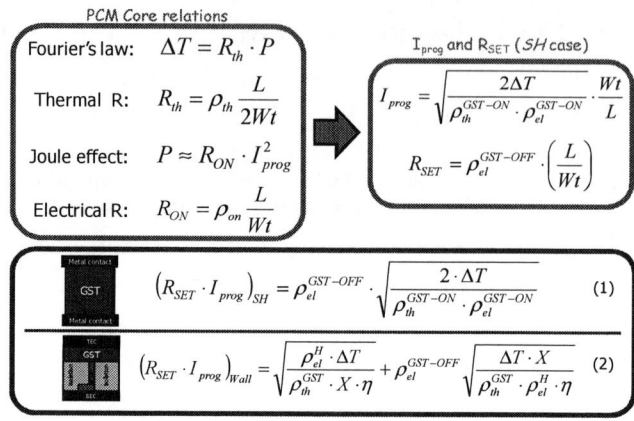

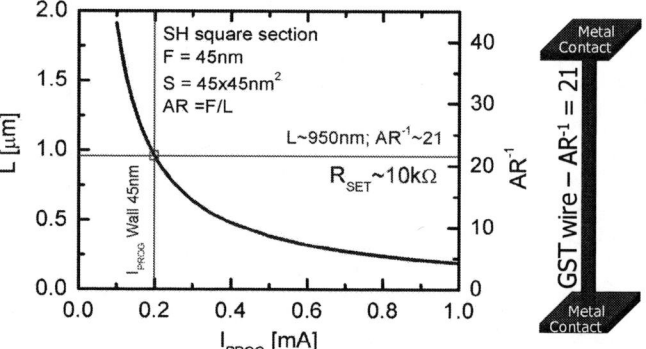

Figure 6. PCM operation equations, leading to R_{SET}·I_{prog} product ruled for the case study of *SH* in Eq. (1) and *Wall* architecture in Eq. (2). In the *Wall* case, X is a W independent factor, $X = f(t_H, L_H, L_{GST})$.

Figure 7. Square section *SH* geometry projection to meet the *Wall* 45nm working point. The ruled *SH* cell height, L, leads to hazardous aspect ratios for technology implementation (see sketch).

state electrical resistance can be combined together to rule the current I_{PROG} and the R_{SET}·I_{PROG} product. Such metrics, for *SH* cells, may be described by using the simple equations in Fig. 6 with thermal parameters taken as *effective* values. As reported by Eq. (1), in a *SH* architecture the studied product depends on the material properties only, *i.e.* on the ON and OFF electrical resistivity, the thermal resistivity and the melting temperature, and keeps constant at different geometries, *i.e.* different L/S where L is the cell height (or length) and S is the cell section. A similar relationship can be derived for the *Wall* architecture, in Eq. (2): here the product R_{SET}·I_{PROG} keeps constant if only the cell width W is changed, while leaving all the other dimensions unchanged, as experimentally done for the *Wall* 30 nm cell in Fig. 3.

Figure 5 highlights that *all the SH architectures*, having different L/S ratios, are aligned on the same line. For example architectures with very different geometrical features, as the vertical *SHW* with W = 60 nm and the planar *Line-Bridge* with section S = 50x100 nm², are almost superimposed at very close nominal L/S values, confirming that R_{SET}·I_{PROG} does not depend on geometry for *SH* (See dashed line in Fig. 5). In the *Wall* case R_{SET}·I_{PROG} shows

978-1-4799-8002-4/14 $31.00 © 2014 IEEE

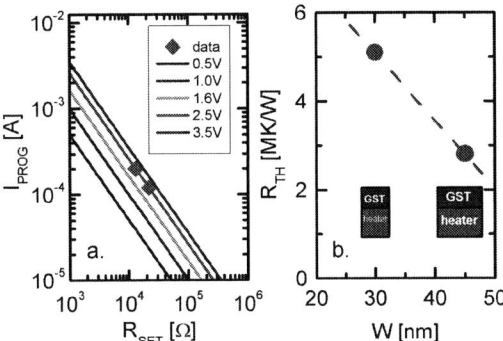

Figure 8. *Wall* architecture width modulation does not impact R_{SET} ·I_{PROG} (a). I_{PROG} reduction coming from W shrink is due to increasing thermal and on-resistances, R_{TH} and R_{ON} ($\Delta T = R_{TH} R_{ON} I^2_{PROG}$) (b).

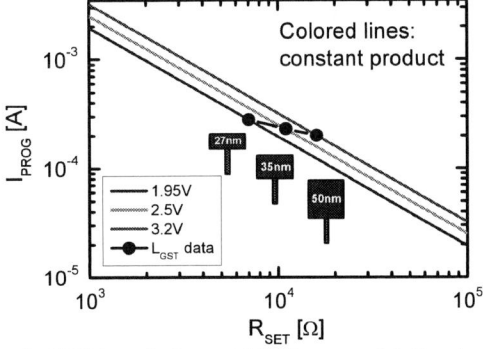

Figure 9. GST length, L_{GST}, trials. L_{GST} modulation towards lower values allows lower R_{SET}·I_{PROG} product, suggesting higher thermal efficiency. Note that higher current injection is required with lower L_{GST}.

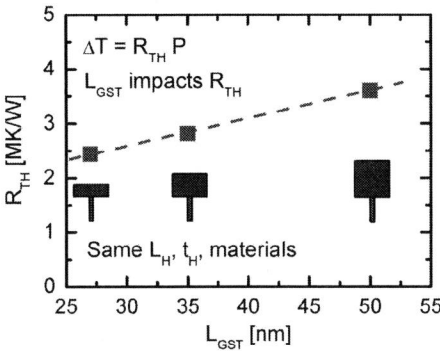

Figure 10. GST length, L_{GST}, trials. Decreasing L_{GST} introduces lower R_{TH}, or lower degree of heat confinement, hence higher power injection requirements, but with better readout performance.

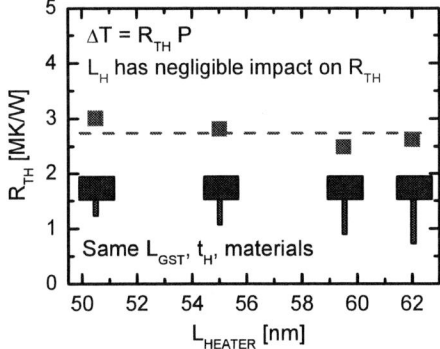

Figure 11. Heater length, L_H, trials. L_H modulation introduces negligible variations in the cell thermal resistance, highlighting that GST represents the dominant path for heat draining.

slightly larger values than *SH*, in fact the 45 nm and the 30 nm points are slightly shifted toward the top-right hand side of the plot. This is ascribed to the presence of the heater electrode as a series element, slightly degrading the cell efficiency. It is thus interesting to design a *SH* cell that meets the *Wall* architecture working point. Figure 7 reports the height, L, of a *SH* cell with square section, S = 45x45 nm^2, calculated to obtain different I_{PROG}, according to Eq. (1) in Fig. 6. To match the proper I_{PROG}, a cell with L = 950 nm and huge aspect ratio would be required (AR^{-1} ≈ 21), resulting in a long GST wire (see sketch in Fig. 7), hence hazardous for technology realization. It is worth noting that the *Wall* architecture shows instead AR^{-1} ≈ 2. Moreover, from the I_{PROG} equation for *SH* (Fig. 6) it is straightforward noticing that a material change may be required to manage *SH* cells from the technological standpoint: the involved material parameters, *i.e.* T_{MELT} and the electrical and thermal resistivities, should be changed with respect to those typical of Ge$_2$Sb$_2$Te$_5$ in order to obtain lower I_{PROG} with more relaxed AR.

Wall optimization

Further guidelines for *Wall* architecture optimization are then proposed. Figure 8a highlights that cell width, W, modulation does not impact the R_{SET}·I_{PROG} product as expected from Eq. (2); reduced W, obtained with the 30 nm

experiment, leads to lower I_{PROG} due to i) increased effective thermal resistance R_{TH} of cell as represented in Fig. 8b, in which R_{TH} is extracted by the method described in Ref. [8], and ii) increased R_{ON} also. In order to further explore the cell efficiency as a function of the *Wall* geometry we modulated both the GST length, L_{GST}, and the heater length, L_H. Cells with different L_{GST}, namely 50 nm, 35 nm and 27 nm, are compared in the I_{PROG} - R_{SET} metrics in Fig. 9. Decreasing L_{GST} results in lower R_{SET}·I_{PROG}, or enhanced efficiency. Despite having both lower R_{SET}·I_{PROG} and lower R_{SET}, a slightly higher I_{PROG} can be measured at decreased L_{GST}. This is due to a decreasing cell thermal resistance, R_{TH}, as depicted in Fig. 10, in agreement with a reduced heat path toward the Top Electrode Contact (TEC) for lower L_{GST}. On the other hand, trials modulating the heater length, L_H, do not introduce significant shift of the R_{SET}·I_{PROG} product far from the standard geometry working point and the R_{TH} extraction also reveals that such trials are pretty aligned, in the range 2.5 - 3 MK/W, as reported in Fig. 11. This suggests that the preferential path for heat dissipation in the *Wall* architecture is mainly through GST towards the TEC, rather than through the heater and towards the BEC. Indeed, decreasing the cell R_{TH} by lowering L_{GST} and/or by optimizing the GST/TEC thermal interface was highlighted in previous work as a strategy for thermal-disturb immune PCM technology [9],

978-1-4799-8002-4/14 $31.00 © 2014 IEEE   683

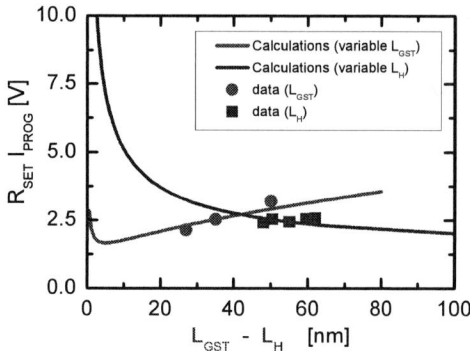

Figure 12. $R_{SET} \cdot I_{PROG}$ calculations with variable L_{GST} (all other parameters fixed) and with variable L_H (all other parameters fixed). Good agreement could be obtained with experimental data.

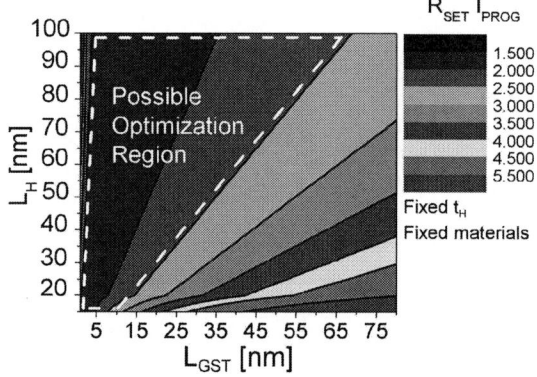

Figure 13. Contour plot of $R_{SET} \cdot I_{PROG}$ as a function of L_{GST} and L_H, combined together from analytical behaviors in Fig. 12. The dashed-line delimited area highlights heater/GST combinations leading to lower $R_{SET} \cdot I_{PROG}$.

owing to more effective heat escape towards the TEC during program. Then a simple model of $R_{SET} \cdot I_{PROG}$, based on the observed R_{TH} functional dependencies, $R_{TH}=f(W, L_{GST}, L_H)$, shown in Figs. 8-11, allows ruling calculations of $R_{SET} \cdot I_{PROG}$ as a function of both L_{GST} and L_H, as represented in Fig. 12. Experimental data are shown to be in good agreement with calculations. Starting from Fig. 12, a matrix combination of both $R_{SET} \cdot I_{PROG}$ dependences on L_{GST} and L_H has been computed, involving larger domains of L_{GST} and L_H. Such combination rules the contour plot of the $R_{SET} \cdot I_{PROG}$ product, represented by the color-map in Fig. 13. It is worth noting that towards high values of L_{GST} (50nm $< L_{GST} <$ 80nm) and low values of L_H ($L_H <$ 50 nm), the product increases, that is, inefficient program/read due to a heater with lower electrical resistance (hence increased I_{PROG}) and higher R_{SET} due to thicker GST. For highly unbalanced lengths, i.e. $L_H \ll L_{GST}$, this model undergoes a validity range violation due to heat dissipation towards the BEC getting comparable to that towards the TEC. Instead, a possible optimization region is found (dashed-line-delimited area) in which $R_{SET} \cdot I_{PROG}$ is ruled in the range $1.5 - 2.5$ V, hence lower than the state-of-the-art 45 nm *Wall* architecture. In fact at lower L_{GST}, e.g. 10nm $< L_{GST} <$ 50nm, efficiency will increase, but an increasing I_{PROG} will be likely found also, due to a R_{TH}

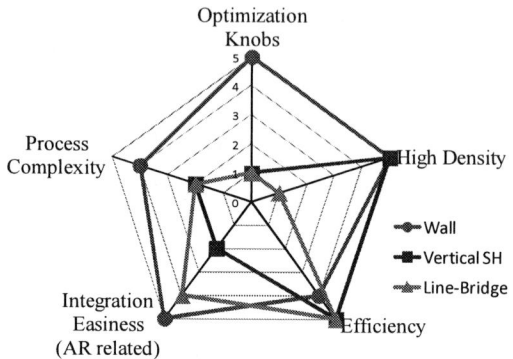

Figure 14. Summary of strengths and drawbacks of the studied architectures: *Wall*, vertical *SH* and *Line-Bridge*.

reduction, even with L_H designed to operate in the optimized area. In this case, in order to make I_{PROG} meet the designed value with no change in the optimized $R_{SET} \cdot I_{PROG}$ product, a cell width W shrink could be applied to balance both R_{ON} and R_{TH}, likely adjusting the required program/read trade-off. Lastly, the product strongly increases for $L_{GST} \ll L_H$ ($L_{GST} <$ 5 nm, $L_H >$ 70 nm), that is, inefficient program/read due on one hand to a heater with very high resistance, detrimental for R_{SET}, and on the other hand very low cell R_{TH}, resulting in an increasing I_{PROG}.

Conclusions

We reported here a comparative study of PCM cell architectures. The developed architectures are considered in a program/read efficiency framework and in an integration context. The *Self-Heating* approach is slightly more efficient, owing to heat generation happening directly within GST, but shows hazardous technology implementation with $Ge_2Sb_2Te_5$, due to high aspect ratios. The *heater-based Wall* architecture represents the best and easiest solution for PCM from the technology standpoint: it features relaxed aspect ratios and benefits from lots of geometry-based knobs for optimization with slightly higher process complexity and slightly lower efficiency. Strengths and drawbacks of the different architectures are schematically reported in Fig. 14.

Acknowledgments

The authors gratefully acknowledge the Micron groups involved in the PCM programs, for their fundamental contribution to the results presented in this work.

References

[1] G. Servalli, *IEDM Tech. Dig.*, p. 113, 2009.
[2] S. Lai, *et al., Symp. of VLSI Tech.*, pp. 132-133, 2013.
[3] D. H. Im, *et al., IEDM Tech. Dig.*, pp. 1-4, 2008.
[4] K. Attenborough, *et al., IEDM Tech. Dig.*, 29.2.1, 2010.
[5] C. Villa, *et al., Proc. ISSCC*, 14.8, 2010.
[6] A. Lacaita, *et al., Microelec. Eng.*, 109, 351-356, 2013.
[7] U. Russo, *et al., Trans. Elec. Dev.*, 55, 2, 506-514, 2008.
[8] M. Boniardi, *et al., Elec. Dev. Lett.*, 33, 4, 594-596, 2012.
[9] A. Redaelli, *et al., IEDM Tech. Dig.*, 2013.

55-μA Ge_xTe_{1-x}/Sb_2Te_3 superlattice topological-switching random access memory (TRAM) and study of atomic arrangement in Ge-Te and Sb-Te structures

N. Takaura[1], T. Ohyanagi[1], M.Tai[1], M.Kinoshita[1], K.Akita[1], T.Morikawa[1],
H.Shirakawa[2], M.Araidai[3], K.Shiraishi[3], Y. Saito[4], and J. Tominaga[4]

[1]Low-power Electronics Association & Project, Onogawa, Tsukuba, Ibaraki, Japan
[2]Graduate School of Pure and Applied Sciences, University of Tsukuba, Tsukuba, Ibaraki, Japan
[3]Graduate School of Engineering, Nagoya University, Nagoya, Aichi, Japan
[4]National Institution of Advanced Industrial Science and Technology, Tsukuba, Ibaraki, Japan
Tel: +81-29-879-8262, FAX: +81-29-856-2622, E-mail: takaura@leap.or.jp

Abstract

Ge_xTe_{1-x}/Sb_2Te_3 superlattice topological-switching random access memory (TRAM) was developed. Set and reset currents of 55 μA, the lowest for an ULSI-grade device, were obtained. TEM analyses of the Ge-Te structures and novel superlattice fabrication enabled us to reveal the retention, endurance, and electrical characteristics of TRAM for the first time.

Introduction

A new type of phase change memory (PCM or PRAM), named topological-switching random access memory (TRAM), has been investigated as a candidate for the next generation of non-volatile memory (Fig. 1) (1). TRAM operates on the basis of Ge atomic movement in the $GeTe/Sb_2Te_3$ superlattice (SL). It is a meta-material made by alternately stacking Sb_2Te_3 and GeTe films. The SL records data by the short-range movement of atoms, while conventional phase change memory (PRAM) records data by order-disorder transition. Topological switching is the generation and extinction of channels for electric currents through changes in the atomic structure. TRAM is characterized by an enhanced atomic motion due to charge injection and non-melting switch (the ability to switch without melting). Therefore, its theoretical programming energy is predicted to be less than 1/20 of that of PRAM (2).

The electrical properties of TRAM can be improved by changing the $GeTe/Sb_2Te_3$ films. We were able to improve the quality of the crystal structure of the $GeTe/Sb_2Te_3$ SL (5). When the sputtering temperature was relatively high, the superlattice structure exhibited some GST defects. A SL with GST defects is a low-quality one. However, we were able to eliminate the GST defects by controlling the substrate temperature and sputtering rate. A SL with an ordered structure and no defects has clear atomic interference fringes. This is a high-quality superlattice.

The atomic structure of the $GeTe/Sb_2Te_3$ SL plays an important role in TRAM low-power operation. The SL consists of a Sb_2Te_3 bottom layer and [$GeTe/Sb_2Te_3$] periodic layers. The results of the material investigation of $GeTe/Sb_2Te_3$ SL we performed are summarized in Table 1. In this work, we show the high-angle-annular-dark-field (HAADF) STEM images of the $GeTe/Sb_2Te_3$ SL and the impact of the Ge-Te and Sb-Te atomic arrangements on TRAM properties. We report on a few new SL structures and their excellent electrical characteristics. In particular, we propose a novel TRAM based on the $Ge_{1-x}Te_x/Sb_2Te_3$ SL with ultra-low set and reset currents for the first time.

Theory

The superlattice resistance is determined by the atomic sequence in the GeTe layer. In the low resistance state (LRS), the atomic sequence is Ge-Te-Te-Ge. On the other hand, in the high resistance state (HRS), the sequence is Te-Ge-Ge-Te. The high resistance of SL is generated by the Te-Te conduction channels between GeTe layers and Sb_2Te_3 layers.

First principle calculations revealed the energy of $GeTe/Sb_2Te_3$ as a function of atomic reaction coordinates (Fig. 2(a)). The minimum energy pathway that represented the change of $GeTe/Sb_2Te_3$ structure was calculated, and the initial, transition, and final structures of $GeTe/Sb_2Te_3$ were obtained as shown in Fig. 2(b). The atomic movement in TRAM was enhanced by charge injection (3). Fig.3 shows the energy of the $GeTe/Sb_2Te_3$ SL in the HRS minus that in the LRS, ΔE_{SL}. The electron injection makes the SL in the LRS unstable, which forces the reset transition to change from the low to the high resistance state. On the other hand, holes make the SL in the HRS unstable, which forces the set transition to change from the high to the low resistance state.

Fig.4 (a) and (b) show the calculated results of negatively and positively charged $GeTe/Sb_2Te_3$ SL. When the SL is negatively charged, the electron density near the Ge and Te atoms increases, thus strengthening the Ge-Te bonds. The Ge atoms start moving in the direction of the increasing electron density. When the SL is positively charged, the electron density near the Ge atoms decreases, which weakens the Ge-Ge bonds. This prompts the Ge atoms to start moving in the direction opposite to that of the decreasing electron density.

During an I-V curve measurement of the set transition, the transition resistance state was clearly observed. In addition, negative resistance was observed as the electric current decreased with increasing voltage (Fig. 25) (3). We considered that the negative resistance was caused by hole generation during the set transition, which enhanced the Ge motion during the set transition (7).

It is assumed that the low-power operation of the SL is achieved when the Ge atoms move easily, which is caused by the existence of atomic vacancies in the Ge-Te layers in the SL (Fig. 13). Of course, the operation of TRAM is affected by the number of periodic layers (Fig. 23(a)) and the quality of the bottom layer of the SL (Fig. 21). However, the thermal properties of the bottom layer in the SL (Fig. 24(a)) are assumed not to change the TRAM operation very much because it is mainly enhanced by charge injection, not Joule heating.

978-1-4799-8002-4/14 $31.00 © 2014 IEEE

Experimental Results

We produced GeTe/Sb$_2$Te$_3$ SL TRAM test element groups (TEGs) with different number of periodic layers and thermal properties of bottom layer. Furthermore, we changed the composition of the Ge-Te layers and fabricated a Ge$_X$Te$_{1-X}$/Sb$_2$Te$_3$ SL TRAM TEG.

(1) Process flow of TRAM TEG fabrication

The process flow of TRAM TEG fabrication is shown in Fig. 5. Some of the GeTe/Sb$_2$Te$_3$ SL TRAM TEGs were fabricated on Si wafer coupons (Figs. 7, 9, 21, 23, and 24), and some of the GeTe/Sb$_2$Te$_3$ SL TRAM TEGs and all of the Ge$_X$Te$_{1-X}$/Sb$_2$Te$_3$ SL TRAM TEGs were developed in the super clean room (SCR) in the national institute of advanced industrial science and technology (AIST) (Figs. 6, 10, 11, 12, and 14-21). The SL films consisted of a Sb$_2$Te$_3$ bottom layer and [GeTe/Sb$_2$Te$_3$] periodic layers. The standard SL structure had 8 layers [GeTe/Sb$_2$Te$_3$ = 1 nm/4 nm]. The SLs were sandwiched between a top electrode (TE) and bottom contact electrode (BCE). The SL TEG fabricated on Si coupons had a BCE with a diameter of 100 nm. The SL TEG fabricated on 300-mm Si wafers had a BCE with a diameter of 50 nm (Figs. 6(a), (b)).

(2) Ge-Te structures

We verified the atomic structures of the SL films with HAADEF-STEM. First, the Ge-Te arrangement in the SL was characterized as a [Ge-Te-Te-Ge], or [5-7-7-5], sequence (Figs. 7 and 8). We observed the change in the atomic sequence from the [Ge-Te-Te-Ge] to a [Ge-Te-Ge-Te] in as-fabricated SL films (Fig. 9). The short-range movement of Ge atoms occurred at Te-Te gaps. The Ge-Te-Ge-Te sequence increased the fluctuation of set resistance, but it was cured by endurance set and reset cycles (Fig. 10). We also observed a Ge-Te sequence in the SL that was named the "Ge$_1$Sb$_4$Te$_7$ defect" (Figs. 11(a)(b)). The defect was considered to deteriorate the retention characteristics of the TRAM (Fig. 12).

(3) Ge$_X$Te$_{1-X}$ /Sb$_2$Te$_3$ superlattice TRAM

The Ge-Te layers in the SL play an important role in low-power TRAM operation. We assumed that the atoms move easily in the [Ge$_X$Te$_{1-X}$ (x < 0.5)], and, therefore, investigated Ge$_X$Te$_{1-X}$/Sb$_2$Te$_3$ SL TRAM. Thin films of Ge$_X$Te$_{1-X}$ and Sb$_2$Te$_3$ layers were deposited with a multi-cathode PVD (Figs. 14 and 15). Figure 16(a) and 16(b) show the reset and set characteristics of the Ge$_X$Te$_{1-X}$/Sb$_2$Te$_3$ TRAM developed in this study. The TRAM TEG with x >> 0.5 did not work because atoms did not move easily. The TRAM TEG with x << 0.5 exhibited set and reset currents that were much less than those with x = 0.5. This was because the atoms in the SL with x << 0.5 moved easily. The number of endurance set and reset cycles of this TRAM was about 100 million (Fig. 17). The reset currents depended on set resistances and the minimum current of 55 µA (Fig. 18(a)). This was achieved by maintaining a resistance ratio of more than 100 (Fig. 19). The minimum set current obtained in this work was 55 µA as well (Fig. 18(b)). The set and reset currents of TRAM in this work were found to be identical (Fig. 20).

(4) Sb-Te structures and periodic/bottom layers of SL

The standard Sb$_2$Te$_3$ layer is a quintuple layer (QL). Nonetheless, non-QLs were generated with a high-temperature process (Figs. 21(a)(b)). However, the fluctuation of resistances and the reset voltage generated by the SL fabricated with a low-temperature process were very small (Fig. 22).

SL memory cells with 6-nm thick Sb$_2$Te$_3$ layers also exhibited set and reset cycles. It was found that the reset voltage of the SL increased as the number of the periodic layers increased (Figs. 23).

In our previous work (1), a bottom layer made of Sb$_2$Te$_3$ was used for stable SL deposition. In this work, we used composite bottom layers. The resistance change of the SL occurred with the composite bottom layers. We found that the reset voltages of the SL were independent on the thermal conductivity of composite bottom layer (Figs. 24(a)(b)). This indicated that the resistance change in the SL was not caused mainly by Joule heating but by charge injection.

Conclusion

An ultra-low current Ge$_{1-x}$Te$_x$/Sb$_2$Te$_3$ superlattice (SL) topological-switching random-access memory (TRAM) was developed. It showed set and reset currents of 55 µA, the lowest for a ULSI-grade device. TEM observations revealed Ge-Te and Sb-Te sequences in the SL, which enabled us to understand the retention and endurance characteristics of TRAM. We also fabricated SLs with 6-nm-thick Sb$_2$Te$_3$ layers and ones with composite bottom layers, and achieved electrical property characteristics in the TRAM.

Acknowledgements

This work was performed as "Ultra-Low Voltage Device Project" funded and supported by METI and NEDO. A part of the device processing was operated by AIST, Japan.

References

(1) M. Tai, T. Ohyanagi, M. Kitamura, M. Kinoshita, T. Morikawa, K. Akita, S. Kato, H. Shirakawa, M. Araidai, KK. Shiraishi, and N. Takaura, *2014 symposium on VLSI technology*, T22.4, June 2014.

(2) R. E. Simpson, P. Fons, A.V. Kolobov, T. Fukaya, M. Krbal, T. Yagi, and J. Tominaga, *Nature Nanotechnology*, **6**, 501-505 (2011)

(3) S. Kato, M. Araidai, K. Kamiya, T. Yamamoto, T. Ohyanagi, N. Takaura, and K. Shiraishi, *2013 International Conference on Solid State Devices and Materials (SSDM2013)*, 544-545, September 2013.

(4) T. Ohyanagi, N. Takaura, M. Kitamura, M. Tai, M. Kinoshita, K. Akita, T. Morikawa, and J. Tominaga, *Jpn. J. Appl. Phys.*, **52**, 05FF01 (2013).

(5) T. Ohyanagi, N.Takaura, M. Kitamura, M. Tai, M. Kinoshita, K. Akita, T. Morikawa, S. Kato, M. Araidai, K. Kamiya, T. Yamamoto, K. Shiraishi, *2013 IEEE International Electron Devices Meeting (IEDM 2012)*, 30.5, December 2013.

(6) T. Morikawa, K. Akita, T. Ohyanagi, M. Kitamura, M. Kinoshita, M. Tai, and N. Takaura, *2012 IEEE International Electron Devices Meeting (IEDM 2012)*, 737-740, December 2012.

(7) N. Takaura, T. Ohyanagi, M. Tai, M. Kitamura, M. Kinoshita, K. Akita, T. Morikawa, S. Kato, M. Araidai, K. Kamiya, T. Yamamoto and K. Shiraishi, *2014 IEEE International Conference on Microelectronic Test Structure(ICMTS2014)*, 2.2, March 2014.

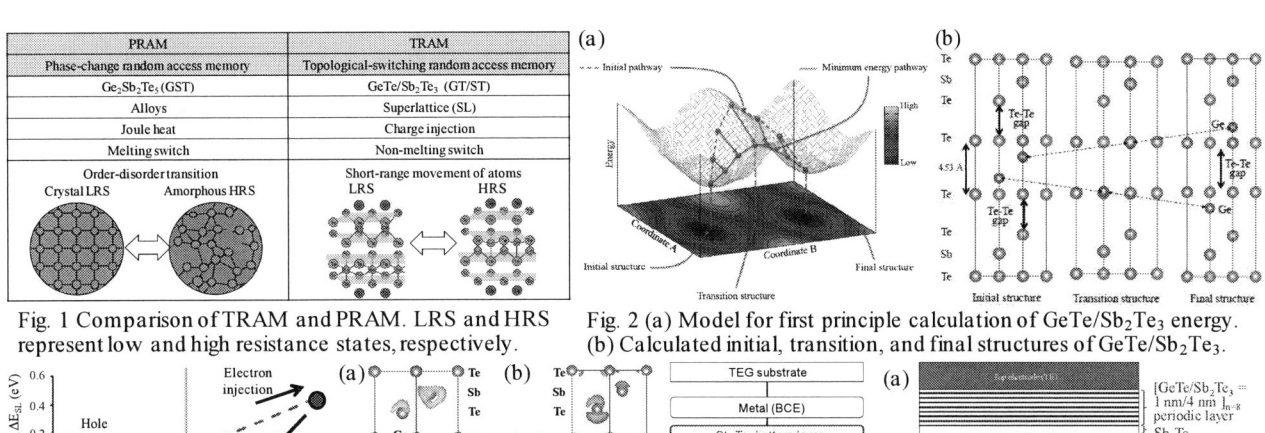

PRAM	TRAM
Phase-change random access memory	Topological-switching random access memory
$Ge_2Sb_2Te_5$ (GST)	$GeTe/Sb_2Te_3$ (GT/ST)
Alloys	Superlattice (SL)
Joule heat	Charge injection
Melting switch	Non-melting switch
Order-disorder transition Crystal LRS Amorphous HRS	Short-range movement of atoms LRS HRS

Fig. 1 Comparison of TRAM and PRAM. LRS and HRS represent low and high resistance states, respectively.

Fig. 2 (a) Model for first principle calculation of $GeTe/Sb_2Te_3$ energy. (b) Calculated initial, transition, and final structures of $GeTe/Sb_2Te_3$.

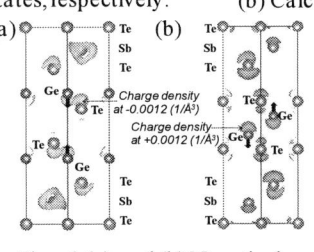

Fig. 3 SL energy difference (ΔE_{SL}) as function of number of charges in SL. ΔE_{SL} defined as the energy of SL in HRS minus that in LRS.

Figs. 4 (a) and (b) Negatively and positively charged SL. Germanium atoms move in direction of bold arrows.

Fig. 5 Process flow of TRAM TEG fabrication

Figs. 6 (a) and (b) Schematic and TEM cross-sectional views of $GeTe/Sb_2Te_3$ SL

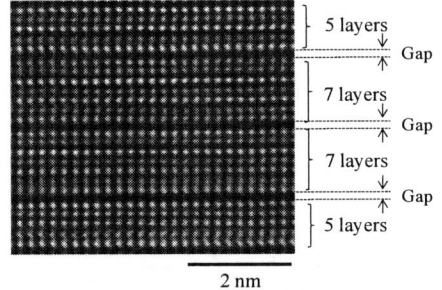

Fig. 7 HAADF-STEM image of $GeTe/Sb_2Te_3$ SL

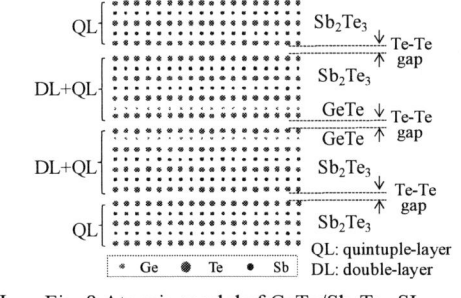

Fig. 8 Atomic model of $GeTe/Sb_2Te_3$ SL

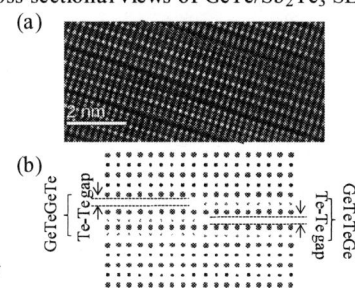

Fig. 9 Ge-Te transition in SL. (a) TEM image. (b) Atomic model.

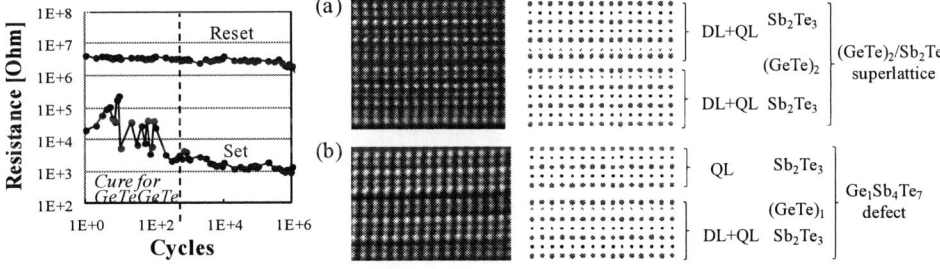

Fig. 10 Endurance of SL including Ge-Te-Ge-Te sequence

Fig. 11 (a) Ge-Te-Te-Ge sequence in SL. (b) Ge-Te sequence in SL, which is $Ge_1Sb_4Te_7$ defect.

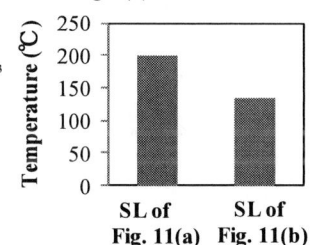

Fig. 12 Temperature that can maintain reset resistance level longer than 2,000 sec.

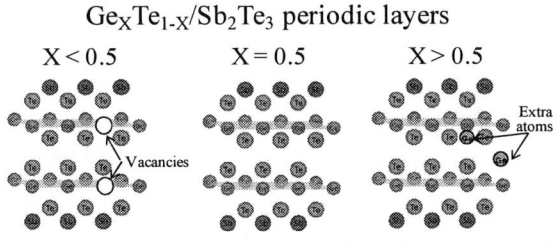

Fig. 13 Atomic models of Ge_xTe_{1-x}/Sb_2Te_3 periodic layers.

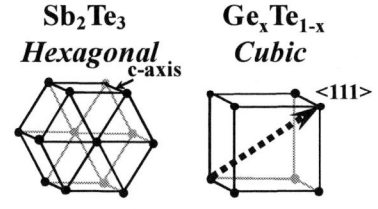

Fig. 14 Crystal structures of Sb_2Te_3 and Ge_xTe_{1-x}. <111> direction of cubic Ge_xTe_{1-x} layer aligned to c-axis of hexagonal Sb_2Te_3 layer.

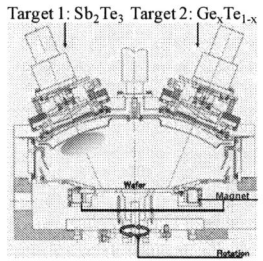

Fig. 15 Multi-cathode PVD of Ge_xTe_{1-x}/Sb_2Te_3 superlattice

978-1-4799-8002-4/14 $31.00 © 2014 IEEE

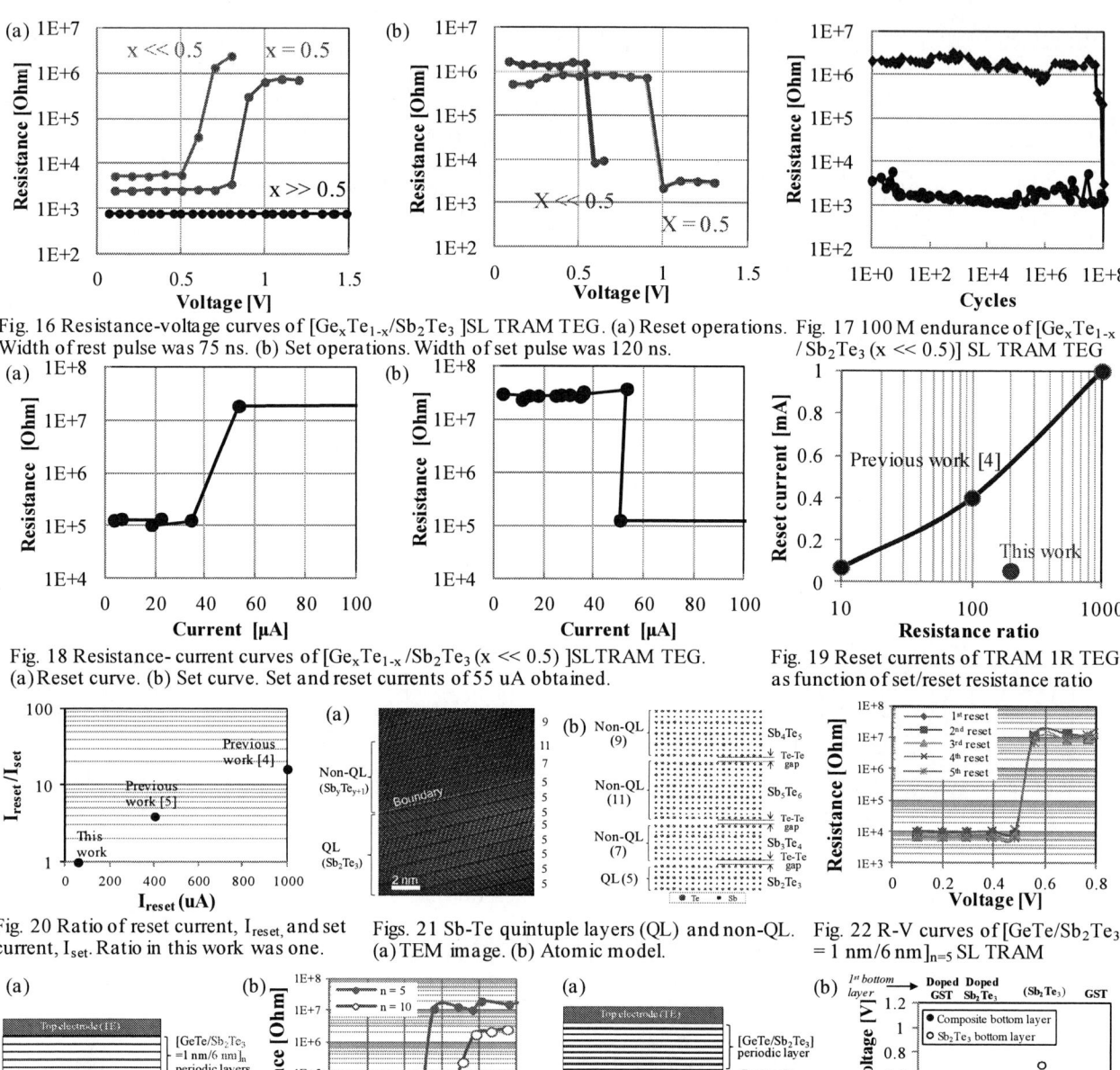

Fig. 16 Resistance-voltage curves of [Ge_xTe_{1-x}/Sb_2Te_3]SL TRAM TEG. (a) Reset operations. Width of rest pulse was 75 ns. (b) Set operations. Width of set pulse was 120 ns.

Fig. 17 100 M endurance of [Ge_xTe_{1-x} /Sb_2Te_3 (x << 0.5)] SL TRAM TEG

Fig. 18 Resistance- current curves of [Ge_xTe_{1-x} /Sb_2Te_3 (x << 0.5)]SLTRAM TEG. (a) Reset curve. (b) Set curve. Set and reset currents of 55 uA obtained.

Fig. 19 Reset currents of TRAM 1R TEG as function of set/reset resistance ratio

Fig. 20 Ratio of reset current, I_{reset}, and set current, I_{set}. Ratio in this work was one.

Figs. 21 Sb-Te quintuple layers (QL) and non-QL. (a) TEM image. (b) Atomic model.

Fig. 22 R-V curves of [$GeTe$/Sb_2Te_3 = 1 nm/6 nm]$_{n=5}$ SL TRAM

Figs. 23 Memory cell with [$GeTe$/Sb_2Te_3 = 1 nm/6 nm] periodic layers. (b) Data of SLs with n = 5 and n = 10.

Fig. 24 (a) Memory cell with composite bottom layer. (b) Reset voltage as function of thermal conductivity, λ, of 1st bottom layer.

Table. 1 Summary of material investigation of GT/ST superlattice

Superlattice		Sequence	Remarks	Figs.
Periodic layer	Ge-Te	Ge-Te-Te-Ge	Standard sequence in our previous work	7, 8
		Ge-Te-Ge-Te	Ge-Te transition and set resistance fluctuation	9, 10
		Ge-Te	$Ge_1Sb_4Te_7$ defects deteriorated TRAM retention.	11, 12
		Ge_xTe_{1-x}	**x << 0.5 ⇒ 55-uA set and reset currents**	16, 17, 18
	Sb-Te	QL	Standard sequence in our works	21
		Non-QL	Not serious. Density can be reduced with proper PVD	21, 22
		Periodic layers	Reset voltage increased as # of periodic layers increased.	23
Bottom layer	Single	Sb_2Te_3	Standard sequence in our works	6
	Composite	1st layer + 2nd Sb_2Te_3	Nominally no change in reset voltage with λ	24

Dangling bonds at the interface between the GeTe and Sb_2Te_3 layers seem to cause the formation of holes, resulting in negative resistance.

Fig. 25 A set behavior of $GeTe$/Sb_2Te_3 SL TRAM TEG.

Phase Change Memory and its intended Applications

Chung H. Lam

IBM Research, T.J. Watson Research Center, 1101 Kitchawan Road, Yorktown Heights, New York 10598, USA.
Phone: 1-914-945-3902, email: clam@us.ibm.com

Abstract

Phase Change Memory (PCM) has been one of the emerging memories for more than a decade. Fundamentals of PCM have been studied in great detail, challenges have been identified, and device structures have been proposed and demonstrated. With this large body of knowledge, it is time to examine probable applications for PCM.

Introduction

PCM is based on a non-volatile reversible transition between crystalline and amorphous phases of a class of materials known as chalcogenides. Crystallization temperatures of chalcogenides are generally between 100°C to 300°C (fig.1) while melting temperatures are above 600°C. The transition from crystalline to amorphous phase is the power-limited melt-quench (reset) operation and the transition from amorphous to crystalline phase is the performance-limited (set) operation (fig.2). Early entries of PCM [1,2] were quickly silenced by the rapid advances of floating gate electrical erasable programmable read only memory (EEPROM). As the critical dimension of semiconductor devices advanced below 1µm, around the turn of the century, the power for the reset operation became manageable with volumetric scaling of the phase change memory element. PCM was reintroduced as Ovonic Unified Memory (OUM) [3], which created much hype for being a replacement of incumbent memory technologies as they reach scaling limits. As it has been more than a decade since the reintroduction, it is evident that the replacement of NAND Flash will not occur; NAND Flash continues to deliver incredible advances [4]. In this talk, we shall examine the current state of PCM development, review requirements for possible memory applications, and identify probable applications for PCM.

Current State of PCM

PCM has been trapped in the trough of disillusionment for the last 5 years according to Gartner's Hype Cycles for Semiconductors (fig.3). It is a rather accurate description of the current state of PCM. PCM vendors originally targeted a $4F^2$ cell [5,6] using a silicon diode as selector (fig.4) to replace the larger NOR Flash cell, which was nearing it scaling limits for much of the rapidly growing mobile market. With the introduction of smart phones, the amount of memories installed in a typical cell phone increased exponentially and NAND Flash quickly took over the market. PCM vendors need to redirect the PCM roadmap [7].

Storage Class Memory (SCM)

With memory latency staying fairly constant, SCM was introduced [8] as a means to alleviate the expanding performance gap between the multi-core processor unit and the system memory. A consequence of the memory latency staying practically constant is that massive quantities of DRAM are required to keep memory bandwidth on par with the increasing parallelism of the processor cores. DRAM is approaching its scaling limit; therefore, massive quantities of DRAM may become too expensive. Without these massive quantities of DRAM, the cost of computing falls upon the frequent memory fetches from the next level of the memory hierarchy, Solid State Drives (SSD) based on NAND Flash, which are about 3 orders of magnitude slower than DRAM. SCM fills this performance difference space.

Challenges in SCM Applications

Critical characteristics required of an SCM technology are low power, high performance, high endurance and high density/cost (fig.5). *Power:* Unlike NOR Flash, which is practically a read-mostly and write-scarcely memory, SCM applications have a very low read to write ratio. In diode selector architecture, unselected word lines in an array need to be biased to turn off the selector. Since the write and read voltage biases are different, frequent switching between read and write operations will result in unacceptable power consumption. PCM vendors with diode selector will need to retool the memory selector. The unanimous choice for SCM selector is the vertical surround gate MOSFET. *Performance:* Data non-volatility is not essential for SCM applications; memory element materials for PCM can be optimized for the performance limiting set operation (fig.6). *Endurance:* PCM vendors have demonstrated [9,10] write endurance better than 10^{10} cycles (fig.7). With wear leveling this is more than adequate for SCM applications. *Density/cost:* Density is the dominant factor in determining the cost of memories. With the vertical surround gate (VSG) MOSFET selector (fig.8), the PCM cell size of $6F^2$ is on par with the DRAM cell size. To achieve a 4x cost reduction compared to DRAM, PCM would have to depend on other techniques. Multi-level cell (MLC) structures (fig.9) immune to resistance drift issues have been demonstrated in PCM [11,12]; however, both of these cell structures are capable of providing only 2 bits/cell. Therefore, PCM needs to scale beyond the scaling limits of DRAM in order to achieve a 4x cost reduction advantage. A concern for PCM to scale beyond 20nm is the thermal disturb of neighbor cells during the reset operation in which the temperature

978-1-4799-8002-4/14 $31.00 © 2014 IEEE

of the selected cell is raised above 600°C in order to melt the phase change material. The metal nitride used in ref. 12 is also shown to act as an effective thermal isolation layer (fig.10).

Ternary Content Addressable Memory (TCAM)

Another probable application for PCM is TCAM. The basic concept of applying PCM to construct high density and low power TCAM was demonstrated using a 2T2R TCAM cell [13]. The 2T2R cell has only 3 terminals and is 10 times smaller than conventional 16-T SRAM-based TCAM cell. However, this compact PCM-based TCAM cell design posts constraints on efficient array layout and write operation. An alternate PCM-based 4T2R TCAM cell is introduced (fig.11) here to alleviate these constraints. In a TCAM array, each match line has its own sense amplifier. In 2T2R cell array, the match line pitch is minimal (fig.12) and is dictated by the width of a pair of the search/word-line transistors. To minimize active search power, the width of search/word-line transistor is minimized. Thus, in order to fit a sense amplifier circuit to the minimal match line pitch, the sense amplifier layout is staggered so that each sense amplifier occupies 2-multiples of match line pitch. As a result the memory array efficiency is much degraded. On the other hand, the match line pitch of the 4T2R cell array is wide enough (fig.12) to accommodate a sense amplifier resulting in a much better array efficiency which more than compensating the larger memory cell size. A more effective memory cell write operation in 4T2R cell array is facilitated with dedicated write bit- and word- lines configured parallel and orthogonal to the match line so that a match word can be written in parallel.

The dedicated write word-line transistor can be optimized for the write operation of the PCM element to maximize the reset to set resistance ratio essential to the match signal. Let n=the length of the search word, R_{match}= resistance of a fully matched n-bit word and $R_{mismatch}$= resistance of a mismatch n-bit word with only 1 mismatched bit. The worse case match line signal is given by:

$$TCAMR = \frac{R_{match} - R_{mismatch}}{R_{mismatch}} = \frac{(\alpha_r - n)}{n} + \frac{(n-1)}{n\alpha_d}$$

where $\alpha_r = R_{resset}/R_{set}$ and $\alpha_d = R_{reset}$ drift factor. While α_d has a diminishing effect on $TCAMR$ with large n, α_r has to be significantly greater than n to maintain an acceptable $TCAMR$ (fig.13). In order to account for the exponential degradation of R_{reset} at elevated operation temperature, α_r must be characterized at 85C or higher dependent on the application environment. For n=40 including sense amplifier circuit operation window, α_r needs to be greater than 200 (fig.13).

Neuromorphic Memory

Much of the recent progress in Artificial Intelligent Machines is based on the applications of the innovative Deep Learning algorithms [14]. These learning algorithms required enormous computer resources to perform brain-like functions such as visual and voice recognitions. Synaptic electronics using non-volatile memory devices such as PCM [15] and RRAM [16] is an emerging field of research aiming to build energy-efficient electronic systems that mimic brain-like functions. A PCM memory crossbar array [17] was demonstrated to exhibit brain-like learning and recognition functions. Robust brain-inspired learning in this small-scale synaptic array is a significant step towards building large-scale computational systems with brain-level computational efficiency.

Conclusion

Although PCM will not replace any incumbent memories, with the successful introduction of PCM-based SCM, TCAM and Neuromorphic memory, PCM will transform the future semiconductor memory landscape. Finally, the author would like to thank the contributions in the works mentioned in this paper from IBM Research PCM teams in Watson Research Center, Zurich Research Center, Tokyo Research Laboratory and IBM PCM research partners.

Reference

[1] Neale, R.G., Nelson D.L. and Moore, G.E., "Nonvolatile and Reprogrammable, the Read-Mostly Memory is Here," Electronics pp.56-60, Sept. 1970.

[2] Shanks, R.R. and Davis, C., "A 1024-Bit Nonvolatile 15ns Bipolar Read-Write Memory," ISSCC pp.112-113, 1978.

[3] Lai, S. and Lowrey. T., "OUM - A 180 nm nonvolatile memory cell element technology for stand alone and embedded applications," IEDM 2001.

[4] Park, K-T et al, "Three-dimensional 128Gb MLC vertical NAND Flash-memory with 24-WL stacked layers and 50MB/s high-speed programming," ISSCC 2014.

[5] Kang, M.J. et al, "PRAM cell technology and characterization in 20nm node size," IEDM 2011.

[6] G. Servalli, "A 45nm Generation Phase Change Memory Technology," IEDM 2009.

[7] Clarke, P., "Exclusive: Micron Drops Phase-Change Memory – for Now," Electronics360.globalspec.com, Jan. 2014.

[8] Freitas, R.F. and Wilcke, W.W., "Storage-class memory: The next storage system technology," IBM J. Res. & Dev., pp.441-447, v.52, 2008.

[9] Lai, S., "Current Status of Phase Change Memory and its Future," IEDM 2003.

[10] Kim, I.SA. et al, "High Performance PRAM Cell Scalable to sub-20nm technology with below 4F² Cell Size, Extendable to DRAM Applications," VLSIT 2010.

[11] Oh, G.H. et al, "Parallel Multi-Confined (PMC) Cell Technology for High Density MLC PRAM," VLSIT 2009.

[12] Kim, S. et al, "A Phase Change Memory Cell with Metallic Surfactant Layer as a Resistance Drift Stabilizer," IEDM 2013.

[13] Rajendran, B. et al, "Demonstration of CAM and TCAM using Phase Change Devices," IMW 2011.

[14] Hinton, G. E., Osindero, S., & Teh, Y., "A fast learning algorithm for deep belief nets," Neural Computation 18 7 2006.

[15] Kuzum, D., Jayasingh, R., Yu, S., Wong, H.-S.P., "Low-Energy Robust Neuromorphic Computation Using Synaptic Devices," TED 59 12 2012.

[16] Park, S. et al, "Nanoscale RRAM-based synaptic electronics: toward a neuromorphic computing device," Nanotechnology 24 38 2013.

[17] Burc Eryilmaz, S. et al, "Experimental Demonstration of Array-level Learning with Phase Change Synaptic Devices," IEDM 2013.

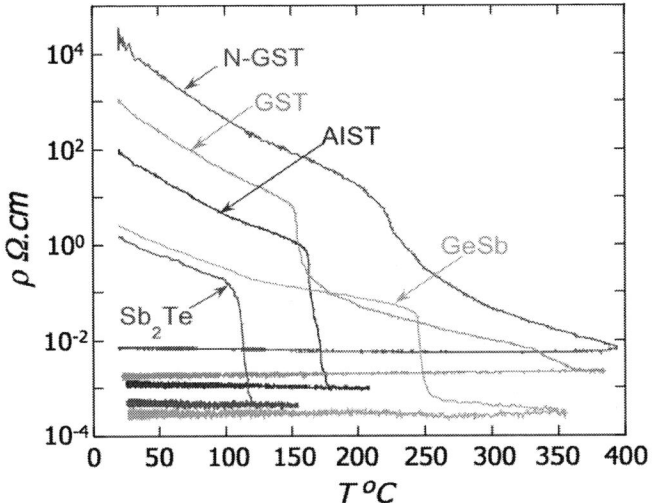

Fig.1 Resistivity transformations upon crystallization of sample amorphous phase change materials.

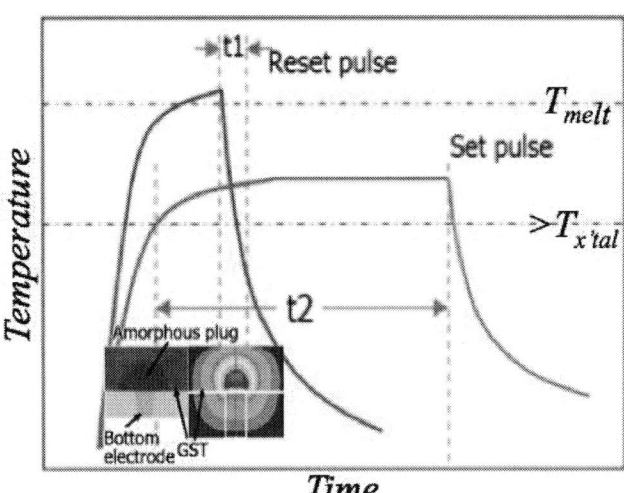

Fig.2 Reset and set operations characteristics of a typical mushroom-type phase change memory element.

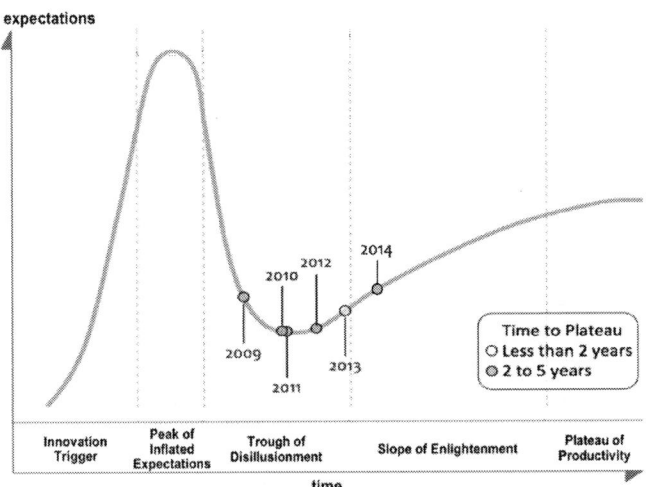

Fig.3 State of Phase Change Memory on the Hype Cycles (adopted from Gartner Hype Cycles 2009-2014).

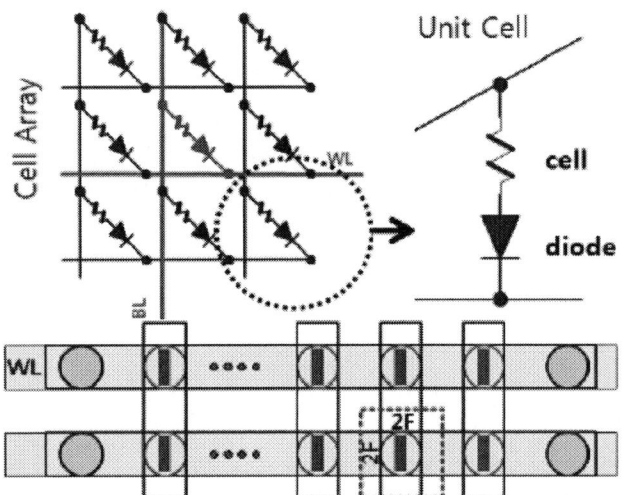

Fig.4 A 4F² PCM cell using a diode as selector. (Source: Kang, M.J. et al, IEDM 2011.)

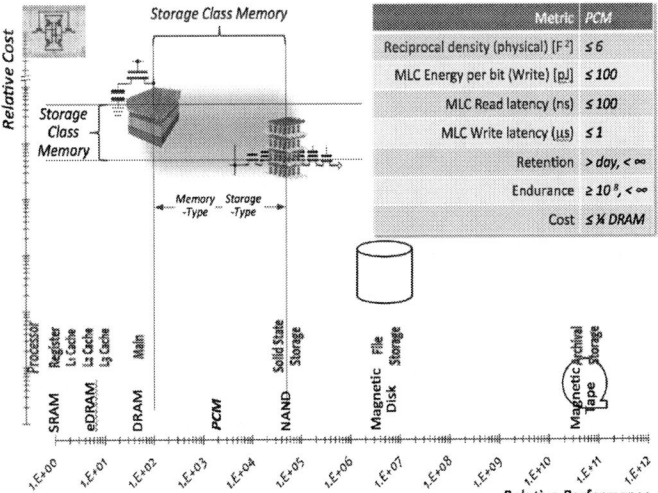

Fig.5 Storage Class Memory in the memory hierarchy and its requirement specifications.

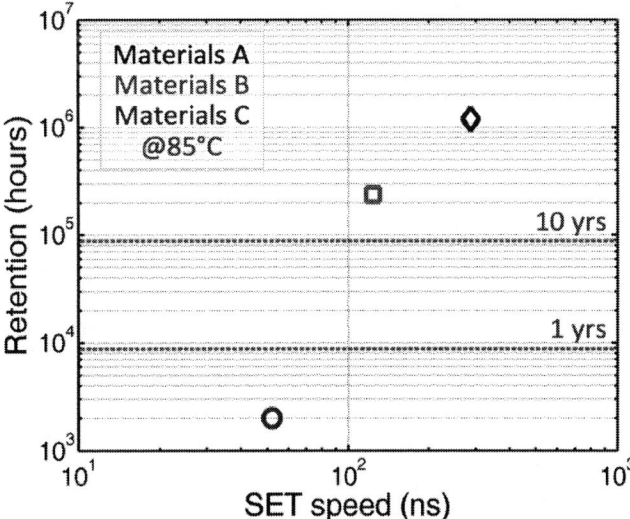

Fig.6 Data retention characteristics of various PCM materials versus performance-limiting set speed.

978-1-4799-8002-4/14 $31.00 © 2014 IEEE

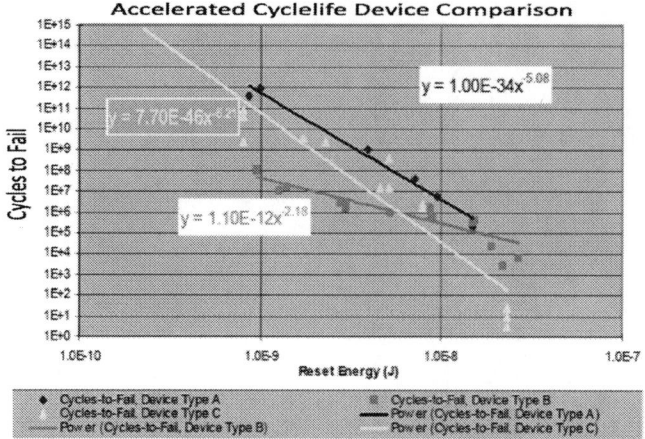

Fig.7 Endurance characteristics of various PCM memory element types. (Source: S. Lai, IEDM 2003.)

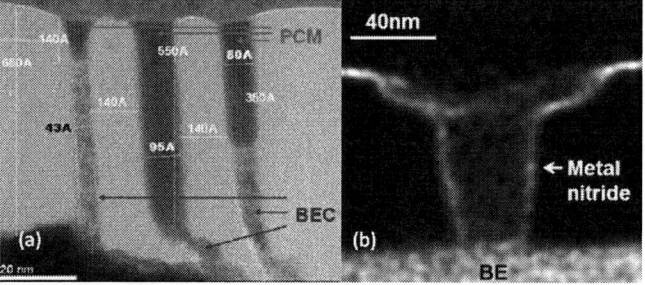

Fig.9 (a) Parallel multi-confined cell (Ref. 11), (b) Metal nitride lined confined cell (Ref. 12).

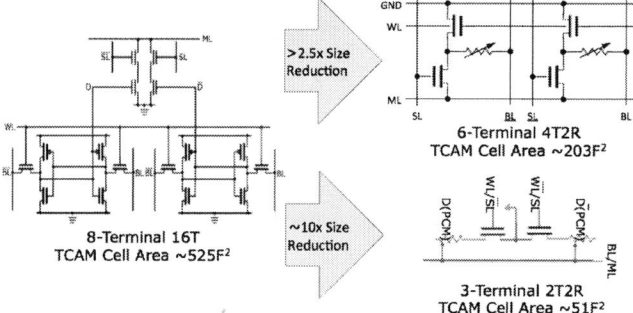

Fig.11 PCM-based TCM cells: 6-Terminal 4T2R cell and 3-Terminal 2T2R cell versus SRAM-based TCAM cell.

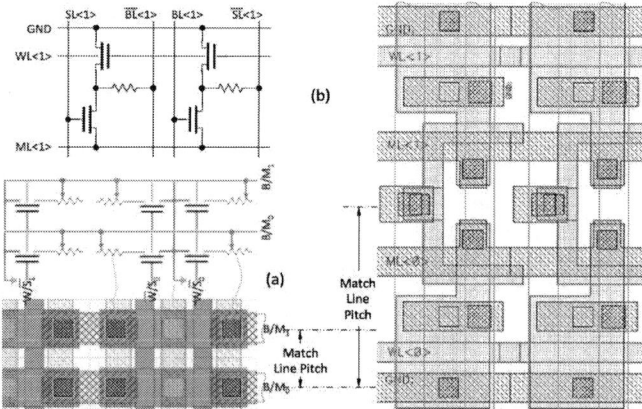

Fig.12 Comparison of match line pitch height in (a) 2T2R TCAM cell layout with (b) 4T2R TCAM cell layout.

Fig.8 VSG MOSFET selector array structure for PCM integration simulated using Coventor®.

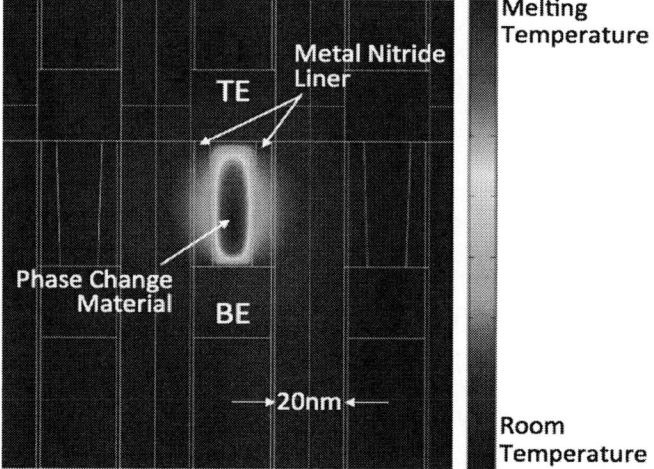

Fig.10 Simulated thermal profile of confined cell with metal nitride liner during reset operation.

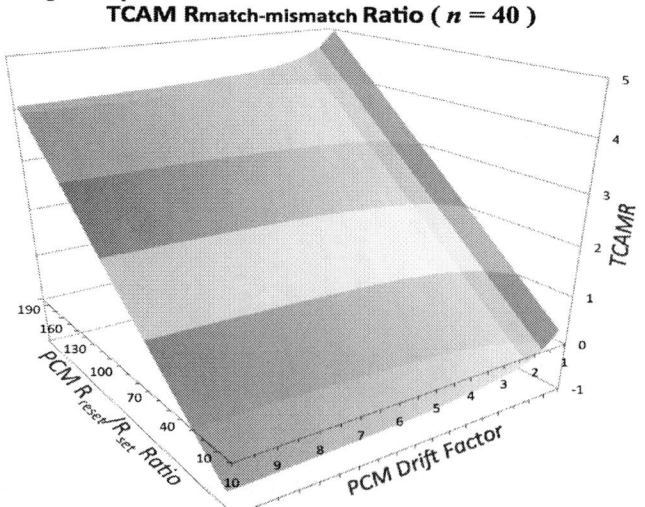

Fig.13 PCM-based TCAM available match-line sense signal.

978-1-4799-8002-4/14 $31.00 © 2014 IEEE

Capacity Optimization of Emerging Memory Systems: A Shannon-Inspired Approach to Device Characterization

Jesse H. Engel[1,2], S. Burc Eryilmaz[2], SangBum Kim[3], Matthew BrightSky[3],
Chung Lam[3], Hsiang-Lan Lung[4], Bruno A. Olshausen[1], and H.-S. Philip Wong[2]

[1]Redwood Center for Theoretical Neuroscience, UC Berkeley, Berkeley, CA, 94720, *E-mail: jhengel@stanford.edu
[2]Dept. of Electrical Engineering and Center for Integrated Systems, Stanford University, Stanford, CA, 94305
[3]IBM Research, T.J. Watson Research Center, Yorktown Heights, NY, 10598
[4]Macronix International Co., Ltd., Emerging Central Lab, 16 Li-Hsin Road, Hsinchu Science Park, Taiwan

Abstract

Traditional approaches to memory characterize the number of distinct states achievable at a given Raw Bit Error Rate (RBER). Using Phase Change Memory (PCM) as an example analog-valued memory, we demonstrate that measuring the mutual information allows optimal design of read-write circuits to increase data storage capacity by 30%. Further, we show the framework can be used for energy efficient memory design by optimizing simulations of a 1Mb memory array to consume 32% less energy/bit. This work provides an information-theoretic framework to guide the design and characterization of other analog-valued emerging memory such as RRAM and CBRAM.

Introduction

The number of data generating devices connected to the internet is expected to grow exponentially to 20 billion by 2020 [1]. Storing and analyzing the data deluge from this 'internet of things' requires new energy-efficient and high-density memory systems. Multi-level storage has successfully achieved high densities in emerging memories such as Flash, PCM, and RRAM, with systems realizing as many as 16 states/cell [2, 3].

Multi-level cells function by modulating an analog-valued property such as cell resistance or threshold voltage and discretizing the output. The dynamics of setting these properties are stochastic due to atomic kinetics and nanoscale fabrication variations, leading to higher RBER with increasing states/cell. Robust memory storage increasingly requires strong Error Correcting Codes (ECCs) to reach low ($\sim10^{-15}$) Uncorrected Bit Error Rates (UBER) [4].

The Shannon Capacity (C) of a memory cell is the theoretical maximum bits/device that can be robustly stored and recalled with the help of an ECC [5] (Fig. 1). As modern coding techniques such as Low-Density Parity Check (LDPC) codes can now approach the Shannon Capacity limit [6], and are employed in noisy MLC-Flash memories [7], the Shannon Capacity is a valuable metric to improve performance of the memory cell/controller/ECC system.

From a Shannon perspective, storage and recall in a memory cell can be thought of as an input write voltage pulse distribution, P(V), traveling through a noisy memory channel, P(R|V), and resulting in an output resistance read distribution, P(R) (Fig. 2). Shannon showed that the capacity of such a channel is given by the maximum mutual information over the possible inputs [5]:

$$C = \max_{P(V)} \sum_V P(V)P(R|V) \, log_2 \frac{P(R|V)}{P(R)}$$

Since P(R)= Σ_V P(V)P(R|V), the capacity of a memory cell is determined entirely by its conditional distribution / transition matrix P(R|V) (Fig. 3).

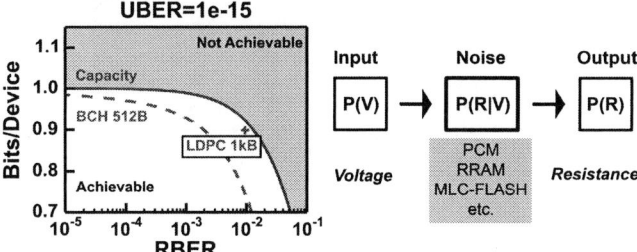

Fig. 1 (left) Capacity sets the frontier between achievable and unachievable rates of reliable storage with error correction. While commonly used Bose Ray-Chaudhuri (BCH) codes perform below the capacity limit, it can be approached by modern LDPC codes with large codewords (ex. 1kB) [4]. **Fig. 2 (right)** Capacity can be measured by viewing information storage as communication through a noisy memory channel. It is determined uniquely by P(R|V).

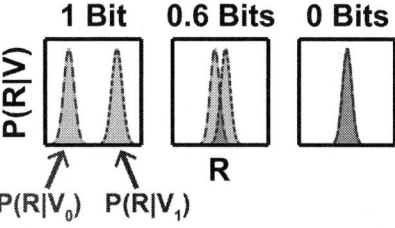

Fig. 3 Overlap of distributions reduces capacity. Capacity is optimized by increasing the number of input states and reducing overlap of output states.

Here, we demonstrate how measuring the full conditional distribution P(R|V), instead of the RBER at a number of distinct states, enables co-optimization of device characteristics and memory controller design to maximize storage capacity (bits/device). Further, we find ECCs that operate without quantization, storing analog signals in analog

978-1-4799-8002-4/14 $31.00 © 2014 IEEE

devices, intrinsically operate at the highest capacity for a given channel.

Device Characterization

For this study we performed pulsed resistance measurements on 100-device 1T1R PCM arrays (Fig. 4). Device fabrication and characterization of these arrays were previously reported in [8-10]. The cell is reset to high resistance states via short current pulses that heat the film above its melting temperature and quickly quench to form a resistive amorphous cap. Slow Joule heating above the crystallization temperature anneals the amorphous cap and sets the low resistance state. Multiple resistance levels are achieved by switching between different volumes of resistive amorphous and conducting crystalline phases [11].

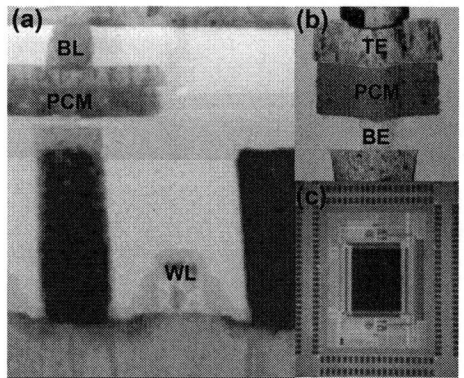

Fig. 4 100-device PCM array (~40nm BE contact diameter) (TEM (a)[8], (b)[9], and (c) optical micrograph) as an example analog-valued memory.

The gate voltage (V_{WL}) on each access transistor controls the current flowing through the cell (Fig. 5). To ensure independent statistics, we apply a pulsing scheme (Fig. 6) where the cell is initialized to a consistent set state, before applying partial reset pulses to the word line (V_{WL}) of varying magnitude and measuring the resulting resistance (R).

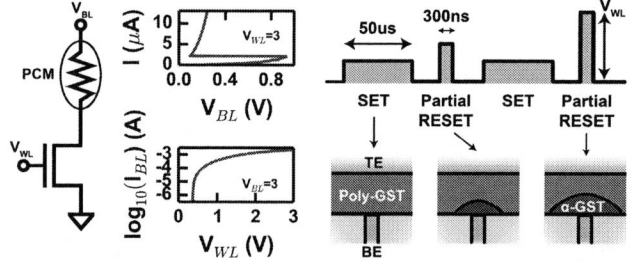

Fig. 5 (left) Current-voltage characteristics for a representative PCM device and access transistor in the array (I_{RESET} ~400µA) **Fig. 6 (right)** Pulsing scheme for partial-RESET operation. Device is SET between each partial-RESET pulse, and R(V_{WL}) is measured.

After collecting 380 trials at each voltage level, we calculate a Gaussian kernel density estimate of the continuous probability density P(R|V_{WL}) (Fig. 7). For simplicity and generality, time-dependent effects of PCM such as resistance drift are not considered here [12]. Similarly, we apply only a single write and read step, even though more robust storage has been demonstrated with read-verify schemes [3]. The results of this investigation can be extended to these modified channel models, and methods for calculating their Shannon capacity have been proposed [13].

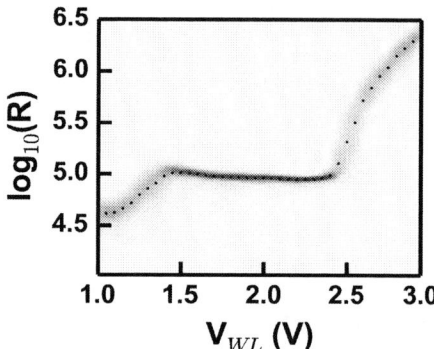

Fig. 7 Kernel density estimation of P(R|V_{WL}) (shaded, darker is higher density), from PCM data over 380 trials. Means of the data at each voltage are represented by dots.

Optimal Discretization

Measuring P(R|V_{WL}) provides a powerful tool to optimize the write pulse and read bin locations of an associated memory controller. We solve for the capacity-achieving input distribution by maximizing the mutual information over P(V) using the Blahut-Arimoto algorithm [14]. While the algorithm only applies to discrete distributions, we can approximate the capacity of analog channels by sufficiently discretizing P(R|V) such that the capacity is not increased by further discretization (ex. >2000 states).

The number of nonzero values in the capacity-achieving input distribution is determined by the balance between using as many input states as possible, and reducing overlap of their outputs (Fig. 8). Interestingly, beyond 12 discrete inputs, no more states are added, as the increased overlap would create a net reduction in mutual information (Fig. 9). The output distributions for these 12 inputs partially overlap, demonstrating higher capacity than achievable with totally distinct states. The unequal input state probabilities are unrealistic for most applications, however, making these probabilities uniform only reduces the channel capacity by approximately 5%.

978-1-4799-8002-4/14 $31.00 © 2014 IEEE

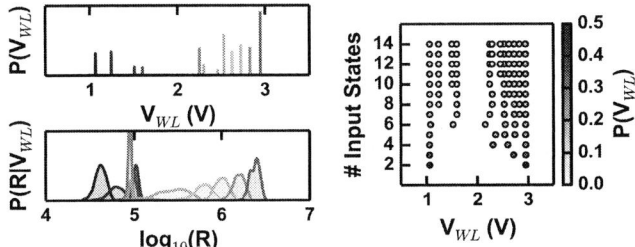

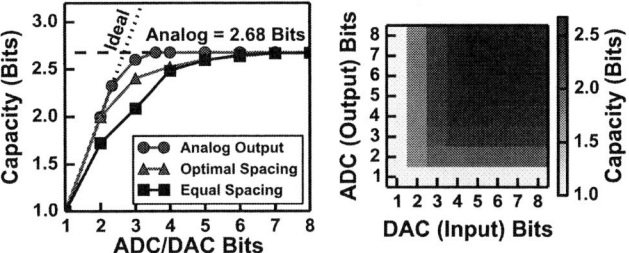

Fig. 8 (left) Optimal 'analog' input distribution P(V_WL), is actually discrete (12 states). Corresponding output P(R|V_WL) achieves the highest storage capacity despite moderate overlap of 12 outputs. **Fig. 9 (right)** Optimal input distributions for increasing number of allowed states. Mutual information is maximized by reducing overlap of outputs. Each row of dots is of the same type as the top graph of figure 8, with height represented by color. For 2 inputs, they are as separate as possible. Beyond 12 discrete inputs, no more states are added, as the increased overlap would create a net reduction in mutual information.

Fig. 11 (left) Higher capacity can be reached for the same number of states if their spacing is optimally chosen. Analog representations, such as in an analog artificial neural network, intrinsically have the same capacity as an infinite bit ADC. **Fig. 12 (right)** The capacity of the PCM device approaches the analog case for increasing numbers of input and output states. Having more output states than input states creates 'soft information', grayscale belief propagation values currently used in MLC-Flash for decoders.

We further consider optimal discretization strategies via employing basin hopping search to find discrete inputs/outputs levels with maximum capacity [15]. Comparing discretization schemes between equally spaced, optimally spaced, and analog-valued input/outputs (Fig. 10), capacities are similar between schemes for low and high bit ADC/DACs (Fig. 11-12), as they are limited by the number of input states and intrinsic overlap of P(R|V) respectively. However, at 3-bits, the optimally spaced scheme gains 20% over the equally spaced case, and analog-valued outputs gain 30% over equal spacing. Having more output states than input states creates 'soft information', grayscale belief about the input given the output, a practice currently used to increase capacity of MLC-Flash decoders that perform variations of belief propagation [16].

Energy Efficient Storage

We calculate the energy consumption of each pulse from oscilloscope measurements (Fig. 13). The current is transistor limited, indicating that the dynamic resistance of the PCM devices is low and pulses are not strongly affected by parasitic capacitances in the 1T1R structure. Additional power consumed due to line losses are included in simulations of a 1Mb square array with 130nm wide, 1:1 aspect ratio, Cu wires [17] (Fig. 14). We then perform constrained optimization to find the input distributions that maximize capacity per unit energy [14] (Fig. 15). Since larger V_WL consume more energy, inputs are constrained to lower voltages in the efficient case (Fig. 16). Although this gives less separable outputs and lower capacities, there is an overall gain in efficiency (nJ/bit). We find an appropriate choice of input pulses can create a 32% reduction in energy/bit for the array.

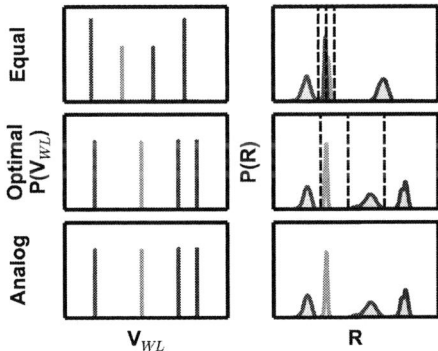

Fig. 10 Quantization schemes for 4 inputs and outputs, drawn from P(R|V) in Fig. 7: equally spaced, optimally spaced, and analog-valued. The equally spaced example has higher probability in the outer states as there is less overlap of the outputs than the inner states.

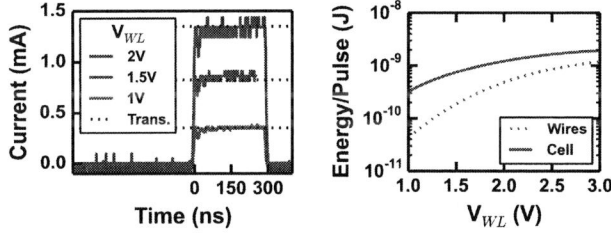

Fig. 13 (left) Current is limited by the access transistor. Current traces measured in oscilloscope match the predicted level (dashed) based on transistor transconductance from Fig. 5. **Fig. 14 (right)** Simulations of square 1Mb array, including I²R losses in the wires and CV² losses from capacitive charging (dashed), and energy dissipated in the memory cell calculated from Fig. 5 (solid).

978-1-4799-8002-4/14 $31.00 © 2014 IEEE

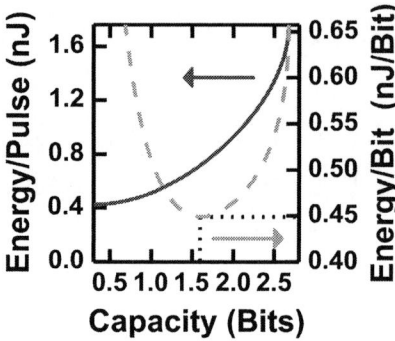

Fig. 15 Simulations of 1Mb array. While capacity (red) decreases with less energy due to less separable outputs, a minima exists for capacity/energy (yellow dashed).

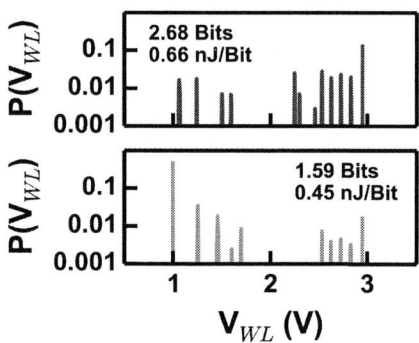

Fig. 16 The ideal input distribution for max capacity (red, 12 states) and max efficiency (nJ/bit) (yellow, 10 states). The efficient case is weighted towards lower V_{WL} that have less current and power per pulse, but also less separable outputs.

Conclusion

We have presented an information theoretic framework for the characterization of analog-valued emerging memory devices. By measuring full device statistics, we have demonstrated the potential for co-optimization of memory device and controller design. Further, the results that analog-valued circuits intrinsically operate at peak capacity are promising for analog artificial neural network designs where emerging memories are proposed as artificial synapses.

Acknowledgements

This work was supported in part by SONIC, one of six centers of STARnet, a Semiconductor Research Corporation program sponsored by MARCO and DARPA, and by member companies of the Stanford Non-Volatile Memory Technology Research Initiative (NMTRI).

References

[1] D. Evans, "The Internet of Things How the Next Evolution of the Internet Is Changing Everything," Cisco, http://www.cisco.com/web/about/ac79/docs/innov/IoT_IBSG_0411 FINAL.pdf, *Internet Business Solutions Group*, , 2011.

[2] C. Trinh et al., "A 5.6MB/s 64Gb 4b/Cell NAND Flash memory in 43nm CMOS," in *2009 IEEE International Solid-State Circuits Conference - Digest of Technical Papers*, 2009, pp. 246–247.

[3] T. Nirschl et al, "Write Strategies for 2 and 4-bit Multi-Level Phase-Change Memory," in *2007 IEEE International Electron Devices Meeting*, 2007, pp. 461–464.

[4] R. Motwani, Z. Kwok, and S. Nelson, "Low Density Parity Check (LDPC) Codes and the Need for Stronger ECC," in *Flash Memory Summit*, Aug. 2011.

[5] C. E. Shannon, "A Mathematical Theory of Communication," *Bell Syst. Tech. J.*, vol. 27, pp. 379–423, 1948.

[6] D. J. C. MacKay and R. M. Neal, "Near Shannon limit performance of low density parity check codes," *Electron. Lett.*, vol. 33, no. 6, p. 457, 1997.

[7] S. Tanakamaru, Y. Yanagihara, and K. Takeuchi, "Error-Prediction LDPC and Error-Recovery Schemes for Highly Reliable Solid-State Drives (SSDs)," *IEEE J. Solid-State Circuits*, vol. 48, no. 11, pp. 2920–2933, Nov. 2013.

[8] M. Breitwisch et al., "Novel Lithography-Independent Pore Phase Change Memory," in *2007 IEEE Symposium on VLSI Technology*, 2007, pp. 100–101.

[9] G. F. Close et al., "Device, circuit and system-level analysis of noise in multi-bit phase-change memory," in *2010 International Electron Devices Meeting*, 2010, pp. 29.5.1–29.5.4.

[10] S. B. Eryilmaz et al., "Experimental Demonstration of Array-level Learning with Phase Change Synaptic Devices," *IEEE Int. Electron Devices Meet.*, vol. 25, no. 5, pp. 1–4, Dec. 2013.

[11] H.-S. P. Wong et al., "Phase Change Memory," *Proc. IEEE*, vol. 98, no. 12, pp. 2201–2227, Dec. 2010.

[12] N. Papandreou, H. Pozidis, T. Mittelholzer, G. F. Close, M. Breitwisch, C. Lam, and E. Eleftheriou, "Drift-Tolerant Multilevel Phase-Change Memory," in *2011 3rd IEEE International Memory Workshop (IMW)*, 2011, pp. 1–4

[13] .L. A. Lastras-Montano, M. Franceschini, T. Mittelholzer, and M. Sharma, "Rewritable storage channels," in *2008 International Symposium on Information Theory and Its Applications*, 2008, pp. 1–6.

[14] S. Arimoto, "An algorithm for computing the capacity of arbitrary discrete memoryless channels," *IEEE Trans. Inf. Theory*, vol. 18, no. 1, pp. 14–20, Jan. 1972.

[15] D. J. Wales and J. P. K. Doye, "Global Optimization by Basin-Hopping and the Lowest Energy Structures of Lennard-Jones Clusters Containing up to 110 Atoms," *J. Phys. Chem. A*, vol. 101, no. 28, pp. 5111–5116, Jul. 1997.

[16] G. Dong, N. Xie, and T. Zhang, "On the Use of Soft-Decision Error-Correction Codes in nand Flash Memory," *IEEE Trans. Circuits Syst. I Regul. Pap.*, vol. 58, no. 2, pp. 429–439, Feb. 2011.

[17] J. Liang, S. Yeh, S. S. Wong, and H.-S. P. Wong, "Effect of Wordline/Bitline Scaling on the Performance, Energy Consumption, and Reliability of Cross-Point Memory Array," *ACM J. Emerg. Technol. Comput. Syst.*, vol. 9, no. 1, pp. 1–14. Feb. 2013.

Device data and Python code for analysis are publicly available at https://github.com/rctn/CapacityOptimization.

Experimental demonstration and tolerancing of a large-scale neural network (165,000 synapses), using phase-change memory as the synaptic weight element

G. W. Burr, R. M. Shelby, C. di Nolfo, J. W. Jang‡, R. S. Shenoy, P. Narayanan,
K. Virwani, E. U. Giacometti, B. Kurdi, and H. Hwang‡

IBM Almaden Research Center, 650 Harry Road, San Jose, CA 95120, Tel: (408) 927–1512, E-mail: *gwburr@us.ibm.com*
‡Pohang University of Science and Technology, Pohang, South Korea

Abstract

Using 2 phase-change memory (PCM) devices per synapse, a 3–layer perceptron network with 164,885 synapses is trained on a subset (5000 examples) of the MNIST database of handwritten digits using a backpropagation variant suitable for NVM+selector crossbar arrays, obtaining a training (generalization) accuracy of 82.2% (82.9%). Using a neural network (NN) simulator matched to the experimental demonstrator, extensive tolerancing is performed with respect to NVM variability, yield, and the stochasticity, linearity and asymmetry of NVM-conductance response.

Introduction

Dense arrays of nonvolatile memory (NVM) and selector device-pairs (Fig. 1) can implement neuro-inspired non-Von Neumann computing [1,2], using pairs [2] of NVM devices as programmable (plastic) bipolar synapses. Work to date has emphasized the Spike-Timing-Dependent-Plasticity (STDP) algorithm [1,2], motivated by synaptic measurements in real brains, yet experimental NVM demonstrations have been limited in size (≤100 synapses).

Unlike STDP, backpropagation [3] is a widely-used, well-studied NN, offering benchmark-able performance on datasets such as handwritten digits (MNIST) [4]. In forward evaluation of a multilayer perceptron, each layer's inputs (x_i) drive the next layer's neurons through weights w_{ij} and a nonlinearity $f()$ (Fig. 2). Supervised learning occurs (Fig. 3) by back-propagating error terms δ_j to adjust each weight w_{ij}. A 3–layer network is capable of accuracies, on previously unseen 'test' images (*generalization*), of ~97% [4] (Fig. 4); even higher accuracy is possible by first "pre-training" the weights in each layer [5]. Like STDP, low-power learning should be achievable by emphasizing brief spikes[7] and local-only clocking.

Considerations for a crossbar implementation

By encoding synaptic weight in conductance difference between paired NVMs, $w_{ij} = G^+ - G^-$ [2], forward propagation simply compares total read signal on bitlines (Fig. 5). However, backpropagation [3] calls for weight updates $\Delta w \propto x_i \delta_j$ (Fig. 6), requiring upstream i and downstream j neurons to exchange information for each synapse. In a crossbar, learning becomes much more efficient when neurons modify weights in parallel, by firing pulses whose overlap at the various NVM devices implements training [1] (Fig. 7). Fig. 8 shows, using a simulation of the NN in Figs. 2,3, that this adaptation for NVM implementation has no effect on accuracy.

However, the conductance response of any real NVM device exhibits imperfections that could still decidedly affect NN performance, including nonlinearity, stochasticity, varying maxima, asymmetry between increasing/decreasing responses, and non-responsive devices at low or high conductance (Fig. 9). **This paper explores the relative importance of each of these factors.**

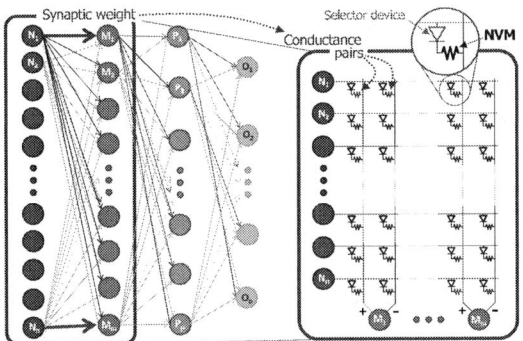

Fig. 1 Neuro-inspired non-Von Neumann computing [1,2], in which neurons activate each other through dense networks of programmable synaptic weights, can be implemented using dense crossbar arrays of nonvolatile memory (NVM) and selector device-pairs.

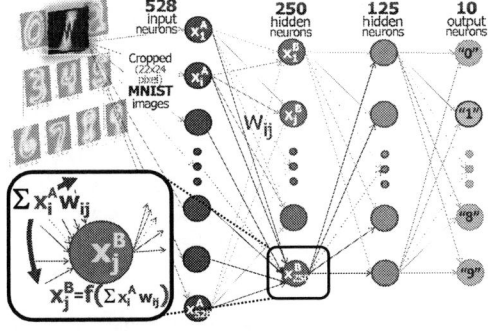

Fig. 2 In forward evaluation of a multilayer perceptron, each layer's neurons drive the next layer through weights w_{ij} and a nonlinearity $f()$. Input neurons are driven by pixels from successive MNIST images (cropped to 22×24); the 10 output neurons identify which digit was presented.

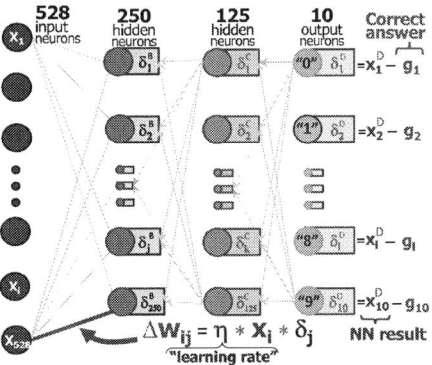

Fig. 3 In supervised learning, error terms δ_j are back-propagated, adjusting each weight w_{ij} to minimize an "energy" function by gradient descent, reducing classification error between computed (x_l^D) and desired output vectors (g_l).

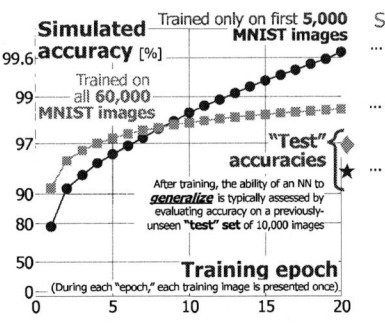

Fig. 4 A 3–layer perceptron network can classify previously unseen ('test') MNIST handwritten digits with up to ~97% accuracy[4]. Training on a subset of the images sacrifices some generalization accuracy but speeds up training.

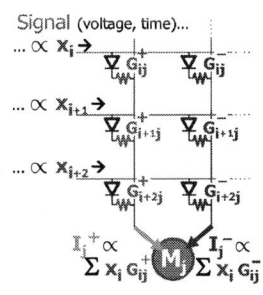

Fig. 5 By comparing total read signal between pairs of bitlines, summation of synaptic weights (encoded as conductance differences, $w_{ij} = G^+ - G^-$) is highly parallel.

978-1-4799-8002-4/14 $31.00 © 2014 IEEE

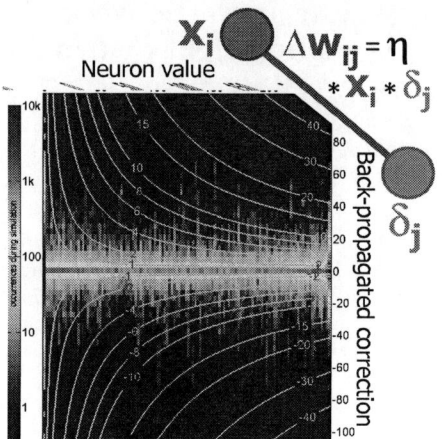

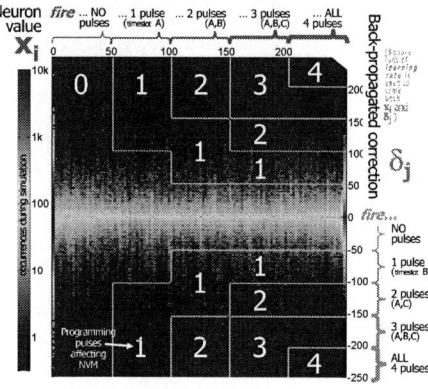

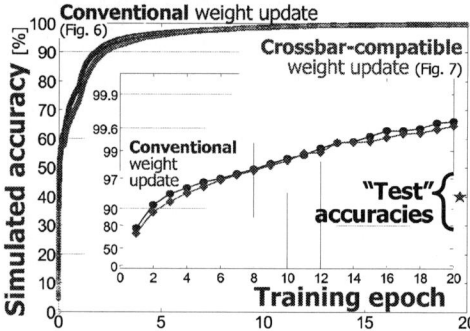

Fig. 6 Back-propagation calls for each weight to be updated by $\Delta w = \eta x_i w_{ij}$, where η is the *learning rate*. Colormap shows log(occurrences), in the 1st layer, during NN training (blue curve, Fig. 4); white contours identify the quantized increase in the integer weight.

Fig. 7 In a crossbar, efficient learning requires neurons to update weights in parallel, firing pulses whose overlap at the various NVM devices implements training.

Fig. 8 Computer NN simulations show that a crossbar-compatible weight-update rule (Fig. 7) is just as effective as the conventional update rule (Fig. 6).

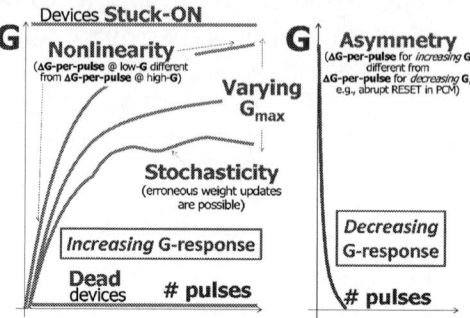

Fig. 9 The conductance response of an NVM device exhibits imperfections, including nonlinearity, stochasticity, varying maxima, asymmetry between increasing/decreasing responses, and non-responsive devices (at low or high G).

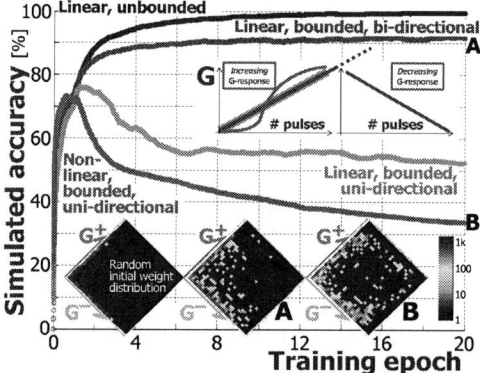

Fig. 10 Bounding G values reduces NN training accuracy slightly, but unidirectionality and nonlinearity in G-response strongly degrade accuracy. Figure insets map NVM-pair synapse states on a diamond-shaped plot of G^+ vs. G^- (weight is vertical position) for a sampled subset of the weights.

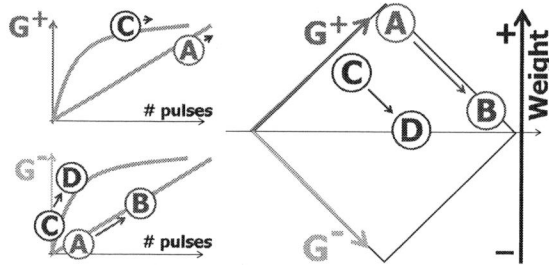

Fig. 11 If G values can only be increased (asymmetric G-response), a synapse at point **A** (G^+ saturated) can only increase G^-, leading to a low weight value (**B**). If response at small G values differs from that at large G (nonlinear G-response), alternating weight updates can no longer cancel. As synapses tend to get herded into the same portion of the G-diamond (**C → D**), the decrease in average weight can lead to network freeze-out.

While bounding G values reduces NN training accuracy slightly (Fig. 10), unidirectionality and nonlinearity in the G-response strongly degrade accuracy. Figure insets (Fig. 10) map NVM-pair synapse states on a diamond-shaped plot of G^+ vs. G^- (weight is vertical position). In this context (Fig. 11), a synapse with a highly *asymmetric* G-response moves only unidirectionally, from left-to-right. Once one G is saturated, subsequent training can only increase the other G value, reducing weight magnitude, deleting trained information, and degrading accuracy. *Nonlinearity* in G-response further encourages weights of low value (Fig. 11), which can lead to network "freeze-out" (no weight changes, Fig. 10 inset). One solution to the highly asymmetric response of PCM devices is occasional RESET [2], moving synapses back to the left edge of the "G-diamond" while preserving weight value (with iterative SETs, Fig. 12 inset). However, if this is not done frequently enough, weight stagnation

will degrade NN accuracy (Fig. 12).

Experimental results

We implemented a 3–layer perceptron of 164,885 synapses (Figs. 2,3) on a 500×661 array of mushroom-cell [6], 1T1R PCM devices (180nm node, Fig. 13). While the update algorithm (Fig. 7) is fully compatible with a crossbar implementation, our hardware allows only sequential access to each PCM device (Fig. 14). For read, a sense amplifier measures G values and thus weights for the software-based neurons, mimicking column- and row-based integrations. Weights are increased (decreased) by identical "partial–SET" pulses (Fig. 7) to increase G^+ (G^-) (Fig. 15). The deviation from true crossbar implementation occurs upon occasional RESET

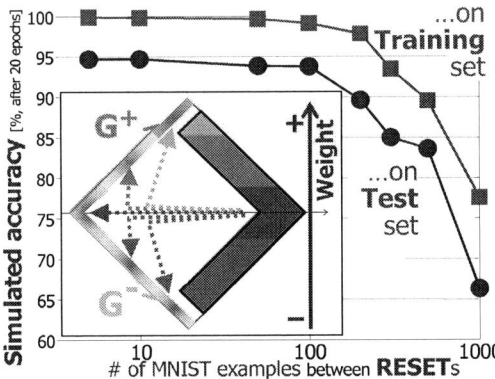

Fig. 12 Synapses with large conductance values (inset, right edge of G-diamond) can be refreshed (moved left) while preserving the weight (to some accuracy), by RESETs to both G followed by a partial SET of one. If such RESETs are too infrequent, weight evolution stagnates and NN accuracy degrades.

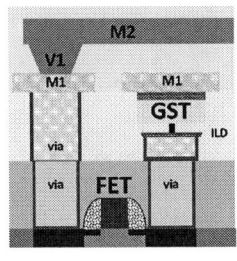

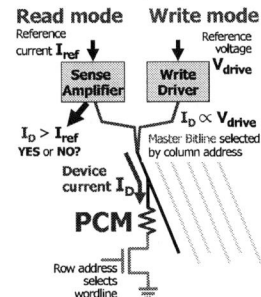

Fig. 13 Mushroom-cell [6], 1T1R PCM devices (180nm node) with 2 metal interconnect layers enable 512×1024 arrays.

Fig. 14 A 1-bit sense amplifier measures G values, passing the data to software-based neurons. Conductances are increased by identical 25ns "partial-SET" pulses to increase G^+ (G^-) (Fig. 7), or by RESETs to both G followed by an iterative SET procedure (Fig. 12).

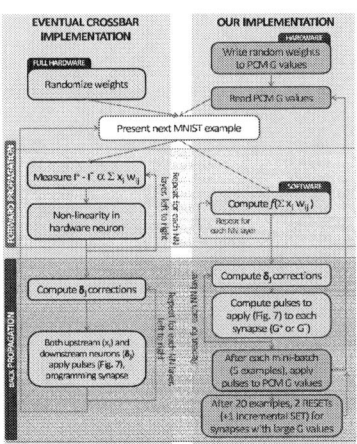

Fig. 15 Although G values are measured sequentially, weight summation and weight update procedures in our software-based neurons closely mimic the column (and row) integrations and pulse-overlap programming needed for parallel operations across a crossbar array. However, since occasional RESET is triggered when both G^+ and G^- are large, serial device access is required to obtain and then re-program individual conductances.

(Fig. 12), triggered when either G^+ or G^- are large, thus requiring both knowledge of and control over individual G values.

Fig. 16 shows measured accuracies for a hardware-synapse NN, **with all weight operations taking place on PCM devices**. To reduce test time, weight updates for each *mini-batch* of 5 MNIST examples were applied together. Fig. 17 plots measured G-response, stochasticity, variability, stuck-ON pixel rate, and RESET accuracy. By matching all parameters including stochasticity (Fig. 18) to those measured during the experiment, a NN computer simulation can

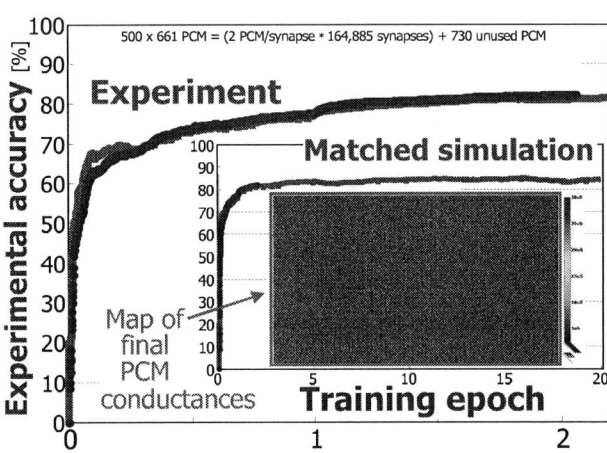

Fig. 16 Training and test accuracy for a 3–layer perceptron of 164,885 hardware-synapses, with all weight operations taking place on a 500 × 661 array of mushroom-cell [6] PCM devices (Fig. 13). Also shown is a matched computer simulation of this NN, using parameters extracted from the experiment.

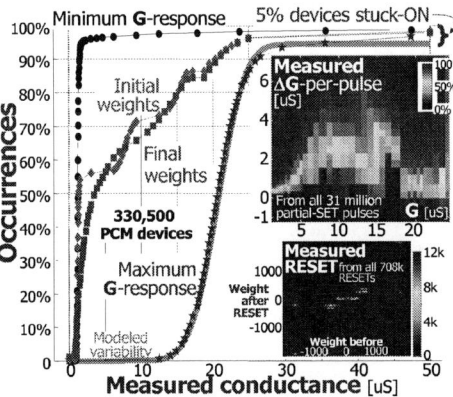

Fig. 17 50-point cumulative distributions of experimentally measured conductances for the 500 × 661 PCM array, showing variability and stuck-ON pixel rate. Insets shows the measured RESET accuracy, and the rate and stochasticity of G-response, plotted as a colormap of ΔG-per-pulse vs. G.

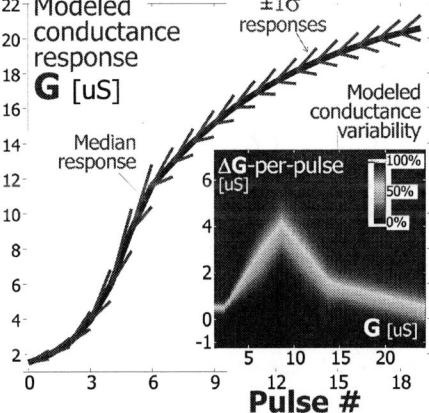

Fig. 18 Fitted G-response vs. # of pulses (blue average, red $\pm 1\sigma$ responses), obtained from our computer model (inset) for the rate and stochasticity of G-response (ΔG-per-pulse vs. G) matched to experiment (Fig. 17).

precisely reproduce the measured accuracy trends (Fig. 16).

Tolerancing and power considerations

We use this matched NN simulation to explore the importance

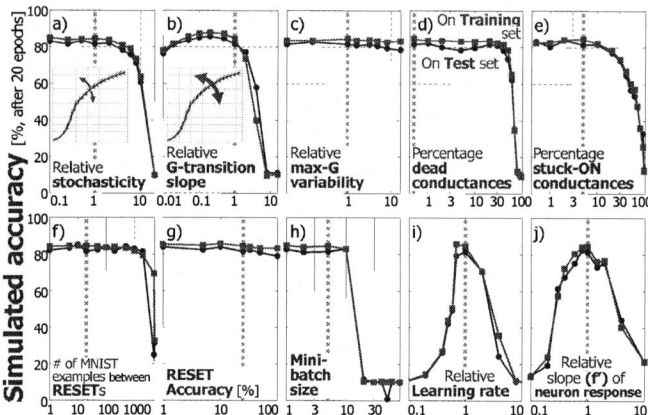

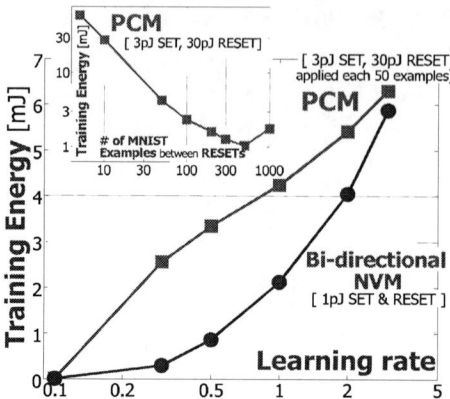

Fig. 19 Matched simulations show that an NVM-based NN is highly robust to stochasticity, variable maxima, the presence of non-responsive devices, and infrequent or inaccurate RESETs. Mini-batch size = 1 avoids the need to accumulate weight updates before applying them. However, the nonlinear and asymmetric G-response limits accuracy to ~85%, and require learning-rate and neuron-response (f') to be precisely tuned.

Fig. 21 Despite the higher power involved in RESET rather than partial-SET (30pJ and 3pJ for highly-scaled PCM [1]), total energy costs of training can be minimized if RESETs are sufficiently infrequent (inset). Low-energy training requires low learning rates, which minimizes the number of synaptic programming pulses. At higher learning rates, even a bi-directional, linear RRAM requiring no RESET and offering low-power (1pJ per pulse) can lead to large training energy.

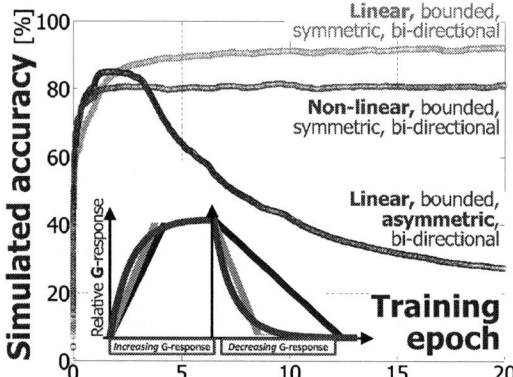

Fig. 20 NN performance is improved if G-response is linear and symmetric (green curve) rather than nonlinear (red). However, asymmetry between the up- and down-going G-responses (blue), if not corrected in the weight-update rule (Fig.7), can strongly degrade performance by favoring particular regions of the G-diamond (Figs.10,11).

Table of conclusions

Large 3-layer network with 2 PCM devices/synapse. (Figs. 1-3, 5) Back-propagation weight update rule compatible with crossbar array. (Figs. 6-8)	Moderately high accuracy (82%) achieved on MNIST handwritten digit recognition with two training epochs. (Fig. 16)
NVM models identified issues for training: Conductance bounds, nonlinearity, and asymmetry must be considered. (Figs. 9-10, 20)	PCM response and asymmetry mitigated by RESET strategy, mapping of response, choice of update pulse. (Figs. 11, 12, 17)
Model of PCM allows well-matched simulation of experiment (Figs. 16-18), variation of network parameters allows tolerancing. (Fig. 19)	NN is resilient to NVM variations (Figs. 19a-e) and RESET strategy (Figs. 19f-g), but sensitive to learning rate and neuron response function. (Figs. 19i-j)
Bidirectional NVM with no special RESET strategy and good performance requires scheme for symmetric response. (Fig. 20)	For PCM, keeping RESET frequency down and learning rate above "freeze-out" threshold allows reasonable training energy. (Fig. 21)

Fig. 22 NN built with NVM-based synapses tend to be highly sensitive to "gradient" effects (nonlinearity and asymmetry in G-response) that "steer" all synaptic weights towards either high or low values, yet are highly resilient to random effects (NVM variability, yield, and stochasticity).

of NVM imperfections. Fig. 19 shows final training (test) accuracy as a function of variations in NVM and NN parameters. NN performance is highly robust to stochasticity, variable maxima, the presence of non-responsive devices, and infrequent RESETs. A mini-batch of size 1 allows weight updates to be applied immediately. However, as mentioned earlier, nonlinearity and asymmetry in G-response limit the maximum possible accuracy (here, to ~85%), and require precise tuning of the learning rate and neuron-response (f'). Too low a learning rate and no weights receive any updates; too high, and the imperfections in the NVM response generate chaos.

NN performance with NVM-based synapses offers high accuracy if G-response is linear and symmetric (Fig. 20, green curve) rather than nonlinear (red curve). Asymmetry in G-response (blue curve) strongly degrades performance. While the asymmetric G-response of PCM makes it necessary to occasionally stop training, measure all conductances, and apply RESETs and iterative SETs, energy usage can be reasonable if RESETs are infrequent (Fig. 21, inset), and learning rate is low (Fig. 21).

Conclusions

Using 2 phase-change memory (PCM) devices per synapse, a 3–layer perceptron with 164,885 synapses was trained with back-

propagation on a subset (5000 examples) of the MNIST database of handwritten digits to high accuracy of (82.2%, 82.9% on test set). A weight-update rule compatible for NVM+selector crossbar arrays was developed; the "G-diamond" concept illustrates issues created by nonlinearity and asymmetry in NVM conductance response. Using a neural network (NN) simulator matched to the experimental demonstrator, extensive tolerancing was performed (Fig.22). NVM-based NN are **highly resilient to random effects** (NVM variability, yield, and stochasticity), but **highly sensitive to "gradient" effects that act to steer all synaptic weights**. A learning-rate just high enough to avoid network "freeze-out" is shown to be advantageous for both high accuracy and low training energy.

References

[1] B. Jackson et al., *ACM J. Emerg. Tech. Comput. Syst.*, **9**(2), 12 (2013).
[2] M. Suri et al., *IEDM Tech. Digest*, 4.4 (2011).
[3] D. Rumelhart et al., *Parallel Distributed Processing*, MIT Press (1986).
[4] Y. LeCun et al., *Proc. IEEE*, **86**(11), 2278 (1998).
[5] G. Hinton et al., *Science*, **313**(5786) 504 (2006).
[6] M. Breitwisch et al., *VLSI Tech. Symp.*, T6B-3 (2007).
[7] B. Rajendran et al., *IEEE Trans. Elect. Dev.*, **60**(1), 246 (2013).

Statistics of set transition in phase change memory (PCM) arrays

M. Rizzi*, N. Ciocchini*, S. Caravati[†], M. Bernasconi[†], P. Fantini[‡] and D. Ielmini*

* DEIB, Politecnico di Milano and IU.NET, 20133 Milano, Italy, e-mail: maurizio1.rizzi@polimi.it

[†] Università degli Studi di Milano-Bicocca, 20126 Milano, Italy

[‡] Micron Technology Inc., Process R&D, 20864 Agrate Brianza (MB), Italy

Abstract

This work presents a statistical characterization of the set operation in phase change memory (PCM) arrays. The set performance was studied in devices programmed with increasing size of the amorphous region by means of suitable reset pulses. The thickness-dependent set time and statistics are explained by the different roles of the nucleation and growth of crystalline grains in the amorphous volume. We then analyzed the set transition kinetics comparing 2 set techniques, namely: i) crystallization in the solid state through rectangular pulses below the reset level and ii) crystallization from the liquid phase through triangular pulses above the reset level. The experimental results provide guiding rules for minimizing the energy consumption and the switching variability in PCM arrays.

Introduction

Phase change memory (PCM) is one of the most promising technologies for memory applications. PCM technology has in fact demonstrated the capability to compete with the mainstream NAND flash [1] in terms of cell-size scaling, with the outlook of achieving extremely high switching speed [2], [3] and low-power operation, thanks to the nanoscaled active volume [4], [5]. The working principle of PCM relies on the property of chalcogenide alloys (typically $Ge_2Sb_2Te_5$, or GST) to reversibly switch between two distinct structural phases, namely a disordered amorphous structure and a polycrystalline phase, each with 2 markedly different electrical resistivities. The GST phase is amorphized in the reset operation consisting of a fast quenching of an electrically-induced molten state. Due to the extremely long crystallization time of amorphous GST at ambient temperature, the retention of the stored data is above 10 years at a typical operating temperature below 85°C. On the other hand, the poly-crystalline phase is recovered in the set operation, where a voltage pulse raises the temperature to several hundred °C thus allowing crystallization in about 100 ns [6]. To develop reliable PCM arrays for future generation memories, the kinetic and statistics of the set/reset program operations must be carefully understood. This work presents a statistical characterization of the electrically-induced crystallization by analyzing 1 Gb PCM arrays at the 45 nm node.

We collected the cell resistance for any given amplitude and duration of the pulses applied to the cells. The measurements

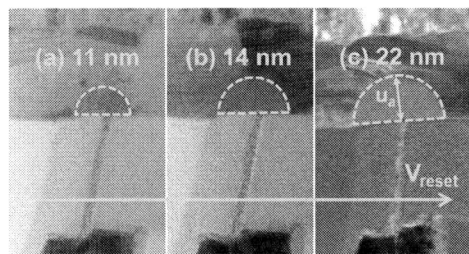

Fig. 1. TEM images of PCM cells programmed at increasing voltage V_{R2} (a), V_{R4} (b) and V_{R6} (c). The thickness u_a of the amorphous cap (yellow dashed line) correspondingly increases from 11 to 22 nm.

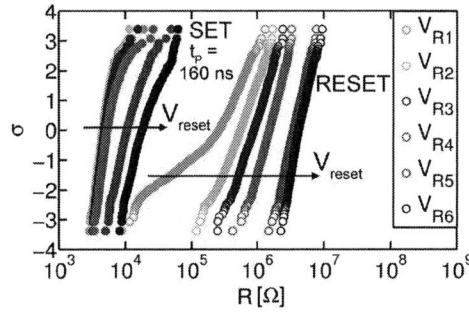

Fig. 2. Cumulative distributions of measured resistance of the set state (filled circles) and reset state (open circles) for increasing reset voltages from V_{R1} to V_{R6}. The set time was 160 ns. Both set and reset resistances increase at increasing reset voltages.

were repeated considering different initial amorphous thicknesses, obtained by modulating the amorphization pulse.

The results allow to address the minimization of the program energy through a careful optimization of the pulse shape.

Set characteristics

Our samples consist of mushroom-type PCM devices in 45 nm technology [7], with confined bottom electrode (BE). Experiments were conducted on PCM arrays of 1 Gb size [8]. To study the dependence of crystallization on the initial reset state, the cells were programmed with box pulses at increasing amplitudes $_{reset}$ from $_{R1}$ to $_{R6}$, all being larger than $_{melt}$ needed to reach the melting temperature in the GST. Fig. 1 shows the TEM pictures of cells programmed at increasing voltage $_{R2}$ (a), $_{R4}$ (b) and $_{R6}$ (c) thus evidencing the strong control of the amorphous GST volume through the electrical

978-1-4799-8002-4/14 $31.00 © 2014 IEEE

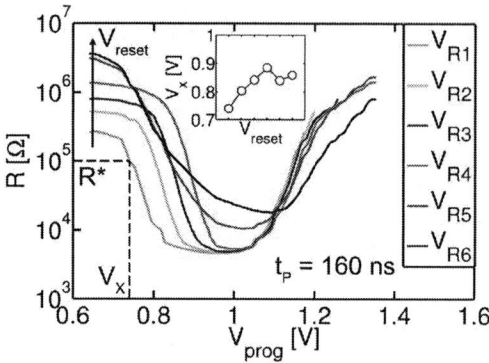

Fig. 3. Measured resistance after the application of rectangular pulses with amplitude V_{prog} and duration $t_P = 160$ ns. The inset shows the crystallization voltage V_x extracted at $R^* = 100$ $k\Omega$ as a function of the reset voltage.

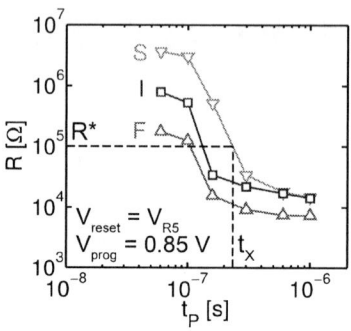

Fig. 5. Resistance as a function of the pulse width t_P for 3 cells with different behaviors, displaying slow (S), intermediate (I) and fast (F) crystallization. Measurements were done at a programming voltage $V_{prog} = 0.85$ V. The crystallization time t_x is defined as the time for which the resistance drops below $R^* = 100$ $k\Omega$.

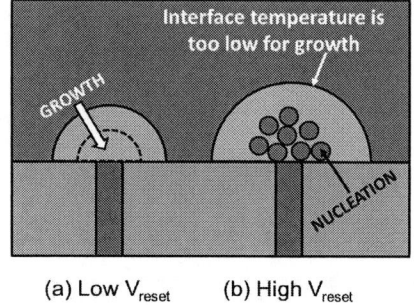

(a) Low V_{reset} (b) High V_{reset}

Fig. 4. Schematic illustration of the different crystallization processes of the amorphous volume at low (a) and high reset voltage (b), corresponding to small and large amorphous volumes, respectively.

pulse. The dashed yellow lines highlight the amorphous regions, which are enclosed between the bottom electrode and the upper crystalline region. Increasing the reset voltage leads to an increase of the amorphous thickness $_a$ from 11 nm (a) to 22 nm (c). It was previously reported that the amorphous thickness is extremely important, since it strongly impacts the statistical variation of the crystallization time [9]. To study the programming statistics in the array, Fig. 2 shows the cumulative distributions of the resistance R for the reset and set states at increasing reset voltage $_{reset}$, considering a population of 1408 cells. Both set and reset R increase with $_{reset}$, as also shown by the R − $_{prog}$ characteristics in Fig. 3. This shows the median R measured after the application of a rectangular pulse of amplitude $_{prog}$ and pulse-width $t_P = 160$ ns. The reset state R increases with $_{reset}$ because of the larger $_a$ in Fig. 1. On the other hand, the set-state R increases because a residual amorphous shell persists after set, thus contributing to the relatively high R in the set state at high $_{reset}$. The inset shows the transition point $_x$ defined as the voltage at which the resistance drops below 100 Ω, as a function of $_{reset}$ showing that, as a general trend, $_x$ increases with $_{reset}$. As the amorphous thickness increases at large $_{reset}$, a larger voltage is needed to increase the temperature at the amorphous-crystalline boundary to promote crystal growth

from the outer crystalline region (Fig. 4a). For large $_{reset}$, instead, $_x$ decreases as the high temperature occurring in the amorphous cap causes crystallization through nucleation, given that the temperature at the amorphous-crystalline boundary is too low for crystalline growth (Fig. 4b) [10].

Set statistics

Fig. 5 shows the cell resistance as a function of the set pulse duration t_P for three selected cells within the array: F and show, fast and slow crystallization respectively, while cell shows an intermediate behavior. The set time t_x is defined as the pulse duration needed to cross the threshold value $R^* = 100$ Ω. The cumulative distribution of t_x is shown in Fig. 6, together with isothermal annealing results at 150°C obtained from previous retention experiments [8]. The comparison reveals that the set time distribution is tighter than the retention time distribution. This is due to the non-Arrhenius crystallization in GST [6], with a relatively low activation energy $_x$ for crystallization at high temperature in the set regime [6]. In fact, the lower $_x$ decreases the variations of t_x in response to $_x$ variations at equal T in the GST, which were recently identified as potential source of retention spread in PCM arrays [10]. Fig. 7 shows the Arrhenius plot for the crystallization time, obtained by measuring the median t_x as a function of $_{prog}$ and extracting the corresponding effective temperature T in the GST volume. The Arrhenius plot clearly indicate the 2 different slopes in the set and retention regimes which are at the basis of the different statistical spread in Fig. 6. Note the markedly lower $_x$ in the set regime at high T [6], compared to the typical $_x = 2.85$ eV in the retention regime (T 250°C) [6].

Fig. 8 shows the median value of t_x (a) and its standard deviation (b) as a function of $_{reset}$ for a fixed programming voltage $_{prog} = 0.75$ V. The median t_x shows a steep increase at low $_{reset}$, which can be explained by the growth regime in Fig. 4a. Growth dominates at low $_{reset}$, thus t_x increases steeply with $_{reset}$ as the temperature at the amorphous-crystalline boundary at $_a$ decreases at the programming

978-1-4799-8002-4/14 $31.00 © 2014 IEEE

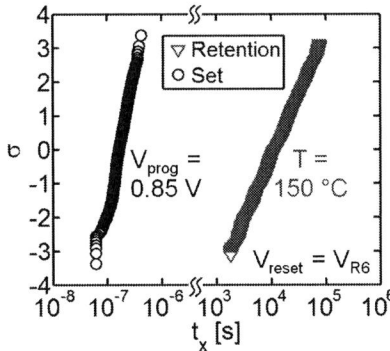

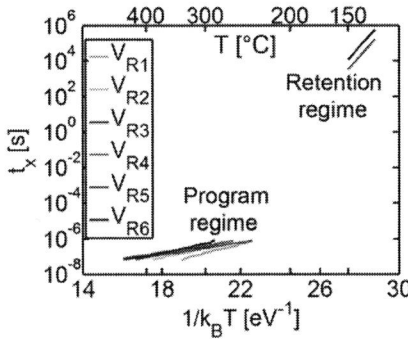

Fig. 6. Cumulative distributions of t_x in the high-temperature program (black dots) and in the low-temperature retention (red triangles) regimes. The comparison reveals a tighter distribution in the program regime.

Fig. 7. Crystallization time t_x as a function of the calculated cell temperature T. The slope E_x is lower in the high-temperature program region, compared to the retention regime, revealing the non-Arrhenius behavior of crystallization.

voltage V_{prog} = 0.75 V in the experiment. At larger V_{reset}, hence larger V_a, nucleation in the bulk of the amorphous phase dominates instead. The statistical spread in Fig. 8b increases for decreasing V_{reset} as the amorphous cap becomes smaller, thus the number of initial defects for crystalline growth decreases with a corresponding larger variation of nucleation barriers [9]. These results suggest that a higher V_{reset} is beneficial to the set/reset characteristics of PCM devices, since it improves the resistance window without severely affecting t_x and ensuring low statistical variability.

Fig. 9a shows the activation energy E_x for crystallization, extracted from the slope of the Arrhenius plot of Fig. 7 in the high-temperature set regime. Note the low value of E_x compared to E_x = 2.85 eV in the low-temperature crystallization (retention) regime, which is due to the non-Arrhenius crystallization behavior of GST [11]. The observed non-Arrhenius behavior was supported by ab-initio molecular dynamics (MD) simulations using a density functional theory (DFT) with Born-Oppenheimer (adiabatic) approximation [12]. From these simulations, E_x was extracted as the activation energy for atomic diffusivity, which dominates the T-dependence of the crystallization dynamics. Fig. 9b shows the computed trajectory of a fast moving Ge atom within a cage of less mobile atoms in GST. The self-diffusion coefficient D was simulated at variable T, thus yielding the Arrhenius plot in Fig. 9c. The calculated D is well captured by an Arrhenius law with an activation energy E_x = 0.33 eV, which compares well with our experimental results in Fig. 9a.

Impact of pulse shape

We compared the set characteristics in PCM arrays for different shapes of the set pulse, including the rectangular pulse (see Fig. 10a) and the triangular pulse (Fig. 10b). In the triangular pulse, the maximum voltage always exceeds the melting point V_{melt}, thus resulting in a transition to the liquid state of the active region. The slow quench allows a direct transition from the liquid to the crystalline phase, if the quenching time is sufficiently long, or to the amorphous phase, if the quenching rate is sufficiently fast.

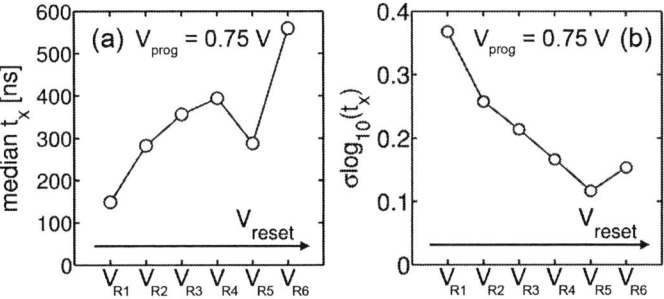

Fig. 8. Median crystallization time t_x (a) and spread of the t_x distribution (b) as a function of the reset voltage V_{reset} for a programming voltage V_{prog} = 0.75 V. t_x increases with V_{reset} due to the increasing size of the amorphous volume. The distribution spread decreases with V_{reset}, as the number of nucleation sites increases with the amorphous volume.

Fig. 10c shows the R distributions for the initial reset states and after the application of either a rectangular or a triangular pulse. By properly choosing the pulses amplitudes and durations (V_{prog} = 0.85 V, t_P = 160 ns for the rectangular pulse and V_{prog} = 150 , t_{quench} = 140 ns for the triangular pulse) we obtained resistance distributions with comparable tails of crystallized cells. In order to optimize the program operation in PCM arrays, we compared the resistance distributions as a function of energy consumption for the different pulse shapes in Fig. 10. Fig. 11a shows the measured resistance as a function of the applied pulse energy, for rectangular and triangular pulses at V_{reset} = V_{R4}. Both the median resistance and the distribution value at +/-3 σ are shown to monitor the distribution spread. The dissipated energy was calculated from the integral of the electrical power during the pulse. Larger (or longer) pulses allow for a lower programmed resistance at the expense of the pulse energy. Looking at the median curve, rectangular and triangular pulses show similar performances. However, triangular pulses show a smaller statistical spread as highlighted by the arrow in Fig. 11a, thus resulting in a wider R read window. Fig. 11b shows the median resistance as a function of the pulse energy, for increasing reset voltage, hence increasing size of the amorphous volume. The energy needed

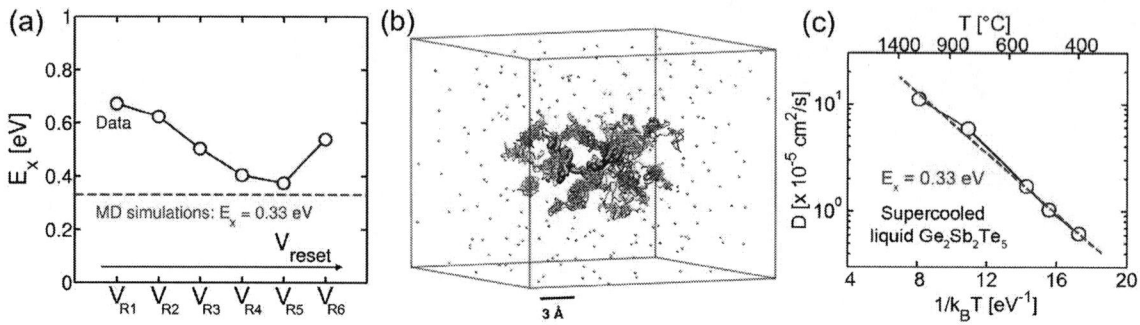

Fig. 9. Activation energy extracted from the Arrhenius plot of Fig. 7 as a function of V_{reset} (symbols) and calculated from MD simulations (dashed line) (a), calculated trajectory of a fast moving Ge atom in GST (b), and calculated diffusion coefficient as a function of 1/kT, obtained from MD calculations (c). The Arrhenius plot of D indicates an activation energy $E_x = 0.33$ eV, thus supporting the non-Arrhenius crystallization of GST.

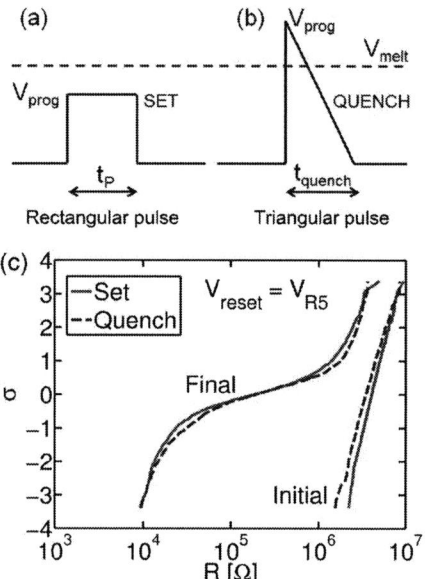

Fig. 10. Schematic illustration of the rectangular (a) and triangular (b) crystallization pulses and resistance distributions for the initial and post-pulse states for a rectangular pulse with $V_{prog} = 0.85$ V and $t_P = 160$ ns and a triangular pulse with $I_{prog} = 150$ μA and $t_{quench} = 140$ ns (c).

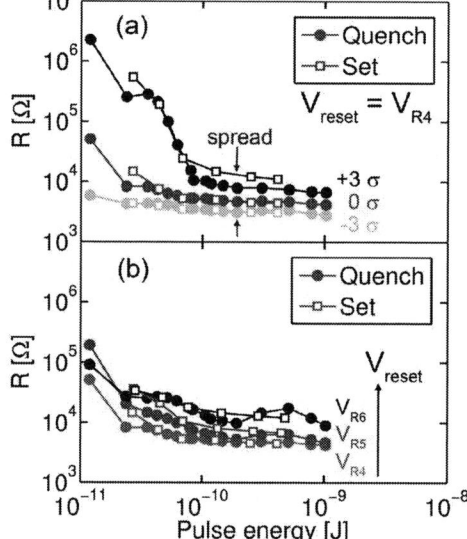

Fig. 11. Measured resistance after an applied crystallization pulse for median and +/-3 σ as a function of the pulse energy (a), and median resistance after set and quench pulses as a function of the pulse energy (b).

to obtain a given resistance value decreases for decreasing volume of the amorphous phase. This shows the beneficial effect of PCM size downscaling for decreasing the energy consumption of the set operation.

Conclusion

We present a statistical study of the set characteristics in PCM arrays. By controlling the amorphous volume through the reset voltage, we find that the crystallization voltage and time increase with the amorphous GST size, which can be explained by the impact of growth and nucleation mechanisms in the amorphous GST. The set time variability decreases at increasing $_{reset}$ as the number of nucleation sites increases in the bigger amorphous volume. Different statistical spread of the crystallization time are found at high and low temperature,

which we attribute to the non-Arrhenius crystallization in GST, as supported by MD simulations. We finally compare the set characteristics of rectangular and triangular pulses with the purpose of optimizing the energy consumption and resistance distribution in the PCM array.

References

[1] Y. Choi, et al., in ISSCC Tech. Dig., pp. 46-48, 2012.
[2] G. Bruns, et al., Appl. Phys. Lett., vol. 95, p. 043108, 2009.
[3] D. Loke, et al., Science, vol. 336, pp. 1566-1569, 2012.
[4] F. Xiong, et al., Science, vol. 332, pp. 568-570, 2011.
[5] J. Liang, et al., in Symp. VLSI Tech. Dig., pp. 100-101, 2011.
[6] N. Ciocchini, et al., in IEDM Tech. Dig., pp. 729-732, 2012.
[7] G. Servalli, in IEDM Tech. Dig., pp. 113-116, 2009.
[8] M. Rizzi, et al., in IEDM Tech. Dig., pp. 578-580, 2013.
[9] U. Russo, et al., IEEE T-ED, pp. 3040-3046, 2006.
[10] M. Rizzi, et al., ESSDERC, 2014.
[11] N. Ciocchini, et al., IEEE T-ED, pp. 3767-3774, 2013.
[12] G. Sosso, et al., J. Phys. Chem. Lett., pp. 4241-4246, 2013.

Circuit-Level Benchmarking of Access Devices for Resistive Nonvolatile Memory Arrays

P. Narayanan, G. W. Burr, R. S. Shenoy, K. Virwani, and B. Kurdi

IBM Research – Almaden, 650 Harry Road, San Jose, CA 95120, Tel: (408) 927–2920, E-mail: *pnaraya@us.ibm.com*

Abstract

Access Devices (1AD) for crossbar resistive (1R) memories are compared via circuit-level analysis. We show that in addition to intrinsic properties, AD suitability for 1AD+1R memories is strongly dependent upon (a) nonvolatile memory (NVM) and (b) circuit parameters. We find that (1) building large arrays ($\geq$1Mb) with $\geq$10uA NVM current would require MIEC ADs and moderate NVM switching voltage ($\leq$1.2V). (2) For all ADs high NVM voltages ($>$2V) are supported only at sub-5uA currents. AD improvements to expand this design space are discussed.

Keywords: 1AD1R, Selector, Access Device, Nonvolatile memory

Introduction

Desirable AD properties for 3D resistive crossbar memories (1AD+1R - Fig. 1) include BEOL compatibility, bipolar operation for RRAM/MRAM and large ON/OFF ratios to support high ON current density through selected cells with ultra-low leakage in unselect and partial select cells. Beyond these basic properties, determining whether an AD is suitable for an NVM requires circuit-level analyses that capture complex interactions between AD, NVM and circuit parameters. We explore capabilities and limitations of ADs in this design space by quantifying key figures of merit such as total power consumption, maximum achievable array size and the ranges of NVM voltages and currents supported.

Simulation Framework

We use a generic NVM model (Fig. 2) that transitions between Low- and High-Resistance States (LRS, HRS – can be Poole-

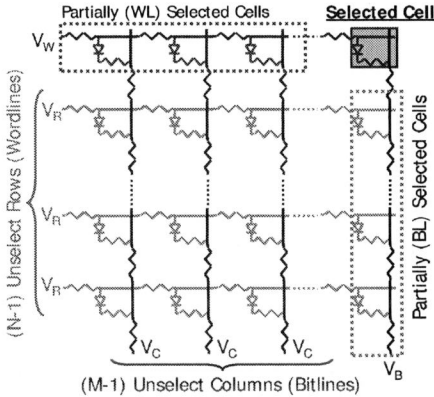

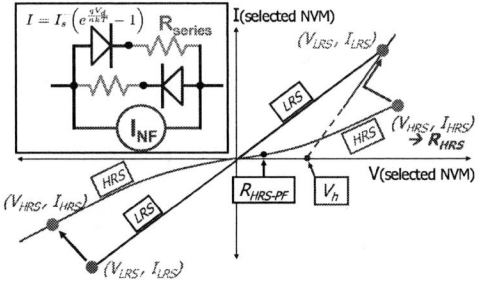

Fig. 1 Crossbar memory array with selected, partially (WL) selected, partially (BL) selected and unselected 1AD+1R cells.

Fig. 2 Generic NVM model for SPICE, with switching between an ohmic LRS and an HRS exhibiting Poole-Frenkel conduction. Inset shows equivalent circuit for SPICE modeling of bipolar diode-type ADs. Total array power to write worst-case selected cell (Fig. 1) must be estimated.

NVM Parameters		
SET Switching	V_{HRS}, I_{HRS}	1.2V, 3uA
RESET Switching	V_{LRS}, I_{LRS}	0.8V, 30uA
Holding V (SET)	V_h	0.35V
Resistance States	R_{LRS}, R_{HRS}	26.7kΩ, 400kΩ
PF HRS@0.1V	R_{HRS-PF}	10MΩ
Circuit Parameters		
Array Size	$N \times M$	1Mb
Interconnect R/cell	R_{int}	2.215Ω

Table. 1 Default NVM and Circuit Simulation parameters.

Frenkel (PF) or Ohmic) as a function of applied voltage (default parameters in Table 1). Inner voltages V_C and V_R for unselected bitlines/wordlines are chosen for an aggregate unselect leakage of 10uA (e.g. for 1Mb array, 10pA/device). Total array power to force a worst-case selected NVM through HRS–to–LRS and LRS–to–HRS transitions is estimated.

We consider 4 bipolar Diode-type ADs (D-ADs - MIEC [1], Varistor [5], Metal-amorphous Si-Metal [3] and Silicon NPN [7] - Fig. 3). D-ADs are modeled as back-to-back diodes with a noise floor (Fig. 2, top left inset). For all ADs, if IV data is not at scale ($\sim$32nm CD) or if only current-density data is available, we estimate currents assuming constant current-density.

Voltage Margin (V_m – defined as voltage range over which current$\leq$10nA), Turn-on Slope (S) and Series Resistance (R_s) parameters are extracted (Table 2). High V_m provides a wide low-leakage zone to accomodate partial-select cells in large arrays. Low S and R_s ensure low Voltage-across-AD (V_{AD}), and consequently low total switching voltage (V_{SW}) to be applied at the edge of the array to induce NVM switching (Fig. 4, inset). AD operating points

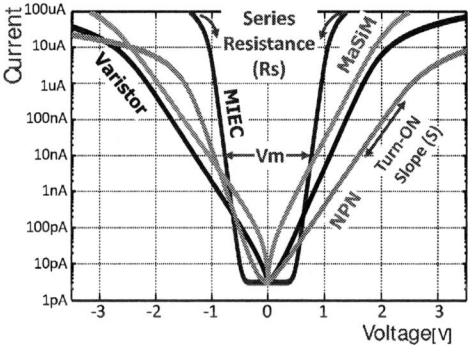

Fig. 3 IV characteristics of Diode-type ADs (D-ADs): AD parameters – Voltage Margin (V_m – voltage range for current$\leq$10nA), turn-on slope S (voltage for 10$\times$ current increase during AD turn-on) and series resistance R_s are indicated.

Diode-Type ADs			
	Vm(V)	Slope(mV/dec)	Rs(kΩ)
MaSiM	1.9	454, 333	3, 17
MIEC	1.54	85, 85	2.8, 2.8
NPN	2.56	219, 430	80, 70
Varistor	2.4	416, 282	19, 14

Threshold Switching ADs			
	Vth (V), Ith (uA)	Vh	Ron
CTS	1.67, 6.2	1.41	1
TVS	1.37, 0.87	0	1.8

Table. 2 Default Access Device Simulation parameters – among D-ADs, MIEC has the best turn-on slope and series resistance. Varistor and NPN have better voltage margin but slope and/or series resistance are also significantly higher and thus high voltages are required to drive high currents (Fig. 3).

978-1-4799-8002-4/14 $31.00 © 2014 IEEE

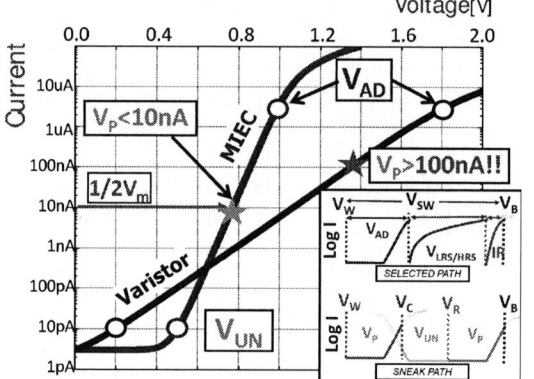

Fig. 4 MIEC, Varistor operating points for selected, unselected and partially selected ADs under default NVM, circuit parameters. Large Varistor V_{AD} causes increased total switching voltage V_{SW}, thereby causing high partial select leakage. MIEC operating conditions are within manageable limits.

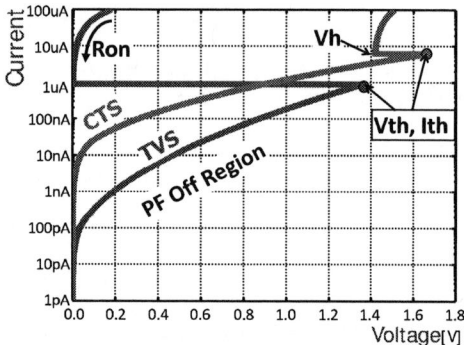

Fig. 5 I–V characteristics of Threshold-switching ADs(T-ADs): Off-state current is modeled as a Poole-Frenkel characteristic. Vth, Ith represents threshold switching condition of AD. V_h represents holding voltage after switching, R_{ON} is threshold device ON-state resistance.

under default NVM and circuit parameters are marked on MIEC and Varistor DC IV curves in Fig. 4.

Threshold switching ADs (T-ADs, e.g. Threshold Vacuum Switch (TVS) [4] and Chalcogenide Threshold Switch (CTS) [6], Fig. 5) can 'snapback' to low holding voltage above a current threshold, thereby reducing V_{AD} and V_{SW} at NVM switching. This may seem an 'unfair' advantage for T-ADs over D-ADs, especially when $V_h \sim 0$ (TVS). However, array design can still be constrained by total switching voltage, power to induce AD thresholding as opposed to NVM switching (Fig. 6), and must be included in circuit analysis.

To evaluate ~Mb arrays, number of circuit nodes is reduced by replacing all unselect cells by a single aggregate device (Fig. 7).

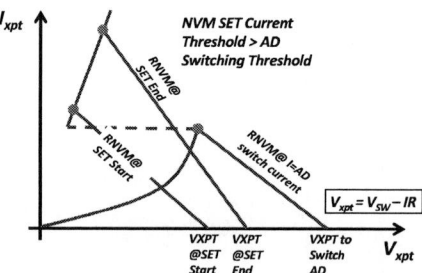

Fig. 6 Maximum voltage required at selected cell, and consequently worst-case power consumption in the array, can occur at the threshold condition of T-ADs (Fig. 5), as opposed to NVM switching conditions – illustrated in this diagram, with a representative T-AD IV and varying load lines representing instantaneous resistances of NVM.

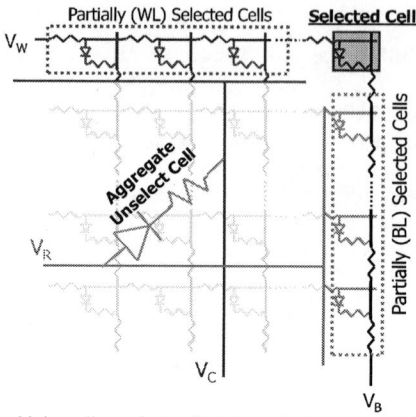

Fig. 7 Combining all unselect cells into a single aggregate Diode+NVM significantly reduces number of nodes, simplifying circuit analysis of large scale cro

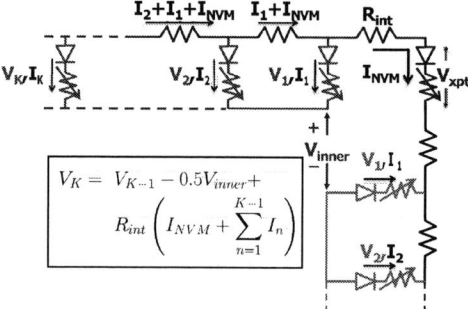

$$V_K = V_{K-1} - 0.5V_{inner} + R_{int}\left(I_{NVM} + \sum_{n=1}^{K-1} I_n\right)$$

Fig. 8 A hybrid circuit-simulation/analytical approach for analysis crossbar array conditions – by iterating currents and voltages outwards from the selected cell, one can estimate the voltage/current/power conditions at the edge of the array to induce NVM switching.

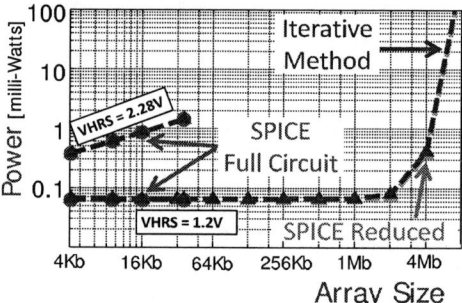

Fig. 9 Power consumption vs. Array size for 1MIEC+1NVM crossbar arrays: plot shows near-identical correspondence between full-SPICE simulations and approximate methods described in Figs. 7 and 8 for 2 different NVM V_{HRS} conditions.

Given operating conditions at selected cell, iterating outwards can determine voltage/current/power at the edge of the array to induce switching (Fig. 8). Fig. 9 shows excellent agreement in power for 1MIEC+1NVM using iterative, reduced SPICE and full simulation.

Design Space Exploration of ADs

Write power poses the most stringent constraint for 1AD+1R designs [2]. Design points become unfavorable if total array power far exceeds baseline power to switch AD+NVM. Fig. 10 plots a color map of total power for MIEC+NVM arrays when varying V_{HRS} and array size. At favorable design points (blue) most of the applied power is consumed at the selected cell; at unfavorable points (red), partial select ADs allow significant sneak path currents that can dominate power consumption.

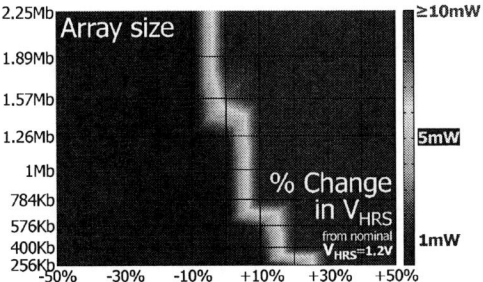

Fig. 10 Colormap of write power vs. V_{HRS} and array size for MIEC ADs from[2]: blue regions represent favorable design points with power consumption dominated by selected cell switching, red regions represent unfavorable design points with extreme sneak path leakage through partial-select cells.

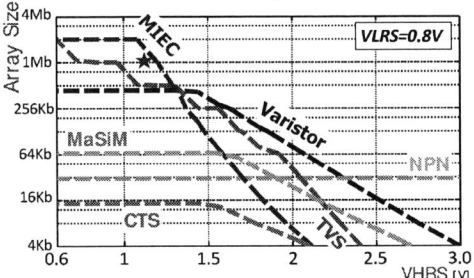

Fig. 11 Write power consumption contours at 1mW vs. NVM V_{HRS} and array size demarcating favorable (left and down) and unfavorable (right and up) design points for D-ADs and T-ADs. MIEC ADs can support array sizes $\geq$1Mb at moderate switching voltages. The star marks MIEC + nominal NVM parameters.

Fig.11 shows 1mW power contours used to delineate favorable (left, down) vs. unfavorable (top, right) regions for all ADs. The graphs validate that AD suitability is coupled strongly to extrinsic parameters - MIEC ADs can support array sizes up to 2Mb at low-to-moderate switching voltages$\leq$1.2V, whereas Varistor is better at $V_{HRS}\geq$1.5V but only on much smaller arrays. No AD can support high NVM switching voltage and large array sizes at default current values. CTS can support only small arrays, since leakage is relatively high, even at low bias. Contour 'plateaus' indicate regions where array power is limited by constant V_{LRS},I_{LRS}. An extreme case is NPN, which shows no dependence on V_{HRS} – high series resistance implies large V_{AD} to deliver 30uA. Fig. 12 assumes V_{LRS} scales as 2/3V_{HRS}. Trends are similar, but MIEC can support larger array sizes (4Mb) at low switching voltage, as V_{AD} reduces. Fig.13 shows that moderate increase in the supported NVM voltage is achieved with LRS non-linearity for MIEC, TVS and Varistor. Other ADs do not appear since array size (1Mb) is larger than what they can support.

Figs.14, 15 plot I vs. V design space for array sizes of 256Kb

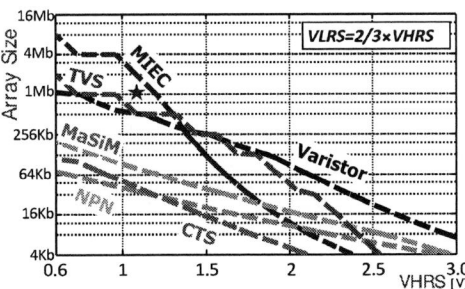

Fig. 12 Varying V_{HRS} and V_{LRS} simultaneously removes 'plateau' contours limited by constant V_{LRS} assumption. Moderate expansion in supported design space observed for MIEC and Varistor.

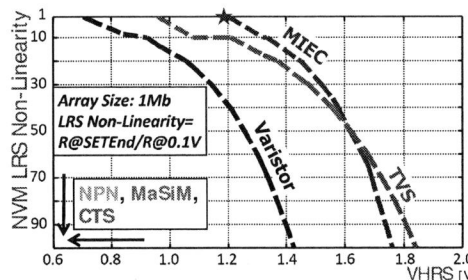

Fig. 13 Increasing NVM non-linearity moderately improves voltage ranges supported by certain ADs since partial select leakage can be reduced. Yet, switching voltages $\geq$2V are not supported. NVM non-linearity also has no impact on ADs that were already unsustainable at 1Mb in Fig.12.

and 1Mb, assuming V_{LRS}=2/3V_{HRS} and I_{LRS}=10$\times I_{HRS}$. D-ADs follow their device I-V (e.g. Varistor contour). In linear segments, D-ADs are limited by R_s – V_{AD} trades off in proportion to V_{HRS}; exponential segments show diode action (large change in current for small change in voltage) which implies that lower NVM switching voltage is an extremely important design consideration. NPN follows a nearly linear contour, given high R_s. At lower I_{LRS}, I_{HRS} (10uA, 1uA - may be required for interconnect scaling), NPN ADs can support 256Kb arrays with a relatively high V_{HRS} of 2.5V.

V_m can be doubled by stacking two ADs in series; V_{HRS} up to 2.5V, and array size of 1Mb can be supported by double MIEC (Fig.16). Stacking is not as effective on other D-ADs with poor S, R_s, since V_m benefits are offset by large increase in V_{AD}.

At $I_{LRS}\leq$10uA, TVS is constrained by NVM switching whereas at higher currents, worst-case power occurs at V_{th}, I_{th} (Fig.17). If AD switching is orders of magnitude faster than NVM, instantaneous power for AD threshold can be ignored and TVS ADs can support a much wider range of NVM V and I (2V at 1Mb in Fig.17). These gains are enabled by low V_h – negligible V_{AD} implies V_{SW} and consequently, power can be low. TVS design space can also

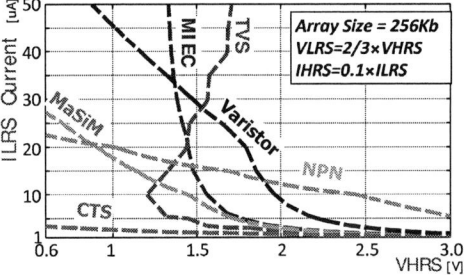

Fig. 14 At constant array size (256Kb), NVM switching voltage supported trades-off against switching current. D-ADs follow their DC IV – linear segments are limited by series resistance: current trades-off in proportion to voltage, exponential segments indicate large increases in supported current for small reduction in voltage.

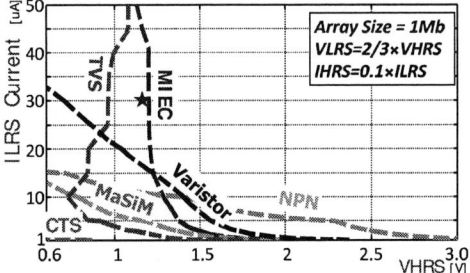

Fig. 15 1mW contours for Write power showing range of Voltages and Currents supported for an array size of 1Mb. MIEC ADs can support large arrays at high currents and moderate switching voltage. NPN can support large switching voltage at extremely low currents.

978-1-4799-8002-4/14 $31.00 © 2014 IEEE

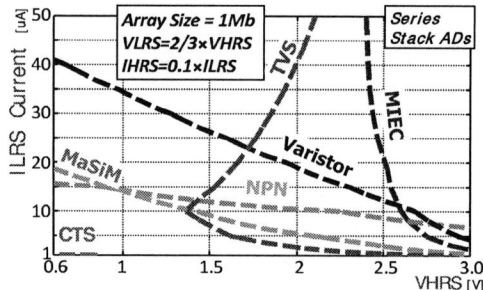

Fig. 16 Voltage-current design space at 1Mb for a composite series stack of two diode-like ADs: Significant gains are seen for MIEC, with a doubling in NVM switching voltage supported at 1Mb array size. Diminished gains for other D-ADs, given doubling in already large slope and series resistance values.

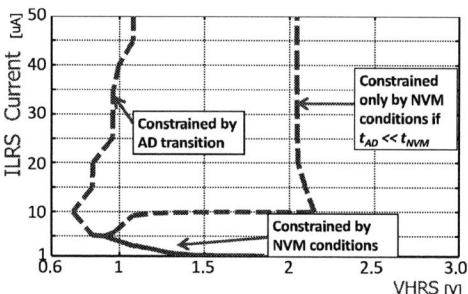

Fig. 17 At currents>10uA, power consumption of TVS+1R arrays is constrained by T-AD threshold point. If AD switching time is much less than NVM switching time, instantaneous power to switch AD can be ignored and larger NVM voltages (2V for 1Mb arrays) supported.

be expanded if I_{th} can be reduced (thereby reducing NVM voltage at I_{th}) while maintaining V_{th} (Fig. 18).

ADs were also compared against low-current, non-linear, 'self-select' RRAM [8]. While series AD increases V_{SW}, this can be offset by improved leakage mitigation on partial select cells – e.g. in Fig. 19 MIEC, NPN show >4× improved array size for the same power, CTS shows degradation and TVS has little impact.

Conclusions

AD suitability was shown to be dependent upon AD, NVM and circuit parameters. MIEC ADs were shown to be the best choice for NVMs with low-to-moderate switching voltages. NPN ADs are most suitable for larger switching voltages and low currents <5uA – MIEC, TVS, Varistor can support low-current/high voltage NVMs if supplemented by NVM non-linearity. Table 3 summarizes the

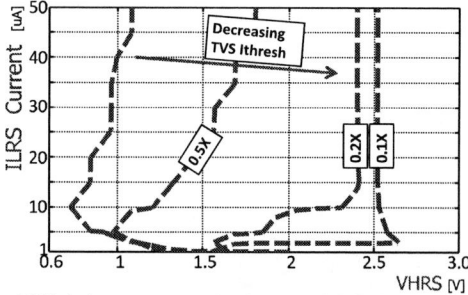

Fig. 18 TVS design space can also be expanded if threshold current can be reduced while maintaining threshold voltage. This considerably reduces voltage across NVM at AD threshold, thereby reducing total crosspoint voltage needed (refer Fig. 6).

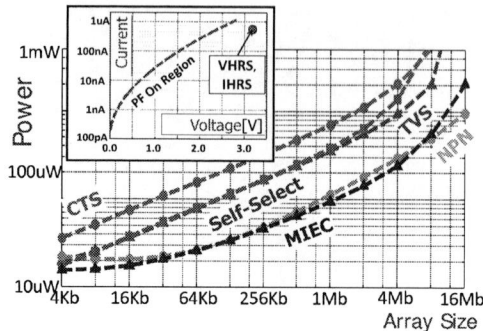

Fig. 19 Integrating D-ADs with low ON-current, high non-linearity NVMs [8] can enable ≥ 4× increase in array size vs. a selector-less array. TVS devices do not show any benefit with this NVM, as switching threshold of the AD is higher than the NVM.

design space. Table 4 identifies key parameters to be improved for the ADs studied.

	256Kb	512Kb	1Mb	2Mb
Low V, Low I	All, CTS<5uA	All, CTS<5uA	All except CTS	MIEC, NPN, Varistor (MaSiM, TVS@1uA)
Low V, High I	TVS, MIEC, Varistor	TVS, MIEC Varistor	TVS, MIEC	MIEC
High V, Low I	All@1uA, NPN@5uA	All@1uA, NPN@5uA	NPN<5uA, others+ non-linear NVM	NPN@1uA, MIEC, Varistor non-linear NVM
High V, High I	None	None	None	None

Table. 3 Design Space Summary

Diode-Type ADs	
MaSiM	Voltage Margin, Slope
MIEC	Voltage Margin
NPN	Slope, Series Resistance
Varistor	Slope
Threshold ADs	
CTS	Low-bias leakage, Ith
TVS	Ith

Table. 4 AD improvements to expand supported design space

References

[1] K. Virwani et al., *IEDM Tech. Digest*, 2.7 (2012).
[2] P. Narayanan et al., *DRC*, V.A-5 (2014).
[3] L. Zhang et al., *IEEE EDL*, 35(2) 2014).
[4] C-H. Ho et al. *IEDM Tech. Digest*, 2.8 (2012).
[5] J. Woo et al. *VLSI Tech. Digest*, 12-4 (2013).
[6] M-J. Lee et al. *IEDM Tech. Digest*, 2.6 (2012).
[7] V.S.S. Srinivasan et al. *IEEE EDL*, 33(10) (2012).
[8] S-G. Park et al., *IEDM Tech. Digest*, 20.8 (2012).
[9] ITRS Interconnect Tables (2011).

A Novel Inspection and Annealing Procedure to Rejuvenate Phase Change Memory from Cycling-Induced Degradations for Storage Class Memory Applications

W.S. Khwa[1,3], J.Y. Wu[1], T.H. Su[1], H.P. Li[1], M. BrightSky[2], T.Y. Wang[1], T.H. Hsu[1], P.Y. Du[1], S. Kim[2], W.C. Chien[1], H.Y. Cheng[1], R. Cheek[2], E.K. Lai[1], Y. Zhu[2], M.H. Lee[1], M. F. Chang[3], H.L. Lung[1], and C. Lam[2]

IBM/Macronix Joint PCRAM Project
Macronix International Co., Ltd., 16 Li-Hsin Road, Hsinchu Science Park, Taiwan
[1]Macronix International Co., Ltd., [2]IBM T.J. Watson Research Center, [3]National Tsing Hua University
E-mail: vincekhwa@mxic.com.tw,   Fax: +886-3-5789087, Phone: +886-3-5786688 #78265

Abstract

A novel Cycle Alarm Point (CAP) inspection is proposed to monitor PCM cycling degradation. The degradation appears in two stages – (1) right shift of R-I during moderate cycling degradation, and (2) left shift of R-I when cycling damage is severe. We further propose an In-Situ-Self-Anneal (ISSA) procedure, such that once a CAP signal is detected, the annealing procedure is issued to rejuvenate the cells. We demonstrate, for the first time, PCM cycling degradation can be recovered repeatedly. This opens a new window to extend PCM endurance and reliability for storage class memory (SCM) applications.

Introduction

The ever-enlarging performance gap between traditional hard disks and the rest of the memory hierarchy cries out for the need of SCM [1]. PCM is a promising candidate for SCM due to its scalability, DRAM-like operation and non-volatility. However, although single PCM cells have demonstrated high endurance up to 1E9 cycles, tail bits caused by various, hard-to-control factors greatly reduce the array cycling endurance. Previously, endurance improvements were proposed based on various materials and structural innovations, such as material doping, confined cell structure, and operation optimization [2-5]. In this work, we study the PCM failure mechanism and cycling induced degradation in detail and propose, for the first time, a novel Cycle Alarm Point (CAP) warning, which can monitor the PCM degradation effectively. We further propose a PCM annealing procedure to refresh the PCM from cycling damage. By detecting and managing the tail bits, the array endurance can be greatly improved for demonstrating a tail-bit free PCM application.

Device Structure and Measurement Platform

Fig.1 shows the PCM device structure with the bottom electrode (BE) diameter ~30nm for the 128Mb PCM test chip. A doped GST is used for improving cycling reliability and reducing operation current [6-7].   In this study, the PCM chip is mounted on a testing platform controlled by a FPGA as shown in **Fig. 2**. The waveforms used to perform RESET and SET operations are shown in **Fig. 3**.

Mechanism of Cycling-Induced Degradation

Fig. 4 illustrates the excellent endurance characteristics of a single PCM cell for >1E8 cycling. However, array endurance is much shorter due to tail bits. Two transition stages can be identified – (A): a downhill reduction of RESET resistance and (B): an uphill recovery of RESET resistance accompanied by increased SET resistance. While the varying SET-RESET window is a key parameter for cycling reliability, the R-I curve is found to be an indicator to monitor the PCM lifetime. **Fig. 5** reveals the detailed R-I curve of a cell from initial to high-cycle states. The right shift of R-I in Fig. 5(a) and the left shift of R-I in **Fig. 5(b)** are representative of RESET resistance degradation and recovery in **Fig. 4**. Note that these shifts are only noticeable in the partial-RESET states. The full-RESET state resistance stays constant throughout the cycling test. The open failure is always observed after the left shift of R-I curve and we found it irreversible (and fatal) in this work.

Another clue showing the relationship between cycling endurance and R-I curve can be found from the TEM images. **Fig. 6(a)** shows the segregation and small voids being generated after 10^5 cycling. **Fig. 6(b)** shows the void agglomeration and accumulation toward the BE after 10^6 cycling. **Fig. 6(c)** shows an open failure when void covers the entire BE area. The right shift or left shift of R-I curve is found to be sensitive to the composition change of GST after cycling [8]. A simple chart explaining the PCM failure mechanism with two steps degradation is shown in **Fig. 7**. We found the early degradation and later degradation can be accurately portrayed by the right and left shift of R-I curve, respectively. If early warning can be detected, then suitable annealing procedures may prevent the device from going into later stage degradation.

Fig. 8 shows the 100K cycled array distribution for SET and RESET states. Although PCM has excellent single cell cycling endurance, as in **Fig. 4** but tail-bits are observed. The tail-bits may be grouped into several categories according to TEM analysis and failure mechanism. Region (I) and region (II) are the good cell and the open cell regions, respectively. Cells in region (III) have lower RESET resistance and they show R-I right shift after cycling. Cells in region (IV) have higher SET resistance and show left R-I shift. These results imply that the cause for cycling induced degradation is complex and it is difficult to control by material or device engineering alone.

Cycle Alarm Point (CAP)

Fig. 9 shows the proposed CAP inspection. A pulse that can cause partial RESET (e.g. 150uA/50ns) is applied and then the cell resistance measured against a reference. As shown in **Fig. 5** earlier, the cell resistance in partial RESET state decreases with increasing cycling. If the resistance falls below reference then a CAP signal is generated. Once a CAP signal is detected, a healing procedure may be issued to refresh the PCM from the cycling damage (**Fig. 10**). In contrast, for the full RESET pulse in **Fig.9**, there will be no CAP signal due to the resistance of full RESET state is always above the reference value thus remains unchanged.

Healing by annealing was first investigated by baking a PCM chip in a furnace. In **Fig. 11(a)**, the cell's R-I curve shifts right due to cycling, the threshold current, which is defined as the minimum current to achieve a cell resistance of 100 kΩ, increased from 80 uA initially to around 120 uA after stressing. With a 350°C annealing for 30 minutes, the R-I curve is restored, the threshold current goes back to 80 uA as in **Fig 11(b)**. **Fig. 11(c)** shows the cell can continue to cycle normally. The healing procedure can be repeated, as shown in **Fig. 11(d).** This provides a possible solution for improving cycling induced degradation on PCM.

In-Situ Self-Anneal (ISSA) Method

The furnace annealing is concept-proving, but not a practical solution in real life. Thus we propose an In-Situ-Self-Anneal (ISSA) method to implement the healing in a practical way. Once the CAP signal is detected, an ISSA pulse is issued as shown in **Fig. 12**. The amplitude of the pulse is critical and should be carefully chosen to provide the necessary healing energy yet not to cause unwanted damage on GST. (Unlike furnace annealing, a strong in-situ pulse can cause GST damage due to electromigration [5].) **Fig. 13** illustrates that the ISSA annealing is repeatable and its effectiveness closely match that of external furnace experiment. **Fig. 14** shows the effective range of proposed ISSA method for PCM to rejuvenate cycling induced degradation. In **Fig. 14(a)**, healing effect is only observed for moderate ISSA current. Higher ISSA current cannot heal the R-I curve because the stressing damage outstrips the benefit of annealing. The upper boundary of ISSA healing current is 175uA, as shown in **Fig.14 (b)**. In addition, longer healing pulse width can achieve better healing effect under optimized ISSA current range.

Fig. 15 shows the simulated temperature distribution for various ISSA current. The simulated temperature range in (b) – (d) is consistent with the experimental data. In the case of (a), the GST temperature at 50 uA cell current is well below the annealing range. For (b), (c), and (d), the GST temperature is high enough to see some healing effect. In general, higher ISSA current helps to raise the GST temperature; but if it is too high, such as (e), the electromigration driven void formation would outstrips the annealing effect.

Fig. 16 shows a tail-bit-free PCM can be achieved after 100K cycling with the proposed ISSA method. As comparing to normal cycling without ISSA, all three kinds of failure mechanisms (right shift of R-I, left shift of R-I and open) can be successfully suppressed. **Fig. 17** provides another evidence for the ISSA healing effect. The SET speed shmoo plots show the cycling induced speed degradation can also be recovered by the proposed ISSA method.

Summary

We investigated PCM cycling-induced degradation in detail and identified the transition stages for PCM failure mechanism after cycling (right shift of R-I, left shift of R-I and then open). A CAP method is proposed to monitor the transitions and identify cells requiring the ISSA procedure for rejuvenation. We have shown that PCM cells are capable of withstanding the required annealing thermal energy and have demonstrated a tail-bit-free, no-degradation PCM array using the proposed ISSA method suitable for SCM applications.

References

[1] R.F. Freitas and W.W. Wilcke, "Storage-class memory: The nextstorage system technology," *IBM J. Res. Dev.*, vol.52, no.4.5, pp.439–447, 2008.

[2] Y.H. Shih, M.H. Lee, M. Breitwisch, R. Cheek, J.Y. Wu, B. Rajendran, Y. Zhu, E.K. Lai, C.F. Chen, H.Y. Cheng, A. Schrott, E. Joseph, R. Dasaka, S. Raoux, H.L. Lung, and C. Lam, "Understanding amorphous states of phase-change memory using Frenkel-Poole model," *IEEE International Electron Devices Meeting (IEDM)*, pp.1,4, 7-9 Dec. 2009.

[3] H.Y. Cheng, T.H. Hsu, S. Raoux, J.Y. Wu, P.Y. Du, M. Breitwisch, Y. Zhu, E.K. Lai, E. Joseph, S. Mittal, R. Cheek, A. Schrott, S.C. Lai, H.L. Lung, and C. Lam, "A high performance phase change memory with fast switching speed and high temperature retention by engineering the GexSbyTez phase change material," *IEEE International Electron Devices Meeting (IEDM)*, pp.3.4.1,3.4.4, 5-7 Dec. 2011.

[4] S.C. Lai, S. Kim, M. BrightSky, Y. Zhu, E. Joseph, R. Bruce, H.Y. Cheng, A. Ray, S. Raoux, J.Y. Wu, T.Y. Wang, N.S. Cortes, C.M. Lin, Y.Y. Lin, R. Cheek, E.K. Lai, M.H. Lee, H.L. Lung, and C. Lam, "A scalable volume-confined phase change memory using physical vapor deposition," *IEEE Symposium on VLSI Technology (VLSIT)*, pp.T132,T133, 11-13 June 2013.

[5] P.Y. Du, J.Y. Wu, T.H. Hsu, M.H. Lee, T.Y. Wang, H.Y. Cheng, E.K. Lai, S.C. Lai, H.L. Lung, S. Kim, M. BrightSky, Y. Zhu, S. Mittal, R. Cheek, S. Raoux, E. Joseph, A. Schrott, L. Jing, and C. Lam, "The impact of melting during reset operation on the reliability of phase change memory," *IEEE International Reliability Physics Symposium (IRPS)*, pp.6C.2.1,6C.2.6, 15-19 April 2012.

[6] M. Breitwisch, T. Nirschl, C.F. Chen, Y. Zhu, M.H. Lee, M. Lamorey, G.W. Burr, E. Joseph, A. Schrott, J. Philipp, R. Cheek, T. Happ, S.H. Chen, S. Zaidi, P. Flaitz, J. Bruley, R. Dasaka, B. Rajendran, S. Rossnage, M. Yang, Y.C. Chen, R. Bergmann, H.L. Lung, and C. Lam, "Novel Lithography-Independent Pore Phase Change Memory," *IEEE Symposium on VLSI Technology (VLSIT)*, pp.100,101, 12-14 June 2007.

[7] C.F. Chen, A. Schrott, M.H. Lee, S. Raoux, Y.H. Shih, M. Breitwisch, F.H. Baumann, E.K. Lai, T.M. Shaw, P. Flaitz, R. Cheek, E. Joseph, S.H. Chen, B. Rajendran, H.L. Lung, and C. Lam, "Endurance Improvement of Ge2Sb2Te5-Based Phase Change Memory," *IEEE International Memory Workshop*, pp.1,2, 10-14 May 2009.

[8] J.Y. Wu, M.H. Lee, W.S. Khwa, H.C. Lu, H.P. Li, Y.Y. Chen, M. BrightSky, T.S. Chen, T.Y. Wang, R. Cheek, H.Y. Cheng, E.K. Lai, Y. Zhu, H.L. Lung, and C. Lam, "A Double-Density Dual-Mode Phase Change Memory Using a Novel Background Storage Scheme," *IEEE Symposium on VLSI Technology (VLSIT)*, pp.116-117, 2014.

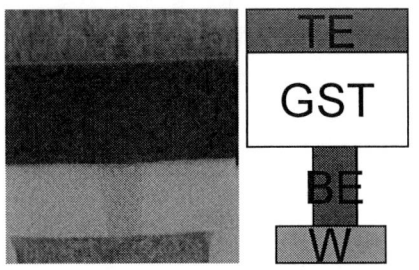

Fig. 1 The device structure from a 128Mb PCM chip.

Fig. 2 The testchip is mounted on a testing platform and controlled by a PC through an FPGA interface.

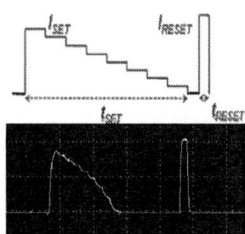

Fig. 3 The SET and RESET waveform from the 128Mb PCM chip. The distortion in the shown wave form is due to RC delay in the test platform and oscilloscope.

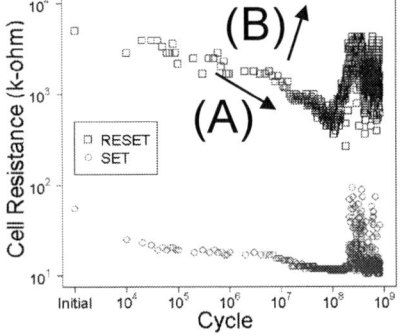

Fig. 4 The cycling endurance up to >1E8 is achieved by a single cell and transitions of cycling window (A: downhill) and (B: uphill) were observed and studied to understand the failure mechanism.

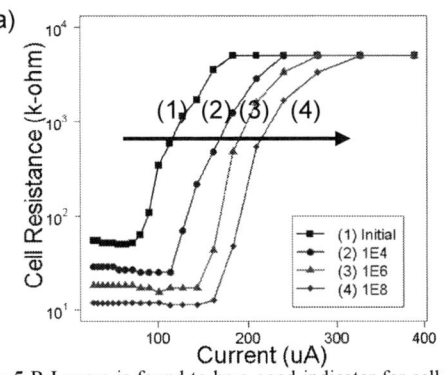

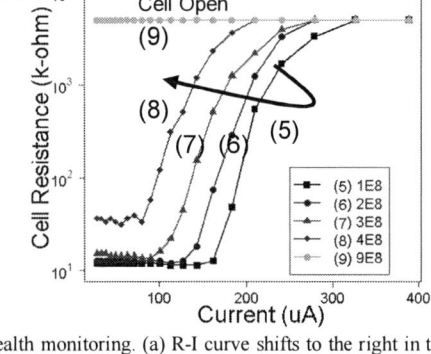

Fig. 5 R-I curve is found to be a good indicator for cell health monitoring. (a) R-I curve shifts to the right in the downhill region of Fig. 4 (indicated by (A)) during cycling. (b) R-I curve shifts to the left after severe cycling ((B) or uphill region in Fig. 4). The fatal open failure is always observed after "left-shift" of the R-I curve. Note that these shifts are only noticeable in the partial-RESET states. The full-RESET state resistance stays constant throughout the cycling test.

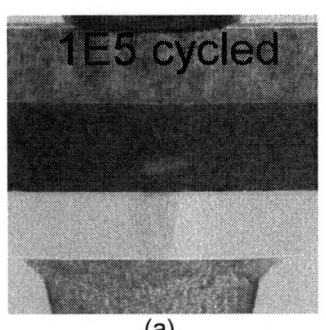

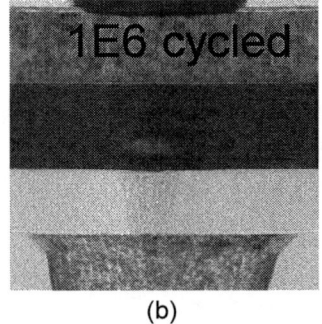

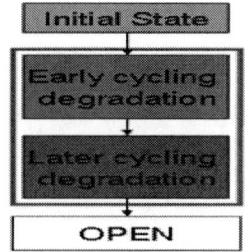

Fig. 7 Cycling degradation may be divided into early (recoverable) and later (not recoverable) stages. If early warning can be detected, then suitable annealing procedures may prevent the device from going into later stage degradation.

Fig. 6 TEM images suggest mechanisms for correlation between cycling endurance and the R-I curve. (a) The 10^5 cycled cell shows phase segregation and small voids are generated, (b) the 10^6 cycled cell shows enlarged void size and the void is closer to the BE, (c) open failure is fatal and occurs when the void covers the entire BE area.

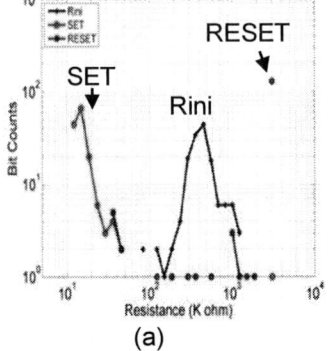

Fig. 8 (a) Resistance distribution of an array after 100K cycles. (b) The tail-bits fall into failure groups that match the failure mechanisms in TEM analysis. The cycling induced degradation has wide variation and it is difficult to control by material and device engineering alone.

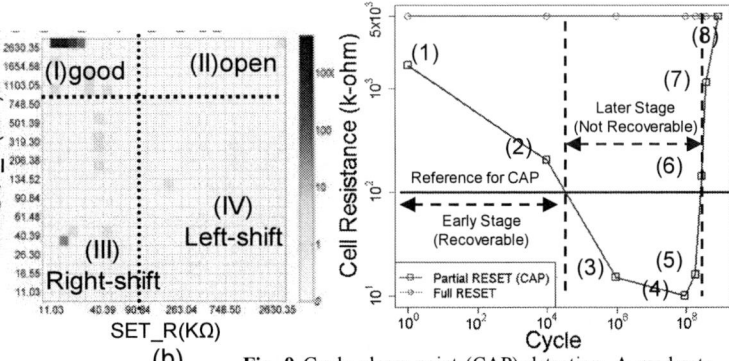

Fig. 9 Cycle alarm point (CAP) detection. A moderate partial RESET pulse (e.g. 150uA/50ns) is issued and the cell resistance measured and compared with a reference resistance. As shown in Fig. 5, partial RESET cell resistance decreases with cycling (points 1 – 4). When the cell resistance falls below the reference, a CAP warning is issued. If the cell is severely damaged then the resistance increases again (points 5 – 8).

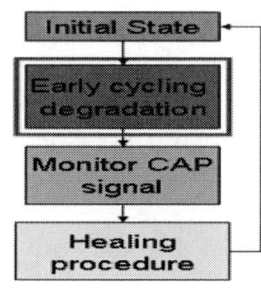

Fig. 10 The proposed healing procedure to further improve PCM reliability. The CAP signal is effective to detect the early (recoverable) cycling degradation stage.

978-1-4799-8002-4/14 $31.00 © 2014 IEEE   711

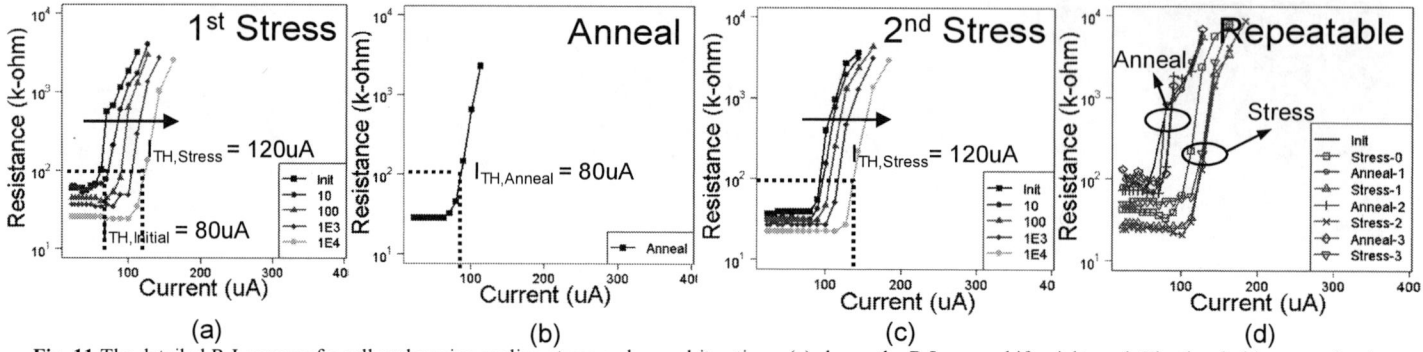

Fig. 11 The detailed R-I curves of a cell undergoing cycling stress and anneal iterations. (a) shows the R-I curve shifts rightward. The threshold current for the cell is 80 uA initially and 120 uA after cycling. (b) illustrates the R-I curve after 350°C anneal for 30 minutes. The threshold current for the cell after anneal is restored back to 80 uA. This demonstrates that the anneal can recover cycling damage. (c) shows the cell's R-I curve can be cycled again. (d) shows the annealing procedure is repeatable.

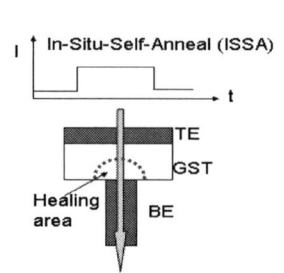

Fig. 12 The proposed electrical ISSA healing procedure. The ISSA current pulses generates localized annealing effect to remove cycling induced damage.

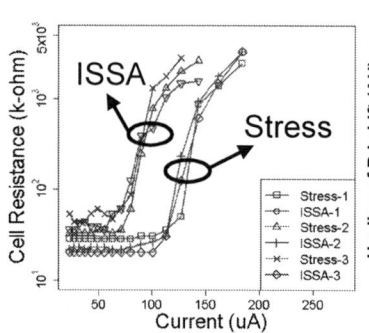

Fig. 13 The healing result from ISSA procedure closely matches that from external furnace experiment in Figure 11 (d).

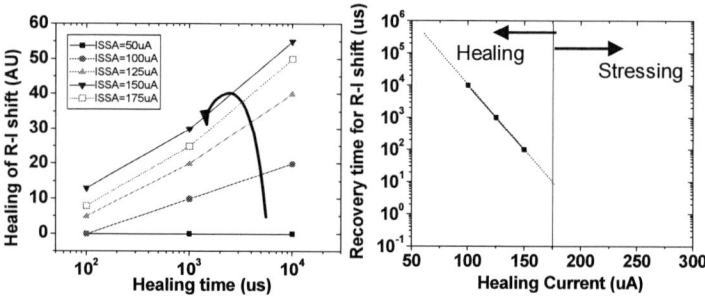

Fig. 14 The ISSA current and pulse width effects. (a) The ISSA current shows an interesting turnaround trend between 50uA to 175uA. (b) larger ISSA current provides faster healing, but too high ISSA current causes electromigration damage and becomes stressing instead of annealing.

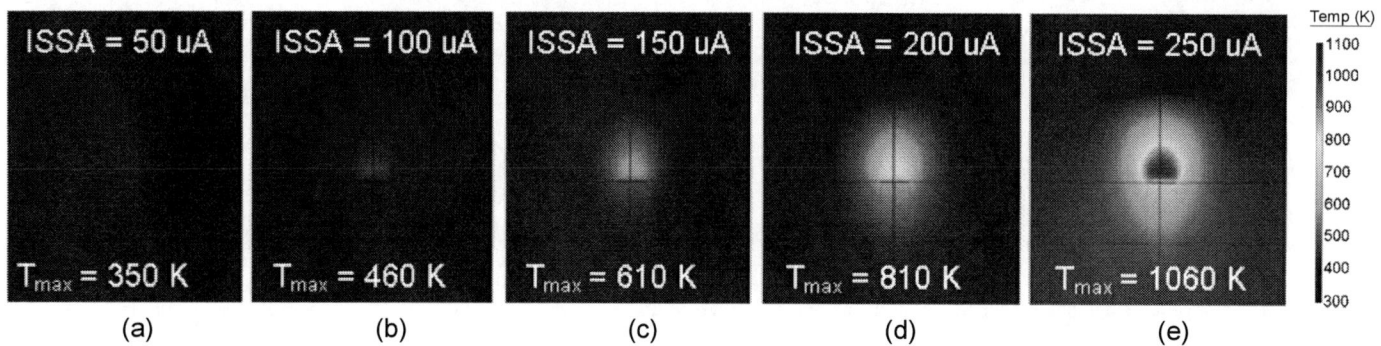

Fig. 15 Thermal simulation for PCM cell temperature under various ISSA current. (a) Similar to electrical data, low ISSA current cannot generate enough heat for annealing, (b/c) optimized ISSA current can generate adequate heat for annealing, (d/e) higher ISSA current cannot anneal even though the temperature is higher. The damage caused by void generation outstrips the benefit of annealing.

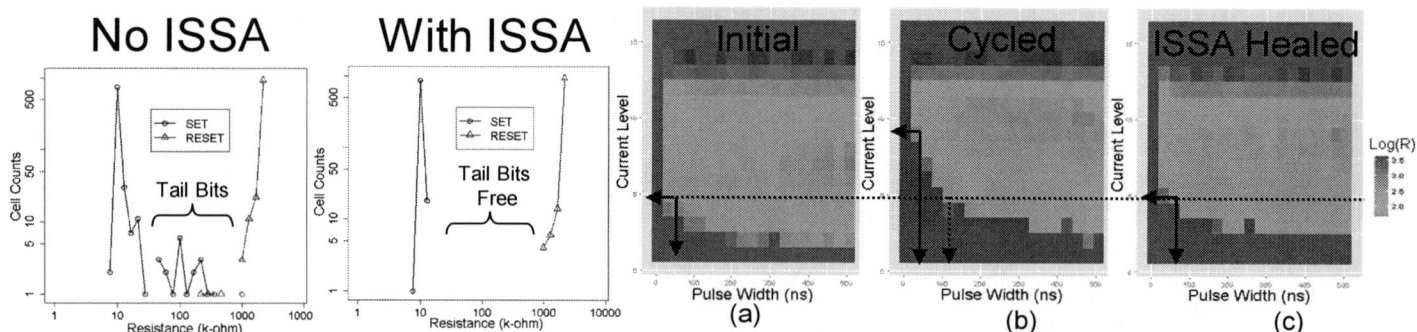

Fig. 16 The SET and RESET distribution for 2048 cells. (a) Without ISSA, significant number of tail bits emerges after 100K cycles. (b) With ISSA, the distribution remains tail bits free after 100K cycles.

Fig. 17 Another evidence for the ISSA healing effect. The SET shmoo shows the degradation of PCM performance can be healed after ISSA pulse. (a) initial state, (b) cycled state and operation current for fastest SET speed is increased, (c) ISSA can heal the degradation.

978-1-4799-8002-4/14 $31.00 © 2014 IEEE

Source-induced RDF Overwhelms RTN in Nanowire Transistor: Statistical Analysis with Full Device EMC/MD Simulation Accelerated by GPU Computing

Akito Suzuki[1], Takefumi Kamioka[2], Yoshinari Kamakura[3], Kenji Ohmori[4], Keisaku Yamada[4], and Takanobu Watanabe[1]

[1]Faculty of Science and Engineering, Waseda University, 3-4-1 Ohkubo, Shinjuku, Tokyo 169-8555, Japan,
[2]Semiconductor Laboratory, Toyota Technological Institute, 2-12-1 Hisakata, Tenpaku, Nagoya 468-8511, Japan,
[3]Graduate School of Engineering, Osaka University, 2-1 Yamada-oka, Suita, Osaka 565-0871, Japan,
[4]Graduate School of Pure and Applied Sciences, University of Tsukuba, 1-1-1 Tennodai, Tsukuba 305-8573, Japan.
Phone and fax: +81-3-5286-1621, Email: suzuki@watanabe.nano.waseda.ac.jp

Abstract

We numerically demonstrate that a random dopant fluctuation (RDF) in a source region causes a noticeable variability in the on-current of Si nanowire (NW) transistors, and its effect is much larger than that of a random telegraph noise (RTN). This work assesses the static and dynamic variability of NW device characteristics using the ensemble Monte Carlo/molecular dynamics (EMC/MD) simulation, which employs parallel computing technique using a graphic processing unit (GPU). The current flow in a one-dimensional NW device is determined by the number of dopants at the source edge, indicating the importance of forming an abrupt source–channel boundary to suppress the variability.

Introduction

The scaling and integration of Si CMOS very-large-scale integration (VLSI) have been impeded by various variability sources, e.g., a random dopant fluctuation (RDF), line-edge roughness, and fluctuation in the metal-gate work function [1–3]. RDF was expected to be suppressed by employing undoped channel in ultra-thin body (UTB) or multi-gate transistors. Contrary to the expectation, recent simulation studies have shown that the RDF in the source and drain (SD) region exerts a critical effect on the variability of fully depleted silicon-on-insulator (SOI) devices [4].

Dynamic current fluctuation induced by a random telegraph noise (RTN) is also a major integration concern because the on-current during one clock cycle varies from time to time depending on the number of interface trap charges (ITCs). However, determining which of the SD-RDF or RTN is the critical bottleneck is important for further scaling and integration. In particular, the current flow through a one-dimensional nanowire (NW) channel, which is a promising structure for extreme scaling, is known to be quite sensitive to a single ITC [5]. Therefore, we must know which of the SD-RDF or RTN dominates the variability of NW channels.

This work compares the static variability induced by SD-RDF and the dynamic variability induced by ITCs in gate-all-around (GAA) NW transistors, as shown in Fig. 1, using the ensemble Monte Carlo/molecular dynamics

(EMC/MD) simulation [6, 7]. EMC/MD allows an intuitive observation of the carrier scattering process and explicitly describes instantaneously the unscreened charges of dopants in the SD regions. So far, statistical analysis based on EMC/MD simulation has been unfeasible because of the huge computational resource requirement. We have overcome this difficulty in previous works [8, 9] by using a graphic processing unit (GPU) parallel computing technique. This work provides the first statistical demonstration of the larger impact of the source-induced RDF than that of RTN by means of EMC/MD.

Simulation Method

Figure 2 shows the GAA n-i-n Si NW model studied in this work. The field effect caused by the surrounding gate electrode was modeled by spreading particles with fractional charges over the gate insulator layer. The fractional charge was varied according to the gate voltage. The number of electrons in the SD regions was adjusted to satisfy the charge neutrality condition of the whole device. To investigate the RTN amplitude caused by the ITCs, fixed charges were randomly placed in the gate oxide layer. The work function of the gate electrode was assumed to be identical to that of the Si channel. Table 1 lists the device parameters determined according to ITRS 2012 [10] and 2013 [11]. The parameter set of ITRS 2012 is employed in the variability analyses, and that of ITRS 2013 is used in the transfer characteristics simulation. Quantum confinement effect in the cross-sectional plane was considered by adopting a confinement potential given by Eq. (1) [5, 12]. Only the conducting electrons were considered as carriers. Figure 3 shows a snapshot of the electron potential landscape. All electrons and impurity ions were randomly placed inside the source and drain region, and no impurity ion was included in the channel region. The single simulation time was 50 ps or 1 ns with a 0.1fs time step.

In the EMC/MD method, the carriers and impurity ions are treated as point charges (Fig. 4). To avoid a singular point in the attractive Coulomb interaction between an electron and a donor ion at zero distance, we employed a softened Coulomb potential (Fig. 5). The softening parameter α was determined to reproduce the experimental low field mobility in an n-type Si bulk [8]. Figure 6 shows the dependence of the low field mobility on the donor

---

978-1-4799-8002-4/14 $31.00 © 2014 IEEE   713

concentration estimated for various α. In this work, α is set to 7×10^{-10} m to reproduce the low field mobility in the SD regions with high donor density.

To enable statistical analysis, we parallelized our EMC/MD simulation code by utilizing a GPU [8, 9]. GPU is an auxiliary processor for real-time rendering of complex 3D models, which is equipped with several thousands of computing cores. It is also effective in scientific computing owing to its large-scale parallel operation ability. Figure 7 shows the speedup rate, which is the ratio of the execution time with the GPU to that with a CPU. The GPU parallel computation speed was enhanced by more than 10 times in the case of several thousand particles. The execution time of the EMC/MD simulation on a CPU is $O(N^2)$ to calculate the Coulomb point-to-point interactions among N particles. The GPU reduces the execution time to $O(N)$. Readers may consult Ref. (8) and (9) for more technical details.

Results and Discussion

The simulated transfer characteristics of the GAA n-i-n Si NW field-effect transistor (FET) is shown in Fig. 8. Here we set the device parameters based on the latest ITRS 2013 [11]. The off-state drain leakage current increases as the gate length (L_G) becomes short. Appreciable deterioration appears when L_G is reduced from 10 to 7 nm, indicating that the NW diameter should be scaled down to more than that of the road map [11] expected for $L_G < 10$ nm.

Figure 9 shows the cumulative probability distribution of the drain currents at the on state and the threshold voltage for various channel lengths. The current variability induced by the SD-RDF increases as the channel length becomes short. Figure 10 shows the cumulative probability distribution of the drain current for $L_G = 10$ nm including the ITCs, which corresponds to the dynamic variability due to the RTN. The number of ITCs ranges from one to four, which corresponds to the trap density of from 4.8×10^{11} to 1.9×10^{12} cm^{-2}. Although the current variability due to RTN increases with the number of ITCs, it is much smaller than that of the SD-RDF.

Figure 11 shows the comparison of the drain current variability due to the SD-RDF and RTN for $L_G = 10$ nm, which shows that the effect of the statistical variability of the SD-RDF is much larger than that of the dynamic fluctuation of the RTN. Here, we show two types of plots for the RTN. The blue plots show the case where the number of ITCs is fixed to four. The green plots show the case where the number of ITCs ranges from zero to four. The deviation from linearity in the green plots does not have a physical meaning because we simply gather 30 data samples for each number of ITCs and a single data without ITC. To obtain a meaningful cumulative probability distribution, we must consider the relative frequency of the number of ITCs. However, the overwhelmingness of the SD-RDF is never changed by the difference in the statistical processing.

The simulation clearly reveals that the drain current is sensitive to the source-induced RDF, and it is quite immune to the drain-induced RDF. Figure 12 shows the dependence of the on-current on the dopant densities in the source and

drain edges within 5 nm from the channel region. The current is strongly correlated to the impurity density in the source edge. Figure 13 shows the comparison of the on-current variability under three different cases with respect to the dopant distribution in the source-edge region: the number and positions are identical, only the number of dopants is fixed, and the position and density of the dopants are varied. The result suggests that the number of dopants at the source edge is a critical factor for the RDF in the NW FET.

Conclusions

We have provided a clear comparison of the effect of the SD-RDF and RTN on the current variability of GAA Si NW FET by performing a full-device EMC/MD simulation powered by GPU computing. The static variability of the source-induced RDF is found to overwhelm the dynamic on-current fluctuation due to RTN. We should note that our study conclusion is applicable to other device configurations such as fully depleted UTB-SOI and FinFET because a thicker channel has better tolerance to RTN [5]. Fabricating an abrupt source–channel boundary is thus the key to suppress the statistical variability in nanoscale devices. Metal SD is potentially an effective solution.

References

[1] X. Wang, A. R. Brown, B. Cheng, and A. Asenov, "Statistical variability and reliability in nanoscale FinFETs," IEDM Tech. Dig., pp. 5.4.1, 2011.

[2] Y. Ye, T. Liu, M. Chen, S. Nassif, and Y. Cao, "Statistical modeling and simulation of threshold variation under random dopant fluctuations and line-edge roughness," VLSI Systems, IEEE Transactions on, 19.6, pp. 987-996, 2011.

[3] R. Huang, R. Wang, J. Zhuge, C. Liu, T. Yu, L. Zhang, X. Huang, Y. Ai, J. Zou, Y. Liu, J. Fan, H. Liao and Y. Wang, "Characterization and analysis of gate-all-around Si nanowire transistors for extreme scaling," Custom Integrated Circuits Conference (CICC), 2011.

[4] S. Markov, A. S. M. Zain, B. Cheng, and A. Asenov, "Statistical variability in scaled generations of n-channel UTB-FD-SOI MOSFETs under the influence of RDF, LER, OTF and MGG," SOI Conference, 2012.

[5] T. Kamioka, H. Imai, Y. Kamakura, K. Ohmori, K. Shiraishi, M. Niwa, K. Yamada, and T. Watanabe, "Current fluctuation in sub-nano second regime in gate-all-around nanowire channels studied with ensemble Monte Carlo/molecular dynamics simulation", IEDM Tech. Dig., pp. 17.2.1, 2012.

[6] C. Jacoboni, and L. Reggiani, "The Monte Carlo method for the solution of charge transport in semiconductors with applications to covalent materials," Reviews of Modern Physics, 55.3, pp. 645, 1983.

[7] Y. Kamakura, H. Ryouke, and K. Taniguchi, "Ensemble Monte Carlo/molecular dynamics simulation of inversion layer mobility in Si MOSFETs—Effects of substrate impurity," IEICE Transactions on Electronics, 86.3, pp. 357-362, 2003.

[8] A. Suzuki, T. Kamioka, H. Imai, Y. Kamakura, and T. Watanabe, "Accelerated parallel computing of carrier transport simulation utilizing graphic processing units," 16th International Workshop on Computational Electronics (IWCE), Book of Abstract, pp. 166-167, 2013.

[9] A. Suzuki, T. Kamioka, Y. Kamakura, and T. Watanabe, "Full-scale whole device EMC/MD simulation of Si nanowire transistor including source and drain regions by utilizing graphic processing units," SISPAD 2014 Proceedings, pp. 357-360, 2014.

[10] ITRS2012, http://www.itrs.net/, 2013.

[11] ITRS2013, http://www.itrs.net/, 2014.

[12] L.Ge, and J. G. Fossum, "Analytical modeling of quantization and volume inversion in thin Si-film DG MOSFETs," Electron Devices, IEEE Transactions on, 49.2, pp. 287-294, 2002.

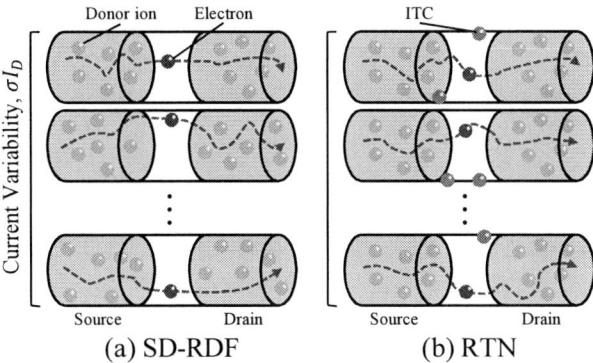

(a) SD-RDF   (b) RTN

Fig. 1 Schematic of the current variability sources in GAA NW FET compared in this work: (a) SD-RDF, in which the distribution of dopant ions in SD regions changes from device to device. (b) RTN, the current varies from time to time due to the temporally trapped charge in the gate oxide film.

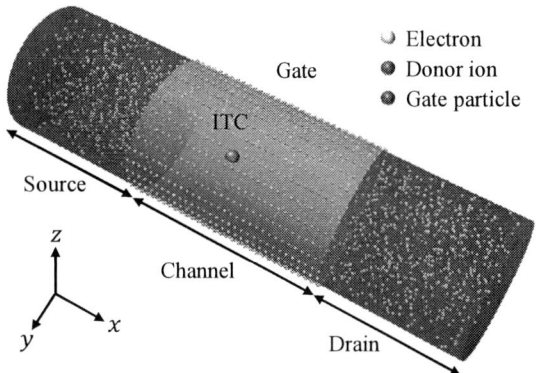

Fig. 2 Simulation model for the GAA n-i-n Si NW FET. One-dimensional periodic boundary condition is adopted in the longitudinal direction. The donor density in the SD region is 10^{20} cm^{-3}.

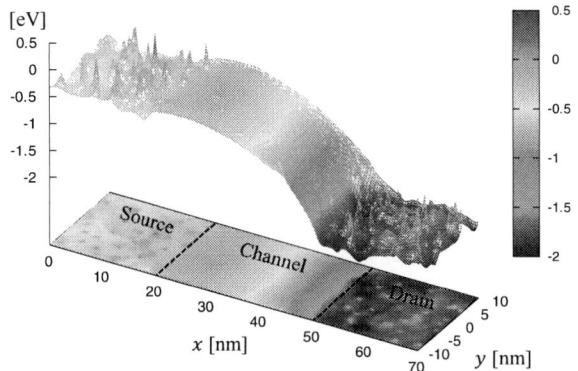

Fig. 3 Electron potential landscape on the cross-sectional plane, including the longitudinal axis, in the GAA NW FET at on state.

Table 1. Device parameters of the GAA n-i-n Si NW FET determined according to ITRS2012 [10] and ITRS2013[11]. L_S and L_D are the length of source and drain regions, respectively. The number of electrons is the total number of them in the source, channel, and drain regions.

ITRS	L_S, L_D [nm]	L_G [nm]	V_{DS} [V]	EOT [nm]	Diameter ϕ [nm]	Number of Electron
2012	20	30	1.0	1.0	20	1280
		20	0.85	0.88	13.3	566
		10	0.65	0.6	6.7	144
		7	0.6	0.48	4.7	72
		5	0.55	0.42	3.3	36
2013	40	20	0.85	0.80	6.4	256
		10	0.75	0.57	3.2	64
		7	0.70	0.48	2.2	32
		5	0.65	0.40	1.6	16

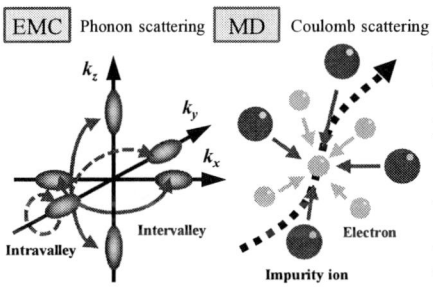

Fig. 4 Scattering processes described in the EMC/MD method. The acoustic and optical phonon scatterings are described as stochastic changes in the momentum of the carriers by EMC algorithm. The real-space trajectories of electrons under the Coulomb interaction are described by the MD algorithm. Bare point-to-point Coulomb potential is assumed between electrons, ITCs, and fractional charges on the gate electrode. A softened Coulomb potential is applied between an electron and a donor ion.

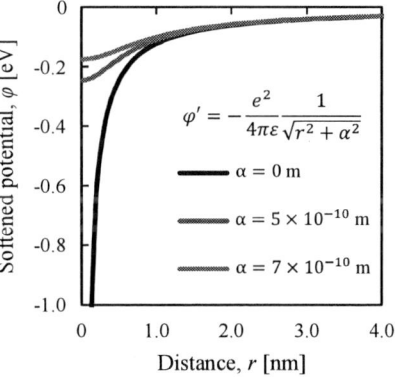

Fig. 5 Softened Coulomb potential to avoid a singular point at zero distance. e is the elementary charge, ε is the permittivity of the semiconductor, and α is the softening factor. The plot for $\alpha = 0$ is the bare point to point Coulomb potential with the singular point at $r = 0$.

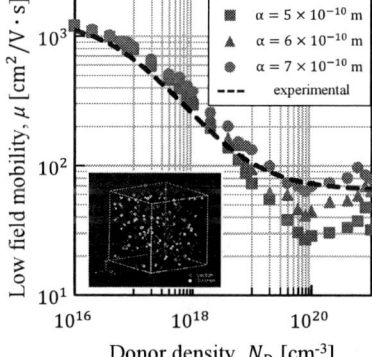

Fig. 6 Low field (1 kV/cm) mobility vs donor density relationship in an n-type bulk Si simulated using the softened Coulomb potential (Part of data are taken from Ref [8]). The unit cell includes 100 electrons and 100 impurity ions. Each data point is the average for 100 ps. In the current work, α is set to $\alpha = 7 \times 10^{-10}$ m.

$$V(r, N_{ele}) = -\frac{2k_b T}{q}\left[\ln\left[\sin\left(\frac{\pi}{\Phi}\left(r + \frac{\Phi}{2}\right)\right)\right] + \ln\left[\exp\left(-\frac{B(N_{ele})}{\Phi}\left(r + \frac{\Phi}{2}\right)\right) + \exp\left(\frac{B(N_{ele})}{\Phi}\left(r - \frac{\Phi}{2}\right)\right)\right] + \ln\left(\frac{A(N_{ele})}{\pi\Phi L_c n_i \sqrt{2\Phi}} N_{ele}\right)\right]$$

Equation (1) Quantum confinement potential in the GAA NW FET [5, 12]. A and B are functions that depend on N_{ele}. N_{ele} is the mean carrier density in the channel, which is self-consistently determined

978-1-4799-8002-4/14 $31.00 © 2014 IEEE

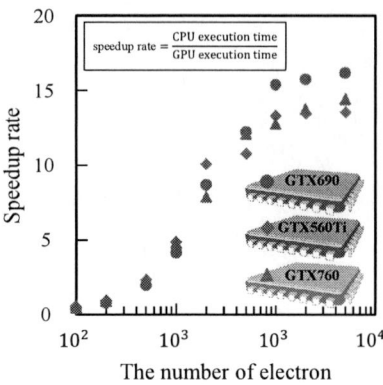

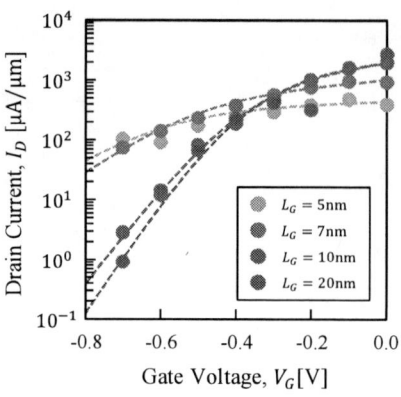

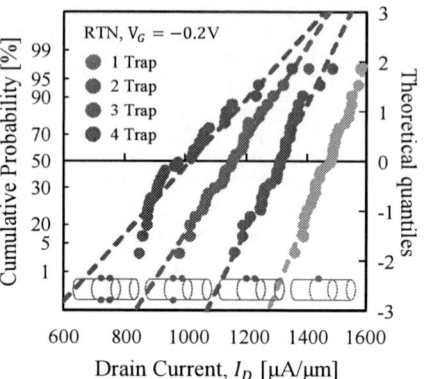

Fig. 7 Speedup rate of the EMC/MD calculation by the GPU. One GPU thread is assigned to the calculation with respect to one electron. We employed an Intel core i7 3930k CPU and NVIDIA GeForce GPUs. (Data of GTX690, GTX560Ti are taken from Ref. [8]).

Fig. 8 Transfer characteristics of the GAA Si NW FETs calculated by EMC/MD based on ITRS2013 [11]. The drain current is normalized by the channel diameter. Each data point is the average on the ensemble of 30 simulations for 50 ps.

Fig. 10 Cumulative probability distribution of the drain currents due to RTN for a 10-nm channel length. The number of ITCs ranges from one to four, which corresponds to the trap density of from 4.8×10^{11} to 1.9×10^{12} cm^{-2}.

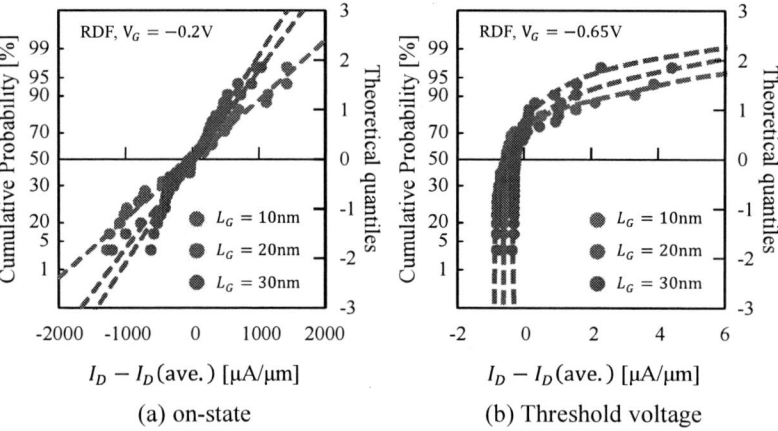

(a) on-state   (b) Threshold voltage

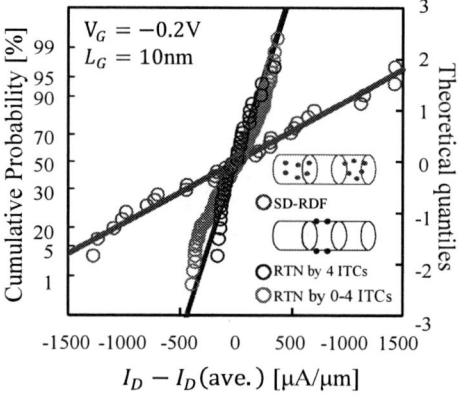

Fig. 9 Cumulative probability distribution of the on-currents induced by the SD-RDF. Each data point is the time average for 1 ns. The current variability increases as the channel length becomes short. The linearity of cumulative probability in on-state indicate that the drain current follows normal distribution. The cumulative probability at the threshold voltage deviates from linearity, because the drain current was almost zero in many cases within the limited simulation duration.

Fig. 11 Comparison of the SD-RDF and RTN. Here, RTN means the on-current variability due to the ITCs. The number of ITCs is 4 (Blue) and ranges between 0-4 (Green). The 2σ variation due to SD-RDF is about 1500 μA/μm, which is more than 3 times compared to the that of RTN. Thus, the effect of the SD-RDF on the on-current is much larger than that of the RTN.

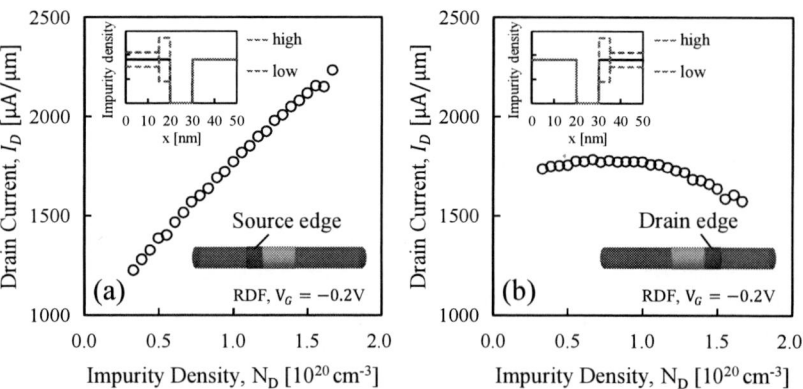

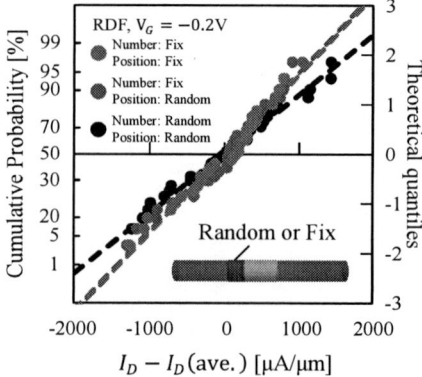

Fig. 12 Relationship between the on-state drain current and the donor densities in the (a) source and (b) drain edges within 5 nm from the channel region. The drain current is well correlated to the donor density at the source edge. Donor ions are randomly placed in the cross-sectional (y–z) plane, and the y–z positions are fixed throughout the simulation. The density at the S/D edge is adjusted by controlling the x coordinate of the donor ions. Each data point is time-averaged for 1 ns.

Fig.13 Cumulative probability distribution of the on-current for three different cases with respect to the dopant distribution in the source-edge region. The dopant number at the source edge is a critical source for the variability.

978-1-4799-8002-4/14 $31.00 © 2014 IEEE   716

Perspective of tunnel-FET for future low-power technology nodes

[1]A.S. Verhulst, [1,2]D. Verreck, [1,2]Q. Smets, [1,2]K-H. Kao, [1,3]M. Van de Put, [1]R. Rooyackers, [1,3]B. Sorée, [1]A. Vandooren, [1,2]K. De Meyer, [1,2]G. Groeseneken, [1,2]M.M. Heyns, [1]A. Mocuta, [1]N. Collaert and [1]A. V-Y. Thean

[1]Imec, Kapeldreef 75, 3001 Leuven, Belgium, [2]KULeuven, 3001 Leuven, Belgium, [3]U Antwerpen, 2020 Wilrijk, Belgium

Contact: anne.verhulst@imec.be

Introduction

Tunnel-FETs (TFETs) promise a subthreshold swing (SS) smaller than 60mV/dec and are considered as interesting candidates to replace MOSFET in future low-power technology nodes. The road ahead is challenging, with a large discrepancy between experiment and prediction, the latter showing extremely promising performance for heterostructure TFET at small supply voltage V_{dd}.

This paper starts with a calibration of the band-to-band tunneling (BTBT) models. It then discusses architecture and material optimizations, highlights the differences between n-TFET and p-TFET, and focuses on unexplored material aspects, like the decrease of dielectric constant with confinement. Parasitic effects are briefly touched upon, with trap-assisted tunneling (TAT) being the most challenging TFET parasitic to overcome. A new metric, V_{TAT}, is defined to capture the TAT impact.

Calibration

To make realistic TFET performance predictions, calibration of the BTBT models is desired, in particular of the prefactor A_{BTBT} and the exponential factor B_{BTBT} of the BTBT rate G_{BTBT} in a uniform electric field [1]:

$$G_{BTBT} = A_{BTBT} (E/E_0)^D \exp(-B_{BTBT}/E) \qquad (1)$$

with E the electric field, $E_0 = 1$V/cm and $D = 2$ (2.5) for direct (indirect) BTBT. We have calibrated the Si, compressively strained Si_xGe_{1-x} and $In_{0.53}Ga_{0.47}As$ BTBT models, based on p-i-n diodes of the respective materials (see **Fig. 1** for InGaAs) [2]. The theoretical predictions for B_{BTBT} are all within 10% of the experimental observations. The prefactor A_{BTBT} of the indirect processes (Si, Si_xGe_{1-x}) needs calibration due to uncertainty in the electron-phonon coupling strength. For $In_{0.53}Ga_{0.47}As$, the good agreement of also A_{BTBT} with experiment supports the model's validity for other direct-bandgap materials. The calibration process reveals the need for a combination of *I-V*, *C-V* and physical characterization to limit the calibration error, a challenge caused by the exponential dependence in the BTBT model.

Architecture and materials

The basic TFET configuration is a gated p-i-n diode. Optimized configurations are shown in **Fig. 2**: the hetero pocketed p-n-i-n TFET and the hetero line-TFET. These are both shown as vertical architectures, which allow for nanowire or fin embodiments. The counter-doped pocket and the line-TFET design promise improvement in both I_{on} and SS [3].

The heterostructure in Fig. 2 is essential to maximize the TFET performance. It allows to have a very small effective bandgap E_g at the source-channel injection side, while keeping a large E_g at the channel-drain side where ambipolar parasitic BTBT currents should be avoided. The ideal heterostructure should fulfill the material specifications shown schematically in **Fig. 3**: small effective tunnel gap to achieve high tunnel rates, high localized doping to create large electric fields to enhance tunnel rates, and large density-of-states (DOS) to avoid

Fermi levels inside the valence or conduction band [4]. This last criterion typically implies large E_g of the individual materials, however E_g cannot be too large else reflection at the hetero-interface will increase [5]. For n-i-p or n-p-i-p p-TFETs with tunneling along the nanowire axis, large DOS as a result of a nearby indirect band edge is ideal (e.g. GaSb).

The performance of promising TFETs is captured by the I_{60} metric, defined as the current at which SS transitions from sub-60mV/dec to super-60mV/dec behavior [6]. High I_{60} values of 1-10 µA/µm, necessary to compete with MOSFET are typically hard to attain experimentally (see **Fig. 4**(a)), because of traps, unwanted gradients in the active dopant profile or too wide body. Simulations, however, predict that TFET has the inherent capability to outperform MOSFET (see **Fig. 4**(b)), provided the TFET has a heterostructure [7-13] with strain promising further performance enhancement [14].

Predictions further indicate that implementations with nanowires or fins should be pursued. For wide body designs (typically 30nm and more), the SS degrades due to lack of electrostatic control of the tunnel junction in the body center. This is illustrated for the p-n-i-n TFET with quantum-mechanical predictions with a 2-dimensional self-consistent Poisson-Schrödinger solver based on the effective mass approximation (see **Fig. 5**) [3]. **Fig. 6** shows a detailed optimization for $Ga_{0.5}As_{0.5}Sb$-$In_{0.53}Ga_{0.47}As$ heterostructure p-n-i-n TFETs with body thickness up to 20nm. The p-source is $Ga_{0.5}As_{0.5}Sb$ and the n-pocket, channel and n-drain are $In_{0.53}Ga_{0.47}As$. Doping in source and pocket are 10^{20} at/cm^3, EOT is 0.6nm, pocket thickness varies from 2 to 4nm and body thickness varies from 5 to 20nm. A quantum-mechanical 2-dimensional solver has been used, calculating ballistic tunneling transitions based on a 15-band k.p material description [15]. Since no scattering mechanisms are included, the predicted currents reach beyond 1mA/µm, which is an overestimation. The simulations illustrate, that with decreasing body thickness, the SS decreases (better gate control) but the on-current also decreases (decreasing cross-sectional area and increasing bandgap due to size confinement). The I_{60} shows an optimal performance for 10nm wide TFETs, with I_{60} values up to 20µA/µm. Variability due to pocket thickness also decreases with decreasing body thickness. The 10nm devices are therefore suited for future technology nodes, when the supply voltage scaling is aggressive, e.g. $V_{dd} = 0.3$V, while today's technology, capable of 20nm wide designs, should be able to realize TFETs which can outperform MOSFET.

For p-TFET, achieving a steep SS is more difficult than for n-TFET. This is partially due to the low DOS in the conduction band of common direct bandgap materials, like most III-V materials [4]. However, also the architecture can significantly affect the TFET's performance. When the tunneling is (partially) oriented towards the gate dielectric, the electrostatic potential profile induces quantum confinement, and the resulting smaller effective E_g for the heavy-hole subband (compared to the ligh-hole subband), results in an earlier onset of phonon-assisted BTBT which

978-1-4799-8002-4/14 $31.00 © 2014 IEEE

degrades I_{60} (**Fig. 7**) [16]. The illustration is made for high-eDOS InGaAs ($In_{0.53}Ga_{0.47}As$ with a "heavy electron" band added to its conduction band), such that a sub-60mV/dec SS is achieved. Configurations whereby tunneling is parallel to the gate should therefore be pursued for p-TFET, like strongly confined n-p-i-p or n-i-p configurations [12], possibly in combination with tensile strain.

Beneficial impact of dielectric constant

A decrease in dielectric constant ε has the same beneficial impact on the electrostatic potential bending $\Delta\Psi$ as an increase in doping N_s, while avoiding the degeneracy penalty:
$$\Delta\Psi \sim N_s/\varepsilon \qquad (2)$$
The dielectric constant of a material decreases upon physical confinement of the material [17]. Hence a thin slab of a material with different atoms, composition or crystal structure inserted at the tunnel junction can enhance the TFET performance (see **Fig. 8**).

Trap-assisted tunneling

A strong temperature-dependent SS is observed in many experimental TFETs. The most straightforward cause of the resulting degraded SS at room temperature is trap-assisted tunneling. No good description of the amount of TAT exists. We propose the following figure of merit for TAT, capturing the TAT impact on the required supply voltage when a given I_{off} and I_{on} are desired: $V_{TAT}@I_{off}$ being the change in voltage at which the desired I_{off} is achieved at room temperature versus 77K (see **Fig. 9**). Introducing heterostructures can increase TAT but further research is required to determine whether the hetero-interface traps have an observable impact on TAT or whether other sources of TAT (interface states, border traps, bulk traps) are dominant.

Conclusions

Theoretically, confined heterostructure p(-n)-i-n (n(-p)-i-p) TFETs are promising candidates for future low-power applications, with n-TFET outperforming p-TFET. An optimal body thickness of about 10nm is predicted for $Ga_{0.5}As_{0.5}Sb$-$In_{0.53}Ga_{0.47}As$ n-TFET with $I_{60}=20\mu A/\mu m$. For p-TFETs, stronger confinement may be required to avoid tunneling to the heavy-hole band. An unexploited domain is the insertion of thin heterostructure slabs offering a locally reduced dielectric constant, enhancing both SS and I_{on}.

Overall, the gap between theory and simulation is decreasing due to calibration efforts. But in-depth knowledge of TAT is needed to unravel the TFET's full potential.

Acknowledgements

This work was supported by imec's Industrial Affiliation Program. D. Verreck and Q. Smets acknowledge the support of a Ph.D. stipend from the IWT-Flanders.

References

[1] E. Kane, J. Phys. Chem. Solids 12, 181 (1959).
[2] Q. Smets et al., J. Appl. Phys. 115, 184503 (2014).
[3] D. Verreck et al., Trans. Elec. Dev. 60, 2128 (2013).
[4] W.G. Vandenberghe et al., Appl. Phys. Lett. 100, 193509 (2012).
[5] M. Van de Put, IEEE Eurocon, 2134 (2013).
[6] W.G. Vandenberghe et al., Appl. Phys. Lett. 102, 013510 (2013).
[7] K. Tomioka et al., VLSI, 48 (2012).
[8] G. Dewey et al., IEDM, 785 (2011).
[9] F. Mayer et al., IEDM, 163 (2008).
[10] U.E. Avci et al., IEDM, 830 (2013).
[11] M. Luisier and G. Klimeck, IEDM, 913 (2009).
[12] B. Rajamohanan et al., J. Appl. Phys. 115, 044502 (2014).
[13] E. Baravelli, Trans. Elec. Dev. 61, 178 (2014).
[14] F. Conzatti, IEDM, 95 (2011).
[15] D. Verreck et al., J. Appl. Phys. 115, 053706 (2014).
[16] A.S. Verhulst et al., Appl. Phys. Lett. 105, 043103 (2014).
[17] G. Zhang et al., Electron Dev. Lett. 29, 1302 (2008).
[18] R. Bijesh et al., IEDM, 687 (2013).
[19] R. Rooyackers et al., IEDM, 92 (2013).

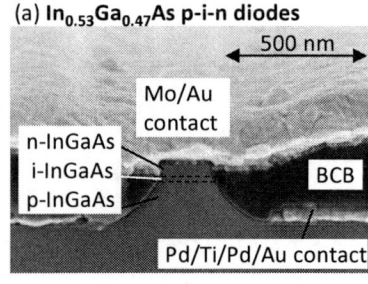

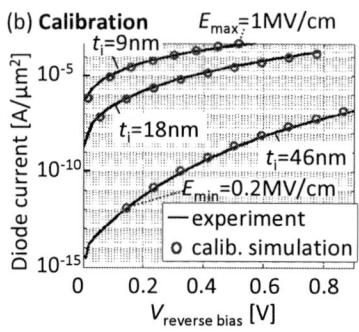

(c) **Semi-classical and quantum-mechical BTBT model parameters**

	lower limit	recommended	upper limit	theory
A_{BTBT} [$cm^{-3}s^{-1}$]	1.1×10^{20}	1.3×10^{20}	1.6×10^{20}	1.6×10^{20}
B_{BTBT} [$V\,cm^{-1}$]	6.0×10^{6}	5.7×10^{6}	5.4×10^{6}	5.6×10^{6}
QM: E_P [eV]	13.5	15	16.5	16.5

Fig. 1. BTBT model calibration: (a) SEM picture of $In_{0.53}Ga_{0.47}As$ p-i-n diode. Three samples with intrinsic region thickness $t_i = \{9nm, 18nm, 46nm\}$ are fabricated. (b) Calibration of the $In_{0.53}Ga_{0.47}As$ BTBT model by simultaneously matching the BTBT current of 3 diodes. (c) Results of the semi-classical (A_{BTBT}, B_{BTBT}) and quantum-mechanical (E_P) BTBT calibration. Comparison with theory confirms the validity of the direct-bandgap BTBT model.

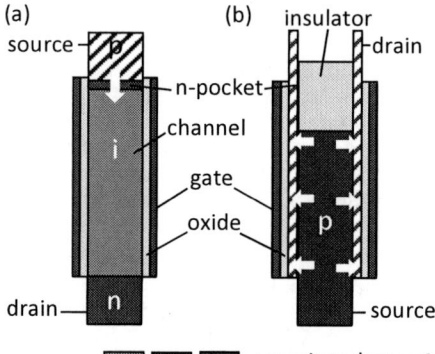

Fig. 2. Optimized TFET architectures: (a) hetero p-n-i-n and (b) hetero line-TFET. The white arrows indicate the location of the BTBT injection. Both designs are suitable for nanowire-based or fin-based fabrication.

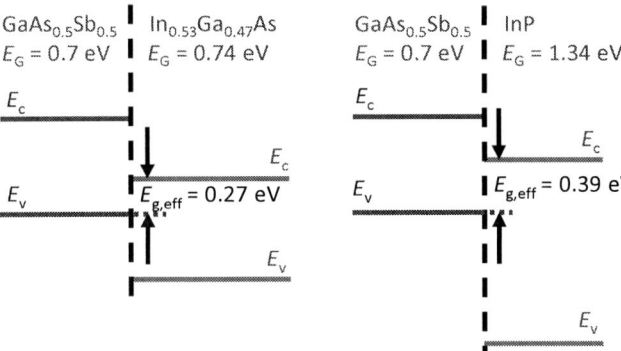

(a) Small effective tunnel gap

$GaAs_{0.5}Sb_{0.5}$ | $In_{0.53}Ga_{0.47}As$
$E_G = 0.7$ eV | $E_G = 0.74$ eV

$GaAs_{0.5}Sb_{0.5}$ | InP
$E_G = 0.7$ eV | $E_G = 1.34$ eV

$E_{g,eff} = 0.27$ eV

$E_{g,eff} = 0.39$ eV

(b) Higher doping → larger E-field → higher BTBT rate

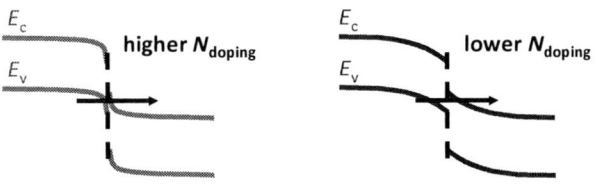

higher N_{doping}

lower N_{doping}

(c) High h-DOS (and n-DOS)

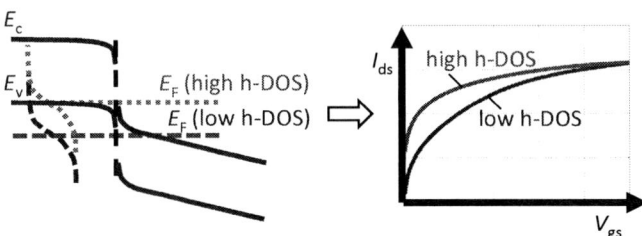

E_F (high h-DOS)

E_F (low h-DOS)

high h-DOS

low h-DOS

Fig. 3. Ideal heterostructure for TFET applications: (a) small effective tunnel gap. The left configuration has the smaller effective E_g and hence larger I_{on}, the right configuration has the larger E_g of the individual materials and hence smaller I_{off}. (b) high doping, (c) high DOS in valence band (n-TFET) and conduction band (p-TFET: not shown) [4].

(a) Experimental I_{60} achievements

n-TFET	I_{60} [A/μm]	p-TFET	I_{60} [A/μm]
Tomioka [7]	6×10^{-9}	Mayer [9]	1×10^{-12}
Dewey [8]	3×10^{-9}		

(b) Predicted I_{60} performance

n-TFET	I_{60} [A/μm]	p-TFET	I_{60} [A/μm]
This work	2×10^{-5}	Rajamoha. [12]	3×10^{-6}
Avci [10]	1×10^{-5}	Baravelli [14]	1×10^{-6}
Luisier [11]	1×10^{-5}	Avci [10]	1×10^{-7}
Rajamoha. [12]	5×10^{-6}		
Baravelli [13]	3×10^{-6}		
Conzatti [14]	3×10^{-6}		

Fig. 4. Highest achieved values for I_{60}, the current up to which the TFET SS beats the MOSFET SS: (a) experimental and (b) simulation. All predictions make use of hetero-TFET, except for Ref. 10 with a p-TFET out of GeSn and Ref. 14 with an n-TFET out of biaxial tensile strained InAs. All predictions make use of nanostructures with body thickness < 20nm.

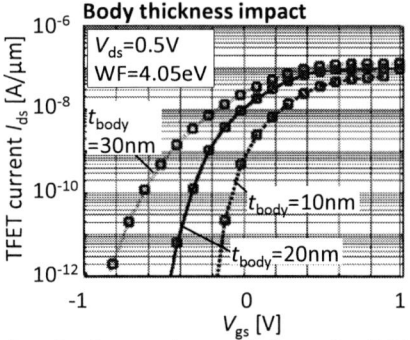

Body thickness impact

$V_{ds}=0.5V$
WF=4.05eV

$t_{body}=30nm$

$t_{body}=10nm$

$t_{body}=20nm$

Fig. 5. Input characteristics of all-Si p-n-i-n TFET simulated with QM solver based on phonon-assisted BTBT [3]. Down to 10nm body thickness, there is no size confinement expected in Si. This figure illustrates that wide-body (30nm for Si) p-n-i-n TFETs have degraded SS.

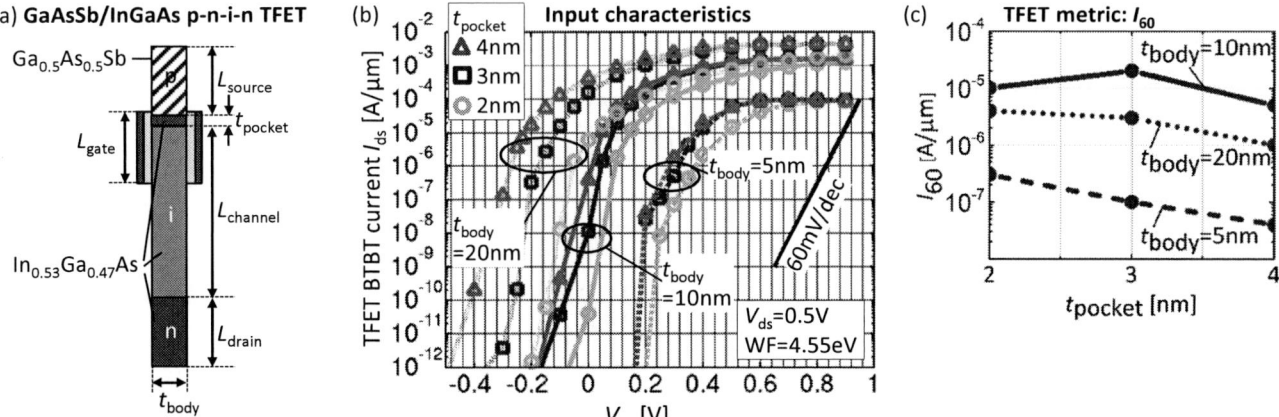

(a) GaAsSb/InGaAs p-n-i-n TFET

$Ga_{0.5}As_{0.5}Sb$

L_{source}
t_{pocket}
L_{gate}
$L_{channel}$
$In_{0.53}Ga_{0.47}As$
L_{drain}
t_{body}

(b) Input characteristics

t_{pocket}: 4nm, 3nm, 2nm

$t_{body}=5nm$
$t_{body}=20nm$
$t_{body}=10nm$

60mV/dec

$V_{ds}=0.5V$
WF=4.55eV

(c) TFET metric: I_{60}

$t_{body}=10nm$
$t_{body}=20nm$
$t_{body}=5nm$

Fig. 6. $Ga_{0.5}As_{0.5}Sb/In_{0.53}Ga_{0.47}As$ p-n-i-n n-TFET optimization: (a) schematic design: $L_{source}=L_{drain}=20nm$, $L_{channel}=50nm$, $L_{gate}=21nm$, EOT=0.6nm, $N_{source}=N_{pocket}=10^{20}$at/cm^3, $t_{pocket}=\{2nm,3nm,4nm\}$, $t_{body}=\{5nm,10nm,20nm\}$; (b) input characteristics, indicating SS steepening but decreasing I_{on} with decreasing t_{body}. This implies that with decreasing supply voltage, narrower body TFETs have to be targeted. (c) I_{60} values up to 20μA/μm, indicating superior performance compared to MOSFET.

978-1-4799-8002-4/14 $31.00 © 2014 IEEE

(a) p-TFET with tunneling orthogonal to gate

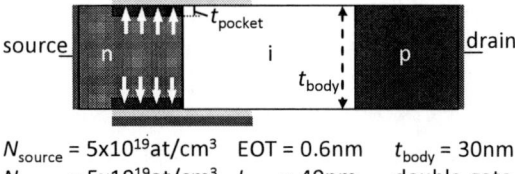

$N_{source} = 5 \times 10^{19} \text{at/cm}^3$ EOT = 0.6nm t_{body} = 30nm
$N_{pocket} = 5 \times 10^{19} \text{at/cm}^3$ L_{gate} = 40nm   double-gate
$L_{source} = L_{drain} = 20$nm $L_{channel} = 50$nm $L_{pocket} = 20$nm

(b) Input characteristics for high-eDOS InGaAs

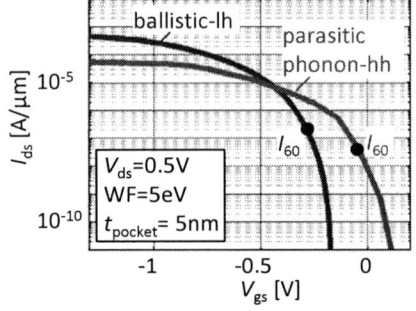

Fig. 7. Impact of hh-band on p-TFET: (a) Schematic of a double-gate pocketed line-TFET, the white arrows indicate the tunneling direction. (b) Corresponding input characteristics for $In_{0.53}Ga_{0.47}As$ with articicially high eDOS. The orientation of the electric field towards the gate-dielectric results in a parasitic phonon-assisted tunneling current to the hh-subband, in addition to the desired ballistic current to the lh-subband. The overall result is a degradation of SS and I_{on} at fixed I_{off} and V_{dd}, reflected in a degradation of I_{60}.

(a) Favorable material combinations

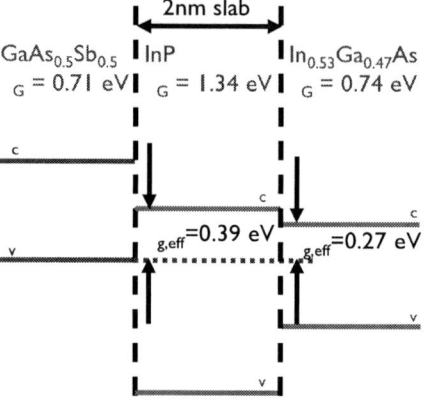

(b) Favorable design

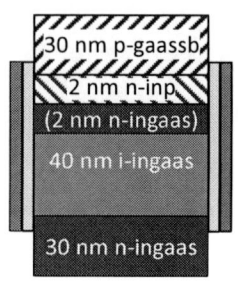

(c) Impact dielectric constant

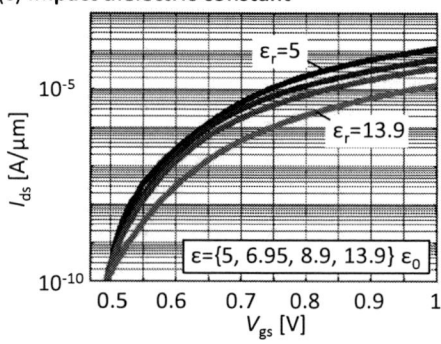

Fig. 8. Beneficial impact of small ε_r on TFET performance: (a) material combination with high transmission probability upon tunneling from E_v of GaAsSb to E_c of InGaAs, (b) TFET design exploiting the decreased ε_r of thin-slab InP, (c) shifted I-V characteristics (coinciding @ $I_{ds} = 10^{-10} A/\mu m$) of all-$In_{0.53}Ga_{0.47}As$ p-n-i-n TFET with default $\varepsilon = 13.9\varepsilon_0$ ($N_{source} = N_{pocket} = 5 \times 10^{19} \text{at/cm}^3$, $t_{pocket} = 2$nm, EOT=0.6nm) compared to the same TFET with artificially decreased ε. Superlinear increase in I_{on} is predicted with decrease in ε_r.

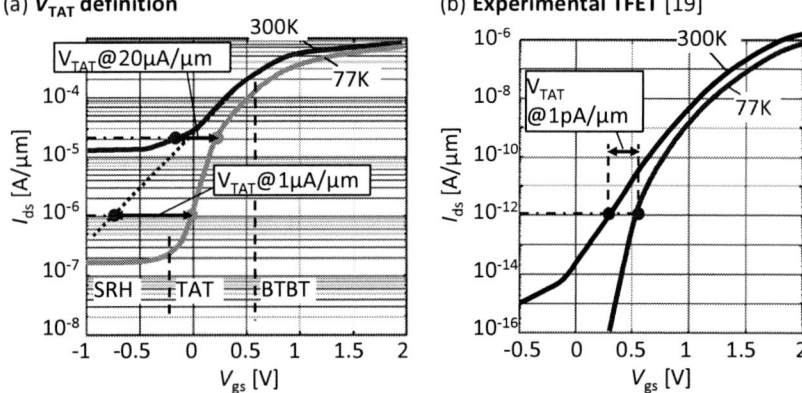

Fig. 9. $V_{TAT}@I_{off}$ defined as ΔV_{gs} between 300K and 77K @I_{off}: the increase in required supply voltage caused by TAT. (a) Illustration of definition with I-V curves modified from [18]: if both I-V curves reach I_{off}, the shift is readily determined: $V_{TAT}@20\mu A/\mu m$=400mV. If SRH limits I_{off} of the 300K I-V curve, the TAT-tail should be extrapolated to estimate ΔV_{gs} due to TAT: $V_{TAT}@1\mu A/\mu m$=750mV. (b) Illustration of V_{TAT} for a vertical Ge-source Si-TFET [19], $V_{TAT}@1pA/\mu m$=300mV.

978-1-4799-8002-4/14 $31.00 © 2014 IEEE   720

Performance evaluation of MoS_2-WTe_2 vertical tunneling transistor using real-space quantum simulator

Kai-Tak Lam, Gyungseon Seol and Jing Guo

Department of Electrical and Computer Engineering, University of Florida, Gainesville, FL 32611, USA
Email: lamkt@ufl.edu

Abstract

Layered two dimensional (2D) semiconductor materials enable vertical interlayer heterojunctions (HJ) without the requirement of lattice matching. Interlayer transport through a MoS_2-WTe_2 vertical HJ transistor is studied by atomistic quantum device simulations. Ultra-steep subthreshold slope (SS) is obtained due to the utilization of band filtering as the switching mechanism. The simulator enables the investigation of the effects of atomic defects and trapped charges on the performance of the atomically thin HT transistor. It is shown that the ultra-steep SS in TMD vertical tunneling FETs is robust against both atomic defects in the TMD layers and charged impurity scattering.

Introduction

Layered two dimensional (2D) semiconductor materials enable vertical interlayer heterojunctions (HJ) without the requirement of lattice matching. Motivated by recent reports of HJ in layered transition metal dichalcogenide (TMD) materials [1,2], we studied the interlayer transport through a MoS_2-WTe_2 PN HJ by atomistic quantum transport simulations based on non-equilibrium Green's function (NEGF) formalism. In the vertical tunneling transistor (vTFET), ultra-steep subthreshold slope (SS) is obtained due to the utilization of band filtering as the switching mechanism. While the roles of atomistic defects and charge impurity scattering are important in 2D semiconductors, they have not been studied in interlayer transport.

Our simulator enables the investigation of atomic defects in the TMD materials, the effect of trapped charges in the gate oxide on the performance of vTFETs, and the issue of lattice mismatch between the layered HJ. Our results show that the ultra-steep SS in TMD vTFETs is robust against both atomic defects in the TMD layers and charged impurity scattering, a dominant extrinsic scattering mechanism in layered 2D semiconductors.

Methodology

Despite of the importance of interlayer transport in 2D semiconductors, only phenomenological non-atomistic treatment has been reported to date [3,4]. Atomistic treatment of interlayer transport has so far been hindered by the unavailability of interlayer binding parameters, and the lattice mismatch, as well as the randomness of stacking order, between the layers. However, device physics and trends insensitive to uncertainties can be obtained by performing atomistic transport simulations with accurate description of electronic structures of each layer at the band edges and reasonable approximations of interlayer binding. Monolayer MoS_2 and WTe_2 are chosen as the bottom and top layers in the vTFETs shown in Fig. 1. The maximum phase difference of 30° is chosen between the layers and the overlapped region is defined as the active region (channel) of the device.

An atomistic nearest neighbor Hamiltonian is used [5], with the on-site potential term in the sublattice A and B being the half band gap, $\pm E_G/2$ and the nearest coupling term being t_1. The material parameters are fitted from *ab inito* band structures as in [6]: $t_1 = 1.10$ and 0.84 eV, $E_G = 1.59$ and 0.75 eV with spin-orbital coupling for MoS_2 and WTe_2 respectively. Interlayer interaction are obtained by identifying nearest WTe_2 sublattice within the loci of MoS_2 sublattice and assigning the coupling according to $t_\perp exp(-8\sqrt{3}r/a_0)$ where $t_\perp$ is the interlayer coupling in eV, r is the planar distance between the neighbors and a_0 is the lattice constant of MoS_2 at 3.18 Å. The NEGF formulism [7] is used for ballistic transport simulation and the self-energies of the doped contacts are calculated numerically [8]. The potential profile is obtained self-consistently with a 2D Poisson solver, taking care of the workfunction difference between the layers (5.07 and 4.05 eV for MoS_2 and WTe_2 respectively), and that of the gates at 4.30 eV. The TMD layers are 6 Å thick with a similarly thick air gap and the insulator material is HfO_2. The length of the layer extensions (lateral junctions) are 10 nm and have a doping concentration of 10^{16} m^{-2}. The length and width of the active region and the insulator thickness are 20, 20 and 3 nm respectively.

978-1-4799-8002-4/14 $31.00 © 2014 IEEE

Discussion

The device switching mechanism is shown in Fig. 2, where a Type-II HJ is formed in the device active region at OFF-state. The E_G of MoS$_2$ effectively filter off carriers from WTe$_2$ (and vice versa) resulting in a very low OFF-state current (I_{OFF}). At ON-state, the conduction band (E_C) of MoS$_2$ is moved below the valence band (E_V) of WTe$_2$, creating a Type-III HJ with a conducting channel at a specific energy range, which results in an ultra-steep SS lower than 60 mV/dec. at room temperature. We first investigate the effect of interlayer coupling to the device performance of vTFETs by varying the values of $t_\perp$ shown in Fig. 3. The current values are normalized with respect to the area of the active region. The ON-state current (I_{ON}) increases with $t_\perp$ while I_{OFF} remains unchanged, which validate the device switching mechanism. This also gives an insight to the effect of different lattice mismatch between the top and bottom layers, where the ultra-steep SS and low I_{OFF} of TMD vTFETs are not expected to be affected, while the I_{ON} will be modulated due to different interlayer interactions.

Next we change the length of the active region from 10 to 40 nm, shown in Fig. 4 and the I_{ON}, normalized with respect to the channel width, increases with a longer active region. This is different from existing FET devices, where the current either reduces, due to higher resistance, or at best remains the same for ballistic devices, as the channel length increases. This effect can only be captured with atomistic transport simulations. Since the current path is vertical between the layers, increasing the channel length would increase the area of the current path, thereby reducing channel resistance.

The n-type and p-type device operations for the same device parameters are demonstrated in Fig. 5 and 6 respectively at different biasing conditions to showcase the versatility of the device. The effect of gate oxide on the device performance is also investigated with different insulator thickness shown in Fig. 7, where a larger gate bias is required to fully switch the device with a thicker gate oxide. The doping concentration of the lateral junctions is examined in Fig. 8, where a very high doping concentration result in a large lateral tunneling current, which greatly degrades the device performance of TMD vTFETs in terms of much larger leakage current.

Finally, the effect of atomic defects on WTe$_2$ layer is simulated with a 0.25% defects shown in Fig. 9(a). The local density of states (LDOS) plots for the perfect and defected devices at OFF-state are shown in Fig. 9(c) and 9(d) where the states in WTe$_2$ layer are highly perturbed. However, the current spectrum shown in Fig. 9(e) is not affected due to the filtering effect of the MoS$_2$ band gap. We have also simulated the effect of a trapped charge in the top gate oxide similar to [9], shown in Fig. 10. Similarly, while states in the WTe$_2$ layers are more perturbed and the current spectrum differs from that of the perfect device, the overall effect of the trapped charge is small and the low I_{OFF} and ultra-steep SS of the TMD vTFETs are not greatly affected, indicating the robustness of the device.

Conclusion

An atomistic transport simulation of TMD vTFET is presented in this work. The ultra-steep SS and low I_{OFF} of the device highlighted the potential of the device for flexible low power electronic applications. Our results show that these features of TMD vTFET are robust against dominant atomistic scale disorders and extrinsic scattering mechanisms in layered semiconductors.

Acknowledgement

This work was supported by NSF DMR-1124894, NSF ECCS-0846563, and ONR.

Reference

[1] C.-H. Lee, G.-H. Lee, A. M. van der Zande, W. Chen, Y. Li, M. Han, et al., "Atomically thin p-n junctions with van der Waals heterointerfaces," Nature Nanotech. 9, pp. 676-681 (2014).

[2] M. Li, D. Esseni, G. Snider, D. Jena and H. G. Xing, "Single particle transport in two-dimensional heterojunction interlayer tunneling field effect transistor," J. Appl. Phys. 115, 074508 (2014).

[3] T. Georgiou, R. Jalil, B. D. Belle, L. Britnell, R. V. Gorbachev, S. V. Morozov, et al., "Vertical field-effect transistor based on graphene-WS$_2$ heterostructures for flexible and transparent electronics," Nature Nanotech. 8, pp. 100-103 (2013).

[4] K.-T. Lam, G. Seol and J. Guo, "Operating principles of vertical transistors based on monolayer two-dimensional semiconductor heterojunctions," Appl. Phys. Lett. 105, 013112 (2014).

[5] D. Xiao, G.-B. Liu, W. Feng, X. Xu and W. Yao, "Coupled spin and valley physics in monolayers of MoS$_2$ and other Group-VI dichalcogenides," Phys. Rev. Lett. 108, 196802 (2012).

[6] K.-T. Lam, X. Cao and J. Guo, "Device performance of heterojunction tunneling field-effect transistors based on transition metal dichalcogenide monolayer," IEEE Electron Device Lett. 34, pp. 1331-1333 (2013).

[7] S. Datta, Quantum Transport: Atom to Transistor, New York, NY: Cambridge University Press (2005).

[8] M. P. López Sancho, J. M. López Sancho and J. Rubio, "Quick iterative scheme for the calculation of transfer matrices: application to Mo(100)," J. Phys. F 14 1205 (1984).

[9] M. G. Pala, D. Esseni, F. Conzatti, "Impact of interface traps on the IV curves of InAs tunnel-FETs and MOSFETs: a full quantum study," IEDM12, 135 (2012).

978-1-4799-8002-4/14 $31.00 © 2014 IEEE

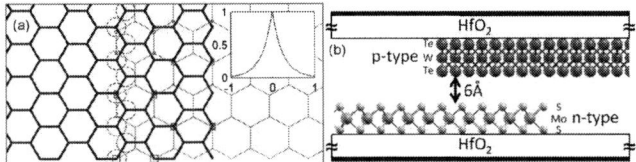

Fig. 1 (a) The top view of the atomic structure for the MoS_2 (thick line) and WTe_2 (thin line) vertical tunneling transistor (MoS_2-WTe_2 vTFET). The MoS_2 and WTe_2 layers are 30° out of phase and are approximately 6 Å apart vertically. The dash circles indicate the loci from the lattice sites of MoS_2 where interlayer couplings are determined. The WTe_2 atoms within these loci are assigned with the coupling strength according to the planar distance as shown in the inset, where unity in the x-axis corresponds to the loci radius and the y-axis is normalized to the maximum interlayer coupling $t_\perp$. The size of square markers is proportional to the interlayer coupling values. (b) The side view of the MoS_2-WTe_2 vTFET showing the double-gated structure. The MoS_2 and WTe_2 are n- and p-type respectively and the dielectric constant of HfO_2 is 20.

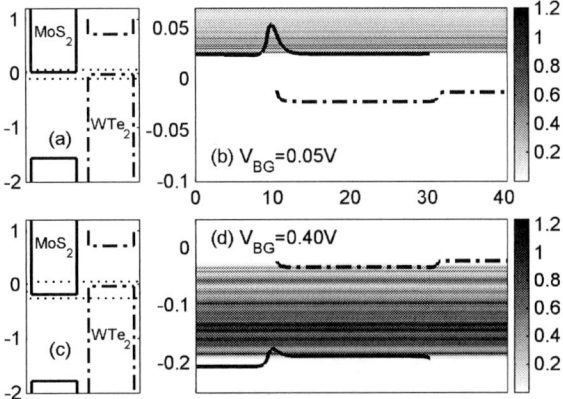

Fig. 2 (a) The band alignment of the overlapped region at OFF-state. The top and bottom solid blocks correspond to the conduction (E_C) and valence band (E_V) of MoS_2 and the dash-dot blocks are for WTe_2. The band gap of MoS_2 and WTe_2 are 1.59 and 0.75 eV respectively. (b) The spatial current spectrum at OFF-state corresponding to the dotted region in (a). The solid line is the E_C of MoS_2 and the dash-dot line is the E_V of WTe_2. The x-axis is the position along the transport direction in [nm] and the unit for the color bar is [nA/eV]. (c) and (d) show the band alignment and spatial current spectrum at ON-state. The unit for the color bar in (d) is [A/eV]. The vertical axes for all figures are energy in [eV].

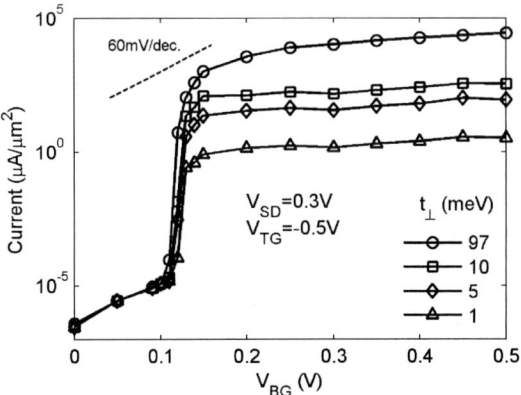

Fig. 3 The current characteristics with different interlayer coupling strengths $t_\perp$. The MoS_2 layer (V_S) is biased at 0.3 V while the WTe_2 layer (V_D) is grounded. The top gate is fixed at -0.5 V while the bottom gate is varied. $t_\perp = 97$ meV, 10% of the averaged intralayer coupling, in all other simulations.

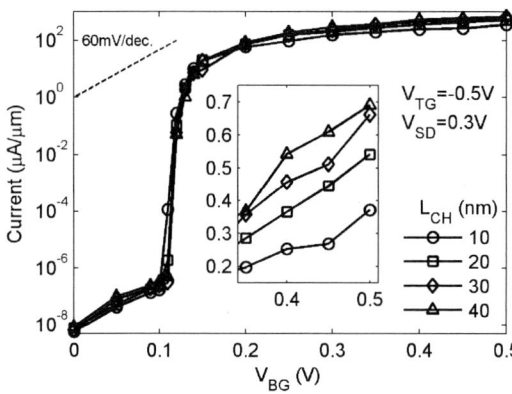

Fig. 4 The current characteristics with different channel lengths L_{CH}, with the linear scale shown in inset. The ON-state current (I_{ON}) increases with the overlapped area while the OFF-state current remains low. The current is normalized to the width of the active region here.

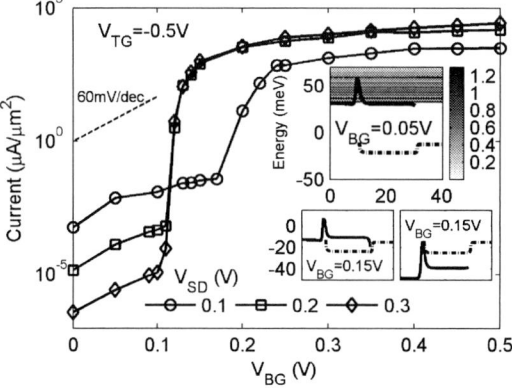

Fig. 5 The current characteristics at different bias conditions for n-type device operations. The top inset shows the spatial current spectrum at OFF-state when $V_{SD} = 0.10$ V and the unit of color bar is [μA/eV]. A larger tunneling current above the E_C of MoS_2 is observed as compared to Fig. 2(b). The bottom insets show the band profiles at $V_{BG} = 0.15$ V for $V_{SD} = 0.10$ (left) and 0.30 V (right), where a larger gate bias is needed for the Type-II to Type-III HJ transition. The vertical axis is energy in [meV] for the insets.

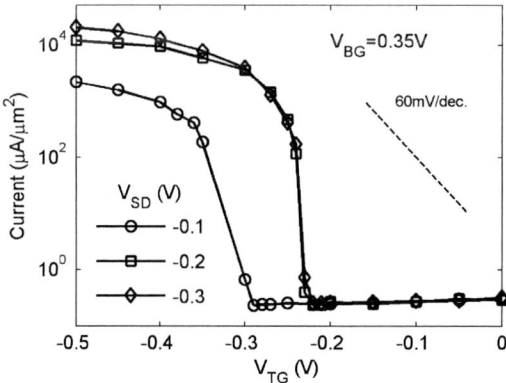

Fig. 6 The current characteristics at different bias conditions for p-type device operations. The V_{BG} is fixed at 0.35V while the MoS_2 layer is grounded and WTe_2 layer is biased from -0.10 to -0.30 V.

978-1-4799-8002-4/14 $31.00 © 2014 IEEE   723

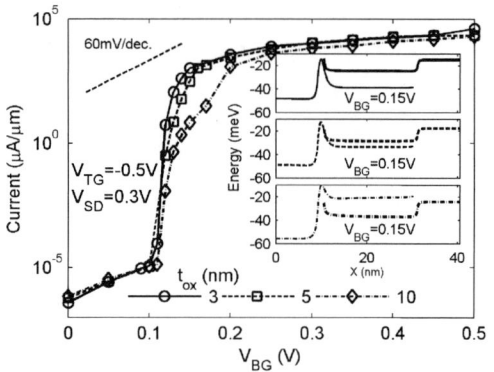

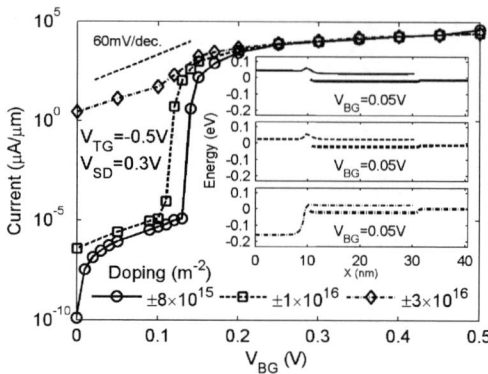

Fig. 7 The current characteristics for devices with different oxide thicknesses. The insets show the band profile of the devices at $V_{BG} = 0.15$ V. The weaker gate control for $t_{ox} = 10$ nm results in a larger bias to fully switch on the device.

Fig. 8 The current characteristics for devices with different doping concentrations at the lateral junctions. The insets show the band profile where the OFF-state current for the higher doped device is large due to lateral junction tunneling.

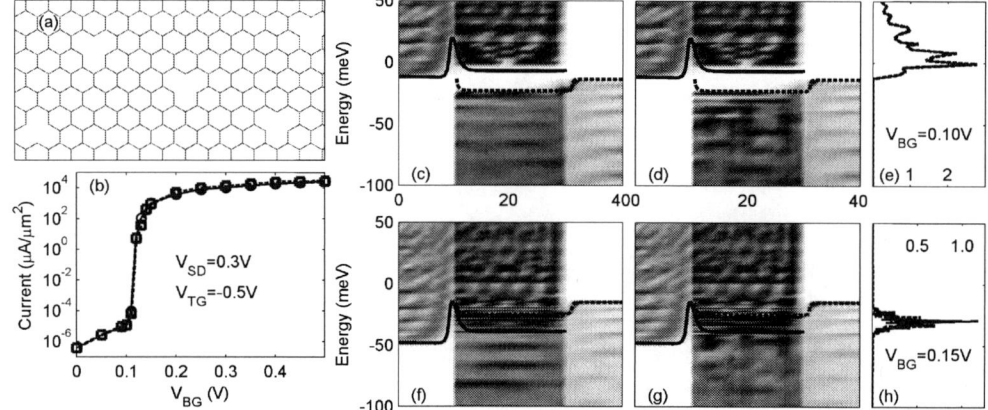

Fig. 9 (a) The atomic schematic of WTe$_2$ layer in the overlapped region with point defects. (b) The current characteristics for the perfect (circle marker, solid line) and defected device (square marker, dash line). The current characteristics of the MoS$_2$-WTe$_2$ vTFET device is not affected by the low point defect of 0.25% in the WTe$_2$ layer. (c) and (d) show the local density of states (LDOS) for the perfect and defected device at OFF-state respectively. The solid and dash lines are the E_C and E_V of MoS$_2$ and WTe$_2$ layers respectively. The LDOS in the valence band of WTe$_2$ in the overlapped region are highly perturbed by the defects. (e) The current spectrum for the perfect (solid line) and defected (dash line) device at OFF-state and the unit for the x-axis is [nA/eV]. (f)-(h) are the LDOS and current spectrum for the perfect and defected devices at ON-state. The unit for the x-axis in (f) is [A/eV]. The perturbation of the LDOS in WTe$_2$ does not have a large effect on the current of the device.

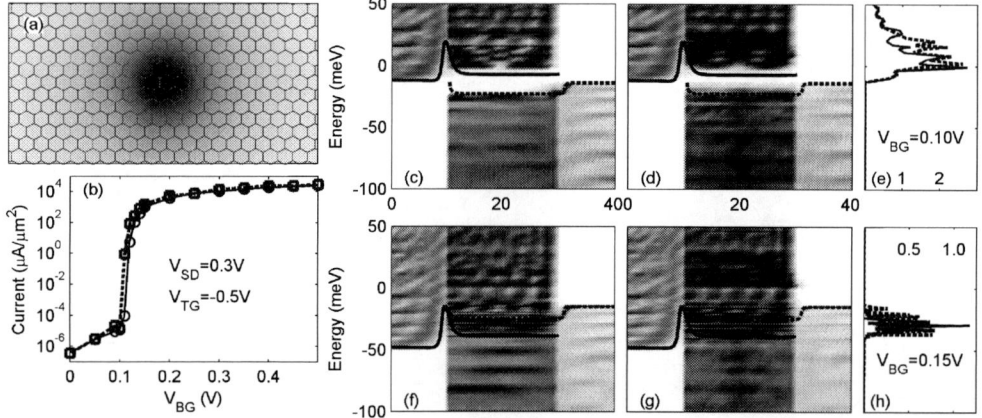

Fig. 10 (a) The atomic schematic of WTe$_2$ layer in the overlapped region overlay with the potential profile due to a single trapped charge in the top gate oxide. The trapped charge has a diameter of 1 nm and the potential decays $1/r$ from the charge. (b) The current characteristics for the devices without (circle marker, solid line) and with (square marker, dash line) single trapped charge. (c) and (d) show the LDOS at OFF-state. The solid and dash lines are the E_C and E_V of MoS$_2$ and WTe$_2$ layers respectively. The LDOS in the valence band of WTe$_2$ in the overlapped region are highly perturbed by the trapped charge potential. (e) The current spectrum for the device without (solid line) and with (dash line) trapped charge at OFF-state and the unit for the x-axis is [nA/eV]. (f)-(h) are the LDOS and current spectrum for devices at ON-state. The unit for the x-axis in (f) is [A/eV]. The effect of trapped charge on device performance is greater than that of the structural defects in our simulations. A maximum potential change of 1 V is assumed due to the trapped charge.

978-1-4799-8002-4/14 $31.00 © 2014 IEEE

Ab-initio simulations of MoS_2 transistors: from mobility calculation to device performance evaluation

Aron Szabo, Reto Rhyner, and Mathieu Luisier

Email: szaboa@iis.ee.ethz.ch, Phone: +41 44 632 5727, Fax: +41 44 632 1194, ETHZ, 8092 Zürich, Switzerland

Abstract—**In this paper we present the first *ab-initio* quantum transport simulations of single-layer MoS_2 field-effect transistors including electron-phonon scattering. It is shown that the relatively high ON-current and the negative differential resistance observed in previous studies are artifacts of the applied models and not physical effects. Despite a relatively high phonon-limited mobility (220 cm^2/Vs) single-layer MoS_2 cannot compete with strained-Si or III-V for potential application as high-performance logic switch. Its electron injection velocity ranging from 1.5e6 to 2.6e6 cm/s is too low for that purpose.**

Introduction

Since the first experimental demonstration of single-layer MoS_2 transistors [1], metal dichalcogenides (MDs) have started to attract a lot of attention as potential replacement for Si at the end of the semiconductor roadmap due to their excellent electrostatic properties resulting from their 2-D nature, their relatively large band gaps, and their promising mobility values. These features make them also well-suited for flexible and transparent electronic applications, while their weakly-bound layered structure is the key to assemble atomically-thin *p-n* junction with defect-free interfaces [2].

Since ultra-scaled MoS_2 transistors have not been fabricated yet, device simulation represents an appealing alternative to explore the characteristics of future MD-based switches and assess their potential in advance. Several theoretical studies have recently focused on this research topic: MoS_2 has been treated in the ballistic limit of transport [3] and in the presence of electron-phonon scattering [4], however in both cases within the effective mass approximation. Full-band calculations have been undertaken too, without elastic or inelastic scattering [5]. Finally, a full-band and dissipative approach has been proposed, but it only evaluates the mobility of MoS_2 based on the Boltzmann Transport Equation [6].

Here, we will present the first *ab-initio* quantum transport simulations of metal dichalcogenide field-effect transistors (FETs) taking electron-phonon scattering into account. Using the device structure sketched in Fig. 1(a) it will be shown that (i) the phonon-limited electron mobility of MoS_2 reaches values up to 220 cm^2/Vs and drops at high carrier concentrations, (ii) the ballistic limit is a ill-conditioned concept in MDs, (iii) the electron injection velocity does not exceed 2.6e6 cm/s, and (iv) strained Si or III-V outperform single-layer MoS_2 as next generation high performance transistor.

Approach

To supply the *ab-initio* components of the simulation approach, the density-functional theory (DFT) tool VASP [7] is

employed. It has been demonstrated earlier that for single-layer MoS_2 the generalized gradient approximation (GGA) of Perdew, Burke, and Ernzerhof (PBE) yields a band gap very close to the experimental one, i.e. E_g=1.8 eV [8]. This requires setting the lattice constant of MoS_2 to it experimental value during the structure relaxation. A $25\times25\times1$ Monkhorst-Pack k-point grid and a 500 eV plane-wave cutoff energy are used in the electronic structure calculation (no spin-orbit coupling). Convergence is reached when the force acting on each ion goes below 10^{-3} eV/Åor the total energy difference between two subsequent iterations gets smaller than 10^{-3} eV.

The plane wave (PW) basis of VASP does not lend itself naturally to transport calculations, but this problem can be circumvented through a unitary and exact transformation of the PW basis into maximally-localized Wannier functions (MLWFs) using the wannier90 code [9]. A tight-binding-like (TB) Hamiltonian matrix is then produced that can be loaded in our Non-equilibrium Green's Function quantum transport (QT) solver [10]. It contains five d-orbitals centered on each Mo atom and three p-orbitals on each S atom. Hopping parameters for distances up to the 6th nearest-neighbor are kept in our TB Hamiltonian in order to reproduce the DFT band structure in the whole energy range of interest with high precision. The accuracy of the method is validated in Fig. 1(b) for single-layer MoS_2. The seven highest valence bands and four lowest conduction bands of MoS_2 are captured by the "VASP+MLWF" technique, which produces a direct band gap at the K point with E_g=1.786 eV, very close to the experimentally determined value.

Using density-functional perturbation theory (DFPT) VASP can also compute the dynamical matrix of MoS_2, from which the phonon frequencies and oscillation amplitudes can be extracted [11]. The results are shown in Fig. 1(c). The electron-phonon coupling constants are calculated based on the small displacement method, where a perturbed and unperturbed Bloch Hamiltonian are transformed into the MLWF basis. Our QT tool can read these DFT data and use them to perform dissipative quantum transport simulations with electron-phonon scattering [12]. A Fröhlich Hamiltonian with the parameters of Ref. [6] is utilized to account for the interactions with polar optical phonons. Note finally that the out-of-plane direction z is modeled via 21 k_z points that are connected to each other via electron-phonon scattering.

Results

First, the low-field, phonon-limited electron mobility μ_{ph} of single-layer MoS_2 is extracted with the dR/dL method [13].

Assuming diffusive transport the channel resistances R_{ch} of MoS$_2$ samples with different lengths L_{tot}=40, 60, and 80 nm are calculated at a very low bias ΔV=1e-3 V. As expected and confirmed in Fig. 2(a) R_{ch} linearly increases with L_{tot}. The corresponding phonon-limited mobilities are reported in Fig. 2(b). Three important observations can be made from this plot. First, polar optical phonon scattering has a much lower influence than acoustic and optical deformations. Secondly, μ_{ph} can be as high as 220 cm^2/Vs, a value larger than in experiments (15-60 cm^2/Vs) [14] due to the neglection of charge impurity scattering [15]. Finally, the mobility rapidly decreases for charges $n_{2D} \geq$5e12 cm^{-2}. This behavior can be explained by the spectral distribution of the electrons, as shown in Fig. 2(c). As a consequence of the carrier concentration increase the satellite band minima along the $K - \Gamma$ direction become more and more populated. Since these valleys have a heavier transport effective mass than the K one (m^*_{trans}=0.87m_0 vs. 0.55m_0), μ_{ph} gets reduced.

The temperature dependence of $\mu_{ph}(T)$ has also been investigated. It follows a $T^{-\gamma}$ law with γ=2.2, which is smaller than in bulk (γ=2.6), but larger than in experiments [14] and other theoretical studies [6], [15] (1.0$\leq \gamma \leq$1.69). Since the latters only consider the K valley and not its satellites, our full-band results suggest that DFT might underestimate the energy separation between them (50 meV). In effect, by increasing the temperature from 200 to 300 K, the carrier concentration in the satellite valleys grows very fast, thus drastically reducing the mobility. With a larger splitting between K and its satellites this effect would be less significant and γ smaller.

As a next step the ballistic transfer characteristics of the MoS$_2$ FET in Fig. 1(a) are analyzed in Fig. 3(a). At large V_{gs}, the current exhibits a negative differential resistance (NDR) due to the transmission probability that decreases as V_{ds} goes from 0.05 to 0.68 V (see Fig. 3(b)). This phenomenon has been reported before [5], but we establish in Fig. 4 that it is an artifact of ballistic simulations. In this regime the same band must be available in the source, channel, and drain region to contribute to the transmission probability [16]. Bands with a small energy width Δ_B, as in MoS$_2$ and other MDs, only propagate if $qV_{ds} \leq \Delta_B$, explaining the observed NDR.

When electron-phonon scattering is turned on the situation goes back to normal, as shown in Fig. 5(a). To understand this evolution the spectral current of the device should be inspected. It is displayed in Fig. 5(b). A high current density region can be found on the drain side of the transistor. It is situated below the conduction band edge of the source and is caused by the absorption/emission of phonons that connect independent sub-bands (or branches) together. This process allows states that would normally be reflected back to the source in the ballistic limit to continue propagating towards the drain. Hence, at high V_{ds} the current increases in the presence of electron-phonon interactions, while it decreases at low V_{ds} due to backscattering (Fig. 5(c)).

Still, if our MoS$_2$ FET is compared to strained-Si or In$_{0.53}$Ga$_{0.47}$As ultra-thin-body (UTB) transistors with similar dimensions [17], a huge performance gap in disfavor of MoS$_2$ is noticed. Although the subthreshold swing does not vary much from one device to the other, MoS$_2$ carries a significantly lower ON-current than s-Si and InGaAs. The electron injection velocity (v_{inj}) is responsible for this rather negative outcome. In MoS$_2$ the bandstructure-limited v_{inj} is equal to 6.8e6 cm/s, but in practical applications, it does not exceed 1.5e6 cm/s at V_{ds}=0.05 V and 2.6e6 cm/s at V_{ds}=0.68 V, in both cases at a carrier density n_{2D}=5e12 cm^{-2} and with electron-phonon scattering.

These results are summarized in the transfer characteristics and table of Fig. 6. Note that the simulations of the strained-Si and In$_{0.53}$Ga$_{0.47}$As UTB FETs are performed in the ballistic limit of transport, not with scattering. Even if these devices operate at only 50% of their ideal performance, which is probably a conservative estimate, their ON-current would be at least two to three times larger than in MoS$_2$. It can therefore be concluded that single-layer MoS$_2$ FETs cannot challenge s-Si and InGaAs as ultimate logic switch.

Conclusion

We have carried out *ab-initio* quantum transport simulations of single-layer MoS$_2$ transistors with electron-phonon scattering, extracting both low-field mobility values and current characteristics. It has been shown that the phonon-limited mobility of MoS$_2$ reaches values up to 220 cm^2/Vs and drops at high carrier concentrations. We have also proved that the negative differential resistance observed in ballistic simulations does not rely on physical principles, but is the consequence of the narrow bands present in metal-dichalcogenides. NDR might eventually manifest itself at very low temperatures. Single-layer MoS$_2$ is certainly not the best-suited candidate as next generation logic device, but it can certainly find other applications in plastic or flexible electronics, for example.

Acknowledgement

This work was supported by SNF Grant PP00P2_133591 and by a grant from the Swiss National Supercomputing Centre (CSCS) under Project ID s503. We also acknowledge PRACE for awarding us access to the Hermit supercomputer at HLRS and financial support from the European Union Seventh Framework Programme (DEEPEN FP7-604416).

References

[1] B. Radisavljevic et al., Nat. Nano. **6**, 147 (2011).
[2] C.-H. Lee et al., Nature Nanotechnology **9**, 676 (2014).
[3] Y. Yoon et al., Nano Lett. **11**, 3768 (2011).
[4] L. Liu et al., IEEE Trans. Elec. Dev. **60**, 4133 (2013).
[5] J. Chang et al., App. Phys. Lett. **103**, 223509 (2013).
[6] K. Kaasbjerg et al., Phys. Rev. B **85**, 115317 (2012).
[7] G. Kresse et al., Phys. Rev. B **54**, 11169 (1996).
[8] K. Liu et al., IEEE Trans. Elec. Dev. **58**, 3042 (2011).
[9] N. Marzari and D. Vanderbilt, Phys. Rev. B **56**, 12847 (1997).
[10] M. Luisier et al., Phys. Rev. B **74**, 205323 (2006).
[11] A. Togo et al., Phys. Rev. B **78**, 134106 (2008).
[12] M. Luisier et al., Phys. Rev. B **80**, 155430 (2009).
[13] K. Rim et al., IEDM Tech. Dig. **2002**, 43-46 (2002).
[14] B. Radisavljeviv and A. Kis, Nature Materials **12**, 815 (2013).
[15] Z.-Y. Ong and M. V. Fischetti, Phys. Rev. B **88**, 165316 (2013).
[16] E. Gnani et al., IEEE Trans. Nano. **7**, 700 (2008).
[17] M. Luisier, IEEE Elec. Dev. Lett. **32**, 1686 (2011).

978-1-4799-8002-4/14 $31.00 © 2014 IEEE

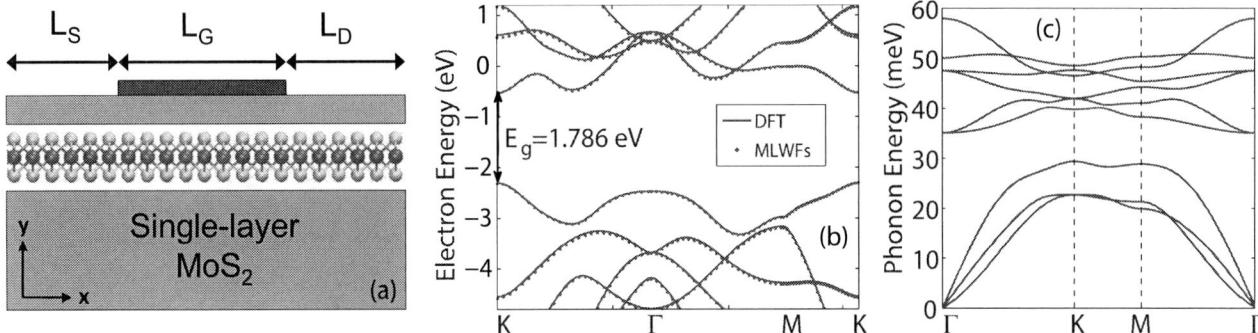

Fig. 1. (a) Schematic view of the n-type single-gate MoS$_2$ field-effect transistor (FET) considered in this work. The source and drain extensions measure L_s=L_d=15 nm each and are doped with a donor concentration N_D=6e13 cm^{-2}. The gate is L_g=10.7 nm long and is separated from the channel by a t_{ox}=3 nm thick HfO$_2$ layer with ϵ_R=20 (EOT=0.58 nm). The single-layer MoS$_2$ is deposited on a SiO$_2$ substrate. All the simulations are done at room temperature. (b) MoS$_2$ electron bandstructure calculated with VASP [7] and with the maximally-localized Wannier functions used in quantum transport. (c) MoS$_2$ phonon bandstructure calculated with VASP. The DFT phonon energies and modes are imported into our quantum transport solver to model electron-phonon scattering.

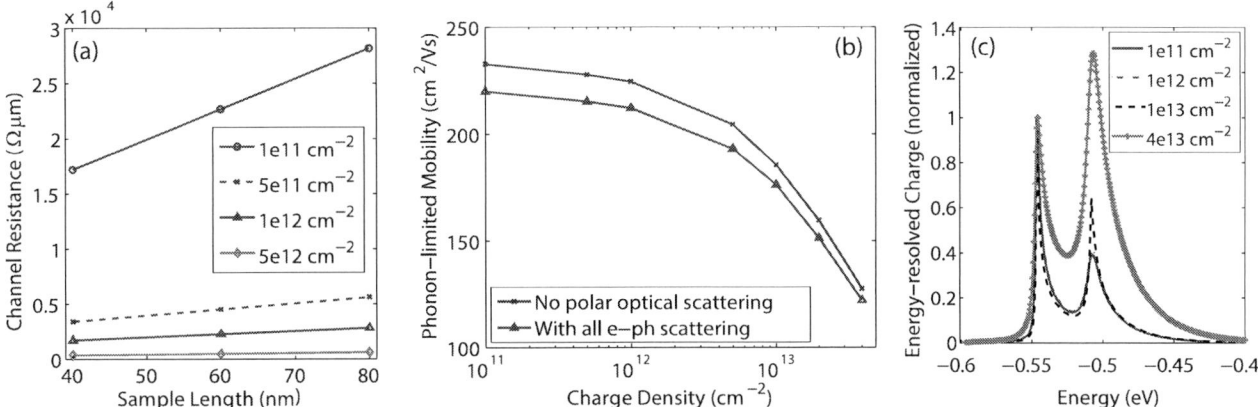

Fig. 2. (a) Channel resistance of single-layer MoS$_2$ as a function of its length at different carrier concentrations. The resistances are extracted from quantum transport calculations performed with a bias difference ΔV=1e-3 V between the two extremities of the MoS$_2$ sample. These simulations include electron-phonon scattering caused by acoustic and optical deformations as well as by polar optical interactions through Fröhlich theory. An almost perfect linear increase of the resistance is observed. (b) Low-field, phonon-limited mobility of single-layer MoS$_2$ as a function of the carrier concentration with (triangles) and without (crosses) polar optical phonon scattering. The "dR/dL" method [13] is employed to calculate the mobilities. (c) Energy-resolved electron density at different concentrations. The height of the first peak at E=-0.55 eV (K valley) is normalized to 1. As the electron population increases, the states around that satellite valley situated at E=-0.5 eV along the K-Γ direction become more and more filled.

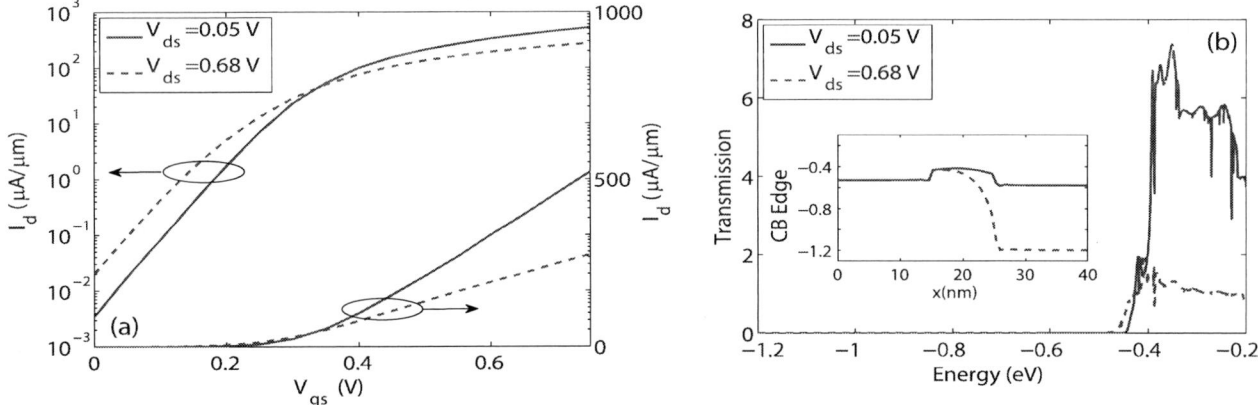

Fig. 3. (a) Ballistic transfer characteristics I_d-V_{gs} at V_{ds}=0.05 and 0.68 V for the single-gate MoS$_2$ transistor in Fig. 1(a). (b) Energy-resolved transmission probability at V_{gs}=0.35 V for the low and high V_{ds} cases. Contrary to the expectations, the two transmission probabilities are not similar: the one at V_{ds}=0.05 V (solid blue line) is much larger than the one at V_{ds}=0.68 V (dashed red line). The inset shows the corresponding conduction band edges.

978-1-4799-8002-4/14 $31.00 © 2014 IEEE

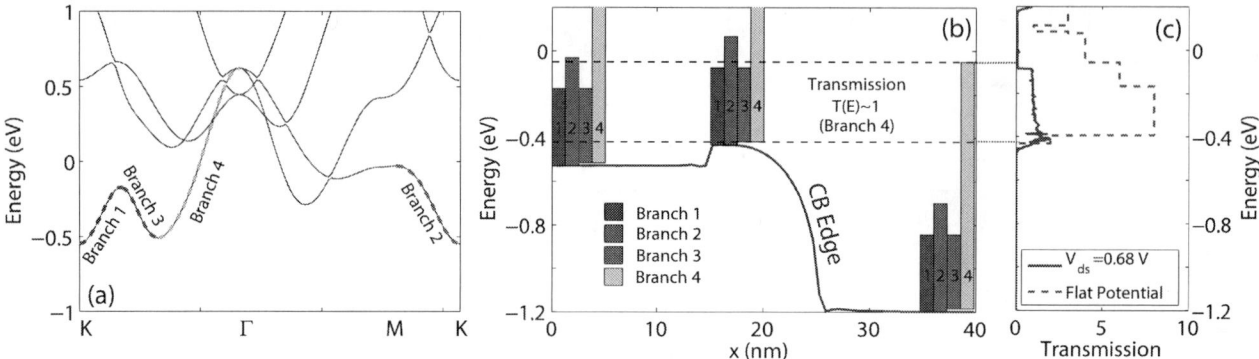

Fig. 4. (a) Conduction bandstructure of single-layer MoS$_2$ highlighting the four lowest branches that contribute to electron transport. Branches 1 and 2 start from the K point, branches 3 and 4 from the minimum situated between K and Γ. (b) Available electron propagation channels from source to drain in the single-layer MoS$_2$ transistor of Fig. 1(a) at V_{gs}=0.35 V and V_{ds}=0.68 V. Each branch in subplot (a) forms a different channel that can be multiply degenerate. A given channel must be available in the source, channel, and drain region so that a reflectionless transmission is possible [16]. Due to the narrow energy width of branches 1, 2, and 3, only a state from branch 4 can propagate from source to drain at high V_{ds}. At low V_{ds}, states in branches 1, 2, and 3 also propagate, explaining the larger transmission at V_{ds}=0.05 V in Fig. 3. (c) Transmission probability at V_{gs}=0.35 V and V_{ds}=0.68 V (solid line) and for a flat potential (dashed line). The solid line is different from 0 only in the energy window delimited by the dashed lines where branch 4 is available.

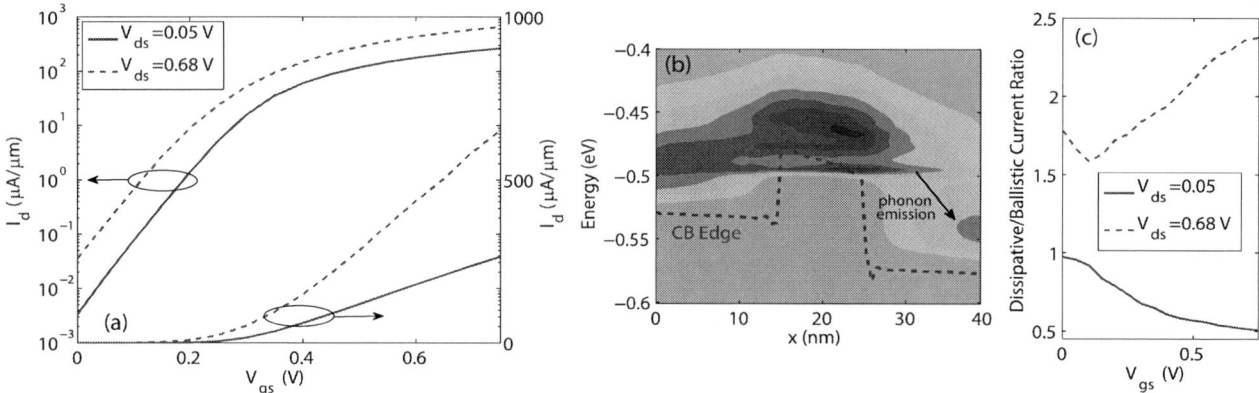

Fig. 5. (a) Transfer characteristics I_d-V_{gs} at V_{ds}=0.05 and 0.68 V with electron-phonon scattering for the same MoS$_2$ transistor as before. Current conservation has been verified in all the cases. (b) Spectral current $I_d(E, x)$ at V_{ds}=0.05 V and V_{gs}=0.7 V. Red indicates high current concentrations, green no current, the dashed line is the conduction band edge. (c) Ratio between the dissipative (electron-phonon) and ballistic currents.

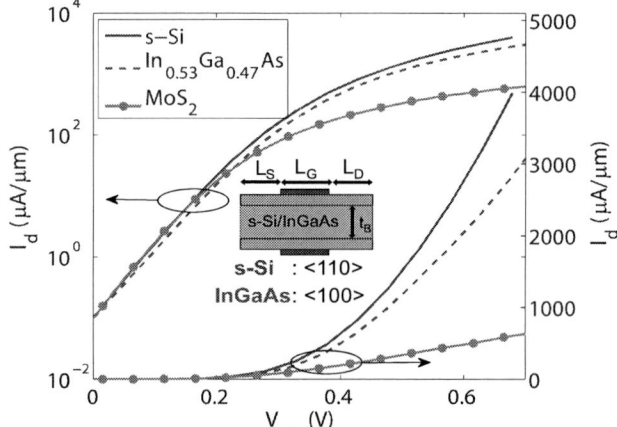

	s-Si	InGaAs	MoS$_2$
V_{DD} (V)	0.68	0.68	0.68
L_g (nm)	10.7	10.7	10.7
$L_s=L_d$ (nm)	15	15	15
t_{body} (nm)	5	5	0.32
Gate Type	Double	Double	Single
Transport	Ballistic	Ballistic	Dissipative
EOT (nm)	0.58	0.58	0.58
m^*_{trans} (m_0)	0.18	0.07	0.55
I_{OFF} ($\mu A/\mu m$)	0.1	0.1	0.1
I_{ON} ($\mu A/\mu m$)	3990	3070	650
$v_{inj,ON}$ (cm/s)			2.6e6
SS (mV/dec)	82	90	78
DIBL (mV/V)	116	109	119

Fig. 6. (Left) Transfer characteristics at V_{ds}=0.68 V for the single-layer MoS$_2$ transistor in Fig. 1(a) and for double-gate ultra-thin-body transistors made of strained Si (transport along <110>) and In$_{0.53}$Ga$_{0.47}$As (<100>) [17] designed according to the ITRS specifications for 2020. (Right) Table summarizing the dimensions and properties of each transistor on the left. Note that m^*_{trans} stands for the transport effective mass of the lowest conduction band.

Performance Evaluation and Design Considerations of 2D Semiconductor Based FETs for Sub-10 nm VLSI

Wei Cao, Jiahao Kang, Deblina Sarkar, Wei Liu and Kaustav Banerjee[+]

Department of Electrical and Computer Engineering, University of California, Santa Barbara, CA, 93106-9560, USA

[+]Contact e-mail: kaustav@ece.ucsb.edu

Abstract

Two-dimensional (2D) crystal semiconductors, such as the well-known molybdenum disulfide (MoS_2), are witnessing an explosion in research activities due to their apparent potential for various electronic and optoelectronic applications. In this paper, dissipative quantum transport simulations using non-equilibrium Green's function (NEGF) formalism are performed to rigorously evaluate the scalability and performance of monolayer/multilayer 2D semiconductor based FETs for sub-10 nm node VLSI technologies. Device design considerations in terms of the choice of prospective 2D materials/structure/technology to fulfill the sub-10 nm ITRS requirements are analyzed for the first time. Firstly, it is found that MoS_2 FETs can meet high-performance (HP) requirement up to 6.6 nm node by employing bilayer MoS_2 as the channel material, while low-standby-power (LSTP) requirements present significant challenges for all sub-10 nm nodes. Secondly, by studying the effects of underlap (UL) FET structures, scattering strength and carrier effective mass, it is found that the high mobility and suitably low effective mass of tungsten diselenide (WSe_2), aided by UL, enable 2D FETs for both HP and LSTP applications at the smallest foreseeable (5.9 nm) node. Finally, possible solutions for sub-5 nm nodes are also proposed based on the effects of critical parameters on device performance.

I. Introduction

Compared to graphene, 2D semiconductors such as MoS_2 and other transition metal dichalcogenides (TMDs), are considered to be more suitable for VLSI applications, due to their intrinsically uniform and non-zero band gaps that significantly lower the device leakage and static power consumption [1]-[5]. The atomic scale thicknesses of these 2D semiconductors offer high scalability to field-effect transistors (FETs) in which they are used as channel materials. *Ballistic transport* simulations were performed in [2],[3] to evaluate the scalability of monolayer/multilayer TMD FETs based on a semiconductor-on-insulator (SOI) FET topology. However, it has been recently reported that MoS_2 suffers from *remote phonon scattering* in high-k dielectric environment and has an intrinsic phonon limited mobility of only ~ 60 cm²/V·s [4]. Moreover, Liu et al. [5] reported, by considering phonon scattering, that the performance of monolayer MoS_2 FET is far from "ballistic" even at sub-10 nm nodes. Additionally, till date, the reported scaling analyses [2],[3],[5] seem generally negative in terms of meeting the ITRS requirements, and no efforts have been made to provide possible solutions for overcoming the apparent challenges. Therefore, a comprehensive performance/scalability evaluation of monolayer/multilayer TMD- and other 2D- FETs based on *dissipative transport theory*, along with practical solutions to meet the ITRS requirements is highly desirable. We present such an evaluation platform and propose solutions for sub-10 nm 2D FETs.

II. Methodology

Fig. 1(a) shows the schematic of an n-type double-gated (DG) 2D FET used in the simulations. HfO_2 is used as both top and bottom gate dielectric. Poisson's equation and transport equations are iteratively solved to obtain self-consistent values of parameters,

as shown in **Fig. 1(b)**. The coupling term in the effective mass Hamiltonian is set to be zero for multilayer devices, due to the weak interlayer interaction [3]. In the transport equations, scattering events are modeled by "Büttiker Probes" (**Fig. 1(c)**) within the NEGF formalism [6]. For each probe (i.e., a scatter), scattering strength or equivalently probe self-energy $\sum_P$ is mapped from experimentally measured or theoretically predicted low-field mobility, which allows us to take into account all types of non-negligible scattering events, such as acoustic/optical phonon scattering from lattice vibrations, remote surface phonon scattering from gate dielectric, and Coulomb scattering from charged defect and/or ionized impurities, etc. [4],[5], in a computationally affordable manner. Fundamentally, scattering events only perturb/change the momentum and energy of mobile carriers, they do not absorb or generate carriers, i.e., the total number of carriers is required to be conserved. This requirement can be achieved by adjusting the probe Fermi level μ_P (assigned to each scatter) to ensure that the net current (including in-scattering and out-scattering currents [6]) through each scatter is zero, i.e., current continuity condition. Detailed information about this approach can be found in [6]. **Figs. 2(a)** and **(b)** show the electron density color map obtained with ballistic transport simulation as well as dissipative transport simulation used in this work, respectively. The difference in essential physics can be clearly observed from the fact that the electrons keep relaxing energy during dissipative transport from source to drain, compared to the conserved energy during ballistic transport. On the other hand, drain current and output resistance (R_{out}) predicted by ballistic and dissipative transport simulations are also different even for gate length as small as 8 nm, as shown in **Figs. 2(c)** and **(d)**, indicating the necessity of dissipative transport simulation in accurately evaluating the performance of MoS_2 FET and other 2D FETs. MoS_2, as the most studied 2D semiconductor, is taken as a representative 2D channel material in this study. Whenever MoS_2 is unable to meet the ITRS requirements, study is extended to other 2D semiconductors by changing critical parameters that affect device performance. Relevant parameters dependent on the number of layers of MoS_2 [1],[3],[4],[7] are listed in **Table I**. Theoretical mobility limits in this table are used for MoS_2, while an optimistic mobility of 200 cm²/V·s is used for Si device in the simulations. A high source/drain doping of 6.5×10^{13} cm⁻² is used to ensure ohmic source/drain contacts [8].

III. Results and Discussion

Device Electrostatics Evaluation: In ultra-short channel FETs, fringing electric field at the two edges of the gate is usually detrimental to device electrostatics. Therefore, the effect of spacer (beside the gate dielectric, as schematically shown in **Fig. 1(a)**) with different dielectric constants is studied at first. **Fig. 3(a)** shows improved subthreshold characteristics with low-k spacer, which is attributed to the fact that fringing electric fields are suppressed by low-k spacer, as shown in **Figs. 3(b)** and **(c)**. The red dashed lines in the insets of **Figs. 3(b)**-**(c)** represent the locations where x and y components of the fringing electric field are extracted, respectively. In **Figs. 4(a),(b)**, subthreshold swing (SS) and drain-induced-barrier-lowing (DIBL) are evaluated for both DG and

978-1-4799-8002-4/14 $31.00 © 2014 IEEE   729

SOI structures from monolayer (1L) to 3 layer- (3L) MoS$_2$ as well as Si based ultra-thin-body (UTB) DG FETs. Values for parameters such as effective-oxide-thickness (EOT), supply voltage V_{dd}, body thickness (only for Si device) at each node (or gate length L_g) are adapted from ITRS [9] and used throughout this work. It is shown that MoS$_2$ FETs have much better electrostatics than Si devices. On the other hand, SOI topology can only sustain good electrostatics for 1L MoS$_2$, while DG topology can do so for up to 3L.

ON-Current and Output Resistance Evaluation: Fig. 5(a) shows the obtained ON-current (I_{on}) for 1L-3L MoS$_2$ FETs for both HP and LSTP applications. It is found that, for HP, 2L case provides the highest I_{on}, and can meet the requirement up to 6.6 nm node. In contrast, for LSTP, 1L is the closest to, although still lower than, the requirement. This is due to the fact that HP relies more on super-threshold performance (current drive capability), while LSTP relies more on the subthreshold performance (electrostatics), as schematically illustrated in **Fig. 5(b)**. Compared to 1L, 2L and 3L have more available states for conduction and higher "mobility" [10] (an extended concept to conveniently quantify the "scattering rate" in sub-10 nm scale, and not identical to the conventional definition for long-channel devices), and thus higher current drive capability, so they have higher I_{on} for HP. 3L shows lower I_{on} than 2L, because its worse electrostatics begins to degrade I_{on} for HP. Therefore, > 3L must be avoided for sub-10 nm nodes. 1L has the best electrostatics, and thus the highest I_{on} for LSTP. The degree of current saturation that can be quantified by R_{out}, is critical for circuit metrics such as noise margin and voltage gain. **Fig. 6** shows the R_{out} at different nodes. It can be observed that device with smaller channel thickness (better electrostatics) displays higher R_{out} or better saturation.

Design Considerations for ITRS: Given the inability of MoS$_2$ FETs in fulfilling the HP requirement at the smallest 5.9 nm node, and LSTP requirement for all sub-10 nm nodes, natural tendencies are to improve the device electrostatics and increase the "mobility". The underlap (UL) structure could be a choice for the former goal. As shown in the upper three subplots of **Fig. 7**, 2L and 3L cases can derive more benefits in lowering SS, compared to the 1L case, because of their larger room for improving electrostatics. Although UL improves device electrostatics by reducing source/drain-to-channel coupling, i.e., lowers SS, it degrades drive current by introducing extra series resistance R_{series}, as shown in the lower three subplots of **Fig. 7**. Therefore, the length of UL should be optimized to maximize I_{on}. **Fig. 8** shows the effect of varying UL on I_{on} for 1L-3L MoS$_2$ FETs. It is found that UL can help 2L and 3L MoS$_2$ FETs achieve higher I_{on} at small nodes for both HP and LSTP, due to greatly reduced SS (see **Fig. 5(b)**). However, it does not help or even degrades 1L MoS$_2$ FET, because the effect of increasing R_{series} prevails over that of reduced SS for 1L. Although the obtained I_{on} values are still lower than the requirement, UL design is demonstrated to be helpful. Based on the results in **Figs. 7** and **8**, it can be observed that 1L MoS$_2$ with 1 nm UL should be used for LSTP, and 2L MoS$_2$ with 0-1 nm UL should be used for HP.

To estimate how much the increased "mobility" can help, ballistic transport simulation (UL is not used) is performed to obtain I_{on} as shown in **Fig. 9**. It can be observed that the HP requirement of all the nodes in the road map can be met, while LSTP requirement beyond 7 nm node cannot, indicating that HP requires high "mobility", while only high "mobility" is not sufficient for LSTP below 7.4 nm nodes. Subsequently, the combined effects of UL and increased "mobility" are studied for the smallest 5.9 nm node in **Figs. 10(a)** and **(b)**, respectively. Note that when varying the "mobility", only scattering rate is changed, effective mass remains the same. For HP, 2L aided by 1 nm UL requires a "mobility" of 138. In comparison, 1L without UL requires a "mobility" of 174. The former is expected to be much easier to achieve than the latter since

carriers in 2L would experience less scattering compared to that in 1L. Such mobility boost can be achieved by engineering the dielectric environment, such as inserting h-BN buffer layer between high-k dielectric and TMD channel [11]. The other solution is to employ high-mobility 2D semiconductors. In fact, 1L WSe$_2$ has been found to exhibit mobility as large as 200 cm^2/V·s [12], and thus, is more promising than MoS$_2$ for HP applications. In contrast, increased "mobility" aided by UL, although helpful, is still not sufficient for LSTP. In order to look for solutions for LSTP, the effect of effective mass aided by 1 nm UL (0 nm UL case is also shown for comparison) and increased "mobility" is studied as shown in **Figs. 11(a)** and **(b)**. It can be observed that smaller effective mass offers higher I_{on} due to higher carrier velocity, but aggressively small effective mass also leads to source-to-drain (S-to-D) tunneling leakage (**Fig. 11(c)**) and thus to severely degraded SS and I_{on}. With the help of UL, devices are more immune to S-to-D tunneling leakage, and thus can derive more benefit from reduced effective mass. For 2D materials with 1 nm UL and an effective mass of 0.3 m_0, the LSTP requirement can be met if 1L "mobility" can reach 190. **Fig. 12** collects currently available experimental/calculation data in the literature [1],[7],[12]-[15] for various 2D materials, where WSe$_2$ is shown to have the desired values. Note that graphene, germanene, and silicene are not included here because of their zero or small band gaps, which make them unsuitable for logic applications.

Possible Solutions for Sub-5 nm Nodes: It has been shown above how hard it is to fulfill the ITRS requirements at the 5.9 nm node. Hence, requirements for sub-5 nm nodes are significantly harder, and the most stringent scaling constraint should be identified to allow appropriate compromise. **Fig. 13** shows the obtained I_{on} from 1L MoS$_2$ FETs (UL=0) with and without V_{dd} scaling. It can be found that if V_{dd} is not scaled, ITRS requirement can be easily met at all nodes in the current roadmap, indicating that for ultra-short channels, such as sub-5 nm nodes, the constraint on V_{dd} can be suitably relaxed to achieve a balance between power consumption and speed. The other direction is to resort to steeper turn-on devices such as tunnel FETs that provide sub-kT/q operation (SS < 60 mV/decade) and thus able to conquer the V_{dd} "barrier" [16]. Alternatively, 3D IC integration with 2D materials [17] could provide an alternative pathway to scaling beyond the 5 nm node.

IV. Summary

Performance and scalability of 2D FETs are comprehensively evaluated for sub-10 nm nodes, through rigorous dissipative quantum transport simulations. As summarized in **Table II**, solutions based on the proper choice of materials/structure/technology for 2D FETs to fulfill the ITRS requirements up to the year 2026 are identified for the first time.

Acknowledgment: This work is being supported by the AFOSR under Grant A9550-14-1-0268 (R18641).

References

[1] B. Radisavljevic, et al., *Nature Nano.* vol. 6, no. 3, pp. 147, 2011. **[2]** Y. Yoon, et al., *Nano Lett.* vol. 11, no. 9, pp. 3768, 2011. **[3]** V. Mishra, et al., *IEEE IEDM Tech. Dig.*, pp. 136, 2013. **[4]** N. Ma, et al., *Phys. Rev. X*, vol. 4, no. 1, pp. 011043, 2014. **[5]** L. Liu, et al., *IEEE Trans. Elect. Dev.*, vol. 60, no. 12, pp. 4133, 2013. **[6]** R. Venugopal, et al., *J. Appl. Phys.*, vol. 93, no. 9, pp. 5613, 2003. **[7]** W. Liu, et al., *IEEE IEDM Tech. Dig.*, pp. 499, 2013. **[8]** J. Kang, et al., *Phys. Rev. X*, vol. 4, no. 3, pp. 031005, 2014. **[9]** Table PIDS 2 and 4, *ITRS* 2012 *update*, online. **[10]** J. Kang, et al., *Appl. Phys. Lett.*, vol. 104, no. 9, pp. 093106, 2014. **[11]** J. Kang, et al., *45th IEEE SISC*, San Diego, CA, Dec 10-13, 2014. **[12]** W. Liu, et al., *Nano Lett.*, vol. 13, no. 5, pp. 1983, 2013. **[13]** J. Qiao, et al., *Nature Comm.*, vol. 5, no. 8, pp. 4475, 2014. **[14]** J. Song, et al., *ACS Nano*, vol. 7, no. 12, pp. 11333, 2013. **[15]** S. Larentis, et al., *Appl. Phys. Lett.*, vol. 101, no. 22, pp. 223104, 2012. **[16]** W. Cao, et al., *AIP Adv.*, vol. 4, pp. 067141, 2014. **[17]** J. Kang, et al., *Proc. SPIE*, vol. 9038, pp. 908305, 2014.

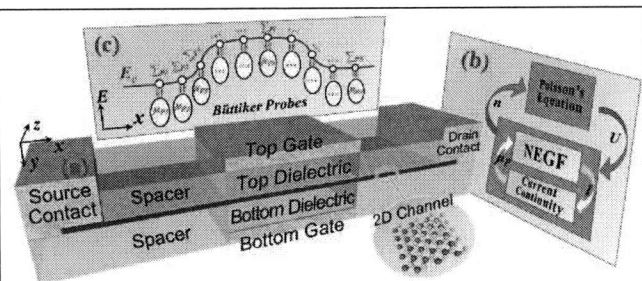

Fig. 1: (a) A schematic of 2D FET with double-gated topology. **(b)** The self-consistent NEGF-Poisson simulation scheme. U, n, and I are self-consistent potential, electron density, and current, respectively. **(c)** Illustration of the concept of "Büttiker probes" to model scattering events along the 2D channel material between Source and Drain contacts. The probe Fermi levels μ_{P1}, μ_{P2}...... μ_{PN}, are adjusted to ensure zero net current at each scatter. $\sum_P$ represent the corresponding probe self-energies.

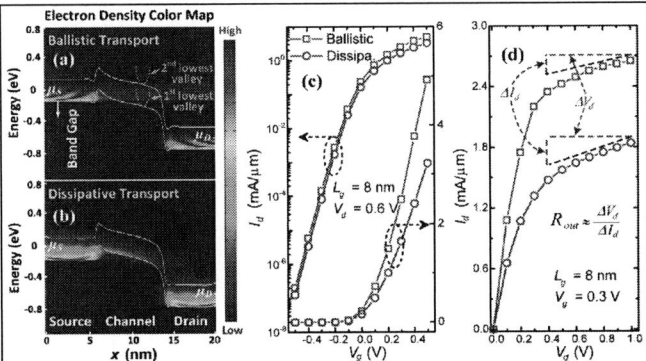

Fig. 2: Energy resolved electron density (brighter color indicates higher density) shows that in **(a)** ballistic transport, electron wave propagates without energy relaxation (flat color contour in the channel), while in **(b)** dissipative transport, carriers keep relaxing energy (bent color contour in the channel); **(c)** Transfer characteristics in linear scale (right) and log scale (left) show that drain current is overestimated by ballistic simulation even for gate length as small as 8 nm; **(d)** Output characteristics show that output resistance (R_{out}) is overestimated by ballistic simulation.

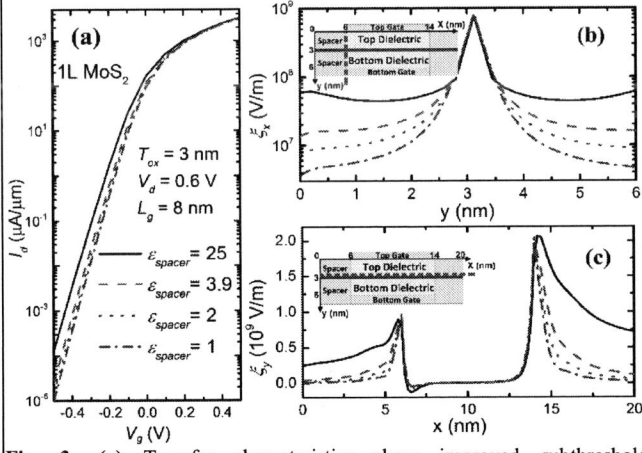

Fig. 3: (a) Transfer characteristics show improved subthreshold characteristics by using low-k spacer. ε_{spacer} is the dielectric constant of the spacer material. **(b)** The parallel component ξ_x (along x direction) of electric field at the source/channel junction; **(c)** the vertical component ξ_y (along y direction) of electric field at the gate dielectric/channel interface. The insets in (b) and (c) show the cross sectional view of the device topology being studied. The dashed red lines in the insets indicate the locations where electric fields are extracted.

Table I: Parameter dependence on number of MoS₂ layers [1],[3],[4],[7].

	# of MoS₂ layers	1	2	3
Mobility (cm²/V·s)	Experimentally obtained at present (under the same recipe)	~10	~20	~30
	Theoretically predicted limits (in high-k environment)	~60	~95 ª	~130 ª
$\varepsilon_{//}$ / $\varepsilon_{\perp}$		2.8 / 4.2	4.2 / 6.5	4.9 / 7.5
$\Delta E_{K-\Lambda}$ (meV)		270	190	130

$\varepsilon_{//}$ / $\varepsilon_{\perp}$: in-plane/out-of-plane dielectric constant; $\Delta E_{K-\Lambda}$ (meV): energy difference between the lowest K valley and the second lowest Λ valley; ª : interpolated values.

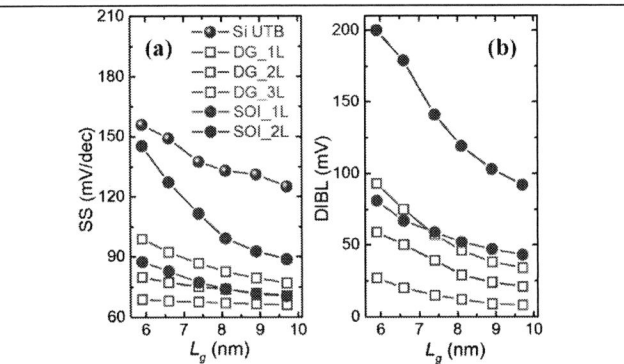

Fig. 4: (a) Subthreshold swing (SS) and **(b)** Drain-induced-barrier-lowering (DIBL) with gate length scaling for 1L-3L MoS₂ FETs, and Si based ultra-thin-body double gated (Si UTB DG) FETs (only SS shown). "DG" and "SOI" represent double-gate and semiconductor-on-insulator structures, respectively. DG has a symmetrical top and bottom gate dielectric, while in SOI the bottom dielectric is 50 nm thick SiO₂ and bottom gate is grounded.

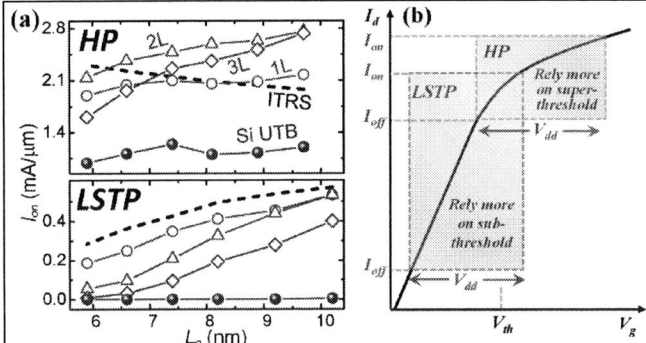

Fig. 5: (a) ON-current (I_{on}) versus gate length for 1L-3L MoS₂ FETs, and Si UTB DG FETs. Black dashed lines represent the ITRS requirement for I_{on}. "HP" and "LSTP" represent high-performance and low-standby-power technologies, respectively; **(b)** A schematic illustration showing the method used for obtaining I_{on} and the design priorities for LSTP and HP technologies.

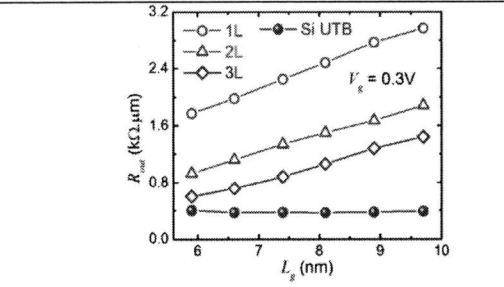

Fig. 6: Output resistance R_{out} extracted from the output characteristics (the method has been explained in Fig. 2(d)) for 1L-3L MoS₂ FETs and Si UTB DG FETs. MoS₂ FETs show much higher R_{out} compared to Si devices, indicating that they exhibit better saturation characteristics. The general observation is that the smaller the channel thickness, the better the electrostatics, and higher the R_{out}.

978-1-4799-8002-4/14 $31.00 © 2014 IEEE

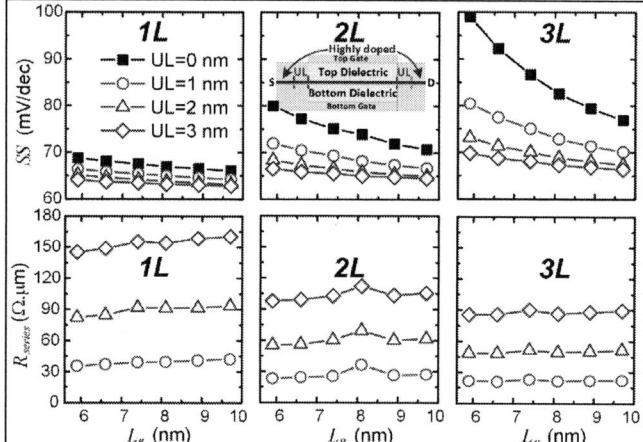

Fig. 7: SS improvement and extra series resistance introduced by using underlap (UL) structure for 1L-3L MoS₂ FETs. The inset in the subplot for 2L is a schematic illustration of the UL structure.

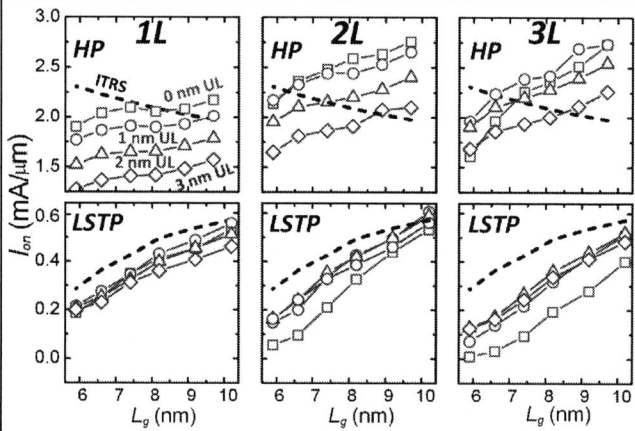

Fig. 8: Top: I_{on} versus gate length for 1L-3L MoS₂ FETs. 2L MoS₂ with 0-1 nm UL can meet the HP requirement except for the smallest 5.9 nm node. **Bottom:** 1L MoS₂ with 1 nm UL is closest to the LSTP requirement. The best choice is 2L with 1 nm UL for HP, and 1L with 1 nm UL for LSTP.

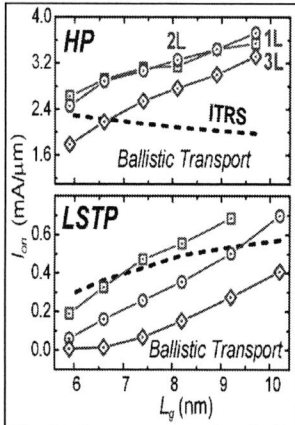
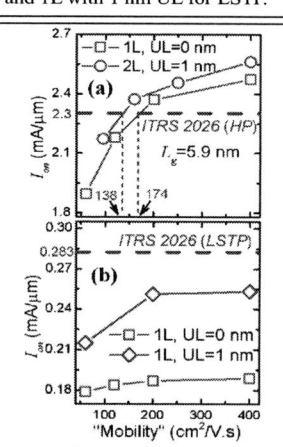

Fig. 9: I_{on} versus gate length for 1L-3L MoS₂ FETs obtained by ballistic simulation. HP requirement can be met, while LSTP requirement for below 7 nm nodes cannot, indicating that only improving "mobility" is not enough for LSTP.

Fig. 10: The "mobility" is varied to meet **(a)** HP and **(b)** LSTP requirements at the 5.9 nm node. "2L, UL=1nm" only requires a "mobility" of 138 for HP, which has been experimentally achieved in 1L WSe₂ [12]. Improved "mobility" with the aid of UL is not enough to meet LSTP requirement.

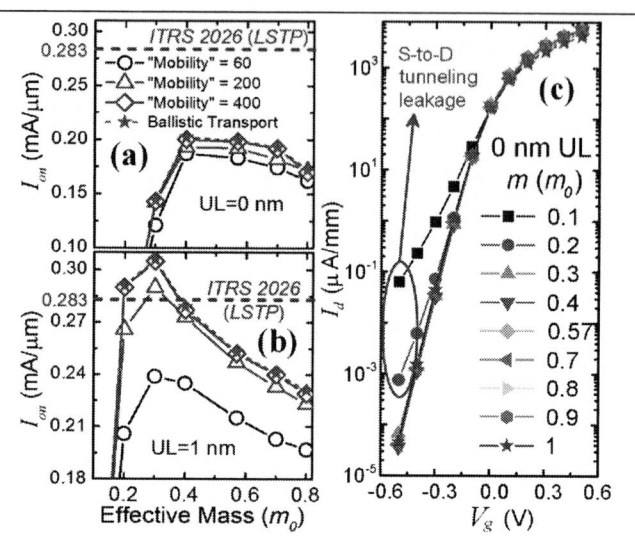

Fig. 11: The effective mass is varied to meet the LSTP requirement at 5.9 nm node in the case of **(a)** UL= 0 nm and **(b)** UL= 1 nm. **(c)** I_d-V_g curves with different effective mass. Comparing (a) and (b), it can be found that with the aid of UL, materials with effective mass around $0.3m_0$, and "mobility" higher than 190 cm²/V·s can meet the LSTP requirement. The reason is that with smaller effective mass, carrier velocity is higher, and thus I_{on} can be higher. However, aggressively small effective mass will lead to increased source-to-drain (S-to-D) tunneling leakage as shown in (c), and significantly degrade I_{on}, which can be observed in (a) and (b). UL can help suppress S-to-D tunneling and derive more benefits from reduced effective mass.

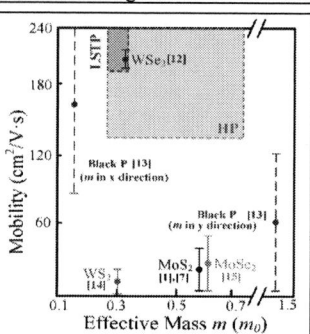
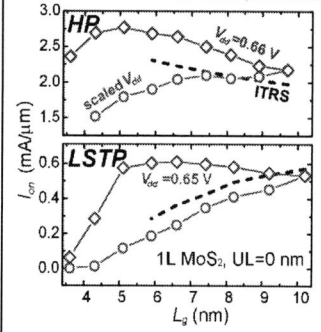

Fig. 12: Collected data of mobility (experiment) and effective mass (first principle calculation) for various 2D semiconductors (1L-2L). Green block is the required range of values for HP, and blue block is for LSTP. Since experimental data is still lacking for the mobility of 1L-2L black phosphorus, a dashed-line error bar is used for an estimation according to an effective mass based rough calculation [13].

Fig. 13: I_{on} versus gate length in the case of scaled and unscaled supply voltage V_{dd}. Without scaling V_{dd}, both HP and LSTP requirement can be easily met, which indicates that we need to consider relaxing constraint on V_{dd} scaling or turn to sub-kT/q devices such as tunnel FETs for below 5 nm nodes. Alternatively, 3D IC integration with 2D materials could provide an alternative pathway to scaling beyond the 5 nm node.

Table II: A summary of possible material/structure/technology choices for sub-10 nm HP and LSTP VLSI applications.

L_g	>6.6 nm	5 nm - 6.6 nm	< 5 nm
HP	• 2L MoS₂ & 1nm UL	• 1 nm UL, & dielectric engineering, such as using h-BN buffer layer to boost the "mobility" of 2L MoS₂ to 138 • 1L or 2L WSe₂ with 1 nm UL	• UL & Slow down V_{dd} scaling • tunnel FETs
LSTP	• 1 nm UL, & dielectric engineering to boost the "mobility" of 1L MoS₂ to >190, & strain engineering to lower the effective mass of MoS₂ to ~$0.3m_0$ • 1L WSe₂ & 1 nm UL		

978-1-4799-8002-4/14 $31.00 © 2014 IEEE

Atomic Disorder Scattering in Emerging Transistors by Parameter-Free First Principle Modeling

Qing Shi[1,*], Lining Zhang[2,*] Yu Zhu[3], Lei Liu[3], Mansun Chan[2], Hong Guo[1]

[1]Dept. of Physics, McGill University, Montréal, Canada, [2]Dept. of ECE, Hong Kong University of Science and Technology, Kowloon, Hong Kong, [3]NanoAcademic Technologies Inc., Brossard, QC Canada.
Tel: (514) 3986530, Fax: (514) 3988434, Email: qing.shi2@mail.mcgill.ca, lnzhang@ieee.org

Abstract

A parameter-free first principle modeling methodology is reported with emphasis on simulating effects of atomistic disorder in nano-scale transistors. The technique is based on the developed theory of nonequilibrium coherent potential approximation and a linear scaling sparse Hamiltonian implementation. Using this technique, effects of disorder scattering to the quantum transport properties of a boron-nitrogen (B-N) co-doped graphene tunnel field effect transistor (TFET) is investigated.

Introduction

Modeling new materials and operation principles of many emerging and nano-scale electronic devices require quantum models beyond the capability of traditional TCAD methods. An important consideration in device physics and modeling is how to deal with the atomic scale disorder. Examples include the random grain boundary scattering in copper and graphene[1] leading to significantly higher resistance of the material; the discrete atomic dopants scattering in transition metal dichalgogenides that changes the contact resistance with metal electrodes [2][3]; as well as the well-known device-to-device variability due to fluctuations of random dopant. Ample experimental results have demonstrated that atomic disorder can seriously affect and may even dominate device performance [1]-[4]. Traditionally there are several theoretical methods in device physics to simulate atomic disorder effects including the density gradient correction in the drift-diffusion model [5], the random-alloy approach in the tight-binding model for computing band structures of semiconductor alloys [6], etc. Microscopically, an atomic disorder interacts with the host material to alter its electronic property so it is desirable to simulate disorder effects from first principles that determine the atomic potential self-consistently and parameter-free. For disorder-free device structures, first principles transport methodology has been realized by carrying out density functional theory (DFT) within the nonequilibrium Green's function (NEGF) formalism [7]. With disorder, however, repetitive computation of many disorder configurations by NEGF-DFT in order to perform disorder averaging, is a bottleneck which has so far prevented realistic devices (e.g. with inevitable disorders) to be simulated by first principles. A methodology that overcomes this bottleneck is in high demand.

To this end, we have recently developed a parameter-free first principles NEGF-DFT methodology to predict atomistic disorder effects in quantum transport [8], and in this work we report further improvements of the algorithm that allows us – for the first time in literature, to realize first principles transistor modeling completely parameter-free. The methodology is described first and then a B-N co-doped graphene TFET is used as an example for demonstration.

Methodology

Fig.1 shows a schematic of a graphene transistor which consists of a heterogeneous material system where the electronic characteristics strongly depend on the atomic configurations in the active region and also the contacts. Multiple disorder scattering of charges must be considered in calculating the electronic structure and quantum transport. The self-consistent NEGF-DFT simulation [7] flow is shown in Fig.2 where a key and novel step is in the third box: the nonequilibrium coherent potential approximation (NECPA) [8] that accounts for multiple disorder scattering effects in both the density matrix and the transmission coefficients, as indicated by the Green's functions with a bar on top. The essence of NECPA [8] is to construct an effective medium by *analytically* carrying out the configurational average over the distribution of random disorder under nonequilibrium transport conditions and the resulting averaged NEGF needs only to be computed once, thereby solving the computation bottleneck mentioned above. Numerically, NECPA adapts perfectly with the DFT implementation by the linear muffin-tin orbital (LMTO) approach which also leads to a very sparse Hamiltonian matrix. This methodology has previously been applied to model magnetic tunnel junctions [9], copper interconnects [10], and Si nano-channels [11] having atomic disorders.

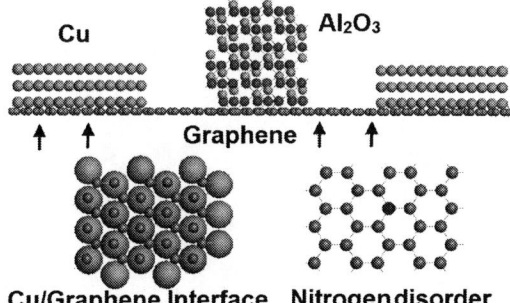

Fig.1 The atomistic view of a graphene transistor model (not drawn in scale). Quantum transport modeling of nanoscale devices requires calculations of homegeneous/heterogeneous materials including the atomic disorder effects.

978-1-4799-8002-4/14 $31.00 © 2014 IEEE

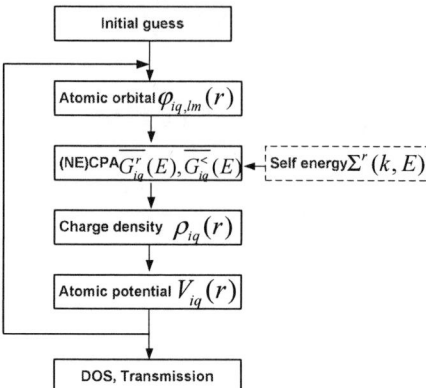

Fig.2. A flow chart of the first principle modeling methodology for disorder scattering. Crystal and disorder atoms are indexed with q, and disorder modeling is carried out in all the steps with highlights of the NECPA for the averaged Green functions (quantities with a bar on top). Disorder material modeling is done in the same way but without the self-energy box indicated by dashed lines. The other quantities are: r, the position along the radius and E, the energy. The algorithm realizes a nearly O(N) scalability along the transport direction.

Disordered Electronic Material

Electronic and atomic structures of the device material are calculated by DFT total energy relaxation, as usual. Again, when there are impurity atoms or other types of disorder, the coherent potential approximation (CPA) is applied [8] and one obtains the averaged properties (e.g. density of states, retarded Green's functions, density matrix elements, etc.).

To see the essence of CPA we consider the B-N co-doped graphene [12] where boron (B) and nitrogen (N) are impurity atoms. Co-doping means the concentration of B and N is the same. Fig.3(a) plots one of many atomic configurations for a given B-N concentration x. The electronic structure varies from one configuration to another, but their *average* is well represented by the effective medium model of CPA in Fig.3(b) where boron/nitrogen occupies the A/B site of graphene with a probability of x%.

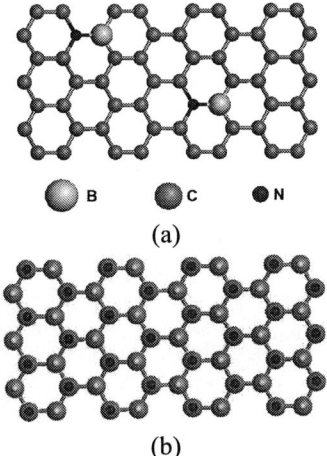

Fig.3 (a) A particular configuration of B-N substitution co-doping in graphene where the co-doping concentration is x%. (b) The effective medium by CPA where a site has x% probability to be impurity.

The calculated band structure of pristine graphene is shown in Fig.4(a), which agrees very well with plane wave methods; the disorder averaged CPA band structure [13] of a 10% B-N co-doped graphene is in Fig.4(b). With B-N co-doping, a band gap of 0.33eV is opened and the band edge electron mass is $0.1m_0$. Note that the CPA bands have a finite width (Fig.4b) which is the disorder broadening. These material properties can also be obtained by brute force DFT calculations of many disorder configurations, as we verified by the VASP program.

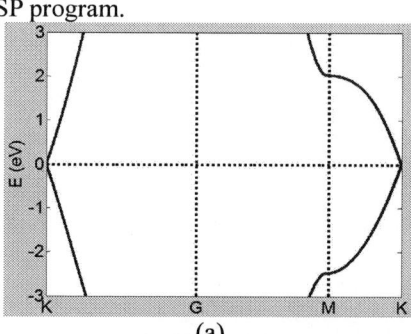

(a)

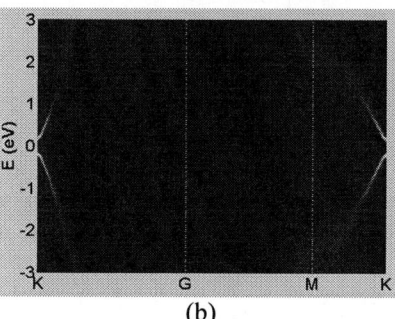

(b)

Fig.4 (a) Band structure of pristine graphene. (b) CPA band of disordered graphene with B-N co-doping at 10%. The CPA bands has a finite width reflecting the disorder broadening. Fermi levels are at "0"eV for both figures.

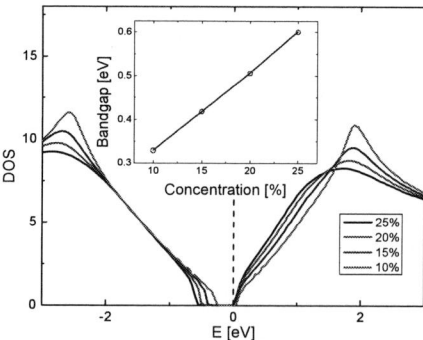

Fig.5. DOS versus energy for several co-doping concentrations. A near linear dependence of the average band gaps on the concentration in the interested range is showed in the inset.

Density of states (DOS) of co-doped graphene is shown in Fig.5. An interesting result is that the band gap linearly increases with co-doping concentration. To make the co-doped graphene p-type or n-type, additional B or N atoms are introduced to the lattice on both the A/B sites, the corresponding DOS at 1% additional doping is shown in Fig.6 where the Fermi level shifts up or down in energy according to the doping type, as expected.

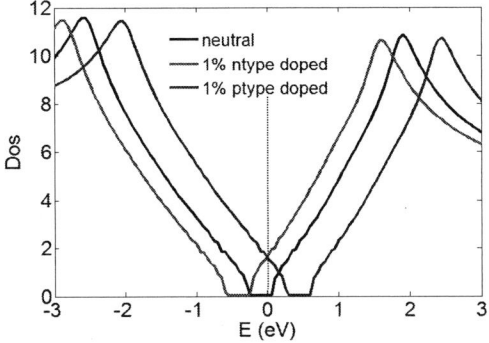

Fig.6 The averaged DOS versus energy of the B-N co-doped graphene with extra individual boron (p-type) or nitrogen (n-type) impurity atoms. Fermi levels move accordingly to the doping type, as expected. The change in band gap is very small with the extra individual dopants.

Modeling Transport with Disorder Scattering

The inputs of our quantum transport modeling are atomic positions of the structure and the applied voltages. We consider double-gate TFET structures whose top view is shown in Fig.7. On the left, the disorder is treated by the widely used virtual crystal approximation (VCA) which essentially provides a shift of the electronic potential due to contributions of impurity atoms. On the right, the disorder is treated by the NECPA methodology presented above that accounts for multiple impurity scattering in addition to the potential change. Thus, a B-N co-doped graphene TFET shown in Fig.7(b) includes the whole channel and part of source/drain treated by NECPA, and we consider practical dimensions where the channel length L_g=10nm, gate dielectric thickness EOT=1.7nm. In the simulation, the width direction is considered infinite (since graphene is a film) which we treat by using periodic conditions in the NEGF-DFT analysis. Devices containing 2160 atoms are calculated with the VCA theme or 4080 atoms with the NECPA theme. The gate modulation on transport is included by solving the real space three-dimensional Poisson equation with the Dirichlet boundary condition at the gate electrodes and floating boundary condition at other surfaces. We emphasize that homogeneous or heterogeneous materials are equally accounted for within our methodology.

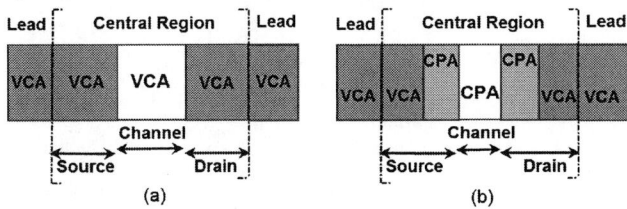

Fig.7 Top views of the double-gate TFET structure (the active region). The leads provide self-energy to the central region in the NEGF-DFT self-consistent analysis. (a) The structure where doping is done by VCA. (b) The structure where CPA accounts for multiple disorder scattering.

The calculated average band profile with V_{ds}=0 is plotted in Fig.8. The potential changes within the VCA and NECPA themes are matched excellently as confirmed by the smooth energy profile. Region I indicates the source depletion while region II is the non-inverted channel of the co-doped graphene TFET which are both treated with NECPA. The interband tunneling process across the source/channel junction in the simulated TFET is thus diffusive and the atomic disorder scattering comes from the co-doping as well as the extra source doping. The transmission spectrum given by NECPA is shown in Fig.9 which clearly confirms the inter-band tunneling physics, for example the transmission peak is achieved at the energy level corresponding almost to the smallest tunneling distance.

As a comparison, the device structure constructed as Fig.7 (a) with VCA doping is also simulated. Our results show that disorder scattering plays a very significant role as indicated by the NECPA curve (black) which represents a much smaller transmission as compared to that of VCA curve where disorder scattering is absent. Microscopically, the decrease in transmissions comes from the diffusive scattering in the tunneling process by the atomic disorders. It is the first evidence that the interband tunneling is affected by the atomic disorders which should be taken into consideration in future TFET modeling. The transmission peak only moves slightly in the tunneling energy window which is not affected by the disorder scattering. The local density of states (LDOS) is plotted in Fig.10 in which the interband tunneling is clearly seen in the band gap. In the 10nm TFET we also observe a direct tunneling current flowing across the whole channel which is also affected by the disorder scattering.

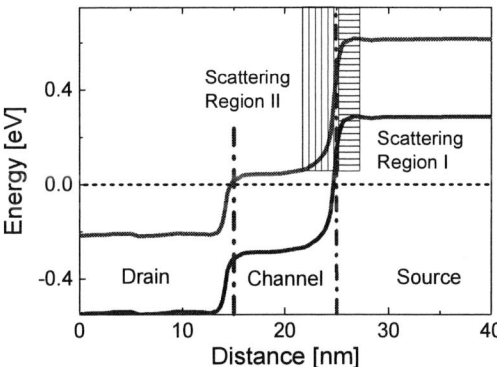

Fig.8 Energy profiles and Fermi level in the (B-N) co-doped graphene TFET obtained from the first principle model with NECPA. Scattering region I includes (B-N) co-doping and extra p-type doping, and region II includes the co-doping.

Non-equilibrium quantum transport of TFETs in Fig.7 (a) or (b) has also been obtained from first principles. Transfer characteristics of the TFETs are calculated and disorder scattering reduces interband tunneling current as seen in Fig.11. Even with scattering the driving current of the co-doped graphene TFET is in the range of mA/µm due to the small band gap and effective mass. We also observe that such disorder scattering has larger influence on off-state

978-1-4799-8002-4/14 $31.00 © 2014 IEEE   735

current than on-state current as shown in Fig.11. It also indicates an increase of ~20% in the subthreshold slope due to disorder scattering.

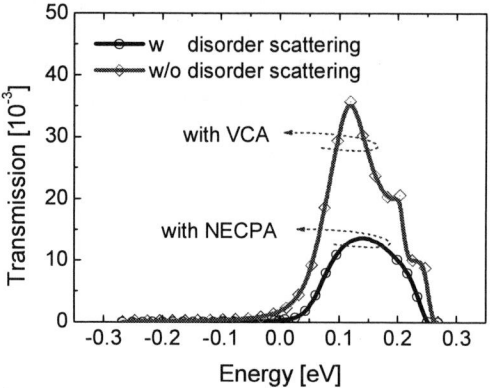

Fig.9 Transmissions of the modeled graphene TFET with VCA doping and CPA doping. The significant decrease in the NECPA results is caused by the diffusive interband tunneling due to the strong disorder scattering.

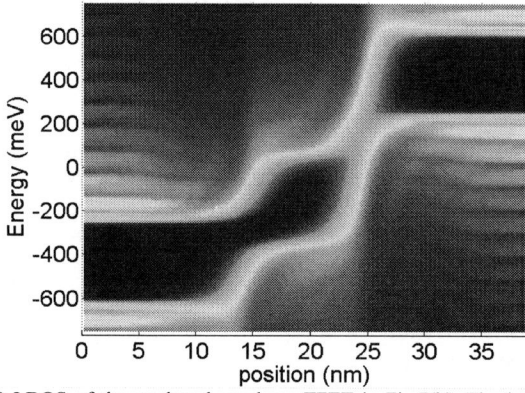

Fig.10 LDOS of the co-doped graphene TFET in Fig.7(b). The interband tunneling is clearly seen from LDOS in the band gap. A direct tunneling current is also observed.

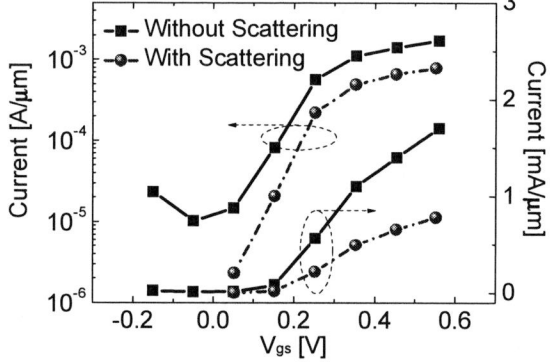

Fig.11 Drain current of the 10nm long graphene TFET with (blue circles) and without (black squares) disorder scattering. Disorder scattering dramatically reduces the inter-band tunneling current.

Conclusion

We reported a parameter-free first principle modeling methodology to handle the important physical effect of atomistic disorder in emerging materials and devices. The method is general and applicable to different materials or heterogeneous systems. Together with a linear scalability of its computational intensity (along the transport direction), the methodology provides a promising TCAD solution to determine disorder induced physical effects.

We applied the methodology to simulate a graphene based TFET by co-doping B-N impurity atoms. Graphene is a zero gap semiconductor in its pristine form, but the co-doping is found to open a substantial gap that linearly scales with the co-doping concentration which is essential for TFET and/or other transistor applications. Another important effect of disorder in the graphene TFET is to reduce the band-to-band tunneling current by a substantial factor, due to the diffusive scattering of the carriers. We note that even if the potential change due to doping is accounted for (e.g. by VCA), it is not enough to obtain correct transport result if multiple impurity scattering is not explicitly included. To this end, the NECPA theory and implementation becomes essential.

Acknowledgments

The work was supported by the Hong Kong's University Grant Committee via the Area of Excellence project AoE-P04-08. H.G. thanks NSERC of Canada for financial support. We thank CalcuQuebec and Compute-Canada for computation allocation which made this work possible. Discussions with Dr. Ferdows Zahid of the University of Hong Kong are gratefully acknowledged.

References

[1] Q. Yu, *et al*, "Control and characterization of individual grains and grain boundaries in graphene grown by chemical vapour deposition," Nature Matter, 10, pp. 443-449, 2011.

[2] H. Fang, *et al*, "Degenerate n-doping of few-layer transition metal dichalcogenides by Potassium," Nano Lett., 13, pp. 1991-1995, 2013.

[3] N. H. Pour, *et al*, "Chemical doping for threshold control and contact resistance reduction in graphene and MoS2 field effect transistors" DRC Tech. Dig., pp. 101-102, 2013.

[4] D. Basu, *et al*, "Effect of edge roughness on electronic transport in graphene nanoribbon channel metal-oxide-semiconductor field-effect transistors," Appl. Phys. Lett. 92, 042114, 2012.

[5] A. Asenov, *et al*, "Problem with the continuous doping TCAD simulations of decananometer CMOS transistors," IEEE Trans. Elec. Dev., 61(2), pp. 2745-2751, 2014.

[6] T. B. Boykin, *et al*, "The electronic structure and transmission characteristics of disordered AlGaAs nanowires," IEEE Trans. Nano., 6(1), pp. 43-47, 2007.

[7] J. Taylor, *et al*, "Ab initio modeling of quantum transport properties of molecular electronic devices," Phys. Rev. B, 63, 245407, 2001.

[8] Y. Zhu, *et al*, "Quantum transport theory with non-equilibrium coherent potentials," Phys. Rev. B, 88, 205415, 2013.

[9] Y. Ke, *et al*, "Oxygen-vacancy-induced diffusive scattering in Fe/MgO/Fe magnetic tunnel junctions," Phys. Rev. Lett, 105, 236801, 2010.

[10] J. S. Chawla, *et al*, "Effect of O2 adsorption on electron scattering at Cu(001) surfaces," Appl. Phys. Lett, 132106, 2010.

[11] J. Maassen, *et al*, "Suppressing Leakage by Localized Doping in Si Nanotransistor Channels," Phys. Rev. Lett. 109, 266803, 2012.

[12] L. Ci, et al, "Atomic layers of hybridized boron nitride and grapheme domains," Nature Materials, 9, pp. 430-435, 2010.

[13] NanoAcademic, Nanodsim Manual, 2013.

Bio-MEMS towards Single-Molecular Characterization

Hiroyuki Fujita

Center for International Research on Micronano Mechatoronics, Institute of Industrial Science,
The University of Tokyo, 4-6-1 Komaba, Meguro-ku, Tokyo 153-8505 Japan

Abstract

MEMS devices, which are capable of handling fluids and objects as small as molecules, enabled bio analysis in the ultimate level of a single molecule. This review includes three recent developments: (1) Micromachined tweezers with sharp tips successfully captured a bundle of DNA molecules and allowed realtime observation of DNA degradation dynamics. (2) A rotational bio molecular motor was encapsulated in a fL-chamber. Exact correlation was obtained between rotational speed and ATP production/consumption. (3) A bio motor system, microtubules and kinesin, was reconstructed on a chip. The chip can distinguish normal and abnormal tau-proteins related to Alzheimer's disease.

Introduction

For almost thirty years, MEMS researchers have investigated the design, fabrication and application of micro/nano mechanisms and actuators of sizes from 10 nanometers to 100 micrometers made by semiconductor-based 3D-micromachining. The MEMS technology enables us to make smart micro systems that have sensing, data processing, communication and actuation capability. Applications include sensors, optics, bio technology, nano technology, and information /communication technology. The combination of MEMS and nano-bio technology is especially beneficial to better quality of life in the future society by providing better understanding of biological processes and developing new medical diagnosis tools.

This paper deals with latest achievements in my group on the MEMS applications to nano-bio technology. Nano tweezers composed of a pair of probes with 10-50 nm tip radius were micromachined with integrated micro actuators for handling DNA molecules. Also, an array of fL-chambers were fabricated on a chip; in each chamber a single enzyme molecule was captured and its enzymatic activity was evaluated. Furthermore, we utilized various functions of bio molecules by integrating them in MEMS. A conveyance device driven by bio motor molecules (microtubules and kinesin) is reconstructed in MEMS fluidic devices. We demonstrated the speed of kinesin is a sensitive index distinguishing the functionality of a mutated-type from a healthy-type of tau, a microtubule-associated-protein. This may lead to the detection of the mutated tau that are bio markers of Alzheimer's disease.

Micromachined Tweezers for DNA handling

A. Basic structure and molecular capturing method

Micromachined tweezers (1,2), named silicon nano tweezers (SNT), comprise two parallel arms ending with sharp tips, designed to trap molecules (Fig. 1a). The sharp tips of SNT were coated with Al and dipped in an aqueous solution of DNA molecules. The high frequency electrostatic field between tips attracted molecules, some of which were captured between them. The mobile arm is displaced by an electrostatic actuator. The motion is acquired by a position sensor, thus the mechanical characteristics of the trapped molecules (stiffness, viscosity) are measured in real time by the change in the resonant frequency characteristics of SNT with DNA bundles (Fig. 1b, c).

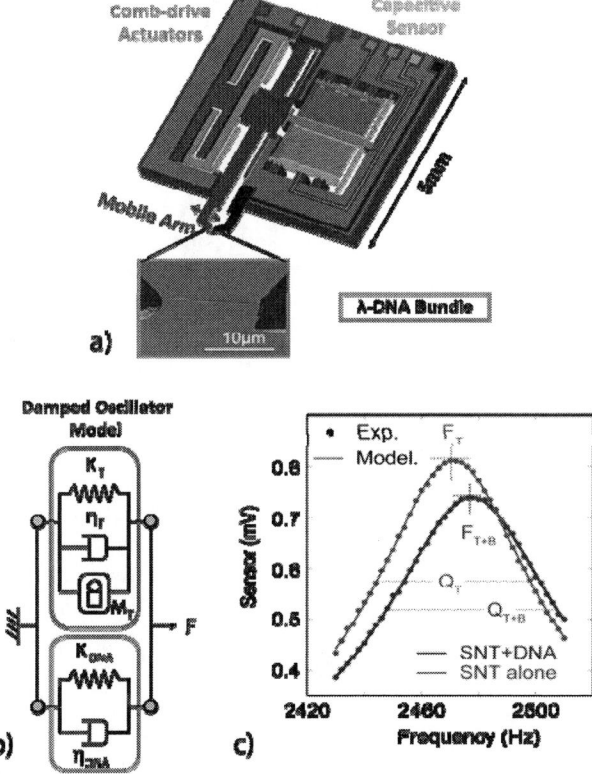

Fig. 1 a) Silicon Nanotweezers, close-up *view on a trap DNA bundle b) Equivalent damped oscillator mode. c) Frequency response of SNT with and without DNA bundle inside a fluidic cavity.*

978-1-4799-8002-4/14 $31.00 © 2014 IEEE   737

B. Application to monitoring DNA damage under X-ray

Tumor cell elimination by X-ray beams in cancer radiotherapy is currently based on a rather empirical understanding of the basic mechanisms and effectiveness of DNA damage by radiation. We conducted real-time biomechanical measurement of the degradation of a DNA bundle in solution when exposed to a therapeutic radiation beam (3). The SNT and associated microfluidic devices can endure the harsh environment of radiation beams and still retains molecular-level accuracy.

The experiments were performed with a Cyberknife, a LINAC accelerator mounted on a robot arm at the Centre Oscar Lambret in Lille, France. All the setup was placed under the Cyberknife. After capturing a DNA bundle, SNT's tips are placed inside a microfluidic cavity; the alignment and the insertion are controlled by a micro-robot. The collimated beam, delivering an intense 6 MeV photon flux, completely encompasses the SNT holding the DNA bundle in the microfluidic cavity.

Once the SNT is locked to its own resonant frequency, the LINAC radiation beam is turned on for 160 s (total dose 30 Gy). The resonant frequencies of the SNT, before, during and after the irradiation are plotted in Fig. 2a. First, SNT without DNA was irradiated to get a reference. The resonant frequency is constant during the irradiation, and the noise level is not significantly modified under irradiation, thus molecular-level accuracy (threshold: 10 ds-DNA) is kept for the measurement. Then DNA bundle was trapped. Two different experiments in the same conditions are compared (two different bundles). For both bundles the resonant frequencies of the SNT decreases in the same proportion. The stiffness of the bundle 1 is calculated from the oscillator model (Fig.1.b) and plotted with the equivalent number of DNA molecule which composed the bundle (Fig. 2b) (4).

Single molecular observation of rotational motor protein

Bio assay in the single-molecular level has mainly been achieved with special optical microscopy, e.g. total reflection near-field microscopy, which limits observation volumes. Physical confinement of a biomolecule in a volume as small as some femtoliters can provide the alternative. In such a small volume, even for a single molecule of common enzymes with a turn-over rate of 10/sec, it is possible to accumulate catalytic product molecules up to µM range within a minute.

Investigation on a single enzyme molecule named adenosine triphosphate synthase (ATPase), a molecular rotary motor of 10 nm in diameter, was carried out in a MEMS fluidic device. Actually, we used one part of the ATPase named

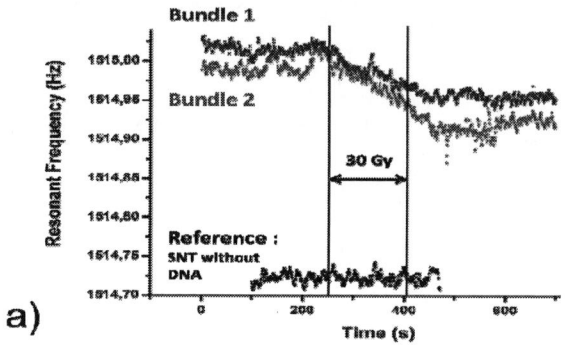

a)

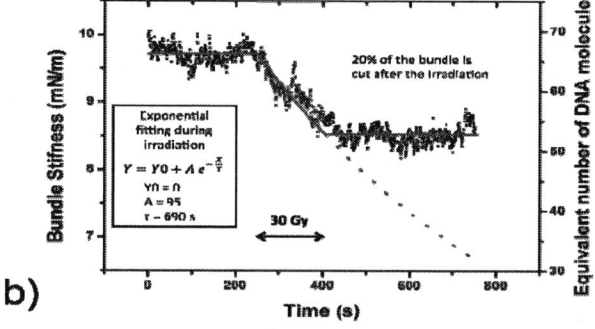

b)

Fig. 2 *a) Locked resonance frequencies recorded during irradiation (total 30Gy with 4 ms pulses at 300 Hz). Same SNT, with 2 different DNA bundles and without DNA. b) Mechanical characterization of the bundle1 damage during irradiation. Decrease of the bundle stiffness (K_{DNA}) with time and the equivalent in term of number of DNA molecule.*

F1-ATPase. The rotation of a single molecule can be observed by attaching a marker on the axis. We encapsulated a single molecule of F1-ATPase in a chamber of a few tens of fL in volume. First, we fabricated a poly-dimethylsiloxane (PDMS) replica of arrayed silicon micro disks on a substrate. The PDMS sheet had many recesses of 1-2 micrometers in depth and around 10 micrometers in diameter. Then, it was pressed against a glass plate on which F1 molecules were immobilized sparsely. We looked for a rotating molecule isolated and encapsulated in a chamber and measured the amount of ATP consumption associated with the rotation of a single molecule in the chamber (5); this concept is depicted in Fig. 3. Please note that F1-ATPase spontaneously rotates in anti-clockwise direction by decomposing ATP in the surrounding solution. If it is forced to rotate in the reversed, i.e. in the clockwise, direction by other molecule or some external force, it synthesizes ATP from adenosine diphosphate (ADP) and phosphate. When the motor is turned at 10 Hz for 1 min, the estimated number of synthesized ATP is calculated to be only 1800 molecules assuming 3 ATP per turn. Such a tiny amount of chemical output gives the concentration of 3 µM in ten-femtoliter chamber in ten minutes. It is high enough to be detected by a conventional fluorescent method for example.

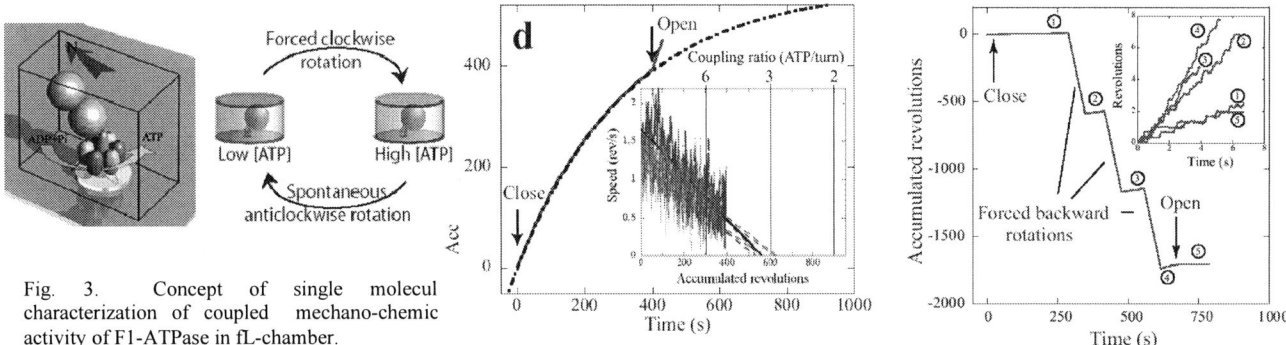

Fig. 3. Concept of single molecul characterization of coupled mechano-chemic activity of F1-ATPase in fL-chamber.

Fig. 4. Decrease of rotational speed of an F1-ATPase molecule due to consumption of ATP in the fL-chamber.

Fig. 5. Increase of rotational speed of an F1-ATPase molecule due to production of ATP by forced inverse rotation in the fL-chamber.

We observed temporal exponential decrease of the speed of a F1 molecule enclosed in a chamber (Fig. 4). This could be attributed to the consumption of ATP in the closed chamber because the rotational speed of F1-ATPase linearly depends on the ATP concentration of surrounding media. From the initial concentration of ATP and the number of turns, we could conclude that three molecules of ATP were consumed in one turn. Furthermore, the motor was forced to rotate in the reversed direction by a magnetic bead attached to the axis of F1-ATPase in a rotating magnetic field. Increase in ATP concentration by the operation was quantitatively determined (Fig. 5) from the spontaneous rotation speed just after the forced rotation. Three molecules of ATP were produced in one (reversed) turn.

Reconstruction of linear protein motor on a chip

The conveyance of bio molecules in cells is conducted by the microtubule (MT) network on which vesicles, filled with cargo molecules and coated with kinesin, move around. We have fabricated the device that imitates the intracellular nanotransport system. Previously, we succeeded in building a molecular sorter by using kinesin motion along

MTs placed on a chip (6-8). As for the medical application of the device, we are investigating the interaction between MTs and microtubule-associated-proteins (MAPs). MAPs' attachment is beneficial to stabilize the supra-molecular structure of MT; if they are mutated and fail to attach, MTs tend to depolymerize, resulting in disruption of intracellular transport. This is one of the causes of neurodegenerative diseases such as Alzheimer's disease. Therefore, MAPs are among candidates for core biomarkers (9). Tau-protein is one of the most common MAPs.

We focused on the kinesin speed decrease caused by tau attachment, known as roadblocking effect. When the sample containing healthy tau is mixed with MT, tau attach to MT hinders kinesin motion and its speed is slower than on MT without tau. If the sample has mutated tau, less molecules bind to MT and the decrease in speed must be smaller.

The effect of different tau-MT incubation times on the kinesin motility was investigated (10). "No tau" case was used as a control. The average bead velocity was reduced with prolonged tau incubation time (Fig. 6a). According to a

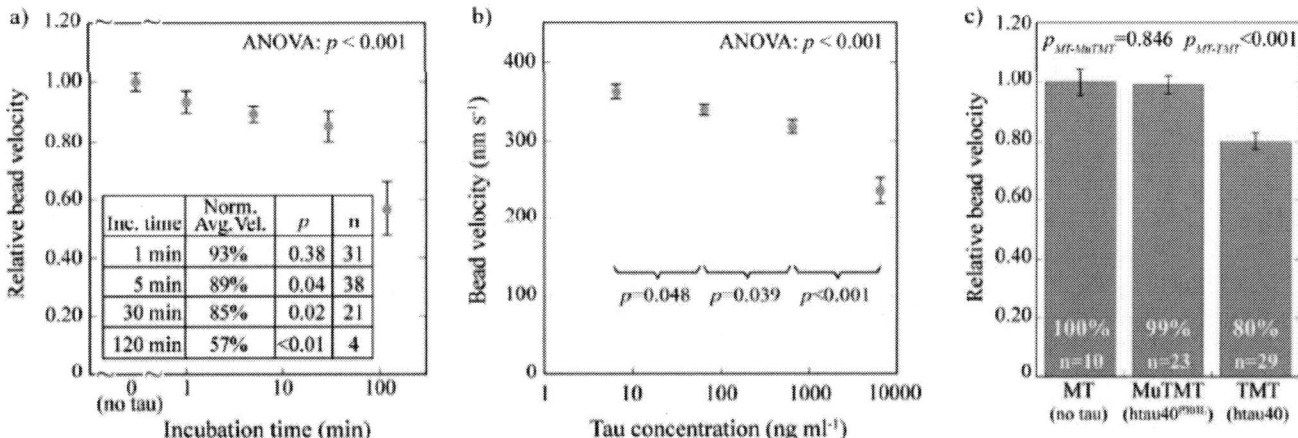

Fig. 6. a) The effect of tau and MTs incubation time on the bead motion. The results were normalized with the value obtained in the control experiment (no tau, n=41). b) Increase in tau concentration results in 5 a significant decrease in bead velocity (n= 57, 75, 52 and 34, respectively). c) While incubation with htau40 results in 20% decrease in bead velocity (*t-test*: *p*<0.001), no significant change was noticed after incubating with htau40P301L (*t-test*: *p*=0.846).

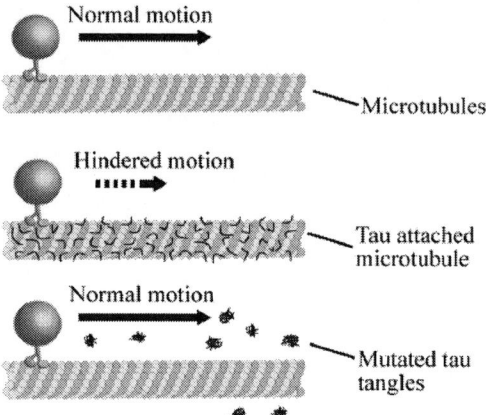

Fig. 7. Schematic view of the kinesin coated bead motion along the MTs. No tau (control) results in unhindered transport of beads. Wild type tau (htau40) binds to the MTs creating "roadblocks" and hinders the motion of the beads. Similar to the control, mutated htau40P301L does not affect kinesin coated bead motion because of its reduced MT-binding capability.

post-hoc Dunnett's test, the difference in the bead velocity between control and 5-min (and longer) incubation times showed significant differences ($p \leq 0.04$) due to the formation of additional "roadblocks" on the MTs. We also confirmed that an increase in the concentration of tau resulted in a significant decrease of average bead velocity (Fig. 6b). Those results indicate that the proposed protocol could accurately detect the effect of tau protein attachment on kinesin motility.

The system was tested to investigate the effect of tau mutation on kinesin-coated bead velocity (Fig. 6c). MTs were co-incubated with either htau40 (healthy tau) or htau40^{P301L} (mutated tau). No tau was used as a control. As expected, htau40 resulted in a significant reduction of bead velocity (*t-test*: $p<0.001$) when compared to the control (Fig. 6c). On the other hand, MTs incubated with htau40^{P301L} showed no significant change (*t-test*: $p=0.846$) in average bead velocity (Fig. 6c). Htau40 binds to MT, hindering kinesin-coated bead motion (Fig. 7). P301L mutation reduces the MT-binding capability, preventing the generation of "roadblocks" and allowing the unhindered movement of kinesin along the MTs.

Conclusion

Latest research activities related to MEMS for nano-bio applications in our group were overviewed; those demonstrate the advantage of MEMS tools for single-molecular bio science. Actuators and sensors integrated in a MEMS device enable us to manipulate molecules and characterize their properties in real-time. The small size of MEMS structure allowed the detection of enzymatic activity of a single molecule. A bio motor system reconstructed on a

chip is useful to determine the effect of slight difference and mutation of molecules on their functionality.

- The SNT associated with microfluidic cavity is able to record the DNA damage rate under irradiation, providing real time data for clinical treatment evaluation.

- Confinement and isolation of a single molecule provided by advanced processing reveal fundamental mechanism often occulted by bulk observation.

- Reconstructed bio molecular system, allow to provide a relevant biological sensing method with in depth caracterization capabilities.

Acknowledgment

The author wishes to acknowledge Prof. G. Hashiguchi with Shizuoka University, Dr. S. Karsten with NuroInDx Co. and Prof. Yokokawa with Kyoto University for their collaboration. Also he appreciates his colleges, Prof. H. Noji, Dr. Y. Rondelez, Prof. D. Collard, Dr. M. Kumemura, and Dr. M.C. Tarhan with The University of Tokyo for their contribution.

References

(1) G. Hashiguchi, T. Goda, M. Hosogi, K. Hirano, N. Kaji, Y. Baba, K. Kakushima, H. Fujita, "DNA manipulation and retrieval from an aqueous solution with micromachined nanotweezers" *Analytical Chemistry*, 75, pp.4347-4350 (2003)

(2) C. Yamahata, D. Collard, B. Legrand, T. Takekawa, M. Kumemura, G. Hashiguchi, and H. Fujita, "Silicon Nanotweezers with Subnanometer Resolution for the Micromanipulation of Biomolecules." *J. Microelectromech. Syst.* 17, 623–631, (2008).

(3) G. Perret, T. Lacornerie, M. Kumemura, N. Lafitte, H. Guillou, L. Jalabert, E. Lartigau, T. Fujii, F Cleri, H. Fujita, D. Collard, "Real Time Bio Mechanical Characterization of DNA Damage under Therapeutic Radiation Beams," *Materials Research Society Fall Meeting*, 14.03, 2013/12/1-6, Boston, Ma, U.S.A

(4) S.B. Smith, Y. Cui, C. Bustamante, "Overstretching B-DNA: The Elastic Response of Individual Double-Stranded and Single-Stranded DNA Molecules, " *Science 271, 795-799* (1996)

(5) Y. Rondelez, G. Tresset, T. Nakashima, Y. Kato-Yamada, H. Fujita, S. Takeuchi, H. Noji, "Highly coupled ATP synthesis by F1-ATPase single molecules", *Nature*, Vol. 433, pp.773-777 (2005)

(6) M.C. Tarhan, R. Yokokawa, F. O. Morin, H. Fujita "Specific Transport of Target Molecules by Motor Proteins in Microfluidic Channels", *ChemPhysChem*, Vol. 14, pp. 1618–1625, (2013)

(7) M.C. Tarhan, Y. Orazov, R. Yokokawa, S.L. Karsten, H. Fujita, "Suspended Microtubules Demonstrate High Sensitivity and Low Experimental Variability in Kinesin Bead Assay", Analyst, Vol.138, No.6, pp.1653-1656, (2013)

(8) M.C. Tarhan, R. Yokokawa, F.O. Morin, H. Fujita, "Specific Transport of Target Molecules by Motor Proteins in Microfluidic Channels", ChemPhysChem 14, 8, 1618–1625, (2013)

(9) A. Seitz, *EMBO Jour.*, Vol. 21, pp. 4896–4905, (2002),.

(10) M. C. Tarhan, Y. Orazov, R. Yokokawa, S. L. Karsten, H. Fujita, "Biosensing MAPs as "roadblocks": kinesin-based functional analysis of tau protein isoforms and mutants using suspended microtubules (sMT)," *Lab Chip*, vol. 13, pp. 3217-3224, (2013).

An AC and Phase Nanowire Sensing for Site-Binding Detection

Marco Tartagni, Marco Crescentini, Michele Rossi, Hywel Morgan[1], Enrico Sangiorgi
Department of Electrical, Electronic and Information Engineering, University of Bologna,
Cesena Campus, I-47521, Cesena, Italy. Tel: +39-0547-339233 - Email: marco.tartagni@unibo.it
[1]Institute for Life Sciences and Electronics and Computer Sciences,
University of Southampton, Southampton, UK

Abstract

Nanowire and nanoribbon sensing is mostly performed by quasi-static measurements (DC) that are unable to detect the time-dependent behavior of charges and their displacement at the interface. In this paper we propose a technique and a model for sensing nanowires in sinusoidal regimes (AC) that is able to capture both magnitude and phase information of the device response. The approach combines the advantages of impedance spectroscopy with the noise reduction performances of lock-in techniques. Experimental results using an integrated circuit (IC) interface reveal how different surface functionalizations give the same DC output but can be discriminated using the AC approach.

Device and DC sensing measurements

The device under test is a comb-shaped set of polysilicon nanowires (Fig.1) that have been fabricated using a very low cost top-down nano-fabrication process based on photolithography, thin film technology, and plasma etching (1). The nanowire lengths are within the 10-50 nm range with typical cross-section of 100 nm. The nanowires are embedded in a microfluidic testing structure as shown in Fig.2, which also shows the geometric parameters. The architecture of Fig.2 could be employed for both DC and AC measurements, where a small sinusoidal voltage signal (100mV) is applied to one end to the nanowires and the current is detected as both a real and imaginary signal. The buffer solution is tied to ground by means of an *AgCl* electrode. The surface of the nanowires have been modified using two different chemistries (2): one is a vapor deposition of 3-aminopropyltri-ethoxysilane (*APTES*), which is known to give smooth, reproducible films and another by exposing the nanowires to a solution of N-(3-Dimethylaminopropyl)-N'-ethylcarbodiimide hydrochloride, N-hydroxysulfosuccinimide sodium salt and succinic acid. The two surfaces have different dissociation constants (pKa) for the functional group so that the functionalization process can be confirmed by DC measurements at a fixed pH 7. Specifically, the nanowire conductivity decreases after *APTES* functionalization due to the amino group (pKa~9) that protonates at pH 7, reducing the local concentration of holes (carriers) in the p-type nanowire. Conversely, it increases after succinic acid treatment, since the terminal carboxyl group (pKa ~ 5) deprotonates at pH 7, resulting in a conductance similar to the bare nanowires. The DC results shown in Fig.3 (B) show a difference between the two surfaces. However, the behavior of the *bare* and *succinic* functionalized wires are similar due to the total static effect of the charges on the potential profile at the nanowire interface (2) influencing the conductance value of the nanowire.

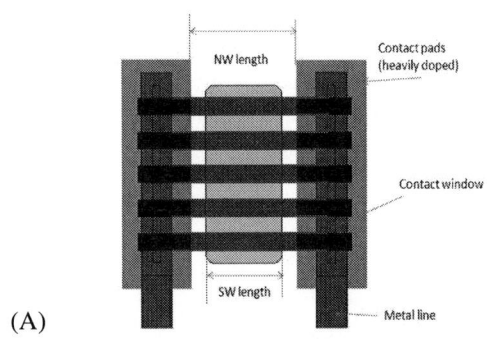

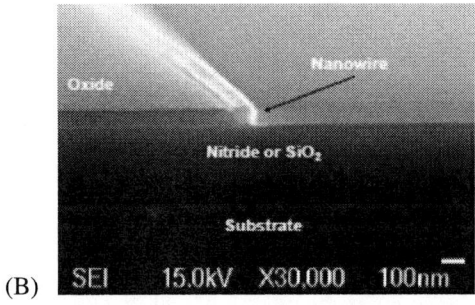

Fig. 1: Top-view of the nanowire structure (A) and cross-sectional SEM image of a fabricated polysilicon nanowire (B). Figures from (1).

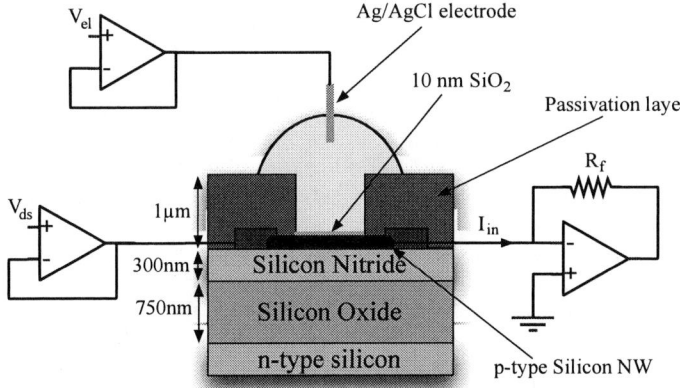

Fig. 2: DC readout operation and cross section of the device.

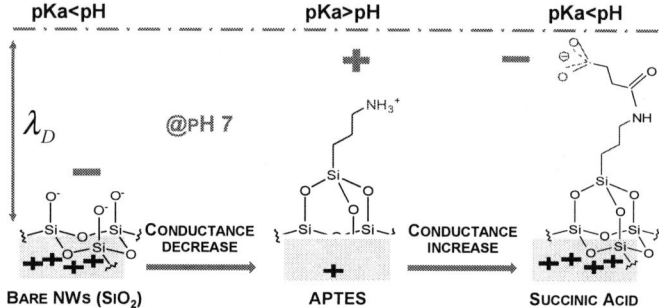

(A)

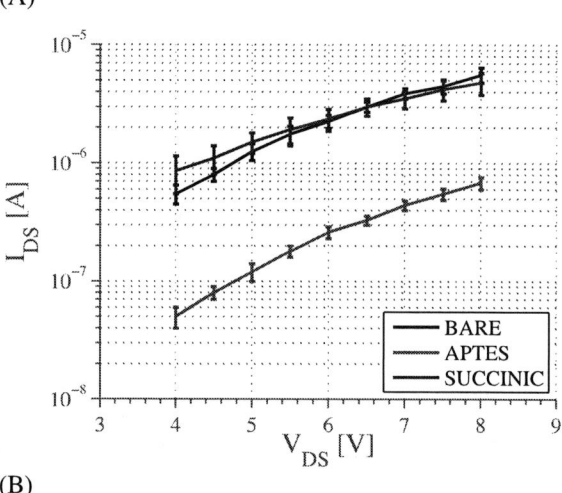

(B)

Fig. 3: Schematic of a nanowire surface functionalization (bare SiO$_2$, APTES and succinic acid) (A). DC measurements from (1) (B).

For AC sensing, an IC interface was developed that is able to measure the complex impedance in real time (3). The chip has been developed using a 0.35 μm CMOS technology embedding 4 cores for multiple concurrent readouts into an area of 1.32 mm^2 (Fig.4). The IC implements a lock-in technique based on a band-pass delta-sigma approach. The architecture is shown in Fig.5 where each core comprises an integrated fully-differential low-noise amplifier (LNA) followed by an anti-aliasing switched-capacitor filter and a band-pass delta-sigma (BPDS) analog-to-digital converter. Demodulation is performed in the digital domain using two XOR gates that avoids the non-linearity of analog multipliers, such as Gilbert cell mixers or switched system. Delta-sigma converters are oversampling converters working at a sampling frequency much higher than the signal bandwidth and are generally composed of an integrator, a comparator and a 1-bit DAC in the feedback loop that allow high resolution due to the "noise-shaping" in the signal bandwidth performed by a digital filter usually implemented in a finite impulse response (FIR) scheme. In our implementation, this is accomplished by using an external field programmable gate array (FPGA) that digitally filters and down-samples the single bit data stream to achieve the desired resolution.

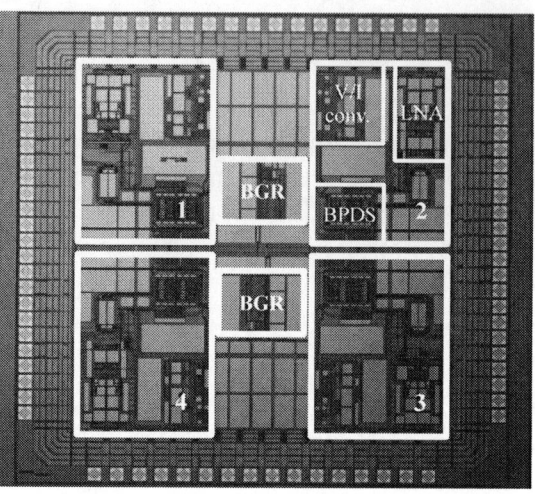

Fig. 4: Photograph of the 4-core integrated circuit (IC) impedance chip used to measure the nanowires.

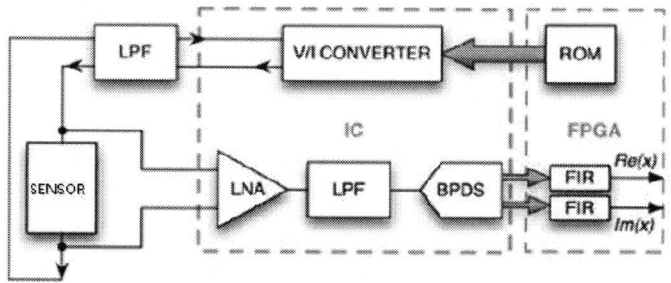

Fig. 5: Architecture of the impedance interface IC. The approach is based on a predominantly digital impedance detection technique based on a lock-in loop.

The reference sinusoidal pattern is modeled using MATLAB® and saved into a read-only-memory (ROM) embedded in the FPGA. The waveform is fed into a voltage-to-current (V/I) converter and then filtered by an off-chip low-pass filter (LPF) to reduce high-frequency quantization noise. The high input impedance LNA acquires and amplifies the voltage drop across the nanowire. The BPDS analog-to-digital converter (ADC) samples LNA output and digitizes it on two concurrent paths for quadrature detection to detect complex components of the impedance. The use of a BPDS converter enables the AC signal to be digitized, thus simplifying the structure. Finally, the demodulation process is performed in the digital domain.

In summary, this is a novel architecture where both the reference signal generation and signal demodulation are performed in the digital domain allowing the system to be easily reprogrammed for the application need.

AC sensing measurements

The previously published DC results (1) were confirmed by performing AC frequency sweeps of nanowires immersed in 1mM phosphate buffer solution (pH 7). Fig.6 shows impedance magnitude and phase results. The magnitude data (Fig.6(A)) are in good agreement with the DC measurements confirming the expected behaviour of the modulation of conductivity in response to changes in the isoelectric point of the nanowire surface over the entire range of frequencies measured. Conversely, the phase plot (Fig.6(B)) shows unexpected and interesting behaviour, where the *APTES* and *succinic* acid treated nanowires are similar but different from *bare* nanowires. Considering only the magnitude, both *bare* and *succinic* acid treated nanowires behave similarly, while the phase data shows that it is possible to distinguish between the two different functionalization steps. This could be explained by a variation of net charges displacements versus voltage (differential capacitance) within the Debye length (λ_D) that is modelled by the double layer capacitance. Therefore, AC readout is sensitive to surface capacitance where DC readout is not, thus giving more information on site-binding effects. The phase behavior could be interpreted by the bypass effect of the surface capacitance on the AC signal.

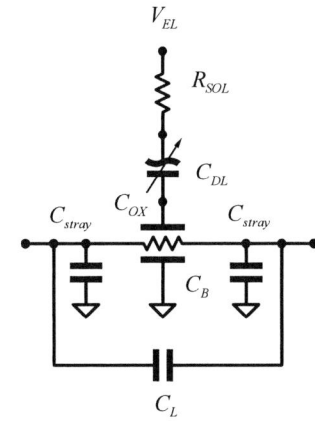

(A)

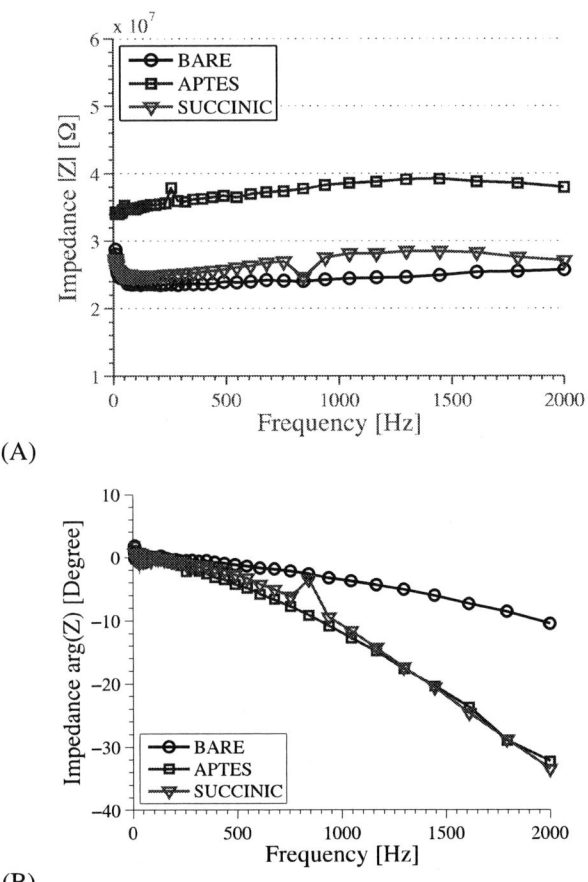

(A)

(B)

Fig. 6: Modulus of the impedance (A) and phase (B) showing how functionalizations that is not distinguishable using modulus appears in the phase.

(B)

(C)

Fig. 7: Distributed model of nanowire AC sensing (A) and comparison of the model with experimental data (bare nanowire) in impedance modulus (B) and phase (C). R_{NW}=3 M , C_L=5 pF, C_{stray}=30 pF, C_S=C_{OX}//C_{DL}=15 pF, C_{DL}=74 pF C_B=100 fF, R_{SOL}=2 M . These parameters refer to 30 parallel Si-nanowires immersed in 1 mM phosphate buffer solution.

Each nanowires is 40 μm long with 100 x 100 nm cross section.

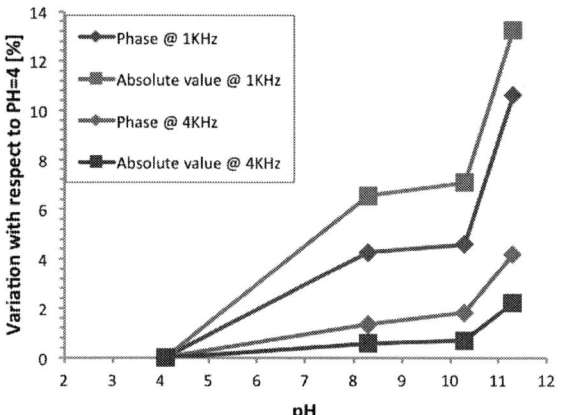

(A)

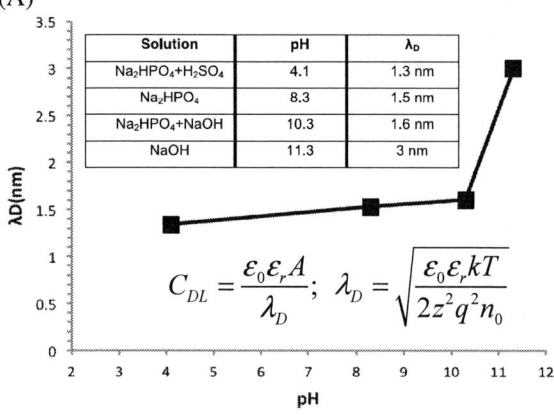

Solution	pH	λ_D
Na₂HPO₄+H₂SO₄	4.1	1.3 nm
Na₂HPO₄	8.3	1.5 nm
Na₂HPO₄+NaOH	10.3	1.6 nm
NaOH	11.3	3 nm

$$C_{DL} = \frac{\varepsilon_0 \varepsilon_r A}{\lambda_D}; \quad \lambda_D = \sqrt{\frac{\varepsilon_0 \varepsilon_r kT}{2z^2 q^2 n_0}}$$

(B)

Fig. 8: AC pH sensing (bare nanowires) (A) showing the correlation of data with the Debye length of the solution (B).

Model

In order to understand the complex impedance behavior of the device, a simple electrical model of the nanowire is proposed, as shown in Fig. 7(A). The values of the components have been measured on the device or calculated from the geometry, while the double layer capacitance has been deduced from experimental measurements. It should be pointed out that the capacitances should be regarded as differential capacitances, with a time-dependent variation of charge with respect to voltages. The nanowire interface has been represented as a distributed RC model, implemented in SPICE with N partition segments; N>5. The model is particularly useful for determining the best AC polarization to trade off the effect of the interface on the resistance (due to charge proximity) with the effect of the interface capacitance (due to charge displacement). The model fits very well with bare nanowire experimental results as shown in Fig.7 (B,C). Furthermore, the model fits well the experimental results of Fig.6 where DC and AC magnitude is mostly affected by R_{NW} value whilst AC phase behavior is mostly affected by $C_s=C_{OX}//C_{DL}$ values. The relationship of the C_s values with a physical model is under investigation and could be referred to dynamic asymmetrical analyte charge distribution similar to models proposed by (4) for conductance. Finally, using the model, we found that with a fixed AC bias current as reference (Kelvin mode), the overall response is more sensitive to the interface capacitance than nanowire resistance. The result is illustrated in Fig. 8 where several buffer solutions have been used to change the pH and Debye length showing how AC detection, either using magnitude or phase, is more sensitive to the double layer capacitance than to pH as in DC mode.

Conclusion

Impedance sensing can be used to measure the electrical properties of nanowires for different surface functionalization. This method has larger sensitivity with respect to Debye length compared to DC methods and could open new perspectives in nanowire biosensing.

Acknowledgments

Authors would like to thank Marta Lombardini for her help and Prof. Peter Ashburn for nanowires samples. The research have been supported by the Engineering and Physical Sciences Research Council (UK) through grant EP/H044795/1, the ENIAC-JTI European project LAB4MEMs and by the grant agreement NANOFUNCTION n°257375 (IT).

References

(1) M. M. A. Hakim, M. Lombardini, K. Sun, F. Giustiniano, P. L. Roach, D. E. Davies, P. H. Howarth, M. R. R. de Planque, H. Morgan, and P. Ashburn, "Thin Film Polycrystalline Silicon Nanowire Biosensors," *Nano Lett.*, (2012).

(2) F. Patolsky, G. Zheng, and C. M. Lieber, "Fabrication of silicon nanowire devices for ultrasensitive, label-free, real-time detection of biological and chemical species.," *Nat. Protoc.*, vol. 1, no. 4, pp. 1711–24, (2006).

(3) M. Crescentini, M. Bennati, and M. Tartagni, "A High Resolution Interface for Kelvin Impedance Sensing," IEEE J. Solid-State Circuits, accepted for publication, 2014.

(4) L. De Vico, L. Iversen, M. H. Sørensen, M. Brandbyge, J. Nygard,• K. L. Martinez•and J. H. Jensen, "Predicting and rationalizing the effect of surface charge distribution and orientation on nano-wire based FET bio-sensors," *Nanoscale*, vol. 3, no. 9, pp. 3635–3640 (2011).

978-1-4799-8002-4/14 $31.00 © 2014 IEEE

MEMS for Cell Mechanobiology

Beth L. Pruitt

Department of Mechanical Engineering, Stanford University

Stanford, CA USA

Living organisms generate and respond to mechanical forces and these forces are sensed and created by specialized cells in the body. Force generation and sensing, or more broadly the mechanobiology coupling tissue (cell) mechanics and biology, are essential in normal development, wound healing, and tissue homeostasis. Our mechanical senses of hearing and touch allow us to navigate our environment and interact with one another, yet they remain the least understood of our perceptive senses. Basic life sustaining functions such as breathing, circulation, and digestion are driven autonomously by coordinated contraction of specialized muscle cells, yet how these functions incorporate active feedback via force sensing at the cellular level is an area of active study. Meanwhile, a variety of specialized stretch activated receptors and mechanically mediated biochemical signaling pathways have been identified in recent years. Importantly, defects in

proteins of these mechanically mediated pathways and receptors have been implicated in disease states spanning cardiovascular disease, cancer growth and metastasis, neuropathy, and deafness. Thus, understanding the mechanical basis of homeostasis (health) and defective cell renewal function (disease) increasingly requires us to consider the role of mechanics. To study how cells and tissues integrate mechanical signals, we and others have developed specialized cell cultures systems and micromachined tools to stimulate and measure forces and displacements at the scale of proteins and cells. A key feature of such experiments is the ability to observe cell outputs such as morphological changes, protein expression, electrophysiological signaling, force generation and transcriptional activity in response to mechanical stimuli. (Figure 1.)

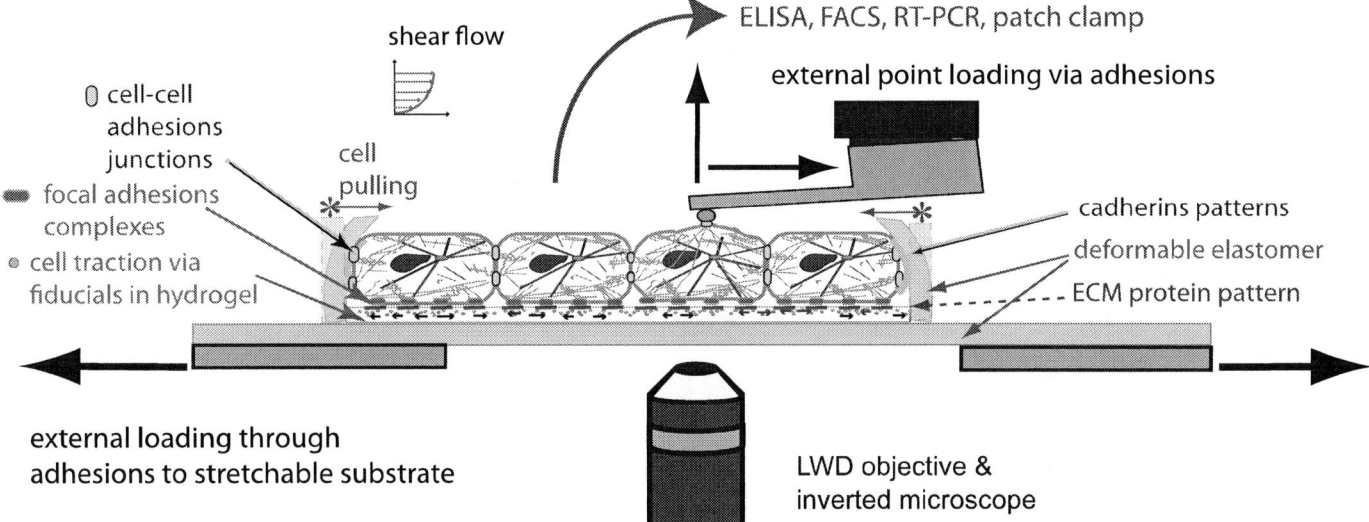

Figure 1. Cells interface with the surrounding extracellular matrix via bound clusters of specialized proteins comprising the focal adhesions complexes and with other cells via different specialized receptors clustered at adhesions junctions. These complexes are linked to cytoskeletal protein structures that give the cell its shape and mechanical properties and which couple mechanical forces to the nucleus. Many of the proteins in the cell-cell and cell-ECM adhesions as well as the cytoskeleton have been shown to take different conformations under physiological loads which results in different binding affinities for biochemical signaling partners and elicits different cell "programs." By deforming cells via pulling, stretching, indenting or shear flow, these different programs can be observed by microscopy, traction force measurements, patch clamp recording of electrophysiological activity, live cell protein expression or downstream gene expression assays.

978-1-4799-8002-4/14 $31.00 © 2014 IEEE

Traditional tools (e.g., tensile testing, Atomic Force Microscopy (AFM), micropipette aspiration, magnetic tweezers, and optical traps) and microfabricated platforms (e.g., silicon force probes, micropillars or micropatterned traction force microscopy substrates) are useful for studying various biological functions related to mechanics such as material properties/cytoskeletal stiffness, traction forces generated by adherent or migratory cells, contractile forces or even the biochemical pathways and electrophysiological signaling and transcriptional activity in response to mechanical stimuli. [2-12] The choice of tool depends on the stiffness and size of the sample and the bandwidth required to observe a response (Figure 2). In our lab, we have focused on the ability to apply precise mechanical stimuli and quantitatively assess the mechanical outputs of cells or Figure 2.

organisms. For example, we have developed comprehensive modeling and optimization routines to design piezoresistive force sensing cantilevers scaled to study the sense of touch and hearing. [1, 13-18] This "force clamp" system is now enabling simultaneous mechanical measurements with patch clamp and imaging studies of *C. elegans*. Further, we have developed advanced cell culture systems to enable measurement platforms integrating micropatterning, stiffness control, external stretch and force measurements using traction force microscopy algorithms. These devices and systems are enabling the study of mechanobiology and cell signaling dependent on cell-cell and cell-matrix adhesion forces. [19-21] My talk will review prior art and opportunities for microfabricated devices in the quantitative study of mechanobiology.

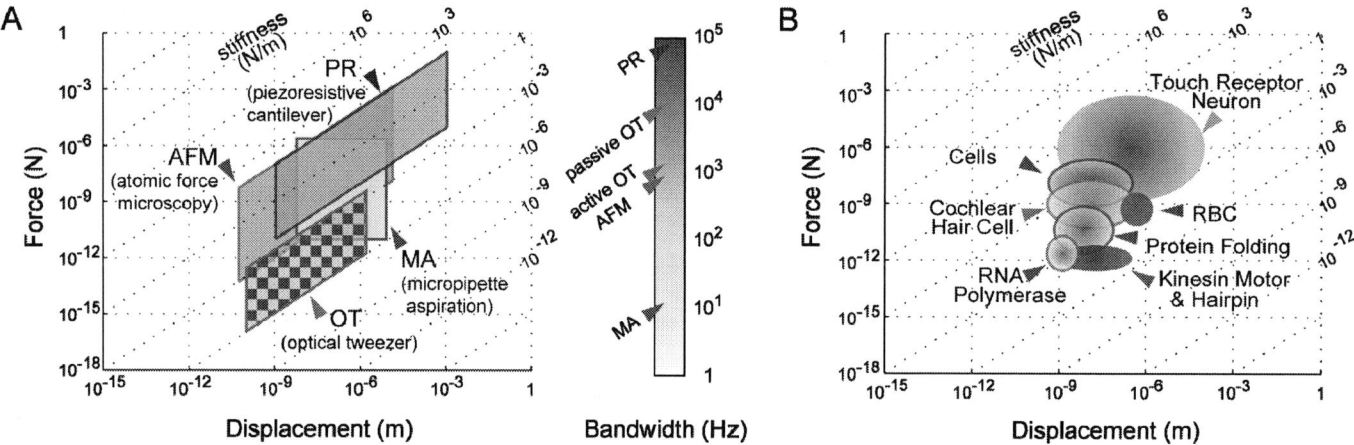

Figure 2. Comparison of the force and displacement range of common instruments for biomechanics (A) and biological materials (B). The color in A indicates the bandwidth of the measurement method, note that custom MEMS piezoresistive sensors can be designed to accommodate many orders of magnitude over the range of measurement space. Adapted with permission from Park 2007.[1]

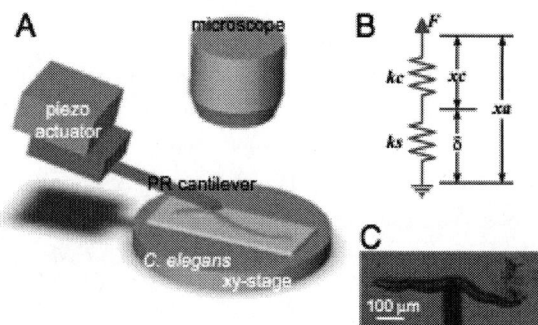

Figure 3. Piezoresistive cantilevers have been deployed in force feedback control to apply precise mechanical stimuli to model organisms like C. elegans (A) Schematic of nematode measurement setup. (B) Mechanical model of sample under indentation is series of cantilever and sample springs. (C) Micrograph of adult nematode under a piezoresistive cantilever. Adapted with permission from Park 2007.[1]

978-1-4799-8002-4/14 $31.00 © 2014 IEEE

[1] S.-J. Park, M. B. Goodman, and B. L. Pruitt, "Analysis of nematode mechanics by piezoresistive displacement clamp," *PNAS,* vol. 104, pp. 17376-17381, 25 October 2007.

[2] D. L. Wang, B.-S. Wung, Y.-J. Shyy, C.-F. Lin, Y.-J. Chao, S. Usami, *et al.*, "Mechanical Strain Induces Monocyte Chemotactic Protein-1 Gene Expression in Endothelial Cells : Effects of Mechanical Strain on Monocyte Adhesion to Endothelial Cells," *Circ Res,* vol. 77, pp. 294-302, August 1, 1995 1995.

[3] R. J. Pelham, Jr. and Y.-l. Wang, "Cell locomotion and focal adhesions are regulated by substrate flexibility," *PNAS,* vol. 94, pp. 13661-13665, December 9, 1997 1997.

[4] K. A. Beningo and Y.-L. Wang, "Flexible substrata for the detection of cellular traction forces," *TRENDS in Cell Biology,* vol. 12, pp. 79-84, February 2002 2002.

[5] K. J. Van Vliet, G. Bao, and S. Suresh, "The biomechanics toolbox: experimental approaches for living cells and biomolecules," *Acta Materialia,* vol. 51, pp. 5881–5905, 2003.

[6] R. Taylor, V. Mukundan, and B. Pruitt, "Tools for Studying Biomechanical Interactions in Cells," in *Mechanobiology of Cell-Cell and Cell-Matrix Interactions*, A. Wagoner Johnson and B. A. C. Harley, Eds., ed: Springer US, 2011, pp. 233-265.

[7] C. S. Simmons, B. C. Petzold, and B. L. Pruitt, "Microsystems for biomimetic stimulation of cardiac cells," *Lab on a Chip,* vol. 12, pp. 3235-3248, 2012.

[8] S. Nishimura, S.-i. Yasuda, M. Katoh, K. P. Yamada, H. Yamashita, Y. Saeki, *et al.*, "Single cell mechanics of rat cardiomyocytes under isometric, unloaded, and physiologically loaded conditions," *Am. J. Physiol. Heart Circ. Physiol.,* vol. 287, pp. H196-202, July 1, 2004 2004.

[9] D. Discher, C. Dong, J. J. Fredberg, F. Guilak, D. Ingber, P. Janmey, *et al.*, "Biomechanics: cell research and applications for the next decade," *Ann Biomed Eng,* vol. 37, pp. 847-59, May 2009.

[10] J. L. Tan, J. Tien, D. M. Pirone, D. S. Gray, K. Bhadriraju, and C. S. Chen*, "Cells lying on a bed of microneedles: An approach to isolate mechanical force," *PNAS,* vol. 100, pp. 1484–1489, February 18, 2003 2003.

[11] C. G. Galbraith and M. P. Sheetz, "A micromachined device provides a new bend on fibroblast traction†forces," *PNAS,* vol. 94, pp. 9114-9118, August 19, 1997 1997.

[12] V. Mukundan, W. Nelson, and B. Pruitt, "Microactuator device for integrated measurement of epithelium mechanics," *Biomedical Microdevices,* vol. 15, pp. 117-123, February 2013 2013.

[13] J. C. Doll, S.-J. Park, and B. L. Pruitt, "Design optimization of piezoresistive cantilevers for force sensing in air and water," *Journal of Applied Physics,* vol. 106, pp. 064310-1-12, 2009.

[14] J. C. Doll, A. W. Peng, A. J. Ricci, and B. L. Pruitt, "Faster than the Speed of Hearing: Nanomechanical Force Probes Enable the Electromechanical Observation of Cochlear Hair Cells," *Nano Letters,* vol. 12, pp. 6107-6111, 2013/01/20 2012.

[15] J. C. Doll and B. L. Pruitt, "Design of piezoresistive versus piezoelectric contact mode scanning probes," *Journal of Micromechanics and Microengineering,* vol. 20, p. 095023, 2010.

[16] S.-J. Park, J. C. Doll, and B. L. Pruitt, "Piezoresistive Cantilever Performance, Part 1: Analytical Model for Sensitivity," *JMEMS,* vol. 19, pp. 137-148, 2010.

[17] S.-J. Park, J. C. Doll, and B. L. Pruitt, "Piezoresistive Cantilever Performance, Part 2: Optimization," *JMEMS,* vol. 19, pp. 149-161, 2010.

[18] S.-J. Park, B. Petzold, M. B. Goodman, and B. L. Pruitt "Piezoresistive cantilever force-clamp system," *Review of Scientific Instruments,* vol. 82, pp. 043703-1-10, 2011.

[19] J. Y. Sim, N. Borghi, K. Kim, C. C. Simmons, A. J. Ribeiro, and B. L. Pruitt, "Changes in Cell Traction Forces in Response to Uniaxial Loading," in *International Conference on Micro and Nano Engineering*, Toulouse, France, 2012.

[20] C. Simmons, J. Sim, P. Baechtold, A. Gonzalez, C. Chung, N. Borghi, *et al.*, "Integrated strain array for high-throughput in situ cellular mechanobiology studies and demonstrated alignment under biaxial strain," *Journal Micromechanics and Microengineering,* vol. 21, p. 054016, 2011.

[21] C. S. Simmons, A. J. S. Ribeiro, and B. L. Pruitt, "Formation of composite polyacrylamide and silicone substrates for independent control of stiffness and strain," *Lab on a Chip,* 2013.

Organic Electrochemical Transistors for BioMEMS Applications

Dimitrios A. Koutsouras, Pierre Leleux, Marc Ramuz, Jonathan Rivnay, and George G. Malliaras

Department of Bioelectronics, Ecole Nationale Supérieure des Mines, CMP-EMSE, MOC

880 route de Mimet, Gardanne, 13541, FRANCE

Abstract

A visible trend over the past few years involves the application of organic electronic materials to the interface with biology, with applications both in sensing and actuation. Examples include biosensors, artificial muscles and neural interface devices. These materials offer an attractive combination of properties, including mechanical flexibility, enhanced biocompatibility, and capability for drug delivery. Most importantly, high ionic mobilities in organic films enable new ways of signal transduction. An example of a device that takes advantage of these properties is the organic electrochemical transistor (OECT). In this device, ions from an electrolyte enter a conducting polymer channel and change its conductivity, hence the drain current. As such OECTs offer a convenient and powerful way to transduce signals of biological origin. Here we report high performance OECTs that are used to record neural activity. As such, they promise to yield a new tool for neuroscience and enhance our understanding on how the brain works.

Introduction

Interest in organic electronic materials stems mostly from a combination of attractive properties that include ease of processing (which can lead to low-cost fabrication), compatibility with a variety of substrates including mechanically flexible ones, and tunability of electronic properties via chemical synthesis[1]. A success story in the field of organic electronics is the organic light emitting diode (OLED). First described in late 80s, this device is currently enjoying wide spread commercialization in small flat panel displays (smartphones), while its production in TVs and solid-state lighting elements is ramping up. For applications in bioelectronics, in particular, the list of attractive properties goes on to include oxide-free interfaces with aqueous media, the ability to transport ions with a relatively high mobility, and the fact that electronic excitations affect molecular structure[2]. The first two properties mean that ions can be exchanged between an aqueous electrolyte and the bulk of organic electronic film, leading to a 3D interaction with the biological milieu, while the last one leads to novel actuation mechanisms[3].

A prototypical system for organic bioelectronics is the conducting polymer poly(3,4-ethylenedioxythiophene) doped with poly(styrene sulfonate) (PEDOT:PSS)[4]. This material, whose structure is shown in Fig. 1, consists of an organic semiconductor (PEDOT), which is degenerately doped *p*-type by the sulfonate groups of the PSS chain. If we wanted to draw an analogy with silicon, PEDOT would be the silicon and the sulfonate ions would be the boron acceptors. The main differences are that doping is not done by substitution (the dopant is not introduced *in* the PEDOT chain, but near it), and the fact that the dopant is introduced in large quantities (there is more PSS than PEDOT in a typical formulation).

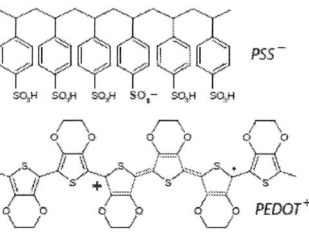

Fig. 1: Chemical structure of the conducting polymer PEDOT:PSS.

The conductivity of a PEDOT:PSS film can be changed through the process of electrochemical doping. According to this process, ions from an electrolyte are injected into the film and change the hole density not just under the surface, but *throughout the entire volume of the film*. Injection of cations, for example, will lead to hole extraction (through a metal electrode) and dedope the film (it is worth noting that the PSS chain "holds" the sulfonate ions in position, so that they do not diffuse into the electrolyte). This process is analogous to compensation doping of silicon, but takes place at room temperature and by applying a small bias. A ramification of this phenomenon is that the effective capacitance at the interface between a conducting polymer film and an electrolyte scales with film volume, rather than area, and can therefore reach very large values.

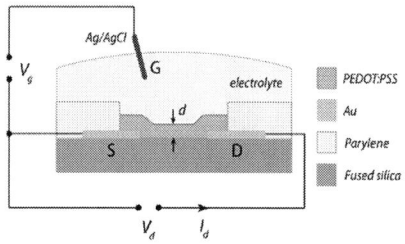

Fig. 2: Schematics of an organic electrochemical transistor with a PEDOT:PSS channel, Au source and drain electrodes, and a Ag/AgCl gate electrode. The substrate (fused silica) can be replaced by a flexible plastic foil, while parylene is used to insulate the contacts from the electrolyte.

OECTs take advantage of this mechanism to deliver a large transconductance. An OECT consist of a PEDOT:PSS channel, in contact with an electrolyte (Fig. 2), and with source and drain electrodes that measure the (hole) drain current. A change in the electrical potential at the interface between the electrolyte and the polymer film drive ions in and out of the channel and changes the conductivity of the latter, thereby modulating the drain current[5]. As the entire volume of the channel participates in the current modulation process, OECTs exhibit a very large transconductance (in the mS range)[6]. As such, they can be very useful for transducing signals of biological origin[7]. For this purpose, the voltage applied at the gate is held constant, and a biological phenomenon is used to modulate the potential at the electrolyte/channel interface. This phenomenon can be, for example, the electrical activity of a neural network in the brain, or an electron transfer reaction due to a redox enzyme[2]. Finally, we should note that simple voltage amplifiers have been fabricated using OECTs, offering >50 dB of power amplification for low frequency signals[8].

Results and discussion

A. OECT fabrication and characteristics
We fabricate OECTs through a combination of solution and vapor deposition and etching processes, and use photolithography to pattern them, mainly to be able to access micron-scale dimensions that are of interest for interfacing with single neurons. The PEDOT:PSS film is deposited from a commercial dispersion using spin coating to a thickness of around 100 nm. It can be patterned either by using an underlying sacrificial layer that forms a contact mask on the substrate, or by protecting parts of it with a photoresist and removing the rest using an oxygen plasma[9]. Au source and drain electrodes are deposited by vacuum evaporation and patterned with photolithography and etching, and then covered with an insulator such as parylene, deposited from vapor. The gate electrode can be held on top of the channel, as shown in Fig. 2, or patterned on the side of the channel (planar configuration) using microfabrication. It is made of Ag/AgCl, PEDOT:PSS, or Pt – the choice of material is known to affect performance[8]. The processes discussed above can be combined in a number of different ways to yield OECTs on glass or plastic substrates. An example of a microfabricated OECT channel is shown in Fig. 3. It should be noted that OECTs can also be fabricated using additive processes such as ink-jet printing[10], and we expect this to be a major advantage for custom-made biosensing applications.

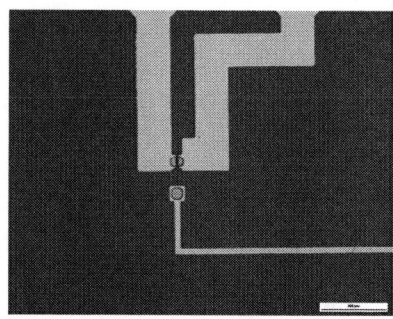

Fig. 3: Micrograph of microfabricated PEDOT:PSS OECT and electrode. A film of parylene coats the entire surface with the exception of the areas where PEDOT:PSS was deposited.

The transfer curve and resulting transconductance are shown in Fig. 4. The transfer curve is typical for depletion operation, where application of positive gate bias causes cations to enter the channel, which decreases hole density and reduces the drain current. The transconductance reaches its highest value around zero gate bias, meaning that the OECT can be used with the gate electrode directly connected with the source (it was designed so intentionally[8]).

The transconductance exceeds 4 mS, which is a very high value for a thin film transistor.

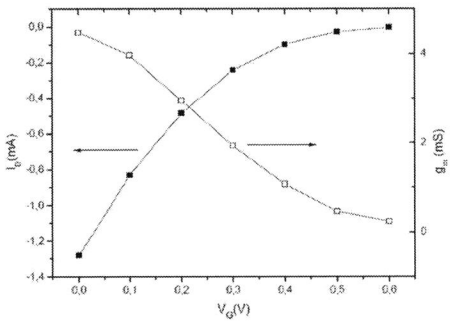

Fig. 4: Transfer curve and resulting transconductance.

B. Applications in recording brain activity

We explored the potential of OECTs in recording brain activity. The first experiment was carried out in a rat, in an electrocorticography configuration, according to which the OECT is in direct contact with the brain. The electrolyte, in this case, is the cerebrospinal fluid, and it is the activity of neurons in the cortex that gates the transistor. We showed that in this configuration the OECT records brain activity with an SNR that is 25 dB higher than that of an electrode of the same size (Fig. 5), due to the local amplification endowed by the transistor[11].

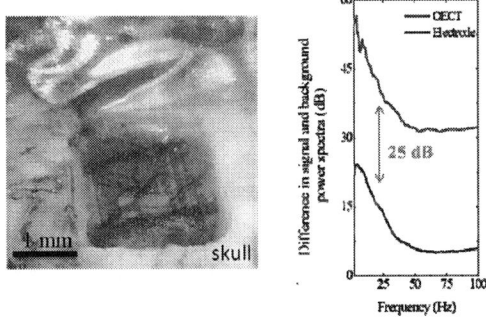

Fig. 5: Flexible array of electrodes and transistors placed on the cortex of a rat. Power spectra plot showing SNR of recordings obtained with electrodes and transistors.

Fig. 6 shows electroencephalography recordings in a human. In this particular experiment the gate electrode of the OECT was in contact with the scalp, eavesdropping on brain activity of underlying neural circuits and transferring it to the OECT channel for transduction. OECTs can, therefore, record brain activity through the scull and recover, for example, the alpha rhythm in a human brain (Fig. 4). This represents, to the best of our knowledge, the smallest signal measured directly with an OECT. The ramification is that OECTs can dramatically simplify the instrumentation currently used for electrophysiology.

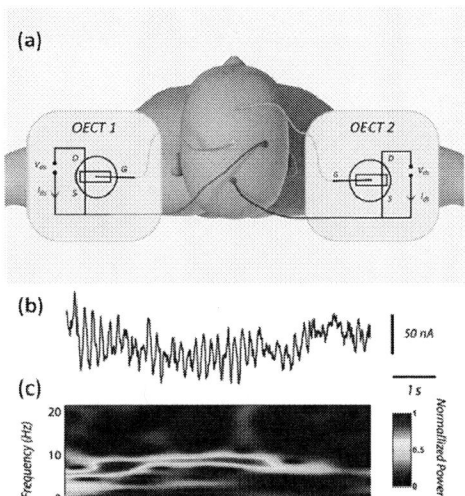

Fig. 6: (a) Wiring configuration used for the measurement of EEG with OECTs, (b) Recording of spontaneous brain activity from one of the OECTs, and (c) associated time frequency spectrogram showing the alpha rhythm around 10 Hz.

Conclusions

Organic electronic materials offer an attractive combination of properties, with the most important one being their 3D interaction with ions in electrolytes. OECTs leverage this property to yield recordings of brain activity with a high signal-to-noise ratio. Examples of recordings in an electrocorticography configuration in a rat and an electroencephalography configuration in a human were discussed.

Acknowledgements

We acknowledge our collaborators Marc Ferro, Thomas Lonjaret, Ilke Uguz, Adel Hama, Esma Ismailova and Róisín Owens, as well as our collaborators from the Institut de Neurosciences des Systèmes, INSERM (France) Adam Williamson, Christophe Bernard, Christian Benar, and Jean-Michel Badier. Funding for this work was provided by the ITN CURIE OLIMPIA and various other regional, national and European projects.

References

1. G. Malliaras and R. Friend, Physics Today **58** (5), 53-58 (2005).

978-1-4799-8002-4/14 $31.00 © 2014 IEEE

2. J. Rivnay, R. M. Owens and G. G. Malliaras, Chemistry of Materials **26** (1), 679-685 (2014).

3. A. M. D. Wan, R. M. Schur, C. K. Ober, C. Fischbach, D. Gourdon and G. G. Malliaras, Adv. Mater. **24** (18), 2501-2505 (2012).

4. A. Elschner, S. Kirchmeyer, W. Lövenich, U. Merker and K. Reuter, in *PEDOT, Principles and Applications of an Intrinsically Conductive Polymer* (CRC Press, 2010), pp. 113-166.

5. D. A. Bernards and G. G. Malliaras, Adv. Funct. Mater. **17** (17), 3538-3544 (2007).

6. D. Khodagholy, J. Rivnay, M. Sessolo, M. Gurfinkel, P. Leleux, L. H. Jimison, E. Stavrinidou, T. Herve, S. Sanaur, R. M. Owens and G. G. Malliaras, Nat Commun **4**, 2133 (2013).

7. P. Lin and F. Yan, Adv. Mater. **24** (1), 34-51 (2012).

8. J. Rivnay, P. Leleux, M. Sessolo, D. Khodagholy, T. Herve, M. Fiocchi and G. G. Malliaras, Adv. Mater. **25** (48), 7010-7014 (2013).

9. J. A. DeFranco, B. S. Schmidt, M. Lipson and G. G. Malliaras, Organic Electronics **7** (1), 22-28 (2006).

10. L. Basiricò, P. Cosseddu, A. Scidà, B. Fraboni, G. G. Malliaras and A. Bonfiglio, Organic Electronics **13** (2), 244-248 (2012).

11. D. Khodagholy, T. Doublet, P. Quilichini, M. Gurfinkel, P. Leleux, A. Ghestem, E. Ismailova, T. Hervé, S. Sanaur, C. Bernard and G. G. Malliaras, Nat. Commun **4**, 1575 (2013).

Multifunctional Smart Lab-on-a-Tube (LOT) Probe
For Monitoring Traumatic Brain Injury (TBI)

Chunyan Li,[1,3] Pei-Ming Wu,[1] Zhizhen Wu,[3] Nirjhar Bhattacharjee,[3] Jed A. Hartings,[2] Raj K. Narayan,[1] and Chong H. Ahn[3]

[1]The Cushing Neuromonitoring Laboratory, Feinstein Institute for Medical Research, Manhasset, NY, USA
[2]Department of Neurosurgery, University of Cincinnati, Cincinnati, OH, USA
[3]Microsystems and BioMEMS Laboratory, University of Cincinnati, Cincinnati, OH, USA

Abstract

A novel multifunctional smart lab-on-a-tube (LOT) is described to continuously and accurately monitor multiple physiological, metabolic and electrophysiological parameters that are vitally important in guiding the care of patients with traumatic brain injury. In addition to measuring various crucial parameters, the newly developed probe allows for drainage of excess cerebrospinal fluid as a strategy to reduce intracranial pressure.

Introduction

Traumatic brain injury (TBI) is a major cause of death and disability in both the civilian and military settings [1]. The mortality from severe TBI is approximately 30% with a substantial proportion of survivors being left with significant disability. The initial brain injury is made worse by secondary events such as ischemia and its many biochemical sequelae [2]. We believe that injured neurons, if provided with the optimal *milieu*, can recover.

The current approach to monitoring patients with any type of acute severe brain injury is based on the insertion of multiple probes into the brain parenchyma [3]. Monitoring of the injured brain has advanced little in recent decades and is limited mainly to intracranial pressure (ICP). Brain oxygen, temperature, blood flow, biochemistry, and electrophysiology are monitored for research and all use separate devices and monitors. The purpose of this research was to develop a novel multimodality lab-on-a-tube (LOT), smart catheter, which could accurately track multiple physiological, metabolic and electrophysiological parameters in the injured brain (Figure 1). In addition, the developed smart catheter would allow for drainage of excess cerebrospinal fluid (CSF) as a strategy to reduce intracranial pressure [4].

Design

Seven microsensors were fabricated on a 6 μm thick polyimide (PI2611) substrate based on the standard Microelectromechanical Systems (MEMS) technology and rolled spirally to form a tube structure (Inner diameter = 0.7

mm; Outer diameter = 0.8 mm; Sensing length = 5 mm). Non-ideal effects can be induced during sensor miniaturization and integration as shown in Figure 2. The types of interference include: (1) electromagnetic interference; (2) electrical feed-through across conductive electrodes; (3) chemical cross-talks among the electrochemical sensors; (4) inductive coupling among the traces; (5) finite traces resistance; and (6) capacitive coupling. The mechanical design and electrical operation of the microsensors were carefully chosen such that potential electronic, thermal and chemical crosstalk among the microsensors was negligible.

First, to counter the electromagnetic interference (EMI), all the sensor cables carrying sensitive analog signals are properly shielded and the corresponding interface circuits are equipped with EMI filters.

Multifunctional Smart Lab-on-a-Tube

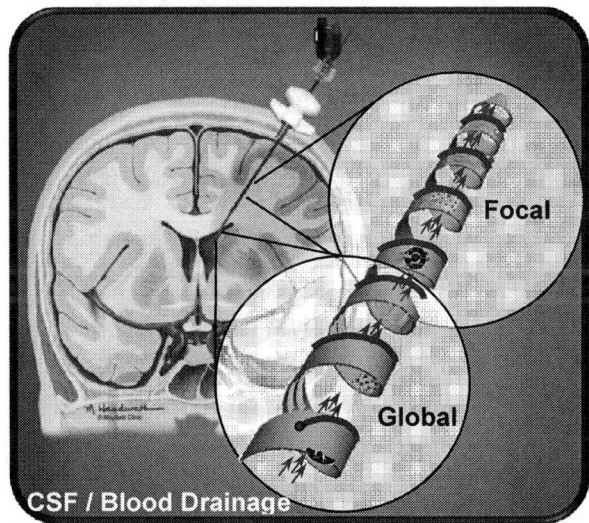

Figure 1: Conceptual drawing of a multifunctional smart lab-on-a-tube (LOT) probe for monitoring traumatic brain injury: It can continuously and accurately monitor multiple physiological, metabolic and electrophysiological parameters in the injured brain. Microsensors outside the tube monitor local brain tissue parameters and the ones outside monitor global cerebrospinal fluid (CSF) parameters. It can also work as a tube to drain excess CSF.

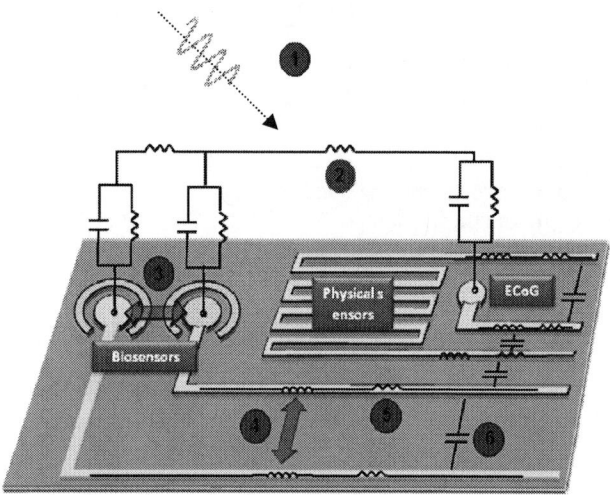

1: Electromagnetic
2: Coupling via solution
3: Chemical crosstalk
4: Inductive coupling
5: Finite trace resistance
6: Capacitive coupling

$E_1 = E_0 \cdot e^{-t/\delta}; H_1 = H_0 \cdot e^{-t/\delta}$

$V_o = V_i \cdot Z_{in} / (Z_{eq} + Z_{in})$

Selectivity coefficient $\neq 0$

$V = L \cdot dI / dt$

$R_c \neq 0$

$I = C \cdot dV / dt$

Figure 2: Challenges of multi-sensor integration: Non-ideal effects are induced during sensor miniaturization and integration.

The output currents of the electrochemical sensors are controlled to generate output currents of less than 10 nA so as to reduce the electrical feed-through between the electrochemical and electrocorticography (ECoG) sensors. Adequate distance is kept among the electrochemical sensors to reduce the chemical crosstalk. For each electrochemical sensor, the counter and the working electrode are positioned in close proximity to confine the fringe electric field to lower the electrical crosstalk among the electrochemical sensors and its influence on ECoG electrodes [5]. The temperature sensor is located outside the "thermal influence" area from flow sensor [6]. The sensor traces are cautiously arranged such that the noisier trace is kept at a distance from the quieter trace to help bring down the impact of the coupling. In addition, the traces are electroplated with copper in order to reduce their inductance and resistance.

Methods

A polysilicon-diaphragm-based pressure sensor was embedded on a flexible polyimide substrate [7]. Temperature and cerebral blood flow microsensors were based on micromachined gold resistance temperature detectors with a 4-wire configuration [8]. The temperature sensor operated with AC excitation current without causing self-heating and the flow sensor employed a periodic heating and cooling technique with a constant-temperature mode [6]. Oxygen

sensor with three-electrode configuration was designed to achieve zero net oxygen consumption. Glucose and lactate sensors were based on amperometric enzyme based electrochemical detection. Heterostructured electroencephalography (EEG) electrode array was developed to achieve a superior signal-to-noise ratio [5].

The entire monitor system is divided into four subsystems, the headstage, the front-end hardware, the real-time hardware, and the monitor display. The headstage, as its name suggests, is installed near the animal's head. It incorporates critical circuits for the ECoG and the electrochemical sensors, which are vulnerable to noise pickup under long sensor cables. Using the headstage minimizes the length of the sensor cable, and thus improves the signal-to-noise ratio (SNR). The front-end subsystem performs functions such as sensor driving, signal conditioning, digitization, and sensor data fusion. Modular system design enables easy assembly and sharing of a single ground bus. All the analog sensor signals are converted into digital ones. The digital signals from various channels are passed to a sequencer so that the 10-channel signals are converted into two serial channels. The serial channels are then sent out to the real-time hardware subsystem via two optical links. The usage of the optical link ensures the interconnection emits no EMI back to the sensitive front-end part. At the real-time hardware subsystem, we employ the Labview-based real-time controller to perform the function of signal processing, calibration, data management, etc. The real-time outcome is displayed on a monitor display mimicking a commercial health monitor. End users not only can see short-terms results but also the past history as well as the correlation among different channels.

Figure 3 shows the developed smart lab-on-a-tube probe and its interface signal conditioning circuits and monitor.

Findings

The working principles and key features of each developed microsensor on smart lab-on-a-tube are shown in Table 1. The pressure sensor was found to have an accuracy of 1 mmHg in the range from 0 to 50 mmHg. The temperature

Smart Lab-on-a-Tube Key Features

Parameter (Unit)	Working Principle	Range	Accuracy
Pressure (mmHg)	Polysilicon diaphragm based	0-50	1
Temperature (°C)	Resistance temperature detector	25-45	0.1
Flow (ml/100g-min)	Thermal diffusion – periodic heating	0-200	5
Oxygen (mmHg)	Electrochemical	0-160	1
Glucose (mM)	Electrochemical	0.03-10	0.05
Lactate (mM)	Electrochemical	0.02-8	0.1
ECoG (mV)	Depth electrode	0-200	0.01

Table 1: Key features for the developed microsensors.

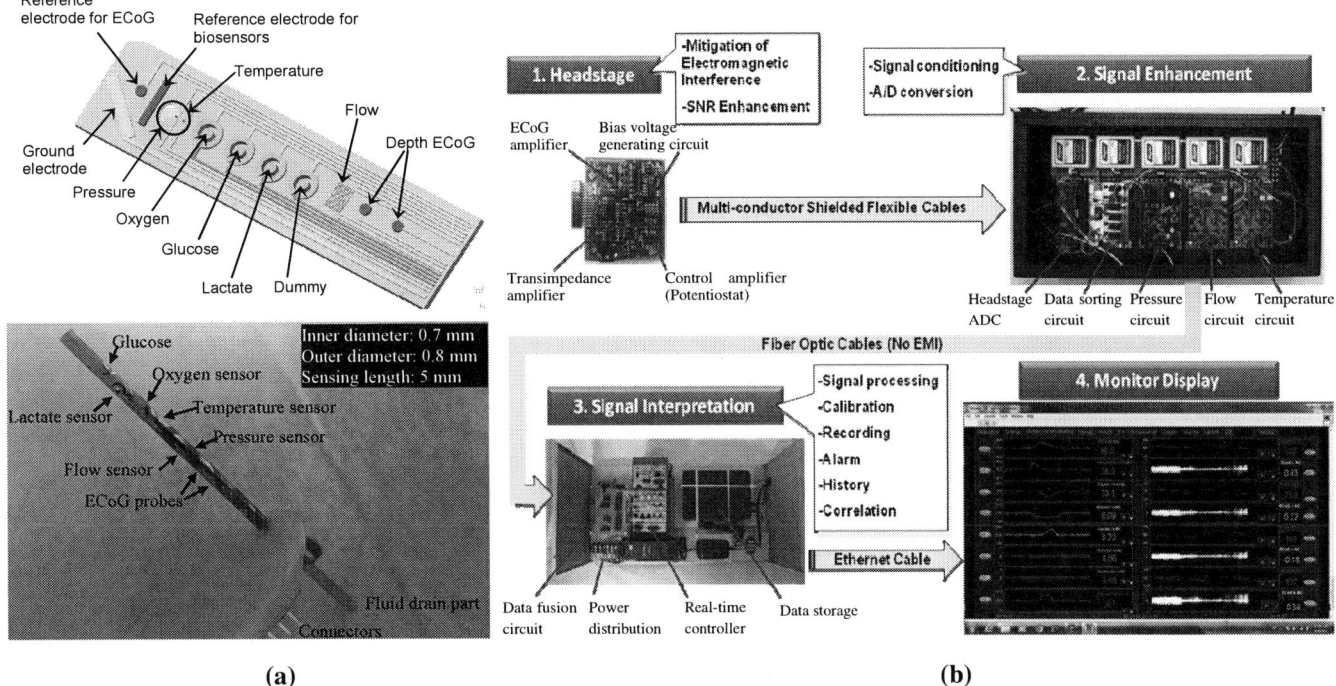

Figure 3: Developed Smart Lab-on-a-Tube (LOT) System: (a) Smart LOT prototype: 7 microsensors were embedded in six layers of PI2611 substrate and (b) Signal conditioning interface system: it consists of headstage front-end circuit, back-end signal enhancement circuit, signal interpretation unit and display.

sensor had a resolution of 0.013 °C and achieved an accuracy of 0.1 °C. The flow sensor had a resolution of 0.18 ml/100g/min and achieved an accuracy of 5 ml/100g/min. The oxygen sensor had an accuracy of 1 mmHg in the range from 0 to 60 mmHg. The glucose sensor had an accuracy of 0.02 mM in the linear range from 0.1 to 10 mM. The lactate sensor had an accuracy of 0.05 mM in the range from 0.05 to 8 mM. No crosstalk was detected among the sensors.

The biocompatibility of the smart catheter was examined by histopathological outcomes. The histological analysis confirms that the smart catheter is minimally toxic to the surrounding tissue. Mild to moderate gliosis in brain parenchyma around the tracts were seen in both the commercial catheter and smart catheter. Neither acute nor chronic inflammation was observed in the samples at day 3 and day 7 post implantation (Figure 4).

To assess the smart catheter, it was inserted into the parietal cortex of anesthetized rats that were subjected to permanent middle cerebral artery occlusion (MCAO) on the same side as the recording device (Figure 5(a)). During the 3 hours after MCA occlusion, spontaneous waves of depolarization appeared in the ischemic penumbra. The first episode appeared 210±13 seconds after occlusion. Spreading depression (SD) consisted of negative DC deflections, followed by positive waves (4.7±0.9 mV). After a delay of 183±11 seconds from the onset of the negative DC shift, the

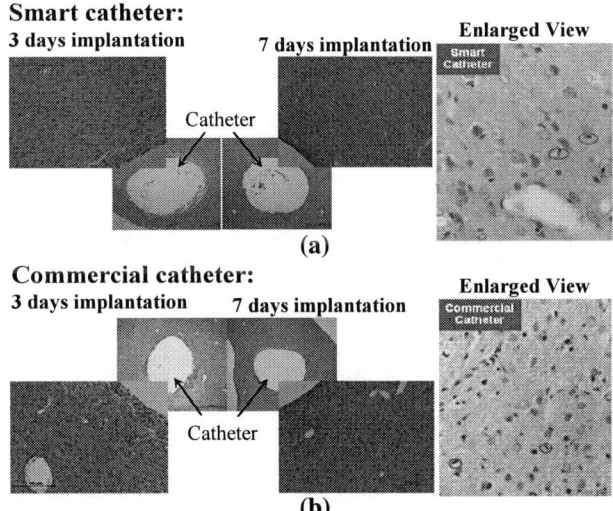

Figure 4: Histopathological examination of smart catheter and commercial catheters with 3 days and 7 days implantation (N=3): There is mild to moderate gliosis in brain parenchyma around tracts. No acute or chronic inflammation was seen.

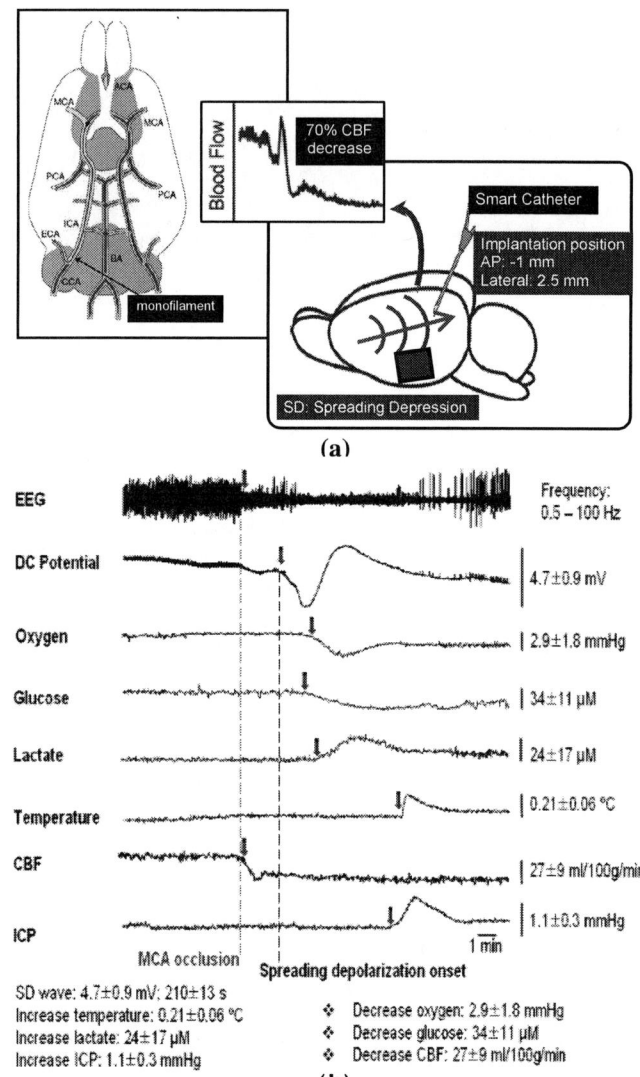

(a)

(b)

SD wave: 4.7±0.9 mV; 210±13 s
Increase temperature: 0.21±0.06 °C
Increase lactate: 24±17 μM
Increase ICP: 1.1±0.3 mmHg

❖ Decrease oxygen: 2.9±1.8 mmHg
❖ Decrease glucose: 34±11 μM
❖ Decrease CBF: 27±9 ml/100g/min

Figure 5: *In vivo* evaluation of smart lab-on-a-tube: (a) permanent middle cerebral artery occlusion (MCAO) model and (b) experimental results after the first episode of spreading depression (SD) appeared after occlusion.

smart catheter recorded increased brain lactate (24±17 μM), intracranial pressure (1.1±0.3mmHg) and temperature (0.21±13°C); and decreased glucose (34±11μM), cerebral blood flow (27±9ml/100g/min) and oxygen (2.9±1.8mmHg).

Conclusions

With a single device, we found changes in cerebral glucose, lactate, oxygen, temperature, local cerebral blood flow and intracranial pressure that correlated with spreading depression. These results demonstrate that the smart LOT is capable of simultaneous and continuous measurement of multiple brain variables, within the pathophysiology ranges observed in

brain injury. The smart catheter will advance the field of neuromonitoring into a completely new era, in which medical decisions will be based on extensive, real-time measures of brain chemistry and physiology during the critical period immediately following a brain injury, when the brain is most vulnerable to secondary insults. Such monitoring has the potential to substantially improve the care of critically brain-injured soldiers and to improve our understanding of brain pathophysiology.

References

(1) V. G. Coronado, L. Xu, S. V. Basavaraju, L. C. McGuire, M. M. Wald, M. D. Faul, B. R. Guzman, and J. D. Hemphill, "Surveillance for Traumatic Brain Injury—Related Deaths --- United States, 1997—2007," *Morbidity and mortality weekly report*, 60(5), pp. 1-32, 2011.

(2) M. W. Greve, and B. J. Zink, "Pathophysiology of traumatic brain injury," *Mt Sinai J Med*, 76, pp. 97-104, 2009.

(3) S. Cecil, P. M. Chen, S. E. Callaway, S. M. Rowland, D. E. Adler, and J. W. Chen, "Traumatic brain injury: advanced multimodal neuromonitoring from theory to clinical practice," *Crit Care Nurse*, 31(2), pp. 25-36, 2011.

(4) C. Li, P. M. Wu, W. Jung, C. H. Ahn, L. A. Shutter, and R. K. Narayan, " A novel lab-on-a-tube for multimodality neuromonitoring of patients with traumatic brain injury (TBI)," *Lab Chip*, 9, pp. 1988-1990, 2009.

(5) C. Li, P. M. Wu, S. J. Prince, Z. Wu, S. Chakraborti, C. H. Ahn, J. A. Hartings, and R. K. Narayan, "Multifunctional lab-on-a-tube (LOT) provbe for simultaneous neurochemical and electrophysiological activity measurements," *The 17th International Conference on Solid-State Sensors, Actuators and Microsystems*, pp. 880-883, 2013.

(6) C. Li, P. M. Wu, J. A. Hartings, Z. Wu, C. H. Ahn, D. LeDoux, L. A. Shutter, and R. K. Narayan, "Smart catheter flow sensor for real-time continuous regional cerebral blood flow monitoring," *Appl Phys Lett*, 99, pp. 233705-1-233705-4, 2011.

(7) Z. Wu, C. Li, N. Bhattacharjee, J. A. Hartings, R. K. Narayan, and C. H. Ahn, "A new intracranial pressure sensor on polyimide lab-on-a-tube using exchanged polysilicon piezoresistors," *The 17th International Conference on Solid-State Sensors, Actuators and Microsystems*, pp. 1779-1782, 2013.

(8) C. Li, P. M. Wu, J. A. Hartings, Z. Wu, C. Cheyuo, P. Wang, D. LeDoux, L. A. Shutter, B. R. Ramaswamy, C. H. Ahn, and R. K. Narayan, "Micromachined lab-on-a-tube sensors for simultaneous brain temperature and cerebral blood flow measurements," *Biomed Microdevices*, 14(4), pp. 759-768, 2012.

Invited

Small Soft Safe Micromachines for Biomedical Applications

Satoshi KONISHI

College of Science and Engineering, BioMedical Devices research Center,

Ritsumeikan University

Kusatsu, Shiga, JAPAN

Abstract

This paper introduces small, soft, and safe (S^3) micromachine and its biomedical applications. MEMS technology has made remarkable development of tiny machines based on LSI technology. Micromachine and MEMS are small in nature. We have developed all polymer pneumatic balloon actuator (PBA) as S^3 actuator. PBA employed pneumatic driving principle. S^3 micromachine is based on its soft and flexible structure and safe driving principle. Polymers such as polyimide and polydimethylsiloxane (PDMS) allow soft and flexible structure. S^3 micromachines are expected to allow minimally invasive medicine. Especially, pneumatic driving method can provide safe operation for medical application. The PBA-based medical tool was applied to a retractor for spacing in front of the endoscope at our early stage. This talk also presents the PBA as a surgical tool for retinal pigment epithelium transplantation. As our recent works on S^3 micromachine, further possibilities will be presented.

Introduction

Micromachines or MEMS technology can provide interfaces with various targets in bio and medical field. Various micro fluidic chips have been developed in the field of uTAS (Micro Total Analysis Systems). uTAS chips uses polymers and glass while MEMS have used Si as primal material because of their technology of origin. Si based MEMS allows an integration of electronics into MEMS. On the other hand, polymer chips provide advantages such as transparency, disposability and simple manufacturing. It is possible to provide fluidic channels by lithography or in-printing of polymers. In addition, soft structure and flexible motion have become important to handle living organisms in biotechnology and medical applications. Polymers are expected to satisfy these requirements. Micromachines are expected to provide precise tools suitable for manipulation of tiny or fragile living organisms in biotechnology and advanced medical tools for minimally invasive medical operations. Medical tools are studied as rather large micromachines for biomedical engineering. Minimally invasive surgery such as an endoscopic surgery requires a

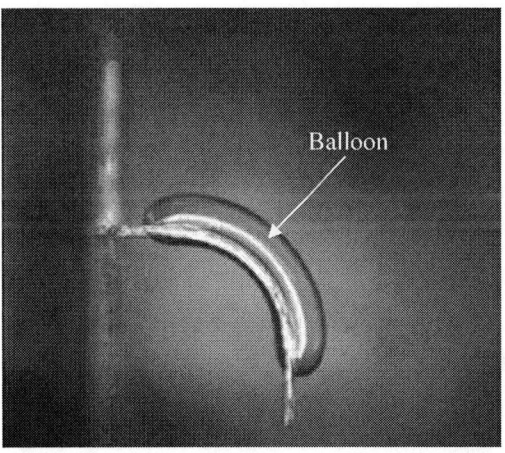

Figure 1 Bending motion of all PDMS PBA. Typical size of PBA is : several mm long, a few mm wide, and a few hundreds μm thick. All PDMS PBA bends due to a tensile stress generated by a swelled balloon.

978-1-4799-8002-4/14 $31.00 © 2014 IEEE

complicated operation in a limited space under indirect view through an endoscope.

This paper introduces small, soft, and safe (S^3) micromachines from our research activities on biomedical micromachines. Micromachines are small in nature. We have developed pneumatic balloon actuator (hereafter PBA) as S^3 actuator (see Fig. 1)[1]. All PDMS PBA was reported in 2005[2]. Bending motion of all PDMS PBA is shown in Fig. 1. Typical size of PBA is: several mm long, a few mm wide, and a few hundreds μm thick. All PDMS PBA is composed of two PDMS layers so as to form cavities as balloons and pneumatic supply channels. PBA can provide out-of-plane bending motion due to a tensile stress generated by a swelled balloon.

The PBA have been applied to various biomedical applications including a retractor for endoscopic surgery, a transplantation tool for eye surgery, and a cellular aggregates manipulator for tissue engineering and so on. This paper describes expanding application of S^3 micromachines in biomedical fields.

S^3 micromachines

S^3 micromachines are featured by their soft and flexible structure and safe driving principle. Polymers such as polyimide and polydimethylsiloxane (PDMS) allow soft and flexible structure. Pneumatic driving method can provide safe operation especially for medical application. PBA can be used as an elemental artificial muscle. The bending type of PBA can provide out-of-plane bending motion due to a tensile stress generated by a swelled balloon. We have developed all-PDMS PBA as the newest generation of PBA. Figure 1 shows a typical bending motion of all-PDMS PBA. We have demonstrated a micro hand composed of five micro fingers with eleven degrees of freedom. We could succeed in demonstrating master-slave operation of the micro hand (see Fig. 2). Examples among our research activities of medical applications will be presented in this paper.

Figure 2 Silicone microhand composed of five micro fingers. Individual fingers are driven by PBA. The operator can control the microhand through the master-slave robot system. The micro hand can grasp and manipulate tiny objects by using its eleven DOF motions.

Biomedical Applications

Biomedical applications of S^3 micromachines are introduced; retractors for spacing in front of the endoscope (see Fig. 3) [3], a surgical tool for the retinal pigment epithelium transplantation (see Fig. 4)[4]. Furthermore, very recent results on an implantable pneumatically actuated stimulating device for in vivo pressure-mediated transfection [5] and manipulating of cellular aggregate by tiny micro fingers [6] are presented on site.

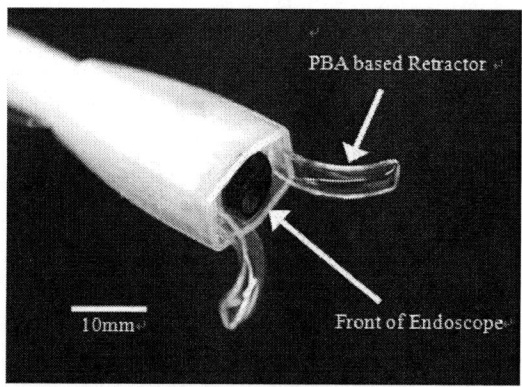

Figure 3 A photograph of developed retractor for the endoscope. The PBA-based retractor is mounted on the head of the endoscope. In the abdominal cavity, pneumatically driven PBA-based retractor pushes away opposing organ.

As one of minimally invasive medicines, we can name laparoscopic surgery. The laparoscopic surgery requires the endoscope which is has been downsized continuously. Many efforts have been made to improve endoscopic tools such as forceps and optical scope. It becomes also important to create a space between a head of an endoscope and objective organs in laparoscopic surgery. We proposed a retractor using PBA to push organs aside [3]. A photograph of developed retractors for the endoscope is shown in Fig. 3. The retractors with the bending PBAs are mounted on the head of the endoscope. The retractor mad of PDMS is flexible enough not to obstruct insertion of the endoscope into the abdominal cavity. In the cavity, the retractor pushes the organ ahead of the endoscope so as to create a space for the medical diagnosis and operation.

We have developed the retractor and presented a novel surgical tool for ESD (Endoscopic Submucosal Dissection) [4]. ESD has recently been established and appears to be a more reliable endoscopic therapy than conventional method. Conventional method enables the resection of affected mucosa, however, it requires more endoscopic skill. We proposed ESD retractor using S^3 micromachine to improve ESD operation. The proposed ESD retractor enables to resect the mucosa under observing the submucosa by holding the resected mucosa to the side. Both insertion and lift operation of the mucosa were carried out by two DOF motion.

We have also proposed a transplantation tool of the retinal pigment epithelium (RPE) sheet (see Fig. 4). The RPE transplantation has been attempted to repair the damaged RPE [3]. It is not easy to carry and transplant the RPE sheet to the bottom of an eye. The RPE sheet transplantation requires minimally invasive operation. The injection method of the RPE sheet with physiological saline by 3-port vitrectmy is currently used while it is hard to manipulate the sheet in the eyeball. PBA was used to fold the tool in a cylindrical shape and open the folded tool in an eye. Figure 6 shows a head of a developed tool. The tool in the photograph is the normal state without pressurization. Head size of tool was designed 3mm × 3mm × 100μm in consideration of the RPE sheet size. Many

PBAs were set in array to roll up the flat structure. The tool also had the bending PBA at its wrist. Generated force by the PBA was measured 3mN. Both in vitro and in vivo test by using the developed tool have been executed successfully.

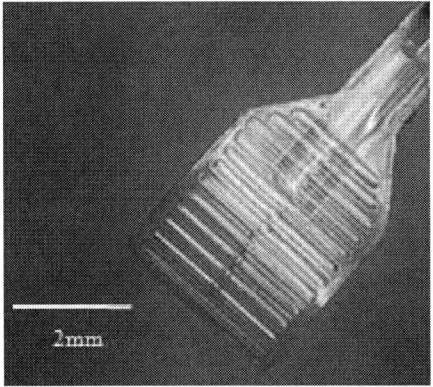

Figure 4 A surgical tool for the retinal pigment epithelium sheet transplantation. The designed head size of tool was 3mm × 3mm × 100μm. The head is retracted into a cylindrical shape with the cell sheet so as to be accommodated into a hollow needle.

Summary

This paper presented pneumatic balloon actuator (PBA) as a small, soft, and safe (S^3) micromachine. PBA is soft and flexible and is driven by pneumatic pressure. All polydimethylsiloxane (PDMS) PBA was mainly discussed in this paper. Various biomedical applications of PBA were reported. First, the retractor for spacing in front of the endoscope was presented. The retractor for ESD was also demonstrated to accomplish more complicated operation. Next, a surgical tool for the retinal pigment epithelium transplantation was reported. The 3mm × 3mm transplantation tool could retract into a cylindrical shape. In addition to these applications, we developed unique applications for biomedical use. Recently, the pneumatically actuated stimulating device for in vivo pressure-mediated transfection and the manipulator of cellular aggregate by tiny micro fingers have been developed.

Acknowledgements

This work was supported by Grant-in-Aid for Scientific Research (A) : JSPS KAKENHI Grant Number 24240075.

References

(1) S. Konishi, F. Kawai and P. Cusin : "Thin Flexible End-Effector Using Pneumatic Balloon Actuator", Sensors and Actuators A: Physical, Vol. 89, pp. 28-35

(2) O. C. Jeong and S. Konishi, IEEE/ASME Journal of Microelectromechanical Systems, Vol. 15, No. 4, pp. 896-902, 2006.

(3) S. Konishi, "Small, Soft and Safe Actuator by Using MEMS-Based Pneumatic Balloon Actuator", APCOT Conf., p.144, Jun., 2006.

(4) Y. Watanabe, M. Maeda, T. Okano, S. Konishi, et al., IEEE MEMS '07, pp. 659-662, Jan, 2007.

(5) Shimizu, K., Kawakami, S., Hashida, M., Konishi, S. et al., Journal of Controlled Release Societyvol. 159,(1)85-91, 2012.

(6) S. Shimomura, S. Konishi et al., *IEEE MEMS'14*, pp.925-926, Jan, 2014

Bio-integrated Systems with Stretchable Designs for Skin-Mounted Wearable Health Monitoring

Milan Raj, Pinghung Wei, Shyamal Patel, Xianyan Wang, Bryan McGrane, Lauren Klinker, Paolo DePetrillo, Roozbeh Ghaffari

Research and Development, MC10 Inc. Cambridge, MA, 02140
Tel: (617) 234-4448, Email: rghaffari@mc10inc.com

Abstract

We present ultrathin, flexible and stretchable epidermal health monitoring devices that contain physiological sensors, an analog front end processor, a core microprocessor, flash memory module, rechargeable battery, and a wireless communication (Bluetooth low energy) module. The system is mechanically matched to the Young's modulus of human skin and designed to record heart rate, electromyography signals, activity, respiration and sleep quality to facilitate continuous (electro-)physiological data recording outside of the hospital setting.

Introduction

Advances in the semiconductor industry have driven important breakthroughs in implantable and non-invasive medical devices, like cochlear implants, pacemakers, implantable cardioverter defibrillators and wearable health monitoring devices. However, there are significant limitations inherent in all standard forms of rigid electronics that comprise these existing classes of devices, in the way they interface with soft biological tissue. These geometrical and mechanical constraints impose unique integration and therapeutic delivery challenges. This paper presents new designs for skin-based systems that incorporate physiological sensors, system on chip, memory, and rechargeable battery configured in stretchable formats at the system level. Quantitative analyses of electromechanical performance and wireless data transmission under mechanical stress highlight the clinical utility of these systems in tracking patients with neuromuscular movement disorders and heart disease. As demonstrations of this technology, we present representative examples of bio-integrated systems that highlight functionality and performance coupled with extreme mechanical flexibility [1-3].

Epidermal Electronics System

Figure 1 shows the system architecture for the epidermal electronics module. The system is fully functional and

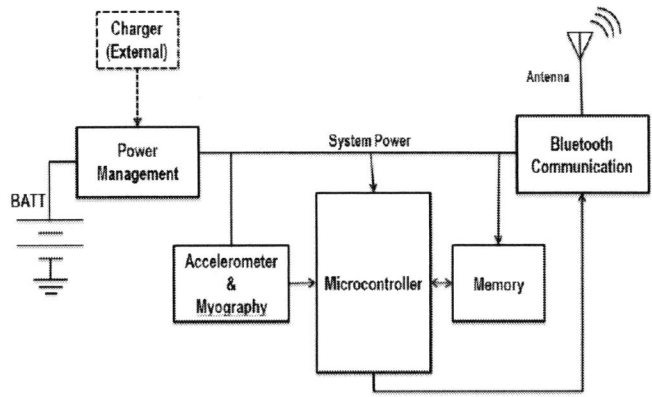

Figure 1— System architecture of epidermal wearable health monitoring device. The device has an onboard rechargeable battery, memory, wireless communication (Bluetooth low energy) and microcontroller, accelerometer, and analog front end coupled to ultrathin dry electrodes.

rechargeable using both wired and wireless power sources. A schematic drawing of a representative epidermal electronics module illustrated in Fig. 2, highlights the stretchable design which consists of interconnects, integrated dry electrodes and adhesive to facilitate skin coupling. The epidermal wearable device can be attached to multiple locations on the human body for localized sensing relevant to specific parts of the anatomy (e.g. resting tremor in the hand).

This fully functional system collects sensor data from multiple locations on the body through direct skin coupling with thin adhesive layers. The mechanical architecture consists of a network of metal interconnects that behave like a matrix of springs. The interconnect structures accommodate strains, and thereby create a highly flexible and stretchable wearable system. Finite element analysis provides support for a high level of flexibility and stretchability, with system-level strains exceeding 20% (Fig. 4). Individual interconnects buckle, twist, and stretch in response to external deformations imposed on the entire system. To prevent damage to individual electronic components, the electronics and sensors are encapsulated with thin low modulus silicone layers on the top and bottom surfaces.

978-1-4799-8002-4/14 $31.00 © 2014 IEEE

Surface-Electromyography and Movement Sensing

A. Surface-electromyography (s-EMG) and movement tracking

S-EMG and movement signals are captured at multiple body locations and stored in flash memory or streamed via Bluetooth low energy through an antenna (for short range data transfer and communication). A 16-bit low-power micro-controller conditions streams of signals from the 3-axis accelerometer and passive electrodes connected to a single channel analog front end [4-5]. The sensor data is high pass filtered (<10 mHz) to remove the DC offset noise from the high-impedance electrodes. The analog to digital output from the sensors is then processed and sampled by the microcontroller, which in turn transmits the data via the Bluetooth communication module or relays the data into the flash memory module (32 MB capacity). The system can operate for about ~12-18 hrs using milliwatts of power in raw data streaming mode. Onboard algorithms reduce the amount of continuous data transmission by up to a factor of 3, thereby enabling multi-day use.

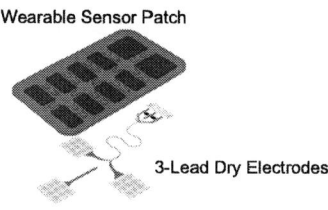

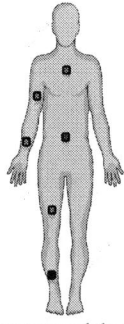

Figure 2— (Top) Wearable sensor module containing dry electrodes and accelerometer sensor in ultrathin, flexible and stretchable form factors. (Bottom) Wearable modules laminate on multiple body locations to track movements and biopotentials.

B. S-EMG and Movement of Limb Muscle Groups

Figure 5 shows EMG measurements from the anterior tibialis muscle in response to vertical movements of the foot. Ultra-thin dry electrodes attached to skin can withstand twisting and stretching while remaining in direct contact with skin. We capture both movement and s-EMG concurrently during prescribed muscle activity. These measurements are comparable in signal to noise ratio (SNR) to the performance of commercial electrodes (Ag/AgCl) electrodes.

C. Comparison to Commercial Systems

Figure 6 shows a comparative analysis of commercial conductive gel encapsulated electrodes versus the ultrathin dry electrodes presented in this study. Both electrodes were located on the bicep muscle. The subject was asked to repeatedly lift flex their bicep muscle by lifting a 1 lb. and 5 lb. weight. The results for commercial and ultrathin electrodes have comparable SNR.

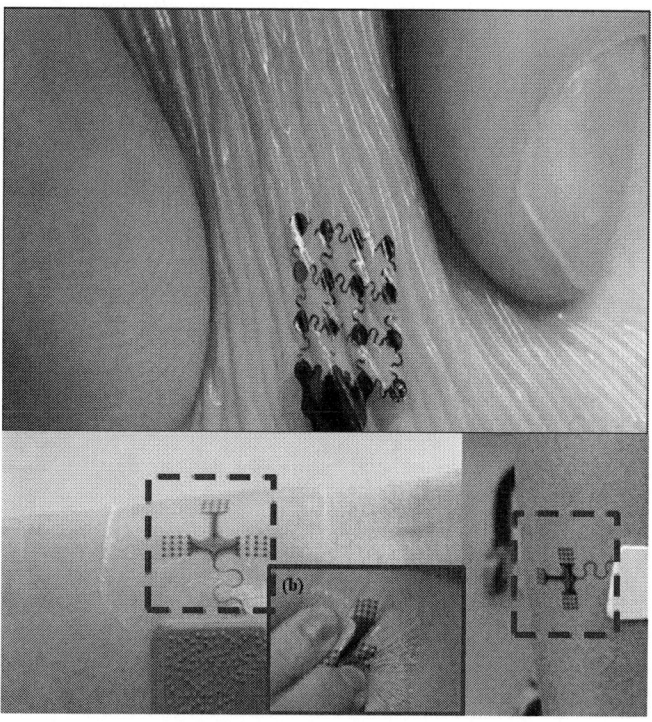

Figure 3 – Ultrathin dry electrode samples laminated on human skin.

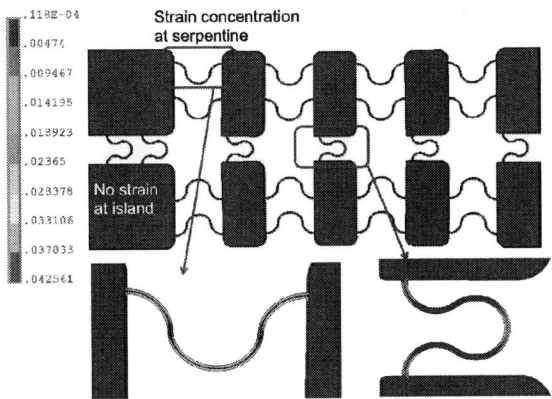

Figure 4 – Finite element analysis of stretchable electronic system mechanical performance in response to tensile stresses. Modeling results show most of the strain is isolated to the serpentine interconnects region.

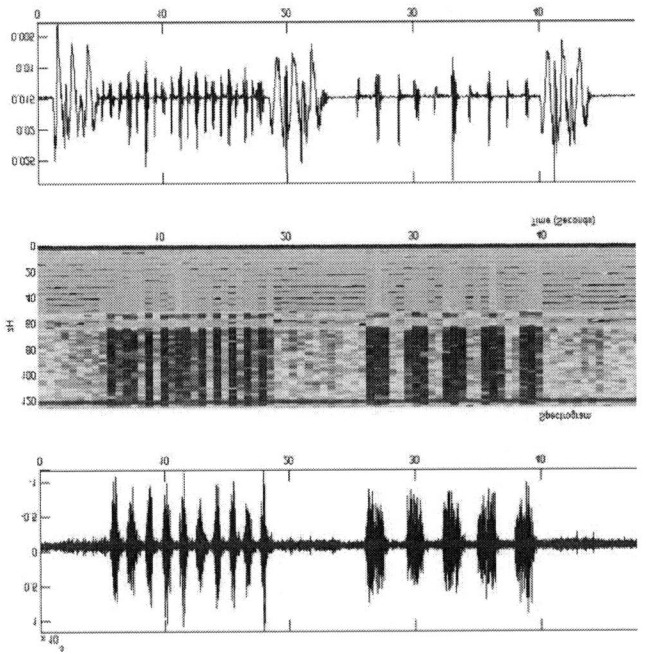

Figure 5 – Accelerometer signals s-EMG spectrogram and s-EMG time waveforms collected from the tibialis anterior muscle.

Conclusions

Our results demonstrate the capabilities of soft, wearable epidermal systems to perform continuous monitoring via direct skin coupling. Functional prototypes of the kind used to measure movement and EMG signals are ideally suited for tracking biometrics from multiple locations on the body in patients with neuromuscular movement diseases. Epidermal flexible/stretchable electronics systems have the potential to provide quantitative feedback to patients and physicians in a continuous manner with limited mechanical load or discomfort on the patient. These results have implications for body-integrated electronics for use in clinical trials and in home-based health monitoring applications.

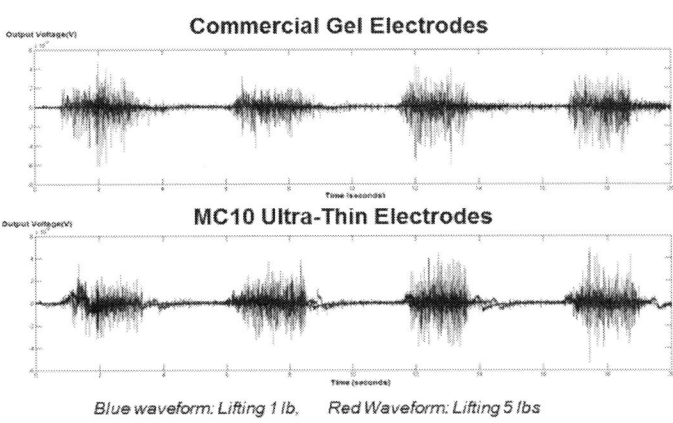

Blue waveform: Lifting 1 lb,   Red Waveform: Lifting 5 lbs

Figure 6 – s-EMG time waveforms using commercial electrodes and ultrathin dry electrodes in response to bicep muscle contractions caused by lifting 1 lb. and 5 lb weights.

References

[1] Rogers, J. A., Someya, T. and Huang, Y., "Materials and Mechanics for Stretchable Electronics," Science 327 pp. 1603-1607 (2010).

[2] Jiang, H., Khang, D.-Y., Song, J., Sun, Y., Huang, Y. and Rogers, J.A., "Finite Deformation Mechanics in Buckled Thin Films on Compliant Supports," Proceedings of the National Academy of Sciences USA 104(40), 15607–15612 (2007).

[3] Kim, D.H., Ahn, J.H., Choi, W.M., Kim, H.S. Kim, T.H., Song, J., Huang, Y.Y., Liu, Z., Lu, C., Rogers, J.A., Stretchable and foldable silicon integrated circuits. Science, Vol. 320, pp. 507-511, (2008).

[4] Hsu, Y. Y., Hoffman, J., Wei, P. H., Klinker, L., Morey, B., Elolampi, B. Davis, D., Rafferty, C., and Dowling, K., "Epidermal electronics: skin sweat patch, " 7th International Microsystems, Packaging, Assembly Conference Taiwan (IMPACT/IEEE), Oct. 24-26, Taipei (2012).

[5] Hsu, Y. Y., Wei, P. H., Elolampi, B., Davis, D., Rafferty, C., and Dowling, K., "A Novel Strain Relief Structure for Stretchable Interconnects, " ASME/IMECE, Nov. 9-15, Houston, TX (2012).

Electrical Characterization of FinFETs with Fins Formed by Directed Self Assembly at 29 nm Fin Pitch Using a Self-Aligned Fin Customization Scheme

Hsinyu Tsai, Hiroyuki Miyazoe, Josephine B. Chang, Jed Pitera, Chi-Chun Liu, Markus Brink, Isaac Lauer, Joy Y. Cheng, Sebastian Engelmann, John Rozen, James J. Bucchignano, David P. Klaus, Simon Dawes, Lynne Gignac, Chris Breslin, Eric A. Joseph, Daniel P. Sanders, Matthew E. Colburn and Michael A. Guillorn

IBM Watson Research Center, 1101 Kitchawan Road, Yorktown Heights, NY 10598, USA
IBM Albany Nanotech Research Center, 257 Fuller Road, Albany, NY 12203, USA
IBM Research Almaden, 650 Harry Road, San Jose, CA 95120, USA
Tel: (914) 945-3622, Email: htsai@us.ibm.com

Abstract

In this work, we report electrical characterization FinFET devices with 29nm-pitch fins patterned using a technique called tone inverted grapho-epitaxy (TIGER). We use a topographic template to direct the self-assembly of block copolymers (BCP) to form small area gratings that are self-aligned to the template. After a tone-inversion operation, blocks of defect free SOI fins bounded by self-aligned exclude regions are formed with the spacing determined by the template line width (LW). This self-aligned customization enables further definition of the active region for FinFETs. Process window and design implications for directed self-assembly (DSA) with TIGER are also discussed.

Introduction

Fin formation is one of the most critical modules in the FinFET device fabrication flow. Given that a typical device is composed of an ensemble of fins, each fin must be nearly identical to avoid performance degradation arising from geometric variation [1]. Techniques for fin patterning must demonstrate the ability to form fins with a high degree of structural precision. Directed self-assembly (DSA) of block copolymers (BCP) is an attractive option to extend 193nm immersion (193i) lithography beyond the 10 nm node. Recent work [2] demonstrated the application of lamellar phase DSA to fabricate fins in electrically testable CMOS FinFET devices using a 42nm-pitch polystyrene-poly(methyl methacrylate) (PS-PMMA) DSA process. The process featured in [2] formed fins in large, grating-like arrays to minimize process variation effects. An additional lithographically aligned exposure followed by an etching process was used to customize the arrays into the discrete groups of active fins. Scaling of the fin pitch below 42nm can be achieved by a BCP with a smaller natural period, λ [3]. However, customization of denser features will require improved overlay accuracy. Recent data for a 28nm-pitch DSA fin patterning process experimentally determined that an overlay of 3nm would be required to accurately customize large fin arrays [4]. 3nm overlay is beyond the capability of state of the art lithography equipment. Next generation 193i scanners may provide suitable overlay performance for this task. However, delivering this level of accuracy in production may prove challenging. In this paper, we discuss a DSA fin patterning process using grapho-epitaxy that incorporates a self-aligned customization strategy. We demonstrate the utility of this approach by showing arrays of functioning FinFETs with uniform groups of SOI fins.

Directed Self-Assembly for Fin Patterning

In grapho-epitaxy DSA, a topographic template pattern is created on a chemically neutral surface. Confinement of the BCP between the sidewalls of the template provides an ordering force that drives the pattern into registry with the surface topography. The gap between two template lines required to form a given number of self-assembled polymer phases in a symmetric lamellar phase BCP can be calculated using a simple analytic model of the free energy per copolymer chain [5]. For reference, the equation describing this phenomenon is given in Fig. 1a. Favorable template gap configurations are plotted in Fig. 1b. Applying a threshold function to this data gives approximate ranges for

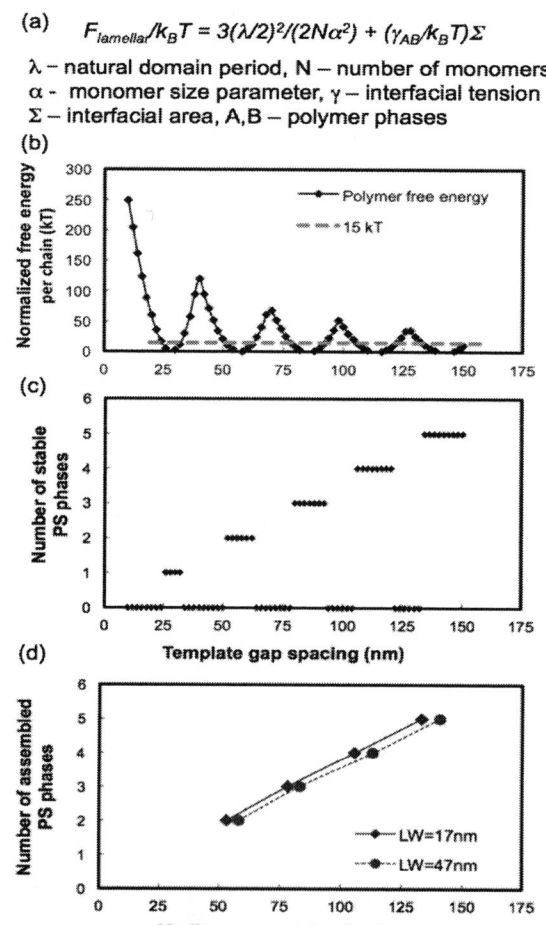

(a) $F_{lamellar}/k_BT = 3(\lambda/2)^2/(2N\alpha^2) + (\gamma_{AB}/k_BT)\Sigma$

λ – natural domain period, N – number of monomers
α – monomer size parameter, γ – interfacial tension
Σ – interfacial area, A,B – polymer phases

Figure 1: (a) analytic model of the free energy per copolymer chain of a symmetric lamellar phase BCP as discussed in [5]. (b) Free energy of a PS-b-PMMA BCP with a natural length, λ, of 29nm confined within a gap of varying size. (c) Thresholding of gap sizes within a 15 k_BT range plotted as a function of the number of stable self-assembled PS phases (d) Experimentally determined median gap spacing for DSA of 2 to 5 lines as a function of template line width (LW).

978-1-4799-8002-4/14 $31.00 © 2014 IEEE

favorable gap spacing conducive to defect free assembly of line space patterns Fig. 1c. Test patterns based on this model were fabricated using electron beam lithography and hydrogen silsesquioxane (HSQ) resist imaged on a neutral polymer surface. Two different line widths (LW), ~17 and ~49 nm, were used for the template pattern with spaces between the template lines varying in 5 nm increments from 50 to 300nm. DSA was performed using a PS-b-PMMA BCP with $\lambda = 29$nm. The results of this study are shown in Fig. 1d. The data demonstrated that template LW played an unexpected role in establishing the process window for the template pattern. In particular, narrower template lines displayed a larger process window over a wider range of template gap spacing. 3D modeling of the DSA process using Monte Carlo techniques [6] was performed to illuminate this phenomenon (Fig. 2). These results illustrate that while template gap spacing is a critical parameter in this process, the BCP, template LW, template height and BCP thickness must also be taken into account. Optimizing these parameters leads to a regime that yields uniform DSA images that are tolerant to a wide range of variations in template gap spacing.

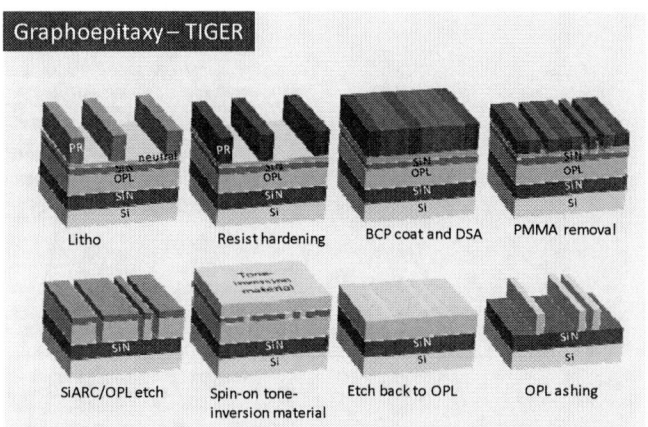

Figure 3: Process flow for forming groups of silicon on insulator (SOI) fins using DSA. A tone inversion operation is applied after etch transferring the PMMA phase of the DSA pattern. The wrapping of the BCP over the template pattern prevents the partial PMMA phases adjacent to the template from participating in image formation during the selective PMMA removal. The tone inversion material allows this image to be used to transfer the image into the hard mask used to define the SOI fins.

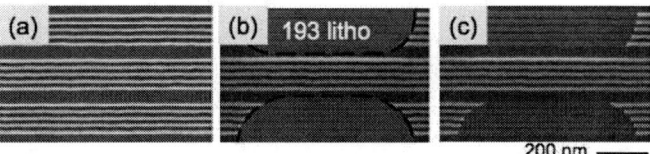

Figure 4: (a) 3 groups of 5 fins formed by the TIGER process (b) illustration of lithographically aligned 193 nm customization exposure (c) single group of fins defined after the customization process. The light gray texture flanking the center group of fins is a result of box texturing.

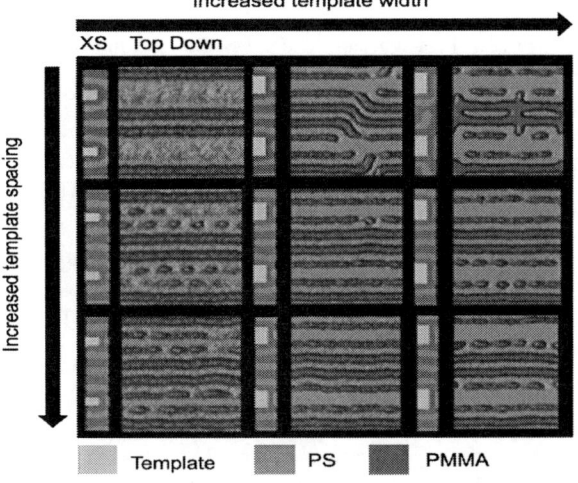

Figure 2: Monte Carlo simulations of the self-assembly process using the single-chain-in-mean-field method described in [6]. Cross section (XS) as well as top down results are shown. Thin templates display defect free phase formation over the displayed range of gap sizes. Wider template lines require a larger template gap to form a defect free phase. This agrees well with the data

In a previous work [7], we demonstrated the ability to form fin patterns from the PMMA phase of DSA patterns using a tone inverted grapho-epitaxy resolution enhancement technique (TIGER, Fig. 3). Use of TIGER combined with the template engineering techniques described above lead to the formation of fin patterns with blocks of defect free SOI fins bounded by self aligned exclude regions (Fig. 4a). The latter feature facilitates circuit patterning by increasing the overlay margin for additional customizing exposures (Fig. 4b-c). An example SRAM pattern formed by the TIGER process is shown in Fig. 5, where the critical cuts were formed by the self-aligned template pattern and the less critical cuts formed with a separate lithography and etch step. Fig. 5a illustrates the type of defects that can occur with a

unoptimized template design. The use of wider template lines simplified the cut mask at a cost of process window reduction. In contrast, Fig. 5c shows defect free DSA formation with a more uniform and narrower template design. This template style is more tolerant to defects arising from pattern density variation.

FinFET Fabrication and Electrical Characterization

Fin width variation can have a detrimental impact on the sub-threshold performance of FinFET devices [8]. There are a number of possible sources of fin width variation. However, these problems are usually pronounced in "cut first" integration schemes where the fins are customized before final transfer into the Si or SOI substrate [4]. TEM images demonstrating the uniformity of fins produced using the TIGER process are shown in Fig. 6. The average fin width in Fig. 6a, where every fin is an end fin, is 8.5nm, and the average fin width in Fig. 6b is only 7.5nm. Nevertheless, since the 1 sigma standard deviation for fin widths is 1.0nm, overall variation is dominated by line width roughness (LWR). Statistical data from SEM measurements on TIGER patterned fins is shown in Fig. 7. From this data, we find that line edge roughness (LER) and LWR are the primary sources of variation. That is to say, the end fins do not look abnormally less uniform than the nested fins in a given group.

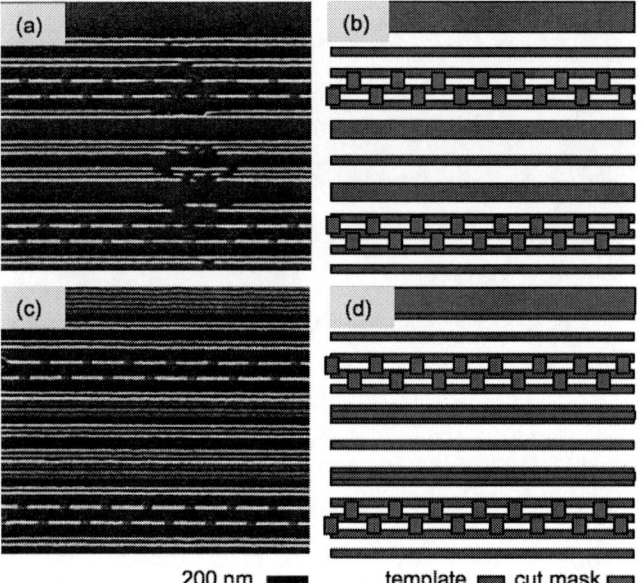

200 nm ▬ template ▬ cut mask ▬

Figure 5: SRAM patterns formed by the TIGER process after additional customization. (a) Results of pattern formed using a combination of template line widths as shown in (b). (c) Pattern formed using only thin template lines as shown in (d). The use of different template line widths in (a) gives rise to different BCP coating behavior over complex patterns. This results in underfilling and overfilling of the BCP leading to defects (broken or missing lines). In contrast, BCP coating over the thin template pattern results in defect free image formation after DSA and tone inversion.

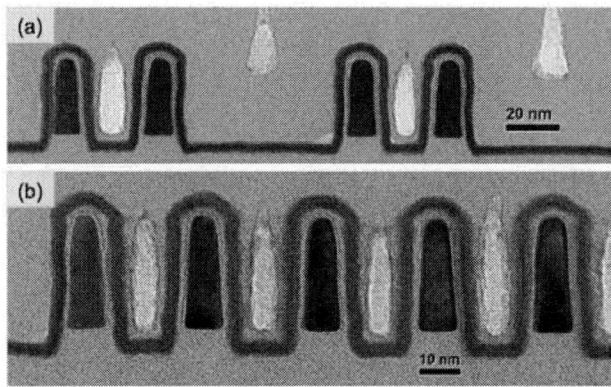

Figure 6: (a) TEM cross sectional image of groups of 2 fins formed by the TIGER process. (b) TEM cross sectional image of a larger group of fin. The dimensions of the fins are well controlled across the group.

Fin type	CD(nm)	LER(nm)	LWR(nm)
edge	15.73	2.09	3.30
nested	14.13	2.04	3.06

Figure 7: critical dimension (CD), line edge roughness (LER) and line width roughness (LWR) measured from SEM images of 5-fin-groups produced using the TIGER process. The CD of the fin is a relative value and should not be construed as the actual physical dimension. The variation in CD of the edge fins compared to the nested fins is contained within the LWR value. While the LWR is higher than desired, a number of process parameters can be tuned to reduce it including improved tone inversion materials and RIE optimization.

To further validate the uniformity of fins produced using the TIGER process, we fabricated NFET FinFET devices using a gate first high-K metal gate integration scheme (Fig. 8). Fig. 9 shows the transfer characteristic of NFET devices with 5 and 10 DSA patterned fins. Sub-threshold slope (STS) and DIBL degradation associated with gross fin width variation was not observed for either groups of 5 or 10 fins [1]. Plots of Id_{SAT} vs. Vt_{SAT} (Fig. 10a-b) and $Ioff_{SAT}$ vs. Vt_{SAT} (Fig. 10c-d) show linear trends demonstrating proper electrostatic behavior and no off-current degradation due to wide end fins. The flat response of DIBL and STS vs. number of fin (Nfin) per device (Fig. 10e-f) further confirms that end

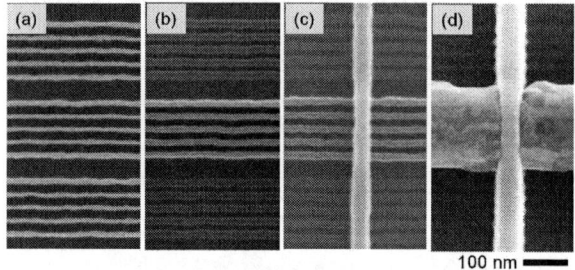

100 nm ▬

Figure 8: fabrication of FinFET devices using the TIGER process showing fins before customization (a) after customization (b) after gate patterning (c) and after spacer formation and epitaxial Si growth. A gate first high-K/metal gate integration scheme was used in this work. The Si epitaxy process includes dopant incorporation of phosphorous during the growth circumventing the need for further ion implantation. The device are completed by performing an activation anneal using a rapid thermal process and forming a Ni-based silicide.

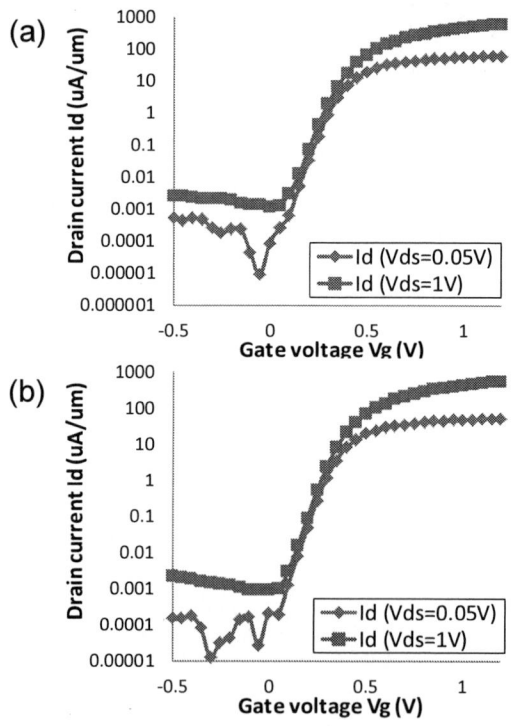

Figure 9: Transfer characteristic of two NFET devices with (a) 5 fins and (b) 10 fins with L_G = 35nm. With a modified definition (reducing the constant current definition from 300nA to 100nA) of Vt_{SAT}, DIBL is 30.5 and 32.6 mV/V; STS_{SAT} is 62 and 66 mV/dec, respectively.

978-1-4799-8002-4/14 $31.00 © 2014 IEEE

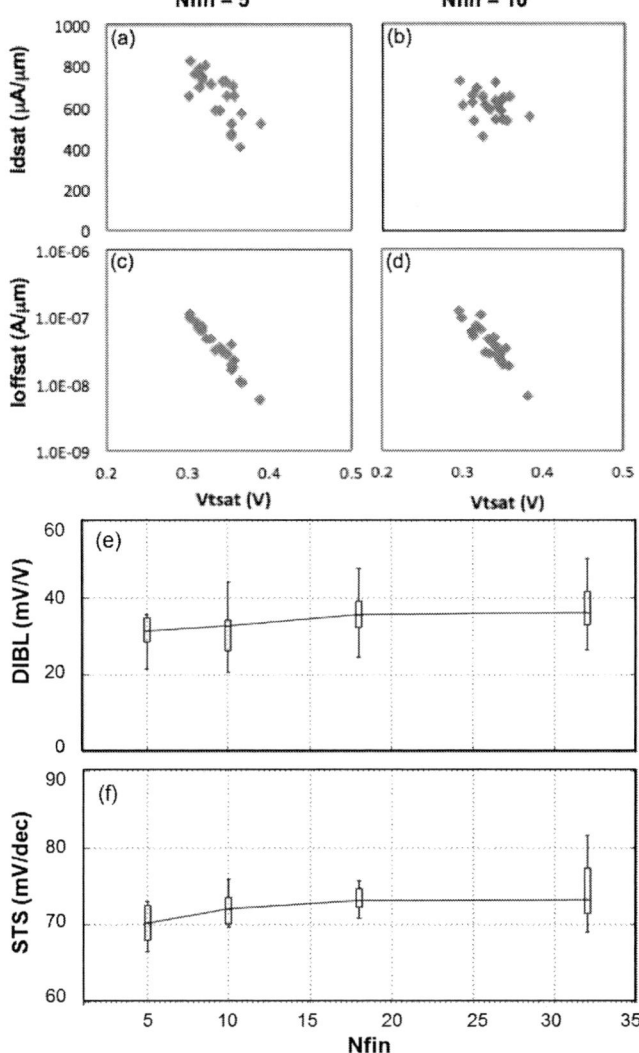

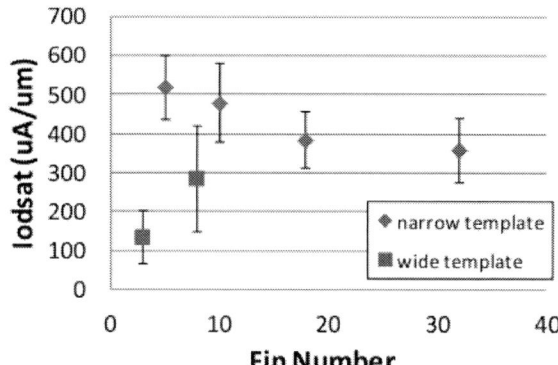

Figure 11: Iod_{SAT} vs. fin number (Nfin) for fins produced using narrow and wide template line width. Ideally, Iod_{SAT} should be a straight line. The increased performance of the 5 and 10 fin device relative to the 18 and 32 fin device is attributable to improved R_{EXT} for lower fin count devices. In contrast, the wide template devices show a strong dependence on Nfin suggesting that a high degree of defectivity is present, eliminating fins from participating in transport leading to misnormlaization of the data.

Conclusion

We have demonstrated a patterning process based on grapho-epitaxy DSA to form the fins of FinFETs using self-aligned customization. This process is compatible with the current generation of 193i scanners and does not require aggressive overlay. It is therefore suitable for scaling beyond the 10 nm node. We showed that DSA interaction with template geometries can affect process window and defectivity. A narrow template design style was identified to be the most robust. Electrical data produced from fins patterned with this approach shows good uniformity and no signs of gross CD variation. These results are also applicable to other DSA-based cut first techniques [9]. Further reduction of LER and LWR is the most critical improvement required to mitigate fin variation and improve device characteristics.

Acknowledgement

The devices in this work were fabricated in the Microelectronics Research Laboratoy (MRL) at the IBM T. J. Watson Research Center. The authors are grateful to the MRL staff for their contributions to the success of this work. The authors would also like to thank C.-H. Lin, L. Liebmann, G. Shahidi, and T.C. Chen for insightful discussions and their support.

References

[1] C. H. Lin, et al., VLSI Technol. Symp. (2011). [2] C. Liu, et al., Proc. SPIE, (2014). [3] H.C. Kim et al, J. Poly Phys (2010) [4] S. Sayan, et al Proce. SPIE (2014) [5] F. S. Bates et al, Phys. Today (1999) [6] Muller, et al, J. Pol. Sci: B (2005) [7] H. Tsai, et al., ACS Nano (2014). [8] T. Yamashita, et al, VLSI Technol Symp. (2011) [9] G. S. Doerk, et al., Proc. SPIE (2013).

Figure 10: (a-b) Id_{SAT} vs. Vt_{SAT} for Nfin = 5 and 10. (c) $Ioff_{SAT}$ vs. Vt_{SAT} for Nfin = 5 and 10. The linear correlation between Id_{SAT}, $Ioff_{SAT}$ and Vt_{SAT} demonstrates that the ensemble of fins is following proper electrostatic scaling theory. Moreover, the off state is not degraded by fins with abnormally wide CD as discussed in [8]. (e) DIBL and (f) STS vs. Nfin in the device for $L_G = 30$ nm. The flat response of DIBL and STS vs. Weff further confirms the uniformity of the TIGER patterning process.

fins are not the primary source of variation. End fin effects should decrease with increasing Nfin in one device. Note that an empirical maximum of 12 DSA lines was used for grapho-epitaxy DSA. As a result, devices with Nfin larger than 12 consist of groups of fins with a small break between, as shown in Fig. 4a.

Yield of fins as a function of template line width was also explored by plotting Iod_{SAT} vs. the Nfin in the device (Fig. 11). Data from devices with a wide template LW shows lower average current and a strong dependence on Nfin due to defects described in Fig. 5. In contrast, data from devices with narrow templates shows good uniformity.

978-1-4799-8002-4/14 $31.00 © 2014 IEEE

Highly Reliable Cu Interconnect Strategy for 10nm Node Logic Technology and Beyond

R.-H. Kim, B.H. Kim, T. Matsuda, J.N. Kim, J.M. Baek, J.J. Lee, J.O. Cha, J.H. Hwang, S.Y. Yoo, K.-M. Chung, K.H. Park, J.K. Choi, E.B. Lee, S.D. Nam, Y.W. Cho, H.J. Choi, J.S. Kim, S.Y. Jung, D.H. Lee, I.S. Kim, D.W. Park, H.B. Lee, S. H. Ahn, S.H. Park, M.-C. Kim, B.U. Yoon , S.S. Paak, N.-I. Lee, J.-H. Ku, J.S. Yoon, H.-K. Kang, and E.S. Jung

Samsung Electronics Co., Ltd.
San #16, Banwol-Dong, Hwasung-City, Gyeonggi-Do 445-701, Korea

Abstract

CVD-Ru represents a critically important class of materials for BEOL interconnects that provides Cu reflow capability. The results reported here include superior gap-fill performance, a solution for plausible integration issues, and robust EM / TDDB properties of CVD-Ru / Cu reflow scheme, by iterative optimization of process parameters, understanding of associated Cu void generation mechanism, and reliability failure analysis, thereby demonstrating SRAM operation at 10 nm node logic device and suggesting its use for future BEOL interconnect scheme.

Introduction

It is increasingly difficult to establish void-free Cu fill beyond 14nm node technology with PVD based TaN Cu-diffusion barrier and Ta adhesion promoter. To enhance Cu gap-fill performance, CVD based Co and Ru liners have been extensively studied (1-3), because of their conformal step coverage characteristics and better Cu wettability compared to their counterpart, PVD-Ta liner. Specifically, Ru liner enables Cu reflow process, mitigating structure related limitations due to its bottom-up growth capability. However, Cu reflow process on Ru liner is a new scheme for Cu back end of line (BEOL), several concerns and issues are brought up, such as gap-fill instability on complicated dual damascene patterns, unproven electro-migration (EM) and time-dependent-dielectric-breakdown (TDDB) characteristics, and lack of relevant integration scheme. In this study, CVD-Ru liner is explored with Cu reflow process, and associated Cu void generation mechanism at various patterns are suggested. Furthermore, its integration and reliability challenges are identified and addressed with major revelation of failure mechanism, providing highly reliable Cu interconnect gap-fill strategy for 10nm node and beyond.

Experimental

Basically, three layers of Cu dual damascene structures with LK (k=2.7) for M1 and ULK (k=2.5) for M2 and M3 have been fabricated to investigate gap-fill performance, electrical

characteristics, and reliability. Cu reflow capability is characterized at 10nm or 7nm equivalent patterns, with and without over-burden electroplating Cu and post chemical mechanical polishing (CMP) procedures, using SEM, TEM, and a defect inspection tool. Electrical properties (i.e., line resistance, via contact resistance, and healthiness of various test element group (TEG)), EM and TDDB performance have been investigated at 10nm node devices, followed by failure analysis.

Results and Discussion

Superior gap-fill capability of Ru liner based Cu reflow scheme over Co liner based seed enhancement layer scheme is clearly shown in Fig. 1, which provides post CMP SEM images at 7nm equivalent via chain and metal line patterns. PVD-TaN/CVD-Ru/Cu-reflow scheme shows void free gap fill performance at both via chain and metal line patterns, while PVD-TaN/CVD-Co/Cu seed scheme shows both slit and pit-type voids. Specifically, coexistence of pit and slit-type Cu voids with seed enhancement layer scheme indicates both insufficient sidewall coverage of seed Cu and top opening before electro-plating, revealing its limitation for use at 7nm node logic technology.

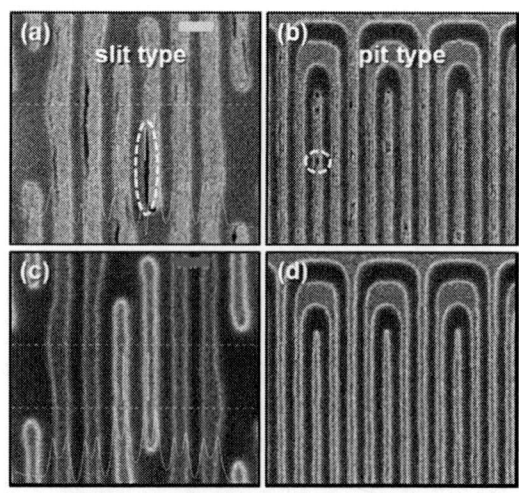

Figure 1. Post CMP top view SEM images of (a), (b) PVD-TaN/CVD-Co/Cu seed, and (c), (d) PVD-TaN/CVD-Ru/Cu-reflow schemes at 7nm equivalent

978-1-4799-8002-4/14 $31.00 © 2014 IEEE

dual damascene patterns.

Fig. 2 shows representative cross sectional TEM image at 7nm node equivalent via chain, confirming no bottom void at via chain when Cu reflow on CVD-Ru is adopted.

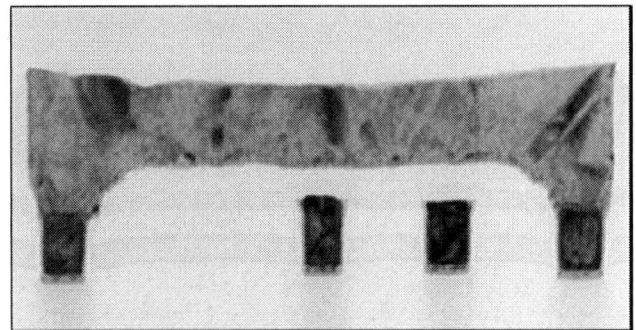

Figure 2. Vertical TEM image of PVD-TaN/CVD-Ru/Cu reflow scheme at 7nm node equivalent dual damascene pattern.

In general, superior gap fill performance of CVD Ru liner is enabled by Cu bottom-up gap-fill property as shown in Fig. 3, induced by reflow of Cu atoms during deposition. Here, the reflow momentum of adsorbed Cu atoms on Ru is provided by thermal and collisional energy transfer from substrate and Ar plasma, respectively, supported by ultralow impurity level Ru surface and good wettability of Cu.

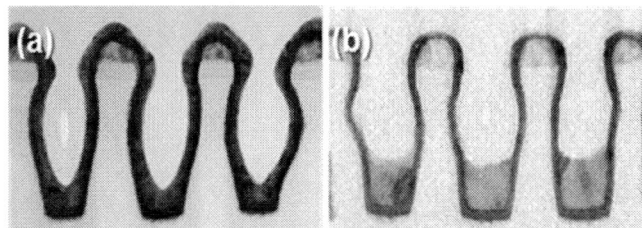

Figure 3. Vertical SEM for (a) PVD-TaN/CVD-Ru/Seed-Cu scheme and (b) PVD-TaN/CVD-Ru/ Cu-reflow scheme. Less Cu overhang and higher bottom-up thickness is clearly shown in Cu reflow scheme.

However, optimization of Cu reflow process is indispensable since it is highly dependent on pattern density / complexity and degree of pattern asymmetry as well as basic feature size and aspect ratio, to have robust gap-fill performance. Specifically, pit-type Cu void generation, associated with pattern asymmetry (Fig. 4a), is found to be one of key issues to be addressed for robust gap-fill performance of CVD-Ru based Cu reflow process. From SEM image after reflow Cu process (i.e., before electro-plating, Fig. 4b), partially filled points at wide trench are observed due to faster bottom-up rate at narrow trench. Even though these partially filled trenches after Cu reflow are supposed to be perfectly filled by subsequent electro-plating process, severe Cu agglomeration and/or overhang profile can degrade bottom-up capability electro-plating, thereby posing a potential risk (Fig. 4c).

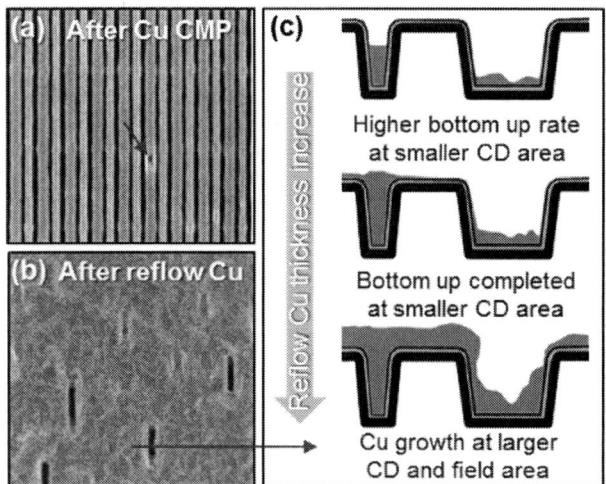

Figure 4. (a) Post CMP SEM image, showing pit type Cu void that is observed at relatively wider pattern rather than adjacent narrow pattern. (b) SEM image after Cu reflow process (before EP-Cu). (c) Schematic illustration as thickness of reflow Cu or deposition time increases (from top to bottom). It represents much faster bottom-up rate at narrow pattern when two different pitched patterns are located side by side.

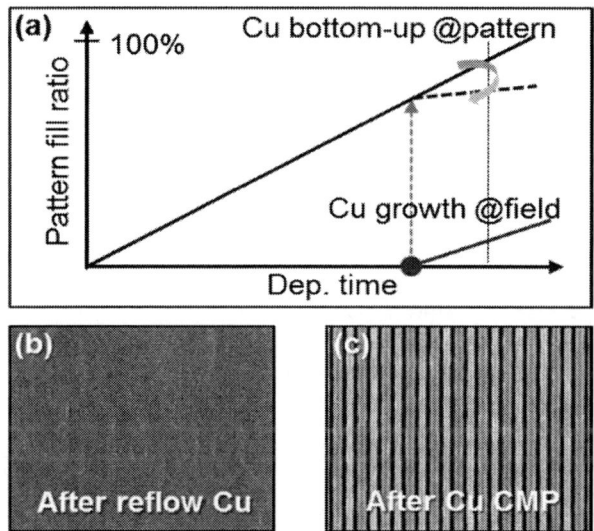

Figure 5. (a) Graphical illustration, suggesting strategy to achieve better gap-fill capability with less pattern dependence; higher bottom-up rate and delayed Cu nucleation and growth at the field area are key factors.

Void generation modeling is provided based on such observations, suggesting higher bottom-up rate inside trench and/or via as well as controlled Cu nucleation and growth at field area are critically important to achieve perfect gap-fill performance (Fig. 5a); Cu can nucleate and grow at field area (i.e., not favorable site) after certain period of reflow time, retarding diffusion process of Cu ad-atoms into trench or via patterns (i.e., favorable site) and causing bad Cu morphology. Specifically, Cu nucleation at the field area becomes

978-1-4799-8002-4/14 $31.00 © 2014 IEEE   769

aggressive when pattern asymmetry exist, since Cu bottom-up process preferentially happens at narrow trench, which can serve as nucleated Cu at the field to adjacent wide trench. With appropriate process optimization, improved Cu morphology and its void-free gap-fill performance are demonstrated as shown in Fig. 5b and 5c, respectively.

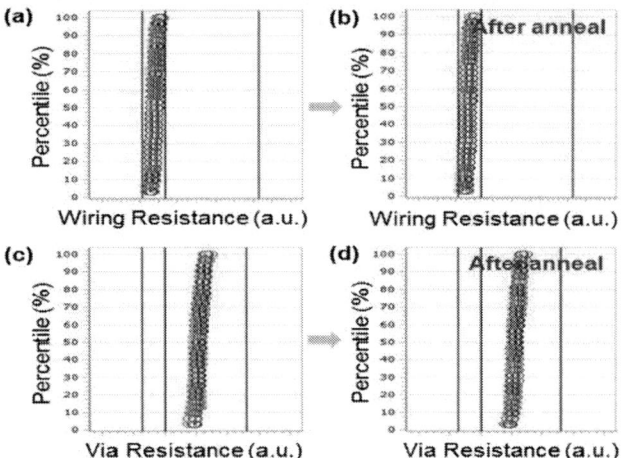

Figure 6. Representative metal line (a and b) and via contact (c and d) resistance at 10nm node technology before and after thermal budget, respectively. Tight distributions in all four graphs are noticeable, demonstrating highly robust wafer-level gap-fill performance.

In addition, very tight distribution of wafer-level metal line and via chain resistance even after application of thermal heat budget (Fig. 6) as well as demonstration of high open/short yield even at 36 million-via chains (Table 1) support highly robust wafer-level gap fill performance of process optimized Ru liner / Cu reflow scheme.

Table 1. Electrical yield summary of M1 and M2 layers. Here, Ru liner / Cu reflow scheme is applied, being characterized at M3 layer

Layer	Item	Yield (%)
M1	Line open yield (120m)	95.2
	Line short yield	100
	Via chain open yield (39.6M)	93.8
	Via chain short yield	95.3
	ISOVIA open	100.0
M2	Line open yield (120m)	95.3
	Line short yield	100.0
	Via chain open yield (39.6M)	93.8
	Via chain short yield	100.0
	ISOVIA open yield	100.0

We reported TDDB failure mechanism in Ru liner / Cu reflow scheme at 14nm node device in previous study (4), demonstrating more hydrophobic ULK application and optimized CMP process significantly reduce early breakdown failure associated with formation of unstable Cu oxide. This

strategy is well adapted to 10nm node device, and over 10 years TDDB lifetime is achieved both for line to line and via to line macros as shown in Fig. 7.

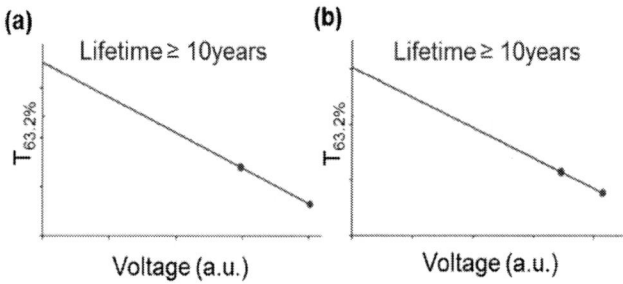

Figure 7. (a) Line-to-Line TDDB and (b) Via-to-Line TDDB results, indicating CVD-Ru liner/Cu-reflow scheme meets TDDB criteria at 10nm node pattern. Here, it is obvious that the same strategy, exploited for 14nm node device, is successfully adapted to 10nm node technology.

The other reliability concern is EM performance, originated from the fact that Cu migrates well on Ru without inter-mixing or formation of intermetallic compound. It can facilitate Cu reflow process as mentioned. However, it may also cause instability problem at Ru/Cu interface under stressing conditions.

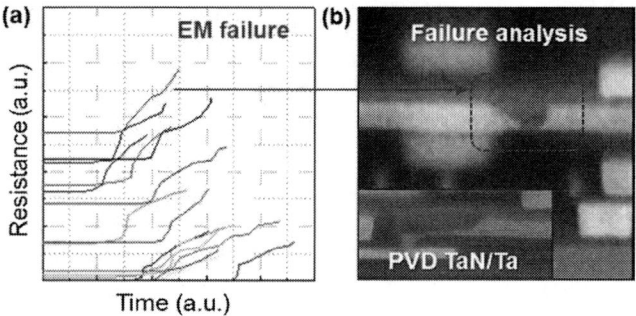

Figure 8. (a) EM test result of Ru liner/reflow-Cu scheme at 10nm node device. (b) Failure analysis of EM failure point. Void is shown at the interface of Cu and capping layer, while no via related voids are observed. Inset image of Fig. 8b represents conventional EM failure point in case of PVD TaN/Ta/Seed-Cu scheme.

Fig. 8a and 8b present representative EM test result of TaN/Ru/Reflow Cu scheme at 10nm node device and its failure analysis result at EM failure point, respectively. Gradual increase in line resistance under stressing condition is noticeable. Failure analysis confirms that Cu void is generated at the interface of Cu and dielectric capping layer, rather than interface of Cu and Ru or at via bottom. This specific EM failure mode, different from that with the conventional TaN/Ta/Seed-Cu scheme (inset of Fig. 8b), indicates robust interface between Ru and Cu. Therefore, simple modifications such as application of CuMn alloy seed with higher Mn content or selective metal capping can boost

978-1-4799-8002-4/14 $31.00 © 2014 IEEE   770

EM life time, from marginal 10 years with the conventional scheme to over 100 years with the optimized scheme, as shown in Fig. 9.

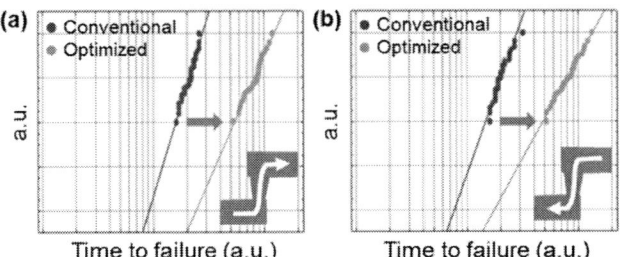

Figure 9. EM result of conventional and optimized CVD-Ru/Cu-reflow scheme at 10nm node device. Figure (a) and (b) correspond to up-stream EM and down-stream EM, respectively. In both cases, much longer time to failure is observed by application of the optimized scheme.

This improvement in EM characteristics is well explained by comparison of EM activation energy between two cases (Fig. 10). It reveals that ~31% increase of activation energy with optimized scheme can be obtained, implying enhanced adhesion between Cu and upper capping layer is critical.

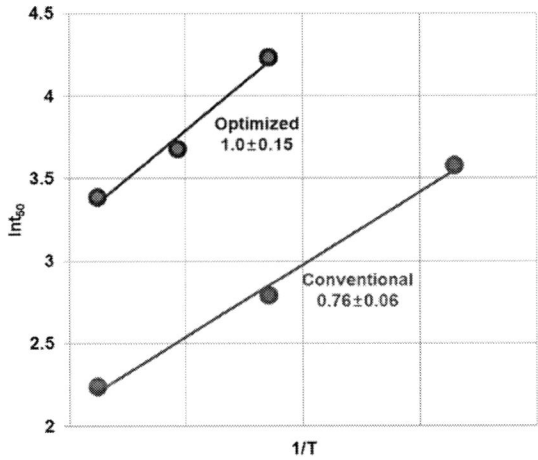

Figure 10. EM activation energy comparison between (a) conventional Ru scheme, and (b) optimized Ru scheme. The normalized activation energy for electro migration increases from 0.76 to 1.0 with optimized scheme.

Lastly, integration challenges associated with Ru liner / Cu reflow scheme are investigated throughout all the necessary process steps. One issue is related to low Ru removal rate with conventional CMP process. Fig. 11a and 11c show post CMP optical microscope images of the conventional and new CMP scheme, respectively. Severe discolor at a monitoring site after conventional CMP due to presence of Ru residue, which is potential risk causing de-focus in photo lithography at upper layer, is clearly solved by using the new CMP scheme. Thus, the optimized Ru CMP enables integration of

Ru liner / reflow-Cu scheme for M1-M3 interconnect layers of logic device without any patterning issues, leading to a successful achievement of 10nm node SRAM yield.

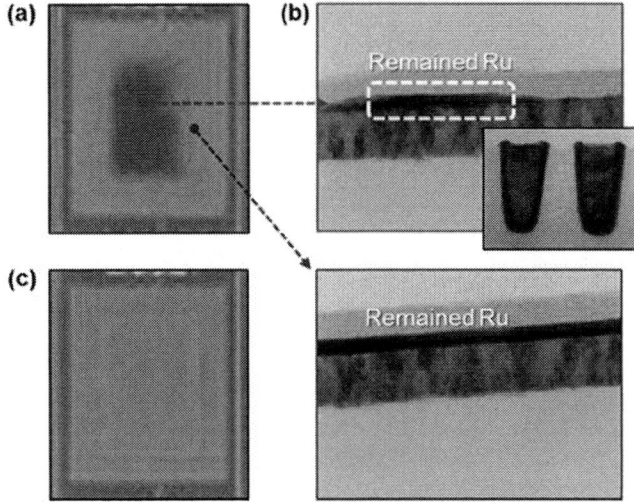

Figure 11. Representative optical images of the conventional CMP (a) and new CMP (c) at monitoring site, respectively. With conventional Ru CMP scheme, severe discolor at a monitoring / align key sites is observed due to remained Ru film even after complete CMP. Vertical TEM images reveal uneven Ru film at two different points as shown in Fig. 12b.

Conclusion

CVD-Ru liner / Reflow-Cu scheme is adopted for 10nm node BEOL technology, demonstrating superior gap-fill performance and robust reliability. Pattern dependency of reflow Cu scheme is tightly controlled, based on the suggested modeling, and the reliability issue associated with Ru liner at 10nm node technology is successfully addressed via failure analysis. In addition, wafer-level yield is successfully demonstrated with optimized CMP, suggesting CVD Ru liner is an appealing candidate for 10nm node and beyond.

References

(1) M. He, X. Zhang, T. Nogami et al., "Mechanism of Co Liner as Enhancement Layer for Cu Interconnect Gap-Fill, Journal of The Electrochemical Society, 160 (12), D3040-3044 (2013).

(2) C.C.Yang, T.Spooner, S.Ponoth et al., "Physical, Electrical, and Reliability Characterization of Ru for Cu Interconnects", IITC2006, pp. 187-189 (2006).

(3) F.Gstrein, R.Akolkar, S.Balakrishnan et al., "Reliability of Cu Interconnects with Gap Fill-Enabling Ruthenium Liners", AMC2008, pp. 13-14 (2008).

(4) T.Matsuda, J.J. Lee, K.H. Han et al., "Superior Cu Fill with Highly Reliable Cu/ULK Integration for 10nm Node and Beyond", IEDM 2013, s29-2.

A new high-k/metal gate CMOS integration scheme (Diffusion and Gate Replacement) suppressing gate height asymmetry and compatible with high-thermal budget memory technologies

R. Ritzenthaler, T. Schram, A. Spessot[1], C. Caillat[1], M. Cho, E. Simoen, M. Aoulaiche, J. Albert, S. A. Chew, K. B. Noh[2], Y. Son[2], P. Fazan[1], N. Horiguchi, and A. Thean

imec, assignee at imec from [1]Micron, [2]SK-Hynix – Kapeldreef 75, B-3001, Leuven, Belgium
Tel: (32) 16 28 77 43 - Fax: (32) 16 28 17 06, Email: romain.ritzenthaler@imec.be

Abstract

A new scheme called in the following "Diffusion and Gate Replacement" (D&GR) MIPS integration is demonstrated. The CMOS flow allows to control the gate height asymmetry between NMOS and PMOS by driving the work function shifter directly into the high-k. Since the threshold voltage (Vth) shifter sources are removed, it is compatible with other processes requiring high-thermal budget such as memory technologies (DRAM periphery).

Introduction

Although Replacement Metal Gate high-k (RMG) has become the popular choice for high-performance sub-22nm CMOS technologies, such gate-stacks are very sensitive to thermal budgets [1]. For applications that require thermal robustness, like control logic for Memory or System-on-Chip coupled with memory elements, gate-first MIPS and variants are preferred. The use of Al and group IIA/IIIB elements as La or Mg for adjusting the threshold voltage of gate first MIPS based devices has been described extensively [2-5]. The dopant atom is added as an oxide cap layer or embedded into the metal gate itself. In most cases the dopant source stays in place during the further processing, rendering it prone to additional diffusion during subsequent thermal processing. Especially in memory applications, control logic devices can be exposed to a high thermal budget required by the fabrication of the DRAM memory elements (storage capacitor, access transistor, Back End of Line (BEOL)). Hence, work-function stability of the control logic devices experiencing higher thermal processing of the memory element is necessary. Moreover, multiple work-function metals stacked in gate-first integration may lead to N-P gate height asymmetry [4,6], complicating gate patterning and further integration steps. The Diffusion and Gate Replacement (D&GR) process module proposed in the present paper solves both issues by diffusing the dopant species from dummy doped N and P metal gates with a single thermal step, followed by the replacement by a common final metal electrode (**Figs. 1-2**).

Process Integration

The new D&GR gate stack module can be inserted in any conventional MIPS transistor flow. This is illustrated in **Fig. 1** that shows its position in our experimental process flow. By diffusing N and P-type shifting elements from doped dummy metal gates or cap layers in a dedicated diffusion step and removing the dopant source afterwards (**Fig. 2**), additional

dopant diffusion and hence unwanted further Vth shift during subsequent thermal processing is avoided.

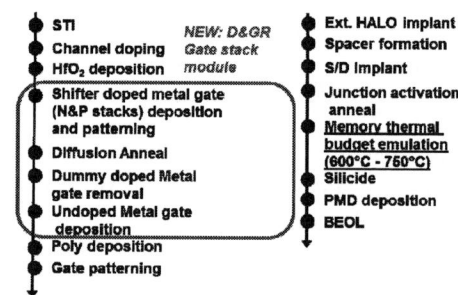

Fig. 1: Schematic representation of the process flow. The position of the new D&GR (Diffusion and Gate Replacement) gate stack module is highlighted.

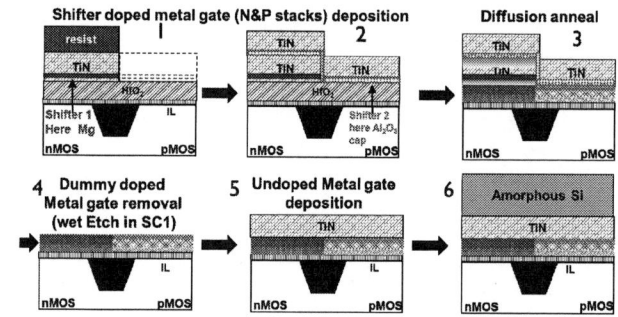

Fig. 2: The proposed D&GR gate stack module. The preferred NMOS first implementation is illustrated here. A Mg based TiN sandwich, followed by an Al_2O_3 cap layer is used for N and PMOS V_{TH} shifting, respectively.

using an SC1 based wet etching, the TiN based dummy electrodes can be removed selectively to HfO_2 and subsequently replaced by an undoped TiN electrode. TEM picture after D&GR process (**Fig. 3**) demonstrates the symmetry of the final stacks. Proof of the diffusion of dopants in the high-k layer is provided by physical analysis (ERD/SIMS, **Figs. 4-5**): after diffusion anneal, dummy electrode removal and final TiN redeposition, dopants are driven into the high-k dielectrics layer. As an additional constraint in a CMOS flow, the dopants in the upper layer must diffuse slower than the dopant underneath in order to avoid cross-contamination (**Fig. 2, step 3**).

978-1-4799-8002-4/14 $31.00 © 2014 IEEE

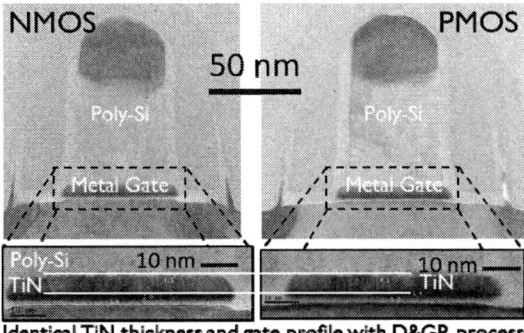

Fig. 3: Final stack (Step 6 of Fig.2) obtained in the CMOS D&GR flow. The gate height symmetry between NMOS and PMOS devices is evidenced, with identical SiO$_2$ Interfacial Layer/HfO$_2$ high-k/TiN Metal gate/Poly-Si gate stack thicknesses and gate foot profile. The gate etch recipe can be optimized simultaneously for NMOS and PMOS.

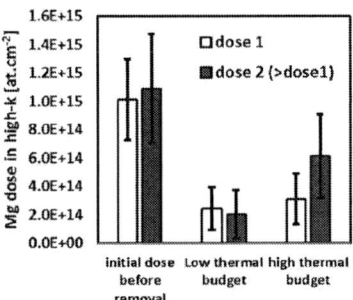

Fig. 4: ERD concentration in HfO$_2$ based stack after TiN/Mg/TiN cap based D&GR processing with different anneal temperatures. The diffusion of Mg into the high-k during D&GR process is evidenced.

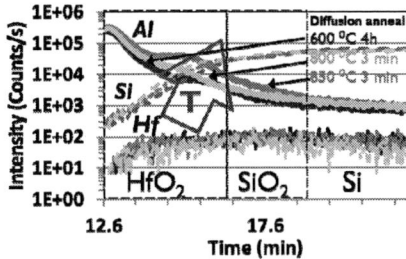

Fig. 5: SIMS profile of a HfO$_2$ based stack after Al$_2$O$_3$ cap based D&GR processing (diffusion anneal and dummy metal removal) with different anneal temperatures. The diffusion of Al into the high-k during D&GR process is evidenced.

The choice of dopants, their removal ability with the wet etch, and their respective process windows (N and P side should be tuned with a single diffusion anneal) must therefore be carefully evaluated. After screening the various options, the most promising species have been defined (**Tab. 1**). Mg is preferred as N-type dopant as it can be removed using the basic SC1 chemistry only, while La in contact with TiN cannot be removed with an SC1 solution [3]. The preferred P-type dopant is Al$_2$O$_3$ cap on the high-k. The driven dopant dose is increased by raising the diffusion anneal thermal budget (duration and temperature) (**Figs. 4-5**), which in turn leads to more aggressive work function (eWF) values (**Fig. 6**). It comes however at the potential cost of increased EOT (**Fig. 7**) and degraded mobility (**Fig. 8**), leading to a necessary trade-off.

	PMOS: Al	PMOS: Al$_2$O$_3$	NMOS: Mg	NMOS: La
Removal with SC1 (cap layer)	Formation of insoluble AlN	OK	OK	Formation of insoluble TiLaN
Removal with SC1 (TiN/dopant/ TiN sandwich)	Formation of insoluble AlN	OK if Al$_2$O$_3$ is thin enough (<1 nm)	OK	Formation of insoluble TiLaN
Residue on removed area → Residual Vt shift	Present in case of cap layer	Residue in case of cap layer on NMOS → use Al$_2$O$_3$ cap layer in an NMOS first flow	None in case of TiN/Mg/TiN sandwich	Present in case of cap layer

Tab. 1: Selection criteria and results for the choice of the NMOS and PMOS dopant species, as well as the dopant location and the integration order.

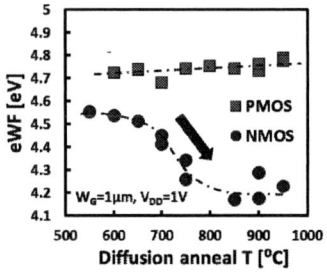

Fig. 6: eWF obtained from the Al$_2$O$_3$ cap based TiN PMOS stack and the TiN/Mg/TiN sandwich NMOS stacks after diffusion anneals in the range 550°C-950°C (<60 s) and subsequent metal gate replacement.

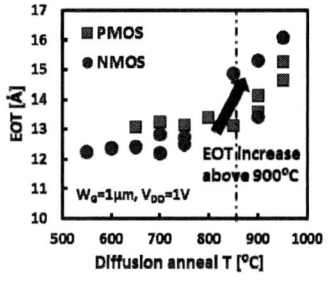

Fig. 7: EOT obtained from the Al$_2$O$_3$ cap based TiN PMOS stack and the TiN/Mg/TiN sandwich NMOS stacks after diffusion anneals in the range 550°C-950°C (<60 s) and subsequent metal gate replacement.

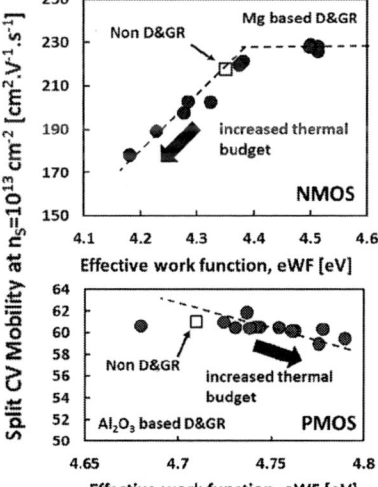

Fig. 8: High field mobility vs. work function for D&GR devices, for various diffusion anneal thermal budget.

Using a single diffusion anneal between 700°C and 900°C (<60 s) both low N- and high P-MOS eWF can be obtained at the same time without significant EOT increase. The location of the dopant within the stack and the integration order (PMOS vs.

978-1-4799-8002-4/14 $31.00 © 2014 IEEE   773

NMOS first) is based on the risk of residual dopant potentially left on the HfO_2 after the initial removal step (**Fig 2, step1**), which will cause a Vth shift opposite to the desired polarity and increased variability fluctuation. Capping layers leave more residues (**Fig. 9.b**) than doping elements sandwiched between two TiN layers (**Fig. 9.a**). The complete process on P-side (experiencing two removals, **Fig. 10**) does not show any eWF penalty. As a consequence, the optimal configuration of the D&GR integration is an 'NMOS first' integration with Mg added in a TiN sandwich, followed by an Al_2O_3 cap layer.

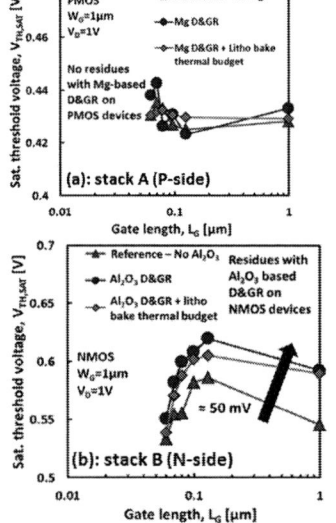

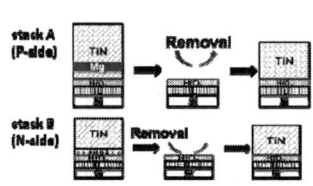

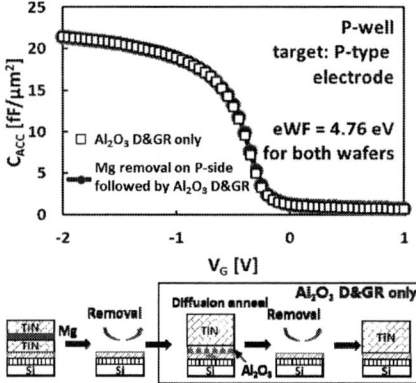

Fig. 9: (a) Vth of a MIPS device without Mg TiN sandwich compared to those submitted to the D&GR process with Mg TiN sandwich. No residual Vth shift is observed. (b) Vth of a MIPS device without Al_2O_3 cap layer compared to those submitted to the D&GR process with an Al_2O_3 cap layer on top of the HfO_2. A residual V_{TH} shift is observed due to the Al_2O_3 residues on the HfO_2.

Fig. 10: eWF extraction from CV measurements on Pwell areas submitted to the full PMOS processing corresponding to the preferred D&GR process flow (**Fig. 2**), as well as for the Al_2O_3 D&GR only case. Identical CV curves are obtained for both cases.

Device Performance

The final gate stacks are evaluated in depth in terms of yield and reliability, and benchmarked to non D&GR high-k/Metal gate (HKMG) integration [4]. Similar N and PMOS gate leakage vs. EOT trends are obtained with the D&GR integration and in absence of the dopant removal (**Fig. 11**), and no failure increase is found for large area capacitors (**Fig. 12**). Reliability-wise, no penalty is measured in BTI lifetime and TDDB with D&GR (**Fig. 13-15**). Low Frequency Noise levels are also very good, and virtually insensitive to the diffusion anneal thermal budget (**Fig. 16**).

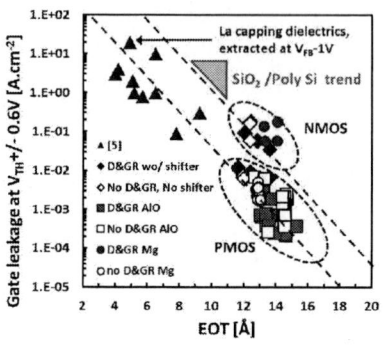

Fig. 11: Gate leakage vs. EOT for Al_2O_3 cap layer (PMOS) and TiN/Mg/TiN sandwich (NMOS) based D&GR devices (filled markers) compared to their equivalent where the Al_2O_3 and Mg sandwich was not removed [4] (empty markers).

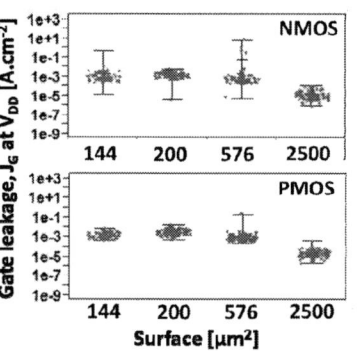

Fig. 12: Gate leakage vs. capacitor surface for Al_2O_3 cap layer (PMOS) and TiN/Mg/TiN sandwich (NMOS) based D&GR devices (170 devices). No gate stack failure increase is measured even for large surfaces, highlighting the good yield of D&GR process.

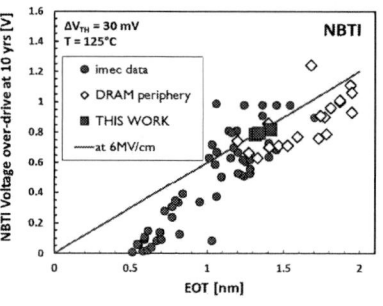

Fig. 13: Benchmark of the NBTI lifetime vs. EOT for Logic ("imec data") and DRAM gate stacks (adapted from [7]). Threshold voltage shift criterion = 30 mV. D&GR process is not yielding a lifetime penalty.

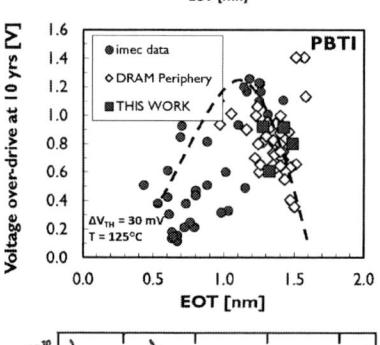

Fig. 14: Benchmark of the PBTI lifetime vs. EOT for Logic and DRAM gate stacks (adapted from [7]). Threshold voltage shift criterion = 30 mV. D&GR process is not yielding a lifetime penalty.

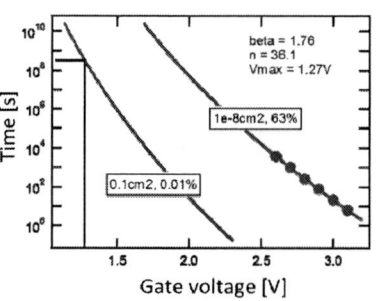

Fig. 15: The TDDB lifetime extrapolation (at T=125°C) for the Mg-based N stack reveals a maximum gate voltage of 1.27V for 10 years without hard breakdown.

978-1-4799-8002-4/14 $31.00 © 2014 IEEE

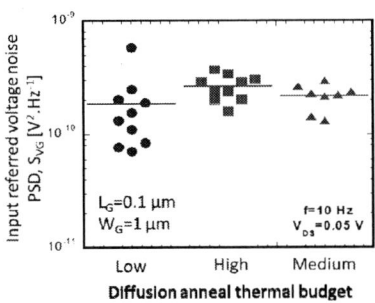

Fig. 16: Input referred Voltage noise Power Spectral Density S_{VG} (W=1µm by L_G=0.1 µm nMOSFET) for low, medium and high diffusion anneal thermal budget. 1/f noise is observed. Gate stack features good noise level, and is not degraded by increasing the diffusion anneal thermal budget.

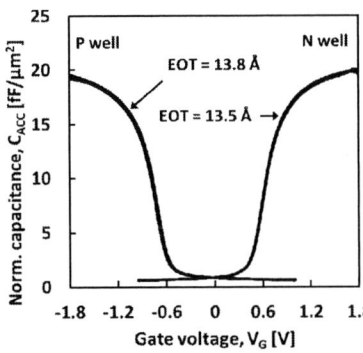

Fig. 17: N and P well gate-to-bulk accumulation capacitance (W=L=50 µm), showing measured gate stacks across wafer.

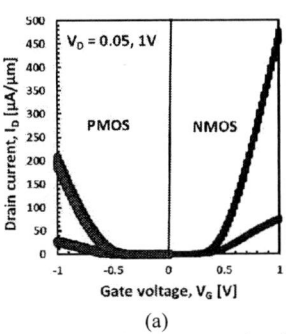 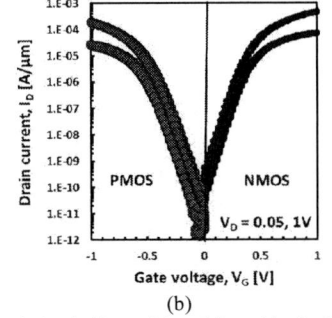

(a)   (b)

Fig. 18: N and PMOS transfer characteristics in linear (a) and logarithmic (b) scale at V_D = +/- 0.05, +/- 1V, illustrating the good control of short channel effects mandatory for Low Power application. W_G = 1 µm.

Final CMOS C-V and transfer characteristics are shown in **Figs.17-18**; intrinsic transistor performance (**Fig. 19**) and ring oscillators gate delay (**Fig. 20**) improvement over the HKMG baseline are demonstrated. Finally, D&GR I/O devices are also successfully integrated, with Vth shift between 150mV and 300mV compared to undoped high-k on thick oxides (5 nm interfacial oxide + HK, **Fig. 21**).

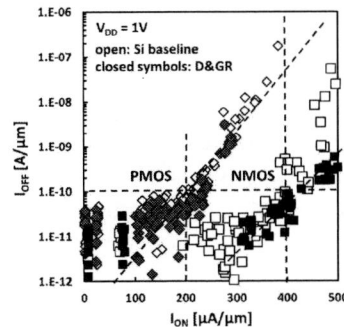

Fig. 19: I_{ON}/I_{OFF} (taken at drain side) obtained for NMOS (resp. PMOS) transistors, featuring identical performance of D&GR transistor over the HKMG baseline [4]. W = 1 µm, V_D = 1 V.

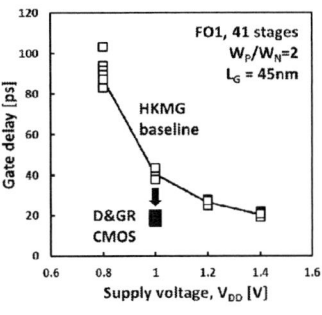

Fig. 20: Gate delay vs. V_{DD}. Values below 20 ps are obtained for V_{DD}=1V, with significant improvements over non D&GR baseline.

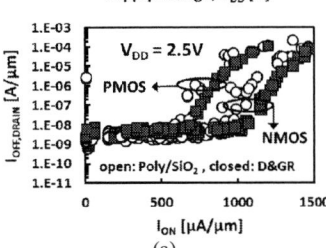 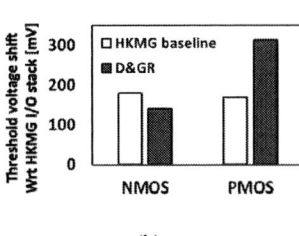

(a)   (b)

Fig. 21: (a) I_{ON}/I_{OFF} (D&GR and PolySi/SiO$_2$ reference) and (b) obtained V_{TH} shift for D&GR I/O transistors.

Conclusions

In this work, a new scheme (Diffusion and Gate Replacement (D&GR)) for MIPS integration is demonstrated. The CMOS flow allows to control the gate height asymmetry between N and P stacks, by driving the work function shifter in the high-k. Since the dopants source is removed, it improves the integration friendliness for high-thermal budget processes, such as control logic for memory technologies, making it an optimal choice for DRAM periphery. A common process window for both NMOS and PMOS is demonstrated, and is carefully optimized with regard to device performance (eWF, EOT, mobility). No penalty is found with D&GR process with regards to gate leakage, reliability (BTI, TDDB) and Low Frequency Noise. Current and ring oscillators performance improvement is obtained over HKMG baseline, and I/O devices successfully demonstrated.

Acknowledgements

This work was performed as part of IMEC's Core Partner Program.

References

[1] R. Ritzenthaler et al., *Solid-State Electronics*, vol. 84, 2013
[2] V. Narayanan et al., *2006 Symposium on VLSI Technology*, vol. 729, no. 2005, pp. 178–179, 2006.
[3] T. Schram et al., *ECS Transactions*, 25(5) 17-28, 2009
[4] R. Ritzenthaler et al., *IEEE Transactions on Electron Devices*, Vol. 61, no. 8, pp. 2935-2943, Aug. 2014
[5] T. Ando, *Materials*, 5(12), pp. 478–500, 2012.
[6] "Samsung 28 nm HKMG: Inside the Apple A7", chipworks blog (jan 20th 2014).
[7] M. Cho et al., *IEEE Transactions on Electron Devices*, vol. 59, no. 8, pp. 2042–2048, 2012.

978-1-4799-8002-4/14 $31.00 © 2014 IEEE

Ultra Low Contact Resistivity (< 1×10^{-8} Ω-cm^2) to In$_{0.53}$Ga$_{0.47}$As Fin Sidewall (110)/(100) Surfaces: Realized with a VLSI Processed III-V Fin TLM Structure Fabricated with III-V on Si Substrates

Rinus T.P. Lee, Y. Ohsawa[1], C. Huffman, Y. Trickett[1], G. Nakamura[1], C. Hatem[2], K.V. Rao[2],
F.Khaja[2], R. Lin[3], K. Matthews, K. Dunn[4], A. Jensen[3], T. Karpowicz[3], Peter F. Nielsen[3],
E. Stinzianni, A. Cordes, P.Y. Hung, D.-H. Kim[5], R.J.W. Hill, W.-Y. Loh and C. Hobbs

SEMATECH, Albany, NY 12203, USA, [1]Tokyo Electron Technology Center, Albany, NY 12203, USA,
[2]Applied Materials, Gloucester, MA 01930, USA, [3]CAPRES A/S, DK-2800 Kongens Lyngby, Denmark,
[4]CNSE, Albany, NY 12203, [5]GLOBALFOUNDRIES assignee at SEMATECH

Phone: +1-518-649-1200, Fax: +1-518-649-1322, E-mail: rinus.lee@sematech.org

ABSTRACT

We report a record low contact resistivity of sub-1.0×10^{-8} $\Omega.cm^2$ realized on n$^+$ In$_{0.53}$Ga$_{0.47}$As fin sidewall surfaces. This is achieved with VLSI processed fin TLM structures on wafer scale III-V on Si substrates. A novel low-damage III-V fin etch was developed and fins down to 35 nm were fabricated. A surface treatment to smoothen the fin sidewall surfaces was proposed, which reduced sidewall surface roughness variation by 90%. Additionally, we show for the first time that implant temperature could be used to eliminate implant damage in III-V fins. This increased activation efficiency (+3.6×) and reduced sheet resistance (-60%).

INTRODUCTION

Recent advances in III-V device technology [1–2] have led to an increasing interest in III-V as a replacement for Si in CMOS. To enable this transition, it is critical that III-V process technologies are developed in parallel with progress in device technology. Furthermore, it is well documented that with channel resistance scaling, series resistance (R_{SERIES}) becomes a limiting factor in I_{Dsat} [3].

In this work, III-V S/D process technologies are proposed to mitigate the issue of increasing R_{SERIES}. Key contributions of this work are the development of a low-damage III-V fin etch and a fin TLM structure for contact resistivity (ρ_{co}) extraction on fin sidewalls. We also show that implant temperature can be used to eliminate amorphization in III-V fins.

RESULTS AND DISCUSSION

A. III-V Fin Etch Development

Fig. 1(a) shows the design of the hetero-buffer used in this work. HRTEM in Fig. 1(b) confirms that a 30 nm lattice-matched In$_{0.53}$Ga$_{0.47}$As device layer was grown on In$_{0.52}$Al$_{0.48}$As. A SiO$_2$/SiN layer was used as the fin etch hardmask (HM). Fins were defined with 193 nm lithography. The HM and III-V fins were etched using a RLSA^{TM} plasma etch system. Table I summarizes the III-V fin etch development. It is evident in Table I that fin sidewall taper angle increases (not ideal) and etch residues decreases (ideal) with increasing pressure [4]. Additionally, etch residues could be reduced further without negative impact on fin sidewall taper angles by optimizing the process gas ratios. It was found that Process D with a combination of medium pressure and increased Ar to Cl$_2$:CH$_4$ gas ratios generates the least amount of etch residue (if any) and provides a near vertical III-V fin profile, which is ideal for this work.

B. III-V Fin TLM Fabrication

III-V fin TLM structures were fabricated with the process shown in Fig. 2. The layers were grown sequentially by MBE on wafer scale Si substrates. In$_{0.53}$Ga$_{0.47}$As layers were doped to 1.0×10^{19} cm^{-3} (μRaman measurement) with Si$^+$ insitu doping or Si$^+$ implant at 25 °C to compare these two techniques for III-V fin S/D doping. Fig. 3 confirmed that a near vertical fin sidewall profile (*sidewall-to-substrate angle of 80°*) without etch-induced damage was realized with Process D. Device isolation was achieved by etching into the In$_{0.52}$Al$_{0.48}$As layer (>100 nm) as shown in Fig. 3. In$_{0.52}$Al$_{0.48}$As is a suitable isolation layer as it remains highly resistive even after Si$^+$ implant (2×10^{15}/cm^2) and activation at 700 °C due to its large bandgap of 1.47 eV (see Fig. 4). A surface treatment process based on an oxidizer/acid for fin smoothing was used in this work, which reduced sidewall surface roughness variation by ~ 90% (see Fig. 5).

Device pitch scaling and the use of self-aligned contact vias will negate the benefits of self-aligned silicides. Furthermore, silicide-like contacts in III-V

978-1-4799-8002-4/14 $31.00 © 2014 IEEE

exhibit $\rho_{co} > 100\times$ higher than non-alloyed metal contacts. For this work, we selected Mo as S/D contact metal for its excellent thermal stability on III-V. Besides, Fig. 6 shows that the electron barrier heights (Φ_e) for non-alloyed metals on $In_{0.53}Ga_{0.47}As$ are pinned at ~ 0.2 eV [5–8]. This indicates that any ρ_{co} scaling based on Φ_e selection would be limited.

C. III-V Fin TLM Characterization

Fig. 7 plots total resistance versus contact spacing for Mo/Si$^+$ insitu doped fins and Mo/Si$^+$ implanted at 25 °C fins (hereafter referred as I/D and I/I–RT, respectively). An excellent linear fit to the experimental data is obtained. Sheet resistance (R_s) is estimated to be 75 Ω/sq for I/D fins and 410 Ω/sq for I/I–RT fins. Qualitatively, this indicates that doping concentration (N_D) for I/D fins are higher than I/I–RT fins. To evaluate the impact of N_D on ρ_{co} and R_s statistically, box plots for I/D fins and I/I–RT fins are shown in Fig. 9 and 10. It was found that ρ_{co} and R_s for I/D fins are significantly lower than I/I–RT fins, with an average ρ_{co} of 1.59×10^{-8} $\Omega.cm^2$. This is the lowest ρ_{co} reported to date on fin sidewalls. It should be noted that a significant percentage (25%) of the extracted ρ_{co} are within 6.0×10^{-9} $\Omega.cm^2$, which meet ITRS requirements for the 7 nm node. This suggests that with further optimization, the average ρ_{co} value can decrease to sub-1.0×10^{-8} $\Omega.cm^2$. Degradation of N_D and the resulting poor ρ_{co} and R_s for I/I–RT fins are due to the incomplete recrystallization of implant damaged III-V fins (inset of Fig. 10). This clearly demonstrates that Si$^+$ implant at 25 °C is not suitable for III-V fin doping due to Si$^+$ implant-induced damage in narrow III-V fins.

D. Hot Implant in III-V Fins

Alternatively, hot implant (I/I–HOT) has been shown to eliminate implant damage in the narrow fins of SOI and Bulk Si FinFETs [9–10]. In this work, we applied this concept to III-V fins for the first time with the process flow summarized in Fig. 11. Fig. 12 shows the dependency of R_s for I/I–RT and I/I-HOT. It is obvious that layers subjected to I/I-HOT have significantly lower R_s, with much tighter distributions than I/I–RT. This is due to enhanced activation efficiency (η_{act}) with I/I-HOT.

Fig. 13 shows that with I/I-HOT, a 3.6× improvement in η_{act} leads to a 60% reduction in R_s. To correlate this improvement with I/I-HOT, a series of TEM images was obtained. Fig. 14(a) shows that I/I–RT forms an amorphous layer around the fin top/sidewalls. This leads to residual defects after activation anneal [see Fig. 14(b)]. In contrast, I/I-HOT does not form an amorphous layer but maintained excellent crystallinity for both as-implanted (I/I-HOT) and after activation annealed III-V fins [see Fig 14(c)–(f)]. This is attributed to enhanced annihilation of defects (dynamic annealing) with elevated temperature (i.e. I/I-HOT).

CONCLUSIONS

Table II summarizes the ρ_{co} and details of leading S/D contact materials on $In_{0.53}Ga_{0.47}As$. Here, we achieved the lowest ρ_{co} ($< 1.0\times10^{-8}$ $\Omega.cm^2$) for Mo on $In_{0.53}Ga_{0.47}As$ fin sidewall surfaces using a III-V fin TLM structure. Additionally, we show that I/I-HOT is a key process knob that can be used to increase η_{act} ($\uparrow 3.6\times$), reduce R_{sheet} ($\downarrow 60\%$) and prevent amorphization in narrow III-V fins. Our results represent significant progress towards the development of III-V fin S/D process technologies.

REFERENCES

[1] M. Radosavljevic, *IEDM*, p.765, 2011
[2] N. Waldron, *VLSI Symp.*, p.82, 2012
[3] Y.C. Yeo, *INEC*, p.129, 2013
[4] Y. Ohsawa, *AVS Symp.*, p.P34, 2012
[5] K. Kajiyama, *APL*, (23), p. 458, 1973
[6] H.J. Lee, *APL*, (63), p.1939, 1993
[7] P. Kordos, *JAP*, (72), p.2347, 1992
[8] H.T. Wang, *APL*, (70) p.2571, 1997
[9] W. Mizubayashi, *IEDM*, p538, 2013
[10] M. Togo, *VLSI Symp.*, p. 196, 2013
[11] L. Czornomaz, *J. SSE*, p.71, 2012
[12] X. Zhang, *J. SSE*, p.82, 2012
[13] R. Dormaier, *JVST B*, p031209, 2012
[14] U. Singisetti, *APL*, p.183502, 2008
[15] W. Lu, *EDL*, p178, 2014

978-1-4799-8002-4/14 $31.00 © 2014 IEEE

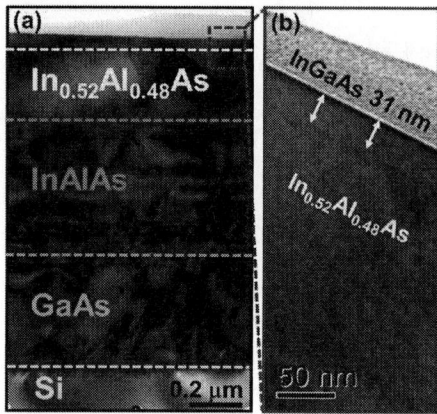

Fig. 1: (a) TEM shows the layer design of the III-V on Si substrate used for III-V fin etch development in this work. **Fig. 1(b)** HRTEM clearly shows that a high quality 30 nm thick $In_{0.53}Ga_{0.47}As$ layer has been successfully grown on $In_{0.52}Al_{0.48}As$.

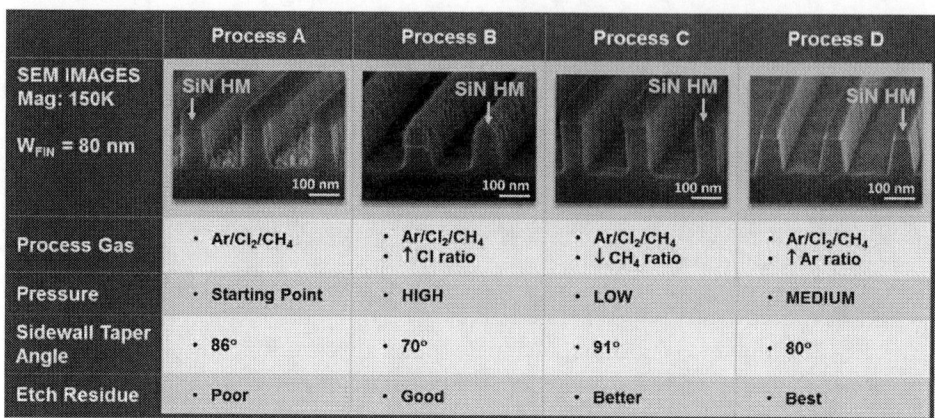

	Process A	Process B	Process C	Process D
SEM IMAGES Mag: 150K W_{FIN} = 80 nm	SiN HM 100 nm	SiN HM 100 nm	SiN HM 100 nm	SiN HM 100 nm
Process Gas	• $Ar/Cl_2/CH_4$	• $Ar/Cl_2/CH_4$ • ↑ Cl ratio	• $Ar/Cl_2/CH_4$ • ↓ CH_4 ratio	• $Ar/Cl_2/CH_4$ • ↑ Ar ratio
Pressure	• Starting Point	• HIGH	• LOW	• MEDIUM
Sidewall Taper Angle	• 86°	• 70°	• 91°	• 80°
Etch Residue	• Poor	• Good	• Better	• Best

Table I: Summary of III-V fin etch process development. The objective is to develop a low-damage III-V fin etch process. Process D with a combination of medium pressure and increased Ar to Cl_2:CH_4 gas ratios generates the least amount of etch residue (if any) and near vertical (80°) III-V fin sidewall profile. To further reduce surface roughness (i.e. striations) on the fin sidewall, a surface treatment process was also developed in this work as shown in Fig. 5.

○ **III-V Epitaxial Growth on Wafer Scale Substrates**
---- Si⁺ insitu doping
---- Undoped (splits for Si⁺ Ion Implantation)
○ **Hard Mask Deposition**
○ **Fin Etch (into $In_{0.52}Al_{0.48}As$ for device isolation)**
○ **Hard Mask Strip**
○ **Surface Treatment (Fin Smoothing)**
● **Si⁺ Ion Implantation / Activation for undoped wafers**
○ **ILD Deposition**
○ **Slot Contact Etch and Cleans**
○ **Mo/Ti/TiN S/D Contact Dep.**
○ **W Contact Via Dep. and CMP**
○ **M1 Metal**

Planview: Slot Contact Etch

Fig. 2: Key steps in the III-V fin TLM structure process flow. Device isolation was realized by etching into the $In_{0.52}Al_{0.48}As$ layer. A post fin etch surface treatment was applied to all wafers to reduce sidewall surface roughness (SWR) variation.

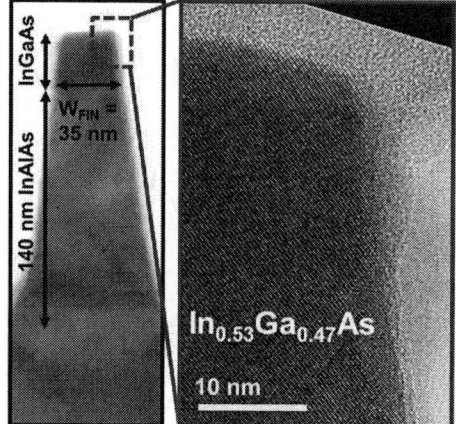

Fig. 3: Cross-sectional TEM image of a 35 nm wide III-V fin fabricated with Process D. HRTEM shows that the etched $In_{0.53}Ga_{0.47}As$ fin top and sidewall surface is free from etch-induced crystalline damage.

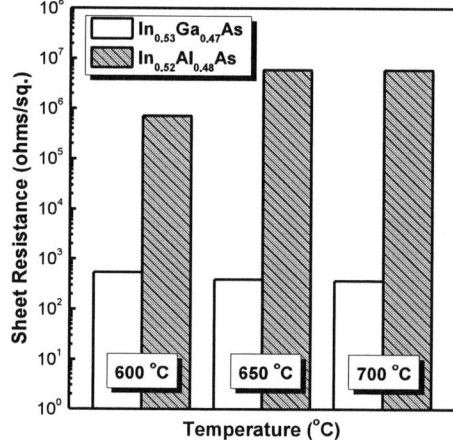

Fig. 4: Plot of sheet resistance for $In_{0.53}Ga_{0.47}As$ and $In_{0.52}Al_{0.48}As$ layers implanted with Si⁺ and annealed. It was found that Si⁺ doped $In_{0.52}Al_{0.48}As$ layers remain highly resistive making it suitable as an isolation layer.

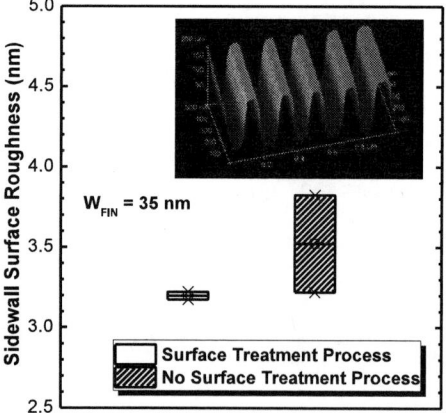

Fig. 5: Comparison of sidewall surface roughness (SWR) for etched III-V fins with and without the surface treatment process. III-V fins that received the surface treatment process show a reduction in SWR variation compared to III-V fins without the surface treatment process.

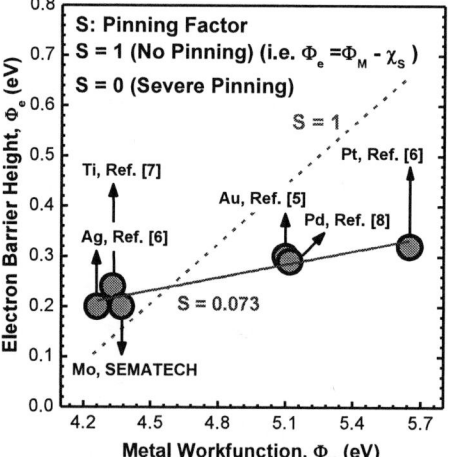

Fig. 6: Plot of electron barrier height (Φ_e) versus metal work function (WF) for non-alloyed metal contacts on $In_{0.53}Ga_{0.47}As$. It becomes evident that Φ_e is independent of its metal WF as S = 0.073.

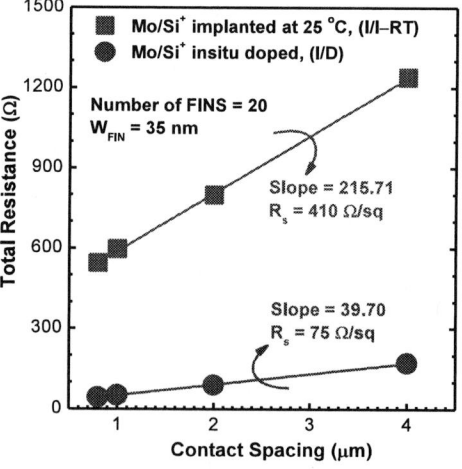

Fig. 7: Plot of total resistance as a function of fin TLM contact spacing for III-V fins doped with Si⁺ insitu doping (I/D) and Si⁺ implantation at 25 °C (I/I–RT) A good linear fit to the data is obtained.

978-1-4799-8002-4/14 $31.00 © 2014 IEEE

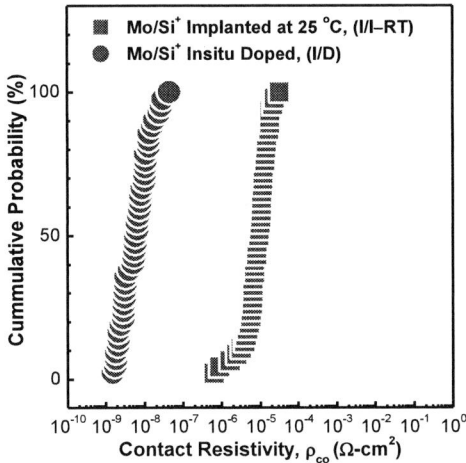

Fig. 8: Plot of cumulative distributions of ρ_{co} for Si$^+$ insitu doped (I/D) fins and Si$^+$ implanted at 25 °C (I/I–RT) fins. Good process uniformity is achieved across wafer scale III-V on Si substrates.

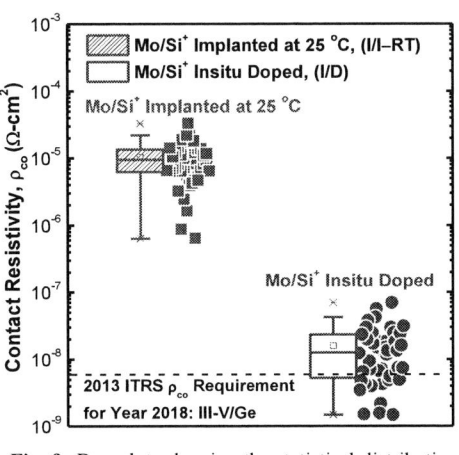

Fig. 9: Box plots showing the statistical distribution in ρ_{co} for Si$^+$ insitu doped (I/D) fins and Si$^+$ implanted at 25 °C (I/I–RT) fins. It is clear that I/D fins have lower ρ_{co} down to sub-1×10^{-8} Ω-cm^2, which meets ITRS ρ_{co} requirements for III-V in year 2018.

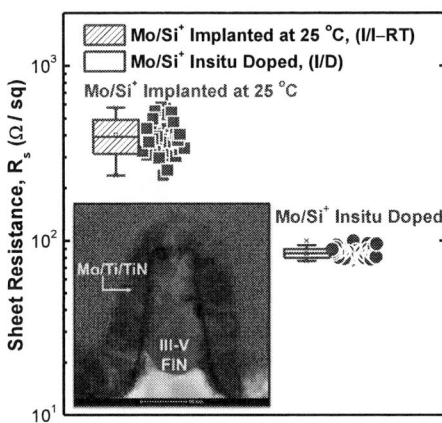

Fig. 10: Box plots showing the statistical distribution in R_s for Mo/Si$^+$ insitu doped (I/D) fins and Mo/Si$^+$ implanted at 25 °C (I/I-RT) fins. The increase in R_s and ρ_{co} for I/I–RT fins is due to implant induced damage in III-V fins (see inset of Fig. 10).

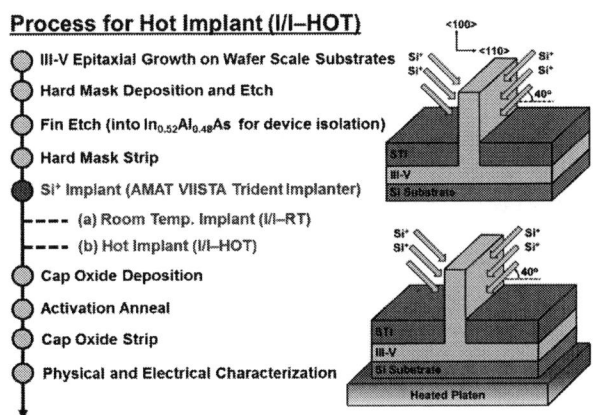

Fig. 11: Key steps in the process flow for the comparison of room temperature implant (I/I–RT) and hot implant (I/I–HOT) into III-V. Schematics show the process configuration used in this work for I/I–RT and I/I–HOT. Sheet resistances and SIMS measurements were obtained from In$_{0.53}$Ga$_{0.47}$As/In$_{0.52}$Al$_{0.48}$As/InP coupon substrates. TEM inspections were performed on III-V fins.

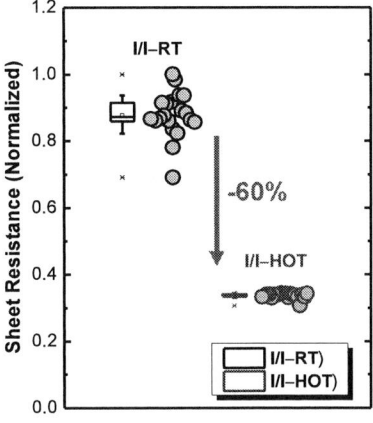

Fig. 12: Plot of normalized sheet resistance as a function of room temperature implant (I/I–RT) and hot implant (I/I– HOT). Lower sheet resistance values with tighter distributions were obtained with I/I–HOT.

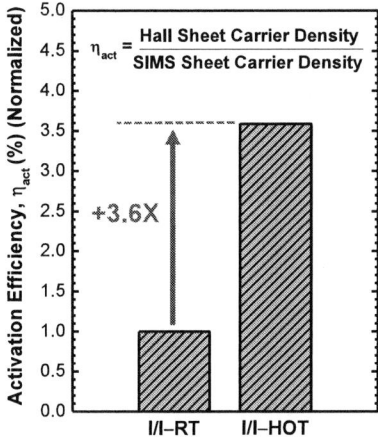

Fig. 13: Plot of activation efficiency as a function of room temperature implant (I/I–RT) and hot implant (I/I–HOT). Activation efficiency for Si$^+$ increases with I/I–HOT.

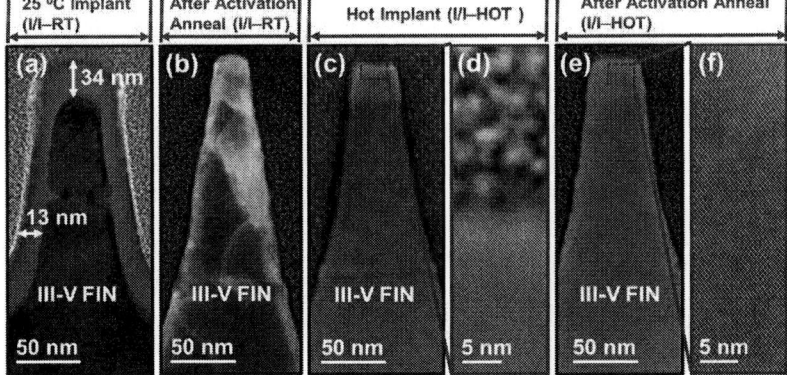

Fig. 14: (a) XTEM image of a III-V fin just after implantation at 25 °C (i.e. I/I–RT). An amorphous layer as thick as 34 nm is formed after I/I–RT. **Fig. 14(b)** XTEM image of the I/I–RT fn after activation anneal. **Fig. 14(c) and (d)** XTEM images of a III-V fin just after hot implant (I/I–HOT). No implant damage is observed and an amorphous layer is not formed after I/I–HOT. **Fig. 14(e) and (f)** XTEM images of the I/I–HOT III-V fin after activation anneal. Excellent crystallinity is maintained.

Contact Materials	Ref.	Metal Dep.	Surface Orientation	Doping (cm^{-3})	N-type ρ_{co} (Ω - cm^2)
Ni-InGaAs	[11]	Metal Last	(100), Planar	1×10^{19}	1.05×10^{-6}
Ni-InGaAs	[12]	Metal Last	(100), Planar	5×10^{19}	2.05×10^{-6}
Ti	[13]	Metal First	(100), Planar	1×10^{19}	2.01×10^{-8}
Pt	[13]	Metal First	(100), Planar	1×10^{19}	1.38×10^{-8}
Pd	[13]	Metal First	(100), Planar	1×10^{19}	7.50×10^{-9}
Mo	[14]	Metal First	(100), Planar	3.6e19	1.30×10^{-8}
Mo	[15]	Metal First	(100), Planar	1×10^{19}	6.90×10^{-9}
Mo	This Work	Metal Last	(110)/(100), Non-Planar	1×10^{19}	1.59×10^{-8}

Table II: Benchmark table comparing ρ_{co} of leading S/D contact materials on In$_{0.53}$Ga$_{0.47}$As reported in literature. We achieved a record low average ρ_{co} of 1.59×10^{-8} Ω-cm^2 for Mo contact on III-V fin sidewall (110)/(100) surfaces using a non-planar structure. This is the first report of ρ_{co} extraction on fin sidewalls, which is representative of the S/D contact/semiconductor interface for a FinFET. In contrast to many of these references a metal last VLSI compatible process was used in this work.

978-1-4799-8002-4/14 $31.00 © 2014 IEEE

Dramatic Effects of Hydrogen-induced Out-diffusion of Oxygen from Ge Surface on Junction Leakage as well as Electron Mobility in n-channel Ge MOSFETs

Choong Hyun Lee, Tomonori Nishimura, Cimang Lu, Shoichi Kabuyanagi, and Akira Toriumi

Department of Materials Engineering, The University of Tokyo
7-3-1, Hongo, Tokyo 113-8656, Japan   Email: lee@adam.t.u-tokyo.ac.jp

Abstract

This paper discusses about effects of oxygen in Ge substrate on MOSFET performance from both viewpoints of advantages and disadvantages. For improvement of electron mobility in Ge n-MOSFETs, oxygen in the channel region should be extracted to suppress additional scattering. On the other hand, oxygen in S/D region is helpful for dramatically reducing junction leakage currents. By understanding these oxygen effects on Ge, high electron mobility Ge n-MOSFETs with the highest I_{on}/I_{off} ratio are demonstrated.

1. Introduction

A significant enhancement of electron mobility and understanding of carrier transport physics in Ge MOSFETs have been made thanks to the elimination of major extrinsic carrier scattering sources such as D_{it} and surface roughness [1,2]. Although we have reported that oxygen-related impurities in Ge substrate could be another origin of electron mobility degradation, it is not obvious whether this is also the case for hole mobility in p-MOSFETs. Furthermore, another big concern for Ge MOSFETs is relatively large junction leakage current, particularly for n-MOSFETs.

In this work, we clarify the role of oxygen in Ge substrate by directly comparing both electron and hole mobility in Ge MOSFETs. Then, we discuss the oxygen effect in Ge substrate on reverse-biased n^+/p junction leakage current. Finally, high electron mobility Ge n-MOSFETs with high I_{on}/I_{off} ratio, designed based upon those understandings, are demonstrated.

2. Oxygen-related Scattering for μ_e and μ_h in MOSFETs

We reported that electron mobility was much improved by extracting oxygen from Ge substrate with H_2 annealing, even though the initial mobility was relatively poor [1]. **Fig. 1** shows the depth profile of oxygen in oxygen-rich ($\sim$1e16 cm^{-3}) Ge wafers measured by SIMS. It clearly shows that the amount of oxygen concentration near Ge surface is reduced by H_2 annealing to make the denuded zone. **Fig. 2(a)** shows the energy distribution of D_{it} at the GeO$_2$/Ge interface with and without H_2 annealing, where GeO$_2$ was thermally grown by low-temperature high-pressure oxidation (LT-HPO) [3] and EOT of GeO$_2$/Ge stack was about 4 nm. It is very clear that there is no interface degradation by H_2 annealing. The V_{FB} shift and D_{it} estimated near the conduction band edge also show no difference of electrical properties in GeO$_2$/Ge stacks as shown in **Fig. 2(b)**. It suggests that oxygen seems to be neutral states in bulk Ge, in spite of oxygen extraction from Ge surface.

To carefully examine the oxygen-related scattering for both electron and hole mobility, we employed oxygen-rich ($\sim$1e16 cm^{-3}) Ge(111) wafers with a resistivity of 0.6 Ωcm for fabricating both n- and p-MOSFETs. After chemical cleaning of Ge, H_2 annealing was carried out at 650 to 850°C to extract the dissolved oxygen in n- and p-Ge substrates. Phosphorus and boron at a dose of 1e15 cm^{-2} were implanted with 50 keV and 20 keV for S/D formation, respectively. Dopant activation was carried out at 600°C for 30 sec in N_2 ambient. 6 nm-thick GeO$_2$ was thermally grown by LT-HPO to have superior GeO$_2$/Ge interface properties, as shown in Fig. 2(a). We note that 4 nm EOT of pure GeO$_2$/Ge stack without any high-k capping layer was employed as a dielectric layer to exclude any additional scattering sources. H_2 annealing temperature dependences of electron and hole mobility as a function of N_s are shown in **Figs. 3(a)** and **(b)**. Interestingly, a significant improvement of electron mobility in a wide range of N_s in n-MOSFETs is observed by oxygen extraction, while very slight effect on hole mobility in p-MOSFETs is shown.

In order to understand asymmetric effect of H_2 annealing between electron and hole mobility in Ge MOSFETs, it is inferred that the neutral states induced by dissolved oxygen may exist in the upper half of Ge band gap as shown in **Fig. 4**. In this configuration of oxygen in Ge, it could work for electron scattering centers because it is likely to be negatively charged in n-MOSFETs operation in over-threshold region. This is quite similar to the border trap consideration in Ge n-MOSFETs [4]. In case of p-MOSFETs, such states should be neutral, and no effect on the hole mobility is reasonably expected.

3. Reduction of Off-state Current by O Incorporation

The improvement of electron mobility directly corresponds to an enhancement of on-state performance in Ge n-MOSFETs as shown in **Fig. 5(a)**. However, it is interesting yet problematic that the off-state current in Ge n-MOSFETs also increases by oxygen extraction with H_2 annealing, while no H_2 annealing effect on off-state current in p-MOSFETs is observed (**Fig. 5(b)**). It indicates that oxygen in Ge substrate has more influence on n-MOSFETs performance such as electron mobility and n^+/p junction leakage current.

In order to understand the role of oxygen in Ge n^+/p junction, oxygen (^{16}O) ion was intentionally implanted to oxygen-less (below 1e15 cm^{-3}) Ge(100) wafer at a dose of 1e13 and 1e14 cm^{-2} with 100 keV, respectively, in which the initial oxygen concentration of Ge wafer was lower than SIMS detection limit. The depth profile of oxygen (as-implanted) is

978-1-4799-8002-4/14 $31.00 © 2014 IEEE

shown in **Fig. 6**. Subsequently, phosphorous ion was implanted at a dose of 1e15 cm^{-2} with 50 keV through 30 nm Y$_2$O$_3$ buffer layer to make n$^+$/p junction. The dopant activation was carried out at 400 to 650°C for 30 sec in N$_2$ ambient. Al was deposited by the vacuum evaporation for the gate electrode and ohmic contact of diodes, respectively. It is worth noting that Y$_2$O$_3$ passivation layer on Ge was employed to reduce peripheral surface-state current due to interface traps in n$^+$/p junction. **Fig. 7(a)** shows *I-V* characteristics of n$^+$/p junctions formed by oxygen, followed by phosphorus ion implantation (I/I), where dopant activation annealing was carried out at 600°C for 30 sec in N$_2$. It shows good rectifying diode characteristics and much lower reverse junction leakage current, compared to conventional phosphorous I/I case. The I_{on}/I_{off} ratio is about 10^6 and reverse junction leakage current is reduced down to 2e-4 A/cm^2, which is the lowest reverse leakage current in Ge n$^+$/p junctions so far reported [5-10]. The annealing temperature dependence of reverse junction leakage current in Ge n$^+$/p junction is shown in **Fig. 7(b)**. It is also possible to reduce S/D formation temperature down to 500°C for Ge n-MOSFETs with a reasonable off-state current with the help of oxygen I/I. In order to clarify the statistical aspects of junction leakage current, the Weibull plot of the leakage current at reverse bias of 1 V for different annealing conditions (500 and 600°C) was shown in **Fig. 8**. Regardless of annealing conditions, n$^+$/p junctions with oxygen I/I exhibit a tight distribution of the low leakage current, compared to conventional I/I case. This effect is more pronounced for the samples with oxygen I/I at a dose of 1e13 cm^{-2}, which might be attributable to less-damage of oxygen I/I.

The temperature dependence of *I-V* characteristics was investigated to see the dominant mechanism of reverse junction leakage current in n$^+$/p junction formed by oxygen, followed by phosphorus I/I. **Fig. 9(a)** shows the wafer temperature dependence of the leakage current at reverse bias of 1 V, plotted as a function of the inverse temperature. The activation energy (E_A) of leakage current in oxygen-implanted n$^+$/p junction is around 0.46 eV, which is much higher than conventional phosphorous I/I one (E_A ~0.25 eV). We note that most of Ge n$^+$/p junctions [5,9,10] have an E_A value around the mid-gap (~0.3 eV) for the leakage current, suggesting that the formation of numerous generation and recombination (G-R) centers near the mid-gap is the dominant mechanism for junction leakage current. Another possible reason is the presence of metallic impurities dissolved deeper inside Ge substrate, which gives rise to electronic defect levels in the band gap of Ge [11]. It can also generate leakage current due to the formation of G-R centers at junction edge.

With respect to oxygen I/I, it might passivate the traps (defects) assisting the tunneling or helping generating electron-hole pairs in the band gap of Ge, so the reverse junction leakage current is significantly suppressed, as schematically depicted in **Fig. 9(b)**. Furthermore, we have found another advantage of oxygen incorporation into n$^+$/p junction, which is an increase of junction breakdown voltage

as shown in **Fig 10**. This fact might be attributed to gettering effects of metallic impurities deeper inside Ge by oxygen, which has been intensively discussed in Si technology [12]. This is a good news because it is expected that the junction leakage current in Ge can be further suppressed.

4. High Mobility Ge n-FETs with High I_{on}/I_{off} Ratio

In order to fully utilize the oxygen effects on Ge MOSFETs, we fabricated high mobility Ge(100) n-MOSFETs with high I_{on}/I_{off} ratio by selective areal profile of oxygen in the active area, as described in **Fig. 11**. Oxygen (1e13 cm^{-2}) was initially implanted to overall Ge substrate for improving S/D junction properties as discussed previously. The oxygen extraction and atomically flat Ge surface in the channel region were achieved at the same time by H$_2$ annealing at 850°C. Note that S/D region was covered by thick SiO$_2$ to prevent the out-diffusion of oxygen by H$_2$ annealing. 6 nm-thick GeO$_2$ was thermally grown by LT-HPO for the formation of atomically flat GeO$_2$/Ge interface. **Fig. 12(a)** shows I_D-V_G curves of Ge(100) n-MOSFETs fabricated with selective areal profile of oxygen. It is worth noting that the record-high I_{on}/I_{off} ratio of 10^5 in Ge n-MOSFETs with a very low subthreshold slope (*SS*=74 mV/dec.) is achieved in spite of the long channel MOSFETs (L$_{ch}$ = 200 μm) thanks to the independent control of carrier scattering from trap-assisted leakage current. **Fig. 12(b)** shows electron mobility as a function of N_s with and without oxygen extraction in the channel region. A significant enhancement of electron mobility together with high I_{on}/I_{off} ratio is achieved by the selective areal profile of oxygen. The performance benchmark of Ge n-MOSFETs is shown in **Table I**. The present work shows an excellent performance in Ge n-MOSFETs in terms of I_{on}/I_{off} ratio, subthreshold slope, and electron mobility. Further enhancement of Ge n-MOSFETs performance can be expected simply with shrinking the device size.

5. Conclusion

We have studied oxygen effects in Ge in terms of carrier scattering and junction leakage current, which have been considered to be incompatible. However, we have demonstrated high electron mobility Ge n-MOSFETs together with high I_{on}/I_{off} ratio by selective areal profile of oxygen for the channel and S/D regions. Although the junction leakage has been considered to be intrinsically poor in Ge, this paper strongly suggests that Ge technology will be further improved for the best with the help of a rich heritage of Si technology.

References

[1] C. H. Lee *et al.*, *IEDM* (2013). [2] C. H. Lee *et al.*, *VLSI*, (2014). [3] C. H. Lee *et al.*, *APEX*, **5**, 114001 (2012). [4] R. Zhang *et al.*, *VLSI*, (2013). [5] K. Morii *et al.*, *IEDM* (2009). [6] W. B. Chen *et al.*, *IEDM* (2010). [7] Y.-C. Fu *et al.*, *IEDM* (2010). [8] Y.-J. Lee *et al.*, *IEDM* (2012). [9] G. Thareja *et al.*, *IEEE EDL*, **32**, 608 (2011). [10] M. Jamil et al., *IEEE EDL*, **32**, 1203 (2011). [11] S. M .Sze and J. C. Irvin, *Solid-State Electron.*, **11**, 599 (1968). [12] F. Shimura, "Semiconductor Silicon Crystal Technology," p. 344, (Academic Press, 1989).

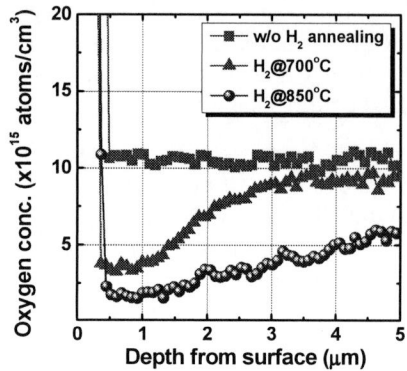

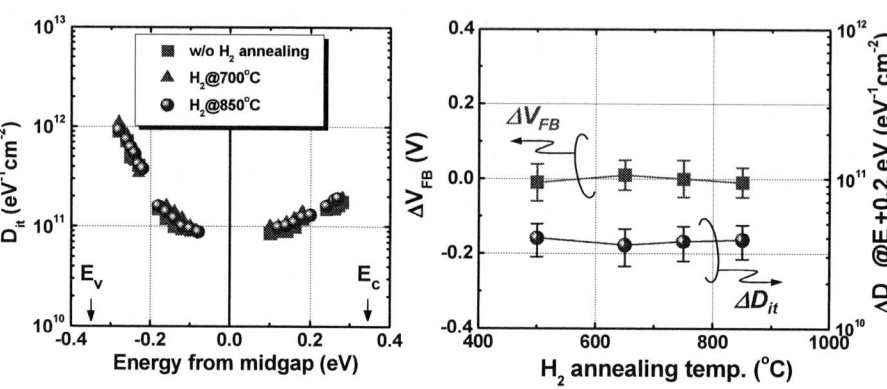

(a)   **(b)**

Fig. 1 Depth profile of oxygen in oxygen-rich Ge wafer measured by SIMS. The amount of oxygen concentration near Ge surface is reduced by H_2 annealing at 700°C and further reduced at 850°C. It clearly shows that the amount of oxygen concentration near Ge surface is reduced by H_2 annealing to make the denuded zone.

Fig. 2 (a) Energy distribution of D_{it} at the GeO$_2$/Ge interface estimated by the low temperature conductance method. The GeO$_2$ was thermally grown by LT-HPO [3] and EOT is about 4 nm. Note that H_2 annealing was carried out before thermal oxidation. It is very clear that there is no interface degradation by H_2 annealing. **(b)** V_{FB} shift and ΔD_{it} estimated near the conduction band edge as a function of H_2 annealing temperature. In spite of oxygen extraction from Ge surface, no differences of electrical properties such as C-V characteristics, V_{FB} shift, and D_{it} spectrum are observed, indicating that oxygen seems to be neutral states in bulk Ge.

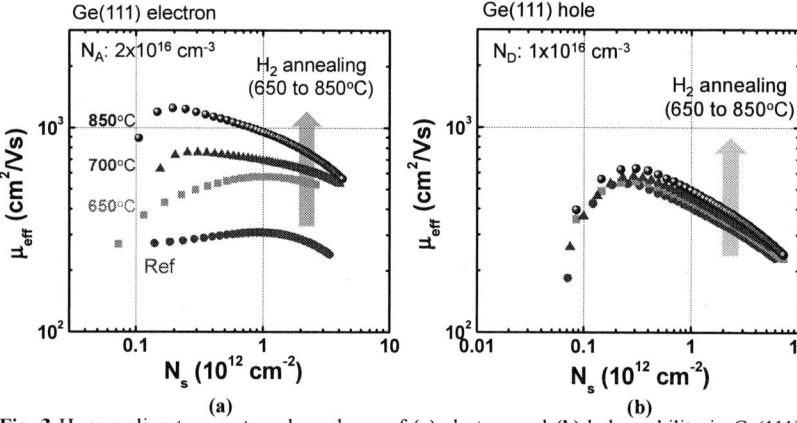

(a)   **(b)**

Fig. 3 H_2 annealing temperature dependence of **(a)** electron and **(b)** hole mobility in Ge(111) MOSFETs as a function of N_s. Pure GeO$_2$/Ge gate stack was used to exclude any additional scattering sources. EOT of gate stack is about 4 nm for all the devices. It is very interesting to see that a significant improvement of electron mobility in Ge n-MOSFETs is observed by H_2 annealing due to oxygen extraction, while very slight oxygen effects on hole mobility in p-MOSFETs is shown. Oxygen in Ge surface works differently for electron and hole mobility.

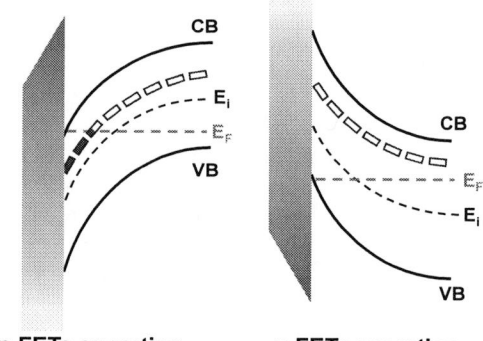

Fig. 4 Schematic of oxygen-related scattering centers for electron and hole mobility in Ge substrates. It is inferred that the neutral states induced by dissolved oxygen may exist in the upper half of Ge band gap, which might work for electron scattering centers because it is likely to be negatively charged in n-MOSFETs operation. However, in p-MOSFETs operation they should be neutral states, and thus no scattering can occur for hole mobility.

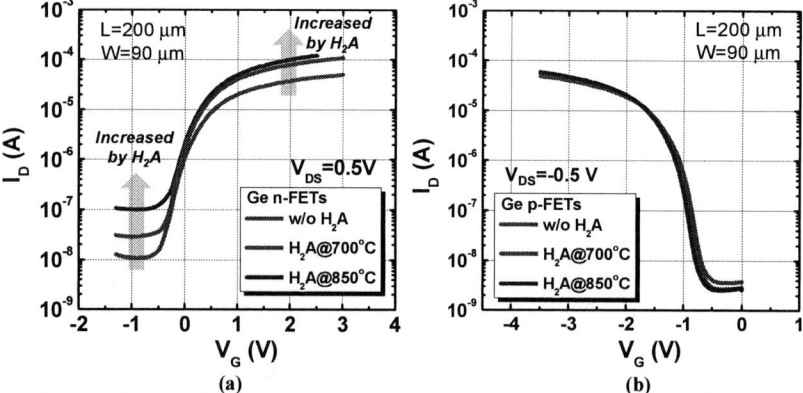

(a)   **(b)**

Fig. 5 (a) Typical transfer characteristics (I_D-V_G) of GeO$_2$/Ge(111) n-MOSFETs with different H_2 annealing temperature. The EOT of GeO$_2$/Ge stack is about 4 nm. The improvement of electron mobility directly corresponds to on-current enhancement in Ge n-MOSFETs due to oxygen extraction, but it is very interesting to see that the off-state current also increases by H_2 annealing. **(b)** Typical transfer characteristics (I_D-V_G) of GeO$_2$/Ge(111) p-MOSFETs with different H_2 annealing temperature. No H_2 annealing dependence of off-state current in p-MOSFETs is observed as well as on-state current. It indicates that oxygen in Ge substrate has more influence on n-MOSFETs performance.

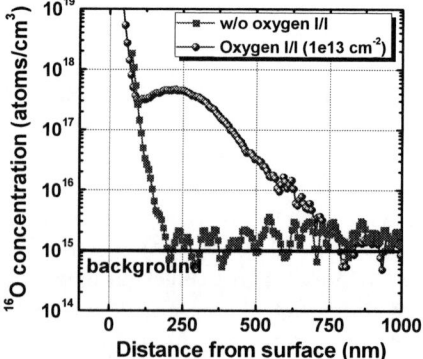

Fig. 6 Depth profile of implanted oxygen (¹⁶O) in Ge n⁺/p junction. Initial oxygen concentration in Ge wafer is below SIMS detection limit (below 1e15 cm⁻³). After oxygen implantation at a dose of 1e13 cm⁻² with 100 keV, 500 nm-thick oxygen implanted-layer in Ge was formed. Subsequently, phosphorous was implanted at a dose of 1e15 cm⁻² with 50 keV to make n⁺/p junction.

978-1-4799-8002-4/14 $31.00 © 2014 IEEE

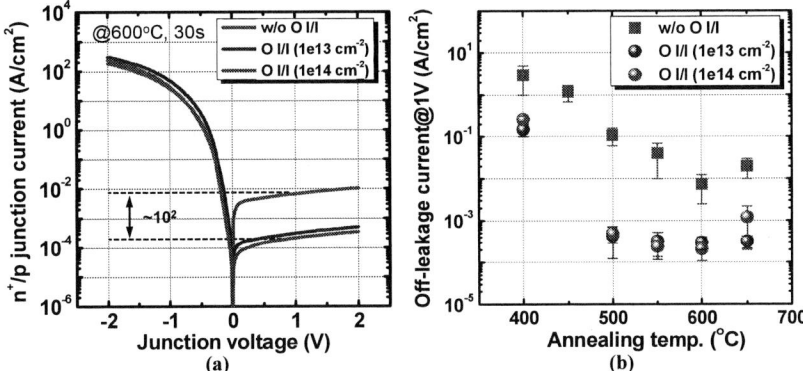

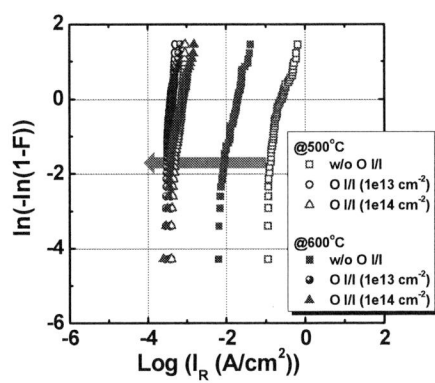

Fig. 7 (a) *I-V* characteristics of Ge n^+/p junctions fabricated by oxygen, followed by phosphorus ion implantation, where dopant activation annealing was carried out at 600°C for 30 sec. The I_{on}/I_{off} ratio is about 10^6 and reverse leakage current is reduced down to 2e-4 A/cm^2 at room temperature. **(b)** Reverse leakage current of n^+/p junction as a function of annealing temperature. It indicates that it is possible to reduce S/D formation temperature down to 500°C for Ge n-MOSFETs with a reasonable off-state current.

Fig. 8 Weibull plot of reverse junction leakage current in Ge n^+/p junctions. Regardless of annealing conditions, n^+/p junctions with oxygen I/I exhibit a tight distribution of the low leakage current, compared to conventional I/I case.

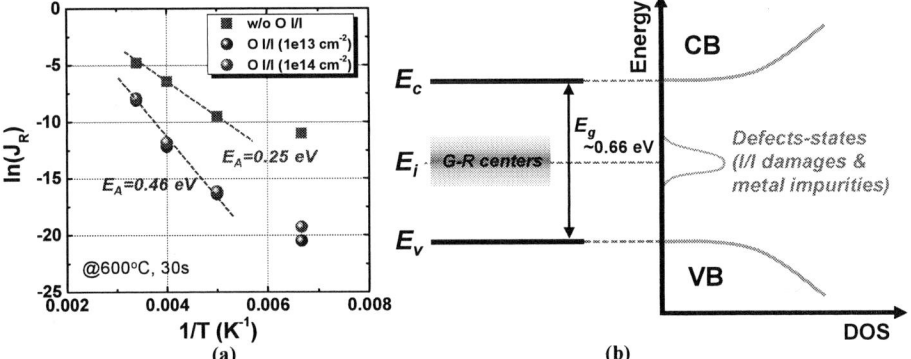

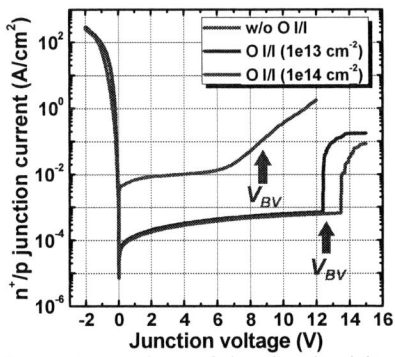

Fig. 9 (a) Temperature dependence of reverse leakage current at reverse junction voltage of 1 V. The activation energy (E_A) of reverse leakage current in oxygen and phosphorus co-implanted junction is 0.46 eV, which is much higher than conventional phosphorus I/I one (E_A ~0.25 eV). **(b)** Schematic of dominant mechanism for reverse leakage current mechanism in Ge n^+/p junction. Oxygen might passivate the defects levels in the band gap of Ge generated by ion implantation or dissolved metallic impurities and significantly suppress the defects-assisted tunneling current.

Fig. 10 Comparison of junction breakdown voltage (V_{BV}) in Ge n^+/p junction. Another advantage of oxygen incorporation into n^+/p junction is an increase of junction breakdown voltage, which might be attributed to the gettering effects of metallic impurities in Ge by oxygen.

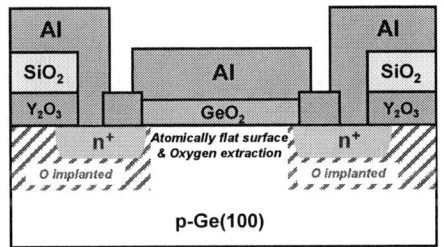

Fig. 11 Schematic of Ge n-MOSFETs with selective areal profile of oxygen for high mobility and high I_{on}/I_{off} ratio. In order to fully utilize the oxygen effects on Ge n-MOSFETs, Oxygen (1e13 cm^{-2}) was initially implanted to overall Ge substrate for improving S/D junction properties. Then, oxygen in the channel region was extracted by H_2 annealing at 850°C. Atomically flat Ge surface can be also achieved at the same time.

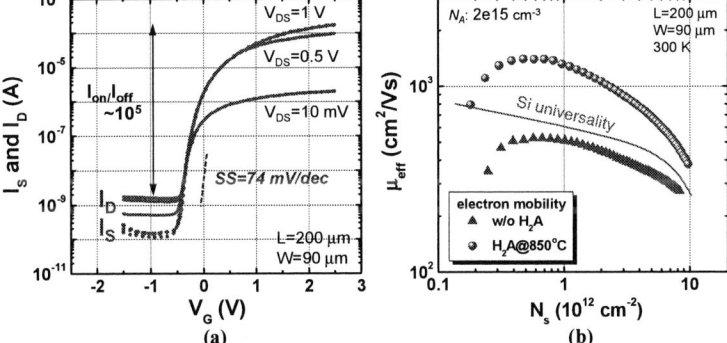

Fig. 12 (a) I_D-V_G curves of the fabricated GeO$_2$/Ge(100) n-MOSFETs with selective areal profile of oxygen, where EOT of GeO$_2$/Ge gate stack is about 4 nm. It shows excellent switching properties with high I_{on}/I_{off} ratio of 10^5 as well as a very low *SS* of 74 mV/dec in spite of the long channel FETs. **(b)** Electron mobility as a function of N_s with and without oxygen extraction in the channel region. A significant enhancement of electron mobility is achieved by oxygen extraction.

Table I Performance benchmark of Ge n-MOSFETs

*Estimated from the literatures

Ref.	Structure	Orientation	EOT (nm)	L_{ch} (μm)	I_{on}/I_{off} (V_D=1V)	S.S. (I_D) (mV/dec.)	Mobility (cm²/Vs)
[1]	Planar	Ge(100)	~20 *	70	10^4	193	804
[2]	Planar	Ge(100)	0.95	10	10^4	106	645
[3]	Planar	Ge(100)	~12 *	45	10^3	150	1050
[4]	Planar	Ge(100)	0.8	50	10^3	~250 *	300
[5]	Planar	GeSn(100)	~3.3 *	5	10^3	135	380
[6]	Planar	GeSn(100)	0.76	5	10^3	80	689
[7]	Fin	Ge(110)	~4 *	0.17	10^4 *	144	-
[8]	GAA	Ge(111)	5.5	0.12	10^4	94	-
This work	**Planar**	**Ge(100)**	**4**	**200**	**10^5**	**74**	**1412**

References

[1] K. Morii *et al.*, *IEDM*, 681 (2009). [2] W. B. Chen *et al.*, *IEDM*, 420 (2010). [3] Y.-C. Fu *et al.*, *IEDM*, 432 (2010). [4] C.-M. Lin *et al.*, *IEDM*, 509 (2012). [5] S. Gupta *et al.*, *IEDM*, 375 (2012). [6] R. Zhang *et al.*, *IEEE ED*, 60, 927 (2013). [7] C.-T. Chung *et al.*, *IEDM*, 383 (2012). [8] S.-H. Hsu *et al.*, *IEDM*, 525 (2012).

Evolution of Directed Ion Beams from Doping to Materials Engineering

Anthony Renau

Applied Materials, Varian Semiconductor Equipment
Gloucester, MA, USA

Abstract

We review recent changes to implanter processing capabilities, including the adoption of cryogenic implants to reduce leakage and contact resistance as well as high temperature implants for finFETs. We discusss some specific 3D challenges and introduce a new process technology for 3D that uses directed ion beams for material modification including implant, etch and deposition.

Introduction

Ion implantation has been used for many years for FEOL doping applications (Fig. 1) and implanters have evolved into highly sophisticated tools such as that shown in Fig 2 [1]. This evolution has been driven by productivity and scaling, which in turn, are driven by dopant placement accuracy [2]. However, only a fraction of today's implants are for electrical doping. The majority use the implanter's ability to accurately and abruptly introduce pretty much any quantity of any atom for material modification. Recently, temperature control has been added to some implanters to govern the substrate crystallinity and, thereby, the incorporation of new materials. This will be discussed further.

Three dimensional structures (e.g. finFETs) have created a number of challenges for implant. For example, we will discuss how temperature control is critical to some aspects of finFET doping. Additionally we will discuss how high aspect ratio structures provide opportunities for a different type of doping process, plasma doping. 3D has also provided a number of

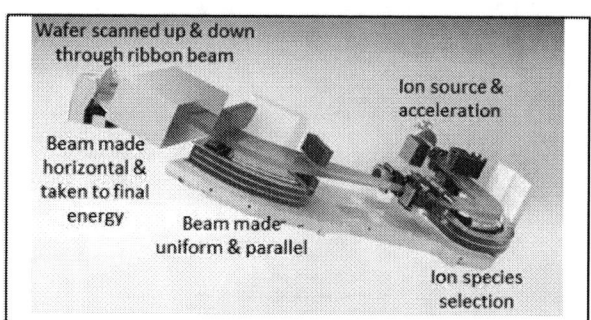

Fig 2. Single wafer tool. Implants ~500WPH by directing a uniform horizontal beam at a vertically scanned wafer

opportunities to extend ion beam processing beyond implant. We discuss a tool that is designed to do this and has directed ion beam processing capability that can enable many new applications.

Evolution of Temperature Control

The rate of lattice disruption caused by implant is greater at low temperatures because the lattice energy is lower. Under these conditions implant amorphizes more rapidly and interstitials tend to remain trapped within that amorphous layer [3]. During anneal, the amorphous layer will re-crystallize readily and fewer residual defects will be left at the end of range (beyond the amorphous / crystal interface). As the implant gets colder there will be less dynamic annealing and fewer interstitial defects. This effect is illustrated in Fig. 3, which shows fluorine decorating the end of range damage and how a -100°C implant stops this. The impact of this on the post-

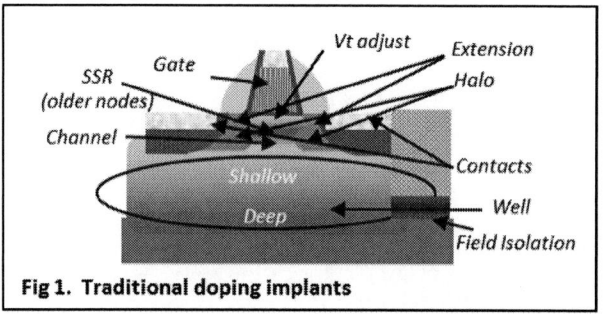

Fig 1. Traditional doping implants

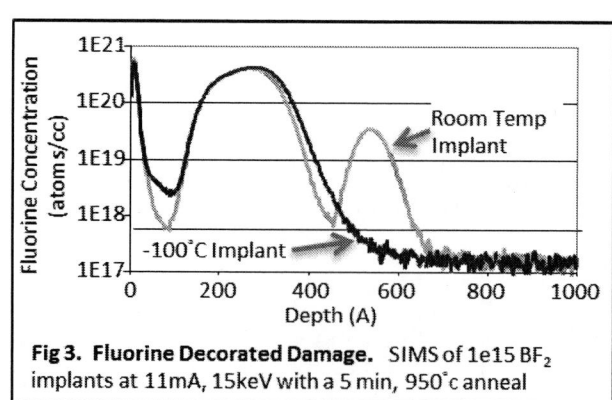

Fig 3. Fluorine Decorated Damage. SIMS of 1e15 BF_2 implants at 11mA, 15keV with a 5 min, 950°c anneal

978-1-4799-8002-4/14 $31.00 © 2014 IEEE

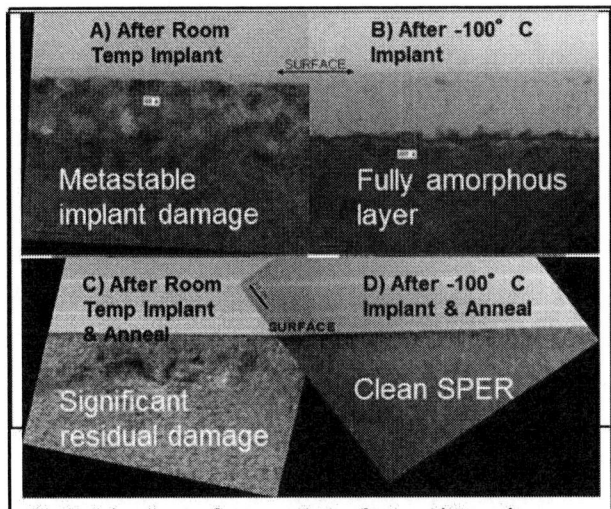

Fig 4. Advantage of cryogenic implants. A) Sample implanted at room temperature showing metastable damage B) The same implant at cryogenic temperatures resulting in fully amorphous layer C) Sample 'A' post anneal showing significant defects D) Sample 'B' post anneal showing clean Solid Phase Epitaxial Regrowth

annealed damage that could lead to device leakage is shown in Fig. 4 [4].

Cryogenic implants have more benefits than leakage current reduction. They can, for example, greatly reduce contact resistance by thermally stabilizing the NiSi formation in a pMOS contact flow [5]. Carbon implants can be used during this process to limit Ni diffusivity. The impact of the temperature of this implant on the thermal stability of the NiSi is shown in Fig 5. The temperature at which NiSi starts to form can be seen clearly, as can the temperature at which it starts to agglomerate (and affect the strained SiGe layer). We believe that the superior stability and resistance with the cryogenic implant is because it makes the implant amorphizing.

High temperature implant capability has also been developed for some processes. It is used when implant

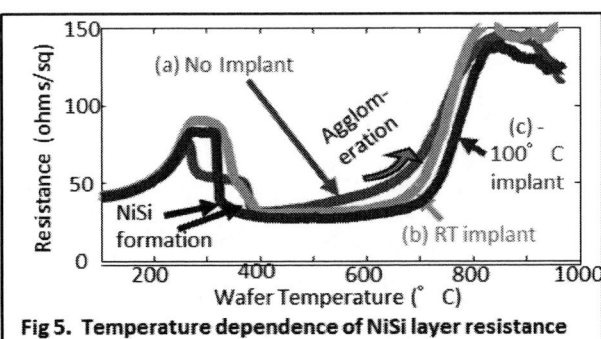

Fig 5. Temperature dependence of NiSi layer resistance Formed (a) without PAI implant, b) with RT PAI C+ and c) -100°C C+ implant. Latter is most thermally stable.

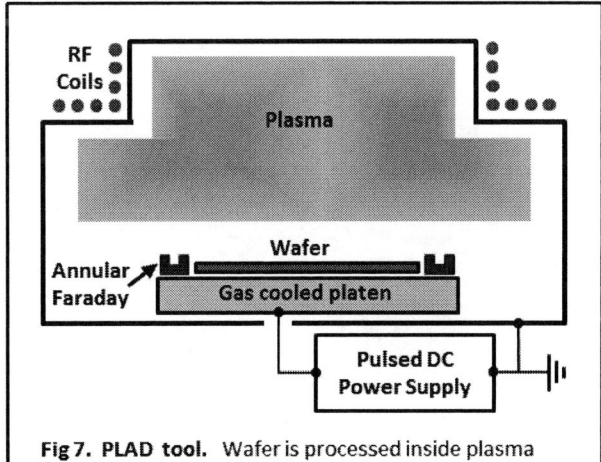

Fig 6. Hot implantation advantage in finFET processing. Arsenic Implant followed by a 1050°C RTA. When implant is at room temp, residual damage can be seen, whereas a 400°C implant enables complete recrystalization.

damage is detrimental and non-recoverable. This might be true of some halo implants (Fig. 1) for which the dose is too low to amorphize. High temperatures are particularly useful for fin implants [6]. Implant damage to a fin can usually not be repaired by solid phase epitaxial regrowth (Fig 6). Oftentimes, chevron or 111 regrowth occurs during anneal.

Challenge of 3D

3D structures introduce other challenges to implant for which new solutions have been developed.

Plasma doping is used for DRAM processes because it is very well suited to extremely high dose applications [7]. Unlike a beam line implanter, the tool processes wafers in the plasma chamber (Fig. 7). The wafer is pulsed negatively with a DC bias power supply. When the pulse is on, ions are implanted. When it is off, the wafer is exposed to the neutrals and radicals from the plasma, which is generated by a sophisticated multi-set-point

Fig 7. PLAD tool. Wafer is processed inside plasma chamber. Ions are implanted when DC pulse is on and are measured by a magnetically suppressed Faraday.

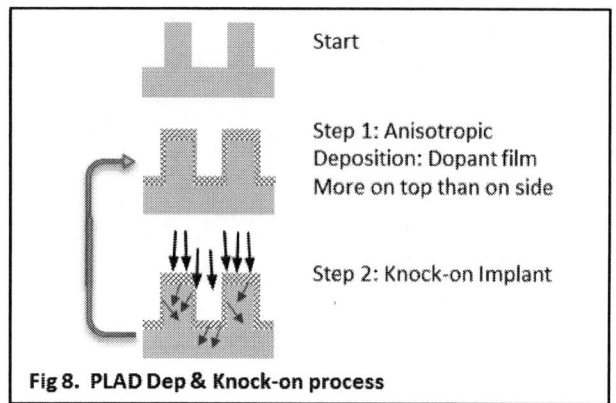

Start

Step 1: Anisotropic
Deposition: Dopant film
More on top than on side

Step 2: Knock-on Implant

Fig 8. PLAD Dep & Knock-on process

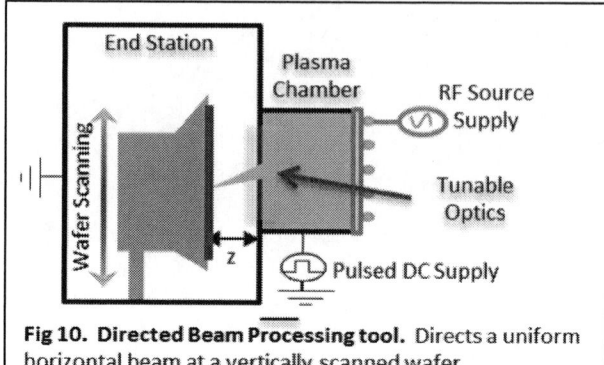

Fig 10. Directed Beam Processing tool. Directs a uniform horizontal beam at a vertically scanned wafer

power supply. Neutral and radical processing can therefore be controlled and enriched or minimized. This capability makes the tool well suited for some key 3D processes. For example:

i) Fins can be doped using a combination (Fig. 8) of isotropic deposition (pulse off) and highly collimated energetic ions (pulse on).

ii) The work function of metal gates can be tuned by using PLAD's highly directional implants to penetrate and modify the high aspect ratio structure (Fig. 9) [8].

Directed ion beam processing

We have shown a little of how beam line and plasma tools have been enhanced for 3D applications. However, we believe that there are many more opportunities for directed ion beam processing particularly if, as PLAD has demonstrated, this is combined with simultaneous control of radicals and neutrals.

Fig. 10 shows a tool that captures this concept for 3D applications. The process chamber has an external ion source and extraction system that allows a ribbon ion

beam to be directed at a scanning wafer. A wide range of angles can be tuned (Fig. 11). This gives the system unique capabilities, some of which are shown in Figs. 12-15.

Since the ribbon ion beam has tunable angles in one direction, but no structure in the other, the tool has directional selectivity (Fig. 12). This means that sidewalls will be processed if they are parallel to the ribbon beam, but not when they are perpendicular. For example, fin sidewall spacers can be etched while gate spacers remain unmodified.

The ion beam direction can be optimized to treat preferentially a particular part of a 3D structure. For example, Fig. 13 shows how directed beam processing can be used to etch the top part of a trench. This could be used for instance as part of high aspect ratio contacting problem, or to improve a replacement gate process.

The beam can also be tuned for more isotropic processing. This might use the beams illustrated in Fig. 11a or Fig. 11d, or a combination of these. If isotropy were required through one cross section only (e.g. though the fins, but not the gates) then the wafer would not rotate between scans. If it were required everywhere,

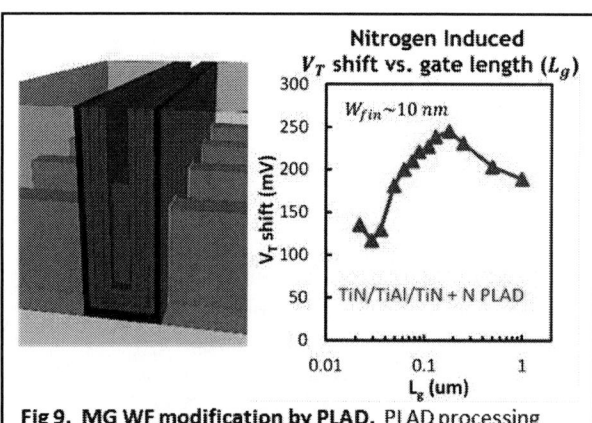

Fig 9. MG WF modification by PLAD. PLAD processing was able to achieve a ~200mV shift in Vt by implanting into the target (red) region of this HAR structure to modify the metal gate work function.

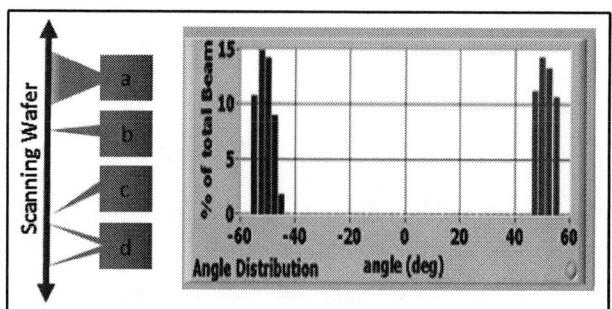

Fig 11. Tunable Angular Distribution. The ion beam can be directed in many ways, e.g. a) broad angle, b) focused, c) glancing, or d) bi-directional. The graph shows the measured beam angular distribution for the latter case.

978-1-4799-8002-4/14 $31.00 © 2014 IEEE   786

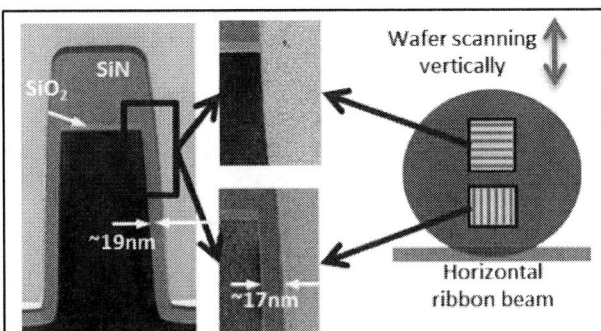

Fig 12. Direction Selectivity. Etching of SiN deposited on fin structures by a directed beam processing tool. The SiN on the side wall of fins parallel to the beam is etched >> 10X more than on the fins perpendicular to the beam.

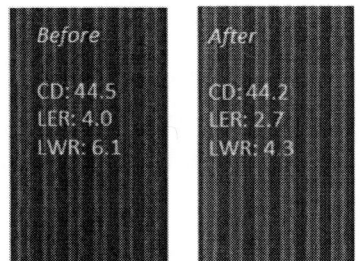

Fig 14. LER Improvement with Directed ion Beam. Over 30% LER improvement with little CD change for optimized mix of angles and process conditions.

then the wafer would be rotated. Photo-resist treatment for line edge roughness reduction is an example of a 3D structure that requires multi directional treatment. This can be optimized to enable, for example, significant improvements in LER for both 193i (figures 14 & 15) and eUV lithography. [9]

Summary

Ion implantation has evolved significantly beyond the tools that were used for the electrical doping of silicon. Today's tools are not only highly automated pieces of equipment that are used to implant at precisely the correct position and angle, but they also control the implanted material temperature and, thereby, the activation and defectivity of the process. A majority of implants are used for materials modification and this is being taken to the next level by a new type of directed ion beam tool that is being developed for a wide variety of materials engineering applications.

References

1) C. Campbell, A. Cucchetti, F. Sinclair, P. Kellerman, S. Radovanov, and S. Falk; "VIISta Trident: new generation high current implant technology." AIP Conference Proceedings 1496, 296-299. (2012)
2) H. L. Gossmann; "Junction formation and its device impact through the nodes", J. Vac. Sci. Technol. B26, 267-272 (2008).
3) A. Murakoshi, K. Suguro, M. Iwase, M. Tomita and K. Okumura, MRS Proceedings, 610, B3.8 doi:10.1557/PROC-610-B3.8 (2000)
4) C. Hatem, A. Renau, N. Variam, H. Maynard, B. Colombeau; Insight Conference, 2009
5) A. Renau, 10th Int. Workshop on Junction Technology, Shanghai, China (2010)
6) M. Togo, Y. Sasaki,G. Zschätzsch,G. Boccardi, R. Ritzenthaler, J. W. Lee, F. Khaja, B. Colombeau, L. Godet, P. Martin, S. Brus, S. E. Altamirano, G. Mannaert, H. Dekkers, G. Hellings, N. Horiguchi, W. Vandervorst, and A. Thean, IEEE VLSI, 14-2, T196 (2013).
7) Y. Jeon et al., in Proceedings IIT2008, edited by E.G. Seebauer et al., AIP Conf. Proceedings 1066, American Inst. of Physics, Melville, NY 2008, pp. 133-136.
8) L. A. Ragnarsson, S. A. Chew, H. Dekkers, M. Toledano Luque, B. Parvais, A. De Keersgieter, K. Devriendt, A. Van Ammel, T. Schram, N. Yoshida, A. Phatak, K. Han, B. Colombeau, A. Brand1, N. Horiguchi, and A. V.-Y. Thean, IEEE VLSI, 46-47 (2014).
9) T. Ma, et al., "Post-litho line edge/width roughness smoothing by ion implantations", Advances in Resist Materials and Processing Technology XXX, Proc. of SPIE, Vol. 8682, 868206, 2013.

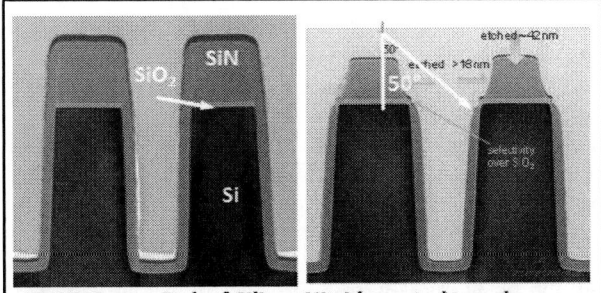

Fig 13. Corner Etch of Silicon Nitride coated trenches. Using a +/- 50° beam to open up the top of a trench.

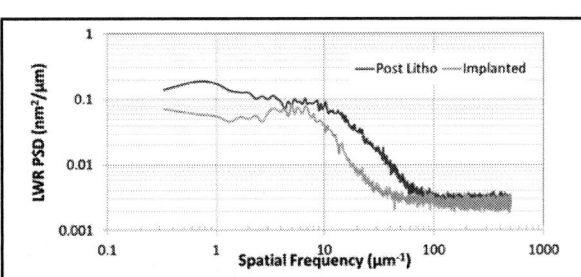

Fig 15. LWR Improvement with Directed ion Beam. Showing significantly improved LWR for 193i resist, particularly at low frequencies

A Novel Junctionless FinFET Structure with Sub-5nm Shell Doping Profile by Molecular Monolayer Doping and Microwave Annealing

Y.-J. Lee[1,*], T.-C. Cho[1,2], K.-H. Kao[3], P.-J. Sung[1,2], F.-K. Hsueh[1,2], P.-C. Huang[2], C.-T. Wu[1], S.-H. Hsu[1], W.-H. Huang[1], H.-C. Chen[1]
Y. Li[4], M. I. Current[5], B. Hengstebeck[6], J. Marino[6], T. Büyüklimanli[6], J.-M. Shieh[1], T.-S. Chao[2], W.-F. Wu[1], W.-K. Yeh[1]

[1]National Nano Device Laboratories, Hsinchu, Taiwan; [2]Dept. of Electrophysics, National Chiao Tung University, Hsinchu, Taiwan.[3]Dept. of Electrical Engineering, National Cheng Kung University, Tainan , Taiwan;[4]Dept. of Electrical and Computer Engineering, National Chiao Tung University, Hsinchu, Taiwan;[5]Current Scientific, San Jose, CA, USA; [6]Evans Analytical Group, Sunnyvale, CA, USA; Tel: +886-3-5726100-7793, Fax: +886-3-5722715, Email: yjlee@narlabs.org.tw

Abstract

For the first time, a novel junctionless (JL) FinFET structure with a shell doping profile (SDP) formed by molecular monolayer doping (MLD) method and microwave annealing (MWA) at low temperature is proposed and studied. Thanks to the ultra thin SDP leading to an easily-depleted channel, the proposed JLFinFET can retain the ideal subthreshold swing (~ 60 mV/dec) at a high doping level according to simulations. Poly Si based JLFinFETs processed with MLD and MWA exhibit superior subthreshold swing (S.S. ~ 67mV/dec) and excellent on-off ratio (>10[6]) for both n and p channel devices. Threshold voltage (V_{TH}) variation due to random dopant fluctuation (RDF) is reduced in MLD-JLFinFETs, which can be attributed to the molecule self-limiting property of MLD on the Si surface and quasi-diffusionless MWA at low temperature. Our results reveal the potential of the proposed SDP enabling a JLFET showing reduced variation and outstanding performance for low power applications.

Introduction

Compared to a conventional MOSFET, a JLFET possesses simplicity of fabrication and immunity of mobility degradation at the channel/oxide interface [1]. For better gate control, a nanoscaled multigate configuration is needed to completely deplete the channel of a JLFET and to achieve better S.S. and a lower leakage current. However, a nanostructure may not only lead to inevitable process difficulties, e.g. RDF and dopant deactivation, it may also degrade the device performance due to quantum confinement (QC) and large series resistance (Rs) [2].Furthermore, RDF is responsible for the variation of electrical performance of nanoscaled devices [3]. With device dimension scaling, junction depth has become a critical parameter in terms of short channel effects. To suppress short channel effects, MLD process has recently attracted much attention due to its characteristics of conformally self-assembled and self-limiting doping and forming ultra shallow junctions without damage [4]. In MLD method, the dopant-containing molecules are covalently bonded to the surface of a semiconductor at low temperature resulting in high doping density [4]. High temperature annealing decomposes the molecules and drives the dopants into the semiconductor at a targeted junction depth by controlling the thermal budget. Furthermore, in order to achieve an ultra shallow junction depth with high doping density and abruptness, high temperature annealing must be avoided. MWA, therefore, may be an ideal technique to activate dopants at low temperature [5].Unlike the conventional JLFET relying on a nanoscaled channel, in this work we propose a novel JLFinFET structure with a size-QC-immune fin width and a SDP (Fig. 1). Based on this structure,

the drawbacks of conventional JLFETs, such as high series resistance and low electronic density of state (DOS), can be relieved and still retain the ideal S.S. and high on-off current ratio (I_{on}/I_{off}). The proposed device is simulated and fabricated adopting MLD and MWA in process.

Results and Discussion

Device Simulation Fig. 2(a) and (b) illustrate the simulated JLFinFET with a SDP (achievable by the MLD method and MWA in reality), which is assumed to be uniform. The electrical characteristics are calculated numerically by a three dimensional simulator, which self-consistently solves the Poisson equation and carrier drift-diffusion transport equations involving quantum potential correction [6]. Since the doping levels are higher than the DOS of the majority band, Fermi statistics is used. For the accuracy of on-current, mobility models (doping-dependent, velocity saturation and interface degradation) and QC model redistributing the carrier at the semiconductor/oxide interface are considered. To estimate the subthreshold leakage currents, doping-dependent bandgap narrowing, Shockley-Read-Hall generation and tunneling models are included as well. Fig. 3 shows transfer characteristics of devices with different SDPs. For the devices with $N = 5 \times 10^{19}$ cm^{-3} and d =5 nm, the gates can hardly deplete the channels and barely modulate the drain currents. However, the devices with the same high doping level (5×10^{19} cm^{-3}) are able to show the ideal S.S.(~60 mV/dec) when d = 2 nm. This can be explained by the carrier distribution in the fin as depicted in Fig. 4. Thanks to the SDP, the hole concentration in the middle of the fin width with d = 2 nm is lower than that with d = 5 nm by almost six orders of magnitude at V_{gs} = 0 V for the same p-type doping level at 5×10^{19} cm^{-3}. Therefore, the shallow junction of the SDP can effectively improve the S.S. and decrease the off-current of a JLFinFET for low power applications.

For a given doping concentration, the device with a shallower doping depth exhibits better gate control as shown in Fig. 3. This is because of the proximity of doped region to the gate. In addition, the shallow junction induces ignorable penalty on the on-state current (compare devices with N = 5×10^{19} cm^{-3} in Fig. 3). This can be understood by viewing the carrier profile along the fin width in Fig. 4 (c). Although SDP is only 2 nm beneath the gate oxide, there is still a significant amount of carriers ($\approx 2 \times 10^{18}$ cm^{-3}) flowing in the middle of the fin contributing the on current. Carriers transporting in the intrinsic region are injected from the source side because the potential of the middle region follows that of the doped surface, especially when the fin width is narrow. Besides the small decrease in the on-current, the proposed SDP relieves the constraint on the

978-1-4799-8002-4/14 $31.00 © 2014 IEEE

depletable nanoscale channel of a conventional JLFET, which suffers from QC and Rs.

Fabricated Device Fig. 5 shows the process flow of fabricating poly-Si JLFinFETs utilizing MLD and MWA. Initially an undoped amorphous Si film is deposited on an oxide-capped Si wafer by using low pressure chemical vapor deposition, followed by a thermal step for recrystallization. The active region is defined by lithography and the fin width is narrowed down to 10 nm. For implanted (imp) JLFinFETs, n and p type doping are formed both with a dose of 5×10^{14} cm^{-2}. For MLD devices, after HF cleaning, Si surfaces were reacted with diethyl 1-propylphosphonate and vinylboronic acid dibutyl ester in mesitylene for n and p doping, respectively. By utilizing the advantage of conformal doping of MLD, a uniform doping profile can be well-defined for a three dimensional structure with better material quality compared to the one obtained by ion implantation. After oxide capping, the samples are treated with rapid thermal annealing (RTA, at 1000 °C) or MWA (~at 550 °C) to drive in and activate the dopants. After the capping oxide removal, an Al_2O_3/TiN gate stack is defined for all devices. Source/Drain(S/D) regions are defined by implantation with a dose of 1×10^{15} cm^{-2} except for imp. JLFinFETs. Subsequently, S/D activation was performed by MWA to suppress dopant diffusion [5].

Fig. 6 displays the successful attachment of B- and P-containing molecules on the silicon surface after the MLD process examined by X-ray photoelectron spectroscopy (XPS). Fig. 7 depicts the SIMS profiles of P and B showing the junction depths of two different annealing conditions. With MLD followed by MWA, ultra-shallow doping profiles of sub 5 nm have been obtained. Although MWA is performed at low temperature, the dopants can still be activated efficiently when annealing time is beyond 150 sec (insets in Fig. 7). Therefore, a combination of MLD and MWA processes is able to achieve an activated and ultrathin junction. Fig.8 shows the results of Raman analysis of a bare Si substrate and a BF_2 implanted sample with a RTA step at 1000 °C. Even after high temperature annealing, the implanted sample still shows a small hump around 490 cm^{-1}, which indicates damage caused by ion implantation. However, the result of the sample with the MLD process almost coincide with the one of the bare Si (inset of Fig. 8), which confirms that MLD is a damage-free process. Capacitance-voltage (CV in Fig. 9) shows invariance and ignorable hysteresis of devices with and without MLD. It indicates that MLD is a contamination-free process for gate oxide interface. Furthermore, the effective oxide thickness (EOT) is about 2.6 nm determined by CV and TEM analysis.

Fig. 10 plots the I_d-V_g of the p-channel imp. JLFinFETs, MLD-JLFinFETs and FinFETs, where I_d has been normalized by the total length of fin width and fin height. The current of imp. JLFinFETs with a dose of 1×10^{15} cm^{-2} cannot be modulated even when the fin width is scaled down to 15 nm. This is because the dopant concentration of channel is too high to be depleted. The gate modulation can be observed by the imp. JLFinFET with W = 15 nm and a dose of 5×10^{14} cm^{-2}. Therefore, the transfer characteristics of conventional JLFinFETs strongly depends on the geometric size of devices and doping concentration. However, MLD-JLFinFETs not only show superior gate control and lower off-current, also exhibit less width dependence influencing S.S. (Fig. 10). For the poly Si MLD-JLFinFET, the one with MWA exhibits excellent S.S. (67mV/dec) and lower off-current compared to the one with RTA as shown in Fig. 11.The later can be explained by a deeper junction depth (Fig. 7) owing to the higher temperature annealing. Both p and n MLD-JLFETs shrinking to W/L= 10/20 nm(TEM image in Fig. 2) still show good transfer characteristics with $I_{on}/I_{off} \sim 10^6$ (inserts in Fig. 11 and13).

V_{TH} variation has been an obstacle for conventional FETs and JLFETs in nanoscale caused by the size-QC and RDFs. The I_d-V_g characteristics of the pFinFETs and imp. pJLFETs show different variation (Fig. 14). However, variability of p-JLFinFETs with a SDP is suppressed due to the uniform and self-limiting doping by MLD and limited diffusion by MWA. Fig. 15 is a plot of the S.S. againsting V_{TH} for imp. JLFETs, MLD-JLFETs and pFinFETs, and MLD-JLFETs show lower V_{TH} variation and better S.S. The temperature-induced variation of I_d-V_g characteristics was shown in Fig. 16. Fig.17 shows the impact of temperature on S.S. of imp. JLFinFETs, MLD-JLFinFETs and FinFETs, presenting that MLD-JLFinFETs can suppress S.S. degradation effectively. MLD-JLFinFETs with SDPs, as a consequence, exhibit better device performance in terms of V_{TH} variation, saturation current and off-current. Finally, the performance benchmark of other state-of-art JLFETs is summarized in Table I.

Conclusion

For the first time, a novel JLFinFET structure with a SDP has been demonstrated with poly-Si adopting the MLD method and MWA at low temperature. Poly-Si n and p JLFinFETs (W/L=10/20 nm) experimentally exhibit superior gate control ($I_{on}/I_{off} > 10^6$) and reduced variability in terms of V_{TH} and S.S., hence variation of device performance due to RDF is suppressed. It is attributed to the conformally self-limiting doping of MLD method and the proximity of proposed SDP to the gate oxide interface for excellent gate control.MLD method poses the potential of reducing device variation induced by RDFs for nanoscaled devices.

Acknowledgement

The authors would like to thank DSG Technologies and Evans Analytical Group for the useful suggestions. And We also thank the support of Ministry of Science and Technology (program number: 103-2221-E-492 -045,103-2218-E-006-018,102-2120-M-002 -001, and 102-2120-M-009 -002).

References

[1] J.-P. Colinge et al., Nature Nanotechnology,p. 225, 2010.
[2] H.-B. Chen et al., VLSI Tech. Dig., p. T232, 2013.
[3] A. Asenovetal., IEEE T-ED, p. 1837, 2003
[4] K.-W. Ang et al., IEDM Dig., p. 35.5.1, 2011.
[5] Y.-J. Lee et al., IEEE T-ED, p. 652, 2014
[6] Sentaurus Device, Synopsys, Version G-2012.
[7] C.-J. Su et al., IEEE EDL, p. 521. 2011.
[8] H.-C. Lin et al., IEEE EDL, p. 53, 2012.
[9] H.-C. Lin et al., IEEE T-ED, p. 1142, 2013.

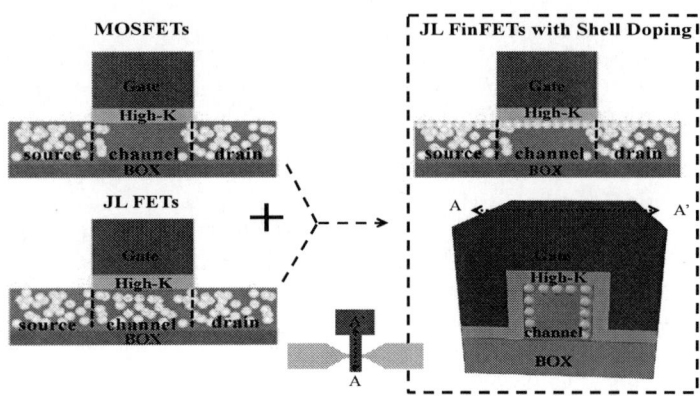

Fig. 1. An illustration of the novel JLFinFETs structure with sub-5nm SDP formed by MLD. Due to the rich self-limiting surface reaction property [3], MLD can improve the S.S. and RDFs-induced variation (Fig. 14).

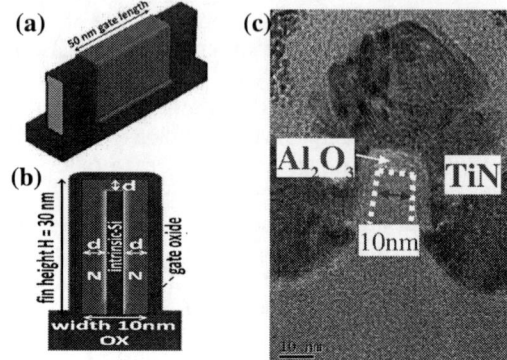

Fig. 2. A simulated FinFET (a) and its cross section along the fin width (b) N and d are the doping concentration and junction depth, respectively. (c) TEM image of a 10-nm channel width with Al_2O_3/TiN gate stack.

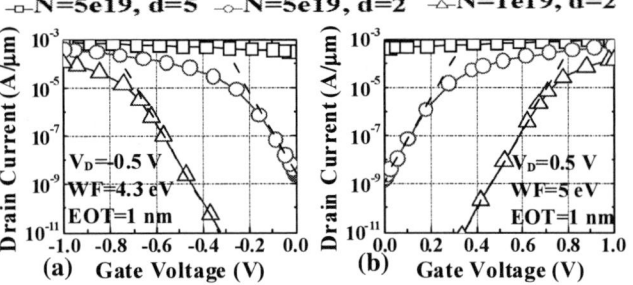

Fig. 3. Simulated transfer characteristics of FinFETs for (a) p- and(b) n-channel for different parameters N and d (Fig.2).Dashed lines indicate an ideal S.S. of 60 mV/dec. N and d are in the units of cm^{-3} and nm, respectively.(WF: workfunction, EOT: effective oxide thickness, V_D: drain bias)

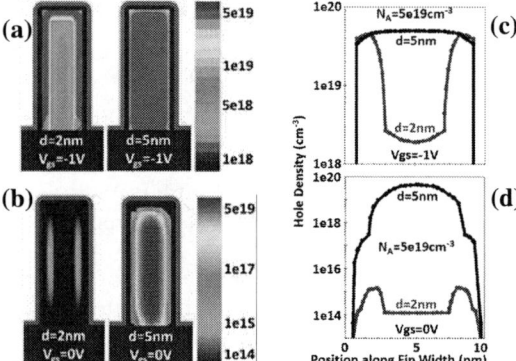

Fig. 4. 2D hole distribution in the fins for V_{gs} = -1/0 V (a/b). 1D hole distribution along the fin for V_{gs} = -1/0 V(c/d). Note that similar results have been observed in n-channel devices (not shown).(V_{ds} = -0.5 V)

- a-Si deposition (60nm)
- SPC and mesa isolation
- (a)Narrowing fin width
- Channeldoping:
  1. BF_2 or ^{31}P (1E15 or 5E14) for imp. JLFinFETs
  2. (b)MLD forSub-5 nm Shell Dopant Profile
  3. Undoped for FinFETs
- (c)Oxide capping
- (d)Annealing (RTA 1000 °C or MWA 550 °C)
- Capping oxide removal
- Gate stack formation and patterning
- Source/Drain Implantation
^{31}P / 1E15/ 10keV for n-channel devices
BF_2 / 1E15/ 10keV for p-channel devices
- MWA Activation (480 °C)

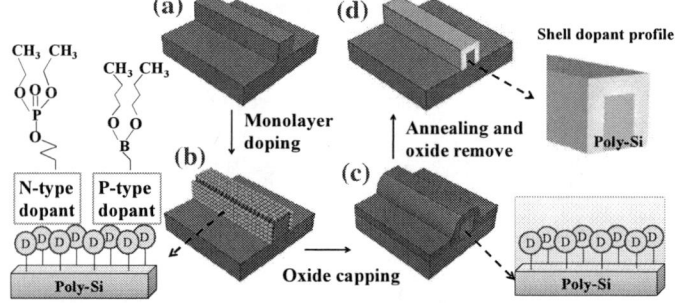

Fig. 5. Process flow for the fabrication of imp. JLFinFETs, MLD-JLFinFETs and conventional FinFETs. The right figure is a schematic illustration of MLD-JLFETs with monolayer doping in narrow fin structure (W=10 nm) resulting in shell doping profile. (SPC：solid phase crystallization)

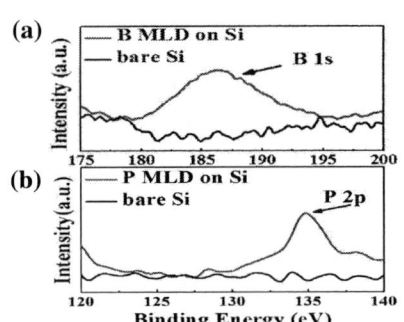

Fig. 6. XPS clearly reveals the covalently bonding of (a) B and (b) P containing monolayer on Si surfaces.

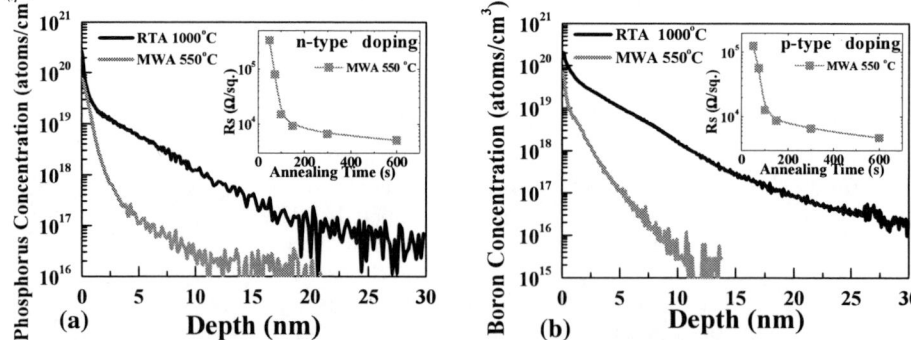

Fig. 7. SIMS profiles of (a) P (b) B showing the junction depth after various anneal splits. When subjected to a MWA anneal, a shallow X_j< 5nm is obtained. The inserts are Rs after microwave annealing splits.

978-1-4799-8002-4/14 $31.00 © 2014 IEEE

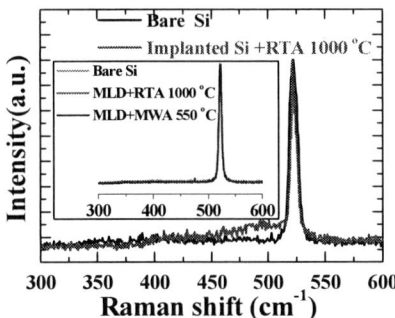

Fig. 8. Raman shows a small hump around 490 cm⁻¹ evident to the damage caused by implantation. Samples with MLD process do not have any damage signal, just like bare Si (inset).

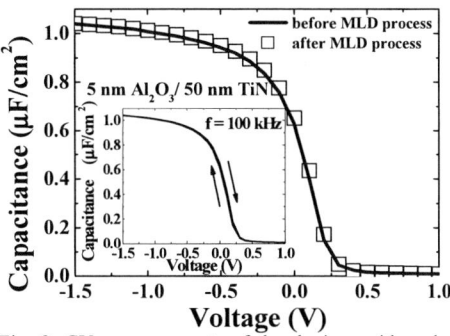

Fig. 9. CV measurements of the devices with and without MLD. The inset shows ignorable hysteresis.

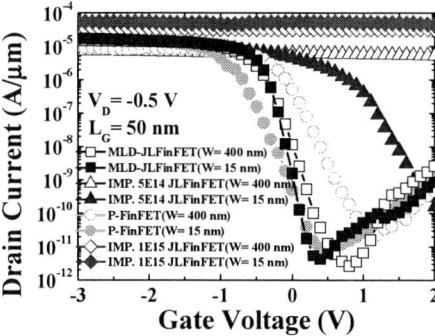

Fig. 10. I_D-V_G of p-channel devices (L=50/W= 400 or 15 nm).

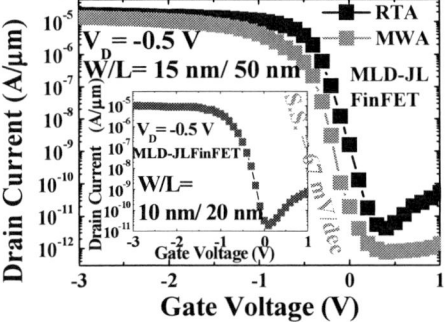

Fig. 11. I_D-V_G curves of p-channel MLD-JLFinFETs with MWA (550ºC) and RTA (1000ºC).The MLD-JLFinFET still reveals good performance even when the device dimension shrinks downto W/L= 10nm/20 nm (inset).

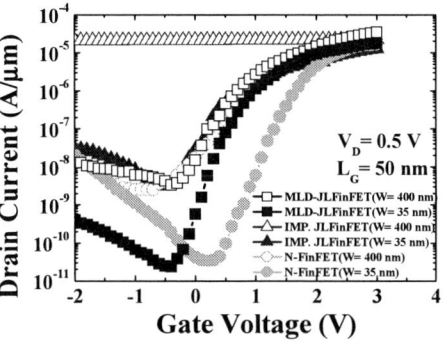

Fig. 12. I_D-V_G curves of n-channel imp. JLFinFETs, MLD-JLFinFETs and conventional FinFETs. (L=50/W= 400 or 35 nm).

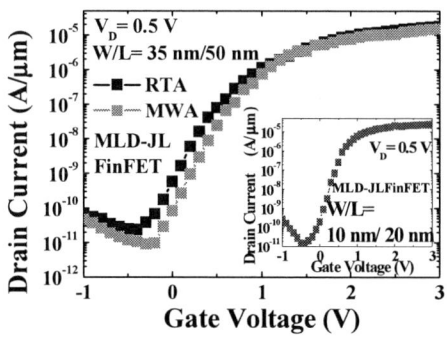

Fig. 13. I_D-V_G curves of n-channel MLD-JLFinFETs with MWA (550ºC) and RTA (1000ºC).The MLD-JLFinFET still reveals good performance even when the device dimension shrinks down to W/L= 10nm/20 nm (inset).

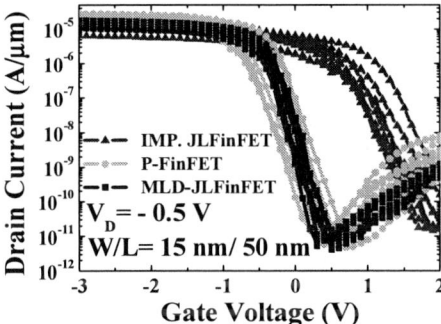

Fig. 14. I_D-V_G curves of p-type devices are compared for variation. MLD-JLFinFETs show tight distribution.

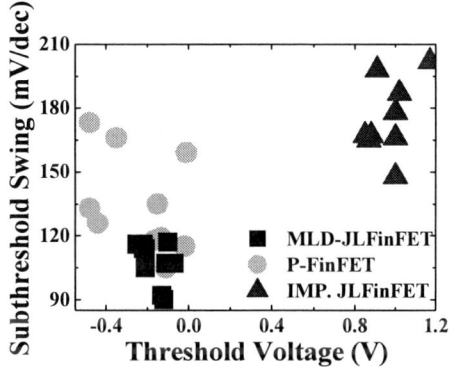

Fig. 15. Subthreshold swing versus threshold voltage for all devices. MLD-JLFinFETs show tight distribution and better swing.

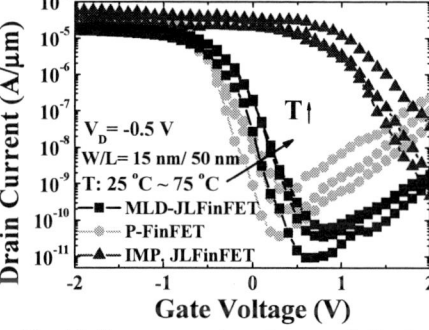

Fig. 16. Temperature dependence on I_D-V_G of p-channel devices. MLD-JLFinFETs show less V_{TH}, on and off current variation due to the SDP.

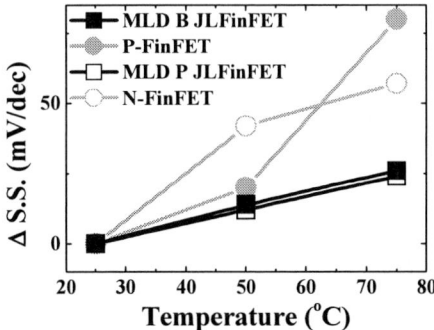

Fig. 17. S.S. deviation from the one at 25ºC as a function of temperature for imp. JLFinFETs, MLD-JLFinFETs and conventional FinFETs. It reveals that MLD-JLFinFETs can suppress temperature-dependent S.S. variation.

Table I. Performance comparison of MLD-JLFinFETs adopting MLD and MWA with previous studies.

JunctionlessFETs	This work	Ref [2]	Ref [7]	Ref [8]	Ref [9]
Cross-section	Narrow Fin +SDP	Nano sheet	Rough rectangular	Flat rectangular	Flat rectangular
Channel Structure	N-SPC JLFinFET	N-SPC JL-GAA	N-SPC JL-GAA	N-SPC JL-planar	N-SPC JL-planar
W/L (nm/nm)	10/20	10*(700/1000)	2*(70/1000)	10000/5000	10000/400
V_{th} (V)	n-type: 0.4 p-type: -0.1	0.25	-0.3	-0.3	0.3
EOT (nm)	2.6	17	15	8	8
I_{ON}/I_{OFF} (V_G;V_D)	>10⁶ (3V;0.5V)	>10⁷ (3V;0.5V)	>10⁶ (5V;1 V)	>10⁷ (3V;0.1 V)	>10⁷ (4V,0.1V)

978-1-4799-8002-4/14 $31.00 © 2014 IEEE

Experimental Demonstration of Four-Terminal Magnetic Logic Device with Separate Read- and Write-Paths

D. M. Bromberg[1], M. T. Moneck[1], V. M. Sokalski[2], J. Zhu[3,4], L. Pileggi[1], J.-G. Zhu[1]

[1]Dept. of Elec. & Comp. Eng., [2]Dept. of Mat. Sci. Eng., Carnegie Mellon University, Pittsburgh, PA, 15213
[3]Sun Yat-Sen University, SYSU-CMU Joint Institute of Engineering, Guangzhou, China
[4]SYSU-CMU Shunde International Joint Research Institute, Guangdong Province, China
Tel: (412) 268-5126, Fax: (412) 268-6662, Email: bromberg@cmu.edu

Abstract

Magnetic logic has recently become an attractive candidate for future electronics. This paper describes the demonstration of a four-terminal spintronic device with distinct read- and write-paths ("mCell"). The mCell enables a non-volatile circuit technology (mLogic) with gain sufficient to drive fanout independent of CMOS [1]. Measured material properties and prototype device results are presented.

Introduction

Spintronics, where the spin polarization of electrons is exploited in computation, has been studied in recent years as a potential platform for logic circuit design. A number of approaches have been put forth that differ in design and implementation, but all tend to share the common thread that electron spin – not charge – is used to represent and transfer data. The magnetization direction of a bistable element generally stores the logic value, making the circuits non-volatile. Proposed technologies range from circuits based solely on dipolar coupling with no electrical signaling between gates [1]-[4] to devices and circuits based on the flow of pure spin currents [5].

Recently, we have proposed a magnetic logic technology ("mLogic") based on a current-driven four terminal device ("mCell") with isolated read- and write- paths [6],[7]. An input current pulse through the write-path switches its magnetic state, which is coupled through an electrically insulating magnetic material to the free layer of a magnetic tunnel junction (MTJ) with two reference layer pillars that constitute the read path. These devices can be configured into circuits based on current steering that require no tight CMOS integration and can operate on low, noisy supply voltages.

mCell Device and Circuits

The mCell (Fig. 1) is programmed by moving a domain wall (DW) with a current through the write-path of a magnetic nanowire by the spin Hall effect (SHE) [9]-[11]. This magnetization state couples to a free layer in the read-path through an electrically insulating magnetic material, causing the magnetization in each layer to align. The free layer in combination with a tunnel barrier and reference layer forms a magnetic tunnel junction (MTJ) in the read path, setting the resistance of the device. The direction of current through the write-path determines which logic state the device enters. Micromagnetic simulation solving the Landau-Lifshitz-Gilbert equation [12] is used to explore device switching characteristics as a function of material properties, device size, and input stimuli. Fig. 2 shows the domain wall speed is

roughly linear with write current density; for scaled devices, these speeds translate to 500 ps – 1 ns switching times. The absolute current scales linearly with device width, and is on the order of tens of microamperes (Fig. 3).

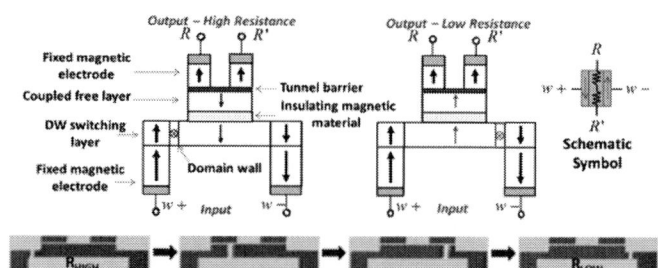

Fig. 1: (top left) Cross-section of mCell device with a write-path between (w⁺,w⁻) and a separate read-path between (R,R'); (top right) schematic symbol of mCell; (bottom) micromagnetic simulation of mCell SHE state switching.

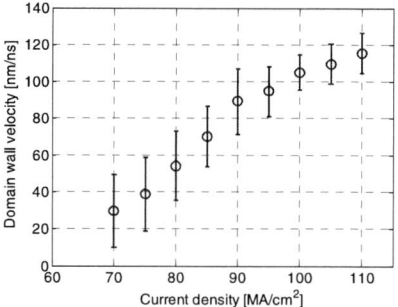

Fig. 2: Wall velocity increases roughly linearly with write current density.

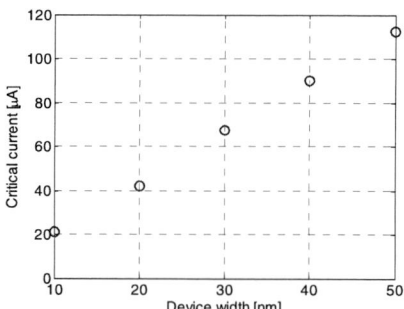

Fig. 3: Critical current scales linearly with device width.

The devices can be configured into circuits based on current steering, where the ratio of read-path resistances in one stage drives a positive or negative output current through fanout mCells connected in series through their write-paths

(Fig. 4). No integrated CMOS is required. Voltages of only ±100 mV or less are required to drive the required currents.

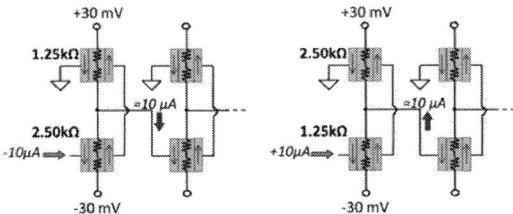

Fig. 4: Ratio of pull-up to pull-down mCell read-path resistance steers current into or out of series-connected write-paths of fanout mCells. The schematics show one inverter driving another.

mCells can also be used to design an all-magnetic MRAM bitcell (Fig. 5). The current flow through the write bitline (WBL) programs the inverter in the bitcell, which is used to isolate the third mCell (storage device) from other bitcells in an array. Driving the write wordlines (WWL+, WWL-) transfers the bit value to the storage cell based on current steering (Fig. 5). The cell is read by sensing the current along the read bitline (RBL) when the read wordline (RWL) is asserted. Current signaling enables low voltage (< 100 mV) operation and minimal energy loss in charging wordline and bitline parasitics. More details on mLogic can be found in [1].

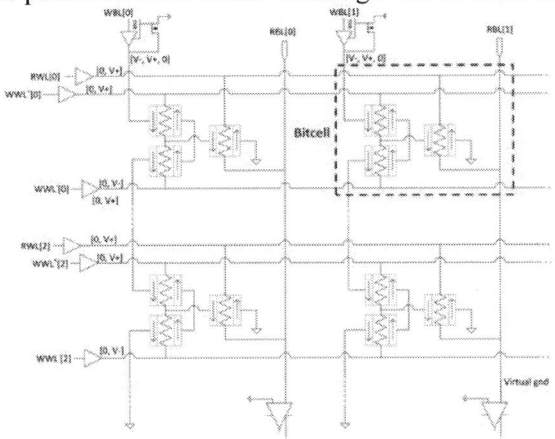

Fig. 5: Schematic of an all-magnetic mCell-based bitcell.

Materials Development

Individual mCell components were developed separately prior to integration in a single device. MTJs based on Ta/FeCoB/MgO/FeCoB/Ta (Fig. 6) were prepared by magnetron sputtering and subsequently annealed in a 4 kOe perpendicular field at 250-350°C resulting in a perpendicular tunneling magnetoresistance (TMR) up to 138% (Fig. 7) [13].

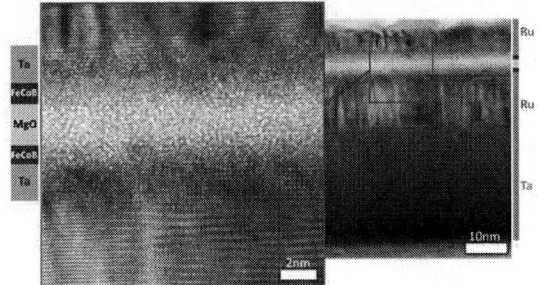

Fig. 6: Cross-sectional TEM image of MTJ films showing a smooth barrier.

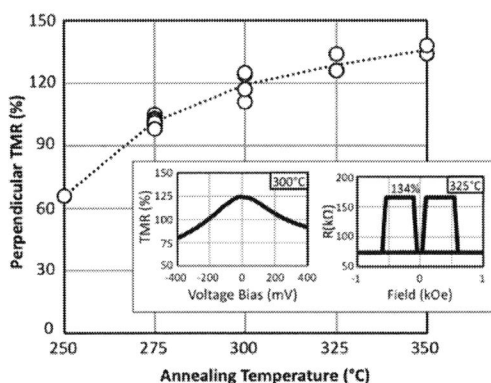

Fig. 7: TMR of 138% achieved with annealing stability up to 350 °C in Ta/FeCoB/MgO/FeCoB/Ta MTJs [13]. Inset: Bias voltage dependence and example R-H loop.

Fig. 8(a) shows Kerr images demonstrating domain wall motion (DWM) in a TaN(3nm) / Pt(2.5) / [Co(0.2) / Ni(0.3)], / Co(0.2) / Ta(0.32) / TaN(6) wire in response to various current pulses. DWM is along the current direction, indicating the driving force is the SHE. No fields were applied. Velocity as a function of current density is plotted in Fig. 8(b). Each point represents an average of five pulses with error bars indicating the standard deviation. Various Ta cap layer thicknesses were tested due to the effect Ta has on the chirality of the DW. SHE driven wall motion requires that the domain wall have some Néel character. This can be realized through the Dzyaloshinskii-Moriya interaction (DMI) from the seed and capping layers. As the DMI term for Ta/Co and Pt/Co have the same sign [9], they will have a canceling effect on the wall character when on opposite sides. As such, we observe that limiting the Ta thickness improves DW velocity, with a maximum value of 125 m/s in the experiment (Fig. 9).

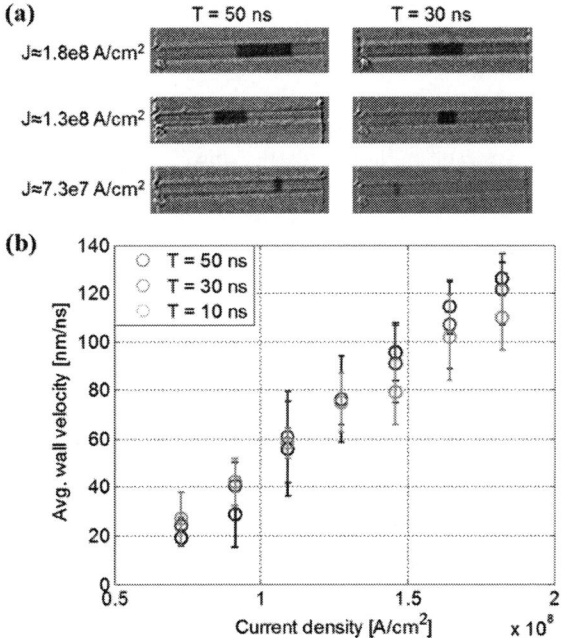

Fig. 8: (a) Differential Kerr images of domain wall displacement for different current densities and pulse widths; (b) Mean wall velocity increases linearly as a function of current density.

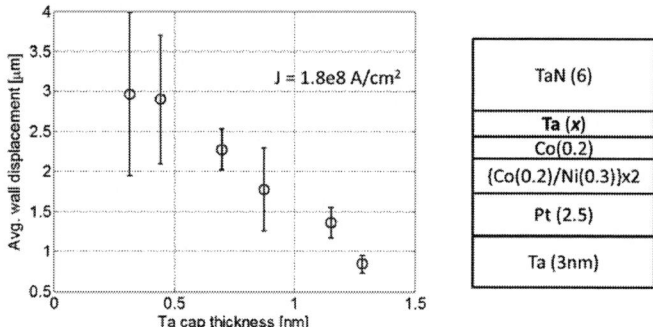

Fig. 9: Average domain wall displacement vs. Ta thickness in TaN(3nm)/Pt(2.5)/[Co(0.2)/Ni(0.3)]₂/Co(0.2)/ Ta(t_{Ta})/TaN(6) nanowires.

The magnetic oxide (coupling, insulating interlayer) was prepared by allowing a native oxide to form on metallic FeCoB in a 0.2 mTorr oxygen atmosphere. The FeCoB passivates, forming an electrically insulating, magnetic oxide layer of ≈1 nm (Fig. 10), similar to that observed in metal FeCo [14]. It has been shown that a 0.5 nm layer of Ta will not break coupling while still supporting DWM in the Co/Ni based write-path and necessary properties in the MTJ [15]. Coupling through the DW nanowire, magnetic oxide, and MTJ free layer was evaluated using the film stack of Fig. 11(a). At least 1 nm of oxide can maintain perpendicular magnetization due to coupling from the MTJ free layer and DW nanowire. This stack includes every layer of a full mCell device except for the top MTJ electrode.

Device Prototype

Based on the results of the developmental work, a prototype mCell with an integrated write- and read-path was fabricated using a combination of e-beam and optical lithography, RIE, and ion milling. The magnetic oxide was replaced by 0.5 nm of Ta in the prototype device to reduce fabrication complexity, meaning read- and write-paths are distinct and coupled but not electrically isolated. The full stack structure is seen in Fig. 12, which was post-annealed at 250°C. The films were annealed at 250°C to preserve DW motion, also resulting in low TMR values.

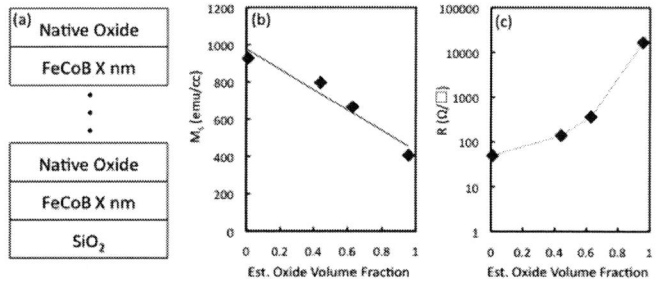

Fig. 10: (a) [FeCoB/NativeOxide]ₙ multi-layers to characterize properties of oxidized FeCoB. Saturation magnetization (b) decays to 400 emu/cm³ and sheet resistance (c) increases by three orders of magnitude as oxide volume fraction approaches 100%.

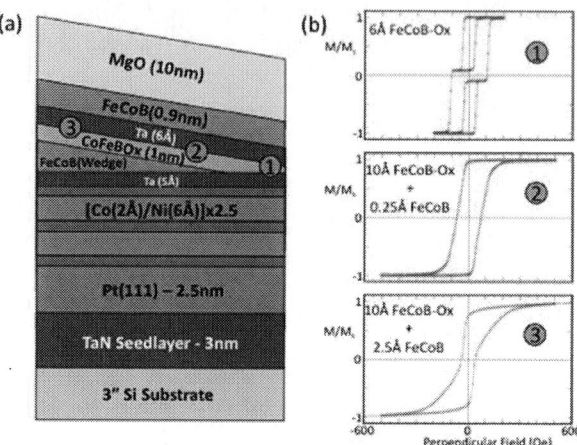

Fig. 11: (a) Film stack used to evaluate coupling within write-path. (b) Variation in M-H loops for different thicknesses of the spacer material (as listed). Full perpendicular remanence is observed for a 1nm FeCoB-oxide spacer layer.

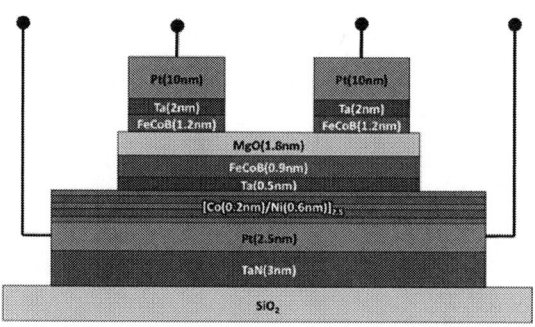

Fig. 12: Schematic stack structure of the prototype mCell.

The devices were patterned to be 1 μm wide and 10 μm long, with 275 nm of space between the two MTJs in the read-path (see plan view SEM and cross-sectional TEM images in Fig. 13(a) and (b)). A high resolution cross-sectional TEM image is given in Fig. 13(c), showing the layer breakdown in the device. A current passed through the side lead of Fig. 13(a) (left of the image) is used to nucleate a DW in the locality of the lead by generation of a circulating Oersted field.

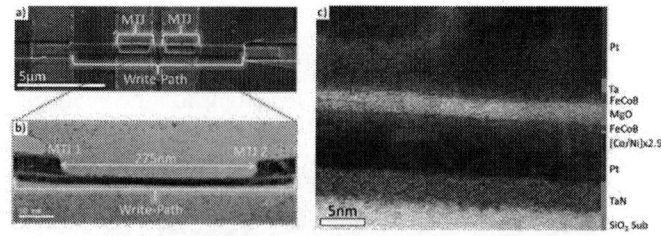

Fig. 13: (a) Plan view SEM image of fabricated device; (b) cross-sectional TEM image of device showing separation of read-path MTJs; (c) high resolution cross-sectional TEM image of device.

Current pulses injected in the write-path move the DW along the current flow by the SHE. Fig. 14 shows the location of the DW and corresponding resistance states in the structure. Polar Kerr microscopy images as well as micromagnetic simulations of the DW location are shown for clarity. Initially, the DW is "outside" the MTJs (region 1, low resistance). A current pulse is injected, bringing the DW to the middle of the device (region 2, midpoint resistance). A bias field of roughly 50 Oe was required to overcome the energy barrier necessary to bring the DW out of its low energy position, due in part to magnetostatic interaction between the read-path free layer and reference layer. A subsequent current pulse brings the DW to the other end of the device (region 3, high resistance). The electrical measurement shown in the figure demonstrates the fundamental concept of the mCell, that by moving a DW with current in the write-path, the read-path resistance can be switched from a low state to a high state (or vice versa) via magnetic coupling.

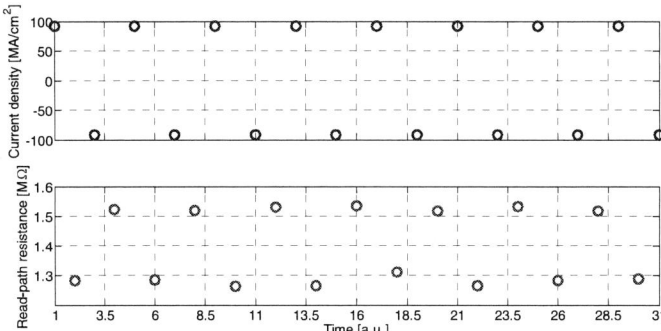

Fig. 16: Positive current pulses bring the device into a low resistance state and negative current pulses bring the device to a high resistance state.

Conclusion

The mCell is a device that could enable all-magnetic circuits and memory for non-volatile electronics. A prototype demonstrates the basic concept of the device, that a write-path current can digitally switch a separate read-path. Measured results of constituent components indicate suitable properties for the devices. Future work will involve integrating the magnetic oxide and improving the MTJ properties as means to enable fanout and the full realization of mLogic circuits.

References

[1] A. Imre, G. Csaba, L. Ji, A. Orlov, G. H. Bernstein, and W. Porod, "Majority logic gate for magnetic quantum-dot cellular automata," *Science* (New York, N.Y.), vol. 311, no. 5758, pp. 205-8, Jan. 2006.
[2] E. Varga, A. Orlov, M. Niemier, X. Sharon Hu, G. H. Bernstein, W. Porod, "Experimental Demonstration of Fanout for Nanomagnetic Logic," *IEEE Transactions on Nanotechnology*, 2010, 9(6).
[3] S. Breitkreutz, J. Kiermaier, I. Eichwald, X. Ju, G. Csaba, D. Schmitt-Landsiedel, M. Becherer, "Majority Gate for Nanomagnetic Logic With Perpendicular Magnetic Anisotropy," *IEEE Trans. Mag.*, 2012, 48(11).
[4] I. Eichwald, A. Bartel, J. Kiermaier, S. Breitkreutz, G. Csaba, D. Schmitt-Landsiedel, M. Becherer, "Nanomagnetic Logic: Error-Free, Directed Signal Transmission by an Inverter Chain," *IEEE Trans. Mag.*, 2012, 48(11).
[5] B. Behin-Aein, D. Datta, S. Salahuddin, and S. Datta, "Proposal for an all-spin logic device with built-in memory," *Nature Nanotechnology*, vol. 5, no. 4, pp. 266–70, Apr. 2010.
[6] D. Morris, D. Bromberg, J. Zhu, L. Pileggi, "mLogic: Ultra-Low Voltage Non-Volatile Logic Circuits Using STT-MTJ Devices", *Proc. of the 49th Annual Design Automation Conference*, 2012.
[7] D. Morris, D. Bromberg, J. Zhu and L. Pileggi, "Magnetic Logic Circuits with Minimal Connections to CMOS," *Proc. of the IEEE CAS-FEST*, 2012.
[8] J.E. Hirsch, "Spin Hall Effect," *Phys. Rev. Lett.*, 83, 1834–1837 (1999).
[9] S. Emori, U. Bauer, S.-M. Ahn, E. Martinez, G. S. D. Beach, "Current-driven dynamics of chiral ferromagnetic domain walls," *Nature Mat.*, vol. 12, 2013.
[10] K.-S. Ryu, L. Thomas, S.-H. Yang, S. Parkin, "Chiral spin torque at magnetic domain walls," *Nature Nanotechnology*, vol. 8, 2013.
[11] K.-S. Ryu, S-H. Yang, L. Thomas, S. S. P. Parkin, "Chiral spin torque arising from proximity-induced magnetization," *Nature Comm.*, **5**, 3910 (2014).
[12] Y. Nakatani, Y. Uesaka, N. Hayashi, "Direct Solution of the Landau-Lifshitz-Gilbert Equation for Micromagnetics," *Japanese Journal of Applied Physics*, vol. 28, no. 12, December, 1989, pp. 2485-2507.
[13] V. Sokalski, D. Bromberg, M. Moneck, E. Yang, and J.G. Zhu, "Increased perpendicular TMR in FeCoB/MgO/FeCoB magnetic tunnel junctions by seedlayer modifications," *IEEE Trans. Mag.*, 2013, 49(7).
[14] G.S.D. Beach and A.E. Berkowitz, "Co-Fe metal/native-oxide multilayers: A new direction in soft magnetic thin film design I. Quasi-static properties and dynamic response," *IEEE Trans. Mag.*, 2005, 41(6).
[15] V. Sokalski, M.T. Moneck, E. Yang, and J.G. Zhu, "Optimization of Ta thickness for perpendicular magnetic tunnel junction applications in the MgO-FeCoB-Ta system," *Appl. Phys. Lett.*, 2012. 101(072411).

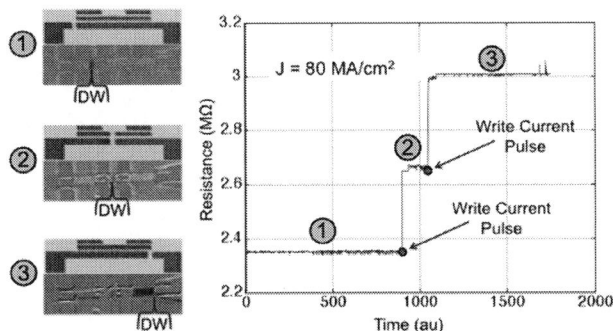

Fig. 14: Read-path resistance changes as the domain wall in the write-path is moved along with current. The resistance hits a midpoint when the wall is between the MTJs and increases to its maximum value when the wall moves under the second MTJ. Micromagnetic and Kerr images shown for domain wall location reference.

Higher current densities can be applied to ensure a single pulse can be used to switch the device (Fig. 15). Fig. 16 shows the device can be reliably switched between high (low) states when negative (positive) current pulses are applied.

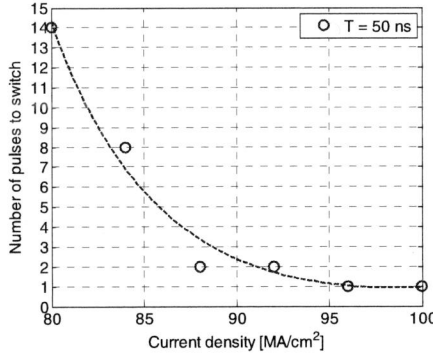

Fig. 15: Fewer pulses are required to switch the device as the write current density increases.

978-1-4799-8002-4/14 $31.00 © 2014 IEEE

Perpendicular-anisotropy CoFeB-MgO based magnetic tunnel junctions scaling down to 1X nm

S. Ikeda[1,2,3], H. Sato[1,2], H. Honjo[1], E. C. I. Enobio[3], S. Ishikawa[3], M. Yamanouchi[2,3], S. Fukami[1,2],

S. Kanai[3], F. Matsukura[1,3,4], T. Endoh[1,2,5] and H. Ohno[1,2,3,4]

[1]Center for Innovative Integrated Electronic Systems, Tohoku University, Sendai 980-0845, Japan

[2]Centerf for Spintronics Integrated Systems, Tohoku University, Sendai 980-8755, Japan

[3]Laboratory for Nanoelectronics and Spintronics, Research Institute of Electrical Communication (RIEC), Tohoku University, Sendai 980-8577, Japan

[4]WPI Advanced Institute for Materials Research (WPI-AIMR), Tohoku University, Sendai 980-8577, Japan

[5]Graduate School of Engineering, Tohoku University, Sendai 980-8579, Japan

+81-22-796-3409, s-ikeda@cies.tohoku.ac.jp

Abstract

CoFeB-MgO based magnetic tunnel junction with perpendicular easy axis (p-MTJ) shows a high potential to be used in spintronics based very large scale integrated circuits and spin-transfer-torque magnetorestive random access memories. In this paper, we review development of p-MTJ using single CoFeB-MgO and double CoFeB-MgO interface structures. The TMR ratio shows 164% after annealing at 400 °C, indicating the CoFeB-MgO p-MTJs have capability for back-end-of-line. Scaling properties of p-MTJs using double CoFeB-MgO interface structure are also reviewed.

Introduction

Nonvolatile very large scale integrated circuits (VLSIs) where magnetic tunnel junctions (MTJs) are integrated with CMOS logic have attracted much attention due to their potential in reduction of both power consumption and interconnection delay [1]. CoFeB-MgO based MTJ is appealing system as a building block in spin-transfer-torque magnetorestive random access memory (STT-MRAM) and spintronics based VLSIs because of their high tunnel magnetoresistance (TMR) ratio [2, 3]. There are two types of MTJs depending on easy axis direction; one has in-plane easy axis (i-MTJ), and the other has perpendicular easy axis (p-MTJ). Early studies focused on i-MTJs to realize STT-MRAM and spintronics based VLSIs [4-10]. However, in principle, p-MTJ has better efficiency in terms of the ratio of thermal stability factor (Δ) to intrinsic critical current (I_{C0}) because demagnetization field along out-of-plane direction increases I_{C0} whereas Δ is not affected by the demagnetization field in i-MTJ, resulting in smaller Δ/I_{C0}. The benefit of p-MTJs has driven intensive research on MTJ using various materials with perpendicular easy axis as summarized in Ref. 11, and STT-MRAMs and several types of spintronics based VLSIs using p-MTJs were demonstrated in recent years [12-20]. Among the explored materials, CoFeB-MgO p-MTJs whose perpendicular magnetic anisotropy originates from interfacial anisotropy at CoFeB-MgO interface [21] are actively studied because of their high potential in meeting major requirements to be used in spintronics based VLISs and STT-MRAMs at a junction diameter (D) of 40 nm [22]. The interfacial anisotropy at ferromagnet-oxide interface was first reported in Pt-Co(FeB)-MO$_x$ (M = Al, Mg, Ta and Ru) trilayer structures [23-25]. Although the earlier studies suggested the presence of interfacial anisotropy, there was always Pt underlayer inserted underneath ferromagnet to stabilize perpendicular easy axis. An effect of interfacial anisotropy at CoFeB-MgO interface on I_{C0} in i-MTJs was also reported [26-28].

Recently, we have shown that higher Δ in double CoFeB-MgO interface recording structure can be obtained while keeping almost the same I_{C0} compared to the single CoFeB-MgO interface recording structure [29]. Thereafter, p-MTJs with the double interface recording structure are being developed [30-32]. Other group also reported high Δ and low I_{C0} in 27 nm-diameter p-MTJ with double interface recording structure [33]. In this paper, we review development in p-MTJs using single and double CoFeB-MgO interface structures. We also review scaling properties of p-MTJ using double CoFeB-MgO interface structure down to 1X nm.

Results and discussion

Figure 1 shows schematic of structure for CoFeB-MgO p-MTJs using CoFeB-MgO interfacial anisotropy which were fabricated on 3 inch thermally oxidized Si substrate, and properties of p-MTJs. High TMR ratio, low I_{C0}, and a relatively high Δ were simultaneously realized at D of 40 nm after annealing at temperature (T_a) 300°C [22]. The high performances of CoFeB-MgO p-MTJ have triggered numerous studies on material and device using CoFeB-MgO system. To realize spintronics based VLSIs and STT-MRAMs, one needs to engineer stack structure and fabrication process for CoFeB-MgO p-MTJs using 300 mm Si wafer process line where capability to withstand annealing at temperature (T_a) up to 350°C at least is required for back-end-of-line (BEOL) process compatibility. For this purpose, T_a dependence of TMR ratio of blanket films for CoFeB-MgO p-MTJs fabricated on 300 mm Si wafer is studied. Figure 2 shows T_a dependence of TMR ratio of the CoFeB-MgO p-MTJ stacks witha Co/Pt multilayer based synthetic ferromagnetic (SyF) reference layer [30, 34]. TMR ratio is measured by current-in-plane tunneling (CIPT) method [35]. TMR ratio exhibits monotonic increase with increasing T_a, and reaches 164% after annealing at T_a = 400°C, indicating that the CoFeB-MgO p-MTJ stack deposited on 300 mm Si wafer have capability to withstand annealing at T_a = 400°C required for standard BEOL.

Although high performances were achieved by CoFeB-MgO p-MTJs, enhancement of Δ is required to ensure 10 years retention time at reduced dimensions. We developed CoFeB-MgO p-MTJ using MgO/CoFeB/Ta/CoFeB/MgO recording structure where Δ increased by a factor of ~2 without increase of I_{C0} at D of 70 nm compared to MgO/CoFeB structure [29]. To study scaling properties of p-MTJs with double CoFeB-MgO interface structure down to 1X nm, we employed stack structure schematically shown in Fig. 3(a). For comparison, we also fabricated p-MTJs with single CoFeB-MgO interface whose structure is shown in Fig. 3(b). Two series of MTJs with double CoFeB-MgO interface

978-1-4799-8002-4/14 $31.00 © 2014 IEEE

structure were fabricated. We determined resistance and area product (RA) for both p-MTJs from the relationship between conductance at parallel configuration and area measured by scanning electron microscope [31], revealing that RA of double CoFeB-MgO interface structure was comparable to that of single CoFeB-MgO interface structure. Both series of MTJs with double CoFeB-MgO interface structure have the same RA value within experimental error. Figures 4(a) and 4(b) show a cross-sectional high resolution transmission electron microscope (HRTEM) image and a high-angle annular dark-field scanning transmission electron microscope (HAADF-STEM) image of stack consisting of sub./Ta(5)/Ru(10)/Ta(5)/CoFeB(0.9)/MgO(0.9)/CoFeB(1.6)/Ta(0.4)/CoFeB(1)/MgO(0.9)/Ta(5)/Ru(5), respectively. As it can be seen, top MgO thickness is reduced to two-thirds of nominal thickness whereas bottom MgO thickness has almost the same thickness as nominal one, which could be a reason for comparable RA of double CoFeB-MgO interface structure to that of single CoFeB-MgO interface structure. Figures 5 show resistance versus magnetic field curves (R-H curves) of the p-MTJs with $D = 11$ nm for double ((a)) and single CoFeB-MgO ((b)) interface structure. Clear parallel and antiparallel states are observed at $H = 0$ for the p-MTJ with double CoFeB-MgO interface structure whereas that with single CoFeB-MgO interface shows no hysteresis, which could be due to smaller Δ of single CoFeB-MgO interface structure. Hereafter we focus on scaling properties of p-MTJs with double CoFeB-MgO interface structure. Figure 6 shows D dependence of TMR ratio. TMR ratio shows virtually the same value within studied D range. We then evaluate Δ of the p-MTJs with double CoFeB-MgO interface structure from the relationship between switching probability (P) and pulse magnetic field with duration of 1 s. Figure 7 shows P plotted against H for the p-MTJ with $D = 56$ nm. From a fit of theoretical equation to the experimental results, one can obtain Δ of 84 for the p-MTJ with $D = 56$ nm. The same measurement results for 11 nmϕ MTJ with double CoFeB-MgO interface structure is shown in Fig. 8, from which Δ of the p-MTJ with $D = 11$ nm was determined to be 29. Figure 9 shows junction size dependence of Δ of p-MTJs with double CoFeB-MgO interface structure. Δ shows almost constant value when D is larger than 31 nm. Below $D = 31$ nm, Δ starts to decrease with reducing D. The results suggest that nucleation type magnetization reversal takes place at D of larger than 31 nm, below which it is expected that single domain type magnetization reversal takes place. We compare the experimental results with expected Δ value calculated from magnetic properties of blanket film taking into account change of demagnetization factors depending on D [32]. In the figure, the calculated Δ value is shown as dotted curve, revealing that the calculated curve explains well the experimental results.

To evaluate I_{C0}, switching probability (P_I) was measured as a function of pulse current amplitude with duration of 0.1 s. Figure 10 shows switching probability (P_I) as a function of pulse current amplitude for double CoFeB-MgO interface structure with $D = 20$ nm, from which average absolute I_{C0} is determined to be 24 μA. The average I_{C0} in 11 nmϕ p-MTJ is 13μA. Figure 11 shows D dependence of I_{C0} for double CoFeB-MgO interface structure. In the figure, a results reported in Ref. 29 is also included for comparison in which the same double CoFeB-MgO interface structure with pseudo-spin-valve-structure was used. I_{C0} shows monotonic reduction with reducing D. Based on macrospin model, I_{C0} is scaled with the product of effective perpendicular magnetic anisotropy energy density (K_{eff}) and volume of recording

layer. K_{eff} of the p-MTJs is again calculated from magnetic properties of blanket film with correction of demagnetization factors [32]. In the Fig. 11, calculated I_{C0} for three different damping constant values are shown for comparison. The behavior of D dependence of I_{C0} cannot be explained by constant damping constant, suggesting that effective damping constant is dependent on D. In order to discuss effective damping constant without an influence of device-to-device variation of Δ, here the ratio of Δ/I_{C0} is considered; Δ/I_{C0} increases from 1.63 to 2.2 with reduction of D from 28 to 11 nm. As shown in Fig. 12, although Δ/I_{C0} is constant in macrospin model, the experimental result increases with reduction of D in junction size range where it is expected that the single domain type magnetization reversal takes place.

Conclusion

We review recent progress on CoFeB-MgO based magnetic tunnel junction with perpendicular easy axis. The CoFeB-MgO p-MTJ with a Co/Pt multilayer based SyF reference layer have capability to withstand annealing at $T_a = 400°C$ required for BEOL. By using double CoFeB-MgO interface structure, we achieved Δ of 58 and I_{C0} of 24 μA at a junction diameter of 20 nm. On the contrary to macrospin model, experimental result for the ratio of Δ to I_{C0} increases with reduction of D in the junction size range where it is expected that single-domain type magnetization reversal takes place, suggesting that effective damping constant is reduced at reduced dimensions.

Acknowledgements

The authors wish to thank C. Igarashi, T. Hirata, H. Iwanuma, Y. Kawato, and K. Goto for technical support. This work was supported in part by JSPS through FIRST program, R&D Project for ICT Key Technology of MEXT, and R&D Subsidiary Program for Promotion of Academia-industry Cooperation of METI.

References

[1] H. Ohno et al., Tech. Dig. - Int. Electron Devices Meet. **2010**, p. 218.
[2] J. Hayakawa etl al., Jpn. J. Appl. Phys. **44**, L587 (2005).
[3] D. D. Dyayaprawira et al., Appl. Phys. Lett. **86**, 092502 (2005).
[4] M. Hosomi et al., Tech. Dig. - Int. Electron Devices Meet. **2005**, p. 459.
[5] T. Kawahara et al., IEEE J. Solid-State Circuits **43**, 109 (2008).
[6] R. Takemura et al., IEEE J. Solid-State Circuits **45**, 869 (2010).
[7] S. Chung et al., Tech. Dig. - Int. Electron Devices Meet. **2010**, p. 304.
[8] S. Matsunaga et al., Appl. Phys. Express **1**, 091301 (2008).
[9] D. Suzuki et al., Dig. Tech. Pap., Symp. VLSI Circuits **2009**, p. 80.
[10] W. Zhao et al., IEEE Trans. Magn. **45**, 3784 (2009).
[11] S. Ikeda et al., SPIN **2**, 124003 (2012).
[12] T. Kishi et al., Tech. Dig. - Int. Electron Devices Meet. **2008**, p. 309.
[13] K. Tsuchida et al., ISSCC Dig. Tech. Pap., (2010) p. 258.
[14] D. C. Worledge et al., Tech. Dig. - Int. Electron Devices Meet. **2010**, p. 296.
[15] W. C. Lim et al., Dig. Tech. Pap., Symp. VLSI Technology, **2013**, p. 64.
[16] T. Ohsawa et al., Dig. Tech. Pap., Symp. VLSI Circuits. **2013**, p. 110.
[17] L. Thomas et al., J. Appl. Phys. **115**, 172615 (2014).
[18] T. Endoh et al., Tech. Dig. - Int. Electron Devices Meet. **2011**, p. 76.
[19] S. Matsunaga et al., Dig. Tech. Pap., Symp. VLSI Circuits. **2011**, p. 298.
[20] M. Natsui et al., Tech. Dig. Pap. –ISSCC. **2013**, p. 194.
[21] M. Endo et al., Appl. Phys. Lett. **96**, 212503 (2010).
[22] S. Ikeda et al., Nature Mater. **9**, 721 (2010).
[23] S. Monso et al., Appl. Phys. Lett. **80**, 4157 (2002).
[24] A. Manchon et al., J. Appl. Phys. **104**, 043914 (2008).
[25] L. E. Nistor et al., Appl. Phys. Lett. **94**, 012512 (2009).
[26] M. Hosomi et al., J. Magn. Soc. Jpn. **2**, 606, (2007).
[27] J. Hayakawa et al., IEEE Trans. Magn. **44**, 1962 (2008).
[28] S. Yakata et al., J. Appl. Phys. **105**, 07D131 (2009).
[29] H. Sato et al., Appl. Phys. Lett. **101**, 022414 (2012).
[30] H. Sato et al., IEEE Trans. Magn. **49**, 4437 (2013).
[31] H. Sato et al., Tech. Dig. -Int. ElectronDevices Meet. 2013, p. 60
[32] H. Sato et al., Appl. Phys. Lett. **105**, 062403 (2014).
[33] L. Thomas et al., J. Appl. Phys. **115**, 172615 (2014).
[34] H. Sato et al., Jpn. J. Appl. Phys. 53, 04EM02 (2014).
[35] D. C. Worledge and P. L. Trouilloud, Appl. Phys. Lett. **83**, 84, (2003).

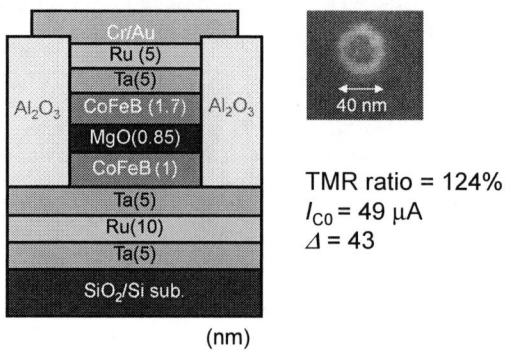

Fig. 1 Schematic of structure of CoFeB-MgO magnetic tunnel junction with perpendicular easy axis (p-MTJ). Properties of p-MTJ with junction diameter of 40 nm after annealing at 300°C are also shown.

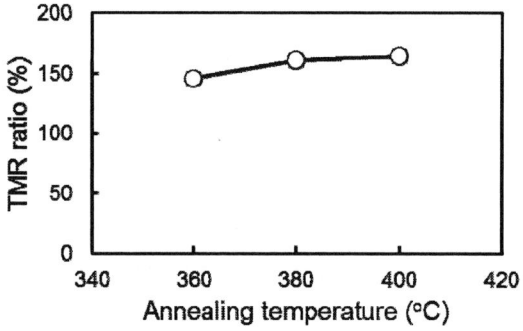

Fig. 2 Annealing temperature dependence of tunnel magnetoresistance (TMR) ratio for CoFeB-MgO p-MTJ stacks with perpendicular easy axis on 300 mm Si wafer where a Co/Pt multilayer based synthetic ferrimagnetic (SyF) reference layer are employed.

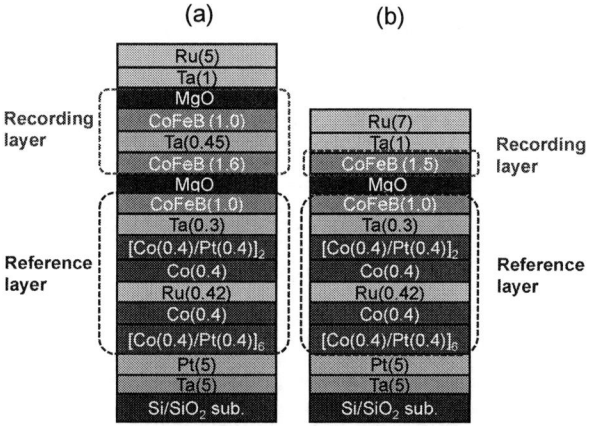

Fig. 3 Stack for magnetic tunnel junction with double CoFeB-MgO interface structure (a), and that with single CoFeB-MgO interface structure.

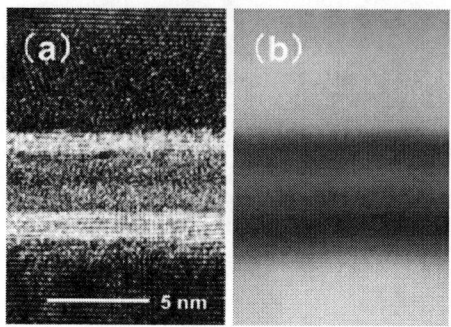

Fig. 4 (a) Cross-sectional high resolution transmission electron microscope (HRTEM) image and (b) high-angle annular dark-field scanning transmission electron microscope (HAADF-STEM) image of stack with structure of Sub./Ta(5)/Ru(10)/Ta(5)/CoFeB(0.9)/MgO(0.9)/CoFeB(1.6)/Ta(0.4)/CoFeB(1.0)/MgO(0.9)/Ta(5)/Ru(5).

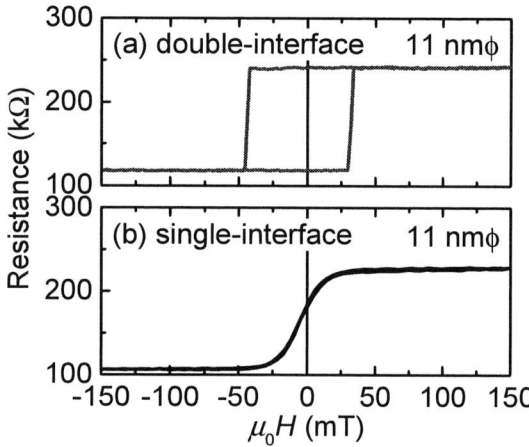

Fig.5 Resistance versus magnetic field curves for magnetic tunnel junctions with junction diameter of 11 nm for double CoFeB-MgO interface structure (a) and single CoFeB-MgO interface structure (b).

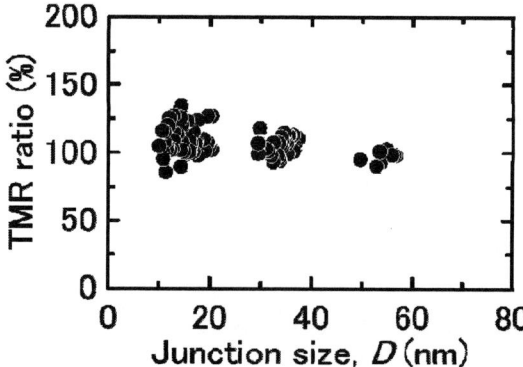

Fig. 6 Tunnel magnetoresistance (TMR) ratio of magnetic tunnel junctions with double CoFeB-MgO interface structure is plotted with respect to the junction size.

978-1-4799-8002-4/14 $31.00 © 2014 IEEE

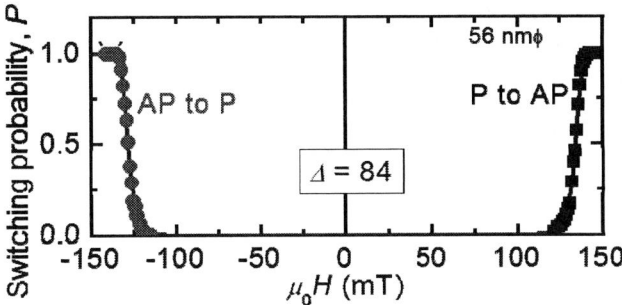

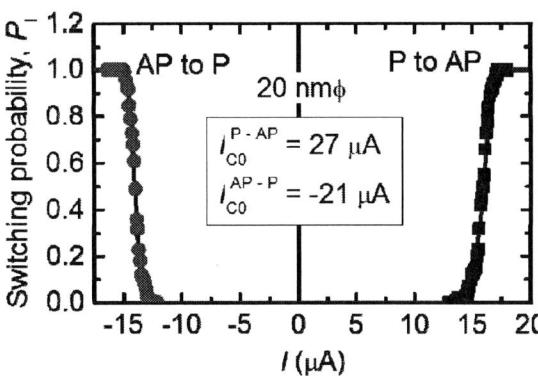

Fig. 7 Switching probability as a function of magnetic field of 56 nmφ magnetic tunnel junction with double CoFeB-MgO interface structure. In the figure, symbols and blue solid curve correspond to experimental results and fitting curve, respectively.

Fig. 10 Switching probability measured by pulse current with duration of 0.1 s as a function of pulse current amplitude for 20 nmφ magnetic tunnel junction with double CoFeB-MgO interface structure.

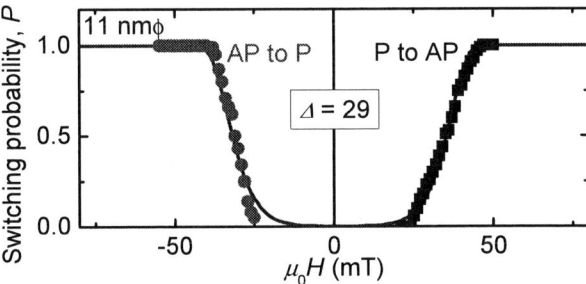

Fig. 8 Switching probability as a function of magnetic field for magnetic tunnel junction using double CoFeB-MgO interface structure with junction diameter of 11 nm.

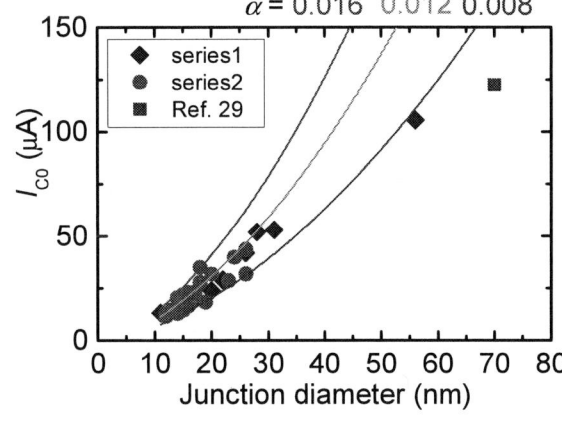

Fig. 11 Average absolute intrinsic critical current of double CoFeB-MgO interface structure is shown with respect to junction diameter.

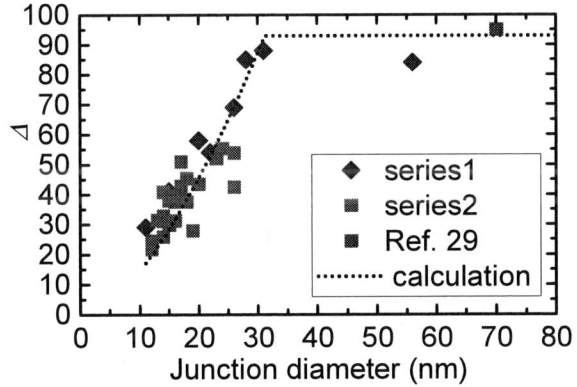

Fig. 9 Thermal stability factor Δ of the magnetic tunnel junctions with double CoFeB-MgO interface structure as a function of junction size.

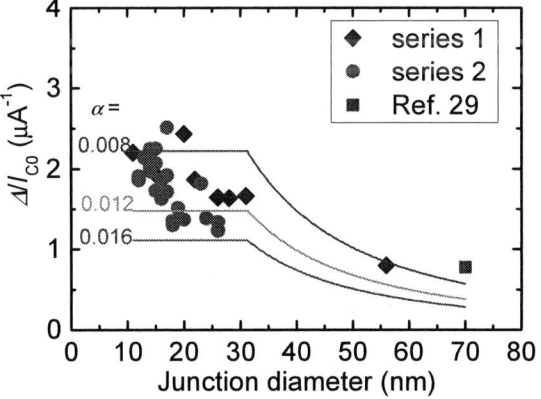

Fig. 12 Ratio of thermal stability factor D to intrinsic critical current I_{C0} of double CoFeB-MgO interface structure is shown with respect to junction diameter.

978-1-4799-8002-4/14 $31.00 © 2014 IEEE

High Performance, Excellent Reliability Multifunctional Graphene Oxide Doped Memristor Achieved by Self-protective Compliance Current Structure

Kuan-Chang Chang[1†], Rui Zhang[2], Ting-Chang Chang[3#], Tsung-Ming Tsai[1*], Tian-Jian Chu[1], Hsin-Lu Chen[4], Chih-Cheng Shih[1], Chih-Hung Pan[1], Yu-Ting Su[3], Pei-Jung Wu[5] and Simon M. Sze[3]

[1]Department of Materials and Optoelectronic Science, National Sun Yat-Sen University, Taiwan; [2]Information Technology Center, BGP, China National Petroleum Corporation, China; [3]Department of Physics, National Sun Yat-Sen University, Taiwan; [4]Department of Mechanical and Electromechnical Engineering, National Sun Yat-Sen University, Taiwan. [5]Department of Chemistry, National Kaohsiung Normal University, Taiwan.

Email: [†]doubleccc@yahoo.com.tw, [#]tcchang3708@gmail.com, [*]tmtsai@faculty.nsysu.edu.tw

Abstract

Double-ended graphene oxide (GO) doped silicon oxide based (SOB) via-structure RRAM with self-protective ability is reported in this paper. The fabricated RRAM exhibits comprehensive outstanding performance including switching speed (~30ns), endurance property ($>10^{12}$ cycles), read disturbance immunity ($>10^{10}$ cycles) and retention ($>10^4$s at 125℃, >144 days at room temperature). Combined with the applicability of complementary resistive switching structure and whole-cycle multi-bit operation, it is quite promising for this RRAM to be applied in future mass productions.

I. Introduction

SOB RRAM is one of the most promising candidates for future non-volatile memory owing to its intrinsic electrical stability, low cost and good compatibility to semiconductor fabrication line[1-2]. Achieving good device performance and high integration density has always been a tough task[3-4]; however, we design double-ended GO doped SOB memory which possesses comprehensive excellent performance and apply it with CRS that has outstanding intrinsic self-current compliance ability, addressing the sneak path problem without the use of any selector and current limiting element. Furthermore, multifunctional property like distinctive multi-bit characteristics in set region and fine multi-bit operation in reset process can be achieved.

II. Experimental Setup

Patterned substrate was fabricated to form 1μm*1μm via holes (Fig. 1), in which switching layers were deposited. The via hole substrate was designed for precise control of the RRAM cell size, by which more accurate and uniform electrical properties can be guaranteed. Switching layers were deposited into the via by RF magnetron co-sputtering with Zr, C, SiO_2 targets respectively constructing four kinds of devices (Fig. 7). The entire electrical measurements were performed using Agilent B1500A, B1530A and Cascade probe station. XPS spectra was applied to analyze the element concentration in $Zr:SiO_2$ and $GO:SiO_2$ films (Fig.2&3) and GO was confirmed by Raman and FTIR spectra (Fig. 4&5). Fig. 6 is the TEM picture and schematic diagram of filament self-aligning formation process of the triple layer device.

III. Electrical Characterization and Discussion

From our previous research we found that finite size material with low k value will concentrate electrical fields [2]. Thus by introducing GO into the silicon oxide film, GO will perform as the media to concentrate electrical fields and lead the filament growth if a metal doped silicon oxide film is stacked. For GO, HRS and LRS are switched by the adsorption and desorption of oxygen-contained chemical groups [3]. Inspired by the special resistive switching of GO and the electrical filed concentration effects, we fabricated double-ended GO RRAM device. The double-ended GO device possesses intrinsic filament searching growth ability as shown in Fig.6. During electro-forming process, electrons tend to conduct through the easiest accessible GO, which also works as the top current limiter, and trigger the accumulation of metal precipitates to form conductive filament. The filament grows in a directional

978-1-4799-8002-4/14 $31.00 © 2014 IEEE

way, approaching to the bottom GO flake to form current conduction path. The double-ended GO structure possesses instinctive ability to restrict multi-filament growth that may lead to multi-switching points. The multi-filament formation reflects on the electrical characteristics as instable current-voltage curves, especially in the reset process and the HRS. Moreover, it is notable to mention the current compliance graphene oxide flake in top GO:SiO$_x$ layer. Due to its finite size which defines the maximal flux of electrons, current is restricted automatically, making it easier to avoid filament over-formation and excessive rupture.

Fig. 7 demonstrates 100-cycle I-V curve comparison of four kinds of devices, which were operated without equipment current compliance. Triple layer device exhibits outstanding self-current-limiting capability and the most stable I-V cycle with HRS/LRS window > 100 (inset of I-V curve). To the AC endurance test, the double-ended GO device can achieve up to more than 10^{12} cycles. It is the first time for SOB RRAM to reach this endurance value. At the same time, triple layer device also owns uniform switching under AC operation, which can be seen in Fig. 8.

To the double-ended GO device, resistive switching happens in the bottom GO:SiO$_x$ layer, which results from the oxygen-contained chemical groups adsorption and desorption, facilitated by filament concentrated electrical field (Fig. 9). Fig.10 shows oxidation and reduction process of GO with stretching and relaxing of carbon-carbon conjugation double bonds, simulated by ChemBio3D Ultra Software. Thus, with the addition of more oxygen-contained chemical groups to GO, its resistance increases not only for its higher oxidation ratio but for the longer hopping distance of electrons. We have also conducted vary amplitude triangular pulse test, as shown in Fig.11. Specially, with the increase of the voltage amplitude, currents around the switching point almost remain the same, which further confirms device's current-limiting ability.

In addition, we also extracted the switching speed of different devices, from which we can find it needs ~30ns to set for triple layer device at 0.7V operation voltage (Fig. 12). The reliability for triple layer device test also includes: $>10^{10}$ read disturbance immunity (Fig. 13), $>10^4$s 125℃ and >144 days room temperature retention (Fig. 14&15). As it is the ultimate aim for device to be integrated, the scalability for RRAM device is an important factor [5]. Owing to the inborn sneak current for cross-bar architecture, CRS memory is a competitive candidate for its simple structure and elimination of any selector element [6]. We successfully fabricated CRS memory by combining two double-ended GO doped SOB structure RRAM and no current degradation observed during $>10^4$ sweeping cycles (Fig. 16). Besides, extraordinary multi-bit property (Fig. 17) in set region and fine multi-state in reset region (Fig. 18) can be achieved, which is extremely different from single set and rough unstable multi-bit reset metal oxide based RRAM.

IV. Conclusion

We first demonstrate double-ended GO SOB RRAM with comprehensive excellent performance and reliability. With its strong intrinsic current self-limiting ability, CRS implementation capability, whole sweeping cycle multi-bit property, double-ended GO SOB structure is a promising technique to the future mass production of RRAM.

Acknowledgement

This work was performed at National Science Council Core Facilities Laboratory for Nano-Science and Nano-Technology in Kaohsiung-Pingtung area and supported by the National Science Council of the Republic of China (Nos. NSC 102-2120-M-110-001 and NSC 101-2221-E-044-MY3).

Reference

[1] K.C. Chang *et al.*, *IEEE EDL*, 34, 399-401, 2013;

[2] T.M. Tsai *et al.*, *Appl. Phys. Lett.*, 102, 253509, 2013;

[3] K.C. Chang *et al.*, *IEEE EDL*, 34, 677-679, 2013;

[4] Y. Deng *et al.*, *Tech. Dig. IEDM*, 629-632, 2013;

[5] H.Y. Cheng *et al.*, *Tech. Dig. IEDM*, 497-500, 2012;

[6] E. Linn *et al.*, *Nature Mater.*, 9, 403-406, 2010.

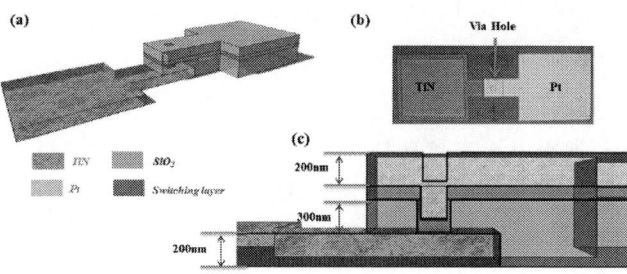

Fig.1 (a) Whole view (b) real picture and
(c) cross-sectional view of the RRAM device.

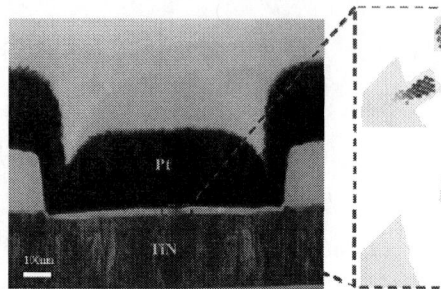

Fig.6 TEM picture and schematic diagram of filament self-aligning formation process
of the double-ended GO SOB RRAM device.

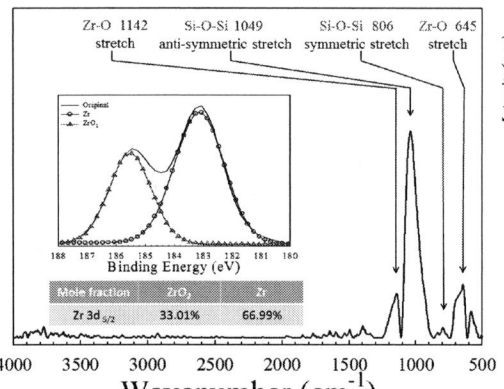

Fig. 2 FTIR spectra of Zr:SiOₓ layer.
Inset is the XPS spectra.

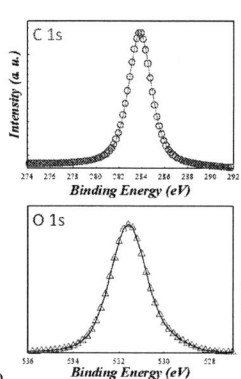

Fig. 3 XPS spectra of GO:SiOₓ layer. The percentage
of carbon element takes 2.75%.

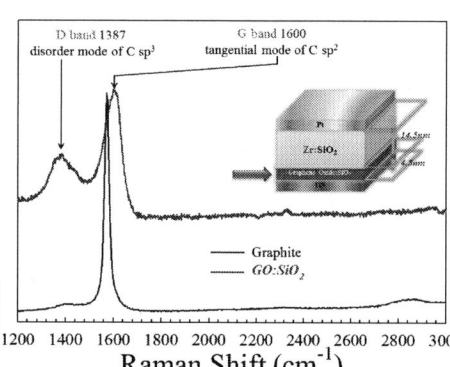

Fig.4 Raman spectra of the inserted GO:SiOₓ layer,
confirming the existence of GO.

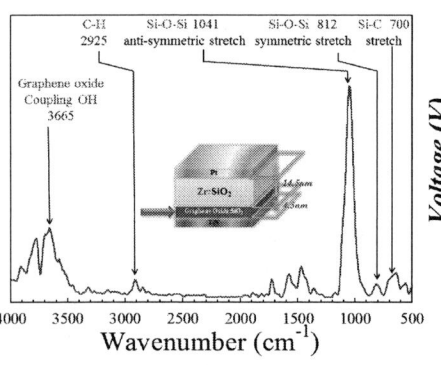

Fig.5 FTIR spectra of the inserted GO:SiOₓ layer.
Graphene oxide coupling OH bond at 3665cm⁻¹
further confirms the existence of graphene oxide.

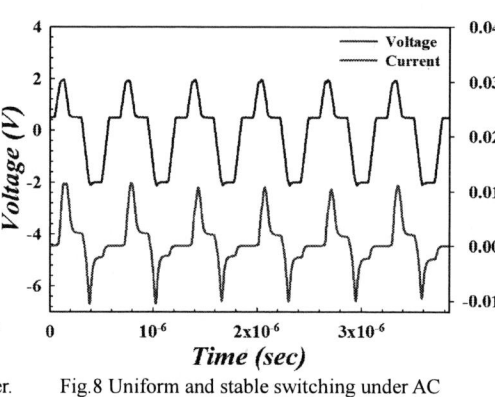

Fig.8 Uniform and stable switching under AC
operation of triple layer device.

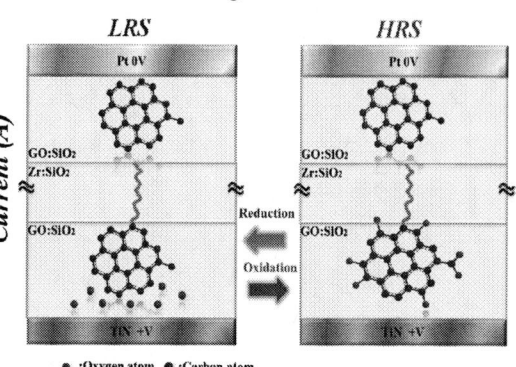

Fig.9 Resistance switching model of triple layer
device.

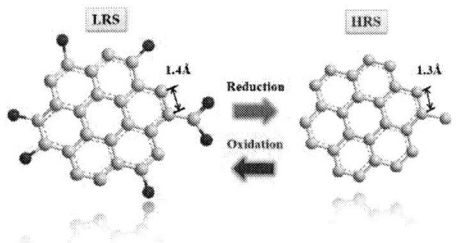

Fig.10 Reduction and oxidation process
simulation of GO. With the addition of
surrounding oxygen-contained chemical groups,
carbon-carbon conjugation double bonds will be
stretched, resulting in longer hopping distance.

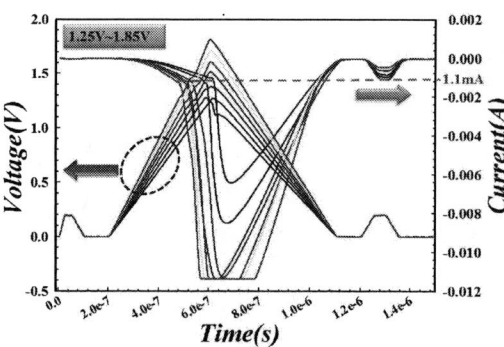

Fig.11 Set process test by vary amplitude triangular
voltage pulse. The abrupt change of current centers on
the point of 1.1mA, confirming the maximum current
compliance ability of two-sided GO structure.

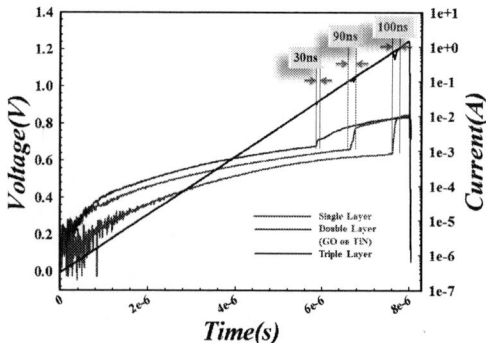

Fig.12 Set process response time comparison of
three kinds of devices.

978-1-4799-8002-4/14 $31.00 © 2014 IEEE   802

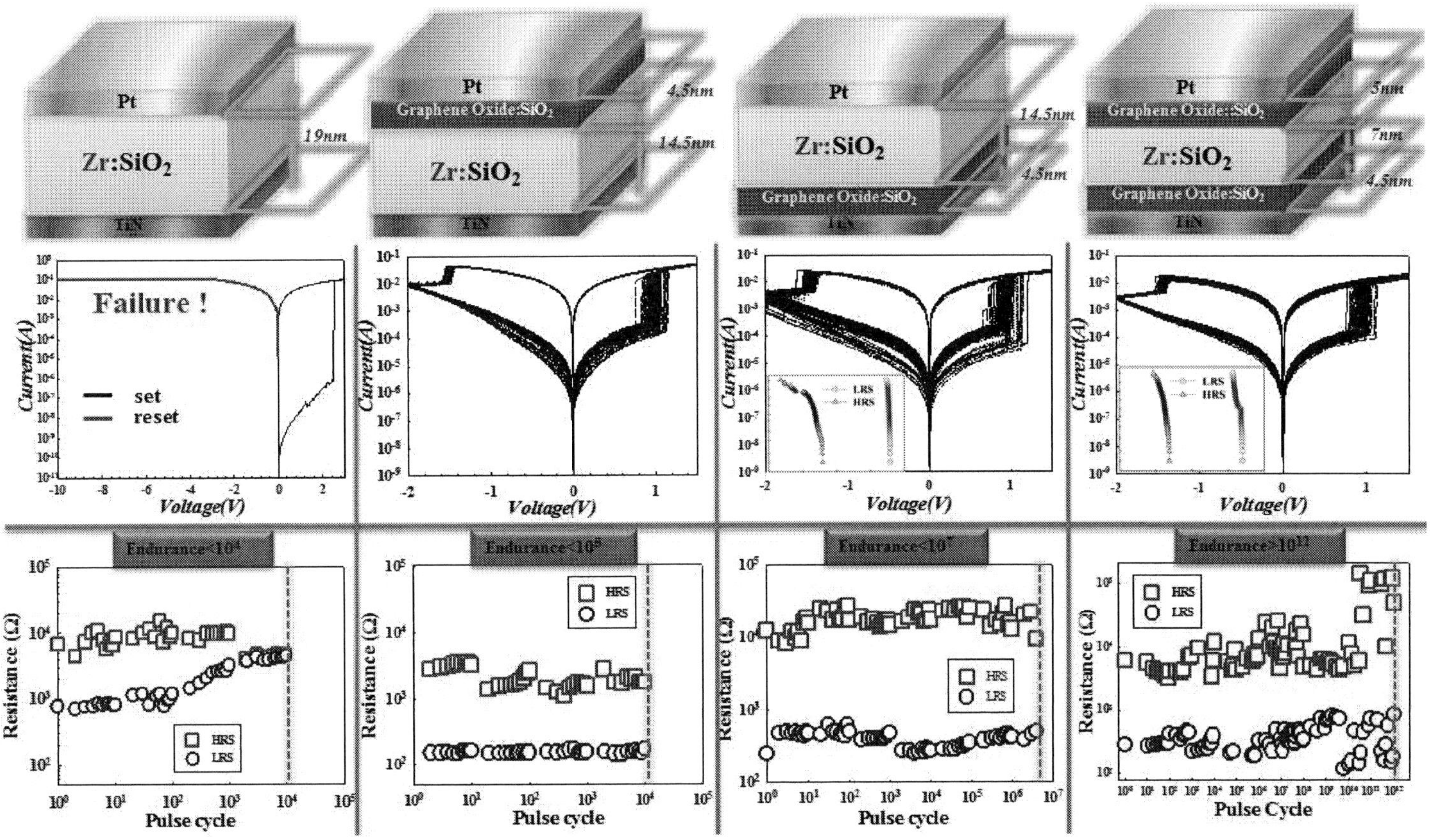

Fig.7 Device structure and the corresponding 100 cycle I-V curve operated without equipment current compliance. The bottom are their endurance performance. Triple layer device exhibits the most stable I-V cycle with HRS/LRS window > 100 (inset of I-V curve) and outstanding endurance properties($>10^{12}$).

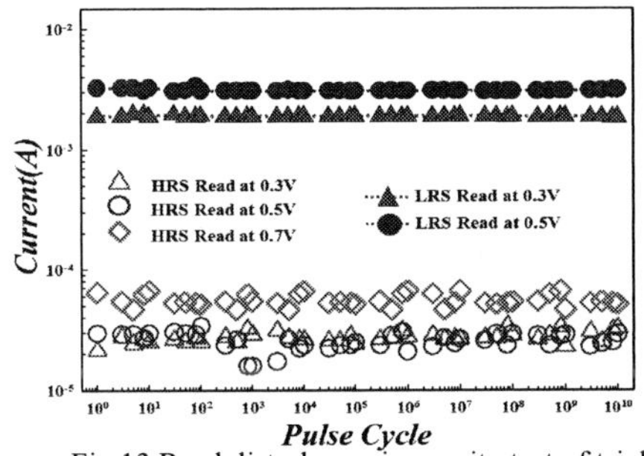

Fig.13 Read disturbance immunity test of triple layer device. The device can maintain its state $>10^{10}$ pulse cycles.

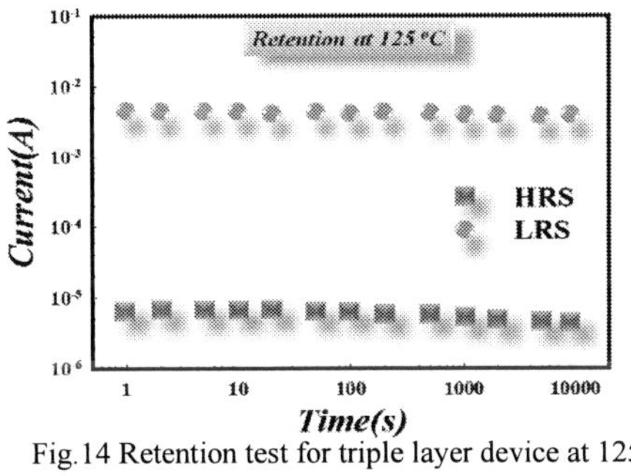

Fig.14 Retention test for triple layer device at 125°C (without equipment compliance). Both HRS and LRS can maintain their states $>10^4$ cycles.

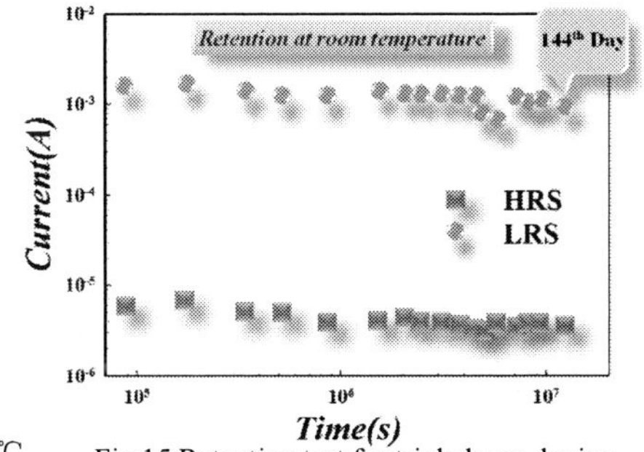

Fig.15 Retention test for triple layer device at room temperature. Both resistance states exhibit no degradation even after 144 days.

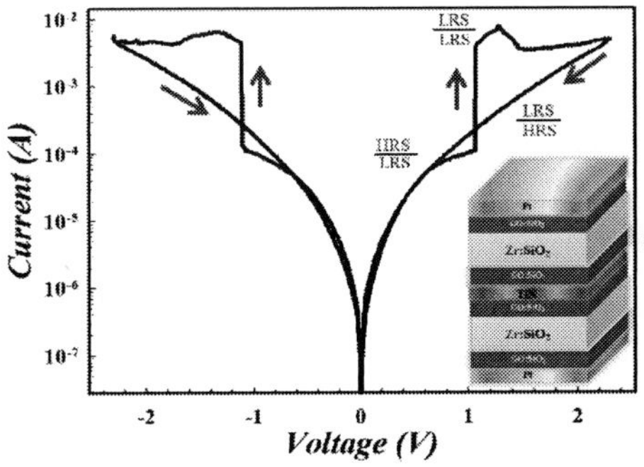

Fig.16 CRS application constructed by stacking two double-ended GO SOB RRAM devices.

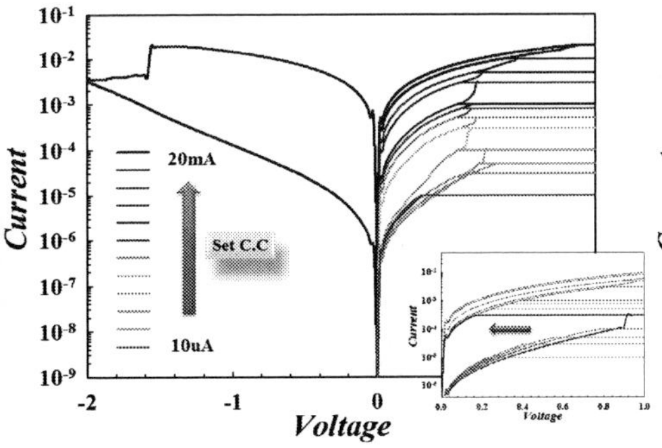

Fig.17 Multi-set process can be achieved for triple layer device. Single layer device exhibits no such property, as shown bottom right.

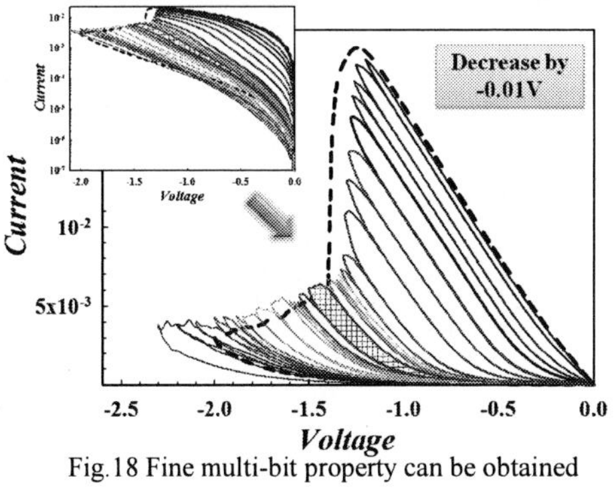

Fig.18 Fine multi-bit property can be obtained from -1.2V to -1.31V with -0.01V successive voltage decrement.

978-1-4799-8002-4/14 $31.00 © 2014 IEEE

Realizing a Topological-Insulator Field-Effect Transistor using Iodostannanane

William G. Vandenberghe[1], Massimo V. Fischetti[1]

[1]Dept. of Materials Science and Engineering, the University of Texas at Dallas, Richardson, TX, USA

Abstract

Monolayer hexagonal tin (stannanane) is a topological insulator and upon functionalization with halogens, such as iodine, a gap exceeding 300 meV is obtained. In a stannanane ribbon the topologically protected edge states lead to very high conductivities and mobilities; moreover the conductivity is strongly dependent on the Fermi level. We show how this property can be exploited to make a topological-insulator field-effect transistor (TIFET). We simulate the input and output characteristics of the TIFET using a drift-diffusion-like approximation and obtain promising transistor characteristics with a high on-current which exceeds the off-current by over three orders of magnitude.

Introduction

As transistor dimensions are decreasing, further scaling is becoming increasingly challenging. The main challenge is to maintain electrostatic control and to reduce power consumption while maintaining or further improving performance. To maintain electrostatic control, a transition to two-dimensional materials will be required at extremely scaled dimensions (1). To further reduce power consumption, the energy per switching cycle QV_{dd} must be reduced and to maintain high performance, a high electron mobility must be retained. Many alternative devices exploiting different switching mechanisms such as BISFETs (2) or TFETs (3) are being investigated. However up to now, none of these devices have proven to be a worthy competitor for Si-based MOSFETs.

When looking at two-dimensional materials, graphene is the first material that comes to mind but the absence of a bandgap makes it unsuited as a channel material. A large amount of research focusing on other two-dimensional materials, such as MoS_2 or other monolayer metal-dichalcogenides, is underway but these materials have much lower mobility. Recently, a new two-dimensional material consisting of a monolayer of tin (Sn, *stannum*) in a buckled hexagonal lattice, similar to that of graphane (4) and germanane (5), has been proposed. To indicate the sp^3 nature of the bond but to distinguish it from stannane (SnH_4), we

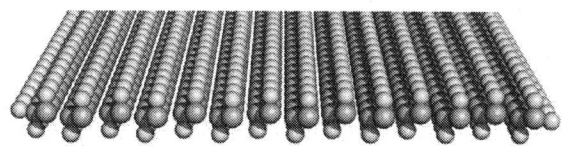

Fig. 1: Illustration of an iodostannanane ribbon

call this monolayer of tin *stannanane* and we call a monolayer of tin functionalized with iodine *iodostannanane*. An interesting property of stannanane and stannanane functionalized with halogens, is that they are two-dimensional semiconductors with an "inverted bandgap" because of strong spin-orbit coupling. More specifically, stannanane is a two-dimensional topological insulator, also known as a quantum spin hall insulator, whose band structure differs fundamentally from most other semiconductors (which are trivial insulators) (6). An interesting property of topological insulators is that at their edge, a set of topologically protected states is present. Moreover, these edge states are spin-polarized, with spin and momentum locked, and in the absence of a spin-breaking interactions, back-scattering is prohibited, conductance is quantized at low temperature and a very high conductivity can be obtained even at room temperature.

We focus on stannanane functionalized with halogens since they are predicted to exhibit a band gap exceeding 300 meV ensuring room temperature operation. For computational convenience, we choose iodine as functionalizing agent since it has a lower energy cut-off in *ab initio* calculations.

In this paper, we first calculate the conductivity and mobility of stannanane functionalized with iodine (iodostannanane, illustrated in Fig. 1) and show that its conductivity is strongly dependent on the Fermi level. Next, we show how this property can be exploited to build a new kind of transistor: the topological-insulator field-effect transistor (TIFET). Employing a drift-diffusion like approximation, we simulate the TIFET input and output characteristics.

Iodostannanane conductivity

In Fig. 2, we show the iodostannanane band structure as calculated by the Vienna *ab-initio* simulation package (VASP) (7) and show that spin-orbit coupling opens a

978-1-4799-8002-4/14 $31.00 © 2014 IEEE

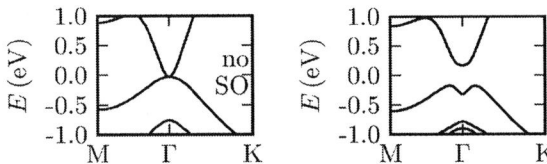

Fig. 2 Band structure of bulk iodostannanane calculated without spin-orbit (SO) coupling (left) and with spin-orbit coupling (right). The opening of the bandgap is indicative of the topological insulating properties of iodostannanane.

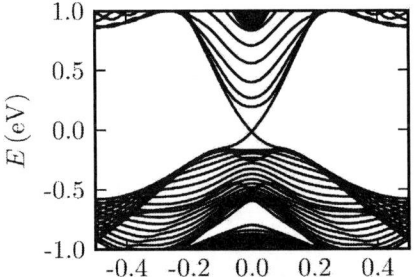

Fig. 3 Bulk band structure of an iodostannanane ribbon. The band structure resembles the bulk iodostannanane folded along the Γ -M direction except for the edge states traversing the bandgap

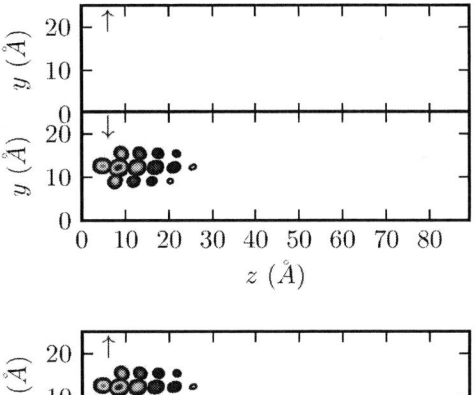

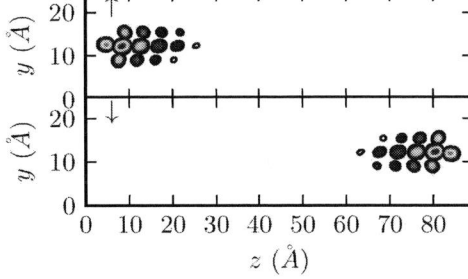

Fig. 4 Wavefunction spin components for the left-edge k>0 state (a) and both edges for the k<0 state (b). Intra-edge backscattering is prohibited because of spin-polarization whereas inter-edge backscattering is unlikely because of the distance between the edge states.

bandgap which is an indication that it is a topological insulator. Additional calculations of the band structure of an iodostannanane ribbon (Fig. 3) reveals the topologically protected edge states with their energy crossing the bulk bandgap. Next, we perform an inverse Fourier transform of the wavefunctions associated with the edge states and show the spin components in Fig. 4. The edge wavefunctions are spin-polarized and this prohibits intra-edge back-scattering. Inter-edge backscattering on the other hand, is allowed but the physical distance between both edges impedes this

process. Edge states with energy in the bulk conduction or valence band, can back-scattering to bulk states.

We calculate the phonon-limited mobility in iodostannanane ribbons based on the Kubo-Greenwood formalism using the equations

$$\sigma_{1\mathrm{D}} = \frac{2e^2}{k_{\mathrm{B}}T} \int \mathrm{d}k\, \tau(k)v(k)^2 f(E(k))(1-f(E(k))) \quad (1)$$

$$\frac{1}{\tau_{\mathrm{TA,LA}}(k)} \approx \frac{2\pi}{\hbar} \frac{D_{\mathrm{TA,LA}}^2 k_{\mathrm{B}}Ta}{2v_{\mathrm{s,TA,LA}}^2 v_{\mathrm{F}} N M \hbar} \mathcal{I}_k \quad (2)$$

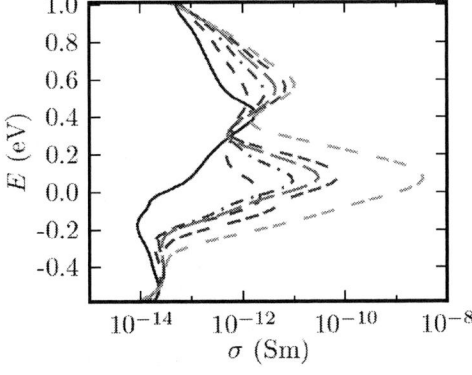

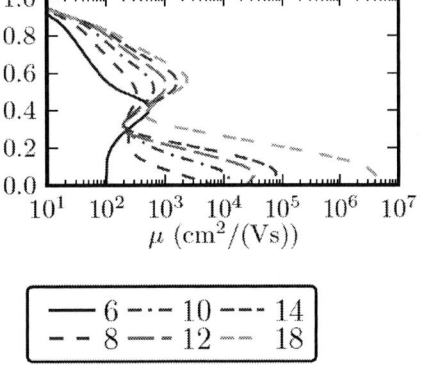

Fig. 5 Calculated iodostannanane conductivity and mobility. The conductivity reaches a maximum when the Fermi level is inside the bulk bandgap and is reduced by up to 4 orders of magnitude when the Fermi level moves into the conduction band or valence band.

where k_B is the Boltzmann constant, T is the temperature (taken as 300 K), τ is the scattering time, $\upsilon(k)$ is the electron velocity, υ_F is the velocity near $k=0$, $f(E)$ is the Fermi-Dirac distributions, $D_{TA,LA}$ and $\upsilon_{TA,LA}$ is the deformation potential and velocity for TA and LA phonons, N is the number of atoms, M is the unit cell mass and $\mathcal{I}_k$ is the overlap integral calculated from the wavefunctions. We consider elastic scattering with the LA and TA phonons whose interaction strength is obtained from first principles and we account for phonon confinement by imposing free-floating boundary conditions on the atom displacement inside the ribbon. The deformation potentials are calculated from the conduction band of the bulk iodostannanane ribbon. The mobility can be computed from the conductivity by using the relation $\sigma = en\mu$.

Fig. 5 shows the conductivity and the mobility of stannanane ribbons. The mobility and conductivity are very high due to the small probability of backscattering and the conductivity can be changed over 3 to 4 orders of magnitude when the Fermi level is changed over a few 100s of meV. The ability to change the conductivity over several orders of magnitude suggests that topological insulators can be used for transistor applications and this property will be exploited in following section.

TIFET

In Fig. 6a we illustrate the TIFET device structure: an iodostannanane ribbon sandwiched between a double gate with source and drain contacts at both ends of the ribbon. The operation principle of the TIFET (illustrated in Figs. 6b-c) is: i) in the on-state, electrons travel from the source to the drain with minimal scattering; ii) in the off-state, the Fermi level is moved into the conduction (or valence) band so that scattering to "bulk" states dominates and strongly reduces the current.

To model the TIFET, we employ a drift-diffusion-like approximation by introducing a quasi-Fermi level ($\varphi(x)$) so the current at each point in the TIFET can be computed as $J = \sigma(\varphi - U) \nabla\varphi$. We further assume the gates perfectly control the potential in the TI and the TIFET can be modeled as a one-dimensional structure. Invoking current continuity, the quasi-Fermi level can be determined by solving an ordinary differential equation given. The equations to model the TIFET are

$$J(x) = \sigma(\varphi - U(x), x)\frac{d\varphi}{dx} \qquad (3)$$

$$\varphi(x) = \int_0^x dx \frac{J(x)}{\sigma(\phi - U(x))} \qquad (4)$$

$$\sigma(E_F, x) = \begin{cases} \sigma_{\text{contact}} & |x - x_0| > \frac{l}{2} \\ \sigma_{\min} + \sigma_{TI} e^{-\alpha E_F^2} & |x - x_0| \leq \frac{l}{2} \end{cases} \qquad (5)$$

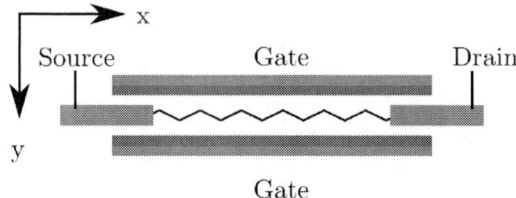

Fig. 6a Schematic side view of the TIFET device structure.

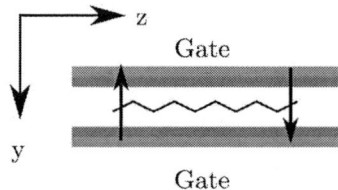

Fig. 6b TIFET in its on-state: The Fermi level is inside the bulk bandgap, only back-scattering to the opposite edge is active and the conductivity is high.

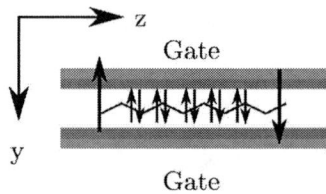

Fig. 6c TIFET in its off-state: The Fermi level is moved into the conduction or valence band, scattering rates strongly increase and the conductivity is low.

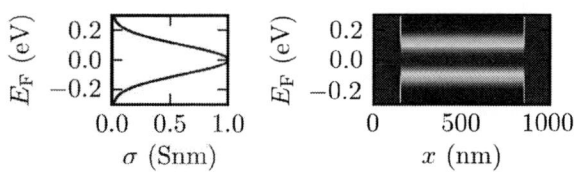

Fig. 7 Simplified model for TI-conductivity with $\sigma_{\min} = 10^{-13}$ Sm and $\sigma_{TI} = 10^{-9}$ Sm used in our simulations (left) and a contour plot of the conductivity in the TIFET device with high-conductivity contacts (right).

where J is the current, $\sigma(E_F, x)$ is the conductivity as a function of Fermi level and position. We take the external potential ($U(x)$) to be perfectly controlled by the gate $U(x)=-qV_{\text{gs}}$. σ_{TI} is the maximum TI conductivity, σ_{contact} is the contact conductivity and we take $\sigma_{\text{contact}} = \sigma_{TI}$ in our example while x_0 is the center of the device and l is the length of the TI. For simplicity, we model the TI conductivity as a Gaussian function with a maximal conductivity (σ_{TI}) inside the bulk bandgap and a minimal conductivity ($\sigma_{\min}$) inside the conduction and valence band while we take the contacts

978-1-4799-8002-4/14 $31.00 © 2014 IEEE   806

to have a high conductivity independent of the Fermi level and equaling the maximal TI conductivity as shown in Fig. 7. In Fig. 8 we show the quasi-Fermi level as a function of position for a given current and gate bias. In Fig. 9, we show the simulated output characteristics of the TIFET respectively and after normalizing to the ribbon width (<10 nm), the current exceeds 10^4 A/μm. The input characteristics are shown in Fig. 10 and an I_{on}/I_{off} exceeding 4 orders of magnitude is observed.

Conclusion

We have shown that TIs exhibit a high conductivity when the Fermi level is inside the bandgap and a lower conductivity when the Fermi level is moved into the conduction band. The TIFET proposed in this paper shows input- and output characteristics similar to those of conventional FETs with very high current. Furthermore, with a minimal charge density and backscattering in the channel, TIFETs could prove to be much more energy-efficient than conventional FETs.

Acknowledgements

We acknowledge the support of Nanoelectronics Research Initiative's (NRI's) Southwest Academy of Nanoelectronics (SWAN).

References

(1) M. V. Fischetti, B. Fu, and W. G. Vandenberghe, "Theoretical Study of the Gate Leakage Current in Sub-10-nm Field-Effect Transistors," *IEEE Trans. Electron Devices*, vol.60, no.11, pp. 3862-3869, November 2013

(2) S. K. Banerjee, L. F. Register, E. Tutuc, D. Reddy, and A. H. MacDonald. "Bilayer pseudospin field-effect transistor (BiSFET): a proposed new logic device." *IEEE Electron Device Letters*, vol. 30, no. 2, pp. 158-160, December 2009.

(3) W.G. Vandenberghe, A. S. Verhulst, B. Sorée, W. Magnus, G. Groeseneken, Q. Smets, M. Heyns, and M. V. Fischetti. "Figure of merit for and identification of sub-60 mV/decade devices." *Applied Physics Letters* 102, no. 1, 013510, January 2013.

(4) D. C. Elias, R. R. Nair, T. M. G. Mohiuddin, S. V. Morozov, P. Blake, M. P. Halsall, A. C. Ferrari et al. "Control of graphene's properties by reversible hydrogenation: evidence for graphane." *Science* 323, no. 5914, pp. 610-613, January 2009.

(5) E. Bianco, S. Butler, S. Jiang, O. D. Restrepo, W. Windl, and J. E. Goldberger. "Stability and exfoliation of germanane: a germanium graphane analogue." *Acs Nano* 7, no. 5, pp. 4414-4421, March 2013.

(6) Xu, Yong, Binghai Yan, Hai-Jun Zhang, Jing Wang, Gang Xu, Peizhe Tang, Wenhui Duan, and Shou-Cheng Zhang. "Large-gap quantum spin Hall insulators in tin films." *Physical review letters* 111, no. 13, 136804, September 2013.

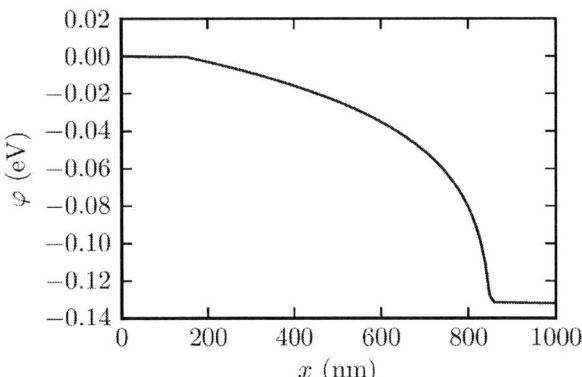

Fig. 8 Quasi-Fermi level as a function of position for V_{gs} =0.24 V and J=2.6 μA.

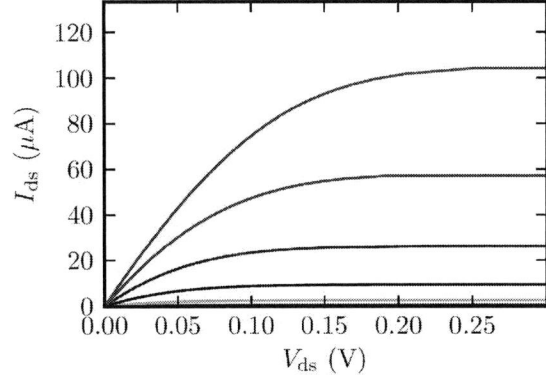

Fig. 9 Simulated output characteristics of the TIFET showing high current and excellent saturation.

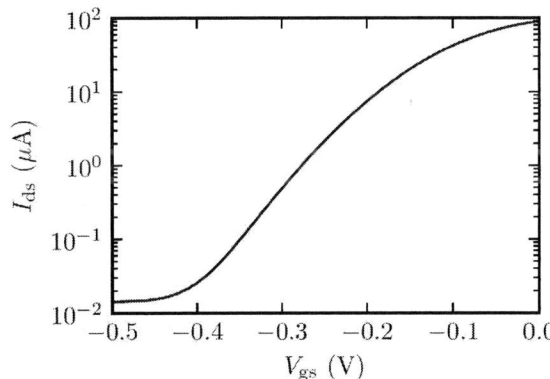

Fig. 10 Simulated input characteristics for I_{ds} = 0.1 V of the TIFET showing an I_{on}/I_{off} exceeding 3 orders of magnitude.

(7) G. Kresse, and J. Hafner. "Ab initio molecular dynamics for liquid metals." *Physical Review B* 47, no. 1, 558, January 1993.

978-1-4799-8002-4/14 $31.00 © 2014 IEEE

Hybrid Si/TMD 2D Electronic Double Channels Fabricated Using Solid CVD Few-Layer-MoS$_2$ Stacking for V$_{th}$ Matching and CMOS-Compatible 3DFETs

Min-Cheng Chen[1], Chia-Yi Lin[1], Kai-Hsin Li[1], Lain-Jong Li[2], Chang-Hsiao Chen[2], Cheng-Hao Chuang[3], Ming-Dao Lee[1], Yi-Ju Chen[1], Yun-Fang Hou[1], Chang-Hsien Lin[1], Chun-Chi Chen[1], Bo-Wei Wu[1], Cheng-San Wu[1], Ivy Yang[1], Yao-Jen Lee[1], Wen-Kuan Yeh[1], Tahui Wang[4], Fu-Liang Yang[5] and Chenming Hu[6]

[1] National Nano Device Laboratories, National Applied Research Laboratories, Taiwan; [2] Physical Sciences and Engineering, King Abdullah University of Science and Technology, Kingdom of Saudi Arabia; [3] Dept. of Phys., Tamkang University, Taiwan; [4] Dept. of Electronics Eng., National Chiao-Tung University, Taiwan; [5] Research Center for Applied Sciences, Academia Sinica, Taiwan; [6] Dept. of Electrical Eng. and Computer Science, University of California, Berkeley, USA; Tel: (886) 357-26100 ext. 7511, Fax: (886) 357-22715, Email: mcchen@narlabs.org.tw

Abstract

Stackable 3DFETs such as FinFET using hybrid Si/MoS$_2$ channels were developed using a fully CMOS-compatible process. Adding several molecular layers (3–16 layers) of the transition-metal dichalcogenide (TMD), MoS$_2$ to Si fin and nanowire resulted in improved (+25%) I$_{on,n}$ of the FinFET and nanowire FET (NWFET). The PFETs also operated effectively and the N/P device V$_{th}$ are low and matched perfectly. The proposed heterogeneous Si/TMD 3DFETs can be useful in future electronics.

I. Introduction

3DFETs can improve sub-20 nm CMOS node performance and substantially reduce supply voltage and short channel effects [1]. However, the traditional silicon channel must be replaced by high-mobility materials in future VLSI applications [2]–[3]. Heterogeneous two-dimensional atomic crystals, namely, transition-metal dichalcogenide (TMD), have atomically smooth surface without dangling bounds and good mobility in CVD deposited films of atomic scale thickness are very attractive enablers of ultimately scaled transistors and 3D ICs [1,4]. However, a manufacturing flow must be realized using low-temperature semiconductor process [5] and TMD by chemical vapor deposition (CVD) [6]. This paper presents the first CMOS process compatible TMD 3D transistor technology using novel hybrid Si/MoS$_2$ channel FinFET and NWFET with improved I$_{on,n}$ and matched V$_{th}$ of N and P devices. In comparison to previously published TMD transistors, this work reports the shortest gate length, thinnest gate dielectric, and first high performance at low voltage (Table 1 and Fig. 1).

II. Process Integration and Device Fabrication

The reported work made use of a previously published low temperature Tri-gate FinFETs technology [7]. The few-layer MoS$_2$ growth step was inserted after blocking oxide deposition and clean (see Figs. 2 and Fig. 4 for FinFET and Fig. 3 and Fig. 5 and for NWFET). The ~5nm SiO$_2$ on the surface of the Si fin is responsible for setting the desirable low and matched V$_{th}$ of N and P devices as shown later. A low-temperature activation process that involved microwave annealing [5] was used after hybrid MoS$_2$ deposition. In Figs. 6 and 7, the TEM images show the hybrid Si/MoS$_2$ channel trigate FinFET and NWFET respectively. Few-layer MoS$_2$ was successfully integrated into 3DFETs technology using low-temperature CVD with the number of MoS$_2$ layers determined by deposition duration (Figs. 8 and 9). The TiN gate over the hybrid Si/MoS$_2$ fins is 50 nm long (Fig. 10).

III. Results and Discussion

A. MoS$_2$ Material Analysis

MoS$_2$ material analysis was performed over flat regions of the wafer and over the hybrid Si/MoS$_2$ fins. Figs. 11(a) and (b) show the S L-edge and Mo M-edge features of the MoS$_2$ clearly observed for both the films deposited on the flat Si substrate and over the hybrid Si/MoS$_2$ fins using X-ray absorption spectroscopy. Fig. 12 shows the X-ray absorption near the edge spectrum, which revealed that the Mo M-edge spectrum exhibited a peak in films deposited on flat Si substrate and on the hybrid Si/MoS$_2$ channel. Strong separated features and high inter-binding hybridization confirmed that few-layer-MoS$_2$ material characteristics were preserved through the hybrid Si/MoS$_2$ channel fabrication process.

B. Device Performance

Few-layer-MoS$_2$ (3–16 layers, 2–10 nm thick) were fully integrated into the hybrid channel 3DFETs. Fig. 13 presents the N/PMOS transfer and output characteristics of the hybrid channel 3DFETs. Both N/P type hybrid channel FETs have highly desirable low V$_{th}$ and perfect V$_{th}$ matching. Fig. 14 illustrates the flat band and the threshold conditions and explains the low and matched V$_{th}$. While the NMOS

978-1-4799-8002-4/14 $31.00 © 2014 IEEE

channel is formed in MoS_2. The PMOS channel is formed in Si. Therefore the Si hole mobility determines the P-type hybrid Si/MoS_2 FinFET performance. The electron mobility of the hybrid Si/MoS_2 channel FinFET is determined by the electron rich MoS_2 and is two times larger than the electron mobility of the Si channel FinFET (Fig. 15). Fig. 16 shows that the hysteretic behavior induced by gate bias sweeping was minimal (delta V_{th} < 50 mV) and exceptional stability was achieved.

Regarding the geometric characteristics of the hybrid Si/MoS_2 electronic double channels, using several molecular layers of MoS_2 improved $I_{on,n}$ by approximately 20% in the hybrid Si/MoS_2 FinFETs device (Fig. 17). The $I_{on,n}$ performance was enhanced (> 25%) by applying the hybrid Si/MoS_2 channels according to the NWFETs (Fig. 18). Table 2 summarizes the demonstrated advantages of the Si/MoS_2 3DFETs over the MoS_2 2DFETs in CMOS operation and V_{th} matching.

IV. Conclusions

In this study, hybrid Si/TMD channel 3DFETs were proposed and demonstrated using a fully CMOS-compatible process. The $I_{on,n}$ improved by more than 25% and N/P device V_{th} matching was achieved. The novel hybrid Si/ TMD channel is a promising technology for high-performance and scaled 2D and 3D FETs in 2D and 3D ICs.

Acknowledgement

This work was performed by the National Nano Device Laboratories facilities and supported by the National Science Council, Taiwan.

References

[1] Chenming Hu, "Thin-Body FinFET as Scalable Low Voltage Transistor" in VLSI-TSA., 2012, pp. 1-4, 2012. C. Hu, 2012 VLSI-TSA., p.1 (2012).

[2] P. Hashemi, K. Balakrishnan, A. Majumdar, A. Khakifirooz, W. Kim, A. Baraskar, Li A. Yang, K. Chan, S. U. Engelmann, John A. Ott, D. A. Antoniadis, E. Leobandung, and D.-G. Park, "Strained $Si_{1-x}Ge_x$-on-Insulator PMOS FinFETs with Excellent Sub-Threshold Leakage, Extremely-High Short-Channel Performance and Source Injection Velocity for 10nm Node and Beyond" in *VLSI Symp. Tech. Dig.*, 2014, pp. 18–19, 2014.

[3] W. Liu, J. Kang, W. Cao, D. Sarkar, Y. Khatami, D. Jena and K. Banerjee "High-Performance Few-Layer-MoS2 Field-Effect-Transistor with Record Low Contact-

Resistance", in *IEDM Tech. Dig.*, 2013, pp. 499–502, 2013.

[4] B. Radisavljevic, A. Radenovic, J. Brivio, V. Giacometti and A. Kis "Single-layer MoS_2 transistors," in Nature Nanotech. vol. 6, pp. 147-150, 2011.

[5] Y.-J. Lee, S.-S. Chuang, C.-I. Liu, F.-K. Hsueh, P.-J. Sung, H.-C. Chen, C.-T. Wu, K.-L. Lin, J.-Y. Yao, Y.-L. Shen, M.-L. Kuo, C.-H. Yang, G.-L. Luo, H.-W. Chen, C.-H. Lai, M. I. Current, C.-Y. Wu, Y.-M. Wan, T.-Y. Tseng, Chenming Hu, and F.-L. Yang "Full Low Temperature Microwave Processed Ge CMOS Achieving Diffusion-Less Junction and Ultrathin 7.5nm Ni Mono-Germanide", in *IEDM Tech. Dig.*, 2012, pp. 514–517, 2012.

[6] H. Wang, L. Yu, Y.-H. Lee, W. Fang, A. Hsu, P. Herring, M. Chin, M. Dubey, L.-J. Li, J. Kong, and T. Palacios "Large-scale 2D Electronics based on Single-layer MoS_2 Grown by Chemical Vapor Deposition", in *IEDM Tech. Dig.*, 2012, pp. 88–91, 2012.

[7] M.-C. Chen, C.-H. Lin, Y.-F. Hou, Y.-J. Chen, C.-Y. Lin, F.-K. Hsueh, H.-L. Liu, C.-T. Liu, B.-W. Wang, H.-C. Chen, C.-C. Chen, S.-H. Chen, C.-T. Wu, T.-Y. Lai, M.-Y. Lee, B.-W. Wu, C.-S. Wu, I. Yang, Y.-P. Hsieh, C.H. Ho, T. Wang, A.B. Sachid, C. Hu and F.-L. Yang "A 10 nm Si-Based Bulk FinFETs 6T SRAM with Multiple Fin Heights Technology for 25% Better Static Noise Margin", in *VLSI Symp. Tech. Dig.*, 2013, pp. 218–219, 2013.

[8] J. Lee, H.-Y. Chang, T.-J. Ha, H. Li, R. S. Ruoff, A. Dodabalapur and D. Akinwande "High-Performance Flexible Nanoelectronics: 2D Atomic Channel Materials for Low-Power Digital and High-Frequency Analog Devices", in IEDM Tech. Dig., 2013, pp. 491–494, 2013.

[9] S. Das and J. Appenzeller "Where does the Current flow in Two Dimensional Layered Systems", in Nano Letters vol. 13, pp.3396-3402, 2013.

[10] L. Yang, K. Majumdar, Y. Du, H. Liu, H. Wu, M. Hatzistergos, P. Y. Hung, R. Tieckelmann, W. Tsai, C. Hobbs and P. D. Ye "High-Performance MoS_2 Field-Effect Transistors Enabled by Chloride Doping: Record Low Contact Resistance (0.5kOmega·μm) and Record High Drain Current (460 μA/μm)" in *VLSI Symp. Tech. Dig.*, 2014, pp. 238–239, 2014.

MoS_2 layer	0.7nm (1L)		0.7nm (1L)	7nm (10L)	4nm (6L)	2-10nm (3-16L)
growth	exfoliation		CVD	CVD	exfoliation	CVD
G_{ox}	30 nm HfO_2	270 nm SiO_2	30 nm HfO_2	10 nm Al_2O_3	90 nm SiO_2	4 nm HfO_2
L_{ch}	0.5 um	1.5 um	1 um	1 um	0.1 um	50 nm
V_g-V_{th}	4 V	30 V	5 V	4 V	> 20 V	1 V
$I_{ds,n}$ (uA/um)	2.5 @ 0.5V	0.09@10mV	16 @ 5V	30 @ 1V	460@1.6V	133 @ 1V
I_{on}/I_{off} (X)	10^8	10^6	10^8	10^7	$6.3x10^5$	10^5
m (cm^2/V-s)	217		190	30	50-60	40
structure	TG planar	BG planar	TG planar	TG planar	BG planar	hybrid 3DFETs
Ref.	2011 nature nano lett. [4]		2012 IEDM [6]	2013 IEDM [8]	2014 VLSI [10]	this work

Table 1. Comparison of prior MoS_2 MOSFET and the this work

top gate
bottom gate
this work
3DFET
[10] 2014 VLSI
V_{bg}=-30 V
[3] 2013 IEDM
V_{bg}=30 V
[9] 2013 nano lett.
V_{bg}=4 V
[8] 2013 IEDM
[6] 2012 IEDM
I_{ds} (μA/μm) @ V_{ds}=1V
L_{ch} (μm)

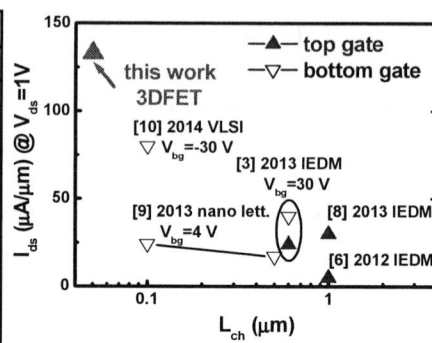

Fig. 1. Comparison of I_{ds} at V_{ds}=1V in prior reports and this work.

- **Fin pattering**
- **Oxide dep. & etch back**
- **OX treatment & clean**
- **MoS_2 & gate stack dep.**
- **Gate pattering**
- **S/D imp. & MWA**

Fig. 2. Key steps of the hybrid Si/ MoS_2 trigate FinFETs.

- **NW pattering**
- **Box etch. & NW trimming**
- **OX treatment & clean**
- **MoS_2 & gate stack dep.**
- **Gate pattering**
- **S/D imp. & MWA**

Fig. 3. Key steps of the hybrid Si/ MoS_2 NWFET.

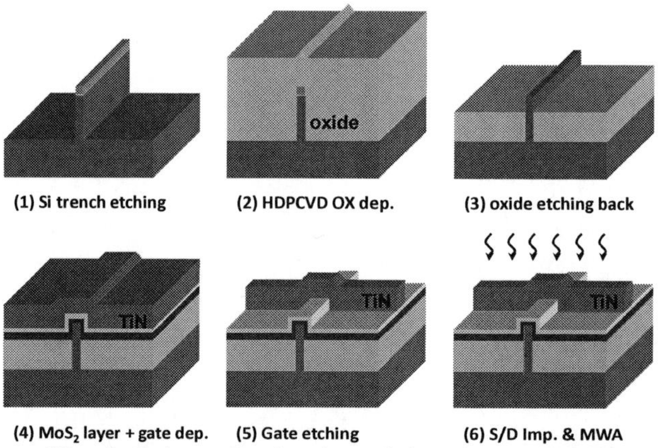

oxide
TiN
(1) Si trench etching
(2) HDPCVD OX dep.
(3) oxide etching back
(4) MoS_2 layer + gate dep.
(5) Gate etching
(6) S/D Imp. & MWA

Fig. 4. Schematic drawing of the trigate FinFETs with the hybrid Si/MoS_2 channels.

MG/HK
MoS_2
oxide
Si Fin
50 nm

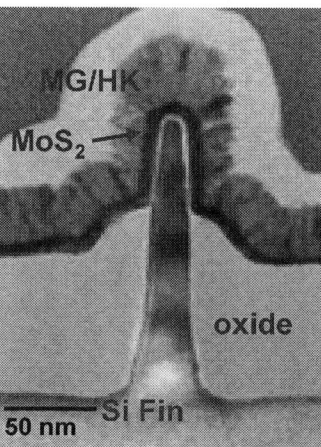

Fig. 6. TEM picture of the hybrid Si/MoS_2 trigate FinFET.

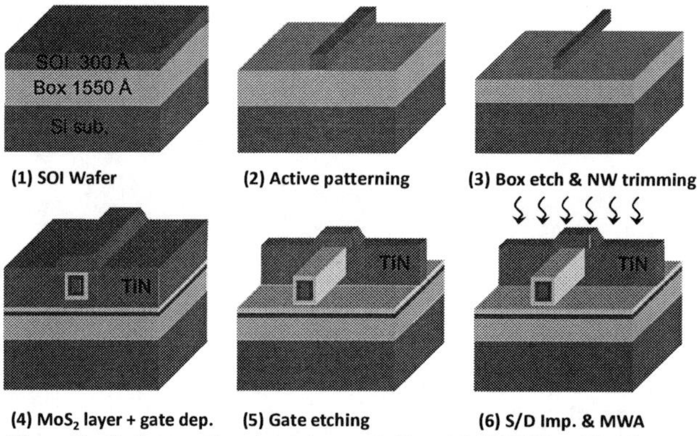

SOI 300 Å
Box 1550 Å
Si sub.
TiN
(1) SOI Wafer
(2) Active patterning
(3) Box etch & NW trimming
(4) MoS_2 layer + gate dep.
(5) Gate etching
(6) S/D Imp. & MWA

Fig. 5. Schematic drawing of the NWFET with the hybrid Si/MoS_2 channels.

2 nm
(a) 3 layers MoS_2
5 nm
(b) 8 layers MoS_2
10 nm
(c) 16 layers MoS_2

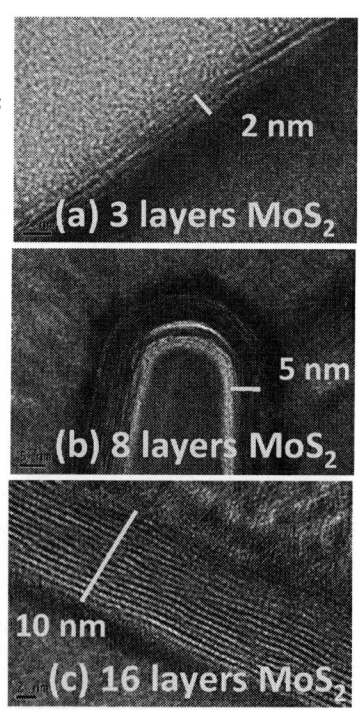

Fig. 8. TEM pictures of the hybrid Si/MoS_2 channels.

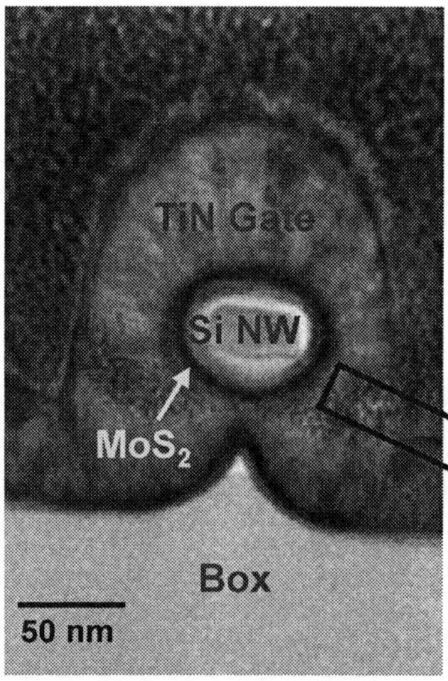

TiN Gate
Si NW
MoS_2
Box
50 nm

Fig. 7. TEM picture of the hybrid Si/MoS_2 NWFET.

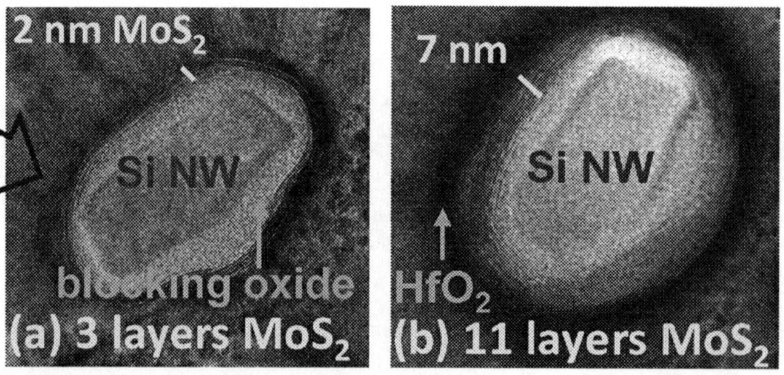

2 nm MoS_2
Si NW
blocking oxide
(a) 3 layers MoS_2
7 nm
Si NW
HfO_2
(b) 11 layers MoS_2

Fig. 9. Enlarge TEM pictures of the hybrid Si/MoS_2 NWFET.

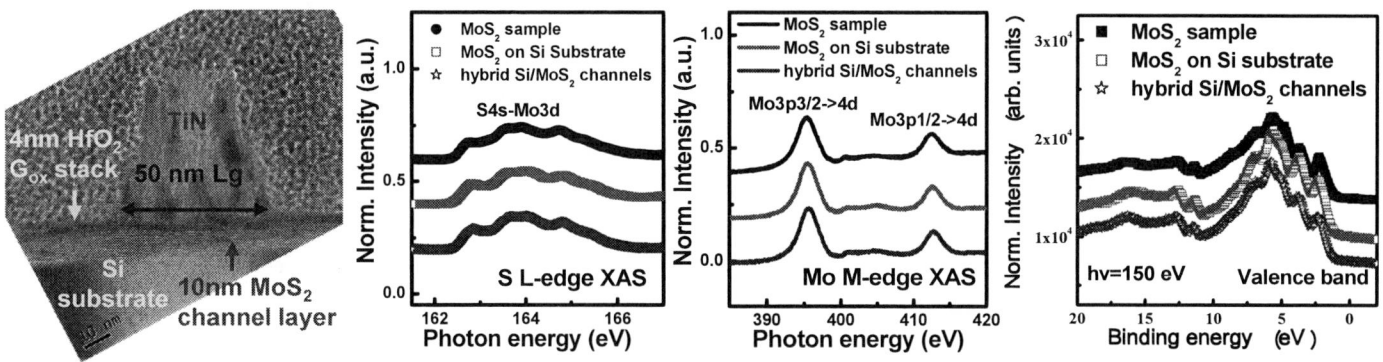

Fig. 10. TEM picture of the hybrid Si/MoS$_2$ channels gate structure.

Fig. 11. (a) S L-edge and (b) Mo M-edge features comparison of MoS$_2$ by using X-ray absorption spectroscopy (XAS).

Fig. 12. X-ray absorption near-edge spectrum (XANES) of MoS$_2$.

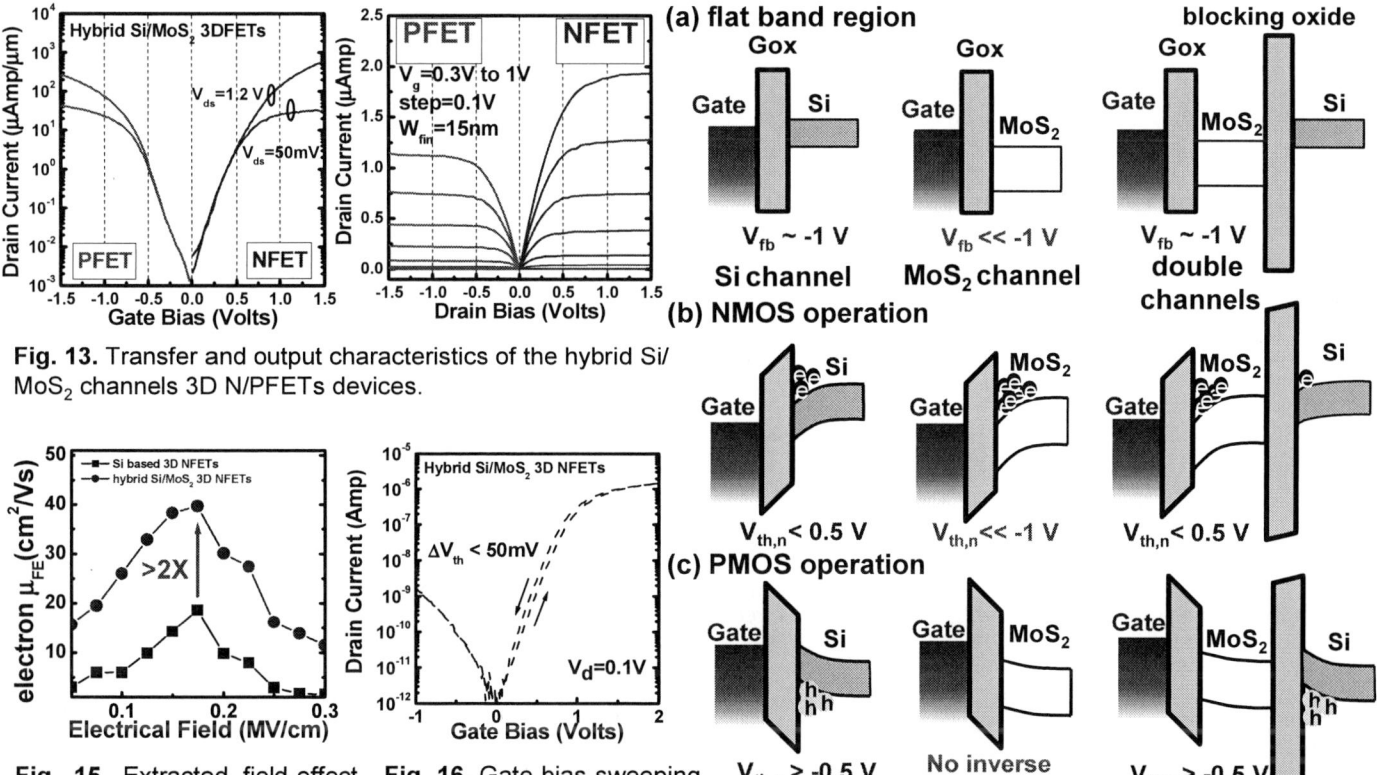

Fig. 13. Transfer and output characteristics of the hybrid Si/MoS$_2$ channels 3D N/PFETs devices.

Fig. 15. Extracted field-effect mobility of N-type Si/MoS$_2$ channel FinFET is twice the mobility of Si channel FinFET.

Fig. 16. Gate bias sweeping of the hybrid Si/MoS$_2$ channels 3D NFETs.

Fig. 14. Band diagram of device operation for Si based, MoS$_2$ FETs and hybrid Si/MoS$_2$ double channels.

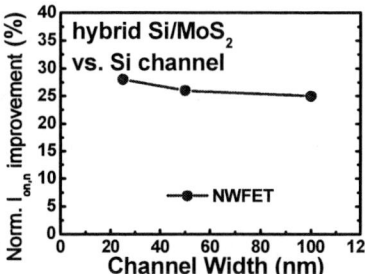

Fig. 17. I$_{on}$ improvement of the hybrid Si/MoS$_2$ channels 3D NFETs.

Fig. 18. I$_{on}$ improvement of the hybrid Si/MoS$_2$ channels 3D NFETs.

	homogeneous 3DFETs	MoS$_2$ channel	hybrid Si/MoS$_2$ 3DFETs (this work)
gate structure	Trigate/GAA NW	bottom/top gate	Trigate / GAA NW
channel material	Si based	MoS$_2$ channel	Si/MoS$_2$ double channels
growth method	Si substrate	exfoliation / CVD	CVD
NMOS operation	Si channel	MoS$_2$ channel	MoS$_2$ channel
I$_{on}$ performance	-	high	~20% / >25%
PMOS operation	Si channel	X	Si channel
I$_{on}$ performance	-	no	comparable
V$_{th}$ balancing	yes	bad	matching
CMOS process	yes	no/hard	compatible

Table 2. Summary table of homogeneous 3DFETs, MoS$_2$ channel and the heterogeneous hybrid Si/MoS$_2$ channels 3DFETs.

High-Performance Carbon Nanotube Field-Effect Transistors

Max M. Shulaker[1]*, Gregory Pitner[1], Gage Hills[1], Marta Giachino[3], H.-S. Philip Wong[1], Subhasish Mitra[1,2]

Department of Electrical Engineering[1], Computer Science[2], and Materials Science and Engineering[3]
Stanford University, Stanford, CA, 94305, U.S.A. *Email: maxms@stanford.edu

Abstract

We demonstrate carbon nanotube (CNT) field-effect transistors (CNFETs) with the highest current drive (per unit layout width)[1] to-date (>100 µA/µm at 400 nm channel length and 1V V_{DS}), while simultaneously achieving high I_{ON}/I_{OFF} (>5,000). This is the first demonstration of CNFETs with CNT density above 100 CNTs/µm consisting of highly-aligned CNTs and achieving both high current drive and high I_{ON}/I_{OFF}. The current drives of the demonstrated CNFETs approach that of similarly-scaled and similarly-biased silicon-based field-effect transistors in production in major semiconductor foundries.

Introduction

Carbon nanotube-based digital systems promise both increased performance and improved energy efficiency beyond the limitations of current silicon-based technologies [1-3]. A typical carbon nanotube (CNT) field-effect transistor (CNFET) is formed by using multiple CNTs in parallel as the channel, with traditional lithographically-defined source, drain, and gate regions (Fig 1). To surpass the speed of silicon-based digital systems and to achieve the promised energy efficiency benefits, the current-drive per unit layout width (I_{ON}) of CNFETs must surpass that of Si-MOSFETs. This requires target CNT density of 100-200 highly aligned CNTs/µm, which translates to an average inter-CNT pitch of 5-10 nm [1,2,4].

Furthermore, high CNT density alone is insufficient for high-performance and highly energy-efficient CNFET digital systems. CNT density variations, caused by variations in inter-CNT spacing (during CNT growth or during subsequent process steps such as CNT transfer), result in variations in CNT counts inside CNFETs. This leads to CNFET I_{ON} variations that can result in reduced yield, increased delay variations, and degraded noise margins of large-scale circuits [5]. As explained in [5], CNT density variations can be quantified using the index of dispersion for count (IDC) metric, which can be expressed as:

$$IDC = \sigma^2(s)/\mu^2(s),$$

where $\sigma(s)$ and $\mu(s)$ are the standard deviation and mean of the inter-CNT spacing distribution, respectively. An ideal IDC value of 0 corresponds to uniform CNT spacing. Our typical CNT growth has IDC value of 0.5 [5].

However, there is a significant gap between the requirement of high CNT density with low IDC and what is achievable today. A typical CNT growth results in 1-10 CNTs/µm, while the best CNT growth techniques achieve <50 CNTs/µm [6]. Efforts to control IDC have resulted in even lower maximum CNT densities of <15 CNTs/µm [7]. Table 1 summarizes previous efforts to achieve high CNT density, all of which have lacked the ability to <u>simultaneously</u> meet the essential requirements of >100 CNTs/µm, high CNFET I_{ON}, high I_{ON}/I_{OFF}, and controlled IDC.

Here we demonstrate a new technique, **C**ontrolled **IDC**, **D**ensity **E**nhancement by **R**epeated Transfers (CIDER), which uses a new low-temperature CNT transfer technique to achieve, for the first time, all of these essential requirements. CIDER achieves these milestones through a new combination of design and processing steps that enables transfer of CNTs from multiple (different) growth substrates onto the same target substrate while maintaining the CNT alignment and IDC of the initial CNT growth (see "Methodology" section for CIDER process and design details). Our key results are:

1) CIDER achieves >100 CNTs/µm, enabled by its capability to allow an arbitrary number (13 demonstrated) of CNT growth sources to be combined.

2) CIDER preserves the "original" IDC (of starting CNT growth) across all CNT transfers while simultaneously achieving > 100 CNTs/µm.

3) CIDER relaxes the need for CNT growth techniques to simultaneously achieve high-density (>100 CNTs/µm) CNTs and controlled low IDC (which has been a *major* obstacle in the field of CNT nanoelectronics). CNT growth techniques with controlled IDC (even at CNT densities of <100 CNTs/µm), when combined with CIDER, can be sufficient to achieve required CNT densities for high-performance and highly energy-efficient CNFET digital VLSI.

4) Highest reported CNFET I_{ON} with >5,000 I_{ON}/I_{OFF} at 400 nm channel length and V_{DS} of 1V, approaching the current drives of similarly-scaled and similarly-biased silicon-based FETs in production.

TABLE 1: SUMMARY OF HIGH CNT DENSITY EFFORTS

Technique	CNT Density (CNTs/µm)	CNFET I_{ON}[1,2]	I_{ON}/I_{OFF}	IDC [3]	Major challenge(s)
Langmuir–Schaefer bi-layer CNT carpet [9]	500	98µA/100nm (V_{DS} = 0.8V, L_C = 120nm)	2 to 600	not enough data reported	2% metallic CNTs, low I_{ON}/I_{OFF}, multi-layered CNTs
CNT placement by trench patterning [7]	15	6.5 µA/µm (V_{DS} = 0.5V, L_C = 100nm)	2 to 50,000	0.3	2% metallic CNTs, CNT density limited by litho. pitch
Repeated CNT growth on same substrate [6]	20-45	10 µA/µm (V_{DS} = 1.0V, L_C = 4 µm)	5	0.4 (after repeated growth)	CNT density saturates at ~45 CNTs/µm
Stacked multiple transfers [10]	55	8.3 µA/µm (V_{DS} = 1.0V, L_C = 500nm)	1,000	>1.0 (greater than initial growth IDC = 0.5)	increases IDC, maximum 4 CNT transfers
Sequential multiple transfers [11]	8	3 µA/µm (V_{DS} = 1.0V, L_C = 1.5µm)	1,000	0.5 (same as initial growth IDC)	maximum 4-5 CNT transfers
CIDER (this work)	100	122 µA/µm (V_{DS} = 1.0V, L_C = 400nm)	5,990	0.5 (same as initial growth IDC)	

[1] All current densities (µA/µm) in this paper are reported per unit layout width (I_{ON} = CNFET current drive / CNFET layout width) with **no** CNT diameter normalization. With diameter normalization, I_{ON} is often expressed: I_{ON} = Current per CNT / CNT diameter.
[2] All p-CNFETs; L_C = channel length.
[3] IDC calculated from figures within the cited papers.

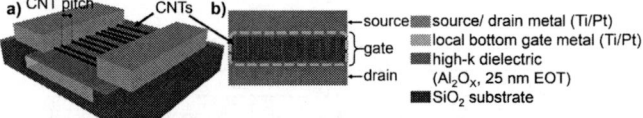

Figure 1. (a) Schematic of a CNFET with a local bottom gate [8]. (b) SEM of a CNFET, showing multiple CNTs bridging the source and drain.

Methodology

CNT Multiple Transfers

CNT transfers, such as those performed in CIDER, offer several key benefits: they decouple the high-temperature CNT growth (~900⁰C) from the final target substrate which undergoes circuit fabrication. As demonstrated in this paper, they can be used to increase the final CNT density above the initial CNT growth density. Moreover, such CNT transfers are silicon-CMOS compatible and enable monolithic 3D integration [12-14].

Ideally, CNT multiple transfers, whereby transfers are performed sequentially onto the same target substrate, should combine CNTs from multiple CNT growth sources to achieve higher CNT density, without degrading IDC (by maintaining the alignment and positioning of the CNTs). However, as Fig. 2a illustrates, simple CNT multiple transfers result in neither increased CNT density nor constant IDC (rather, IDC degrades drastically after a few CNT transfers). Substantial obstacles for CNT multiple transfers can be summarized as:

1) Limited number of CNT transfers: After 2 or 3 CNT transfers, further transfers fail to adhere to the target substrate, due to previously transferred CNTs acting as surface contamination. This places significant burden on the starting CNT growth: it must achieve <u>both</u> CNT density of ~50 CNTs/μm and controlled low IDC values, which is very difficult today. Ideally, an arbitrary number of CNT multiple transfers can be performed, relaxing the requirements for CNT growth to achieve only controlled low IDC (at a reasonable CNT density).

2) Increased IDC: Fig. 2a shows CNTs bundling (with other CNTs) after 2-3 CNT transfers. This bundling has many negative consequences: a) drastically increased IDC, b) reduced gate control due to CNT-CNT electrostatic shielding, c) increased contact resistance (CNTs under other CNTs cannot make direct contact with the source and drain metal), and d) it makes it very difficult to remove an overwhelming majority of metallic CNTs. These render the CNTs unusable for high-performance, highly energy-efficient digital systems.

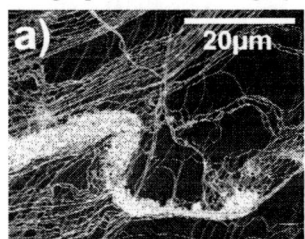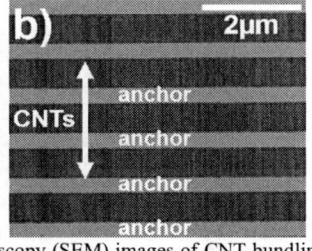

Figure 2. (a) Scanning electron microscopy (SEM) images of CNT bundling after performing 3 sequential transfers without CIDER. (b) SEM after performing 5 transfers with CIDER. SEM shows rows of anchors (CNFET source/drain contacts, see "CIDER Process Flow" for details), with highly-aligned CNTs bridging them.

CIDER Process Flow

CIDER overcomes all of the above obstacles, enabling high-density and highly-aligned CNTs with controlled IDC for high current-drive CNFETs. We first describe and characterize CIDER's ability to overcome CNT multiple transfer obstacles (discussed above), followed by CNFET electrical measurements and benchmarking versus silicon-based FETs in production in major semiconductor foundries.

CIDER processing and design steps are shown in Fig. 3. CNTs are grown wafer-scale on a crystalline ST-cut quartz substrate, yielding >99.5% highly aligned CNTs [15], with an average CNT density of 8 CNTs/μm and IDC value of 0.5.

To perform CIDER CNT transfer, the target substrate is first prepared by depositing an ultra-thin (~7 nm) sacrificial polymer layer (PMGI-based DUV photoresist) [16]. A CNT single transfer is performed onto this polymer surface, using a gold/thermal release tape stack to transfer the CNTs from the growth substrate to the target substrate (Fig. 3b, details in [15]). These process steps (polymer deposition followed by CNT transfer) are repeated for as many transfers as required to reach the target CNT density. Following these CNT multiple transfers, the polymer is lithographically patterned with horizontal lines at the minimum lithographic pitch, perpendicular to the direction of the CNTs (Fig 3a, step IV). This lithographic patterning of the polymer defines the regions where metal "anchors" (ultimately the source and drain CNFET contact metal, in this instance titanium/platinum) are electron-beam evaporated. The remaining polymer is used to lift-off the unwanted metal (Fig. 3a, step V). These patterning steps (Fig. 3a, steps IV-V) are essential for controlling IDC, as first the patterned polymer and then the metal contacts act as anchors for the CNTs, preventing CNT bundling (detailed discussion in "CIDER Characterization" section). To form the final circuit layout, the unneeded sections of these metal anchors are removed (along with the unneeded CNTs), leaving the remaining metal as the source and drain contacts for the CNFETs. There is no area cost associated with these metal anchors as these are the source and drain contacts, and it is compatible with the imperfection-immune paradigm (a set of process and design steps used to overcome the inherent imperfections of CNTs, such as mis-positioned and metallic CNTs) [5].

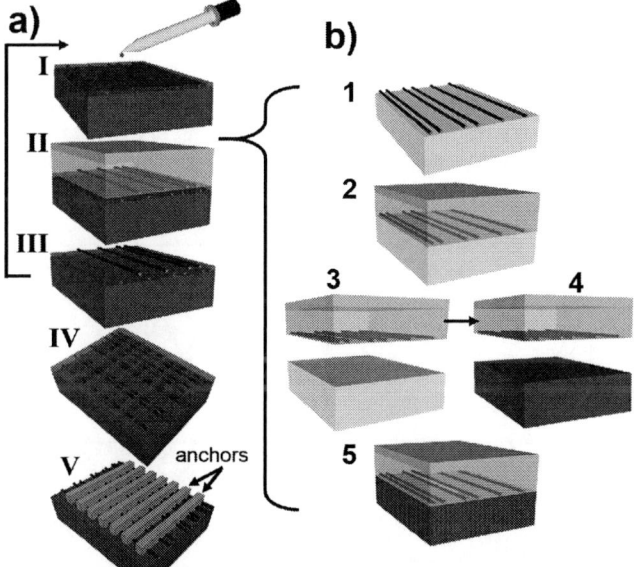

Figure 3. CIDER process flow. (a) CIDER process (steps I-V). Steps I-III are repeated as many times as the number of transfers are performed. Step II, drawn as a single step, is the process shown in (b), encompassing steps 1-5. After all transfers are performed, steps IV-V show two-step CNT anchor process. (b) Single CNT transfer process. Quartz growth substrate is coated in 150 nm gold and thermal release tape (TRT). TRT is removed from the growth wafer, along with the gold with embedded CNTs, and placed on the target substrate.

CIDER Characterization: Overcoming Obstacles

1) CIDER enables a large (arbitrary for practical purposes) number of CNT multiple transfers. CNT transfers adhere to the target substrate due to weak van der Waals interactions, whose strength is determined by the adhesion energy attraction between the two contacting materials and by the distance separating them [17]. The polymer layer applied before each CNT transfer specifically addresses both of these challenges, and is therefore essential to enabling a large number of CNT multiple transfers. First, the polymer acts as an adhesive for the gold transfer film, increasing the adhesion energy between the gold and the target oxide substrate by >6x (Fig. 4) [18].

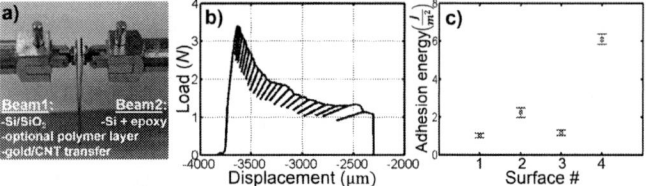

Figure 4. Adhesion energy measurements. (a) Double cantilever beam (DCB) test set-up used to measure adhesion energy [18]. The fracture interface is between the gold and either a blank SiO_2 surface or an optional interfacial layer (such as the polymer layer used in CIDER). (b) Load versus displacement curve generated from the test shown in (a), from which the adhesion energy can be calculated. (c) Adhesion energy for different interfaces. Surface #1: gold-SiO_2, #2: gold-commercial gold adhesion promoter, #3: evaporated gold on the PMGI-polymer, #4: transferred gold film onto the PMGI-polymer. The evaporated gold on the polymer acts to completely remove the solvent from the resist (due to the low (<5e-7 Torr) evaporation pressure), which is equivalent to fully drying an epoxy. This shows the polymer is acting as an epoxy for the transfer to adhere to, as the transferred gold transfer film has ~6X adhesion energy to the wafer with polymer compared to without polymer.

As noted earlier, previously transferred CNTs contribute to increased surface roughness, decreasing the necessary intimate contact between the gold transfer film and the substrate. The polymer coating applied before each transfer planarizes the surface, resulting in a near atomically-smooth transfer surface. Fig. 5 shows that the RMS surface roughness remains <6Å after every transfer for all 13 transfers. As a new polymer surface is deposited before every transfer, the adhesion energy and surface roughness remain constant regardless of transfer count, allowing for a large (arbitrary) number of CNT transfers.

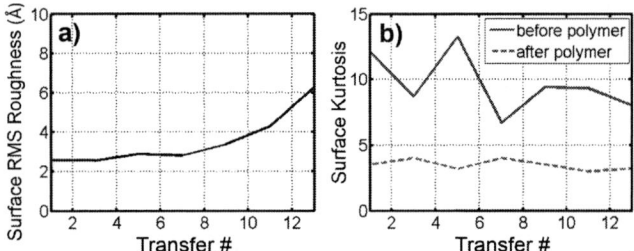

Figure 5. (a) After 13 transfers, surface roughness continues to maintain <6Å RMS, allowing CNT transfers to maintain intimate contact with the substrate surface and thus adhere to the wafer. (b) Surface kurtosis ("peakiness") also improves from the application of the polymer layer.

2) CIDER maintains IDC of the original CNT growth across all CNT multiple transfers.
As shown in Fig. 2a, CNTs become misaligned and "bundle" together after simple repeated CNT transfers. CNTs bundle when immersed in a liquid and later dried, due to capillary forces "zipping" CNTs together [11]. This occurs when the wafer is submerged in gold etchant to remove the gold transfer film after each transfer, leaving CNTs on the wafer substrate. Without CIDER, the CNTs currently being transferred and any previously transferred CNTs on the substrate are free to bundle.

CIDER avoids these CNT-CNT interactions by relying on combined processing and design solutions. First, by covering previously transferred CNTs with the polymer film (described earlier to increase the adhesion energy of the gold transfer film and planarize the surface), all previously transferred CNTs are immobile, and thus have no interactions with any subsequently transferred CNTs. Second, after all CNT transfers have been performed, the sacrificial polymer layers must be removed. To avoid CNT bundling during this polymer removal step, the polymer is removed using a two-step anchor process. As discussed in Fig. 3, the polymer photoresist is lithographically patterned to expose regions that will become CNFET source and drain contacts. During the removal of the polymer from the patterned regions, the CNTs are prevented from bundling due to the remaining polymer sections anchoring the CNTs in place. When the remaining polymer is removed (which in the process also patterns the metal in a lift-off process), the metal source and drain CNFET contacts then act as the CNT anchors. Thus, during the entire process, the CNTs are always anchored in place by two different sets of anchors: first the polymer that remains after lithographic patterning (Fig. 3a, step IV), then the metal source and drain contacts (Fig. 3a, step V, Fig. 6).

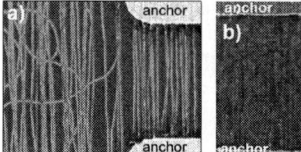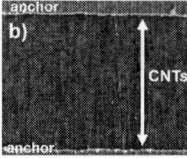

Figure 6. Controlled experiment demonstrating anchors maintaining CNT alignment. (a) CNTs between anchors remain aligned, while CNTs outside anchors are free to bundle. (b) SEM of CNTs after performing 13 CIDER multiple transfers, maintaining overall alignment between the metal anchors (i.e. CNFET source and drain contacts). Channel length = 400 nm.

The design of these two sets of anchors (patterned polymer followed by metal CNFET source and drain contacts) is critical to avoid increased IDC during CNT multiple transfers. Fig. 7 illustrates that the pitch of the anchors (determined by the width and spacing of the lithographically patterned polymer, shown in Fig. 7a) determines their effectiveness, with the smallest anchor pitch resulting in the highest degree of CNT alignment. The width of the initially-patterned polymer anchors translates to the CNFET channel length while the width of the second set of metal anchors translates to the CNFET source and drain contact length. Therefore, the minimum anchor pitch corresponds to a scaled CNFET channel length and contact length, which are ideally scaled. Fig. 7c shows that the IDC for all 13 transfers maintain the starting growth IDC of ~0.5 (using our 400 nm minimum-pitch anchors, which also sets our CNFET channel length to 400 nm). In Fig. 8, atomic force microscopy (AFM) measurements of the CNFET channel show the CNTs are in contact with the substrate and not suspended above.

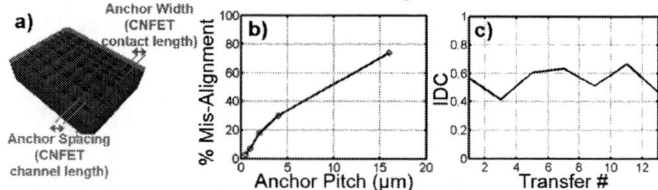

Figure 7. (a) Schematic of anchor process, after polymer has been selectively patterned for CNFET source and drain contacts. (b) Percent CNT misalignment (percentage of CNTs that cross other CNTs within the CNFET channel) versus anchor pitch. (c) Each transfer maintains the initial IDC growth value of ~0.5.

978-1-4799-8002-4/14 $31.00 © 2014 IEEE

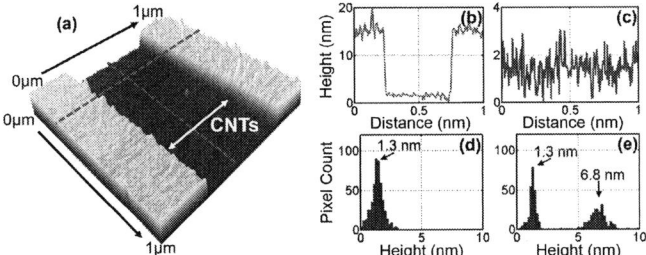

Figure 8. (a) AFM of a CNFET following CIDER process. (b) Height profile of red line in (a). (c) Height profile of blue line in (a). Elevation of CNTs in the channel is <3 nm while the metal height is ~15 nm, so CNTs are not suspended. (d) Histogram of (c), with unimodal distribution with a peak at 1.3 nm, ~diameter of the as-grown CNTs. (e) Histogram of height in a controlled experiment in which CNTs are transferred across a 5 nm step to "emulate" suspended CNTs; shows bimodal distribution with a peak at ~1.3 nm and another at ~6.8 nm, resulting from the two layers of CNTs. (d-e) further confirm CNTs are not suspended.

Electrical Measurements: High Current-Drive CNFETs

Using CIDER, we perform 13 CNT transfers with starting CNT growth density of ~8 CNTs/µm, resulting in CNT density >100 CNTs/µm. Fig. 9 shows 50 p-type CNFET I_D-V_{GS} curves. As fabricated, the CNFETs achieve 347 µA/µm with an I_{ON}/I_{OFF} ratio of ~2, due to the presence of metallic CNTs. This is an 11.1x increase (close to the ideal 13x for the 13 transfers performed) in drive current versus a set of 50 control CNFETs with only a single CNT transfer. We remove metallic CNTs using electrical breakdown [19], whereby the gates of CNFETs are used to turn off semiconducting CNTs, and a large source-drain bias is applied. With a sufficient source-drain bias, the metallic CNTs pass enough current that they breakdown due to a Joule self-heating process. Electrical breakdown has been shown to selectively remove >99.99% of all metallic CNTs with inadvertent removal of <4% semiconducting CNTs [20]. It can be applied wafer-scale, rendering the process compatible for VLSI applications [19,20]. After electrical breakdown, the average CNFET I_{ON} is 84 µA/µm with an I_{ON}/I_{OFF} of 5,660. From the 50 CNFETs, the best I_{ON} is 122 µA/µm, with an I_{ON}/I_{OFF} of 5,990 (Fig. 9).

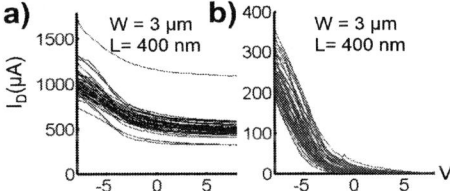

Figure 9. I_D-V_{GS} curves for 50 CNFETs. (a) Before electrical breakdown. (b) After electrical breakdown.

Electrical breakdown is invariant to the number of transfers performed, requiring the same average breakdown voltage and retaining the same proportion of the initial semiconducting drive current (taken as I_{ON}-I_{OFF}) as CNFETs fabricated with a single CNT transfer (47% versus 54%, respectively, Fig. 10).

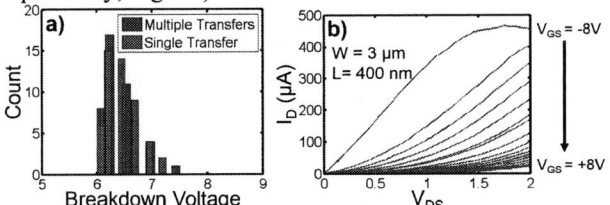

Figure 10. (a) Distributions of required breakdown voltages for CNFETs fabricated after single transfer versus 13 CIDER CNT multiple transfers. (b) Typical I_D-V_{DS} curve for a CNFET after 13 CIDER CNT transfers.

Benchmarking against silicon-FETs

We used foundry PDK models for 350, 130, 65, and 28 nm node Si-FETs, to extract the drive current for a Si-FET with similar scaling (400 nm channel length) and biasing (V_{DS}=1V, V_t chosen for the same I_{OFF} as our CNFETs (~10 nA/µm), and I_{ON} taken at the same maximum gate field). Our average CNFET I_{ON} achieves 65% of a Si-FET I_{ON}; the best CNFET I_{ON} achieves 82-94% of the Si-FET I_{ON} (Fig. 11).

Device	Channel Length (nm)	I_D/µm (µA) @ 1 V_{DS}	I_D/µm (µA) @ 1 V_{DS}	I_D/µm (µA) @ 0.85 V_{DS}	I_D/µm (µA) @ 0.5 V_{DS}
		pre-electrical breakdown	post-electrical breakdown	post-electrical breakdown	post-electrical breakdown
CNFET (average)	400	347	84	68	38
CNFET (best)	400	587	122	100	59

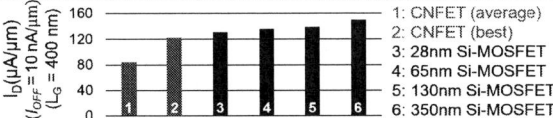

Figure 11. (top) CNFET drive currents before and after electrical breakdown, under different biasing conditions for the same CNFETs. Both the average and best CNFET data are given. (bottom): Comparing drive currents between both the average and best CNFETs and commercial Si-FETs in use today. The technology node is the node of the foundry PDK model used to extract the current. The Si-FET current is extracted under similar scaling and biasing conditions as our CNFETs (details in text) from the foundry PDK models.

Conclusion

We demonstrate carbon nanotube field-effect transistors (CNFETs) with the highest current drive per unit layout width to-date with high I_{ON}/I_{OFF} (>100 µA/µm at 400 nm channel length with >5000 I_{ON}/I_{OFF} and 1V V_{DS}). This is the first demonstration of CNFETs approaching the performance necessary to realize CNFET logic circuits for high-performance and highly energy-efficient digital systems.

Acknowledgements

This research was supported in part by STARnet SONIC, NSF, & Hertz/SGF for MMS. We thank Prof. Eric Pop, Prof. Zhenan Bao, Dr. J. Provine of Stanford for discussions.

References

[1] L. Chang, et al., "IEDM short course," *IEDM*, 2012.
[2] L. Wei, et al., "A non-iterative compact model for carbon nanotube FETs incorporating source exhaustion effects," *IEDM*, pp. 917-920, 2009.
[3] A. Franklin, et al., "Sub-10 nm carbon nanotube transistor," *Nano Letters*, vol. 12(2), pp. 758-762, 2012.
[4] J. Deng, et al., "Carbon nanotube transistor circuits: circuit-level performance benchmarking and design options for living with imperfections," *ISSCC*, pp.587-8, 2007
[5] J. Zhang, et al., "Robust digital VLSI using carbon nanotubes," *TCAD*, vol. 31, pp. 453-471, 2012.
[6] S. Hong, et al., "Improved density in aligned arrays of single-walled carbon nanotubes by sequential chemical vapor deposition on quartz," *Advanced Materials*, vol. 22(16), pp. 1862-1830, 2010.
[7] H. Park, et al., "High-density integration of carbon nanotubes via chemical self-assembly," *Nature Nanotech.*, vol. 7, pp. 787-791, 2012.
[8] A. Franklin, et al., "Current scaling in aligned carbon nanotube array transistors with local bottom gating," *Electron Device Lett.*, vol. 31(7), pp. 644-646, 2010.
[9] Q. Cao, et al., "Arrays of single-walled carbon nanotubes with full surface coverage for high-performance electronics," *Nature Nanotech.*, vol. 8, pp. 180-186, 2013.
[10] C. Wang, et al., "Synthesis and device applications of high-density aligned carbon nanotubes using low-pressure chemical vapor deposition and stacked multiple transfer," *Nano Research*, vol. 3, pp. 831-842, 2010.
[11] M. Shulaker, et al., "Linear increases in carbon nanotube density through multiple transfer technique," *Nano Letters*, vol. 11, pp. 1881-1886, 2011.
[12] H. Wei, et al., "Monolithic three-dimensional integration of carbon nanotube FET complementary logic circuits," *IEDM*, pp. 511-514, 2013.
[13] H. Wei, et al., "Monolithic three-dimensional integrated circuits using carbon nanotube FETs and interconnects," *IEDM*, pp.577-580, 2009.
[14] M. Shulaker, et al., "Monolithic three-dimensional integration of carbon nanotube FETs with silicon CMOS," *VLSI Symp.*, pp. 172-173, 2014.
[15] N. Patil, et al., "Wafer-scale growth and transfer of aligned single-walled carbon nanotubes," *Nanotech.*, vol. 8, pp.498-504, 2009.
[16] A. McCullough, et al., "Polydimethylglutarimide (PMGI) Resist-A Progress Report," *Microlitho. Conf.*, pp. 316-320, 1986.
[17] F. DelRio, et al., "The role of van der Waals forces in adhesion of micromachined surfaces," *Nature Materials*, vol. 4, pp. 629-634, 2005.
[18] R. Dauskardt, "Adhesion and debonding of multi-layer thin film structures," *Engineering Fracture Mechanics*, vol. 61, pp. 141-162, 1998.
[19] N. Patil, et al., "VMR: VLSI-compatible metallic carbon nanotube removal for imperfection-immune cascaded multi-stage digital logic circuits using carbon nanotube FETs," *IEDM*, pp. 573-576, 2009.
[20] M. Shulaker, et al., "Sensor-to-digital interface built entirely with carbon nanotube FETs," *JSSC*, vol. 49, pp. 190-201, 2014.

New Insights into the Design for End-of-life Variability of NBTI in Scaled High-κ/Metal-gate Technology for the nano-Reliability Era

Pengpeng Ren[1], Runsheng Wang[1,*], Zhigang Ji[2,#], Peng Hao[1], Xiaobo Jiang[1], Shaofeng Guo[1], Mulong Luo[1],
Meng Duan[2], Jian F. Zhang[2,#], Jianping Wang[3], Jinhua Liu[3], Weihai Bu[3], Jingang Wu[3],
Waisum Wong[3,5], Shaofeng Yu[3,5], Hanming Wu[3,5], Shiuh-Wuu Lee[3,5], Nuo Xu[4], Ru Huang[1,5,*]

[1]Institute of Microelectronics, Peking University, Beijing 100871, China (*Emails: ruhuang@pku.edu.cn, r.wang@pku.edu.cn)
[2]School of Engineering, Liverpool John Moores University, Liverpool L3 3AF, UK (#Emails: z.ji@ljmu.ac.uk, j.f.zhang@ljmu.ac.uk)
[3]Semiconductor Manufacturing International Corporation (SMIC), Shanghai 201203 and Beijing 100176, China
[4]Department of EECS, University of California, Berkeley, CA 94720, USA
[5]Innovation Center for MicroNanoelectronics and Integrated System, Beijing 100871, China

Abstract

In this paper, a new methodology for the assessment of end-of-life variability of NBTI is proposed for the first time. By introducing the concept of characteristic failure probability, the uncertainty in the predicted 10-year VDD is addressed. Based on this, variability resulted from NBTI degradation at end of life under specific VDD is extensively studied with a novel characterization technique. With the further circuit level analysis based on this new methodology, the timing margin can be relaxed. The new methodology has also been extended to FinFET in this work. The wide applicability of this methodology is helpful to future reliability/variability-aware circuit design in nano-CMOS technology.

Introduction

As CMOS devices downscaling into nanoscale region, dynamic variability induced by NBTI effects has been paid growing attention [1-10]. With device aging, the induced variations will directly degrade the circuit stability [2-5], which is especially significant at end of life in more aggressive technology nodes (Fig. 1). Therefore, assessment of the end-of-life variability emerges as a big necessity for practical circuit design in the nano-reliability era, including both the impacts of device-to-device variation (DDV) [1, 3, 7-9] and the cycle-to-cycle variation (CCV) recently found in our work [9-10]. However, it faces great challenges: The conventional assessment methodology is not suitable for characterizing NBTI reliability and the resulted variability in nano-devices, due to the stochastic nature of trapping/detrapping within the gate oxide [6-10]. New methodology is thus intensively required.

In this paper, a new methodology is proposed to address the challenges in characterizing the end-of-life variability for nanoscale devices, with demonstrations on high-κ/metal-gate (HKMG) and FinFET technology for the first time. By introducing the novel characterization technique, variability induced by the NBTI degradation at end of life is experimentally studied. The impacts on circuit and yield analysis are also investigated.

New Assessment Methodology for End-of-life Variability

To assess the end-of-life variability, the 10-year VDD should be predicted at first. However, nano-devices brought big challenges into 10-year VDD prediction with conventional constant voltage stress (CVS) procedure [11] in two aspects: (1) Requirements of multiple identical devices in CVS method cannot be met by the severe DDV effect. (2) Time-dependent

variability during degradation within a single device (CCV effect) makes the conventional power factor extraction unreliable (Fig. 2). In addition, the method using large devices to predict 10-year VDD also fails, since the degradation of large devices is not consistent with the average degradation of nano-devices (Fig. 3). A new assessment methodology with novel characterization technique is thus required. The underlying physics and models should be also explored, with extension to circuit level analysis (Fig.4).

As shown in Fig. 5, the NBTI degradation in nano-devices manifests large fluctuations. To evaluate the extent of device degradation at end of life for specific VDD, the concept of failure probability is introduced in the new methodology, which includes not only the impacts of DDV, but also the transient failure caused by CCV effect. Due to the large variations of the degradation, the conventional definition of 10-year VDD needs to be improved. Rather than a single value, the 10-year VDD becomes probabilistic in nano-devices, and each one is with a characteristic value of failure probability. The 10-year VDD will be determined by the target failure probability in practice. In other words, the end-of-life variability becomes a 2-D problem at specific VDD (Fig. 6): one dimension is the DDV of the mean degradation; the other is the DDV of the fluctuations in the degradation within one device (CCV effect). In order to transform the degradation under accelerated stress to normal VDD for end-of-life characterization, novel characterization technique is required, which will be shown next. Devices measured in this work are scaled HKMG planar pFETs and SOI FinFETs.

Novel Fast Voltage Step Stress (FVSS) Technique

A new characterization technique named FVSS is proposed to address the challenges in 10-year VDD prediction for nanoscale devices. By introducing the concept of stepped stressing [12], the 10-year VDD can be predicted on a single device, rather than multiple devices with the impact of DDV. V_{th} sensing method is based on the modified ultra-fast technique [13] to fully capture the CCV effect during degradation. As shown in Fig. 7, the device is continuously stressed by stepped stress V_{Gstr} for the same time Δt until reaching pre-specified step N. The stress bias is interrupted quickly and periodically to monitor the V_{th} shift (ΔV_{th}) within 5μs under the same V_{Gmea} ($<V_{Gstr}$). With ultra-fast measurement, ΔV_{th} can be obtained with higher sampling rate, which enables the extraction of n in the first step of FVSS technique. Due to the gradual increase of V_{Gstr}, the degradation is

978-1-4799-8002-4/14 $31.00 © 2014 IEEE

accelerated (Fig. 8). With Eq. 2, the stress time under high V_{Gstr} can be equivalently transformed to effective stress time under the first V_{Gstr} ($=V_1$) with the optimal parameters m and A to restore the power law statistics under V_1. Thus the 10-year VDD can be predicted on a single device with the extracted parameters. In addition, the effective stress time can be transformed under any V_G bias (e.g., VDD) with Eq. 3. The FVSS method is firstly verified with large devices. Good agreement is achieved between the FVSS and conventional CVS methods (Fig. 9). Thus it can be applied to nano-devices (Fig. 10&11). As expected, the ΔV_{th} manifests large fluctuations against stress time due to CCV effect.

Experimental Results and Discussions

a) Statistics of failure probability: As shown in Fig. 12, the failure probability at specific VDD after transformation can be extracted as the probability that ΔV_{th} larger than the failure criterion (50mV as an example here) around 10-year lifetime. The extracted failure probability (Fig. 13) presents a large dispersion with varying VDD for nano-device compared with large device, due to CCV effect. With DDV effect further taken into account (Fig. 14), the failure probabilities of different devices have a wider distribution. It can be observed in Fig. 14(c) that DDV is much larger than CCV, mostly due to the fact that the measured devices (30×300 nm^2) are not sufficiently small. Thus, for a given VDD, the effective occupation probability will be 1 or 0 for most of the devices. As a direct result, the distribution of the failure probability among different devices should be U-shape like, which is consistent with the experimental results in Fig. 15. This interesting behavior will be discussed later in more details with correlation to trap energy distribution.

b) HK process, FinFET and AC NBTI: The mean failure probability is extracted and compared between two HK processes (Fig. 15). HK process #1 presents less degradation and variation due to process optimization. The VDD corresponding to 100% and 0% failure are further extracted and compared in Fig. 16. With the scaling of the gate area, the dispersion of failure probability becomes larger, indicating the more severe impacts of CCV and DDV. Fig. 17 shows the results of FVSS technique applied to FinFETs (Fig. 17), confirming its applicability beyond planar technology. The methodology can also be extended to AC NBTI characterization (Fig. 18). The impact of CCV is also non-negligible under practical AC circuit operation conditions.

c) Degradation and variation at end of life: Once the degradation is transformed under specific VDD, the average and deviation values of $\mu(\Delta V_{th})$ and $\sigma(\Delta V_{th})$ among different devices at end of life can be extracted around 10-year lifetime in terms of $\mu(\mu)$, $\sigma(\mu)$, $\mu(\sigma)$ and $\sigma(\sigma)$ respectively, as shown in Fig. 19. Large variations can be observed at end of life, which will directly degrade the parametric yield. With the increase of VDD, more traps are generated, contributing to the increase of dynamic variations.

d) Distributions of 10-year VDD: From another perspective of statistics, the dispersion of failure probability can be evaluated alternatively by the distribution of 10-year VDD among different devices with the same mean failure probability. The distribution of 10-year VDD well fits Weibull distribution [Fig. 20(a)]. The shape factor of the Weibull distribution keeps the same with the varying failure probability, and decreases with the shrinking gate area [Fig. 20(b)].

Physical Model of Failure Probability

The U-shape like distribution of failure probability is fundamentally correlated with the energy distribution of oxide traps. Fig. 21 shows the trap charge density N_{ot} and energy density D_{ot} extracted in large device with the method in [14]. Since the occupation probability is directly determined by the differences between Fermi level and trap energy level, the distribution of the occupation probability is correlated with the energy distribution of oxide traps (Fig. 22). With the proposed physical model (Eq. 4), the theoretical distribution of failure probability is well consistent with the experiment results. The impact of different energy distribution of oxide traps on failure probability is simulated (Fig. 23). Therefore, the new methodology could also be applied to new materials (e.g. Ge, III-V) which have different trap energy distributions from Si.

Impacts on Circuits and Yield Analysis

Based on the above results, the new methodology is extended to circuit and yield analysis. Considering the severe impacts of DDV and CCV at end of life, the frequency shift of ring oscillator (RO) presents wide distribution with the varying VDD (Fig. 24). The individual RO circuit fails stochastically among the operation cycles due to CCV effect (Fig. 25). In other words, it does not fail in some operation cycles within the 'dying' part (Fig. 26), contradicted to the view that the 'dying' part is regarded as totally failed in conventional assessment methodology. Therefore, the additional timing margin can be relaxed with the new methodology. On the other hand, the impact on end-of-life parametric yield is also evaluated. With specific failure criterions, the 10-year VDD of RO is determined by the target yield (Fig. 27).

Summary

We have proposed a new methodology for assessing end-of-life variability of NBTI in this paper. The uncertainty in the predicted 10-year VDD is addressed by introducing the concept of characteristic failure probability. At specific VDD, the induced variations from NBTI degradation at end of life are extensively studied for the first time with the proposed novel technique. With the further analysis on circuit level, timing margin can be relaxed with the new methodology. It is thus helpful to the variability-aware circuit design in the nano-reliability era.

Acknowledgement

This work was partly supported by the 973 Projects (2011CBA00601), NSFC (61106085), National S&T Major Project (2009ZX02035-001).

References

[1] B. Kaczer, *et al.*, *IRPS*, p.26, 2010; [2] T. Naphade, *et al.*, *IEDM*, p.838, 2013; [3] Y. Mitani, *et al.*, *IRPS*, p.857, 2011; [4] R. Wang, *et al.*, *IEDM*, p.978, 2013; [5] M. Nafria, *et al.*, *IEDM*, p.127, 2011; [6] T. Grasser, *et al.*, *IEDM*, p.85, 2010; [7] M. Toledano-Luque, *et al.*, *VLSI*, p.152, 2011; [8] M. Duan, *et al.*, *IEDM*, p.774, 2013; [9] C. Liu, *et al.*, *IEDM*, p.571, 2011; [10] C. Liu, *et al.*, *IEDM*, p.466, 2012; [11] JEDEC-JEP122E; [12] Z. Ji, *et al.*, *IRPS*, GD.2.1, 2014; [13] A. E. Islam, *et al.*, *IEDM*, p.805, 2007; [14] Z. Ji, *et al.*, *TED*, p. 228, 2010.

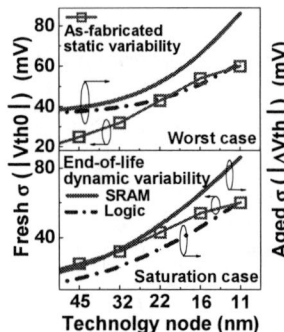

Fig. 1 Comparison between the predicted as-fabricated static variability and the total dynamic variability at end of life by the proposed compact model (Eq. 1).

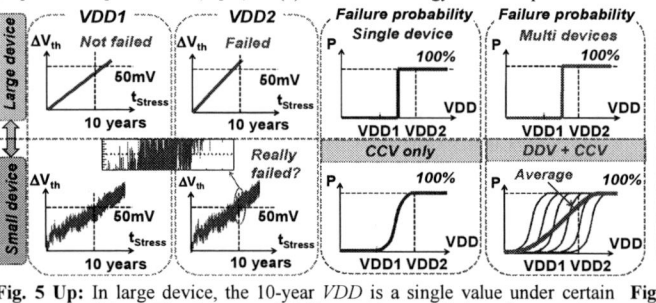

Fig. 2 (a) BTI degradation under CVS test in large devices. (b) The 10-year VDD can be extracted with extrapolation. (c) When CVS method is applied to small devices, fluctuations in the degradation make the conventional power law fitting unreliable. (d) New methodology is thus required in nanoscale devices.

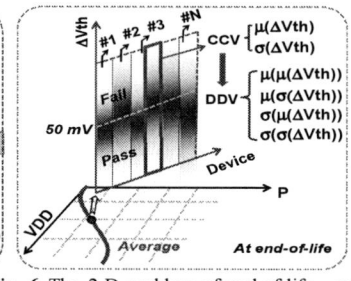

Fig. 3 Comparison of NBTI degradation between large and small devices under the same stress conditions. The average degradation of 1000 cycles of 10 small devices is considered for fair comparison.

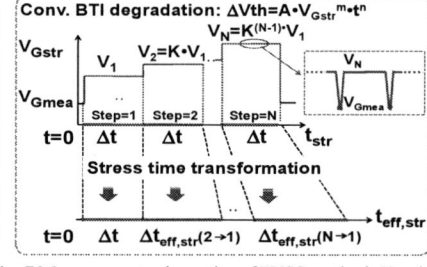

Fig. 4 The characterization scheme for end-of-life variability of NBTI proposed in this work.

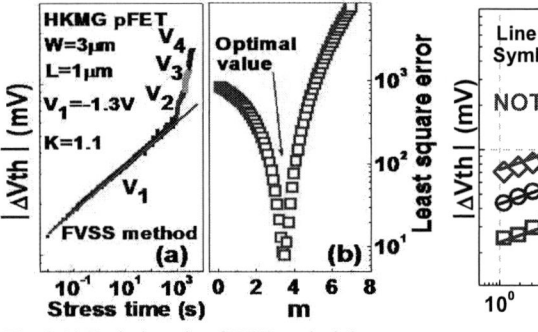

Fig. 5 Up: In large device, the 10-year VDD is a single value under certain process conditions. The failure probability at 10-year lifetime shows a sharp trend from 0% to 100% with the varying VDD. **Below:** In nanoscale devices, the V_{th} shift (ΔV_{th}) manifests large fluctuations with aging time. The failure probability shows dispersion with the varying VDD due to CCV effect. When DDV is also taken into account, wider dispersion of failure probability can be observed.

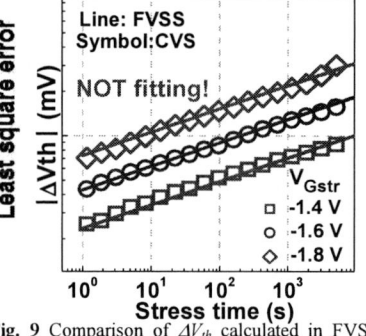

Fig. 6 The 2-D problem of end-of-life variability of NBTI at specific VDD. The degradation and the resulted variation can be evaluated as $\mu(\mu(\Delta V_{th}))$, $\sigma(\mu(\Delta V_{th}))$, $\mu(\sigma(\Delta V_{th}))$ and $\sigma(\sigma(\Delta V_{th}))$ respectively, considering the impacts of CCV and DDV.

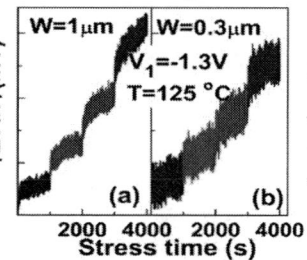

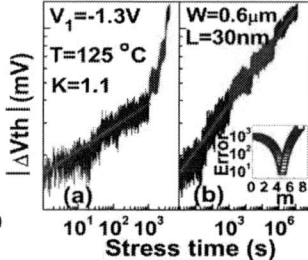

Fig. 7 Measurement schematics of FVSS method. V_{Gstr} is stepped increased with a constant multiple K and interrupted quickly and periodically to monitor the ΔV_{th}. The stress time under high V_{Gstr} can be equivalently transformed to effective stress time under V_1 with optimal m and A to restore the power law statistics under V_1.

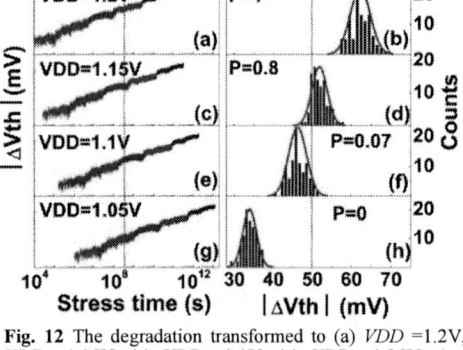

Fig. 8 (a) Typical results of FVSS method (large device). (b) Least square error between the ΔV_{th} before and after the stress time transformation. At the minimum point, the optimal value of m can be obtained.

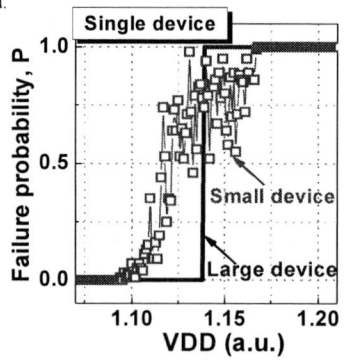

Fig. 9 Comparison of ΔV_{th} calculated in FVSS method with CVS measurement data. With the values of n, m and A extracted from Fig. 8, ΔV_{th}, and 10-year VDD can be calculated in FVSS method.

Fig. 10 Time evolutions of ΔV_{th} with FVSS method, (a) HKMG device with W=1μm, L=30nm, (b) HKMG device with W=0.3μm, L=30nm.

Fig. 11 Time evolutions of ΔV_{th} with FVSS method (a) before and (b) after effective stress time transformation. The insert is to extract the optimal value of m.

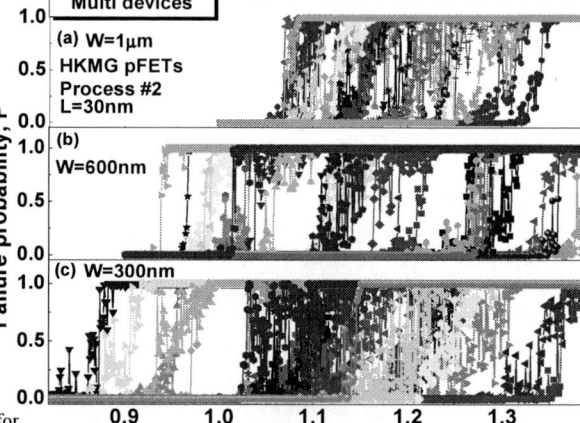

Fig. 12 The degradation transformed to (a) VDD =1.2V, (c) VDD =1.15V, (e) VDD =1.1V, (g) VDD =1.05V. Around 10-year lifetime, the ΔV_{th} follows the normal distribution. The corresponding failure probability is extracted as the probability that ΔV_{th} larger than the failure criterion (50mV as an example here). The results are shown in (b), (d), (f) and (h) respectively.

Fig. 13 The extracted failure probability for a single device with varying VDD, compared with that of a large device. The 10-year VDD is no more a single value, but becomes probabilistic and each with one characteristic value of failure probability.

Fig. 14 For multi devices with (a) W=1μm, (b) W=600nm, and (c) W=300nm, the failure probabilities show a wide dispersion with varying VDD, especially in smaller devices. Each curve corresponds to one device. L=30nm for all the devices under test.

978-1-4799-8002-4/14 $31.00 © 2014 IEEE

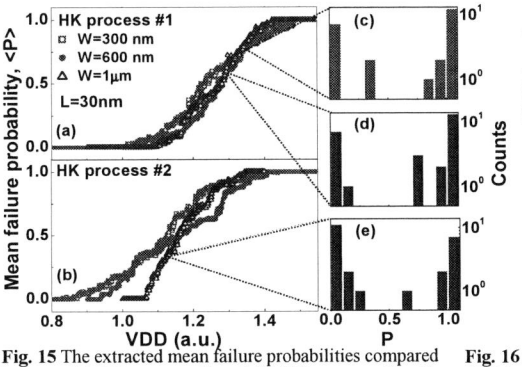
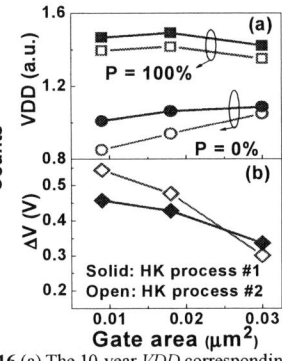
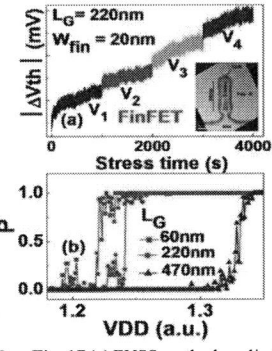
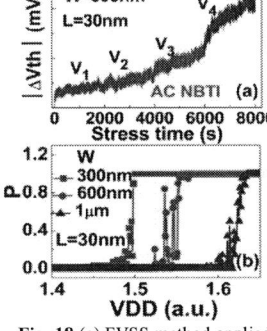

Fig. 15 The extracted mean failure probabilities compared between (a) HK process #1 and (b) HK process #2 with different gate areas. The distributions of the failure probabilities of different devices turn to be U-shape like and are shown in (c), (d) and (e) respectively.

Fig. 16 (a) The 10-year VDD corresponding to 100% failure, 0% failure and (b) their difference $\Delta V = VDD(P=100\%) - VDD(P=0\%)$ with different gate areas. Large dispersion ($\Delta V \approx 0.5V$) can be observed.

Fig. 17 (a) FVSS method applied to FinFETs. (b) The extracted failure probabilities compared among devices with different gate lengths.

Fig. 18 (a) FVSS method applied to AC NBTI. (b) The extracted failure probabilities compared among devices with different gate widths.

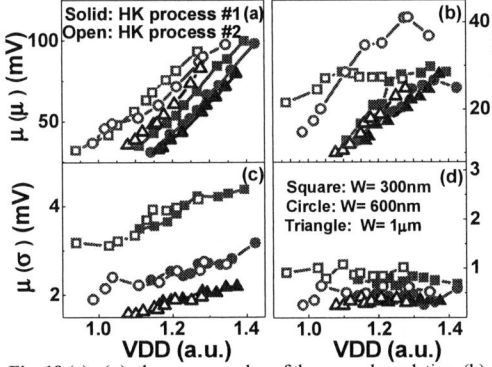
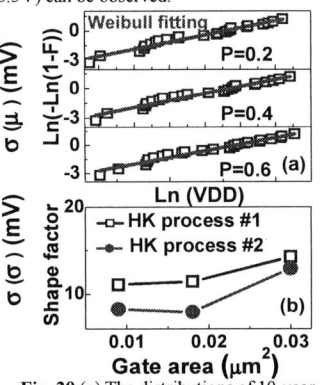
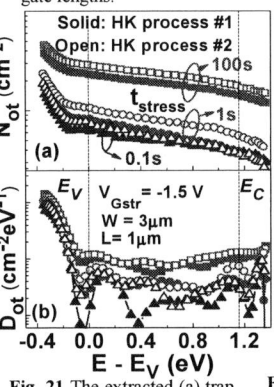
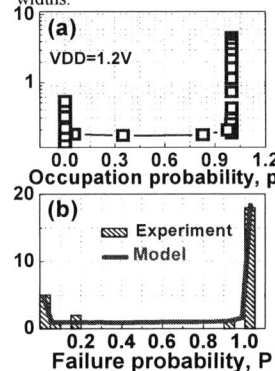

Fig. 19 (a) $\mu(\mu)$: the average value of the mean degradation, (b) $\sigma(\mu)$: the deviation value of the mean degradation, (c) $\mu(\sigma)$: the average value of the variation, and (d) $\sigma(\sigma)$: the deviation values of the variation in individual devices at end of life for two HK processes.

Fig. 20 (a) The distributions of 10-year VDD of different devices with the same failure probability, fitted by Weibull distribution with the shape factor in (b).

Fig. 21 The extracted (a) trap charge density N_{ot} and (b) energy density D_{ot} with $E - E_V$ for two HK processes.

Fig. 22 (a) Distributions of the occupation probability of oxide traps. (b) Comparison of the distribution of failure probability between modeling and experiment.

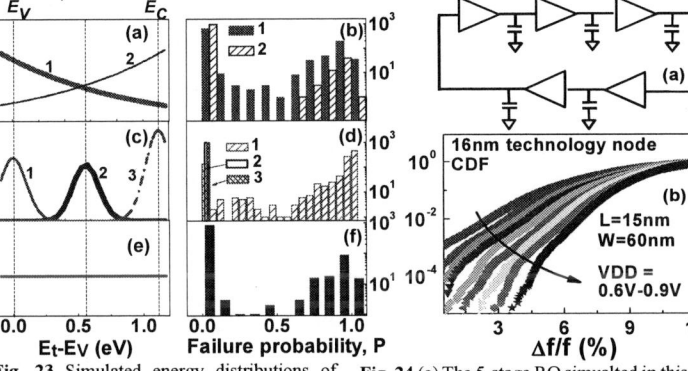
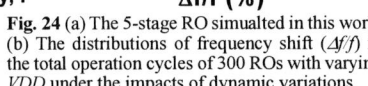
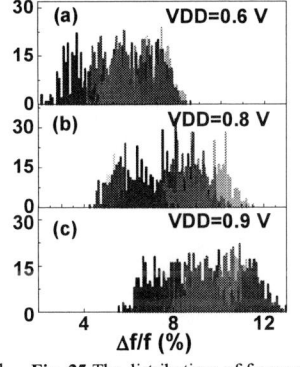
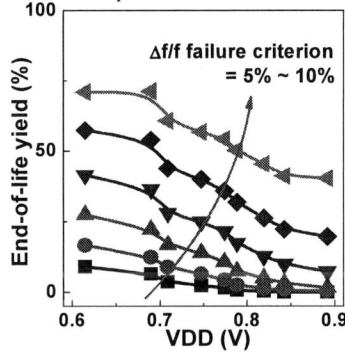

Fig. 23 Simulated energy distributions of oxide traps in (a), (c) and (e), corresponding to the modeled distributions of failure probability in (b), (d) and (f), respectively.

Fig. 24 (a) The 5-stage RO simulated in this work. (b) The distributions of frequency shift ($\Delta f/f$) in the total operation cycles of 300 ROs with varying VDD under the impacts of dynamic variations.

Fig. 25 The distributions of frequency shift ($\Delta f/f$) among 500 operation cycles for individual ROs under different VDD. 12 ROs are shown examples for each VDD.

Fig. 27 The end-of-life parametric yield of ROs under varying VDD with different failure criterions.

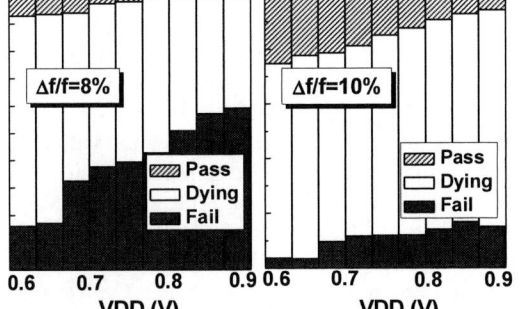
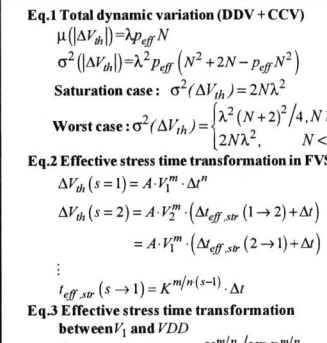
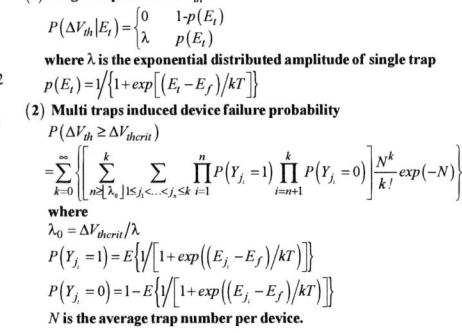

Fig. 26 With the failure criterion (left) $\Delta f/f=6\%$, and (right) $\Delta f/f=8\%$, the proportion compared between *Pass* (all 'healthy'), *Fail* (all 'dead') and *Dying* (fail only in some operation cycles while in others not).

Eq.1 Total dynamic variation (DDV + CCV)

$$\mu(|\Delta V_{th}|) = \lambda p_{eff} N$$

$$\sigma^2(|\Delta V_{th}|) = \lambda^2 p_{eff}\left(N^2 + 2N - p_{eff} N^2\right)$$

Saturation case: $\sigma^2(\Delta V_{th}) = 2N\lambda^2$

Worst case: $\sigma^2(\Delta V_{th}) = \begin{cases} \lambda^2(N+2)^2/4, N \geq 2 \\ 2N\lambda^2, \quad N < 2 \end{cases}$

Eq.2 Effective stress time transformation in FVSS

$$\Delta V_{th}(s=1) = A \cdot V_1^m \cdot \Delta t^n$$

$$\Delta V_{th}(s=2) = A \cdot V_2^m \cdot \left(\Delta t_{eff,str}(1 \to 2) + \Delta t\right)$$

$$= A \cdot V_1^m \cdot \left(\Delta t_{eff,str}(2 \to 1) + \Delta t\right)$$

$$\vdots$$

$$t_{eff,str}(s \to 1) = K^{m/n}(s-1) \cdot \Delta t$$

Eq.3 Effective stress time transformation between V_1 and VDD

$$t_{eff,str,VDD} = t_{eff,str,V_1} \cdot V_1^{m/n}/VDD^{m/n}$$

Eq.4 Physical model of failure probability

(1) Single trap induced ΔV_{th}

$$P(\Delta V_{th}|E_t) = \begin{cases} 0 & 1-p(E_t) \\ \lambda & p(E_t) \end{cases}$$

where λ is the exponential distributed amplitude of single trap

$$p(E_t) = 1/\left\{1 + exp\left[(E_t - E_f)/kT\right]\right\}$$

(2) Multi traps induced device failure probability

$$P(\Delta V_{th} \geq \Delta V_{thcrit})$$

$$= \sum_{k=0}^{\infty}\left[\left\{\sum_{n \geq \lambda_o}\sum_{1 \leq j_1 < ... < j_k \leq k}\prod_{i=1}^{n}P(Y_{j_i}=1)\prod_{i=n+1}^{k}P(Y_{j_i}=0)\right\}\frac{N^k}{k!}exp(-N)\right\}$$

where

$$\lambda_0 = \Delta V_{thcrit}/\lambda$$

$$P(Y_{j_i}=1) = E\left\{1/\left[1 + exp\left((E_j - E_f)/kT\right)\right]\right\}$$

$$P(Y_{j_i}=0) = 1 - E\left\{1/\left[1 + exp\left((E_j - E_f)/kT\right)\right]\right\}$$

N is the average trap number per device.

NBTI of Ge pMOSFETs: understanding defects and enabling lifetime prediction

J. Ma[1], W. Zhang[1], J. F. Zhang[1], B. Benbakhti[1], Z. Ji[1], J. Mitard[2], J. Franco[2],
B. Kaczer[2], and G. Groeseneken[2]

[1]School of Engineering, Liverpool John Moores University, Liverpool L3 3AF, UK
[2]IMEC, Leuven B3001, Belgium

Abstract

Conventional lifetime prediction method developed for Si is inapplicable to Ge devices. This work demonstrates that the defects are different in Ge and Si devices. Based on the investigation of defect difference, for the first time, a method is developed for Ge devices to restore the power law for NBTI kinetics, enabling lifetime prediction. This method is applicable for both GeO_2/Ge and Si-cap/Ge devices, assisting in further Ge process/device optimization.

Introduction

As the downscaling of CMOS technology reaches its end, Ge pMOSFETs, have attracted many attentions and been considered as a strong candidate for next technology nodes, because of its high hole mobility. The record hole mobility has been reported for two advanced fabrication approaches, Al_2O_3/GeO_2/Ge and HfO_2/SiO_2/Si-cap/Ge structures [1-3]. Reliability, however, is still problematic and currently impedes the progress [4-5]. Large NBTI degradation exists in GeO_2/Ge, but little information is available on the defects. Ge device, with Si-cap on top of the Ge channel, has been demonstrated with superior reliability, but its lifetime, τ, cannot be predicted by conventional power law extrapolation [3, 4]. *This work demonstrates that the defects behave differently in Ge and Si devices. For the first time, a new method is developed for Ge devices, based on the understanding of defects, to restore the power law for NBTI kinetics, which enables lifetime, τ, prediction and process/device optimization* (**Figs. 1a-d**).

Current issues

It was reported that NBTI degradation in Si-cap Ge devices by DC measurement cannot be described by the conventional power law, $\Delta V_{th} = C \cdot V_{ov}^{\gamma} \cdot t^n$ [3]. The GeO_2/Ge has higher NBTI [5, 6] and the degradation does not follow the power law, either. The NBTI measured by fast pulse technique is examined here, and the power law is also inapplicable for both Si-cap/Ge (**Fig. 1a**) and Al_2O_3/GeO_2/Ge (**Fig. 1b**), preventing reliable lifetime (τ) prediction (**Fig. 1c**). For Si devices, our latest results show that the power law can be restored for the generated defects (GD), after removing the as-grown hole traps (AHT) [7].

Although this Si method is able to predict NBTI lifetime for Si device with various gate stacks, it does not work for Ge devices, since it fails to give a Vg-independent power exponent (**Fig. 2**) that is essential for reliable lifetime prediction. There is a pressing need to develop a method for Ge devices to restore the power law, enable NBTI lifetime prediction, and assist in further development of Ge devices and process. *The key advance of this work is to meet this need (**Fig. 1d**), based on an in-depth understanding of defects in Ge and their differences from Si devices.*

Understanding of defects

A. Defect Differences

The details of Si and Ge devices used in this work are summarized in Table 1. Their defects behave differently in four aspects:

1) **Recovery**: NBTI in GeO_2/Ge devices is fully recoverable, but not in Si device (**Fig. 3a and 4a**);

2) **2nd stress:** After the recovery, the 2nd stress in Ge follows the same kinetics as the 1st one (**Fig. 3b**), indicating all defects returned to their fresh states after recovery. For Si device, however, the 2nd stress deviates from the 1st one after AHTs are filled (**Fig. 4b**);

3) **Recharge:** Following discharge through which the energy profiles (**Fig. 2a**) are obtained [5, 7, 8], the traps in Ge cannot be recharged until charging energy level (EL) is swept back near Ge Ev (**Fig. 3c**). For Si, recharge starts once energy level is swept lower than Ec (**Fig. 4c**);

4) **Temperature:** For Ge devices, the recharge does not appear in the upper half of its band gap and is independent of temperature (**Fig. 3d**). For Si devices, recharge happens near Si Ec, and it clearly rises when the temperature is lowered from 125 °C to room temperature (**Fig. 4d**).

B. Energy Alternating Defects (EAD)

The above differences are caused by the presence of EAD in Ge, which is absent in Si devices. As illustrated in **Figs. 5a and b,** the energy level of EAD alternates with its charge status: it shifts above Ev when charged, and back below Ev when neutralized. In contrast, the energy levels of generated defects (GD) in Si devices do not alternate and are above Ev even after neutralization (**Fig. 5b**). Since EADs in Ge return to their fresh states once neutralized, the 2nd stress in Ge has the same kinetics as 1st one (**Fig.3b**). The recharging of EAD can only take place when the bias is below ~Ev, the same as in a fresh device (**Fig. 3c**), so that it cannot happen when biased at ~Ec at either room temperature (RT) or 125 °C (**Fig. 3d**). On the other hand, the generated defects in Si devices keep their high energy level after neutralization and do not return to their fresh states, resulting in different kinetics during the 1st and 2nd stresses (**Fig. 4b**). The neutralized GDs at high energy level recharge once they are above Ef (**Fig. 4c**). They also recharge when the temperature switch from 125 °C to RT, as there are less electrons at RT that can reach and neutralize them (**Fig. 4d and 5b**). The energy alternation with charge status change is further supported by first-principle calculations (**Fig. 6**) [9-11], suggesting that EADs are intrinsic defects in Al_2O_3/GeO_2/Ge. The absence of 'permanent' component in Ge (**Fig. 3a**) is because the charged EADs are sufficiently close to Ge Ec and they can be fully neutralized, as the Ec offset at GeO_2/Ge interface is smaller than that at SiON/Si [12].

978-1-4799-8002-4/14 $31.00 © 2014 IEEE

C. As-grown hole traps (AHT)

All the AHTs in Si sample are below Ev and measured by sweeping energy level from high to low [7, 8]. When this Si method is applied to GeO$_2$/Ge devices, it appears that AHTs in Ge were also below Ev ('■', **Fig. 7a**). This, however, is an artifact and the AHTs above Ev in Ge sample (Grey triangle, **Fig. 7a**) were not detected by the Si method because of insufficient charging during the sweeping due to lower hole density above Ev. By sweeping energy level from low to high, an AHT 'tail' was observed above Ge Ev, which is independent of temperature (125 °C / RT).

D. Separating EAD from AHT

To support that AHTs and EADs are two different types of defects in Ge device, **Fig. 7b** shows that EADs increase with stress time, but AHTs do not (as indicated by the marked parallel shift), since they are 'as-grown'. The initial degradation is dominated by the filling of AHTs, which is insensitive to temperature (**Fig. 7c**), supporting **Fig. 7a**. In contrast, the charging of EADs is a thermally accelerated process and does not saturate with stress time (**Figs. 7b and c**). In order to separate EADs from AHTs, we obtain the saturation level of AHT for a given stress Eox from **Fig. 7a and b**. EADs is then extracted by subtracting these saturated AHTs from the total ΔVth (**Fig. 7d**).

Restore power law and enable lifetime prediction in Ge

When EADs were extracted by evaluating AHTs with Si-method, power-law was restored (**Fig. 2b**), but the time exponent 'n' varies substantially with Eox ('▲' **Fig. 8a**), which prevents reliable prediction. In contrast, 'n' is a constant when AHTs were evaluated by the new Ge-method, demonstrating that the AHT-tail above Ev plays a crucial role. This tail does not scale with Eox and impacts more on the raw 'n' at lower Eox. After taking it into account, the variation of both 'n' and the lifetime power exponent, m=γ/n (**Fig. 1c**) disappears, enabling lifetime prediction (**Fig. 8b**). It works for both RT and 125°C. The exponents (m, n, γ) are summarized in Table 1.

Si-cap/Ge devices and process optimization

Fig. 9a compares the AHTs in optimized and non-optimized Si-cap Ge devices. The optimized one does not have a tail above Ev, but the non-optimized one does. The AHTs saturate with stress time clearly for both (**Figs. 9b and c**). The energy level of Si-cap/Ge (optimized) is further below Ev (~0.4eV) than that of Si-cap/Ge (non-optimized). In general, the non- and optimized Si-cap devices behave like GeO$_2$/Ge and Si devices, respectively. When the Ge-method is applied, power law was restored for both the non- and optimized Si-cap devices (**Figs. 10a and b**). A constant time-to-failure power exponent, m, is restored in both cases, enabling reliable lifetime prediction. For the purpose of comparison, the symbols '●' represent results under similar Eox in **Figs. 10a and b**. As expected, for a given ΔVth, the optimized one can survive much longer than the non-optimized one. The processing temperature for the non-optimized one is higher and Ge can diffuse through Si-cap, making it like GeO$_2$/Ge. **Fig. 11** compares the lifetime of different devices/processes by using the new Ge-method at both RT and 125 °C. Si-cap Ge shows superior reliability, even better than SiON/Si and optimization is clearly needed for GeO$_2$/Ge, agreeing with [3]. For the optimized Si-cap device, an overdrive voltage of 1.77 V (Eox = -12.6 MV/cm) can be used to keep ΔVth within 100 mV for 10 years. The Si-cap (optimized) has a large m (see table1) and longer life time/higher maximum operational voltage than Si technology at 125 °C. Power law lifetime prediction is restored for both GeO$_2$/Ge and Si-cap Ge devices, enabling device/process evaluation.

Conclusions

Conventional lifetime prediction method developed for Si is inapplicable to Ge devices. Defects behave differently in Ge devices and their energy level alternates with charging status. The as-grown hole traps have a tail above Ev for Ge devices, but not Si. EAD is experimentally separated from AHT in Ge. For the first time, the importance of the AHT tail is demonstrated for restoring power law of NBTI degradation with a constant power exponent. The newly developed Ge method enables lifetime prediction for GeO$_2$/Ge devices and it is also applicable to the Si-cap/Ge devices, assisting in Ge process/device development and optimization.

References

(1) B. Kaczer, J. Franco, J. Mitard, P. J. Roussel, A. Veloso, and G. Groeseneken, "Improvement in NBTI reliability of Si-passivated Ge/high-k/metal-gate pFETs," *Microelectron. Eng.*, vol. 86, no. 7–9, pp. 1582-1584, 2009.

(2) R. Zhang, P. C. Huang, N. Taoka, M. Takenaka, and S. Takagi, "High mobility Ge pMOSFETs with 0.7 nm ultrathin EOT using HfO$_2$/Al$_2$O$_3$/GeO$_x$/Ge gate stacks fabricated by plasma post oxidation," in *VLSI Symp. Tech. Dig.*, 2012, pp. 161-162.

(3) J. Franco, B. Kaczer, P. J. Roussel, J. Mitard, S. Sioncke, L. Witters, H. Mertens, T. Grasser, and G. Groeseneken, "Understanding the suppressed charge trapping in relaxed- and strained-Ge/SiO$_2$/HfO$_2$ pMOSFETs and implications for the screening of alternative high-mobility substrate/dielectric CMOS gate stacks," in *IEDM Tech. Dig.*, 2013, pp. 15.12.11-15.12.14.

(4) G. Groeseneken, M. Aoulaiche, M. Cho, J. Franco, B. Kaczer, T. Kauerauf, J. Mitard, L. A. Ragnarsson, P. Roussel, and M. Toledano-Luque, "Bias-temperature instability of Si and Si(Ge)-channel sub-1nm EOT p-MOS devices: Challenges and solutions," in *Proc. IPFA*, 2013, pp. 41-50.

(5) J. Ma, J. F. Zhang, Z. Ji, B. Benbakhti, W. Zhang, J. Mitard, B. Kaczer, G. Groeseneken, S. Hall, J. Robertson, and P. Chalker, "Energy Distribution of Positive Charges in Al$_2$O$_3$/GeO$_2$/Ge pMOSFETs," *IEEE Electron Device Lett.*, vol. 35, no. 2, pp. 160-162, 2014.

(6) J. Ma, J. F. Zhang, Z. Ji, B. Benbakhti, W. Zhang, X. F. Zheng, J. Mitard, B. Kaczer, G. Groeseneken, S. Hall, J. Robertson, and P. Chalker, "Characterization of negative-bias temperature instability of Ge MOSFETs with GeO$_2$/Al$_2$O$_3$ stack," *IEEE Trans. Elec. Dev.*, vol. 61, pp. 1307-1315, 2014.

(7) S. W. M. Hatta, Z. Ji, J. F. Zhang, M. Duan, W. Zhang, N. Soin, B. Kaczer, S. De Gendt, and G. Groeseneken, "Energy distribution of positive charges in gate dielectric: probing technique and impacts of different defects," *IEEE Trans. Elec. Dev.*, vol. 60, pp. 1745-1753, 2013.

(8) Z. Ji, S. F. W. M. Hatta, J. F. Zhang, J. G. Ma, W. Zhang, N. Soin, B. Kaczer, S. De Gendt, and G. Groeseneken, "Negative bias temperature instability lifetime prediction: Problems and solutions," in *IEDM Tech. Dig.*, 2013, pp. 15.16.11-15.16.14.

(9) J. R. Weber, A. Janotti, and C. G. Van de Walle, "Native defects in Al$_2$O$_3$ and their impact on III-V/Al$_2$O$_3$ metal-oxide-semiconductor-based devices," *J. Appl. Phys.*, vol. 109, no. 3, pp. 033715-1 - 033715-7, 2011.

(10) D. Liu, Y. Guo, L. Lin, and J. Robertson, "First-principles calculations of the electronic structure and defects of Al$_2$O$_3$," *J. Appl. Phys.*, vol. 114, no. 8, pp. 083704-1 - 083704-5, 2013.

(11) J. F. Binder, P. Broqvist, and A. Pasquarello, "Charge trapping in substoichiometric germanium oxide," *Microelectron. Eng.*, vol. 88, no. 7, pp. 1428-1431, 2011.

(12) L. Lin, K. Xiong, and J. Robertson, "Atomic structure, electronic structure, and band offsets at Ge:GeO:GeO$_2$ interfaces," *Appl. Phys. Lett.*, vol. 97, no. 24, pp. 242902-1 - 242902-3, 2010.

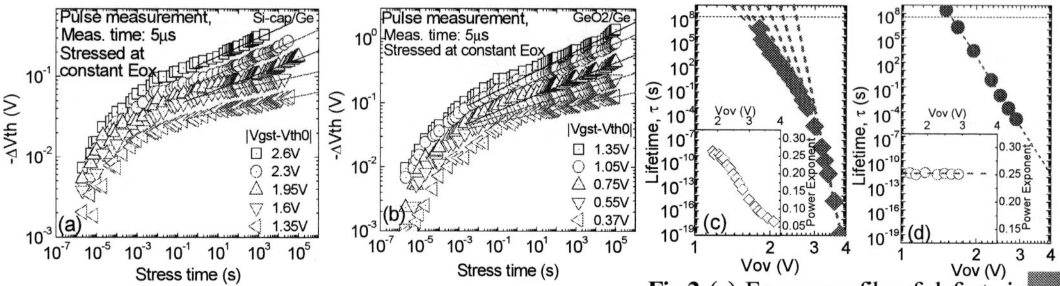

Fig.1 NBTI in (a) Si-cap and (b) GeO$_2$/Ge devices does not follow a power law. (c) Power law extrapolation failed for Si-cap devices, as the exponent (inset) is not a constant [3]. (d) Power law is restored by the new technique developed in this work with a constant exponent (inset).

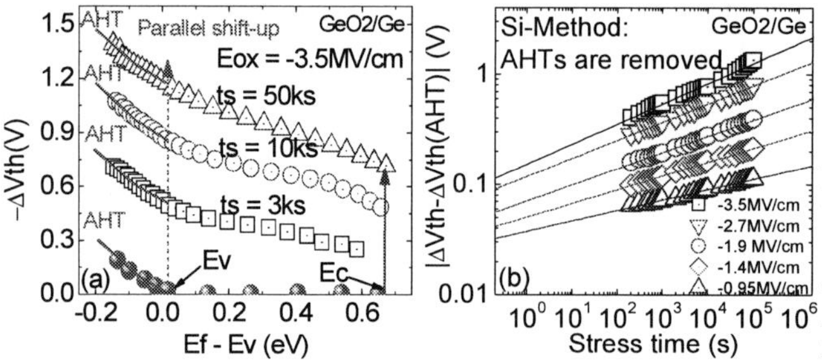

Fig.2 (a) Energy profile of defects in GeO$_2$/Ge are obtained by discharging defects against energy levels from low-to-high. AHTs, obtained with the Si-method by sweep-charging from high-to-low on fresh device [6], are below Ev and do not increase with stress time. (b) Removing AHT leads to a varying power exponent (slope), preventing reliable extrapolation from high stress Vg to low operation Vg.

Table 1: Gate stack and exponents
a) 2.3nm plasma-N **SiON/Si** (125°C: n=0.20, m=16.1, γ=3.22)
b) 4nm Al$_2$O$_3$/1.2nm **GeO$_2$/Ge** (RT: n=0.20, m=14.4, γ=2.88; 125°C: n=0.24, m=10.9, γ=2.62)
c) 4nmHfO$_2$/~0.5nmSiO$_2$/**Si-cap/Ge(non-optimized)** (RT: n=0.19, m=25.3, γ=2.92)
d) 2nmHfO$_2$/~0.4nmSiO$_2$/**Si-cap/Ge(optimized)** (thick Si-cap: RT: n=0.25, m=46.0, γ=11.5 ; 125°C: n=0.28, m=34.4, γ= 9.63; thin Si-cap: 125°C: n=0.19, m=34.0, γ=6.46)

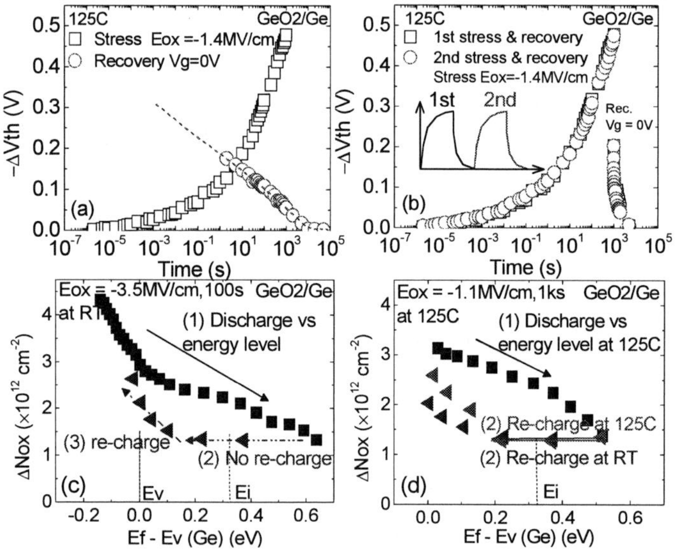

Fig.3 Defects in GeO$_2$/Ge device: (a) Degradation is fully recoverable without a permanent component. (b) The 2nd stress after recovery follows the same kinetics as the 1st one. All defects returned to their fresh states after recovery. (c) Negligible recharge when biased in the upper half of bandgap. (d) Recharge does not increase when switching from 125°C to room temperature (RT).

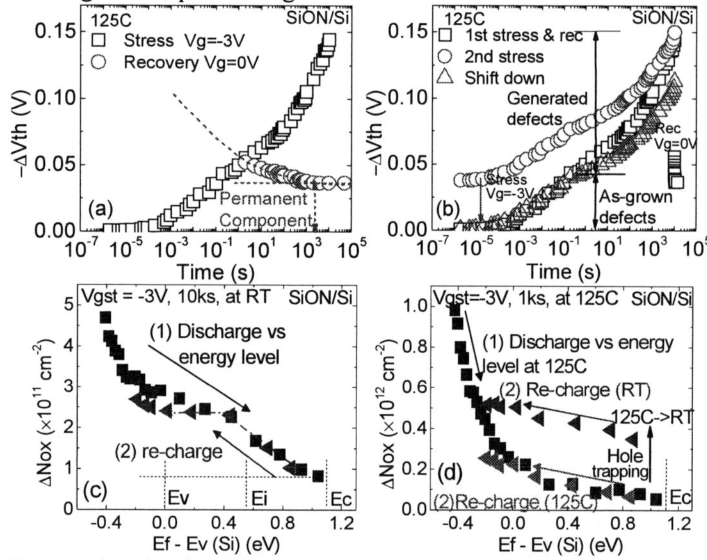

Fig.4 Defects in SiO$_2$/Si device: (a) NBTI is not fully recovered due to permanent component. (b) 2nd stress after recovery follows the same kinetics for AHTs, but different kinetics for generated defects (GD). 'Δ' is a parallel downward shift of 'O'. (c) Recharge occurs in the upper half of band gap. (d) Recharge increases when switching from 125°C and RT. The hole traps neutralized at 125°C at high energy levels are recharged at RT due to lower electron energy at RT.

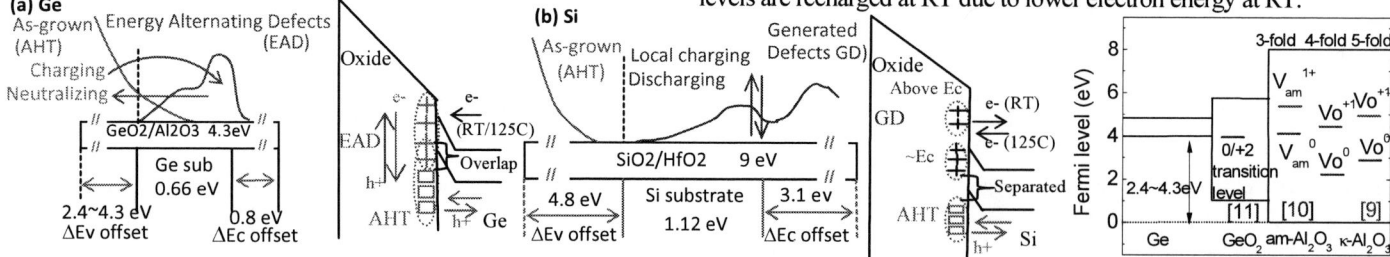

Fig. 5 Illustration of defect differences in Ge and Si devices. (a) GeO$_2$/Ge: **AHTs**, either charged or neutral, are mainly below Ev with a tail above Ev. **EADs** are below Ev when neutral, shift to above Ev once charged, return to fresh states below Ev after neutralization. (b) SiO$_2$/Si: **AHTs** are below Ev without the tail. **GDs** are generated and have high energy levels, either charged or neutral. **GDs** above Ec cannot be fully neutralized, leading to the permanent component. **GDs** neutralized at 125°C can be recharged at RT by e-tunneling to Si conduction band. Small Ec offset at oxide/Ge allows full neutralization of **EADs**.

Fig. 6 First principle calculations show intrinsic energy alternating defects in Al$_2$O$_3$ [9, 10]. For GeO$_2$, the charge transition level is reported for hole traps [11].

978-1-4799-8002-4/14 $31.00 © 2014 IEEE

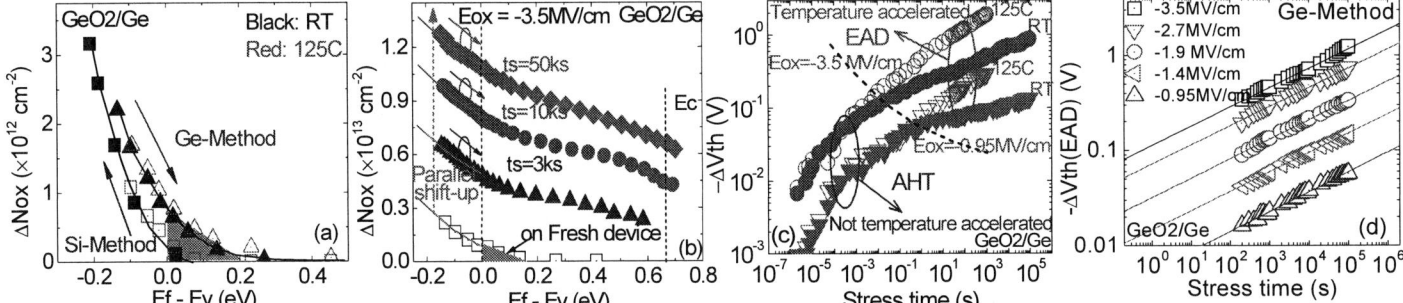

Fig. 7 Restoring power law extrapolation for GeO₂/Ge devices (a) A comparison of AHTs extracted using the Ge- and Si-Method. The Ge method detects a tail above Ev (Grey triangle). (b) AHTs do not increase with stress time, resulting in the marked parallel shift. (c) AHTs are filled first during stress and is temperature independent, whilst EADs is the opposite. (d) Power law is restored after removing AHT extracted with Ge-Method in (a), during which the filling time is kept short enough so that EADs are negligible.

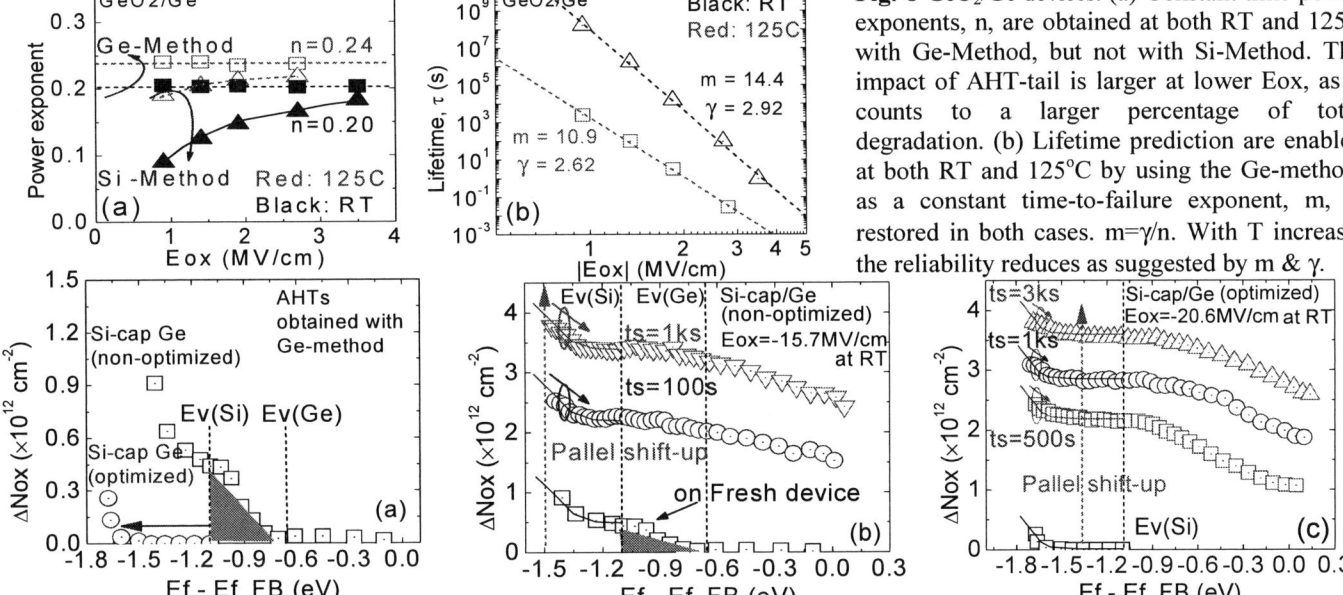

Fig. 8 GeO₂/Ge devices: (a) Constant time power exponents, n, are obtained at both RT and 125C with Ge-Method, but not with Si-Method. The impact of AHT-tail is larger at lower Eox, as it counts to a larger percentage of total degradation. (b) Lifetime prediction are enabled at both RT and 125°C by using the Ge-method, as a constant time-to-failure exponent, m, is restored in both cases. m=γ/n. With T increase, the reliability reduces as suggested by m & γ.

Fig. 9 (a) The energy profile of AHT in Si-cap/Ge (optimized) is further away (~0.4eV) from Ev than Si-cap/Ge (non-optimized). AHT tail is observable in fresh non-optimized device inside Si bandgap, but not in the optimized one.

(b) Degradation of a non-optimized Si-cap/Ge device. Like GeO₂/Ge device: AHTs have a tail above Ev(Si) and do not increase with stress time.

(c) Degradation of an optimized Si-cap/Ge device. Like SiON/Si device: AHTs do not have a tail above Ev and do not increase with stress time (also in the optimized thin Si-cap/Ge, not shown).

Fig. 10 Power law with constant time power exponent, n, is restored with Ge-Method for Si-cap/Ge (a) non-optimized and (b) optimized device. Both tests are at RT. Constant time-to-failure power exponent, m, is also restored for both cases, as shown in Fig.11, enabling reliable lifetime prediction. Eox of '●' is similar for (a)&(b).

Fig. 11 A comparison of lifetime prediction on different CMOS processes by the new Ge-method developed in this work at (a) RT and (b) 125°C. Si-cap/Ge MOSFETs shows superior reliability. Si-cap/Ge (optimized) show process improvement over the non-optimized one (a). Thick&thin Si-cap (optimized) leads to a larger m & γ (table1) and longer lifetime/higher maximum operational voltage than Si technology at 125 °C (b). Power law is restored in all cases, enabling process evaluation.

978-1-4799-8002-4/14 $31.00 © 2014 IEEE

Accurate Prediction of PBTI Lifetime for N-type Fin-Channel Tunnel FETs

W. Mizubayashi, T. Mori, K. Fukuda, Y. X. Liu, T. Matsukawa, Y. Ishikawa, K. Endo,
S. O'uchi, J. Tsukada, H. Yamauchi, Y. Morita, S. Migita, H. Ota, and M. Masahara

Nanoelectronics Research Institute (NeRI), National Institute of Advanced Industrial Science and Technology (AIST),
1-1-1 Umezono, Tsukuba, Ibaraki 305-8568, Japan
Phone: +81-29-849-1629, Fax: +81-29-861-5170, E-mail: w.mizubayashi@aist.go.jp

Abstract

The positive bias temperature instability (PBTI) characteristics for n-type fin-channel tunnel FETs (TFETs) with high-k gate stacks have been thoroughly investigated and compared with conventional FinFETs. The subthreshold slope (SS) is not degraded at all while the threshold voltage (V_{th}) shifts in the positive direction by the PBTI stress. The activation energy of ΔV_{th} for TFETs is almost the same as FinFETs, indicating that the PBTI mechanism for TFETs is almost the same as FinFETs. It was found that, by applying a positive bias to the n^+-drain (normal operation condition), the PBTI lifetime is dramatically improved as compared with that in the conventional stress test (both the p^+-source and n^+-drain are grounded). This is because carrier injection from the n^+-drain is the main cause of the PBTI, especially for n-type TFETs. Thus, the realistic impact of the PBTI is significantly mitigated for n-type TFETs.

Introduction

The increasing power consumption of CMOS devices is a critical issue. To reduce the power consumption, it is necessary to lower the operation voltage. To reduce the operation voltage, TFETs are considered as a promising candidate because of their steeper SS than the limitation for conventional MOSFETs (60 mV/dec) [1-4]. Since TFETs are assumed to operate at an ultralow voltage below 0.5 V, the allowable BTI lifetime becomes severer. Nevertheless, PBTI for the TFETs has scarcely been studied so far [5, 6]. The detailed mechanisms of PBTI degradation for TFETs have not been comprehensively understood yet. This is because the asymmetry source-drain doping for TFETs makes it difficult to understand the PBTI mechanism as shown in Fig. 1.

In this work, we systematically investigated the PBTI characteristics of n-type TFETs with high-k gate stacks.

Experimental

We fabricated both n-type fin-channel TFETs and n-type FinFETs. (110) fin channel was formed on (100) SOI substrate. n^+poly-Si/HfAlO$_x$/SiO$_2$ (equivalent oxide thickness

(EOT) = 2.2 and 2.4 nm) gate stacks were formed. BF$_2^+$ was implanted at 5 keV with doses of 1.0×10^{15} cm^{-2} in the source region, while As$^+$ was implanted at 5 keV with doses of 1.0×10^{15} cm^{-2} in the drain region [7]. To suppress the spread of the implanted dopants by activation annealing, we performed flash lamp annealing (1200 °C, 3 ms) [2]. Finally, the back-end process was carried out.

V_{th} was defined as the gate voltage (V_g) at the drain current (I_d) = 10^{-11} A/ m. The simulation was performed utilizing HyENEXSS, ver. 5.5 [8, 9].

Results and Discussion

A. EOT Dependence of V_{th} in TFETs (Initial State)

We investigated the impact of the EOT on V_{fb} and V_{th} in n-type TFETs (Figs. 2-5). V_{fb} is identical regardless of the EOT (Fig. 2). On the other hand, V_{th} shifts in the negative direction with thinning EOT (Figs. 3 and 5). In the case of TFETs, the drain current is determined by the band-to-band tunneling probability. V_{th} can be described as

$$V_{th} \propto \left| -B \sqrt{\frac{\varepsilon_{Si}}{\varepsilon_{ox}}} t_{ox} t_{Si} \right|, \qquad (1)$$

where B is Kane's parameter [10], ε_{Si} is the dielectric constant of Si, ε_{ox} is the dielectric constant of SiO$_2$, t_{ox} is the EOT, and t_{Si} is the Si thickness (Fig. 6). The experimental and simulated V_{th} can be reproduced by our proposed model (Fig. 5). As a result, V_{th} for TFETs significantly depends on the EOT. Thus, for TFETs, it is necessary to consider the V_{th} shift by EOT in the V_{th} design.

B. PBTI Characteristics for n-Type TFETs
PBTI Mechanism

In order to understand the PBTI mechanism for n-type TFETs, we systematically measured and analyzed their characteristics. V_{th} shifts in the positive direction by the PBTI stress. ΔV_{th} follows a power law of the stress time ($\Delta V_{th} = \alpha t^n$) regardless of the measurement temperature (Fig. 7). Figure 8 shows the exponent n in the power law as a function of the stress voltage for the TFETs. The exponent n is 0.1–0.2 regardless of the EOT, the stress voltage, and the measurement temperature. On the other hand, the SS is hardly degraded by the PBTI stress (Fig. 9).

The activation energy (E_a) of ΔV_{th} for the TFETs is estimated to be 0.026 eV, which is almost the same as that for the FinFETs (E_a = 0.022 eV) (Fig. 10). This result clearly indicates that the PBTI mechanism for n-type TFETs is almost the same as that for n-type FinFETs.

We measured the recovery characteristics for the n-type TFETs (Fig. 11). The recovery test was performed at 25 °C for suppressing the impact of trap assisted tunneling (TAT). The following PBTI tests were carried out at 25 °C. Fig. 11(a) shows ΔV_{th} as a function of stress time before and after the PBTI stress. The recovery phenomenon is observed in the n-type TFETs. The recovery ratios are about 60% for both n-type TFETs and n-type FinFETs (Fig. 11(b)).

Impact of Electric Field Concentration at Source/Gate Edge

In the case of n-type TFETs, since they have p⁺-source, the electric field (lateral E_x and vertical E_y) is concentrated at the source/gate edge (Fig. 1). We investigated the impact of the E_y concentration at the source/gate edge on PBTI degradation (Fig. 12). In order to separate the effects on the source/gate and the gate/drain edge, the PBTI test was performed under gate/source or gate/drain stress. ΔV_{th} at 1000 s under gate/drain stress is very small (about 5 mV) as compared with that under gate or gate/source stress (Fig. 12). Thus, this result clearly reveals that the E_y concentration at the source/gate edge has no impact on PBTI degradation, but the carrier injection from the n⁺-drain is the main cause of PBTI for n-type TFETs.

Accurate Lifetime Prediction in PBTI

In order to accurately predict the realistic PBTI lifetime for n-type TFETs, we investigated the impact of the drain bias on the PBTI characteristics. ΔV_{th} follows a power law of stress time regardless of the drain bias condition (Fig. 13(a)). The exponent n in the power law is about 0.1 regardless of the drain bias (Fig. 13(b)). As shown in Fig. 14, ΔV_{th} at 1000 s decreases with increasing V_d for the n-type TFETs. By applying a positive bias to the n⁺-drain, the electric field in the gate dielectric at the gate/drain edge decreases, as shown in Fig. 15, and thus carrier injection from the drain is suppressed. On the other hand, carrier injection scarcely occurs from the p⁺-source even under the GND condition. This is the main reason for the PBTI improvement by the drain bias. Furthermore, the measured ΔV_{th} is fitted to $\exp(-\beta V_d)$ (Fig. 14), where β is defined as the PBTI improvement factor. Since the carrier injection from the drain

to the gate follows the tunneling mechanism through the gate dielectrics, the relational equation can be described as $\exp(-\beta V_d)$. β for n-type TFETs is about three times that for n-type FinFETs and becomes large with thinning EOT (Fig. 16). Fig. 17 shows the lifetime of PBTI at ΔV_{th} = 50 mV as a function of the stress voltage in n-type TFETs. The lifetime follows $\exp(-\gamma V_g)$ regardless of the drain bias. The voltage to guarantee the 10-years-operation is improved from 0.9 V to 1.5 V by applying a positive drain bias (+0.5 V). This means that n-type TFETs can be guaranteed the 10 years lifetime under normal operation conditions.

Conclusion

We systematically investigated the PBTI characteristics for n-type TFETs with high-k gate stacks (Table 1). The PBTI mechanism for n-type TFETs is almost the same as n-type FinFETs. In the case of TFETs, since the polarity of the carriers is asymmetry in the source and drain, the PBTI lifetime can be accurately estimated by applying drain bias. It was found that n-type TFETs can be guaranteed the 10 years lifetime under normal operation conditions.

Acknowledgements

The research is granted by the Japan Society for the Promotion of Science (JSPS) through the "Funding Program for World-Leading Innovative R&D on Science and Technology (First Program)," initiated by the Council for Science and Technology Policy (CSTP).

References

(1) A. Villalon et al., "Strained tunnel FETs with record I_{ON}: first demonstration of ETSOI TFETs with SiGe channel and RSD," in VLSI Symp. Tech. Dig., 2012, pp. 49-50.

(2) T. Mori et al., "EOT scaling in tunnel field-effect transistors: trade-off between subthreshold steepness and gate leakage," in Ext. Abst. of the 2012 SSDM, 2012, pp. 74-75.

(3) Y. Morita et al., "Synthetic electric field tunnel FETs: drain current multiplication demonstrated by wrapped gate electrode around ultrathin epitaxial channel," in VLSI Symp. Tech. Dig., 2013, pp. T236-T237.

(4) R. Rooyackers et al., "A new complementary hetero-junction vertical tunnel-FET integration scheme," in IEDM Tech. Dig., 2013, pp. 92-95.

(5) G. F. Jiao et al., "New degradation mechanisms and reliability performance in tunneling field effect transistors," in IEDM Tech. Dig., 2009, pp. 741-744.

(6) G. Han et al., "PBTI characteristics of n-channel tunneling field effect transistor with HfO₂ gate dielectric: new insights and physical model," in 2012 Symp. on VLSI-TSA Proc., 2012, p. T82.

(7) T. Mori et al., "Tunnel field-effect transistors with extremely low off-current using shadowing effect in drain implantation," Jpn. J. Appl. Phys., vol. 50, pp. 06GF14-1-3, 2011.

(8) HyENEXSS, ver. 5.5.

(9) K. Fukuda et al., "On the nonlocal modeling of tunnel-FETs - device and compact models -," in Proc. SISPAD, 2012, pp. 284-287.

(10) E. O. Kane, "Zener tunneling in semiconductors," J. Phys. Chem. Solids, vol. 12, pp. 181-188, 1959.

(11) J. Knoch and J. Appenzeller, "A novel concept for field-effect transistors - the tunneling carbon nanotube FET," in 63rd DRC Proc., 2005, pp. 153-156.

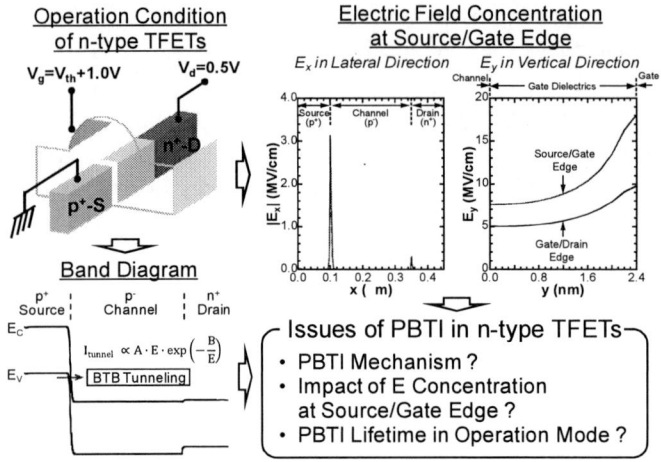

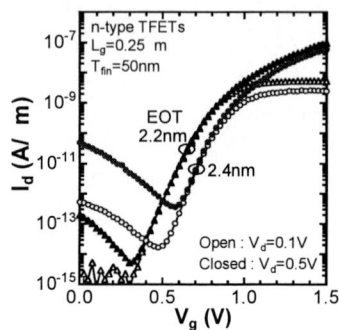

Fig. 1. Schematic illustration of issues of PBTI in n-type TFETs.

Fig. 2. C-V_g characteristics of n-type TFETs. V_{fb} is identical regardless of the EOT.

Fig. 3. I_d-V_g characteristics of n-type TFETs with EOT = 2.2 and 2.4 nm. The I_d-V_g curves shift in the negative direction with thinning EOT.

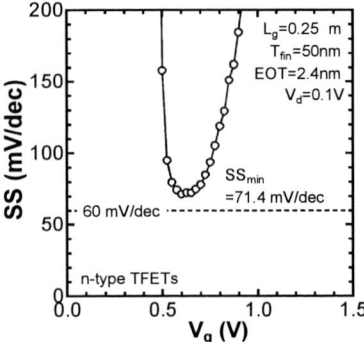

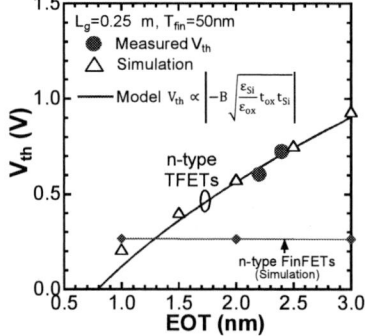

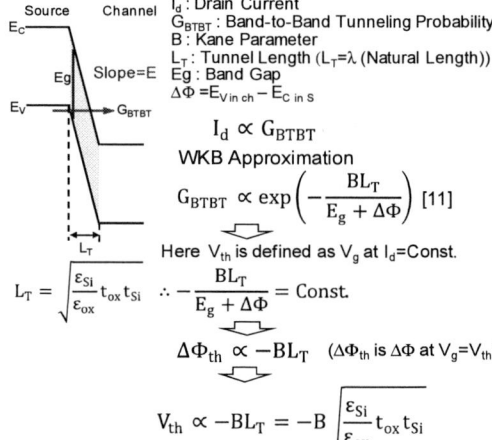

Fig. 4. Typical SS as a function of V_g in n-type TFETs. SS_{min} is 71.4 mV/dec.

Fig. 5. V_{th} as a function of EOT in n-type TFETs and n-type FinFETs. The V_{th} behavior in n-type TFETs is different from that in n-type FinFETs. V_{th} shifts in the negative direction with thinning EOT in n-type TFETs. The measured and simulated data can be reproduced well by our model.

Fig. 6. Schematic illustration of our proposed V_{th} model for TFETs.

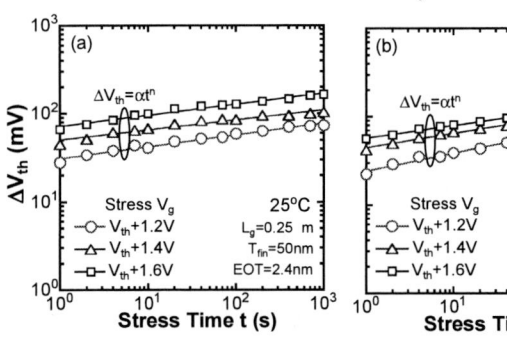

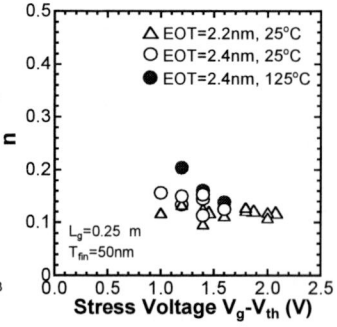

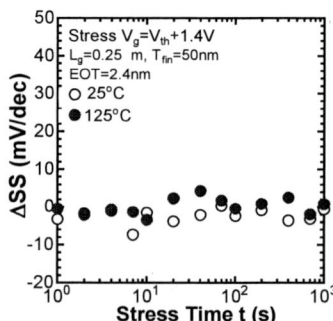

Fig. 7. ΔV_{th} as a function of stress time in n-type TFETs. The measurement temperatures are (a) 25 °C and (b) 125 °C. The PBTI stress conditions are as follows: the stress bias was only applied at the gate, and the source, drain, and substrate were grounded. ΔV_{th} follows a power law of stress time regardless of the measurement temperature.

Fig. 8. Exponent n as a function of stress voltage for n-type TFETs. n is about 0.1–0.2 regardless of the EOT and measurement temperature.

Fig. 9. ΔSS as a function of stress time for n-type TFETs. ΔSS hardly changes with the stress time. This means that the SS is not degraded by the PBTI stress.

978-1-4799-8002-4/14 $31.00 © 2014 IEEE

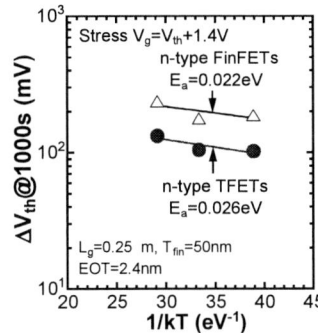

Fig. 10. Arrhenius plots of ΔV_{th} at 1000 s at stress $V_g = V_{th}+1.4V$ for n-type TFETs and n-type FinFETs. The activation energy of ΔV_{th} in the n-type TFETs is almost the same as that in the n-type FinFETs.

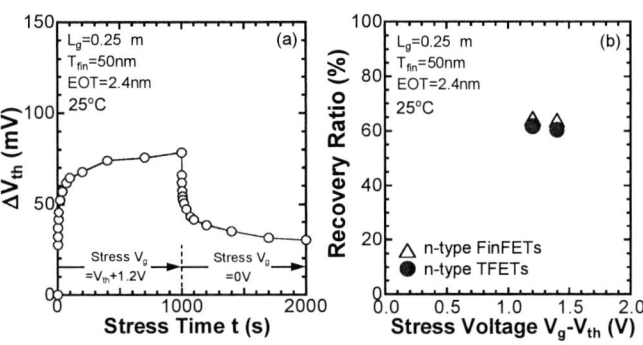

Fig. 11. (a) ΔV_{th} as a function of stress time for n-type TFETs before and after the PBTI stress. (b) Recovery ratio as a function of stress voltage for n-type TFETs and n-type FinFETs. (a) The recovery phenomenon is observed in the n-type TFETs. (b) The recovery ratio in the n-type TFETs is almost the same as that in the n-type FinFETs regardless of the stress voltage.

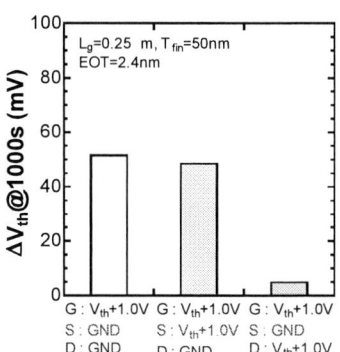

Fig. 12. ΔV_{th} at 1000 s for n-type TFETs under various stress conditions.

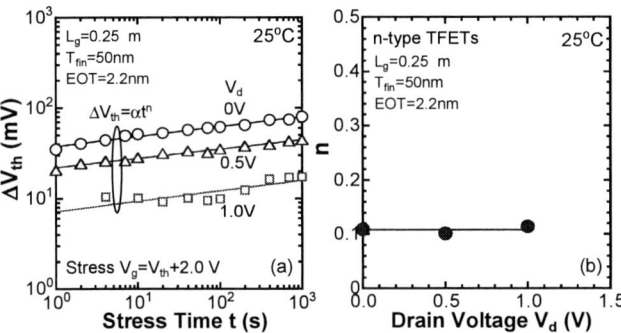

Fig. 13. ΔV_{th} as a function of stress time at stress $V_g = V_{th}+2.0$ V and $V_d = 0, 0.5,$ and 1.0 V for n-type TFETs. (b) Exponent n as a function of drain voltage.

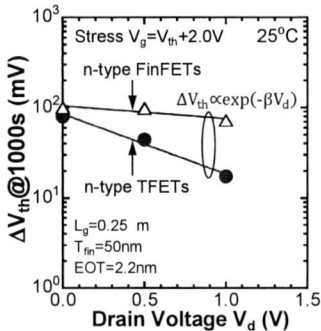

Fig. 14. ΔV_{th} at 1000 s as a function of drain voltage for n-type TFETs and n-type FinFETs. ΔV_{th} at 1000 s decreases with increasing V_d for n-type TFETs.

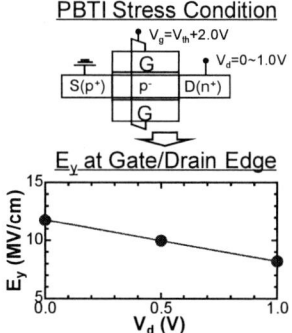

Fig. 15. Schematic illustration of E_y at gate/drain edge in n-type TFETs (simulation). E_y at the gate/drain edge decreases with increasing V_d.

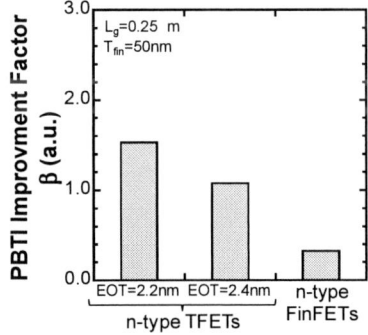

Fig. 16. PBTI improvement factor β for n-type TFETs and n-type FinFETs. β for n-type TFETs increases with thinning EOT and is about three times that for n-type FinFETs.

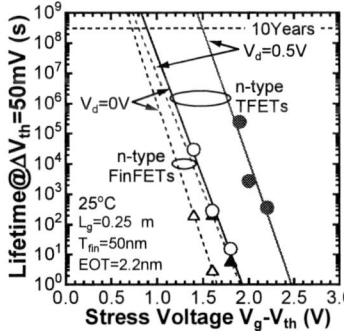

Fig. 17. Lifetime at $\Delta V_{th} = 50$ mV as a function of stress voltage for n-type TFETs and n-type FinFETs. The lifetime longer than 1000 s was estimated from the extrapolation of the measured data. The lifetime is dramatically improved by applying drain bias.

Table 1. Summary of PBTI characteristics for n-type TFETs and n-type FinFETs.

	n-type TFETs	n-type FinFETs
Exponent n	0.1~0.2	0.1~0.18
ΔSS	No Degradation	No Degradation
Activation Energy of ΔV_{th}	0.026eV	0.022eV
E_y Concentration at Source/Gate Edge	No Impact	No E_y Concentration at S/G Edge
PBTI Improvement Factor	1.53 (EOT=2.2nm) 1.08 (EOT=2.4nm)	0.32 (EOT=2.2nm)
PBTI Lifetime in Operation Mode	Dramatic Improvement by Drain Bias	Slight Improvement by Drain Bias

BTI reliability of advanced gate stacks for Beyond-Silicon devices: challenges and opportunities

G. Groeseneken[*], J. Franco, M. Cho, B. Kaczer, M. Toledano-Luque, Ph. Roussel, T. Kauerauf, A. Alian,
J. Mitard, H. Arimura, D. Lin, N. Waldron, S. Sioncke, L. Witters, H. Mertens, L.-Å. Ragnarsson,
M. Heyns, N. Collaert, A. Thean and A. Steegen

*imec, Kapeldreef 75, B3001, Leuven, Belgium; *also with Dept. of Electrical Engineering (ESAT), KU Leuven, Belgium;*
[†] *Phone: +32-16-281269, E-mail: Guido.groeseneken@imec.be*

Abstract

Our present understanding of BTI in Si and (Si)Ge based sub 1-nanometer EOT MOSFET devices is reviewed and extended to benchmark other Beyond-Si based devices. We discuss the evolution of NBTI for Si-based pMOS devices as a possible showstopper for further scaling below 1nm EOT. Then we present the BTI reliability framework which was developed for SiGe based MOSFET devices, showing strongly improved BTI reliability, explained by carrier-defect decoupling. Also the important issue of increasing stochastic behavior and time dependent variability is discussed. Based on the presented framework developed for SiGe stacks we benchmark alternative Beyond-Si gate stacks using a metric for carrier-defect decoupling, allowing to screen stacks for acceptable reliability.

1. Introduction

Bias Temperature Instability (BTI) is one of the most critical issues limiting the device reliability for sub 1-nanometer equivalent oxide thickness Si-MOSFET's [1-2]. At the same time, the use of high-mobility channels is being considered for further device performance enhancement in future CMOS technology nodes. (Si)Ge and III-V based quantum well devices are the first candidates for p- and n-type channels, respectively. Although promising drive current performance has been reported for these devices by several groups [3-5], much less information on their BTI reliability is available in literature. Nevertheless, it is mandatory that, besides good performance and mobility, also sufficient BTI reliability can be guaranteed before these Beyond-Silicon devices can be considered for production.

In this paper we review our present understanding of BTI in Si and (Si)Ge based sub 1-nanometer EOT MOSFET devices and extend it to benchmark other Beyond-Si based devices. In section 2 we discuss the evolution of NBTI for Si-based pMOS devices as a possible showstopper for further scaling below 1nm EOT. In section 3 we present the BTI reliability framework which was developed for SiGe based MOSFET devices, showing strongly improved BTI reliability. In section 4 the important issue of increasing stochastic behavior and time dependent variability is discussed. Based on the framework model we developed for SiGe stacks we then benchmark in section 5 alternative Beyond-Si gate stacks using a metric for carrier-defect decoupling, allowing to screen stacks for acceptable reliability.

2. BTI of sub-1nm EOT Si-based p-MOSFET's

Fig. 1 shows the voltage overdrive (V_{ov}) at 10 year NBTI lifetime as function of EOT for 70 Hf-based high-k metal gated Si p-MOSFET devices [6]. The 10 year V_{ov} is decreasing when reducing EOT, and the devices below 1nm EOT show an even accelerated decrease in maximum V_{ov} and operating fields (inset) resulting in predicted lifetimes far below the specifications. We have found the mechanism for this severe degradation in sub-nanometer EOT devices to be the increased bulk charge trapping effect, enhanced by the reduced interfacial layer thickness (Fig. 2) [7]. It can be concluded that at least 0.4nm of interfacial layer is needed to reduce massive hole direct tunneling into the high-k oxide defects for sub-1nm EOT devices (Fig. 3) [6]. As a result BTI is a severe showstopper for reliability of sub-1nm EOT Si-based devices.

3. BTI framework of (Si)Ge based MOSFET's

Contrary to the Si-based devices, Si-Ge quantum well devices using a Si passivation layer to enable the use of a standard SiO_2/HfO_2 dielectric stack [3] show strongly improved BTI reliability together with a 1.5-2.4 x improvement in mobility. **Fig. 4** shows the impact of Ge-content, quantum well (QW) and Si cap thickness [8]. The main gate-stack parameter affecting the NBTI robustness is the Si passivation layer thickness, with thinner Si caps yielding improved reliability and reduced equivalent thickness in inversion (T_{inv}) of the gate stack (**Fig. 5**). Combining the effects of high Ge-fraction, thicker QW and thinner Si-cap the lifetime can be optimized above the ITRS spec and this for both gate first (MIPS) or RMG process flows (**Fig. 6**). This reliability optimization can be done without jeopardizing the drive current performance (**Fig. 7**). The BTI pre-factors show a strong reduction and a stronger field acceleration (higher γ-factor) for SiGe devices, particularly for a reduced Si-cap thickness, also confirmed by data from IBM [1] and NUS [9] (**Fig. 8**). This effect is crucial in understanding the superior reliability of SiGe technology, and is caused by a favorable energy decoupling between channel holes and dielectric defect levels (**Fig. 9**). This is quantitatively confirmed by representing the defect bands as Gaussian energy distributions centered at 0.95eV and 1.4eV below the Si valence band in the SiO_2 IL and the HfO_2, respectively (**Fig.10**). The model is calibrated for the Si-

978-1-4799-8002-4/14 $31.00 © 2014 IEEE

reference device, and then applied to the SiGe devices using fixed defect band parameters. The model matches excellently the BTI data: both the reduced degradation and the stronger field dependence are readily captured (**Fig. 11**). The framework also explains the other experimental observations concerning the Ge fraction and the QW thickness [8]. In order to minimize the fraction of accessible defects, by pushing up the Fermi level in the channel with respect to the defect band, the valence band offset between SiGe and Si has to be maximized: a higher Ge fraction (reduced bandgap and higher ΔE_v) and thick QW (to reduce quantization) are therefore beneficial.

4. Stochastic behavior and time-dependent variability

An important complication in nanoscaled CMOS technologies is that only a handful of defects is present in each device, while their relative impact on the device characteristics is strong. As a result the behavior of these defects is stochastic and widely distributed in time (**Fig. 12**) [10]. Similarly to large area devices, nanoscaled SiGe channel pMOSFETs with a reduced Si cap thickness show a reduced average number of charging defects per device $\langle N_T \rangle$, and a stronger field acceleration but also significantly lower average ΔV_{th} step height η (**Fig. 13**), ascribed to a reduced electrostatic impact of the accessible traps located further from the channel. The knowledge of the single-defect impact distribution, combined with the assumption of Poisson-distributed number of defects per device, allows predicting the distribution of the total degradation per device and projecting the fraction of failing devices at 10 years(**Fig. 14**) [11, 12]. A dramatic improvement of the time-dependent variability distributions is apparent for SiGe devices with reduced Si cap thickness. The reliability improvement at median percentiles, observed in large area devices, is strongly magnified at the more relevant high percentiles of working devices, giving an even higher benefit for SiGe devices.

5. Benchmarking BTI of Beyond-Si gate stacks

Based on the BTI framework developed for SiGe devices we can now analyze and benchmark alternative gate stacks and substrate materials using a unified picture of channel carrier-defect energy coupling. **Fig. 15** shows a calculation of our defect band model for various (Si)Ge/HfO$_2$ based gate stacks [13], where the distribution parameters have been fitted to yield the typically observed voltage acceleration $\gamma \sim 3$ for Si, in the measurable ΔV_{th} range. For different energy injection levels, representing SiGe ($\Delta E \sim 0.3 eV$) and r-Ge ($\Delta E \sim 0.5 eV$) valence bands, the same defect band is expected to induce different $\Delta V_{th}(V_{ov})$ curves. The field acceleration factor γ can be considered as a measure for the defect-energy decoupling, with a higher γ leading to an expected better BTI reliability at low operating voltages. Following a similar approach for InGaAs/Al$_2$O$_3$ gate stacks, accounting for the higher energy level populated in the thinner quantum wells and including an exponential attenuation factor to account for inversion charge centroid displacement, we obtained a wide distribution of electron trap levels with median energy only 1eV above the InGaAs conduction band (**Fig. 16**) [14]. Using this framework we can benchmark all investigated nMOS and pMOS Beyond-Si gate stacks in a T_{inv}-independent $\Delta N_{eff}(E_{ox})$ plot (**Fig. 17**). Similar shifts are observed for InGaAs/Al$_2$O$_3$ and Ge/GeO$_x$/Al$_2$O$_3$ devices, showing a very flat voltage dependence ($\gamma \sim 1.5$) ascribed to the wide energy distribution and hence strong carrier-defect coupling of the Al$_2$O$_3$ oxide defects. The more reliable stacks on the other hand show a steeper E_{ox} dependence ($\gamma \sim 7$-12). **Fig. 18** shows the benchmark plot of maximum operating overdrive for all Beyond-Si devices investigated. Si-passivation of Ge-based channels with SiO$_2$/HfO$_2$ dielectric stack yields good to excellent pMOS reliability and sufficient nMOS reliability. Alternative passivation schemes based on a thin GeO$_x$ interfacial layer obtained by plasma oxidation of the Ge surface through a thin Al$_2$O$_3$ capping layer [3] yield extremely poor CMOS reliability, similar to InGaAs n-channel devices with Al$_2$O$_3$ gate dielectric [14].

6. Conclusions

When developing novel Beyond-Si high-mobility channel device technologies, next to the drive current and mobility, the BTI reliability is one of the important performance parameters that should be accounted for when selecting the optimal stack. We developed a framework to analyze, benchmark and screen such Beyond-Si gate stacks based on the carrier-defect coupling of oxide defects. The field acceleration factor γ can be used as a powerful metric for this defect-energy decoupling, with a higher γ leading to an expected better BTI reliability (**Fig. 19**). Higher γ-values and thus good carrier-defect energy decoupling are obtained for (Si)Ge/Si-cap/HfO$_2$ pMOS and nMOS devices, whereas poor decoupling is found for all the GeO$_x$ and for the InGaAs/Al$_2$O$_3$ gate stacks.

Acknowledgement:

This work is supported by the IMEC Core Logic CMOS Program.

References

(1) S.A. Krishnan et al, *IEEE IRPS Proceedings*, p. 387, 2011.
(2) M. Sato et al. Jap. J. of App. Phys., Vol. 47, p. 2354, 2008.
(3) J. Mitard et al, *IEDM Tech Dig.*, p. 249, 2010.
(4) S. Krishnan et al, *IEDM. Tech Dig.*, p. 634, 2011.
(5) R. Zhang et al, *IEDM Tech Dig.*, p. 642, 2011.
(6) M. Cho et al. *IEEE Trans El. Dev.*, Vol. 59, p. 2042, 2012.
(7) M. Cho et al, *IEEE Trans El. Dev.*, Vol. 58, p. 3342, 2011.
(8) J. Franco et al, *IEEE Trans El. Dev.*, Vol. 60, p. 396, 2013.
(9) X. Gong et al., ECS Trans., vol. 50, p. 949. 2012.
(10) B. Kaczer et al, *IEEE IRPS Proceedings*, p. 26, 2010.
(11) M. Toledano-Luque et al, *IEEE VLSI Techn. Symp Tech Dig*, p. 152, 2011.
(12) J. Franco et al, , *IEEE Trans El. Dev.*, Vol. 60, p. 405, 2013.
(13) J. Franco et al, *IEDM Tech Dig.*, p. 397, 2013.
(14) J Franco et al, *IEEE IRPS Proceedings*, p. 6.A 2.1, 2014

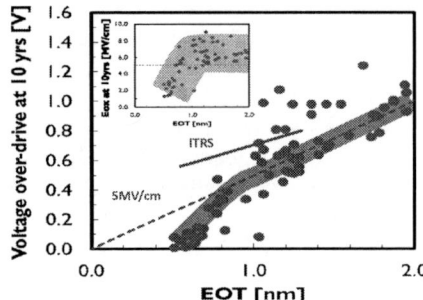

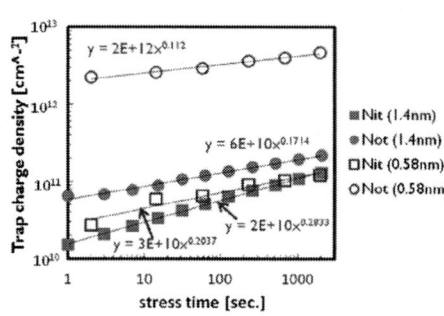

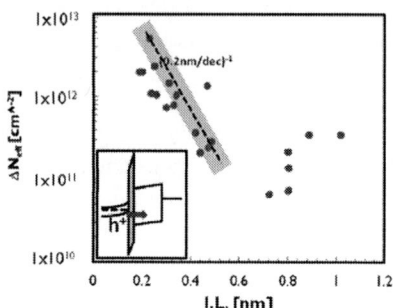

Fig. 1. Over-drive voltage V_{ov} at 10 years for 30mV V_{th}-shift, extracted from NBTI on 70 different p-MOSFET devices. Below 1nm EOT, V_{ov} decreases more than the iso-electric field at 5MV/cm. ITRS roadmap data is also added. The inset shows electric field at 10 years.

Fig. 2. The interface trap density (ΔN_{it}) and the bulk trap density (ΔN_{ot}) increase as a function of stress time for 1.4nm, and 0.58nm EOT devices. The ΔN_{ot} largely increases in the 0.58nm EOT device, showing that the bulk defect dominates the NBTI degradation in sub-nanometer EOT regime.

Fig. 3. The effective trap density (ΔN_{eff}) as a function of the interfacial layer (I.L.) thickness. The inset shows a band diagram indicating high hole tunneling current into the oxide bulk defect through a very thin interfacial layer.

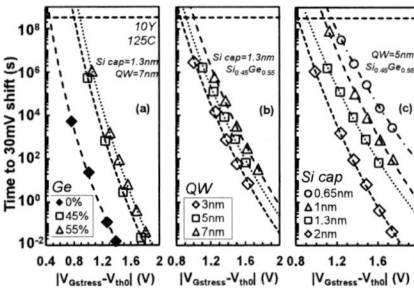

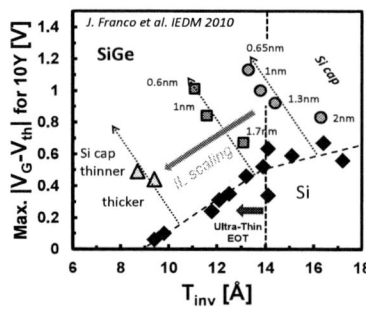

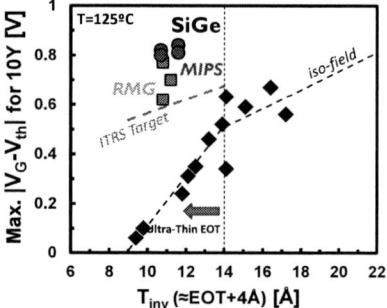

Fig. 4. Extrapolated lifetimes as a function of gate voltage overdrive for varying (a) Ge content, (b) QW thickness and (c) Si Cap thickness. The NBTI performance is improved for increasing Ge fraction, increasing QW thickness and reducing Si Cap thickness. The latter also enables T_{inv} reduction.

Fig. 5. Maximum operating overdrive for 10 year lifetime (T=125°C, failure criterion ΔV_{th}=30mV) vs. T_{inv}. SiGe devices with a thin Si cap offer improved NBTI reliability, i.e. higher maximum operating overdrive.

Fig. 6. A high Ge fraction (55%) in a 6.5nm thick QW, combined with a thin Si cap (0.8nm) boosts the maximum operating overdrive ($|V_G-V_{th0}|$) to meet the target V_{DD} at ultra-thin EOT in a MIPS flow or an a RMG flow.

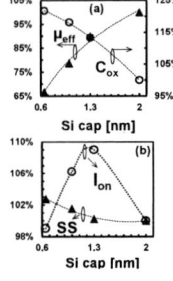

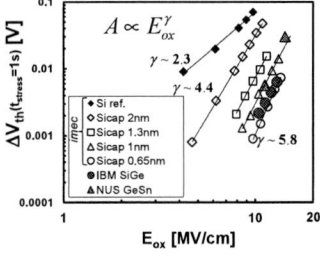

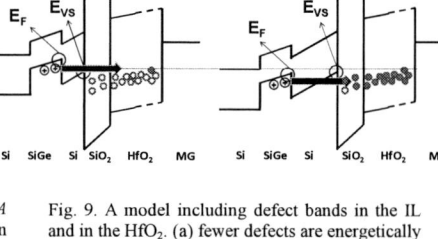

Fig. 7. (a) SiGe devices with thinner Si-cap show reduced mobility (poor interface passi-vation) but higher C_{ox} (reduced hole displace-ment). (b) This yields an optimum I_{on} for a medium Si cap. But the I_{on} dependence on the Si-cap thickness is weak (~±5%).

Fig. 8. NBTI ΔV_{th} power-law pre-factors A ($\equiv\Delta V_{th}(t_{stress}=1s)$]: a lower NBTI is found in SiGe devices, particularly for a reduced Si cap thickness. The stronger E_{ox}-acceleration (higher γ) for SiGe w.r.t. the Si ref. device is a key signature of a favourable energy misalignment between channel holes and dielectric defect levels (cf. Fig 9). NBTI data published by IBM [1] for SiGe pMOSFETs and by NUS [9] for GeSn pMOSFETs with SiO_2/HfO_2 dielectric stack compare well.

Fig. 9. A model including defect bands in the IL and in the HfO_2. (a) fewer defects are energetically favorable for trapping channel holes thanks to the higher Fermi level in SiGe as compared to Si. (b) The additional voltage drop on a thicker Si cap 'pushes' down the Fermi level in the channel (when benchmarking at constant electric field or constant gate overdrive) and therefore more oxide defects become energetically favorable for hole trapping.

Fig. 10. Defect band for SiO_2/HfO_2 stacks, centered at 0.95eV and 1.4eV below the Si valence band in the SiO_2 IL and in the HfO_2, respectively, used to model the SiGe BTI experiments (see Fig 10). The defect band is modeled as a Gaussian distribution over energy. Charged defects at different spatial positions contribute differently to the total ΔV_{th} due to electrostatics

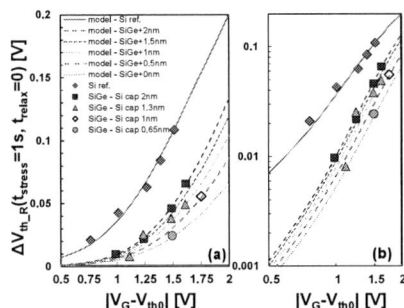

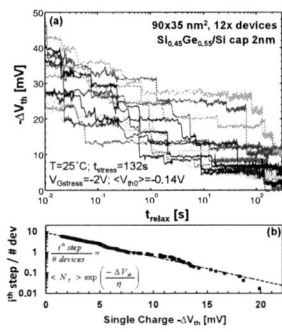

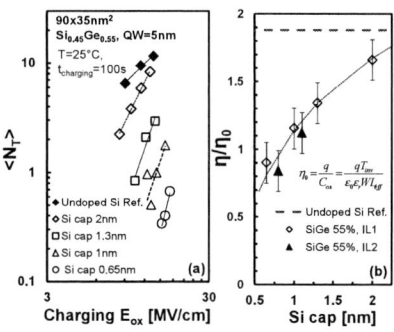

Fig. 11. Calculated ΔV_{th} vs. experimental BTI data. The model was first calibrated on the Si ref. data, then the same defect band parameters were used to calculate the expected ΔV_{th} for SiGe devices (including the valence band offset between the SiGe and the Si cap, and the voltage drop on Si caps of varying thickness). The simple model matches the experimental data remarkably well. (a) Lin-lin, (b) log-log scales.

Fig. 12. (a) NBTI relaxation transients on nanoscaled SiGe devices. For each device, multiple single defect discharge events are visible. (b) Weighted complementary Cumulative Distribution Function of the individual ΔV_{TH} step heights observed on 41 devices, showing an exponential distribution [10], with an average value $\eta \approx 3.9mV$. The average number of defects per device, $\langle N_T \rangle$, can be read in this plot from the intersection with the y-axis [11]

Fig. 13. (a) Nanoscaled SiGe pMOSFETs with a reduced Si cap thickness show reduced average number of charging defects per device $\langle N_T \rangle$, and a stronger field acceleration. (b) Extracted average ΔV_{th} step heights η for SiGe devices with different Si cap and for undoped Si channel devices, after a charging phase at $E_{ox} \approx 12MV/cm$. The values of η are normalized to the single charge electrostatic ($\eta_0 = q/C_{ox}$). Devices with the thinnest Si cap show a significantly lower η, confirmed for two different SiO_2 interfacial layer thicknesses. The red dashed line is the benchmark value experimentally estimated on undoped Si channel. devices.

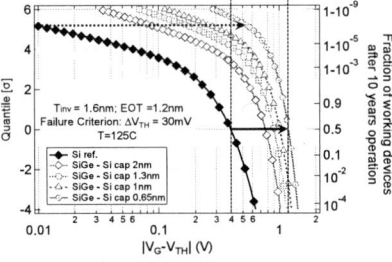

Fig. 14. Calculated fractions of working devices after 10 year operation at varying operating voltages [11] for various gate stacks. A strong improvement of the distributions is apparent for SiGe devices with reduced Si cap thickness. The improvement observed in large area SiGe devices (shown by the solid arrow at median percentile) is expected to be magnified at high percentiles (shown by the dotted arrow, at ~1 ppb).

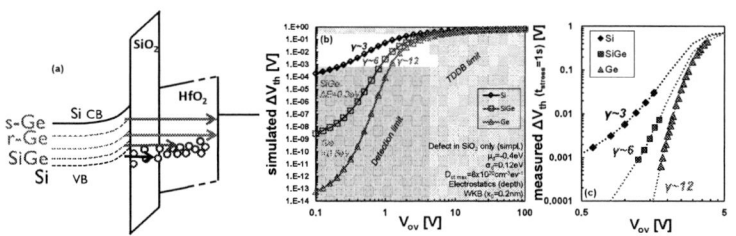

Fig. 15. (a) Sketch of a defect band in the dielectric bandgap: more defects become energetically favorable for channel carrier with increasing V_{ov}; different channel materials yield different carrier energy injection levels (Si→SiGe→Ge→strained Ge). (b) Calculated $\Delta V_{th}(V_{ov})$ for a Normal distribution of defect energy levels [13]. The defect band parameters were fitted to yield a typically observed voltage acceleration exponent $\gamma \sim 3$ for Si devices. Then the same parameters were used to calculate the expected ΔV_{th} for SiGe devices (valence band offset $\Delta E_v \sim 0.3eV$) and for r-Ge devices ($\Delta E_v \sim 0.5eV$). The simulated trends well describe (c) the experimental $\Delta V_{th}(V_{ov})$ data for Si, SiGe, r-Ge devices with identical SiO_2/HfO_2 gate stacks (lines are guide to the eye).

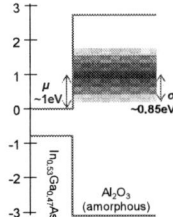

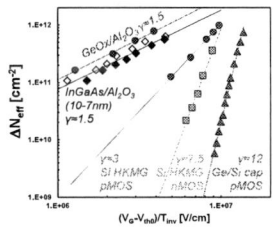

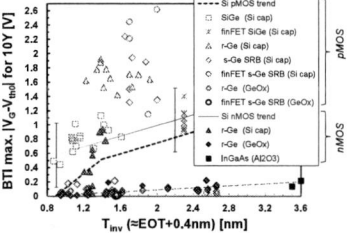

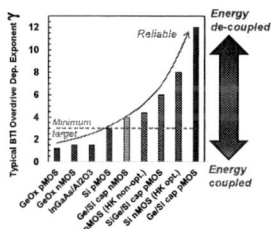

Fig. 16. A wide distribution of electron trap levels ($\sigma \sim 0.85eV$) is found in the Al_2O_3 oxide, with median energy only ~1eV above the InGaAs conduction band, and only ~0.38eV above the higher energy level populated in the thinnest quantum well [14]. Such a wide distribution of defect levels aligned with the channel Fermi level and accessible at low oxide fields jeopardize the device reliability.

Fig. 17. BTI shifts plotted in a T_{inv}-independent benchmark $\Delta N_{eff}(E_{ox})$. Similar shifts are observed for InGaAs/Al$_2$O$_3$ and Ge/GeO$_x$/Al$_2$O$_3$ devices . Note the very flat voltage dependence ($\gamma \sim 1.5$) ascribed to a wide energy distribution of oxide defects (Fig. 15). For comparison, data of Si hk/MG (p and nMOS) and Ge/Si cap/hk/MG pMOS are also shown. Note the steeper voltage acceleration for the more reliable technologies

Fig. 18. Benchmark plot of max. overdrive for advanced CMOS technologies. NBTI limits Si pMOS reliability at CET<1.4nm (EOT<1nm), PBTI can be mitigated by high-k optimization (error bars on the Si nMOS trendline represents optimization range). Si-passivation of Ge-based channels with SiO$_2$/HfO$_2$ dielectric stack yields good pMOS reliability (SiGe→relaxed-Ge→strained-Ge, planar/FinFETs) and sufficient nMOS reliability (relaxed-Ge). Alternative passivation schemes based on a thin GeO$_x$ interfacial layer by plasma oxidation of the Ge surface through a thin Al$_2$O$_3$ cap layer [3] yields poor CMOS reliability (relaxed-Ge), similar to InGaAs n-channel devices with~Al$_2$O$_3$ gate dielectric [14].

Fig. 19. Benchmarking the γ-factor extracted on various Beyond-Si technologies. Higher γ-values are observed for stacks with good carrier-defect energy decoupling (e.g. SiGe or Ge with Si-cap pMOS, Si or Ge/Si cap nMOS with optimized high k). Note the very low γ-values for the GeO$_x$ and InGaAs/Al$_2$O$_3$ gate stacks. γ is a solid metric for carrier-defect coupling when screening substrate/ dielectric systems. A minimum reliability target can be set to the level of Si pMOS, i.e. $\gamma \sim 3$.

New Understanding of State-Loss in Complex RTN:
Statistical Experimental Study, Trap Interaction Models, and Impact on Circuits

Jibin Zou[1], Runsheng Wang[1,*], Shaofeng Guo[1], Mulong Luo[1], Zhuoqing Yu[1], Xiaobo Jiang[1], Pengpeng Ren[1], Jianping Wang[2],
Jinhua Liu[2], Jingang Wu[2], Waisum Wong[2,3], Shaofeng Yu[2,3], Hanming Wu[2,3], Shiuh-Wuu Lee[2,3], Yangyuan Wang[1,3], Ru Huang[1,3,*]

[1]Key Laboratory of Microelectronic Devices and Circuits (MOE), Institute of Microelectronics, Peking University, Beijing 100871, China.
[2]Semiconductor Manufacturing International Corporation (SMIC), Shanghai 201203 and Beijing 100176, China.
[3]Innovation Center for MicroNanoelectronics and Integrated System, Beijing 100871, China
*Email: ruhuang@pku.edu.cn; r.wang@pku.edu.cn

Abstract

In this paper, the statistical characteristics of complex RTN (both DC and AC) are experimentally studied for the first time, rather than limited case-by-case studies. It is found that, over 50% of RTN-states predicted by conventional theory are lost in actual complex RTN statistics. Based on the mechanisms of non-negligible trap interactions, new models are proposed, which successfully interpret this state-loss behavior, as well as the different complex RTN characteristics in SiON and high-κ devices. The circuit-level study also indicates that, predicting circuit stability would have large errors if not taking into account the trap interactions and RTN state-loss. The results are helpful for the robust circuit design against RTN.

Introduction

The random telegraph noise (RTN) has emerged as a critical concern for modern VLSI design [1-8] due to its increasing amplitude as device scaling, which leads to severe time-dependent variations in nanoscale technology nodes. It is understood that single oxide trap behavior causes RTN with 2 distinct current-state variations; and multiple traps accounts for the more-frequently-observed cases of complex RTN with more states [5-7]. Conventionally, complex RTN is treated as the superposition of individual traps independently (i.e. n traps lead to 2^n RTN states). On the other hand, the trap interaction can be significant in complex RTN, as we have found in 2-trap (4-state) RTN experiments recently [7]. However, rather than limited case-by-case studies of individual RTN, the characteristics of complex RTN with more states needs to be further investigated from statistical experiments, for deep understanding of the trap interaction behavior and its practical impact on circuit stability.

Therefore, in this paper, the statistical characteristics of complex RTN are experimentally studied for the first time. It is observed that lots of RTN states predicted by the conventional theory are lost, due to the non-negligible trap interaction behaviors. Process dependence (SiON or high-κ) and frequency dependence (DC or AC RTN [8]) of the RTN state-loss are also discussed. According to the balancing degree of trap interactions, theoretical models based on inheritance hierarchical and non-hierarchical systems are proposed for multiple traps, which can well explain the experiments and provide guidelines for circuit analysis. The results are helpful for the understanding of trap behaviors and the future robust design against RTN.

Devices and RTN Characterization

Devices used in this work are with high-κ/metal-gate (HKMG) or SiON/Poly-Si gate stacks. The statistics of all the device drain current with RTN (2-state and complex cases) and without RTN (i.e., 1-state) were characterized. More than 270 devices were measured. Fig. 1(a)&(d) show two examples, with histogram as Fig. 1(b)&(e). With Gaussian Mixture Model and EM algorithm [9] as RTN/current state extraction method, these two examples show 4 distinct RTN states respectively, which is believed to be caused by 2 discrete traps by common understanding. Current variations under both DC and AC RTN cases can be analyzed in the same way, as shown in Fig. 2. Fig. 2 (c)&(d) represents current-state statistics of (a)&(b), which indicates that the RTN time constant statistics change largely under digital circuit operations [8]. More interestingly, Fig. 2 (e)-(h) shows that AC signals can stimulate additional RTN state. This is because that slow trap can be activated with increased frequency [7, 8].

Results and Discussions
A. New observations on complex RTN statistics

Trap number per device is believed to be Poisson distributed [10] (as the inset of Fig. 3). The conventional understanding of trap-number and RTN-state-number relationships predicts that n traps induce 2^n state, which results in the ideal RTN state number distribution with peaks on particular positions of 2, 4, 8, etc., as shown in Fig. 3. However, the statistical experimental results exhibit irregular RTN state distribution for both SiON and high-κ cases in Fig. 4, with only one clear peak on distribution and sharp drop towards the higher state number. Besides, HKMG devices show more states, i.e. more traps per device than SiON case. The most likely reason of the mismatch between experiments and the ideal cases is that the activities of some traps are impacted by the occupation of other certain traps, i.e., n traps in one device cannot lead to all possible 2^n variation states in drain current, due to trap interactions that change the trap physical properties. There are two microscopic

978-1-4799-8002-4/14 $31.00 © 2014 IEEE

mechanisms of trap interaction: one is the Coulomb repulsion effect between each traps (Fig. 5), the other is the channel percolation effect induced local carrier density perturbation [11] beneath each trap (Fig. 6). By studying a 4-state (2-trap) RTN case, strong trap interactions can be observed in Fig. 7: one filled trap degrades the occupancy rate of the other one by 10%~60%. With the physical property of certain traps being largely affected by some filled traps, it is reasonable to understand the RTN state-loss in Fig. 4. It is worth noticing that, channel percolation effect can cause unbalanced trap interactions between each other, as shown in Fig. 6; while the interactions by Coulomb repulsion effect are balanced or "equal" for each trap. Thus, the trap interaction behavior should be modeled with regards to different balancing degrees of interactions, for deeper understanding of complex RTN statistics and RTN state-loss.

B. Theoretical Modeling of Trap Interaction Mechanism

In single nanoscale device with few oxide traps, where dominant trap interaction mechanism is channel percolation effect, multi-trap exhibits an unbalanced hierarchical priority. As shown in Fig. 8, an inheritance hierarchical system (model HS) is proposed to describe trap interaction behavior with an average degradation rate of q, to represent the observed trap interactions (e.g., as in Fig. 7). On the other hand, for the case of much more traps in one device, Coulomb repulsion effect prevails as dominant interaction mechanism, due to the fact that traps are closer and more traps can average the percolation effects to some extent. In this case, trap interactions are balanced due to the nature of Coulomb repulsion, and interactions will be stronger with reduced trap distance. Thus, the more-trap case suits a non-hierarchical system (model NHS) considering all traps equally as shown in Fig. 9. Fig. 10 gives the results of Model HS with various trap number and q. Only a few states can be frequently observed, and the state-loss is more than 50%, especially for the stronger trap interaction cases. If further including Poisson distribution of trap number, the correlation between experimental results and theoretical calculations are plotted in Fig. 11, as a function of average trap number (λ) and q. Model NHS results in Fig. 12 reveals that under AC operations, some additional traps per device (i.e., larger λ) are activated. Fig. 13 compares the experimental statistics of complex RTN with both models. Note that, λ and q are self-consistently solved and have unique solution in each model. It can be observed that model HS fits the SiON devices well with a lower λ (~1.4) and weaker trap interactions (q~0.7), while model NHS better describes the HKMG devices with a higher λ (~3.5) and stronger trap interactions (q~0.3). This is exactly consistent with the expectations above that the process with more traps per device has balanced but stronger trap interactions.

C. Impacts on Digital Circuits

With the above new understanding of trap interactions, the complex RTN impact on circuit can be precisely predicted. As shown in Fig. 14, A 5-stage ring oscillator (RO) is adopted in the study as typical digital circuit. RTN states with particular trap-filling combinations are applied as a result of the above trap interaction models. Transient circuit simulations of RTN are performed on our recently-developed platform [12]. Monte Carlo simulations with random RTN amplitude (ΔV_{TH}) are repeated many times for average, as shown in Fig. 15. It is found that, the RTN state-loss results in the non-Gaussian distribution of RO frequency (f) degradation (Fig. 16), which distinctly differs from the Gaussian distributions predicted by conventional theory. The overall distribution of $\Delta f/f$ from the conventional prediction and the proposed new models are shown in Fig. 17. Although they all present exponential trends at the tail, new models show steeper probability drops at higher degradation range, due to the fact that larger variations caused by many traps being filled at the same time will not frequently appear in the presence of strong trap interaction. For the same set of experimental data extraction, Model NHS (thus high-k technology) tends to have smaller variation than HS (thus SiON technology) at high-σ, due to more balanced trap interactions. Fig. 18 shows Weibull plot of ideal non-interaction case and interaction cases. The 3-σ and 6-σ of RO f degradation are extracted in Fig. 19. Without considering trap interactions, circuit instability due to complex RTN will be largely overestimated by 48%~72%.

Summary

The state-loss in complex RTN statistics are observed for the first time, which cannot be interpreted by conventional theory. New models based on non-negligible trap interactions are proposed. The process (SiON vs. high-κ) and frequency (DC vs. AC) dependence of complex RTN statistics are also found. The results indicate that, precise prediction of practical RTN impacts on circuit stability should be with the new understanding of state-loss induced by trap interactions.

Acknowledgement

This work was partly supported by the 973 Projects (2011CBA00601), NSFC (61106085), and National S&T Major Project (2009ZX02035-001).

References: [1]. H. Miki, et al., ***VLSI 2012***, p. 137. [2]. A.P. van der Wel, et al., ***APL***, 183507, 2005. [3]. K. Ohmori, et al., ***VLSI 2011***, p. 202. [4]. N. Tega, et al., ***VLSI 2009***, p. 50. [5]. T. Nagumo, et al., ***IEDM 2009***, p. 759. [6]. E. R. Hsieh, et al., ***IEDM 2012***, p. 454. [7]. P. Ren, et al, ***IEDM 2013***, p. 778. [8]. J. Zou, et al., ***VLSI 2013***, p. 186. [9]. A. Dempster, et al., ***J. R. Stat. Soc. B***, p. 1, 1977. [10]. B. Kaczer, et al., ***EDL***, p. 411, 2010. [11]. A. Asenov, et al., ***TED***, p. 839, 2003. [12]. R. Wang, et al., ***IEDM 2013***, p. 834.

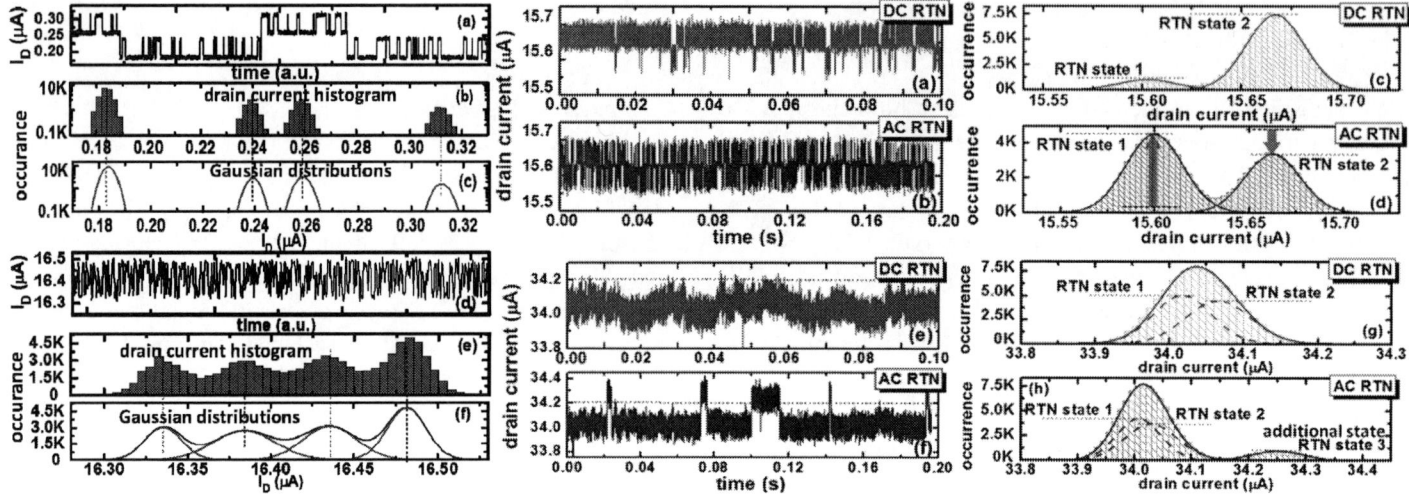

Fig.1. (a)(b)(d)&(e) Device currents and the histogram plots. (c)&(f) Current-variation or RTN-state extraction by Gaussian Mixed Model based on EM algorithm.

Fig.2. (a)~(d) and (e)~(h) are two sets of experimental results. Under AC operating condition (b), RTN time constants can be largely different from the DC case (a). AC signals can also stimulate additional RTN state (f) compared to (e), indicating more traps are activated.

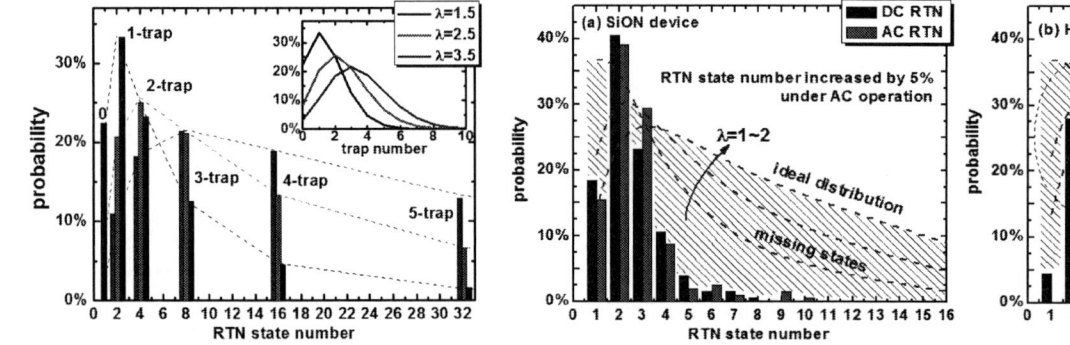

Fig.3. Ideal RTN state distribution follows discrete distribution with peaks at 2^n positions, where trap number n follows Poisson distribution as inset figure.

Fig.4. Measured RTN-state distributions show sharp trends for both SiON and HKMG devices. No discrete peaks (like Fig. 3) are observed. Lots of RTN states predicted by conventional theory are lost under both DC and AC operations. HKMG devices show more states, indicating more traps per device, as expected.

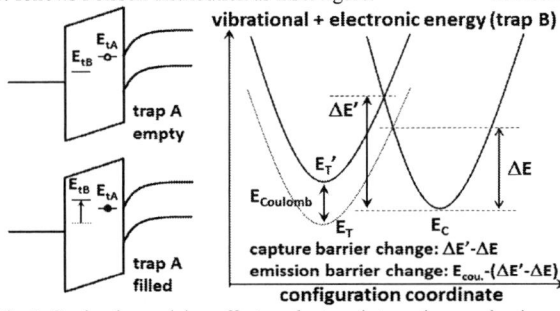

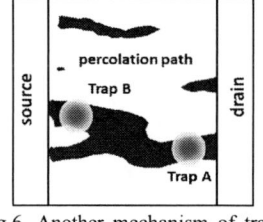

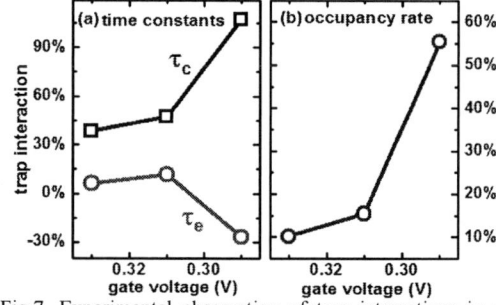

Fig.5. Coulomb repulsion effect as the trap interaction mechanism: capture of trap A changes capture & emission barrier of trap B due to additional Coulomb barrier.

Fig.6. Another mechanism of trap interaction behaviors (by atomistic simulation): traps on the same percolation path will have different and unbalanced impact on each other.

Fig.7. Experimental observation of trap interactions in a 4-sate RTN case-study: capture of one trap degrades the other's occupancy rate by 10%~60%.

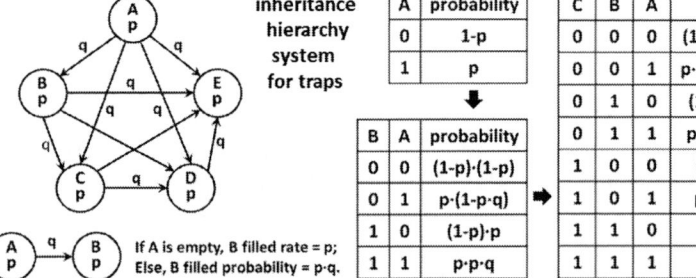

Fig.8. Inheritance hierarchy system model for traps (model HS): multiple traps are considered to be interacting with a hierarchical priority map (left figure). If one higher-level trap (A) is trapped, the occupancy rate of the lower-level ones (B, C, D, E) adopts a degeneration rate of q. Lower q means stronger interactions. Lower-level trap inherits all impact from higher ones. Right charts show an example of occupancy probability for a three-trap system.

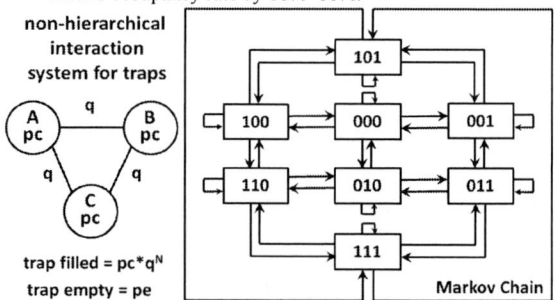

Fig.9. Non-hierarchical interaction system for traps (model NHS): each trap is considered to be equal and share the mutual influence. The RTN states transition map could be described by Markov Chain as shown on right. The filled traps (number of N) give a q^N impact on all other ones.

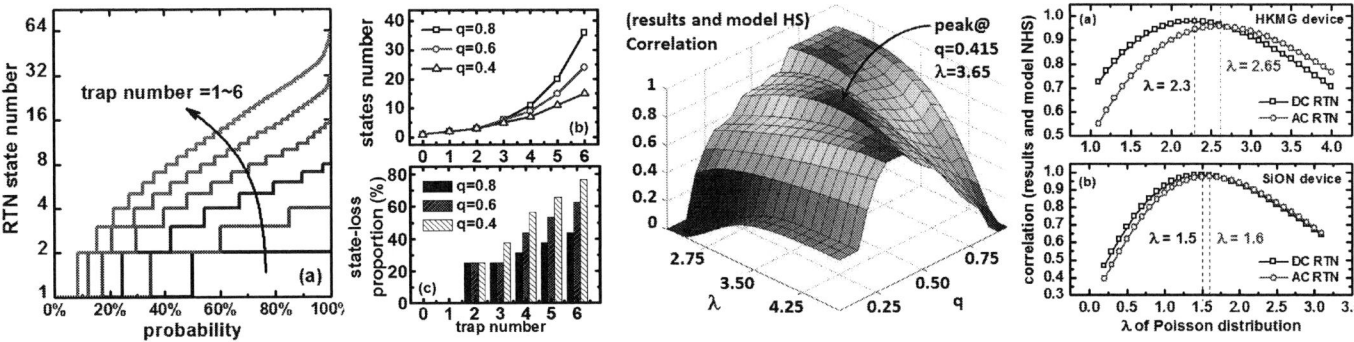

Fig.10. (a) Practical statistics of trap-number and RTN state-number relationships calculated by model HS. (b) The stronger interaction between traps (lower q), the less states will be observed. (c) The proportion of the RTN state-loss.

Fig.11. Correlation between model HS and experimental results as a function of q and λ: for a particular set of results, a medium degeneration rate of q is found.

Fig.12. Correlation between model NHS and experimental results as a function of λ: in AC RTN case, average trap number is increased.

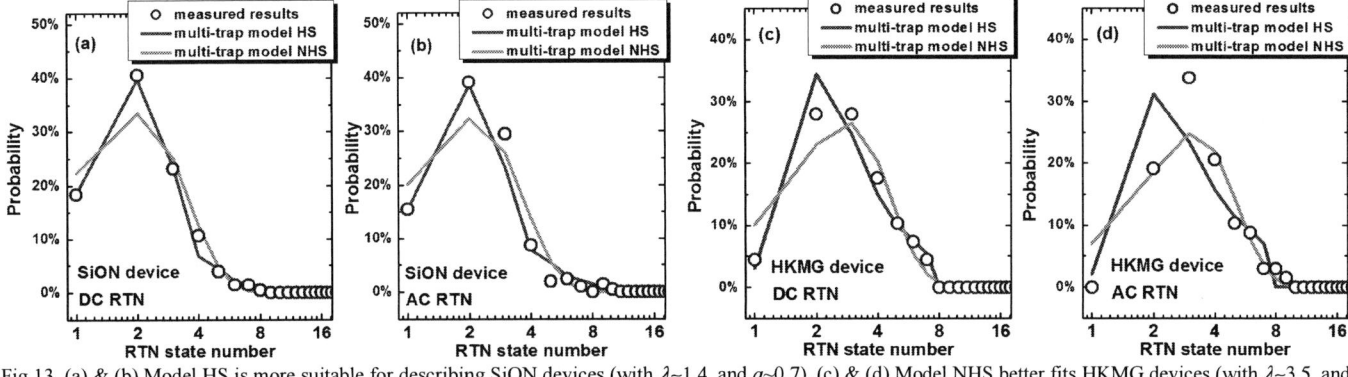

Fig.13. (a) & (b) Model HS is more suitable for describing SiON devices (with $\lambda \sim 1.4$ and $q \sim 0.7$). (c) & (d) Model NHS better fits HKMG devices (with $\lambda \sim 3.5$ and $q \sim 0.3$), which means more traps & stronger trap interactions. Note that, λ and q are self-consistently solved and have unique solution in each model.

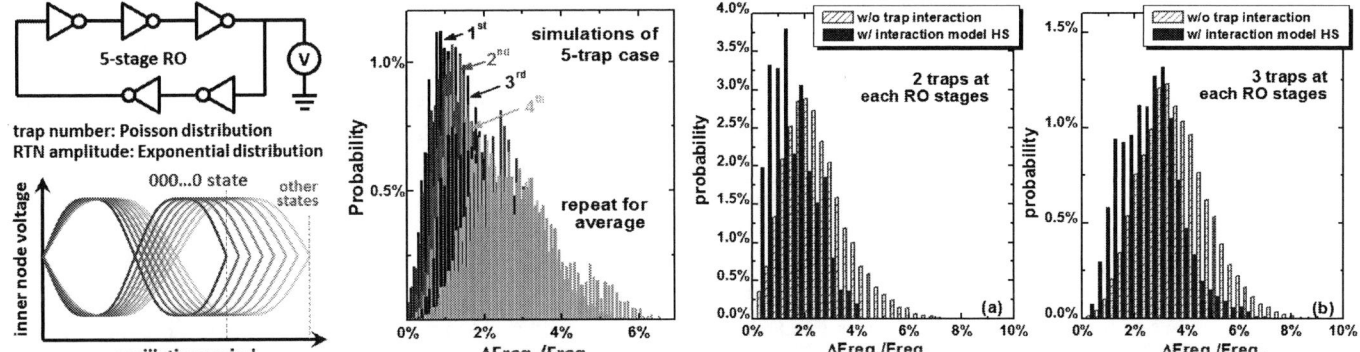

Fig.14. 5-stage RO & RTN-induced jitter noise schematics. Practical trap filling combinations are applied.

Fig.15. ΔV_{TH} follows exponential distribution. Simulations are repeated many times to average f degradation.

Fig.16. 2-trap & 3-trap at each stage of RO cases: w/o consideration of trap interaction, f degradation and jitter variation is larger; while w/ trap interaction case shows non-Gaussian distributions.

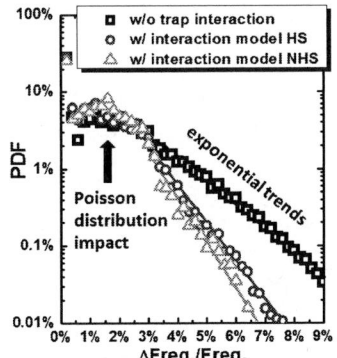

Fig.17. With trap number following Poisson distribution & ΔV_{TH} as exponential distribution, PDF of f degradation can be obtained.

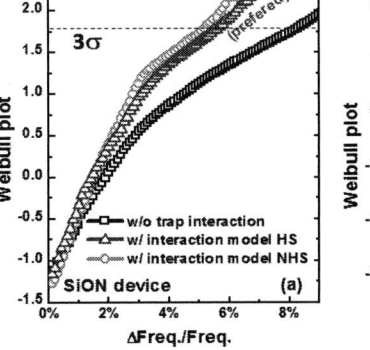

Fig.18. Weibull plot of f degradations for (a) SiON and (b) HKMG cases. Compared to w/o trap interaction, trap interaction models shows less degradation due to that many possible RTN states are concealed by trap interaction behaviors.

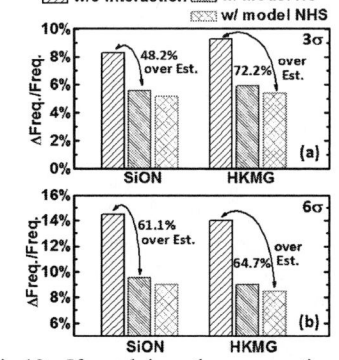

Fig.19. If applying the conventional prediction (w/o interaction), there is an overestimation of 48%~72% for RO frequency variation.

978-1-4799-8002-4/14 $31.00 © 2014 IEEE   835

New Observations on Hot Carrier induced Dynamic Variation in Nano-scaled SiON/Poly, HK/MG and FinFET devices based on On-the-fly HCI Technique: The Role of Single Trap induced Degradation

Changze Liu[*], Kyong Taek Lee, Sangwoo Pae and Jongwoo Park

Technology Reliability, System LSI division, Samsung Electronics Co. Ltd., Yongin-City, Gyeonggi-Do, Korea 446-771
Phone: 82-31-209-4566, Fax: 82-31-209-4132, [*]E-mail: changze.liu@samsung.com

Abstract

In this paper, HCI induced dynamic variation in nano-scaled MOSFETs is systematically studied. Based on the proposed on-the-fly HCI technique, individual defect related HCI variation in small area device is observed for the first time. The fundamental properties of HCI variation sources (single trap induced degradation and trap number) are further investigated. The results show universal scaling trend for all the SiON/Poly, HK/MG and FinFET devices which confirms that the device dimension scaling is the dominant factor for the enhanced individual trap effect. Based on the new observations, HCI variation model is further discussed for the accurate prediction for design. Moreover, HCI variation is compared with BTI and RTN in terms of individual trap. The results show that HCI effect has the largest single trap impacts, which implies the defects responsible for HCI could be closer to dielectric-silicon interface than that for BTI and RTN.

Introduction

In nano-scaled MOSFETs nowadays, variability is considered as a critical issue that all kinds of variations together shaving away the design margin [1]. Recently, dynamic variation caused by NBTI has attracted increasing attention due to its impacts dramatically increase with the device dimension scaling down [2-8]. Moreover, this NBTI variation is further found to be originated from a handful of individual defects, which has a stochastic nature in the atomic level [3]. On the other hand, same as NBTI effect, the channel hot carrier injection (HCI) effect is also a serious reliability issue, especially in NMOS devices. A nature question could arise that whether HCI degradation also shows the fundamental random behavior due to individual traps as the cold carrier induced NBTI, which implies the origin of HCI induced variation in nano-scaled devices. However, the above question has not been investigated so far. In this paper, based on the proposed on-the-fly HCI technique, the fundamental variation sources of HCI variability are experimentally investigated for the first time. The statistical distribution and scaling trend of single trap induced HCI degradation are further extracted and compared in nano-scaled SiON/Poly, HK/MG and FinFET devices. Based on the new observations, the variation model for HCI variability is also discussed as well as the comparison of HCI, BTI and RTN variations in terms of individual defects.

Device and Characterization Method

In this work, small-scaled SiON/Poly, HK/MG and FinFET devices are adopted for the comparison of HCI variations between different processes. **Fig.1** shows the measurement schematics for HCI characterization in this work, which combines the measurement stress measurement (MSM) with on-the-fly HCI methods. For the exclusion of RTN effects, an I_d-t sampling test is performed before and after stress in the MSM stage as that in NBTI test [6]. From the results of MSM test shown in **Fig.2**, no clear recovery is observed after HCI stress, which is different from the single trap induced step-like recovery in NBTI. Due to the lack of detail information, on-the-fly HCI measurement is also adopted here to check the impacts of single trap induced degradation (STID). **Fig.3** shows the results of on-the-fly HCI. Large device shows the continuous degradation as expected, while step-like degradation related to individual defect can be clearly seen in small area device of all three processes, which confirms the enhancement of stochastic mechanism in HCI degradation. As shown in **Fig.4**, HCI degradation by using MSM method and on-the-fly method are further compared, which shows a clear linear correlation. Due to the larger current enhances the Coulomb screening effect of STID, smaller degradation is found in on-the-fly method. In the following part, this factor will be taken into account when calculating the STID. **Fig.5** further shows the average line of all the on-the-fly traces (for all the same devices on the wafer) in the small area devices, which confirms that the intrinsic HCI effect is the same while the large variation is the main property for nano-scaled device HCI.

Results and Mechanism Analysis of HCI Variation

A. Variation sources for nano-scaled device HCI variation: Based on the results of on-the-fly HCI traces, STID (RTN related trap is not included) can be further extracted. As shown in **Fig.6**, the STID distribution follows the Weibull distribution with Weibull slope $\beta>2$. Although the mean HCI degradation is tripled from stress bias 1.7V to 1.9V, the STID distribution does not show an evident stress bias dependence (mean value of STID increases only 14%) especially for large degradation region. This indicates that the STID in HCI degradation also has similar intrinsic

978-1-4799-8002-4/14 $31.00 © 2014 IEEE

behavior which is mainly determined by the device dimension as the case of NBTI [3] and RTN [9]. This can be further confirmed by the non-correlated single trap capture time (the time to find step-like jump) and the STID (**Fig.7**). Increasing the stress bias, the trap generation rate can be increased, which results in a reduced capture time (**Fig.8**), yet the average STID still keeps the same value. For trap number distribution at the fixed stress time, the trap number per device is found to agree with Poisson distribution, as shown in **Fig.9**. In addition, the average trap number under different stress bias is further extracted (**Fig.10**), which shows an increase of 130%. The above results confirm that for nano-scaled device HCI effect, STID performed like a relatively fixed value and the increased HCI degradation by longer stress time and higher bias is mainly attributive to more and faster increasing of the trap number. For accurate prediction of HCI variation, device dimension and process dependences of the above two variation sources are further investigated. **Fig. 11** (planar+SiON/Poly) and **Fig. 12** (planar+HK/MG) show the scaling trend of STID distribution. Larger STID mean value and smaller Weibull slope (means the variation of STID is larger) can be found in the smaller devices. The STID mean value shows a universal 1/WL trend that the same model line can fit well with all the data based on SiON/Poly, HK/MG and FinFET devices, as shown in **Fig.13**. This confirms that, fundamentally, the STID of HCI follows the ratio of defect charge screening length over device dimension to the first order, as that was found in RTN effect [9]. Seen from **Fig. 14**, Weibull slope scaling trend is further extracted (to consider the different stress condition for different processes, the results are normalized to the effective lateral field), which does not depend on processes either. Moreover, the average trap number N shows to be proportional to WL for all devices, shown in **Fig.15**. Only about 2 traps are responsible for HCI in the smallest device studied in this work.

B. Modeling of HCI variation: Due to the similar variation sources, HCI variation model can be also built based on the frameworks for NBTI or RTN. However, the existed compact model [2-3] uses the exponential (for STID) + Poisson distribution (for N) framework (Eq.3), which cannot be right for HCI variation since its STID follows the Weibull distribution. For simplicity, a correction factor $f(\beta)$ related to Weibull slope is considered here (Eq.4). To verify this model, HCI variations with different device area and bias conditions are further tested here. **Fig.16** shows the HCI variation with varying stress biases in SiON/Poly device while **Fig.17** and **Fig.18** show the device area dependence in planar SiON/Poly and HK/MG devices respectively. $f(\beta)$ is further extracted from the data, which shows a value of 0.1 to 0.3 for all conditions. The results indicate that the model may overestimate the HCI variation by several times if not considering the Weibull distribution related factor. Furthermore, $f(\beta)$ is found to have a trend of $f(\beta) \sim 1/\beta$ empirically, as shown in **Fig. 19**. As the β value decreases to 1 (Weibull distribution evolves into exponential distribution in this case), $f(\beta)$ also approaches to 1, which further confirms that the assumption of $f(\beta)$ is valid for the approximate calculation empirically.

Comparison between HCI, BTI and RTN Variations

In nano-scaled devices, HCI, BTI and RTN are all induced by the individual traps. However, the defects related to these mechanisms could be different from each other. For the further investigation of this difference, the STID distributions of the above effects are further compared, noting that the STID is found to be not sensitive to bias, temperature and time in BTI [3-4] and HCI variation. Using the BTI MSM method shown in **Fig.20**, STID is then extracted from the steps in the recovery traces (**Fig.21**) in all same devices. The comparison of STID distributions in HCI (NMOS), PBTI (NMOS), RTN (NMOS) and NBTI (PMOS) effects for HK/MG devices is shown in **Fig.22**. The results indicate HCI > NBTI > PBTI ≈ RTN in terms of single trap, which implies the location of the defect related to these effects since larger impacts could be related to the defect charge closer to the channel, according to Coulomb scattering theory. A possible location could be HCI trap at the interface, NBTI trap in the interfacial layer (IL) and PBTI / RTN trap in the bulk HK layer. For multiple-trap complex RTN, the impact of HCI stress should be also taken into account. Therefore, **Fig.23** and **Fig.24** further plots the RTN amplitude (worst impact is considered for multiple trap RTN) before and after HCI stress for SiON/Poly and HK/MG devices. RTN amplitude distribution does not increase obviously due to HCI stress. In addition, **Fig.25** compares the scaling trend of HCI and RTN variations in NMOS device. HCI variation shows larger value and slope (scaling factor with area) which confirms the results of the larger STID distribution in HCI effect.

Summary

In this paper, using on-the-fly HCI technique, single defect induced HCI variation in small area device is observed for the first time. The variation sources (STID and average trap number) show universal scaling trend for SiON/Poly, HK/MG and FinFET devices, which indicates that new material and structure do not change the fundamental behavior of HCI variation for the first order. In addition, HCI variation model is further discussed based on the Weibull (STID) + Possion (trap number) distribution assumptions. Interestingly, the STID distribution of HCI shows larger impacts than BTI and RTN, which implies that the defects responsible for HCI could be different and may locate more closely to the dielectric-silicon interface.

References:

[1] K.V. Aadithya et. al., Tran. on CADICS ,p. 73, 2013; [2] C. Prasad et.al., *IEEE IRPS*, 6A.5, 2014; [3] B. Kaczer, et al., *IEEE EDL*, p. 411, 2010; [4] B. Kaczer, et al., *IIRW*, p.69 2013; [5] T. Grasser, et al., *IEEE IEDM*, p. 82, 2010; [6] C. Liu et.al., *IEEE IEDM*, p.571, 2011 [7] C. Liu et.al., *IEEE IEDM*, p.466, 2012 [8] M. Duan, et al., *IEEE IEDM*, p.774, 2013.[9] H. Miki et.al., *IEEE IEDM*, p.450, 2012

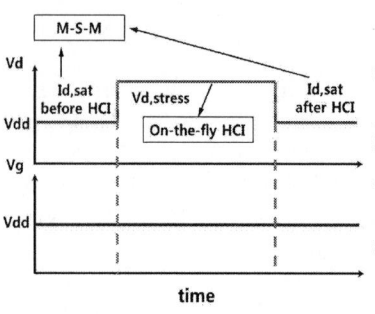

Fig. 1. The measurement schematics for HCI characterization in this work, which combines the measurement - stress - measurement (MSM) method with on-the-fly method.

Fig. 2. The experimental results of the MSM method in HK/MG planar device, a) for the device does not have RTN, b) for the device has RTN at time=0. The definition of HCI degradation is also shown here. After HCI stress no clear recovery was found even the recovery time is over 200s.

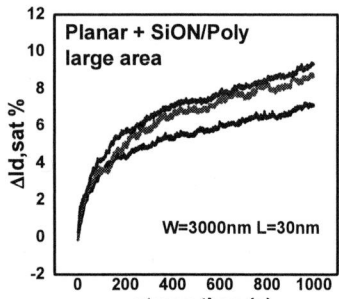 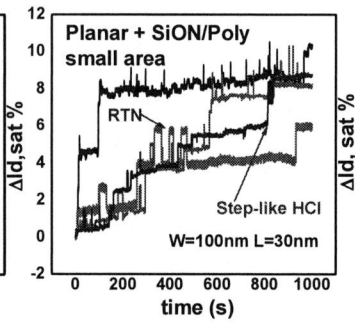 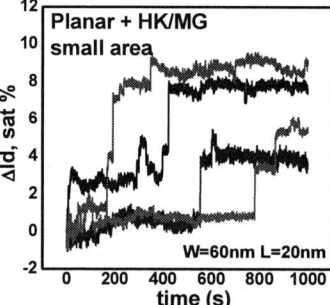 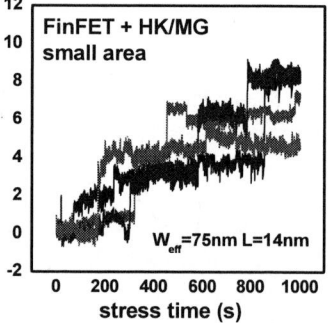

Fig. 3. The experimental results of on-the-fly HCI method for large area planar SiON/Poly device, small area planar SiON/Poly device , small area planar HK/MG device and small area FinFET HK/MG device. The large area device shows the conventional continuous HCI degradation. The small area devices show the step-like degradation due to the enhanced single trap impacts. New gate material and new device structure show the similar properties for HCI in nano-scaled devices. The RTN effect can be also observed in the results which will be excluded when extraction the single trap induced HCI.

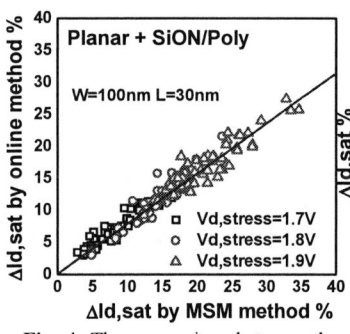 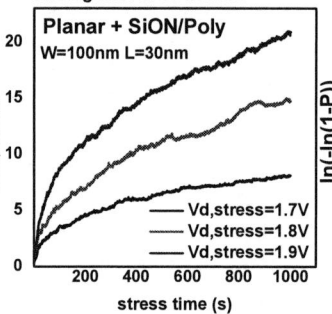 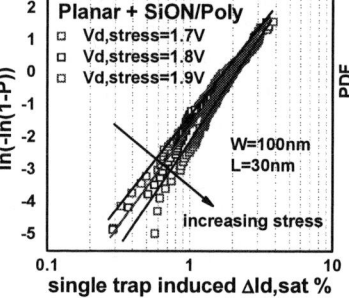 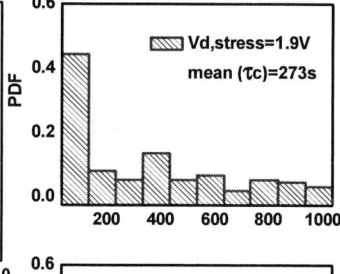

Fig. 4. The comparison between the results of MSM method and that of on-the-fly HCI method. Clear correlation can be found here. On-the-fly method shows smaller degradation.

Fig. 5. The average line of the on-the-fly HCI results. Each curve is averaged based on the same devices in the whole wafer (over 60 devices for each condition).

Fig. 6. The distribution of single trap induced degradation (STID) in planar SiON/Poly devices under different stress biases.

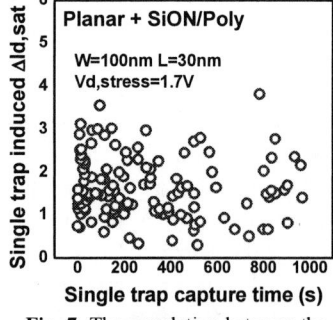 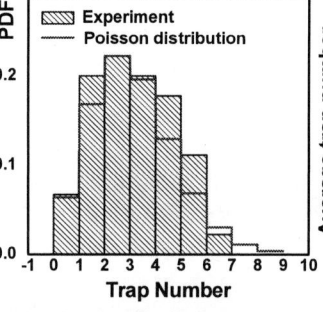 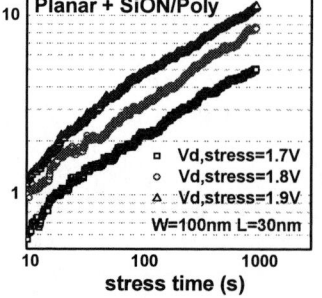 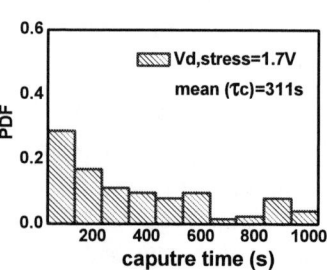

Fig. 7. The correlation between the amplitude of STID and the capture time for the same trap. No clear correlation was found here.

Fig. 9. The distribution of trap number in a single device at a fixed stress time. The result shows good agreement with the Poisson distribution.

Fig. 10. The extracted average trap number in a single device as a function of stress time. More trap was generated under higher stress.

Fig. 8. The distribution of capture time of single trap in HCI effect with the varying stress biases. Shorter mean value was found under higher stress.

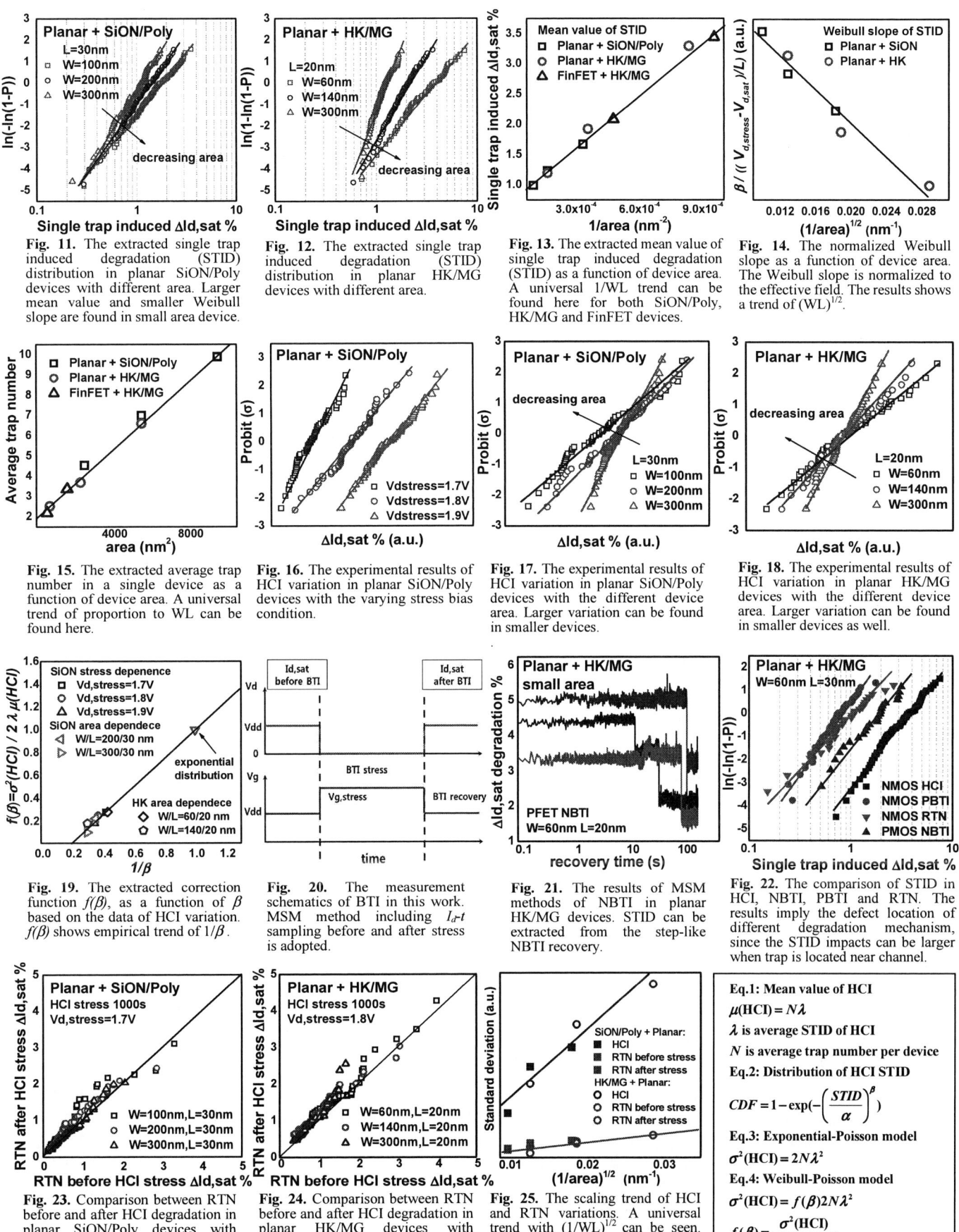

Fig. 11. The extracted single trap induced degradation (STID) distribution in planar SiON/Poly devices with different area. Larger mean value and smaller Weibull slope are found in small area device.

Fig. 12. The extracted single trap induced degradation (STID) distribution in planar HK/MG devices with different area.

Fig. 13. The extracted mean value of single trap induced degradation (STID) as a function of device area. A universal 1/WL trend can be found here for both SiON/Poly, HK/MG and FinFET devices.

Fig. 14. The normalized Weibull slope as a function of device area. The Weibull slope is normalized to the effective field. The results shows a trend of $(WL)^{1/2}$.

Fig. 15. The extracted average trap number in a single device as a function of device area. A universal trend of proportion to WL can be found here.

Fig. 16. The experimental results of HCI variation in planar SiON/Poly devices with the varying stress bias condition.

Fig. 17. The experimental results of HCI variation in planar SiON/Poly devices with the different device area. Larger variation can be found in smaller devices.

Fig. 18. The experimental results of HCI variation in planar HK/MG devices with the different device area. Larger variation can be found in smaller devices as well.

Fig. 19. The extracted correction function $f(\beta)$, as a function of β based on the data of HCI variation. $f(\beta)$ shows empirical trend of $1/\beta$.

Fig. 20. The measurement schematics of BTI in this work. MSM method including I_d-t sampling before and after stress is adopted.

Fig. 21. The results of MSM methods of NBTI in planar HK/MG devices. STID can be extracted from the step-like NBTI recovery.

Fig. 22. The comparison of STID in HCI, NBTI, PBTI and RTN. The results imply the defect location of different degradation mechanism, since the STID impacts can be larger when trap is located near channel.

Fig. 23. Comparison between RTN before and after HCI degradation in planar SiON/Poly devices with different area. RTN amplitude distribution keeps the same range.

Fig. 24. Comparison between RTN before and after HCI degradation in planar HK/MG devices with different area.

Fig. 25. The scaling trend of HCI and RTN variations. A universal trend with $(1/WL)^{1/2}$ can be seen. HCI shows to be dominant in NMOS device dynamic variation.

Eq.1: Mean value of HCI

$\mu(HCI) = N\lambda$

λ is average STID of HCI

N is average trap number per device

Eq.2: Distribution of HCI STID

$$CDF = 1 - \exp(-\left(\frac{STID}{\alpha}\right)^{\beta})$$

Eq.3: Exponential-Poisson model

$\sigma^2(HCI) = 2N\lambda^2$

Eq.4: Weibull-Poisson model

$\sigma^2(HCI) = f(\beta)2N\lambda^2$

$f(\beta) = \dfrac{\sigma^2(HCI)}{2\lambda \mu(HCI)}$

Multiple Breakdown Phenomena and Modeling for Non-Uniform Dielectric Systems

Ernest Wu, Baozhen Li, James. H. Stathis*, and Ravi Achanta**

IBM Co. SRDC, Essex Junction, VT, USA; * IBM Research Div. Yorktown Height, NY, USA; **IBM SRDC, Hopewell Junction, NY, USA

Abstract

We report a wide range of experimental observations of multiple breakdown (BD) phenomena in BEOL/FEOL/MOL dielectric systems with large variability (non-uniformity). Newly developed successive breakdown theory of time-dependent clustering model can well capture these multiple BD events with and without correlation. The understanding of these effects can potentially lead to much improved and realistic projection for future technology nodes.

Introduction

It has been recently reported that dielectric breakdown with high variability exhibits non-uniform characteristics so that classical Weibull/Poisson approach is no longer valid for BEOL /MOL/FEOL in modern technology nodes [1-4]. As a result, the use of classic Weibull/Poisson model can yield pessimistic reliability projection [1]. It is shown that this non-uniformity can arise from spatial clustering of BD spots in some regions versus others [5]. In addition, multiple BD modes can also contribute to BD non-uniformity [2,3]. To correctly perform TDDB reliability projection, a time-dependent clustering model with the classic Weibull/Poisson model as a special case ($\alpha \to \infty$) is reported recently [1]. On the other hand, direct observations of multiple BD phenomena in non-uniform dielectric systems such as BEOL/FEOL have not been reported yet. For FEOL gate dielectrics, multiple BD effect in a single-mode with very small positive correlation (0.3-0.5) was reported [6,7]. In this work, we report successive BD theory of time-dependent clustering model and its verification with several sets of experimental data using grouping methodology [8-10]. More importantly, we will present a wide range of direct experimental observations of multiple BD phenomena involving clustering and bimodality effects with or without BD correlation for several dielectric systems. These results provide much needed insights for BD statistics of non-uniform dielectric systems and for future understanding of BD defect clustering and correlation effects.

Multiple Breakdown Phenomena and Modeling

1). Successive Breakdown Theory of Clustering Model

Table 1 summarizes the equations of successive BD theory of time-dependent clustering model (Eq. 1) and Weibull/Poisson model (Eq. 1a). A comparison of two models is given in Fig. 1 without correlation. At low percentiles, the differences at small K values between various α values are small since the differences between two models are relatively small. For K$\geq$2, a cross-over effect occurs from high to low percentiles. As expected, at high percentiles smaller α values tend to cause failure rate to saturate towards long times [1]. At low percentiles, this trend reverses as seen in Fig. 1. At a fixed time within a K value, the BD probability for small α is much higher than those of higher α values. These differences

become increasingly larger towards larger K values. Based on a phenomenological approach using a macroscopic Markov technique [6,7], the successive BD theory for time-dependent clustering model can be generalized to include the correlation effect as given in Eq. (3). Nevertheless, this integral can only be numerically solved due to the lack of analytic solutions as we will show the results in comparison with T_{BD}/V_{BD} data.

Table 1. Model Equations

Reliability function of successive BD for clustering model:

$$R_i(\lambda) = \frac{\Gamma(i+\alpha)}{i!\Gamma(\alpha)} \frac{(\lambda/\alpha)^i}{(1+(\lambda/\alpha))^{i+\alpha}} \quad (1) \qquad R_i(\lambda) \underset{\alpha \to \infty}{=} \frac{\lambda^i e^{-\lambda}}{i!} \quad (1a)$$

where α is the clustering factor and $\lambda \equiv (t/t_{63})^\beta$

Reliability function of successive BD for grouping n devices [8]:

$$R^G(j,n) = \left(\frac{n!}{j!(n-j)!}\right) R_0(\lambda)^j (1-R_0(\lambda))^{n-j} \quad (2)$$

Reliability function of successive BD with correlation [6]:

$$R_{i+1}(\lambda, \eta) = k_i \int_0^\lambda R_i(\lambda') \exp[-k_{i+1}(\lambda-\lambda')]d\lambda' \quad (3)$$

where

$$k_i = 1 + i\eta \quad (3a)$$

and η is the correlation parameter.

CDF of successive BD theory:

$$F_K(t) = 1 - \sum_{i=0}^{K-1} R_i(\lambda) \quad (4)$$

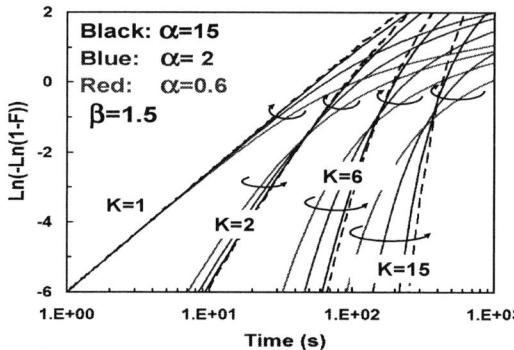

Fig. 1. Theoretical successive BD events of clustering model with different α values and Weibull/Poisson model (dashed lines).

2). Multiple BD events via grouping method

For model verification, it is useful to compare theoretical results with T_{BD} or V_{BD} data using grouping method [8-10]. The comparison of successive BD theory (Eqs. 1 and 2) with grouping experimental V_{BD} data is given in Fig. 2 for poly gate (PC) and diffusion-contact (CA) of mid-of-line (MOL) with α=0.7 and T_{BD} data of process-damaged oxides [11] with α=0.5. Note that the same set of parameters was used for both 1st BD distribution (Figs. 2a and 2c) are used to generate successive BD distributions (Figs. 2b and 2d) which are in good agreement with experimental data and validate the clustering model. In grouping methodology, the binomial selection statistics should be properly considered [8]. Fig. 3a shows the effect of K_{max} (10 vs. 100) on successive BD events

978-1-4799-8002-4/14 $31.00 © 2014 IEEE

for Weibull/Poisson model (Eq. 1a) in which the grouping effect diminishes for $K_{max}=100$ [8]. In contrast, the successive BD events of clustering model with increasing K_{max} do not converge to those of Eq. 1, but rather converge to those of Eq. 1a as shown in Fig. 3b.

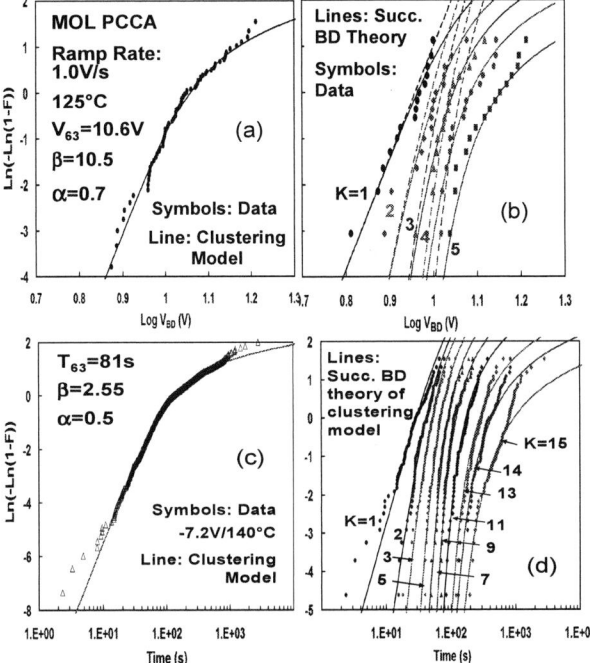

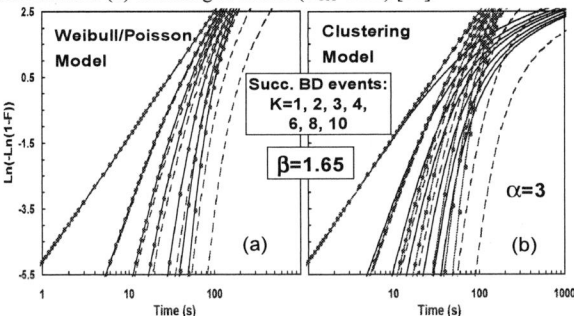

Fig. 2. MOL PC-CA V_{BD} data with its original distribution (a) and the grouped distributions (b) in comparison with successive BD theory of clustering model. Similarly, the original distribution (c) and grouped distributions (d) of damaged oxides ($T_{OX}=62$Å) [11].

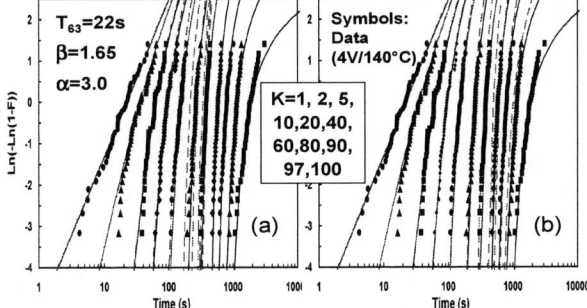

Fig. 3. Theoretical results of successive BD theory. Black lines in (a) and (b) are the results of Eq. 1a and Eq. 1. The dashed blue lines in (a) and (b) are the successive BD distributions of Eq. 2 [8] in respective combination with Eq. 1a and Eq. 1 for $K_{max}=10$ while the red dots are for $K_{max}=100$. The red lines in (b) are the same as the black lines of (a).

Fig. 4. Comparison of grouped T_{BD} data ($T_{OX}=27$Å) with successive BD theory ($K_{max}=100$) of clustering model (black lines). Red dashed lines: The use of Eq. 1a in (a) and the use of Eq. 1a and Eq. 2 in (b).

This is because the grouping method ($K_{max}=100$) yields an effectively larger area ($A'=K_{max}A_0$). This is consistent with the fact that larger area represents the low-percentile portion of the distribution in which clustering effect is expected to vanish. These predictions are confirmed experimentally as seen in Fig. 4a. Nevertheless, the differences remain between Weibull/Poisson model and clustering model at higher K values as seen in Figs. 4, revealing the usefulness of grouping method to distinguish different models in comparison with experimental data.

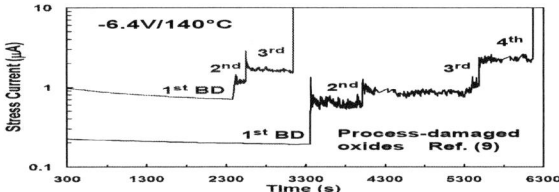

Fig. 5. Current transients of process-damaged oxides [11] similar to Figs. 2c and 2d showing several jumps as multiple BD events

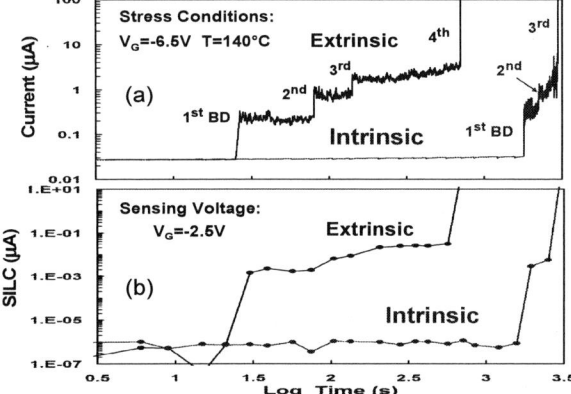

Fig. 6. Time-evolution of stress currents (a) and stress-induced leakage currents (SILC) (b) for a bimodal BD distribution with extrinsic and intrinsic modes in competition within the same samples. The stair-case plateaus indicate the occurrence of multiple BD events.

3). Direct observation and modeling of multiple BD events

The grouping method is an indirect demonstration of successive BD events and useful for verification of theoretical models [8-10]. In this method, the successive BD events are taken from first BD events. Naturally, one would not expect any correlation effect. To directly measure multiple BD events, we need to directly determine multiple BD events from I-t traces as shown in Figs. 5 and 6 for SiO_2 gate oxides. All these current traces reveal several stair-case jumps in currents, indicating the occurrence of multiple BD events. Fig. 7 shows that measured multiple BD distributions corresponding to those I-t traces in Fig. 5 in agreement with successive BD theory of time-dependent clustering model while Weibull/Poisson model does not give a good fit to the data. Note the modeling parameters are consistent with those used in the grouping experiments (Figs. 2c and 2d). In Fig. 8, we show the measured 1st and 2nd BD distributions of oxides showing bimodal behavior. In this case, the non-uniformity arises from the competition of extrinsic and intrinsic modes within same samples as modeled according to the following equation:

$$F_K(t) = 1 - (1 - F_{K,Ext})(1 - F_{K,Int}). \qquad (5)$$

978-1-4799-8002-4/14 $31.00 © 2014 IEEE   841

The modeling results without correlation show a large discrepancy with experimental data for 2nd BD events by ~10X in time for extrinsic mode. A positive correlation parameter of η=2 in successive BD theory (Eq. 2) for both extrinsic and intrinsic modes is required to achieve an agreement with experimental data. This contrasts with previous report of a very small correlation factor of 0.3-0.5 for single intrinsic mode [6,7]. This means trap-generation process is highly enhanced with strong correlation [6,7]. In principle in the case of competition between extrinsic and intrinsic modes, once a 1st BD event of extrinsic mode occurs, the 2nd BD statistics of this bimodal system is determined by the competition among 2nd BD of both extrinsic mode and intrinsic mode, but also the contribution from the 1st BD of intrinsic mode. We found the latter contribution is small as compared to that of Eq. 4. Its effect can be compensated by using two different η values for extrinsic and intrinsic modes.

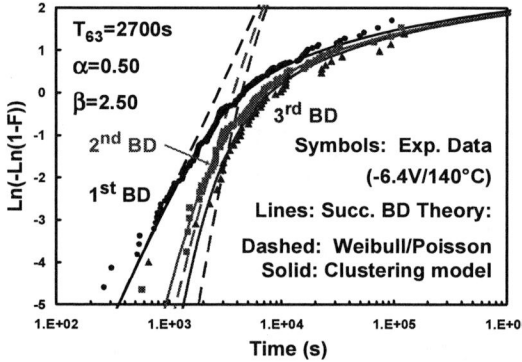

Fig. 7. Multiple BD distributions of process-damaged oxides directly measured from I-t traces in Fig. 5 in comparison with successive BD theory (Eq. 1) without correlation, showing good agreement.

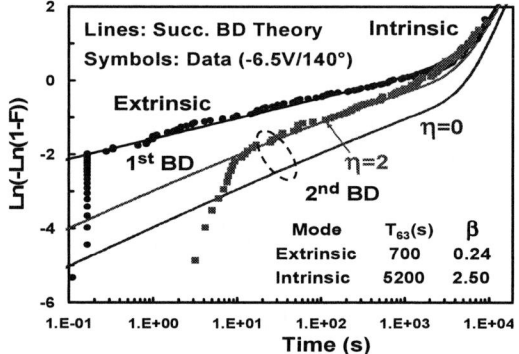

Fig. 8. Measured multiple BD distributions from I-t traces (Fig. 6) in comparison with successive BD theory with bimodal competition (Eqs. 4 and 5), R_K of each mode is from Eq. 1a without clustering ($\alpha \to \infty$). A very large positive correlation (η=2) needed to fit well with the data.

Recently, multiple BD phenomena up to 10th successive BD events were reported for a nanolaminated HfO$_2$/Al$_2$O$_3$ stack insulator by atomic-layer deposition [12]. The corresponding successive BD distributions reveal a bending of failure distributions at longer times (Fig. 9). The authors performed a correlation study indicating that lower initial leakage currents are responsible for the samples with longer T$_{BD}$ [12]. The authors found it is necessary to screen the data to recover the successive BD characteristics of Weibull/Poisson model for low-percentile portion of the data. In Fig. 9, we compare their

multiple BD data with the successive BD theory (Eq. 1) of clustering model. The model can capture both low- and high-percentile portion of these successive BD events without screening any data, thus providing a more natural mathematical framework while classic Weibull/Poisson model fail to explain the data at high-percentiles.

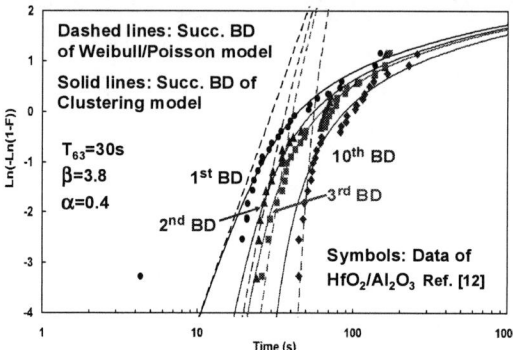

Fig. 9. A comparison of multiple BD data with successive BD theory of time-dependent clustering model for nanolaminated HfO$_2$/Al$_2$O$_3$ stack insulators by ALD deposition.

Figs. 10 show the typical time-evolution of stress currents and stress-induced leakage current (SILC) of BEOL low-κ samples. The stress is interrupted to measure the SILC at low voltage to ensure the detection of successive BD. In this study, we found the 1st and 2nd BD events primarily dominate during the stress. The corresponding T$_{BD}$ data of 1st and 2nd BD events are given in Fig. 11a for BEOL low-κ samples. It can be seen that the successive BD theory (Eq. 1) without correlation cannot explain experimental data for up to 4th BD events as shown in Fig. 11a.

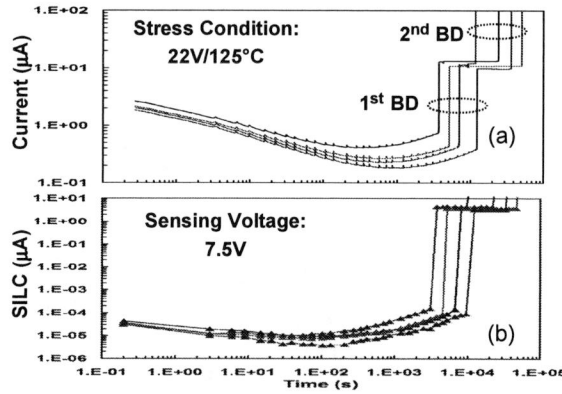

Fig. 10. Time-evolution of stress currents (a) and SILC (b) for some BEOL low-k samples, showing strong trapping behaviors.

In order to explain this discrepancy with the T$_{BD}$ data, we consider several positive values of correlation parameter to compare with the data. The results of successive BD theory with positive correlation further deviate from the experimental data in opposite direction. On the other hand, when a negative correlation parameter (anti-correlation) is introduced, an excellent agreement is achieved between experimental data and successive BD theory (Eq. 2). The results of negative correlation sharply contrast with those of gate dielectrics in which a positive correlation is observed either in single intrinsic BD mode [6,7] or in competing extrinsic and intrinsic

978-1-4799-8002-4/14 $31.00 © 2014 IEEE   842

modes (Fig. 8). This negative correlation may be caused by decreasing rate of BD path generation due to voltage-drop as a result of BD current increase after 1st BD events [7]. To investigate this possibility, we plot a second set of BEOL low-κ T_{BD} data from a different fabrication process but using the same test structures as shown in Fig. 12. In this case, the agreement is excellent between experimental data and successive BD theory of Weibull/Poisson model ($\alpha \rightarrow \infty$) without any correlation effect. This result may suggest that voltage-drop does not play a major role as the root-cause for negative correlation although the root-cause of fabrication process remains unknown at this moment.

One possible cause of this negative correlation effect is that BD propagation effect [13] after first BD events happens rapidly so that some BD precursors might be quenched so that the occurrence of some successive BD spots is suppressed. Since Fig. 10b shows the SILCs remain less than 10μA after 1st BD events for circuit failure [14,15], this means negative-correlation effect can lead to an improved reliability projection. The lifetime enhancement factor of 2nd BD defined in Fig. 13a can be as large as 10^2-10^5 for several β values (Fig. 13b). It should be cautioned that we found for large K values or small α values, non-physical results (*i.e.* F_{Cum} decrease with increasing time) can occur in the case of negative correlation. A new model to replace the linear expression of Eq. (3a) may be needed for general applications of Eq. (3). Our results provide new frontiers for investigation and understanding of BD defect generation process in BEOL/MOL/FEOL dielectrics.

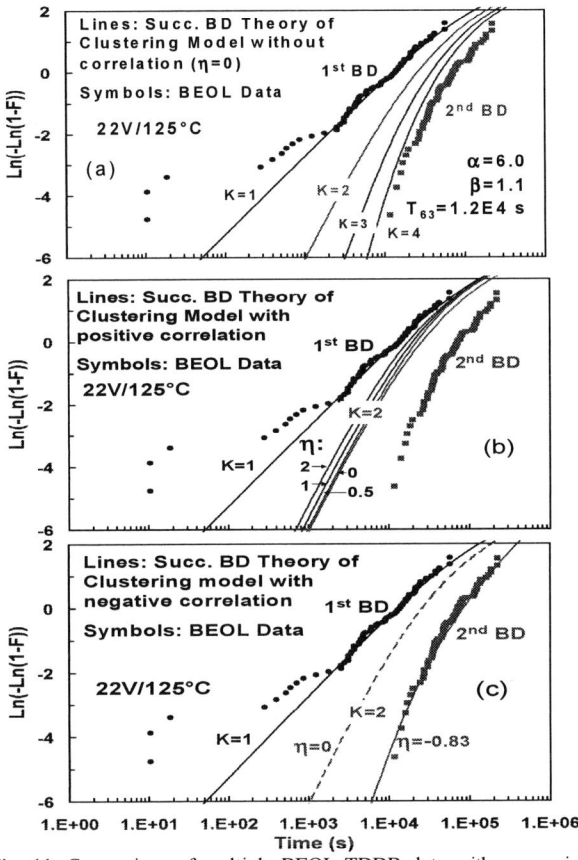

Fig. 11. Comparison of multiple BEOL TDDB data with successive BD theory of clustering model (Eqs. 1 and 2): (a) without any correlation. (b) with positive correlation. (c) with negative correlation.

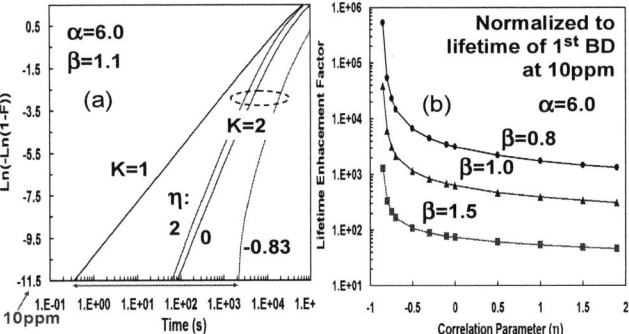

Fig. 13. (a) The 1st and 2nd BD distributions for various η values. (b) Lifetime enhancement factor up to 10^6 vs. correlation parameter (η).

Conclusions

We report a wide variety of multiple BD phenomena for non-uniform dielectric systems (MOL/BEOL/FEOL) as compared to successive BD theory of clustering model. The grouping experiments are shown as the useful tools to distinguish model choices. We found strong positive correlation of bimodal modes of oxides in agreement with successive BD theory. For the first time, a strong negative-correlation is reported for BEOL low-κ dielectrics. Successive BD theory of clustering model can provide realistic and accurate lifetime projection for dielectric systems with high variability in future technology nodes.

References

[1] Wu *et al.* IEDM, p.301, 2013; Wu *et al.* APL,103,152907, 2013.
[2] Stucchi, *et al*, *TDMR*, V.11, p. 278, 2011.
[3] Lee, *et al.* *TDMR*, V.10, p,307, 2010.
[4] Yokogawa *et al.* IRPS, p.149, 2011.
[5] Wu *et al.* VLSI, p.126, 2014.
[6] Alam *et al.* IRPS, p. 406, 2003; IEDM, p. 151, 2002.
[7] Suñé *et al.* IEEE Trans. Elec. Device, V.51, p.1584, 2004.
[8] Suñé, *et al.* IEDM, p.147, 2002
[9] Wu & Suñé, IRPS, p. 6A.2.1, 2012.
[10] Achanta *et al.* IITC, p.219, 2014.
[11] Wu *et al.* IRPS, p. 5.b.2.1, 2014.
[12] Martinez-Domingo, *et al.* Microelec. Eng. V88, p.380, 2011.
[13] Lombardo *et al.* J. Appl. Phys, V.98, p.121301, 1999.
[14] Rodriguez *et al.* IRPS, p.11, 2003.
[15] Stathis, *et al.* Micro. Relia, V.43, p.1193, 2003

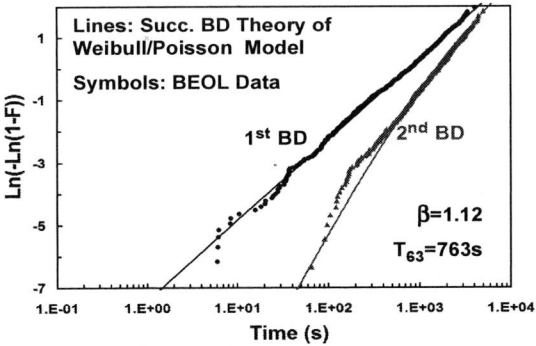

Fig. 12. The 1st and 2nd BD distributions of BEOL low-κ samples from a second set of T_{BD} data with a different fabrication process in comparison with successive BD theory of Weibull/Poisson model without correlation.

A physics-based compact model for FETs from diffusive to ballistic carrier transport regimes

Shaloo Rakheja[1], Mark Lundstrom[2], and Dimitri Antoniadis[1]

[1] Microsystems Technology Laboratories, Massachusetts Institute of Technology, Cambridge MA 02139

[2] Network for Computational Nanotechnology, Purdue University, West Lafayette, IN 47907

Abstract

This paper discusses a new emission-diffusion-based compact model for FETs to describe carrier transport in both short and long channel devices. The new model provides a description of the current at any drain bias without empirical fitting and predicts the injection velocity (device on-current). The new model is fully consistent with the widely used virtual-source model for describing transport in quasi-ballistic transistors. The accuracy of the new model is demonstrated by comparison with measured I-V data of III-V HEMTs and ETSOI silicon MOSFETs.

I.   Introduction

The basic virtual source (VS) model provides a simple, physical description of transistors that operate in the quasi-ballistic regime [1-2]. Through comparisons to experimental data, key device parameters in the VS model can be extracted. However, the VS model suffers from three limitations: i) it is restricted to short channels, ii) the transition between linear and saturation regime is treated empirically, and iii) the injection velocity cannot be predicted, it must be extracted by fitting the model to measured data. The enhanced, VS emission-diffusion (VS ED) model discussed in this paper not only overcomes these limitations of the basic VS model, but is also applicable from drift-diffusive to ballistic transport regimes.

The remainder of the paper is organized as follows. In Section II, the physics of the VSED and VS models is briefly discussed, and the link between the two models is established. In Section III, the VSED model is calibrated with experimental I-V data for III-V HEMTs and Si ETSOI MOSFETs. Section IV concludes the paper.

II.   Virtual Source Emission-Diffusion (VSED) and Basic Virtual Source (VS) Models

The VSED model for MOSFETs[1] is similar to the ED model for Schottky barriers [3] but incorporates MOS electrostatics. The current, I_D, is given as

$$I_D = W Q_n(0) v_T F_{sat,VSED},\qquad(1)$$

$$Q_n(0) = \qquad(2)$$
$$C_{inv} n\phi_t \ln\left(1 + exp\left(\frac{V_{gsi}-(V_{t0}-\delta V_{dsi}-FF\phi_t)}{n\phi_t}\right)\right),$$

$$FF = \frac{1}{1+exp\left(\left(V_{gsi}-(V_{t0}-\delta V_{dsi}-0.5\alpha\phi_t)\right)/n\phi_t\right)},\qquad(3)$$

where W is the device width, v_T is the thermal velocity, $Q_n(0)$ is the charge at the virtual source point or top of the barrier (ToB), V_{gsi} and V_{dsi} are the intrinsic gate-source and drain-source voltages, respectively; C_{inv} is the per-unit-area gate-channel capacitance in strong inversion, V_{t0} is the device threshold voltage, α is an empirical parameter associated with threshold voltage shift in units of thermal voltage, ϕ_t, between strong and weak inversion (typically fixed as 3.5), δ is the drain-induced barrier lowering (DIBL), and $n=n_0+n_d V_{dsi}$ is the non-ideality factor related to sub-threshold slope $S = 2.3n\phi_t$; n_d is the punch-through factor. Eq. (2) for inversion charge captures the channel charge behavior in both strong and weak inversion operation regimes.

The function $F_{sat,VSED}$ in (1) describes the output current transition from the linear regime to the saturation regime and is given as

$$F_{sat,VSED} = \left(\frac{\mathcal{T}}{2-\mathcal{T}}\right)\left(\frac{1-exp\left(-\frac{V_{dsi}}{\phi_t}\right)}{1+\frac{\mathcal{T}}{2-\mathcal{T}}exp\left(-\frac{V_{dsi}}{\phi_t}\right)}\right).\qquad(4)$$

$F_{sat,VSED}$ is physically derived and not an empirical expression as in the VS model. The parameter $\mathcal{T}$ in (4) is the transmission coefficient and is related to the carrier mean free path, λ, and the critical length, l, for scattering at the ToB.

$$\mathcal{T}=\frac{\lambda}{\lambda+l};\ \lambda = 2\frac{\phi_t}{v_T}\mu_{eff};\ v_T = 2\sqrt{\frac{k_B T}{2\pi m^*}},\qquad(5)$$

where μ_{eff} is the carrier mobility in a long-channel device, and m^* is the effective transport mass of carriers in the device (single parabolic band and non-degenerate conditions assumed). The critical length l is a function of the channel potential profile, $V(x)$, along the transport direction x, and is mathematically expressed as

$$l = \int_0^{L_G} e^{-\frac{V(x)-V(0)}{\phi_t}}\,dx,\qquad(6)$$

where L_G is the channel length assumed equal to the gate length. The spatial profile of $V(x)$ is controlled by 2D electrostatics and transport in the channel. If $V(x)$ is known, the model is complete; if not, approximations must be made.

For long channel devices with $L_G > \lambda$, $V(x)$ can be obtained using the drift-diffusion (DD) approach with gradual channel approximation (GCA) [4] as

$$V(x) = V_{gt1}\left(1 - \sqrt{1 - \frac{x}{L_G}(1-\eta^2)}\right),\qquad(7)$$

$$V_{gt1} = \frac{2V_{gt}}{1+\sqrt{\frac{2V_{gt}}{L_G E_{crit}}}},\qquad(8)$$

$$V_{gt} = \frac{Q_n(0)}{C_{inv}},\qquad(9)$$

where E_{crit} is the critical electric field for velocity saturation, and the parameter η is given as

$$\eta = 1 - F_v\left(\frac{V_{dsi}}{V_{gt1}}\right),\qquad(10)$$

---

[1] The derivation of the equations in the VSED model will be presented elsewhere.

$$F_v(x) = \frac{x}{(1+x^\beta)^{\frac{1}{\beta}}}, \qquad (11)$$

where β is an empirical parameter[2]. The corresponding critical length in the DD regime is given as

$$l(DD) = \frac{2L_G}{\left(\frac{V_{gt1}}{\phi_t}\right)^2 (1-\eta^2)} \left(\exp\left(\frac{V_{gt1}}{\phi_t}(\eta - 1)\right) \left(1 - \frac{V_{gt1}}{\phi_t}\eta\right) - \left(1 - \frac{V_{gt1}}{\phi_t}\right) \right). \qquad (12)$$

For short channel devices that are more ballistic, we use a simplified potential profile, $V(x) = V_{dsi,eff} x/L_G$, where $V_{dsi,eff}$ is the effective intrinsic drain-source voltage drop. While $V_{dsi,eff}$ is equal to V_{dsi} in the linear regime of transport, numerical simulations as in [5-6] show that V_{dsi} saturates to V_{dsat} for high V_{dsi}. V_{dsieff} can, therefore, be given as

$$V_{dsi,eff} = V_{dsat,VSED} F_v\left(\frac{V_{dsi}}{V_{dsat,VSED}}\right), \qquad (13)$$

$$V_{dsat,VSED} = \theta \phi_t \frac{\lambda}{\lambda+L_G}, \qquad (14)$$

where $F_v(x)$ is given in (11); $V_{dsat,VSED}$ is the saturation voltage in the VSED model; θ is fitted for a given device technology (independent of gate length). The corresponding critical length for ballistic transport is given as

$$l(Ballistic) = L_G \frac{\phi_t}{V_{dsi,eff}} \left(1 - \exp\left(-\frac{V_{dsi,eff}}{\phi_t}\right)\right). \qquad (15)$$

As shown in **Figure 1**, the F_{sat} functions in both the VSED and VS models (Eq. (1) in Ref. [1]) are remarkably similar. The inset plot shows ξ, which is the ratio of the critical length, l, and the gate length, L_G, as a function of V_{ds} normalized to the saturation voltage, $V_{dsat,VSED}$, in the VSED model.

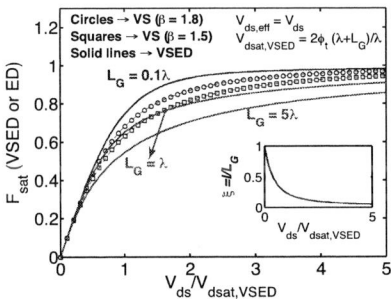

Figure 1: F_{sat} functions either in the VSED or the VS model versus the drain-source bias normalized to the saturation voltage as defined in the VSED model. Two values of β are chosen in the VS model. The inset shows ξ, which is the ratio of the critical length, l, and the gate length, L_G, as a function of $V_{ds}/V_{dsat,VSED}$. As $V_{ds} \to 0$, l approaches L_G (i.e. $\xi \to 1$.)

A comparison of output characteristics obtained from the VSED model and the standard DD model (see Chap. 4 in [3]) for a long-channel device in **Figure 2** demonstrates that the

VSED model is able to reproduce $(V_{gs}-V_t)^2$ relationship in saturation.

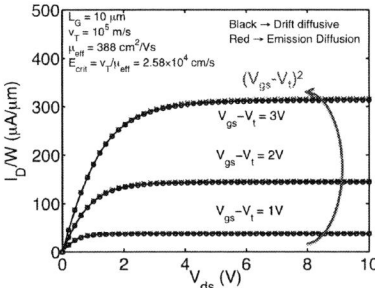

Figure 2: Comparison of output characteristics from the drift diffusion and VSED models and a demonstration of the capability of the VSED model to generate $(V_{gs}-V_t)^2$ relation in saturation for a long-channel device.

a. Link between VSED and VS model

The link between the VS and the VSED models is illustrated in **Figure 3** for the *low-V_{ds}* regime and in **Figure 4** for the *high-V_{ds}* regime. The VS model parameters (mobility and VS injection velocity) that are extracted from calibration with experimental data are related to the physical parameters in the VSED model, all of which are known either experimentally or from band-structure. Thus, the VSED and the VS models are completely consistent with each other, but the VSED model also describes long-channel devices.

Low V_{ds} regime

VSED model
$$I_D = WQ_n(0)\mathcal{T}\frac{v_T}{2\phi_t}V_{ds}$$

$$\mu = L_G \mathcal{T}\frac{v_T}{2\phi_t}$$

$$\mathcal{T}(V_{ds} \to 0) = \frac{\lambda}{\lambda+L_G}$$

$$\mu = \frac{\lambda v_T}{2\phi_t}\left(\frac{L_G}{L_G+\lambda}\right)$$

VS model
$$I_D = \mu\frac{W}{L_G}Q_n(0)V_{ds}$$

$$\mu = \mu_{eff}\left(\frac{L_G}{L_G+\lambda}\right)$$

Figure 3: Low-V_{ds} relationship between the VS and VSED models. The mobility, μ, extracted from VS model is related to the device parameters from the VSED model. It can be seen that the mobility in the VS model scales with gate length. For devices much longer than the MFP, μ becomes limited by the effective mobility, μ_{eff}, which can be experimentally obtained.

High V_{ds} regime

VSED model
$$I_D = WQ_n(0)\left(\frac{\mathcal{T}}{2-\mathcal{T}}\right)v_T$$

$$I_D = WQ_n(0)\left(\frac{\mathcal{T}}{2-\mathcal{T}}\right)v_T$$

$$\mathcal{T} = \frac{\lambda}{\lambda+l} = \frac{\lambda}{\lambda+2\xi L_G}$$

VS model
$$I_D = WQ_n(0)v_{x0}$$

$$v_{x0} = v_T\frac{\lambda}{\lambda+2\xi L_G}$$

Figure 4: High-V_{ds} relationship between the VS and VSED models. The VS velocity, v_{x0}, extracted from the VS model is related to thermal velocity, v_T, MFP, λ, and gate length, L_G. The parameter $\xi = l/L_G << 1$ in high V_{ds} conditions as shown in the inset of Fig. 1. The dependence of ξ on V_{ds} is weak in the high-V_{ds} regime, since l varies slowly with V_{ds} in this voltage regime. ξ is fitted in the VS model.

---

[2] Limiting V_{dsi} using empirical function $F_v(x)$ may be avoided if $V(x)$ is known a priori from self-consistent solution of transport and Poisson's equation.

III. Comparison to Experimental Data

In the VSED model, there are only a few parameters that are fixed for a device class irrespective of the device length. These are the carrier effective mass m^* (or thermal velocity) and experimentally obtained long-channel effective mobility, μ_{eff}. The determination of m^* requires the knowledge of band-structure, strain, and degeneracy. Here, we use a value that best fits all the devices in a given device class. Fitting the VSED model to both short and long channel devices as given below demonstrates that (i) the injection velocity (on-current) can be predicted (and not fitted as in the VS model), and (ii) the model can be applied to both long and short channel devices using an appropriate $V(x)$.

a. Silicon ETSOI MOSFETs

The VSED model is verified against ETSOI MOSFETs with an SOI thickness of 6 nm and a poly-Si/SiN gate stack with an EOT of 1.1 nm ($C_{inv} = 2\ \mu F/cm^2$) fabricated at IBM [8]. The effective mass is taken to be $0.22m_0$, which gives a thermal velocity of 1.14×10^7 cm/s. The measured low-field, long-channel mobility of these devices is 350 cm²/Vs [8], which gives a MFP of 15.8 nm. Fits to the experimental data for the 40 nm and 980 nm devices using the VSED model with critical length as in (12) (corresponding to classical DD $V(x)$) are shown in **Figures 5 and 6**, respectively. The fitting parameters for all of the devices are shown in **Table I**.

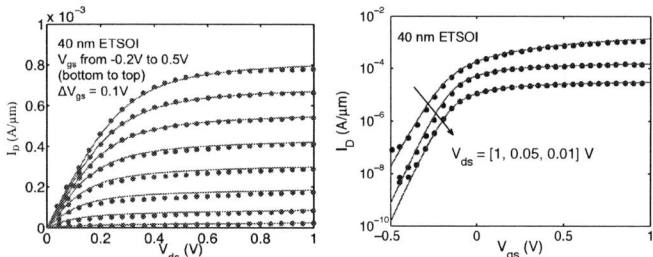

Figure 5: Output (left) and transfer (right) curves for the 40 nm ETSOI MOSFET from [8]. Symbols represent experimental data, while solid lines correspond to VSED model fits.

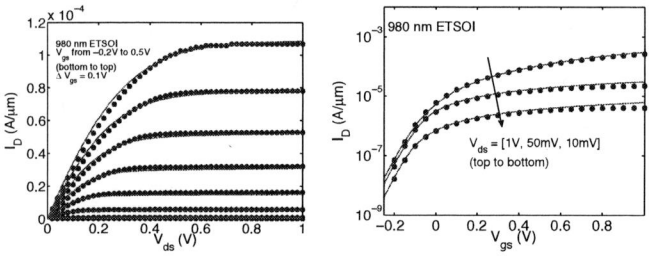

Figure 6: Output (left) and transfer (right) curves for the 980 nm ETSOI MOSFET [data courtesy Dr. Amlan Majumdar, IBM T.J. Watson]. Symbols represent experimental data, while solid lines correspond to VSED model fits.

b. III-V HEMTs

The VSED model is also verified against III-V InGaAs HEMTs with L_G ranging from 30 nm to 130 nm [9]. The effective mass in this set of devices is taken to be $0.022m_0$ [10], which gives a thermal velocity of 3.62×10^7 cm/s. The measured low-field, long-channel mobility is 12,500 cm²/Vs giving a MFP of 153 nm.

To obtain C_{inv} for the HEMT devices, $\partial Q_{inv}/\partial V_g$ obtained by differentiating (2) is fitted to the experimental data on gate capacitance as shown in Figure 7. The values of n_0, α, and V_{t0} are obtained from the transfer curve measurement of the device with $L_G = 30$ nm. A value of $C_{inv} = 1.08\ \mu F/cm^2$ provides a good match with experimental data as shown in the figure below.

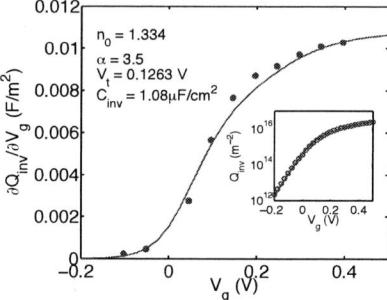

Figure 7: Inversion capacitance versus V_g for III-V HEMTs [9]. Symbols represent experimental data, while solid lines are model fits using (2)-(4). Parameters n_0 and V_{t0} are obtained from transfer curve data of the 30 nm HEMT device. Here, $Q_{inv} = Q_n(0)$. The fit to the inversion charge obtained by integrating the experimental inversion capacitance data is shown in the inset.

As shown in Figures 8 and 9, excellent fit to the I-V data of HEMTs is obtained using the VSED model with critical length as in (15) (corresponding to ballistic $V(x)$). The fitting parameters for all of the devices are shown in **Table I**.

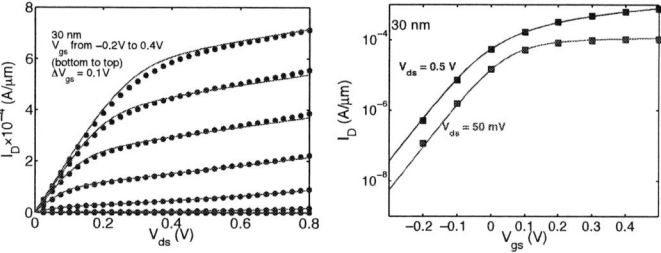

Figure 8: Output (left) and transfer (right) curves for the 30 nm HEMT device from [9]. Symbols represent experimental data, while solid lines correspond to VSED model fits.

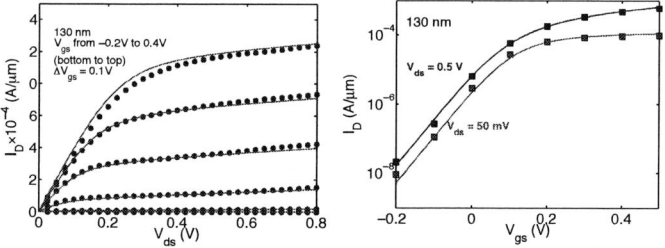

Figure 9: Output (left) and transfer (right) curves for the 130 nm HEMT device from [9]. Symbols represent experimental data, while solid lines correspond to VSED model fits.

The I-V data of HEMTs is also fitted using the basic VS model (fits not shown here). The inverse of the extracted mobility from the VS model is plotted against the inverse of the gate length in **Figure 10**, and the data is fit with a straight line. The y-axis intercept of the plot provides information on effective mobility, μ_{eff}, while the slope of the straight line equals λ/μ_{eff} (see **Figure 3**). From the best-fit line, we find μ_{eff} = 12,195 cm^2/Vs and λ=171 nm, which are very close to the corresponding values in the VSED model. Furthermore, the inverse of the VS velocity extracted from the VS model fits is plotted versus L_G in **Figure 11**. The data points are fit with a straight line, where the y-axis intercept provides information on the thermal velocity (see **Figure 4**). The thermal velocity extracted from this procedure is 3.57×10^7 cm/s, which is very close to the corresponding value in the VSED model, further establishing the equivalence between the two models.

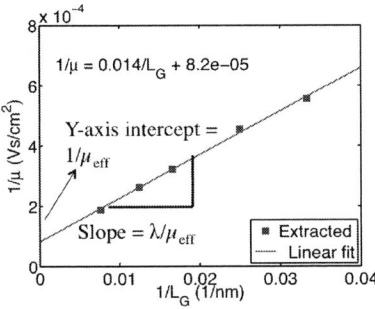

Figure 10: Inverse of extracted mobility versus inverse of gate length for III-V HEMT devices using the VS model. The extracted mobility is fit with a straight line. μ_{eff} = 12,195 cm^2/Vs, λ=171 nm.

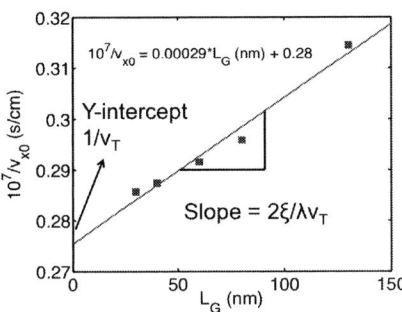

Figure 11: Extraction of ballistic velocity by fitting $1/v_{x0}$ versus L_G with a straight line. v_{x0} is extracted by fitting the III-V HEMTs data using the VS model. v_T = 3.57×10^7 cm/s and $\xi=l/L_G$ = 0.09.

IV. Conclusion

A new, emission-diffusion model of the MOSFET, which not only reduces the empiricism of the VS model but is also applicable to both long and short channel devices, is described and illustrated in this paper. The paper introduces a new approach for predicting the I-V characteristics of novel channel material MOSFETs using only a few physical parameter inputs, such as long channel mobility and thermal velocity.

Acknowledgement:

The authors acknowledge the support of NSF and SRC through the NCN-NEEDS program under contract 1227020-EEC and MIT SMART/LEES program. The authors would also like to thank Dr. Amlan Majumdar from IBM Thomas J. Watson Research Center for providing ETSOI experimental data and for insightful discussions on the VS model.

References:

[1] A. Khakifirooz et al., IEEE Trans. on Electron Devices, vol. 56, no. 8, pp. 1674-1680, Aug. 2009.

[2] S. Rakheja and D.A. Antoniadis, https://nanohub.org/resources/19684

[3] C.R. Crowell and S.M. Sze, Solid-State Electron., vol. 9, pp. 1035-1048, 1966.

[4] Y. Tsividis, Operation and Modeling of the MOS Transistor, McGraw Hill, 2nd Ed., 1999.

[5] Y. Liu et al, IEEE Trans. on Electron Devices, vol. 59, no. 4, pp. 994-1001, 2012.

[6] A. Rahman and M.S. Lundstrom, IEEE Trans. on Electron Devices, vol. 49, no. 3, pp. 481-489, 2002.

[7] Y. Liu et al., IEEE Trans. on Electron Devices, vol. 55, no. 3, pp. 866-871, 2008.

[8] A. Majumdar and D.A. Antoniadis, IEEE Trans. on Electron Devices, vol. 61, no. 2, pp. 351-358, 2014.

[9] D.-H. Kim et al., IEEE Int'l Electron Devices Meeting (IEDM), 9-9 Dec, 2009.

[10] U. Vergaftman et al., Journal of Applied Physics, vol. 89, p. 5815, 2001.

[11] F.E. Stern and W.E. Howard, Physical Review B, vol. 163, pp. 816- 835, 1996.

Table I: Extracted parameters in the VSED model for HEMT [9] and ETSOI [8] devices. Universally fitted parameters for III-V HEMT devices: v_T = 3.62×10^7 cm/s (m* = 0.022m$_0$), μ_{eff} = 12,500 cm^2/Vs, θ = 5.0; For ETSOI devices, universally fitted parameters are: v_T = 1.14×10^7 cm/s (m* = 0.22m0), μ_{eff} = 350 cm^2/Vs, and E$_{crit}$ = 1.14×10^5 V/cm. In the table, R$_s$ and R$_d$ are the source-drain access resistances such that V$_{dsi}$ = V$_{ds}$-I$_D$(R$_s$+R$_d$) and V$_{gsi}$ = V$_{gs}$-I$_D$×R$_s$; where V$_{ds}$ and V$_{gs}$ are the external drain-source and gate-source voltages, respectively.

	III-V HEMT devices from [9]					ETSOI devices from [8]			
Parameter	**30 nm**	**40 nm**	**60 nm**	**80 nm**	**130 nm**	**30 nm**	**40 nm**	**50 nm**	**980 nm**
$R_s = R_d$ [$\Omega.\mu$m]	200	200	195	190	190	130.0	125.0	128.5	160.0
n_0	1.334	1.313	1.279	1.261	1.200	1.24	1.13	1.05	1.0
δ [mV/V]	110.0	103.8	79.9	72.1	43.0	80.0	70.0	60.0	10.0
n_d [1/V]	0	0	0	0	0	0.36	0.24	0.17	0.0
V_{t0} [mV]	126.3	132.2	148.8	157.7	180.0	-125.0	-125.0	-130.0	-85.0
β	1.20	1.25	1.30	1.35	1.40	1.8	1.8	2.5	2.5

A Physics-Based RTN Variability Model for MOSFETs

Maurício Banaszeski da Silva[1,2], Hans Tuinhout[1], Adrie Zegers-van Duijnhoven[1],

Gilson I. Wirth[2] and Andries Scholten[1]

[1]NXP Semiconductors – Eindhoven, Netherlands, [2]UFRGS, PGMicro – Porto Alegre, RS, Brazil

Tel: + 55 51 3308-4443, Fax: + 55 51 3308-3293   Email: mbsilva@inf.ufrgs.br

Abstract

Low Frequency Noise (LFN) and Random Telegraph Noise (RTN) are performance limiters in many analog and digital circuits. For small area devices the noise PSD can easily vary by more than 4 orders of magnitude, imposing serious threat in circuit performance and possibly reliability. In this paper we propose a new RTN/LFN variability area scaling model. The model is validated through numerous experimental results for n-channel and p-channel devices from different CMOS nodes. Using this model we demonstrate that the variability found in our measurements can be explained using reasonable physical quantities and we clarify why variability, $\sigma[\log(S_{Id})]$, of RTN/LFN doesn't follow a $1/\sqrt{area}$ dependency.

Introduction

Low Frequency Noise (LFN) is a serious performance limiter in mixed signal CMOS circuits such as RF mixers, VCOs, and time-to-digital A/D converters. Moreover, Random Telegraph Noise (RTN) is emerging as a potential yield hazard in the most advanced CMOS nodes, as it can for instance appear as a time dependent SNM limiter in SRAMs [1]. Therefore, it is essential to provide adequate models for LFN/RTN and its variability. LFN variability studies [e.g., 2, 3, and 4] generally propose an area scaling of the noise variability, $\sigma[\log(S_{Id})]$, based on a $1/\sqrt{area}$ dependency. The work in [5] reasons that LF noise variability deviates from the $1/\sqrt{area}$ dependency based on the statistics of sums of lognormal distributions, but eventually it relies on a rather arbitrary empirical function to fit the variability area scaling.

In this work, we explain why LF noise variability indeed should **not** follow a $1/\sqrt{area}$ dependency. Moreover, through a statistical derivation, we show that this is the natural outcome of the lognormal distribution of RTN based LFN. We then derive a physics-based model for the variability of RTN and LF noise. The model holds for n- as well as p-channel devices. It is validated on a wide range of contemporary mixed-signal bulk CMOS device types. The physical interpretation of the elementary noise mechanisms on which the model is based explains the device length and bias dependency of the variability found in our LFN measurements.

RTN and LF noise variability

Variability of the LFN power spectral density (PSD) depends strongly on the gate area in MOSFETs [2]. As shown in Figs. 1 and 7, the noise PSD can easily vary by more than 4 orders of magnitude for small area devices. For such devices, the LFN spectra are generally dominated by a few Lorentzians, each of which in the time domain correspond to random telegraph signals associated with single traps [6]. Fig. 2 demonstrates that the statistical distributions of LF noise variability for different transistor sizes are well described by lognormal distributions. Fig. 1 shows that the noise spectra of small area devices are composed mainly of Lorentzians. This figure also shows that the average of all noise spectra has the $(1/f^\alpha)$ dependency that should be expected from averaging over such a large number of spectra. This observation, combined with studies as for instance reported in [7], supports our fundamental assumption that RTN effects dominate the LFN behavior. Therefore RTN and LFN should be treated in a single model to describe LFN variability.

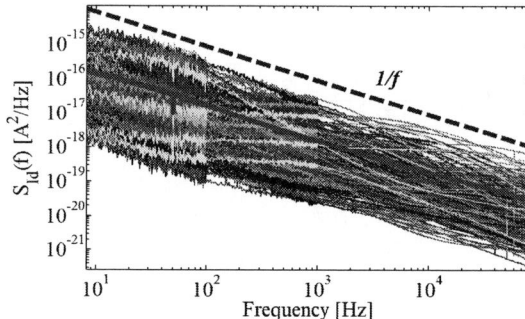

Figure 1. PSD of 282 NMOS devices with $W \times L = 1\mu m \times 0.06\mu m$. The red line is the calculated average of the 282 spectra. Note the abundance of Lorentzians (Random Telegraph Signals) even in a population of these relatively large transistors. Example spectra of four arbitrary devices are highlighted in yellow.

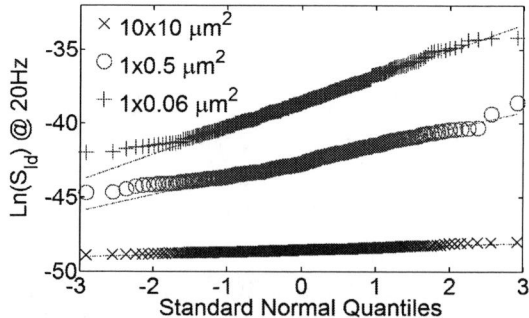

Figure 2. Q-Q plot of $\ln(S_{Id})$ at 20 Hz for three populations of 282 NMOS transistors demonstrating the lognormal nature of the noise distribution for these 3 device areas.

RTN Variability Model

When the statistical distribution of the LFN PSD is assumed lognormal, we can use the lognormal distribution properties [8] to calculate the standard deviation of the natural logarithm of the noise PSD

$$\sigma\left[\ln\left(S_{Id}(f)\right)\right] = \sqrt{\ln\left(1 + \frac{\text{Var}\left[S_{I_d}(f)\right]}{\text{E}\left[S_{I_d}(f)\right]^2}\right)}, \qquad (1)$$

where $\text{E}[S_{Id}(f)]$ and $\text{Var}[S_{Id}(f)]$ are the expectation and variance of the noise PSD respectively.

Following [6], the noise PSD that results from all traps in a transistor can be calculated by

$$S_{Id}(\omega) = 4\sum_{i=1}^{N_{tot}} \Delta I_{d,i}^2 \frac{\beta_i}{(1+\beta_i)^2} \frac{\overline{\tau}_i}{1 + \overline{\tau}_i^2 \omega^2}, \qquad (2)$$

where N_{tot}, ΔI_{di}, τ_i are the number of traps in the transistor, the fluctuation in the current caused by trap i, the geometric mean of the emission and capture times $(1/\tau = 1/\tau_e + 1/\tau_c)$ respectively and β given by $\beta = \tau_e/\tau_c = e^{(E_F - E_T)/kT}$, with E_T the trap energy and E_F the Fermi energy.

The expectation and variance of the noise PSD can be derived when E_T, ΔI_d, N_{tot} and τ, in the above equation, are treated as random variables, as proposed in [7]. When assuming N_{tot} Poisson distributed, τ log-uniform distributed and that the traps are uniformly distributed from source to drain, we derived that

$$\text{E}\left[S_{Id}(f)\right] = \frac{kT}{\gamma f} \frac{I_d^2}{WL^2} \int_0^L E[\Delta \tilde{I}_d^2(x)]N_{tr}(E_F)dx \qquad (3)$$

$$\text{Var}\left[S_{Id}(f)\right] = \frac{kT}{3\pi^2 \gamma f^2} \frac{I_d^4}{W^3 L^4} \int_0^L E[\Delta \tilde{I}_d^4(x)]N_{tr}(E_F)dx, \qquad (4)$$

where L, W, N_{tr} and $\Delta\tilde{I}_d$ are the channel length, channel width, the trap density (cm^{-2}.eV^{-1}) and the normalized ΔI_d ($\Delta\tilde{I}_d = \Delta I_d WL/I_d$) respectively. $E[\Delta\tilde{I}_d^n(x)]$ represents the expectation of $\Delta\tilde{I}_d^n$ for position x along the channel, with $x=0$ corresponding to the source side and $x=L$ to the drain side of the channel.

The expectation given by (3) converges to the commonly used model by Hung et al. [9] when $\Delta\tilde{I}_d = R/N + \alpha\mu_{\text{eff}}$, in which $R = C_{\text{inv}}/(C_{\text{inv}} + C_{\text{ox}} + C_d)$, N the inversion charge density and α the scattering coefficient. In the derivation used by Hung, τ is assumed to be associated with the depth location of the trap inside the oxide, and γ represents the attenuation coefficient of the electron wave function ($1/\gamma \approx 10^{-8}$cm). The log-uniform distribution of τ is attributed to a uniform distribution of traps inside the oxide. Campbell et al. [10] demonstrated that the elastic tunneling model underpinning this assumption is not realistic. In our approach τ is simply treated as a log-uniform random variable, implying that γ represents the density of τ in the log scale, hence $\gamma = \ln(\tau_{\max}/\tau_{\min})$. It should be noted that despite these differences in the assumptions, the resulting N_{tr}/γ (cm^{-2}.eV^{-1}) is the same in both works.

The key quantity of the proposed variability model in (1), the so called *normalized variance* (Var/E^2), is determined by

$$\frac{\text{Var}\left[S_{Id}(f)\right]}{\text{E}\left[S_{Id}(f)\right]^2} = \frac{1}{3\pi^2 kT} \frac{\gamma}{W} \frac{\int_0^L E[\Delta I_d^4(x)]N_{tr}(E_F)dx}{\left(\int_0^L E[\Delta I_d^2(x)]N_{tr}(E_F)dx\right)^2}. \qquad (5)$$

This equation hence describes how noise variability depends on the device dimensions, the trap density and on the bias condition. The bias dependence must consequently be attributed to the trap density at E_F ($N_{tr}(E_F)$) and how the current fluctuation ΔI_d depends on the inversion charge density and electric fields at position x.

Results: Uniformly Charged Channel

For strong inversion and small V_{ds}, the channel can be assumed to be uniformly charged (inverted) from source to drain. In this case ΔI_d and N_{tr} can be assumed to be independent of trap position x, and (5) can be simplified into

$$\frac{\text{Var}\left[S_{Id}(f)\right]}{\text{E}\left[S_{Id}(f)\right]^2} = \frac{1}{3\pi^2 kT} \frac{\gamma}{N_{tr}(E_F)} \frac{1}{WL} \frac{E\left[\Delta I_d^4\right]}{E\left[\Delta I_d^2\right]^2}. \qquad (6)$$

In this equation, $E[\Delta I_d^4]/E[\Delta I_d^2]^2$ is the factor that accounts for the statistical variation of ΔI_d caused by e.g. random dopants and the distribution of traps into the oxide depth. If ΔI_d is equal for all traps, as in Hung's model, $E[\Delta I_d^4]/E[\Delta I_d^2]^2 = 1$. If ΔI_d is exponentially distributed [11], $E[\Delta I_d^2] = 2E[\Delta I_d]^2$ and $E[\Delta I_d^4] = 24E[\Delta I_d]^4$ resulting in $E[\Delta I_d^4]/E[\Delta I_d^2]^2 = 6$.

If K is defined as $WL\text{Var}[S_{Id}]/E[S_{Id}]^2$, equations (1) and (6) can be simplified into

$$\sigma\left[\ln\left(S_{Id}(f)\right)\right] = \sqrt{\ln\left(1 + \frac{K}{WL}\right)}, \text{ and } K = \frac{1}{3\pi^2 kT} \frac{\gamma}{N_{tr}(E_F)} \frac{E\left[\Delta I_d^4\right]}{E\left[\Delta I_d^2\right]^2}. \qquad (7)$$

Fig. 3 shows an example of the measured K for a wide range of device areas. This demonstrates that for a uniformly charged channel (small V_{ds}) the noise variability can be described for all channel lengths using one single value for K.

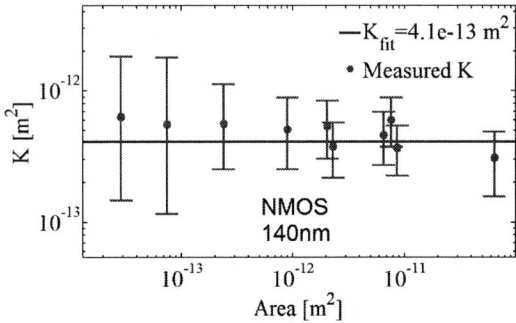

Figure 3. Measured $K = WL\text{Var}[S_{Id}]/E[S_{Id}]^2$ and fitted K. Population size for each geometry: 43, V_{ds}=100mV and V_{gs}=1.4V. Uncertainty bars represent the 0.02 and 0.98 quantiles from a bootstrap analysis of the measured K's.

Figs. 4 and 5 then show how in practice the area scaling of the noise variability follows the predicted behavior of (7). The dashed lines represent the conventional $1/\sqrt{\text{area}}$

dependency when calculated using the (smallest statistical uncertainty) large geometry devices. They clearly show the overestimation of the variability for small devices when the $1/\sqrt{\text{area}}$ relation is used. Moreover, Fig. 5 also demonstrates that the variability does converge to the conventional $1/\sqrt{\text{area}}$ model for large devices.

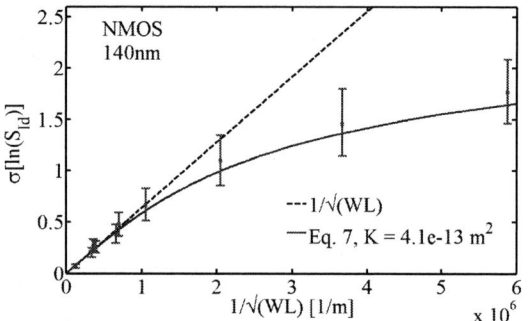

Figure 4. Example of the new variability area scaling model. Note the large deviation from the conventional $1/\sqrt{\text{area}}$ scaling model (dashed line) for small devices.

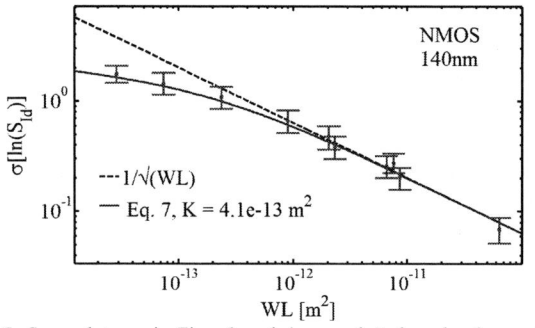

Figure 5. Same data as in Figs. 3 and 4 now plotted on log-log scale, to demonstrate that the model converges to $1/\sqrt{\text{area}}$ for large devices and that no noise saturation occurs near the origin of Fig. 4.

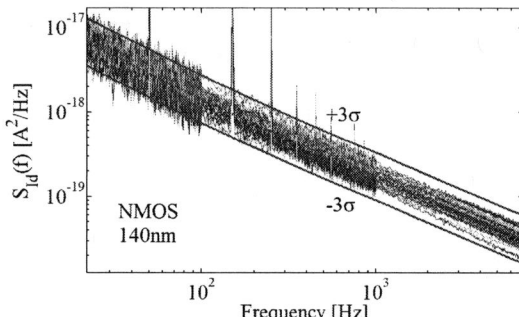

Figure 6. Example of the Noise Variability model prediction for a population of large geometry devices plotted back into the original spectra using (7) with K=4.1e-13m^2. Measured: 43 devices (WL=30×0.336μm^2) with V_{ds}=100mV and V_{gs}=1.4V. For these large area devices the many individual Lorentzians are summed, implying that they are no longer discernible as individual humps. The apparent larger spectrum noise in the lower frequency bands (>1kHz and <100Hz) is due to the lower number over time traces used for the FFT averaging at lower frequencies.

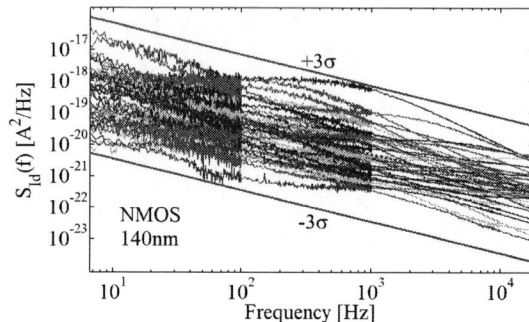

Figure 7. Example of the Noise Variability model prediction for a population of small geometry devices plotted back into the original spectra. K=4.1e-13m^2. Measured: 43 devices (WL=0.232×0.16μm^2) with V_{ds}=100mV and V_{gs}=1.4V.

Figs. 6 and 7 demonstrate, for two of the populations (a large and a small device geometry), how well the 3-sigma PSD predictions based on our Noise Variability model correspond with the observed variability of the LF noise spectra.

Figs. 8 and 9 show applications of the model for N- and PMOS devices from two additional contemporary mixed-signal CMOS technologies. For all cases the simplified model allows an excellent fit of the noise variability with area. Fig. 9 extends the low V_{ds} simplification to V_{ds}=0.5V. The fitting still holds for this case when the V_{gs} is large (hence implying a similar gradient of N for different device lengths).

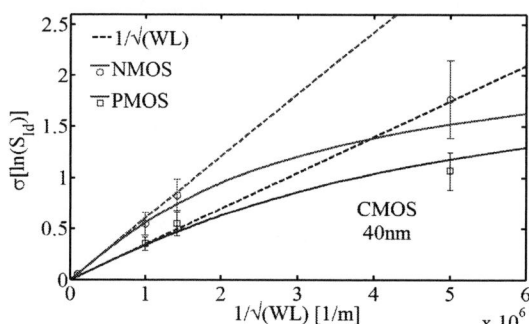

Figure 8. LF noise PSD variability area scaling in a CMOS 40nm node. Fitted using (7) with K=3.7e-13m^2 for NMOS and 1.2e-13 m^2 for PMOS. Measured 53 devices using V_{ds}=±50mV and V_{gs}=±1.1V.

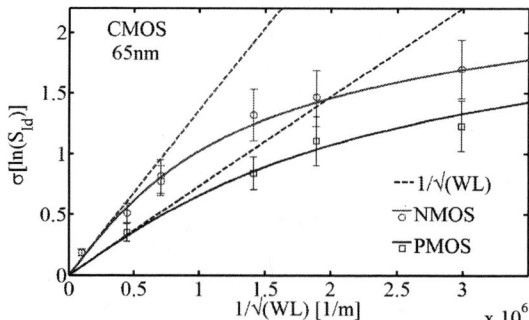

Figure 9. LF noise PSD variability area scaling in a CMOS 65nm node. Fitted using (7) with K=1.9e-12m^2 for NMOS and K=5.4e-13m^2 for PMOS. Measured 68 devices using V_{ds}=±500mV and V_{gs}=±2.5V.

Results: Non-uniformly Charged Channel

When the inversion layer charge density (N) and electrical fields are a strongly varying function of channel position x, the dependence of ΔI_d on trap position x will become appreciable. This makes K bias dependent. Equation (5) predicts however, that when changing the W with a fixed L, the model should scale for any bias combination. This is confirmed in Fig. 10. The fitted K's from this experiment (varying W, fixed L), for each bias combination, are summarized in Fig. 11. Following our model, for the short device (and for small V_{ds} on the large device), the change in K with V_{gs} (Fig. 11) could either be attributable to a gate bias dependency of $N_{tr}(E_F)$, or to an increase of $E[\Delta I_d^4]/E[\Delta I_d^2]^2$ associated with a non-homogeneity of the channel (e.g. random carrier concentration fluctuations -percolation effects- or non-uniform doping). Fig. 11 shows that for the short device there is practically no dependence of the variability with V_{ds}, whereas for the long channel device there is a large change of the variability with V_{ds}. This can be understood, at least qualitatively, from (5) by assuming a different impact of V_{ds} on the carrier density at the drain side for short and long channels and assuming an enhancement of the contribution of the halo regions to the noise [12].

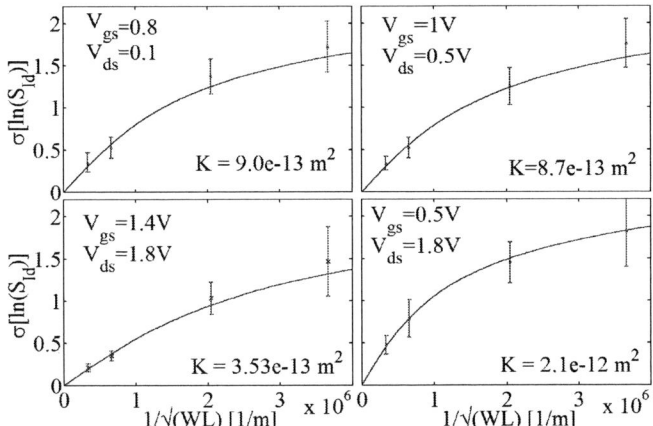

Figure 10. Bias dependence of K for NMOS devices in the 140nm technology. L is fixed (0.336µm), width is varied. Data fitted using (7).

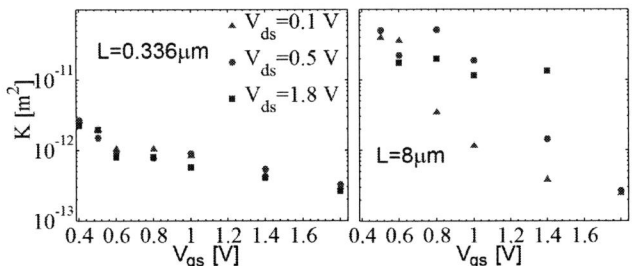

Figure 11. $K=WLVar[S_{id}(f)]/E[S_{id}(f)]^2$ as a function of V_{gs} for L=0.336µm (left) and L=8µm (right). Red (diamonds), black (triangles) and blue (circles) are, respectively for, V_{ds}=0.1V, V_{ds}=0.5V and V_{ds}=1.8V.

Fig. 12 shows the measurements of the expectation and variance of S_{Id}, for a population of transistors with area of 8x0.16µm^2 under varying bias conditions. The dashed lines in

these graphs are based on (3) and (4). It proves possible to replicate the measurements using a constant trap density of N_{tr}/γ=3e9 eV^{-1}cm^{-2} and an exponential distribution of $\Delta \tilde{I}_d$ with mean equal to R/N. We obtained realistic values for R and N as a function of the channel position, using the Medici® TCAD simulator on a tuned transistor.

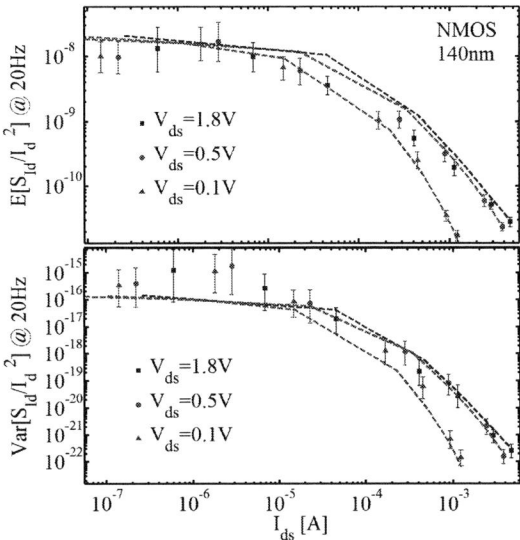

Figure 12. Measured expectation and variance of the noise of 43 devices with WxL = 8x0.16µm^2. Dashed lines represented the values calculated using (3) and (4) and physical values of R and N extracted from TCAD simulations.

Conclusion

In this work we propose a new physics-based RTN/LFN variability area scaling model. We explain why variability of RTN/LF noise does not follow a 1/√area dependency, as it is still commonly assumed in literature. The applicability of the model is demonstrated through numerous results for n-channel and p-channel devices from different mixed-signal CMOS technology nodes. We also show that the LFN variability depends on bias, and we provide the qualitative physical interpretations of these observations

References

[1] Seng Oon Toh et al., IEDM, pp 1-4, 2009.
[2] Ghibaudo and Roux-dit-Buisson, ESSDERC, pp.693-700, 1994.
[3] E. G. Ioannidis et al., IEDM, pp 449-452, 2011.
[4] P. Srinivasan et al., IEDM, pp 458-461, 2012.
[5] Bo Yu et al., CICC, pp 1-4, 2012
[6] Kirton, M. J., Uren, M. J., Adv. Phys., vol. 38, no. 4, pp 367-468, 1989.
[7] Gilson I. Wirth et al., IEEE TED, vol. 52, no. 7, pp 1575-1588, 2005.
[8] Norman L. Johnson, Samuel Kotz and N. Balakrishnan, Continuous Univariate Distributions, Vol. 1, Wiley, 1994.
[9] K K. Hung et al., IEEE TED, vol. 37, no.5, pp.1323-1333, 1990
[10] Campbell et al. IRPS, pp 382-388, 2009.
[11] Bukhori et al., IIRW, pp. 76-79, 2010.
[12] Paydavosi et al., ESSDERC, pp. 238-241, 2013.

Experiment and Model for Deviation from Pelgrom Scaling Relation in Device Width

Ning Lu, Jeffrey S. Brown, Rainer Thoma, Pooja M. Kotecha[*] and Richard A. Wachnik[*]

IBM Semiconductor Research and Development Center, Essex Junction, VT 05452
[*]IBM Semiconductor Research and Development Center, Hopewell Junction, NY 12533
Email: lun@us.ibm.com

Abstract

Through modeling, simulations, and experimental data, we show that FETs exhibit several width dependent characteristics purely due to un-correlated random variations among sub-threshold currents in different width segments. They include unit-width median sub-threshold current and constant-current threshold voltage. The width scaling relation for threshold voltage mismatch is different from Pelgrom scaling relation for sufficiently large variation when compared to the thermal voltage.

Introduction

Total chip leakage is a very real concern in FET device design, device selection, and in the design and modeling of circuits. A deep understanding of the width scaling relation of sub-threshold drain current and an accurate model of sub-threshold current (including both median and mean values) are thus very important and useful. A closely related FET characteristic is the constant-current threshold voltage (designated as Vt in this paper). It has been reported that average Vt decreases with FET width [1, 2], number of fingers in a planar FET [3], and number of fins in a finFET [4]. The Vt mismatch scaling relation in FET length has been reported [5] to deviate from the Pelgrom [6] scaling relation. On the other hand, Vt mismatch scaling relation in FET width is still said to follow the Pelgrom scaling relation [1, 4, 5, 7]. This paper shows that unit-width median sub-threshold drain current increases with total device width, resulting in lowering average Vt. We furthermore show and verify with experimental data that Vt mismatch scaling relation in FET width deviates from the known Pelgrom relation in general.

Width scaling relation of leakage currents

There are several kinds of leakage currents in semiconductor devices. Examples include sub-threshold drain current, gate leakage current, diode current, etc. A common feature for them is that each of them is an exponential function of a random variation. The exponent may arise for leakage from emission over a barrier as described by Maxwell-Boltzmann statistics or from other physical mechanisms. The random variation itself is symmetric about its mean value, which is also its median value. A fin FET device often contains several fins (Fig. 1), and thus is an ensemble FET. Other examples of an ensemble FET include multi-finger FETs in planar FET technologies. For an ensemble FET consisting of m sub-devices (e.g., $m = N_f$ for a planar FET with N_f fingers and $m = N_{fin}$ for a single-finger finFET with N_{fin} fins), sub-threshold drain current in each sub-device can be written as

$$i_{ds,k} = f(V_{ds}, L_1, w_1, \cdots) \exp[(V_{gs} - v_{th,k})/nv_t], \quad k = 1, 2, \cdots, m,$$

where V_{gs} is gate bias, V_{ds} is drain bias, L_1 is median/mean channel length, w_1 is single sub-device's median/mean channel width, $v_t = k_B T/q$ is the thermal voltage, and nv_t is related to sub-threshold slope S through $S = 2.3 nv_t$. In the long channel case, n is sub-threshold swing/body-effect. Among multiple sub-devices of the ensemble FET, Vt random variations of each sub-device contain both a correlated (σ_c) and an un-correlated (σ_u) component,

$$v_{th,k} = V_{th,ave} + \sigma_c G + \sigma_u g_k, \quad k = 1, 2, \cdots, m,$$

where each of G, g_1, g_2, ..., g_m is an independent random variable of mean zero and standard variation one. The correlation coefficient between the threshold voltages of any two sub-devices is $r = \sigma_c^2/(\sigma_c^2 + \sigma_u^2)$. Random variations in Vt's short channel effect are included in σ_c and σ_u, but random variations in channel length, width, mobility, etc. inside the multiplier function f are ignored. For each sub-device, its median sub-threshold leakage current is independent of sub-device index k,

$$i_{ds}(\text{median}) = f(V_{ds}, L_1, w_1, \cdots) \exp[(V_{gs} - V_{th,ave})/nv_t], \quad (1)$$

and its mean sub-threshold leakage current is also independent of sub-device index k,

$$i_{ds}(\text{mean}) = \langle i_{ds,k} \rangle = U_c U_u i_{ds}(\text{median}). \quad (2)$$

In Eq. (2),

$$U_c = \exp[\sigma_c^2/(2n^2 v_t^2)],$$
$$U_u = \exp[\sigma_u^2/(2n^2 v_t^2)] \quad (3)$$

are uplift factors of a single sub-device for correlated and uncorrelated random variations, respectively. Both correlated and uncorrelated random variations contribute to the uplift. The total sub-threshold drain current of the ensemble FET is

$$I_{ds} = \sum_{k=1}^{m} i_{ds,k}. \quad (4)$$

The mean leakage of the ensemble FET is simply m times the mean leakage current of single sub-device,

$$\langle I_{ds} \rangle = m i_{ds}(\text{mean}). \quad (5)$$

Intuitively, Eq. (5) can be understood by noticing that total leakage current remains the same no matter how summation is done. The statistical averages in Eqs. (2) and (5) are for all FETs of a same type and same channel length and width on a chip (circuit designer's view), for kerf data measured over a long period (fab's view), or for all Monte Carlo samples in a SPICE simulation (modeler's view). We emphasize that per-device mean leakage, $\langle I_{ds} \rangle/m$, of an ensemble FET with m individual sub-devices stays the same when m increases (Fig. 2), independent of the ratio of correlated component (σ_c) to uncorrelated component (σ_u). To find the median leakage of the ensemble FET, we need first to replace the sum of multiple

978-1-4799-8002-4/14 $31.00 © 2014 IEEE

uncorrelated log-normal distributions in an ensemble FET by a single log-normal distribution multiplied by m. One form of a statistical model is

$$I_{ds} = m f(V_{ds}, L_1, w_1, \cdots) U_{unc}(m) \exp[(V_{gs} - V_{th})/nv_t], \quad (6)$$

with

$$V_{th} = V_{th,ave} + \sigma_c G + \sigma_{unc}(m) g_1 ,$$

where each of G and g_1 is an independent random variable of mean zero and standard deviation one. The values of a new uplift factor $U_{unc}(m)$ and standard deviation $\sigma_{unc}(m)$ are obtained by matching two expressions of the mean of I_{ds} [one from Eq. (4) and the other from Eq. (6)] and also matching two expressions of the variance of I_{ds},

$$U_{unc}(m) = U_u / \sqrt{1 + (U_u^2 - 1)/m} , \quad (7)$$

$$\sigma_{unc}(m) = nv_t \sqrt{\ln[1 + (U_u^2 - 1)/m]} . \quad (8)$$

They have desired properties,
$U_{unc}(1) = 1, \qquad U_{unc}(m; \sigma_u = 0) = 1, \qquad U_{unc}(\text{large } m) = U_u ;$
$\sigma_{unc}(1) = \sigma_u , \qquad \sigma_{unc}(m; \sigma_u = 0) = 0 .$
We can now find the median leakage of the ensemble FET easily from Eqs. (6) and (1),

$$I_{ds}(\text{median}) = m U_{unc}(m) i_{ds}(\text{median}) . \quad (9)$$

Should random variations be completely correlated (i.e., when $\sigma_u = 0$), per-device median leakage would also stay the same when sub-device number m is increased, $I_{ds}(\text{median})/m$ $= i_{ds}(\text{median})$. Since there are un-correlated random variations (e.g., RDF), the ratio of per-device median leakage to single-device median leakage increases monotonically with sub-device number m,

$$\frac{I_{ds}(\text{median})/m}{i_{ds}(\text{median})} = U_{unc}(m) . \quad (10)$$

Only uncorrelated random variation (σ_u) contributes to this increase. It follows from Eqs. (5), (9), and (2) that, for an ensemble FET, the ratio of mean leakage to median leakage (ensemble FET's uplift factor) also depends on sub-device number m,

$$\langle I_{ds} \rangle / I_{ds}(\text{median}) = U_c U_u / U_{unc}(m) = U_c \sqrt{1 + (U_u^2 - 1)/m} . \quad (11)$$

Again, both correlated and uncorrelated random variations contribute to the uplift. Fig. 3 plots Eq. (10) and the ratio $U_u / U_{unc}(m)$ vs. sub-device number m.

Width scaling relation of average threshold voltage
Threshold voltage is often defined as the gate bias V_{gs} required to achieve a drain current of, say, $(300 \text{ nA}) W/L_1$ for nFET and $(70 \text{ nA}) W/L_1$ for pFET (constant current definition; $W = mw_1$). For an ensemble FET, the mean of its threshold voltage is approximately the same as the median of its threshold voltage (and the two values are equal under the approximation used in this paper), even when the mean of sub-threshold current is substantially larger than the median of sub-threshold current. The median of threshold voltage is determined by median sub-threshold drain current. Since per-device median sub-threshold drain current increases with sub-device number m (towards per-device mean sub-threshold current), Vt decreases with

increasing sub-device number m. To find the amount of this Vt drop, we recast statistical model (6) to a form that is Vt-friendly,

$$I_{ds} = m f(V_{ds}, L_1, w_1, \cdots) \exp\{[V_{gs} - V_{t\_shift}(m) - V_{th}]/nv_t\} . \quad (12)$$

A comparison between statistical models (6) and (12) shows that V_{t_shift} is related to $U_{unc}(m)$ through the following relation,

$$V_{t\_shift}(m) = -nv_t \ln U_{unc}(m) .$$

As expected, there is no Vt change when $m = 1$, $V_{t\_shift}(1) = 0$. As a statistical variable, Vt is found to be

$$V_t(m) = V_{t\_shift}(m) + V_{th,ave} + \sigma_c G + \sigma_{unc}(m) g_1 . \quad (13)$$

The median and mean of Vt are identical and are equal to

$$\langle V_t(m) \rangle = V_{t\_shift}(m) + V_{th,ave} .$$

The mean of an ensemble FET's Vt is different from the mean of its sub-device's Vt and the difference is

$$\langle V_t(m) \rangle - \langle V_t(1) \rangle = V_{t\_shift}(m) = -\frac{nv_t}{2} \ln\left(\frac{m}{1 + (m-1)/U_u^2} \right) . \quad (14)$$

Equation (14) clearly shows that correlated random variation (σ_c) does not contribute to the lowering of average Vt. This differs from [2]. Fig. 4 compares average Vt from Monte Carlo simulation using multiple FET instances [Fig. 1(b)] with Vt shift given by Eq. (14). Fig. 5 compares average Vt from Monte Carlo simulation using multiple FET instances [Fig. 1(b)] with nominal Vt from a single run using a single FET instance [Fig. 1(c)] after improving BSIMSOI model. In the large m limit, Vt change reaches a plateau,

$$V_{t\_shift}(m) \approx -\sigma_u^2 /(2nv_t) , \quad m >> U_u^2 .$$

Width scaling relation of threshold voltage mismatch
The standard deviation of un-correlated variation in the Vt of the ensemble FET is given in Eq. (8). Threshold voltage mismatch between two adjacent ensemble FETs (each with m individual sub-devices) is found from Eq. (13) as

$$\Delta V_t = \sigma_{unc}(m)(g_1 - g_2) ,$$

and its standard deviation is simply

$$\sigma(\Delta V_t) = \sqrt{2} \sigma_{unc}(m), \quad (15)$$

which differs from Pelgrom scaling in general. When the sub-device number m is large, mismatch scaling relation becomes proportional to the inverse of the square root of m, $\sigma(\Delta V_t) =$ $nv_t \sqrt{2(U_u^2 - 1)/m}$. In the small variation limit ($\sigma_u^2 << n^2 v_t^2$), mismatch relation (15) reduces to the Pelgrom scaling relation [6], $\sigma(\Delta V_t) = \sqrt{2} \sigma_u / \sqrt{m}$. For a wide planar FET, integer m in Eqs. (5)–(15) is replaced by a real value W/w_{min}, where w_{min} is the minimum FET width, and i_{ds} and σ_u are for a FET with the minimum width $w_1 = w_{min}$. In Fig. 6, we plot relation (8) for several different ratios of σ_u / nv_t. In Fig. 7 and Fig. 8, we show Vt mismatch hardware data from two different SOI technology nodes. Deviation from the Pelgrom scaling relation can be clearly seen.

Summary

We have shown several interesting scaling relations with respect to FET sub-device number/width. They include increases in unit-width median leakage currents (e.g., sub-threshold drain current), a lowering of average Vt, and a non-Pelgrom width scaling relation for Vt mismatch. The non-Pelgrom width scaling relation for Vt mismatch has been confirmed experimentally from two different planar FET technologies.

References

[1] A. Asenov, IEEE TED, Vol. 45, p. 2505, 1998.
[2] J. Gu et al., DAC, p. 87, 2007.
[3] J. Watts et al., NSTI-Nanotech, Vol. 3, p. 703, 2006.
[4] C.-H. Lin et al., VLSI Symp., p. 16, 2011.
[5] J. Johnson et al., El. Dev. Lett., Vol. 29, p. 802, 2008.
[6] M. J. M. Pelgrom et al., IEEE J. Solid State Circuits, p. 1433, 1989.
[7] Q. Zhang et al., IEEE TED, Vol. 61, p. 643, 2014.

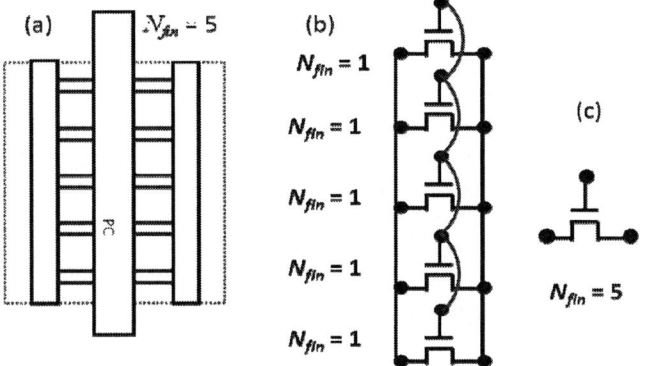

Figure 1. (a) A finFET layout with 5 fins. (b). A corresponding netlist that uses five separate finFET instances (each with N_{fin} = 1). They are connected in parallel. (c) Another netlist that uses a single finFET instance with N_{fin} = 5. Without an improvement to a compact FET model, only Monte Carlo simulations using netlist (b) can give correct results on median sub-threshold drain current, average constant-current threshold voltage (Vt), and Vt mismatch.

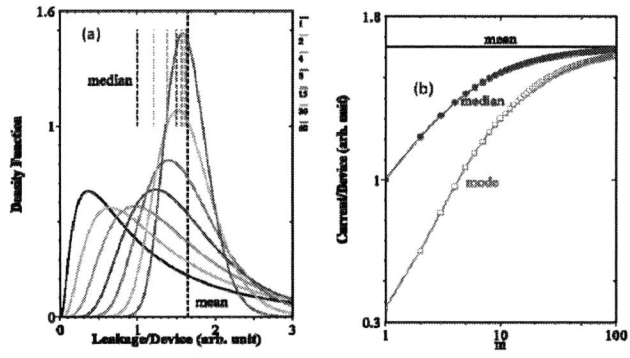

Figure 2. (a) Density functions of leakage current for sub-device number m = 1, 2, 4, 8, 15, 30, and 60. (b) Mean, median, and mode of the density function of leakage current as

a function of sub-device number m. When the sub-device number m of an ensemble device increases, per-device median leakage current [short vertical dashed lines in (a)] increases but per-device mean leakage current [long vertical dashed line in (a)] stays the same.

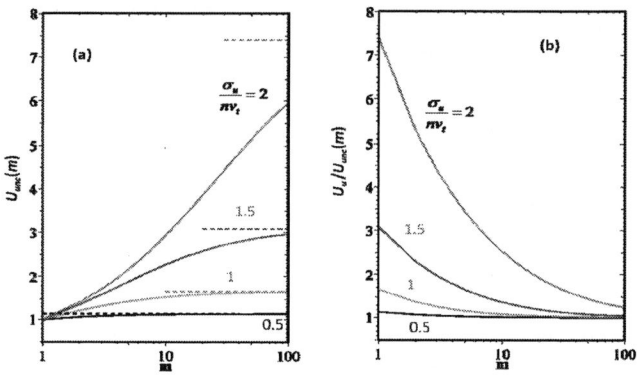

Figure 3. For the leakage current of an ensemble device containing m sub-devices, (a) shows the ratio $U_{unc}(m)$ of per-device median current to single-device median current [Eq. (10)] as a function of sub-device number m, and (b) shows the ratio $U_u/U_{unc}(m)$ [the ratio of mean sub-threshold current to median sub-threshold current when $\sigma_c = 0$; see Eq. (11)] as a function of sub-device number m. Each dashed line in (a) is the asymptotic value of $U_{unc}(m)$ for the given ratio of σ_u/nv_t in large m limit, which is U_u.

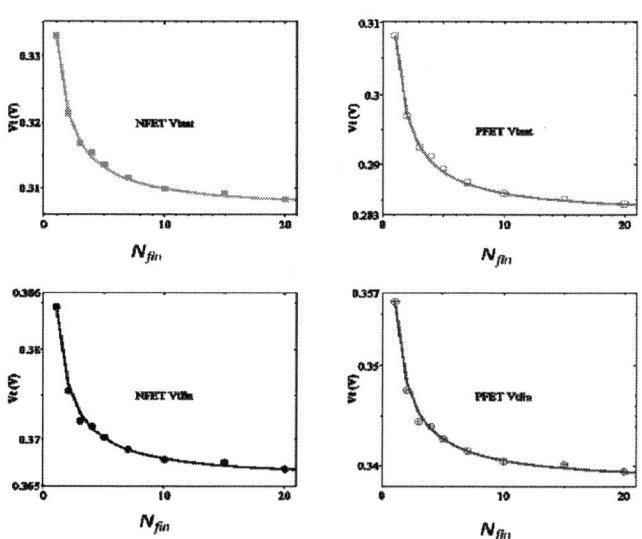

Figure 4. Constant-current threshold voltage (Vt) vs. fin number N_{fin} in a 14nm finFET technology. Symbols are average Vt obtained by netlisting N_{fin} FET instances in parallel and run Monte Carlo simulations of 5,000 runs. Curves are direct fitting of Eq. (14). Each ratio of σ_u to nv_t is between 0.5 and 2. NFET Vtsat: nv_t = 112 mV, σ_u = 77 mV; NFET Vtlin: nv_t = 81 mV, σ_u = 55 mV; PFET Vtsat: nv_t = 73 mV, σ_u = 51 mV; PFET Vtlin: nv_t = 76 mV, σ_u = 114 mV.

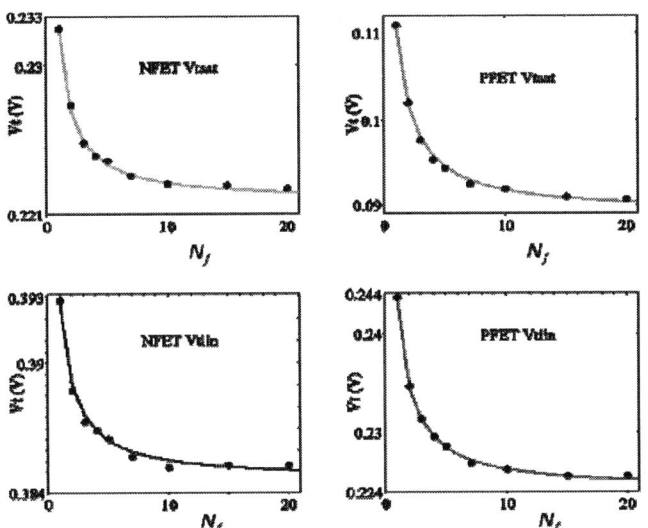

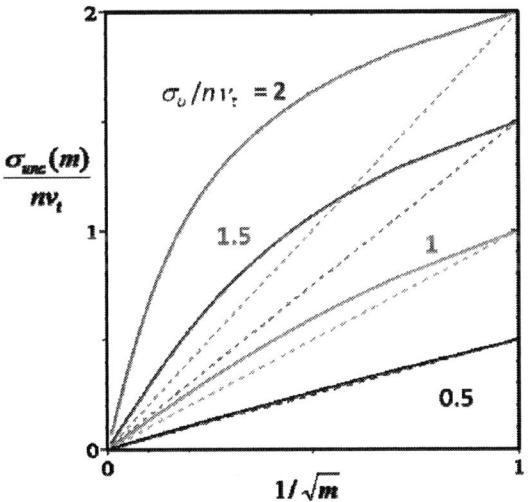

Figure 5. Constant-current threshold voltage (Vt) vs. finger number N_f of a mult-finger FET in a 22nm SOI technology. Symbols are average Vt obtained by netlisting N_f FET instances in parallel and run Monte Carlo simulations of 10,000 runs. Each curve is from a single model run of using only one FET instance (with $N_f = 1, 2, \ldots, 20$, respectively) after setting VOFF parameter in BSIMSOI model equal to $V_{t\_shift}(N_f)$ in Eq. (14). Each ratio of σ_u to nv_t is between 0.5 and 1.

Figure 6. Standard deviation [Eq. (8)] of un-correlated component of Vt variation in an ensemble FET vs. $1/\sqrt{m}$, where m is sub-device number (solid curves). Dashed straight lines are $\sigma_{unc}(m) = \sigma_u / \sqrt{m}$.

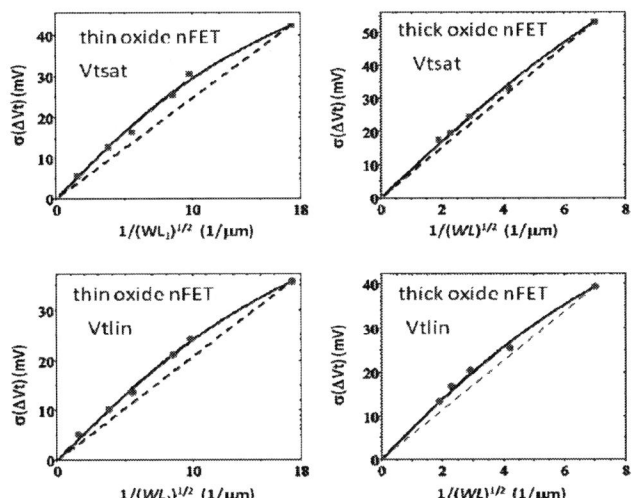

Figure 7. Vt mismatch as a function of total FET width W. Symbols are measured data from a 32nm SOI technology [7]. Each dashed line is a straight line that connects the data point of the narrowest FET ($W = w_{min}$) and the origin. Solid curves are fitting results using Eqs. (15) and (8). Each ratio of σ_u to nv_t is between 0.5 and 2. Thin-oxide nFET: Vtsat, $nv_t = 27$ mV, $\sigma_u = 30$ mV; Vtlin, $nv_t = 25$ mV, $\sigma_u = 25$ mV; Thick-oxide nFET: Vtsat: $nv_t = 54$ mV, $\sigma_u = 37$ mV; Vtlin: $nv_t = 31$ mV, $\sigma_u = 28$ mV. The experimental data clearly deviate from a straight line. Excellent model-hardware correspondence is demonstrated.

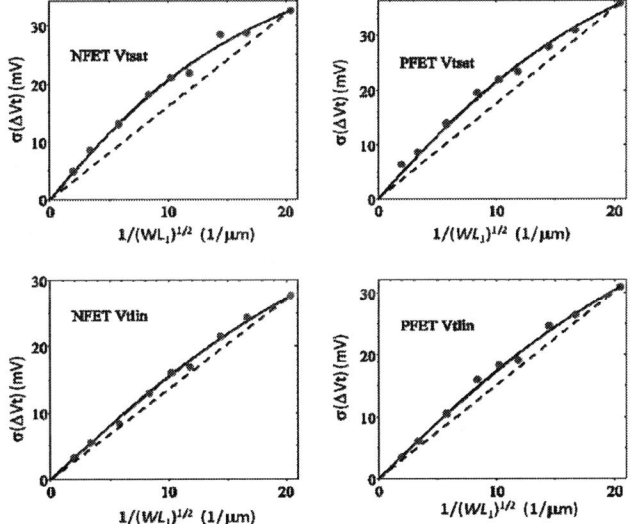

Figure 8. Vt mismatch as a function of $1/\sqrt{WL_1}$. Symbols are measured data from a 22nm SOI technology. All data have a same nominal channel length L_1. Each dashed line is a straight line that connects the data point of the narrowest FET ($W = w_{min}$) and the origin. Solid curves are fitting results using Eqs. (15) and (8). Each ratio of σ_u to nv_t is between 0.5 and 2. The experimental data clearly deviate from a straight line. Excellent model-hardware correspondence is demonstrated.

A Comprehensive Platform for Thermal Studies in TSV-based 3D Integrated Circuits

P. M. Souaré[1234], P. Coudrain[1], J.P. Colonna[23], V. Fiori[1], A. Farcy[1], F. de Crécy[23], A. Borbely[4], H. Ben-Jamaa[23], C. Laviron[23], S. Gallois-Garreignot[1], B. Giraud[23], N. Hotellier[1], R. Franiatte[23], S. Dumas[23], C. Chancel[23], J.-M. Rivière[1♣], J. Pruvost[1♣], S. Chéramy[23], C. Tavernier[1], J. Michailos[1], L. Le Pailleur[1]

[1] STMicroelectronics, 850 rue Jean Monnet 38926 Crolles, ♣Grenoble, France
[2] Université Grenoble Alpes, F-38000 Grenoble, France
[3] CEA, Leti, Minatec Campus, F-38054 Grenoble, France
[4] Ecole des Mines de Saint Etienne, 158 Cours Fauriel 42023 Saint-Étienne, France
(Email: papa-momar.souare@st.com)

Abstract

We present an advanced and comprehensive platform for thermal dissipation studies in TSV-based 3D ICs. A 2-tier 3D test chip with through silicon via (TSV) and μ-bump is used for thermal characterization with unprecedented precision and design exploration capabilities. A comprehensive calibrated 3D finite element model is associated to provide a predictive tool that is able to simulate the thermal mapping in any given 3D interconnect configuration with minimal error. Guidelines are finally provided for thermal optimization of 3D designs with a precision far beyond the prior art.

Keywords : Thermal, TSV, 3D ICs, self heating, sensor, FEM simalution, thermoelectric measurement.

Thermal dissipation in 3D IC

Thermal dissipation is a major concern in 3D circuits where the dense stacking of thin silicon layers leads to a dramatic increase of heat fluxes [1]. Moreover, the thermal impact of 3D interconnects such as TSV and μ-bump still has to be properly characterized in realistic 3D environments. Simplistic studies have led to incorrect assumptions, especially on the benefit of TSV on the dissipation, which depends on heating configurations. Developing a comprehensive strategy for 3D thermal characterization and modeling becomes mandatory to support realistic 3D design. In this paper we present an advanced 3D thermal test chip with the highest thermal mapping precision reported [2][3][4][5] (Tab.1). Embedded heaters and numerous sensors enable ultra-precise mapping with a large set of 3D interconnect configurations associated with various heating scenarios. A finite element modeling strategy is conducted in parallel to provide a calibrated and predictive tool able to simulate any scenario of 3D configurations with minimal error. A large focus is made on thermal effects of 3D-technological parameters, such as TSV and 3D interconnects.

3D Thermal Test Chip

The 2-tier 3D-thermal test chip is presented in Fig.1,2&3. It is composed of a 7.5 mm² top die on a 31 mm² bottom die connected to a BGA by μ-bumps. TSV-middle integration is used in a Face-to-Back configuration (Fig.2) [6]. 3D packages are either molded or non-molded to allow IR thermography. The chips were designed as a combination of parallel test circuits: 8 central in each die (Fig.1). A redistribution layer (RDL) is used to connect these circuits to the BGA (Fig.1a). Each circuit is composed of 8 MOS-based heaters (60x200 μm²) and diode temperature sensors (Fig.1b), with a total of 128 sensors in each circuit after stacking (64 in each tier). A counter addresses sensor responses through decoders (Fig.4) 3D stacking efficiency was first checked by elementary device measurements on the top die through the 3D interconnect stack from the BGA (Fig.5). The 3D test chip is monitored with a test PCB (Fig.6). Heaters and sensors were then calibrated, with a sensitivity of -1.64 mV/°C for the diodes (Fig.7). The thermal test chip can perform multi-pattern thermal excitation with a combination of top and bottom heaters, and simultaneously provides ultra-precise thermal map in tiers (Fig.8, 9 & 10).

FEM Model

A numerical model covering the 3D package and the PCB (Fig.11) was built with Ansys with homogeneous materials for BEOL, μ-bump, TSV and large bump layers (Tab.2). Homogenization was performed to simplify sub-assemblies in blocks. Proper orthotropic properties were computed to account the real thermal behavior of the layers. Using this finite element model, a 10880-point DOE was performed using Optimal Design of Response Surface Modeling. This results in a model of the temperature field as a function of input technological parameters (Tab.2) and boundary conditions which allows fitting the simulations to measurements. Nearly perfect agreement is obtained between simulations and measurements, under the different heating scenarios (Fig.8, 9 & 10). The validation of the numerical model enabled to forecast the temperature with 1°C accuracy in the 3D stack, and to evaluate different solutions for optimizing heat dissipation. This ultra-precise model is now used for 3D interconnects layout and heating profile exploration.

Thermal Guidelines

Thermal recommendations are presented in the scope of three important concerns for 3D circuit design and fabrication: design partitioning, inter-tier connections and package parameters. They are established from measurements on different heating scenarios and interconnect configurations, as well as explorations with the FEM model.

Design partitioning on the tiers controls the power pattern of the chip and may lead to generation of localized hotspots. The effect of hotspot is illustrated in Fig.12 where 300 mW are either spread over 8 heaters or localized on a single heater: the maximal temperature in the bottom die is increased by 5 °C. This hotspot scenario (300 mW on single heater on bottom chip) is used in the following developments.

Package parameters have a major influence. Despite relatively poor thermal conductivity, the molding allows to dissipate heat over the top chip and reduce the temperature by about 10 °C in bottom die (Fig.13). In 3D ICs, the thinning of silicon allows shorter interconnects and leads to reasonable aspect ratio for TSV process, but this thinning is detrimental to thermal dissipation [7]. Maximal temperature is increased by 12 °C in the bottom and 6 °C in the top dies when the silicon is thinned as specified in Fig.14 & 15, due to reduced spreading efficiency in thin silicon.

3D interconnects such as TSV and bumps can exhibit a strong influence on the thermal diffusion depending on the layout. The presence of TSVs around hotspots leads to an increase of the bottom die temperature. Actually, lateral spreading in the silicon is reduced by the SiO_2 via electrical insulation (Fig.16): TSVs must be kept away from hotspots. Addition of μ-bumps and RDL above TSVs however changes these results. It creates a thermal path between the dies so that the top die can act as a heat spreader (Fig.17). A design flow with a global approach for thermal optimization should therefore take into account these three components and their interactions. Thereby, our model is now used for 3D interconnect layouting and heating profile exploration, with application-dependent guidelines for designers. For example, an interposer can be used to spread heat from top dies if μ-bumps are located close to hotspots and TSV are moved away using RDL. Moreover, any application and related heating pattern can benefit from this approach, as soon as the interaction of hotspots and interconnects is properly analyzed.

Conclusion

We have presented a unique platform for thermal studies in 3D IC. A dedicated test chip presents the largest functionality reported and allows investigating thermal performances of 3D interconnects. A comprehensive FEM model has been calibrated by electrical measurements with a minimal error. Based on these results, along with a large exploration of 3D configurations and several heating scenarios, we are able to provide thermal optimization for 3D circuits with a fine understanding of the interconnect layout impact.

Acknowledgement

This work was funded thanks to the French national program "Programme d'Investissements d'Avenir, IRT Nanoelec" ANR-10-AIRT-05.

The authors would like to thank STMicroelectronics Crolles and Cea-LETI Minatec teams for wafer level processing and characterization, CPA team for 3D chips assembly and packaging, and LETI-DACLE team for the design of test PCB and system level characterization.

Reference

[1]. M. Rencz, et al., « Thermal Issues in Stacked Die Packages », SEMI-THERM, pp 307-312, 2005.

[2]. H. Oprins, et al., « Steady State and Transient Thermal Analysis of Hot Spots in 3D Stacked ICs using Dedicated Test Chips », SEMI-THERM, pp 1081-1088, 2011.

[3]. A. Jain, et al., « Analytical and Numerical Modeling of the Thermal Performance of Three-Dimensional Integrated Circuits », IEEE Transaction on CPMT, vol. 30, pp. 56-63, 2010.

[4]. K. R. Vaddina, et al., « Thermal Analysis of Advanced 3D Stacked Systems », Symposium on VLSI, pp 371-372, 2011.

[5]. C. Santos, et al., « System-Level Thermal Modeling for 3D Circuits: Characterization with a 65nm Memory on Logic Circuit », 3DIC, pp 1-6, 2013.

[6]. D. Dutoit, et al., « A 0.9 pJ/bit, 12.8 GByte/s wideIO memory interface in a 3D-IC NoC-based MPSoC », Symposium on VLSI Tech., pp. 22-23, 2013.

[7]. P. M. Souaré, et al., « Thermal Effects of Silicon Thickness in 3D ICs: Measurements & Simulations », IEEE Transaction on CPMT, pp. 1-9, 2014.

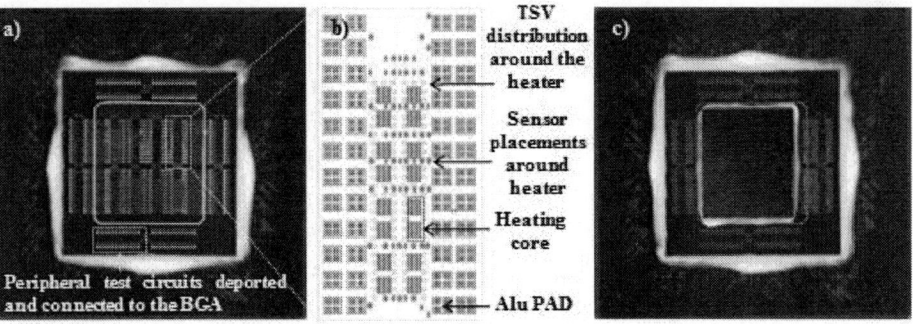

Figure 1: Optical top view a) bottom die on BGA b) Placement of sensors and TSVs around heating core on scribe c) top and bottom die on BGA

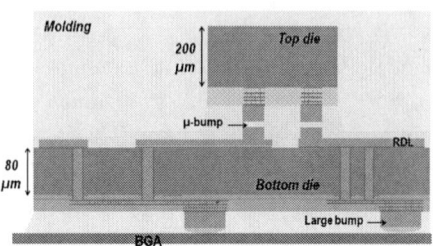

Figure 2: 3D stack of the thermal chip

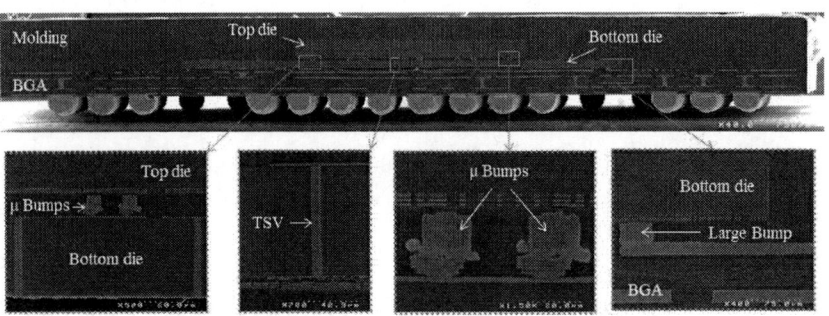

Figure 3: SEM cross section of thermal chip

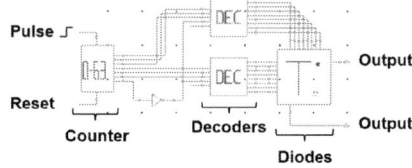

Figure 4: Electrical schematic of the logic circuitry

Table 2: Material properties

Materials		Conductivity (W/m*K)	
		XY	Z
Silicon		150	150
TSVs	Ø10 µm, pitch 40 µm	138.04	160.88
	SiO2 0.3 µm		
BEOL	M1 to M7	2.36	2.54
Large bump	Ø55 µm, pitch 120 µm	2.66	7.32
RDL + Passiv		3.75	260.42
µ-bump	Ø20 µm, pitch 40 µm	3.81	15.58
Underfill		1.5	1.5
BGA	4 layers	92.09	0.6
Molding		0.88	0.88

Table 1: Thermal benchmarking

	3D Thermal Test Chip	Wide IO by STE [5]	IMEC 3D Test Chip [2]
Sensors	6x128	14	30
Heaters	96 in chip	8 in chip	6 in chip
Sensing localization	Top and bottom die	Top and bottom die	Top die
Heating localization	Top and bottom die	Bottom die only	Top die only
Heating configuration	Hotspot & Average Temp.	Hotspot & Average Temp.	Hotspot
3D Model	FEM	Compact	FEM
Sensors addressing	Auto (Logic circuitry)	Manual	Manual
Accuracy / Model	± 1°C => 99%	NC => 95%	± 1°C to 4°C => 88 to 93%

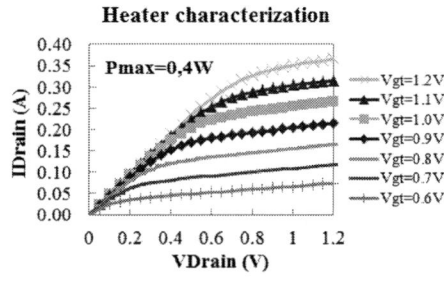

Heater characterization

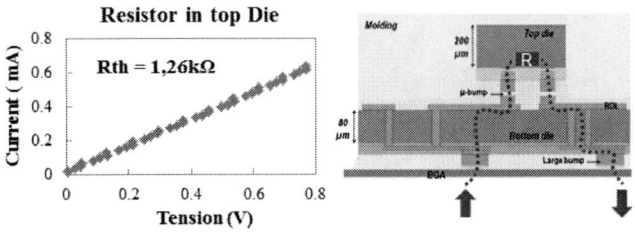

Figure 5: Validation of 3D integration by measuring a resistance on the top die from PCB

Figure 6: Test card PCB controlled with Labview scripts

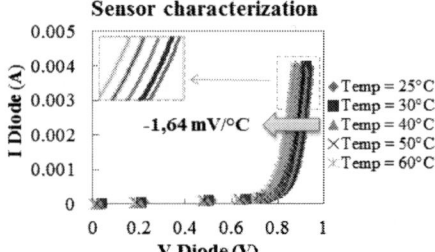

Sensor characterization

Figure 7: Calibration of Diode sensors and MOS heaters

978-1-4799-8002-4/14 $31.00 © 2014 IEEE

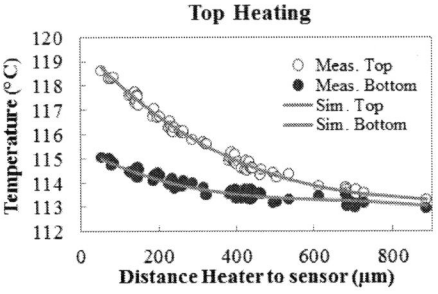

Figure 8: Correlation measurement vs. simulation in top heating configuration.

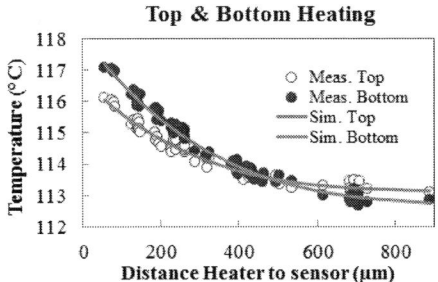

Figure 9: Correlation measurement vs. simulation bottom heating configuration.

Figure 10: Correlation measurement vs. simulation in top and bottom heating configuration.

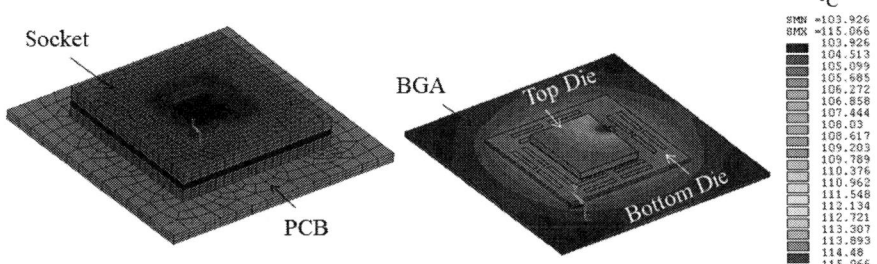

Figure 11: FEM 3D model: thermal mapping + meshing

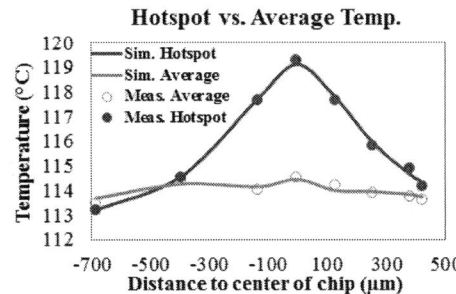

Figure 12: Effect of hotspot surface: temperature profile where the power is applied onto a single hotspot vs. in 8 hotspots.

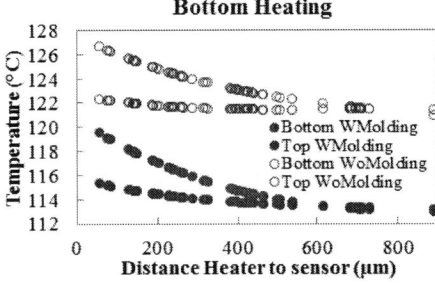

Figure 13: Temperature profile in two dice with vs. without molding.

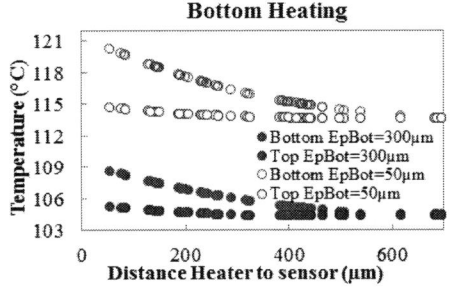

Figure 14: Temperature profile in two dice according the bottom die thickness

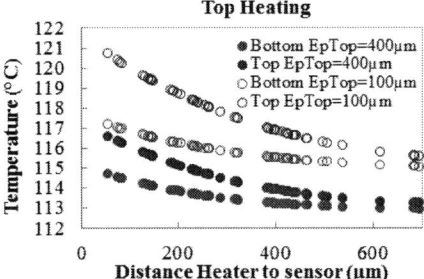

Figure 15: Temperature profile in two dice according the top die thickness

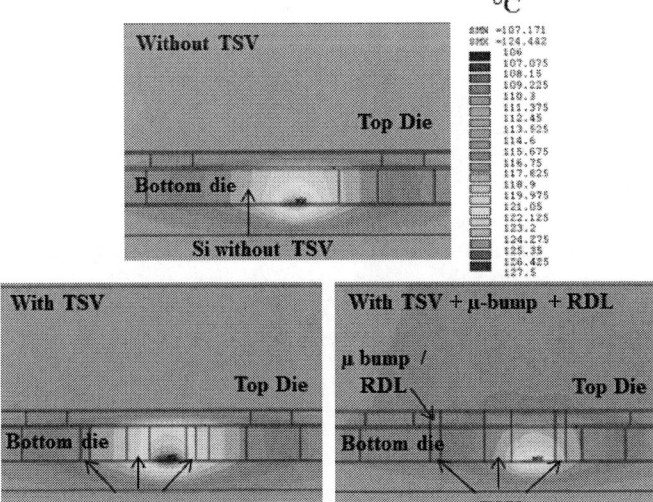

Figure 16: Temperature mapping between with/without TSVs and with TSVs +μCP + RDL configuration

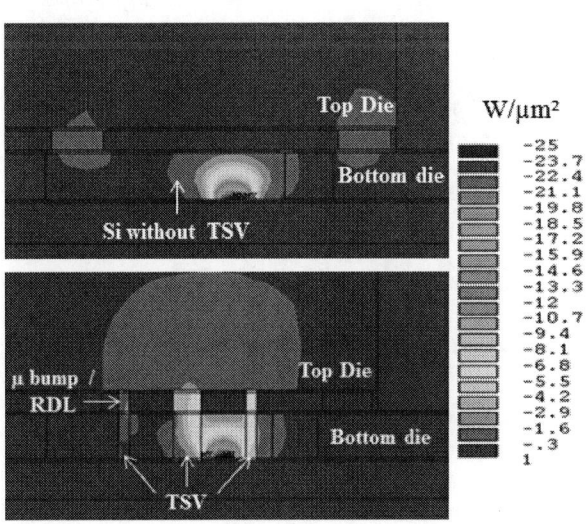

Figure 17: Heat Fluxes mapping between without TSVs and with TSVs +μCP + RDL configuration

978-1-4799-8002-4/14 $31.00 © 2014 IEEE   859

A New Surface Potential-Based Compact Model for a-IGZO TFTs in RFID Applications

Zhiwei Zong[1], Ling Li[1*], Jin Jang[2], Zhigang Li[1], Nianduan Lu[1], Liwei Shang[1], Zhuoyu Ji[1], Ming Liu[1*]

[1]Institute of Microelectronics of Chinese Academy of Sciences, China

[2]Kyung Hee University, Seoul, Korea

E-mail[*]: lingli@ime.ac.cn; liuming@ime.ac.cn

Abstract

For the first time, we present a surface potential-based compact model for a-IGZO TFTs based on multiple trapping-release theory and benchmark our work against device measurements. This model does not require time-consuming calculation. Meanwhile, we have developed the automatic parameter extraction program, which can extract the parameters rapidly and accurately. Moreover, the compact model is coded in Verilog-A, and implemented in a vendor CAD environment. This model provides physics-based consistent description of DC and AC device characteristics and enables accurate circuit-level performance prediction and RFID circuit design of a-IGZO TFTs.

Introduction

Amorphous In-Ga-Zn-O thin film transistors (a-IGZO TFTs) have been under active research and development due to the higher mobility and stability than amorphous silicon thin films and the good uniformity compared with polycrystalline silicon thin film transistors [1-4], and becoming increasingly important for the circuit application such as flexible display and transparent TFTs. Oxide semiconductor TFTs, such as a-IGZO TFTs are expected to be a promising candidate constructing RFID tags [5-7]. Due to the different charge transport mechanisms, compact model for a-IGZO cannot use organic TFT or amorphous TFT model directly. A circuit friendly compact model of a-IGZO is therefore required.

Device Modeling

Fig. 1 shows the geometric definition for a-IGZO TFT compact model and optical image of a fabricated a-IGZO TFT. Schematic view of the multiple trapping and release transport mechanism is illustrated in Fig. 2(a). All parameters are listed in Table 1.

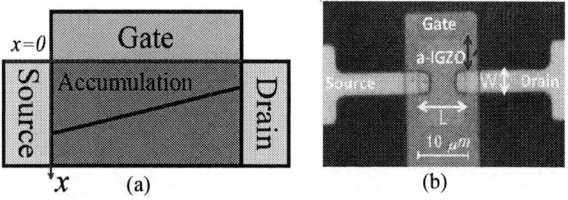

Fig.1 (a) Geometric definition for a-IGZO TFT compact model. (b) Optical image of a fabricated a-IGZO TFT with channel width W of 4μm and channel length L of 10μm.

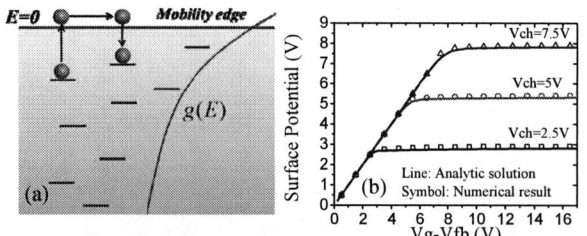

Fig.2 (a) Schematic view of the multiple trapping and release transport mechanism. (b) Comparison of the calculated surface potential between analytic solution and numerical results for different channel voltages.

Table 1. Parameters used in this work

$n(x)$	Total carrier concentration
$g(E)$	Trap density of states
$n_e(x)$	Extended states
$E_F(x)$	Quasi-Fermi level
V_g	Gate voltage
C_{OX}	Insulator capacitance per unit area
$\varphi(x)$	Surface potential
V_{fb}	Flat band voltage
v_0	Attempt-to-escape frequency
τ_0	Lifetime of carriers
μ_{eff}	Effective mobility
N_t	Total concentration of localized states
T_{TA}	Characteristic temperature of the exponential DOS
V_{CH}	Channel voltage

A. Multiple trapping-release theory

The total carrier concentration $n(x)$ is given by the sum of the carrier density in extended states $n_e(x)$ and carrier density in localized states as Eq. (1), where $g(E)$ is the trap density of states (DOS) energy distribution and $E_F(x)$ is the quasi-Fermi level [8,9]. In equilibrium state, the carrier trapping rate equals the releasing rate, and the Fermi energy is determined by Eq. (2).

$$n(x) = n_e(x) + \int_{-\infty}^{0} \frac{g(E)}{1+exp\left(\frac{E-E_F(x)}{k_B T}\right)} \, dE \tag{1}$$

$$E_F(x) = k_B T \ln\left(\frac{n_e(x)}{v_0 \tau_0 N_t}\right) \tag{2}$$

Connecting Eq. (1) and Eq. (2), the electric field $F(0)$ at the interface can be expressed in Eq. (3) and Eq. (4).

$$F(x) = \sqrt{\frac{2q}{\varepsilon_s}\left(v_0 \tau_0 N_t \frac{k_B T}{q} exp\frac{E_{F0}+q\varphi(x)}{k_B T} + \int_0^{\varphi(x)}\int_{-\infty}^0 \frac{g(E)}{1+exp\frac{E-(E_{F0}+q\varphi(x))}{k_B T}}dEd\varphi\right)} \tag{3}$$

$$C_{ox}(V_g - V_{fb} - \varphi_s) = \varepsilon_s F(0) \tag{4}$$

Based on exponential DOS in a-IGZO TFTs, the field effective mobility increases with gate voltage while approachs to a saturation value at high gate voltage as Eq. (5) and Eq. (6).

$$\mu = \frac{n_e(\varphi_s)\mu_0}{N_t v_0 \tau_0 exp\frac{E_{f0}+q\varphi_s}{k_B T} + \int_{-\infty}^0 \frac{g(E)}{1+exp\frac{E-(E_{f0}+q\varphi_s)}{k_B T}}dE} \tag{5}$$

$$\mu_{eff} = \sqrt[m]{\frac{\mu^m \mu_0^m}{\mu^m + \mu_0^m}} \tag{6}$$

B. Calculation of surface potentials

To calculate the surface potential accurately without using the numerical computation, Eq. (4) should be transformed to Eq. (7) at first.

$$V_g - V_{fb} - \varphi_S = G_T \exp(\frac{q\varphi_s - qV_{CH}}{2k_BT}) + G_{TA} \exp(\frac{q\varphi_s - qV_{CH}}{2k_BT_{TA}}) \quad (7)$$

where G_T and G_{TA} are:

$$G_T = \frac{1}{C_i}\sqrt{q\varepsilon_s \nu_0 \tau_0 N_t \frac{k_BT}{q} \exp(\frac{E_{F0}}{k_BT})} \quad (8)$$

$$G_{TA} = \frac{1}{C_i}\sqrt{q\varepsilon_s N_t \frac{k_BT_{TA}}{q} \Gamma(1+\frac{T}{T_{TA}})\Gamma(1-\frac{T}{T_{TA}}) \exp(\frac{E_{F0}}{k_BT_{TA}})} \quad (9)$$

By using two-order Taylor expansion and Schroder series, we can obtain:

$$x_i = xg\{[(xg+1)^2 + 2xn + 2\log(\frac{xg}{G_T})]^{\frac{1}{2}} - xg - 1\} \quad (10)$$

where

$$x = \frac{\varphi_s}{2k_BT_{TA}/q}, xg = \frac{V_g - V_{fb}}{2k_BT_{TA}/q}, xn = \frac{V_{CH}}{2k_BT_{TA}/q} \quad (11)$$

Finally, the analytical solution of surface potential can be written as Eq. (12), Fig. 2(b) shows the comparison between calculated surface potential and numerical result.

$$\varphi_s = \left(x_i - \frac{f}{\partial f}\left[1 + \frac{\partial^2 f}{2\partial f}\frac{f}{\partial f}\right]\right) \cdot 2\frac{k_BT_{TA}}{q} \quad (12)$$

$$f = (xg - x) - G_{TA}\exp(x - xn) - G_T[\exp(x-xn)]^{\frac{T_{TA}}{T}} \quad (13)$$

C. Current model

Using the gradual channel approximation, the current equation is given in Eq. (14). Integrating equation (14) from $\Phi_s = \Phi_{ss}$ to $\Phi_s = \Phi_{sd}$, the static current of a-IGZO TFTs is obtained in Eq. (15). Based on Eq. (15), a-IGZO TFTs characteristics can be described by a uniform formula without threshold voltage.

$$I_{ds} = -\mu_{eff}WQ_i\frac{dV}{dy} = -\mu_{eff}WQ_i\left(2\Phi_t\frac{C_{ox}}{Q_i}+1\right)\frac{d\Phi}{dy} \quad (14)$$

$$I_{ds0} = \mu_{eff}\frac{W}{L}\Big(2\Phi_t C_{ox}(\Phi_{sd} - \Phi_{ss}) - 0.5 \\ * \left((V_g - V_{fb} - \Phi_{sd})^2 - (V_g - V_{fb} - \Phi_{ss})^2\right)\Big) \quad (15)$$

D. Charge model

In addition to the drain current behavior, correct compact model of terminal charge and trans-capacitance are also required for transient mode simulation. The surface potential based gate charge can be expressed in Eq. (16).

$$Q_g = WLC_{ox}\frac{2/3[A^2+B^2+AB]}{A+B} \quad (16)$$

where $A = V_g - V_{fb} - \Phi_{ss}$ and $B = V_g - V_{fb} - \Phi_{sd}$.

According to Ward's charge-partitioning scheme [10] and total charge neutrality requirement, the drain charge Q_d and source charge Q_s can be expressed as Eq. (17). Finally, the trans-capacitances are now evaluated as Eq. (18).

$$Q_d = WLC_{ox}\frac{2}{15}\frac{[2A^3+4A^2B+6AB^2+3B^3]}{(A+B)^2}; \quad Q_s = -Q_d - Q_g \quad (17)$$

$$C_{i,j} = \pm\frac{\partial Q_i}{\partial V_j} \quad (18)$$

where i and j denote the transistor terminals G, D, or S. The negative sign is used when $i \neq j$.

Model Validation

Output characteristics curves and transfer characteristics curves are shown in Fig. 3(a) and Fig. 4(a) which demonstrate good agreement with the experimental data. Drain conductance and transconductance curves are shown in Fig. 3(b) and Fig. 4(b). In addition, comparison between the calculated and experimental data for output characteristics under different temperatures is shown in Fig. 5(a), from which we can find that the model is accurate in different temperatures. Meanwhile, we have compared the model capacitance with the experimental data in Fig. 5(b).

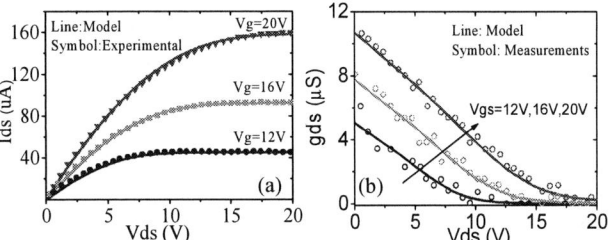

Fig.3 Comparison between the calculation and experimental data for output characteristics of IGZO under different gate voltages (a) and g_{ds}-V_{ds} curves (b). Good agreement has been achieved between the model and measurement.

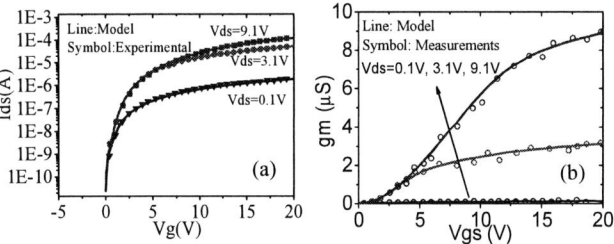

Fig.4 Comparison between the calculation and experimental data for transfer characteristics of IGZO under different drain-source voltages (a) and transconductance curves (b).

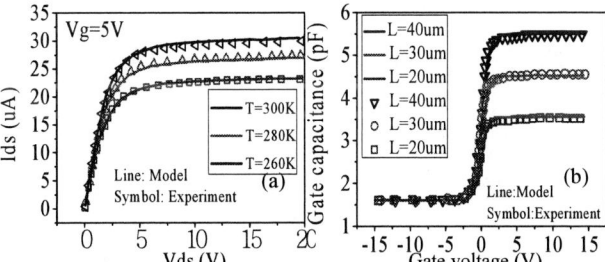

Fig.5 (a) Comparison between the calculation and experimental data for output characteristics of IGZO TFT at different temperatures. (b) Comparison between the calculation and experimental data for gate capacitance of IGZO TFT under different channel lengths.

In this model, there are three main physical parameters include maximum mobility μ_0, characteristic temperature T_a, and the product of escape frequency ν_0 and lifetime of carriers τ_0 need to be extracted before fitting parameter extraction. According to the flow in Fig. 6, the three parameters can be extracted sequentially. Fitting parameter extraction flow is shown in Fig. 7. Table 2 lists the key parameters and fitting parameters of the model.

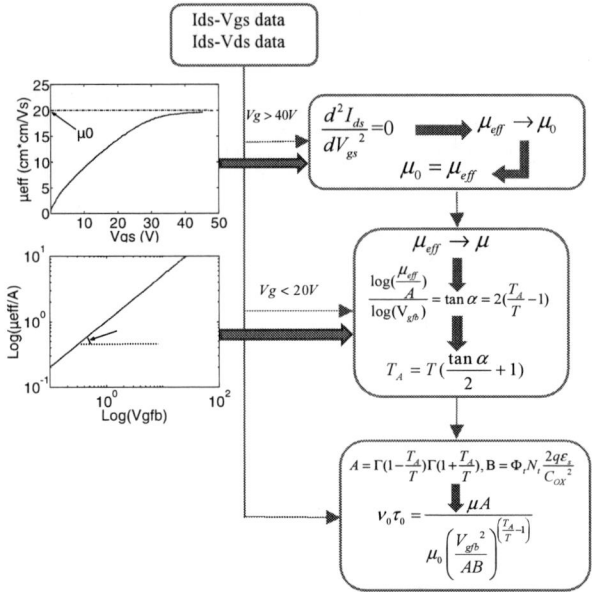

Fig.6 Extraction of key physical parameters of the model. The effective mobility tends to be saturated when V_g>40V. Hence we can obtain the μ_0 in this situation. Also, the characteristic temperature can be obtained in double logarithmic diagram of V_{gfb}-μ_{eff} when V_g<20V. Finally, the product of v_0 and τ_0 can be obtained by formula-transformation.

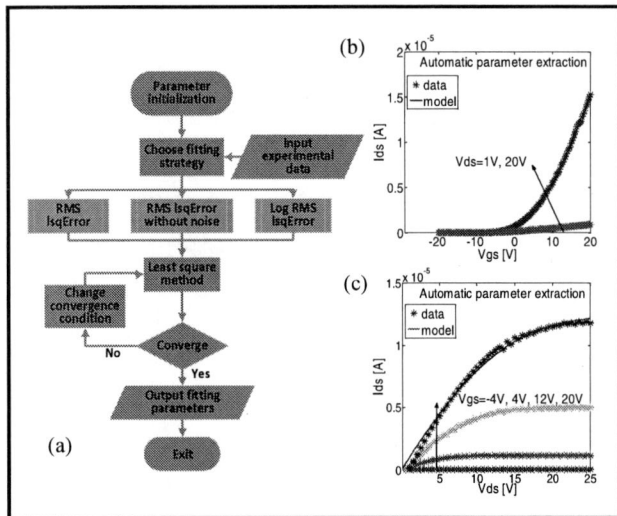

Fig.7 (a) Extraction flow of fitting parameters. The model fitting with experimental data are shown in (b) and (c).

Table 2. Extracted parameter values

Parameter	Extracted value	Parameter	Extracted value
T_A (K)	405	$v_0\tau_0$	1
n	5	C_{ox} (F/cm²)	8.85×10^{-8}
m	1	V_{fb} (V)	0.5
ss	2.75	μ_0 (cm²/Vs)	19.7

Gummel symmetry test is shown in Fig. 8 which displays

good symmetry and continuity characteristics of this model.

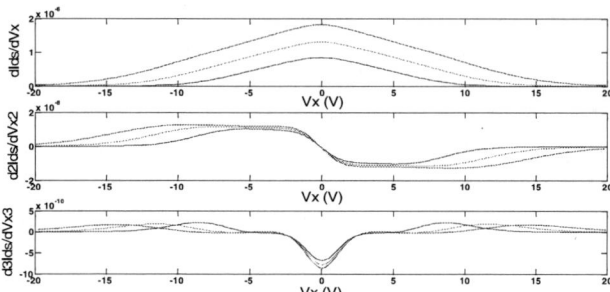

Fig.8 Gummel symmetry test for the 1, 2, 3-order derivative of the drain current under different gate voltages (Vg=12V, 16V, 20V in blue, green and red lines, respectively).

RFID Simulation Results and Discussion

RFID digital block is shown in Fig. 9. Based on the proposed model, the simulated waveforms of RFID logic circuit is shown in Fig. 10, where the first line is the clock signal waveform, the second line is ROM data output waveform, and the third line is Manchester encoded data waveform. The result shows that the logic circuit constructed by our model can achieve a certain function.

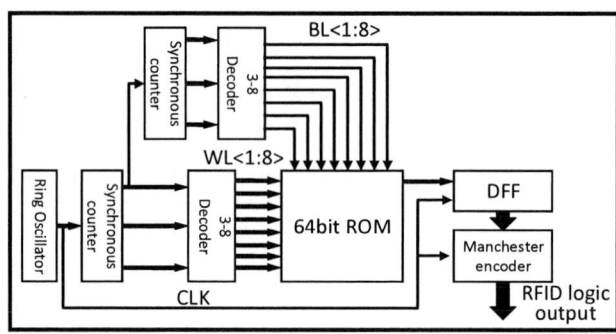

Fig.9 RFID digital block includes ring oscillator, synchronous counter, 3-8 decoder, 64bit ROM, D flip-flop and Manchester encoder.

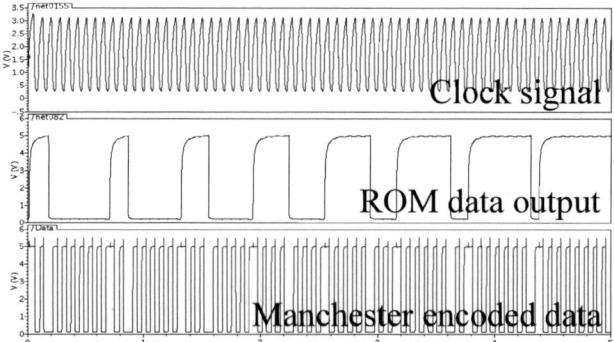

Fig.10 The simulated waveforms of RFID logic circuit, including clock signal generated by ring oscillator, ROM data read by WL and BL sequentially, then the output data is encoded by Manchester encoder.

Vin-Vout curves of inverter at different W/L show that the switch speed increases with W/L (Fig. 11). Fig. 12 shows clock frequency increases with driver transistor's

W/L and V_{DD}. It is found that W/L plays a little role in the low V_{DD} region. Besides, the circuit cannot work when W/L is lower than 37/1 because logic delay is too large to make the latch realize correct function, which determines the lowest operating frequency in a particular supply voltage. Meanwhile, we estimate the power consumption and operating frequency with V_{DD}. The results show a power consumption-frequency law, as shown in Fig. 13. In addition, we found that for each supply voltage V_{DD}, there is a maximum value of driver transistor's W/L, as shown in Fig.14. In other words, an upper limit frequency is always exists for each supply voltage. From what has been discussed above, the minimum power consumption and maximum frequency in particular condition can be obtained.

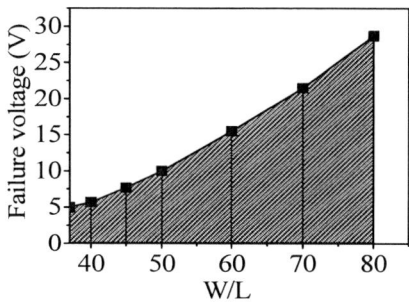

Fig.14 Failure voltage at different W/L of driver transistor. The shaded area shows available region for V_{DD} and W/L which is instructive to IGZO-based RFID circuit design.

Conclusion

In this work, we developed a compact model of a-IGZO TFTs for digital applications. Based on our model, the simulation result can match the experimental data very well. To verify this compact model in a systematic optimization of the circuit design more clearly, we use this model to simulate the RFID digital circuit. Furthermore, the maximum operating frequency and the minimum power consumption can be obtained, which is instructive to IGZO-based circuit design and performance prediction.

References

[1] K. Hoshino, et al., "*Constant-Voltage-Bias Stress Testing of a-IGZO Thin-Film Transistors*", Electron Devices, IEEE Transactions on , vol.56, no.7, pp.1365,1370, July 2009
[2] M. K. Kim, et al., "*High mobility bottom gate InGaZnO thin film transistors with SiOx etch stopper*", Applied Physics Letters, vol.90, no.21, pp.212114,212114-3, May 2007
[3] Y. W. Jeon, et al., "*Subgap Density-of-States-Based Amorphous Oxide Thin Film Transistor Simulator (DeAOTS)*", Electron Devices, IEEE Transactions on, vol.57, no.11, pp.2988, 3000, Nov. 2010
[4] L. Li, et al., "*Field Effect Mobility Model in Oxide Semiconductor Thin Film Transistors With Arbitrary Energy Distribution of Traps*", Electron Device Letters, IEEE. 2014
[5] T. Kawamura, et al., "*Oxide TFT rectifier achieving 13.56-MHz wireless operation with DC output up to 12V*", IEDM, 2010
[6] H. Ozaki, et al.,"*20-μW operation of an a-IGZO TFT-based RFID chip using purely NMOS "active" load logic gates with ultra-low-consumption power*", VLSI Circuits (VLSIC), 2011
[7] A. Chasin, et al., "*Integrated UHF a-IGZO energy harvester for passive RFID tags*", IEDM, 2013
[8] V. I. Arkhipov, et al., "*Space-charge-limited currents in materials with Gaussian energy distributions of localized states*", Appl. Phys. Lett., 79, no.25, pp. 4154-416, Dec. 2001.
[9] L. Li, et al., "*Diffusion-controlled charge injection model for organic light-emitting diodes*", Appl. Phys. Lett. 91,172111 (2007).
[10] D. E. Ward, "*Charge-based modeling of capacitance in MOS transistors*", Ph.D. dissertation, Stanford Univ., Stanford, CA, 1981.

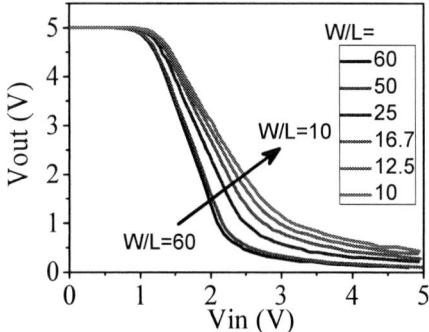

Fig.11 V_{in}-V_{out} curves of inverter at different W/L of driver transistor (the load transistor's W/L is fixed at 50/30).

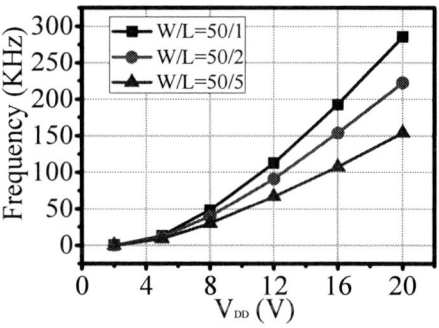

Fig.12 Frequency of ring oscillator vs. V_{DD} at different W/L of driver transistor.

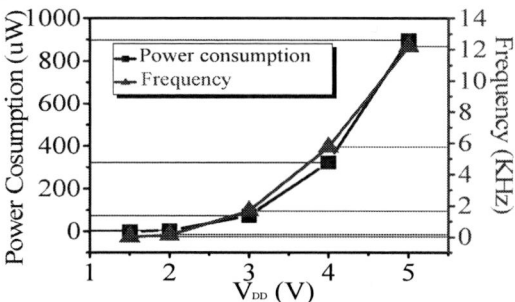

Fig.13 The circuit power consumption and operating frequency vs. V_{DD}. The W/L of driver transistor we choose is 37/1 which is the least workable W/L for circuit.

978-1-4799-8002-4/14 $31.00 © 2014 IEEE

Physics-Based Factorization of Magnetic Tunnel Junctions for Modeling and Circuit Simulation

Kerem Yunus Camsari[1], Samiran Ganguly[2], Deepanjan Datta[3], Supriyo Datta[4]

School of Electrical & Computer Engineering, Purdue University, West Lafayette, IN, 47907, USA[1,2,4]
GLOBALFOUNDRIES Singapore, 738406, Singapore[3]

Abstract

We present a physics-based factorization of Magnetic Tunnel Junctions (MTJ) in terms of a minimal number of experimentally and theoretically accessible parameters that can be used to optimize existing MTJ designs as well as to probe emerging MTJ devices. Our model fully captures angular/voltage dependence of state-of-the-art MTJs and Spin Valves (SV) and is compatible with existing circuit simulation frameworks such as Verilog-A and SPICE.

Introduction

The growing practical interest in Magnetic Tunnel Junction (MTJ)-based devices requires flexible as well as robust device models to be used in circuit simulators. We introduce such a robust, physics-based device model that **(1)** factorizes MTJs into a combination of two ferromagnet/interface conductance matrices that are described by four experimentally and theoretically accessible parameters (G_0, P, a, b), **(2)** captures angular as well as voltage dependence of charge and spin-transfer-torque conductances of MTJs with a single set of these 4 parameters, **(3)** similarly factorizes spin-valves into two conductance matrices reproducing established spin valve physics as a demonstration of the generality of the factorization approach. This suggests that the physics-based factorization can be used to model emerging devices, such as double-interface MTJ stacks (FM/I/FM/I/NM), by factorizing them into a combination of smaller conductance matrices, for exploratory design and circuit modeling in Verilog-A/SPICE framework. These features represent key advantages over (1) dedicated Non-equilibrium Green's Function (NEGF) or Scattering Theory based models ([1]–[3]) due to rapid development time and over (2) compact models that only capture partial device physics [4].

Building Blocks: FM||NM Interface Conductances

Our starting point to describe spin valves and MTJs is in terms of a generalized [4×4] conductance matrix that describes charge/spin current flow through a thin FM layer interfacing a NM (normal metal, non-magnetic) layer with the following parameters:

$$G_{FM} = G_0 \begin{bmatrix} 1 & P & 0 & 0 \\ P & 1 & 0 & 0 \\ 0 & 0 & a & b \\ 0 & 0 & -b & a \end{bmatrix}$$

Charge conductance (G_0), interfacial polarization (P), and spin-mixing conductances for Slonczewski (a) and field-like (b) spin-torque currents. The collinear parameters P and G_0 can either be theoretically (S-Matrix, NEGF) or experimentally obtained. Similarly, the non-collinear parameters (a) and (b), also known as the spin-mixing interface conductances, can be predicted from first-principles or directly measured. Recent experiments [5] have proved the validity of this way of representing the FM/NM interfaces by testing the model against a wide variety of physical phenomena. The conductance matrix is written in (c,z,x,y) basis where z denotes the magnetization axis of the magnet. The voltages in the NM and FM side correspondingly, are 4-component vectors $(V_c, V_z, V_x, V_y)^T$ described in terms of spin and charge quasi-Fermi levels (T denotes transpose). It is important to note that even though the voltage vector at the NM side can have any arbitrary value, the spin-voltage in the FM side is assumed to have only a charge voltage $(V_c, 0, 0, 0)^T$ since the ordinary leads attached to the outer terminals do not have any spin accumulation. This assumption may break down for spin-driven MTJs where spin-current is injected to the free layer from an external spin source like a heavy metal exhibiting GSHE, however this is beyond our scope. Under these assumptions for the spin voltages and the conductance matrix of the interface, the 4-component spin current can then be written as:

$$I_{FM||NM} = G_{FM} (V_{NM} - V_{FM}) \qquad (1)$$

with the current direction into the FM defined as positive.

Spin-Valve (SV)

We demonstrate the power of the physics-based factorization technique by first showing that it can be applied to describe angular magnetoresistance of CPP-type spin-valve devices. Postulating that the spin-valve is composed of two FM||NM interfaces that are attached back to back, the combined conductance of the SV can be written as:

$$G_{SV} = (G_{FM1}^{-1} + G_{FM2}^{-1})^{-1} \qquad (2)$$

where G_{FM1} and G_{FM2} represent two FM conductances making an angle with respect to each other. In order to describe the conductance matrix of a magnet pointing in a direction other than +z direction, a 4×4 rotation matrix is used for the second magnet such that ($G_{FM2} = U\, G'_{FM2}\, U^T$), where G'_{FM2} is the intrinsic conductance of FM_2 in its own magnetization axis. The rotation matrix U is obtained from

978-1-4799-8002-4/14 $31.00 © 2014 IEEE

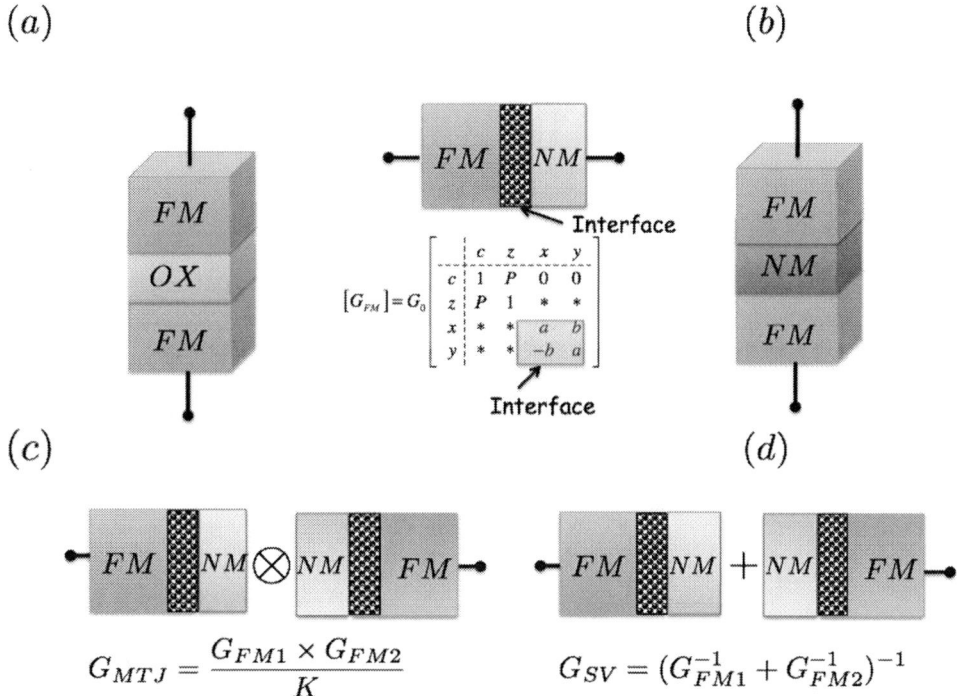

FIG.1: A Magnetic Tunnel Junction (MTJ) and an All-Metallic Spin Valve (SV) (a)-(b) are decomposed as two individual ferromagnet/interface units that are described by simple [4×4] conductance matrices. (c) In the case of MTJs, the "tunneling" physics of require these matrix conductances to be multiplied in contrast to the physics of (d) spin valves where they are added serially in accordance with basic circuit theory, to account for the highly diffusive nature of transport in these structures. The different interface properties corresponding to spacer materials of SVs and MTJs are adjusted in the conductance matrix through spin-mixing conductances which can either be calculated from *ab-initio* theories or obtained directly from experiment [5].

the standard Rodrigues' formula, where the rotation is defined to be within the z-x plane. When the combined conductance of the Spin Valve is computed from this formula, the angle dependent charge conductance (the (1,1) entry of the combined matrix) reads:

$$\left(\frac{R(\theta) - R(0)}{R(\pi) - R(0)} \right)_{SV} = \frac{1 - \cos\theta}{\chi(1 + \cos\theta) + 2} \qquad (3)$$

where χ is a parameter in terms of the interface mixing conductance and polarizations of the interfaces, $\chi = a/(1 - P^2) - 1$ that exactly matches established SV theory based on Boltzmann equation (Eq. (2) in [6]) as well as explains related experiments quantitatively [7] (Fig.3). Note that the parameter "b" is assumed to be zero, since this component is known to vanish for Ohmic interfaces such as a typical Py ‖ Cu interface.

Magnetic Tunnel Junction (MTJ)

The combined [4×4] conductance of the MTJ that provides the spacer current can be written in terms of the matrix product of the same building blocks:

$$[G_{MTJ}]_{12} = (G_{FM1} \times G_{FM2})/K \qquad (4)$$

where K is a normalization factor given in terms of $\sqrt{G_{0_1} G_{0_2}}$, the charge conductance G_0 of the individual FM blocks, and G_{FM2} makes an angle with G_{FM1}, similarly obtained from a rotation matrix. The multiplication of the FM matrix conductances is a generalized version of Julliere's original treatment [8] of tunneling between FM films, and can be intuitively understood by considering the charge resistance of a tunneling junction (R_J) where cascading tunnel junctions with length d1 and d2, results in an exponential increase in $R_J = e^{-\alpha d_1} e^{-\alpha d_2}$. The first column (charge-driven currents) of the matrix obtained from combining the interface conductances in this way, reproduces all of the angle dependent spin-currents in the spacer *exactly* as they are *analytically* given by NEGF formalism in the linear response regime [1] (Fig. 4).

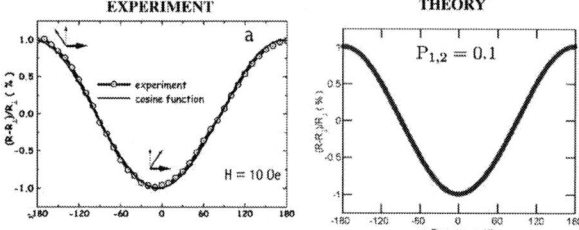

FIG. 2: We factorize the MTJ into distinct conductance matrices generalizing Julliere's original treatment [12] by multiplying [4×4] G_{FM} matrices to account for tunneling physics. The (c,c) element of the analytically obtained [4×4] matrix for the composite structure matches established MTJ conductance ($G(1+P_1P_2 \cos\theta)$) [[1], [9]] and reproduces the well-known experimental angular dependence of low-field TMR [10].

In general the MTJ conductance for the spacer currents can be written as (Fig.4) :

$$I_2 = G_{12}V_2 - G_{21}V_1 \qquad (5)$$

where G_{12} and G_{21} are the conductance matrices associated with different orders of multiplication. The order of multiplication in the MTJ conductance is significant and it predicts a sign difference (unlike Spin-Valves) for the out-of-plane spin-current components, ($[G_{21}]^{m\times M}=-[G_{12}]^{m\times M}$) implying a quadratic bias dependence for out-of-plane currents [6]. For in-plane spin-currents and charge currents, there is no such sign difference in the charge-driven columns of G_{12} and G_{21}. The common assumption that the current through the spacer of the MTJ conserves all spin and charge currents requires the columns of the 2-port conductance matrix (Fig. 4b) to add to zero, separately.

A.   Voltage Dependence of MTJs

The combined interface conductance matrix presented so far, works only in the linear response regime. To capture the high bias features of MTJs, we phenomenologically lump the bias dependence of the MTJ into the interface polarizations making energy dependent polarizations voltage dependent, $P(E) \rightarrow P(V)$ and assuming a symmetric junction so that P1(V)=P2(—V)). A typical polarization dependence of voltage is illustrated in Fig. 5. We show that lumping all the voltage dependence to polarizations of the interfaces captures effects such as voltage dependence of tunneling magnetoresistance (TMR), voltage asymmetry and quadratic dependence of in-plane and out-of-plane spin currents, respectively. These findings are shown in (Fig. 5-6-7) detailing the used parameters obtained from matched experiments.

Conclusion

We have introduced a physics-based model that factorizes MTJs into simpler units that are described by 4 experimentally and theoretically accessible parameters. The model produces established analytical results for MTJs and SVs, and is shown to match experiments.

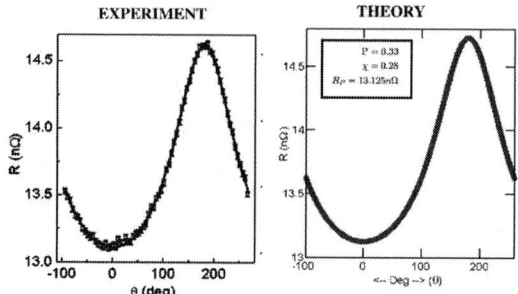

FIG. 3: A spin valve (SV) structure is obtained by adding the resistance matrices showing the robustness of the physics-based factorization. The combined conductance G_{SV} matches established SV theory (Eq. 2 in [6]) by fully reproducing the non-collinear angle dependence of GMR. Experimental features of [7] are also reproduced quantitatively within our model with the Slonczewski type mixing conductance "a" while field-like mixing conductance "b" is known to be 0, for such FM/NM interfaces [6].

The power of the factorization approach suggests this model can be used in an exploratory mode for emerging devices or to optimize existing challenges in MTJ designs.

Acknowledgments

This work was supported in part by C-SPIN, one of six centers of STARnet, a Semiconductor Research Corporation program, sponsored by MARCO and DARPA, and in part by the National Science Foundation through the NCN-NEEDS program, contract 1227020-EEC.

References

[1]   D. Datta, B. Behin-Aein, S. Datta, and S. Salahuddin, "Voltage Asymmetry of Spin-Transfer Torques," *IEEE Trans. Nanotechnol.*, vol. 11, no. 2, pp. 261–272, 2012.

[2]   J. Xiao, G. E. W. Bauer, and A. Brataas, "Spin-transfer torque in magnetic tunnel junctions: Scattering theory," *Phys. Rev. B*, vol. 77, no. 22, p. 224419, Jun. 2008.

[3]   I. Theodonis, N. Kioussis, A. Kalitsov, M. Chshiev, and W. H. Butler, "Anomalous Bias Dependence of Spin Torque in Magnetic Tunnel Junctions," *Phys. Rev. Lett.*, vol. 97, no. 23, p. 237205, Dec. 2006.

[4]   A. Nigam, K. Munira, A. Ghosh, S. Wolf, E. Chen, and M. R. Stan, "Self consistent parameterized physical MTJ compact model for STT-RAM," in *Semiconductor Conference (CAS), 2010 International*, 2010, vol. 02, pp. 423–426.

[5]   M. Weiler et. al., "Experimental Test of the Spin Mixing Interface Conductivity Concept," *Phys. Rev. Lett.*, vol. 111, no. 17, p. 176601, Oct. 2013.

[6]   G. E. W. Bauer, Y. Tserkovnyak, D. Huertas-Hernando, and A. Brataas, "Universal angular magnetoresistance and spin torque in ferromagnetic/normal metal hybrids," *Phys. Rev. B*, vol. 67, no. 9, p. 094421, Mar. 2003.

[7]   S. Urazhdin, R. Loloee, and W. P. Pratt, "Noncollinear spin transport in magnetic multilayers," *Phys. Rev. B*, vol. 71, no. 10, p. 100401, Mar. 2005.

[8]   M. Julliere, "Tunneling between ferromagnetic films," *Phys. Lett. A*, vol. 54, pp. 225–226, Sep. 1975.

978-1-4799-8002-4/14 $31.00 © 2014 IEEE

[9] J. C. Slonczewski, "Currents, torques, and polarization factors in magnetic tunnel junctions," *Phys. Rev. B*, vol. 71, no. 2, p. 024411, Jan. 2005.

[10] H. Jaffrès et. al., "Angular dependence of the tunnel magnetoresistance in transition-metal-based junctions," *Phys. Rev. B*, vol. 64, no. 6, p. 064427, Jul. 2001.

[11] J. C. Sankey et. al., "Measurement of the spin-transfer-torque vector in magnetic tunnel junctions," *Nat. Phys.*, vol. 4, no. 1, pp. 67–71, Jan. 2008.

[12] H. Kubota et. al., "Quantitative measurement of voltage dependence of spin-transfer torque in MgO-based magnetic tunnel junctions," *Nat. Phys.*, vol. 4, no. 1, pp. 37–41, Jan. 2008.

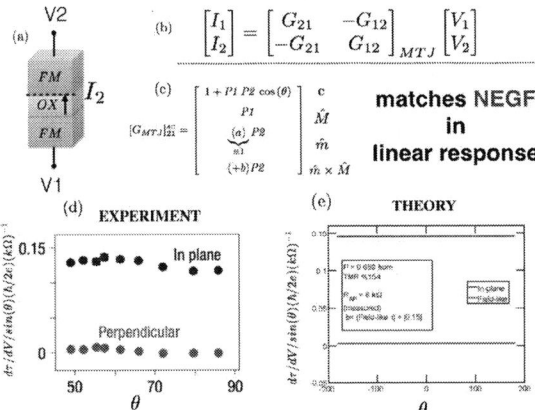

FIG. 4: (a-b) Assuming no scattering within the short spacer [1], the 2-port MTJ conductances can be written in terms of [4x4] G_{12} and G_{21} matrices. This implies the sum rule $G_{11}+G_{21}=0$ and $G_{22}+G_{12}=0$ hold, leaving only 2-independent conductances, G_{12} and G_{21} that are given in terms of ($G_{FM1} \times G_{FM2}$ and $G_{FM2} \times G_{FM1}$ respectively ($G_{12} \neq G_{21}$ since $[G_{12}]^{m\times M} = -[G_{21}]^{m\times M}$). (c) The charge-driven spin current column of the combined conductance matrix of the MTJ, where theta is the angle between the free layer and the reference layer, exactly matches results of NEGF [1] in linear response, assuming a=1 (Ballistic limit) (d-e). Experimental angular [11] dependence of zero-bias torkances showing sin (θ) dependence is reproduced by P= 0.659 obtained from reported TMR=154% and R_{AP}=8 kΩ with a=1 and b=15% (for field-like torque 15% of in-plane torque), assuming they are both voltage independent.

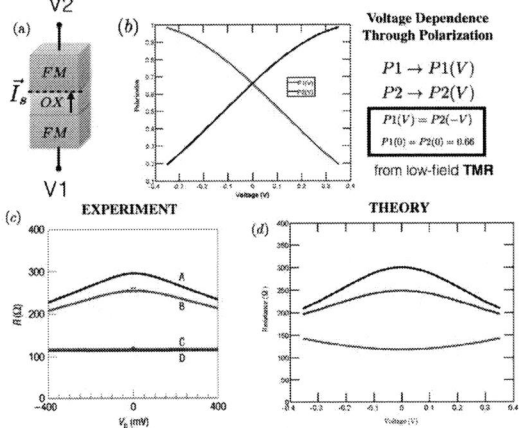

FIG. 5: (a-b) The voltage dependent features of the MTJ devices

are approximated by turning the energy dependent polarizations into voltage dependent polarizations, P (E)→ P (V) as shown in (b) similar to the treatment described in [9]. (c) Experimental voltage dependence [11] of charge conductance of the MTJ is reproduced for different free/fixed layer configurations by these voltage dependent polarizations, where $(R_{MTJ})^{-1}$=G_{MTJ0} (1+P1(V) P2(V) cos(θ), P1(V=0)=P2(V=0)=0.659 obtained from low-field TMR=154%. Note that the voltage dependent polarizations are equal to 0.659 at V=0. (d) The conductance, $G_{MTJ0} = (169\,\Omega)^{-1}$ is obtained from the experimental R_{AP}=294 Ω and assumed to be voltage independent, possibly the reason for the apparent discrepancy between theory and experiment between (c)-(d).

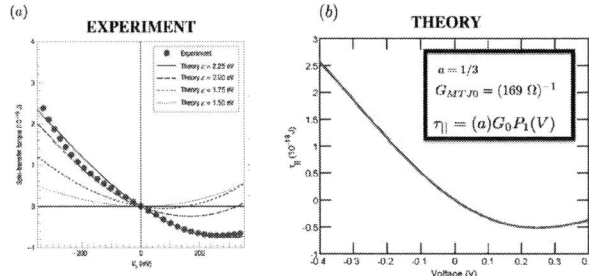

FIG. 6: (a) Using the same voltage dependent polarization factors for the two interfaces described in (Fig. 5), the unified model reproduces the in-plane spin-torque incident to the free layer (FM2). (b) For this experiment, the in-plane spin conductance is assumed to be a=0.33 (compared to a=1 for Fig.4) and voltage independent and $G_{MTJ0} = (169\,\Omega)^{-1}$ as in Fig. 5 (Measured in Ref. [12]).

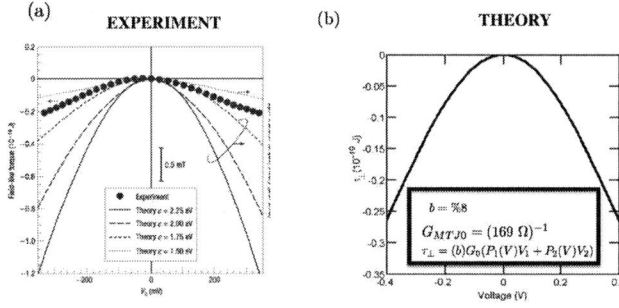

FIG. 7: (a) The form of the out-of-plane spin torque (inset-b) is obtained from $I_2 = (G_{12}V_2 - G_{21}V_1)$ (Fig.4), where $V_2 = +V/2$ and V_1=-V/2, and the non-reciprocity between G_{12} and G_{21} leads to the quadratic bias dependence of the out-of-plane spin current, since $G_{12} = G_{21}$ for all directions (c,M,m) except the out-of-plane direction (m×M) where $[G_{12}]^{m\times M} = -[G_{21}]^{m\times M}$. (b) Using the same voltage dependent interface polarization factors for the two interfaces (Fig. 5), and $G_{MTJ0} = (169\,\Omega)^{-1}$, and the out-of-plane spin-mixing conductance 'b' is assumed to be 8%, proposed model quantitatively reproduces the experimentally measured out-of-plane spin torque [12].

978-1-4799-8002-4/14 $31.00 © 2014 IEEE

AUTHOR INDEX

Abbate, F. 77
Abe, K. 650
Abisset, A. 518
Absil, P. 128, 168
Achanta, Ravi 840
Adelmann, Christoph 164
Adisusilo, I. N. 176
Afanas'Ev, V. V. 530
Afzalian, A. 235
Afzali-Ardakani, Ali 602
Agnello, P. 74, 490
Agostinelli, M. 71
Agrawal, A. 414
Aguilera, M. 40
Ahmad, Zeshan 92
Ahmed, N. 80
Ahn, Chong H. 753
Ahn, S. H. 768
Aiyar, A. 74
Ajoy, Arvind 347
Akbar, S. 71
Akinwande, Deji 216
Akita, K. 685
Akiyama, S. 40
Alain, F. 406
Alam, M. A. 502, 510
Alapati, R. 462
Alayan, M. 144
Albert, J. 772
Alessandri, C. 128
Alian, A. 506, 828
Allibert, F. 172, 219
Alpetkin, E. 74
Amano, M. 650
Ambrogio, S. 359, 363
An, J.-J. 74
Ando, T. 582
Andrieu, F. 172, 223
Anthopoulos, T. D. 610
Antoniadis, D. A. 578
Antoniadis, Dimitri 295, 844
Antoniadis, Dimitri A. 574
Aoki, M. 486
Aoulaiche, M. 772
Aquilino, M. 74, 490
Araidai, M. 538, 685
Aratani, K. 140
Arcamone, J. 223, 550
Arimura, H. 418, 828
Arndt, G. 550
Arnold, N. 490
Arulkumaran, S. 594
Asai, Y. 40
Asano, K. 36
Asselberghs, I. 128

Augendre, E. 172, 219
Augur, R. 462
Aussenac, F. 642
Avenier, G. 77
Baek, J. M. 768
Baek, J. U. 482
Baek, R. H. 578
Baets, R. 128
Baghini, Maryam Shojaei 63
Bai, Y. 657
Baillin, X. 223
Balakrishnan, Karthik 402
Balakrishnan, M. 140
Balan, V. 152
Balatti, S. 359, 363
Baliga, B. J. 20
Banerjee, Kaustav 120, 729
Bao, R. 74
Baraskar, Ashish 402
Barci, M. 136
Barge, D. 642
Barla, K. 418
Barnes, J.-P. 642
Barnola, S. 223
Barraud, S. 152, 172, 406, 518
Barth, M. 414, 673
Bashir, R. 390
Basker, V. 74
Basu, A. 582
Bataillon, J.-L. 219
Bateman, B. 657
Batude, P. 223, 642
Bea, J. C. 374
Bea, J.-C. 669
Beece, A. 462
Beigne, E. 223
Belmonte, A. 351
Benbakhti, B. 820
Bender, H. 351, 418
Benedetto, K. 562
Benevent, V. 642
Beneyton, R. 77
Ben-Jamaa, H. 856
Benoist, T. 144
Bernard, M. 136, 152
Bernasconi, M. 701
Bernier, N. 406
Berthier, L. 77
Bertin, F. 223
Besson, P. 642
Beyer, G. 168
Beyne, E. 168
Bhattacharjee, Nirjhar 753
Bianchini, R. 77
Bichler, O. 661

AUTHOR INDEX

Birkhahn, R. 40
Blackwell, D. 462
Blaise, P. 136, 152, 546
Boeuf, F. 219
Boniardi, M. 681
Bonilla, G. 522
Bonnevialle, A. 546
Borbely, A. 856
Borot, B. 77
Boschke, R. 514
Bost, M. 71
Bottacchi, F. 610
Bouche, G. 462
Boulenc, Pierre 100
Bowonder, A. 71
Bowyer, S. 657
Boyd, K. 74
Brar, B. 578
Breil, N. 74
Brems, S. 128
Breslin, C. 764
Brianceau, P. 152
Brightsky, M. 709
Brightsky, Matthew 693
Brink, M. 764
Brodsky, M. 74
Bromberg, D. M. 792
Bronner, G. 657
Brookes, S. 414
Brown, Jeffrey S. 852
Bruce, R. L. 582
Brun, P. 77
Brunet, L. 642
Bryant, A. 74
Bu, Weihai 319, 816
Bucchignano, J. J. 764
Büchele, Patric 259
Buczko, M. 77
Burr, G. W. 697, 705
Büthe, L. 610
Büyüklimanli, T. 788
Cai, H. 398
Cai, J. 74
Caillat, C. 772
Calarrudo, M. 657
Calderoni, A. 140, 359
Campanella, Humberto 566
Campbell, Michael 271
Campidelli, Y. 77
Campolini, L. 59
Camsari, Kerem Yunus 864
Canderle, E. 77
Cantarella, G. 610
Cao, Wei 729
Carabasse, C. 152

Caravati, S. 701
Cardon, C. 140
Carminati, Y. 77
Carrere, J.-P. 80
Caruso, Enrico 188
Carver, Alexander G. 267
Cassé, M. 172, 406, 518, 642
Celano, U. 351
Céli, D. 77
Celik, M. 172, 219
Cha, J. O. 768
Chae, K. S. 482
Chafik, F. 172, 219
Chan, Kevin K. 402
Chan, Mansun 311, 733
Chancel, C. 856
Chandorkar, A. N. 63
Chandra, B. 490
Chandramoorthy, N. 673
Chanemougame, D. 172, 219
Chang, C. H. 48, 634
Chang, Chao-Yuan 132
Chang, Chu-En 96
Chang, Chun-Yen 622
Chang, J. 764
Chang, Ken 454
Chang, Kuan-Chang 800
Chang, Kuo-Ping 498
Chang, Leland 307
Chang, M. F. 709
Chang, M.-H. 450
Chang, Meng-Fan 67
Chang, Pengying 192
Chang, S. Y. 48
Chang, S. Z. 48
Chang, T. 48
Chang, Ting-Chang 622, 800
Chang, Tzong-Sheng 148, 494
Chang, V. S. 48
Chang, W. 48, 74, 490
Chang, Wen-Hao 132
Chang, Y.-C. 450
Chang, Yao-Feng 216
Chang-Liao, K. S. 438
Chang-Liao, Kuei-Shu 410
Chao, T.-S. 788
Chapon, J. D. 77
Charbon, E. 255
Charbon, Edoardo 251
Charpin, C. 144
Charvet, A. M. 562
Chasin, Adrian 598
Cheam, Daw D. 570
Cheek, R. 709
Chen, Bangtao 570

AUTHOR INDEX

Chen, Bo-Yuan 67
Chen, C. 200
Chen, C. C. 48
Chen, C. Y. 355
Chen, C.-C. 808
Chen, C.-H. 808
Chen, C.-M. 450
Chen, Chang-Hsiao 132
Chen, Chao-Yu 558
Chen, Cheng 335
Chen, Cheng-Chi 554
Chen, Chien-Fu 67
Chen, Chih-Ping 498
Chen, Chun-Yuan 410
Chen, H. F. 48
Chen, H.-C. 788
Chen, Hong-Yu 116, 156
Chen, Hsin-Lu 800
Chen, Hung-Bin 622
Chen, I-Ru 327, 677
Chen, J. H. 48
Chen, Jim 494
Chen, K. J. 442
Chen, K. S. 48
Chen, Kevin J. 287, 434
Chen, L. C. 450
Chen, M. 390
Chen, M.-C. 808
Chen, P. C. 450
Chen, R. 48, 200
Chen, S.-H. 514
Chen, Shu-En 148
Chen, Shu-Han 231
Chen, T.-T. 390
Chen, X. 490
Chen, Y. Y. 355
Chen, Y.-J. 808
Chen, Yang-Yin 164
Chen, Yenhao 677
Chen, Yen-Ting 239
Chen, Yu-Sheng 239
Chen, Z. 200
Cheng, C.-W. 390, 582
Cheng, Chao-Ching 231
Cheng, H. Y. 709
Cheng, J. Y. 48, 764
Cheng, K. 172, 219
Cheng, L.-H. 390
Cheng, Samuel L. 267
Cheng, Ya-Chi 622
Chéramy, S. 856
Chevalier, P. 77
Chevallier, C. 657
Chew, S. A. 772
Chhajed, S. 140

Chhun, S. 642
Chi, Li-Jen 132
Chiang, M. C. 48
Chiang, W. C. 634
Chidambarrao, D. 74
Chien, W. C. 709
Chikarmane, V. 71
Chin, Chi-Hang 558
Chin, Yung-Wen 148
Chini, Alessandro 446
Chiu, H.-C. 450
Chiu, Y. H. 48
Cho, Chunhum 108
Cho, M. 772, 828
Cho, T.-C. 788
Cho, Y. W. 768
Cho, Y.-D. 578
Chobpattana, V. 586
Choe, Sonbeak 275
Choi, H. J. 768
Choi, J. K. 768
Choi, Jeong-Hyuk 367
Choi, Ji-Min 646
Choi, M. 168
Choi, Munkang 180
Choi, Pilsoon 295
Choi, Won Ho 315
Choi, Woosung 184
Choi, Yang-Kyu 204, 646
Chokawa, K. 538
Chouksey, S. 71
Chowdhury, S. 40
Christiansen, C. 74
Chu, H. C. 634
Chu, Tian-Jian 800
Chua, Geng L. 570
Chung, C.-H. 808
Chung, K.-M. 768
Chung, Ming-Hsien 618
Chung, Steve S. 303
Ciocchini, N. 701
Clarke, R. Chris 291
Clement, H. 40
Clendenning, S. B. 200
Clermidy, F. 144, 152
Clevenger, L. 74
Clima, S. 355
Clima, Sergiu 164
Cluzel, J. 152
Cohen, S. 371
Coignus, J. 144
Colburn, M. E. 764
Colinet, E. 550
Collaert, N. 168, 418, 506, 717, 828
Collins, C. 371

AUTHOR INDEX

Colonna, J. P. 856
Conklin, D. 74
Conrad, N. 502
Conrad, Nathan 227
Conrad, Nathan J. 112
Cook, B. 140
Cooper, D. 223
Cordes, A. 776
Cosemans, Stefan 164
Cossalter, J. 77
Cotter, M. 673
Coudrain, P. 856
Courouble, K. 77
Cramer, Harlan 291
Crescentini, Marco 741
Crespi, L. 681
Croon, Jeroen 430
Crotti, Davide 164
Cueto, O. 152, 223
Cupeta, C. 681
Current, M. I. 788
Da Silva, Maurício Banaszeski 848
Dahmani, F. 136
Damarla, G. 140
Dansas, H. 642
Dao, V. T. S. 255
D'Arrigo, G. D. 681
Dasgupta, A. 71
Datta, Deepanjan 864
Datta, S. 414, 673
Datta, Supriyo 864
Davies, W. 490
Dawes, S. 764
De Crécy, F. 856
De Marchi, Michele 339
De Meyer, K. 717
De Micheli, Giovanni 339
De Moor, Piet 100
De Munck, Koen 100
De Salvo, B. 136, 152, 172, 219, 223, 546, 661
De-Buttet, C. 77
Deglise, C. 77
Degraeve, R. 355
Del Alamo, J. A. 578
Del Alamo, Jesús A. 574, 590
Delaye, V. 152, 562
Deleonibus, S. 223
Delprat, D. 406
Demkov, Alexander A. 216
Deng, Yexin 112
Depetrillo, Paolo 761
Deprat, F. 642
Derrier, N 77
Deshpande, Abhishek 315
Detavernier, C. 351

Dewan, C. 74
Di Nolfo, C. 697
Diaz, C. H. 235
Dicioccio, L. 223
Divakaruni, R. 74
Dolmanan, S. B. 594
Dong, H. 74
Doornbos, G. 235
Doris, B. 172, 219
Druais, G. 642
Du, Gang 192
Du, P. Y. 709
Du, Pei-Ying 498
Duan, Meng 816
Dubois, E. 212
Duchemin, I. 518
Duijnhoven, Adrie Zegers-Van 848
Dumas, S. 856
Dunn, D. 40
Dunn, K. 776
Dupré, C. 550
Duraffourg, L. 223, 550, 562
Durand, C. 77
Duriez, B. 235
Eberhart, H. 458
Economikos, L. 74
Eichfeld, C. 414
Ekerdt, John G. 216
El-Sayed, A.-M. 530
Endo, K. 243, 538, 824
Endo, Kazuhiko 299
Endoh, T. 796
Endoh, Tetsuo 654
Eneman, G. 168, 418
Engbrecht, E. 74
Engel, B. 74
Engel, Jesse H. 693
Engel-Herbert, R. 414
Engelmann, S. 764
Engelmann, Sebastian U. 402
Enobio, E. C. I. 796
Ercan, Alper 100
Ernst, T. 223, 550, 562
Eryilmaz, S. Burc 693
Escarabajal, Y. 219
Esseni, David 188
Etoh, T. G. 255
Faber, H. 610
Fanet, H. 223
Fang, S. 74
Fantini, A. 355
Fantini, P. 701
Farcy, A. 856
Farmer, D. B. 582
Farooq, M. 371

AUTHOR INDEX

Favennec, L. 77
Faynot, O. 172, 219, 223, 406
Fazan, P. 772
Feng, Philip X.-L. 196
Fenouillet-Beranger, C. 642
Fernando, Sanchitha N. 570
Ferrer, D. 74
Fiori, V. 856
Fischer, K. 71
Fischer, M. 140
Fischer, Rene 259
Fischetti, Massimo V 804
Flatresse, P. 59
Foong, Fatt 606
Fossum, Eric R. 247
Foussadier, F. 77
Franco, J. 506, 820, 828
Franiatte, R. 856
Frank, M. M. 582
Frans, Yohan 454
Frederick Jr., Marlin 323
Freeman, G. 490
Freitag, Ron 291
Friedman, A. 74
Fu, Q. 71
Fujiki, Jun 327, 677
Fujita, Hiroyuki 737
Fujita, S. 650
Fukami, S. 796
Fukuda, K. 36, 243, 824
Fukuda, Koichi 299
Fukuda, S. 486
Fukushima, T. 374, 669
Furchi, Marco M. 124
Gaben, L. 406
Gabor, A. 74
Gaillardon, Pierre-Emmanuel 339
Gallagher, William J. 307
Gallois-Garreignot, S. 856
Gamrat, C. 661
Ganguly, Samiran 864
Gao, S. 462
Garbin, D. 144, 661
Gardner, D. S. 200
Garros, X. 642
Gasparutto, D. 223
Geelen, Bert 261
Geisler, M. 128
Ghaffari, Roozbeh 761
Ghani, T. 71
Ghibaudo, G. 152, 172, 661
Giachino, Marta 812
Giacometti, E. U. 697
Giammaria, G. 351
Gignac, L. 764

Giles, M. 71
Giraud, B. 144, 856
Gloria, D. 77
Glowacki, F. 406
Goes, W. 530
Golz, J. 371
Gong, Xiao 231
Gonzalez, Pilar 261
Goodsall, Timothy 267
Goren, David 307
Gossard, A. C. 586
Gossner, Harald 63
Goswami, Sushmit 295
Goto, Masahide 84
Gourhant, O. 77
Gourvest, E. 77
Goux, L. 351, 355
Govindaraju, S. 71
Govoreanu, Bogdan 164
Grampeix, H. 152, 546
Grasser, T. 506, 530
Graves-Abe, T. L. 371
Greene, B. 74
Grenouillet, L. 172, 219
Gritters, J. 40
Groeseneken, G. 355, 506, 717, 820, 828
Groeseneken, Guido 164, 514
Gros-Jean, M. 77
Grot, Annette 265
Grover, R. 71
Gu, Alex 566
Gu, Alex Y. D. 570
Gu, J. J. 502, 510
Gu, Y. D. 208, 398
Guan, X. 74
Guarin, F. 74
Guedj, C. 406
Guillaumet, S. 219
Guillermet, M. 77
Guillorn, M. 764
Guillot, N. 642
Guo, Hong 733
Guo, Jing 721
Guo, Lingyi 335
Guo, Shaofeng 319, 542, 816, 832
Guo, W. 168
Gupta, Ankur 63
Gupta, Shalini 291
Gustafson, J. L. 200
Guy, J. 136, 152
Häberlen, O. 28
Haendler, S. 77
Hagiwara, Kei 84
Han, Jin-Woo 204, 646

AUTHOR INDEX

Han, W.	71
Handa, Hiroyuki	283
Hang, P.	462
Hanken, D.	71
Hannah, E.	200
Hanyu, Takahiro	654
Hao, Peng	319, 542, 816
Hao, Y.	442
Haond, M.	219
Hara, H.	650
Haralson, E.	71
Haran, B.	219
Haran, M.	71
Haras, Maciej	212
Hartings, Jed A.	753
Hartman, Jeffrey	291
Hartmann, J.-M.	406
Hasanuzzaman, M.	74
Hashemi, Pouya	402
Hashimoto, H.	374, 669
Haspeslagh, Luc	100
Hatada, A.	486
Hatem, C.	776
Haukness, B.	657
Hawkins, K.	490
Haxaire, K.	77
He, H.	219
He, Keliang	196
Heckscher, M.	71
Heidel, Timothy D.	44
Hekmatshoar, Bahman	602
Hellings, G.	514
Hendrickx, Paul	164
Hengstebeck, B.	788
Hennessy, John	267
Henry, H. George	291
Henson, K.	74
Hentz, S.	550
Heremans, Paul	598
Herget, Philipp	307
Hertel, T.	610
Heussner, R.	71
Heylen, N.	478
Heyns, M.	828
Heyns, M. M.	717
Higurashi, Eiji	84
Hikavyy, A.	418
Hill, R. J. W.	776
Hills, Gage	812
Hiraiwa, A.	279
Hiraiwa, Miori	275
Hiramoto, Toshiro	84
Hirose, Y.	88
Ho, Chong Pei	566
Ho, H.	74

Hobbs, C.	776
Hody, Hubert	164
Hoenk, Michael E.	267
Holland, M.	235
Holzwarth III, C. W.	200
Honda, M.	140
Hondongwa, Donald	247
Honea, J.	40
Hong, J.	74
Hong, J. P.	482
Honjo, H.	796
Hopstaken, M.	582
Hopstaken, Marinus	402
Horiguchi, N.	418, 772
Hosoda, T.	40
Hotellier, N.	856
Hou, S. Y.	634
Hou, Tuo-Hung	132, 665
Hou, Y.-F.	808
Howe, Roger T.	96
Howell, Robert S.	291
Hoyos, D.	74
Hsieh, Chih-Chang	498
Hsieh, E. R.	303
Hsieh, Min-Che	148
Hsieh, Tung-Ying	67, 410
Hsieh, Y.-J.	390
Hsiung, C.-W.	450
Hsu, Chung-Wei	665
Hsu, S.-H.	788
Hsu, T. H.	709
Hsu, Tzu-Hsuan	498
Hsu, Wei-Ting	132
Hsueh, F.-K.	788
Hu, C.	808
Hu, H. P.	634
Hu, Y. T.	128
Huang, C. Y.	586
Huang, Guo-Wei	67
Huang, H. T.	48
Huang, J.-C.	390
Huang, Joanne	180
Huang, P.-C.	788
Huang, Qianqian	335
Huang, R. M.	303
Huang, Ru	319, 335, 542, 816, 832
Huang, S.	442
Huang, S. W.	634
Huang, Shih-Hsien	239
Huang, W.-H.	788
Huang, Wen-Hsien	67, 410
Huet, K.	642
Huffman, C.	776
Hung, C. M.	303
Hung, Chih-Ming	470

AUTHOR INDEX

Hung, Min-Feng 618
Hung, P. Y. 776
Hur, Sunghoi 367
Hurkx, Fred 430
Hutin, L. 223
Hutin, Louis 677
Huyghebaert, C. 128
Hwang, H. 697
Hwang, In Seol 108
Hwang, J. H. 768
Iba, Y. 486
Idrissi-Eloudrhiri, A. 518
Ielmini, D. 359, 363, 701
Iguchi, Yoshinori 84
Ikeda, S. 796
Ikeda, Shoji 654
Ikegami, K. 650
Imanishi, K. 40
Imbert, G. 77
Imura, S. 88
Inoue, Hideo 13
Irrera, F. 351
Ishida, Hidetoshi 283
Ishida, Masahiro 283
Ishihara, R. 614
Ishikawa, S. 796
Ishikawa, Tsuyoshi 24
Ishikawa, Y. 824
Ishikawa, Yuki 299
Itabashi, K. 40
Itai, S. 650
Ito, A. 458
Ito, J. 650
Iwai, H. 518
Iyer, S. S. 371, 490
Izumi, Shigeto 630
Jaeger, D. 490
Jagannathan, B. 74
Jain, P. 71
Jain, S. 74
Jalaguier, E. 546
James, R. 71
Jana, Raj K. 347
Jang, J. W. 697
Jang, Jin 860
Jang, S. M. 48
Jayapala, Murali 261
Jeannot, S. 546
Jehl, X. 223
Jena, Debdeep 347
Jeng, S. P. 634
Jeng, S.-J. 74
Jenny, C. 77
Jensen, A. 776
Jeon, M. S. 482

Jewell, April D. 267
Jhan, Yi-Ruei 618
Jhaveri, R. 71
Ji, H. M. 208
Ji, Jingwei 542
Ji, Li 216
Ji, Z. 506, 820
Ji, Zhigang 816
Ji, Zhuoyu 860
Jiang, Q. 442
Jiang, Qimeng 287
Jiang, Xiaobo 816, 832
Jiang, Zizhen 116, 156
Jin, I. 71
Jin, Seonghoon 184
Jin, W. 200
Jin, Z. 442
Jo, Sung Hyun 160
Joblot, S. 77
Johnson, A. 140
Johnson, J. 74
Johnson, J. B. 490
Johnson, R. 219
Jones, Todd J. 267
Joseph, E. A. 764
Josse, E. 219
Jovanovic, N. 144
Julien, C. 77
Jung, E. S. 768
Jung, Eun-Seung 184
Jung, R. 219
Jung, S. Y. 768
Jung, Ukjin 108
Jurczak, M. 351, 355
Jurczak, Malgorzata 164
Kabuyanagi, Shoichi 780
Kachtouli, R. 642
Kaczer, B. 506, 820, 828
Kajitani, Ryo 283
Kalnitsky, A. 450
Kam, H. 71
Kamakura, Y. 176, 255
Kamakura, Yoshinari 713
Kamata, C. 650
Kamata, Y. 422
Kamimuta, Y. 422
Kamioka, Takefumi 713
Kamiya, K. 538
Kamiyama, M. 40
Kanagala, V. 462
Kanai, S. 796
Kanamura, M. 40
Kang, Daewoong 367
Kang, Gyuseong 315
Kang, H.-K. 768

AUTHOR INDEX

Kang, Jiahao.....................................120, 729
Kannan, B..74
Kanyandekwe, J...219
Kao, K.-H..717, 788
Kao, Ming-Hsuan...67
Kar, G. S..478
Karda, K...140
Karl, E...71
Karpowicz, T...776
Kaste, E..74
Kato, Y...88
Kauerauf, T......................................506, 828
Kawa, Jamil..180
Kawarada, H..279
Kawasaki, S...40
Kawasumi, A..650
Ke, Tung-Huei..598
Ke, Y...74
Kee, Jack Sheng..394
Kellam, M..657
Kenney, Christopher J.....................................96
Kenyon, C...71
Kerber, P..582
Kerdiles, S..642
Khaja, F...776
Khakifirooz, Ali...402
Khan, B...74
Khare, M...219
Khwa, W. S...709
Ki, Wing-Hung..311
Kikkawa, T..40
Kikuchi, K..88
Kim, B..74
Kim, B. H..768
Kim, Chris H...315
Kim, D.-H..578, 776
Kim, I. S..768
Kim, J. N..768
Kim, J. S..768
Kim, Jee-Yeon..204
Kim, Jongyeon..315
Kim, M.-C..768
Kim, Minsoo..331
Kim, R.-H..768
Kim, S...709
Kim, Sangbum...693
Kim, T.-W..578
Kim, W...478
Kim, Yonghun...108
Kim, Young-Tae...184
Kim, Yun Ji..108
Kimoto, T...36
King, Matthew..291
King, Ya-Chin....................................148, 494
Kinoshita, M...685

Kinoshita, Yusuke..283
Kirahata, T..490
Kirsch, P. D...578
Kitagawa, E..650
Kiuchi, K...40
Klaus, D. P..764
Kleemeier, W.....................................172, 219
Klinker, Lauren..761
Ko, C.-H...414
Ko, Chih-Hsin..231
Ko, D. H...578
Koba, S..176
Kobayashi, Masaharu.......................................84
Koike, M...422
Kong, W..490
Konishi, Satoshi...757
Konno, S...669
Kopta, A..32
Korber, M..140
Koswatta, S...74
Kotecha, Pooja M...852
Kothandaraman, C...371
Kothari, H.........................172, 219, 522
Koutsouras, Dimitrios A..................................749
Koyama, M..518
Koyanagi, M......................................374, 669
Kreupl, F..657
Krishnan, S...74
Krishnasamy, R...490
Ku, J.-H...768
Ku, Y...48
Kubota, M...88
Kukita, K..176
Kumar, A..74
Kumar, Tanmay..160
Kunihiro, T..140
Kuo, Yue...104
Kurdi, B...697, 705
Kuroda, Tadahiro...474
Kushida, K...650
Kuss, J..219
Kwan, Man Ho...450
Kwon, H. M...578
Kwon, O.-J...490
Kwon, T...74
Kwon, U...74
Kwong, D. L......................................208, 398
Kwong, Dim-Lee...570
Kye, J...462
Lacaita, A. L..681
Lacatena, V..212
Ladner, C..562
Lafond, D..642
Lagrain, P...418
Lai, E. K..709

AUTHOR INDEX

Lai, F.-L. 390
Lai, H.-C. 390
Lai, Y.-A. 450
Lal, R. 40
Lam, C. 709
Lam, Chung 693
Lam, Chung H. 689
Lam, Kai-Tak 721
Lambrechts, Andy 261
Lamy, Y. 223
Lan, Huang-Siang 239
Landers, W. 371
Lansbergen, G. P. 450
Lanzerotti, L. 74
Laska, T. 28
Lauer, I. 764
Laurent, A. 642
Laurent, L. 562
Lauwers, Anne 100
Lavallee, G. 414
Laviron, C. 856
Le Carval, G. 152
Le Royer, C. 223
Lebeau, J. 414
Lee, Byoung Hun 108
Lee, C. H. 48
Lee, Changsub 367
Lee, Chengkuo 566
Lee, Choong Hyun 780
Lee, D. H. 768
Lee, D. Y. 482
Lee, E. B. 768
Lee, H. B. 768
Lee, H. M. 48
Lee, H.-K. 74
Lee, Han-Bo-Ram 108
Lee, J. J. 768
Lee, J.-W. 514
Lee, Jack C. 216
Lee, Jaesung 196
Lee, K. W. 374
Lee, K.-T. 582
Lee, K.-W. 669
Lee, Keun-Ho 184
Lee, Kyong Taek 836
Lee, M. H. 709
Lee, M.-D. 808
Lee, N.-I. 768
Lee, Rinus T. P. 776
Lee, S. 586
Lee, S. E. 482
Lee, Sang Kyung 108
Lee, Seunghyun 116
Lee, Shiuh-Wuu 319, 816, 832
Lee, T. L. 48

Lee, W.-H. 74
Lee, Y.-J. 414, 788, 808
Lei, Dian 231
Leleux, Pierre 749
Leobandung, E. 582
Leobandung, Effendi 402
Le-Pailleur, L. 856
Leverd, F. 77, 80
Levesque, A. 74
Levin, T. 219
Levinson, H. 462
Lherron, B. 219
Li, Baikui 287
Li, Baozhen 840
Li, Chao 394
Li, Cheng 378, 386
Li, Cheng-Syun 558
Li, Chunyan 753
Li, H. 140
Li, H. P. 709
Li, K.-H. 808
Li, Kuan-Hsien 554
Li, L.-J. 808
Li, Lain-Jong 132
Li, Ling 860
Li, Ming-Huang 554, 558
Li, Sheng-Shian 554, 558
Li, W. 74
Li, X. 490, 673
Li, Xiang 120
Li, Xiaohan 216
Li, Xing 311
Li, Xiuyan 534
Li, Y. 458, 788
Li, Z. 74
Li, Zhigang 860
Liang, M. 48
Liao, T.-C. 390
Liao, W. S. 634
Liaw, J. J. 48
Licitra, C. 136
Lim, J. 140
Lim, Sung Kwan 108
Lin, Angel T . H. 570
Lin, C. Y. 48
Lin, C.-H. 74, 808
Lin, C.-Y. 808
Lin, Chein-Din 67
Lin, Chih-Pin 132
Lin, Chrong Jung 148, 494
Lin, D. 828
Lin, H. L. 634
Lin, H. S. 450
Lin, I. 462
Lin, Jianqiang 574

AUTHOR INDEX

Lin, M.-C. 450
Lin, Po-Yen 494
Lin, R. 776
Lin, Tzu-Ping 132
Lin, Tzu-Yao 239
Lin, Xi-Wei 180
Lin, Y. S. 450
Lin, Y.-R. 414
Lin, Yen-Chuan 665
Lin, You-Ru 231
Linder, B. 74
Linder, B. P. 522
Liniger, E. G. 522
Linten, D. 514
Lisauskas, Alvydas 92
Liu, A. Q. 398
Liu, B 219
Liu, C. 442
Liu, C. W. 239
Liu, C.-C. 764
Liu, Changze 836
Liu, Cheng 434
Liu, G. 442
Liu, H. 442
Liu, Han 112
Liu, J. 371, 490
Liu, Jie 180
Liu, Jinhua 816, 832
Liu, Lei 733
Liu, M. 71
Liu, Ming 860
Liu, P.-C. 450
Liu, Pang-Shiuan 132
Liu, Q. 172, 219
Liu, S. 390, 442
Liu, Shenghou 434
Liu, T. H. 634
Liu, Tsu-Jae King 327, 677
Liu, W. 74
Liu, Wei 120, 729
Liu, X. 442
Liu, Xiaoyan 192
Liu, Y. 200, 243
Liu, Y. X. 824
Liu, Y.-S. 390
Liu, Yongxun 299
Lizzit, Daniel 188
Lo, Guo-Qiang 394
Lo, Roger 498
Loh, W.-Y. 776
Longobardi, Giorgia 430
Loo, R. 418
Loubet, N. 172, 219
Lu, Chih-Yuan 498
Lu, Cimang 780

Lu, G.-N. 80
Lu, Nianduan 860
Lu, Ning 852
Lu, V. 642
Lu, Wei D. 160
Lu, Y. 442
Lu, Yan 311
Lu, Yunyou 287, 434
Ludurczak, W. 562
Lue, Hang-Ting 498
Luisier, Mathieu 725
Lundsrom, Mark 844
Lung, H. L. 709
Lung, Hsiang-Lan 693
Luo, Mulong 319, 816, 832
Luo, Wei 227, 426
Luo, Xianshu 394
Luo, Y. 71
Luo, Zhe 112
Lv, Yang 315
Ma, J. 820
Ma, Jiaju 247
Ma, T. P. 438
Ma, Terry 180
Maciejewski, E. 74, 490
Maeda, Hidehiro 630
Maeda, T. 40
Maffini-Alvaro, V. 406
Magyari-Köpe, Blanka 156
Mahajan, S. 74
Maitrejean, S. 172, 219
Maize, K. 510
Majumdar, A. 582
Mäkilä, E. 200
Malik, R. 490
Malliaras, George G. 749
Mamdy, B. 80
Manojkumar, C. M. 594
Mao, Junfa 120
Marino, J. 788
Martin, A. J. 371
Martin, F. 223
Martin, R. M. 582
Martin-Fernandez, Iñigo 382
Martini, S. 172
Martinie, S. 518
Masahara, M. 243, 824
Masahara, Meishoku 299
Masarotto, L. 562
Masuduzzaman, M. 510
Maszara, W. 578
Mathieu, B. 642
Matsuda, T. 768
Matsukawa, T. 243, 824
Matsukawa, Takashi 299

AUTHOR INDEX

Matsukura, F.	796
Matsuoka, T.	255
Matthews, K.	776
Maury, P.	77
Mayuzumi, S.	140
Mazuré, C.	406
McCaffrey, V.	363
McCarthy, L.	40
McDaniel, Martin D.	216
McGrane, Bryan	761
McKay, J.	40
McStay, K.	74
Mehandru, R.	71
Meneghesso, Gaudenzio	446
Menegini, Matteo	446
Mertens, H.	418, 828
Mertens, S.	478
Messenger, B.	490
Meyer, R.	657
Michailos, J.	856
Migita, S.	243, 824
Migita, Shinji	299
Milenin, A. P.	418
Min, T.	478
Minoglou, Kyriaki	100
Mishra, N.	657
Mishra, U.	40
Mitard, J.	168, 418, 820, 828
Mitchell, W.	586
Mitra, Subhasish	638, 812
Mitsuishi, Hajime	630
Miyakawa, K.	88
Miyashita, T.	48
Miyata, K.	140
Miyazaki, Y.	40
Miyazoe, H.	764
Mizubayashi, W.	243, 824
Mizubayashi, Wataru	299
Mo, R. T.	582
Mo, Renee T.	402
Mochizuki, A.	40
Mochizuki, Akira	654
Mocuta, A.	418, 717
Mohammad, Mohammad Ali	378, 386
Mohney, S. E.	414
Molas, G.	136, 144, 152, 546
Moneck, M. T.	792
Monfray, S.	212
Montagné, A.	77
Moon, B. K.	200
Moon, Dong-Il	204
Morarka, S.	71
Morgan, Hywel	741
Mori, N.	176
Mori, T.	243, 422, 824
Moriceau, H.	223
Morikawa, T.	685
Morimoto, Masao	56
Morin, P.	172, 219
Morini, F.	212
Morita, Tatsuo	283
Morita, Y.	243, 824
Morita, Yukinori	299
Moroz, V.	168
Moroz, Victor	180
Mounet, C.	406
Mu, X. J.	208
Mueller, Thomas	124
Müller, Robert	598
Münzenrieder, N.	610
Murugesan, M.	374, 669
Musolf, Juergen	606
Myny, Kris	598
Nagai, C.	669
Nagai, Shuichi	275
Nagashio, Kosuke	534
Nakabayashi, M.	486
Nakada, T.	88
Nakahira, Hosei	630
Nakamura, G.	776
Nakane, Ryosho	331
Nakazawa, K.	140
Nam, S. D.	768
Namal, I.	610
Narasimha, S.	74, 490
Narayan, Raj K.	753
Narayanan, P.	697, 705
Narayanan, Sundar	160
Narayanan, V.	74, 582, 673
Narducci, Margarita	566
Natarajan, S.	71
Natsui, Masanori	654
Nayfeh, H.	74
Nazarian, Hagop	160
Nechay, Bettina	291
Neiberg, L.	71
Nelson, E.	490
Nemouchi, F.	223, 642
Newbury, Joseph S.	402
Ney, D.	77
Ng, G. I.	594
Nguyen, B.-Y.	406
Nguyen, C.	144
Nguyen, P.	406, 518
Nguyen, S.	657
Nicoll, W.	74
Nielsen, Peter F.	776
Nii, Koji	56
Nikolic, B.	59
Nikzad, Shouleh	267

AUTHOR INDEX

Niquet, Y.-M. 518
Nishi, Y. 546
Nishi, Yoshio 156, 638
Nishimura, Tomonori 534, 780
Nishizawa, Yutaka 184
Noguchi, H. 650
Noh, K. B. 772
Northrop, G. 74
Norum, J. 490
Noshiro, H. 486
Nummy, K. 490
O, Kenneth K. 92
Oakley, J. 371
Oates, Anthony S. 526
Ocak, Ilker E. 570
Ochiai, T. 650
Ogawa, Masahiro 283
Ogino, A. 74
Ogino, T. 40
Ohmori, Kenji 713
Ohno, H. 796
Ohno, Hideo 654
Ohsawa, Y. 776
Ohtake, H. 88
Ohtake, Hiroshi 84
Ohyanagi, T. 685
Okabe, Kye 156
Okada, Masashi 630
Okamoto, Kazuya 630
Okino, T. 88
Okumura, H. 36
Okuno, H. 223
Okuno, J. 140
Ollier, E. 550
Olshausen, Bruno A. 693
Onishi, K. 490
Onishi, M. 669
Onizawa, Naoya 654
Opsomer, K. 351
Orlando, B. 80
Osgnach, Patrik 188
O'Sullivan, Eugene 307
Ota, H. 243, 824
Ota, Hiroyuki 299
Otsuka, Nobuyuki 275
Otsuka, W. 140
Ott, John A. 402
Ottosson, Kristoffer 606
O'Uchi, S. 243, 824
O'Uchi, Shin-Ichi 299
Ouerghi, I. 562
Owen, Man Hon Samuel 231
Oxland, R. 235
Paak, S. S. 768
Packan, P. 71

Pae, Sangwoo 836
Pai, C. Y. 634
Paik, H. 673
Pain, L. 223
Pal, Asish 638
Pala, M. 172
Palacios, T. 438
Palacios, Tomás 295
Palestri, Pierpaolo 188
Paliwal, A. 71
Palma, G. 144
Palmour, John W. 1
Pan, Chih-Hung 800
Pan, K. H. 48
Pantouvaki, M. 128
Paraschiv, Vasile 164
Parihar, A. 673
Parikh, P. 40
Park, D. W. 768
Park, D.-G. 582
Park, Dae-Gyu 402
Park, J. G. 482
Park, Jongwoo 836
Park, K. H. 768
Park, S. H. 768
Park, S. O. 482
Park, W.-K. 578
Park, Woojin 108
Park, Youngkwan 184
Parke, Justin 291
Parker, C. 71
Parks, C. 371
Parries, P. 490
Pasini, L. 642
Passlack, M. 235
Patel, P. 71
Patel, R. 71
Patel, Shyamal 761
Peh, Li-Shiuan 295
Pei, C. 74, 490
Pellissier-Tanon, D. 80
Pellizzer, F. 681
Pelloux-Prayer, J. 518
Pelto, C. 71
Peng, Y.-C. 390
Perniola, L. 136, 144, 152, 223, 546, 661
Perreau, P. 642
Perrot, C. 80
Petitdidier, S. 77
Petitprez, E. 642
Petrarca, K. 371
Petti, L. 610
Pham, Anh-Tuan 184
Philippe, J. 550, 562
Pierre, F. 136

AUTHOR INDEX

Pileggi, L. 792
Pilorget, S. 172, 219
Pint, C. 200
Pipes, L. 71
Pitera, J. 764
Pitner, Gregory 812
Planes, N. 59
Plekhanov, P. 71
Pofelski, A. 172
Poindexter, D. 490
Poiroux, T. 172
Pollet, O. 152
Polvino, S. 74
Polyushkin, Dmitry K. 124
Posadas, Agham B. 216
Pospischil, Andreas 124
Posseme, N. 223
Poulain, C. 223
Poupon, G. 223
Pourghaderi, M. A. 506
Prall, K. 140
Prati, Enrico 9
Previtali, B. 642
Prince, M. 71
Pruitt, Beth L. 745
Pruvost, J. 856
Qian, Chuang 327, 677
Qian, H. 657
Qiu, Yingxin 542
Quémerais, T. 77
Radens, C. 74
Radhakrishna, Ujwal 295
Radhakrishnan, K. 398
Rafhay, Q. 661
Raghuram, U. 657
Ragnarsson, L.-å. 828
Raj, Milan 761
Rajamani, S. 71
Rajan, S. 594
Raju, Salahuddin 311
Rakheja, Shaloo 844
Ralston, Parrish 291
Ramachandran, R. 74
Ramadout, B. 77
Ramaswamy, N. 140, 359
Rambal, N. 642
Ramuz, Marc 749
Ranica, R. 59
Ranjan, K. 594
Rao, K. V. 776
Rao, V. Ramgopal 63
Rashed, M. 462
Raychowdhury, A. 673
Raynaud, C. 466
Rayner Jr., G. B. 414

Redaelli, A. 681
Reddy, B. 390
Redolfi, A. 355
Redolfi, Augusto 164
Reimbold, G. 518
Reisinger, H. 530
Reita, C. 562, 642
Ren, M. 398
Ren, Pengpeng 319, 542, 816, 832
Ren, Tian-Ling 378, 382, 386
Ren, Z. 74
Renaldo, Karen 291
Renau, Anthony 784
Reyboz, M. 144
Rhyner, Reto 725
Ribes, G. 77
Rice, J. 74
Richard, C. 77
Richard, E. 642
Richard, O. 351
Ritzenthaler, R. 772
Rivière, J.-M. 856
Rivnay, Jonathan 749
Rivoire, M. 80
Rizzi, M. 701
Robbelein, Jo 100
Robillard, J.-F. 212
Robin, O. 77
Robison, R. 74
Robson, N. 371, 490
Rodwell, M. 586
Rombawa, S. 490
Roodman, Aaron J. 96
Rooyackers, R. 717
Rosa, J. 77
Rosenblatt, S. 371
Roskos, Hartmut G. 92
Rosmeulen, Maarten 100
Rossi, Michele 741
Rott, K. 530
Rouhi, A. 168
Roule, A. 136, 152
Roussel, Ph. 828
Roussel, Ph. J. 506
Rouvière, J.-L. 518
Roy, F. 80
Rozeau, O. 172, 406
Rozen, J. 582, 764
Ruoff, Rodney S. 216
Rupp, R. 28
Rzepa, G. 530
Saad, Yves 180
Sadana, D. K. 582
Safran, J. 371
Saida, D. 650

AUTHOR INDEX

Saito, T. 279
Saito, Y. 685
Salahuddin, Sayeef 343
Salm, E. 390
Salot, R. 223
Samavedam, S. 462
Sampson, R. 172, 219
Samson, M.-P. 406, 642
Sanders, D. P. 764
Sandford, J. 71
Sang, X. 414
Sangiorgi, Enrico 741
Sankarapandian, M. 219
Sanquer, M. 223
Saraf, I. 74
Saraswat, Krishna 638
Saraya, Takuya 84
Sardesai, V. 74
Sarkar, Deblina 120, 729
Sasaki, Y. 418
Sato, H. 796
Sato, Y. 374, 669
Saudari, S. 74
Savoj, Jafar 454
Savoye, M. 550
Saxod, O. 77
Schanovsky, F. 530
Scheer, P. 172
Scheiblin, P. 562
Schepis, D. 74
Schlom, D. G. 673
Schmidt, Oliver 259
Scholten, Andries 848
Scholz, M. 514
Schram, T. 772
See, Y. 514
See, Y. C. 235
Segal, Julie D. 96
Seignard, A. 80
Sekar, D. C. 657
Sell, B. 71
Selmi, Luca 188
Senapati, B. 462
Seo, K.-S. 578
Seol, Gyungseon 721
Seol, Myeong-Lok 204
Servalli, G. 681
Shakouri, A. 510
Shan, Jie 196
Shang, H. 74
Shang, Liwei 860
Shao, Chi-Shen 622
Shapiro, Charles 267
Sharma, Dinesh Kumar 63
Sharma, Jaibir 570

Shelby, R. M. 697
Shen, Chang-Hong 67, 410
Shen, L. 40
Shenoy, R. S. 697, 705
Sheraw, C. 74, 490
Shi, Qing 733
Shieh, Chan-Long 606
Shieh, J.-M. 788
Shieh, Jia-Min 67, 410
Shieh, Tai-Cheng 239
Shih, Chih-Cheng 800
Shih, Yen-Hao 498
Shim, T. H. 482
Shimoda, Toshimasa 630
Shimomura, N. 650
Shimonomura, K. 255
Shin, C.-S. 578
Shin, S. H. 502, 510, 578
Shinada, Takahiro 9
Shiraishi, K. 538, 685
Shirakawa, H. 538, 685
Shiu, Jung-Hau 410
Shiu, K.-T. 582
Shluger, A. L. 530
Shono, K. 40
Shoron, O. F. 594
Shrivastava, Mayank 63
Shu, Yi 378
Shukla, N. 673
Shulaker, Max M. 638, 812
Si, M. 502, 510
Si, Mengwei 426
Sibaja-Hernandez, A. 514
Siddiqui, S. 74
Signamarcheix, T. 223
Sills, S. 140, 359
Simoen, E. 772
Singh, J. 462
Singh, Navab 570
Singh, Pushpapraj 570
Sinha, Ashutosh 52
Sioncke, S. 418, 506, 828
Siow, L. Y. 208
Sivakumar, S. 71
Skotnicki, T. 212, 219
Smets, Q. 717
Smith, K. 40
Smith, Lee 180
Smith, P. 40, 71
Snider, Gregory 347
Snook, Megan 291
Sohn, Joon 116, 156
Sokalski, V. M. 792
Son, Y. 772
Song, B. 71

AUTHOR INDEX

Song, Duheon................367
Song, Junfeng................394
Song, L................74
Song, M.-H................514
Sonský, Jan................430
Soon, Jeffrey Bo Woon................566
Sorée, B................717
Souaré, P. M................856
Sousa, V................152
Soussan, Philippe................261
Southwick, R. G................172, 522
Späth, F................610
Spessot, A................772
Sque, Stephen................430
Stathis, James H................522, 840
Steegen, A................828
Stein, K................74
Steinbacher, Frank................259
Stemmer, S................586
Stephens, J................462
Stesmans, A................530
Steudel, Soeren................598
Stewart, Eric J................291
Stiffler, S................74
Stinzianni, E................776
Su, Jun-Ji................622
Su, R. Y................450
Su, S.-H................390
Su, T. H................709
Su, Yu-Ting................800
Suda, J................36
Sugaya, Isao................630
Sugibayashi, Tadahiko................654
Sugii, T................486
Sumino, J................140
Sun, B................442
Sun, C. L................208
Sun, Jeff................677
Sun, X................438
Sun, Y................582
Sung, P.-J................788
Sung, Yuh-Te................494
Suzuki, Akito................713
Suzuki, Asamira................275
Suzuki, Daisuke................654
Swab, P................657
Swenson, B................40
Swerts, J................478
Szabo, Aron................725
Sze, Simon M................800
T., Arun V................414
Tabone, C................406
Tack, Klaas................261
Tahmasebi, T................478
Tai, M................685

Takagi, Shinichi................331
Takahashi, A................486
Takaura, N................685
Takemura, Y................482
Takenaka, Mitsuru................331
Tamura, Satoshi................283
Tanaka, C................650
Tanaka, Kenichiro................283
Tanaka, Koji................56
Tanaka, Miki................56
Tanaka, Shinji................56
Tanaka, T................669
Tanaka, Yasunori................24
Tang, Xi................287
Tang, Z................442
Tang, Zhikai................434
Tao, J. F................208, 398
Tao, Li................216
Tartagni, Marco................741
Tavernier, C................77, 856
Tedde, Sandro F................259
Tedesco, S................223
Teehan, S................219
Teo, K. L................594
Teranishi, N................88
Testini, A................562
Tezuka, T................422
Thangaraju, S................462
Thean, A................168, 506, 514, 772, 828
Thean, A. V.-Y................418, 717
Thibeault, B................586
Thoma, Rainer................852
Thomas, O................59, 144
Thor, D................40
Tian, He................378, 382, 386
Ting, K. C................48
Tiron, R................223
Toffoli, A................136, 144, 152, 406
Toh, S. O................59
Toledano-Luque, M................828
Tominaga, J................685
Tone, K................71
Ton-That, V................462
Toqué-Trésonne, I................642
Toriumi, Akira................534, 780
Torsi, A................140
Toshiyoshi, Hiroshi................84
Tournier, A................80
Tran, C................74
Tran, H................140
Traoré, B................546
Treu, M................28
Trickett, Y................776
Trifunovic, M................614
Triozon, F................518

AUTHOR INDEX

Tripathy, S. 594
Troeger, T. 71
Tröster, G. 610
Tsai, C. H. 48
Tsai, C. S. 634
Tsai, C. T. 303
Tsai, C. Y. 634
Tsai, Gene 231
Tsai, H. 764
Tsai, J. L. 450
Tsai, M. H. 48
Tsai, M. W. 450
Tsai, Tsung-Ming 800
Tsuboi, H. 279
Tsuchiya, H. 176
Tsui, Chi-Ying 311
Tsui, F. 390
Tsui, R. F. 48
Tsukada, J. 824
Tsukada, Junichi 299
Tsukamoto, M. 140
Tsukamoto, Yasumasa 56
Tsunoda, K. 486
Tsuzaki, Y. 486
Tu, Wen-Hsien 239
Tuan, H. C. 450
Tuinhout, Hans 848
Tunga, K. 371
Turkyilmaz, O. 144
Tzou, T. 490
Udrea, Florin 430
Ueda, Daisuke 275
Ueda, Tetsuzo 283
Ujita, Shinji 283
Umeda, Hidekazu 283
Upadhyaya, Parag 454
Urteaga, M. 578
Usuda, K. 422
Utomo, H. 74
Van Campenhout, J. 128
Van Dal, M. J. H. 235
Van De Put, M. 717
Van Der Plas, G. 168
Van Der Zwan, M. 614
Van Thourhout, D. 128
Vandenberghe, William G. 804
Vandervorst, W. 351
Vandooren, A. 717
Vardi, Alon 590
Vega, R. 74
Vellianitis, G. 235
Venigalla, R. 219
Verhulst, A. S. 717
Verreck, D. 717
Verzellesi, Giovanni 446

Vianello, E. 136, 144, 223, 546, 661
Villalon, A. 223
Vinet, M. 172, 219, 223, 406, 642
Virollet, N. 80
Virwani, K. 697, 705
Vizioz, C. 406
Vogt, C. 610
Vrancken, C. 418
Wachnik, Richard A. 852
Wahab, M. A. 510
Wakabayashi, Yuki 331
Wakimura, G. 176
Waldron, N. 506, 828
Waltl, M. 530
Wang, B. 140
Wang, C. P. 200
Wang, D. 363
Wang, Debin 382
Wang, G. 74, 490
Wang, Guangming 606
Wang, H. 74, 442
Wang, Hanxing 287
Wang, Hsing-Hsiang 67, 410
Wang, I-Ting 665
Wang, Jiannong 287
Wang, Jianping 319, 816, 832
Wang, Jian-Ping 315
Wang, Jiaxin 335
Wang, M. 219, 522
Wang, Naigang 307
Wang, Nan 566
Wang, Runsheng 319, 335, 542, 816, 832
Wang, S.-P. 450
Wang, T. 808
Wang, T. Y. 303, 709
Wang, W. 74
Wang, X. 74, 442, 490
Wang, Xianyan 761
Wang, Xue-Feng 378
Wang, Yangyuan 319, 335, 832
Wang, Yu-Fen 665
Wang, Z.-Q. 359
Wang, Zenghui 196
Watanabe, H. 538
Watanabe, Takanobu 713
Wathuthanthri, Ishan 291
Weber, O. 219, 223
Wehbi, A. 462
Wehelle-Gamage, D. 74
Wei, A. 462
Wei, K. 442
Wei, Pinghung 761
Wells, D. 140
Wen, C.-H. 390
Wiedemer, J. 71

AUTHOR INDEX

Wimmer, Y. 530
Wirth, Gilson I. 848
Witters, L. 168, 418, 828
Witters, Thomas 164
Wong, H.-S. Philip 116, 156, 638, 693, 812
Wong, I-Hsieh 239
Wong, Justin C. 343
Wong, K.-Y. 450
Wong, Waisum 319, 816, 832
Woo, A. 458
Woo, Jong-Ho 204
Woo, Y. 462
Woodard, E. 74
Wouters, Dirk J. 164
Wu, B.-W. 808
Wu, C. B. 450
Wu, C.-H. 414, 450
Wu, C.-S. 808
Wu, C.-T. 788
Wu, Cheng-Hsien 231
Wu, Chunlei 335
Wu, E. Y. 522
Wu, Ernest 840
Wu, H. 657
Wu, H.-Y. 450
Wu, Hanming 319, 816, 832
Wu, Heng 227, 426
Wu, J. 657
Wu, J. Y. 709
Wu, Jingang 319, 816, 832
Wu, Meng-Chyi 410
Wu, Min-Hsin 618
Wu, Pei-Jung 800
Wu, Pei-Ming 753
Wu, S.-Y. 48
Wu, Shih-Chieh 132
Wu, Tony F. 638
Wu, Tsung-Ta 67, 410
Wu, W.-F. 788
Wu, Y. 40
Wu, Y. K. 48
Wu, Yung-Chun 618, 622
Wu, Zhizhen 753
Xiao, Tian 606
Xie, Xuejun 120
Xu, D. 279
Xu, Nuo 327, 677, 816
Xu, Xianfan 112
Xu, Xin 231
Xu, Y. 74
Yabuuchi, Makoto 56
Yajima, Takeaki 534
Yamada, Keisaku 713
Yamada, T. 279
Yamada, Yasuhiro 275

Yamamoto, T. 48
Yamanouchi, M. 796
Yamashita, T. 219
Yamauchi, H. 824
Yamauchi, Hiromi 299
Yamazaki, Y. 486
Yang, C.-K. 390
Yang, Chih-Chao 67, 410
Yang, F. J. 450
Yang, F.-L. 808
Yang, Fu-Liang 67, 410
Yang, I. 808
Yang, J.-H. 390
Yang, Jin Ho 108
Yang, M. 71
Yang, Rui 196
Yang, S. 442
Yang, S. H. 48
Yang, Shu 434
Yang, Y. 74
Yang, Yi 378, 382, 386
Yang, Z. C. 398
Yano, Koji 24
Yao, C. H. 48
Yao, F. W. 450
Yasuda, S. 140
Yatsuo, Tsutomu 24
Yau, J.-B. 582
Yckache, K. 562
Ye, Le 335
Ye, P. D. 502, 510
Ye, Peide D. 112, 227, 426
Yea, S. 40
Yeh, Mu-Shih 618
Yeh, W.-K. 788, 808
Yeo, Y.-C. 414
Yeo, Yee-Chia 231
Yew, T. R. 303
Yin, M. 490
Yokoyama, Masafumi 331
Yonezawa, Y. 36
Yoo, S. Y. 768
Yoon, B. U. 768
Yoon, J. S. 768
Yoshida, C. 486
Yu, C. Y. 450
Yu, D. 462
Yu, Doug 634
Yu, Douglas C. H. 626
Yu, Edward T. 216
Yu, Gang 606
Yu, H. B. 208
Yu, J. L. 450
Yu, S.-Y. 414
Yu, Scott 319

AUTHOR INDEX

Yu, Shaofeng 816, 832
Yu, Zhuoqing 832
Yuan, L. 462
Yue, C. Patrick 311
Zahurak, J. 140
Zaleski, M. 462
Zamdmer, N. 74
Zanoni, Enrico 446
Zeng, J. 462
Zhan, N. 74
Zhang, C. 255
Zhang, J. 442, 442, 502, 614
Zhang, J. F. 820
Zhang, Jian 339
Zhang, Jian F. 816
Zhang, Jingyun 426
Zhang, K. 71
Zhang, Leqi 164
Zhang, Lining 733
Zhang, Q. X. 208
Zhang, Rui 800
Zhang, W. 820
Zhang, Xin 307
Zhang, Xing 192
Zhang, Y. 438
Zhang, Yuegang 382
Zhang, Yunqiang 180
Zhao, C. 442
Zhao, K. 74
Zhao, Liang 156, 638
Zhao, Xin 590
Zheng, Xuqian 196
Zheng, Y. 414, 442
Zhong, Freeman 52
Zhou, Hong 426
Zhou, Qian 231
Zhu, C. 74
Zhu, Hao 335
Zhu, J. 792
Zhu, J.-G. 792
Zhu, Y. 208, 582, 709
Zhu, Yu 733
Zimmer, B. 59
Zong, Zhiwei 860
Zou, Jibin 832